Eleanor Jones
5466/65

Wilson and Gisvold's textbook of organic medicinal and pharmaceutical chemistry

Wilson and Gisvold's textbook of organic medicinal and pharmaceutical chemistry

13 contributors

edited by

Robert F. Doerge, Ph.D.

*Professor Emeritus of
Pharmaceutical Chemistry
School of Pharmacy
Oregon State University, Corvallis, Oregon*

eighth edition

Philadelphia
London · Mexico City · New York
St. Louis · São Paulo · Sydney

J. B. Lippincott Company

Printed in the United States of America

Eighth Edition

Copyright © 1982 by J. B. Lippincott Company
Copyright © 1977, 1971, 1966, 1962, 1956 by J. B. Lippincott Company
Copyright 1954 by J. B. Lippincott Company
First edition Copyright 1949 by J. B. Lippincott Company under the title Organic Chemistry in Pharmacy
For information address J. B. Lippincott Company, East Washington Square,
Philadelphia, Pennsylvania 19105.

5 6

Library of Congress Cataloging in Publication Data

Textbook of organic medicinal and pharmaceutical chemistry.
    Wilson and Gisvold's Textbook of organic medicinal and
pharmaceutical chemistry.

    Includes bibliographies and index.
    1. Chemistry, Pharmaceutical. 2. Chemistry, Organic.
I. Wilson, Charles Owens, date.
II. Gisvold, Ole, date. III. Doerge, Robert F. IV. Title.
V. Title: Wilson and Gisvold's Textbook of organic medicinal and
pharmaceutical chemistry. [DNLM: 1. Chemistry, Pharmaceutical.
QV 744 W754]
RS403.T43 1982   615′.3   81-12385
ISBN 0-397-52092-1   AACR2

contents

5 sulfonamides, sulfones, and folate reductase inhibitors with antibacterial action   189
DWIGHT S. FULLERTON

6 antimalarials   205
DWIGHT S. FULLERTON

7 antibiotics   225
ARNOLD R. MARTIN

8 antineoplastic agents 295

WILLIAM A. REMERS

9 central nervous system depressants 335

T. C. DANIELS and E. C. JORGENSEN

10 central nervous system stimulants 383

T. C. DANIELS and E. C. JORGENSEN

11 adrenergic agents 401

PATRICK E. HANNA

12 cholinergic drugs and related agents 433

GEORGE H. COCOLAS

contributors

Neal Castagnoli, Jr., Ph.D.

Professor of Chemistry and Pharmaceutical Chemistry, School of Pharmacy
University of California, San Francisco, California

George H. Cocolas, Ph.D.

Professor of Medicinal Chemistry, School of Pharmacy
University of North Carolina at Chapel Hill,
Chapel Hill, North Carolina

T.C. Daniels, Ph.D

Professor Emeritus of Pharmaceutical Chemistry, School of Pharmacy
University of California, San Francisco, California

Charles M. Darling, Ph.D

Professor of Pharmaceutical Chemistry, School of Pharmacy
Auburn University, Auburn, Alabama

Jaime N. Delgado, Ph.D.

Professor of Medicinal Chemistry, College of Pharmacy
The University of Texas at Austin, Austin, Texas

Robert F. Doerge, Ph.D.

Professor Emeritus of Pharmaceutical Chemistry
School of Pharmacy, Oregon State University, Corvallis, Oregon

Dwight S. Fullerton, Ph.D.

Professor of Pharmaceutical Chemistry, School of Pharmacy
Oregon State University, Corvallis, Oregon

Patrick E. Hanna, Ph.D.

Associate Professor of Medicinal Chemistry, College of Pharmacy and School of Medicine
University of Minnesota, Minneapolis, Minnesota

E.C. Jorgensen, Ph.D. *

Formerly, Associate Dean and Professor of Chemistry and Pharmaceutical Chemistry
School of Pharmacy, University of California, San Francisco, California

Lawrence K. Low, Ph.D.

Research Associate in Medicinal Chemistry, Department of Pharmaceutical Sciences
School of Pharmacy, University of Washington, Seattle, Washington

Arnold R. Martin, Ph.D.

Professor of Medicinal Chemistry, College of Pharmacy
The University of Arizona, Tucson, Arizona

William A. Remers, Ph.D.

Professor of Medicinal Chemistry
Head, Department of Pharmaceutical Sciences, College of Pharmacy
The University of Arizona, Tucson, Arizona

Robert E. Willette, Ph.D.

Director, Duo Research Division
Research Designs, Inc., Annapolis, Maryland

* Deceased

preface

The eighth edition of *Textbook of Organic Medicinal and Pharmaceutical Chemistry,* like the previous seven, is written for the undergraduate pharmacy student who has previously completed a regular year's course in the fundamentals of organic chemistry. Knowledge of physiology, physiologic chemistry, and pharmacology are also essential for maximal understanding of the presentations on therapeutic agents. The information assembled is that which is of practical value to present-day pharmacists in their everyday communication with patients and physicians. Pharmacists should have a thorough knowledge of what a drug is, its limitations, applications, stability, forms and uses, as well as of those characteristics that pertain strictly to its dose formulation.

Products used in medicine and in pharmacy have been discussed under chapter headings that are understood readily by those in the pharmaceutical profession. In order to present the topics in a logical manner for student comprehension, a combination chemical, pharmacologic, and therapeutic classification has been used.

Chapters 2 and 3 are independent; the information presented here, in most cases, should be reckoned with before a proper appreciation of the other chapters can be obtained. A final chapter seemed necessary to include pharmaceutic products designated as Miscellaneous Organic Pharmaceuticals.

As each class of agents is introduced, there is a discussion of the basic principles that relate to absorption, stability, ionization, salt formation, and other physicochemical properties that affect the therapeutic application such as dose form, excretion, or duration of action. Structure-to-activity relationships are dealt with in a manner relevant to undergraduate educa-

tion; these discussions aid greatly in understanding drug design and drug action.

To clarify the selection of certain organic compounds as pharmaceuticals and the importance of physical properties, we include a chapter on Physicochemical Properties in Relation to Biological Action. Usually, the chemical properties of a compound are responsible for the method of detoxification in the body. This area is covered in a chapter on Metabolic Changes of Drugs and Related Compounds.

A pharmacist serves primarily as an expert on drugs to the patient, the physician, and all other health professionals. This was true in the past and, with advances in the knowledge of drugs, is even more true today. The product-oriented member of the health team is the pharmacist. In this respect there is no competition with any other group of practitioners in the health professions: physicians, dentists, nurses, medical technicians, or medical social workers. In fact, these health professionals expect the pharmacist to be the expert in this area. The pharmacist's knowledge encompasses drug reactions of all types, and the presentation in this textbook forms the basis for understanding much of the information on drugs used in the daily practice of pharmacy.

We have made extensive use of tables in an effort to focus information upon groups of therapeutic agents for better retention of knowledge. Tables are intended as a guide to names, dose forms, dosages, applications, and categories. In all cases the *United States Pharmacopeia Dispensatory Information (U.S.P. D.I.)* should be consulted for complete dosage information. Separate appendices are included for many of the pharmaceutic aids and necessities. A comprehensive listing of pka's of drug molecules is included in an appendix.

The authors of this textbook have based their discussions primarily on drugs that are in the present *U.S.P./N.F.* In addition, drugs that are important from a historical standpoint, as well as promising new leads from the current literature, have been included.

This edition reflects in the title the outstanding teaching tool that Professors Wilson and Gisvold admirably pioneered. Since the second edition, I have been a contributing author; beginning with the fifth edition, I shared the editorship with Professors Wilson and Gisvold.

With the death of Professor Gisvold and the retirement of Professor Wilson from active academic life, I will serve as the sole editor and will attempt, along with the contributing authors, to continue the tradition of excellence established by these two pharmaceutical educators.

Robert F. Doerge, Ph.D.

1

introduction

Robert F. Doerge

In the development of organic therapeutic agents, pharmaceutical scientists have explored numerous approaches to finding and developing organic compounds that are now available to us in dosage forms suitable for the treatment of our ills and often for the maintenance of our health.

Pure organic compounds, natural or synthetic, are the chief source of agents for the cure, the mitigation or the prevention of disease today. These remedial agents have had their origin in a number of ways, (1) from naturally occurring materials of both plant and animal origin, and (2) from the synthesis of organic compounds whose structures are closely related to those of naturally occurring compounds (e.g., morphine, atropine, steroids and cocaine) that have been shown to possess useful medicinal properties. Although these first two approaches have led to the development of a great many of our useful medicinal agents, a third approach (3), that of pure synthesis, has provided significant discoveries of medicinal agents; i.e., historically, Ehrlich's outstanding synthetic efforts to develop antiprotozoal drugs which yielded the useful organoarsenicals and various antimicrobial dyes; the development of the active sulfanilamide as a study of the metabolic products of the azo dye Prontosil; the discovery of the diuretic and antidiabetic properties of certain analogs of sulfanilamide during a study of their biological properties other than antimicrobial; the discovery of the outstanding analgesic properties of Demerol® as an observation in connection with its biological testing as an antispasmodic agent. This discovery gave an outstanding lead to the development of other important analgesics. Other routes together with those found by serendipity could be cited. It so often happens that it is the keen observation of a research scientist who has had a broad pharmaceutical training that identifies research responses as having significant medicinal application.

Many of the compounds from the first group are prepared synthetically today as a financially expedient measure. Examples of these are most of the vitamins, some sex hormones, corticometric principles, methyl salicylate, amino acids, camphor and menthol. On the other hand, cardiac glycosides, quinine, atropine, antibiotics and insulin either cannot be synthesized or can be isolated from natural sources at a cost that can compete with synthetic methods. Examples of compounds found in the second group are the numerous sympathomimetic drugs and local anesthetics, antispasmodics, mydriatic and myotic drugs and recently the prostaglandins. Examples of the third group include the synthetic antimalarials, dyes, some analgesics, diuretics, phenols, antihistamines, barbiturates and surface-active agents.

Even though isolation and synthesis of many active constituents from animal and plant sources have been accomplished, there are new problems yet to be solved. These accomplishments, together with others, will add to the present large number of useful organic medicinal compounds and bring about a more nearly complete complement of agents for therapeutic use.

In many cases, at present, there is no simple and direct correlation between the activity of organic compounds and their chemical structure beyond the broad generalization that compounds similarly constituted may be expected to have similar activities. This is not always true, for often a fine shade of difference in chemical structure may lie between a very active compound on the one hand and a completely inactive one on the other or even one whose activity may be antagonistic to the original model. The last-mentioned compounds have received intensive research attention

1

in recent years and usually are referred to as metabolic antagonists. The very useful sulfa drugs were shown to be metabolic antagonists. These studies are useful to the biochemist and the physiological chemist, and it is hoped that new and useful medicinal agents also will be developed as a result of the extension of these studies. In other cases, each member of a whole series of compounds more or less related in structure to one another may have some activity. This can well be illustrated by sympathomimetic drugs, which include a series of compounds from the simple 2-aminoheptane to epinephrine. The fact that a series of compounds, the members of which are structurally related to one another, exhibits a similarity in activity does not preclude the possibility that some other compounds, unrelated structurally, can have similar activity. For example, anesthetic properties are present not only in the cocaine or procaine type of molecule but also in benzyl alcohol, quinine, Nupercaine, phenacaine, plasmochin and other compounds. Nevertheless, a convenient method for the study of organic medicinal agents according to a hybrid chemical classification is, in part, a desirable approach, because it allows the student to become familiar with the chemical, the physical and the biochemical properties of such groups. It is well to remember that the chemical, the physical and conformational and, now, the biochemical properties of organic compounds are functions of their structures. Therefore, much can be gained by studying medicinal agents from these points, noting the changes in activities that are effected by the changes in these factors.

Sometimes the activity of a drug is dependent chiefly on its physical and chemical properties, whereas in other instances the arrangement, the position and the size of the groups in a given molecule also are important and lead to a high degree of specificity. In the latter case, this high degree of specificity is usually associated with the mode of action of the drug involving enzymes or enzyme systems.

When the inhibiting activity or properties of a compound cannot rationally be explained on a structure-activity basis involving some limited active center on the perimeter of an enzyme molecule, other explanations have been offered. In some cases such an activity might be quite specific for a given enzyme and the assumption is proposed that there is a specific direct interaction between the enzyme and the compound in question. Such types of specificity might be expected when one considers the great number of primary, secondary and tertiary structures possible in the polypeptide chains of which enzymes are composed. These features contribute not only to the specificity of the "active center" but also significantly to the other important essential properties of such enzymes necessary

for their normal function. Non-"active-center" interactions could lead to the alteration of one or more of the features necessary for the activity of the enzyme in question. Some of such interactions may be described as allosteric in nature and still be quite specific in nature.

Intercalation interactions of certain drugs with enzymes usually also produce an alteration in enzyme activity that may be analogous to those produced by "allosterically" active drugs. The recognition of "receptive centers" or sites of physiologic action is also better understood and is contributing to our knowledge of structure-activity relationship.

As our knowledge of enzyme systems increases, the development of more nearly perfect organic drugs will be made possible. Furthermore, this increased knowledge will help us to explain the mode of action of a number of valuable organic drugs now in use.

In the case of the steroid hormones a tremendous number of modifications have been reported in the literature. These modifications were prepared in order to develop more potent steroids, particularly those that emphasize one activity at the expense of another—for example, the accentuation of the anti-inflammatory activity of the corticoids, the anabolic activity of the androgens, etc. An attempt has been made in this revision to include the important changes that have been made in the steroid structures and to correlate these changes with their effects upon the biological activities.

Many modifications of the purines, the pyrimidines, the nucleosides and the nucleotides have been reported in the literature in the quest for antimetabolites that might prove to be useful in the treatment of cancer. Very few of these antimetabolites have survived clinical trial, and even these are not curative agents. No attempt has been made to discuss the changes in the structures of this group of metabolites that have proved to be effective in the production of antimetabolites. Suffice it to say that not only has success been encountered with the preparation of reversible competitive antimetabolites but some of these modified compounds are actually incorporated in vivo into nucleic acids. The latter phenomenon is called lethal synthesis and probably would preclude the use of such antimetabolites as medicinal agents.

The chemical properties frequently determine the locale of absorption. For example, weakly acidic, feebly basic and neutral drugs are absorbed from both the stomach and the intestines. Basic drugs of the order $pKa = 9$ are not absorbed significantly from the stomach but are well absorbed from the intestines provided that the other factors are favorable. On the other hand, strong bases and strong acids are poorly absorbed, if at all, from the gastrointestinal tract. Although strong bases such as streptomycin, curare, etc., are not absorbed orally, many such drugs are absorbed when they are administered by injection.

Among many other factors that no doubt influence the absorption and the distribution of drugs, particle size can play a significant role in the rate of absorption of drugs from the site of administration. In the case of medicinal agents that have a high solubility in both water and lipids, particle size is much less significant than it is in those cases in which the partition coefficient is very high. Between these extremes an intermediate situation exists.

Advancement in the synthesis of polypeptides has proceeded at such a rapid pace that some very low molecular weight polypeptide hormones such as oxytocin, vasopressin, etc., have been synthesized. In addition, some analogs of these polypeptides also have been synthesized in order to delineate the contribution to activity of some of the amino acid residues. The amino acid sequence also has been determined in the complex polypeptide chains A and B of insulin, corticotropin, glucagon, etc. In some cases it has been shown that the biological activity of these protein hormones is not a function of the molecule as a whole; rather, the activity is due to a portion of the molecule (active center). In these instances fragments of the parent polypeptide chain are active. This poses the possibility that these fragments may be synthesized and analogous synthetic studies carried out as with oxytocin, vasopressin, etc.

In order to appreciate and understand organic medicinal products more fully, the student should be well grounded in such fields as organic, physical, biological and physiologic chemistry, bacteriology, physiology and zoology.

It is not possible in a textbook of reasonable size to include a complete discussion of all subjects mentioned. Furthermore, the inclusion of commercially available forms of the medicinal compounds per se or in combination is not considered because of their usually nonfundamental nature. The material presented here should provide a basis for a more detailed study of the scientific literature. The student should use the references and selected readings for this purpose.

2

physicochemical properties in relation to biological action

T. C. Daniels

E. C. Jorgensen*

During the past century, a period which completely encompasses the era of development and growth of the field of synthetic drugs, much consideration has been given to the possible relationships existing between chemical constitution, physical properties and biological action. During the 19th century, medicinally useful agents were developed most often by isolation from the natural sources known and tested by folk medicine. As the science of chemistry developed, chemical structures were elucidated for these natural products, and it was observed that similar structural units were sometimes present in compounds possessing related biological activity. For example, in 1869, Crum-Brown and Fraser[1] observed that tertiary amines with varying pharmacologic properties tended to show similar pharmacologic properties when quaternized.

Considerable research in recent years has been focused on the development of general principles which would guide the rational development of useful new agents with select and discrete physiologic properties. Early attempts to relate a particular single functional group to a specific biological response were largely unsuccessful.[2] More recently, many pharmacologic classes of drugs have been recognized as being describable as collections of functional groups, arranged in particular patterns in space. Compounds with similar pharmacologic properties may be made up of similar regions of high or low electron densities, or of charged or polar groups, separated by specific distances from nonpolar or hydrophobic residues, such as aromatic rings. Examples where such a generalized structural formula may be used to describe a class of drugs are found in the adrenergic agents (Chap. 11), cholinergic

agents (Chap. 12) and narcotic analgesic agents (Chap. 17).

There are also many compounds unrelated chemically which show the same general pharmacologic properties. As examples may be mentioned the relatively large number of different chemical classes of compounds that have local anesthetic properties or may serve as hypnotics or as antipyretic analgesics. In some cases, such an apparently heterogeneous pharmacologic class may be described in terms of a set of physicochemical properties, such as a limited range of partition coefficients between lipoidal and aqueous phases. In other cases, failure to find a simple relationship between chemical structure, physical properties and biological action may be accounted for in terms of the complex nature of biological systems.

Many competing events take place between the introduction of a drug and its final interaction with a specific receptor or organized tissues in which the desired response is to be initiated. Structural features which contribute the proper physical properties must be present to shepherd the drug through these devious pathways, and an adequate number of molecules must survive the passage to bring about a significant reaction with the receptor, or to disrupt the order within organized tissues. In order to exert their biological effects, drugs must be soluble in and transported by the body fluids, pass various membrane barriers, escape excessive distribution into inert body depots, endure metabolic attack, penetrate to the sites of action and there orient and interact in a specific fashion, causing the alteration of function termed the action of the drug.

The physicochemical properties of a compound are measurable characteristics by which the compound may interact with other systems. Biological response

* Deceased

to a drug is a consequence of the interaction of that drug with the living system, causing some change in the biological processes present before the drug was administered. Since physicochemical properties determine the processes by which drugs reach and interact with their sites of action, it is important to examine the extent to which any one property correlates with the observed biological activity. The possible importance of such properties as solubility, partition coefficients, surface activity, degree of dissociation at the pH of the body fluids, interatomic distances between functional groups, redox-potentials (reduction-oxidation), hydrogen bonding, dimensional factors, chelation, and the spatial configuration of the molecule are worthy of consideration. Moreover, the physicochemical properties of the cellular components with which the drug interacts are of interest. The physical properties of drugs can be varied through chemical modification more or less at will, but for the most part the properties of the cellular constituents remain constant.

COMPLEX OF EVENTS BETWEEN DRUG ADMINISTRATION AND DRUG ACTION

Following introduction into the body, a drug must pass many barriers and survive competing alternate pathways before it finally reaches the site of action (cell receptors) where the useful biological response is developed. As a basic requirement, in order to reach cellular sites of action, all biologically active substances must have or acquire by binding to transport protein, or by chemical or enzymatic modification, some minimum solubility in the polar extracellular fluids. Before consideration of factors and forces important to the interaction of the drug and the receptor, physicochemical properties affecting the absorption, the distribution and the metabolism of a drug

should be considered. A simplified diagram of this system is presented in Figure 2-1.

INFLUENCE OF ROUTE OF ADMINISTRATION

Parenteral administration of a drug involves no absorption complications, if carried out by the intravenous, the intra-arterial, the intraspinal or the intracerebral routes, for these place the drug directly into the body fluids. However, the subcutaneous, the intramuscular, the intradermal, the intraperitoneal, etc., routes produce a depot from which the drug must reach the blood or the lymph in order to produce systemic effects. Factors of importance in determining the rate at which this takes place are those which determine the dissolution rate of the drug and its transfer from one phase to another.[3] These factors are similar to those which have been studied in far greater detail for absorption of a drug from the gastrointestinal tract.

Following oral administration of a drug, dissolution rate is of primary importance in determining eventual levels attained in the blood and the tissues.[4,5] If the drug is too insoluble in the environment of the gastrointestinal tract to dissolve at an appreciable rate, it cannot diffuse to the gastrointestinal wall and be absorbed. It will be excreted unchanged in the feces. Variation in particle size and surface area, coating, chemical modification, etc., producing differences in rates of dissolution, provide an important approach to "prolonged action" medication by slowing dissolution rates, and thus absorption.[6]

Following dissolution of a drug, its action in tissues outside of the gastrointestinal tract must be preceded by passage through the membranes separating the lumen of the stomach and the intestines from the mucosal blood supply. The same factors influencing the penetration of the gastrointestinal-plasma barrier are

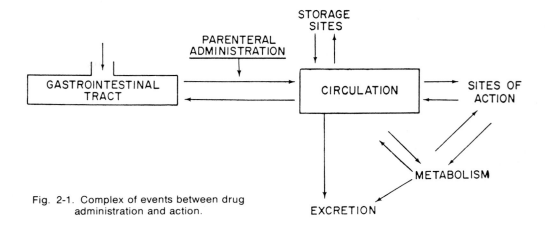

Fig. 2-1. Complex of events between drug administration and action.

important in the penetration of other membranes such as the blood-brain barrier and select cells and tissues.[7] The major differences are in variations in acidity for various body compartments: the high acidity of the stomach (pH 1 to 3.5), the far less acidic environment of the lumen of the intestine (duodenal contents, pH 5 to 7; duodenum to ileo-cecal valve, pH 6 to 7; lower ileum, pH about 8), and the essentially neutral environment of the circulating fluids of the body and of the tissues and organs supplied by these fluids (plasma, cerebrospinal fluid, pH 7.4).

Absorption from the gastrointestinal tract, as well as penetration of other membrane barriers, may be *passive* or *active*. Substances which are normal cellular metabolites, or close chemical relatives, may pass the membrane by a process of active absorption, energy being used by the body to effect the transfer of such normal food stuff as glucose and amino acids from the gastrointestinal tract to the plasma. The preferential cellular uptake of potassium ion over sodium ion is assumed to involve such a carrier system. Some lipid-insoluble substances penetrate cell membranes by a passive diffusion process, so that the cell wall is thought of as sievelike with a connection via the pores between aqueous phases on both sides of the membrane.[8] The rate of such passive diffusion through aqueous channels depends on the size of the pores, the molecular volume of the solute, and the differences in transmembrane concentrations. Penetration through pores has been shown to be important for absorption from the intestine only for those drugs with molecular weights less than 100.[9] Pores of other sizes must be present in other tissues; for example, the glomerulus of Bowman's capsule in the kidney is permeable to molecules smaller in size than albumin (molecular weight 70,000).

However, although there are specialized transport mechanisms for natural cell substances and diffusion through pores for small polar molecules, most organic compounds foreign to the body penetrate tissue cells as though the boundaries were lipid in nature, with passage across these barriers predictable from the lipid-solubilities of the molecules. This concept was advanced in 1901 by Overton[10] following a study of the lipid-solubility of organic compounds and the relationship between this property and ease of penetration of cells. The fatlike nature of cellular membranes has since stood the test of considerable experimentation.

The relative lipid/water-solubility of drugs (partition coefficient) has been shown to be an important physical property in governing the rate of passage through a variety of membrane barriers.[11] Examples include passage across the mucosal membranes of the oral cavity (*buccal* and *sublingual* absorption), the gastrointestinal membranes (stomach, small intestine,

colon), through the skin, across the renal tubule epithelium, into the bile, the central nervous system and tissue cells. Most drugs are weak acids or bases, and the degree of their ionization, as determined by the dissociation constant (pKa) of the drug and pH of the environment, influences their lipid/water-solubilities. The nonionized molecule possesses the higher lipid-solubility and passes most membrane barriers more readily than does the ionized molecule. The highly charged nature of the lipoprotein making up the cell wall accounts for this difference in the ability of undissociated molecules and their ions to penetrate the cell. The electrostatic forces interacting between the ion and the cell wall serve to repel or bind the ion, thus decreasing cell penetration. Also, hydration of the ion results in a species larger in size than the undissociated molecule, and this may interfere with diffusion through pores.

For an understanding of the effect of the acidic or basic character of a drug on its passage through membranes separating compartments of differing pH, the Henderson-Hasselbach equation is useful. The dissociation constant (pKa) is the negative log of the acidic dissociation constant and is the preferred expression for both acids and bases.

for acids
$$R—COOH = R—COO^- + H^+$$
$$pKa = pH + \log \frac{\text{non-ionized acid}}{\text{ionized acid}}$$
$$= pH + \log \frac{[RCOOH]}{[RCOO^-]}$$

for bases
$$RNH_3^+ = R—NH_2 + H^+$$
$$pKa = pH + \log \frac{\text{ionized base}}{\text{non-ionized base}}$$
$$= pH + \log \frac{[RNH_3^+]}{[RNH_2]}$$

An acid with a small pKa (pKa = 1) placed in an environment with pH 7 would be almost completely ionized:

$$(R—COO^-/RCOOH = 10^6/1);$$

it would be classed as a strong acid. Weak acids and strong bases have a large pKa; weak bases and strong acids, a small pKa.

ABSORPTION FROM THE STOMACH

It was noted in 1940[12] that large doses of alkaloids were not toxic in the presence of a highly acidic gastric content. When the gastric contents were made alkaline, the animals rapidly died. This indicated that the gastric epithelium is selectively permeable to the undissociated alkaloidal bases, which produce their toxic effects upon absorption. Later, the observation by Shore[13] that certain parenterally administered weak bases concentrated in the gastric juice led to the development of a general pH—partition hypothesis explaining the rate and the extent of absorption of ionizable drugs from the gastrointestinal tract. A simplified example of this process is presented in Figure 2-2. When a lipid-soluble, moderately weak base such as an aromatic amine ($ArNH_2$; pKa = 4.0) is administered orally and passes into the strongly acidic environment of the stomach (pH = 1), the base will exist largely (1,000/1) in the poorly lipid-soluble ionic form ($ArNH_3^+$) and will be only slowly absorbed through the gastric epithelium. As a specific example, absorption of aniline (pKa = 4.6) from the rat stomach during a 1-hour exposure was measured as 6 percent of the administered dose.[14] Weaker bases (pKa <2.5), such as acetanilide (pKa = 0.3), caffeine (pKa = 0.8) and antipyrine (pKa = 1.4), are absorbed better (36 to 14%), since they are significantly nonionized even in this strongly acidic environment. By decreasing the acidity of the rat stomach to pH 8 with sodium

bicarbonate, even aniline is absorbed to the extent of 56 percent. This approaches a maximum value under the experimental conditions used, limited by the rate of blood flow in the gastric mucosa. These results are consistent with the concept of the gastric mucosa being selectively permeable to the lipid-soluble undissociated form of drugs.

Weak acids (pKa 2.5 to 10) exist largely in the nonionized form in the stomach and are well absorbed. For example, salicylic acid (pKa = 3.0) is 61 percent absorbed, benzoic acid (pKa = 4.2) is 55 percent absorbed from the rat stomach following 1-hour exposure. That lipid-solubility is the physical property governing the passage of uncharged molecules across membrane barriers is supported by the observation that 3 barbiturates with similar pKa values were absorbed at rates proportional to the lipid/water partition coefficients (K = chloroform/water) of the nonionized forms.[15] Thiopental (pKa = 7.6; K = >100) was absorbed very rapidly, the less lipid-soluble secobarbital (pKa = 7.9; K = 23.3) less rapidly, and barbital with its poor lipid-solubility (pKa = 7.8; K = 0.7) was absorbed very slowly. The same pattern of absorption has been observed in man.

Substances completely ionized, and therefore

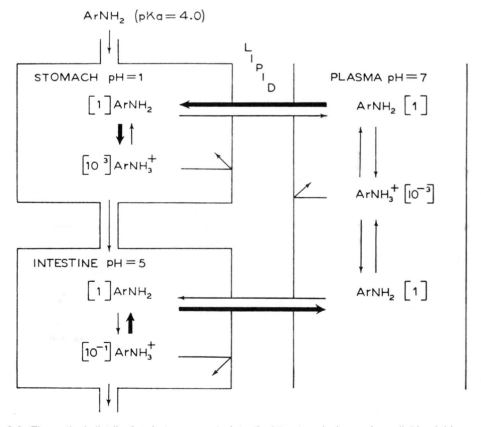

Fig. 2-2. Theoretical distribution between gastrointestinal tract and plasma for a lipid-soluble aromatic amine with pKa = 4.0. Data from Brodie and Hogben.[14]

poorly lipid-soluble, at the pH of the stomach (or the intestines), such as the strongly acidic sulfonic acids ($R—SO_3H$) and completely ionized quaternary ammonium compounds (R_4N^+), are not well absorbed.

The stomach also serves as a "site of loss" for weak bases administered intravenously. Since these exist largely in the nonionized form in the blood (Fig. 2-2), they penetrate cellular membranes readily, including the lipid barrier between the mucosal blood supply and the stomach, where they may be trapped as the ions. This site of loss has been confirmed for many basic drugs,[13] since they have been found concentrated in gastric juice following intravenous administration.

ABSORPTION FROM THE INTESTINES

Due to its large surface area, the small intestine is the major site of absorption for most drugs. As is true for the stomach, the nonionized form of a drug is absorbed from the intestine faster than the ionized form, although ions appear to be appreciably absorbed as well.[16] When weak bases pass from the strongly acidic environment of the stomach into the less-acidic intestinal lumen, the extent of ionization decreases as shown in Figure 2-2. The concentration of nonionized species for a base with pKa = 4.0, is about 10 times that of the ionized species, and, since the neutral molecule freely diffuses through the intestinal mucosa, the drug is well absorbed.

When aniline (pKa = 4.6) was perfused through the small intestine of the rat over a wide concentration range, and at the fairly rapid rate of 1.5 ml. per minute, 53 to 59 percent of the administered dose was absorbed, even though the time of drug contact was only about 7 minutes. Although the measured pH of the content of the lumen of the intestine is 6.5, absorption studies are more consistent with a "virtual" pH of 5.3 for the absorbing surface of the intestinal mucosa.[17] This is derived principally from the observation that a drastic reduction in the extent of absorption from the intestine occurred for acids with a pKa <2.5 (strong acids), and for bases with a pKa >8.5 (strong bases). The rat colon has been shown[18] to follow the same pattern as the intestine, with lipid-solubility being the primary physical property governing the rate of drug absorption.

Studies of physicochemical factors related to absorption following oral administration of quaternary ammonium ions illustrate additional complexities in the problem of absorption from the gastrointestinal tract. Some appear to be unabsorbed, e.g., pyrvinium pamoate (Povan) and dithiazanine iodide, anthelmintic drugs whose lack of absorption by the host pre-

vents undesirable systemic effects and preserves high gastrointestinal concentrations for toxic effects against intestinal parasites. Others, in spite of their permanent ionic character, cross the intestinal epithelium but at a very slow rate as compared with most uncharged molecules. This rate falls with time following administration, suggesting the formation of nonabsorbable complexes with the charged carboxyl and sulfonic acid residues of intestinal mucosa. Mucin added to the intestinal loop has been shown to decrease the rate of absorption of quaternary ammonium compounds.[19]

When the relatively inactive trimethylene-bis (trimethylammonium) dichloride was administered orally

$$(CH_3)_3 \overset{+}{N} - (CH_2)_3 - \overset{+}{N}(CH_3)_3 \cdot 2\,Cl^-$$

Trimethylene-bis(trimethylammonium) Dichloride

IN 292

together with an active hypotensive bis-quaternary compound, IN 292, a marked enhancement of effect has been shown, although none was shown following concomitant intravenous administration.[20] It is postulated that the inactive quaternary ammonium competes with the active quaternary for mucosal binding sites, allowing enhanced absorption of the active molecule.

ABSORPTION OF DRUGS INTO THE EYE

When a drug is applied topically to the conjunctival sac, a portion will pass directly through the conjunctiva membrane into the blood and the remainder passes through the cornea at rates dependent on the degree of ionization and the partition coefficient of the drug.[21] Since the nonionized molecule possesses the higher lipid-solubility, weak acids penetrate more rapidly from solutions having a low pH and weak bases from solutions buffered at high pH values.

Drugs may pass from the bloodstream into the ocular fluid by two general routes: (1) through the epithelium of the ciliary body, and (2) through the capillary walls and connective tissue of the iris. The rate of absorption into the aqueous humor appears to parallel closely the partition coefficient of the drug.

SITES OF LOSS

Relatively few drug molecules will survive to reach the site of action in a complex biological system. The sites to which a drug may be lost may be reversible storage depots, or enzyme systems which produce metabolic alteration to a more or less active form, or the drug may be excreted before or after metabolism (Fig. 2-1). All of these ways in which the drug may be lost rather than react at the normal site have been collectively described as "sites of loss." The distribution between the sites of loss and of action is largely dependent on the physicochemical properties of the drug, including solubility, degree of ionization, and the nature and the strength of the forces binding the drug at these sites.

Storage sites

Body compartments exist in a variety of types, each characterized by the nature of the physicochemical factors which retain the drug in competition with other sites. The gastrointestinal tract has already been mentioned as such a site, retaining molecules which lack adequate appropriate lipid/water partition characteristics, small size or special transport systems.

Protein binding

Binding of drugs by plasma protein is usually readily reversible, with most drugs bound to proteins in the albumin fraction.[14] Binding resembles salt formation, for generally the ionic form of a drug interacts with the charged residues of plasma protein, aided by secondary binding by nonionic polar and nonpolar portions of the molecule. The latter forces may be adequate alone, for drugs which are not electrolytes are also protein bound, e.g., hydrocortisone. The resulting protein binding may act as a transport system for the drug, which, while bound, is hindered in its access to the sites of metabolism, action and excretion. The drug-protein complex is too large to pass through the renal glomerular membranes and therefore remains in the circulating blood, thereby prolonging the duration of action. Protein binding not only may prevent rapid excretion of the drug; also, it limits the amount of free drug available for metabolism and for interaction with specific receptor sites. For example, the trypanocide, suramin, remains in the body in the protein-bound form for several months following a single intravenous injection. Its slow dissociation releases enough free drug for protection against sleeping sickness.

Protein binding may also limit access to certain body compartments. The placenta is able to block passage of proteins from the maternal to the fetal circulation. Hormones, such as thyroxine, which are firmly bound to maternal transport proteins and which are not required or desirable in the developing fetus before the appearance of a functional fetal endocrine gland, are thus excluded by this placental barrier.[22]

There exists a high degree of structural specificity for the interaction between plasma proteins and many small molecules. Drug binding by protein is generally more dependent on detailed chemical structure than are the competing events of drug absorption and localization in lipoidal tissues, which are most dependent on partition character between polar and nonpolar solvents. For example, specific structural requirements for binding of thyroxine analogs to a thyroxine-binding fraction of serum albumin have been established.[23] For maximal binding, the molecular features of thyroxine are most favorable: a diphenyl ether nucleus, 4 iodine atoms, a free phenolic hydroxyl group and an alanine side chain, or an anionic group separated by 3 carbon atoms from the aromatic nucleus. The resulting association constant of 500,000 is significantly reduced if any of these structural features are altered (see Table 2-1). Compounds possessing single aromatic rings have very low association constants ranging from 8 to 11,000, the strongest of these being salicylic acid and 2,4-dinitrophenol.

In general, structural requirements for plasma protein binding are related, but not as specific as those for the biological receptor. Thus, salicylate may give some thyroxinelike effects in high doses by displacing thyroxine from serum protein, but there is no indication that salicylate will elicit a thyroxinelike effect at the cell receptor.

Other drugs may exert an indirect biological effect by displacing active substances from protein binding. Thus, the sulfonylurea antidiabetic agents are thought to exert some activity by displacing insulin from its complex with plasma protein, as well as by releasing insulin from pancreatic β-cells.

The protein-bound anticoagulants dicumarol and warfarin (Coumadin) are displaced from protein by many other drugs, including phenylbutazone (Butazolidin), clofibrate (Atromid-S), norethandrolone (Nilevar), sulfonamides, etc. This displacement of the anticoagulants potentiates their action by increasing the amount of free drug available for competitive inhibition of vitamin K in the clotting process. The resulting increase in clotting time may lead to hemorrhaging. Displacement of a protein-bound drug by administration of another drug is more generally a therapeutic hazard than has been commonly recognized.

Tissue proteins or related tissue constituents may also bind drugs, thus providing depots outside of the

TABLE 2-1

RELATIONSHIP BETWEEN STRUCTURE OF THYROXINE ANALOGS AND BINDING BY THYROXINE-BINDING ALBUMIN

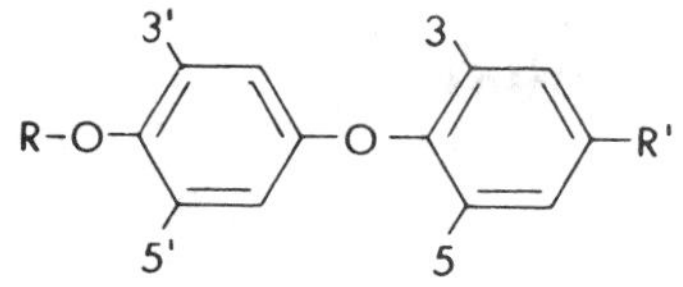

	Substituents			Association
R	3′,5′	3,5	R′	Constant
H	I_2	I_2	$CH_2CH(NH_2)COOH$	500,000
CH_3	I_2	I_2	$CH_2CH(NH_2)COOH$	20,000
H	I,H	I_2	$CH_2CH(NH_2)COOH$	24,600
H	I_2	I_2	CH_2CH_2COOH	160,000
H	I_2	I_2	CH_2COOH	100,000
H	I_2	I_2	COOH	72,000
H	I_2	I_2	$CH_2CH_2NH_2$	32,000
H	Cl_2	Cl_2	$CH_2CH(NH_2)COOH$	23,400
H	$(NO_2)_2$	$(NO_2)_2$	$CH_2CH(NH_2)COOH$	6,600
H	H_2	I_2	$CH_2CH(NH_2)COOH$	6,400
H	H_2	I, H	$CH_2CH(NH_2)COOH$	5,060
H	H_2	H_2	$CH_2CH(NH_2)COOH$	660

Data from Sterling.[23]

plasma. For example, the antimalarial drug quinacrine (Atabrine) shows a 2,000-fold concentration in liver over plasma 4 hours after administration; in 14 days of daily administration, the concentration in liver is 20,000 times that in plasma. Similar concentration of drug occurs in other body tissues such as lung, spleen and muscle.

Neutral fat

Neutral fat constitutes some 20 to 50 percent of body weight and as such makes up a depot of considerable importance. Drugs with high partition coefficients (lipid/water) are concentrated in these inert depots. Physical solution in lipid has been suggested[24] as the principal reason for the rapid disappearance of ultra-short-acting barbiturates from the plasma (see Chap. 9). Thiopental, a thiobarbiturate with pKa = 7.6, is approximately 50 percent ionized in the plasma (pH = 7.4). However, the high lipid-solubility for the undissociated molecule as compared with its oxygen analog causes this to partition rapidly into neutral fat, thus decreasing blood levels below those adequate for the maintenance of anesthesia. Some N-methyl barbituric acid analogs (e.g., hexobarbital) also have a short duration of anesthetic activity. This has been attributed to the influence of the N-methyl group on acid strength; these have a pKa of about 8.4 compared with 7.6 for those without the N-methyl substituent. At physiologic pH hexobarbital exists largely in the undissociated form which would distribute rapidly into the lipid depots.

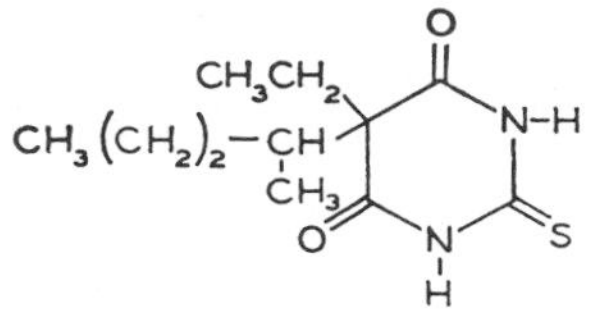

Thiopental pKa = 7.6

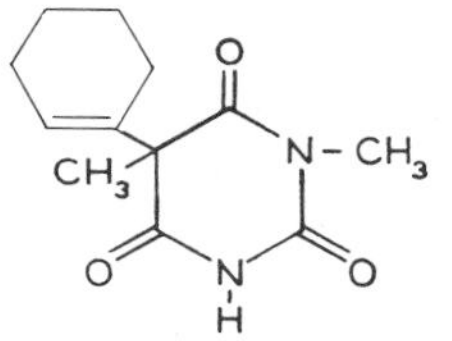

Hexobarbital pKa = 8.4

An alternate explanation has been proposed,[25] since the blood supply to neutral fat is too poor to account for the initial depletion from the plasma which takes place within a few minutes. The lean body tissues such as viscera and muscle are well perfused with blood, and their cells possess the lipid-permeable membrane which allows these to serve as the initial depot. Redistribution to body fat, metabolism and excretion appear to occur more slowly, and these are related to prolongation of depression rather than rapid recovery from anesthesia.

Lipid accumulation has also been implicated in the long duration of action of the adrenergic blocking agents (see Chap. 11), dibenamine and dibenzyline.[26,27]

Some 20 percent of the initial dose is rapidly deposited in fat, followed by a slow return to the bloodstream. The strength of the covalent bond formed at the active sites with these alkylating agents also appears to play a role in their long duration of action.[28]

As an example of a different type relating to either neutral fat depots or "lipophilic receptor sites," the neuromuscular blocking agent hexafluorenium (Mylaxen) is relatively ineffective in the unanesthetized animal.[29] However, in the presence of anesthesia with cyclopropane, ether and related lipophilic anesthetics, potent blockade of muscular function occurs. It was assumed that the lipophilic fluorene groups were absorbed by biologically inert sites of loss which are lipid in nature, and that concomitant administration of a nonpolar anesthetic would result in greater saturation of these lipophilic receptors. This would result in a decrease in available sites of loss with a resulting increase in the hexafluorenium available at the neuromuscular junction. As a test for this hypothesis, relatively inert compounds more closely related in structure to the lipophilic portion of the drug were administered: dibenzylamine, 9-dimethylaminofluorene and its quaternary derivative. All caused significant potentiation of the hexafluorenium effect in the order:

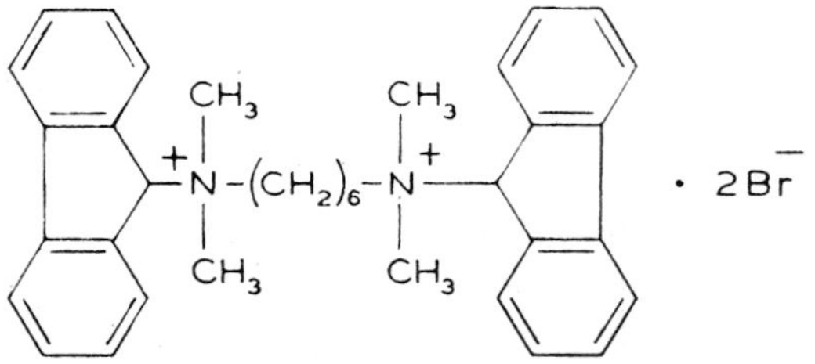

Hexafluorenium

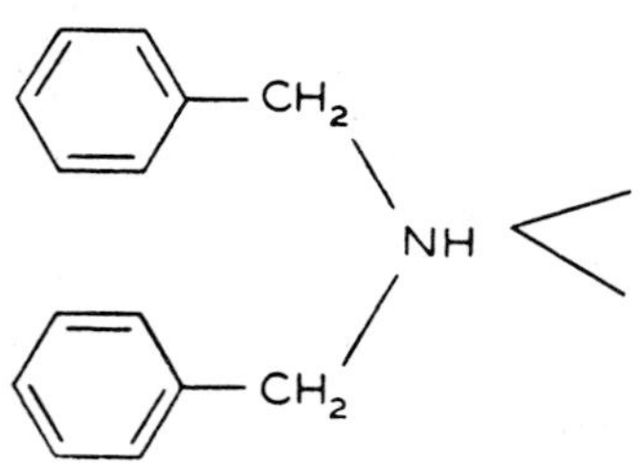

Dibenzylamine

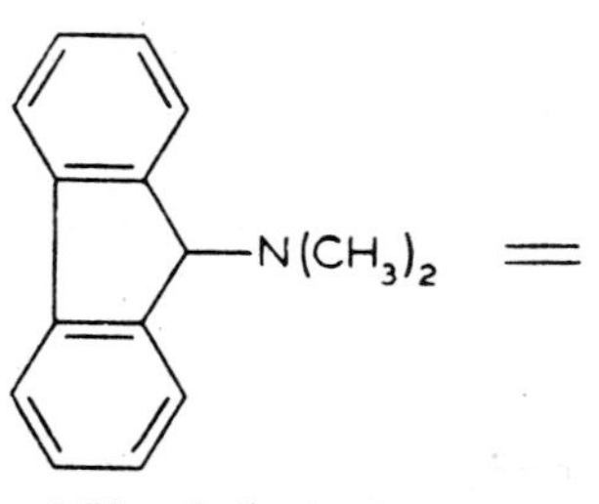

9-Dimethylaminofluorene

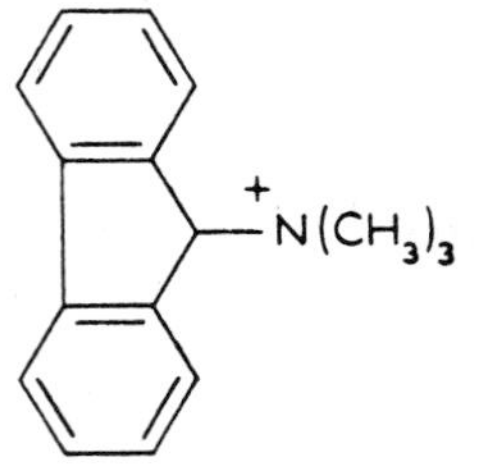

Quaternary Ammonium Derivative

The coplanarity of the fluorene rings could also permit enhanced binding by van der Waals' forces and could account for an increased effect over dibenzylamine. The equal effect for the tertiary and the quaternary bases demonstrates that the lipid portion, rather than the ionic, is involved.

This property of synergistic activity for biologically inert substances, involving the blockage of sites of loss and conservation of drugs, has further examples at other sites.

Metabolism and excretion

The nature of the processes involved in drug metabolism and excretion is discussed in detail in Chapter 3. Excretion, either of the unaltered drug or its metabolites, is an irreversible site of loss. However, metabolic alteration of the drug may lead to a metabolite with enhanced, reduced or essentially unchanged biological activity.

One of the major routes of excretion is by way of the kidney, which implies the presence or the formation of a water-soluble substance. Following glomerular filtration, tubular reabsorption into plasma is virtually complete for substances with a high partition coefficient (lipid/water). Since most active drugs (by virtue of their ability to penetrate lipid cellular membranes) are lipid-soluble, metabolic conversion, usually in the liver, to a more polar form is essential for their excretion. Presumably, a lipid membrane surrounds the liver microsomes in which are found the nonspecific enzyme systems responsible for most metabolic conversions. This membrane is readily penetrated by the lipophilic drug, and metabolism to a more polar form results, followed by increased excretion during the next passage through the kidney.

The potentiation of the action of a wide variety of drugs, such as analgesics, central nervous system stimulants and depressants, etc., by the compound SKF 525 (β-diethylaminoethyl 2,2-diphenylvalerate) has been accounted for on the basis of its inhibition of many metabolic reactions.[30] This is another example of a synergistic effect by blocking a site of loss.

SITES OF ACTION

After a drug reaches the bloodstream, and a portion of it survives distribution to sites of loss, other cell boundaries must be crossed before it reaches its site of action.

Penetration of drugs into tissue cells

The capillary wall is sufficiently porous to permit the passage of water-soluble molecules of relatively large size. The boundaries of organ tissue cells present a barrier of lipid character to the passage of foreign substances. A completely ionized molecule such as hexamethonium does not enter tissue cells,[31] but lipid-soluble, nonionized molecules possessing high partition coefficients readily penetrate a variety of cells and tissues. Thus, in dogs, phenobarbital has been shown[32] to be increased in concentration in body tissues (brain, fat, liver and muscle) when the plasma pH was lowered. Presumably, the increased concentration of undissociated molecules facilitated cell membrane penetration, providing a shift of drug from extracellular to intracellular fluids.

A wide variation occurs in the rate at which various drugs penetrate cerebrospinal fluid and the brain. Here, too, lipid-solubility of the nonionized molecule is the physical property largely governing the rate of entry.[33] The dissociation constant of a weak acid or base is of importance insofar as this determines the concentration of lipid-soluble undissociated drug in the plasma. Alteration of the plasma pH has been shown to produce the expected increase or decrease in penetration of cerebrospinal fluid by weak acids and bases.[33,34] Changes which produce a higher concentration of undissociated molecules lead to increased penetration; a higher concentration of ions leads to decreased penetration. Sulfonic acids and quaternary ammonium compounds do not penetrate the cerebrospinal fluid in any significant amount.

SKF 525

Some ionic species, such as the amino acids, use special transport systems to cross membrane barriers. Levodopa (Chap. 20), the 3-hydroxy derivative of L-tyrosine, readily crosses the blood-brain barrier. Subsequent metabolic decarboxylation within the central nervous system produces the active anti-parkinson agent, dopamine. Dopamine, which exists largely in the protonated form at physiologic pH, is itself too polar to cross the lipidlike blood-brain barrier.

The same characteristic lipid barrier is found in a variety of other cells, including those of some bacteria. The toxic effects of certain chelating agents, such as 8-hydroxy-quinoline (oxine), are best explained in terms of bacteria cell penetration by the lipid-soluble saturated Fe (oxine)$_3$ species (see this chapter, chelating agents).

SOLUBILITY AND PARTITION COEFFICIENTS

The absolute and relative solubilities of drugs in aqueous and lipid phases of the body are physical properties of primary importance in providing and maintaining effective concentrations of drugs at their sites of action. The regular changes in biological activity which often occur within homologous series provide useful examples in understanding the correlation of solubility and partition properties with drug action.

BIOLOGICAL ACTIVITIES OF HOMOLOGOUS SERIES

In homologous series of undissociated or slightly dissociated compounds[35] in which the change in structure involves only an increase in the length of the carbon chain, gradations in the intensity of action have been observed for a number of unrelated pharmacologic groups of compounds, e.g., normal alcohols, alkyl resorcinols, alkyl hydrocupreines, alkyl phenols and cresols[36] (antibacterial), esters of *p*-aminobenzoic acid (anesthetics), alkyl 4,4′-stilbenediols[37] (estrogenic). Frequently, the lower members of a homologous series show a low order of biological activity; with the increasing length of the carbon chain (nonpolar portion of the molecule), the activity increases, passing through a maximum. Further increase in the length of the carbon chain results in a rapid decrease in the activity. The increase in activity roughly parallels the decrease in water-solubility and the increase in lipid-solubility (partition coefficient), which may be associated with the availability of the compound for the cell where the action occurs. The observed decrease in activity with further increase in length of the chain may be due to the diminishing solubility of the compounds in the extracellular fluid which serves as a medium of transport to the cell surface.

This has been illustrated graphically by Ferguson in Figure 2-3, in which is plotted the log of toxic concentration *v.* the log of solubility of the normal alcohols for two organisms. Also given on the graph is the "saturation line." A compound falling on this line would have to be present at the concentration of its saturated solution to show the bactericidal effect. If a line for a series crosses this saturation line, then the series will have a sharp cutoff of activity at the point of intersection as the series is ascended, because those compounds beyond the crossover point which would appear on the dotted line will not have enough solubility to give a bactericidal concentration. This neatly accounts for the observation made with a number of substances that the biological activity increases on ascending a series and then abruptly falls off in going to the next higher homolog. This cutoff point will depend on the resistance of the particular organism. The more resistant the organism, the higher the concentration necessary for killing and the earlier the cutoff will appear in the series.

The 4-*n*-alkylresorcinols also illustrate the relationship of biological activity in homologous series differing only in the length of the carbon chain. The phenol coefficients of 4-*n*-alkylresorcinols against *B. typhosus* are shown in Figure 2-4. Against this organism, a maximum is reached when the alkyl side chain has 6 carbons. Schaffer and Tilley[38] studied the same series of resorcinols and observed that the phenol coefficients for *Staph. aureus* continued to increase through the 4-*n*-nonylresorcinol, indicating a difference in sensitivity of different organisms to members of a homologous series of compounds.

The normal aliphatic alcohols show a regular increase in antibacterial activity as the homologous series is ascended from methyl through octyl alcohols.[38]

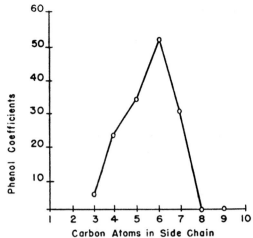

Fig. 2-4 Phenol coefficients of 4-*n*-alkylresorcinols against *B. typhosus.*

(See also Fig. 2-3.) The branched chain alcohols are more water-soluble and have lower partition coefficients than the corresponding primary alcohols; the branched chain alcohols are also less active as antibacterial agents. *n*-Hexyl alcohol is more than twice as active as the secondary hexyl alcohol and 5 times as active as the tertiary hexyl alcohol. The higher molecular weight alcohols (e.g., cetyl) are inactive as antibacterial agents; therefore, the alcohols have the same type of curve as the alkylresorcinols against *B. typhosus,* i.e., activity increasing with increase in the length of the carbon sidechain through a maximum.

Esters of a number of aromatic carboxylic acids have been studied, and it has been observed that as the carbon length of the alcohol is increased the antibacterial activity rapidly increases.[39] The phenol coefficients of the esters of vanillic acid are reported as follows: methyl, 1.7; ethyl, 7.3; *n*-propyl, 33.4. The isopropyl ester, which has a lower partition coefficient than the *n*-propyl ester, has a phenol coefficient of only 11.2.

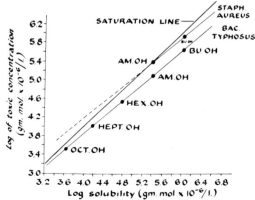

Fig. 2-3. Bactericidal concentration vs. solubility for normal primary alcohols. (Ferguson, J.: Proc. Roy. Soc. (Ser. B) 127:387, 1939)

TABLE 2-2

PHENOL AND PARTITION COEFFICIENTS OF ESTERS OF *p*-HYDROXYBENZOIC ACID

Ester	Inhibition of Fermentation	Staph. aureus Bactericidal	Partition Coefficient*
Methyl	3.7	2.6	1.2
Ethyl	5.3	7.1	3.4
n-Propyl	25.0	15.0	13.0
i-Propyl	15.0	13.0	7.3
n-Butyl	40.0	37.0	17.0
Amyl	53.0	. . .	150.0
Allyl	15.0	12.0	7.6
Benzyl	69.0	83.0	119.0
Phenol	1.0	1.0	. . .

* Lipid/water.

The partition coefficients and observed antibacterial activity in a series of esters of *p*-hydroxybenzoic acid are quite parallel, as shown in Table 2-2.[40]

PARTITION COEFFICIENTS AND GENERAL ANESTHESIA

The discussion of partition coefficients thus far has dealt largely with antibacterial agents. Historically, partition coefficients were first correlated with the biological activity of hypnotic and narcotic drugs by Overton[10] and Meyer.[41]

The theory of narcosis, as expressed by Meyer in 1899, may be summarized as follows: chemically unreactive compounds such as ethers, hydrocarbons, halogenated hydrocarbons, etc., exert a narcotic action on living tissue in proportion to their ability to concentrate in those cells, such as nerve cells, in which lipid substances predominate. Their efficiency as hypnotics or anesthetics is therefore dependent on the *partition coefficient* which determines the distribution of the compound between the aqueous phase and lipid phase of the tissue.

This concept that anesthesia is produced by the disruptive presence of substances in the lipid phase of cells has been supported in that excellent correlation between hypnotic activity and the partition coefficient has been observed[42,43] for many compounds. However, this correlation is not proof that the proposed mechanism is correct, since it relates only to the availability of the compound for the site of action.

Alternate theories of general anesthesia

All substances with high lipid/water partition coefficients are not effective as general anesthetics. A possible explanation may be related to molecular size. The recognition that the chemically unreactive rare gas xenon was capable of producing general anesthesia[44] led Wulf and Featherstone[45] to call attention to the relationship between some fundamental properties of molecules and their depressant effects. They pointed out that a correlation exists between the constants "a" and "b" in the van der Waals' equation (these terms measure the sphere of influence of a molecule) and the presence or the absence of anesthetic potency. In general, a critical "size" (van der Waals' "b," relating to molecular volume) was found necessary for the anesthetic molecule. This was larger than that for substances (such as H_2O, "b" = 3.05; O_2, "b" = 3.18; N_2, "b" = 3.91) which might normally occupy the lateral space separating lipid and protein molecules of the cell. Molecular volumes ("b" values) for some of the anesthetic agents are: N_2O, 4.4; Xe, 5.1; ethylene, 5.7; cyclopropane, 7.5; chloroform, 10.2; ethyl ether, 13.4. None had a value lower than 4.4. Wulf and Featherstone suggest that the anesthetic agents may occupy the space between lipid layers normally occupied by water, oxygen and nitrogen, causing a separation of these layers which would be dependent upon the molecular volume of the anesthetic. This alteration in cell structure could produce a depression of function leading to anesthesia.

Pauling[46] has proposed a theory of anesthesia which focuses attention on the aqueous phase, rather than the lipid phase of the central nervous system. The formation in the brain fluid of hydrate microcrystals, such as those known in vitro for chloroform, xenon and other anesthetic agents, is suggested. The anesthetic agents, together with side chains of proteins and other solutes in the encephalonic fluid, could occupy and stabilize by van der Waals' forces chambers made up of water molecules. The resulting microcrystalline hydrates could alter the conductivity of electrical impulses necessary for maintenance of mental alertness, leading to narcosis or anesthesia.

Most of the current theories of general anesthesia are based on positive correlations obtained between the partial pressures of agents required to produce anesthesia and such physical properties as solubility in oil, the distribution between oil and water, the vapor pressure ("thermodynamic activity") of the pure liquid (see Ferguson principle, below) or the partial pressure of hydrate crystals. All of these physical properties which correlate with anesthetic activity are related to the van der Waals' attraction of the molecules of anesthetic agent for other molecules, and all are interrelated, since the energy of intermolecular attraction is approximately proportional to the polarizability (mole refraction) of the molecules of anesthetic agent. As yet, no direct experimental evidence uniquely supports one theory of mode of anesthetic action at the molecular level.

Ferguson principle

The observation that many compounds containing diverse chemical groups show narcotic or anesthetic action is indicative that mainly physical rather than chemical properties are involved. The fact that narcotic action is attained rapidly and remains at the same level as long as a reservoir or critical concentration of the drug is maintained but quickly disappears when the supply of drug is removed suggests that an equilibrium exists between the external phase and the phase at the site of action in the organism designated the *biophase.*

In many homologous series the toxicity increases as the series is ascended. Fühner,[47] in 1904, found the decrease in concentration required for an equitoxic effect proceeded according to a geometric progression, 1, 1/3, $1/3^2$, $1/3^3$, . . ., as the number of carbon atoms increases arithmetically. This finding holds for a number of series of cellular depressants, including alcohols, ketones, amines, esters, urethanes and hydrocarbons. Certain, but not all, physical properties change according to a geometric progression in ascending a homologous series. These include vapor pressure, water-solubility, surface activity and distribution between immiscible phases. Since logarithms represent a geometric progression, a plot of the logarithms of the value of these various properties against the number of carbon atoms gives straight lines (Fig. 2-5). The attribute these physical properties have in common is that they involve a distribution between heterogeneous phases, e.g., solubility involves distribution between solid or liquid and saturated solution; surface activity the distribution between solution and surface; vapor pressure the distribution between liquid and vapor, etc. Toxicity, or cell depressant action resulting from such physical properties, also must involve such an equilibrium between the agent in the biophase and the agent in the extracellular fluids.

This logarithmic change in distribution coefficiency, which is the common denominator involved in each of these properties, results from the relation

$$\log k = (\bar{F}^\circ_2 - \bar{F}^\circ_1)/RT$$

in which the distribution coefficient k is a log function of the difference in the partial molal free energies $\bar{F}^\circ_1$ and $\bar{F}^\circ_2$ of the substance in its standard states in phases 1 and 2. In a homologous series, each additional CH_2 group gives rise to a constant increment in the difference between the partial molal free energies.

The fact that the biological effect parallels some physical property, such as the oil/water distribution ratio, is in itself not evidence that a particular mechanism is involved; for example, that narcosis takes place in a lipid medium. On the contrary, it may relate only to the fact that both the biological effect and the oil/water distribution ratio have in common a heterogeneous phase distribution.

Ferguson[48] advanced the concept that it is unnecessary to define the nature of the biophase, or receptor, nor is it necessary to measure the concentration at this site. If equilibrium conditions exist between the drug in the biophase and that in the extracellular fluids, although the concentration in each phase is different, the tendency for the drug to escape from each phase is the same. In such a system the partial molal free energy for the substance must be equal in each

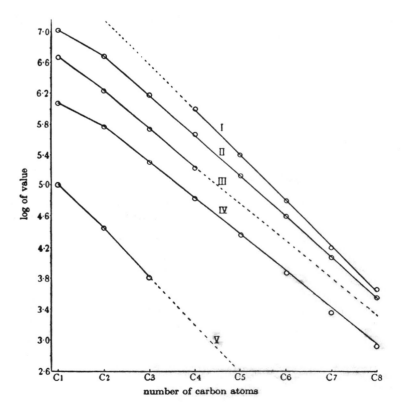

Fig. 2-5 Properties of normal primary alcohols. I. Solubility (mole $\times$ 10^{-6}/liter). II. Toxic concentration for *B. typhosus* (mole $\times$ 10^{-6}/liter). III. Concentrations reducing surface tension of water to 50 dynes/cm. (mole $\times$ 10^{-6}/liter). IV. Vapor pressure at 25° (mm. $\times 10^4$). V. Partition coefficient between water and cottonseed oil ($\times$ 10^3). (Ferguson, J.: Proc. Roy. Soc. (Ser. B) 127:387, 1939)

phase ($\bar{F}_1 = \bar{F}_2$), since this serves as a quantitative measure of the escaping tendency from that phase. The degree of saturation of each phase is a reasonable approximation of the tendency to escape from that phase, and this may be called the *thermodynamic activity*. Since this thermodynamic activity is the same in both the biophase and the extracellular phase, measurements made in the latter, which is accessible, may be directly equated with the former, which is inaccessible.

For a variety of depressant gases and vapors (Table 2-3), the isonarcotic concentrations varied from 100 to 0.5 percent by volume, but the ratio of the partial vapor pressure to saturation pressure, which gives the approximate thermodynamic activity, varied from only 0.01 to 0.07.

Similar data may be compiled, using the data for partial pressures of gases and vapors required to pro-

duce anesthesia in man. As shown in Table 2-4, the approximate thermodynamic activities (degree of saturation of the vapors) range between 0.01 and 0.05, while the concentration range was 200-fold.

Where the biological activity parallels thermodynamic activity, the compounds are said to be *structurally nonspecific*. In general, structurally nonspecific substances which are present in the same proportional saturation have the same thermodynamic activity and the same degree of biological action.

In contrast with those compounds, such as a variety of cellular depressants, which are structurally nonspecific and are characterized by a wide variety of chemical types giving a like biological response, dependent only on their thermodynamic activity, there are those compounds that are said to be *structurally specific*.

Compounds that are structurally specific usually

TABLE 2-3

ISONARCOTIC CONCENTRATIONS OF GASES AND VAPORS FOR MICE AT 37° C.*

Substance	Vapor Pressure (mm.) p_s	Narcotic Concentration % by Volume c	Partial Pressure (mm.) at Narcotic Concentration $(760xc/100) = p_t$	Approximate Thermodynamic Activity p_t/p_s
Nitrous oxide	59,300	100	760	0.01
Acetylene	51,700	65	494	0.01
Methyl ether	6,100	12	91	0.02
Methyl chloride	5,900	14	106	0.01
Ethylene oxide	1,900	5.8	44	0.02
Ethyl chloride	1,780	5.0	39	0.02
Diethyl ether	830	3.4	26	0.03
Methylal	630	2.8	21	0.03
Ethyl bromide	725	1.9	14	0.02
Dimethylacetal	288	1.9	14	0.05
Diethylformal	110	1.0	8	0.07
Dichlorethylene	450	0.95	7	0.02
Carbon disulfide	560	1.1	8	0.02
Chloroform	324	0.5	4	0.01

* Adapted from a table by Ferguson.[48]

TABLE 2-4

ISOANESTHETIC CONCENTRATION OF GASES AND VAPORS IN MAN AT 37°*

Substance	Vapor Pressure* mm. p_s	Anesthetic† Conc. in Vol. % c	Partial Pressure at Anesthetic Conc. $(760xc/100) = p_t$	Approximate Thermodynamic Activity p_t/p_s
Nitrous oxide	59,300	100	760	0.01
Ethylene	49,500	80	610	0.01
Acetylene	51,700	65	495	0.01
Ethyl chloride	1,780	5	38	0.02
Ethyl ether	830	5	38	0.05
Vinyl ether	760	4	30	0.04
Ethyl bromide	725	1.9	14	0.02
1,2-Dichloroethylene	450	0.95	7	0.02
Chloroform	324	0.5	4	0.01

* From data in Table 2-3 and the Handbook of Chemistry and Physics, Chemical Rubber Company.
† From data in Goodman, L. S., and Gilman, A.: The Pharmacological Basis of Therapeutics, New York, Macmillan, 1965.

are effective in lower concentrations than those that are nonspecific. However, equilibria are involved with the former as well as with the latter. This may involve equilibria between an external phase and the biophase, or equilibria between the drug and the receptors or the enzymes on or within the cell. The bonds involved may be any of the known types, including covalent, ionic, ion-dipole, dipole-dipole, hydrogen, van der Waals' and hydrophobic bonds. In cases of the structurally specific agents the bonds are likely to be stronger and the equilibrium shifted over to the side favoring maximum biological activity. The law of mass action and its equations are applicable to such situations. However, it should be realized that physical properties may be important in determining the action of both structurally specific and structurally nonspecific compounds.

Hansch quantitative structure-activity relationships

A quantitative measure of the importance of partition behavior on drug action has been introduced which has been equally applicable to both structurally specific and nonspecific drugs. Hansch[49] has determined the effect of substituent groups on distribution between water and the nonpolar solvent, 1-octanol. The distribution coefficients for the parent compound, e.g., phenoxyacetic acid ($C_6H_5OCH_2COOH$) and a derivative, e.g., 3-trifluoromethylphenoxyacetic acid (3-CF_3-$C_6H_4OCH_2COOH$) are measured, and a value, pi (π), for the substituent trifluoromethyl group is determined by the difference between the logarithm of the distribution coefficients:

$$\pi_{CF_3} = \log P_{CF_3} - \log P_H,$$

where P_{CF_3} is the partition coefficient of the 3-trifluoromethyl derivative, and P_H that of the unsubstituted parent compound. The π values, taken in conjunction with the Hammett sigma (σ) values[50] (measures of the electronic contributions of substituents relative to hydrogen) have been used effectively in correlating chemical structure, physical properties and biological activities.

Hansch has assumed that a rate-limiting condition for many biological responses involves the movement of the drug through a large number of cellular compartments made up of essentially aqueous or organic phases. The molecule possessing solubility and structural characteristics such that the sum of the free energy changes is minimal for the many partitionings made between phases, including adsorption-desorption steps at solid surfaces, will have ideal lipohydrophilic character and will most easily reach its site of action. The π value is a measure of the substituent's

TABLE 2-5

CONSTANTS FOR SOLUBILITY (π) AND ELECTRONIC (σ) EFFECTS OF 3-SUBSTITUENTS IN PHENOXYACETIC ACID*

R	π†	σ‡
n-C_4H_9	+1.90	−0.15
SCF_3	+1.58	+0.51
SF_5	+1.50	+0.68
n-C_3H_7	+1.43	−0.15
OCF_3	+1.21	+0.35
I	+1.15	+0.28
CF_3	+1.07	+0.55
C_2H_5	+0.97	−0.15
Br	+0.94	+0.23
SO_2CF_3	+0.93	+0.93
Cl	+0.76	+0.23
SCH_3	+0.62	−0.05
CH_3	+0.51	−0.17
OCH_3	+0.12	−0.27
NO_2	+0.11	+0.78
H	0	0
COOH	−0.15	+0.27
$COCH_3$	−0.28	+0.52
CN	−0.30	+0.63
OH	−0.49	−0.36
$NHCOCH_3$	−0.79	−0.02
SO_2CH_3	−1.26	+0.73

* Data from Hansch.[51,53,55]
† $\pi = \log P_x - \log P_H$, where P_x and P_H are the partition coefficients between 1-octanol and water.
‡ σ = Hammett sigma constant for 4-substituents.

contribution to solubility behavior in such a series of partitions.

Table 2-5 lists some typical substituent constants[51] arranged in order of decreasing contribution to lipophilic character when substituted in the 3 position of phenoxyacetic acid. Values of π and σ are approximately constant and additive in a variety of different aromatic systems, as long as no strong group interactions occur. Therefore, the substituent constants for a polysubstituted aromatic compound are approximately equal to the sum of the π and σ values for individual substituents. The additive character of these constants has been demonstrated by good correlations obtained from the action of polysubstituted phenols on gram-negative and gram-positive organisms, the action of thyroxine analogs on rodents and the carcinogenic activity of derivatives of dimethylaminoazobenzene and of aromatic hydrocarbons and benzacridines.[49]

A different set of π values has been obtained for substituents not attached to an aromatic nucleus.[52] In a homologous series, if functional groups are separated by two or more methylene (CH_2) groups, interaction is small and values may usually be determined additively. Both the methyl and the methylene (—CH₂—) groups have an additive π value of about $+0.50$; thus π values for a homologous series substituted in the 3 position of phenoxyacetic acid are: H = 0; CH₃ = 0.51; C₂H₅ = 0.97; n-C₃H₇ = 1.43; n-C₄H₉ = 1.90.

Relative to hydrogen = 0, a positive value for π means that the group enhances solubility in nonpolar solvents, a negative value that solubility in polar solvents is enhanced. A positive value for σ denotes an electron-attracting effect; a negative value denotes electron-donation by the group. Thus, the methyl group is typical of alkyl groups, in enhancing nonpolar solubility ($\pi = +0.51$) and is electron donating ($\sigma = -0.17$). By contrast, the acetamido group (CH₃CONH—) as a substituent strongly enhances water-solubility ($\pi = -0.79$) and is a weak electron acceptor ($\sigma = +0.10$).

Particularly noteworthy is the exceptionally strong lipophilic character of fluoro-substituted groups as compared with the hydrogen-substituted analog; e.g., CF₃ > CH₃; SCF₃ > SCH₃; OCF₃ > OCH₃; SO₂CF₃ > SO₂CH₃. The frequent enhancement of biological activity when a hydrogen, a methyl or a halogen is replaced by the trifluoromethyl group may be related to the significant contribution to lipophilic character.

In relating the application of the pi (π) and sigma (σ) substituent constants to biological activity, Hansch has derived the equation:[53]

$$\log (1/C) = -k\pi^2 + k'\pi + \rho\sigma + k''$$

where C is the concentration of drug necessary to produce the biological response (log A, the logarithm of relative biological activities is equally applicable), k, k' and k'' are constants for the system being studied, ρ (rho) is a reaction constant, π is the substituent constant for solubility contribution, and σ is the substituent constant for electronic contributions. In this form, contributions by steric factors are assumed to be constant as substituents are varied.

As an example of an application of these substituent constants, the relative antibacterial activities of chloromycetin (R = NO₂) and a series of its analogs have been compared,[53] in which the 4-nitro group has been varied in its nature.

When substituent constants, π and σ, and relative observed antibacterial activities are substituted in the equation log A = $-k\pi^2 + k'\pi + \rho\sigma + k''$, the system and the reaction constants which best fit the experimental data are: log A = $-0.54\pi^2 + 0.48\pi + 2.14\sigma + 0.22$. A comparison of observed antibacterial activities and those calculated from the derived equation (Table 2-6) shows excellent correlation. From these data, it is concluded that a strong electron-attracting group enhances activity ($\sigma_{NO_2} = +0.71$), as does a moderately lipophilic group ($\pi_{NO_2} = +0.06$). The great potential such correlations hold for directing the course of structure-activity studies is apparent.

Benzeneboronic acids (X—C₆H₄—B(OH)₂) are carriers of boron, which, if localized in tumor tissue, could be useful in the treatment of cancer. Radiation with neutrons would lead to neutron capture by boron and release local high concentrations of high-energy alpha radiation capable of destroying the tumor. The problem of structural factors leading to selective localization of compounds in tumor tissue has been evaluated

TABLE 2-6

ANALOGS OF CHLOROMYCETIN TESTED AGAINST STAPHYLOCOCCUS AUREUS*

R—⟨C₆H₄⟩—CH–CH–CH₂OH (with OH on first CH and NHCOCHCl₂ on second CH)

Substituent R	Electronic σ†	Solubility π	Log A‡ Calculated	Log A‡ Observed
NO₂	0.71	0.06	1.77	2.00
CN	0.68	−0.31	1.47	1.40
SO₂CH₃	0.65	−0.47	1.27	1.04
COOCH₃	0.32	−0.04	0.89	1.00
Cl	0.37	0.70	1.08	1.00
N–N—C₆H₅	0.58	1.72	0.69	0.78
OCH₃	0.12	−0.04	0.46	0.74
NHCOC₆H₅	0.22	0.72	0.76	0.40
NHCOCH₃	0.10	−0.79	−0.28	−0.30
OH	0	−0.62	−0.29	<−0.40
COOH	0.36	−0.16	0.90	<−0.40

* Data from Hansch.[53]
† σ = Hammett sigma constant for 3-substituents.
‡ A = activity relative to chloromycetin = 100.

by the Hansch method,[54] and it has been found that penetration of the brain is highly dependent on π, while localization of boronic acids in tumor tissue is dependent on electron-releasing substituents $(-\sigma)$. Since the compounds are not significantly ionized at physiologic pH, it is suggested that an electron-releasing group, which would facilitate cleavage to boric acid, might release this polar molecule inside the tumor, where it would be trapped by lipophilic barriers. Alternatively, electron release might enhance binding of the boronic acid with an electron-deficient component of the tumor tissue.

The hypnotic activities of a variety of drugs, including barbiturates, tertiary alcohols, carbamates, amides, and N,N-diacylureas (see Chap. 9) have been correlated with their distribution behavior using the model nonpolar-polar system, octanol-water.[55] The most active depressant drugs, of all classes, have partition coefficients of about 100/1 (log P = 2) in the octanol/water system. All effective hypnotics contain a very polar nonionic portion of the molecule, as illustrated by their large negative π values: 5,5-unsubstituted barbituric acid, −1.35; hydroxyl (—OH), −1.16; carbamate (—OCO-NH₂), −1.16; carboxamide (—CONH₂), −1.71; N,N-diacylurea (—CONHCONH-CO—), −1.68. In addition, they possess hydrocarbon or halogenated hydrocarbon residues which are sufficiently lipophilic to provide the intact molecule with nonionic surface-active character, and a distribution coefficient (log P) in the usual range of 1 to 3.

Examples of the additive nature of the Hansch substituent constants (π) in estimating the partition coefficient (log P), and the closeness of this value to the ideal coefficient for hypnotics (log P = 2), are illustrated with calculations for the hypnotic-sedative drugs amobarbital, a barbiturate, and ethchlorvynol, an acetylenic tertiary alcohol. More accurate methods for calculating log P values from substituent constants, π, have been developed which take into account the hydrophobic contribution of the hydrogen atom.[56]

In addition to correlations based upon electronic and solubility constants for substituents, parameters for steric contributions of substituents have been applied.[57] Steric constants (Es) derived from substituent effects on the rates of hydrolysis of aliphatic esters or *ortho*-substituted benzoic acid esters, or calculated values based upon van der Waals' radii, have been used to correlate structure-activity relationships in substituted phenoxy-ethylcyclopropylamine $[R—C_6H_4OCH_2CH_2—NH—CH(CH_2)_2]$ monoamine oxidase inhibitors. The reduced activity produced by *meta* substitution was best correlated with steric inhibition of fit to the enzyme surface. Molar refraction (MR) and molecular weight (MW) are also useful and

Ethchlorvynol
(Placidyl)

Substituent	π
C — OH	-1.16
C≡CH	0.84
CH₃CH₂	1.00
ClHC≡CH	1.32
	$\Sigma\pi = 2.00 =$ log P

Amobarbital
(Amytal)

Substituent	π
-ĊCONHCONHĊO	-1.35
CH₃CH₂	1.00
(CH₃)₂CHCH₂CH₂	2.30
	$\Sigma\pi = 1.95 =$ log P

readily calculated measures of size and steric effects of substituent groups.[58]

Although partition coefficients may frequently be correlated with the observed biological activity, many other molecular characteristics are involved in the initiation of drug effects at drug receptors.

DRUG-RECEPTOR INTERACTIONS

CHARACTERISTICS OF THE DRUG

Most drugs that belong to the same pharmacologic class have certain structural features in common. These frequently include, for example, a basic nitrogen atom, an aromatic ring, an ester or amide group, a phenolic or alcoholic hydroxyl group, or an aliphatic or alicyclic portion of the molecule. Structural features usually are present in the molecule which permit these "functional" groups to be oriented in a similar pattern in space.

Paul Ehrlich's introduction of the receptor concept provided the basis for relating structural similarities in molecules with similarities in biological activity.

The drug receptor is conceived as a relatively small region of a macromolecule, which may be an isolable enzyme, a structural and functional component of a cell membrane, or a specific intracellular substance, such as a protein or a nucleic acid. Specific regions of these macromolecules are visualized as being oriented in space in a manner which permits their functional groups to interact with the complementary functional groups of the drug, this interaction initiating changes in structure and function of the macromolecule which lead ultimately to the observable biological response. The concept of specifically oriented functional areas forming a receptor leads directly to specific structural requirements for functional groups of a drug which must be complementary to the receptor.

Isosterism

In the search for novel, more potent, less toxic and more selectively acting drugs, those associated with pharmaceutical research have developed considerable intuition, based on a large body of experimental knowledge, in selecting appropriate structural modifications of pharmacologically active compounds. As understanding of the stereochemical and physicochemical nature of molecular features has increased, intuition has been strengthened by the application of modern structural theory and by the techniques of quantitative structure-activity relationship studies. The term *isosterism* has been widely used to describe the selection of structural components whose steric, electronic and solubility characteristics make them interchangeable in drugs of the same pharmacologic class.

The concept of isosterism has evolved and changed significantly in the years since its introduction by Langmuir[59] in 1919. Langmuir, while seeking a correlation which would explain similarities in physical properties for nonisomeric molecules, defined *isosteres* as compounds or groups of atoms having the same number and arrangement of electrons. Those isosteres which were isoelectric, i.e., with the same total charge as well as same number of electrons, would possess similar physical properties. For example, the molecules N_2 and CO, both possessing 14 total electrons and no charge, show similar physical properties. Related examples described by Langmuir were CO_2 and N_2O, and N_3^- and NCO^-.

With increased understanding of the structures of molecules, less emphasis has been placed on the number of electrons involved, for variations in hybridization during bond formation may lead to considerable differences in the angles, the lengths and the polarities of bonds formed by atoms with the same number of peripheral electrons. Even the same atom may vary widely in its structural and electronic characteristics when it forms a part of a different functional group. Thus, nitrogen is part of a planar structure in the nitro group but forms the apex of a pyramidal structure in ammonia and the amines.

Groups of atoms which impart similar physical or chemical properties to a molecule, due to similarities in size, electronegativity or stereochemistry, are now frequently referred to under the general term of *isostere*. The early recognition that benzene and thiophene were alike in many of their properties led to the term "ring equivalents" for the vinylene group (—CH=CH—) and divalent sulfur (—S—). This concept has led to replacement of the sulfur atom in the phenothiazine ring system of tranquilizing agents with the vinylene group to produce the dibenzazepine class of antidepressant drugs (see Chap. 10). The vinylene group in an aromatic ring system may be replaced by other atoms isosteric to sulfur, such as oxygen (furan) or NH (pyrrole); however, in such cases, aromatic character is significantly decreased.

Examples of isosteric pairs which possess similar steric and electronic configurations are: the carboxylate (COO^-) and sulfonamido (SO_2NR^-) ions; ketone (CO) and sulfone (SO_2) groups; chloride (Cl) and trifluoromethyl (CF_3) groups. Divalent ether (—O—), sulfide (—S—), amine (—NH—) and methylene (—CH_2—) groups, although dissimilar electronically, are sufficiently alike in their steric nature to be frequently interchangeable in drugs.

Compounds may be altered by isosteric replacements of atoms or groups, in order to develop analogs with select biological effects, or to act as antagonists to normal metabolites. Each series of compounds showing a specific biological effect must be considered separately, for there are no general rules which will predict whether biological activity will be increased or decreased. It appears that when isosteric replacement involves the bridge connecting groups necessary for a given response, a gradation of like effects results, with steric factors (bond angles) and relative polar character being important. Some examples of this type are:

Antibacterial: X = S, Se, O, NH, CH_2

Thyroid Hormone Analogs: X = O, S, CH_2

Antihistamines: X = O, NH, CH_2.
Cholinergic Blocking Agents: X = —COO—, —CONH—, —COS—.

When a group is present in a part of a molecule where it may be involved in an essential interaction or may influence the reactions of neighboring groups, isosteric replacement sometimes produces analogs which act as antagonists. Some examples from the field of cancer chemotherapy are:

Adenine	NH_2 } Metabolites
Hypoxanthine	OH
6-Mercaptopurine	SH — Antimetabolite

The 6-NH_2 and 6-OH groups appear to play essential roles in the hydrogen-bonding interactions of base pairs during nucleic acid replication in cells. The substitution of the significantly weaker hydrogen-bonding isosteric sulfhydryl groups results in a partial blockage of this interaction, and a decrease in the rate of cellular synthesis.

In a similar fashion, replacement of the hydroxyl group of pteroylglutamic acid (folic acid) by the amino group leads to aminopterin, an antagonist useful in the treatment of certain types of cancer.

As a better understanding develops of the nature of the interactions between drug, metabolizing enzymes and biological receptor, selection of isosteric groups with particular electronic, solubility and steric properties should permit the rational preparation of more selectively acting drugs. But in the meanwhile, results obtained by the systematic application of the principles of isosteric replacement are aiding in the understanding of the nature of these receptors.

The Hansch[53] approach, discussed previously, which correlates the contributions to biological activity made by selected physicochemical properties of substituents, has greatly facilitated the systematic alteration of groups in the design of more useful drugs.

Steric features of drugs

Regardless of the ultimate mechanism by which the drug and the receptor interact, the drug must approach the receptor and fit closely to its surface. Steric factors determined by the stereochemistry of the receptor site surface and that of the drug molecules are therefore of primary importance in determining the nature and the efficiency of the drug-receptor interaction. Unless the drug is of the structurally nonspecific cellular depressant type discussed under the Ferguson principle, it must possess a high degree of structural

specificity to initiate a response at a particular receptor.

Some structural features contribute a high degree of structural rigidity to the molecule. For example, aromatic rings are planar, and the atoms attached directly to these rings are held in the plane of the aromatic ring. Thus, the quaternary nitrogen and carbamate oxygen attached directly to the benzene ring in the cholinesterase inhibitor neostigmine are restricted to the plane of the ring, and, consequently, the spatial arrangement of at least these atoms is established.

Neostigmine

The relative positions of atoms attached directly to multiple bonds are also fixed. In the case of the double bond, *cis* and *trans* isomers result. For example, diethylstilbestrol exists in two fixed stereoisomeric forms. *Trans*-diethylstilbestrol is estrogenic, while the *cis*-isomer is only 7 percent as active. In *trans*-diethylstilbestrol, resonance interactions and minimal steric interference tend to hold the two aromatic rings and connecting ethylene carbon atoms in the same plane.

trans-Diethylstilbestrol

cis-Diethylstilbestrol

Geometric isomers, such as the *cis* and the *trans* isomers, hold structural features at different relative positions in space. These isomers also have significantly different physical and chemical properties. Therefore, their distributions in the biological medium are different, as well as their capabilities for interacting with a biological receptor in a structurally specific manner.

More subtle differences exist for *conformational* isomers. Like geometric isomers, these exist as different arrangements in space for the atoms or groups in a single classic structure. Rotation about bonds allows interconversion of conformational isomers; however, an energy barrier between isomers is often sufficiently high for their independent existence and reaction. Differences in reactivity of functional groups, or interaction with biological receptors, may be due to differences in steric requirements. In certain semirigid ring systems, such as the steroids, conformational isomers show significant differences in biological activities (see Chap. 18).

The principles of conformational analysis have established some generalizations in regard to the more stable structures for reduced (nonaromatic) ring systems. In the case of cyclohexane derivatives, bulky groups tend to be held approximately in the plane of the ring, the *equatorial* position. Substituents attached to bonds perpendicular to the general plane of the ring (*axial* position) are particularly susceptible to steric crowding. Thus, 1,3-diaxial substituents larger than hydrogen may repel each other, twisting the flexible ring and placing the substituents in the less crowded equatorial conformation.

Equatorial (e) and *axial (a)* substitution in the chair form of cyclohexane.

Similar calculations may be made for reduced heterocyclic ring systems, such as substituted piperidines. Generally, an equilibrium mixture of conformers may exist. For example, the potent analgesic trimeperidine (see Chap. 17) has been calculated to exist largely in the form in which the bulky phenyl group is in the *equatorial* position, this form being favored by 7 kcal./mole over the *axial* species. The ability of a molecule to produce potent analgesia has been related to the relative spatial positioning of a flat aromatic nucleus, a connecting aliphatic or alicyclic chain, and a nitrogen atom which exists largely in the ionized form at physiologic pH.[60] It might be expected that one of the conformers would be responsible for the analgesic activity; however, in this case it appears that both the *axially* and the *equatorially* oriented phenyl group may contribute. In structurally related isomers whose conformations are fixed by the fusion of an additional ring, both compounds in which the phenyl group is the *axial* and those in which it is in the *equatorial* position have equal analgesic potency.[61]

In a related study of conformationally rigid diaster-

Trimeperidine (*equatorial*-phenyl)

Trimeperidine (*axial*-phenyl)

Equatorial-phenyl (analgesic E.D._{50} 18.4 mg./kg.)

Axial-phenyl analgesic E.D._{50} 18.7 mg./kg.)

Ring-fused Analgesics

eoisomeric analogs of meperidine, the *endo*-phenyl epimer was found to be more potent than was the *exo*-isomer.[62] However, the *endo*-isomer was shown to penetrate brain tissue more effectively due to slight differences in pKa values and partition coefficients between the isomers. This emphasizes the importance of considering differences in physical properties of closely related compounds before interpreting differences in biological activities solely on steric grounds and relative spatial positioning of functional groups.

Open chains of atoms, which form an important part of many drug molecules, are not equally free to assume all possible conformations, there being some which are sterically preferred.[63] Energy barriers to free rotation of the chains are present, due to interactions of nonbonded atoms. For example, the atoms tend to position themselves in space so as to occupy staggered positions, with no two atoms directly facing (eclipsed). Thus, for butane at 37°, the calculated relative probabilities for four possible conformations show that the maximally extended *trans* form is favored 2-to-1 over the two equivalent bent (skew) forms. The *cis* form, in which all of the atoms are facing or *eclipsed,* is much hindered, and only about 1 molecule in 1,000 may be expected to be in this conformation at normal temperatures.

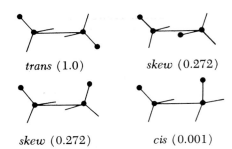

trans (1.0)   *skew* (0.272)

skew (0.272)   *cis* (0.001)

Relative probabilities for the existence of conformations of butane

Nonbonded interactions in polymethylene chains tend to favor the most extended *trans* conformations, although some of the partially extended *skew* conformations also exist. A branched methyl group reduces somewhat the preference for the *trans* form in that portion of the chain, and therefore the probability distribution for the length of the chain is shifted toward the shorter distances. This situation is present in substituted chains which contain the elements of many drugs, such as the β-phenylethylamines. It should be noted that such amines are largely protonated at physiologic pH, and exist in a charged tetra-covalent form. Thus, their stereochemistry closely resembles that of carbon, although in the diagrams below, the hydrogen atoms attached to nitrogen are not shown. As may be expected, the fully extended *trans* form, with maximal separation of the phenyl ring and the nitrogen atom, is favored and a smaller population of the two equivalent *skew* forms, in which the ring and the nitrogen are closer together, exists in solution. Introduction of an α-methyl group alters the favored position of the *trans* form, since positioning of the bulky methyl group away from the phenyl group (*skew* form 2) also results in a decrease in nonbonded interactions. Clearly, *skew* form 1 with both the methyl and the amine group close to phenyl is less favorable. The over-all result is a reduction in the average distance between the aromatic group and the basic nitrogen atom in α-methyl-substituted β-arylethylamines. This

trans

skew   *skew*

Conformations of β-phenylethylamines

steric factor influences the strength of the binding interaction with a biological receptor required to produce a given pharmacologic effect. It is possible that the altered stereochemistry of α-methyl-β-arylethylamines may partially account for their slow rate of metabolic deamination (see Chap. 11).

trans

skew form 1   *skew* form 2

Conformations of α-Methyl-β-phenylethylamines

The introduction of atoms other than carbon into a chain strongly influences the conformation of the chain. Due to resonance contributions of forms in which a double bond occupies the central bonds of esters and amides, a planar configuration is favored, in which minimal steric interference of bulky substituents occurs. Thus, an ester is mainly in the *trans,* rather than the *cis* form. For the same reason, the amide linkage is essentially planar, with the more bulky substituents occupying the *trans* position. Therefore, ester and amide linkages in a chain tend to hold bulky groups in a plane and to separate them as far as possible. As components of the side chains of drugs, ester and amide groups favor fully extended chains and, also, add polar character to that segment of the chain.

trans-planar form   resonance form   *cis*-planar form

Stabilizing planar structure of esters

trans-planar form   resonance form   *cis*-planar form

Stabilizing planar structure of amides

The above considerations make it clear that the ester linkages in succinyl choline provide both a polar segment which is readily hydrolyzed by plasma cholinesterase (see Chap. 12), and additional stabilization to the fully extended form. This form is also favored by repulsion of the positive charges at the ends of the chain.

Extended form of succinyl choline

The conformations favored by stereochemical considerations may be further influenced by *intramolecular interactions* between specific groups in the molecule. *Electrostatic forces,* involving attractions by groups of opposite charge, or repulsion by groups of like charge, may alter molecular size and shape. Thus, the terminal positive charges on the polymethylene bis-quaternary ganglionic blocking agent hexamethonium and the neuromuscular blocking agent decamethonium make it most likely that the ends of these molecules are maximally separated in solution.

$$(CH_3)_3\overset{+}{N}-(CH_2)_n-\overset{+}{N}(CH_3)_3$$
Hexamethonium n = 6
Decamethonium n = 10

In some cases *dipole-dipole interactions* appear to influence structure in solution. Methadone may exist partially in a cyclic form in solution, due to dipolar attractive forces between the basic nitrogen and carbonyl group.[64] In such a conformation, it closely resembles the conformationally more rigid potent analgesics, morphine, meperidine and their analogs (see Chap. 17), and it may be this form which interacts with the analgesic receptor.

Ring conformation of methadone by dipolar interactions

An intramolecular *hydrogen bond,* usually formed between donor —OH and =NH groups, and acceptor oxygen (:Ö=) and nitrogen (:N≡) atoms, might be expected to add stability to a particular conformation of a drug in solution. However, in aqueous solution donor and acceptor groups tend to be bonded to water, and little gain in free energy would be achieved by the formation of an intramolecular hydrogen bond, particularly if unfavorable steric factors involving nonbonded interactions were introduced in the process. Therefore, it is likely that internal hydrogen bonds play only a secondary role to steric factors in determining the conformational distribution of flexible drug molecules.

Conformational flexibility and multiple modes of action

It has been proposed that the conformational flexibility of most open-chain neurohormones, such as acetylcholine, epinephrine, serotonin, and related physiologically active biomolecules, such as histamine, permits multiple biological effects to be produced by each molecule, by virtue of the ability to interact in a different and unique conformation with different biological receptors. Thus, it has been suggested that acetylcholine may interact with the muscarinic receptor of postganglionic parasympathetic nerves and with acetylcholinesterase in the fully extended conformation, and in a different, more folded structure, with the nicotinic receptors at ganglia and at neuromuscular

Quasi-ring form of acetylcholine

Extended conformation of acetylcholine

junctions.[65,66] Acetylcholine bromide exists in a quasi-ring form in the crystal, with an N-methyl hydrogen atom close to, and perhaps forming a hydrogen bond with, the backbone oxygen.[67] In solution, however, it is able to assume a continuous series of conformations, some of which are energetically favored over others.[66]

Conformationally rigid acetylcholine-like molecules have been used to study the relationships between these various possible conformations of acetylcholine and their biological effects. (+)-*trans*-2-acetoxycyclopropyl trimethylammonium iodide, in which the quaternary nitrogen atom and acetoxyl groups are held apart in a conformation approximating that of the extended conformation of acetylcholine, was about 5 times as active as acetylcholine in its muscarinic effect on dog blood pressure, and equiactive to acetylcholine in its muscarinic effect on the guinea pig ileum.[68] The (+)-*trans*-isomer was hydrolyzed by acetylcholinesterase at a rate equal to the rate of hydrolysis of acetylcholine. It was inactive as a nicotinic agonist. In contrast, the (−)-*trans*-isomer and the mixed (+),(−)-*cis*-isomers were 1/500 and 1/10,000 as active as acetylcholine in muscarinic tests on guinea-pig ileum and were inactive as nicotinic agonists. Similarly, the *trans*-diaxial relationship between the quaternary nitrogen and acetoxyl group led to maximal muscarinic response and rate of hydrolysis by true acetylcholinesterase in a series of isomeric 3-trimethylammonium-2-acetoxyl decalins.[69] These results could be interpreted that either acetylcholine was acting in a *trans* conformation at the muscarinic receptor, and was not acting in a *cisoid* conformation at the nicotinic receptor, or that the nicotinic response is highly sensitive to steric effects of substituents being used to orient the molecule.

In contrast to the concept of acetylcholine reacting with muscarinic and nicotinic receptors in different conformations, Chothia[70] has proposed that acetylcholine interacts in the same conformation, but in a different manner, with each receptor. The conformations of acetylcholine (Fig. 2-6) are primarily defined by rotations about the C_α—C_β and C_β—O_1 bonds, since the C_1—N—C_α—C_β sequence and the O_2—C_4—O_1—C_β ester group exist largely in planar conformations due to steric and resonance factors. Acetylcholine and several selective muscarinic and nicotinic agents have been shown to be in closely similar conformations in the crystal state.[71] In these compounds the C_α—C_β bond, or its equivalent, is rotated so that the N and O_1 (ether oxygen) are about 60 to 75° from the *cis* coplanar conformation. The C_α—C_β—O_1—C_4 atoms are essentially in a *trans* planar extended chain. This conformation presents a methyl side, defined by a plane close to C_2, O_1 and C_5 (methyl carbon), and a carbonyl side, defined by a plane close to C_3, C_β and O_2 (car-

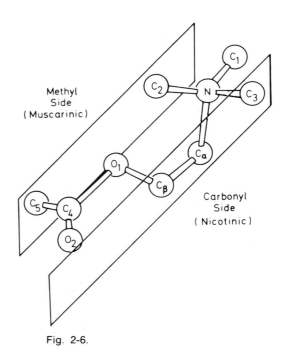

Fig. 2-6.

bonyl oxygen). Compounds with high muscarinic and low nicotinic activity, such as *trans*-2-acetoxycyclopropyl trimethylammonium iodide, L-(+)-acetyl-β-methylcholine and muscarine, show structures in the crystal state which have free access to their methyl sides, while their carbonyl sides are blocked by the spatial position occupied by the extra methyl or methylene groups. In preferential nicotinic agonists, such as L-(+)-acetyl-α-methylcholine, the carbonyl side is exposed and access to the methyl side is blocked. Chothia[70] has proposed that the methyl sides of acetylcholine and its predominantly muscarinic analogs interact with the muscarine receptor, while it is the interaction with groups on the carbonyl side of acetylcholine and its nicotinic analogs which activates the nicotinic receptor.

Using an approach which focuses on the parent molecule, rather than on conformationally fixed analogs, molecular orbital calculations have indicated that histamine may exist in two extended conformations (A,B) of nearly equal and minimal energy[72] rather than the earlier predicted coiled form (C) involving intramolecular hydrogen bonds.[73] In one extended conformation (A), one imidazole ring nitrogen atom is about 4.55 Å from the side chain nitrogen, while in conformation B this distance is about 3.60 Å. Histamine receptors have been differentiated into at least two classes, there being different structural requirements for stimulation of smooth muscle, such as the guinea-pig ileum (histamine H_1-receptor, blocked by classical antihistamines), and for the stimulation of secretion of gastric acid (histamine H_2-receptor, not

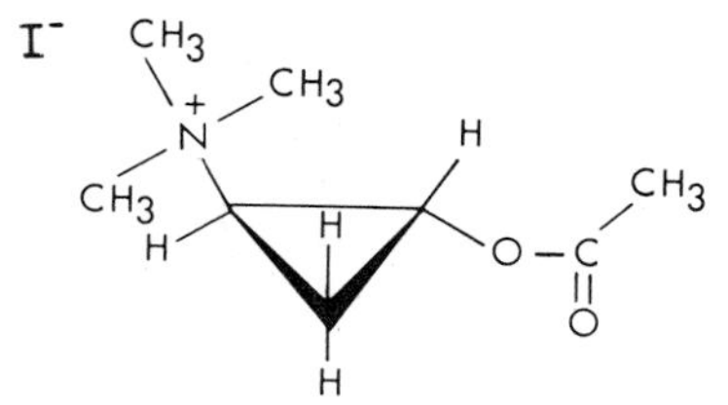

trans-2-Acetoxycyclopropyl Trimethylammonium
Iodide

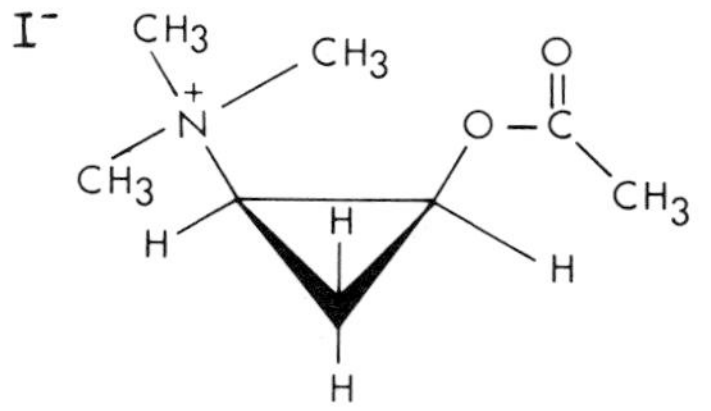

cis-2-Acetoxycyclopropyl Trimethylammonium
Iodide

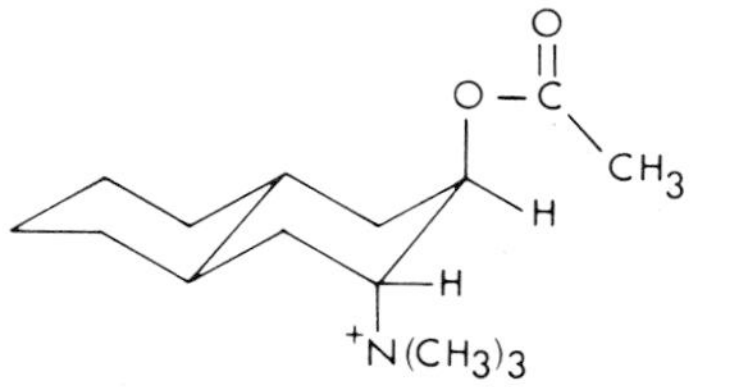

trans-diaxial 3-Trimethylammonium-2-acetoxy-
decalin

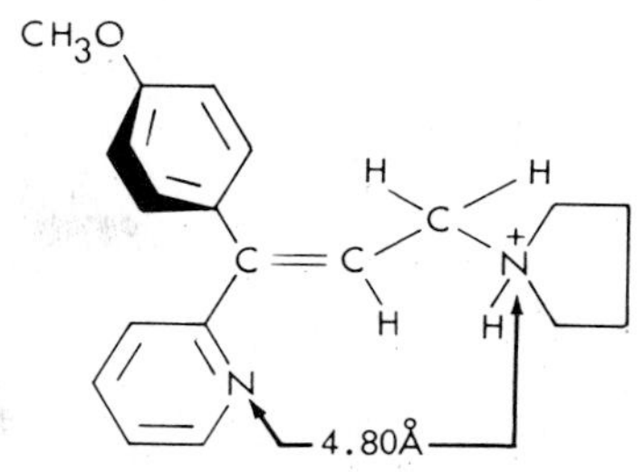

Triprolidine (antihistamine)

ring by chains four atoms long, two atoms longer than the dimethylene side chain of histamine. It appears likely that H_2-antagonist activity results from the interaction of the side chain and its polar residues with a receptor region distinct from that with which the positively charged side chain of histamine interacts.

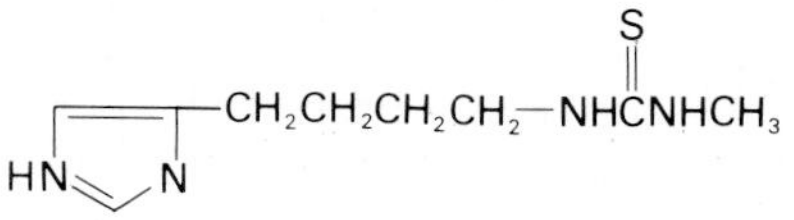

Burimamide

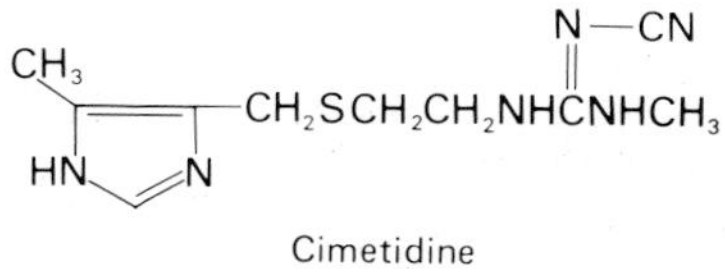

Cimetidine

blocked by classical antihistamines). It is proposed, on the basis of the internitrogen distance of closest approach of 4.8 ± 0.2 Å for the relatively rigid antihistamine triprolidine, that histamine acts on smooth muscle (H_1-receptor) in conformation A, in which the internitrogen distance of 4.55 Å closely approximates the spacing found in the specific antagonist. It is further presumed that the histamine-induced release of gastric acid may be brought about by a histamine H_2-receptor interaction in an alternate conformation of closer internitrogen spacing, such as conformation B.

The recently discovered histamine H_2-receptor antagonists,[74] burimamide and cimetidine (see Chap. 16) contain uncharged polar residues on the side chain, such as the thiourea (—NH—CS—NH—) or N-cyanoguanidine [—NH—C(=NCN)NH—] groups. These polar residues are separated from the imidazole

Optical isomerism and biological activity

The widespread occurrence of differences in biological activities for *optical isomers* has been of particular importance in the development of theories in regard to the nature of drug-receptor interactions. *Diastereoisomers*, compounds with two or more asymmetric centers, have the same functional groups and, therefore, can undergo the same types of chemical reactions. However, the diastereoisomers (e.g., ephedrine, *pseudo*-ephedrine, see Chap. 11) have different physical properties, undergo different rates of reactions, have substituent groups which occupy different relative positions in space, and the different biological properties shown by such isomers may be accounted

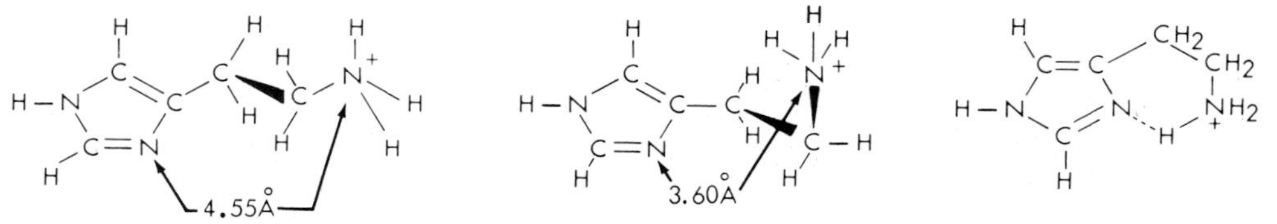

Conformations of Histamine

for by the influence of any of these factors on drug distribution, metabolism or interaction with the drug receptor.

However, *optical enantiomers,* also called *optical antipodes* (mirror images) present a very different case, for they are compounds whose physical and chemical properties are usually considered identical except for their ability to rotate the plane of polarized light. Here one might expect the compounds to have the same biological activity. However, such is not the case with many of the enantiomers that have been investigated.

As examples of compounds whose optical isomers show different activities may be cited the following: (−)-hyoscyamine is 15 to 20 times more active as a mydriatic than (+)-hyoscyamine; (−)-hyoscine is 16 to 18 times as active as (+)-hyoscine; (−)-epinephrine is 12 to 15 times more active as a vasoconstrictor than (+)-epinephrine; (+)-norhomoepinephrine is 160 times more active as a pressor than (−)-norhomoepinephrine; (−)-synephrine has 60 times the pressor activity of (+)-synephrine; (−)-amino acids are either tasteless or bitter, while (+)-amino acids are sweet; (+)-ascorbic acid has good antiscorbutic properties, while (−)-ascorbic acid has none.

Although it is well established that optical antipodes have different physiologic activities, there are different interpretations as to why this is so. Differences in distribution of isomers, without considering differences in action at the receptor site, could account for different activities for optical isomers. Diastereoisomer formation with optically active components of the body fluids (e.g., plasma proteins) could lead to differences in absorption, distribution and metabolism. Distribution could also be affected by preferential metabolism of one of the optical antipodes by a stereospecific enzyme (e.g., D-amino acid oxidase). Preferential adsorption could also occur at a stereospecific site of loss (e.g., protein binding). Cushny[75] accounted for this difference by assuming that the optical antipodes reacted with an optically active receptor site to produce diastereoisomers with different physical and chemical properties. Easson and Stedman,[76] taking a somewhat different view, point out that optical antipodes can in theory have different physiologic effects for the same reason that structural isomers can have different effects, i.e., because of different molecular arrangements, one antipode can react with a hypothetical receptor while the other cannot. Assuming a receptor in tissues to which a drug can be attached and have activity only if the complementary parts B, D, C are superimposed, it is apparent that of the two enantiomers, only I can be so superimposed. Under these conditions, I therefore would be active, and II would show no activity. This interpretation in

a sense is not greatly different from that given by Cushny, because the receptor has a unique configuration not much different from that of an optically active compound. Both theories demand a structure of unique configuration in the body, but in the one theory only one enantiomer reacts, while in the other they both react, with one combination having greater biological activity than the other.

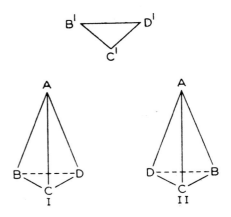

Easson and Stedman[76] have also postulated that the optical antipodes of epinephrine owe their differences in activity to a difference in ease of attachment to the receptor surface. This is illustrated below for the pressor activity of (−)- and (+)-epinephrine.[77]

Thus, only in (−)-epinephrine can the three groups essential for maximal pressor activity in sympathomimetic amines—the positively charged nitrogen, the aromatic ring and the alcoholic hydroxyl group—attach to the complementary receptor surface. In the (+)-isomer, any two binding groups may orient to attach, but not all three. This is consistent with the observation[78] that desoxyepinephrine, which lacks the alcoholic hydroxyl and therefore may only bind in two positions, has about the same pressor effect as (+)-epinephrine.

In the field of potent analgesics (morphine, meperidine, methadone, etc.) Beckett[77] has described a receptor surface made up of three regions (see Chap. 17). These are: (1) a flat surface providing binding to an aromatic ring, (2) a cavity into which the connecting chain between the aromatic group and nitrogen may fit and be held by van der Waals' forces and (3) an anionic site which binds the cationic nitrogen. Configurational studies have shown that all potent analgesics thus far studied either possess or may adopt the conformations which allow ready association with this receptor. Optical antipodes to the potent analgesics are less active as analgesics, although some, e.g., dextromethorphan (Romilar), retain antitussive properties.

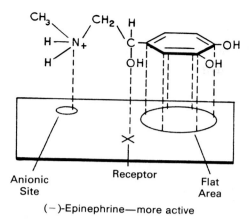

(−)-Epinephrine—more active

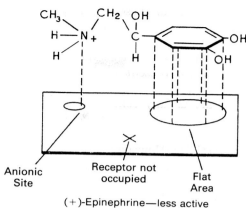

(+)-Epinephrine—less active

THE DRUG RECEPTOR

The drug receptor is a component of the cell whose interaction with the drug initiates a chain of events leading to an observable biological response. Primarily by analogy with the well-studied substrate-enzyme interactions, it has been usually assumed that those drug receptors which are not enzymes resemble enzymes in their general nature but are, in contrast, an integral part of the organized structure of the cell and, therefore, may not be isolable by presently available techniques. There are a number of specific examples of drug-enzyme interactions which are related to pharmacologic effects. The best established of these are inhibitors of acetylcholinesterase acting as cholinergic agents (e.g., physostigmine, isoflurophate; see Chap. 12), inhibitors of carbonic anhydrase acting as diuretics (e.g., acetazolamide; see Chap. 13), inhibitors of monoamine oxidase acting as central nervous system stimulants (e.g., tranylcypromine; see Chap. 10), and the aldehyde oxidase inhibitor disulfiram, used to discourage the chronic consumption of alcohol. However, most drug actions appear to take place on or within the cell in regions which have not been isolated and characterized as enzymes.

Many drug effects are believed to take place at the receptors for hormones or neurotransmitter substances. Mammalian organs use these two types of chemical signals to transmit messages in the absence of direct nerve connections. The hormones (amino acids, peptides, steroids, glycoproteins; see Chaps. 18 and 20) are widely distributed in the circulation, and their abilities to carry specific messages reside in their selective interactions with tissue-specific receptors on cell membranes, or within the cell. In contrast, the neurotransmitters (e.g., acetylcholine) are distributed to very few cells. They are released from storage sites in nerves to stimulate the receptors of nearby postsynaptic effector cells.

Certain hormone and neurotransmitter receptors are associated with structural and functional elements of the cell by relatively weak noncovalent bonds, and may be solubilized and isolated. The nicotinic cholinergic receptor occurs in high concentrations in the electric organs of the eels, *Electrophorus electricus* and *Torpedo californica*. This membrane-bound receptor has been solubilized by treatment with nonionic detergents, without loss of binding affinity for cholinergic agonists or antagonists. The receptor has been purified by affinity chromatography and characterized as a glycoprotein[79,80] of molecular weight 255,000, composed of five subunits of molecular weights ranging from 40,000 to 65,000. Two of the subunits are identical and all are structurally related in their amino acid sequences. The subunits associate to form a hollow cylindrical structure which acts as a transmembrane channel about 25 Å in diameter. Interaction of acetylcholine with specific elements on the outer (synaptic) face of the receptor stimulates ion movements through the channel, resulting in alteration of the transmembrane electrical potential. When combined with lipids from the same organism, the purified cholinergic receptor from *Torpedo* forms a sealed vesicle which is chemically excitable by acetylcholine as measured by enhanced radioactive sodium ion efflux from the vesicle.[81] These results show that the isolated receptor retains both a specific cholinergic binding site, as well as the molecular elements for ion translocation, a role which the receptor plays in carrying out depolarization of the intact postsynaptic membrane. Using fluorescent probes, changes in physicochemical state of the cholinergic receptor have been demonstrated upon desensitization with local anesthetics.[82]

Like the nicotinic cholinergic receptor, many other neurotransmitter receptors and receptors for peptide and protein hormones appear to be localized in the cell membrane. Some of these, such as the β-adrenergic receptor, act by activating intracellular adenylate cyclase, producing the "second messenger," cyclic adenosine monophosphate. Techniques and receptor-rich sources of tissues have not yet been developed for the isolation and characterization of these membrane-bound receptors. In contrast, intracellular receptors play a dominant role in the actions of the steroid and thyroid hormones, and much is known about the nature and functions of these hormone receptors.

The lipophilic steroid hormones are solubilized and transported attached to plasma proteins. The small amount of free steroid in equilibrium with the protein-bound steroid enters the cell where it binds with high affinity to a receptor protein in the cytoplasm.[83] The steroid-receptor complex moves rapidly into the cell nucleus where it binds to the chromosomes. This combination of steroid-receptor-chromatin-DNA initiates specific RNA synthesis, leading to the stimulation of new protein synthesis. The progesterone receptor has been partially purified by affinity chromatography, and is characterized as a dimeric protein, made up of two nonidentical cigar-shaped subunits, each having a molecular weight of about 100,000.

Intracellular receptor proteins with high affinity for the thyroid hormones, and which mediate thyroid hormone action, appear to be permanent residents in the nuclear chromatin. They are nonhistone proteins which are extracted by high-salt concentrations (0.4 M KCl), have been partially purified, and show a molecular weight of about 65,000.[84] There is an excellent correlation between binding affinities to the nuclear receptors of rat liver cells, and the in vivo hormonal effectiveness of a wide variety of thyroid hormone analogs.[85] Like the steroid receptors, the thyroid hormone nuclear receptors appear to mediate the principal effects of the thyroid hormones by stimulating synthesis of specific proteins.

These recent and rapidly expanding studies of hormone and neurotransmitter receptors have established that drug receptors exist as discrete entities. At least in some cases these are capable of being separated from structural elements of the cell without loss of binding affinity for their ligands, and with retention of function, when recombined with other requisite cellular constituents. Many of these receptors require a membranelike structure for their functional activity.

The cell membrane is one region of the cell which contains organized components which can interact with small molecules in a specific manner. The structural unit of the cell membrane is thought to consist of a bimolecular layer of lipid molecules about 25 Å thick, held between two layers, each about 25 Å thick, which are at least partially protein. The lipid layers may consist of cholesterol and phospholipids, with the nonpolar hydrocarbon chains held together at the center by van der Waals' attraction, the polar heads being oriented outward, and associated by polar bonds with the protein sheaths. High molecular weight, charged mucopolysaccharides, with their constituent carboxylic and sulfate ester groups acting as solvated anions, may be associated with the protein in one or both of the outer layers. Water-filled pores, lined by the polar side chains of protein molecules, are assumed to permit passage of small polar molecules. The proteins constitute a potentially highly organized region of the cell membrane. Molecular specificity is well known in such proteins as enzymes and antibodies, and it is generally believed that proteins are an important component of the drug receptor. The nature of the amide link in proteins provides a unique opportunity for the formation of multiple internal hydrogen bonds, as well as internal formation of hydrophobic, van der Waals' and ionic bonds by side chain groups, leading to such organized structures as the α-helix, which contains about four amino acid residues for each turn of the helix. An organized protein structure would hold the amino acid side chains at relatively fixed positions in space and available for specific interactions with a small molecule.

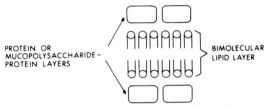

PROTEIN OR MUCOPOLYSACCHARIDE-PROTEIN LAYERS / BIMOLECULAR LIPID LAYER

Schematic representation of the cell membrane

Proteins have the potential to adopt many different conformations in space without breaking their covalent amide linkages. They may shift from highly coiled structures to partially disorganized structures, with parts of the molecule existing in random chain, or to folded sheet structures, depending on the environment. In the monolayer of a cell membrane, the interaction of a foreign small molecule with an organized protein may lead to a drastic change in the structural and physical properties of the membrane. Such changes could well be the initiating events in the production of a tissue or organ response to a drug, such as the ion-translocating effects produced by interaction of acetylcholine and the cholinergic receptor.[81]

The large body of information now available on relationships between chemical structure and biological activity strongly supports the concept of flexible receptors. The fit of drugs onto or into macromolecules is only rarely an all-or-none process as pictured by the earlier "lock and key" concept of a receptor. Rather, the binding or partial insertion of groups of moderate size onto or into a macromolecular pouch appears to be a continuous process, even though over a limited range, as indicated by the frequently occurring regular increase and decrease in biological activity as one ascends a homologous series of drugs. A range of productive associations between drug and receptor may be pictured, which lead to agonist responses, such as those produced by cholinergic drugs. Similarly, strong associations may lead to unproductive changes in the configuration of the macromolecule, leading to an antagonistic or blocking response, such as that produced by anticholinergic agents. The fundamental structural unit of the drug receptor is generally considered to be protein in nature, although this may be supplemented by its associations with other units, such as mucopolysaccharides and nucleic acids.

In the maximally extended protein, the distance between peptide bonds ("identity distance") is 3.61 Å. For many types of biological activity, the distance between functional groups leading to maximal activity approximates this identity distance or some whole number multiple of it. Many parasympathomimetic (acetylcholinelike) and parasympatholytic (cholinergic blocking) agents have a separation of 7.2 Å (2 × 3.6) between the ester carbonyl group and nitrogen.[86] This distance is doubled between quaternary nitrogens of curarelike drugs; 14.5 Å (4 × 3.61).[87] The preferred separation of hydrogen bonding groups in estrogenic compounds (e.g., hydroxyls of diethylstilbestrol) is 14.5 Å (4 × 3.61).[88]

Identity Distance in Extended Protein

A related spacing of 5.5 Å, which corresponds to two turns of the α-helical structure common to proteins, is found between functional groups of many drugs. The most frequently occurring of these is the R—X—CH$_2$—CH$_2$—NR$_2'$ (X = N; X = O, or X = C) structure which is present in local anesthetics, antihistamines, adrenergic blocking agents and others.[89]

Studies involving the relative effectiveness of various molecules of well-defined structural and functional types have contributed to an understanding of the stereochemical and physicochemical properties of their biological receptors. Pfeiffer[86] concluded that parasympathomimetic stimulant action depends on two adjacent oxygen atoms at distances of approximately 5.0 Å and 7.0 Å from a methyl group or groups attached to nitrogen. Since these compounds (acetyl-

choline, methacholine, urecholine, etc.) do not have rigid structures, the actual distance between the oxygen and the methyl groups varies; however, the more extended conformations would be favored in solution.

Welsh and Taub[90] have concluded that a carbonyl group at a maximum distance of 7 Å from the quaternary nitrogen is an important linking group with the acetylcholine receptor protein of the *Venus* heart. They suggest that some type of bond forms between the carbonyl carbon or ketone oxygen and an appropriate group in the protein molecule.

The nature of the acetylcholinesterase receptor site probably has been investigated more thoroughly than the reactive site of any other enzyme. On the basis of studies with enzyme inhibitors, Nachmansohn and Wilson[91] suggested two functional sites: a center of high electron density which binds the cationic nitrogen, and an esteratic site which interacts with the carbonyl carbon atom. Friess and his co-workers[92] attempted to define the distance between the anionic and the esteratic sites by studying enzyme inhibition with cyclic aminoalcohols (e.g., *cis*-2-dimethylaminocyclohexanol) and their esters. The *cis*-isomers were more active than the *trans*, and a distance of about 2.5 Å was indicated as separating the nitrogen and the oxygen and, by inference, the receptors which bind these on the enzyme. Krupka and Laidler[93] correlated previous stereochemical studies with kinetic data and described a complex esteratic site made up of three components: a basic site (imidazole nitrogen, 5 Å from the anionic site), an acid site, 2.5 Å from the anionic site, and a serine hydroxyl group. Following stereospecific binding of acetylcholine, the serine hydroxyl is acetylated to effect ester cleavage. Subsequently, a

Anionic Site
Acetylcholinesterase

Esteratic Site
(Nachmansohn
and Wilson)

Acetylcholinesterase
(Krupka and Laidler)

water molecule is held in the proper position through hydrogen bonds with imidazole, serine is deacetylated (hydrolyzed) and the reactive enzyme is regenerated.

THE DRUG-RECEPTOR INTERACTION; FORCES INVOLVED

A biological response is produced by the interaction of a drug with a functional or organized group of molecules which may be called the biological receptor site. This interaction would be expected to take place by utilizing the same bonding forces involved as when simple molecules interact. These, together with typical examples, are collected in Table 2-7.

Most drugs do not possess functional groups of a type which would lead to ready formation of the strong and essentially irreversible covalent bonds between drug and biological receptors. In most cases it is desirable that the drug leave the receptor site when the concentration decreases in the extracellular fluids; therefore, most useful drugs are held to their receptors by ionic or weaker bonds. However, in a few cases where relatively long-lasting or irreversible effects are desired (e.g., antibacterial, anticancer), drugs which form covalent bonds are effective and useful.

The alkylating agents, such as the nitrogen mustards (e.g., mechlorethamine) used in cancer chemotherapy, furnish an example of drugs which act by formation of covalent bonds. These are believed to form the reactive immonium ion intermediates, which alkylate and thus link together proteins or nucleic acids, preventing their normal participation in cell division.

Covalent bond formation between drug and receptor is the basis of Baker's[94] concept of *"active-site-directed irreversible inhibition."* Considerable experimental evidence on the nature of enzyme inhibitors has supported this concept. Compounds studied possess appropriate structural features for reversible and

TABLE 2-7

TYPES OF CHEMICAL BONDS*

Bond Type	Bond Strength kcal./mole	Example
Covalent	40–140	CH_3-OH
Reinforced ionic	10	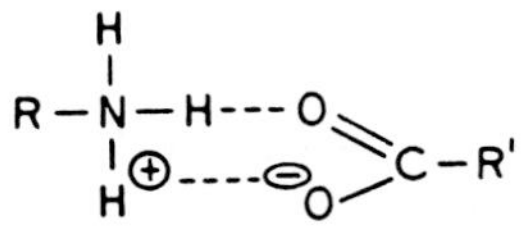
Ionic	5	
Hydrogen	1–7	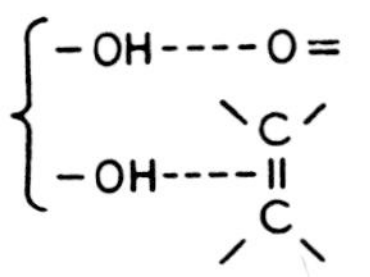
Ion-dipole	1–7	
Dipole-dipole	1–7	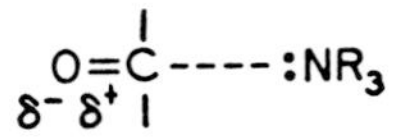
van der Waals'	0.5–1	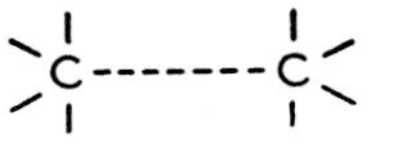
Hydrophobic	1	See Text

* Adapted from a table *in* Albert, A.: Selective Toxicity, p. 183, New York, Wiley, 1968.

highly selective association with an enzyme. If, in addition, the compounds carry reactive groups capable of forming covalent bonds, the substrate may be irreversibly bound to the drug-receptor complex by covalent bond formation with reactive groups adjacent to the active site. In studies with reversibly binding antimetabolites that carried additional alkylating and acylating groups of varying reactivities, selective irre-

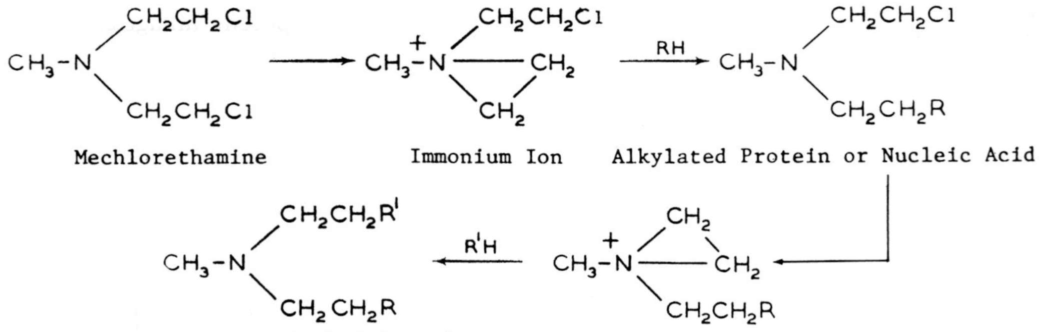

Mechlorethamine   Immonium Ion   Alkylated Protein or Nucleic Acid

Cross-linked Protein or
Nucleic Acid

R, R′ = free amino groups of proteins, adenyl or phosphate groups of nucleic *acids*.

versible binding by the related enzymes lactic dehydrogenase and glutamic dehydrogenase has been demonstrated. The selectivity of response has been attributed to the formation of a covalent bond between the carbophenoxyamino substituent of 5-(carbophenoxyamino)salicylic acid and a primary amino group in glutamic dehydrogenase[95] and between the maleamyl substituent of 4-(maleamyl)salicylic acid and a sulfhydryl group in lactic dehydrogenase.[96] Assignments of covalent bond formation with specific groups in the enzymes are based on the fact that the α,β-unsaturated carbonyl system of maleamyl groups reacts most rapidly with sulfhydryl groups, much more slowly with amino groups and extremely slowly with hydroxyl groups. In contrast, the carbophenoxy group will react only with a primary amino group on a protein. The diuretic drug, ethacrynic acid (see Chap. 13), is an α,β-unsaturated ketone, thought to act by covalent bond formation with sulfhydryl groups of ion-transport systems in the renal tubules.

5-(Carbophenoxyamino)salicylic acid

4-(Maleamyl)salicylic acid

In the purine series, similar studies[97] have led to the rational development of an active-site-directed inhibitor of adenosine deaminase. Studies on 9-alkyladenines showed that hydrophobic interactions between the 9-alkyl substituent and a nonpolar region of the enzyme were important in the formation of the reversible drug-inhibitor complex. A nonpolar aromatic group, containing the active but nonselective bromoacetamido group, was substituted in the 9 position, and the resulting 9-(p-bromoacetamidobenzyl)adenine was shown to form initially a reversible enzyme-inhibitor complex, followed by formation of an irreversible complex, presumably by alkylation.

Other examples of covalent bond formation between drug and biological receptor site include the reaction of arsenicals and mercurials with essential sulfhydryl groups, the acylation of bacterial cell-wall constituents by penicillin and the inhibition of cholinesterase by the organic phosphates.

It is desirable that most drug effects be reversible. For this to occur, relatively weak forces must be involved in the drug-receptor complex, and yet strong enough so that other binding sites of loss will not competitively deplete the site of action. Compounds with a high degree of structural specificity may orient several weak binding groups, so that the summation of their interactions with specifically oriented complementary groups on the receptor will provide the total bond strength sufficient for a stable combination.

Thus, for drugs acting by virtue of their structural specificity, binding to the receptor site will be carried out by hydrogen bonds, ionic bonds, ion-dipole and dipole-dipole interactions, van der Waals' and hydrophobic forces. Ionization at physiologic pH would normally occur with the carboxyl, sulfonamido and aliphatic amino groups, as well as the quaternary ammonium group at any pH. These sources of potential ionic bonds are frequently found in active drugs. Differences in electronegativity between carbon and other atoms such as oxygen and nitrogen lead to an unsymmetrical distribution of electrons (dipoles) which are also capable of forming weak bonds with

9-(p-Bromoacetamidobenzyl)adenine

regions of high or low electron density, such as ions or other dipoles. Carbonyl, ester, amide, ether, nitrile and related groups which contain such dipolar functions are frequently found in equivalent locations in structurally specific drugs. Many examples may be found among the potent analgesics, the cholinergic blocking agents and local anesthetics.

The relative importance of the *hydrogen bond* in the formation of a drug-receptor complex is difficult to assess. Many drugs possess groups, such as carbonyl, hydroxyl, amino and imino, with the structural capabilities of acting as acceptors or donors in the formation of hydrogen bonds. However, such groups would usually be solvated by water, as would the corresponding groups on a biological receptor. Relatively little net change in free energy would be expected in exchanging a hydrogen bond with a water molecule for one between drug and receptor. However, in a drug-receptor combination, a number of forces could be involved, including the hydrogen bond which would contribute to the stability of the interaction. Where multiple hydrogen bonds may be formed, the total effect may be a sizeable one, such as that demonstrated by the stability of the protein α-helix, and by the stabilizing influence of hydrogen bonds between specific base pairs in the double helical structure of deoxyribonucleic acid.

Van der Waals' forces are attractive forces created by the polarizability of molecules and are exerted when any two uncharged atoms approach very closely. Their strength is inversely proportional to the seventh power of the distance. Although individually weak, the summation of their forces provides a significant bonding factor in higher molecular weight compounds. For example, it is not possible to distil normal alkanes with more than 80 carbon atoms, since the energy of about 80 kcal. per mole required to separate the molecules is approximately equal to the energy required to break a carbon-carbon covalent bond. Flat structures, such as aromatic rings, permit close approach of atoms. With van der Waals' force approximately 0.5 to 1.0 kcal./mole for each atom, about 6 carbons (a benzene ring) would be necessary to match the strength of a hydrogen bond. The aromatic ring is frequently

found in active drugs, and a reasonable explanation for its requirement for many types of biological activity may be derived from the contributions of this flat surface to van der Waals' binding to a correspondingly flat receptor area.

The hydrophobic nature of structural elements which may participate in van der Waals' interactions provides additional binding energy.

The *hydrophobic bond* appears to be one of the more important forces of association between nonpolar regions of drug molecules and biological receptors. A nonpolar region of a molecule cannot be solvated by water, and, as a consequence, the water molecules in that region associate through hydrogen bonds to form quasi-crystalline structures ("icebergs"). Thus, a nonpolar segment of a molecule produces a higher degree of order in surrounding water molecules than is present in the bulk phase. If two nonpolar regions, such as hydrocarbon chains of a drug and a receptor, should come close together, these regions would be shielded to a greater extent from interaction with water molecules. As a result some of the quasi-crystalline water structures would collapse, producing a gain in entropy relative to the isolated nonpolar structures. The gain in free energy achieved through a decrease in the ordered state of many water molecules stabilizes the

Isolated nonpolar chains in an ordered aqueous environment

Association of nonpolar chains displacing ordered water structures

Schematic representation of hydrophobic bond formation

close contact of nonpolar regions, this association being called "hydrophobic bonding."

THE DRUG-RECEPTOR INTERACTION AND SUBSEQUENT EVENTS

Once bound at the receptor site, drugs may act, either to initiate a response (*stimulant* or *agonist* action), or to decrease the activity potential of that receptor (*antagonist* action) by blocking access to it by active molecules. The chain of events leading to an observable biological response must be initiated in some fashion by either the process of formation or the nature of the drug-receptor complex. Current theories in regard to the mechanism of action of drugs at the receptor level are based primarily on the studies of Clark[98] and Gaddum,[99] whose work supports the assumption that the tissue response is proportional to the number of receptors occupied. The "occupancy theory" of drug action has been modified by Ariëns[100] and Stephenson,[101] who have divided the drug-receptor interaction into two steps: (a) combination of drug and receptor, and (b) production of effect. Thus, any drug may have structural features which contribute independently to the *affinity* for the receptor, and to the efficiency with which the drug-receptor combination initiates the response (*intrinsic activity* or *efficacy*). The Ariëns-Stephenson concept retains the assumption that the response is related to the number of drug-receptor complexes.

In the Ariëns-Stephenson theory, both agonist and antagonist molecules possess structural features which would enable formation of a drug-receptor complex (strong affinity). However, only the agonist possesses the ability to cause a stimulant action, i.e., possesses intrinsic activity. The affinity of a drug may be estimated by comparison of the dose required to produce a pharmacologic response with the dose required by a standard drug. Thus, acetylcholine produces a normal "S"-shaped curve if the logarithm of the dose is plot-

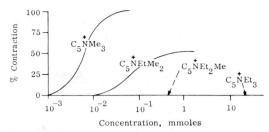

Fig. 2-8. Dose-response curves for contraction produced by pentyl trialkylammonium salts on the rat jejunum. (Modified from Ariëns[100])

ted against the percent contraction of the rat jejunum (a segment of the small intestine). A series of related alkyl trimethylammonium salts (ethoxyethyl trimethylammonium, pentyl trimethylammonium, propyl trimethylammonium; Fig. 2-7) are able to produce the same degree of contraction of the tissue as can acetylcholine, but higher doses are required. The shape of the dose-response curve is the same, but the series of parallel curves are shifted to higher dose levels. Therefore, the alkyl trimethylammonium compounds are said to possess the same intrinsic activity as acetylcholine, being able to produce the same maximal response, but to show a lower affinity for the receptor, since larger amounts of drug are required.

By contrast, structural change of a molecule can lead to a gradual decline in the maximal height and slope of the log dose-response curves (Fig. 2-8), in which case the loss in activity may be attributed to a decline in intrinsic activity. For example, pentyl trimethylammonium ion is able to produce a full acetylcholinelike contraction. Successive substitution of methyl by ethyl groups (pentyl ethyl dimethylammonium, pentyl diethyl methylammonium, pentyl triethylammonium) leads to successive decreases in the maximal effect obtainable, with pentyl triethylammonium ion producing no observable contraction. The loss in acetylcholinelike activity for pentyl triethylammonium ion is apparently due to a loss in intrinsic activity, without a significant decrease in the affinity for the receptor, since the compound acts as a competitive inhibitor (antagonist) for active derivatives of the same series.

In the case of an antagonist, it is desirable to have high affinity and low or zero intrinsic activity—that is, to bind firmly to the receptor, but to be devoid of activity. Many examples are available where structural modifications of an agonist molecule lead successively to compounds with decreasing agonist and increasing antagonist activity. Such modifications on acetylcholinelike structures, usually by addition of bulky non-polar groups to either end (or both ends) of the molecule, may lead to the complete antagonistic activity found in the parasympatholytic compounds (e.g., atropine) discussed in Chapter 12.

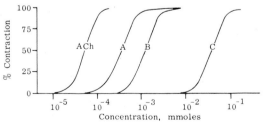

Fig. 2-7. Dose-response curves for contraction produced by acetylcholine (ACh) and alkyltrimethylammonium salts on the rat jejunum.

A. $CH_3CH_2OCH_2CH_2\overset{+}{N}Me_3$.

B. $CH_3CH_2CH_2CH_2CH_2\overset{+}{N}Me_3$. C. $CH_3CH_2CH_2\overset{+}{N}Me_3$.
(Modified from Ariëns, E. J., and Simonis, A. M.[100])

In contrast to the occupancy theory, Croxatto[102] and Paton[103] have proposed that excitation by a stimulant drug is proportional to the *rate* of drug-receptor combination rather than to the number of receptors occupied. The *rate theory* of drug action proposes that the rate of association and dissociation of an agonist is rapid, and this leads to the production of numerous impulses per unit time. An antagonist, with strong receptor-binding properties, would have a high rate of association but a low rate of dissociation. The occupancy of receptors by antagonists, assumed to be a nonproductive situation, prevents the productive events of association by other molecules. This concept is supported by the fact that even blocking molecules are known to cause a brief stimulatory effect before blocking action develops. During the initial period of drug-receptor contact when few receptors are occupied, the rate of association would be at a maximum. When a significant number of sites are occupied, the rate of association would fall below the level necessary to evoke a biological response.

The *occupation* and the *rate* theories of drug action do not provide specific models at the molecular level to account for a drug acting as agonist or antagonist. The *induced-fit* theory of enzyme-substrate interaction,[104] in which combination with the substrate induces a change in conformation of the enzyme, leading to an enzymatically active orientation of groups, provides the basis for similar explanations of mechanisms of drug action at receptors. Assuming that protein constituents of membranes play a role in regulating ion flow, it has been proposed[105] that acetylcholine may interact with the protein and alter the normal forces which stabilize the structure of the protein, thereby producing a transient rearrangement in the membrane structure and a consequent change in its ion-regulating properties. If the structural change of the protein led to a configuration in which the stimulant drug was bound less firmly and dissociated, the conditions of the *rate* theory would be satisfied. A drug-protein combination which did not lead to a structural change would result in a stable binding of the drug and a blocking action.

A related hypothesis (the *macromolecular perturbation theory*) of the mode of acetylcholine action at the muscarinic (postganglionic parasympathetic) receptor has been advanced by Belleau.[106] It is proposed that interaction of small molecules (substrate or drug) with a macromolecule (such as the protein of a drug receptor) may lead either to *specific conformational perturbations* (SCP) or to *nonspecific conformational perturbations* (NSCP). A SCP (specific change in structure or conformation of a protein molecule) would result in the specific response of an agonist (i.e.,

the drug receptor would possess intrinsic activity). If a NSCP occurs, no stimulant response would be obtained, and an antagonistic or blocking action may be produced. If a drug possesses features which contribute to formation of both a SCP and a NSCP, an equilibrium mixture of the two complexes may result, which would account for a partial stimulant action.

The alkyl trimethylammonium ions ($R\overset{+}{-}NMe_3$), in which the alkyl group, R, is varied from 1 to 12 carbon atoms, provide a homologous series of muscarinic drugs which serve as models for the macromolecular perturbation theory of events which may occur at the drug receptor. With these simple analogs, hydrophobic forces, in addition to ion-pair formation, are considered to be the most important in contributing to receptor binding. Lower alkyl trimethylammonium ions (C_1 to C_6) stimulate the muscarinic receptor and are considered to possess a chain length which is able to form a hydrophobic bond with nonpolar regions of the receptor, altering receptor structure in a specific perturbation (Fig. 2-9; e.g., $C_5\overset{+}{N}Me_3$). With a chain of 8 to 12 carbon atoms, the antagonistic action observed is considered to result from a nonspecific conformational perturbation (NSCP) of a network of nonpolar residues at the periphery of the catalytic surface (Fig. 2-9; e.g., $C_9\overset{+}{N}Me_3$). The intermediate heptyl and octyl derivatives act as partial agonists, and it is considered that they may form an equilibrium mixture of drug-receptor combinations, with both active SCP forms and inactive NSCP forms present (Fig. 2-9; e.g., $C_7\overset{+}{N}Me_3$).

The events initiated by specific conformational changes of the receptor are unknown. However, it is suggested[106] that water freed of its ordered structure during the formation of a hydrophobic bond may be available for hydration of Na^+ and K^+ during transport. An alternate concept is that a certain chain length (C_1-C_6) or hydrophobic character for a portion of the drug is effective in bringing order to protein strands, so that groups whose interaction contributes to the energy necessary for ion transport are brought into proximity. A longer hydrophobic chain (C_9-C_{12}) could bring about disorder by associating with an additional segment of protein, either altering the ordered state required or screening the site from a necessary interaction with other molecules. A specific conformational perturbation of the enzymatic drug receptor monoamine oxidase during association with its substrates also has been proposed by Belleau.[107]

As discussed earlier (see The Drug Receptor) a variety of specific operational effects are known to be produced when a receptor is occupied. These include changes in membrane potential and ion flux, alterations in intracellular cyclic nucleotide levels (produc-

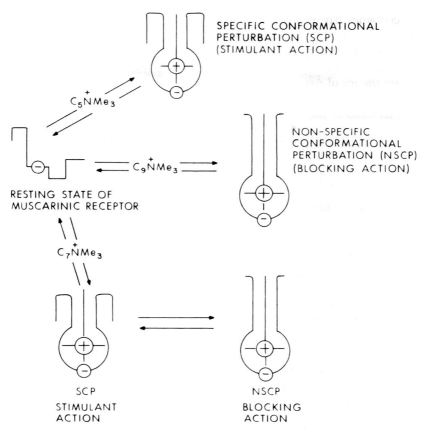

SPECIFIC CONFORMATIONAL
PERTURBATION (SCP)
(STIMULANT ACTION)

$C_5\overset{+}{N}Me_3$

NON-SPECIFIC
CONFORMATIONAL
PERTURBATION (NSCP)
(BLOCKING ACTION)

$C_9\overset{+}{N}Me_3$

RESTING STATE OF
MUSCARINIC RECEPTOR

$C_7\overset{+}{N}Me_3$

SCP
STIMULANT
ACTION

NSCP
BLOCKING
ACTION

Fig. 2-9. Schematic representation of alkyl trimethylammonium ions reacting with the muscarinic receptor. (Modified from Belleau[106])

tion of "second messenger," cyclic AMP), and induction of enzyme synthesis.

In general, drugs which reduce the activity of other drugs, neurotransmitters or hormones (such as acetylcholine, epinephrine, serotonin, various steroids and histamine) are called *antagonists* or *blocking agents*. If the substrate is a compound normally required in the metabolism of the organism (e.g., vitamins, coenzymes), the drug that blocks its use is called an *antimetabolite*. The best example of an effective and useful antimetabolite is the antibacterial sulfonamides (see Chap. 5). These drugs, close structural relatives to *p*-aminobenzoic acid, interfere with the incorporation of the latter into pteroylglutamic acid (folic acid) required by some bacteria. Since mammals obtain their folic acid preformed from blood sources, a toxic effect selective for *p*-aminobenzoic-acid-requiring bacteria is achieved by the sulfonamide antimetabolites.

RATIONAL DRUG DESIGN

The increased understanding of the interrelationships between drug-receptor interactions and the physicochemical properties of drug molecules has greatly enhanced the ability of medicinal chemists to design new drug entities. It has also helped to clarify observed structure-activity relationships and to provide model receptors as templates in a search for more effective and less toxic agents. The following example illustrates this process.

HISTAMINE H₂-RECEPTOR ANTAGONISTS

Histamine stimulates contraction of smooth muscle, and this effect is suppressed by classical antihistamines containing aromatic lipophilic groups connected by a chain of about three atoms to a basic nitrogen atom (see Chap. 16). The receptor that mediates this effect has been defined as the histamine H_1-receptor. In contrast, the histamine stimulation of gastric acid secretion is not blocked by H_1-antagonists, and this effect was postulated to be mediated by a receptor with different characteristics, defined as the histamine H_2-receptor.[108] Systematic efforts to design a specific H_2-antagonist were initiated in 1964 and culminated in the introduction in 1977 of the novel drug cimetidine.

It has long been recognized that two classes of receptors exist for epinephrine, called alpha- and beta-adrenergic receptors (see Chap. 11). This classification is based on the existence of selective agonists and antagonists for different pharmacological effects. Using a simple analogy between the aromatic ring structure and side chain groups present in both epinephrine and histamine, initial systematic alterations of the imidazole ring of histamine were carried out. These studies showed that the 4-methyl derivative was a selective H_2-agonist with only weak H_1-agonist properties. This separation of agonist effects established the validity of the dual receptor concept and showed that simple ring alterations did not produce an antagonist.

Histamine: R=H
4-Methyl histamine: R=CH$_3$

Conversion of the side chain ammonium group of histamine to the strongly basic guanidine group [$-NH(C=NH)NH_2$] gave the first, although weak, H_2-antagonist. Increasing the length of the methylene side chain of the guanidine analog increased antagonistic potency, but the compounds retained an undesirable partial agonist activity. Replacement of the positively charged guanidine group with the still polar but uncharged thiourea residue removed agonist activity and gave the potent antagonist burimamide.

Burimamide

Because of the low potency of burimamide on oral use, due to poor absorption from the gastrointestinal tract, a more lipophilic derivative, metiamide, was prepared. This involved addition of a 4-methyl substituent to the imidazole ring, and replacement with a sulfur atom of one methylene group in the side chain of burimamide.

Metiamide

Although this was orally effective, observation of a low incidence of the blood disorder of agranulocytosis

during clinical trials led to further alteration of the thiourea moiety, which was implicated in this side reaction. The residue first found to produce an antagonist effect, guanidine, was modified by the addition of the N-cyano substitution to form the uncharged but still polar N-cyanoguanidine moiety. The strongly electronegative cyano group reduces the basic character of the guanidine residue. The resulting compound, cimetidine, is the first clinically useful H_2-receptor blocking agent, and it is widely used to inhibit gastric acid secretion in the treatment of duodenal ulcer (see Chap. 16).

Cyanoguanidine

Cimetidine

SELECTED PHYSICOCHEMICAL PROPERTIES

Factors which influence the passage of a drug from its site of administration to its site of action have been described in the preceding sections. Some specific physicochemical properties which are important to drug action will now be discussed in greater detail.

IONIZATION

Acids and bases may be responsible for biological action as undissociated molecules or in the form of their respective ions. A great many compounds, particularly the weak acids and bases, appear to act as undissociated molecules. It is probable that when action takes place inside the cell or within cell membranes, molecules gain entrance to the cell in the undissociated form and after reaching the site of action may then function as ions. Numerous examples of this type are known, and several have been described in previous sections. A general relationship between biological activity and pH for weak acids and bases which require a high concentration of undissociated molecules for maximal effect is shown in Figure 2-10.

For example, the antibacterial activity of benzoic acid, salicylic acid, mandelic acid and other acids of this type is greatest in acid media. The efficiency of these acids as antibacterial agents may be increased as much as 100 times in going from neutral to acid solutions (pH 3).[109]

The antibacterial action of the phenols is greatest

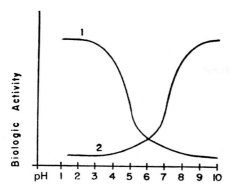

Fig. 2-10. Relationship between biological activity and pH of weak acids and bases (1 = acids; 2 = bases.)

at a pH below 4.5, and the activity again increases at a pH of 10 and above. This increase at high pH has been attributed to partial oxidation of the phenol to a more active quinone.

The solubility and partition coefficients of acids and bases may be altered greatly by changes in pH. Thus, cocaine hydrochloride is freely soluble in water (1:0.4), while the free base has only low water-solubility (1:600). On the other hand, the solubility of the free base in chloroform, ether and vegetable oils is rather high, and the partition between these solvents and water will favor the nonpolar solvents (high partition coefficients). The partition coefficients of the salt between water and all of the nonpolar solvents will, of course, be extremely low. The same is true in general for the salts of other acids and bases. The salts of acids or bases that are absorbed as undissociated molecules will develop their biological activity in proportion to the concentration of free undissociated molecules in the solution.

Minor changes in structure can produce significant changes in the degree of ionization of a weak acid or base. This change may be the primary reason for the presence or absence of biological activity in closely related compounds. Thus, all barbituric acid derivatives that are useful central nervous system depressants are 5,5-disubstituted, whereas barbituric acid and its 5-monosubstituted derivatives are inactive. It is most likely that the relatively high acidity of barbituric acid (pKa 4.0) and of its 5-monosubstituted derivatives (e.g., 5-ethylbarbituric acid; pKa 4.4) is responsible for the lack of hypnotic-sedative properties. These compounds are stronger acids, since they are able to assume a completely aromatic structure which can stabilize the barbiturate ion by delocalization of the extra pair of electrons formed. At physiologic pH of 7.4, these compounds are about 99.9 percent in the polar ionic form and, therefore, do not effectively penetrate the lipoidal barriers to the central nervous

system. In contrast, 5,5-disubstituted barbituric acid derivatives cannot assume fully aromatic character and are much weaker acids, usually ranging in pKa from 7.0 to 8.5 (e.g., 5,5-diethylbarbituric acid, barbital; pKa 7.4). Therefore, at physiologic pH, these compounds exist about 50 percent or higher in the nonpolar, nonionized form, capable of ready passage into the lipoidal tissues of the central nervous system.

Barbiturate Ion

R = H: Barbituric Acid
R = C_2H_5:   5-Ethylbarbituric Acid

5,5-Diethylbarbituric Acid   5,5-Diethylbarbiturate Ion
(Barbital)

In addition to modifying the physical properties of solutions, changes in pH may also affect the reactivity of acidic and basic groups on the cell surface or within the cell. At the isoelectric point, potential anions and cations in a protein or cell exist as "zwitterions." The effect of modifying the pH above or below the isoelectric point may be shown as follows:

Alanine
(Zwitterion)

Cation

Anion

Increasing the pH of the medium will increase the concentration of the anions on the cell, thereby increasing the activity for biologically active cations. Decreasing the pH of the medium will increase the concentration of cations on the cell and thereby increase the activity for biologically active anions.

Active ions

Some compounds show increased biological activity when their degree of ionization is increased. Because of the difficulty with which ions penetrate membranes, it is most likely that compounds of this type exert their effect on the outside of the cell.

The percent of ionization at the body pH (7.3) for a large number of acridine compounds has been recorded by Albert and co-workers.[110] It was observed that a basicity sufficient to induce at least 75 percent ionization at pH 7.3 at 20° (or 60% at 37°) is necessary for effective antibacterial action in the series. It also was shown that the acridine cations are largely responsible for the activity. The undissociated molecules, anions or zwitterions have an insignificant effect on activity. Representative members of the group are shown in Table 2-8.

Amino group substitution has a marked influence on the base strength of the heterocyclic nitrogen. Resonance stabilization of the ion by an amino group in the 3, 6 and the 9 positions will increase base strength, leading to a higher concentration of ion at pH 7.3 and increased bacteriostatic activity.

3-Aminoacridinium ions

Substitution of an amino group in the 4 position leads to base-weakening intramolecular hydrogen bonding; substitution in the 1 and the 2 positions permits no resonance stabilization of the ion, and base strength remains low.

4-Aminoacridine

Thus, substitution to produce a biologically active cation at physiologic pH is the most important structural feature of the aminoacridines. Other features of the molecule are also of importance, for if the total flat surface of the molecule is reduced below about 38 square angstroms, antibacterial activity is largely lost. Examples are 9-aminoacridine with one ring reduced (9-aminotetrahydroacridine; antibact. conc. 1/5,000); and 4-aminoquinoline (antibact. conc. <1/5,000).

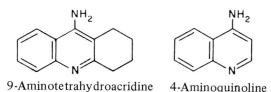

9-Aminotetrahydroacridine   4-Aminoquinoline

It is postulated that a sufficiently flat surface is necessary to supplement, by van der Waals' forces, the ionic bond between the drug cation and the receptor anion.

The basic dyes (e.g., triphenylmethanes, acridines, etc.) appear to function as antibacterial cations by reacting with essential anions (acid groups) of the bacterial cell to give slightly dissociated compounds. The slightly dissociated salt would have a relatively high stability constant, and the larger this stability constant, the better the compound can compete with hydrogen ions for the essential anionic groups on the cell. In this manner, functional groups of the organism can be blocked and cellular metabolism inhibited (bacteriostasis). This means that active compounds of this type must be relatively highly ionized at body pH. A large number of substances show marked antibacterial

TABLE 2-8

IONIZATION AND BACTERIOSTATIC EFFECTS OF AMINOACRIDINES*

Acridine	Min. Bacteriostatic Conc., Strept. Pyog.	Percent Ionized (pH 7.3; 37°)
3-NH₂	1/80,000	73
9-NH₂	1/160,000	99
3,6-diNH₂	1/160,000	99
3,7-diNH₂	1/160,000	76
3,9-diNH₂	1/160,000	100
4,9-diNH₂	1/80,000	98
4-NH₂	1/5,000	<1
2-NH₂	1/10,000	2
1-NH₂	1/10,000	2
4,5-diNH₂	<1/5,000	<1
2,7-diNH₂	1/20,000	4

* Adapted from Albert, A.: Selective Toxicity, p. 262, New York, Wiley, 1968.

activity and yet have little else in common beyond possessing cations of high molecular weight (150 or more) and being highly ionized at pH of 7. Among these may be listed the aliphatic amines, quaternary ammonium compounds, diamines, amidines, diamidines, guanidines, biguanidines, pyridinium compounds, etc.

The relatively high molecular weight lipophilic residues joined to the cationic head must contribute hydrophobic or van der Waals' binding to complementary nonpolar regions of the bacteria, in addition to the interaction between the drug cation and anionic or polar regions of the bacteria. Sufficient total associative forces must be present to prevent displacement by competing cations, such as hydrogen ion, so that the cationic antiseptic will be retained and will produce a toxic disruption of normal bacterial function.

In the same manner that the quaternary ammonium compounds (invert soaps) function as biologically active cations, the ordinary soaps may function as anions. However, their antibacterial activity is extremely low. Less is known concerning the biologically active anions than of the corresponding cationic compounds, yet it may be assumed that they differ only in their mode of action by competing for hydroxyl ions, rather than hydrogen ions, on a cationic group of an essential enzyme. Highly ionized anionic compounds normally show no significant biological activity, and this may be attributed to the predominantly anionic nature of living cells. For example, most bacteria have an isoelectric point of approximately 4 and, at a pH of 7 or more, are anionic in character.

It has been pointed out that in certain cases biological activity increases with increased ionization. In other cases, where undissociated molecules are responsible for the biological effect, the activity decreases with increased ionization. However, it should be kept in mind that in the case of biologically active acids and bases the concentration of ions and undissociated molecules is determined by the pKa of the acid or the base and the pH value of the environment in which action occurs.

HYDROGEN BONDING AND BIOLOGICAL ACTION

The hydrogen bond is a bond in which a hydrogen atom serves to hold two other atoms together. Atoms capable of forming hydrogen bonds are electronegative atoms with at least one unshared electron pair together with a complete octet, and these include F, O, N, and to a lesser degree Cl and S.

The strength of the hydrogen bond varies from 1 to 10 kcal. per mole and usually is about 5 kcal. per mole.

Thus it is only about one tenth as strong as most covalent bonds, which range from about 35 to 110 kcal. per mole. In spite of the relative weakness of the bond, it may have a profound effect on the properties of substances which relate to biological action.

Proteins are held in specific configurations by hydrogen bonds and the denaturation of proteins involves the breaking of some of these bonds. It is significant that virtually all reagents that denature proteins are reagents capable of breaking hydrogen bonds.

The most common hydrogen bonds are the following: O—H · · · O, N—H · · · O, N—H · · · N, F—H · · · F, O—H · · · N and N—H · · · F. If such bonds occur within a molecule they are termed *intramolecular*; if they occur between two molecules they are called *intermolecular* hydrogen bonds. Molecules are known that form both intramolecular and intermolecular hydrogen bonds simultaneously, an example being salicylic acid. *o*-Nitrophenol is an example of a molecule that forms an intramolecular hydrogen bond, while *p*-nitrophenol can form only intermolecular hydrogen bonds. Intermolecular bonds are frequently much weaker than the intramolecular bonds. Strong intramolecular hydrogen bonds usually are found in 5-membered rings (6-membered, counting the hydrogen atom).

o-Nitrophenol

p-Nitrophenol

Since the physical and chemical properties of a compound may be greatly altered by hydrogen bonding, it is reasonable to expect that this may also have a significant effect and show some correlation with biological properties. In a number of cases, such a correlation is present.

1-Phenyl-3-methyl-5-pyrazolone shows no analgesic properties; on the other hand, 1-phenyl-2,3-dimethyl-5-pyrazolone (antipyrine) is a well-known analgesic agent. The former has a melting point of 127° and is comparatively insoluble at ordinary temperatures in water and only slightly soluble in ether. The latter has a lower melting point (112°), is soluble in water (1:1) and moderately soluble in ether (1:43). It is unusual for a methyl group to bring about such large changes.

The effect appears to be best explained by the fact that the first compound through intermolecular hydrogen bonding forms a linear polymer.

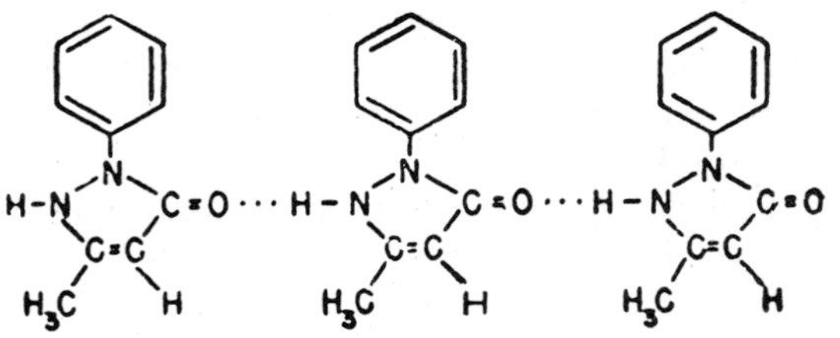

Intermolecular hydrogen bonded
1-Phenyl-3-methyl-5-pyrazolone

The resulting large attractive force between the molecules raises the melting point and lowers the solubility, especially in the nonpolar solvents which are not capable of breaking the hydrogen bonds. On the other hand, the methyl compound (antipyrine) cannot form hydrogen bonds and has only comparatively weak attractive forces between its molecules.

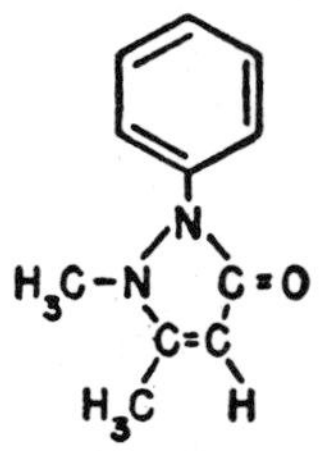

1-Phenyl-2,3-dimethyl-5-pyrazolone
(Antipyrine)

Its melting point is lower in spite of its having a higher molecular weight, and it is freely soluble in nonpolar solvents. Thus antipyrine is adequately soluble in both polar and nonpolar solvents, and has the proper partition characteristics to penetrate the central nervous system.

Salicylic acid (o-hydroxybenzoic acid) has quite an appreciable antibacterial activity, but the para isomer (p-hydroxybenzoic acid) is inactive. The reverse is true for the esters. Methyl salicylate has an extremely weak antibacterial action, but methyl p-hydroxybenzoate shows good action. A number of the esters of p-hydroxybenzoic acid (especially methyl and propyl) are used as preservatives in various pharmaceutical and cosmetic preparations. The difference in antibacterial action of the free acids and their esters may be

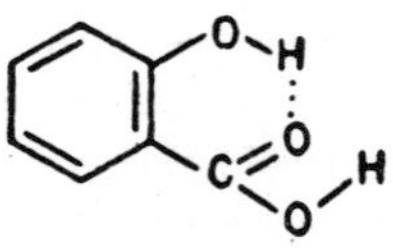

accounted for through hydrogen bond formation. Only the ortho isomer (salicylic acid) shows analgesic and antipyretic properties. Likewise, salicylic acid is the only one of the three isomers that can form intramolecular hydrogen bonds. The m- and the p-isomers can form only intermolecular hydrogen bonds.

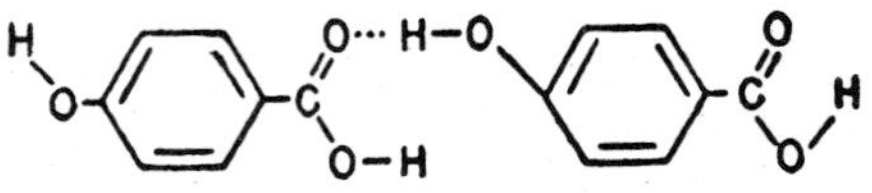

p-Hydroxybenzoic Acid (dimer)

Salicylic acid is a much stronger acid (pKa = 3.0) than p-hydroxybenzoic acid (pKa = 4.5). Salicylic acid is less soluble in water than the p-isomer, but its partition coefficient (benzene/water) is approximately 300 times greater. The higher melting point of p-hydroxybenzoic acid may be associated with intermolecular hydrogen bonding, which can also account for the low partition coefficient and low bactericidal action. It should be noted that salicylic acid with an intramolecular hydrogen bond has the phenolic hydroxyl masked, but the carboxylic acid group is free and can function as an antibacterial agent similar to benzoic acid. p-Hydroxybenzoic acid, on the other hand, must form intermolecular hydrogen bonds leading to a high degree of association and thereby lowering its antibacterial activity. In the case of the esters of salicylic and p-hydroxybenzoic acid, the opposite effect on bactericidal power is observed. Methyl salicylate is without significant antibacterial activity, but the esters of p-hydroxybenzoic acid show useful antibacterial properties. Methyl salicylate, through intramolecular hydrogen bond formation, has the phenolic hydroxyl group masked.

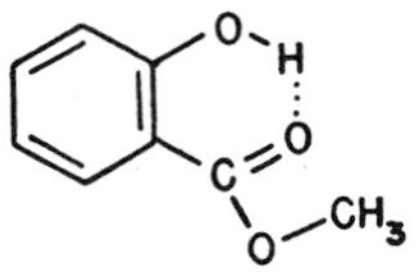

Methyl Salicylate

Methyl p-hydroxybenzoate and other esters of p-hydroxybenzoic acid can form only intermolecular hydrogen bonds, which may be illustrated with the following structure:

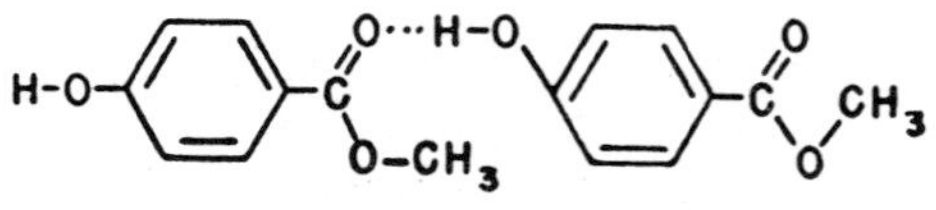

Salicylic Acid

Methyl p-Hydroxybenzoate (dimer)

Association through hydrogen bonding to form the dimer or higher polymers may occur, but the partition coefficient and the antibacterial activity data suggest that the esters of *p*-hydroxybenzoic acid are not highly associated and that they may function as substituted phenols.

The nucleic acids, fundamental reproductive units of cells, provide an important example of molecules held together by specific hydrogen bonds.[111] Nucleic acids are composed of purines and pyrimidines in glycosidic combination with ribose or 2-deoxyribose forming nucleosides, which are in turn phosphorylated to form nucleotides. The nucleotides are linked together through phosphate bridges forming long chains, and these chains associated with other chains through hydrogen bonding form the nucleic acids of molecular weight 200,000 to 2,000,000. The nucleic acids are in turn associated through weak saltlike linkages with proteins. The nucleoproteins have in common the ability to duplicate themselves in the proper environment, and hydrogen bonds play an important role in this process of reproduction. Examples of essentially pure nucleoproteins are the chromosomes and the plant viruses.

The genetic code of the cell, which constitutes the instructions for the synthesis of the cell's proteins, is contained in the cell nucleus in the form of a double-chain molecular helix of deoxyribonucleic acid (DNA). The code consists of sequences of 4 purine and pyrimidine bases. The purine-pyrimidine pairs, adenine-thymine and guanine-cytosine, occupy adjacent positions

Adenine   Thymine

Guanine   Cytosine

on neighboring nucleic acid strands, and are held together by specific hydrogen bonds. A triplet code is involved: a sequence of 3 purine or pyrimidine bases is needed to specify which amino acid will be incorporated in a specific location in the protein.

The large DNA molecule does not take part directly in protein synthesis; instead the genetic code is transcribed into shorter single chains of ribonucleic acid (RNA), called "messenger RNA," which is presumed to be a copy of the bases in one strand of DNA. Still smaller units of RNA, called "transfer RNA," which are specific for each amino acid, pick up and activate the amino acid by acylation of the ribose portion of the RNA. The activated amino acid is deposited at a position in the polypeptide chain specified by messenger RNA, the site of protein synthesis within the cell being a particle called the ribosome.

Hydrogen bonds play a key role in maintaining the structural integrity of the base pairs of DNA. Similar weak chemical bonds must be responsible for the interactions between amino acids, messenger RNA and transfer RNA which ultimately result in the synthesis of specific polypeptide chains. A more detailed understanding of the specific interactions involved in these processes will aid in understanding and attacking many problems of abnormal cellular function.

Mutagenic agents may cause rupture of the nucleic acid chain, thus disrupting the self-duplicating sequence. Many current attempts at chemotherapeutic control of cancer are directed at this level of cellular function. The *alkylating agents* (nitrogen mustards) are thought to act by replacing the weak and reversible hydrogen bonds between adjacent nucleic acid strands with strong and relatively irreversible covalent bonds. In this way nucleic acid regeneration and cell division in the rapidly proliferating cancer cells may be inhibited. Some *antimetabolites* are thought to act by their structural similarity to the purine and pyrimidine constituents of nucleic acids. An example is mercaptopurine, an analog of adenine in which the 6-amino group, normally responsible for specific hydrogen bonding, is replaced by a 6-thiol group. This is believed to be a close enough analog to fit into pathways leading to normal nucleic acid formation but incapable of replacing the functions of adenine which contribute to cellular reproduction.

CHELATION AND BIOLOGICAL ACTION

The term *chelate* is applied to those compounds that result from a combination of an electron donor with a metal ion to form a ring structure. The compounds capable of forming a ring structure with a metal are designated as *ligands*. If the metal is bonded to carbon, the ring structure is not a chelate but an organometallic compound which has different properties. If the metal is not in a ring, the compound is called simply a metal *complex*. Nearly all the metals can form chelates and complexes. However, the electron donor atoms in the chelating agent are limited almost entirely to N, O and S. If the complex-forming ligand supplies both electrons for chelation, then the bond is classified as a co-ordinate covalent bond and by convention is represented as M←X, where M is the metal and X the ligand. If one electron is supplied by the metal and one by the ligand (normal covalent bond), the bond is shown as M—X.

A ligand molecule containing only 2 electron-donating groups is designated as "bidentate" and is able to form only a single ring; if it contains 3 electron-donating groups, it is "tridentate" and may form 2 rings in an interlocked complex; the porphyrins are able to form a number of interlocked ring systems with metals and are designated as "polydentate" structures.

The size of the rings in chelate compounds is of interest with respect both to stability and occurrence. Three-membered rings have not been identified, but 4-membered rings are known, and 4-membered rings containing sulfur may be quite stable. The 5- and the 6-membered chelate rings are most common and usually show the greatest stability.

The "normal" chemical reactions of a metal ion in solution disappear if a chelate is formed. The chelate may serve to prevent precipitation of an ion which normally would precipitate. For example, the cupric ion in basic Fehling's solution normally would be precipitated as cupric hydroxide but is prevented from doing so by the formation of the copper tartrate chelate. In this case, the formation of the chelate results in an increased water-solubility. In other cases, the chelate may be insoluble in water and soluble in organic solvents. Water-soluble chelating agents, called *sequestering agents*, often are used to remove objectionable metal ions by combining with them to form stable water-soluble chelates.

A number of naturally occurring chelates are present in biological systems. The amino acids, proteins and acids of the tricarboxylic acid cycle are the principal *ligands*; and the metals involved are iron, magnesium, manganese, copper, cobalt and zinc, among others. A group with iron present is the heme-proteins, such as hemoglobin, which is found in the red blood cells of vertebrates and is involved in oxygen transport. In the heme-proteins, an iron atom is covalently bound to a porphine derivative. Other heme-proteins are myoglobin, an intracellular pigment involved in oxygen storage; catalase, a compound present in the tissues of plants and animals which catalyzes the decomposition of H_2O_2; peroxidases, which are enzymes involved in the oxidation of a substrate with peroxides; cytochromes, which play a part in cellular oxidation.

Copper-containing enzymes include the oxidases, ascorbic acid oxidase, tyrosinase, polyphenoloxidase and laccase. Magnesium is present in chlorophyll, but it is involved also in the action of some proteolytic enzymes, phosphatases and carboxylases. Manganese activates most carboxylases and some proteolytic enzymes. Zinc, which is present in insulin, has an activating effect on some carboxylases, proteolytic enzymes and phosphatases. Cobalt activates some enzymes belonging to each of the above classes and is present in vitamin B_{12}.

The fact that a number of biologically important compounds are chelates opens up other approaches to chemotherapy. One such approach depends on the use of an unnatural chelating agent to reduce or eliminate the toxic effects of a metal. To serve in this capacity the chelating agent (ligand) must effectively compete with the chemical systems in the body to which the excess metal is bound.

The agent, because of its greater affinity for the metal, forms a more stable chelate, thereby decreasing the concentration of the toxic metal ion in the tissues by binding it as a soluble chelate for excretion by the kidneys. The stability of chelates is expressed as a constant (log K_s) which represents the over-all equilibrium between the metal and the chelates which it forms with the ligand. In the case of a trivalent metal (e.g., Fe^{+++}), log K_1 represents the stability constant for the 1:1 chelate; log K_2 the 2:1 chelate; log K_3 the 3:1 chelate, and the over-all constant (log K_s) is the product of the individual constants,[112] that is,

$$\log K_s = \log K_1 + \log K_2 + \log K_3.$$

A chelate varies with the ligand and the metal to which it is bound. For example, the *stability constants* (log K_s) of glycine 2:1 chelates with several divalent metals are as follows: Cu^{++} 15; Ni^{++} 11; Co^{++} 9; Fe^{++} 8; and Mn^{++} 5.5.

Dimercaprol was first introduced in 1945 under the name BAL (British anti-lewisite) as an antidote for the organic arsenical "lewisite." Subsequent studies have revealed its effectiveness for the treatment of poisoning due to antimony, gold and mercury, as well as arsenic.[113]

($\pm$)Penicillamine, a hydrolysis product of the various penicillins, is an effective antidote for the treatment of poisoning by copper.[114] Hepatolenticular degeneration (Wilson's disease), a familial disorder in which there is decreased excretion of copper and, sometimes, increased excretion of amino acids, may be treated with penicillamine to promote the removal of protein-bound copper and its excretion in the urine.

Penicillamine also has been used with some success as an antidote for the treatment of mercury and lead (divalent metals) poisoning. The 2:1 copper chelate, possessing two ionizable and solubilizing carboxyl groups, acts as a sequestering agent.

Penicillamine   Penicillamine   2:1 Copper Chelate
                        1:1 Copper Chelate   Forms water-soluble salts

Deferoxamine mesylate (Desferal), a trihydroxamic acid compound made up of the elements of 3 moles of a 1,5-pentamethylene diamine, 2 moles of succinic acid, and one of acetic acid, is isolated from *Streptomyces pilosus*.[115] The compound combines with Fe^{+++} to form a water-soluble chelate which is excreted by the kidneys. The agent removes excess iron from the tissues but does not displace it from essential proteins (e.g., transferrin) involved in the iron transport mechanism. The compound is reported to be selective for iron with little or no affinity for calcium, copper and other metals.[116,117] It is nontoxic and has been used successfully in the treatment of primary (hereditary) and secondary hemochromatosis and as an effective antidote for the treatment of acute iron poisoning in children.

The reddish-colored iron chelate of deferoxamine has a high stability constant ($\log K_s = 30.7$) which may be attributed to its unique chemical structure in which the iron is octahedrally bound by the hydroxamic acid oxygen atoms and carbonyl oxygens of the ligand.

The compound ethylenediaminetetraacetic acid (EDTA) forms water-soluble stable chelates with many metals and is an important sequestering agent

which has received wide application. Among these may be mentioned: its use as an antioxidant for the stabilization of drugs which rapidly deteriorate in the presence of trace metals (e.g., ascorbic acid, epinephrine and penicillin); prevention of rancidity in detergents; purifying oils; clarifying wines and soap solutions; removal of lead arsenate spray residues from fruits; removal of radioactive contaminants; titration of metals; and as an antidote for heavy metal poisoning.

Like penicillamine, EDTA is able to form water-soluble, stable metal chelates in the body which may be excreted readily. The free acid and sodium salts of EDTA, when administered to mammals, produce an excessive loss of essential body calcium and are quite toxic. The calcium-disodium salt (edathamil) is comparatively nontoxic and serves as an effective antidote for the treatment of lead poisoning. It is also reported to be effective as an antidote for other heavy metals, including copper, chromium, iron and nickel.[118]

The general structure given below is assigned to the water-soluble metal chelates of EDTA. It should be noted that the metal (M) is bound by 2 coordinate-covalent bonds ($\rightarrow$) and 2 normal covalent bonds (—). The strength of the metal binding (stability constant)

Deferoxamine
(Chelating groups circled)

Deferoxamine–iron chelate

varies with each metal. For the divalent ions the order is as follows:

$$Cu^{++} > Ni^{++} > Pb^{++} > Co^{++}, Zn^{++}$$
$$> Fe^{++} > Mn^{++} > Mg^{++}, > Ca^{++}.[119]$$

Disodium salt of metal chelates with
ethylenediamine tetraacetic acid
M = Bound metal

We are indebted to Albert and co-workers[120,121] for much of our present understanding of the structure-activity relationships of the chelates as antibacterial agents. 8-Hydroxyquinoline (oxine) was observed to precipitate a number of the heavy metals under physiologic conditions of temperature and pH, and initially it was suggested that the antifungal and antibacterial properties may be due to the removal of trace metals essential for metabolism of the organisms.[122] A study of the 7 isomeric monohydroxyquinolines demonstrated that only the 8-hydroxy isomer was active in inhibiting the growth of microorganisms and that the same isomer was the only one to form metal chelates. This observation stimulated a study of derivatives of 8-hydroxyquinoline and related analogs which led to the following generalizations on structure-activity relationship: (1) Both the 8-methyl ether and the 1-methyl derivatives of 8-hydroxyquinoline which are unable to form chelates show no antibacterial effect. (2) Substitution of a mercapto for the hydroxy group in oxine gives an active chelating agent which is also active as an antibacterial. (3) The substitution of a methyl group in the 2 position of oxine gives an active chelating agent in vitro, but the compound is relatively inactive as an antibacterial. This decreased activity is attributed to lack of penetration of the cell, or interaction with the cell receptor, due to steric hindrance. (4) The introduction of a highly ionizable group in oxine (e.g., 8-hydroxyquinoline-5-sulfonic acid) does not alter the chelating property in vitro, but the antibacterial activity is lost, presumably due to the inability of the ion to penetrate the cell wall. A high partition coefficient appears to be essential for antibacterial activity.

It has been well established that oxine and its analogs act as antibacterial and antifungal agents by complexing with iron or copper. Oxine, in the absence of these metals, is nontoxic to microorganisms. The site of action (within the cell or on the cell surface) has not been established. However, Albert and co-workers,[123] in a study of a series of mono-aza and alkylated mono-aza-oxines, observed that the compounds chelated with metals as effectively as oxines and that the antibacterial activity paralleled the oil/water partition coefficients. This observation suggests that the site of action of oxine and its analogs is inside the bacterial cell. It has also been suggested that the site of action may be on the cell surface.[124]

Since ferrous iron is easily oxidized after chelating with oxine, it is reasonable to believe that ferric chelate may predominate, although the toxic action of oxine is equally developed by the addition of ferrous or ferric salts. The addition of an excess of either iron or oxine inhibits the antibacterial action.[125] Thus, the

1:1-oxine-ferric chelate
unsaturated: active

2:1-oxine-ferric chelate
unsaturated: active

3:1-oxine-ferric chelate
saturated: inactive

growth of *Staph. aureus* in untreated meat broth is completely inhibited by oxine (M/100,000), but this concentration has no effect on the organism when suspended in distilled water. The toxic effect of oxine is due to its combination with trace amounts of iron in the meat broth; when the concentration of oxine was increased (M/800) the inhibition of growth (antibacterial effect) disappears, due to a "concentration quenching" effect which is attributed to shifting of the equilibrium from the unsaturated (1:1- and 2:1-complexes) to the saturated, nontoxic 3:1-oxine-iron complex; and inhibition of growth again occurs when the concentration of iron is increased (M/800), since the equilibrium is shifted from the saturated 3:1-complex to the unsaturated 1:1- and 2:1-complexes which are toxic.

If the site of action is within the cell, it is reasonable to assume that only the saturated (3:1-oxine-ferric complex) will be able to penetrate the cell membrane. The unsaturated 1:1- and 2:1-complexes as cations cannot penetrate. It is postulated that the nontoxic 3:1-complex which is able to penetrate the cell membrane breaks down inside the cell to form the toxic unsaturated 1:1- or 2:1-complexes.[123] Although it appears less likely that the site of action is outside the cell membrane, it must be assumed that, if such is the case, the unsaturated 2:1-complex would be responsible for the toxic effect. Since excess iron or oxine decreases this toxic effect, the equilibrium would be shifted to form primarily the 1:1- or the 3:1-complexes, respectively, which would bring about a concomitant decrease in toxicity.

The antibacterial properties of oxine-iron complexes are antagonized by metals that form more stable complexes. The addition of low concentrations of cobaltous sulfate (M/25,000) completely inactivates the antibacterial action of (M/100,000) oxine-iron solutions.[125]

The structures below are representative of the biologically active and inactive analogs of oxine. It should

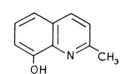
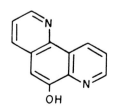

8-Hydroxyquinoline (Oxine)
Chelates: active

8-Methoxyquinoline
Nonchelating: inactive

Oxine Methochloride
Nonchelating: inactive

8-Mercaptoquinoline
Chelates: active

8-Hydroxyquinoline-5-
sulfonic acid
Chelates: inactive

7-Chloro-8-hydroxy-
quinoline
Chelates: active

2-Methyl-oxine
Chelates: decreased activity

4-Aza-oxine
Chelates: active

4-Hydroxyacridine
Chelates: active

6-Hydroxy-*m*-phenanthroline
Chelates: active

5,6 Benzo-oxine
Chelates: active

be noted that all active compounds (antibacterial) form metal chelates, but not all chelating structures are biologically active.

Numerous drugs unrelated to oxine form chelate complexes with metals; and although the complex formation may have no direct relation with the major action of the drug, it may be responsible for significant side-effects. Thus, the antitubercular agent, thiacetazone, may produce an onset of *diabetes mellitus,* and it has been suggested that this may be due to its ability to chelate with zinc in the *beta* cells of the pancreas, thereby inhibiting the production of insulin. Diphenyldithiocarbazone, oxine and alloxan are believed to react in the same manner to produce a diabetogenic effect. The anemia produced by administration of the hypotensive agent hydralazine (Apresoline) has been attributed to its ability to complex with iron.[126] Dimercaprol and the antitubercular drug isonicotonic acid hydrazide (INH) tend to induce histaminelike actions, and it has been suggested that this may be due to complexing with a copper-catalyzed enzyme responsible for the destruction of histamine.[127] INH may function as a chelating agent in inhibiting the growth of *Mycobacterium tuberculosis,* but the evidence for this mode of action is not conclusive. The drug is an active chelating agent (see below), and derivatives, such as 1-methyl-1-isonicotinoyl hydrazine, which are unable to chelate are inactive.[128] The salicylates, catechol amines, biguanides, tetracyclines and many other commonly used drugs form metal chelates. Boric acid chelates with the 3,4-hydroxyl groups of epinephrine and related catechol amines without altering the pharmacologic properties.[129]

In summary, chelation may be used for a variety of purposes, including (1) sequestration of metals to control the concentration of metal ions (e.g., buffer systems); (2) stabilization of drugs (e.g., epinephrine); (3) elimination of toxic metals from intact organisms (e.g.,

EDTA as an antidote for treatment of lead poisoning); (4) improvement of metal absorption, which has been demonstrated in plants and, by analogy, may also be true in mammals (e.g., EDTA-iron complex increases uptake of iron in plants); and (5) increasing the toxic effects of a metal (e.g., antibacterial activity of the unsaturated oxine-iron chelates).

The full significance and importance of chelation in biology and medicine remains to be established, but from the knowledge available it is evident that the chelates represent an extremely important group of naturally occurring compounds and that foreign ligands may play an increasingly important role in chemotherapy. A review of the role of metal-binding in the biological activities of drugs has been presented by Foye (see Selected Reading).

OXIDATION-REDUCTION POTENTIALS AND BIOLOGICAL ACTION

The oxidation-reduction potential (redox potential) may be defined as a quantitative expression of the tendency that a compound has to give or to receive electrons. The oxidation-reduction potential may be compared with an acid-base reaction. The latter case may be regarded as the transfer of a proton from an atom in one molecule to the atom in another, while in the case of an oxidation-reduction reaction, there is an electron transfer. Since living organisms function at an optimum redox potential range which varies with the organism, it might be assumed that the oxidation-reduction potentials of compounds of a certain type would correlate with the observed biological effect.[130] However, there are a number of reasons why few satisfactory correlations have been observed. The oxidation-reduction potential applies to a single reversible ionic equilibrium which does not exist in a living orga-

Isonicotinic Acid
Hydrazide (INH)

Enol Form

1:1-INH-Ferric
Chelate

1:1-Catecholamine-
Boron Chelate

2:1-INH-Ferric
Chelate

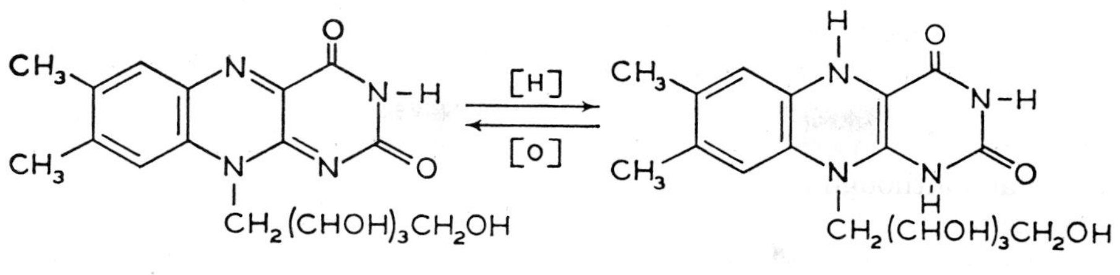

Riboflavin   Dihydroriboflavin

nism. The living cell is obliged to carry on a great many reactions simultaneously, involving oxidations of an ionic and a non-ionic character, some of which are reversible and others irreversible. The access of a drug to the sites of oxidation-reduction reactions in the intact animal is hindered by the complex competing events occurring during absorption, distribution, metabolism and excretion. Included among these may be multiple competing biological redox systems. Therefore, it is to be expected that correlations between redox potential and biological activity generally hold only for compounds of very similar structure and physical properties. In such series, variations in routes of distribution and in steric factors which might modify the drug redox system interaction would be minimized. Only a few series studied have met these criteria. Page and Robinson[131] have studied the relation between the bacteriostatic activity and the normal redox potentials of substituted quinones. They find no simple relation between the reduction potentials (E_o') of 20 substituted quinones and their observed bacteriostatic activity. The quinones showing marked activity against *Staph. aureus* gave reduction potentials falling between -0.10 and $+0.15$ volt, with optimum activity associated with a potential of approximately $+0.03$ volt. The same authors failed to find a similar relation between oxidation-reduction potentials of 18 commercially distributed oxidation-reduction indicators of varied structure and their activity against *Staph. aureus*.

The reduction potentials of a number of acridines have been studied.[132] The more active compounds have an $E_{1/2h}$ of less than -0.4, but no detailed correlation with the antibacterial activity was observed.

Riboflavin, in its cofactor form, owes its biological activity to its ability to accept electrons and be reduced to the dihydro form. This reaction has a poten-

tial of $E_o = -0.185$ volt. Recognizing that retention of most structural features, but alteration of this redox system, could lead to compounds antagonistic to riboflavin, Kuhn[133] prepared the analog in which the two methyl groups of riboflavin were replaced by chlorines. This compound had a potential of $E_o = -0.095$ volt, and its antagonistic properties were suggested as being due to the dichloro-dihydro form being a weaker reducing agent than the dihydro form of riboflavin. It

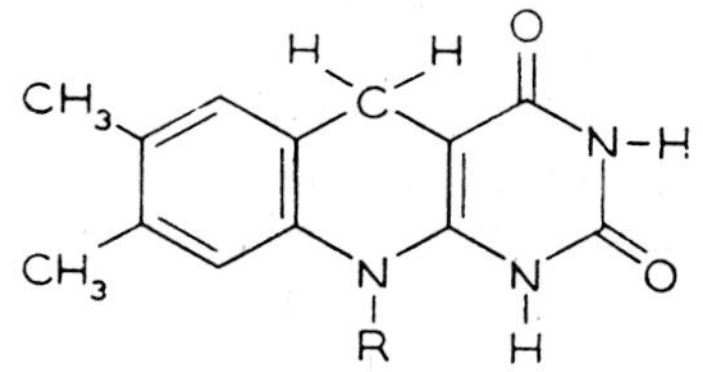

Altered redox potential

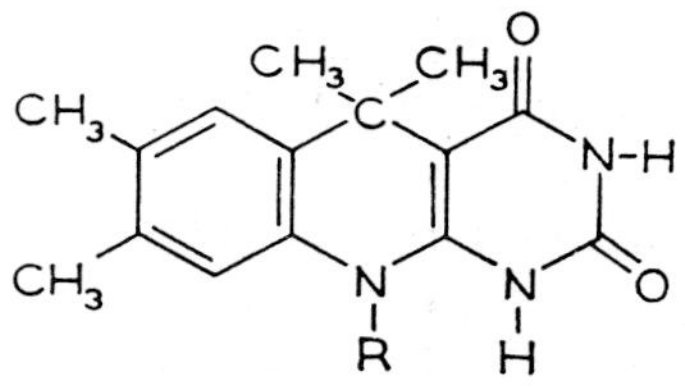

Fixed in dihydro form

Riboflavin analogs

may be absorbed at the specific receptor site but not have a negative enough potential to carry out the biological reductions of riboflavin. Also, "nonredox analogs of riboflavin" have been proposed as potential anticancer agents,[134] and compounds have been prepared which alter the redox potential or fix the molecule in the nonoxidizable dihydro form.

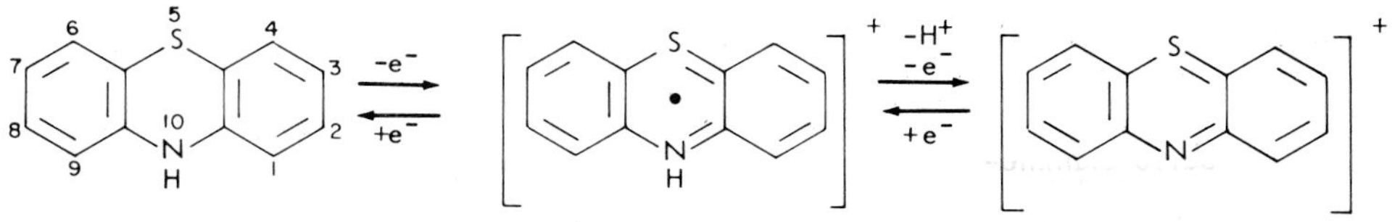

Phenothiazine   Semiquinone Ion
                        (Active species)   Phenozothionium Ion

The anthelmintic activities of a series of substituted phenothiazines have been correlated with the possession of a redox potential which could lead to maximal formation of semiquinone ion (a radical ion) at physiologic pH.[135] Against mixed infestations of *Syphacia obvelata* and *Aspirculurus tetraptera* in mice, anthelmintic activity was present in unsubstituted phenothiazine and those substituted derivatives (3-EtO—; 3-MeO—; 3-Me; 2-Cl—7-MeO—; 4-Cl—7-MeO—; 3-F—; 3-Cl—; 3-Br—) with E_m values* which were within 0.1 volts of the value 0.583 V (acetic acid-water). At a potential similar to that of the biological oxidation-reduction system involved, semiquinone concentration would be maximal and thus facilitate or compete with an essential biological electron transfer reaction, producing a toxic or paralyzing effect. When corrected for solvent effects, the active potential is in the range of that of isolated cytochromes.

The additional requirement of a free 3 or 7 position in the phenothiazine nucleus for significant anthelmintic activity, and the inactivity of phenothiazine tranquilizing drugs (2-substituted 10-dimethylaminopropylphenothiazines) again point up the difficulty of correlating redox potential and activity for compounds with differing structural and solubility characteristics.

SURFACE ACTIVITY: ADSORPTION AND ORIENTATION AT SURFACES

The orientation of surface-active molecules at the surface of water or at the interface of polar and nonpolar liquids takes place with the nonpolar (e.g., hydrocarbon) portion of the molecule oriented toward the va-

* E_m = bivalent midpoint electrode potential.

por phase or nonpolar liquid and the polar groups (e.g., —COOH, —OH, —NH$_2$, —NO$_2$, etc.) toward the polar liquid.

A surfactant molecule exhibits two distinct regions of lipophilic and hydrophilic character and such compounds are commonly categorized as *amphiphilic*, or as *amphiphils*. Molecules of this type may vary markedly from predominantly hydrophilic to predominantly lipophilic, depending on the relative ratio of polar to nonpolar groups present. The polar or hydrophilic groups differ widely in their degree of polarity and, as might be expected, the more polar groups (e.g., $-OSO_3^-$, $-SO_3^-$, $-NR_3^+$ etc.) are able to increase the hydrophilic character of a molecule to a greater extent than the weaker polarizing groups such as $-NH_2$, $-OH$, $-COOH$. Thus, water-solubility is lost, and lipid-solubility is gained in the normal alcohols, amines and carboxylic acids at C_4 to C_6. The higher molecular weight members are miscible with oil and immiscible with water. However, the more active *n*-alkylated anions and cations (e.g., $-OSO_3^-$ and $-NR_3^+$) increase the hydrophilic character, and water-solubility is not lost until the number of carbon atoms reaches C_{14} or higher.

There are four general classes of surface-active agents: (1) *anionic* compounds, such as the ordinary soaps, salts of bile acids, salts of the sulfate or phosphate esters of alcohols, and salts of sulfonic acids; (2) *cationic* compounds, such as the high molecular weight aliphatic amines, and quaternary ammonium derivatives; (3) *nonionic* compounds (e.g., polyoxyethylene ethers and glycol esters of fatty acids); and (4) *amphoteric* surfactants.

The surface-active ions of intermediate to high molecular weight (ca. 150 to 300) show the same electrical and osmotic properties in dilute solutions as equivalent concentrations of inorganic electrolytes. Therefore, the ions in dilute solution are distributed in the

Phenol + Sodium oleate

1:1 association molecular dispersion

Sodium oleate / water

Spherical micelle

water layer / phenol layer / water layer

More stable layered micelle

NP = nonpolar hydrocarbon chain
P = polar carboxyl group

monomeric state. However, with increasing concentration of the surfactant, a critical point is reached, at which the molecules associate (become polymeric) in an oriented fashion to form *micelles.* The concentration at which the polymeric species develops is commonly designated as the *critical micelle concentration* (CMC) and differs for each surfactant.

At the critical micelle concentration large polymers (macromolecules) begin to form, and the solution becomes colloidal in nature. This is a reversible process; therefore, a micelle on dilution will revert to the monomeric state. The *solubilization* of organic compounds (insoluble in water) begins at the CMC and increases rapidly with increasing concentration of the solubilizing agent. The simplified structures above show the micelles of sodium oleate solubilizing an insoluble phenol. When phenols are solubilized by soap, mixed micelles (soap and phenol) are formed, and the activity of the phenol may be enhanced or reduced depending on the ratio of soap to phenol used.

The anthelmintic activity of hexylresorcinol is reported to be increased by low concentrations and decreased by high concentrations of soap. If the soap concentration is kept below the CMC, a 1:1 association of the phenol and soap occurs which facilitates the penetration of phenol through the surface of the worm. If the CMC is exceeded the micelle competes favorably with the worms for the phenol and there is decreased activity.[136]

Compounds showing pronounced surface activity usually are unsuited for use in the animal body. Such compounds are lost through adsorption on proteins, and they also have the undesirable feature of disorganizing the cell membrane and producing hemolysis of red blood cells. In general, highly surface-active agents are limited in use to topical application as skin disinfectants or for the sterilization of inanimate objects, such as instruments. The surface-active cations of the quaternary ammonium type are used for this purpose. They are characterized by having a hydrophilic cation attached to a long nonpolar group. Compounds of this type are nonspecific antibacterial agents, since they are adsorbed on all tissues as well as on bacteria. Their activity is greatly reduced by body fluids and by high molecular weight anions, such as the ordinary soaps.

The antibacterial activity of high molecular weight quaternary ammonium compounds (cationic soaps) appears to be dependent on two or more factors, such as: (1) the charge density on the nitrogen atom; (2) the size and the length of the nonpolar groups attached to the nitrogen; and (3) the lipophilic-hydrophilic balance. The fact that the lower and excessively high molecular weight compounds are inactive indicates that more than a charged nitrogen is necessary for antibacterial activity. The active compounds are those in which the charged nitrogen is unsymmetrically positioned in the molecule; therefore, long nonpolar chains appear to be a necessary structural feature. This suggests that the most active compounds will be those having a maximum charge on an unsymmetrically positioned nitrogen and lipophilic-hydrophilic balance to impart optimum surface activity. Molecules of this type will be attracted and held comparatively firmly to the bacterial cell wall by the cation reacting with anionic cellular groups to form reinforced ionic bonds and the nonpolar portion of the molecule associating through van der Waals' bonds. This action occurs below the CMC.

Like the soaps, the quaternary ammonium compounds can form mixed micelles. Thus, the bactericidal concentration of dodecyl (C_{12}) dimethyl benzyl ammonium chloride must be doubled in the presence of 25 percent of hexadecyl (C_{16}) dimethyl benzyl ammonium chloride.[137] Albert[138] interpreted this finding in terms of mixed micelle formation. His reasoning may be stated briefly as follows: Drugs act in their monomolecular form, but at the CMC and above the micelle competes with the microorganism for the monomers, thereby reducing the effective antibacterial concentration.

Surface-active agents can be expected to have a pronounced effect on the permeability of the cell. Mildly surface-active agents may be adsorbed as a monolayer on the cell membrane and thereby interfere with the absorption of other compounds through this membrane or may alter membrane structure and function. Many central nervous system depressant drugs, such as the hypnotic-sedative, anticonvulsant and central relaxant agents possess the general structure of nonionic surface-active compounds.

REFERENCES

1. Crum-Brown, A., and Fraser, T.: Tr. Roy. Soc. Edinburgh 25:151, 1868–9.
2. Burger, A.: J. Chem. Ed. 35:142, 1958.
3. Wagner, J. B.: J. Pharm. Sci. 50:359, 1961.
4. Wurster, D. E., and Taylor, P. W.: J. Pharm. Sci. 54:169, 1965.
5. Levy, G.: J. Pharm. Sci. 50:388, 1961.
6. Lazarus, J., and Cooper, J.: J. Pharm. Pharmacol. 11:257, 1959.
7. Wagner, J. C.: Biopharmaceutics and Relevant Pharmacokinetics, Hamilton, Ill., Drug Intelligence Publication, 1971.
8. Collander, R., and Bärlund, H.: Acta bot. fenn. 11:1, 1933.
9. Schanker, L. S., *et al.:* J. Pharmacol. Exp. Ther. 123:81, 1958.
10. Overton, E.: Studien über Narkose, Jena, Fischer, 1901.
11. Schanker, L. S.: J. Med. Pharm. Chem. 2:343, 1960.
12. Travell, J.: J. Pharmacol. Exp. Ther. 69:21, 1940.
13. Shore, P. A., Brodie, B. B., and Hogben, C. A. M.: J. Pharmacol. Exp. Ther. 119:361, 1957.
14. Brodie, B. B., and Hogben, C. A. M.: J. Pharm. Pharmacol. 9:345, 1957.
15. Schanker, L. S., *et al.:* J. Pharmacol. Exp. Ther. 120:528, 1957.
16. Benet, L. Z.: Biopharmaceutics as a Basis for the Design of Drug Products, *in* Ariëns, E. J. (ed.): Drug Design, vol. 4, New York, Academic Press, 1973.

17. Hogben, C. A. M., *et al.*: J. Pharmacol. Exp. Ther. 125:275, 1959.
18. Schanker, L. S.: J. Pharmacol. Exp. Ther. 126:283, 1959.
19. Levine, R., Blaire, M., and Clark, B.: J. Pharmacol. Exp. Ther. 121:63, 1957.
20. Cavallito, C. J., and O'Dell, T. B.: J. Am. Pharm. A. (Sci. Ed.) 47:169, 1958.
21. Schanker, L. W.: Physiological Transport of Drugs, *in* Harper, N. J., and Simmonds, A. B. (eds.): Advances in Drug Research, pp. 71–106, London, Academic Press, 1964.
22. Erenberg, A., and Fisher, D. A.: Thyroid Hormone Metabolism in the Foetus, *in* Foetal and Neonatal Physiology, Cambridge, Cambridge University Press, 1973.
23. Sterling, K. L.: J. Clin. Invest. 43:1721, 1964.
24. Brodie, B. B., Bernstein, E., and Mark, L. C.: J. Pharmacol, Exp. Ther. 105:421, 1952.
25. Price, H. L., *et al.*: J. Clin. Pharmacol. Ther. 1:16, 1960.
26. Axelrod, J., Aronow, L., and Brodie, B. B.: J. Pharmacol. Exp. Ther. 106:166, 1952.
27. Brodie, B. B., Aronow, L., and Axelrod, J.: J. Pharmacol. Exp. Ther. 111:21, 1954.
28. Agarwal, S. L., and Harvey, S. C.: J. Pharmacol. Exp. Ther. 117:106, 1956.
29. Cavallito, C., *et al.*: Anesthesiology, 17:547, 1956.
30. Fouts, J. R., and Brodie, B. B.: J. Pharmacol. Exp. Ther. 115:68, 1955.
31. Paton, W. D. M., and Zaimis, E. J.: Pharmacol. Rev. 4:219, 1952.
32. Waddell, W. J., and Butler, T. C.: J. Clin. Invest. 36:1217, 1957.
33. Brodie, B. B., Kurz, H., and Schanker, L. S.: J. Pharmacol. Exp. Ther. 130:20, 1960.
34. Roll, D. P., Stabenau, J. R., and Zubrod, C. G.: J. Pharmacol, Exp. Ther. 125:185, 1959.
35. Daniels, T. C.: Ann. Rev. Biochem. 12:447, 1943.
36. Coulthard, C. E., Marshall, J., and Pyman, F. L.: J. Chem. Soc. 280, 1930.
37. Dodds, E. C., *et al.*: Proc. Roy. Soc. London, (Ser. B) 127:140, 1939.
38. Schaffer, J. M., and Tilley, F. W.: J. Bact. 12:303, 1926; 14:259, 1927.
39. Sabalitschka, Th., and Tietz, H.: Arch. Pharm. 269:545, 1931.
40. Sabalitschka, Th., and Tietz, H.: Pharm. Acta Helv. 5:286, 1930.
41. Meyer, H.: Arch. Exp. Path. Pharmakol. 42:109, 1899; 46:338, 1901.
42. Winterstein, H.: Die Narkose, ed. 2, Berlin, Springer, 1926.
43. Meyer, K. H., and Gottlieb-Billroth, H.: Z. Physiol. Chem. 112:6, 1920.
44. Cullen, S. C., and Gross, E. G.: Science 113:580, 1951.
45. Wulf, R. J., and Featherstone, R. M.: Anesthesiology 18:97, 1957.
46. Pauling, L.: Science 134:15, 1961.
47. Führer, H.: Arch. Exp. Path. Pharmakol. 51:1, 52:69, 1904.
48. Ferguson, J.: Proc. Roy. Soc. London (Ser. B) 127:387, 1939.
49. Hansch, C., and Fujita, T.: J. Am. Chem. Soc. 86:1616, 1964.
50. Jaffé, H. H.: Chem. Rev. 53:191, 1953.
51. Fujita, T., Iwasa, J., and Hansch, C.: J. Am. Chem. Soc. 86:5175, 1964.
52. Iwasa, J., Fujita, T., and Hansch, C.: J. Med. Chem. 8:150, 1965.
53. Hansch, C., *et al.*: J. Am. Chem. Soc. 85:2817, 1963.
54. Hansch, C., Steward, A. R., and Iwasa, J.: Mol. Pharmacol. 1:87, 1965.
55. Hansch, C., Steward, A. R., Anderson, S. M., and Bentley, D.: J. Med. Chem. 11:1, 1968.
56. Leo, A., Jow, P. Y. C., Silipo, C., and Hansch, C.: J. Med. Chem. 18:865, 1975.
57. Kutter, E., and Hansch, C.: J. Med. Chem. 12:647, 1969.
58. Hansch, C., *et al.*: J. Med. Chem. 16:1207, 1973.
59. Langmuir, I.: J. Am. Chem. Soc. 41:1543, 1919.
60. Beckett, A. H., and Casy, A. F.: J. Pharm. Pharmacol. 6:986, 1954.
61. Eddy, N. B.: Chem. & Ind. 1959, 1462.
62. Portoghese, P. S., Mikhail, A. A., and Kupferberg, H. J.: J. Med. Chem. 11:219, 1968.
63. Gill, E. W.: Prog. Med. Chem. 4:39, 1965.
64. Beckett, A. H.: J. Pharm. Pharmacol. 8:848, 1956.
65. Martin-Smith, M., Smail, G. A., and Stenlake, J. B.: J. Pharm. Pharmacol. 19:561, 1967.
66. Kier, L. B.: Mol. Pharmacol. 3:487, 1967; 4:70, 1968.
67. Chothia, C., and Pauling, P.: Nature 219:1156, 1968.
68. Chiou, C. Y., Long, J. P., Cannon, J. G., and Armstrong, P. D.: J. Pharmacol. Exp. Ther. 166:243, 1969.
69. Smissman, E., Nelson, W., Day, J., and LaPidus, J.: J. Med. Chem. 9:458, 1966.
70. Chothia, C.: Nature 225:36, 1970.
71. Chothia, C., and Pauling, P.: Nature 226:541, 1970.
72. Kier, L. B.: J. Med. Chem. 11:441, 1968.
73. Niemann, C. C., and Hayes, J. T.: J. Am. Chem. Soc. 64:2288, 1942.
74. Black, J. W., *et al.*: Nature 236:385, 1972.
75. Cushny, A. R.: Biological Relations of Optically Active Isomeric Substances, Baltimore, Williams & Wilkins, 1926.
76. Easson, L. H., and Steadman, E.: Biochem. J. 27:1257, 1933.
77. Beckett, A.: Prog. Drug. Res. 1:455–530, 1959.
78. Blaschko, H.: Proc. Roy. Soc. London (Ser. B) 137:307, 1950.
79. Raftery, M. A., Hunkapiller, M. W., Strader, C. D., and Hood, L. E.: Science 208:1454, 1980.
80. Meunier, J. C., Sealock, R., Olsen, R., and Changeux, J. P.: Eur. J. Biochem. 45:371, 1974.
81. Michaelson, D. M., and Raftery, M. A.: Proc. Nat. Acad. Sci. 71:4768, 1974.
82. Cohen, J. B., Weber, M., and Changeux, J. B.: Mol. Pharmacol. 10:904, 1974.
83. O'Malley, B. W., and Schrader, W. T.: Sci. Am. 234:32, 1976.
84. Surks, M. I., Koerner, D., Dillman, W., and Oppenheimer, J. H.: J. Biol. Chem. 248:7066, 1973.
85. Koerner, D., Schwartz, H. L., Surks, M. I., Oppenheimer, J. H., and Jorgensen, E. C.: J. Biol. Chem. 250:6417, 1975.
86. Pfeiffer, C.: Science 107:94, 1948.
87. Barlow, R. B., and Ing, H. R.: Brit. J. Pharmacol. 3:298, 1948.
88. Fisher, A., Keasling, H., and Schueler, F.: Proc. Soc. Exp. Biol. Med. 81:439, 1952; *see* Schueler, Selected Reading, p. 410.
89. Gero, A., and Reese, V. J.: Science 123:100, 1956.
90. Welsh, J. H., and Taub, R.: J. Pharmacol. Exp. Ther. 103:62, 1951.
91. Nachmansohn, D., and Wilson, I. B.: Advances Enzym. 12:259, 1951.
92. Friess, S. L., *et al.*: J. Am. Chem. Soc. 76:1363, 1954; 78:199, 1956; 79:3269, 1957; 80:5687, 1958.
93. Krupka, R. M., and Laidler, K. J.: J. Am. Chem. Soc. 83:1458, 1961.
94. Baker, B. R.: J. Pharm. Sci. 53:347, 1964.
95. Baker, B. R., and Patel, R. P.: J. Pharm. Sci. 52:927, 1963.
96. Baker, B. R., and Alumaula, P. I.: J. Pharm. Sci. 52:915, 1963.
97. Schaeffer, H. J.: J. Pharm. Sci. 54:1223, 1965.
98. Clark, A. J.: J. Physiol. 61:530, 547, 1926.
99. Gaddum, J. H.: J. Physiol. 61:141, 1926; 89:7P, 1937.
100. Ariëns, E. J., and Simonis, A. M.: J. Pharm. Pharmacol. 16:137, 289, 1964.
101. Stephenson, R. P.: Brit. J. Pharmacol. 11:379, 1956.
102. Croxatto, R., and Huidobro, F.: Arch. Int. Pharmacodyn. 106:207, 1956.
103. Paton, W. D. M.: Proc. Roy. Soc. London (Ser. B) 154:21, 1961.
104. Koshland, D. E.: Proc. Nat. Acad. Sci. 44:98, 1958.
105. Nachmansohn, D.: Chemical and Molecular Basis of Nerve Activity, New York, Academic Press, 1959.
106. Belleau, B.: J. Med. Chem. 7:776, 1964.
107. Belleau, B., and Moran, J.: Ann. N.Y. Acad. Sci. 107:822, 1963.
108. Black, J. W., *et al.*: Nature 236:385, 1972.
109. Rahn, O., and Conn, J. E.: Ind. & Eng. Chem. 36:185, 1944.
110. Albert, A., *et al.*: Brit. J. Exp. Path. 26:160, 1945.

111. Overend, W., and Peacocke, A.: Endeavor 16:90, 1957.
112. Albert, A.: Biochem. J. 47:531, 1950; 50:690, 1952.
113. Stocken, L. A., and Thompson, R. H. S.: Physiol. Rev. 29:168, 1949.
114. Walshe, J.: Am. J. Med. 21:487, 1956.
115. Bickel, H.: Experientia 16:129, 1960.
116. Moeschlin, S., and Schnider, U.: New Eng. J. Med. 269:57, 1963.
117. Brannerman, R. M., *et al.*: Brit. Med. J. 1573, 1962.
118. Chenoweth, M. B.: Pharmacol. Rev. 8:57, 1956.
119. Mellor, D. P., and Maley, L.: Nature 161:436, 1948.
120. Albert, A., and Magrath, D.: Biochem. J. 41:534, 1947.
121. Albert, A., Gibson, M. I., and Rubbo, S. D.: Brit. J. Exp. Path. 34:119, 1953.
122. Albert, A.: M. J. Australia 1:245, 1944.
123. Albert, A., *et al.*: Brit. J. Exp. Path. 35:75, 1954.
124. Beckett, A. H., *et al.*: J. Pharm. Pharmacol. 10:160T, 1958.
125. Rubbo, S., Albert, A., and Gibson, M.: Brit. J. Exp. Path. 31:425, 1950.
126. Perry, H. M., and Schroeder, H. A.: Am. J. Med. 16:606, 1954.
127. Bruns, F., and Stüttgen, G.: Biochem. Z. 322:68, 1951.
128. Cymerman-Craig, J., *et al.*: Nature 176:34, 1955.
129. Tautner, E. M., and Messer, M.: Nature 169:31, 1952.
130. Goldacre, R. J.: Australian J. Sc. 6:112, 1944.
131. Page, J. E., and Robinson, F. A.: Brit. J. Exp. Path. 24:89, 1943.
132. Breyer, B., Buchanan, G. S., and Duewell, H.: J. Chem. Soc. 360, 1944.
133. Kuhn, R., Weygand, F., Möller, E.: Chem. Ber. 76(2):1044, 1943.
134. Reist, E. J., *et al.*: J. Org. Chem. 25:1368, 1455, 1960.
135. Tozer, T. N., Tuck, L. D., and Craig, J. C.: J. Med. Chem. 12:294, 1969.
136. Alexander, A. E., and Trim, A. R.: Proc. Roy. Soc. London (Ser. B) 133:220, 1946.
137. Valko, E., and Dubois, A.: J. Bact. 47:15, 1944.
138. Albert, A.: Selective Toxicity, p. 97, New York, Wiley, 1965.

SELECTED READINGS

Albert, A.: Selective Toxicity, ed. 4, London, Chapman and Hall, 1973.
Ariëns, E. J.: Molecular Pharmacology, vol. 1, New York, Academic Press, 1964.
Barlow, R. B.: Introduction to Chemical Pharmacology, ed. 2, New York, Wiley, 1964.
Beckett, A.: Stereochemical factors in biological activity, Prog. Drug Res. 1:455–530, 1959.
Bloom, B. M., and Laubach, G. D.: The relationship between chemical structure and pharmacological activity, Ann. Rev. Pharmacol. 2:62, 1962.
Brodie, Bernard B., and Hogben, Adrian M.: Some physico-chemical factors in drug action, J. Pharm. Pharmacol. 9:345, 1957.
Foye, W. O.: Role of metal-binding in the biological activities of drugs, J. Pharm. Sci. 50:93, 1961.
Gill, E. W.: Drug receptor interactions, Prog. Med. Chem. 4:39, 1965.
Gourley, D. R. H.: Basic mechanisms of drug action, Prog. Drug Res. 7:11, 1964.
Kollman, P. A.: The Nature of the Drug-Receptor Bond, *in* Wolff, M. (ed.): Burger's Medicinal Chemistry, ed. 4, part I, p. 313, New York, Wiley-Interscience, 1980.
Kuntz, I. D., Jr.: Drug-Receptor Geometry, *in* Wolff, M. (ed.): Burger's Medicinal Chemistry, ed. 4, part I, p. 285, New York, Wiley-Interscience, 1980.
Lehmann F., P. A.: Stereoselective Molecular Recognition in Biology, *in* Cuatrecases, P., and Greaves, M. F. (eds.): Receptors and Recognition, Series A, Vol. 5, p. 3, London, Chapman and Hall, 1978.
Martin, Y. C.: Quantitative Drug Design, *in* Grunewald, G. (ed.): Medicinal Research, Vol. 8, New York, Dekker, 1978.
Schueler, F. W.: Chemobiodynamics and Drug Design, New York, Blakiston, 1960.
Wooley, D. W.: A Study of Antimetabolites, New York, Wiley, 1952.

3

metabolic changes of drugs and related organic compounds

Lawrence K. Low

Neal Castagnoli, Jr.

Metabolism plays a central role in the elimination of drugs and other foreign compounds (xenobiotics) from the body. Most organic compounds entering the body are relatively lipid-soluble (lipophilic). Therefore, in order to be absorbed, they must traverse the lipoprotein membranes of the lumen walls of the gastrointestinal (GI) tract. Once in the bloodstream, these molecules can diffuse passively through other membranes to reach various target organs to effect their pharmacological actions. Owing to their reabsorption in the renal tubules, lipophilic compounds are not excreted to any significant extent in the urine.

If lipophilic drugs or xenobiotics were not metabolized to polar, water-soluble products that are readily excretable, they would remain indefinitely in the body, eliciting their biological effects. Thus, the formation of water-soluble metabolites not only enhances drug elimination but also leads to compounds that are generally pharmacologically inactive and relatively nontoxic. For this reason, drug metabolism reactions have been traditionally regarded as *detoxication* (or *detoxification*) processes.[1] However, it is not correct to assume that drug metabolism reactions are always detoxifying. Many drugs are biotransformed to pharmacologically active metabolites. Some metabolites have significant activity that contributes substantially to the pharmacological effect ascribed to the parent drug. In some cases, the parent compound is inactive and must be converted to a biologically active metabolite.[2,3] In addition, it is becoming increasingly clear that not all metabolites are nontoxic. Indeed many toxic side effects (e.g., tissue necrosis, carcinogenicity, teratogenicity) of drugs and environmental contaminants can be directly attributable to the formation of

chemically reactive metabolites that are highly detrimental to the body.[4,5,6]

GENERAL PATHWAYS OF DRUG METABOLISM

Drug metabolism reactions have been divided into two categories: *phase I (functionalization)* and *phase II (conjugation)* reactions.[1,7] Phase I, or functionalization reactions, includes oxidative, reductive, and hydrolytic biotransformations (see below).[8] The purpose of these reactions is to introduce a polar functional group (e.g., OH, COOH, NH_2, SH) into the xenobiotic molecule. This can be achieved by direct introduction of the functional group (e.g., aromatic and aliphatic hydroxylation) or by modifying or "unmasking" existing functionalities (e.g., reduction of ketones and aldehydes to alcohols; oxidation of alcohols to acids; hydrolysis of ester and amides to yield COOH, NH_2, and OH groups; reduction of azo and nitro compounds to give NH_2 moieties; oxidative N- , O- , and S-dealkylation to give NH_2, OH, and SH groups). While phase I reactions may not produce sufficiently hydrophilic or inactive metabolites, they generally tend to provide a functional group or "handle" in the molecule that can undergo subsequent phase II reactions.

The purpose of phase II reactions is to attach small, polar, and ionizable endogenous compounds such as glucuronic acid, sulfate, glycine, and other amino acids to the functional "handles" of phase I metabolites to form water-soluble conjugated products. Parent compounds that already have existing

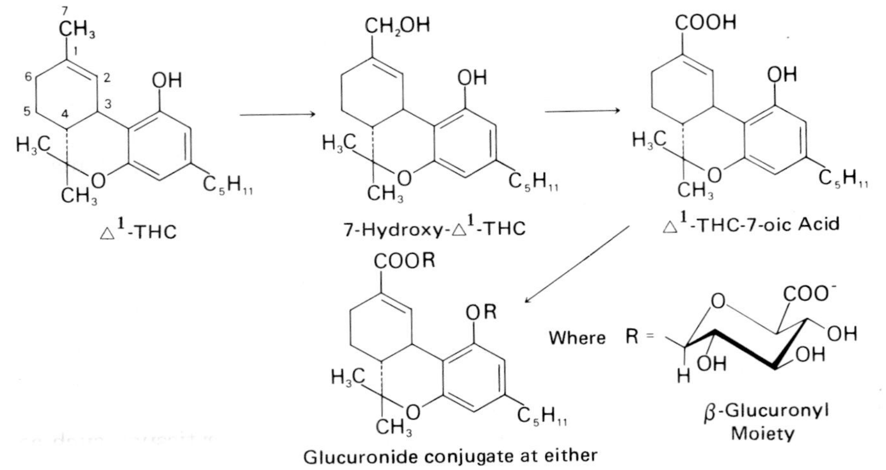

Glucuronide conjugate at either
COOH or phenolic OH group

functional groups, such as OH, COOH, and NH_2, are often directly conjugated by phase II enzymes. Conjugated metabolites are readily excreted in the urine and are generally devoid of pharmacological activity and toxicity. Other phase II pathways, such as methylation and acetylation, serve to terminate or attenuate biological activity, while glutathione conjugation serves to protect the body against chemically reactive compounds or metabolites. Thus, it is apparent that phase I and phase II reactions complement one another in detoxifying and facilitating the elimination of drugs and xenobiotics.

To illustrate, consider the principal psychoactive constituent of marijuana, Δ^1-tetrahydrocannabinol (Δ^1-THC). This lipophilic molecule (octanol/water partition coefficient ca. 6000)[9] undergoes allylic hydroxylation to give 7-hydroxy-Δ^1-THC in man.[10] More polar than its parent compound, the 7-hydroxy me-

tabolite is further oxidized to the corresponding carboxylic acid derivative Δ^1-THC-7-oic acid, which is ionized (pk$_a$ COOH $\sim$ 5) at physiological pH. Subsequent conjugation of this metabolite (either at the COOH or phenolic OH) with glucuronic acid leads to water-soluble products which are readily eliminated in the urine.[11]

In the above series of biotransformations, the parent Δ^1-THC molecule is made increasingly more polar, ionizable and hydrophilic. The attachment of the glucuronyl moiety (with its ionized carboxylate group and three polar hydroxyl groups; see structure above) to the Δ^1-THC metabolites notably favors partitioning of the conjugated metabolites into aqueous media.

The purpose of this chapter is to provide the student with a broad overview of drug metabolism. Various phase I and phase II biotransformation pathways (see above) will be outlined. Representative drug ex-
(see below) will be outlined. Representative drug ex-

GENERAL SUMMARY OF PHASE I AND PHASE II METABOLIC PATHWAYS

PHASE I OR FUNCTIONALIZATION REACTIONS

Oxidative Reactions

Oxidation of aromatic moieties

Oxidation of olefins

Oxidation at benzylic, allylic carbon atoms and carbon atoms alpha to carbonyl and imines

Oxidation at aliphatic and alicyclic carbon atoms

Oxidation involving carbon-heteroatom systems:

  Carbon-nitrogen systems (aliphatic and aromatic amines; includes N-dealkylation, oxidative deamination, N-oxide formation, N-hydroxylation).

  Carbon-oxygen systems (O-dealkylation)

  Carbon-sulfur systems (S-dealkylation, S-oxidation and desulfuration)

Oxidation of alcohols and aldehydes

Other miscellaneous oxidative reactions

Reductive Reactions

Reduction of aldehydes and ketones

Reduction of nitro and azo compounds

Miscellaneous reductive reactions

Hydrolytic Reactions

Hydrolysis of esters and amides

Hydration of epoxides and arene oxides by epoxide hydrase

PHASE II OR CONJUGATION REACTIONS

Glucuronic acid conjugation

Sulfate conjugation

Conjugation with glycine, glutamine and other amino acids

Glutathione or mercapturic acid conjugation

Acetylation

Methylation

amples for each pathway will be presented. Drug metabolism examples in man will be emphasized, although discussion of metabolism in other mammalian systems is necessary. The central role of the cytochrome P-450 monooxygenase system in oxidative drug biotransformation will be elaborated. Discussion of other enzyme systems involved in phase I and phase II reactions will be presented in their respective sections. In addition to stereochemical factors that may affect drug metabolism, biological factors such as age, sex, heredity, disease state, and species variation will be considered. The effects of enzyme induction and inhibition on drug metabolism, as well as a section on pharmacologically active metabolites, will be included.

SITES OF DRUG BIOTRANSFORMATION

Although biotransformation reactions may occur in many tissues, the liver by far is the most important organ in drug metabolism.[12] It is particularly rich in almost all the drug metabolizing enzymes to be discussed in this chapter. The liver is a well-perfused organ and plays a paramount role in the detoxification and metabolism of endogenous and exogenous compounds present in the bloodstream. Orally administered drugs which are absorbed through the gastrointestinal tract must first pass through the liver. Therefore, they are susceptible to hepatic metabolism (first-pass effect) prior to reaching the systemic circulation. Depending on the drug, this metabolism can sometimes be quite significant and, as a result, decrease oral bioavailability. For example, in man, a number of drugs are extensively metabolized by the first-pass effect.[13] The following list includes some of those drugs.

Isoproterenol
Lidocaine
Meperidine
Morphine
Nitroglycerin
Pentazocine
Propoxyphene
Propranolol
Salicylamide

Some drugs (e.g., lidocaine) are so effectively removed by first-pass metabolism that they are ineffective when given orally.[14] Thus, one can appreciate the enormous metabolizing capability of the liver.

Since a majority of drugs are administered orally, the intestine appears to play an important role in the extrahepatic metabolism of xenobiotics. For example, in man, orally administered isoproterenol has been demonstrated to undergo considerable sulfate conjugation in the intestinal wall.[15] A number of other drugs (e.g., levodopa, chlorpromazine and diethylstilbestrol)[16] have also been reported to be metabolized in the gastrointestinal tract. Esterases and lipases present in the intestine may be particularly important in carrying out hydrolysis of many ester prodrugs (see section on hydrolysis below).[17] Bacterial flora present in the intestine and colon appear to play an important role in the reduction of many aromatic azo and nitro drugs (e.g., sulfasalazine).[18] Intestinal β-glucuronidase enzymes are capable of hydrolyzing glucuronide conjugates excreted in the bile, thus liberating the free drug or its metabolite for possible reabsorption (enterohepatic circulation or recycling).[19]

Although other tissues, such as kidney, lungs, adrenals, placenta, brain, and skin, have some degree of drug metabolizing capability, the biotransformations that they carry out oftentimes are more substrate selective and more limited to particular types of reactions (e.g., oxidation, glucuronidation).[20] In many instances, the full metabolic capabilities of these tissues have not been fully explored.

ROLE OF CYTOCHROME P-450 MONOOXYGENASES IN OXIDATIVE BIOTRANSFORMATIONS

Of the various phase I reactions that will be considered, oxidative biotransformation processes are by far the most common and important in drug metabolism. The general stoichiometry which describes the oxidation of many xenobiotics (R-H) to their corresponding oxidized metabolites (R-OH) is given by the equation below:[21]

$$RH + NADPH + O_2 + H^+ \rightarrow ROH + NADP^+ + H_2O$$

The enzyme systems carrying out this biotransformation are referred to as mixed function oxidases or monooxygenases.[22,23] The reaction requires both molecular oxygen and the reducing agent NADPH (reduced form of nicotinamide adenosine dinucleotide phosphate). It should be emphasized that during this oxidative process, one atom of molecular oxygen (O_2) is introduced into the substrate R-H to form R-OH, while the other oxygen atom is incorporated into water. The mixed function oxidase system[24] is actually made up of several components, the most important being an enzyme called cytochrome P-450, which is responsible for transferring *an oxygen atom* to the substrate R-H. Other important components of this system include the NADPH-dependent cytochrome P-450 reductase and a NADH-linked cytochrome b_5. The latter two components along with the cofactors

NADPH and NADH supply the reducing equivalents (namely electrons) needed in the overall metabolic oxidation of foreign compounds. The proposed mechanistic scheme by which the cytochrome P-450 monooxygenase system catalyzes the conversion of molecular oxygen to an "activated oxygen" species will be elaborated below.

The cytochrome P-450 enzyme is a heme-protein.[25] The heme portion is an iron-containing porphyrin called protoporphyrin IX, while the protein portion is called the apoprotein. Cytochrome P-450 is found in high concentrations in the liver, the major organ involved in the metabolism of xenobiotics. The presence of this enzyme in many other tissues (e.g., lung, kidney, intestine, skin, placenta, adrenal cortex) is a reflection that these tissues have drug oxidizing capability too. The name cytochrome P-450 is derived from the fact that the reduced (Fe^{+2}) form of this enzyme binds with carbon monoxide to form a complex that has a distinguishing spectroscopic absorption maximum at 450 nm.[26]

An important feature of the hepatic cytochrome P-450 mixed function oxidase system is its ability to metabolize an almost unlimited number of diverse substrates by a variety of oxidative transformations.[27] This versatility is believed to be attributable to the substrate nonspecificity of cytochrome P-450 as well as to the presence of multiple forms of the enzyme.[28] Some of these P-450 enzymes are selectively inducible by various chemicals (e.g., phenobarbital, benzo[a]pyrene, 3-methylcholanthrene).[29] One of these inducible forms of the enzyme (cytochrome P-448) is of particular interest and will be discussed later.

The cytochrome P-450 monooxygenases are located in the endoplasmic reticulum, a highly organized and complex network of intracellular membranes which is particularly abundant in tissues like the liver.[30] When these tissues are disrupted by homogenization, the endoplasmic reticulum loses its structure and is converted into small vesicular bodies known as microsomes.

Microsomes isolated from hepatic tissue appear to retain all of the mixed function oxidase capabilities of intact hepatocytes; because of this, microsomal preparations (with the necessary cofactors, e.g., NADPH, Mg^{+2}) are frequently utilized for in vitro drug metabolism studies. Because of its membrane-bound nature, the cytochrome P-450 monooxygenase system appears to be housed in a lipoidal environment. This may explain in part why lipophilic xenobiotics are generally good substrates for the monooxygenase system.[31]

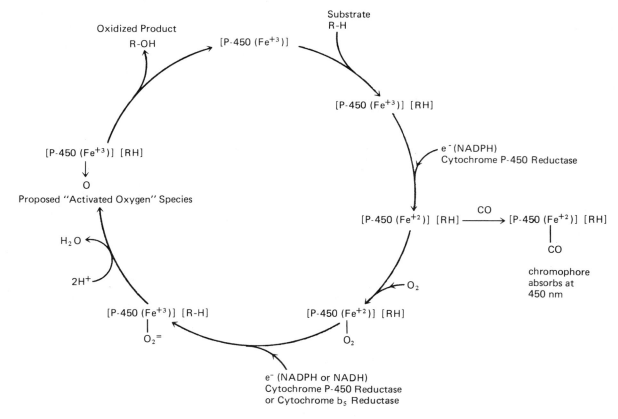

FIG. 3-1. *Proposed catalytic reaction cycle involving cytochrome P-450 in the oxidation of xenobiotics.*

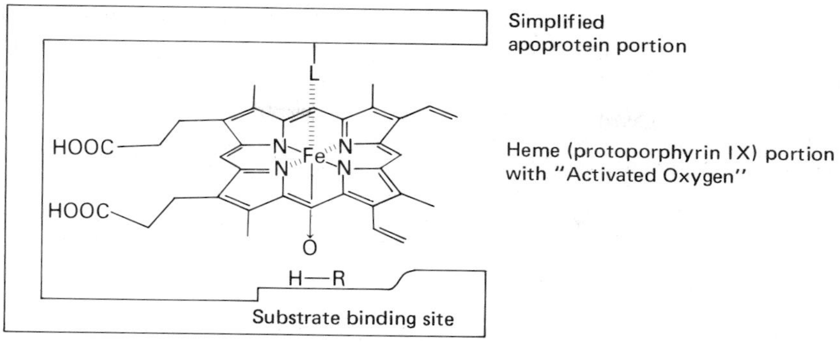

Simplified apoprotein portion

Heme (protoporphyrin IX) portion with "Activated Oxygen"

Substrate binding site

FIG. 3-2. *A simplified depiction of the proposed "activated oxygen"- cytochrome P-450-substrate complex. Note the simplified apoprotein portion and the heme (protoporphyrin IX) portion of cytochrome P-450 and the close proximity of the substrate RH undergoing oxidation.*

The catalytic role that the cytochrome P-450 monooxygenase system plays in the oxidation of xenobiotics is summarized in the cycle shown in Figure 3-1.[32,33] The initial step of this catalytic reaction cycle starts with the binding of the substrate to the oxidized (Fe^{+3}) resting state of cytochrome P-450 to form a P-450-substrate complex. The next step involves the transfer of one-electron from NADPH-dependent cytochrome P-450 reductase to the P-450-substrate complex. This one-electron transfer reduces Fe^{+3} to Fe^{+2}. It is this reduced (Fe^{+2}) P-450-substrate complex that is capable of binding dioxygen (O_2). The dioxygen-P-450-substrate complex which is formed then undergoes another one-electron reduction (by cytochrome P-450 reductase-NADPH and/or cytochrome b_5 reductase-NADH) to yield what is believed to be a peroxide dianion-P-450 (Fe^{+3})-substrate complex. Water (containing one of the oxygen atoms from the original dioxygen molecule) is released from the latter intermediate to form an "activated oxygen"-P-450-substrate complex (Fig. 3-2). The "activated oxygen" in this complex resembles oxene ($\ddot{O}$) and is therefore highly electron-deficient and a potent oxidizing agent. The "activated oxygen" is transferred to the substrate (RH) and the oxidized substrate product (ROH) released from the enzyme complex to regenerate the oxidized form of cytochrome P-450.

It is important to recognize that the key sequence of events appears to center around the alteration of a dioxygen-P-450-substrate complex to an "activated oxygen"-P-450-substrate complex, which is then capable of effecting the critical transfer of oxygen from P-450 to the substrate.[34,35] In view of the potent oxidizing nature of the "activated oxygen" (i.e., oxene) being transferred, it is not surprising that a large number of substrates are capable of being oxidized by cytochrome P-450.

The many types of oxidative reactions carried out by cytochrome P-450 will be enumerated in the sections to follow. Many of these oxidative pathways are summarized schematically in Figure 3-3 (see also Phase I and Phase II Metabolic Pathways above).[34]

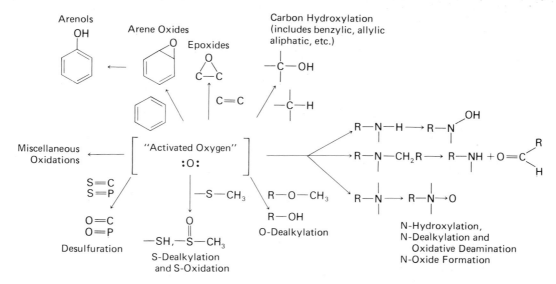

FIG. 3-3. *Schematic summary of cytochrome P-450 catalyzed oxidation reactions (Adapted from Ullrich, ref. 34).*

The versatility of cytochrome P-450 in carrying out a variety of oxidation reactions on a vast number of substrates may be attributable to the multiple forms of the enzyme. Thus it is important for the student to realize that the biotransformation of a parent xenobiotic to several oxidized metabolites is carried out not just by one form of P-450 but more likely by several different forms.[36] Extensive studies indicate that the apoprotein portions of various cytochrome P-450's appear to differ from one another in their tertiary three-dimensional structure (owing to differences in amino acid sequence or the make-up of the polypeptide chain).[25,28] Since the apoprotein portion is important in substrate binding and catalytic transfer of "activated oxygen," these structural differences may account for some substrates being preferentially or more efficiently oxidized by one particular form of cytochrome P-450.

OXIDATIVE REACTIONS

OXIDATION OF AROMATIC MOIETIES

Aromatic hydroxylation refers to the mixed function oxidation of aromatic compounds (arenes) to their corresponding phenolic metabolites (arenols).[37] Almost all aromatic hydroxylation reactions are believed to

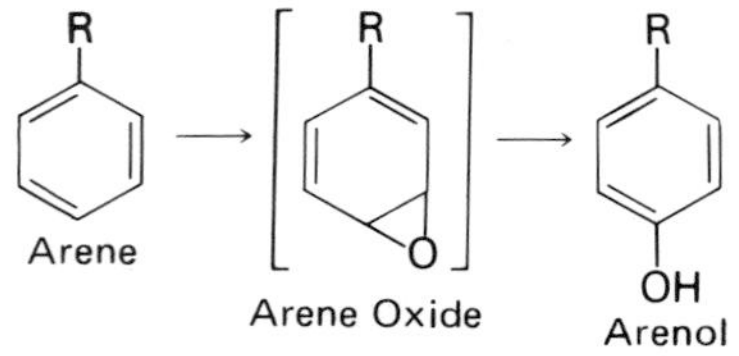

proceed initially through an epoxide intermediate called an arene oxide, which rearranges rapidly and spontaneously to the arenol product in most instances. The importance of arene oxides in the formation of arenols and in other metabolic and toxicological reactions will be discussed shortly.[38] Attention will now focus on the aromatic hydroxylation of a number of drugs and xenobiotics.

Most foreign compounds containing aromatic moieties are susceptible to aromatic oxidation. In man, aromatic hydroxylation is a major route of metabolism for a number of drugs containing phenyl groups. Important therapeutic agents such as propranolol,[39] phenobarbital,[40] phenytoin,[41] phenylbutazone,[42] phenformin,[43] 17α-ethinylestradiol,[44] S(-)-warfarin,[45] among others, undergo extensive aromatic oxidation (see Fig. 3-4 for structure and site of hydroxylation). In most of the drugs just mentioned, hydroxylation occurs at the *para*-position.[47] Most phenolic metabolites formed from aromatic oxidation undergo further conversion to polar and water-soluble glucuronide or

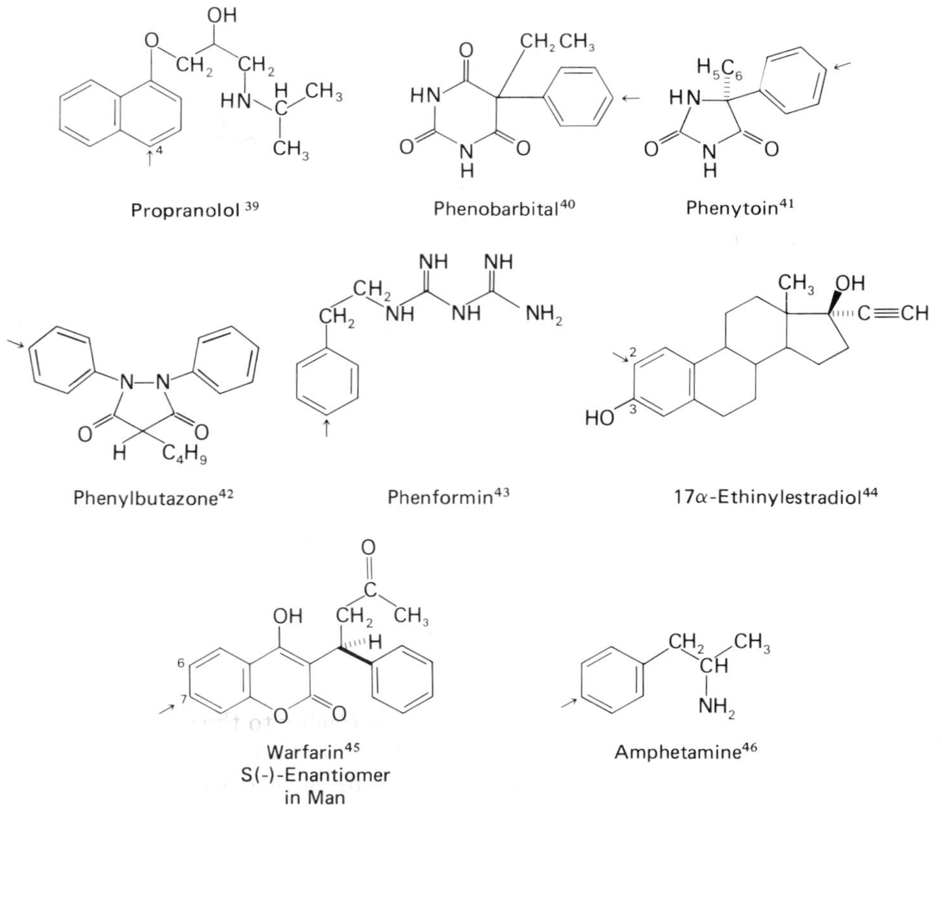

FIG. 3-4. *Examples of drugs and xenobiotics that undergo aromatic hydroxylation in man. Arrow indicates site of aromatic hydroxylation.*

Phenytoin → *p*-Hydroxyphenytoin → O-Glucuronide Conjugate

sulfate conjugates, which are readily excreted in the urine. For example, the major urinary metabolite of phenytoin found in man is the O-glucuronide conjugate of *p*-hydroxyphenytoin.[41] Interestingly, the *para*-hydroxylated metabolite of phenylbutazone, oxyphenbutazone, is pharmacologically active and has been marketed itself as an antiinflammatory agent (Tandearil®, Oxalid®).[42] Of the two enantiomeric forms of the oral anticoagulant warfarin (Coumadin®), only the more active S(-)-enantiomer has been shown to undergo significant aromatic hydroxylation to 7-hydroxywarfarin in man.[45] In contrast, the R(+)-enantiomer is metabolized by keto reduction (see section on stereochemical aspects of drug metabolism below).[45]

Often the substituents attached to the aromatic ring may influence the ease of hydroxylation.[47] As a general rule, microsomal aromatic hydroxylation reactions appear to proceed most readily in activated (electron-rich) rings, whereas deactivated aromatic rings (e.g., those containing electron-withdrawing groups Cl, $-\overset{+}{N}R_3$, COOH, SO_2NHR) are generally slow or resistant to hydroxylation. The deactivating groups (Cl, $-\overset{+}{N}H=C$) present in the antihypertensive clonidine (Catapres®) may explain why this drug undergoes little aromatic hydroxylation in man.[48] The uricosuric agent probenecid (Benemid®) with its electron-withdrawing carboxy and sulfamido groups has not been reported to undergo any aromatic hydroxylation.[49]

Clonidine Hydrochloride   Probenecid

In compounds where two aromatic rings are present, hydroxylation occurs preferentially in the more electron-rich ring. For example, aromatic hydroxylation of diazepam (Valium®) occurs primarily in the more activated ring to yield 4'-hydroxydiazepam.[50] A similar situation is seen in the 7-hydroxylation of the antipsychotic agent chlorpromazine (Thorazine®)[51]

and in the *para*-hydroxylation of *p*-chlorobiphenyl to *p'*-chloro-*p'*-hydroxybiphenyl.[52]

Diazepam   Chlorpromazine

p-Chlorobiphenyl

Recent environmental pollutants, such as polychlorinated biphenyls (PCBs) and 2,3,7,8-tetrachlorodibenzo-*p*-dioxin (TCDD), have attracted considerable public concern over their toxicity and health hazards.

Polychlorinated Biphenyl Mixtures
The number of chlorine atoms (m, n) present in two aromatic rings varies considerably

2,3,7,8-Tetrachlorodibenzo-*p*-dioxin (TCDD)

These compounds appear to be resistant to aromatic oxidation because of the numerous electronegative chlorine atoms present in the aromatic rings comprising their structures. The metabolic stability coupled to the lipophilicity of these environmental contaminants probably explains their long persistence in the body once absorbed.[53,54]

Arene oxide intermediates are formed when a double bond in aromatic moieties is epoxidized. Arene oxides are of significant toxicological concern, since it is known that these intermediates are electrophilic and chemically reactive (owing to the strained three-membered epoxide ring). Detoxification of arene oxides occurs mainly by spontaneous rearrangement to arenols,

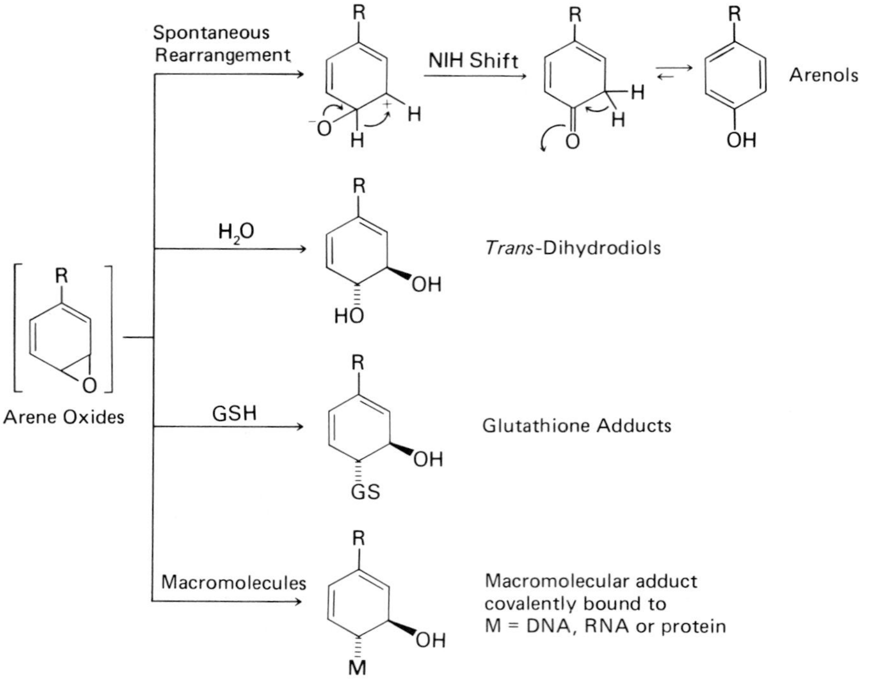

FIG. 3-5. *Possible reaction pathways for arene oxides.*[37, 38]

but enzymatic hydration to *trans*-dihydrodiols and enzymatic conjugation with glutathione (GSH) also play very important roles (Fig. 3-5).[37,38] If not effectively detoxified by the three pathways above, arene oxides will covalently bind with nucleophilic groups present on proteins, DNA, and RNA, thus leading to serious cellular toxicity.[5,37] A few examples of reactive arene oxides causing carcinogenicity and cytotoxicity are examined above.

Quantitatively, the most important detoxification reaction for arene oxides is the spontaneous rearrangement to the corresponding arenols. In many cases, this rearrangement is accompanied by a novel intramolecular hydride (deuteride) migration called the "NIH shift".[55] It was named after the National Institute of Health (NIH) laboratory in Bethesda, Maryland, where this process was discovered. The general features of the NIH shift are illustrated with the mixed function aromatic oxidation of 4-deuterioanisole to 3-deuterio-4-hydroxyanisole as shown in Figure 3-6.[56]

After its metabolic formation, the arene oxide ring opens in the direction which generates the most resonance stabilized carbocation (positive charge on C-3 carbon is resonance stabilized by the OCH_3 group). The zwitterionic species (positive charge on the C-3 carbon atom and negative charge on the oxygen atom) then undergoes a 1,2-deuteride shift (NIH shift) to form the dienone. Final transformation of the dienone to 3-deuterio-4-hydroxyanisole occurs with the preferential loss of a proton because of the weaker bond energy of the C-H bond (as compared to the C-D bond). Thus, the deuterium is retained in the molecule

by undergoing this intramolecular NIH shift. The experimental observation of an NIH shift for aromatic hydroxylation of a drug or xenobiotic is taken as indirect evidence for the involvement of an arene oxide.

In addition to the NIH shift, the zwitterionic species may also undergo direct loss of D^+ to generate 4-hydroxyanisole, in which there is no retention of deuterium (Fig. 3-6). The alternative pathway (direct loss of D^+) may be more favorable than the NIH shift in some aromatic oxidation reactions. Therefore, depending on the substituent group on the arene, some aromatic hydroxylation reactions do not display any NIH shift.

Two extremely important enzymatic reactions also aid in neutralizing the reactivity of arene oxides. The first of these involves the hydration (i.e., nucleophilic attack of water on the epoxide) of arene oxides to yield inactive *trans*-dihydrodiol metabolites (see Fig. 3-5). This reaction is catalyzed by microsomal enzymes called epoxide hydrases.[57] Often epoxide hydrase inhibitors such as cyclohexene oxide and 1,1,1-trichloropropene-2, 3-oxide have been utilized to demonstrate the detoxification role of these enzymes. Addition of these inhibitors is frequently accompanied by

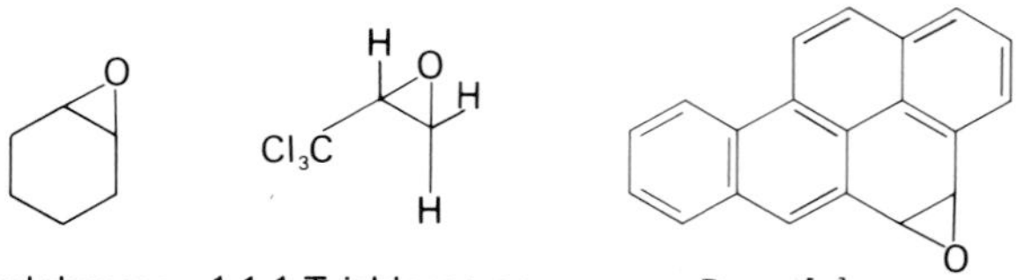

Cyclohexene oxide   1,1,1-Trichloropropene 2,3-oxide   Benzo[a]pyrene 4,5-oxide

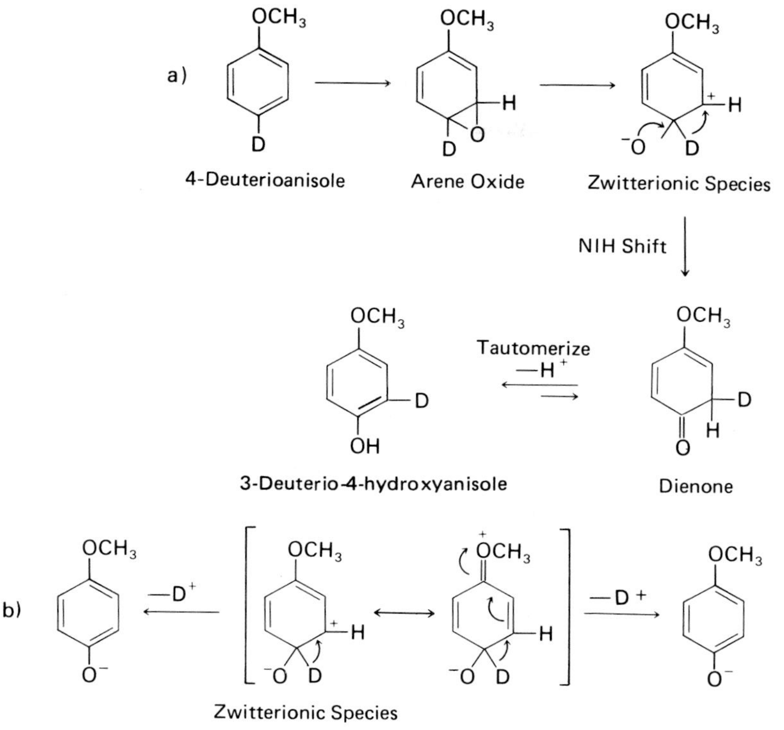

FIG. 3-6. a) *General features of the NIH shift or 1,2-hydride (deuteride) shift in the the mixed function oxidation of 4-deuterioanisole to 3-deuterio-4-hydroxyanisole;*

b) *Direct loss of D⁺ from Zwitterionic species leading to no retention of deuterium in 4-hydroxyanisole.*

an increase in toxicity of the arene oxide being tested, since formation of nontoxic dihydrodiols is blocked. For example, the mutagenicity of benzo[a]pyrene-4,5-oxide as measured by the Ames *Salmonella typhimurium* test system is potentiated when cyclohexene oxide is added.[58] Dihydrodiol metabolites have been reported in the metabolism of a number of aromatic hydrocarbons (e.g., naphthalene, benzo[a]pyrene and other related polycyclic aromatic hydrocarbons).[37] A few drugs (e.g., phenytoin,[59] phenobarbital,[60] glutethimide[61]) have also been observed to yield dihydrodiol

products as minor metabolites in man. Dihydrodiol products are susceptible to conjugation with glucuronic acid as well as an enzymatic dehydrogenation to the corresponding catechol metabolite, as exemplified by the metabolism of phenytoin below.[59]

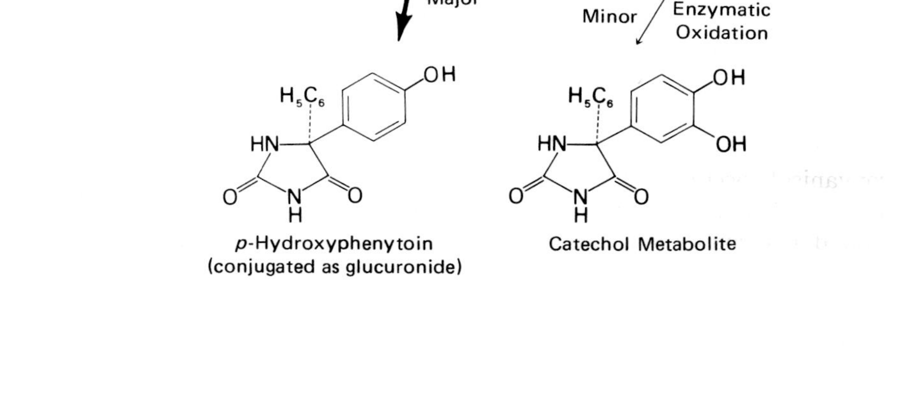

A second enzymatic reaction involves nucleophilic ring opening of the arene oxide by the sulfhydryl group present in glutathione to yield the corresponding *trans*-1,2-dihydro-1-S-glutathionyl-2-hydroxy adduct or glutathione adduct (see Fig. 3-5).[37] The reaction is catalyzed by various glutathione S-transferases.[62] Because glutathione is found in practically all mammalian tissues, it plays an important role not only in the detoxification of arene oxides but also in the detoxification of a variety of other chemically reactive and potentially toxic intermediates. Initially formed glutathione adducts from arene oxides are modified in a series of reactions to yield "premercapturic acid" or mercapturic acid metabolites.[63] Since it is classified as a phase II pathway, glutathione conjugation will be covered in greater detail in a later section.

Because of their electrophilic and reactive nature, arene oxides may also undergo spontaneous reactions with nucleophilic functionalities present on biomacromolecules.[38] Such reactions lead to modified protein, DNA, and RNA structures and often cause dramatic alterations in how these macromolecules function. Much of the cytotoxicity and irreversible lesions caused by arene oxides are presumed to be the result of their covalent binding to cellular components. Several well-established examples of reactive arene oxides that cause serious toxicity are presented below.

The administration of bromobenzene to rats causes severe liver necrosis.[64] Extensive in vivo and in vitro studies indicate that the liver damage results from the interaction of a chemically reactive metabolite, namely 4-bromobenzene oxide, with hepatocytes.[65] Extensive covalent binding to hepatic tissue was confirmed by using radiolabeled bromobenzene. The severity of necrosis correlated well with the amount of covalent binding to hepatic tissue. It was demonstrated that the depletion of hepatic glutathione stores using diethyl maleate or large doses of bromobenzene in rats led to more severe liver necrosis.

Polycyclic aromatic hydrocarbons are ubiquitous environmental contaminants that are formed from auto emission, refuse burning, industrial processes, cigarette smoke, and other combustion processes. Benzo[a]pyrene, a potent carcinogenic agent, is perhaps the most extensively studied of these polycyclic aromatic hydrocarbons.[66] From inspection of its structure, aromatic hydroxylation of benzo[a]pyrene can obviously occur at a number of positions. The identification of a number of dihydrodiol metabolites is viewed as indirect evidence for the formation and involvement of arene oxides in the metabolism of benzo[a]pyrene. Although certain arene oxides of benzo[a]pyrene (e.g., 4,5-oxide, 7,8-oxide, 9,10-oxide) appear to display some mutagenic and tumorigenic activity, it does not appear that they represent the ultimate reactive species responsible for benzo[a]pyrene's carcinogenicity. In recent years, extensive studies have led to the characterization of a specific sequence of metabolic reactions (Fig. 3-7) that give rise to a highly reactive intermediate that covalently binds to DNA. Metabolic activation of benzo[a]pyrene to the ultimate carcinogenic species involves an initial epoxidation reaction to form the 7,8-oxide, which is then converted by epoxide hydrase to (-)-7R, 8R- dihydroxy -7, 8-dihydrobenzo[a]pyrene.[67] The two-step enzymatic formation of this *trans*-dihydrodiol is stereospecific. Subsequent epoxidation at the 9,10-double bond of the latter metabolite generates predominantly (+)-7R, 8S-dihydroxy-9R,10R-oxy-7,8,9,10-tetrahydrobenzo[a]pyrene or (+)7,8-diol-9,10-epoxide. It is this key electrophilic diol epoxide metabolite that readily reacts with DNA to form a number of covalently bound adducts.[68,69] Careful degradation studies have shown that the principal adduct involves attack of the C-2 amino group of deoxyguanosine at C-10 of the diol epoxide. It is now clear that these reactions are responsible for genetic code alterations that ultimately lead to the malignant transformations. Covalent binding of the diol epoxide metabolite to deoxyadenosine and to deoxycytidine have also been established.[69]

Another carcinogenic polycyclic aromatic hydrocarbon, 7,12-dimethylbenz[a]anthracene, also has been demonstrated to form covalent adducts to nucleic acids (RNA).[70] The ultimate carcinogenic reactive species appears to be the 5,6-oxide resulting from epoxidation of the 5,6-double bond in this aromatic hy-

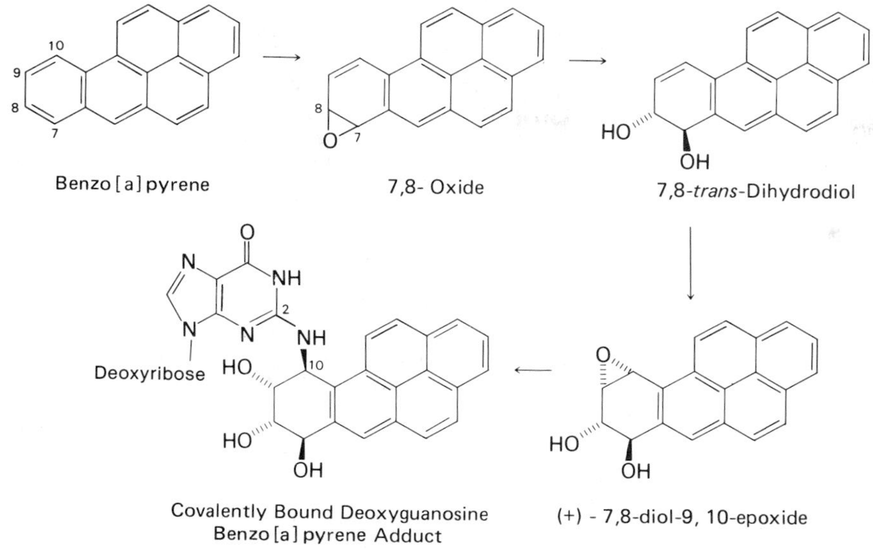

Benzo[a]pyrene   7,8-Oxide   7,8-*trans*-Dihydrodiol

Covalently Bound Deoxyguanosine Benzo[a]pyrene Adduct

(+)-7,8-diol-9,10-epoxide

FIG. 3-7. *Metabolic sequence leading to the formation of the ultimate carcinogenic species of benzo[a]pyrene, (+)-7R,8S-dihydroxy-9R,10R-oxy-7,8,9,10-tetrahydrobenzo[a]pyrene or (+)-7,8-diol-9,10-epoxide.*

drocarbon. The arene oxide intermediate binds covalently to guanosine residues of RNA to yield the two adducts shown below.

OXIDATION OF OLEFINS

The metabolic oxidation of olefinic carbon-carbon double bonds leads to the corresponding epoxide (or oxirane). Epoxides derived from olefins generally tend to be somewhat more stable than the arene oxides formed from aromatic compounds. A few epoxides are stable enough that they are directly measurable in biological fluids (e.g., plasma, urine). Like their arene oxide counterparts, epoxides are susceptible to enzymatic hydration by epoxide hydrase to form *trans*-1,2-dihydrodiols (also called 1,2-diols or 1,2-dihydroxy compounds).[57] In addition, a number of epoxides undergo glutathione conjugation.[71]

A well-known example of olefinic epoxidation is the metabolism of the anticonvulsant drug carbamazepine (Tegretol®) to carbamazepine-10,11-epoxide in man.[72] The epoxide is reasonably stable and can be quantitatively measured in the plasma of patients receiving

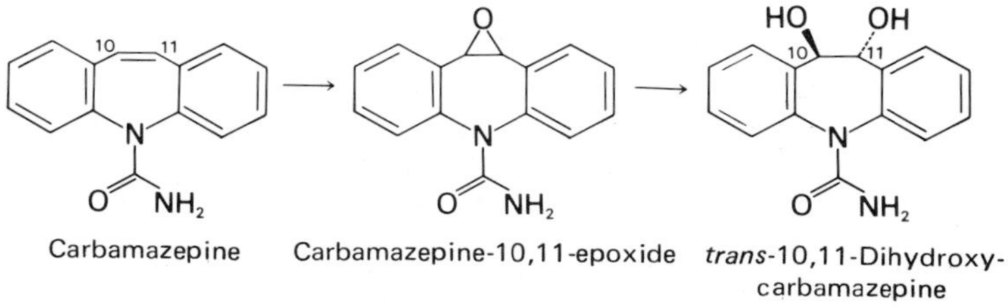

Carbamazepine   Carbamazepine-10,11-epoxide   *trans*-10,11-Dihydroxy-carbamazepine

the parent drug. Recent evidence suggests that the epoxide metabolite has marked anticonvulsant activity and may therefore contribute significantly to the therapeutic effect of the parent drug.[73] Subsequent hydration of the epoxide produces 10,11-dihy-

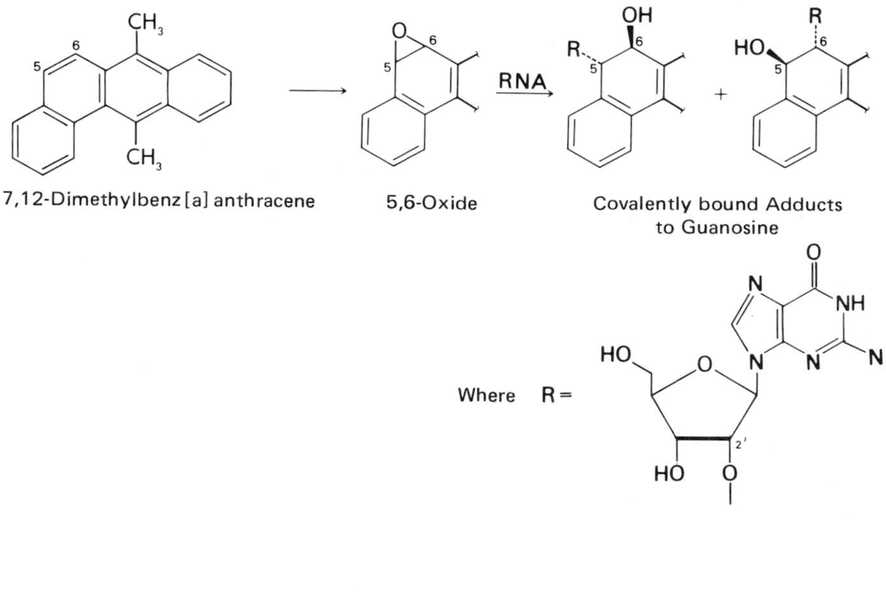

7,12-Dimethylbenz[a]anthracene   5,6-Oxide   Covalently bound Adducts to Guanosine

Where R =

droxycarbamazepine, a significant urinary metabolite (10–30%) in man.[72]

Epoxidation of the olefinic 10,11-double bond in the antipsychotic agent protriptyline (Vivactil®)[74] and in the H$_1$-histamine antagonist cyproheptadine (Periactin®)[75] has also been demonstrated. Frequently, the

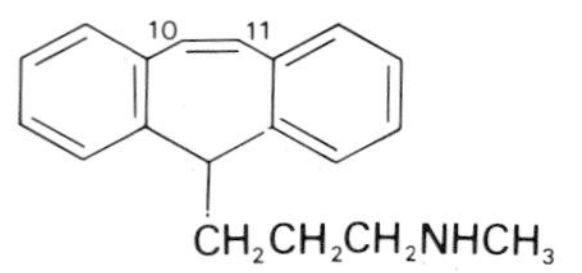

Protriptyline

Cyproheptadine

epoxides formed from the biotransformation of an olefinic compound are minor products, owing to their further conversion to the corresponding 1,2-diols. For example, dihydroxyalcofenac is a major human urinary metabolite of the antiinflammatory agent alcofenac.[76] However, the epoxide metabolite from which it is de-

rived is present in minute amounts. The presence of the dihydroxy metabolite (called secodiol) of secobarbital but not the epoxide product has also been reported in man.[77]

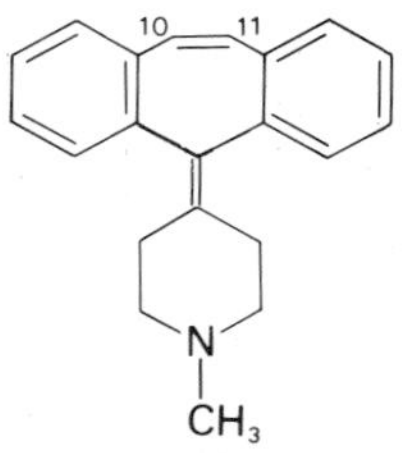

Indirect evidence for the formation of epoxides comes also from the isolation of glutathione or mercapturic acid metabolites. After administration of styrene to rats, two urinary metabolites have been identified as the isomeric mercapturic acid derivatives resulting from nucleophilic attack of glutathione on the intermediate epoxide.[78] In addition, styrene oxide has been found to covalently bind to rat liver microsomal proteins and nucleic acids.[79] These results indicate that styrene oxide is relatively reactive toward nucleophiles (e.g., glutathione and nucleophilic groups on protein and nucleic acids).

There appear to be a number of metabolically generated epoxides that display similar chemical reactivity toward nucleophilic functionalities. Thus, it has been suggested that the toxicity of some olefinic compounds may be a consequence of their metabolic conversion to chemically reactive epoxides.[80] One case

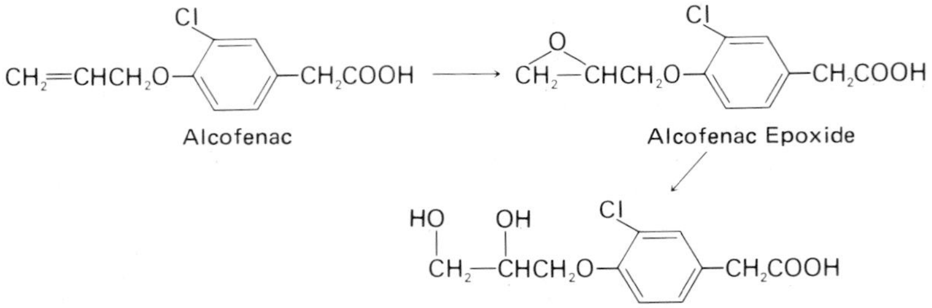

Alcofenac

Alcofenac Epoxide

Dihydroxyalcofenac

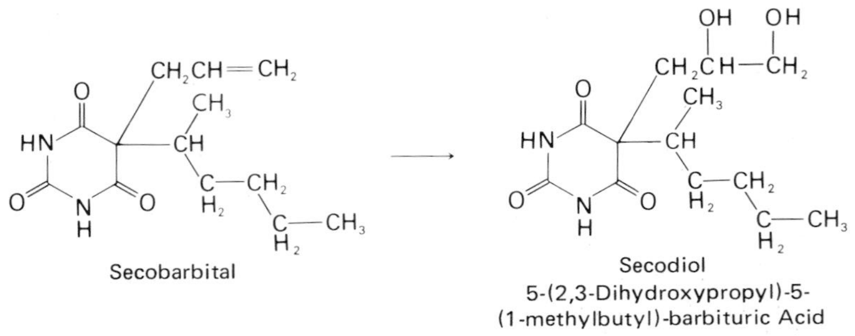

Secobarbital

Secodiol
5-(2,3-Dihydroxypropyl)-5-
(1-methylbutyl)-barbituric Acid

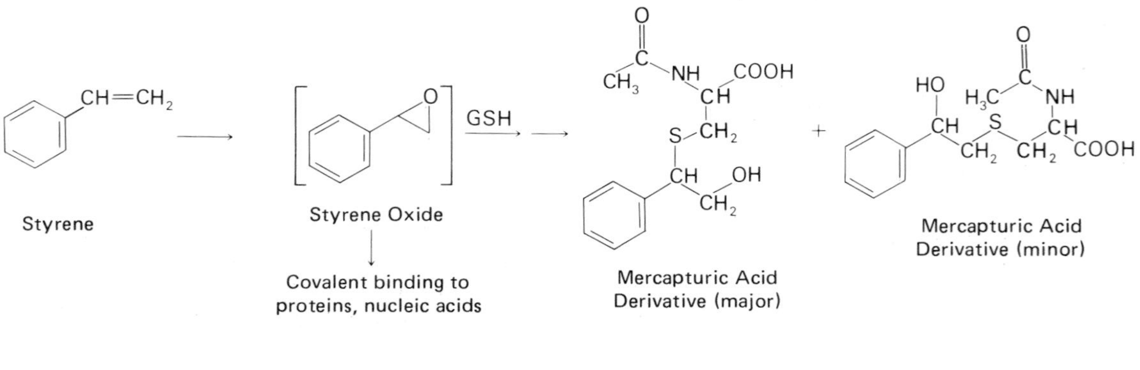

Styrene

Styrene Oxide

Covalent binding to
proteins, nucleic acids

Mercapturic Acid
Derivative (major)

Mercapturic Acid
Derivative (minor)

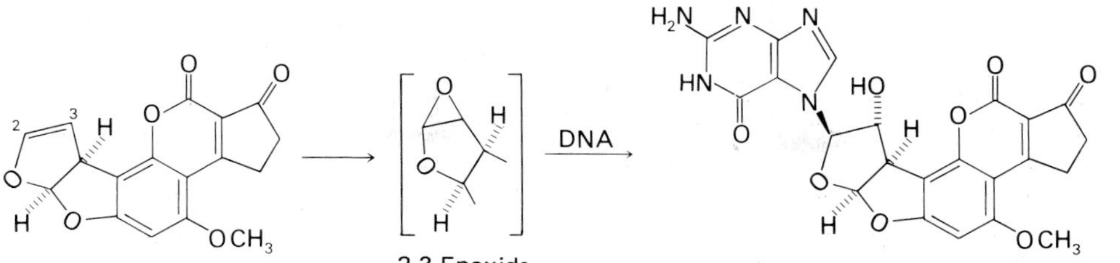

Aflatoxin B$_1$ → 2,3-Epoxide → DNA → 2,3-Dihydro-2-(N^7-guanyl)-3-hydroxyaflatoxin B$_1$

that clearly links metabolic epoxidation as a biotoxification pathway involves aflatoxin B$_1$. This naturally occurring carcinogenic agent contains an olefinic (C2-C3) double bond adjacent to a cyclic ether oxygen. The hepatocarcinogenicity of aflatoxin B$_1$ has been clearly linked to its metabolic oxidation to the corresponding 2,3-oxide, which is extremely reactive.[81] Extensive in vitro and in vivo metabolic studies indicate that this 2,3-oxide binds covalently to DNA, RNA, and proteins. A major DNA adduct has been isolated and characterized as 2,3-dihydro-2-(N^7-guanyl)-3-hydroxyaflatoxin B$_1$.[82]

Other olefinic compounds, such as vinyl chloride,[83] stilbene,[84] and the carcinogenic estrogenic agent diethylstilbestrol (DES),[85] have been observed to undergo metabolic epoxidation. It has been suggested that the corresponding epoxide metabolites may be the reactive species responsible for mediating the cellular toxicity seen with these compounds.

Vinyl chloride

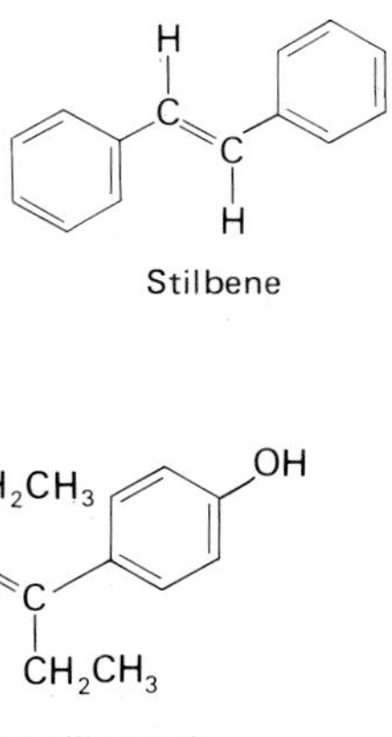

Stilbene

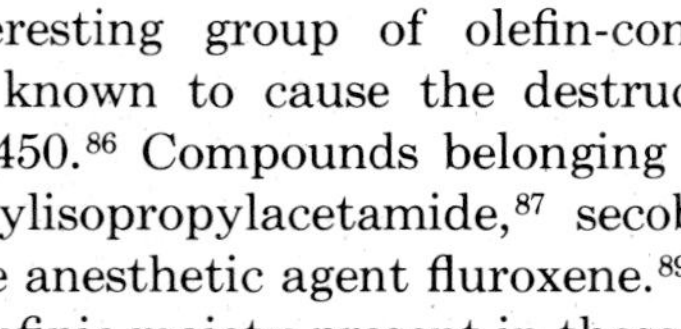

Diethylstilbestrol
(DES)

An interesting group of olefin-containing compounds is known to cause the destruction of cytochrome P-450.[86] Compounds belonging to this group include allylisopropylacetamide,[87] secobarbital[88] and the volatile anesthetic agent fluroxene.[89] It is believed that the olefinic moiety present in these compounds is

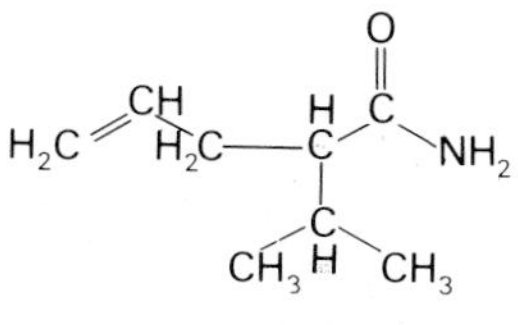

Allylisopropylacetamide

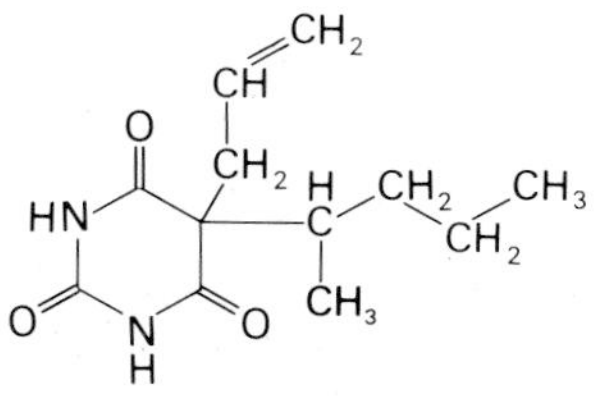

Secobarbital

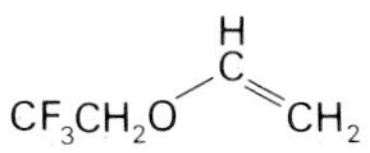

Fluroxene

metabolically activated by cytochrome P-450 to form a very reactive intermediate that covalently binds to the heme portion of cytochrome P-450.[90] As a result of this interaction, the cytochrome P-450 enzyme is functionally inactivated or "destroyed." The altered heme moiety that accumulates from this destructive process appears to have a "greenish" pigmentation.[86] Although there is some evidence that the epoxide metabolites are not the reactive intermediates involved in the destruction process, the exact chemical nature of the species involved in binding covalently to the heme portion of cytochrome P-450 remains unclear.[90] Prolonged administration of the above three agents is expected to lead to some inhibition of oxidative drug metabolism.

OXIDATION AT BENZYLIC CARBON ATOMS

Carbon atoms attached to aromatic rings (benzylic position) are susceptible to oxidation forming the corresponding alcohol (or carbinol) metabolite.[47] Primary alcohol metabolites are often oxidized further to aldehydes and carboxylic acids (CH$_2$OH→CHO→COOH), and secondary alcohols are converted to ketones by soluble alcohol and aldehyde dehydrogenases.[91] Alternatively, the alcohol may be directly conjugated with glucuronic acid.[92] The benzylic carbon atom present in the oral hypoglycemic agent tolbutamide (Orinase®) is extensively oxidized to the corresponding alcohol

Tolbutamide   Alcohol Metabolite   Carboxylic Acid Metabolite

Tolmetin   Dicarboxylic Acid Metabolite

and carboxylic acid. Both metabolites have been isolated in the urine of man.[93] Similarly, the "benzylic" methyl group in the antiinflammatory agent tolmetin (Tolectin®) undergoes oxidation to yield the dicarboxylic acid product as the major metabolite in man.[94] The sedative hypnotic agent methaqualone has been observed to undergo benzylic oxidation at its C-2′ methyl group to give 2′-hydroxymethylmethaqualone as a minor metabolite.[95] Benzylic hydroxylation occurs to a significant extent in the metabolism of the β-adrenergic blocker metoprolol (Lopressor®) to yield α-hydroxymetoprolol.[96] Additional examples of drugs and xenobiotics undergoing benzylic oxidation are shown in Figure 3-8.

OXIDATION AT ALLYLIC CARBON ATOMS

Microsomal hydroxylation at allylic carbon atoms is commonly observed in drug metabolism. An illustrative example of allylic oxidation is given by the psychoactive component of marijuana, Δ^1-tetrahydrocannabinol (Δ^1-THC). This molecule contains three allylic

carbon centers (C-7, C-6, and C-3). Allylic hydroxylation occurs extensively at C-7 to yield 7-hydroxy-Δ^1-THC as the major plasma metabolite in man.[10] Pharmacological studies show that this 7-hydroxy metabolite is as active or even more active than Δ^1-THC itself and may contribute significantly to the overall CNS psychotomimetic effects of the parent compound.[103] Hydroxylation also occurs to a minor extent at the allylic C-6 position to give both the epimeric 6α- and 6β-hydroxy metabolites.[10] Metabolism does not occur at C-3, presumably because of steric hindrance.

The antiarrhythmic agent quinidine is metabolized via allylic hydroxylation to 3-hydroxyquinidine, the principal plasma metabolite found in man.[104] Recent reports indicate that this metabolite shows significant antiarrhythmic activity in animals and possibly in man as well.[105]

Other examples of allylic oxidation include the sedative hypnotic hexobarbital (Sombulex®) and the analgesic pentazocine (Talwin®). The 3′-hydroxylated metabolite formed from hexobarbital is susceptible to glucuronide conjugation as well as further oxidation to the 3′-oxo compound.[106] Hexobarbital is a chiral

Methaqualone   2′-Hydroxymethylmethaqualone

Metroprolol   α-Hydroxymetroprolol

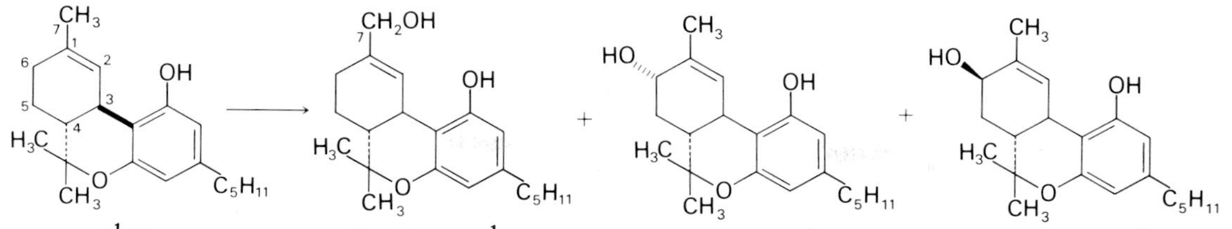

Δ¹-THC   7-Hydroxy-Δ¹-THC   6α-Hydroxy-Δ¹-THC   6β-Hydroxy-Δ¹-THC

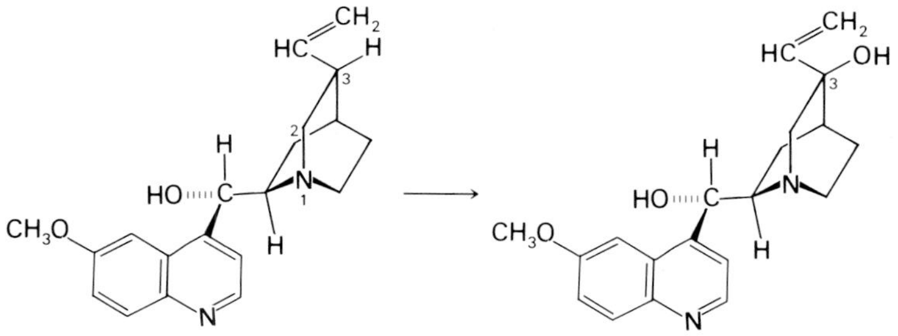

Quinidine   3-Hydroxyquinidine

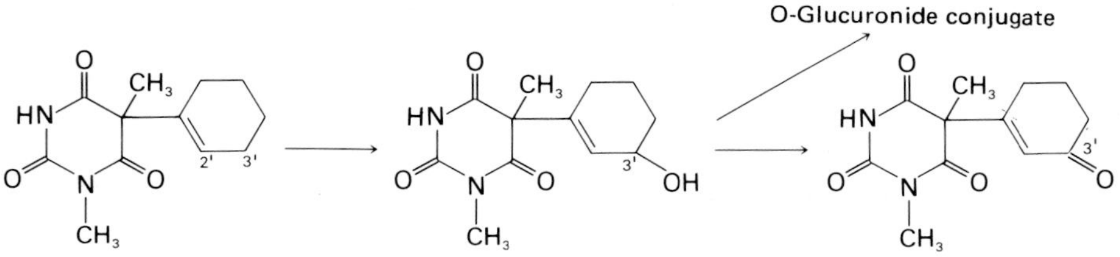

O-Glucuronide conjugate

Hexobarbital   3'-Hydroxyhexobarbital   3'-Oxohexobarbital

barbiturate derivative which exists in two enantiomeric forms. Studies in man indicate that the pharmacologically less active R(-)-enantiomer is more rapidly metabolized than its S(+)-isomer.[107] Pentazocine undergoes allylic hydroxylation at the two terminal methyl groups of its N-butenyl side chain to yield either the *cis* or *trans* alcohol metabolites shown below. In man, greater amounts of the *trans* alcohol are formed.[108]

In the case of the hepatocarcinogenic agent safrole, allylic hydroxylation is known to be involved in a bioactivation pathway leading to the formation of chemically reactive metabolites.[109] This process involves initial hydroxylation at the C-1' carbon of safrole. It should be noted that this center is both allylic and benzylic. The hydroxylated metabolite then undergoes further conjugation to form a sulfate ester. This chemically reactive ester intermediate presumably undergoes nucleophilic displacement reactions with DNA or RNA in vitro to form covalently bound adducts.[110] As shown in the scheme below, nucleophilic attack by DNA, RNA, or other nucleophiles is facilitated by a good leaving group (e.g., $SO_4^=$) at the C-1' position. The leaving group tendency of the alcohol OH group itself is not great enough to facilitate displacement reactions. It should be stressed that allylic hydroxylation generally is not a pathway that leads to the generation of reactive intermediates. Its involvement in the biotoxification of safrole appears to be an exception.

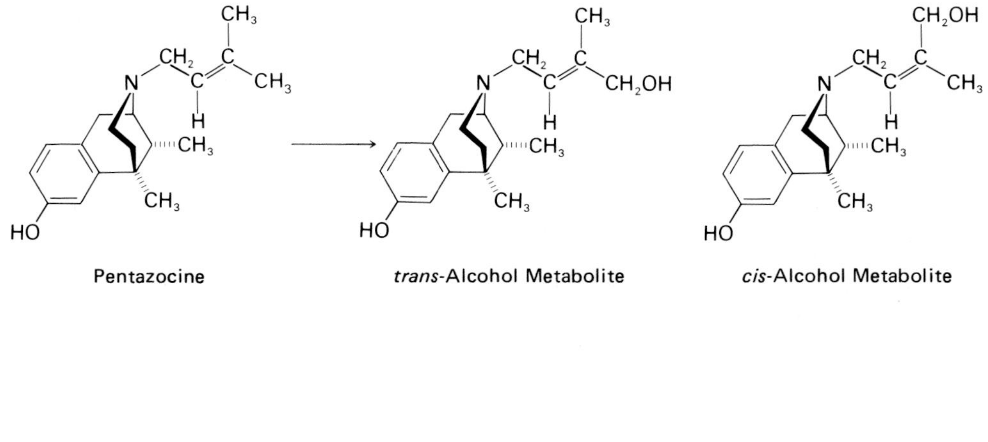

Pentazocine   *trans*-Alcohol Metabolite   *cis*-Alcohol Metabolite

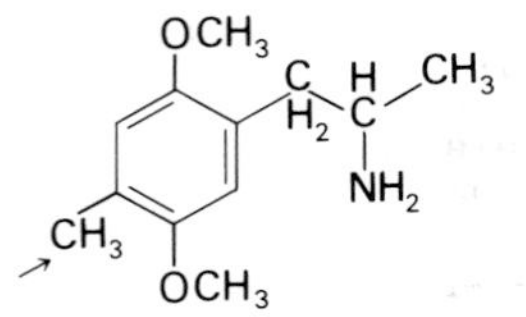

"STP"
1-(2,5-Dimethoxy-4-methylphenyl)
-2-aminopropane (DOM)[97]

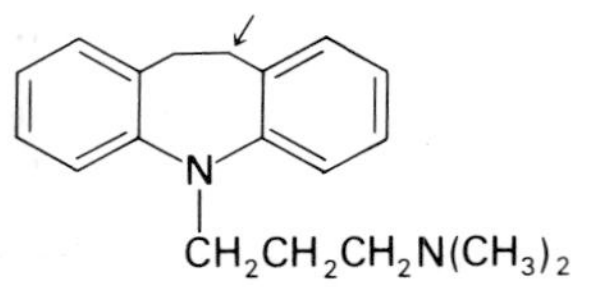

Imipramine[98]

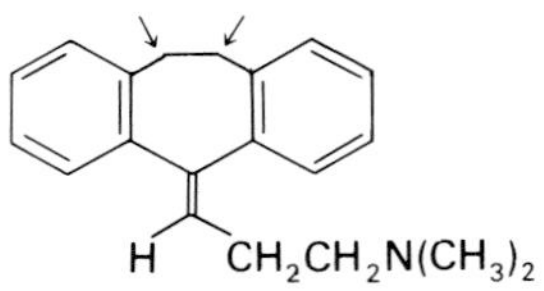

Amitriptyline[99]

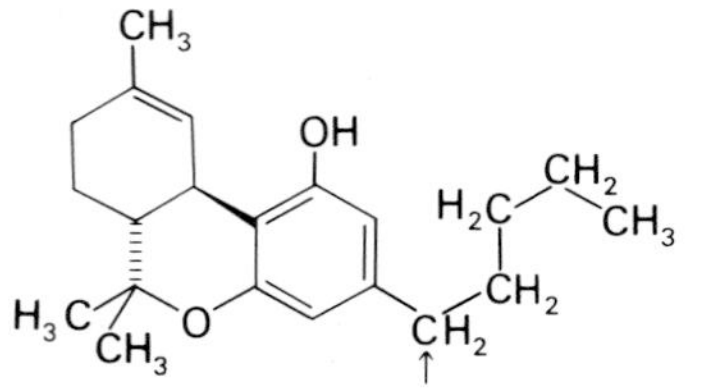

Δ^1-Tetrahydrocannabinol[100]

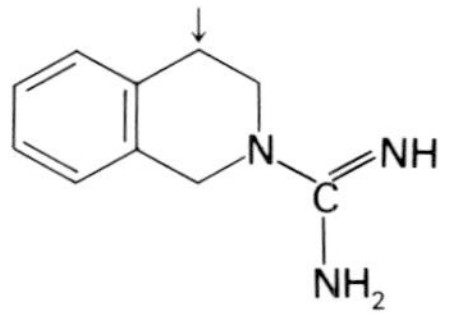

Debrisoquin[101]

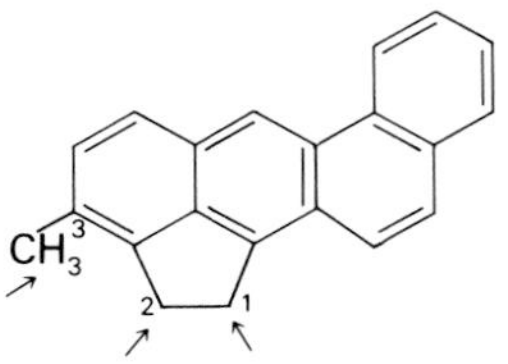

3-Methylcholanthrene[102]

FIG. 3-8. Examples of drugs and xenobiotics undergoing benzylic hydroxylation. Arrow indicates site of hydroxylation.

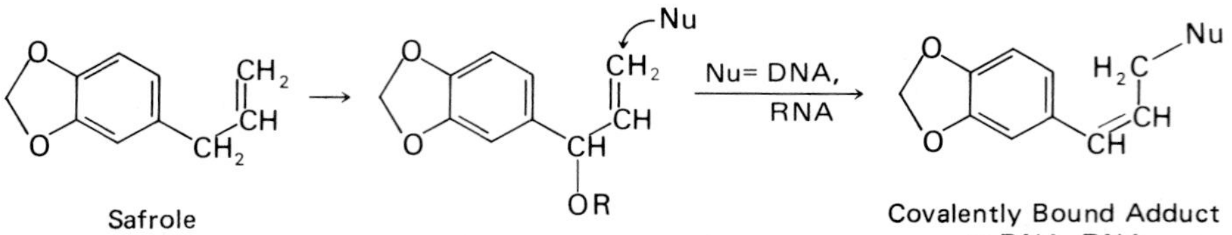

Safrole

1'—Hydroxysafrole, R=H
1'—Hydroxysafrole, R=SO$_3^-$
O—Sulfate ester, R=SO$_3^-$

Nu= DNA,
RNA

Covalently Bound Adduct
to DNA, RNA

OXIDATION AT CARBON ATOMS ALPHA TO CARBONYLS AND IMINES

The mixed function oxidase system also oxidizes carbon atoms adjacent (i.e., alpha) to carbonyl and imino (C=N) functionalities. An important class of drugs undergoing this type of oxidation is the benzodiazepines. For example, diazepam (Valium®), flurazepam

(Dalmane®), and nimetazepam are all oxidized to their corresponding 3-hydroxy metabolites.[111] The C-3 carbon atom undergoing hydroxylation is alpha to both a lactam carbonyl and an imino functionality.

In the case of diazepam, the hydroxylation reaction proceeds with remarkable stereoselectivity to form primarily (90%) 3-hydroxydiazepam (also called N-methyloxazepam) having the S absolute configuration

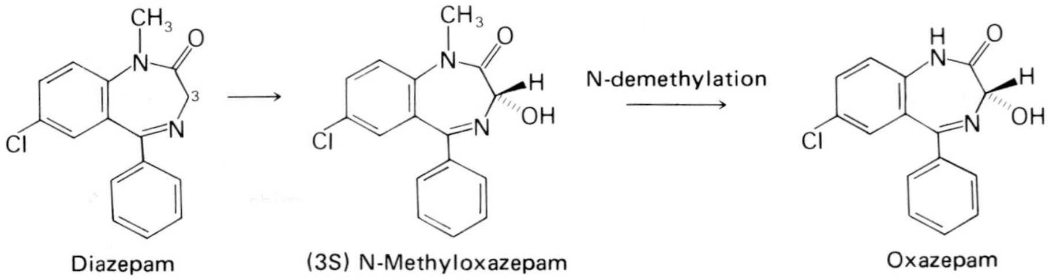

Diazepam

(3S) N-Methyloxazepam
or 3-Hydroxydiazepam

N-demethylation

Oxazepam

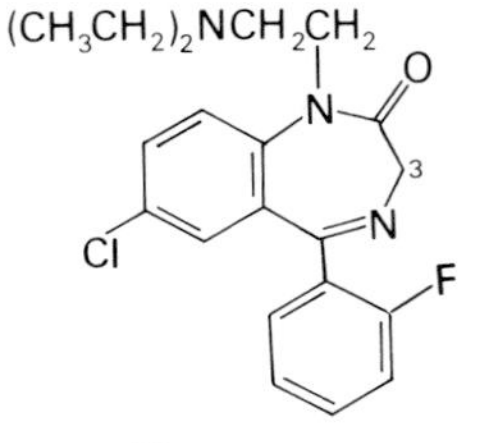

Flurazepam

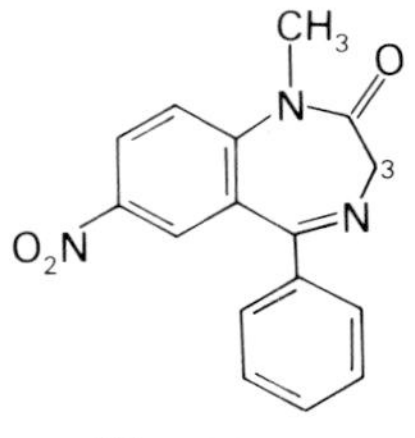

Nimetazepam

at C-3.[112] Further N-demethylation of the latter metabolite gives rise to the pharmacologically active 3S(+)-oxazepam.

Hydroxylation of carbon atom alpha to carbonyl functionalities generally occurs only to a limited extent in drug metabolism. An illustrative example involves the hydroxylation of the sedative hypnotic glutethimide (Doriden®) to 4-hydroxyglutethimide.[113]

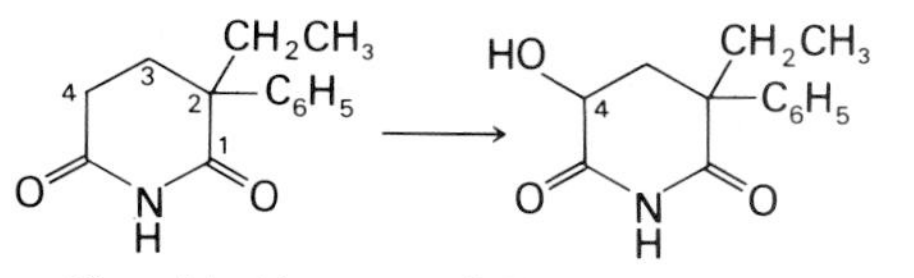

Glutethimide   4-Hydroxyglutethimide

OXIDATION AT ALIPHATIC AND ALICYCLIC CARBON ATOMS

Alkyl or aliphatic carbon centers are subject to mixed function oxidation. Metabolic oxidation at the terminal methyl group is often referred to as ω-oxidation while oxidation of the penultimate carbon atom (i.e., next-to-the-last carbon) is called *ω−1 oxidation*.[47] The initial alcohol metabolites formed from these enzymatic ω- and ω−1 oxidations are susceptible to further oxidation to yield aldehyde, ketones, or carboxylic acids. Alternatively, the alcohol metabolites may undergo glucuronide conjugation.

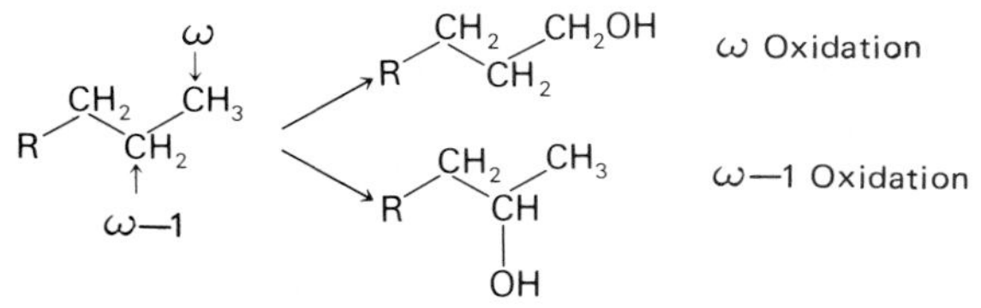

ω Oxidation

ω−1 Oxidation

Aliphatic ω- and ω−1 hydroxylations commonly take place in drug molecules having straight or branched alkyl chains. For example, the antiepileptic agent valproic acid (Depakene®) undergoes both ω- and ω−1 oxidation to the 5-hydroxy and 4-hydroxy metabolites, respectively.[114] Further oxidation of the 5-hydroxy metabolite yields 2-n-propylglutaric acid.

Numerous barbiturates and oral hypoglycemic sulfonylureas also have aliphatic side chains that are susceptible to oxidation. For example, the sedative hypnotic amobarbital (Amytal®) undergoes extensive ω−1 oxidation to the corresponding 3'-hydroxylated metabolite.[115] Other barbiturates, such as pentobarbital,[116] thioamylal,[117] and secobarbital,[77] have also been reported to be metabolized by way of ω and ω−1 oxidation.

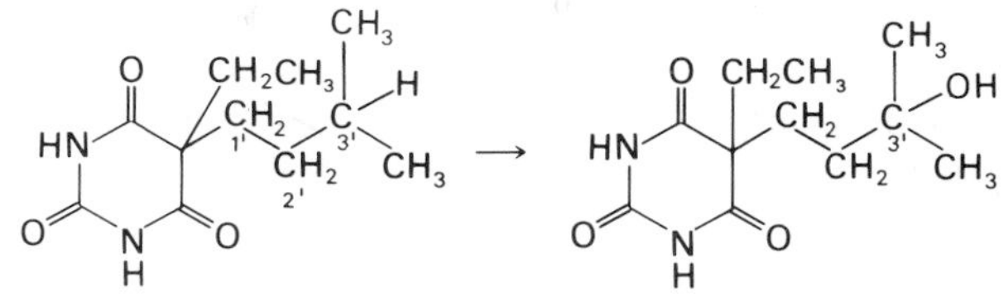

Amobarbital   3'-Hydroxyamobarbital

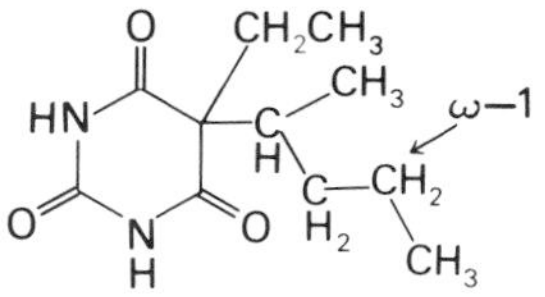

Pentobarbital

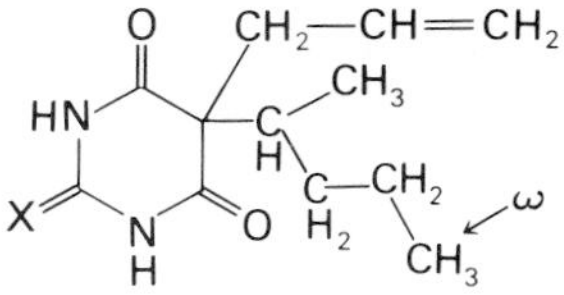

Thioamylal X = S
Secobarbital X = O

The n-propyl side chain attached to the oral hypoglycemic agent chlorpropamide (Diabinese®) undergoes extensive ω−1 hydroxylation to yield the secondary alcohol 2'-hydroxychlorpropamide as a major urinary metabolite in man.[118]

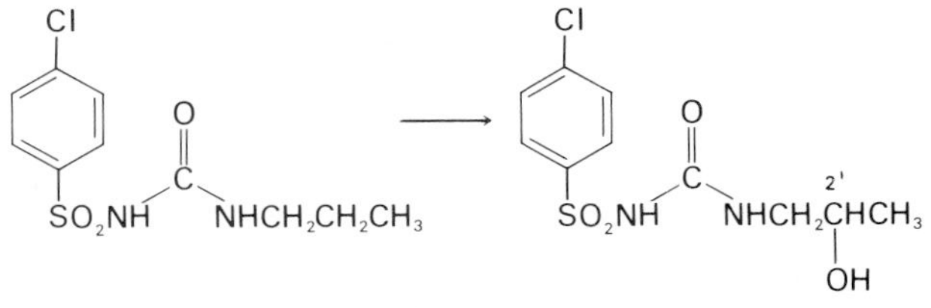

Chlorpropamide   2'-Hydroxychlorpropamide

nC_3H_7
$HOCH_2CH_2CH_2CHCOOH$ $\longrightarrow$ $HOOCCHCH_2CHCOOH$
5-Hydroxyvalproic Acid   2-n-Propylglutaric Acid

ω Oxidation

nC_3H_7
$CH_3CH_2CH_2CHCOOH$ ω−1 Oxidation
Valproic Acid

$CH_3CHCH_2CHCOOH$
4-Hydroxyvalproic Acid

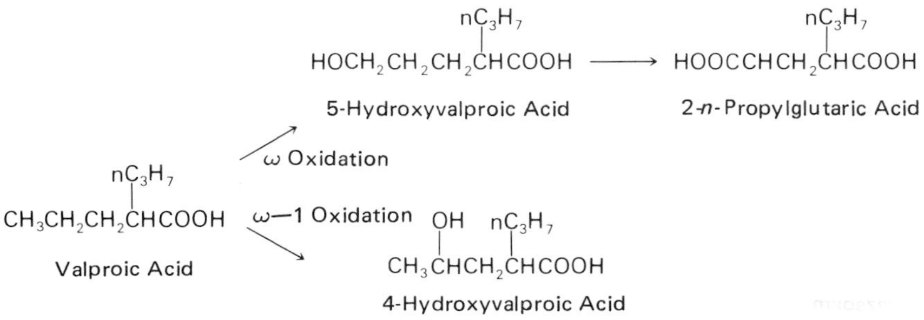

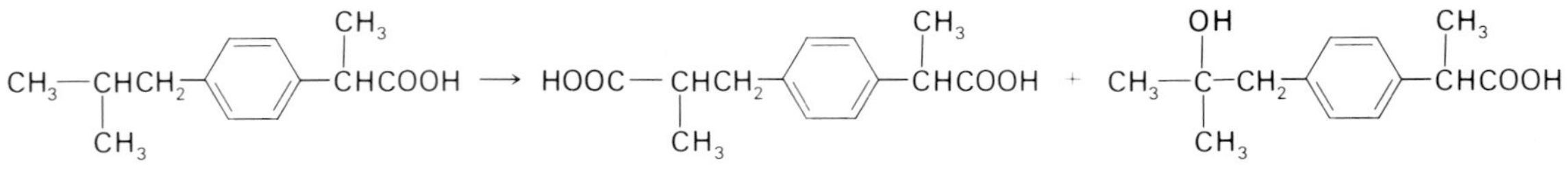

Ibuprofen → Carboxylic Acid Metabolite + Tertiary Alcohol Metabolite

Omega and $\omega-1$ oxidation of the isobutyl moiety present in the antiinflammatory agent ibuprofen (Motrin®) yield the corresponding carboxylic acid and tertiary alcohol metabolites shown below.[119] Additional examples of drugs reported to undergo aliphatic hydroxylation include meprobamate,[120] glutethimide,[113] ethosuximide,[121] and phenylbutazone.[122]

The cyclohexyl group is commonly found in many medicinal agents and is susceptible also to mixed function oxidation (alicyclic hydroxylation).[47] Enzymatic introduction of a hydroxyl group into a monosubstituted cyclohexane ring generally occurs at C-3 or C-4 and can lead to *cis* and *trans* conformational stereoisomers, as shown in the scheme below.

An example illustrating this hydroxylation pathway is seen in the metabolism of the oral hypoglyce-mic agent acetohexamide (Dymelor®). In man, the *trans*-4-hydroxycyclohexyl product has been reported as a significant metabolite.[123] Small amounts of the other possible stereoisomers, namely, the *cis*-4- and *cis*-3- and *trans*-3-hydroxycyclohexyl derivatives, have also been detected. Another related oral hypoglycemic agent glipizide is oxidized in man to the *trans*-4- and *cis*-3-hydroxylcyclohexyl metabolites in about a 6:1 ratio.[124]

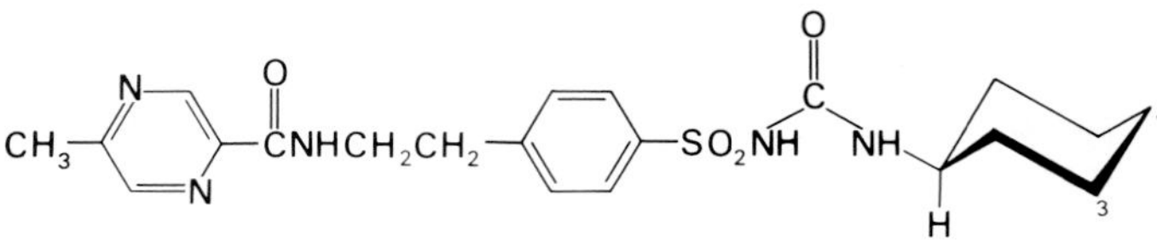

Glipizide

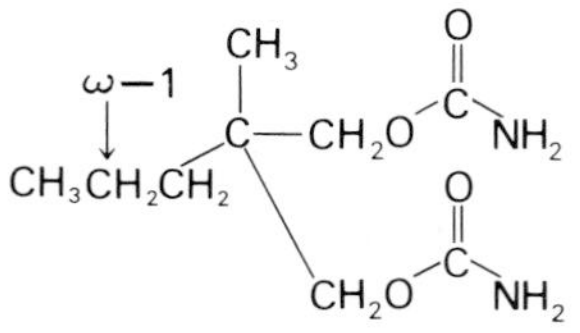

Meprobamate 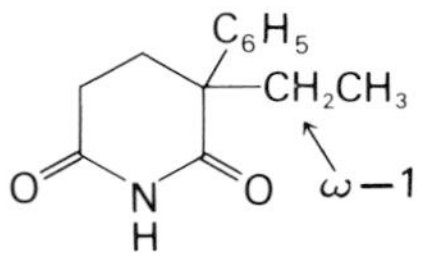 Glutethimide 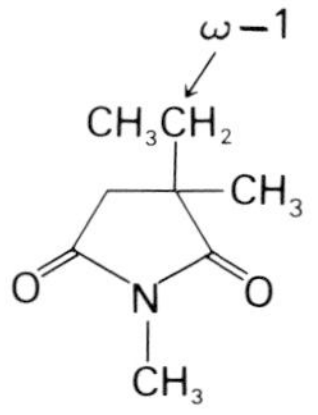 Ethosuximide 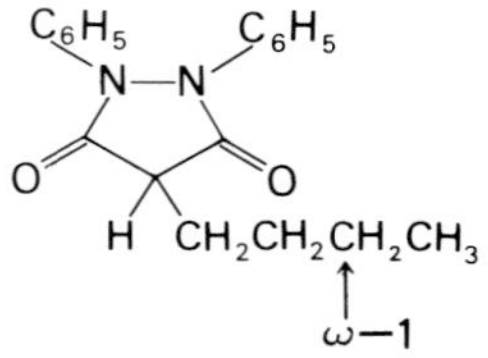 Phenylbutazone

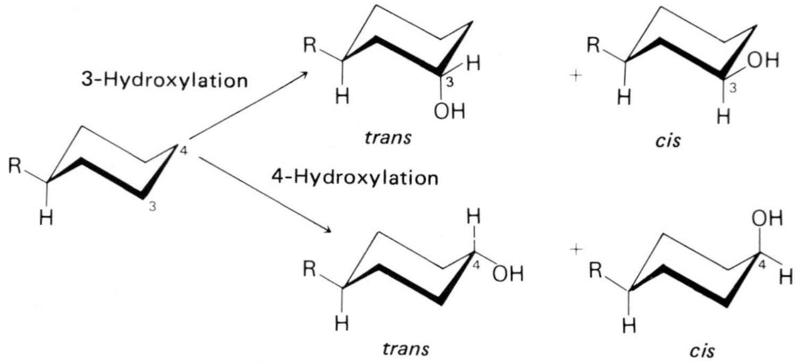

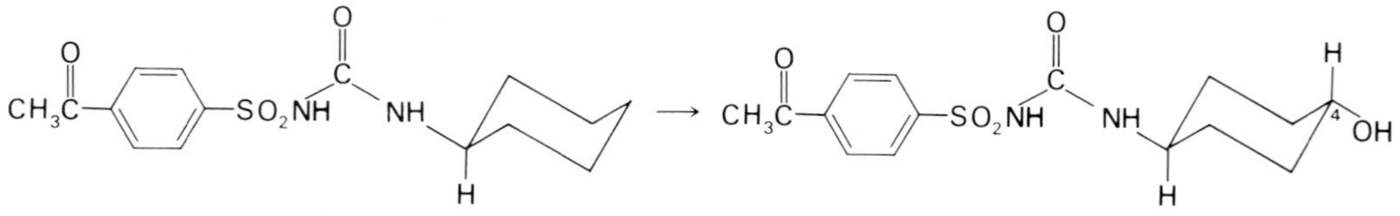

3-Hydroxylation *trans* *cis*

4-Hydroxylation *trans* *cis*

Acetohexamide → *trans*-4-Hydroxyacetohexamide

Phencyclidine   4-Hydroxycyclohexyl Metabolite   4-Hydroxypiperidyl Metabolite

Minoxidil   4'-Hydroxyminoxidil

Two human urinary metabolites of phencyclidine (PCP) have been identified as the 4-hydroxypiperidyl and 4-hydroxycyclohexyl derivatives of the parent compound.[125a] Thus, from these results, it appears that "alicyclic" hydroxylation of the six-membered piperidyl moiety may closely parallel the hydroxylation pattern of the cyclohexyl moiety. The stereochemistry about the hydroxylated centers in the two metabolites has not been clearly established. Biotransformation of the antihypertensive agent minoxidil (Loniten®) yields the 4'-hydroxypiperidyl metabolite. In the dog this product is a major urinary metabolite (29–47%), whereas in man it is detected in small amounts (approximately 3%).[125b]

OXIDATION INVOLVING CARBON-HETEROATOM SYSTEMS

Nitrogen and oxygen functionalities are commonly found in a majority of drugs and foreign compounds, whereas sulfur functionalities occur only occasionally. Metabolic oxidation of carbon–nitrogen, carbon–oxygen and carbon–sulfur systems involves principally two basic types of biotransformation processes:

1. Hydroxylation of the α-carbon atom attached directly to the heteroatom (N,O,S). The resulting intermediate is often unstable and decomposes with the cleavage of the carbon-heteroatom bond:

Where X=N,O,S   Usually Unstable

Oxidative N-, O-, and S-dealkylation as well as oxidative deamination reactions fall under this mechanistic pathway.

2. Hydroxylation or oxidation of the heteroatom (N, S only) (e.g., N-hydroxylation, N-oxide formation, sulfoxide, and sulfone formation).

A number of structural features frequently determine which pathway will predominate, especially in the case of carbon–nitrogen systems. Metabolism of some nitrogen-containing compounds is complicated by the fact that carbon- or nitrogen-hydroxylated products may undergo secondary reactions to form other, more complex, metabolic products (e.g., oxime, nitrone, nitroso, imino). Other oxidative processes that do not fall under the two basic categories above will be discussed individually in the appropriate carbon-heteroatom section. The metabolism of carbon-nitrogen systems will be discussed first, followed by the metabolism of carbon–oxygen and carbon–sulfur systems.

OXIDATION INVOLVING CARBON-NITROGEN SYSTEMS

Metabolism of nitrogen functionalities (e.g., amines, amides) is of importance, since such functional groups are found in many natural products (e.g., morphine, cocaine, nicotine) and in a large number of important drugs (e.g., phenothiazines, antihistamines, tricyclic antidepressants, β-adrenergic agents, sympathomimetic phenylethylamines, barbiturates, benzodiazepines).[126] The discussion to follow divides nitrogen-containing compounds into three basic classes:

1. Aliphatic (tertiary, secondary and primary) and alicyclic (tertiary, secondary) amines

2. Aromatic and heterocyclic nitrogen compounds
3. Amides

The susceptibility of each class of these nitrogen compounds to either α-carbon hydroxylation or N-oxidation and the metabolic products that are formed will be discussed.

The hepatic enzymes responsible for carrying out α-carbon hydroxylation reactions are the cytochrome P-450 mixed function oxidases. However, the N-hydroxylation or N-oxidation reactions appear to be catalyzed not only by cytochrome P-450 but also by a second class of hepatic mixed function oxidases called amine oxidases (sometimes called N-oxidases).[127] These enzymes are NADPH-dependent flavoproteins and do not contain cytochrome P-450.[128] They require NADPH and molecular oxygen to carry out N-oxidation.

Tertiary aliphatic and alicyclic amines

The oxidative removal of alkyl groups (particularly methyl groups) from tertiary aliphatic and alicyclic amines is carried out by hepatic cytochrome P-450 mixed function oxidase enzymes. This reaction is commonly referred to as oxidative N-dealkylation.[129] The initial step involves α-carbon hydroxylation to form a carbinolamine intermediate, which is unstable and undergoes spontaneous heterolytic cleavage of the C-N bond to give a secondary amine and a carbonyl moiety

(aldehyde or ketone).[130] In general, small alkyl groups like methyl, ethyl, and isopropyl are rapidly removed.[129] N-Dealkylation of the *t*-butyl group is not possible by the carbinolamine pathway, since α-carbon hydroxylation cannot occur. Removal of the first alkyl group from a tertiary amine occurs more rapidly than the removal of the second alkyl group. In some instances, bisdealkylation of the tertiary aliphatic amine to the corresponding primary aliphatic amine occurs very slowly.[129] For example, the tertiary amine imipramine (Tofranil®) is monodemethylated to desmethylimipramine (desipramine).[98,131] This major plasma metabolite in man is pharmacologically active and contributes significantly to the antidepressant activity of the parent drug.[132] Very little of the bisdemethylated metabolite of imipramine is detected. In contrast, the local anesthetic and antiarrhythmic agent lidocaine is extensively metabolized by N-deethylation to both monoethylglycylxylidine and glycyl-2,6-xylidine in man.[133]

Numerous other tertiary aliphatic amine drugs are metabolized principally by oxidative N-dealkylation. Some of these include the antiarrhythmic disopyramide (Norpace®),[134] the antiestrogenic agent tamoxifen (Nolvadex®),[135] diphenhydramine (Benadryl®),[136] chlorpromazine (Thorazine®)[137] and (+)-α-propoxyphene (Darvon®).[138] In cases in which the tertiary amine contains several different substituents capable of undergoing dealkylation, the smaller alkyl group is preferentially and more rapidly removed. For exam-

Tertiary Amine   Carbinolamine   Secondary Amine   Carbonyl Moiety (aldehyde or ketone)

Imipramine   Desmethylimipramine (desipramine)   Bisdesmethylimipramine

Lidocaine   Monoethylglycylxylidine (MEGX)   Glycyl-2, 6-xylidine

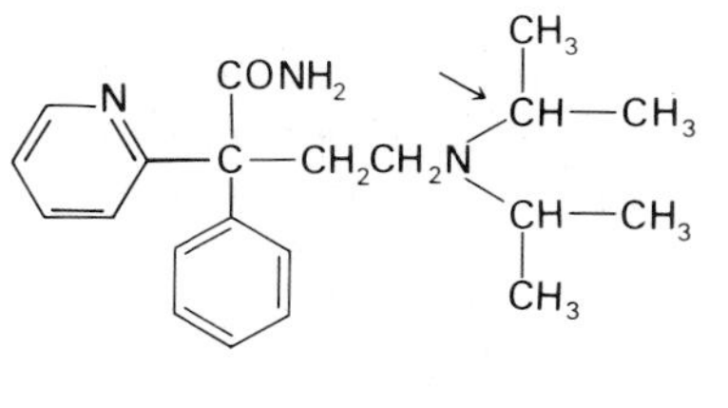

Disopyramide

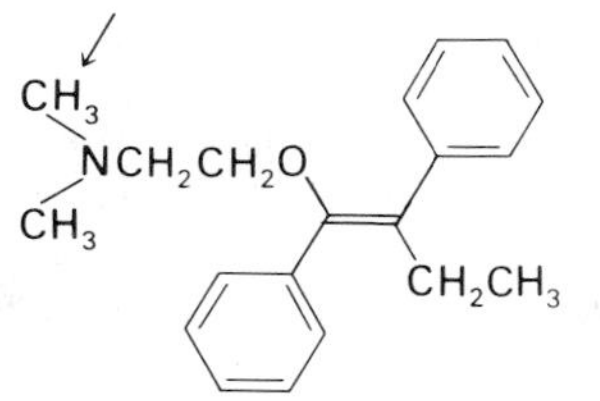

Tamoxifen

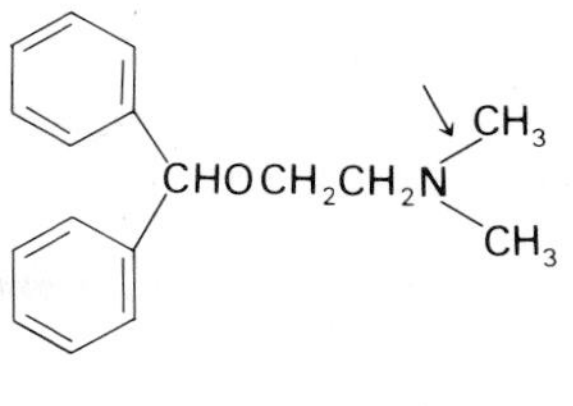

Diphenhydramine

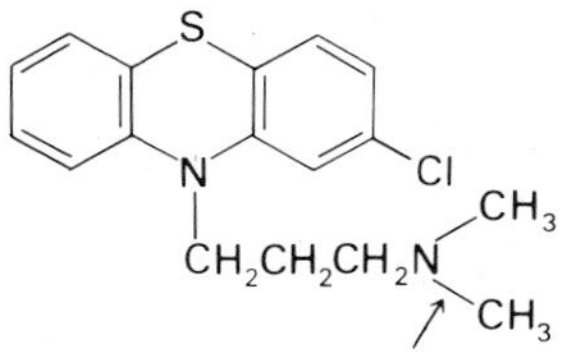

Chlorpromazine

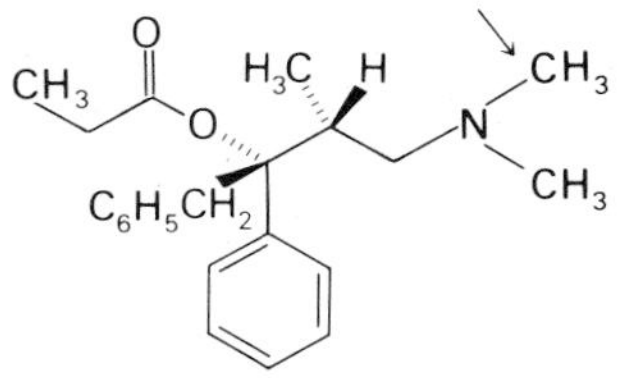

(+)-α-Propoxyphene

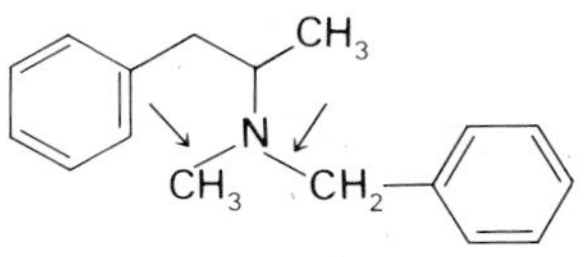

Benzphetamine
(N-demethylation
and N-debenzylation)

ple, in benzphetamine (Didrex®), the methyl group is removed much more rapidly than the benzyl moiety.[139]

An interesting cyclization reaction occurs with methadone upon N-demethylation. The demethylated metabolite, normethadone, undergoes a spontaneous cyclization reaction to form the enamine metabolite, 2-ethylidene-1,5-dimethyl-3,3-diphenylpyrrolidine (EDDP).[140] Subsequent N-demethylation of EDDP and isomerization of the double bond leads to 2-ethyl-5-methyl-3,3-diphenyl-1-pyrroline (EMDP).

In many cases, bisdealkylation of a tertiary amine leads to the corresponding primary aliphatic amine metabolite, which is susceptible to further oxidation. For example, the bisdesmethyl metabolite of the H_1-histamine antagonist brompheniramine (Dimetane®) undergoes oxidative deamination and further oxidation to the corresponding propionic acid metabolite.[141] Oxidative deamination will be discussed in greater detail when we examine the metabolic reactions of secondary and primary amines.

Like their aliphatic counterparts, alicyclic tertiary

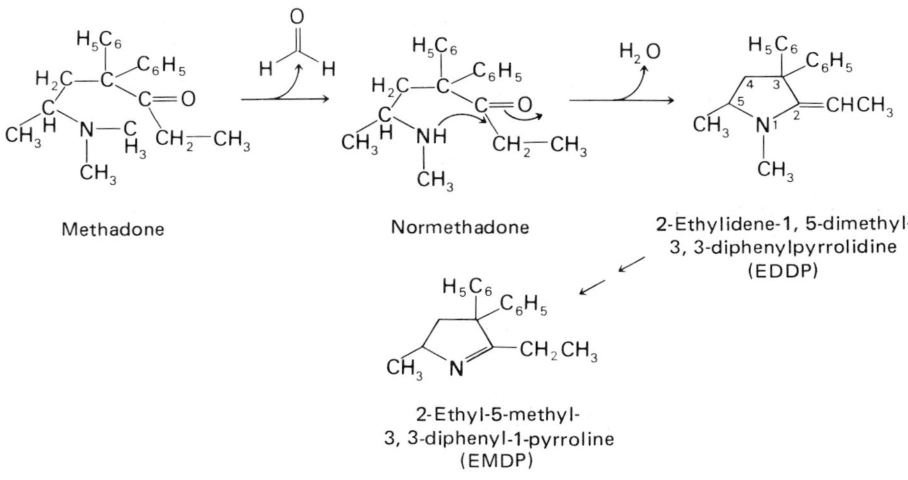

Methadone

Normethadone

2-Ethylidene-1, 5-dimethyl-3, 3-diphenylpyrrolidine (EDDP)

2-Ethyl-5-methyl-3, 3-diphenyl-1-pyrroline (EMDP)

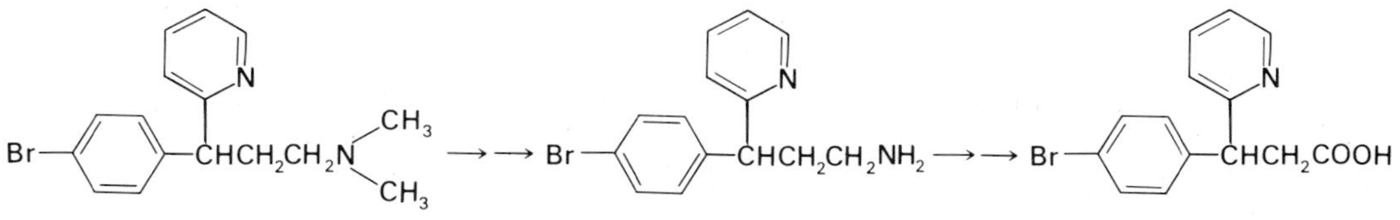

Brompheniramine

Bisdesmethyl Metabolite

3-(p-Bromophenyl)-3-pyridyl-propionic Acid

amines are susceptible to oxidative N-dealkylation reactions. For example, the analgesic meperidine (Demerol®) is metabolized principally by this pathway to yield normeperidine as a major plasma metabolite in

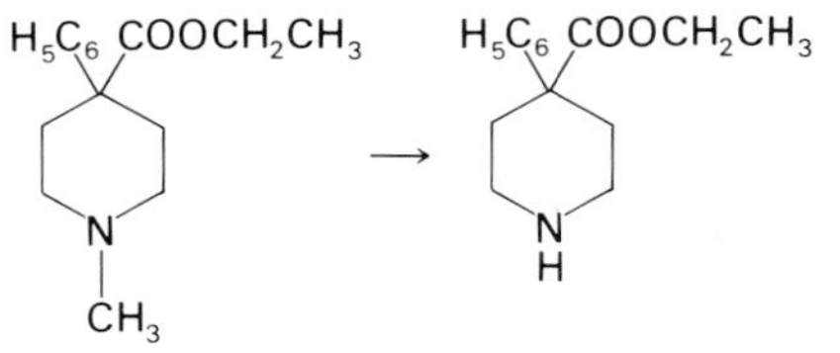

Meperidine   Normeperidine

man.[142] Morphine, N-ethylnormorphine, and dextromethorphan undergo N-dealkylation to some extent too.[143]

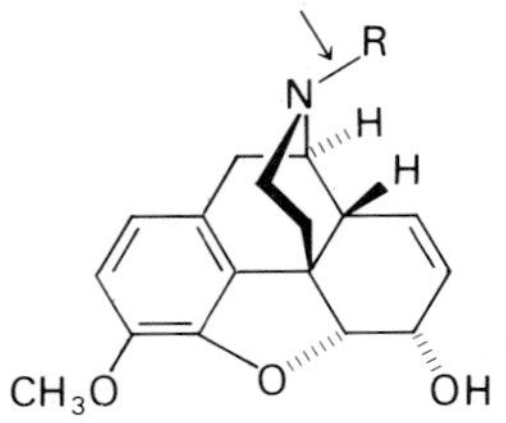

Morphine   R= CH₃
N-Ethylnormorphine R=CH₂CH₃

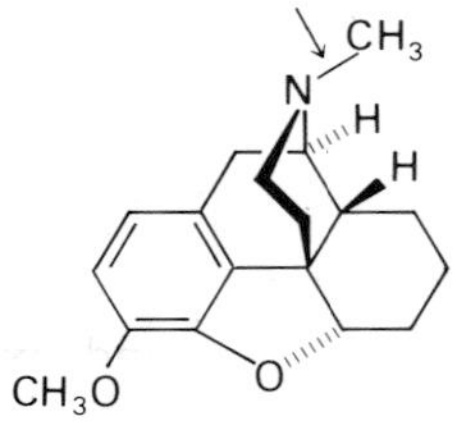

Dextromethorphan

Direct N-dealkylation of *t*-butyl groups as discussed earlier is not possible by the α-carbon hydroxylation pathway. However, in vitro studies indicate that N-*t*-butylnorchlorocyclizine indeed is metabolized to significant amounts of norchlorocyclizine, whereby the *t*-butyl group is lost.[144] Careful studies showed that the *t*-butyl group is removed via initial hydroxylation of one of the methyl groups of the *t*-butyl moiety to the carbinol or alcohol product.[145] Further oxidation generates the corresponding carboxylic acid, which upon decarboxylation forms the N-

isopropyl derivative. The N-isopropyl intermediate is dealkylated by the normal α-carbon hydroxylation (i.e., carbinolamine) pathway to give norchlorocyclizine and acetone. Whether this is a general method for the loss of *t*-butyl groups from amines is still unclear. It appears that indirect N-dealkylation of *t*-butyl groups is not significantly observed. The N-*t*-butyl group present in many β-adrenergic antagonists such as terbutaline and salbutamol remains intact and does not appear to undergo any significant metabolism.[146]

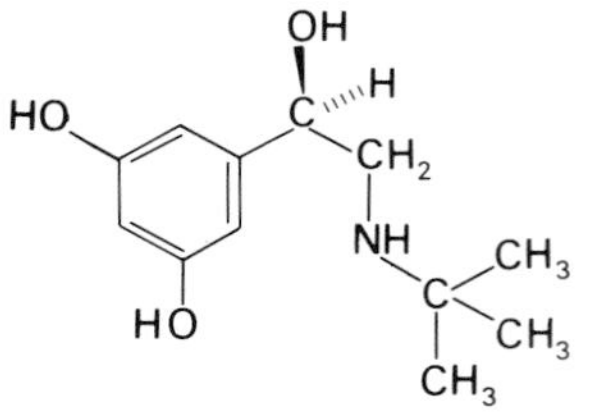

Terbutaline 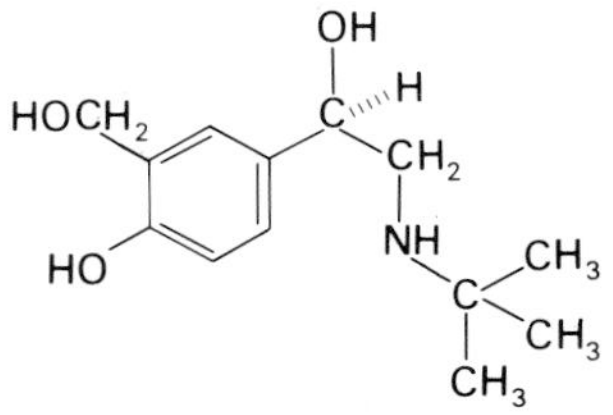   Salbutamol

Alicyclic tertiary amines often generate lactam metabolites via α-carbon hydroxylation reactions. For example, the tobacco alkaloid nicotine is initially hydroxylated at the ring carbon atom α to the nitrogen to yield a carbinolamine intermediate. Furthermore, enzymatic oxidation of this cyclic carbinolamine generates the lactam metabolite, cotinine.[147]

Formation of lactam metabolites has been also reported to occur to a minor extent for the antihistamine cyproheptadine (Periactin®)[148] and the antiemetic diphenidol (Vontrol®).[149]

N-Oxidation of tertiary amines occurs with a number of drugs.[150] The true extent of N-oxide formation is often complicated by the susceptibility of N-oxides to undergo in vivo reduction back to the parent tertiary amine. Tertiary amines such as H₁-histamine antagonists (e.g., orphenadrine, tripelenamine), phenothiazines (e.g., chlorpromazine), tricyclic antidepressants (e.g., imipramine), and narcotic analgesics (e.g., morphine, codeine, and meperidine) have been

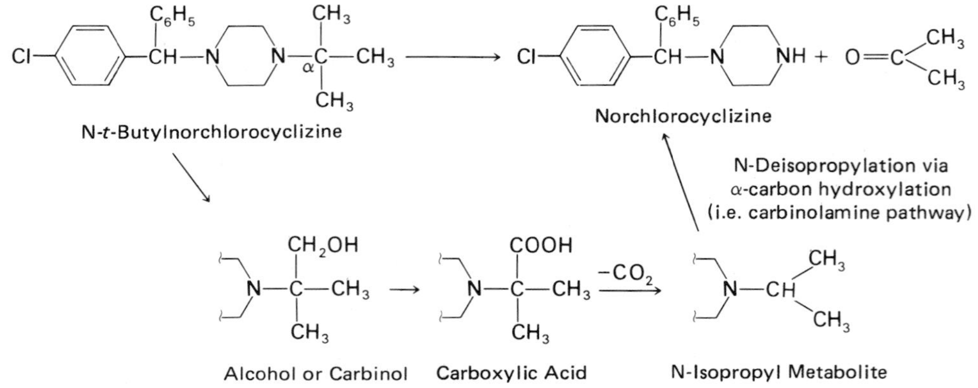

N-*t*-Butylnorchlorocyclizine

Norchlorocyclizine

N-Deisopropylation via
α-carbon hydroxylation
(i.e. carbinolamine pathway)

Alcohol or Carbinol   Carboxylic Acid   N-Isopropyl Metabolite

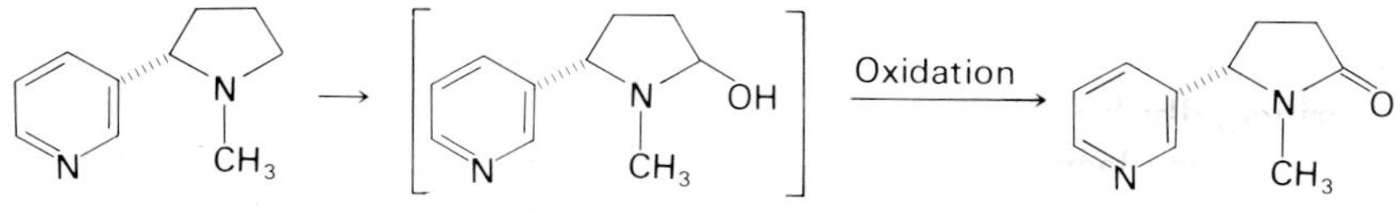

Nicotine → Carbinolamine —Oxidation→ Cotinine

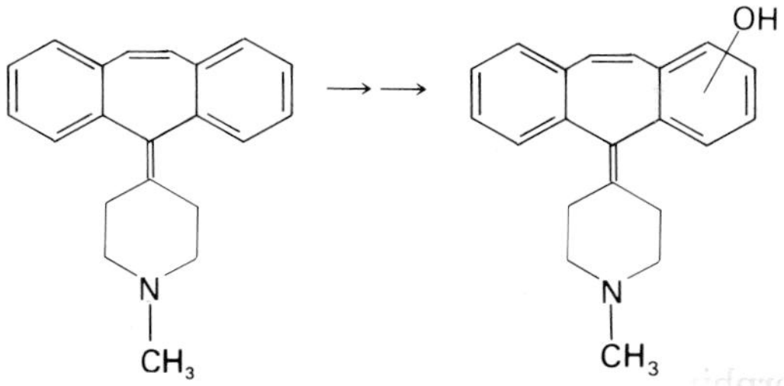

Cyproheptadine

Lactam Metabolite

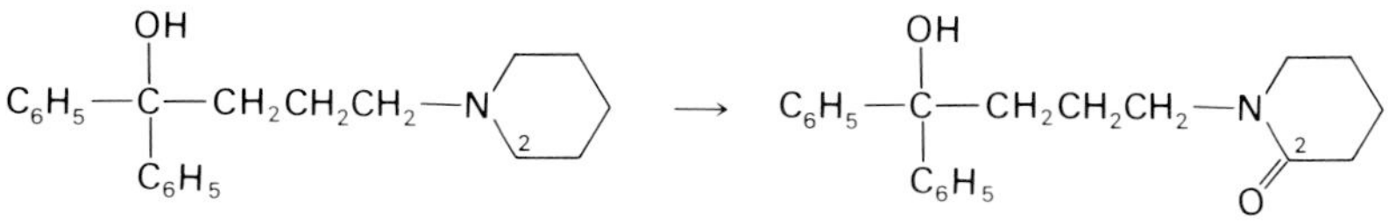

Diphenidol

2-Oxodiphenidol

reported to form N-oxides products. In some instances, N-oxides have been shown to possess pharmacological activity.[151] For example, comparison of imipramine N-oxide with imipramine indicates that the N-oxide itself possesses antidepressant and cardiovascular activity similar to the parent drug.[152]

Secondary and primary amines

Secondary amines (either parent compounds or metabolites) are susceptible to oxidative N-dealkylation, oxidative deamination, and N-oxidation reactions.[129,153] As in the case of tertiary amines, N-dealkylation of secondary amines proceeds via the carbinolamine pathway. Dealkylation of secondary amines gives rise to the corresponding primary amine metabolite. For example, the β-adrenergic blockers

propranolol[39] and oxprenolol[154] undergo N-deisopropylation to the corresponding primary amines. N-Dealkylation appears to be a significant biotransformation pathway for the secondary amine drugs, meth-

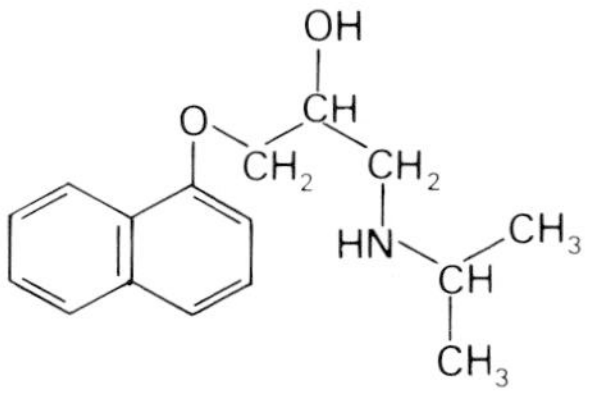

Propranolol

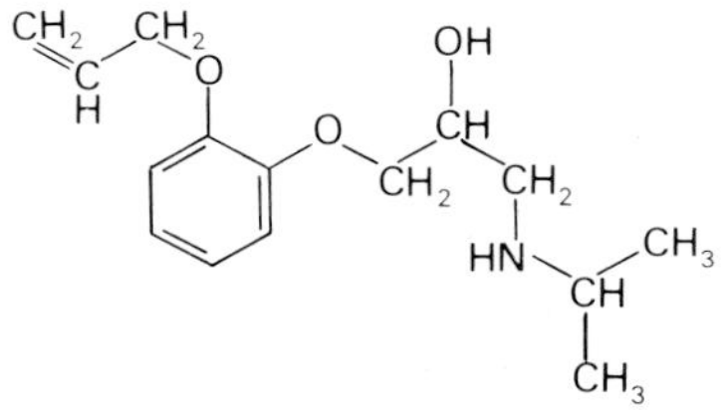

Oxprenolol

amphetamine[155] and ketamine,[156] yielding amphetamine and norketamine, respectively.

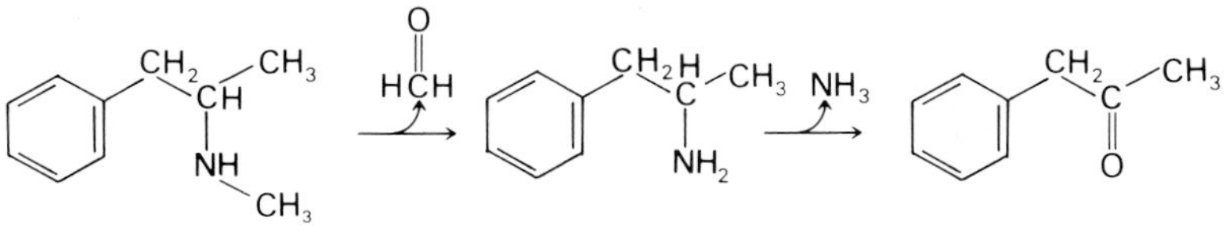

Methamphetamine

Amphetamine

Phenylacetone

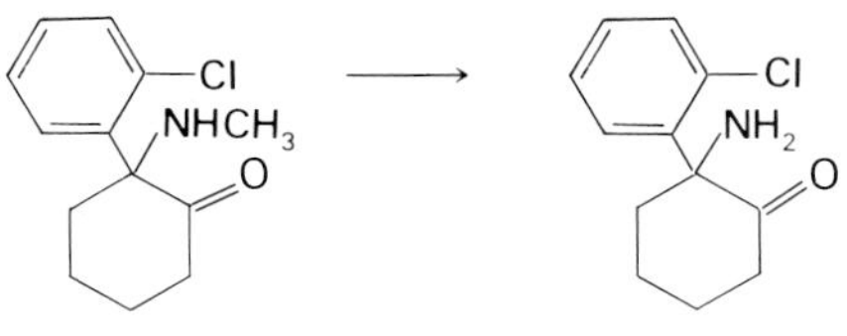

Ketamine

Norketamine

The primary amine metabolites formed from oxidative dealkylation are susceptible to *oxidative deamination*. This process is similar to N-dealkylation in that it involves an initial α-carbon hydroxylation reaction to form a carbinolamine intermediate which then undergoes subsequent carbon–nitrogen cleavage to the carbonyl metabolite and ammonia. If α-carbon hydroxylation cannot occur, then oxidative deamination

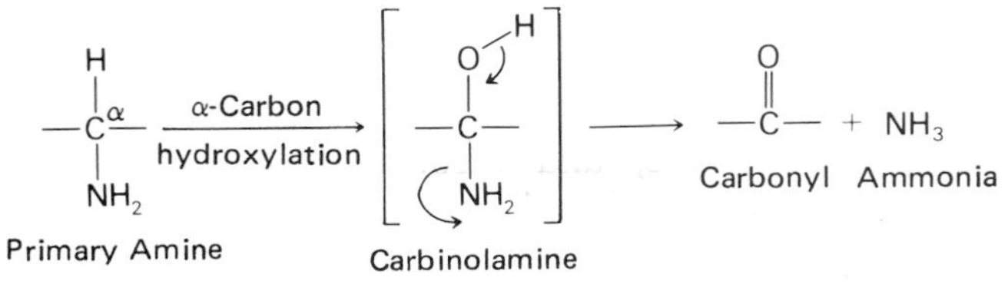

is not possible. For example, deamination does not occur for norketamine, since α-carbon hydroxylation cannot take place.[156] In the case of methamphetamine, oxidative deamination of primary amine metabolite amphetamine produces phenylacetone (see above).[155]

In general, dealkylation of secondary amines is believed to take place before oxidative deamination occurs. However, there is some evidence that this may not always be the case. Direct deamination of the secondary amine has also been observed to occur. For example, in addition to undergoing deamination via

its desisopropyl primary amine metabolite, propranolol can undergo a direct oxidative deamination reaction (also via α-carbon hydroxylation) to yield the aldehyde metabolite and isopropylamine (Fig. 3-9).[157] How significantly direct oxidative deamination contributes to the metabolism of secondary amines remains unclear.

Some secondary alicyclic amines like their tertiary amine analogs are metabolized to their corresponding lactam derivatives. For example, the anorectic agent phenmetrazine (Preludin®) is principally metabolized

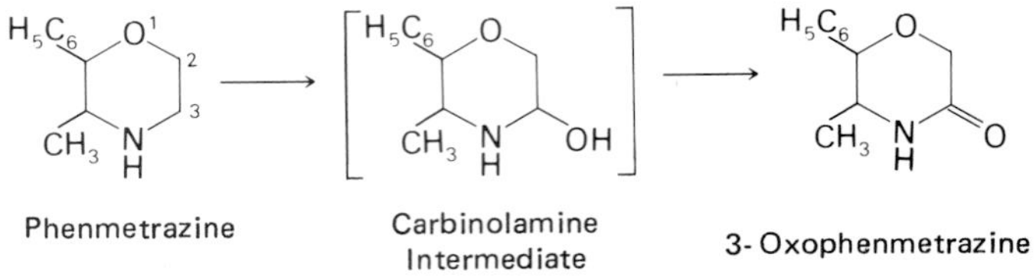

to the lactam product 3-oxophenmetrazine.[158] In man, this lactam metabolite is a major urinary product. Methylphenidate (Ritalin®) has also been reported to yield a lactam metabolite, 6-oxoritalinic acid, via oxidation of its hydrolyzed metabolite ritalinic acid in man.[159]

Metabolic N-oxidation of secondary aliphatic and alicyclic amines leads to a number of N-oxygenated

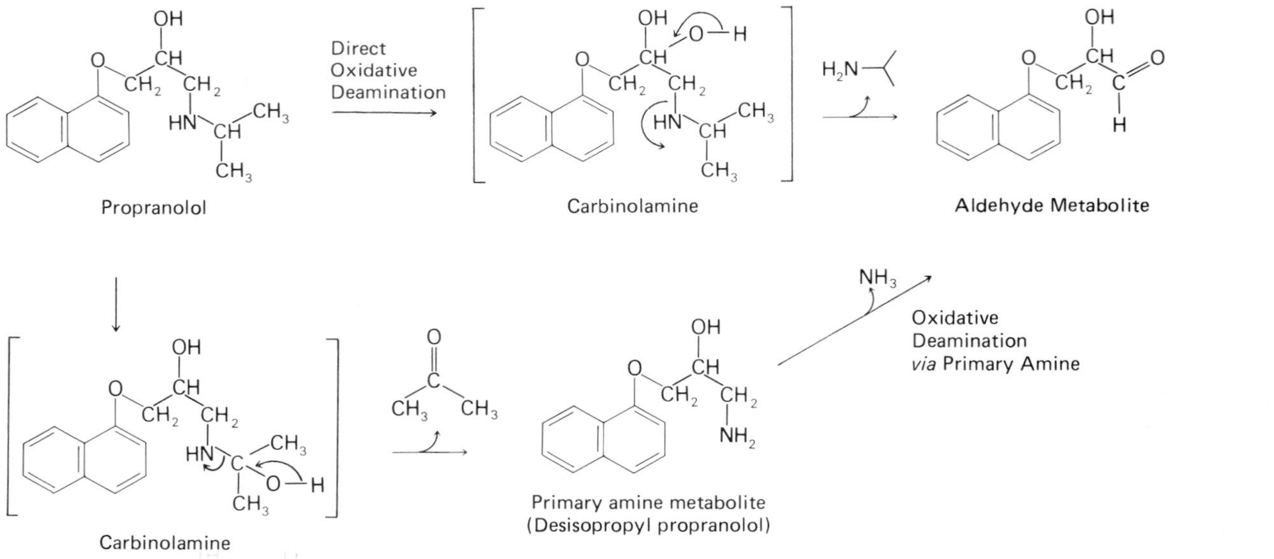

FIG. 3-9. *The metabolism of propranolol to its aldehyde metabolite* via *direct deamination of the parent compound and* via *deamination of its primary amine metabolite desisopropyl propranolol.*

Methylphenidate   Ritalinic Acid   6-Oxoritalinic Acid

products.[153] N-Hydroxylation of secondary amines generates the corresponding N-hydroxylamine metabolites. Often these hydroxylamine products are susceptible to further oxidation (either spontaneous or enzymatic) to the corresponding nitrone derivatives.

Secondary amine   Hydroxylamine   Nitrone

For example, N-benzylamphetamine has been observed to undergo metabolism to both the corresponding N-hydroxylamine and the nitrone metabolites.[160] In man, the nitrone metabolite of phenmetrazine (Preludin®) found in the urine is believed to be formed via

Phenmetrazine   N-Hydroxyphenmetrazine   Nitrone Metabolite

further oxidation of the N-hydroxylamine intermediate, N-hydroxyphenmetrazine.[158] It should be emphasized that in comparison to oxidative dealkylation and deamination, N-oxidation occurs to a much lesser extent for secondary amines.

Primary aliphatic amines (whether parent drugs or metabolites) are biotransformed by oxidative deamination (via carbinolamine pathway) or by N-oxida-tion. In general, oxidative deamination of most exogenous primary amines is carried out by the mixed function oxidases discussed earlier. However, endogenous primary amines, such as dopamine, norepinephrine, tryptamine, and serotonin, are metabolized through oxidative deamination by a specialized family of enzymes called monoamine oxidases (MAO).[161] These enzymes are primarily involved in inactivating the above neurotransmitter amines. It does not appear that MAO plays any significant role in the metabolism of xenobiotic primary amines.

Structural features, especially the α-substituents of the primary amine, often determine whether carbon or nitrogen oxidation will occur. For example, compare amphetamine with its α-methyl homolog, phentermine. In amphetamine, α-carbon hydroxylation can occur to form the carbinolamine intermediate, which is converted to the oxidatively deaminated product, phenylacetone.[46] In the case of phentermine, α-carbon hydroxylation is not possible and precludes oxidative deamination for this drug. Consequently, phentermine would be expected to undergo N-oxidation readily. In man, p-hydroxylation and N-oxidation are the main pathways for biotransformation of phentermine.[162]

Indeed, N-hydroxyphentermine is a significant (5%) urinary metabolite in man.[162] As shall be discussed shortly, N-hydroxylamine metabolites are susceptible to further oxidation to yield other N-oxygenated products.

Xenobiotics, such as the hallucinogenic agents mescaline[163] and 1-(2,5-dimethoxy-4-methylphenyl)-2-aminopropane (DOM or "STP"),[97] are oxidatively deaminated. Primary amine metabolites arising from N-dealkylated or decarboxylation reactions also undergo deamination. The example of the bisdesmethyl

N-Benzylamphetamine   Hydroxylamine Metabolite   Nitrone Metabolite

Amphetamine   α-Carbon Hydroxylation   Carbinolamine   Phenylacetone

Phentermine   α-Carbon hydroxylation not possible, hence do not see oxidative deamination

N-Hydroxylation

p-Hydroxyphentermine   N-Hydroxyphentermine

primary amine metabolite derived from bromopheniramine was discussed earlier (see section on tertiary aliphatic and alicyclic amines).[141] In addition, many

Mescaline   1-(2,5-Dimethoxy -4-methylphenyl)-2-aminopropane DOM or "STP"

tertiary aliphatic amines (e.g., antihistamines) and secondary aliphatic amines (e.g., propranolol) are

dealkylated to their corresponding primary amine metabolites, which are amenable to oxidative deamination. S(+)-α-Methyldopamine resulting from decarboxylation of the antihypertensive agent S(−)-α-methyldopa (Aldomet®) is deaminated to the corresponding ketone metabolite, 3,4-dihydroxyphenylacetone.[164] In man, this ketone is a significant urinary metabolite.

The N-hydroxylation reaction is not restricted to α-substituted primary amines like phentermine. Amphetamine has been observed to undergo some N-hydroxylation in vitro to N-hydroxyamphetamine.[165] However, N-hydroxyamphetamine is susceptible to further conversion to the imine or oxidation to the oxime intermediate. It is interesting to note that the

S(-)-α-Methyldopa   Enzymatic   S(+)-α-Methyldopamine   Oxidative Deamination   3,4-Dihydroxyphenylacetone

Amphetamine   N-Hydroxyamphetamine   Imine   Oxidation   Phenylacetone   Oxime

oxime intermediate arising from this N-oxidation pathway can undergo hydrolytic cleavage to yield phenylacetone, the same product obtained via the α-carbon hydroxylation (carbinolamine) pathway.[166] Thus, it is apparent that conversion of amphetamine to phenylacetone may arise via either the α-carbon hydroxylation or N-oxidation pathway. The debate concerning the relative importance of the two pathways remains ongoing.[166,167] The general consensus however is that both metabolic pathways (carbon and nitrogen oxidation) are probably operative. Whether α-carbon or nitrogen-oxidation predominates in the metabolism of amphetamine appears to be species dependent.

In primary aliphatic amines like phentermine,[162] chlorphentermine (*p*-chlorophentermine),[168] and amantadine,[169] N-oxidation appears to be the major biotransformation pathway, since α-carbon hydroxylation cannot occur. In man, chlorphentermine is

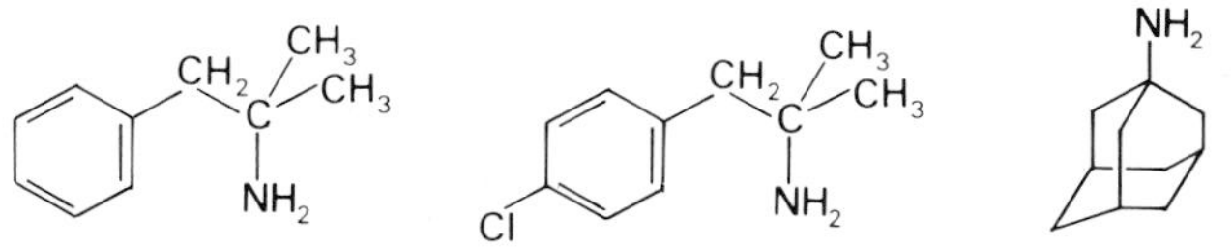

Phentermine   Chlorphentermine   Amantadine

extensively N-hydroxylated. About 30 percent of a dose of chlorphentermine is found in the urine (48 hours) as N-hydroxychlorphentermine (free and conjugated) and an additional 18 percent as other products of N-oxidation (presumably the nitroso and nitro

metabolites).[168] In general, N-hyroxylamines are chemically unstable and susceptible to spontaneous or enzymatic oxidation to the nitroso and nitro derivatives. For example, the N-hydroxylamine metabolite

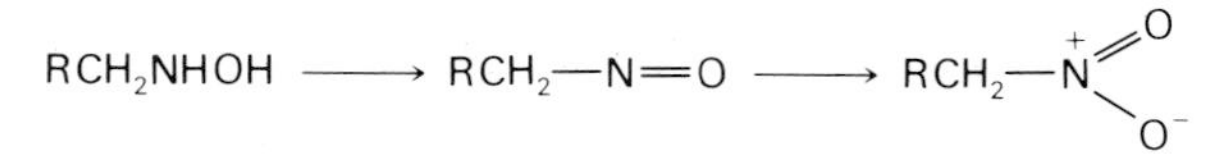

Hydroxylamine   Nitroso   Nitro

of phentermine undergoes further oxidation to the nitroso and nitro products.[162] The antiviral and antiparkinson agent amantadine (Symmetrel®) has been reported to undergo N-oxidation to yield the corresponding N-hydroxy and nitroso metabolites in vitro.[169]

Aromatic amines and heterocyclic nitrogen compounds

The biotransformation of aromatic amines parallels the carbon and nitrogen oxidation reactions seen for aliphatic amines.[170,171] In the case of tertiary aromatic amines such as N,N-dimethylaniline, oxidative N-dealkylation as well as N-oxide formation takes place.[172] Secondary aromatic amines may undergo N-dealkylation or N-hydroxylation to give the corresponding N-hydroxylamines. Further oxidation of the N-hydroxylamine leads to nitrone products, which in turn may be hydrolyzed to primary hydroxylamines.[173] Tertiary

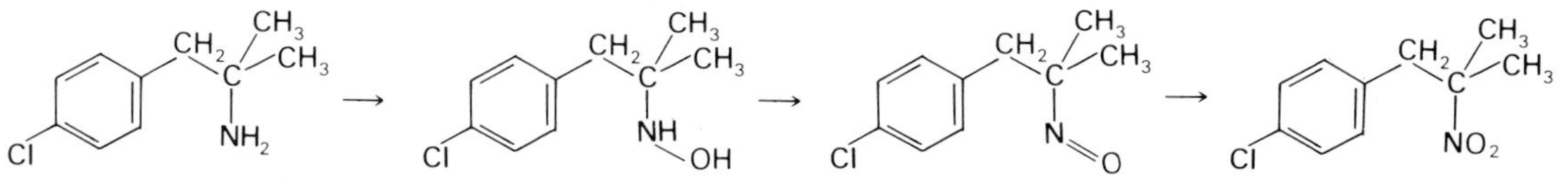

Chlorphentermine   N-Hydroxychlorphentermine   Nitroso Metabolite   Nitro Metabolite

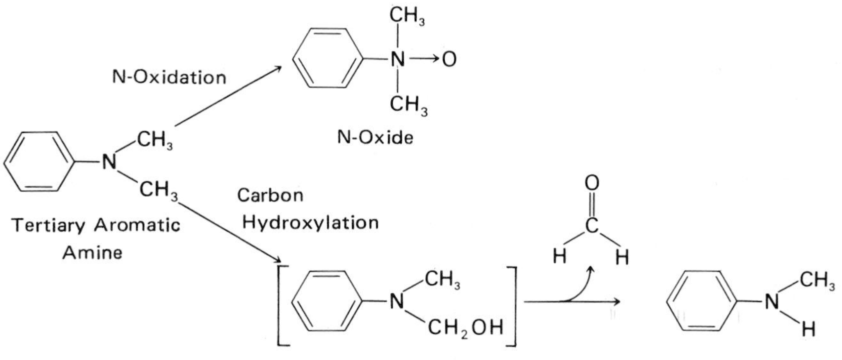

N-Oxidation

N-Oxide

Tertiary Aromatic Amine

Carbon Hydroxylation

Carbinolamine

Secondary Aromatic Amines → Hydroxylamine (Secondary) → (Oxidation) → Nitrone → (H₂O) → Hydroxylamine (primary)

and secondary aromatic amines are rarely encountered in medicinal agents. In contrast, primary aromatic amines are found in many drugs and often are generated from enzymatic reduction of aromatic nitro compounds, reductive cleavage of azo compounds and hydrolysis of aromatic amides.

N-Oxidation of primary aromatic amines generates the N-hydroxylamine metabolite. For example, aniline is metabolized to the corresponding N-hydroxy product.[171] Oxidation of the hydroxylamine derivative to

Aniline (primary aromatic amine) → Hydroxylamine → Nitroso

the nitroso derivative can also occur. When considering primary aromatic amine drugs or metabolites, N-oxidation constitutes only a minor pathway in comparison with other biotransformation pathways like N-acetylation and aromatic hydroxylation in man. However, some N-oxygenated metabolites have been reported. For example, the antileprotic agent dapsone and its N-acetylated metabolite are significantly metabolized to their corresponding N-hydroxylamine derivatives.[174] The N-hydroxy metabolites are further conjugated with glucuronic acid.

It has been established that methemoglobinemia toxicity caused by a number of aromatic amines including aniline and dapsone is a result of the bioconversion of the aromatic amine to its N-hydroxy derivative. Apparently the N-hydroxylamine oxidizes the Fe^{+2} form of hemoglobin to its Fe^{+3} form. This oxidized (Fe^{+3}) state of hemoglobin (called methemoglobin or ferrihemoglobin) is no longer capable of transporting oxygen and leads to serious hypoxia or anemia.[175]

A number of aromatic amines (especially azo amino dyes) are known to be carcinogenic. It is believed that N-oxidation plays an important role in bioactivating these aromatic amines to potentially reactive electrophilic species that covalently bind to cellular protein, DNA, or RNA. A well-studied example is the carcinogenic agent N-methyl-4-aminoazobenzene.[176] N-Oxidation of this compound leads to the corresponding hydroxylamine, which has been shown to undergo sulfate conjugation. Owing to the good leaving group ability

Dapsone R = H
N-Acetyldapsone R = CCH₃ (O)

N-Hydroxydapsone R = H

N-Acetyl-N'-hydroxydapsone R = CCH₃ (O)

N-Methyl-4-aminoazobenzene → Hydroxylamine → Sulfate Conjugate

Covalently bound adducts ← DNA, RNA and protein — Nitrenium Ion

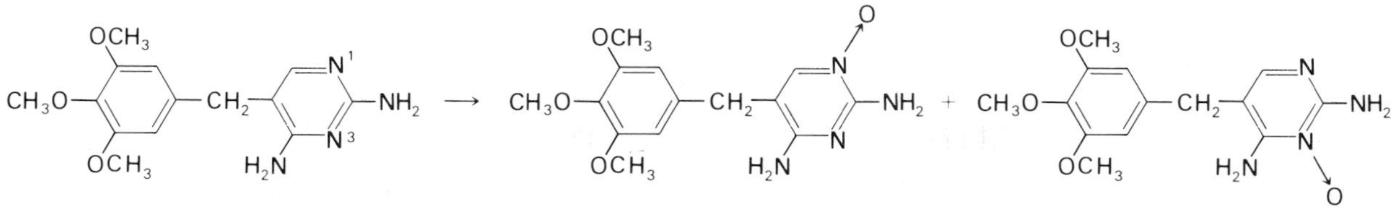

Trimethoprim → 1-N-Oxide + 3-N-Oxide

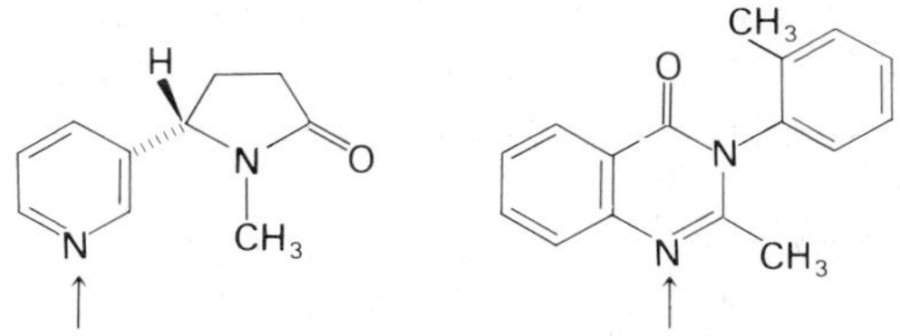

Cotinine   Methaqualone

of the sulfate ($SO_4^=$) anion, this sulfate conjugate can spontaneously ionize to form a highly reactive, resonance stabilized nitrenium species. Covalent adducts between this species and DNA, RNA, and protein have been characterized.[177] The sulfate ester is believed to be the ultimate carcinogenic species. Thus, the above example indicates that certain aromatic amines can be bioactivated to reactive intermediates by N-hydroxylation and O-sulfate conjugation. Whether primary hydroxylamines can be similarly bioactivated is not clear. In addition, it is not known if this biotoxification pathway plays any significant role in the toxicity of aromatic amine drugs.

N-Oxidation of the nitrogen atoms present in aromatic heterocyclic moieties of many drugs occurs to a minor extent. For example, N-oxidation of the folic acid antagonist trimethoprim (Proloprim®, Trimpex®) has been reported in man to yield approximately equal amounts of the isomeric 1-N-oxide and 3-N-oxide as minor metabolites.[178] The pyridinyl nitrogen atom present in cotinine (the major metabolite of nicotine) undergoes oxidation to yield the corresponding N-oxide metabolite.[179] Formation of an N-oxide metabolite of methaqualone (Quaalude®, Parest®) has been observed in man as well.[180]

Amides

Amide functionalities are susceptible to oxidative carbon–nitrogen bond cleavage (by way of α-carbon hydroxylation) and N-hydroxylation reactions. Oxidative dealkylation of many N-substituted amide drugs and xenobiotics has been reported. Mechanistically oxidative dealkylation proceeds by way of an initially formed carbinolamide, which is unstable and fragments to form the N-dealkylated product. For example, diazepam undergoes extensive N-demethylation to the pharmacologically active metabolite, desmethyldiazepam.[181]

Various other N-alkyl substituents present in benzodiazepines (e.g., flurazepam)[111] and in barbiturates (e.g., hexobarbital and mephobarbital)[106] are similarly oxidatively N-dealkylated. Alkyl groups attached to

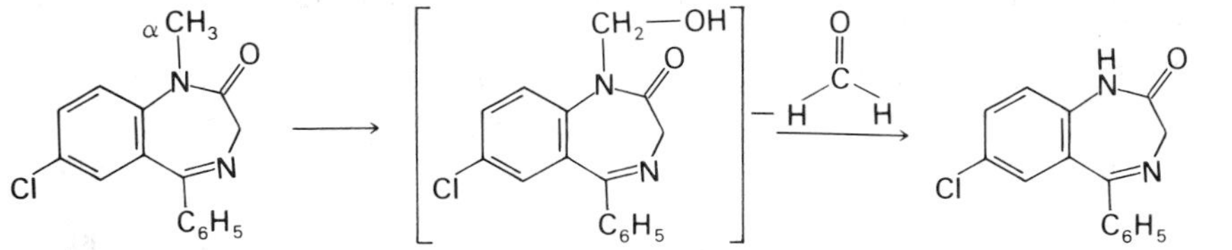

Diazepam → Carbinolamide → Desmethyldiazepam

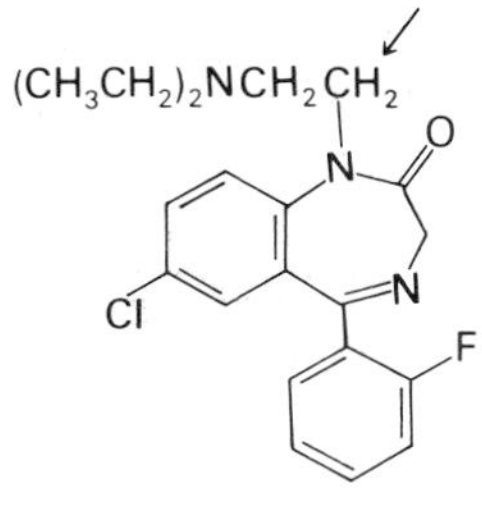

Flurazepam

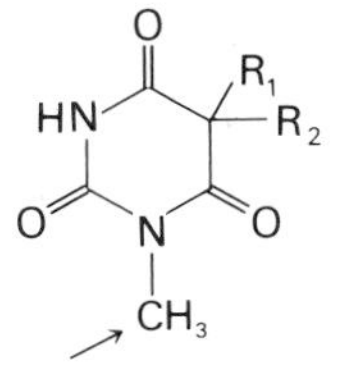

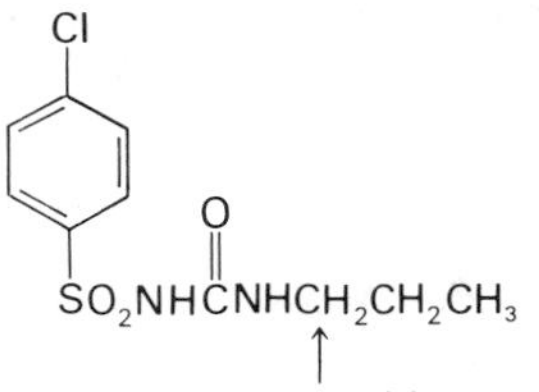

Hexobarbital $R_1=$ cyclohexenyl, $R_2=CH_3$
Mephobarbital $R_1=C_6H_5$, $R_2=CH_2CH_3$

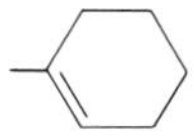

Chlorpropamide

Cotinine   5-Hydroxycotinine   γ-(3-Pyridyl)-γ-oxo-N-
                                          methylbutyramide

the amide moiety of some sulfonylureas such as the oral hypoglycemic chlorpropamide[182] also are subject to dealkylation to a minor extent.

In the case of cyclic amides or lactams, hydroxylation of the alicyclic carbon alpha to the nitrogen atom leads also to carbinolamides. An example of this pathway is the conversion of cotinine to 5-hydroxycotinine. Interestingly, the latter carbinolamide intermediate is in tautomeric equilibrium with the ring-opened metabolite, γ-(3-pyridyl)-γ-oxo-N-methyl butyramide.[183]

Metabolism of the important cancer chemotherapeutic agent cyclophosphamide (Cytoxan®) follows a similar hydroxylation pathway described above for cyclic amides. This drug is a cyclic phosphoramide derivative and is for the most part the phosphorous counterpart of a cyclic amide. Since cyclophosphamide itself is pharmacologically inactive,[184] metabolic bioactivation is required in order for the drug to mediate its antitumorigenic or cytotoxic effects. The key biotransformation pathway leading to the active metabolite involves an initial carbon hydroxylation reaction at C-4 to form the carbinolamide, 4-hydroxycyclophosphamide.[185] 4-Hydroxycyclophosphamide is in equilibrium with the ring-opened dealkylated metabolite, aldophosphamide. Although it has potent cytotoxic properties, aldophosphamide undergoes a further elimination reaction (reverse Michael reaction) to generate acrolein and the phosphoramide mustard

[N,N-bis(2-chloroethyl)phosphorodiamidic acid]. The latter metabolite is the principal species responsible for cyclophosphamide's antitumorigenic properties and chemotherapeutic effect. Enzymatic oxidation of 4-hydroxycyclophosphamide and aldophosphamide leads to the relatively non-toxic metabolites 4-ketocyclophosphamide and carboxycyclophosphamide, respectively.

N-Hydroxylation of aromatic amides, which occurs to a minor extent, is of some toxicological interest, since this biotransformation pathway may lead to the formation of chemically reactive intermediates. Several examples of cytotoxicity or carcinogenicity having been clearly associated with metabolic N-hydroxylation of the parent aromatic amide have been reported. For example, the well-known hepatocarcinogenic 2-acetylaminofluorene (AAF) undergoes an N-hydroxylation reaction catalyzed by cytochrome P-450 to form the corresponding N-hydroxy metabolite (also called a hydroxamic acid).[186] Further conjugation of this hydroxamic acid produces the corresponding O-sulfate ester, which ionizes to generate the electrophilic nitrenium species (see below). The covalent binding of this reactive intermediate to DNA is known to occur and is likely to be the initial event that ultimately leads to malignant tumor formation.[187] Sulfate conjugation plays an important role in this biotoxification pathway (see section on sulfate conjugation for further discussion).

Cyclophosphamide   4-Hydroxycyclophosphamide   4-Ketocyclophosphamide

Phosphoramide Mustard   Acrolein   Aldophosphamide   Carboxyphosphamide
N,N-bis (2-Chloroethyl)-
phosphorodiamidic Acid

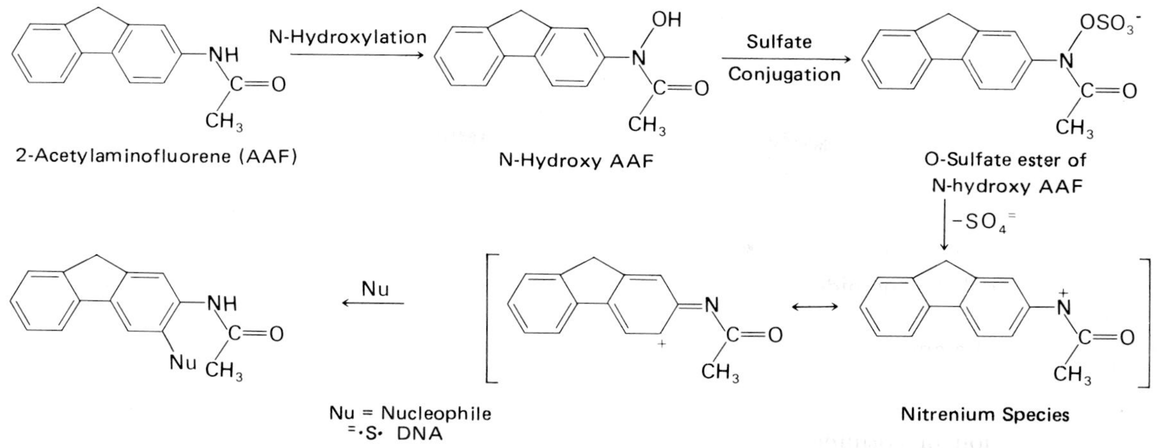

2-Acetylaminofluorene (AAF)

N-Hydroxy AAF

O-Sulfate ester of
N-hydroxy AAF

Nu = Nucleophile
= ·S· DNA

Nitrenium Species

Acetaminaphen is a relatively safe and nontoxic analgesic agent if used at therapeutic doses. When large doses of this drug are ingested, extensive liver necrosis is produced in man and animals.[188] Considerable evidence argues that this hepatotoxicity is dependent upon the formation of a metabolically generated reactive intermediate.[189] Until recently,[190] the accepted bioactivation pathway was believed to involve an initial N-hydroxylation reaction to form N-hydroxyacetaminophen.[191] Spontaneous dehydration of this N-hydroxyamide produces N-acetylimidoquinone, the proposed reactive metabolite. Usually the glutathione present in the liver combines with this reactive metabolite to form the corresponding glutathione conjugate. If glutathione levels are sufficiently depleted by large doses of acetaminophen, covalent binding of the reactive intermediate occurs with macromolecules present in the liver, thus leading to cellular necrosis.

However, very recent studies indicate that the reactive N-acetylimidoquinone intermediate is not formed from N-hydroxyacetaminophen.[189,190] It probably arises via some other oxidative process. Therefore, the mechanistic formation of the reactive metabolite of acetaminophen remains unclear.

OXIDATION INVOLVING CARBON-OXYGEN SYSTEMS

Oxidative O-dealkylation of carbon–oxygen systems (principally ethers) is catalyzed by microsomal mixed function oxidases.[129] Mechanistically, the biotransformation involves an initial α-carbon hydroxylation to form either a hemiacetal or hemiketal, which undergoes spontaneous carbon–oxygen bond cleavage to yield the dealkylated oxygen species (phenol or alco-

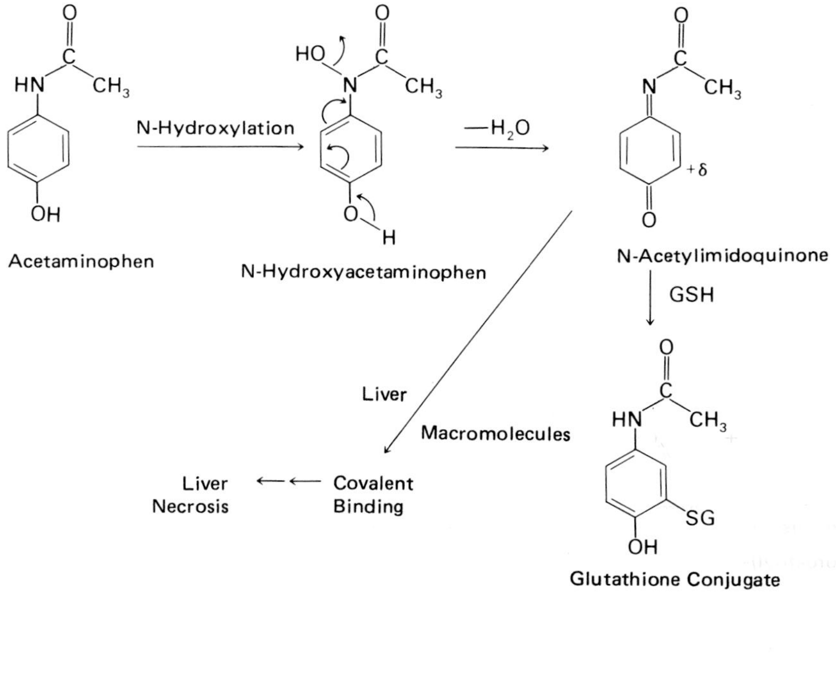

Acetaminophen

N-Hydroxyacetaminophen

N-Acetylimidoquinone

Liver

Macromolecules

Liver
Necrosis ← ← Covalent
Binding

Glutathione Conjugate

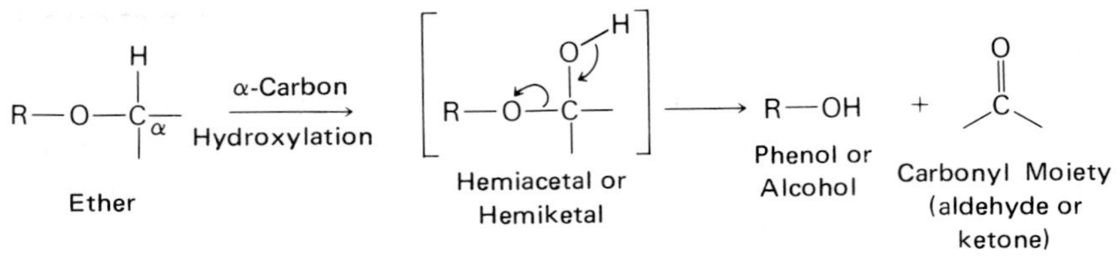

hol) and a carbon moiety (aldehyde or ketone). Small alkyl groups (e.g., methyl or ethyl) attached to oxygen are rapidly O-dealkylated. For example, morphine is the metabolic product resulting from O-demethyla-

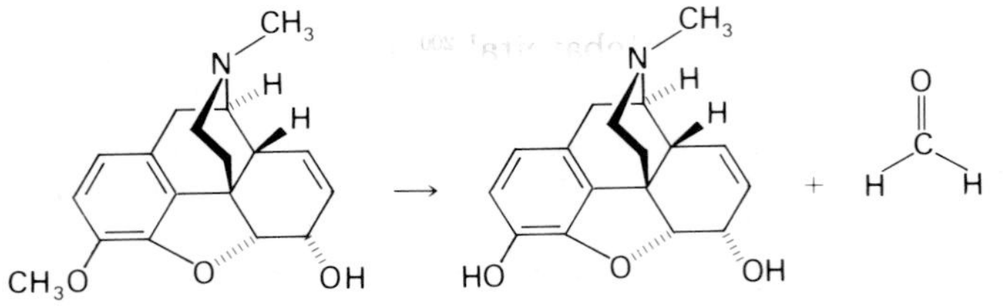

Codeine Morphine

tion of codeine.[192] The antipyretic and analgesic activities of phenacetin in man appear to be a consequence of O-deethylation to the active metabolite

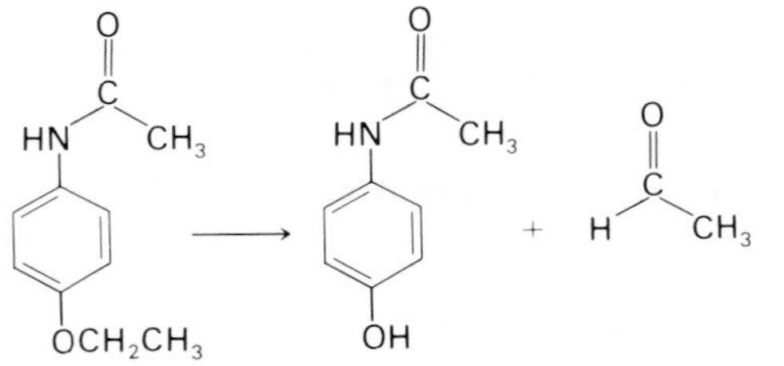

Phenacetin Acetaminophen

acetaminophen.[193] A number of other drugs containing ether groups, such as indomethacin (Indocin®),[194] prazosin (Minipress®)[195] and metoprolol (Lopressor®),[96] have been reported to undergo significant O-demethylation to their corresponding phenolic or alcoholic metabolites, which are further conjugated. In many drugs which have several non-equivalent methoxy groups, one particular methoxy group often appears to be selectively or preferentially O-demethylated. For example, the 3,4,5-trimethoxyphenyl moiety in both mescaline[196] and trimethoprim[178] undergoes O-demethylation to yield predominantly the corresponding 3-0-demethylated metabolites. 4-0-Demethylation also occurs to a minor extent for both drugs. The phenolic and alcoholic metabolites formed from oxidative O-demethylation are susceptible to conjugation, particularly glucuronidation.

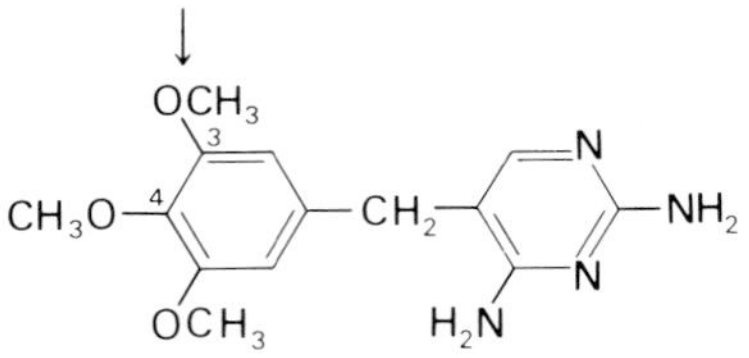

Trimethoprim

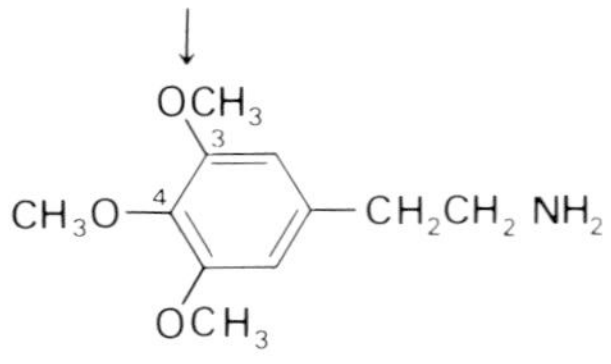

Mescaline

OXIDATION INVOLVING CARBON-SULFUR SYSTEMS

Carbon-sulfur functional groups are susceptible to metabolic S-dealkylation, desulfuration, and S-oxidation reactions. The first two processes involve oxidative carbon-sulfur bond cleavage. S-Dealkylation is

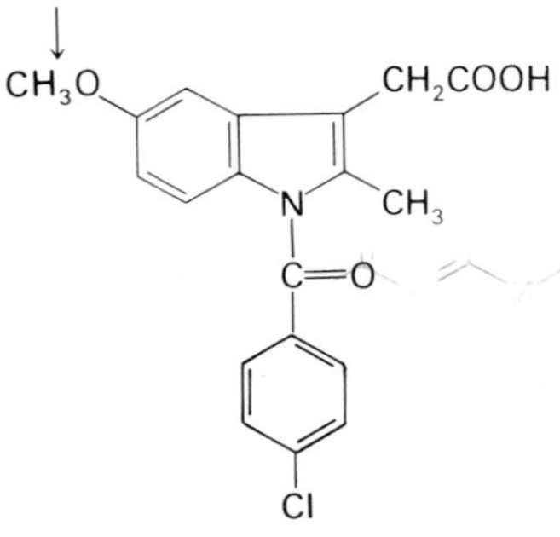

Indomethacin

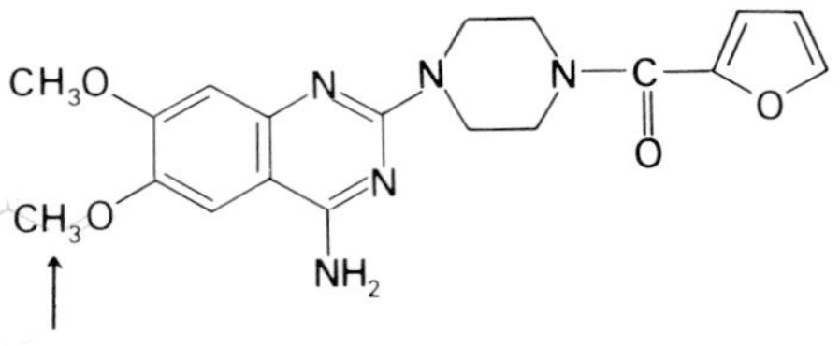

Prazosin

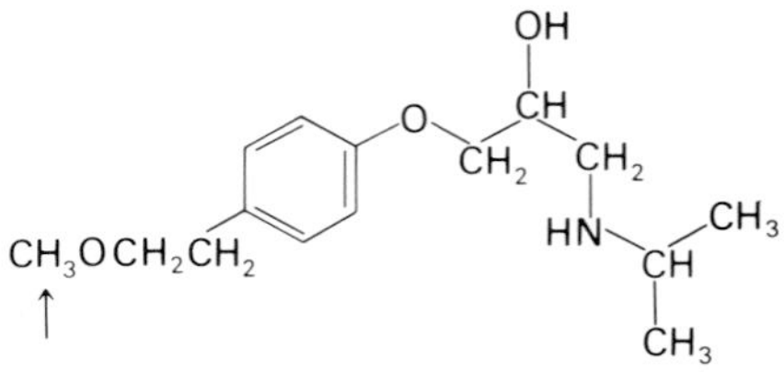

Metoprolol

analogous to O- and N-dealkylation mechanistically (i.e., involves α-carbon hydroxylation) and has been observed for various sulfur xenobiotics.[129,197] For example, 6-(methylthio)purine is oxidatively demethylated in rats to 6-mercaptopurine.[198] S-Demethylation of

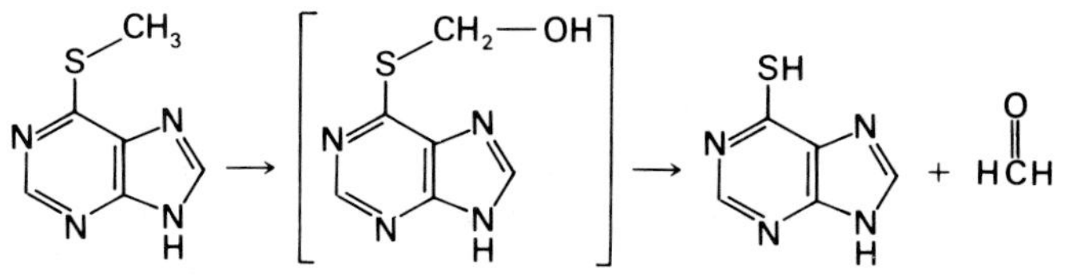

6-(Methylthio)-purine

6-Mercaptopurine

methitural[197] and S-debenzylation of 2-benzylthio-5-trifluoromethylbenzoic acid have been reported too. In contrast to O- and N-dealkylation, examples of drugs undergoing S-dealkylation in man are limited, owing to the small number of sulfur-containing medicinals and to competing metabolic S-oxidation processes (see below.)

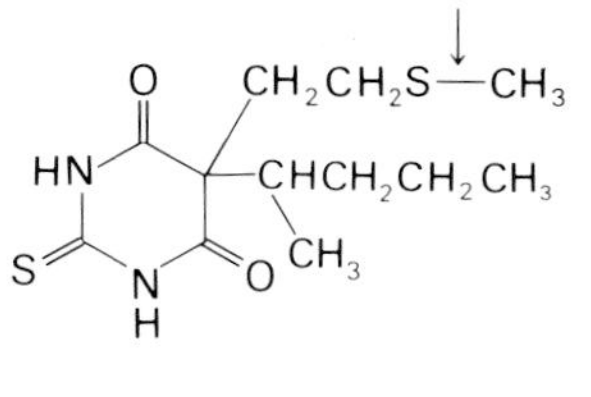

Methitural

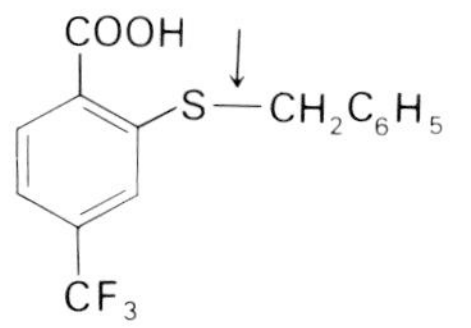

2-Benzylthio-5-trifluoromethylbenzoic Acid

Oxidative conversion of carbon–sulfur double bonds (C=S) (thiono) to the corresponding carbon–oxygen double bond (C=O) is called desulfuration. A well-known drug example of this metabolic process is the biotransformation of thiopental to its corresponding

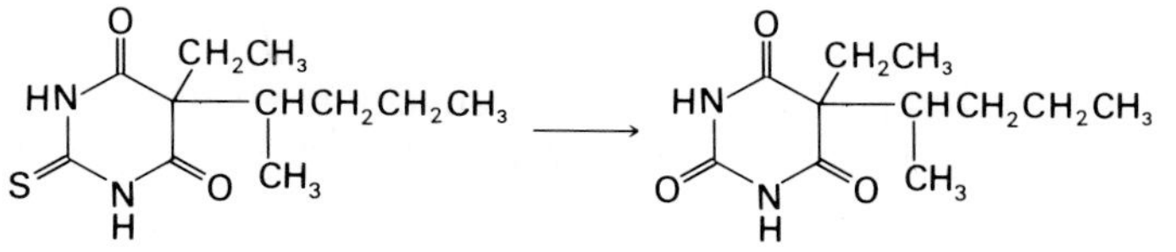

Thiopental

Pentobarbital

oxygen analog pentobarbital.[200] An analogous desulfuration reaction also occurs with the P=S moiety present in a number of organophosphate insecticides such as parathion.[201] Desulfuration of parathion leads to the formation of paraoxon, which is the active metabolite responsible for the anticholinesterase activity of the parent drug. The mechanistic details of desulfuration are poorly understood, but it appears to involve microsomal oxidation of the C=S or P=S double bond.[202]

Organosulfur xenobiotics commonly undergo S-oxidation to yield sulfoxide derivatives. A number of phenothiazine derivatives are metabolized by this pathway. For example, both sulfur atoms present in thioridazine (Mellaril®)[203] are susceptible to S-oxidation. Oxidation of the 2-methylthio group yields the active sulfoxide metabolite, mesoridazine. Interest-

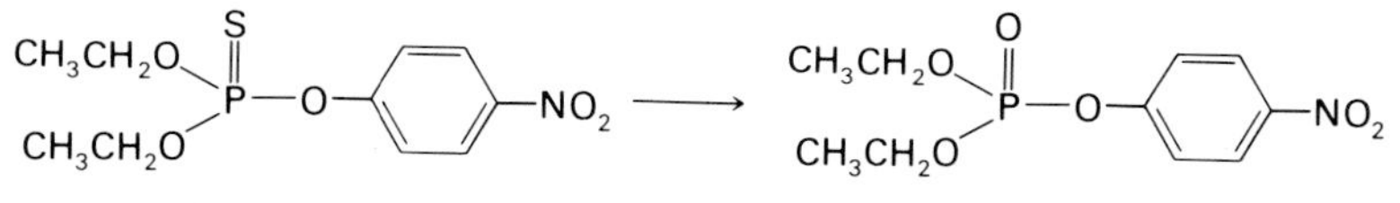

Parathion

Paraoxon

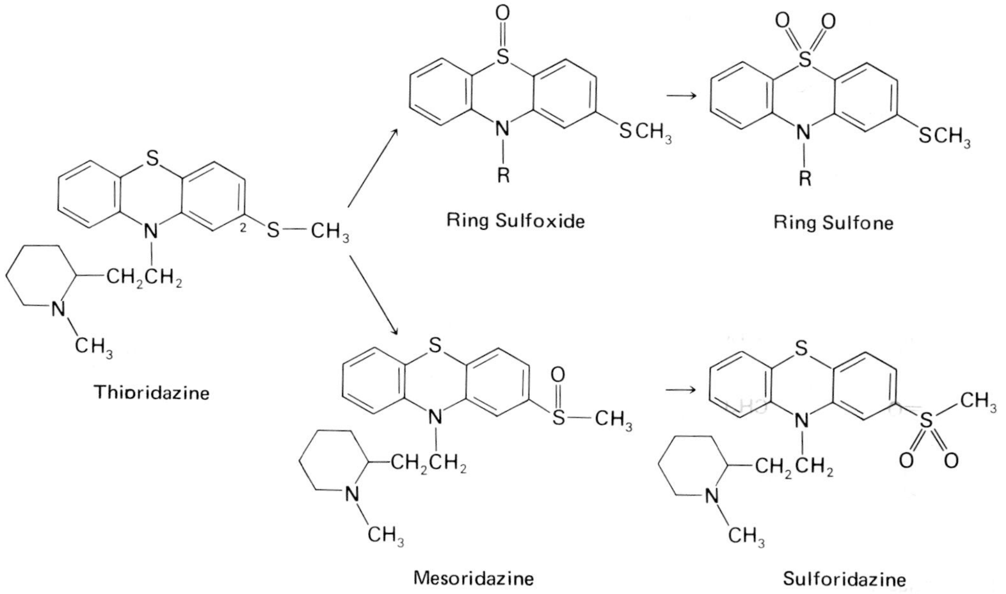

Thioridazine

Ring Sulfoxide

Ring Sulfone

Mesoridazine

Sulforidazine

Cimetidine   X = N—C ≡ N
Metiamide   X = S

Sulfoxide Metabolite

ingly, mesoridazine is twice as potent an antipsychotic agent as thioridazine in man and has been introduced into clinical use as Serentil®.[204]

S-Oxidation constitutes an important pathway in the metabolism of the H_2-histamine antagonists cimetidine (Tagamet®)[205] and metiamide.[206] The corresponding sulfoxide derivatives are the major urinary metabolites found in man.

Sulfoxide drugs and metabolites may be further oxidized to sulfones ($-SO_2-$). The sulfoxide group present in the immunosuppressive agent oxisuran is metabolized to a sulfone moiety.[207] In man, dimethyl-

Oxisuran   Sulfone Metabolite

sulfoxide is primarily found in the urine as the oxidized product dimethyl sulfone. Sulfoxide metabolites such as those of thioridazine have been reported to

Dimethyl Sulfoxide   Dimethyl Sulfone

undergo further oxidation to their sulfone derivatives (see above).[203]

OXIDATION OF ALCOHOLS AND ALDEHYDES

Many oxidative processes (e.g., benzylic, allylic, alicyclic, or aliphatic hydroxylation) generate alcohol or carbinol metabolites as intermediate products. If not conjugated, these alcohol products are further oxidized to aldehydes (if primary alcohols) or to ketones (if secondary alcohols). Aldehyde metabolites resulting from oxidation of primary alcohols or from oxidative deamination of primary aliphatic amines often undergo facile oxidation to generate polar carboxylic acid derivatives.[91] As a general rule, primary alcoholic groups and aldehyde functionalities are quite vulnerable to oxidation. A number of drug examples in which primary alcohol metabolites and aldehyde metabolites are oxidized to carboxylic acid products were cited in earlier sections.

$$RCH_2OH \rightleftharpoons RCHO \rightarrow RCOOH$$

Primary Alcohol   Aldehyde   Acid

Although secondary alcohols are susceptible to oxidation, this reaction is not often significant, since the reverse reaction, namely, reduction of the ketone back to the secondary alcohol, occurs quite readily. In addition, the secondary alcohol group being polar and functionalized is more likely to be conjugated than the ketone moiety.

The bioconversion of alcohols to aldehydes and ketones is catalyzed by soluble alcohol dehydrogenases present in the liver and other tissues. NAD^+ is required as a coenzyme, although $NADP^+$ may also serve as a coenzyme. The reaction catalyzed by alcohol dehydrogenase is reversible but often proceeds to the right, since the aldehyde formed is further oxidized to the acid. Several aldehyde dehydrogenases, including aldehyde oxidase and xanthine oxidase, carry out the oxidation of aldehydes to their corresponding acids.[91, 209]

The metabolism of cyclic amines to their lactam metabolites has been observed for a number of drugs (e.g., nicotine, phenmetrazine, methylphenidate). It appears that soluble or microsomal dehydrogenase and oxidases are involved in oxidizing the carbinol group of the intermediate carbinolamine to a carbonyl moiety.[209] For example, in the metabolism of medazepam to diazepam the intermediate carbinolamine (2-hydroxymedazepam) undergoes oxidation of its 2-hydroxy group to a carbonyl moiety. A microsomal dehydrogenase has been demonstrated to carry out this oxidation.[210]

Medazepam   2-Hydroxymedazepam   Diazepam

OTHER OXIDATIVE BIOTRANSFORMATION PATHWAYS

In addition to the many oxidative biotransformations discussed already, oxidative aromatization or dehydrogenation and oxidative dehalogenation reactions also occur. Metabolic aromatization has been reported in the case of norgestrel. Aromatization or dehydrogenation of the A ring present in this steroid leads to the corresponding phenolic product, 17α-ethinyl-18-homo-

Norgestrel   17α-Ethinyl-18-homoestradiol

estradiol as a minor metabolite in women.[211] In mice, the terpene ring of Δ^1-THC or $\Delta^{1,6}$-THC undergoes aromatization to give cannabinol.[212]

A number of halogen-containing drugs and xenobiotics are metabolized via oxidative dehalogenation. For example, the volatile anesthetic agent halothane is metabolized principally to trifluoroacetic acid in man.[213] It has been postulated that this metabolite arises from cytochrome P-450 mediated hydroxylation of halothane to form an initial carbinol intermediate that spontaneously eliminates hydrogen bromide (dehalogenation) to yield trifluoroacetyl chloride. The latter acyl chloride is chemically reactive and reacts rapidly with water to form trifluoroacetic acid. Alternatively, it can acylate tissue nucleophiles. Indeed, in vitro studies indicate that halothane is metabolized to a reactive intermediate (presumably trifluoroacetyl chloride) that covalently binds to liver microsomal proteins. Chloroform also appears to be oxidatively metabolized by a similar dehalogenation pathway to yield the chemically reactive species, phosgene. It has been postulated that phosgene may be responsible for the hepato- and nephrotoxicity associated with chloroform.[215]

A final example of oxidative dehalogenation con-

Δ^1-THC $\Delta^{1,6}$-THC   Cannabinol

Halothane   Carbinol Intermediate   Trifluoroacetyl Chloride   Trifluoroacetic Acid

Chloroform   Phosgene   Tissue nucleophiles   Covalent Binding cerns the antibiotic chloramphenicol. In vitro studies have demonstrated that the dichloroacetamide portion of the molecule undergoes oxidative dechlorination to yield a chemically reactive oxamyl chloride intermediate that is capable of reacting with water to form the corresponding oxamic acid metabolite or capable of acylating microsomal proteins.[216] Thus, it appears that in several instances oxidative dehalogenation can lead to the formation of toxic and reactive acyl halide intermediates.

REDUCTIVE REACTIONS

Reductive processes play an important role in the metabolism of many compounds containing carbonyl, nitro, and azo groups. Bioreduction of carbonyl compounds generates alcohol derivatives,[91,217] while nitro and azo reduction lead to amino derivatives.[218] The hydroxyl and amino moieties of the metabolites are much more susceptible to conjugation than the functional groups of the parent compounds. Thus, reductive processes as such facilitate drug elimination.

Reductive pathways that are less frequently encountered in drug metabolism include reduction of N-oxides to their corresponding tertiary amines and reduction of sulfoxides to sulfides. Reductive cleavage of disulfide linkages and reduction of carbon–carbon double bonds also occur but constitute only minor pathways in drug metabolism.

REDUCTION OF ALDEHYDE AND KETONE CARBONYLS

The carbonyl moiety, particularly the ketone group, is frequently encountered in many drugs. In addition, metabolites containing ketone and aldehyde functionalities often arise from oxidative deamination of xenobiotics (e.g., propranolol, chlorpheniramine, amphetamine). Owing to their ease of oxidation, aldehydes are mainly metabolized to carboxylic acids. Occasion-

ally, aldehydes are reduced to primary alcohols. Ketones, on the other hand, are generally resistant to oxidation and are primarily reduced to secondary alcohols. Alcohol metabolites arising from reduction of carbonyl compounds generally undergo further conjugation (e.g., glucuronidation).

A number of soluble enzymes called aldo-keto reductases carry out bioreduction of aldehydes and ketones.[91,219] They are found in the liver and other tissues (e.g., kidney). As a general class these soluble enzymes have similar physiochemical properties and broad substrate specificities and require NADPH as a cofactor. Oxidoreductase enzymes that carry out both oxidation and reduction reactions also are capable of reducing aldehydes and ketones.[219] For example, the important liver alcohol dehydrogenase is an NAD^+-dependent oxidoreductase that oxidizes ethanol and other aliphatic alcohols to aldehydes and ketones. However, in the presence of NADH or NADPH, the same enzyme system is capable of reducing carbonyl derivatives to their corresponding alcohols.[91]

The number of aldehydes undergoing bioreduction is small because of the relative ease of oxidation of aldehydes to carboxylic acids. However, one fre-

Propranolol — N-Dealkylation → N-Desisopropyl propranolol — Oxidative Deamination → Aldehyde Intermediate — Oxidation → Naphthoxylactic Acid; Aldehyde Intermediate — Reduction → Propranolol Glycol (conjugated)

quently cited example of a parent aldehyde drug undergoing extensive enzymatic reduction is the sedative-hypnotic chloral hydrate. Bioreduction of this hydrated aldehyde yields trichloroethanol as the major metabolite in man.[220] Interestingly, this alcohol metabolite is pharmacologically active. Further glucuronidation of the alcohol leads to an inactive conjugated product that is readily excreted in the urine.

Aldehyde metabolites resulting from oxidative deamination of drugs also have been observed to undergo reduction to a minor extent. For example, in man the β-adrenergic blocker propranolol is converted to an intermediate aldehyde via N-dealkylation and oxidative deamination. Although the aldehyde is primarily oxidized to the corresponding carboxylic acid (naphthoxylactic acid), a small fraction is also reduced to the alcohol derivative (propranolol glycol).[221]

Two major polar urinary metabolites of the histamine H$_1$-antagonist chlorpheniramine have recently been identified in dogs, as the alcohol and carboxylic acid products (conjugated) derived respectively from reduction and oxidation of an aldehyde metabolite. The aldehyde precursor arises from bis-N-demethylation and oxidative deamination of chlorpheniramine.[222]

Bioreduction of ketones often leads to the creation of an asymmetric center and thus two possible stereoisomeric alcohols.[91,223] For example, reduction of acetophenone by a soluble rabbit kidney reductase leads to the enantiomeric alcohols, S(-)- and R(+)-methylphenylcarbinol with the S(-)-isomer predominating (3:1 ratio).[224] The preferential formation of one stereoisomer over the other is termed *product stereoselectivity* in drug metabolism.[223] Mechanistically, ketone re-

Chlorpheniramine — 1) bis-N-demethylation 2) Oxidative Deamination → Aldehyde Metabolite; Aldehyde Metabolite — Reduction → 3-(p-Chlorobenzyl)-3-(2-pyridyl)-propan-1-ol; Aldehyde Metabolite — Oxidation → 3-(p-Chlorobenzyl)-3-(2-pyridyl)-propanoic Acid

Acetophenone → S(−)-Methyl Phenyl Carbinol (75%) + R(+)-Methyl Phenyl Carbinol (25%)

duction involves a "hydride" transfer from the reduced nicotinamide moiety of the cofactor NADPH or NADH to the carbonyl carbon atom of the ketone. It is generally agreed that this step proceeds with considerable *stereoselectivity*.[91,223] For this reason, it is not surprising to find many reports of xenobiotic ketones that are preferentially reduced to a predominant stereoisomer. Oftentimes, ketone reduction yields alcohol metabolites that are pharmacologically active.

Although a large number of ketone-containing drugs undergo significant reduction, only a few selected examples will be presented in detail here. Those xenobiotics that are not discussed in the text have been structurally tabulated in Figure 3-10. The keto group undergoing reduction has been designated with an arrow.

Ketones lacking asymmetric centers in their molecules, such as acetophenone or the oral hypoglycemic acetohexamide, usually give rise to predominantly one enantiomer upon reduction. In man, acetohexamide is rapidly metabolized in the liver to give principally

S(−)-hydroxyhexamide.[225] This metabolite is as active a hypoglycemic agent as its parent compound and is further eliminated via the kidney.[226] Acetohexamide is usually not recommended in diabetic patients with renal failure owing to the possible accumulation of its active metabolite, hydroxyhexamide.

When chiral ketones are reduced, they yield two possible diastereomeric or epimeric alcohols. For example, the R(+)-enantiomer of the oral anticoagulant warfarin undergoes extensive reduction of its side chain keto group to generate the R,S-(+)-alcohol as the major plasma metabolite in man.[45,234] Small amounts of the R,R-(+)-diastereomer are also formed. In contrast, the S(−)-enantiomer undergoes little ketone reduction and is primarily 7-hydroxylated (i.e., aromatic hydroxylation) in man.

Reduction of the 6-keto functionality in the narcotic antagonist naltrexone can lead to either the epimeric 6α- or 6β-hydroxy metabolites depending on the animal species.[231] In man and rabbit, bioreduction of naltrexone is highly stereoselective and generates only 6β-naltrexol, whereas in chicken, reduction occurs to yield only 6α-naltrexol.[231,235] However, in monkey and guinea pig, both epimeric alcohols are formed (predominantly 6β-naltrexol).[236] It appears that in the latter two species, reduction of naltrexone to the epimeric 6α- and 6β-alcohols is carried out by two distinctly different reductases found in the liver.[235,236]

Reduction of oxisuran appears to be an important pathway by which the parent drug mediates its immu-

Acetophenone + Reduced Nicotinamide Moiety of NADPH or NADH → S(−)-Methyl Phenyl Carbinol + Oxidized Nicotamide Moiety of NADP⁺ or NAD⁺

Acetohexamide → S(−)-Hydroxyhexamide

R(+)-Warfarin → R,S-(+)-Alcohol Major Diastereomer + R,R-(+)-Alcohol Minor Diastereomer

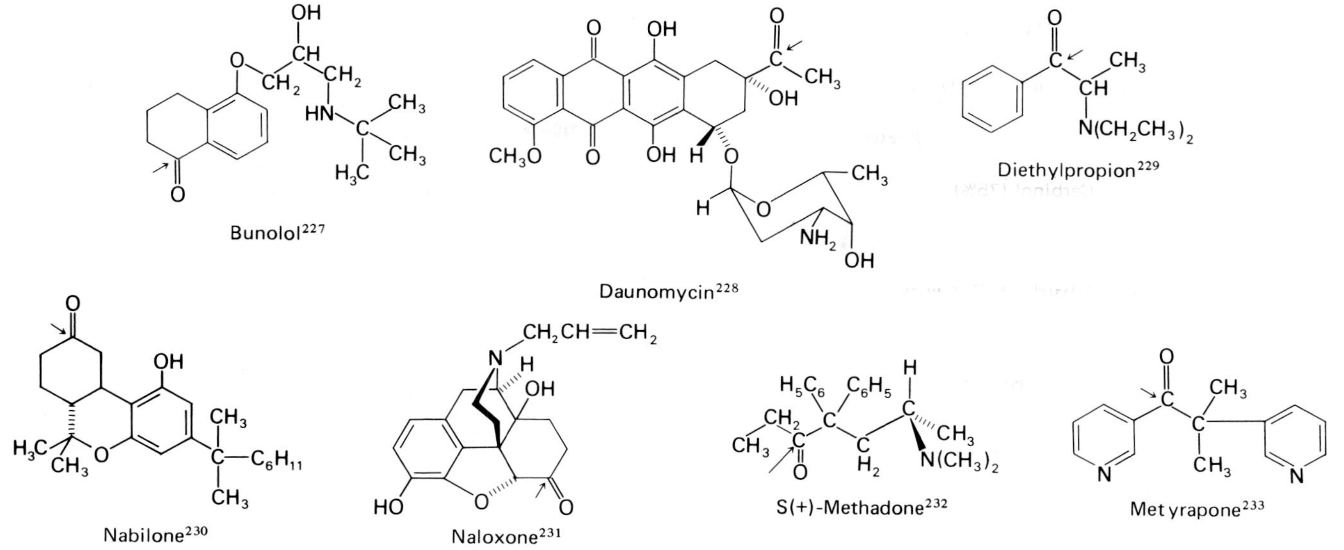

Bunolol[227]

Daunomycin[228]

Diethylpropion[229]

Nabilone[230]

Naloxone[231]

S(+)-Methadone[232]

Metyrapone[233]

FIG. 3-10. *Additional examples of xenobiotics that undergo extensive ketone reduction, which were not covered in the text. Arrow indicates the keto group undergoing reduction.*

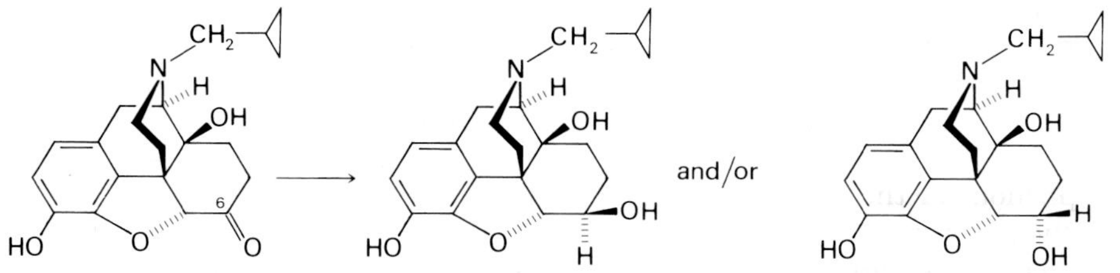

Naltrexone

6β-Naltrexol

and/or

6α-Naltrexol

nosuppressive effects. Studies indicate that oxisuran has its greatest immunosuppressive effects in those species that form alcohols as their major metabolic products (e.g., man, rat).[237,238] In species in which reduction is a minor pathway (e.g., dog), oxisuran shows little immunosuppressive activity.[238] These findings tend to indicate that the oxisuran alcohols (oxisuranols) are pharmacologically active and contribute significantly to the overall immunosuppressive effect of the parent drug. The sulfoxide group in oxisuran is chiral by virtue of the lone pair of electrons on sulfur. Therefore, reduction of oxisuran leads to diastereomeric alcohols.

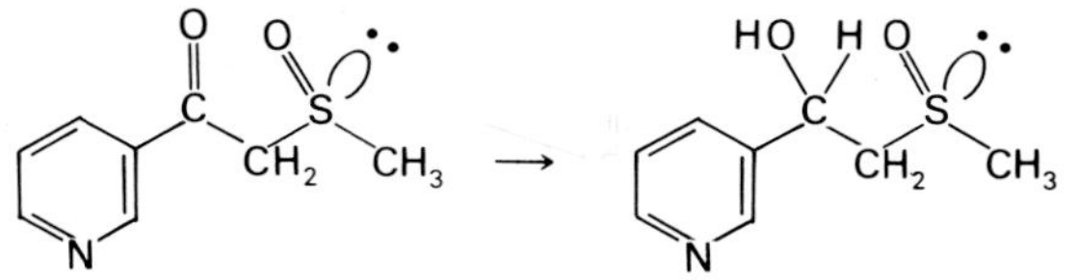

Oxisuran

Oxisuranols (diastereomeric mixture)

Reduction of α,β-unsaturated ketones results not only in reduction of the ketone group but also in reduction of the carbon–carbon double bond as well. Steroidal drugs often fall into this class, including norethindrone, a synthetic progestin found in many oral

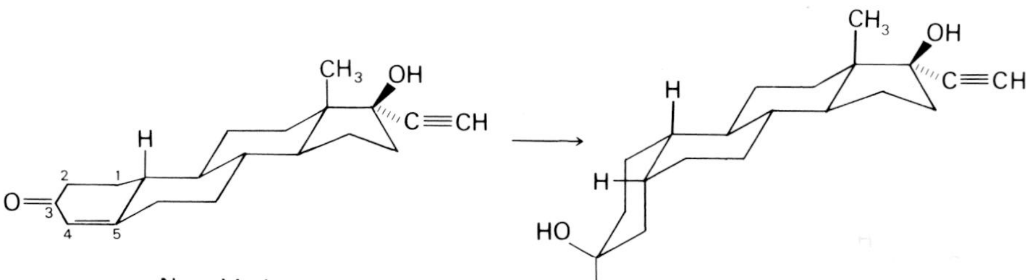

Norethindrone

3β, 5β-Tetrahydronorethindrone

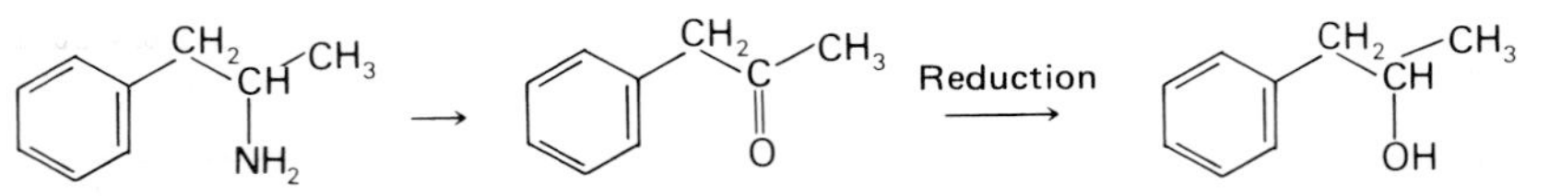

Amphetamine — Phenylacetone — 1-Phenyl-2-propanol

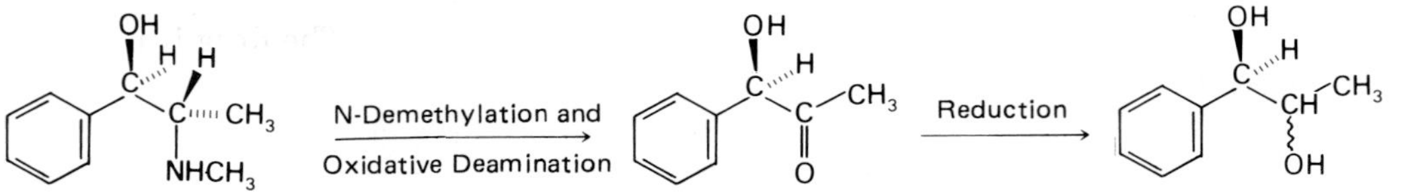

(-)-Ephedrine — 1-Hydroxy-1-phenylpropan-2-one — 1-Phenyl-1, 2-propanediol (as glucuronide conjugate)

contraceptive drug combinations. In women, the major plasma and urinary metabolite of norethindrone is the 3β,5β-tetrahydro derivative.[239]

Ketones resulting from metabolic oxidative deamination processes are also susceptible to reduction. For example, rabbit liver microsomal preparations have been shown to metabolize amphetamine to phenylacetone, which is subsequently reduced to 1-phenyl-2-propanol.[240] In man, a minor urinary metabolite of (-)-ephedrine has been identified as the diol derivative formed from keto reduction of the oxidatively deaminated product, 1-hydroxy-1-phenylpropan-2-one.[241]

REDUCTION OF NITRO AND AZO COMPOUNDS

The reduction of aromatic nitro and azo xenobiotics leads to aromatic primary amine metabolites.[218] Aromatic nitro compounds are initially reduced to the nitroso and hydroxylamine intermediates as shown in the metabolic sequence below:

$$Ar{-}\overset{+}{N}(=O)O^{-} \longrightarrow Ar{-}N{=}O \longrightarrow Ar{-}NHOH \longrightarrow Ar{-}NH_2$$

Nitro — Nitroso — Hydroxylamine — Amine

Azo reduction on the other hand is believed to proceed via a hydrazo intermediate (-NH-NH-) that is subsequently reductively cleaved to yield the corresponding aromatic amines:

$$Ar{-}N{=}N{-}Ar' \longrightarrow Ar{-}NH{-}NH{-}Ar' \longrightarrow Ar{-}NH_2 + H_2N{-}Ar'$$

Azo — Hydrazo — Amines

Bioreduction of nitro compounds is carried out by NADPH-dependent microsomal and soluble nitro reductases present in the liver. A multi-component hepatic microsomal reductase system requiring NADPH appears to be responsible for azo reduction. In addition, bacterial reductases present in the intestine are also capable of reducing nitro and azo compounds, especially those which are poorly absorbed or which are excreted mainly in the bile.[244]

Various aromatic nitro drugs undergo enzymatic reduction to the corresponding aromatic amines. For example, the 7-nitro benzodiazepine derivatives clonazepam and nitrazepam are extensively metabolized to their respective 7-amino metabolites in man.[245,246] The skeletal muscle relaxant dantrolene (Dantrium®) also has been reported to undergo reduction to aminodantrolene in humans.[247]

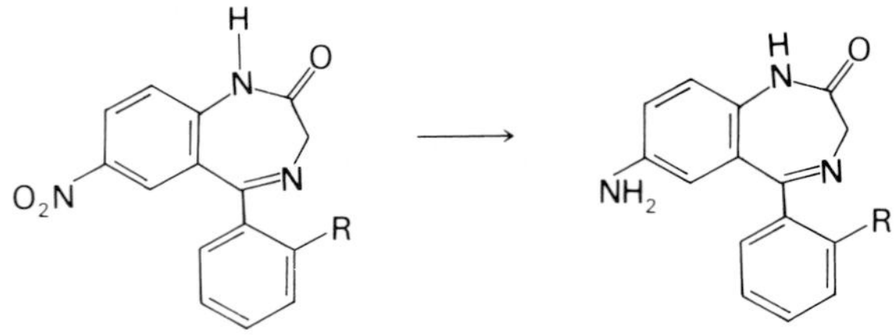

Clonazepam, R=Cl
Nitrazepam, R=H

7-Amino Metabolite

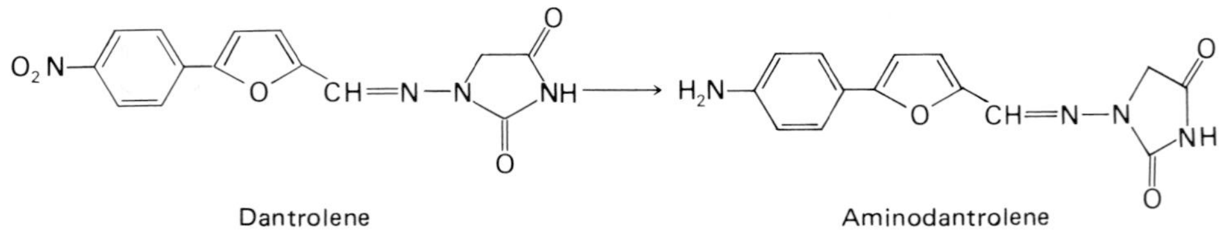

Dantrolene — Aminodantrolene

For some nitro xenobiotics, bioreduction appears to be a minor metabolic pathway in vivo owing to competing oxidative and conjugative reactions. However, under artificial anaerobic in vitro incubation conditions, these same nitro xenobiotics are rapidly enzymatically reduced. For example, most of the urinary metabolites of metronidazole found in man are either oxidation or conjugation products. Reduced metabolites of metronidazole have not been detected.[248] However, when incubated anaerobically with guinea pig liver preparations, metronidazole undergoes considerable nitro reduction.[249]

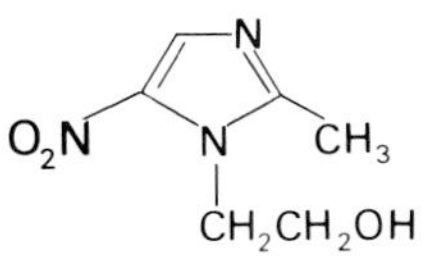

Metronidazole

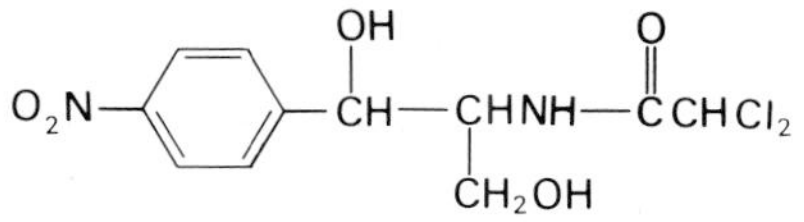

Chloramphenicol

Bacterial reductase present in the intestine also tends to complicate in vivo interpretations of nitro reduction. For example, in rats the antibiotic chloramphenicol is not reduced in vivo by the liver but is excreted in the bile and subsequently reduced by intestinal flora to form the amino metabolite.[250]

The enzymatic reduction of azo compounds is best exemplified by the conversion of Prontosil to the active sulfanilamide metabolite in the liver.[251] This reaction has historical significance, since it led to the discovery of sulfanilamide as an antibiotic and eventually to the development of a large number of therapeutic sulfonamide drugs. Bacterial reductases present in the intestine play a significant role in reducing azo xenobiotics, particularly those which are poorly absorbed.[244] For example, the two azo dyes tartrazine[252] and amaranth[253] have poor oral absorption as a result of the many polar and ionized sulfonic acid

groups present in their structures. Therefore, these two azo compounds are primarily metabolized by bacterial reductases present in the intestine. The importance of intestinal reduction is further revealed in the metabolism of sulfasalazine (formerly salicylazosulfapyridine) (Azulfidine®), a drug used in the treatment of ulcerative colitis. The drug is poorly absorbed and

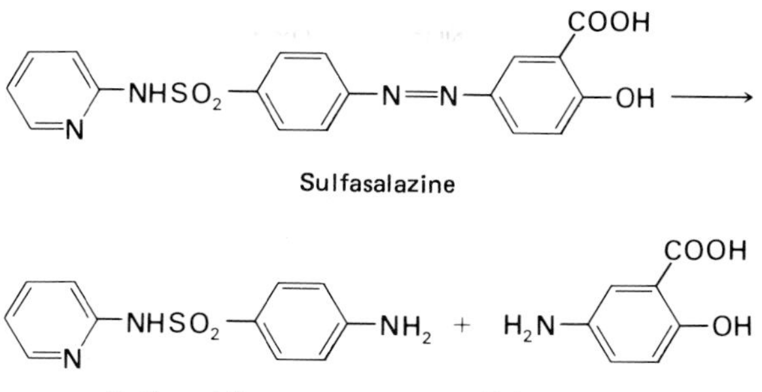

Sulfasalazine

Sulfapyridine   5-Aminosalicylic Acid

undergoes reductive cleavage of the azo linkage to yield sulfapyridine and 5-aminosalicylic acid.[254] The reaction occurs primarily in the colon and is carried out principally by intestinal bacteria. Studies in germ-free rats, lacking intestinal flora, have demonstrated that sulfasalazine is not reduced to any appreciable extent.[255]

MISCELLANEOUS REDUCTIONS

Several minor reductive reactions also have been demonstrated to occur. Reduction of N-oxides to the corresponding tertiary amine occurs to some extent. This reductive pathway is of interest, since a number of tertiary amines are oxidized to form polar and water-soluble N-oxide metabolites. If reduction of N-oxide metabolites occurs to a significant extent, drug elimi-

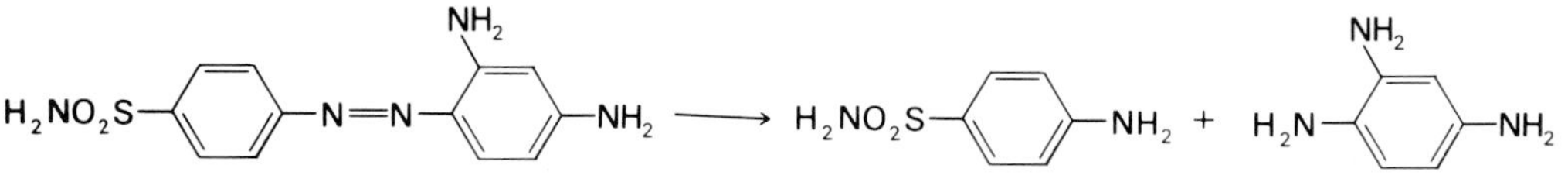

Prontosil   Sulfanilamide   1,2,4-Triaminobenzene

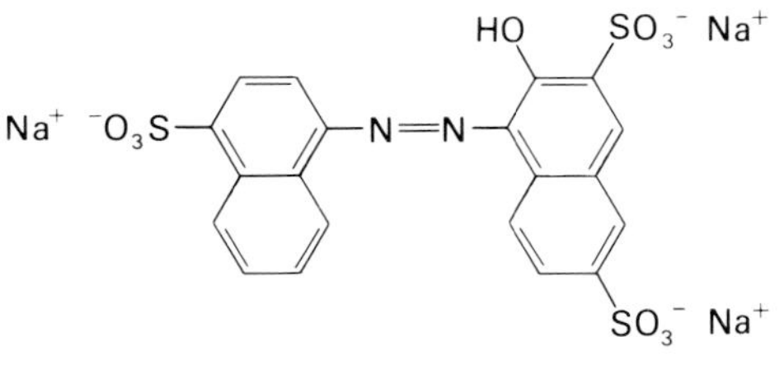

Tartrazine

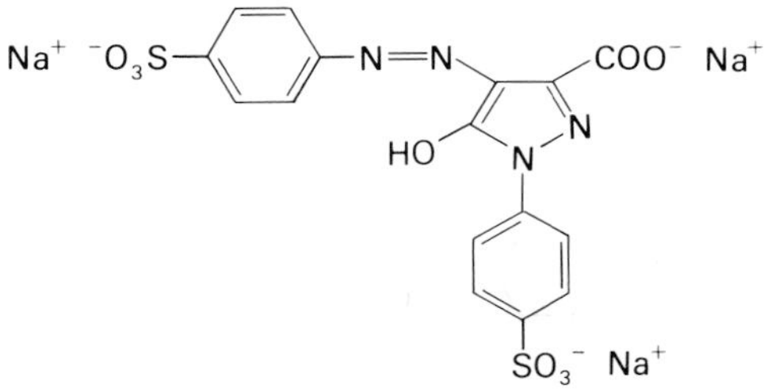

Amaranth

nation of the parent tertiary amine would be impeded. N-Oxide reduction is often assessed by administering the pure synthetic N-oxide in vitro or in vivo and then attempting to detect the formation of the tertiary amine. For example, imipramine N-oxide has been demonstrated to undergo reduction in rat liver preparations.[256]

Imipramine N-Oxide → Imipramine

Reduction of sulfur-containing functional groups, such as the disulfide and sulfoxide moieties, also constitutes minor reductive pathways. Reductive cleavage of the disulfide bond in disulfiram (Antabuse®) yields N,N-diethyldithiocarbamic acid (free or glucuronidated) as a major metabolite in man.[257] Although sulfoxide functionalities are mainly oxidized to sul-

Disulfiram → N,N-Diethylthiocarbamic Acid

fones ($-SO_2-$), they sometimes undergo reduction to sulfides. A recent example demonstrating the importance of this reductive pathway is seen in the metabolism of the antiinflammatory agent sulindac (Clinoril®). Studies in man show that sulindac undergoes reduction to an active sulfide that is responsible for

Sulindac → Sulindac Sulfide Metabolite

the overall antiinflammatory effect of the parent drug.[258] Sulindac or its sulfone metabolite exhibits little antiinflammatory activity. Another example of sulfide formation involves the reduction of dimethyl sulf-

oxide (DMSO) to dimethyl sulfide. In man, DMSO is metabolized to a minor extent via this pathway.

Dimethyl Sulfoxide → Dimethyl Sulfide

The characteristic unpleasant odor of dimethyl sulfide is evident on the breath of patients using this agent.[259]

HYDROLYTIC REACTIONS

HYDROLYSIS OF ESTERS AND AMIDES

The metabolism of ester and amide linkages in many drugs is catalyzed by hydrolytic enzymes present in various tissues and in plasma. The metabolic products formed, namely carboxylic acids, alcohols, phenols, and amines, generally are polar and functionally more susceptible to conjugation and excretion than the parent ester or amide drugs. The enzymes carrying out ester hydrolysis include a number of nonspecific esterases found in the liver, kidney, and intestine as well as the pseudocholinesterases present in plasma.[260,261] Amide hydrolysis appears to be mediated by liver microsomal amidases, esterases and deacylases.[261]

Hydrolysis is a major biotransformation pathway for drugs containing the ester functionality. This is because of the relative ease of hydrolyzing the ester linkage. A classical example of ester hydrolysis is the metabolic conversion of aspirin (acetylsalicylic acid)

Aspirin (Acetylsalicylic Acid) → Salicylic Acid + Acetic Acid

to salicylic acid.[262] Of the two ester moieties present in cocaine, it appears that in general the methyl group is preferentially hydrolyzed to yield benzoylecgonine as the major urinary metabolite in man.[263] However, the hydrolysis of cocaine to methylecgonine has also recently been demonstrated to occur in plasma and blood to a minor extent.[264] Methylphenidate (Ritalin®) is rapidly biotransformed by hydrolysis to yield ritalinic acid as the major urinary metabolite in man.[265] Often ester hydrolysis of the parent drug leads to pharmacologically active metabolites. For example,

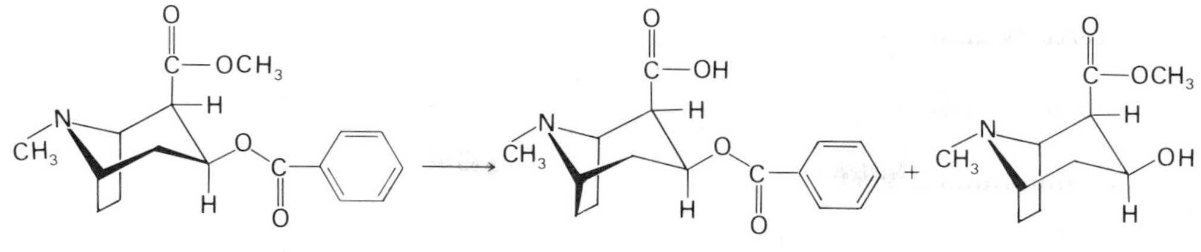

Cocaine → Benzoylecgonine + Methylecgonine

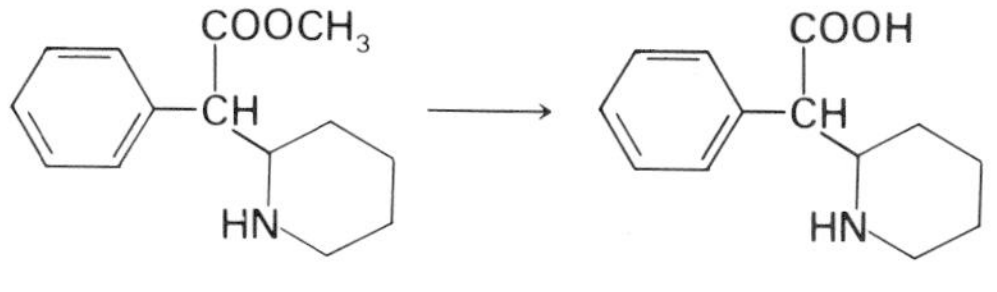

Methylphenidate → Ritalinic Acid

hydrolysis of diphenoxylate in man leads to diphenoxylic acid (difenoxin), which is apparently five times more potent an antidiarrheal agent than the parent ester.[266] The rapid metabolism of clofibrate (Atromid-S®) yields *p*-chlorophenoxyisobutyric acid (CPIB) as the major plasma metabolite in man.[267] Studies in rats

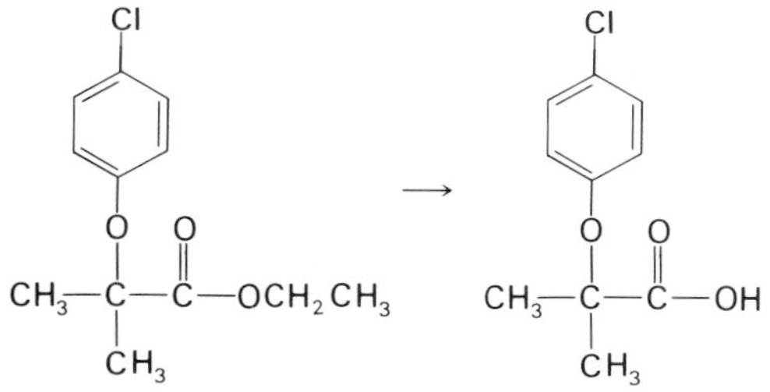

Clofibrate → p-Chlorophenoxyisobutyric Acid

indicate that the free acid CPIB is responsible for clofibrate's hypolipidemic effect.[268]

In recent years, many parent drugs have been chemically modified or derivatized to generate so-called *prodrugs* in order to overcome some undesirable property (e.g., bitter taste, poor absorption, poor solubility, irritation at site of injection). The rationale behind the prodrug concept was to develop an agent which once inside the biological system will be biotransformed to the active parent drug.[17] The presence

of esterases in many tissues and in plasma makes ester derivatives logical prodrug candidates, since hydrolysis would cause the ester prodrug to revert to the parent compound. For example, antibiotics such as chloramphenicol and clindamycin have been derivatized as their palmitate esters in order to minimize their bitter taste and to improve their palatability in pediatric liquid suspensions.[269] After oral administration, intestinal esterases and lipases hydrolyze the palmitate esters to the free antibiotics. In order to improve the

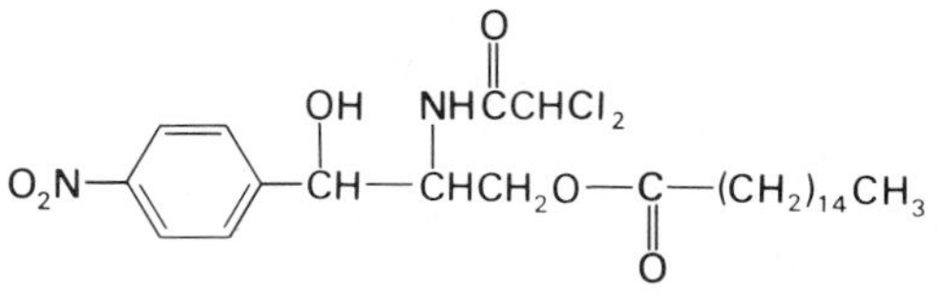

Chloramphenicol
Palmitate

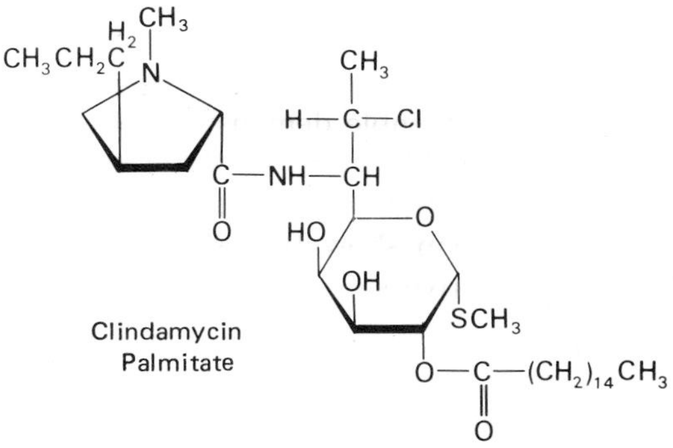

Clindamycin
Palmitate

poor oral absorption of carbenicillin, a lipophilic indanyl ester has been formulated (Geocillin®).[270] Once orally absorbed, the ester is rapidly hydrolyzed to the parent drug. A final example involves derivatization of

Diphenoxylate → Diphenoxylic Acid (Difenoxin)

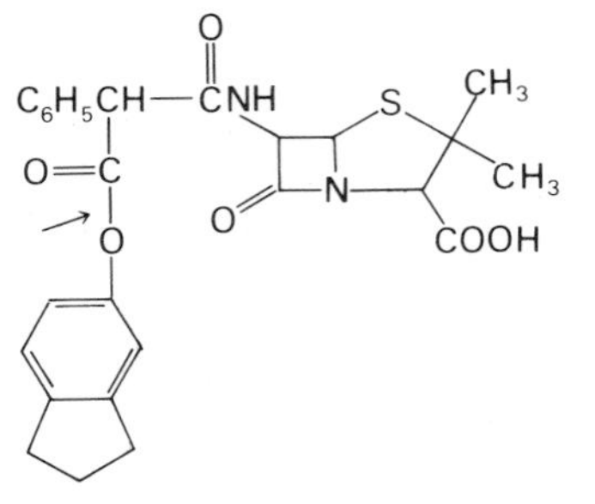

Carbenicillin Indanyl Ester

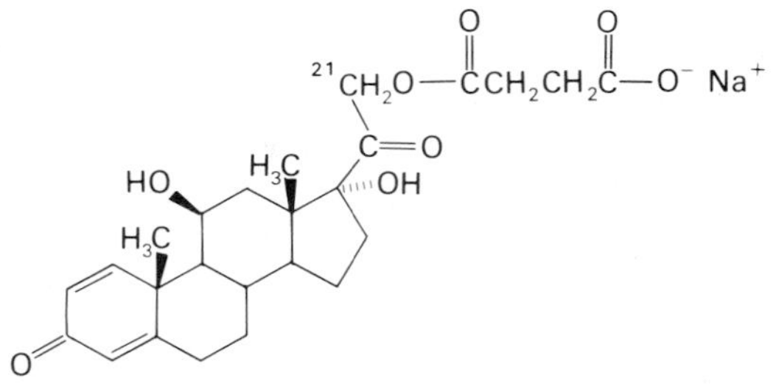

Prednisolone Hemisuccinate
Sodium Salt

prednisolone to its C-21 hemisuccinate sodium salt. This water-soluble derivative is extremely useful for parenteral administration and is metabolized to the parent steroid drug by plasma and tissue esterases.[271]

Amides are slowly hydrolyzed in comparison to esters.[261] For example, hydrolysis of the amide bond of procainamide is relatively slow compared with the hydrolysis of the ester linkage in procaine.[260,272] Drugs in

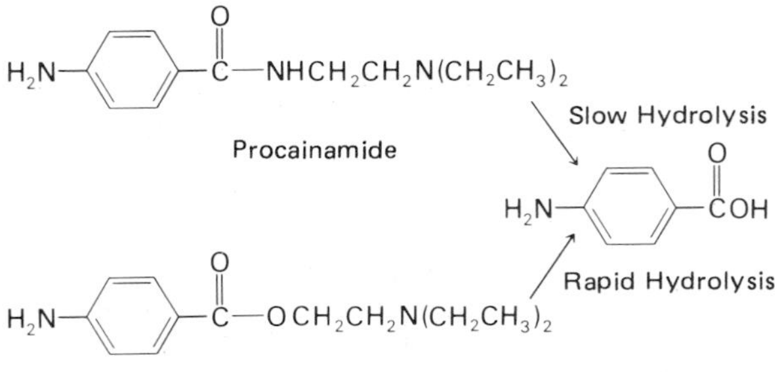

Procainamide

Slow Hydrolysis

Rapid Hydrolysis

Procaine

which amide cleavage has been reported to occur to some extent include lidocaine,[273] carbamazepine,[72] indomethacin,[194] and prazosin (Minipress®).[195] Amide linkages present in barbiturates (e.g., hexobarbital)[274] as well as in hydantoins (e.g., 5-phenylhydantoin)[275]

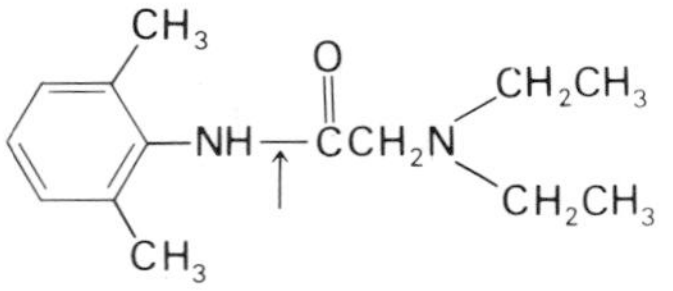

Lidocaine

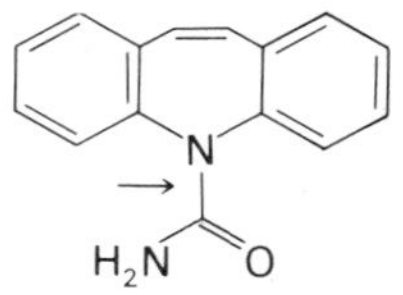

Carbamazepine

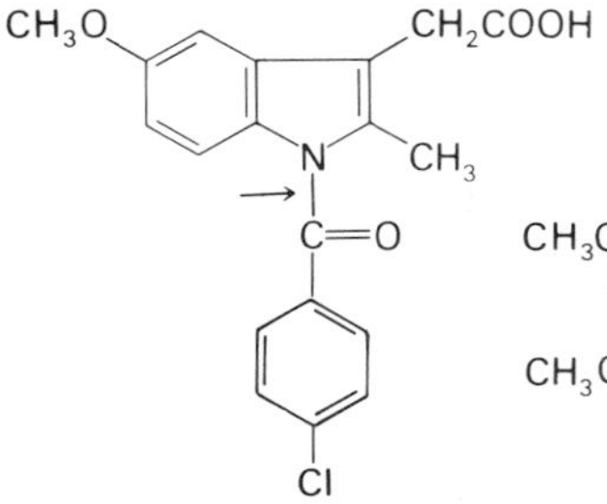

Indomethacin

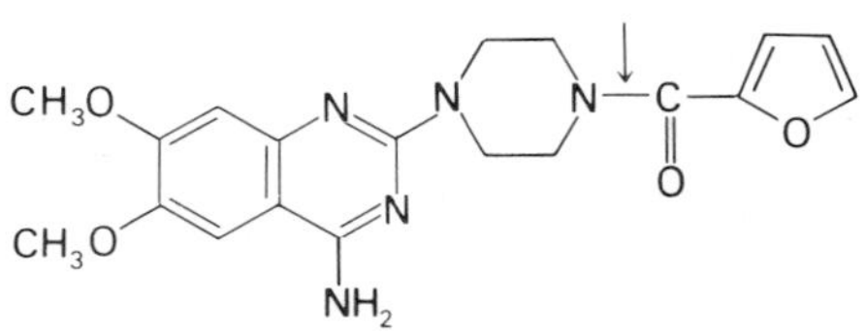

Prazosin

and succinimides (phensuximide)[275] are also susceptible to hydrolysis.

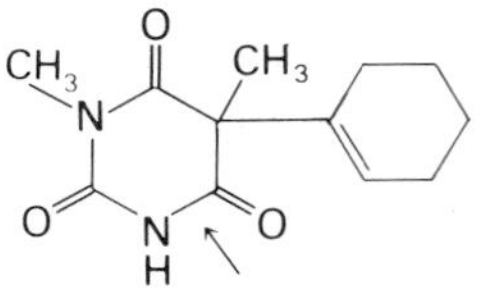

Hexobarbital

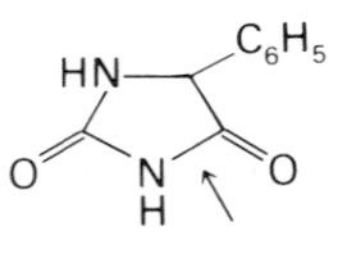

5-Phenylhydantoin

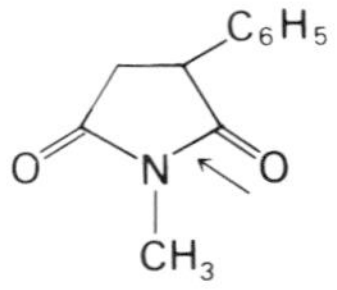

Phensuximide

MISCELLANEOUS HYDROLYTIC REACTIONS

In addition to hydrolysis of amides and esters, hydrolytic cleavage of other moieties also occur to a minor extent in drug metabolism.[8] These include the hydrolysis of phosphate esters (e.g., diethylstilbestrol diphosphate), sulfonylureas, cardiac glycosides, carbamate esters, and organophosphate compounds. Glucuronide and sulfate conjugates are also capable of undergoing hydrolytic cleavage by β-glucuronidase and sulfatase enzymes. These hydrolytic reactions will be discussed later under conjugation reactions. Finally, the hydration or hydrolytic cleavage of epoxides and arene oxides by epoxide hydrase is sometimes considered under hydrolysis reactions.

PHASE II OR CONJUGATION REACTIONS

Phase I or functionalization reactions do not always produce hydrophilic or pharmacologically inactive metabolites. However, various phase II or conjugation reactions are capable of converting these metabolites to more polar and water-soluble products. Many conjugative enzymes accomplish this objective by attaching small, polar, and ionizable endogenous molecules such as glucuronic acid, sulfate, glycine and glutamine to the phase I metabolite or parent xenobiotic. The resulting conjugated products are relatively water-soluble and readily excretable. In addition, they generally are biologically inactive and nontoxic. Other phase II reactions such as methylation and acetylation do not generally increase water solubility but serve mainly to terminate or attenuate pharmacological activity. The role of glutathione is to combine with chemically reactive compounds to prevent damage to important biomacromolecules such as DNA, RNA, and proteins. Thus, phase II conjugation reactions can be regarded as truly detoxifying pathways in drug metabolism with a few exceptions.

A distinguishing feature of most phase II reactions is that the conjugating group (glucuronic acid, sulfate, methyl and acetyl) is initially activated in the form of a coenzyme before transfer or attachment of the group is made to the accepting substrate by the appropriate transferase enzyme. In other cases, such as glycine and glutamine conjugation, the substrate is initially activated. A large number of endogenous compounds, such as bilirubin, steroids, catecholamines, histamine, also undergo conjugation reactions and utilize the same coenzymes, although they appear to be mediated by more specific transferase enzymes. The phase II conjugative pathways to be discussed include those listed earlier in this chapter. Although other conjugative pathways (e.g., conjugation with glycosides, phosphate, other amino acids, conversion of cyanide to thiocyanate) exist, they are only of minor importance in drug metabolism and will not be covered in this chapter.

GLUCURONIC ACID CONJUGATION

Glucuronidation is the most common conjugative pathway in drug metabolism for several reasons: (1) a readily available supply of D-glucuronic acid (derived form D-glucose); (2) a large number of functional groups that can combine enzymatically with glucuronic acid; and (3) the glucuronyl moiety (with its ionized carboxylate [pK_a 3.2] and polar hydroxyl groups) when attached to xenobiotic substrates, greatly increases the water solubility of the conjugated product.[92,276,277] Formation of β-glucuronides involves two steps, synthesis of an activated coenzyme, uridine-5'-diphospho-α-D-glucuronic acid (UDPGA) and subsequent transfer of the glucuronyl group from UDPGA to an appropriate substrate.[92,277] The transfer step is catalyzed by microsomal enzymes called UDP-glucuronyltransferases. They are found primarily in the liver but also occur in many other tissues, including kidney, intestine, skin, lung, and brain.[277]

The sequence of events involved in glucuronidation is summarized in Figure 3-11.[92,277] The synthesis of the coenzyme UDPGA utilizes α-D-glucose-1-phosphate as its initial precursor. It should be noted that all glucuronide conjugates have the β-configuration or β-linkage at C-1 (hence the term β-glucuronides). In contrast, the coenzyme UDPGA has an α-linkage. In the enzymatic transfer step, it appears that nucleophilic displacement of the α-linked UDP moiety from UDPGA by the substrate RXH proceeds with complete inversion of configuration at C-1 to give the β-glucuronide. Glucuronidation of one functional group is usually sufficient to effect excretion of the conjugated metabolite; diglucuronide conjugates usually do not occur.

The diversity of functional groups undergoing glucuronidation is illustrated by the examples given below.[8] Metabolic products are classified as oxygen-, nitrogen-, sulfur-, or carbon-glucuronide according to the heteroatom attached to the C-1 atom of the glucuronyl group. Two important functionalities, the hydroxy and carboxy, form O-glucuronides. Phenolic and alcoholic hydroxyls are the most common functional groups undergoing glucuronidation in drug metabo-

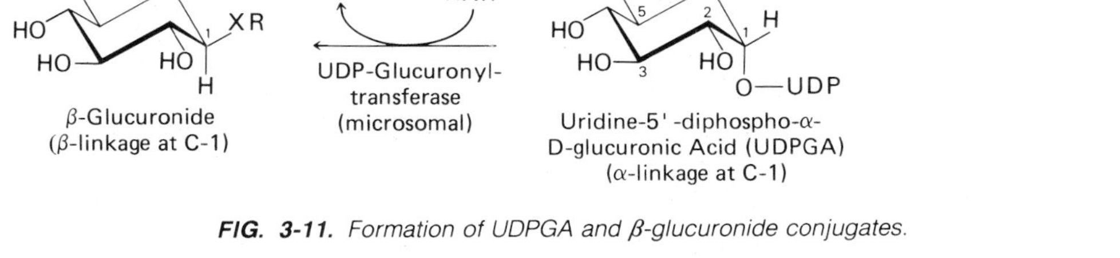

FIG. 3-11. *Formation of UDPGA and β-glucuronide conjugates.*

lism. As we have seen, phenolic and alcoholic hydroxyl groups are present in many parent compounds and arise via various phase I metabolic pathways. Morphine,[278] acetaminophen,[279] and *p*-hydroxyphenytoin (the major metabolite of phenytoin)[41] represent a few examples of phenolic compounds that undergo considerable glucuronidation. Alcoholic hydroxyls such as those present in trichloroethanol (major metabolite of chloral hydrate),[220] chloramphenicol,[280] and propranolol[281] are also commonly glucuronidated. Occurring less frequently is glucuronidation of other hydroxyl groups such as enols (—C=C—OH),[282] N-hydroxylamines (RNHOH),[174] and N-hydroxylamides (RCNHOH).[186] For examples refer to the list of glucuronides (below).

TYPES OF COMPOUNDS FORMING OXYGEN-, NITROGEN-, SULFUR-, AND CARBON-GLUCURONIDES

Oxygen-Glucuronides
Hydroxyl compounds
  Phenols: Morphine, Acetaminophen, *p*-Hydroxyphenytoin
  Alcohols: Trichloroethanol, Chloramphenicol, Propranolol
  Enols: 4-Hydroxycoumarin
  N-Hydroxylamines: N-Hydroxydapsone
  N-Hydroxylamides: N-Hydroxy-2-acetylaminofluorene
Carboxyl Compounds
  Aryl acids: Benzoic acid, Salicylic acid
  Arylalkyl acids: Naproxen, Fenoprofen

Nitrogen-Glucuronides
Arylamines: 7-Amino-5-nitroindazole
Alkylamines: Desipramine
Amides: Meprobamate
Sulfonamides: Sulfisoxazole
Tertiary amines: Cyproheptadine, Tripelennamine

Sulfur-Glucuronides
Sulfhydryl groups: Methimazole, Propylthiouracil, Diethylthiocarbamic acid

Carbon-Glucuronides
3,5–Pyrazolidinedione: Phenylbutazone, Sulfinpyrazone
  (For structures and site of β-glucuronide attachment, see Figure 3-12.)

The carboxy group is also subject to conjugation with glucuronic acid. For example, arylaliphatic acids, such as the antiinflammatory agents naproxen[283] and fenoprofen,[284] are primarily excreted as their O-glucuronide derivatives in man. Carboxylic acid metabolites such as those arising from chlorpheniramine[222] and propranolol[221] (see section on reduction of aldehyde

and ketone carbonyls) have been observed to form O-glucuronide conjugates. Aryl acids (e.g., benzoic acid,[285] salicylic acid[286]) also undergo conjugation with glucuronic acid, but a more important pathway for these compounds appears to be conjugation with glycine.

The formation of N-glucuronides with aromatic amines, aliphatic amines, amides, and sulfonamides occurs occasionally. Representative examples are found in the list of glucuronides given above. Glucuronidation of aromatic and aliphatic amines is generally a minor pathway in comparison with N-acetylation or oxidative processes (e.g., oxidative deamination). More recently, tertiary amines such as the antihistaminic agents cyproheptadine (Periactin®)[291] and tripelennamine[292] have been observed to form interesting quaternary ammonium glucuronide metabolites.

Since the thiol group (SH) does not commonly occur in xenobiotics, S-glucuronide products have been reported for only a few drugs. For instance, the thiol groups present in methimazole (Tapazole®),[293] propylthiouracil,[294] and N,N-diethyldithiocarbamic acid (major reduced metabolite of disulfiram, Antabuse®)[295] have been demonstrated to undergo conjugation with glucuronic acid.

The formation of glucuronides attached directly to a carbon atom is relatively novel in drug metabolism. Recent studies in man have shown that conjugation of phenylbutazone (Butazolidin®)[296] and sulfinpyrazone (Anturane®)[297] yield the corresponding C-glucuronide metabolites:

Phenylbutazone, R = CH₂CH₂CH₂CH₃
Sulfinpyrazone, R = CH₂CH₂SC₆H₅

C-Glucuronide Metabolite

Besides xenobiotics, a number of endogenous substrates, notably bilirubin[298] and steroids,[299] are eliminated as glucuronide conjugates. Glucuronide conjugates are primarily excreted in the urine. However as the molecular weight of the conjugate exceeds 300 daltons, biliary excretion may become an important route of elimination.[300] Glucuronides that are excreted in the bile are susceptible to hydrolysis by β-glucuronidase enzymes present in the intestine. The hydrolyzed

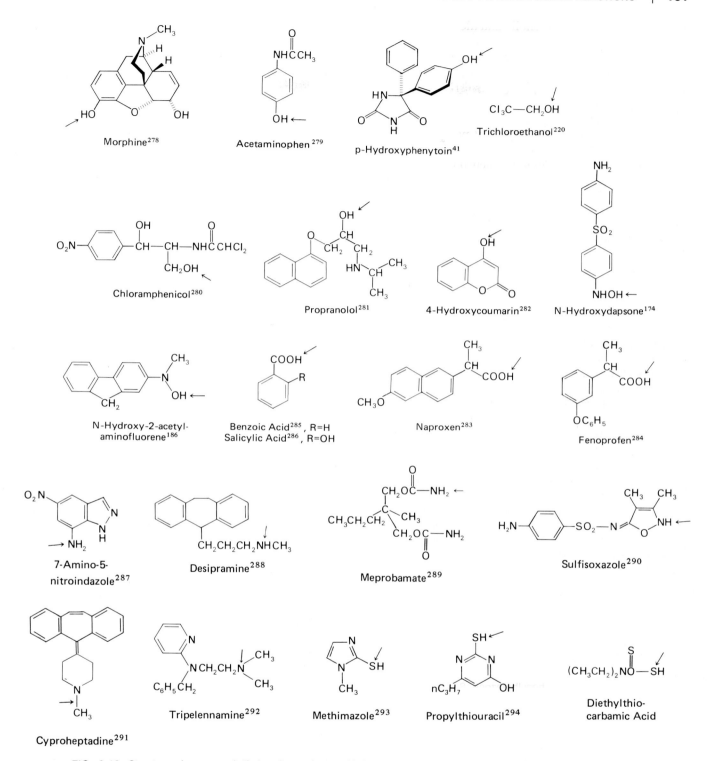

FIG. 3-12. *Structure of compounds that undergo glucuronidation (arrow indicates the site of β- glucuronide attachment).*

product may be reabsorbed in the intestine, thus leading to enterohepatic recycling.[19] β-Glucuronidases are also present in many other tissues, including the liver, the endocrine system, and the reproductive organs. While the function of these hydrolytic enzymes in drug metabolism is unclear, it appears that, in terms of hormonal and endocrine regulation, β-glucuronidases may be functioning to liberate active hormones (e.g., steroids) from their inactive glucuronide conjugates.[19]

In neonates and children, glucuronidating processes are often not fully developed. In such subjects, drugs and endogenous compounds (e.g., bilirubin) which are normally metabolized by glucuronidation may accumulate and cause serious toxicity. For example, neonatal hyperbilirubinemia may be attributable to the inability of newborns to conjugate bilirubin with glucuronic acid.[301] Similarly, the inability of infants to glucuronidate chloramphenicol has been suggested to be responsible for the "gray baby syndrome," which results from accumulation of toxic levels of the free antibiotic.[302]

SULFATE CONJUGATION

Conjugation of xenobiotics with sulfate occurs primarily with phenols and occasionally with alcohols, aromatic amines, and N-hydroxy compounds.[303-305] In contrast to glucuronic acid, the amount of available sulfate is rather limited. A significant portion of the sulfate pool is utilized by the body to conjugate numerous endogenous compounds such as steroids, heparin, chondroitin, catecholamines, and thyroxine. The sulfate conjugation process involves activation of inorganic sulfate to the coenzyme 3'-phosphoadenosine-5'-phosphosulfate (PAPS). Subsequent transfer of the sulfate group from PAPS to the accepting substrate is catalyzed by various soluble sulfotransferases present in the liver and other tissues (e.g., kidney, intestine).[305] The sequence of events involved in sulfoconjugation is depicted in Figure 3-13. Sulfate conjugation generally leads to water-soluble and inactive metabolites. However, it appears that the O-sulfate conjugates of some N-hydroxy compounds give rise to chemically reactive intermediates that are toxic.[186]

Phenols comprise the main group of substrates undergoing sulfate conjugation. Thus, drugs containing phenolic moieties are often susceptible to sulfate formation. For example, the antihypertensive agent α-methyldopa (Aldomet®) is extensively metabolized to its 3-O-sulfate ester in man.[306] The β-adrenergic bron-

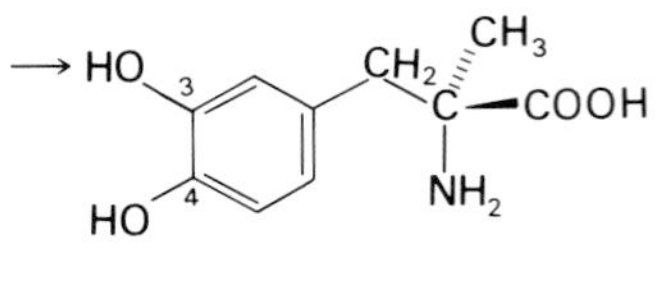

α-Methyldopa

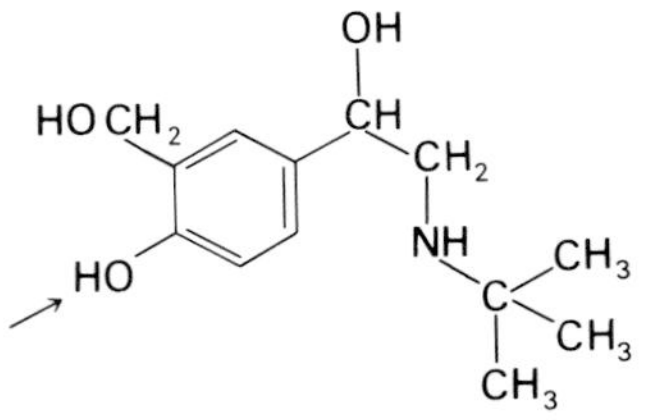

Salbutamol
(Albuterol)

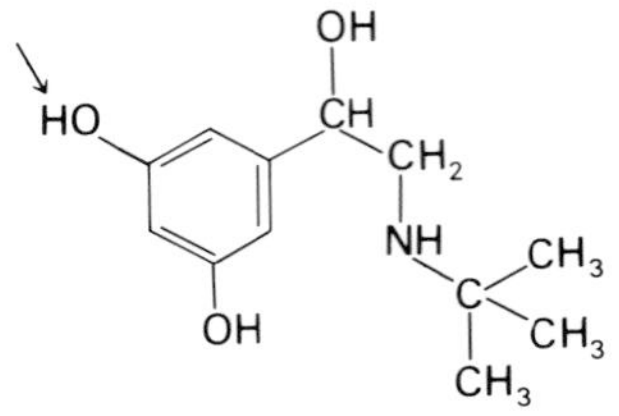

Terbutaline

chodilators salbutamol (albuterol)[307] and terbutaline (Brethine®, Bricanyl®)[308] also undergo sulfate conjugation as their principal route of metabolism in humans. However, for many phenols sulfoconjugation may represent only a minor pathway. Glucuronidation of phenols is frequently a competing reaction and

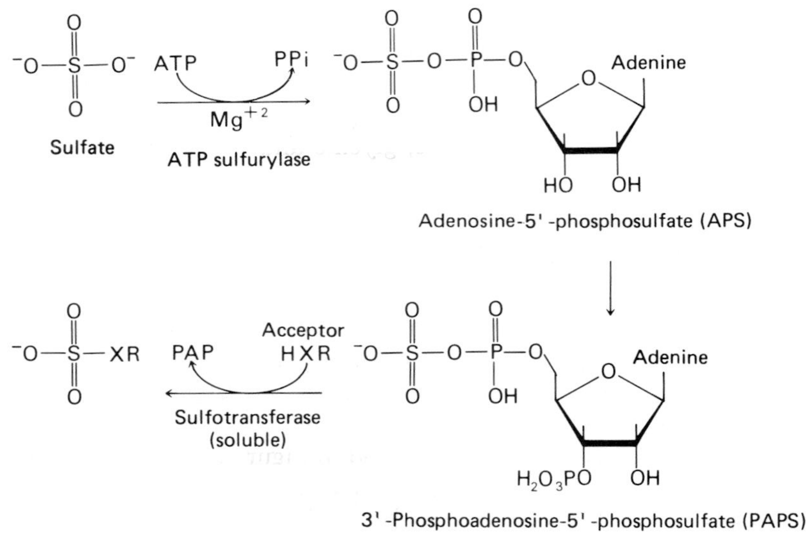

Sulfate

ATP sulfurylase

Adenosine-5'-phosphosulfate (APS)

Acceptor

Sulfotransferase
(soluble)

3'-Phosphoadenosine-5'-phosphosulfate (PAPS)

FIG. 3-13. *Formation of PAPS and sulfate conjugates.*

may predominate as the conjugative route for some phenolic drugs. In adults, the major urinary metabolite of the analgesic acetaminophen is the O-glucuronide conjugate with the O-sulfate conjugate being formed in small amounts.[279] Interestingly, in infants

Acetaminophen   O-Glucuronide Conjugate   O-Sulfate Conjugate

and young children (ages 3 to 9 years) a different urinary excretion pattern is observed: the O-sulfate conjugate is the main urinary product.[309] The explanation for this reversal stems from the fact that neonates and young children have a decreased glucuronidating capacity owing to undeveloped glucuronyltransferases or low levels of glucuronyltransferases. Sulfate conjugation however is well developed and becomes the main route by which acetaminophen is conjugated in this pediatric age group.

Other functionalities such as alcohols (e.g., aliphatic C_1 to C_5 alcohols, diethylene glycol)[310] and aromatic amines (e.g., aniline, 2-naphthylamine)[311] are also capable of forming sulfate conjugates. These reactions, however, have only minor importance in drug metabolism. The sulfate conjugation of N-hydroxylamines and N-hydroxylamides takes place occasionally as well. O-Sulfate ester conjugates of N-hydroxy compounds are of considerable toxicological concern, since they can lead to reactive intermediates that are responsible for cellular toxicity. The carcinogenic agents N-methyl-4-aminoazobenzene and 2-acetylaminofluorene are believed to mediate their toxicity via N-hydroxylation to the corresponding N-hydroxy compounds (see earlier section on N-hydroxylation of amines and amides). Sulfoconjugation of the N-hydroxy metabolites yields O-sulfate esters, which presumably are the ultimate carcinogenic species. Loss of $SO_4^=$ from the above sulfate conjugates generates electrophilic nitrenium species, which may react with nucleophilic groups (e.g., NH_2, OH, SH) present in proteins, DNA, and RNA to form covalent linkages that lead to structural and functional alteration of these crucial biomacromolecules. The consequences of this are cellular toxicity (tissue necrosis) or alteration of genetic code material leading eventually to cancer. Some evidence supporting the role of sulfate conjugation in metabolic activation of N-hydroxy compounds

to reactive intermediates comes from the observation that the degree of hepatotoxicity and hepatocarcinogenicity of N-hydroxy-2-acetylaminofluorene is markedly dependent on the level of sulfotransferase activity in the liver.[313]

The analgesic phenacetin is metabolized to N-hydroxyphenacetin and subsequently conjugated with sulfate.[314] The O-sulfate conjugate of N-hydroxyphenacetin has been demonstrated to covalently bind to microsomal proteins.[315] It has been suggested that this pathway may represent one possible route leading to reactive intermediates that are responsible for the hepatotoxicity and nephrotoxicity associated with phenacetin. Other pathways (e.g., arene oxides) leading to reactive electrophilic intermediates are also possible.[6]

Phenacetin   N-Hydroxyphenacetin   O-Sulfate conjugate of N-Hydroxyphenacetin

CONJUGATION WITH GLYCINE, GLUTAMINE AND OTHER AMINO ACIDS

The amino acids glycine and glutamine are utilized by mammalian systems to conjugate carboxylic acids, particularly aromatic acids and arylalkyl acids.[316,317] Glycine conjugation is common to most mammals, while glutamine conjugation appears mainly confined to man and other primates. The quantity of amino acid conjugates formed from xenobiotics is quite small because of the limited availability of amino acids in the body and competition with glucuronidation for carboxylic acid substrates. In contrast to glucuronic acid and sulfate, glycine and glutamine are not converted to activated coenzymes. Instead the carboxylic acid substrate is activated with ATP and coenzyme A to form an acyl coenzyme A complex. The latter intermediate in turn acylates glycine or glutamine under the influence of specific glycine or glutamine N-acyltransferase enzymes. The activation and acylation steps take place in the mitochondria of liver and kidney cells. The sequence of metabolic events associated with glycine and glutamine conjugation of phenylacetic acid is summarized in Figure 3-14. Amino acid conjugates, being polar and water-soluble, are mainly excreted renally and sometimes in the bile.

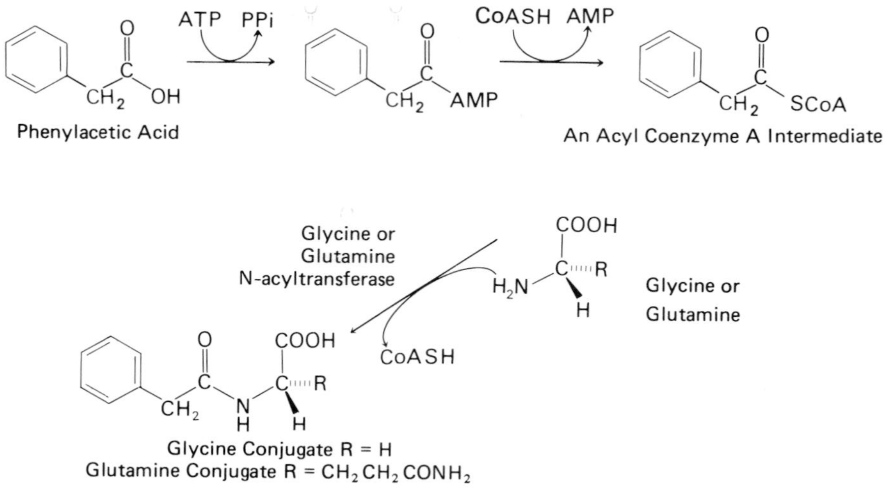

FIG. 3-14. *Formation of glycine and glutamine conjugates of phenylacetic acid.*

Aromatic acids and arylalkyl acids are the major substrates undergoing glycine conjugation. The conversion of benzoic acid to its glycine conjugate, hippuric acid, is a well-known metabolic reaction in many mammalian systems.[319] The extensive metabolism of salicylic acid (75 percent of dose) to salicyluric acid in

Benzoic Acid, R=H   Hippuric Acid, R=H
Salicylic Acid, R=OH   Salicyluric Acid, R=OH

man is another illustrative example.[320] Carboxylic acid metabolites resulting from oxidation or hydrolysis of

many drugs are also susceptible to glycine conjugation. For example, the H_1-histamine antagonist brompheniramine is oxidized to a propionic acid metabolite that is conjugated with glycine in both man and dog.[141] Similarly, *p*-fluorophenylacetic acid derived from the metabolism of the antipsychotic agent haloperidol (Haldol®) is found as the glycine conjugate in the urine of rats.[321] Phenylacetic acid and isonicotinic acid resulting respectively from the hydrolysis of the anticonvulsant phenacemide (Phenurone®)[322] and the antituberculosis agent isoniazid[323] are also conjugated with glycine to some extent.

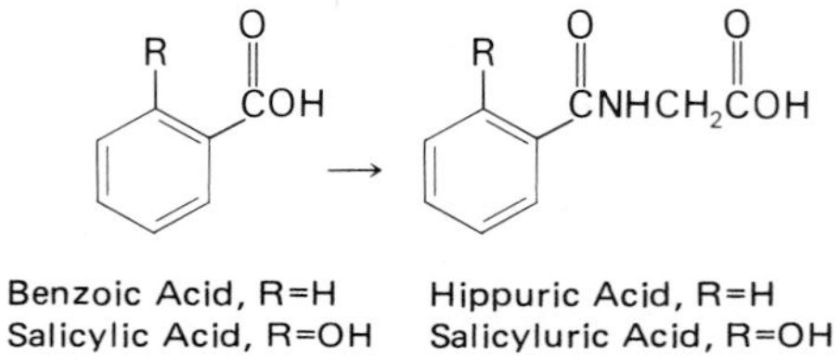

Glutamine conjugation occurs mainly with arylacetic acids, including endogenous phenylacetic[318] and indolylacetic acid.[324] A few glutamine conjugates of drug metabolites have been reported. For example, in man the 3,4-dihydroxy-5-methoxyphenylacetic acid metabolite of mescaline is found as a conjugate of glu-

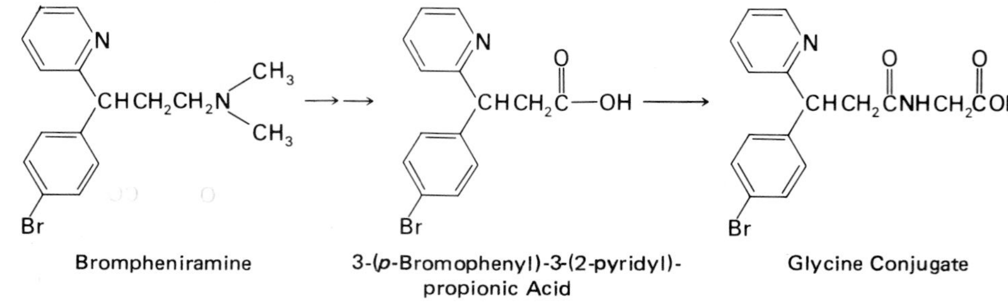

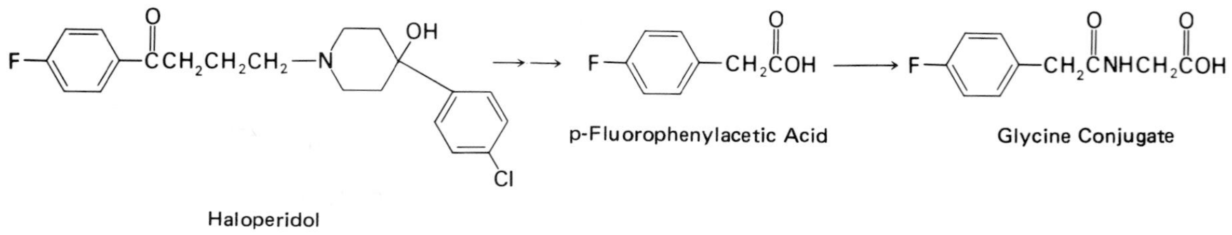

Brompheniramine   3-(*p*-Bromophenyl)-3-(2-pyridyl)-propionic Acid   Glycine Conjugate

Haloperidol   p-Fluorophenylacetic Acid   Glycine Conjugate

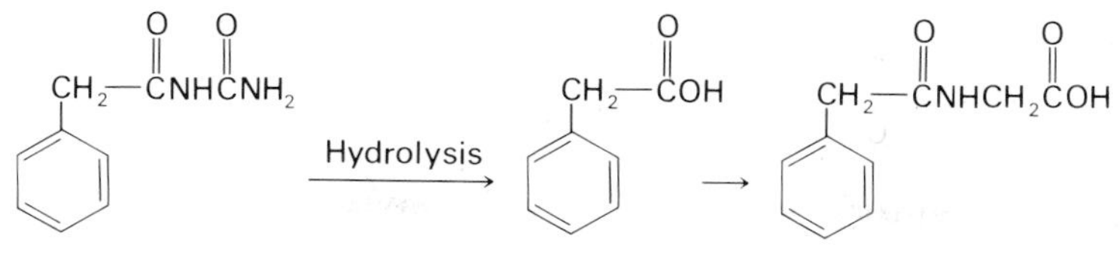

Phenacemide   Phenylacetic Acid   Glycine Conjugate

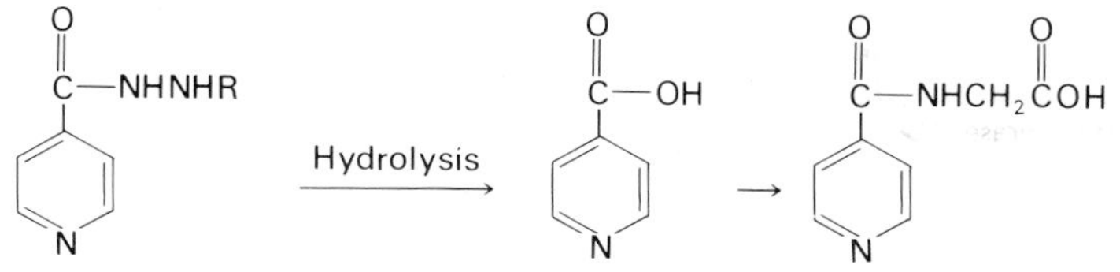

Isoniazid (R=H)
or
N-Acetylisoniazid
(R = COH$_3$)

Isonicotinic Acid   Glycine Conjugate

tamine.[325] Diphenylmethoxyacetic acid, a metabolite of the antihistamine diphenhydramine (Benadryl®), is biotransformed further to the corresponding glutamine derivative in the rhesus monkey.[326]

A number of other amino acids are known to be involved in the conjugation of carboxylic acids, but these reactions occur only occasionally and appear to be highly substrate and species dependent.[317,327] Ornithine (in birds), aspartic acid and serine (in rats), alanine (in mouse and hamster), taurine ($H_2NCH_2CH_2SO_3H$) (in mammals and pigeons) and histidine (in African bats) are among these amino acids.[327]

GLUTATHIONE OR MERCAPTURIC ACID CONJUGATES

Glutathione conjugation is an important pathway by which chemically reactive electrophilic compounds are detoxified.[328,329] It is now generally accepted that reactive electrophilic species manifest their toxicity (e.g.,

tissue necrosis, carcinogenicity, mutagenicity, teratogenicity) by combining covalently with nucleophilic groups present in vital cellular proteins and nucleic acids.[4,330] Many serious drug toxicities may be explainable also in terms of covalent interaction of metabolically generated electrophilic intermediates with cellular nucleophiles.[5,6] Glutathione protects vital cellular constituents against chemically reactive species by virtue of its nucleophilic sulfhydryl group. It is the sulfhydryl (SH) group that reacts with electron-deficient compounds to form S-substituted glutathione adducts (Fig. 3-15).[328,329]

Glutathione (GSH) is a tripeptide (γ-glutamylcysteinylglycine) found in most tissues. Xenobiotics conjugated with GSH usually are not excreted as such but undergo further biotransformation to give S-substituted N-acetylcysteine products called mercapturic acids.[63,71,329] This process involves enzymatic cleavage of two amino acids (namely glutamic acid and glycine) from the initially formed glutathione adduct and subsequent N-acetylation of the remaining S-substituted cysteine residue. The formation of glutathione conju-

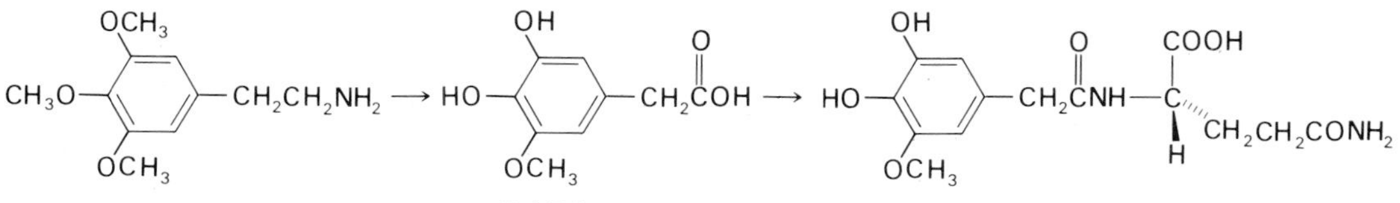

Mescaline   3,4-Dihydroxy-5-methoxyphenylacetic Acid   Glutamine Conjugate

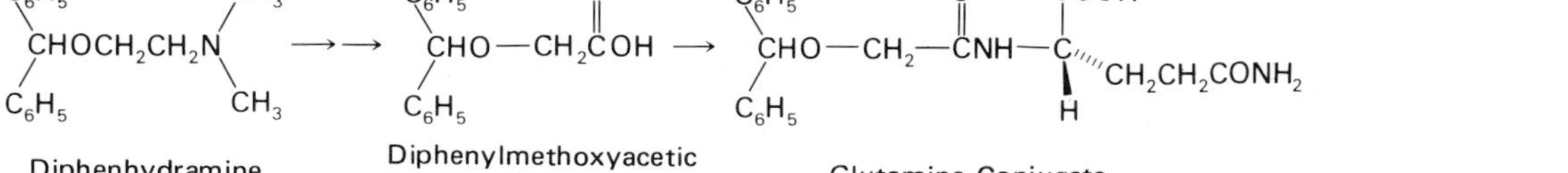

Diphenhydramine   Diphenylmethoxyacetic Acid   Glutamine Conjugate

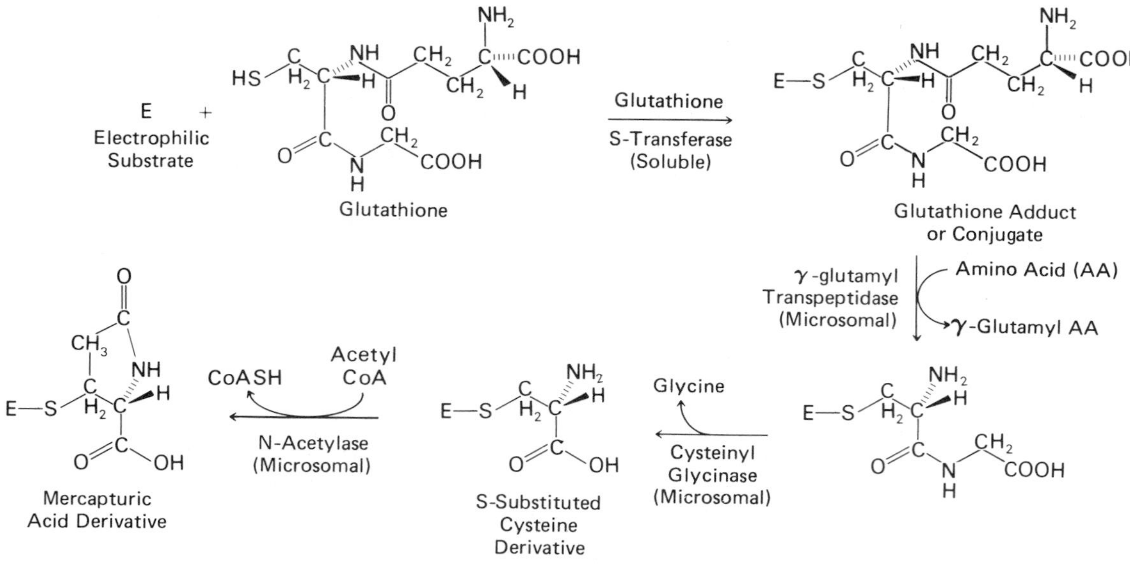

FIG. 3-15. *Formation of glutathione conjugates of electrophilic xenobiotics or metabolites (E) and their conversion to mercapturic acids.*

gates and their conversion to mercapturic acid derivatives are outlined in Figure 3-15.

Conjugation of a wide spectrum of substrates with GSH is catalyzed by a family of cytoplasmic enzymes known as glutathione S-transferases.[62] These enzymes are found in most tissues, particularly the liver and kidney. Degradation of GSH conjugates to mercapturic acids is carried out principally by renal and hepatic microsomal enzymes (Fig. 3-15).[63] Unlike other conjugative phase II reactions, GSH conjugation does not require the initial formation of an activated coenzyme or substrate. The inherent reactivity of the nucleophilic GSH toward an electrophilic substrate usually provides sufficient driving force. The substrates susceptible to GSH conjugation are quite varied and encompass many chemically different classes of compounds. A major prerequisite is that the substrate be sufficiently electrophilic in nature. Compounds that react with glutathione do so by two general mechanisms: (1) nucleophilic displacement at an electron-deficient carbon or heteroatom, or (2) nucleophilic addition to an electron-deficient double bond.[328]

Many aliphatic and arylalkyl halides (Cl, Br, I), sulfates (OSO_3^-), sulfonates (OSO_2R), nitrates (NO_2), and organophosphates (O-P[OR]$_2$) possess electron-deficient carbon atoms that react with GSH (via aliphatic nucleophilic displacement) to form glutathione conjugates, as shown below:

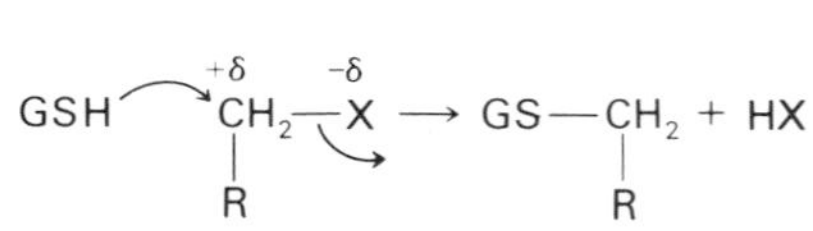

R= Alkyl, Aryl, Benzylic, Allylic
X= Br, Cl, I, OSO_3^-, OSO_2R, $OPO(OR)_2$

The carbon center is rendered electrophilic as a result of the electron-withdrawing group (e.g., halide, sulfate, phosphate) attached to it. Nucleophilic displacement is often facilitated when the carbon atom is benzylic or allylic or when X is a good leaving group (e.g., halide, sulfate). Many industrial chemicals such as benzyl chloride ($C_6H_5CH_2Cl$), allyl chloride

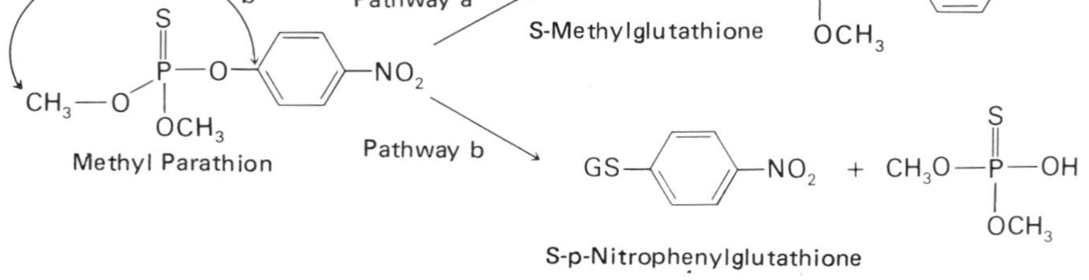

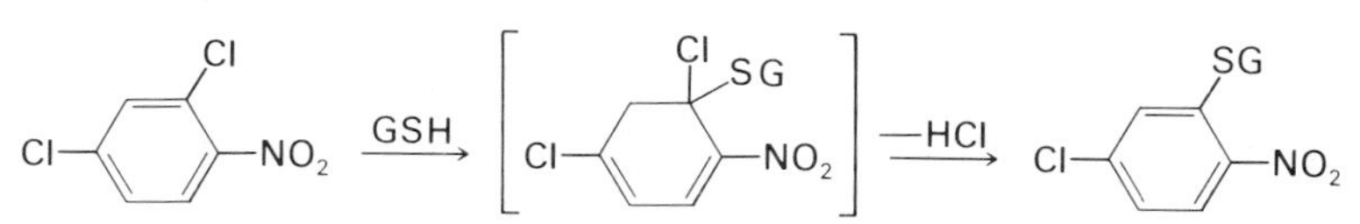

2,4-Dichloronitrobenzene

(CH$_2$=CHCH$_2$Cl), and methyl iodide are known to be toxic and carcinogenic. The reactivity of these three halides toward GSH conjugation in mammalian systems is demonstrated by the formation of the corresponding mercapturic acid derivatives.[329] Organophosphate insecticides such as methyl parathion are detoxified by two different glutathione pathways.[331] Pathway a involves aliphatic nucleophilic substitution and yields S-methylglutathione. Pathway b, on the other hand, involves aromatic nucleophilic substitution and produces S-p-nitrophenylglutathione. Aromatic or heteroaromatic nucleophilic substitution reactions with GSH occur only when the ring is rendered sufficiently electron-deficient by the presence of one or more strongly electron-withdrawing substituents (e.g., NO$_2$, Cl). For example, 2,4-dichloronitrobenzene is susceptible to nucleophilic substitution by glutathione, whereas chlorobenzene is not.[332]

The metabolism of the immunosuppressive drug azathioprine (Imuran®) to 1-methyl-4-nitro-5-(S-glutathionyl)imidazole and 6-mercaptopurine is an example of heteroaromatic nucleophilic substitution reaction involving glutathione.[333] Interestingly, 6-mercaptopurine formed in this reaction appears to be responsible for azathioprine's immunosuppressive activity.[334]

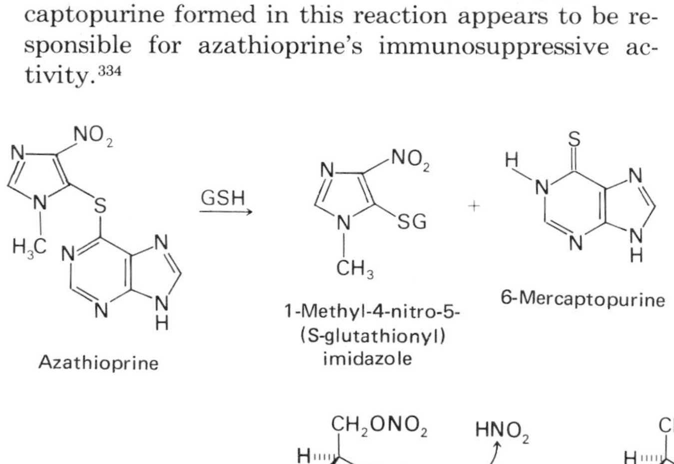

Arene oxides and aliphatic epoxides (or oxiranes) represent a very important class of substrates that are conjugated and detoxified by glutathione.[335] The three-membered oxygen-containing ring in these compounds is highly strained and therefore is reactive toward ring cleavage by nucleophiles (e.g., GSH, H$_2$O or nucleophilic groups present on cellular macromolecules). As discussed previously, arene oxides and epoxides are intermediary products formed from cytochrome P-450 oxidation of aromatic compounds (arenes) and olefins, respectively. If reactive arene oxides (e.g., benzo[a]pyrene-4,5-oxide, 4-bromobenzene oxide) and aliphatic epoxides (e.g., styrene oxide) are not "neutralized" or detoxified by glutathione S-transferase, epoxide hydrase or other pathways, they ultimately covalently bind to cellular macromolecules to cause serious cytotoxicity and carcinogenicity. The isolation of glutathione or mercapturic acid adducts from benzo[a]pyrene, bromobenzene and styrene clearly demonstrates the importance of GSH in reacting with the reactive epoxide metabolites generated from these compounds.

Glutathione conjugation involving substitution at heteroatoms such as oxygen is often seen with organic nitrates. For example, nitroglycerin (Nitrostat®) and isosorbide dinitrate (Isordil®) are metabolized by a pathway involving an initial glutathione conjugation reaction. However, the GSH conjugate products are not metabolized to mercapturic acids but instead are converted enzymatically to the corresponding alcohol derivatives and glutathione disulfide (GSSG).[336]

The nucleophilic addition of glutathione to electron-deficient carbon–carbon double bonds occurs mainly in compounds with α,β-unsaturated double bonds. In most instances, the double bond is rendered

Azathioprine — 1-Methyl-4-nitro-5-(S-glutathionyl)imidazole + 6-Mercaptopurine

Nitroglycerin

Isosorbide

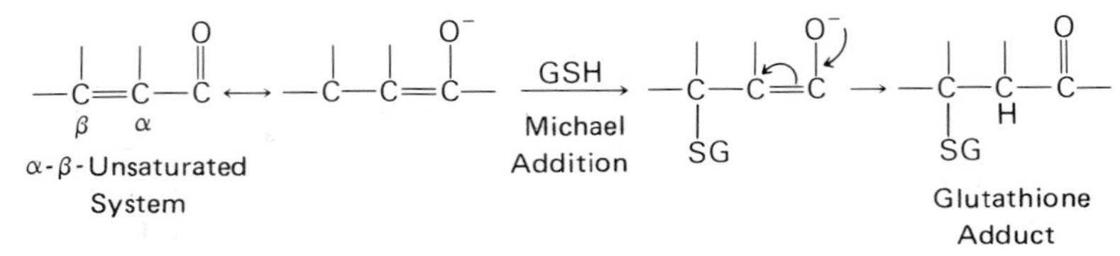

electron deficient by resonance or conjugation with a carbonyl group (ketone or aldehyde), ester, nitrile, etc. Such α,β-unsaturated systems undergo so-called Michael addition reactions with GSH to yield the corresponding glutathione adduct.[328,329] For example, in rats and dogs the diuretic agent ethacrynic acid (Edecrin®) reacts with GSH to form the corresponding glutathione or mercapturic acid derivatives.[337] The compound diethyl maleate is readily conjugated with glutathione and has been used experimentally to de-

plete hepatic GSH stores in laboratory animals.[338] A number of other α,β-unsaturated compounds such as acrolein, crotonaldehyde, o-chlorobenzylidenemalononitrile, and arecoline have been demonstrated to form mercapturic acid or glutathione derivatives.[329] It should be emphasized that not all α,β-unsaturated compounds are conjugated with glutathione. Many steroidal agents possessing α,β-unsaturated carbonyl moieties such as prednisone and digitoxigenin have not been observed to undergo any significant conjuga-

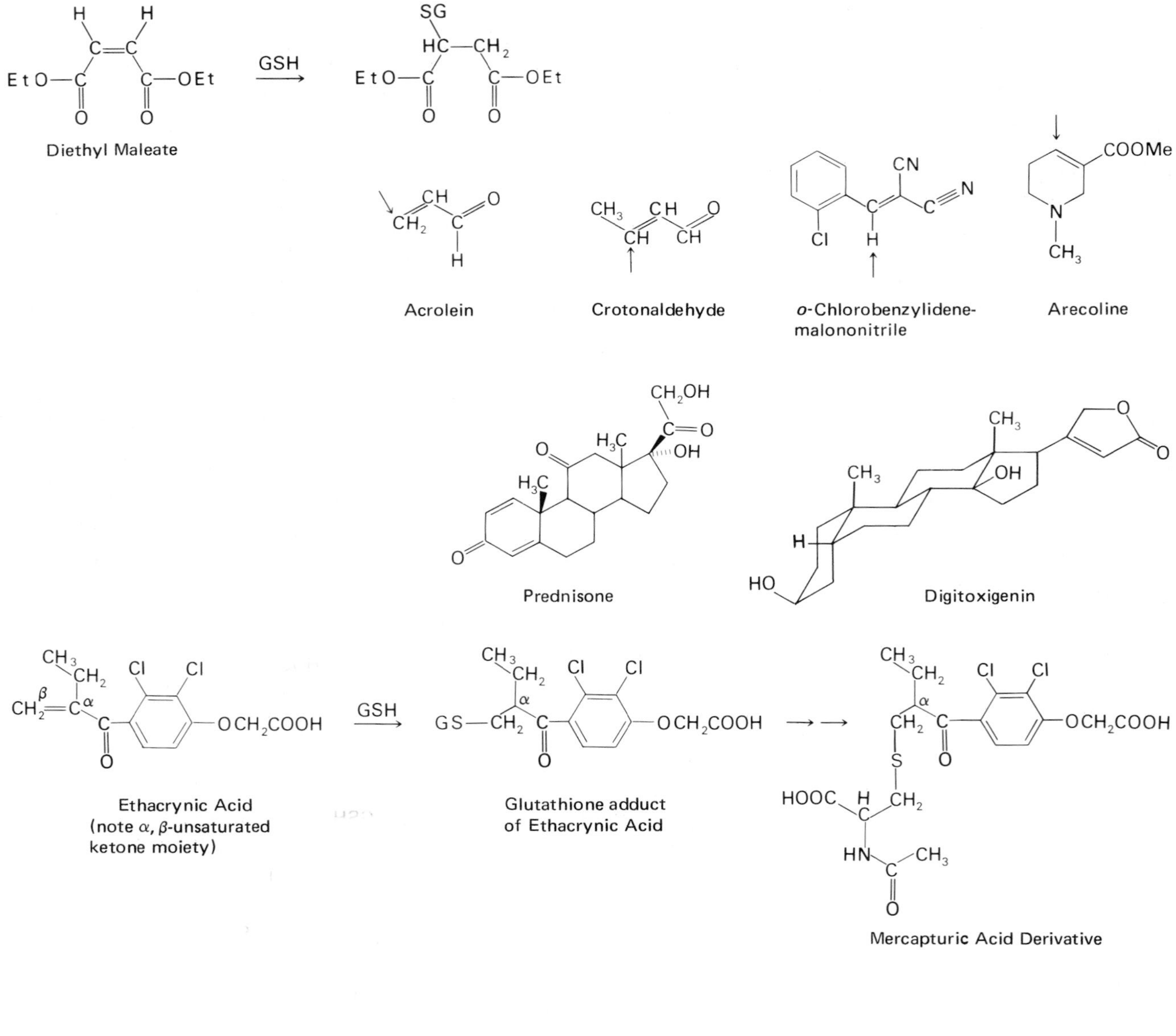

Diethyl Maleate

Acrolein

Crotonaldehyde

o-Chlorobenzylidene-malononitrile

Arecoline

Prednisone

Digitoxigenin

Ethacrynic Acid
(note α,β-unsaturated ketone moiety)

Glutathione adduct of Ethacrynic Acid

Mercapturic Acid Derivative

Acetaminophen   N-Acetylimidoquinone   Mercapturic Acid Derivative

tion with glutathione. Steric factors, decreased reactivity of the double bond as well as other factors (e.g., susceptibility to metabolic reduction of the ketone or of the C=C double bond) may account for these observations.

In some cases, metabolic oxidative biotransformation reactions may generate chemically reactive α,β-unsaturated systems that react with glutathione. For example, metabolic oxidation of acetaminophen presumably generates the chemically reactive intermediate, N-acetylimidoquinone. Michael addition of GSH to the imidoquinone leads to the corresponding mercapturic acid derivative in both animals and man.[189,191] 2-Hydroxyestrogens such as 2-hydroxy-17β-estradiol undergo conjugation with glutathione to yield the two isomeric mercapturic acid or glutathione derivatives. Although the exact mechanism is unclear, it appears that the 2-hydroxyestrogen is oxidized to a chemically reactive orthoquinone or semiquinone intermediate that reacts with GSH at either the electrophilic C-1 or C-4 position.[339]

ACETYLATION

Acetylation constitutes an important metabolic route for drugs containing primary amino groups.[316, 340] This encompasses primary aromatic amines (ArNH$_2$), sulfonamides (H$_2$NC$_6$H$_4$SO$_2$NHR), hydrazines (-NHNH$_2$), hydrazides (-CONHNH$_2$), and primary aliphatic amines. The amide derivatives formed from

acetylation of these amino functionalities are generally inactive and non-toxic. Since water solubility is not greatly enhanced by N-acetylation, it appears that the primary function of acetylation is one of termination of pharmacological activity and detoxification. However, a few recent reports indicate that acetylated metabolites may be as active (e.g., N-acetylprocainamide)[341] or more toxic (e.g., N-acetylisoniazid)[342] than their corresponding parent compounds.

The acetyl group utilized in N-acetylation of xenobiotics is supplied by acetylcoenzyme A.[316] Transfer of the acetyl group from this cofactor to the accepting amino substrate is carried out by soluble N-acetyltransferases present in hepatic reticuloendothelial cells. Other extrahepatic tissues such as the lung, spleen, gastric mucosa, red blood cells, and lymphocytes also show acetylation capability. N-Acetyltransferase enzymes display broad substrate specificity and catalyze the acetylation of a number of drugs and xenobiotics (Fig. 3-16).[340] Aromatic compounds possessing a primary amino group such as aniline,[316] p-aminobenzoic acid,[343] p-aminosalicylic acid,[325] procainamide (Pronestyl®)[341, 343] and dapsone (Avlosulfon®)[344] are especially susceptible to N-acetylation. Aromatic amine metabolites resulting from the reduction of aryl nitro compounds are also N-acetylated. For example, the anticonvulsant clonazepam (Clonopin®) undergoes nitro reduction to its 7-amino metabolite, which in turn is N-acetylated.[245] Another related benzodiazepam analog, nitrazepam, follows a similar pathway.[246]

2-Hydroxy-17β-estradiol   Orthoquinone   Semiquinone

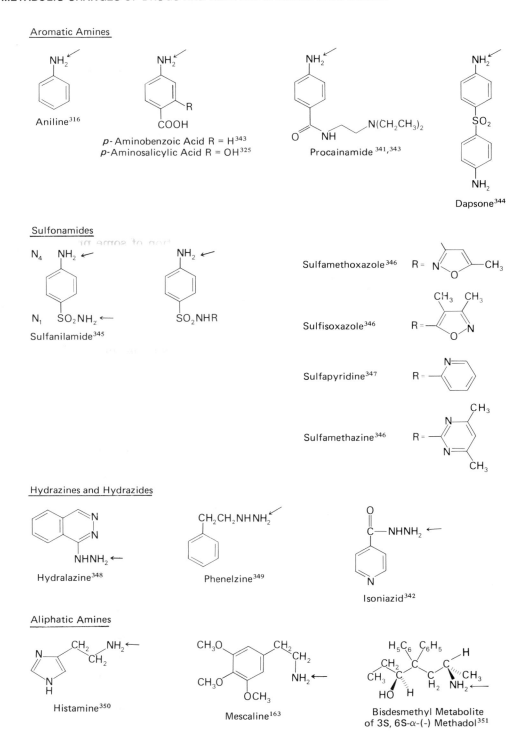

FIG. 3-16. *Examples of different types of compounds undergoing N-acetylation (arrow indicates the site of N-acetylation).*

The metabolism of a number of sulfonamides such as sulfanilamide,[345] sulfamethoxazole (Gantanol®),[346] sulfisoxazole (Gantrisin®),[346] sulfapyridine[347] (major metabolite from azo reduction of sulfasalazine [Azulfidine®]) and sulfamethazine[316] occurs mainly by acetylation at the N_4 position. In the case of sulfanilamide, acetylation also takes place at the sulfamido N_1-position.[345] N-Acetylated metabolites of sulfonamides tend to be less water soluble than their parent compounds and have the potential of crystallizing out in

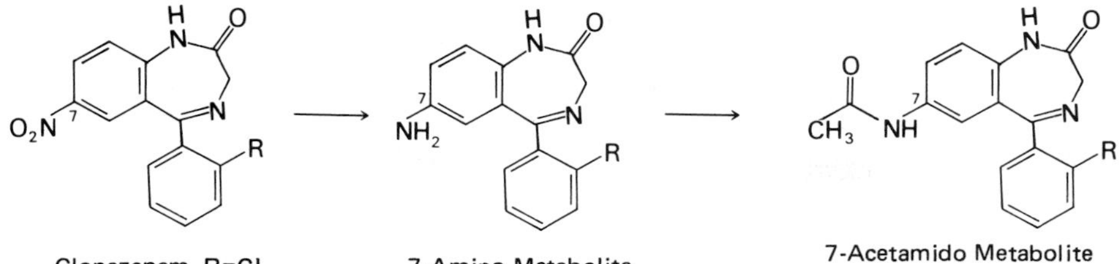

Clonazepam, R=Cl
Nitrazepam, R=H

7-Amino Metabolite

7-Acetamido Metabolite
or
N-Acetylated Metabolite

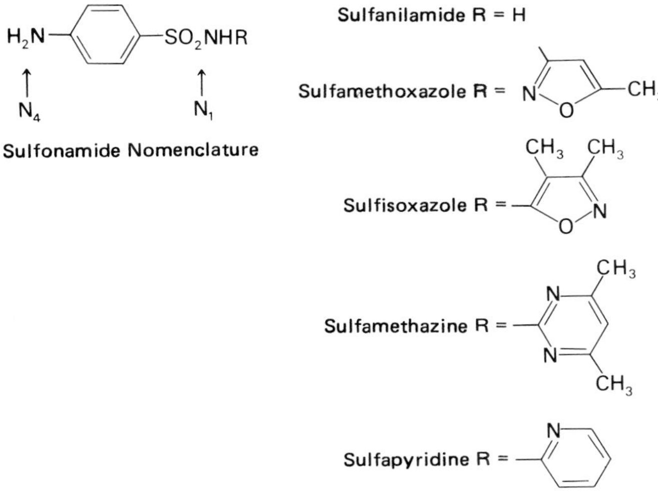

Sulfanilamide R = H

Sulfamethoxazole R =

Sulfisoxazole R =

Sulfamethazine R =

Sulfapyridine R =

Sulfonamide Nomenclature

renal tubules (crystalluria), thereby causing kidney damage. The frequency of crystalluria and renal toxicity is especially prominent with older sulfonamide derivatives such as sulfathiazole.[1, 327] However, newer sulfonamides such as sulfisoxazole and sulfamethoxazole are metabolized to relatively water-soluble acetylated derivatives which are less likely to precipitate out.

The biotransformation of hydrazine and hydrazide derivatives also proceeds by acetylation. The antihypertensive hydralazine (Apresoline®)[348] and the monoamine oxidase (MAO) inhibitor phenelzine (Nardil®)[349] are two representative hydrazine compounds that are metabolized by this pathway. It should be noted that the initially formed N-acetyl derivative of hydralazine is unstable and cyclizes intramolecularly to form 3-methyl-s-triazolo[3,4-a]phthalazine as the major isolable hydralazine metabolite in man.[348] The

antituberculosis drug isoniazid or isonicotinic acid hydrazide (INH) is extensively metabolized to N-acetylisoniazid.[342]

The acetylation of some primary aliphatic amines such as histamine,[350] mescaline,[163] and the *bis*-N-demethylated metabolite of α(-)-methadol[351] also has been reported. In comparison to oxidative deamination processes, N-acetylation is only a minor pathway in the metabolism of this class of compounds.

The acetylation pattern of a number of drugs (e.g., isoniazid, hydralazine, procainamide) in the human population displays a bimodal character in which the drug is either rapidly or slowly conjugated with acetyl coenzyme A.[352,353] This phenomenon is termed acetylation polymorphism. Individuals are classified as being either slow or rapid acetylator phenotypes. This variation in acetylating ability is genetic in origin and is mainly caused by differences in N-acetyltransferase activity. The proportion of rapid and slow acetylators varies widely among different ethnic groups throughout the world. For example, a high proportion of Eskimos and Orientals are rapid acetylators, whereas Egyptians and some Western European groups are mainly slow acetylators.[353] Other populations are intermediate between these two extremes. Because of the bimodal distribution of the human population into rapid and slow acetylators, there appears to be significant individual variations in therapeutic and toxicological responses to drugs displaying acetylation polymorphism.[316,352,353] It appears that slow acetylators are more likely to develop adverse reactions, while rapid acetylators seem more likely to show an inadequate therapeutic response to standard doses of the drug. The antituberculosis drug isoniazid illustrates many of these points. The plasma half-life of isoniazid

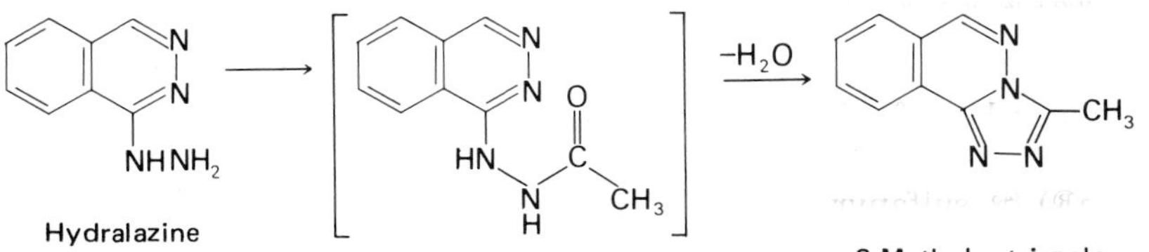

Hydralazine

N-Acetylhydralazine

3-Methyl-s-triazolo-
[3,4-a]phthalazine

in rapid acetylators ranges from 45 to 80 minutes, whereas in slow acetylators the half-life is about 140 to 200 minutes.[354] Thus, for a given fixed dosing regimen, slow acetylators tend to accumulate higher plasma concentrations of isoniazid than do rapid acetylators. Higher concentrations of isoniazid may explain the greater therapeutic response (i.e., higher cure rate) among slow acetylators, but they probably also account for the greater incidence of adverse effects (e.g., peripheral neuritis and drug-induced systemic lupus erythematosus syndrome) observed among slow acetylators.[353] Slow acetylators of isoniazid also appear to be more susceptible to certain drug interactions involving drug metabolism. For example, phenytoin toxicity associated with concomitant use with isoniazid appears to be more prevalent in slow acetylators than in rapid acetylators.[355] Isoniazid inhibits the metabolism of phenytoin, thus leading to an accumulation of high and toxic plasma levels of phenytoin.

Interestingly, patients who are rapid acetylators appear to be more likely to develop isoniazid associated hepatitis.[342] This liver toxicity presumably arises from initial hydrolysis of the N-acetylated metabolite N-acetylisoniazid to acetylhydrazine. The latter metabolite is further converted (by cytochrome P-450 enzyme systems) to chemically reactive acylating intermediates that covalently bind to hepatic tissue, causing necrosis. Pathological and biochemical studies in experimental animals appear to support this hypothesis. Therefore, rapid acetylators run a greater risk of incurring liver injury by virtue of producing more acetylhydrazine.

It appears that the tendency of drugs such as hydralazine and procainamide to cause lupus erythematosus syndrome and to elicit formation of antinuclear antibodies (ANA) is related to acetylator phenotype with greater prevalence in slow acetylators.[356] Rapid acetylation may be preventing the immunological trig-

gering of ANA formation and the lupus syndrome. It is of interest to note that the N-acetylated metabolite of procainamide has recently been shown to be as active an antiarrhythmic agent as the parent drug[341] and has a half-life twice as long in man.[357] These findings indicate that N-acetylprocainamide may be a promising alternative to procainamide as an antiarrhythmic agent with decreased lupus-inducing potential.

METHYLATION

Methylation reactions play an important role in the biosynthesis of many endogenous compounds (e.g., epinephrine and melatonin) and in the inactivation of numerous physiologically active biogenic amines (e.g., norepinephrine, dopamine, serotonin, and histamine).[358] However, methylation constitutes only a minor pathway for conjugating drugs and xenobiotics. Methylation generally does not lead to polar or water-soluble metabolites except when it creates a quaternary ammonium derivative. Most methylated products tend to be pharmacologically inactive, although there are a few exceptions.

The coenzyme involved in methylation reactions is S-adenosylmethionine (SAM). The transfer of the activated methyl group from this coenzyme to the acceptor substrate is catalyzed by various cytoplasmic and microsomal methyltransferases (Fig. 3-17).[358,359] Methyltransferases having particular importance in the metabolism of foreign compounds include catechol-O-methyltransferase (COMT), phenol-O-methyltransferase, and non-specific N-methyltransferases and S-methyltransferases.[358] One of these enzymes, COMT, should be familiar, since it carries out O-methylation of such important neurotransmitters as norepinephrine and dopamine, the consequence of which is termination of their activity. Besides being

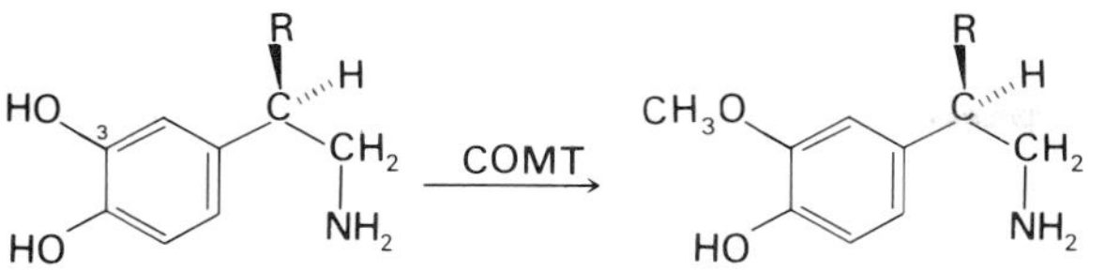

Norepinephrine, R=OH
Dopamine, R=H

Normetanephrine, R=OH
3-Methoxytyramine, R=H

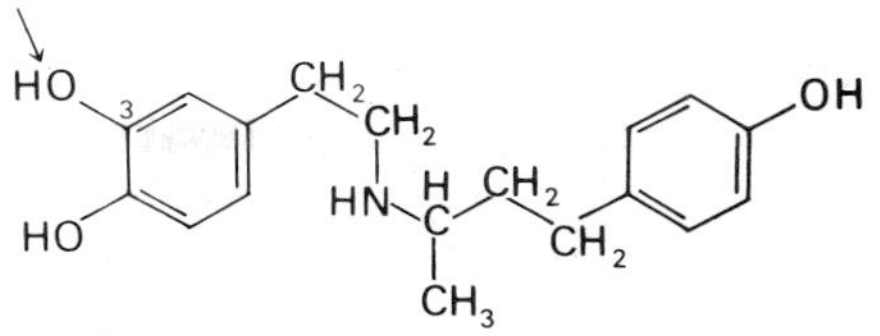

Dobutamine

present in the CNS and peripheral nerves, COMT is widely distributed in other mammalian tissue, particularly the liver and kidney. The other methyltransferases mentioned above are primarily located in the liver, kidney, or lungs. Transferases that specifically methylate histamine, serotonin, and epinephrine usually are not involved in the metabolism of xenobiotics.[358]

Foreign compounds undergoing methylation include catechols, phenols, amines, N-heterocyclic, and thiol compounds. Catechol and catecholamine-like drugs are metabolized by COMT to inactive monomethylated catechol products. Examples of drugs undergoing significant O-methylation by COMT in man include the antihypertensive S(-)α-methyldopa (Aldomet®),[360] the antiparkinsonism agent S(-)-dopa (Levo-

dopa®),[361] isoproterenol (Isuprel®),[362] and dobutamine (Dobutrex®).[363] The student should note the marked structural similarities between these drugs and the endogenous catecholamines such as norepinephrine and dopamine. In the above four drugs, COMT selectively O-methylates only the phenolic OH at C-3. Bismethylation does not occur. Catechol metabolites arising from aromatic hydroxylation of phenols (e.g., 2-hydroxylation of 17α-ethinylestradiol)[44] and from the arene oxide dihydrodiol-catechol pathway (see section on oxidation of aromatic moieties above, e.g., the catechol metabolite of phenytoin)[364] also undergo O-methylation. Substrates undergoing

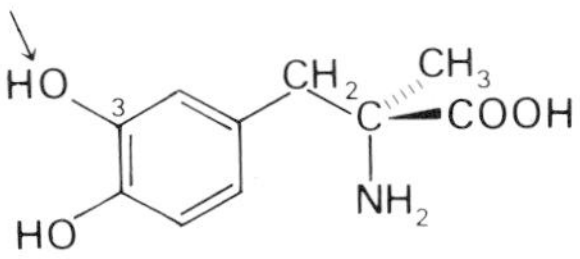

S(-)-α-Methyldopa

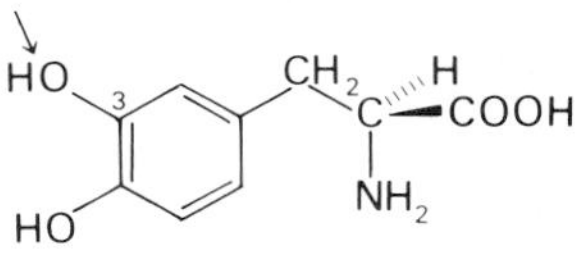

S(-)-Dopa

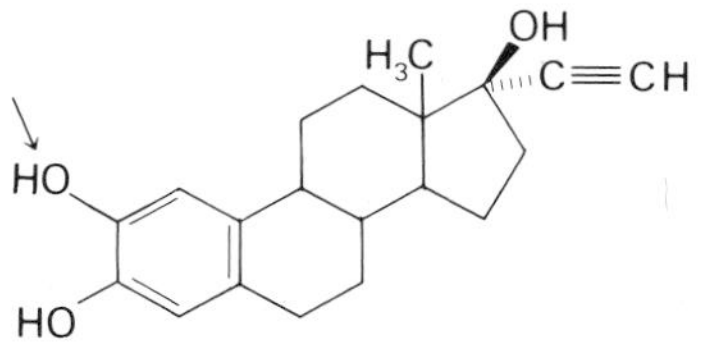

2-Hydroxy-17α-ethinylestradiol

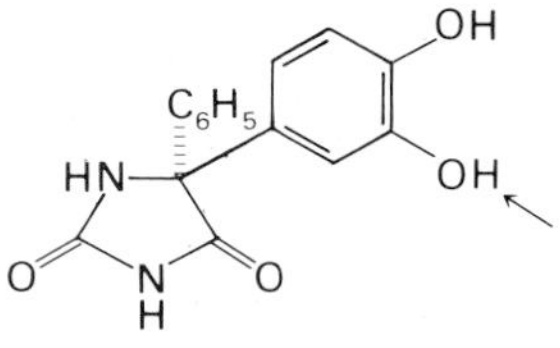

Catechol Metabolite of Phenytoin

Isoproterenol

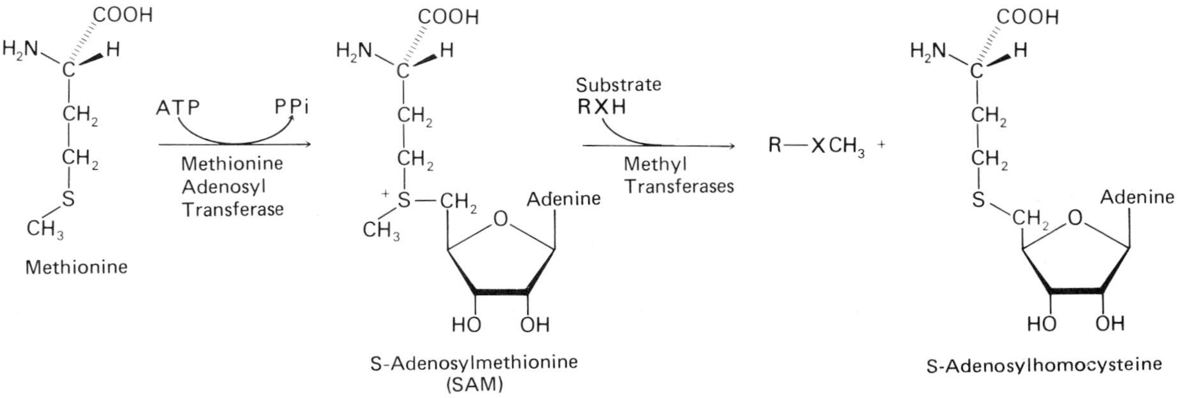

O-methylation by COMT are required to contain an aromatic 1,2-dihydroxy group (i.e., catechol group). Resorcinol (1,3-dihydroxybenzene) or p-hydroquinone (1,4-dihydroxybenzene) derivatives are not substrates for COMT. This would explain why isoproterenol undergoes extensive O-methylation[362] but terbutaline (which contains a resorcinol moiety) does not.[308]

Methionine

S-Adenosylmethionine (SAM)

S-Adenosylhomocysteine

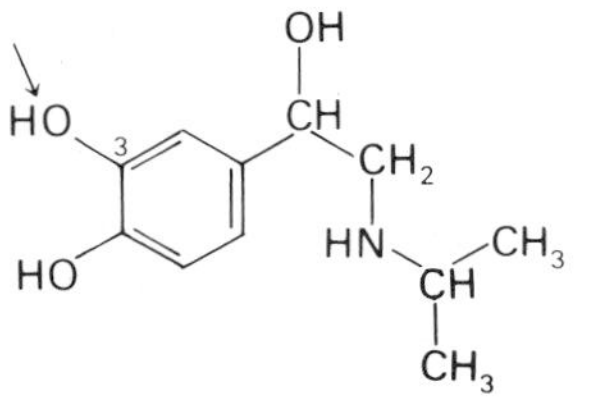

FIG. 3-17. Conjugation of exogenous and endogenous substrates (RXH) by methylation.

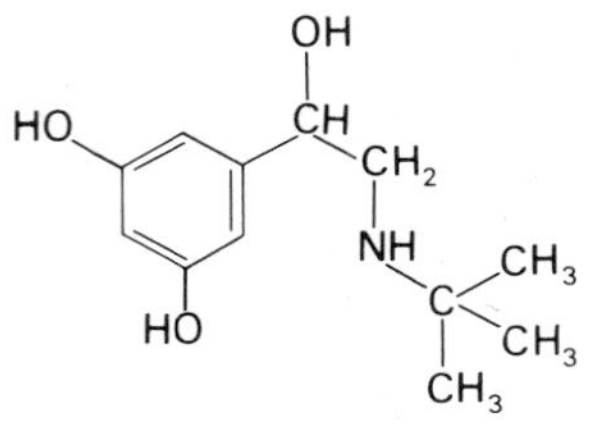

Terbutaline
(not a substrate for COMT)

Phenols have also been occasionally reported to undergo O-methylation, but only to a minor extent.[358] One interesting example involves the conversion of morphine to its O-methylated derivative, codeine, in man. This metabolite is formed in significant amounts in tolerant subjects and may account for up to 10 percent of the morphine dose.[365]

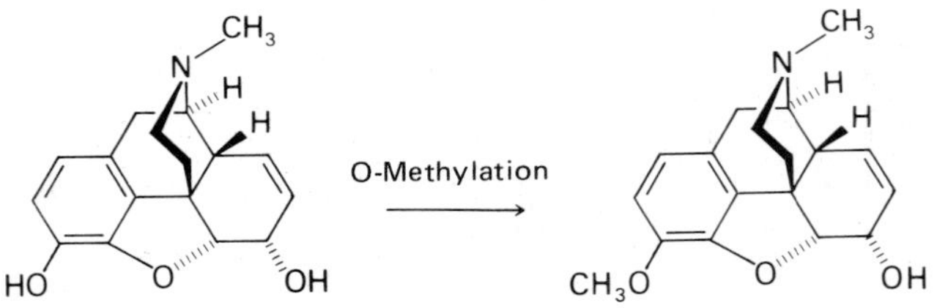

Morphine —O-Methylation→ Codeine

Although N-methylation of endogenous amines (e.g., histamine, norepinephrine) occurs commonly, biotransformation of nitrogen containing xenobiotics to N-methylated metabolites occurs only to a limited extent. Some examples reported include the N-methylation of the antiviral and antiparkinsonism agent amantadine (Symmetrel®) in dogs[366] and the in vitro

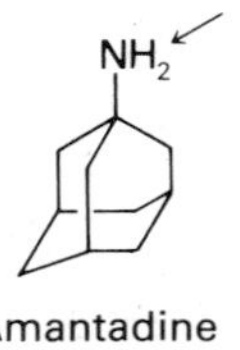

Amantadine 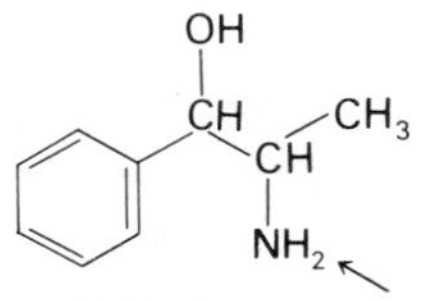   Norephedrine

N-methylation of norephedrine using rabbit lung preparations.[358] N-Methylation of nitrogen atoms present in heterocyclic compounds (e.g., pyridine derivatives) also takes place. For example, the pyridinyl nitrogens of nicotine[147] and nicotinic acid[367] are N-methylated to yield quaternary ammonium products.

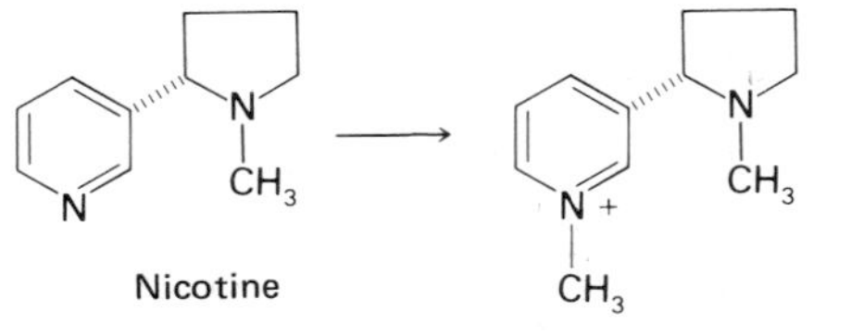

Nicotine

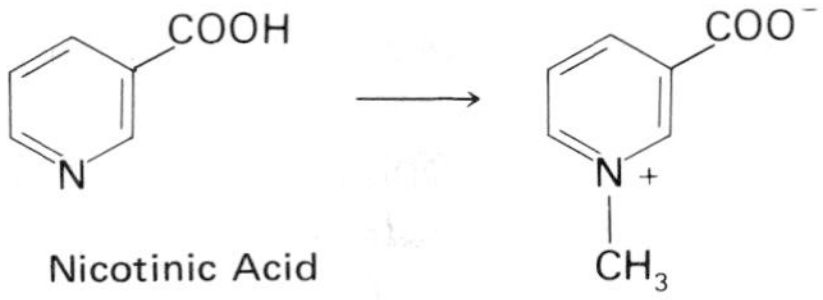

Nicotinic Acid   Trigonelline

Thiol containing drugs such as propylthiouracil,[368] 2,3-dimercapto-1-propanol (BAL),[369] and 6-mercaptopurine[370] have also been reported to undergo S-methylation.

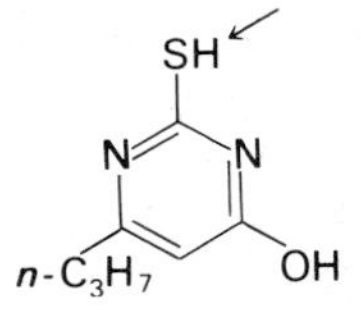

Propylthiouracil 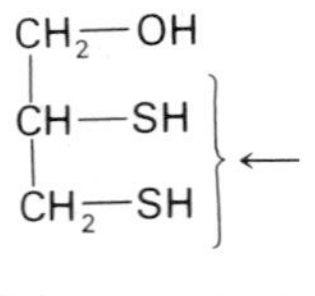   2,3-Dimercapto-1-Propanol (BAL) 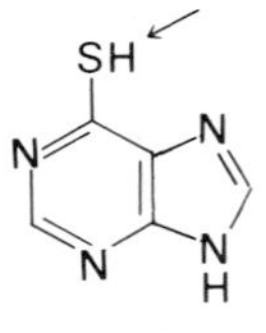   6-Mercaptopurine

FACTORS AFFECTING DRUG METABOLISM

Drugs and xenobiotics often are metabolized by several different phase I and phase II pathways to give a number of metabolites. The relative amount of any particular metabolite is determined by the concentration and activity of the enzyme(s) responsible for the biotransformation. The rate of metabolism of a drug is particularly important with regard to pharmacological action as well as to toxicity. For example, if the rate of metabolism of a drug is decreased, this generally leads to an increase in the intensity and duration of the drug. In addition, decreased metabolic elimination may also lead to accumulation of toxic levels of the drug. Conversely, an increase in the rate of metabolism leads to decreases in intensity and duration of action as well as to decreased efficacy. A number of factors may affect drug metabolism and will be discussed below. These include age, species and strain, genetic or hereditary factors, sex, enzyme induction, and enzyme inhibition.[29, 371]

AGE DIFFERENCES

Age related differences in drug metabolism are generally quite apparent in the newborn.[372,373] In most fetal and newborn animals, undeveloped or deficient oxidative and conjugative enzymes are chiefly responsible for the reduced metabolic capability seen. In general, the ability to carry out metabolic reactions increases rapidly after birth and approaches adult levels in about one to two months. An illustration of the influence of age on drug metabolism is seen in the duration of action (sleep time) of hexobarbital in newborn and adult mice.[374] When given a dose of 10 mg./kg. body weight, the newborn mouse sleeps more than 6 hours. In contrast, the adult mouse sleeps for less than 5 minutes when given the same dose.

In humans, oxidative and conjugative (e.g., glucuronidation) capabilities of newborns are also low compared with adults. For example, the oxidative (cytochrome P-450) metabolism of tolbutamide appears to be markedly decreased in newborns.[375] In comparison to the half-life of 8 hours in adults, the plasma half-life of tolbutamide in infants is greater than 40 hours. As discussed earlier, infants possess poor glucuronidating ability due to a deficiency in glucuronyltransferase activity. The inability of infants to conjugate chloramphenicol with glucuronic acid appears to be responsible for the accumulation of toxic levels of this antibiotic, resulting in the so-called "gray baby syndrome."[302] Similarly, neonatal hyperbilirubinemia (or kernicterus) results from the inability of newborn babies to glucuronidate bilirubin.[301]

The effect of old age on drug metabolism has not been as well studied as in the young. There is some evidence in animals and in man that suggests that drug metabolism diminishes with old age.[376] However, much of the evidence is based on prolonged plasma half-life of drugs that are totally or mainly metabolized by hepatic microsomal enzymes (e.g., antipyrine, phenobarbital, acetaminophen). At this time, the quantitative importance of old age on drug metabolism is not known.

SPECIES AND STRAIN DIFFERENCES

The metabolism of many drugs and foreign compounds is often species-dependent. Different animal species may biotransform a particular xenobiotic by similar or markedly different metabolic pathways. Even within the same species, there may be individual variations (strain differences) that may result in significant differences in a specific metabolic pathway.[377,378]

Species variation has been observed in many oxidative biotransformation reactions. For example, metabolism of amphetamine occurs by two main pathways, oxidative deamination or aromatic hydroxylation. In man, rabbit and guinea pig, oxidative deamination appears to be the predominant pathway, while in the rat aromatic hydroxylation appears to be the more important route.[379] Phenytoin is another drug example showing marked species difference in

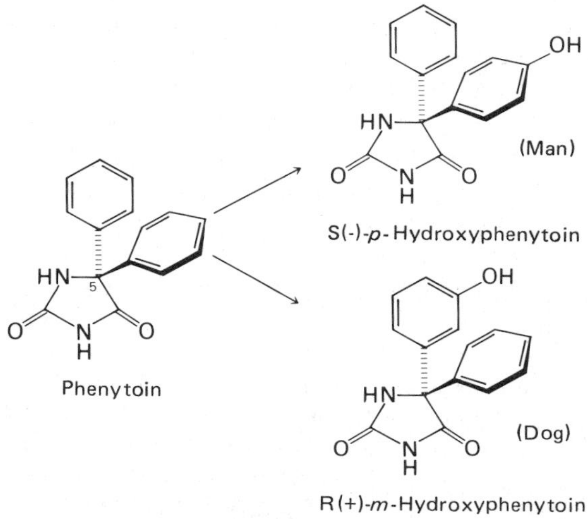

Phenytoin

S(-)-*p*-Hydroxyphenytoin (Man)

R(+)-*m*-Hydroxyphenytoin (Dog)

metabolism. In man, phenytoin undergoes aromatic oxidation to yield primarily S(-)-*p*-hydroxyphenytoin, whereas in dogs, oxidation occurs mainly to give R(+)-*m*-hydroxyphenytoin.[380] There is not only a dramatic difference in the position (i.e., meta or para) of aromatic hydroxylation but also a pronounced difference in which of the two phenyl rings (at C-5 of phenytoin) undergoes aromatic oxidation.

Species differences in many conjugation reactions have also been observed. Often these differences are caused by the presence or absence of transferase en-

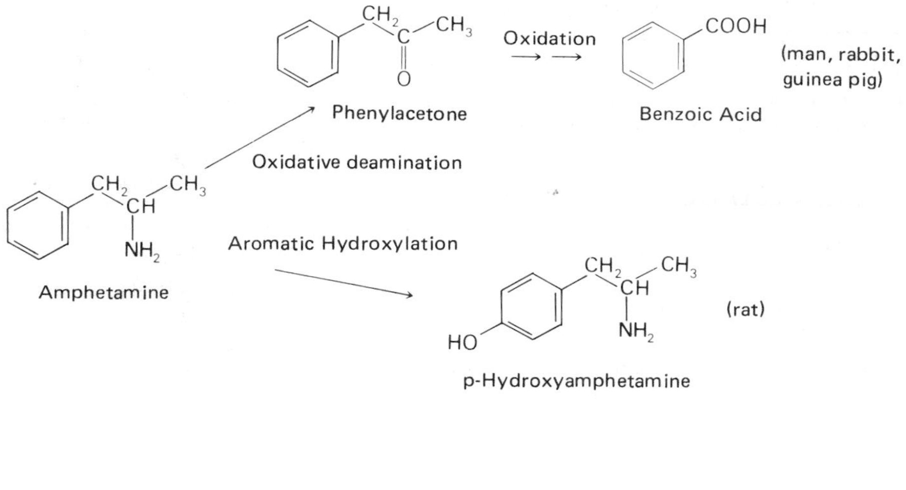

Phenylacetone

Benzoic Acid (man, rabbit, guinea pig)

Amphetamine

Oxidative deamination

Aromatic Hydroxylation

p-Hydroxyamphetamine (rat)

zymes involved in the conjugative process. For example, cats lack glucuronyltransferase enzymes and therefore tend to conjugate phenolic xenobiotics by sulfation instead.[381] In pigs, the situation is reversed. Pigs are not able to conjugate phenols with sulfate (owing to lack of sulfotransferase enzymes) but appear to have good glucuronidation capability.[381] The conjugation of aromatic acids with amino acids (e.g., glycine, glutamine) has been noted to be dependent on the animal species as well as on the substrate. For example, glycine conjugation is a common conjugation pathway for benzoic acid in many animals. However, in certain birds (e.g., duck, goose, turkey), glycine is replaced by the amino acid ornithine.[382] Phenylacetic acid is a substrate for both glycine and glutamine conjugation in man and other primates. However, in nonprimates such as the rabbit and rat, phenylacetic acid is excreted only as the glycine conjugate.[383]

Strain differences in drug metabolism particularly in inbred mice and rabbits have been noted. These differences appear to be caused by genetic variations in the amount of metabolizing enzyme present among the different strains. For example, in vitro studies indicate that Cottontail rabbit liver microsomes metabolize hexobarbital about ten times more rapidly than New Zealand rabbit liver microsomes.[384] In man, interindividual differences in drug metabolism will be considered under hereditary or genetic factors.

HEREDITARY OR GENETIC FACTORS

Marked individual differences in the metabolism of a number of drugs exist in man.[354] It appears that genetic or hereditary factors are mainly responsible for the large differences seen in the rate of metabolism of these drugs. One frequently cited example that dramatically illustrates the influence of genetic factors on drug metabolism concerns the biotransformation of the antituberculosis agent isoniazid or isonicotinic acid hydrazide (INH). Metabolism of isoniazid occurs mainly by N-acetylation.[342] Studies indicate that individuals differ markedly in their ability to acetylate the drug, either slowly or rapidly. Rapid acetylators appear to have more hepatic N-acetyltransferase enzymes than do slow acetylators. In addition, the level of N-acetyltransferase has been shown to be genetically determined and is transmitted as an autosomal recessive trait in man. As discussed earlier in the section on acetylation, the proportion of rapid and slow acetylators varies widely among different ethnic groups. For instance, a high proportion (90%) of Eskimos and Orientals are rapid acetylators, whereas Egyptians and Mediterranean Jews are mainly slow acetylators.[353]

The rate at which isoniazid is acetylated is clinically significant in terms of therapeutic response and toxicity. In general, it seems that rapid acetylators are more likely to show an inadequate therapeutic response (lower cure rate against tuberculosis) while slow acetylators are more likely to develop a greater incidence of adverse effects (e.g., peripheral neuritis, lupus erythematosus syndrome).[353] Other drugs such as hydralazine, procainamide, and dapsone also show similar bimodal distribution in the rate at which they are acetylated.[353]

Genetic factors also appear to influence the rate of oxidation of drugs like phenytoin, phenylbutazone, dicumarol and nortriptyline.[385,386] The rate of oxidation of these drugs varies widely among different individuals; however, these differences do not appear to be distributed bimodally, as in the case of acetylation. In general, individuals who tend to oxidize one drug rapidly are also likely to oxidize other drugs rapidly. Numerous studies in twins (identical and fraternal) and in families tend to indicate that oxidation of these drugs is under genetic control.[386]

SEX DIFFERENCES

The rate of metabolism of xenobiotics also varies according to sex in some animal species. For example, marked difference is observed in female and male rats. Adult male rats metabolize a number of foreign compounds at a much faster rate than female rats (e.g., N-demethylation of aminopyrine, hexobarbital oxidation, glucuronidation of o-aminophenol). It appears that this sex difference is also dependent on substrate, since some xenobiotics are metabolized at the same rate in both female and male rats. Differences in microsomal oxidation have been shown to be under the control of sex hormones, particularly androgens. The anabolic action of androgens appears to increase metabolism.[387]

Sex differences in drug metabolism appear to be species-dependent. Rabbits and mice, for example, do not show significant sex difference in drug metabolism.[387] In man, there have been a few reports of sex difference in metabolism. For instance, nicotine and aspirin appear to be metabolized differently in women and men.[388]

ENZYME INDUCTION

The activity of hepatic microsomal enzymes such as the cytochrome P-450 mixed function oxidase system can be markedly increased upon exposure to a number of drugs, pesticides, polycyclic aromatic hydrocarbons, and environmental xenobiotics. The process by which the activity of these drug metabolizing enzymes is increased is referred to as enzyme induction.[389-391] The increase in activity appears to be caused by an increase in the amount of newly synthesized enzyme. Enzyme induction often leads to an increase in the rate of drug metabolism and to a decrease in the duration of drug action.

Inducing agents may increase the rate of their own metabolism as well as those of other unrelated drugs or foreign compounds (see Table 3-1).[29] Concomitant administration of two or more drugs may often lead to serious drug interactions as a result of enzyme induction. For instance, a clinically significant drug interaction occurs with phenobarbital and warfarin.[392] Induction of microsomal enzymes by phenobarbital causes an increase in the metabolism of warfarin and consequently a marked decrease in the anticoagulant effect. Therefore, if a patient is on warfarin anticoagulant therapy and begins taking phenobarbital, careful attention must be paid to readjustment of the warfarin dose. Dosage readjustment must also be made if a patient on both warfarin and phenobarbital therapy suddenly stops taking the barbiturate. The ineffectiveness of oral contraceptives in women on concurrent phenobarbital or rifampin therapy has been attributed to the enhanced metabolism of estrogens (e.g., 17α-ethinylestradiol) caused by phenobarbital[393] and rifampin[394] induction.

Inducers of microsomal enzymes may also enhance the metabolism of endogenous compounds such as steroidal hormones and bilirubin. For example, phenobarbital has been observed to increase the metabolism of cortisol, testosterone, vitamin D, and bilirubin in man.[389] The enhanced metabolism of vitamin D_3 induced by phenobarbital and phenytoin appears to be responsible for the osteomalacia seen in patients on long-term use of these two anticonvulsant drugs.[395] Interestingly, phenobarbital causes the induction of glucuronyltransferase enzymes, thereby enhancing the conjugation of bilirubin with glucuronic acid. Phenobarbital has been used in some cases to treat hyperbilirubinemia in neonates.[396]

In addition to drugs, other chemicals such as polycyclic aromatic hydrocarbons (e.g., benzo[a]pyrene, 3-methylcholanthrene) and environmental pollutants (e.g., pesticides, polychlorinated biphenyls, TCDD) also may induce certain oxidative pathways and thereby alter drug response.[389,391] Cigarette smoke contains minute amounts of polycyclic aromatic hydrocarbons such as benzo[a]pyrene, which are potent inducers of microsomal cytochrome P-450 enzymes. This induction has been noted to increase the oxidation of some drugs in smokers. For example, theophylline is metabolized more rapidly in smokers than in nonsmokers. This difference is reflected in the marked difference in the plasma half-life of theophylline between smokers ($t_{1/2}$ 4.1 hrs) and nonsmokers ($t_{1/2}$ 7.2 hrs).[397] Other drugs, such as phenacetin, pentazocine, and propoxyphene, have also been reported to undergo more rapid metabolism in smokers than in nonsmokers.[398]

Occupational and accidental exposures to chlorinated pesticides and insecticides have also been shown to stimulate drug metabolism. For instance the half-life of antipyrine in workmen occupationally exposed to the insecticides lindane and DDT has been reported to be significantly shorter (7.7 hrs versus 11.7 hrs) in contrast to control subjects.[399] A case has also been reported in which a worker exposed to chlorinated insecticides showed a lack of response (i.e., decreased anticoagulant effect) to a therapeutic dose of warfarin.[400]

Multiple forms of cytochrome P-450 have been demonstrated to exist.[28] Many chemicals have been shown to selectively induce one or more distinct forms of cytochrome P-450.[28] These inducers fall into two categories: "phenobarbital-like" inducers or "polycyclic aromatic hydrocarbon-like" inducers (e.g., benzo[a]pyrene, 3-methylcholanthrene). Phenobarbital-like compounds generally induce one or more forms of cytochrome P-450, in which the spectral maximum of the reduced cytochrome P-450 carbon monoxide complex occurs at 450 nm. In contrast, polycyclic aromatic hydrocarbon-like chemicals induce a different form of cytochrome P-450 in which the reduced cytochrome P-450-carbon monoxide complex occurs at 448 nm.[25,28] This distinct enzyme form is often referred to as cytochrome P-448. Xenobiotics such as benzo[a]pyrene, 3-methylcholanthrene and TCDD have been shown to

TABLE 3-1

DRUGS THAT INDUCE METABOLISM IN MAN

Inducing Agent	Enhances Metabolism of:
Phenobarbital and other barbiturates	Coumarin anticoagulants, phenytoin, cortisol, testosterone, bilirubin, vitamin D, acetaminophen, oral contraceptives
Glutethimide	Glutethimide, warfarin
Phenylbutazone	Aminopyrine, cortisol
Meprobamate	Meprobamate
Ethanol	Pentobarbital, tolbutamide
Phenytoin	Cortisol, nortriptyline, oral contraceptives
Rifampin	Rifampin, hexobarbital, tolbutamide, coumarin anticoagulants, oral contraceptives, methadone, digitoxin, cortisol
Griseofulvin	Warfarin
Carbamazepine	Carbamazepine, warfarin, phenytoin

(From Nelson, S.D., *in* Wolff, M.E. (*ed.*): Burger's Medicinal Chemistry, ed. 4, part I, p. 227, New York, Wiley-Interscience, 1980. Reprinted with permission.)

induce cytochrome P-448.[389] Cytochrome P-448 is particularly interesting in that it shows a greater selectivity for the oxidation of polycyclic aromatic hydrocarbons.

Enzyme induction may also affect toxicity of some drugs by enhancing the metabolic formation of chemically reactive metabolities. Particularly important is the induction of cytochrome P-450 enzymes involved in the oxidation of drugs to reactive intermediates. For example, the oxidation of acetaminophen to a reactive imidoquinone metabolite appears to be carried out by a phenobarbital-inducible form of cytochrome P-450 in rats and mice. Numerous studies in these two animals indicate that phenobarbital pretreatment leads to an increase in vivo hepatotoxicity and covalent binding as well as to an increase in the formation of reactive metabolite in microsomal incubation mixtures.[188,189,191] Induction of cytochrome P-448 is of toxicological concern, since it is now well-established that this particular enzyme is involved in the metabolism of polycyclic aromatic hydrocarbons to reactive and carcinogenic intermediates.[67,401] For example, the metabolic bioactivation of benzo[a]pyrene to its ultimate carcinogenic diol epoxide intermediate is carried out by cytochrome P-448 (see earlier section on aromatic oxidation for the bioactivation pathway of benzo[a]pyrene to its diol epoxide).[401] Thus, it is becoming increasingly apparent that enzyme induction in a number of cases may enhance the toxicity of some xenobiotics by increasing the rate of formation of reactive metabolites.

ENZYME INHIBITION

A number of drugs and xenobiotics are capable of inhibiting drug metabolism.[29,371] With metabolism decreased, drug accumulation often occurs, thus leading to prolonged drug action and serious adverse effects. Enzyme inhibition can occur by a number of mechanisms, including substrate competition, interference with protein synthesis, inactivation of drug metabolizing enzymes, hepatotoxicity leading to impairment of enzyme activity, etc. Some drug interactions resulting from enzyme inhibition have been reported in man.[402] For example, phenylbutazone has been noted to stereoselectively inhibit the metabolism of the more potent S(-)-enantiomer of warfarin. This inhibition may explain the excessive hypoprothrombinemia (increased anticoagulant effect) and many instances of hemorrhaging seen in patients on both warfarin and phenylbutazone therapy.[45] The metabolism of phenytoin has been shown to be inhibited by drugs such as chloramphenicol, disulfiram, and isoniazid.[392] Interestingly, phenytoin toxicity as a result of enzyme inhibi-

tion by isoniazid appears to occur primarily in slow acetylators.[355] Several drugs such as dicumarol, chloramphenicol, and phenylbutazone[392] have been observed to inhibit the biotransformation of tolbutamide, which may lead to a possible hypoglycemic response.

Other compounds, such as SKF-525A, metyrapone, piperonyl butoxide, and cobaltous chloride, have been used experimentally in animals as general inhibitors of miscrosomal enzymes.[29,403]

MISCELLANEOUS FACTORS AFFECTING DRUG METABOLISM[29,371]

Other factors also may influence drug metabolism. Dietary factors such as the protein:carbohydrate ratio in the diet have been found to affect the metabolism of a few drugs. Indoles present in vegetables such as brussel sprouts, cabbage, and cauliflower and polycyclic aromatic hydrocarbons present in charcoal-broiled beef have been shown to cause enzyme induction and to stimulate the metabolism of some drugs. Vitamins, minerals, starvation, and malnutrition also appear to have an influence on drug metabolism. Finally, physiological factors such as the pathological state of the liver (e.g., hepatic cancer, cirrhosis, hepatitis), pregnancy, hormonal disturbances (e.g., thyroxine, steroids), and circadian rhythm may also markedly affect drug metabolism.

STEREOCHEMICAL ASPECTS OF DRUG METABOLISM

Many drugs (e.g., warfarin, propranolol, hexobarbital, glutethimide, cyclophosphamide, ketamine, and ibuprofen) are often administered as racemic mixtures in man. The two enantiomers present in a racemic mixture may differ from one another in pharmacological activity. In most cases, one enantiomer tends to be much more active than the other. For example, the S(-)-enantiomer of warfarin is five times more potent as an oral anticoagulant than is the R(+)-enantiomer.[404] In some instances, the two enantiomers may have totally different pharmacological activities. For example, (+)-α-propoxyphene (Darvon®) is an analgesic, whereas (-)-α-propoxyphene (Novrad®) is an antitussive.[405] Such differences in activity between stereoisomers should not be surprising, since in Chapter 2 we learned that stereochemical factors generally have a dramatic influence on how the drug molecule interacts with the target receptors to elicit its pharmacological response. By the same token, the preferential interaction of one stereoisomer with drug metabolizing enzymes may lead one to anticipate differences in metabolism for the two enantiomers of a racemic mixture. Indeed, it is well established that individual enantiomers of a racemic drug are often metabolized at different rates. For example, studies in man indicate that the less active (+)-enantiomer of proproanolol undergoes more rapid metabolism than the corresponding (-)-enantiomer.[396] Allylic hydroxylation of hexobarbital has been observed to occur more rapidly with the R(-)-enantiomer in man.[407] The term *substrate stereoselectivity* is frequently used to denote the preference of one stereoisomer as a substrate for a metabolizing enzyme or metabolic process.[223]

Individual enantiomers of a racemic mixture may also be metabolized by different pathways. For instance, in dogs the (+)-enantiomer of the sedative hypnotic glutethimide (Doriden®) is primarily hy-

Warfarin

R (+)-Enantiomer

R, S(+)-Alcohol

S(-)-

7-Hydroxywarfarin

CH₂CH₃
C₆H₅

Glutethimide

4-Hydroxyglutethimide
(+)-Enantiomer

(-)-Enantiomer

droxylated alpha to the carbonyl to yield 4-hydroxyglutethimide, whereas the (-)-enantiomer undergoes aliphatic $\omega-1$ hydroxylation of its C-2 ethyl group.[113] Dramatic differences in the metabolic profile of two enantiomers of warfarin have also been noted. In man, the more active S(-)-isomer is 7-hydroxylated (aromatic hydroxylation), whereas the R(+)-isomer undergoes keto reduction to yield primarily the RS warfarin alcohol as the major plasma metabolite.[45, 234] While a large number of other examples of substrate stereoselectivity or enantioselectivity in drug metabo-

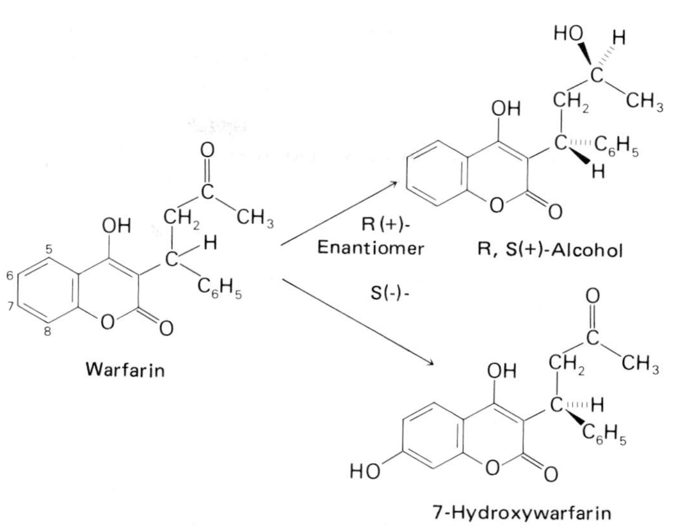

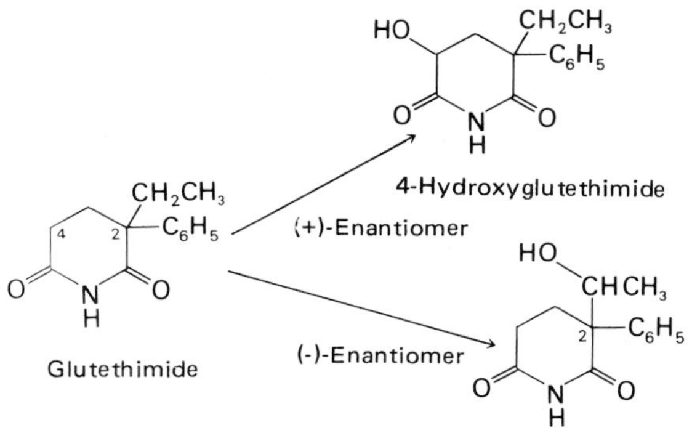

lism exist, the cases presented above will suffice to emphasize the point.[223,408]

Drug biotransformation processes often lead to the creation of a new asymmetric center in the metabolite (i.e., stereoisomeric or enantiomeric products). The preferential metabolic formation of a stereoisomeric product is called *product stereoselectivity*.[223] For example, bioreduction of ketone xenobiotics as a general rule produces predominantly one stereoisomeric alcohol (see section on reduction of keto carbonyls).[91, 223] The preferential formation of S(-)-hydroxyhexamide from the hypoglycemic agent acetohexamide[225] and the exclusive generation of 6β-naltrexol from naltrexone[231] (see section on reduction of keto carbonyls for structure) are two examples of highly stereoselective bioreduction processes in man.

Oxidative biotransformations have been noted to display product stereoselectivity, too. For example, phenytoin contains two phenyl rings in its structure both of which *a priori* should be susceptible to aromatic hydroxylation. However, in man, *p*-hydroxylation occurs preferentially (approximately 90%) at the pro-S phenyl ring to give primarily S(-)-5-(4-hydroxyphenyl)-5-phenylhydantoin. Although the other phenyl ring is also *p*-hydroxylated, it occurs to only to a minor extent (10%).[380] Microsomal hydroxylation of

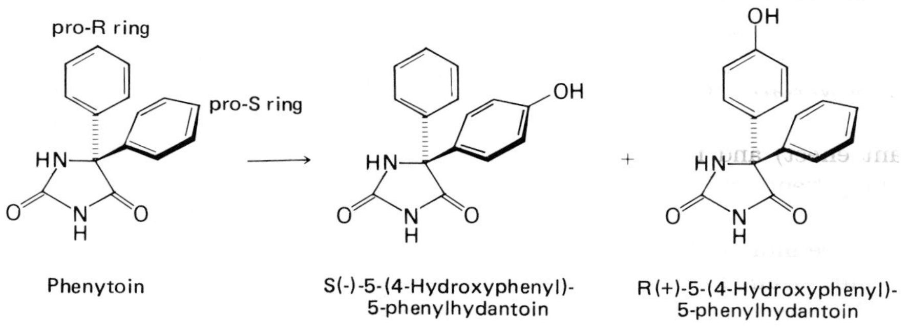

pro-R ring

pro-S ring

Phenytoin

S(-)-5-(4-Hydroxyphenyl)-5-phenylhydantoin

R(+)-5-(4-Hydroxyphenyl)-5-phenylhydantoin

the C-3 carbon of diazepam and desmethyldiazepam (using mouse liver preparations) has been reported to proceed with remarkable stereoselectivity to yield optically active metabolites having the 3S absolute configuration.[112] Interestingly, these two metabolites are also pharmacologically active and one of them,

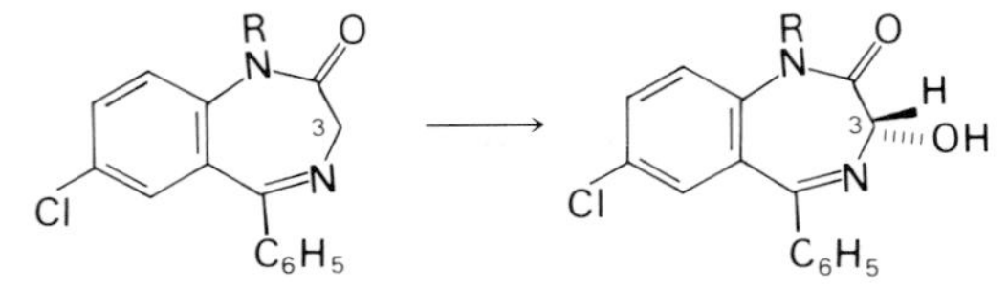

Diazepam, R=CH$_3$
Desmethyldiazepam, R=H

(3S) N-Methyloxazepam, R=CH$_3$
S(+)-Oxazepam, R=H

oxazepam, is marketed as a drug (Serax®). The allylic hydroxylation of the N-butenyl side group of the analgesic pentazocine (Talwin®) leads to two possible alcohols (*cis* and *trans* alcohols). In man, mouse, and monkey, pentazocine is metabolized predominantly to the *trans* alcohol metabolite, whereas the rat primarily tends to form the *cis* alcohol.[108] The product stereoselectivity observed in this biotransformation involves *cis* and *trans* geometric stereoisomers.

More recently, the term *regioselectivity*[409] has been introduced in drug metabolism to denote the selective metabolism of two or more similar functional groups (e.g., OCH$_3$, OH, NO$_2$) or two or more similar atoms which are positioned in different regions of a molecule. The examples described below should make this concept clear. For example, of the four methoxy groups present in papaverine, the 4'-OCH$_3$ group is regioselectively O-demethylated in several species (e.g., rat,

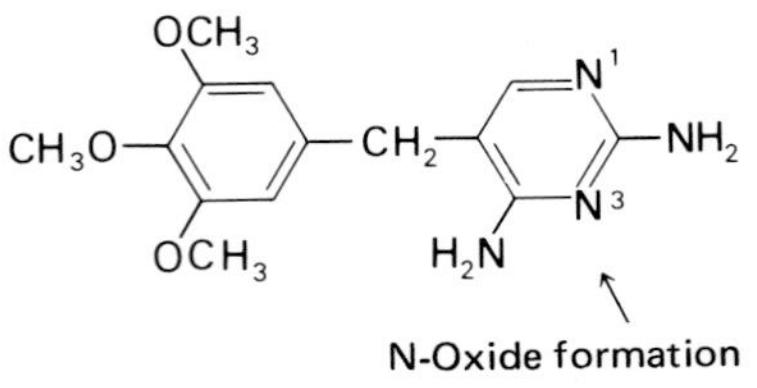

N-Oxide formation

Trimethoprim

guinea pig, rabbit, and dog).[410] Trimethoprim (Trimpex®, Proloprim®) has two heterocyclic sp^2 nitrogen atoms (N^1 and N^3) in its structure. In dogs, it appears that oxidation occurs regioselectively at N^3 to give the corresponding 3-N-oxide.[178] Nitroreduction of the 7-nitro group in 5,7-dinitroindazole to yield the 7-amino derivative in the mouse and rat has been shown to occur with a high degree of regioselectivity.[287] Substrates amenable to O-methylation by COMT appear to proceed with remarkable regioselectivity, as typified by the cardiotonic agent dobutamine (Dobutrex®). O-Methylation occurs exclusively with the phenolic hydroxy group at C-3.[363]

5,7-Dinitroindazole

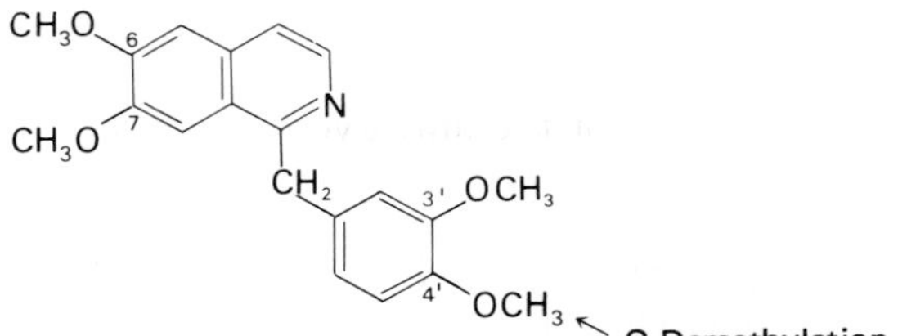

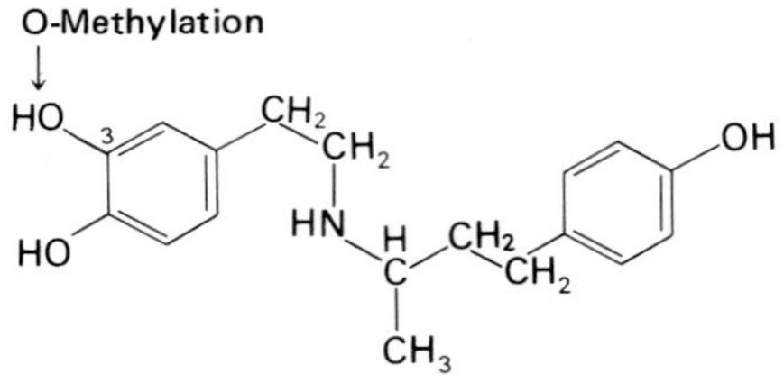

Dobutamine

Papaverine

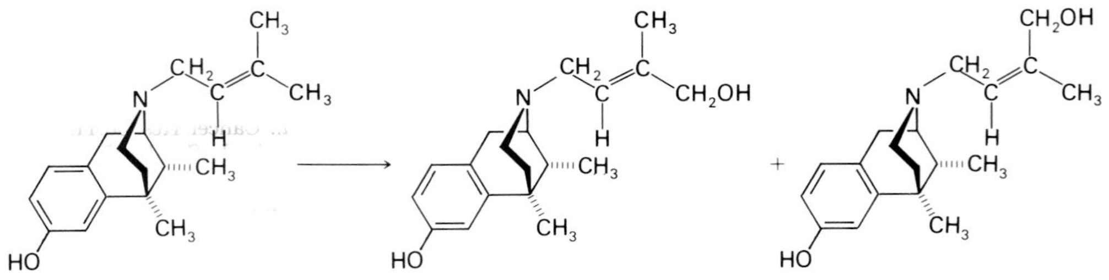

Pentazocine   *trans*-Alcohol   *cis*-Alcohol

PHARMACOLOGICALLY ACTIVE METABOLITES

The traditional notion that drug metabolites are inactive and are insignificant in drug therapy has dramatically changed in recent years. There is now increasing evidence to indicate that many drugs are biotransformed to pharmacologically active metabolites that contribute to the therapeutic as well as toxic effects of the parent compound. Metabolites that have been shown to have significant therapeutic activity in humans are listed in Table 3-2.[2,411] The parent drug from which the metabolite is derived and the biotransformation process involved are also given.

How significantly an active metabolite contributes to the therapeutic or toxic effects ascribed to the parent drug will be dependent on its relative activity and its quantitative importance (e.g., plasma concentration). In addition, whether the metabolite accumulates after repeated administration (e.g., desmethyldiazepam in geriatric patients) or in patients with renal failure are also determinants.

From a clinical standpoint, active metabolites are especially important in patients with decreased renal function. If renal excretion is the major pathway for elimination of the active metabolite, then accumulation is likely to occur in patients with renal failure. Especially in the case of drugs like procainamide, clofibrate and digitoxin, caution should be exercised in treating patients with renal failure.[2,411] Many of the toxic effects seen for these drugs have been attributed to high plasma levels of their active metabolites. For example, the severe muscle weakness and tenderness (myopathy) seen with clofibrate in renal failure patients is believed to be caused by high levels of the active metabolite, chlorophenoxyisobutyric acid.[412] Cardiovascular toxicity due to digitoxin and procainamide in anephric subjects has been attributed to high plasma levels of digoxin and N-acetylprocainamide, respectively. In such situations, appropriate reduction in dosage and careful monitoring of plasma levels of the parent drug and its active metabolite are often recommended.

The pharmacological activity of some metabolites has led many manufacturers to synthesize these metabolites and to market them as separate drug entities. For example, oxyphenbutazone (Tandearil®, Oxalid®) is the *p*-hydroxylated metabolite of the antiinflammatory agent phenylbutazone (Butazolidin®, Azolid®); notriptyline (Aventyl®) is the N-demethylated metabolite of the tricyclic antidepressant amitriptyline (Elavil®); oxazepam (Serax®) is the N-demethylated and 3-hydroxylated metabolite of diazepam (Valium®); mesoridazine (Serentil®) is the sulfoxide metabolite of the antipsychotic agent thioridazine (Mellaril®).

Acknowledgment

The authors would like to kindly thank Professor William F. Trager of the University of Washington for his review of the entire chapter and for his many criticisms and suggestions.

TABLE 3–2

PHARMACOLOGICALLY ACTIVE METABOLITES IN HUMANS

Parent Drug	Metabolite	Biotransformation process
Acetohexamide	Hydroxyhexamide	Ketone reduction
Acetylmethadol	Noracetylmethadol	N-demethylation
Amitriptyline	Nortriptyline	N-demethylation
Azathioprine	6-Mercaptopurine	Glutathione conjugation
Carbamazepine	Carbamazepine-9, 10-epoxide	Epoxidation
Chloral Hydrate	Trichloroethanol	Aldehyde reduction
Clofibrate	Chlorophenoxyisobutyric acid	Ester hydrolysis
Chlorpromazine	7-Hydroxychlorpromazine	Aromatic hydroxylation
Cortisone	Hydrocortisone	Ketone reduction
Diazepam	Desmethyldiazepam and Oxazepam	N-demethylation and 3-hydroxylation
Digitoxin	Digoxin	Alicyclic hydroxylation
Diphenoxylate	Diphenoxylic Acid	Ester hydrolysis
Imipramine	Desipramine	N-demethylation
Mephobarbital	Phenobarbital	N-demethylation
Metoprolol	α-Hydroxymethyl-metoprolol	Benzylic hydroxylation
Phenacetin	Acetaminophen	O-Deethylation
Phenylbutazone	Oxybutazone	Aromatic hydroxylation
Prednisone	Prednisolone	Ketone reduction
Primidone	Phenobarbital	Hydroxylation and oxidation to ketone
Procainamide	N-Acetylprocainamide	N-acetylation
Propranolol	4-Hydroxypropranolol	Aromatic hydroxylation
Quinidine	3-Hydroxyquinidine	Allylic hydroxylation
Sulindac	Sulfide Metabolite of Sulindac	Sulfoxide reduction
Thioridazine	Mesoridazine	S-oxidation
Warfarin	Warfarin Alcohols	Ketone reduction

REFERENCES

1. Williams, R.T.: Detoxication Mechanisms, ed. 2, New York, Wiley, 1959.
2. Drayer, D.E.: Clin. Pharmacokin, 1:426, 1976.
3. Colvin, M., *et al.*: Cancer Res. 36:1121, 1976.
4. Jollow, D.J., Kocsis, J., Snyder, R., and Vainio, H.: Biological Reactive Intermediates, New York, Plenum Press, 1977.
5. Gillette, J.R., *et al.*: Ann. Rev. Pharmacol. 14:271, 1974.
6. Nelson, S.D., *et al.*: in Jerina, D.M. (ed.): Drug Metabolism Concepts, p. 155, Washington, D.C., American Chemical Society, 1977.
7. Testa, B., and Jenner, P.: Drug Metab. Rev. 7:325, 1978.

8. Low, L.K., and Castagnoli, N., Jr.: *in* Wolff, M.E. (ed.): Burger's Medicinal Chemistry, ed. 4, Part 1, p. 107, New York, Wiley-Interscience, 1980.
9. Gill, E.W., *et al.:* Biochem. Pharmacol. 22: 175, 1973.
10. Wall, M.E., *et al.:* J. Am. Chem. Soc. 94: 8579, 1972; Lemberger, L: Drug Metab. Disposit 1:641, 1973.
11. Green, D.E., *et al.: in* Vinson, J.A. (ed.): Cannabinoid Analysis in Physiological Fluids, p. 93, Washington, D.C., American Chemical Society, 1979.
12. Williams, R.T.: *in* Brodie, B.B., and Gillette, J.R. (eds): Concepts in Biochemical Pharmacology, Part 2, p. 226, Berlin, Springer-Verlag, 1971.
13. Rowland, M.: *in* Melmon, K.L., and Morelli, H.F. (eds.): Clinical Pharmacology: Basic Principles in Therapeutics, ed. 2, p. 25, New York, Macmillan, 1978.
14. Benowitz, N.L., and Meister, W.: Clin. Pharmacokin. 3:177, 1978.
15. Connolly, M.E., *et al.:* Br. J. Pharmacol. 46: 458, 1972.
16. Gibaldi, M., and Perrier, D.: Drug Metab. Rev. 3: 185, 1974.
17. Sinkula, A.A., and Yalkowsky, S.H.: J. Pharm. Sci. 64: 181, 1975.
18. Scheline, R.R.: Pharmacol. Rev. 25:451, 1973; Peppercorn, M.A., and Goldman, P.: J. Pharmacol. Exp. Ther. 181:555, 1972.
19. Levvy, G.A., and Conchie, J.: *in* Dutton, G.J. (ed.): Glucuronic Acid, Free and Combined, p. 301, New York, Academic Press, 1966.
20. Testa, B., and Jenner, P.: Drug Metabolism: Chemical and Biochemical Aspects, p. 419, New York, Marcel Dekker, 1976.
21. Powis, G., and Jansson, I.: Pharmacol. Ther. 7:297, 1979.
22. Mason, H.S.: Ann. Rev. Biochem. 34:595, 1965.
23. Hayaishi, O.: *in* Hayaishi, O. (ed.): Oxygenases, p. 1, New York, Academic Press, 1962.
24. Mannering, G.J.: *in* LaDu, B.N., *et al.* (eds.): Fundamentals of Drug Metabolism and Disposition, p. 206, Baltimore, Williams & Wilkins, 1971.
25. Sato, R., and Omura, T. (eds.): Cytochrome P-450, Tokyo, Kodansha, and New York, Academic Press, 1978.
26. Omura, T., and Sato, R.: J. Biol. Chem. 239:2370, 1964.
27. Gillette, J.R.: Adv. Pharmacol. 4:219, 1966.
28. Johnson, E.F.: Rev. Biochem. Toxicol. 1:1. 1979; Guengerich, F.P.: Pharmacol. Ther. 6:99, 1979.
29. Nelson, S.D.: *in* Wolff, M.E. (ed.): Burger's Medicinal Chemistry, ed. 4, Part I, p. 227, New York, Wiley-Interscience, 1980.
30. Claude, A.: *in* Gillette, J.R., *et al.* (eds): Microsomes and Drug Oxidations, p. 3, New York, Academic Press, 1969.
31. Hansch, C.: Drug Metab. Rev. 1:1, 1972.
32. Estabrook, R.W., and Werringloer, J.: *in* Jerina, D.M. (ed.): Drug Metabolism Concepts, p. 1, Washington, D.C., American Chemical Society, 1977.
33. Trager, W.F.: *in* Jenner, P., and Testa, B. (eds.): Concepts in Drug Metabolism, Part A, p. 177, New York, Marcel Dekker, 1980.
34. Ullrich, V.: Topics Current Chem. 83:68, 1979.
35. White, R.E., and Coon, M.J.: Ann. Rev. Biochem. 49:315, 1980.
36. Frommer, U., *et al.:* FEBS Lett. 41:14, 1974.
37. Daly, J.W., *et al.:* Experientia 28:1129, 1972.
38. Jerina, D.M., and Daly, J.W.: Science 185:573, 1974.
39. Walle, T., and Gaffney, T.E.: J. Pharmacol. Exp. Ther. 182:83, 1972; Bond, P.: Nature 213:721, 1967.
40. Whyte, M.P., and Dekaban, A.S.: Drug Metab. Disposit. 5:63, 1977.
41. Witkin, K.M., *et al.:* Ther. Drug Monitoring 1:11, 1979; Richens, A.: Clin. Pharmacokin. 4:153, 1979.
42. Burns, J.J., *et al.:* J. Pharmacol. Exp. Ther. 113:481, 1955; Yü, T.F., *et al.:* J. Pharmacol. Exp. Ther. 123:63, 1958.
43. Murphy, P.J., and Wick, A.N.: J. Pharm. Sci. 57:1125, 1968; Beckmann, R.: Ann. N.Y. Acad. Sci. 148:820, 1968.
44. Williams, M.C., *et al.:* Steroids 25:229, 1975; Ranney, R.E.: J. Toxicol. Environ. Health 3:139, 1977.
45. Lewis, R.J., *et al.:* J. Clin. Invest. 53:1697, 1974.
46. Beckett, A.H., and Rowland, M.: J. Pharm. Pharmacol. 17:628, 1965; Dring, L.G., *et al.:* Biochem. J. 116:425, 1970.
47. Daly, J.: *in* Brodie, B.B., and Gillette, J.R. (eds.): Concepts in Biochemical Pharmacology, Part 2, p. 285, Berlin, Springer-Verlag, 1971.
48. Lowenthal, D.T.: J. Cardiovas. Pharmacol. 2(Suppl. 1):S29, 1980; Davies, D.S., *et al.:* Adv. Pharmacol. Ther. 7:215, 1979.
49. Dayton, P.G., *et al.:* Drug Metab. Disposit. 1:742, 1973.
50. Schreiber, E.E.: Ann. Rev. Pharmacol. 10:77, 1970.
51. Hollister, E.E.: Res. Comm. Chem. Pathol. Pharmacol. 2:330, 1971.
52. Safe, S., *et al.:* J. Agric. Food Chem. 23:851, 1975.
53. Allen, J.R., *et al.:* Food Cosmet. Toxicol. 13:501, 1975; Vinopal, J.H., *et al.:* Arch. Environ. Contamin. Toxicol. 1:122, 1973.
54. Hathway, D.E. (Sr. Reporter): Foreign Compound Metabolism in Mammals, vol. 4, p. 234, London, The Chemical Society, 1977.
55. Guroff, G., *et al.:* Science, 157:1524, 1967.
56. Daly, J., *et al.:* Arch. Biochem. Biophys. 128:517, 1968.
57. Oesch, F.: Progr. Drug Metab. 3:253, 1978; Lu, A.A.H., and Miwa, G.T.: Ann. Rev. Pharmacol. Toxicol. 20:513, 1980.
58. Ames, B.N., *et al.:* Science 176:47, 1972.
59. Maguire, J.H., *et al.:* Ther. Drug Monitoring 1:359, 1979.
60. Harvey, D.G., *et al.:* Res. Comm. Chem. Pathol. Pharmacol. 3:557, 1972.
61. Stillwell, W.G.: Res. Comm. Chem. Pathol. Pharmacol. 12:25, 1975.
62. Jakoby, W.B., *et al.: in* Arias, I.M., and Jakoby, W.B. (eds.): Glutathione, Metabolism and Function, p. 189, New York, Raven Press, 1976.
63. Boyland, E.: *in* Brodie, B.B., and Gillette, J.R. (eds.): Concepts in Biochemical Pharmacology, Part 2, p. 584, Berlin, Springer-Verlag, 1971.
64. Brodie, B.B., *et al.:* Proc. Nat. Acad. Sci. 68:160, 1971.
65. Jollow, D.J., *et al.:* Pharmacology 11:151, 1974.
66. Gelboin, H.V., *et al.: in* Jollow, D.J., *et al.* (eds.): Biological Reactive Intermediates, p. 98, New York, Plenum Press, 1977.
67. Thakker, D.R., *et al.:* Chem.-Biol. Interact. 16:281, 1977.
68. Weinstein, I.B., *et al.:* Science 193:592, 1976; Jeffrey, A.M., *et al.:* J. Am. Chem. Soc. 98:5714, 1976; Koreeda, M., *et al.:* J. Am. Chem. Soc. 98:6720, 1976.
69. Straub, K.M., *et al.:* Proc. Nat. Acad. Sci. 74:5285, 1977.
70. Kasai, H., *et al.:* J. Am. Chem. Soc. 99:8500, 1977.
71. Chausseaud, L.F.: Drug Metab. Rev. 2:185, 1973.
72. Pynnönen, S.: Ther. Drug Monitoring 1:409, 1979.
73. Eadie, M.J., and Tyrer, J.H.: Anticonvulsant Therapy, ed. 2, p. 142, Edinburgh, London, Churchill Livingston, 1980.
74. Hucker, H.B., *et al.:* Drug Metab. Disposit. 3:80, 1975.
75. Hintze, K.L., *et al.:* Drug Metab. Disposit. 3:1, 1975.
76. Slack, J.A., and Ford-Hutchinson, A.W.: Drug Metab. Disposit. 8:84, 1980.
77. Waddell, W.J.: J. Pharmacol. Exp. Ther. 149:23, 1965.
78. Seutter-Berlage, F., *et al.:* Xenobiotica 8:413, 1978.
79. Marniemi, J., *et al.:* Ullrich, V., *et al.* (eds.): Microsomes and Drug Oxidations, p. 698, Oxford, Pergamon Press, 1977.
80. Garner, R.C.: Prog. Drug Metab. 1:77, 1976.
81. Swenson, D.H., *et al.:* Biochem. Biophys. Res. Comm. 60:1036, 1974; *ibid.,* 53:1260, 1973.
82. Essigman, J.M., *et al.:* Proc. Nat. Acad. Sci. 74:1870, 1977; Croy, R.G., *et al.: ibid.,* 75:1745, 1978.
83. Henschler, D., and Bonser, G.: Adv. Pharmacol. Ther. 9:123, 1979.
84. Watabe, T., and Akamatsu, K.: Biochem. Pharmacol. 24:442, 1975.
85. Metzler, M.: J. Toxicol. Environ. Health 1(Suppl.):21, 1976; Neuman, H.G., and Metzler, M.: Adv. Pharmacol. Ther. 9:113, 1979.
86. Unseld, A., and DeMatteis, F.: *in* Doss, M. (ed.): Porphyrins in Human Disease, p. 71, Basel, S. Karger, 1976.
87. DeMatteis, F.: Biochem. J. 124:767, 1971; Levin, W., *et al.:* Arch. Biochem. Biophys. 148:262, 1972.
88. Levin, W., *et al.:* Science 176:1341, 1972; Levin, W., *et al.:* Drug Metab. Disposit. 1:275, 1973.
89. Ivanetich, K.M., *et al.: in* Ullrich, V., *et al.* (eds.): Microsomes and Drug Oxidations, p. 76, Oxford, Pergamon Press, 1977.
90. Ortiz de Montellano, P.R., *et al.:* Biochem. Biophys. Res. Comm. 83:132, 1978.

91. McMahon, R.E.: *in* Brodie, B.B., and Gillette, J.R. (eds.): Concepts in Biochemical Pharmacology, Part 2, p. 500, Berlin, Springer-Verlag, 1971.
92. Dutton, G.: *in* Brodie, B.B., and Gillette, J.R. (eds.): Concepts in Biochemical Pharmacology, Part 2, p. 378, Berlin, Springer-Verlag, 1971.
93. Thomas, R.C., and Ikeda, G.J.: J. Med. Chem. 9:507, 1966.
94. Selley, M.L., *et al.*: Clin. Pharmacol. Ther. 17:599, 1975; Sumner, D.D., *et al.*: Drug Metab. Disposit. 3:283, 1975.
95. Permisohn, R.C., *et al.*: J. Forensic Sci. 21:98, 1976.
96. Borg, K.O., *et al.*: Acta Pharmacol. Toxicol. 36(Suppl. V):125, 1975; Hoffmann, K.J. Clin. Pharmacokin. 5:181, 1980.
97. Ho, B.T., *et al.*: J. Med. Chem. 14:158, 1971; Matin, S., *et al.*: J. Med. Chem. 17:877, 1974.
98. Crammer, J.L., and Scott, B.: Psychopharmacologia 8:461, 1966.
99. Hucker, H.B.: Pharmacologist 4:171, 1962; Kraak, J.C., and Bijster, P.: J. Chromatogr. 143:499, 1977.
100. Harvey, D.J., *et al.*: *in* Frigerio, A., and Ghisalberti, E.L. (eds.): Mass Spectrometry in Drug Metabolism, p. 403, New York, Plenum Press, 1977.
101. Allen, J.G., *et al.*: Drug Metab. Disposit. 3:332, 1975.
102. Sims, P.: Biochem. Pharmacol. 19:795, 1970; Takahashi, G., and Yasuhria, K.: Cancer Res. 32:710, 1972.
103. Lemberger, L., *et al.*: Science 173:72, 1971; *Ibid.*, 177:62, 1972.
104. Carroll, F.I., *et al.*: J. Med. Chem. 17:985, 1974; Drayer, D.E., *et al.*: Clin. Pharmacol. Ther. 27:72, 1980.
105. Drayer, D.E., *et al.*: Clin. Pharmacol. Ther. 24:31, 1978.
106. Bush, M.T., and Weller, W.L.: Drug Metab. Rev. 1:249, 1972; Thompson, R.M., *et al.*: Drug Metab. Disposit. 1:489, 1973.
107. Breimer, D.D., and Van Rossum, J.M.: J. Pharm. Pharmacol. 25:762, 1973.
108. Pittman, K.A., *et al.*: Biochem. Pharmacol. 18:1673, 1969; *Ibid.*, 19:1833, 1970.
109. Miller, J.A., and Miller, E.C.: *in* Jollow, D.J., *et al.* (eds.): Biological Reactive Intermediates, p. 14, New York, Plenum Press, 1977.
110. Wislocki, P.G., *et al.*: Cancer Res. 36:1686, 1976.
111. Garattini, S., *et al.*: *in* Usdin, E., and Forrest, I. (eds.): Psychotherapeutic Drugs, Part II, p. 1039, New York, Marcel Dekker, 1977; Greenblatt, D.J., *et al.*: Clin. Pharmacol. Ther. 17:1, 1975; Yanagi, Y., *et al.*: Xenobiotica 5:245, 1975.
112. Corbella, A., *et al.*: J. Chem. Soc. Chem. Comm. 721, 1973.
113. Keberle, H., *et al.*: Arch. Int. Pharmacodyn. 142:117, 1963; Keberle, H., *et al.*: Experientia 18:105, 1962.
114. Ferrandes, B., and Eymark, P.: Epilepsia, 18:169, 1977; Kuhara, T., and Matsumoto, J.: Biomed. Mass. Spectrom. 1:291, 1974.
115. Maynert, E.W.: J. Pharmacol. Exp. Ther. 150:117, 1965.
116. Palmer, K.H., *et al.*: J. Pharmacol. Exp. Ther. 175:38, 1970; Holtzmann, J.L., and Thompson, J.A.: Drug Metab. Disposit. 3:113, 1975.
117. Carroll, F.J., *et al.*: Drug Metab. Disposit. 5:343, 1977.
118. Thomas, R.C., and Judy, R.W.: J. Med. Chem. 15:964, 1972.
119. Adams, S.S., and Buckler, J.W.: Clin. Rheum. Dis. 5:359, 1979.
120. Ludwig, B.J., *et al.*: J. Med. Pharm. Chem. 3:53, 1961.
121. Horning, M.G., *et al.*: Drug Metab. Disposit. 1:569, 1973.
122. Dieterle, W., *et al.*: Arzneim. Forsch. 26:572, 1976.
123. McMahon, R.E., *et al.*: J. Pharmacol. Exp. Ther. 149:272, 1965.
124. Fuccella, L.M., *et al.*: J. Clin. Pharmacol. 13:68, 1973.
125a. Lin, D.C.K., *et al.*: Biomed. Mass Spectrom. 2:206, 1975; Wong, L.K., and Biemann, K.: Clin. Toxicol. 9:583, 1976.
125b. Thomas, R.C., *et al.*: J. Pharm. Sci. 64:1360, 1366, 1975; Gottlieb, T.B., *et al.*: Clin. Pharmacol. Ther. 13:436, 1972.
126. Gorrod, J.W. (ed.): Biological Oxidation of Nitrogen, Amsterdam, Elsevier-North Holland, 1978.
127. Gorrod, J.W.: Chem.-Biol. Interact. 7:289, 1973.
128. Ziegler, D.M., *et al.*: Drug Metab. Disposit. 1:314, 1973; Arch. Biochem. Biophys. 150:116, 1972.
129. Gram, T.E.: *in* Brodie, B.B., and Gillette, J.R. (eds.): Concepts in Biochemical Pharmacology, Part 2, p. 334, Berlin, Springer-Verlag, 1971.
130. Brodie, B.B., *et al.*: Ann. Rev. Biochem. 27:427, 1958; McMahon, R.E.: J. Pharm. Sci. 55:457, 1966.
131. Nagy, A., and Johansson, R.: Arch. Pharm. 290:145, 1975.
132. Gram, L.F., *et al.*: Psychopharmacologia 54:255, 1977.
133. Collinsworth, K.A., *et al.*: Circulation 50:1217, 1974; Narang, P.K., *et al.*: Clin. Pharmacol. Ther. 24:654, 1978.
134. Hutsell, T.C., and Kraychy, S.J.: J. Chromatogr. 106:151, 1975; Heel, R.C., *et al.*: Drugs 15:331, 1978.
135. Adam, H.K., *et al.*: Biochem. Pharmacol. 27:145, 1979.
136. Chang, T.K., *et al.*: Res. Comm. Chem. Pathol. Pharmacol. 9:391, 1974; Glazko, A.J., *et al.*: Clin. Pharmacol. Ther. 16:1066, 1974.
137. Hammar, C.G., *et al.*: Anal. Biochem. 25:532, 1968; Beckett, A.H., *et al.*: J. Pharm. Pharmacol. 25:188, 1973.
138. Due, S.L., *et al.*: Biomed. Mass Spectrom. 3:217, 1976.
139. Beckett, A.H., *et al.*: J. Pharm. Pharmacol. 23:812, 1971.
140. Pohland, A., *et al.*: J. Med. Chem. 14:194, 1971.
141. Bruce, R.B., *et al.*: J. Med. Chem. 11:1031, 1968.
142. Szeto, H.H., and Inturri, C.E.: J. Chromatogr. 125:503, 1976.
143. Misra, A.L.: *in* Adler, M.L., *et al.* (eds.): Factors Affecting the Action of Narcotics, p. 297, New York, Raven Press, 1978.
144. Kamm, J.J., *et al.*: J. Pharmacol. Exp. Ther. 182:507, 1972.
145. Kamm, J.J., *et al.*: J. Pharmacol. Exp. Ther. 184:729, 1973.
146. Goldberg, M.E., (ed.): Pharmacological and Biochemical Properties of Drug Substances, vol. 1, p. 257, 311, 1977.
147. Gorrod, J.W., and Jenner, P.: Essays Toxicol. 6:35, 1975; Beckett, A.H., and Triggs, E.J.: Nature (London) 211:1415, 1966.
148. Hucker, H.B., *et al.*: Drug Metab. Disposit. 2:406, 1974; Wold, J.S., and Fischer, L.J.: J. Pharmacol. Exp. Ther. 183:188, 1972.
149. Kaiser, C., *et al.*: J. Med. Chem. 15:1146, 1972.
150. Bickel, M.H.: Pharmacol. Rev. 21:325, 1969.
151. Jenner, P.: *in* Gorrod, J.W. (ed.): Biological Oxidation of Nitrogen, p. 383, Amsterdam, Elsevier-North Holland, 1978.
152. Faurbye, A., *et al.*: Am. J. Psychiatr. 120:277, 1963; Theobald, W., *et al.*: Med. Pharmacol. Exp. 15:187, 1966.
153. Coutts, R., and Beckett, A.H.: Drug Metab. Rev. 6:51, 1977.
154. Leinweber, F.-J., *et al.*: J. Pharm. Sci. 66:1570, 1977.
155. Beckett, A.H., and Rowland, M.: J. Pharm. Pharmacol. 17:109S, 1965; Caldwell, J., *et al.*: Biochem. J. 129:11, 1972.
156. Wieber, J., *et al.*: Anaesthol. 24:260, 1975; Chang, T., and Glazko, A.J.: Anaesthol. 36:401, 1972.
157. Tindell, G.L., *et al.*: Life Sci. 11:1029, 1972.
158. Franklin, R.B., *et al.*: Drug Metab. Disposit. 5:223, 1977.
159. Barlett, M.F., and Egger, H.P.: Fed. Proc. 31:537, 1972.
160. Beckett, A.H., and Gibson, G.G.: Xenobiotica 8:73, 1978.
161. Tipton, K.F., *et al.*: *in* Wolsterholme, G.E.W., and Knight, J. (eds.): Monoamine Oxidase and Its Inhibition, p. 5, Amsterdam, Elsevier-North Holland, 1976.
162. Beckett, A.H., and Brookes, L.G.: J. Pharm. Pharmacol. 23:288, 1971.
163. Charalampous, K.D., *et al.*: J. Pharmacol. Exp. Ther. 145:242, 1964; Psychopharmacologia 9:48, 1966.
164. Au, W.Y.W., *et al.*: Biochem. J. 129:110, 1972.
165. Parli, C.J., *et al.*: Biochem. Biophys. Res. Comm. 43:1204, 1971; Lindeke, B., *et al.*: Acta Pharm. Suecica 10:493, 1973.
166. Hucker, H.B.: Drug Metab. Disposit. 1:332, 1973; Parli, C.H., *et al.*: Drug Metab. Disposit. 3:337, 1973.
167. Wright, J., *et al.*: Xenobiotica 7:257, 1977; Gal, J., *et al.*: Res. Comm. Chem. Pathol. Pharmacol. 15:525, 1976.
168. Beckett, A.H., and Bélanger, P.M.: J. Pharm. Pharmacol. 26:205, 1974.
169. Bélanger, P.M., and Grech-Bélanger, O.: Can. J. Pharm. Sci. 12:99, 1977.
170. Weisburger, J.H., and Weisburger, E.K.: Pharmacol. Rev. 25:1, 1973; Miller, E.C., and Miller, J.A.: Pharmacol. Rev. 18:805, 1966.
171. Weisburger, J.H., and Weisburger, E.K.: *in* Brodie, B.B., and Gillette, J.R. (eds.): Concepts in Biochemical Pharmacology, Part 2, p. 312, Berlin, Springer-Verlag, 1971.
172. Uehleke, H.: Xenobiotica 1:327, 1971.
173. Beckett, A.H., and Bélanger, P.M.: Biochem. Pharmacol. 25:211, 1976.
174. Israili, Z.H., *et al.*: J. Pharmacol. Exp. Ther. 187:138, 1973.
175. Kiese, M.: Pharmacol. Rev. 18:1091, 1966.
176. Lin, J.-K., *et al.*: Cancer Res. 35:844, 1975; Poirer, L.A., *et al.*: Cancer Res. 27:1600, 1967.

177. Lin, J.-K., *et al.*: Biochemistry 7:1889, 1968; Biochemistry 8:1573, 1969.
178. Schwartz, D.E., *et al.*: Arzneim-Forsch. 20:1867, 1970.
179. Dagne, E., and Castagnol, N., Jr.: J. Med. Chem. 15:840, 1972.
180. Ericsson, O., and Danielsson, B.: Drug Metab. Disposit. 5:497, 1977.
181. Garrattini, S., *et al.*: Drug Metab. Rev. 1:291, 1972.
182. Brotherton, P.M., *et al.*: Clin. Pharmacol. Ther. 10:505, 1969.
183. Langone, J.J., *et al.*: Biochemistry 12:5025, 1973.
184. Grochow, L.B., and Colvin, M.: Clin. Pharmacokin. 4:380, 1979.
185. Colvin, M., *et al.*: Cancer Res. 33:915, 1973; Connors, T.A., *et al.*: Biochem. Pharmacol. 23:115, 1974.
186. Irving, C.C.: *in* Fishman, W.H. (ed.): Metabolic Conjugation and Metabolic Hydrolysis, Vol. 1, p. 53, New York, Academic Press, 1970.
187. Miller, J., and Miller, E.C.: *in* Jollow, D.J., *et al.* (eds.): Biological Reactive Intermediates, p. 6, New York, Plenum Press, 1970.
188. Jollow, D.J., *et al.*: J. Pharmacol. Exp. Ther. 187:195, 1973; Prescott, L.F., *et al.*: Lancet 1:519, 1971.
189. Hinson, J.A.: Rev. Biochem. Toxicol. 2:103, 1980.
190. Hinson, J.A., *et al.*: Life Sci. 24:2133, 1979; Nelson, S.D., *et al.*: Biochem. Pharmacol. 29:1617, 1980.
191. Potter, W.Z., *et al.*: J. Pharmacol. Exp. Ther. 187:203, 1973.
192. Adler, T.K., *et al.*: J. Pharmacol. Exp. Ther. 114:251, 1955.
193. Brodie, B.B., and Axelrod, J.: J. Pharmacol. Exp. Ther. 97:58, 1949.
194. Duggan, D.E., *et al.*: J. Pharmacol. Exp. Ther. 181:563, 1972; Kwan, K.C., *et al.*: J. Pharmacokin. Biopharm. 4:255, 1976.
195. Brogden, R.N., *et al.*: Drugs 14:163, 1977; Taylor, J.A., *et al.*: Xenobiotica 7:357, 1977.
196. Daly, J., *et al.*: Ann. N.Y. Acad. Sci. 96:37, 1962.
197. Mazel, P., *et al.*: J. Pharmacol. Exp. Ther. 143:1, 1964.
198. Sarcione, E.J., and Stutzman, L.: Cancer Res. 20:387, 1960; Elion, G.B., *et al.*: Proc. Am. Assoc. Cancer Res. 3:316, 1962.
199. Taylor, J.A.: Xenobiotica 3:151, 1973.
200. Brodie, B.B., *et al.*: J. Pharmacol. Exp. Ther. 98:85, 1950; Spector, E., and Shideman, F.E.: Biochem. Pharmacol. 2:182, 1959.
201. Neal, R.A.: Arch. Int. Med. 128:118, 1971; Neal, R.A.: Biochem. J. 103:183, 1967.
202. Neal, R.A.: Rev. Biochem. Toxicol. 2:131, 1980.
203. Gruenke, L., *et al.*: Res. Comm. Chem. Pathol. Pharmacol. 10:221, 1975; Zehnder, K., *et al.*: Biochem. Pharmacol. 11:535, 1962.
204. Aguilar, S.J.: Dis. Nerv. System 36:484, 1975.
205. Taylor, D.C., *et al.*: Drug Metab. Disposit. 6:21, 1978.
206. Taylor, D.C.: *in* Wood, C.J., and Simkins, M.A. (eds.): International Symposium on Histamine H₂-Receptor Antagonists, p. 45, Welwyn Garden City, V.K., Smith, Kline and French, 1973.
207. Crew, M.C., *et al.*: Xenobiotica 2:431, 1972.
208. Hucker, H.B., *et al.*: J. Pharmacol. Exp. Ther. 154:176, 1966; *Ibid.*, 155:309, 1967.
209. Hathway, D.E., (Sr. Reporter): Foreign Compound Metabolism in Mammals, Vol. 3, p. 512, London, The Chemical Society, 1975.
210. Schwartz, M.A., and Kolis, S.J.: Drug Metab. Disposit. 1:322, 1973.
211. Sisenwine, S.F., *et al.*: Drug Metab. Disposit. 3:180, 1975.
212. McCallum, N.K.: Experientia 31:957, 1975; *Ibid.*, 31:520, 1975.
213. Cohen, E.N., and Van Dyke, R.A.: Metabolism of Volatile Anesthetics, Reading, MA, Addison-Wesley, 1977; Cohen, E.N., *et al.*: Anesthesiology 43:392, 1975.
214. Van Dyke, R.A., *et al.*: Drug Metab. Disposit. 3:51, 1975; *Ibid.*, 4:40, 1976.
215. Pohl, L.: Rev. Biochem. Toxicol. 1:79, 1979.
216. Pohl, L., *et al.*: Biochem. Pharmacol. 27:335, 1978; *Ibid.*, 27:491, 1978.
217. Parke, D.V.: The Biochemistry of Foreign Compounds, p. 218, Oxford, Pergamon Press, 1968.
218. Gillette, J.R.: *in* Brodie, B.B., and Gillette, J.R. (eds.): Concepts in Biochemical Pharmacology, Part 2, p. 349, Berlin, Springer-Verlag, 1971.
219. Bachur, N.R.: Science 193:595, 1976.
220. Sellers, E.M., *et al.*: Clin. Pharmacol. Ther. 13:37, 1972.
221. Pritchard, J.F., *et al.*: J. Chromatogr. 162:47, 1979.
222. Osterloh, J.D., *et al.*: Drug Metab. Disposit. 8:12, 1980.
223. Jenner, P., and Testa, B.: Drug Metab. Rev. 2:117, 1973.
224. Culp, H.W., and McMahon, R.E.: J. Biol. Chem. 243:848, 1968.
225. McMahon, R.E., *et al.*: J. Pharmacol. Exp. Ther. 149:272, 1965; Galloway, J.A., *et al.*: Diabetes 16:118, 1967.
226. Yü, T.F., *et al.*: Metabolism 17:309, 1968.
227. DiCarlo, F.J., *et al.*: Clin. Pharmacol. Ther. 22:858, 1977.
228. Bachur, N.R., and Gee, M.: J. Pharmacol. Exp. Ther. 177:567, 1971; Huffmann, D.H., *et al.*: Clin. Pharmacol. Ther. 13:895, 1972.
229. Testa, B., and Beckett, A.H.: J. Pharm. Pharmacol. 25:119, 1973; Testa, B.: Acta Pharm. Suecica 10:441, 1973.
230. Billing, R.E., *et al.*: Xenobiotica 10:33, 1980.
231. Pollock, S.H., and Blum, K., *et al.* (eds.): Alcohol and Opiates, p. 359, New York, Academic Press, 1977; Chatterjie, N., *et al.*: Drug Metab. Disposit. 2:401, 1974.
232. Sullivan, H.R., and Due, S.L.: J. Med. Chem. 16:909, 1973.
233. Hollands, T.R., and Johnson, W.J.: Biochem. Med. 7:288, 1973.
234. Chan, K.K., *et al.*: J. Med. Chem. 15:1265, 1972.
235. Dayton, H., and Inturrisi, C.E.: Drug Metab. Disposit. 4:474, 1974.
236. Roerig, S., *et al.*: Drug Metab. Disposit. 4:53, 1976; Malsperis, L., *et al.*: Res. Comm. Chem. Pathol. Pharmacol. 14:393, 1976.
237. Bachur, N.R., and Felsted, R.L.: Drug Metab. Disposit. 4:239, 1976; Crew, M.C., *et al.*: Clin. Pharamcol. Ther. 14:1013, 1973.
238. DiCarlo, F.J., *et al.*: Xenobiotica 2:159, 1972; Crew, M.C., *et al.*: Xenobiotica 2:431, 1972; DiCarlo, F.J., *et al.*: J. Reticulo-endothel. Soc. 14:387, 1973.
239. Gerhards, E., *et al.*: Acta Endocrinol. 68:219, 1971.
240. Wright, J., *et al.*: Xenobiotica 7:257, 1977.
241. Kawai, K., and Baba, S.: Chem. Pharm. Bull. (Japan) 24:2728, 1976.
242. Gillette, J.R., *et al.*: Mol. Pharmacol. 4:541, 1968; Fouts, J.R., and Brodie, B.B.: J. Pharmacol. Exp. Ther. 119:197, 1957.
243. Hernandez, P.H., *et al.*: Biochem. Pharmacol. 16:1877, 1967.
244. Scheline, R.R.: Pharmacol. Rev. 25:451, 1973; Walker, R.: Food Cosmet. Toxicol. 8:659, 1970.
245. Min, B.H., and Garland, W.A.: J. Chromatog. 139:121, 1977.
246. Rieder, J., and Wendt, G.: *in* Garattini, S., *et al.* (eds.): The Benzodiazepines, p. 99, New York, Raven Press, 1973.
247. Conklin, J.D., *et al.*: J. Pharm. Sci. 62:1024, 1973; Cox, P.L., *et al.*: J. Pharm. Sci. 58:987, 1969.
248. Stambaugh, J.E., *et al.*: J. Pharmacol. Exp. Ther. 161:373, 1968.
249. Mitchard, M.: Xenobiotica 1:469, 1971.
250. Glazko, A.J., *et al.*: J. Pharmacol. Exp. Ther. 96:445, 1949; Smith, G.N., and Worrel, G.S.: Arch. Biochem. Biophys. 24:216, 1949.
251. Tréfouël, J., *et al.*: C.R. Seances Soc. Biol., Paris 190:756, 1935.
252. Jones, R., *et al.*: Food Cosmet. Toxicol. 2:447, 1966; *Ibid.*, 4:419, 1966.
253. Ikeda, M., and Uesugi, T.: Biochem. Pharmacol. 22:2743, 1973.
254. Peppercorn, M.A., and Goldman, P.: J. Pharmacol. Exp. Ther. 181:555, 1972; Das, E.M.: Scand. J. Gastroenterol. 9:137, 1974.
255. Schröder, H., and Gustafsson, B.E.: Xenobiotica 3:225, 1973.
256. Bickel, M.H., and Gigon, P.L.: Xenobiotica 1:631, 1971.
257. Eldjarn, L.: Scand. J. Clin. Lab. Invest. 2:202, 1950; Staub, H.: Helv. Physiol. Acta 13:141, 1955.
258. Duggan, D.E., *et al.*: Clin. Pharmacol. Ther. 21:326, 1977; Duggan, D.E., *et al.*: J. Pharmacol. Exp. Ther. 20:8, 1977.
259. Kolb, H.K., *et al.*: Arzneim.-Fosch. 15:1292, 1965.
260. LaDu, B.N., and Snady, H.: *in* Brodie, B.B., and Gillette, J.R. (eds.): Concepts in Biochemical Pharmacology, Part 2, p. 477, Berlin, Springer-Verlag, 1971.
261. Junge, W., and Krisch, K.: CRC Crit. Rev. Toxicol. 3:371, 1975.
262. Davison, C.: Ann. N.Y. Acad. Sci. 179:249, 1971.

263. Kogan, M.J., et al.: Anal. Chem. 49:1965, 1977.
264. Inaba, T., et al.: Clin. Pharmacol. Ther. 23:547, 1978; Stewart, D.J., et al.: Life Sci. 1557, 1977.
265. Wells, R., et al.: Clin. Chem. 20:440, 1974.
266. Rubens, R., et al.: Arzneim.-Forsch. 22:256, 1972.
267. Gugler, L., and Jensen, C.: J. Chromatogr. 117:175, 1976.
268. Houin, G., et al.: Eur. J. Clin. Pharmacol. 8:433, 1975.
269. Sinkula, A.A., et al.: J. Pharm. Sci. 62:1106, 1973; Martin, A.R.: in Wilson, C.O., et al. (eds.): Textbook of Organic Medicinal and Pharmaceutical Chemistry, ed. 7, p. 304, Philadelphia, Lippincott, 1977.
270. Knirsch, A.K., et al.: J. Infect. Dis. 127:S105, 1973.
271. Stella, V.: in Higuchi, T., and Stella, V. (eds.): Prodrugs as Novel Drug Delivery Systems, p. 1, Washington, D.C., American Chemical Society, 1975.
272. Mark, L.C., et al.: J. Pharmacol. Exp. Ther. 102:5, 1951.
273. Nelson, S.D., et al.: J. Pharm. Sci. 66:1180, 1977.
274. Tsukamoto, H., et al.: Pharm. Bull. (Tokyo) 3:459, 1955; Ibid., 3:397, 1955.
275. Dudley, K.H., et al.: Drug Metab. Disposit. 6:133, 1978; Ibid., 2:103, 1974.
276. Dutton, G.J., et al.: in Parke, D.V., and Smith R.L. (eds.): Drug Metabolism: From Microbe to Man, p. 71, London, Taylor & Francis, 1977.
277. Dutton, G.J.: in Dutton, G.J. (ed.): Glucuronic Acid, Free and Combined, p. 186, New York, Academic Press, 1966; Dutton, G.D., et al.: Progr. Drug Metab. 1:2, 1977.
278. Brunk, S.F., and Delle, M.: Clin. Pharmacol. Exp. Ther. 16:51, 1974; Berkowitz, B.A., et al.: Clin. Pharmacol. Exp. Ther. 17:629, 1975.
279. Andrews, R.S., et al.: J. Int. Med. Res. 4(Suppl. 4):34, 1976.
280. Thies, R.L., and Fischer, L.J.: Clin. Chem. 24:778, 1978.
281. Walle, T., et al.: Fed. Proc. 35:665, 1976; Walle, T., et al.: Clin. Pharmacol. Ther. 26:167, 1979.
282. Roseman, S., et al.: J. Am. Chem. Soc. 76:1650, 1954.
283. Segre, E.J.: J. Clin. Pharmacol. 15:316, 1975.
284. Rubin, A., et al.: J. Pharm. Sci. 61:739, 1972; Rubin, A., et al.: J. Pharmacol. Exp. Ther. 183:449, 1972.
285. Bridges, J.W., et al.: Biochem. J. 118:47, 1970.
286. Gibson, T., et al.: Br. J. Clin. Pharmacol. 2:233, 1975; Tsuichiya, T., and Levy, G.: J. Pharm. Sci. 61:800, 1972.
287. Woolhouse, N.M., et al.: Xenobiotica 3:511, 1973.
288. Bickel, M.H., et al.: Experientia, 29:960, 1973.
289. Tskamoto, H., et al.: Chem. Pharm. Bull. (Tokyo) 11:421, 1963.
290. Uno, T., and Kono, M.: J. Pharm. Soc. (Japan) 82:1660, 1962.
291. Porter, C.C., et al.: Drug Metab. Disposit. 3:189, 1975.
292. Chaundhuri, N.K., et al.: Drug Metab. Disposit. 4:372, 1976.
293. Sitar, D.S., and Thornhill, D.P.: J. Pharmacol. Exp. Ther. 184:432, 1973.
294. Lindsay, R.H., et al.: Pharmacologist 18:113, 1976; Sitar, D.S., and Thornhill, D.P.: J. Pharmacol. Exp. Ther. 183:440, 1972.
295. Dutton, G.J., and Illing, H.P.A.: Biochem. J. 129:539, 1972.
296. Dieterle, W., et al.: Arzneim.-Forsch. 26:572, 1976; Richter, J.W., et al.: Helv. Chim. Acta 58:2512, 1975.
297. Dieterle, W., et al.: Eur. J. Clin. Pharmacol. 9:135, 1975.
298. Schmid, R., and Lester, R.: in Dutton, G.J. (ed.): Glucuronic Acid, Free and Combined, p. 493, New York, Academic Press, 1966.
299. Hadd, H.E., and Blickenstaff, R.T.: Conjugates of Steroid Hormones, New York, Academic Press, 1969.
300. Smith, R.L.: The Excretory Function of Bile: The Elimination of Drugs and Toxic Substance in Bile. London, Chapman & Hall, 1973.
301. Stern, L., et al.: Am. J. Dis. Children 120:26, 1970.
302. Weiss, C.F., et al.: New England J. Med. 262:787, 1960.
303. Dodgson, K.S.: in Parke, D.V., and Smith, R.L. (eds.): Drug Metabolism: From Microbe to Man, p. 91, London, Taylor & Francis, 1977; Roy, A.B.: in Brodie, B.B., and Gillette, J.R. (eds.): Concepts in Biochemical Pharmacology, Part 2, p. 536, Berlin, Springer-Verlag, 1971.
304. Williams, R.T.: in Bernfeld, P. (ed.): Biogenesis of Natural Products, ed. 2, p. 611, Oxford, Pergamon Press, 1967.
305. Roy, A.B.: Adv. Enzymol. 22:205, 1960.
306. Kwan, K.C., et al.: J. Pharmacol. Exp. Ther. 198:264, 1976; Stenback, O., et al.: Eur. J. Clin. Pharmacol. 12:117, 1977.
307. Lin, C., et al.: Drug Metab. Disposit. 5:234, 1977.
308. Nilsson, H.T., et al.: Xenobiotica 2:363, 1972.
309. Miller, R.P., et al.: Clin. Pharmacol. Ther. 19:284, 1976; Levy, G., et al.: Pediatrics 55:818, 1975.
310. James, S.P., and Waring, R.H.: Xenobiotica 1:572, 1971; Bostrum, H., and Vestermark, A.: Acta Physiol. Scand. 48:88, 1960.
311. Boyland, E., et al.: Biochem. J. 65:417, 1957; Roy, A.B.: Biochem. J. 74:49, 1960.
312. Irving, C.C.: in Gorrod, J.W. (ed.): Biological Oxidation of Nitrogen, p. 325, Amsterdam, Elsevier-North Holland, 1978.
313. Irving, C.C.: Cancer Res. 35:2959, 1975; Jackson, C.D., and Irving, C.C.: Cancer Res. 32:1590, 1972.
314. Hinson, J.A., and Mitchell, J.R.: Drug Metab. Disposit. 4:430, 1975.
315. Mulder, G.J., et al.: Biochem. Pharmacol. 26:189, 1977.
316. Weber, W.: in Brodie, B.B., and Gillette, J.R. (eds.): Concepts in Biochemical Pharmacology, Part 2, p. 564, Berlin, Springer-Verlag, 1971.
317. Williams, R.T., and Millburn, P.: in Blaschko, H.K.F. (ed.): MTP International Review of Science, Biochemistry Series One; vol. 12, Physiological and Pharmacological Biochemistry, p. 211, Baltimore, University Park Press, 1975.
318. James, M.O., et al.: Proc. Roy. Soc., Ser. B 182:25, 1972.
319. Bridges, J.W., et al.: Biochem. J. 118:47, 1970.
320. Wan, S.H., and Riegelman, S.: J. Pharm. Sci. 61:1284, 1972; VonLehmann, B., et al.: J. Pharm. Sci. 62:1483, 1973.
321. Braun, G.A., et al.: Eur. J. Pharmacol. 1:58, 1967.
322. Tatsumi, K., et al.: Biochem. Pharmacol. 16:1941, 1967.
323. Weber, W.W., and Hein, D.W.: Clin. Pharmacokin. 4:401, 1979.
324. Smith, R.L., and Caldwell, J.: in Parke, D.V., and Smith, R.L. (eds.): Drug Metabolism: From Microbe to Man, p. 331, London, Taylor & Francis, 1977.
325. Williams, R.T.: Clin. Pharmacol. Ther. 4:234, 1963.
326. Drach, J.C., et al.: Proc. Soc. Exp. Biol. Med. 135:849, 1970.
327. Caldwell, J.: in Jenner, P., and Testa, B. (eds.): Concepts in Drug Metabolism, Part A, p. 211, New York, Marcel Dekker, 1980.
328. Jerina, D.M., and Bend, J.R.: in Jollow, D.J., et al. (eds.): Biological Reactive Intermediates, p. 207, New York, Plenum Press, 1977.
329. Chasseaud, L.F.: in Arias, I.M., and Jakoby, W.B. (eds.): Glutathione: Metabolism and Function, p. 77, New York, Raven Press, 1976.
330. Weisburger, E.K.: Ann. Rev. Pharmacol. Toxicol. 18:395, 1978.
331. Hollingworth, R.M., et al.: Life Sci. 13:191, 1973; Benke, G.M., and Murphy, S.D.: Toxicol. Appl. Pharmacol. 31:254, 1975.
332. Bray, H.G., et al.: Biochem. J. 67:607, 1957.
333. Chalmers, A.H.: Biochem. Pharmacol. 23:1891, 1974; de Miranda, P., et al.: J. Pharmacol. Exp. Ther. 187:588, 1973; Ibid., 195:50, 1975.
334. Elion, G.B.: Fed. Proc. 26:898, 1967.
335. Jerina, D.M.: in Arias, I.M., and Jakoby, W.B. (eds.): Glutathione: Metabolism and Function, p. 267, New York, Raven Press, 1976.
336. Needleman, P.: in Needleman, P. (ed.): Organic Nitrates, p. 57, Berlin, Springer-Verlag, 1975.
337. Klaasen, C.D., and Fitzgerald, T.J.: J. Pharmacol. Exp. Ther. 191:548, 1974.
338. Boyland, E., and Chasseaud, L.F.: Biochem. J. 109:651, 1968.
339. Kuss, E., et al.: Hoppe-Seylers Z. Physiol. Chem. 352:817, 1971; Nelson, S.D., et al.: Biochem. Biophys. Res. Comm. 70:1157, 1976.
340. Weber, W.: in Fishman, W.H. (ed.): Metabolic Conjugation and Metabolic Hydrolysis, vol. 3, p. 250, New York, Academic Press, 1973; Williams, R.T.: Fed. Proc. 26:1029, 1967.
341. Elson, J., et al.: Clin. Pharmacol. Ther. 17:134, 1975; Drayer, D.E., et al.: Proc. Soc. Exp. Biol. Med. 146:358, 1974.
342. Mitchell, J.R., et al.: Ann. Intern. Med. 84:181, 1976; Nelson, S.D., et al.: Science 193:193, 1976.

343. Giardina, E.G., *et al.*: Clin. Pharmacol. Ther. 19:339, 1976; *Ibid.*, 17:722, 1975.
344. Peters, J.H., and Levy, L.: Ann. N.Y. Acad. Sci. 179:660, 1971.
345. Reimerdes, E., and Thumim, J.H.: Arzneim.-Forsch. 20:1171, 1970.
346. Vree, T.B., *et al.*: in Merkus, F.W.H.M. (ed.): The Serum Concentration of Drugs, p. 205, Amsterdam, Excerpta Medica, 1980.
347. Garrett, E.R.: Int. J. Clin. Pharmacol. 16:155, 1978.
348. Reidenberg, M.W., *et al.*: Clin. Pharmacol. Ther. 14:970, 1973; Israili, Z.H., *et al.*: Drug Metab. Rev. 6:283, 1977.
349. Evans, D.A.P., *et al.*: Clin. Pharmacol. Ther. 6:430, 1965.
350. Tabor, H., *et al.*: J. Biol. Chem. 204:127, 1953.
351. Sullivan, H.R., and Due, S.L.: J. Med. Chem. 16:909, 1973; Sullivan, H.R., *et al.*: J. Am. Chem. Soc. 94:4050, 1972; Life Sci. 11:1093, 1972.
352. Drayer, D.E., and Reidenberg, M.M.: Clin. Pharmacol. Ther. 22:251, 1977.
353. Lunde, P.K.M., *et al.*: Clin. Pharmacokin. 2:182, 1977.
354. Kalow, W.: Pharmacogenetics: Heredity and the Response to Drugs, Philadelphia, Saunders, 1962.
355. Kutt, H., *et al.*: Am. Rev. Resp. Dis. 101:377, 1970.
356. Alarcon-Segovia, D.: Drugs, 12:69, 1976; Reidenberg, M.M., and Martin, J.H.: Drug Metab. Disposit. 2:71, 1974.
357. Strong, J.M., *et al.*: J. Pharmacokin. Biopharm. 3:233, 1975.
358. Axelrod, J.: in Brodie, B.B., and Gillette, J.R. (eds.): Concepts in Biochemical Pharmacology, Part 2, p. 609, Berlin, Springer-Verlag, 1971.
359. Mudd, S.H.: in Fishman, W.H. (ed.): Metabolic Conjugation and Metabolic Hydrolysis, vol. 3, p. 297, New York, Academic Press.
360. Young, J.A., and Edwards, K.D.G.: Med. Res. 1:53, 1962; J. Pharmacol. Exp. Ther. 145:102, 1964.
361. Shindo, H., *et al.*: Chem. Pharm. Bull. (Tokyo) 21:826, 1973.
362. Morgan, C.D., *et al.*: Biochem. J. 114:8P, 1969.
363. Weber, R., and Tuttle, R.R.: in Goldberg, M.E. (ed.): Pharmacological and Biochemical Properties of Drug Substances, vol. 1, p. 109, Washington, D.C., American Pharmaceutical Association, 1977.
364. Glazko, A.J.: Drug Metab. Disposit. 1:711, 1973.
365. Börner, U., and Abbott, S.: Experientia 29:180, 1973.
366. Bleidner, W.E., *et al.*: J. Pharmacol. Exp. Ther. 150:484, 1965.
367. Komori, Y., and Sendju, Y.: J. Biochem. 6:163, 1926.
368. Lindsay, R.H., *et al.*: Biochem. Pharmacol. 24:463, 1975.
369. Bremer, J., and Greenberg, D.M.: Biochim. Biophys. Acta 46:217, 1961.
370. Allan, P.W., *et al.*: Biochim. Biophys. Acta 114:647, 1966; Elion, G.B.: Fed. Proc. 26:898, 1967.
371. Testa, B., and Jenner, P.: Drug Metabolism: Chemical and Biochemical Aspects, p. 329–418, New York, Marcel Dekker.
372. Ward, R.M., *et al.*: in Avery, G.S. (ed.): Drug Treatment, ed. 2, p. 76, Sydney, ADIS Press, 1980.
373. Morselli, P.L.: Drug Disposition During Development, New York, Spectrum, 1977.
374. Jondorf, W.R., *et al.*: Biochem. Pharmacol. 1:352, 1958.
375. Nitowsky, H.M., *et al.*: J. Pediatr. 69:1139, 1966.
376. Crooks, J., *et al.*: Clin. Pharmacokin. 1:280, 1976; Crooks, J., and Stevenson, I.H. (eds.): Drugs and the Elderly, London, Macmillan, 1979.
377. Williams, R.T.: Ann. N.Y. Acad. Sci. 179:141, 1971.
378. Williams, R.T.: in LaDu, B.N., *et al.* (eds.): Fundamentals of Drug Metabolism and Disposition, p. 187, Baltimore, Williams & Wilkins, 1971.
379. Williams, R.T., *et al.*: in Snyder, S.H., and Usdin, E. (eds.): Frontiers in Catecholamine Research, p. 927, New York, Pergamon Press, 1973.
380. Butler, T.C., *et al.*: J. Pharmacol. Exp. Ther. 199:82, 1976.
381. Williams, R.T.: Biochem. Soc. Trans. 2:359, 1974.
382. Bridges, J.W., *et al.*: Biochem. J. 118:47, 1970.
383. Williams, R.T.: Fed. Proc. 26:1029, 1967.
384. Cram, R.L., *et al.*: Proc. Soc. Exp. Biol. Med. 118:872, 1965.
385. Kutt, H., *et al.*: Neurology 14:542, 1962.
386. Vesell, E.S.: Progr. Med. Genet. 9:291, 1973.
387. Kato, R.: Drug Metab. Rev. 3:1, 1974.
388. Beckett, A.H., *et al.*: J. Pharm. Pharmacol. 23:62S, 1971; Menguy, R., *et al.*: Nature 239:102, 1972.
389. Conney, A.H.: Pharmacol. Rev. 19:317, 1967; Snyder, R., and Remmer, H.: Pharmacol. Ther. 7:203, 1979.
390. Parke, D.V.: in Parke, D.V. (ed.): Enzyme Induction, p. 207, London, Plenum Press, 1975.
391. Estabrook, R.W., and Lindenlaub, E. (eds.): The Induction of Drug Metabolism, Stuttgart, Schattauer Verlag, 1979.
392. Hansten, P.D.: Drug Interactions, ed. 4, p. 38, Philadelphia, Lea & Febiger, 1979.
393. Laenger, H., and Detering, K.: Lancet 600, 1974.
394. Skolnick, J.L., *et al.*: J. Am. Med. Assoc. 236:1382, 1976.
395. Dent, C.E., *et al.*: Br. Med. J. 4:69 1970.
396. Yeung, C.Y., and Field, C.E.: Lancet 135, 1969.
397. Jenne, J., *et al.*: Life Sci. 17:195, 1975.
398. Pantuck, E.J., *et al.*: Science 175:1248, 1972; Clin. Pharmacol. Ther. 14:259, 1973; Vaughan, D.P., *et al.*: Br. J. Clin. Pharmacol. 3:279, 1976.
399. Kolmodin, B., *et al.*: Clin. Pharmacol. Ther. 10:638, 1969.
400. Jeffrey, W.H., *et al.*: J. Am. Med. Assoc. 236:2881, 1976.
401. Gelboin, H.V., and TS'o P.O.P. (eds.): Polycyclic Hydrocarbons and Cancer: Environment, Chemistry, Molecular and Cell Biology, New York, Academic Press, 1978.
402. Vesell, E.S., and Passananti, G.T.: Drug Metab. Disposit. 1:402, 1973; Anders, M.W.: Ann. Rev. Pharmacol. 11:37, 1971.
403. Mannering, G.J.: in Brodie, B.B., and Gillette, J.R. (eds.): Concepts in Biochemical Pharmacology, Part 2, p. 452, Berlin, Springer-Verlag, 1971.
404. Hewick, D., and McEwen, J.: J. Pharm. Pharmacol. 25:458, 1973.
405. Casy, A.F.: in Burger, A. (ed.): Medicinal Chemistry, ed. 3, Part I, p. 81, New York, Wiley-Interscience, 1970.
406. George, C.F., *et al.*: Eur. J. Clin. Pharmacol. 4:74, 1972.
407. Breimer, D.D., and Van Rossum, J.M.: J. Pharm. Pharmacol. 25:762, 1973.
408. Low, L.K., and Castagnoli, N., Jr.: Ann. Rep. Med. Chem. 13:304, 1978.
409. Testa, B., and Jenner, P.: J. Pharm. Pharmacol. 28:731, 1976.
410. Belpaire, F.M., *et al.*: Xenobiotica 5:413, 1975.
411. Drayer, D.E.: U.S. Pharmacist (Hosp. Ed.) 5:H15, 1980.
412. Pierides, A.M., *et al.*: Lancet 2:1279, 1975; Gabriel, R., and Pearce, J.M.S.: Lancet 2:906, 1976.

SELECTED READINGS

Aitio, A. (ed.): Conjugation Reactions in Drug Biotransformation, Amsterdam, Elsevier, 1978.
Brodie, B.B., and Gillette, J.R. (eds.): Concepts in Biochemical Pharmcology, Part 2, Berlin, Springer-Verlag, 1971.
Doull, J., Klaassen, C.D., and Amdur, M.O. (eds.): Casarett and Doull's Toxicology, ed. 2, New York, Macmillan, 1980.
Drayer, D.E.: Pharmacologically Active Drug Metabolites, Clin. Pharmacokin. 1:426, 1976.
Estabrook, R.W., and Lindenlaub, E. (eds.): The Induction of Drug Metabolism, Stuttgart, Schattauer Verlag, 1979.
Gillette, J.R., Mitchell, J.R., and Brodie, B.B.: Biochemical Mechanisms of Drug Toxicity, Ann. Rev. Pharmacol. 14:271, 1974.
Gorrod, J.W. (ed.): Drug Toxicity, London, Taylor & Francis, 1979.
Gorrod, J.W., and Beckett, A.H. (eds.): Drug Metabolism in Man, London, Taylor & Francis, 1978.
Hodgson, E., and Guthrie, F.E. (eds.): Introduction to Biochemical Toxicology, New York, Elsevier, 1980.
Jakoby, W.B. (ed.): Enzymatic Basis of Detoxification, Vols. I and II, New York, Academic Press, 1980.
Jenner, P., and Testa, B.: The Influence of Stereochemical Factors on Drug Disposition, Drug Metab. Rev. 2:117, 1973.
Jenner, P., and Testa, B. (eds.): Concepts in Drug Metabolism, Part A, New York, Marcel Dekker, 1980.

Jerina, D.M. (ed.): Drug Metabolism Concepts, Washington, D.C., American Chemical Society, 1977.

La Du, B.N., Mandel, H.G., and Way, E.L. (eds.): Fundamentals of Drug Metabolism and Drug Disposition, Baltimore, Williams & Wilkins, 1971.

Low, L.K., and Castagnoli, N.: Drug Biotransformations, *in* Wolff, M.E. (ed.): Burger's Medicinal Chemistry, ed. 4, Part I, p. 107, New York, Wiley-Interscience, 1980.

Nelson, S.D.: Chemical and Biological Factors Influencing Drug Biotransformation, *in* Wolff, M.E. (ed.): Burger's Medicinal Chemistry, ed. 4, Part I, p. 227, New York, Wiley-Interscience, 1980.

Parke, D.V.: The Biochemistry of Foreign Compounds, New York, Pergamon Press, 1968.

Parke, D.V., and Smith, R.L. (eds.): Drug Metabolism: From Microbe to Man, London, Taylor & Francis, 1977.

Sato, R., and Omura, T. (eds.): Cytochrome P-450, New York, Academic Press, and Tokyo, Kadansha, 1978.

Testa, B., and Jenner, P.: Drug Metabolism: Chemical and Biochemical Aspects, New York, Marcel Dekker, 1976.

4

anti-infective agents

Arnold R. Martin

Chemotherapy may be defined as the study and the use of agents which are selectively more toxic to the invading organisms than to the host. Paul Ehrlich, the father of chemotherapy, was more absolute in his concept and used the term to describe the cure of an infectious disease *without* injury to the host. This ideal has been rather closely approached by the antibiotic, penicillin. The scientific principles of chemotherapy were established chiefly during the period 1919–1935, but only since this time and especially with the advent of the sulfonamides and the antibiotics have the material benefits in terms of useful medicinal products been realized. The only chemotherapeutic agents known before the time of Ehrlich were cinchona for malaria, ipecac for amebic dysentery and mercury for treating the symptoms of syphilis.

The first 30 years of the 20th century saw the development of useful chemotherapeutic agents, among which were organic compounds containing heavy metals such as arsenic, mercury and antimony, dyes, and a few modifications of the quinine molecule. These agents represented extremely important advances but even so had many drawbacks. The next 30 years of the 20th century comprise the period of greatest advance in the area of chemotherapy. During this time the sulfonamides and sulfones (see Chap. 5), many phenols and their derivatives, the antimalarial agents (see Chap. 6), the surfactants and, of great importance, the antibiotics (see Chap. 7) were studied and introduced into medical practice. The development of these newer drugs has relegated some of the older drugs to positions of minor importance or historical interest only.

The knowledge and the use of chemotherapeutic agents can be classified according to the diseases and the infestations for which they are used; or they can be classified according to separate compounds or groups of related compounds. In this book the chapters covering chemotherapeutic agents are organized by an amalgamation of the two systems. When the knowledge is best expressed and interrelated by the chemical classification this method is used, but where several classes of drugs may be rather specific for a single disease or group of related diseases the medical classification is used. Those groups of antibiotics employed for the specific treatment of tuberculosis (cycloserine, viomycin, capreomycin and rifampin) and for the treatment of fungal infections (griseofulvin and the polyene antibiotics) are covered in this chapter.

LOCAL ANTI-INFECTIVE AGENTS

Local anti-infectives are also known as antiseptics and disinfectants and constitute a widely used group of drugs. Generally, the term *antiseptic* includes those agents applied to living tissues; antiseptics are bacteriostatic and do not necessarily sterilize the surface under treatment. The ideal antiseptic would destroy bacteria, spores, fungi, viruses and other infective agents without harming the tissues of the host; however, most have a limited spectrum of activity and many have an adverse effect on tissues. Disinfectants, on the other hand, are applied to inanimate objects, are bactericidal and rapidly produce an irreversibly lethal effect.

While there is extensive use of antibiotics for systemic infections, their topical use is limited because of their high degree of antigenicity. In addition to the allergic reactions that may result, the sensitivity that may develop during the treatment of a minor or suspected infection may seriously jeopardize the patient during the treatment of a later, more severe systemic infection.

Several chemical classes of compounds possess activity as local anti-infective agents.

ALCOHOLS AND RELATED COMPOUNDS

Various alcohols and alcohol derivatives have been used as antiseptics. Ethyl alcohol and isopropyl alcohol are widely used for this purpose.

Antibacterial action and chemical structure

The antibacterial values of the straight chain alcohols increase with an increase in molecular weight, but as the molecular weight increases the water-solubility decreases so that beyond C_8 the activity begins to fall off. The isomeric alcohols show a drop in activity from primary to secondary to tertiary. Thus, n-propyl alcohol has a phenol coefficient against *Staph. aureus* of 0.082 as compared with 0.054 for isopropyl alcohol. Of course, because the latter is commercially available at a lower price it is more widely used than n-propyl alcohol. Isopropyl alcohol is slightly more effective than ethyl alcohol against the vegetative phase, but both are rather ineffective against the spore phase.

Products

Alcohol U.S.P., ethanol, ethyl alcohol, spiritus vini rectificatus (cologne spirit, wine spirit). Ethanol has been known since earliest times as a fermentation product of carbohydrates. An important source today is from the fermentation of molasses. A synthetic method of preparation using acetylene or ethylene has

$$CH_2=CH_2 + HOSO_2OH$$
$$\downarrow$$
$$CH_3CH_2OSO_2OH$$
Ethyl Sulfuric Acid

$$2\ CH_3CH_2OSO_2OH \xrightarrow[H_2SO_4]{} (CH_3CH_2O)_2SO_2$$
Diethyl Sulfate

$$C_2H_5OSO_2OH + H_2O$$
$$\downarrow$$
$$C_2H_5OH + H_2SO_4$$

$$(C_2H_5O)_2SO_2 + H_2O$$
$$\downarrow$$
$$C_2H_5OH + C_2H_5OSO_2OH$$
Alcohol

been employed, although only the ethylene procedure has shown commercial possibilities. By using sulfuric acid on ethylene to form ethyl sulfuric acid and diethyl sulfate, which are diluted with an equal volume of water, alcohol is formed and removed by distillation.

The commercial product is about 95 percent alcohol by volume, because this concentration of alcohol (92.3% w/w) and water forms a constant-boiling mixture at 78.2°. Pure alcohol boils at 78.3° and cannot be obtained by direct distillation.

Ethanol is a clear, colorless, volatile liquid having a burning taste and a characteristic odor. It is flammable and miscible with water, ether, chloroform and most alcohols. Its chemical properties are characteristic of primary alcohols. Most incompatibilities associated with it are due to solubility characteristics. Ethanol does not dissolve most inorganic salts, gums or proteins. Due to the aldehydes sometimes present in alcohol, the following chemical changes are often observed: the reduction of mercuric chloride to mercurous chloride, the formation of explosive mixtures with silver salts in the presence of nitric acid and the development of a dark color with alkalies.

Ethanol suspected of containing methanol is treated with resorcinol and concentrated sulfuric acid. A pink color denotes presence of methanol. Detection of 2-propanol in ethanol is facilitated by a 1 percent solution of p-dimethylaminobenzaldehyde in concentrated sulfuric acid. Positive test is a brilliant red-violet ring which slowly decomposes. Similar red-brown color is given by 1-propanol.

The Treasury Department of the U. S. Government oversees the use of alcohol and provides definitions and information pertaining thereto.*

"The term 'alcohol' means that substance known as ethyl alcohol, hydrated oxide of ethyl, or spirit of wine, from whatever source or whatever process produced, having a proof of 160 or more, and not including the substances commonly known as whisky, brandy, rum, or gin."

Besides alcohol available as ethyl alcohol, there are two other forms: (1) completely denatured alcohol and (2) specially denatured alcohol. Denatured alcohol is ethyl alcohol to which has been added such denaturing materials as render the alcohol unfit for use as an intoxicating beverage. It is free of tax and is solely for use in the arts and industries.

Completely denatured alcohol is prepared according to one of two formulas:

A. Contains ethyl alcohol, wood alcohol and benzene. This is not suitable even for external use.

---

* Regulation No. 3, Industrial and Denatured Alcohol, published by U. S. Treasury Department 1927, 1938.

B. Contains ethyl alcohol, methanol, aldehol† and benzene. This mixture is usually used as an antifreeze.

Specially denatured alcohol is ethyl alcohol treated with one or more acceptable denaturants so that its use may be permitted for special purposes in the arts and industries. Examples are: menthol in alcohol intended for use in dentifrices or mouthwashes; iodine in alcohol intended for preparation of tincture of iodine; phenol, methyl salicylate or sucrose octaacetate in alcohol intended for bathing or as an antiseptic, and methanol in alcohol to be used in the preparation of solid drug extracts.

Ethyl alcohol has a low narcotic potency. It seldom is used in medical practice as a therapeutic agent but almost always is employed as a solvent, preservative, mild counterirritant or antiseptic. It may be injected near nerves and ganglia to alleviate pain or ingested as a source of food energy, for hypnotic effect, as a carminative or as a mild vasodilator. The body readily oxidizes ethanol, first to acetaldehyde and then to carbon dioxide and water. (See disulfiram.)

Externally, it is refrigerant, astringent, rubefacient and slight anesthetic (Rubbing Alcohol U.S.P.).

The specific uses of alcohol in pharmacy are extremely varied and numerous. Spirits are a class of pharmaceuticals using alcohol exclusively as the solvent, whereas elixirs are hydroalcoholic preparations. Most fluid extracts contain a small percentage of alcohol as a preservative and solvent.

A concentration of 70 percent has long been held to be optimal for bactericidal action, but there is little evidence to support it. The rate of kill of organisms suspended in alcohol concentration between 60 and 95 percent is always so rapid that it is difficult to establish a significant difference.[1] Lower concentrations are also effective, but longer contact times are necessary, e.g., a period of 24 hours is required for a 15 percent solution to kill *Staph. albus*.[2] It has been reported that concentrations over 70 percent can be used safely for preoperative treatment of the skin.[3]

It also is the initial material used for the production of other medicinal agents, such as chloroform, ether and iodoform.

Dehydrated Alcohol U.S.P., dehydrated ethanol, absolute alcohol. Absolute or dehydrated alcohol is ethyl hydroxide in a form as pure as it is possible to obtain. It contains not less than 99 percent by weight of C_2H_5OH.

There are many laboratory procedures available for the preparation of anhydrous ethanol. Some of the compounds used in these methods are calcium oxide,

† Aldehol is an oxidation product of kerosene (b.p., 340° to 370°), having a boiling point of 200° to 240°, composed of glycols, aldehydes and acids.

anhydrous calcium sulfate, anhydrous sodium sulfate, aluminum ethoxide, diethyl phthalate and diethyl succinate. Commercially, absolute alcohol is prepared by azeotropic distillation of an ethanol and benzene mixture. Because the ethanol contains about 5 percent water, the resultant combination, ethyl alcohol-water-benzene, first distills at 64.8° (C_6H_6 74%, water 7.5% and C_2H_5OH 18.5%). All of the water is removed at this temperature, and then the remaining ethyl alcohol and benzene distill at 68.2°. The ethyl alcohol is always in great excess; thus, when all the benzene has been removed, pure ethyl alcohol is collected at 78.3°.

Dehydrated alcohol has a great affinity for water and must be stored in tightly closed containers. It is used primarily as a chemical agent but has been injected for the relief of pain in carcinoma and in other conditions where pain is local.

Isopropyl Alcohol U.S.P., 2-propanol. Isopropyl alcohol[2] became recognized about 1935 as a suitable substitute for ethyl alcohol in many external uses, but it must not be taken internally.

Most of the isopropyl alcohol used in the United States is made by hydration of propylene, using sulfuric acid as a catalyst.

$$CH_3CH{=}CH_2 \xrightarrow[H_2SO_4]{H_2O} CH_3CHOHCH_3$$

Isopropyl Alcohol

Isopropyl alcohol is a colorless, clear, volatile liquid having a slightly bitter taste and a characteristic odor. It is miscible with water, ether and chloroform.

It is used to remove creosote from the skin and as a disinfectant for the skin and surgical instruments. A 40 percent solution is approximately equal in antiseptic power to a 60 percent solution of ethyl alcohol. The effective concentrations are between 50 and 95 percent by weight. A 91 percent solution in water forms a constant boiling mixture and is thus the most economical concentration. It is frequently used by diabetics for cold sterilization of their syringes and needles. This is Azotropic Isopropyl Alcohol U.S.P.

Full-strength isopropyl alcohol is a skin irritant, due primarily to its defatting properties. If splashed in the eyes, it must be washed out at once with water. It possesses none of the effects of ethyl alcohol when used as a beverage; in fact, even very dilute aqueous solutions are not palatable.

In recent years it has been used in many toiletries and pharmaceuticals as a solvent and preservative and to replace, in some cases, ethyl alcohol.

Ethylene Oxide has been used for many years to sterilize temperature-sensitive medical equipment and

TABLE 4-1

ALCOHOL PRODUCTS

	Approximate Percentage of Alcohol Content, by Volume	Category	Application
Alcohol U.S.P.	95	Topical anti-infective; pharmaceutic aid (solvent)	Topically to the skin, as a 70 percent solution
Rubbing Alcohol U.S.P.	70	Rubefacient	
Diluted Alcohol U.S.P.	50	Pharmaceutic aid (solvent)	
Isopropyl Alcohol U.S.P.	100	Local anti-infective; pharmaceutic aid (solvent)	
Azeotropic Isopropyl Alcohol U.S.P.	91	Local anti-infective	
Isopropyl Rubbing Alcohol U.S.P.	70	Rubefacient	

more recently has been found to be of value in the sterilization of certain thermolabile pharmaceuticals. As a gas it will diffuse through porous material, it is readily removed by aeration following treatment and effectively destroys all forms of microorganisms at ordinary temperatures.[4] It is a colorless, flammable gas at ordinary room temperature and pressure but can be liquefied at 12°. The gas in air forms explosive mixtures in all proportions from 3 to 80 percent by volume. The explosion hazard is eliminated when the ethylene oxide is mixed with more than 7.15 times its volume of CO_2. Carboxide is a commercially available product which is 10 percent ethylene oxide and 90 percent CO_2; it can be released in the air in any quantity without forming an explosive mixture. Water vapor is a factor in ethylene oxide sterilization. The amount of water vapor which must be added to the gas appears to depend on the amount of moisture absorbed by the material to be sterilized.[5] Plastic intravenous injection equipment can be sterilized in the shipping carton, using ethylene oxide.[6]

Formaldehyde Solution U.S.P., formalin, formol. Formaldehyde solution is a colorless, aqueous solution containing not less than 37 percent of formaldehyde (CH_2O) with methanol added to prevent polymerization. It is miscible with water or alcohol and has the pungent odor that is typical of the lower members of the aliphatic aldehyde series.

Owing to the ease with which oxidation and polymerization can take place, the chief impurities to be found in the solution are formic acid and paraformaldehyde.

On long standing, especially in the cold, the solution may become cloudy. Therefore, it should be preserved in tightly closed containers at temperatures not below 15°.

Formaldehyde was prepared first by Hofmann (1868) by passing a hot mixture of methyl alcohol and air over platinum. It still is obtained commercially in the same way, although a variety of catalysts have been utilized, including copper, silver, oxides of iron and molybdenum and vanadium pentoxide. It also can be produced by the oxidation of methane in natural gas.[7] Formaldehyde does not occur naturally in significant amounts. It is frequently found in the aqueous distillate during the preparation of volatile oils from plants.

Formaldehyde differs from typical aliphatic aldehydes in some important reactions. When evaporated with a solution of ammonia, it forms methenamine by condensation.

$$6\,HCHO + 4\,NH_3 \longrightarrow (CH_2)_6\,N_4 + 6\,H_2O$$

Methenamine

It shows a remarkable tendency to polymerize, since evaporation of the solution yields a white, friable mass of paraformaldehyde $(CH_2O)_n$. If a strong solution is distilled with 2 percent sulfuric acid and the vapors are condensed quickly, a crystalline trimer known as trioxane or trioxymethylene is formed. Either one of these products can be depolymerized by heat, thus giving a convenient source of formaldehyde for synthetic and other processes.

Trioxymethylene

Formaldehyde, either as a gas or in solution, has a powerful effect on all kinds of tissue; it is irritating to mucous membranes, hardens the skin and kills bacteria or inhibits their growth. It is an excellent germicide, probably equal to phenol or mercury, and its volatility renders it more penetrating. A dilution of 1:5,000 inhibits the growth of any organism, and, in many cases, 1:20,000 will retard any multiplication. Large doses by mouth cause the usual symptoms of gastroenteritis and ultimate collapse. The gas is very irritating when inhaled.

The gas has been employed to disinfect rooms, excreta, instruments and clothing but is little used at present.

Usually, applications to the body are not to be recommended, but, diluted with water or alcohol, the solution has been applied as a hardener of the skin, to prevent excessive perspiration and, also, to disinfect the hands or the site of an operation.

Paraformaldehyde, obtained by evaporating a formaldehyde solution, also is known as paraform, triformol and, erroneously, trioxymethylene. It is a white powder that is slowly soluble in cold water and more readily soluble in hot water, but with some decomposition, to produce an odor of formaldehyde. Because it can be converted completely to the gas by heating, it is used largely as a convenient form of transportation. It has been used as the active ingredient of contraceptive creams.

Glutaral Concentrate U.S.P., Glutaraldehyde, Cidex®. This aldehyde is used as a sterilizing solution for equipment and surgical instruments which cannot be heat-sterilized. A 2 percent aqueous solution is moderately bacteriacidal against a broad spectrum of bacteria and spores when adjusted to pH 7.5 to 8.5. The acidity of aqueous solutions is probably caused by the cyclic hydrated (hemiacetal) form. Such acidic solutions are stable for at least 2 years but lack sporicidal activity. At high pH's, glutaraldehyde polymerizes quite rapidly, but at pH 7.5 to 8.5 the rate is slow enough so that the activity is maintained for about 2 weeks.

$$\underset{\text{Glutaraldehyde}}{H-\overset{\displaystyle O}{\overset{\displaystyle \|}{C}}-CH_2CH_2CH_2-\overset{\displaystyle O}{\overset{\displaystyle \|}{C}}-H}$$

The commercial product (Cidex®) is a stabilized alkaline glutaraldehyde solution, which actually consists of two components that are mixed together immediately prior to use. The activated solution thus prepared contains 2 percent glutaraldehyde buffered to pH 7.5–8.0. Stabilized solutions retain 86 percent of their original activity 30 days after preparation,[8] while the non-stabilized alkaline solutions lose 44 percent of their activity after 15 days.

PHENOLS AND THEIR DERIVATIVES

Physiologic properties

The bactericidal activity of most substances has been compared with that of Phenol U.S.P. as a standard, and this activity is reported as the phenol coefficient.[9,10] The phenol coefficient is defined as the ratio of the dilution of a disinfectant to the dilution of phenol required to kill a given strain of a definite microorganism, *Eberthella typhosa*, under carefully controlled conditions in a specified length of time. For example, if the dilution of the substance undergoing the test is 10 times as great as that of phenol, which is the compound used and taken as unity, then the phenol coefficient (P.C.) is 10. This method of testing contains variables that do not permit easy duplication of results by different laboratories; also, for another organism the coefficient may be considerably different. The phenol coefficient of many phenols is very temperature-dependent.[11]

The pH at which phenolics are tested or used can markedly alter their effectiveness. Thus, lower pH conditions can enhance the activities of the more acidic phenols through a suppression of their ionization. The reason for this is that the nonionized (neutral), more lipid-soluble form can penetrate cell membranes more readily than the ionized, less lipid-soluble form. For the less acidic phenols this is not the case, i.e., the activity of *p-tert*-amylphenol-C^{14} is not affected by pH changes from 4.8 to 10 as measured by its uptake. The active form of the phenol may be either the neutral or ionized form or both. The anion ARO^- can accept some proton at some key enzymatic or biologically important site. At pH 6.8 and at physiologic pH, most phenols are very little ionized and are assumed to be active in this state.

Almost all phenolic compounds exhibit some antibacterial properties, and this activity is not too specific, although in some cases the phenol coefficients of a given phenol for *Eberthella typhosa* and *Staphylococcus aureus* may differ quite widely.

The cell walls of gram-negative bacteria contain more lipids than the mucopeptide nature of the cell walls of gram-positive organisms. This may account for the greater antibacterial activity of some of the more lipid-soluble phenolics for gram-negative bacteria than for gram-positive bacteria. This is not true in all cases, i.e., *n*-octyl resorcinol has a phenol coefficient of 680 for *Staph. aureus* and is quite inert for *E. typhosa*. Also, *n*-heptyl phenol is extremely effective against *S. typhosa*.

Cell membranes of microbes are networks of highly organized structures to or in which many enzymes are fixed. Some of these are cytochromes, Na^+, K^+, activated ATPases, NAD-oxidase and acid phosphatase. One would suspect that sufficient disruption of such highly organized structures by phenols would indeed lead to inhibition of some vital processes. In some cases phenolics may interact with DNA to exert some supplementary effects. The antimicrobial activity of phenols may be due to structural damage and alteration of permeability mechanisms of microsomes, lysosomes and cell walls.

Although this type of activity is characteristic of some antibiotics, the general antibacterial effects of many phenols are irreversible by dilution with water. Furthermore, bacteria cannot acquire immunity to an initial inhibitory concentration of a phenol. Therefore, phenols have great value as economically useful antimicrobial agents.

Because phenol itself is antiseptic, early workers (Ehrlich, 1906, and others) sought to improve its activity by modifications of its structure. The introduction of the halogens, chlorine or bromine, into the nuclei of phenols increases their antiseptic activities. This activity increases with the increase in the number of halogens introduced; however, the solubility decreases, thus rendering the polyhalogenated phenols much less useful. Furthermore, phenols, as well as their halogen derivatives, are too toxic for internal use. The introduction of nitro groups increases the antiseptic activity to a moderate degree, whereas carboxyl and sulfonic acid groups are ineffective or moderately effective. The introduction of alkyl groups into phenol, cresols and so on causes a marked increase in antiseptic activity. The structure and the size of the alkyl chain exert marked differences in their effects. Normal alkyl chains are more effective than isoalkyl chains, which in turn are more effective than secondary chains, and the tertiary chains are the least effective; however, the latter do exert considerable activity. Alkoxyl groups also increase the activity of phenols. It is noteworthy that in some cases increased antimicrobial activity is not accompanied by an equal increase in toxicity, i.e., *n*-amyl phenol is one tenth as toxic as phenol and *p-n*-amyl-*o*-chlorophenol is about one thirtieth as toxic as *o*-chlorophenol.

Table 4–2 gives the phenol coefficients of some of the substituted phenols and of some of the better-known antiseptics.[12]

The alkyl phenols, although powerful antiseptics, are too toxic for internal use and are used for skin sterilization, skin antiseptics and oral antiseptics.

Phenols exert a definite vermicidal activity which is enhanced by the presence of alkyl groups. The most effective anthelmintics in this group must have a solubility in water of a relatively low order (1:1,000 to 1:2,000) so as to prevent too great absorption from the stomach and intestines.

Phenols and their derivates have antiseptic, anthelmintic, anesthetic, keratolytic, caustic, vesicant and protein precipitant properties. The extent of activity, in any one or more of the above properties, varies with the type of phenol, i.e., mono-, di- and trihydroxy substitution, and with the type and extent of substitution, i.e., alkyl, alkoxyl, acetoxy, halogen, nitro, sulfonic, and other groups.

As a rule, phenols are inactive in the presence of serum, possibly because they combine with serum albumin and serum globulin and, thus, are not free to act upon the bacteria. A second undesirable feature of phenols for bloodstream infections and other infections involving blood is the inhibitory effect upon leukocytic activity as compared with their activity to inhibit bacterial growth.

Because phenols have a well-known tendency to bind to proteins, a limited number of substituted phenols have been examined for their serum albumin and mitochondrial protein-binding properties. The binding properties depend on the lipophilic character of the substituent, and a linear free-energy relationship exists between the logarithm of the binding constants and substitutent π ($\pi = \log P_x/P_H$, where P_H is the partition coefficient of a parent compound between octanol and water, and P_x is that of the derivative).[13]

TABLE 4-2

PHENOL COEFFICIENTS OF SOME SUBSTITUTED PHENOLS AND SOME ANTISEPTICS*

Compound	Organism		
	E. typhosa	Staph. aureus	Strep. hemolyticus
Phenol	1.0	. . .	. . .
2-Chlorophenol	3.6	3.8	. . .
3-Chlorophenol	7.4	5.8	. . .
4-Chlorophenol	3.9	3.9	. . .
2,4-Dichlorophenol	13.0	13.0	. . .
2,4,6-Trichlorophenol	23.0	25.0	. . .
p-Methylphenol†	2.5§	. . .	. . .
p-Ethylphenol	7.5§	10.0	. . .
p-n-Propylphenol	20.0§	14.0	. . .
p-n-Butylphenol	70.0§	21.0	. . .
p-n-Amylphenol	104.0§	20.0	. . .
Thymol	. . .	28.0‖	. . .
Chlorothymol	. . .	61.3	
4-Ethylmetacresol‡	12.5§	. . .	. . .
4-*n*-Propylcresol	34.0§	. . .	. . .
4-*n*-Butylcresol	100.0§	. . .	. . .
4-*n*-Amylcresol	280.0§	. . .	. . .
p-Methyl-*o*-chlorophenol	6.3	7.5	5.6
p-Ethyl-*o*-chlorophenol	17.3	15.7	15.0
p-n-Propyl-*o*-chlorophenol	38.0	32.0	35.0
p-n-Butyl-*o*-chlorophenol	87.0	94.0	89.0
p-n-Amyl-*o*-chlorophenol	80.0	286.0	222.0

*After Suter, C. M.: Chem. Rev. 28:269, 1941.
†Position shown to be unimportant.
‡The 4-alkylorthocresols and the 2-alkylparacresols assay much like the 4-alkylmetacresols.
§20°.
‖25°.

PRODUCTS

Phenol U.S.P., carbolic acid. Phenol is monohydroxybenzene obtained from coal tar in 0.7 percent yield by extraction with alkali. The phenolates are decomposed, and the phenolic fraction subjected to fractional distillation for separation and purification of the individual phenolic fractions. This source of phenol was not sufficient to meet the demand, and phenol is now synthesized on a commercial scale.[14]

Phenol is often called carbolic acid, and this terminology is derived from its weakly acidic properties. Its sodium and potassium salts (sodium and potassium phenolates) are soluble in water.

Phenol occurs as colorless to light pink, interlaced, or separate, needle-shaped crystals or as a white or light pink, crystalline mass that has a characteristic odor. It is soluble 1:15 in water, very soluble in alcohol, glycerin, fixed and volatile oils and is soluble in petrolatum and liquid petrolatums 1:70. Water is soluble 10 percent in phenol.

Phenol is the most stable member of the group of phenols, although slight oxidation does take place upon exposure to air. It can be sterilized by heat and readily forms eutectic mixtures with a number of compounds, such as thymol, menthol and salol. Phenol is substituted readily by bromine to form the insoluble tribromo derivative.

Phenol is one of the oldest antiseptics, having been introduced in surgery by Sir Joseph Lister in 1867. In addition to its bactericidal activity, which is not very strong, it has a caustic and slight anesthetic action. Phenol is, in general, a protoplasmic poison and is toxic to all types of cells. High concentrations will precipitate proteins, whereas low concentrations denature proteins without coagulating them. This denaturing activity does not firmly bind phenol and, thus, it is free to penetrate the tissues. The action on tissues is a toxic one, and pure phenol is corrosive to the skin, destroying much tissue, and may lead to gangrene. Even the prolonged use of weak solutions of phenol in the form of lotions is apt to cause tissue damage and dermatitis.

Phenol is used commonly in 0.1 to 1 percent concentrations as an antipruritic in phenolated calamine lotion or as an ointment or simple aqueous solution. Aqueous solutions stronger than 2 percent should not be applied to the surface of the body. Pure phenol in very small amounts may be used to cauterize small wounds. A 4 percent solution in glycerin may be used if necessary. Crude phenol is cheap enough to be used for a disinfectant. Phenol is too soluble and too readily absorbed to be of value as an intestinal antiseptic.

Liquefied Phenol U.S.P. Liquefied carbolic acid is phenol maintained in a liquid state by the presence of 10 percent of water. Liquefied phenol is a solution of water in phenol and is a convenient way in which to use it in most pharmaceutic applications. However, its water content precludes its use in fixed oils, petrolatum and liquid petrolatum.

Parachlorophenol U.S.P., 4-chlorophenol. Mono-p-chlorophenol is prepared by the chlorination of phenol under conditions which will yield predominantly the para isomer, which can be separated very easily from the ortho isomer by fractional distillation.

Phenol   p-Chlorophenol   o-Chlorophenol

Parachlorophenol has a phenol coefficient of about 4 and is used in combination with camphor in liquid petrolatum.

The introduction of chlorine, although it markedly increases the antiseptic value of the parent compound, also decreases the solubility in water. Therefore, the polychlorinated phenols have little application in pharmacy. However, it is worthy of note that pentachlorophenol is an outstanding commercial wood preservative by virtue of its powerful fungicidal properties. It has a phenol coefficient of 50.

Hexachlorophene U.S.P., Gamophen®; Surgi-Cen®, pHisoHex®, Hex-O-San®, Germa-Medica®, 2,2' - methylene - bis(3,4,6 - trichlorophenol), 2,2'dihy - droxy-3,5,6,3',5',6'-hexachlorodiphenylmethane. It is synthesized as follows:

2,4,5-Trichlorophenol   Hexachlorophene

It is a white to light tan, crystalline powder that is insoluble in water, soluble in alcohol, acetone, the lipid solvents and is stable in air.

Most biphenolic compounds are far more effective than the monomers; moreover, their chlorine content further increases the antiseptic activity (phenol coefficient, 40 for *S. Aureus* and 15 for *Salm. typhi*).[15] This type of compound is deposited on the skin either through a combination with the epidermis or the sebaceous glands or both. Its usefulness as an antiseptic in low concentrations, is, therefore, prolonged. Hexachlorophene is incorporated in soaps, detergent creams, oils and other suitable vehicles for topical application in 2 to 3 percent concentrations. It is effective against gram-positive bacteria, whereas gram-negative organisms are much more resistant to its action.

Although the systemic toxicity of hexachlorophene in animals[16] following oral and parenteral administration had been known for some time, it was not until the late 1960's and the early 1970's that reports of neurological toxicity in infants and burn patients prompted the Food and Drug Administration to ban its use in OTC antiseptic and cosmetic preparations.[17]

Many bisphenols have pronounced fungicidal and bactericidal activities.[18,19] Additionally, two bisphenols, bithional and dichlorophen, are employed as anthelmintic agents.

Some surface-active agents, such as Tween 80, markedly decrease the activity of hexachlorophene.[20]

Cresol. Cresol, also called cresylic acid and tricresol, is a mixture of three isomeric cresols obtained from coal tar or petroleum. The alkali-soluble fraction of coal tar is subjected to fractional distillation, and the three isomeric hydroxytoluenes are obtained as one fraction because they are not readily resolved into pure entities.

o-Cresol   *m*-Cresol   *p*-Cresol

Cresol is a yellowish to brownish-yellow or pinkish, highly refractive liquid that has a phenol-like, sometimes empyreumatic odor. It is soluble in alcohol or glycerin, and 1 ml. is soluble in 50 ml. of water.

By virtue of the methyl groups, the cresols have a phenol coefficient for *E. typhosa* of about 2.5. Cresol supplies the need of a cheap antiseptic and disinfectant.

Thymol N.F. Thymol or thyme camphor is isopropyl *m*-cresol. It is prepared from the oil of thyme (*Thymus vulgaris*) by extraction with alkali and subsequent acidulation of the alkaline extract to liberate

Thymol

the thymol. It can be distilled to effect further purification if necessary.

Thymol occurs as colorless and, at times, large crystals, or as a white crystalline powder that has an aromatic, thymelike odor and a pungent taste. It is soluble 1:1,000 in water, 1:1 in alcohol and is soluble in volatile and vegetable oils.

Although thymol is affected by light, it is quite stable and can be sterilized by heat. It forms eutectic mixtures (see Phenol).

Thymol is used in Trichloroethylene N.F. at a level of about 0.01 percent as an antimicrobial agent.

Thymol has fungicidal properties and is effective in controlling dermatitis caused by pathogenic yeasts. It is effective in a 1 percent alcoholic solution for the treatment of epidermophytosis and in a 2 percent concentration in dusting powders for ringworm.

Thymol   Chlorothymol

It is an effective antiseptic. It has a phenol coefficient of 61.

Eugenol U.S.P., 4-allyl-2-methoxyphenol, is a phenol obtained from clove oil and from other sources. Clove oil, which contains not less than 82 percent of eugenol, is extracted with alkali, and the eugenol is freed subsequently from the separated alkaline extract by acidulation. Further purification can be effected by distillation. A number of other volatile oils also contain eugenol.

Eugenol is a colorless or pale yellow liquid, having a strongly aromatic odor of clove and a pungent, spicy taste. It is slightly soluble in water, soluble in twice its volume of 70 percent alcohol and is miscible with alcohol and with fixed and volatile oils.

Eugenol

The para-allyl and ortho-methoxy groups contribute to the antiseptic and anesthetic activity of the phenolic group, so much that eugenol is used for toothaches, and for its antiseptic activity in mouthwashes. It has a phenol coefficient of 14.4

Resorcinol U.S.P., *m*-dihydroxybenzene, resorcin, is prepared synthetically.

Resorcinol occurs as white or nearly white, needle-shaped crystals or powder. It has a faint, characteristic odor and a sweetish, followed by a bitter taste. One gram is soluble in 1 ml. of water and in 1 ml. of alcohol, and it is freely soluble in glycerin. Resorcinol should be stored in dark-colored or light-resistant containers. It is much less stable in solution, particularly in the presence of alkaline substances.

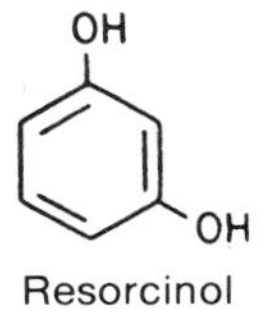

Resorcinol

Although resorcinol is feebly antiseptic (phenol coefficient = 0.4 for both *E. typhosa* and *Staph. aureus*), it is used in 1 to 3 percent solutions and in ointments and pastes in 10 to 20 percent concentrations for its antiseptic and keratolytic action in skin diseases, such as ringworm, parasitic infections, eczema, psoriasis and seborrheic dermatitis. Resorcinol has some fungicidal properties. It is very poorly bound by protein.

Resorcinol Monoacetate U.S.P., euresol. This compound is prepared by partial acetylation of resorcinol.

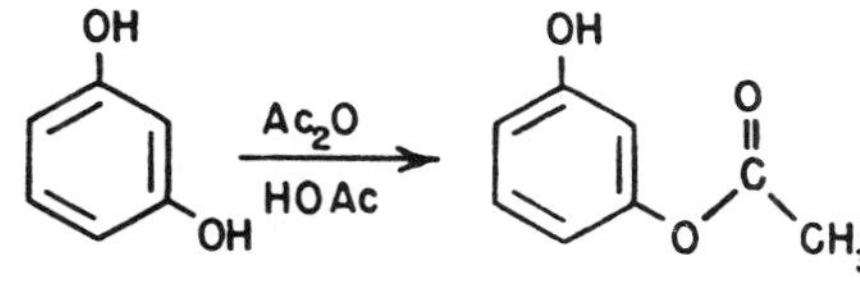

Resorcinol   Resorcinol Monoacetate

Resorcinol monacetate is a viscous, pale yellow or amber liquid with a faint characteristic odor and a burning taste. It is soluble in alcohol and sparingly soluble in water. Although partial acetylation of resorcinol has increased its stability, it should be stored in tight, light-resistant containers.

As a general rule, the esters of phenols are not as stable as other organic esters. They are hydrolyzed

very easily by alkalies or alkaline solutions. They will even hydrolyze very slowly in the solid state in the presence of moisture.

Resorcinol is partially acetylated to produce a milder product with a longer-lasting action. Prior to hydrolysis, the ester group contributes properties similar to an alkyl group. The resorcinol monoacetate is hydrolyzed slowly to liberate the resorcinol. It is used in skin conditions such as alopecia, seborrhea, acne, sycosis and chilblains.

Resorcinol monoacetate is equal or superior to undecylenic acid for *Candida albicans* and *Microsporum gypseum*. Resorcinol monbenzoate is fungistatic at 0.1 to 0.01 percent concentrations and is less toxic and more active than resorcinol monoacetate.

It is used in 5 to 20 percent concentrations in ointments and in 3 to 5 percent alcoholic solutions for scalp lotions.

Hexylresorcinol U.S.P., Crystoids®, 4-hexylresorcinol, is prepared as shown below. The first step takes place in the presence of anhydrous zinc chloride, and the condensation is of the Friedel-Crafts type. Resorcinol is substituted so readily that an acid, rather than an acid chloride, can be used. Also, because of this ease of substitution, the moderately active zinc chloride provides adequate catalysis. In the second step, a typical Clemmensen reduction, using zinc amalgam and dilute hydrochloric acid, will reduce the ketone to the hydrocarbon.

Hexylresorcinol occurs as white, needle-shaped crystals. It has a faint odor and a sharp astringent taste; it produces a sensation of numbness when placed on the tongue. It is freely soluble in alcohol, glycerin and vegetable oils and is soluble 1:2,000 in water. It is sensitive to light.

Although resorcinol is feebly antiseptic, it is less toxic than phenol. This is the basis for the preparation of numerous alkylated resorcinols, the most effect of which was the 4-*n*-hexyl, which has a phenol coefficient of 46 to 56 against *E. typhosus* and 98 against *Staph. aureus*.

Hexylresorcinol was introduced by Leonard and marketed as a 1:1,000 solution under the name of S.T. 37. It was recommended as a general skin antiseptic, being effective for both gram-positive and gram-negative organisms. It can be administered dissolved in olive oil in capsules to be used for an effective urinary

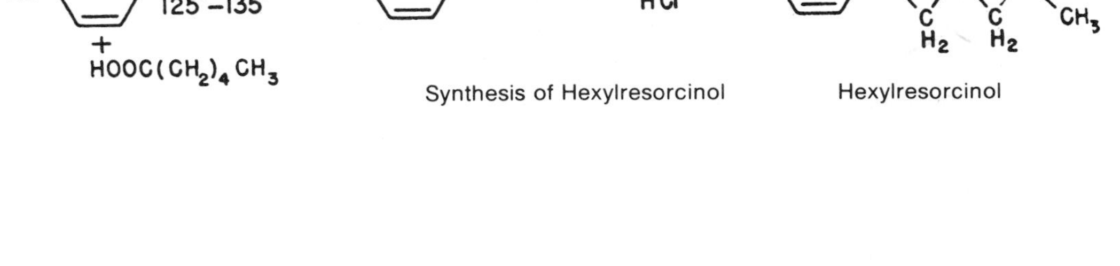

Synthesis of Hexylresorcinol   Hexylresorcinol

TABLE 4–3

PHENOL COEFFICIENTS OF 4-ALKYLRESORCINOLS*

| 4-Alkylresorcinol | Phenol Coefficient | |
	E. typhosa	Staph. aureus
n-Propyl	5	3.7
n-Butyl	22	10.0
Isobutyl	15	. .
n-Amyl	33	30
Isoamyl	24	. .
n-Hexyl	46-56	98
Isohexyl	27	. .
n-Heptyl	30	280
n-Octyl	0†	680
n-Nonyl	. .	980

*After Suter, C.M.: Chem. Rev. 28:269, 1941.
†At 45° C., more active than hexyl, heptyl and octyl resorcinols.[3]

antiseptic and an anthelmintic for Ascaris and hookworms. Its low solubility in water makes it an effective anthelmintic; however, sufficient amounts are absorbed to be of value in urinary tract infections.

Hexylresorcinol is irritating to the respiratory tract and to the skin, and an alcoholic solution has vesicant properties. This vesicating effect is a general property of alkylated phenols and reaches a very high degree in urushiol.

The alkyl substituted phenols and resorcinols possess the ability to reduce surface tension. It is believed that these compounds may owe at least part of their increased bactericidal activity to this ability to lower surface tension, because many surface-active agents are very effective germicides. Hexylresorcinol, like many of the alkyl phenols, exhibits some local anesthetic activity.

Anthralin U.S.P., cignolin, dithranol, 1,8,9-anthracenetriol, 1,8-dihydroxyanthranol, is prepared as follows:

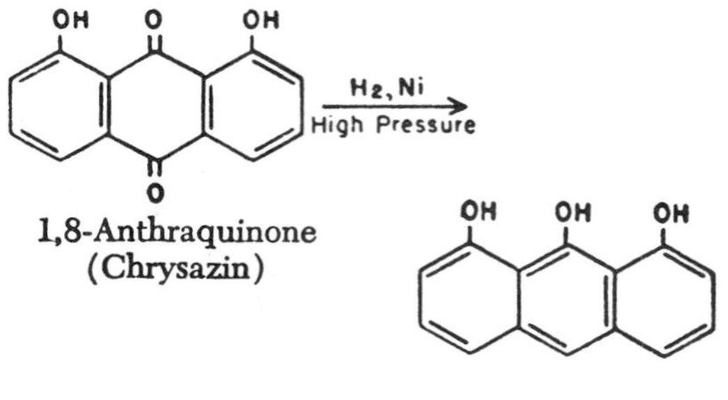

1,8-Anthraquinone
(Chrysazin)

Anthralin

TABLE 4–4

PHENOL PRODUCTS

Name Proprietary Name	Preparations	Category	Application*
Phenol U.S.P.		Pharmaceutic aid (preservative)	
Liquefied Phenol U.S.P.		Topical antipruritic	Topically to the skin, as a 0.5 to 2 percent lotion or ointment
Parachlorophenol U.S.P.	Camphorated Parachlorophenol U.S.P.	Anti-infective (dental)	Topically to root canals and the periapical region
Hexachlorophene U.S.P.	Hexachlorophene cleansing emulsion U.S.P.	Topical anti-infective; detergent	Topically to the skin, as the sole detergent, followed by thorough rinsing
pHisoHex, Presulin Cleanser	Hexachlorophene Liquid Soap U.S.P.		
Thymol N.F.		Pharmaceutic aid (stabilizer)	
Anthralin U.S.P.	Anthralin Ointment U.S.P.	Topical antipsoriatic	Topical, to the skin, as a 0.1 to 1 percent ointment once or twice daily
Eugenol, U.S.P.	Zinc-Eugenol Cement U.S.P.	Dental Protective	Topical to dental cavities
Resorcinol U.S.P.		Keratolytic	Topically to the skin, as a 2 to 20 percent or ointment
	Resorcinol Ointment U.S.P.	Local antifungal; keratolytic	Topical, to the skin as a 2 to 20 percent ointment or lotion
Resorcinol Monoacetate U.S.P. *Euresol*	Resorcinol Lotion U.S.P.	Antiseborrheic; keratolytic	Topical, for application to the scalp
Hexylresorcinol U.S.P. *Crystoids*	Hexylresorcinol Pills U.S.P.	Anthelmintic (intestinal round-worms and trematodes)	1 g. May be repeated at weekly intervals if necessary

*See U.S.P. D.I. for complete dosage information.

Anthralin occurs as an odorless, tasteless, crystalline, yellowish-brown powder that is insoluble in water, slightly soluble in alcohol and soluble in most lipoid solvents. Anthralin has antiseptic, irritant and proliferating properties which indicate its use as a substitute for chrysarobin in the treatment of psoriasis, chronic dermatomycosis and chronic dermatoses.

Chloroxine

Chloroxine, 5,7-dichloro-8-quinclinol., this halogenated 8-hydroxyquinoline is used in a 2 percent cream as a keratolytic agent for application to the scalp.

OXIDIZING AGENTS

Oxidizing agents which are of value as antiseptics depend on the liberation of oxygen, and many are in the inorganic class. Included are such compounds as hydrogen peroxide, other metal peroxides, potassium permanganate and sodium perborate.

Carbamide Peroxide Solution U.S.P. This is a solution of about 12.6 percent carbamide peroxide in anhydrous glycerin. Carbamide peroxide is a stable complex of urea and hydrogen peroxide, $H_2NCONH_2 \cdot H_2O_2$. Hydrogen peroxide is released when the glycerin solution is mixed with water. Several drops are applied to the affected area and then removed after 2 to 3 minutes.

Hydrous Benzoyl Peroxide U.S.P., Benoxyl®, Oxy-5®, Persadox®, Vanoxide®, is a white, granular powder with a characteristic odor. It contains about 30 percent water to make it safer to handle.

Benzoyl Peroxide

Benzoyl peroxide is employed in concentrations of 5 percent to 10 percent as a keratolytic and keratogenic agent for the control of acne. Like other peroxides, it is chemically unstable and can explode when heated. Nonstabilized aqueous solutions slowly decompose to hydrogen peroxide and benzoic acid. Lotions containing benzoyl peroxide are stabilized with the addition of 2 parts of dicalcium phosphate.

The value of benzoyl peroxide in acne treatment is believed to derive from its irritant properties.[21] It induces proliferation of epithelial cells, leading to sloughing and repair.

HALOGEN-CONTAINING COMPOUNDS

Iodophors

Various surfactants will act as solubilizers or carriers for iodine with the resulting complex possessing antibacterial properties. In practice, the nonionic surfactants along with the addition of an acid to stabilize the product and to enhance the antibacterial properties have been most successful.[22] About 80 percent of the iodine which dissolves in the carrier remains as bacteriologically active or available iodine. Phosphoric acid is used because of its buffering action in the pH range of 3 to 4.[23] Iodophors have been found to be fungicidal, active against tubercle bacilli and effective in moderate concentrations against *Bacillus subtilis;* they show some loss of activity in the presence of serum.[24]

Povidone-Iodine U.S.P., Betadine®, Isodine®, is a complex of iodine with poly(1-vinyl-2-pyrrolidinone). It is water-soluble and releases iodine slowly, providing a nontoxic, nonvolatile and nonstaining antiseptic. It contains about 10 percent of available iodine.

As an aqueous solution it is useful for skin preparation prior to surgery and injections, for the treatment of wounds and lacerations and for bacterial and mycotic infections of the skin.

Chlorine-containing compounds

N-Chlorocompounds are represented by amides, imides and amidines in which one or more of the hydrogen atoms attached to nitrogen have been replaced by chlorine. All of these products are designed to liberate hypochlorous acid (HClO) and, therefore, simulate the antiseptic action of hypochlorites, such as Sodium Hypochlorite Solution U.S.P.

In contact with water, the N-chlorocompounds slowly liberate hypochlorous acid. The antiseptic property is greatest at pH 7 and decreases as the solution becomes more alkaline or acidic. It is known that hypochlorous acid will chlorinate amide nitrogen, and it is assumed to attack bacterial protein by this route. Proteins are chlorinated as follows:

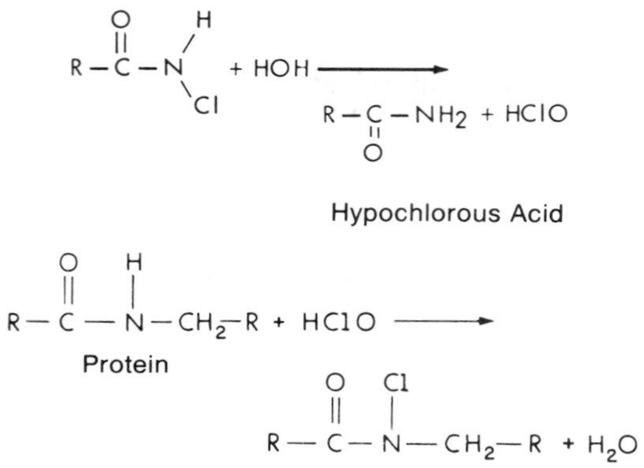

Hypochlorous Acid

Protein

The term "active chlorine" is associated with these N-chlorocompounds and hypochlorites, which means the chlorine that is liberated from a substance when treated with an acid.

Products

Halazone U.S.P., *p*-dichlorosulfamoylbenzoic acid, *p*-sulfondichloramidobenzoic acid, *p*-carboxysulfondichloramide, is a white crystalline powder with a chlorinelike odor. It is affected by light. It is slightly soluble in water and chloroform and is soluble in dilute alkalines. The sodium salt of the compound is used in sterilizing drinking water.

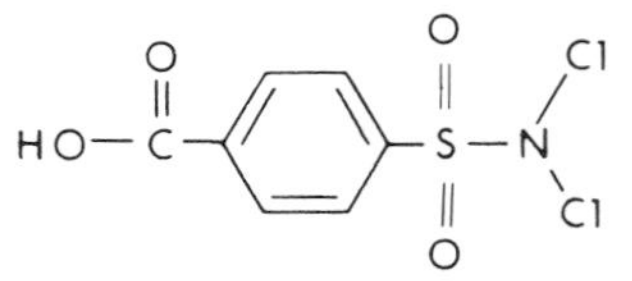

Halazone

Chloroazodin U.S.P., Azochloramid®, N,N'-dichlorodicarbonamidine, contains the equivalent of not less than 37.5 percent and not more than 39.5 percent of active chlorine (Cl). It is prepared by treating a solution of guanidine nitrate in dilute acetic acid and sodium acetate with a solution of sodium hypochlorite at 0°.

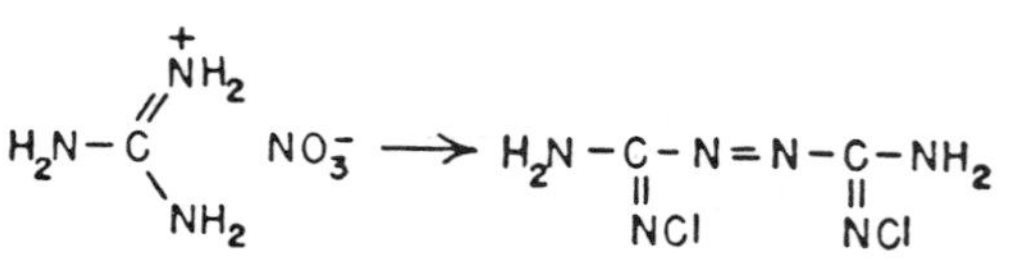

Chloroazodin

It consists of bright yellow needles or flakes with a faint odor of chlorine and a slightly burning taste, and it is explosive at about 155°. It is not very soluble in water or other solvents, including glyceryl triacetate (triacetin), and the solutions decompose on warming or exposure to light.

Chloroazodin is similar to the chloramines and to sodium hypochlorite, but it does not react rapidly with water or reagents, and its action in use is relatively prolonged. Solutions are used on wounds (1:3,300), as a packing for cavities and for lavage and irrigation; dilutions up to 1:13,200 have been proposed for mucous membranes. It often is used in isotonic solutions buffered at pH 7.4. For dressing and packing the stable solution in glyceryl triacetate is employed, and a dilution of this in a vegetable oil (1:2,000) is claimed to be nonirritating to mucous membranes. Tablets of a saline mixture with buffer are available for making solutions.

CATIONIC SURFACTANTS

These agents in aqueous media undergo ionization to form cations having surface-active properties. As an example, the surface activity of lauryl triethyl ammonium chloride resides in the cation,

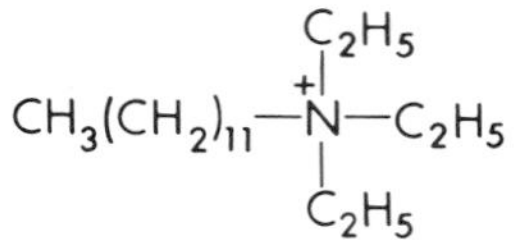

which forms upon ionization. Pharmaceutically important cationic surfactants are the quaternary ammonium compounds. Other derivatives of lesser interest include amine salts (primary, secondary and tertiary), sulfonium and phosphonium compounds.

Major interest in these agents centers on their antimicrobial activity. Although such activity in this class of compounds was reported as early as 1908, it was not until a report by Domagk in 1935 that attention was directed to their use as antiseptics, disinfectants and preservatives.

The cationic surfactants have bactericidal action in high dilutions against a broad range of organisms, both gram-negative and gram-positive. They are active against a number of fungus and protozoal organisms, including several pathogenic varieties. Aqueous solutions of these agents are not active against such organisms as spore-forming bacteria, *Mycobacterium tuberculosis* and viruses.

Various mechanisms have been proposed for the antimicrobial action of the cationic agents. In general, these surfactants are considered to be adsorbed upon the surface of the bacteria. Probably the action most responsible for activity is the inactivation of certain enzymes which follows adsorption. Also contributing to the ultimate death of the bacterial cell is the occurrence of a change in the cell wall integrity and a lysis of intracellular components.

In addition to the potent and wide range of antimicrobial activity, these agents possess several other advantages which make them useful germicides. Included are their properties of low toxicity, high water-solubility, nonstaining, high stability in aqueous solutions and noncorrosiveness to metallic instruments. Also, they possess the surfactant properties of wetting and detergency which increase their usefulness as germicides and disinfectants.

In considering the limitations of these agents, attention is directed toward their numerous incompatibilities. They are inactivated by soaps and other anionic surfactants. All traces of soap should be removed from the skin or other surfaces before using these agents. Other anionic agents, including dyes[25] and various drugs,[26] are similarly incompatible with the cationic surfactants. The presence of calcium and magnesium ions in hard water has been reported to reduce antibacterial activity.[27,28] Nonionic surfactants have also been reported to reduce this activity.[29,30] Inactivation occurs when they are used in the presence of blood, serum, food residues and other complex organic substances. Cationic agents are adsorbed on the glass surfaces of containers; the greatest loss occurs in small containers where a relatively large surface area/volume of liquid ratio exists.[31] In the presence of talc and kaolin they are adsorbed upon the surface of the solid particles and are inactivated.[32] Temperature and pH are additional factors which influence their action. Activity is greater in basic solution than in neutral or acid solution and greater activity occurs as temperature is increased. The effect of these various inactivating factors is frequently reduced through the use of concentrations much greater than required for antimicrobial activity under ideal conditions.

Products

Benzalkonium Chloride N.F., Zephiran® Chloride, Benasept®, Germicin®, Pheneen®, Alkylbenzyldimethylammonium chloride. Benzalkonium chloride is a mixture of alkyldimethylbenzylammonium chlorides of the general formula $[C_6H_5CH_2N(CH_3)_2R]Cl$, in which R represents a mixture of alkyls, including all or some of the group beginning with n-C_8H_{17} and extending to higher homologs, with n-$C_{12}H_{25}$, n-$C_{14}H_{29}$, and n-$C_{16}H_{33}$ comprising the major portion. Variations in properties occur among the members in this mixture. The relationship of physical and antimicrobial properties for a series of alkyl homologs[33] may be seen in Table 4-5.

It is a white, bitter-tasting gel, freely soluble in benzene and very soluble in water and alcohol. Aque-

TABLE 4–5

PROPERTIES OF ALKYLDIMETHYLBENZYLAMMONIUM CHLORIDE DERIVATIVES

R	S.T.* of 0.01% Sol. (100 p.p.m.)	CMC† × 10 (moles/l)	Minimum Bacteriocidal Conc. (p.p.m.)		Minimum Fungicidal Conc. (p.p.m.) C. albicans
			S. aureus	Ps. aeruginosa	
C_8H_{17}	72.3	220	250	>1000	>1000
C_9H_{19}	72.2	84	250	1000	750
$C_{10}H_{21}$	71.9	37	50	750	250
$C_{11}H_{23}$	70.9	14.0	7.5	250	75
$C_{12}H_{25}$	68.7	6.9	7.5	250	25
$C_{13}H_{27}$	67.1	2.7	5	100	7.5
$C_{14}H_{29}$	62.4	1.2	0.75	250	5
$C_{15}H_{31}$	53.9	0.60	2.5	500	2.5
$C_{16}H_{33}$	43.7	0.24	5	500	10
$C_{17}H_{35}$	43.2	0.10	5	500	25
$C_{18}H_{37}$	43.4	0.033	5	500	100
$C_{19}H_{39}$	43.6	0.018	10	750	100

*Surface tension (S.T.) as dynes/cm., measured at room temperature (about 25°C.). Surface tension of water at room temperature is 72.0 dynes/cm.
†Critical micelle concentration (CMC) at room temperature (about 25°C.).

ous solutions are colorless, alkaline to litmus and foam strongly.

Benzalkonium chloride possesses wetting, detergent and emulsifying actions. It is used as a surface antiseptic for intact skin and mucosa at 1:750 to 1:20,000 concentrations. Above 1:10,000, it has proved to be irritant on prolonged contact. It is effective against many pathogenic nonsporulating bacteria and fungi after several minutes exposure. For irrigation, 1:20,000 to 1:40,000 solutions are employed. For storage of surgical instruments, 1:750 to 1:5,000 solutions are used, with 0.5 percent sodium nitrite being added as an anticorrosive agent. For presurgical antisepsis, all traces of soap used in preliminary scrubbing must be removed, or inactivation of the cationic detergent will ensue.

Benzethonium Chloride U.S.P., Phemerol® Chloride, benzyldimethyl[2-[2-[p-(1,1,3,3-tetramethylbutyl)phenoxy]ethoxy]ethyl]ammonium chloride.

The structure of this agent and its relationship to the other available analogs of dimethylbenzylammonium chloride may be seen in Table 4-6.

Benzethonium chloride is a colorless, odorless, bitter crystalline powder, soluble in water and in chloroform.

Its actions and uses are similar to those of benzalkonium chloride. It is employed at 1:750 concentration for general antisepsis of the skin, and for irrigation of the eye, nose or mucous membranes, a 1:5,000 solution is used. Also, a 1:500 alcoholic tincture is available.

Methylbenzethonium Chloride U.S.P., Diaparene®, benzyldimethyl[2-[2-[[4-(1,1,3,3-tetramethylbutyl)tolyl]oxy]ethoxy]ethyl]ammonium chloride.

This is a colorless, crystalline compound, bitter in taste, soluble in water, alcohol and chloroform.

It has a specific use in the bacteriostasis of the intestinal saprophyte, *Bacterium ammoniagenes,* which produces ammonia in decomposed urine and is responsible for diaper dermatitis in infants. The agent is marketed for topical use in the treatment of diaper dermatitis and as a general antiseptic.

Cetylpyridinium Chloride U.S.P., Ceepryn®, 1-hexadecylpyridinium chloride.

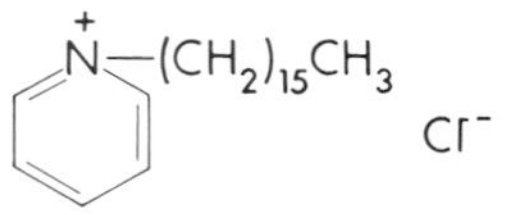

Cetylpyridinium Chloride

In this compound the quaternary nitrogen is part of a heterocyclic nucleus. The cetyl derivative has been selected in preference to other alkyl derivatives studied because of its maximal activity. Also, it is believed that the absence of a benzyl group reduces the toxicity of the compound.

Cetylpyridinium chloride is a white powder which is very soluble in water and alcohol.

It is available for use as a general antiseptic in 1:100 to 1:1,000 aqueous solution on intact skin, 1:1,000 for minor lacerations, and 1:2,000 to 1:10,000 on mucous membranes. The agent is available in the form of throat lozenges and 1:2,000 phosphate-buffered mouthwash/gargle.

Chlorhexidine Gluconate, Hibiclens®, is 1,6-di-(4′-chlorophenyldiguanidino)hexane gluconate, the

TABLE 4-6

ANALOGS OF DIMETHYLBENZYLAMMONIUM CHLORIDE

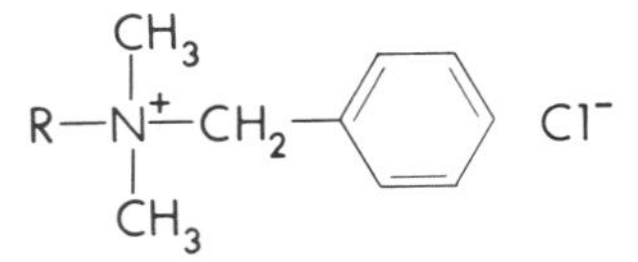

Compound	R
Benzalkonium Chloride	$n - C_8H_{17}$ to $C_{16}H_{33}$
Benzethonium Chloride	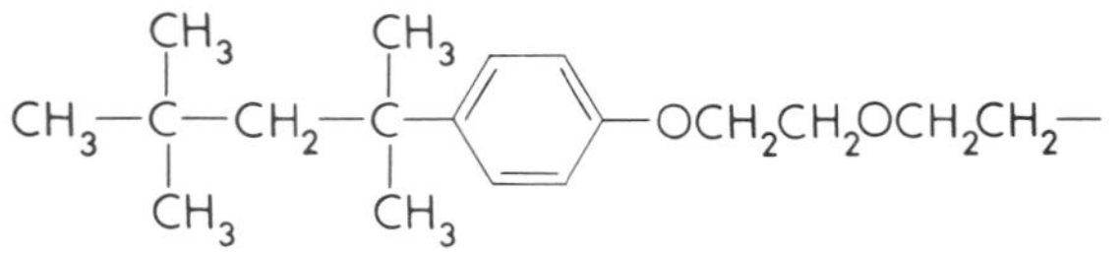
Methylbenzethonium Chloride	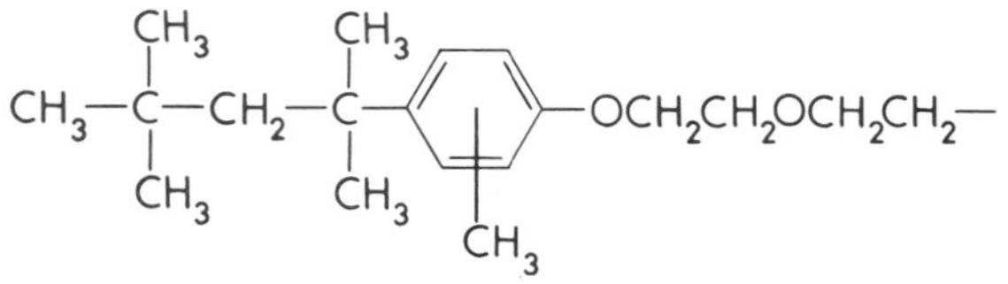

TABLE 4-7

CATIONIC SURFACTANTS

Name Proprietary Name	Preparations	Category	Application	Usual Dose
Benzalkonium Chloride N.F. *Zephiran Chloride, Benasept, Germicin, Pheneen*	Benzalkonium Chloride Solution N.F.	Pharmaceutic aid (antimicrobial preservative)		
Benzethonium Chloride U.S.P. *Phemerol*	Benzethonium Chloride Solution U.S.P. Benzethonium Chloride Tincture U.S.P.	Local anti-infective; pharmaceutic aid (preservative)	Topical, 1:750 solution or 0.2 percent tincture to the skin or 1:5000 solution nasally	
Methylbenzethonium Chloride U.S.P. *Diaparene*	Methylbenzethonium Chloride Lotion U.S.P. Methylbenzethonium Chloride Ointment U.S.P. Methylbenzethonium Chloride Powder U.S.P.	Local anti-infective	Topical, 0.067 percent lotion, 0.1 percent ointment, or 0.055 percent powder	
Cetylpyridinium Chloride U.S.P. *Ceepryn*	Cetylpyridinium Chloride Lozenges U.S.P. Cetylpyridinium Chloride Solution U.S.P.	Local anti-infective; pharmaceutic aid (preservative)	Topical, 1:100 to 1:1,000 solution to intact skin, 1:1,000 solution for minor lacerations, and 1:2,000 to 1:10,000 solution to mucous membranes	Sublingual, 1:1500 lozenge

most effective of a series of antibacterial biguanides originally developed in Great Britain.[34] The antimicrobial properties of the biguanides were discovered as a result of earlier investigations of these substances as potential antimalarial agents (see Chap. 6). Although the biguanides are technically not bis-quaternary ammonium compounds and perhaps should therefore be separately classified, they share many of the same physical, chemical and antimicrobial properties with the cationic surfactants. The biguanides are strongly basic compounds which exist as dications at physiological pH. Like the cationic surfactants, they are inactivated by anionic detergents, and complex anions such as phosphate, carbonate, and silicate.

Chlorhexidine has broad spectrum antibacterial activity, but it is not active against acid-fast bacteria, spores, or viruses. It has been employed for such topical antiseptic uses as preoperative skin disinfection, wound irrigation, bladder irrigation, mouthwashes, and general sanitation. Chlorhexidine is not absorbed through skin or mucous membranes and does not cause systemic toxicity.

DYES

The discovery that some dyes would stain certain tissues and not others led Ehrlich to the idea that dyes might be found that would selectively stain, combine with and destroy pathogenic organisms without causing appreciable harm to the host. He and other workers studied a number of dyes with this idea in view and, as a result of these studies, some azo, thiazine, triphenylmethane and acridine dyes came into use as antiseptics and trypanocides and for other medicinal purposes. However, there appears to be no correlation between the dyeing properties of a series of compounds and their antiseptic or bacteriostatic properties.

Prior to the advent of the sulfonamides and the antibiotics the organic dyes were used more extensively as antibacterial agents than they are today. They were used topically for various skin infections. Their chief disadvantage is that they stain the skin and clothing.

The dyes considered in this chapter as well as many

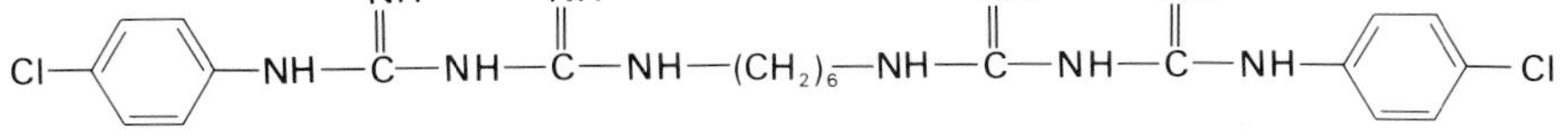

Chlorhexidine

of the certified dyes belong to 4 classes: the azo dyes, the acridine dyes, the triphenylmethane dyes and the thiazine dye, methylene blue. They can be further subdivided on the basis of the charge on the color nucleus when in aqueous solution. Those that ionize with a negative charge are "acid dyes" and are anionic, while those that ionize with a positive charge are called "basic dyes" in contrast with the acid dyes and are cationic.

The acid dyes are usually sulfonic acids and in the salt form are water-soluble and are generally insoluble in hydrocarbons. They all tend to form slightly water-soluble complexes with the basic or cationic dyes. This may also occur with high molecular weight amine salts. The basic dyes, being cationic, do not combine with metal ions. Metal ions such as Mg^{++}, Ba^{++}, Ca^{++}, Cu^{++} and Fe^{++} will discolor some dyes and may form insoluble precipitates with the acidic dyes. As a general rule, the basic dyes are more resistant to reducing conditions than other dyes. They are considered to be light-sensitive, yet in some cases they may be relatively stable. Light-stability of dyes used for coloring sugar-coated tablets is often a problem. The use of insoluble pigments incorporated in a titanium dioxide and syrup suspension will, in many cases, obviate this problem.[35]

Some dyes change color rapidly with the pH and can be used as indicators, while others discolor more slowly and are a stability problem when used as a colorant. Some of the acid dyes may even precipitate at low pH.

Commercial dyes are frequently impure; some of them are mixed with diluents, such as inorganic salts or dextrose. Others may be mixtures of several different colored compounds rather than being composed of one specific compound. Dyes with the same name may vary considerably, depending on the manufacturer.

Some of the confusion in regard to dyes has been removed by standards set up by several official bodies. All dyes used in coloring pharmaceutical products and foods must conform to the Coal Tar Color Regulations established by the United States Food, Drug and Cosmetic Act. Standards for medicinal dyes and food colors also are sanctioned by the *United States Pharmacopeia,* the *National Formulary* and the Dye Certification Division of the United States Department of Agriculture.

Certified colors that are used as colorants in foods and drugs are analyzed and approved by the Food and Drug Administration. To be of certifiable purity each batch of colorant must be virtually free of undesirable by-products and metallic impurities, particularly lead, arsenic and copper. The tolerance for lead is 10 p.p.m. and for arsenic is 1.4 p.p.m. Color certification is controlled to the extent that neither the producer nor the seller may open a container without losing the right to call it certified. If certified dyes are mixed to get a particular shade, the mixture must be recertified. This also applies when repackaged in smaller units without mixing or diluting.

The Food and Drug Administration has classified the certifiable dyes under 3 groups: Group I—Food, Drug and Cosmetic dyes (F. D. & C. dyes), these may be used for coloring foods, drugs and cosmetics; Group II—Drug and Cosmetic dyes (D. & C. dyes), these are designated for use in drugs and cosmetics but not for use in foods; Group III—External Drug and Cosmetic dyes (Ext. D. & C. dyes), these are restricted to use in preparations that will not come in contact with the lips or other mucous membranes and are strictly for use only in externally applied drugs and cosmetics.

Gentian Violet U.S.P., Pyoktannin®, N,N,N',N',N'',N''-hexamethylpararosaniline chloride, crystal violet, methyl violet, methylrosaniline chloride. The commercial product usually contains small amounts of the closely related compounds, penta- and tetramethylpararosaniline chlorides. Some of the methyl violets of commerce have methyl groups substituted in the ring, and there is considerable lack of uniformity in composition of those being distributed commercially. The pure synthetic crystal violet is presumably free of nuclear methyl groups.

Gentian violet occurs as a green powder or as green particles with a metallic luster. The commercial dye frequently contains dextrose and other diluents and should not be used medicinally. It is soluble in water (1:35), in alcohol (1:10) and in glycerin (1:15), but it is insoluble in ether. The dye is much more effective against gram-positive organisms than against gram-negative organisms. It is used topically as a 1 to 3 percent solution in the treatment of *Monilia albicans* infections, vaginal yeast infections, impetigo and Vincent's angina.

In addition to its use as an antibacterial agent, gentian violet is employed as the dye in indelible pencils. Copying leads contain about 33 percent of the dye. Eye injuries from indelible pencils are complicated by the toxic effect of the dye which causes local necrosis that may lead to blindness. In making routine examination of such injuries, employing sodium fluorescein, it was observed that the dye surrounding the injured membrane precipitated and could be removed by flushing with the anionic fluorescein solution. By repeated washings, most of the dye can be removed in this manner. At the present time, sodium fluorescein is the agent of choice for treatment of such injuries, although it should be recognized that other anionic agents of high molecular weight may be superior. The mechanism probably involves precipitation of the dye first, followed by solubilization with excess of the anionic agent.

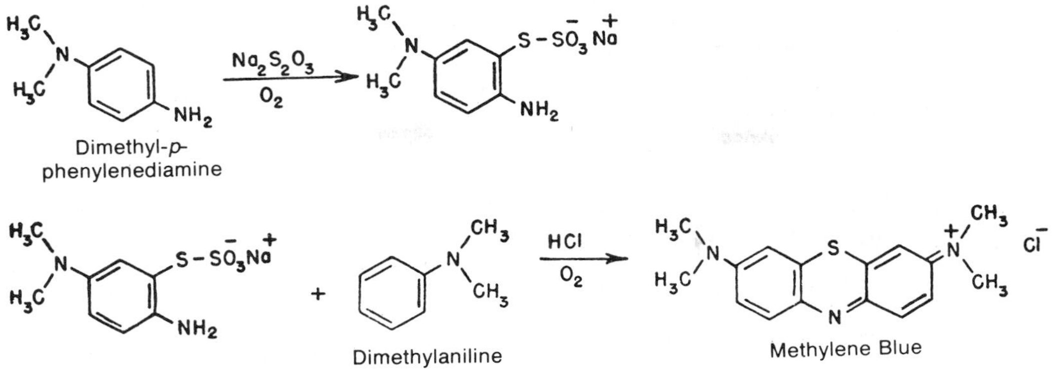

Gentian violet is also used systemically for strongyloidiasis and oxyuriasis.

Basic Fuchsin U.S.P. is a mixture of the hydrochlorides of rosaniline and pararosaniline. It is a metallic-green powder or crystalline solid, soluble in alcohol, with the solution being a carmine red. It is also soluble in water but is insoluble in ether.

Basic fuchsin is an ingredient of carbol-fuchsin solution (Castellani's Paint), which is used topically in the treatment of various fungous infections, including ringworm and "athlete's foot."

It is employed also as Schiff's reagent in testing for aldehydes. This reagent is fuchsin decolorized with sulfur dioxide.

Methylene Blue U.S.P., 3,7-bis(dimethylamino)-phenazathionium chloride, occurs as dark green crystals or powder with a bronze luster. It is soluble in chloroform, in water (1:25) and in alcohol (1:65). Its solutions may be sterilized by autoclaving.

Methylene blue may be synthesized as shown opposite.

It has a comparatively low toxicity and is used to test the renal function of the kidneys and also as a dye in vital nerve staining. It has some action against ma-

laria but is inferior to the cinchona alkaloids, quinacrine and some of the new synthetics in this respect. It is a weak antiseptic that has been used in treating skin diseases and some urinary conditions. Methylene blue also is employed in the treatment of cyanosis resulting from the sulfonamide drugs and as an antidote for cyanide and nitrate poisoning. In proper concentrations, it has been shown to increase the rate of conversion of methemoglobin to hemoglobin. It is used to test for the presence of anaerobic bacteria in milk by the Thundberg technique. Methylene blue is only fairly fast to light, shows moderate stability to oxidizing and reducing agents and good stability to ferrous ions.

MERCURY COMPOUNDS

Mercury and its compounds have been used since early times in the treatment of various diseases. Metallic mercury incorporated in ointment bases was applied locally for the treatment of skin infections and syphilis. A few inorganic mercury compounds have been used orally but are no longer commonly used because of the gastrointestinal disturbances and other toxic manifestations resulting therefrom. A number of

TABLE 4-8

PHARMACEUTIC DYES

Name	Preparations	Category	Application*	Usual Dose*	Usual Pediatric Dose*
Gentian Violet U.S.P.	Gentian Violet Vaginal Suppositories U.S.P.	Topical anti-infective	Topically to the vagina, as a 1.35 percent cream once every 2 days		
	Gentian Violet Solution U.S.P.		Topically to the skin and mucous membranes, as a 0.5 to 2 percent solution 2 or 3 times daily for 3 days		
Methylene Blue U.S.P.	Methylene Blue Injection U.S.P.	Antidote to cyanide poisoning; antidote to methemoglobinemia		I.V., 1 to 2 mg. per kg. of body weight	1 to 2 mg. per kg. of body weight or 25 to 50 mg. per square meter of body surface

*See U.S.P. D.I. for complete dosage information.

organic mercury compounds are now in use mainly as antiseptics, disinfectants and diuretics. In some of these, the mercury is attached to carbon and is held rather firmly to the organic portion of the molecule; in others the mercury is attached to oxygen or nitrogen and may be ionized almost completely or partially.

It appears[36] that the antibacterial action of mercury compounds is explained best on the basis of their interfering with SH (sulfhydryl) compounds that are essential cellular metabolites. Large concentrations of SH compounds will inactivate mercury compounds completely as far as their bactericidal or bacteriostatic action is concerned. This reaction is reversible. Apparently, mercury compounds inhibit the growth of bacteria because the mercury combines with SH groups to form a complex of the type R—S—Hg—R', thus depriving the cell of the SH groups necessary for its metabolism. However, if other SH-containing compounds are introduced which take the mercury away from the bacteria, the latter can grow again.

Experiments have been carried out in which bacteria that have been rendered inactive by mercury compounds have resumed growth when treated with hydrogen sulfide, thioglycollic acid and other sulfhydryl compounds. Thus, apparently, the mercury compounds are not bactericidal but only bacteriostatic in character.

The antibacterial activity of mercurial antiseptics is reduced greatly in the presence of serum and other proteins because the proteins supply SH groups which inactivate the mercury compounds by combining with mercury as they do with arsenic (see BAL). Thus, mercurial antiseptics are more effective on relatively unabraded skin than on highly abraded areas or mucous membranes. Mercurial antiseptics do not kill spores effectively.

The disadvantages of mercurials for antiseptic and disinfectant use far outweigh any possible advantages they might have. Hence, other more effective and less potentially toxic agents are now preferred.

Products

Nitromersol U.S.P., Metaphen®, 6-(hydroxymercuri)-5-nitro-o-cresol inner salt, occurs as a yellow powder that is practically insoluble in water and has a low solubility in alcohol, acetone and ether. It dissolves in alkalies due to the formation of a salt. Two of the structures for the compound given in the literature are shown below. The *U.S.P.* gives formula (1). It is very probable that neither of these structures is correct. In (2) there is too great a distance between the Hg and the O for the formation of a bond, and in (1) the normal valence angle of 180° for mercury would

have to be distorted greatly to form the 4-membered ring. The forms in which it is supplied most commonly are a 1:500 aqueous solution and a 1:200 alcohol-acetone-aqueous solution, in both of which the compound is present as the sodium salt.

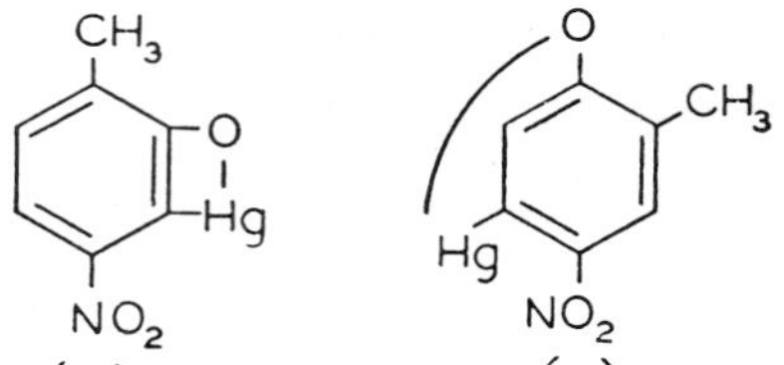

Nitromersol

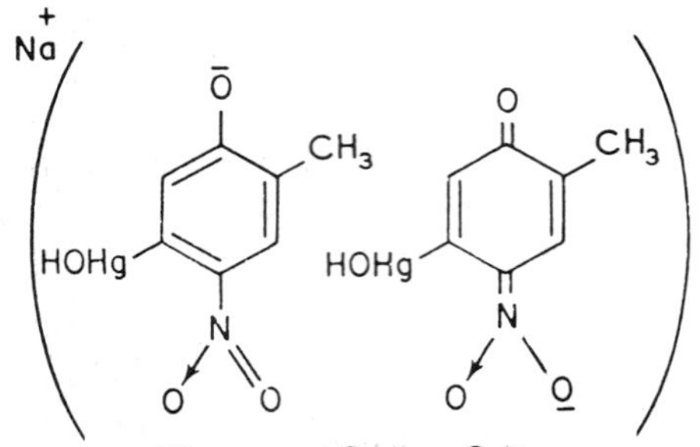

Nitromersol Sodium Salt

Thimerosal U.S.P., Merthiolate®, sodium [(o-carboxyphenyl)thio]ethylmercury. This occurs as a cream-colored powder. It is soluble in water and is compatible with alcohol, soaps and physiologic salt solution. It does not stain fabric or tissues. It is used as an antiseptic in various ways: 1:1,000 tincture for skin disinfection, 1:1,000 aqueous solution for wounds and denuded surfaces, 1:5,000 in ophthalmic ointment, 1:20,000 to 1:5,000 aqueous for urethral irrigation, 1:5,000 to 1:2,000 aqueous for nasal mucous membranes.

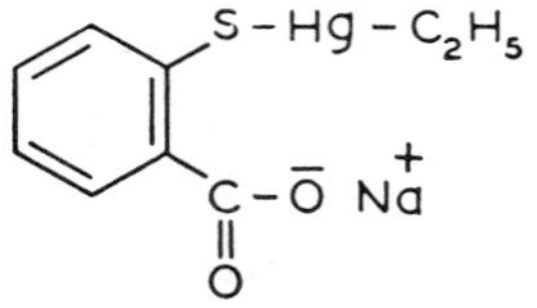

Thimerosal

NITROFURAN DERIVATIVES

The nitrofuran compounds used in medicine have resulted primarily from the extensive efforts of a single laboratory. The essential features are a nitro group in the 5 position and an enamine group in the 2 position. Several hundred members of the series have been studied but at present only four are used in the United States.

O$_2$N — CH=N—R

Nitrofurazone   R = —N—C—NH$_2$ (with H and O above)

Nifuroxime   R = —OH

Furazolidone   R = —N (with cyclic O—C=O ring)

Nitrofurantoin   R = —N (with cyclic NH, C=O ring)

Products

Nitrofurazone U.S.P., Furacin®, is 5-nitro-2-furalde-hyde semicarbazone. It is an odorless, tasteless, lemon-yellow crystalline solid that is stable at autoclave temperatures for 15 minutes. It decomposes above 227°. In crystalline form or in solution, it darkens on long exposure to light; however, there is no loss in antibacterial activity. Nitrofurazone is very slightly soluble in water and practically insoluble in ether, chloroform and benzene. It is slightly soluble in propylene glycol (1:300), acetone and alcohol. The best solubility is in the polyethylene glycols. There is no deterioration, either in solution or the dry state. In dispensing preparations of nitrofurazone, light-resistant containers should be used.

Nitrofurazone was first studied in 1944 and was found to possess good bacteriostatic and bactericidal properties. It is effective against a very wide range of both gram-positive and gram-negative organisms but is not fungistatic. Its action is inhibited by organic matter, such as blood, serum or pus, as well as *p*-aminobenzoic acid.

Studies on related compounds reveal that no other substitution, either in the 5 or in the 2 position of furan, will reproduce the activity of nitrofurazone. Even analogs of thiophene or pyrrole are inactive. Nitrofurazone is unique, even among other 5-nitrofuran derivatives, in its effect on bacteria. No functional group or specific property has been identified as the key to its activity.

The mode of action of nitrofurazone on the bacterial cell is still obscure. Indications are that it temporarily blocks an energy transfer by the organism necessary for cell division. It is known that the nitro group is reduced, presumably to the 5-hydroxylamine (HOHN-) derivative, with total loss of color. The anti-bacterial action may result from its inhibition of bacterial respiratory enzymes. Since it can be reduced, it may act as a hydrogen acceptor.

Nitrofurazone is available in solutions, ointments and suppositories (usually 0.2%). Water-soluble bases are used which are composed of a mixture of glycols. The compound primarily is used topically for mixed infections associated with burns, ulcers, wounds and some skin diseases.

PRESERVATIVES

Preservatives are added to various liquid dosage forms to prevent microbial spoilage. They are used also for the same purpose in many cosmetic preparations. Their use in oral or external preparations is to prevent growth of microorganisms. Parenteral and ophthalmic preparations are sterile products, so the use of a preservative here is to maintain sterility in the case of contamination during use.

The ideal preservative is one that would be effective in low concentrations against all possible invading microorganisms; it would be nontoxic even when used over protracted periods; its taste, odor and color would be imperceptible; it would be compatible with other constituents which may be included in a formulation; and, finally, it would be adequately stable under conditions of use so that its activity would be maintained during the shelf-life of the formulation. The ideal preservative does not exist, so that often combinations are used. Some preservatives, even though not ideal, are used because there has been extensive experience in their use.

PARAHYDROXYBENZOIC ACID DERIVATIVES

p-Hydroxybenzoic acid possesses only slight antiseptic action, but when esterified (for example, with the alcohols methyl, ethyl, propyl or butyl) the resulting compound is very active. Esters of *p*-hydroxybenzoic acid have a low order of acute toxicity and are less toxic than the corresponding esters of salicylic acid or benzoic acid. The toxicity increases as the molecular weight increases, the butyl ester being about 3 times as toxic as the methyl ester. Hydrolysis in vivo would yield *p*-hydroxybenzoic acid, which has a low order of toxicity.[37]

The preservative effect of these esters also increases with the molecular weight; the methyl ester is more effective against molds and the propyl ester more effective against yeasts. The ester grouping apparently behaves like an alkyl group (see hexylresorcinol) in its

TABLE 4–9

ANTISEPTICS AND DISINFECTANTS

Name Proprietary Name	Preparations	Category	Application
Alcohol U.S.P.		Topical anti-infective	Topically to the skin, as a 70 percent solution
Isopropyl Alcohol U.S.P.		Local anti-infective	
Ethylene Oxide		Disinfectant	
Formaldehyde Solution U.S.P.		Disinfectant	Full strength or as a 10 per cent solution to inanimate objects
Glutaral Concentrate U.S.P. *Cidex*		Disinfectant	
Hexachlorophene U.S.P. *pHisoHex, Presulin Cleanser*	Hexachlorophene Detergent Lotion U.S.P. Hexachlorophene Liquid Soap U.S.P.	Topical anti-infective; detergent	Topically to the skin, as the sole detergent, followed by thorough rinsing
Carbamide Peroxide Solution U.S.P.		Local anti-infective (dental)	Several drops onto affected area; expectorate after 2 to 3 minutes
Povidone-Iodine U.S.P. *Betadine*	Povidone-Iodine Solution U.S.P.	Topical anti-infective	Topically to the skin and mucous membranes, as the equivalent of a 0.75 to 1 percent solution of iodine
Halazone U.S.P.	Halazone Tablets for Solution U.S.P.	Disinfectant	2 to 5 p.p.m. in drinking water
Benzethonium Chloride U.S.P. *Phemerol*	Benzethonium Chloride Solution U.S.P. Benzethonium Chloride Tincture U.S.P.	Local anti-infective; pharmaceutic aid (preservative)	Topical, 1:750 solution or 0.2 percent tincture to the skin or 1:5000 solution nasally
Methylbenzethonium Chloride U.S.P. *Diaparene*	Methylbenzethonium Chloride Lotion U.S.P. Methylbenzethonium Chloride Ointment U.S.P. Methylbenzethonium Chloride Powder U.S.P.	Local anti-infective	Topical, 0.067 percent lotion, 0.1 percent ointment, or 0.055 percent powder
Cetylpyridinium Chloride U.S.P. *Ceepryn*	Cetylpyridinium Chloride Lozenges U.S.P. Cetylpyridinium Chloride Solution U.S.P.	Local anti-infective; pharmaceutic aid (preservative)	Topical, 1:100 to 1:1,000 solution to intact skin, 1:1,000 solution for minor lacerations, and 1:2,000 to 1:10,000 solution to mucous membranes
Nitromersol U.S.P. *Metaphen*	Nitromersol Solution U.S.P. Nitromersol Tincture U.S.P.	Local anti-infective	Topical, solution or tincture
Thimerosal U.S.P. *Merthiolate*	Thimerosal Aerosol U.S.P. Thimerosal Solution U.S.P. Thimerosal Tincture U.S.P.	Local anti-infective	Topical, a 0.1 percent aerosol, solution or tincture

effect on the antiseptic action. Oil-solubility increases along with the size of the ester group, thus making the propyl ester better than the methyl for oils and fats. The esters may be used to preserve almost any pharmaceutical.[38] Clinical research has indicated that methyl- and propylparaben prevent the overgrowth of monilia, the most frequently occurring fungus infection associated with antibiotic therapy.

Methylparaben N.F., Methylben, methyl *p*-hydroxybenzoate. Methyl *p*-hydroxybenzoate occurs as small, white crystals or as a crystalline powder which has a slightly burning taste, and a faint, characteristic odor or none at all. It is soluble in water, alcohol (1:2) and ether (1:10) and slightly soluble in benzene and in carbon tetrachloride. It is used as a preservative primarily to protect against yeasts.

Methylparaben

Propylparaben N.F., Propylben, propyl *p*-hydroxybenzoate. This ester is prepared and used in the same manner as methylparaben. It occurs as a white powder or as small, colorless crystals which are only slightly soluble in water and soluble in most organic solvents. It is used as a preservative, primarily against yeast (see methylparaben).

Ethylparaben N.F., ethyl *p*-hydroxybenzoate. Ethylparaben occurs as small, colorless crystals or white powder. It is slightly soluble in water and in glycerin, and freely soluble in acetone, in alcohol, in ether, and in propylene glycol.

Butylparaben N.F., butyl *p*-hydroxybenzoate, occurs as small, colorless crystals or white powder. It is very slightly soluble in water or glycerol but is very soluble in alcohols and in propylene glycol.

OTHER PRESERVATIVES

Products

Chlorobutanol N.F., Chloretone®, 1,1,1-trichloro-2-methyl-2-propanol. Chlorobutanol is tertiary trichlorobutyl alcohol which may be synthesized from acetone and chloroform.

$$CH_3COCH_3 + CHCl_3 \xrightarrow{KOH} CH_3-\underset{\underset{CCl_3}{|}}{\overset{\overset{CH_3}{|}}{C}}-OH$$

Chlorobutanol

It is a white, crystalline solid having a characteristic camphorlike odor and taste. It is available in two forms: the anhydrous form, and the hydrated form containing not over one-half molecule of water of hydration. The anhydrous form is used in preparing oil solutions. Because it volatilizes readily at room temperatures, chlorobutanol is difficult to dry and must be stored carefully. The compound dissolves in water (1:125), alcohol (1:1), glycerin (1:10), all oils or in organic solvents.

Chlorobutanol is widely used as a bacteriostatic agent in pharmaceuticals for injection, ophthalmic use or intranasal administration. When used in aqueous preparations, it has the distinct disadvantage of being only slowly soluble.[39] It is more soluble in boiling water, but when heated the compound hydrolyzes and is lost by volatilization as well. Solutions which are buffered below pH 5 and in closed systems can be auto-

claved at 121° for 20 minutes with only slight loss due to hydrolysis.[40]

As part of a thorough kinetic study of the degradation of chlorobutanol in aqueous solution, it was calculated that solutions at pH 5 would lose 13 percent when heated at 115° for 30 minutes.[41] The solution could then be stored at 25° for well over 5 years before showing a further 10 percent loss. The hydrolysis of chlorobutanol can be represented as follows:

$$CH_3\underset{\underset{CCl_3}{|}}{\overset{\overset{CH_3}{|}}{C}}OH + H_2O \longrightarrow$$

$$CH_3CCH_3 + CO + 3\ HCl$$
$$\underset{O}{\overset{||}{}}$$

Hydrolysis of Chlorobutanol

When chlorobutanol is used in oil solutions these problems of hydrolysis and slow rate of solubility are not met.

Benzyl Alcohol N.F., phenylcarbinol, phenylmethanol. Benzyl alcohol occurs free in nature (Oil of Jasmine, 6%) and is found as an ester of acetic, cinnamic and benzoic acids in gum benzoin, storax resin, Peru balsam and tolu balsam and in some volatile oils (jasmine and hyacinth). In maize, a glucoside of benzyl alcohol is found. It is synthesized readily from toluene (1) and by the Cannizzaro reaction from benzaldehyde (2).

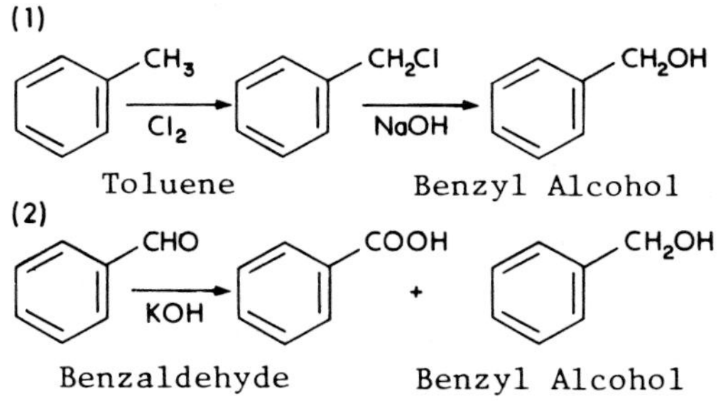

TABLE 4–10

PARABENS

Name	Preparations	Category	Solubility in Water
Methylparaben N.F.	Hydrophilic Ointment U.S.P.	Pharmaceutic aid (antifungal preservative)	1:400
Ethylparaben N.F.		Pharmaceutic aid (antifungal preservative)	1:600
Propylparaben N.F.	Hydrophilic Ointment U.S.P.	Pharmaceutic aid (antifungal preservative)	1:2,500
Butylparaben N.F.		Pharmaceutic aid (antifungal preservative)	1:5,000

The alcohol is soluble in water (1:25) and in 50 percent alcohol (1:15). It is miscible with fixed and volatile oils, ether, alcohol or chloroform. Benzyl alcohol is a clear liquid with a faint aromatic odor. It can be boiled without decomposition.

The chemical properties of benzyl alcohol are much the same as those of primary alcohols, since it is phenylmethanol. On oxidation it first yields benzaldehyde and then benzoic acid. It differs from the aliphatic alcohols in being resinified by sulfuric acid, and it does not form the corresponding sulfuric ester.

Benzyl alcohol commonly is incorporated as a preservative in vials of injectible drugs and also because it exerts a local anesthetic[42] effect when injected or applied on mucous membranes. The concentrations usually employed are 1 to 4 percent (maximum solubility) in water or saline solution. In such small doses it is nonirritating and nontoxic. Since it is also strongly antiseptic, ointments containing benzyl alcohol up to 10 percent are useful in preventing secondary infection in the itching of pruritus and other skin conditions. A suitable lotion may be prepared with equal parts of benzyl alcohol, water and alcohol. A saturated piece of cotton is effective for toothache when used in the same manner as clove oil.

There are a few other aromatic alcohols of minor importance. For a pharmacologic study of some of these see Hirschfelder.[43]

Phenylethyl Alcohol U.S.P., phenethyl alcohol, 2-phenylethanol, orange oil or rose oil, $C_6H_5CH_2$-CH_2OH. This compound is useful in perfumery, occurs in oils of rose, orange flowers, pine needles and Neroli. It is prepared by the reduction of ethyl phenylacetate or with phenylmagnesium chloride and ethylene oxide.

This alcohol is soluble in water (2%). It may be sterilized by boiling, since it boils at 220°.

Hjort[44] found it to be slightly more anesthetic than benzyl alcohol and of the same order of toxicity.

Benzoic Acid U.S.P. Benzoic acid and its esters occur in nature as constituents in gum benzoin, in Peru and tolu balsams and in cranberries. As hippuric acid, it occurs in combination with glycine in the urine of herbivorous animals.

The acid may be obtained by distillation from a natural product, such as benzoin, or prepared synthetically by several procedures.

Benzoic acid forms white crystals, scales or needles, that are odorless or may have a slight odor of benzoin or benzaldehyde. It sublimes at ordinary temperature and distills with steam. It is slightly soluble in water (0.3%), benzene (1%) and benzin, but it is more soluble in alcohol (30%), chloroform (20%), acetone (30%), ether (30%) and volatile and fixed oils.

The acid is more strongly acidic (pKa 4.2) than acetic acid (pKa 4.8), and most of the common electron-attracting substituents increase the acidity. Solutions containing the ions of iron, silver, lead or mercury form a precipitate of the respective salt with benzoic acid. The iron salt is a reddish-tan or salmon-colored precipitate.

Benzoic acid is used externally as an antiseptic[45] and is employed in lotions, ointments and mouthwashes. In concentrations over 0.1 percent, it may produce local irritation. It is employed as a food preservative, especially in the form of its salts (e.g., sodium benzoate). When used as a preservative in foods and in pharmaceutical products, benzoic acid and its salts are more effective as the pH is lowered; thus it is the undissociated benzoic acid molecule which is the effective agent. The pKa of benzoic acid is 4.2, so that at this pH only 50 percent would be in the undissociated form, while at pH 3.5 over 80 percent would be in the undissociated form.[46,47] When benzoic acid or its salts are used in emulsions, the effectiveness as a preservative depends on the distribution between the oil phase and the water phase as well as the pH of the system.[48]

Sodium Benzoate N.F., Sodium benzoate is prepared by adding sodium bicarbonate to an aqueous suspension of benzoic acid. The product has a sweet, astringent taste; it is stable in air and is a white, odorless, crystalline substance or a granular powder of 99 percent purity. It is soluble in water or alcohol.

Sodium benzoate has chemical properties similar to those of all benzoates and the incompatibilities are similar to those of benzoic acid. It is used primarily

(0.1%) as a preservative in acid media for the antiseptic effect of benzoic acid. It is not effective in preserving nonacid products.

Sodium Propionate N.F., Mycoban®. In 1943, sodium propionate became important as an effective agent in the treatment of fungus infections.[49] It is now widely used as an antifungal preservative. The salt occurs as transparent, colorless crystals that have a faint odor resembling acetic and butyric acids and are deliquescent in moist air. It is soluble in water (1:1) or alcohol (1:24). Sodium propionate is most effective at pH 5.5 and usually is used in a 10 percent ointment, powder or solution. Other fatty acid salts containing an odd number of carbon atoms are effective as fungicides, but they are more toxic and not so readily available. Undecylenic acid is one that is being used. By 1944 propionate-propionic acid mixtures had been found to be superior, not only as fungicides but also as bactericides. However, they are only slightly better than the undecylenate-undecylenic acid mixture.

Propionic acid salts of sodium, calcium, zinc, potassium and copper are used in preparations for the treatment of fungus infections, such as athlete's foot (tinea pedis).

Sorbic Acid N.F., 2,4-hexadienoic acid, has been found to be effective for inhibiting the growth of molds and yeasts. It is soluble to the extent of 0.15 percent in water. The pKa is 4.8. In a test of its fungistatic properties, concentrations as low as 0.05 percent were found to be effective.

$$CH_3CH=CHCH=CHCOOH$$
Sorbic Acid

Sorbic acid has been found to be useful as a mold inhibitor in various medicinal syrups, elixirs, ointments and lotions containing sugars and other components that support mold growth. It is used in films and other food-packaging materials.

Potassium Sorbate N.F., 2,4-hexadienoic acid, potassium salt; potassium 2,4-hexadienoate. This compound occurs as white crystalline powder and has a characteristic odor. It is freely soluble in water and soluble in alcohol. It is used like sorbic acid, especially where greater solubility in water is required. The suggested experimental levels are 0.025 to 0.1 percent by total weight.

Phenylmercuric Nitrate N.F., merphenyl nitrate, occurs as a white crystalline powder and is a mixture of phenylmercuric nitrate and phenylmercuric hydroxide. It is very slightly soluble in water and slightly soluble in alcohol and glycerin. It is used in 1:10,000 to 1:50,000 concentrations in injections.

There is a tendency to avoid the use of organic mercurials as preservatives in new products, because the preservative action is greatly diminished in the presence of serum proteins.

Phenylmercuric Nitrate   Phenylmercuric Hydroxide

Phenylmercuric Acetate N.F., acetoxyphenylmercury, occurs in the form of white prisms that are soluble in alcohol and benzene but only slightly soluble in water. It is used for its bacteriostatic properties. It has been used as a herbicide, also as a trichomonicide in the preparation Nylmerate Jelly, used as a vaginal antiseptic.

ANTIFUNGAL AGENTS

Many remedies have been used against fungus infections, and research still continues, which would lead one to conclude that the ideal antifungal agent has not yet been found. The majority of fungal infections (mycoses) involve superficial invasion of the skin or mucous membranes of body orifices. These diseases, which can usually be controlled by local application of an antifungal agent, are conveniently divided into two etiological groups: (1) the dermatophytoses (tinea infections), which are contagious superficial epidermal infections caused by various *Epidermophyton, Microsporum* and *Trichophyton* spp.; and (2) mycoses caused by pathogenic saphrophytic yeasts, which are contagious and usually superficial infections involving the skin and mucous membranes. Some species of saphrophytic yeasts (*Aspergillus, Blastomyces, Can-*

TABLE 4-11

PRESERVATIVES

Name	Use
Chlorobutanol N.F.	Antimicrobial
Benzyl Alcohol N.F.	Bacteriostatic (injections)
Phenylethyl Alcohol U.S.P.	Bacteriostatic
Sodium Benzoate N.F.	Antifungal
Sodium Propionate N.F.	Antifungal
Sorbic Acid N.F.	Antimicrobial
Potassium Sorbate N.F.	Antimicrobial
Phenylmercuric Nitrate N.F.	Bacteriostatic
Phenylmercuric Acetate N.F.	Bacteriostatic

dida, Coccidioides, Cryptococcus, Histoplasma) under certain conditions are capable of invading deeper body cavities and causing systemic mycoses. Such infections may become serious and occasionally life theatening, and they are frequently difficult to treat.

Fatty acids in perspiration have been found to be fungistatic, and this discovery has led to the introduction of fatty acids in therapy. The use of copper and zinc salts provides the added antifungal activity of the metal ion. Aromatic acids, especially salicylic acid, which also has a useful keratolytic action, and its derivatives are employed for their topical fungistatic effect. A variety of alkylated and/or halogenated phenols and their derivatives are useful for the treatment of local fungal infections. The antifungal activity of the aforementioned compounds is largely confined to local dermatophytic infections. Deep dermatophytic infections, resistant to topical therapy, may be treated systemically with the antibiotic griseofulvin. Several years ago, it was discovered that local and gastrointestinal yeast infections, which became prevalent as superinfections resulting from the misuse of broad spectrum antibiotics such as the tetracyclines, could be combated effectively with the polyene antibiotic nystatin. More recently, two additional polyene antibiotics, first candicidin and later pimaricin, have been introduced for the topical treatment of yeast infections. Research in Europe led to the discovery that certain highly substituted imidazole derivatives possessed broad spectrum antifungal activity. Two such agents, miconidazole and clotrimoxazole, with activity against both dermatophytes and pathogenic yeasts, are now available.

Very few compounds have as yet been found which combine the properties required for the treatment of systemic yeast infections, namely, effectiveness against the causative organisms and a reasonable margin of safety, Amphoteracin B (a polyene antibiotic) and flucytosine are the only two agents that have been approved for such use in the United States.

SYNTHETIC ANTIFUNGAL AGENTS

Propionic Acid has become an important fungicide because it is nontoxic, nonirritant and readily available.

It is a clear, corrosive liquid with a characteristic odor and is soluble in water or alcohol. In 1939 Peck,[50] in his studies on perspiration, observed that it was not the pH of perspiration that was responsible for the fungicidal and fungistatic effect but the presence of fatty acids and their salts. Previous to this, Bruce had found that fatty acids of odd-numbered carbon atoms were bacteriostatic while acids of even-numbered car-

bon atoms were not. Note, however, that caprylic acid is active. Chemical analysis of sweat showed it to contain, among other ingredients, 0.0091 percent of propionic acid. The fungicidal action of propionic acid salts, such as those of sodium, ammonium, calcium, zinc and potassium, was found to be the same as that of the free acid. The free acid may be used to treat fungus infections, such as athlete's foot, but usually is employed in the form of its salts because they are more easily handled and are odorless.

A number of fatty acids and their salts are efficient fungicides, but propionic, caprylic and undecylenic acids are used because of availability.

Zinc Propionate occurs as plates or as needles in the case of the monohydrate. It is freely soluble in water and sparingly soluble in alcohol. It decomposes in a moist atmosphere, giving off propionic acid. Therefore, it should be kept in well-closed containers. It is used as a fungicide, particularly on adhesive tape to reduce irritation caused by fungi and bacterial action.

Sodium Caprylate. Caprylic acid, found in several oils such as coconut and palm-kernel, is the acid from which the sodium salt is prepared. Like propionic acid, caprylic acid is an ingredient of perspiration, where it contributes to the antifungal properties. The sodium salt is soluble in water, sparingly soluble in alcohol and occurs as cream-colored granules.

It is an antifungal agent similar to propionates and undecylenates, being effective against infections due to trichophytons, microsporons and *Candida albicans*. There appears to be no skin sensitivity produced by continuous or repeated use. Sodium caprylate is available as a solution, powder or ointment.

$$CH_3 (CH_2)_5 CH_2COO^- Na^+$$

Sodium Caprylate

Free caprylic acid is a light amber, oily liquid possessing a disagreeable odor. It is insoluble in water and only slightly soluble in alcohol, as would be expected.

Zinc Caprylate is a fine, white powder that is practically insoluble in water and alcohol. It decomposes on exposure to moist atmosphere, liberating caprylic acid, and, therefore, the container should be kept well closed. It is used as a fungicide as is zinc propionate. Aluminum and copper salts also are used in proprietaries.

Undecylenic Acid U.S.P., 10-undecenoic acid, may be represented as $CH_2=CH(CH_2)_8COOH$. It may be obtained by the destructive distillation of castor oil. The ricinoleic acid, present in castor oil as the glyceride, is the source of undecylenic acid.

$$CH_3(CH_2)_5CHOHCH_2CH=CH(CH_2)_7COOH$$

Ricinoleic Acid

↓ Vacuum

$$CH_3(CH_2)_5CHO \quad + \quad CH_2=CHCH_2(CH_2)_7COOH$$

n-Heptyl Aldehyde · · · · · · Undecylenic Acid

It occurs as a yellow liquid having a characteristic odor and a persistent bitter or acrid taste. At lower temperatures (between 21° and 22°) it congeals and at 24° melts. The acid is practically insoluble in water and miscible with alcohol, chloroform, ether, benzene and with both fixed and volatile oils. It possesses the properties of a double bond and is a very weak organic acid.

The higher fatty acids (heptylic, caprylic, pelargonic, capric and undecylenic) have been found to be effective antifungal agents.[51] Undecylenic acid is one of the best fatty acids available as a topical fungistatic agent.[52] It may be used in up to 10 percent strength in solutions, emulsions, adsorbed on powders or in ointments. Application to eyes, ears, nose or other areas of mucous membrane is not advisable. Even local use as a fungicide may be irritating. Internally, a very pure form is used (dose 7.5 to 10 g. daily) in capsules for the treatment of psoriasis and neurodermatitis.

There are in use a number of undecylenic acid salts, such as zinc undecylenate, copper undecylenate (Undesilin® and Decupryl®), sodium and potassium. Mixtures of the acid and salts are also used.

Zinc Undecylenate U.S.P., zinc 10-undecenoate, is a fine white powder practically insoluble in water and alcohol. It is used as a fungicide in connection with the free acid and other compounds.

Triacetin U.S.P., Enzactin®, Fungacetin®, glyceryl triacetate, is an ester of glycerin and acetic acid, prepared by heating a mixture of the two.

It is a colorless, oily liquid having a slight odor and a bitter taste. It is soluble in water (6:100), soluble in organic solvents and miscible with alcohol.

Triacetin acts as a topical antifungal agent by virtue of the acetic acid which is formed by slow enzymatic hydrolysis by esterases in the skin. The rate of release is self-limited, because as the pH drops to 4 the esterases are inactivated. It is nonirritating to the skin.

Salicylic Acid U.S.P., o-hydroxybenzoic acid. This acid has been known for over 135 years, having been discovered in 1839. It is found free in nature and in the form of salts and esters. A very common ester is methyl salicylate (oil of wintergreen). Salicylic acid may be obtained from oil of wintergreen by saponification with sodium hydroxide and then neutralization with hydrochloric acid. This is referred to as "natural salicylic acid" and is used to prepare salts which are preferred by some. The natural acid usually is tinted pink or yellow and has a faint wintergreenlike odor. At one time it was believed that the synthetic salicylic acid was contaminated with some cresotinic acid $[C_6H_3 \cdot CH_3(OH)(COOH)]$ and was thus more toxic, its salts less desirable. It has since been shown, not only that cresotinic acid is absent, but also that cresotinic acid is nontoxic.

In 1859, Kolbe introduced a method for the synthetic preparation of salicylic acid, and, with slight changes, this is still used. Sodium phenolate is prepared and saturated under pressure with carbon dioxide; the resulting product then is acidified and salicylic acid is isolated. When the reaction is carried out at 200°, the para structural isomer (p-hydroxybenzoic acid) is obtained preferentially.

Salicylic acid usually occurs as white, needlelike crystals or as a fluffy, crystalline powder. The synthetic acid is stable in air and is odorless. It is slightly soluble in water (1:460) and is soluble in most organic solvents.

The chemical properties of this acid are due to the phenolic hydroxyl group and to the carboxyl group. Since it is also a phenol, it responds with the reactions of phenols, such as the producing of a violet color with

Salicylic Acid

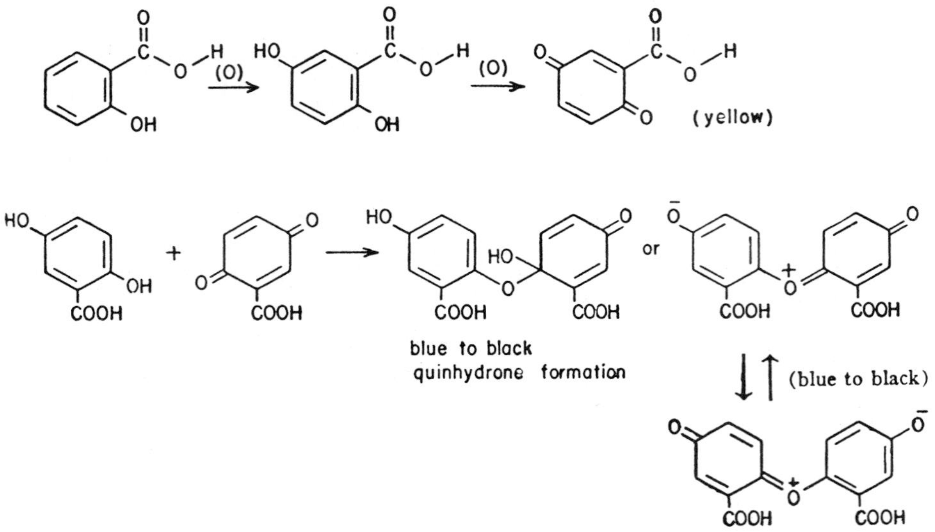

blue to black
quinhydrone formation

(blue to black)

ferric salts, halogenation and oxidation. Oxidizing agents form colored compounds, perhaps of a quinoid type, and destroy the molecule. The colored compounds produced on standing in alkaline solution are due to quinhydrone formation.

Insoluble salts are formed with ions of the heavy metals, such as silver, mercury, lead, bismuth and zinc. Reducing agents break down salicylic acid to pimelic acid. Boric acid and salicylic acid combine to form borosalicylic acid.

Salicylic acid has strong antiseptic and germicidal properties because it is a carboxylated phenol. The presence of the carboxyl group appears to enhance the antiseptic property and to decrease the destructive, escharotic effect. It is used externally as a mild escharotic and antiseptic in ointments and solutions. Many hair tonics and remedies for athlete's foot, corns and warts employ the keratolytic action of salicylic acid.

Tolnaftate U.S.P., Tinactin®, is O-2-naphthyl *m*,N-dimethylthiocarbanilate.

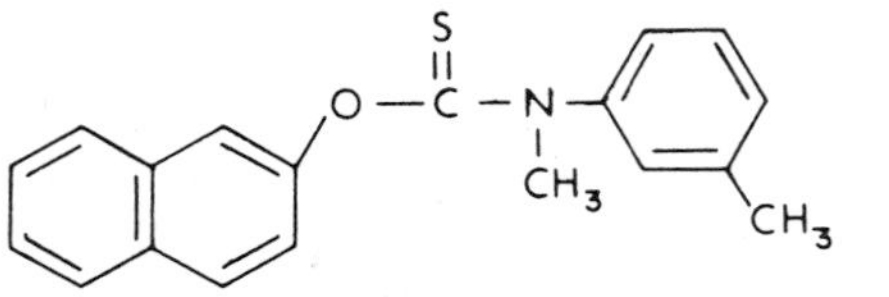

Tolnaftate

This compound, which is essentially an ester of β-naphthol, is reported to be a potent antifungal agent. Only one or two drops of a 1 percent solution in a polyethylene glycol is adequate for areas as large as the hand.

Parachlorometaxylenol, PCMX®, *p*-chloro-*m*-xylenol or 4-chloro-3,5-dimethylphenol is a relatively

nonirritating antiseptic agent with useful antifungal properties. It is available in a variety of topical dosage forms for the treatment of tinea infections such as athlete's foot and jock itch.

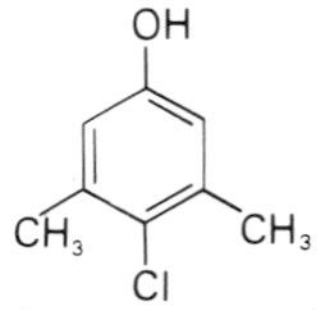

Parachlorometaxylenol

Acrisorcin U.S.P., Akrinol®, 9-aminoacridinium 4-hexylresorcinolate. This compound is prepared from 9-aminoacridine and 4-hexylresorcinol and occurs as yellow crystals slightly soluble in water and soluble in alcohol.

Acrisorcin is used in the treatment of tinea versicolor (caused by the fungus *Malassezia furfur*). Treatment is usually for at least 6 weeks.

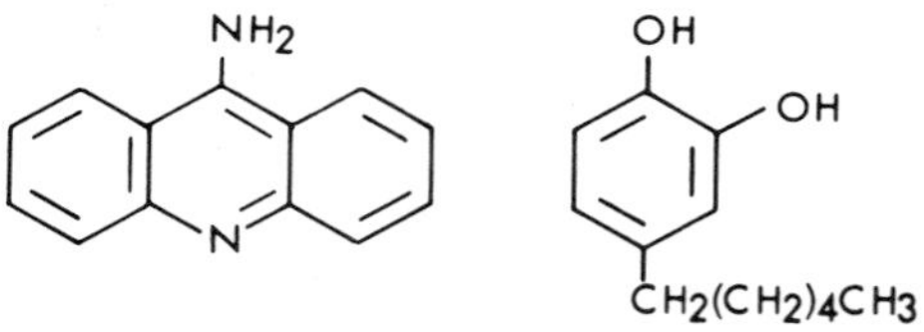

Acrisorcin

Haloprogin, Halotex®, is 3-iodo-2-propynyl 2,4,5-trichlorophenyl ether. It is used generally as a 1 percent cream or solution for the treatment of superficial fungal infections of the skin. Haloprogin is light-sensitive, thus the formulations should be protected from strong light. Haloprogin is reactive with metals, but is compatible with the aluminum tube in which the cream is supplied.

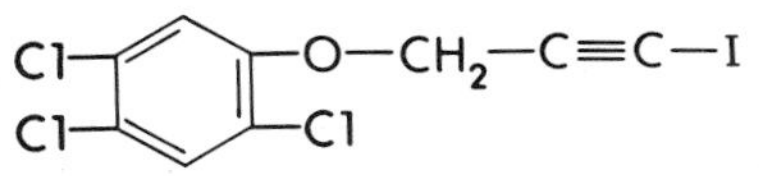

Haloprogin

Clotrimazole U.S.P., Lotrimin®, 1-(o-chloro-α, α-diphenylbenzyl)imidazole, is a topical antifungal agent that has been shown to be effective for tinea infections and for candidiasis caused by *Candida albicans*. It is supplied as a 1 percent solution in polyethylene glycol 400. The chemical is stable at room temperature for at least 5 years. Although clotrimazole is effective against a variety of pathogenic yeasts and is well absorbed orally, it causes severe gastrointestinal disturbances and is thus not considered suitable for the treatment of systemic infections.

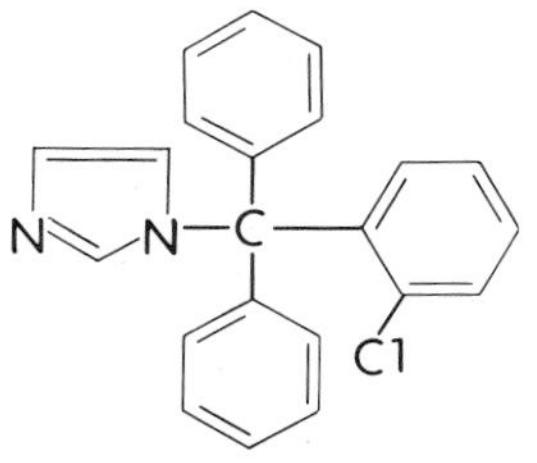

Clotrimazole

Miconazole Nitrate U.S.P., Monistat®, Mica-Tin®, is 1-[2-(2,4-dichlorophenyl)-2-[(2,4-dichlorophenyl)methoxy]ethyl]-1H-imidazole mononitrate, the most effective of a large number of antifungal phenylimidazole compounds developed at Janssen laboratories in Belgium.[53] Miconazole nitrate is used topically in the treatment of tinea infections and vaginally in the treatment of moniliasis. It is supplied as a 2 percent topical cream and as a 2 percent vaginal cream.

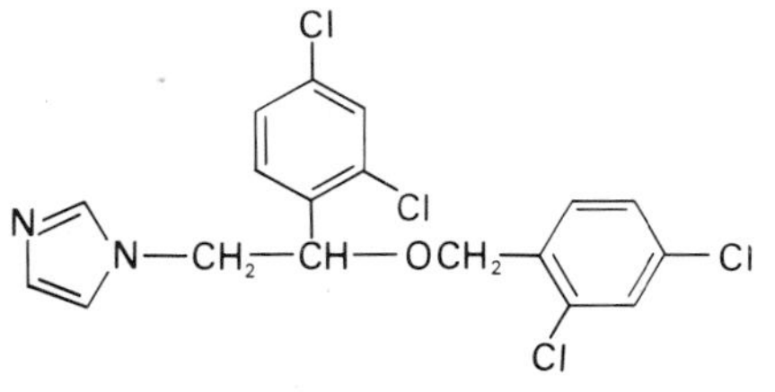

Miconazole

TABLE 4–12

SYNTHETIC ANTIFUNGAL AGENTS

Name / Proprietary Name	Preparations	Application*
Undecylenic Acid U.S.P. / Zinc Undecylenate U.S.P.	Compound Undecylenic Acid Ointment U.S.P.	Topically to the skin, once a day at bedtime, as required
Triacetin U.S.P. / *Enzactin, Fungacetin*	Powder Ointment Aerosol	Topical, 3 to 5 percent ointment
Salicylic Acid U.S.P.	Salicylic Acid Ointment U.S.P.	Keratalytic topically to the skin, as a 3 to 10 percent ointment
Toinaftate U.S.P. / *Tinactin*	Toinaftate Cream U.S.P.	Topically to the skin, as a 1 percent cream twice daily
	Toinaftate Topical Solution U.S.P.	Topically to the skin, as a 1 percent solution twice daily
Parachloro-metaxylenol PCMX	Cream Solution	Topical, 3 percent cream to the affected area as required
Acrisorcin U.S.P. / *Akrinol*	Acrisorcin Cream U.S.P.	Topical, 0.2 percent cream, to the affected area twice daily
Compound Undecylenic Acid Powder U.S.P. Compound Undecylenic Acid Topical Aerosol Powder U.S.P.	Cream	
Clotrimazole U.S.P. / *Lotromin* / *Lotromin*	Clotrimazole Topical Solution U.S.P. Clotrimazole Cream U.S.P.	Topical, 1 percent solution twice daily for tinea / Topical 1 percent cream twice daily for vaginal candidiasis
Miconidazole Nitrate U.S.P. / *Monistat* / *MicaTin*	Miconidazole Nitrate Cream U.S.P. Miconidazole Nitrate Lotion U.S.P.	Topical, 2 percent cream or lotion twice a day
Flucytosine U.S.P.	Flucytosine Capsules U.S.P.	Orally, 12.5 to 37.5 mg. per kg. of body weight every 6 hours
Clotrimazole	Vaginal Cream U.S.P.	Intravaginal, 5 g. once daily, preferably at bedtime, for 7 to 14 consecutive days
Clotrimazole	Vaginal Tablets U.S.P.	Intravaginal, 100 mg. once daily, preferably at bedtime, for 7 consecutive days

*See U.S.P. D.I. for complete dosage information.

Flucytosine U.S.P., Ancobon®, 5-Fluorocytosine, 5-FC, is an orally active antifungal agent. It is indicated only in the treatment of serious systemic infections caused by susceptible strains of pathogenic yeasts, especially *Candida albicans* and *Cryptococcus neoformans*. The mode of antifungal action of 5-FC in

Flucytosine

susceptible fungi, and mechanisms of resistance to the drug in nonsusceptible strains have been studied in some detail.[54] Incorporation of fluorinated pyrimidine into RNA following selective deamination to 5-fluorouracil (5-FU) in the fungus appears to be required for antifungal activity. Resistant strains appear to either be deficient in enzymes required for the bioactivation of 5-FC (e.g., uridinemonophosphate pyrophosphorylase or cytosine deaminase) or have a surplus of de novo pyrimidine synthesizing capacity. The comparative lack of toxicity of 5-FC in man is apparently due to the fact that it is not deaminated to the toxic antimetabolite 5-FU in cells of the host following oral administration. The half-life of flucytosine, which is excreted largely unchanged, is 4 to 8 hours.

ANTIFUNGAL ANTIBIOTICS

Griseofulvin U.S.P., Fulvicin®, Grisactin®, Grifulvin®. Although griseofulvin was reported in 1939 by Oxford *et al.*[55] as an antibiotic obtained from *Penicillin griseofulvum Dierckx,* it was not until 1958 that its use for the treatment of fungal infections in man was demonstrated successfully. Previously, it had been used for its antifungal action in plants and animals. Its release in the United States in 1959, 20 years after its discovery, as a potent agent for the treatment of ringworm infections re-emphasizes the need for the broad screening of drugs to find their potential uses.

Griseofulvin

The structure of griseofulvin was determined by Grove *et al.*[56] to be 7-chloro-2′,4,6-trimethoxy-6′β-methylspiro[benzofuran-2(3H),1′-[2]cyclohexene]3,4′-dione. It is a white, bitter, thermostable powder that may occur also as needlelike crystals. It is relatively soluble in alcohol, chloroform and acetone. In the dry state it is stable for at least 20 months.

Since its introduction, griseofulvin has provided startling cures for infections due to trichophytons and microspora resulting in refractory ringworm infections of the body, the nails and the scalp (tinea corporis, tinea unguium and tinea capitis) and athlete's foot (tinea pedis). In the treatment of these infections it is administered orally and is absorbed from the gastrointestinal tract. Following systemic circulation, it is concentrated in the keratin of growing skin, nails and hair. As new tissue develops, the fungistatic action of the griseofulvin prevents the growth of the organism in it. The old tissue continues to support viable fungi, so the drug must be continued until exfoliation of the old tissue is complete. In the case of infected nails, therapy may need to be continued for months because of the slow rate of growth. Griseofulvin does not cause many adverse side-effects, but careful observation of patients receiving it is indicated. It is not active against bacteria and other fungi or yeasts.

A number of methods for the synthesis of griseofulvin have been developed that have permitted the synthesis of some structural analogs. None of these has shown activity superior to that of griseofulvin. The mode of action of griseofulvin is unknown, and little fundamental work has been published concerning possible mechanisms of its inhibitory effects. Of interest is the effect crystal size has on absorption of the orally administered powder. "Microsize" griseofulvin may be administered in significantly smaller doses than the conventional size powder to obtain the same effect. The *U.S.P.* specifies that the official product is the "Microsize" powder.

THE POLYENES

A number of antibiotics are known to contain a conjugated polyene system as a characteristic chemical grouping. Rather surprisingly, such antibiotics often show similar antifungal activity, which suggests a structure-activity relationship for which there is not yet a satisfactory explanation. Among the polyenes are a group of macrocyclic lactones that show some degree of chemical relationship. They differ from the macrolide antibiotics of the erythromycin type (see Chap. 7) by having a larger lactone ring in which there is a conjugated polyene system. Many of them contain a glycosidically linked sugar such as the aminodesoxyhexose, mycosamine, that is present in amphotericin B, nystatin, pimaricin and some others. The macrolide polyenes are sometimes classified by the number of double bonds present in the conjugated group, into tetraenes, pentaenes, hexaenes and heptaenes. Characteristic ultraviolet absorption spectra are used as the basis for the classification determination.

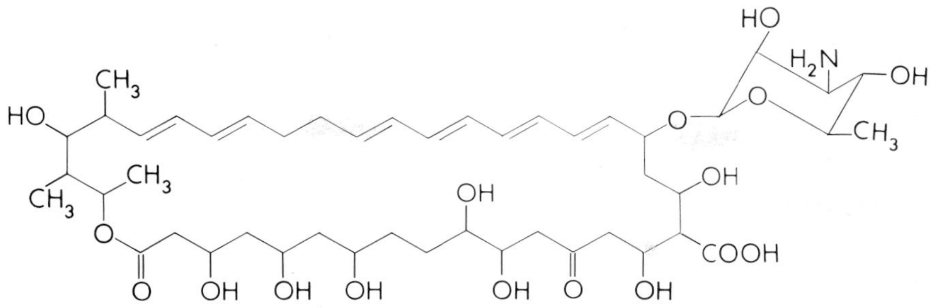

Nystatin

The macrolide polyenes include four antibiotics that are used as antifungal agents in the United States. They are amphotericin B, natamycin (pimaricin), candicidin and nystatin. The first complete structure for one of these, pimaricin, was reported by Golding *et al.*[57] More recently, the complete structures of amphotericin B[58,59] and nystatin[60,61] have been elucidated by x-ray crystallographic and chemical degradation procedures. Their general lack of water-solubility, their poor stability and their rather toxic properties have contributed to their failure to achieve a more important place in therapy. To improve their usefulness, their amphoteric characteristics have been overcome by acylating the amino group of the sugar function and then forming water-soluble salts of the free carboxyl group on the macrolide ring with bases.[62] However, none of these derivatives has yet been marketed.

Nystatin U.S.P., Mycostatin®. In 1951, Hazen and Brown[63] reported the isolation of nystatin from a strain of *Streptomyces noursei.* It has become established in human therapy as a valuable agent for the treatment of both gastrointestinal and local infections of *Candida albicans.* However, amphotericin B, which may be administered parenterally, has replaced nystatin for the treatment of systemic yeast infections. There is divided opinion among clinicians whether or when nystatin should be given with tetracyclines to prevent monilial overgrowth. Perhaps the majority now favor treatment of intestinal candidiasis only after it occurs as a result of tetracycline therapy. Its success against other monilial infections is less impressive, but it shows in vitro activity against many yeasts and molds. Its dosage is expressed in terms of units. One mg. of nystatin contains not less than 2,000 U.S.P. Units.

Nystatinolide, the aglycon portion of nystatin, consists of a 38-membered lactone ring with single tetraene and diene chromophores isolated from each other by a methylene group, one carboxyl, one keto and eight hydroxyl groups. It is glycosidically linked to the amino sugar mycosamine (3-amino-3,6-dideoxypyranose). The structure[60,61] of nystatin is given below.

Nystatin is a yellow to light tan powder that has a cereal-like odor. It is very slightly soluble in water and only sparingly soluble in nonpolar solvents. It is unstable to moisture, heat, light and air, and its solutions are inactivated rapidly by acids and bases.

Amphotericin B U.S.P., Fungizone®. A polyene antibiotic having potent antifungal action was reported in 1956 by Gold *et al.*[64] to be produced from a Streptomyces species isolated from a sample of soil obtained from the Orinoco River in Venezuela. The species name *Streptomyces nodosus* has been given to this organism. The antibiotic material was shown to contain two closely related substances that were given the names amphotericins A and B. The B compound is the more active and in purified form is being used for its broad-spectrum activity against a number of deep-seated and systemic infections caused by yeastlike fungi. It does not exhibit any activity against bacteria, protozoa or viruses.

As its name indicates, this compound is an amphoteric substance that at its isoelectric point is water-

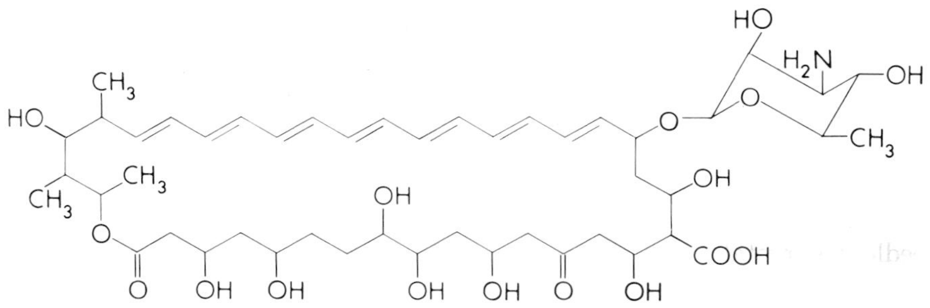

Amphotericin B

insoluble. For the treatment of systemic yeast infections it is administered intravenously in the form of a colloidal suspension with desoxycholate. Like nystatin its aglycon portion consists of a 38-membered lactone-containing ring with the same substituents, including the amino sugar, mycosamine. However, the polyene chromophore, in contrast to nystatin's, is a fully conjugated heptaene. The structure of amphotericin B[58,59] is given on the previous page.

Amphotericin B and other polyenes exert a fungicidal action, apparently as a result of altering the permeability of yeast cell membranes to promote the loss of essential cell constituents. They are known to have a high affinity for sterols in yeast cell membranes.[65]

Amphotericin B is very poorly absorbed from the gastrointestinal tract, and so its preferred route of administration is intravenous infusion. Since aqueous solutions deteriorate rapidly and should not be used af-

Natamycin

ter 24 hours, it is available only as the dry powder that is to be dissolved in 5 percent dextrose solution just before use. The dry powder, as well as any solution made for a day's use, should be stored in a refrigerator and protected from light.

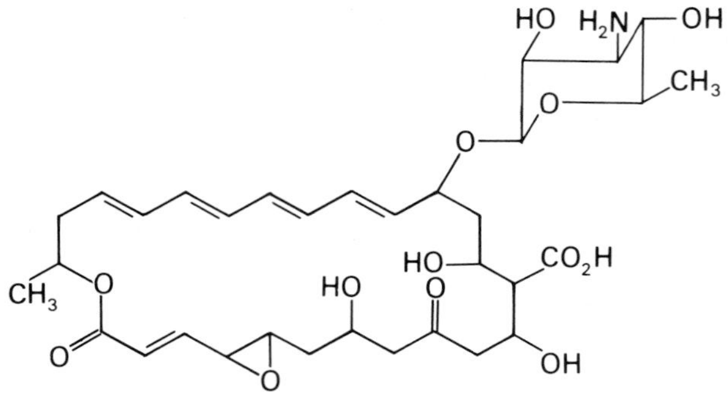

TABLE 4–13

ANTIFUNGAL ANTIBIOTICS

Name Proprietary Name	Preparations	Application*	Usual Adult Dose*	Usual Pediatric Dose*
Griseofulvin U.S.P. *Fulvicin U/F,* *Grifulvin V,* *Grisactin*	Griseofulvin Tablets U.S.P. Griseofulvin Capsules U.S.P. Griseofulvin Oral Suspension U.S.P.		Tinea corporis, tinea cruris, or tinea capitis— oral, 500 mg. daily, as a single dose or in divided doses; tinea pedis or tinea unguium—1 g. daily in divided doses	Oral, 10 mg. per kg. of body weight or 300 mg. per square meter of body surface daily, as a single dose or in divided doses, or for children 14 to 23 kg.—oral, 125 to 250 mg. daily as a single dose or in divided doses; children 23 kg. and over—250 to 500 mg. daily
Nystatin	Nystatin Vaginal Tablets U.S.P.		100,000 Units once or twice daily for 2 weeks	
	Nystatin Topical Powder U.S.P. Nystatin Cream U.S.P. Nystatin Lotion U.S.P.	Topical, to the skin, as a 100,000 units-per- gram powder, cream, or lotion, 2 to 3 times daily		
Nystatin U.S.P. *Mycostatin, Nilstat*	Nystatin Ointment U.S.P.	Topically to the skin, as 100,000 Units per g. ointment 2 or 3 times daily		
	Nystatin Oral Suspension U.S.P.		400,000 to 600,000 Units 4 times daily	Premature and low birth-weight infants—100,000 Units 4 times daily; older infants—200,000 Units 4 times daily; children—see Usual Adult Dose
	Nystatin Tablets U.S.P.		500,000 to 1,000,000 Units 3 times daily	
Amphotericin B U.S.P. *Fungizone*	Amphotericin B for injection U.S.P.		I.V. infusion, 250 μg. per kg. of body weight in 500 ml. of 5 percent Dextrose Injection, over a period of 6 hours	
Candicidin U.S.P. *Candeptin, Vanobid*	Candicidin Ointment U.S.P.	Vaginal, 0.06 percent ointment twice daily for 14 days		
	Candicidin Vaginal Tablets U.S.P.	Vaginal, 3-mg. suppository twice daily for 14 days		

*See *U.S.P. D.I.* for complete dosage information.

Candicidin U.S.P., Candeptin®. The macrolide polyene antibiotic candicidin was isolated in 1953 by Lechevalier et al.[66] from a strain of *Streptomyces griseus*. Although its potent antifungal property had been known for some time, it was not until 1964 that it became available for medicinal use in the United States. It is recommended for use in the treatment of monilia infections of the vaginal tract. Its chemistry is not yet well known but it is a heptaene macrolide closely related to amphotericin B. It is available as a 3-mg. vaginal tablet and as a vaginal ointment containing 3 mg. of candicidin per 5 g. of ointment.

Natamycin U.S.P., pimaricin, is a polyene macrolide antibiotic obtained from *Streptomyces natalensis*. It was first isolated in 1958 by Struyk et al.[67] and its structure was fully elucidated eight years later.[57] The pimaricin structure (shown below) consists of a 26-membered lactone ring containing a tetraene chromophore, a double bond conjugated with the lactone carbonyl group, three hydroxyl groups, one keto group, an epoxide and a carboxyl group. It is, of course, an amphoteric substance.

The antifungal properties of pimaricin have been known for some time. It was, however, only recently introduced specifically for the treatment of mycotic keratosis. It is currently available as Natamycin (Ophthalmic Suspension U.S.P.).

ANTIVIRAL AGENTS

The chemotherapy of viral disease is today at about the same stage of development as was the chemotherapy of bacterial infections prior to the discovery and development of the sulfonamides. Viral diseases such as smallpox and poliomyelitis are at present controlled by public health measures and immunization. With few exceptions, treatment of viral diseases consists of making the condition tolerable for the patient and ensuring that a secondary bacterial infection does not develop.

The two major obstacles to effective antiviral chemotherapy are, first, the close relationship that exists between the multiplying virus and the host cell and, second, the fact that many viral-caused diseases can be diagnosed and recognized only after it is too late for effective treatment. In the first case, an effective antiviral agent must prevent completion of the viral growth cycle in the infected cells without being toxic to the surrounding normal cells. One encouraging development is the discovery that some virus-specific enzymes are elaborated during multiplication of the virus particles and this may be a point of attack by a specific enzyme inhibitor. However, recognition of the

disease state too late for effective treatment would render antiviral drugs useless, even if they were available. Thus, until early recognition of the impending disease state is provided, most antiviral chemotherapeutic agents will have their greatest value as prophylactic agents.

Two approaches to the development of prophylactic antiviral agents that show promise involve the discovery of substances which either prevent penetration of virus particles into host cells or induce the in vivo synthesis of interferon, an antiviral protein produced by host cells in response to viral infection. The first approach has led to the introduction into therapy of amantadine for the prevention of certain forms of influenza. Some interferon inducers with broad spectrum antiviral activity (such as the synthetic RNA copolymer, poly I.C.[68] and the tricyclic compound tilorone[69]) have been discovered, but none of these substances has achieved clinical acceptance to date.

From a chemotherapeutic standpoint, viruses that infect animals can be divided into 2 groups. The first and smaller group is made up of the rickettsias and the large viruses (both held by some investigators not to be true viruses) which are more or less effectively controlled by some of the sulfonamides (see Chap. 5) and antibiotics (see Chap. 7). The larger group (true viruses), with the notable exception of the herpes simplex virus, cannot be controlled by chemotherapeutic agents. Otherwise, only when secondary bacterial infection is present can the course of disease caused by true viruses be shortened or modified. Most attempts to inhibit virus multiplication without causing damage to the host have been unsuccessful, probably because virus multiplication is so intimately dependent on host cell metabolism.[70]

In 1963 the first antiviral agent which acts as an antimetabolite became available commercially when idoxuridine was marketed for the treatment of herpes keratitis. It was not until fifteen years later, however, that a second antiviral antimetabolite, the purine nucleoside, vidarabine, was introduced. Vidarabine represents a significant advance in viral chemotherapy because of its utility for intravenous administration in the treatment of herpes encephalitis. The antitumor antimetabolite, Cytarabine U.S.P. (cytosine arabinoside), is also employed for the treatment of herpes infections (see Chap. 8).

Idoxuridine U.S.P., Stoxil®, Dendrid®, Herplex®, is 2′-deoxy-5-iodouridine. It is slightly soluble in water and insoluble in chloroform or ether. This product has been found to be an effective antiviral agent in the treatment of dendritic keratitis caused by herpes simplex. Until this discovery was made there was no satisfactory chemotherapy for this infection.

Idoxuridine

Aqueous solutions are slightly acidic in reaction and are stable for one year if refrigerated.[71] At room temperature there is up to 10 percent loss in 1 year. Solutions may not be sterilized by autoclaving and must be dispensed in amber bottles to protect from light. The ointment does not require refrigeration and is stable for at least 2 years.

Trifluridine, Viroptic®, is 2′-deoxy-5-trifluoromethyluridine. This product is closely related in structure to idoxuridine, having a trifluoromethyl group at the 5-position rather than an iodo group. Trifluridine is used as a sterile, buffered, 1 percent aqueous solution in the treatment of keratoconjunctivitis caused by herpes simplex virus. It is reported to be effective in cases that are resistant to idoxuridine treatment.

Trifluridine

Trifluridine solutions are heat sensitive, so the product should be kept refrigerated until dispensed to the patient. At 25°C the product will lose 10 percent of its potency in approximately 30 days. When refrigerated at 4 to 8°C, the product shows less than 10 percent loss of potency for up to 36 months.

Sterile Vidarabine U.S.P., Vira A®, ara-A, is 9-β-D-arabinofuranosyladenine or adenine arabinoside. Originally synthesized in 1960 as a potential anticancer agent, ara-A was later found to have broad spectrum activity against DNA viruses.[72] First marketed in 1977 as an alternative to idoxuridine for the treatment of herpes keratitis, it received FDA approval one year later for the treatment of herpes encephalitis. Vidarabine thus became the first antiviral drug to be introduced for the systemic treatment of a viral infec-

Vidarabine

tion. In cases of viral encephalitis, the drug must be administered by continuous intravenous infusion because of its poor water solubility and relatively rapid metabolic conversion in vivo. Vidarabine undergoes deamination catalyzed by adenosine deaminase to form the considerably less active hypoxanthine derivative, hypoxanthine arabinoside (ara-H).

Amantadine Hydrochloride U.S.P., Symmetrel®, 1-adamantanamine hydrochloride, is a white crystalline powder, freely soluble in water and insoluble in alcohol or chloroform; it has a bitter taste. It is useful in the prevention but not the treatment of influenza caused by the A_2 strains of the Asian influenza virus. Aside from vaccination, it is the only prophylactic presently available against any strain of Asian influenza. It appears to exert its effect by preventing penetration of the adsorbed virus into the host cell. It has no therapeutic value, and is ineffective once the virus has penetrated the host cell. Compared to vaccination, amantadine has two advantages: it is oral medication, and it provides immediate protection. Its

Amantadine Hydrochloride

chief drawbacks are that protection stops shortly after daily dosage stops and the protection is against only the A_2 strains of the virus.

Amantadine also finds occasional use in the management of Parkinson's disease in patients who do not tolerate full therapeutic doses of levodopa.

Investigational Drugs, which are currently undergoing extensive clinical evaluation in man and which represent potentially significant advances in antiviral chemotherapy, include acycloguanosine, inosiplex and ribavirin.

Acycloguanosine, Zovirax®, is 9-(2-hydroxy-ethoxymethyl)guanosine. This compound, which can be

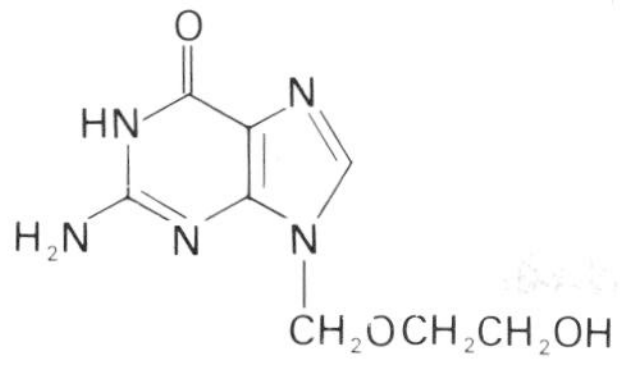

Acycloguanosine

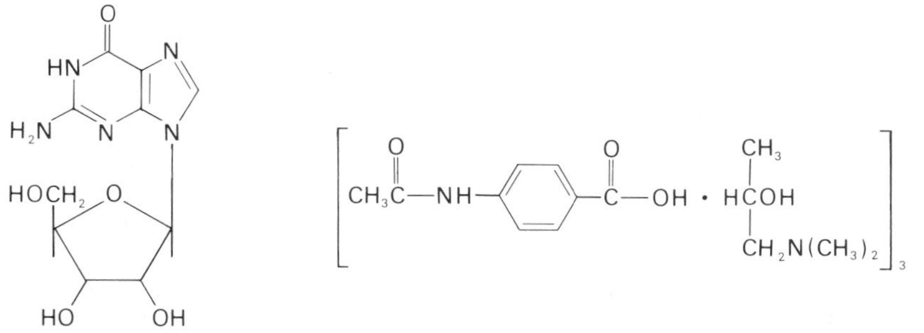

Inosiplex

considered to be a purine nucleoside with an incomplete sugar moiety, is significantly more active than vidarabine against herpes viruses in vitro. It is believed to be converted to the triphosphate which in turn specifically inhibits viral DNA synthesis.[73] Clinical studies have demonstrated its effectiveness against cutaneous and ocular herpes infections. Effective blood levels and an apparent lack of toxicity following oral administration in animals suggest that the drug may be useful in the treatment of herpes encephalitis.

Ribavirin, Virazole®, is 1-β-D-ribofuranosyltriazole-3-carboxamide. This nucleoside is isosteric with 1-β-D-ribofuranosylimidazole carboxamide, a precursor in the de novo synthesis of purine nucleotides. Ribavirin has broad spectrum antiviral activity in vitro against both DNA and RNA viruses.[74] Its clinical efficacy against cutaneous herpes infections has been clearly demonstrated, but its activity against hepatitis (type A) and influenza infections is more doubtful.

Inosiplex, Isoprinosine®, is a 1:3 complex of inosine and the 1-(dimethylamino)-2-propanol salt of 4-acetamidobenzoic acid. This agent has been shown to exhibit activity against a wide variety of both DNA and RNA viruses. Its broad spectrum of activity is thought to result from a two-pronged mode of action, namely, the simultaneous stimulation of host T-cell mediated immunity and the direct inhibition of viral replication. Clinical studies in humans indicate potentially useful activity against herpes, rhino and influenza viruses. The drug is marketed in several foreign countries.

Methisazone, Marboran®, is 1-methylisatin-3-thiosemicarbazone. The activity of this type of heterocyclic thiosemicarbazone against various DNA viruses of the orthopox group (e.g., pox, vaccinia and variola) has been known for more than two decades. Despite its effectiveness in the treatment of smallpox, methisazone has not been widely used for this purpose because of the prophylactic nature of its action and the greater desirability of prevention of the disease through public health measures such as vaccination. Methisazone may continue to find use in the control of vaccinia infections that occasionally occur as a result of smallpox vaccinations, but since mass vaccination for the prevention of smallpox is no longer advocated by public health officials in the United States and many foreign countries, the frequency of even this use of the drug is likely to decrease.

The mechanism of antiviral action of methisazone

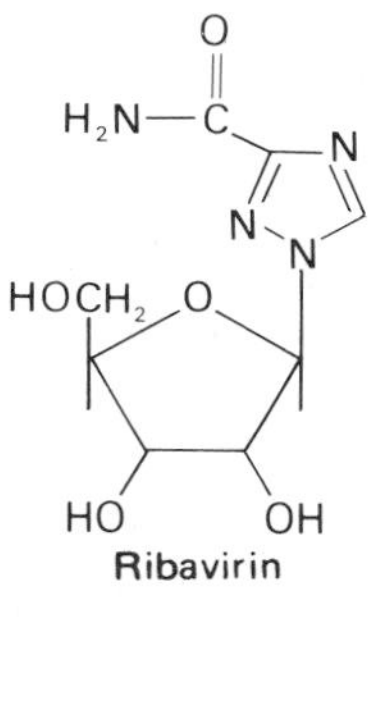

Ribavirin

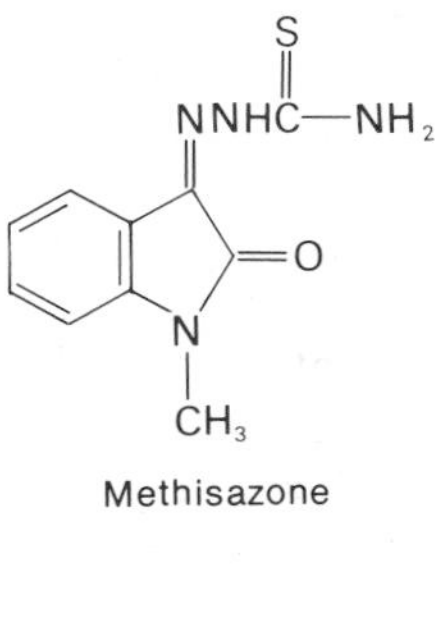

Methisazone

TABLE 4–14

ANTIVIRAL AGENTS

Name Proprietary Name	Preparations	Application	Usual Dose
Idoxuridine U.S.P. *Dendrid, Herplex, Stoxil*	Idoxuridine Ophthalmic Ointment U.S.P. Idoxuridine Ophthalmic Solution U.S.P.	Topically to the conjunctiva, as an 0.5 percent ointment 5 times daily or 0.1 ml. of a 0.1 percent solution 10 to 20 times daily	
	Sterile Vidarabine U.S.P. Vidarabine Ophthalmic Ointment U.S.P.	Topically into the lower conjunctival sac as a 5 percent ointment 5 times daily at 3 hour intervals	
Amantadine Hydrochloride U.S.P. *Symmetrel*	Amantadine Hydrochloride Capsules U.S.P. Amantadine Hydrochloride Syrup U.S.P.		200 mg. daily, given in a single dose or in 2 divided doses

remains to be clarified. It appears to inhibit viral protein synthesis in some manner, perhaps through the formation of a metal chelate. The drug, which is available only for investigational use in America, is administered orally in the form of tablets or an an oral suspension. It is insoluble in water.

URINARY TRACT ANTI-INFECTIVES

Certain antibacterial agents that, by virtue of their solubility properties, concentrate in the urine are effective in the treatment of infections of the urinary tract. Generally speaking, the more typical first time infections such as acute cystitis can be eradicated with use of either a short-acting sulfonamide (see Chap. 5) or the antibiotic tetracycline (see Chap. 7). Acute renal tissue infections such as glomerulonephritis require use of an antibiotic that achieves high concentrations in the kidneys and to which the causative organism is highly susceptible. A β-lactam (penicillin or cephalosporin) or an aminoglycoside (see Chap. 7) may be used, depending on the causative organism. Chronic, recurring infections tend to be less responsive to therapy and often require long term treatment and/or the use of several drugs. Enteric bacterial strains resistant to particular chemotherapeutic agents are frequently the cause. Cotrimoxazole, a combination of the dihydrofolate reductase inhibitor trimethoprim and the sulfonamide sulfamethiazole (see Chap. 5), is a treatment of choice for chronic urinary tract infections. β-Lactam and aminoglycoside antibiotics are also frequently employed. In addition, there is a group of antibacterial agents described in this chapter which, because they achieve significantly higher concentrations in the urine and in the kidneys as compared with other body fluids and tissues, are well suited for the treatment of chronic urinary tract

infections. They are usually employed when antibiotics or sulfonamides are either ineffective owing to the emergence of bacterial resistance or contraindicated because the patient is allergic to them.

Methenamine U.S.P., Urotropin®, Uritone®, hexamethylenetetramine, depends upon the liberation of formaldehyde for its activity. It is manufactured by evaporating a solution of formaldehyde to dryness with strong ammonia water.

Methenamine

The compound consists of colorless crystals or a white crystalline powder without odor. It sublimes at about 260° without melting and burns readily with a smokeless flame. It dissolves in 1.5 ml. of water to make an alkaline solution and in 12.5 ml. of alcohol. Warm acids will liberate formaldehyde, which may be recognized by its odor, and the subsequent addition of alkalies will give an odor of ammonia. The assay of the compound depends upon decomposition with volumetric solution of sulfuric acid and titration of the excess with sodium hydroxide.

Methenamine is used internally as an antiseptic, especially in the urinary tract. In itself it has practically no bacteriostatic power and can be efficacious only when it is acidified to produce formaldehyde. Because concentration in the kidney and the bladder never can become very high, the success in treating infections of the urinary tract usually has not been great. In order to obtain a maximum effect, the administration of the compound generally is accompanied by ascorbic acid, sodium biphosphate, ammonium chloride or a similar acidifying agent. A recent clinical study[75] casts doubt upon the effectiveness of ascorbic acid as a urinary acidifier, since the maximal change in urinary pH brought about by this agent is very small (less than 0.24 of a pH unit).

Methenamine Mandelate U.S.P., Mandelamine®, hexamethylenetetramine mandelate, is a white crystalline powder with a sour taste and practically no odor. It is very soluble in water and has the advantage of furnishing its own acidity, although in its use the custom is to carry out a preliminary acidification of the urine for 24 to 36 hours before administration. It is effective with smaller amounts of mandelic acid and thus avoids the gastric disturbances attributed to the acid when used alone.

Methenamine Hippurate U.S.P., Hiprex®, is the hippuric acid salt of methenamine. It is readily absorbed after oral administration and is concentrated in the urinary bladder, where it exerts its antibacterial activity. Its activity is increased in acid urine.

Nitrofurantoin U.S.P., Furadantin®, 1-[(5-nitrofurfurylidene)amino]hydantoin, is a nitrofuran derivative that is suitable for oral use. The compound has been used successfully in treating infections of the urinary tract. It has been effective for infections that were resistant to antibiotics. Few side-effects, such as diarrhea, pruritus or crystalluria, have been observed.

Nitrofurantoin

Nalidixic Acid U.S.P., NegGram®, is 1-ethyl-1,4-dihydro-7-methyl-4-oxo-1,8-naphthyridine-3-carboxylic acid.

Nalidixic Acid

Nalidixic acid is useful in the treatment of infections of the urinary tract in which gram negative bacteria are predominant. The activity against indole

positive *Proteus* spp. is particularly noteworthy, and nalidixic acid and its congeners represent important alternatives for the treatment of urinary tract infections caused by strains of these bacteria resistant to other agents. The mode of action of nalidixic acid appears to involve the selective inhibition of bacterial DNA synthesis[76] by a mechanism that remains to be clarified. Gram positive bacteria are much less sensitive to the drug.

Nalidixic acid is rapidly absorbed, extensively metabolized, and rapidly excreted following oral administration. The recommended dosage regimen is 1 g. 4 times daily for 2 weeks.

X = CH Oxolinic Acid
X = N Cinoxacin

Cinoxacin, Cinobac®, is 1-ethyl-1,4-dihyro-4-oxo[1,3]dioxolo[4,5g] cinnoline-3-carboxylic acid. This close congener (isostere) of oxolinic acid has similar antibacterial properties to those of nalidixic and oxolinic acids and is recommended for the treatment of urinary tract infections caused by strains of gram negative bacteria susceptible to these agents. Early clinical studies indicate that the drug possesses pharmacokinetic properties superior to those of either of its predecessors. Thus, higher urinary concentrations of cinoxacin are achieved following oral administration as compared with nalidixic acid or oxolinic acid. Cinoxacin appears to be more completely absorbed and less protein bound than is nalidixic acid and significantly less metabolized to inactive metabolites than is oxolinic acid (which has been subsequently withdrawn from the market in the United States). The recommended dosage schedule for cinoxacin, which remains on an investigational status, is 500 mg. every 12 hours.

Phenazopyridine Hydrochloride U.S.P., Pyridium®, 2,6-diamino-3-(phenylazo)pyridine monohydrochloride, is a brick-red fine crystalline powder. It is slightly soluble in alcohol, in chloroform, and in water.

Phenazopyridine Hydrochloride

TABLE 4–15

URINARY TRACT ANTIBACTERIAL AGENTS

Name Proprietary Name	*Preparations*	*Usual Adult Dose**	*Usual Dose Range**	*Usual Pediatric Dose**
Methenamine U.S.P.	Methenamine Elixir U.S.P. Methenamine Tablets U.S.P.	1 g. every 6 hours	Up to 12 g. daily	
Methenamine Mandelate U.S.P. *Mandelamine*	Methenamine Mandelate Oral Suspension U.S.P. Methenamine Mandelate Tablets U.S.P.	1 g. every 6 hours		Under 6 years of age—18.3 mg. per kg. of body weight every 6 hours; 6 to 12 years—500 mg. every 6 hours
	Methenamine Hippurate Tablets U.S.P. *Hiprex, Urex*	1 g. every 12 hours	Up to 4 g. daily	Children 6 to 12 years—500 mg. to 1 g. every 12 hours
Nalidixic Acid U.S.P. *NegGram*	Nalidixic Acid Tablets U.S.P.	1 g. 4 times daily for 1 to 2 weeks. Thereafter, for prolonged treatment, the dose may be reduced to 500 mg. 4 times daily		
Nitrofurantoin U.S.P. *Furadantin, N-Toin*	Nitrofurantoin Oral Suspension U.S.P. Nitrofurantoin Tablets U.S.P. Nitrofurantoin Capsules U.S.P.	50 to 100 mg. every 6 hours; or 1.25 to 1.75 mg. per kg. of body weight every 6 hours	Up to 600 mg. daily; or up to 10 mg. per kg. of body weight daily	Use for infants below 1 month of age is not recommended. Older infants and children—1.25 to 1.75 mg. per kg. of body weight every 6 hours

*See U.S.P. D.I. for complete dosage information.

Phenazopyridine hydrochloride was formerly used as a urinary antiseptic. Although it is active in vitro against staphylococci, streptococci, gonococci and *E. coli,* it has no useful antibacterial activity in the urine. Thus, its present utility lies in its local analgesic effect on the mucosa of the urinary tract. It is now usually given in combination with urinary antiseptics. For example, it is available as Azo-Gantrisin, a fixed dose combination with the sulfonamide antibacterial sulfisoxazole. The drug is rapidly excreted in the urine, to which it gives an orange-red color. Stains in fabrics may be removed by soaking in a 0.25 percent solution of sodium dithionate.

Category—analgesic (urinary tract).

Usual dose—100 mg. 3 or 4 times daily.

Occurrence

Phenazopyridine Hydrochloride Tablets U.S.P.

ANTITUBERCULAR AGENTS

The development of effective chemotherapeutic agents for tuberculosis began in 1938, when it was observed that sulfanilamide had a slight inhibitory effect on the course of experimental tuberculosis in guinea pigs. Later, the activity of the sulfones was discovered. Dapsone, 4,4'-diaminodiphenylsulfone, was investigated clinically, but was considered to be too toxic. Later evidence indicates this was probably due to the use of too large doses. Dapsone is now considered one of the most effective drugs for the treatment of leprosy (see Chap. 5). It also appears to be of value in treating certain resistant forms of malaria. Major early advances in the chemotherapy of tuberculosis were, first, the discovery of the antitubercular activity of streptomycin by Waksman and his associates (in 1944); next, of the usefulness of *p*-aminosalicylic acid; and, finally, of the activity of isoniazid in 1952. Later, the antitubercular properties of first the synthetic agent, ethambutol, and then the semisynthetic antibiotic rifampin, were discovered.

Combination therapy, with the use of two or more antitubercular drugs, has been well documented to reduce the emergence of strains of *Mycobacterium tuberculosis* resistant to individual agents and has become standard medical practice. The choice of antitubercular combination is dependent on a variety of factors including: the location of the disease (pulmonary, urogenital, gastrointestinal or neural); the results of susceptibility tests and the pattern of resistance in the locality; the physical condition and age of the patient; and the toxicities of the individual agents. A combination of isoniazid and ethambutol, with or without streptomycin, has become the preferred choice of treatment among clinicians in this country. However, one or more of a relatively large group of compounds may be substituted for ethambutol or streptomycin. Antibiotics in this group include: rifampin, cycloserine, kanamycin, viomycin and capreomycin. Therapy with the antibiotic streptomycin or a suitable substitute is usually discontinued when the sputum becomes negative so that its toxic effects may be minimized.

A major advance in the treatment of tuberculosis was signalled by the introduction into therapy of the antibiotic rifampin. Recent clinical studies indicate that when rifampin is included in the regimen, particularly in combination with isoniazid and ethambutol, a significant shortening of the period required for successful therapy is possible. Previous treatment schedules without rifampin required maintenance therapy for at least 2 years, whereas those based on the isoniazid–rifampin combination achieve equal or better results in 6 to 9 months.

SYNTHETIC ANTITUBERCULAR AGENTS

Aminosalicylic Acid U.S.P., PAS, Parasal®, Pamisyl®, 4-aminosalicylic acid. The acid is available as practically odorless, white or yellowish-white crystals which darken on exposure to light and air. It is slightly soluble in water (0.1%) but is more soluble in alcohol, methanol and isopropyl alcohol. Solubility is increased with alkaline salts of alkali metals (sodium bicarbonate) and in weak nitric acid. The amine salts of hydrochloric and sulfuric acids are insoluble. Aqueous solutions have a pH of about 3.2, and, when heated, the acid decomposes.

Until recently, p-aminosalicylic acid was considered a first line drug for the chemotherapy of tuberculosis and was generally included in combination regimens with isoniazid and streptomycin. However, the introduction of the more effective and generally better tolerated agents, ethambutol and rifampin, have relegated it to an alternative drug status. The acid is taken orally, usually in tablet form. Often, severe gastrointestinal irritation accompanies the use of PAS or its sodium salt. To overcome this disadvantage, coated tablets, capsules and granules are used. Often, an antacid, such as aluminum hydroxide, is prescribed concurrently.

Studies of structural modifications have shown that the maximum activity is obtained when the hydroxyl group is in the 2 position and the free amino group in the 4 position. However, esters and acylation of the amino group, if labile enough to be hydrolyzed in vivo to p-aminosalicylic acid, may be used. In fact, advantages of less gastric irritation are claimed for some of these derivatives.

p-Aminosalicylic acid is rapidly and almost completely absorbed after oral administration. It is distributed freely and equally to most tissues and fluids with the exception of the cerebrospinal fluid, where levels are lower and less consistently obtained. After an oral dose of 4 g. in man, a maximum plasma level of about 7.5 mg. percent is reached in about an hour. It is excreted in the urine, both unchanged and as metabolites and has a biological half-life of about 2 hours. Up to one-third of the dose is excreted unchanged, up to two-thirds as acetyl p-aminosalicylic acid and up to about one-fourth is conjugated with glycine and excreted as p-aminosalicyluric acid.

Aminosalicylate Sodium U.S.P., Parasal® Sodium, Pasara® Sodium, Pasem® Sodium, Parapas® Sodium, Paraminose®, sodium 4-aminosalicylate. This compound is the dihydrate salt, occurring as a yellow-white, odorless powder or in crystals. It is soluble in alcohol and very soluble in water, provided that the solution has a pH of 7.25. Aqueous solutions decompose quite readily, the rate depending on the pH and the temperature. The pH of maximum stability is in the range of 7 to 7.5. Two types of reactions are involved in the decomposition process. The first is decarboxylation to yield m-aminophenol. The second involves the oxidation of p-aminosalicylic acid or of the m-aminophenol, or both, with the formation of brown to black pigments. Freshly prepared solutions of pure sodium p-aminosalicylate are nearly colorless, but on standing they develop an amber and eventually a dark brown to black color. The presence of the amber color is not necessarily a sign of extensive decomposition; however, the U.S.P. cautions that solutions should be prepared within 24 hours of administration and that in no case should a solution be used if its color is darker than that of a freshly prepared solution. It is generally agreed that solutions for parenteral or topical use should be sterilized by filtration.

Using 4.8 percent solutions suitable for intravenous infusion, it was found that the following amounts of m-aminophenol formed when stored at the conditions indicated[77] in the table below. The addition of 0.1 percent of sodium sulfite will prevent discoloration (oxidation) but not decarboxylation.[78]

Temperature, °C.	Time, Days	Mg./100 ml.
20	1	None
20	2	10
0	7	7
0	30	11
0	45	15
−5	30	9
−5	60	11
−5	90	14
−5	120	15

Aminosalicylate Potassium U.S.P., Paskalium®, Paskate®, potassium 4-aminosalicylate. This salt has properties similar to those of sodium *p*-aminosalicy-late. It is reported to cause less gastric irritation than the free acid or the sodium salt. Of course, its use is indicated when the sodium ion intake must be kept at low levels.

Aminosalicylate Calcium U.S.P., Parasa® Calcium, calcium 4-aminosalicylate, is available in three forms: powder, granules and capsules. It exhibits all the desirable actions of *p*-aminosalicylic acid but materially reduces gastrointestinal irritation.

Benzoylpas Calcium U.S.P., Benzapas®, is the calcium salt of N-benzoylaminosalicylic acid. This derivative of PAS is practically insoluble in water and, when completely hydrolyzed, yields 47.4 percent of PAS. This modification has been made to decrease the incidence of gastric irritation which may occur when the free acid or the inorganic salts must be administered for long periods.

Isoniazid U.S.P., Rimifon®, INH, isonicotinic acid hydrazide, isonicotinyl hydrazide, occurs as nearly colorless crystals which are very soluble in water. Hydrazides are prepared readily by refluxing a methyl or ethyl ester with a hydrazine.

Antitubercular drugs have been studied ever since Koch identified the tubercle bacillus, *Mycobacterium tuberculosis*. Up to 1952, the primary compounds used to treat tuberculosis were sulfonamides, various sulfones, *p*-aminosalicylic acid, streptomycin, dihydro-streptomycin and tibione.[79] In 1945, and again in 1948, it was pointed out that nicotinamide had tuberculo-static activity equal to that of *p*-aminosalicylic acid. In view of this and the fact that tibione is a thiosemi-carbazone of *p*-acetamidobenzaldehyde, the thiosemi-carbazones of alpha, beta and gamma nicotinaldehyde were prepared and studied. Of these pyridine analogs of tibione, the alpha is inactive, and the beta and the gamma are superior to tibione.

In the method used for synthesizing[80] gamma-nico-tinaldehyde thiosemicarbazone, isonicotinylhydrazine (isoniazid) was an intermediate. Since the product was available, it was subjected routinely to study on tuberculosis. Experiments on animals and humans re-

Isoniazid

vealed no serious or irreversible toxic effects. Reactions included central nervous system stimulation (leg twitching and insomnia) or autonomic activity (dryness of secretions) and dizziness. There is a wide margin of safety between the therapeutic and lethal doses in animals, the oral L.D.$_{50}$ being about twenty times the oral therapeutic dose.

Isoniazid is a remarkably effective drug and is now considered one of the primary drugs (along with rifampin and streptomycin) for chemotherapy of tuberculosis. But, even so, it is not completely effective in all types of the disease. Isoniazid is well absorbed after oral administration and is rather rapidly excreted, with between 50 and 70 percent of a dose being eliminated in the urine within 24 hours. There was no evidence of elevated plasma levels in patients receiving a dose of 1.5 mg. per kg. twice daily for several weeks. It is excreted unchanged and in several metabolically modified forms, the principal metabolites being pyruvic acid isonicotinylhydrazone, α-ketoglutaric acid isonicotinylhydrazone, acetylisoniazid, isonicotinic acid and isonicotinuric acid.[81] Although the metabolism of isoniazid is relatively complex, the principal path of inactivation involves acetylation of the primary hydrazine nitrogen. The capacity to inactivate by acetylation in humans is an inherited characteristic. Approximately half of the population are fast acetylators (plasma half life 45 to 80 min.) and the other half slow acetylators (plasma half life 140 to 200 min.). Isoniazid is freely distributed to all the tissues and fluids of the body, including the cerebrospinal fluid and the placental fluid in the pregnant woman.

The activity of the drug is only on the growing bacilli and not on the resting forms. At present there is no completely satisfactory explanation of the mechanism of action of isoniazid. There is evidence to indicate that it may exert its effect by interference with enzyme systems requiring pyridoxal phosphate as a coenzyme.

The principal toxic reactions are peripheral neuritis and gastrointestinal disturbances such as loss of appetite and constipation. The side-effects are dosage-related, and the incidence may be expected to increase as the dose is increased. The peripheral neuritis resembles that caused by pyridoxine deficiency and it is now current practice for many physicians treating tuberculosis to give the patients fairly large doses of pyridoxine. The mechanism by which isoniazid produces the peripheral neuropathy is not well understood. The pyridoxine does not seem to interfere with the antibacterial action of isoniazid.

The problem often occurs that resistant strains of the tubercle bacillus develop during therapy. For this reason isoniazid is seldom used as the sole chemotherapeutic agent but is usually administered with rifampin given orally or with streptomycin administered intramuscularly. In some dosage regimens all three drugs are used.

None of the derivatives of isoniazid is more useful in therapy than the parent compound. Any change in structure leads to a decrease in potency and, in most cases, to a loss of potency. The isopropyl derivative 1-isonicotinyl-2-isopropylhydrazine shows good activity, but clinical trial proved that it was too toxic for use considering that other safer drugs were available. However, the isopropyl derivative, iproniazid, is worthy of special mention because it has led to the development of a group of psychomotor stimulants useful in drug therapy of certain kinds of depression.

Pyrazinamide U.S.P., Aldinamide®, pyrazinecarboxamide, is the pyrazine analog of nicotinamide. It occurs as a white crystalline powder, practically insoluble in water, slightly soluble in acetone, in alcohol and in chloroform. It is a fairly active drug, but it causes a rather significant incidence of liver damage. Because of its hepatotoxic potential, the drug is generally reserved for the treatment of hospitalized patients when the primary drugs and other secondary drugs cannot be used because of bacterial resistance or because the patients cannot tolerate them. Pyrazinamide increases reabsorption of urates and should thus be used with care in patients with a history of gout.

Pyrazinamide

Ethionamide U.S.P., Trecator S.C.®, 2-ethylthioisonicotinamide. This drug occurs as a yellow, crystalline substance, very sparingly soluble in water or ether, and soluble in hot acetone or in dichloroethane. In contrast to ring modifications in the isoniazid series, the 2-alkyl substituted thioisonicotinamides were more active than the parent compound. The 2-ethyl derivative was the most interesting of the group studied.

Ethionamide is a secondary drug in the chemotherapy of tuberculosis and is intended mainly for use

Ethionamide

in the treatment of pulmonary tuberculosis resistant to isoniazid or when the patient is intolerant to other drugs. It is administered orally and the highest tolerated dosage is generally recommended.

Ethambutol U.S.P., Myambutol®, (+)-2,2'-(ethylene diimino)di-1-butanol dihydrochloride, EBM, is a white crystalline powder freely soluble in water and slightly soluble in alcohol.

Ethambutol Dihydrochloride

This compound is remarkably stereospecific. Tests have shown that although the toxicities of the dextro, levo, and meso isomers are about equal, their activities vary considerably. The dextro isomer is 16 times as active as the meso isomer, and the levo isomer is even less active than the meso isomer. In addition, the length of the alkylene chain, the nature of the branching of the alkyl substituents on the nitrogens, and the extent of N-alkylation all have a pronounced effect on the activity.

Ethambutol is rapidly absorbed after oral administration, and peak serum levels occur in about 2 hours. It is rapidly excreted, mainly in the urine. Up to 80 percent is excreted unchanged, with the balance being metabolized and excreted as 2,2'-(ethylenediimino)dibutyric acid and as the corresponding di-aldehyde.

It is recommended not for use alone but in conjunction with other antitubercular drugs in the chemotherapy of pulmonary tuberculosis.

ANTITUBERCULAR ANTIBIOTICS

Cycloserine U.S.P., Seromycin®, D-(+)-4-amino-3-isoxazolidinone. One of the simplest structures to possess antibiotic action is the antitubercular substance, cycloserine. It has been isolated from three different species of Streptomyces: *S. orchidaceus, S. garyphalus* and *S. lavendulus*. Its structure has been determined by Kuehl *et al.*[82] and Hidy *et al.*[83] to be D-4-amino-3-isoxazolidinone. No doubt the compound exists in equilibrium with its enol form.

Cycloserine

TABLE 4–16

TUBERCULOSTATIC AGENTS

Name Proprietary Name	Preparations	Usual Adult Dose*	Usual Dose Range*	Usual Pediatric Dose*
Aminosalicylic Acid U.S.P. *Pamisyl, Parasal, Rezipas, NatriPas, Pamisyl Sodium, Pasara Sodium, Pasna*	Aminosalicylic Acid Tablets U.S.P.	3 g. 4 times daily	10 to 20 g. daily	
Aminosalicylate Sodium U.S.P.	Aminosalicylate Sodium Tablets U.S.P.	4 to 5 g. 3 times daily	8 to 15 g. daily	100 mg. per kg. of body weight or 2.7 g. per square meter of body surface, 3 times daily
Aminosalicylate Potassium U.S.P. *Parasal Potassium, Paskalium*	Aminosalicylate Potassium Tablets U.S.P.	3 g. 4 times daily	10 to 20 g. daily	
Aminosalicylate Calcium U.S.P. *Parasal Calcium*	Aminosalicylate Calcium Capsules U.S.P. Aminosalicylate Calcium Tablets U.S.P.	4 g. 4 times daily	10 to 25 g. daily in 4 divided doses	
Benzoylpas Calcium U.S.P. *Benzapas*	Benzoylpas Calcium Tablets U.S.P.		10 to 15 g. daily in 2 or 3 divided doses	
Isoniazid U.S.P. *Hyzyd, Niconyl, Nydrazid*	Isoniazid Injection U.S.P.	I.M., 5 mg. per kg. of body weight once daily, up to 300 mg. daily		
	Isoniazid Syrup U.S.P. Isoniazid Tablets U.S.P.	Prophylaxis—oral, 300 mg. once daily; treatment, in combination with other tuberculostatics—5 mg. per kg. of body weight up to 300 mg. once daily	Up to 20 mg. per kg. of body weight, not to exceed 600 mg. daily	Prophylaxis—oral, 10 mg. per kg. of body weight, up to 300 mg. once daily; treatment, in combination with other tuberculostatics—10 to 20 mg. per kg. of body weight, up to 500 mg. once daily
Pyrazinamide U.S.P. *Aldinamide*	Pyrazinamide Tablets U.S.P.	5 to 8.75 mg. per kg. of body weight 4 times daily	1 to a maximum of 3 g. daily	
Ethionamide U.S.P. *Trecator S.C.*	Ethionamide Tablets U.S.P.	250 mg. 2 to 4 times daily	500 mg. to 1 g. daily	4 to 5 mg. per kg. of body weight, up to a maximum of 250 mg. 3 times daily
	Ethambutol Hydrochloride Tablets U.S.P.	In combination with other tuberculostatics—initial treatment, oral, 15 mg. per kg. of body weight once daily; retreatment—25 mg. per kg. of body weight once daily	Initial treatment—500 mg. to 1.5 g. daily; retreatment—900 mg. to 2.5 g. daily	Children under 13 years of age—use is not recommended; children 13 years of age and over—see Usual Adult Dose

*See U.S.P. D.I. for complete dosage information.

In aqueous solutions, cycloserine will form a dipolar ion that, on standing will dimerize to 2,5-bis(aminoxymethyl)-3,6-diketopiperazine:

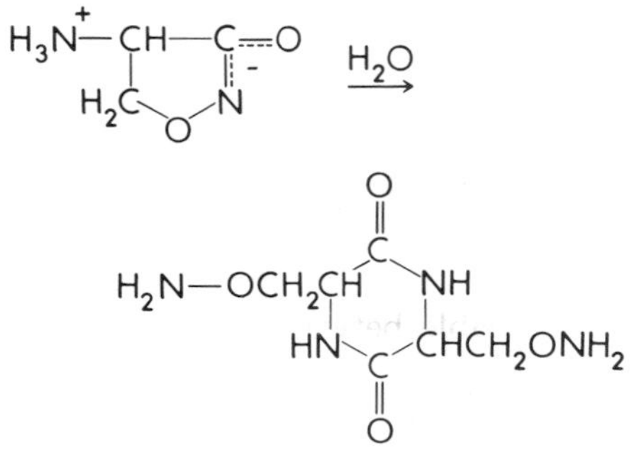

Cycloserine is a white to pale yellow, crystalline material that is soluble in water. It is quite stable in alkali but is unstable in acid. It has been synthesized from serine by Stammer *et al.*[84] and by Smrt *et al.*[85] Configurationally, cycloserine resembles D-serine, but the L-form has similar antibiotic activity. Most interesting is the observation that the racemic mixture is more active than either enantiomorph, indicating that the isomeric pair act on each other synergistically.

Cycloserine is presumed to exert its antibacterial action by preventing the synthesis of crosslinking peptide in the formation of bacterial cell walls.[86] Rando[87] has recently suggested that it is an antimetabolite for alanine which acts as a suicide substrate for the pyridoxal phosphate requiring enzyme, alanine racemase. Irreversible inactivation of the enzyme thus deprives the cell of D-alanine required for the synthesis of the crosslinking peptide.

Although cycloserine exhibits antibiotic activity in vitro against a wide spectrum of both gram-negative and gram-positive organisms, its relatively weak potency and frequent toxic reactions limit its use to the treatment of tuberculosis. It is recommended for cases which fail to respond to other tuberculostatic drugs or are known to be infected with organisms resistant to other agents. It is usually administered orally in combination with other drugs, commonly isoniazid.

Sterile Viomycin Sulfate U.S.P., Viocin® Sulfate. Viomycin is a cyclic peptide isolated from a number of Streptomyces species. Its use is confined to the treatment of tuberculosis for which it is a second-line agent occasionally substituted for streptomycin in infections resistant to that antibiotic. Viomycin exerts a bacteriostatic action against the tubercle bacillus by a mechanism that has not been determined. It is significantly less potent than streptomycin and its toxicity is greater. Toxic effects of viomycin are primarily associated with damage to the eighth cranial nerve and to the kidney.

Viomycins are strongly basic peptides. At least two components have been obtained from S. vinaceus and have been named vinactins A and B. A closely related substance, identified as vinactin C, has also been found to be present. Vinactin A appears to be the ma-

Vinactin A

jor component of viomycin. Some disagreement remains concerning details of the chemical structure of the viomycins. Early work by Haskell et al.[88] and Mayer et al.[89] showed that vinactin A had no free α-amino groups and, on vigorous acid hydrolysis, yielded carbon dioxide, ammonia, urea, L-serine, α,β-diaminopropionic acid, β-lysine and a guanidino compound. Based on additional chemical and spectroscopic evidence at least three different structures have been proposed for vinactin A.[90-92] Doubt about the peptide se-

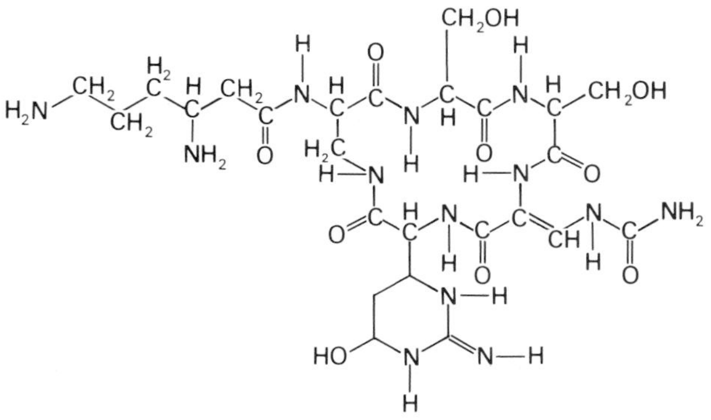

quence of the antibiotic, raised as a result of x-ray crystallographic studies on a closely related antibiotic, tuberactinomycin N,[93] appears to have been resolved by the chemical studies of Noda et al. who have suggested the structure of vinactin A shown.[92] It is perhaps noteworthy that the more recently proposed structures lack the fused hetero aromatic ring system of structures suggested earlier[90] to explain the ultraviolet spectrum of the antibiotic. A possible explanation for this could be the existence of the antibiotic in a different chemical form in solution as compared to the solid state.

Viomycin sulfate is an odorless powder that varies in color from white to slightly yellow. It is freely soluble in water, forming solutions ranging in pH from 4.5 to 7.0. It is insoluble in alcohol and other organic solvents. Since it is slightly hygroscopic it should be stored in tightly closed containers. It is administered in aqueous solutions intramuscularly.

Sterile Capreomycin Sulfate U.S.P., Capastat® Sulfate. Capreomycin is a strongly basic cyclic peptide isolated from Streptomyces capreolus in 1960 by Herr et al.[94] It was released in the United States in 1971 exclusively as a tuberculostatic drug. Capreomycin,

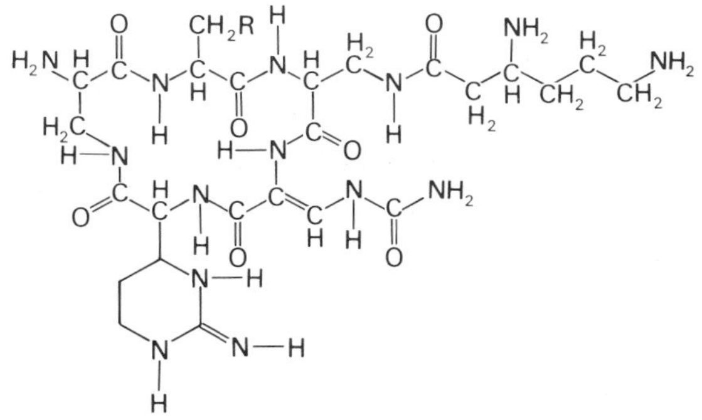

Capreomycin 1A   R=OH
           1B   R=H

which resembles viomycin chemically and pharmacologically, is a second-line agent employed in combination with other antitubercular drugs. In particular, it may be used in place of streptomycin where either the patient is sensitive to, or the strain of M. tuberculosis is resistant to, streptomycin. Like viomycin, capreomycin is a potentially toxic drug. Damage to the eighth cranial nerve and renal damage, as with viomycin, are the more serious toxic effects associated with capreomycin therapy. There is, as yet, insufficient clinical data on which to reliably compare the relative toxic potential of capreomycin with either viomycin or streptomycin. Cross-resistance among strains of tubercle bacilli is probable between capreomycin and viomycin, but rare between either of these antibiotics and streptomycin.

Four capreomycins, designated, 1A, 1B, IIA and IIB have been isolated from *S. capreolus*. The clinical agent contains primarily 1A and 1B. The close chemical relationship between capreomycins 1A and 1B and viomycin was established[95] and the total synthesis and proof of structure of the capreomycins later accomplished.[96] The structures of capreomycins II_a and II_b correspond to those of I_a and I_b, but lack the β-lysyl residue. The sulfate salts are freely soluble in water.

Rifampin U.S.P., Rifadin®, Rimactane®, rifampicin. The rifamycins are a group of chemically related antibiotics obtained from *Streptomyces mediterrani*. They belong to a new class of antibiotics that contain a macrocyclic ring bridged across two nonadjacent (ansa) positions of an aromatic nucleus and called ansamycins. The rifamycins and many of their semisynthetic derivatives have a broad spectrum of antimicrobial activity. They are most notably active against gram-positive bacteria and *Mycobacterium tuberculosis*. However, they are also active against some gram-negative bacteria and many viruses. Rifampin, a semisynthetic derivative of rifamycin B, was released as an antitubercular agent in the United States in 1971. Its structure is shown below.

The chemistry of rifamycins and other ansamycins has been reviewed recently by Rinehart.[97] All of the rifamycins (A, B, C, D and E) are biologically active. Some of the semisynthetic derivatives of rifamycin B are the most potent known inhibitors of DNA-directed RNA-polymerase in bacteria[98] and their action is bactericidal. They have no activity against the mammalian enzyme. The mechanism of action of rifamycins as inhibitors of viral replication appears to be different from that for their bactericidal action. Their net effect is to inhibit the formation of the virus particle, apparently by the prevention of a specific polypeptide conversion.[99] Rifamycin B (which lacks a substituent at C-4 and has a glycolic acid attached by an ether linkage at C-3), rifamycin SV (which lacks a C-4 substituent and the glycolic acid linked at C-3) and rifamide (the amide of rifamycin B) have antibacterial activity. However, only rifampin is well absorbed orally and finds clinical use in the United States. Rifamide is available in Europe for the treatment of hepatobiliary infections. It is 80 percent excreted in the bile following parenteral administration (I.M.). Some derivatives of 4-formylrifamycin SV are active against RNA-dependent DNA-polymerase in several RNA tumor viruses.[100] N-Demethylrifampin, N-demethyl-N-benzylrifampin and 2,6-dimethyl-N-demethyl-N-benzylrifampin were very active in this system, while rifampin was ineffective. The clinical utility of these agents as antitumor agents has not been established.

Rifampin occurs as an orange to reddish-brown crystalline powder that is soluble in alcohol, but only sparingly soluble in water. It is unstable to moisture and a dessicant (silica gel) should be included with rifampin capsule containers. The expiration date for capsules thus stored is two years. Rifampin is stable in the solid state, but undergoes a variety of chemical changes in solution, the rates and nature of which are pH and temperature dependent.[101] In alkaline pH, it oxidizes to the quinone in the presence of oxygen; in acidic solutions, it hydrolyzes to 3-formylrifamycin SV. Slow hydrolysis of ester functions also occurs, even at neutral pH. Rifampin is well absorbed following oral administration to provide effective blood levels for 8 hours or more. However, food markedly reduces its oral absorption and rifampin should be administered on an empty stomach. It is distributed in effective concentrations to all body fluids and tissues except the brain, despite the fact that it is 70 to 80 percent protein bound in the plasma. The principal excretory route is via the bile and feces, and high concentrations of rifampin and its primary metabolite, deacetylrifampin, are found in the liver and biliary system. Deacetylrifampin is also microbiologically active. Equally high concentrations of rifampin are found in the kidney, and although significant amounts of the drug are passively reabsorbed in the renal tubules, its urinary excretion is significant.

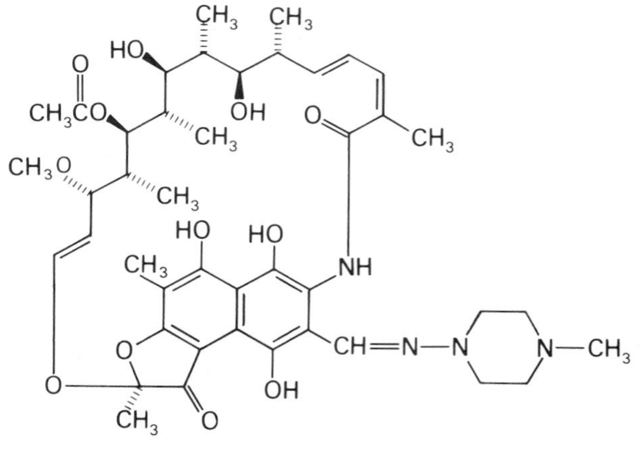

Rifampin

Rifampin is the most active agent in clinical use for the treatment of tuberculosis. As little as 5 µg. per ml. are effective against sensitive strains of *Mycobacterium tuberculosis*. However, resistance to it develops rapidly in most species of bacteria, including the tubercle bacillus. For this reason rifampin is used only in combination with other antitubercular drugs, and it is ordinarily not recommended for the treatment of other bacterial infections where other antibacterial agents are available. Toxic effects associated with rifampin are relatively infrequent. It may, however, in-

TABLE 4–17

ANTITUBERCULAR ANTIBIOTICS

Name Proprietary Name	Preparations	Usual Adult Dose*	Usual Dose Range*	Usual Pediatric Dose*
Cycloserine U.S.P. *Seromycin*	Cycloserine Capsules U.S.P.	250 mg. 2 to 4 times daily	250 mg. to 1 g. daily	5 mg. per kg. of body weight or 150 mg. per square meter of body surface, twice daily initially, then titrate the dose to yield a blood level of 20 to 30 μg. per ml
Viomycin Sulfate U.S.P. *Viocin*	Sterile Viomycin Sulfate U.S.P.	I.M., the equivalent of 1 g. of viomycin twice a day, twice weekly	4 to a maximum of 14 g. weekly	Use in children is not recommended unless crucial to therapy. The equivalent of 20 mg. per kg. of body weight or 600 mg. of viomycin per square meter of body surface twice a day, twice weekly
Capreomycin Sulfate U.S.P. *Capastat*	Sterile Capreomycin Sulfate U.S.P.	I.M., the equivalent of 1 g. of capreomycin once daily for 2 to 4 months, then 1 g. 2 or 3 times weekly		
Rifampin U.S.P. *Rifadin, Rimactane*	Rifampin Capsules U.S.P.	In combination with other tuberculostatics: oral— 600 mg. once daily		Children 5 years of age and over—in combination with other tuberculostatics—oral, 10 to 20 mg. per kg. of body weight once daily

*See *U.S.P. D.I.* for complete dosage information.

terfere with liver function in some patients and should not be combined with other potentially hepatotoxic drugs, nor employed in patients with impaired hepatic function (e.g., chronic alcoholics). The incidence of hepatotoxicity was found to be significantly higher when rifampin was combined with isoniazid than it was when either agent is combined with ethambutol. Allergic and sensitivity reactions to rifampin have been reported, but they are infrequent and usually not serious.

Rifampin is also employed to eradicate the carrier state in asymptomatic carriers of *Neisseria meningitidis* to prevent outbreaks of meningitis in high-risk areas such as military camps. Serotyping and sensitivity tests should be performed prior to its use, since resistance develops rapidly. However, a daily dose of 600 mg. of rifampin for four days is sufficient to eradicate sensitive strains of *N. meningitidis*. Rifampin has also been shown to be very effective against *Mycobacterium leprae* in experimental animals. Its utility in the treatment of human leprosy remains to be established.

ANTIPROTOZOAL AGENTS

Diseases caused by protozoa, especially in the United States and other countries in the temperate zone, are not as widespread as bacterial and viral diseases. Protozoal diseases are more prevalent in the tropical countries of the world, where they occur both in man and in livestock, causing suffering, death and great

economic loss. The main protozoal diseases in humans in the United States include malaria, amebiasis, trichomoniasis and trypanosomiasis. The antimalarial agents are covered in Chapter 6.

Amebiasis, usually thought of as a tropical disease, is actually worldwide in occurrence; in some areas in temperate climates, where sanitary conditions are poor, the incidence may be 20 percent or more. An ideal chemotherapeutic agent against amebiasis would be effective against the causative organism, *Entamoeba histolytica,* irrespective of whether it occurs in the lumen of the colon, the wall of the colon, or extraintestinally, in the liver, lung or other organs, or in the skin.

Amebicides have traditionally been divided into two groups: agents effective against the extraintestinal forms of the parasite, and those effective only against intestinal forms.

The first group includes emetine (which was first described by Pelletier in 1817 and reported to be of value in the chemotherapy of acute amebic dysentery in 1912), and the antimalarial drugs, chloroquine and amodiaquine. After the 1912 report, emetine was quickly taken into use, but the extent of its use has fluctuated because of the relatively narrow margin between effective and toxic doses. A great deal of research has been done in efforts to develop a substitute that would be free from the serious toxic effects associated with emetine. Such efforts have apparently been rewarded with the discovery of the in vivo amebicidal activity of metronidazole and related nitroheterocyclic compounds.

The second group of amebicides, effective against

hepatic abscesses, include bialamicol, the antibiotic Paromomycin Sulfate U.S.P. (see Chap. 7); Metronidazole U.S.P.; the arsenicals Carbarsone U.S.P. and Glycobiarsol U.S.P.; and three derivatives of 8-hydroxy-7-iodoquinoline–chiniofon, Iodoquinol U.S.P. and Iodochlohydroxyquin U.S.P.

Trichomoniasis, caused by *Trichomonas vaginalis,* is common in the United States. Although it is often considered to be a relatively unimportant affliction, it causes serious physical discomfort and, sometimes, marital problems because of its disruptive effect on sexual relations. Chemotherapeutic research in the past 20 years has led to marked progress in knowledge of the biological properties of the causative organism. Accurate testing methods have been devised and highly effective compounds with systemic activity have been discovered.

Heterocyclic nitro compounds, such as the nitrofurans, furazolidone, and nifuroxime and the nitroimidazole, Metronidazole U.S.P., provide effective treatment for trichomoniasis. Other trichomonacides include arsenicals and the 8-hydroxyquinolines, Clioquinol U.S.P. and Iodoquinol U.S.P.

Various forms of trypanosomiasis, chronic tropical diseases caused by pathogenic members of the family *Trypanosonidae,* occur both in man and in livestock. The principal disease in man, sleeping sickness, can be broadly classified into two main geographical and etiological groups: African sleeping sickness caused by *Trypanosoma gambiense* (West African), *T. rhodesiense* (East African), or *T. congolense,* and South American sleeping sickness (Chagas' disease) caused by *Trypanosoma cruzi.* Of the various forms of trypanosomiasis, Chagas' disease is the more serious and generally the more resistant to chemotherapy. Leishmaniasis is a chronic tropical disease caused by various flagellate protozoa of the genus *Leishmania.* The more common visceral form caused by *Leishmania donovani,* called Kala-azar, is similar to Chagas' disease. Fortunately, although these diseases are widespread in tropical areas of Africa and South and Central America, they are of minor importance in the United States, Europe and Asia.

The successful chemotherapy of trypanosomiasis and leishmaniasis remains somewhat primitive and often less than effective. In fact, it is doubtful that these diseases can be controlled by chemotherapeutic measures alone, without successful control of the intermediate hosts and vectors that transmit them. Heavy metal compounds, such as the arsenicals and antimonials, are sometimes effective but frequently toxic. The old standby suramin, the newer bisamidines (Hydroxystilbamidine U.S.P.), and pentamidine appear to be of some value in long- and short-term prophylaxis. The recently introduced nitrofuran derivative, nifurti-

mox, may be a major breakthrough in the control of these diseases. However, its potential toxicity remains to be completely evaluated.

PRODUCTS

Emetine Hydrochloride U.S.P. The nonphenolic alkaloid, emetine, is obtained either by isolation from natural sources or synthetically by methylating naturally occurring cephaëline (phenolic). Emetine is obtained from the crude drug by first extracting the total alkaloids with a suitable solvent and then separating them by a method similar to that outlined for the separation of phenolic and nonphenolic bases.

The free base is levorotatory and occurs as a water-insoluble, light-sensitive, white powder. It is soluble in alcohol or the immiscible solvents. It contains 2 basic nitrogens and forms salts quite readily. The *U.S.P.* sets limits for the water of hydration content (8 to 15%). Other than the hydrochloride, the hydrobromide and the camphosulfonate (as a solution) sometimes are used.

Emetine Hydrochloride

The hydrochloride occurs as a white or very slightly yellowish, odorless, crystalline powder. The salt is freely soluble in water (1:4) and in alcohol. Its solutions have an approximate pH of 5.5 but, when prepared for injection, they should be adjusted to a pH of 3.5 Sterilization of solutions may be effected by bacteriologic filtration. Solutions are light-sensitive and should be preserved in light-resistant containers.

As the name implies, emetine possesses emetic action, due to its marked irritation of mucous membranes when ingested orally. However, it is used principally for its amebicidal qualities. Considerable research has shown that, while emetine causes prompt recession of the symptoms of acute intestinal amebiasis, it cures only 10 to 15 percent of the cases, and is now considered to be the least valuable agent for curing the disease.

The recession of symptoms quite often leads patients to believe that they are cured, although they are still carriers. Therefore, emetine probably is used best for symptomatic control of acute amebic dysentery and should be supplemented by other more effective drugs.

Metronidazole U.S.P., Flagyl® is 2-methyl-5-nitroimidazole-1-ethanol. It is the most useful of a vast number of antiprotozoal nitroimidazole derivatives that have been synthesized in various laboratories throughout the world. Metronidazole was first marketed for the topical treatment of *Trichomonas vaginalis* vaginitis. It has since been shown to be effective orally against both the acute and carrier states of the disease. The drug also possesses useful amebicidal activity and is, in fact, effective against both intestinal and hepatic amebiasis. It has also found use in the treatment of such other protozoal diseases as giardiasis and balantidiasis.

Metronidazole is a pale yellow, crystalline compound which has limited solubility in water yet is adequately absorbed after oral administration. It is stable in air but darkens on exposure to light. After oral administration, the serum and urine levels reach their peaks in about 2 to 3 hours. The known metabolites formed from oxidation of the 2-methyl and 1-ethanol groups are all inactive. Darkened urine may occur when the drug is given in doses higher than those generally recommended. The pigment responsible for the darkened urine has not been positively identified, but is probably a metabolite of metronidazole. The darkened urine appears to have no clinical significance.

Metronidazole

8-Hydroxyquinoline, oxine, quinophenol, oxyquinoline is the parent compound from which the antiprotozoal oxyquinolines have been derived. The antibacterial and antifungal properties of oxine and its derivatives, which are believed to result from the ability to chelate metal ions (see Chap. 2), are well known. Aqueous solutions of acid salts of oxine, particularly the sulfate (Chinosol, quinosol), in concentrations of 1:3000 to 1:1000, have been used as topical antiseptics. The substitution of an iodine atom at the 7-position of 8-hydroxyquinolines produces compounds with broad spectrum antimicrobial properties. The chief therapeutic application of such compounds, however, is in the treatment of intestinal amebiasis.

Iodochlorhydroxyquin, Vioform®, 5-chloro-7-iodo-8-quinolinol, 5-chloro-8-hydroxy-7-iodoquinoline, consists of a spongy, voluminous, yellowish-white powder that has a slight, characteristic odor and is affected by light. It is practically insoluble in water or alcohol but dissolves in hot ethyl acetate and in hot acetic acid. It is prepared by iodinating 5-chloro-8-hydroxyquinoline, a direct product of a Skraup synthesis.

The compound originally was introduced as an odorless substitute for iodoform and acts by slow liberation of iodine. It is used as an undiluted powder in surgery, in atopic dermatitis, in eczema of the external auditory canal, in chronic dermatitis, in oil dermatitis and in acute psoriasis and impetigo. It also may be applied as a 2 to 3 percent ointment, emulsion or paste or used in suppositories for the vagina in the treatment of trichomonas vaginitis.

However, the chief use at present is as a remedy in amebic dysentery. For this purpose, it is more toxic than chiniofon but also more potent, and it can bring about apparent cures without any unpleasant symptoms, although in a few cases there may be abdominal distress, and the possibility of iodism is always present.

Iodoquinol U.S.P., diiodohydroxyquin, Diodoquin®, 5,7-diiodo-8-quinolinol, 8-hydroxy-5,7-diiodoquinoline, is a light yellowish to tan, microcrystalline, odorless powder that is insoluble in water. It is recommended in the treatment of amebic dysentery and the infestation by *Trichomonas hominis (intestinalis)* and is claimed to be just as effective as chiniofon and much less toxic. However, several cases of acute dermatitis have followed its use.

Diloxanide Furoate, Furamide®, is the 2-furoate ester of 2,2-dichloro-4-hydroxy-N-methylacetanilide. It was developed as a result of the discovery that various α,α-dichloroacetamides possessed amebicidal activity in vitro. Diloxanide itself and many of its esters are also active, and drug metabolism studies indicate that hydrolysis of the ester is required for the amebicidal effect. Nonpolar esters of diloxanide are more potent than polar ones. Diloxanide furoate has been used in the treatment of asymptomatic carriers of *E. histolytica*. Its effectiveness against acute intestinal amebiasis or hepatic abscesses, however, has not been established. Diloxanide furoate is a white crystalline powder. It is administered only orally as 500 mg. tablets and may be obtained in the United States from the Center for Disease Control in Atlanta, Georgia.

Diloxanide Furoate

Hydroxystilbamidine Isethionate U.S.P., 2-hydroxy-4,4'-stilbenedicarboxamidine diisethionate, 2-hydroxy-4,4'-diamidinostilbene. This consists of yellow crystals which are stable in air but are light-sensitive. The pH of a 1 percent aqueous solution is about 4. Solutions for medicinal use should be freshly prepared and free of any cloudiness. The solution when given by intravenous infusion should be carefully protected from light.

Hydroxystilbamidine is considered a drug of choice for the prophylaxis and treatment of African trypanosomiasis. The drug does not penetrate the central nervous system and is therefore of doubtful value for the treatment of late stages of the disease. Hydroxystilbamidine has also been used as an antifungal agent in the treatment of systemic and pulmonary North American blastomycosis.

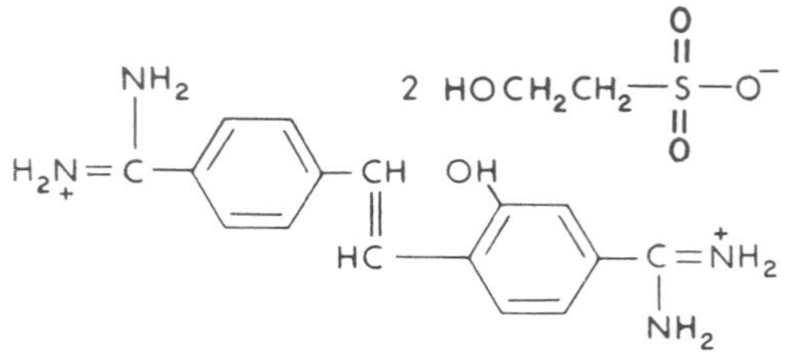

Hydroxystilbamidine Isethionate

Pentamidine Isoethionate, Lomidine®, is 4,4'-(pentamethylenedioxy)-dibenzamidine diisethionate. This water soluble, crystalline salt is stable to light and air. It is available from the Center for Disease Control for the prophylaxis and treatment of African trypanosomiasis. It is also of some value for treating visceral leishmaniasis. Pentamidine rapidly disappears from the plasma following intravenous injection and is distributed to the tissues, where it is stored for long periods of time. This property probably contributes to the utility of the drug as a prophylactic agent.

Nifurtimox, Bayer 2502, Lampit®, is 4-[(5-nitrofurfurylidene)amino]-3-methylthiomorpholine-1,1-dioxide. The observation that various derivatives of 5-nitrofuraldehyde possessed, in addition to their antibacterial and antifungal properties, significant and potentially useful antiprotozoal activity eventually led to discovery of particular nitrofurans with antitrypanosomal activity. The most important of such compounds is nifurtimox because of its demonstrated effectiveness against *T. cruzi,* the parasite responsible for South American trypanosomiasis. In fact, use of this drug represents the only clinically proven treatment for both acute and chronic forms of the disease. Nifurtimox is available in the United States from the Center for Disease Control.

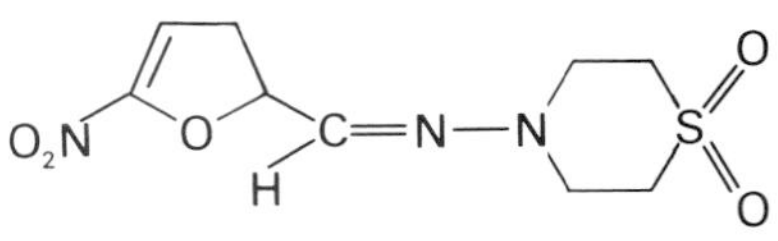

Nifurtimox

Suramin Sodium is a high molecular weight bisurea derivative containing six sulfonic acid groups as their sodium salts. It was developed in Germany shortly after World War I as a by-product of research efforts directed toward the development of potential antiparasitic agents from dyestuffs. The drug has been used for more than half a century for the treatment of early cases of trypanosomiasis. It was not until several decades later, however, that suramin was discovered to be a long-term prophylactic agent whose effectiveness after a single intravenous injection is maintained for periods of up to three months. The drug is tightly bound to plasma proteins, causing its excretion in the urine to be almost negligible.

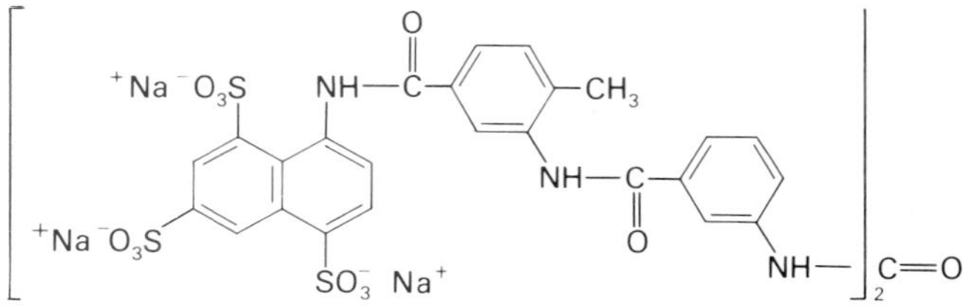

Suramin Sodium

Tissue penetration of the drug does not occur, apparently because of its high molecular weight and highly ionic character. Thus, an injected dose remains in the plasma for a very long period of time. Newer, more effective drugs are now available for short-term treatment and prophylaxis of African sleeping sickness. Suramin is also used for prophylaxis of onchocerciasis. It is available from the Center for Disease Control.

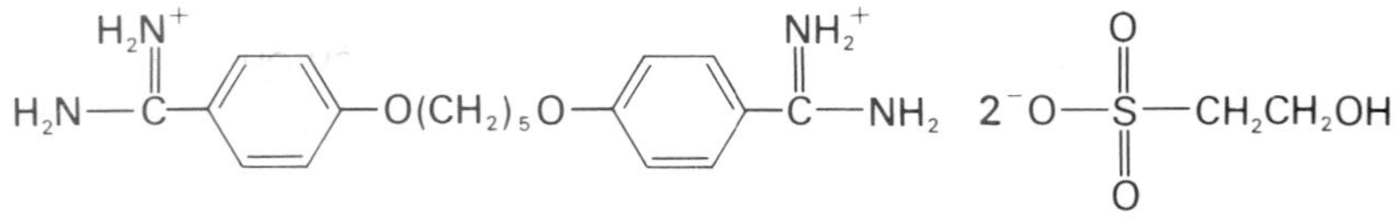

Pentamidine Isoethionate

Carbarsone U.S.P., N-carbamoylarsanilic acid, *p*-ureidobenzenearsonic acid. This compound occurs as a white, crystalline powder. It is slightly soluble in water and alcohol, nearly insoluble in ether but soluble in basic aqueous solutions. It is synthesized from urethane and arsanilic acid.

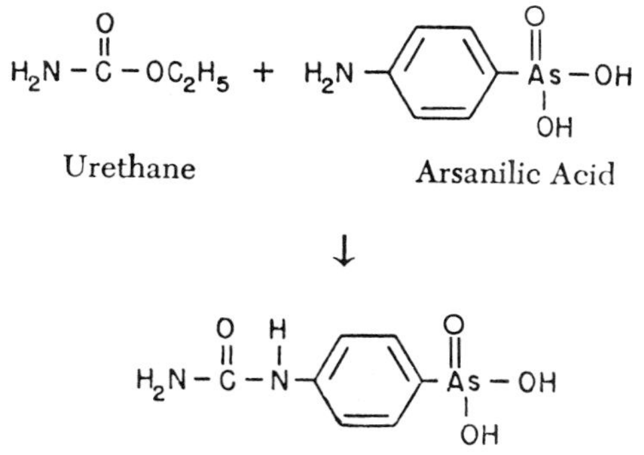

Urethane   Arsanilic Acid

↓

Carbarsone

Carbarsone is used orally in the chemotherapy of intestinal amebiasis and intravaginally as suppositories in the treatment of vaginitis by *Trichomonas vaginalis*. When given orally it is readily absorbed from the gastrointestinal tract and is then slowly excreted in the urine. For this reason, rest periods between treatment periods are needed to prevent cumulative poisoning. It is one of the safest arsenicals in use. Dimercaprol is a useful antidote.

Glycobiarsol U.S.P., Milibis®, (hydrogen N-glycoloylarsanilato)oxobismuth, bismuthyl N-glycoloylarsanilate. The compound is a yellow to pink powder that decomposes on heating and is very slightly soluble in water or alcohol. The saturated aqueous solution is acidic, the pH being in the range 2.8 to 3.5.

The compound is made from bismuth nitrate and sodium *p*-N-glycoloylarsanilate and is used in the treatment of intestinal amebiasis. It is reported to have low toxicity which may be due to low solubility. It imparts a black color to feces, as a result of the formation of bismuth sulfide. Glycobiarsol is also used as vaginal suppositories in the treatment of trichomonal and monilial vaginitis.

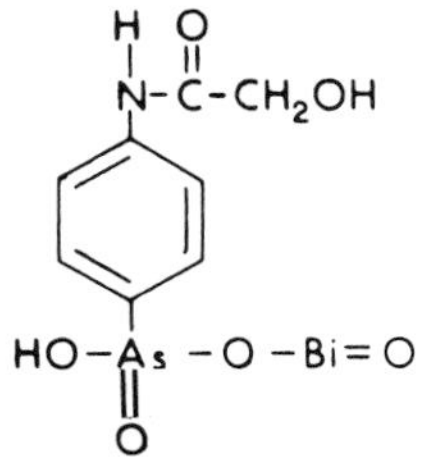

Bismuth Glycoloylarsanilate

Melarsoprol, Mel B®, Arsobal®, 2-*p*-(4,6-diamino-*s*-triazin-2-ylamino)phenyl-4-hydroxymethyl-1,3,2-dithiarsoline, is prepared by reduction of the corresponding pentavalent arsanilate to the trivalent arsenoxide followed by reaction of the latter with 2,3-dimercaptopropanol (BAL). It has become the drug of choice for the treatment of the latter stages of both forms of African trypanosomiasis. Melarsoprol has the advantage of excellent penetration into the central nervous system and is therefore effective against mengioencephalitic forms of *T. gambiense* and *T. rhodesiense*. Trivalent arsenicals tend to be more toxic to the host (as well as the parasites) than the corresponding pentavalent compounds. The bonding of arsenic with sulfur atoms tends to reduce host toxicity, increase chemical stability (to oxidation), and improve distribution of the compound to the arsenoxide. However, melarsoprol shares the toxic properties of other arsenicals and its use must be monitored for signs of arsenic toxicity.

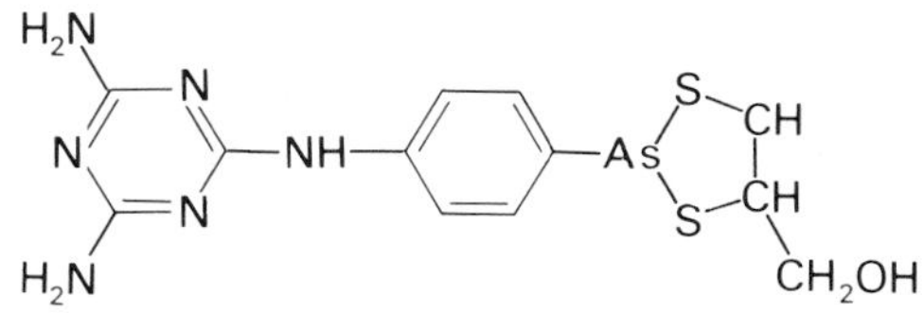

Melarsoprol

Sodium Stibogluconate, Pentostam®. Sodium antimony gluconate is a pentavalent antimonial compound intended primarily for the treatment of various forms of leishmaniasis. It is available from the Center for Disease Control as the disodium salt which is chemically stable and freely soluble in water. The 10 percent aqueous solution used for either intramuscular or intravenous injection has a pH of about 5.5. Like all antimonial drugs, this drug has a low therapeutic index, and patients undergoing therapy with it should be monitored carefully for signs of heavy metal poisoning. Other organic antimonial compounds are employed primarily for the treatment of schistosomiasis and other flukes.

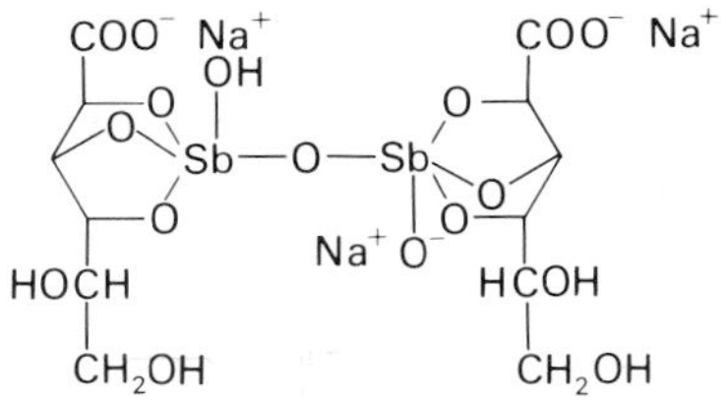

Sodium Stibogluconate

Dimercaprol U.S.P., 2,3-dimercapto-1-propanol, BAL (British Anti-Lewisite), dithioglycerol, is a color-

TABLE 4-18

ANTIPROTOZOAL AGENTS

| Name | | | | | | |
Proprietary Name	Preparations	Category	Application*	Usual Adult Dose*	Usual Dose Range*	Usual Pediatric Dose*
Emetine Hydrochloride U.S.P.	Emetine Hydrochloride Injection U.S.P.	Amebicide		I.M. or S.C., 1 mg. per kg. of body weight, but not exceeding a total of 65 mg. once daily for 3 to 10 days	Not exceeding 65 mg. daily or a total dose of 650 mg. in 10 days	S.C. 500 μg per kg. of body weight or 15 mg. per square meter of body surface, twice daily for 4 to 6 days, but not exceeding a total of 65 mg. per day
Metronidazole U.S.P. *Flagyl*	Metronidazole Tablets	Amebicide, Trichomonacide		Antiamebic—500 to 750 mg. 3 times daily for 5 to 10 days Antitrichomonal—250 mg. 3 times daily		
Iodochlorhydroxyquin U.S.P. *Vioform*	Iodochlorhydroxyquin Cream U.S.P. Iodochlorhydroxyquin Ointment U.S.P. Compound Iodochlorhydroxyquin Powder U.S.P. Iodochlorhydroxyquin Tablets U.S.P.	Amebicide, local anti-infective	Topical, to the skin, 3 percent cream or ointment 2 or 4 times daily or 25 percent powder, as required	Oral, 250 mg. 3 times daily for 10 days	250 to 500 mg. daily	
Iodoquinol U.S.P. *Diodoquin, Yodoxin*	Diiodohydroxyquin Tablets U.S.P.	Amebicide		650 mg. 3 times daily for 20 days		13.3 mg. per kg. of body weight 3 times daily, not to exceed 1.95 g. in 24 hours
Hydroxystilbamidine Isethionate U.S.P.	Sterile Hydroxystilbamidine	Leishmanicide Trypanocide		I.M., 225 mg. every 24 hours, administered in 10 ml. of 5 percent Dextrose Injection or 0.9 percent Sodium Chloride Injection; I.V. infusion, 225 mg. every 24 hours, administered in 200 ml. of 5 percent Dextrose Injection or 0.9 percent Sodium Chloride Injection over a period of 2 to 3 hours		I.V. infusion, 3 to 4.5 mg. per kg. of body weight every 24 hours, administered in an appropriate amount of 5 percent Dextrose Injection or 0.9 percent Sodium Chloride Injection over a period of 2 to 3 hours

Drug	Classification	Dose	Dose
Carbarsone U.S.P. / Carbarsone Capsules U.S.P.	Amebicide	250 mg. 2 or 3 times daily for 10 days	100 to 250 mg.
Glycobiarsol U.S.P. / Glycobiarsol Tablets U.S.P. *Milibis*	Amebicide	500 mg. 3 times daily for 7 to 10 days	
Nifurtimox	Trypanocide	Complicated dosage schedule available from the Center for Disease Control	
Diloxanide Furoate	Amebicide	500 mg. orally 3 times daily for 10 days	500 mg. orally 3 times daily for 10 days
Suramin	Trypanocide	100 to 200 mg. I.V. test dose weekly, then 1 g. I.V. on days 1,3,7,14, and 21	20 mg. per kg. I.V. on days 1,3,7,14, and 21
Pentamidine Isethionate	Trypanocide	4 mg. per kg. I.V. per day for 10 days	4 mg. per kg. I.V. per day for 10 days
Melarsoprol	Trypanocide	2 to 3.6 mg. per kg. I.V. per day in 3 doses; after 1 week 3.6 mg. per kg. per day for 3 doses. Repeat after 10 to 21 days	18 to 25 mg. per kg. I.V. over 1 month. Initial dose of 0.36 mg. per kg. I.V.; intervals of 1 to 5 days depending on reactions for a total of 9 to 10 doses
Stibogluconate Sodium	Leishmanicide	600 mg. I.M. or I.V. 1 to 6 times daily	10 mg. per kg. I.M. or I.V. 1 to 6 times daily

* See U.S.P. D.I. for complete dosage information.

less liquid with a mercaptanlike odor. BAL is soluble in water (1:20), in benzyl benzoate and in methanol. It was developed during World War II by the British as an antidote for "Lewisite." The name BAL is an abbreviation for British Anti-Lewisite. It is an effective antidote for poisoning with arsenic, gold, antimony, mercury and perhaps other heavy metals. The skin damage resulting from arsenical vesicant agents can be prevented by a previous application of BAL preparations. The damage to the skin by the same agents also can be arrested and perhaps reversed by application of BAL shortly after exposure. In systemic poisoning resulting from various arsenical agents, parenteral administration of BAL in oil has been demonstrated to be quite effective.

$$CH_2-CHCH_2OH$$
$$\underset{SH}{|} \quad \underset{SH}{|}$$

Dimercaprol

The antidote properties of BAL for the metals are associated with the fact that the heavy metal ions tie up the —SH groups in the tissues and thus interfere with the pyruvate oxidase and perhaps other enzyme systems which are dependent on the —SH groups for their activity. The synthetic dithiol compounds, such as BAL, compete effectively with the tissues for the metal, removing the metal by forming a ring compound of the type

$$\underset{-C-S}{\overset{|}{\underset{|}{}}}\diagdown_{\diagup}As-R$$
$$-\underset{|}{C}-S\diagup$$

which is relatively nontoxic and is excreted fairly rapidly. To exhibit the detoxifying effect, it is apparently necessary for the compound to have 2 thiol groups on adjacent carbon atoms or on atoms separated by one other atom so stable 5- or 6-membered ring compounds can be formed. Monothiol compounds are much less effective.

BAL may be applied locally as an ointment, 5 percent W/V in a base of lanolin, Lanette wax and diethylphthalate. It is injected intramuscularly as a 5 or 10 percent solution in peanut oil, to which 2 g. of benzyl benzoate is added for each gram of BAL to make the latter miscible with the peanut oil in all proportions. Solutions of this type can be sterilized in nitrogen-filled ampuls by heating to 170° for 1 hour without having more than 1.5 percent of the BAL destroyed in the process. Solutions of BAL in water or propylene glycol are reported to be unstable. 1,2,3-Tri-

mercaptopropane has been reported to occur as an impurity in varying amounts in the commercial product.[102]

Category—antidote to arsenic, gold and mercury poisoning; metal-complexing agent.

Usual adult dose—I.M., 2.5 mg. per kg. of body weight 4 to 6 times daily on the first 2 days, then twice daily for the next 8 days, if necessary.

Usual dose range—2.5 to 5 mg. per kg.

Usual pediatric dose—I.M., 2.5 to 3 mg. per kg. of body weight, 6 times daily on the first day, 4 times daily on the second day, twice daily on the third day, then once daily for the next 10 days, if necessary.

Occurrence	Percent Dimercaprol
Dimercaprol Injection U.S.P.	10

ANTHELMINTICS

Anthelmintics are drugs which have the capability of ridding the body of parasitic worms or helminths. The prevalence of human helminthic infestations is widespread throughout the globe and represents a major world health problem, particularly in third world countries. Helminths parasitic to man and other animals are derived from two families: *Platyhelminthes* and *Nemanthelminthes*. Cestodes (tapeworms) and trematodes (flukes) belong to the former, and the nematodes or true roundworms to the latter. The helminth infestations of major concern on the North American continent are caused by roundworms, i.e., hookworm, pinworm, and *Ascaris*. Human tapeworm and fluke infestations are rarely seen in the United States.

Several classes of chemicals are used as anthelmintics and include (1) chlorinated hydrocarbons, (2) phenols and derivatives, (3) dyes, (4) piperazine and related compounds, (5) antimalarial compounds (see Chap. 6), (6) various heterocyclic compounds, (7) alkaloids and other natural products, and (8) antimonial compounds.

Tetrachloroethylene U.S.P., perchloroethylene, tetrachloroethene, $Cl_2C{=}CCl_2$.

Tetrachloroethylene may be synthesized from dry hydrogen chloride and carbon monoxide at 300° and 200 atmospheres pressure in the presence of a nickelous oxide catalyst or by passing symmetrical ethylene dichloride and chlorine over heated pumice at 400°.

Tetrachloroethylene is a colorless, mobile liquid of ethereal odor with a specific gravity of 1.61 and a boiling point of 122°. It is miscible with an equal volume of alcohol and with most organic solvents. Like trichloroethylene, it is unstable to air, moisture and light, decomposing in part to phosgene and hydrochloric acid. The *U.S.P.* permits up to 1 percent alcohol as a preservative. It is noninflammable. It is used industrially as a solvent and as a cleaner of textiles and metals.

Although it is a potent anesthetic, it is a skin and respiratory irritant and difficult to vaporize. Its specific use in medicine is as an anthelmintic in hookworm infestation. Wright and Schaffer,[103] in their attempt to correlate anthelmintic efficacy of chlorinated alkyl hydrocarbons and chemical structure, observed the following trends: in any one homologous series anthelmintic efficacy increases with the lengthening of the carbon chain; correspondingly, there is a decrease in water-solubility, about 1:1,000 to 1:5,000, for most effective anthelmintic compounds; the optimum range varies for different homologous series. Substitution of bromine or iodine for chlorine makes less difference in anthelmintic efficacy than does change in water-solubility; the optimum solubility range for these halogenated hydrocarbons is 1:1,000 to 1:1,700.

All of these compounds are irritant to the gastrointestinal tract and produce varying degrees of liver and kidney degeneration. Tetrachloroethylene is about equally as efficient as carbon tetrachloride and is preferred in hookworm treatment because it is less toxic and does not raise the guanidine content of the blood, a criterion of importance where calcium deficiency exists.

It may be given on sugar or in gelatin capsules after first emptying the gastrointestinal tract. It is followed by a saline cathartic. Oils, fats and alcohol favor absorption and toxic side-effects and, therefore, should be avoided.

Piperazine U.S.P., Arthriticin®, diethylenediamine, dispermine, hexahydropyrazine, occurs as colorless, volatile crystals that are freely soluble in water or glycerol. It crystallizes as a hexahydrate from water. It can be made by warming ethylene chloride with ammonia in alcoholic solution.

$$2\ CH_2Cl\!-\!CH_2Cl + 6\ NH_3 \rightarrow$$
$$NH(CH_2\!-\!CH_2)_3NH + 4\ NH_4Cl$$
Piperazine

It was introduced into medicine because it will dissolve uric acid in a test tube, and it was hoped that this might be of service in gout and other rheumatic diseases. The clinical results, however, have been almost nil because the distribution of therapeutic doses could not be expected to furnish sufficient concentration. The claim that piperazine is a powerful diuretic has not been confirmed. Piperazine is used as a stabi-

lizing buffer for the estrone sulfate ester (see Ogen®, a piperazine estrone sulfate).

After the discovery of the activity of the piperazine derivative, diethylcarbamazine, it was established that piperazine itself was active and is used commonly today as an anthelmintic for the treatment of pinworms (*Enterobius vermicularis; Oxyuris v.*) and roundworms (*Ascaris lumbricoides*) in children and adults. A number of salts of piperazine are available by brand names, usually in the form of a syrup or tablet. It appears to function by inducing a state of narcosis in the worms. An important aspect of successful treatment is that the worms be voided before the effects of the drug have worn off. Piperazine and its salts have generally replaced gentian violet as the drug of choice in the treatment of human pinworm infections.

Piperazine Citrate U.S.P., Antepar® Citrate, Multifuge® Citrate, Parazine® Citrate, Pipazin® Citrate, tripiperazine dicitrate, occurs as a white crystalline powder with a slight odor. It is insoluble in alcohol and soluble in water, a 10 percent solution having a pH of 5 to 6.

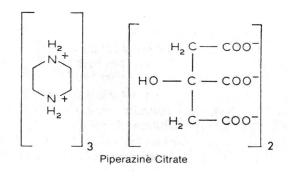
Piperazine Citrate

Piperazine citrate is administered orally. In some commercial products, the dose is expressed in terms of the equivalent amount of piperazine hexahydrate, i.e., 550 mg. anhydrous piperazine citrate equivalent to 500 mg. piperazine hexahydrate.

Piperazine Phosphate U.S.P., Antepar®, Vermizine®. This is formed from 1 mole each of piperazine and phosphoric acid; the pH of a 1:100 solution is between 6.0 and 6.5. Like the other salts of piperazine the dose is expressed in terms of the equivalent amount of piperazine hexahydrate.

Diethylcarbamazine Citrate U.S.P., Hetrazan®, N,N-diethyl-4-methyl-1-piperazine-carboxamide dihydrogen citrate, 1-diethyl-carbamyl-4-methylpiperazine dihydrogen citrate, has been introduced for the treatment of filariasis. It is highly specific for certain parasites, including filariae and ascaris.

It is a colorless, crystalline solid, highly soluble in water, alcohol and chloroform but insoluble in most

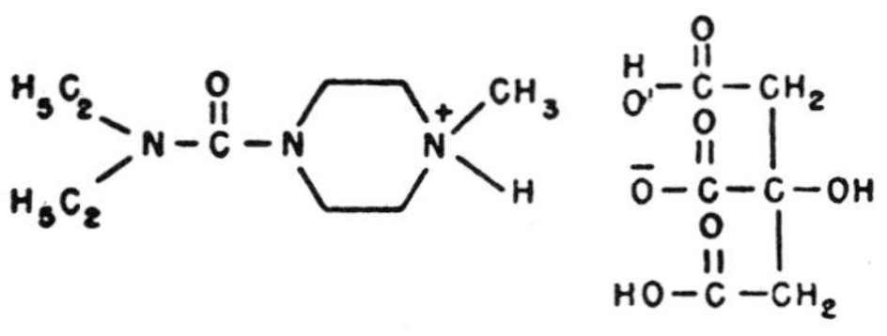

Diethylcarbamazine Citrate

organic solvents. A 1 percent solution has a pH of 4.1. The drug is stable under varied conditions of climate and moisture.

Gentian Violet U.S.P. is used in the treatment of pinworm, but has been largely replaced by piperazine and its salts. It is one of the few drugs effective in strongyloides infestations. For pinworm infestations, it is administered as enteric-coated tablets before or with meals 3 times daily for 8 to 10 days, and for 16 to 18 days for strongyloides infestations. The adult dosage is 60 mg. 3 times daily; the dose for children should not exceed 90 mg. total daily dose. The usual size tablets are 10 and 30 mg.

Pyrvinium Pamoate U.S.P., Povan®, 6-(dimethylamino)-2-[2-(2,5-dimethyl-1-phenylpyrrol-3-yl)vinyl]-1-methylquinolinium 4,4'-methylenebis[3-hydroxy-2-naphthoate], is a red cyanine dye. It is used in the chemotherapy of pinworm infestation. The drug is sparingly soluble and poorly absorbed from the intestinal tract, and because of its local irritant action it may cause nausea and vomiting. If vomiting occurs before the drug has left the stomach the vomitus may be red colored; in addition, during treatment the feces will be frequently stained reddish-brown.

Single dose treatment is usually highly effective in eradicating pinworm infestation in children and adults.

Pyrantel Pamoate U.S.P., Antiminth®, is *trans*-1,4,5,6,-tetrahydro-1-methyl-2-[2-(2-thienyl)vinyl]pyrimidine pamoate. This drug has shown activity against pinworm and roundworm infestations. The anthelmintic action may be due to a neuromuscular blocking action. Over half of a dose is excreted in the feces unchanged, while only about 7 percent is ex-

creted in the urine as the intact molecule or as metabolites. It is not a dye, thus it does not discolor the feces or urine. Purging is not necessary before or after use of the drug.

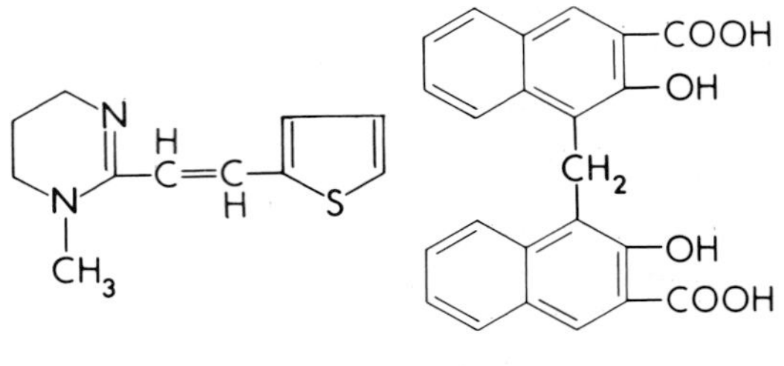

Pyrantel Pamoate

Thiabendazole U.S.P., Mintezol®, is 2-(4-thiazolyl)benzimidazole. Thiabendazole is a stable compound, both as a solid and in solution. It forms colored complexes with metal ions, such as iron. It has a basic pKa of 4.7 and is only slightly soluble in water but becomes more soluble as the pH is raised or lowered; its maximum solubility is at pH 2.5, at which it will give a 1.5 percent solution.[104]

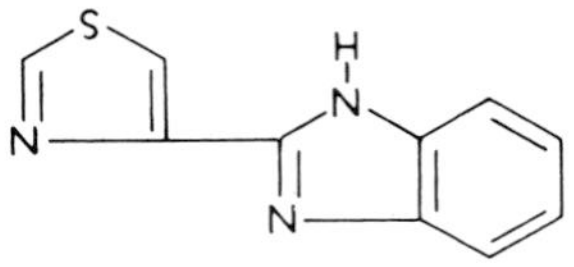

Thiabendazole

Thiabendazole is effective in the treatment of several helminthic diseases. It has shown a high degree of efficacy against threadworm and pinworm, moderate effectiveness against large roundworm and hookworm, and less activity against whipworm. It has been used successfully in the treatment of cutaneous larva migrans (creeping eruption). There have been reports that, in several cases of trichinosis, relief of symptoms and fever have followed its use, but there is no evi-

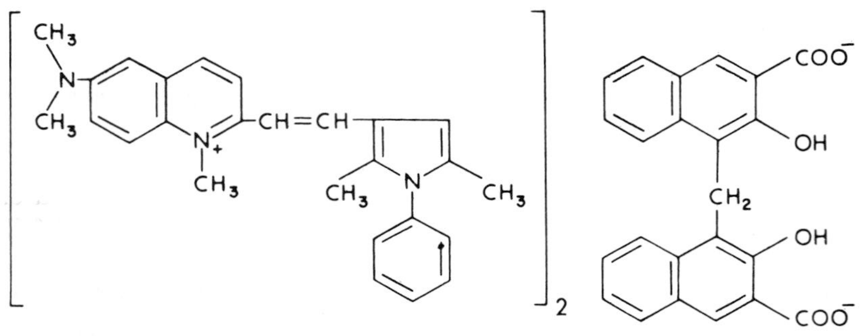

Pyrvinium Pamoate

dence that it will eliminate the adult *Trichinella spiralis*. It is an odorless, tasteless, nonstaining compound and generally is administered as a suspension given after meals.

In addition to its use in human medication, it has been widely accepted for controlling gastrointestinal parasites in livestock. It is also highly active as a fungicide, and a wettable powder has been marketed to control stem-end rot and fruit spoilage in citrus fruit.

Mebendazole U.S.P., Vermox®, methyl 5-benzoylimidazole-2-carbamate. This is a broad-spectrum anthelmintic and is especially useful against whipworm infestations. It is stable under normal conditions of temperature, light and moisture.

Mebendazole

Mebendazole is contraindicated in pregnant women because it has shown teratogenic activity in pregnant rats. It has not been studied extensively in children under 2 years of age.

The drug blocks the uptake of glucose by the susceptible helminths.

It is supplied as 100-mg. tablets for trichuriasis.

Tetramisole, Ripercol®, Nilverm®, Anthelvet® and *Levamisole*, Nemicide®, Tamisol®, Keterax®, are (±)- and (-)-2,3,5,6-tetrahydro-6-phenylimidazo(2,1-b)thiazole, respectively. These broad spectrum anthelmintic agents, which were developed at Janssen Pharmaceutica in Belgium, are effective against a wide range of parasitic nematodes in man and other animals.[105] Their use in the United States is restricted to veterinary practice. Virtually all of the anthelmintic activity resides in the levo isomer, which apparently exerts its action through the stereospecific inhibitor of fumarate reductase in various nematodes.[106] Perhaps of greater potential importance to human medicine, however, is the observed ability of levamisole to re-

Tetramisole

store host defense mechanisms. Favorable clinical results have been obtained using it as an immunostimulant for such conditions as rheumatoid arthritis, systemic lupus erythematosis, Crohn's disease and various neoplastic diseases. It is available only for investigational use as an immunosuppressant in the United States at the present time.

Bephenium Hydroxynaphthoate U.S.P., Alcopara®, benzyldimethyl(2-phenoxyethyl)ammonium 3-hydroxy-2-naphthoate, is a pale yellow, crystalline powder, with a bitter taste, and is sparingly soluble in water. It is useful in the treatment of hookworm infestation and in mixed infestations which include hookworm and large roundworm. Because of the bitter taste, the drug is generally mixed with milk, fruit juice, or carbonated beverage just prior to administration; no food should be taken for at least 2 hours afterwards.

Bephenium Hydroxynaphthoate

Niridazole, Ambilhar®, 1-(5-nitro-2-thiazolyl)-2-oxotetrahydroimidazole is an antischistosomal drug that was synthesized as part of a systematic investigation of heterocyclic nitro compounds as potential antiparasitic agents. Its principal therapeutic application is in the treatment of infections caused by *Schistosoma haematobium* (urinary schistosomiasis), but it is also moderately effective against intestinal schistosomiasis (*S. mansoni*) and guinea worm (*Dracunculus medinensis*). Niridazole has antiprotozoal activity as

Niridazole

well, and has been used effectively for intestinal amebiasis. The drug is generally well tolerated and appears to be relatively nontoxic. However, occasional neuropsychiatric reactions (i.e., mental disorientation, mania, and convulsions) and electroencephalographic changes have been reported following oral administration in man, and animal studies indicate that it may be both mutagenic and carcinogenic. Its use in the mass chemotherapy of schistosomiasis, therefore, requires further evaluation.

TABLE 4–19

ANTHELMINTICS

Name / Proprietary Name	Preparations	Effective Against	Usual Dose*	Usual Pediatric Dose*
Tetrachloroethylene U.S.P.	Tetrachloroethylene Capsules U.S.P.	Hookworms and some trematodes	0.12 ml. per kg. of body weight as a single dose, up to a maximum of 5 ml.	0.1 ml. per kg. of body weight or 3 ml. per square meter of body surface, as a single dose, up to a maximum of 5 ml.
Piperazine Citrate U.S.P. *Antepar Citrate, Multifuge Citrate, Ta-Verm, Vermidole*	Piperazine Citrate Syrup U.S.P.	Intestinal pinworms and roundworms	Against *Enterobius*—the equivalent of 2 g. of piperazine hexahydrate once daily for 7 days. Against *Ascaris*—3.5 g. once daily for 2 days	Against *Enterobius*—the following amounts, or 1 g. per square meter of body surface, are usually given once daily for 7 days: up to 7 kg. of body weight—250 mg.; 7 to 14 kg.—500 mg.; 14 to 27 kg.—1 g.; over 27 kg.—2 g. Against *Ascaris*—the following amounts, or 2 g. per square meter of body surface, and usually given once daily for 2 days: up to 14 kg. of body weight—1 g.; 14 to 23 kg.—2 g.; 23 to 45 kg.—3 g.; over 45 kg.—3.5 g.
	Piperazine Citrate Tablets U.S.P.	Intestinal pinworms and roundworms	Against *Enterobius*—the equivalent of 2 g. of piperazine hexahydrate once daily for 7 days. Against *Ascaris*—3.5 g. once daily for 2 days	Against *Enterobius*—the following amounts, or 1 g. per square meter of body surface, and usually given once daily for 7 days: up to 7 kg. of body weight—250 mg.; 7 to 14 kg.—500 mg.; 14 to 27 kg.—1 g.; over 27 kg.—2 g. Against *Ascaris*—the following amounts, or 2 g. per square meter of body surface, are usually given once daily for 2 days: up to 14 kg. of body weight—1 g.; 14 to 23 kg.—2 g.; 23 to 45 kg.—3 g.; over 45 kg.—3.5 g.
Piperazine Phosphate U.S.P. *Antepar, Vermizine*	Piperazine Phosphate Tablets U.S.P.	Intestinal roundworms and trematodes	Antienterobiasis, an amount of piperazine phosphate equivalent to 2 g. of piperazine hexahydrate daily for 7 days; antiascariasis, an amount of piperazine phosphate equivalent to 3.5 g. of piperazine hexahydrate daily for 2 days	
Pyrvinium Pamoate U.S.P. *Povan*	Pyrvinium Pamoate Oral Suspension U.S.P. Pyrvinium Pamoate Tablets U.S.P.	Intestinal pinworms	The equivalent of 5 mg. of pyrvinium per kg. of body weight, as a single dose.	See under Usual Dose, or the equivalent of 150 mg. of pyrvinium per square meter of body surface, as a single dose
Pyrantel Pamoate U.S.P. *Antiminth*	Pyrantel Pamoate Oral Suspension U.S.P.	Intestinal pinworms and roundworms	11 mg. per kg. of body weight	Children—11 mg. per kg. of body weight
Thiabendazole U.S.P. *Mintezol*	Thiabendazole Oral Suspension U.S.P.	Pinworms, threadworms, whipworms, roundworms, hookworms, and in cutaneous larva migrans	Adults under 68 kg.—25 mg. per kg. of body weight twice daily for 1 to 4 days; adults 68 kg. and over—1.5 g. twice daily for 1 to 4 days; up to a maximum of 3 g. daily for 1 to 4 days	22 mg. per kg. of body weight, or 650 mg. per square meter of body surface twice daily, for 1 to 4 days
Mebendazole U.S.P. *Vermox*	Mebendazole Tablets U.S.P.		100 mg. morning and evening for 3 consecutive days	

Generic Name / Brand	Dosage Form	Use	Usual Dose	Usual Pediatric Dose
Bephenium Hydroxynaphthoate U.S.P. *Alcopara*	Bephenium Hydroxynaphthoate for Oral Suspension U.S.P.	Hookworms	Against *Ancylostoma duodenale*—the equivalent of 2.5 g. of bephenium twice daily for 1 day; against *Necator americanus*—2.5 g. twice daily for 3 days	Under 23 kg. of body weight—500 mg. to 1.25 g. twice daily for one day; over 23 kg.—see Usual Dose
Niclosamide	Tablets	Tapeworms (fish, beef, pork, dwarf)	A single oral dose of 2 g. (4 tablets) chewed thoroughly	A single oral dose of 1 g. (2 tablets) for 11 to 34 kg.
Bithionol		Flukes (lung, liver)	30 to 50 mg. per kg orally on alternate days for 10 to 15 doses	30 to 50 mg. per kg orally on alternate days for 10 to 15 doses
Niridazole	Tablets	Antischistosomal	25 mg. per kg. orally per day	25 mg. per kg. orally per day
Antimony Potassium Tartrate U.S.P.		Schistosomicide	I.V., as a 0.5 percent solution given once every other day, the first dose 40 mg., each succeeding dose increased by 20 mg. until 140 mg. is reached, then 140 mg. every other day to a total of 2 g.	
Stibophen Injection *Fuadin*		Schistosomicide	I.M. or I.V., 100 mg. on the first day, then 300 mg. every other day to a total of 2.5 to 4.6 g.	

*See U.S.P. D.I. for complete dosage information.

Niridazole occurs as a yellow crystalline powder that is sparingly soluble in water. It is supplied as 500-mg. tablets and can be obtained from the Center for Disease Control.

Bithionol, Actamer®, Bitin®, is bis(2-hydroxy-3,5-dichlorophenyl)sulfide. This chlorinated bisphenol was formerly used in soaps and cosmetics for its antimicrobial properties but was removed from the market for topical use because of reports of contact photodermititis. Bithionol has useful anthelmintic properties and has been employed as a fasciolicide and taeniacide. It is still considered the agent of choice for the treatment of infestations caused by the liver fluke, *Fasciola hepatica* and the lung fluke, *Paragonimus westermani*. Niclosamide is believed to be superior to it for the treatment of tapeworm infestations.

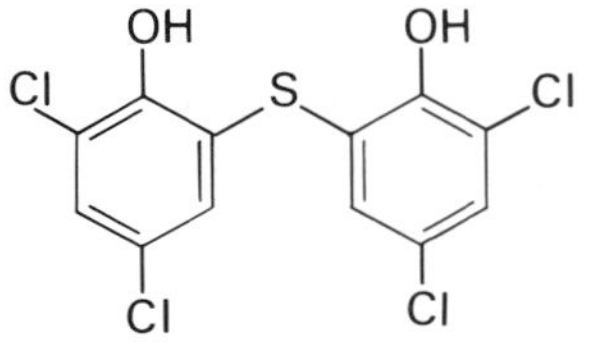

Bithionol

Dichlorophen, Anthiphen®, is 2,2′-methylene-bis(4-chlorophenol). This biphenol occurs as a cream colored powder with a slightly phenolic odor and is only slightly soluble in water. It was synthesized as part of a series of investigations that eventually led to the development of hexachlorophene. It has broad spectrum antimicrobial activity and has been widely employed for its antifungal properties as a mildew preventing agent. Effectiveness as a taeniacide in veterinary practice eventually led to the introduction of dichlorophen into human medicine for the treatment of tapeworm infestations. A relatively high incidence of gastrointestinal disturbances has been reported with dichlorophen, and consequently niclosamide has largely replaced it as a taeniacide.

Niclosamide, Cestocide®, Mansonil®, Yomesan®, 2,5′-dichloro-4′-nitrosalicylanilide, occurs as a yellowish-white, water insoluble powder. It is a potent tae-

niacide that causes rapid disintegration of worm segments and the scolex. Penetration of the drug into various cestodes appears to be facilitated by the digestive juices of the host, since very little of the drug is absorbed by the worms in vitro. Niclosamide is well tolerated following oral administration and little or no systemic absorption of it occurs. A saline purge 1 to 2 hours after the ingestion of the taenicide is recommended to remove the damaged scolex and worm segments. This procedure is mandatory in the treatment of pork tapeworm infestations to prevent possible cystecercosis resulting from release of live ova from worm segments damaged by the drug.

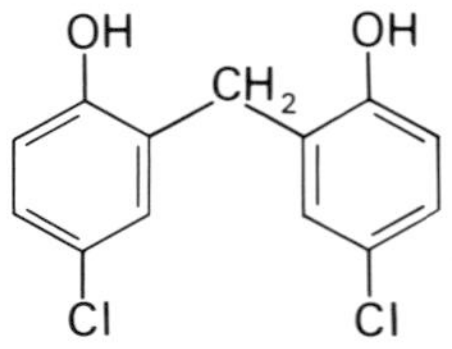

Dichlorophen

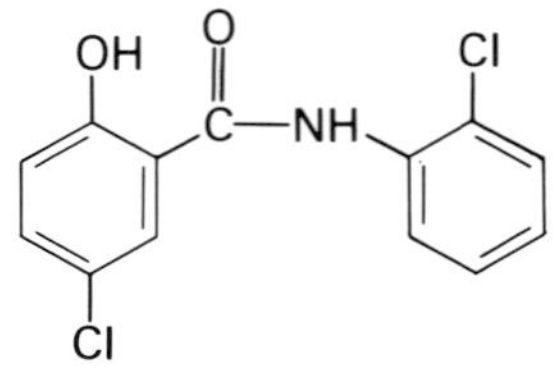

Niclosamide

Antimony Potassium Tartrate U.S.P., antimonyl potassium tartrate, tartar emetic, occurs either as colorless, odorless, transparent crystals or as a white powder, depending on whether or not the compound contains water of crystallization. The crystals effloresce when exposed to air. It is soluble in water (1:12), in glycerol (1:15) and is insoluble in alcohol.

The precise structure of antimony potassium tartrate has been the subject of considerable controversy through the years. X-ray crystallographic[107] and chemical[108] techniques have both been applied to this problem. The monomeric structure shown below has been suggested for tartar emetic in the solid state

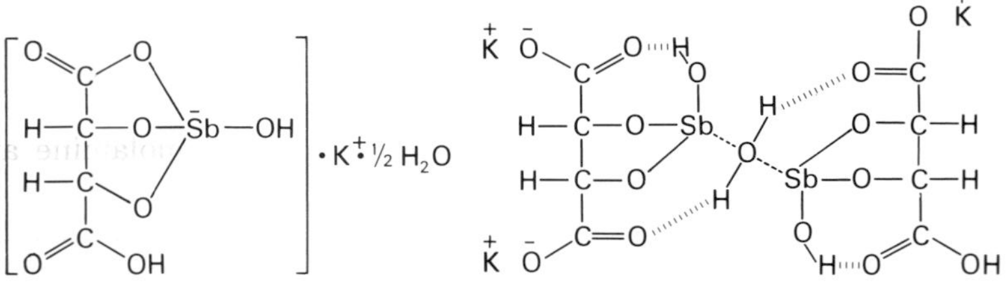

Antimony Potassium Tartrate

based on x-ray crystallography.[107] This representation suffers from two major shortcomings: first, the bicyclic ring system is highly strained, and second, the position of the water molecule is not specified. Chemical studies indicate that tartar emetic readily dimerizes in solution.[108] The dimeric structure recently proposed by Steck[109] is nonstrained, defines the position of the water molecule of crystallization, and satisfies the bonding properties of antimony.

The compound is used orally as an expectorant and an emetic. It is employed intravenously in the treatment of a number of tropical diseases, including leishmaniasis and schistosomiasis. It is considered the drug of choice against *Schistosoma japonicum.* The average oral dose as an expectorant is 3 mg.

Stibophen, Fuadin®, pentasodium antimony-*bis*-[catechol-2,4-disulfonate]. This compound is a white, odorless, crystalline powder. It is freely soluble in water and nearly insoluble in alcohol and ether. It is used in the treatment of schistosomiasis and granuloma inguinale. The compound is sensitive to light and should not be brought into contact with iron. Unused portions of opened ampules should be discarded because the compound is subject to oxidation. The ampuls usually contain about 0.1 percent sodium bisulfite to protect the solution during processing and storage.

Stibophen

Sodium Stibocaptate, Astiban®, Sodium Antimony Dimercaptosuccinate is a trivalent antimonial derivative of 2,3-dimercaptosuccinic acid in the form of the hexasodium salt. It provides a water-soluble derivative that is effective against all three forms of schistosomiasis following intramuscular administration. Because of the toxicity commonly associated with antimonial compounds, this drug is not suitable for mass chemotherapy of schistosomiasis, and its use must be carefully monitored. The drug is available as

Sodium Antimony Dimercaptosuccinate

a 3.6 percent sterile solution in propylene glycol for injection from the Center for Disease Control. It is only sparingly soluble in water.

ANTISCABIOUS AND ANTIPEDICULAR AGENTS

Antiscabious agents or scabicides are drugs used against the mite, *Sarcoptes scabiei,* which thrives when personal hygiene is neglected. The incidence of scabies is believed to be increasing both in the United States and worldwide; it has, in fact, reached pandemic proportions.[110] The ideal scabicide must kill both the parasites and their eggs. Sulfur preparations have been used for many years but are now being supplanted by more effective and less offensive agents. Antipedicular agents or pediculicides are used to eliminate head, body and crab lice. Like the ideal scabicide, the ideal pediculicide must kill both the parasites and their eggs.

PRODUCTS

Benzyl Benzoate U.S.P. This ester occurs naturally in Peru balsam and in some resins. It is prepared synthetically from benzyl alcohol and benzoic acid by several methods, such as that using benzyl alcohol and benzoyl chloride.

$$C_6H_5COCl + C_6H_5CH_2OH \longrightarrow$$
$$C_6H_5OOCH_2C_6H_5$$
Benzyl Benzoate

The ester is a clear, oily, colorless liquid, having a faint aromatic odor and a sharp, burning taste. The liquid is insoluble in glycerin and water but is miscible in all proportions with chloroform, alcohol or ether. It congeals at about 18° to 20°. Benzyl benzoate is neutral to litmus and with potassium hydroxide is readily saponified. It is used as a solvent with a vegetable oil for Dimercaprol Injection U.S.P.

It was introduced into medicine several years ago as an antispasmodic because of the benzyl group, but in 1937[111] it was found to be an effective parasiticide of especial value for the treatment of scabies by local application. For scabies, a 25 percent emulsion with the aid of triethanolamine and oleic acid usually is used. Due to some local anesthetic effect, there is instantaneous relief from itching. A single treatment often produces a complete cure.[112] Other advantages are absence of odor, no staining of clothes and no skin irritation. It is used topically, as a lotion over previously dampened skin of the entire body, except the face.

Lindane U.S.P., gamma benzene hexachloride, γ-1,2,3,4,5,6-hexachlorocyclohexane, benzene hexachloride (666, Gamex, B.H.C., Gammexane). This halogenated compound was prepared first in 1825 and has been a subject of research since that time. Bender,[113] in 1935, reported the value of lindane as an insecticide in his patent dealing with its preparation by the addition of benzene to liquid chlorine.

Lindane

Lindane is a mixture of a number of isomers, 5 of which have been isolated: alpha, beta, gamma, delta and epsilon. The gamma isomer, which is present to the extent of 10 to 13 percent in the commercial product, is by far the most active form and is responsible for the insecticidal property. The gamma isomer is extracted with organic solvents and obtained in 99 percent purity. This is known as lindane. The formula given above corresponds to one of the optically active inositols which would, of course, have OH groups substituted for the chlorine.

In powder form, it has a light buff to tan color, a persistent musty odor and a bitter taste. It is insoluble in water but readily soluble in many organic solvents[114] such as xylene, carbon tetrachloride, methanol, benzene or kerosene. The compound is unusually stable in neutral or acid environments. It withstands the effect of hot water and may be recrystallized with hot concentrated nitric acid. In the presence of alkalis, such as dry lime or lime water, however, hydrogen chloride is split out readily, leaving a mixture of the isomers of trichlorobenzene.

Lindane exhibits three modes of action against insects: (1) contact, (2) fumigant and (3) stomach poison. Insects which are susceptible to it are affected most by the first two modes of action, since the effect is rapid and the contact or fumigant action is felt before enough material can be eaten to be lethal. Physiologically, the effect upon insects seems to be the same as with DDT, that is, the nervous system is affected first. It is widely used in the destruction of cotton insects, aphids of fruit and vegetables.

Toxicity of this compound to warm-blooded animals is about the same as, or lower than, DDT, and, although it may be irritating to some people, recent studies indicate that a considerable quantity would have to be ingested before any ill effects would be produced.

Pharmaceutically, it is used externally as a parasiticide in the form of lotions and ointments in the treatment of scabies and pediculosis.

Crotamiton U.S.P., Eurax®, N-ethyl-*o*-crotonotoluide. This compound is a colorless, odorless oily liquid. It is practically insoluble in water, but dissolves in oils, fats, alcohol, acetone and ether. It is stable to light and air.

Crotamiton

Crotamiton, in the form of a lotion or in a washable ointment base, is used in the prevention and treatment of scabies. It also has an antipruritic action.

TABLE 4–20

SCABICIDES AND PEDICULICIDES

Name Proprietary Name	Preparations	Category	Application
Benzyl Benzoate U.S.P.	Benzyl Benzoate Lotion U.S.P.	Scabicide	Topical, as lotion over previously dampened skin of entire body, except face
Lindane U.S.P. *Kwell*	Lindane Cream U.S.P.	Pediculicide; scabicide	Topically to the skin, as a 1 percent cream once or twice weekly
	Lindane Lotion U.S.P.	Pediculicide; scabicide	Topically to the skin, as a 1 percent lotion once or twice weekly
	Lindane Shampoo U.S.P.		Topical, to the scalp, as a 1 percent shampoo for one application, repeated after 7 days if necessary

Diethyltoluamide U.S.P., N,N-diethyl-*m*-toluamide, is useful as a repellant for various kinds of insects, especially mosquitoes. It occurs as a colorless liquid with a faint, pleasant odor. Only the *meta*-isomer has activity as repellant.

Diethyltoluamide

It is practically insoluble in water but is miscible with alcohol, isopropyl alcohol and solvents such as ether and chloroform.

Category—arthropod repellant.

Application—topical, to skin and clothing, 15 percent ointment.

REFERENCES

1. DuMez, A. G.: J. Am. Pharm. A. 28:416, 1939.
2. Smyth, H. F.: J. Indust. Hyg. & Toxicol. 23:259, 1941.
3. Leech, P. N.: J.A.M.A. 109:1531, 1937.
4. Gilbert, G. L., et al.: Appl. Microbiol. 12:496, 1964.
5. Opfell, J. B., et al.: J. Am. Pharm. A. (Sci. Ed.) 48:617, 1959.
6. Grundy, W. E., et al.: J. Am. Pharm. A. (Sci. Ed.) 46:439, 1957.
7. Berl, E.: U.S. Patent 2,270,779, Jan. 20, 1942; through Chem. Abstr. 36:3187⁹, 1942.
8. Miner, N. A., et al.: Am. J. Pharm. 34:376, 1977.
9. U.S. Department of Agriculture: Circular 198, Dec., 1931.
10. J. Am. Pharm. A. 36:129, 134, 1947.
11. Reddish, G. F.: Antiseptics, Disinfectants, Fungicides, and Physical Sterilization, ed. 2, p. 537, Philadelphia, Lea & Febiger, 1957.
12. Suter, C. M.: Chem. Rev. 28:269, 1941.
13. Hansch, C., et al.: J. Am. Chem. Soc. 87:5770, 1965.
14. Weiss, J. M.: Chem. & Eng. News 30:4715, 1952.
15. Sykes, G.: Disinfection and Sterilization, ed. 2, p. 320, London, Spon, 1965.
16. Gaines, T. B., et al.: Toxicol. Appl. Pharmacol. 25:332, 1973.
17. U.S. Food and Drug Administration, "Hexachlorophene and Newborns", bulletin, December 1971.
18. Marsh, P. B., et al.: Ind. & Eng. Chem. 41:2176, 1949.
19. Florestano, H. J., and Bahler, M. E.: J. Am. Pharm. A. (Sci. Ed.) 42:576, 1953.
20. Erlandson, A. L., and Lawrence, C. A.: Science 118:274, 1953.
21. Vasarenish, A., Arch. Dermatol. 98:183, 1968.
22. Gershenfeld, L.: J. Milk & Food Tech. 18:223, 1955.
23. Brost, G. A., and Krupin, F.: Soap Chem. Specialties 33:93, 1957.
24. Lawrence, C. A., et al.: J. Am. Pharm. A. (Sci. Ed.) 46:500, 1957.
25. Lachman, L., Kuramoto, R., and Cooper, J.: J. Am. Pharm. A. 47:871, 1958.
26. Miller, O. H.: J. Am. Pharm. A. (Pract. Ed.) 13:657, 1952.
27. Mueller, W. S., and Seeley, D. B.: Soap Sanit. Chem. 27:131, 1951 (Nov.).
28. Ridenour, G. M., and Armbruster, E. H.: Am. J. Public Health 38:504, 1948.
29. DeLuca, P. P. and Kostenbauder, H. B.: J. Am. Pharm. A. 49:430, 1960.
30. Bradshaw, J. W., Rhodes, C. T., and Richardson, G.: J. Pharm. Sci. 61:1163, 1972.
31. Pivnick, H., Tracy, J. M., and Glass, D. G.: J. Pharm. Sci. 52:883, 1963.
32. Batuyios, N. H., and Brecht, E. A.: J. Am. Pharm. A. 46:524, 1957.
33. Cutler, R. A., et al.: Soap Chem. Specialties 43:84, 1967 (Mar).
34. Rose, F. L., and Swain, G., J. Chem. Soc. 422, 1956.
35. Tucker, S. J., et al.: J. Am. Pharm. A. (Sci. Ed.) 47:849, 1958.
36. Fildes, P.: Brit. J. Exp. Path. 21:67, 1940.
37. Richardson, A., et al.: J. Am. Pharm. A. 45:268, 1956.
38. Neidig, C. P., and Burrell, H.: Drug Cosmet. Ind. 54:408, 1944.
39. Deeb, E. N., and Boenigk, J. W.: J. Am. Pharm. A. (Sci. Ed.) 47:807, 1958.
40. Murphy, J. T., et al.: Arch. Ophthal. 53:63, 1955.
41. Nair, A. D., and Lach, J. L.: J. Am. Pharm. A. (Sci. Ed.) 48:390, 1959.
42. Macht, D. I.: J. Pharmacol. Exp. Ther. 11:263, 1918.
43. Hirschfelder, A. D., et al.: J. Pharmacol. Exp. Ther. 15:237, 1920.
44. Hjort, A. M., and Eagan, F. T.: J. Pharmacol. Exp. Ther. 14:211, 1919.
45. Goshorn, R. H., and Degering, E. F.: Ind. & Eng. Chem. 30:646, 1938.
46. Bandelin, F. J.: J. Am. Pharm. A. (Sci. Ed.) 47:691, 1958.
47. Rahn, O., and Conn, J. E.: Ind. & Eng. Chem. 36:185, 1944.
48. Garrett, E. R., and Woods, O. R.: J. Am. Pharm. A. (Sci. Ed.) 42:736, 1953.
49. Kenney, E. L.: Bull. Johns Hopkins Hosp. 73:379, 1943.
50. Peck, S. M., et al.: Arch. Derm. 39:126, 1939.
51. Keeney, E. L., et al.: Bull. Johns Hopkins Hosp. 75:417, 1944.
52. Schwartz, L.: Am. Prof. Pharm. 13:157, 1947.
53. Godefroi, E. F., et al.: J. Med. Chem. 12:784, 1969.
54. Polak, A., and Scholer, H. J. Chemotherapy 21:113, 1975.
55. Oxford, A. E., et al.: Biochem. J. 33:240, 1939.
56. Grove, J. F., et al.: J. Chem. Soc., p. 3977, 1952.
57. Golding, B. T., et al.: Tetrahedron Letters 3551, 1966.
58. Mechlinski, W., et al.: Tetrahedron Letters 3873, 1970.
59. Borowski, E., et al.: Tetrahedron Letters 3909, 1970.
60. Chong, C. N., and Richards, R. W.: Tetrahedron Letters 5145, 1970.
61. Borowski, E., et al.: Tetrahedron Letters 685, 1971.
62. Lechevalier, H. A., et al.: Antibiotics & Chemother. 11:640, 1961.
63. Hazen, E. L., and Brown, R.: Proc. Soc. Exp. Biol. Med. 76:93, 1951.
64. Gold, W., et al.: Antibiotics Annual 1955–1956, p. 579, New York, Medical Encyclopedia, 1956.
65. Norman, A. W., et al.: J. Biol. Chem. 247:1918, 1972.
66. Lechevalier, H. A., et al.: Mycologia 45:155, 1953.
67. Struyk, A. P., et al.: Antibiot. Ann. 1957–58, 857, 1958.
68. Field, A. K., Proc. Nat. Acad. Sci. U. S. 58:1004, 1967.
69. Mayer, G. D., and Krueger, R. F., Science 169:1213, 1970.
70. Tamm, I.: Yale J. Biol. Med. 29:33, 1956.
71. Ravin, L. J., and Gulesich, J. J.: J. Am. Pharm. A. NS4:122, 1964.
72. Pavan-Langston, D., et al.: "Adenosine Arabinoside: An Antiviral Agent," Raven Press, N.Y., 1975.
73. Schaeffer, H. J., et al.: Nature 272:583, 1978.
74. Sidwell, R. W., et al.: Science 177:705, 1972.
75. Nahata, M. C., et al.: Am. J. Hosp. Pharm. 34:1234, 1977.
76. Dietz, W. H., et al.: J. Bacteriol. 91:768, 1966.
77. Külling, E.: Pharm. Acta Helvet. 34:430, 1959.
78. Schneller, G. H.: Am. Prof. Pharm. 18:148, 1952.
79. Fox, H. H.: J. Chem. Ed. 29:29, 1952.
80. ———: Science 116:131, 1952.
81. Boxenbaum, H. G., and Riegelman, S.: J. Pharm. Sci. 63:1191, 1974.

82. Kuehl, F. A., Jr., *et al.*: J. Am. Chem. Soc. 77:2344, 1955.
83. Hildy, P. H., *et al.*: J. Am. Chem. Soc. 77:2345, 1955.
84. Stammer, C. H., *et al.*: J. Am. Chem. Soc. 77:2346, 1955.
85. Smrt, J.: Experientia 13:291, 1957.
86. Neuhaus, F. C. and Lynch, J. L., Biochemistry, 3:471, 1964.
87. Rando, R. R., Biochem. Pharmacol., 24:1153, 1975.
88. Haskell, T. H., *et al.*: J. Am. Chem. Soc. 74:599, 1952.
89. Mayer, R. L.: Experientia 10:335, 1954.
90. Bowie, J. H., *et al.*: Tetrahedron Letters 3305, 1964.
91. Bancroft, B. W., *et al.*: Experientia 27:501, 1971.
92. Noda, T., *et al.*: J. Antibiot. 25:427, 1971.
93. Yoshioka, H., *et al.*: Tetrahedron Letters 2043, 1971.
94. Herr, E. B., *et al.*: Indiana Acad. Sci. 69:134, 1960.
95. Bancroft, B. W., *et al.*: Nature 231:301, 1971.
96. Nomoto, S., *et al.*: J. Antibiotics 30:955, 1977.
97. Rinehart, K. L.: Acc. Chem. Res. 5:57, 1972.
98. Hartmann, G., *et al.*: Angew. chem. 80:710, 1968.
99. Katz, E., and Moss, B.: Proc. Nat. Acad. Sci. U.S. 66:677, 1970.
100. Gurgo, C., *et al.*: Nature (New Biol.) 229:111, 1971.
101. Gallo, G. G., and Radaelli, P. *In* Analyt. Profiles of Drug Substances, Florey, K. (ed.), Vol. 5, 1976, p. 491.
102. Ellin, R. I., and Kondritzer, A. A.: J. Am. Pharm. A. (Sci. Ed.) 47:12, 1958.
103. Wright, W. H., and Schaffer, J. M.: Am. J. Hyg. 16:325, 1932.
104. Robinson, H. J., *et al.*: Tox. Appl. Pharmacol. 7:53, 1965.
105. Janssen, P. A. J., Fortschr. Arzneimittel Forsch. 20:307, 1976.
106. Van den Bossche, H., and Janssen, P. A. J., Life Sci., 6:1781, 1967.
107. Grdenic, D., and Kamenar, B.: Acta Crystallogr. 19:192, 1965.
108. Banerjee, A. K., and Chari, K. V. R.: J. Inorg. Nucl. Chem. 31:2958, 1969.
109. Steck, E. A.: Progr. Drug Res. 18:304, 1974.
110. Orkin, M. and Maibach, H. I., N. Eng. J. Med. 298:496, 1978.
111. Kissmeyer, A.: Lancet 1:21, 1941.
112. MacKenzie, I. F.: Brit. Med. J. 2:403, 1941.
113. Bender, H.: U.S. Patent 2,010,841, Aug. 13, 1935; through Chem. Abstr. 29:6607[7], 1935.
114. Chamlin, G. R.: J. Chem. Ed. 23:283, 1945.

SELECTED READING

Bambury, R. E.: Synthetic Antibacterial Agents, in Burger's Medicinal Chemistry, Wolff, M. E. (ed.), Part II, ed. 4, p. 41, Wiley-Interscience, N. Y., 1979.
D'Arcy, P. F. and Scott, E. M.: Antifungal Agents. *In* Prog. Drug Res. 22:267, 1978.
Davis, A.: Drug Treatment in Intestinal Helminthiasis, World Health Organization, Geneva, 1973.
Gutteridge, W. E. and Coombs, G. H.: Biochemistry of Parasitic Protozoa, McMillan Press, Ltd., London, 1977.
Islip, P. J.: Anthelmintic Agents. *In* Burger's Medicinal Chemistry, Wolff, M. E. (ed.), Part II. ed. 4, p. 481, Wiley-Interscience, N. Y., 1979.
Krogstad, D. J., *et al.*: Current Concepts in Parasitology: Amebiasis, N. Engl. J. Med. 298:262, 1978.
Lawrence, C. A. and Block, S. S. (eds.): Disinfection, Sterilization and Preservation, 2nd ed., Lea and Febiger, Philadelphia, 1977.
Lawrence, C. A.: Quaternary Ammonium Compounds, Academic Press, New York, 1950.
Marvis, M. (ed.): Development of Chemotherapeutic Agents for Parasitic Diseases, North Holland, Amsterdam, 1975.
Miller, M. J.: Protozoan and Helminth Parasites: A Review of Current Treatment, Prog. Drug Res. 20:433, 1976.
Ostrolenk, M., and Brewer, C. M.: A Bactericidal Spectrum of Some Common Organisms, J. Am. Pharm. A. 38:95, 1949.
Ross, W. J.: Antiamebic Agents. *In* Burger's Medicinal Chemistry, Wolff, M. E. (ed.), Part II, ed. 4, p. 415, Wiley-Interscience, New York, 1979.
Ross, W. J.: Chemotherapy of Trypanosomiasis and Other Protozoan Diseases. *In* Burger's Medicinal Chemistry, Wolff, M. E. (ed.), Part II, ed. 4, p. 439, Wiley-Interscience, New York, 1979.
Sensi, P., and Gialdroni-Grassi, G.: Antimycobacterial Agents. *In* Burger's Medicinal Chemistry, Wolff, M. E. (ed.), Part II, ed. 4, p. 289, Wiley-Interscience, New York, 1979.
Sidwell, R. W., and Wilkowski, J. T.: Antiviral Agents. *In* Burger's Medicinal Chemistry, Wolff, M. E. (ed.), Part II, ed. 4, p. 543, Wiley-Interscience, New York, 1979.
Suter, C. M.: The Relationship Between the Structure and Bactericidal Properties of Phenols, Chem. Rev. 28:269, 1941.

5

sulfonamides, sulfones, and folate reductase inhibitors with antibacterial action

Dwight S. Fullerton

SULFONAMIDES AND FOLATE REDUCTASE INHIBITORS

PRONTOSIL AND GERHARD DOMAGK (1895–1964)

The founding of chemotherapy, drug design, and medicinal chemistry by Paul Ehrlich (1854–1915) and the antisyphilitic drug Salvarsan in 1908 will be discussed in Chapter 6. Ehrlich's discovery led to intensive investigations of dyes as antimicrobial agents, especially in Germany. Although Salvarsan and related drugs were revolutionary in treating some protozoal infections and syphilis, they were not useful in treating a major killer of the times—streptococcal and staphylococcal infections.

Fritz Mietzsch and Joseph Klarer of the I.G. Farbenindustrie (Bayer) laboratories began a systematic synthesis of azo dyes as possible antimicrobials. Sulfonamide azo dyes were included because they were relatively easy to synthesize and had improved staining properties. The Bayer pathologist-bacteriologist who evaluated the new Mietzsch-Klarer dyes was Gerhard Domagk, who, like Ehrlich, was a physician by training.[1-3] In 1932, Domagk began a study of a bright red dye later to be named Prontosil and found that it caused remarkable cures of streptococcal infections of mice.[1] However, Prontosil was inactive on bacterial cultures. Domagk's studies on Prontosil continued, and in 1933 the first of many human cures of severe staphyloccal septicemias was reported.[4] Domagk even

saved the life of his own daughter from a severe streptococcal infection. For his pioneering efforts in chemotherapy, Gerhard Domagk was awarded the Nobel prize for medicine and physiology in 1939. The Gestapo prevented him from actually accepting the award, but he received it in Stockholm in 1947.

Prontosil's inactivity in vitro but excellent activity in vivo attracted much attention. In 1935, Trefouel, Trefouel, Nitti and Bovet[5] reported their conclusion from a structure–activity study of sulfonamide azo dyes that the azo linkage was metabolically broken to release the active ingredient, sulfanilamide. Their reported finding was confirmed in 1937 when Fuller[6] isolated sulfanilamide from the blood and urine of patients being treated with Prontosil. Modern chemotherapy and the concept of the prodrug (see Chap. 2) were firmly established.

THE MODERN ERA

Following Prontosil's dramatic successes, a cascade of sulfanilamide derivatives began to be synthesized and tested—over 4500 by 1948 alone.[7] From these only about two dozen have actually been used in clinical practice. In the late 1940's, penicillins began to replace the sulfanilamides in chemotherapy. This was largely because of the sulfanilamides' toxicity for some patients and because sulfanilamide-resistant bacterial strains were becoming an increasing problem—the result of indiscriminant use worldwide.

Prontosil   Sulfanilamide

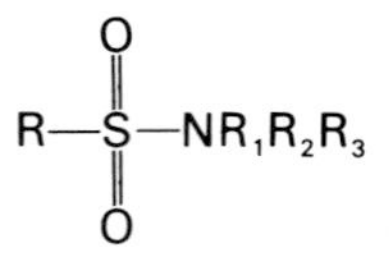

General Sulfonamide Structure

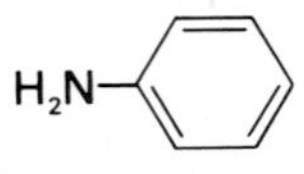

Aniline

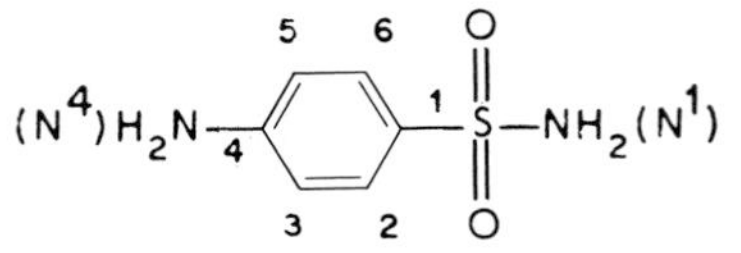

Sulfanilamide

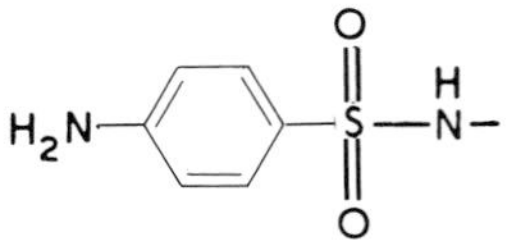

Sulfanilamido-

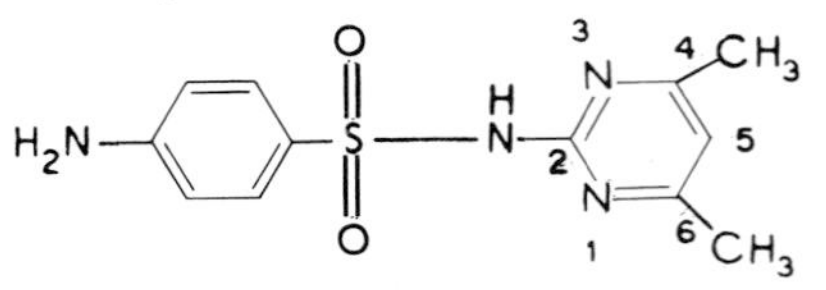

N^1-(4,6-Dimethyl-2-pyrimidyl)sulfanilamide
Sulfamethazine

FIG. 5-1. *Nomenclature and numbering.*

Today a few sulfonamides and especially sulfonamide-trimethoprim combinations are extensively used for urinary tract infections or for burn therapy.[8-12] They are also the drugs of choice or alternates for a few other types of infections (Table 5-1), but their overall use is otherwise quite limited in modern antimicrobial chemotherapy,[8-12] having been largely replaced by antibiotics.

CHEMISTRY AND NOMENCLATURE

The term *sulfonamide* is commonly used to refer to antibacterials which are (1) aniline-substituted sulfonamides, the "sulfanilamides" (Fig. 5-1); (2) prodrugs that produce sulfanilamides, e.g., sulfasalazine; and (3) nonaniline sulfonamides, e.g., mafenide. However, a number of other widely used drugs are also sulfonamides or sulfanilamides. Included among these nonantibacterial sulfonamides are tolbutamide (an oral diabetic drug, see Chap. 14), furosemide (a potent diu-

retic, see Chap. 13), and chlorthalidone (also a diuretic, see Chap. 13).

As reviewed in Chapter 2, pKb's are not used in pharmaceutical chemistry to compare compounds. If a pKa of an amine is given, it refers to its salt acting as the conjugate acid, e.g.:

Aniline, pKa 4.6 refers to:

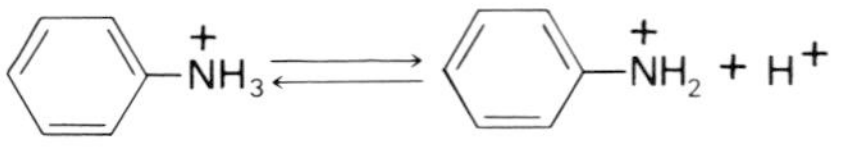

not

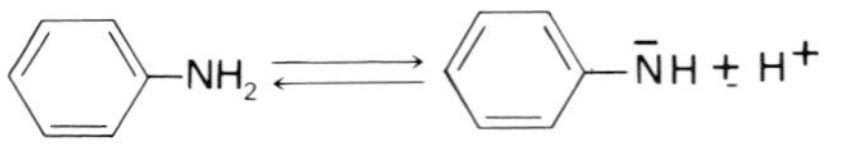

A minus charge on a nitrogen atom is normally not very stable, unless the minus charge can be greatly delocalized by resonance. This is exactly the case with the sulfanilamides. Thus, the single pKa usually given with sulfanilamides refers to loss of an amide H+, e.g.:

Sulfanilamide, pKa 10.4
refers to:

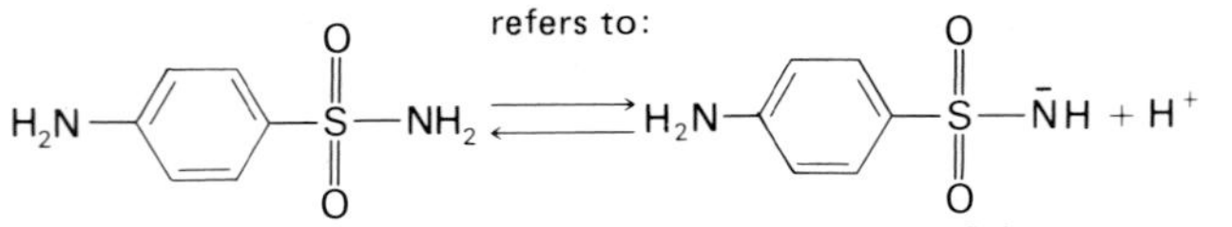

and Sulfisoxazole, pKa 5.0
refers to:

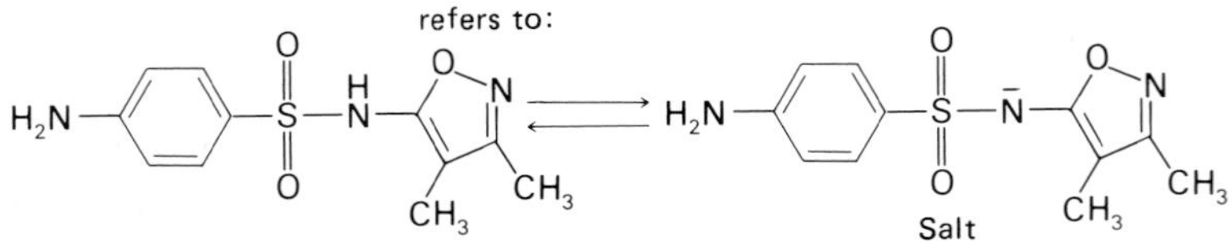

TABLE 5–1

CURRENT THERAPY WITH SULFONAMIDE ANTIBACTERIALS[4,5,6]

Disease/Infection	Sulfonamides Commonly Used
WIDE USE	
Previously untreated urinary-tract infections	Sulfamethoxazole and Trimethoprim Sulfisoxazole Sulfadiazine Trisulfapyrimidines
Burn therapy—prevention and treatment of bacterial infection	Silver sulfadiazine Mafenide
Conjunctivitis and related superficial ocular infections	Sodium sulfacetamide
Chloroquine-resistant malaria (Chapter 7)	Combinations with quinine, others. Sulfadoxine Sulfalene
LESS COMMON INFECTIONS/DISEASES—DRUGS OF CHOICE OR ALTERNATES	
Nocardiosis	Sulfisoxazole Sulfadiazine
Toxoplasmosis	Sulfisoxazole
Severe travelers' diarrhea[13]	Trimethoprim–Sulfamethoxazole
Meningococcal infections	Only if proved to be sulfonamide sensitive; otherwise penicillin G or ampicillin should be used. Sulfisoxazole, sulfadiazine.
LESS CERTAIN USES	
Streptococcal infections	Most are resistant to sulfonamides
Prophylaxis of rheumatic fever recurrences	Most are resistant to sulfonamides
Other bacterial infections	Penicillins' low cost and bacterial resistance to sulfonamides have decreased sulfonamide use world wide—but still widely used in a few countries.
Vaginal infections	FDA Bulletin[8] finds no evidence of effectiveness.
Reduction of bowel flora	Effectiveness not established.[9] Phthalylsulfathiazole
Ulcerative colitis	Corticosteroid therapy often preferred. Relapses common with sulfanilamides. Phthalylsulfathiazole Salicylazosulfapyridine Side effects of the sulfanilamides sometimes seem like the ulcerative colitis (Werlin and Grand[14])
FUTURE USES	

Many new uses for trimethoprim–sulfanilamide combinations have been proposed[15,16]. It is likely that some of these will be approved in the near future.

Thus, sulfisoxazole (pKa 5.0) is slightly weaker an acid than acetic acid (pKa 4.8).

REDUCING CRYSTALLURIA BY LOWERING pKa

Sulfanilamide, although revolutionary in the early 1930's, often caused severe kidney damage from crystals of sulfanilamide forming in the kidneys. Sulfanilamides and their metabolites (usually acetylated at N^4) are excreted almost entirely in the urine. Unfortunately, sulfanilamide is not very water soluble. Unless the pH is above the pKa (that is, above pH 10.4), little of its water-soluble salt is present. Since urine pH is typically about 6, and often slightly lower during bacterial infections, essentially all sulfanilamide is in the relatively insoluble nonionized form in the kidneys.

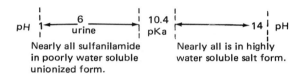

At pH=pKa, that is at pH 10.4 for sulfanilamide, there will be a 1:1 mixture of nonionized and salt forms.

How can a sulfanilamide be made "more soluble" in the urine? There are several options:

1. Greatly increase urine flow. Thus, during the early days of sulfanilamide and sulfanilamide derivative use, patients were warned to "force fluids."
2. Raise the pH of the urine. The closer that the pH of the urine gets to 10.4 (in the case of sulfanilamide itself), the more of the highly water-soluble salt form will be present. Thus, sometimes oral sodium bicarbonate was and occasionally still is given to raise urine pH.
3. Make derivatives of sulfanilamide which have lower pKa's, closer to the pH of urine. This has been the approach taken with virtually all sulforamides clinically used today. e.g.:

Sulfanilamide	pKa
Sulfadiazine	6.5
Sulfamerazine	7.1
Sulfamethazine	7.4
Sulfisoxazole	5.0
Sulfamethoxazole	6.1

4. Mix sulfonamides to reach the total dose. Since solubilities of sulfanilamides are independent, more of a *mixture* of sulfanilamides can stay in water

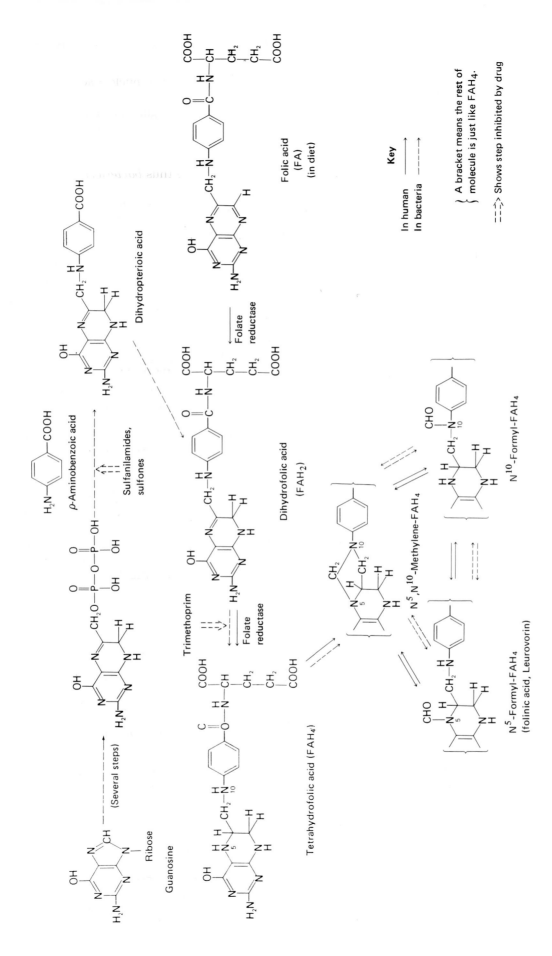

FIG. 5-2. *Sites of action of sulfanilamides and trimethoprim.*

192

solution at a particular pH than a single sulfonamide. Thus, Trisulfapyrimidines U.S.P. ("triple sulfas") contains a mixture of Sulfadiazine, Sulfamerazine, and Sulfamethazine. However, such mixtures are little used today because the individual agents have sufficiently low pKa's to be adequately urine soluble *providing at least normal urine flow is maintained.* Patients must still be cautioned to maintain a normal fluid intake, even if they don't feel like drinking during the illness. "Forcing fluids," however, is no longer necessary.

It would be reasonable to ask, "Why do the modern sulfonamides have such low pKa's?" The answer is that the heterocyclic rings attached to N^1 are electron withdrawing—providing additional stability for the salt form. Therefore the nonionized forms can more easily given up a H^+, so the pKa's are lower. Why aren't simpler electron withdrawing groups used, such as *p*-nitrophenyl—the usual example studied in introductory organic chemistry classes? Such compounds were in fact extensively investigated, as were thousands of others. However, the other sulfonamides generally were too toxic, not sufficiently active, or both.

MECHANISM OF ACTION

Folinic acid (N^5-formyltetrahydrofolic acid, Fig. 5-2), N^5, N^{10}-methylenetetrahydrofolic acid, and N^{10}-formyltetrahydrofolic acid are indispensable for several biosynthetic pathways in humans, bacteria, animals, and plants (Fig. 5-3). Without these folate coenzymes, for example, deoxythymidine monophosphate will not be available to produce nucleic acids needed for cell division. Other reactions requiring the folate coenzymes are shown in Figure 5-3. The result of any drug blocking the biosynthesis of folate coenzymes in bacteria, for example, is that growth and cell division are stopped. Such drugs—including the sulfonamides, and trimethoprim—are thus *bacteriostatic.*

As shown in the biosynthetic pathway in Figure 5-2, folate coenzymes are formed from dietary folic acid in humans (and other animals). However bacteria (and protozoa, see Chap. 6) must form them from para-aminobenzoic acid (PABA). The microbes cannot use dietary folic acid from the host, for reasons not yet completely understood,[3] but it may be that folic acid cannot penetrate the cell wall.

The sulfonamide and sulfone antibacterials act as competitive inhibitors for the incorporation of PABA to form dihydropteroic acid (Fig. 5-2). Trimethoprim is an inhibitor of folate reductase, needed to convert dihydrofolic acid (FAH_2) into tetrahydrofolic acid (FAH_4) in bacteria. Evidence for these inhibitions has been summarized in detail by Anand.[3]

Although trimethoprim does not have a very high affinity for malaria protozoa's folate reductase (see Chap. 6), it does have high affinity for the bacterial folate reductase. The reverse situation exists for the antimalarial drug pyrimethamine.[17] Trimethoprim still does have some affinity for human folate reductase—the cause of some toxic effects discussed later in this chapter and in Chapter 6.

Drugs with additional or modified mechanisms of action, e.g., silver sulfadiazine, will be discussed with individual descriptions which follow later in the chapter.

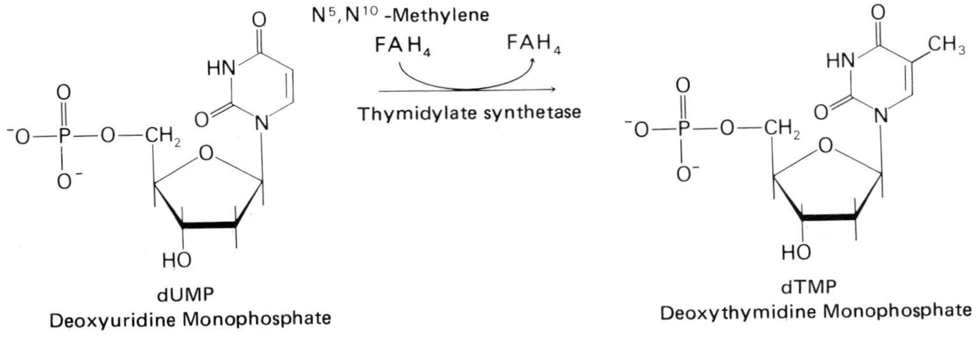

Other examples

Coenzyme	Reaction
N^{10}-Formyl FAH_4	Met-t-RNA → Formyl Met-t-RNA
N^5,N^{10}-Methylene FAH_4	Glycine → Serine
N^5-Formyl FAH_2	Homocysteine → Methionine

FIG. 5-3. Reactions requiring the folate coenzymes.

SYNERGISM OF SULFONAMIDES AND FOLATE REDUCTASE INHIBITORS

Blocking the biosynthesis of folate coenzymes at more than one point in the biosynthetic pathway of bacteria (or protozoa, see Chap. 6) will result in a synergistic antimicrobial effect. An additional benefit is that the microbe will not be able to develop resistance as quickly as with a single pathway blocker. This synergistic approach is widely used in antibacterial therapy with the combination of sulfamethoxazole and trimethoprim[15,16] and in antimalarial therapy with pyrimethamine plus a sulfanilamide or quinine (Table 5-1). Additional explanations for the synergistic antimicrobial actions of trimethoprim and sulfamethoxazole have been at times vigorously debated.[3,15,16,18]

Other combinations of trimethoprim have also been investigated, e.g., with rifampicin.[19,20]

MECHANISMS OF RESISTANCE

As noted earlier in this chapter, wide and unselective use of sulfonamides led to the emergence of many drug resistant strains of bacteria. The cause of the resistance is probably increased production of PABA by the resistant bacteria,[21] although other mechanisms may account for resistance in some cases.[3,16] If a microbe is resistant to one sulfonamide, it is generally resistant to all. A special concern is the finding that sulfonamide resistance can be transferred from a resistant bacterial species to a previously sulfanilamide-sensitive species. The transferring substances have been called "R-factors."[22]

Several explanations have been presented to account for bacterial resistance to trimethoprim,[16] including natural (intrinsic) resistance, development of an ability by the bacteria to use the host's d-TMP (Fig. 5-3), and R-factor transmission.

TOXICITY AND SIDE-EFFECTS

A variety of serious toxic and hypersensitivity problems have been reported with the sulfonamide and sulfonamide–trimethoprim combinations. Weinstein[9] notes that these occur in about 5 percent of all patients. Hypersensitivity reactions include drug fever, Stevens–Johnson syndrome, skin eruptions, allergic myocarditis, photosensitization and related conditions. Hematologic side-effects also sometimes occur, especially hemolytic anemia in individuals with a deficiency of glucose 6-phosphate (discussed further in Chap. 6), Other hematologic side-effects that have been reported include agranulocytosis, aplastic ane-

mias, and others. Crystalluria may still occur, even with modern sulfonamides, when the patient does not maintain a normal fluid intake. Nausea and related gastrointestinal side-effects are sometimes noted.

Detailed summaries of incidences of side-effects with trimethoprim–sulfamethoxazole have been published by Wormser and co-workers[15] and by Gleckman and associates.[16]

METABOLISM, PROTEIN BINDING, AND DISTRIBUTION

With the exception of the poorly absorbed sulfonamides used for ulcerative colitis and reduction of bowel flora, and the topical burn preparations (e.g., mafenide), sulfonamides and trimethoprim tend to be quickly absorbed and well distributed. As Weinstein noted, sulfonamides can be found in the urine "within 30 minutes after oral ingestion."

The sulfonamides vary widely in plasma protein binding—e.g., sulfisoxazole 76 percent, sulfamethoxazole 60 percent, sulfamethosypyridazine 77 percent, and sulfadiazine 38 percent. (An excellent table comparing the percent of protein bound, lipid solubility, plasma half-life, and percent N^4-metabolites has been published by Anand.[3]) The fraction that is protein bound is not active as an antibacterial, but because the binding is reversible, free, and therefore active, sulfonamide eventually becomes available. Generally, the more lipid soluble a sulfonamide is, at physiological pH, the more it will be protein bound. Fujita and Hansch have found that among sulfonamides with similar pKa, the lipophilicity of the N^1 group has the largest effect on protein binding.[23] N^4-Acetate metabolites of the sulfonamides are more lipid soluble and therefore better protein bound than the starting drugs themselves (which have a free 4-NH_2 group that decreases lipid solubility). Surprisingly, the N^4-acetylated metabolites, although more strongly protein bound, are more rapidly excreted than the starting drugs.

At present, the relationship between plasma protein binding and biological half-life is not clear. Many competing factors are involved, as reflected in sulfadiazine with a serum half-life of 17 hours being much less protein bound than sulfamethoxazole with a serum half-life of 11 hours.[3]

Sulfonamides are excreted primarily as mixtures of unmetabolized drugs, N^4-acetates, and glucuronides.[24] The N^4-acetates and glucuronides are inactive. Sulfisoxazole, for example, is excreted about 80 percent unchanged, and sulfamethoxazole, 20 percent unchanged. Sulfadimethoxine is about 80 percent excreted as the glucuronides. The correlation between structure and route of metabolism has not been delineated yet, although progress has been made by Fujita.[25] Vree and co-workers, however, have described the excretion kinetics and pKa's of N^1 and N^4 acetyl-sulfamethoxazole and other sulfonamides.[26]

Trimethoprim and sulfamethoxazole are both partially plasma protein bound—about 45 percent of trimethoprim, and about 66 percent of sulfamethoxazole. Whereas about 80 percent of excreted trimethoprim and its metabolites are active as antibacterials, only 20 percent of sulfamethoxazole's are active, mostly unmetabolized sulfamethoxazole. Six metabolites of trimethoprim are known.[27] It is likely therefore that sulfonamide–trimethoprim combinations using a sulfonamide with a higher active urine concentration will be developed in the future for urinary-tract infections. Sulfamethoxazole and trimethoprim have a similar half-life, about 10 to 12 hours, but the half-life of the active fraction of sulfamethoxazole is less—about 9 hours.[27] (Ranges of half-lives have been summarized by Gleckman,[16] and a detailed summary of pharmacokinetics has been made by Hansen.[27]) In patients with impaired renal function, sulfamethoxazole and its metabolites may greatly increase in the plasma. The *Hospital Formulary* warns that the fixed combination of sulfamethoxazole and trimethoprim should not be used with patients with low creatinine clearances.[8]

STRUCTURE-ACTIVITY RELATIONSHIPS

As noted earlier in this chapter, several thousand sulfonamides have been investigated as antibacterials (and many as antimalarials, see Chap. 6). From these efforts several structure–activity relationships have been proposed, elegantly summarized by Anand.[3] The aniline (N^4) amino group is very important for activity because any modification of it other than to make prodrugs results in a loss of activity. As noted earlier, for example, the N^4-acetylated metabolites of sulfonamides are all inactive.

A variety of studies have shown that the active form of sulfonamides is the N^1-ionized salt form. Thus, although many modern sulfonamides are much more active than unsubstituted sulfanilamide, they are only 2 to 6 times more active when comparing amounts of N^1-ionized forms.[28] Maximal activity seems to be exhibited by sulfonamides between pKa's 6.6 and 7.4.[28–31] This reflects in part the need for enough nonionized (i.e., more lipid soluble) drug to be present at physiological pH to be able to pass through bacterial cell walls.[32,33] Fujita and Hansch[23] also related pKa, partition coefficients, and electronic (Hammett) parameters with sulfonamide activity.

WELL-ABSORBED, SHORT-, AND INTERMEDIATE-ACTING SULFONAMIDES

Sulfamethizole U.S.P., N^1-(5-methyl-1,3,4-thiadiazol-2-yl)sulfanilamide, 5-methyl-2-sulfanilamido-1,3,4-thiadiazole. Plasma half-life = 2.5 hours. This compound is a white, crystalline powder soluble 1:2,000 in water.

Sulfamethizole

Sulfisoxazole U.S.P., N^1-(3,4-dimethyl-5-isoxazolyl)sulfanilamide, 5-sulfanilamido-3,4-dimethylisoxazole. Plasma half-life = 6 hours. This compound is a white, odorless, slightly bitter, crystalline power. Its pKa is 5.0. At pH 6 this sulfonamide has a water-solubility of 350 mg. in 100 ml., and its acetyl derivative has a solubility of 110 mg. in 100 ml. of water.

Sulfisoxazole possesses the action and the uses of other sulfonamides and is used for infections involving sulfonamide-sensitive bacteria. It is claimed to be effective in treatment of gram-negative urinary infections.

Sulfisoxazole

Sulfisoxazole Acetyl U.S.P., N-(3,4-dimethyl-5-isoxazolyl)-N-sulfanilylacetamide, N^1-acetyl-N^1-(3,4-dimethyl-5-isoxazolyl)sulfanilamide, shares the actions and the uses of the parent compound, sulfisoxazole. The acetyl derivative is tasteless and, therefore, suitable for oral administration, especially in liquid preparations of the drug. The acetyl compound is split in the intestinal tract and absorbed as sulfisoxazole. That is, it is a *prodrug* for sulfisoxazole.

Sulfisoxazole Acetyl

Sulfisoxazole Diolamine U.S.P., 2,2'-iminodi-ethanol salt of N^1-(3,4-dimethyl-5-isoxazolyl)sulfanil-amide.

This salt is prepared by adding enough diethanola-mine to a solution of sulfisoxazole to bring the pH to about 7.5. It is used as a salt to make the drug more soluble at physiologic pH range of 6.0 to 7.5 and is used in solution for systemic administration of the drug by slow intravenous, intramuscular or subcuta-neous injection when sufficient blood levels cannot be maintained by oral administration alone. It also is used for instillation of drops or ointment in the eye for the local treatment of susceptible infections.

Sulfamethazine U.S.P., N^1-(4,6-dimethyl-2-py-rimidinyl)sulfanilamide, 2-sulfanilamido-4,6-dimeth-ylpyrimidine. Plasma half-life = 7 hours. This com-pound is similar in chemical properties to sulfamerazine and sulfadiazine but does have greater water-solubility than either of them. Its pKa is 7.2. Because it is more soluble in acid urine than is sulfa-merazine, the possibility of kidney damage from use of the drug is decreased. The human body appears to handle the drug unpredictably; hence, there is some disfavor to its use in this country except in combina-tion sulfa therapy (in Trisulfapyrimidines, U.S.P.) and in veterinary medicine (structure in Fig. 5-1).

Sulfacetamide, N-sulfanilylacetamide, N^1-acetyl-sulfanilamide. Plasma half-life = 7 hours. This com-pound is a white, crystalline powder, soluble in water (1:62.5 at 37°) and in alcohol. It is very soluble in hot water, and its water solution is acidic. It has a pKa of 5.4.

Sulfacetamide

Sulfachloropyridazine, N^1-(6-chloro-3-pyrida-zinyl)sulfanilamide. Plasma half-life = 8 hours.

Sulfachloropyridazine

Sulfapyridine U.S.P., N^1-2-pyridylsulfanilamide. Plasma half-life = 9 hours. This compound is a white, crystalline, odorless and tasteless substance. It is sta-ble in air but slowly darkens on exposure to light. It is soluble in water (1:3,500), in alcohol (1:440) and in

acetone (1:65) at 25°. It is freely soluble in dilute min-eral acids and aqueous solutions of sodium and potas-sium hydroxide. The pKa is 8.4. Its outstanding effect in curing pneumonia was first recognized by Whitby; however, because of its relatively high toxicity it has been supplanted largely by sulfadiazine and sulfamer-azine. A number of cases of kidney damage have re-sulted from acetylsulfapyridine crystals deposited in the kidneys. It also causes severe nausea in a majority of patients. Because of its toxicity it is only used for dermatitis herpetiformis.

Sulfapyridine

Sulfapyridine was the first drug to have an out-standing curative action on pneumonia. It gave impe-tus to the study of the whole class of N^1-heterocyclic-ally substituted derivatives of sulfanilamide.

Sulfamethoxazole U.S.P., Gantanol®, N^1-(5-methyl-3-isoxazolyl)sulfanilamide. Plasma half-life = 11 hours.

Sulfamethoxazole

Sulfamethoxazole is a sulfonamide drug closely re-lated to sulfisoxazole in chemical structure and anti-microbial activity. It occurs as a tasteless, odorless, almost white, crystalline power. The solubility of sul-famethoxazole at the pH range of 5.5 to 7.4 is slightly less than that of sulfisoxazole but greater than that of sulfadiazine, sulfamerazine, or sulfamethazine.

Following oral administration, sulfamethoxazole is not as completely or as rapidly absorbed as sulfisoxa-zole and its peak blood level is only about 50 percent as high.

Sulfadiazine U.S.P., N^1-2-pyrimidinylsulfanila-mide, 2-sulfanilamidopyrimidine. Plasma half-life =

Sulfadiazine

17 hours. Sulfadiazine is a white, odorless, crystalline powder soluble in water to the extent of 1:8,100 at 37°, 1:13,000 at 25°, in human serum to the extent of 1:620 at 37°, and sparingly soluble in alcohol and acetone. It is readily soluble in dilute mineral acids and bases. Its pKa is 6.3.

Sulfadiazine Sodium U.S.P., soluble sulfadiazine. This compound is an anhydrous, white, colorless, crystalline powder soluble in water (1:2) and slightly soluble in alcohol. Its water solutions are alkaline (pH 9-10) and absorb carbon dioxide from the air with precipitation of sulfadiazine. It is administered as a 5 percent solution in sterile water intravenously for patients requiring an immediate high blood level of the sulfonamide.

MIXED SULFONAMIDES

The danger of crystal formation in the kidneys on administration of sulfonamides has been reduced greatly through the use of the more soluble sulfonamides such as sulfisoxazole. This danger may be diminished still further by administering mixtures of sulfonamides. When several sulfonamides are administered together, the antibacterial action of the mixture is the summation of the activity of the total sulfonamide concentration present, but the solubilities are independent of the presence of similar compounds. Thus, by giving a mixture of sulfadiazine, sulfamerazine and sulfacetamide, the same therapeutic level can be maintained with much less danger of crystalluria, since only one third the amount of any one compound is present. Some of the mixtures employed are the following:

Trisulfapyrimidines Oral Suspension U.S.P. This mixture contains equal weights of Sulfadiazine U.S.P., Sulfamerazine U.S.P., and Sulfamethazine U.S.P., either with or without an agent to increase the pH of the urine.

Trisulfapyrimidines Tablets U.S.P. These tablets contain essentially equal quantities of sulfadiazine, sulfamerazine, and sulfamethazine.

Sulfacetamide, Sulfadiazine and Sulfamerazine Tablets, Buffonamide®, Cetazine®, Incorposal®. These are a mixture of equal weights of these sulfonamides either with or without an agent to increase the pH of the urine.

Sulfacetamide, Sulfadiazine and Sulfamerazine Oral Suspension. The suspension usually available contains 167 mg. of each sulfonamide in each 4 ml.

Sulfadiazine and Sulfamerazine Tablets, Duosulf®, Duozine®, Merdisul®, Sulfonamide Duplex. These are a mixture of equal weights of sulfadiazine and sulfamerazine either with or without an agent to increase the pH of the urine.

WELL-ABSORBED, LONG-ACTING SULFONAMIDES

In view of the well-known toxic and hypersensitivity problems of all sulfonamides, use of long-acting sulfonamides is rare except as part of mixed sulfonamide preparations. Many have been dropped from the United States market, and now only two remain—sulfamerazine (which some authors classify as "intermediate") and sulfameter.

Sulfamerazine U.S.P., N^1-(4-methyl-2-pyrimidinyl)sulfanilamide, 2-sulfanilamido-4-methylpyrimidine. Plasma half-life = 24 hours. This compound is a white, crystalline compound with a slightly bitter taste. It slowly darkens on exposure to light. It dissolves in water at 20° (1:6,250) and at 37° (1:3,300). It is readily soluble in dilute acids and bases, sparingly so in acetone, slightly soluble in alcohol, and very slightly soluble in chloroform and ether. Its pKa is 7.1.

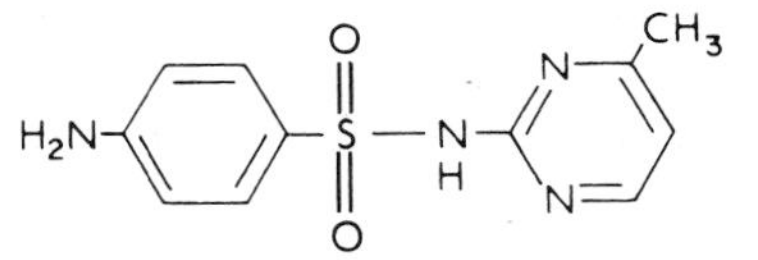

Sulfamerazine

It has a higher incidence of toxic reactions than sulfadiazine, including renal complications, drug fever and rashes.

Sulfameter, N^1-(5-methoxy-2-pyrimidinyl)sulfanilamide, is a long-acting sulfonamide. Plasma half-life = 48 hours. It occurs as a white, crystalline powder and is insoluble in water and slightly soluble in alcohol. It should be protected from light.

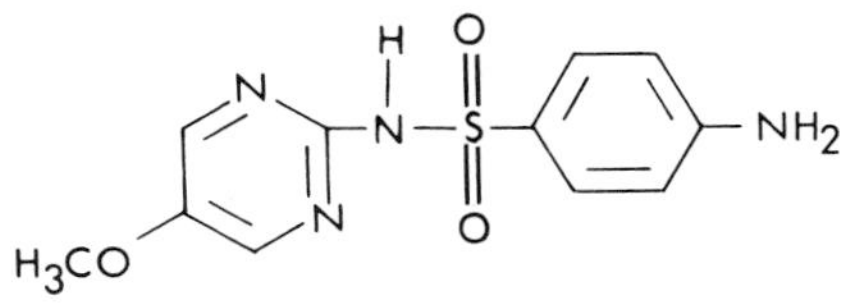

Sulfameter

Sulfameter is readily absorbed from the gastrointestinal tract. Following oral administration, measurable levels of the drug are reached in approximately 2 hours and peak serum levels occur within 4 to 8 hours. Limited data suggest that measurable amounts of the drug are still present in the plasma 96 hours after the

TABLE 5–2

SHORT- AND INTERMEDIATE-ACTING SULFONAMIDES

Name Proprietary Name	Preparations	Usual Adult Dose*	Usual Dose Range*	Usual Pediatric Dose*
Sulfamethizole U.S.P. *Thiosulfil, Utrasul*	Sulfamethizole Oral Suspension U.S.P. Sulfamethizole Tablets U.S.P.	500 mg. 4 times daily	500 mg. to 1 g. daily	
Sulfisoxazole U.S.P. *Gantrisin, Sulfalar*	Sulfisoxazole Tablets U.S.P.	2 to 4 g. initially, then 750 mg. to 1.5 g. every 4 hours; or 1 to 2 g. every 6 hours	2 to 12 g. daily	Use in infants under 1 month of age is not recommended; over 1 month—75 mg. per kg. of body weight or 2 g. per square meter of body surface initially, followed by 25 mg. per kg. or 667 mg. per square meter every 4 hours not exceeding 6 g. daily
Sulfisoxazole Acetyl U.S.P. *Gantrisin Acetyl, Lipo Gantrisin*	Sulfisoxazole Oral Suspension U.S.P.	The equivalent of 2 to 4 g. of sulfisoxazole initially, then 750 mg. to 1.5 g. every 4 hours; or 1 to 2 g. every 6 hours	2 to 12 g. daily	Use in infants under 1 month of age is not recommended; over 1 month—the equivalent of 75 mg. per kg. of body weight or 2 g. per square meter of body surface initially, followed by 25 mg. per kg. or 667 mg. per square meter, every 4 hours, not exceeding 6 g. daily
Sulfisoxazole Diolamine U.S.P. *Gantrisin Diethanolamine*	Sulfisoxazole Diolamine Injection U.S.P.	I.M. or I.V., 4 g. initially, then 1 to 2 g. every 4 to 6 hours		
Sulfamethazine U.S.P.	(see Mixed Sulfonamides)			
Sulfacetamide *Sulamyd*	Tablets	1 g. 3 times daily		Children over 2 months of age—60 mg. per kg. of body weight daily in 3 or 4 divided doses
Sulfapyridine U.S.P.	Sulfapyridine Tablets U.S.P.			
Sulfamethoxazole U.S.P. *Gantanol*	Sulfamethoxazole Oral Suspension U.S.P. Sulfamethoxazole Tablets U.S.P.		Initial, 2 g., then 1 g. 2 or 3 times daily	
Sulfadiazine U.S.P. *Coco-Diazine*	Sulfadiazine Tablets U.S.P.	2 to 4 g. initially, then 500 mg. to 1 g. 4 times daily	2 to 8 g. daily	Use in infants under 2 months of age is not recommended; over 2 months—75 mg. per kg. of body weight or 2 g. per square meter of body surface initially, followed by 37.5 mg. per kg. or 1 g. per square meter, 4 times daily, not exceeding 6 g. daily
Sulfadiazine Sodium U.S.P.	Sulfadiazine Sodium Injection U.S.P.	I.V. initially, 50 mg. per kg. of body weight or 1.125 g. per square meter of body surface as a 5 percent solution, then 25 mg. per kg. or 563 mg. per square meter, 4 times daily; S.C., initially, 50 mg. per kg. or 1.125 g. per square meter as a 5 percent solution, then 33 mg. per kg. or 750 mg. per square meter, 3 times daily		Use in infants under 2 months of age is not recommended; over 2 months—see Usual Dose

*See U.S.P. D.I. for complete dosage information.

TABLE 5–2 (Cont.)

SHORT- AND INTERMEDIATE-ACTING SULFONAMIDES

Name Proprietary Name	Preparations	Usual Adult Dose*	Usual Dose Range*	Usual Pediatric Dose*
MIXED SULFONAMIDES				
Sulfamerazine U.S.P. Sulfamethazine U.S.P. Sulfadiazine U.S.P. *Terfonyl, Sulfalose, Trisulfazine, Trionamide, Neotrizine, Sulfose, Truozine*	Trisulfapyrimidines Oral Suspension U.S.P. Trisulfapyrimidines Tablets U.S.P.	2 to 4 g. initially, then 500 mg. to 1 g. 4 times daily	2 to 7 g. daily	Use in infants under 2 months of age is not recommended; over 2 months—75 mg. per kg. of body weight or 2 g. per square meter of body surface initially, followed by 37.5 mg. per kg. or 1 g. per square meter, 4 times daily, not exceeding 6 g. daily
Sulfacetamide U.S.P. Sulfadiazine U.S.P. Sulfamerazine U.S.P.	Sulfacetamide, Sulfadiazine, and Sulfamerazine Tablets	4 g. initially, then 500 mg. to 1 g. every 4 hours		
	Sulfacetamide, Sulfadiazine, and Sulfamerazine Oral Suspension	4 g. initially then 500 mg. to 1 g. every 4 hours		
Sulfadiazine U.S.P. Sulfamerazine, U.S.P.	Sulfadiazine and Sulfamerazine Tablets	4 g. initially then 2 g. every 4 hours		

*See *U.S.P. D.I.* for complete dosage information.

drug is administered; approximately 90 percent of the sulfonamide in the plasma is nonacetylated.

Sulfameter is administered orally as a single daily dose, preferably after breakfast.

SULFONAMIDES FOR OPHTHALMIC INFECTIONS

Sulfacetamide Sodium U.S.P., Sodium Sulamyd®, N-sulfanilylacetamide monosodium salt.

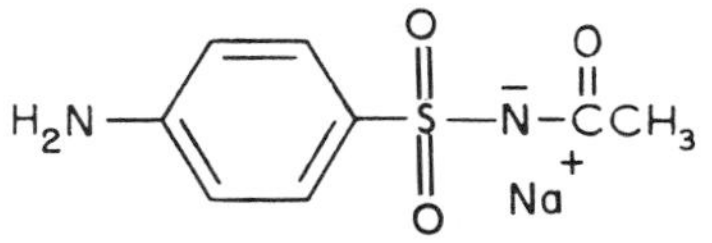

This compound is obtained as the monohydrate and is a white, odorless, bitter, crystalline powder which is very soluble (1:2.5) in water. Because the so-

Sulfacetamide Sodium

dium salt is highly soluble at the physiologic pH of 7.4, it is especially suited, as a solution, for repeated topical applications in the local management of ophthalmic infections susceptible to sulfonamide therapy.

Sulfisoxazole Diolamine U.S.P. Also used in I.V. and I.M. preparations, this salt of sulfisoxazole was described along with the short and intermediate acting sulfonamides.

TABLE 5–3

LONG-ACTING SULFONAMIDES

Name Proprietary Name	Preparation	Usual Adult Dose*
Sulfomerazine U.S.P.	Sulfamerazine Tablets, U.S.P.	Initial 4 g. of sulfamerazine; maintenance, 1 g. of sulfamerazine every 6 hours
Sulfameter *Sulla*	Sulfameter Tablets	For adults weighing more than 45 kg.—1.5 g. the first day, followed by 500 mg. daily thereafter

*See *U.S.P. D.I.* for complete dosage information.

SULFONAMIDES FOR BURN THERAPY

Mafenide Acetate U.S.P., Sulfamylon®, *p*-aminomethylbenzenesulfonamide acetate. This compound is a homolog of the sulfanilamide molecule. It is not a true sulfanilamide-type compound, as it is not inhibited by *p*-aminobenzoic acid. Its antibacterial action involves a mechanism that is different from that of true sulfanilamide-type compounds. This compound is particularly effective against *Clostridium welchii* in topical application and was used during World War II by the German army for prophylaxis of wounds. It is not effective by mouth. It is employed currently alone

TABLE 5–4

SULFONAMIDES FOR OPHTHALMIC INFECTIONS

Name Proprietary Name	Preparations	Application*
Sulfacetamide Sodium U.S.P. *Blefcon, Bleph-10 Liquifilm, Bufopto Sulfacel-15, Cetamide, Isopto Cetamide, Sodium Sulamyd, Sulf-30*	Sulfacetamide Sodium Ophthalmic Ointment U.S.P.	Topically to the conjunctiva, as a 10 to 30 percent ointment 5 times daily
	Sulfacetamide Sodium Ophthalmic Solution U.S.P.	Topically to the conjunctiva, 0.05 to 0.1 ml. of a 10 to 30 percent solution 6 to 12 times daily
Sulfisoxazole Diolamine U.S.P. *Gantrisin Diethanolamine*	Sulfisoxazole Diolamine Ophthalmic Ointment U.S.P. Sulfisoxasole Diolamine Ophthalmic Solution U.S.P.	Topical, to the conjunctiva, 4 percent ointment or solution

*See *U.S.P. D.I.* for complete dosage information.

or with antibiotics in the treatment of slow-healing infected wounds.

Some patients treated for burns with large quantities of this drug developed metabolic acidosis. In order to overcome this side-effect a series of new organic salts was prepared.[15] The acetate in an ointment base proved to be the most efficacious.

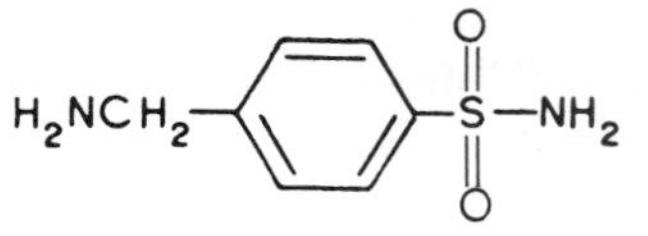

Mafenide

Silver Sulfadiazine, Silvadene®. The silver salt of sulfadiazine applied in a water-miscible cream base has proved to be an effective topical antimicrobial agent, especially against *Pseudomonas sp.* This is of particular significance in burn therapy because Pseudomonas is often responsible for failures in therapy.

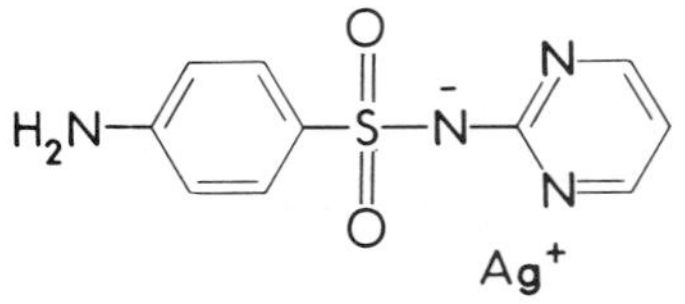

Silver Sulfadiazine

TABLE 5–5

SULFONAMIDES FOR BURN THERAPY

Name Proprietary Name	Preparations	Application*
Mafenide Acetate U.S.P. *Sulfamylon*	Mafenide Acetate Cream U.S.P.	Topically to the skin, as the equivalent of an 8.5 percent cream of mafenide, 1 or 2 times daily in a 2-mm. thickness, repeated whenever necessary to keep affected areas covered at all times
Silver Sulfadiazine *Silvadene*	Silver Sulfadiazine Cream	Topical to burns by sterile application

*See *U.S.P. D.I.* for complete dosage information.

The salt is only very slightly soluble and does not penetrate the cell wall but acts on the external cell structure. Studies using radioactive silver have shown essentially no absorption into body fluids. Sulfadiazine levels in the serum were found to be of the order of 0.5 to 2 mg. percent.

This preparation is reported to be simpler and easier to use than other standard burn treatments such as application of freshly prepared dilute silver nitrate solutions or mafenide ointment.

SULFONAMIDES FOR INTESTINAL INFECTIONS, ULCERATIVE COLITIS, OR REDUCTION OF BOWEL FLORA

Each of the sulfonamides in this group is a prodrug, which is designed to be poorly absorbable, although in practice a little usually is absorbed. Therefore, usual precautions with sulfonamide therapy should be observed. In the large intestine the N^4-protecting groups are cleaved, releasing the free sulfonamide antibacterial agent.

Phthalylsulfathiazole U.S.P., 4′-(2-thiazolylsulfamoyl)phthalanilic acid, 2-(N^4-phthalylsulfanilamido) thiazole. This compound is an odorless, white, crystalline powder with a slightly bitter taste. It slowly darkens on exposure to light. It is insoluble in water and chloroform and slightly soluble in alcohol. It is readily

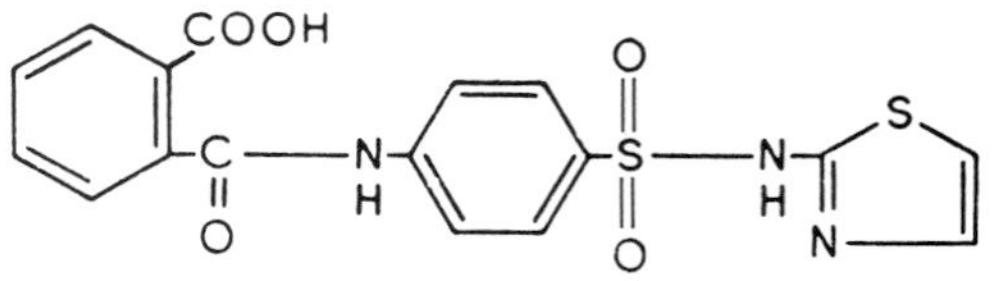

Phthalylsulfathiazole

soluble in strong acids and bases and liberates carbon dioxide from a solution of sodium bicarbonate.

This compound is poorly absorbed from the intestinal tract and has properties similar to those of its succinic acid analog, although it is considered to be somewhat more potent.

Phthalylsulfacetamide, 4′-(acetylsulfamoyl)-phthalanilic acid, N^1-acetyl-N^4-phthaloylsulfanilamide. This compound occurs as a white, crystalline solid that is very sparingly soluble in water. It possesses the property of diffusing into the intestinal wall but is absorbed into the bloodstream in amounts too small to give a systemic effect.

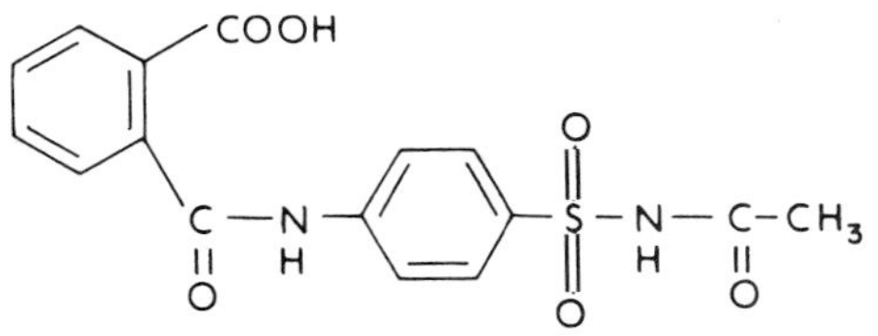

Phthalylsulfacetamide

Its main use is as an intestinal antibacterial agent in gastrointestinal infections and preoperative "sterilization" of the gastrointestinal tract. It may be used after abdominal surgery.

Sulfasalazine U.S.P., 5-[p-(2-pyridylsulfamoyl)-phenylazo]salicylic acid. This compound is a brownish-yellow, odorless powder, slightly soluble in alcohol but practically insoluble in water, ether and benzene.

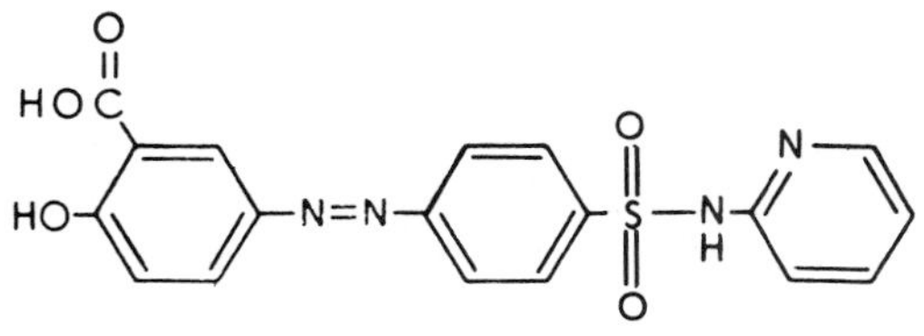

It is broken down in the body to m-aminosalicylic acid and sulfapyridine. The drug is excreted through the kidneys and is colorimetrically detectable in the urine, producing an orange-yellow color when the urine is alkaline and no color when the urine is acid.

FOLATE REDUCTASE INHIBITORS

Trimethoprim U.S.P., 2,4-diamino-5-(3,4,5-trimethoxybenzyl)pyrimidine. Trimethoprim is closely related to a number of antimalarials, but it does not have good antimalarial activity by itself. However it is a potent antibacterial. Originally introduced in combination with sulfamethoxazole, it is now available as a single agent. Approved by the FDA on May 30, 1980, trimethoprim as a single agent is used only for the treatment of uncomplicated urinary tract infections.

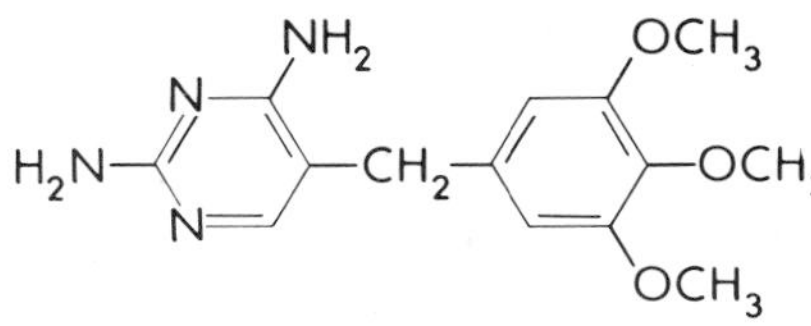

Trimethoprim

Sulfasalazine

The case for trimethoprim to be a single agent was summarized in 1979 by Wormser and Deutsch.[15] They point out for example that several studies comparing trimethoprim with trimethoprim–sulfamethoxazole for treatment of chronic urinary tract infections found no statistically relevant difference between the two treatments. Furthermore, some patients cannot take sulfonamide products for the reasons discussed previ-

TABLE 5–6

SULFONAMIDES FOR INTESTINAL INFECTIONS, ULCERATIVE COLITIS, OR REDUCTION OF BOWEL FLORA

Name Proprietary Name	Preparations	Usual Adult Dose*	Usual Dose Range*

Phthalylsulfathiazole U.S.P.
  Sulfathalidine, Cremo-thalidine
  Rothalid
	Phthalylsulfathiazole Tablets U.S.P.	1 g. every 4 hours	4 to 12 g. daily

Phthalylsulfacetamide *Enterosulfon*	Phthalylsulfacetamide Tablets	2 g. 3 times daily	1.5 to 4 g. daily
Sulfasalazine U.S.P.	Sulfasalazine Tablets U.S.P.	Initial, 4 to 8 g. daily; maintenance, 500 mg. 4 times daily 10 ml. of a 10 percent suspension as a retention enema after each stool and at bedtime	

*See *U.S.P. D.I.* for complete dosage information.

ously in this chapter, especially hypersensitivity. In contrast to these and similar arguments, the concern is that, when used as a single agent, bacteria now susceptible to trimethoprim will rapidly develop resistance. However, in combination with a sulfonamide, the bacteria will be less likely to do so. That is, they won't survive long enough to easily develop resistance to both drugs.

Trimethoprim–Sulfonamide Combinations

Sulfamethoxazole and Trimethoprim. The synergistic action of the combination of these two drugs has been previously discussed in this chapter. The combination is available only as 160 mg. of trimethoprim and 800 mg. of sulfamethoxazole. The combination has been widely used for urinary tract infections as well as for acute otitis media, meningococcal infections, chronic bacterial prostatitis, and other bacterial infections. The sulfamethoxazole–trimethoprim literature is extensive, with hundreds of publications related to their use, pharmacokinetics, and chemistry by mid-1980 alone. Several excellent reviews have been published, which are recommended for further reading.

SULFONES

The sulfones are primarily of interest as antibacterial agents, although there are some reports of their use in the treatment of malarial and rickettsial infections. They are less effective agents than are the sulfonamides. *p*-Aminobenzoic acid partially antagonizes the action of many of the sulfones, suggesting that the mechanism of action is similar to that of the sulfona-

mides. It also has been observed that infections which arise in patients being treated with sulfones are cross-resistant to sulfonamides. Some sulfones have found use in the treatment of leprosy.

The search for antileprotic drugs has been hampered by the inability to cultivate *Mycobacterium leprae* on artificial media and by the lack of experimental animals susceptible to human leprosy. Recently a method of isolating and growing *M. leprae* in the foot pads of mice has been reported and may allow for the screening of possible antileprotic agents. Sulfones were introduced into the treatment of leprosy after it was found that sodium glucosulfone was effective in experimental tuberculosis in guinea pigs.

Dapsone

The parent sulfone in the clinically useful area is dapsone (4,4'-sulfonyldianiline). Four types of variations on this structure have given useful compounds.

1. Substitution on both the 4 and 4' amino- functions
2. Monosubstitution on only one of the amino- functions
3. Nuclear substitution on one of the benzenoid rings
4. Replacement of one of the phenyl rings with a heterocyclic ring

The antibacterial activity and the toxicity of the disubstituted sulfones are thought to be due chiefly to the formation in vivo of dapsone. Hydrolysis of disubstituted derivatives to the parent sulfone apparently occurs readily in the acid medium of the stomach, but

TABLE 5–7

FOLATE REDUCTASE INHIBITORS

Name Proprietary Name	Preparations	Usual Adult Dose*	Usual Pediatric Dose*
Trimethoprim U.S.P. *Proloprim Trimpex*		100 mg. two times a day for 10 days. Approved (May 30, 1980) only for uncomplicated urinary-tract infections	
	Sulfamethoxazole and Trimethoprim Tablets U.S.P. *Bactrina, Septra, Bactrim-DS, Septra-DS*	Two DS tablets a day—each tablet containing 160 mg. of trimethoprim and 800 mg. of sulfamethoxazole—for 10 to 14 days	8 mg. per kg. per day of trimethoprim and 40 mg. per kg. per day of sulfamethoxazole in two divided doses per day for 10 to 14 days. The oral suspension contains 40 mg. of trimethoprim and 200 mg. of sulfamethoxazole per 5 ml.
	Sulfamethoxazole and Trimethoprim Oral Suspension U.S.P.		

*See U.S.P. D.I. for complete dosage information.

TABLE 5–8

LEPROSTATIC SULFONES

Name Proprietary Name	Preparations	Usual Adult Dose†	Usual Dose Range†	Usual Pediatric Dose†
Dapsone* U.S.P. *Avlosulfon*	Dapsone Tablets U.S.P.	25 mg. 2 times a week for 1 month, then increased by 25 mg. per dose at monthly intervals to a maximum of 100 mg. 4 times a week		6 to 12.5 mg. 2 times a week for 1 month, then increased by 6 to 12.5 mg. per dose at monthly intervals to a maximum of 50 mg. 4 times a week
Sulfoxone Sodium U.S.P. *Diasone Sodium*	Sulfoxone Sodium Tablets U.S.P.	300 mg. 1 or 2 times daily	300 mg. to 1 g. daily	
Glucosulfone Sodium *Promin*	Glucosulfone Sodium Injection	I.V., 2 g. daily for 6 days of each week	2 to 5 g.	
Acetosulfone Sodium** *Promacetin*	Acetosulfone Sodium Tablets	500 mg. daily for the first 2 weeks; thereafter increase every 2 weeks by increments of 500 mg. to 1.5 g. until a maximal daily dose of 3 to 4 g. is reached		Children—7.1 mg. per kg. of body weight administered in the same schedule as for adults

*Additional use, dermatitis herpetiformis suppressant, 100 to 200 mg. once daily; usual dose range—50 to 400 mg. daily.
**Additional use, dermatitis herpetiformis suppressant.
†See U.S.P. D.I. for complete dosage information.

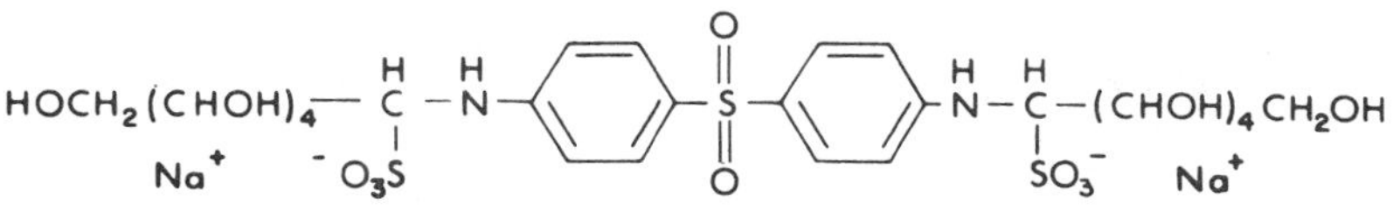

Glucosulfone Sodium

only to a very limited extent following parenteral administration. Mono-substituted and nuclear-substituted derivatives are believed to act as entire molecules.

PRODUCTS

Dapsone U.S.P., Avlosulfon®, DDS, 4,4′-sulfonyldianiline, *p,p*′-diaminodiphenyl sulfone. This compound occurs as an odorless, white crystalline powder which is very slightly soluble in water and sparingly soluble in alcohol. The pure compound is light-stable, but the presence of traces of impurities, including water, makes it photosensitive and thus susceptible to discoloration in light. Although no chemical change is detectable following discoloration, the drug should be protected from light.

Dapsone is used in the treatment of both lepromatous and tuberculoid types of leprosy.

Sulfoxone Sodium U.S.P., Diasone® Sodium, disodium [sulfonylbis(*p*-phenyleneimino)]dimethanesul-

Sodium Sulfoxone

finate. This compound is a white to pale yellow powder with a characteristic odor. It is slightly soluble in alcohol and very soluble in water. It is affected by light.

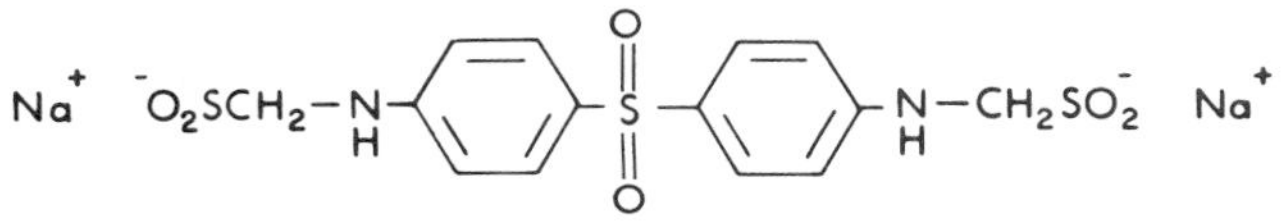

Sulfoxone sodium is used in the treatment of leprosy. Lesions usually do not progress under therapy, although not all respond favorably.

Acetosulfone Sodium, Promacetin®, N-(6-sulfanilylmetanilyl)acetamide sodium derivative. This agent is used in the treatment of both lepromatous

Acetosulfone Sodium

and tuberculoid types of leprosy. It is also effective in controlling the symptoms of dermatitis herpetiformis.

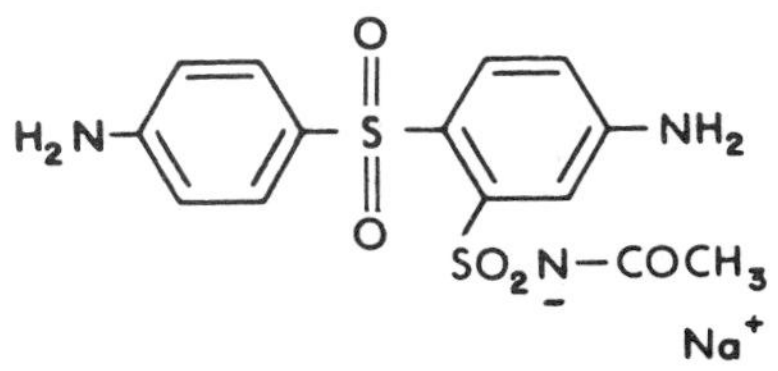

REFERENCES

1. Domagk, G.: Deut. Med. Wochenschr. 61:250, 1935.
2. Baumler, E.: In Search of the Magic Bullet, Thames and Hudson, London, 1965.

3. Anand, N.: Sulfonamides and Sulfones, Chapter 13 *in* Wolff, M. E. (ed.): Burger's Medicinal Chemistry, Part II, ed. 4, New York. Wiley, 1979.
4. Forester, J.: Z. Haut. Geschlechtsk, 45:459, 1933.
5. Trefouel, J., et al.: C. R. Seanc. Soc. Biol. 120:756, 1935.
6. Fuller, A. T.: Lancet 1:194, 1937.
7. Northey, E. H.: The Sulfonamides and Allied Compounds, American Chemical Society Monograph Series, American Chemical Society, Washington, D.C., 1948.
8. The Hospital Formulary, American Society of Hospital Pharmacists, 1980.
9. Weinstein, L., *in* Goodman, L. S. and Gilman, A. (eds.): The Pharmacological Basis of Therapeutics, ed. 6, New York, Macmillan, 1980.
10. Krupp, M. A., and Chatton, M. J.: Current Medical Diagnosis and Treatment, Lange Medical Publications, Los Altos, Calif., 1980.
11. AMA Drug Evaluations, Third Edition, American Medical Association, Publishing Sciences Group, Littelton, Mass., 1977.
12. FDA Drug Bulletin, U.S. Department of Health, Education and Welfare, Food and Drug Administration, February, 1980.
13. The Medical Letter 21:42, 1979.
14. Werlin, S. L., and Grand, R. J.: J. Pediatrics 92:450, 1978.
15. Wormser, G. P., and Deutsch, G. T.: Ann. Internal Med. 91:420, 1979.
16. Gleckman, R., et al.: Am. J. Hosp. Pharm. 36:893, 1979.
17. Bushby, S. R., and Hitchings, G. H.: Brit. J. Pharmacol. Chemotherap. 33:72, 1968.
18. Letters of Burchall, J.J., Then, R., and Poe, M.: Science 197:1300–1301, 1977.
19. Palminteri, R., and Sassella, D.: Chemotherapy 25:181, 1979.
20. Harvey, R. J.: J. Antimicrob. Chemotherap. 4:315, 1978.
21. White, P. J., and Woods D. D.: J. Gen. Microbiol, 40:243, 1965.
22. Watanabe, T.: Bacteriol. Rev. 27:87, 1963.
23. Fujita, T., and Hansch, C.: J. Med. Chem. 10:991, 1967.
24. Zbinden, G. *in* Gould, R. F. (ed.); Molecular Modification in Drug Design, American Chemical Society, Washington, D.C., 1964.
25. Fujita, T., *In* Gould. R.F. (ed.): Molecular Modification in Drug Design, American Chemical Society, Washington, D.C., 1964.
26. Vree, T.B., *et al.:* Clin. Pharmacokin. 4: 310, 1979.
27. Hansen, I.: Antibiotics Chemother. 25:217, 1978.
28. Fox, C. L. and Ross, H. M.: Proc. Soc. Exp. Biol. Med. 50:142, 1942.
29. Yamazaki, M., *et al.:* Chem. Pharm. Bull. 18:702, 1970.
30. Bell, R. H., and Roblin, R. O.: J. Am. Chem. Soc. 64:2905, 1942.
31. Cowles, P. B.: Yale J. Biol. Med. 14:599, 1942.
32. Brueckner, A. H.: J. Biol. Med. 15:813. 1943

SELECTED READINGS

Anand, N.: Sulfonamides and Sulfones, Chapter 13 *in* Wolff, M. E. (ed.); Burger's Medicinal Chemistry, Part II," ed. 4, New York, Wiley, 1979.

Baumler, E.: In Search of the Magic Bullet, London, Thames and Hudson, 1965.

Northey, E. H.: The Sulfonamides and Allied Compounds, American Chemical Society Monograph Series, American Chemical Society, Washington, D.C., 1948.

6

Dwight S. Fullerton

antimalarials

"Few people realize that there are far more kinds of parasitic than nonparasitic organisms in the world. Even if we exclude viruses, rickettsias . . . and the many kinds of parasitic bacteria and fungi, the parasites are still in the majority. The parasitic way of life . . . is a highly successful one. Humans are hosts to over 100 kinds of parasites, again not counting viruses, bacteria and fungi. . . . At least 45,000 species of protozoa have been described to date, many of which are parasitic. Parasitic protozoa still kill, mutilate, and debilitate more people in the world than any other group of disease organisms. . . .

G.D. Schmidt and L.S. Roberts
"Foundations of Parasitology"[1]

Malaria, African sleeping sickness, leishmaniasis, Chagas' disease, and other protozoal disease (Table 6-1) of humans and their livestock continue to have a devastating impact worldwide.[1-15] Over 12 million South Americans suffer from Chagas' disease alone,[1] and millions in Africa and the Mediterranean areas from African sleeping sickness or the disfigurement of leishmaniasis.[8,13]

However, none of these protozoal diseases has had the enormous effect upon civilization either historically or in modern times as has malaria. One million African children alone still die each year from the disease. Loss of productivity from the debilitating and cyclic clinical stages of malaria is enormous. It has been noted by Schmidt and Roberts[1] that a single day of malaria fever requires the caloric equivalent of two days of hard labor (and thus of food). With malaria protozoa in some areas becoming resistant to commonly used antimalarial drugs,[10,14] and with insecticide-related problems increasing (resistance by the

mosquito vector, human and environmental harm), the adverse impact of malaria upon the world is likely to continue.

Although some protozal infections of farm animals (e.g., coccidiosis in chickens) have been at times a serious problem in the United States, and *Trichomonas vaginitis* in humans is common, malaria and other life-threatening protozoal infections are not. However, the ease of international travel has caused increased awareness of the prevention and treatment of protozoal infections by American physicians and pharmacists. During the Viet Nam war, of course, several thousand cases of malaria were reported in the United States—largely attributed to returning servicemen.

The epidemiology, diagnosis, microbiology, medicinal chemistry, and chemotherapy of malaria and other parasitic disease have recently been reviewed.[1-15] Current approaches to prevention and treatment of parasitic infections including malaria have been summarized[1,2,6-8,13] and drugs of choice and doses given in a 1979 *Medical Letter*.[15]

ETIOLOGY

Malaria in humans is caused by four species of *Plasmodium* protozoa, which, as shown in Figure 6-1 spend half their life cycle in female *Anopheles* mosquitos. (Male *Anopheles* mosquitos do not feed on vertebrate blood.) Several hundred *Anopheles* species are known, several of which are commonly found in the United States. Resistance to DDT, dieldrin, and other insecti-

TABLE 6-1

MALARIA AND OTHER COMMON PROTOZOAL INFECTIONS IN HUMANS AND FARM ANIMALS

Disease	Protozoa	Insect Vector	Primary Occurrence	Clinical Notes
Malaria	*Plasmodium vivax*, others	Mosquitoes	Tropical	High fever and chills, cyclical
African Sleeping Sickness	*Trypanosoma rhodesiense, gambiense*, others	Tsetse flies	Tropical Africa	*Rhodesiense* usually causes death before CNS depression, but commonly seen with *gambiense*
Chagas' Disease	*Trypanosoma cruzi*	Common "bedbug"	South America	Bug usually bites victim close to mouth, so is called the "kissing bug." Protozoa invade many tissues including the heart. Disfiguring edema
Leishmaniasis (Kala-azar)	*Leishmania donovani*, others	Sand fly	Middle East, tropical Africa, tropical South America	Progressive wasting and anemia, severely enlarged spleen and liver
Amoebiasis (amoebic dysentery)	*Entamoeba histolytica*, vaginalis	None—transmitted by human and animal wastes	Tropical regions	
Trichomonal vaginitis	*Trichomonas vaginalis*	None—usually transmitted sexually	Worldwide	Can be serious in women and men (see discussion by Kreier).[2]
Coccidiosis in farm animals	*Eimeria* species	None—usually by animal wastes	Worldwide	Great economic losses even in USA
Toxoplasmosis	*Toxoplasma* species	None—usually by contact with infected cats	Worldwide	
Babesiasis in cattle	*Babesia* species	Ticks	Worldwide	

cides has been reported for an increasing number of *Anopheles* species—making malaria control more difficult. Once infected, the mosquito carries sporozoites for life.

DDT

Dieldrin

As can be seen in the simplified life-cycle illustrated in Figure 6-1, the malaria protozoa undergo a number of morphological changes in the human host. (Detailed descriptions of these changes are contained in several texts.[1,3]) The rates at which these changes take place vary among the *Plasmodium* species. The patient generally has no adverse symptoms until erythrocytes rupture (generally 1½ to 2 weeks after the initial bite), releasing antigenic cell residues and protozoal wastes—leading to (1) a recurring (every 3 to 4 days) attack of nausea, vomiting, severe chills, delirium, and high fevers (38–40°C); (2) severe anemia (and thrombocytopenia) from hemolysis of erythrocytes as more and more merozoites are produced; and (3) jaundice from excess bilirubin (a metabolite of hemoglobin) production.

The four *Plasmodium* species are as follows:

P. falciparum ("malignant tertian, subtertian"). Of the four, only *P. falciparum* does not have a secondary schizont (secondary exo-erythrocytic) stage. However, it is the most lethal. Enormously high concentrations of the protozoa in the hosts' blood are often found—with over 65 percent[1] of the erythrocytes affected in some cases. Unlike malaria from the other three species, the patient may feel quite ill in between acute attacks. Because all the merozoites are released from the primary schizont stage at the same time, reinfection from a secondary schizont stage is not a problem. About half of all malaria is caused by this species.

P. vivax ("benign tertian malaria"). This form of malaria is called tertian because clinical symptoms usually recur every 48 hours (i.e., if the patient is sick on day 1, sickness will occur again on day 3, the *tertiary* day). Not as many merozoites are produced as with *P. falciparum*, but many reenter new liver cells to form secondary schizonts, which can cause relapses for several years. Drugs specific for this stage (site 3) must be given for the patient to be truly cured. Only young erythrocytes are attacked, thus limiting the total erythrocytic involvement. About 40 percent of all malaria is caused by *P. vivax*.

P. malariae ("Quartan malaria"). This species has a life-cycle like that of *P. vivax*. Relapses may occur for decades.

FEMALE ANOPHELES MOSQUITO   HUMAN

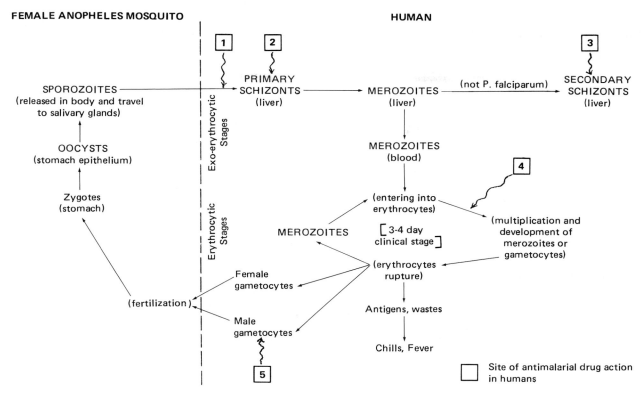

FIG. 6–1. *Life cycle of Plasmodium: (1) sporozoitocide—no drugs effective; (2) primary (exo-erythrocytic) schizonticide—primaquine, pyrimethamine, chloroguanide, cycloguanil pamoate; (3) secondary (exo-erythrocytic) schizonticide—primaquine; (4) erythrocytic schizonticide (fast-acting—chloroquine, quinine, amodiaquine; slow-acting—chloroguanide, pyrimethamine, sulfonamides, cycloguanil pamoate); (5) gametocytocide—primaquine.*

P. ovale ("mild tertian malaria"). *P. ovale* is the least common of the four types of malaria, and, like *P. vivax* and *P. malariae,* has a long-lasting secondary schizont stage.

BIOCHEMICAL DEPENDENCE ON HOST ERYTHROCYTES

Malaria protozoa are very species specific, some infecting only birds, and others only humans. Only *Anopholes* mosquitos harbor the protozoa in half their life cycles—none of the several thousand other mosquitos of other genera. This specificity reflects specialized biochemical dependence upon vertebrate host and vector host. The biochemistry of this specialized metabolic and biochemical dependence has been intensively studied,[5,6] especially with the goal of discovering biochemical (drug-attackable) vulnerability of the malaria protozoa.

Most research has focused on the biochemical dependence of malaria protozoa on erythrocytes. The reasons are primarily that parasite–erythrocyte biochemistry can be studied with fewer technical problems than that of other tissues, and drug treatment and study at this stage is direct (it's easy to get drugs into blood, to obtain blood samples, etc.).

Malaria protozoa need host erythrocytes to make (replicate) their own DNA and RNA, needed for the rapid production of large numbers of merozoites. Although the malaria parasites can synthesize their own pyrimidines (cytosine, uracil, thymine), they must obtain purines (adenine and guanine) from erythrocytes of the host. Phosphate is also used from the host.

Host hemoglobin and plasma have been found to be digested by the malaria protozoa and used as sources of several amino acids. The protozoa also can synthesize some amino acids of their own. (It is not yet certain which amino acids must be obtained from the host.) It also appears that the malaria protozoa are dependent upon the host as a source of pentoses (for DNA and RNA synthesis)—that they do not have an operable pentose phosphate pathway.

HISTORY

EARLY CINCHONA USE

The general antifebrile properties of the bark of the cinchona undoubtedly were known to the Incas before the arrival of the Spaniards early in the 16th century. However, it was probably the observations of early Jesuit missionaries that led to the discovery that infusions of cinchona bark were effective for the treatment of the tertian "ague" that was common in tropical Central and South America even then, and to the introduction of the crude drug into Western Europe. The first recorded use in South America was about 1630, and, in Europe, in 1639. Thus began the first era in the chemotherapy of malaria. Cinchona and the purified alkaloids obtained from it were to remain the only drugs of significance in the treatment of malaria for three centuries.

PAUL EHRLICH AND THE FIRST SYNTHETIC ANTIMALARIALS

Much has been written about Paul Ehrlich (1854–1915), who can be considered the founder of modern medicinal chemistry, chemotherapy, and molecular pharmacology.[16–19] (Films about his life are also available for classroom use.[19]) His Nobel laureate for medicine in 1908 (with Ilya Mechnikov), however, was for his research on immunology and the development of diptheria and other antitoxins.

In the late 1800's, the German dye industry was producing hundreds of new dyes. Ehrlich's first work on the biological properties of these dyes was on blood cells, demonstrating the specific staining of parts of certain leucocytes—the basis of modern hematology. Could pathogenic microorganisms also be specifically stained and perhaps killed without harm to the host ("magic bullet" concept)? Ehrlich's laboratory set about to find out, focusing first on malaria protozoa using methylene blue and related dyes. (Methylene blue, synthesized by Hoechst in 1885, was known to stain nerve tissues selectively without toxic effect.) For five years, Ehrlich and his associates studied hundreds of dyes—many of which they synthesized—leading to the first synthetic antimalarial, trypan red (Fig. 6-2) in 1904. However, trypan red did not have the antimalarial potency needed for an effective human cure. Perhaps a potent poison such as arsenic could be combined with an effective protozoal stain to carry the poison selectively to the malaria protozoa:

one part of the drug for good affinity, one part of the drug for good intrinsic activity. Ehrlich had simultaneously conceived the basis of modern drug design and chemotherapy of the live receptor (see the discussion by Albert[18]).

Atoxyl was Ehrlich's first antimalarial success using the concept, but it was still too neurotoxic to the host. In 1906, he began focusing on spirochetes, specifically on *S. pallida,* identified in 1905 as the cause of syphilis. His organo-arsenicals seemed to be as toxic to these spirochetes as to malaria protozoa, and syphilis was a far greater cause of death in Europe than was malaria. In 1909, his #606 (marketed in 1910 as Salvarsan) was the world's first successful antisyphilitic drug.[19] Ehrlich's neosalvarsan with greater water solubility (permitting larger IV doses) and less toxicity followed in 1912. In the subsequent five years, the incidence of syphilis dropped by 50 to 80 percent from its pre-1910 levels in Europe.

Erhlich's early success with drug design and chemotherapy with organic dyes and organo-arsenicals prompted an explosion of research in drug design and chemotherapy: Suramin in 1917 (still an alternative drug of choice[15] for African sleeping sickness), the antimalarial pamaquin in 1926 (Table 6-2), and the first sulfonamide antimalarial–antibacterial, prontosil, in 1932. The development of other clinically useful organo-arsenic, and antimony antiprotozoal agents is also a direct outcome of Ehrlich's drug design success with Salvarsan.

PROTOZOAL RESISTANCE AND STIMULATION OF RESEARCH BY WAR

Following the pioneering work of Ehrlich and his immediate successors in antimalarial drug design, the necessities of war provided the greatest stimulus to the development of new antimalarials. During 1941 to 1946, for example, over 15,000 substances were synthesized and screened as possible antimalarial agents by the United States, Australia, and Great Britain. Activity increased again during the Vietnam war, especially because of the increasing problem of resistance to commonly use antimalarials. The recent review of Van den Bossche[13] notes that, during the decade 1968–1978, over 250,000 compounds were investigated as part of a U.S. Army search program.

MODERN MALARIA CHEMOTHERAPY— AN OVERVIEW

Most drugs used in modern malaria chemotherapy (Table 6-2)—chloroquine, amodiaquine, pyrimethamine, quinine, sulfonamides and pyrimethamine—act

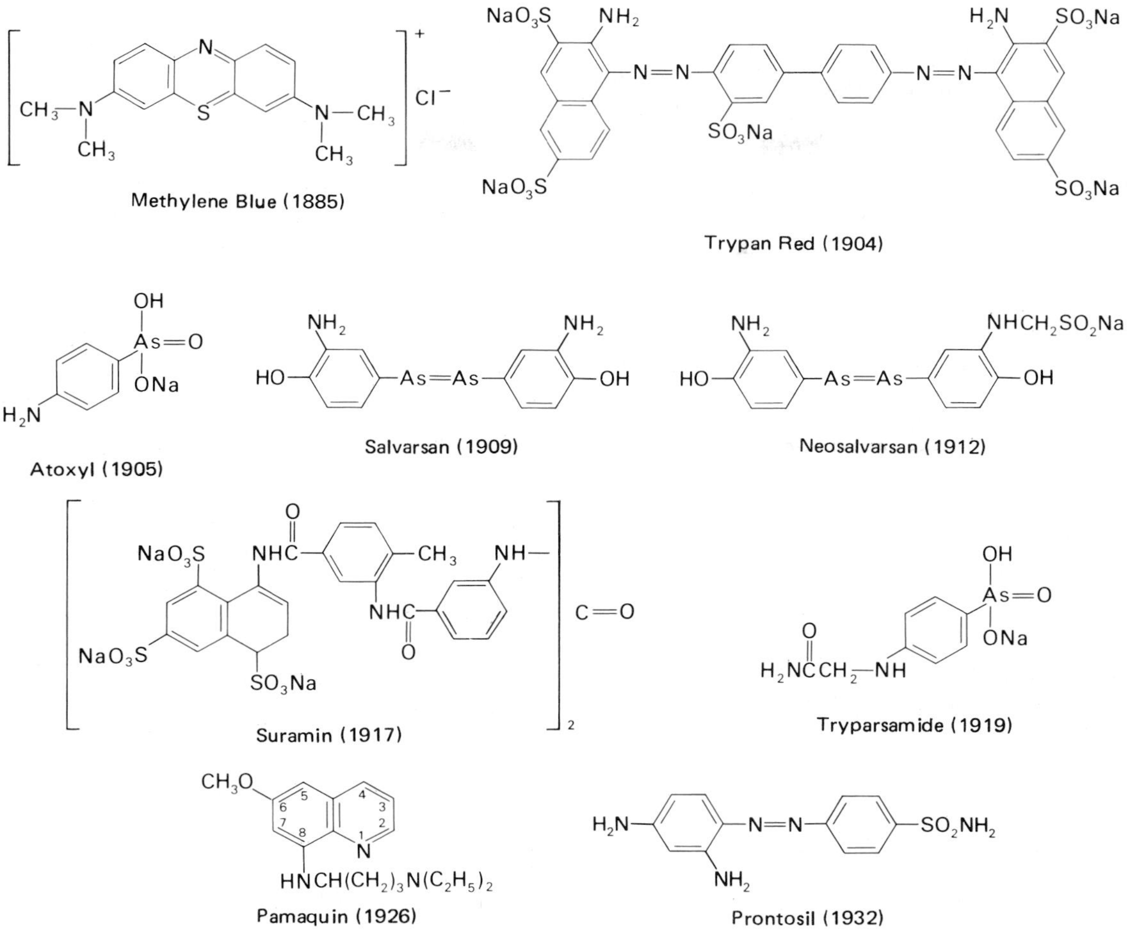

FIG. 6–2. *Antimalarial and related drugs of Paul Ehrlich and other early medicinal chemists.*

TABLE 6–2

DRUGS OF CHOICE FOR HUMAN MALARIA

Malaria Species	Therapeutic Goal	Drug of Choice	Alternative
	Treatment of Species Which Are Not Drug-Resistant		
All	Suppression while visiting endemic area (continue for 6 weeks after leaving)	Chloroquine phosphate	Amodiaquine hydrochloride
All except *P. falciparum**	Elimination of Secondary Schizonts (take for 2 weeks—the last two weeks of chloroquine/amodiaquine therapy above)	Primaquine phosphate	
All	Treatment of uncomplicated attack (clinical stage)	Chloroquine phosphate	Amodiaquine hydrochloride
All	Treatment of severe attack where patient cannot take an oral dose	Chloroquine hydrochloride	Quinine dihydrochloride
	Treatment of a Drug-Resistant Species, e.g., Chloroquine-Resistant *P. falciparum*		
P. falciparum	Suppression while visiting endemic area	Pyrimethamine plus sulfadoxine	
P. falciparum	Treatment of uncomplicated attack (clinical stage)	Quinine sulfate and pyrimethamine or sulfadiazine	Quinine sulfate plus tetracycline
P. falciparum	Treatment of severe illness—where patient cannot take an oral dose	Quinine dihydrochloride	

(Modified from *The Medical Letter*, December 28, 1979 to be consistent with Figure 6–3.)
(*P. falciparum* does not have secondary schizonts, so treatment in this case is not necessary)

primarily at the erythrocytic stage in the malaria life-cycle (site 4 on Fig. 6-1). Since the severe and life-threatening clinical symptoms of malaria occur at this stage, these drugs are very useful in (1) treating all four human malarias, and (2) in preventing clinical symptoms of all four human malarias. However, cures from these "site four drugs" can result only with *P. falciparum.* The other three species, *P. vivax, P. malariae,* and *P. ovale,* have a "secondary exo-erythrocytic" (secondary schizont) stage which can periodically release new merizoites for years or decades. An additional drug which is effective at the "site three" stage—usually primquine—must be given when the patient leaves the endemic area to result finally in a cure for these three species.

It would obviously be desirable to have drugs available that would protect humans from initial infection by the mosquito, at "site one" (Fig. 6-1). Unfortunately, no drugs are yet available which are effective at this site.

Primaquine is quite active at "site two," so it could be used as a prophylactic against all forms of human malaria. However, its toxicity generally precludes its chronic prophylactic use. Pyrimethamine and proguanil are not as effective. Primaquine is also effective as a gametocytocide ("site five"). However, in most countries the best means for controlling the spread of the disease is through community sanitation and the use of insecticides.

A complicating factor in modern malaria chemotherapy is that drug-resistant strains of *Plasmodia* (especially *falciparum*) have been reported in many geographical areas—e.g., chloroquine resistant *P. falciparum* or *P. vivax* in some areas; quinine-resistant *P. falciparum* in another. Similarly, resistance of certain Anopheles species to insecticides is a growing problem. (Tables in Thompson and Werbel[14] and in the review of Sweeney and Strube[6] vividly illustrate the two problems.)

QUINOLINES AND ANALOGS

All the antimalarial agents in Table 6-2 have one common structural feature—a quinoline ring, or a "quinoline with an additional benzene added" (an *acridine* ring). Some but not all have quinine's CH_3O- group on the quinoline ring. None except the cinchona alkaloids has a quinuclidine ring.

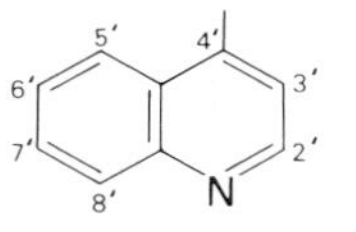

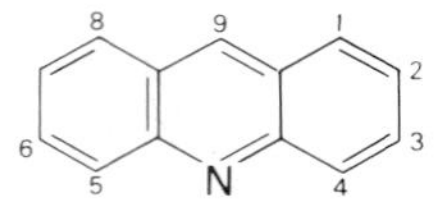

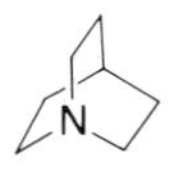

Quinine's structure (Table 6-3) has been known ever since 1908[20] (and proved by total synthesis by Woodward and Doering in 1945).[21] It was the structural model for all the other quinoline antimalarials shown in Table 6-3, along with the methylene blue–trypan red structures. It was found that changing methylene blue's two amino groups from

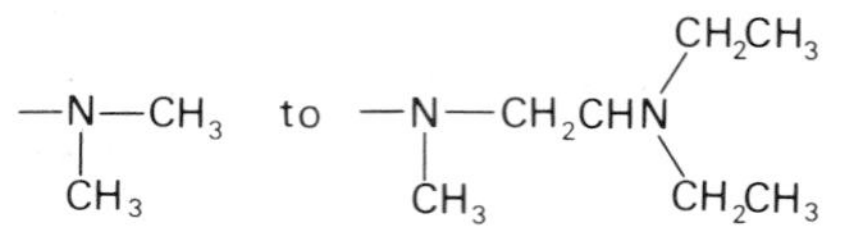

improved the antimalarial activity.[6] The approach used by Schulemann and co-workers in developing Pamaquine in 1926[22] (Fig. 6-2 and Table 6-3) was to synthesize combinations of the 6-methyoxy quinoline moiety of quinine with variations of the "improved methylene blue side group" shown above. Later followed the development of all the other *8-aminoquinoline* antimalarials in Table 6-2.

The same general approach with the acridine structure (since methylene blue has three rings and acridine is a quinoline analog), led to the less toxic quinacrine (Table 6-3) in 1932.[23,24] Including a chlorine atom in the structure was entirely fortuitous, but quinacrine's good activity stimulated the synthesis of many chloroquinolines in the decades that followed. Many other 9-aminoacridines were synthesized in later years, but none was superior to quinacrine.

The next logical step in drug design taken by German chemists was to "divide" the quinacrine structure into its two *4-aminoquinoline* "parts," one of which is chloroquine itself.

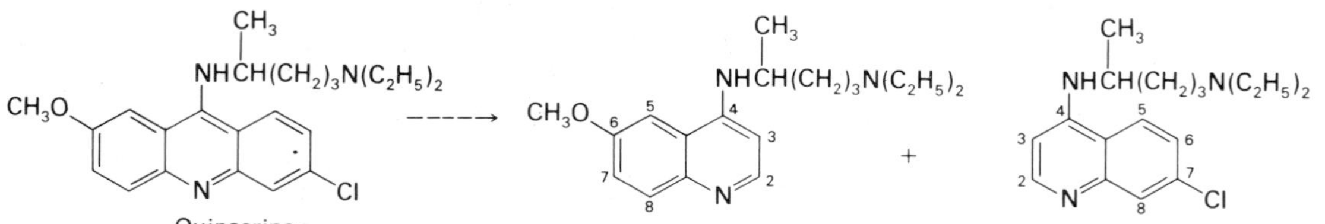

TABLE 6-3

QUINOLINES AND ANALOGS

CINCHONA ALKALOIDS AND OTHER 4-QUINOLINEMETHANOLS

Rubane Quinine Quinidine Mefloquine

Cinchonidine Epiquinine Epiquinidine Cinchonine

8-AMINOQUINOLINES

Compound	R
Pamaquine	$-CH(CH_3)-CH_2CH_2CH_2-N(C_2H_5)_2$
Primaquine	$-CH(CH_3)-CH_2CH_2CH_2-NH_2$
Pentaquine	$-CH_2CH_2CH_2CH_2CH_2-NH-CH(CH_3)_2$
Isopentaquine	$-CH(CH_3)CH_2CH_2CH_2-NH-CH(CH_3)_2$

9-AMINOACRIDINES

Quinacrine Hydrochloride

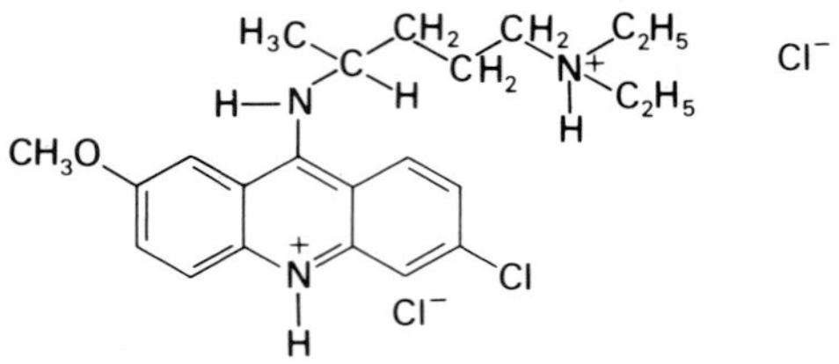

7-CHLORO-4-AMINOQUINOLINES

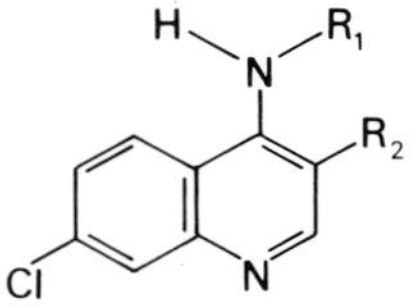

Compound	R_1	R_2
Chloroquine	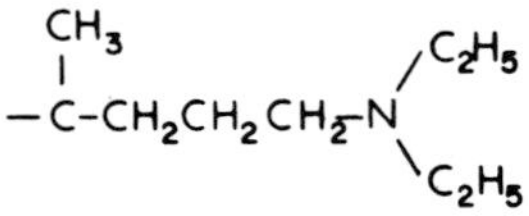 $-\overset{\underset{\textstyle CH_3}{\mid}}{C}-CH_2CH_2CH_2-N\overset{\textstyle C_2H_5}{\underset{\textstyle C_2H_5}{}}$	H
Hydroxychloroquine	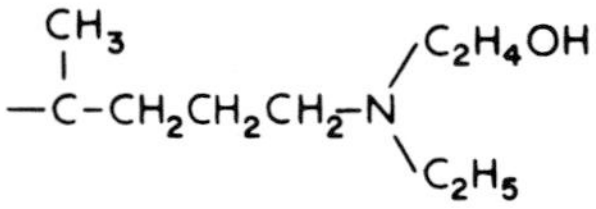 $-\overset{\underset{\textstyle CH_3}{\mid}}{C}-CH_2CH_2CH_2-N\overset{\textstyle C_2H_4OH}{\underset{\textstyle C_2H_5}{}}$	H
Sontoquine	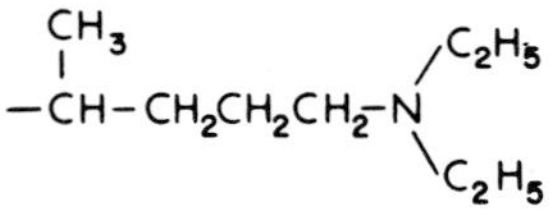 $-\overset{\underset{\textstyle CH_3}{\mid}}{CH}-CH_2CH_2CH_2-N\overset{\textstyle C_2H_5}{\underset{\textstyle C_2H_5}{}}$	CH_3
Amodiaquine	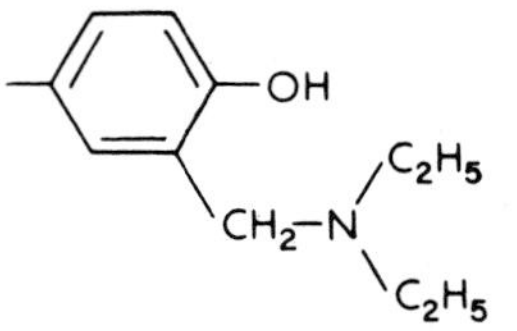 (2-hydroxy-phenyl)$-CH_2-N\overset{\textstyle C_2H_5}{\underset{\textstyle C_2H_5}{}}$, OH	H

Although chloroquine was active, it was considered too toxic, and many other *7-chloro-4-aminoquinolines* (Table 6-3) like chloroquine were investigated by the Germans in the late 1930's. Sontoquine (Table 6-3) was actually used by German soldiers during World War II. After samples of sontoquine were captured by the Allied forces,[25] over 200 7-chloro-4-aminoquinolines were then synthesized in the United States. Hundreds of 9-aminoacridines analogs were also synthesized.

Another drug design approach begun in the late 1930's was to make derivatives of *4-quinolinemethanol*—the central part of the quinine structure:

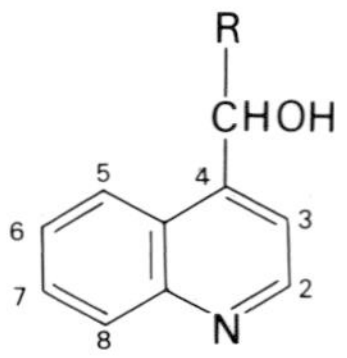

Many antimalarials of this structure have been synthesized, with the most promising being mefloquine (WR 142,490) (Table 6-3) in 1971.[27-29] The excellent reviews of antimalarial design by Sweeney and Strube,[6] and by Thompson and Werbel,[14] are recommended for further reading.

STRUCTURE–ACTIVITY RELATIONSHIPS

Quinine has four asymmetric centers—C8, C9, C3, and C4. Quinidine is the diastereomer formed by "inverting" C8 and C9. (Inversion is used only in the discussion sense. The biosyntheses of the cinchona alkaloids has been elegantly elicidated by Leete.[30]) Cinchonidine is the desmethoxy derivative of quinine, and cinchonine of quinidine. All four of these naturally occurring cinchona alkaloids are active antimalarials (activity varying depending upon the malarial species used in biological testing). Thus the 6-CH$_3$O group is not essential for activity, as illustrated also by mefloquine[27-29] and the 7-chloro-4-aminoquinolines (Table 6-3).

Quinine is found in highest concentration in cinchona bark (about 5%; quinidine, 0.1%; conchonine, 0.3%; cinchonidine, 0.4%), and is the commercially used antimalarial of the group. Quinidine's cardiac effects (see Chap. 14) also preclude its use as an antimalarial. Inversion of only C9 gives epiquinine and epiquinidine, both of which are inactive. Presumably this structural selectivity against malarial protozoa may result from the epi-alkaloids' inability to intercalate between base pairs of the protozoa DNA.

The good antimalarial activity of the 8-aminoquinolines, quinacrine, and the 7-chloro-4-aminoquinolines shows that the CH$_2$OH (aminoalcohol) of the cinchona alkaloids is also not essential for activity. Similarly, a chlorine atom on the quinoline ring is not a necessity, but it significantly increases activity for some analogs.

What are the exact structural requirements for good antimalarial activity of the quinoline–quinoline analog group? Although thousands of quinoline analogs have been synthesized in the more than 50 years since the development of pamaquine in 1926, the question cannot yet be satisfactorily answered. Major progress in development of a quantitative structural model was reported by Kim, Hansch, and co-workers in 1979.[31] The difficulty is compounded by the fact that a variety of different malaria protozoa species—many species which do not attack humans—have been used for evaluation of new antimalarials. Furthermore, as discussed in the next section, some quinoline analogs' antimalarial action may not involve a specific receptor.

It is also important to note that other aromatic ring systems besides quinoline and acridine have been used to make active antimalarials, e.g.:

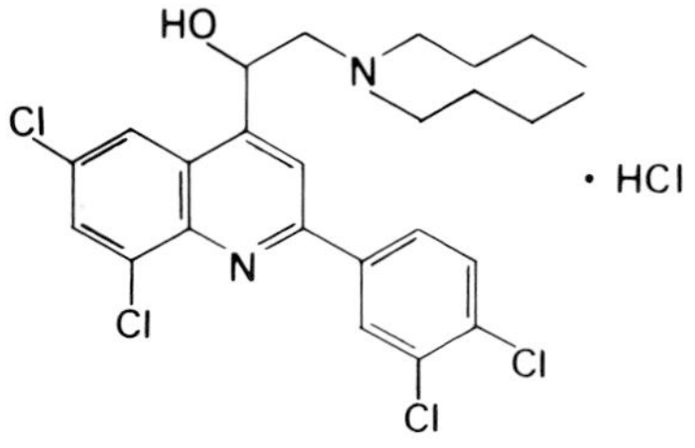

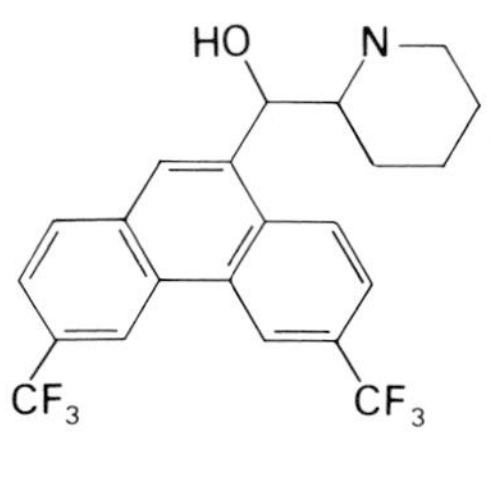

A number of other antimalarials with unusual aromatic rings now being investigated were reviewed in 1979. (Aromatic compounds which are tetrahydrofolate reductase inhibitors are discussed in the next section.)

The intensive investigations of quinolines and acridines have revealed useful structure–activity guidelines. The quinoline ring system has been found to be more active for a given set of substituents than for the acridine ring system. The dialkylaminoalkylamino $-N(CH_2)_nNR_2$ side group seen throughout

$$-\underset{\underset{R}{|}}{N}(CH_2)_nNR_2$$

Table 6-3 has been found to provide maximal activity, particularly when n = 2 to 5 or 6. However, as illustrated with the active analogs below, considerable structural variation is possible:

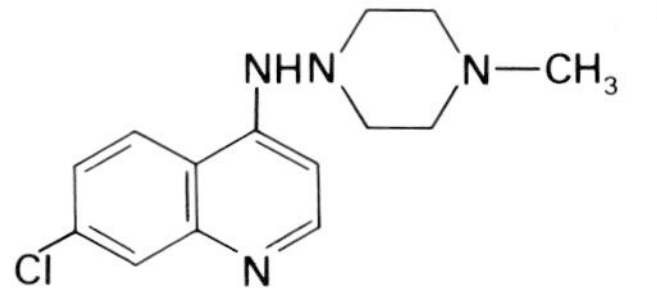

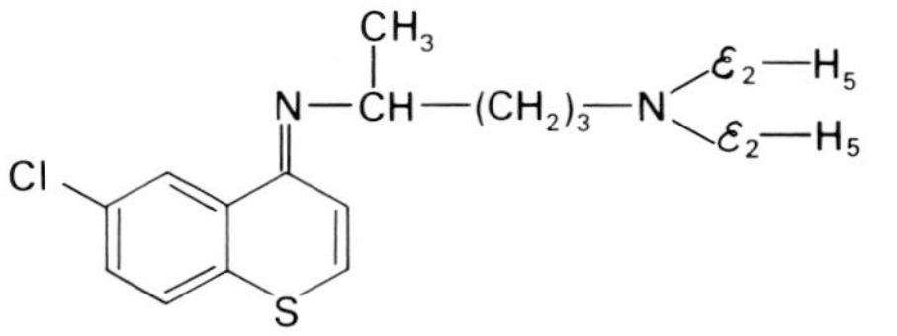

A chlorine group at C7 generally provides maximal activity (and toxicity) with chloro-4-aminoquinolines, but in some cases C6-chloro analogs are equally active. Other relationships have also been found, e.g., with 8-aminoquinolines, 6-CH$_3$O analogs have been found to be more active (and toxic) than 2- or 4-CH$_3$O analogs. Interested readers are encouraged to consult recently published reviews[6,14,27] for further information on these and additional structure–activity relationships.

MECHANISMS OF ACTION

The quinoline and quinoline analog antimalarials are known to exert a number of effects upon protozoal cells. These effects[6] include the following: (1) inhibition of protein synthesis,[34] DNA replication,[35] and RNA transcription;[6] (2) clumping of pigment into autophagic vacuoles,[35]—the pigment being fragments of partially digested hemoglobin; (3) binding of the quinoline drug to DNA;[37,38] and (4) degradation of protozoal RNA.[39] It has been shown also that erythrocytes containing chloroquine-sensitive malaria protozoa concentrate chloroquine to high levels, but much less with chloroquine-resistant malaria. The quinolines may also raise the pH of the malarial food vacuoles.[40] (Most studies have been on chloroquine, but to the extent that the other quinolines have been studied, these effects appear to be found to varying degrees with all the quinoline antimalarials, with the exception of mefloquine.)

An important insight into the molecular basis of quinoline's antimalarial action has been the finding by several research groups that these drugs intercalate between base pairs of adjacent strands of DNA,[38,41] causing the DNA to begin unwinding. The 7-chloro of chloroquine is believed to bind specifically to the amino group of guanines in the DNA.[42]

Significantly, mefloquine does not appear to bind to DNA—and it may be that mefloquine's potent antimalarial activity may be from raising the protozoa's intracellular pH[9]. Mefloquine also inhibits the pigment clumping of chloroquine.

CINCHONA ALKALOIDS

ABSORPTION, DISTRIBUTION, AND EXCRETION

After oral administration, the cinchona alkaloids are absorbed rapidly and nearly completely, with peak blood level concentrations occurring in 1 to 4 hours. About 70 percent is protein bound. Blood levels fall off very quickly after administration is stopped. A single dose of quinine is disposed of in about 24 hours. Various tissues contain enzymes capable of metabolizing the cinchona alkaloids, but the principal action appears to take place in the liver, where an oxidative process results in the addition of a hydroxyl group to the 2' position of the quinoline ring. The resulting degradation products, called carbostyrils, are much less toxic, are eliminated more rapidly and possess lower antimalarial activity than the parent compounds. The carbostyrils may be further oxidized to dihydroxy compounds. Excretion is mainly in the urine.

TOXICITY

The toxic reactions to the cinchona alkaloids have been studied extensively. These drugs are not as effective as chloroquine and are more toxic, so are not generally used except as noted in Table 6-2 (and in some poor countries). Acute poisoning with quinine is not common. In one case, a death was reported after administration of 18 g.; in another case, it was reported that the patient recovered after administration of 19.8 g. of quinine. A fatality resulted after the intravenous

administration of 1 g. of quinine. The toxic manifestations that are most common are due to hypersensitivity to the alkaloids and are referred to collectively as cinchonism. Frequent reactions are allergic skin reactions, tinnitus, slight deafness, vertigo and slight mental depression. The most serious is amblyopia, which may follow the administration of very large doses of quinine but is not common; usual therapeutic regimens do not produce this effect.

OTHER ROUTES OF ADMINISTRATION AND DOSAGE FORMS

In addition to antimalarial action, cinchona alkaloids are antipyretic. The action of quinine on the central temperature-regulating mechanism causes peripheral vasodilation. This effect accounts for the traditional use of quinine in cold remedies and fever treatments. Quinine has been used as a diagnostic agent for myasthenia gravis (by accentuating the symptoms). Also, it has been used for the treatment of night cramps or "restless legs." The antifibrillating effect of quinidine in the treatment of cardiac arrhythmias is discussed in Chapter 14.

The antimalarial action of cinchona alkaloids may be obtained by oral, intravenous or intramuscular administration. Administration by injection, particularly intravenous injection, is not without hazard and should be used cautiously. For intramuscular injection, quinine dihydrochloride is usually used.

Crude extract preparations containing the alkaloids of cinchona have been used widely as economical antimalarials for oral administration. During World War II, a mixture known as quinetum containing a large amount of quinine, was used. As the interest in pure quinine increased, another crude mixture ("cinchona fibrifuge"), composed of the alkaloids remaining after quinine removal, was introduced to replace quinetum. Subsequently, the Malaria Commission of the League of Nations introduced "totaquina." The *N.F. X* defined it as containing 7 to 12 percent of anhydrous crystallizable cinchona alkaloids. Totaquine now is the most widely used of inexpensive antimalarial drugs. The usual dose is 600 mg.

Quinine. Quinine is obtained from quinine sulfate prepared by extraction from the crude drug. To obtain it from solutions of quinine sulfate, a solution of the sulfate is alkalinized with ammonia or sodium hydroxide. Another method is to pour an aqueous solution of quinine bisulfate into excess ammonia water, with stirring. In either procedure the precipitated base is washed and recrystallized. The pure alkaloid crystallizes with 3 molecules of water of crystallization. It is efflorescent, losing 1 molecule of water at 20° under normal conditions and losing 2 molecules in a dry atmosphere. All water is removed at 100°.

It occurs as a levorotatory, odorless, white crystalline powder possessing an intensely bitter taste. It is only slightly soluble in water (1:1,500), but it is quite soluble in alcohol (1:1), chloroform (1:1) or ether.

It behaves as a diacidic base and forms salts readily. These may be of two types, the *acid* or *bi-salts* and the *neutral* salts. The neutral salts are formed by involvement of only the tertiary nitrogen in the quinuclidine nucleus, and the acid salts are the result of involvement of both basic nitrogens. Inasmuch as the quinoline nitrogen is very much less basic than the quinuclidine nitrogen, involvement of both nitrogens results in a definitely acidic compound.

PRODUCTS

Quinine Sulfate U.S.P., quininium sulfate. Quinine sulfate is the usual salt of quinine and is ordinarily the "quinine" asked for by the layman.

It is prepared in one of two ways, i.e., from the crude bark or from the free base. When prepared from the crude bark, the powdered cinchona is alkalinized and then extracted with a hot, high-boiling petroleum fraction to remove the alkaloids. By carefully adding diluted sulfuric acid to the extract, the alkaloids are converted to sulfates, the sulfate of quinine crystallizing out first. The crude alkaloidal sulfate is decolorized and recrystallized to obtain the article of commerce. Commercial quinine sulfate is not pure but contains from 2 to 3 percent of impurities, which consist mainly of hydroquinine and cinchonidine.

To obtain quinine sulfate from the free base, it is neutralized with dilute sulfuric acid. The resulting sulfate, when recrystallized from hot water, forms masses of crystals with the approximate formula $(C_{20}H_{24}O_2N_2)_2 \cdot H_2SO_4 \cdot 8H_2O$. This compound readily effloresces in dry air to the official dihydrate, which occurs as fine, white needles of a somewhat bulky nature.

Quinine sulfate often is prescribed in liquid mixtures. From a taste standpoint, it is better to suspend the salt rather than to dissolve it. However, in the event that a solution is desired, it may be accomplished by the use of alcohol or, more commonly, by addition of a small amount of sulfuric acid to convert it to the more soluble bisulfate. The capsule form of administration is the most satisfactory for masking the taste of quinine when it is to be administered orally.

The sulfate salts of cinchonidine and cinchonine may be used as antimalarials. The dextrorotatory cinchonine salt is of value in the treatment of patients who display a sensitivity to the levorotatory cinchona alkaloids.

7-CHLORO-4-AMINOQUINOLINES

ABSORPTION, DISTRIBUTION AND EXCRETION

Chloroquine is absorbed readily from the gastrointestinal tract, but amodiaquin gives lower plasma levels than others in the group. Peak plasma concentrations are reached in 1 to 3 hours, with blood levels falling off rather rapidly after administration is stopped. About half the drug in the plasma is protein bound. These drugs tend to concentrate in the liver, the spleen, the heart, the kidney and the brain. Half-life of chloroquine is about 3 days from a single dose, and a week or more following daily dosage for 2 weeks. Small amounts of 4-aminoquinolines have been found in the skin but probably not in sufficient quantity to account for their suppressant action on polymorphous light dermatoses. These compounds are excreted rapidly, with most of the unmetabolized drug being accounted for in the urine.

TOXICITY

The toxicity of 4-aminoquinolines is quite low in the usual antimalarial regimen. Side-effects can include nausea, vomiting, anorexia, abdominal cramps, diarrhea, headache, dizziness, pruritus and urticaria. Long-term administration in high doses (for other uses than malaria) may have serious effects on the eyes, and ophthalmologic examinations should be carefully carried out. Also, periodic blood examinations should be made. Patients with liver diseases particularly should be watched when 4-aminioquinolines are used.

OTHER USES, ROUTES OF ADMINISTRATION AND DOSAGE FORMS

The 4-aminoquinolines, particularly chloroquine and hydroxychloroquine, have been used in the treatment of extraintestinal amebiasis. They are of value in the treatment of chronic discoid lupus erythematosus but are of questionable value in the treatment of the systemic form of the disease. Symptomatic relief has been secured through the use of 4-aminoquinolines in the treatment of rheumatoid arthritis. A 1979 paper by Marks and Power[43] suggests chloroquine use for this purpose is *not* obsolete. Although the mechanism for their effect in collagen diseases has not been established, these drugs appear to suppress the formation of antigens that may be responsible for hypersensitivity reactions which cause the symptoms to develop. Long-term therapy of at least 4 to 5 weeks is usually required before beneficial results are obtained in the treatment of collagen diseases.

For the treatment of malaria, these drugs usually are given orally as salts of the amines in tablet form. In case of nausea or vomiting after oral administration, intramuscular injection may be used. For prophylactic treatment, the drugs may be incorporated into table salt. To protect the drugs from the high humidity of tropical climates, coating of the granules with a combination of cetyl and stearyl alcohols has been employed. These drugs are sometimes combined with other drugs such as proguanil or pyrimethamine to obtain a broader spectrum of activity (see Table 6-2).

PRODUCTS

Chloroquine U.S.P., CQ, 7-chloro-4-[[4-(diethylamino)-1-methylbutyl]amino]quinoline. Chloroquine occurs as a white or slightly yellow crystalline powder that is odorless and has a bitter taste. It is usually partly hydrated, very slightly soluble in water and soluble in dilute acids, chloroform and ether.

Chloroquine Phosphate U.S.P., Aralen®, Resochin®, 7-chloro-4-[[4-(diethylamino)-1-methylbutyl]amino]quinoline phosphate. Chloroquine phosphate occurs as a white, crystalline powder that is odorless, has a bitter taste and slowly discolors on exposure to light. It is freely soluble in water, and aqueous solutions have a pH of about 4.5. It is almost insoluble in alcohol, ether and chloroform. It exists in two polymorphic forms, either of which (or a mixture of both) may be used medicinally.

Hydroxychloroquine Sulfate U.S.P., Plaquenil® Sulfate, 2-[[4-[(7-chloro-4-quinolyl)amino]pentyl]ethylamino]ethanol sulfate (1:1).

Hydroxychloroquine sulfate occurs as a white or nearly white, crystalline powder that is odorless but has a bitter taste. It is freely soluble in water, producing solutions with a pH of about 4.5. It is practically insoluble in alcohol, ether and chloroform.

While successful as an antimalarial, hydroxychloroquine has achieved greater use than chloroquine in the control and the treatment of collagen diseases because it is somewhat less toxic.

Amodiaquine Hydrochloride U.S.P., Camoquin® Hydrochloride, 4-[(7-chloro-4-quinolyl)amino]-α-(diethylamino)-o-cresol dihydrochloride dihydrate. Amodiaquine dihydrochloride occurs as a yellow,

odorless, crystalline powder having a bitter taste. It is soluble in water, sparingly soluble in alcohol and very slightly soluble in ether, chloroform and benzene. The pH of a 1 percent solution is between 4 and 4.8. The synthesis of amodiaquine is more expensive than that of chloroquine.

This compound is an economically important antimalarial. Amodiaquine is highly suppressive in *P. vivax* and *P. falciparum* infections, being three to four times as active as quinine. However, it has no curative activity except against *P. falciparum*. Amodiaquine is altered rapidly in vivo to yield products which appear to be excreted slowly and have a prolonged suppressive activity.

8-AMINOQUINOLINES

ABSORPTION, DISTRIBUTION AND EXCRETION

The 8-aminoquinolines are absorbed rapidly from the gastrointestinal tract, to the extent of 85 to 95 percent within 2 hours after oral administration. Peak plasma concentration is reached within 2 hours after ingestion, after which the drug rapidly disappears from the blood. The drugs are localized mainly in liver, lung, brain, heart and muscle tissue. Metabolic changes in the drug are produced very rapidly, and, on excretion, metabolic products account for nearly all of the drug. Only about 1 percent of the drug is eliminated unchanged through the urine. It may be that the antiplasmodial and the toxic properties of these drugs are produced by metabolic transformation products. To maintain therapeutic blood level concentrations, frequent administration of 8-aminoquinolines may be necessary.

TOXICITY

The toxic effects of the 8-aminoquinolines are found principally in the central nervous system and the hematopoietic system. Occasionally, anorexia, abdominal pain, vomiting and cyanosis may be produced. The toxic effects related to the blood system are more common; hemolytic anemia (particularly in dark-skinned people), leukopenia and methemoglobinemia are the usual findings. A genetic deficiency of glucose-6-phosphate dehydrogenase—found in up to 100 million people[44] but rarely in Caucasians—weakens erythrocytes and makes them more easily damaged by drugs like the 8-aminoquinolines. For most Caucasians, primaquine is quite nontoxic. (The discussion of Rollo[11] is recommended for further clinical information.) Toxicity is increased by quinacrine; therefore, the simultaneous use of quinacrine and 8-aminoquinolines must be avoided.

USES, ROUTES OF ADMINISTRATION AND DOSAGE FORMS

Primaquine is used mainly to prevent relapses due to the exoerythrocytic forms of the parasites (Table 6-1). Primaquine is usually administered orally, in tablet form, as salts such as hydrochlorides or phosphates. Pamaquine is used as the methylene-bis-β-hydroxynaphthoate (naphthoate or pamoate), because this salt is of low solubility and is absorbed slowly, and thus, blood levels are maintained for longer periods and are more uniform.

PRODUCTS

Primaquine Phosphate U.S.P., primaquinium phosphate; 8-[(4-amino-1-methylbutyl)-amino]-6-methoxyquinoline phosphate. Primaquine phosphate is an orange-red, crystalline substance having a bitter taste. It is soluble in water and insoluble in chloroform and ether. Its aqueous solutions are acid to litmus. It may be noted that it is the primary amine homolog of pamaquine.

Primaquine has been found to be the most effective and the best tolerated of the 8-aminoquinolines. Against *P. vivax,* it is 4 to 6 times as active an exoerythrocytic schizontocide as pamaquine and about one half as toxic. When 15 mg. of the base are administered daily for 14 days, radical cure is achieved in most *P. vivax* infections. Success has been achieved against some very resistant strains of *P. vivax* by administering 45 mg. of the base once a week for 8 weeks, with simultaneous administration of 300 mg. of chloroquine base. This regimen also tends to lessen the toxic hemolytic effects produced in primaquine-sensitive individuals.

9-AMINOACRIDINES

Quinacrine, found to be the most useful compound in this class, is used rarely for malaria today. A brief discussion on its use has been consolidated below.

PRODUCTS

Quinacrine Hydrochloride U.S.P., Atabrine®, mepacrine hydrochloride, Atebrin®, 6-chloro-9-[[4-(diethylamino)-1-methylbutyl]amino]-2-methoxyacridine dihydrochloride.

Quinacrine Hydrochloride

The wide use of this compound during the early 1940's resulted in a large number of synonyms for quinacrine in various countries throughout the world.

The dihydrochloride salt is a yellow crystalline powder that has a bitter taste. It is sparingly soluble (1:35) in water and soluble in alcohol. A 1:100 aqueous solution has a pH of about 4.5 and shows a fluorescence. Solutions of the dihydrochloride are not stable and should not be stored. A dimethanesulfonate salt produces somewhat more stable solutions, but they too should not be kept for any length of time.

The yellow color that quinacrine imparts to the urine and the skin is temporary and should not be mistaken for jaundice. Quinacrine may produce toxic effects in the central nervous system, such as headaches, epileptiform convulsions and transient psychoses that may be accompanied by nausea and vomiting. Hematopoietic disturbances such as aplastic anemia may occur. Skin reactions and hepatitis are other symptoms of toxicity. Deaths have occurred from exfoliative dermatitis caused by quinacrine.

As an antimalarial, quinacrine acts as an erythrocytic schizontocide in all kinds of human malaria. It has some effectiveness as a gametocytocide in *P. vivax* and *P. malariae* infections. It may be employed in the treatment of blackwater fever when the use of quinine is contraindicated. It is also an effective curative agent for the treatment of giardiasis due to *Giardia lamblia,* eliminating the parasite from the intestinal tract. It is an important drug for use in the elimination of intestinal cestodes such as *Taenia saginata* (beef tapeworm), *T. solium* (pork tapeworm) and *Hymenolepis nana* (dwarf tapeworm). Like the 4-aminoquinolines, quinacrine may also be used to treat light-sensitive dermatoses such as chronic discoid lupus erythematosus.

MEFLOQUINE

Mefloquine (WR 142,490)[27-30] is the most promising of a group of new experimental antimalarial drugs. It has been found to be effective in curing multi-drug resistant *falciparum* malaria, apparently at the erythrocytic stage in the life cycle. (For a review, see the discussion of Roxman and Canfield.[27]) Side-effects are slight, as reported in Phase I clinical trials in 1975 and

1976. It also appears that mefloquine provides suppressive prophylactic activity against both *falciparum* and *vivax* malaria. Single doses of 1 g. are often used for *falciparum* malaria cures, and 250 mg. to 500. mg as weekly doses for *falciparum* or *vivax* malaria suppression. Pharmacokinetics of mefloquine have recently been reported by Desjardins and co-workers.[30]

Mefloquine (racemic mixture)

TETRAHYDROFOLATE SYNTHESIS INHIBITORS

USES AND MECHANISMS OF ACTION

The biosyntheses and roles of p-aminobenzoic acid (PABA), folic acid (FA), dihydrofolic acid (FAH_2), tetrahydrofolic acid (FAH_4), and FAH_4 cofactors (such as folinic acid, FAH_4-CHO) have been described in Chapter 5. The FAH_4 cofactors are essential as one carbon donors on several biosynthetic pathways, especially the conversion of uridine to thymidine nucleic acids needed for DNA synthesis. Malaria protozoa cannot convert FA to FAH_4, but can convert FAH_2 to FAH_4.[45]

Ferone and co-workers[46,47] have found that malarial dihydrofolate reductase is structurally different from mammalian dihydrofolate reductase—and up to 2000 times more sensitive to the antimalarial drugs discussed in this section (much less so with trimethoprim). The malaria protozoa are also unable to use the host's pyrimidine nucleosides (but can use purine nucleosides), and must synthesize their own.[48,49] Synthesizing thymidine nucleotides requires FAH_4-CHO. Thus, any drug that can inhibit the malarial protozoa's biosynthesis of FAH_2 or can selectively inhibit the protozoa's dihydrofolate reductase can inhibit the growth of and kill the protozoa.

A variety of structurally different small molecules

have been found to be very effective in inhibiting competitively one of the steps involved in the formation of FAH_4 in malaria protozoa (Fig. 6-3). An examination of the drugs of choice (Table 6-2), however, shows that these drugs are usually reserved for treatment of malaria strains resistant to one or more of the quinoline type antimalarials. Since the FAH_4 synthesis inhibitors tend to be slow acting, they are often given in combination with a quinoline antimalarial for treatment of acute clinical attacks. Combinations of FAH_4 synthesis inhibitors are often used, but it should be noted that indiscriminate use may lead to resistant strains (as with any antimalarial) of malaria and resistant bacteria strains as well.

The biguanides, diaminopyrimidines, and dihydrotriazines are selective inhibitors of malarial protozoa dihydrofolate reductase. The sulfonamides and sulfones apparently competitively block the incorporation of PABA into the malaria's dihydropteroic acid (FAH_2)—the same mechanism as with bacteria.

As shown in the malaria life cycle (Fig. 6-1), pyrimethamine and chloroguanide are effective against both the primary (exoerythrocytic) schizont (site 2) and erythrocytic (site 4) stages. The sulfonamides and sulfones are effective only against the erythrocytic (site 4) stage. Thus, all can cure *falciparum* infections, but not *vivax*, *malarie*, or *ovale*. They can suppress clinical symptoms of all four species. Pyrimethamine and chloroguanide both could theoretically be effective as prophylactic agents (since they act at the primary schizont stage), but proguanil's slow onset of action and short half-life limit its use for this purpose. Pyrimethamine, however, is an effective prophylactic and excellent clinical suppressive.

The number of FAH_4-synthesis inhibitors in actual clinical use is small, but extensive structure–activity studies have been completed. Interested readers are encouraged to consult recent reviews[6,10] for detailed information.

DIAMINOPYRIMIDINES

Following the observations made in the late 1940's that some 2,4-diaminopyrimidines were capable of interfering with the utilization of folic acid by *Lactobacillus casei*, a property also shown by proguanil, these

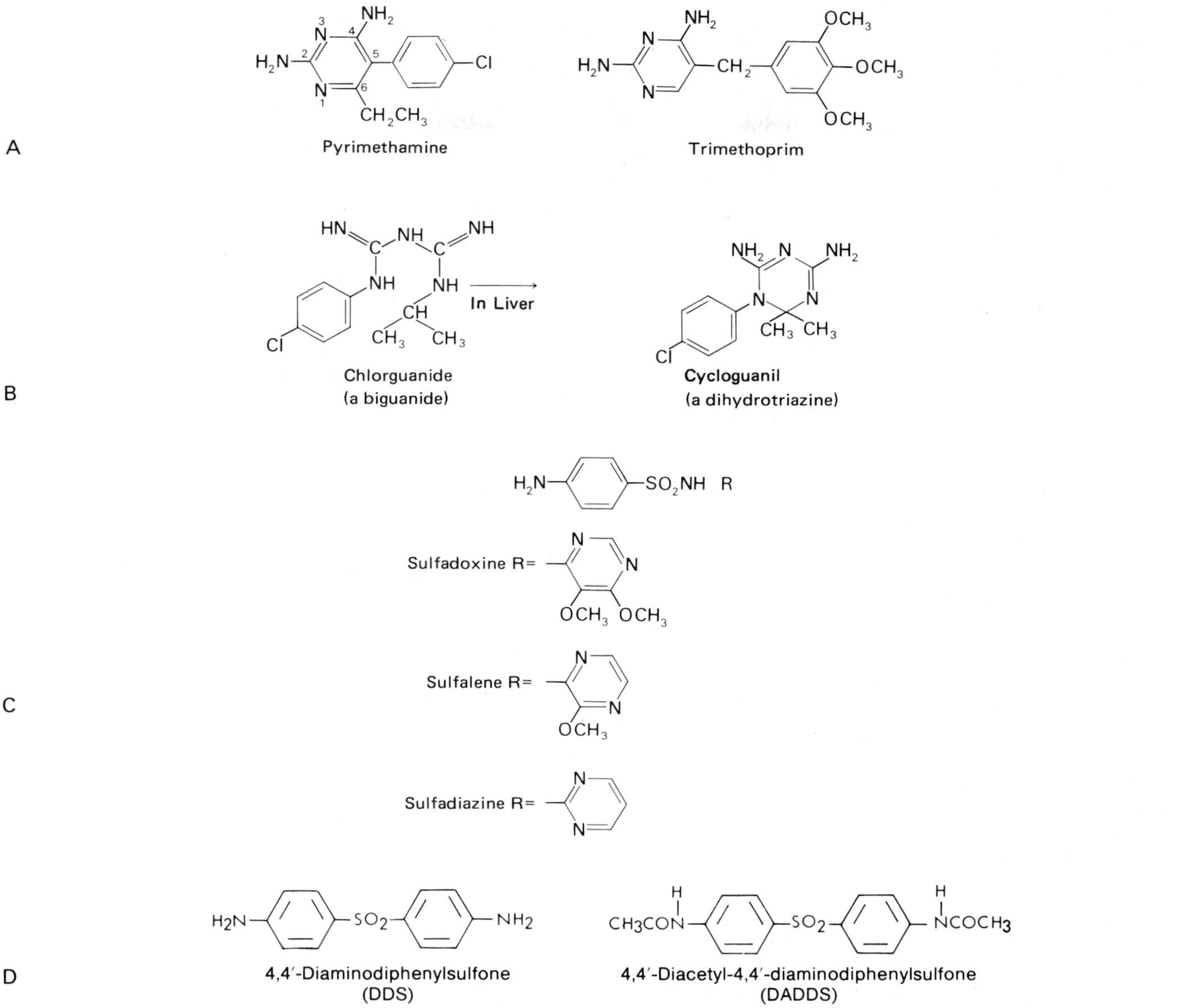

FIG. 6–3. *Tetrahydrofolate synthesis inhibitors. A. Diamino pyrimidines. B. Biguanides and dihydrotriazines. C. Sulfonamides most commonly used for malaria. D. Sulfones.*

compounds received intensive study as potential antimalarials. It was noted that certain 2,4-diamino-5-phenoxypyrimidines possessed a structural resemblance to proguanil, and a series of such compounds was synthesized and found to possess good antimalarial action. Subsequently, a large series of 2,4-diamino-5-phenylpyrimidines was prepared and tested for activity. Maximum activity was obtained when an electron-attracting group was present in the 6 position of the pyrimidine ring and when a chlorine atom was present in the *para* position of the phenyl ring. If the two rings were separated by either an oxygen atom or a carbon atom, antimalarial action decreased. The best in the series of compounds was the one that became known as pyrimethamine.

PRODUCTS

Pyrimethamine U.S.P., Daraprim®, 2,4-diamino-5-(*p*-chlorophenyl)-6-ethylpyrimidine. Pyrimethamine is an effective erythrocytic schizontocide against all human malarias. It also acts as a primary exoerythrocytic schizontocide in most infections.

Pyrimethamine is slowly but completely absorbed from the gastrointestinal tract. It is localized in the liver, the lungs, the kidney and the spleen and is slowly excreted through the urine, chiefly in metabolized form. A single weekly dose of 25 mg. is sufficient for suppression. It is relatively nontoxic, but overdoses may lead to depression of cell growth by inhibition of folic acid activity.

It is administered in the form of the free base, a relatively tasteless powder.

Trimethoprim U.S.P., 2,4-Diamino-5-(3,4,5-trimethoxybenzyl)pyrimidine. Trimethoprim is marketed by itself and in combination with sulfonamides (e.g., in the combination products Bactrim®, Septra®, and others) primarily as antibacterial products (see Chap. 5). It was developed as an antibacterial agent, but subsequent tests as an antimalarial showed it to be active. Studies by Ferone and co-workers[45–47] have shown trimethroprim not to be as selective for protozoal dihydrofolate reductase as is pyrimethamine. As a result, when used as an antimalarial, it is usually combined with another drug. Its antimalarial effectiveness for human malarias can best be described as mixed. (For a review of the early studies, see the discussion of Thompson.[14]) An additional limiting factor in its antimalarial use is trimethoprim's much shorter half-life (about 24 hours) than pyrimethamine.

BIGUANIDES AND DIHYDROTRIAZINES

Although in several malaria species the biguanides have some antimalarial activity, they are largely prodrugs for their active metabolites, the dihydrotriazines. For example, as shown in Figure 6-3, choroquanide is rapidly metabolized to the potent antimalarial dihydrotriazine, cycloquanil. The dihydrotriazines such as cycloquanil are so rapidly metabolized that, with one exception (cycloquanil pamoate), they are not used for human infections. Cycloquanil pamoate is an I.M. injectable depot preparation that can provide antimalarial protection for several months from a single dose.

A large number of biguanide and dihydrodriazine antimalarial agents have been synthesized. Useful structure–activity relationships have been found, for example, substitution of a halogen on the para position of the phenyl ring significantly increases activity. Chlorine is used in chloroguanide, but the bromine analog also is very active. Later, it was observed that a second chlorine added to the 3 position of the phenyl ring of proguanil further enhanced activity. However, the dichloro compound, chlorproguanil, is more toxic than chloroguanide itself.

The biguanides are absorbed from the gastrointestinal tract very quickly, but not as rapidly as quinine or chloroquine. They concentrate in the liver, the lungs, the spleen and the kidney but appear not to cross the bloodbrain barrier. Of the amount in plasma, about 75 percent is protein bound. They are metabolized in large part in the body and are eliminated very rapidly,

principally in the urine. As a result, frequent administration of these drugs is necessary.

The toxic manifestations of biguanides are very mild in man. Some gastrointestinal disturbances may occur if the drugs are taken on an empty stomach but not if they are taken after meals. With excessive doses (1 g. of chlorguanide), some renal disorders such as hematuria and albuminuria may develop.

PRODUCTS

Chloroguanide Hydrochloride, Proguanil hydrochloride, Paludrine®, 1-(p-chlorophenyl)-5-isopropylbiguanide hydrochloride. Occurs as a white, crystalline powder or as colorless crystals that are soluble in water (1:75) and alcohol (1:30). It is odorless, has a bitter taste and is stable in air but slowly darkens on exposure to light.

Cycloguanil Pamoate, Camolar®, CI-501, cycloguanil embonate, 4,6-diamino-1-(p-chlorophenyl)-1,2-dihydro-2,2-dimethyl-s-triazine (2:1) with 4,4′-methylenebis[3-hydroxy-2-naphthoic acid]. A single I.M. injection of cycloquanil pamoate can provide protection against all four human malarials for several months. As with all antimalarials, effectiveness is dependent upon the malaria strain not being resistant to the drug or its structural analogs, e.g., chloroguanide. Unfortunately, resistance develops quickly to cycloguanil pamoate, and injection sites can be painful. The problems do not seem to be significantly improved by combinations with other antimalarials.[50]

SULFONAMIDES

As can be seen in Table 6-2, sulfonamides such as sulfadoxine are used in antimalarial therapy against drug-resistant malarial strains. They are effective against erythrocytic stages of the malaria protozoa (Fig. 6-1, site 4). The azo dye prontozil (see discussion in Chap. 5) was found to have antimalarial activity in vivo against both *falciparum* and *P. vivax* in the late 1930's. Later when it was discovered that prontosil was a prodrug for sulfanilamide, many other related sulfonamides were investigated as antibacterials and antimalarials.

Medium- or long-acting aulfonamides have been used clinically as antimalarials, particularly sulfadiazine, sulfadoxine and sulfalene (see Chap. 5). However, each is much more effective when given in combination with pyrimethamine. Trimethoprim combinations have also been investigated.

SULFONES

It has been known for some time that 4,4'-diaminodiphenylsulfone, Dapsone U.S.P. (DDS), was active against a number of the plasmodium species causing malaria.[52] However, it was considered to be an inferior antimalarial drug until it was discovered that it served effectively as a chemoprophylactic agent against chloroquine-resistant *P. falciparum* infections in southeast Asia.

The effectiveness of DDS has prompted the development of programs seeking the synthesis of sulfone compounds of superior activity and with longer duration of action.[6,14] Among the compounds tested N,N-diacetyl-4,4'-diaminodiphenylsulfone (DADDS) has been found to be promising. Its more prolonged activity and lower toxicity as compared to DDS are probably related to its slow conversion to either the monoacetyl derivative or DDS itself, both of which act as the antimalarial agents. Long-acting depot combinations of DADDS and cycloguanil pamoate are also under clinical investigation.

TABLE 6–4

ANTIMALARIALS

Name Proprietary Name	Preparations	Usual Adult Dose*	Usual Dose Range*	Usual Pediatric Dose*
Quinine Sulfate U.S.P.	Quinine Sulfate Capsules U.S.P. Quinine Sulfate Tablets U.S.P.	Therapeutic—200 mg. to 1 g. 3 times daily for 6 to 12 days For chloroquine—resistant *P. falciparum:* 650 mg. every 8 hr. for at least 3 days with concurrent or consecutive administration of tetracycline or pyrimethamine and sulfadiazine		The following dose is to be divided into 2 or 3 portions and continued for 7 to 10 days: up to 1 year of age—100 to 200 mg.; 1 to 3 years of age—200 to 300 mg.; 4 to 6 years of age—300 to 500 mg.; 7 to 11 years of age—500 mg. to 1 g.; 12 to 15 years of age—1 to 2 g.
Chloroquine U.S.P.	Chloroquine Hydrochloride Injection U.S.P.	Antiamebic—I.M., 200 to 250 mg. once daily for 10 to 12 days	160 to 800 mg. daily	Extraintestinal antiamebic—I.M., 15 mg. per kg. of body weight per day for 2 days, followed by 7.5 mg. per kg. for at least 2 to 3 weeks
		Antimalarial—I.M., initially 200 to 250 mg., repeated in 6 hours if necessary; not to exceed 1 g. in the first 24 hours		Antimalarial—6.25 per kg. of body weight, repeated in 6 hours if necessary. In no instance should a single dose exceed 6.25 mg. per kg. or the total daily dose exceed 12.5 mg. per kg.
Chloroquine Phosphate U.S.P. *Aralen, Resochin*	Chloroquine Phosphate Tablets U.S.P.†	Antiamebic—250 mg. 4 times a day for 2 days, followed by 250 mg. 2 times a day for at least 2 to 3 weeks		Extraintestinal antiamebic—initial, the equivalent of 6 mg. per kg. of body weight twice daily for 2 days; maintenance, the equivalent of 6 mg. per kg. of body weight once daily
		Antimalarial, suppressive—500 mg. once weekly; therapeutic—1 g. initially followed by 500 mg. once a day on the second and third days Lupus erythematosus suppressant—Dosage not established		Antimalarial, suppressive—8.3 mg. per kg. of body weight up to 500 mg., once weekly; therapeutic—41.7 mg. per kg. of body weight administered over 3 days as follows: 16.7 mg. per kg. not to exceed a single dose of 1 g., then 8.3 mg. per kg. not to exceed a single dose of 500 mg. 6, 24, and 48 hours later
Hydroxychloroquine Surfate U.S.P. *Plaquenil Sulfate*	Hydroxychloroquine Sulfate Tablets U.S.P.‡	Antimalarial, suppressive—400 mg. once every 7 days; therapeutic—initially 800 mg. followed by 400 mg. in 6 to 8 hours, and 400 mg. once daily on the 2nd and 3rd days Lupus erythematosus suppressant—400 mg. 1 or 2 times a day		Antimalarial, suppressive—6.4 mg. per kg. not to exceed the adult dose, once every 7 days; therapeutic—32 mg. per kg. of body weight is administered over a 3-day period as follows: 12.9 mg. per kg. not to exceed a single dose of 800 mg.; then 6.4 mg. per kg. not to exceed a single dose of 400 mg. 6, 24, and 48 hours later

*Refer to the *U.S.P. D.I.* for complete dosage information, along with "Antimalarial Drugs of Choice" with doses from *The Medical Letter.*[15]
†Also used as an antiamebic and as a lupus erythematosus suppressant.
‡Also used as a lupus erythematosus suppressant.
§ Also used as an anthelmintic (intestinal tapeworms) and as an antiprotozoal (giardiasis).

TABLE 6–4

ANTIMALARIALS (CONT.)

Name Proprietary Name	Preparations	Usual Adult Dose*	Usual Dose Range*	Usual Pediatric Dose*
Amodiaquine Hydrochloride U.S.P. *Camoquin Hydrochloride*	Amodiaquine Hydrochloride Tablets U.S.P.	Suppressive—the equivalent of amodiaquine—300 to 600 mg. as a single dose once every 7 days; therapeutic—the equivalent of 600 mg. of amodiaquine initially, then the equivalent of 300 mg. 6, 24, and 48 hours later		
Primaquine Phosphate U.S.P.	Primaquine Phosphate Tablets U.S.P.	26.3 mg. once daily for 14 days		
Quinacrine Hydrochloride U.S.P. *Atabrine Hydrochloride*	Quinacrine Hydrochloride Tablets U.S.P.§	Antiprotozoal (giardiasis)—100 mg. 3 times daily for 5 to 7 days Anthelmintic (tapeworms)—200 mg. of quinacrine hydrochloride with 650 mg. of sodium bicarbonate every 10 minutes for 4 doses Antimalarial, suppressive—100 mg. once daily; therapeutic—200 mg. of quinacrine hydrochloride with 1 g. of sodium bicarbonate every 6 hours for 5 doses, then 100 mg. 3 times daily, up to a maximum total dose of 2.8 g. in 7 days	300 to 900 mg. daily	Antiprotozoal (giardiasis)—2.7 mg. per kg. of body weight or 83.3 mg. per square meter of body surface, 3 times daily, up to a maximum of 300 mg. per day Anthelmintic (tapeworms)—7.5 mg. per kg. of body weight or 250 mg. per square meter of body surface every hour for 2 doses, up to a maximum of 800 mg. Antimalarial, suppressive—50 mg. once daily; therapeutic—children 1 to 4 years of age—100 mg. 3 times daily on the 1st day, then 100 mg. once daily for 6 days; children 4 to 8 years of age—200 mg. 3 times daily on the 1st day, then 100 mg. twice daily for 6 days
Chloroguanide Hydrochloride *Paludrine*	Tablets	100 mg. daily		Children less than 2 years of age—25 to 50 mg. daily; 2 to 6 years—50 to 75 mg. daily; 6 to 10 years—100 mg. daily
Pyrimethamine U.S.P. *Daraprim*	Pyrimethamine Tablets U.S.P.	Suppressive—25 mg. once weekly; therapeutic—25 to 50 mg. once daily for 2 days	25 mg. weekly to 75 mg. daily	Suppressive—infants and children under 4 years of age—6.25 mg. once weekly; 4 to 10 years of age—12.5 mg. once weekly; over 10 years of age—see Usual Dose. Therapeutic—4 to 10 years of age—25 mg. once daily for 2 days

*Refer to the *U.S.P. D.I.* for complete dosage information, along with "Antimalarial Drugs of Choice" with doses from *The Medical Letter.*[15]
†Also used as an antiamebic and as a lupus erythematosus suppressant.
‡Also used as a lupus erythematosus suppressant.
§ Also used as an anthelmintic (intestinal tapeworms) and as an antiprotozoal (giardiasis).

OTHER ANTIMALARIALS

The emergence of drug-resistant strains of malaria has prompted a reinvestigation of antibiotics and an intensive investigation into new types of antimalarials. (For excellent reviews of experimental new antimalarials, see Roxman and Canfield,[27] and Sweeney and Strube.[6]) Tetracyclines in combination with other antimalarials have been found to be effective against chloroquine-resistant strains of *P. falciparum.*[53] However, the World Health Organization has warned against the long-term use of tetracyclines as clinical suppressive or prophylactic agents because of the high risk of resistant bacterial strains.[51]

Clindamycin in combination with quinine has also been found to have some effectiveness against some human species, but its side effects and slow activity make it unlikely that it will be generally clinically used. Other antimalarials that have been investigated include quinones and a variety of heterocycles.[6,27]

REFERENCES

1. Schmidt, G. D., and Roberts, L. S.: Foundation of Parasitology, Mosby, Saint Louis, 1977.
2. Kreier, J. P., (ed.): Parasitic Protozoa, I-IV, Academic Press, New York, 1977.
3. Rieckmann, K. H., and Silverman, M.: Plasmodia of Man, *in* Kreier, J. P. (ed.): Parasitic Protozoa, vol. III, Academic Press, New York, 1977.
4. Van den Bossche, H., (ed.): Biochemistry of Parasites and Host–Parasite Relationships, North-Holland, New York, 1976.
5. Gutteridge, W. E., and Coombs, G. H.: Biochemistry of Parasitic Protozoa, Baltimore, University Park Press, 1977.

6. Sweeney, T. R., and Strube, R. E.: Antimalarials *in* Wolff, M. E. (ed): Burger's Medicinal Chemistry, vol. 2, ed. 4, New York, Wiley, 1979.
7. Ross, W. J.: Antiamebic Agents *in* Wolff, M. E. (ed.): Burger's Medicinal Chemistry, volume 2, ed. 4, New York, Wiley, 1979.
8. Ross, W. J.: Chemotherapy of Trypanosomiasis, and Other Protozoan Disease *in* Wolff, M. E. (ed.): Burger's Medicinal Chemistry, vol. 2, ed. 4, New York, Wiley, 1979.
9. Fisher, M. H., and Wang, C. C.: Antiparasitic Agents, *in* Clark, F.H. (ed.): Annual Reports in Medicinal Chemistry, vol. 13 New York, Academic Press, 1978.
10. Werbel, L. M., Worth, D. F. and Weitzel, Antiparasitic Agents *in* Hess, H. J. (ed.): Annual Reports in Medicinal Chemistry, vol. 14, New York, Academic Press, 1979.
11. Rollo, I. M., Chemotherapy of Parasitic Disease, *in* Goodman, L. S., and Gilman, A. (ed.): The Pharmacological Basis of Therapeutics, ed. 5, New York, MacMillian, 1975.
12. Kier, L. B.: Parasite Chemotherapy, *in* Foye, W.O. (ed.): Principles of Medicinal Chemistry, ed. 2, Philadelphia, Lea and Febiger, 1974.
13. Van den Bossche, H.: Chemotherapy of Parasitic Infections, Nature 273:626, 1978.
14. Thompson, P. E., and Werbel, L. M.: Antimalarial Agents—Chemistry and Pharmacology, New York, Academic Press, 1972.
15. Drugs For Parasitic Infections, The Medical Letter, 21:105 (December 28, 1979).
16. Browning, C. H.: Nature, 1975:570, 616, 1955.
17. Baumler, E.: In Search of the Magic Bullet, pp. 15–39, London, Thames and Hudson, 1965.
18. Albert, A.: Selective Toxicity, pp. 130–137, London, Chapman and Hall, 1973.
19. "Dr. Erhlich's Magic Bullett" 103-minute Warner Brothers, film 1940. Available from: United Artists 16, 8901 Beverly Blvd., Los Angeles, California, 90048. A 33 minute version, "Magic Bullets," is available from Utah State University, Audio Visual Services, Logan, Utah 84321.
20. Rabe, P.: Chem. Ber. 41:62, 1908.
21. Woodward, R. B., and Von E. Doering, W. J. Am. Chem. Soc. 67:860, 1945.
22. Schulemann, W.: Proc. Royal Soc. Med. 25:897, 1932.
23. Kikuth, W.: Deut. Med. Wochenschr. 58:530, 1932.
24. Mauss, H., and Mietzsch, F.: Klin. Wochenschr, 12:1276, 1933.
25. Kikuth, W., *in* Sweeney, T. R., and Strube, R. E.: Antimalarials, p. 151, *in* Wolff, M. E. (ed.): Burger's Medicinal Chemistry, vol. 2, ed. 4, New York, Wiley, 1979.
26. Ainley, A. D., and King, H.: Proc. Roy. Soc., Ser. B. 125:60, 1938.
27. Roxman, R. S., and Canfield, C. J.: Advances in Pharmacology, 16:1, 1979.
28. Trenholme, G. M., Williams, R. L., Desjardins, R. E., Frischer, H. Carson, P. E., and Rieckmann, K. H.: Science, 190:792, 1975.
29. Ohnmacht, C. J., Patel, A. R. and Lutz, R. E.: J. Med. Chem. 14:926, 1971.
30. Leete, E.: Account. Chem. Research 1:59, 1969.
31. Kim, K. H., Hansch, C., Fukunaga, J. Y., Steller, E. E., Jow, P. Y. C., Craig, P. M., and Page, J.: J. Med. Chem. 22:366, 1979.
32. Raxdan, R. K., Bruni, R. J., Mehta, A. C., Weinhardt, K. K., and Papanastassiou, J. Med. Chem. 21:643, 1978.
33. Elslager, E. F., *et al.*: J. Med. Chem. 12:600, 1969.
34. Gutteridge, *et al.*: Parasitology 64:37, 1972.
35. Schellenberg, K., and Coatney, G. R.: Biochem. Pharmacology 6:143, 1961.
36. Warhurst, D. C., and Hockley, D. J.: Nature, 214:935, 1967.
37. Cohen, S. M., and Yielding, K. L.: J. Biol. Chem. 240:3123, 1965.
38. Cohen, S. M., and Yielding, K. L.: Proc. Nat. Acad. Sci. 54:521, 1965.
39. Warhurst, D. C., and Williamson, J.: Chem. Biol. Interactions 2:89, 1970.
40. Homewood, C. A., *et al.*: Nature 235:50, 1972.
41. O'Brien, R. L., *et al.*: Nature 55:1511, 1966.
42. O'Brien, R. L., and Hahn, F. E.: Antimicrob. Agents Chemother. 315, 1966.
43. Marks, J. S., and Power, B. J.: Lancet 371, 1979.
44. Marks, P. A., and Banks, J.: Ann. New York Acad. Sci. 123:198, 1965.
45. Ferone, R., and Hitchings, G. H.: J. Protozoal. 13:504, 1966.
46. Ferone, R., *et al.*: Mol. Pharmacol. 5:49, 1969.
47. Ferone, R.: J. Biol. Chem. 245:850, 1970.
48. Krooth, R. S., *et al.*: Science 164;1073, 1968.
49. Neame, K. D., *et al.*: Parisitology 69:329, 1974.
50. Clyde, D. F.: J. Tro. Med. Hyg. 72:81, 1969.
51. World Health Organization, Chemotherapy of Malaria and Resistance to Antimalarials, Report No. 529, World Health Organization, Geneva, 1973.
52. Powell, R. D., *et al.*: Int. J. Lepr. 35:590, 1967.
53. Rieckmann, K. H., *et al.*: Am. J. Trop. Med. Hyg. 20:811, 1971.

SELECTED READINGS

Drugs for Parasitic Infections, The Medical Letter, 21:105 (December 28, 1979).

Sweeney, R. R., and Strube, R. E.: Antimalarials *in* Wolff, M.E. (ed.): Burger's Medicinal Chemistry, vol. 2, ed. 4, New York, Wiley, 1979.

Thompson, P. E., and Werbel, L. M.: Antimalarial Agents—Chemistry and Pharmacology, New York, Academic Press, 1972.

7
antibiotics

Arnold R. Martin

The accidental discovery of penicillin by Sir Alexander Fleming[1] in 1929 was the prime factor in starting the fascinating and fruitful research activities that have produced the amazingly effective anti-infective agents commonly known as antibiotics. However, it was not until Florey and Chain and their associates at Oxford (1940) undertook to apply antibiotics in therapy that Fleming's discovery became meaningful to practical medicine. Long before this, man had learned to use empirically as anti-infective material a number of crude substances which we now assume were effective because of antibiotic substances contained in them. As early as 500 to 600 B.C., the Chinese used a molded curd of soybean to treat boils, carbuncles and similar infections. Vuillemin[2] in 1889 used the term *antibiosis* (literally, against life) to apply to the biological concept of survival of the fittest in which one organism destroys another to preserve itself. It is from this root that the widely used word *antibiotic* has evolved. So broad has its use become, not only by the lay public but also by the medical professions and science in general, that the term is almost impossible to define satisfactorily. There is no knowledge today that can relate either chemically or biologically all the various substances designated as antibiotics other than by their abilities to antagonize the same or similar microorganisms.

Waksman[3] proposed the widely cited definition that "an antibiotic or an antibiotic substance is a substance produced by microorganisms, which has the capacity of inhibiting the growth and even of destroying other microorganisms." However, the restriction that an antibiotic must be a product of a microorganism is not in keeping with common use. The definition of Benedict and Langlykke[4] more aptly describes the use of the term today. They state that an antibiotic is ". . . a chemical compound derived from or produced by a living organism, which is capable, in small concentrations, of inhibiting the life processes of microorganisms." In this chapter, only those substances of importance to modern medical practice and those that meet the requirements proposed by Baron[5] (points 1, 3, and 4 below) plus one additional provision (point 2) will be included. With the present-day activity of medicinal chemists in synthesizing structural analogs of important naturally occurring medicinal agents, it has become necessary to add the qualification that permits the inclusion of synthetically obtained compounds not known to be products of metabolism. Therefore, a substance is classified as an antibiotic if:

1. It is a product of metabolism (although it may be duplicated or even have been anticipated by chemical synthesis).
2. It is a synthetic product produced as a structural analog of a naturally occurring antibiotic.
3. It antagonizes the growth and/or the survival of one or more species of microorganisms.
4. It is effective in low concentrations.

The possibility that Nature held the secret to many antibiotic substances in addition to penicillin became a driving force in the search for new compounds with the discovery by Dubos in 1939 that *Bacillus brevis* produced tyrothricin. Under the direction of S. A. Waksman, who later became a Nobel Laureate for his contributions, work leading to the isolation (1944) of streptomycin from *Streptomyces griseus* was undertaken. The discovery that this antibiotic possessed in vivo activity against *Mycobacterium tuberculosis* as well as gram-negative organisms was electrifying. Evidence was now ample that antibiotics were produced widely in nature. Broad screening programs were set up to find agents that would be effective in the treatment of infections that hitherto had been resistant to chemotherapeutic agents, as well as to provide safer and more rapid therapy for infections for which the previously available treatment had various shortcomings. The development of the broad-spectrum antibiotics such as chloramphenicol and the tetracyclines, the isolation of antifungal antibiotics such as nystatin and griseofulvin and the production of an ever-increasing number of antibiotics that may be used to treat infections that have developed resistance to some of the older antibiotics attest to the success of the many research programs on antibiotics throughout the world.

The natural scientific interest in the field of antibiotics, as well as the commercial success of antibiotics used in therapy, has led to the isolation of antibiotic substances that may now be numbered in the thousands. Of course, only a few of these have been made available for use in medical practice, because, to be useful as a drug, a substance must possess not only the ability to combat the disease process but other attributes as well. For an antibiotic to be successful in therapy, it should be decisively effective against a pathogen without producing significant toxic side-effects. In addition it should be sufficiently stable so that it can be isolated and processed and then stored for a reasonable length of time without appreciable loss in activity. It is important that it be amenable to processing into desirable dosage forms from which it may be absorbed readily. Finally, the rate of detoxification and elimination from the body should be such as to require relatively infrequent dosage to maintain proper concentration levels, yet be sufficiently rapid and complete that the removal of the drug from the body is accomplished soon after administration has been discontinued.

Relatively few substances that have shown promise as antibiotics have been able to fulfill these requirements to the extent that their commercial production has been warranted. Although the antibiotic substances that have shown sufficient promise to be named may be numbered in the hundreds, few of them have been produced in large enough quantities to place them on clinical trial, and only a few more than 5 dozen antibiotics are now released for general medical practice in the United States. To pharmacists and physicians faced with an array of dosage forms and sizes of each antibiotic, not to mention combinations, the number of antibiotics may loom large. When viewed from the standpoint of the number of microorganisms and other living organisms investigated for antibiotic activity, when considered from the standpoint of research activity and cost, and when evaluated from the standpoint of the needs yet remaining for agents that will successfully combat infectious diseases for which there are no satisfactory cures, the number of antibiotics successfully developed to date is not large.

Some groups of antibiotics, because of certain unique properties, have been designated for specialized uses such as the treatment of tuberculosis or fungal infections. Others are employed for cancer chemotherapy. These antibiotics are described along with other drugs of the same therapeutic class. Thus, antifungal and antitubercular antibiotics are discussed in Chapter 4 and antineoplastic antibiotics appear in Chapter 8.

The spectacular success of antibiotics in the treatment of the diseases of man has prompted the expansion of their use into a number of related fields. Extensive use of their antimicrobial power is made in veterinary medicine. The discovery that low-level administration of antibiotics to meat-producing animals resulted in faster growth, lower mortality rates and better quality has led to use of these products as feed supplements. A number of antibiotics are being used to control bacterial and fungal diseases of plants. Their use in food preservation is being studied carefully. Indeed, such uses of antibiotics have made necessary careful studies of their chronic effects on man and their effect on various commercial processes. For example, foods having low-level amounts of antibiotics may be capable of producing allergic reactions in hypersensitive persons, or the presence of antibiotics in milk may interfere in the manufacture of cheese.

The success of antibiotics in therapy and related fields has made them one of the most important products of the drug industry today. The quantity of antibiotics produced in the United States each year may now be measured in several millions of pounds and valued at billions of dollars. With research activity stimulated to find new substances to treat viral infections so far combated with limited success, with the promising discovery that some antibiotics are active against cancers that may be viral in origin, the future development of more antibiotics and increase in the amounts produced seems to be assured.

The commercial production of antibiotics for medicinal use follows a general pattern, differing in detail for each antibiotic. The general scheme may be divided into 6 steps: (1) preparation of a pure culture of the desired organism for use in inoculation of the fermentation medium; (2) fermentation during which the antibiotic is formed; (3) isolation of the antibiotic from the culture media; (4) purification; (5) assay for potency, tests for sterility, absence of pyrogens, other neccessary data; (6) formulation into acceptable and stable dosage forms.

The ability of some antibiotics such as chloramphenicol and the tetracyclines to antagonize the growth of a large number of pathogens has resulted in their being designated as "broad-spectrum" antibiotics. Others such as bacitracin and nystatin have a high degree of specificity and are classifed as "narrow-spectrum" antibiotics. Designations of spectrum of activity are of somewhat limited utility to the physician unless they are based on clinical effectiveness of the antibiotic against specific microorganisms. Many of the "broad-spectrum" antibiotics are active only in relatively high concentrations against some of the species of microorganisms often included in the "spectrum."

The manner in which antibiotics exert their actions against susceptible organisms is varied. The mechanisms of action of some of the more common antibiotics are summarized in Table 7-1. In many instances, the mechanism of action is not fully known; in a few cases, penicillins, for example, the site of action is known, but precise details of the mechanism are still under investigation. The biochemical processes of microorganisms are lively subjects for research, since an understanding of those mechanisms that are peculiar to the metabolic systems of infectious organisms is the basis for the future development of modern chemotherapeutic agents. Antibiotics that interfere with those metabolic systems found in microorganisms and not in mammalian cells are the most successful anti-infective agents. For example, those antibiotics that interfere with the synthesis of bacterial cell walls have

a high potential for selective toxicity. The fact that some antibiotics structurally resemble some essential metabolites of microorganisms has suggested that competitive antagonism may be the mechanism by which they exert their effects. Thus, cycloserine is believed to be an antimetabolite for D-alanine, a constituent of bacterial cell walls. Many antibiotics selectively interfere with microbial protein (e.g., the aminoglycosides, the tetracyclines, the macrolides, chloramphenicol and lincomycin) or nucleic acid synthesis (e.g., rifampin). Others, such as the polymyxins and the polyenes, are believed to interfere with the integrity and function of cell membranes of microorganisms. The mechanism of action of an antibiotic determines, in general, whether the agent exerts a *cidal* or a *static* action. The distinction may be important for the treatment of serious, life-threatening infections, particularly if the natural defense mechanisms of the host are either deficient or overwhelmed by the infection. In such situations, a cidal agent is obviously indicated. Much work remains to be done in this area, and, as mechanisms of actions are revealed, the development of improved structural analogs of effective antibiotics will probably continue to increase.

The chemistry of antibiotics is so varied that a chemical classification is of limited value. However, it is worthy of note that some similarities can be found, indicating, perhaps, that some antibiotics are the products of similar mechanisms in different organisms and that these structurally similar products may exert their activities in a similar manner. For example, a number of important antibiotics have in common a macrolide structure, that is, a large lactone ring. In this group are erythromycin and oleandomycin. The

TABLE 7-1

MECHANISMS OF ANTIBIOTIC ACTION

Site of Action	Antibiotic	Process Interrupted	Type of Activity
Cell wall	Bacitracin	Mucopeptide synthesis	Bactericidal
	Cephalosporins	Cell wall cross-linking	Bactericidal
	Cycloserine	Synthesis of cell wall peptides	Bactericidal
	Penicillin	Cell wall cross-linking	Bactericidal
	Vancomycin	Mucopeptide synthesis	Bactericidal
Cell membrane	Amphotericin B	Membrane function	Fungicidal
	Nystatin	Membrane function	Fungicidal
	Polymyxins	Membrane integrity	Bactericidal
Ribosomes			
50S subunit	Chloramphenicol	Protein synthesis	Bacteriostatic
	Erythromycin	Protein synthesis	Bacteriostatic
	Lincomycins	Protein synthesis	Bacteriostatic
30S subunit	Aminoglycosides	Protein synthesis and fidelity	Bactericidal
	Tetracyclines	Protein synthesis	Bacteriostatic
Nucleic acids	Actinomycin	DNA and m-RNA synthesis	Pancidal
	Griseofulvin	DNA and m-RNA synthesis	Fungicidal
DNA and/or RNA	Mitomycin C	DNA synthesis	Pancidal
	Rifampin	m-RNA synthesis	Bactericidal

tetracycline family presents a group of compounds very closely related chemically. A number of compounds contain closely related amino sugar moieties such as are found in streptomycins, kanamycins, neomycins, paromomycins and gentamicins. The antifungal antibiotics nystatin and the amphotericins (see Chap. 4) are examples of a group of conjugated polyene compounds. The bacitracins, tyrothricin and polymyxin are among a large group of polypeptides that exhibit antibiotic action. The penicillins and cephalosporins are β-lactam ring containing antibiotics derived from amino acids.

The normal biological processes of microbial pathogens are varied and complex. Thus, it seems reasonable to assume that there are many ways in which they may be inhibited and that different microorganisms that elaborate antibiotics antagonistic to a common "foe" produce compounds that are chemically dissimilar and that act on different processes. In fact, Nature has produced many chemically different antibiotics that are capable of attacking the same microorganism by different pathways. The diversity of structure in antibiotics has proved to be of real value clinically. As the pathogenic cell is called on to combat the effect of one antibiotic and, thus, develops drug resistance, another antibiotic, attacking another metabolic process of the resisting cell, will deal it a crippling blow. The development of new and different antibiotics has been a very important step in providing the means for treating resistant strains of organisms which previously had been susceptible to an older antibiotic. More recently the elucidation of biochemical mechanisms of microbial resistance to antibiotics, such as the inactivation of penicillins and cephalosporins by β-lactamase-producing bacteria, has stimulated research in the development of semisynthetic analogs that resist microbial biotransformation. The evolution of nosocomial (hospital-acquired) strains of staphylococci resistant to penicillin and gram-negative bacilli (e.g., *Pseudomonas* and *Klebsiella* sp., *Escherichia coli,* etc.) resistant often to several antibiotics has become a serious medical problem. No doubt the promiscuous and improper use of antibiotics has contributed to the emergence of resistant bacterial strains. The successful control of diseases caused by resistant strains of bacteria will require not only the development of new and improved antibiotics, but the rational use of the agents currently available, as well.

β-LACTAM ANTIBIOTICS

Antibiotics containing the β-lactam (a 4-membered cyclic amide) ring structure constitute the dominant class of agents currently employed for the chemo-

therapy of bacterial infections. The first antibiotic to be used in therapy, penicillin (penicillin G or benzyl penicillin) and a close biosynthetic relative, phenoxymethyl penicillin (penicillin V), remain the agents of choice for the treatment of infections caused by most species of gram-positive bacteria and gram-negative cocci. The discovery of a second major group of β-lactam antibiotics, the cephalosporins, and chemical modifications of naturally occurring penicillins and cephalosporins have provided semisynthetic derivatives that are variously effective against bacterial species known to be resistant to penicillin, in particular, penicillinase-producing staphylococci and gram-negative bacilli. Thus, apart from a few strains that have either inherent or acquired resistance, the vast majority of bacterial species are sensitive to one or more of the β-lactam antibiotics currently available.

In addition to a broad spectrum of antibacterial action, two additional properties contribute to the unequaled importance of β-lactam antibiotics in chemotherapy, namely, a potent and rapid cidal action against bacteria in the growth phase and very low incidence of toxic and other adverse reactions in the host. The uniquely lethal antibacterial action of these agents has been attributed to a selective inhibition of bacterial cell-wall synthesis.[6] Specifically, inhibition of the biosynthesis of the dipeptidoglycan that is needed to provide strength and rigidity to the cell wall is the basic mechanism involved. Penicillins and apparently also cephalosporins,[7] acylate a specific bacterial D-alanine transpeptidase, thus rendering it inactive for its role in forming a peptide cross-link of two linear peptidoglycan strands by transpeptidation and loss of D-alanine. A fully rigid cell wall does not form, and lysis of the developing bacterial cell eventually occurs owing to the increasing internal osmotic pressure. Nondividing bacterial cells are resistant to the action of β-lactam antibiotics, thus explaining the antagonism observed with attempted combination therapy with certain bacteristatic agents that act by inhibiting bacterial protein synthesis.

THE PENICILLINS

Until 1944, it was assumed that the active principle in penicillin was a single substance and that variation in activity of different products was due to the amount of inert materials in the samples. Now it is known that, during the biological elaboration of the antibiotic, a number of closely related compounds may be produced. These compounds differ chemically in the acid moiety of the amide side chain. Variations in this moiety produce differences in antibiotic effect and in chemical-physicial properties, including stability. Thus, it has become proper to speak of penicillins, referring to a group of compounds, and to identify each of the penicillins specifically. As each of the different penicillins was first isolated, letter designations were used in America; the British used Roman numerals.

Over 30 penicillins have been isolated from fermentation mixtures. Some of these occur naturally; others have been biosynthesized by altering the culture media so as to provide certain precursors that may be incorporated as acyl groups. Commercial production of biosynthetic penicillins today depends chiefly on various strains of *P. notatum* and *P. chrysogenum.* In recent years, many more penicillins have been prepared semisynthetically, and undoubtedly many more will be added to the list in attempts to find superior products.

Because penicillin, when it was first used in chemotherapy, was not a pure compound and exhibited varying activity among samples, it was necessary to evaluate it by microbiological assay. The procedure for assay was developed at Oxford, England, and the value became known as the Oxford Unit. One Oxford Unit is defined as the smallest amount of penicillin that will inhibit, in vitro, the growth of a strain of Staphylococcus in 50 ml. of culture media under specified conditions. Now that pure crystalline penicillin is available, the *U.S.P.* defines "Unit" as the antibiotic activity of 0.6 microgram of U.S.P. Penicillin G Sodium Reference Standard. The weight-unit relationship of the penicillins will vary with the nature of the acyl substituent and with the salt formed of the free acid. One milligram of penicillin G sodium is equivalent to 1,667 units. One milligram of penicillin G procaine is equivalent to 1,009 units. One milligram of penicillin potassium is equivalent to 1,530 units.

The commercial production of penicillin has increased markedly since its introduction. As production increased, the cost of penicillin dropped correspondingly. When penicillin was first available, 100,000 units of it sold for $20. Fluctuations in the production of penicillins through the years have reflected changes in the popularity of broad-spectrum antibiotics as compared with penicillins, the development of penicillin-resistant strains of a number of pathogens, the more recent introduction of semisynthetic penicillins, the use of penicillins in animal feeds and for veterinary purposes, and the increase in marketing problems in a highly competitive sales area.

Table 7-2 shows the general structure of the penicillins and relates the structures of the more familiar ones to their various designations.

TABLE 7-2

STRUCTURE OF PENICILLINS

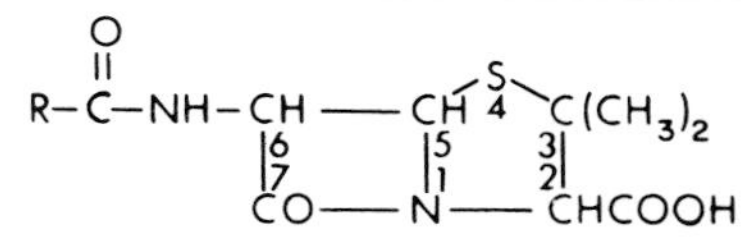

$$R-\overset{\overset{O}{\|}}{C}-NH-\underset{\underset{|7}{|6}}{CH} - \underset{\underset{|1}{|5}}{CH} \overset{S}{\underset{4}{\diagup}} \underset{\underset{2|}{3|}}{C}(CH_3)_2$$
$$CO - N - CHCOOH$$

Generic Name	Chemical Name	R Group
Penicillin G	Benzylpenicillin	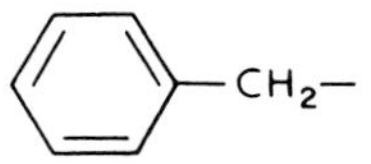 ⬡—CH₂—
Penicillin V	Phenoxymethylpenicillin	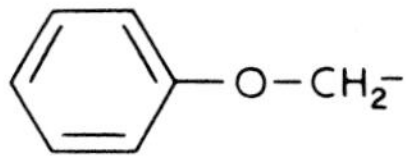 ⬡—O—CH₂⁻
Phenethicillin	Phenoxyethylpenicillin	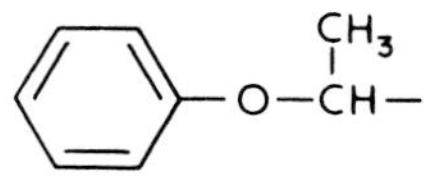 ⬡—O—CH— (CH_3)
Propicillin	(−)-Phenoxypropylpenicillin	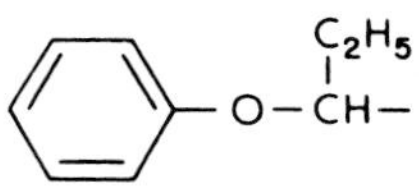 ⬡—O—CH— (C_2H_5)

(Continued)

Methicillin	2,6-Dimethoxyphenylpenicillin	
Nafcillin	2-Ethoxy-l-naphthylpenicillin	
Oxacillin	5-Methyl-3-phenyl-4-isoxazolylpenicillin	
Cloxacillin	5-Methyl-3-(2-chlorophenyl)-4-isoxazolylpenicillin	
Dicloxacillin	5-Methyl-3-(2,6-dichlorophenyl)-4-isoxazolylpenicillin	
Ampicillin	D-α-Aminobenzylpenicillin	
Amoxicillin	D-α-Amino-p-hydroxybenzyl-penicillin	
Carbenicillin	α-Carboxybenzylpenicillin	
Ticarcillin	α-Carboxy-3-thienylpenicillin	

The systematic nomenclature of penicillins is somewhat complex and very cumbersome. Two numbering systems for the fused bicyclic heterocyclic system are in existence. The *Chemical Abstracts* system initiates the numbering with the sulfur atom and assigns the ring nitrogen as the four position. Thus, penicillins are named as 4-thia-1-azabicyclo[3.2.0]heptanes according to this system. The numbering system adopted by the U.S.P. is the reverse of the *Chemical Abstracts* procedure, assigning the nitrogen atom as atom number one and the sulfur atom as number four. Three simplified forms of penicillin nomenclauture have been adopted for general use. One utilizes the name *penam* for the unsubstituted bicyclic system, including the amide carbonyl group, with one of the numbering systems described above. Thus, penicillins are generally designated according to the *Chemical Abstracts* system as 5-acylamino-2,2-dimethylpenam-3-carboxylic acids. The second, seen more frequently in the medical literature, uses the name penicillanic acid to describe the ring system with substituents that are generally present, i.e., 2,2-dimethyl and 3-carboxyl. A third form followed in this chapter uses trivial nomenclature to name the entire 6-carbonylamino-penicillanic acid portion of the molecule as penicillin and then distinguishes compounds on the basis of the R group of the acyl portion of the molecule. Penicillin G is thus named benzyl penicillin, penicillin V is phenoxymethylpenicillin, methacillin is 2,6-dimethoxyphenylpenicillin and so on. For the most part, the latter two systems serve well for naming and comparing closely similar penicillin structures, but they are too restrictive to be applied to compounds with unusual substituents or to ring-modified derivatives.

The penicillin molecule contains three asymmetric carbon atoms, i.e., C-3, C-5 and C-6. All naturally occurring and microbiologically active synthetic and semisynthetic penicillins have the same absolute configuration about these three centers. The carbon atom bearing the acylamino group (C-6) has the L-configuration, while the carbon to which the carboxyl group is attached has the D-configuration. Thus, the acylamino and carboxyl groups are *trans* to each other with the former in the α- and the latter in the β-orientation with respect to the penam ring system. The atoms comprising the 6-aminopenicillanic acid portion of the structure are biosynthetically derived from two amino acids, L-cysteine (S-1, C-5, C-6, C-7 and 6-amino) and L-valine (2,2-dimethyl, C-2, C-3 and 3-carboxyl). The absolute stereochemistry of the penicillins is designated as 3S:5R:6R as shown below.

Examination of the structure of the penicillin molecule shows it to contain a fused ring sytem of unusual design, the β-lactam thiazolidine structure. The nature of the β-lactam ring delayed the elucidation of the structure of penicillin, but its determination was reached as a result of a collaborative research program involving research groups in Great Britain and the United States during the years 1943 to 1945.[8] Attempts to synthesize these compounds resulted at best

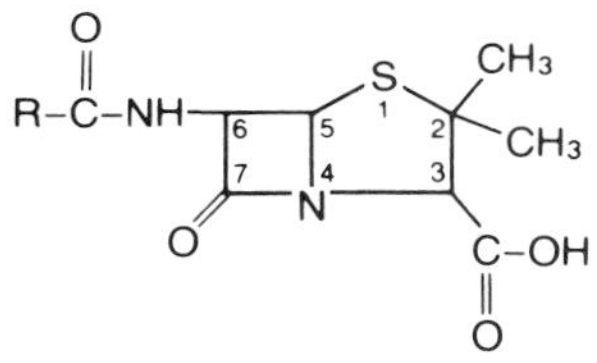

Chemical Abstracts

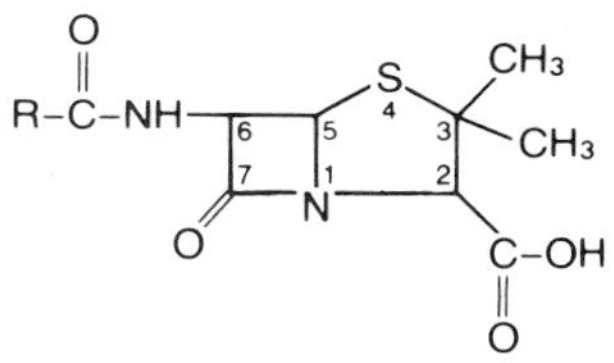

U.S.P.

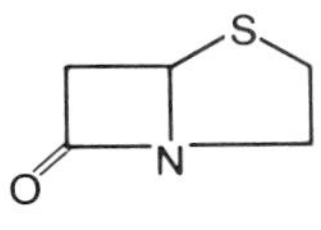

Penam

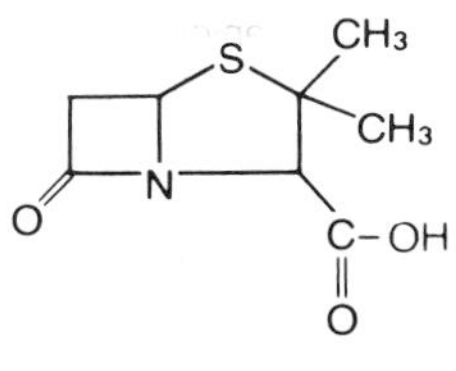

Penicillanic Acid

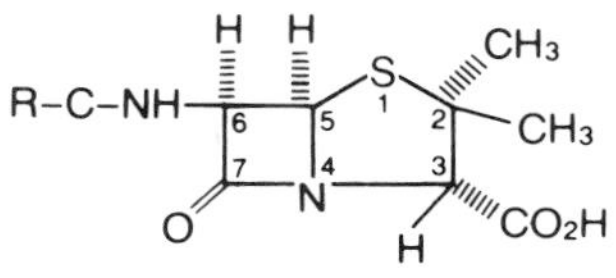

Synthesis of Phenoxymethylpenicillin

only in trace amounts until Sheehan and Henery-Logan[9] adapted techniques developed in peptide syntheses to the synthesis of penicillin V. This procedure is not likely to replace the established fermentation processes, because the last step in the reaction series develops only 10 to 12 percent of penicillin. It is of advantage in research because it provides a means of obtaining many new amide chains hitherto not possible to achieve by biosynthetic procedures.

Two other developments have provided additional means for making new penicillins. A group of British scientists, Batchelor *et al.,*[10] have reported the isolation of 6-aminopenicillanic acid from a culture of *P. chrysogenum.* This compound can be converted to

penicillins by acylation of the 6-amino group. Sheehan and Ferris[11] provided another route to synthetic penicillins by converting a natural penicillin such as penicillin G potassium to an intermediate from which the acyl side chain has been removed, which then can be treated to form biologically active penicillins with a variety of new side chains. By these procedures, new penicillins superior in activity and stability to those formerly in wide use have been found and, no doubt, others will be produced. The first commercial products of these research activities were phenoxyethylpenicillin (phenethicillin) and dimethoxyphenylpenicillin (methicillin).

The early commercial penicillin was a yellow-to-

Conversion of Natural Penicillin to Synthetic Penicillin

brown amorphous powder which was so unstable that refrigeration was required to maintain a reasonable level of activity for a short period of time. Improved procedures for purification provide the white crystalline material in use today. The crystalline penicillin must be protected from moisture, but, when kept dry, the salts will remain stable for years without refrigeration. Many penicillins have an unpleasant taste, which must be overcome in the formation of pediatric dosage forms. All of the natural penicillins are strongly dextrorotatory. The solubility and other physicochemical properties of the penicillins are affected by the nature of the acyl side chain and by the cations used to make salts of the acid. Most penicillins are acids with pKa's in the range of 2.5 to 3.0, but some are amphoteric. The free acids are not suitable for oral or parenteral administration. However, the sodium and potassium salts of most penicillins are soluble in water and are readily absorbed orally or parenterally. Salts of penicillins with organic bases, such as benzathine, procaine and hydrabamine, have limited water-solubility and are therefore useful as depot forms to provide effective blood levels over a long period in the treatment of chronic infections. Some of the crystalline salts of the penicillins are hygroscopic, making it necesary to store them in sealed containers.

The main cause of deterioration of penicillin is the reactivity of the strained β-lactam ring, particularly to hydrolysis. The course of the hydrolysis and the nature of the degradation products are influenced by the pH of the solution.[12,13] Thus, the β-lactam carbonyl group of penicillin readily undergoes nucleophilic attack by water or (especially) hydroxide ion to form the inactive penicilloic acid, which is reasonably stable in neutral to alkaline solutions but readily undergoes decarboxylation and further hydrolysis reactions in acidic solutions. Other nucleophiles, such as hydroxylamine, alkylamines and alcohols react to open the β-lactam ring to form the corresponding hydroxyamic acids, amides and esters. It has been speculated[14] that one of the causes of penicillin allergy may be the formation of antigenic penicilloyl proteins formed in vivo by the reaction of nucleophilic groups (e.g., ε-amino) on specific body proteins with the β-lactam carbonyl group. In strongly acidic solutions (pH<3) penicillin undergoes a complex series of reactions leading to a variety of inactive degradation products.[13] The first step appears to involve rearrangement to the penicillenic acid. This process is initiated by protonation of the β-lactam nitrogen followed by nucleophilic attack of the acyl oxygen atom on the β-lactam carbonyl carbon. The subsequent opening of the β-lactam ring destabilizes the thiazo-line ring which then also suffers acid catalyzed ring opening to form the penicillenic acid. The latter is very unstable and experiences two major degradation pathways. The most easily understood path involves the hydrolysis of the oxazolone ring to form the unstable penamaldic acid. Since it is an enamine, penamaldic acid easily hydrolyzes to penicillamine (a major degradation product) and the penaldic acid. The second path involves a complex rearrangement of the penicillenic acid to a penillic acid through a series of intramolecular processes that remain to be completely elucidated. Penillic acid (an imidazoline-2-carboxylic acid) readily decarboxylates and also suffers hydrolytic ring opening under acidic conditions to form a second major end product of acid catalyzed penicillin degradation, the penilloic acid. The penicilloic acid, the major product formed under weakly acidic to alkaline (as well as enzymatic) hydrolytic conditions, can not be detected as an intermediate under strongly acidic conditions. However, it is known to exist in equilibrium with the penamaldic acid and to undergo decarboxylation to form the penilloic acid in acid. The third major final product of the degradation is penicilloaldehyde formed by decarboxylation of penaldic acid (a derivative of malonaldehyde).

By controlling the pH of aqueous solutions within a range of 6.0 to 6.8, and by refrigeration of the solutions, aqueous preparations of the soluble penicillins may be stored for periods up to several weeks. The relationship of these properties to the pharmaceutics of penicillins has been reviewed by Schwartz and Buckwalter.[15] It has been noted that some buffer systems, particularly phosphates and citrates, exert a favorable effect on penicillin stability independent of the pH effect. However, Finholt *et al.*[16] have shown that these buffers may catalyze penicillin degradation if the pH is adjusted to obtain the requisite ions. Hydroalcoholic solutions of penicillin G potassium show about the same degree of instability as do aqueous solutions.[17] Since penicillins are inactivated by metal ions such as zinc and copper, it has been suggested that the phosphates and the citrates combine with these metals so as to prevent their existing as ions in solution.

Oxidizing agents also inactivate penicillins, but reducing agents have little effect on them. Temperature affects the rate of deterioration; although the dry salts are stable at room temperature and do not require refrigeration, prolonged heating will inactivate the penicillins.

Acid catalyzed degradation in the stomach contributes in a major way to the poor oral absorption of penicillin. Thus, efforts to obtain penicillins with improved pharmacokinetic and microbiological proper-

ties have sought to find acyl functionalities that would minimize sensitivity of the β-lactam ring to acid hydrolysis and at the same time maintain antibacterial activity. Substitution of an electron withdrawing group in the α-position of benzylpenicillin has been shown to markedly stabilize the penicillin to acid catalyzed hydrolysis.[18] Thus, the phenoxymethyl-, α-aminobenzyl- and α-halobenzylpenicillins are significantly more stable than benzylpenicillin in acidic solutions. The increased stability imparted by such electron withdrawing groups has been attributed to a decrease in reactivity (nucleophilicity) of the side chain amide carbonyl oxygen atom toward participation in β-lactam ring opening to form the penicillenic acid. Obviously, α-aminobenzylpenicillin (ampicillin) exists as the protonated form in acidic (as well as neutral) solutions and the ammonium group is known to be powerfully electron withdrawing.

Some bacteria, in particular most species of gram negative bacilli, are naturally resistant to the action of penicillins. Other normally sensitive species are capable of developing penicillin resistance (either through natural selection of resistant individuals or through mutation). The best understood, and probably the most important, biochemical mechanism of penicillin resistance is the bacterial elaboration of enzymes which inactivate penicillins. Such enzymes, which have been given the nonspecific name penicillinases, are of two general types: β-lactamases and acylases. By far the most important of these are the β-lactamases, enzymes which catalyze the hydrolytic opening of the β-lactam ring of penicillins to produce inactive penicilloic acids. Synthesis of bacterial β-lactamases may be under chromosomal or plasmid R-factor control and may be either constitutive or inducible (stimulated by the presence of the substrate) depending upon the bacterial species. The well known resistance among strains of *Staphylococcus aureus* is apparently due entirely to the production of a β-lactamase. Resistance among gram-negative bacilli, on the other hand, may be a result of other, poorly characterized, "resistance factors" or of β-lactamase elaboration. β-Lactamases produced by gram-negative bacilli appear to be cytoplasmic enzymes that remain in the bacterial cell, while those elaborated by *S. aureus* are synthesized in the cell wall and released extracellularly. The medical significance of these differences remains to be thoroughly examined.

Specific acylases, enzymes that are capable of hydrolyzing the acylamino side chain of penicillins, have been obtained from several species of gram-negative bacteria, but their possible role in bacterial resistance has not been well defined. These enzymes find some commercial use in the preparation of 6-aminopenicillanic acid (6-APA) for the preparation of semisynthetic penicillins. 6-APA is less active and more rapidly hydrolyzed (enzymatically and nonenzymatically) than is penicillin.

The availability of 6-APA on a commercial scale made possible the synthesis of numerous semisynthetic penicillins modified at the acyl amino side chain. Much of the early work done in the 1960s was directed toward the preparation of derivatives that would resist destruction by β-lactamases, particularly those produced by penicillin-resistant strains of *Staphylococcus aureus,* which constituted a very serious health problem at that time. In general, it was found that increasing the steric hindrance at the α-carbon of the acyl group increased resistance to staphylococcal β-lactamase, with maximal resistance being observed with quaternary substitution.[19] More fruitful from the standpoint of antibacterial potency, however, was the observation that the α-acyl carbon could be part of an aromatic (e.g., phenyl or naphthyl) or heteroaromatic (e.g., 4-isoxazoyl) system.[20] Substitutions at the ortho positions of a phenyl ring, e.g., 2,6-dimethoxyl (methacillin), or the 2-position of a 1-naphthyl system, e.g., 2-ethoxyl (nafcillin), increase the steric hindrance of the acyl group and confer increased β-lactamase resistance over the unsubstituted compounds or those substituted at positions more distant from the α-carbon. Bulkier substituents are required to confer effective β-lactamase resistance among 5-membered ring heterocyclic derivatives.[21] Thus, members of the 4-isoxazoylpenicillin family (e.g., oxacillin, cloxacillin, dicloxacillin and fluoxicillin) require both the 3-aryl and 5-methyl (or 3-methyl and 5-aryl) substituents for effectiveness against β-lactamase-producing *S. aureus.* Increasing the bulkiness of the acyl group is not without its price, however, since all of the clinically available penicillinase-resistant penicillins are significantly less active than either penicillin G or penicillin V against most non-β-lactamase-producing bacteria normally sensitive to the penicillins. The isoxazoyl penicillins, particularly those with an electronegative substituent in the 3-phenyl group (cloxacillin, dicloxacillin and fluoxicillin), are also resistant to an acid catalyzed hydrolysis of the β-lactam for the reasons described earlier. However, steric factors that confer β-lactamase resistance do not necessarily also confer stability to acid. Thus methacillin, which has electron donating groups (by resonance) ortho to the carbonyl carbon, is even more labile to acid catalyzed hydrolysis than is penicillin G.

Another highly significant advance arising from the preparation of semisynthetic penicillins has been the discovery that the introduction of an ionized or polar group into the α-position of the side chain benzyl carbon atom of penicillin G confers activity against gram-negative bacilli. Thus, derivatives with an ionized α-

amino group, such as ampicillin and amoxicillin, are generally effective against such gram-negative genera as *Escherichia, Klebsiella, Hemophilis, Salmonella, Shigella* and nonindole *Proteus.* Activity against penicillin G-sensitive, gram-positive species is, furthermore, largely retained. The introduction of an α-amino group in ampicillin (or amoxicillin) creates an additional asymmetric center. It is noteworthy that the extension of antibacterial spectrum brought about by the substituent applies only to the D-isomer, which is 2 to 8 times more active than either the L-isomer or benzylpenicillin (which are equiactive) against various species of the genera of gram-negative bacilli listed above. The basis for the expanded spectrum of activity associated with the ampicillin group is not well understood. It is apparent that the activity is not related to β-lactamase inhibition since ampicillin (and amoxicillin) are even more labile than is penicillin G to the action of β-lactamases elaborated by both *S. aureus* and by various species of gram-negative bacilli, including strains among the "ampicillin-sensitive" group. Improved penetration into gram-negative bacteria by the molecule imparted by the α-amino group has been suggested as a possible mechanism, but there is little hard evidence available to date in support of this theory. α-Hydroxy substitution also yields "expanded spectrum" penicillins with similar activity and stereoselectivity to that of the ampicillin group. However, the α-hydroxybenzylpenicillins are about 2 to 5 times less active than their corresponding α-aminobenzyl counterparts, and unlike the latter, are not very stable under acidic conditions.

Stereoselectivity of antibacterial action also extends to the phenoxymethylpenicillins derived from asymmetric phenoxyacetic acids, e.g., phenoxyethylpenicillin (phenethicillin). Thus, L-phenethicillin is 2 to 5 times more active than the corresponding D-isomer against susceptible bacterial species. It has been noted previously that greater antibacterial activity is observed for the D-isomers derived from α-aminophenylacetic acid (ampicillin and amoxicillin) and α-hydroxyphenylacetic acid (α-hydroxybenzylpenicillin). This apparent anomaly arises from the Fischer-Rosanoff (D- and L-) convention for the assignment of stereochemistry. Actually, each of the most active isomers (L-phenethicillin, D-ampicillin, D-amoxicillin and D-α-hydroxybenzylpenicillin) have the same spatial arrangements.[22]

The incorporation of an acidic substituent at the α-benzyl carbon atom of penicillin G also imparts clinical effectiveness against gram-negative bacilli and, furthermore, extends the spectrum of activity to include organisms that are resistant to ampicillin. Thus α-carboxybenzylpenicillin (carbenicillin) is active against ampicillin-sensitive, gram-negative species and additional gram-negative bacilli of the genera *Pseudomonas, Klebsiella, Enterobacter,* indole-producing *Proteus, Serratia* and *Providencia.* The potency of carbenicillin against most species of penicillin G sensitive, gram-positive bacteria is several orders of magnitude lower than that of either penicillin G or ampicillin, presumably because of poorer penetration of a more highly ionized molecule into these bacteria. (It will be noted that α-aminobenzylpenicillins exist as zwitterions over a broad pH range, and as such, are considerably less polar than is carbenicillin.) Carbenicillin is active against both β-lactamase-producing and non-β-lactamase-producing strains of gram-negative bacteria. It is known to be resistant to many of the β-lactamases produced by gram-negative bacteria, especially members of the Enterobacteriaceae family.[23] Resistance to β-lactamases elaborated by gram-negative bacteria is, therefore, probably an important component of carbenicillin's activity against some ampicillin-resistant organisms. However, β-lactamases produced by *Pseudomonas* species readily hydrolyze carbenicillin. Although carbenicillin is also somewhat resistant to staphylococcal β-lactamase, it is considerably less so than methacillin or the isoxazoyl penicillins and its inherent antistaphylococcal activity is also less impressive as compared with the β-lactamase resistant penicillins. When compared with the aminoglycoside antibiotics, the potency of carbenicillin against such gram-negative bacilli as *Pseudomonas aeruginosa, Proteus vulgaris* and *Klebsiella pneumoniae* is much less impressive. Large parenteral doses are thus required to achieve bactericidal concentrations in the plasma and in the tissues. However, the low toxicity of carbenicillin (and the penicillins, in general) usually permits (in the absence of allergy) the use of such high doses without untoward effects. Furthermore, carbenicillin (and other penicillins), when combined with aminoglycosides, exerts a synergistic cidal action against bacterial species sensitive to both agents, frequently allowing the use of a lower dose of the more toxic aminoglycoside than normally required for treatment of a life-threatening infection. Unlike the situation with ampicillin, the introduction of asymmetry at the α-benzyl carbon in carbenicillin imparts little or no stereoselectivity of antibacterial action; the individual enantiomers are nearly equally active and, furthermore, are readily epimerized to the racemate in aqueous solution. Since it is a derivative of phenylmalonic acid, carbenicillin readily decarboxylates to benzylpenicillin in the presence of acid. It is, therefore, not active (as carbenicillin) orally and must be administered parenterally. Esterification of the α-carboxyl group, e.g., as the 5-indanyl ester, partially protects the compound from acid catalyzed destruction and provides an orally active derivative that is

hydrolyzed to carbenicillin in the plasma. However, the blood levels of free carbenicillin achievable with oral administration of such esters may not be sufficiently high to effectively treat serious infections caused by some species of gram-negative bacilli such as *Pseudomonas aeruginosa.*

The nature of the acylamino side chain also determines the extent to which penicillins are plasma protein bound. Quantitative structure activity (QSAR) studies of the binding of penicillins to human serum[24,25] indicate that hydrophobic groups (positive π dependence) in the side chain appear to be largely responsible for increased binding to serum proteins. Penicillins with polar or ionized substituents in the side chain exhibit low to intermediate fractions of protein binding. Thus, ampicillin, amoxicillin and cyclacillin experience 25 to 30 percent protein binding; carbenicillin and ticarcillin show 45 to 55 percent. Those with nonpolar, lipophilic substituents (nafcillin and the isoxazoylpenicillins) are highly protein bound with fractions exceeding 90 percent. The penicillins with less complex acyl groups (benzylpenicillin, phenoxymethylpenicillin, and methacillin) fall in the range of 35 to 80 percent. Protein binding is thought to restrict the tissue availability of drugs if the fraction of binding is sufficiently high. Thus, the tissue distribution of the penicillins in the highly bound group may be inferior to that of other penicillins. The similarity of biological half-lives for various penicillins, however, indicate that plasma protein binding has little effect on their durations of action. All of the commercially available penicillins are actively secreted by the renal active transport system for anions. The reversible nature of protein binding does not compete very effectively with the active tubular secretion process.

Allergic reactions to various penicillins, ranging in severity from a variety of skin and mucous membrane rashes to drug fever and anaphylaxis, constitute the major problem associated with the use of this class of antibiotics. Estimates place the incidence of hypersensitivity to penicillin G throughout the world at between 1 and 10 percent of the population. In the United States and other industrialized countries, it is nearer to the higher figure, ranking penicillin as the most common cause of drug-induced allergy. The penicillins most frequently implicated as causes of allergic reactions are penicillin G and ampicillin. However, virtually all commercially available penicillins have been reported to cause such reactions, and, in fact, cross-sensitivity among most chemical classes of 6-acylaminopenicillanic acid derivatives has been demonstrated.[26] The chemical mechanisms by which penicillin preparations become antigenic have been studied extensively.[14] Evidence suggests that penicillins, or their rearrangement products formed in vivo (e.g., penicillenic acids), react with lysine-ε-amino groups of proteins to form penicilloyl proteins, which are major antigenic determinants.[27,28] Early clinical observations with the biosynthetic penicillins, penicillin G and penicillin V, indicated a higher incidence of allergic reactions with unpurified amorphous preparations as compared with highly purified crystalline forms, giving rise to the suggestion that small amounts of highly antigenic penicilloyl proteins present in unpurified samples were a cause. Polymeric impurities in ampicillin dosage forms have been implicated as possible antigenic determinants and as a possible explanation for the high incidence of allergic reactions with this particular semisynthetic penicillin. Ampicillin is known to undergo pH-dependent polymerization reactions (especially in concentrated solutions) involving nucleophilic attack of the side chain amino group of one molecule on the β-lactam carbonyl carbon atom of a second molecule, etc.[29] The high incidence of antigenicity shown by ampicillin polymers together with their isolation and characterization in some ampicillin preparations supports the theory that they contribute to ampicillin-induced allergy.[30]

A variety of designations have been used for classifying penicillins, based on their sources, chemistry, pharmacokinetic properties, resistance to enzymatic inactivation, antibacterial spectrum of activity, and clinical uses (Table 7-3). Thus, penicillins may be biosynthetic, semisynthetic or (potentially) synthetic; acid resistant or not; orally or (only) parenterally active; and resistant to β-lactamases (penicillinases) or not. They may have a narrow, intermediate, or broad spectrum of antibacterial activity and may be in-tended for multipurpose or limited clinical use. In the latter two connections, it is important to emphasize that designations of spectrum of activity as narrow, intermediate and broad are relative and, furthermore, do not necessarily imply the breadth of therapeutic applications of a particular antibiotic. Indeed, the classification of penicillin G as a "narrow spectrum" antibiotic has meaning only relative to other penicillins. Although the β-lactamase resistant penicillins have a spectrum of activity similar to that of penicillin G, they are generally reserved for the treatment of infections caused by penicillin G-resistant, β-lactamase-producing S. aureus, because their activity against most penicillin G sensitive bacteria is significantly inferior. Similarly, carbenicillin and ticarcillin are usually reserved for the treatment of infections caused by ampicillin-resistant, gram-negative bacilli, since they offer no advantage (and have some disadvantages) to ampicillin or penicillin G in infections sensitive to them.

PRODUCTS

Penicillin G, benzyl penicillin. For years, the most popular penicillin has been benzyl penicillin. In fact, with the exception of patients allergic to it, penicillin G remains the agent of choice for the treatment of more different kinds of bacterial infections than any other antibiotic. It first was made available in the form of the water-soluble salts of potassium, sodium and calcium. These salts of penicillin are inactivated by the gastric juice, and are not effective when administered orally unless antacids such as calcium carbonate, aluminum hydroxide and magnesium trisilicate or

TABLE 7-3

CLASSIFICATION AND PROPERTIES OF PENICILLINS

Penicillin	Source	Acid Resistance	Oral Absorption	Plasma Protein Binding %	β-Lactamase Resistant (S. aureus)	Spectrum of Activity	Clinical Use
Benzylpenicillin	Biosynthetic	Poor	Poor	50–60	No	Narrow	Multipurpose
Phenoxymethyl-penicillin	Biosynthetic	Good	Good	55–80	No	Narrow	Multipurpose
Phenethicillin	Semisynthetic	Good	Good	75–80	No	Narrow	Multipurpose
Methicillin	Semisynthetic	Poor	Poor	30–40	Yes	Narrow	Limited Use
Nafcillin	Semisynthetic	Fair	Variable	90	Yes	Narrow	Limited Use
Oxacillin	Semisynthetic	Good	Good	85–94	Yes	Narrow	Limited Use
Cloxacillin	Semisynthetic	Good	Good	88–96	Yes	Narrow	Limited Use
Dicloxacillin	Semisynthetic	Good	Good	95–98	Yes	Narrow	Limited Use
Ampicillin	Semisynthetic	Good	Fair	20–25	No	Intermediate	Multipurpose
Amoxicillin	Semisynthetic	Good	Good	20–25	No	Intermediate	Multipurpose
Cyclacillin	Semisynthetic	Good	Good	20–25	No	Intermediate	Multipurpose
Carbenicillin	Semisynthetic	Poor	Poor	50–60	No	Broad	Limited Use
Ticarcillin	Semisynthetic	Poor	Poor	45	No	Broad	Limited Use

Penicillin G Procaine

a strong buffer such as sodium citrate are added. Also, because penicillin is poorly absorbed from the intestinal tract oral doses must be very large—about 5 times the amount necessary with parenteral administration. Only after the production of penicillin had increased sufficiently so that low-priced penicillin was available did the oral dosage forms become popular. The water-soluble potassium and sodium salts are used orally and parenterally to achieve rapid high blood level concentrations of penicillin G. The more water-soluble potassium salt is usually preferred when large doses are required. However, situations in which hyperkalemia is a danger, as in renal failure, require use of the sodium salt. Of course, the potassium salt is preferred for patients on "salt-free" diets or with congestive heart conditions.

The rapid elimination of penicillin from the blood-stream through the kidneys by active tubular secretion and the need for maintaining an effective blood level concentration have led to the development of "repository" forms of this drug. Suspensions of penicillin in peanut oil or sesame oil with white beeswax added were first employed for prolonging the duration of injected forms of penicillin. This dosage form was replaced by a suspension in vegetable oil to which aluminum monostearate or aluminum distearate was added. Today, most repository forms are suspensions of high molecular weight amine salts of penicillin in a similar base.

Penicillin G Procaine U.S.P., Crysticillin®, Duracillin®, Wycillin®. The first widely used amine salt of penicillin G was made from procaine. It can be made readily from penicillin G sodium by treatment with procaine hydrochloride. This salt is considerably less soluble in water than are the alkaline metal salts, requiring about 250 ml. to dissolve 1 g. The free penicillin is released only as the compound dissolves and

dissociates. It has an activity of 1,009 units per mg. A large number of preparations for injection of penicillin G procaine are commercially available. Most of these are either suspensions in water to which a suitable dispersing or suspending agent, a buffer and a preservative have been added, or suspensions in peanut oil or sesame oil that have been gelled by the addition of 2 percent aluminum monostearate. Some of the commercial products are mixtures of penicillin G potassium or sodium with penicillin G procaine to provide a rapid development of a high blood level concentration of penicillin through the use of the water-soluble salt plus the prolonged duration of effect obtained from the insoluble salt. In addition to the injectable forms, penicillin G procaine is available in oral dosage forms. It is claimed that the rate of absorption of this salt is as rapid as that of other forms usually administered orally.

Penicillin G Benzathine U.S.P., Bicillin®, Permapen®, is N,N'-dibenzylethylenediamine dipenicillin G. Since it is the salt of a diamine, two moles of penicillin are available from each molecule of the salt. It is very insoluble in water, requiring about 5,000 ml. to dissolve 1 g. This property gives the compound great stability and prolonged duration of effect. At the pH of gastric juice it is quite stable, and food intake does not interfere with its absorption. It is available in tablet form and in a number of parenteral preparations. The activity of penicillin G benzathine is equivalent to 1,211 units per mg.

A number of other amines have been used to make penicillin salts, and research is continuing to investigate this subject. Other amines that have been used include 2-chloroprocaine, L-N-methyl-1,2-diphenyl-2-hydroxyethylamine (L-ephenamine), dibenzylamine, tripelennamine (Pyribenzamine®), and N,N'-bis-(dehydroabiethyl)ethylenediamine (hydrabamine).

Penicillin G Benzathine

Penicillin V U.S.P., Pen Vee®, V-Cillin®. Phenoxymethyl penicillin was reported by Behrens *et al.*[31] in 1948 as a biosynthetic product. However, it was not until 1953 that its clinical value was recognized by some European scientists. Since then it has enjoyed wide use because of its resistance to hydrolysis by gastric juice and its ability to produce uniform concentrations in blood (when administered orally). The free acid requires about 1,200 ml. of water to dissolve 1 g., and it has an activity of 1,695 units per mg. For parenteral solutions, the potassium salt usually is employed. This salt is very soluble in water. Solutions of it are made from the dry salt at the time of administration. Oral dosage forms of the potassium salt are also available, providing rapid blood level concentrations of this penicillin. The salt of phenoxymethyl penicillin with N,N'-bis(dehydroabietyl)ethylenediamine (hydrabamine) (Compocillin-V) provides a very long-acting form of this compound. Its high degree of water-insolubility makes it a desirable compound for aqueous suspensions used as liquid oral dosage forms.

Penicillin V

Phenethicillin Potassium U.S.P., Chemipen®, Darcil®, Syncillin®, Potassium (1-phenoxyethyl)penicillin. Late in 1959, the first of the penicillins to be produced as a result of synthetic procedures was placed on the market. It is a close structural analog of penicillin and has similar properties.

It is interesting to note that the methylene carbon between the carbonyl group and the ether oxygen of the acyl moiety in phenethicillin is asymmetric. The optical isomers have been isolated and tests have shown (−)-α-phenoxyethyl penicillin is somewhat more active than the (+)-form. However, the small difference in activity is of no clinical significance and the racemic mixture is the material made available for medical practice.

Phenethicillin Potassium

The advantages claimed for this product, which differs from penicillin V only by a methyl group on the acyl moiety, include high stability in acidic solutions, high resistance to degradation by penicillinase and unusually high blood level concentrations when given by oral administration. Observations indicate that phenethicillin yields a blood level concentration higher than that obtained by intramuscular injection of penicillin G and about twice the level obtained by an equivalent oral dose of penicillin. However, phenethicillin is intrinsically less active than penicillin V, in vitro, against most strains of penicillin-sensitive bacteria.

Like penicillin G, phenethicillin is used as the potassium salt. It is recommended for its effect against Streptococci, *Diplococcus pneumoniae,* Neisseria and *Staphylococcus aureus.* Most gram-negative organisms, the Rickettsiae, syphilis and infections resulting in endocarditis or meningitis are resistant to phenethicillin. Of interest is the report that some strains of staphylococci that are resistant to other penicillins have been inhibited by this penicillin in vitro. It appears to have the ability to produce some of the allergic reactions that develop in the use of other penicillins. One mg. is approximately equivalent to 1,600 U.S.P. Units.

Methicillin Sodium U.S.P., Staphcillin®, 2,6-dimethoxyphenyl penicillin sodium. During 1960, the second penicillin produced as a result of the research that developed synthetic analogs was introduced for medicinal use. By reacting 2,6-dimethoxybenzoyl chloride with 6-aminopenicillanic acid, 6-(2,6-dimethoxybenzamido)penicillanic acid forms. The sodium salt is a white crystalline solid that is extremely soluble in water, forming clear neutral solutions. Like other penicillins, it is very sensitive to moisture, losing about half of its activity in 5 days at room temperature. Refrigeration at 5° reduces the loss in activity to about 20 percent in the same period. Solutions prepared for parenteral use may be kept as long as 24 hours if refrigerated. It is extremely sensitive to acid, a pH of 2 causing a 50 percent loss in activity in 20 minutes; thus it cannot be used orally.

Methicillin Sodium

Methicillin sodium is particularly resistant to inactivation by penicillinase found in staphylococcal organisms and somewhat more resistant than penicillin G to penicillinase from *B. cereus.* Methicillin and many other penicillinase-resistant penicillins are inducers of penicillinase, an observation that has implications against the use of these agents in the treatment of penicillin-G-sensitive infections. Clearly the use of a penicillinase-resistant penicillin should not be followed by penicillin G.

It may be assumed that the absence of the benzyl methylene group of penicillin G and the steric protection afforded by the 2- and 6-methoxy groups makes this compound particularly resistant to enzyme hydrolysis.

Methicillin sodium has been introduced for use in the treatment of staphylococcal infections due to strains found resistant to other penicillins. It is recommended that it not be used in general therapy to avoid the possible widespread development of organisms resistant to it.

Oxacillin Sodium U.S.P., Prostaphlin®, (5-methyl-3-phenyl-4-isoxazolyl)penicillin sodium monohydrate. Oxacillin sodium is the salt of a semisynthetic penicillin that is highly resistant to inactivation by penicillinase. Apparently, the steric effects of the 3-phenyl and 5-methyl groups of the isoxazolyl ring prevent the binding of this penicillin to the β-lactamase active site and thus protect the lactam ring from degradation in much the same way as has been suggested for methicillin. It is also relatively resistant to acid hydrolysis and, therefore, may be administered orally with good effect.

Oxacillins

X, Y = H: Sodium Oxacillin
X = Cl; Y = H: Sodium Cloxacillin
X, Y = Cl: Sodium Dicloxacillin
X = Cl; Y = F: Sodium Floxicillin

Oxacillin sodium, which is available in capsule form, is well absorbed from the gastrointestinal tract, particularly in fasting patients. Effective blood levels of oxacillin are obtained in about 1 hour, but despite extensive plasma protein binding, it is rapidly excreted through the kidneys.

The use of oxacillin and other isoxazolyl penicillins should be restricted to the treatment of infections caused by staphylococci resistant to penicillin G. Although their spectrum of activity is similar to that of penicillin G, the isoxazolyl penicillins are, in general, inferior to it and the phenoxymethyl penicillins for the treatment of infections caused by penicillin G-sensitive bacteria. Since they cause allergic reactions similar to those produced by other penicillins, the isoxazolyl penicillins should be used with great caution in patients that are penicillin-sensitive.

Cloxacillin Sodium U.S.P., Tegopen®, [3-(o-chlorophenyl)-5-methyl-4-isoxazolyl]penicillin sodium monohydrate. The chlorine atom ortho to the position of attachment of the phenyl ring to the isoxazole ring enhances the activity of this compound over that of oxacillin, not by an increase in intrinsic activity or absorption, but by achieving higher blood plasma levels. In almost all other respects it resembles oxacillin.

Dicloxacillin Sodium U.S.P., Dynapen®, Pathocil®, Veracillin®, [3-(2,6-dichlorophenyl)-5-methyl-4-isoxazolyl]penicillin sodium monohydrate. The substitution of chlorine atoms on both carbons ortho to the position of attachment of the phenyl ring to the isoxazole ring is presumed to further enhance the stability of this oxacillin congener and to produce high plasma concentrations of dicloxacillin. Its medicinal properties and use are like those of cloxacillin sodium. However, progressive halogen substitution also increases the fraction of protein binding in the plasma, potentially reducing the concentration of free antibiotic in the plasma and in the tissues. Its medicinal properties and use are like those of cloxacillin sodium.

Nafcillin Sodium U.S.P. Unipen®, 6-(2-ethoxyl-1-naphthyl)penicillin sodium. Nafcillin sodium is another semisynthetic penicillin produced as a result of the search for penicillinase-resistant compounds. Like oxacillin, it is resistant to acid hydrolysis also. Like methicillin and oxacillin, nafcillin has substituents in positions ortho to the point of attachment of the aromatic ring to the carboxamide group of penicillin. No doubt, the ethoxy group and the second ring of the naphthalene group play steric roles in stabilizing nafcillin against penicillinase. Very similar structures have been reported to produce similar results in some substituted 2-biphenylpenicillins.[20]

Nafcillin Sodium

Nafcillin sodium may be used in infections caused solely by penicillin-G-resistant staphylococci or when streptococci are present also. Although it is recommended that it be used exclusively for such resistant infections, it is effective also against pneumococci and Group A beta-hemolytic streptococci. Since, like other penicillins, it may cause allergic side-effects, it should be administered with care. When given orally, it is absorbed somewhat slowly from the intestine, but satisfactory blood levels are obtained in about 1 hour.

Relatively small amounts are excreted through the kidneys, with the major portion excreted in the bile. Even though some cyclic reabsorption from the gut may thus occur, nafcillin should be readministered every 4 to 6 hours when given orally. This salt is readily soluble in water and may be administered intramuscularly or intravenously to obtain high blood level concentrations quickly for the treatment of serious infections.

Ampicillin U.S.P., Penbritin®, Polycillin®, Omnipen®, Amcil®, Principen®, 6-[D-α-aminophenylacetamido]penicillanic acid, D-α-aminobenzylpenicillin.

With ampicillin another goal in the research on semisynthetic penicillins—an antibacterial spectrum broader than that of penicillin G—has been attained. This product is active against the same gram-positive organisms that are susceptible to other penicillins, and it is more active against some gram-negative bacteria and enterococcal infections. Obviously, the α-amino group plays a significant role in the broader activity, but the mechanism for its action is not known. It has been suggested that the amino group confers an ability to cross cell wall barriers that are impenetrable to other penicillins. It is noteworthy that D-(−)-ampicillin, prepared from D-(−)-phenylalanine, is significantly more active than L-(+)-ampicillin.

Ampicillin

Ampicillin is not resistant to penicillinase, and it produces the allergic reactions and other untoward effects that are found in penicillin-sensitive patients. However, because such reactions are relatively few, it may be used in some infections caused by gram-negative bacilli for which a broad spectrum antibiotic such as a tetracycline or chloramphenicol may be indicated but not preferred because of undesirable reactions or lack of bactericidal effect. However, ampicillin is not so widely active that it should be used as a broad-spectrum antibiotic in the same manner as the tetracyclines. It is particularly useful for the treatment of acute urinary tract infections caused by *Escherichia coli* or *Proteus mirabilis* and is the agent of choice against *Haemophilus influenzae* infections. Ampicillin, together with probenecid to inhibit its active tubular excretion, has also become a treatment of choice for gonorrhea in recent years. However, β-lactamase-producing strains of gram-negative bacteria that are highly resistant to ampicillin appear to be increasing in the world population. The threat from such resistant strains is particularly great with *Hemophilus in-*

fluenzae and *Neisseria gonorrhea* because few alternative therapies for infections caused by these organisms are available. Incomplete absorption together with excretion of effective concentrations in the bile may contribute to the effectiveness of ampicillin in the treatment of salmonellosis and shigellosis.

Ampicillin is water-soluble and stable to acid. The protonated α-amino group of ampicillin has a pKa of 7.3[32] and is thus extensively protonated in acidic media, which explains ampicillin's stability toward acid hydrolysis and instability toward alkaline hydrolysis. It is administered orally and is absorbed from the intestinal tract to produce peak blood level concentrations in about 2 hours. Oral doses must be repeated about every 6 hours, because it is rapidly excreted unchanged through the kidneys. It is available as a white, crystalline, anhydrous powder that is sparingly soluble in water or as the colorless or slightly buff-colored crystalline trihydrate that is soluble in water. Either form may be used for oral administration either in capsules or as a suspension. Earlier claims of higher blood levels for the anhydrous form as compared with the trihydrate following oral administration have recently been disputed.[33,34] The white, crystalline sodium salt is very soluble in water and solutions for injections should be administered within one hour after being made.

Hetacillin U.S.P., Versapen®, is prepared by the reaction of ampicillin with acetone. In aqueous solution it is rapidly converted back to ampicillin and acetone. The spectrum of antibacterial action is identical with that of ampicillin and probably is due to the hydrolysis product, ampicillin. Although hetacillin is more slowly excreted than ampicillin, initial blood levels are lower following equivalent oral doses. Thus, it appears that hetacillin represents only another form in which to administer ampicillin and offers no advantages over it. Hetacillin occurs as a fine, off-white powder that is freely soluble in water and in alcohol. It is available for intramuscular administration in a preparation together with lidocaine for patients unable to take it orally. The water-soluble potassium salt is used for intravenous administration.

Hetacillin

Pivampicillin, Bacampicillin, and **Talampicillin** are prodrug derivatives of ampicillin that show improved absorption over the parent drug following oral administration. Each of these prodrug forms of ampicillin are acid-stable ester derivatives, which undergo hydrolysis of the ester function in the plasma to release ampicillin. Pivampicillin is the pivaloyloxymethyl; bacampicillin, the ethoxycarbonylethyl; and talampicillin, the phthalidyl ester of ampicillin. Each is employed as the hydrochloride salt. Clinical studies indicate greater oral effectiveness of each of these agents as compared to ampicillin. Of the three prodrugs, talampicillin reportedly releases ampicillin more rapidly in the plasma.[35] Bacampicillin has recently been approved for clinical use in the United States and is marketed as Spectrobid®.

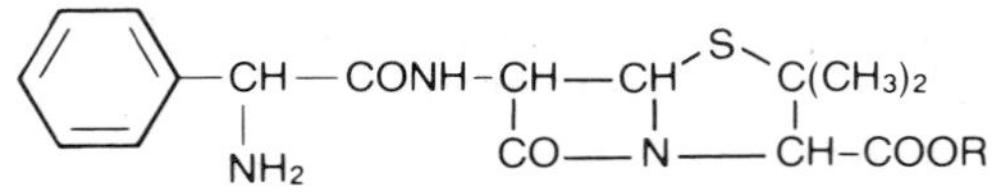

Pivampicillin R = $(CH_3)_3COOCH_2-$

Bacampicillin R = $C_2H_5COOCH(CH_3)-$

Talampicillin R =

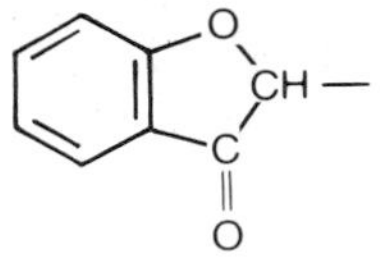

Amoxicillin U.S.P., Amoxil®, Larotid®, Polymox®, 6-[D-(−)-α-amino-p-hydroxyphenyl-acetamido]penicillanic acid. Amoxicillin, a semisynthetic penicillin introduced in 1974, is simply the p-hydroxy analog of ampicillin prepared by the acylation of 6-APA with D-tyrosine. Its antibacterial spectrum is nearly identical to that of ampicillin and it, like ampicillin, is also resistant to acid, susceptible to alkaline and β-lactamase hydrolysis and weakly protein bound. Early clinical reports[36] indicated that orally administered amoxicillin possesses significant advantages over ampicillin, including: more complete gastrointestinal absorption to give higher plasma and urine levels, less diarrhea, and little or no effect of food on absorption. Thus, it appears that amoxicillin may replace ampicillin for the treatment of certain systemic and urinary tract infections wherein oral administration is desirable, particularly if relative costs become more competitive. Amoxicillin is reported to be less effective than ampicillin in the treatment of bacillary dysentery presumably because of its greater gastrointestinal absorption.

Amoxicillin is a fine, white to off-white crystalline powder that is sparingly soluble in water. It is available in a variety of oral dosage forms. Aqueous suspensions are stable for one week at room temperature.

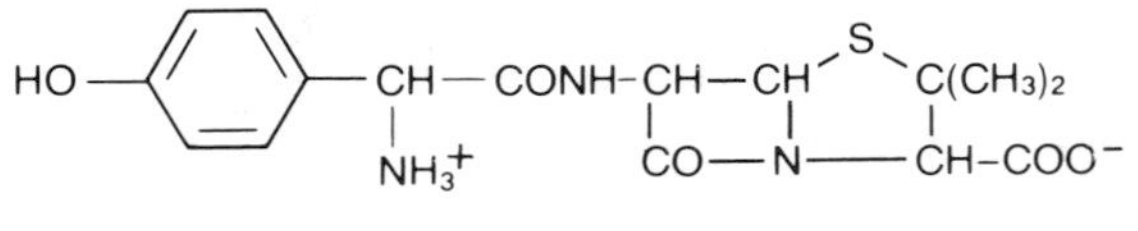

Amoxicillin

Cyclacillin U.S.P., Cyclapen®, 1-aminocyclohexylpenicillin was approved for introduction into the American market in 1979. This agent has properties that are very similar to those of ampicillin. Its spectrum of antibacterial activity is virtually identical. Although it is somewhat resistant to most β-lactamases, cyclacillin is not very active against β-lactamase-producing strains of *S. aureus* or gram-negative bacilli. Furthermore, its potency against most species of ampicillin-sensitive bacteria is 25 to 50 percent lower than ampicillin's. The major advantages of cyclacillin appear to derive from faster and more complete oral absorption and less tendency to form potentially antigenic polymers. Indeed, higher blood levels following oral administration and a significantly lower incidence of skin rashes for cyclacillin as compared with ampicillin have been reported in early clinical studies.

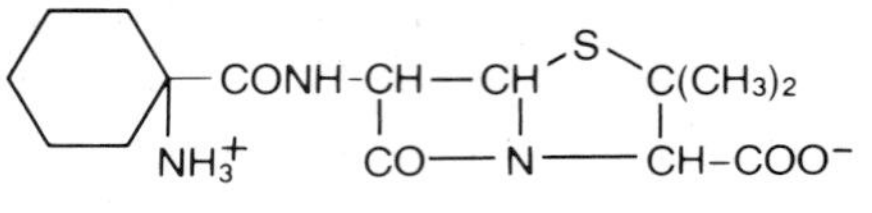

Cyclacillin

Carbenicillin Disodium, Sterile U.S.P., Geopen®, Pyopen®, disodium α-carboxybenzylpenicillin. A new semisynthetic penicillin released in the United States in 1970 is carbenicillin, a product introduced in England and first reported by Acred *et al.*[37] in 1967. Examination of its structure shows that it differs from ampicillin by having an ionizable carboxyl group substituted on the *alpha* carbon atom of the benzyl side chain rather than an amino group. Carbenicillin has a broad range of antimicrobial activity, broader than any other known penicillins, a property attributed to the unique carboxyl group. It has been proposed that the carboxyl group confers improved penetration of the molecule through cell wall barriers of gram-negative bacilli as compared with other penicillins.

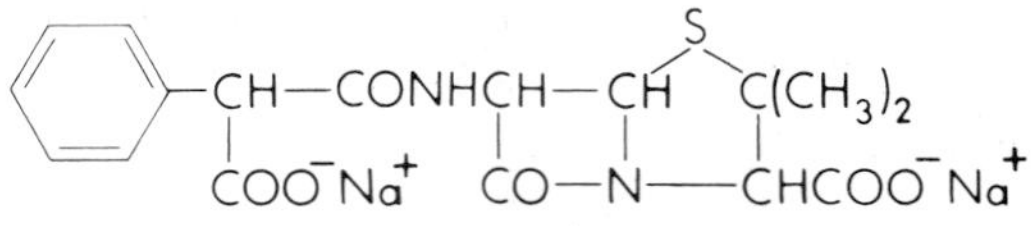

Carbenicillin

Carbenicillin is not stable in acids and is inactivated by penicillinase. It is a malonic acid derivative and thus decarboxylates readily to penicillin G, which is acid-labile. Solutions of the disodium salt should be freshly prepared, but may be kept for 2 weeks when refrigerated. It must be administered by injection and is usually given intravenously.

Carbenicillin has been effective in the treatment of systemic and urinary tract infections due to *Pseudomonas aeruginosa,* indole-producing *Proteus* and *Pro-*

videncia species, all of which are resistant to ampicillin. The low toxicity of carbenicillin, with the exception of allergic sensitivity, permits the use of large dosages in serious infections. Most clinicians prefer to use a combination of carbenicillin and gentamicin for serious *Pseudomonas* and mixed coliform infections. However, the two antibiotics are chemically incompatible and should never be combined in the same intravenous solution.

Carbenicillin Indanyl Sodium U.S.P. Geocillin®, 6-[2-phenyl-2-(5-indanyloxycarbonyl)acetamido]penicillanic acid. Efforts to obtain orally active forms of carbenicillin led to the eventual release of the 5-indanylester in 1972. Approximately 40 percent of the usual oral dose of indanyl carbenicillin is absorbed. Following absorption the ester is rapidly hydrolyzed by plasma and tissue esterases to yield carbenicillin. Thus, despite the fact that the highly lipophilic and highly protein-bound ester has in vitro activity comparable to carbenicillin, its activity in vivo is due to carbenicillin. Indanyl carbenicillin thus provides an orally active alternative for the treatment of carbenicillin-sensitive systemic and urinary tract infections caused by *Pseudomonas* indole-positive *Proteus* species and selected species of gram-negative bacilli.

In clinical trials with indanyl carbenicillin, a relatively high incidence of gastrointestinal symptoms (nausea, occasional vomiting and diarrhea) was reported. It seems doubtful the high doses required for the treatment of serious systemic infections could be tolerated by most patients. Indanyl carbenicillin occurs as the sodium salt, an off-white, bitter tasting powder that is freely soluble in water. It is stable to acid and resistant to penicillinase. It should be protected from moisture to prevent hydrolysis of the ester.

Carbenicillin Indanyl Sodium

Ticarcillin Disodium, Sterile U.S.P., Ticar®, α-carboxy-3-thienylpenicillin, is an isostere of carbenicillin wherein the phenyl group is replaced by a thienyl group. This semisynthetic penicillin derivative, like carbenicillin, is unstable in acid and must therefore be administered parenterally. It is similar to carbenicillin in antibacterial spectrum and pharmacokinetic properties. Two advantages for ticarcillin are claimed: (1) slightly better pharmacokinetic properties, including higher serum levels and a longer duration of action, and (2) greater in vitro potencies against several species of gram-negative bacilli, most notably *Pseudomonas aeruginosa* and *Bacteroides fragilis*. Whether these theoretical advantages will be of importance clinically remains to be firmly established.

Ticarcillin Disodium

Piperacillin is the more promising of a series of semisynthetic α-amino-substituted acyl derivatives of ampicillin developed in Japan. It has been discovered that N-ureido and N-guanyl substituents (e.g., 4-ethyl-2,3-dioxo-1-piperazinylcarbonyl in the case of piperacillin) on the α-amino group of ampicillin create penicillins that combine the antimicrobial properties of ampicillin and carbenicillin. Thus piperacillin and other ureido ampicillins (e.g., azlocillin and mezlocillin) exhibit broad spectrum activity in vitro against gram-positive and gram-negative bacteria.[38,39] More significant, however, is the fact that piperacillin is several times more potent than either carbenicillin or ticarcillin against *Pseudomonas aeruginosa*. Its activity against most other ampicillin and carbenicillin-sensitive bacteria is comparable to, or slightly greater than, that of the standard penicillins. Additionally, it is active against some β-lactamase-producing strains, including *Hemophilus influenzae* and *Enterobacter*. Most β-lactamase-producing strains of Enterobacteriaceae and *Staphylococcus aureus* are resistant, however, and a significant inoculum effect is observed with it when tested against some gram-negative bacilli including *Ps. aeruginosa*. The ultimate fate of piperacillin and other ureido and guanyl derivatives of ampicillin in therapy awaits the results of their clinical evaluation.

Piperacillin

Mecillinam, 6β-(hexahydro-1H-azepin-1-yl)-methyleneaminopenicillanic acid and its analogues differ structurally from other penicillins in that they are not acyl derivatives, but rather alkylideneamino (or amidino) derivatives of 6-APA. This structural difference confers unique biochemical and antimicrobial properties to the 6β-amidinopenicillins which may prove to be of medical significance. In contrast to most other penicillins, mecillinam has significantly greater gram-negative, as compared with gram-positive, antibacterial activity.[40] It is particularly active against the *Enterobacteriaceae,* including some ampicillin-resistant strains. Unfortunately, the activity of mecillinam against *Hemophilus* and *Neisseria* species is low and *Pseudomonas* species are resistant to it. Of perhaps greater interest is the observation that, although mecillinam inhibits cell wall synthesis in gram-negative bacteria, it does so by a mechanism that differs from the action of the penicillins.[41] In fact, mecillinam does not inhibit peptide cross linking and, under certain circumstances, it exerts a synergistic action when combined with conventional penicillins and cephalosporins.

Mecillinam, itself, is poorly absorbed from the gastrointestinal tract, but the acid stable pivaloyloxymethyl ester (as the hydrochloride salt) provides a form that is well absorbed and efficiently converted to mecillinam in the plasma. Mecillinam is inactivated by β-lactamases and its in vitro activity, particularly against ampicillin-resistant bacteria, is reduced by increased inoculum size.

Mecillinam

TABLE 7-4

PENICILLINS

Name Proprietary Name	Preparations	Usual Adult Dose*	Usual Dose Range*	Usual Pediatric Dose*
Penicillin G U.S.P.	Penicillin G Potassium for Injection U.S.P. Sterile Penicillin G Potassium U.S.P. Sterile Penicillin G Sodium U.S.P.	I.M. or I.V., 1,000,000 to 5,000,000 Units every 4 to 6 hours	Up to 160,000,000 Units daily	Premature and full-term newborn infants—I.M. or I.V., 30,000 Units per kg. of body weight every 12 hours; older infants and children—I.M. or I.V., 4,167 to 16,667 Units per kg. of body weight every 4 hours; 6,250 to 25,000 Units per kg. of body weight every 6 hours
	Penicillin G Sodium for Injection U.S.P.	I.M., 400,000 Units 4 times daily; I.V., 10,000,000 Units daily		
	Penicillin G Potassium Tablets U.S.P. Penicillin G Potassium Tablets for Oral Solution U.S.P.	Oral, 200,000 to 500,000 Units, every 6 to 8 hours	Up to 2,000,000 Units daily	Infants and children under 12 years of age—4,167 to 15,000 Units per kg. of body weight, every 6 hours, or 8,333 to 30,000 Units per kg. of body weight every 8 hours; children 12 years and older—see Usual Adult Dose
Crysticillin, Duracillin, Wycillin	Sterile Penicillin G Procaine Suspension U.S.P.	I.M., 600,000 to 1,200,000 Units daily	Up to 4,800,000 Units daily	Infants and children up to 32 kg. of body weight—I.M., 10,000 Units per kg. of body weight daily for 10 days
Bicillin, Permapen	Sterile Penicillin G Benzathine Suspension U.S.P.	I.M., 1,200,000 to 2,400,000 Units as a single dose; 1,200,000 Units once a month or 600,000 Units every 2 weeks	Up to 2,400,000 Units daily	Up to 2 years of age—50,000 Units per kg. of body weight as a single dose
	Penicillin G Benzathine Tablets U.S.P. Penicillin G Benzathine Oral Suspension U.S.P.	400,000 to 600,000 Units every 4 to 6 hours	Up to 12,600,000 Units daily	Infants and children up to 12 years of age—4,167 to 15,000 Units per kg. of body weight every 4 hours; 6,230 to 28,500 Units per kg. of body weight every 6 hours; or 6,333 to 30,000 Units per kg. of body weight every 8 hours

(Continued)

*See U.S.P. D.I. for complete dosage information.

TABLE 7-4

PENICILLINS

Name / Proprietary Name	Preparations	Usual Adult Dose*	Usual Dose Range*	Usual Pediatric Dose*
Penicillin V U.S.P. *Pen-Vee, V-Cillin*	Penicillin V for Oral Suspension U.S.P.	The equivalent of 125 to 500 mg. (200,000 to 800,000 Units) of penicillin every 6 to 8 hours	Up to 7.2 g. daily	Infants and children up to 12 years of age—2.5 to 4 mg. (4,167 to 16,667 Units) per kg. of body weight every 4 hours; 3.75 to 15.6 mg. (6,250 to 25,000 Units) per kg. of body weight every 6 hours; or 5 to 20.8 mg. (8,333 to 33,333 Units) per kg. of body weight every 8 hours; children 12 years and older—see Usual Adult Dose
Penicillin V Potassium U.S.P. *Pen-Vee-K, V-Cillin-K*	Penicillin V Potassium for Oral Solution U.S.P. Penicillin V Potassium Tablets U.S.P.			
Penicillin V Benzathine U.S.P. *Pen-Vee*	Penicillin V Benzathine Oral Suspension U.S.P.	125 to 250 mg. every 6 to 8 hours	125 to 375 mg. every 6 to 8 hours	
Penicillin V Hydrabamine U.S.P. *Compocillin-V*	Penicillin V Hydrabamine Oral Suspension U.S.P. Penicillin V Hydrabamine Tablets U.S.P.	125 to 250 mg. every 6 to 8 hours	125 to 375 mg. every 6 to 8 hours	
Phenethicillin Potassium U.S.P. *Chemipen, Darcil, Maxipen, Ro-Cillin, Syncillin*	Phenethicillin Potassium for Oral Solution U.S.P. Phenethicillin Potassium Tablets U.S.P.	125 to 250 mg. 3 times daily	125 to 500 mg.	
Methicillin Sodium U.S.P. *Staphcillin*	Methicillin Sodium for Injection U.S.P.	I.M., 1 g. every 4 to 6 hours; I.V., 1 g. every 6 hours	Up to 24 g. daily	I.M., 25 mg. per kg. of body weight every 6 hours; I.V., 16.7 to 33.3 mg. per kg. of body weight every 4 hours; or 25 to 50 mg. per kg. of body weight every 6 hours
Oxacillin Sodium U.S.P. *Prostaphlin*	Oxacillin Sodium Capsules U.S.P. Oxacillin Sodium for Oral Solution U.S.P.	Oral, the equivalent of 500 mg. to 1 g. of oxacillin every 4 to 6 hours	Up to 6 g. daily	Children up to 40 kg. of body weight—12.5 to 25 mg. per kg. of body weight every 6 hours daily; children 40 kg. and over—see Usual Adult Dose
	Oxacillin Sodium for Injection U.S.P.	I.M. or I.V., the equivalent of 250 mg. to 1 g. of oxacillin every 4 to 6 hours	Up to 20 g. daily	I.M. or I.V., up to 40 kg. of body weight—the equivalent of 12.5 to 25 mg. of oxacillin per kg. of body weight every 4 hours; 40 kg. of body weight and over—see Usual Adult Dose
Cloxacillin Sodium U.S.P. *Tegopen*	Cloxacillin Sodium Capsules U.S.P. Cloxacillin Sodium for Oral Solution U.S.P.	The equivalent of 250 to 500 mg. every 6 hours	Up to 6 g. daily	Up to 20 kg. of body weight—12.5 to 25 mg. per kg. of body weight every 6 hours; 20 kg. of body weight and over—See Usual Adult Dose
Dicloxacillin Sodium U.S.P. *Dynapen, Pathocil, Veracillin*	Dicloxacillin Sodium Capsules U.S.P. Dicloxacillin Sodium for Oral Suspension U.S.P. Sterile Dicloxacillin Sodium U.S.P.	The equivalent of 125 to 250 mg. of dicloxacillin every 6 hours	Up to 6 g. daily	Up to 40 kg. of body weight—3.125 to 6.25 mg. per kg. of body weight every 6 hours; 40 kg. and over—see Usual Adult Dose
Nafcillin Sodium U.S.P. *Unipen*	Nafcillin Sodium Capsules U.S.P. Nafcillin Sodium for Oral Solution U.S.P. Nafcillin Sodium Tablets U.S.P.	The equivalent of 250 mg. to 1 g. of nafcillin every 4 to 6 hours	Up to 6 g. daily	Newborn infants—the equivalent of 10 mg. of nafcillin per kg. of body weight every 4 to 6 hours; older infants and children—the equivalent of 6.25 to 12.5 mg. of nafcillin per kg. of body weight every 6 hours

(Continued)

*See U.S.P. D.I. for complete dosage information.

TABLE 7–4

PENICILLINS

Name Proprietary Name	Preparations	Usual Adult Dose*	Usual Dose Range*	Usual Pediatric Dose*
	Nafcillin Sodium for Injection U.S.P.	I.M., the equivalent of 500 mg. of nafcillin every 4 to 6 hours; I.V., 500 mg. to 1 g. every 4 hours	I.M., up to 12 g. daily; I.V., up to 20 g. daily	Newborn infants—I.M., the equivalent of 10 mg. of nafcillin per kg. of body weight twice daily; older infants and children—I.M., the equivalent of 25 mg. of nafcillin per kg. of body weight or 750 mg. per square meter of body surface, twice daily
Ampicillin U.S.P. *Penbritin, Polycillin, Omnipen, Amcill, Principen*	Ampicillin Capsules U.S.P. Ampicillin Chewable Tablets U.S.P. Ampicillin for Oral Suspension U.S.P. Sterile Ampicillin for Suspension U.S.P.	250 to 500 mg. every 6 hours	Up to 6 g. daily	Infants and children up to 20 kg. of body weight—12.5 to 25 mg. per kg. of body weight every 6 hours, or 16.7 to 33.3 mg. per kg. of body weight every 8 hours; children 20 kg. of body weight and over—see Usual Adult Dose
	Sterile Ampicillin Sodium U.S.P. *Penbritin-S, Polycillin-N, Omnipen-N*	I.M., or I.V., the equivalent of ampicillin, 250 to 500 mg. every 6 hours	Up to 16 g. daily; or up to 300 mg. per kg. of body weight daily	I.M., or I.V., the equivalent of ampicillin, infants up to 20 kg. of body weight—6.25 to 25 mg. per kg. of body weight every 6 hours; infants and children 20 kg. of body weight and over—see Usual Adult Dose
Hetacillin U.S.P. *Versapen*	Hetacillin for Oral Suspension U.S.P. Hetacillin Tablets U.S.P.	Oral, the equivalent of ampicillin, 225 to 450 mg. of every 6 hours	Up to 6 g. daily	Oral, the equivalent of ampicillin, infants and children up to 40 kg. of body weight—5.625 to 11.25 mg. per kg. of body weight every 6 hours; children 40 kg. of body weight and over—see Usual Adult Dose
Hetacillin Potassium U.S.P. *Versapen-K*	Hetacillin Potassium Capsules U.S.P.	See Dose for Oral Suspension	See Dose for Oral Suspension	See Dose for Oral Suspension
Amoxicillin U.S.P. *Amoxil, Larotid*	Amoxicillin Capsules U.S.P. Amoxicillin for Oral Suspension U.S.P.	The equivalent of anhydrous amoxicillin, 250 to 500 mg. every 8 hours	The equivalent of anhydrous amoxicillin, up to 4.5 g. daily	Oral, the equivalent of anhydrous amoxicillin, infants up to 6 kg. of body weight—25 to 50 mg. every 8 hours; infants 6 to 8 kg. of body weight—50 to 100 mg. every 8 hours; infants and children 8 to 20 kg. of body weight—6.7 to 13.3 mg. per kg. of body weight every 8 hours; children 20 kg. of body weight and over—see Usual Adult Dose
Cyclacillin *Cyclapen*	Tablets Oral Suspension	Oral, 250 mg. 4 times daily	1 to 2 g. daily	50 to 100 mg. per kg. per day
Sterile Ticarcillin Disodium U.S.P. *Ticar*	Sterile Ticarcillin Disodium U.S.P.	Septicemia, respiratory-tract, skin, and soft tissue infections—I.V. infusion, the equivalent of ticarcillin, 3 g. every 3 to 6 hours; 25 to 37.5 mg. per kg. of body weight every 3 hours; 33.5 to 50 mg. per kg. of body weight every 4 hours; or 50 to 75 mg. per kg. of body weight every 6 hours	The equivalent of ticarcillin, up to 500 mg. per kg. of body weight daily	
Carbenicillin Indanyl Sodium U.S.P. *Geocillin*	Carbenicillin Indanyl Sodium Tablets U.S.P.	Oral, the equivalent of carbenicillin, 382 to 764 mg. every 6 hours		

*See U.S.P. D.I. for complete dosage information.

(Continued)

TABLE 7–4

PENICILLINS

Name Proprietary Name	Preparations	Usual Adult Dose*	Usual Dose Range*	Usual Pediatric Dose*
	Sterile Carbenicillin Disodium U.S.P.	Septicemia, meningitis, respiratory-tract or soft-tissue infection—I.M., or I.V., the equivalent of carbenicillin, 50 to 83.3 mg. per kg. of body weight every 4 hours; urinary-tract infections—I.M., or I.V., the equivalent of carbenicillin, 1 to 2 g. every 6 hours or up to 50 mg. per kg. of body weight every 6 hours	The equivalent of carbenicillin, up to 42 g. daily	Neonates up to 2 kg. of body weight, septicemia, meningitis, respiratory-tract or soft-tissue infections—I.M., or I.V., the equivalent of carbenicillin, 100 mg. per kg. of body weight initially, then 75 mg. per kg. of body weight every 8 hours during the first week of life; 100 mg. per kg. of body weight every 6 hours thereafter; neonates 2 kg. of body weight and over, septicemia, meningitis, respiratory-tract or soft-tissue infections—I.M., or I.V., the equivalent of carbenicillin, 100 mg. per kg. of body weight initially, then 75 mg. per kg. of body weight every 6 hours during the first 3 days of life; 100 mg. per kg. of body weight every 6 hours thereafter; older infants and children, septicemia, meningitis, respiratory-tract or soft-tissue infections—see Usual Adult Dose; urinary-tract infections—I.M., or I.V., the equivalent of carbenicillin, 12.5 to 50 mg. per kg. of body weight every 6 hours, or 8.3 to 33.3 mg. per kg. of body weight every 4 hours

*See U.S.P. D.I. for complete dosage information.

(Continued)

CEPHALOSPORINS

The cephalosporins are antibiotics obtained from species of the fungus *Cephalosporium* and from semisynthetic processes. Although work began on this group of antibiotics in 1945, it has been only since 1964 that they have gained a place in therapy. The earlier developments pertaining to the isolation, the chemistry and the antibacterial properties of the cephalosporins and their relationships to the penicillins have been reviewed by Hou and Poole[12] and by Van Heyningen.[42] Compounds having three different chemical structures have been isolated from *Cephalosporium*. One of these, cephalosporin P₁, has a steroid structure. It possesses low antibacterial properties and has not been employed in clinical medicine.

Of greater interest is the antibiotic cephalosporin N that was first isolated from *C. salmosynnematum* and was given the name synnematin and then synnematin B. Its structure was determined to be D-(4-amino-4-carboxybutyl)penicillin and it is now frequently referred to as penicillin N.

Its structure shows it to be an acyl derivative of 6-APA and D-α-aminoadipic acid. The unusual zwitterionic side chain produces a compound less effective

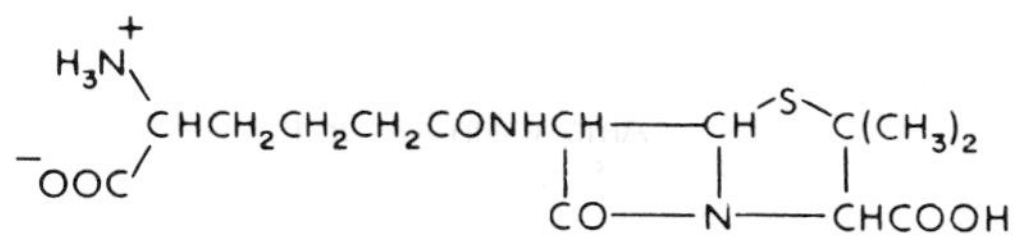

Penicillin N
(Cephalosporin N, Synnematin B)

against gram-positive organisms than are other penicillins. However, it is more active than penicillin G against a number of gram-negative organisms and particularly some of the salmonellae. It has been employed successfully in clinical trials for the treatment of typhoid fever but it has not been released as an approved drug.

The third antibiotic isolated from Cephalosporia is cephalosporin C. Its structure shows it to be a congener of penicillin N, containing a dihydrothiazine ring instead of the thiazolidine ring of the penicillins. Because early studies of the antibacterial properties of cephalosporin C showed it to be similar in spectrum to penicillin N but less active, interest in it was not great in spite of its resistance to degradation by penicillinase. However, the discovery that the α-aminoadipoyl side chain could be removed to efficiently produce 7-

aminocephalosporanic acid (7-ACA)[43,44] prompted investigations that have led to semisynthetic cephalosporins of medicinal value. The relationship of 7-ACA and its acyl derivatives to 6-APA and the semisynthetic penicillins is obvious. Woodward *et al.*[45] have prepared both cephalosporin C and the clinically useful cephalothin by an elegant synthetic procedure, but the commercially available drugs are obtained as semisynthetic products from 7-ACA.

The systematic chemical nomenclature of the cephalosporins is slightly more complex than even that of the penicillins, because of the presence of a double bond in the dihydrothiazine ring. The fused ring system is designated by *Chemical Abstracts* as 5-thia-1-azabicyclo[4.2.0]oct-2-ene. Using this system, cephalothin is 3-(acetoxymethyl)-7-(2-thienyl)-8-oxo-5-thia-1-azabicyclo[4.2.0]oct-2-ene-2-carboxylic acid. A simplification that retains some of the systematic nature of the *C.A.* procedure is to name the saturated bicyclic ring system with the lactam carbonyl oxygen as *cepham* (cf. penam for penicillins). According to this system all of the commercially available cephalosporins and cephamycins are named as 3-*cephems* (or Δ^3-cephems) to designate the position of the double bond. (Interestingly, all known 2-cephems are inactive, presumably because the β-lactam lacks the necessary ring strain to be sufficiently reactive. The trivialized forms of nomenclature of the type that have been applied to the penicillins are not consistently applicable to the naming of cephalosporins because of variations in the substituent at the 3-position. Thus, although some cephalosporins have been named as derivatives of cephalosporanic acids, this practice applies only to the derivatives that have a 3-acetoxymethyl group.

In the preparation of semisynthetic cephalosporins the following improvements are sought: (1) increased acid stability; (2) improved pharmacokinetic properties, particularly better oral absorption; (3) broadened antimicrobial spectrum; (4) increased activity against resistant micro-organisms (as a result of resistance to enzymatic destruction, improved penetration, increased receptor affinity, etc.); (5) decreased allergenicity; and (6) increased tolerance following parenteral administration.

Structures of cephalosporins currently marketed in the United States are represented in Table 7-5.

TABLE 7-5

STRUCTURE OF CEPHALOSPORINS

Generic Name	R_1	R_2
1. ORAL CEPHALOSPORINS		
Cephaloglycin	(phenyl)—CH(NH₂)—	$-CH_2OCOCH_3$
Cephalexin	(phenyl)—CH(NH₂)—	$-CH_3$
Cephradine	(cyclohexadienyl)—CH(NH₂)—	$-CH_3$
Cefadroxil	HO—(phenyl)—CH(NH₂)—	$-CH_3$
Cefaclor	(phenyl)—CH(NH₂)—	$-Cl$

Table 7–5 (Cont.)

STRUCTURE OF CEPHALOSPORINS

2. PARENTERAL CEPHALOSPORINS

Cephalothin

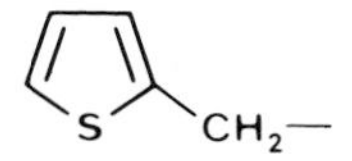

(thiophene ring)–CH_2–   –CH_2OCOCH_3

Cephaloridine

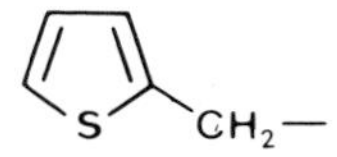

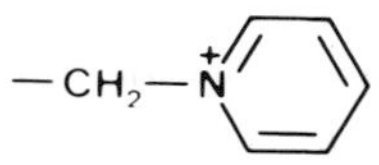

(thiophene ring)–CH_2–   –CH_2–N$^+$(pyridinium)

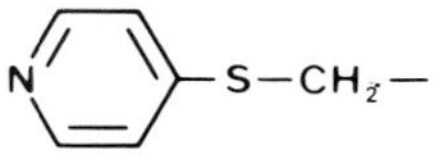

Cephapirin

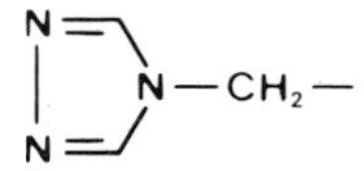

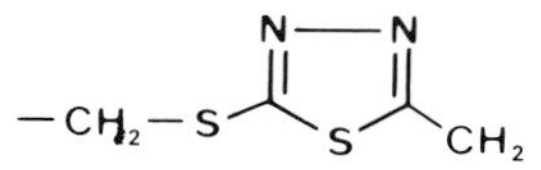

(pyridine ring)–S–CH_2–   –CH_2OCOCH_3

Cefazolin

(tetrazole ring)N–CH_2–   –CH_2–S–(thiadiazole ring)–CH_2

Cefamandole

(phenyl ring)–$\underset{\underset{OH}{|}}{CH}$–   –$CH_2$–$S$–(triazole ring with N–$CH_3$)

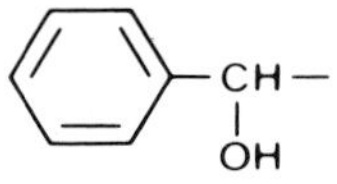

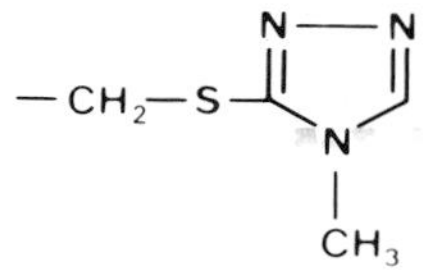

3. PARENTERAL CEPHAMYCIN

Cefoxitin

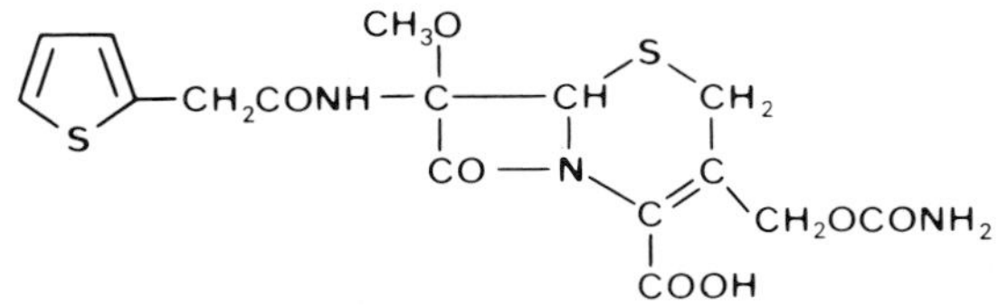

(thiophene ring)–CH_2CONH–$\underset{\underset{CO-N}{|}}{\overset{\overset{CH_3O}{|}}{C}}$ ... –CH_2OCONH_2, $COOH$

To date the more useful semisynthetic modifications of the basic 7-ACA nucleus have resulted from acylations of the 7-amino group with different acids or nucleophilic substitution or reduction or the acetoxyl group. Structure activity relationships among the cephalosporins appear to parallel those among the penicillins insofar as the acyl group is concerned. However, the presence of an allylic acetoxyl function in the 3-position provides a reactive site at which various 7-acylamino cephalosporanic acid structures can easily be varied by nucleophilic displacement reactions. Reduction of the 3-acetoxymethyl to a 3-methyl substituent to prepare 7-ACDA (7-aminodesacetyl-cephalosporanic acid) derivatives can be accomplished by catalytic hydrogenation, but the process currently employed for the commercial synthesis of 7-ACDA derivatives involves the rearrangement of the corresponding penicillin sulfoxide.[46] Perhaps the most noteworthy development thus far is the discovery that 7-phenylglycyl derivatives of 7-ACA and especially 7-ACDA are active orally.

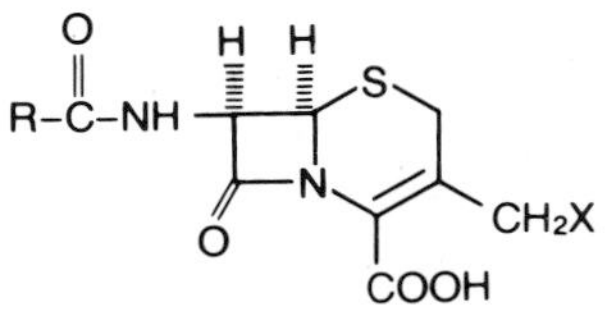

Cephalosporins

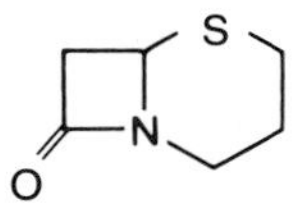

Cepham

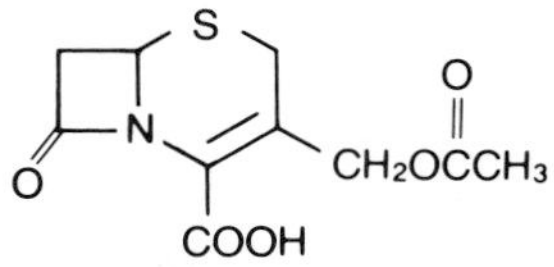

Cephalosporanic Acid

The oral activity conferred by the phenylglycyl substituent is attributed to increased acid stability of the lactam ring resulting from the presence of a protonated amino group on the 7-acylamino portion of the molecule. The situation, then, is analogous to that of the α-aminobenzylpenicillins (e.g., ampicillin). Also important for high acid stability (and, therefore, good oral activity) of the cephalosporins is the absence of the leaving group at the 3-position. Thus, despite the presence of the phenylglycyl side chain in its structure, the cephalosporanic acid derivative, cephaloglycin, is poorly absorbed orally, presumably because of solvolysis of the 3-acetoxyl group in the low pH of the stomach. The resulting 3-hydroxyl derivative is known to undergo lactonization under acidic conditions. The 3-hydroxyl derivatives and especially the corresponding lactones are considerably less active in vitro than the parent cephalosporins. Generally, acyl derivatives of 7-ACDA show lower in vitro antibacterial potencies than the corresponding 7-ACA analogues.

Hydrolysis of the ester function, catalyzed by hepatic and renal esterases, is responsible for some in vivo inactivation of parenteral cephalosporins containing a 3-acetoxymethyl substituent (e.g., cephalothin and cephapirin). The extent of such inactivation (20–35%) is not sufficiently great to compromise seriously the in vivo effectiveness of acetoxyl cephalosporins. Parenteral cephalosporins lacking a hydrolyzable group at the 3-position (cephaloridine, cefazolin and cephamandole) are, of course, not subject to hydrolysis by hepatic esterases.

The cephalosporins are considered broad spectrum antibiotics with patterns of antibacterial effectiveness comparable to ampicillin's. Several significant differences exist, however. Cephalosporins are much more resistant to inactivation by β-lactamases, particularly those produced by gram-positive bacteria, than is ampicillin. However, ampicillin is generally more active against non-β-lactamase-producing strains of gram-positive and gram-negative bacteria sensitive to both it and the cephalosporins. Cephalosporins, among β-lactam antibiotics, exhibit uniquely potent activity against most species of *Klebsiella*. Differential potencies of cephalosporins as compared to penicillins against different species of bacteria have been attributed to several variable characteristics of individual bacterial species and strains, the most important of which probably are: (1) resistance to inactivation by β-lactamases, (2) permeability into bacterial cells, and (3) intrinsic activity against bacterial enzymes involved in cell wall synthesis and cross-linking.

The susceptibility of cephalosporins to various β-lactamases varies considerably with the source and properties of these enzymes. Cephalosporins are significantly less sensitive than all but the β-lactamase-resistant penicillins to hydrolysis by the enzymes from *Staphylococcus aureus* and *Bacillus subtilis*. The "penicillinase" resistance of cephalosporins appears to be a property of the bicyclic cephem ring system rather than of the acyl group. Despite natural resistance to staphylococcal β-lactamase, the different cephalosporins exhibit considerable variation in rates of hydrolysis by the enzyme.[47] Thus, cephalothin and cefoxitin are the more resistant, and cephaloridine and cefazolin are the least resistant of several cephalosporins tested in vitro. The same acyl functionalities that impart β-lactamase resistance in the penicillins unfortunately render cephalosporins virtually inactive against *S. aureus* and other gram-positive bacteria.

β-Lactamases elaborated by gram-negative bacteria present an exceedingly complex picture. Nearly a dozen distinct enzymes from various species of gram-negative bacilli have been identified and characterized,[23] differing widely in specificity for various β-lactam antibiotics. Most of these enzymes hydrolyze penicillin G and ampicillin at faster rates than they do the cephalosporins. Nonetheless, inactivation by β-lactamases is an important factor in determining resistance to cephalosporins in many strains of gram-negative bacilli. Recently, cephalosporins resistant to some of these enzymes have been developed. These newer agents are of two distinct structural types. One type consists of 7-ACA derivatives with acyl groups containing polar substituents on the carbon atom alpha to the carbonyl, e.g., cefamandole, cefoperazone, cefuroxime and cefotaxime. Of these newer cephalosporins only cefamandole has been approved for clinical use in the United States. Agents of the second type, represented by cefoxitin, are semisynthetic derivatives obtained from cephamycin C, a cephamycin antibiotic obtained from actinomycetes. A 7α-methoxyl substituent in cephamycins is the structural feature that distinguishes them from the cephalosporins and apparently also confers resistance to β-lactamases.

Cephalosporins experience a variety of hydrolytic degradation reactions, the specific nature of which depends on the individual structure (see Table 7-5).[48] Among 7-acylaminocephalosporanic acid derivatives the 3-acetoxylmethyl group is the most reactive site. In addition to its reactivity to nucleophilic displacement reactions, the acetoxyl function of this group of cephalosporins also readily undergoes solvolysis in strongly acidic solutions to form the desacetylcephalosporin derivatives. The latter lactonize to form the desacetylcephalosporin lactones, which are virtually inactive. The 7-acylamino group of some cephalosporins can also be hydrolyzed under enzymatic (acylases), and possibly nonenzymatic conditions to give 7-ACA (or 7-ADCA) derivatives. 7-Aminocephalosporanic acid, following hydrolysis or solvolysis of the 3-

acetoxylmethyl group, also lactonizes under acidic conditions.

The reactive functionality common to all cephalosporins is, of course, the β-lactam. Hydrolysis of the β-lactam of cephalosporins is believed initially to give cephalosporoic acids (where the R′ group is stable, e.g., R′ = H or S-heterocycle) or possibly anhydrodesacetylcephalosporoic acids (in the case of the 7-acylaminocephalosporanic acids). It has not been possible to isolate either of these initial hydrolysis products in aqueous systems, however. Apparently, both types of cephalosporanic acids undergo fragmentation reactions that have not been fully characterized. However, studies of the in vivo metabolism[49] of orally administered cephalosporins have demonstrated the formation of arylacetylglycines and arylacetamidoethanols, which are believed to be formed from the corresponding arylacetylaminoacetaldehydes by metabolic oxidation and reduction respectively. The aldehydes no doubt arise from nonenzymatic hydrolysis of the corresponding cephalosporoic acids. No evidence for the intramolecular opening of the β-lactam ring by the 7-acylamino oxygen to form oxazolones of the penicillenic acid type has been found in the cephalosporins. The formation of dimers and possibly also polymers from 7-ADCA derivatives containing an α-amino group in the acylamino side chain may also occur, especially in concentrated solutions and at alkaline pH's.

$$R\text{-}CONHCH_2CH_2OH \leftarrow R\text{-}CONHCH_2CHO \rightarrow R\text{-}CONHCH_2CO_2H$$

PRODUCTS

Cephalothin Sodium U.S.P., Keflin®, sodium cephalosporin C.

Cephalothin sodium occurs as a white to off-white crystalline powder that is practically odorless. It is freely soluble in water and is insoluble in most organic solvents. Although it has been described as a broad-spectrum antibacterial compound, it is not in the same class as the tetracyclines. Its spectrum of activity is broader than that of penicillin G, and more similar to that of ampicillin. Unlike ampicillin, cephalothin is resistant to penicillinase produced by *Staphylococcus aureus* and provides an alternative to the use of penicillinase-resistant penicillins for the treatment of infections caused by such strains.

Cephalothin Sodium

Cephalothin is poorly absorbed from the gastrointestinal tract and must be administered parenterally for systemic infections. It is relatively nontoxic and is acid-stable. It is excreted rapidly through the kidneys, about 60 percent being lost within 6 hours of administration. Pain at the site of I.M. injection and thrombophlebitis following I.V. injection of cephalothin have been reported. Hypersensitivity reactions from cephalothin have been observed and there is some evidence of cross-sensitivity in patients noted previously to be penicillin-sensitive.

Sterile Cephaloridine U.S.P., Loridine®, pyridinomethyl-7-(2-thiophene-2-acetamido)-3-cephem-4-carboxylate, 3-pyridinomethyl-7-(2-thienylacetamido)-desacetylcephalosporanic acid.

When cephalosporin C or 7-ACA are treated with organic bases such as pyridine, a nucleophilic displacement of the acetoxyl group occurs. The pyridinium compound thus produced is more potent than the acetoxyl analog. Among a series of 7-acetamido-3-pyridinomethyl-3-cephem-4-carboxylates, the 2-thiophene-2-acetamido compound was the best. It is active against gram-negative organisms.

Cephaloridine

Cephaloridine occurs as a white crystalline powder that discolors when exposed to light. It is somewhat unstable and should be stored in a refrigerator. It is very soluble in water and deteriorates rapidly in aqueous solutions which should be used within 24 hours of their preparation and then only if stored at 2° to 15° C.

The intramuscular injection of cephaloridine is less painful than I.M. injection of sodium cephalothin and it is not excreted as rapidly. Furthermore, it is more stable to biotransformation and much less protein bound than is cephalothin. Thus, cephaloridine is preferred for tissue infections. However, in elevated doses it may produce a nephrotoxicity that makes the control of dosage necessary. It is capable of causing hypersensitivity reactions.

Sterile Cefazolin Sodium U.S.P., Ancef®, Kefzol®. Cefazolin is one of a series of semisynthetic cephalosporins in which the C-3 acetoxy function has been replaced by a thiol-containing heterocycle, in this case, 5-methyl-2-thio-1,3,4-thiadiazole. It also contains the somewhat unusual tetrazolylacetyl acylating group. Cefazolin was released in 1973 as the water-soluble sodium salt. It is active only by parenteral administration.

In comparison with other currently available cephalosporins, cefazolin provides higher serum levels, slower renal clearance and a longer half-life. It is ap-

proximately 75 percent protein bound in the plasma, a value that is higher than for other cephalosporins. Early in vitro and clinical studies suggest that cefazolin is more active against gram-negative bacilli but less active against gram-positive cocci than either cephalothin or cephaloridine. Thrombophlebitis following I.V. injection and pain at the site of I.M. injection of cefazolin appear to be the lowest of the parenteral cephalosporins.

Cephazolin Sodium

Sterile Cephapirin Sodium U.S.P., Cefadyl®. Cephapirin is a semisynthetic 7-ACA derivative released in the United States in 1974. It closely resembles cephalothin in chemical and pharmacokinetic properties. Like cephalothin, cephapirin is unstable to acid and must be administered parenterally in the form of an aqueous solution of the sodium salt. It is moderately protein bound (45 to 50%) in the plasma and is rapidly cleared by the kidney. Cephapirin and cephalothin are very similar in antimicrobial spectrum and potency. Conflicting reports concerning the relative incidence of pain at the site of injection and thrombophlebitis after I.V. injection of cephapirin and cephalothin are difficult to assess on the basis of available clinical data.

Cephapirin Sodium

Cefamandole Nafate U.S.P., Mandol®, is the formate ester of cefamandole, a new semisynthetic cephalosporin that incorporates D-mandelic acid as the acyl portion and a thiol containing heterocycle (5-thio-1,2,3,4-tetrazole) in place of the acetoxyl function on the C-3 methylene carbon atom. Esterification of the α-hydroxyl group of the D-mandeloyl function overcomes the instability of cefamandole in solid state dosage forms[50] and provides satisfactory concentrations of the parent antibiotic in vivo through the action of plasma and tissue esterases.

Cefamandole Nafate

The D-mandeloyl moiety of cefamandole appears to confer resistance to β-lactamases, since some β-lactamase-producing, gram-negative bacteria (particularly *Enterobacterceae*) that show resistance to cefazolin and the "older" cephalosporins are sensitive to cefamandole. Additionally, it is active against some ampicillin-resistant strains of *Neisseria* and *Hemophilus*. Although resistance to β-lactamases may be a factor in determining sensitivity of individual bacterial strains to cefamandole, a recent study[51] indicates that other factors, such as permeability and intrinsic activity, are frequently more important. It should be noted that the L-mandelolyl isomer is significantly less active than the D-isomer.

Sterile Cefoxitin Sodium U.S.P. Recently, cephalosporin antibiotics and penicillin N were isolated from *Streptomyces* species.[52] These include close relatives of cephalosporin C and four 7 α-methoxy substituted cephalosporins, called cephamycins, including cephamycin C. Cefoxitin is a semisynthetic derivative prepared from cephamycin C which exhibits a broader spectrum of antibacterial activity than other cephalosporins. Although it is less active than either cephalothin or cephaloridine against gram-positive bacteria, cefoxitin is effective against certain gram-negative bacilli (for example, *Enterobacter, Serratin marcescens*, indole-producing *Proteus* and *Bacteroides*) resistant to these antibiotics.[53]

Cephamycin C: R = H₃N⁺—CH—CH₂CH₂CH₂—
                         |
                         COO⁻

Cefoxitin: R = (thiophene)—CH₂—

It has been proposed that the broader spectrum of activity of cefoxitin is related to its observed resistance to β-lactamases. Higher blood levels and less pain following I.M. injection compared with cephalothin

are claimed for cefoxitin, but additional clinical studies are needed to assess its importance in chemotherapy. Cefoxitin is unstable in acid and must, therefore, be administered parenterally.

Cephaloglycin U.S.P., Kafocin®, 7-[D-2-amino-2-phenyl)acetamido]-3-methyl-3-cephem-4-carboxylic acid, 7-(D-α-aminophenylacetamido)cephalosporanic acid.

Cephaloglycin is a congener of ampicillin introduced during 1970. It differs from cephalothin by having a phenylglycine group instead of the 2-thiophenyl-2-acetamido function. It occurs as a white to off-white powdered dihydrate that is acid-stable and absorbed after oral administration, an advantage over the earlier cephalosporin compounds. It is recommended for the treatment of acute and chronic infections of the urinary tract, particularly those due to susceptible strains of *Escherichia coli, Proteus* species, *Klebsiella-Aerobacter*, enterococci and staphylococci. However, oral absorption of cephaloglycin is significantly lower than that of ampicillin, which it closely resembles in antibacterial spectrum, and is, therefore, not recommended for systemic infections. Newer cephalosporins with improved absorption and distribution properties will no doubt replace it, even for the treatment of urinary tract infections.

Cephaloglycin

Cephalexin U.S.P., Keflex®, Keforal®, 7α-(D-amino-α-phenylacetamido)-3-methylcephemcarboxylic acid.

Cephalexin was purposely designed as an orally active semisynthetic cephalosporin. The oral inactivation of cephalosporins has been attributed to two causes: instability of the β-lactam ring to acid hydrolysis (cephalothin and cephaloridine) and solvolysis or microbial transformation of the 3-methylacetoxy group (cephalothin, cephaloglycin). The α-amino group of cephalexin renders it acid-stable (like cephaloglycin) and reduction of the 3-acetoxymethyl to a methyl group circumvents reaction at that site.

Cephalexin

Cephalexin occurs as the white crystalline monohydrate. It is freely soluble in water, resistant to acid, and well absorbed orally. Food does not interfere with its absorption. Because of minimal protein binding and nearly exclusive renal excretion, cephalexin is particularly recommended for the treatment of urinary tract infections. It is also sometimes employed for upper respiratory tract infections. Its spectrum of activity is very similar to those of cephalothin and cephaloridine. Cephalexin is somewhat less potent than these two agents following parenteral administration and is, therefore, inferior to them for the treatment of serious systemic infections.

Cephradine U.S.P., Anspor®, Velosef®. Cephradine is the only cephalosporin derivative that is available in both oral and parenteral dosage forms. It closely resembles cephalexin chemically (since it may be regarded as a partially hydrogenated derivative of cephalexin), and has very similar antibacterial and pharmacokinetic properties. It occurs as the crystalline hydrate which is readily soluble in water. Cephradine is stable to acid and almost completely absorbed following oral administration. It is minimally protein bound and is excreted almost exclusively via the kidney. It is recommended for the treatment of uncomplicated urinary tract infections and upper respiratory tract infections caused by susceptible organisms.

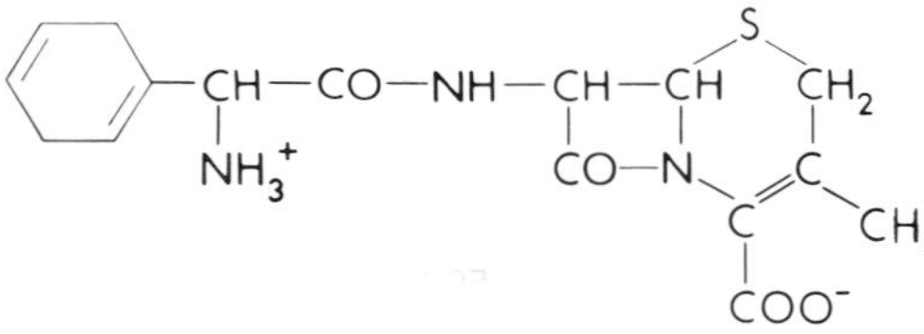

Cephradine

Cefadroxil U.S.P., Duricef®, is a new, orally active semisynthetic derivative of 7-ACDA wherein the 7-acyl group is the D-tyrosyl (or hydroxylphenylglycyl) moiety. This compound is well absorbed following oral administration to give plasma levels that reach 75 to 80 percent of those of an equal dose of its close structural analogue cephalexin. The main advantage claimed for cephadroxil results from its somewhat prolonged duration of action which permits once-a-day dosing. The prolonged duration of action of this compound is related to relatively slow urinary excretion of

the drug as compared with other cephalosporins, but the basis for the latter effect remains to be completely explained. The antibacterial spectrum of action and therapeutic indications of cefadroxil are very similar to those of cephalexin and cephradine. The D-tyrosyl isomer is much more active than the L-isomer.

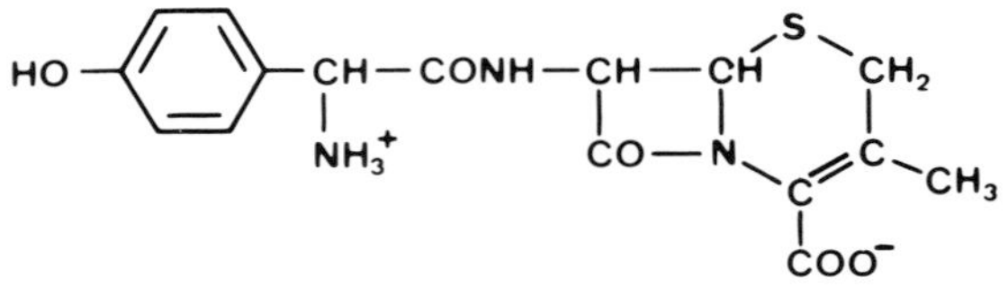

Cefadroxil

Cefaclor U.S.P., Ceclor®, is an orally active semisynthetic cephalosporin that was introduced in the American market in 1979. It differs structurally from cephalexin in that the 3-methyl group has been replaced by a chlorine atom. It is synthesized from the corresponding 3-methylenecepham sulfoxide ester by ozonolysis followed by halogenation of the resulting β-ketoester.[54] The 3-methylenecepham sulfoxide esters are prepared, in turn, by rearrangement of the corresponding 6-acylaminopenicillanic acid derivatives. Cefaclor is moderately stable in acid and achieves sufficient oral absorption to provide effective plasma levels (equal to about 2/3 of those obtained with cephalexin). The compound is apparently unstable in solution since about 50 percent of its antimicrobial activity is lost in two hours in serum at 37° C.[55] The antibacterial spectrum of activity is similar to cephalexin's, but it is claimed to be more potent against some species sensitive to both agents. The drug is currently being recommended for the treatment of non-life-threatening infections caused by *Hemophilus influenzae*, particularly strains resistant to ampicillin.

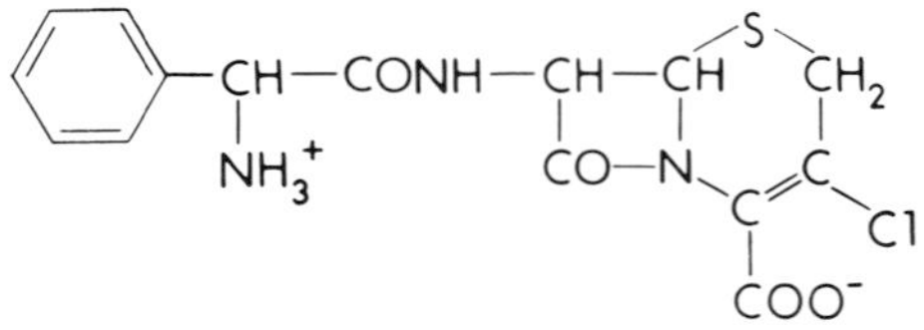

Cefaclor

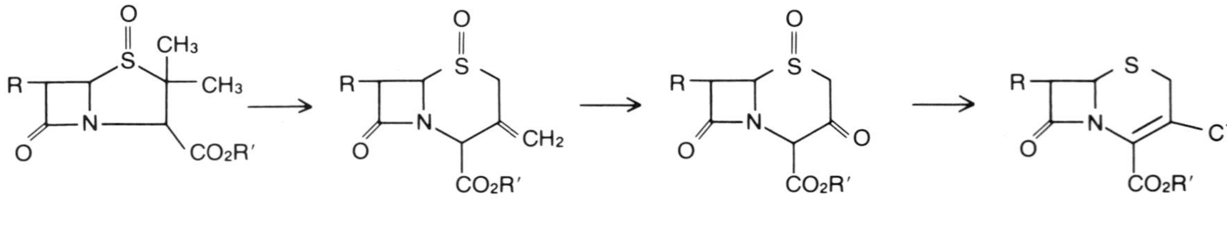

TABLE 7-6

CEPHALOSPORINS

Name Proprietary Name	Preparations	Usual Adult Dose*	Usual Dose Range*	Usual Pediatric Dose*
Cephalothin Sodium U.S.P. *Keflin*	Cephalothin Sodium for Injection U.S.P.	I.M. or I.V., the equivalent of 500 mg. to 1 g. of cephalothin every 4 to 6 hours	Up to 12 g. daily	I.M. or I.V. the equivalent of cephalothin—13.3 to 26.6 mg. per kg. of body weight every 4 hours, or 20 to 40 mg. per kg. of body weight every 6 hours
	Sterile Cephaloridine U.S.P. *Loridine*	I.M. and I.V., 250 mg. to 1 g. every 8 hours	Up to 4 g. daily	Infants and children 1 month of age and over—I.M. or I.V., 10 to 33.3 mg. per kg. of body weight every 8 hours, not to exceed 4 g. daily
Sterile Cefazolin Sodium U.S.P. *Amcef, Kelzol*		I.M. or I.V., the equivalent of cefazolin, 250 mg. to 1 g. every 6 to 8 hours	The equivalent of cephazolin, up to 12 g. daily	Infants and children 1 month of age and over—I.M. or I.V., the equivalent of cefazolin, 6.25 to 25 mg. per kg. of body weight every 6 hours, or 8.3 to 33.3 mg. per kg. of body weight every 8 hours
Sterile Cephapirin Sodium U.S.P.		I.M. or I.V., the equivalent of cephapirin, 500 mg. to 1 g. every 4 to 6 hours	The equivalent of cephapirin, up to 12 g. daily	Infants and children 3 months of age and over—I.M. or I.V., the equivalent of cephapirin, 10 to 20 mg. per kg. of body weight every 6 hours
Cephaloglycin U.S.P. *Kafocin*	Cephaloglycin Capsules N.F.	250 to 500 mg. every 6 hours	Up to 2 g. or more daily	Infants and children 1 year of age and over—oral, the equivalent of anhydrous cephaloglycin, 6.25 to 12.5 mg. per kg. of body weight every 6 hours
Cephalexin U.S.P. *Keflex*	Cephalexin Capsules U.S.P.	The equivalent of 250 mg. of cephalexin 4 times daily	1 to 4 g. daily	6 to 12 mg. per kg. of body weight 4 times daily
	Cephalexin Capsules U.S.P. Cephalexin for Oral Suspension U.S.P. Cephalexin Tablets U.S.P.	Oral, the equivalent of anhydrous cephalexin, 250 to 500 mg. every 6 hours	Up to 4 g. or more daily	Oral, the equivalent of anhydrous cephalexin, 6.25 to 25 mg. per kg. of body weight every 6 hours
Cephradine U.S.P. *Anspor, Velosef*	Sterile Cephradine U.S.P. Cephradine Capsules U.S.P. Cephradine for Oral Suspension U.S.P.	I.M. or I.V., 2 equally divided doses 4 times daily	2 to 4 g. in equally divided doses 4 times daily	50 to 100 mg. per kg. per day up to 300 mg. per kg. per day
	Cephradine Tablets U.S.P.	Oral, 250 to 500 mg. every 6 hours or 500 mg. every 12 hours	Up to 6 g. daily	
	Cephradine for Injection U.S.P.	I.M. or I.V., 500 mg. to 1 g. every 6 hours	Up to 8 g. daily	Infants and children 1 year of age and over—I.M. or I.V., 12.5 to 25 mg. per kg. of body weight every 6 hours
Cefadroxil U.S.P. *Duricef*	Cefadroxil Capsules U.S.P.	Oral, the equivalent of cefadroxil, skin and skin-structure infections—500 mg. every 12 hours, or 1 g. once daily; urinary-tract infections—1 g. every 12 hours	Up to 6 g. daily	
Cefaclor U.S.P. *Ceclor*	Cefaclor Capsules U.S.P.			
	Cefaclor for Oral Suspension U.S.P.	Oral, the equivalent of anhydrous cefaclor, 250 to 500 mg. every 8 hours	Up to 4 g. daily	Infants 1 month of age and over—oral, the equivalent of anhydrous cefaclor, 6.7 to 13.4 mg. per kg. of body weight every 8 hours

*See U.S.P. D.I. for complete dosage information.

1-Oxacephalosporins. As a consequence of the application of new synthetic methods for the modification of penam and cepham ring systems and for total syntheses inspired by Woodward's synthesis of cephalosporins[45] several classes of isosteres of the bicyclic ring systems of penicillins and cephalosporins are now available. At the present time, the more promising group of such isosteric compounds is the 1-oxacephalosporin class. These compounds have been prepared by semisynthetic procedures involving the thiazolidine ring opening of penicillin-S-oxides[56] and also by total synthesis methods.[57] Thus, oxygen analogues of 7-ACA, 7-ADCA and the cephamycins have been prepared and evaluated for antimicrobial properties.

In general, the 1-oxacephalosporins parallel the corresponding cephalosporin and cephamycin derivatives in antimicrobial spectrum of activity and in pharmacokinetic properties. The antibacterial potencies of the 1-oxa derivatives, however, tend to be consistently higher than those of their sulfur isosteres. The "1-oxacephalosporin" that has received the most attention to date is moxalactam, a parenterally active compound whose structure is given below. Examination of the structure of this compound reveals three interesting features: (1) a 7α-methoxyl group (note the analogy to the cephamycins); (2) an α-carboxy-*p*-hydroxyphenylacyl group (note the analogy to carbenicillin); and (3) the same 5-thio heterocyclic moiety on the C-3 methylene carbon atom that is present in cefamandole. Moxalactam exhibits significantly greater activity against most species of *Enterobacteriaceae* and anaerobic gram-negative bacilli than the cephalosporins and cefoxitin. Furthermore, it is 4 times more active than carbenicillin against *Pseudomonas aeruginosa*, a property possessed by very few cephalosporins. The expanded spectrum of activity of LY 127935 compared with conventional cephalosporins, like cefoxitin's, appears to be in part related to greater β-lactamase stability, although other factors evidently are also involved.[58] If the results of currently ongoing clinical trials prove to be as impressive as the in vitro results, moxalactam should find an important place in antibacterial chemotherapy.

OTHER β-LACTAM ANTIBIOTICS

Research efforts will no doubt continue to be directed toward attempts to improve upon penicillins and especially cephalosporins through structural modifications made possible by new techniques for semisynthesis and total synthesis. It is likely, however, that efforts in this already highly developed area will reach a point of diminishing returns, perhaps in the not too distant future. Recent discoveries of several novel, naturally occurring β-lactam-containing structures with (in some cases) unique biochemical and antimicrobial properties, have created new possibilities for development of potentially useful chemotherapeutic agents. Three new classes of β-lactam antibiotics obtained from fermentation are of particular interest currently: they are the thienamycins, nocardicins, and clavulanic acid.

Thienamycin is a novel β-lactam antibiotic first isolated and identified by researchers at Merck[59] from fermentation of cultures of *Streptomyces cattleya*. Its structure and absolute configuration were established by both spectroscopic and total synthesis procedures.[60,61] Two structural features of thienamycin are shared with the penicillins and cephalosporins: a fused bicyclic ring system containing a β-lactam and an equivalently attached 3-carboxyl group. In other respects the thienamycins represent a significant departure from the established β-lactam antibiotics. The bicyclic system consists of a carbapenam containing a double bond between carbons 2 and 3, i.e., it is a 2-carbapenem (or Δ^2-carbapenem) system. The presence of the double bond in the bicyclic structure creates considerable ring strain and increases the reactivity of the β-lactam to ring opening reactions. The side chain is unique in two respects: it is a simple-1-hydroxyethyl group, instead of the familiar acylamino side chain; and it is oriented α to the bicyclic ring system rather than having the usual β orientation of the penicillins and cephalosporins. The remaining feature is a 2-aminoethylthioether function at C-2. The absolute stereochemistry of thienamycin has been determined to be 5R: 6S: 8S. Several additional structurally related an-

LY 127935

tibiotics have been isolated from various *Streptomyces* species, including the four epithienamycins which are isomeric to thienamycin at C-5, C-6 or C-8 and derivatives wherein the 2-aminoethylthio side chain is modified.

Thienamycin displays outstanding broad spectrum antibacterial properties in vitro.[62] It is highly active against most aerobic and anaerobic gram-positive and gram-negative bacteria, including *Staphylococcus aureus, Pseudomonas aeruginosa* and *Bacteroides fragilis*. Furthermore, thienamycin is resistant to inactivation by most β-lactamases elaborated by gram-negative and gram-positive bacteria and is therefore effective against many strains resistant to penicillins and cephalosporins. Resistance to lactamases appears to be a function of the 1-hydroxyethyl side chain, since this property is lost in the 7-nor derivative and the epithienamycins show variable resistance to the different β-lactamases.

The pharmacokinetic properties of thienamycin compare favorably with those of the injectable cephalosporins and the compound appears to cause few and infrequent adverse reactions in experimental animals. A crucial shortcoming of the antibiotic, however, is that it is very unstable in solution. It suffers some hydrolysis of the β-lactam in both dilute acidic and alkaline solutions and has an optimum pH between 6 and 7. The rate of inactivation markedly increases as the concentration increases, indicating possible formation of an inactive dimer or polymers. Lack of solution stability may therefore prevent thienamycin from becoming a clinically useful antibiotic. It is quite likely, however, that stable derivatives with retained antibacterial properties, such as N-formimidoylthienamycin, will overcome these problems.

Nocardicins are β-lactam ring-containing antibiotics obtained from various *Nocardia* species. A total of seven nocardicins have been isolated from fermentation broths,[63] the most extensively studied of which is nocardicin A. Nocardicins resemble the conventional β-lactam antibiotics structurally and stereochemically, having R-acylamino and S-carboxyl functions in positions corresponding to those of the penicillins and cephalosporins. They differ from the latter in that the β-lactam is not fused in a bicyclic system. Nocardicin A has a very narrow spectrum of antibacterial activity. It exhibits significant in vitro activity against only a few species of gram-negative bacteria, notably *Pseudomonas aeruginosa, Proteus vulgaris,* and *Neisseria gonorrhea*. Although it inhibits bacterial cell wall synthesis, it apparently does not prevent cell wall cross-linking and its mechanism is clearly different from that of penicillin G.[64] The potential for the development of nocardicin derivatives with clinical utility remains to be demonstrated.

Clavulanic Acid is a β-lactam antibiotic produced by the same actinomycete that produces cephamycin C, *Streptomyces clavuligerus*. Structurally, it is a l-oxopenam lacking the 6-acylamino side chain of penicillins, but possessing a 2-hydroxyethylidene moiety at C-2. Clavulanic acid exhibits weak broad spectrum antibacterial activity comparable to 6-APA and is, therefore, not likely to become useful as an antibiotic.

Clavulanic Acid

It is, however, a potent irreversible inhibitor of many β-lactamases produced by gram-positive and gram-negative bacteria, especially those mediated by plasmids. Not surprisingly, combination of clavulanic acid with convential β-lactam antibiotics potentiates the actions of the latter against β-lactamase-producing bacteria in vitro. Whether this and similar observa-

Thienamycin

Nocardicin A

tions with other β-lactamase inhibitors (including some β-lactamase-resistant penicillins) will lead to useful clinical applications remains to be seen. At present, clavulanic acid and other β-lactamase inhibitors are laboratory curiosities with experimental usefulness.

THE AMINOGLYCOSIDES

The discovery of streptomycin, the first aminoglycoside antibiotic to be used in chemotherapy, was the result of a planned and deliberate search begun in 1939 and brought to fruition in 1944 by Waksman and his associates.[65] This success stimulated world-wide searches for antibiotics from the actinomycetes and particularly from the genus *Streptomyces*. Among the many antibiotics isolated from that genus, a number are compounds closely related in structure to streptomycin. Five of them, kanamycin, neomycin, paromomycin, gentamicin, and tobramycin are currently marketed in the United States. Amikacin, a semi-synthetic derivative of kanamycin A, has recently been added and it is likely that additional aminoglycosides will be introduced in the future. All aminoglycoside antibiotics are very poorly absorbed (less than 1% under normal circumstances) following oral administration and some of them (kanamycin, neomycin and paromomycin) are administered by that route for the treatment of gastrointestinal infections. Because of the potent broad spectrum nature of their antimicrobial activity, they are also used for the treatment of systemic infections. However, their undesirable side-effects, particularly oto- and nephrotoxicity, have led to restrictions in their systemic use to serious infections or infections caused by bacterial strains resistant to other agents. When administered for systemic infections, aminoglycosides must be given parenterally, usually by intramuscular injection. An additional antibiotic obtained from *Streptomyces,* spectinomycin, is also an aminoglycoside, but differs chemically and microbiologically from other members of the group. It is employed exclusively for the treatment of uncomplicated gonorrhea.

Aminoglycosides are so named because their structures consist of aminosugars linked glycosidically. All have at least one aminohexose and some have a pentose lacking an aminogroup (e.g., streptomycin, neomycin and paromomycin). Additionally, each of the clinically useful aminoglycosides contains a highly substituted 1,3-diaminocyclohexane central ring: in kanamycin, neomycin, gentamicin and tobramycin it is deoxystreptamine; and in streptamycin it is streptamycin.

dine. The aminoglycosides are thus strongly basic compounds that exist as polycations at physiological pH. Their inorganic acid salts are very soluble in water. All are available as sulfates. Solutions of the aminoglycoside salts are stable to autoclaving. The high water solubility of the aminoglycosides no doubt contributes to their pharmacokinetic properties. They distribute well into most body fluids, but not into the central nervous system, bone or fatty or connective tissues. They tend to concentrate in the kidney and are excreted by glomerular filtration. Metabolism of aminoglycosides in vivo apparently does not occur.

Although the aminoglycosides are classified as broad spectrum antibiotics, their greatest utility lies in the treatment of serious systemic infections caused by aerobic gram-negative bacilli. In this regard the choice of agent is generally between kanamycin, gentamicin, tobramycin and amikacin. Aerobic gram-negative and gram-positive cocci (with the exception of staphylococci) tend to be less sensitive and thus the β-lactam and other antibiotics tend to be preferred for the treatment of infections caused by these organisms. Anaerobic bacteria are generally resistant to the aminoglycosides. Streptomycin is the most effective of the group for the chemotherapy of tuberculosis, brucellosis, tularemia and yersina infections. Paromomycin is used primarily in the chemotherapy of amebic dysentery. Under certain circumstances, aminoglycoside and β-lactam antibiotics are known to exert a synergistic action in vivo against certain bacterial strains when the two are administered jointly. Thus, carbenicillin and gentamicin are synergistic against gentamicin-sensitive strains of *Pseudomonas aeruginosa* and several other species of gram-negative bacilli; penicillin G and streptomycin (or gentamicin or kanamycin) tend to be more effective than either agent alone in the treatment of enterococcal endocarditis; and so on. The two antibiotic types should not be combined in the same solution because they are chemically incompatible.

Most of the studies concerning the mechanism of antibacterial action of the aminoglycosides have been carried out with steptomycin. However, the specific actions of other aminoglycosides are thought to be qualitatively similar. The aminoglycosides act directly on the bacterial ribosome to inhibit the initiation of protein synthesis and to interfere with the fidelity of translation of the genetic message. They bind to the 30S ribosomal subunit to form a complex that is unable to initiate proper amino acid polymerization.[66] The binding of streptomycin and other aminoglycosides to the ribosome also causes misreading mutations of the genetic code apparently resulting from failure of specific aminoacyl-RNAs to recognize the proper codons on m-RNA and the incorporation of improper amino acids into the peptide chain.[67] Evidence suggests[68] that the deoxystreptamine containing aminoglycosides differ quantitatively from streptomycin in causing misreading at lower concentrations than are required to prevent initiation of protein synthesis, while streptomycin inhibits initiation and causes misreading equally effectively. Spectinomycin, on the other hand, prevents the initiation of protein synthesis, but apparently does not cause misreading. All of the commercially available aminoglycoside antibiotics are bactericidal in their action, except for spectinomycin. The mechanistic basis for the bactericidal action of the aminoglycosides, however, has not been elucidated.

The development of strains of *Enterobacteriaceae* resistant to antibiotics has become well recognized as a serious medical problem. Nosocomial (hospital-acquired) infections caused by these organisms are often resistant to antibiotic therapy. Research has clearly established that multiple resistance among gram-negative bacilli to a variety of antibiotics occurs and can be transmitted to previously nonresistant strains of the same species and, indeed, to different species of bacteria. The mechanism of transfer of resistance from one bacterium to another has been directly attributed to extrachromosomal R-factors (DNA) which are self-replicative and transferable by conjugation (direct contact). The aminoglycoside antibiotics, because of their potent bactericidal action against gram-negative bacilli, are now preferred for the treatment of many serious infections caused by coliform bacteria. However, a pattern of bacterial resistance has developed to each of the aminoglycoside antibiotics as the clinical use of them has become more widespread. Consequently, there are bacterial strains resistant to streptomycin, kanamycin and gentamicin. Strains carrying R-factors for resistance to these antibiotics synthesize enzymes capable of acetylating, phosphorylating and/or adenylating key amino or hydroxyl groups of the aminoglycosides. Much of the recent effort in aminoglycoside research is directed toward identification of new, or modification of existing, antibiotics that are resistant to inactivation by bacterial enzymes.

Resistance of individual aminoglycosides to specific inactivating enzymes can, in large measure, be understood on the basis of chemical principles. As a first principle, it can be assumed that if the target functional group is absent in a position of the structure normally attacked by an inactivating enzyme, then the antibiotic will be resistant to the enzyme. Secondly, steric factors may confer resistance to attack at functionalities otherwise susceptible to enzymatic attack. For example, the conversion of a primary amino group to a secondary amine has been shown to inhibit N-acetylation by certain aminoglycoside acetyl transferases. At least nine different aminoglycoside inactivating enzymes have been identified and partially characterized.[69] The sites of attack of these enzymes and the biochemistry of the inactivation reactions are briefly described below using the kanamycin B structure (which holds the dubious distinction of being a substrate for all of the enzymes described) for illustrative purposes. Aminoglycoside inactivating enzymes include: (1) aminoacetyltransferases that acetylate the 6'-NH_2 of ring I, the 3-NH_2 of ring II or the 2'-NH_2 of ring I; (2) phosphotransferases that phosphorylate the 3'-OH of ring I; and nucleotidyltransferases that adenylate the 2''-OH of ring III, the 4'-OH of ring I or the 4''-OH of ring III.

The gentamicins and tobramycin lack a 3'-hydroxyl group in ring I (see the section on the individual products for structures) and thus are not inactivated by the phosphotransferase enzymes that phosphorylate that group in the kanamycins. Gentamicin C_1, but not gentamicins C_{1a} or C_2 or tobramycin, is resistant to the acetyltransferase which acetylates the 6'-amino group in ring I of kanamycin B. All gentamicins are resistant to the nucleotidyltransferase enzyme that adenylates the secondary *equatorial* 4''-hydroxyl group of kanamycin B, since the 4''-hydroxyl group in the gentamicins is tertiary and has an *axial* orientation. Removal of functional groups susceptible to attack in an aminoglycoside can, in some cases, lead to derivatives that resist enzymatic inactivation and retain activity. For example, the 3'-deoxy-, 4'-deoxy- and 3, 4'-dideoxy-kanamycins are more similar to the gentamicins and tobramycin in their patterns of activity against clinical isolates that resist one or more of the aminoglycosides. The most significant breakthrough yet achieved in the search for aminoglycosides resistant to bacterial enzymes has been amikacin, the l-N-L-γ-amino-α-butyryl (L-AHBA) derivative of kanamycin A. This remarkable compound retains most of the intrinsic potency of kanamycin A and is resistant to all aminoglycoside-inactivating enzymes known except the aminoacetyltransferase that acetylates the 6'-amino group of ring I.[70] The cause of resistance of amikacin to enzymatic inactivation is not known, but it has been suggested that the introduction of the L-AHBA group into kanamycin A markedly decreases its affinity for the inactivating enzymes. The importance of amikacin's resistance to enzymatic inactivation is reflected in the results of an investigation of the comparative effectiveness of amikacin and other aminoglycosides against clinical isolates of bacterial strains known to be resistant to one or more of the aminoglycosides.[71] In this study amikacin was effective against 91 percent of the isolates (with a range of 87 to 100% depending on the species). The strains found susceptible to other systemically useful amino-

Kanamycin B

glycosides were: kanamycin, 18 percent; gentamicin, 36 percent; and tobramycin, 41 percent.

Despite the complexity inherent in various aminoglycoside structures, some conclusions regarding structure–activity relationships in this antibiotic class have been made.[72] Such conclusions have been formulated on the basis of comparisons of naturally-occurring aminoglycoside structures, the results of selective semisynthetic modifications, and the elucidation of sites of inactivation by bacterial enzymes. It is convenient to discuss aminoglycoside SAR in terms of substituents in rings I, II, and III sequentially.

Ring I is crucially important for characteristic broad spectrum antibacterial activity, and it is the primary target for bacterial inactivating enzymes. Amino functions at 6' and 2' are particularly important, since kanamycin A (6'-amino, 2'-amino) is more active than kanamycin B (6'-amino, 2'-hydroxyl) which, in turn, is more active than kanamycin C (6'-hydroxyl, 2'-amino). Methylation at either the 6'-carbon or the 6'-amino positions does not appreciably lower antibacterial activity and confers resistance to enzymatic acetylation of the 6'-amino group. Removal of the 3'-hydroxyl or the 4'-hydroxyl group or both in the kanamycins (e.g., 3',4'-dideoxykanamycin B or dibekacin) does not reduce antibacterial potency. The gentamicins also lack oxygen functions at these positions, as do sisomicin and netilmicin, which also have a 4',5'-double bond. None of these derivatives is inactivated by phosphotransferase enzymes that phosphorylate the 3'-hydroxyl group. Evidently the 3'-phosphorylated derivatives have very low affinity for aminoglycoside-binding sites in bacterial ribosomes.

Few modifications of ring-II (deoxystreptamine) functional groups are possible without appreciable loss of activity in most of the aminoglycosides. How-ever, the 1-amino group of kanamycin A can be acylated (amikacin, *vide supra*), and activity is largely retained. Netilmicin (1-N-ethylsisomicin) retains the antibacterial potency of sisomicin and is resistant to several additional bacterial inactivating enzymes. 2-Hydroxysisomicin is claimed to be resistant to bacterial strains that adenylate the 2''-hydroxyl group of ring III, while 5-deoxysisomicin exhibits good activity against bacterial strains that elaborate 3-acetylating enzymes.

Ring III functional groups appear to be somewhat less sensitive than those of either ring I or ring II to structural changes. Although the 2''-deoxygentimicins are significantly less active than their 2''-hydroxyl counterparts, the 2''-amino derivatives (seldomycins) are highly active. The 3''-amino group of gentamicins may be primary or secondary with high antibacterial potency. Furthermore, the 4''-hydroxyl group may be *axial* or *equatorial* with little change in potency.

Despite improvements in antibacterial potency and spectrum among newer naturally occurring and semisynthetic aminoglycoside antibiotics, efforts to find agents with improved margins of safety over the earlier discovered antibiotics have clearly been disappointing. The potential for toxicity of these important chemotherapeutic agents continues to restrict their use largely to the hospital environment. The discovery of agents with higher potency toxicity ratios remains an important goal of aminoglycoside research.

PRODUCTS

Streptomycin Sulfate, Sterile U.S.P. Streptomycin sulfate is a white, odorless powder that is hygroscopic but stable toward light and air. It is freely soluble in water, forming solutions that are slightly acidic or nearly neutral. It is very slightly soluble in alcohol and is insoluble in most other organic solvents. Acid hydrolysis of streptomycin yields streptidine and streptobiosamine, the compound that is a combination of L-streptose and N-methly-L-glucosamine.

Streptomycin acts as a triacidic base through the effect of its two strongly basic guanidino groups and the more weakly basic methylamino group. Aqueous solutions may be stored at room temperature for 1 week without any loss of potency, but they are most stable if the pH is adjusted between 4.5 and 7.0. The solutions decompose if sterilized by heating, so sterile solutions are prepared by adding sterile distilled water to the sterile powder. The early salts of streptomycin contained impurities that were difficult to remove and caused a histaminelike reaction. By forming a complex with calcium chloride, it was possible to free the streptomycin from these impurities and to obtain a product that was generally well tolerated.

The organism that produces streptomycin, *Streptomyces griseus,* also produces a number of other antibiotic compounds, hydroxystreptomycin, mannisidostreptomycin and cycloheximide (q.v.). None of these has achieved importance as a medicinally useful substance. The term streptomycin A has been used to refer to what is commonly called streptomycin, and mannisidostreptomycin has been called streptomycin B. Hydroxystreptomycin differs from streptomycin in having a hydroxyl group in place of one of the hydrogens of the streptose methyl group. Mannisidostreptomycin has a mannose residue attached by glycosidic linkage through the hydroxyl group at carbon four of the N-methyl-L-glucosamine moiety. The work of Dyer[73,74] to establish the complete stereostructure of streptomycin has been completed with the total synthesis of streptomycin and dihydrostreptomycin[75] by Japanese scientists.

A clinical problem which develops sometimes with the use of streptomycin is the early development of resistant strains of bacteria, making necessary a change in therapy of the disease. Another factor which limits its therapeutic efficacy is its chronic toxicity.

Certain neurotoxic reactions have been observed after the use of streptomycin. They are characterized by vertigo, disturbance of equilibrium and diminished auditory acuity. Minor toxic effects include skin rashes, mild malaise, muscular pains and drug fever.

As a chemotherapeutic agent, the drug is active against a great number of gram-negative and gram-positive bacteria. One of the greatest virtues of streptomycin is its effectiveness against the tubercle bacillus. It is not a cure in itself but is a valuable adjunct to the standard treatment of tuberculosis. The greatest drawback to the use of this antibiotic is the rather rapid development of resistant strains of microorganisms. In infections that may be due to both streptomycin- and penicillin-sensitive bacteria, the combined administration of the two antibiotics has been advocated. The possible development of damage to the optic nerve by the continued use of streptomycin-containing preparations has led to the discouragement of the use of such products. There is an increasing tendency to reserve the use of streptomycin products for the treatment of tuberculosis. However, it remains one of the agents of choice for the treatment of certain "occupational" bacterial infections such as brucellosis, tularemia, bubonic plague and glanders. The fact that streptomycin is not absorbed when given orally and is not significantly destroyed in the gastrointestinal tract accounts for the fact that at one time it was rather widely used in the treatment of infections of the intestinal tract. For systemic action, streptomycin usually is given by intramuscular injection.

Neomycin Sulfate U.S.P., Mycifradin®, Neobiotic®. In a search for less toxic antibiotics than streptomycin, Waksman and Lechevalier[76] obtained neomycin in 1949 from *Streptomyces fradiae.* Since that time neomycin has increased steadily in importance, and today it is considered to be one of the most useful antibiotics in the treatment of gastrointestinal infections, dermatologic infections and acute bacterial peritonitis. Also, it is employed in abdominal surgery to reduce or avoid complications due to infections from bacterial flora of the bowel. It has a broad–spectrum

Streptomycin

activity against a variety of organisms. It shows a low incidence of toxic and hypersensitive reactions. It is very slightly absorbed from the digestive tract, so its oral use does not ordinarily produce any systemic effect. Neomycin-resistant strains of pathogens have seldom been reported to develop from those organisms against which neomycin is effective. A complete review on neomycin has been edited by Waksman.[77]

Neomycin as the sulfate salt is a white to slightly yellow crystalline powder that is very soluble in water. It is hygroscopic and photosensitive (but stable over a wide pH range and to autoclaving). Neomycin sulfate contains the equivalent of 60 percent of the free base.

Neomycin, as produced by *S. fradiae,* is a mixture of closely related substances. Included in the "neomycin complex" is neamine (originally designated neomycin A) and neomycins B and C. *S. fradiae* also elaborates another antibiotic called fradicin that has some antifungal properties but no antibacterial activity. This substance is not present in "pure" neomycin.

The structures of neamine and neomycin B and C are known and the absolute configurational structures of neamine and neomycin have been reported by Hichens and Rinehart.[78] Neamine may be obtained by methanolysis of neomycins B and C during which the glycosidic link between the deoxystreptamine and D-ribose is broken. Therefore, neamine is a combination of deoxystreptamine and neosamine C linked glycosidically (alpha) at the 4-position of deoxystreptamine. According to Hichens and Rinehart, neomycin B differs from neomycin C by the nature of the sugar attached terminally to D-ribose. That sugar, called neosamine B, differs from neosamine C in its stereochemistry. It has been suggested by Rinehart *et al.*[79] that in neosamine B the configuration is that of 2,6-diamino-2,6-dideoxy-L-idose in which the orientation of the 6-aminomethyl group is inverted to that of the 6-amino-6-deoxy-D-glucosamine in neosamine C. In both instances the glycosidic links are assumed to

be *alpha*. However, Huettenrauch[80] more recently has suggested that both of the diamino sugars in neomycin C have the L-idose configuration and that the glycosidic link is beta in the one attached to D-ribose. The proof of these details concerning the absolute configuration of neomycin B is dependent upon further evidence. The combination of neosamine B with D-ribose is called neobiosamine B, and the combination of neosamine C with D-ribose is called neobiosamine C. In both molecules, the glycosidic links at the D-ribose fragment are *beta* oriented.

Paromomycin Sulfate U.S.P., Humatin®. The isolation of paromomycin was reported in 1956 as an antibiotic obtained from a Streptomyces species (P-D 04998) that is said to resemble closely *S. rimosus.* The parent organism had been obtained from soil samples collected in Colombia. However, paromomycin more closely resembles neomycin and streptomycin, in antibiotic activity, than oxytetracycline, the antibiotic obtained from *S. rimosus.*

The general structure of paromomycin was first reported by Haskell *et al.*[81] as one compound. Subsequently, chromatographic determinations have shown paromomycin to consist of two fractions which have been named paromomycin I and paromomycin II. The absolute configurational structures for the paromomycins were suggested by Hichens and Rinehart[78] as shown in the structural formula, and have been confirmed by DeJongh *et al.*[82] by mass spectrometric studies. It may be noted that the structure of paromomycin is the same as that of neomycin B except that paromomycin contains D-glucosamine instead of the 6-amino-6-deoxy-D-glucosamine found in neomycin B. The same relationship in structures is found between paromomycin II and neomycin C. The combination of D-glucosamine with deoxystreptamine is obtained by partial hydrolysis of both paromomycins and is called paromamine [4-(2-amino-2-deoxy-α-4-glucosyl)deoxy-streptamine].

Neomycin C

CH$_2$OH

HO

HO
D-GLUCOSAMINE NH$_2$

NH$_2$

NH$_2$

OH
DEOXYSTREPTAMINE

CH$_2$OH

D-RIBOSE

OH

R$_1$

HO

HO R$_2$ NH$_2$ O OH

NEOSAMINE B OR C

Paromomycin I : R$_1$ = H; R$_2$ = CH$_2$NH$_2$
Paromomycin II: R$_1$ = CH$_2$NH$_2$; R$_2$ = H

Paromomycin has broad-spectrum antibacterial activity and has been employed for the treatment of gastrointestinal infections due to *Salmonella, Shigella* and enteropathogenic *Escherichia coli*. However, its use is largely restricted to the treatment of intestinal amebiasis at the present time. Paromomycin is soluble in water and stable to heat over a wide pH range.

Kanamycin Sulfate U.S.P., Kantrex®. Kanamycin was isolated in 1957 in Japan by Umezawa and co-workers[83] from *Streptomyces kanamyceticus*. Its activity against mycobacteria and many intestinal bacteria, as well as a number of pathogens that show resistance to other antibiotics, brought a great deal of attention to this antibiotic. As a result, kanamycin was tested and released for medical use in a very short time.

Research activity has been focused intensively on the determination of the structures of the kanamycins. It has been determined by chromatography that *S. kanamyceticus* elaborates three closely related structures that have been designated kanamycins A, B and C. Commercially available kanamycin is almost pure kanamycin A, the least toxic of the three forms. The kanamycins differ only by the nature of the sugar moieties attached to the glycosidic oxygen on the 4 position of the central deoxystreptamine. The absolute configuration of the deoxystreptamine in kanamycins has been reported as represented below by Tatsuoka *et al.*[84] The chemical relationships among the kanamycins, the neomycins and the paromomycins have been reported by Hichens and Rinehart.[78] It may be noted that the kanamycins do not have the D-ribose molecule that is present in neomycins and paromomycins. Perhaps this structural difference is significant in the lower toxicity observed with kanamycins. The kanosamine fragment linked glycosidically

to the 6-position of deoxystreptamine is 3-amino-3-deoxy-D-glucose (3-D-glucosamine) in all three kanamycins. The structures of the kanamycins have been proved by total synthesis.[85,86] It may be seen that they differ in the nature of the substituted D-glucoses attached glycosidically to the 4-position of the deoxystreptamine ring. Kanamycin A contains 6-amino-6-deoxy-D-glucose; kanamycin B contains 2,6-diamino-2,6-dideoxy-D-glucose; and kanamycin C contains 2-amino-2-deoxy-D-glucose (see below).

Kanamycin is basic and forms salts of acids through its amino groups. It is water-soluble as the free base but is used in therapy as the sulfate salt, which is very soluble. It is very stable to both heat and chemicals. Solutions resist both acids and alkali within the pH range of 2.0 to 11.0. Because of possible inactivation of either agent, kanamycin and penicillin salts should not be combined in the same solution.

The use of kanamycin in the United States is usually restricted to infections of the intestinal tract (such as bacillary dysentery) and to systemic infections arising from gram-negative bacilli (e.g., *Klebsiella, Proteus, Enterobacter* and *Serratia*) that have developed resistance to other antibiotics. It has been recommended also for antisepsis of the bowel preoperatively. It is poorly absorbed from the intestinal tract, so systemic infections must be treated by intramuscular or, in the case of serious infections, intravenous injections. Injections of it are rather painful, and the concomitant use of a local anesthetic is indicated. The use of kanamycin in treatment of tuberculosis has not been widely advocated, since the discovery that mycobacteria develop resistance to it very rapidly. Aoki, Hayashi and Ito[87] have found kanamycin to inhibit oxidative mechanisms in the tubercle bacilli as does streptomycin. Their tests indicated that the interference of kanamycin is not identical with that of streptomycin in the oxidation of benzoic acid, niacin and malonic acid. However, clinical experience as well as experimental work of Morikubo[88] indicate that kanamycin does develop cross-resistance in the tubercle bacilli with dihydrostreptomycin, viomycin and other antituberculars. Like streptomycin, kanamycin may cause a decrease in or complete loss of hearing. Upon development of such symptoms, its use should be stopped immediately. Umezawa et al.[89] have reported that the N-methanesulfonate salts of kanamycin are considerably less toxic than the monosulfate.

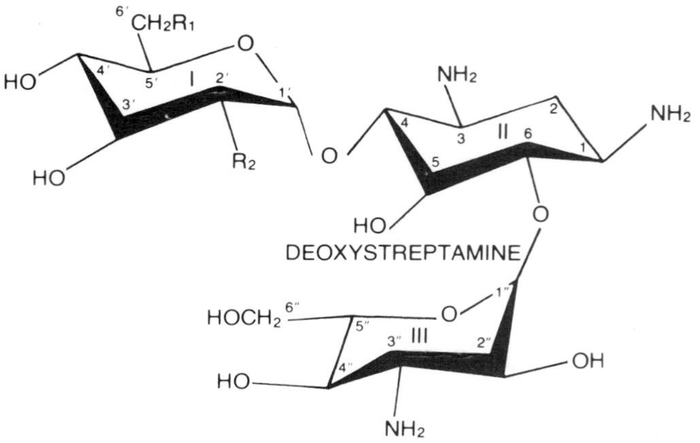

KANOSAMINE

Kanamycin A: $R_1 = NH_2$; $R_2 = OH$
Kanamycin B: $R_1 = NH_2$; $R_2 = NH_2$
Kanamycin C: $R_1 = OH$; $R_2 = NH_2$

Amikacin U.S.P., Amikin®, 1-N-γ-amino-α-hydroxybutyrylkanamycin A, is a semisynthetic aminoglycoside first prepared in Japan. The synthesis formally involves simple acylation of the 1-amino group of the deoxystreptamine ring of kanamycin A with L-

γ-amino-α-hydroxybutyric acid (L-AHBA). This particular acyl derivative retains about 50 percent of the original activity of kanamycin A against sensitive strains of gram-negative bacilli. The L-AHBA derivative is much more active than the D-isomer.[90] The remarkable feature of amikacin is that it resists attack by most bacterial inactivating enzymes and is therefore effective against strains of bacteria that are resistant to other aminoglycosides[71] including gentamicin and tobramycin. In fact, it is resistant to all known aminoglycoside-inactivating enzymes, except the aminotransferase that acetylates the 6'-amino group of the aminoglycosides.[70]

Preliminary studies indicate that amikacin may be less ototoxic than either kanamycin or gentamicin.[91] It should be noted, however, that higher dosages of amikacin are generally required for the treatment of most gram-negative bacillary infections. For this reason, and to discourage the proliferation of bacterial strains resistant to it, amikacin is currently recommended for the treatment of serious infections caused by bacterial strains resistant to other aminoglycosides.

Gentamicin Sulfate U.S.P., Garamycin®. Gentamicin was isolated in 1958 and reported in 1963 by Weinstein et al.[92] to belong to the streptomycinoid (aminocyclitol) group of antibiotics. It is obtained commercially from *Micromonospora purpurea*. Like the other members of its group, it has a broad spectrum of activity against many common pathogens of both gram-positive and gram-negative types. Of particular interest is its high degree of activity against *Pseudomonas aeruginosa* and other gram-negative enteric bacilli.

Gentamicin is effective in the treatment of a variety of skin infections for which a topical cream or ointment may be used. However, since it offers no real advantage over topical neomycin in the treatment of

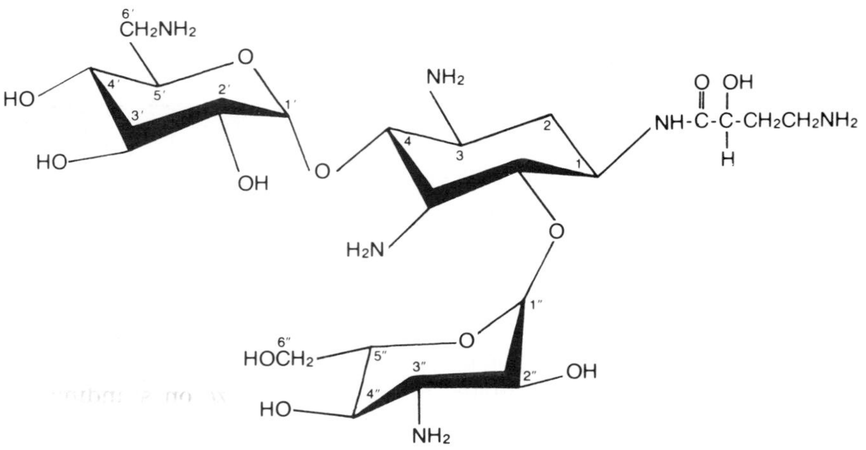

Amikacin

all but pseudomonal infections, it is recommended that topical gentamicin be reserved for use in such infections and in the treatment of burns complicated by pseudomonemia. An injectable solution containing 40 mg. of gentamicin sulfate per ml. may be used for serious systemic and genitourinary tract infections caused by gram-negative bacteria, particularly *Pseudomonas, Enterobacter* and *Serratia* sp. Because of the development of strains of these bacterial species resistant to previously effective broad-spectrum antibiotics, gentamicin is being employed with increasing frequency for the treatment of hospital-acquired infections caused by such organisms.

Gentamicin sulfate is a mixture of the salts of compounds identified as gentamicins C_1, C_2 and C_{1a}. The structures of these gentamicins have been reported by Cooper *et al.*[93] to have the structures shown below. Furthermore, the absolute stereochemistries of the sugar components and the geometries of the glycosidic linkages have been established.[94]

Co-produced but not a part of the commercial product are gentamicins A and B. Their structures have been reported by Maehr and Schaffner[95] and are closely related to the gentamicins C. Although the gentamicin molecules are similar in a number of respects to other aminocyclitols such as streptomycins, they are sufficiently different so that their medical effectiveness is significantly greater.

Gentamicin sulfate is a white to buff-colored substance that is soluble in water and insoluble in alcohol, acetone and benzene. Its solutions are stable over a wide pH range and may be autoclaved. It is chemically incompatible with carbenicillin and the two should not be combined in the same I.V. solution.

Tobramycin U.S.P., Nebcin®, introduced in 1976, is the most active of chemically related aminoglycosides called nebramycins obtained from a strain of *Streptomyces tenebrarius.* Five members of the nebramycin complex have been identified chemically.[96] Factors 4 and 4′ are 6″-O-carbamoyl-kanamycin B and kanamycin B, respectively; factors 5′ and 6 are 6″-O-carbamoyl-tobramycin and tobramycin; and factor 2 is apramycin, a tetracyclic aminoglycoside with an unusual bicyclic central ring structure. Kanamycin B and tobramycin probably do not occur in fermentation broths per se but are formed by hydrolysis of the 6-O″-carbamoyl derivatives in the isolation procedure.

The most important property of tobramycin is an activity against most strains of *Psuedomonas aeruginosa* exceeding that of gentamicin by two- to fourfold. Some gentamicin-resistant strains of this troublesome organism are sensitive to tobramycin, but others are resistant to both antibiotics.[97] Other gram-negative bacilli and staphylococci are generally more sensitive to gentamicin. Tobramycin more closely resembles kanamycin B in structure (it is 3′-deoxy kanamycin B). Further clinical studies are required to determine if tobramycin should replace gentamicin for the general treatment of psuedomonal infections.

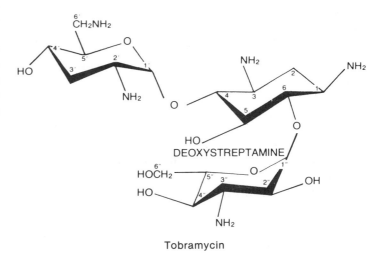

Tobramycin

Spectinomycin Hydrochloride, Sterile U.S.P., Trobicin®. This aminocyclitol antibiotic, isolated from *Streptomyces spectabilis* and once called actinospectocin, was first described by Lewis and Clapp.[98] Its structure and absolute stereochemistry have recently been confirmed by x-ray crystallography.[99] It occurs as the white crystalline dihydrochloride pentahydrate, which is stable in the dry form and very soluble in water. Solutions of spectinomycin, a hemiacetal, slowly hydrolyze on standing and should be freshly prepared and used within 24 hours. It is administered by deep intramuscular injection.

R_1-CHNHR$_2$

DEOXYSTREPTAMINE

PAROSAMINE

Gentamicin C_1: $R_1 = R_2 = CH_3$
Gentamicin C_2: $R_1 = CH_3$; $R_2 = H$
Gentamicin C_{1a}: $R_1 = R_2 = H$

Spectinomycin

Spectinomycin is a broad-spectrum antibiotic with moderate activity against many gram-positive and gram-negative bacteria. It differs from streptomycin and the streptamine-containing aminoglycosides in chemical and antibacterial properties. Like streptomycin, spectinomycin interferes with the binding of t-RNA to the ribosomes and thus interferes with the initiation of protein synthesis. Unlike streptomycin or the streptamine-containing antibiotics, however, it does not cause misreading of the messenger. Spectinomycin exerts a bacteriostatic action and is inferior to other aminoglycosides for most systemic infections. At present it is recommended as an alternative to penicil-

TABLE 7–7

AMINOGLYCOSIDE ANTIBIOTICS

Name Proprietary Name	Preparations	Usual Adult Dose*	Usual Dose Range*	Usual Pediatric Dose*
Streptomycin Sulfate U.S.P.	Streptomycin Sulfate Injection U.S.P. Sterile Streptomycin Sulfate U.S.P.	I.M., the equivalent of streptomycin, in combination with other antibacterials, 250 mg. to 1 g. every 6 hours, or 500 mg. to 2 g. every 12 hours	Tuberculosis—1 g. twice weekly to 2 g. daily; other infections—up to 4 g. daily	I.M., the equivalent of streptomycin, in combination with other antibacterials, 5 to 10 mg. per kg. of body weight every 6 hours; or 10 to 20 mg. per kg. of body weight every 12 hours
Neomycin Sulfate U.S.P. *Mycifradin, Neobiotic*	Neomycin Sulfate Ointment U.S.P. Neomycin Sulfate Cream U.S.P.	For external use, topically to the skin, as the equivalent of a 0.35 percent ointment or cream of neomycin 1 to 3 times daily		
	Neomycin Sulfate Ophthalmic Ointment U.S.P.	For external use, topically to the conjunctiva, as the equivalent of a 0.35 percent ointment of neomycin every 8 to 24 hours		
	Neomycin Sulfate Oral Solution U.S.P. Neomycin Sulfate Tablets U.S.P.	Hepatic coma—the equivalent 700 mg. to 2.1 g. of neomycin every 6 hours for 5 to 6 days; infectious diarrhea—8.75 mg. per kg. of body weight every 6 hours for 2 to 3 days; preoperative preparation—700 mg. every hour for 4 doses, then 700 mg. every 4 hours for the balance of 24 hours; or 10.3 mg. per kg. of body weight every 4 hours for 2 to 3 days	Antibacterial, the equivalent of neomycin, up to 8.4 g. daily for 24 to 48 hours in preoperative bowel preparation	Adjunct in hepatic coma—437.5 mg. to 1.225 g. per square meter of body surface every 6 hours for 5 to 6 days
	Sterile Neomycin Sulfate U.S.P.	I.M., the equivalent of neomycin, 1.3 to 2.6 g. per kg. of body weight every 6 hours	Up to a maximum of 10.5 mg. per kg. of body weight daily, but not to exceed 700 mg. daily for more than 10 days	
Paromomycin Sulfate U.S.P. *Humatin*	Paromomycin Sulfate Capsules U.S.P. Paromomycin Sulfate Syrup U.S.P.	The equivalent of 500 mg. of paromomycin every 6 hours taken with meals	500 mg. to 1 g. of paromomycin	
Kanamycin Sulfate U.S.P. *Kantrex, Klebcil*	Kanamycin Sulfate Capsules U.S.P.	Intestinal infections—the equivalent of 1 g. of kanamycin every 6 to 8 hours for 5 to 7 days; preoperative preparation—the equivalent of 1 g. of kanamycin every hour for 4 hours; then 1 g. every 6 hours for 36 to 72 hours	3 to 12 g. daily	Intestinal infections—12.5 mg. per kg. of body weight or 375 mg. per square meter of body surface, 4 times daily

*See *U.S.P. D.I.* for complete dosage information.

TABLE 7–7

AMINOGLYCOSIDE ANTIBIOTICS

Name Proprietary Name	Preparations	Usual Adult Dose*	Usual Dose Range*	Usual Pediatric Dose*
	Kanamycin Sulfate Injection U.S.P.	I.M., the equivalent of kanamycin—3.75 mg. per kg. of body weight every 6 hours; 5 mg. per kg. of body weight every 6 hours; 5 mg. per kg. of body weight every 8 hours; or 7.5 mg. per kg. of body weight every 12 hours	Up to 15 mg. per kg. of body weight daily, but not to exceed 1.5 g. daily	I.M., or I.V., the equivalent of kanamycin, premature and full-term neonates up to 1 year of age—7.5 mg. per kg. of body weight every 12 hours; older infants and children—3 to 7.5 mg. per kg. of body weight every 12 hours
Gentamicin Sulfate U.S.P. *Garamycin*	Gentamicin Sulfate Cream U.S.P. Gentamicin Sulfate Ointment U.S.P.	For external use, topically to the skin, the equivalent of 0.1 percent of gentamicin as a cream or ointment 3 or 4 times daily		
	Gentamicin Sulfate Injection U.S.P.	I.M., or I.V., the equivalent of gentamicin, 1 to 1.7 mg. per kg. of body weight every 8 hours; or 750 μg to 1.25 mg. per kg. of body weight every 6 hours	The equivalent of gentamicin up to 8 mg. per kg. of body weight daily in severe, life-threatening infections	I.M., or I.V., the equivalent of gentamicin, premature or full-term neonates 1 week of age or less—2.5 to 3 mg. per kg. of body weight every 12 hours; older neonates and infants—2 to 2.5 mg. per kg. of body weight every 8 hours; children—1 to 2.5 mg. per kg. of body weight every 8 hours
	Gentamicin Sulfate Ophthalmic Ointment U.S.P.	Topical, to the conjunctiva, the equivalent of gentamicin, a thin strip (approximately 1 cm.) of a 0.3 percent ointment every 6 to 12 hours		
	Gentamicin Sulfate Ophthalmic Solution U.S.P.	Topical, to the conjunctiva, the equivalent of gentamicin, 1 drop of a 0.3 percent solution every 4 to 8 hours; topical to the ear canal, 3 or 4 drops of a 0.3 percent solution every 4 to 8 hours		
Amikacin U.S.P. *Amakin*	Amikacin Sulfate Injection U.S.P.	I.M., or I.V., the equivalent of amikacin, 5 mg. per kg. of body weight every 8 hours, or 7.5 mg. per kg. of body weight every 12 hours	The equivalent of amikacin, up to 15 mg. per kg. of body weight daily, but not to exceed 1.5 g. daily for more than 10 days	I.M., or I.V., the equivalent of amikacin, neonates—initially, 10 mg. per kg. of body weight, then 7.5 mg. per kg. of body weight every 12 hours
Tobramycin U.S.P. *Nebcin*	Tobramycin Sulfate Injection U.S.P.	I.M., or I.V., the equivalent of tobramycin, 750 μg. to 1.25 mg. per kg. of body weight every 6 hours, or 1 to 1.7 mg. per kg. of body weight every 8 hours	Up to 8 mg. per kg. of body weight daily in severe, life-threatening infections	I.M., or I.V., the equivalent of tobramyacin, neonates one week of age or less—up to 2 mg. per kg. of body weight every 12 hours
	Sterile Spectinomycin Hydrochloride U.S.P. *Trobicin*	I.M., the equivalent of 2 to 4 g. of spectinomycin		Dosage in infants and children not established

*See U.S.P. D.I. for complete dosage information.

lin G salts for the treatment of uncomplicated gonorrhea. A cure rate of greater than 90 percent has been observed in clinical studies for this indication. Many physicians prefer to use a tetracycline or erythromycin for prevention or treatment of suspected gonorrhea in penicillin-sensitive patients since, unlike these agents, spectinomycin is ineffective against syphillis. Furthermore, it is considerably more expensive than erythromycin and most of the tetracyclines.

THE TETRACYCLINES

Among the most important broad-spectrum antibiotics are the members of the tetracycline family. Eight such compounds—tetracycline, oxytetracycline, chlortetracycline, demeclocycline, methacycline, doxycycline, minocycline and rolitetracycline—have been introduced into medical use. A number of others have been shown to possess antibiotic activity. The tetracyclines are obtained by fermentation procedures from *Streptomyces* species or by chemical transformations of the natural products. Their chemical identities have been established by degradation studies and confirmed by the synthesis of three members of the group, oxytetracycline,[100,101] 6-demethyl-6-deoxytetracycline,[102] and anhydrochlortetracycline[103] in their (±) forms. The important members of the group are derivatives of an octahydronaphthacene, a hydrocarbon system comprised of four annelated six-membered rings. It is from this tetracyclic system that the group name is derived. The antibiotic spectra and chemical properties of these compounds are very similar but not identical. Their structural and stereochemical relationships are shown in Table 7-8.

The stereochemistry of the tetracyclines is very complex. Carbon atoms 4, 4a, 5, 5a, 6 and 12a are potentially asymmetric depending on substitution. Oxytetracycline and doxycycline, each with a 6 α-hydroxyl substituent, have six asymmetric centers, while the others lacking asymmetry at C-6 have only five. Determination of the complete absolute stereochemistry of the tetracyclines proved to be a very difficult problem. By detailed X-ray diffraction analysis[104,105,106] it was established that the stereochemical formula shown in Table 7-8 represents the orientations found in the natural and semisynthetic tetracyclines. These studies also confirmed that conjugated systems exist in the structure from C-10 through C-12 and from C-1 through C-3 and that the formula represents only one of a number of canonical forms existing in those portions of the molecule.

The tetracyclines are amphoteric compounds, forming salts with either acids or bases. In neutral solu-

TABLE 7-8

STRUCTURE OF TETRACYCLINES

	R_1	R_2	R_3	R_4
Tetracycline	H	CH_3	OH	H
Chlortetracycline	Cl	CH_3	OH	H
Oxytetracycline	H	CH_3	OH	OH
Demeclocycline	Cl	H	OH	H
Methacycline	H	CH_2	—	OH
Doxycycline	H	H	CH_3	OH
Minocycline	$N(CH_3)_2$	H	H	H

tions these substances exist mainly as zwitterions. The acid salts, which are formed through protonation of the enol group on carbon atom 2, exist as crystalline compounds that are very soluble in water. However, these amphoteric antibiotics will crystallize out of aqueous solutions of their salts unless stabilized by an excess of acid. The hydrochloride salts are used most commonly for oral administration and are usually encapsulated because of their bitter taste. Water-soluble salts may be obtained also from bases such as sodium or potassium hydroxides but are not stable in aqueous solutions. Water-insoluble salts are formed with divalent and polyvalent metals.

The unusual structural groupings in the tetracyclines produce three acidity constants in aqueous solutions of the acid salts. The particular functional groups responsible for each of the thermodynamic pKa values have been determined by Leeson *et al.*[107] to be as shown in the formula below. These groupings had been identified by Stephens *et al.*[108] previously as the sites for protonation, but their earlier assignments as to which produced the values responsible for pKa_2 and pKa_3 were opposite to those of Leeson *et al.* This latter assignment has been substantiated by Rigler *et al.*[109]

TABLE 7-9

pKa VALUES (OF HYDROCHLORIDES) IN AQUEOUS SOLUTION AT 25%

	pKa_1	pKa_2	pKa_3
Tetracycline	3.3	7.7	9.5
Chlortetracycline	3.3	7.4	9.3
Demeclocycline	3.3	7.2	9.3
Oxytetracycline	3.3	7.3	9.1
Doxacycline	3.4	7.7	9.7
Minocycline	2.8	7.8	9.3

The approximate pKa values for each of these groups in the six tetracycline salts in common use are shown in Table 7-9. The values are taken from Stephens et al.[108,] from Benet and Goyan[110] and from Barringer et al.[111] The pKa of the 7-dimethylamino group of minocycline (not listed) is 5.0.

An interesting property of the tetracyclines is their ability to undergo epimerization at carbon atom 4 in solutions of intermediate pH range. These isomers are called *epi*tetracyclines. Under the influence of the acidic conditions, an equilibrium is established in about a day and consists of approximately equal amounts of the isomers. The partial structures below indicate the two forms of the epimeric pair. The 4-*epi*tetracyclines have been isolated and characterized. They exhibit much less activity than the "natural' isomers, thus accounting for a decrease in therapeutic value of aged solutions.

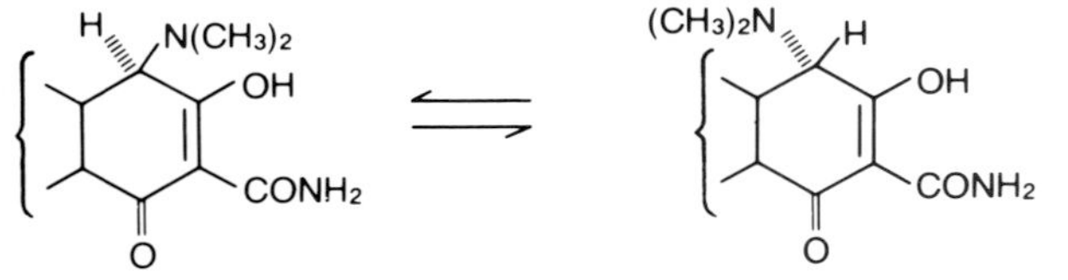

epi   natural

Strong acids and strong bases attack the tetracyclines having a hydroxyl group on the number 6 carbon atom, causing a loss in activity through modification of the C ring. Strong acids produce a dehydration through a reaction involving the 6-hydroxyl group and the 5a-hydrogen. The double bond thus formed between positions 5a and 6 induces a shift in the position of the double bond between carbon atoms 11a and 12 to a position between carbon atoms 11 and 11a, forming the more energetically favored resonant system of the naphthalene group found in the inactive anhydrotetracyclines. Bases promote a reaction between the 6-hydroxyl group and the ketone group at the 11 position, causing the bond between the 11 and 11a atoms to cleave and to form the lactone ring found in the inactive isotetracyclines. These two unfavorable reac-

tions stimulated the research that has led to the development of the more stable and longer-acting compounds, 6-deoxytetracycline, methacycline, doxycycline and minocycline.

Stable chelate complexes are formed by the tetracyclines with many metals including calcium, magnesium and iron. Such chelates are usually very insoluble in water, accounting for the impairment in absorption of most (if not all) tetracyclines in the presence of milk, calcium-, magnesium- and aluminum-containing antacids, and iron salts. Soluble alkalinizers, such as sodium bicarbonate, also decrease the gastrointestinal absorption of the tetracylines.[112] Deprotonation of tetracyclines to more ionic species, and their observed instability in alkaline solutions, may account for this observation. The affinity of tetracyclines for calcium causes them to be laid down in newly formed bones and teeth as tetracycline-calcium orthophosphate complexes. Deposits of these antibiotics in teeth cause a yellow discoloration which darkens (a photochemical reaction) over a period of time. Tetracyclines are distributed into the milk of lactating mothers and also cross the placental barrier into the fetus. The possible effects of these agents on bones and teeth of the child should be taken into consideration before their use in pregnancy or in children under 8 years of age is instituted.

The strong binding properties of the tetracyclines with metal ions caused Albert[113] to suggest that their antibacterial properties may be due to an ability to remove essential metal ions as chelated compounds. Elucidation of details of the mechanism of action of the tetracyclines,[114] however, has more clearly defined more specific roles for magnesium ion in molecular processes affected by these antibiotics in bacteria. Tetracyclines are specific inhibitors of bacterial protein synthesis. They bind to the 3OS ribosomal subunit and thereby prevent the binding of aminoacyl transfer RNA to the messenger RNA-ribosome complex. The binding of aminoacyl transfer RNA and the binding of tetracyclines at the ribosomal binding site both require magnesium ion.[115] Tetracyclines also bind to mammalian ribosomes, but with lower affinities and they apparently do not achieve sufficient intracellular concentrations to interfere with protein synthesis. The selective toxicity of the tetracyclines against bacteria is strongly dependent upon the self-destructive capacity of bacterial cells to concentrate these agents in the cell. The active uptake of tetracyclines by bacterial cells has been found to be an energy–dependent process that requires ATP and magnesium ions.[116] Bacterial resistance to the action of the tetracyclines appears to result primarily, if not exclusively, from inability of the compounds to penetrate the bacterial cell wall. Loss of the capacity to transport tetracy-

5,6-Anhydrotetracycline   Isotetracycline

clines actively[117] and the presence of tetracycline-binding proteins at the cell surface[118] have both been implicated as mechanisms preventing cell wall penetration in resistant bacterial strains. Inactivation of tetracyclines by bacterial enzymes has not been observed to occur.

The tetracyclines are truly broad-spectrum antibiotics with the broadest spectrum of any known antibacterial agents. They are active against a wide range of gram-positive and gram-negative bacteria, spirochetes, mycoplasmas, rickettsiae and clamydia. Their potential indications are, therefore, numerous. However, their bacteriostatic action is a disadvantage in the treatment of life-threatening infections such as septicemia, endocarditis and meningitis wherein the aminoglycosides are usually preferred for gram-negative, and the penicillins for gram-positive, infections. Because of incomplete absorption and effectiveness against the natural bacterial flora of the intestine, tetracyclines may induce superinfections caused by the pathogenic yeast, *Candida albicans*. Resistance to tetracyclines among both gram-positive and gram-negative bacteria is relatively common. Superinfections due to resistant *Staphylococcus aureus* and *Pseudomonas aeruginosa* have resulted from the use of these agents over a period of time. Parenteral tetracyclines may cause severe liver damage, especially when given in excessive dosage to pregnant women or to patients with impaired renal function.

As a result of the large amount of research carried out to prepare semisynthetic modifications of the tetracyclines and to obtain individual compounds by total synthesis, several interesting structure–activity relationships have emerged. Recent reviews are available that discuss structure–activity relationships among the tetracyclines in detail,[119-121] as well as their synthesis and chemical properties.[120-122] Only a brief

review of the salient structure–activity features will be presented here. All derivatives containing fewer than four rings are inactive or nearly inactive. The simplest tetracycline derivative that retains the characteristic broad-spectrum activity associated with this antibiotic class is 6-demethyl-6-deoxytetracycline. It has become evident that many of the precise structural features present in this molecule must remain unmodified in order for derivatives to retain activity. Thus the integrity of substituents at carbon atoms 1,2,3,4,10,11,11a, and 12 cannot be drastically violated without deleterious effects on the antimicrobial properties of the resulting derivatives.

Only very slight modifications of A-ring substituents can be made without dramatic loss of antibacterial potency. The enolized tricarbonylmethane system at carbons 1 to 3 must be intact for good activity. Replacement of the amide at C-2 with other functions such as aldehyde, nitrile, etc., reduces or abolishes activity. Monoalkylation of the amide nitrogen reduces activity proportionate to the size of the alkyl group. Aminoalkylation of the amide nitrogen accomplished by the Mannich reaction, yields derivatives that are significantly more water soluble than the parent tetracycline and are hydrolyzed to it in vivo (e.g., rolitetracycline). The dimethylamino group at the 4-position must have the α-orientation: 4-epitetracyclines are very much less active than the natural isomers. Removal of the 4-dimethylamino group reduces activity even further. Activity is largely retained in the primary and N-methyl-secondary amines, but rapidly diminishes in the higher alkylamines. A *cis* A/B ring fusion with an α-hydroxyl group at 12a is apparently also essential. Esters of the 12a hydroxyl group are all inactive with the exception of the formyl ester which readily hydrolyzes in aqueous solutions. Alkylation at 11a also leads to inactive compounds demonstrating the importance of an enolizable β-diketone functionality at C-11 and C-12. The importance of the shape of the tetracyclic ring system is further illustrated by a substantial loss in antibacterial potency resulting from epimerization at C-5a. Dehydrogenation to form a double bond between C-5a and C-11a markedly decreases activity as does aromatization of ring C to form anhydrotetracyclines.

In contrast, substituents at positions 5, 5a, 6, 7, 8, and 9 can be modified with varying degrees of impunity resulting in retention and, in some cases, improvement of antibiotic activity. A 5-hydroxyl group, as in oxytetracycline and doxycycline, may influence pharmacokinetic properties, but does not change antimicrobial activity as compared to the 5-deoxy compounds. 5a-Epitetracyclines (prepared by total synthesis) although highly active in vitro, are unfortunately much less impressive in vivo. Acid-stable 6-deoxytetracyclines and 6-demethyl-6-deoxytetracylines have been utilized to prepare a variety of mono- and disubstituted derivatives by electrophilic substitution reactions at C-7 and C-9 of the D ring. The more useful results have been achieved with the introduction of substituents at C-7. Oddly, both strongly electron withdrawing groups, e.g., chloro (chlortetracycline) and nitro, and strongly electron donating groups e.g., dimethylamino (miocycline), enhance activity. This unusual circumstance is reflected in quantitative structure activity (QSAR) studies of 7- and 9-substituted tetracyclines[120,123] which indicated a squared dependence on σ, Hammett's electronic substituent constant, and in vitro inhibition of an *Escherichia coli* strain. The effect of introducing substituents at C-8 has not been studied because this position cannot be directly substituted using classical electrophilic aromatic substitution reactions and thus 8-substituted derivatives are available only recently through total synthesis.[124]

The most fruitful site for semisynthetic modification of the tetracyclines has been the 6-position. Neither the 6α-methyl nor the 6β-hydroxyl group is essential for antibacterial activity. In fact, doxycycline and methacycline are more active in vitro against most bacterial strains than their parent oxytetracycline. The conversion of oxytetracycline to doxycycline, which can be accomplished by reduction of methacycline,[125] gives a 1:1 mixture of doxycycline and epidoxycycline (which has a β-oriented methyl group), whereas if the C-11a-α-fluoro derivative of methacycline is employed, the β-methyl epimer is exclusively formed.[126] 6-Epidoxycycline is much less active than doxycycline, 6-Demethyl-6-deoxytetracycline, synthesized commercially by catalytic hydrogenolysis of the 7-chloro and 6-hydroxyl groups of 7-chloro-6-demethyltetracycline obtained by fermentation of a mutant strain of *Streptomyces aureofaciens*,[127] is slightly more potent than tetracyline. More successful from a clinical standpoint, however, is 6-demethyl-6-deoxy-7-dimethylaminotetracycline (minocycline),[128] because of its activity against tetracyline-resistant bacterial strains.

6-Deoxytetracyclines also possess important chemical and pharmacokinetic advantages over their 6-oxy counterparts. Unlike the latter, they are incapable of forming anhydrotetracyclines under acidic conditions, because they can not dehydrate at C-5a and C-6. They are also more stable in base, because they do not readily undergo β-ketone cleavage, followed by lactonization, to form isotetracyclines. Despite the fact that it lacks a 6-hydroxyl group, methacycline shares the instability of the 6-oxytetracyclines. It suffers prototropic rearrangement to the anhydrotetracycline in acid and β-ketone cleavage followed by lactonization to the isotetracycline in base. Reduction of the 6-hydroxyl group also brings about a dramatic change in the solubility properties of tetracyclines. This effect is reflected in the significantly higher oil/water partition coefficients of the 6-deoxytetracyclines as compared with the tetracyclines.[129,130] The greater lipid solubility of the 6-deoxy compounds, in turn, has important pharmacokinetic consequences.[120] Hence, doxycycline and minocycline are more completely absorbed following oral absorption, exhibit higher fractions of plasma protein binding, and have higher volumes of distribution and lower renal clearances than the corresponding 6-deoxytetracyclines. Partition coefficients and pharmacokinetic data for six commercially available tetracyclines are listed in Table 7-10.

Polar substituents, i.e., hydroxyl groups, at C-5 and C-6 contribute decreased lipid versus water solubility to the tetracyclines. However, the 6-position is considerably more sensitive than is the 5-position to this effect. Thus, doxycycline (6-deoxy-5-oxytetracycline) has a much higher partition coefficient than either tetracycline or oxytetracycline. Nonpolar substituents (those with positive π values, see Ch. 2), e.g., 7-dimethylamino, 7-chloro, and 6-methyl, have the opposite effect. Thus, the partition coefficient of chlortetracycline is substantially greater than tetracycline's and slightly greater than demeclocycline's. Interestingly, minocycline (5-demethyl-6-deoxy-7-dimethylaminotetracycline) has the highest partition coefficient of the commonly used tetracyclines.

The poorer oral absorption of the more water–soluble compounds, tetracycline and oxytetracycline, can be attributed to several factors. In addition to their comparative difficulty in penetrating lipid membranes, the polar tetracyclines probably experience more complexation with metal ions in the gut and also undergo some acid–catalyzed destruction in the stom-

TABLE 7-10

PHARMACOKINETIC PROPERTIES* OF TETRACYCLINES

Tetracycline	Substituents				$^Kp.c.$† Octanol/ Water pH 5.6	Percent Absorbed Orally	Percent Excreted Feces	Percent Excreted Urine	Percent Protein Bound	Volume of Distribution (percent of body weight)	Renal Clearance (ml./min./ 1.73 M²)	Half- life (hrs)
	C-5α	C-6α	C-6β	C-7								
Tetracycline	H	CH₃	OH	H	0.056	58	20–50	60	24–65	156–306	50–80	10
Oxytetracycline	OH	CH₃	OH	H	0.075	77–80	~50	70	20–35	189–305	99–102	9
Chlortetracycline	H	CH₃	OH	Cl	0.41	25–30	>50	18	42–54	149	32	7
Demeclocycline	H	H	OH	Cl	0.25	66	23-72	42	68–77	179	35	15
Doxycycline	OH	CH₃	H	H	0.95	93	20–40	27–39	60–91	63	18–28	15
Minocycline	H	H	H	N(CH₃)₂	1.10	~100	40	5–11	55–76	74	5–15	19

* Values taken from Brown and Ireland[120] and references cited therein.
† Values taken from Colazzi and Klink.[129]

ach. Poorer oral absorption, coupled with biliary excretion of some tetracyclines, is also thought to cause a higher incidence of superinfections caused by resistant microbial strains. On the other hand, the more polar tetracyclines are excreted in higher concentrations in the urine (e.g., 70% for tetracycline and 60% for oxytetracycline) than the more lipid soluble compounds (e.g., 33% for doxycycline and only 11% for minocycline). Significant passive renal tubular reabsorption, coupled with higher fractions of protein binding, contribute to the lower renal clearance and more prolonged durations of action of doxycycline and minocycline compared with the other tetracyclines, especially tetracycline and oxytetracycline. Although all tetracyclines are widely distributed into tissues, the more polar ones have larger volumes of distribution than do the nonpolar compounds. The more lipid-soluble tetracyclines, however, distribute better to poorly vascularized tissue. It is also claimed that the distribution of doxycycline and minocycline into bone is less than with other tetracyclines.[131]

Nearly one decade has passed since the last tetracycline derivative, minocycline, was introduced into medical practice, and it might appear that the possibilities for discovery of improved compounds prepared by structural modification or total synthesis have been largely exhausted. As was mentioned earlier, however, the structure–activity relationships of 8-substituted compounds have not been explored. Furthermore, 6-thiatetracycline (actually, the 6-thia isostere of 6-deoxy-6-demethyltetracycline) has recently been claimed in a preliminary report[132] to be superior to all known tetracyclines. The total synthesis of 6-thiatetracycline has been reported recently.[133] Because the ideal tetracycline from the pharmacokinetic point of view has not yet been found, it is possible that one of the new leads will produce compounds with improved clinical properties over those currently available.

PRODUCTS

Tetracycline U.S.P., Achromycin®, Cyclopar®, Panmycin®, Tetracyn®. During the chemical studies on chlortetracycline, it was discovered that controlled catalytic hydrogenolysis would selectively remove the 7-chloro atom and thus produce tetracycline. This process was patented by Conover[134] in 1955. Later, tetracycline was obtained from fermentations of *Streptomyces* species but the commercial supply is still chiefly dependent upon the hydrogenolysis of chlortetracycline.

Tetracycline is 4-dimethylamino-1,4,4a,5,5a,6,11,12a-octahydro-3,6,10,12,12a-pentahydroxy-6-methyl-1,11-dioxo-2-naphthacenecarboxamide. It is a bright-yellow crystalline salt that is stable in air but darkens in color upon exposure to strong sunlight. Tetracycline is stable in acid solutions having a pH higher than 2. It is somewhat more stable in alkaline solutions than chlortetracycline, but, like those of the other tetracyclines, such solutions rapidly lose their potencies. One gram of the base requires 2,500 ml. of water and 50 ml. of alcohol to dissolve it. The hydrochloride salt is most commonly used in medicine, although the free base is absorbed from the gastrointestinal tract about equally well. One gram of the hydrochloride salt dissolves in about 10 ml. of water and in 100 ml. of alcohol. Tetracycline has become the most popular antibiotic of its group, largely because its blood level concentration appears to be higher and more enduring than that of either oxytetracycline or chlortetracycline. Also, it is found in higher concentration in the spinal fluid than are the other two compounds.

Tetracycline

A number of combinations of tetracycline with agents that increase the rate and the height of blood level concentrations are on the market. One such adjuvant is magnesium chloride hexahydrate (Panmycin). Also, an insoluble tetracycline phosphate complex (Tetrex) is made by mixing a solution of tetracycline, usually as the hydrochloride, with a solution of sodium metaphosphate. A variety of claims concerning the efficacy of these adjuvants have been made. The mechanisms of their actions are not clear but it has been reported[135,136] that these agents enhance blood level concentrations over those obtained when tetracycline hydrochloride alone is administered orally. Remmers et al.[137,138] have reported on the effects that selected aluminum-calcium gluconates complexed with some tetracyclines have on the blood level concentrations when administered orally, intramuscularly or intravenously. Such complexes enhanced blood levels in dogs when injected but not when given orally. They have also observed enhanced blood levels in experimental animals when complexes of tetracy-

clines with aluminum metaphosphate, with aluminum pyrophosphate and aluminum-calcium phosphinicodilactates were administered orally. As has been noted previously, the tetracyclines are capable of forming stable chelate complexes with metal ions such as calcium and magnesium that would retard absorption from the gastrointestinal tract. The complexity of the systems involved has not permitted unequivocal substantiation of the idea that these adjuvants act by competing with the tetracyclines for substances in the alimentary tract that would otherwise be free to complex with these antibiotics and thus retard their absorption. Certainly, there is no evidence that they act by any virtue they possess as buffers, an idea alluded to sometimes in the literature.

Rolitetracycline U.S.P., Syntetrin®, N-(pyrrolidinomethyl)tetracycline, has been introduced for use by intramuscular and intravenous injection. This derivative is made by condensing tetracycline with pyrrolidine and formaldehyde in the presence of t-butyl alcohol. It is very soluble in water, 1 g. dissolving in about 1 ml., and provides a means of injecting the antibiotic in a small volume of solution. It is recommended in cases for which the oral dosage forms are not suitable.

N-(pyrrolidinomethyl)tetracycline

Chlortetracycline Hydrochloride U.S.P., Aureomycin® Hydrochloride. Chlortetracycline was isolated by Duggar[139] in 1948 from *Streptomyces aureofaciens*. This compound, which was produced in an extensive search for new antibiotics, was the first of the group of highly successful tetracyclines. It soon became established as a valuable antibiotic with broad-spectrum activities. It is used in medicine chiefly as the acid salt of the compound whose systematic chemical designation is 7-chloro-4-(dimethylamino)-1,4,4a,5,5a,6,11,12a-octahydro-3,6,10,12,12a-pentahydroxy-6-methyl-1,11-dioxo-2-naphthacenecarboxamide. The hydrochloride salt is a crystalline powder having a bright yellow color that suggested its brand name Aureomycin. It is stable in air, but is slightly photosensitive and should be protected from light. It is odorless and has a bitter taste. One gram of the hydrochloride salt will dissolve in about 75 ml. of water, producing a pH of about 3. It is only slightly soluble in alcohol and practically insoluble in other organic solvents.

Chlortetracycline

Chlortetracycline hydrochloride is most generally administered orally in capsules to avoid its bitter taste. It may also be administered parenterally (I.V.).

The 7-bromo analog of chlortetracycline has been isolated from Streptomyces species grown on special media rich in bromide ion. Bromtetracycline has antibiotic properties very similar to those of chlortetracycline.

Oxytetracycline Hydrochloride U.S.P., Terramycin®. Early in 1950, Finlay et al.[140] reported the isolation of oxytetracycline from *Streptomyces rimosus*. It was soon established that this compound was a chemical analog of chlortetracycline and showed similar antibiotic properties. The structure of oxytetracycline was elucidated by Hochstein et al.,[141] and this work provided the basis for the confirmation of the structure of the other tetracyclines.

Oxytetracycline

Oxytetracycline hydrochloride is a crystalline compound having a pale-yellow color and a bitter taste. The amphoteric base is only very slightly soluble in water and slightly soluble in alcohol. It is an odorless substance with a slightly bitter taste. It is stable in air but darkens upon exposure to strong sunlight. The hydrochloride salt is a stable yellow powder having a more bitter taste than the free base. It is much more soluble in water, 1 g. dissolving in 2 ml., and also is more soluble in alcohol. Both compounds are inactivated rapidly by alkali hydroxides and by acid solutions below pH 2. Both forms of oxytetracycline are absorbed from the digestive tract rapidly and equally well, so that the only real advantage the free base offers over the hydrochloride salt is its less bitter

taste. Oxytetracycline hydrochloride also is used for parenteral administration (I.V. and I.M.).

Methacycline Hydrochloride U.S.P., Rondomycin®, 6-deoxy-6-demethyl-6-methylene-5-oxytetracycline hydrochloride. The synthesis of methacycline, reported by Blackwood et al.[142] in 1961, was accomplished by chemical modification of oxytetracycline. It has an antibiotic spectrum similar to that of the other tetracyclines but has a greater potency; about 600 mg. of methacycline is equivalent to 1 g. of tetracycline. Its particular value lies in its longer serum half-life, doses of 300 mg. producing continuous serum antibacterial activity for 12 hours. Its toxic manifestations and contraindications are similar to those of the other tetracyclines.

Methacycline

The greater stability of methacycline, both in vivo and in vitro, is a result of the modification at carbon atom 6. The removal of the 6-hydroxy group markedly increases the stability of ring C to both acids and bases, preventing the formation of anhydrotetracyclines by acids and of isotetracyclines by bases. Methacycline hydrochloride is a yellow to dark yellow crystalline powder that is slightly soluble in water and insoluble in nonpolar solvents. It should be stored in tight, light-resistant containers in a cool place.

Demeclocycline U.S.P., Declomycin®, 7-chloro-6-demethyltetracycline, was isolated in 1957 by McCormick et al.[127] from a mutant strain of *Streptomyces aureofaciens*. Chemically, it is 7-chloro-4-(dimethylamino) - 1,4,4a,5,5a,6,11,12a - octahydro - 3,6,10,12,12a - pentahydroxy-1,11-dioxo-2-naphthacenecarboxamide. Thus, it differs from chlortetracycline only by the absence of the methyl group on carbon atom 6. The absence of this methyl group enhances the stability of ring C to both acid and alkali.

Demeclocycline

Demeclocycline is a yellow, crystalline powder that is odorless and has a bitter taste. It is sparingly soluble in water. A 1 percent solution has a pH of about 4.8. It has an antibiotic spectrum similar to that of other tetracyclines, but it is slightly more active than the others against most of the microorganisms for which they are used. This, together with its slower rate of elimination through the kidneys, gives demeclocycline an effectiveness comparable with that of the other tetracyclines, at about three-fifths of the dose. Like the other tetracyclines, it may cause infrequent photosensitivity reactions that produce erythema after exposure to sunlight. It appears that demeclocyline may produce the reaction somewhat more frequently than the other tetracyclines. The incidence of discoloration and mottling of the teeth in youths found with demeclocyline appears to be as low as with the other tetracyclines.

Doxycycline U.S.P., Vibramycin®, α-6-deoxy-5-oxytetracycline. A more recent addition to the tetracycline group of antibiotics available for antibacterial therapy is doxycycline, first reported by Stephens *et al.*[143] in 1958. It was first obtained in small yields by a chemical transformation of oxytetracycline but it is now produced by catalytic hydrogenation of methacycline or by reduction of a benzyl mercaptan derivative of methacycline with Raney nickel. In the latter process a nearly pure form of the 6α-methyl epimer is produced. It is worthy of note that this isomer has the 6-methyl group oriented differently from that in the tetracyclines bearing also a 6-hydroxy group and that the 6α-methyl epimer is more than three times as active as its β-epimer.[125] Apparently the difference in orientation of the methyl groups, slightly affecting the shapes of the molecules, causes a significant difference in biological effect. Also, as in methacycline, the absence of the 6-hydroxyl group produces a compound that is very stable to acids and bases and that has a long biological half-life. In addition, it is very well absorbed from the gastrointestinal tract, thus allowing a smaller dose to be administered. High tissue levels are obtained with it and, unlike other tetracyclines, doxycycline apparently does not accumulate in patients with impaired renal function. It is therefore preferred for uremic patients with infections outside the urinary tract. However, its low renal clearance may limit its effectiveness in urinary tract infections.

Doxycycline

Doxycycline is available as the hyclate salt, a hydrochloride salt solvated as the hemiethanolate hemihydrate, and as the monohydrate. The hyclate form is sparingly soluble in water and is used in the capsule dosage form; the monohydrate is water-insoluble and is used for aqueous suspensions which are stable for periods up to 2 weeks when kept in a cool place.

Minocycline Hydrochloride U.S.P., Minocin®, Vectrin®, 7-dimethylamino-6-demethyl-6-deoxytetracycline. Minocycline, the most potent tetracycline currently employed in therapy, is obtained by reductive methylation of 7-nitro-6-demethyl-6-deoxytetracycline.[128] It was released for use in the United States in 1971. Since minocycline, like doxycycline, lacks the 6-hydroxyl group it is stable to acids and does not dehydrate or rearrange to anhydro or lactone forms. Minocycline is well absorbed orally to give high blood and tissue levels. It has a very long serum half-life resulting from slow urinary excretion and moderate protein binding. Doxycycline and minocycline, along with oxytetracycline, show the least in vitro calcium binding of the clinically available tetracyclines. The improved distribution properties of the 6-deoxytetracyclines has been attributed to a greater degree of lipid-solubility.

Minocycline

Perhaps the most outstanding property of minocycline is its activity toward gram-positive bacteria, especially staphylococci and streptococci. In fact, minocycline has been found to be effective against staphylococcal strains that are resistant to methacillin and all other tetracyclines, including doxycycline.[144] While it is doubtful that minocycline will replace bactericidal agents for the treatment of life-threatening staphylococcal infections, it may become a useful alternative for the treatment of less serious tissue infections. Minocycline has been recommended for the treatment of chronic bronchitis and other upper respiratory tract infections. Despite its relatively low renal clearance, partially compensated for by high serum and tissue levels, it has also been recommended for the treatment of urinary tract infections. It has been shown to be effective in the eradication of *Neisseria meningitidis* in asymptomatic carriers.

TABLE 7–11

TETRACYCLINES

Name Proprietary Name	Preparations	Usual Adult Dose*	Usual Dose Range*	Usual Pediatric Dose*
Tetracycline U.S.P. *Achromycin, Cyclopar, Panmycin, Steclin, Tetracyn, Robitet, Bristacycline*	Tetracycline Oral Suspension U.S.P. Tetracycline for Oral Suspension U.S.P.	The equivalent of 250 to 500 mg. of tetracycline hydrochloride every 6 hours or 500 mg. to 1 g. every 12 hours	Up to 4 g. daily	Oral, the equivalent of tetracycline hydrochloride, children 8 years of age and over—6.25 to 12.5 mg. per kg. of body weight every 6 hours; or 12.5 to 25 mg. per kg. of body weight every 12 hours
Tetracycline Hydrochloride U.S.P. *Achromycin, Bristacycline, Panmycin, Steclin, Sumycin, Tetracyn*	Tetracycline Hydrochloride Capsules U.S.P.	Oral, the equivalent of tetracycline hydrochloride, 250 to 500 mg. every 6 hours; or 500 mg. to 1 mg. every 12 hours	Up to 4 g. daily	Oral, the equivalent of tetracycline hydrochloride, children 8 years of age and over—6.25 to 12.5 mg. per kg. of body weight every 6 hours; or 12.5 to 25 mg. per kg. of body weight every 12 hours
	Tetracycline Hydrochloride Tablets U.S.P.			
	Tetracycline Hydrochloride Ointment U.S.P.	Topical to the skin, as a 3 percent ointment once or twice daily		
	Tetracycline Hydrochloride for Topical Solution U.S.P.	Topical, to the skin, as a 0.22 percent solution twice daily, morning and evening		
	Tetracycline Hydrochloride Ophthalmic Ointment U.S.P.	Topical, to the conjunctiva, a thin strip (approximately 1 cm.) of a 1 percent ointment every 2 to 4 hours or more frequently		
	Tetracycline Hydrochloride Ophthalmic Suspension U.S.P.	Topical, to the conjunctiva, 1 drop of a 1 percent suspension every 6 to 12 hours or more frequently		
	Tetracycline Hydrochloride for Intramuscular Injection U.S.P.	IM, 100 mg. every 8 hours; 150 mg. every 12 hours; or 250 mg. once daily	Up to 1 g. daily	Children 8 years of age and over—I.M., 5 to 8.3 mg. per kg. of body weight every 8 hours; or 7.5 to 12.5 mg. per kg. of body weight every 12 hours; maximal dose should not exceed 250 mg.
	Tetracycline Hydrochloride for Intravenous Injection U.S.P.	250 to 500 mg. every 12 hours	Up to 2 g. daily	Children 8 years of age and over—I.V., 5 to 10 mg. per kg. of body weight every 12 hours
	Tetracycline Phosphate Complex U.S.P.			
	Tetracycline Phosphate Complex Capsules U.S.P.	See Tetracycline Oral Suspension U.S.P.	See Tetracycline Oral Suspension U.S.P.	See Tetracycline Oral Suspension
Rolitetracycline U.S.P. *Syntetrin*	Rolitetracycline for Injection U.S.P.		I.M., 150 to 350 mg. every 12 hours; I.V. infusion, 350 to 700 mg. every 12 hours	Children 8 years of age and over—oral, 6.25 to 12.5 mg. per kg. of body weight every 6 hours
Chlortetracycline Hydrochloride U.S.P. *Aureomycin*	Chlortetracycline Hydrochloride Capsules U.S.P.	Oral 250 to 500 mg. every 6 hours	Up to 4 g. daily	

*See U.S.P. D.I. for complete dosage information.

TABLE 7–11

TETRACYCLINES

Name Proprietary Name	Preparations	Usual Adult Dose*	Usual Dose Range*	Usual Pediatric Dose*
	Chlortetracycline Hydrochloride Ophthalmic Ointment U.S.P.	Topical, to the conjunctiva, a thin strip (approximately 1 cm.) of a 1 percent ointment every 2 to 4 hours or more often		See Usual Adult Dose
	Chlortetracycline Hydrochloride Ointment U.S.P.	Topical, to the skin, as a 3 percent ointment once or twice daily		See Usual Adult Dose
Oxytetracycline U.S.P. *Terramycin*	Oxytetracycline Tablets U.S.P.	Oral, the equivalent of oxytetracycline, 250 to 500 mg. every 6 hours	Up to 4 g. daily	Children 8 years of age and over—oral, the equivalent of oxytetracycline, 6.25 to 12.5 mg. per kg. of body weight every 6 hours
	Oxytetracycline Injection U.S.P.	IM, 100 mg. every 8 hours; 150 mg. every 12 hours; or 250 mg. once daily	Up to 500 mg. daily	Children 8 years of age and over—I.M., 5 to 8.3 mg. per kg. of body weight every 8 hours; or 7.5 to 12.5 mg. per kg. of body weight every 12 hours; maximal daily dose should not exceed 250 mg. for single injection
Oxytetracycline Calcium U.S.P. *Terramycin*	Oxytetracycline Calcium Oral Suspension U.S.P.	See Oxytetracycline Tablets U.S.P.	See Oxytetracycline Tablets U.S.P.	See Oxytetracycline Tablets U.S.P.
Oxytetracycline Hydrochloride U.S.P. *Terramycin*	Oxytetracycline Hydrochloride Capsules U.S.P.	See Oxytetracycline Tablets U.S.P.	See Oxytetracycline Tablets U.S.P.	See Oxytetracycline Tablets U.S.P.
	Oxtetracycline Hydrochloride for Injection U.S.P.	IV, the equivalent of oxytetracycline, 250 to 500 mg. every 12 hours	Up to 2 g. daily	8 years of age and over—I.V., the equivalent of oxytetracycline, 5 to 10 mg. per kg. of body weight every 12 hours
Methacycline Hydrochloride U.S.P. *Rondomycin*	Methacycline Hydrochloride Capsules U.S.P.	Oral, 150 mg. every 6 hours; or 300 mg. every 12 hours	Up to 2.4 g. daily	Children 8 years of age and over—oral, 1.65 to 3.3 mg. per kg. of body weight every 6 hours; or 3.3 to 6.6 mg. per kg. of body weight every 12 hours
	Methacycline Hydrochloride Oral Suspension U.S.P.	See Methacycline Hydrochloride Capsules U.S.P.	See Methacycline Hydrochloride Capsules U.S.P.	See Methacycline Hydrochloride Capsules U.S.P.
Demeclocycline U.S.P. *Declomycin*	Demeclocycline Oral Suspension U.S.P.	Oral, the equivalent of demeclocycline hydrochloride, 150 mg. every 6 hours; or 300 mg. every 12 hours	Up to 2.4 g. daily	Children 8 years of age and over—oral, the equivalent of demeclocycline hydrochloride, 1.65 to 3.3 mg. per kg. of body weight every 6 hours; or 3.3 to 6.6 mg. per kg. of body weight every 12 hours
Demeclocycline Hydrochloride U.S.P. *Declomycin*	Demeclocycline Hydrochloride Capsules U.S.P.	See Demeclocycline Oral Suspension U.S.P.	See Demeclocycline Oral Suspension U.S.P.	See Demeclocycline Oral Suspension U.S.P.
Doxycycline U.S.P. *Vibramycin*	Doxycycline for Oral Suspension U.S.P.	Oral, the equivalent of anhydrous doxycycline, 100 mg. every 12 hours the first day, then 100 to 200 mg. once daily; or 50 to 100 mg. every 12 hours	Up to 300 mg. daily; or up to 600 mg. daily for 1 day in acute gonococcal infections	Oral, the equivalent of anhydrous doxycycline, children 45 kg. of body weight and under—2.2 mg. per kg. of body weight every 12 hours the first day, then 2.2 to 4.4 mg. per kg. of body weight once daily; or 1.1 to 2.2 mg. per kg. of body weight every 12 hours

*See *U.S.P. D.I.* for complete dosage information.

TABLE 7–11

TETRACYCLINES

Name Proprietary Name	Preparations	Usual Adult Dose*	Usual Dose Range*	Usual Pediatric Dose*
Doxycycline Hyclate U.S.P. *Vibramycin*	Doxycycline Hyclate for Injection U.S.P. Doxycycline Hyclate Tablets U.S.P.	I.V. infusion, the equivalent of doxycycline, 200 mg. once daily or 100 mg. every 12 hours the first day, then 100 to 200 mg. once daily; or 50 to 100 mg. every 12 hours	Up to 300 mg. daily	I.V. infusion, the equivalent of doxycycline, children 45 kg. of body weight and under—4.4 mg. per kg. of body weight once daily or 2.2 mg. per kg. of body weight every 12 hours the first day; then 2.2 to 4.4 mg. per kg. of body weight once daily or 1.1 to 2.2 mg. per kg. of body weight every 12 hours
	Doxycycline Calcium Oral Suspension U.S.P.	See Doxycycline for Oral Suspension U.S.P.	See Doxycycline for Oral Suspension U.S.P.	See Doxycycline for Oral Suspension U.S.P.
Minocycline Hydroxchloride U.S.P. *Minocin, Vectrin*	Minocycline Hydrochloride Capsules U.S.P.	Oral, the equivalent of minocycline, 200 mg. initially, then 100 mg. every 12 hours; or 100 to 200 mg. initially, then 50 mg. every 6 hours	Up to 150 mg. the first day; then up to 200 mg. daily	Oral, the equivalent of minocycline, children 8 years of age and over—4 mg. per kg. of body weight initially, then 2 mg. per kg. of body weight every 12 hours
	Minocycline Hydrochloride Tablets U.S.P.	See Minocycline Hydroxhloride Capsules U.S.P.	See Minocycline Hydrochloride Capsules U.S.P.	See Minocycline Hydrochloride Capsules U.S.P.

*See *U.S.P. D.I.* for complete dosage information.

THE MACROLIDES

Among the many antibiotics isolated from the actinomycetes is the group of chemically related compounds called the macrolides. It was in 1950 that picromycin, the first of this group to be identified as a macrolide compound, was first reported. In 1952, erythromycin and carbomycin were reported as new antibiotics and these were followed in subsequent years by other macrolides. At present more than three dozen such compounds are known, and new ones are likely to appear in the future. Of all of these, only two, erythromycin and oleandomycin, have been consistently available for medical use in the United States. One other, carbomycin, has been available, but, because of its poor and irregular absorption from the gastrointestinal tract and its inferior antibacterial activity when compared to erythromycin, it never enjoyed wide use and was withdrawn from the market. Oleandomycin has been withdrawn. Spiramycin is used in Europe and other parts of the world, but its activity in vitro is inferior to that of erythromycin, and it is difficult to account for its reputed therapeutic success. Various members of the leucomycin group have been used clinically in various parts of the world. Josamycin (leucomycin A₃), isolated from *Streptomyces narbonensis* variety *josamyceticus,* has been marketed in Japan and in Europe primarily for the treatment of respiratory and genitourinary infections caused by gram-positive bacteria. It has recently been subjected to clinical evaluations in the United States and compares favorably with erythromycin for many indications. Rosamicin, a macrolide antibiotic obtained from *Micromonospora rosaria,* has shown some promise for the treatment of genitourinary infections in in vitro tests. Both josamycin and rosamicin remain on investigational status in the United States.

The macrolide antibiotics have three common chemical characteristics: (1) a large lactone ring (which prompted the name *macrolide*), (2) a ketone group, and (3) a glycosidically linked amino sugar. Usually the lactone ring has 12, 14 or 16 atoms in it and is often partially unsaturated with an olefinic group conjugated with the ketone function. (The polyene macrocyclic lactones, such as pimaricin, and the polypeptide lactones generally are not included among the macrolide antibiotics.) They may have, in addition to the amino sugar, a neutral sugar that is glycosidically linked to the lactone ring (see erythromycin). Because of the presence of the dimethylamino group on the sugar moiety, the macrolides are bases which form salts with pKa values between 6.0 and 9.0. This feature has been employed to make clinically useful salts. The free bases are only slightly soluble in water but dissolve in the somewhat polar organic solvents. They are stable in aqueous solutions at or below room temperature but are inactivated by acids, bases and heat.

The chemistry of the macrolide antibiotics has been reviewed by Wiley,[145] Miller,[146] and Morin and Gorman.[147]

The antibacterial spectrum of activity of the more potent macrolides resembles that of penicillin. They are frequently active against bacterial strains that are resistant to the penicillins. The macrolides are generally effective against most species of gram-positive bacteria, both cocci and bacilli, and also exhibit useful effectiveness against gram-negative cocci, especially *Neisseria* sp. Many of the macrolides are also effective against *Treponema pallidum* (like penicillin) and *Mycobacterium pneumoniae* (unlike penicillin). Their activity against most species of gram-negative bacilli is generally low and often unpredictable, although some strains of *Hemophilus influenzae* and *Brucella* sp. are sensitive.

PRODUCTS

Erythromycin U.S.P., E-Mycin®, Erythrocin®, Ilotycin®. Early in 1952, McGuire *et al.*[148] reported the isolation of erythromycin from *Streptomyces erythreus*. It achieved a rapid early acceptance as a well-tolerated antibiotic of value for the treatment of a variety of upper respiratory and soft-tissue infections caused by gram-positive bacteria. It is also effective against many venereal diseases including gonorrhea and syphilis and provides a useful alternative to penicillin for the treatment of many infections in patients allergic to penicillins. More recently, erythromycin has been shown to be effective therapy for Eaton Agent pneumonia (*M. pneumoniae*) and Legionnaires' disease.

The commercial product is actually erythromycin A, which differs from its biosynthetic precurser erythromycin B in having a hydroxyl group at the 12-position of the aglycone. The chemical structure of erythromycin A was reported by Wiley *et al.*[149] in 1957 and its stereochemistry by Celmer[150] in 1965. An elegant synthesis of erythronolide A, the aglycone present in erythromycin A, has recently been described by Corey and his associates.[151]

The amino sugar attached through a glycosidic link to the number 5 carbon atom is desosamine, a structure found in a number of other macrolide antibiotics. The tertiary amine of desosamine (3,4,6-trideoxy-3-dimethylamino-D-*xylo*-hexose) confers a basic character to erythromycin and provides the means by which acid salts may be prepared. The other carbohydrate structure linked as a glycoside to carbon atom 3 is called cladinose (2,3,6-trideoxy-3-methoxy-3-C-methyl-L-*ribo*-hexose) and is unique to the erythromycin molecule.

Erythromycin A

As is common with other macrolide antibiotics, compounds closely related to erythromycin have been obtained from culture filtrates of *S. erythreus*. Two such analogs have been found and are designated as erythromycins B and C. Erythromycin B differs from erythromycin A only at the number 12 carbon atom where a hydrogen has replaced the hydroxyl group. The B analog is more acid-stable but has only about 80 percent of the activity of erythromycin. The C analog differs from erythromycin by the replacement of the methoxyl group on the cladinose moiety by a hydrogen atom. It appears to be as active as erythromycin but is present in very small amounts in fermentation liquors.

Erythromycin is a very bitter, white or yellowish-white crystalline powder. It is soluble in alcohol and in the other common organic solvents but only slightly soluble in water. Saturated aqueous solutions develop an alkaline pH in the range of 8.0 to 10.5. It is extremely unstable at a pH of 4 or lower. The optimum pH for stability of erythromycin is at or near neutrality.

Erythromycin may be used as the free base in oral dosage forms and for topical administration. To attempt to overcome its bitter taste and irregular oral absorption (resulting from acid destruction and adsorption onto food), however, a variety of enteric coated and delayed-release dose forms of erythromycin base have been developed. These forms have been fully successful in overcoming the bitter taste, but have only marginally solved problems of oral absorption of the antibiotic. Chemical modifications of erythromycin have been made with two different goals primarily in mind: (1) to increase either its water or lipid solubility for parenteral dosage forms; and (2) to increase its acid stability (and possibly increase its lipid solubility) for improved oral absorption. Modified derivatives of the antibiotic are of two types: acid salts of the dimethylamino group of the desosamine moiety such as the glucoheptonate, the lactobionate and the stearate; and esters of the 2'-hydroxyl group of the desosamine such as the ethylsuccinate and the propionate (available as the laurylsulfate salt and known as the estolate).

The stearate salt and the ethylsuccinate and propionate esters are used in oral dose forms intended to improve absorption of the antibiotic. The stearate releases erythromycin base in the intestinal tract which is then absorbed. The ethylsuccinate and the estolate are absorbed largely intact and are partially hydrolyzed by plama and tissue esterases to give free erythromycin. The question of bioavailability of the antibiotic from its various oral dosage and chemical forms has been the subject of considerable concern and dispute over the past decade.[152-154] It is generally believed that the 2'-esters per se have little or no intrinsic antibacterial activity[155] and therefore must be hydrolyzed to the parent antibiotic in vivo. Although the ethylsuccinate is hydrolyzed more efficiently than the estolate in vivo and, in fact, provides higher levels of erythromycin following intramuscular administration, an equal oral dose of the estolate gives higher levels of the free antibiotic following oral administration.[153] Superior oral absorption of the estolate is attributed to both its greater acid stability and higher intrinsic absorption compared to the ethylsuccinate. Also the oral absorption of the estolate, unlike that of both the stearate and the ethylsuccinate, is not affected by food or fluid volume content of the gut. Superior bioavailability of active antibiotic from oral administration of the estolate over the ethylsuccinate, stearate or erythromycin base cannot necessarily be assumed, however, since the estolate is more extensively protein bound than erythromycin itself.[156,157] Measured fractions of plasma protein binding for erythromycin-2'-propionate and erythromycin base range from 0.94 to 0.98 for the former and 0.73 to 0.90 for the latter, indicating a much higher level of free erythromycin in the plasma. Furthermore, studies comparing the clinical effectiveness of recommended doses of the stearate, estolate, ethylsuccinate or free base in the treatment of respiratory tract infections have failed to demonstrate substantial differences between them.[158] It should be noted, however, that the recommended oral dose of the base is higher than that of both the stearate and the estolate.

The water-insoluble ethylsuccinate ester is also available as a suspension for intramuscular injection. The glucoheptonate and lactobionate salts, on the other hand, are highly water soluble derivatives that provide high blood levels of the active antibiotic immediately following intravenous injection. Aqueous solutions of these salts may also be administered by I.M. injection, but this is not a common practice.

Some details of the mechanism of antibacterial action of erythromycin are known. It selectively binds to a specific site on the 50S ribosomal subunit to prevent the translocation step of bacterial protein synthesis.[159]

Erythromycin does not bind to mammalian ribosomes. Broadly based nonspecific resistance to the antibacterial action of erythromycin among many species of gram-negative bacilli appears to be related in large part to the inability of the antibiotic to penetrate effectively the cell walls of these organisms.[160] In fact, the sensitivities of members of the *Enterobacteriaceae* family are pH dependent, with MICs decreasing as a function of increasing pH. Furthermore, protoplasts from gram-negative bacilli, lacking cell walls, are sensitive to erythromycin. A highly specific resistance mechanism to the macrolide antibiotics occurs in erythromycin-resistant strains of *Straphylococcus aureus*.[161] Such strains have been shown to produce an enzyme that methylates a specific adenine residue at the erythromycin binding site of the bacterial 50S ribosomal subunit. The thus methylated ribosomal RNA remains active in protein synthesis but no longer binds erythromycin. Bacterial resistance to the lincomycins apparently also occurs by this mechanism.

Troleandomycin, TAO®, triacetlyoleandomycin. Oleandomycin, as its triacetyl derivative troleandomycin, remains available as an alternative to erythromycin for limited indications permitting use of an oral dosage form. Oleandomycin was originally isolated by Sobin, English, and Celmer.[162] The structure of oleandomycin was first proposed by Hochstein et al.[163] and its absolute stereochemistry elucidated by Celmer.[164] The oleandomycin structure (below) consists of 2 sugars and a 14-member lactone ring designated *oleandolide*. One of the sugars is deosamine, also present in erythromycin, the other is L-oleandrose. The sugars are glycosidically linked to the 5- and 3-positions, respectively, of oleandolide.

Oleandomycin

Oleandomycin contains 3 hydroxyl groups that are subject to acylation, one in each of the sugars and one in the oleandolide. The triacetyl derivative retains the in vivo antibacterial activity of the parent antibiotic but possesses superior pharmacokinetic properties. It is hydrolyzed in vivo to oleandomycin. Troleandomycin has been found to achieve more rapid and higher blood-level concentrations following oral administration than oleandomycin phosphate, and it has the additional advantage of being practically tasteless. Troleandomycin occurs as a white crystalline solid that is nearly insoluble in water. It is relatively stable in the solid state but undergoes chemical degradation in either aqueous acidic or alkaline conditions.

Approved medical indications for troleandomycin are currently limited to the treatment of upper-respiratory infections caused by such organisms as *Streptococcus pyogenes* and *Streptococcus pneumoniae*. In this regard, it may be considered as an alternative drug to oral forms of erythromycin. It is available in capsules and as a suspension.

TABLE 7-12

MACROLIDE ANTIBIOTICS

Name Proprietary Name	Preparations	Usual Adult Dose*	Usual Dose Range*	Usual Pediatric Dose*
Erythromycin U.S.P. *Erythrocin, Iotycin, E-Mycin*	Erythromycin Ointment U.S.P.	For external use, topically to the skin, as a 1 percent ointment 3 or 4 times daily		
	Erythromycin Ophthalmic Ointment U.S.P.	For external use, topically to the conjunctiva, as a 0.5 percent ointment one or more times daily		
	Erythromycin Tablets U.S.P.	Antibacterial, oral, 250 mg. every 6 hours or 500 mg. every 12 hours	Antibacterial, up to 4 g. or more daily	Antibacterial, oral, 7.5 to 25 mg. per kg. of body weight every 6 hours; or 15 to 50 mg. per kg. of body weight every 12 hours
Erythromycin Ethylsuccinate U.S.P. *Erythrocin Ethylsuccinate, Pediamycin*	Erythromycin Ethylsuccinate Oral Suspension U.S.P.	Antibacterial, oral, the equivalent of erythromycin, 400 Mg. every 6 hours, or 800 mg. every 12 hours	Antibacterial, the equivalent of erythromycin, up to 4 g. daily	Antibacterial, oral, the equivalent of erythromycin, 7.5 to 25 mg. per kg. of body weight every 6 hours; or 15 to 50 mg. per kg. of body weight every 12 hours
	Erythromycin Ethylsuccinate for Oral Suspension U.S.P.	See Erythromycin Ethylsuccinate Oral Suspension U.S.P.	See Erythromycin Ethylsuccinate Oral Suspension U.S.P.	See Erythromycin Ethylsuccinate Oral Suspension U.S.P.
	Erythromycin Ethylsuccinate Tablets U.S.P. Erythromycin Ethylsuccinate Chewable Tablets U.S.P.			
	Sterile Erythromycin Gluceptate U.S.P. *Ilotycin Gluceptate*	I.V. infusion, the equivalent of 250 to 500 mg. of erythromycin every 6 hours, or 3.75 to 5 mg. per kg. of body weight every 6 hours	Up to 4 g. daily	I.V. infusion, the equivalent of 3.75 to 5 mg. per kg. of body weight every 6 hours
	Erythromycin Lactobionate for Injection U.S.P. *Erythrocin Lactobionate*	See Sterile Erythromycin Gluceptate U.S.P.	See Sterile Erythromycin Gluceptate U.S.P.	See Sterile Erythromycin Gluceptate U.S.P.
Erythromycin Stearate U.S.P. *Erythrocin Stearate, Bristamycin, Ethril*	Erythromycin Stearate Tablets U.S.P. Erythromycin Stearate for Oral Suspension U.S.P.	See Erythromycin Tablets U.S.P.	See Erythromycin Tablets U.S.P.	See Erythromycin Tablets U.S.P.
Erythromycin Estolate U.S.P. *Ilosone*	Erythromycin Estolate Capsules Erythromycin Estolate Tablets U.S.P. Erythromycin Estolate for Oral Suspension U.S.P. Erythromycin Estolate Oral Suspension U.S.P.			
	Erythromycin Estolate Chewable Tablets U.S.P.	See Erythromycin Tablets U.S.P.		
Troleandomycin Tao®	Troleandomycin Capsules Troleandomycin Oral Suspension	Antibacterial, oral, 250 to 500 mg. 4 times a day		Antibacterial, oral, 125 to 250 mg. 4 times a day

*See U.S.P. D.I. for complete dosage information.

THE LINCOMYCINS

The lincomycins are sulfur-containing antibiotics isolated from *Streptomyces lincolnensis.* Lincomycin is the most active and medically useful of the compounds obtained from fermentation. Extensive efforts to modify the lincomycin structure in order to improve its antibacterial and pharmacological properties resulted in the preparation of the 7-chloro-7-deoxy derivative, clindamycin. Of the two antibiotics, clindamycin appears to have the greater antibacterial potency and better pharmacokinetic properties as well. Lincomycins resemble the macrolides in antibacterial spectrum and biochemical mechanisms of action. They are primarily active against gram-positive bacteria, particularly the cocci, but are also effective against non-spore-forming anaerobic bacteria, actinomycetes, mycoplasma and some species of *Plasmodium.* Lincomycin binds to the 50S ribosomal subunit to inhibit protein synthesis. Its action may be bacteriostatic or bactericidal depending on a variety of factors, which include the concentration of the antibiotic. A pattern of bacterial resistance and cross-resistance to lincomycins similar to that observed with the macrolides has been emerging.

PRODUCTS

Lincomycin Hydrochloride U.S.P., Lincocin®. This antibiotic, which differs chemically from other major antibiotic classes, was first isolated by Mason *et al.*[165] Its chemistry has been described by Hoeksema and his coworkers[166] who assigned the structure, later confirmed by Slomp and MacKellar,[167] given below. Total syntheses of the antibiotic were independently accomplished in 1970 through research efforts in England and in the United States.[168,169] The structure contains a basic function, the pyrrolidine nitrogen, by which water-soluble salts having an apparent pKa of 7.6 may be formed. When subjected to hydrazinolysis lincomycin is cleaved at its amide bond into *trans*-L-4-*n*-propylhygric acid (the pyrrolidine moiety) and methyl α-thiolincosamide (the sugar moiety). Lincomycin-related antibiotics have been reported by Argoudelis[170] to be produced by *S. lincolnensis.* These antibiotics differ in structure at one or more of three positions of the lincomycin structure: (1) the N-methyl of the hygric acid moiety is substituted by a hydrogen; (2) the *n*-propyl group of the hygric acid moiety is substituted by an ethyl group; and (3) the thiomethyl ether of the α-thiolincosamide moiety is substituted by a thioethyl ether.

Lincomycin

Lincomycin is employed for the treatment of infections caused by gram-positive organisms, notably staphylococci, β-hemolytic streptococci, and pneumococci. It is moderately well absorbed orally and is widely distributed in the tissues. Effective concentrations are achieved in bone for the treatment of staphylococcal osteomyelitis, but not in the cerebral spinal fluid for the treatment of meningitis. Lincomycin was at one time thought to be a very nontoxic compound, with a low incidence of allergy (skin rashes) and occasional gastrointestinal complaints (nausea, vomiting and diarrhea) as the only adverse effects. However, recent reports of severe diarrhea and the development of pseudomembranous colitis in patients treated with lincomycin (or clindamycin) have brought about the need for reappraisal of the position these antibiotics should have in therapy. In any event, clindamycin is superior to lincomycin for the treatment of most infections for which these antibiotics are indicated.

Lincomycin hydrochloride occurs as the monohydrate, a white crystalline solid that is stable in the dry state. It is readily soluble in water and alcohol and its aqueous solutions are stable at room temperature. It is slowly degraded in acid solutions but is well absorbed from the gastrointestinal tract. Lincomycin diffuses well into peritoneal and pleural fluids and into bone. It is excreted in the urine and the bile. It is available in capsule form for oral administration and in ampules and vials for parenteral administration.

Clindamycin Hydrochloride U.S.P., Cleocin®, 7S-chloro-7S-deoxy-lincomycin. In 1967 Magerlein *et al.*[171] reported that replacement of the 7R-hydroxy group of lincomycin by chlorine with inversion of configuration resulted in a compound with enhanced antibacterial activity in vitro. Clinical experience with this semisynthetic derivative, called clindamycin and released in 1970, has established that its superiority over lincomycin is even greater in vivo. Improved absorption and higher tissue levels of clindamycin, and its greater penetration into bacteria, have been attributed to its higher partition coefficient compared to that of lincomycin. Structural modifications at C-7, for example 7S-chloro and 7R-OCH$_3$, and of the C-4 alkyl group of the hygric acid moiety,[172] appear to influence activity of congeners more through an effect on the partition coefficient of the molecule than through a stereospecific binding role. On the other hand, changes in the α-thiolincosamide portion of the molecule appear to markedly decrease activity, as is evidenced by the marginal activity of 2-deoxy-lincomycin, its β-anomer and 2-0-methyllincomycin.[172,173] Exceptions to this are fatty acid and phosphate esters of the 2-hydroxyl group of lincomycin and clindamycin, which are rapidly hydrolyzed in vivo to the parent antibiotics.

Clindamycin

Clindamycin is recommended by the manufacturer for the treatment of a wide variety of upper respiratory, skin and tissue infections caused by susceptible bacteria. Certainly, its activity against streptococci, staphylococci and pneumococci is undisputably high; and it is one of the most potent agents available against some non-spore-forming anaerobic bacteria, the *Bacteriodes* species in particular. However, an ever-increasing number of reports of clindamycin-associated gastrointestinal toxicity, which range in severity from diarrhea to an occasionally serious pseudomembranous colitis, have caused some clinical experts to call for a reappraisal of the appropriate position of this antibiotic in therapy. Clindamycin- (or lincomycin-) associated colitis may be particularly dangerous in elderly or debilitated patients and has caused deaths in such individuals. This condition, which is usually reversible when the drug is withdrawn, is now believed to result from an overgrowth of a clindamycin-resistant strain of the anaerobic intestinal bacterium, *Clostridium difficile*.[174] Damage to the intestinal lining is caused by a glycoprotein endotoxin released by lysis of this organism. Vancomycin has been found to be effective in the treatment of clindamycin-induced pseudomembranous colitis and in the control of the experimentally induced bacterial condition in animals. Clindamycin should be reserved for staphylococcal tissue infections such as cellulitis and osteomyelitis in penicillin-allergic patients and for severe anaerobic infections outside the central nervous system. It should not ordinarily be used to treat upper respiratory tract infections caused by bacteria sensitive to other, safer antibiotics or in prophylaxis.

Clindamycin is rapidly absorbed from the gastrointestinal tract, even in the presence of food. It is available as the crystalline, water-soluble hydrochloride hydrate (hyclate) and the 2-palmitate ester hydrochloride salts in oral dosage forms, and as the 2-phosphate ester in solutions for I.M. and I.V. injection. All forms are chemically very stable in solution and in the dry state. See Table 7–13 for structured relationships.

THE POLYPEPTIDES

Among the most powerful bactericidal antibiotics are those possessing a polypeptide structure. Many of them have been isolated but, unfortunately, their clinical use has been limited by their undesirable side-reactions, particularly renal toxicity. The chief source of the medicinally important members of this class has been various species of the genus *Bacillus*. A few have been isolated from other bacteria but have not gained a place in medical practice. Three medicinally useful polypeptide antibiotics have been isolated from a *Streptomyces* species.

Polypeptide antibiotics are of three main types: neutral, acidic and basic. It had been presumed that the neutral compounds such as the gramicidins possessed cyclopeptide structures and thus had no free amino or carboxyl groups. It has been shown that the neutrality is due to the formylation of a terminal group and that the neutral gramicidins are linear rather than cyclic. The acidic compounds have free carboxyl[175] groups, indicating that at least part of the structure is noncyclic. The basic compounds have free amino groups and, similarly, are noncyclic at least in part. Some, like the gramicidins, are active against gram-positive organisms only; others, like the polymyxins, are active against gram-negative organisms and thus have achieved a special place in antibacterial therapy. Significant comments about the biosynthesis and structure-activity relationships of peptide antibiotics have been published by Bodanszky and Perlman.[176]

Gramicidin U.S.P. Gramicidin is obtained from tyrothricin, a mixture of polypeptides usually obtained by extraction of cultures of *Bacillus brevis*. Tyrothricin was isolated in 1939 by Dubos[177] in a planned search to find an organism growing in soil that would have antibiotic activity against human pathogens. Having only limited use in therapy now, it is of his-

TABLE 7–13

LINCOMYCINS

Name Proprietary Name	Preparations	Usual Adult Dose*	Usual Dose Range*	Usual Pediatric Dose*
Lincomycin Hydrochloride U.S.P. *Lincocin*	Lincomycin Hydrochloride Injection U.S.P.	I.M., the equivalent of 600 mg. of lincomycin once or twice daily; I.V. infusion, the equivalent of 600 mg. to 1 g. of lincomycin over a period of not less than 1 hour, 2 or 3 times daily	600 mg. to 8 g. daily	Dosage is not established in children under 1 month of age. Over 1 month—I.M., 10 mg. per kg. of body weight or 300 mg. per square meter of body surface, 1 or 2 times daily; I.V. infusion, 5 to 10 mg. per kg. or 150 to 300 mg. per square meter over a period of not less than 1 hour, 2 times daily
	Lincomycin Hydrochloride Capsules U.S.P. Lincomycin Hydrochloride Syrup	The equivalent of 500 mg. of lincomycin every 6 to 8 hours		Infants 1 month of age and over—oral, the equivalent of lincomycin, 7.5 to 15 mg. per kg. of body weight every 6 hours; or 10 to 20 mg. per kg. of body weight every 8 hours
Clindamycin Hydrochloride U.S.P. *Cleocin*	Clindamycin Hydrochloride Capsules U.S.P.	The equivalent of 150 to 450 mg. of clindamycin every 6 hours		Infants under 1 month of age—use with caution; infants 1 month of age and over—oral, the equivalent of clindamycin, 2 to 6.3 mg. per kg. of body weight every 6 hours; or 2.7 to 8.3 mg. per kg. of body weight every 8 hours
Clindamycin Palmitate Hydrochloride U.S.P. *Cleocin Palmitate*	Clindamycin Palmitate Hydrochloride for Oral Solution U.S.P.	12 mg. of clindamycin, as clindamycin palmitate hydrochloride, per kg. 3 or 4 times daily	8 to 25 mg. of clindamycin, present as clindamycin palmitate hydrochloride, per kg. of body weight, divided into 3 or 4 equal doses. In children weighing 10 kg. or less, 37.5 mg. of clindamycin 3 times daily is the minimum recommended dose	
Clindamycin Phosphate U.S.P. *Cleocin Phosphate*	Clindamycin Phosphate Injection U.S.P.	I.M. or I.V., 300 mg. of clindamycin, as the phosphate, 2 to 4 times daily	600 mg. to 2.7 g. of clindamycin, as the phosphate, daily, divided into 2, 3 or 4 equal doses; in children over 1 month of age, 10 to 40 mg. of clindamycin per kg. of body weight daily, divided into 3 or 4 equal doses	

*See U.S.P. D.I. for complete dosage information.

torical interest as the first in the series of modern antibiotics. Tyrothricin is a white to slightly gray or brownish-white powder with little or no odor or taste. It is practically insoluble in water and is soluble in alcohol and in dilute acids. Suspensions for clinical use can be prepared by adding an alcoholic solution to calculated amounts of distilled water or isotonic saline solutions.

Tyrothricin is a mixture of two groups of antibiotic compounds, the gramicidins and the tyrocidines. Gramicidins are the more active components of tyrothricin, and this fraction, occurring in 10 to 20 percent quantities in the mixture, may be separated and used in topical preparations for the antibiotic effect. Five gramicidins, A_3, A_2, B_1, B_2, and C, have been identified. Their structures have been proposed and confirmed through synthesis by Sarges and Witkop.[175] It may be noted that the gramicidins A differ from the gramicidins B by having a tryptophan moiety substituted by an L-phenylalanine moiety. In gramicidin C, a tyrosine moiety substitutes for a tryptophan moiety. In both of the gramicidin A and B pairs, the only difference is the amino acid located at the end of the chain having the neutral formyl group on it. If that

OH
|
(CH$_2$)$_2$
|
HC=O
|
L-Val-Gly-L-Ala-D-Leu-L-Ala-D-Val-L-Val-D-Val-L-Try-D-Leu-L-Try-D-Leu-L-Try-D-Leu-L-Try-NH

Valine-gramicidin A

OH
|
(CH$_2$)$_2$
|
HC=O
|
L-Ileu-Gly-L-Ala-D-Leu-L-Ala-D-Val-L-Val-D-Val-L-Try-D-Leu-L-Try-D-Leu-L-Try-D-Leu-L-Try-NH

Isoleucine-gramicidin A

OH
|
(CH$_2$)$_2$
|
HC=O
|
L-Val-Gly-L-Ala-D-Leu-L-Ala-D-Val-L-Val-D-Val-L-Try-D-Leu-L-Phel-D-Leu-L-Try-D-Leu-L-Try-NH

Valine-gramicidin B

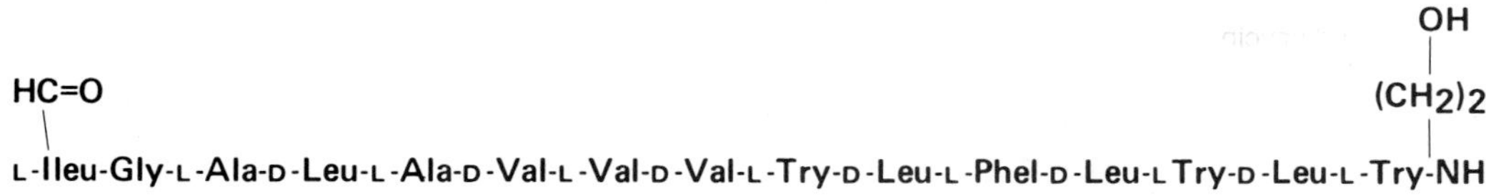

OH
|
(CH$_2$)$_2$
|
HC=O
|
L-Ileu-Gly-L-Ala-D-Leu-L-Ala-D-Val-L-Val-D-Val-L-Try-D-Leu-L-Phel-D-Leu-L-Try-D-Leu-L-Try-NH

Isoleucine-gramicidin B

amino acid is valine, the compound is either valine-gramicidin A or valine-gramicidin B. If that amino acid is isoleucine, the compound is isoleucine-gramicidin, either A or B.

Tyrocidine is a mixture of tyrocidines A, B, C and D whose structures have been determined by Craig and co-workers.[178]

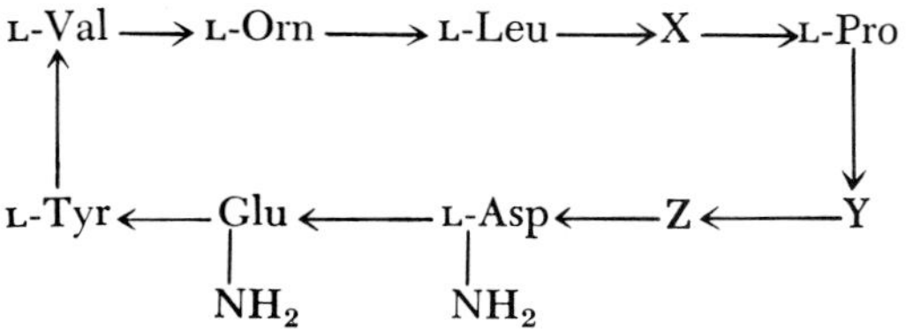

L-Val → L-Orn → L-Leu → X → L-Pro
↑   ↓
L-Tyr ← Glu ← L-Asp ← Z ← Y
        | |
       NH$_2$ NH$_2$

	X	Y	Z
Tyrocidine A:	D-Phe	D-Phe	D-Phe
Tyrocidine B:	D-Phe	L-Try	D-Phe
Tyrocidine C:	D-Try	L-Try	D-Phe
Tyrocidine D:	D-Try	L-Try	D-Try

The synthesis of tyrocidine A has been reported by Ohno *et al.*[179]

Tyrothricin and gramicidin are effective primarily against gram-positive organisms. Their use is restricted to local applications. The ability of tyrothricin to cause lysis of erythrocytes makes it unsuitable for the treatment of systemic infections. Its applications should avoid direct contact with the bloodstream through open wounds or abrasions. It is ordi-narily safe to use tyrothricin in troches for throat infections, as it is not absorbed from the gastrointestinal tract.

Bacitracin U.S.P. The organism from which Johnson, Anker and Meleney[180] produced bacitracin in 1945 is a strain of *Bacillus subtilis*. The organism had been isolated from debrided tissue from a compound fracture in 7-year-old Margaret Tracy, hence the name bacitracin. Production of bacitracin is now accomplished from the licheniformis group (Sp. *Bacillus subtilis*). Like tyrothricin, the first useful antibiotic obtained from bacterial cultures, bacitracin is a complex mixture of polypeptides. So far, at least 10 polypeptides have been isolated by countercurrent distribution techniques: A,A′,B,C,D,E,F$_1$,F$_2$,F$_3$ and G. It appears that the commercial product known as bacitracin is a mixture principally of A with smaller amounts of B, D, E and F.

The official product is a white to pale-buff powder that is odorless or nearly so. In the dry state, bacitracin is stable, but it rapidly deteriorates in aqueous solutions at room temperature. Because of its hygroscopic nature, it must be stored in tight containers, preferably under refrigeration. The stability of aqueous solutions of bacitracin is affected by pH and temperature. Slightly acidic or neutral solutions are stable for as long as 1 year if kept at a temperature of 0 to 5°. If the pH rises above 9, inactivation occurs very rapidly. For greatest stability, the pH of a bacitracin solution is best adjusted at 4 to 5 by the simple addition of acid. The salts of heavy metals precipitate bacitracin from its solutions, with resulting inactivation. However, EDTA also inactivates bacitracin, leading to the discovery that a divalent ion, i.e., Zn^{++}, is required for activity. In addition to being soluble in water, bacitracin is soluble in low molecular weight alcohols but is insoluble in many other organic solvents, including acetone, chloroform and ether.

The principal work on the chemistry of the bacitracins has been directed toward bacitracin A, the component in which most of the antibacterial activity of crude bacitracin resides. The structure shown below is that proposed by Stoffel and Craig[181] but it has not yet been confirmed by synthesis.

The chemistry of the other bacitracins has been worked on only to a limited extent. While there is evidence of considerable similarities in structure to bacitracin A among the other members of the group, there is considerable difficulty in fixing the dissimilarities that do exist.

The activity of bacitracin is measured in units. The potency per mg. is not less than 40 U.S.P. Units except for material prepared for parenteral use which has a potency of not less than 50 Units per mg. It is a bactericidal antibiotic that is active against a wide variety of gram-positive organisms, very few gram-negative organisms and some others. It is believed to exert its bactericidal effect through an inhibition of mucopeptide cell wall synthesis. Its action is enhanced by zinc. Although bacitracin has found its widest use in topical preparations for local infections, it is quite effective in a number of systemic and local infections when administered parenterally. It is not absorbed from the gastrointestinal tract, so oral administration is without effect except for the treatment of amebic infections within the alimentary canal.

Polymyxin B Sulfate U.S.P., Aerosporin®. Polymyxin was discovered in 1947 almost simultaneously in three separate laboratories in America and Great Britain.[182-184] As often happens when similar discoveries are made in widely separated laboratories, differences in nomenclature referring both to the antibiotic-producing organism and the antibiotic itself appeared in references to the polymyxins. Since it now has been shown that the organisms first designated as *Bacillus polymyxa* and *B. aerosporus Greer* are identical species, the one name, *B. polymyxa,* is used to refer to all of the strains that produce the closely related polypeptides called polymyxins. Other organisms (see colistin, for example) also produce polymyxins. Identified so far are polymyxins A, B_1, B_2, C, D_1, D_2, M, colistin A (polymyxin E_1), colistin B (polymyxin E_2), circulins A and B, and polypeptin. The known structures of this group and their properties have been reviewed by Vogler and Studer.[185] Of these, polymyxin B as the sulfate is usually used in medicine because, when used systemically, it causes less kidney damage than the others.

Polymyxin B sulfate is a nearly odorless, white to buff-colored powder. It is freely soluble in water and slightly soluble in alcohol. Its aqueous solutions are slightly acidic or nearly neutral (pH 5 to 7.5) and, when refrigerated, are stable for at least 6 months.

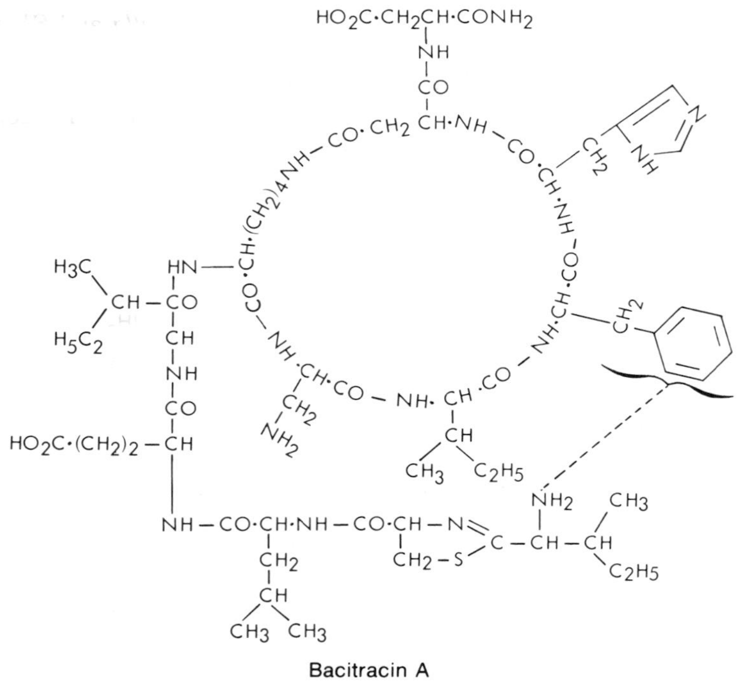

Bacitracin A

$$
\begin{array}{c}
\text{NH}\underline{\quad}\text{CO} \\
| \qquad | \\
C_6H_5CH_2-\text{CH} \quad \text{CH}-CH_2CH_2NH_2 \\
| \qquad | \\
\text{CO} \qquad \text{NH} \\
| \qquad | \\
\text{NH} \qquad \text{CO} \qquad CH_2CH_2NH_2 \qquad CH_2CH_2NH_2 \qquad CH_3 \\
| \qquad | \qquad\qquad | \qquad\qquad | \\
(H_3C)_2CHCH_2-\text{CH} \quad \text{CH}-NHCO-C-NH-CO-CH-NH-CO-CH-NH-CO-(CH_2)_4-CHCH_2CH_3 \\
| \qquad | \qquad | \qquad\qquad | \\
\text{CO} \qquad CH_2 \qquad H \qquad CHOHCH_3 \\
| \qquad | \\
\text{NH} \qquad CH_2 \\
| \qquad | \\
H_2NCH_2CH_2-\text{CH} \quad \text{NH} \\
| \qquad | \\
\text{CO} \qquad \text{CO} \\
| \qquad | \\
\text{NH} \qquad \text{CH}-CHOHCH_3 \\
| \qquad | \\
H_2NCH_2CH_2-\text{CH} \quad \text{NH} \\
\diagdown \text{CO}
\end{array}
$$

Polymyxin B₁

Alkaline solutions are unstable. Polymyxin B has been shown by Hausmann and Craig,[186] who used countercurrent distribution techniques, to contain two fractions that differ in structure only by one fatty acid component. Polymyxin B₁ contains (+)-6-methyloctan-1-oic acid (isopelargonic acid), a fatty acid isolated from all of the other polymyxins. The B₂ component contains an isooctanoic acid, $C_8H_{16}O_2$, of undetermined structure. The structural formula for polymyxin B has been proved by the synthesis accomplished by Vogler et al.[187]

Polymyxin B sulfate is useful against many gram-negative organisms. Its main use in medicine has been in topical applications for local infections in wounds and burns. For such use it is frequently combined with bacitracin, which is effective against gram-positive organisms. Polymyxin B sulfate is poorly absorbed from the gastrointestinal tract, so oral administration of it is of value only in the treatment of intestinal infec-tions such as pseudomonas enteritis or those due to *Shigella*. It may be given parenterally by intramuscular or intrathecal injection for systemic infections. The dosage of polymyxin is measured in U.S.P. Units. One mg. contains not less than 6,000 U.S.P. Units.

Colistin Sulfate U.S.P., Coly-Mycin S®. In 1950, Koyama and co-workers[188] isolated an antibiotic from *Aerobacillus colistinus* (*B. polymyxa* var. *colistinus*) that has been given the name colistin. It had been used in Japan and in some European countries for a number of years before it was made available for medicinal use in the United States. It is especially recommended for the treatment of refractory urinary tract infections caused by gram-negative organisms such as *Aerobacter, Bordetella, Escherichia, Klebsiella, Pseudomonas, Salmonella* and *Shigella*.

Chemically, colistin is a polypeptide that has been reported by Suzuki et al.[189] to be heterogeneous with the major component being colistin A. They proposed

$$
\begin{array}{c}
\text{NH}\underline{\quad}\text{CO} \\
| \qquad | \\
(H_3C)_2CHCH_2-\text{CH} \quad \text{CH}-CH_2CH_2NH_2 \\
| \qquad | \\
\text{CO} \qquad \text{NH} \\
| \qquad | \\
\text{NH} \qquad \text{CO} \qquad CH_2CH_2NH_2 \qquad CH_2CH_2NH_2 \qquad CH_3 \\
| \qquad | \qquad\qquad | \qquad\qquad | \\
(H_3C)_2CHCH_2-\text{CH} \quad \text{CH}-NHCO-CH-NH-CO-CH-NH-CO-CH-NH-CO-(CH_2)_4-CHCH_2CH_3 \\
| \qquad | \qquad\qquad | \\
\text{CO} \qquad CH_2 \qquad CHOHCH_3 \\
| \qquad | \\
\text{NH} \qquad CH_2 \\
| \qquad | \\
H_2NCH_2CH_2-\text{CH} \quad \text{NH} \\
| \qquad | \\
\text{CO} \qquad \text{CO} \\
| \qquad | \\
\text{NH} \qquad \text{CH}-CHOHCH_3 \\
| \qquad | \\
H_2NCH_2CH_2-\text{CH} \quad \text{NH} \\
\diagdown \text{CO}
\end{array}
$$

Colistin A (Polymyxin E₁)

the structure shown here for colistin A, which may be noted to differ from polymyxin B_1 only by the substitution of D-leucine for D-phenylalanine as one of the amino-acid fragments in the cyclic portion of the structure. Wilkinson and Lowe[190] have corroborated the structure and have shown colistin A to be identical with polymyxin E_1. Some additional confusion on nomenclature for this antibiotic exists, as Koyama *et al.* originally named the product colimycin, and that name is still used. Particularly, it has been the basis

for variants used as brand names as Coly-Mycin®, Colomycin®, Colimycine® and Colimicina®.

Two forms of colistin have been made, the sulfate and methanesulfonate, and both forms are available for use in the United States. The sulfate is used to make an oral pediatric suspension; the methanesulfonate is used to make an intramuscular injection. In the dry state, the salts are stable, and their aqueous solutions are relatively stable at acid pH from 2 to 6. Above pH 6, solutions of the salts are much less stable.

TABLE 7–14

POLYPEPTIDE ANTIBIOTICS

Name Proprietary Name	Preparations	Application	Usual Adult Dose*	Usual Dose Range*	Usual Pediatric Dose*
Gramicidin U.S.P.		Topical, 0.05 percent solution			
Bacitracin U.S.P. *Baciguent*	Bacitracin Ointment U.S.P.	Topically to the skin, 2 or 3 times daily			
	Bacitracin Ophthalmic Ointment U.S.P.	Topically to the conjunctiva, 2 or 3 times daily			
	Sterile Bacitracin U.S.P.		I.M., 10,000 to 20,000 Units 3 to 4 times daily	30,000 to 100,000 Units daily	Premature infants—300 Units per kg. of body weight 3 times daily; full-term newborn infants to 1 year of age—330 Units per kg. 3 times daily; older infants and children—500 Units per kg. or 15,000 Units per square meter of body surface, 4 times daily
Bacitracin Zinc U.S.P.	Bacitracin Ointment U.S.P.	Topically to the skin, 2 or 3 times daily			
Polymyxin B Sulfate U.S.P. *Aerosporin*	Sterile Polymyxin B Sulfate U.S.P.		I.M., 6250 to 7500 Units per kg. of body weight 4 times daily; intrathecal, 50,000 Units once daily for 3 or 4 days, then 50,000 Units once every 2 days; I.V. infusion, 7500 to 12,500 Units per kg. of body weight in 300 to 500 ml. of 5 percent Dextrose Injection as a continuous infusion, twice daily. The total daily dose must not exceed 25,000 Units per kg. daily		I.M. see Usual Dose. Intrathecal, children under 2 years of age—20,000 Units once daily for 3 or 4 days or 25,000 Units once every 2 days; children over 2 years of age—see Usual Dose. I.V. infusion 7500 to 12,500 Units per kg. of body weight in 300 to 500 ml. of 5 percent Dextrose Injection over a period of 60 to 90 minutes, twice daily. The total daily dose must not exceed 25,000 Units per kg. daily
	Polymixin B Sulfate Otic Solution U.S.P.				
Colistin Sulfate U.S.P. *Coly-Mycin S*	Colistin Sulfate for Oral Suspension U.S.P.			3 to 15 mg. per kg. daily	The equivalent of 2 to 5 mg. of colistin per kg. of body weight 3 times daily
	Sterile Colistimethate Sodium U.S.P. *Coly-Mycin M*		I.M. or I.V., the equivalent of 1.25 mg. of colistin per kg. of body weight 2 to 4 times daily	1.5 to 5 mg. per kg. daily	See Usual Dose

* See *U.S.P. D.I.* for complete dosage information.

Sterile Colistimethate Sodium U.S.P., Coly-Mycin M®, pentasodium colistinmethanesulfonate, sodium colistimethanesulfonate. In colistin, five of the terminal amino groups of the α, γ-aminobutyric acid fragment may be readily alkylated. In colistimethate sodium, the methanesulfonate radical is the attached alkyl group and, through each of them, a sodium salt may be made. This provides a highly water-soluble compound that is very suitable for injection. In the injectable form, it is given intramuscularly and is surprisingly free from toxic reactions as compared with polymyxin B. Colistimethate sodium does not readily induce the development of resistant strains of microorganisms, and no evidence of cross-resistance with the common broad-spectrum antibiotics has been shown. It is used for the same conditions as those mentioned for colistin.

UNCLASSIFIED ANTIBIOTICS

Among the many hundreds of antibiotics that have been evaluated for activity are a number that have gained significant clinical attention but which do not fall into any of the previously considered groups. Some of these have quite specific activities against a narrow spectrum of microorganisms. Some have found a useful place in therapy as substitutes for other antibiotics to which resistance has developed.

CHLORAMPHENICOL

Chloramphenicol U.S.P., Chloromycetin®, Amphicol®. The first of the widely used broad-spectrum antibiotics, chloramphenicol, was isolated by Ehrlich et al.[191] in 1947. They obtained it from *Streptomyces venezuelae,* an organism that was found in a sample of soil collected in Venezuela. Since that time, chloramphenicol has been isolated as a product of a number of organisms found in soil samples from widely separated places. More important, its chemical structure was soon established and in 1949, Controulis, Rebstock and Crooks[192] reported its synthesis. This opened the way for the commercial production of chloramphenicol by a totally synthetic route. It was the first and is still the only therapeutically important antibiotic to be so produced in competition with microbiological processes. A number of synthetic procedures have been developed for chloramphenicol. The commercial process most generally used has started with p-nitroacetophenone.[193]

Chloramphenicol is a white crystalline compound that is very stable. It is very soluble in alcohol and other polar organic solvents but is only slightly soluble in water. It has no odor but has a very bitter taste.

Chloramphenicol

It may be noted that chloramphenicol possesses two asymmetric carbon atoms in the acylamidopropanediol chain. Biological activity resides almost exclusively in the D-*threo* isomer; the L-*threo* and the D- and L-*erythro* isomers are virtually inactive.

Chloramphenicol is very stable in the bulk state and in solid dose forms. In solution, however, it slowly undergoes a variety of hydrolytic and light-induced reactions.[194] The rates of these reactions are dependent on pH, heat, and light. Hydrolytic reactions include general acid–base catalyzed hydrolysis of the amide to give 1-(p-nitrophenyl)-2-aminopropan-1,3-diol and dichloroacetic acid and alkaline hydrolysis (above pH 7) of the α-chloro groups to form the corresponding α,α-dihydroxy derivative.

The metabolism of chloramphenicol has been investigated thoroughly.[195] The main path involves formation of the 3-O-glucuronide. Minor reactions include reduction of the p-nitro group to the aromatic amine, hydrolysis of the amide, and hydrolysis of the α-chloroacetamido group followed by reduction to give the corresponding α-hydroxyacetyl derivative.

Strains of certain bacterial species are resistant to chloramphenicol by virtue of the ability to produce chloramphenicol acetylase, an enzyme that acetylates the hydroxyl groups at the 1- and 3-positions. Both the 3-acetoxy and the 1,3-diacetoxy metabolites are devoid of antibacterial activity.

A large number of structural analogs of chloramphenicol have been synthesized to provide a basis for correlation of structure to antibiotic action. It appears that the p-nitrophenyl group may be replaced by other aryl structures without appreciable loss in activity. Substitution on the phenyl ring with several different types of groups for the nitro group, a very unusual structure in biological products, does not cause a great decrease in activity. However, all such compounds tested to date are less active than chloramphenicol. Recently, as part of a quantitative SAR study, Hansch et al.[196] reported that the 2-NHCOCF$_3$ derivative is 1.7 times as active as chloramphenicol against *Escherichia coli*. Modifications of the side chain show it to possess a high degree of specificity in structure for antibiotic action. A conversion of the alcohol group on carbon atom 1 of the side chain to a keto group causes an appreciable loss in activity. The relationship of the structure of chloramphenicol to its antibiotic activity will not be clearly seen until the mode of action of this compound is known. The review article by Brock[197] reports on the large amount of research that has been devoted to this problem. It has been established that chloramphenicol exerts its bacteriostatic action by a strong inhibition of protein synthesis. The details of such inhibition are as yet undetermined, and the precise point of action is unknown. Some process lying between the attachment of amino acids to soluble RNA and the final formation of protein appears to be involved.

The broad-spectrum activity of chloramphenicol and its singular effectiveness in the treatment of a number of infections not amenable to treatment by other drugs has made it an extremely popular antibiotic. Unfortunately, instances of serious blood dyscrasias and other toxic reactions have resulted from the promiscuous and widespread use of chloramphenicol in the past. Because of these reactions, it is now recommended that it not be used in the treatment of infections for which other antibiotics are as effective and not as hazardous. When properly used with careful observation for untoward reactions, chloramphenicol provides some of the very best therapy for the treatment of serious infections.

It is specifically recommended for the treatment of serious infections caused by strains of gram-positive and gram-negative bacteria that have developed resistance to penicillin G and ampicillin such as: *Hemophilus influenzae, Salmonella typhi, Streptococcus pneumoniae,* and *Neisseria meningitidis.* Because of its penetration into the central nervous system, chloramphenicol represents a particularly important alternative therapy for meningitis. It is not recommended for the treatment of urinary tract infections, since only 5 to 10 percent of the unconjugated form is excreted in the urine. Chloramphenicol is also employed for the treatment of rickettsial infections, such as Rocky Mountain Spotted Fever.

Because of its bitter taste, this antibiotic is administered orally either in capsules or as the palmitate ester. Chloramphenicol Palmitate U.S.P. is insoluble in water and may be suspended in aqueous vehicles for liquid dosage forms. The ester forms by reaction with the hydroxyl group on the number 3 carbon atom. In the alimentary tract it is slowly hydrolyzed to the active antibiotic. Parenteral administration of chloramphenicol is made by use of an aqueous suspension of very fine crystals or by use of a solution of the sodium salt of the succinate ester of chloramphenicol. Sterile chloramphenicol sodium succinate has been used to prepare aqueous solutions for intravenous injections.

Category—antibacterial.
Usual adult dose—oral, 12.5 mg. of chloramphenicol per kg. of body weight every 6 hours; intravenous, 12.5 mg. of chloramphenicol per kg. every 6 hours.*
Usual dose range—up to 100 mg. per kg. daily.*
For external use—topically to the conjunctiva, as a 1 percent ointment every 3 hours 1 drop of a 0.5 percent solution every 1 to 4 hours.*

Occurrence
Chloramphenicol Capsules U.S.P.
Chloramphenicol Ophthalmic Ointment U.S.P.
Chloramphenicol Ophthalmic Solution U.S.P.
Chloramphenicol for Ophthalmic Solution U.S.P.
Chloramphycol Injection U.S.P
Chloramphenicol Palmitate U.S.P.
Chloramphenicol Palmitate Oral Suspension U.S.P.
Sterile Chloramphenicol Sodium Succinate U.S.P.

Vancomycin Hydrochloride U.S.P., Vancocin®. The isolation of vancomycin from *Streptomyces orientalis* was described in 1956 by McCormick *et al.*[198] The organism was originally obtained from cultures of an Indonesian soil sample and subsequently has been obtained from Indian soil. It was introduced in 1958 as an antibiotic active against gram-positive cocci, particularly streptococci, staphylococci and pneumococci. It is recommended for use when infections have not responded to treatment with the more common antibiotics or when the infection is known to be caused by a resistant organism. It is particularly effective for the treatment of endocarditis caused by gram-positive bacteria.

Vancomycin has not exhibited cross-resistance with any other known antibiotic. Perkins[199] states that vancomycin interferes with mucopeptide biosythesis, perhaps in a manner similar to that of penicillin.

Vancomycin hydrochloride is a free-flowing, tan to brown powder that is relatively stable in the dry state. It is very soluble in water and insoluble in organic solvents. The salt is quite stable in acidic solutions. The free base is an amphoteric substance, the structure of which is undetermined. The presence of carboxyl, amino and phenolic groups has been determined. The purification of vancomycin by utilization of its chelating property to form a copper complex has been reported by Marshal.[200] Recent chemical investigations by scientists at Cambridge[201] have accounted for all, or nearly all, of the carbon skeleton of the antibiotic. Thus, it is now known that vancomycin contains 5 benzene rings. In a 3-ring unit connected through ether linkages two sugars, glucose and vancosamine, are sequentially attached to a central pyrogallol system. The other two rings consist of a biphenyl system that contains three phenolic groups. It is as-

---

*See *U.S.P. D.I.* for complete dosage information.

Novobiocin

sumed that the aromatic residues are connected by amide bonds to the two aspartic acid and one N-terminal N-methyl-leucine residues isolated previously. Removal of the glucose unit (by mild acid hydrolysis) produces a compound, aglucovancomycin, that retains about three-fourths the activity of vancomycin.

Vancomycin hydrochloride is always administered intravenously (never intramuscularly), either by slow injection or by continuous infusion, for the treatment of systemic infections. In short-term therapy, the toxic side-reactions are usually slight, but continued use may lead to impairment of auditory acuity, renal damage, and to phlebitis and skin rashes. Because it is not absorbed, vancomycin may be administered orally for the treatment of staphylococcal enterocolitis and for pseudomembranous colitis associated with clindamycin therapy. It is likely that some conversion to aglucovancomycin occurs in the low pH of the stomach.

Category—antibacterial.

Usual dose—I.V. infusion, the equivalent of 500 mg. of vancomycin in 100 to 200 ml. of 5 percent Dextrose Injection or Sodium Chloride Injection over a period of 20 to 30 minutes, 4 times daily.*

Usual dose range—1 to 2 g. daily.

Usual pediatric dose—premature and full-term newborn infants, the equivalent of 5 mg. of vancomycin per kg. of body weight twice daily; older infants and children; 10 mg. per kg. or 300 mg. per square meter of body surface, 4 times daily.

Occurrence
Sterile Vancomycin Hydrochloride U.S.P.
  Vancomycin Hydrochloride for Oral Solution
  U.S.P.

Novobiocin, Albamycin®, streptonivicin. In the search for new antibiotics, three different research groups independently isolated novobiocin from *Streptomyces* species. It was first reported in 1955 as a product from *S. spheroides* and from *S. niveus*. It is currently produced from cultures of both species. Until the common identity of the products obtained by the different research groups was ascertained, confusion in the naming of this compound existed. Its chemical

*See *U.S.P. D.I.* for complete dosage information.

identity has been established as 7-[4-(carbamoyloxy)-tetrahydro-3-hydroxy-5-methoxy-6,6-dimenthylpyran-2-yloxy]-4-hydroxy-3-[4-hydroxy-3(3-methyl-2-butenyl)benzamido]-8-methylcoumarin by Shunk *et al.*[202] and Hoeksema, Caron and Hinman[203] and confirmed by Spencer *et al.*[204,205]

Chemically novobiocin has a unique structure among antibiotics although, like a number of others, it possesses a glycosidic sugar moiety. The sugar in novobiocin, devoid of its carbamate ester, has been named noviose and is an aldose having the configuration of L-lyxose. The aglycon moiety has been termed novobiocic acid.

Novobiocin is a pale-yellow, somewhat photosensitive compound that crystallizes in two chemically identical forms having different melting points. It is soluble in methanol, ethanol and acetone but is quite insoluble in less polar solvents. Its solubility in water is affected by pH. It is readily soluble in basic solutions, in which it deteriorates, and is precipitated from acidic solutions. It behaves as a diacid, forming two series of salts. The enolic hydroxyl group on the coumarin moiety behaves as a rather strong acid and is the group by which the commercially available sodium and calcium salts are formed. The phenolic—OH group on the benzamido moiety also behaves as an acid but is weaker than the former. Disodium salts of novobiocin have been prepared. The sodium salt is stable in dry air but decreases in activity in the presence of moisture. The calcium salt is quite water-insoluble and is used to make aqueous oral suspensions. Because of its acidic characteristics, novobiocin combines to form salt complexes with basic antibiotics. Some of these salts have been investigated for their combined antibiotic effect, but none has been placed on the market, as no advantage is offered by them.

The antibiotic activity of novobiocin is exhibited chiefly against gram-positive organisms and *Proteus vulgaris*. Because of its unique structure, it appears to exert its action in a manner (still unknown) different from other anti-infectives. It may be that its ability to bind magnesium causes an intracellular deficiency of that ion which is necessary for the maintenance of the integrity of the cell membrane.[206,207] Although resistance to novobiocin can be developed in microorganisms, cross-resistance with other antibiotics is not developed. For this reason, the medical use of novobiocin is reserved for the treatment of infections, particularly staphylococcal, resistant to other antibiotics and the sulfas and for patients who are allergic to the other drugs.

A syrup or suspension of the calcium salt is available for pediatric use. The sodium salt is used for injection and in oral capsules. The suggested dosage for adults is 250 to 500 mg. every 6 hours or 500 mg. to 1 g. every 12 hours, continued for 48 hours after the temperature becomes normal. Parenteral dosage for adults is 500 mg. every 12 hours to be changed to oral treatment as soon as possible.

REFERENCES

1. Fleming, A.: Brit. J. Exp. Path. 10:226, 1929.
2. Vuillemin, P.: Assoc. franc avance sc. Part 2:525–543, 1889.
3. Waksmann, S. A.: Science 110:27, 1949.
4. Benedict, R. G., and Langlykke, A. F.: Ann. Rev. Microbiol. 1:193, 1947.
5. Baron, A. L.: Handbook of Antibiotics, p. 5. New York, Reinhold, 1950.
6. Strominger, J. L., et al.: Pencillin-sensitive Enzymatic Reactions, in Perlman, D. (ed.): Topics in Pharmaceutical Sciences, vol. 1, p. 53, New York, Interscience Publ. 1968.
7. Edwards, J. R., and Park, J. T.: J. Bacteriol. 99:459, 1969.
8. Clarke, H. T., et al.: The Chemistry of Penicillin, p. 454, Princeton, N. J., Princeton Univ. Press, 1949.
9. Sheehan, J. C., and Henery-Logan, K. R.: J. Am. Chem. Soc. 81:3089, 1959.
10. Batchelor, F. R., et al.: Nature 183:257, 1959.
11. Sheehan, J. C., and Ferris, J. P.: J. Am. Chem. Soc. 81:2912, 1959.
12. Hou, J. P., and Poole, J. W.: J. Pharm. Sci. 60:503, 1971.
13. Blaha, J. M., et al.: J. Pharm. Sci. 65:1165, 1976.
14. Schwartz, M.: J. Pharm. Sci. 58:643, 1969.
15. Schwartz, M. A., and Buckwalter, F. H.: J. Pharm. Sci. 51:1119, 1962.
16. Finholt, P., Jurgensen, G., and Kristiansen, H.: J. Pharm. Sci. 54:387, 1965.
17. Segelman, A. B., and Farnsworth, N. R.: J. Pharm. Sci. 59:725, 1970.
18. Doyle, F. P., et al.: Nature 191:1091, 1961.
19. Brain, E. G., et al.: J. Chem. Soc. 1445, 1962.
20. Stedmen, R. J., et al.: J. Med. Chem. 7:251, 1964.
21. Nayler, J. H. C.: Adv. Drug Res. 7:52, 1973.
22. Gourevitch, A., et al.: Antimicrob. Ag. Chemo. 576, 1961.
23. Matthew, M.: J. Antimicrob. Chemother. 5:349, 1979.
24. Hansch, C., and Deutsch, E. W.: J. Med. Chem. 8:705, 1965.
25. Bird, A. E., and Marshall, A. C.: Biochem. Pharmacol. 16:2275, 1967.
26. Stewart, G. W.: The Pencillin Group of Drugs, Amsterdam, Elsevier, 1965.
27. Batchelor, F. R., et al.: Nature 206:362, 1965.
28. DeWeck, A. L.: Int. Arch. Allergy 21:20, 1962.
29. Smith, H., and Marshall, A. C.: Nature 232:45, 1974.
30. Monro, A. C., et al.: Int. Arch. Appl. Immunol. 50:192, 1976.
31. Behrens, O. K., et al.: J. Biol. Chem. 175:793, 1948.
32. Hou, J. P., and Poole, J. W.: J. Pharm. Sci. 58:1150, 1969.
33. Mayersohn, M., and Endrenyi, L: Can. Med. Assoc. J. 109:989, 1973.
34. Hill, S. A., et al.: J. Pharm. Pharmacol. 27:594, 1975.
35. Clayton J. P., et al.: Antimicrob. Ag. Chemother. 5:670, 1974.
36. Neu, H. C.: J. Infect. Dis. 12S;1, 1974.
37. Ancred, P., et al.: Nature 215:25, 1967.
38. Verbist, L.: Antimicrob. Ag. Chemother. 13:349, 1978.
39. Fu, K. P., and Neu, H. C.: Antimicrob. Ag. Chemother. 13:358, 1978.
40. Gedes, A. M., et al.: J. Antimicrob. Chemother. 3(Suppl. B): 1, 1977.
41. Spratt, B. G.: J. Antimicrob. Chemother. 3:13, 1977.
42. Van Heyningen, E.: Cephalosporins, in Harper, N. J., and Simmonds, A. B. (eds.): Advances in Drug Research, vol. 4, p. 1, New York, Academic Press, 1967.
43. Morin, R. B., et al.: J. Am. Chem. Soc. 84:3400, 1962.
44. Fechtig, B., et al.: Helv. Chim. Acta 51:1108, 1968.
45. Woodward, R. B., et al.: J. Am. Chem. Soc. 88:852, 1966.
46. Morin, R. B., et al.: J. Am. Chem. Soc. 85:1896, 1963.
47. Fong, I., et al.: Antimicrob. Ag. Chemother. 9:939, 1976.
48. Yamana, T., and Tsuji, A.: J. Pharm. Sci. 65:1563, 1976.
49. Sullivan, H. R., and McMahon, R. E.: Biochem. J. 102:976, 1967.
50. Indelicato, J. M., et al.: J. Pharm. Sci. 65:1175, 1976.
51. Ott, J. L., et al.: Antimicrob. Ag. Chemother. 15:14, 1979.
52. Nagarajan, R., et al.: J. Am. Chem. Soc. 93:2308, 1971.
53. Moellering, R. C., et al.: Antimicrob. Agents Chemother. 6:320, 1974.
54. Kukolja, S., In Elks, J. (ed.): Recent Advances in the Chemistry of Beta-Lactam Antibiotics, p. 181, Chichester, England, The Chemical Society (London), Burlington House, 1977.
55. Gillett, A. P., et al.: Postgrad. Med. 55 (Suppl. 4): 9, 1979.
56. Uyeo, S., et al.: J. Am. Chem. Soc. 101:4403, 1979.
57. Cama, L. D., and Christensen, B. G.: J. Am. Chem. Soc. 96:7583, 1974.
58. Neu, H. C., et al.: Antimicrob. Ag. Chemother. 16:141, 1979.
59. Merck & Co., Inc.: U. S. Patent 3,950,357 (April 12, 1976).
60. Johnston, D. B. R., et al.: J. Am. Chem. Soc. 100:313, 1978.
61. Albers-Schonberg, G., et al.: J. Am. Chem. Soc. 100:6491, 1978.
62. Kahan, J. S., et al.: Abstract 227, 16th Conference on Antimicrobial Agents and Chemotherapy, Chicago, 1976.
63. Hashimoto, M., et al.: J. Antibiot. 29:890, 1976.
64. Aoki, H., et al.: J. Antibiot. 29:492, 1976.
65. Schatz, A., et al.: Proc. Soc. Exp. Biol. Med. 55:66, 1944.
66. Weisblum, B., and Davies, J.: Bacteriol. Rev 32:493, 1968.
67. Davies, J., and Davis, B. D.: J. Biol. Chem. 243:3312, 1968.
68. Lando, D., et al.: Biochem. 12:4528, 1973.
69. Benveniste, R., and Davies, J.: Ann. Rev. Biochem. 42:471, 1973.
70. Chevereau, P. J. L., et al.: Biochemistry 13:598, 1974.
71. Price, K. E., et al.: Antimicrob. Ag. Chemother. 5:143, 1974.
72. Cox, D. A., et al.: The Aminoglycosides in Topics in Sammes, P. G. (ed.): Antiobiotic Chemistry, Vol 1, p. 44, Chichester, England, Ellis Harwood, Ltd., 1977.
73. Dyer, J. R., and Todd, A. W.: J. Am. Chem. Soc. 85:3896, 1963.
74. Dyer, J. R., et al.: J. Am. Chem. Soc. 87:654, 1965.
75. Umezawa, S., et al.: J. Antibiot. 27:997, 1974.
76. Waksman, S. A., and Lechevalier, H. A.: Science 109:305, 1949.
77. Waksman, S. A. (ed.): Neomycin, Its Nature and Practical Applications, Baltimore, Williams & Wilkins, 1958.
78. Hichens, M., and Rinehart, K. L., Jr.: J. Am. Chem. Soc. 85:1547, 1963.
79. Rinehart, K. L., Jr., et al.: J. Am. Chem. Soc. 84:3218, 1962.
80. Huettenrauch, R.: Pharmazie 19:697, 1964.
81. Haskell. T. H., et al.: J. Am. Chem. Soc. 81:3482, 1959.
82. DeJongh, D. C., et al.: J. Am. Chem. Soc. 89:3364, 1967.
83. Umezawa, H., et al.: J. Antibiot. [A] 10:181, 1957.
84. Tatsuoka, S., et al.: J. Antibiot. [A] 17:88, 1964.
85. Nakajima, M.: Tetrahedron Letters 623, 1968.
86. Umezawa, S., et al.: J. Antibiot. 21:162, 367, 424, 1968.
87. Aoki, T., et al.: J. Antibiot. [A] 12:98, 1959.
88. Morikubo, Y.: J. Antibiot. [A] 12:90, 1959.
89. Umezawa, S., et al.: J. Antibiot. [A] 12:114, 1959.
90. Kawaguchi, H., et al.: J. Antibiot. 25:695, 1972.
91. Paradelis, A. G., et al.: Antimicrob. Ag. Chemother. 14:514, 1978.
92. Weinstein, M. J., et al.: J. Med. Chem. 6:463, 1963.

93. Cooper, D. J., *et al.*: J. Infect. Dis. 119:342, 1969.
94. ———: J. Chem. Soc. C. 3126, 1971.
95. Maehr, H., and Schaffner, C. P.: J. Am. Chem. Soc. 89:6788, 1968.
96. Koch, K. F., *et al.*: J. Antibiot. 26:745, 1973.
97. Lockwood, W., *et al.*: Antimicrob. Agents Chemother. 4:281, 1973.
98. Lewis, C., and Clapp, H.: Antibiotics & Chemother. 11:127, 1961.
99. Cochran, T. G., and Abraham, D. J.: J. Chem. Soc. Chem. Commun. 494, 1972.
100. Muxfeldt, H., *et al.*: J. Am. Chem. Soc. 90:6534, 1968.
101. Muxfeldt, H., *et al.*, J. Am. Chem. Soc. 101:689, 1979.
102. Korst, J. J., *et al.*: J. Am Chem. Soc. 90:439, 1968.
103. Muxfeldt, H. *et al.*: Angew. Chem. (Internat. Ed.) 12:497, 1973.
104. Hirokawa, S., *et al.*: Z. Krist. 112:439, 1959.
105. Takeuchi, Y., and Buerger, M. J.: Proc. Nat. Acad. Sci. U.S. 46:1366, 1960.
106. Cid-Dresdner, H.: Z. Krist. 121:170, 1965.
107. Leeson, L. J., Krueger, J. E., and Nash, R. A.: Tetrahedron Letters, No. 18:1155, 1963.
108. Stephens, C. R., *et al.*: J. Am. Chem. Soc. 78:4155, 1956.
109. Rigler, N. E., *et al.*: Anal. Chem. 37:872, 1965.
110. Benet, L. Z., and Goyan, J. E.: J. Pharm. Sci. 55:983, 1965.
111. Barringer, W., *et al.*: Am. J. Pharm. 146:179, 1974.
112. Barr, W. H., *et al.*: Clin. Pharmacol. Therap. 12:779, 1971.
113. Albert, A.: Nature 172:201, 1953.
114. Jackson, F. L.: Mode of Action of Tetracyclines, *in* Schnitzer, R. J., and Hawking, F. (eds.): Experimental Chemotherapy, vol. 3. p. 103, New York, Academic Press, 1964.
115. Bodley, J. W., and Zieve, P. J.: Biochem. Biophys. Res. Commun. 36:463, 1969.
116. Dockter, M. E., and Magnuson, J. A.: Biochem. Biophys. Res. Commun. 42:471, 1973.
117. Izaka, K., and Arima, K.: Nature 200:384, 1963.
118. Sompolinsky, D., and Krausz, J.: Antimicrob. Ag. Chemother. 4:237, 1973.
119. Durckheimer, W.: Angew. Chem. Int. Ed. 14:721, 1975.
120. Brown, J. R., and Ireland, D. S.: Adv. Pharmacol. Chemother. 15:161, 1978.
121. Mitscher, L. A.: The Chemistry of the Tetracycline Antibiotics, New York, Marcel-Dekker, 1978.
122. Cline, D. L. J.: Chemistry of Tetracyclines, Quart, Rev. 22:435, 1968.
123. Cammarata, A., and Yau, S. J., J. Med. Chem. 13:93, 1970.
124. Glatz, B., *et al.*: J. Am. Chem. Soc. 101:2171, 1979.
125. Schach von Wittenau, M., *et al.*: J. Am. Chem. Soc. 84:2645, 1962.
126. Stephens, C. R., *et al.*: J. Am. Chem. Soc. 85:2643, 1963.
127. McCormick, J. R. D., *et al.*: J. Am. Chem. Soc. 79:4561, 1957.
128. Martell, M. J., Jr., and Booth, J. H.: J. Med. Chem. 10:44, 1967.
129. Colazzi, J. L., and Klink, P. R.: J. Pharm. Sci. 58:158, 1969.
130. Schumacher, G. E., and Linn, E. E.: J. Pharm. Sci. 67:1717, 1978.
131. Schach von Wittenau, M.: Chemother. 13S:41, 1968.
132. Dingeldein, E., and Wahlig, H.: Seventeenth Interscience Conference on Antimicrobial Agents and Chemotherapy, New York, Abstract 71, 1977.
133. Kirchlechner, R., and Rogalalaski, W.: Tetrahedron Letters 247, 251, 1979.
134. Conover, L. H.: U.S. Patent 2,699,054, Jan. 11, 1955.
135. Bunn. P. A., and Cronk, G. A.: Antibiot. Med. 5:379, 1958.
136. Gittinger, W. C., and Weinger, H.: Antibiot. Med. 7:22, 1960.
137. Remmers, E. G., *et al.*: Pharm. Sci. 53:1452, 1534, 1964.
138. ———: J. Pharm. Sci. 54:49, 1965.
139. Duggar, B. B.: Ann. N. Y. Acad. Sci. 51:177, 1948.
140. Finlay, A. C., *et al.*: Science 111:85, 1950.
141. Hochstein, F. A. *et al.*: J. Am. Chem. Soc. 75:5455, 1953.
142. Blackwood, R. K., *et al.*: J. Am. Chem. Soc. 83:2773, 1961.
143. Stephens, C. R., *et al.*: J. Am. Chem. Soc. 80:5324, 1958.
144. Minuth, J. N.: Antimicrob. Agents Chemother. 6:411, 1964.
145. Wiley, P. F.: Research Today (Eli Lily & Co.) 16:3, 1960.
146. Miller, M. W.: The Pfizer Handbook of Microbial Metabolites, New York, McGraw-Hill, 1961.
147. Morin, R., and Gorman, M.: Kirk-Othmer Encyl. Chem. Technol., ed. 2, 12:637, 1967.
148. McGuire, J. M., *et al.*: Antibiotics & Chemother. 2:821, 1952.
149. Wiley, P. F., *et al.*: J. Am. Chem. Soc. 79:6062, 1957.
150. Celmer, W. D.: J. Am. Chem. Soc. 87:1801, 1965.
151. Corey, E. J., *et al.*: J. Am. Chem. Soc. 7131, 1979.
152. Stephens, C. V., *et al.*: J. Antibiot. 22:551, 1969.
153. Bechtol, L. D., *et al.*: Cur. Ther. Res. 20:610, 1976.
154. Welling, P. G., *et al.*: J. Pharm. Sci. 68:150, 1979.
155. Tardew, P. L., *et al.*: Appl. Microbiol. 18:159, 1969.
156. Wiegand, R. G., and Chun, A. H.: J. Pharm. Sci. 61:425, 1972.
157. Janicki, R. S., *et al.*: Clin. Pediat. 14:1098, 1975.
158. Nicholas, P.: N. Y. State J. Med. 77:2088, 1977.
159. Wilhelm, J. M., *et al.*: Antimicrob. Agents Chemother. 1967:236.
160. Gutman, L. T., *et al.*: Lancet 1:464, 1967.
161. Lai, C. J., and Weisblum, B.: Proc. Nat. Acad. Sci. U. S. 68:856, 1971.
162. Sobin, B. A., *et al.*: Antibiotics Annual 1954–1955, p. 827, New York, Medical Encyclopedia, 1955.
163. Hochstein, F. A., *et al.*: J. Am. Chem. Soc. 82:3227, 1960.
164. Celmer, W. D.: J. Am. Chem. Soc. 87:1797, 1965.
165. Mason, D. J., *et al.*: Antimicrob. Against Chemother. 1962:554.
166. Hoeksema, H., *et al.*: J. Am. Chem. Soc. 86:4223, 1964.
167. Slomp, G., and MacKellar, F. A.: J. Am. Chem. Soc. 89:2454, 1967.
168. Howarth, G. B. *et al.*: J. Chem. Soc. C. 2218, 1970.
169. Majerlein, B. J.: Tetrahedron Letters 685, 1970.
170. Argoudelis, A. D., *et al.*: J. Am. Chem. Soc. 86:5044, 1964.
171. Magerlein, B. J., *et al.*: J. Med. Chem. 10:355, 1967.
172. Bannister, B.: J. Chem. Soc. Perkin 1:1676, 1973.
173. ———: J. Chem. Soc. Perkin 1:3025, 1972.
174. Bartlett, J. G.: Rev. Infect. Dis. 1:370, 1979.
175. Sarges, R., and Witkop, B.: J. Am. Chem. Soc. 86:1861, 1964.
176. Bodanszky, M., and Perlman, D.: Science 163:352, 1969.
177. Dubos, R. J.: J. Exp. Med. 70:1, 1939.
178. Paladini, A., and Craig, L. C.: J. Am. Chem. Soc. 76:688, 1954; King, T. P., and Craig, L. C.: J. Am. Chem. Soc. 77:6627, 1955.
179. Ohno, M., *et al.*: Bull. Soc. Chem. Japan 39:1738, 1966.
180. Johnson, B. A., *et al.*: Science 102:376, 1945.
181. Stoffel, W., and Craig, L. C.: J. Am. Chem. Soc. 83:145, 1961.
182. Benedict, R. G., and Langlykke, A. F.: J. Bact. 54:24, 1947.
183. Stansly, P. J., *et al.*: Bull. Johns Hopkins Hosp. 81:43, 1947.
184. Ainsworth, G. C., *et al.*: Nature 160:263, 1947.
185. Vogler, K., and Studer, R. O.: Experientia 22:345, 1966.
186. Hausmann, W., and Craig, L. C.: J. Am. Chem. Soc. 76:4892, 1954.
187. Volger, K., *et al.*: Experientia 20:365, 1964.
188. Koyama, Y., *et al.*: J. Antibiot. [A] 3:457, 1950.
189. Suzuki, T., *et al.*: J. Biochem. 54:414, 1963.
190. Wilkinson, S., and Lowe, L. A.: J. Chem. Soc. 1964:4107.
191. Ehrlich, J., *et al.*: Science 106:417, 1947.
192. Controulis, J., *et al.*: J. Am. Chem. Soc. 71:2463, 1949.
193. Long, L. M., and Troutman, H. D.: J. Am. Chem. Soc. 71:2473, 1949.
194. Szulcewski, D., and Eng, F.: Analytical Profiles of Drug Substances, 4:47, 1972.
195. Glazko, A.: Antimicrob. Ag. Chemother. 655, 1966.
196. Hansch. C., *et al.*: J. Med. Chem. 16:917, 1973.
197. Brock, T. D.: Chloramphenicol, *in* Schnitzer, R. J., and Hawking, F. (eds.): Experimental Chemotherapy, vol. 3. p. 119, New York, Academic Press, 1964.
198. McCormick, M. H., *et al.*: Antibiotics Annual 1955–1956, p. 606, New York, Medical Encyclopedia, 1956.
199. Perkins, H. R.: Biochem. J. 111:195, 1969.
200. Marshall, F. J.: J. Med. Chem. 8:18, 1965.
201. Smith, K. A., Williams, D. H., and Smith, G. A.: J. Chem. Soc. Perkin 1:2371, 1974.
202. Shunk, C. H., *et al.*: J. Am. Chem. Soc. 78:1770, 1956.

203. Hoeksema, H., *et al.:* J. Am. Chem. Soc. 78:2019, 1956.
204. Spencer, C. H., *et al.:* J. Am. Chem. Soc. 78:2655, 1956.
205. ———: J. Am. Chem. Soc. 80:140, 1958
206. Brock, T. D.: Science 136:316, 1962.
207. ———: J. Bacteriol. 84:679, 1962.

SELECTED READINGS

Barza, M.: Drug Therapy Reviews: Spectrum, Pharmacology and Therapeutic Use of Antibiotics. Part 2. Penicillins, Am. J. Hosp. Pharm. 34:57, 1977.

Barza, M., and Miao, P. V. W.: Drug Therapy Reviews: Antimicrobial Spectrum, Pharmacology and Therapeutic Use of Antibiotics. Part 3. Cephalosporins, Am. J. Hosp. Pharm. 43:621, 1977.

Barza, M., and Schiefe, R. T.: Drug Therapy Reviews: Antimicrobial Spectrum, Pharmacology and Therapeutic Use of Antibiotics. Part 4. Aminoglycosides, Am. J. Hosp. Pharm. 34:723, 1977.

Barza, M., and Schiefe, R. T.: Drug Therapy Reviews: Spectrum, Pharmacology and Therapeutic Use of Antibiotics. Tetracyclines, Am. J. Hosp. Pharm. 34:49, 1977.

Beniveniste, R., and Davies, J.: Mechanisms of Antibiotic Resistance in Bacteria. Ann. Rev. Biochem. 42:471, 1973.

Bird, A. E., and Naylo, J. H. C.: Design of Penicillins, *in* Ariens, E. J. (ed.): Drug Design, Vol. 2, pp. 277–318, New York, Acamedic Press, 1971.

Brown J. R., and Ireland, D. S.: Structural Requirements for Tetracycline Activity, Advan. Pharmacol. Chemother, 15:161, 1978.

Corcoran, J. W., and Hahn, F. E. (eds.): Antibiotics, Vol. 3, New York, Springer-Verlag, 1975.

Cox, D. A., *et al.:* The Aminoglycosides, *in* Sammes, P. G. (ed.): Topics in Antibiotic Chemistry, p. 5, Chichester, England, Ellis Harwood Ltd, 1977.

Davis, J. E., and Rownd, R.: Transmissible Multiple Drug Resistance in *Enterobacteriaceae,* Science 176:758, 1972.

Durckheimer, W.: Tetracyclines: Chemistry, Biochemistry and Structure–Activity Relationships, Angew. Chem. Int. Ed. 14:721, 1975.

Elks, J. (ed.): Recent Advances in the Chemistry of β-Lactam Antibiotics, Chichester, England, The Chemical Society, Burlington House, 1977.

Flynn, E. H. (ed.): Caphalosporin and Penicillins: Chemistry and Biology, New York, Academic Press, 1972.

Gale, E. F., Cundliffe, E., and Reynolds, P. E.: The Molecular Basis of Antibiotic Action, London, John Wiley & Sons, 1972.

Gottlieb, D., and Shaw, P. D. (eds.): Antibiotics, Vols. 1 and 2, New York, Springer-Verlag, 1967.

Hollstein, U.: Nonlactam Antibiotics, *in:* Wolff, M.E., (ed.): Burger's Medicinal Chemistry, Part II, ed. 4, p. 173, 1979.

Hoover, J. R. E., and Dunn, G. L.: The β-Lactam Antibiotics, *in* Wolff, M. E. (ed.): Burger's Medicinal Chemistry Part II, ed. 4, 83, 1979.

Hou, J. P., and Poole, J. W.: β-Lactam Antibiotics: Their Physicochemical Properties in Relation to Structure, J. Pharm. Sci. 60:503, 1971.

Masamune, S., Bates, G. S., and Corcoran, J. W.: Macrolides. Recent Progress in Chemistry and Biochemistry. Angew. Chem. Int. Ed. 16:585, 1977.

Mitscher, L. A.: The Chemistry of the Tetracycline Antibiotics, Medical Research Series Vol. 9, New York, Marcel-Dekker, 1978.

Nayler, J. H. C.: Advances in Penicillin Research, Adv. Drug Res. 7:1, 1973.

Nicholas, P.: Erythromycin: Clinical Review, N.Y. State J. Med. 77:2088, 1977.

O'Callaghan, C. H.: Description and Classification of the Newer Cephalosporins and Their Relationships with Established and Compounds, J. Antimicrob. Chemother. 5:635, 1979.

Owens, D. R., *et al.:* The Cephalosporin Group of Antibiotics, Adv. Pharmacol. Chemother. 13:83, 1975.

Perlman, D. (ed.): Structure–Activity Relationships Among the Somisynthetic Antibiotics, New York, Academic Press, 1977.

Schonfeld, H. (ed.): Antibiotics and Chemotherapy: Pharmacokinetics, Vol. 25, Basel, S. Karger AG, 1978.

Smith, H.: Antibiotics in Clinical Practice, Kent, England, Pitman Medical Pub. Co. Ltd., 1977.

Thrupp, L. D.: Newer Cephalosporins and Expanded Spectrum Penicillins, Ann. Rev. Pharmacol. 14:435, 1974.

8

antineoplastic agents

William A. Remers

The chemotherapy of neoplastic disease has become increasingly important in recent years. An indication of this importance is the establishment of a medical specialty in oncology, wherein the physician practices various protocols of adjuvant therapy. Most cancer patients now receive some form of chemotherapy, even though it is merely palliative in many cases.

Cancer chemotherapy has received no spectacular breakthrough of the kind that the discovery of penicillin provided for antibacterial chemotherapy. However, there has been substantial progress in many aspects of cancer research. In particular, an increased understanding of tumor biology has led to elucidation of the mechanisms of action for antineoplastic agents. It also has provided a basis for the more rational design of new agents. Recent advances in clinical techniques, including large cooperative studies, are allowing more rapid and reliable evaluation of new drugs. The combination of these advantages with improved preliminary screening systems is enhancing the emergence of newer and more potent compounds.

At present at least ten different neoplasms can be "cured" by chemotherapy in the majority of patients. Cure is defined here as an expectation of normal longevity. These neoplasms are acute leukemia in children, Burkitt's lymphoma, choriocarcinoma in women, Ewing's sarcoma, Hodgkin's disease, lymphosarcoma, mycosis fungoides, rhabdomyosarcoma, retinoblastoma in children, and testicular carcinoma.[1] Unfortunately, only these relatively rare neoplasms are readily curable. Considerable progress is being made in the treatment of breast cancer by combination drug therapy. However, for carcinoma of the pancreas, colon, liver, or lung (except small cell carci-

noma) the outlook is bleak. Short-term remissions are the best that can be expected for most patients with these diseases.

There are cogent reasons why cancer is more difficult to cure than bacterial infections. One is that there are qualitative differences between human and bacterial cells. For example, bacterial cells have distinctive cell walls and their ribosomes are different from those of human cells. In contrast, the differences between normal and neoplastic human cells are merely quantitative. Another difference is that immune mechanisms and other host defenses are very important in killing bacteria and other foreign cells, whereas they play a negligible role in killing cancer cells. By their very nature, the cancer cells have eluded or overcome the immune surveillance system of the body. Thus, it is necessary for chemotherapeutic agents to kill every single clonogenic malignant cell, because even one can reestablish the tumor. This kind of kill is extremely difficult to effect because antineoplastic agents kill cells by first-order kinetics. That is, they kill a constant fraction of cells. Suppose that a patient had a trillion leukemia cells. This amount would cause a serious debilitation. A potent anticancer drug might reduce this population 10,000-fold, in which case the symptoms would be alleviated and the patient would be in a state of remission. However, the remaining hundred million leukemia cells could readily increase to the original number after cessation of therapy. Furthermore, a higher proportion of resistant cells would be present, which would mean that retreatment with the same agent would achieve a lesser response than before. For this reason, multiple drug regimens are used to reduce drastically the number of neoplastic cells.

295

Typical protocols for leukemia contain four different anticancer drugs, usually with different modes of action. The addition of immunostimulants to the therapeutic regimen helps the body's natural defense mechanisms to identify and eliminate the remaining few cancer cells.

A further complication to chemotherapy is the relative unresponsiveness of slow-growing solid tumors. Current antineoplastic agents are most effective against cells with a high growth fraction. They act to block the biosynthesis or transcription of nucleic acids or to prevent cell division by interfering with mitotic spindles. Cells in the phases of synthesis or mitosis are highly susceptible to these agents. In contrast, cells in the resting state are resistant to many agents. Slow-growing tumors characteristically have many cells in the resting state.[2]

Most antineoplastic drugs are highly toxic to the patient and must be administered with extreme caution. Some of them require a clinical setting where supportive care is available. The toxicity usually involves rapidly proliferating tissues such as bone marrow and the intestinal epithelium. However, individual drugs produce distinctive toxic effects on the heart, lungs, kidneys, and other organs. Chemotherapy is seldom the initial treatment used against cancer. If the cancer is well defined and accessible, surgery is the preferred method. Skin cancers and certain localized tumors are treated by radiotherapy. Even some widely disseminated tumors such as Hodgkin's disease are treated by radiation, although chemotherapy might be equally effective. Generally, chemotherapy is important where the tumor is inoperable or where metastasis has occurred. Chemotherapy is finding increasing use after surgery to ensure that no cells remain to regenerate the parent tumor.

The era of chemotherapy of malignant disease was born in 1941 when Huggins demonstrated that the administration of estrogens produced regressions of metastatic prostate cancer.[3] In the following year, Gilman and others began clinical studies on the nitrogen mustards and discovered that mechlorethamine was effective against Hodgkin's disease and lymphosarcoma.[4] These same two diseases were treated with cortisone acetate in 1949 and dramatic, although temporary, remissions were observed.[5] The next decade was marked by the design and discovery of antimetabolites: methotrexate in 1949, 6-mercaptopurine in 1952, and 5-fluorouracil in 1957. Additional alkylating agents such as melphalan and cyclophosphamide were developed during this period, and the activity of natural products such as actinomycin, mitomycin C, and the vinca alkaloids was discovered. During the 1960's, progress continued in all of these areas with the discovery of cytosine arabinoside, bleomycin, doxorubi-

cin, and carmustine. Novel structures such as procarbazine, dacarbazine, and *cis*-platinum complexes were found to be highly active. In 1965 Kennedy reported that remissions occurred in 30 percent of postmenopausal women with metastatic breast cancer upon treatment with high doses of estrogen.[6]

Much of the leadership and financial support for the development of antineoplastic drugs derives from the National Cancer Institute. In 1955 this organization established the Cancer Chemotherapy National Service Center (now the Division of Cancer Treatment) to coordinate a national voluntary cooperative cancer chemotherapy program. By 1958 this effort had evolved into a targeted drug development program. A massive screening system was established to discover new lead compounds, and thousands of samples have been submitted to it. At the present time the primary screen is P388 lymphocytic leukemia in mice. Compounds active in this screen are tested further against lymphoid leukemia, melanoma, and lung carcinoma in mice. A panel of human tumor xenographs in mice is available for additional preliminary screening.[7] Compounds of significant interest are subjected to preclinical pharmacology and toxicology evaluation in mice and dogs. Clinical trials are generally underwritten by the National Cancer Institute (NCI). They involve three discrete phases. Phase I is the clinical pharmacology stage. The dosage schedule is developed and toxicity parameters are established in it. Phase II involves the determination of activity against a "signal" tumor panel, which includes both solid and hematological types.[8] A broad-based multicenter study usually is undertaken in Phase III. It features randomization schemes designed to validate the efficacy of the new drug statistically in comparison to alternative modalities of therapy. As might be anticipated, the design of clinical trials for antineoplastic agents is very complicated, especially in the matter of controls. Ethical considerations do not permit patients to be left untreated if any reasonable therapy is possible.

It should be mentioned that a number of pharmaceutical industry laboratories and foreign institutions have made significant contributions to the development of anticancer drugs. Frequently their research is in collaboration with the NCI Division of Cancer Treatment.

ALKYLATING AGENTS

Toxic effects of sulfur mustard and ethylenimine on animals were described in the nineteenth century.[9] The powerful vesicant action of sulfur mustard led to its use in World War I, and medical examination of the victims revealed that tissues were damaged at sites distant from the area of contact.[10] Systemic effects included leukopenia, bone marrow aplasia, lymphoid tissue suppression, and ulceration of the gastrointestinal tract. Sulfur mustard was shown to be active against animal tumors, but it was too nonspecific for clinical use. A variety of nitrogen mustards were synthesized between the two world wars. Some of these compounds, for example mechlorethamine, showed selective toxicity, especially to lymphoid tissue. This observation led to the crucial suggestion that nitrogen mustards be tested against tumors of the lymphoid system in animals. Success in this area was followed by cautious human trials that showed mechlorethamine to be useful against Hodgkin's disease and certain lymphomas. This work was classified during World War II but was finally published in a classical paper by Gilman and Philips in 1946.[4] In this paper the chemical transformation of nitrogen and sulfur mustards to cyclic "onium" cations was described and the locus of their interaction with cancer cells was established to be the nucleus. The now familiar pattern of toxicity to rapidly proliferating cells in bone marrow and the gastrointestinal tract was established.

Alkylation is defined as the replacement of hydrogen on an atom by an alkyl group. The alkylation of nucleic acids or proteins involves a substitution reaction in which a nucleophilic atom (nu) of the biopolymer displaces a leaving group from the alkylating agent.

$$\text{nu-H} + \text{alkyl-Y} \rightarrow \text{alkyl-nu} + H^+ + Y^-$$

The reaction rate depends on the nucleophilicity of the atom (S,N,O), which is greatly enhanced if the nucleophile is ionized. A hypothetical order of reactivity at physiological pH would be ionized thiol, amine, ionized phosphate, and ionized carboxylic acid.[11] Rate differences among various amines would depend on the degree to which they are protonated and their conjugation with other functional groups. The N-7 position of guanine in DNA (Scheme 4) is strongly nucleophilic. Reaction orders depend on the structure of the alkylating agent. Methanesulfonates, epoxides, and aziridines give second-order reactions that depend on concentrations of the alkylating agent and nucleophile. The situation is more complex with β-haloalkylamines (nitrogen mustards) and β-haloalkylsulfides (sulfur mustards) because these molecules undergo neighboring-group reactions in which the nitrogen or sulfur atom displaces the halide to give strained, 3-membered "onium" intermediates. These "onium" ions react with nucleophiles in second-order processes. However, the overall reaction kinetics depend on the relative rates of the two steps. In the case of mechlorethamine, the aziridinium ion is rapidly formed in wa-

ter, but reaction with biological nucleophiles is slower. Thus, the kinetics will be second order.[12]

In contrast, sulfur mustard forms the less stable episulfonium ion more slowly than this ion reacts with biological nucleophiles. Thus, the neighboring-group reaction is rate limiting and the kinetics are first order.[13]

Aryl-substituted nitrogen mustards such as chlorambucil

are relatively stable toward aziridinium ion formation, because the aromatic ring decreases the nucleophilicity of the nitrogen atom. These mustards react according to first-order kinetics.[13] The stability of chlorambucil allows it to be taken orally, whereas mechlorethamine is given by intravenous administration of freshly prepared solutions. The requirement for freshly prepared solutions is based on the gradual decomposition of the aziridinum ion by interaction with water.

Ethylenimines and epoxides are strained ring systems, but they do not react as readily as aziridinium or episulfonium ions with nucleophiles. Their reactions are second order and enhanced by the presence of acid.[11]

Examples of antitumor agents containing ethylenimine groups are triethylene melamine and thiotepa.

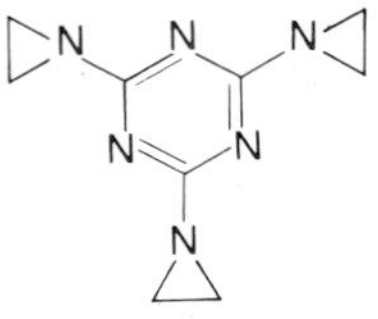

Triethylene Melamine 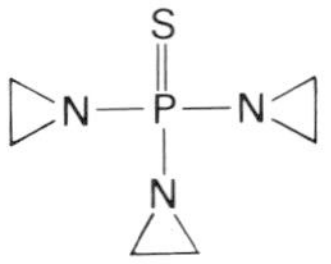   Thiotepa

The use of epoxides as cross-linking agents in textile chemistry suggested that they be tried in cancer chemotherapy. Simple diepoxides such as 1,2:3,4-diepoxybutane showed clinical activity against Hodgkin's disease,[14] but none of these compounds became an established drug. Dibromomannitol is presently under clinical study. It gives the corresponding diepoxide upon continuous titration at pH 8. This diepoxide (1,2:5,6-dianhydro-D-mannitol) shows potent alkylating activity against experimental tumors.[15] Thus, it is supposed that dibromomannitol and related compounds such as dibromodulcitol act by way of the diepoxides. However, this reaction sequence has not yet been verified in vivo.

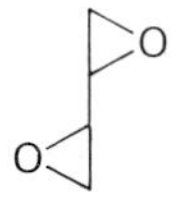

1,2:3,4-Diepoxybutane

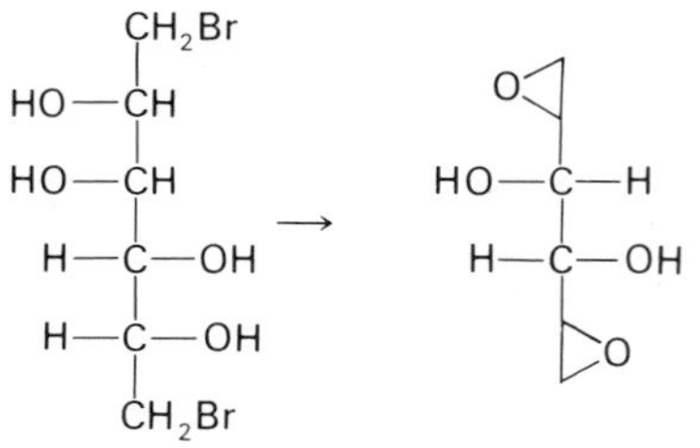

Dibromomannitol   Dianhydro-D-mannitol

A somewhat different type of alkylating agent is the N-alkyl-N-nitrosourea. Compounds of this class are unstable in aqueous solution under physiological conditions. They produce carbonium ions (also called carbenium ions) that can alkylate, and isocyanates that can carbamoylate. For example, methylnitrosourea decomposes initially to form isocyanic acid and methyldiazohydroxide. The latter species decomposes further to methyldiazonium ion and finally to methyl carbonium ion, the ultimate alkylating species.[16]

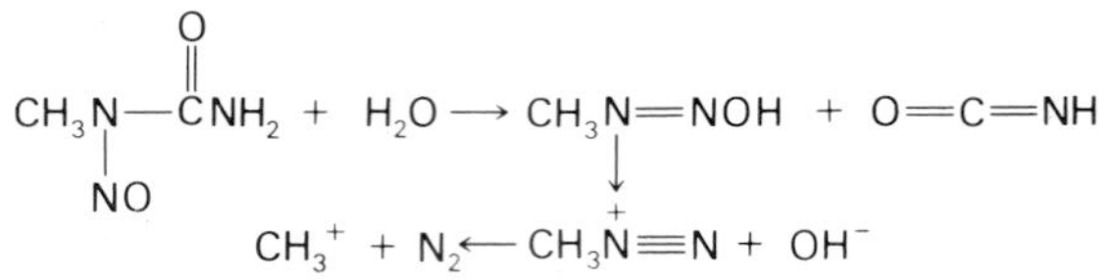

Substituents on the nitrogen atoms of the nitrosourea influence the mechanism of decomposition in water, which determines the species generated and controls the biological effects. Carmustine (BCNU) undergoes an abnormal decomposition in which the urea oxygen displaces a chlorine to give a cyclic intermediate (Scheme 1). This intermediate decomposes to vinyl diazohydroxide, the precursor to vinylcarbonium ion, and 2-chloroethylisocyanate. The latter species gives 2-chloroethylamine, an additional alkylating agent.[16]

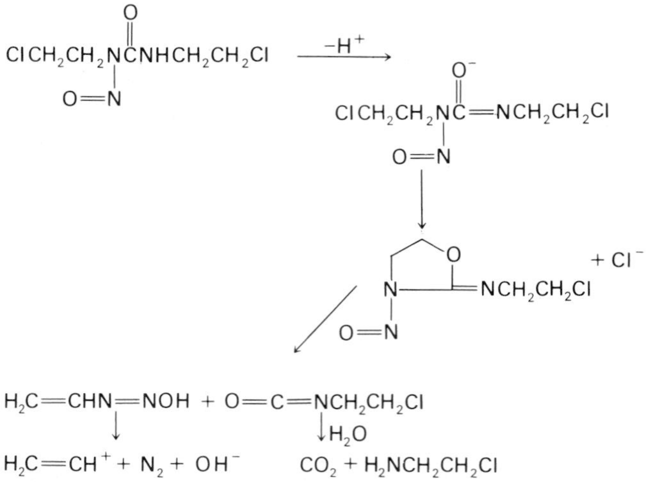

Scheme 1. *Decomposition of Carmustine*

Some clinically important alkylating agents are not active until they have been transformed by metabolic processes. The leading example of this group is cyclophosphamide, which is converted by hepatic cytochrome P-450 into the corresponding 4-hydroxy derivative by way of the 4-hydroperoxy intermediate (Scheme 2). The 4-hydroxy derivative is a carbinolamine in equilibrium with the open-chain aminoaldehyde form. Nonenzymatic decomposition of the latter form generates phosphoramide mustard and acrolein. Recent studies[17] based on ^{31}P NMR have shown that the conjugate base of phosphoramide mustard cyclizes to the aziridinium ion,[18] which is the principal cross-linking alkylator formed from cyclophosphamide. The maximal rate of cyclization occurs at pH 7.4. It was suggested that selective toxicity toward certain neoplastic cells might be based on their abnormally low pH. This would afford a slower formation of aziridinium ions and they would persist longer because of decreased inactivation by hydroxide ions.[17]

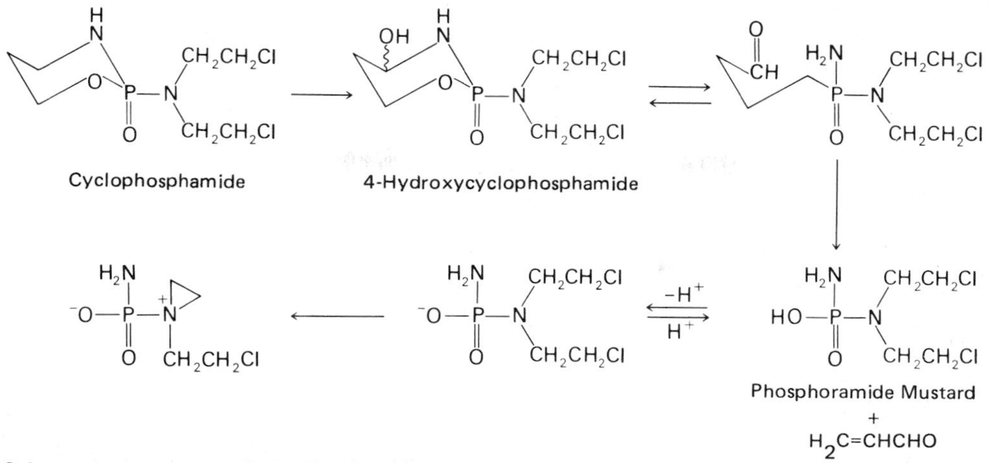

Scheme 2. *Activation of Cyclophosphamide*

The phosphorus atom of cyclophosphamide is substituted asymmetrically, which means that it is capable of being resolved into optical antipodes. This resolution has been made and the antipodes have been tested against tumors. The levorotatory form has twice the therapeutic index of the dextrorotatory form.[19]

Iphosphamide, an isomer of cyclophosphamide in which one of the 2-chloroethyl substituents is on the ring nitrogen, also has potent antitumor activity. It must be activated by hepatic enzymes, but its metabolism is slower than that of cyclophosphamide.[20]

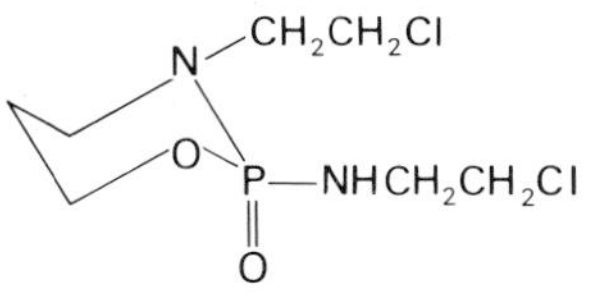

Iphosphamide

Other examples of alkylating species are afforded by carbinolamines as found in maytansine, and vinylogous carbinolamines as found in certain pyrrolizine diesters.[21]

When mitomycin C is reduced enzymatically to its hydroquinone, the spontaneous elimination of methanol affords the vinylogous carbinolamine system. Loss of the carbamoyloxy group from this system gives a stabilized carbonium ion that is capable of alkylating DNA. The aziridine ring of mitomycin C provides a second alkylating group and, together with the vinylogous carbinolamine, it allows mitomycin C to crosslink double helical DNA.[22] Molecules like mitomycin C are said to act by "bioreductive alkylation."[23]

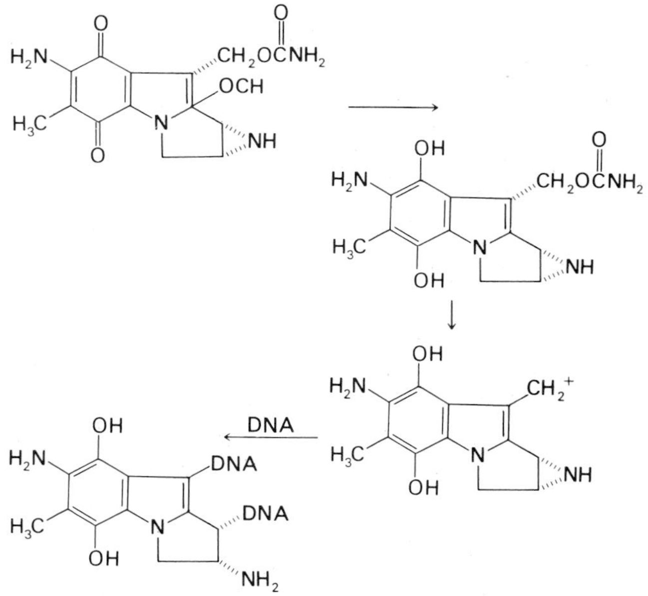

Scheme 3. *Mitomycin C Activation and DNA Alkylation*

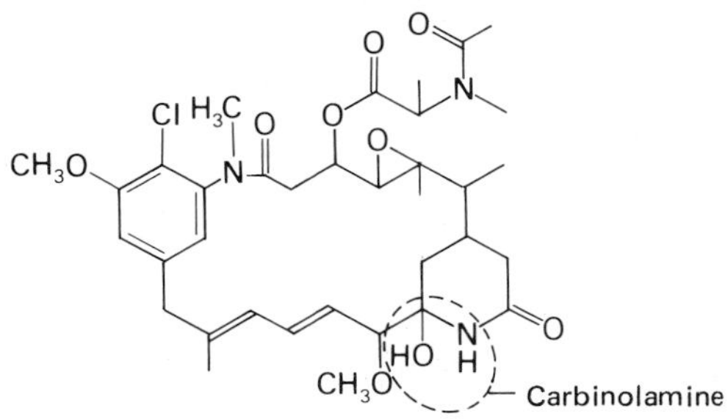

Maytansine

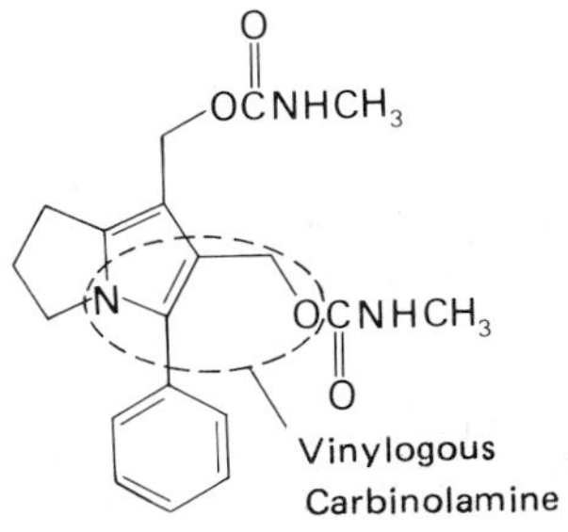

Pyrrolizine Diester

Another type of alkylating species occurs in α, β-unsaturated carbonyl compounds. These compounds can alkylate nucleophiles by conjugate addition.

$$nu-H + H_2C=CHCR \longrightarrow nuCH_2CH_2CR$$

Although there are no established clinical agents of this type, many natural products active against experimental tumors contain α-methylene lactone or α, β-unsaturated ketone functionalities. For example, the sesquiterpene helenalin has both of these systems.[24]

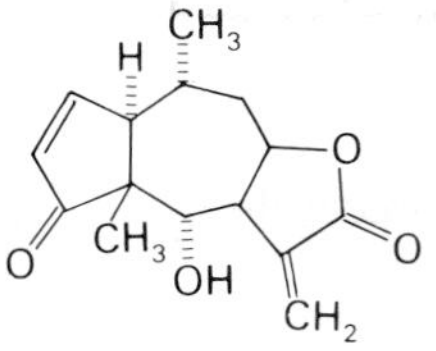

Helenalin

Alkylation can also occur by free radial reactions. The methylhydrazines are a chemical class prone to decomposition in this manner. These compounds were tested as antitumor agents in 1963 and one of them, procarbazine, was found to have a pronounced, but rather specific, effect on Hodgkin's disease.[25] Procarbazine is relatively stable at pH 7, but air oxidation to azoprocarbazine occurs readily in the presence of metalloproteins. Isomerization of this azo compound to the corresponding hydrazone, followed by hydrolysis, gives methylhydrazine and p-formyl-N-isopropylbenzamide. The formation of methylhydrazine from procarbazine has been demonstrated in living organisms.[26] Methylhydrazine is known to be oxidized to methyldiazine, which can decompose to nitrogen, methyl radical, and hydrogen radical.[27]

The methyl group of procarbazine is incorporated intact into cytoplasmic RNA.[28] However, the methylating species has not been conclusively established. Formation of methyl radical seems certain because methane is generated. The metabolism of procarbazine is a complex process involving more than one pathway. In humans, the conversion to azoprocarbazine is very rapid, with procarbazine having a half-life of 7 to 10 minutes. The major metabolite is N-isopropylterephthalamic acid and the N-methyl group appears as both carbon dioxide and methane.[29]

Dacarbazine was originally thought to be an antimetabolite because of its close resemblance to 5-aminoimidazole-4-carboxamide, an intermediate in purine biosynthesis (Scheme 5). However, it now appears to be an alkylating agent.[30] The isolation of an N-demethyl metabolite suggested that there might be a sequence in which this metabolite was hydrolyzed to methyldiazohydroxide, a precursor to methylcarbonium ion,[31] but it was found that this metabolite was less active than starting material against the Lewis lung tumor. An alternative mode of action was proposed in which dacarbazine undergoes acid-catalyzed hydrolysis to a diazonium ion, which can react in this form or decompose to the corresponding carbonium ion. Support for the latter mechanism was afforded by a correlation between the hydrolysis rates of phenyl-substituted dimethyltriazines and their antitumor activities.[32]

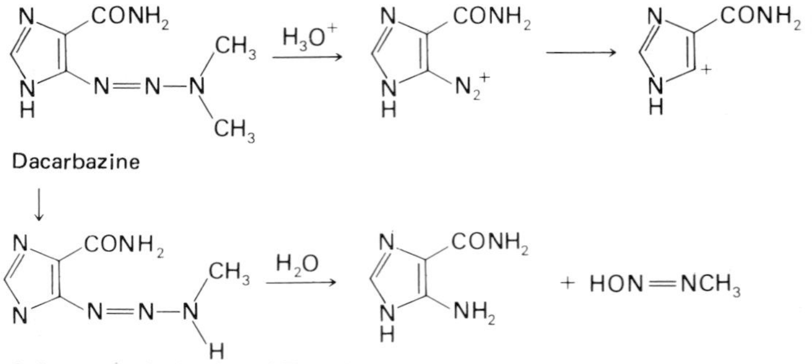

Scheme 5. *Activation of Dacarbazine*

The interaction of alkylating agents with biopolymers has been studied extensively. However, no mode of action for the lethality to cancer cells has been conclusively established. A good working model has been developed for the alkylation of bacteria and viruses, but there are uncertainties in extrapolating it to mammalian cells. The present working hypothesis is that most alkylating agents produce cytotoxic, mutagenic, and carcinogenic effects by reacting with cellular DNA. They also react with RNA and proteins, but these effects are thought to be less significant.[33] The

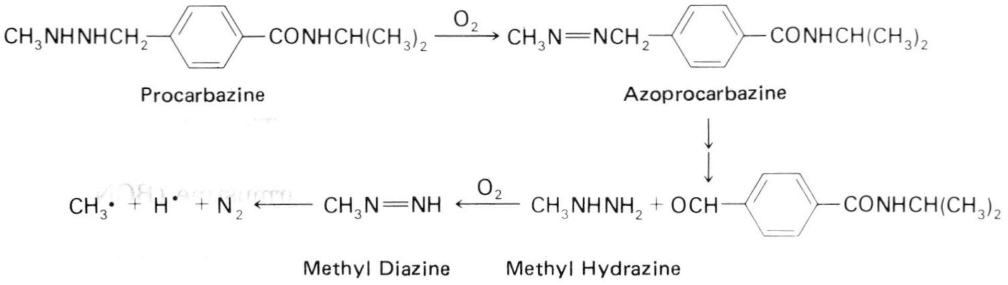

Procarbazine

Azoprocarbazine

$$CH_3 \cdot + H \cdot + N_2 \longleftarrow CH_3N=NH \xleftarrow{O_2} CH_3NHNH_2 + OCH \text{—} \bigcirc \text{—} CONHCH(CH_3)_2$$

Methyl Diazine   Methyl Hydrazine

most active clinical alkylating agents are bifunctional compounds capable of cross-linking DNA. Agents such as methylnitrosourea that give simple alkylation are highly mutagenic relative to their cytotoxicity. The cross-linking process can be either interstrand or intrastrand. Interstrand links can be verified by a test based on the thermal denaturation and renaturation of DNA. When double helical DNA is heated in water, it unwinds and the strands separate. Renaturation, in which the strands recombine in the double helix, is slow and difficult. In contrast, if the two strands are cross-linked they cannot separate. Hence, they renaturate rapidly on cooling. Interstrand cross-linking occurs with mechlorethamine and other "two-armed" mustards, but busulfan appears to give intrastrand links according to this test.[34]

In DNA the 7-position (nitrogen) of guanine is especially susceptible to alkylation by mechlorethamine and other nitrogen mustards (Scheme 4).[35] The alkylated structure has a positive charge in its imidazole ring, which renders the guanine-ribose linkage susceptible to cleavage. This cleavage results in the deletion of guanine, and the resulting "apurinic acid" ribose–phosphate link is readily hydrolyzable. Alkylation of the imidazole ring also activates it to cleavage of the 8, 9-bond.[11]

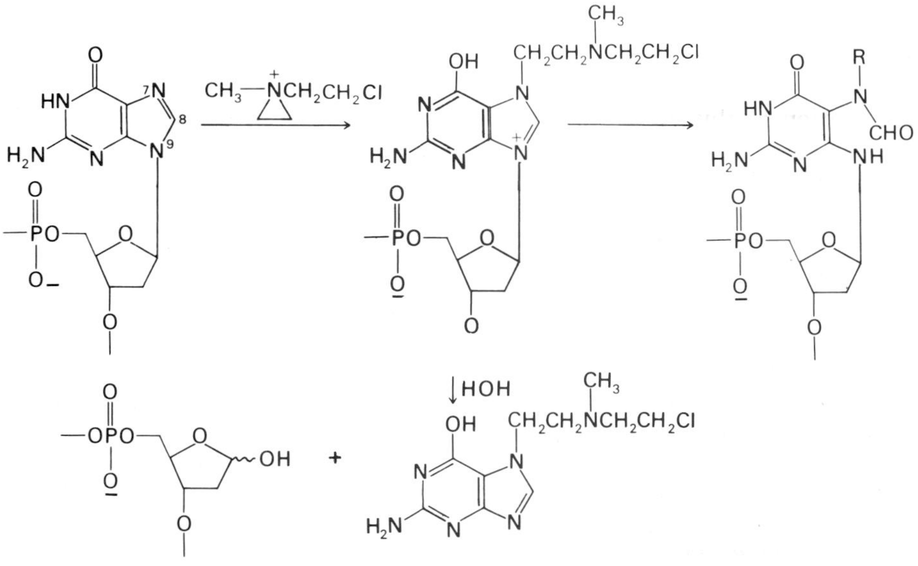

methanesulfonate. This ethyl derivative pairs with thymine, whereas guanine normally pairs with cytosine.[37]

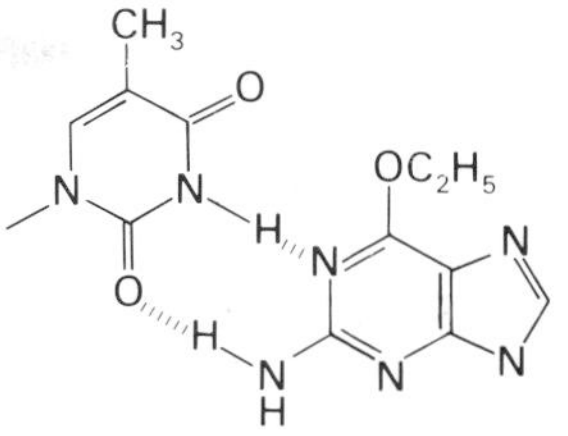

Other base positions of DNA attacked by alkylating agents are N-3 of guanine, N-3, N-1, and N-7 of adenine, 0-6 of thymine, and N-3 of cytosine. The importance of these minor alkylation reactions is difficult to assess. The phosphate oxygens of DNA are alkylated to an appreciable extent, but the significance of this feature is unknown.[38]

Guanine is also implicated in the cross-linking of double helical DNA. Di(guanin-7-yl) derivatives have been identified among the products of reaction with mechlorethamine.[39] Busulfan alkylation has given 1', 4'-di(guanin-7-yl)butane, but this product is considered to have resulted from intrastrand linking.[34] Because DNA alkylated with mitomycin C does not have

Scheme 4. Alkylation of Guanine in DNA

Other consequences of the positively charged purine structure are facile exchange of the 8-hydrogen, which can be used as a probe for 7-alkylation,[36] and a shift to the enolized pyrimidine ring as the preferred tautomer. The latter effect has been cited as a possible basis for abnormal base pairing in DNA replication, but this has not been substantiated. One example in which alkylation of guanine does lead to abnormal base pairing is the 0-6-ethylation produced by ethyl

a readily exchangeable 8-hydrogen on guanine, it has been suggested that the crosslink occurs at the 0-6 of guanine, rather than at N-7.[36]

Alkylating agents also interact with enzymes and other proteins. Thus, the repair enzyme, DNA nucleotidyltransferase of L1210 leukemia cells, was inhibited strongly by carmustine (BCNU), lomustine (CCNU), and 2-chloroethyl isocyanate. Because 1-(2-chloroethyl)-1-nitrosourea was a poor inhibitor of this enzyme, it was concluded that the main interaction with the enzyme was carbamoylation by the alkyl isocyanates generated in the decomposition of carmustine and lomustine.[40]

Alkylating agents can damage tissues with low mitotic indices, but they are most cytotoxic to rapidly proliferating tissues that have large proportions of cells in cycle. Nucleic acids are especially susceptible to alkylation when their structures are changed or unpaired in the process of replication. Thus, alkylating agents are most effective in the late G_1 or S phases. Alkylation may occur to some degree at any stage in the cell cycle, but the resulting toxicity is usually expressed when cells enter the S phase (Fig. 8-1). Progression through the cycle is blocked at G_2, the premitotic phase, and cell division fails.[41,42]

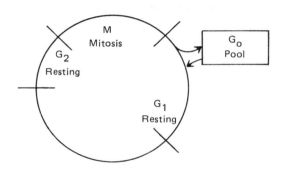

FIG. 8-1. *The cell life cycle.*

If cells can repair damage to their DNA before the next cell division, the effects of alkylation will not be lethal. Cells have developed a complex mechanism to accomplish this repair. First a recognition enzyme discovers an abnormal region in the DNA. This recognition brings about the operation of an endonuclease, which makes a single-strand breakage in the DNA. An exonuclease then removes a small segment of DNA containing the damaged bases. Finally, the DNA is restored to its original structure by replacing the bases and rejoining the strand.[43,44] It is evident that tumor cells with efficient repair mechanisms will be relatively resistant to alkylating agents. Tumor cells outside the cell cycle in the resting phase (G_0) will have a rather long time to repair their DNA. Thus, slow-growing tumors should not respond well to alkylating agents. This limitation is observed clinically.

PRODUCTS

Mechlorethamine Hydrochloride U.S.P, Mustargen®, nitrogen mustard, HN_2, NSC-762, 2,2-dichloro-N-methyldiethylamine hydrochloride. This compound is prepared by treating 2,2'-(methylimino)-diethanol with thionyl chloride.[45] It occurs as hygro-

scopic leaflets that are very soluble in water. The dry crystals are stable at temperatures up to 40°C. They are very irritating to mucous membranes and harmful to eyes. The compound is supplied in rubber-stoppered vials containing a mixture of 10 mg. mechlorethamine hydrochloride and 90 mg. of sodium chloride. It is diluted with 10 ml. of sterile water immediately prior to injection into a rapidly flowing intravenous infusion.

Mechlorethamine is effective in Hodgkin's disease. Current practice is to give it in combination with other agents. The combination with vincristine (Oncovin®), procarbazine, and prednisone, known as the MOPP regimen, is considered the treatment of choice. Other lymphomas and mycosis fungoides can be treated with mechlorethamine. The most serious toxic reaction is bone-marrow depression, which results in leukopenia and thrombocytopenia. Emesis is prevalent and lasts about 8 hours. Nausea and anorexia persist longer. These gastrointestinal effects may be prevented by administering a short-acting barbiturate and prochlorperazine. Extravasation produces intense local reactions at the site of injection. If it occurs, the immediate application of sodium thiosulfate solution can protect the tissues. This is because thiosulfate ion reacts very rapidly with the aziridinium ion formed from mechlorethamine.

Cyclophosphamide U.S.P., Cytoxan®, NSC-26271, N,N-bis(2-chloroethyl)tetrahydro-2H-1,3,2-oxazaphosphorin-2-amine-2-oxide. It is prepared by treating bis(2-chloroethyl)phosphoramide dichloride with propanolamine.[46] The monohydrate is a low-melting solid that is very soluble in water. It is supplied as 25 and 50 mg. white tablets, 50 mg. unit-dose cartons, and as a powder (100, 200, or 500 mg.) in sterile vials.

Cyclophosphamide has advantages over other alkylating agents in that it is active orally and parenterally and can be given in fractionated doses over prolonged periods of time. It is active against multiple myeloma, chronic lymphocytic leukemia, and acute leukemia of children. In combination with other chemotherapeutic agents, it has given complete remissions and even cures in Burkitt's lymphoma and acute lymphoblastic leukemia in children.[47] The most frequently encountered toxic effects are alopecia, nausea, and vomiting. Leukopenia occurs, but thrombocytopenia is less frequent than with other alkylating agents. Sterile hemorrhagic cystitis may result and even be fatal. Gonadal suppression has been reported in a number of patients.

Melphalan U.S.P., Alkeran®, L-sarcolysin, L-PAM, L-phenylalanine mustard, NSC-8808, 4-[bis(2-chloroethyl)amino]-L-phenylalanine. This compound is prepared by treating L-N-phthalimido-*p*-amino-phenylalanine ethyl ester with ethylene oxide, followed by phosphorus oxychloride, and finally hydrolysis with hydrochloric acid.[48] It is soluble in alcohols, but poorly soluble in water. Oral absorption is good and the drug is equally effective whether given orally or intravenously. Scored 2 mg. tablets are available.

Melphalan is active against multiple myeloma. Recent clinical studies have shown it to be highly effective in preventing the recurrence of cancer in premenopausal women who undergo radical mastectomy.[49] The clinical toxicity is mainly hematological, which means that the blood count must be carefully followed. Nausea and vomiting are infrequent and alopecia does not occur.

Chlorambucil U.S.P., Leukeran®, chloraminophene, NSC-3088, p-(di-2-chlorethyl)aminophenylbutyric acid. It is prepared by treating p-aminophenyl butyric acid with ethylene oxide, followed by thionyl chloride.[50] Chlorambucil is soluble in ether and aqueous alkali. Its oral absorption is efficient and reliable. Sugar-coated 2 mg. tablets are supplied.

Chlorambucil acts most slowly and is the least toxic of any nitrogen mustard derivative in use. It is indicated especially in treatment of chronic lymphocytic leukemia and primary macroglobulinemia. Other indications are lymphosarcoma and Hodgkin's disease.[51] Many patients develop progressive, but reversible, lymphopenia during treatment. Most patients also develop a dose-related and rapidly reversible neutropenia. For these reasons weekly blood counts are made to determine the total and differential leukocyte levels. The hemoglobin levels are also determined.

Busulfan U.S.P., Myleran®, NSC-750, 1,4-di(methanesulfonyloxy)butane. This compound is synthesized by treating 1,4-butanediol with methanesulfonyl chloride in the presence of pyridine.[52] It is obtained as crystals that are soluble in acetone and alcohol. Although practically insoluble in water, it dissolves slowly upon hydrolysis. However, it is stable in dry form. It is supplied as scored 2 mg. tablets.

Busulfan is well absorbed orally and metabolized rapidly. Much of the drug undergoes a process known as "sulfur stripping" in which its interaction with thiol compounds such as glutathione or cysteine results in loss of two equivalents of methanesulfonic acid and formation of a cyclic sulfonium intermediate involving the sulfur atom of the thiol.[53] Such sulfonium intermediates are stable in vitro, but in vivo they are readily converted into the metabolite 3-hydroxy thiolane-1,1-dioxide.[54] That the sulfur atom of this thiolane does not come from a methanesulfonyl group was shown by the nearly quantitative isolation of labelled methanesulfonic acid in the urine when busulfan ^{35}S is administered to animals.[55]

The main therapeutic use of busulfan is in chronic granulocytic leukemia. Remissions in 85 to 90 percent of patients are observed after the first course of therapy. However, it is not curative. Toxic effects are mostly limited to myelosuppression in which the depletion of thrombocytes may lead to hemorrhage. Blood counts should be taken not less than weekly. The rapid destruction of granulocytes can cause hyperuricemia that might result in kidney damage. This complication is prevented by using allopurinol, a xanthine oxidase inhibitor.[56]

Carmustine, BiCNU®, BCNU, NSC-409,962, 1,3-bis(2-chloroethyl)-1-nitrosourea. This compound was synthesized at the Southern Research Institute by treating 1,3-bis(2-chloroethyl)urea with sodium nitrite and formic acid.[57] It is a low-melting white powder that changes to an oily liquid at 27°C. This change is considered a sign of decomposition, and such samples should be discarded. Carmustine is most stable in petroleum ether or water at pH 4. It is administered intravenously because metabolism is very rapid. However, some of the degradation products have prolonged half-lives in plasma. Carmustine is supplied as a lyophilized powder. When it is diluted with 3 ml. of the supplied sterile diluent, ethanol, and further diluted with 27 ml. of sterile water, a 10 percent ethanolic solution containing 3.3 mg. per ml. is obtained.

Because of its ability to cross the blood–brain barrier, carmustine is used against brain tumors and other tumors, such as leukemias, that have metastasized to the brain.[58] It also is used as secondary therapy in combination with other agents for Hodgkin's disease and other lymphomas. Multiple myeloma responds to a combination of carmustine and prednisone. Delayed myelosuppression is the most frequent and serious toxicity. This condition usually occurs 4 to

6 weeks after treatment. Thrombocytopenia is the most pronounced effect, followed by leukopenia. Nausea and vomiting frequently occur about 2 hours after treatment.

Carmustine is given as a single dose by intravenous injection at 100 to 200 mg. per square meter. A repeat course is not given until the blood elements return to normal levels. This process requires about 6 weeks.

Lomustine, CeeNU®, CCNU, NSC-79037, [1-(2-chlorethyl)]-3-cyclohexyl-1-nitrosourea. This compound also was synthesized at the Southern Research Institute. The procedure involved treating ethyl 5-(2-chloroethyl)-3-nitrosohydantoate with cyclohexylamine, followed by renitrosation of the resulting intermediate, [1-(2-chloroethyl)]-3-cyclohexylurea.[59] Lomustine is a yellow powder that is soluble in absolute or 10 percent ethanol but is relatively insoluble in water. It is sufficiently stable to metabolism to be administered orally. The high lipid solubility of lomustine allows it to cross the blood–brain barrier rapidly. Levels in the CSF are 50 percent higher than those in plasma. Lomustine is supplied in dose packs that contain two each of color-coded 100, 40, and 10 mg. capsules. The total dose prescribed is obtained by appropriate combination of these capsules.

Lomustine is used against both primary and metastatic brain tumors, and as secondary therapy in Hodgkin's disease. The most common adverse reactions are nausea and vomiting, thrombocytopenia, and leukopenia. As in the case of carmustine, the myelosuppression caused by lomustine is delayed.[60]

The recommended dosage of lomustine is 130 mg. per square meter orally every 6 weeks. A reduced dose is given to patients with compromised bone marrow function.

Thiotepa U.S.P., TSPA, NSC-6396, N,N′,N″-triethylenethiophosphoramide, tris(1-aziridinyl)phosphine sulfide. This compound is prepared by treating trichlorophosphine sulfide with aziridine[61] and is obtained as a white powder that is soluble in water. It is supplied in vials containing 15 mg. of thiotepa, 80 mg. of sodium chloride and 50 mg. of sodium bicarbonate. Sterile water is added to make an isotonic solution. Both the vials and solutions must be stored at 2° to 8°C. The solutions may be stored 5 days without loss of potency.

Thiotepa has been tried against a wide variety of tumors and it has given palliation in many types, although with varying frequency. The most consistent results have been obtained in breast, ovarian, and bronchogenic carcinomas and malignant lymphomas. It also is used to control intracavity effusions resulting from neoplasms. Thiotepa is highly toxic to bone marrow, and blood counts must be made during therapy.

Procarbazine Hydrochloride U.S.P., Matulane®, MIH, NSC-77213, N-isopropyl-α-(2-methylhydrazino)-p-toluamide. It is prepared from N-isopropyl-p-toluamide in a process involving condensation with diethyl azodicarboxylate, methylation with methyl iodide and base, and acid hydrolysis.[62] Although soluble in water, it is unstable in solution. Capsules containing the equivalent of 50 mg. of procarbazine as its hydrochloride are supplied.

Procarbazine has demonstrated activity against Hodgkin's disease. For this condition it is used in combination with other agents such as mechlorethamine, vincristine, and prednisone (MOPP program). Toxic effects such as leukopenia, thrombocytopenia, nausea, and vomiting occur in a majority of patients. Neurological and dermatological effects also occur. Concurrent intake of alcohol or certain amines are contraindicated. The weak monoamine oxidase-inhibiting properties of procarbazine may potentiate amines to produce hypertension.

Dacarbazine, DTIC-Dome®, DIC, DTIC, NSC-45388, 5-(3,3-dimethyl-1-triazenyl)-1H-imidazole-4-carboxamide. This compound is prepared by treating the diazonium salt, prepared from 5-aminoimidazole-4-carboxamide, with dimethylamine in methanol.[63] It is obtained as a colorless to ivory-colored solid that is very sensitive to light. It does not melt but decomposes explosively when heated above 250°C. Water solubility is good, but solutions must be protected from light. Dacarbazine is supplied in vials containing either 100 or 200 mg. When reconstituted with 9.9 and 19.7 ml., respectively, of sterile water these samples give solutions containing 10 mg. per ml. at pH 3.0 to 4.0. Such solutions may be stored at 4°C for 72 hours.

Dacarbazine is indicated for the treatment of metastatic malignant melanoma. Combination with other antineoplastic drugs is superior to its use as a single agent. Anorexia, nausea, and vomiting are the most frequent toxic reactions. However, leukopenia and thrombocytopenia are the most serious effects. Blood counts should be made and therapy temporarily suspended if the counts are too low.

The recommended daily dosage is 2 to 4.5 mg. per kg. for 10 days with repetition at 4 week intervals. Extravasation of the drug during injection may result in tissue damage and severe pain.

Uracil Mustard U.S.P., Uramustine®, NSC-34462, 5-[bis(2-chloroethyl)amino]uracil, 2,6-dihydroxy-5-bis(2-chloroethyl)aminopyrimidine, is prepared by treating 5-aminouracil with ethylene oxide, followed by thionyl chloride.[64] This crystalline solid is sparingly soluble in water. It is supplied as capsules containing 1 mg. of uracil mustard.

Uracil mustard is an analog of nitrogen mustard in which the uracil moiety was designed to serve as a carrier group. However, it shows little selectivity in this respect. It is used in the palliative treatment of chronic lymphocytic leukemia, lymphosarcoma, giant follicular lymphoma, and Hodgkin's disease. Nausea, vomiting, and diarrhea are the most common untoward effects. They are dose related. Complete blood counts should be made once or twice weekly during the first month of therapy. Cumulative, irreversible bone-marrow damage may occur as the total dose reaches 1 mg. per kg. of body weight.

ANTIMETABOLITES

Antimetabolites are compounds that prevent the biosynthesis or utilization of normal cellular metabolites. They usually are closely related in structure to the metabolite that is antagonized. Many antimetabolites are enzyme inhibitors. In this capacity they may combine with the active site as if they are the substrate or cofactor. Alternatively, they may bind to an allosteric regulatory site, especially when they resemble the end product of a biosynthetic pathway under feedback control.[11] Sometimes the antimetabolite must be transformed biosynthetically (anabolized) into the active inhibitor. For example 6-mercaptopurine is converted into the corresponding ribonucleotide, which is a potent inhibitor of the conversion of 5-phosphoribosyl pyrophosphate into 5-phosphoribosylamine, a rate controlling step in the de novo synthesis of purines[5] (Scheme 5). An antimetabolite and its transformation products may inhibit a number of different enzymes. Thus, 6-mercaptopurine and its anabolites interact with more than 20 enzymes. This multiplicity of effects makes it difficult to decide which ones are crucial to the antitumor activity. The anabolites of purines and pyrimidine antagonists may be incorporated into nucleic acids. In this event, part of their antitumor effect might result from malfunction of the nucleic acids.[66] Although antimetabolites of every type have been tested against neoplasms, nearly all of the clinically useful agents are related to metabolites and cofactors in the biosynthesis of nucleic acids.

TABLE 8–1

ALKYLATING AGENTS

Name Proprietary Name	Preparations	Usual Adult Dose*	Usual Dose Range*	Usual Pediatric Dose*
Mechlorethamine Hydrochloride U.S.P. *Mustargen*	Mechlorethamine Hydrochloride for Injection U.S.P.	I.V., 400 µg. per kg. of body weight as a single dose; or divided into 2 or 4 daily doses		See Usual Adult Dose
Cyclophosphamide U.S.P. *Cytoxan, Endoxan*	Cyclophosphamide for Injection U.S.P. Cyclophosphamide Tablets U.S.P.	Initial, I.V., 40 to 50 mg. per kg. of body weight in divided doses given over a period of 2 to 5 days; maintenance, 10 to 15 mg. per kg. every 7 to 10 days or 3 to 5 mg. per kg. twice weekly		
Melphalan U.S.P. *Alkeran*	Melphalan Tablets U.S.P.	150 µg. per kg. of body weight for 7 days, followed by a rest period of at least 3 weeks		
Chlorambucil U.S.P. *Leukeran*	Chlorambucil Tablets U.S.P.	Initial, 100 to 200 µg. per kg. of body weight once daily		100 to 200 µg. per kg. of body weight or 4.5 mg. per square meter of body surface daily
Uracil Mustard U.S.P.	Uracil Mustard Capsules U.S.P.	1 to 2 mg. daily until bone-marrow depression or clinical improvement occurs		
Busulfan U.S.P. *Myleran*	Busulfan Tablets U.S.P.	4 to 6 mg. once daily	4 to 12 mg. daily	1.8 to 4.6 mg. per square meter of body surface daily
Thiotepa U.S.P.	Thiotepa for injection U.S.P.	I.V., 30 mg. at 1- to 4-week intervals, or 800 µg. per kg. of body weight every 4 weeks, or 200 to 400 µg. per kg. of body weight every 2 weeks	I.V., 60 mg. per dose	
Procarbazine Hydrochloride U.S.P. *Matulane*	Procarbazine Hydrochloride Capsules U.S.P.	Initial, the equivalent of procarbazine, 2 to 4 mg. per kg. of body weight, followed by 4 to 6 mg. per kg.		Initial, the equivalent of procarbazine, 50 mg. daily, followed by 100 mg. per square meter of body surface

*See *U.S.P. D.I.* for complete dosage information.

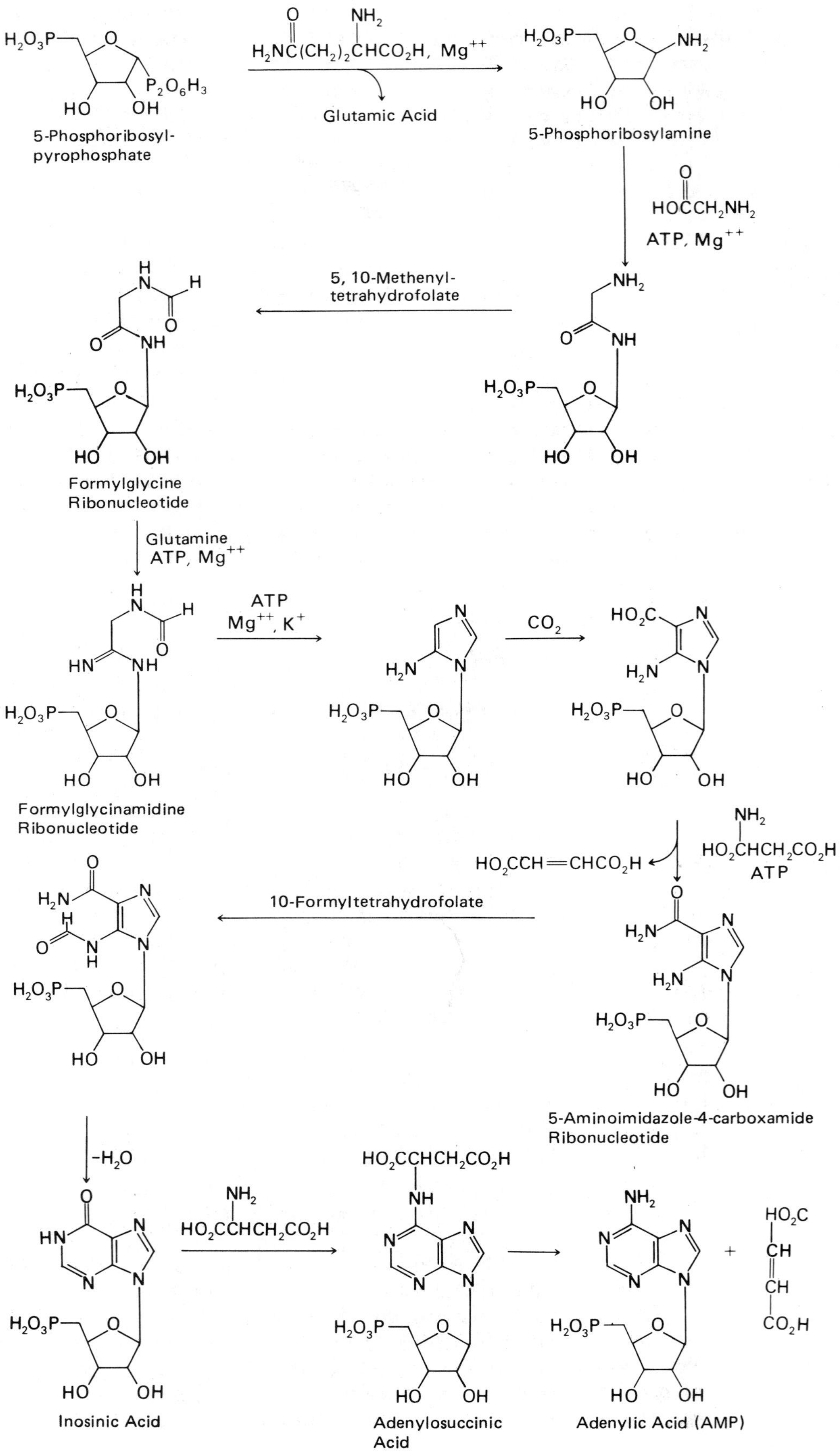

5-Phosphoribosyl-
pyrophosphate

$H_2NC(CH_2)_2CHCO_2H$, Mg^{++}

Glutamic Acid

5-Phosphoribosylamine

$HOCCH_2NH_2$
ATP, Mg^{++}

Formylglycine
Ribonucleotide

5, 10-Methenyl-
tetrahydrofolate

Glutamine
ATP, Mg^{++}

Formylglycinamidine
Ribonucleotide

ATP
Mg^{++}, K^+

CO_2

5-Aminoimidazole-4-carboxamide
Ribonucleotide

$HO_2CCH=CHCO_2H$

$HO_2CHCH_2CO_2H$
ATP

10-Formyltetrahydrofolate

$-H_2O$

Inosinic Acid

$HO_2CHCH_2CO_2H$

Adenylosuccinic
Acid

Adenylic Acid (AMP)

HO_2C
CH
CH
CO_2H

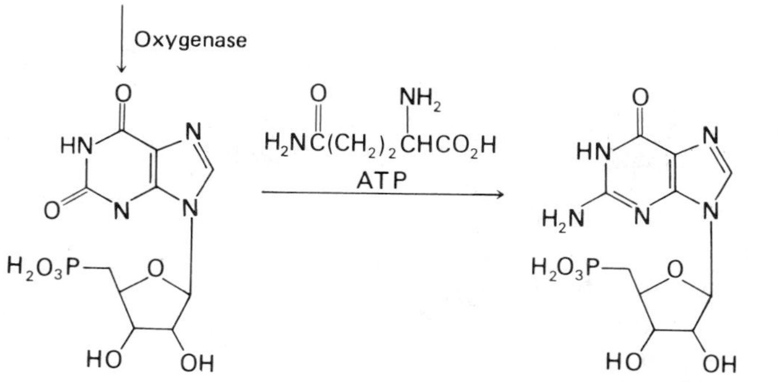

Oxygenase

Xanthylic Acid → Guanylic Acid (GMP)

6-Thioinosinate R = H
6-Methylthioinosinate R = CH₃

Azathioprene R =

Following the formulation of the antimetabolite theory by Woods and Fildes in 1940,[67] antimetabolites based on a variety of known nutrients were prepared. The first purine analog to show antitumor activity in mice, 8-azaguanine, was synthesized by Roblin in 1945.[68] This compound was introduced into clinical trials but was abandoned in favor of newer and more effective agents such as 6-mercaptopurine and 6-thioguanine developed by Elion and Hitchings.[69] 6-Mercaptopurine was synthesized in 1952[70] and was shown to be active against human leukemia in the following year.

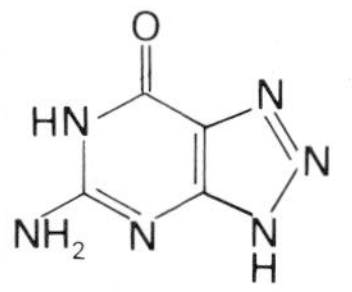

8-Azaguanine 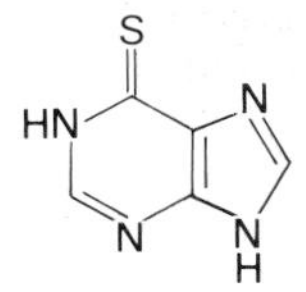   6-Mercaptopurine

In order to be active against neoplasms, 6-mercaptopurine must be converted into its ribonucleotide, 6-thioinosinate, by the enzyme hypoxanthine–guanine phosphoribosyltransferase. Neoplasms that lack this enzyme are resistant to the drug.[71] 6-Thioinosinate is a potent inhibitor of the conversion of 5-phosphoribosyl pyrophosphate into 5-phosphoribosylamine, as mentioned above. It also inhibits the conversion of inosinic acid to adenylic acid at two stages: (1) the reaction of inosinic acid with aspirate to give adenylosuccinic acid, and (2) the loss of fumaric acid from adenylosuccinic acid to give adenylic acid.[72] Furthermore, it inhibits the oxidation of inosinic acid to xanthylic acid.[73] The mode of action of 6-mercaptopurine is further complicated by the fact that its ribose diphosphate and triphosphate anabolites are also active enzyme inhibitors.[72] Still more complex is the ability of 6-thioinosinate to act as a substrate for a methyl transferase, requiring S-adenosylmethionine, that converts it into 6-methylthioinosinate. The latter compound is responsible for certain of the antimetabolite activities of 6-mercaptopurine.[74]

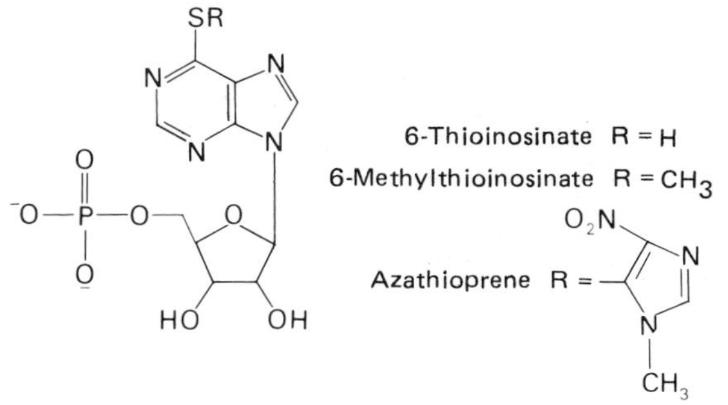

Metabolic degradation (catabolism) of 6-mercaptopurine by guanase gives 6-thioxanthine, which is oxidized by xanthine oxidase to afford 6-thiouric acid.[75] Allopurinol, an inhibitor of xanthine oxidase, increases both the potency and the toxicity of 6-mercaptopurine. However, its main importance is as an adjuvant to chemotherapy because it prevents the uric acid toxicity caused by the release of purines from destroyed cancer cells. Heterocyclic derivatives of 6-mercaptopurine such as azathioprine (Imuran®) were designed to protect it from catabolic reactions.[76] Although azathioprine has antitumor activity, it is not significantly better than 6-mercaptopurine. However, it has found an important role as an immunosuppressive agent in organ transplants.[69]

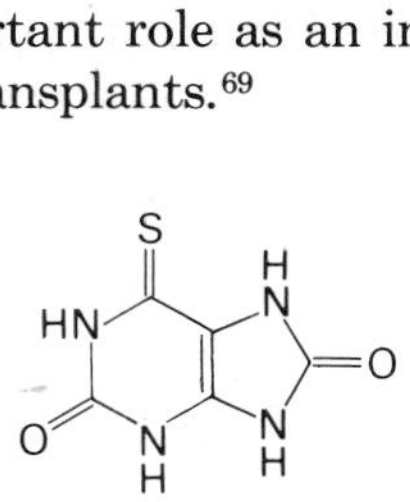

6-Thiouric Acid 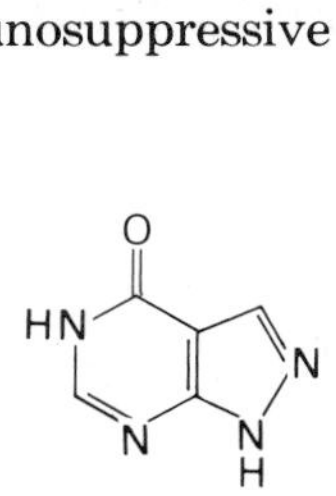   Allopurinol

Thioguanine is converted into its ribonucleotide by the same enzyme that acts on 6-mercaptopurine. It is converted further into the di- and triphosphates.[77] These species inhibit most of the same enzymes that are inhibited by 6-mercaptopurine. Thioguanine is incorporated into RNA and its 2'-deoxy metabolite is incorporated into DNA. The significance of these "fraudulent" nucleic acids in lethality to neoplasms is uncertain.[78]

Adenine arabinoside was first prepared by chemical synthesis[79] and later isolated from cultures of *Streptomyces antibioticus*.[80] It has a sugar, D-arabinose, that is epimeric with D-ribose at the 2'-position. This structural change makes it a competitive inhibitor of DNA polymerase.[81] In addition to its antineoplastic activity, adenosine arabinoside has potent antiviral action. The 5'-phosphate of adenosine arabinoside is included in the NCI listing of compounds in development. Adenine arabinoside and its derivatives are limited in their antitumor effect by susceptibility to adenosine deaminase. This enzyme converts them into hypoxanthine arabinoside derivatives. The resistance of certain tumors correlates with their levels of adenosine deaminase.[82]

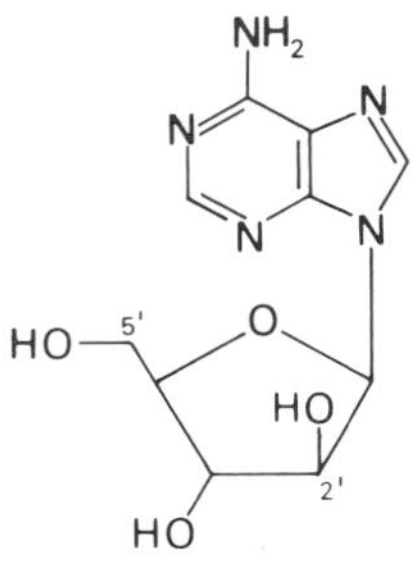

Adenosine
Arabinoside

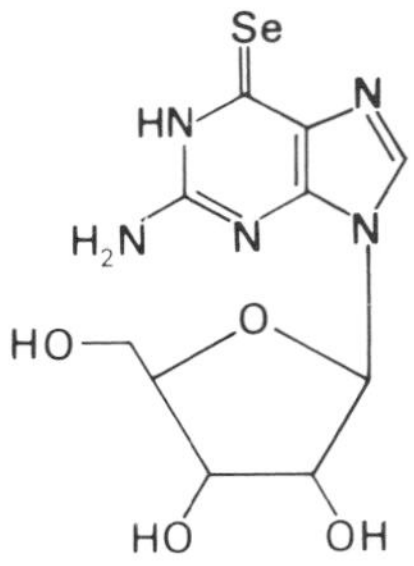

Selenoguanosine

Many other purine analogs have been found active against neoplasms. Among the compounds currently undergoing clinical study are selenoguanosine, α-2'-deoxythioguanosine, and "tricyclic nucleoside".[83]

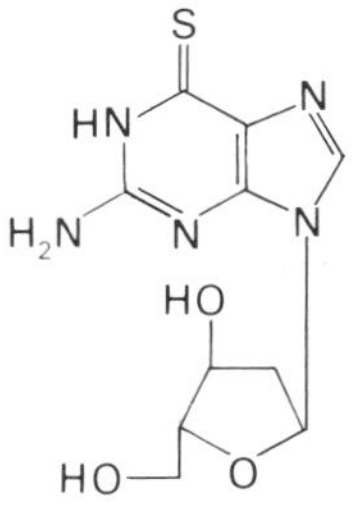

α-2'-Deoxythioguanosine

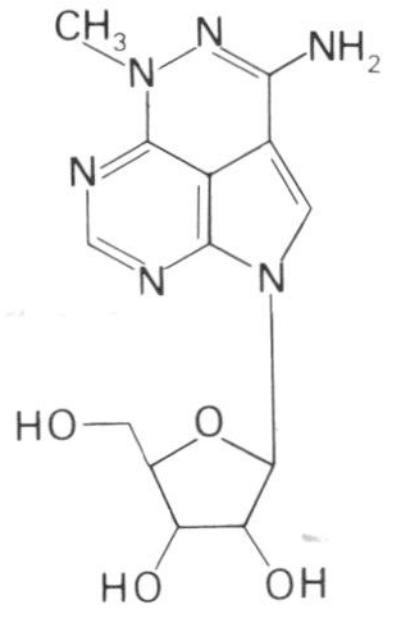

Tricyclic Nucleoside

The invention of 5-fluorouracil as an antimetabolite of uracil by Heidelberger in 1957 provided one of our foremost examples of rational drug design.[84] Starting with the observation that in certain tumors uracil was used more than orotic acid for nucleic-acid pyrimidine biosynthesis, the major precursor in normal tissue, he decided to synthesize an antimetabolite of uracil with only one modification in the structure. The 5-position

was chosen for a substituent in order to block the conversion of uridylate to thymidylate (Scheme 6) thus diminishing DNA biosynthesis. Fluorine was chosen as the substituent because the increased acidity caused by its inductive effect was expected to cause the molecule to bind strongly to enzymes. These choices were well founded, as 5-fluorouracil soon became one of the most widely used antineoplastic agents.

5-Fluorouracil is activated by anabolism to 5-fluoro-2'-deoxyuridylic acid. This conversion may proceed by two routes. In one route, 5-fluorouracil reacts with ribose-1-phosphate to give its riboside, which is phosphorylated by uridine kinase.[85] The resulting compound, 5-fluorouridylic acid, is converted into its 2'-deoxy derivative by ribonucleotide reductase. 5-Fluorouracil also may be transformed directly into 5-fluorouridylic acid by a phosphoribosyltransferase, which is present in certain tumors.[86] An alternative pharmaceutical based on 5-fluorouracil is its 2-deoxyriboside.[84] This compound is phosphorylated by 2'-deoxyuridine kinase. The resulting 5-fluoro-2'-deoxyuridylic acid is a powerful competitive inhibitor of thymidylate synthetase, the enzyme that converts 2'-deoxyuridylic acid to thymdylic acid. This blockade is probably the main lethal effect of 5-fluorouracil and its metabolites.[87] In the inhibiting reaction, the sulfhy-

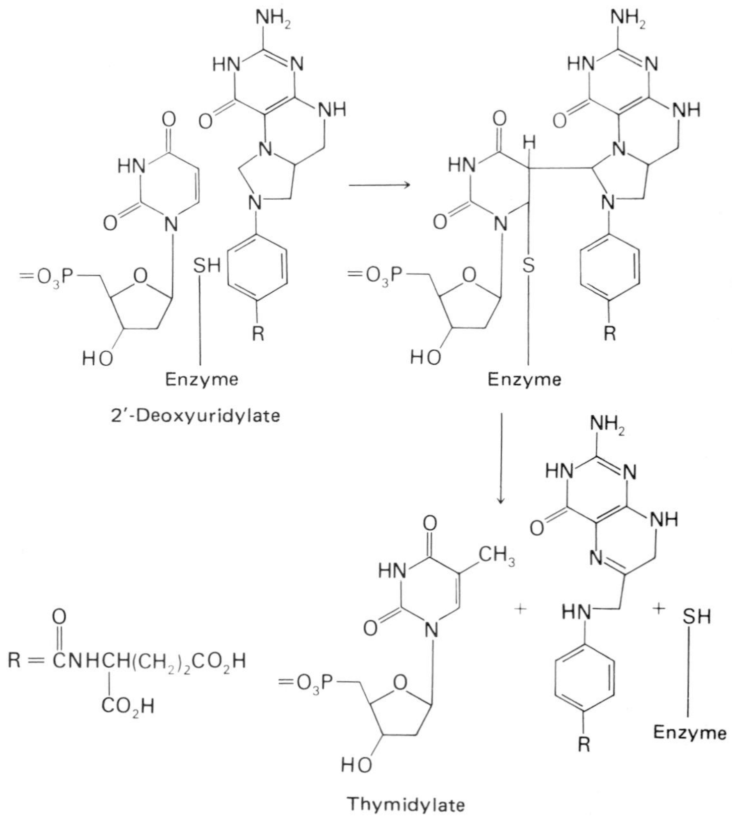

$$R = \overset{O}{\overset{\|}{C}}NHCH(CH_2)_2CO_2H$$
$$CO_2H$$

Scheme 6. *Conversion of Uridylate into Thymidylate*

dryl group of a cysteine residue in the enzyme adds to the 6-position of the fluorouracil moiety. The 5-position then binds to the methylene group of 5,10-methylenetetrahydrofolate. Ordinarily, this step would be followed by the transfer of the 5-hydrogen of uracil to the methylene group, resulting in the formation of thymidylate and dihydrofolate. However, the 5-fluorine is stable to transfer, and a terminal product results involving the enzyme, cofactor, and substrate, all covalently bonded. Thus, 5-fluoro-2'-deoxyuridylic acid would be classified as a Kcat inhibitor.[88]

Trifluorothymidine was designed by Heidelberger as an antimetabolite of thymine.[84] The riboside is essential because mammalian cells are unable to convert thymine and certain analogs into thymidine and its analogs. Thymidine kinase converts trifluorothymidine into trifluorothymidylic acid, which is a potent inhibitor of thymidylate synthetase.[84] In contrast to the stability of most trifluoromethyl groups, the one of trifluorothymidylic acid is extraordinarily labile. It reacts with glycine to give an amide at neutral pH.[91]

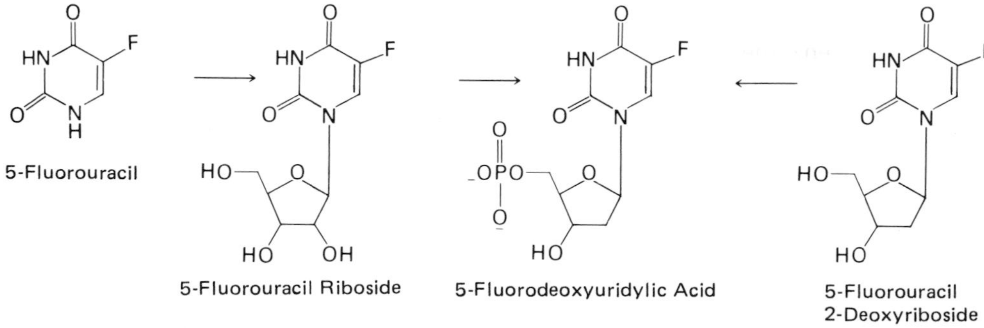

5-Fluorouracil

5-Fluorouracil Riboside

5-Fluorodeoxyuridylic Acid

5-Fluorouracil 2-Deoxyriboside

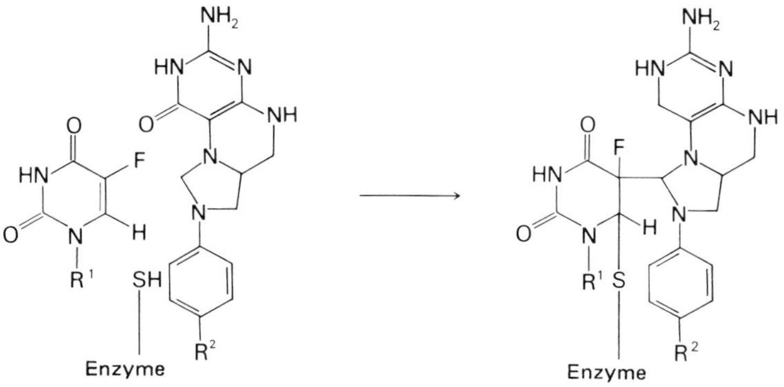

Enzyme

Enzyme

Recently the tetrahydrofuranyl derivative of 5-fluorouracil, known as ftorafur, was prepared in the Soviet Union.[89] It is active in clinical cancer and is less myelosuppressive than 5-fluorouracil. However, it has gastrointestinal and CNS toxicity. Ftorafur is slowly metabolized to 5-fluorouracil; thus, it may be considered a prodrug.[90]

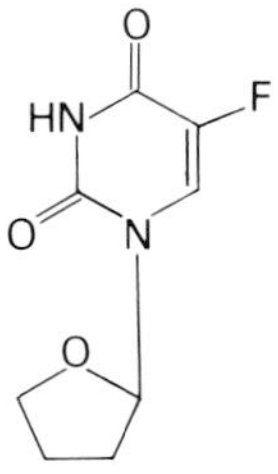

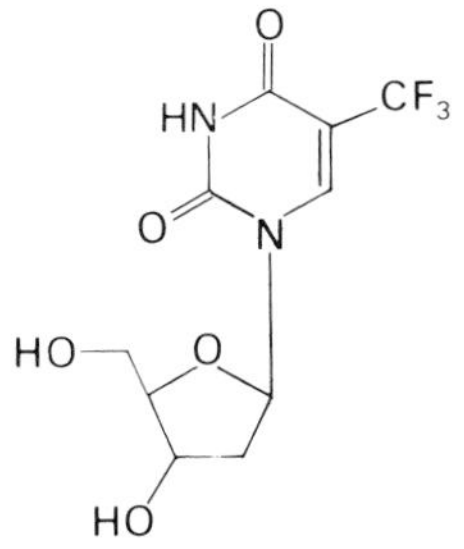

Kinetic studies have shown that this reaction involves initial nucleophilic attack at position-6 followed by loss of HF to give the highly reactive difluoromethylene group.[92] Glycine then adds to this group and hydrolysis of the remaining two fluorine atoms follows (Scheme 7). The interaction of trifluorothymidylic acid with thymidylate synthetase apparently follows a similar course. Thus, after preincubation it becomes irreversibly bound to the enzyme, and the kinetics are noncompetitive.[84]

Cytosine arabinoside was synthesized in 1959[93] and later found as a fermentation product.[94] It is noteworthy structurally in that its arabinose moiety is epimeric at the 2'-position with ribose. This modification, after anabolism to the triphosphate, causes it to inhibit the conversion of cytidylic acid to 2'-deoxycytidylic acid.[95] For a number of years this inhibition was

Ftorafur

Trifluorothymidine

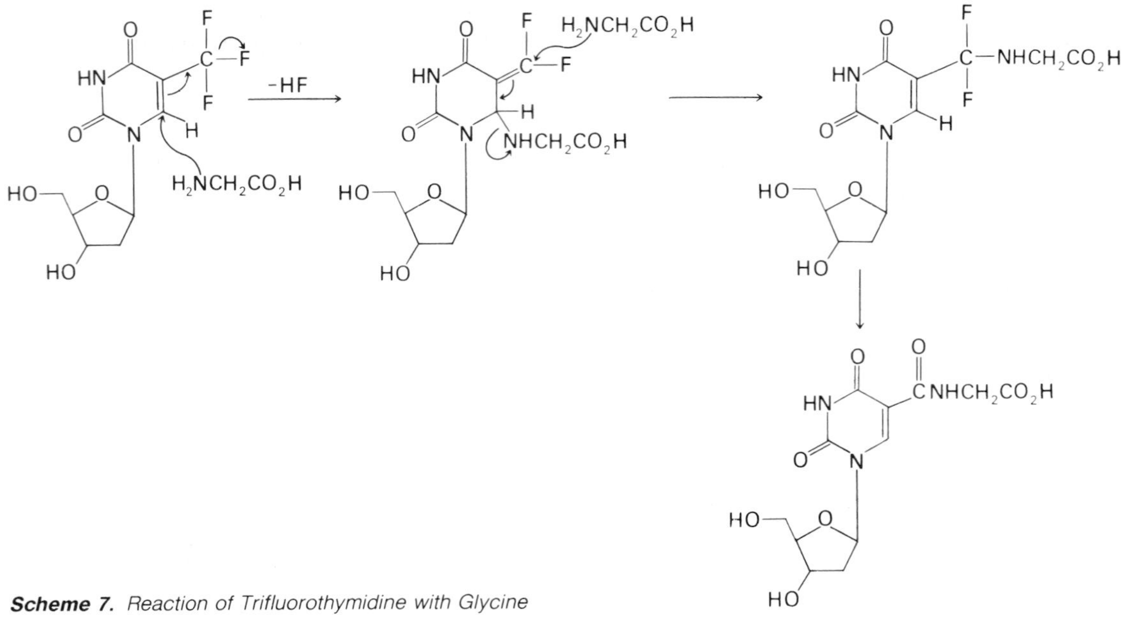

Scheme 7. *Reaction of Trifluorothymidine with Glycine*

believed to be the main mode of action of cytosine arabinoside triphosphate; however, it was shown recently that various deoxyribonucleosides were just as effective as cytosine arabinoside in reducing cellular levels of 2′-deoxycytidylic acid.[96] Other modes of action include the inhibition of DNA-dependent DNA polymerase[97] and miscoding following incorporation into DNA and RNA.[98] Cytosine arabinoside is readily transported into cells and phosphorylated by deoxycytidine kinase. It acts predominantly in the S phase of the cell cycle. Tumor-cell resistance is based on low levels of deoxycytidine kinase and the elaboration of deaminases that convert cytosine arabinoside into uridine arabinoside.[99] Cytidine deaminase has been partially purified and found to be inhibited by tetrahydrouridine.[100] A combination of cytosine arabinoside and tetrahydrouridine is presently in clinical trial.

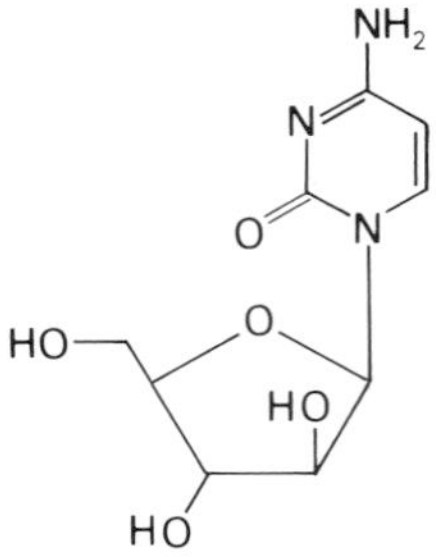

Cytosine arabinoside 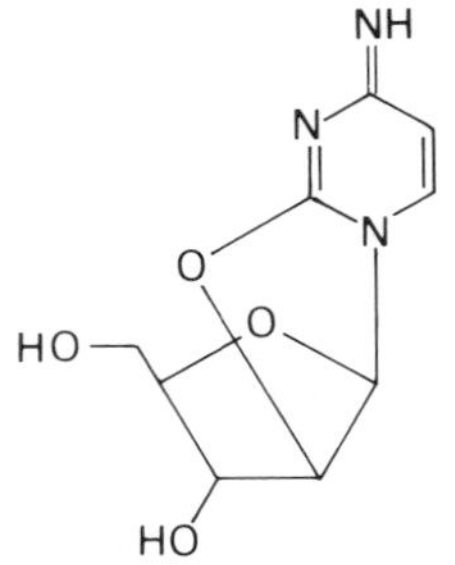   Cyclocytidine

A new analog of cytosine arabinoside is cyclocytidine. This analog apparently is a prodrug that is slowly converted into cytosine arabinoside. It is reported to be resistant to deamination and to have a

therapeutic index superior to that of the parent compound.[101]

A number of pyrimidine nucleoside analogs have one more or one less nitrogen in the heterocyclic ring. They are known as azapyrimidine or deazapyrimidine nucleosides. One of these antimetabolites is 5-azacytidine. It was synthesized in 1964 by Šorm in Czechoslovakia[102] and later was isolated as an antibiotic by Hanka.[103] The mode of action of this compound is complex, involving anabolism to phosphate derivatives and deamination to 5-azauridine. In certain tumor systems it is incorporated into nucleic acids that possibly cause misreading.[104] One of its main effects is the inhibition of orotidylate decarboxylase (Scheme 8), which prevents the new synthesis of pyrimidine nucleotides.[105] Tumor resistance is based on decreased phosphorylation of the nucleoside, decreased incorporation into nucleic acids, and increase in RNA and DNA polymerase activity.[106]

Pyrimidine nucleoside antagonists active against experimental tumors and presently in clinical study include dihydro-5-azacytidine and 3-deazauridine.[83]

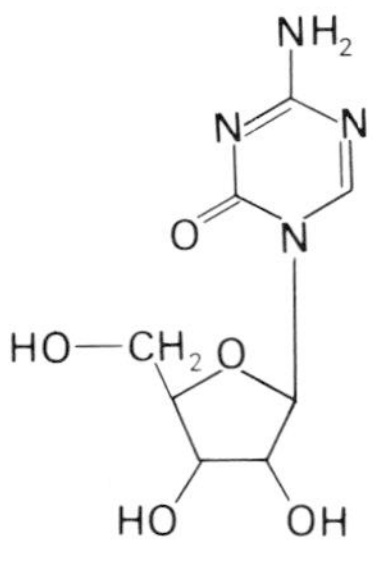

5-Azacytidine 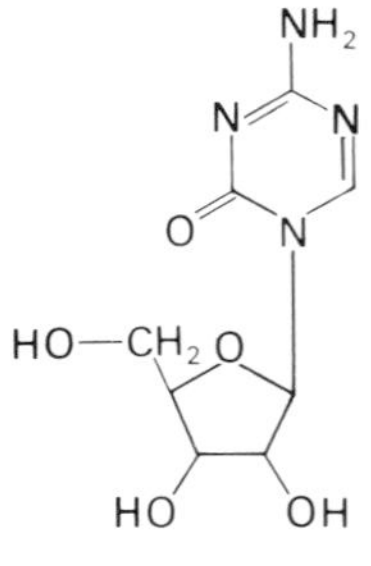   Dihydro-5-azacytidine 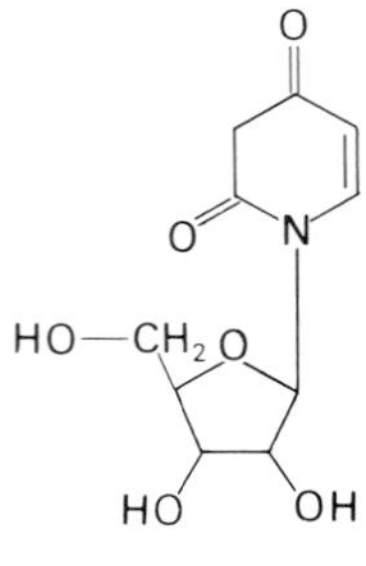   3-Deazauridine

The resistance to purine and pyrimidine antimetabolites such as adenosine arabinoside and cytosine arabinoside by neoplastic cells that produce deaminases has stimulated a search for compounds that might inhibit these deaminases. In principle, a potent deaminase inhibitor would produce a synergistic effect on the antitumor activity of the antimetabolite, even though it might not be active itself. Two types of deaminase inhibitors have emerged recently. One type is the purine analog in which the pyrimidine ring has been expanded to a seven-membered ring. The first example of this type was coformycin, an unusual nucleoside produced in the same cultures as the antibiotic formycin.[107] It strongly synergized the action of formycin against organisms that produce deaminases. In clinical trials it showed a synergistic effect on the activities of adenine arabinoside and cytosine arabinoside. At the present time, 2′-deoxycoformycin is undergoing clinical evaluation.[83] A second type of adenosine deaminase inhibitor has the adenine portion unchanged, but is modified in the ribose moiety. Such modifications have been designed to probe the active site of the enzyme and take advantage of strong binding to adjacent lipophilic regions.[108] The compound EHNA is an example of a rationally designed inhibitor.

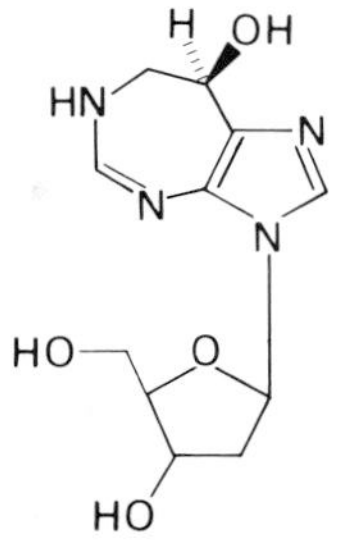
H
OH
HN
N
N
N
HO
O
HO

2′-Deoxycoformycin

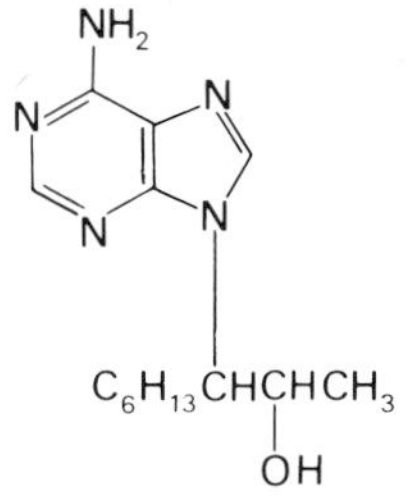
NH2
N
N
N
N
C6H13CHCHCH3
OH

EHNA

Following the discovery of folic acid, a number of analogs based on its structure were synthesized and tested as antimetabolites. The N^{10}-methyl derivative of folic acid was found to be an antagonist, but it had no antitumor activity. Antitumor activity finally was found for the 4-amino-4-deoxy derivative, aminopterin and its N^{10}-methyl homolog, methotrexate (amethopterin).[109]

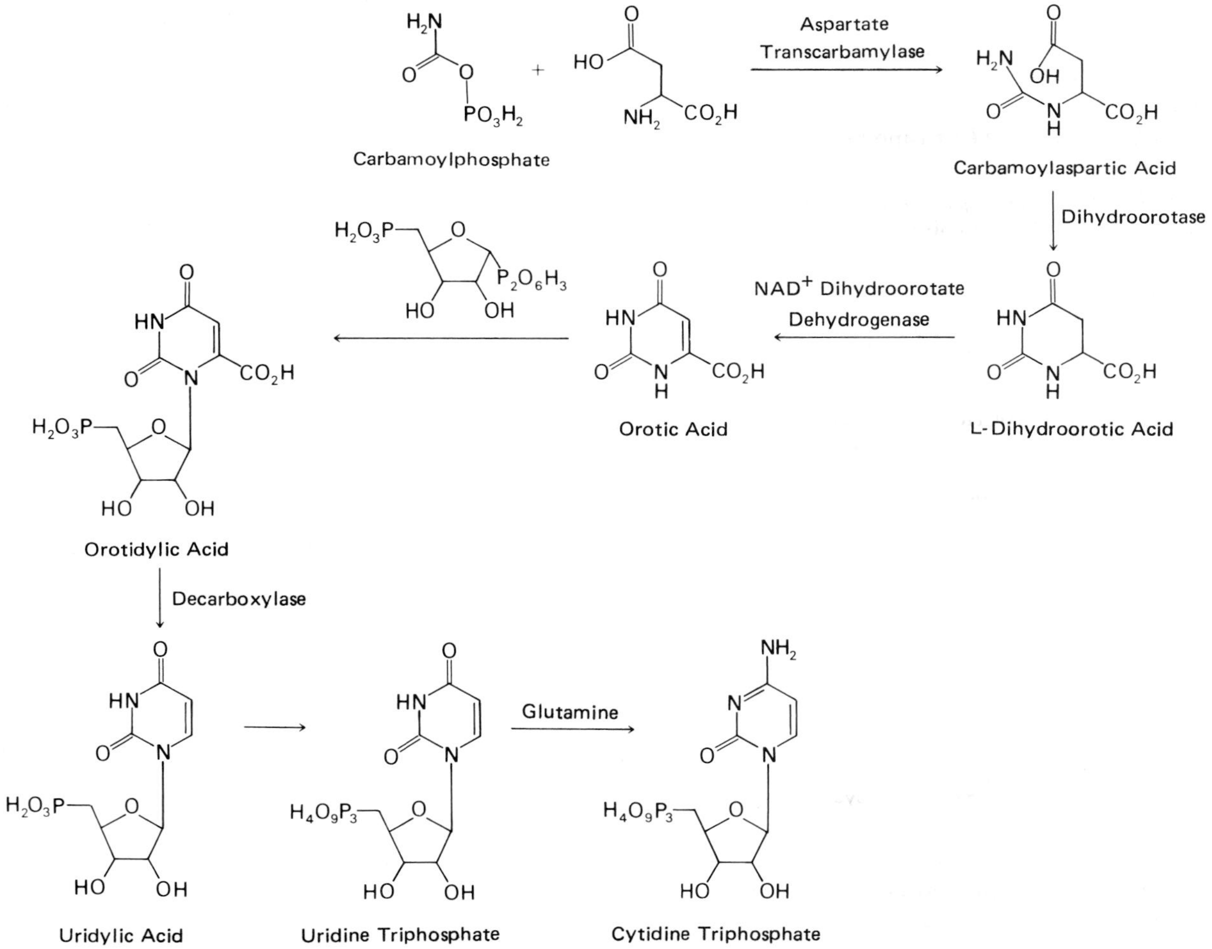
Carbamoylphosphate
Aspartate
Transcarbamylase
Carbamoylaspartic Acid
Dihydroorotase
L-Dihydroorotic Acid
NAD+ Dihydroorotate
Dehydrogenase
Orotic Acid
Orotidylic Acid
Decarboxylase
Glutamine
Uridylic Acid
Uridine Triphosphate
Cytidine Triphosphate

Scheme 8. *De Novo Synthesis of Pyrimidine Nucleotides (Simplified)*

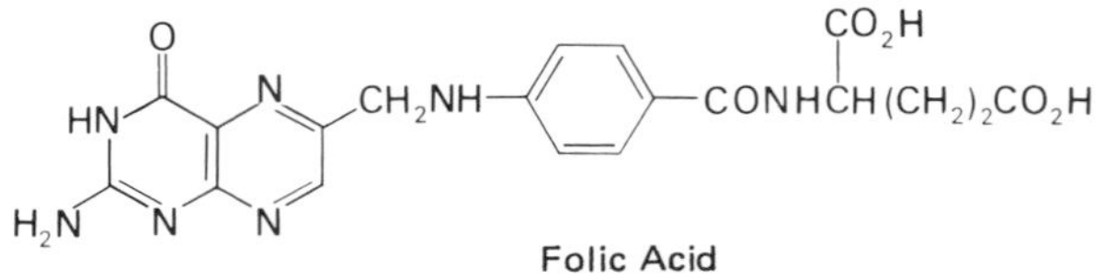

Folic Acid

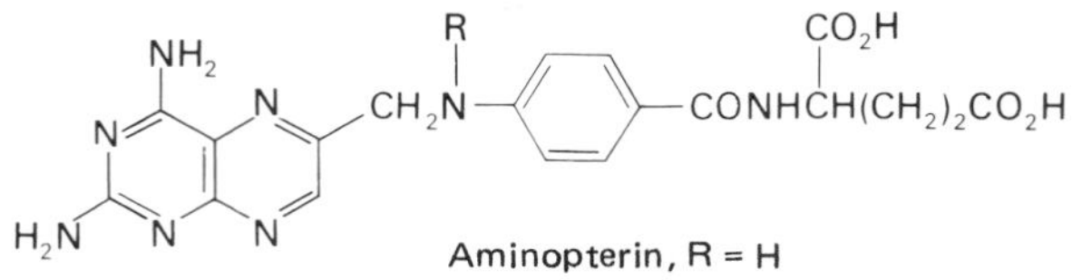

Aminopterin, R = H
Methotrexate, R = CH$_3$

Methotrexate and related compounds inhibit the enzyme dihydrofolic reductase. They bind so tightly to it that their inhibition has been termed "pseudoirreversible". The basis of this binding strength is in the diaminopyrimidine ring, which is protonated at physiological pH. At pH 6, methotrexate binds stoichiometrically with dihydrofolate reductase ($K_I \approx 10^{-10}$M), but at higher pH the binding is weaker and competitive with the substrate.[110]

Folic acid antagonists kill cells by inhibiting DNA synthesis in the S phase of the cell cycle. Thus, they are most effective in the log growth phase.[111] Their effect on DNA synthesis results partially from the in-

hibition of dihydrofolic reductase, which depletes the pool of tetrahydrofolic acid. Folic acid is reduced stepwise to dihydrofolic acid and tetrahydrofolic acid, with dihydrofolic reductase thought to catalyze both steps.[112] As shown in Scheme 9, tetrahydrofolic acid accepts the β-carbon atom of serine, in a reaction requiring pyridoxal phosphate, to give N^5,N^{10}-methylenetetrahydrofolic acid. The last compound transfers a methyl group to 2'-deoxyuridylate to give thymidylate in a reaction catalyzed by thymidylate synthetase. Dihydrofolic acid is generated in this reaction and it must be reduced back to tetrahydrofolic acid before another molecule of thymidylate can be synthesized. It is partly by their effect in limiting thymidylate synthesis that folic acid analogs prevent DNA synthesis and kill cells. This effect has been termed the "thymineless death."[113]

The inhibition of dihydrofolic reductase produces other limitations on nucleic acid biosynthesis. Thus, N^5,N^{10}-methylenetetrahydrofolic acid is oxidized to the corresponding methenyl derivative, which gives N^{10}-formyltetrahydrofolic acid on hydrolysis (Scheme 9). The latter compound is a formyl donor to 5-aminoimidazole-4-carboxamide ribonucleotide in the biosynthesis of purines.[114] N^5-Formyltetrahydrofolic acid,

Dihydrofolic Acid

Tetrahydrofolic Acid

5,10-Methenyltetrahydrofolic Acid

5,10-Methylenetetrahydrofolic Acid

10-Formyltetrahydrofolic Acid

5-Formyltetrahydrofolic Acid

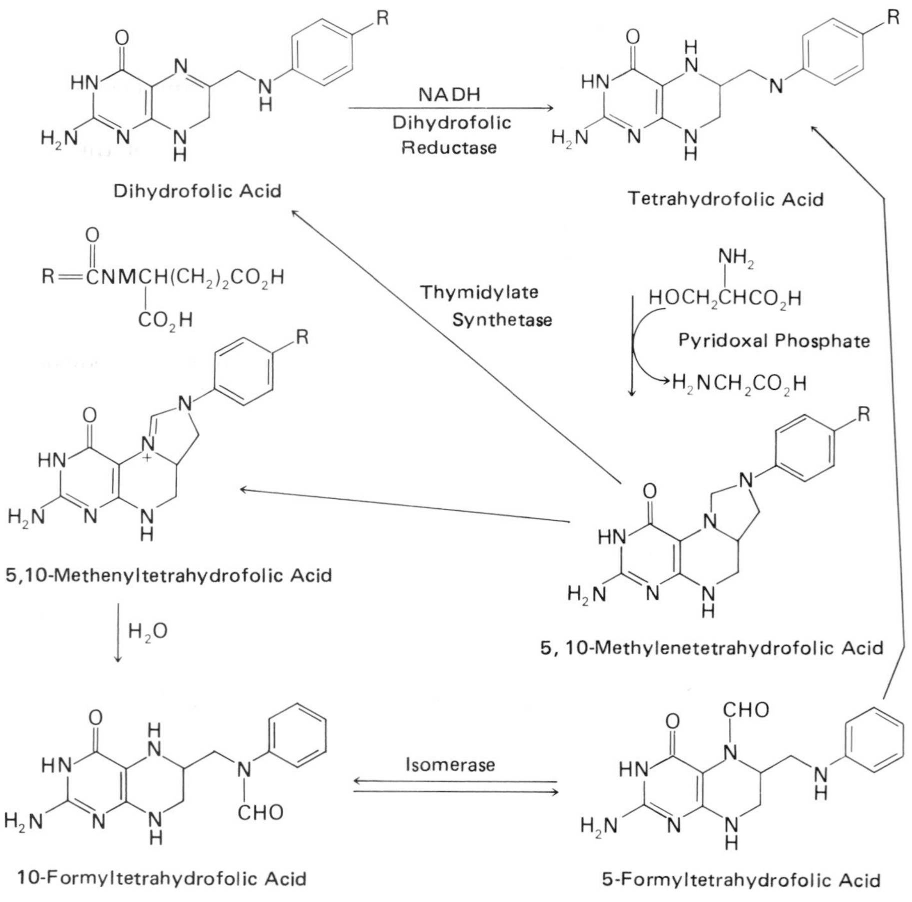

Scheme 9. *Interconversions of Folic Acid Derivatives*

also known as leucovorin and citrovorum factor, is interconvertible with the N^{10}-formyl analog by way of an isomerase-catalyzed reaction. It carries the formimino group for the biosynthesis of formiminoglycine, a precursor of purines (Scheme 5). Leucovorin is utilized in "rescue therapy" with methotrexate. It prevents the lethal effects of methotrexate on normal cells by overcoming the blockade of tetrahydrofolic acid production. In addition, it inhibits the active transport of methotrexate into cells and stimulates its efflux.[115]

Recently it has been shown that giving thymidine with methotrexate to mice bearing L1210 leukemia increased their survival time. This finding contradicts the idea that thymine deficiency is the most lethal effect of methotrexate on tumors. It suggests that the blockade of purine biosynthesis might have greater effects on tumor cells than normal cells.[116] Consequently, the administration of thymidine might protect the normal cells relative to the tumor cells. This reasoning has been applied to the clinical situation with the result that "thymidine rescue" has shown promise in humans receiving methotrexate.[117]

Because the binding affinities of methotrexate and close analogs for the dihydrofolic reductases of normal and neoplastic cells are nearly identical, selectivity is based on other factors, such as growth rate, enzyme levels, coenzyme pools, active transport, and salvage pathways for nucleotides. Methotrexate is actively transported into cells, and one mechanism of resistance to it involves impaired transport.[118] Another mode of resistance is a substantial increase in the number of gene copies for dihydrofolic reductase in certain tumor cells.[119]

Numerous analogs closely related to methotrexate have been prepared and tested against neoplasms. Most structural variations, such as alkylation of the amino groups, partial reduction, and removal or relocation of heterocyclic nitrogens lead to decreased activity. 3′,5′-Dichloroamethopterin is one of the more active inhibitors of dihydrofolic reductase. The analog with homologation in the side chain, homofolic acid, is a potent inhibitor of thymidylate synthetase. It is presently in clinical studies.[83]

Although the active sites of dihydrofolic reductases from normal and neoplastic cells are identical, Baker proposed that regions adjacent to the active sites of these enzymes might be different. He designed inhibitors to take advantage of these differences, thus affording species specificity. One of these inhibitors, known as "Baker's antifol," shows activity against experimental tumors that are resistant to methotrexate.[120] It is being developed by the NCI.

Glutamine and glutamate are the donors of the 3- and 9-nitrogen atoms of purines and the 2-amino

Homofolic Acid

Baker's Antifol

group of guanine.[121] They also contribute the 3-nitrogen atom and the amino group of cytosine[122] (Schemes 5 and 8). Thus, they are involved at five different sites of nucleic-acid biosynthesis. Although glutamine is not an essential nutrient for normal cells, many tumors are dependent upon exogenous sources of it. This provides a rationale for the selective action of agents that interfere with the uptake, biosynthesis or functions of glutamine.

In 1954, azaserine was isolated from a streptomyces species.[123] It was found to antagonize many of the metabolic processes involving glutamine, with the most important effect being the conversion of formylglycine ribonucleotide into formylglycinamidine ribonucleotide (Scheme 5).[124] A related compound, 6-diazo-5-oxo-L-norleucine (DON) was isolated in 1956 and found to possess similar antagonism.[125] Inhibition by DON and azaserine of the enzyme that helps to produce formylglycinamidine ribonucleotide is reversible at first, but upon incubation in the absence of glutamine it becomes irreversible.[126] A study involving incubation with azaserine-^{14}C followed by digestion with proteolytic enzymes and acid hydrolysis produced S-carboxymethylcysteine-^{14}C, which showed that azaserine had reacted covalently with a sulfhydryl group of cysteine on the enzyme.[127]

Azaserine

314 ANTINEOPLASTIC AGENTS

DON is a more potent inhibitor than azaserine of this enzyme and of the enzyme that converts uridine nucleosides into cytidine nucleosides.[128] Although both compounds show good antitumor activity in animal models, they have been generally disappointing in clinical trials. DON is classified as a compound in development by the National Cancer Institute.[83] Three other antibiotics containing DON in conjugated form show antitumor activity in animals. They appear to exert their effects only after cleavage to DON.[129]

PRODUCTS

Mercaptopurine U.S.P., Purinethol®, 6-mercaptopurine, MP, Lukerin, Mercaleukin, NSC-755, 6-purinethiol. This compound is prepared by treating hypoxanthine with phosphorus pentasulfide[130] and is obtained as yellow crystals of the monohydrate. Solubility in water is poor. It dissolves in dilute alkali but undergoes slow decomposition. Scored 50 mg. tablets are supplied.

Mercaptopurine is used primarily for treating acute leukemia. A higher proportion of children than adults respond.[131] The chief toxic effect is leukopenia. Thrombocytopenia and bleeding occur in high doses. Because the leukopenia is delayed, it is important to discontinue the drug temporarily at the first sign of an abnormally large drop in the white cell count.

The tolerated dose varies with the individual patient. Allopurinol potentiates the effect of mercaptopurine by inhibiting its metabolism. However, it also increases its toxicity. If allopurinol is given for potentiation or reduction of hyperuricemia resulting from the killing of leukemia cells, the doses of mercaptopurine must be decreased.[132]

Thioguanine U.S.P., Thioguanine Tabloid®, 6-thioguanine, TG, NSC-752, 2-aminopurine-6-thiol. The preparation of this compound is by treating gua-

nine with phosphorus pentasulfide in pyridine.[133] Scored 40 mg. tablets are supplied.

Thioguanine is used in treating acute leukemia, especially in combination with cytarabine.[134] Cross-resistance exists between thioguanine and mercaptopurine. The chief toxic effect is delayed bone-marrow depression, resulting in leukopenia and eventually thrombocytopenia and bleeding.

The usual initial dose is 2 mg. per kg. daily by the oral route. If there is no clinical improvement or leukopenia after 4 weeks the dosage is increased to 3 mg. per kg. daily. In contrast to mercaptopurine, thioguanine may be continued in the usual dose when allopurinol is used to inhibit uric acid formation.

Fluorouracil U.S.P., Fluorouracil Ampuls®, Fluoroplex®, Efudex®, 5-FU, 5-fluoro-2,4(1H,3H)-pyrimidinedione, 2,4-dioxo-5-fluoropyrimidine. This compound is prepared by condensing S-ethylisothiouronium bromide with the potassium salt (enolate) of ethyl 2-fluoro-2-formylacetate.[135] Recently the preparation of fluorouracil by direct fluorination of uracil has been demonstrated.[136] Fluorouracil is obtained as white crystals. It is supplied in 10 ml. ampuls containing 500 mg. of fluorouracil in a colorless to faint-yellow water solution at pH 9. These ampuls should be stored at room temperature and protected from light. Topical formulations of fluorouracil are Efudex® Solution, which contains 2 per cent or 5 per cent fluorouracil compounded with propylene glycol, tris(hydroxymethyl)aminomethane, hydroxypropyl cellulose, methyl and propyl parabens and disodium edetate, and Efudex® Cream, which contains 5 per cent fluorouracil in a vanishing-cream base consisting of white petrolatum, stearyl alcohol, propylene glycol, polysorbate 60 and methyl and propyl parabens.[137]

Fluorouracil is effective in the palliative management of carcinoma of the breast, colon, pancreas, rectum, and stomach in patients who cannot be cured by surgery or other means.[138] The topical formulations are used with favorable results for the treatment of premalignant keratoses of the skin and superficial basal-cell carcinomas.[139] Parenteral administration almost invariably produces toxic effects. Leukopenia usually follows every course of therapy, with the lowest white blood cell counts occurring between days 9 and 14 after the first course. Gastrointestinal hemorrhage may occur and may even be fatal. Stomatitis, esophagopharyngitis, diarrhea, nausea, and vomiting are commonly seen; alopecia and dermatitis also occur. Therapy must be discontinued if leukopenia or gastrointestinal toxicity becomes too severe. Topical administration is contraindicated in patients who develop hypersensitivity. Prolonged exposure to ultraviolet radiation may increase the intensity of topical inflammatory reactions.

Floxuridine U.S.P., FUDR, fluorodeoxyuridine, NSC-27640, 2'-deoxy-5-fluorouridine, 1-(2-deoxy-β-D-ribofuranosyl)-5-fluorouracil. This compound is prepared by condensing monomercuri-5-fluorouracil with 3,5-di-0-p-toluyl-2-deoxyribosyl-1-chloride followed by alkaline hydrolysis.[140] It is supplied in 5 ml. vials containing 500 mg. of floxuridine as sterile powder. Reconstitution is by the addition of 5 ml. of sterile water. The resulting solutions should be stored under refrigeration for not more than 2 weeks.

Floxuridine is used for palliation of gastrointestinal adenocarcinoma metastatic to the liver in patients who are considered incurable by surgery or other means.[141] It is administered by continuous regional intra-arterial infusion. Because floxuridine is rapidly catabolized to fluorouracil, it gives the same toxic reactions as fluorouracil.

Cytarabine U.S.P., Cytosar-U®, Ara-c, cytosine arabinoside, NSC-63878, 1-β-D-aribinofuranosylcytosine. It is synthesized from uracil arabinoside in a route involving acetylation, treatment with phosphorus pentasulfide, and heating with ammonia.[142] It is supplied as the freeze-dried solid in vials containing 100 mg. or 500 mg. The 100 mg. sample is reconstituted with 5 ml. of sterile water containing 0.9 per cent benzyl alcohol to give 20 mg. cytarabine per ml., whereas the 500 mg. sample is reconstituted with 10 ml. of sterile water containing 0.9 per cent benzyl alcohol to give 50 mg. cytarabine per ml. These solutions may be stored at room temperature for 48 hours.

Cytarabine is indicated primarily for inducing the remission of acute granulocytic leukemia of adults. It also is used for other acute leukemias of adults and children.[143] Remissions have been brief unless followed by maintenance therapy or given in combination with other antineoplastic agents.[144] Side-effects include severe leukopenia, thrombocytopenia, and anemia. Gastrointestinal disturbances also are relatively frequent.

Methotrexate U.S.P., amethopterin, methylaminopterin, NSC-740, 4-amino-N[10]-methylpteroylglutamic acid, L-(+)-N-[p-[[(2,4-diamino-6-pteridinyl)methyl]methylamino]benzoyl]glutamic acid. This compound is prepared by combining 2,4,5,6-tetrahydropyrimidine, 2,3-dibromopropionaldehyde, disodium p-(methylamino)benzoylglutamate, iodine, and potassium iodide, followed by heating with lime water.[112] It is isolated as the monohydrate, a yellow solid. Recent studies indicate that the commercial preparation contains a number of impurities including 4-amino-N[10]-methylpteroic acid and N[10]-methylfolic acid.[145] Methotrexate is soluble in alkaline solutions but decomposes in them. It is supplied as 25 mg. tablets and in vials containing either 5 mg. or 50 mg. of methotrexate sodium in 2 ml. of solution. The 5 mg. sample contains 0.90 per cent of benzyl alcohol as preservative, 0.63 per cent of sodium chloride, and sodium hydroxide to give pH 8.5. The 50 mg. sample contains 0.90 per cent of benzyl alcohol, 0.26 per cent of sodium chloride, and sodium hydroxide to give pH 8.5.

Methotrexate was the first drug to produce substantial (although temporary) remissions in leukemia.[146] It is still used for this purpose against acute lymphocytic leukemia and acute lymphoblastic leukemia. Because it has some ability to enter the CNS, it is used in the treatment and prophylaxis of meningeal leukemia. The discovery that methotrexate afforded a high percentage of apparently permanent remissions in choriocarcinoma in women justified the use of the term "cure" in cancer chemotherapy.[147] Methotrexate is used in combination chemotherapy for palliative management of breast cancer, epidermoid cancers of the head and neck, and lung cancer. It also is used against severe, disabling psoriasis. The most common toxic reactions are ulcerative stomatitis, leukopenia, and abdominal distress. A high dose of methotrexate combined with leucovorin "rescue" produces renal failure in some patients. This condition is thought to result from crystallization of the drug or its metabolites in acidic urine and it is countered by hydration and alkalinization.[148]

Azathioprine U.S.P., Imuran®, 6-[(1-methyl-4-nitroimidazole-5-yl)thio]purine. This compound is prepared from 6-mercaptopurine and 5-chloro-1-methyl-4-nitroimidazole. It is supplied as 50 mg. scored tablets. The injectable sodium salt is available in 20 ml. vials containing 100 mg. of azathioprine.

Azathioprine is well absorbed when taken orally. It is extensively converted to 6-mercaptopurine. The main indication for azathioprine is as an adjunct to prevent the rejection of renal homotransplants. It is contraindicated in patients who show hypersensitivity to it. The chief toxic effects are hematological, expressed as leukopenia, anemia, and thrombocytopenia. Complete blood counts should be performed at least weekly and the drug should be discontinued if there is a rapid fall or persistent decrease in leukocytes. Patients with impaired renal function might have slower elimination of the drug, which requires appropriate reduction of the dose.

ANTIBIOTICS

Seven different antibiotics now are established clinical anticancer agents, and a number of other antibiotics are undergoing clinical development. Most of these agents have been approved within the last few years. Thus, the recognition of antibiotics as an important class of antineoplastic drugs is very recent. However,

TABLE 8–2

ANTIMETABOLITES

Name Proprietary Name	Preparation	Usual Adult Dose*	Usual Dose Range*	Usual Pediatric Dose*
Mercaptopurine U.S.P. *Purinethol*	Mercaptopurine Tablets U.S.P.	2.5 mg. per kg. of body weight daily in single or divided doses		Children 5 years of age and older—2.5 mg. per kg. of body weight or 50 mg. per square meter of body surface daily in single or divided doses
Azathioprine U.S.P. *Imuran*	Azathioprine Tablets U.S.P.	Initial, 3 to 5 mg. per kg. of body weight once daily; maintenance, 1 to 4 mg. per kg. once daily		
	Azathioprine Sodium for Injection U.S.P.	I.V., the equivalent of azathioprine, 3 to 5 mg. per kg. of body weight daily		
Thioguanine U.S.P.	Thioguanine Tablets U.S.P.	Initial, 2 mg. per kg. of body weight once daily, increase to 3 mg. per kg. if no toxicity or improvement after 4 weeks		See usual adult dose
Fluorouracil U.S.P. *Efudex*	Fluorouracil Cream U.S.P.			
	Fluorouracil Injection U.S.P.	Initial, I.V., 12 mg. per kg. of body weight once daily for 4 days; if no toxicity occurs, 7 to 10 mg. per kg. is given every 3 or 4 days for a total course of 4 weeks	Up to 800 mg. daily	
	Fluorouracil Topical Solution U.S.P.	Superficial basal-cell carcinomas—topical, to the skin, as a 5 percent cream or solution twice daily in a sufficient amount to cover the lesions, for at least 2 to 6 weeks		
Floxuridine U.S.P.	Sterile Floxuridine U.S.P.	Intra-arterial, 100 to 600 μg. per kg. of body weight continuously over 24 hours, continued until toxicity or a response occurs		
Cytarabine U.S.P. *Cytosar-U*	Sterile Cytarabine U.S.P.	I.V. infusion, 0.5 to 1 mg. per kg. of body weight daily over one to 24 hours, for 10 days, increasing to 2 mg. per kg. of body weight daily thereafter, if neither response nor toxicity occurs		See Usual Adult Dose
Methotrexate U.S.P.	Methotrexate Tablets U.S.P.	Antineoplastic—oral, 15 to 50 mg. per square meter of body surface once or twice a week, depending on condition and other agents being used concurrently		Antineoplastic—oral, 20 to 30 mg. per square meter of body surface once a week
	Methotrexate Sodium Injection U.S.P.	Antineoplastic—I.M. or I.V., 15 to 50 mg. per square meter of body surface once or twice a week		Antineoplastic—I.M., the equivalent of methotraxate—20 to 30 mg. per square meter of body surface once a week

*See *U.S.P. D.I.* for complete dosage information.

some of the compounds in this class have been known for a long time. For example, dactinomycin (actinomycin D) was first isolated in 1940 by Waksman and Woodruf,[150] although its activity against neoplasms was not described until 1958. Furthermore, mithramycin, originally discovered as aureolic acid in 1953, had to be rediscovered twice before its antitumor activity was established in 1962.[151] These compounds were originally rejected as antibacterial agents because of their cytotoxicity. Only later was it found that this toxicity could be turned to an advantage in the chemotherapy of cancer. The discovery of antitumor activity is much simpler today, and some laboratories routinely screen extracts of microorganism cultures for cytotoxicity and lysogenic phage induction in bacteria (a predictor of potential antitumor activity). Such assays can be performed on small quantities of the extracts.[152]

The production of antitumor agents from microbial fermentations has some special advantages and disadvantages over chemical synthesis. In some cases, the biosynthesis can be controlled to afford novel analogs. This has been true for actinomycins[153] and bleomycins.[154] Strain selection and fermentation conditions can optimize the formation of a particular component of an antibiotic mixture. Thus, *Streptomyces parvullus* produces dactinomycin almost exclusively, in contrast to other species that form complex mixtures of actinomycins.[152] The fermentation in *Streptomyces caespitosus* has been developed similarly, to produce almost all mitomycin C. In some cases, such as with doxorubicin, improvement of the antibiotic yield has been difficult. This results in an expensive product and intensive research on chemical synthesis.

The actinomycins comprise a large number of closely related structures. All of them contain the same chromophore, a substituted 3-phenoxazone-1,9-dicarboxylic acid known as *actinocin*. Each of the carboxyl groups is bonded to a pentapeptide lactone by way of the amino group of an L-threonine unit of this pentapeptide. The hydroxyl group of the L-threonine forms part of the lactone along with L-methylvaline, the fifth amino acid from the chromophore. D-Valine or D-alloisoleucine is the second amino acid and the fourth amino acid usually is sarcosine. The third amino acid is more variable, consisting of L-proline, L-hydroxyproline, L-oxoproline, or others produced by controlled biosynthesis. Actinomycins that have two identical pentapeptide lactones are called isoactinomycins, whereas those with different pentapeptide lactones are called anisoactinomycins. The individual pentapeptide lactones are designated α and β, depending on their attachment to the 9- or 1-carboxylic acids, respectively.[155] Dactinomycin (actinomycin D, actinomycin C$_1$) is an isoactinomycin with an amino acid sequence of L-threonine, D-valine, L-proline, sarcosine, and L-N-methylvaline. Actinomycin C$_3$, which is used in Germany, differs from actinomycin D by a D-alloisoleucine unit instead of D-valine in both the α and β chains.[155]

Dactinomycin

The mode of action of actinomycins has been studied extensively and it now is generally accepted that they intercalate into double helical DNA. In the intercalation process, the helix unwinds partially to permit the flat phenoxazone chromophore to fit in between successive base pairs. Adjacent G-C pairs are especially suitable because the 2-amino groups of the guanines can hydrogen bond with the carbonyl groups of

threonines in the actinomycin. This bonding reinforces the π-bonding between the heterocyclic chromophores. Additional stability is conferred by the interaction between the pentapeptide lactone chains and DNA. These chains lie in the minor groove of the double helix, running in opposite directions to each other, and they make numerous van der Waal's interactions with the DNA.[156]

Intercalation into DNA changes its physical properties in characteristic ways. Thus the length, viscosity, and melting temperature increase, whereas the sedimentation coefficient decreases.[157,158] Changes in substituents on the actinomycins influence their binding to DNA, usually by making it less effective. Opening of a lactone ring or changing the stereochemistry of an amino acid abolishes activity, and replacement of the 4- and 6-methyl groups by other substituents reduces it. Replacement of the 2-amino group also reduces activity.[152]

The main biochemical consequence of the intercalation of actinomycins into DNA is the inhibition of DNA and RNA synthesis. This inhibition eventually results in depletion of RNA and proteins and leads to cell death.[159]

Anthracyclines are another large and complex family of antibiotics. Many members of this family were isolated and identified before a useful antitumor agent, daunorubicin, was found. This significant discovery was made independently in France and Italy in 1963.[160,161] Daunorubicin proved to be active against acute leukemias and it was developed as a clinical agent. However, it was pushed into the background by the discovery of doxorubicin (Adriamycin®) in 1969.[162] Doxorubicin is active against a broad spectrum of tumors including both solid and hematological types. It is presently the most widely used antineoplastic agent.[163]

Many anthracyclines, including all of those with antitumor activity, occur as glycosides of the anthracyclinones. The glycosidic linkage usually involves the 7-hydroxyl group of the anthracyclinone and the β-anomer of a sugar with L-configuration. Anthracyclinone refers to an aglycone containing the anthraquinone chromophore within a linear hydrocarbon skeleton related to that of the tetracyclines.[164] The anthracyclinones differ from each other in the number and location of phenolic hydroxyl groups, the degree of oxidation of the two-carbon side chain at position 9, and the presence of a carboxylic acid ester at position 10. Thus, daunorubicin is a glycoside formed between daunomycinone and L-daunosamine, whereas doxorubicin is its 14-hydroxy analog.[165] In contrast, aclacinomycin A, a compound currently under development, has aklavinone in combination with a trisaccharide chain.[166]

Daunorubicin, R = H
Doxorubicin, R = OH

Aclacinomycin A

Daunomycinol, R = H
Adriamycinol, R = OH

Daunorubicin and doxorubicin exhibit biological effects similar to those of actinomycin, and they are thought to intercalate into double helical DNA.[167] However, the significance of this intercalation to their antineoplastic activity is uncertain. Thus, the N-trifluoroacetyl derivative of doxorubicin 14-valerate does not bind to DNA or even penetrate the nuclei of cells, yet it has antitumor activity.[168] It has been suggested that it either reacts by a different mechanism or is metabolized to an active form. Reduction of doxorubicin followed by intercalation causes DNA-strand scission. This scission is thought to result from the attack of hydroxyl radicals generated from redox cycles involving doxorubicin.[169] In contrast to daunorubicin, aclacinomycin and related compounds do not induce lysogenic phage in bacteria. They are believed to interfere with RNA synthesis more than with DNA synthesis. Aclacinomycin also lacks the cardiotoxicity shown by daunorubicin and doxorubicin.[166]

In contrast to the actinomycins, anthracyclines are metabolized in the liver. Daunorubicin is readily converted into its 13-hydroxy analog, daunorubicinol, which is further cleaved to the aglycone.[170] The 14-hydroxyl group of doxorubicin makes it less susceptible to reduction of the 13-carbonyl group. However, the 13-hydroxy derivative, adriamycinol, is found among the metabolites, along with the 4-demethyl-4-sulfate. Both daunomycinol and adriamycinol are active against neoplastic cells, but their rates of uptake are low.[171]

Many derivatives and analogs of daunorubicin and doxorubicin have been prepared. The 4-hydroxy analog of daunorubicin, known as carminomycin, has been developed in the Soviet Union,[172] whereas daunorubicin benzoylhydrazone (rubidazone) received clinical trials in France.[173] Transformation in the sugar moiety

can maintain antitumor activity. Thus the 4'-deoxy and 4'-epi analogs of daunorubicin and doxorubicin are highly active against experimental tumors.[174] N-Trifluoroacetyldoxorubicin 14-valerate (AD-32) is in development at the National Cancer Institute.[175] More recently, the 4-deoxy analog of doxorubicin has shown potent oral activity in mice.[176]

4'-Deoxydoxorubicin, R = H
4'-Epidoxorubicin, R = OH

The aureolic acid group of antitumor antibiotics includes aureolic acid (mithramycin), the olivomycins, the chromomycins, variamycin, and related compounds. Mithramycin is the only member approved for clinical use in the United States. It is restricted to testicular carcinoma and hypercalcemia that is resistant to other drugs. However, chromomycin A_3 is used in Japan and olivomycin A is used in the Soviet Union.[152] Aureolic acid group compounds have complex structures consisting of an aglycone and two carbohydrate chains. The aglycones are tetrahydroanthracene derivatives with phenolic hydroxyl groups at positions 6, 8, and 9 and a pentanyl side chain that is highly oxygenated. The carbohydrate chains contain either two or three 2,6-dideoxy sugars of novel structures.[177]

Mithramycin and related compounds are weakly acidic owing to the phenolic groups (pKa = 5). They readily form sodium salts that show brilliant yellow fluorescence.[178] The chromophore is responsible for complex formation with divalent metals such as mag-

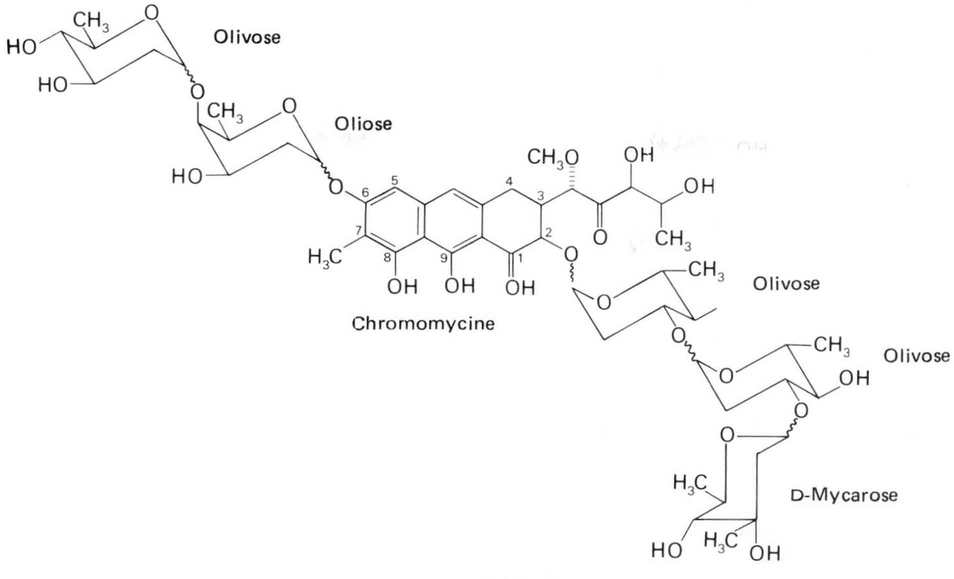

Olivose

Oliose

CH₃O OH OH

CH₃

Olivose

Olivose

H₃C

Chromomycine

D-Mycarose

Mithramycin

nesium and calcium. Such complex formation is required before aureolic acids can bind with DNA.[179] The nature of this DNA binding is uncertain at the present time. Intercalation has been suggested, but the evidence for this process is incomplete.[180] Whatever the exact nature of the binding, mithramycin and other aureolic acids inhibit DNA-dependent RNA polymerase, and this effect leads to cell death.[181]

The discovery of bleomycin in 1966 resulted from the establishment by H. Umezawa of a program for screening microbial-culture filtrates against experimental tumors.[182] Bleomycin is a mixture of closely related compounds that is partly resolved before formulation for clinical use.[183] The presently used commercial product, Blenoxane®, contains bleomycins A_2 and B_2. A variety of other antibiotics have structures similar to those of the bleomycins. They include the phleomycins (which differ from bleomycins in having one thiazole ring partly reduced), zorbamycin and the zorbonamycins, antibiotic YA-56, victomycin, the tallysomycins, and the platomycins.[152] New bleomycin analogs also have been prepared by controlled biosynthesis from bleomycinic acid.

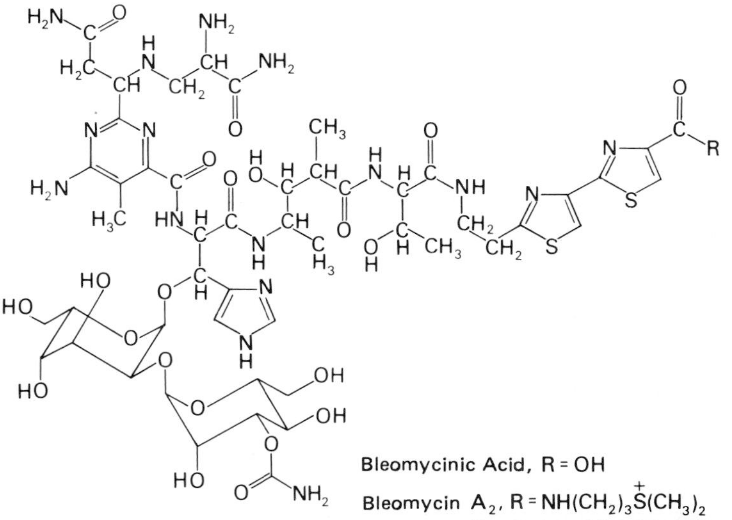

Bleomycins and their analogs occur naturally as blue copper chelates. Removal of the copper by chemical reduction or complexing agents affords the antibiotics as white solids.[184,185] Copper-free bleomycin is preferred for chemotherapy because of its decreased toxicity. The complexation of bleomycin with metal ions occurs readily and is a key factor in its mode of action. Inside the cell, bleomycin forms a chelate with Fe(II) that has square pyramidal geometry.[186] Nitrogen atoms from bleomycin occupy five of the positions in this structure. The sixth position may be occupied by the carbonyl group of the carbamate function, but this group is readily displaced by molecular oxygen. The resulting complex may give rise to hydroxyl radicals and superoxide radicals. Because the bithiazole portion of bleomycin can intercalate partially into DNA, these highly reactive radicals are generated close to the double helix, and they cause cleavage of the phosphodiester bonds. This degradation of DNA strands is thought to be the lethal event in cells.[187]

Bleomycin is inactivated by an intracellular enzyme named bleomycin hydrolase, an aminopeptidase that hydrolyzes the carboxamide group of the β-aminoalanine carboxamide residue to the corresponding carboxylate. This structural change increases the pKa of the α-amino group from 7.3 to 9.4, which results in

Bleomycinic Acid, R = OH

Bleomycin A_2, R = NH(CH₂)₃S⁺(CH₃)₂

Bleomycin B_2, R = NH(CH₂)₄NHCNH₂

Bleomycin A$_2$ Fe(II) Chelate

poorer binding to DNA.[185] Chelation with Fe(II) still occurs, but the production of hydroxyl radicals is drastically reduced.[187] Bleomycin-hydrolase levels in tumor cells help to determine their resistance to bleomycin. Thus, squamous-cell carcinoma is characterized by ready uptake of bleomycin and low levels of the hydrolase. It is especially sensitive to bleomycin.[183]

Bleomycins undergo two different inactivating reactions under mildly alkaline conditions. One is migration of the carbamoyl group to an adjacent hydroxyl group of the mannose residue. The resulting product is called an isobleomycin.[189] Copper-chelated bleomycins do not undergo this reaction. However, they are slowly transformed into epibleomycins, which are racemized at the carbon atom substituted at the 2-position of the pyrimidine ring.[190] Epibleomycins retain about 25 percent of the antitumor activity of the parent bleomycins.

Bleomycinic acid is obtained by chemical degradation of bleomycin A$_2$ or enzymatic degradation of bleomycin B$_2$. It can be readily transformed into semisynthetic bleomycins such as PEP-bleomycin, which possesses reduced pulmonary toxicity.[183]

The mitomycins were discovered in Japan in the late 1950's and one of them, mitomycin C, was rapidly developed as an anticancer drug.[191] However, the initial clinical experience with this compound in the United States was disappointing. It was not approved until 1974, following extensive studies and the establishment of satisfactory dosage schedules. Porfiromycin, the N-methyl homolog of mitomycin C, was discovered at the Upjohn Company.[192] It has received clinical study, but it is not yet an approved agent.

Structures of the mitomycins were elucidated at Lederle Laboratories. These compounds have an unusual combination of three different carcinostatic functions: quinone, carbamate, and aziridine.[193] They

are arranged in such a way that the molecule is relatively unreactive in its natural state. However, chemical or enzymatic reduction to the corresponding hydroquinone is followed by the loss of methanol (water from mitomycin B), and the resulting indolohydroquinone becomes a bifunctional alkylating agent capable of cross linking double helical DNA (Scheme 3).[194] Mitomycins bound to DNA may undergo successive redox cycles, each of which results in the generation of hydrogen peroxide. This potent oxidizing agent can cause single-strand cleavage of the DNA.[195]

Mitomycins are unstable in both acids and bases. Mild acid hydrolysis results in opening of the aziridine ring and loss of methanol or water to give mitosenes such as 2,7-diamino-1-hydroxymitosene.[196] Catalytic

Mitomycin A; X = CH$_3$O, Y = H
Mitomycin C; X = H$_2$N, Y = H
Porfiromycin; X = H$_2$N, Y = CH$_3$
7-Hydroxymitosane; X = OH, Y = H

Mitomycin B; X = CH$_3$O
Mitomycin D; X = H$_2$N

hydrogenation followed by reoxidation gives aziridinomitosenes, which retain a significant amount of antitumor activity in animals.[197] Mild alkaline hydrolysis of mitomycin C gives the corresponding 7-hydroxy mitosane, a compound receiving clinical study in Japan.[198]

The mitomycins have recently been totally synthesized by Kishi.[199] Many analogs have been prepared by chemical transformation or partial synthesis, and a number of them show potent activity in animals. Control of the quinone reduction potential is especially

2,7-Diamino-1-hydroxymitosene; $R^1 = OH$, $R^2 = NH_2$

1,2-Aziridino-7-aminomitosene; R^1, $R^2 = {>}NH$

stressed in analog studies, because reduction is the key step in bioactivation of these molecules.[200]

Streptozotocin was isolated from *Streptomyces achromogenes* in 1960.[201] It is the nitrosomethylurea derivative of 2-deoxyglucose.[202] The simplicity of its structure and the cost of preparing it by fermentation have led to the development of practical syntheses from 2-amino-2-deoxyglucose.[203] Streptozotocin is an alkylating agent similar in reactivity to other nitrosomethylureas, except that its glucose moiety causes it to be especially taken up in the pancreas. This effect is detrimental in that it produces diabetes, but it makes the molecule especially effective against malignant insulinomas.[204] The chloroethyl analog of streptozotocin, called chlorozotocin, shows good antitumor activity in animals and is not diabetogenic.[205]

Streptozotocin; $R = CH_3$

Chlorozotocin; $R = CH_2CH_2Cl$

PRODUCTS

Dactinomycin U.S.P., Cosmegen®, Actinomycin D, Actinomycin C₁, actinomycin IV, NSC-3053. This compound is obtained from the fermentation of selected strains of *Streptomyces parvullus*.[206] It is soluble in alcohols and alcohol–water mixtures; however, these solutions are very sensitive to light. Vials containing 0.5 mg. of lyophilized powder are supplied.

Dactinomycin is used against rhabdomyosarcoma and Wilms' tumor in children.[207] It can be lifesaving for women with choriocarcinoma resistant to methotrexate. In combination with vincristine and cyclophosphamide, it has received some use in solid tumors in children. Toxic reactions include anorexia, nausea, and vomiting. Bone-marrow depression resulting in pancytopenia may occur within a week after therapy.

Alopecia, erythema, and tissue injury at the site of injection may occur.

Daunorubicin Hydrochloride U.S.P., Cerubidin®, daunomycin, rubidomycin, NSC-82151. It is obtained from the fermentation of *Streptomyces peucetius*.[161] The hydrochloride salt is a red crystalline compound that is soluble in water and alcohols. The pink solution turns blue at alkaline pH. Daunorubicin hydrochloride is available as lyophilized powder in 20-mg. vials. In this form it is stable at room temperature, but after reconstitution with 5 to 10 ml. of sterile water, it should be used within 6 hours.

Daunorubicin is used in the treatment of acute lymphocytic and granulocytic leukemias.[208] Toxic effects include bone-marrow depression, stomatitis, alopecia, and gastrointestinal disturbances. At higher doses cardiac toxicity may develop. Severe and progressive congestive heart failure may follow initial tachycardia and arrhythmias.

The usual dose of daunorubicin is 30 to 60 mg. per square meter daily for 3 days. It is administered intravenously, taking care to prevent extravasation.

Doxorubicin Hydrochloride U.S.P., Adriamycin®, NSC-123127, 14-hydroxydaunomycin. This compound is obtained from cultures of *Streptomyces peucetius* variety *caesius*.[209] The orange-red needles are soluble in water and alcohols. Above pH 9 the orange solutions turn blue-violet. Doxorubicin hydrochloride is supplied as a freeze-dried powder in two different sizes: 10 mg. plus 50 mg. of Lactose U.S.P., and 50 mg. plus 250 mg. of Lactose U.S.P. These amounts are reconstituted with 5 ml. and 25 ml., respectively, of Sodium Chloride Injection U.S.P.

Adriamycin is one of the most effective antitumor agents. It has been used successfully to produce regressions in acute leukemias, Hodgkin's disease and other lymphomas, Wilms' tumor, neuroblastoma, soft-tissue and bone sarcomas, breast carcinoma, ovarian carcinoma, transitional-cell bladder carcinoma, thyroid carcinoma, and small-cell bronchogenic carcinoma.[210] Combination chemotherapy with a variety of other agents is being developed for specific tumors. The dose-limiting toxicities are myelosuppression and cardiotoxicity. There is a high incidence of bone-marrow depression, primarily of leukocytes, that usually reaches its nadir at 10 to 14 days. Red blood cells and platelets also may be depressed. Thus careful blood counts are essential. Acute left ventricular failure has occurred, particularly in patients receiving a total dose exceeding the currently recommended 550 mg. per square meter. Cardiomyopathy and congestive heart failure may be encountered several weeks after discontinuing Adriamycin. Toxicity is augmented by impaired liver function, because this is the site of metabolism. Thus, evaluation of liver function by conventional laboratory tests is recommended before individual dosing.

The recommended dosage schedule is 60 to 75 mg. per square meter intravenously at 21-day intervals. This dose is decreased if liver function or bone-marrow reserves are inadequate. Care must be taken to avoid extravasation.

Bleomycin Sulfate, Sterile, U.S.P., Blenoxane®, NSC-125066. This product is a mixture of cytotoxic glycopeptides isolated from a strain of *Streptomyces verticillus.*[211] The main component is bleomycin A_2 and bleomycin B_2 also is present. Bleomycin is a whitish powder that is readily soluble in water. It occurs naturally as a blue copper complex, but the copper is removed from the pharmaceutical form. It is supplied in ampuls containing 15 units of sterile bleomycin sulfate.

Bleomycin is used for the palliative treatment of squamous-cell carcinomas of the head and neck, esophagus, skin, and genitourinary tract including penis, cervix, and vulva.[212] It also is used against testicular carcinoma, especially in combination with cisplatin and vinblastine.[213] The principal toxicities of bleomycin are in skin and lungs. Other tissues contain an aminopeptidase that rapidly inactivates it. Bleomycin has very little bone-marrow toxicity, thus it may be used in combination with myelosuppressive agents. Pulmonary toxicity is induced in about 10 percent of treated patients, with pulmonary fibrosis and death occurring in about 1 percent. Skin or mucous-membrane toxicity occurs in about half of the patients. Anaphylactoid reactions are possible in lymphoma patients.

The recommended dosage is 0.25 to 0.50 units per kg. (10 to 20 units per square meter) given intravenously, intramuscularly, or subcutaneously once or twice weekly. For maintenance of Hodgkin's disease patients in remission, a dose of 1 unit daily or 5 units weekly is given. Blenoxane is stable for 24 hours at room temperature in sodium chloride or 5 percent dextrose solutions for injection.

Mitomycin U.S.P., Mutamycin®, mitomycin C, NSC-26980. This compound is obtained from cultures of *Streptomyces caespitosus* as blue-violet crystals.[214]

TABLE 8–3

OTHER ANTINEOPLASTIC AGENTS

Name Proprietary Name	Preparation	Usual Adult Dose*	Usual Dose Range*	Usual Pediatric Dose*
Dactinomycin U.S.P. *Cosmegen*	Dactinomycin for Injection U.S.P.	I.V., 10 to 15 μg. per kg. of body weight daily for a minimum of 5 days every 4 to 6 weeks		
Mithramycin U.S.P. *Mithracin*	Mithramycin for Injection U.S.P.	I.V. infusion, 25 to 30 μg. per kg. of body weight daily for 8 to 10 days	Up to 30 μg. per kg. of body weight daily for 10 days per course of therapy	
Hydroxyurea U.S.P. *Hydrea*	Hydroxyurea Capsules U.S.P.	For solid tumors, 80 mg. per kg. of body weight as a single dose every third day, or 20 to 30 mg. per kg. of body weight daily as a single dose		
Pipobroman U.S.P. *Vercyte*	Pipobroman Tablets U.S.P.	Initial, 1 to 2.5 mg. per kg. of body weight; maintenance, 100 to 200 μg. per kg. or 7 to 175 mg.		
Vinblastine Sulfate U.S.P. *Velban*	Sterile Vinblastine Sulfate U.S.P.	I.V., 3.7 mg. per square meter of body surface initially, followed at weekly intervals by doses increased gradually to 18.5 mg. per square meter of body surface		Initial 2.5 mg. per square meter of body surface, gradually increasing until a maximum of 7.5 mg. per square meter of body surface
Vincristine Sulfate U.S.P. *Oncovin*	Vincristine Sulfate for Injection U.S.P.	I.V., 1.4 mg. per square meter of body surface given weekly as a single dose	2 mg. per dose	1.5 to 2 mg. per square meter body surface, given weekly as a single dose
Testolactone U.S.P. *Teslac*	Testolactone Tablets U.S.P.	250 mg. four times daily		
	Sterile Testolactone Suspension U.S.P.	I.M., 100 mg. 3 times per week		
Dromostanolone Propionate U.S.P. *Drolban*	Dromostanolone Propionate Injection U.S.P.	Parenteral, 100 mg. 3 times weekly		

*See *U.S.P. D.I.* for complete dosage information.

It is soluble in water and polar organic solvents. Vials containing either 5 mg. of mitomycin and 10 mg. of mannitol or 20 mg. of mitomycin and 40 mg. of mannitol are supplied. The unreconstituted product is stable at room temperature for at least 2 years.

Mitomycin is useful in treating disseminated breast, gastric, pancreatic, or colorectal adenocarcinomas in combination with fluorouracil and Adriamycin (FAM program). It is used in combination with cyclophosphamide and Adriamycin for lung cancer. Recently, complete remissions of superficial transitional-cell carcinomas of the bladder have been obtained in 60 percent of patients.[215] The dose-limiting toxicity is myelosuppression, characterized by delayed, cumulative pancytopenia. Fever, anorexia, nausea and vomiting also occur.

Mitomycin at 10 to 20 mg. per square meter is given as a single dose by intravenous catheter. No repeat dose should be given until the leukocyte and platelet counts have recovered (about 8 weeks).

Mithramycin U.S.P., Mithracin®, aureolic acid, mitramycin, NSC-24559. This antibiotic is obtained from *Streptomyces plicatus*[178] or *Streptomyces argillaceus* as a yellow solid that melts at 180 to 183°C. It is soluble in polar organic solvents and aqueous alkali; however, it is susceptible to air oxidation in alkali. Mithramycin readily forms complexes with magnesium and other divalent metal ions, and these complexes have drastically altered optical rotations. Vials containing 2.5 mg. of mithramycin as a freeze-dried powder are supplied.

Mithramycin was used formerly in the treatment of advanced embryonal tumors of the testes.[204] However, it has been superseded by newer agents such as bleomycin and cisplatin. The main present use of mithramycin is in Paget's disease, in which it gives reduction of alkaline-phosphatase activity and relief of bone pain.[216] It also is useful in treating patients with severe hypercalcemia or hypercalciuria resulting from advanced metastatic cancer involving bones. Mithramycin may produce severe hemorrhaging. Bone-marrow, liver, and kidney toxicity also occur. The lower total dose used for hypercalcemia results in less toxicity.

MISCELLANEOUS COMPOUNDS

In 1965 Rosenberg investigated the effects of electrical fields on bacteria and found that *Escherichia coli* formed long filaments instead of dividing.[231] He subsequently discovered that this effect was caused not by the electrical current, but by a complex, $[Pt(Cl_4)(NH_3)_2]^\circ$, formed from the platinum electrode in the presence of ammonium and chloride ions.[232] This discovery was followed by testing a variety of platinum neutral complexes against tumors, with the result that *cis*-dichlorodiammineplatinum II (cisplatin) eventually became established as a clinical agent.[233]

Cisplatin

This platinum complex is a potent inhibitor of DNA polymerase. Its activity and toxicity resemble those of the alkylating agents. Considerable evidence has been obtained for DNA cross-linking by the platinum complex, wherein the two chlorides are displaced by nitrogen or oxygen atoms of purines. This evidence includes facilitated renaturation, increased sedimentation coefficient, hyperchromicity of the DNA ultraviolet spectrum, and selective reaction of the complex with guanine over other bases.[234]

Many other platinum complexes have been found active against tumors. Generally, they fall into the classification of *cis*-isomers in which one pair of ligands are monodentate anions of intermediate leaving ability (such as chloride) or bidentate anions (such as malonate), and the other pair are mono- or bidentate amines.[235] Among the more significant analogs are *cis*-dichloro(dipyridine)platinum(II), *cis*-dichloro*bis*(cyclohexylamine)platinum(II), and *cis*-di(cyclohexylamine)platinum(II) malonate.[236]

Hydroxyurea has been known for more than 100 years, but its antitumor activity was not discovered until 1963.[237] It is active against rapidly proliferating cells in the synthesis phase, during which it prevents the formation of deoxyribonucleotides from ribonucleotides. The mode of its action is inhibition of ribonucleotide diphosphate reductase, an enzyme consisting of two protein subunits.[238] It does this by interfering with the iron-containing portion of one of these subunits.[239]

Hydroxyurea   Guanazole

Another very old compound recently found active against tumors is guanazole.[240] This diaminotriazole resembles hydroxyurea in its ability to limit DNA synthesis by inhibiting the reduction of ribonucleotides. It is clinically active in inducing remissions of acute adult leukemia.[241]

In 1953 Kidd found that injections of guinea-pig serum caused regressions of certain transplanted tumors in mice and rats.[242] Subsequent investigation revealed that these tumors required L-asparagine as a nutrient, but the presence of the enzyme L-asparaginase in the guinea-pig serum created a deficiency in this amino acid.[243] The practical preparation of L-asparaginase for clinical trials follows the discovery that *Esherichia coli* produces a form of it that has antineoplastic activity.[244] Thus, mass cultures are harvested and treated with ammonium sulfate to rupture the cells, and the liberated enzyme is isolated by solvent extraction and chromatography. Very pure material is obtained by gel filtration or affinity chromatography, followed by crystallization. The *E. coli* enzyme has a molecular weight of 120,000 to 141,000 daltons, an isoelectric point of 4.9 to 5.2, and a K_m of 1.2×10^{-5}.[245]

Earlier preparations of L-asparaginase contained endotoxicins from *E. coli,* but these are absent in the purer new preparations. Clearance of the enzyme from plasma is due to an immunological reaction in which it combines with protein. This reaction may lead to sensitization in some patients. Patients who cannot tolerate L-asparaginase from *E. coli* might be treated by the preparation from *Erwinia carotovora.*[246] Tumor resistance is based on the development of asparagine synthetase by the tumor cells.[247]

A variety of thiosemicarbazones and guanylhydrazones have antitumor activity, although none is an established clinical agent at this time. 5-Hydroxypyridine-2-carboxaldehyde thiosemicarbazone and related heterocyclic compounds are powerful chelating agents for transition metals including iron.[248] There is a direct correlation between chelating ability and antitumor activity. It has been suggested that these compounds inhibit ribonucleoside diphosphate reductase by coordinating with the iron that it contains.[249]

5-Hydroxypyridine-2-
carboxaldehyde Thiosemicarbazone

Bis(thiosemicarbazones) of α-ketoaldehydes and α-diketones also have antineoplastic activity. Among the most thoroughly studied compounds of this type are the derivatives of methylglyoxal and 3-ethoxy-2-oxobutyraldehyde.[250] These compounds form strong chelates with copper and zinc, but the function of these species in causing cytotoxicity is unknown.[251]

Methylglyoxal *bis*(guanylhydrazone) has antitumor activity in humans. It interferes with nuclear and mitochondrial metabolism.[252] Many of its actions are re-

Methylglyoxal *Bis*(thiosemicarbazone), R = CH₃
3-Ethoxy-2-oxobutyraldehyde
Bis(thiosemicarbazone), R = CH₃CHOC₂H₅

Copper chelate

lated to the functions of spermidine, which it resembles in structure. Thus, it competes with spermidine for the transport carrier and intracellular binding site. It also inhibits spermidine biosynthesis. Its antiproliferative effects on cells can be prevented by administering spermidine.[253] Many other *bis*(guanylhydrazones) have been prepared, but none has proven superior to the methylglyoxal derivative.

Methylglyoxal *Bis*(guanylhydrazone)

H₂NCH₂CH₂CH₂NHCH₂CH₂CH₂CH₂NH₂

Spermidine

Phthalanilides such as NSC 60,339 showed antitumor activity, but they were highly toxic in clinical trials. They inhibit oxidative phosphorylation in mitochondria and are thought to influence the phospholipid-histone-DNA equilibrium in the nucleus and mitochondria.[254]

NSC 60,339

The antischistosomal drug hycanthone has been shown to have antitumor activity in animals and it is relatively nontoxic.[255] It is an intercalating agent that inhibits DNA and RNA synthesis.[256]

Unusual antitumor properties are shown by the

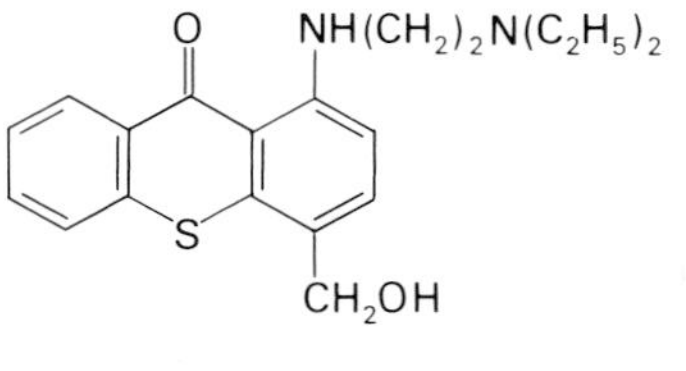

Hycanthone

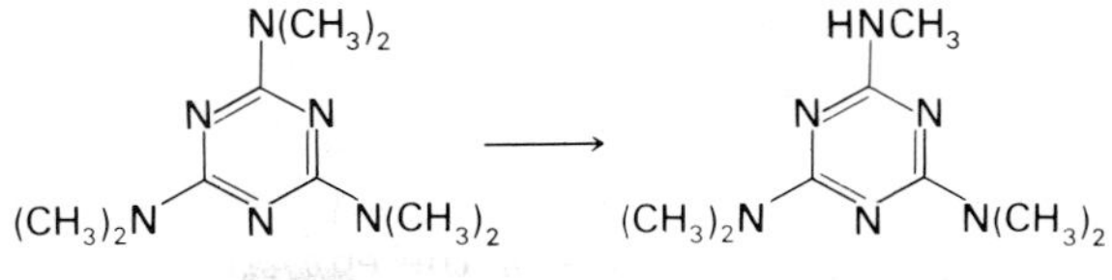

Hexamethylmelamine   Pentamethylmelamine

compound known as ICRF-159. It inhibits the metastases of Lewis lung tumor in mice without affecting the primary tumor.[257] This activity is thought to be caused by normalization of the developing blood vessels at the invading margins of the primary tumor. Another antineoplastic action of ICRF-159 is potent inhibition of DNA synthesis.[258] Phase-II clinical trials indicated activity against lymphoma, Kaposi's sarcoma, osteosarcoma, and advanced head and neck cancer. Toxic effects included leukopenia, nausea and vomiting, and alopecia.[259] A recent study on cyclopropyl analogs of ICRF-159 revealed that, in one tumor model, controlling the drug geometry resulted in a separation of antitumor activity (*cis*-isomer) from tumor potentiating activity (*trans*-isomer).[260]

Among the new antineoplastic drugs, 4-[(9-acridinyl)amino]methanesulfon-*m*-anisidide (*m*-AMSA) appears to be valuable because it showed a wide spectrum of activity in early clinical trials. Phase-II trials are underway in a panel of signal tumors, with some remissions already shown in refractory cases of breast cancer, malignant melanoma, and acute myelocytic leukemia. Leukopenia is the limiting toxicity.[259]

m-AMSA is an acridine derivative that is thought to bind to DNA through intercalation. However, it does not affect DNA synthesis.[263] This compound was rationally designed as one member of a group of acridinylaminomethanesulfonamides.[264] Previously, a number of acridine derivatives had shown antitumor activity.

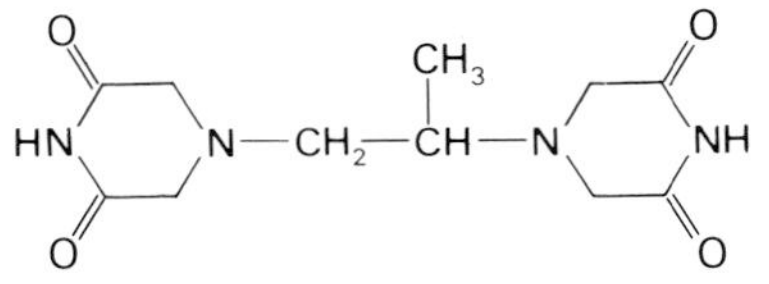

ICRF-159

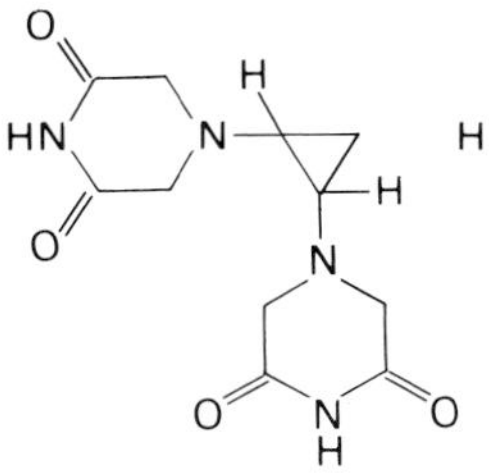

cis-isomer

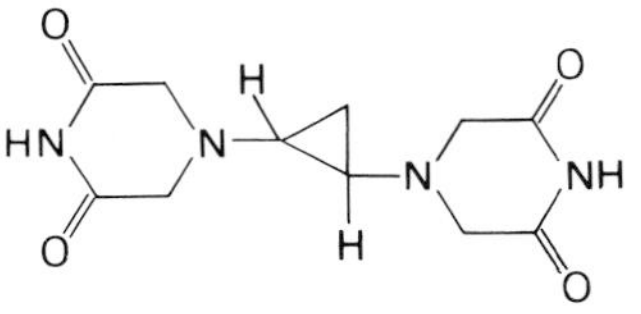

trans-isomer

Pipobroman is used against hematological cancers, especially polycythemia vera and chronic granulocytic leukemia.[261] Its mode of action is unknown, although it has sometimes been included among the alkylating agents. It has not been possible to demonstrate alkylating properties in vitro under physiological conditions.

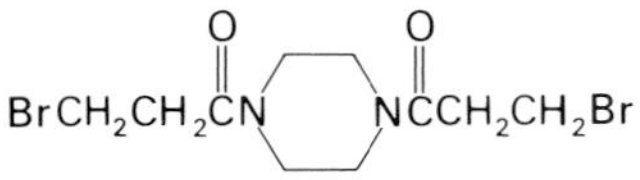

Pipobroman

Hexamethylmelamine is clinically active against melanoma. It is metabolized to the water-soluble compound pentamethylmelamine and eight other compounds. Pentamethylmelamine also is active against melanoma, but it is unclear if its formation represents a necessary conversion to activate hexamethylmelamine.[262]

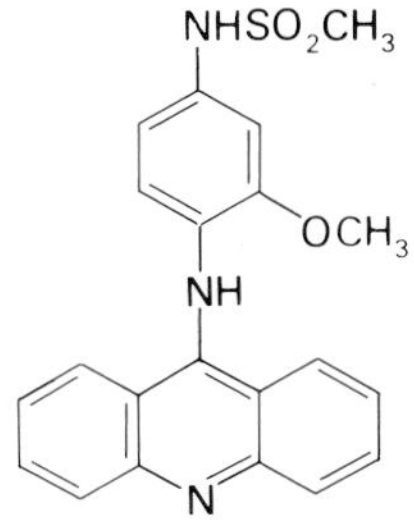

m-AMSA

Cisplatin, Platinol®, NSC-119875. This compound is prepared by treating potassium chloroplatinite with ammonia.[233] It is a water-soluble (1mg. / 1 ml.) white solid. Amber vials containing 10 mg. of cisplatin as a lyophilized powder are supplied. For reconstitution, 10 ml. of sterile water is added and the resulting solution is diluted in 2 liters of 5 percent dextrose in 0.5 or 0.33N saline containing 37.5 g. of mannitol.[265]

Cisplatin is used in combination with bleomycin and vinblastine for metastatic testicular tumors. This combination represents a significant improvement over previous treatments.[266] As a single agent or in combination with doxorubicin, cisplatin is used for the remission of metastatic ovarian tumors. Other tumors that have shown sensitivity to cisplatin include penile cancer, bladder cancer, cervical cancer, head and neck cancer, and small-cell cancer of the lung. The major dose-limiting toxicity is cumulative renal insufficiency associated with renal tubular damage. Hydrating patients with intravenous fluids prior to and during cisplatin treatment significantly reduces the incidence of renal toxicity.[267] Myelosuppression, nausea and vomiting, and ototoxicity also occur frequently.

The usual dosage for metastatic testicular tumors is 20 mg. per square meter intravenously daily for five days, once every three weeks for three courses. Metastatic ovarian tumors are treated with 50 mg. per square meter intravenously once every three weeks. Pretreatment hydration is recommended for both regimens.[265]

Hydroxyurea U.S.P., Hydrea®, hydroxycarbamide, NSC-32065. This compound is prepared from hydroxylamine hydrochloride and potassium cyanide.[268] It is a crystalline solid with good solubility in water. Capsules containing 500 mg. of hydroxyurea are supplied.

Hydroxyurea is active against melanoma, chronic myelocytic leukemia, and metastatic ovarian carcinoma. It is used in combination with radiotherapy for head and neck cancer. The main toxicity is bone-marrow depression expressed as leukopenia, anemia, and, occasionally, thrombocytopenia. Gastrointestinal toxicity and dermatological reactions also occur.

Pipobroman, U.S.P., Vercyte®, NSC-25154, 1,4-*bis*(3-bromopropionyl)piperazine. This compound is prepared from piperazine and 3-bromopropionyl bromide.[269] It is supplied as 10 mg. or 25 mg. grooved tablets.

Pipobroman is used primarily for treating polycythemia vera. It also is used in patients with chronic granulocytic leukemia refractory to busulfan. Nausea, vomiting, abdominal cramping, diarrhea, and skin rash occur.

PLANT PRODUCTS

The use of higher plants in treating neoplastic disease dates to antiquity. Dioscorides described the use of colchicine for this purpose in the first century. In more recent years scientists have attempted to select and screen systematically plants reputed to have antitumor activity. If the presence of activity is established for one member of a plant family, other members of this family are selected and tested. A major impetus to this research was given by Hartwell at the National Cancer Institute, who established an extensive system of plant collection, screening, and isolation.[218] Approximately 100,000 plants have been screened under this program.

Resin of the may apple, *Podophyllum peltatum,* has long been used as a remedy for warts. One of its constituents, podophyllotoxin, has antineoplastic activity, but it is highly toxic.[219] This lignin inhibits mitosis by destroying the structural organization of the mitotic apparatus.[220] Early derivatives of podophyllotoxin showed poor clinical activity, but newer analogs, such as the epipodophyllotoxin derivatives VP-16, 213 (also designated VP-16) and VM-26, appear promising. The former compound has especially good activity against small-cell lung carcinoma. Both of these analogs differ from podophyllotoxin in that they are not inhibitors of microtubule assembly.[221]

VP-16, 213 R = CH₃
VM-26 R =

The most important antitumor agents from plants are the vinca alkaloids. These compounds were isolated from the periwinkle *Catharanthus rosea* at the Eli Lilly Company.[222] They have complex structures composed of an indole-containing moiety named catharanthine and an indoline-containing moiety named vindoline.[223] Four closely related compounds have antitumor activity: vincristine, vinblastine, vinrosidine, and vinleurosine. Among this group, vincristine and vinblastine are proven clinical agents. These two compounds are used against different types of tumors, despite the similarity of their structures. A number of semisynthetic compounds have been prepared. Among them, vinglycinate and 6,7-dihydrovinblastine have clinical potential.[224] Vindesine has undergone Phase-II clinical studies. It is considered to resemble vincristine but to be less neurotoxic.[225]

Vinca alkaloids cause mitotic arrest by promoting the dissolution of microtubules in cells. Microtubule crystals containing the alkaloids are formed in the cytoplasm.[226] Vinblastine is the most active compound, whereas vincristine is the only compound to cause irreversible inhibition of mitosis.[227] Cells can resume mi-

TABLE 8–4

VINCA ALKALOIDS AND THEIR ANALOGS

Catharanthine

Vindoline

	R	R₁	R₂	R₃	R₄
Vincristine	CH_3CO	CHO	H	OH	OCH_3
Vinblastine	CH_3CO	CH_3	H	OH	OCH_3
Vinrosidine	CH_3CO	CH_3	OH	H	OCH_3
Vinleurosine	CH_3CO	CH_3	H	?	OCH_3
Vinglycinate	$(CH_3)_2NCH_2CO$	CH_3	H	OH	OCH_3
Vindesine	H	CH_3	H	OH	NH_2

tosis following brief exposure to other vinca alkaloids after these compounds are withdrawn.[228]

As noted above, colchicine, obtained from the crocus *Colchicum autumnale,* has long been known for its antitumor activity. However, it is not now used clinically for this purpose. Its main use is in terminating acute attacks of gout. Among colchicine derivatives, demecolcine (colcemid) is active against myelocytic leukemia, but only at near-toxic doses. Colchicines have an unusual tricyclic structure containing a tropolone ring. They inhibit mitosis at metaphase by disorienting the organization of the spindle and asters.[229]

Colchicine, R = $COCH_3$
Colcemid, R = CH_3

Many other plant constituents show significant antitumor activity in animals and some of them have been given clinical evaluation. The more important compounds are anguidine, bouvardin, bruceantin, camptothecin, indicene-N-oxide lapachol, maytansine, and thalicarpine.[83]

PRODUCTS

Vinblastine Sulfate U.S.P., Velban®, vincaleucoblastine, VLB, NSC-49842. This antitumor alkaloid is isolated from *Vinca rosea* Linnaeus, the periwinkle plant.[222] It is soluble in water and alcohol. Vials containing 10 mg. of vinblastine sulfate as a lyophilized plug are supplied. It is reconstituted by the addition of sodium chloride solution for injection preserved with phenol or benzyl alcohol.

Vinblastine has been used for the palliation of a variety of neoplastic diseases. It is one of the most effective single agents against Hodgkin's disease and it may be used in combination chemotherapy for patients who have relapses after treatment by the MOPP program. Advanced testicular germinal-cell tumors respond to vinblastine alone or in combination.[208] Beneficial effects are also obtained against lymphocytic lymphoma, histiocytic lymphoma, mycosis fungoides, Kaposi's sarcoma, Letterer-Siwe disease, resistant choriocarcinoma, and carcinoma of the breast. The limiting toxicity is leukopenia, which reaches its nadir in five to ten days after the last dose. Gastrointestinal and neurological symptoms occur and are dose dependent. Extravasation during injection can lead to cellulitis and phlebitis.

Vincristine Sulfate U.S.P., Oncovin®, leurocristine, VCR, LCR, NSC-67574. This alkaloid is isolated from *Vinca rosea* Linnaeus.[222] The sulfate is a crystalline solid that is soluble in water. It is supplied in vials containing either 1 mg. of vincristine sulfate and 10 mg. of lactose or 5 mg. of vincristine and 50 mg. of lactose. Each size has an accompanying vial of 10 ml. of bacteriostatic sodium chloride solution containing 90 mg. of sodium chloride and 0.9 percent benzyl alcohol. The reconstituted pharmaceutical may be stored 14 days in a refrigerator.

Vincristine is effective against acute leukemia. In combination with prednisone it produces complete remission in 90 percent of children with acute lymphoblastic leukemia.[143] It is used in the MOPP program of combination chemotherapy for Hodgkin's disease.[230] Other tumors that respond to vincristine in combination with other antineoplastic agents include lymphosarcoma, reticulum-cell sarcoma, rhabdomyosarcoma, neuroblastoma, and Wilms' tumor. Although the tumor spectra of vinblastine and vincristine are similar, there is a lack of cross-resistance between the two. Because vincristine is less myelosuppressive than vinblastine, it is preferred in combination with myelotoxic agents. The most serious clinical toxicity of vincristine is neurological, with paresthesia, loss of deep-tendon reflexes, pain, and muscle weakness occurring. These symptoms can usually be reversed by lowering the dose or suspending therapy. Constipation and alopecia also occur. The rapid action of vincristine in destroying cancer cells may result in hyperuricemia. This complication can be prevented by administering allopurinol.

HORMONES

Steroid hormones, including estrogen, androgens, progestins, and glucocorticoids act on the appropriate target tissues at the level of transcription. Generally the effect is derepression of genetic template operation, which stimulates the cellular processes. However, glucocorticoids act in lymphatic tissues to impair glucose uptake and protein synthesis. Target cells contain in their cytoplasm specific protein receptors with very high affinities for the hormones. Binding of the hormone to the receptor causes a transformation in the receptor structure, which is followed by migration of the resulting complex into the nucleus. In the nucleus, the complex interacts with an acceptor site to influence transcription.[270]

Normal and well-differentiated neoplastic target cells have a number of hormone receptors and they are dependent upon the hormones for stimulation.[271-274] Less differentiated neoplastic cells become independent of hormonal control and lose their specific receptors. Thus, some neoplasms are hormone dependent and responsive to hormone-based therapy, whereas others are independent and unresponsive. Assays of the number of hormone receptors present in the neoplastic cells should be valuable in predicting the probability of a favorable response.

Hormonal effects in breast cancer are complex and not completely understood. The hormone dependency of breast cancer has been known since 1889,[275] and removal of the ovaries of premenopausal women, which results in decreased estrogens, is an established treatment. Some patients who do not respond to this procedure do respond to adrenalectomy, which suggests that the hormone dependence is not simply related to estrogens.[276] It has been shown that remission after adrenalectomy occurs more often in patients with estrogen receptors than in those lacking receptors. Administration of estrogens to postmenopausal women with metastatic breast cancer resulted in objective remissions in about 30 percent of the cases.[277] This response appears paradoxical, but the estrogen levels resulting from drug treatment are much greater than physiological levels. A recent suggestion is that high estrogen levels interfere with the peripheral action of prolactin, a pituitary hormone that also stimulates breast tissue.[278] Ethinyl estradiol is given orally in the treatment of breast cancer in postmenopausal women

and estradiol dipropionate or benzoate are used parenterally. The experimental compounds MER-25 and tamoxifen are antiestrogens that are used in the treatment of premenopausal women.

MER-25

Tamoxifen

Androgens are active against metastatic breast cancer in about 20 percent of postmenopausal women. Their mode of action is not completely understood. Inhibition of the release of pituitary gonadotrophins has been suggested, but the situation must be more complicated than this because certain androgens are active in hypophysectomized patients.[279] Other useful effects of androgens in advanced breast cancer are stimulation of the hematopoietic system and reversal of bone demineralization. Testosterone propionate is the androgen most frequently used against breast cancer. Other compounds are 2α-methyltestosterone, fluoxymesterone, and 19-nor-17α-methyltestosterone. Testolactone is preferred in some cases because it has no androgenic side-effects.

Estrogens can be used to induce remissions of disseminated prostatic cancer. It is not certain whether their effect is due to direct interference with peripheral androgens, inhibition of pituitary gonadotrophin, or both.[280] Diethylstilbestrol is the compound most widely used for advanced prostatic cancer and it benefits over 60 percent of patients. Chlorotrianisine also is used.

Progesterone and its analogs are active against certain neoplasms that are stimulated by estrogens. They appear to exert antiestrogenic effects of uncertain mechanism. The neoplasms treated by progestins are metastatic endometrial carcinoma and advanced renal-cell carcinoma.[281] Progesterone suspension in oil, megesterol acetate and medroxyprogesterone acetate are used against endometrial cancer. They provide regressions of several months to 3 years in about 30 percent of women.[281] Medroxyprogesterone acetate causes regression of renal-cell carcinoma in about 20 percent of men and 8 percent of women.

Glucocorticoids cause pronounced acute changes in lymphoid tissues. Lymphocytes in the thymus and lymph nodes are dissolved and lymphopenia occurs in peripheral blood.[282] This property is used to advantage in the treatment of leukemia and Hodgkin's disease, wherein profound temporary regressions are observed following the administration of cortisone derivatives or ACTH.[283] Prednisone is usually the corticoid chosen for this purpose and it is almost always used in combination with other chemotherapeutic agents such as mechlorethamine, vincristine, and procarbazine. Such combinations are effective in maintaining the remissions in many cases. Glucocorticoids also are useful in treating metastatic prostate cancer of patients who have relapsed after castration. The rationale for this use is that they inhibit release of ACTH from the pituitary, which leads to adrenal atrophy and decreased adrenal production of androgens.[284] Prednisone and cortisone acetate are used in the treatment of metastatic breast cancer. Their value in this condition derives not from an antineoplastic effect, but in alleviating specific complications such as hypercalcemia and anemia.[285]

Mitotane is unique among antitumor agents in its highly selective effect on one gland, the adrenal cortex. It has a direct cytotoxic action on adrenal cortical cells, in which it extensively damages the mitochondria.[286] This leads to cell death and atrophy of the gland. Mitotane is used specifically against adrenocortical carcinoma.[287]

Mitotane

The role of prolactin in stimulating hormone-dependent breast cancer was described above, and it was noted that high estrogen levels are thought to interfere with the peripheral activity of prolactin. Other types of compounds are under investigation as potential prolactin inhibitors. The ergoline derivatives lergotrile and 8-cyanomethyl-8-methylergoline are examples of this approach. They inhibit the release of prolactin from the pituitary.[288]

Lergotrile   8-Cyanomethyl-8-methylergoline

A thorough discussion of the structures, nomenclature, properties, and dose forms of the steroid hormones is presented in Chapter 18. Only the products not included in that chapter are described below.

PRODUCT

Mitotane U.S.P., Lysodren®, *o,p'*-DDD, CB133, 1,1-dichloro-2-(*o*-chlorphenyl)-2-(*p*-chlorophenyl)ethane. This compound is obtained as a constituent of commercial DDD, which is prepared from 2,2-dichloro-1-(*o*-chlorophenyl)ethanol, chlorobenzene, and sulfuric acid.[289] Isolation from commercial DDD gives mitotane as crystals that are soluble in alcohol and other organic solvents.[290] Scored 500 mg. tablets are supplied.

Mitotane is indicated only for treating inoperable adrenal cortical carcinoma. Frequently occurring side-effects include gastrointestinal disturbances, CNS depression, and skin toxicity.

The usual regimen is 8 to 10 g. daily, divided into three or four doses.

Dromostanolone Propionate U.S.P., Drolban®, 17β-hydroxy-2α-methyl-5α-androstan-3-one propionate, 2α-methyldihydrotestosterone propionate. This semi-synthetic androgen is prepared from dihydrotestosterone in a route involving condensation with ethyl formate followed by hydrogenation to give the 2α-methyl derivative, and then reaction with propionic anhydride.[291] The compound is supplied in rubber-stoppered vials containing 500 mg. of dromostanolone propionate in 10 ml. of sesame oil, with 0.5 percent phenol as a preservative.

Dromostanolone propionate is used in the palliative treatment of metastatic breast carcinoma in postmenopausal women. It is contraindicated in premenopausal women and in carcinoma of the male breast. The most usual side-effect is virilism, although this is less intense than that afforded by testosterone propionate. Edema occurs occasionally.

Testolactone U.S.P., Teslac®, D-homo-17α-oxa-androsta-1,4-dien-3,17-dione, 1-dehydrotestololactone, is prepared by microbial transformation of progesterone.[292] It is soluble in alcohol and slightly soluble in water. The compound is supplied as a sterile aqueous suspension providing 100 mg. of testolactone per ml. in multiple-dose vials of 5 ml. Tablets containing 50 mg. or 250 mg. of testolactone also are supplied.

Testolactone is used in the palliative treatment of advanced or disseminated breast cancer in postmenopausal women. It is contraindicated in breast cancer in men. Testolactone is devoid of androgenic activity in the commonly used doses.

Megestrol Acetate, Megace®, 17α-acetoxy-6-methylpregna-4,6-dien-3,20-dione. This compound is prepared by a multi-step synthesis from 17α-hydroxy-pregnadienolone.[293] It is supplied as light blue scored tablets containing 20 or 40 mg. of megestrol acetate.

Megestrol acetate is indicated for the palliative treatment of advanced breast or endometrial carcinoma when other methods of treatment are inappropriate. No serious side-effects or adverse reactions have been reported. However, there is an increased risk of birth defects in children whose mothers take the drug during the first four months of pregnancy. The usual doses are 160 mg. per day in four equal doses for breast cancer and 40 to 320 mg. per day in divided doses for endometrial cancer.

Tamoxifen Citrate, Nolvadex®, (Z)-2-[4-(1,2-diphenyl-1-butenyl)phenoxy]-N,N-dimethylethanamine citrate is prepared by treating 2-ethyldeoxybenzoin with 4-[(2-N,N-dimethylamino)ethoxy]phenylmagnesium bromide,[294] followed by dehydration and separation of the E and Z isomers.[295] The citrate salt of the Z isomer is soluble in water. Tablets containing 15.2 mg. of tamoxifen citrate, which is equivalent to 10 mg. of tamoxifen, are supplied. They should be protected from heat and light.

Tamoxifen is a nonsteroidal agent that has shown potent antiestrogenic properties in animals. In the rat model, it appears to exert its antitumor effects by binding to estrogen receptors.[296] Tamoxifen is useful in the palliative treatment of advanced breast cancer in postmenopausal women. There are no known contraindications. The most frequent side-effects are hot flashes, nausea, and vomiting. They are rarely severe enough to require dose reduction. The usual dose is one or two 10 mg. tablets twice daily.

IMMUNOTHERAPY

It is now generally accepted that cells of neoplastic potential are continually produced in the human body and that our immune surveillance system destroys them. The development of tumors implies that this system is not functioning properly. Evidence for this factor in carcinogenesis includes a high rate of cancer in organ-transplant patients whose immune systems are suppressed by drugs such as azathioprine, and a high correlation between cancer and immunodeficiency diseases such as bacterial and viral infections.[297] Stimulation of the body's immune system should provide a valuable method of cancer treatment, because it is capable of eradicating the neoplastic cells completely. Research in this area is expanding rapidly and some promising leads are emerging.

The first attempt at immunotherapy was made in the 1890's by Coley, who injected bacterial toxins in cancer patients. His results were generally unaccepted because of rather extravagant claims. However, his techniques have been revived in recent years. Most oncologists now use a live bacteria tuberculosis vaccine named bacillus Calmette-Guerin (BCG).[297] This vaccine is given to certain patients who show a functioning immune system as determined by sensitivity toward dinitrochlorobenzene.[298] Remissions have been obtained in malignant melanoma, breast cancer, and leukemia. Unfortunately, BCG causes a number of undesirable effects including fever, hypersensitivity, and liver disorders. Other immunostimulants currently under investigation as anticancer agents are the methanol-extracted residue (MER) of BCG, *Corynebacterium parvulum, Bordetella pertussis* vaccine, and synthetic polynucleotides.[299]

One approach to overcoming the difficulties of BCG therapy is to develop simpler chemical structures with immunostimulant properties. Two such compounds are presently under clinical investigation as potential anticancer drugs. One of them is levamisole, an anthelmintic agent found to be an immunostimulant by Renoux in 1972. It appears to be most effective in patients with small-tumor burdens and it acts by stimulating the responsiveness of lymphocytes to tumor antigens. Advantages of levamisole include oral activity and few adverse reactions. Tilorone is one member of a large family of synthetic compounds that show immunostimulant activity. It stimulates the production of interferon, which affords antiviral activity. However, its effect is more general. The main cause of its antitumor action appears to be an effect on "T" cells originating in the thymus. Tilorone causes gastrointestinal disorders, dizziness, and headaches in many patients.[300]

Levamisole

Tilorone

REFERENCES

1. Zubrod, C. G.: Agents of Choice in Neoplastic Disease, *in* Sartorelli, A. C., and Johns, D. J. (eds.): Handbook of Experimental Pharmacology, vol. 38, part 1, p. 7, New York, Springer-Verlag, 1974.
2. Clarkson, B.: Clinical Applications of Cell Cycle Kinetics, *in* Sartorelli, A. C., and Johns, D. J. (eds.): Handbook of Experimental Pharmacology, vol. 38, part 1, p. 156, New York, Springer-Verlag, 1974.
3. Huggins, C., and Hodges, C. V.: Cancer Res. 1:293, 1941.
4. Gilman, A., and Philips, F. S.: Science 103:409, 1946.
5. Pearson, O. H.: Cancer 2:943, 1949.
6. Kennedy, B. J.: Cancer 18:1551, 1965.

7. Wood, H. B., Jr.: Cancer Chemother. Rep. 2:9, 1971.
8. Goldin, A., *et al.*: Evaluation of Antineoplastic Activity: Requirements of Test Systems, *in* Sartorelli, A. C., and Johns, D. J. (eds.): Handbook of Experimental Pharmacology, vol. 38, part 1, p. 12, New York, Springer-Verlag, 1974.
9. Himmelweit, F. (ed.): The Collected Papers of Paul Ehrlich, vol. 1, pp. 596–618, London, Pergamon Press, 1956.
10. Lynch, V., *et al.*: J. Pharmacol. Exp. Therap. 12:265, 1918.
11. Montgomery, J. A., *et al.*: Drugs for Neoplastic Diseases, *in* Burger, A. (ed.): Medicinal Chemistry, Third Edition, p. 680, New York, Wiley, 1970.
12. Connors, T. A.: Mechanism of Action of 2-Chloroethylamine Derivatives, Sulfur Mustards, Epoxides and Aziridines, *in* Sartorelli, A. C., and Johns, D. J. (eds.): Handbook of Experimental Pharmacology, vol. 38, part 2, p. 19, New York, Springer-Verlag, 1975.
13. Price, C. C.: Chemistry of Alkylation, *in* Sartorelli, A. C., and Johns, D. J. (eds.): Handbook of Experimental Pharmacology, vol. 38, part 2, p. 4, New York, Springer-Verlag, 1975.
14. White, F. R.: Cancer Chemother. Rep. 4:55, 1959.
15. Jarmon, M., and Ross, W. C. J.: Chem. Ind. (London): 1789, 1967.
16. Montgomery, J. A., *et al.*: J. Med. Chem. 10:668, 1967.
17. Engle, T. W., *et al.*: J. Med. Chem. 22:897, 1979.
18. Colvin, M., *et al.*: Cancer Res. 36:1121, 1976.
19. Karle, I. L., *et al.*: J. Am. Chem. Soc. 99: 4803, 1977.
20. Ahmann, D. L., *et al.*: Cancer Chemother, Rep. 58:861, 1974.
21. Anderson, W. K., and Corey, P. F.: J. Med. Chem. 20:812, 1977.
22. Szybalski, W., and Iyer, V. N.: The Mitomycins and Porfiromysins, *in* Gottlieb, D., and Shaw, P. D. (eds.): Antibiotics, vol. 1, p. 221, New York, Springer-Verlag, 1967.
23. Lin, A. J., *et al.*: Cancer Chemother. Rep. 4:23, 1974.
24. Buchi, G., and Rosenthal, D.: J. Am. Chem. Soc. 78:3860, 1956.
25. Bollag, W.: Cancer Chemother. Rep. 33:1, 1963.
26. Chabner, B. A., *et al.*: Proc. Soc. Exp. Biol. 132:1169, 1969.
27. Tsuji, T., and Kosower, E. M.: J. Am. Chem. Soc. 93:1992, 1971.
28. Kreis, W., and Yen, W.: Experientia 21:284, 1965.
29. Schwartz, D. E., *et al.*: J. Labelled Compounds 111:487, 1967.
30. Preussmann, R., and Von Hodenberg, A.: Biochem. Pharmacol. 19:1505, 1970.
31. Skibba, J. L., *et al.*: Cancer Res. 30:147, 1970.
32. Sava, G., *et al.*: Cancer Treatment Rep. 63:93, 1979.
33. Ludlum, D. B.: Molecular Biology of Alkylation: An Overview, *in* Sartorelli, A. C., and Johns, D. J. (eds.): Handbook of Experimental Pharmacology, vol. 38, part 2, p. 7, New York, Springer-Verlag, 1975.
34. Kohn, K. W., *et al.*: J. Molec, Biol. 19:226, 1966.
35. Ross, W. C. J.: Biological Aklylating Agents, London, Butterworths, 1962.
36. Thomasz, M.: Biochem. Biophys. Acta 213:288, 1970.
37. Lawley, P. D., and Martin, C. M.: Biochem. J. 145:85, 1975.
38. Ludlum, D. B.: Biochem. Biophys. Acta 142:282, 1967.
39. Brooks, P., and Lawley, P. D.: Biochem. J. 80:496, 1961.
40. Wheeler, G. P.: Mechanism of Action of Nitrosoureas, in Sartorelli, A. C., and Johns, D. J., (eds.): Handbook of Experimental Pharmacology, vol. 38, part 2, p. 75, New York, Springer-Verlag, 1975.
41. Levis, A. G., *et al.*: Nature 207:608, 1965.
42. Ludlum, D. B.: Biochem. Biophys. Acta 95:674, 1965.
43. Boyce, R. P., and Howard-Flanders, P.: Proc. Nat. Acad. Sci. 51:293, 1964.
44. Setlow, R. B., and Carrier, W. L.: Proc. Nat. Acad. Sci. 51:226, 1964.
45. Prelog, V., and Stepan, V.: Coll. Czech. Chem. Commun. 7:93, 1935.
46. Arnold, H., *et al.*: Nature 181:931, 1958.
47. Zubrod, C. G.: Cancer 21:553, 1968.
48. Bergel, F., and Stock, J. A.: J. Chem. Soc. 1954:2409.
49. Fisher, B., *et al.*: New Engl. J. Med. 292:117, 1975.
50. Phillips, A. P., and Mentha, J. W.: U.S. Patent 3,046,301, Oct. 29, 1959.
51. Calabresi, P., and Welch, A. D.: Ann. Rev. Med. 13:147, 1962.

52. Timmis, G. M.: U.S. Patent 2,917,432, Dec. 15, 1959.
53. Parham, W. E., and Wilbur, J. M., Jr.: J. Org. Chem. 26:1569, 1961.
54. Roberts, J. J., and Warwick, G. P.: Biochem. Pharmacol. 6:217, 1961.
55. Warwick, G. P.: Cancer Res. 23:1315, 1963.
56. Physicians' Desk Reference, 33rd. Edition, p. 746, Oradell, N.J. Medical Economics Co., 1979.
57. Johnston, T. P., *et al.*: J. Med. Chem. 6:669, 1963.
58. Walker, M. D.: Cancer Chemother. Rep. 4:21, 1973.
59. Johnston, T. P., *et al.*: J. Med. Chem. 9:892, 1966.
60. Moertel, C. G.: Cancer Chemother. Rep. 4:27, 1973.
61. Kuh, E., and Seeger, D. R.: U.S. Patent 2,670,347, Feb. 23, 1954.
62. Hoffman-LaRoche & Co., A.-G.: Belg. Patent 618,638, Dec. 7, 1962.
63. Shealy, Y. F., *et al.*: J. Org. Chem. 27:2150, 1962.
64. Lyttle, D. A., and Petering, H. F.: J. Am. Chem. Soc. 80:6459, 1958.
65. Lukens, L. N., and Herrington, K. A.: Biochim. Biophys. Acta 24:432, 1957.
66. Patterson, A. R. P., and Tidd, D. M.: 6-Thiopurines, *in* Sartorelli, A. C., and Johns, D. J. (eds.): Handbook of Experimental Pharmacology, vol. 38, part 2, p. 384, New York, Springer-Verlag, 1975.
67. Woods, D. D., and Fildes, P.: J. Chem. Soc. 59:133, 1960.
68. Roblin, R. O., *et al.*: J. Am. Chem. Soc. 67:290, 1945.
69. Hitchings, G. H., and Elion, G. B.: Accounts Chem. Res. 2:202, 1969.
70. Elion, G. B., *et al.*: J. Am. Chem. Soc. 74:411, 1952.
71. Brockman, R. W.: Advanc. Cancer Res. 7:129, 1963.
72. Atkison, M. R., *et al.*: Biochem. J. 92:398, 1964.
73. Salser, J. S., *et al.*: J. Biol. Chem. 235:429, 1960.
74. Bennett, L. L., Jr., and Allan, P. W.: Cancer Res. 31:152, 1971.
75. Currie, R., *et al.*: Biochem. J. 104:634, 1967.
76. Elion, G. B.: Fed. Proc. 26:898, 1967.
77. Moor, E. C., and LePage, G. A.: Cancer Res. 18:1075, 1958.
78. LePage, G. A., *et al.*: Cancer Res. 24:835, 1964.
79. Lee, W. W., *et al.*: J. Am. Chem. Soc. 82:2648, 1960.
80. Parke-Davis and Co.: Belg. Patent 671,557, 1967.
81. Furth, J. J., and Cohen, S. S.: Cancer Res. 27:1528, 1967.
82. Brink, J. J., and LePage, G. A.: Cancer Res. 24:312, 1964.
83. Lomax, N. R., and Narayanan, V. L.: Chemical Structures of Interest to the Division of Cancer Treatment, NCI, Jan., 1979.
84. Heidelberger, C.: Flourinated Pyrimidines and Their Nucleosides, *in* Sartorelli, A. C., and Johns, D. J. (eds.): Handbook of Experimental Pharmacology, vol. 38, part 2, p. 193, New York, Springer-Verlag, 1975.
85. Sköld, O.: Biochim. Biophys. Acta 29:651, 1958.
86. Reyes, P.: Biochemistry 8:2057, 1969.
87. Cohen, S. S., *et al.*: Proc. Nat. Acad. Sci. 44:1004, 1958.
88. Santi, D. V., and McHenry, C. S.: Proc. Natl. Acad. Sci. 69:1855, 1972.
89. Hiller, S. A., *et al.*: Dokl. Akad. Nauk S.S.R. 176:332, 1967.
90. Benvenuto, J., *et al.*: Cancer Res. 38:3867, 1978.
91. Heidelberger, C., *et al.*: J. Med. Chem. 7:1, 1964.
92. Santi, D. V., and Sakai, T. T.: Biochemistry 10:3598, 1971.
93. Cohen, S. S.: Progr. Nucleic Acid Res. 5:1, 1966.
94. Bergmann, W., and Feeney, R. J.: J. Org. Chem. 16:981, 1951.
95. Chu, M. Y., and Fischer, G. G.: Biochem. Pharmacol. 11:423, 1962.
96. Larsson, A., and Reichard, P.: J. Biol. Chem. 241:2540, 1966.
97. Creasey, W. A., *et al.*: Cancer Res. 28:1074, 1968.
98. Borun, T. W., *et al.*: Proc. Nat. Acad. Sci. 58:1977, 1967.
99. Creasey, W. A.: Arabinosylcytosine, *in* Sartorelli, A. C., and Johns, D. J. (eds): Handbook of Experimental Pharmacology, vol. 38, part 2, p. 245, New York, Springer-Verlag, 1975.
100. Stoller, R. G., *et al.*: Biochem. Pharmacol. 27:53, 1978.
101. Hoshi, A., *et al.*: Gann 62:145, 1971.
102. Sörm, F., *et al.*: Experientia 20:202, 1964.
103. Hanka, L. J., *et al.*: Antimicrob. Agents Chemother.: 619, 1966.
104. Paces, V., *et al.*: Biochem. Biophys. Acta 161:352, 1968.
105. Vesely, J., *et al.*: Biochem. Pharmacol. 17:519, 1968.

106. Vesely, J., *et al.*: Cancer Res. 30:2180, 1970.
107. Nakamura, H., *et al.*: J. Am. Chem. Soc. 96:4327, 1974.
108. Schaeffer, H. J., and Schwender, C.F.: J. Med. Chem. 17:6, 1974.
109. Seeger, D. R., *et al.*: J. Am. Chem. Soc. 71:1753, 1949.
110. Werkheiser, W.: J. Biol. Chem. 236:888, 1961.
111. Hryniuk, W. M., *et al.*: Molec. Pharmacol. 5:557, 1969.
112. Zakrzewski, S. F., *et al.*: Molec. Pharmacol. 2:423, 1969.
113. Cohen, S.: Ann. N.Y. Acad. Sci. 186:292, 1971.
114. Li, M. C., *et al.*: Proc. Soc. Exp. Biol. 97:29, 1958.
115. Bertino, J. R., *et al.*: Proc. 5th Int. Congr. Pharmacol. 3:376, 1973.
116. Semon, J. H., and Grindley, G. B.: Cancer Res. 38:2905, 1978.
117. Howell, S. B., *et al.*: Cancer Res. 38:325, 1978.
118. Goldin, A., *et al.*: J. Nat. Cancer Inst. 22:811, 1959.
119. Alt, F. W., *et al.*: J. Biol. Chem. 253:1357, 1978.
120. Baker, B. R.: Accts. Chem. Res. 2:129, 1969.
121. Hartman, S. C.: Purines and Pyrimidines, *in* Greenberg, D.M. (ed.): Metabolic Pathways, Vol. 4, p. 1, New York, Academic Press, 1970.
122. Eidinoff, M. L., *et al.*: Cancer Res. 18:105, 1958.
123. Fusari, S. A., *et al.*: J. Am. Chem. Soc. 76:2881, 1954.
124. DeWald, H. A., and Moore, A. M.: J. Am. Chem. Soc. 80:3941, 1958.
125. Dion, H. W., *et al.*: J. Am. Chem. Soc. 78:3075, 1956.
126. French, T. C., *et al.*: J. Biol. Chem. 238:2186, 1963.
127. Dawid, I. B., *et al.*: J. Biol. Chem. 238:2187, 1963.
128. Levenberg, B., *et al.*: J. Biol. Chem. 225:163, 1957.
129. Bennett, L. L., Jr.: Glutamine Antagonists, *in* Sartorelli, A. C., and Johns, D. J. (eds.): Handbook of Experimental Pharmacology, vol. 38, part 2, p. 496, New York, Springer-Verlag, 1975.
130. Elion, G. B., *et al.*: J. Am. Chem. Soc. 74:411, 1952; Beaman, A. G., and Robbins, R. K.: J. Am. Chem. Soc. 83:4042, 1961.
131. Burchenal, J. H., *et al.*: Blood 8:965, 1953.
132. Physicians' Desk Reference, 33rd. Edition, p. 749, Oradell, N.J., Medical Economics Inc., 1979.
133. Elion, G. B., and Hitchings, G. H.: J. Am. Chem. Soc. 77:1676, 1955.
134. Clarkson, B. D.: Cancer 5:227, 1970.
135. Duschinsky, R., *et al.*: J. Am. Chem. Soc. 79:4559, 1957.
136. Earl, R. A., and Townsend, L. B.: J. Heterocyclic Chem. 9:1141, 1972.
137. Physicians' Desk Reference, 33rd. Edition, p. 749, Oradell, N.J., Medical Economics, Inc., 1979.
138. Moore, G. E., *et al.*: Cancer Chemother. Rep. 52:641, 1968.
139. Klein, E., *et al.*: Topical 5-Fluorouracil Chemotherapy for Premalignant and Malignant Epidermal Neoplasms, *in* Brodsky, I., and Kahn, S. B. (eds.): Cancer Chemotherapy II., p. 147, New York, Grune & Stratton, Inc., 1972.
140. Hoffer, M., *et al.*: J. Am. Chem. Soc. 81:4112, 1959.
141. Sullivan, R. D., and Miller, E.: Cancer Res. 25:1025, 1965.
142. Hunter, J. H.: U.S. Patent 3,116,282, Dec. 31, 1963.
143. Greenwald, E. S.: Cancer Chemotherapy, Flushing, N.Y., Medical Examination Publishing Co., Inc., 1973.
144. Clarkson, B. D.: Cancer 30: 1572, 1972.
145. Hignite, C. E., *et al.*: Cancer Treatment Rep. 62:13, 1978.
146. Farber, S., *et al.*: New Engl. J. Med. 238:787, 1948.
147. Hertz, R.: Ann. Intern. Med. 59:931, 1963.
148. Stoller, R. G., *et al.*: New Engl. J. Med. 297:630, 1977.
149. Hitchings, G. H., and Elion, G.B.: U.S. Patent 3,056,785, Oct. 2, 1962.
150. Waksman, S. A., and Woodruff, H.B.: Proc. Soc. Exp. Biol. Med. 45:609, 1940.
151. Rao, K. V., *et al.*: Antibiot. Chemother. 12:182, 1962.
152. Remers, W. A.: The Chemistry of Antitumor Antibiotics, vol. 1, New York, Wiley, 1979.
153. Schmidt-Kastner, G.: Naturwissenschaften 43:131, 1956.
154. Umezawa, H.: Bleomycin: Discovery, Chemistry, and Action, *in* Umezawa, H. (ed.): Bleomycin, Fundamental and Clinical Studies, p. 3, Tokyo, Gann Monograph on Cancer Research, 1976.
155. Brockman, H.: Fortschr. Chem. Org. Naturst. 18:1, 1960.
156. Sobell, H. M., and Jain, S. C.: J. Mol. Biol. 68:21, 1972.
157. Goldberg, I. H., and Friedman, P. A.: Pure Appl. Chem. 28:499, 1971.
158. Wells, R. D., and Larson, J. E.: J. Mol. Biol. 49:319, 1970.
159. Reich, E., *et al.*: Science 134:556, 1961.
160. DuBost, N., *et al.*: C. R. Acad. Sci., Paris 257:1813, 1963.
161. Grein, A., *et al.*: Giorn. Microbiol. 11:109, 1963.
162. DiMarco, A., *et al.*: Cancer Chemother. Rep. 53:33, 1969.
163. Rauscher, F. J., Jr.: Special Communication from the Director, National Cancer Program, June 20, 1975.
164. Brockman, H.: Fortschr. Chem. Org. Naturst. 21:1, 1963.
165. Arcamone, F., *et al.*: Tetrahedron Lett. 1968:3349.
166. Oki, T., *et al.*: J. Antibiot. 28:830, 1975.
167. DiMarco, A.: Daunomycin and Related Antibiotics, *in* Gottlieb, D., and Shaw, P. D. (eds.): Antibiotics, I, pp. 190–210, New York, Springer-Verlag, 1967.
168. Krishan, A., *et al.*: Cancer Res. 36:2108, 1976.
169. Lown, J. W., *et al.*: Biochem. Biophys. Res. Commun. 76:705, 1979.
170. Bachur, N. R., and Gee, M.: J. Pharmacol. Exp. Ther. 177:567, 1971.
171. Bachur, N. R., *et al.*: J. Med. Chem. 19:651, 1976.
172. Gause, G. F., *et al.*: Antibiotiki 18:675, 1973.
173. Jacquillat, C.: National Cancer Institute Conference on Anthracyclines, Bethesda, MD, April, 1976.
174. Arcamone, F., *et al.*: J. Med. Chem. 19:1424, 1976.
175. Israel, M., *et al.*: 5th Internat. Symp. on Med. Chem., Paris, 1976, Abstracts, p. 63.
176. DiMarco, A., *et al.*: Cancer Treat. Rep. 62:375, 1978.
177. Berlin, Yu. A.: Nature 218:193, 1968.
178. Rao, K. V., *et al.*: Antibiot. Chemother. 12:182, 1962.
179. Nayak, R., *et al.*: FEBS Lett. 30:157, 1973.
180. Gauze, G. F.: Olivomycin, Chromomycin, and Mithramycin, *in* Corcoran, J. W., and Hahn, F. E. (eds.): Antibiotics, III., pp. 197–202, New York, Springer-Verlag, 1975.
181. Kersten, W.: Abh. Deut. Akad, Wiss. Berlin, Kl. Med. 1968:593.
182. Umezawa, H., *et al.*: J. Antibiot., Ser. A. 19:260, 1966.
183. Umezawa, H., *et al.*: J. Antibiot., Ser. A. 19:210, 1966.
184. Ikekawa, T., *et al.*: J. Antibiot., Ser. A. 17:194, 1964.
185. Argoudelis, A. A., *et al.*: J. Antibiot. 24:543, 1971.
186. Takita, T., *et al.*: J. Antibiot. 31:1073, 1978.
187. Sugiura, Y., and Kikuchi, T.: J. Antibiot. 31:1310, 1978.
188. Umezawa, H., *et al.*: J. Antibiot. 25:409, 1972.
189. Nakayama, Y., *et al.*: J. Antibiot. 26:400, 1973.
190. Muraoka, Y., *et al.*: J. Antibiot. 29:853, 1976.
191. Wakaki, S., *et al.*: Antibiot. Chemother. 8:288, 1958.
192. DeBoer, C., *et al.*: Antimicrobial Agents Annual 1960:17, 1961.
193. Webb, J. S., *et al.*: J. Am. Chem. Soc. 84:3185, 1962.
194. Iyer, V. N., and Szybalski, W.: Science 145:55, 1964.
195. Tomasz, M.: Chem.-Biol. Interactions 13:89, 1976.
196. Taylor, W. G., and Remers, W. A.: J. Med. Chem. 18:307, 1975.
197. Patrick, J. B., *et al.*: J. Am. Chem. Soc. 86:1889, 1964.
198. Imai, R.: 24th. Meeting of the Japan Chemotherapy Society, June, 1976, Abstracts.
199. Nakatsubo, F., *et al.*: J. Am. Chem. Soc. 99:8115, 1977.
200. Kinoshita, S., *et al.*: J. Med. Chem. 14:103, 1971.
201. Vavra, J. J., *et al.*: Antibiot. Ann. 1960:230, 1960.
202. Herr, R. R., *et al.*: J. Am. Chem. Soc. 89:4808, 1967.
203. Hessler, E. J., and Jahnke, H. K.: J. Org. Chem. 35:245, 1970.
204. Kennedy, B. J.: Cancer 26:755, 1970.
205. Johnston, T. P., *et al.*: J. Med. Chem. 18:104, 1975.
206. Manaker, R. A.: Antibiot. Ann. 55:853, 1954.
207. Farber, S.: J. Am. Med. Assoc. 198:826, 1966.
208. Livingston, R. B., and Carter, S. K.: Single Agents in Cancer Chemotherapy, New York, Plenum Press, 1970.
209. Arcamone, F., *et al.*: Tetrahedron Lett. 1969:1007.
210. Chabner, B. A., *et al.*: New Engl. J. Med. 292:1107, 1975.
211. Umezawa, H., *et al.*: J. Antibiot., Ser. A., 19:200, 1966.
212. Blum. R. H., *et al.*: Cancer 31:903, 1973.
213. Einhorn, L. H., and Donohue, J.: Ann. Int. Med. 87:293, 1977.
214. Wakaki, S., *et al.*: Antibiot. Chemother. 8:288, 1958.
215. Baker, L. H.: The Development of an Acute Intermittent Schedule—Mitomycin C., *in* Carter, S. K., and Crooke, S. T.

(eds.): Mitomycin C: Current Status and New Developments, p. 77, New York, Academic Press, 1969.

216. Veldhius, J. D.: Lancet 1:1152, 1978.
217. Moertel, C. G., et al.: Cancer Chemother. Rep. 55:303, 1971.
218. Hartwell, L. J.: Lloydia 31: 71, 1968.
219. Kelley, M. G., and Hartwell, J. L.: J. Nat. Cancer Inst. 14:967, 1953.
220. Sartorelli, A. C., and Creasey, W. A.: Ann. Rev. Pharmacol. 9:51, 1969.
221. Loike, J. D., et al.: Cancer Res. 38:2688, 1978.
222. Svoboda, G.: Lloydia 24:173, 1961.
223. Neuss, N., et al.: J. Am. Chem. Soc. 86:1440, 1964.
224. Creasey, W. A.: Vinca Alkaloids and Colchicine, in Sartorelli, A. C., and Johns, D. J. (eds.): Handbook of Experimental Pharmacology, vol. 38, part 2, pp. 670–694, New York, Springer-Verlag, 1975.
225. Barnett, C. J., et al.: J. Med. Chem. 21:88, 1978.
226. Bensch, K. G., and Malawista, S. E.: J. Cell Biol. 40:95, 1969.
227. Journey, L. J., et al.: Cancer Chemother. Rep. 52:509, 1968.
228. Krishan, A.: J. Nat. Cancer Inst. 41:581, 1968.
229. Taylor, E. W.: J. Cell Biol. 25:145, 1965.
230. DeVita, V. T., et al.: Cancer 30:1495, 1972.
231. Rosenberg, B., et al.: Nature 205:698, 1965.
232. Rosenberg, B., et al.: J. Bacteriol. 93:716, 1967.
233. Rosenberg, B., et al.: Nature 222:385, 1969.
234. Gale, G. R.: Platinum Compounds, in Sartorelli, A. C., and Johns, D. J. (eds.): Handbook of Experimental Pharmacology, vol. 38, part 2, pp. 829–838, New York, Springer-Verlag, 1975.
235. Rozencweig, M., et al.: Cisplatin, in Pinedo, H.M. (ed.): Cancer Chemotherapy 1979, p. 107, Amsterdam, Excerpta Medica, 1979.
236. Burchenal, J. H., et al.: Cancer Treatm. Rep. 63:9, 1979.
237. Stearns, B., et al.: J. Med. Chem. 6:201, 1963.
238. Krakoff, I. H., et al.: Cancer Res. 28:1559, 1968.
239. Brown, N. C., et al.: Biochem. Biophys. Res. Commun. 30:522, 1968.
240. Brockman, R. W., et al.: Cancer Res. 30:2358, 1970.
241. Hewlett, J. S.: Proc. Amer. Ass. Cancer Res. 13:119, 1972.
242. Kidd, J. G.: J. Exp. Med. 98:565, 1953.
243. McCoy, T. A., et al.: Cancer Res. 19:591, 1959.
244. Mashburn, L. T. and Wriston, J. C., Jr.: Arch. Biochem. Biophys. 105:450, 1964.
245. Wriston, J. C., Jr.: Enzymes 4:101, 1971.
246. Hrushesky, W. J., et al.: Med. Ped. Oncol. 2:441, 1976.
247. Broome, J. D., and Schwartz, J. H.: Biochem. Biophys. Acta 138:637, 1967.
248. Brockman, R. W., et al.: Proc. Soc. Exp. Biol. 133:609, 1970.
249. Michaud, R. L., and Sartorelli, A. C.: 155th Amer. Chem. Soc. National Meeting, San Francisco, April, 1968. Abstract No. 54.
250. French, F. A., and Freedlander, B. L.: Cancer Res. 18:1298, 1958.
251. Booth, B. A., and Sartorelli, A. C.: Molec. Pharmacol. 3:290, 1967.
252. Pressman, B. C.: J. Biol. Chem. 238:401, 1963.
253. Mihich, E.: Pharmacologist 5:270, 1963.
254. Yesair, D. W., and Kensler, C. J.: The Phthalanilides, in Sartorelli, A. C., and Johns, D. J. (eds.): Handbook of Experimental Pharmacology, vol. 38, part 2, pp. 820–828, New York, Springer-Verlag, 1975.
255. Hirschberg, E., et al.: J. Nat. Cancer Inst. 20:567, 1959.
256. Waring, M. J.: Humangenetik 9:234, 1970.
257. Creighton, A. M., et al.: Nature 222:384, 1969.
258. Creighton, A. M., and Birnie, G. D.: Int. J. Cancer 5:47, 1970.
259. Von Hoff, D. D., et al.: New Anticancer Drugs, in Pinedo, H. M. (ed.): Cancer Chemotherapy 1979, pp. 126–166, Amsterdam, Excerpta Medica, 1979.
260. Witiak, D. T., et al.: J. Med. Chem. 21:1194, 1978.
261. Bond, J. V.: Proc. Amer. Ass. Cancer Res. 3:306, 1962.
262. Legha, S. S., et al.: Cancer 38:27, 1976.
263. Wilson, W. R.: Chem. N.Z. 37:148, 1973.
264. Atwell, G. J., et al.: J. Med. Chem. 15:611, 1972.
265. Bristol Laboratories: Platinol Product Monograph; Syracuse, N.Y.

266. Einhorn, L. H., and Donohue, J.: Ann. Int. Med. 87:293, 1977.
267. Einhorn, L. H.: Combination Chemotherapy with Cis-Diammindichloroplatium, Vinblastine, and Bleomycin in Disseminated Testicular Cancer, in Carter, S. K., et al. (eds.): Bleomycins-Current Status and New Developments, New York, Academic Press, 1978.
268. Hantzsch, A.: Ann. 299:99, 1898.
269. Horrom, B. W., and Carbon, J. A.: Ger. Pat. 1,138,781 (Oct. 31, 1962).
270. Gorski, J., et al.: J. Cell Comp. Physiol. 66:91, 1965.
271. Jensen, E. V., and Jacobson, H. I.: Rec. Progr. Hormone Res. 18:387, 1962.
272. Bruchovsky, N., and Wilson, J.D.: J. Biol. Chem. 243:2012, 1968.
273. Sherman, M. R., et al.: J. Biol. Chem. 245:6085, 1970.
274. Wira, C., and Munck, A.: J. Biol. Chem. 245:3436, 1970.
275. Schinzinger, A.: 18th Kongress als Beilage zum Centralblatl f. Chir. 29:5, 1889.
276. Dao, T. L.: Some Current Thoughts on Adrenalectomy, in Segaloff, A., et al. (eds.): Current Concepts of Breast Cancer, pp. 189–199, Baltimore, Williams & Wilkins, 1967.
277. Kennedy, B. J.: Cancer 18:1551, 1965.
278. Pearson, O. H., et al., in Dao, T.L. (ed.): Estrogen Target Tissues and Neoplasia, pp. 287–305, Chicago, University of Chicago Press, 1972.
279. Beckett, V. L., and Brennan, M. J.: Surg. Gynecol. Obstet. 109:235, 1959.
280. Dao, T. L.: Pharmacology and Clinical Utility of Hormones in Hormone Related Neoplasma, in Sartorelli, A. C., and Johns, D. J. (eds.): Handbook of Experimental Pharmacology, vol. 38, part 2, p. 172, New York, Springer-Verlag, 1975.
281. Bloom, H. J. G.: Brit. J. Cancer 25:250, 1971.
282. Kelley, R., and Baker, W.: Clinical Observations on the Effect of Progesterone in the Treatment of Metastatic Endometrial Carcinoma, in Pincus, G., and Vollmer, E. P. (eds.): Biological Activities of Steroids in Relation to Cancer, pp. 427–443, New York, Academic Press, 1960.
283. Dougherty, T. F., and White, A.: Amer. J. Anat. 77:81, 1965.
284. Heilman, F. R., and Kendall, E. C.: Endocrinology 34:416, 1944.
285. Dao, T. L.: Third National Cancer Conference Proceedings, pp. 292–296, Philadelphia, Lippincott, 1957.
286. Hart, M. M., and Straw, J. A.; Steroids 17:559, 1971.
287. Bergenstal, D. M., et al.: Ann. Int. Med. 53:672, 1960.
288. Floos, H. J., et al.: J. Pharm. Sci. 62: 699, 1973.
289. Haller, H. L., et al.: J. Am. Chem. Soc. 67:1600, 1945.
290. Cueto, C., and Brown, J. H. V.: Endocrinology 62:326, 1958.
291. Ringold, H. E., Batres, E., Halpern, O., and Necoecha, E.: J. Am. Chem. Soc. 81:427, 1959.
292. Fried, J., Thoma, R. W., and Klingsberg, A.: J. Am. Chem. Soc. 75:5764, 1953.
293. Ringold, H. E., Ruelas, J. P., Batres, E., and Djerassi, C.: J. Am. Chem. Soc. 81:3712, (1959).
294. Imperial Chemical Industries: Belg. Patent 637,389, Mar. 13, 1964.
295. Bedford, G. R., and Richardson, D. N.: Nature 212:733, 1966.
296. Jordan, V. C., and Jaspan, T.: J. Endocrinology 68:453, 1976.
297. Morton, D.: Report to the American Association for the Advancement of Science, San Francisco, February, 1974.
298. O'Brien, P. H.: J. South Carolina Med. Assoc. 68:466, 1972.
299. Goodnight, J. E., Jr., and Morton, D. L.: Ann. Rev. Med. 29:231, 1978.
300. Sanders, H. J.: Chem. and Eng. News, p. 74, Dec. 23, 1974.

SELECTED READINGS

Baker, B. R.: Design of Active-Site-Directed Irreversible Enzyme Inhibitors, New York, Wiley, 1967.

Baserga, R. (ed.): The Cell Cycle and Cancer, New York, Marcel Dekker, 1971.

Brodsky, I., and Kahn, S. B. (eds.): Cancer Chemotherapy II, New York, Grune & Stratton, 1972.

Calabresi, P., and Parks, R. E., Jr.: Chemotherapy of Neoplastic Diseases, *in* Goodman, L. S., and Gilman, A. (eds.): The Pharmacological Basis of Therapeutics, ed. 5, pp. 1248–1307, New York, Macmillan, 1975.

Cline, M. J., and Haskell, C. M.: Cancer Chemotherapy, Philadelphia, W.B. Saunders, 1980.

Devita, V. T., Jr., and Busch, H. (eds.): Methods in Cancer Research, vol. 16, New York, Academic Press, 1979.

Ferguson, L. N.: Cancer and Chemicals, Chem. Rev. 75: 289, 1975.

Greenspan, E. M. (ed.): Clinical Cancer Chemotherapy, New York, Raven Press, 1975.

Greenwald, E. S.: Cancer Chemotherapy, Flushing, N.Y., Medical Examination Publishing Co., 1973.

Holland, J. F., and Frei, E., III (eds.): Cancer Medicine, Philadelphia, Lea & Febiger, 1973.

Livingston, R. B., and Carter, S. K.: Single Agents in Cancer Chemotherapy, New York, Plenum Press, 1970.

McKearns, K. W. (ed.): Hormones and Cancer, New York, Academic Press, 1974.

Mongomery, J. A.: *et al.:* Drugs for Neoplastic Disease, *in* Wolff, M. E. (ed.): Burgers' Medicinal Chemistry, ed. 4, part 2, pp. 595–670, New York, Wiley, 1979.

Pinedo, H. M. (ed.): Cancer Chemotherapy 1979, Amsterdam-Oxford, Excerpta Medica, 1979.

Plants and Cancer, *in* Proceedings of the 16th Annual Meeting of the Society for Economic Botany: Cancer Treat. Rep. 60: 973, 1976.

Remers, W. A.: The Chemistry of Antitumor Antibiotics, vol. 1, New York, Wiley, 1979.

Ross, W. C. J.: Biological Alkylating Agents, London, Butterworths, 1962.

Sartorelli, A. C., and Johns, D. J., (eds.): Antineoplastic and Immunosuppressive Agents: Handbook of Experimental Pharmacology, vol. 38, parts 1 and 2, New York, Springer-Verlag, 1974 and 1975.

Schnitzer, R. J., and Hawking, F. (eds.): Experimental Chemotherapy, vols. 4 and 5, New York, Academic Press, 1967.

Stoll, B. A., (ed.): Endocrine Therapy in Malignant Disease, Philadelphia, W. B. Saunders, 1972.

Suhadolnik, R. J.: Nucleoside Antibiotics, New York, Wiley, 1970.

9
central nervous system depressants

T. C. Daniels

E. C. Jorgensen

The agents described in this chapter produce depressant effects on the central nervous system as their principal pharmacologic action. These include the general anesthetics, hypnotic-sedatives, central nervous system depressants with skeletal-muscle-relaxant properties, tranquilizing agents and anticonvulsants. The general anesthetics (e.g., ether) and hypnotic-sedatives (e.g., phenobarbital) produce a generalized or nonselective depression of central nervous system function and overlap considerably in their depressant properties. Many sedatives, if given in large enough doses, produce anesthesia. The central depressant properties of a group of skeletal-muscle-relaxing agents (e.g., meprobamate) resemble closely the depressant properties of the hypnotic-sedatives. The tranquilizing agents (e.g., reserpine, chlorpromazine) exert a more selective action and, even in high doses, are incapable of producing anesthesia. The anticonvulsants also act in a more selective fashion, modifying the brain's ability to respond to seizure-evoking stimuli.

The analgesics, another group of agents which selectively depress central nervous system function, are discussed separately in Chapter 17.

The brain possesses a unique and specialized mechanism for excluding many substances presented to it by the circulation.[1] This *blood-brain barrier* appears to involve a complex interplay of anatomic, physiologic and biochemical factors. The capillary endothelium and surrounding glial cells play an important role among the anatomic components that may selectively bar or admit substances to the functional areas of the brain. There are few metabolic reserves in the central nervous system, and a substantial flow of nutrients and oxygen must be supplied continuously.

Cerebral circulation is large, but, in spite of this, only minute amounts of exogenous substances are accepted by the brain. The concept of the blood-brain barrier is used comprehensively to describe all phenomena which either hinder *or* facilitate the penetration of substances into the central nervous system. Penetration may occur by many different mechanisms: dialysis, ultrafiltration, osmosis, Donnan equilibrium, lipid-soluble, active transport, or diffusion due to concentration differences created by special tissue affinities or metabolic activity. In general, however, lipid-soluble, nonionized molecules pass most readily into the central nervous system, whether their ultimate pharmacologic effect is depression or stimulation. Except for the relatively few active transport systems involving ionic molecules, weak acids or weak bases pass into the brain when their acid or base strengths are such that a high proportion exists as the nonionized lipid-soluble form at the pH of the plasma (pH 7.4). For this reason, metabolically induced changes in plasma pH, such as those produced by respiratory acidosis or alkalosis, may strongly influence the effects of drugs on the central nervous system. Penetration of the brain by weak acids such as phenobarbital and acetazolamide is increased under conditions of hypercapnia (plasma pH 6.8), and decreased with hypocapnia (plasma pH 7.8). The converse would be true of weak bases, such as amphetamine.

GENERAL ANESTHETICS

General anesthesia is the controlled, reversible depression of the functional activity of the central nervous system, producing loss of sensation and consciousness. The relief of pain through general anesthesia during surgery was first carried out by Crawford Long (Georgia, 1841), who used ether during the removal of a cyst. However, it was the use of nitrous oxide anesthesia by Horace Wells (Connecticut, 1844) during extraction of a tooth that excited in William Morton, a dental associate, awareness of the possibilities of anesthesia during surgery. Morton, while a student at Harvard Medical School, learned of the anesthetizing properties of ethyl ether from his chemistry instructor, Professor Charles Jackson. Morton then persuaded the professor of surgery, J. C. Warren, to allow him to administer ether as a general anesthetic during surgery. The success attending this led to the rapid introduction of ether anesthesia for surgical operations. The word anesthesia, signifying insensibility, was coined by Oliver Wendell Holmes in a letter to Morton shortly after his successful demonstration. Chloroform was introduced in Edinburgh in 1847, and the search for new and better anesthetics has continued to this day.

The stages of anesthesia developed are related to functional levels of the central nervous system successively depressed and are present to varying degrees for all agents capable of producing general anesthesia.

Stage I (Cortical Stage): Analgesia is produced, consciousness remains, but the patient is sleepy as the higher cortical centers are depressed.

Stage II (Excitement): Loss of consciousness results, but depression of higher motor centers involving the brain stem and the cerebellum leads to excitement and delirium.

Stage III (Surgical Anesthesia): Spinal cord reflexes are diminished in activity, and skeletal-muscle relaxation is obtained. This stage, in which most operative procedures are performed, is further divided into Planes i–iv, mainly differentiated on the basis of increasing somatic-muscle relaxation and decreased respiration.

Stage IV (Medullary Paralysis): Respiratory failure and vasomotor collapse occur, due to depression of vital functions of the medulla and the brain stem.

General anesthesia may be produced by a variety of chemical types and routes of administration. Inhalation of gases or the vapors from volatile liquids is by far the most frequently used, although the intravenous and the rectal routes ("fixed anesthetics") are also used. The manner in which the general anesthetics of widely varying structure act to depress central nervous system function is unknown. Theories derived from the relatively high lipid-solubility of most members of this class, the size of anesthetic molecules and their participation in the formation of hydrate microcrystals are discussed in Chapter 2.

HYDROCARBONS

The saturated hydrocarbons possess an anesthetic effect which increases from methane to octane and then decreases. However, toxicity is too high in these to be useful, and only hydrocarbons with unsaturated character are used.

Cyclopropane U.S.P., trimethylene. Cyclopropane was introduced for use as a general anesthetic in 1934 and is the most potent gaseous anesthetic agent currently in use. Following premedication with depressant drugs such as morphine or barbiturates, surgical anesthesia may be obtained with concentrations of about 15 volume percent cyclopropane and 85 volume percent oxygen. Induction of anesthesia is rapid, requiring only 2 to 3 minutes. Cyclopropane is rapidly eliminated by the lungs. The high potency of the anesthetic, which allows use of high oxygen concentrations, is particularly advantageous in providing adequate tissue oxygenation.

Cyclopropane

Cyclopropane is a colorless gas, b.p. $-33°$, which liquefies at 4 to 6 atmospheres pressure. It is flammable and forms an explosive mixture with air in a concentration range from 3.0 to 8.5 percent; in oxygen, 2.5 to 50 percent. The solubility is about 1 volume of gas in 2.7 volumes of water; it is freely soluble in alcohol.

HALOGENATED HYDROCARBONS

The anesthetic potency of the lower molecular weight hydrocarbons is increased as hydrogen is successively replaced by halogen. Thus, anesthetic potency increases in the order methane, methyl chloride, dichloromethane, chloroform, carbon tetrachloride. However, the favorable factors of general increase in anesthetic potency and decrease in flammability are counterbalanced by a general increase in toxicity which has limited the anesthetic applications of the halogenated hydrocarbons.

Ethyl Chloride U.S.P., chloroethane, CH_3CH_2Cl. Ethyl chloride is capable of producing rapid induction of anesthesia, followed by rapid recovery after administration ceases. Surgical anesthesia may be obtained with 4 volume percent of the vapor. However, its potential for producing liver damage and cardiac arrhythmias has limited its application in anesthesia to occasional use as an induction anesthetic before another agent is used.

It is also used to produce local anesthesia of short duration. When sprayed on the unbroken skin, rapid evaporation freezes the tissues and allows short, minor operations to be performed.

Ethyl chloride is a gas, b.p. 12°, available under pressure in the liquid form. It is flammable, and explosive when mixed with air.

Halothane U.S.P., Fluothane®, 2-bromo-2-chloro-1,1,1-trifluoroethane. Halothane, a general anesthetic with potency estimated at 4 times that of ether, was introduced in 1956. Experience with the chemical inertness and low toxicity of fluorinated hydrocarbons containing the CF_3 or CF_2 groupings, and use as refrigerants, led to the development of halothane as an anesthetic.[2] The presence of a trifluoromethyl group, and bromine, chlorine and hydrogen atoms on a single carbon atom, produced an asymmetric molecule, not yet separated into its diasteriomeric forms, with physical properties and anesthetic potency close to those of chloroform ($CHCl_3$), but with much lower toxicity. The solubility and vapor pressure of halothane were found to be in a desirable range for the potential production of anesthesia, as proposed by Ferguson's principle (see Chap. 2). In addition, the high electronegativity of the fluorine atom stabilizes the C—F bonds of CF_3, but tends to weaken the adjacent C—C and C—halogen bonds. As a result, the major metabolic products are chloride and bromide ions, and trifluoroacetic acid (CF_3COOH).

$$\begin{array}{ccc} F & Cl & \\ | & | & \\ F-C-C-H \\ | & | & \\ F & Br & \end{array}$$
Halothane

The induction period is very rapid, surgical anesthesia being produced in 2 to 10 minutes. Recovery is equally rapid following removal of anesthetic. Side-effects produced by halothane are hypotension and bradycardia. Liver necrosis, in some cases similar to that induced by chloroform and carbon tetrachloride, has been observed. An impurity, 2,3-dichloro-1,1,1,4,-4,4-hexafluorobutene-2, has been found in low concentrations in freshly opened bottles of halothane.[3] Its concentration is increased in the presence of copper, oxygen and heat. This impurity has proved to be acutely toxic to dogs in anesthetic concentrations and produces degenerative changes in the lungs, the liver and the kidney of rats. It is possible that the degradation product may be responsible for liver damage in the United States where copper vaporizers are widely used. Halothane is a volatile, nonflammable liquid, b.p. 50°, which is given by inhalation. Anesthesia may be induced with 2 to 2.5 percent, vaporized by a flow of oxygen. The compound is sensitive to light and is distributed in brown bottles, stabilized by the addition of 0.01 percent thymol.

ETHERS

Ether U.S.P., ethyl ether, diethyl ether, $CH_3CH_2OCH_2CH_3$. Ether was the first (1842) of the general anesthetics used in surgical anesthesia, and because of the wealth of knowledge and experience concerning its effects in each plane of anesthesia, it is still one of the safest. The prolonged induction time may be avoided by the initial use of a more rapidly acting agent (e.g., vinyl ether, nitrous oxide), followed by a gradual change to ether at the proper concentration for maintenance.

Ether is flammable and forms explosive mixtures with air and oxygen. It occurs as a colorless, mobile liquid having a burning, sweetish taste and a characteristic odor. Ether U.S.P. is intended for anesthetic use and thus has rigid specifications as to content and method of handling. It may contain up to 4 percent of alcohol and water. The alcohol has little value as a preservative but does raise the boiling point and prevent frosting on the anesthetic mask. In the *U.S.P.*, a caution limits the size of containers to 3 kilos and permits the ether to be used only up to 24 hours after the container has been opened.

The ether must be free of acids, aldehydes and peroxides. Acids are tested for by using 0.02N sodium hydroxide. A test for aldehydes, sensitive to 1 part in 1,000,000, using Nessler's solution (alkaline mercuric-potassium iodide T.S.) shows no yellow color when the test is negative. The peroxide test is carried out on 10 ml. of ether. One ml. of potassium iodide T.S. is added, and the mixture is shaken for 1 hour. If peroxides are present, the potassium iodide, as a reducing agent, has its iodide ion oxidized to free elemental iodine (colored).

Vinyl Ether U.S.P., divinyl oxide, $CH_2=CH-O-CH=CH_2$. Vinyl ether alone is useful for induction anesthesia, with production of onset in about 30 seconds, or for short operative procedures. It produces extensive liver damage on prolonged use.

Vinyl ether is a volatile liquid, b.p. 28°, with about the same explosive hazard as ether.

Methoxyflurane U.S.P., Penthrane®, 2,2-dichloro-1,1-difluoroethyl methyl ether, methyl β-dichloro-α-difluoroethyl ether, $CHCl_2-CF_2-O-CH_3$. Methoxyflurane is a volatile liquid, b.p. 101°, whose vapors produce general anesthesia with a slow onset and a fairly long duration of action. It was introduced in 1962. It is nonflammable in any concentration in air or oxygen and produces anesthesia at concentrations of 1.5 to 3 percent when vaporized by a rapid flow of oxygen. Methoxyflurane is a stable compound, even in the presence of bases. It is metabolized in the liver to produce inorganic fluoride ion which is responsible for a "fluoride diabetes insipidus" seen in most patients treated with methoxyflurane. This is characterized by unresponsiveness to vasopressin (ADH), leading to polyuria, dehydration, thirst and plasma hyperosmolality.

Enflurane U.S.P., Ethrane®, 2-chloro-1,1,2-trifluoroethyl difluoromethyl ether, HF_2COCF_2CHFCl. Enflurane is a stable, colorless, nonflammable halogenated ether with physical and anesthetic properties similar to those of halothane. The compound may be vaporized for inhalation anesthesia in oxygen or nitrous oxide and oxygen, and in a 2 to 5 percent concentration gives an induction time of from 4 to 6 minutes. It is available as a liquid in 125- and 250-ml. containers.

ALCOHOLS

Alcohols, most notably ethanol, have been long known and used for their ability to depress certain higher centers of the central nervous system. As the series is ascended, hypnotic activity of the normal alcohols reaches a maximum at 6 or 8 carbons and then declines as the alkyl chain is further lengthened. None of the unsubstituted alcohols possesses sufficient potency for use as a general anesthetic. Some halogenated alcohols, particularly those bearing 3 bromine or chlorine atoms on a single carbon atom (e.g., tribromoethanol, trichloroethanol), are potent hypnotics and are capable of producing basal anesthesia.

ULTRASHORT-ACTING BARBITURATES

The ultrashort-acting barbiturates (Table 9-2), as their sodium salts, may be administered intravenously or by retention enema for the production of surgical anesthesia. Anesthesia begins rapidly (in less than 1 minute) and is usually of short duration. Intravenous barbiturate anesthesia provides smooth induction, with rapid passage through the excitement stage, absence of salivary secretions, fair muscular relaxation, nonexplosive properties, and rapid depletion of the agent from the central nervous system, leading to a short and uncomplicated period of postoperative recovery. Disadvantages include potent respiratory depression, tissue-irritating properties on extravasation, laryngospasm, and a precipitous fall in blood pressure if the intravenous barbiturates are administered too rapidly.

The rapid onset and brief duration of action of the ultrashort-acting barbiturate anesthetics have been considered to be due to their high lipid-solubility, enabling free passage through the lipoid cellular membrane of the blood-brain barrier, followed by rapid loss to the peripheral lipoidal storage areas.[4] Metabolism of the barbiturates occurs too slowly to have any significant influence on the onset and duration of anesthesia. Thiopental, for example, is metabolized at the rate of 10 to 15 percent per hour. However, the rate at which fat concentrates thiopental following its intravenous injection is too slow to account for the rate at which the central nervous system is depleted.[5] Instead, the lean body tissues (e.g., muscle), which are well perfused with blood, provide the initial pool and rapidly take up most of the thiopental lost by the brain. Redistribution to body fat and metabolic degradation appear to occur more slowly and do not account for the rapid recovery from thiopental anesthesia.

The production of surgical anesthesia by barbiturates for short operations is not without dangers, chiefly because of variations in individual susceptibility to respiratory depression and to other side-effects. In general, the ultrashort-acting barbiturates produce a greater degree of respiratory depression for a given degree of skeletal muscular relaxation than do the inhalation anesthetics.

Methohexital Sodium, Brevital® Sodium, sodium α-($\pm$)-1-methyl-5-allyl-5-(1-methyl-2-pentynyl)barbiturate. Methohexital sodium was introduced in 1960

TABLE 9-1

GASEOUS AND LIQUID ANESTHETICS

Name Proprietary Name	Physical State	Category	Application
Cyclopropane U.S.P.	Colorless gas	General anesthetic (inhalation)	By inhalation as required
Ethyl Chloride U.S.P.	Colorless gas	Local anesthetic	Topical, spray on intact skin
Halothane U.S.P. *Fluothane*	Liquid, b.p. 49–51°	General anesthetic (inhalation)	By inhalation as required
Ether U.S.P.	Liquid, b.p. 35°	General anesthetic (inhalation)	By inhalation as required
Vinyl Ether U.S.P. *Vinethene*	Liquid, b.p. 28–31°	General anesthetic (inhalation)	By inhalation as required
Enflurane, U.S.P. *Ethrane*	Liquid, b.p. 55.5–57.5°	General anesthetic (inhalation)	By inhalation as required
Methoxyflurane U.S.P. *Penthrane*	Liquid, b.p. 105°	General anesthetic (inhalation)	By inhalation as required

TABLE 9-2

ULTRASHORT-ACTING BARBITURATES USED TO PRODUCE GENERAL ANESTHESIA

General Structure

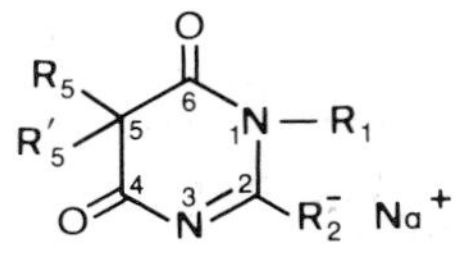

Generic Name Proprietary Name	Substituents			
	R_5	R'_5	R_1	R_2
Methohexital Sodium *Brevital Sodium*	$CH_2{=}CH{-}CH_2{-}$	$CH_3CH_2C{\equiv}C{-}\overset{\displaystyle CH_3}{\underset{\textstyle \vert}{CH}}{-}$	CH_3	O
Thiamylal Sodium *Surital Sodium*	$CH_2{=}CH{-}CH_2{-}$	$CH_3CH_2CH_2\overset{\displaystyle CH_3}{\underset{\textstyle \vert}{CH}}{-}$	H	S
Thiopental Sodium *Pentothal Sodium*	$CH_3CH_2{-}$	$CH_3CH_2CH_2\overset{\displaystyle CH_3}{\underset{\textstyle \vert}{CH}}{-}$	H	S

as an intravenously administered ultrashort-acting barbiturate. Induction of anesthesia with methohexital is as rapid as with thiopental, and recovery is more rapid, perhaps due to a faster metabolism. The drug has no muscle-relaxant properties; for surgical procedures requiring muscle relaxation, it requires supplementation with a gaseous anesthetic and a muscle relaxant. Methohexital sodium made from methohexital, sodium hydroxide and sodium carbonate, is supplied in crystalline form together with anhydrous sodium carbonate and is administered only by the intravenous route.

Hexobarbital U.S.P., Evipan Sodium, N-methyl-5-cyclohexenyl-5-methylbarbiturate. Hexobarbital is a nonthiobarbiturate similar in action to thiopental, but is used primarily as a short-acting sedative. Its ultra-short duration of action has been attributed to the influence of the N-methyl group on acid strength. The N-methyl-substituted barbiturates have a pK_a of approximately 8.4 compared with 7.6 for those without the N-methyl group. At physiologic pH, hexobarbital exists largely in the undissociated form. It has been suggested that its relatively high lipid solubility causes it to partition rapidly into neutral fatty depots, thus decreasing the blood levels below that required for sedation. The drug is no longer used as a general anesthetic but is used as a sedative–hypnotic. The usual adult dose is from 200 to 500 mg. by mouth.

Thiamylal Sodium, Surital® Sodium, sodium 5-allyl-5-(1-methylbutyl)-2-thiobarbiturate. Thiamylal sodium was introduced in 1952 as an ultrashort-acting intravenous anesthetic. It is available in vials as a

TABLE 9-3

ULTRASHORT-ACTING BARBITURATES

Name Proprietary Name	Preparations	Category	Usual Adult Dose*
Methohexital Sodium *Brevital Sodium*	Methohexital Sodium for Injection U.S.P.	General anesthetic (intravenous)	I.V., 5 to 12 ml. of a 1 percent solution at the rate of 1 ml. every 5 seconds for induction, then 2 to 4 ml. every 4 to 7 minutes as required
Hexobarbital U.S.P. *Sombulex*	Hexobarbital Tablets U.S.P.	Sedative	Hypnotic—oral, 250 to 500 mg. at bedtime
Thiamylal Sodium *Surital Sodium*	Thiamylal Sodium for Injection U.S.P.	General anesthetic (systemic)	I.V., induction, 3 to 6 ml. of a 2.5 percent solution at the rate of 1 ml. every 5 seconds; maintenance, 500 μl. to 1 ml. as required
Thiopental Sodium U.S.P. *Pentothal Sodium*	Thiopental Sodium for Injection U.S.P.	Anticonvulsant; general anesthetic (intravenous)	Anticonvulsant—I.V., 3 to 10 ml. of a 2.5 percent solution over a 10-minute period; anesthetic (induction)—I.V., 2 to 3 ml. of a 2.5 percent solution at intervals of 30 to 60 seconds as necessary

* See *U.S.P. D.I.* for complete dosage information.

sterile powder admixed with anhydrous sodium carbonate as a buffer. Onset of anesthesia occurs in 20 to 60 seconds and the effect lasts for 10 to 30 minutes after the last injection.

A nonsterile form of thiamylal sodium, characterized by the green dye which it contains, is available for rectal instillation. The dose of a 5 to 10 percent solution is determined by the physician.

Thiopental Sodium U.S.P., Pentothal® Sodium, sodium 5-ethyl-5-(1-methylbutyl)-2-thiobarbiturate. Thiopental sodium has been the most widely used of the intravenous barbiturate anesthetics. Onset is rapid (about 30 seconds) and duration brief (10 to 30 minutes). Ampules containing the sterile white to yellowish-white powder also contain anhydrous sodium carbonate as a buffer.

Thiopental sodium is available as a nonsterile powder containing a green dye and is intended for rectal application by enema; 45 mg. per kg. of body weight, in 10 percent solution; range, 25 to 45 mg. per kg.

MISCELLANEOUS

Ketamine Hydrochloride U.S.P., Ketalar®, (±)-2-(o-chlorophenyl)-2-methylaminocyclohexanone hydrochloride. Ketamine hydrochloride was introduced in 1970 as an anesthetic agent, with rapid onset and short duration of action on parenteral administration. Unlike the ultrashort-acting barbiturates, which are sodium salts of acids, ketamine is solubilized for parenteral administration as the hydrochloride of a weakly basic amine. Anesthesia is produced within 30 seconds after intravenous administration, with the effect lasting for 5 to 10 minutes. Intramuscular doses bring on surgical anesthesia within 3 to 4 minutes, with a duration of action of 12 to 25 minutes. It may also be used as an induction anesthetic, prior to the use of other anesthetics, or administered together with volatile anesthetics.

Ketamine Hydrochloride

Ketamine was developed as a structural analog of phencyclidine [1-(1-phenylcyclohexyl)piperidine], a parenteral anesthetic. Like the parent compound, ketamine has produced disagreeable dreams or hallucinations during the brief period of awakening and reorientation. Other untoward effects, including moderate increase in blood pressure, are minimal.

Ketamine hydrochloride is a water-soluble white crystalline powder. The drug is available at concentrations of 10 mg. per ml. or 50 mg. per ml.

Category—general anesthetic (systemic).

Usual dose—I.V., 1 to 4.5 mg. per kg. of body weight for induction, administered slowly produces anesthesia in 1 minute (lasting 5 to 10 minutes); I.M., 6.5 to 13 mg. per kg. of body weight for induction, then one half to full induction dose for maintenance, as required.*

* See *U.S.P. D.I.* for complete dosage information.

Occurrence
Ketamine Hydrochloride Injection U.S.P.

Nitrous Oxide U.S.P., nitrogen monoxide, N_2O, is useful for the rapid induction of anesthesia. For surgical anesthesia, a concentration of 80 to 85 percent is required. The 15 to 20 percent oxygen which may be used with nitrous oxide provides borderline oxygenation of tissues, and the gas is not recommended for prolonged administration.

Category—general anesthetic (inhalation).

Application—by inhalation, 60 to 80 percent, with oxygen 20 to 40 percent, as required.

SEDATIVES AND HYPNOTICS

Historically, the first sedative-hypnotic was ethanol, obtained by fermentation of a variety of carbohydrates. The opium poppy provided the limited armamentarium of the early physician with a second source of a depressant drug. The introduction of inorganic bromides as sedative-hypnotics and anticonvulsants in the 1850's was followed shortly by the development of the effective depressants chloral, paraldehyde, sulfonal and urethan. The recognition of the depressant properties of the barbiturates in 1903 was followed by a variety of related sedative-hypnotics which possess many properties in common.

A characteristic shared by all of the sedative-hypnotic drugs is the general type of depressant action on the cerebrospinal axis. In their clinical applications they differ mainly in the time required for onset of depression and in the duration of the effect produced. The degree of depression depends largely upon the potency of the agent selected, the dose used and the route of administration. All sedative-hypnotic drugs are capable of producing depression ranging from slight sedation, a condition in which the patient is awake but possesses decreased excitability, to sleep. In sufficiently high doses, depression of the central nervous system continues, and many sedative-hypnotic agents may produce surgical anesthesia which resembles that brought about by the volatile anesthetics.

The same sequelae of events, including a stage of excitement due to depression of the higher cortical centers, proceed into surgical anesthesia with both sedative-hypnotics and anesthetics. However, the dangers attending use of anesthetic doses of these drugs largely limit their use to the production of sedation and sleep. The longer-acting central nervous system depressants are usually selected for the production of sedation. The situations in which such sedation is useful include[6]: (1) sudden, limited stressful situations involving great emotional strain, (2) chronic tension states created by disease or sociologic factors, (3) hypertension, (4) potentiation of analgesic drugs, (5) the control of convulsions, (6) adjuncts to anesthesia, (7) narcoanalysis in psychiatry.

The hypnotic dose is used to overcome insomnia of many types. Sedative-hypnotics with a short to moderate duration of action are useful in relieving the insomnia of individuals whose high level of activity during the day makes it difficult for them to decrease their activities as a prelude to sleep. Once asleep, they have no more need for the drug, and the shorter-acting agents provide little after-depression. For others, who for reasons of health, external disturbances or psychic abnormalities awake frequently during the night, the longer-acting agents are more useful.

The use of hypnotic drugs for the treatment of insomnia should normally be restricted to short-term use because there is little evidence that the sedative–hypnotic drugs continue to be effective when used nightly over long periods. Sleep laboratory research on most hypnotics has found them to lose their effectiveness within 3 to 14 days of continuous use.

STRUCTURE-ACTIVITY RELATIONSHIPS

Although the sedative-hypnotic drugs include many chemical types, they have certain common physicochemical and structural features. The polar portion of the molecule is one of the most water-solubilizing of the nonionic functional groups. These include, with their Hansch π values[7] as measures of polar character (see Chap. 2), the unsubstituted barbituric acid nucleus

$$-\overset{|}{\underset{|}{C}}-\underset{CONHCO,}{CONH} \quad \pi = -1.35,$$

acyclic diureides ($-CONHCONHCO-$, $\pi = -1.68$), amides ($-CONH_2$, $\pi = -1.71$), alcohols ($-OH$, $\pi = -1.16$), carbamates ($-OCONH_2$, $\pi = -1.16$), and sulfones ($-SO_2CH_3$, $\pi = -1.26$). These polar groups are attached to a nonpolar moiety, usually alkyl, aryl

or haloalkyl, so that the partition coefficient between a lipid and an aqueous phase (octanol-water) for the resulting molecule is close to 100 (log P = 2). In general, the potency of many classes of sedative-hypnotic drugs varies in a parabolic fashion, with a maximum close to a partition coefficient value of log P = 2.0.

It appears that these molecules have the proper solubility characteristics to be absorbed from the gastrointestinal tract, to be transported in the aqueous body fluids and to be sufficiently lipophilic so that they readily penetrate the central nervous system where their nonionic surfactant characteristics may serve to distort essential lipoprotein matrices, thus depressing function.

In addition to their solubility characteristics, most of the useful sedative-hypnotic drugs possess structural features which resist the rapid metabolic attack which their partition behavior would normally facilitate (see Chap. 3).

Thus, tertiary alcohols, which are resistant to metabolic oxidation, are generally more effective than primary or secondary alcohols, which are rapidly oxidized, conjugated and excreted. Amides and carbamates are generally hydrolyzed slowly as compared with esters. Sulfones require cleavage of a carbon-sulfur bond for further metabolic oxidation to occur, and they are generally excreted unchanged.

BARBITURATES

The barbiturates are widely used as sedative-hypnotic drugs. Barbital, the first member of the class, was introduced in 1903; the method of synthesis of the thousands of analogs prepared since has undergone little change.

Diethylmalonate reacts with alkyl halides in the presence of sodium alkoxides to form the intermediate monoalkyl malonic ester. This may be allowed to react with a different alkyl halide to form a dialkyl malonic ester, which may be condensed with urea in the presence of a sodium alkoxide to form the sodium salt of 5,5-dialkylbarbituric acid. In an acid environment, the free 5,5-dialkylbarbituric acid is formed.

If thiourea is used in place of urea in the condensation, thiobarbiturates which contain a sulfur atom attached to the 2-carbon atom are formed. The use of N-methylurea in this condensation leads to the 1-methylbarbiturates.

All of the barbiturates are colorless, crystalline solids that melt at from 96° to 205°. They are not very soluble in water but form sodium salts that are quite soluble. Solutions of the latter are usually rather strongly alkaline and often hydrolyze enough to give precipitates of the barbiturate. Any admixture with

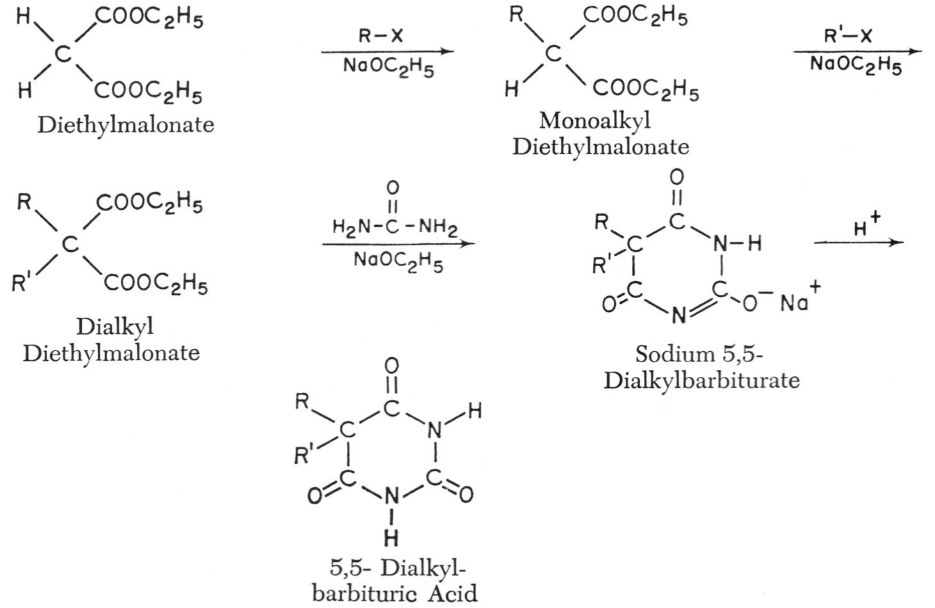

$$\text{H}_2\text{C}(\text{COOC}_2\text{H}_5)_2 \xrightarrow[\text{NaOC}_2\text{H}_5]{\text{R}-\text{X}} \text{RHC}(\text{COOC}_2\text{H}_5)_2 \xrightarrow[\text{NaOC}_2\text{H}_5]{\text{R}'-\text{X}}$$

Diethylmalonate · Monoalkyl Diethylmalonate

Dialkyl Diethylmalonate

$$\xrightarrow[\text{NaOC}_2\text{H}_5]{\text{H}_2\text{N}-\text{CO}-\text{NH}_2}$$

Sodium 5,5-Dialkylbarbiturate $\xrightarrow{\text{H}^+}$

5,5-Dialkyl-barbituric Acid

acidic substances will be almost certain to give such a precipitate, an incompatibility that frequently must be considered in the dispensing laboratory. The alkalinity of the solutions for parenteral injections can be overcome largely by appropriate buffering, principally by the use of sodium carbonate.

Many of the names of the barbituric acid derivatives end with the suffix "al." This has been used to denote hypnotics since the introduction of chloral hydrate, in 1896, and has been applied to a wide variety of chemical types (e.g., Carbromal, Veronal, Luminal).

The action of an ideal hypnotic would be exerted only on cells in the psychic center of the brain and on the centers of pain perception, with no effect on those of motor control, of the automatic process, such as respiration and circulation, or on any other functions or glands. Therefore, the ideal agent would bring about sleep and freedom from pain without interfering with other normal processes; the effect should be of sufficient duration for the purpose intended, and there should be no undesirable secondary reactions. While no such agent has yet been discovered, the barbiturates appear to approach closest to these criteria, although chloral, paraldehyde, codeine and a few others seem to be of advantage under differing circumstances.

The mechanism by which hypnotics bring about the desired selective depression is not well understood, but the barbiturates are believed to act on the brain stem reticular formation to reduce the number of nerve impulses ascending to the cerebral cortex. No one knows how narcotics in general affect the cell activities unselectively or why some of them act only on particular cells. The Meyer-Overton law, that the de-

pressant efficiency of any agent is measured by lipid-solubility, is accepted generally as applying to hypnotics. However, it is readily apparent from a knowledge of the physiologic action of thousands of compounds that other factors also must be considered. For a general discussion of these factors relating to narcosis, see Chapter 2.

Variations in properties among the barbiturates involve chiefly the dose required, the length of time after administration before the effects are observed, duration of action, ratio of therapeutic to toxic or fatal dose and extent of accumulation. Since the dosage can be regulated and the margin of safety is usually satisfactory, the main considerations are promptness and duration of action, chiefly the latter. These factors, together with the structures and the usual doses of the currently distributed barbiturates, are compiled in Table 9-4.

Structure-activity relationships

Major findings in regard to structure-activity relationships are as follows:

Both hydrogen atoms in position 5 of barbituric acid must be replaced for maximal activity. This is likely due to the susceptibility to rapid metabolic attack[7] and to the high acidity and ionization of C—H bonds in such a position.

Increasing the length of an alkyl chain in the 5 position enhances the potency up to 5 or 6 carbon atoms; beyond that, depressant action decreases and convulsant action may result. This is probably due to the excess over ideal lipophilic character.

TABLE 9-4

BARBITURATES USED AS SEDATIVES AND HYPNOTICS

General Structure

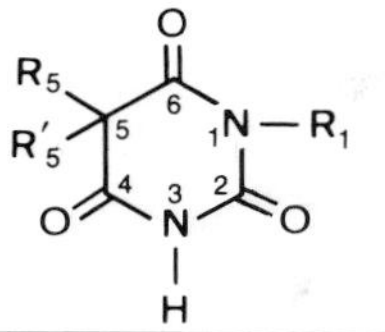

A. LONG DURATION OF ACTION (6 OR MORE HOURS)

| Generic Name / Proprietary Name | Substituents | | | Sedative Dose (in mg.) | Hypnotic Dose (in. mg.) | Usual Onset of Action (in min.) |
	R_5	R'_5	R_1			
Barbital / *Veronal*	C_2H_5	C_2H_5	H	—	300	30–60
Mephobarbital U.S.P. / *Mebaral*	C_2H_5	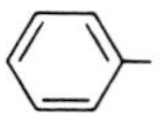 (phenyl)	CH_3	30–100*	100	30–60
Metharbital U.S.P. / *Gemonil*	C_2H_5	C_2H_5	CH_3	50–100*	—	30–60
Phenobarbital U.S.P. / *Luminal*	C_2H_5	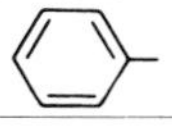 (phenyl)	H	15–30*	100	20–40

*Daytime sedative and anticonvulsant.

B. INTERMEDIATE DURATION OF ACTION (3–6 HOURS)

| Generic Name / Proprietary Name | Substituents | | | Sedative Dose (in mg.) | Hypnotic Dose (in mg.) | Usual Onset of Action (in min.) |
	R_5	R'_5	R_1			
Butalbital U.S.P. / *Sandoptal*	$CH_2{=}CHCH_2{-}$	$(CH_3)_2CHCH_2{-}$	H	—	200–600	20–30
Amobarbital U.S.P. / *Amytal*	$CH_3CH_2{-}$	$(CH_3)_2CHCH_2CH_2{-}$	H	20–40	100	20–30
Aprobarbital / *Alurate*	$CH_2{=}CHCH_2{-}$	$(CH_3)_2CH{-}$	H	20–40	40–160	—
Butabarbital Sodium U.S.P. / *Butisol Sodium*	$CH_3CH_2{-}$	$CH_3CH_2\overset{CH_3}{\underset{\vert}{C}}H{-}$	H	15–30	100	20–30
Butallylonal / *Pernocton*	$CH_2{=}\overset{}{\underset{\vert}{C}}{-}CH_2{-}$ with Br	$CH_3CH_2\overset{CH_3}{\underset{\vert}{C}}H{-}$	H	—	200	—
Butethal	$CH_3CH_2{-}$	$CH_3CH_2CH_2CH_2{-}$	H	—	100–200	30–60
Allobarbital / *Dialog*	$CH_2{=}CHCH_2{-}$	$CH_2{=}CHCH_2{-}$	H	30	100–300	15–30
Probarbital	$CH_3CH_2{-}$	$(CH_3)_2CH{-}$	H	50	130–390	20–30
Talbutal U.S.P. / *Lotusate*	$CH_2{=}CHCH_2{-}$	$CH_3CH_2\overset{CH_3}{\underset{\vert}{C}}H{-}$	H	50	120	20–30
Vinbarbital	$CH_3CH_2{-}$	$CH_3CH_2CH{=}\overset{CH_3}{\underset{\vert}{C}}{-}$	H	30	100–200	20–30

C. SHORT DURATION OF ACTION (LESS THAN 3 HOURS)

| Generic Name / Proprietary Name | Substituents | | | Sedative Dose (in mg.) | Hypnotic Dose (in mg.) | Usual Onset of Action (in min.) |
	R_5	R'_5	R_1			
Cyclobarbital / *Phanodorn*	$CH_3CH_2{-}$	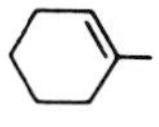 (cyclohexenyl)	H	—	100–300	15–30
Cyclopentenylallyl-barbituric Acid / *Cyclopal*	$CH_2{=}CHCH_2{-}$	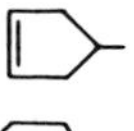 (cyclopentenyl)	H	50–100	100–400	15–30
Heptabarbital / *Medomin*	$CH_3CH_2{-}$	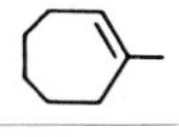 (cycloheptenyl)	H	50–100	200–400	20–40

(Continued)

C. SHORT DURATION OF ACTION (LESS THAN 3 HOURS) *(Continued)*

Generic Name Proprietary Name	Substituents			Sedative Dose (in mg.)	Hypnotic Dose (in mg.)	Usual Onset of Action (in min.)
	R_5	R'_5	R_1			
Hexethal *Ortal*	CH_3CH_2-	$CH_3(CH_2)_5-$	H	50	200–400	15–30
Pentobarbital Sodium U.S.P. *Nembutal Sodium*	CH_3CH_2-	$CH_3CH_2CH_2\overset{\displaystyle CH_3}{\overset{\displaystyle \vert}{CH}}-$	H	30	100	20–30
Secobarbital U.S.P. *Seconal*	$CH_2{=}CHCH_2-$	$CH_3CH_2CH_2\overset{\displaystyle CH_3}{\overset{\displaystyle \vert}{CH}}-$	H	15–30	100	20–30

Branched, cyclic or unsaturated chains in the 5 position generally produce a briefer duration of action than do normal saturated chains containing the same number of carbon atoms. This appears to be due to a combination of decreased lipophilic character and increased ease of metabolic conversion to a more polar, inactive metabolite.

Compounds with alkyl groups in the 1 or 3 position may have a shorter onset and duration of action. The N-methyl group results in a barbiturate which is a weaker acid (e.g., hexobarbital, $pK_a = 8.4$) compared with the usual $pK_a = 7.6$ for barbiturates without the N-methyl substituent. The weaker acid is largely in the nonionic lipid-soluble form (plasma pH 7.4), which readily enters the central nervous system and rapidly equilibrates into peripheral fatty stores.

Replacement of oxygen by sulfur on the 2-carbon shortens the onset and duration of action. Thiobarbiturates, although little different from barbiturates in acid strength (e.g., thiopental, $pK_a = 7.4$), are much more lipid-soluble in the nonionized form than are the corresponding oxygen analogs. Rapid movement into and out of the central nervous system, as well as ease of metabolic attack, accounts for the rapid onset and short duration of action.

It may be noted from Table 9-4, that the total number of carbon atoms contained in the groups substituted in the 5 position of barbituric acid is closely related to the duration of action. The compounds with the most rapid onset and shortest duration of action, following oral administration, are those with the most lipophilic substituents, totaling 7 to 9 carbon atoms. This may be related to rapid absorption and distribution to the central nervous system, followed by rapid loss to neutral storage sites, such as to lean body tissue and to peripheral fat. Conversely, the barbiturates with the slowest onset and longest duration of action contain the most polar side chains—either the 4 carbon atoms contributed by 2 ethyl groups, or an ethyl and a phenyl group (e.g., phenobarbital). The phenyl group attached to a polar substituent has a water-solubility greater than that expected of its 6-carbon content, apparently due to the polarizability of its *pi* electrons. In a barbiturate, for example, the lipophilic character of the benzene ring ($\pi = +1.77$) is between that of a 3- and 4-carbon aliphatic chain (*n*-propyl, $\pi = +1.5$; *n*-butyl, $\pi = +2.0$).[7] The barbiturates with an intermediate duration of action have 5 alkyl substituents of intermediate polarity (5 to 7 carbon atoms total).

Lipophilia is somewhat reduced by branched chains and unsaturation, but the total carbon content of the groups in the 5 position provides a good first approximation of duration of action. The long-acting, relatively polar phenobarbital, for example, both enters and leaves the central nervous system very slowly as compared with the more lipophilic thiopental. In addition, the lipoidal barriers to drug-metabolizing enzymes lead to a slower metabolism for the more polar barbiturates, phenobarbital being metabolized to the extent of only about 10 percent per day.

Pharmacologic properties

The effects following administration of the barbiturates pursue about the same course, regardless of the compound used. Therapeutic doses in the smaller amounts calm nervous conditions of any origin and in larger amounts cause a dreamless sleep in from 20 to 60 minutes after oral administration and almost immediately if given intravenously. In some patients, there may be considerable excitement before sedation is initiated. Still larger doses produce a form of anesthesia, and in all cases, up to this stage, there is little disturbance of other functions, such as respiration, circulation, metabolism or the action of smooth muscle. The effects last for one half to 12 hours, depending on the compound, the dosage and the stage to which the narcosis has been carried. The patient awakens refreshed but may not be as alert as usual and may experience some lassitude for a time. In the presence of pain, these drugs have very little analgesic action but may potentiate other compounds that do have such effect. Excretion is principally by the kidneys, where they appear partly unchanged, partly oxidized in the side chain, and partly conjugated. Some of the compounds, especially those containing sulfur, are destroyed almost entirely in the body, probably in the liver. For a more detailed description of barbiturate metabolism, see Chapter 3.

Untoward reactions are uncommon except with very large doses. The chief one of inconvenience is the appearance in some individuals of delayed effects, extreme depression, excitement or even mania. Occasionally, some persons are hypersensitive and experience dermatologic lesions as manifested by wheals, angioneurotic edema or scarletinal-like rashes. The respiration and the circulation are depressed slightly by anesthetic doses, but the temperature and basal metabolism are scarcely affected. Relatively large amounts cause profound and prolonged coma, a marked fall in blood pressure and eventual paralysis of the respiratory center. The remedial measures used are persistent administration of central nervous system stimulants (see Chap. 10), various supportive procedures and artificial respiration or oxygen if necessary. Even in therapeutic doses, there is an occasional fatal collapse due to peripheral paralysis of the blood vessels. The margin between therapeutic and toxic doses is comparatively large, and poisoning would be of little importance were it not for the fact that the barbiturates are widely used, and their ready availability leads to frequent accidental or deliberate overdosage.

Habituation to the barbiturates is widespread and well recognized. It is not so well known that these widely used drugs are capable of producing a primary addiction.[8] Tolerance to increased doses develops slowly, but physical dependence may develop fairly rapidly. Oral ingestion of about 800 mg. daily of the potent, short-acting barbiturates for a period of 8 weeks will result in mild to moderate withdrawal symptoms in most individuals. The average daily dose for the barbiturate addict is about 1.5 g. Abrupt withdrawal of the drug from an addicted individual will frequently result in delirium and grand-mal-like convulsions. Severe withdrawal symptoms, including insomnia, nausea, cramps, vomiting, orthostatic hypotension, convulsive seizures and visual and auditory hallucinations, may continue for days. The extent of mental, emotional and neurologic impairment together with the severity of the withdrawal reactions have caused barbiturate addiction to be classed by some as a public health and medical problem more serious than morphine addiction.

The barbiturates are used chiefly as sedatives and hypnotics in a wide variety of conditions. Selection of a barbiturate with the appropriate onset time and duration of action is desirable.[6] The disorders for which sedation is indicated vary from a state of "overwrought nerves" to a violent mania. The barbiturates have the advantage over the bromides in that the action may be enhanced to more profound states by increasing the dose. They are indicated in any type of insomnia that is not due to pain and, even in cases where pain is present, may advantageously be combined with analgesics. They also are applied to suppress a variety of convulsions with origins in the central nervous system, including those from tetanus, meningitis, chorea, epilepsy, eclampsia, insulin overdosage and poisoning by strychnine and similar drugs. In epilepsy, phenobarbital or a similar long-acting barbiturate will diminish the number and the severity of the attacks.

Barbiturates also are employed to produce anesthesia, either as premedication before other agents, or to act as the sole anesthetic. In providing preliminary sedation, the short- or the intermediate-acting barbiturates are superior in some respects to morphine, and it often may be of advantage to combine the two. In somewhat larger doses, the ultrashort-acting barbiturates contribute to the narcosis, thus diminishing the amount of volatile anesthetic required and reducing the undesirable side-effects of the latter. In combination with morphine and scopolamine, the barbiturates are used in producing obstetric amnesia.

Some of the more frequently used barbiturates are described briefly in the following sections. For the structures, the usual dosages required to produce sedation and hypnosis, the times of onset and the duration of action see Table 9-4.

Barbiturates with a long duration of action (six hours or more)

Barbital, Veronal®, 5,5-diethylbarbituric acid. Barbital is used orally as a hypnotic-sedative with a duration of about 8 to 12 hours. It is also available as the more water-soluble salt, barbital sodium (Veronal® Sodium).

Mephobarbital U.S.P., Mebaral®, 5-ethyl-1-methyl-5-phenylbarbituric acid. Mephobarbital produces sedation of long duration, but is a relatively weak hypnotic. It is used orally in the prevention of grand mal and petit mal epileptic seizures.

Metharbital U.S.P., Gemonil®, 5,5-diethyl-1-methylbarbituric acid. Metharbital produces less se-

TABLE 9–5

SEDATIVE-HYPNOTIC BARBITURATES

Name Proprietary Name	Preparations	Category	Usual Adult Dose*	Usual Dose Range*	Usual Pediatric Dose*
Mephobarbital U.S.P. *Mebaral*	Mephobarbital Tablets U.S.P.	Anticonvulsant; sedative		Anticonvulsant, 400 to 600 mg. daily; sedative, 32 to 100 mg. 3 or 4 times daily	
Metharbital N.F. *Gemonil*	Metharbital Tablets N.F.	Anticonvulsant	Initial, 100 mg. 1 to 3 times daily	100 to 800 mg. daily	
Phenobarbital U.S.P. *Luminal*	Phenobarbital Elixir U.S.P. Phenobarbital Tablets U.S.P.	Anticonvulsant; hypnotic; sedative	Anticonvulsant, 50 to 100 mg. 2 or 3 times daily; hypnotic, 100 to 200 mg. at bedtime; sedative, 15 to 30 mg. 2 or 3 times daily	30 to 600 mg. daily	Anticonvulsant, 15 to 50 mg. 2 or 3 times daily; sedative, 2 mg. per kg. of body weight or 60 mg. per square meter of body surface, 3 times daily
Phenobarbital Sodium U.S.P. *Luminal Sodium* *Phenalix*	Phenobarbital Sodium Injection U.S.P. Sterile Phenobarbital Sodium U.S.P.	Anticonvulsant; hypnotic; sedative	Anticonvulsant, I.M. or I.V., 200 to 320 mg.; may repeat in 6 hours as necessary; hypnotic, I.M. or I.V., 130 to 200 mg.; sedative, I.M. or I.V., 100 to 130 mg.; may repeat in 6 hours as necessary	30 to 600 mg. daily	Anticonvulsant, I.M., 3 to 5 mg. per kg. of body weight or 125 mg. per square meter of body surface; sedative, 2 mg. per kg. of body weight, or 60 mg. per square meter of body surface, 3 times daily
	Phenobarbital Sodium Tablets U.S.P.		Anticonvulsant, 50 to 100 mg. 2 or 3 times daily as necessary; hypnotic, 100 to 200 mg. at bedtime; sedative, 15 to 30 mg. 2 or 3 times daily	30 to 600 mg. daily	Sedative, 2 mg. per kg. of body weight or 60 mg. per square meter of body surface, 3 times daily
Amobarbital U.S.P. *Amytal*	Amobarbital Elixir U.S.P. Amobarbital Tablets U.S.P.	Sedative	Sedative, 25 mg.; hypnotic 100 mg.	25 to 200 mg.	
Amobarbital Sodium U.S.P. *Amytal Sodium*	Amobarbital Sodium Capsules U.S.P. Sterile Amobarbital Sodium U.S.P.	Hypnotic and sedative	Hypnotic, oral, I.M. or I.V., 65 to 200 mg. at bedtime; sedative, oral, I.M. or I.V., 30 to 50 mg. 2 or 3 times daily	Hypnotic, oral, I.M. or I.V., 50 to 200 mg. daily; sedative, oral, I.M. or I.V., 15 mg. to 1 g. daily	Sedative, oral, 2 mg. per kg. of body weight or 60 mg. per square meter of body surface, 3 times daily
Butabarbital Sodium U.S.P. *Butisol Sodium*	Butabarbital Sodium Capsules U.S.P. Butabarbital Sodium Elixir U.S.P. Butabarbital Sodium Tablets U.S.P.	Sedative; hypnotic	Sedative, 15 to 30 mg. 3 or 4 times daily; hypnotic, 100 mg.	Sedative, 7.5 to 60 mg.; hypnotic, 100 to 200 mg.	
Talbutal U.S.P. *Lotusate*	Talbutal Tablets U.S.P.	Sedative	Hypnotic, 120 mg. from 15 to 30 minutes before bedtime		
Pentobarbital Sodium U.S.P. *Nembutal*	Pentobarbital Sodium Capsules U.S.P.	Hypnotic, sedative	Hypnotic, 100 mg. at bedtime; sedative, 30 mg. 3 or 4 times daily	50 to 200 mg. daily	Sedative, 2 mg. per kg. of body weight or 60 mg. per square meter of body surface, 3 times daily
	Pentobarbital Sodium Elixir U.S.P.		Hypnotic, 100 mg. at bedtime; sedative, 20 mg. 3 or 4 times daily	50 to 200 mg. daily	Same as for Pentobarbital Sodium Capsules U.S.P.
	Pentobarbital Sodium Injection U.S.P.		Hypnotic, I.M., 150 to 200 mg.; I.V., 100 mg. repeated as necessary; sedative, I.M., 30 mg. 3 or 4 times daily	50 to 500 mg. daily	Same as for Pentobarbital Sodium Capsules U.S.P.

* See *U.S.P. D.I.* for complete dosage information.
(Continued)

TABLE 9–5 (Continued)

SEDATIVE-HYPNOTIC BARBITURATES

Name Proprietary Name	Preparations	Category	Usual Adult Dose*	Usual Dose Range*	Usual Pediatric Dose*
Secobarbital U.S.P. *Seconal*	Secobarbital Elixir U.S.P.	Hypnotic; sedative	Hypnotic, 100 mg. at bedtime; sedative, 30 to 50 mg. 3 or 4 times daily	90 to 300 mg. daily	Sedative, 2 mg. per kg. of body weight or 60 mg. per square meter of body sur- face, 3 times daily
Secobarbital Sodium U.S.P. *Seconal Sodium*	Secobarbital Sodium Capsules U.S.P.	Hypnotic; sedative	Hypnotic, 100 mg. at bedtime; sedative, 30 to 50 mg. 3 or 4 times daily	90 to 300 mg. daily	Sedative, 2 mg. per kg. of body weight or 60 mg. per square meter of body sur- face, 3 times daily
	Secobarbital Sodium Injection U.S.P. Sterile Secobarbital Sodium U.S.P.		Hypnotic, I.M. or I.V., 2.2 mg. per kg. of body weight; sedative, I.M. or I.V., 1.1 to 1.65 mg. per kg. of body weight	1.1 to 4.4 mg. per kg.	Sedative, see Usual Dose.

*See *U.S.P. D.I.* for complete dosage information.

dation than phenobarbital and is most often used in the control of epileptic seizures of the grand mal, the petit mal, the myoclonic or the mixed type.

Phenobarbital U.S.P., Luminal®, 5-ethyl-5-phenylbarbituric acid. Phenobarbital is a long-acting hypnotic-sedative, more potent than barbital but slower in onset of action, requiring about 1 hour. The duration of action is 10 to 16 hours. The plasma half-life ranges from 72 to 96 hours. It is also effective in the prevention of epileptic seizures, being more effective in the grand mal than in the petit mal types. **Phenobarbital Sodium U.S.P.,** a salt that is more water-soluble, is available for either oral use or parenteral use by the subcutaneous, the intramuscular and the intravenous routes.

Barbiturates with an intermediate duration of action (3 to 6 hours)

Amobarbital U.S.P., Amytal®, 5-ethyl-5-isopentylbarbituric acid. The doses of amobarbital range as follows: sedation, 16 to 50 mg.; hypnosis or preanesthetic medication, 100 to 200 mg.; anticonvulsant, 200 to 400 mg. The plasma half-life ranges from 16 to 24 hours. **Amobarbital Sodium U.S.P.** is the water-soluble salt used for oral, rectal, intramuscular or subcutaneous administration; the usual dose for hypnosis is 100 mg.

Aprobarbital, Alurate®, 5-allyl-5-isopropylbarbituric acid. Aprobarbital is used orally in a dose that ranges from 20 to 160 mg., as a sedative and hypnotic.

Butabarbital Sodium U.S.P., Butisol® Sodium, sodium 5-*sec*-butyl-5-ethylbarbiturate. Butabarbital sodium is used orally as a sedative at a dose of 8 to 60

mg.; as a hypnotic, 100 to 200 mg. Sedation is sustained for about 5 to 6 hours.

Talbutal U.S.P., Lotusate®, 5-allyl-5-*sec*-butylbarbituric acid. Talbutal was introduced in 1955 as a hypnotic-sedative with intermediate duration of action. The sedative dose is 30 to 50 mg., the hypnotic dose 120 mg.

Barbiturates with a short duration of action (less than 3 hours)

Pentobarbital Sodium U.S.P., Nembutal®, sodium 5-ethyl-5-(1-methylbutyl)barbiturate. Pentobarbital sodium is a short-acting barbiturate, used as a hypnotic at a usual oral or intravenous dose of 100 mg. The usual dose range for oral administration is 15 to 200 mg. daily, for intravenous administration 50 to 200 mg. daily.

Secobarbital U.S.P., Seconal®, 5-allyl-5-(1-methylbutyl)barbituric acid. Secobarbital is the free acid form, used for its hypnotic effect in a usual adult oral dose of 100 mg. at bedtime.

Secobarbital Sodium U.S.P. is used for hypnosis in either an oral or rectal dose of 100 to 200 mg. The sodium salt is also used to produce hypnosis or as an adjunct to anesthesia in a usual parenteral dose of 100 mg.

NONBARBITURATE SEDATIVE-HYPNOTICS

Many drugs varying widely in their chemical structures are capable of producing sedation and hypnosis that closely resembles that of the barbiturates. The same factors are important in selecting either a non-barbiturate or a barbiturate sedative-hypnotic, and these are principally the time required for onset of the depressant effect, the duration of the effect, and the incidence and the nature of undesirable side-effects.

In all ureides at least one of the nitrogen atoms is flanked by two carbonyl groups, resulting in an acidic hydrogen atom (—CO—NH—CO—) which will form a water-soluble salt in the presence of alkali hydroxides.

The acyclic ureides are prepared from the corresponding alkyl or dialkylmalonic acids which are decarboxylated to form the substituted acetic acids. The acid halides of these, or their α-halo derivatives may be condensed with urea to form the desired ureide.

Amides and imides

A group of cyclic amides and imides, some bearing a close structural relationship to the barbiturates, have proved to be effective as sedative-hypnotic drugs.

Glutethimide U.S.P., Doriden®, 2-ethyl-2-phenyl-glutarimide. Glutethimide was introduced in 1954 as a sedative-hypnotic drug, closely related in action to the barbiturates. Its hypnotic effects begin about 30 minutes after administration and last for about 4 to 8 hours. The drug is useful for the induction of sleep in cases of simple insomnia and has been used as a daytime sedative to relieve anxiety-tension states. Glutethimide is a white powder, soluble in alcohol but practically insoluble in water.

Glutethimide

Numerous reports of addiction to glutethimide have been published,[9,10] including epileptic seizures during withdrawal.

Methyprylon U.S.P., Noludar®, 3,3-diethyl-5-methyl-2,4-piperidinedione. Methyprylon is a sedative-hypnotic, structurally related to the barbiturates and similar in its actions. The drug, introduced in 1955, is useful for the induction of sleep within 15 to 30 minutes in patients with simple insomnia. It is intermediate in its duration of action.

Methyprylon is a white powder, moderately soluble in water and very soluble in alcohol.

Methyprylon

Methaqualone Hydrochloride U.S.P., Quaalude®, Parest®, Sopor®, 2-methyl-3-o-tolyl-4(3H)-quinazolinone. Methaqualone, which is also distributed as the hydrochloride salt, is a sedative-hypnotic drug, introduced in 1965. The drug is contraindicated in pregnant women, and caution is recommended in its use in anxiety states where mental depression and suicidal tendencies may exist. Long-term use may lead to psychological or physical dependence.

Methaqualone Hydrochloride

Alcohols and their carbamates

Ethanol has played a prominent role as a sedative and hypnotic for centuries. However, the feeling of stimulation which precedes that of depression has been recognized more widely. Because of the many problems associated with the use of alcohol, such as the development of chronic alcoholism on continued use, other depressant drugs have been favored for sedative-hypnotic use.

The hypnotic activity of the normal alcohols increases as the molecular weight and the lipid-solubility increase, reaching a maximum depressant effect at 8 carbons (see Chap. 2). Branching of the alkyl chain increases activity, and the order of potency in an isomeric series of alcohols is tertiary > secondary > primary. This may be due to a greater resistance to metabolic inactivation for the more highly branched compounds (see Chap. 3). Replacement of a hydrogen by a halogen has an effect equivalent to increasing the alkyl chain and, for the lower molecular weight alcohols, results in increased potency.

Ethanol is a depressant drug whose apparent stimulation is produced as a result of the increased activity of lower centers freed from control by the depression of higher inhibitory mechanisms.

n-Butyl alcohol has been used clinically for the relief of pain, presumably taking advantage of its weak sedative-hypnotic properties. Some higher alcohols and their derivatives used for their central depressant effects are described below.

Ethchlorvynol U.S.P., Placidyl®, 1-chloro-3-ethyl-1-penten-4-yn-3-ol. Ethchlorvynol is a colorless to yellow liquid with a pungent odor. It darkens on exposure to light and to air.

Ethchlorvynol

Ethchlorvynol was introduced as a mild hypnotic-sedative in 1955. It has a fairly rapid onset of action and a duration of about 5 hours. The drug is most useful in the induction of sleep for patients with simple insomnia and for use as a daytime sedative. Physical dependence has been reported following excessive intake.

Ethinamate U.S.P., Valmid®, 1-ethynylcyclohexanol carbamate. Ethinamate was introduced as a sedative-hypnotic in 1955. The onset of its depressant effects requires about 20 to 30 minutes. The drug is metabolized rapidly and the duration of action is short, lasting less than 4 hours. Tolerance and physical dependence have been observed on prolonged use of large doses. Ethinamate is not effective for more than 7 days, so prolonged use is inadvisable.

Ethinamate

Aldehydes and derivatives

Members of this group were among the first of the organic hypnotics. Chloral was introduced in 1869 in the mistaken belief that it would be converted to the anesthetic chloroform in the body. However, instead of undergoing the haloform reaction of the test tube, chloral is reduced in the body to trichloroethanol, which may be responsible for most of chloral's depressant properties (see Chap. 3). Analogs and derivatives, such as chloral betaine and petrichloral, have been introduced to reduce or eliminate the disadvantages of chloral.

Chloral Hydrate U.S.P., Noctec®, chloral, trichloroacetaldehyde monohydrate, $CCl_3CH(OH)_2$. Chloral is a reliable and safe hypnotic, useful in inducing sleep where insomnia is not due to pain, for the drug is a poor analgesic. The parent aldehyde, chloral (CCl_3CHO, note that the synonym for the hydrate is a misnomer), is an oily liquid which in the presence of water yields the crystalline hydrate.

Chloral hydrate occurs as colorless or white crystals having an aromatic, penetrating and slightly acrid odor and a bitter caustic taste. It is very soluble in water (1:0.25) and in alcohol (1:1.3).

In the usual oral dose (500 mg. to 2 g.) chloral hydrate causes sedation in 10 to 15 minutes. Sleep occurs within an hour and lasts for 5 to 8 hours. The sleep is light, and the patient is readily aroused. Complete anesthesia is possible with doses of 6 g. or more, but this approaches the dose causing marked respiratory depression, and chloral hydrate is not safely used for anesthesia. Alcohol synergistically increases the depressant effect of chloral and a mixture of the two ("knockout drops," "Mickey Finn") is a very potent depressant, although the chloral alcoholate formed (CCl_3—$CHOH$—O-C_2H_5, a hemiacetal) is no more hypnotic than is the hydrate. Chloral hydrate causes local irritation and may cause nausea, vomiting and diarrhea, particularly if taken with inadequate fluids.

Chloral Betaine, Beta-Chlor®. Chloral betaine is a chemical complex of chloral hydrate and betaine with a hypnotic and sedative potency equal to that of the chloral hydrate which it contains. The 870-mg. tablets contain 500 mg. of chloral hydrate. The complex is tasteless, and is said to produce gastric irritation infrequently.

$$CCl_3CH(OH)_2 \cdot (CH_3)_3\overset{+}{N}CH_2COO^-$$

Petrichloral, pentaerythritol chloral. Petrichloral, the hemiacetal between pentaerythritol and chloral, was introduced in 1955 as a sedative-hypnotic. It lacks the acrid odor and the bitter taste of chloral and is said to be free of gastric upset and after-taste. The recommended oral dose for daytime sedation is 300 mg., the hypnotic dose 600 mg. to 1.2 g.

Petrichloral

Chlorhexadol, 2-methyl-4-(2′,2′,2′-trichloro-1′-hydroxyethoxy)-2-pentanol.

$$Cl_3C-CH-O-CH-CH_2-\underset{\underset{OH}{|}}{\overset{\overset{CH_3}{|}}{C}}-CH_3$$

Chlorhexadol

Chlorhexadol, like petrichloral, is a hemiacetal formed from chloral and an alcohol. It hydrolyzes in the stomach to produce chloral, but provides better patient acceptability due to improved taste and odor characteristics. The amount of 1.6 g. of chlorhexadol provides 1 g. of chloral hydrate.

Triclofos Sodium, Triclos®, 2,2,2-trichloroethyl dihydrogen sodium phosphate.

$$Cl_3C-CH_2O-\overset{\overset{O}{\|}}{\underset{\underset{OH}{|}}{P}}-O^-\ Na^+$$

Triclofos Sodium

Triclofos sodium is a phosphate ester of trichloroethanol which is rapidly hydrolyzed in the body to release the active sedative-hypnotic component, trichloroethanol. The serum level of trichloroethanol peaks in about one hour, and has a half-life of about eleven hours. Triclofos is free of the unpleasant odor and taste of chloral hydrate and shares the common active intermediary metabolite, trichloroethanol. It is available in liquid and tablet form, and for its hypnotic effect is given orally at a usual dose of 1.5 g. 15 to 30 minutes before bedtime.

Paraldehyde U.S.P., 2,4,6-trimethyl-*s*-trioxane, paracetaldehyde.

Paraldehyde

Paraldehyde, in 1882, was the second of the synthetic organic compounds to be introduced for use as a sedative-hypnotic.

Paraldehyde is a colorless liquid with an odor which is not pungent or unpleasant but has a disagreeable taste. The drug is more potent and toxic than ethanol but less so than chloral hydrate. With the usual oral or rectal dose of 10 ml., sleep is induced within 10 to 15 minutes. The sleep is a natural one and is accompanied by little change in respiration or circulation.

The chief objections to the use of paraldehyde are its disagreeable taste, which is difficult to mask, and the potent odor which appears on the breath of patients within a few minutes following its ingestion. Although an excellent and safe depressant drug, its odor prevents use as a daytime sedative, and the more acceptable barbiturates have largely replaced paraldehyde in routine use as a soporific. It is used most frequently in delirium tremens and in the treatment of psychiatric states characterized by excitement, where drugs must be given over long periods.

Paraldehyde may oxidize to form acetic acid on storage. Stored samples have been found to consist of up to 98 percent of acetic acid. Since oxidation occurs more rapidly in opened, partially filled containers, the drug should not be dispensed from a container that has been opened for longer than 24 hours. Refrigeration is indicated to retard oxidation. The U.S. Public Health Service has recommended that its hospitals stock the drug only in single-dose ampules for injection and in the oral capsule form.

Miscellaneous

Many of the antihistaminic drugs (see Chap. 16) possess a significant degree of central nervous system depression as their principal side-effect. In certain cases, this has been selected as the therapeutic effect. For example, promethazine (Phenergan®) has been used as a preoperative sedative. Several of the antihistaminic drugs, principally methapyrilene, make up the depressant component of a large number of proprietary sleeping preparations.

CENTRAL RELAXANTS (CENTRAL NERVOUS SYSTEM DEPRESSANTS WITH SKELETAL-MUSCLE-RELAXANT PROPERTIES)

The search for drugs capable of diminishing skeletal muscle tone and involuntary movement has led to the introduction during recent years of a variety of agents capable of a relatively weak, centrally mediated muscle relaxation. However, these agents have been of more significance in the therapeutic application of their mild depressant properties on the central nervous system. This depressant effect has been variously described as "tranquilization," "ataraxia," "calming" and "neurosedation." At therapeutic levels, this has

TABLE 9-6

NONBARBITURATE SEDATIVES AND HYPNOTICS

Name Proprietary Name	Preparations	Category	Usual Adult Dose*	Usual Dose Range*	Usual Pediatric Dose*
Glutethimide U.S.P. *Doriden*	Glutethimide Capsules U.S.P. Glutethimide Tablets U.S.P.	Sedative; hypnotic		Sedative, 125 to 250 mg. 1 to 3 times daily; hypnotic, 500 mg. to 1 g.	
Methyprylon U.S.P. *Noludar*	Methyprylon Capsules U.S.P. Methyprylon Tablets U.S.P.	Hypnotic		50 to 400 mg. at bedtime	
Methaqualone U.S.P. *Quaalude, Sopor*	Methaqualone Tablets U.S.P.	Sedative	Sedative, 75 mg. 3 or 4 times daily; hypnotic, 150 to 300 mg. at bedtime		
Methaqualone Hydrochloride U.S.P. *Optimil, Parest, Somnafac*	Methaqualone Hydrochloride Capsules U.S.P.	Sedative	Sedative, 100 mg. after each meal and at bedtime; hypnotic, 200 or 400 mg. before bedtime	100 to 400 mg.	
Ethchlorvynol U.S.P. *Placidyl*	Ethchlorvynol Capsules U.S.P.	Sedative	Sedative, 100 mg. 2 or 3 times daily; hypnotic, 500 mg.	Sedative, 100 to 200 mg.; hypnotic, 500 mg. to 1 g.	
Ethinamate U.S.P. *Valmid*	Ethinamate Capsules U.S.P.	Hypnotic	500 mg.	500 mg. to 1 g.	
Chloral Hydrate U.S.P. *Noctec, Felsules, Rectules, Aquachloral*	Chloral Hydrate Capsules U.S.P.	Hypnotic and sedative	Hypnotic, 500 mg. to 1 g. at bedtime; sedative, 250 mg. 3 times daily	250 mg. to 2 g. daily	
	Chloral Hydrate Syrup, U.S.P.		Hypnotic, 500 mg. to 1 g. at bedtime; sedative, 250 mg. 3 times daily	250 mg. to 2 g. daily	Hypnotic, 50 mg. per kg. of body weight or 1.5 g. per square meter of body surface, up to 1 g. per dose, at bedtime; sedative, 8 mg. per kg. or 250 mg. per square meter, up to 500 mg. per dose, 3 times daily
Chloral Betaine *Beta-Chlor*	Chloral Betaine Tablets	Hypnotic	870 mg. to 1.74 g. 15 to 30 minutes before bedtime		
Paraldehyde U.S.P. *Paral*		Hypnotic; sedative	Hypnotic, 10 to 30 ml.; sedative, 5 to 10 ml.	3 to 30 ml.	Hypnotic, 0.3 ml. per kg. of body weight or 12 ml. per square meter of body surface, per dose; sedative, 0.15 ml. per kg. or 6 ml. per square meter of body surface, per dose
	Sterile Paraldehyde U.S.P.	Hypnotic; sedative	Hypnotic, I.M., 10 ml.; I.V. infusion, diluted with several volumes of Sodium Chloride Injection, 10 ml.; sedative, I.M., 5 ml; I.V. infusion, diluted with several volumes of Sodium Chloride Injection, 5 ml.	3 to 10 ml.	Hypnotic, I.M., 0.3 ml. per kg. of body weight or 12 ml. per square meter of body surface, per dose; sedative, I.M., 0.15 ml. per kg. of body weight or 6 ml. per square meter of body surface, per dose.

* See *U.S.P. D.I.* for complete dosage information

been shown to resemble more closely the well-known sedation produced by the sedative-hypnotics (e.g., amobarbital), and to be quite different from the depressant effects on the central nervous system produced by the tranquilizing agents (e.g., reserpine, chlorpromazine) (see Tables 9-10, 9-11).

The skeletal-muscle relaxation is produced by drugs of this group in a manner completely different from that of curare and its analogs, which act at the neuromuscular junction. These centrally acting muscle relaxants block impulses at the interneurons of polysynaptic reflex arcs, mainly at the level of the spinal cord. This is demonstrated by the abolishment or the diminution of the flexor and crossed extensor reflexes which possess one or more interneurons between the afferent (sensory) and the efferent (motor) fibers.

The knee-jerk response, which acts through a monosynaptic reflex system and therefore possesses no interneurons, is unaffected by these drugs.

The skeletal-muscle relaxation produced by this centrally mediated mechanism can be applied therapeutically by employing those members of this class which produce muscle relaxation without excessive sedation. The therapeutic applications of the skeletal-muscle-relaxant effect include relief in the variety of conditions in which painful muscle spasm may be present, such as bursitis, spondylitis, disk syndromes, sprains, strains and low back pain.

The major sites of the sedative effects of these drugs are the brain stem and subcortical areas. The ascending reticular formation, which receives and transmits some sensory stimuli, transmits and maintains a state of arousal. When the passage of stimuli is blocked at the level of the ascending reticular formation, response to sensory stimuli is reduced, and depression, ranging from sedation to anesthesia, may occur. The barbiturates[11] and other sedative-hypnotics, as well as meprobamate[12] and its analogs, are capable of producing inhibition of this arousal system. Suppression of polysynaptic reflexes at the spinal level is not sufficient to account for depression of the arousal system.

The depressant effect of members of this class has been applied in producing mild hypnosis in case of simple insomnia, or as an adjunct to psychotherapy in the management of anxiety and tension states associated or unassociated with physical ills, such as hypertension and cardiovascular disorders, in which excessive excitation should be avoided.

Meprobamate, a member typical of this class, has been shown[13] to produce withdrawal symptoms similar to those of the barbiturates. Prolonged administration of overdosages of any drug of this class with significant sedative action may lead to withdrawal symptoms (convulsions, tremor, abdominal and muscle cramps, vomiting or sweating) if medication is stopped abruptly.

As a general class, members of this group may be described as mild sedatives with moderate to weak skeletal-muscle-relaxing properties.

GLYCOLS AND DERIVATIVES

In 1945, during a study on potential preservatives for penicillin during production and processing, F. M. Berger observed the muscle-relaxant effects of aryl glycerol ethers in experimental animals. Following a study of a series of structural analogs, 3-o-toloxy-1,2-propanediol (mephenesin) was introduced in 1948 as a skeletal-muscle relaxant. However, the duration of action was too brief. Since metabolic attack of the terminal hydroxy group occurred in part (see Chap. 3), structural analogs which protected this functional group, including the carbamate (mephenesin carbamate), were prepared. Due to their limited effectiveness as compared with newer agents, mephenesin and mephenesin carbamate are no longer distributed. Further structural alterations demonstrated that the aromatic nucleus was not a requisite for activity, and, in an attempt to prolong muscle-relaxant activity in the aliphatic series, 2-methyl-2-propyl-1,3-propanediol dicarbamate (from which the generic name meprobamate was derived) was synthesized.

The drug not only showed the desired longer-acting skeletal-muscle-relaxant properties but was shown to be an effective central nervous system depressant, producing a sedative effect in experimental animals and in man. The drug was marketed in 1955 under the trade name Miltown (its development originated in Milltown, a village in New Jersey) and was widely promoted for its "tranquilizing" properties. Since then a number of related glycols and their carbamate derivatives, as well as structurally unrelated compounds, have been shown to possess in varying degrees the properties of skeletal-muscle relaxation mediated through the blockade of polysynaptic reflexes and a mild depression of the central nervous system.

Mephenesin: R = H

Mephenesin Carbamate: R = —C—NH$_2$ (with O double bond)

Guaifenesin U.S.P., Robitussin®, 3-(o-methoxyphenoxy)-1,2-propanediol. Guaifenesin is rarely used alone for its sedative action. It is most often used in combination with antihistamines, analgesics and vasoconstrictors in cough medicines for its expectorant action.

Guaifenesin: R = H

Methocarbamol: R = —C—NH$_2$ (with O double bond)

Category—expectorant.
Usual dose—100 mg. every 3 or 4 hours.*

Occurrence
Guaifenesin Capsules U.S.P.
Guaifenesin Syrup U.S.P.
Guaifenesin Tablets U.S.P.

Methocarbamol U.S.P., Robaxin®, 3-(o-methoxy-phenoxy)-1,2-propanediol 1-carbamate. Methocarbamol, the carbamate derivative of glyceryl guaiacolate, was introduced in 1957 for the relief of skeletal-muscle spasm. Peak plasma concentrations of methocarbamol are reached more slowly (1 hour) than for mephenesin (30 minutes) but are more sustained. Preparations of methocarbamol for parenteral use contain polyethylene glycol as a solvent. This is contraindicated in patients with impaired renal function, since it increases urea retention and acidosis in such patients.

Chlorphenesin Carbamate, Maolate®, 3-p-chlorophenoxy-2-hydroxypropyl carbamate. This close analog of mephenesin carbamate and of methocarbamol was introduced in 1967 as a muscle relaxant. Chlorphenesin carbamate is used in the short-term relief of the discomfort from a variety of traumatic or inflammatory disorders of skeletal muscle, such as strains or sprains. The drug is rapidly absorbed, maximum serum concentrations being reached in 1 to 3 hours after oral administration. The biological half-life in man is 3.5 hours. It is rapidly excreted in the urine as a glucuronide conjugate along with traces of phenolic and acidic metabolites.

$$Cl-\langle\ \rangle-O-CH_2-CH-CH_2-O-\overset{O}{\overset{\|}{C}}-NH_2$$
$$\underset{OH}{|}$$

Chlorphenesin Carbamate

Meprobamate U.S.P., Equanil®, Miltown®, 2-methyl-2-propyltrimethylene dicarbamate, 2-methyl-2-propyl-1,3-propanediol dicarbamate. Meprobamate produces skeletal-muscle relaxation by interneuronal blockade at the spinal level. The duration of its muscle-relaxant effect is 8 to 10 times longer than that produced by mephenesin, but of the same type. Certain types of abnormal motor activity and muscle spasm may be reduced by meprobamate. However, the major application of the drug has been in the treatment of excessive central nervous system stimulation (e.g., simple insomnia) and in psychoneurotic anxiety and tension states. Meprobamate is also effective in the prevention of attacks of the petit mal form of epilepsy. Meprobamate Injection U.S.P. may be used as an adjunct in the treatment of tetanus.

The drug is a white powder, possessing a bitter taste. It is relatively insoluble in water (0.34% at 20°) and freely soluble in alcohol and most organic sol-

TABLE 9-7

CENTRALLY ACTING SKELETAL-MUSCLE RELAXANTS

Name Proprietary Name	Preparations	Category	Usual Adult Dose*	Usual Dose Range*	Usual Pediatric Dose*
Methocarbamol U.S.P. *Robaxin*	Methocarbamol Injection U.S.P. Methocarbamol Tablets U.S.P.	Skeletal-muscle relaxant		Oral, 1.5 to 2 g. 4 times daily for the first 2 or 3 days, then 1 g. 4 times daily	
Chlorphenesin Carbamate *Maolate*	Chlorphenesin Carbamate Tablets	Skeletal-muscle relaxant	800 mg. 3 times daily; maintenance, 400 mg. 4 times daily		
Meprobamate U.S.P. *Equanil, Miltown, Bamate, Arcoban, Kesso-Bamate, Meprospan, Tranmep*	Meprobamate Injection U.S.P.	Adjunct in tetanus (sedative)	I.M., 400 mg. 6 to 8 times daily	1.2 to 3.2 g. daily	Infants—125 mg. 4 times daily; children—200 mg. 6 to 8 times daily
	Meprobamate Tablets U.S.P.	Sedative	400 mg. 3 or 4 times daily	1.2 to 2.4 g. daily	Dosage is not established in children under 6 years of age; 6 to 12 years of age—100 to 200 mg. 2 or 3 times daily
	Meprobamate Oral Suspension U.S.P.	Tranquilizer	400 mg. of meprobamate 3 or 4 times daily		

*See *U.S.P. D.I.* for complete dosage information.

vents. The drug is stable in the presence of dilute acid or alkali.

The usual oral dose is 400 mg. 3 times daily.

Meprobamate: R = H
Carisoprodol: R = —CH(CH₃)₂
Tybamate: R = —(CH₂)₃CH₃

Carisoprodol, Soma®, Rela®, N-isopropyl-2-methyl-2-*n*-propyl-1,3-propanediol dicarbamate. This N-isopropyl derivative of meprobamate was introduced in 1959 for therapeutic application of its centrally mediated skeletal-muscle-relaxant properties. It is recommended for use in acute skeletomuscular conditions characterized by pain, stiffness and spasticity. Drowsiness is the principal side-effect of the drug.

The usual oral dose of carisoprodol is 350 mg., 4 times daily. Peak blood levels are reached 1 to 2 hours after ingestion. The compound is a bitter, odorless, white crystalline powder. It is soluble in water to the extent of 30 mg. in 100 ml. at 25°.

Tybamate, Tybatran®, N-butyl-2-methyl-2-*n*-propyl-1,3-propanediol dicarbamate, 2-methyl-2-propyl-trimethylene butylcarbamate carbamate. Tybamate is an N-butyl substituted analog of meprobamate introduced in 1965 for the treatment of psychoneurotic anxiety and tension states. The usual oral dose is 250 to 500 mg., 3 or 4 times daily.

BENZODIAZEPINE DERIVATIVES

The benzodiazepine derivatives have become increasingly important as antianxiety, anticonvulsant, antipsychotic, and sedative–hypnotic drugs. Owing to their wide acceptance, they have displaced the barbiturates as the drugs most frequently prescribed.

The initial synthesis of the 1,4-benzodiazepines resulted from an attempt to prepare 2-methylamino-6-chloro-4-phenylquinazoline-3-oxide by the treatment of 6-chloro-2-chlormethyl-4-phenylquinazoline-3-oxide with methylamine. The unexpected ring enlargement reaction instead produced the benzodiazepine chlordiazepoxide, which was shown to possess sedative, muscle-relaxant, and anticonvulsant properties much like those of the barbiturates. The drug and its congeners have been widely used for the relief of anxiety, tension, apprehension and related neuroses, and of agitation during withdrawal from alcohol. Long-term treatment with larger than usual doses of the benzodi-

azepines should be avoided since this regimen may lead to psychic and physical dependence.

6-Chloro-2-chloromethyl-4-phenylquinazoline-3-oxide

Chlordiazepoxide

Chlordiazepoxide Hydrochloride U.S.P., Librium®, 7-chloro-2-(methylamino)-5-phenyl-3H-1,4-benzodiazepine 4-oxide hydrochloride. Chlordiazepoxide was introduced in 1960 for use in the treatment of anxiety and tension. It has been shown to block spinal reflexes at one-tenth the dose required of meprobamate and is therefore a moderately effective skeletal-muscle relaxant. Doses larger than those necessary to block the spinal reflexes depress the reticular activating system in the same manner as meprobamate. These factors, together with the lack of appreciable effect in the conditioned response test, and the ability to elevate the convulsant threshold, place chlordiazepoxide in the class of mild central depressants, with a centrally mediated skeletal-muscle-relaxant effect. The drug is absorbed rapidly from the gastrointestinal tract, and peak blood levels are reached in 2 to 4 hours. The drug is excreted slowly, and its plasma half-life is 20 to 24 hours.

Chlordiazepoxide Hydrochloride

The major metabolites of chlordiazepoxide are oxazepam, a lactam, an amino acid resulting from ring opening of the lactam, and a small amount of a conjugate of the amino acid, all primarily excreted in the urine.

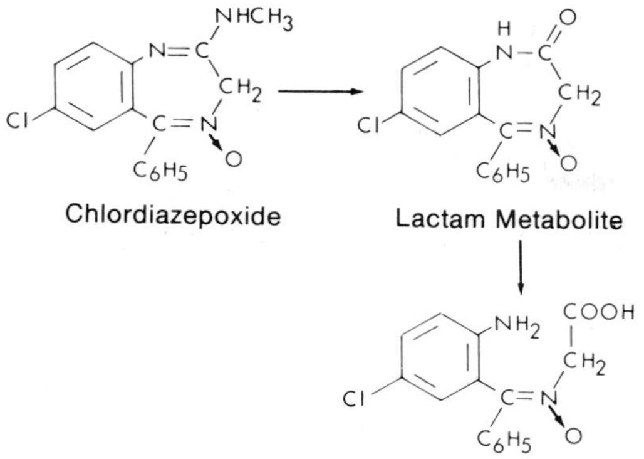

Chlordiazepoxide

Lactam Metabolite

Amino Acid Metabolite

Oxazepam

Chlordiazepoxide hydrochloride is a colorless crystalline substance, light-sensitive and highly soluble in water but unstable in aqueous solution.

Diazepam U.S.P., Valium®, 7-chloro-1,3-dihydro-1-methyl-5-phenyl-2H-1,4-benzodiazepin-2-one. Diazepam is a substituted benzodiazepine, introduced in 1964, which is related in structure and pharmacology to chlordiazepoxide. Diazepam is used for the control of anxiety and tension states, the relief of muscle spasm, and for the management of acute agitation during withdrawal from alcohol. However, it should be used with caution in the long-term treatment of alcoholism, since habituation and dependence on the drug may result. Diazepam also has significant anticonvulsant properties. The parenteral administration of diazepam is considered to be the most effective treatment of status epilepticus. Diazepam is metabolized in the liver, one of the metabolites being the 3-hydroxy derivative, oxazepam, which is also active as a sedative and muscle relaxant.

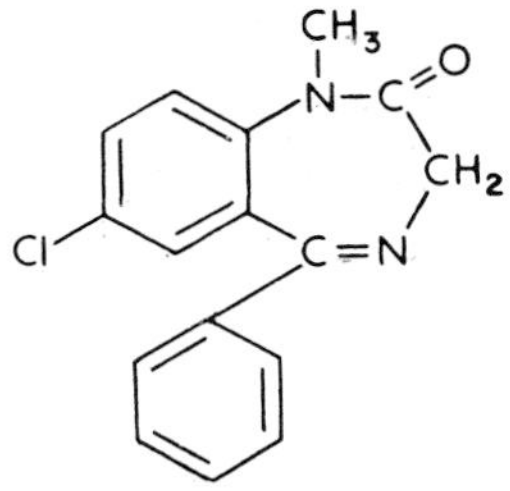

Diazepam

Oxazepam U.S.P., Serax®, 7-chloro-1,3-dihydro-3-hydroxy-5-phenyl-2H-1,4-benzodiazepin-2-one. Oxazepam is a benzodiazepine derivative introduced in 1965 for use in the relief of psychoneuroses characterized by anxiety and tension. It is said to show a lower incidence of side-effects and reduced toxicity, perhaps due to the ease of conjugation of the 3-hydroxy group, and elimination as the glucuronide which is the major metabolite.

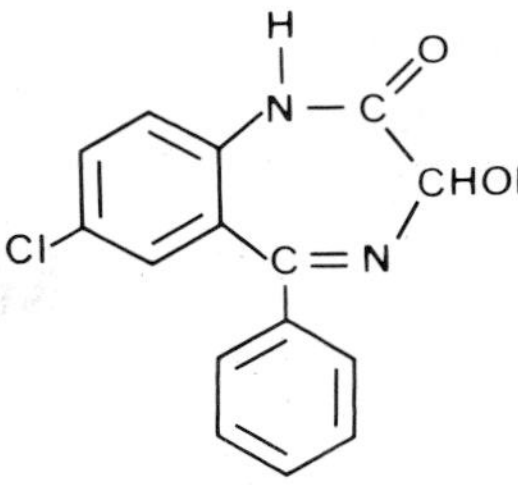

Flurazepam Hydrochloride U.S.P., Dalmane®, 7-chloro-1-(2-diethylaminoethyl)-5-(2-fluorophenyl)-1,3-dihydro-2H-1,4-benzodiazepin-2-one dihydrochloride. Flurazepam is a pale-yellow, crystalline compound, freely soluble in alcohol and in water.

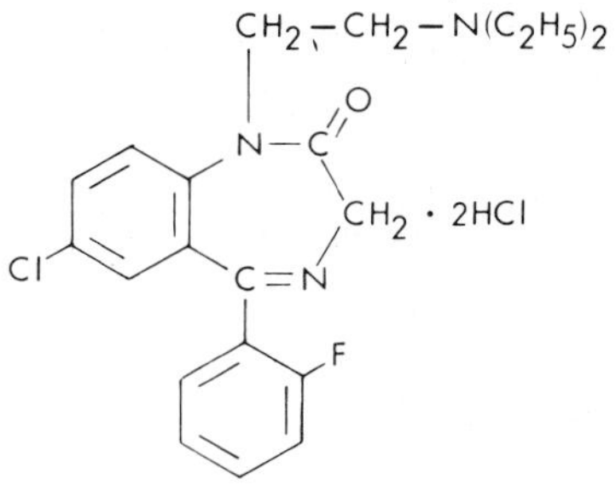

Flurazepam Hydrochloride

Flurazepam was introduced in 1970 as a hypnotic drug, useful in types of insomnia characterized by difficulty in falling asleep, and by early awakening. Although flurazepam is chemically related to the other benzodiazepines which show antianxiety properties, the potent hypnotic effect appears to be unique to flurazepam. The drug is reported to provide 7 to 8 hours of restful sleep, during which normal dreaming activity, as characterized by rapid eye movements, is maintained.

The usual oral dose of 15 to 30 mg. is rapidly absorbed, and induces sleep in about 20 minutes. The major urinary metabolite of flurazepam is the conjugated N-hydroxyethyl derivative. A second major metabolite N-dealkyl flurazepam has a long half-life (47 to 100 hours) and remains in the body for several days.

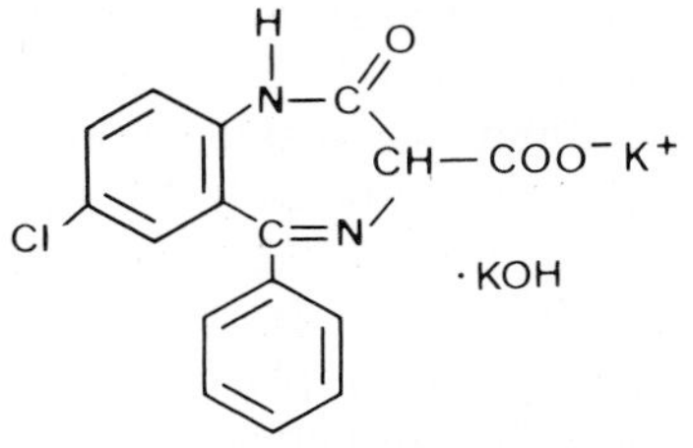

Clorazepate Dipotassium

TABLE 9-8

BENZODIAZEPINE DERIVATIVES

Name Proprietary Name	Preparations	Category	Usual Adult Dose*	Usual Dose Range*	Usual Pediatric Dose*
Chlordiazepoxide U.S.P. *Libritabs*	Chlordiazepoxide Tablets U.S.P.	Tranquilizer	5 or 10 mg. 3 or 4 times daily	5 to 25 mg.	
Chlordiazepoxide Hydrochloride U.S.P. *Librium*	Chlordiazepoxide Hydrochloride Capsules U.S.P.	Sedative	5 to 25 mg. 3 or 4 times daily	10 to 300 mg. daily	Children under 6 years of age—dosage is not established; children 6 years of age and over—5 to 10 mg. 2 to 4 times daily
	Sterile Chlordiazepoxide Hydrochloride U.S.P.	Sedative (alcohol withdrawal)	I.M. or I.V., 50 to 100 mg., repeated in 2 or 4 hours if necessary	25 to 300 mg. during a 6-hour period, but not to exceed 300 mg. daily	
Diazepam U.S.P. *Valium*	Diazepam Injection U.S.P.	Sedative	I.M. or I.V., 2 to 10 mg., repeated in 2 to 4 hours if necessary	2 to 15 mg.; do not exceed 30 mg. in an 8-hour period	40 to 200 µg. per kg. of body weight or 1.2 to 6 mg. per square meter of body surface; dose may be repeated in 2 to 4 hours. Do not exceed 18 mg. per square meter in an 8-hour period
	Diazepam Tablets U.S.P.		2 to 10 mg. 2 to 4 times daily.	2 to 40 mg. daily.	Infants under 6 months of age: use is not recommended; over 6 months of age: 1 to 2.5 mg. 3 or 4 times daily.
Oxazepam U.S.P. *Serax*	Oxazepam Capsules U.S.P. Oxazepam Tablets U.S.P.	Sedative	10 to 15 mg. 3 or 4 times daily	10 to 30 mg.	
Flurazepam Hydrochloride U.S.P. *Dalmane*	Flurazepam Hydrochloride Capsules U.S.P.	Hypnotic	15 to 30 mg. at bedtime		
Clorazepate Dipotassium *Tranxene*	Capsules	Antianxiety agent	15 to 60 mg. daily in divided doses; elderly or debilitated patients, 7.5 to 15 mg. daily		Over 6 years—7.5 to 60 mg. daily in divided doses
Lorazepam *Ativan*	Lorazepam Tablets	Antianxiety agent	2 to 6 mg. daily in divided doses; elderly or debilitated patients—1 to 2 mg. daily	Use not established	
Prazepam U.S.P. *Verstran*	Prazepam Tablets U.S.P.	Antianxiety agent	20 to 60 mg. daily in divided doses; elderly or debilitated patients—10 to 15 mg. daily	Use not established	

*See *U.S.P. D.I.* for complete dosage information.

Clorazepate Dipotassium, Tranxene®, 7-chloro-2,3-dihydro-2-oxo-5-phenyl-1H-1,4-benzodiazepine-3-carboxylic acid dipotassium salt monohydrate.

Clorazepate was introduced in 1972 and has uses as a sedative and antianxiety agent similar to those of the other benzodiazepines. It is metabolically decarboxylated to produce peak plasma levels of nordiazepam in one hour. This principal active metabolite has a half-life of 24 hours. The most common side-effects are drowsiness and ataxia. Phenothiazines, alcohol and other central nervous system depressant drugs may enhance the effects of clorazepate.

Lorazepam, Activan®, 7-chloro-5-(2-chlorophenyl)-1,3-dihydro-3-hydroxy-2H-1,4-benzodiazepin-2-one.

Lorazepam is related to oxazepam, being a dichloro rather than a monochloro derivative. Owing to its higher lipophilicity, lorazepam requires a lower dose

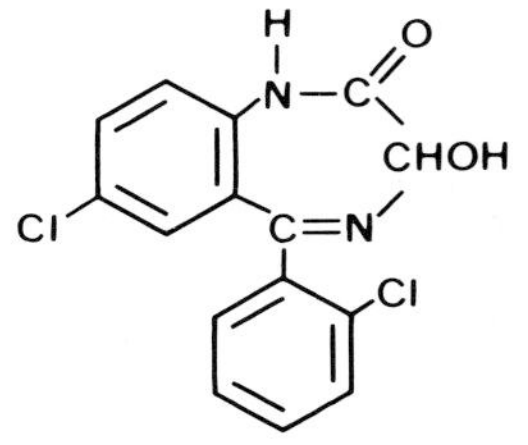

Lorazepam

than does oxazepam to produce its antianxiety effects. Peak plasma levels of lorazepam are found 2 hours after oral administration. The drug is rapidly conjugated at its 3-hydroxy group to form its major metabolite, lorazepam glucuronide, which shows no CNS activity. The mean half-life of unconjugated lorazepam is about 12 hours, and of the glucuronide, about 18 hours. The drug is about 85 percent bound to plasma proteins.

Prazepam U.S.P., Verstran®, 7-chloro-1-(cyclopropylmethyl)-1,3-dihydro-5-phenyl-2H-1,4-benzodiazepin-2-one.

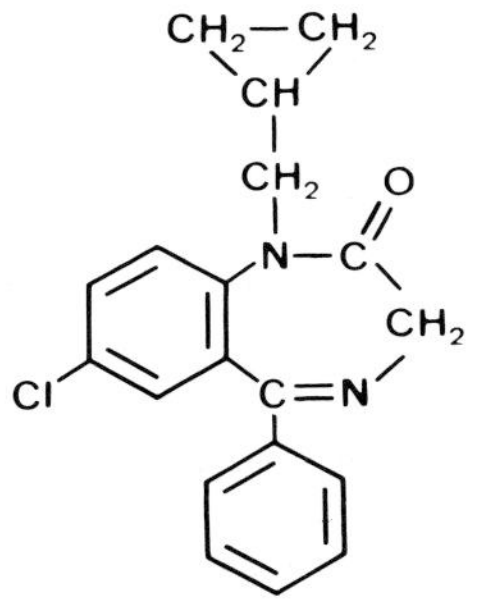

Prazepam

Prazepam is a sedative and antianxiety drug. The mean half-life of the drug is 63 hours. The major metabolite present 6 hours postadministration is the N-dealkylated derivative norprazepam. The drug is largely excreted as the metabolic oxidation products 3-hydroxyprazepam and oxazepam.

MISCELLANEOUS

Several compounds of unique structure have been found to inhibit polysynaptic reflexes of the spinal cord, as well as to depress higher centers. Like previous central relaxants, their application as skeletal-muscle relaxants or sedative drugs has depended upon which effect predominates.

Certain derivatives of benzoxazole have been shown to inhibit transmission of nervous impulses through polysynaptic reflex arcs, thus acting as skeletal-muscle-relaxing agents in the same manner as mephenesin and related compounds.

The first compound of this type, zoxazolamine (Flexin), was introduced as a muscle relaxant in 1956. The uricosuric effect of the drug was recognized, and, in 1958, zoxazolamine was recommended for treatment of gout. A high frequency of side-effects which included, in a small percentage of cases, serious hepatic toxicity, led to its withdrawal in 1962. A related benzoxazole (chlorzoxazone) with a lower incidence of side-effects is currently used for its muscle-relaxing properties.

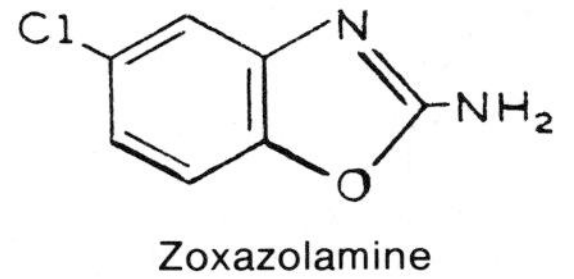

Zoxazolamine

Chlorzoxazone, Paraflex®, 5-chloro-2-benzoxazolinone. Chlorzoxazone was introduced in 1958 as a skeletal-muscle relaxant used for reduction of painful muscle spasms in medical and orthopedic disorders such as bursitis, myositis, sprains and strains, and acute or chronic back pain. A low incidence of liver damage and gastrointestinal disturbances, as side-effects, has been reported.

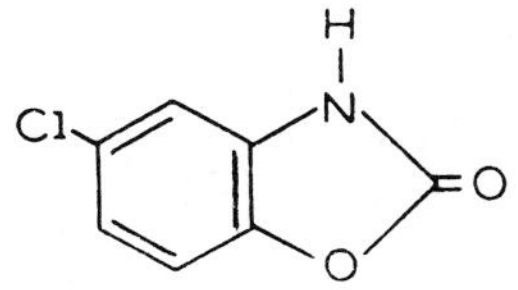

Chlorzoxazone

The usual oral dose of chlorzoxazone is 250 mg., 3 or 4 times a day.

Chlormezanone, Trancopal®, 2-(4-chlorophenyl)-3-methyl-4-metathiazanone 1,1-dioxide. Chlormezanone was introduced in 1958 as a skeletal-muscle relaxant, useful in the treatment of conditions characterized by muscle spasm. Its mild depressant effect on the central nervous system resembles that of meprobamate, and the drug is recommended for use in anxiety and tension states.

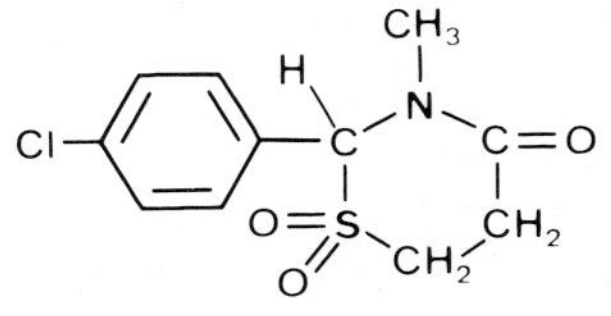

Chlormezanone

Chlormezanone is a white crystalline powder with solubility in water of less than 0.25 percent and less than 1 percent in alcohol. The usual oral dose is 100 to 200 mg., 3 or 4 times daily.

Dantrolene Sodium, Dantrium®, 1-[5-(*p*-nitrophenyl)furfurylideneamino]hydantoin sodium salt.

Dantrolene Sodium

Dantrolene sodium was introduced in 1974 as a muscle relaxant, recommended for control of the spasticity resulting from a variety of disorders, including spinal cord injury, stroke and multiple sclerosis. Although this substituted hydantoin has been shown to produce relaxation of the contractile state of isolated skeletal muscles, it also acts as a central nervous system depressant producing side-effects of drowsiness and generalized weakness. The extent of involvement of central nervous system centers in the muscle-relaxant effect is not known. The absorption after oral administration is slow and incomplete, but dose-related blood levels are obtained. The mean biological half-life in man is 8.7 hours following a single 100-mg. oral dose. The recommended initial oral dose of 25 mg. twice a day is increased gradually to a usual maximum dose of 100 mg. 4 times a day.

Cyclobenzaprine Hydrochloride U.S.P., Flexeril®, 3-(5-H-dibenzo[a,d]cyclohepten-5-ylidene)-N,N-dimethyl-1-propanamine hydrochloride.

Cyclobenzaprine

Cyclobenzaprine hydrochloride is a centrally acting skeletal-muscle relaxant that relieves acute and painful muscle spasm. It is a close structural relative to the tricyclic antidepressants (Chap. 10) and shows similar side-effects, including sedative and anticholinergic properties. Cyclobenzaprine is absorbed well on oral administration and is eliminated slowly with a plasma half-life of one to three days. It is highly bound to plasma proteins. It is extensively metabolized and its metabolites are excreted primarily by the kidneys. The usual oral dose is 30 mg. three times a day, with a range of 60 to 90 mg. a day in divided doses.

A large number of compounds which act as peripheral cholinergic blocking (parasympatholytic) agents possess a central depressant effect which produces a reduction of voluntary muscle spasm. They produce no inhibition of transmission through peripheral neuromuscular pathways (no blockade of polysynaptic reflexes) and therefore the muscle relaxation produced is unlike that produced by the agents discussed as "central relaxants" or central nervous system depressants with skeletal-muscle-relaxant properties. The inhibition by the "centrally acting cholinergic blocking agents" is apparently on the extrapyramidal system. These drugs are used primarily in the relief of rigidity and spasticity of paralysis agitans (Parkinson's disease). Examples of this group of drugs are benztropine, cycrimine, procyclidine and trihexyphenidyl. For a detailed discussion of this class, see Chapter 12.

TRANQUILIZING AGENTS

The introduction of two drugs, one an alkaloid from a small Asian shrub, the other a synthetic compound related to the antihistamines, has led to revolutionary

TABLE 9–9

A PHARMACOLOGIC COMPARISON OF CENTRAL RELAXANTS, SEDATIVE-HYPNOTICS AND TRANQUILIZERS[14,15,16]

Action	Central Relaxants — Meprobamate	Sedative-Hypnotics — Barbiturates	Tranquilizers — Reserpine	Tranquilizers — Chlorpromazine
Adrenergic blocking (central)	No	No	Yes	Yes
Cholinergic blocking (peripheral)	No	No	No	Yes
Antihistaminic	No	No	No	Yes
Anesthesia	Yes	Yes	No	No
Arousal	Difficult	Difficult	Easy	Easy
Addiction liability	Yes	Yes	No	No
Ataxia	Yes	Yes	No	No
Convulsant threshold	Raised	Raised	Lowered	Lowered
Excitement	Present	Present	Absent	Absent
Lethal dose	Respiratory depression	Respiratory depression	Convulsions (muscle spasticity)	Convulsions

advances in the understanding and the treatment of certain types of mental disease. The Rauwolfia alkaloids were first made generally available to Western medicine in 1953 and were followed shortly by chlorpromazine in 1954. These drugs possess a variety of pharmacologic properties, but their unique depressant effects on the central nervous system have been widely used in the treatment of serious mental and emotional disorders which are characterized by varying degrees of excitation. The Rauwolfia alkaloids, chlorpromazine and its analogs, and other chemically and pharmacologically dissimilar compounds whose use in mental disorders followed, have been described by a variety of names, among those "ataraxic," "neurosedative," "calming." The designation most widely used for the depressant drugs has been "tranquilizing agent."

This term may be applied more specifically to represent those agents capable of exerting a unique type of selective central nervous system depression. They act differently from the barbiturates and other sedatives which act by producing a general central nervous system depression. The action of the tranquilizers is believed to take place primarily in the paleocortex and the subcortical areas of the brain. They give strong sedation without producing sleep and produce a state of indifference and disinterest. They are effective in reducing excitation, agitation, aggressiveness and impulsiveness which are not controlled by the ordinary sedative-hypnotics (e.g., phenobarbital) and central relaxants (central depressant drugs with skeletal-muscle-relaxing properties, e.g., meprobamate).

A comparison of some pharmacologic properties of the central relaxants, sedative-hypnotics and tranquilizers displays close similarity for the depressant effects of the central relaxants and the sedative-hypnotics, and a unique set of properties for the tranquilizers (see Table 9-9).

In addition to the analogs of the Rauwolfia alkaloids and chlorpromazine, a group of diphenylmethane derivatives and miscellaneous compounds structurally related to chlorpromazine have been shown to be best described as tranquilizing agents. Members of this group do not produce true anesthesia, even in high

TABLE 9-10

RAUWOLFIA ALKALOIDS AND SYNTHETIC ANALOGS

Generic Name Proprietary Name	Substituents			Usual Oral Dose (mg./day)
	R_1		R_2	Psychoses
Reserpine U.S.P. *Sandril, Serpasil*	(3,4,5-trimethoxybenzoyl ester)		—OCH₃	*3–5*
Rescinnamine *Moderil*	(3,4,5-trimethoxycinnamoyl ester)		—OCH₃	3–12
Deserpidine *Harmonyl*	(3,4,5-trimethoxybenzoyl ester)		H	2–3

dose, and arousal is easy from the sleep which may be induced. They possess no demonstrated addiction liability, produce no ataxia, muscle tone is increased at high doses, and the convulsant threshold is lowered. No excitement stage precedes hypnosis, and toxic doses produce convulsions. In contrast, most of these properties are opposite or significantly different from both central relaxants and sedative-hypnotics.

Certain biogenic amines, the catecholamines norepinephrine and dopamine, and the indoleamine serotonin, are known to occur in appreciable quantities in both the central and the peripheral nervous systems. Much is known about their roles as neurotransmitting substances in the peripheral nervous system (see Chap. 11), but very little is known of their function in the central nervous system. By analogy to their peripheral function, it is presumed that imbalances in the normal pattern of synthesis, distribution and metabolism of these amines in the central nervous system may lead to changes in brain function resulting in marked alterations of mood and behavior.[17]

Norepinephrine is present in many parts of the brain, with the highest concentration in the hypothalamic area. In contrast, epinephrine occurs in the brain in very low concentration compared to norepinephrine. Dopamine is present in highest concentration in the basal ganglia, and in lower concentrations elsewhere in the brain. Serotonin occurs both in the brain and in peripheral tissues in appreciable amount. Norepinephrine, about which most is known in terms of peripheral function, serves as a general model for theories of behavioral changes induced by drug-neurohormonal interactions. In general, drugs that produce high levels of available norepinephrine in the central nervous system produce excitation or stimulation. Drugs that enhance the depletion and inactivation of norepinephrine in the CNS produce sedation or depression.

Norepinephrine is synthesized from tyrosine, with the intermediate formation of 3,4-dihydroxyphenyl-alanine (dopa) and 3,4-dihydroxyphenethylamine (dopamine). Hydroxylation of dopamine at the β-carbon produces norepinephrine. A wide variety of structural analogs can serve as substrates for one or more steps in the synthetic pathway, e.g., L-α-methyl-m-tyrosine, L-α-methyldopa. These yield "false transmitters" which are generally much weaker in their neurohormonal actions. Norepinephrine is stored within the nerve in intraneuronal granules (see Fig. 9-1), and may be released intracellularly by the action of the sedative-hypotensive reserpine alkaloids or the related synthetic benzoquinolizine derivatives. The released intracellular norepinephrine may be inactivated, mainly by mitochondrial monoamine oxidase, forming deaminated catechol metabolites, such as 3,4-dihydroxymandelic acid, before leaving the cell. This depletion of a physiologically active form of norepinephrine is associated with the sedative-hypotensive properties of the reserpine alkaloids. Monoamine oxidase inhibitors (see Chap. 10), which block the intracellular inactivation of norepinephrine, act as stimulant drugs.

Norepinephrine is discharged from neuronal endings in its physiologically active form, either by nerve impulses or by the action of some sympathomimetic drugs. It is presumed to produce its stimulant effect by direct action either as a neurohormone on central adrenergic receptors, or as a regulator of synaptic transmission by mediating the release of other chemical transmitters, such as acetylcholine. Some centrally acting sympathomimetic drugs may exert a direct effect on such receptors. The phenothiazine tranquilizers are thought to act by blocking the effective interaction of norepinephrine with its receptors. Norepinephrine released to the synaptic cleft is inactivated by cellular re-uptake, or by enzymatic methylation of the 3-hydroxyl group by catechol-O-methyl transferase, to form the less active normetanephrine. The tricyclic antidepressants (see Chap. 10) are thought to elevate mood by inhibition of the cellular

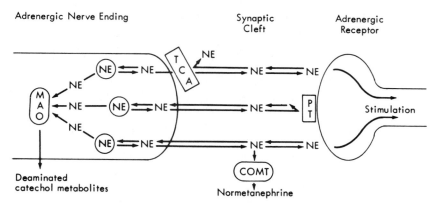

FIG. 9-1. Key: COMT = catechol O-methyl transferase, MAO = monoamine oxidase, NE = norepinephrine, PT = phenothiazine, TCA = tricyclic antidepressant.

re-uptake of norepinephrine, thus prolonging its existence and action within the synaptic cleft.

The central roles of dopamine and serotonin are less well defined. Serotonin is synthesized by decarboxylation of 5-hydroxytryptophan and, like norepinephrine, exists in the neuron in free and bound forms. Serotonin is metabolized by monamine oxidases, forming 5-hydroxyindoleacetic acid.

THE RAUWOLFIA ALKALOIDS AND SYNTHETIC ANALOGS

Rauwolfia serpentina and other Rauwolfia species have been widely used in India for centuries in a variety of ailments including snakebite, dysentery, cholera, fevers, insomnia and insanity. A gradually increasing literature from India that emphasized the effectiveness of plant extracts and dried root powder in the reduction of elevated blood pressure culminated in a publication by Vakil in 1949.[18] This led, in 1950, to trial of the crude drug in the United States for treatment of hypertension. Hypotensive and sedative effects which developed slowly were observed.

Concurrent with the medical interest, Swiss chemists during 1947 to 1951 studied the structures of the crystalline alkaloids from *Rauwolfia serpentina* reported by Indian chemists in 1931[19] but found in these only moderate sedative and hypotensive activity. However, pharmacologic tests revealed the potent activity of the crude drug was concentrated in the noncrystalline "oleoresin fraction," and from this was isolated reserpine, the major active constituent. Animal studies demonstrated the unique sedative effects of reserpine, a quiet and subdued state being gradually developed, often leading to sleep from which the animals could be aroused readily. Unlike the sedative-hypnotics, large doses did not cause deep hypnosis and anesthesia. Together with this sedative effect, blood pressure was gradually lowered.

At the same time (1952) that enthusiastic clinical reports on the hypotensive effect and the unique sedation produced by oral use of the powdered root were being presented, the crystalline alkaloid, reserpine, was made available. Clinical reports on the antihypertensive effects of reserpine noted the sedative effect and suggested use in the treatment of psychiatric states of agitation and anxiety. Although used in India for at least 5 centuries in treatment of the mentally disturbed (called in some areas pagal-kadawa, or "insanity remedy"), trial in psychotherapy outside of India was delayed until 1954 when the powdered whole root was used with moderate success in a wide variety of mental disorders characterized by excitement (mania) rather than depression.[20,21] A better un-

derstanding of the type of patient responsive to Rauwolfia therapy has led to its favorable application as a psychotherapeutic-sedative in the management of patients with anxiety or tension psychoneuroses and in those chronic psychoses involving anxiety, compulsive aggressive behavior and hyperactivity. The introduction of the first of the phenothiazine tranquilizing agents, chlorpromazine, in 1954, has limited the application of the Rauwolfia alkaloids for psychotherapeutic treatment. The alkaloids and their synthetic analogs are widely used for their hypotensive effects, either alone in cases of mild or labile hypertension or in combination with the more potent hypotensive agents (e.g., ganglionic blocking agents) for the management of essential hypertension (see Chap. 14 for detailed descriptions of these drugs).

The Rauwolfia alkaloids and their synthetic analogs (see Table 9-10) are presently used in relief of symptoms of agitated psychotic states (e.g., schizophrenia), primarily in those individuals unable to tolerate phenothiazines or related derivatives, or in those who also require antihypertensive medication.

It has been proposed that reserpine and related Rauwolfia alkaloids produce both their sedative and hypotensive effects by the release and the depletion of body amines such as serotonin, norepinephrine and hydroxytyramine.[22] In support of this concept, it has been shown that only those Rauwolfia alkaloids which produce a sedative response affect brain serotonin levels.[23] A reserpine analog, the *m*-dimethylaminobenzoic acid ester of methyl reserpate, which has a more prolonged effect on levels of brain norepinephrine than on brain serotonin, demonstrates depressant effects which closely parallel changes in the level of serotonin but not norepinephrine.

BENZOQUINOLIZINE DERIVATIVES

The slow onset of action, the prolonged duration, and the significant side-effects produced by the Rauwolfia alkaloids and their derivatives encouraged a search for compounds with similar tranquilizing properties. Many partial structures of reserpine have been synthesized and tested, but the most promising lead has come from an unrelated program.

Emetine is a potent but toxic amebicidal alkaloid not present in Rauwolfia species. Approaches to its chemical synthesis yielded a substituted benzoquinolizine, *tetrabenazine*, as an intermediate which was found to have pharmacologic properties closely resembling those of reserpine; it caused release and depletion of brain amines and was effective in depressing the conditioned avoidance response in experimental animals. It is less potent than reserpine. However, it has a briefer onset and duration of action than reserpine, being rapidly metabolized.

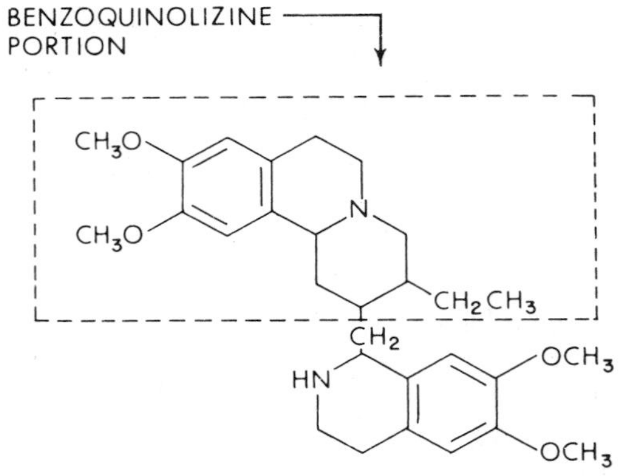

BENZOQUINOLIZINE PORTION

Emetine

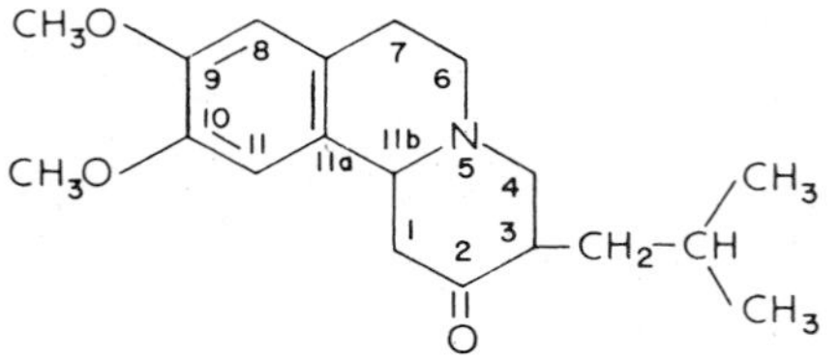

Tetrabenazine

Benzquinamide, Emete-con®, N,N-diethyl-2-acetoxy-9,10-dimethoxy-1,2,3,4,6,7-hexahydro-11bH-benzo[a]quinolizine-3-carboxamide hydrochloride. Benzquinamide, like reserpine and tetrabenazine, disrupts the conditioned avoidance response in experimental animals. However, unlike the other analogs it produces no measurable changes in the level of brain amines. This indicates the possibility that the tranquilizing effect may be separated from the amine-depletion effect and from resulting side-effects.

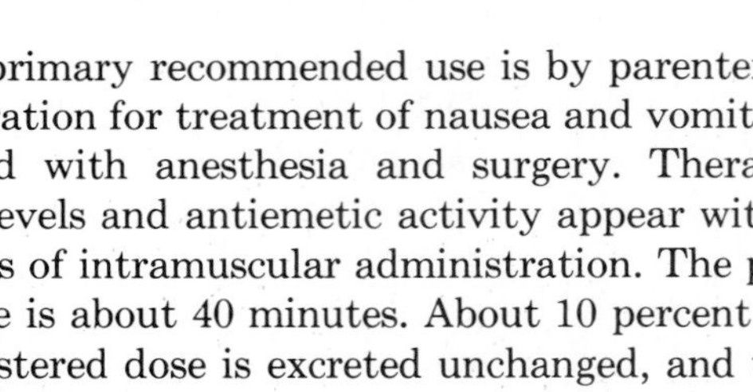

Benzquinamide Hydrochloride

Its primary recommended use is by parenteral administration for treatment of nausea and vomiting associated with anesthesia and surgery. Therapeutic blood levels and antiemetic activity appear within 15 minutes of intramuscular administration. The plasma half-life is about 40 minutes. About 10 percent of the administered dose is excreted unchanged, and the remainder is converted into more polar metabolites which are excreted in the urine and, via biliary secretion, in the feces.

The usual intramuscular dose is 50 mg., which may be repeated after one hour with subsequent doses at 3- to 4-hour intervals. A single intravenous dose of 25 mg. should be administered slowly. Subsequent doses should be given intramuscularly. The usual oral dose is 200 mg. 3 times daily.

PHENOTHIAZINE DERIVATIVES

During World War II a number of phenothiazine derivatives were prepared in the Paris laboratories of the French pharmaceutical manufacturer Rhone Poulenc. Among these was a series of 10-(2-dimethylaminoalkyl)phenothiazines which on pharmacologic screening were found to possess strong antihistaminic properties (see Chap. 16). 10-(2-Dimethylamino)propylphenothiazine (promethazine) was studied extensively and among its diverse pharmacologic properties were a sedative effect alone and a potentiating effect on the sedative action of the barbiturates. Structural analogs of promethazine were prepared in an attempt to develop derivatives with a more marked central depressant action; among these chlorpromazine was synthesized by Charpentier in 1950.

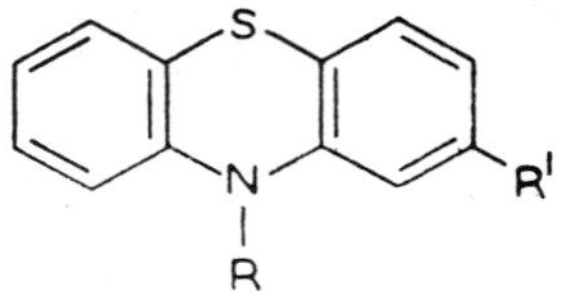

Promethazine: R = $CH_2CH(CH_3)N(CH_3)_2$
R' = H
Chlorpromazine: R = $CH_2CH_2CH_2N(CH_3)_2$
R' = Cl

A general synthesis for phenothiazine derivatives of this type is shown below.

Between 1951 and 1954 chlorpromazine was used extensively in Europe under the name of Largactil. The drug was first used in 1951 in combination with meperidine and promethazine and external cooling to lower body temperature to produce "artificial hibernation" for surgery. Its synergistic effect with analgesics was applied in 1952 and in the same year it was first reported to be useful in the quieting and the control of hyperactive psychotic patients. Chlorpromazine was introduced in the United States in 1954, gaining rapid and widespread acceptance for its use as an antiemetic, for the potentiation of anesthetics, analgesics and sedatives and in the treatment of major mental and emotional disorders. Since 1954 a number of analogs varying principally in the nature of the aminopropyl side chain and the substituent in the 2-position of the phenothiazine ring have been introduced (see Table 9-11).

Like the Rauwolfia alkaloids, chlorpromazine and its analogs produce a central reduction of sympathetic outflow at the level of the hypothalamus. The resulting decrease in peripheral sympathetic tone which would lead to reserpinelike side-effects, if unopposed, is counterbalanced by the pronounced atropinelike (cholinergic-blocking) properties of the phenothiazine derivatives. Therefore, side-effects are a mixture of those produced by central adrenergic and peripheral cholinergic blockade. These include dry mouth, dizziness, blurred vision, orthostatic hypotension and tachycardia.

Central effects of sedation and drowsiness, desirable in the treatment of anxiety and agitation, are undesirable in other applications. At high doses, side-effects occur frequently, including extrapyramidal symptoms resembling parkinsonism. These include tremor, spasticity and contraction of muscles of the head and the shoulders. The parkinsonism-like symptoms are reversible on lowered dosage or temporary discontinuation of the drug. More rapid reversal may be achieved by administration of antiparkinsonism drugs (see Chap. 12), or use of intravenous Caffeine and Sodium Benzoate Injection U.S.P. (see Chap. 10). Severe hypotension may also occur, calling for immediate supportive measures, including the use of intravenous vasopressor drugs, such as Levarterenol Bitartrate U.S.P. Epinephrine should not be used, since phenothiazine derivatives reverse its action, resulting in a further lowering of blood pressure. Jaundice has been reported in about 1 percent of the patients receiving chlorpromazine, and a much smaller percentage have developed agranulocytosis. Ocular changes including opacities of the lens and cornea have been observed in patients receiving large doses of the phenothiazines for an extended period. This appears to be due to deposition of fine particulate matter, which is mobilized upon withdrawal of the drug. Large doses of phenothiazines also lead to deposits in the skin. These react with light, producing a dark purple-brown coloration.

The therapeutic applications of the phenothiazine tranquilizing agents may be grouped into three major areas:

1. *Antiemetic Effect.* These agents are the best available for the treatment or the prevention of emesis which is drug-induced (e.g., nitrogen mustards), due to infections or toxicoses, or postoperative. They are generally less effective in the prevention of motion sickness. Certain derivatives, but not all, are recommended for the relief of nausea and vomiting during pregnancy.

2. *Potentiation of the Effects of Anesthetics, Analgesics and Sedatives.* As an adjuvant in surgical procedures, the phenothiazine tranquilizing agents reduce apprehension by their sedative effects. They also potentiate the anesthetics, potent (narcotic) analgesics and sedatives, permitting their use in a smaller dose which results in decreased respiratory depression.

3. *Treatment of Moderate and Severe Mental and Emotional States.* The phenothiazine tranquilizing drugs are used most widely in the treatment of mental and emotional disorders. Anxiety, tension and agitation are reduced in both psychoneurotics and psychotics. Selected cases of schizophrenia, mania, toxic and senile psychoses respond.

The antiemetic and tranquilizing effects of the phenothiazines combine to relieve the acute syndrome following withdrawal from addicting drugs. Miscellaneous somatic disorders which are relieved by these drugs include refractory hiccups and severe asthmatic attacks.

Structure-activity relationships for the phenothiazine tranquilizing agents

The large number of structural variations carried out in the phenothiazine series have resulted in the establishment of some fairly consistent patterns of relationships between structure and activity.[24]

meta-Substituted Diphenylamine

(small amount)

TABLE 9–11

PHENOTHIAZINE DERIVATIVES (AMINOPROPYL SIDE CHAIN)

Generic Name Proprietary Name	R_{10}	R_2	Year of Intro-duction
PROPYL DIALKYLAMINO SIDE CHAIN			
Promazine Hydrochloride U.S.P. *Sparine*	$-(CH_2)_3N(CH_3)_2 \cdot HCl$	H	1955
Chlorpromazine Hydrochloride U.S.P. *Thorazine*	$-(CH_2)_3N(CH_3)_2 \cdot HCl$	Cl	1954
Triflupromazine Hydrochloride U.S.P. *Vesprin*	$-(CH_2)_3N(CH_3)_2 \cdot HCl$	CF_3	1957
ALKYL PIPERIDYL AND PYRROLIDINYL SIDE CHAIN			
Thioridazine Hydrochloride U.S.P. *Mellaril*		SCH_3	1959
Mesoridazine Besylate U.S.P. *Serentil*		$\overset{O}{\underset{}{\uparrow}}$ SCH_3	1970
PROPYL PIPERAZINE SIDE CHAIN			
Prochlorperazine Maleate U.S.P. *Compazine*		Cl	1957
Trifluoperazine Hydrochloride U.S.P. *Stelazine*		CF_3	1959
Thiethylperazine Maleate U.S.P. *Torecan*		SCH_2CH_3	1961
Butaperazine Maleate *Repoise*		$CO(CH_2)_3CH_3$	1968
Perphenazine U.S.P. *Trilafon*		Cl	1957
Fluphenazine Hydrochloride U.S.P. *Permitil, Prolixin*		CF_3	1960
Acetophenazine Maleate U.S.P. *Tindal*		$\overset{O}{\overset{\|}{-C}}-CH_3$	1961
Carphenazine Maleate U.S.P. *Proketazine*		$\overset{O}{\overset{\|}{-C}}-CH_2CH_3$	1963

Phenothiazine Tranquilizing
Agents—General Structure

The nature and the position of substituents on the phenothiazine nucleus strongly influence activity. Replacement of the hydrogen in position 2 (R_2) by chlorine (chlorpromazine), trifluoromethyl (triflupromazine) or a dimethylsulfonamido group results in increased activity. The trifluoromethyl analog is generally more potent than the chloro compound, but this is accompanied usually by an increase in extrapyramidal symptoms. Tranquilizing activity is retained with a variety of 2-substituents such as thioalkyl (thioridazine, thiethylperazine) and acyl groups (butaperazine, acetophenazine, carphenazine). The 2-thioalkyl derivatives are said to produce fewer extrapyramidal side-effects. A ring substituent in positions 1,3,4 or simultaneous substitution in both aromatic rings results in loss of tranquilizing activity.

The three-carbon side chain connecting the nitrogen of the phenothiazine ring and the more basic side chain nitrogen is optimal for tranquilizing activity. Compounds with a two-carbon side chain (aminoethyl side chain) still possess a moderate central depressant activity, but their antihistaminic and antiparkinsonism effects generally predominate. If the side chain is altered significantly in its length and polarity, tranquilizing activity is lost, although compounds of this type show antitussive properties.

Branching at the β-position of the side chain (R_3) with a small group such as methyl produces variable changes in tranquilizing activity. Depending upon the nature of R_2, either enhanced or decreased tranquilizing activity is found. In general β-methyl substitution enhances the antihistaminic and antipruritic effects (trimeprazine). This has been attributed to steric repulsion between the methyl group at the β-position, and the 1,9-*peri* hydrogens of the phenothiazine rings, resulting in a decrease in the coplanarity of the benzene rings.[25] It would also be expected that branching at the β-position would slow the rate of metabolic attack on the side chain. This substitution would also exert an effect on the conformational preference of the aminopropyl side chain. The relative importance of these conformational or steric effects in determining the level of tranquilizing, antihistaminic, or antipruritic activity has not been established. Antipruritic activity is retained if the branching on the β-carbon is

part of a ring (methdilazine). Side-chain substitution with a large or polar group such as phenyl, dimethylamino or hydroxyl results in loss of tranquilizing activity. The importance of the side chain is further emphasized by the fact that stereospecificity exists, the levo isomer being far more active than the dextro isomer for β-methyl derivatives.

Substitution of the piperazine group (prochlorperazine, trifluoperazine) in place of the terminal dimethylamino moiety on the side chain increases potency but usually results in an increase in extrapyramidal symptoms. Substitution of $—CH_2CH_2OH$ for the terminal methyl group on piperazine (perphenazine, fluphenazine) results in a slight increase in potency. In general, the dimethylamino compounds seem more likely to produce a parkinsonlike syndrome (tremors, rigidity, salivation), while the piperazine derivatives produce, in addition, dyskinetic reactions, generally involving the muscles of the face and the neck. Skin and liver disorders and blood dyscrasias have been associated with the dimethylamino and alkylpiperidyl types to a greater extent than with the piperazine derivatives.

Quaternization of the side chain nitrogen in any of the phenothiazine derivatives results in a decrease in lipid-solubility, leading to decreased penetration of the central nervous system and virtual loss of central effects.

Products

The structures of the phenothiazine derivatives containing the aminopropyl side chain attached to the nitrogen atom of the phenothiazine ring are presented in Table 9-11.

Chlorpromazine Hydrochloride U.S.P., Thorazine® Hydrochloride, 2-chloro-10-[3-(dimethylamino)-propyl]phenothiazine hydrochloride. Chlorpromazine hydrochloride is used orally or parenterally in the treatment of nausea and vomiting, to potentiate the effects of anesthetics, analgesics, hypnotics and sedatives, and in a variety of mental and emotional disturbances. The free base (Chlorpromazine U.S.P.) is available in suppository form. Chlorpromazine hydrochloride is a white crystalline powder, very soluble in alcohol or water.

Thioridazine Hydrochloride U.S.P., Mellaril®, 10-[2-(1-methyl-2-piperidyl)ethyl]-2-(methylthio)-phenothiazine monohydrochloride. Thioridazine inhibits psychomotor function, and its use ranges from the treatment of minor conditions of anxiety and tension, to the more severe psychoneuroses and psychoses. It shows minimal antiemetic activity and minimal extrapyramidal stimulation. Retinal pigmentary degeneration with diminished visual acuity, which does not regress when the drug is discontinued, has been observed in patients taking large doses of thioridazine (more than 800 mg. daily).

Mesoridazine Besylate U.S.P., Serentil®, 10-[2-(1-methyl-2-piperidyl)ethyl]-2-(methylsulfinyl)phenothiazine monobenzenesulfonate. Mesoridazine, a sulfoxide metabolite of thioridazine, is useful in treating major psychoses, including schizophrenia. Indications and adverse reactions are like those of thioridazine.

Prochlorperazine Maleate U.S.P., Compazine® Dimaleate, 2-chloro-10-[3-(4-methyl-1-piperazinyl)-propyl]phenothiazine dimaleate. Prochlorperazine is capable of producing the same effects as chlorpromazine in a much smaller dose. Except for a very low potentiating effect on other central depressant drugs, the two drugs are similar in their applications and side-effects.

Prochlorperazine maleate (a white crystalline powder, practically insoluble in water and alcohol) is administered orally. The water-soluble ethanedisulfonate salt **Prochlorperazine Edisylate U.S.P.** (Compazine® Edisylate) is used by the oral or intramuscular routes. The free base **Prochlorperazine U.S.P.,** (Compazine®) is available in suppository form.

Promazine Hydrochloride U.S.P., Sparine® Hydrochloride, 10-[3-(dimethylamino)propyl]phenothiazine hydrochloride. Promazine is used in the management of acute neuropsychiatric agitation but is less potent than chlorpromazine. The hydrochloride salt is a white to slightly yellow crystalline solid that oxidizes on prolonged exposure to air and acquires a blue or pink color. It is available for oral, intramuscular or intravenous administration; however, the oral route is preferred because of the production of orthostatic hypotension on parenteral administration.

Triflupromazine Hydrochloride U.S.P., Vesprin®, 10-[3-(dimethylamino)propyl]-2-(trifluoromethyl)phenothiazine hydrochloride. Triflupromazine is a potent tranquilizing agent used for the treatment of anxiety and tension, and for the management of psychotic disorders. It is also employed for the control of nausea and vomiting, and as an adjunct to the potent analgesics and general anesthetics. Adverse reactions are the same as for other phenothiazine tranquilizers, and include parkinsonism, hypotension, liver damage and blood dyscrasias. Liquid forms of the drug should be protected from exposure to light.

Trifluoperazine Hydrochloride U.S.P., Stelazine® Hydrochloride, 10-[3-(4-methyl-1-piperazinyl)propyl]-2-(trifluoromethyl)phenothiazine dihydrochloride. Trifluoperazine is a relatively highly potent drug used in the control of acute and chronic psychoses marked by hyperactivity. Extrapyramidal symp-

toms occur frequently at doses required for control of psychoses. Intramuscular injections are given usually when rapid control of symptoms is necessary.

Thiethylperazine Maleate U.S.P., Torecan®, 2-(ethylthio)-10-[3-(4-methyl-1-piperazinyl)propyl]-phenothiazine maleate. Thiethylperazine is a potent tranquilizing agent which may also be used as an antiemetic and for the treatment of vertigo. The compound may give a higher incidence of extrapyramidal reactions but less agranulocytosis and jaundice than chlorpromazine. The salt with malic acid, **Thiethylperazine Maleate U.S.P.,** is used for intramuscular application, while the maleate salt is used for oral or rectal application. The initial dose may be by intramuscular injection or by suppository in the semiconscious or actively vomiting patient.

Butaperazine Maleate, Repoise®, 1-{10-[3-(4-methyl-1-piperazinyl)propyl]phenothiazin-2-yl}-1-butanone dimaleate. Butaperazine is reported to be effective in the management of chronic schizophrenic patients who are under close psychiatric supervision. It is also used for the control of all forms of psychomotor agitans, mania, hallucinations, anxiety and tension. Adverse reactions are similar to those of other major tranquilizers but it is said to give a higher incidence of extrapyramidal reactions and a lower incidence of undesirable sedation.

Perphenazine U.S.P., Trilafon®, 2-chloro-10-{3-[4-(2-hydroxyethyl)piperazinyl]propyl}phenothiazine, 4-[3-(2-chlorophenothiazin-10-yl)propyl]-1-piperazineethanol. Perphenazine is a major tranquilizing agent of relatively high potency. Its uses include acute and chronic schizophrenia and the manic phase of manic-depressive psychoses. The drug is also used as an antiemetic. Adverse reactions are similar to those of the other phenothiazine tranquilizers. Significant autonomic side-effects, such as blurred or double vision, nasal congestion, dryness of the mouth and constipation, are infrequent at doses below 24 mg. per day.

Fluphenazine Hydrochloride U.S.P., Permitil®, Prolixin®, 4-{3-[2-(trifluoromethyl)phenothiazin-10-yl]propyl}-1-piperazineethanol dihydrochloride, 10-{3-[4-(2-hydroxyethyl)piperazinyl]propyl}-2-trifluoromethylphenothiazine dihydrochloride. Fluphenazine is the most potent of the currently available phenothiazine tranquilizers on a milligram basis. It is effective in the control of major psychotic states marked by hyperactivity but displays a high incidence of extrapyramidal side-effects at the dose required.

A solution of the enanthic acid (heptanoic acid) ester of fluphenazine, **Fluphenazine Enanthate U.S.P.,** in sesame oil containing 1.5 percent of benzyl alcohol is useful in the treatment of chronic schizophrenia. A single dose of 12.5 to 25 mg. may be given by the parenteral route (subcutaneous or intramuscular), with the therapeutic effect lasting for 1 to 3 weeks. The decanoic acid ester, **Fluphenazine Decanoate,** is also available for use by injection as a suspension in oil, 25 mg. per ml.

Acetophenazine Maleate U.S.P., Tindal®, 10{3-[4-(2-hydroxyethyl)-1-piperazinyl]propyl}phenothiazin-2-yl methyl ketone dimaleate. Acetophenazine, a 2-acyl phenothiazine derivative, is a homolog of carphenazine, and is also similar in structure to butaperazine. The compound is a relatively potent tranquilizing agent and, like others of its class, may be preferable for treatment of nonhospitalized patients because of its low incidence of agranulocytosis and jaundice.

Carphenazine Maleate U.S.P., Proketazine®, 1-{10-(3-[4-(2-hydroxyethyl)-1-piperazinyl]propyl)phenothiazin-2-yl}-1-propanone dimaleate. Proketazine is a comparatively short-acting antipsychotic agent used in the management of schizophrenic psychotic reactions. Like other phenothiazine tranquilizers, it may produce adverse reactions including extrapyramidal symptoms and blood dyscrasias.

Piperacetazine U.S.P., Quide®, 2-acetyl-10-{3-[4-(2-hydroxyethyl)piperidino]propyl}phenothiazine. Piperacetazine differs from acetophenazine by the presence of a piperidine ring, rather than a piperazine

ring, in the side chain. Its uses and side-effects are similar to those of the other phenothiazine tranquilizers.

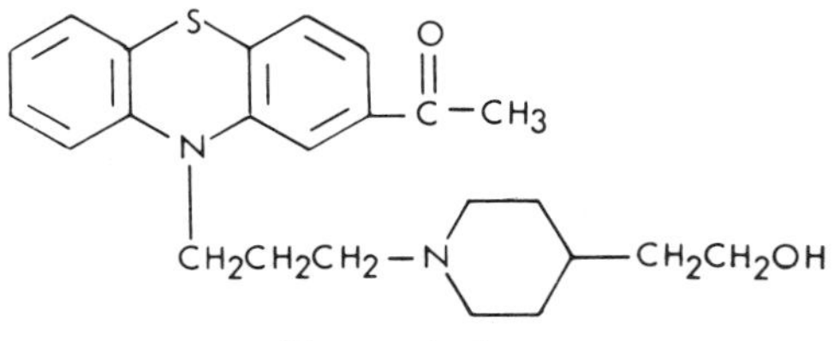

Aminoethyl side chain phenothiazine derivatives

In addition to the chlorpromazinelike compounds which have 3 carbon atoms separating the heterocyclic and aliphatic or alicyclic nitrogen atoms, derivatives in which 2 carbon atoms separate the nitrogens (e.g., promethazine) show marked central depressant properties. Some of these compounds are used therapeutically in other ways (e.g., the antihistamine promethazine, the antiparkinson agent, ethopropazine), but newer analogs (e.g., the sedative propiomazine) reflect the recognition that the sedation produced by this group is like that of chlorpromazine and unlike that of the sedative-hypnotics (e.g., phenobarbital, meprobamate). The sedative side-effects of members of this series are frequently applied therapeutically in the relief of excited emotional states. In contrast to the phenothiazine derivatives with aminopropyl (C_3) side chains, tranquilizing potency in the aminoethyl series is generally *reduced* by substitution in the 2 position of the phenothiazine nucleus. For example, 2-chloropromethazine is less effective than promethazine.

The structures of representative members of this class are presented in Table 9-12. These drugs are discussed in more detail in Chapter 12, Cholinergic Agents and Related Drugs, and Chapter 16, Histamine and Antihistaminic Agents.

Promethazine Hydrochloride U.S.P., Phenergan®, 10-(2-dimethylaminopropyl)phenothiazine hydrochloride. Promethazine is employed primarily for

Piperacetazine

TABLE 9-12

PHENOTHIAZINE DERIVATIVES (AMINO-ETHYL SIDE CHAIN)

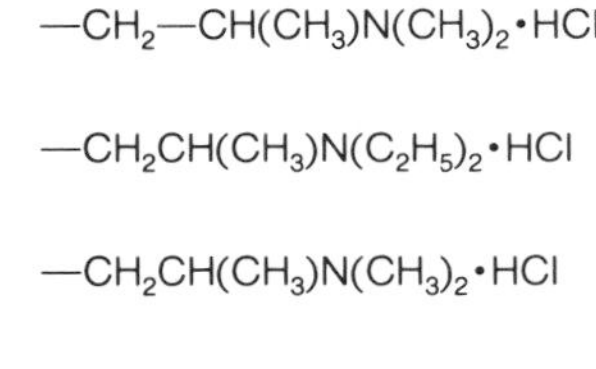

Generic Name Proprietary Name	R	R'	Date of Introduction
Promethazine Hydrochloride U.S.P. *Phenergan*	—CH₂—CH(CH₃)N(CH₃)₂·HCl	H	1951
Ethopropazine Hydrochloride U.S.P. *Parsidol*	—CH₂CH(CH₃)N(C₂H₅)₂·HCl	H	1954
Propiomazine Hydrochloride U.S.P. *Largon*	—CH₂CH(CH₃)N(CH₃)₂·HCl	—C—CH₂CH₃ ‖ O	1960

its antihistaminic effect (see Chap. 16) but is also used for its sedative and antiemetic properties, and for its potentiating effect on analgesics and other central nervous system depressants. The average adult oral dose for sedation is 25 mg. taken before retiring.

Propiomazine Hydrochloride U.S.P., Largon®, (±)-1-(10-[2-(dimethylamino)propyl]phenothiazin-2-yl)-1-propanone hydrochloride. The sedative effect of propiomazine is utilized to provide night-time, presurgical or obstetrical sedation of short duration and

(Text continues on p. 371)

TABLE 9-13

PHENOTHIAZINE TRANQUILIZERS

Name Proprietary Name	Preparations	Category	Usual Adult Dose*	Usual Dose Range*	Usual Pediatric Dose*
Chlorpromazine U.S.P. *Thorazine*	Chlorpromazine Suppositories U.S.P.	Antiemetic; tranquilizer	Antiemetic—rectal, 50 to 100 mg. 3 or 4 times daily as necessary; tranquilizer—rectal, 100 mg. 3 or 4 times daily as necessary	50 to 400 mg. daily	Antiemetic or tranquilizer—use not recommended in infants under 6 months of age; children 6 months and older—1 mg. per kg. of body weight 3 or 4 times daily as necessary
Chlorpromazine Hydrochloride U.S.P. *Thorazine, Promapar*	Chlorpromazine Hydrochloride Injection U.S.P.	Antiemetic; tranquilizer	Antiemetic—I.M., 25 to 50 mg., repeated 6 to 8 times daily as necessary; tranquilizer—I.M., 25 to 50 mg. repeated in 1 hour if necessary	Tranquilizer, 25 mg. to 1 g. daily	Antiemetic—use not recommended in infants under 6 months of age; children under 5 years of age—550 μg. per kg. of body weight 3 or 4 times daily as necessary, up to 40 mg. a day; 5 to 12 years of age—550 μg. per kg. 3 or 4 times daily as necessary, up to 75 mg. daily; over 12 years of age, see Usual Dose; tranquilizer—use not recommended in infants under 6 months of age; over 6 months of age—550 μg. per kg. of body weight or 15 mg. per square meter of body surface, 4 times daily up to 40 mg. per day for children 6 months to 5 years of age, and up to 75 mg. per day for children 5 to 12 years of age
	Chlorpromazine Hydrochloride Syrup U.S.P. Chlorpromazine Hydrochloride Tablets U.S.P.		Antiemetic, 10 to 25 mg. 4 to 6 times daily as necessary; tranquilizer, 10 to 50 mg. 2 or 3 times daily	Antiemetic, 10 to 300 mg. daily; tranquilizer, 10 mg. to 1 g. daily	Antiemetic—use not recommended in infants under 6 months of age; over 6 months of age, 550 μg. per kg. of body weight 4 to 6 times daily as necessary; tranquilizer—use not recommended in infants under 6 months of age; over 6 months of age—550 μg. per kg. of body weight or 15 mg. per square meter of body surface, 4 times daily
Thioridazine Hydrochloride U.S.P. *Mellaril*	Thioridazine Hydrochloride Oral Solution U.S.P. Thioridazine Hydrochloride Tablets U.S.P.	Tranquilizer	Initial, 25 to 100 mg. 3 times daily; maintenance, 10 to 200 mg. 2 to 4 times daily	20 to a maximum of 800 mg. daily	Use in children under 2 years of age is not recommended; 2 years of age and over—250 μg. per kg. of body weight or 7.5 mg. per square meter of body surface, 4 times daily
Mesoridazine Besylate U.S.P. *Serentil*	Mesoridazine Besylate Injection U.S.P. Mesoridazine Besylate Oral Solution U.S.P. Mesoridazine Besylate Tablets U.S.P.	Antipsychotic agent	Oral, 50 to 400 mg. of mesoridazine, as the besylate, daily; I.M., 25 to 200 mg. of mesoridazine, as the besylate, daily	25 to 400 mg. of mesoridazine, as the besylate, daily	

* See *U.S.P. D.I.* for complete dosage information.
(Continued)

TABLE 9–13 *(Continued)*

PHENOTHIAZINE TRANQUILIZERS

Name Proprietary Name	Preparations	Category	Usual Adult Dose*	Usual Dose Range*	Usual Pediatric Dose*
Prochlorperazine U.S.P. *Compazine*	Prochlorperazine Suppositories U.S.P.	Antiemetic	Rectal, 25 mg. 2 times daily		Use in children under 9 kg. of body weight or 2 years of age is not recommended; 9 to 13 kg.—2.5 mg. once or twice daily, not exceeding 7.5 mg. daily; 14 to 17 kg.—2.5 mg. 2 or 3 times daily, not exceeding 10 mg. daily; 18 to 39 kg.—2.5 mg. 3 times a day or 5 mg. twice daily, not exceeding 15 mg. daily
Prochlorperazine Edisylate U.S.P. *Compazine*	Prochlorperazine Edisylate Injection U.S.P.	Antiemetic; tranquilizer	Antiemetic—I.M. or I.V., the equivalent of 5 to 10 mg. of prochlorperazine 6 to 8 times daily as necessary; tranquilizer—I.M. or I.V., the equivalent of 10 to 20 mg. of prochlorperazine 4 to 6 times daily	Antiemetic, 5 mg. to not more than 40 mg. daily; tranquilizer, 10 to 200 mg. daily	Use in children under 9 kg. of body weight or 2 years of age is not recommended; antiemetic—I.M., the equivalent of 132 µg. of prochlorperazine per kg. of body weight; tranquilizer—I.M., the equivalent of 132 µg. of prochlorperazine per kg.
	Prochlorperazine Edisylate Oral Solution U.S.P. Prochlorperazine Edisylate Syrup U.S.P.		Antiemetic, the equivalent of 5 to 10 mg. of prochlorperazine 3 or 4 times daily as necessary; tranquilizer, the equivalent of 5 to 35 mg. of prochlorperazine 3 or 4 times daily	5 to 150 mg. daily	Use in children under 9 kg. of body weight or 2 years of age is not recommended; antiemetic, 9 to 13 kg.—the equivalent of 2.5 mg. of prochlorperazine once or twice daily, not exceeding 7.5 mg. daily; 14 to 17 kg.—the equivalent of 2.5 mg. of prochlorperazine 2 or 3 times daily, not exceeding 10 mg. daily; 18 to 39 kg.—the equivalent of 2.5 mg. of prochlorperazine 3 times daily or 5 mg. twice daily, not exceeding 15 mg. daily; tranquilizer, the equivalent of 100 µg. per kg. of body weight or 2.5 mg. of prochlorperazine per square meter of body surface, 4 times daily
Prochlorperazine Maleate U.S.P. *Compazine*	Prochlorperazine Maleate Tablets U.S.P.	Antiemetic; tranquilizer	Antiemetic, the equivalent of 5 to 10 mg. of prochlorperazine 3 or 4 times daily as necessary; tranquilizer, the equivalent of 5 to 35 mg. of prochlorperazine 3 or 4 times daily	5 to 150 mg. daily	Use in children under 9 kg. of body weight or 2 years of age is not recommended; antiemetic, 9 to 13 kg.—the equivalent of 2.5 mg. of prochlorperazine once or twice daily, not exceeding 7.5 mg. daily; 14 to 17 kg.—the equivalent of 2.5 mg. of prochlorperzine 2 or 3 times daily, not exceeding 10 mg. daily; 18 to 39 kg.—the equivalent of 2.5 mg. of prochlorperazine 3 times daily or 5 mg. twice daily, not exceeding 15 mg. daily; tranquilizer, the equivalent of 100 µg. per kg. of body weight or 2.5 mg. of prochlorperazine per square meter of body surface, 4 times daily

*See U.S.P. D.I. for complete dosage information.

(Continued)

TABLE 9–13 *(Continued)*

PHENOTHIAZINE TRANQUILIZERS

Name Proprietary Name	Preparations	Category	Usual Adult Dose*	Usual Dose Range*	Usual Pediatric Dose*
Promazine Hydrochloride U.S.P. *Sparine*	Promazine Hydrochloride Injection U.S.P. Promazine Hydrochloride Oral Solution U.S.P. Promazine Hydrochloride Syrup U.S.P. Promazine Hydrochloride Tablets U.S.P.	Antipsychotic agent		Oral, I.M. or I.V., 10 to 200 mg. every 4 to 6 hours	
Triflupromazine U.S.P. *Vesprin*	Triflupromazine Oral Suspension U.S.P.	Antipsychotic agent		The equivalent of 30 to 150 mg. of triflupromazine hydrochloride daily	
Triflupromazine Hydrochloride U.S.P. *Vesprin*	Triflupromazine Hydrochloride Injection U.S.P. Triflupromazine Hydrochloride Tablets U.S.P.	Antipsychotic agent		Oral, 30 to 150 mg. daily; I.M., 5 to 10 mg., repeated every 4 hours, if necessary; I.V., 1 to 3 mg., repeated in 4 hours, if necessary	
Trifluoperazine Hydrochloride U.S.P. *Stelazine*	Trifluoperazine Hydrochloride Injection U.S.P. Trifluoperazine Hydrochloride Syrup U.S.P. Trifluoperazine Hydrochloride Tablets U.S.P.	Antipsychotic agent		Oral, non-hospitalized patients, 1 to 2 mg. twice daily; hospitalized patients, 2 to 5 mg. daily initially, gradually increasing to the optimum level of 15 to 20 mg. daily, although a few patients may require 40 mg. or more daily; I.M., 1 to 2 mg. every 4 to 6 hours as required	
Thiethylperazine Maleate U.S.P. *Torecan*	Thiethylperazine Maleate Suppositories U.S.P. Thiethylperazine Maleate Tablets U.S.P.	Antiemetic		Oral or rectal, 10 to 30 mg. daily	
Thiethylperazine Maleate U.S.P. *Torecan*	Thiethylperazine Maleate Injection U.S.P.	Antiemetic	I.M., 10 to 30 mg. daily	I.M., 10 to 30 mg. daily	
Butaperazine Maleate *Repoise*	Butaperazine Maleate Tablets	Antipsychotic agent	15 to 30 mg. daily in 3 divided doses; may be increased gradually to a maximum daily dose of 100 mg.		
Perphenazine U.S.P. *Trilafon*	Perphenazine Injection U.S.P. Perphenazine Oral Solution U.S.P. Perphenazine Syrup U.S.P. Perphenazine Tablets U.S.P.	Antipsychotic agent		Oral, non-hospitalized patients, 2 to 8 mg. 3 times daily, hospital-ized patients, 8 to 16 mg. 2 to 4 times daily; I.M., 5 to 10 mg. initially, followed by 5 mg. in 6 hours	

*See U.S.P. D.I. for complete dosage information.
(Continued)

TABLE 9-13 *(Continued)*

PHENOTHIAZINE TRANQUILIZERS

Name Proprietary Name	Preparations	Category	Usual Adult Dose*	Usual Dose Range*	Usual Pediatric Dose*
Fluphenazine Hydrochloride U.S.P. *Permitil, Prolixin*	Fluphenazine Hydrochloride Elixir U.S.P. Fluphenazine Hydrochloride Injection U.S.P. Fluphenazine Hydrochloride Oral Solution U.S.P. Fluphenazine Hydrochloride Tablets U.S.P.	Tranquilizer	I.M., 312.5 μg. to 2.5 mg. 4 times daily Initial, 625 μg. to 2.5 mg. 4 times daily; maintenance, 1 to 5 mg. once daily	1.25 to 10 mg. daily 500 μg. to 20 mg. daily	250 to 750 μg. 1 to 4 times daily, up to 10 mg. daily in older children
Fluphenazine Enanthate U.S.P. *Prolixin*	Fluphenazine Enanthate Injection U.S.P.	Tranquilizer	I.M. or S.C., 25 mg. every 2 weeks	12.5 to 100 mg. every 1 to 3 weeks	Dosage is not established in children under 12 years of age
Fluphenazine Decanoate *Prolixin*	Fluphenazine, Decanoate Injection	Tranquilizer		I.M. or S.C., 6.25 to 50 mg. every 2 to 4 weeks	
Acetophenazine Maleate U.S.P. *Tindal*	Acetophenazine Maleate Tablets U.S.P.	Antipsychotic agent	20 mg. 3 times daily	40 to 80 mg. daily	
Carphenazine Maleate U.S.P. *Proketazine*	Carphenazine Maleate Solution U.S.P. Carphenazine Tablets U.S.P.	Antipsychotic agent		12.5 to 50 mg. 3 times daily; increased by 12.5 to 50 mg. daily at intervals of from 4 days to 1 week; the maximum daily dose recommended is 400 mg.	
Piperacetazine U.S.P. *Quide*	Piperacetazine Tablets U.S.P.	Antipsychotic agent	Initial, 10 mg. 2 to 4 times daily, may be increased up to 160 mg. daily within 3 to 5 days; maintenance, up to 160 mg. daily in divided doses		
Propiomazine Hydrochloride U.S.P. *Largon*	Propiomazine Hydrochloride Injection U.S.P.	Sedative	I.M. or I.V., 20 mg.	10 to 40 mg.	

*See *U.S.P. D.I.* for complete dosage information.

ready arousal. The drug enhances the effect of other central nervous system depressants; therefore, the dose of such agents should be reduced in the presence of propiomazine. Propiomazine is given by either the intravenous or intramuscular route; the solution contains 20 mg. per ml., together with preservatives and buffer salts.

DIPHENYLMETHANE DERIVATIVES

The diphenylmethane derivatives are a group of drugs with diverse pharmacologic actions. However, the sedative properties shown by many members of this series more closely resemble the sedation produced by the tranquilizing agents than by the sedative-hypnotics. Table 9-14 lists those compounds whose sedative or tranquilizing properties have led to their use as psychotherapeutic or calming agents in the treatment of a variety of emotional or mental disorders characterized by tension, anxiety and agitation. In general, this group of drugs is not effective against the major psychoses. Some possess significant antihistaminic properties (hydroxyzine, buclizine, chlorcyclizine), while others (benactyzine) show pharmacologic properties and side-effects which are primarily anticholinergic. This combination of anticholinergic, antihistaminic

TABLE 9–14

STRUCTURAL RELATIONSHIPS OF DIPHENYLMETHANE DERIVATIVES

Generic Name Proprietary Name	R_1	R_2	R_3	Year of Intro- duction
Hydroxyzine Hydrochloride U.S.P. *Atarax, Vistaril* Hydroxyzine Pamoate U.S.P.	H	$-N$⟨⟩$N-CH_2-CH_2-O-CH_2-CH_2-OH$ $\cdot 2\ HCl$	Cl	1956
Buclizine Hydrochloride *Bucladin-S*	H	$-N$⟨⟩$N-CH_2-$⟨⟩$-C(CH_3)_3$ $\cdot HCl$	Cl	1956
Chlorcyclizine Hydrochloride U.S.P.	H	$-N$⟨⟩$N-CH_3 \cdot HCl$	Cl	1950
Benactyzine Hydrochloride *Deprol*	OH	$\overset{O}{\overset{\|}{-C}}-O-(CH_2)_2N(C_2H_5)_2 \cdot HCl$	H	1957

and tranquilizing properties has also been noted for the phenothiazine derivatives.

Products

Hydroxyzine Hydrochloride U.S.P. Atarax® Hydrochloride, 1-(*p*-chlorobenzhydryl)-4-[2-(2-hydroxyethoxy)ethyl]piperazine dihydrochloride, 2-{2-[4-(*p*-chloro-α-phenylbenzyl)-1-piperazinyl]ethoxy}ethanol dihydrochloride. Hydroxyzine hydrochloride is useful for the management of neuroses with agitation and anxiety as characterizing features. It is of little use in frank psychoses or depressive states. In addition to its sedative effects, the drug possesses antihistaminic properties useful in the management of acute and chronic urticaria and other allergic states (see Chap. 16). Anticholinergic properties have been demonstrated pharmacologically.

Hydroxyzine hydrochloride is a white solid, very soluble in water and in ethanol. Intramuscular or intravenous administration may be used for emergencies where rapid onset of response is necessary.

The salt of hydroxyzine with pamoic acid (1,1'-methylene bis[2-hydroxy-3-naphthalene carboxylic acid]) is **Hydroxyzine Pamoate U.S.P.** (Vistaril®), used orally for the central depressant and antihistaminic properties of hydroxyzine.

Buclizine and *chlorcyclizine* resemble hydroxyzine in indications, side-reactions and potency.

Benactyzine Hydrochloride, Deprol®, 2-diethylaminoethyl benzilate hydrochloride. Benactyzine is an anticholinergic compound with about one-fourth the peripheral activity of atropine. In a dose which produces little peripheral effect, benactyzine is useful in the management of psychoneurotic disorders characterized by anxiety and tension. The usual oral dose of benactyzine hydrochloride ranges from 1 to 3 mg., three times daily.

RING ANALOGS OF PHENOTHIAZINES

There is a group of tranquilizing agents derived by isosteric replacement of one or more groups or atoms in the structure of the phenothiazine tranquilizing agents. The compounds thus derived possess many clinically useful pharmacologic properties in common with phenothiazine tranquilizers.

Chlorprothixene U.S.P., Taractan®, *cis*-2-chloro-9-(3-dimethylaminopropylidene)thioxanthene. Chlorprothixene, an isostere of chloropromazine in which nitrogen is replaced with a methylene group, was released in 1961 for use as a psychotherapeutic drug. It appears to be effective in the treatment of schizophrenia, and in psychotic and severe neurotic conditions characterized by anxiety and agitation, thus resembling the phenothiazines. In addition, it is claimed to exert some benefit in depressive states. Chlorprothixene potentiates the effect of sedatives, has a hypotensive effect and shows antihistaminic and antiemetic properties.

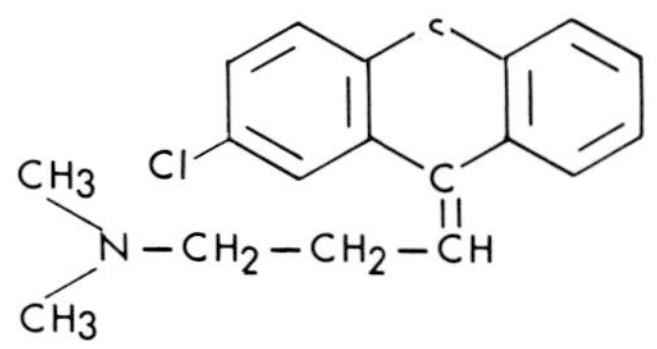

Chlorprothixene

Thiothixene U.S.P., Navane®, *cis*-N,N-dimethyl-9-[3-(4-methyl-1-piperazinyl)propylidene]thioxanthene-2-sulfonamide. Thiothixene, a thioxanthene derivative related to chlorprothixene, was introduced in 1967 as an antipsychotic agent useful in the management of schizophrenia and other psychotic states. It also shows antidepressant properties. Thiothixene is similar in its actions to the phenothiazine tranquilizers and may potentiate the actions of the central nervous system depressants including anesthetics, hypnotics, and alcohol. At higher dosage levels it may produce extrapyramidal symptoms and orthostatic hypotension.

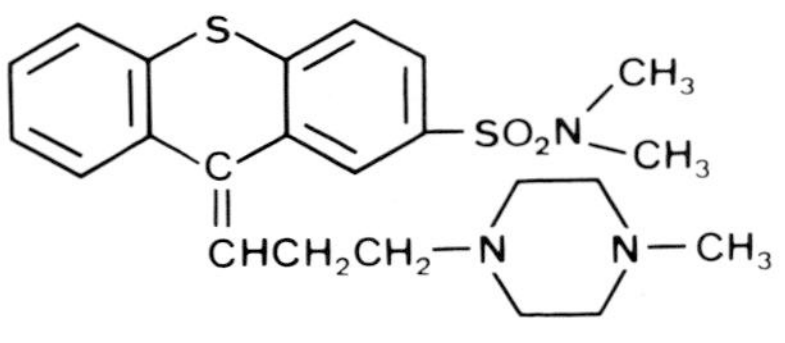

Thiothixene

The salt, **Thiothixene Hydrochloride U.S.P.,** is available for oral or parenteral use.

Loxapine Succinate, Loxitane®, 2-chloro-11-(4-methyl-1-piperazinyl)dibenz[b,f][1,4]oxazepine succinate.

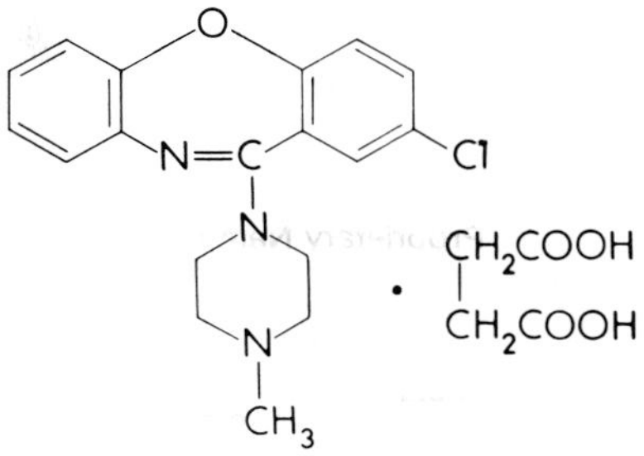

Loxapine Succinate

Loxapine is a tricyclic antipsychotic agent, in which the central ring contains the unique oxazepine structure. It is used in the treatment of acute and chronic schizophrenia. Side-effects are similar to those of other antipsychotic tranquilizers, including parkinson-like symptoms and anticholinergic effects.

Fluorobutyrophenones

A series of related fluorobutyrophenones, derived from studies in Europe on potential analgesics,[26] was found to be effective in the management of major psychoses. The first compound of the series, haloperidol, was introduced in the United States in 1967.

Haloperidol U.S.P., Haldol®, 4-[4-(p-chlorophenyl)-4-hydroxypiperidino]-4'-fluorobutyrophenone. Haloperidol is a major tranquilizer. It is used in the management of the agitated states, as well as mania, aggressiveness, and hallucinations associated with acute and chronic psychoses including schizophrenia

TABLE 9-15

DIPHENYLMETHANE DERIVATIVES

Name Proprietary Name	Preparations	Category	Usual Adult Dose*	Usual Dose Range*
Hydroxyzine Hydrochloride U.S.P. *Atarax*	Hydroxyzine Hydrochloride Injection U.S.P. Hydroxyzine Hydrochloride Syrup U.S.P. Hydroxyzine Hydrochloride Tablets U.S.P.	Tranquilizer; antihistaminic	Oral, 25 mg. 3 times daily; I.M., 50 to 100 mg. every 4 to 6 hours	25 to 100 mg.
Hydroxyzine Pamoate U.S.P. *Vistaril*	Hydroxyzine Pamoate Capsules U.S.P. Hydroxyzine Pamoate Oral Suspension U.S.P.	Tranquilizer; antihistaminic		The equivalent of 25 mg. of hydroxyzine hydrochloride 3 times daily to the equivalent of 100 mg. of hydroxyzine hydrochloride 4 times daily

* See *U.S.P. D.I.* for complete dosage information.

and psychotic reactions in adults with organic brain damage. Haloperidol is also used for control of facial tics and vocal utterances of Gilles de la Tourette's syndrome.

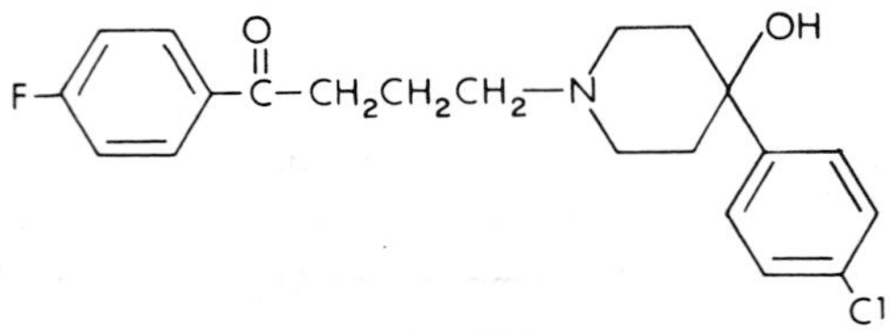

Haloperidol

The extrapyramidal reactions, parkinsonlike symptoms, impaired liver function and blood dyscrasias observed in the phenothiazine tranquilizers have been reported to occur with haloperidol also. The drug also potentiates the actions of central nervous system depressant drugs such as analgesics, anesthetics, barbiturates and alcohol.

Droperidol U.S.P., Inapsine®, 1-{1-[3-(p-fluorobenzoyl)propyl]-1,2,3,6-tetrahydro-4-pyridyl}-2-benzimidazolinone. Droperidol, a fluorobutyrophenone tranquilizer, is used alone or together with the potent narcotic analgesic, fentanyl (Sublimaze®) (see Chap. 17). The combination (Innovar®) is administered by the intramuscular or intravenous routes for preanesthetic sedation and analgesia, and as an adjunct to the induction of anesthesia. Like the phenothiazines, droperidol may be used alone as an antipsychotic or antiemetic agent. Droperidol is available for intravenous or intramuscular use only. The parenteral solution contains 2.5 mg. of droperidol per ml., together with lactic acid to hold the pH at 3.4 ± 0.4. For sedation and analgesia, the usual intramuscular dose is 0.5 to 2 ml., each ml. containing 50 μg. of fentanyl and 2.5 mg. of droperidol. As an adjunct to the induction of anesthesia, the usual intravenous dose is 1 ml. for each 20 to 25 pounds of body weight.

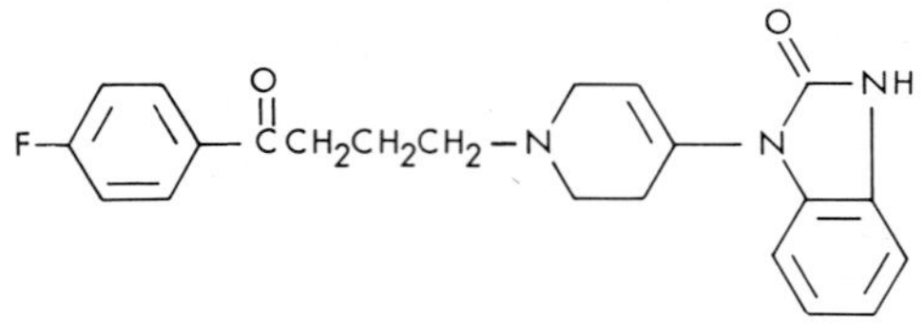

Droperidol

TABLE 9–16

MISCELLANEOUS TRANQUILIZING AGENTS

Name Proprietary Name	Preparations	Category	Usual Adult Dose*	Usual Dose Range*	Usual Pediatric Dose*
Chlorprothixene U.S.P. *Taractan*	Chlorprothixene Injection U.S.P. Chlorprothixene Oral Suspension U.S.P. Chlorprothixene Tablets U.S.P.	Antipsychotic agent		Oral: moderate anxiety, 10 mg. 3 or 4 times daily (up to 60 mg. daily); severe neurotic and psychotic states, 25 to 50 mg. 3 or 4 times daily (up to 600 mg. daily). I.M.: moderate anxiety, 12.5 to 25 mg.; severe neurotic and psychotic states, 75 to 200 mg. daily	
Thiothixene U.S.P. *Navane*	Thiothixene Capsules U.S.P.	Antipsychotic agent	20 to 30 mg. daily in divided doses	6 to 60 mg. daily, in divided doses	
Thiothixene Hydrochloride U.S.P. *Navane*	Thiothixene Hydrochloride Injection U.S.P. Thiothixene Hydrochloride Oral Solution U.S.P.	Antipsychotic agent	Oral, 6 to 60 mg. daily, in divided doses; I.M., 4 mg. 2 to 4 times daily	Oral, 20 to 60 mg. daily, in divided doses; I.M., 16 to 30 mg. daily, in divided doses	
Loxapine Succinate *Loxitane*	Loxapine Succinate Capsules	Antipsychotic agent	Initial, 10 mg. twice daily; maintenance, 60 to 100 mg. daily; total daily doses over 250 mg. are not recommended		
Haloperidol U.S.P. *Haldol*	Haloperidol Oral Solution U.S.P. Haloperidol Tablets U.S.P.	Tranquilizer	500 μg. to 5 mg. 2 or 3 times daily	1 to 100 mg. daily	Dosage is not established in infants and children
Droperidol U.S.P. *Inapsine*	Droperidol Injection U.S.P.	Antipsychotic agent	I.M. or I.V., 1.25 to 10 mg.	1.25 to 10 mg.	

*See *U.S.P. D.I.* for complete dosage information.

MISCELLANEOUS TRANQUILIZING AGENTS

Chemical types distinctly different from the phenothiazines and their ring analogs, and from the fluorobutyrophenones, have been found useful in the treatment of major psychoses.

Molindone Hydrochloride, Lidone®, 3-ethyl-6,7-dihydro-2-methyl-5-morpholinomethylindole-4(5H)one hydrochloride.

Molindone Hydrochloride

Molindone, a unique antipsychotic indoleamine, is useful in the treatment of chronic and acute schizophrenia. Principal side-effects are like those of the phenothiazines: extrapyramidal parkinsonlike symptoms, anticholinergic effects and drowsiness.

The usual initial oral dose is 10 mg. per day, being increased gradually until symptoms are controlled. In severe schizophrenia this may require up to 225 mg. per day. Peak blood levels are reached within one hour after administration, and the drug is excreted rapidly.

Lithium Carbonate U.S.P., Eskalith®, Lithane®, Li_2CO_3. Lithium carbonate is used to treat the mildly active patient in the manic phase of manic-depressive psychoses. The drug has a slow onset of action requiring seven or more days to develop its maximal effect. It is rapidly absorbed and also rapidly excreted. Side-effects are usually mild if the blood-level concentration is kept below 1.7 mEq. per liter. Major adverse reactions occur if blood levels of lithium exceed this concentration. Mild side-effects include nausea, thirst, fatigue, and a fine tremor of the hands or jaw. Moderate side-effects include anorexia, vomiting, diarrhea, and coarse tremor.

Lithium, owing to its ability to mimic the body's natural electrolytes sodium and potassium, exerts widespread effects on the body. Lithium is believed to act in manic-depressive disorders by its effects on membrane transport of choline that result in reduced choline turnover in the brain.

The usual initial oral dose of 600 mg. three times a day is reduced to a maintenance dose of 300 mg. three times a day, in order to stabilize the serum lithium level at 0.6 to 1.2 mEq. per liter, measured 8 to 12 hours after administration.

ANTICONVULSANT DRUGS

The primary use of anticonvulsant drugs is in the prevention and the control of epileptic seizures. In most cases treatment is symptomatic; only in a few cases suitable for surgery is a cure possible.

The disease affects approximately from 0.5 to 1.0 percent of the population. Since symptomatic treatment is frequently lifelong, and a feeling of inferiority and self-consciousness often causes a withdrawal from society, this disease constitutes a major public health problem.

Until recently, only two useful drugs were available which could depress the motor cortex (prevent convulsions) as well as the sensory cortex (produce sleep). These were the bromides, which were introduced in about 1857, and phenobarbital, which has been used since 1912. Since the introduction of phenytoin (Dilantin®) in 1938, a number of anticonvulsant drugs have followed which are better able to control seizures; they demonstrate that sedation may be dissociated from anticonvulsant activity in the various types of epilepsy. Each type of epilepsy may be distinguished by clinical and electroencephalographic patterns, and each responds differently to the various classes of anticonvulsant drugs. The major types of epileptic seizures are:

1. *Grand Mal.* Sudden loss of consciousness followed by general muscle spasms lasting for an average of 2 to 5 minutes. The frequency and the severity of attacks are variable.

Structure common to anticonvulsant drugs.

2. *Petit Mal.* Sudden, brief loss of consciousness with minor movements of the head, the eyes and the extremities, lasting for about 5 to 30 seconds. The patient is immediately alert, ready to continue normal activity. There may be many episodes in a day; the highest incidence is found in children.

3. *Psychomotor Seizures.* Automatic, patterned movements lasting from 2 to 3 minutes occur. Amnesia is common, with often no memory of the incident remaining. This state is sometimes confused with psychotic behavior.

Compounds being tested in the laboratory for anticonvulsant activity are assayed for protection against convulsions induced both chemically and electrically. Clinically useful drugs are usually effective in elevating the threshold to seizures produced by the central nervous system stimulant pentylenetetrazol (Metrazol, see Chap. 10) or by electroshock.

The drugs acting as selective depressants of convulsant activity have a common structural feature, as shown above. The single exception to this structural pattern among currently useful anticonvulsant drugs is primidone (Mysoline®). However, it is known[27,28] that primidone is metabolically oxidized in man to form the barbiturate phenobarbital, which may account for the structural uniqueness of the drug.

One of the most important factors for persons susceptible to seizures is the control of living conditions. States favorable for prevention of seizures include dehydration, systemic acidosis, adequate oxygen and freedom from stress. Therefore, adjuncts to anticonvulsant therapy include drugs that produce acidosis, such as glutamic acid, and those that produce both acidosis and dehydration, such as the carbonic anhydrase inhibiting diuretics, e.g., acetazolamide (Diamox®), ethoxzolamide (Cardrase®) (see Chap. 13).

Table 9-17 lists the names and the types of seizure for which the drugs are most effective.

BARBITURATES

Of the commonly employed barbiturates, only phenobarbital, mephobarbital (Mebaral®) and metharbital (Gemonil®) show the selective anticonvulsant activity which makes them useful in the symptomatic treatment of epilepsy. The mechanism by which these drugs reduce the excitability of the motor cortex is unknown. The structures of these barbiturates, all members of the long-acting class, are listed in Table 9-4.

The anticonvulsant activity of the barbiturates is not related to the sedation they produce, for protection from convulsions is often shown at nonsedating dose levels. The three anticonvulsant barbiturates show a high degree of effectiveness in grand mal but are of less benefit in petit mal and psychomotor epilepsy.

HYDANTOINS

As cyclic ureides, related in structure to the barbiturates, many hydantoins were synthesized as potential hypnotics following the introduction of the barbiturates in 1903. The first of the hydantoins, nirvanol, was introduced as a hypnotic and anticonvulsant in 1916 but has since been replaced by less toxic analogs. Following systematic pharmacologic and clinical studies, 5,5-diphenylhydantoin was reported[29] in 1938 to be the least hypnotic and most strongly anticonvulsant of the related compounds studied. The hydantoins are most effective against grand mal; psychomotor attacks are sometimes controlled. These drugs are ineffective against petit mal.

The cyclic ureide structure exists in equilibrium with its enolic form, 2,4-dihydroxy-5,5-disubstituted imidazole. Sodium, or other metal salts of the acidic 2-hydroxyl group may be formed in alkaline solution; the insoluble free-acid form is formed in the presence of acid.

The names, the structures and the years of introduction for the anticonvulsant hydantoins are listed in Table 9-18.

TABLE 9–17

DRUGS USED IN THE TREATMENT OF EPILEPSY

Drug	Types of Seizure
I. Barbiturates	
Phenobarbital	Grand mal
Mephobarbital	Grand mal
Metharbital	Grand mal
II. Hydantoins	
Phenytoin	Grand mal*
Mephenytoin	Grand mal*
Ethotoin	Grand mal*
III. Oxazolidinediones	
Trimethadione	Petit mal
Paramethadione	Petit mal
IV. Succinimides	
Phensuximide	Petit mal
Methsuximide	Petit mal*
Ethosuximide	Petit mal
V. Benzodiazepines	
Clonazepam	Petit mal
Diazepam IV	Status epilepticus
VI. Miscellaneous	
Primidone	Grand mal*
Carbamazepine	Grand mal*
Phenacemide	Psychomotor seizures
Valproic acid	Petit mal

*Some effectiveness against psychomotor seizures.

Keto   5,5-Disubstituted   Enol
Hydantoin

Phenytoin U.S.P., Dilantin®, 5,5-diphenyl-2,4-im-idazolidinedione, 5,5-diphenylhydantoin. Phenytoin, formerly named diphenylhydantoin, is an anticonvulsant with little or no sedative properties. It is most effective in controlling grand mal seizures when used alone or in combination with phenobarbital. Various untoward effects have been observed; these include dizziness, skin rashes, itching, tremors, fever, vomiting, blurred vision, difficult breathing and hyperplasia of the gums.

Phenytoin is a white powder, practically insoluble in water, and slightly soluble in alcohol. The water-soluble and somewhat hygroscopic sodium salt is available as **Phenytoin Sodium U.S.P.** Phenytoin sodium capsules are available in both prompt and extended forms, characterized by rapid or slow rates of dissolution.

Mephenytoin U.S.P., Mesantoin®, 3-methyl-5-ethyl-5-phenylhydantoin. Mephenytoin, like phenytoin, is an anticonvulsant with little or no sedative

effect. It is used primarily for the control of grand mal seizures, but may also be used in conjunction with other anticonvulsants for the control of psychomotor and jacksonian seizures. Significant adverse reactions include blood dyscrasias and skin rashes.

Ethotoin, Peganone®, 3-ethyl-5-phenylhydantoin. Ethotoin is used primarily for the control of grand mal epilepsy. Adverse reactions are like those of related hydantoins: blood dyscrasias, skin rash, ataxia and gum hypertrophy. The drug may be used alone, but it is most frequently used in combination with other anticonvulsants.

OXAZOLIDINEDIONES

The oxazolidine-2-4-diones, compounds isosterically related to the hydantoins by substitution of an oxygen for nitrogen, were first tested as hypnotics in 1938.[30] The most active anticonvulsant drugs of this type (3,5,5-trialkyloxazolidine-2,4-diones) may be synthesized by condensation of an ester of dialkylglycollic acid with urea in the presence of sodium ethylate, followed by N-alkylation with alkylsulfates.[31]

The oxazolidinediones are effective in the treatment of petit mal and were uniquely so when first introduced in 1946. They are ineffective against grand mal but are used in conjunction with other drugs in the treatment of mixed types of seizures, such as combined petit mal and grand mal epilepsy.

TABLE 9–18

THE ANTICONVULSANT HYDANTOIN DERIVATIVES

Generic Name Proprietary Name	Substituents			Year of Introduction
	R_5	R'_5	R_3	
Phenylethylhydantoin *Nirvanol*	(phenyl)	$CH_3—CH_2—$	H	1916
Phenytoin U.S.P. *Dilantin, Diphentoin*	(phenyl)	(phenyl)	H	1938
Mephenytoin U.S.P. *Mesantoin*	(phenyl)	$CH_3—CH_2—$	$CH_3—$	1947
Ethotoin *Peganone*	(phenyl)	H	$CH_3—CH_2—$	1957

Trimethadione U.S.P., Tridione®, 3,5,5-trimethyl-2,4-oxazolidinedione. Trimethadione was introduced in 1946 for use in the treatment and the prevention of epileptic seizures of the petit mal type. Although trimethadione is among the more effective agents for this purpose, it is recommended that it be reserved for refractory cases because of toxicity. Toxic effects include gastric irritation, nausea, skin eruptions, sensitivity to light, disturbances of vision and dizziness and drowsiness. Serious reactions include aplastic anemia and nephrosis, indicating the need for routine blood and urine examinations.

Trimethadione $R_5 = R_5' = CH_3$
Paramethadione $R_5 = CH_3; R_5' = C_2H_5$

Trimethadione is a white, granular substance with a weak camphorlike odor. It is soluble in water or alcohol, giving a slightly acidic solution.

Paramethadione U.S.P., Paradione®, 5-ethyl-3,5-dimethyl-2,4-oxazolidinedione. Paramethadione, introduced in 1949, has the same use and side-effects as trimethadione, although individual variation may show one to be effective in patients in which the other is ineffective.

The compound is an oily liquid, slightly soluble in water but readily soluble in alcohol.

SUCCINIMIDES

Extensive screening for anticonvulsant activity among aliphatic and heterocyclic amides revealed high activity within a series of α,N-disubstituted derivatives of succinimide. The discovery of their usefulness in the treatment of petit mal seizures led to the introduction of phensuximide (Milontin®) in 1953 as a therapeutic companion to the oxazolidinediones. Methsuximide (Celontin®) followed in 1958, and ethosuximide (Zarontin®) in 1960.

The succinimides appear to be less potent than the oxazolidinediones and to possess less significant side-effects. Periodic blood and urine studies are advisable during treatment. These drugs are moderately effective in the control of petit mal seizures but are ineffective against grand mal. They are administered orally.

Ethosuximide U.S.P., Zarontin®, 2-ethyl-2-methylsuccinimide. Ethosuximide has been shown to be effective in pure petit mal but less effective in mixed petit mal seizures. It is considered the drug of choice for treatment of petit mal due to its lower incidence of major adverse reactions. Rare cases of agranulocytosis, cytopenia and bone marrow depression have been reported. The oral dose of 250-mg. capsules must be adjusted according to patient response.

Phensuximide $R = $ ⟨phenyl⟩ , $R' = H$, $R'' = CH_3$

Methsuximide $R = $ ⟨phenyl⟩ , $R' = CH_3$, $R'' = CH_3$

Ethosuximide $R = C_2H_5-$, $R' = CH_3$, $R'' = H$

Phensuximide U.S.P., Milontin®, N-methyl-2-phenylsuccinimide. Phensuximide is a crystalline solid, slightly soluble in water (0.4%), readily soluble in alcohol. Aqueous solutions are fairly stable at pH 2 to 8, but hydrolysis occurs under more alkaline conditions.

Methsuximide U.S.P., Celontin®, N,2-dimethyl-2-phenylsuccinimide. Methsuximide, when used in treatment of petit mal seizures, has been found to produce undesirable side-effects in about 30 percent of the patients taking the drug. Among the more serious were psychic disturbances (ranging from alteration of mood to acute psychoses), hepatic dysfunction and bone marrow aplasia.

The drug has shown usefulness in a significant number of cases of psychomotor seizures. Physical properties are like those of phensuximide.

BENZODIAZEPINES

The benzodiazepines are CNS depressants and are used primarily as sedative–hypnotics and in anxiety and tension states. However, some members of this class (diazepam, clonazepam) are used as anticonvulsant drugs in the treatment of specific kinds of seizures. Their effectiveness in the treatment of epilepsy is limited by the rapid development of tolerance.

Diazepam U.S.P., Valium®. In addition to its use as a sedative and antianxiety drug (earlier in this chapter) diazepam is an effective anticonvulsant agent. Given parenterally, it is useful in the acute treatment of status epilepticus and in the interruption of severe recurrent convulsive seizures. Owing to the short-lived effect of diazepam after intravenous administration, long-term control of seizures is best effected by other drugs.

Clonazepam U.S.P., Clonopin®, 5-(2-chlorophenyl)-1,3-dihydro-7-nitro-2H-1,4-benzodiazepin-2-one.

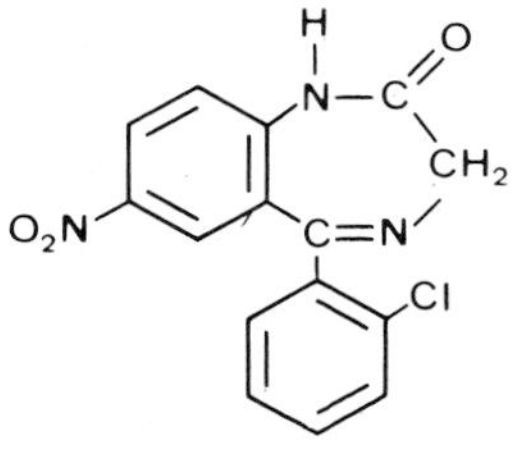

Clonazepam

Clonazepam is a long-acting CNS depressant used solely as an anticonvulsant for treatment of petit mal and myoclonic seizures in patients uncontrolled by oxazolidinediones or the succinimides. Following single oral doses of clonazepam, maximal blood levels are achieved in one to two hours. The half-life of the parent drug varies from 18 to 50 hours. Metabolism involves oxidative hydroxylation of the 3-position methylene group and reduction of the 7-nitro group to the amino, which is partially acetylated before excretion.

MISCELLANEOUS

Several compounds of miscellaneous structure show useful anticonvulsant properties. All possess structural features common to the class, but each chemical type, as yet, is represented by a single member.

Primidone U.S.P., Mysoline®, 5-ethyldihydro-5-phenyl-4,6-(1H,5H)-pyrimidinedione, 5-phenyl-5-ethylhexahydropyrimidine-4,6-dione. Primidone, a 2-deoxy analog of phenobarbital, was synthesized in 1949[32] and introduced in 1954 for use in the control of grand mal and psychomotor epilepsy. It is prepared by the reductive desulfurization of 5-ethyl-5-phenylthiobarbituric acid. The high incidence (20%) of drowsiness, and the anticonvulsant effects of primidone may be due to its oxidation to phenobarbital. This has been shown to occur to the extent of about 15 percent in man.[29]

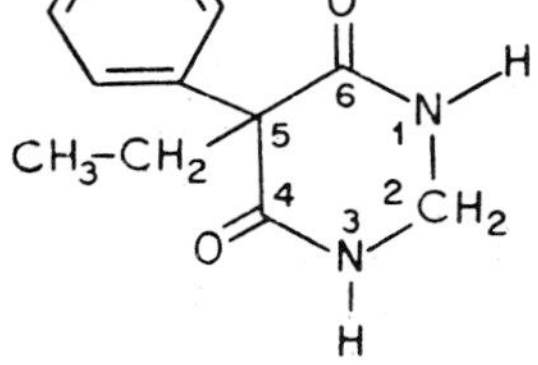

Primidone

Although, in general, the toxicity of primidone is low, a few cases of megaloblastic anemia have been associated with its use.

Primidone is a white, odorless, crystalline powder. It has low solubility in water (1:2,000) and in alcohol (1:200).

Phenacemide U.S.P., Phenurone®, phenylacetylurea. Phenacemide was introduced in 1951 for the treatment of psychomotor, grand mal and petit mal epilepsies and in mixed seizures. Serious side-effects associated with the use of phenacemide include personality changes (suicide attempts and toxic psychoses), fatalities attributed to liver damage, and bone marrow depression. Its unique effectiveness in the control of psychomotor seizures may indicate its use in spite of these hazards, but only after other drugs have been found to be ineffective.

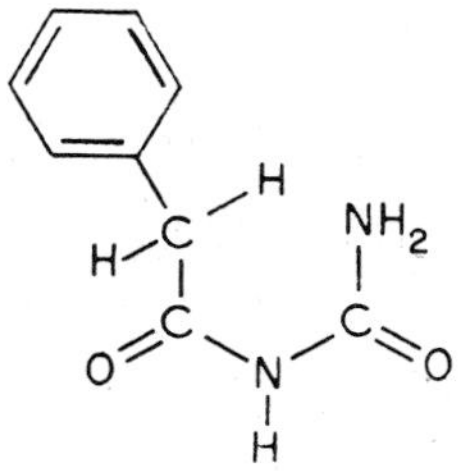

Phenacemide

Phenacemide is an odorless and tasteless, white, crystalline solid. It is very slightly soluble in water and slightly soluble in alcohol.

Carbamazepine U.S.P., Tegretol®, 5H-dibenz[b,f]azepine-5-carboxamide. Carbamazepine contains the dibenzazepine ring system of the psychotherapeutic drug imipramine. It differs from imipramine in having the double bond in the 10,11 position of the ring, and in having the carboxamide side chain rather than the dimethylaminopropyl group.

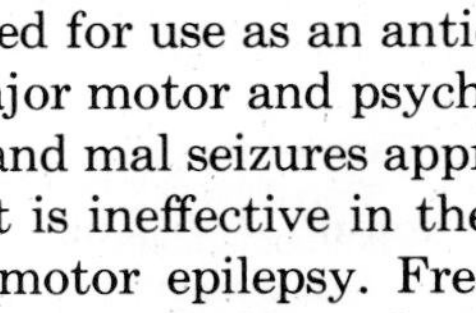

Carbamazepine

Carbamazepine was introduced in 1968 for relief of pain of trigeminal neuralgia (*tic douloureux*) and in 1974 was approved for use as an anticonvulsant agent for control of major motor and psychomotor epilepsy. Its efficacy in grand mal seizures approximates that of phenobarbital. It is ineffective in the control of petit mal and minor motor epilepsy. Frequent minor adverse reactions include dizziness, drowsiness, ataxia, or dermatologic reactions. Serious and sometimes fatal blood-cell abnormalities have been reported.

Valproic Acid, Depakene®, 2-*n*-propylpentanoic acid.

$$CH_3CH_2CH_2 \diagdown$$
$$CH\!-\!COOH$$
$$CH_3CH_2CH_2 \diagup$$

Valproic Acid

Valproic acid and its sodium salt are used in the treatment of absence seizures (very brief loss of sensory contact or consciousness), including petit mal.

The anticonvulsant activity may be related to increased brain levels of gamma-aminobutyric acid. Valproic acid is rapidly absorbed and peak serum levels occur between one and four hours following a single oral dose. The serum half-life is between eight and twelve hours. The drug is largely metabolized in the liver and is excreted in the urine as the glucuronide and as products of beta and omega oxidation, 2-propyl-3-ketopentanoic acid, 2-propyl-5-hydroxypentanoic acid, and 2-propylglutaric acid. Valproic acid (pK_a 4.8) is a colorless liquid with a characteristic odor. Capsules of valproic acid contain the free acid; the syrup contains the sodium salt.

TABLE 9–19

ANTICONVULSANT DRUGS

Name / Proprietary Name	Preparations	Category	Usual Adult Dose*	Usual Dose Range*	Usual Pediatric Dose*
Phenytoin U.S.P. / *Dilantin*	Phenytoin Oral Suspension U.S.P. Phenytoin Tablets U.S.P.	Anticonvulsant; cardiac depressant (anti-arrhythmic)	Anticonvulsant, initial, 100 mg. 3 times daily; cardiac depressant, 100 mg. 2 to 4 times daily	200 to 600 mg. daily	Anticonvulsant, 1.5 to 4 mg. per kg. of body weight or 125 mg. per square meter of body surface, twice daily, not to exceed 300 mg. daily
Phenytoin Sodium U.S.P. / *Dilantin Sodium*	Phenytoin Sodium Capsules U.S.P.	Anticonvulsant; cardiac depressant (anti-arrhythmic)	Anticonvulsant, initial, 100 mg. 3 times daily; cardiac depressant, 100 mg. 2 to 4 times daily	200 to 600 mg. daily	Anticonvulsant, 1.5 to 4 mg. per kg. of body weight or 125 mg. per square meter of body surface, twice daily, not to exceed 300 mg. daily
	Sterile Phenytoin Sodium U.S.P.		Anticonvulsant, I.V., 150 to 250 mg. then 100 to 150 mg. repeated in 30 minutes as necessary, at a rate not exceeding 50 mg. per minute; cardiac depressant, I.V., 50 to 100 mg., repeated every 10 to 15 minutes as necessary, up to a maximum total dose of 10 to 15 mg. per kg. of body weight	50 to 800 mg. daily	Anticonvulsant, I.V., 1.5 to 4 mg. per kg. of body weight or 125 mg. per square meter of body surface, twice daily
Mephenytoin U.S.P. / *Mesantoin*	Mephenytoin Tablets U.S.P.	Anticonvulsant	100 mg.	200 to 600 mg. daily	
Ethotoin / *Peganone*	Ethotoin Tablets	Anticonvulsant	Initially, 1 g. daily in divided doses; maintenance, 2 to 3 g. daily in 4 to 6 divided doses		Children, initially, 750 mg. daily; maintenance, 500 mg. to 1 g. daily in divided doses
Trimethadione U.S.P. / *Tridione*	Trimethadione Capsules U.S.P. Trimethadione Oral Solution U.S.P. Trimethadione Tablets U.S.P.	Anticonvulsant	Initial, 300 mg. 3 times daily	900 mg. to 2.4 g. daily	13 mg. per kg. of body weight or 335 mg. per square meter of body surface, or the following amounts, are usually given 3 times daily: infants—100 mg.; 2 years—200 mg.; 6 years—300 mg.; 13 years—400 mg.
Paramethadione U.S.P. / *Paradione*	Paramethadione Capsules U.S.P. Paramethadione Oral Solution U.S.P.	Anticonvulsant	Initial, 300 mg. 3 times daily	900 mg. to 2.4 g. daily	Under 2 years of age—100 mg. 3 times daily; 2 to 6 years—200 mg. 3 times daily; over 6 years—see Usual Adult Dose

*See U.S.P. D.I. for complete dosage information.

(Continued)

TABLE 9–19 (Continued)

ANTICONVULSANT DRUGS

Name Proprietary Name	Preparations	Category	Usual Adult Dose*	Usual Dose Range*	Usual Pediatric Dose*
Ethosuximide U.S.P. *Zarontin*	Ethosuximide Capsules U.S.P.	Anticonvulsant	250 mg. twice daily initially, increased as necessary every 4 to 7 days in increments of 250 mg.	500 mg. to 1.5 g. daily	Children under 6 years of age—250 mg. once daily; over 6 years—250 mg. twice daily initially, increased as necessary every 4 to 7 days in increments of 250 mg.
Phensuximide U.S.P. *Milontin*	Phensuximide Capsules U.S.P. Phensuximide Oral Suspension U.S.P.	Anticonvulsant		500 mg. to 1 g. 2 or 3 times daily, irrespective of age	
Methsuximide U.S.P. *Celontin*	Methsuximide Capsules U.S.P.	Anticonvulsant	Initial, 300 mg. daily; maintenance, 300 mg. to 1.2 g. daily		
Diazepam U.S.P. *Valium*	Diazepam Injection U.S.P.	Anticonvulsant (Status epilepticus)	Initial, 5 to 10 mg. I.V. If necessary repeated at 10 to 15 min. intervals to a maximum of 30 mg.		Infants, 30 days to 5 years of age—0.2 to 0.5 mg. I.V. every 2 to 5 min. to a maximum of 5 mg. Children, 5 years or older—1 mg. I.V. every 2 to 5 min. to a maximum of 10 mg.
Clonazepam *Clonopin*	Tablets	Anticonvulsant	Initial, 0.5 mg. 3 times daily	Dose increased in increments of 0.5 to 1 mg. every 3 days to a maximum of 20 mg. daily	Initial, 0.01 to 0.03 mg. per kg. of body weight per day, increased not to exceed 0.1 to 0.2 mg. per kg. body weight per day.
Primidone U.S.P. *Mysoline*	Primidone Oral Suspension U.S.P. Primidone Tablets U.S.P.	Anticonvulsant	Initial—week one, 250 mg. once daily at bedtime; week two, 250 mg. twice daily; week three, 250 mg. 3 times daily; week four, 250 mg. 4 times daily; maintenance, 250 to 500 mg. 3 times daily	250 mg. to not more than 2 g. daily	Children under 8 years of age—initial, week one, 125 mg. once daily at bedtime; week two, 125 mg. twice daily; week three, 125 mg. 3 times daily; week four, 125 mg. 4 times daily; maintenance, 250 mg. 2 or 3 times daily; children over 8 years—see Usual Adult Dose
Phenacemide U.S.P. *Phenurone*	Phenacemide Tablets U.S.P.	Anticonvulsant	Initial, 250 to 500 mg. 3 times daily; maintenance, 250 to 500 mg. 3 to 5 times daily	2 to 5 g. daily, in divided doses	
Carbamazepine U.S.P. *Tegretol*	Carbamazepine Tablets U.S.P.	Anticonvulsant	Anticonvulsant, initial 200 mg. 2 times daily; trigeminal neuralgia, 100 mg. 2 times daily	Anticonvulsant, 800 to 1200 mg. daily. Trigeminal neuralgia, 400 to 800 mg. daily	Anticonvulsant, 6 to 12 years initial 100 mg. 2 times daily, increased not to exceed 1000 mg. daily
Valproic Acid *Depakene*	Valproic Acid Capsules Valproate Sodium Syrup	Anticonvulsant	Initial, 15 mg. per kg. per day, increased weekly by 5 to 10 mg. per kg. per day to a maximum of 60 mg. per kg. per day		Children weighing over 10 kg. initial dose 15 mg. per kg. per day, increased weekly by 5 to 10 mg. per kg. per day to a maximum of 60 mg. per kg. per day

*See *U.S.P. D.I.* for complete dosage information.

REFERENCES

1. Roth, L. J., and Barlow, C. F.: Science 134:22, 1961.
2. Suckling, C.W.: Brit. J. Anaesth. 29:466, 1957.
3. Cohen, E. N., *et al.*: Science 141:899, 1963.
4. Brodie, B. B., Bernstein, E., and Mark, L. C.: J. Pharmacol. Exp. Ther. 105:421, 1952.
5. Price, H. L., *et al.*: J. Clin. Pharmacol. Ther. 1:16, 1960.
6. Friend, D.: J. Clin. Pharmacol. Ther. 1 (6):5, 1960.
7. Hansch, C., *et al.*: J. Med. Chem. 11:1, 1968.
8. Goodman, L. S., and Gilman, A.: The Pharmacological Basis of Therapeutics, p. 290, New York, Macmillan, 1970.
9. Rogers, G. A.: Am. J. Psychiat. 115:551, 1958.
10. Bonnet, H., *et al.*: J. Med. Lyon 39:924, 1958.
11. Domino, E. F.: J. Pharmacol. Exp. Ther. 115:449, 1955.
12. Schallek, W., Kuehn, A., and Seppelin, D. K.: J. Pharmacol. Exp. Ther. 118:139, 1956.
13. Swinyard, E. A., and Chin, L.: Science 125:739, 1957.
14. Berger, F. M.: Ann. N.Y. Acad. Sci. 67:685, 1957.
15. Jacobsen, E.: J. Pharm. Pharmacol. 10:282, 1958.
16. Burbridge, T. N.: A Pharmacologic Approach to the Study of the Mind, Springfield, Ill., Charles C. Thomas, 1959.
17. Schildkraut, J. J., and Kety, S. S.: Science 156:21, 1967.
18. Vakil, R. J.: Brit. Heart J. 11:350, 1949.
19. Siddiqui, S., and Siddiqui, R. H.: J. Indian Chem. Soc. 8:667, 1931.
20. Noce, R. H., Williams, D. B., and Rapaport, W.: J.A.M.A. 156:821; 1954; 158:11, 1955.
21. Kline, N. S.: Ann. N. Y. Acad. Sci. 59:107, 1954.
22. Burns, J. J., and Shore, P. A.: Ann. Rev. Pharmacol. 1:79, 1961.
23. Brodie, B. B., Shore, P. A., and Pletscher, A.: Science 123:992, 1956.
24. Gordon, M., Craig, R. N., and Zirkle, C. L.: Molecular Modification in the Development of Phenothiazine Drugs, Molecular Modification in Drug Design, Advances in Chemistry Series 45, Am. Chem. Soc., Washington, D.C., 1964.
25. Bloom, B. M., and Laubach, G. D.: Ann. Rev. Pharmacol. 2:69, 1962.
26. Janssen, P. A. J., *et al.*: J. Med. Pharm. Chem. 1:281, 1959.
27. Butler, T. C., and Waddell, W. S.: Proc. Soc. Exp. Biol. Med. 93:544, 1956.
28. Plaa, G. L., Fujimoto, J. M., and Hine, C. H.: J.A.M.A. 168:1769, 1958.
29. Merritt, H. M., and Putnam, T. J.: Arch. Neurol. Psychiat. 39:1003, 1938; Epilepsia 3:51, 1945.
30. Erlenmeyer, H.: Helv. chim. acta 21:1013, 1938.
31. Spielman, M. A.: J. Am. Chem. Soc. 66:1244, 1944.
32. Bogue, J. Y., and Carrington, H. C.: Brit. J. Pharmacol. 8:230, 1953.

SELECTED READINGS

Burger, A. (ed.): Drugs Affecting the Central Nervous System, Medicinal Research, Vol. 2, New York, Dekker, 1968.
Delgado, J. N., and Isaacson, E. I.: Anticonvulsants, *in* Burger, A. (ed.): Medicinal Chemistry, ed. 3, p. 1886, New York, Wiley-Interscience, 1970.
Gordon, M.: Psychopharmacological Agents, New York, Academic Press, vols. 1 and 2, 1964 and 1965.
Jucker, E.: Some new developments in the chemistry of psychotherapeutic agents, Angewandte Chemie, Internat. Ed. 2:492, 1963.
Mautner, H. G., and Clemson, H. C.: Hypnotics and Sedatives, *in* Burger, A. (ed.): Medicinal Chemistry, ed. 3, p. 1365, New York, Wiley-Interscience, 1970.
Patel, A. R.: General Anesthetics, *in* Burger, A. (ed.): Medicinal Chemistry, ed. 3, p. 1314, New York, Wiley-Interscience, 1970.
Simpson, L. L. (ed.): Drug Treatment of Mental Disorders, New York, Raven Press, 1975.
Toman, J. E. P., and Goodman, H. S.: Anticonvulsants, Pharmacol. Rev. 28:409, 1948.
Zirkle, C. L., and Kaiser, C.: Antipsychotic Agents, *in* Burger, A. (ed.): Medicinal Chemistry, ed. 3, p. 1410, New York, Wiley-Interscience, 1970.

10

central nervous system stimulants

T. C. Daniels

E. C. Jorgensen

The central nervous system is a complex network of subunits which act as conducting pathways between peripheral receptors and effectors, enabling man to respond to his environment. It also adjusts behavior to the quality and the intensity of stimuli and coordinates activities, providing a unified set of actions. Drugs which have in common the property of increasing the activity of various portions of the central nervous system are called central nervous system stimulants.

Until recently, the major therapeutic applications of central nervous system stimulants were their use as respiratory stimulants and analeptics. Respiratory stimulation may be brought about not only by the action of drugs directly upon the respiratory center of the medulla but by pH changes in the blood which supplies the center. Carbonic acid is most effective in this manner, and carbon dioxide, in some cases, is the respiratory stimulant of choice. Stimulation of the chemoreceptor of the carotid body (cyanide), afferent impulses from sensory stimuli (ammonia inhalation) and higher centers (visual stimuli) may affect respiration through the respiratory center.

Analeptics, agents used to lessen narcosis brought about by excess of depressant drugs, often stimulate a variety of other centers as well. The vasomotor center, which maintains the constriction of the blood vessel walls, is frequently affected. Many analeptic drugs are also pressor drugs because their stimulation of the vasomotor center produces an increase in vasoconstriction. The resulting increased peripheral resistance to blood flow causes an elevation of blood pressure.

Stimulation of the emetic center, also located in the medulla, is not infrequent with therapeutic doses of many drugs. A few drugs, such as apomorphine, apparently exert a selective effect on the emetic chemoreceptor trigger zone of the medulla and may be used as emetics in the treatment of poisoning.

In the sense that an effect on the "appetite control center" may be classed as a central stimulant effect, drugs classed as anorexigenic agents and used to decrease appetite in the control of obesity are included in this discussion.

There are many drugs of varying pharmacologic classes which, in addition to their desired effects, elicit a pronounced stimulatory effect on the central nervous system. Examples are found in the local anesthetics (cocaine), parasympatholytics (atropine), sympathomimetics (many, including ephedrine, amphetamine, etc.). Important toxic effects occur in high doses of salicylates, local anesthetics and many others.

In 1955, Goodman and Gilman[1] summarized the status of central stimulants.

> Although the central nervous system stimulants are sometimes dramatic in their pharmacological effects, they are relatively unimportant from a therapeutic point of view. It is not possible to stimulate the nervous system over a long period of time, for heightened nervous activity is followed by depression, proportional in degree to the intensity and duration of the stimulation. Consequently, therapeutic excitation of the central nervous system is usually of brief duration and is reserved for emergencies characterized by severe central depression.

This statement was based on the respiratory stimulant and analeptic properties of available agents such as picrotoxin, pentylenetetrazol, nikethamide, and caffeine and the related xanthines. This viewpoint has been altered by the recognition that a more moderate and prolonged degree of central stimulation may be achieved in the treatment of patients with mental depression. This group of drugs is now called *psychomotor stimulants* and may be subdivided into the *central stimulant sympathomimetics,* the *antidepressants* (*monoamine oxidase inhibitors, tricyclic antidepressants*), and a group of compounds of *miscellaneous* chemical and pharmacologic class.

ANALEPTICS

The following drugs are used chiefly as analeptics to counteract respiratory depression and coma resulting from overdosage of central depressant agents.

Picrotoxin is the active principle from the seed ("fishberries") of the shrub *Anamirta cocculus.* It is a molecular compound easily separated into the component dilactones, the active picrotoxinin[2] and inactive picrotin.[3]

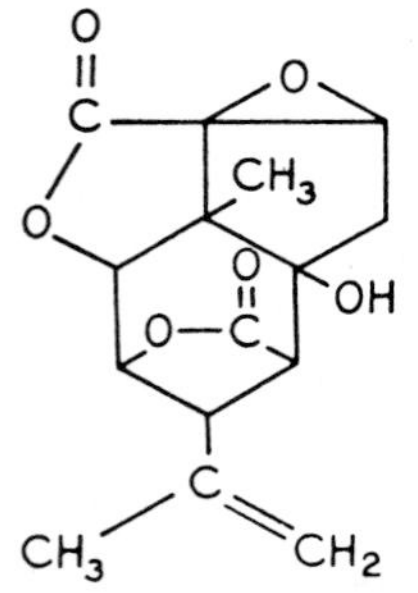

Picrotoxinin 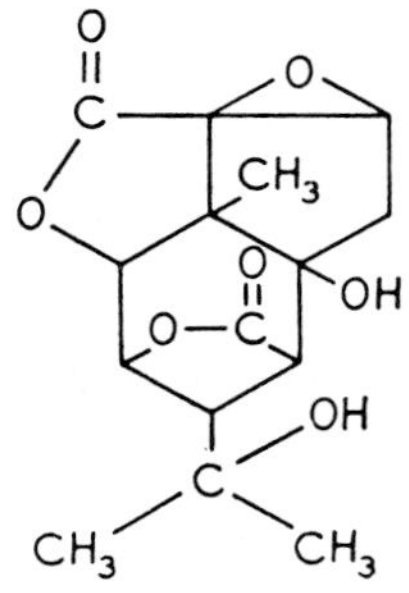   Picrotin

Picrotoxin is a powerful central nervous stimulant, which was once used to treat the adverse effects resulting from overdoses of barbiturate and other central depressant drugs. The use of picrotoxin did not produce an earlier arousal time, and the margin between analeptic and convulsant dose is narrow. Injected doses of 20 mg. may produce severe clonic and tonic convulsions. Accordingly, picrotoxin is no longer used.

Pentylenetetrazol, Metrazol®, 6,7,8,9-tetrahydro-5H-tetrazoloazepine, 1,5-pentamethylenetetrazole. Pentylenetetrazol was formerly widely used by injection for treatment of drug-induced coma, and as a convulsant for shock therapy. Its use has been largely supplanted by more effective and less hazardous agents. Although currently suggested for oral use as a cerebral stimulant for treatment of senility or mental depression, its effectiveness for these uses is not well documented.

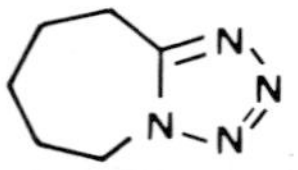

Pentylenetetrazol

Usual dose—for drug-induced coma, 500 mg., intravenous, followed by 1 g. every 30 minutes as needed. For senility, oral dose 100 mg. 3 or 4 times daily.

Nikethamide, Coramine®, N,N-diethylnicotinamide. Nikethamide is a respiratory stimulant. It has an intermediate central stimulant effect, resembling that of the amphetamines rather than the more potent picrotoxin or pentylenetetrazol. However, since the margin between the analeptic and convulsant dose is narrow, it is considered to be of questionable merit in treatment of drug-induced coma.

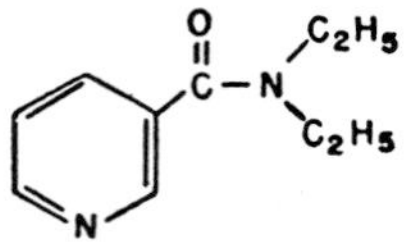

Nikethamide

It is a viscous, high-boiling oil which is miscible with water, alcohol and ether.

Usual dose range—intramuscular and intravenous, 1.25 g. repeated as needed at 5-minute intervals.

Ethamivan U.S.P., N,N-diethylvanillamide, 3-methoxy-4-hydroxybenzoic acid diethylamide. Ethamivan is an analeptic drug useful as an adjunctive agent in the treatment of severe respiratory depression. Continuous intravenous infusion maintains an increase in both rate and depth of respiration in such patients. The drug produces general stimulation of the central nervous system, and excessively high doses may lead to convulsions.

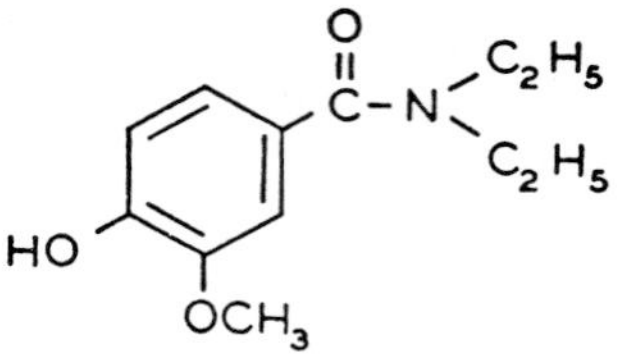

Ethamivan

Category—central and respiratory stimulant.

Usual dose range—I.V., 500 μg. to 5 mg. per kg. of body weight, given slowly as a single injection; may be followed with continuous intravenous infusion at the rate of 10 mg. per minute, as determined by response of the patient.

Occurrence
Ethamivan Injection U.S.P.

Doxapram Hydrochloride U.S.P., Dopram®, 1-ethyl-3,3-diphenyl-4-(2-morpholinoethyl)-2-pyrrolidinone hydrochloride hydrate. Doxapram is used to

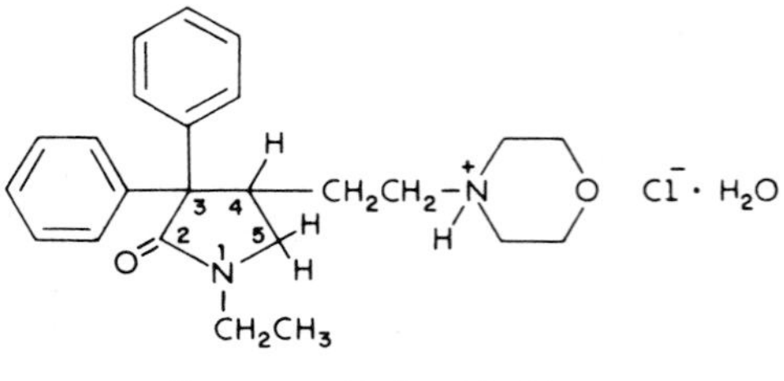

Doxapram Hydrochloride

stimulate respiration in patients with postanesthetic respiratory depression, and to hasten arousal during this period. However, its use for this purpose is considered less effective than adequate airway management and support of ventilation.

Category—respiratory stimulant.

Usual dose range—I.V., 1 to 1.5 mg. per kg. of body weight.

Occurrence

Doxapram Hydrochloride Injection U.S.P.

Flurothyl U.S.P., Indoklon®, bis(2,2,2-trifluoroethyl)ether, CF_3CH_2—O—CH_2CF_3. Flurothyl is an inhalant which may be used in place of electroshock therapy in depressive disorders. Convulsions usually occur after 4 to 6 inhalations of the vapor.

Flurothyl is a colorless flammable liquid, b.p. 63.9°, with a mild ethereal odor. It is slightly soluble in water. The drug is supplied in 2-ml. and 10-ml. ampules.

Category—central stimulant (convulsant).

Application—1 ml. by special inhalation.

PURINES

Purines occur widely distributed among natural products (e.g., in uric acid, coffee, tea, cocoa, nucleic acids and enzymes). The 2,6-dihydroxylated purines, or xanthine derivatives, are caffeine, theobromine and theophylline. The worldwide use of stimulating drinks containing one or more of these principles causes them to assume added significance.

Table 10-1 summarizes the structural relationships of xanthine alkaloids. The relative pharmacologic potencies of the xanthines are summarized[1] in Table 10-2.

TABLE 10-1

XANTHINE ALKALOIDS

Xanthine

(R, R′ & R″ = H)

Compound	R	R′	R″	Common Source
Caffeine	CH_3	CH_3	CH_3	Coffee, Tea
Theophylline	CH_3	CH_3	H	Tea
Theobromine	H	CH_3	CH_3	Cocoa

TABLE 10-2

RELATIVE PHARMACOLOGIC POTENCIES OF THE XANTHINES

Xanthine	C.N.S. Stimulation	Respiratory Stimulation	Diuresis	Coronary Dilatation	Cardiac Stimulation	Skeletal-Muscle Stimulation
Caffeine	1*	1	3	3	3	1
Theophylline	2	2	1	1	1	2
Theobromine	3	3	2	2	2	3

*1 = most potent.

In therapeutics, caffeine is the drug of choice among the three xanthines for obtaining a *stimulating effect* on the *central nervous system*. This stimulant action is almost physiologic in nature and helps to combat fatigue and sleepiness. Apparently, little tolerance is built up toward caffeine stimulation; therefore, habitual coffee drinkers continue to experience stimulation from day to day. Ordinarily, caffeine is not of value in other conditions, in spite of its other pharmacologic actions, because of excessive stimulation at the dose necessary to elicit other effects.

The xanthine alkaloids have poor water-solubility, and this has prompted the use of numerous solubilizers. Alkali salts of organic acids (sodium acetate, sodium benzoate and sodium salicylate) are often used to solubilize caffeine.

These combinations usually are referred to as double salts, mixtures, combinations or complexes, indicating that their true nature is not well understood. Studies by Blake and Harris[4] showed that there is no chemical compound formed between caffeine and citric acid, sodium benzoate or sodium acetate. With sodium salicylate, some hydrogen bonding between the hydroxyl of the salicylate and the carbonyl of caffeine was observed.

Caffeine U.S.P., 1,3,7-trimethylxanthine. Caffeine is a very weak base and does not form salts which are stable in aqueous or alcoholic solutions. Caffeine occurs as a white powder, or as white, glistening needles. The alkaloid is odorless and possesses a bitter taste. Caffeine is soluble in water (1:50), alcohol (1:75) or chloroform (1:6) but is less soluble in ether. Its solubility is increased in hot water (1:6 at 80°) or hot alcohol (1:25 at 60°).

A cup of coffee or tea (the prepared beverage) contains about 60 mg. of caffeine.

In physiologic action, caffeine and its relatives theobromine and theophylline are qualitatively alike. The primary effect from therapeutic doses is a stimulation of the central nervous system, beginning in the psychic center and progressing downward, with little or no reversal by continued or larger doses. There is also a direct stimulation of all muscles, partly central and partly peripheral, an increase in diuresis and vaso-dilation by direct action on peripheral vessels. The dominant effect on the psychic center causes increased flow of thought, lessens drowsiness and mental fatigue, relieves headache and gives a sense of comfort and well-being. In combination with the action on muscles, it brings about a condition in which more work can be done before fatigue sets in, and this is performed more rapidly and accurately, although later there may be impairment of these qualities for some time.

Caffeine often is employed in headaches of certain kinds, such as in neuralgia, rheumatism, migraine and in those due to fatigue, frequently combined with other analgesics, such as phenacetin and aspirin. It may be given as a diuretic in cardiac edema, but is usually inactive or harmful in the presence of renal disease. Large doses cause insomnia, restlessness, excitement, mild delirium, tinnitus, tachycardia and diuresis. Caffeine, as a cranial vasoconstrictor, is used together with ergotamine tartrate for the treatment of migraine.

Occurrence	Percent Caffeine
Caffeine and Sodium Benzoate Injection U.S.P.	45–52
Ergotamine Tartrate and Caffeine Suppositories U.S.P.	
Ergotamine Tartrate and Caffeine Tablets U.S.P.	

PSYCHOMOTOR STIMULANTS

The psychomotor stimulants are used to elevate the mood or improve the outlook of patients with mental depression. They may be divided into the general classes: *central stimulant sympathomimetics,* the *antidepressants* (*monoamine oxidase inhibitors, tricyclic antidepressants*), and *miscellaneous.*

CENTRAL STIMULANT SYMPATHOMIMETICS

The sympathomimetic agents are discussed in Chapter 11 in terms of peripheral or autonomic effects. Certain of the sympathomimetics, by virtue of structural features and physical properties, exert a significant stimulant effect on the central nervous system. The structures of these compounds are presented in Table 10-3.

These agents vary in their intensity of central stimulant activity.[5] The most potent stimulating drugs are amphetamine and its N-methyl analog, methamphetamine. Steric differences are of considerable importance, for the *dextro* isomers are 10 to 20 times more stimulating than the *levo* isomers. The branched methyl group (amphetamine, methamphetamine) or similar substitution, such as incorporation of the nitrogen in a ring system (methylphenidate, phenmetrazine), is an important feature of these central stimulants, presumably providing resistance to enzymatic inactivation by steric protection of the amino group. These compounds are attacked by monoamine oxidase at a slower rate than those which do not possess branching in the chain connecting the aromatic nucleus and amino group. The parent β-phenylethylamine is not useful as a central stimulant. An increase in the number of carbons in the branching side chain also decreases activity; 1-phenyl-2-aminobutane and 1-phenyl-2-aminopentane show a low order of central stimulation as compared with amphetamine, as does the α,α-dimethyl analog, mephentermine. N-alkyl substitution by groups larger than methyl also decreases activity. Substitutions which increase the hydrophilic character decrease central stimulant activity. A hydroxyl group in the 2-position (phenylpropanolamine, ephedrine) or in the aromatic nucleus (hydroxyamphetamine) results in sympathomimetic amines with distinctly less central stimulant activity than their nonhydroxylated analogs. Reduction of the aromatic ring, or its replacement by an alkyl group, produces compounds with little or no stimulating action on the central nervous system, although many retain peripheral activity and serve as vasoconstrictors, useful as nasal decongestants.

Those stimulants which are most potent (amphetamine, methamphetamine) may be used as analeptics in reversing the profound central depression due to anesthetic, narcotic and hypnotic drugs, although supportive therapy without drug intervention is currently recommended practice. Narcoleptic patients show considerable relief from attacks of sleep and cataplexy and are often improved by the potent amphetamine analogs. A centrally mediated decrease in appetite brought about by amphetamine and its analogs has caused these to be used as anorexigenic agents, as adjuncts to dietary control in the management of obesity. In addition to the potent central stimulants amphetamine and methamphetamine, several related sympathomimetic amines are advocated for use as anorexigenic agents and are said to decrease appetite, with a lowered degree of the central stimulant effects leading to restlessness and insomnia. These include benzphetamine, diethylpropion, phenmetrazine, phendimetrazine, phentermine and chlorphentermine (Table 10-3).

The use of amphetamine and methamphetamine for fatigue and as "pep pills" has been long recognized and has led to frequent abuse.[6] The amphetamines have not been shown to produce true addiction, al-

TABLE 10–3

SYMPATHOMIMETICS WITH SIGNIFICANT CENTRAL STIMULANT ACTIVITY

Generic Name	Base Structure		
Amphetamine	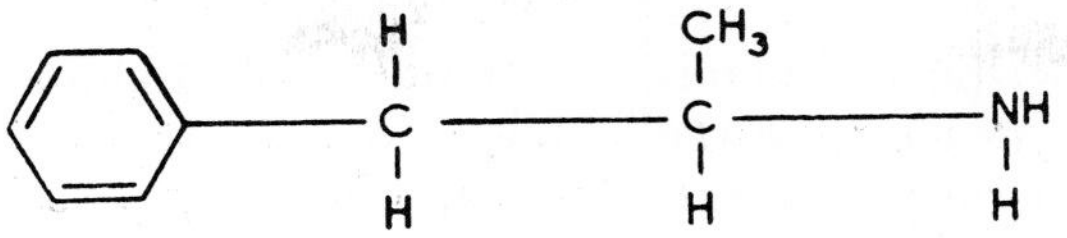		
Methamphetamine	H	H	CH_3
Phentermine	H	CH_3	H
Benzphetamine	H	H	CH_3 $CH_2C_6H_5$
Diethylpropion	O*	H	C_2H_5 C_2H_5

Fenfluramine

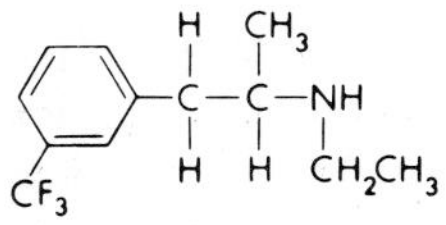

Chlorphentermine

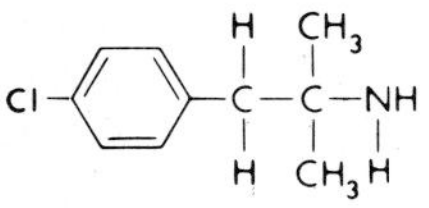

Clortermine

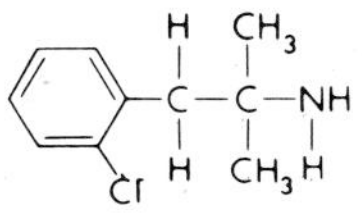

Phenmetrazine

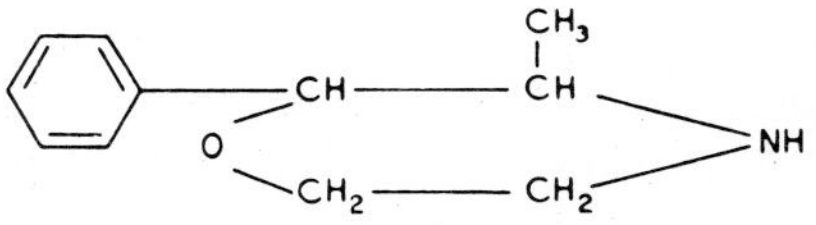

Phendimetrazine

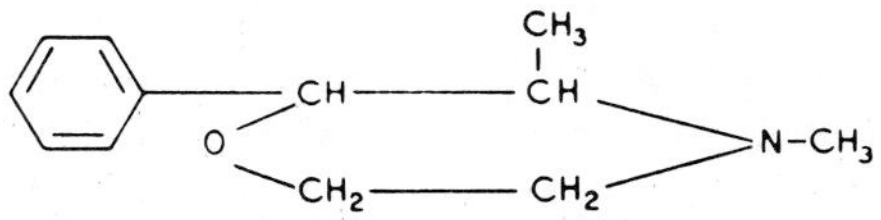

Methylphenidate

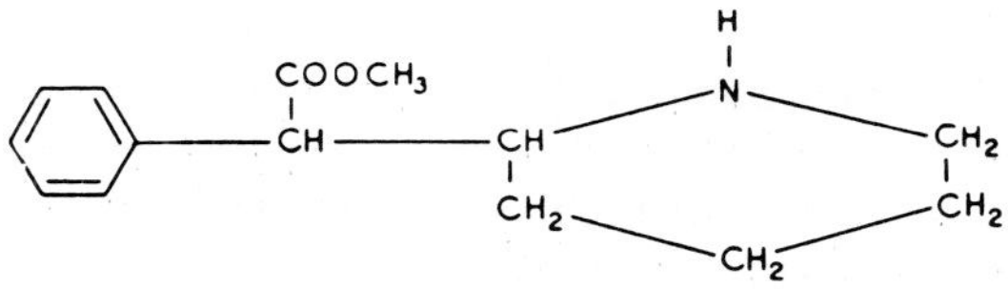

*Carbonyl.

though tolerance to larger doses, without increased effect, and habituation occur.

The sympathomimetic amines with central stimulant properties have been used in the treatment of those psychogenic disorders related to depressive states. However, they have largely been replaced in treatment of depression by the tricyclic antidepressants and monoamine oxidase inhibitors.

Amphetamine shows little or no contracting effect on chronically denervated tissue, nor does it produce a pressor response in animals in which pretreatment with reserpine has depleted stores of catecholamines. Therefore, amphetamine is thought to act primarily by releasing catecholamines such as norepinephrine and not to exert a direct adrenergic effect, at least in peripheral tissues.

Products

For the structures of the following sympathomimetics with significant central stimulant activity, see Table 10-3.

Amphetamine, Benzedrine®, (±)-1-phenyl-2-aminopropane, is a colorless liquid. The free base and the carbonate salt have been used as nasal decongestants. The generic name was derived from one chemical designation *a*lpha *m*ethyl *ph*enyl*et*hyl *amine;* the proprietary name from an alternate chemical name: *benz*yl *m*ethyl carbinam*ine.* Two salts, Amphetamine Sulfate U.S.P. (Benzedrine® Sulfate) and *amphetamine phosphate,* are used as analeptic agents, in the treatment of narcolepsy, as adjuncts to treatment of alcoholism, to decrease appetite in the management of obesity, and in depressive conditions characterized by apathy and psychomotor retardation. Amphetamine is a fairly strong base, pKa = 9.77, which is slightly soluble in water, and readily soluble in alcohol, ether and aqueous acids. The sulfate salt is given orally, the usual dosage being 5 mg. 3 times a day. The phosphate salt is more soluble and is used both orally and by intramuscular or intravenous injection.

Dextroamphetamine Sulfate U.S.P., Dexedrine® Sulfate, (+)-α-methylphenethylamine sulfate, is the *dextro* isomer with the same actions and uses as the racemic amphetamine sulfate but possessing a greater stimulant activity. It was introduced in 1944.

Dextroamphetamine Phosphate U.S.P., monobasic (+)-α-methylphenethylamine phosphate. The phosphate salt is used for the same purpose as the sulfate, for its central stimulant effects. Aqueous solutions have a pH between 4 and 5.

Methamphetamine Hydrochloride, Desoxyn®, desoxyephedrine hydrochloride, (+)-N,α-dimethylphenethylamine hydrochloride. This N-methyl analog of amphetamine was introduced in 1944, in part for treatment of depression, but it has largely been replaced by the tricyclic antidepressants and monoamine oxidase inhibitors. The drug may be used for treatment of narcolepsy and in the management of hyperactive children. Its appetite-depressant properties are used to supplement dietary control in the treatment of obesity. However, methamphetamine, as the street drug "speed" or "crystal," is one of the most widely abused of the sympathomimetics. It is taken orally for its euphoretic or antidepressant effect, to temporarily improve performance or defer fatigue. A compulsive pattern of use (psychic dependence) often develops when methamphetamine is used by injection, and prolonged use frequently leads to major psychotic states of suspicion, anxiety and paranoia. The drug occurs as colorless or white crystals with a bitter taste, m.p. 170° to 175°. The hydrochloride is soluble in water (1:2), alcohol (1:3) and chloroform (1:5), and is insoluble in ether. The free base is a fairly strong base, pKa = 9.86, which is readily soluble in ether.

Phentermine, Ionamin®, 1-phenyl-2-methyl-2-aminopropane, α,α-dimethylphenethylamine. Phentermine was introduced in 1959 as an agent to lessen appetite in the management of obesity. The free base is bound to an ion-exchange resin for delayed release into the gastrointestinal tract. Phentermine is also available as the hydrochloride salt.

The usual oral dose of phentermine resin is 15 to 30 mg. before breakfast. The usual dose of the hydrochloride salt is 8 mg. taken before mealtime.

Chlorphentermine Hydrochloride, Pre-Sate®, 1-(4-chlorophenyl)-2-methyl-2-aminopropane hydrochloride, *p*-chloro-α,α-dimethyl-β-phenylethylamine hydrochloride. Chlorphentermine is the *p*-chloro analog of phentermine, introduced in 1965 for treatment of obesity. As with other sympathomimetic amines, chlorphentermine should not be taken by patients with glaucoma or those receiving monoamine oxidase inhibitors. The recommended adult daily dose is 1 tablet (65 mg. as the free base, 75 mg. as the hydrochloride) after the morning meal.

Clortermine Hydrochloride, Voranil®, *o*-chloro-α,α-dimethyl-β-phenylethylamine hydrochloride. Clortermine, an isomer of chlorphentermine, is an appetite-depressant drug introduced in 1973. As is typical of most sympathomimetics of this type, it produces elevation of blood pressure, and has the potential for abuse of its central stimulatory effects. The recommended oral dose is 50 mg. daily, taken as a single dose mid-morning.

Benzphetamine Hydrochloride, Didrex®, (+)-1-phenyl-2-(N-methyl-N-benzylamino)propane hydrochloride, (+)-N-benzyl-N,α-dimethylphenethylamine hydrochloride. Benzphetamine is an anorexigenic agent introduced in 1960. At the oral dose of about 75 mg. a day in divided doses, little restlessness, anxiety, insomnia and other symptoms of excess central stimulation are said to occur.

Diethylpropion Hydrochloride U.S.P., Tenuate®, Tepanil®, 2-(diethylamino)propiophenone hydrochloride, 1-phenyl-2-diethylaminopropanone-1 hydrochloride, was introduced in 1959 for the suppression of appetite in the management of obesity. Central and cardiovascular stimulation appear to be minimal at the recommended doses.

Fenfluramine Hydrochloride, Pondimin®, N-ethyl-α-methyl-*m*-trifluoromethyl-β-phenylethylamine hydrochloride. Fenfluramine was introduced in 1973 for use as an appetite depressant as an adjunct to a restricted diet in the management of obesity. Although fenfluramine is an amphetamine analog, it appears to differ from other members of the class by side-effects more related to central nervous system depression than to stimulation, at therapeutic dose levels. The most common side-effects are drowsiness, diarrhea and dryness of mouth, although anxiety and nervousness have been noted at higher dose levels. The recommended oral dose ranges from 20 to 40 mg. three times daily before meals.

Phenmetrazine Hydrochloride U.S.P., Preludin®, (±)-3-methyl-2-phenylmorpholine hydrochloride, was introduced in 1956 as an appetite suppressant with the side-effects of nervousness, euphoria and insomnia attributable to central nervous stimulation being much less than with amphetamine.

Phendimetrazine Tartrate, Plegine®, (+)-3,4-dimethyl-2-phenylmorpholine bitartrate, was introduced in 1961 as an anorexigenic agent. It appears to possess the same degree of effectiveness and the same order of central stimulation as its close analog, phenmetrazine. The usual oral dose is 35 mg. taken 1 hour before meals.

Methylphenidate Hydrochloride U.S.P., Ritalin® Hydrochloride, methyl α-phenyl-2-piperidineacetate hydrochloride. Methylphenidate is a mild cortical stimulant used in the treatment of depressive states since 1956. It is considered to be effective in the treatment of narcolepsy, and as adjunctive therapy in the syndrome described as "minimal brain dysfunction" in children. This state is characterized by a history of short attention span, emotional lability and hyperactivity. Typical central nervous system stimulant side-effects, such as nervousness, insomnia and anorexia may occur. Psychic dependence has occurred after long-term use of large doses. The usual dose is 10 mg. 3 times daily.

Mazindol, Sanorex®, 5-p-chlorophenyl-5-hydroxy-2,3-dihydro-5H-imidazo(2,1a)isoindole.

Mazindol

Although structurally unique, mazindol produces typical amphetaminelike central nervous system stimulation, and was introduced in 1973 for its appetite-depressant effects as an adjunct to dietary restriction in the management of obesity. The isoindole form is a ring-closed tautomer of the substituted imidazoline, 2-[2'-(p-chlorobenzoyl)phenyl]-2-imidazoline.

Since mazindol appears to inhibit storage-site uptake of norepinephrine, it may potentiate the pressor effects of exogenous catecholamines and blood pressure should be monitored if a pressor amine is administered concurrently. The recommended dose is 1 mg. three times daily, 1 hour before meals, or 2 mg. taken 1 hour before lunch in a single daily dose.

Pemoline, Cylert®, 2-amino-5-phenyl-2-oxazolin-4-one and magnesium hydroxide.

Pemoline

Pemoline, an equimolar mixture of the oxazolinone and magnesium hydroxide, is a central nervous system stimulant, introduced in 1975. Like certain amphetamine analogs, it is recommended for use as an adjunct to social therapy in children with "minimal brain dysfunction." This poorly characterized state may include hyperactivity, short attention span and learning disability. The mechanism of its central stimulant effect is not known, but studies in rats show an increased rate of dopamine synthesis in the brain. The major side-effects are insomnia and anorexia.

Pemoline is administered as a single daily oral dose ranging from 37.5 to 75 mg. Peak blood levels are reached in 2 to 4 hours, and the serum half-life is approximately 12 hours. About 75 percent of an oral dose appears in the urine in 24 hours, 43 percent as unchanged pemoline. Other metabolites include the 2,4-dione, conjugated pemoline, and the ring-cleaved hydrolytic product, mandelic acid.

MONOAMINE OXIDASE INHIBITORS

In 1952 the independent observations were made that iproniazid produced central stimulation in patients being treated for tuberculosis[7] and also inhibited the enzyme monoamine oxidase.[8] These properties were not shared by the related antitubercular agent isoniazid. In 1957 it was noted[9] that pretreatment of animals with iproniazid reversed the usual depressant effect of reserpine, producing instead central stimulation. At the same time it was observed that the reserpine-induced depletion of serotonin and norepinephrine in the brain was prevented by pretreatment with iproniazid.

These observations prompted the successful clinical re-examination of iproniazid as a central stimulant in the treatment of mental depression.[10] It has been proposed that the clinical antidepressant actions of iproniazid and related compounds which inhibit mono-amine oxidase are due to the decreased metabolic destruction of brain amines such as norepinephrine and serotonin. In experimental animals (rabbit, rat, mice, monkey) both serotonin and norepinephrine levels increase after treatment with monoamine oxidase inhibitors, and in a variety of species concentration of brain norepinephrine seems to be best correlated with excitation.[11] However, the metabolic destruction of other amines whose physiologic function in the central nervous system has not been as well explored is prevented by the monoamine oxidase inhibitors. Examples include phenylethylamine, tyramine and its ortho and meta isomers, 3,4-dihydroxyphenylethylamine, and tryptamine. One or more of these or others not yet discovered may contribute to the pharmacologic effects of the monoamine oxidase inhibitors.[12]

Iproniazid: R = CH (CH$_3$)$_2$
Isoniazid: R = H

Serotonin

Norepinephrine

The hypothesis[13] that oxidative deamination of the catecholamines by monoamine oxidase represents the major metabolic pathway in the brain was a result of the above observations. It has been demonstrated[14] subsequently that the major route of metabolism for the catecholamines in peripheral tissue is via methylation of the 3-hydroxyl group by the enzyme catechol-O-methyl transferase (see Chap. 3). However, brain and heart tissues, where potentiation and protection of catecholamines have been demonstrated, possess relatively low concentrations of O-methyl transferase, and it seems possible that, in the blood and most peripheral tissues, O-methylation is the more important reaction and monoamine-oxidase-mediated oxidative demethylation more important in the brain and the heart.

The monoamine oxidase inhibitors are used in the treatment of psychotic patients with mild to severe depression. Responsive patients show an increased sense of well-being, increased desire and ability to communicate, elevation of mood, increased physical activity and mental alertness as well as improvement in appetite. These drugs are used in the milder depressive states in place of electroshock. Because of their slow onset of action, they are of no value in psychiatric emergencies.

The monoamine oxidase inhibitors, by an unknown mechanism, reduce the frequency and the severity of migraine attacks.

Because of their enzyme-inhibiting properties, they potentiate and prolong the actions of many drugs such as amphetamines, caffeine, barbiturates and local anesthetics.

Toxic side-effects to the monoamine oxidase inhibitors are large in number, and some are of a serious nature. Such side-effects, including hepatic toxicities and visual disturbances, have resulted in the removal from distribution of some of the earlier agents (e.g., iproniazid, pheniprazine, etryptamine) as less toxic drugs were developed.

Occasional hypertensive crises, severe occipital headache, palpitation, nausea, vomiting and intracranial bleeding, sometimes resulting in death, have been reported with patients using monoamine oxidase inhibitors. These side-effects have been related to the long-lasting inhibitory properties of monoamine oxidase inhibitor drugs, which permit a strong pressor response to a variety of amines from exogenous sources. Tyramine and related amines present in high concentration in certain cheeses, wines, beer, liver, etc., have been implicated and patients taking monoamine oxidase inhibitors should be warned to avoid foods and beverages with a high tyramine content.

Because of the toxicity of the monoamine oxidase inhibitors, the tricyclic antidepressants are presently considered the drugs of choice in the treatment of depression.

In addition to the hydrazine derivatives, a number of nonhydrazines have been shown to possess potent monoamine-oxidase inhibiting properties. The hydrazines (e.g., phenelzine, isocarboxazid) have a slow onset of response, 2 or 3 weeks often being required before any degree of improvement in the mentally depressed state is noted. Tranylcypromine, a nonhydrazine monoamine oxidase inhibitor, frequently produces a response within several days.

Some currently available monoamine oxidase inhibitors are listed by name and structure in Table 10-4.

Structure-activity relationships

The in vitro potency of a large series of hydrazine derivatives in inhibiting the metabolism of serotonin

TABLE 10–4

MONOAMINE OXIDASE INHIBITORS

Generic Name Proprietary Name	Structure
Phenelzine Sulfate U.S.P. *Nardil*	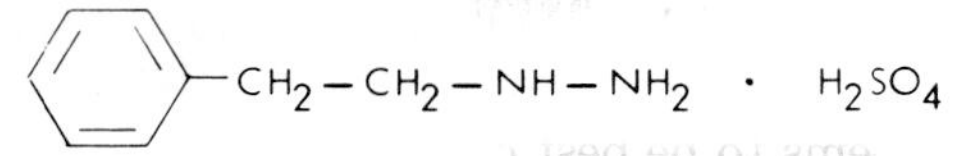 $\langle\text{phenyl}\rangle$—$CH_2$—$CH_2$—$NH$—$NH_2$ · H_2SO_4
Isocarboxazid U.S.P. *Marplan*	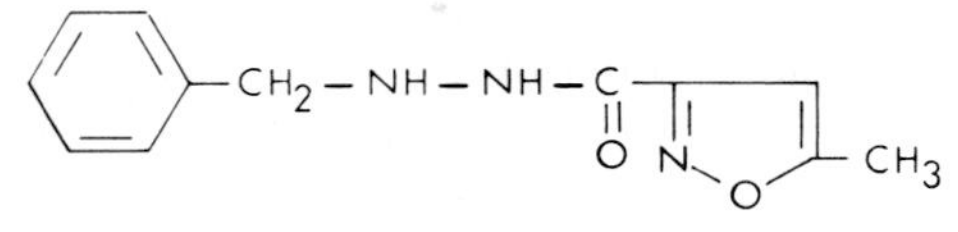 $\langle\text{phenyl}\rangle$—$CH_2$—$NH$—$NH$—$C$(=$O$)— (5-methyl-3-isoxazolyl, CH_3, N, O)
Tranylcypromine Sulfate U.S.P. *Parnate*	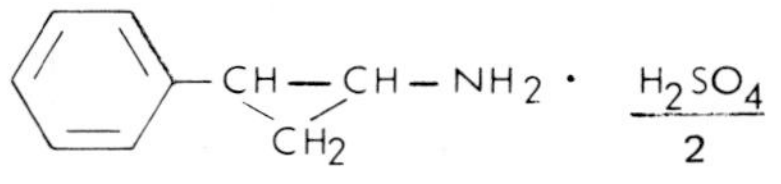 $\langle\text{phenyl}\rangle$—$CH$—$CH$—$NH_2$ · $\dfrac{H_2SO_4}{2}$ (with CH_2 bridge)
Pargyline Hydrochloride U.S.P. *Eutonyl*	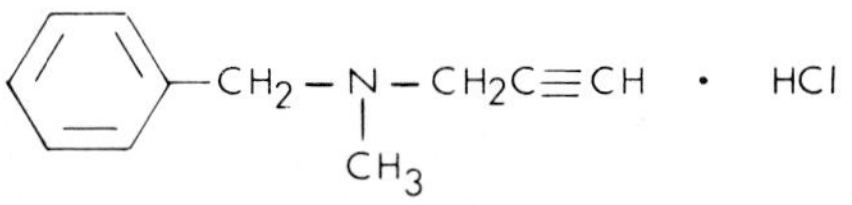 $\langle\text{phenyl}\rangle$—$CH_2$—$N(CH_3)$—$CH_2C\equiv CH$ · HCl

by rat-liver homogenate has been used to establish structure-activity relationships for the property of monoamine oxidase inhibition.[15] Analeptic properties were tested by measuring arousal of mice from a reserpine-induced stupor. Maximum monoamine oxidase and analeptic activities were shown by compounds with the amphetaminelike structure. For example, pheniprazine, a nitrogen isostere of methamphetamine, was one of the most potent agents tested, showing strong enzyme-inhibiting properties, and analeptic activity comparable with that of amphetamine.

Nuclear substitution (methoxy, methyl, hydrogenation) reduced both enzyme-inhibitory and analeptic properties of pheniprazine, just as it does with amphetamine. Both N-acylation and N-alkylation of the hydrazines that have been tested decreased enzyme-inhibitory and analeptic potency. Replacement of the phenyl ring by several heterocyclic ring systems reduced enzyme-inhibitory properties; analeptic properties were absent. An increase or a decrease in chain length between aryl and hydrazinyl groups caused

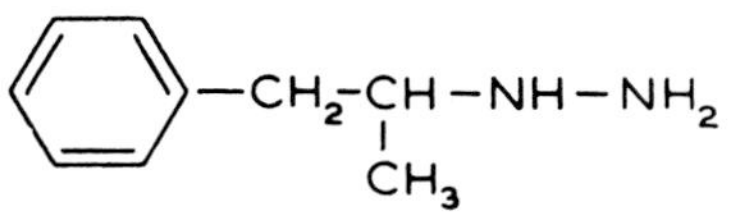

$\langle\text{phenyl}\rangle$—$CH_2$—$CH(CH_3)$—$NH$—$NH_2$

Pheniprazine

$\langle\text{phenyl}\rangle$—$CH_2$—$CH(CH_3)$—$NH$—$CH_3$

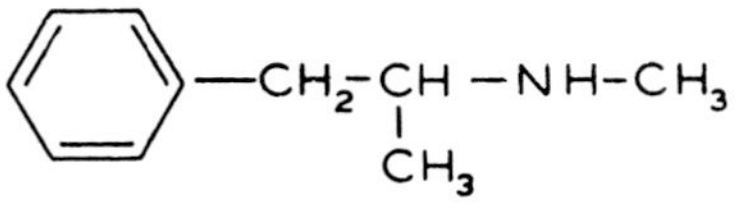

Methamphetamine

variations in analeptic and enzyme-inhibiting properties which demonstrated that these were separable. For example, iproniazid, benzylhydrazine, α-phenylethylhydrazine and γ-phenylisobutylhydrazine showed significant monoamine-oxidase-inhibiting properties, both in vivo and in vitro, but were without significant analeptic effect in mice.

Products

Phenelzine Sulfate U.S.P., Nardil®, β-phenylethylhydrazine dihydrogen sulfate. Phenelzine was introduced in 1959 as a potent monoamine oxidase inhibitor used in the treatment of depression. Side-effects include postural hypotension, constipation and edema. Phenothiazine tranquilizers are recommended in the treatment of accidental overdosage of monoamine oxidase inhibitors.

Isocarboxazid U.S.P., Marplan®, 5-methyl-3-isoxazolecarboxylic acid 2-benzylhydrazide, 1-benzyl-2-(5-methyl-3-isoxazolylcarbonyl)hydrazine. Isocarboxazid was introduced in 1959 for the treatment of mental depression and is used in moderate to severe depressive states in adults. Side-effects include orthostatic hypotension, constipation, and the more serious potential for hypertensive crises due to the potentiation of the pressor effects of sympathomimetics.

Tranylcypromine Sulfate U.S.P., Parnate® Sulfate, ($\pm$)-*trans*-2-phenylcyclopropylamine sulfate. Tranylcypromine was synthesized[16] in 1948 as an amphetamine analog. Following the introduction of the hydrazine derivatives, tranylcypromine was retested and found to be a potent monoamine oxidase inhibitor. The drug was introduced in 1961 for the treatment of patients with psychoneurotic and psychotic depression.

Due to hypertensive crises from its use in the presence of sympathomimetic amines from drug or food sources, tranylcypromine was withdrawn briefly from distribution in 1964. Because of its effectiveness in the treatment of depression, the drug was returned for restricted use in the treatment of hospitalized cases of severe depression or in closely supervised cases outside the hospital in which other medication has been found ineffective. It is not to be used in patients over 60 years of age or with a history of hypertension or other cardiovascular disease.

Tranylcypromine sulfate is available as 10-mg. tablets.

Pargyline Hydrochloride U.S.P., Eutonyl®, N-methyl-N-(2-propynyl)benzylamine hydrochloride. Pargyline is a monoamine oxidase inhibitor which possesses hypotensive and stimulant properties. The postural hypotension common as a side-effect in the monoamine oxidase inhibitors is emphasized in pargyline, and it is recommended for the treatment of hypertension rather than for use in depressed states. (See Chap. 14.)

As with other monoamine oxidase inhibitors, patients receiving pargyline should not receive sympathomimetic amines, such as amphetamine, ephedrine and their analogs; foods that contain pressor amines, such as aged cheese containing tyramine;

drugs that cause a sudden release of catecholamines, such as parenteral reserpine; tricyclic antidepressants, such as imipramine, or other monoamine oxidase inhibitors.

TRICYCLIC ANTIDEPRESSANTS

Following the discovery of the therapeutic value of the phenothiazine derivative chlorpromazine in the treatment of psychiatric disorders, many structural analogs were tested. Among these, the dibenzazepine derivative imipramine was found to be of therapeutic value in the treatment of depressive states, a condition in which chlorpromazine is not effective. A group of compounds, related in structure and pharmacologic effects, is now available for the treatment of depression: imipramine (Tofranil®), desipramine (Norpramin®, Pertofrane®), amitriptyline (Elavil®), nortriptyline (Aventyl®), and protriptyline (Vivactil®). Because the initial compounds introduced as antidepressant drugs were related in structure by their similar three-ring systems, these drugs are frequently designated the *tricyclic antidepressants*. Unlike the hydrazine derivatives, they do not inhibit monoamine oxidase. They rarely produce stimulation and excitement and may produce mild sedation like that of the phenothiazine tranquilizers. However, unlike the phenothiazine tranquilizers, they are effective in the treatment of emotional and psychiatric disorders in which the major symptom is depression.

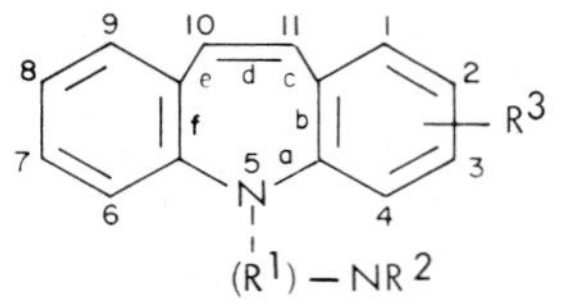

5H-Dibenz[*b,f*]azepine

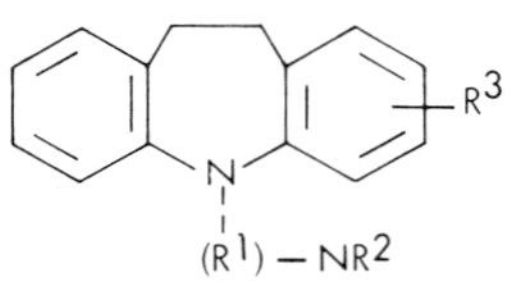

10,11-Dihydro-5H-dibenz[*b,f*]azepine

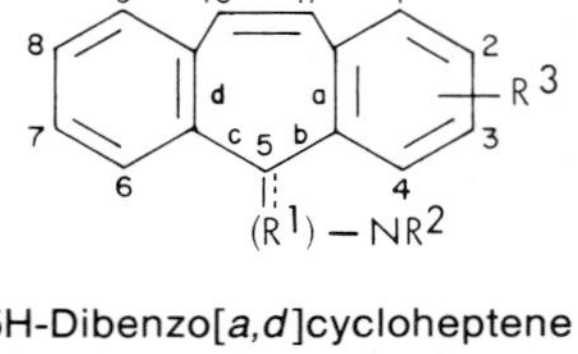

5H-Dibenzo[*a,d*]cycloheptene

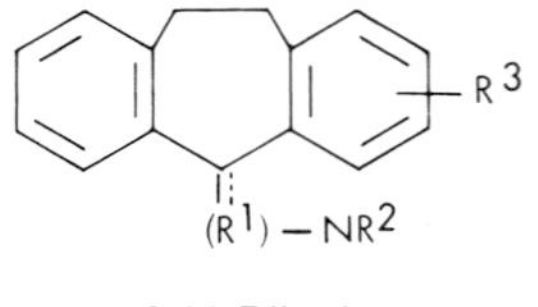

10,11-Dihydro-
5H-dibenzo[*a,d*]cycloheptene

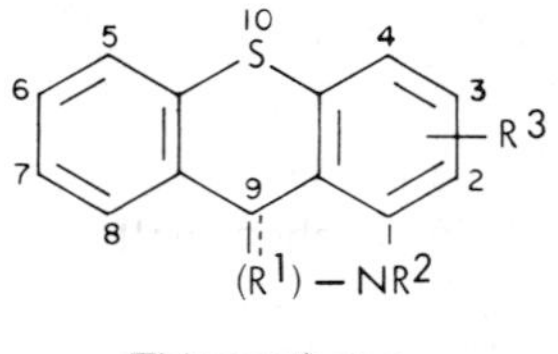

Thioxanthene

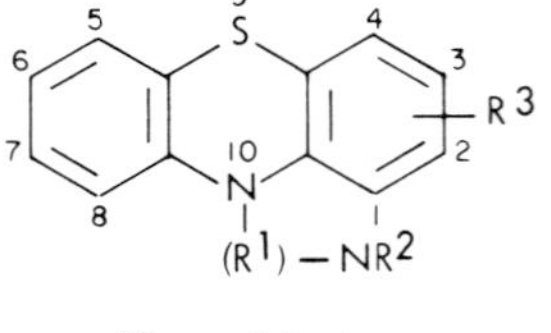

Phenothiazine

Depressed individuals, particularly those with endogenous depression rather than exogenous or reactive depressions, may respond with an elevation of mood, increased physical activity, mental alertness and an improved appetite. All of the tricyclic antidepressants have a slow onset of action. There may be a delay of several weeks after starting therapy before the clinical effects are apparent. They also produce numerous undesirable side-effects, including the following: atropinelike anticholinergic effects consisting of dry mouth, blurred vision, tachycardia, urinary retention, constipation, and aggravation of glaucoma; increased tension and agitation; postural hypotension, which is dose related and sometimes a limiting factor in their use. Dangerous synergistic effects may occur when monoamine oxidase inhibitors are administered with imipramine and related compounds. For this reason, it is recommended that a period of at least 2 weeks should be allowed before changing from a monoamine oxidase inhibitor to an imipramine-type compound, or vice versa.

Although the mechanism of the antidepressant action in man is unknown, it is postulated that this may be related to the inhibition of re-uptake of norepinephrine into adrenergic neurons by tricyclic antidepressants. (See Fig. 9-1, Chap. 9.)

Structure-activity relationships

A number of tricyclic ring systems, if appropriately substituted, may possess antidepressant properties. These include the 5H-dibenz[*b,f*]azepine (dibenzazepine, iminostilbene); the related ring system with the 10,11 double bond reduced, 10,11-dihydro-5H-dibenz[*b,f*]azepine (dihydrodibenzazepine, iminodibenzyl); the ring system without a heteroatom, 5H-dibenzo[*a,d*]cycloheptene (dibenzocycloheptene); the corresponding reduced system, 10,11-dihydro-5H-dibenzo-[*a,d*]cycloheptene (dihydrodibenzocycloheptene); the sulfur-bridged analog, thioxanthene; and the parent structure, originally related only to tranquilizing action, the phenothiazine.

Relationships between structure and antidepressant activity have been developed in a pharmacologic screening test based on the reversal of depression produced in the rat by reserpinelike benzoquinolizine compounds.[17] The following generalizations relate to derivatives of the ring structures: dibenzazepines, dibenzocycloheptenes, thioxanthenes, and phenothiazines.

Variations in R¹ (side chain). Activity is restricted to compounds having two or three carbons in the side chain. Compounds lacking the side chain, or with branched chains and chains containing more than four carbons are inactive.

Variations in R² (N-substituents). Activity is confined to methyl-substituted or unsubstituted amines. Ethyl or higher alkyl groups on the side-chain nitrogen result in compounds that are inactive, and show toxicity that increases with increasing length of the side chain. Almost all antidepressant compounds are primary and secondary amines. The antidepressant action of some tertiary amines has been attributed to the rapid formation of their secondary analogs in the body. Generally, the tertiary amines show sedative properties. Thus, imipramine exerts a weak tranquilizing action. Amitriptyline is even more pronounced in this respect, and triflupromazine is a potent tranquilizer. Imipramine and amitriptyline were revealed as antidepressants in pharmacologic tests, only if the secondary amine active metabolites were allowed to accumulate by repeated administration of the parent drug.

Variations in R³ (ring substituents). A number of ring-substituted compounds are active (e.g., 3-chloro, 10-methyl, 10,11-dimethyl) provided that they contain the aminoethyl or aminopropyl side chain.

Variations in the 10,11-Bridge. The bridge in the 10,11-position may be formed by —CH₂CH₂—(dihydrodibenzazepine) or by —CH=CH—(dibenzazepine). Thus, when a dibenzazepine is active, the corresponding 10,11-dihydro compound is also active. In the case of desipramine, activity is also preserved if the C-2 bridge is replaced by an S-bridge (the related phenothiazine derivative, desmethylpromazine), or is left out altogether (a diphenylamine). This suggests that the 10,11-bridge is not vital for antidepressant activity.

Variations in Ring Systems. The ring nitrogen of desipramine can be replaced by carbon to yield the active dihydrodibenzocycloheptene, nortriptyline. Of 20 phenothiazines tested, all were inactive except desmethylpromazine and desmethyltriflupromazine. Several appropriately substituted thioxanthenes were active, as were a number of related tricyclic ring structures. Removal of one benzene ring (bicyclic ring structures) resulted in loss of activity.

Products

Imipramine Hydrochloride U.S.P., Tofranil®, Presamine®, 5-[3-(dimethylamino)propyl]-10,11-dihydro-5H-dibenz[*b,f*]azepine hydrochloride. Imipramine was introduced in 1959 for the treatment of mental depression. In its clinical effect and pharmacology it shows some similarity to the phenothiazine derivatives to which it is chemically related in that it has mild tranquilizing properties. Unlike the phenothiazines, it is effective as an antidepressant agent. The drug is a potent parasympatholytic and displays prominent atropinelike side-effects.

Imipramine is most useful in treating endogenous depression. In treating depressions accompanied by anxiety, the drug is sometimes used together with a phenothiazine tranquilizing agent. Severe toxic reactions have occurred when imipramine was taken concurrently or immediately after the administration of monoamine oxidase inhibitors.

Imipramine undergoes metabolic N-demethylation to form the antidepressant metabolite desmethylimipramine (desipramine), which is then slowly demethylated to form the primary amine desdimethylimipramine. Both imipramine and desipramine are hydroxylated in the 2-position, followed by O-glucuronide formation.

Imipramine: R = CH₃
Desipramine: R = H

Imipramine hydrochloride is available as 10-,25- and 50-mg. tablets for oral use or in ampules containing 25 mg. for intramuscular administration. Small crystals may form in some ampules, but this has no influence on therapeutic effectiveness. The crystals redissolve when the ampules are immersed in hot tap water for 1 minute. Imipramine pamoate (Tofranil-PM®) is also available for oral administration.

Desipramine Hydrochloride U.S.P., Norpramin®, Pertofrane®, 5-(3-methylaminopropyl)-10,11-dihydro-5H-dibenz[*b,f*]azepine hydrochloride. Desipramine is a metabolite of imipramine which demonstrates similar antidepressant activity and was introduced in 1964. Although it is produced relatively slowly in the body by N-demethylation of imipramine, its subsequent metabolism and excretion is even slower, permitting accumulation. The slow onset of action of imipramine encouraged the theory that the parent compound might be exerting its antidepressant effect through a metabolite. Desipramine appears to have a somewhat shorter onset of action than imipramine, but it is somewhat less potent. The therapeutic range of effectiveness in the treatment of depressive states is the same as for imipramine, as are the side-effects and the precautions for use. Atropine-like side-effects are common, and the concomitant or prior use of monoamine-oxidase-inhibiting compounds is not recommended.

Studies of ring analogs of the phenothiazine tranquilizing drugs included a series in which the sulfur bridge of phenothiazine was replaced by an ethylene bridge, and the ring nitrogen was replaced by carbon. Antidepressant activity was noted as well as retention of tranquilizing properties to a slight extent in a member of the series, amitriptyline, and the compound was introduced as an antidepressant drug in 1961. Its metabolite, nortriptyline, was introduced in 1964.

Alternate methods have been developed for addition of the dimethylaminopropylidine side chain to 2,3:6,7-dibenzosuberone in syntheses of amitriptyline.[18,19]

Amitriptyline: R = CH₃
Nortriptyline: R = H

Amitriptyline Hydrochloride U.S.P., Elavil®, 5-(3-dimethylaminopropylidene)-10,11-dihydro-5H-dibenzo[*a,d*]cycloheptene hydrochloride. Amitriptyline is recommended for the treatment of mental depression. It also has a tranquilizing component of action which is useful in cases in which anxiety accompanies depression. The sedative effect of amitriptyline is manifested quickly; however, the antidepressant effect may vary in onset from about 4 days to 6 weeks. Generally, improvement in mood and behavior is seen in 2 to 3 weeks after the start of medication.

Minor side-effects reflecting amitriptyline's anticholinergic activity are common. These include dryness of mouth, blurred vision, tachycardia and urinary retention. Amitriptyline is contraindicated in the presence of glaucoma and in patients with cardiovascular complications. The drug should not be administered with a monoamine oxidase inhibitor, since serious potentiation of side-effects may occur. Such combinations have caused cardiovascular collapse, impaired consciousness, hyperpyrexia, convulsions, and death. The drug is not recommended for use in children under 12 years of age.

Metabolic alteration of amitriptyline occurs by monodemethylation of the side chain nitrogen, and hydroxylation of the 10-position, forming *cis* and *trans* isomers of 10-hydroxyamitriptyline. Aromatic hydroxylation, rupture of the ethylene bridge, and oxidative deamination also occur. Nortriptyline is not excreted in the urine in appreciable amounts after administration of amitriptyline.

Amitriptyline is available as 10-, 25- and 50-mg. coated tablets, and as an injection for intramuscular use, containing 10 mg. per ml.

Nortriptyline Hydrochloride U.S.P., Aventyl®, Pamelor®, 5-(3-methylaminopropylidene)-10,11-dihydro-5H-dibenzo[*a,d*]cycloheptene hydrochloride. Nortriptyline is the N-demethylated metabolite of amitriptyline. It possesses antidepressant and tranquilizing properties like those of the parent drug. The anticholinergic side-effects of nortriptyline are reported to be less than those of amitriptyline; however, they are still significant enough to preclude use in patients with glaucoma and urinary retention. The drug should not be used concurrently with monoamine oxidase inhibitors or before an interval of 1 to 2 weeks following termination of monoamine oxidase inhibitor therapy. The drug is not recommended for use in children.

Nortriptyline undergoes N-demethylation, as well as hydroxylation in the 10 position, forming *cis* and *trans* isomers of 10-hydroxynortriptyline, which are excreted in the urine as conjugates.

Nortriptyline is administered orally in capsule or liquid form for the treatment of mental depression, anxiety-tension states and psychosomatic disorders.

Nortriptyline is available as 10-mg. or 25-mg. capsules or in a liquid preparation containing 10 mg. in 5 ml.

Protriptyline Hydrochloride U.S.P., Vivactil®, 5-(3-methylaminopropyl)-5H-dibenzo[*a,d*]cycloheptene hydrochloride. Protriptyline, an isomer of nortriptyline containing an endocyclic rather than exocyclic double bond, was introduced in 1967 as a selective antidepressant agent. It is used for the treatment of mental depression in patients under close medical supervision, and is said to produce little sedation. Because of potentially serious drug interactions, protriptyline should not be used in patients receiving monoamine oxidase inhibitors, guanethidine, or other hypotensive agents. The drug is not recommended for use in children.

Protriptyline

Trimipramine Maleate, Surmontil®, was introduced in the United States in 1979. It has utility as an antidepressant. Its actions are similar to those of related compounds in this group.

Trimipramine

It is used as the racemic mixture.

Doxepin Hydrochloride U.S.P, Sinequan®, Adapin®, N,N-dimethyl-3-(dibenz[*b,e*]oxepin-11(6H)-ylidene)propylamine hydrochloride. Doxepin was introduced in 1969 as an antidepressant drug useful in the treatment of mild to moderate endogenous depression. It differs in structure from amitriptyline by the presence of an oxygen atom in the central ring, which leads to the formation of *cis* and *trans* isomers. The *cis* isomer is more active than the *trans*. Doxepin is a mixture of the isomers. As with related tricyclic antidepressant agents, the drug-induced elevation of mood may be accompanied by atropine-like anticholinergic side-effects such as dryness of mouth, and by sedation. The drug is not recommended for use in children.

Doxepin Hydrochloride

Doxepin is available in capsule or concentrated solution form for oral use.

MISCELLANEOUS PSYCHOMOTOR STIMULANTS

Deanol Acetamidobenzoate, Deaner®, the *p*-acetamidobenzoic acid salt of 2-dimethylaminoethanol. Deanol base, dimethylaminoethanol, is the nonquaternized precursor to choline. The salt was introduced in 1958 for use in the treatment of a variety of mild depressive states and for alleviation of behavior problems and learning difficulties of school-age children. It has been proposed that deanol penetrates the central nervous system, there serving as a precursor to choline and acetylcholine. The drug is of low toxicity, and side-effects are relatively mild. These include headache, constipation, muscle tenseness and twitching, insomnia and postural hypotension. The oral dose is 10 to 50 mg. daily.

TABLE 10–5

PSYCHOMOTOR STIMULANTS

Name / Proprietary Name	Preparations	Category	Usual Adult Dose*	Usual Dose Range*	Usual Pediatric Dose*
Amphetamine Phosphate, Dextro Sulfate U.S.P. *Dexedrine*	Amphetamine Phosphate, Dextro Sulfate Elixir U.S.P. Amphetamine Phosphate, Dextro Sulfate Tablets U.S.P.	Central stimulant	Narcolepsy—5 to 20 mg. 1 to 3 times daily	2.5 to 60 mg. daily	Hyperkinesia: children under 3 years of age—use is not recommended; 3 to 5 years of age—2.5 mg. once daily, increased by 2.5 mg. at weekly intervals; 6 years of age and over—5 mg. once or twice daily, increased by 5 mg. at weekly intervals Narcolepsy: 6 to 12 years of age— 2.5 mg. twice daily, increased by 5 mg. at weekly intervals; 12 years of age and over—5 mg. twice daily, increased by 10 mg. at weekly intervals
Amphetamine Phosphate, Dextro U.S.P. *Dextro-Profetamine*	Amphetamine Phosphate, Dextro Tablets U.S.P.	Central stimulant	5 mg. every 4 to 6 hours	5 to 10 mg.	
Methamphetamine Hydrochloride *Desoxyn*	Methamphetamine Hydrochloride Tablets	Central stimulant	Narcolepsy—5 to 60 mg. daily in divided doses		
Diethylpropion Hydrochloride U.S.P. *Tenuate, Tepanil*	Diethylpropion Hydrochloride Tablets U.S.P.	Anorexic	25 mg. 3 times daily		
Phenmetrazine Hydrochloride U.S.P. *Preludin*	Phenmetrazine Hydrochloride Tablets U.S.P.	Anorexic	25 to 75 mg. daily, in divided doses, 1 hour before meals		
Methylphenidate Hydrochloride U.S.P. *Ritalin*	Methylphenidate Hydrochloride Tablets U.S.P.	Central stimulant	Narcolepsy—10 mg. 2 or 3 times daily	10 to 60 mg. daily	Hyperkinesia: use in children under 6 years of age is not recommended; over 6 years—5 mg. twice daily, increased by 5 to 10 mg. at weekly intervals
Phenelzine Sulfate U.S.P. *Nardil*	Phenelzine Sulfate Tablets U.S.P.	Antidepressant	The equivalent of 15 mg. of phenelzine once daily or every other day	7.5 to 75 mg. daily	
Isocarboxazid U.S.P. *Marplan*	Isocarboxazid Tablets U.S.P.	Antidepressant	Initial, 30 mg. daily as a single dose or in divided doses; maintenance, 10 to 20 mg. daily		
Tranylcypromine Sulfate U.S.P. *Parnate*	Tranylcypromine Sulfate Tablets U.S.P.	Antidepressant		Initial, 10 mg. in the morning and afternoon daily for 2 weeks; if no response appears, increase dosage to 20 mg. in the morning and 10 mg. in the afternoon daily for another week; maintenance, 10 to 20 mg. daily	

(Continued)

* See *U.S.P. D.I.* for complete dosage information.

TABLE 10–5

PSYCHOMOTOR STIMULANTS (CONT.)

Name Proprietary Name	*Preparations*	*Category*	*Usual Adult Dose**	*Usual Dose Range**	*Usual Pediatric Dose**
Imipramine Hydrochloride U.S.P. *Tofranil, Presamine*	Imipramine Hydrochloride Injection U.S.P.	Antidepressant	I.M., 25 to 50 mg. 3 or 4 times daily	50 to 300 mg. daily	375 μg. per kg. of body weight or 11 mg. per square meter of body surface 4 times daily. Dose is not established in children under 12 years of age.
	Imipramine Hydrochloride Tablets U.S.P.		25 to 50 mg. 3 or 4 times daily	50 to 300 mg. daily	375 μg. per kg. of body weight or 11 mg. per square meter of body surface 4 times daily. Dosage is not established in children under 12 years of age
Desipramine Hydrochloride U.S.P. *Norpramin, Pertofrane*	Desipramine Hydrochloride Capsules U.S.P. Desipramine Hydrochloride Tablets U.S.P.	Antidepressant	150 mg. daily in divided doses	50 to 200 mg. daily	
Amitriptyline Hydrochloride U.S.P. *Elavil*	Amitriptyline Hydrochloride Injection U.S.P.	Antidepressant	I.M., 20 to 30 mg. 4 times daily	80 to 120 mg. daily	Dosage is not established in children under 12 years of age
	Amitriptyline Hydrochloride Tablets U.S.P.		25 mg. 2 to 4 times daily	30 to 300 mg. daily	Dosage is not established in children under 12 years of age
Nortriptyline Hydrochloride U.S.P. *Aventyl*	Nortriptyline Hydrochloride Capsules U.S.P. Nortriptyline Hydrochloride Oral Solution U.S.P.	Antidepressant		An amount of nortriptyline hydrochloride equivalent to 20 to 100 mg. of nortriptyline daily in divided doses	
Protriptyline Hydrochloride U.S.P. *Vivactil*	Protriptyline Hydrochloride Tablets U.S.P.	Antidepressant	15 to 40 mg. daily in 3 or 4 divided doses	15 to 60 mg. daily in divided doses	
Doxepin Hydrochloride U.S.P. *Sinequan*	Doxepin Hydrochloride Capsules U.S.P. Doxepin Hydrochloride Oral Solution U.S.P.	Antidepressant	Initial, 75 mg. daily	75 to 150 mg. daily	Not recommended for use in children under 12 years of age

* See *U.S.P. D.I.* for complete dosage information.

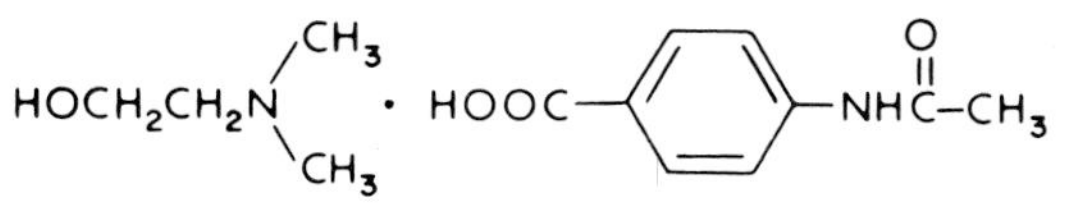

Deanol Acetamidobenzoate

HALLUCINOGENS (PSYCHODELICS, PSYCHOTOMIMETICS)

A wide variety of drugs are capable of stimulating the central nervous system to afford alterations of mood and perception, illusions, or bizarre hallucinations which resemble naturally occurring psychotic states. These hallucinogenic effects may be produced as toxic side-effects of drugs used therapeutically, such as bromides, amphetamines, cocaine and certain anticholinergics. These effects may occur when the drugs are used at high doses, or when they are administered by routes which produce high concentrations in the central nervous system. In addition, there exists a group of agents which have no currently accepted medical use, which are highly effective in altering mood and perception, and which are used illegally for this purpose. The incentive for their use is derived to a lesser extent from the hallucinogenic component of action, and to a greater extent from the "mind-expanding" experience generated by the drugs. The term "psychodelic" is preferred by some to better describe the drug class and its use. A state of heightened awareness and perception to sensory input is induced, and external stimuli are perceived in ways which are not part of normal experience; for example, sounds may be seen as waves of color. The drug user may feel a sense of dissociation, being both impartial observer and participant in the experience. The surroundings may appear strikingly beautiful, thoughts are perceived as being profound and clear. In contrast to these pleasurable events, the mood may shift into one of anxiety, fear, and panic—a "bad trip." Prolonged psychotic episodes may occur following sustained use of hallucinogenic drugs.

The major hallucinogenic drugs are related to the neurotransmitter substances, 5-hydroxytryptamine (serotonin), an *indole ethylamine,* and norepinephrine, a *β-phenylethylamine.*

INDOLE ETHYLAMINES

Serotonin does not cause behavioral changes after ingestion or injection, since it does not effectively penetrate the central nervous system. A precursor of serotonin, 5-hydroxytryptophan, is capable of central nervous system penetration; particularly in the presence of a monoamine oxidase inhibitor, it produces elevated brain serotonin levels and excited behavior.

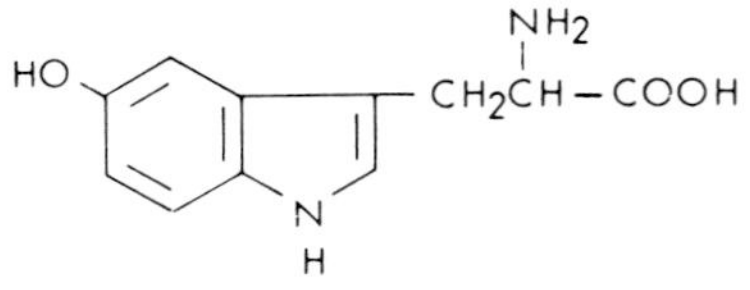

5-Hydroxytryptophan

Aromatic L-Amino Acid Decarboxylase

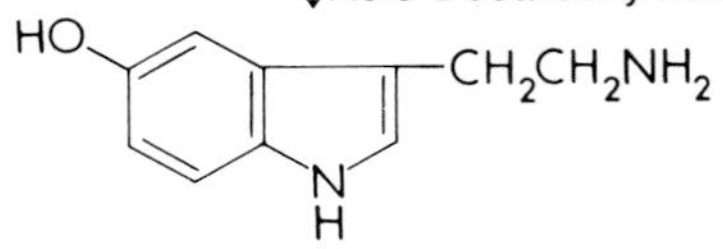

5-Hydroxytryptamine
(Serotonin, 5-HT)

A number of hallucinogens possess the 3-(β-aminoethyl)indole structure in common with serotonin but, in addition, are N,N-dimethyl derivatives. This tertiary amine structure facilitates penetration of the central nervous system by its increased lipophilic character and appears to delay metabolic oxidative deamination reactions, thus prolonging the existence of active amines in the body.

Dimethyltryptamine, DMT, N,N-dimethyltryptamine, and N,N-diethyltryptamine (DET) are hallucinogenic if smoked or given by injection. Their psy-

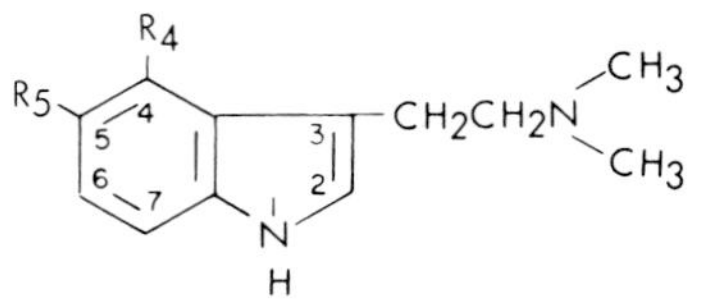

Dimethyltryptamine: $R_4 = R_5 = H$
Bufotenine: $R_4 = H; R_5 = OH$
Psilocybin: $R_4 = OPO(OH)_2; R_5 = H$
Psilocyn: $R_4 = OH; R_5 = H$

chotomimetic effects have a duration of less than one hour, and are accompanied by pronounced sympathomimetic side-effects.

Bufotenin, 5-hydroxy-3-(β-dimethylaminoethyl)-indole, is the N,N-dimethyl homolog of serotonin. It occurs naturally in the secretions of the skin of a toad (L. *bufo*) and in the seeds of a plant, *Piptadenia peregrina,* used in the form of a snuff by some South American Indians. Bufotenine is hallucinogenic after injection.

Psilocybin, 4-phosphoryloxy-N,N-dimethyltryptamine, is the active hallucinogenic principle of the mushroom *Psilocybe mexicana.* The mushroom is used in religious ceremonies by Mexican Indians. Psilocybin resembles mescaline and lysergic acid diethylamide in its pharmacologic properties. It has a short onset and duration of action, the peak effect being reached about 2 minutes after administration of the usual 5- to 10-mg. oral dose. Psilocyn, 4-hydroxy-N,N-dimethyltryptamine, which is the hydrolysis product of psilocybin, is also hallucinogenic.

Lysergic Acid Diethylamide, LSD, a potent hallucinogen, contains both the indole ethylamine (indole nucleus, C_4, C_5, N—CH_3) and the phenylethylamine (benzene ring, C_{10}, C_5, N—CH_3) units within its structure.

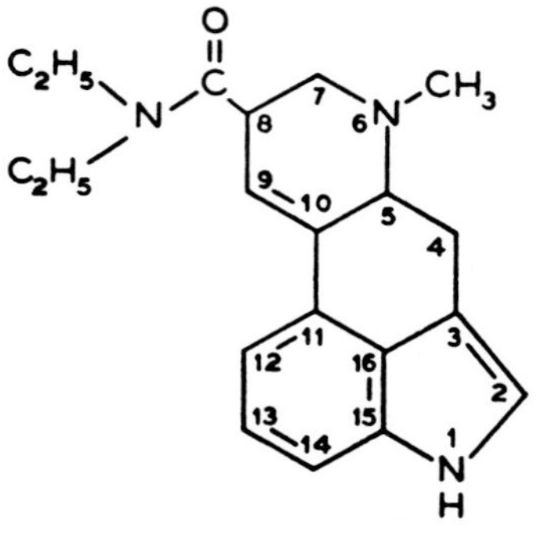

Lysergic Acid Diethylamide

The usual hallucinogenic dose of LSD ranges from 100 to 400 micrograms, taken orally, usually absorbed on an inert support such as sucrose in sugar-cube form, or lactose in tablets or capsules. Clammy skin, anxiety and a slight clouding of consciousness occur about 20 to 45 minutes after ingestion, followed in about 15 minutes by the major psychic effects, which last for about 6 hours. With intensities varying widely depending on the individual and the surroundings, effects include disturbances of perception, hallucinations characterized by vivid patterns of color, excitation, euphoria, and loss of personal identity. A phenothiazine tranquilizer, such as chlorpromazine, may be used orally or intramuscularly to terminate the effects of LSD. Prolonged psychic disturbances may occur after use of the drug, frequently including flashes of color. Prolonged psychotic episodes, including schizophrenia, have been reported after discontinuance of use of LSD.

β-PHENYLETHYLAMINES

Several β-phenethylamine derivatives, structurally related to norepinephrine and to amphetamine, act rapidly and intensely on the central nervous system. Sympathomimetic side-effects, such as elevated blood pressure and pupillary dilatation, are greater than with LSD.

Mescaline, 3,4,5-trimethoxyphenethylamine, isolated from the stem parts (mescal buttons, peyote) of the cactus *Lophophora williamsii,* has long been used for its hallucinogenic effect during religious ceremonies of certain Indian tribes in Mexico and the southwestern United States. In contrast to the amphetamine-like phenyl-2-aminopropanes, mescaline is susceptible to rapid metabolic attack by monoamine oxidase and a relatively high oral dose (250 to 500 mg.) is required to produce the hallucinogenic effect.

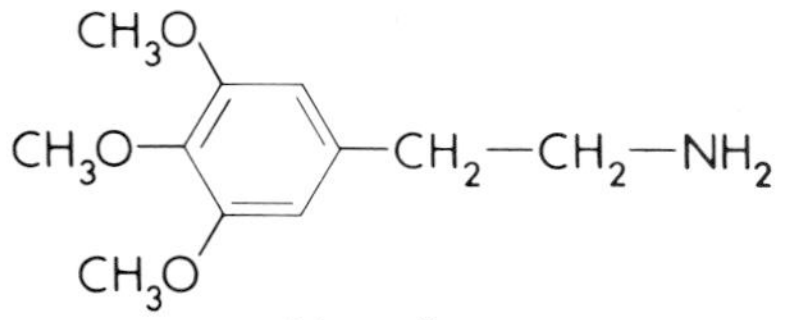

Mescaline

Synthetic analogs of mescaline that possess the branched methyl side chain of amphetamine show greatly enhanced potency as central nervous system stimulants and hallucinogens and are used illegally for these effects. Thus, 1-(2,5-dimethoxy-4-methylphenyl)-2-aminopropane (DOM), also called STP (Serenity, Tranquility, Peace), is about 100 times as potent as mescaline, although still only one thirtieth as potent as LSD. The usual hallucinogenic oral dose of DOM is 5 to 10 mg. A related synthetic analog, 3,4-methylenedioxyamphetamine (MDA), has also been used illegally as a substitute for LSD in producing its hallucinatory and so-called mind-expanding properties.

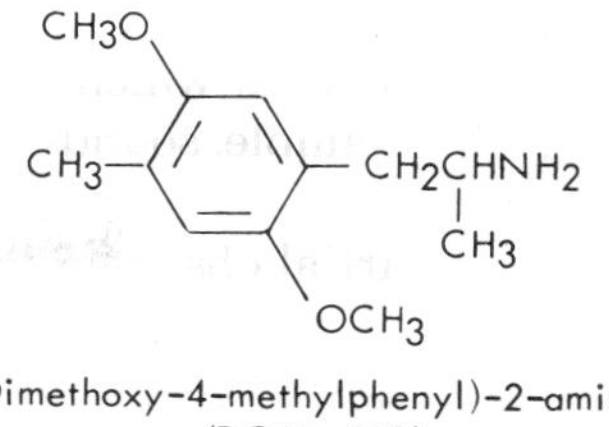

1-(2,5-Dimethoxy-4-methylphenyl)-2-aminopropane
(DOM, STP)

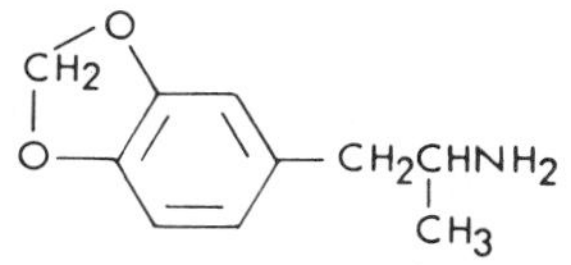

3,4-Methylenedioxyamphetamine
(MDA)

Cocaine (see Chap. 15), although not a phenylethylamine, produces central nervous system arousal or stimulant effects, as well as acute and chronic toxic effects which closely resemble those of the amphetamines. This may be due to the inhibition by cocaine of re-uptake of the norepinephrine released by adrenergic nerve terminals, leading to an enhanced adrenergic stimulation of norepinephrine receptors. The increased sense of well-being, mood elevation and intense but short-lived euphoric state produced by cocaine requires frequent administration, usually by the intranasal ("sniffing") or intravenous routes.

MISCELLANEOUS HALLUCINOGENS

Phencyclidine Hydrochloride, 1(1-phenylcyclohexyl) piperidine hydrochloride. Phencyclidine is an analgesic anesthetic agent which was first used primarily in veterinary practice to immobilize primates. However, since it produces hallucinations, euphoria, and alterations of mood and perception, it has been widely used illegally under the name of "PCP" (derived from the chemical name) or "angel dust". It is used both by injection and by smoking, routes that deliver a maximal concentration to the central nervous system. Action of the drug may terminate in a variable period of amnesia.

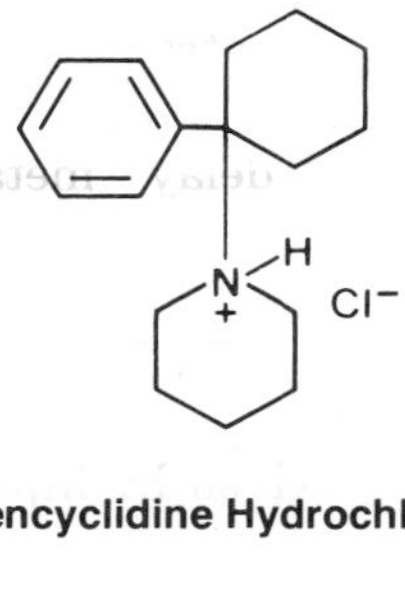

Phencyclidine Hydrochloride

Cannabis (Marihuana, Hashish). The products derived from the hemp plant, *Cannabis sativa,* are not central nervous system stimulants, but rather depressants. Their euphoretic properties, stemming from depression of higher centers, have caused them to be associated with the more potent hallucinogenic agents. For this reason, they are included in the present discussion, although they resemble alcohol and related depressants drugs rather than the arylethylamines.

Marihuana is widely (although illegally) used. The active components are present in the leaves and especially concentrated in the resin exuded from the flowering tops of the female Indian hemp plant, *Cannabis sativa L.* The highly potent resin is known as hashish in the Middle East, or charas in India. The term marihuana generally refers to the mixed leaves and flowering tops of cannabis; it is usually smoked. The major compound, $(-)$-Δ^1-*trans*-tetrahydrocannabinol (THC), has been synthesized in the racemic form [20] and produces psychotomimetic effects at a dose of about 200 micrograms per kg. of body weight when smoked.[21] When smoked or injected, any cannabis preparation rapidly produces a maximal excitatory effect as the THC rapidly enters the central nervous system; the latency period for onset of effects may be 45 to 60 minutes after ingestion. The effect quickly passes from excitation to sedation as concentrations decrease in the central nervous system, owing to redistribution of the highly lipophilic drug to peripheral lipoidal tissues. Current evidence indicates that chronic marihuana use is less hazardous to health than is excessive use of alcohol. Marihuana has been advocated in the relief of pain in the terminal cancer patient and also for its anticonvulsant effects.

$(-)$-Δ^1-*trans*-Tetrahydrocannabinol

Several different numbering systems for the cannabinoids have been used, thus leading to some confusion. A system that regards these compounds as substituted monoterpenes has been widely adopted[22] and is used here.

Since most of the agents included in this section (Hallucinogens) have a high potential for abuse and have no currently accepted medical use in treatment in the United States, they have been designated as "Schedule 1" drugs under the Comprehensive Drug Abuse Prevention and Control Act of 1970 (see Federal Regulations, April 24, 1971). These substances are not available for prescription use.

REFERENCES

1. Goodman, A., and Gilman, L.: The Pharmacological Basis of Therapeutics, ed. 3, pp. 324 and 340, New York, Macmillan, 1965.
2. Conroy, H.: J. Am. Chem. Soc. 79:5551, 1957.
3. Holker, J. S. E., Robertsen, A., and Taylor, J. H.: J. Am. Chem. Soc. 80:2987, 1958.
4. Blake, M., and Harris, H. E.: J. Am. Pharm. A. (Sci. Ed.) 41:521, 1952.
5. Lands, A. M.: First Symposium on Chemical-Biological Correlation, pp. 73-119, National Academy of Sciences, Washington, D.C., 1951.
6. Leake, C. D.: The Amphetamines, Their Actions and Uses, Springfield, Ill., Charles C Thomas, 1958.
7. Selikoff, I. J., Robitzek, E. H., and Ornstein, G. G.: Am. Rev. Tuberc. 67:212, 1953.
8. Zeller, E. A., *et al.*: Experientia 8:349, 1952.
9. Shore, P. A., and Brodie, B. B.: Proc. Soc. Exp. Biol. Med. 94:433, 1957.
10. Loomer, H. P., Saunders, J. C., and Kline, N. S.: Psychiat. Res. Rep. 8:129, 1958.
11. Spector, S., Shore, P. A., and Brodie, B. B.: J. Pharmacol. Exp. Ther. 128:15, 1960.
12. Jepson, J. B., *et al.*: Biochem. J. (London) 74:5P, 1960.
13. Shore, P. A., *et al.*: Science 126:1063, 1957.
14. Axelrod, J., Senohi, S., and Witkop, B.: J. Biol. Chem. 233:697, 1958.
15. Biel, J. H., Nuhfer, P. A., and Conway, A. C.: Ann. N. Y. Acad. Sci. 80:568, 1959.
16. Burger, A., and Yost, W.: J. Am. Chem. Soc. 70:2198, 1948.
17. Bickel, M. H., and Brodie, B. B.: Int. J. Neuropharmacol. 3:611, 1964.
18. Jucker, E.: Chimia 15:267, 1961.
19. Hoffsommer, R. D., Taub, D., and Wendler, N. L.: J. Org. Chem. 27:4134, 1962.
20. Mechoulam, R., and Gaoni, Y.: J. Am. Chem. Soc. 87:3273, 1965.
21. Isbell, H.: Psychopharmacologia 11:184, 1967.
22. Mechoulam, R., and Gaoni, Y.: Progr. Chem. Org. Nat. Prod. 25:175, 1967.

SELECTED READINGS

Biel, J. H.: Some Rationales for the Development of Antidepressant Drugs, *in* Molecular Modification in Drug Design, Advances in Chemistry Series, no. 45, Am. Chem. Soc. Applied Pub., Washington, D.C., 1964.

Burger, A.: Hallucinogenic Agents, *in* Burger, A. (ed.): Medicinal Chemistry, ed. 3, p. 1511, New York, Wiley-Interscience, 1970.

Burns, J. J., and Shore, P. A.: Biochemical effects of drugs, Ann. Rev. Pharmacol. 1:79, 1961.

Hoffer, A., and Osmond, H.: The Hallucinogens, New York, Academic Press, 1967.

Kaiser, C., and Zirkle, C. L.: Antidepressant Drugs, *in* Burger, A. (ed.): Medicinal Chemistry, ed. 3, p. 1476, New York, Wiley-Interscience, 1970.

Nieforth, K. A., and Cohen, M. L.: Central Nervous System Stimulants *in* Faye, W. O. (ed.): Principles of Medicinal Chemistry, p. 275, Philadelphia, Lea & Febiger, 1974.

Rice, L. M., and Dobbs, E. C.: Analeptics, *in* Burger, A. (ed.): Medicinal Chemistry, ed. 3, p. 1402, New York, Wiley-Interscience, 1970.

Whitelock, O. V. (ed.): Amine oxidase inhibitors, Ann, N. Y. Acad. Sci. 80:551, 1959.

11
adrenergic agents

Patrick E. Hanna

Adrenergic drugs are those chemical agents that exert their principal pharmacologic and therapeutic effects by acting at peripheral sites to either enhance or reduce the activity of components of the sympathetic division of the autonomic nervous system. In general, those substances that produce effects similar to stimulation of sympathetic nervous activity are known as *sympathomimetics, adrenomimetics,* or *adrenergic stimulants.* Those that decrease sympathetic activity are referred to as *sympatholytics, antiadrenergics,* or *adrenergic blocking agents.* In addition to their effects on sympathetic nerve activity, a number of adrenergic agents produce important effects on the central nervous system.

This chapter includes discussion of both adrenergic stimulants and adrenergic blocking agents. We assume that the reader has a basic understanding of both the principal anatomical features and the principal functions of the autonomic nervous system, including the underlying concepts of neurochemical transmission. Both Mayer[1] and Day[2] have published useful reviews of these topics and we refer the reader to them for supplemental background information.

ADRENERGIC NEUROTRANSMITTERS

FUNCTION

The adrenergic nerves in the autonomic nervous system are the postganglionic sympathetic fibers; preganglionic fibers and postganglionic parasympathetic fibers are cholinergic. The neurotransmitter, or neuro-

hormone, which is released from all preganglionic nerves and from postganglionic parasympathetic neurons is acetylcholine. The adrenergic neurotransmitter that is liberated from postganglionic sympathetic neurons as a result of sympathetic nerve stimulation is norepinephrine (NE).

Norepinephrine, after its release from the sympathetic nerve ending into the synaptic cleft, interacts with specific postsynaptic receptors on cells of the effector organ (Fig. 11-1). The effector organ is the organ or tissue (usually a gland, smooth muscle or cardiac muscle) that is innervated by the postganglionic nerve. Interaction of norepinephrine with the adrenergic receptors of the effector cells ultimately results in the production of a physiological response (muscle contraction or relaxation, glandular secretion, etc.) that is characteristic of that organ or tissue.

The action of NE at adrenergic receptors is terminated by a combination of processes, including uptake into the neuron and extraneuronal tissues, diffusion away from the synapse, and metabolism. In most cases the primary mechanism for termination of the action of NE appears to be reuptake (uptake$_1$) of the catecholamine into the nerve terminal. This is an energy-requiring process; it involves a membrane pump system that has a high affinity for norepinephrine. The uptake system also transports certain amines other than NE into the nerve terminal. Much of the NE that reenters the sympathetic neuron is then transported by a second active uptake process into the storage granules where it is held in a stable complex with ATP and protein until sympathetic nerve activity or some other stimulus causes it to be released into the synaptic cleft.

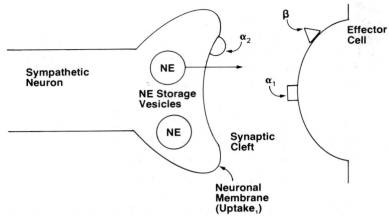

FIG. 11-1. *Postganglionic sympathetic neuron, synaptic cleft, and effector cell.*

In addition to the intraneuronal uptake of NE, there exists an extraneuronal uptake process that is commonly referred to as uptake$_2$. This uptake process, which was first discovered in cardiac muscle cells, is present in a variety of tissues. Its physiological significance is not known, but it may serve to help regulate the concentration of NE and other catecholamines in the vicinity of adrenergic receptors.[3] Since catecholamines that are taken up into extraneuronal tissues are rapidly metabolized, the uptake$_2$ process does not serve as a catecholamine-conservation mechanism in the fashion of the uptake$_1$ pump, which helps conduct NE safely to its storage granules.

Epinephrine is also an adrenergic neurotransmitter, but it is not released from sympathetic nerve endings in the fashion described for norepinephrine. Epinephrine is synthesized and stored in the adrenal medulla from which it is released into the circulation. Until the 1950's it was generally believed, however, that epinephrine not only was a product of adrenal medullary cells but also was the principal sympathetic neurotransmitter released at postganglionic adrenergic receptors. It is now known that epinephrine is distributed by means of the circulation to various organs and tissues where it exerts its effects at adrenergic receptor sites.

The termination of the action of epinephrine appears not to be highly dependent on the neuronal-membrane pump (uptake$_1$), for which it has less affinity than NE. Current evidence indicates that the physiological activity of epinephrine may, in large part, be terminated by metabolism in extraneuronal tissues following its removal from the vicinity of the adrenergic receptors by the uptake$_2$ mechanism.

STRUCTURE AND PHYSICOCHEMICAL PROPERTIES

The neurotransmitters, NE and epinephrine, belong to the chemical class of substances known as *catecholamines*. This name arises from the fact that the two hydroxyl groups present on the aromatic ring of each of these compounds are situated *ortho* to each other; the same arrangement of aromatic hydroxyl groups is found in catechol.

Norepinephrine and epinephrine are relatively small, polar substances with a high degree of water solubility. These physicochemical characteristics are typical of a number of endogenous agonists, such as acetylcholine, histamine, serotonin, and dopamine. The catecholamines contain both acidic (the aromatic hydroxyls) and basic (the aliphatic amine) functional groups. The magnitude of the dissociation constants of the phenolic and amino groups of the catecholamines has been the subject of considerable investigation and some controversy.[4] Ganellin has pointed out that the pKa values of 9.9 and 8.7 (which are frequently attributed to a phenolic hydroxyl and the protonated amino group, respectively, of the epinephrine cation) have been shown by several research groups to be incorrectly assigned.[4] The lower pKa value of the protonated form of epinephrine is due to ionization of a phenolic hydroxyl group. Ganellin has calculated the relative populations of the various ionized and nonionized species of NE and epinephrine at pH 7.4 and has found that the cation (shown below) is present to an extent slightly greater than 95 percent for both catecholamines. The zwitterionic form in which both the

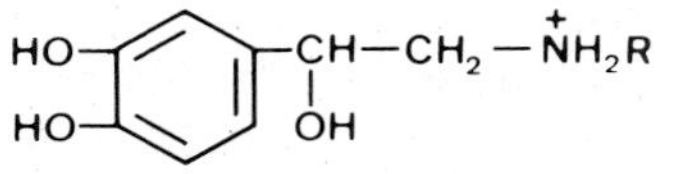

R = H or CH₃; Cationic Form of Norepinephrine and Epinephrine

aliphatic amine is protonated and one of the hydroxyl groups is ionized is present to the extent of approximately 3 percent. Thus, at physiological pH, less than 2 percent of either epinephrine or NE exists in the nonionized form. This largely accounts for the high degree of water solubility of these compounds as well as of other catecholamines such as isoproterenol and dopamine.

Aromatic compounds that contain *ortho*-hydroxyl substituents are highly susceptible to oxidation. Therefore, catecholamines such as epinephrine and

NE undergo oxidation in the presence of oxygen (air) or other oxidizing agents to produce a mixture of colored products. For this reason, solutions of catecholamine drugs are usually stabilized by the addition of an antioxidant (reducing agent) such as ascorbic acid or sodium bisulfite.

BIOSYNTHESIS

Catecholamine biosynthesis takes place in adrenergic and dopaminergic neurons in the central nervous system, in sympathetic neurons in the autonomic nervous system, and in the adrenal medulla. Norepinephrine biosynthesis takes place by a three-step process beginning with the conversion of tyrosine to L-dihydroxyphenylalanine (L-DOPA) (Fig. 11-2). The tyrosine hydroxylase-catalyzed formation of DOPA takes place in the cytoplasm of the neuron and is the rate-limiting process in catecholamine biosynthesis. End-product inhibition of the tyrosine hydroxylase reaction is believed to be a key mechanism for the regulation of catecholamine biosynthesis, because norepinephrine reduces tyrosine hydroxylase activity markedly.

The second enzymatic process in NE biosynthesis is the decarboxylation of L-DOPA to produce dopamine. This reaction is catalyzed by L-aromatic amino acid decarboxylase (DOPA decarboxylase), an enzyme that resides in the cell cytoplasm but which, unlike tyrosine hydroxylase, exhibits broad substrate specificity. Thus, a variety of aromatic amino acids that have the L-configuration may serve as substrates for this en-

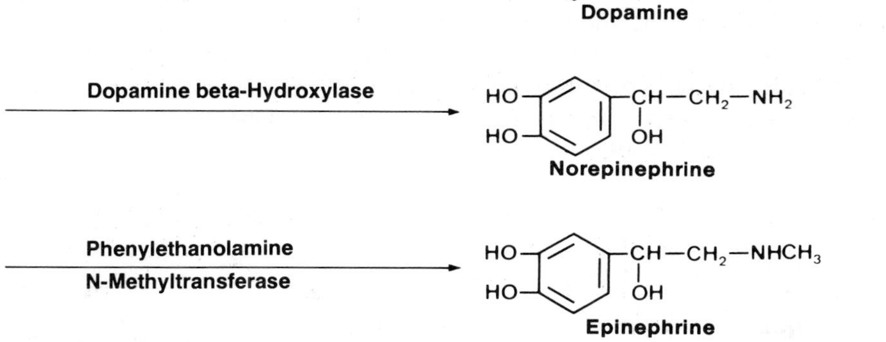

FIG. 11-2. *Catecholamine biosynthesis.*

zyme. DOPA-decarboxylase exhibits a high rate of activity compared to tyrosine hydroxylase.

The dopamine that is formed in the cytoplasm of the neuron is transported by an uptake process into the storage vesicles where it is stereospecifically hydroxylated by another enzyme of rather wide substrate selectivity, dopamine β-hydroxylase. The NE thus formed is stored in the vesicles until it is released into the synaptic cleft to interact with the adrenergic receptors.

In the adrenal medulla, a fourth biosynthetic reaction takes place by which NE is converted to epinephrine. This reaction is catalyzed by phenylethanolamine-N-methyltransferase (PNMT), a cytoplasmic enzyme that is specific for phenylethanolamine derivatives. PNMT is highly localized in the adrenal medulla, but small amounts are present in other tissues such as heart and brain. The epinephrine is held in the storage granules of the chromaffin cells in the adrenal medulla.

There are two drugs in clinical use that were designed to produce their therapeutic effects by inhibition of norepinephrine and epinephrine biosynthesis. The first of these, methyldopa (α-methyldopa, L-α-methyl-3,4-dihydroxyphenylalanine, Aldomet®) is an inhibitor of DOPA decarboxylase and is used as an antihypertensive agent. Although methyldopa does inhibit this enzyme, apparently by serving as an alternative substrate, its mechanism of antihypertensive activity is rather complex. (This drug is discussed in Chap. 14.)

The second catecholamine biosynthesis inhibitor in clinical use is metyrosine (α-methyl-p-tyrosine, Demser®). Metyrosine differs structurally from tyrosine only by the presence of the α-methyl group. This is the same structural feature that was incorporated into DOPA to produce methyldopa. However, because methyltyrosine inhibits the first, and rate-limiting, step in catecholamine biosynthesis, it is a much more effective inhibitor of epinephrine and NE production than is methyldopa. Metyrosine has been approved for use in the United States in patients with pheochromocytoma. This condition involves the presence of chromaffin-cell tumors that produce large amounts of NE and epinephrine. Although these tumors, which can occur in the adrenal medulla, are often benign, patients frequently suffer hypertensive episodes. Metyrosine reduces the frequency and severity of these episodes by lowering catecholamine production.

Metyrosine, which is given orally in doses as high as 2 to 3 g. per day, is particularly useful for preoperative management of pheochromocytoma.

METABOLISM

As discussed earlier, removal of NE from the synaptic cleft by uptake into the sympathetic neuron is believed to be the primary mode of termination of its action on the effector organ receptors. This is in contrast to the mechanism of termination of the cholinergic activity of acetylcholine, which involves metabolism of the neurotransmitter by acetylcholinesterase (Chap. 12). However, both NE and epinephrine are extensively metabolized prior to excretion. Little NE or epinephrine is excreted unchanged under ordinary circumstances.

The two principal enzymes involved in catecholamine metabolism are monoamine oxidase (MAO) and catechol 0-methyltransferase (COMT). Both of these enzymes are widely distributed throughout the body. MAO is associated primarily with the outer membrane of mitochondria, and COMT is found in the cellular cytoplasm. It appears that COMT is not present in sympathetic neurons while the neuronal mitochondria do contain MAO. MAO has a role in the metabolism of intraneuronal catecholamines. COMT acts primarily upon catecholamines that enter the circulation and the extraneuronal tissues after being released from nerves or from the adrenal medulla, or after being administered exogenously.

Neither COMT nor MAO exhibits a high degree of substrate specificity. MAO oxidatively deaminates a variety of compounds that contain an amino group attached to a terminal carbon. In addition, mammalian tissues contain more than one type of MAO, and the various types of MAO exhibit different substrate selectivity.[5] Similarly, COMT catalyzes the methylation of a variety of catechol-containing molecules. The lack of substrate specificity of COMT and MAO is manifested in the metabolic disposition of NE and epinephrine (Fig. 11-3). MAO and COMT not only use NE and epinephrine as substrates but also they each can act upon the metabolites produced by the other.

As illustrated in Figure 11-3, both NE and epinephrine are oxidized by MAO to form 3,4-dihydroxymandelic acid. Although the initial product formed in this reaction is an aldehyde, the aldehyde is so rapidly oxi-

Methyldopa

Metyrosine

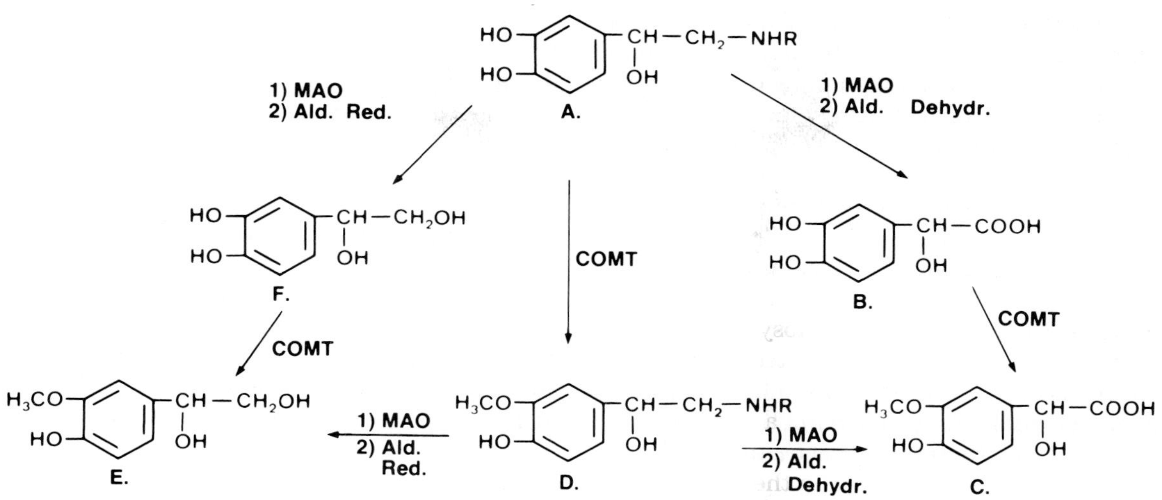

FIG. 11-3. *Norepinephrine and epinephrine metabolism. A. R = H = norepinephrine; R = CH₃ = epinephrine. B. 3,4-Dihydroxymandelic acid. C. 3-Methoxy-4-hydroxymandelic acid. D. R = H = normetanephrine; R = CH₃ = metanephrine. E. 3-Methoxy-4-hydroxyphenylglycol. F. 3,4-Dihydroxyphenylglycol. MAO = monoamine oxidase; COMT = catechol 0-methyltransferase; Ald. Dehydr. = aldehyde dehydrogenase; Ald. Red. = aldehyde reductase.*

dized in most tissues that the carboxylic acid is the main product. However, in the human central nervous system, as well as (to a limited extent) in certain peripheral tissues, substantial quantities of the aldehyde are reduced to the primary alcohol, resulting in the excretion of 3-methoxy-4-hydroxyphenylglycol (Fig. 11-3).

In vivo, COMT methylates almost exclusively the *meta*-hydroxyl group of catechols, regardless of whether the catechol is NE, epinephrine, or one of the oxidation products formed by MAO. Therefore, a converging pattern of metabolism of NE and epinephrine occurs in which 3-methoxy-4-hydroxymandelic acid (VMA) and 3-methoxy-4-hydroxyphenylglycol are common end products regardless of whether the compound metabolized is NE or epinephrine and regardless of whether the first metabolic step is oxidation by MAO or methylation by COMT.

Under normal circumstances, 3-methoxy-4-hydroxymandelic acid is the principal urinary metabolite of NE or epinephrine. However, varying amounts of the other metabolites are also found in the urine, both in the free forms and as sulfate or glucuronide conjugates.

ADRENERGIC RECEPTORS

In recent years a great deal of drug research activity has focused upon the development of selective adrenergic agonists and antagonists. This high level of effort and the progress that has stemmed from it have,

in large part, been the result of the acceptance of Ahlquist's proposal that there exists more than one type of adrenergic receptor.[6] Ahlquist proposed that two general types of adrenergic receptors exist in mammalian tissues and he designated them α and β. The postsynaptic α-adrenergic receptors are involved primarily in contraction of smooth muscle while β-receptors are associated with the relaxation of smooth muscle and the stimulation of cardiac muscle. There are also α and β receptors that mediate the effects of catecholamines on carbohydrate and lipid metabolism. Both α and β receptors mediate the relaxation of intestinal smooth muscle.

The key factor in the classification of adrenergic receptors is the grouping together of certain receptors on the basis of the order of potency of agonists that activate them. A second important factor that supports the concept of multiple types of adrenergic receptors is the existence of selective antagonists. Careful analysis of the potency order of agonists and the selectivity of antagonists has led to the further subclassification of both α and β receptors into groupings that have been designated α_1, α_2, β_1, and β_2.

ALPHA-ADRENERGIC RECEPTORS

For a number of years, the discussion and definition of α-adrenergic receptors were considered a rather simple matter. Alpha receptors were designated as those postsynaptic receptors that exhibited the following order of agonist potency: epinephrine > norepinephrine > > isoproterenol. Additionally, α-adrenergic receptors are blocked by antagonists such as phentolamine and phenoxybenzamine. This definition is a most useful one for characterizing such prominent α-receptors as those associated with contraction of vascular smooth muscle and contraction of the radial muscle of the iris. The physiological and therapeutic importance of these types of α-receptors, now designated as "α_1," is apparent.

In recent years, a second subclass of α-adrenergic receptors, known as α_2, has been discovered.[7,8] The α_2-receptors that have been studied to date are all inhibitory receptors in terms of their functions, but there is no reason to believe that certain excitatory α-receptors will not eventually be included in the α_2 classification. A typical representative of α_2-receptors is the presynaptic α-receptor found on the terminal of the sympathetic neuron (see Fig. 11-1). Interaction of this receptor with an agonist such as NE results in inhibition of norepinephrine release. Thus, α_2-receptors may play a role in the regulation of NE release. Although this α_2-receptor is presynaptic, others are postsynaptic and some are associated with nonneural tissue. The potency order of agonists and the binding affinity of antagonists for α_2-receptors are different than for the α_1-type.[7,8] Although α_2-receptors may be involved in the antihypertensive action of drugs such as clonidine, the extent to which α_2-receptors are involved in the therapeutic actions of adrenergic drugs or in the production of their side-effects is not known.

Epinephrine, NE and isoproterenol produce metabolic responses such as glycogenolysis and lipolysis. Recent studies have indicated that both α- and β-receptors may play a role in the mediation of the catecholamine-induced glycogenolytic response in the liver.[9] This may, in part, account for the difficulty encountered by investigators who, in the past, attempted to classify the metabolic adrenergic receptors as α- or β-.

An important question that has not been resolved concerns the mechanism, or mechanisms, by which activation of α-adrenergic receptors produces the tissue responses with which they are associated. There is, however, considerable evidence that Ca^{++} ions play a role in both the smooth-muscle effects and in the metabolic effects mediated by α-receptors. Alpha receptor stimulation results in increased Ca^{++} uptake by liver cells as well as in enhanced transmembrane Ca^{++} fluxes in other tissues. It is believed that the resulting rise in cellular Ca^{++} concentrations results in stimulation or inhibition of enzymes that participate in the mediation of the α-adrenergic response.

Research in the solubilization and purification of receptors has led to an enhanced understanding of the molecular mechanisms of action of several classes of therapeutically important drugs. However, in contrast to the large amount of progress that has been made in the characterization of β-adrenergic receptors, relatively little has been accomplished with α-receptors. At the time of writing this chapter there appears to be only a single report that describes the successful solubilization of an α-receptor.[10]

BETA-ADRENERGIC RECEPTORS

In 1967, almost 20 years after Ahlquist's landmark paper proposed the existence of α- and β-adrenergic receptors, Lands and co-workers suggested that β-receptors could be designated as β_1- and β_2-subtypes.[11] Beta$_1$ receptors are those that exhibit the agonist potency order: isoproterenol > epinephrine = NE. The agonist potency order for β_2-receptors is as follows: isoproterenol > epinephrine > NE. Some researchers have proposed further subclassifications of β-receptors, but until more definitive evidence is developed, the distinction between β_1- and β_2-subtypes appears to be the most useful approach to classification. Further evidence for the existence of more than one type of β-receptor is derived from recent research on selective β_1- and β_2-receptor antagonists. This topic is discussed later in this chapter.

Cardiac stimulation and lipolysis are typical physiological effects associated with β_1-receptors; bronchodilatation and vasodilatation are primarily β_2-responses. The development of agonists and antagonists that selectively modulate these β-receptor controlled responses has led to potentially important advances in drug therapy that will be discussed below.

In contrast to the paucity of information regarding the biochemical mechanisms by which α-receptors produce their effects, an abundance of data indicates that the β-adrenergic effects of catecholamines and other β-receptor agonists are mediated by stimulation of the enzyme adenylate cyclase. This enzyme catalyzes the conversion of adenosine triphosphate (ATP) to cyclic 3',5'-adenosine monophosphate (cyclic AMP). Cyclic AMP is released from the membrane-bound enzyme into the cell where it functions as a "second messenger." According to this concept, the β-adrenergic agonist acts as the "first messenger" in that it carries information to the cell by interacting with β-receptors on the cell membrane. The receptor–agonist interaction results in stimulation of adenylate cyclase and production of cyclic AMP, which acts as a "second messenger" by carrying the information to intracellular sites.[12] It is now known that a variety of hormones and drugs exert their effects by stimulating or depressing adenylate cyclase activity subsequent to their interaction with membrane receptors. The intracellular function of cyclic AMP, the second messenger, appears to be the activation of a group of enzymes called *protein kinases* which phosphorylate specific proteins.

Thus, the phosphorylated proteins mediate the actions of cyclic AMP. Cyclic AMP functions as the mediator of the action of the drug or neurotransmitter that originally interacted with the β-adrenergic receptor.[13] The action of cyclic AMP is terminated by a class of enzymes known as phosphodiesterases. Although cyclic AMP is not the only substance that functions as a second messenger, it is the one that has been studied most extensively, and it appears to be the one involved in mediation of β-adrenergic responses. It should also be noted that, in certain instances, a β-adrenergic response may take place without a measurable increase in cyclic AMP levels. The possible implications of such apparent exceptions to the general relationship between β-adrenergic and adenylate-cyclase activities have been discussed by Kunos.[14]

At one time it was proposed that adenylate cyclase itself was the β-adrenergic receptor. According to this concept, β-adrenergic agonists and antagonists interact directly with the catalytic site of adenylate cyclase to either activate or inhibit the enzyme.[15] However, it is now well established that the β-adrenergic receptor site that binds β-agonists and antagonists resides on a portion of the cell membrane that is separate from adenylate cyclase. The β-adrenergic receptor and adenylate cyclase are separate, membrane-bound macromolecules that appear to interact through a "coupling system" involving membrane lipids.[16,17,18] In addition, guanyl nucleotides are involved in modulation of the binding of agonists and antagonists to the receptor.[19]

A full understanding of the molecular characteristics of biological receptors requires their purification in a way that allows them to retain their ability to bind in a highly specific fashion to agonists and antagonists. In the case of membrane-bound macromolecules, such as the adrenergic receptor proteins, this is a formidable task. However, a great deal of progress has been made in the purification of β-adrenergic receptors. It is reasonable to expect that information derived from studying these purified receptors will contribute both to the development of more selective and less toxic adrenergic drugs and to a more comprehensive understanding of the molecular mechanism of drug action.[20]

SYMPATHOMIMETIC AGENTS

MECHANISM OF ACTION AND STRUCTURE–ACTIVITY CONSIDERATIONS

In terms of mechanism of action, sympathomimetic agents may be classified as producing their effects by direct, indirect, or mixed mechanisms. Direct-acting agents elicit a sympathomimetic response by interacting directly with adrenergic receptors. Indirect-acting agents produce their effects primarily by causing the release of NE from adrenergic nerve terminals; the norepinephrine that is released by the indirect-acting agent then activates the receptors to produce the response. Those compounds with a mixed mechanism of action both interact directly with adrenergic receptors and cause the release of NE. Most sympathomimetics appear to have direct and indirect actions, although one mechanism or the other often is predominant. Unfortunately, complete mechanistic data are not available for many compounds.

Direct-acting Sympathomimetics

The mechanism by which an agent produces a sympathomimetic effect is, in most cases, intimately related to its chemical structure. For example, the prototypical direct-acting compounds are NE, epinephrine, and isoproterenol. Each of these three substances is a catecholamine. More fundamentally, they are phenylethylamine derivatives that contain the appropriate substituents to impart direct receptor-activating capabilities. These "appropriate" substituents are the catechol hydroxyl groups in the *meta*- and *para*-positions of the aromatic ring and the β-hydroxyl group on the ethylamine portion of the molecule. It was previously believed that, as a general rule, a potent, direct-acting phenylethylamine sym-

Phenylethylamine

pathomimetic agent should contain at least the *meta*- and *para*-hydroxyl groups or, alternatively, the *meta*-hydroxyl and the β-hydroxyl. However, exceptions to the generalizations can readily be found. For example, it is well established that β-receptor agonist activity is often retained in phenylethanolamines when certain groups are substituted for the *meta*-hydroxyl, as will be discussed later in this chapter.[21,22]

The presence of the amino group in phenylethylamines is important for direct agonist activity. Both primary and secondary amines are found among the potent direct-acting agonists, but tertiary amines tend to be poor direct agonists. The amino group should be separated from the aromatic ring by two carbon atoms for optimal activity.

Direct receptor-agonist activity is enhanced by the presence of a hydroxyl group, of the correct stereochemical configuration, on the β-carbon, but is reduced by the presence of a methyl group on the α-carbon. However, it is important to note that an α-

methyl group increases the duration of action of the phenylethylamine agonist by making the compound resistant to metabolic deamination by monoamine oxidase. Such compounds often exhibit enhanced oral effectiveness and greater central nervous system activity than their counterparts that do not contain an α-methyl group.

A highly critical factor in the interaction of adrenergic agonists with their receptors is that of stereoselectivity. Those direct-acting sympathomimetics that exhibit chirality by virtue of the presence of a β-hydroxyl group (phenylethanolamines) invariably exhibit a high degree of stereoselectivity in producing their agonist effects. That is, one enantiomeric form of the drug has greater affinity for the receptor than the other. This is true for both α- and β-agonists. For epinephrine, NE, and related compounds, the more potent enantiomer has the R(-), or D(-), configuration. It appears that for all direct-acting agonists and antagonists that are structurally similar to NE, the more

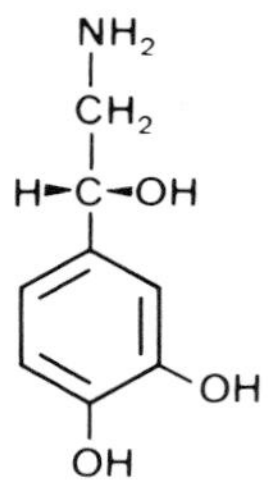

R (-) Norepinephrine

potent enantiomer is capable of assuming a conformation that results in the arrangement in space of the aromatic group, the amino group, and the β-hydroxyl group in a fashion resembling that of R(-) NE. This explanation of stereoselectivity is based upon the presumed interaction of these three critical pharmacophoric groups with three complementary binding areas on the receptor.

Indirect-acting Sympathomimetics

Certain structural characteristics tend to impart indirect sympathomimetic activity to phenylethylamines. As in the case of direct-acting agents, the presence of the catechol hydroxyls enhances the potency of indirect-acting phenethylamines. In contrast to the direct-acting agents, the presence of a β-hydroxyl group decreases, and an α-methyl group increases, the effectiveness of indirect-acting agents. The presence of nitrogen substituents decreases indirect activity, with substituents larger than methyl rendering the compound virtually inactive. Phenyl-

ethylamines that contain a tertiary amino group are also ineffective as NE-releasing agents. Given the above structure–activity considerations, it is easy to understand why amphetamine and p-tyramine are often cited as the prototype indirect-acting sympathomimetics.

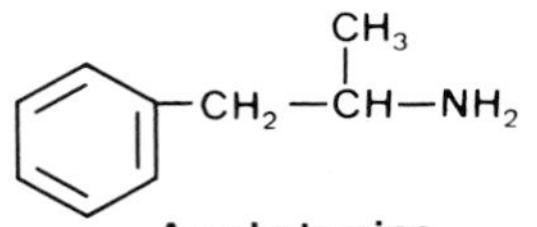

Amphetamine

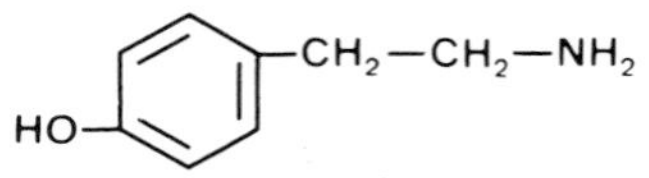

p-Tyramine

Although p-tyramine is not a clinically useful agent, its α-methylated derivative, hydroxyamphetamine, is an effective, indirect-acting sympathomimetic drug. Amphetamine-type drugs are discussed in more detail in the chapter on central nervous system stimulants (Chap. 10).

Sympathomimetic Agents with a Mixed Mechanism of Action

Those phenylethylamines that are considered to have a mixed mechanism of action usually have no hydroxyls on the aromatic ring but usually do have a β-hydroxyl group. Thus, D(-) ephedrine, which is discussed later in this chapter, is the classic example of a sympathomimetic with a mixed mechanism of action.

α- and β-Receptor Agonists

If a phenylethylamine has the structural prerequisites that cause it to act primarily by a direct-receptor activation mechanism, then it is the nature of the nitrogen substituent that determines whether it will act primarily at α- or β-receptors. In general, as the bulk of the nitrogen substituent increases, α-receptor agonist activity decreases and β-receptor activity increases. Thus, isoproterenol is a potent β-receptor agonist, but has little affinity for α-receptors. However, the converse is not true; NE, which is an effective β_1-receptor agonist, is also a potent α-agonist. Epinephrine is a potent agonist at α-, β_1-, and β_2-receptors.

The fact that isoproterenol activates β-receptors and not α-receptors has resulted in its being widely used for the management of bronchial asthma. Isoproterenol is an effective bronchodilator but it has several deficiencies as a therapeutic agent. It stimulates both β_1- and β_2-receptors. The β_1-component of its action imparts to it an undesirable cardiac stimulatory effect. After oral administration, its aborption is rather erratic and undependable, and it has a duration of action of only a few minutes, regardless of the route of administration. The principal reason for its poor absorption characteristics and short duration of action is its facile metabolic transformation by sulfate and glucuronide conjugation of the ring hydroxyls, and methylation by COMT. Unlike epinephrine and NE, isoproterenol does not appear to undergo oxidative deamination by MAO.

The problems of lack of β-receptor selectivity and rapid metabolic inactivation associated with isoproterenol have been at least partially overcome by the design and development of selective β_2-adrenoceptor stimulants. In several instances it has been determined that an N-*tert*-butyl group enhances β_2-selectivity. For example, N-*tert*-butyl NE, which is not in clinical use, is 9 to 10 times as potent an agonist at tracheal β_2-receptors than at cardiac β_1-receptors.

Isoproterenol

N-*tert*-Butylnorepinephrine

Several modifications of the catechol portion of β-agonists have been useful in the development of selective drugs. The resorcinol structure has served as a successful replacement for catechol. Resorcinol is not a substrate for COMT. Therefore, β-agonists that contain this ring structure tend to have better absorption characteristics and a longer duration of action than their catechol-containing counterparts. Terbutaline is an example of a drug that contains both the N-*tert*-butyl substituent and the resorcinol structure. Unfortunately, terbutaline does not have as much β_2-adrenoceptor selectivity as is desired for optimal therapeutic use.

As mentioned earlier, a variety of functional groups can replace the *meta*-hydroxyl of the catechol structure in β_2-agonists. Advantage of this fact was taken in the design of salbutamol, a β_2-selective agent that is

not yet available for general clinical use in the United States. Salbutamol, which is a saligenin derivative, is β_2-receptor selective, orally effective, and has an acceptable duration of action. It is not metabolized by COMT.

Resorcinol   **Terbutaline**

Salbutamol

Another molecular modification that enhances β_2-agonist selectivity is the presence of an ethyl group on the carbon adjacent to the nitrogen. For example, the α-ethyl derivative of isoproterenol, which is known as isoetharine, is a β_2-selective agonist used in the treatment of bronchial asthma.

Isoetharine

Although numerous clinical studies have shown that β_2-agonists are useful in the management of bronchial asthma and that they produce less cardiac stimulation than nonselective agents, they should not be used indiscriminately or excessively. In sufficient doses, they can produce cardiovascular effects. Also, tachyphylaxis may develop in patients who use β_2-adrenergic agonists, resulting in diminished responsiveness to the drugs. Finally, an annoying but apparently harmless problem is the tremor that occurs as the result of the stimulation of β_2-receptors in skeletal muscle.

Extensive structure–activity investigations have shown that large N-substituents other than isopropyl and *tert*-butyl are useful and effective in the development of both β_1- and β_2-adrenoreceptor agonists. This fact, combined with the knowledge that dopamine has a direct stimulant effect on cardiac β-receptors, provides at least a partial explanation for the effectiveness of dobutamine as an adrenergic β_1-receptor agonist. Dobutamine, like dopamine, lacks a β-hydroxyl group on the side chain.

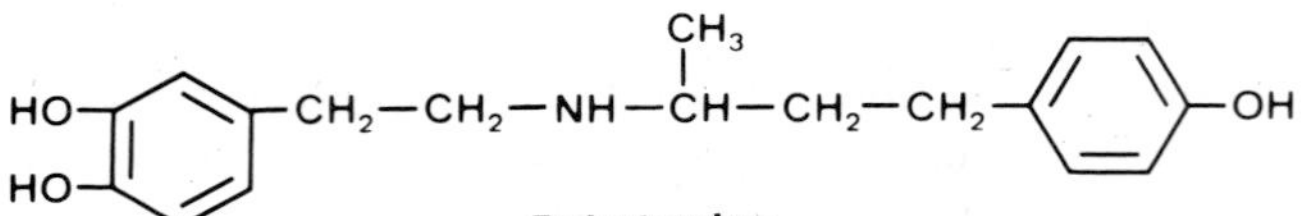

Dobutamine

PRODUCTS

Catecholamines, Phenylethanolamines, Phenylethylamines, and Related Agents

Epinephrine U.S.P., Adrenalin®, (−)-3,4-dihydroxy-α-[(methylamino)methyl]benzyl alcohol. This compound is a white, odorless, crystalline substance which is light-sensitive. In the official product norepinephrine is present. Initially, epinephrine was isolated from the medulla of the adrenal glands of animals used for food. Although synthetic epinephrine became available soon after the structure of the hormone had been elucidated, the synthetic (±)-base has not been used widely in medicine because the natural levorotatory form is about 15 times as active as the racemic mixture.

Because of its catechol nucleus, it is oxidized easily and darkens slowly on exposure to air. Dilute solutions are partially stabilized by the addition of chlorobutanol and by reducing agents, e.g., sodium bisulfite or ascorbic acid. As the free amine, it is available in oil solution for intramuscular injection and in aqueous solution for inhalation. Like other amines, it forms salts with acids; for example, those now used include the hydrochloride, the borate and the bitartrate. The bitartrate has the advantage of being less acid and, therefore, is used in the eye because its solutions have a pH close to that of lacrimal fluid. Epinephrine is destroyed readily in alkaline solutions, by aldehydes, weak oxidizing agents and oxygen of the air.

Although an intravenous infusion of epinephrine has pronounced effects on the cardiovascular system, its use in the treatment of heart block or circulatory collapse is limited because of its tendency to induce cardiac arrhythmias. It increases systolic pressure by increasing cardiac output, and it lowers diastolic pressure by causing an overall decrease in peripheral resistance; the net result is little change in mean blood pressure.

It is sometimes useful in the treatment of glaucoma, because it apparently reduces the formation of aqueous humor, which results in a lowering of intraocular pressure.

Local application is limited but is of value as a constrictor in hemorrhage or nasal congestion. One of its major uses is to enhance the activity of local anesthetics. It is used by injection to relax the bronchial mus-

cle in asthma and in anaphylactic reactions. Forms of administration are aqueous or oil solutions, ointment, suppositories, and inhalation.

Epinephrine has the following disadvantages: short duration of action; decomposition of its salts in solution; vasoconstrictive action frequently followed by vasodilation; and inactivity on oral administration.

Epinephryl Borate, Ophthalmic Solution U.S.P. Epinephrine forms a soluble epinephryl borate complex at a neutral or slightly alkaline pH. The buffered solution has a pH of about 7.4 and the complex probably has the following structure:

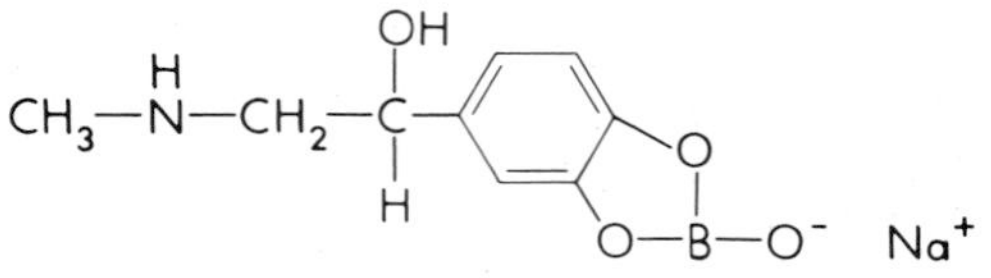

Epinephryl Borate

It is used like other epinephrine preparations by topical application in the treatment of primary open-angle glaucoma. It possesses the same limitations as the other preparations but it is claimed to cause less stinging upon application. In the lacrimal fluid it immediately dissociates to yield free epinephrine.

Norepinephrine Bitartrate U.S.P., Levophed® Bitartrate, (−)-α-(aminomethyl)-3,4-dihydroxybenzyl alcohol bitartrate, (−)norepinephrine bitartrate. Norepinephrine differs from epinephrine in that it is a primary amine rather than a secondary amine.

The bitartrate is a white, crystalline powder which is soluble in water (1:2.5) and in alcohol (1:300). Solutions of the hydrochloride of norepinephrine are comparable with those of epinephrine hydrochloride with regard to stability. The bitartrate salt is available as a more stable injectable solution. It has a pH of 3 to 4 and is preserved by using sodium bisulfite. It is used to maintain blood pressure in acute hypotensive states resulting from surgical or nonsurgical trauma, central vasomotor depression and hemorrhage.

Isoproterenol Hydrochloride U.S.P., Aludrine® Hydrochloride, Isuprel® Hydrochloride, 3,4,-dihydroxy-α-[(isopropylamino)methyl]benzyl alcohol hydrochloride, isopropylarterenol hydrochloride, isoproterenolium chloride.

This compound is a white, odorless, slightly bitter, crystalline powder. It is soluble in water (1:3) and in alcohol (1:50). A 1 percent solution in water is slightly acidic (pH 4.5 to 5.5). It gradually darkens on exposure to air and light. Its aqueous solutions become pink on standing.

Isoproterenol is a potent β-adrenergic agonist that has virtually no effect on α-receptors. Because it is not selective for either β_1- or β_2-receptors, it causes an increase in cardiac output by stimulating cardiac β_1-receptors and brings about bronchodilatation by stimulating β_2-receptors in the respiratory tract. It also produces the metabolic effects expected of a potent β-agonist.

It is available for use by inhalation, or injection, in liquid form, and as sublingual tablets. Its principal clinical use is for the relief of bronchospasm associated with bronchial asthma. Cardiac stimulation is an undesirable and occasionally dangerous side-effect. On the other hand, advantage is sometimes taken of isoproterenol's effect on the heart by using it for the treatment of heart block.

Isoproterenol Sulfate U.S.P., Medihaler-Iso®, Norisodrine® Sulfate, 3,4-dihydroxy-α-[(isopropylamino)methyl]benzyl alcohol sulfate, 1-(3',4'-dihydroxyphenyl)-2-isopropylaminoethanol sulfate. This compound is a white, odorless, slightly bitter crystalline powder. It is slightly soluble in alcohol and freely soluble in water and is hygroscopic. A 1 percent solution in water is acidic (pH 3.5 to 4.5). Its aqueous solutions become pink on standing. It is used for the same purpose as is the corresponding hydrochloride.

Phenylephrine Hydrochloride U.S.P., Neo-Synephrine® Hydrochloride, Isophrin® Hydrochloride, ($-$)-m-hydroxy-α-[(methylamino)methyl]benzyl alcohol hydrochloride, phenylephrinium chloride. This compound is a white, odorless, crystalline, slightly bitter powder which is freely soluble in water and in alcohol. It is relatively stable in alkaline solution and is unharmed by boiling for sterilization.

The duration of action is about twice that of epinephrine. It is a vasoconstrictor and is active when given orally. It is relatively nontoxic and, when applied to mucous membrane, reduces congestion and swelling by constricting the blood vessels of the mucous membranes. It has little central nervous stimulation and finds its main use in the relief of nasal congestion. It is also used as a mydriatic agent, as an agent to prolong the action of local anesthetics and to prevent a drop in blood pressure during spinal anesthesia. Phenylephrine is a direct-acting, α-receptor agonist.

Metaproterenol, Alupent®, Metaprel®, 3,5-dihydroxy-α-[(isopropylamino)methyl]benzyl alcohol. Metaproterenol is the resorcinol counterpart of isoproterenol. It is used as the sulfate salt and, like isoproterenol, as the racemic mixture. It is more effective when given orally, and has a longer duration of action than isoproterenol. Metaproterenol does not exhibit significant β_2-adrenoreceptor selectivity; therefore, it produces cardiovascular effects similar to those caused by isoproterenol. It is not metabolized by COMT, its principal metabolite being the glucuronide conjugate. Metaproterenol is used in tablet, syrup, and inhalation forms for the management of bronchial asthma.

Metaproterenol

Terbutaline Sulfate U.S.P., Bricanyl®, Brethine®, 1-(3,5-dihydroxyphenyl)-2-*tert*-butylaminoethanol sulfate. This drug has been introduced with the implication that it acts preferentially at β_2-receptor sites, thus making it useful in the treatment of bronchial asthma and related conditions. However, the common cardiovascular effects that are associated with other adrenergic agents are also seen in the use of terbutaline sulfate. The drug is administered orally and is not metabolized by catechol-O-methyltransferase. Evidence indicates that it is excreted primarily as a conjugate.

Terbutaline Sulfate

Dopamine Hydrochloride, Intropin®. Dopamine is the precursor in the biosynthesis of norepinephrine. It is used in the treatment of shock and, in contrast to the usual catecholamines, increases blood flow to the kidney in doses that have no chronotropic effect on the heart or cause no increase in blood pressure.

Dopamine Hydrochloride

The increased blood flow to the kidneys enhances the glomerular filtration rate, Na^+ excretion and, in turn, urinary output. This is accomplished with doses of 1 to 2 μg. per kg. per minute. The infusion is made using solutions that are neutral or slightly acidic in reaction.

In doses slightly higher than those required to increase renal blood flow, dopamine stimulates the β-receptors of the heart to increase cardiac output. Some of dopamine's effect on the heart is due to NE release. It is known that the dilatation of renal blood vessels produced by dopamine is the result of its agonist action on a specific dopaminergic receptor, rather than on β-receptors. Infusion at a rate greater than 10 μg. per kg. per minute results in a stimulation of α-receptors leading to vasoconstriction and an increase in arterial blood pressure. See Table 11-1 for a summary of use and dosage information.

Ephedrine U.S.P., (−)-*erythro*-α-[(1-methylamino)ethyl]benzyl alcohol. Ephedrine is an alkaloid which can be obtained from the stems of various species of *Ephedra*. The drug Ma Huang, containing ephedrine, was known to the Chinese in 2800 B.C., but the active principle, ephedrine, was not isolated until 1885.

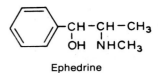

Ephedrine

Ephedrine has 2 asymmetric carbon atoms; thus there are 4 optically active forms. The *erythro* racemate is called ephedrine, and the *threo* racemate is known as *pseudo*ephedrine (ψ-ephedrine). Natural

TABLE 11–1

CATECHOLAMINES, PHENYLETHANOLAMINES, PHENYLETHYLAMINES, AND RELATED AGENTS

Name Proprietary Name	Preparations	Category	Application	Usual Adult Dose*	Usual Dose Range*	Pediatric Dose*
Epinephrine U.S.P.	Epinephrine Inhalation U.S.P.	Adrenergic (bronchodilator)		Oral inhalation, the equivalent of a 1 percent solution of epinephrine applied as a fine mist as required.		Children up to 6 years of age—dosage must be individualized by physician
	Epinephrine Injection U.S.P.	Adrenergic		I.M. or S.C., the equivalent of 200 to 500 μg. of epinephrine, repeated as necessary	I.M. or S.C., 100 μg. to 1 mg., I.V., 25 to 400 μg.	S.C., 10 μg. per kg. of body weight or 300 μg. per square meter of body surface, up to 500 μg. per dose, repeated as necessary up to 6 times daily
	Epinephrine Nasal Solution U.S.P.	Adrenergic (nasal)				
	Sterile Epinephrine Oil Suspension U.S.P.	Adrenergic (bronchodilator)		S.C., 500 μg. initially, then 500 μg. to 1.5 mg. not more often than every 6 hours, as needed	400 μg. to 6 mg. daily	S.C., 25 μg. per kg. of body weight or 625 μg. per square meter of body surface, repeat not more often than every 6 hours
	Epinephrine, Ophthalmic Solution U.S.P.					
Epinephrine Bitartrate U.S.P. *Asmatane, Medihaler-Epi, Epitrate, Lyophrin*	Epinephrine Bitartrate Inhalation Aerosol U.S.P.		Oral inhalation, the equivalent of 160 μg. of epinephrine			
Epinephryl Borate	Epinephryl Borate Ophthalmic Solution U.S.P.	Adrenergic (ophthalmic)	1 to 2 drops in each eye as directed. The frequency of instillation should be titrated tonometrically to the individual response of each patient			
Norepinephrine Bitartrate U.S.P. *Norepinephrine Bitartrate*	Norepinephrine Bitartrate Injection U.S.P.	Adrenergic (vasopressor)		I.V. infusion, the equivalent of 4 mg. of levarterenol, in 1000 ml. of 5 percent Dextrose Injection or 5 percent Dextrose and Sodium Chloride Injection, at a rate adjusted to maintain blood pressure at the desired level		2 g. per square meter of body surface per minute

*See *U.S.P. D.I.* for complete dosage information.

TABLE 11-1

CATECHOLAMINES, PHENYLETHANOLAMINES, PHENYLETHYLAMINES, AND RELATED AGENTS *(Continued)*

Name Proprietary Name	*Preparations*	*Category*	*Application*	*Usual Adult Dose**	*Usual Dose Range**	*Pediatric Dose**
Isoproterenol Hydrochloride U.S.P. *Isuprel Hydrochloride, Norisodrine, Aerotrol, Iprenol, Proternol*	Isoproterenol Hydrochloride Inhalation U.S.P.	Adrenergic (bronchodilator)		Oral inhalation, 125 to 250 µg. as a 0.5 to 1 percent solution repeated at 5- to 10-minute intervals as necessary, up to a maximum of 750 µg. per attack	Oral inhalation, 125 to 750 µg. per attack	
	Isoproterenol Hydrochloride Injection U.S.P.			I.V., 10 to 20 µg., repeated as necessary		
	Isoproterenol Hydrochloride Tablets U.S.P.			Sublingual, 10 to 15 mg. 3 times daily		Sublingual, 5 to 10 mg. 3 times daily
Isoproterenol Sulfate U.S.P. *Medihaler-Iso, Norisodrine Sulfate, Iso-Autohaler, Metermatic*	Isoproterenol Sulfate Aerosol N.F.	Adrenergic (bronchodilator)		Oral inhalation, 80 to 160 µg. in an aerosol, allowing at least 2 minutes to elapse between inhalations		
Phenylephrine Hydrochloride U.S.P. *Neo-Synephrine Hydrochloride, Isopto Frin*	Phenylephrine Hydrochloride Injection U.S.P.	Adrenergic (vasopressor)		I.M. or S.C., 2 to 5 mg., repeated in 1 or 2 hours as necessary; I.V., 200 µg. repeated in 10 to 15 minutes as necessary; I.V. infusion, 10 mg. in 500 ml. of Dextrose Injection or Sodium Chloride Injection, at a rate adjusted to maintain blood pressure at the desired level	I.M. or S.C., 1 to 10 mg.; I.V., 100 to 500 µg.; I.V. infusion, 10 to 20 mg. or more	I.M. or S.C., 100 µg. per kg. of body weight or 3 mg. per square meter of body surface
Alconefrin, Isohalent Improved, Isophrin, Synasal	Phenylephrine Hydrochloride Nasal Solution U.S.P.	Adrenergic (nasal)	Intranasal, 2 or 3 drops of a 0.25 to 0.5 percent solution in each nostril every 3 to 4 hours as necessary			2 or 3 drops of a 0.125 percent solution in each nostril every 3 to 4 hours as necessary
BufOpto, Efricel, Degest, Eye-Gene, Isopto Frin, Prefrin, Liquifilm, Tear-Efrin	Phenylephrine Hydrochloride Ophthalmic Solution U.S.P.	Adrenergic (ophthalmic)	Topically to the conjunctiva, 0.05 ml. of a 0.120 to 10 percent solution, repeated as necessary			
Metaproterenol *Alupent, Metaprel*	Metaproterenol Inhalation Metaproterenol Tablets	Adrenergic (bronchodilator)		20 mg. 3 or 4 times daily	Oral Inhalation, 650 µg. to 1.95 mg. every 3 to 4 hours	Children up to 12 years of age—use is not recommended
Terbutaline Sulfate *Brethine, Bricanyl*	Terbutaline Sulfate Tablets			5 mg. at approximately 6-hour intervals, 3 times daily		
Ephedrine Hydrochloride U.S.P.		Adrenergic (bronchodilator)				
Ephedrine Sulfate U.S.P. *Isofedrol*	Ephedrine Sulfate Capsules U.S.P.	Adrenergic (bronchodilator)		25 to 50 mg. every 3 or 4 hours, as necessary		500 µg. per kg. of body weight or 16.7 mg. per square meter of body surface, every 4 to 6 hours

* See *U.S.P. D.I.* for complete dosage information.

TABLE 11–1

CATECHOLAMINES, PHENYLETHANOLAMINES, PHENYLETHYLAMINES, AND RELATED AGENTS (Continued)

Name Proprietary Name	Preparations	Category	Application	Usual Adult Dose*	Usual Dose Range*	Pediatric Dose*
	Ephedrine Sulfate Injection U.S.P.	Adrenergic		I.M. or S.C., 25 to 50 mg., repeated if needed		I.V. or S.C., 500 µg. per kg. of body weight or 16.7 mg. per square meter of body surface, every 4 to 6 hours
	Ephedrine Sulfate Nasal Solution U.S.P.	Adrenergic (nasal)	Intranasal, 0.1 to 0.15 ml. of a 1 to 3 percent solution 2 or 3 times daily			
	Ephedrine Sulfate Syrup U.S.P. Ephedrine Sulfate Tablets U.S.P.	Adrenergic (bronchodilator)		25 to 50 mg. every 3 or 4 hours, as necessary	25 to 300 mg. daily	500 µg. per kg. of body weight or 16.7 mg. per square meter of body surface, every 4 to 6 hours
Pseudoephedrine Hydrochloride U.S.P. *Sudafed*	Pseudoephedrine Hydrochloride Syrup U.S.P. Pseudoephedrine Hydrochloride Tablets U.S.P.	Adrenergic		60 mg. every 4 hours	Not over 240 mg. in 24 hours	Oral, 4 mg. per kg. of body weight or 125 mg. per square meter of body surface per day, in 4 divided doses
Phenylpropanol-amine Hydrochloride U.S.P. *Propadrine Hydrochloride*		Adrenergic (vasoconstrictor)				
Mephentermine Sulfate U.S.P. *Wyamine Sulfate*	Mephentermine Sulfate Injection U.S.P. Mephentermine Sulfate Tablets U.S.P.	Adrenergic (vasopressor)		Oral, 12.5 to 25 mg. once or twice daily; I.M. or I.V., the equivalent of 15 to 30 mg. of mephentermine; infusion, 150 mg. in 500 ml. of an isotonic solution at a rate adjusted to maintain blood pressure	The equivalent of 12.5 to 80 mg. of mephentermine or mephentermine sulfate, repeated as necessary	
Metaraminol Bitartrate U.S.P. *Aramine Bitartrate*	Metaraminol Bitartrate Injection U.S.P.	Adrenergic (vasopressor)		I.M. or S.C., the equivalent of 2 to 10 mg. of metaraminol; I.V., the equivalent of 500 µg. to 5 mg. of metaraminol; I.V. infusion, the equivalent of 15 to 100 mg. of metaraminol in 500 ml. of 5 percent Dextrose Injection or Sodium Chloride Injection at a rate adjusted to maintain blood pressure at the desired level	I.V. infusion, 15 to 500 mg.	I.M. or S.C., 100 µg. per kg. of body weight or 3 mg. per square meter of body surface; I.V., 10 µg. per kg. or 300 µg. per square meter; I.V. infusion, 400 µg. per kg. or 12 mg. per square meter as a 0.004 percent solution at a rate adjusted to maintain blood pressure at the desired level
Hydroxyamphet-amine Hydrobromide U.S.P. *Paredrine Hydrobromide*	Hydroxyamphet-amine Hydrobromide Ophthalmic Solution U.S.P.	Adrenergic (ophthalmic)	Topical, to the conjunctiva, 100 µl. of a 0.25 to 1 percent solution, repeated as necessary			
Methoxamine Hydrochloride U.S.P. *Vasoxyl Hydrochloride*	Methoxamine Hydrochloride Injection U.S.P.	Adrenergic (vasopressor)		I.M., 10 to 15 mg.; I.V., 3 to 5 mg.	I.M., 5 to 20 mg.; I.V., 3 to 10 mg.	I.M., 250 µg. per kg. of body weight or 7.5 mg. per square meter of body surface; I.V., 80 µg. per kg. or 2.5 mg. per square meter, given slowly

*See U.S.P. D.I. for complete dosage information.

TABLE 11–1

CATECHOLAMINES, PHENYLETHANOLAMINES, PHENYLETHYLAMINES, AND RELATED AGENTS (Continued)

Name / Proprietary Name	Preparations	Category	Application	Usual Adult Dose*	Usual Dose Range*	Pediatric Dose*
Ethylnorepinephrine Hydrochloride U.S.P. *Bronkephrine*	Ethylnorepinephrine Hydrochloride Injection U.S.P.	Adrenergic (bronchodilator)	I.M. or S.C., 1 to 2 mg.		I.M. or S.C., 200 μg. to 1 mg.	
Methoxyphenamine Hydrochloride U.S.P. *Orthoxine Hydrochloride*		Adrenergic (bronchodilator)			50 to 100 mg. every 4 hours as necessary	
Nylidrin Hydrochloride U.S.P. *Arlidin*	Nylidrin Hydrochloride Injection U.S.P. Nylidrin Hydrochloride Tablets U.S.P.	Peripheral vasodilator		Oral, 6 mg. 3 times daily; I.M. or S.C., 5 mg. 1 or more times daily	Oral, 3 to 12 mg.; I.M. or S.C., 2.5 to 5 mg.	
Isoxsuprine Hydrochloride U.S.P. *Vasodilan*	Isoxsuprine Hydrochloride Injection U.S.P. Isoxsuprine Hydrochloride Tablets U.S.P.	Peripheral vasodilator		Oral, 10 to 20 mg. 3 or 4 times daily; I.M., 5 to 10 mg. 2 or 3 times daily		

* See *U.S.P. D.I.* for complete dosage information.

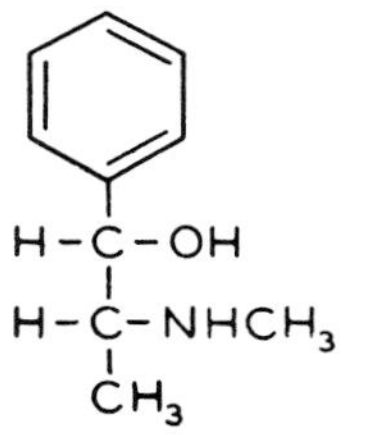

Erythro form Ephedrine — *Threo* form ψ-Ephedrine

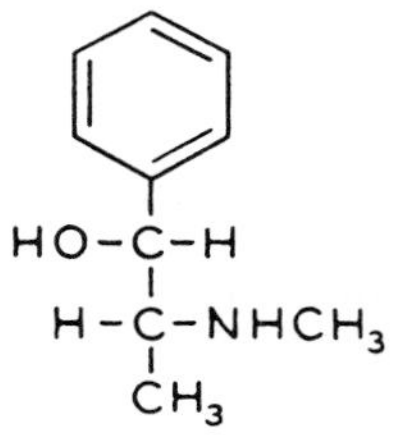

ephedrine is D(−) and is the most active of the 4 isomers as a pressor amine. Table 11–2 lists the relative pressor activity of isomers of ephedrine. Racemic ephedrine, racephedrine, is used for the same purpose as the optically active alkaloids.

The ephedrine alkaloid occurs as a waxy solid and as crystals or granules and has a characteristic pronounced odor. Because of its instability in light, it de-

TABLE 11–2

Isomer		Relative Pressor Activity
D	(−) Ephedrine	36
DL	(±) Ephedrine	26
L	(+) Ephedrine	11
L	(+) *Pseudo*ephedrine	7
DL	(±) *Pseudo*ephedrine	4
D	(−) *Pseudo*ephedrine	1

composes gradually and darkens. It may contain up to one-half molecule of water of hydration. It is soluble in alcohol, water (5%), some organic solvents and liquid petrolatum. The free alkaloid is a strong base and an aqueous solution of the free alkaloid has a pH above 10. The salt form has a pKa of 9.6.

Ephedrine simulates epinephrine in physiologic effects but its pressor action and local vasoconstrictor action are of greater duration. It causes more pronounced stimulation of the central nervous system than does epinephrine, and it is effective when given orally or systemically.

Ephedrine and its salts are used orally, intravenously, intramuscularly and topically in a variety of conditions such as allergic disorders, colds, hypotensive conditions and narcolepsy. It is employed locally to constrict the nasal mucosa and cause decongestion, to dilate the pupil or the bronchi and to diminish hyperemia. Systemically, it is effective for asthma, hay fever, urticaria, low blood pressure and the alleviation of muscle weakness in myasthenia gravis.

Pseudoephedrine Hydrochloride U.S.P., Sudafed®, (+)-*threo*-α-[(1-methylamino)ethyl]benzyl alcohol hydrochloride, isoephedrine hydrochloride. The hydrochloride salt is a white, crystalline material, soluble in water, in alcohol, and in chloroform. Pseudoephedrine, like ephedrine, is a useful bronchodilator and nasal decongestant, but is much less active in increasing blood pressure. However, it should be used with caution in hypertensive individuals.

Phenylpropanolamine Hydrochloride U.S.P., Propadrine® Hydrochloride, (±)-1-phenyl-2-amino-1-propanol hydrochloride, (±)-norephedrine hydrochloride. Propadrine is the primary amine corresponding to ephedrine, and this modification gives an agent which has slightly higher vasopressor action and lower toxicity and central stimulation action than has ephedrine. It can be used in place of ephedrine for most purposes and is used widely as a nasal decongestion agent. For the latter purpose it is applied locally to shrink swollen mucous membranes; its action is more prolonged than that of ephedrine. It also is stable when given orally.

Phenylpropanolamine appears to have a mixed mechanism of action. Phenylpropanolamine is commonly used as the active component of over-the-counter appetite suppressants, but its efficacy as an anorectic agent is questionable. No sympathomimetic amines have been found to be effective for long-term use in weight-reduction programs.

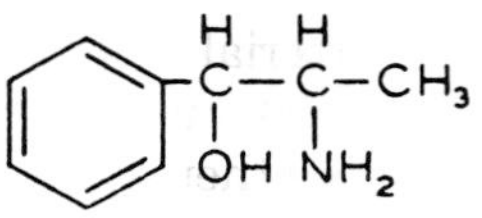

Phenylpropanolamine

Mephentermine Sulfate U.S.P., Wyamine® Sulfate, N,α,α-trimethylphenethylamine sulfate. The sulfate is a white crystalline powder with a faint fishy odor. It is soluble 1:20 in water and 1:50 in alcohol. A 1 percent solution in water is acidic, pH 5.5 to 6.2.

It exhibits pressor amine properties and is used topically as a nasal decongestant. It may be injected parenterally as a vasopressor agent in acute hypotensive states. Mephentermine is an indirect-acting agent with a prolonged duration of action.

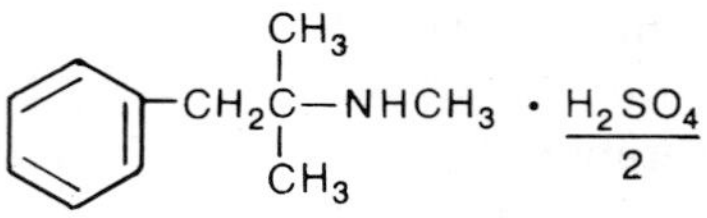

Mephentermine Sulfate

Metaraminol Bitartrate U.S.P., Aramine® Bitartrate, (−)-α-(1-aminoethyl)-*m*-hydroxybenzyl alcohol tartrate, (−)-*m*-hydroxynorephedrine bitartrate. Met-

araminol bitartrate is freely soluble in water and 1:100 in alcohol. This compound is a potent vasopressor with prolonged duration of action. It is employed as a nasal decongestant in the symptomatic relief of nasal edema accompanying the common cold, rhinitis, sinusitis and nasopharyngitis.

Metaraminol bitartrate is useful for parenteral administration in hypotensive episodes during surgery, for sustaining blood pressure in patients under general or spinal anesthesia and for the treatment of shock associated with trauma, septicemia, infectious diseases and adverse reactions to medication. It does not produce central nervous system stimulation.

Hydroxyamphetamine Hydrobromide U.S.P., Paredrine® Hydrobromide, (±)-*p*-(2-aminopropyl)-phenol hydrobromide, 1-(*p*-hydroxyphenyl)-2-amino-propane hydrobromide. This compound is a white, crystalline material which is very soluble in water (1:1) and in alcohol (1:2.5).

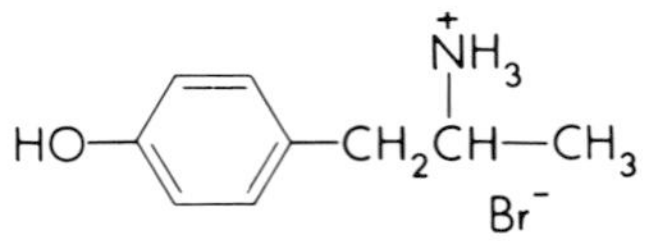

Hydroxyamphetamine Hydrobromide

It has been found useful for its synergistic action with atropine in producing mydriasis. A more rapid onset, more complete dilation and more rapid recovery are observed with a mixture of atropine and hydroxyamphetamine hydrobromide than with atropine alone.

Hydroxyamphetamine has little or no ephedrine-like central-nervous-system-stimulating action but retains the ability to shrink the nasal mucosa. Its actions as a bronchodilator or as an appetite-reducing agent are too weak to make it useful in these fields.

Levonordefrin U.S.P., Cobefrin®, (−)-α-(1-aminoethyl)-3,4-dihydroxybenzyl alcohol. This compound is a strong vasoconstrictor and has been recommended for use with local anesthetics.

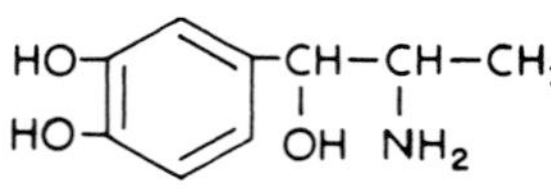

Levonordefrin

The structure has the catechol nucleus of epinephrine but the side chain of norephedrine.

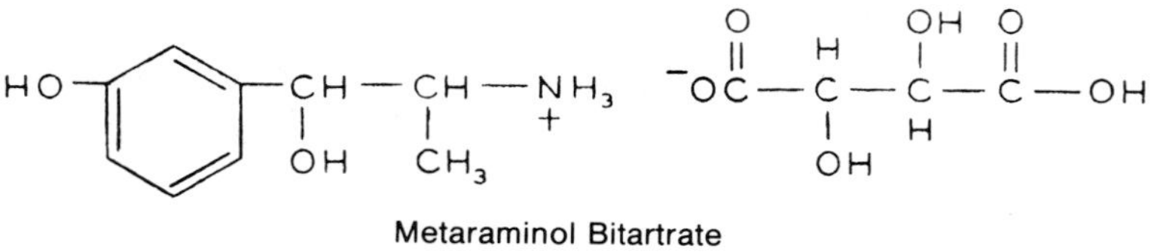

Metaraminol Bitartrate

Isoetharine Hydrochloride U.S.P., Bronkosol®. Isoetharine, which is also available as the mesylate salt, is a β_2-selective agent used as an inhalation aerosol for the treatment of bronchial asthma. It is metabolized by COMT.

Ethylnorepinephrine Hydrochloride U.S.P., Bronkephrine®. This compound is a hybrid of the epinephrine and ephedrine structures. The side chain differs from that of ephedrine by the presence of an ethyl rather than a methyl group. There are 2 asymmetric centers, thus giving rise to 4 possible isomers. No information is available on the relative activities of the individual isomers. The predominant action of the drug is as a β-adrenergic stimulant. It is weaker than isoproterenol in this regard. Although not the drug of choice, it does have utility in the treatment of asthma. Bronkephrine sulfate is administered by subcutaneous or intramuscular injection.

Ethylnorepinephrine Hydrochloride

Methoxyphenamine Hydrochloride U.S.P., Orthoxine® Hydrochloride, 2-(o-methoxyphenyl)isopropylmethylamine hydrochloride. Methoxyphenamine hydrochloride is a bitter, odorless, white, crystalline powder which is freely soluble in alcohol and water. A 5 percent solution is slightly acidic (pH 5.3 to 5.7).

Methoxyphenamine Hydrochloride

It is a sympathomimetic compound whose predominant actions are bronchodilation and inhibition of smooth muscle. Its effects on blood vessels are slight, its pressor effect being considerably less than that of ephedrine or epinephrine. Methoxyphenamine is useful as a bronchodilator in the treatment of asthma and also is effective in allergic rhinitis, acute urticaria and gastrointestinal allergy.

Administration of this drug produces no alterations in blood pressure and only slight cardiac stimulation. The actions on the central nervous system are minor.

Methoxamine Hydrochloride U.S.P., Vasoxyl® Hydrochloride, α-(1-aminoethyl)-2,5-dimethoxybenzyl alcohol hydrochloride, 2-amino-1-(2,5-dimethoxyphenyl)propanol hydrochloride. This compound is a white, platelike crystalline substance with a bitter taste. It is odorless or has only a slight odor. It is soluble in water

1:2.5 and in alcohol 1:12. A 2 percent solution in water is slightly acidic (pH 4.0 to 5.0), and it is affected by light.

Methoxamine Hydrochloride

Methoxamine hydrochloride is a sympathomimetic amine that exhibits the vasopressor action characteristic of other agents of this class, but is unlike most pressor amines in that the cardiac rate decreases as the blood pressure increases when this agent is used. The drug tends to slow the ventricular rate; it does not produce ventricular tachycardia, fibrillation or an increased sinoatrial rate. It is free of cerebral-stimulating action.

It is used primarily during surgery to maintain adequately or to restore arterial blood pressure, especially in conjunction with spinal anesthesia. It is also used in myocardial shock and other hypotensive conditions associated with hemorrhage, trauma and surgery. It is applied topically for relief of nasal congestion.

Dobutamine Hydrochloride, Dobutrex®. Dobutamine is a catecholamine that exerts its principal agonist action on cardiac β_1-receptors. It is a considerably less potent agonist at β_2-receptor sites. Its cardiac stimulant effect does not involve an indirect NE-releasing action. In vivo, the inotropic action of dobutamine is predominant over its chronotropic effects. It does not act as an agonist at the dopaminergic receptors that mediate renal vasodilatation.

Dobutamine is used as an intravenous infusion in the treatment of heart failure. Its uses and adverse effects, such as precipitation of arrhythmias, have been thoroughly reviewed.[23]

Dobutamine has a plasma half-life of about 2 minutes. It is metabolized by COMT and by conjugation.

Nylidrin Hydrochloride U.S.P., Arlidin®, p-hydroxy - α - {1 - [(1 - methyl - 3 - phenylpropyl)-amino]ethyl}benzyl alcohol hydrochloride. This compound is a white, odorless, practically tasteless, crystalline powder. It is soluble in water 1:65 and in alcohol 1:40. A 1 percent solution in water is acidic (pH 4.5 to 6.5).

Nylidrin Hydrochloride

Nylidrin acts as a peripheral vasodilator. It is indicated in vascular disorders of the extremities that may be benefited as the result of increased blood flow. It is administered orally.

Isoxsuprine Hydrochloride U.S.P., Vasodilan®, *p*-hydroxy-*α*-{1-[(1-methyl-2-phenoxyethyl)amino]-ethyl}benzyl alcohol hydrochloride.

Isoxsuprine Hydrochloride

This compound is used as a vasodilator for symptomatic relief in peripheral vascular disease and cerebrovascular insufficiency. Isoxsuprine appears to activate the β_2-receptors in small blood vessels with a resulting dilatation of those vessels.

Ritodrine Hydrochloride, Yutopar®, 1-(4-hydroxyphenyl)-2-[2-(4-hydroxyphenyl)ethylamino]propanol hydrochloride. Ritodrine is a selective β_2-receptor agonist that is used to control premature labor and to reverse fetal distress caused by excessive uterine activity. Its uterine inhibitory effects are more sustained than its effects on the cardiovascular system, which are minimal compared to those caused by nonselective β-agonists. The cardiovascular effects usually associated with its administration are mild tachycardia and slight diastolic-pressure decrease. It is available as an injection and in tablet form for oral administration. It is usually administered initially by intravenous infusion in order to stop premature labor. Subsequently, it may be given by the oral route.

Ritodrine

Aliphatic Amines. In the classic work by Barger and Dale[24] in 1910, the pressor action of the aliphatic amines was described, but only since the early 1940's have these agents become of pharmaceutical importance. An investigation in 1944[25] determined the influence of the location of an amino group on an aliphatic carbon chain and the effect of branching of the carbon chain carrying an amino group on pressor action. Optimal conditions were found in compounds of 7 to 8 carbon atoms with a primary amino group in the 2-position. Branching of the chain increases pressor activity.

A series of secondary β-cyclohexylethyl- and β-cyclopentylethylamines have been shown to have sympathomimetic activity.[26]

Tuaminoheptane U.S.P., Tuamine®, 2-aminoheptane. This compound is a colorless to pale-yellow liquid. It is freely soluble in alcohol and sparingly soluble in water. A 1 percent solution in water is alkaline (pH 11.5). It is a vasoconstrictor and a sympathomimetic amine. Inhalation of the vapors is an effective treatment of acute rhinologic conditions and is very useful when prolonged and repeated medication is required. It should be used with caution by those who have cardiovascular disease.

$$CH_3CHCH_2CH_2CH_2CH_2CH_3$$
$$|$$
$$NH_2$$

Tuaminoheptane

Tuaminoheptane Sulfate U.S.P., Tuamine® Sulfate, 1-methylhexylamine sulfate. This compound is a white, odorless powder. It is soluble in alcohol and freely soluble in water. A 1 percent solution in water is slightly acidic (pH 5.4).

The vasoconstrictive effects of a 1 percent solution of tuaminoheptane sulfate exceed those of a similar concentration of ephedrine, and the duration of effect is greater than that of ephedrine. For topical use a 1 percent solution of tuaminoheptane sulfate may be applied to the mucous membranes of infants and adults and usually is adequate for routine treatment.

Cyclopentamine Hydrochloride U.S.P., Clopane® Hydrochloride, N,*α*-dimethylcyclopentaneethylamine hydrochloride, 1-cyclopentyl-2-methylaminopropane hydrochloride. This white, bitter, crystalline powder is a sympathomimetic agent with uses and actions characteristic of other pressor amines. It is soluble in water 1:1 and in alcohol 1:2. Its effects are similar to those of ephedrine, but it produces only slight cerebral excitation. Orally, it is more effective than ephedrine. Presently, cyclopentamine is used by topical application for the temporary relief of nasal congestion.

Cyclopentamine Hydrochloride

Too frequent application topically should be avoided to prevent side-effects such as increased blood pressure, nervousness, nausea and dizziness.

Propylhexedrine U.S.P., Benzedrex®, N,α-dimethylcyclohexaneethylamine. This material is a clear, colorless liquid, with a characteristic fishy odor. Propylhexedrine is very soluble in alcohol and very slightly soluble in water. It volatilizes slowly at room temperature and absorbs carbon dioxide from air. Its

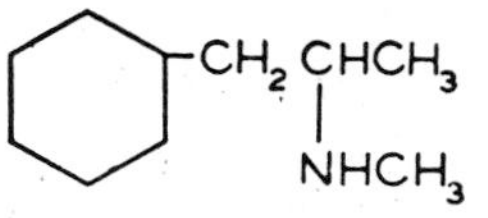

Propylhexedrine

uses and actions are similar to those of other volatile sympathomimetic amines. It produces vasoconstriction and a decongestant effect on the nasal membranes but has only about one half the pressor effect of amphetamine and produces decidedly less effect on the nervous system. Therefore its major use is for local shrinking effect on nasal mucosa in the symptomatic relief of nasal congestion caused by the common cold, allergic rhinitis or sinusitis.

Isometheptene Mucate, 2-methylamino-6-methyl-5-heptene mucate. Isometheptene is a sympathomimetic amine that has limited use as a compo-

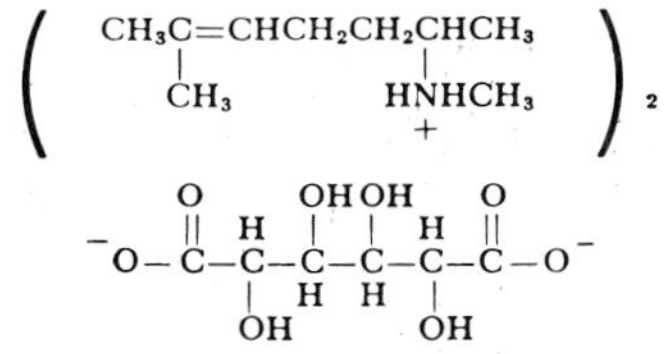

Isometheptene Mucate

nent of medications used to treat vascular and tension headaches. It is claimed to constrict cranial and cerebral arterioles, which reduces the stimuli that lead to vascular headaches.

Imidazoline Derivatives. A number of important adrenergic agents are derivatives of imidazoline. While 2-benzylimidazoline (Priscoline) is a vasodilator and sympatholytic agent, introduction of a hydroxyl group into the *para* position of the benzenoid ring converts the compound into a potent pressor agent.

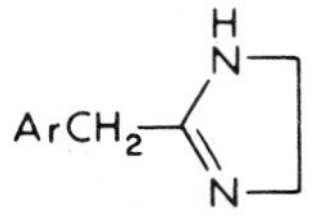

2–Aralkylimidazoline

Naphazoline Hydrochloride U.S.P., Privine® Hydrochloride, 2-(1-naphthylmethyl)-2-imidazoline monohydrochloride. This compound is a bitter, odorless, white, crystalline powder which is a potent vasoconstrictor, similar to ephedrine in its action. It is freely soluble in water and in alcohol. When applied to nasal and ocular mucous membranes it causes a prolonged reduction of local swelling and congestion. It is of value in the symptomatic relief of disorders of the upper respiratory tract. In acute nasal congestion, excessive use of vasoconstrictors may delay recovery. A rebound congestion of the mucosa is sometimes caused by naphazoline hydrochloride but can be alleviated by discontinuing all nasal medication.

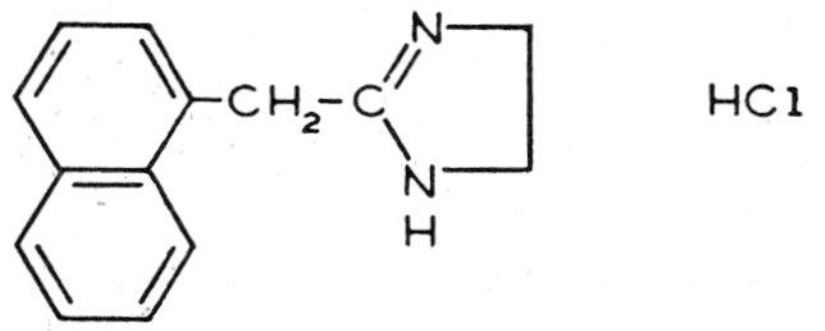

Naphazoline Hydrochloride

Tetrahydrozoline Hydrochloride U.S.P., Tyzine® Hydrochloride, Visine®, 2-(1,2,3,4-tetrahydro-1-naphthyl)-2-imidazoline monohydrochloride. This

TABLE 11–3

ALIPHATIC ADRENERGIC AMINES USED AS VASOCONSTRICTORS

Name Proprietary Name	Preparations	Application	Usual Adult Dose*
Tuaminoheptane U.S.P. *Tuamine*	Tuaminoheptane Inhalant U.S.P.	By inhalation, no more frequently than twice an hour	
Tuaminoheptane Sulfate U.S.D. *Tuamine Sulfate*	Tuaminoheptane Sulfate Solution U.S.P.	To the nasal mucosa, 0.5 to 2 percent solution	
Cyclopentamine Hydrochloride U.S.P. *Clopane Hydrochloride*	Cyclopentamine Hydrochloride Solution U.S.P.		Intranasal, 1 or 2 drops of a 0.5 or 1 percent solution every 3 or 4 hours
Propylhexedrine U.S.P. *Benzedrex*	Propylhexedrine Inhalant U.S.P.		Inhalation, 2 inhalations (about 500 μg.) through each nostril as required

*See *U.S.P. D.I.* for complete dosage information.

compound is closely related to naphazoline hydrochloride in its pharmacologic action. When applied topically to the nasal mucosa, the drug causes vasoconstriction, which results in reduction of local swelling and congestion. It is also useful in a 0.05 percent solution (Visine®) as an ocular decongestant. When used 2 or 3 times daily, there is no influence on pupil

Tetrahydrozoline Hydrochloride

size. It does not appear to increase intraocular pressure; however, its use in the presence of glaucoma is not recommended.

Xylometazoline Hydrochloride, U.S.P., Otrivin® Hydrochloride, 2-(4-*t*-butyl-2,6-dimethylbenzyl)-2-imidazoline hydrochloride. This compound is used as a nasal vasoconstrictor. Its duration of action is approximately 4 to 6 hours.

Xylometazoline Hydrochloride

Oxymetazoline Hydrochloride U.S.P., Afrin® Hydrochloride, 6-*t*-butyl-3-(2-imidazolin-2-ylmethyl)-2,4-dimethylphenol monohydrochloride, 2-(4-*t*-butyl-2,6-dimethyl-3-hydroxybenzyl)-2-imidazoline hydrochloride. This compound, closely related to xylometazoline, is a long-acting vasoconstrictor. It is used as a topical aqueous nasal decongestant in a wide variety of disorders of the upper respiratory tract.

ADRENERGIC BLOCKING AGENTS

NEURONAL BLOCKING AGENTS

Neuronal blocking agents are drugs which produce their pharmacological effects primarily by preventing the release of NE from sympathetic nerve terminals. The drugs of this type enter the adrenergic neuron by way of the uptake$_1$ process and they cause the release of some of the stored NE. However, NE depletion does not appear to account for their anti-adrenergic activity. Currently, it is believed that neuronal blocking agents produce their effects by stabilization of the neuronal membrane or the membranes of the storage vesicles. This stabilization makes the membranes less responsive to nerve impulses, thus inhibiting the release of NE into the synaptic cleft. The two neuronal blockers that are currently available for clinical use in the United States are bretylium tosylate, which

TABLE 11-4

IMIDAZOLINE ADRENERGIC AMINES USED AS VASOCONSTRICTORS

Name Proprietary Name	Preparations	Application*	Usual Pediatric Dose*
Naphazoline Hydrochloride U.S.P. *Privine Hydrochloride*	Naphazoline Hydrochloride Nasal Solution U.S.P.	Intranasal, 0.05 to 0.1 ml. of a 0.05 to 0.1 percent solution 4 to 8 times daily	
Albalon, Clear Eyes, Naphcon, Privine Hydrochloride, Vasocon	Naphazoline Hydrochloride Ophthalmic Solution U.S.P.	Topically to the conjunctiva, 0.05 to 0.15 ml. of a 0.012 or 0.1 percent solution 8 to 12 times daily as necessary	
Tetrahydrozoline Hydrochloride U.S.P. *Tyzine Hydrochloride*	Tetrahydrozoline Hydrochloride Nasal Solution U.S.P.	Intranasal, 0.1 to 0.2 ml. of a 0.1 percent solution up to 8 times daily as necessary	Children 2 to 6 years of age—intranasal, 0.1 to 0.15 ml. of a 0.05 percent solution up to 8 times daily as necessary
Visine	Tetrahydrozoline Hydrochloride Ophthalmic Solution U.S.P.	Topically to the conjunctiva, 0.05 to 0.1 ml. of a 0.05 percent solution 2 or 3 times daily	Children over 6 years—use adult dosage
Xylometazoline Hydrochloride U.S.P. *Otrivin Hydrochloride*	Xylometazoline Hydrochloride Solution U.S.P.	Nasal, 2 or 3 drops of a 0.1 percent solution every 4 to 10 hours as needed	Children 6 months to 12 years of age—intranasal, 2 or 3 drops of a 0.05 percent solution into each nostril every 4 to 10 hours as needed
Oxymetazoline Hydrochloride U.S.P. *Afrin Hydrochloride*	Oxymetazoline Hydrochloride Nasal Solution U.S.P.	Intranasal, 2 or 3 drops or sprays of a 0.05 percent solution into each nostril every 12 hours	Children 2 to 6 years of age—2 or 3 drops of a 0.025 percent solution into each nostril every 12 hours; children over 6 years of age—use adult dosage.

*See *U.S.P. D.I.* for complete dosage information.

is administered parenterally to control cardiac arrhythmias, and guanethidine sulfate, which is used as an antihypertensive agent. These compounds are discussed in the chapter on cardiovascular drugs (Chap. 14).

Guanethidine Sulfate

Bretylium Tosylate

α-ADRENERGIC BLOCKING AGENTS

Theoretically, adrenergic blocking agents that exert their antagonist effect at postsynaptic α-receptors should be useful as antihypertensive agents. Unfortunately, this class of compounds has not proven to be very useful for the management of hypertension. The α-antagonist prazosin appears to constitute the single exception to this generally disappointing performance.[27,28,29]

There appear to be several reasons for the general lack of usefulness of α-antagonists in hypertension. These drugs lower blood pressure by blocking vascular α-receptors, thereby inhibiting peripheral vasoconstriction. However, the reduction in blood pressure is accompanied by a reflex tachycardia and an increased cardiac output. Because the chronotropic response is mediated by β-receptors, α-antagonists do not prevent this effect, which is the indirect result of their vasodilator action. It is also believed that certain α-antagonists, such as phentolamine, tolazoline, and phenoxybenzamine, have the ability to block presynaptic α_2-receptors as well as postsynaptic α_1-receptors. Because blockade of the presynaptic α_2-adrenoceptors would be expected to enhance NE release, this would also contribute to the tachycardia and enhanced cardiac output that occur subsequent to the administration of α-adrenergic blocking agents. Among the other problems associated with the use of α-antagonists are the following: short duration of action (phentolamine); poor absorption from the GI tract (phentolamine); weak α-blocking activity (tolazoline), agonist action at other receptors (tolazoline); and postural hypotension (phenoxybenzamine). Clinical uses of α-antagonists are mentioned under the descriptions of the individual drug products.

Structure–activity relationships among the various types of α-antagonists are difficult to perceive, if they indeed do exist. Unlike the β-antagonists, which bear clear structural similarities to the adrenergic agonists NE, epinephrine, and isoproterenol, the α-antagonists comprise a mixture of chemical classes that bears little obvious resemblance to the agonists.

Ergot Alkaloids. The oxytocic action of ergot was recognized as early as the 16th century, and it was used by midwives for years prior to its acceptance by the medical profession. Modern acceptance is based largely on the extensive research conducted during the past half century. Ergotoxine, isolated in 1906, and ergotamine, isolated in 1920, for many years were thought to be the principal alkaloids present. Since then, the former has been shown to be nonhomogeneous and composed of equal parts of three bases: ergocornine, ergocristine and ergocryptine. In 1933, sensibamine was reported as a new base, only to be shown later to be a mixture of equal parts of ergotamine and ergotaminine. Similarly, ergoclavine (1934) has been shown to be a mixture of ergosine and

TABLE 11–5

ERGOT ALKALOIDS*

	R_1	R_2
Ergotamine Group		
Ergotamine	—CH$_3$	-CH$_2$⟨phenyl⟩
Ergosine	—CH$_3$	—CH$_2$CH(CH$_3$)$_2$
Ergotoxine Group		
Ergocristine	—CH(CH$_3$)$_2$	-CH$_2$⟨phenyl⟩
Ergocryptine	—CH(CH$_3$)$_2$	—CH$_2$CH(CH$_3$)$_2$
Ergocornine	—CH(CH$_3$)$_2$	—CH(CH$_3$)$_2$

* Each of the listed alkaloids has an inactive diastereoisomer derived from *iso*-lysergic acid which, in the above formulas, differs only in that the configuration of the hydrogen and the carboxyl groups at position 8 is interchanged. The nomenclature also differs, in that the suffix "*in*" is added to the name, e.g., ergotaminine instead of ergotamine. However, in the case of ergonovine, the diastereoisomer is named "ergometrinine" because this derives from the name of ergonovine commonly used in England, i.e., ergometrine.

ergosinine. In 1935, an active water-soluble alkaloid was reported simultaneously by four research groups and is the alkaloid now known as ergonovine (ergometrine in Great Britain).

The active alkaloids are all amides of lysergic acid, whereas the inactive diastereoisomeric counterparts are similarly derived from *iso*-lysergic acid. The only difference between the two acids is the configuration of the substituents at position 8 of the molecule. The structure of ergonovine is the simplest of these alkaloids, being the amide of lysergic acid derived from (+)-2-aminopropanol. The other alkaloids are of a more complex polypeptidelike structure.

The isomers of ergonovine (A) (Table 11-6) have been prepared for pharmacologic study. Only the propanolamides of (+)-lysergic acid were found to be active. The optical configuration of the amino alcohol did not seem to be important to pharmacologic activity. Other partially synthetic derivatives of (+)-lysergic acid have been prepared, with two of them showing notable activity: methylergonovine, the amide formed from (+)-lysergyl chloride and 2-aminobutanol (B), and the N-diethyl amide of (+)-lysergic acid (C). The latter compound, also known as LSD, has an oxytocic action comparable with that of ergonovine and, in addition, is known to cause, in very small doses (100 to 400 μg.), marked psychic changes combined with hallucinations and colored visions. The most recent active synthetic derivative of lysergic acid is the 1-methyl butanolamide (D), known generically as methysergide. The hydrogenation of the C-9 to C-10 double bond in the lysergic acid portion of the ergot alkaloids, other than ergonovine, enhances the adrenergic blocking activity as assayed against the con-

strictive action of circulating epinephrine upon the seminal vesicle of an adult guinea pig. Comparative activities are demonstrated in the following results recorded by Brugger.[30]

Pharmacologically, the naturally occurring ergot alkaloids may be placed in two classes: (1) the water-insoluble, polypeptidelike group comprising ergocryptine, ergocornine, ergocristine (ergotoxine group), ergosine and ergotamine and (2) the water-soluble alkaloid ergonovine. The members of the water-insoluble group are typical adrenergic blocking agents in that they inhibit all responses to the stimulation of adrenergic nerves and block the effects of circulating epinephrine. In addition, they cause a rise in blood pressure by constriction of the peripheral blood vessels due to a direct action on the smooth muscle of the vessels. The most important action of these alkaloids however, is their strongly stimulating action on the smooth muscle of the uterus, especially the gravid or puerperal uterus. This activity develops more slowly and lasts longer when the water-insoluble alkaloids are used than when ergonovine is administered. Toxic doses or the too frequent use of these alkaloids in small doses are responsible for the symptoms of ergotism. These alkaloids are rendered water-soluble by preparing salts of them with such organic acids as tartaric, maleic, ethylsulfonic or methylsulfonic.

Ergonovine has little or no activity as an adrenergic blocking agent and, indeed, has many of the pharmacologic properties (produces mydriasis in the rabbit's eye, relaxes isolated strips of gut, constricts blood vessels) of a sympathomimetic drug. It does not raise the blood pressure when injected intravenously into an anesthetized animal. It possesses a strong, prompt, oxytocic action. Although ergonovine exerts a constrictive effect upon peripheral blood vessels, no cases have been reported yet of ergotism due to its use. It is highly active orally and causes little nausea or vomiting. It usually is dispensed as a salt of an organic acid, such as maleic, tartaric or hydracrylic acid.

TABLE 11-6

	R_1	R_2	R_3
A. Ergonovine	—H	—CH(CH₃)CH₂OH	—H
B. Methylergonovine	—H	—CH(C₂H₅)CH₂OH	—H
C. LSD	—C₂H₅	—C₂H₅	—H
D. Methysergide	—H	—CH(C₂H₅)CH₂OH	—CH₃

Ergotamine	1	Dihydroergotamine	7
Ergocornine	2	Dihydroergocornine	25
Ergocristine	4	Dihydroergocristine	35
Ergocryptine	4	Dihydroergocryptine	35

PRODUCTS

Ergonovine Maleate U.S.P., Ergotrate® Maleate. This water-soluble alkaloid was isolated, as indicated, from ergot, in which it occurs to the extent of 200 μg. per gram of ergot. The several research groups which isolated the alkaloid almost simultaneously named the alkaloid according to the dictates of each. Thus, the names ergometrine, ergotocin, ergostetrine and ergobasine were assigned to this alkaloid. To clarify the confusion, the Council on Pharmacy and Chemistry of the American Medical Association adopted a new name, ergonovine, which is in general use today. Of course, commercial names differ from the Council-accepted name, the principal USA-one being Ergotrate® (ergonovine maleate).

The free base occurs as white crystals which are quite soluble in water or alcohol and levorotatory in solution. It readily forms crystalline, water-soluble salts, behaving in this respect as a mono-acidic base. The nitrogen involved in salt formation obviously is not the one in the indole nucleus, since it is far less basic than the other nitrogen. The official salt is the maleate. It is said to be a convenient form in which to crystallize the alkaloid and is also quite stable.

Ergonovine maleate occurs in the form of a light-sensitive, white or nearly white, odorless, crystalline powder. It is soluble in water (1:36) and in alcohol (1:20) but is insoluble in ether and in chloroform.

Ergonovine has a powerful stimulating action on the uterus and is used for this effect. Since it seems to exercise a much greater effect on the gravid uterus than on the nongravid one, it is used safely in small doses with ample effect.

During the third stage of labor, these drugs should not be used until at least after presentation of the head and preferably after passage of the placenta. Ordinarily, 200 μg. of ergonovine is injected at this stage to bring about prompt and sustained contraction of the uterus. The effect lasts about 5 hours and prevents excessive blood loss. It also lowers the incidence of uterine infection. A continued effect may be obtained by further administration of the alkaloid, either orally or parenterally.

Ergotamine Tartrate U.S.P., Gynergen®, ergotamine tartrate (2:1) (salt). Ergotamine, one of the insoluble ergot alkaloids, is obtained from the crude drug by the usual isolation methods.

It occurs as colorless crystals or as a white to yellowish-white crystalline powder. It is not especially soluble in water (1:500) or in alcohol (1:500) although the aqueous solubility is increased with a slight excess of tartaric acid. Ergotamine has both direct vasoconstrictor properties and an α-receptor antagonist action.

Previous to the discovery of ergonovine, ergotamine was the ergot drug of choice as a uterine stimulant, either orally or parenterally. Because it offered no advantage over ergonovine except for a more sustained action and, in addition, was more toxic, it fell into disuse. However, it has been employed as a specific analgesic in the treatment of migraine headache, in which capacity it is reasonably effective. Cafergot®, a combination of ergotamine tartrate and caffeine, is an available product. It is of no value in other types of headaches and sometimes fails to abort migraine headaches. It has no prophylactic value. It is customary to administer 250 μg. subcutaneously to determine whether idiosyncrasy to the drug exists. In the event that no sensitivity is shown, the full dose is injected. Oral or sublingual administration may be resorted to, but they are much less effective than the parenteral route. Care should be exercised in its continued use to prevent signs of ergotism.

Dihydroergotamine Mesylate U.S.P., D.H.E.-45®, dihydroergotamine monomethanesulfonate. This compound is produced by the hydrogenation of the easily reducible C-9 to C-10 double bond in the lysergic acid portion of the ergotamine molecule. It occurs as a white, yellowish, or faintly red powder which is only slightly soluble in water and chloroform but soluble in alcohol.

Dihydroergotamine, although very closely related to ergotamine, differs significantly from the latter in its action. For all practical purposes, the uterine action is lacking. However, the adrenergic blocking action is stronger. Nausea and vomiting are at a minimum, as is its cardiovascular action. One of its principal uses has been in the relief of migraine headache in a manner similar to ergotamine, over which it excels not only in decreased toxicity but also in a higher percentage of favorable results. Because the drug is not very effective orally, it usually is administered subcutaneously, intravenously or intramuscularly.

Methylergonovine Maleate U.S.P., Methergine®, N-[α-(hydroxymethyl)propyl]-D-lysergamide. This compound occurs as a white to pinkish-tan microcrystalline powder which is odorless and has a bitter taste. It is only slightly soluble in water and alcohol and very slightly soluble in chloroform and ether.

It is very similar to ergonovine in its pharmacologic actions. It is said to be about one to three times as powerful as ergonovine in its action. The action of methylergonovine is quicker and more prolonged than that of ergonovine. It has been shown to be relatively nontoxic in the doses used. It is marketed as 200-μg. tablets and as ampules.

Methysergide Maleate U.S.P., Sansert®. This drug was introduced in 1962. It occurs as a white to yellowish-white, crystalline powder that is practically odorless. It is only slightly soluble in water and alcohol and very slightly soluble in chloroform and ether.

Although it is closely related in structure to methylergonovine it does not possess the potent oxytocic action of the latter. It has been shown to be a potent serotonin antagonist and has found its principal utility in the prevention of migraine headache, but the exact mechanism of prevention has not been elucidated. Methylsergide is a weak α-antagonist and a weak vasoconstrictor.

Methysergide produces a variety of untoward side-effects although most of them are mild and will disappear with continued use. Some of the most common of these effects are nausea, epigastric pain, dizziness, restlessness, drowsiness, leg cramps and psychic effects. However, it has become increasingly evident that this drug must be carefully administered under a physician's and pharmacist's watchful eye. The reason for this is that, when administered on a long-term uninterrupted basis, it appears to be prone to induce retroperitoneal fibrosis, pleuropulmonary fibrosis and fibrotic thickening of cardiac valves. As a consequence of these potential fibrotic manifestations, the drug has been reserved for "prophylaxis in patients whose vascular headaches are frequent and/or severe and uncontrollable and who are under close medical supervision." Because of its side-effects it should not be continuously administered for over a 6-month period without a drug-free interval of 3 to 4 weeks between each 6-month course of treatment. Furthermore, the dosage should be reduced gradually during the last 2 to 3 weeks of the 6-month treatment period to avoid "headache rebound." The drug is not recommended for children.

Methysergide is an effective blocker of the effects of serotonin, which may be involved in the mechanism of vascular headaches. The complete mechanism and the involvement of the drug have not been completely clarified as yet. However, it is used as indicated above for the prevention and reduction of intensity as well as frequency of vascular headaches in patients (1) suffering from one or more severe vascular headaches per week, or (2) suffering from vascular headaches that are uncontrollable or so severe that preventive therapy is indicated regardless of the frequency of the attack.

β-Haloalkylamines. Although dibenamine (N,N-dibenzyl-β-chloroethylamine), the prototype of these compounds, was characterized in 1934 by Eisleb[31] incidental to a description of some other synthetic intermediates, it was the report of Nickerson and Goodman[32] in 1947 on the pharmacology of the compound that revealed the powerful adrenergic blocking properties. The blockade produced by this group of compounds seems to be the most complete of the entire group of blocking agents. The differences in activity of the members of this group differ only quantitatively, being qualitatively the same. When given in adequate doses, they produce a slowly developing prolonged adrenergic blockade which is not overcome by epinephrine. Much of the early work done with this group was confined to dibenamine but this has been largely supplanted by the more orally useful and potent phenoxybenzamine. The mass of pharmacologic data accumulated with respect to this class of compounds has led to the establishment of certain structural requirements that are necessary for activity. Ullyot and Kerwin[33] state that most of the presently known effective compounds may be defined broadly by the following formula:

$$C_6H_5-CH_2$$
$$\diagdown$$
$$N-CH_2CH_2Cl$$
$$\diagup$$
$$C_6H_5-CH_2$$

Dibenamine

TABLE 11-7

ERGOT ALKALOID PRODUCTS

Name Proprietary Name	Preparations	Category	Usual Adult Dose*
Ergonovine Maleate U.S.P. *Ergotrate Maleate*	Ergonovine Maleate Injection U.S.P. Ergonovine Maleate Tablets U.S.P.	Oxytocic	Oral, 200 μg. 3 or 4 times daily; I.M. or I.V., 200 μg. repeated after 2 to 4 hours, if necessary
Ergotamine Tartrate U.S.P. *Gynergen, Ergomar*	Ergotamine Tartrate Injection U.S.P.	Analgesic (specific in migraine)	I.M. or S.C., 250 to 500 μg., repeated in 40 minutes if necessary
	Ergotamine Tartrate Tablets U.S.P.		Oral or sublingual, 1 to 2 mg., then 1 to 2 mg. every 30 minutes, if necessary, to a total of 6 mg. per attack
Dihydroergotamine Mesylate U.S.P *D.H.E. 45*	Dihydroergotamine Mesylate Injection U.S.P.	Antiadrenergic	Parenteral, 1 mg., may be repeated at 1-hour intervals to 3 mg.
Methylergonovine Maleate U.S.P. *Methergine*	Methylergonovine Maleate Injection U.S.P. Methylergonovine Maleate Tablets U.S.P.	Oxytocic	Oral, 200 μg. 3 or 4 times daily; I.M. or I.V., 200 μg., repeated after 2 to 4 hours, if necessary
Methysergide Maleate U.S.P. *Sansert*	Methysergide Maleate Tablets U.S.P.	Analgesic (specific in migraine)	2 mg. 2 to 4 times daily

*See *U.S.P. D.I.* for complete dosage information.

R' = Aralkyl (benzyl, phenethyl, etc.)
= Phenoxyalkyl (β-phenoxyethyl, etc.)
R'' = Alkyl, alkenyl, dialkylamino-alkyl, aralkyl,
β-phenoxylethyl, etc.
X = Halogen, sulfonic acid ester.

The mechanism whereby β-haloalkylamines produce a long-lasting, α-adrenoreceptor blockade involves the formation of an intermediate aziridinium ion (ethylene iminium ion, immonium ion). This positively charged electrophile reacts with a nucleophilic group on the receptor, resulting in the formation of a covalent bond between the drug and the receptor.

This alkylation of the receptor through the formation of a covalent bond results in prolonged α-receptor blockade. Although the aziridinium ion intermediate has long been believed to be the active receptor alkylating species, only in recent years has it been demonstrated unequivocally that the aziridinium ions derived from dibenamine and phenoxybenamine are capable of a α-receptor alkylation.[34]

Phenoxybenzamine Hydrochloride U.S.P., Dibenzyline® Hydrochloride, N-(2-chloroethyl)-N-(1-methyl-2-phenoxyethyl)benzylamine hydrochloride. The compound exists in the form of colorless crystals which are soluble in water, freely soluble in alcohol and chloroform and insoluble in ether. It slowly hydrolyzes in neutral and basic solutions but is stable in acid solutions and suspensions.

Phenoxybenzamine Hydrochloride

The action of phenoxybenzamine has been described as representing a "chemical sympathectomy" because of its selective blockade of the excitatory responses of smooth muscle and of the heart muscle.

Although phenoxybenzamine is capable of blocking acetylcholine, histamine, and serotonin receptors, its primary pharmacological effects, especially vasodilatation, may be attributed to its α-adrenergic blocking capability. As would be expected of a drug that produces such a profound α-blockade, the administration of phenoxybenzamine is frequently associated with reflex tachycardia, increased cardiac output, and postural hypotension. There is also evidence indicating that blockade of presynaptic α_2-receptors contributes to the increased heart rate produced by phenoxybenzamine.

The onset of action is slow, but the effects of a single dose of phenoxybenzamine may last 3 to 4 days. The principal effects following administration are an increase in peripheral blood flow, increase in skin temperature and a lowering of blood pressure. It has no effect on the parasympathetic system and has little effect on the gastrointestinal tract. The most common side-effects are miosis, tachycardia, nasal stuffiness and postural hypotension, which are all related to the production of adrenergic blockade.

Oral phenoxybenzamine is the drug of choice for the preoperative management of patients with pheochromocytoma and in the chronic management of patients whose tumors are not amendable to surgery.[27] In some patients, the administration of a β-blocking agent after phenoxybenzamine blockade has been established as useful.

Phenoxybenzamine is used to treat peripheral vascular disease such as Raynaud's syndrome. It has also been used in shock and frostbite to improve blood flow to peripheral tissues. In low doses, it has been used in hypertensive patients who have developed vascular supersensitivity to sympathomimetics during treatment with adrenergic neuronal blocking agents.

Imidazolines. Like the ergot alkaloids, the imidazoline α-antagonists are competitive (reversible) blocking agents. They are structurally similar to the imidazoline α-agonists such as naphazoline, tetrahydrozoline, and xylometazoline. For reasons discussed earlier, these agents have not proven to be useful for the treatment of hypertension, and their clinical applications have been limited largely to treatment of conditions involving peripheral vasospasm. Phentolamine has found some use in the treatment of congestive heart failure.

Tolazoline Hydrochloride U.S.P., Priscoline® Hydrochloride, 2-benzyl-2-imidazoline monohydrochloride.

Tolazoline Hydrochloride

Tolazoline is well absorbed after administration by the oral route, and it is largely excreted unmetabolized in the urine. Its α-antagonistic action is relatively weak, but its histaminelike and acetylcholinelike agonist actions probably contribute to its vasodilator activity. Its histaminelike effects include stimulation of gastric-acid secretion, rendering it inappropriate for administration to patients who have gastric or peptic ulcers. It has been used to treat Raynaud's syndrome and other conditions involving peripheral vasospasm.

The drug occurs as a white or creamy white, bitter, crystalline powder possessing a slight aromatic odor. It is freely soluble in water and alcohol. A 2.5 percent aqueous solution is slightly acidic (pH 4.9 to 5.3). It is only slightly soluble in ether and ethyl acetate but is soluble in chloroform.

Phentolamine Mesylate U.S.P., Regitine® Methanesulfonate, m-[N-(2-imidazolin-2-ylmethyl)-p-toluidino]phenol monomethanesulfonate. This compound may be made by the procedure of Urech and co-workers.[113] It occurs as a white, odorless, bitter powder which is freely soluble in alcohol and very soluble in water. Aqueous solutions are slightly acidic (pH 4.5 to 5.5) and deteriorate slowly. However, the chemical itself is stable when protected from moisture and light. The stability and the solubility of this salt of phentolamine are superior to those of the hydrochloride and account for the use of the methanesulfonate (mesylate) rather than the hydrochloride for parenteral injection.

Phentolamine Mesylate

Phentolamine Hydrochloride U.S.P., Regitine® Hydrochloride, m-[N-(2-imidazolin-2-ylmethyl)-p-toluidino]phenol monohydrochloride. It occurs as a white or slightly grayish, odorless, bitter powder. It is slightly soluble in alcohol and sparingly soluble in water. Its solutions in water are slightly acidic (pH 4.5 to 5.5) and foam when shaken. It is affected by light, and its aqueous solutions are unstable.

Although phentolamine was once used for the diagnosis of pheochromocytoma, it has been replaced largely by the chemical analysis of catecholamines and catecholamine-metabolite levels in the urine. It is used to prevent hypertension during pheochromocytoma surgery and to counteract the pressor effects of sympathomimetic overdose.

Phentolamine hydrochloride is available in tablet form for oral administration, although it is much less effective than parenteral administration of phentolamine mesylate. Like tolazoline, which is also more effective after parenteral administration than after oral administration, phentolamine produces side-effects associated with cardiac and gastrointestinal stimulation. It is used to increase blood flow to the extremities and it has found some application in the preoperative management of pheochromocytoma.

Prazosin Hydrochloride (Minipress®) is a quinazoline derivative that has been found to be of value in the management of hypertension.

Prazosin

Prazosin is an α-adrenergic antagonist that offers distinct advantages over the other α-blockers. It

TABLE 11–8

IMIDAZOLINE PRODUCTS

Name Proprietary Name	Preparations	Category	Usual Adult Dose*	Usual Pediatric Dose*
Tolazoline Hydrochloride U.S.P. *Priscoline Hydrochloride*	Tolazoline Hydrochloride Injection U.S.P. Tolazoline Hydrochloride Tablets U.S.P.	Peripheral vasodilator	Oral and parenteral, 50 mg. 4 times daily	
Phentolamine Mesylate U.S.P. *Regitine Methanesulfonate*	Phentolamine Mesylate for Injection U.S.P.	Antiadrenergic; diagnostic aid (pheochromocytoma)	Antiadrenergic—I.M. or I.V., 5 to 10 mg.; diagnostic I.M. or I.V., 5 mg.	Diagnostic—I.V., 100 µg. per kg. of body weight or 3 mg. per square meter of body surface
Phentolamine Hydrochloride U.S.P. *Regitine Hydrochloride*	Phentolamine Hydrochloride Tablets U.S.P.	Antiadrenergic	50 mg. 4 to 6 times daily	

*See *U.S.P. D.I.* for complete dosage information.

causes peripheral vasodilatation without an increase in heart rate or cardiac output. This advantage is attributed, at least in part, to the fact that prazosin blocks postjunctional α_1-receptors selectively without blocking presynaptic α_2-receptors. Prazosin is effective after oral administration and has a plasma half-life of approximately 4 hours. It is extensively metabolized and is excreted mainly in the bile.

Although its side-effects are usually minimal, the most frequent one, known as the *first dose phenomenon,* is sometimes severe. This is a dose-dependent effect characterized by dizziness, palpitations, and syncope. The cause of the "first dose effect" is unknown.

Prazosin, in addition to its use in treatment of hypertension, has been found to be of value in the treatment of heart failure. Its pharmacology and therapeutic efficacy have been reviewed.[28]

β-ADRENERGIC BLOCKING AGENTS

Although the α-receptor antagonist action of agents such as the ergot alkaloids was discovered many years ago, the first β-blocker was not reported until 1958 when Powell and Slater described the activity of dichloroisoproterenol (DCI).[35] Shortly thereafter, Moran and Perkins reported that DCI blocked the effects of sympathomimetics on the heart.[36] The structure of DCI is identical to that of isoproterenol, with the exception that the catechol hydroxyl groups have been replaced by two chloro groups. This simple structural modification, involving the replacement of the aromatic hydroxyl groups, has provided the basis for nearly all of the approaches employed in subsequent efforts to design and synthesize therapeutically useful β-receptor antagonists. Unfortunately, DCI did not prove to be a pure antagonist, but rather, it was found to be a partial agonist. The substantial, direct sympathomimetic action of DCI precluded its development as a clinically useful drug.

Dichloroisoproterenol

Pronethalol, whose structural similarities to isoproterenol and DCI are obvious, was the next important β-antagonist to be described.[37] Although pronethalol had much less intrinsic sympathomimetic activity than DCI, it was withdrawn from clinical testing because of reports that it caused thymic tumors in

mice. However, within two years of the report on pronethalol, Black and co-workers described the β-

Pronethalol

blocking actions of propranolol, a close structural relative to pronethalol. Propranolol has become one of the most thoroughly studied and widely used drugs in the therapeutic armamentarium. It is the standard against which all other β-antagonists are compared.

Propranolol

Propranolol belongs to the group of β-blocking agents known as *aryloxypropranolamines.* This term reflects the fact that an -OCH$_2$- group has been incorporated into the molecule between the aromatic ring and the ethylamino side chain. Because this structural feature is so frequently found in β-antagonists, the assumption is often made that the -OCH$_2$- group is responsible for the antagonist properties of the molecules. However, this is not the case; in fact, the -OCH$_2$- group is present in a number of experimental compounds that are potent β-agonists.[38] This latter fact again leads to the conclusion that it is the nature of the aromatic ring and its substituents which is the primary determinant of β-antagonist activity.

Propranolol, like the other β-receptor antagonists that will be discussed, is a competitive antagonist whose receptor-blocking actions can be reversed with sufficient concentrations of β-agonists. Currently, propranolol is approved for use in the United States for hypertension, cardiac arrythmias, angina pectoris due to coronary atherosclerosis, hypertrophic subaortic stenosis, and prophylaxis of migraine headache. Propranolol is under investigation for the treatment of a variety of other conditions, including anxiety and schizophrenia.

Although some degree of controversy exists, it appears that nearly all of the pharmacological effects of propranolol may be attributed to β-receptor blockade. By blocking the β-receptors of the heart, it slows the heart, reduces the force of contraction and reduces cardiac output. Because of reflex sympathetic activity and blockade of vascular β_2-receptors, administration of propranolol may result in increased peripheral resistance. The antihypertensive action of propranolol may be attributed, at least in part, to its ability to reduce cardiac output as well as to its suppression of renin release from the kidney.

Since propranolol exhibits no selectivity for β_1-receptors, it blocks β_2-receptors in the respiratory tract. Therefore, propranolol is contraindicated in the presence of conditions such as asthma or bronchitis.

The rationale for the use of propranolol or other β-antagonists in the prophylaxis of migraine largely resides in the assumption that the drugs cause a vasoconstriction that inhibits the vasodilating effects of chemical mediators. However, since little is known about the causal factors and processes in migraine, it is difficult to specify the precise reason for the effectiveness of β-blockers. Likewise, it is uncertain whether propranolol's effectiveness in anxiety states is due to an action in the central nervous system, a peripheral β-blocking effect, or a combination of these.

The use of propranolol in cardiovascular diseases is covered in the chapter on cardiovascular drugs (Chap. 14). In addition, the clinical pharmacology of propranolol and other β-adrenoceptor antagonists has been thoroughly reviewed by other authors.[2,27,39]

A facet of the pharmacological action of propranolol which has received a good deal of attention is its so-called *membrane stabilizing activity*. This is a nonspecific effect (i.e., not mediated by a specific receptor), which is also referred to as a *local anesthetic* effect or a *quinidinelike* effect. Although various authors have implied that the membrane stabilizing effect may be therapeutically important, it now seems clear that the concentrations required to produce this effect far exceed those obtained with normal therapeutic doses of propranolol and related β-blocking drugs. It is most unlikely that the nonspecific-membrane stabilizing activity plays any role in the clinical efficacy of β-blocking agents.[27,40]

The β-blocking agents exhibit a high degree of stereoselectivity in the production of their β-blocking effects. As discussed earlier with regard to sympathomimetic agents, the configuration of the hydroxyl-bearing carbon of the side chain plays a critical role in the interaction of β-antagonist drugs with β-receptors. The available data indicate that the pharmacologically more active enantiomer interacts with the receptor recognition site in a manner analogous to that of the agonists. However, the structural features of the aromatic portion of the antagonist appear to perturb the receptor, or to interact with it in a manner that inhibits activation. In spite of the fact that nearly all of the β-antagonist activity resides in one enantiomer, propranolol and most other β-blocking agents, except timolol, are available for clinical use as racemic mixtures. Both enantiomers of propranolol have mem-brane stabilizing activity, but only the levorotatory isomer is a potent β-receptor antagonist, the dextrorotatory isomer being approximately 100 times less effective.[40]

The metabolism of propranolol is a topic that has received intense study during the past several years. Propranolol is well absorbed after oral administration, but it undergoes extensive "first pass" metabolism before it reaches the systemic circulation. The term *first pass metabolism* refers to the fact that the compound is efficiently extracted from the portal vein by the liver where it undergoes biotransformation. Lower doses of propranolol are more efficiently extracted than higher doses; this indicates that the extraction process may become "saturated" at higher doses. The outcome of this saturation of the extraction process is that a larger percentage of a high oral dose than of a lower dose of propranolol reaches the systemic circulation. The extensive first-pass metabolism of propranolol accounts for the fact that, when the drug is given intravenously, much smaller doses are required to achieve a therapeutic effect than when it is given by the oral route.

Numerous metabolites of propranolol have been identified, but the major metabolite in people, after a single oral dose, is naphthoxylactic acid, which is

Naphthoxylactic Acid

formed by a series of metabolic reactions involving N-dealkylation, deamination, and oxidation of the resultant aldehyde.[41,42] A propranolol metabolite of particular interest is 4-hydroxypropranolol. This compound is a potent β-antagonist that has some intrinsic sympathomimetic activity.[43] Interestingly, 4-hydroxypropranolol has been detected in plasma following intravenous administration of propranolol, although the plasma levels of the metabolite are considerably higher after oral administration.[44] It is not known what contribution, if any, 4-hydroxypro-

4-Hydroxypropranolol

pranolol makes to the pharmacologic effects seen after administration of propranolol. It has been suggested that 4-hydroxypropranolol contributes to propranolol's pharmacologic effects after low single doses or long-term oral administration.[44]

The half-life of propranolol after a single oral dose is 3 to 4 hours, and it increases to 4 to 6 hours after long-term therapy. This relatively short half-life seems inconsistent with the observations that the pharmacological effects of propranolol may persist for 2 to 3 days after discontinuation of the drug and that some patients can be treated effectively with one or two doses of propranolol per day. It has been proposed that propranolol glucuronide, which is the principal metabolite formed during long-term therapy, may serve as a storage pool for propranolol and thereby be at least partially responsible for the slow accumulation of propranolol.[45] According to this mechanism, propranolol glucuronide undergoes deconjugation in various tissues as well as during enterohepatic circulation, thus providing for a slow release of propranolol from the stored glucuronide. Another mechanism that may contribute to the slow accumulation of propranolol is the storage of propranolol in sympathetic neurons from which it is released during sympathetic nerve stimulation.[46] Although a number of authors have commented on the interindividual variation of metabolism of propranolol, and upon the lack of correlation of propranolol plasma levels and interindividual therapeutic effects, there does not appear to be a high degree of patient variation in propranolol disposition during long-term therapy.[44] This is somewhat surprising in view of the fact that, after oral administration, propranolol is almost completely metabolized. Very little is excreted unchanged.

The discovery that β-blocking agents are useful in the treatment of cardiovascular disease such as hypertension stimulated a search for cardioselective β-blockers. Cardioselective β-antagonists are drugs that have a greater affinity for the β_1-receptors of the heart than for β_2-receptors in other tissues. Such cardioselective agents should provide two important therapeutic advantages. The first advantage would be the lack of an antagonist effect on the β_2-receptors in the bronchi. Theoretically, this would make β_1-blockers safe for use in patients who have bronchitis or bronchial asthma. The second major therapeutic advantage of cardioselective agents would be the absence of blockade of the vascular β_2-receptors, which mediate vasodilatation. This would be expected to reduce or eliminate the increase in peripheral resistance that sometimes occurs after the administration of nonselective β-antagonists. Unfortunately, cardioselectivity is usually observed with β_1-antagonists only at relatively low doses. At normal therapeutic doses, much of the selectivity is lost.[27,40]

The goal of complete cardioselectivity is one that

may never be achieved, even if a drug is developed that retains its β_1-selectivity at therapeutic doses. One obstacle to the attainment of complete cardioselectivity may be the presence of both β_1- and β_2-receptors in cardiac and lung tissue. Although it is true that β_1-receptors are predominant in the heart and that the β-receptors of the lung are primarily β_2-, recent evidence indicates that both types of receptors are present in both tissues.[47,48] Thus, theoretically, any β-blocker would block at least some of the receptors in either tissue.

The prototype β_1-antagonist is practolol. Although it was not released for use in the United States, it was the first cardioselective β_1-antagonist to be extensively used in humans. However, because it produced several toxic effects, it is no longer in general use in most countries. In Great Britain it is used only in hospitals; its principal application is in the management of cardiac arrythmias.[2]

Although medicinal chemists have been successful in synthesizing a number of β_1-selective antagonists, it is not clear why some compounds exhibit selectivity and others do not. One common structural feature of a number of cardioselective antagonists is the presence of a *para*-substituent on the aromatic ring along with the absence of *meta*-substituents. Practolol and metoprolol are examples of this structural type. However, there are a sufficient number of exceptions

Practolol

Metoprolol

to show that this type of ring substitution is not the sole determinant of β_1-receptor selectivity.[49,50]

Metoprolol (Lopressor®) was released in 1978 for clinical use in the United States as an antihypertensive agent. Like propranolol, metoprolol undergoes extensive first-pass metabolism after oral administration. Metoprolol has a bioavailability of approximately 50 percent of an oral dose compared to approximately 30 percent for propranolol. The elimination half-life of metoprolol is 3 to 4 hours, which is very similar to that of propranolol.[40] But unlike propranolol, metoprolol does not appear to be converted to any pharmacologically important metabolites. Although metoprolol is considered to be a β_1-receptor- selective antagonist, it is important to be aware that most of its cardioselectivity is usually lost at full therapeutic dose. Therefore, in spite of its cardioselectivity at low doses, it is not normally indicated for patients with bronchial asthma.

Nadolol (Corgard®) is a nonselective β-blocker approved late in 1979 for general use in the United States as an antihypertensive agent and in the management of angina pectoris. Unlike propranolol, nadolol has no membrane-stabilizing properties. It also has no intrinsic sympathomimetic activity, a characteristic it shares with both propranolol and metoprolol.

Nadolol

Nadolol exhibits profound differences from propranolol and metoprolol with regard to its absorption and disposition. Whereas propranolol and metoprolol are almost completely absorbed after oral administration, only about 30 percent of an oral dose of nadolol is absorbed. In contrast to the extensive first-pass metabolism undergone by propranolol and metoprolol, nadolol appears to be excreted unchanged after either oral or intravenous administration.[51] It is excreted primarily by the kidneys and, because of its lack of metabolism, has a long duration of action. The serum half-life of nadolol is 12 to 20 hours, a property that permits the drug to be administered in single daily doses.

Timolol is a nonselective β-blocker approved in 1978 for general use in the United States for the management of glaucoma. Timolol has no significant membrane-stabilizing (local-anesthetic) action or intrinsic sympathomimetic activity. Although the chemical structure of the "aromatic ring" portion of timolol is significantly different from that found in other β-blockers, the side chain is structurally identical to the ones found in numerous other β-antagonists. It is supplied for clinical use as the more potent S-isomer.

Timolol

Although timolol is used topically to lower intra-ocular pressure in the treatment of glaucoma, it is an effective antihypertensive agent and it is marketed in Great Britain for use in the treatment of hypertension.[2] The mechanism whereby β-blockers lower intraocular pressure is not known with certainty, but it appears to be at least partially due to a β-receptor antagonism that results in a reduction in the formation and secretion of aqueous humor. It seems somewhat paradoxical that sympathomimetic agents such as epinephrine, phenylephrine, norepinephrine, and isoproterenol also are used to lower intraocular pressure because of their ability to inhibit aqueous-humor production. β-blockers offer an advantage over many other drugs used in the treatment of glaucoma, because they do not have an effect on pupil size and they do not cause spasm of the ciliary muscle. Timolol also offers an advantage over β-blockers such as propranolol in that it does not have a local anesthetic effect. The treatment of open-angle glaucoma with timolol and other β-blockers has been thoroughly reviewed.[52,53]

Two other drugs of interest that have β-receptor blocking activities are labetalol and butoxamine.

Labetalol

Butoxamine

Labetalol has both α-receptor and β-receptor antagonist properties. It is a more potent β-antagonist than α-antagonist, and it has membrane-stabilizing activity but no intrinsic sympathomimetic action. Labetalol is a clinically useful antihypertensive agent, although it has not been released for general use in the United States. The rationale for its use in the management of hypertension is that its α-receptor blocking effects produce vasodilatation and that its β-receptor blocking effects prevent the reflex tachycardia that is usually associated with vasodilatation.[2,54]

Butoxamine is of interest because it is a selective β_2-receptor antagonist. It blocks the β_2-receptors in uterine smooth muscle, in bronchial smooth muscle, and in skeletal muscle. Because of its β_2-receptor selectivity, butoxamine is a useful research tool, but it does not, at present, have clinical use.

PRODUCTS

Propranolol Hydrochloride U.S.P., Inderal®, 1-(isopropylamino)-3-(1-naphthyloxy)-2-propanol hydrochloride. Propranolol hydrochloride is a white to off-white crystalline solid, soluble in water or ethanol and insoluble in nonpolar solvents. Its therapeutic indications were discussed earlier in this chapter. Propranolol is contraindicated in the following instances: bronchial asthma; allergic rhinitis during the pollen season; sinus bradycardia and greater than first-degree block; cardiogenic shock; right ventricular failure secondary to pulmonary hypertension; congestive heart failure unless the failure is secondary to a tachyarrhythmia treatable with propranolol hydrochloride; patients who are taking adrenergic-augmenting psychotropic drugs. In patients with angina pectoris, the sudden withdrawal of propranolol or other β-blockers should be avoided in order to prevent exacerbation of the angina. This so-called *propranolol-withdrawal rebound* has been associated with fatal myocardial infarctions, and it is recommended that withdrawal from the drug be achieved by a gradual reduction of the dose.[55] Propranolol hydrochloride is supplied in both tablet and injectable form. For most therapeutic purposes, the dosage of propranolol must be individualized. For the management of hypertension, the usual initial dose is 40 mg. twice daily; for angina pectoris the usual initial dosage is 10 to 20 mg. three or four times daily.

Metoprolol Tartrate, Lopressor®, 1-isopropylamino-3-[*p*-(2-methoxyethyl)phenoxy]-2-propanol dextro-tartrate. Metoprolol tartrate is a water-soluble, white crystalline solid. It is a selective β_1-receptor antagonist that is indicated for use in the management of hypertension. Metoprolol tartrate is contraindicated in sinus bradycardia, heart block greater than first degree, cardiogenic shock, and overt cardiac failure. Abrupt withdrawal of the drug should be avoided and it should not be used by patients who have bronchospastic diseases unless the patients do not respond to, or cannot tolerate other antihypertensive drugs. In such cases, the lowest dose possible of metoprolol tartrate should be used, and a β_2-agonist should be administered concomitantly. Although the dose should be individualized, the usual initial dose is 50 mg. twice daily.[56] It is supplied only in tablet form.

Nadolol, Corgard®, 2,3-*cis*-5-[3-(1,1-dimethylethyl)amino-2-hydroxypropoxy]-1,2,3,4-tetrahydro-2,3-naphthalenediol. Nadolol is a nonselective β-blocker that has been approved for the management of hypertension and angina pectoris. Like propranolol and metoprolol, it is often used in combination with other antihypertensive agents, such as diuretics. It is contraindicated in the presence of bronchial asthma, sinus bradycardia and greater than first-degree conduction block, cardiogenic shock, and overt cardiac failure. Sudden cessation of nadolol therapy may exacerbate ischemic heart disease. Because it is excreted unchanged, it must be used with caution and in lower dose in patients with renal impairment. The usual dose in hypertension is 80 to 320 mg. daily, and in angina pectoris it is 80 to 240 mg. daily. Nadolol is supplied in tablet form.

Timolol Maleate U.S.P., Timoptic®, (S)-1(*tert*-butylamino)-3-[(4-morpholino-1,2,5-thiadiazol-3-yl)oxy]-2-propanol maleate. Timolol maleate is a nonselective β-blocker used in the management of chronic open-angle glaucoma and ocular hypertension. It is available as an ophthalmic solution in 0.25 and 0.5 percent concentrations. The usual starting dose is one drop of 0.25 percent solution in each eye twice a day. The systemic effects of ophthalmically administered timolol are usually minor and may include slight bradycardia or acute bronchospasm in patients with bronchospastic disease.

REFERENCES

1. Mayer, S. E., *in* Gilman, A. G., Goodman, L. S. and Gilman A. (eds.): The Pharmacological Basis of Therapeutics, 6th ed. Chap. 4, New York, Macmillan, 1980.
2. Day, M. D.: Autonomic Pharmacology, New York, Longman, Inc., 1979.
3. Trendelenburg, U.: Trends in Pharmacol. Sci. 1:4, 1979.
4. Ganellin, C. R.: J. Med. Chem. 20:579, 1977.
5. Fowler, C. J., *et al.*: Biochem. Pharmacol. 27:97, 1978.
6. Ahlquist, R. P.: Am. J. Physiol. 154:586, 1948.
7. Berthelsen, S. and Pettinger, W. A.: Life Sci. 21:595, 1977.
8. Wood, C. L., *et al.*: Biochem. Pharmacol. 28:1277, 1979.
9. Exton, J. H.: Biochem. Pharmacol. 28:2237, 1979.
10. Wood, C. L., *et al.*: Biochem. Biophys. Res. Comm. 88:1, 1979.
11. Lands, A. M., *et al.*: Nature 214:597, 1967.
12. Robison, G. A., *et al.*: Cyclic AMP, Chap. 2, New York, Academic Press, 1971.
13. Greengard, P.: Science 199:146, 1978.
14. Kunos, G.: Ann. Rev. Pharmacol. Toxicol. 18:291, 1978.
15. Belleau, B.: Ann. N. Y. Acad. Sci. 139:580, 1967.
16. Limbird, L. E., and Lefkowitz, R. J.: J. Biol. Chem. 252:799, 1977.
17. Limbird L. E., and Lefkowitz, R. J.: Mol. Pharmacol. 12:559, 1976.
18. Ross, E. M. and Gilman, A. G.: Proc. Natl. Acad. Sci. U.S.A. 74:3715, 1977.
19. Kent, R. S., DeLean A. and Lefkowitz, R. J.: Mol. Pharmacol. 17:14, 1980.
20. Caron, M. G., *et al.*: J. Biol. Chem. 254:2923, 1979.
21. Triggle, D. J., and Triggle, C. R.: Chemical Pharmacology of the Synapse, Chap. 3, New York, Academic Press, 1976.
22. Kaiser, C., *et al.*: J. Med. Chem. 18:674, 1975.
23. Sonnenblick, E. H., *et al.*: New Eng. J. Med. 300:17, 1979.
24. Barger, and Dale, H. H.: J. Physiol. 41:19, 1910.
25. Rohrmann, E. and Shonle, H.: J. Am. Chem. Soc. 66:1517, 1944.
26. Lands, A. M., *et al.*: J. Pharmacol. Exp. Ther. 89:271, 1947.
27. McDevitt, D. G.: Drugs 17:267, 1979.
28. Brogden, R. N., *et al.*: Drugs 14:163, 1977.
29. Hoffman, B. B., and Lefkowitz, R. J.: New Eng. J. Med. 302:1390, 1980.
30. Brugger, J.: Helv. Physiol. Pharmacol. Acta 3:117, 1945.
31. Eisleb, O.: U.S. Patent 1,949,247. Chem Abstr. 28:2850, 1934.
32. Nickerson, M., and Goodman, L. S.: J. Pharmacol. Exp. Ther. 89:167, 1947.

33. Ullyot, G. E., and Kerwin, J. F.: Medicinal Chemistry, vol. 2, p. 234, New York, Wiley, 1956.
34. Henkel, J. G., *et al.:* J. Med. Chem. 19:6, 1976.
35. Powell, C. E., and Slater, I. H.: J. Pharmacol. Exp. Ther. 122:480, 1958.
36. Moran, N. C., and Perkins, M. E.: J. Pharmacol. Exp. Ther. 124:223, 1958.
37. Black, J. W., and Stephenson, J. S.: Lancet 2:311, 1962.
38. Kaiser, C., *et al.:* J. Med. Chem. 20:687, 1977.
39. Shand, D. G.: Circulation 52:6, 1975.
40. Frishman, W.: Am. Heart J. 97:663, 1979.
41. Walle, T., *et al.:* Clin. Pharmacol. Ther. 26:548, 1979.
42. Walle, T., and Gaffney, T. E.: J. Pharmacol. Exp. Ther. 182:83, 1972.
43. Fitzgerald, J. D., and O'Donnell, S. R.: Br. J. Pharmacol. 43:222, 1971.
44. Walle, T., *et al.:* Clin. Pharmacol. Ther. 27:23, 1980.
45. Walle, T., *et al.:* Clin. Pharmacol. Ther. 26:686, 1979.
46. Daniell, H. B.: *et al.:* J. Pharmacol. Exp. Ther. 208:354, 1979.
47. Daly, M. J., and Levy, G. P.: The subclassification of β-adrenoceptors: Evidence in support of the dual β-adrenoceptor hypothesis, *in* Kalsner, S. (ed.): Trends in Autonomic Pharmacology, Vol. 3, p. 347, Baltimore, Urban and Schwarzenberg, 1979.
48. Barnett, D. B., *et al.:* Nature 273:167, 1978.
49. Clarkson, R., *et al.:* Ann. Reports Med. Chem. 10:51, 1975.
50. Evans, D. B., *et al.:* Ann. Reports Med. Chem. 14:81, 1979.
51. Dreyfuss, J., *et al.:* J. Clin. Pharmacol. 17:300, 1977.
52. Boger, W.: Drugs 18:25, 1979.
53. Heel, R. C., *et al.:* Drugs 17:38, 1979.
54. Brogden, R. N., *et al.:* Drugs 15:251, 1978.
55. Miller, R. R., *et al.:* New Eng. J. Med. 293:416, 1978.
56. Koch-Weser, J.: New Eng. J. Med. 301:698, 1979.

SELECTED READINGS

Baselt, R. C.: Decongestants, *in* Disposition of Toxic Drugs and Chemicals in Man, Vol. 2, p. 70, Canton, Connecticut, Biomedical Publications, 1978.

Dring, L. G., and Millburn, P.: Sympathomimetics and Bronchodilators, *in* Hathway, D. E. (ed.): Foreign Compound Metabolism in Mammals, Vol. 5, Chap. 8, London, Burlington House, 1979.

Lefkowitz, R. J.: Recent Developments in Adrenergic Receptor Research, Ann. Reports Med. Chem. 15:217, 1980.

Modell, W., Schild, H. O., and Wilson, A.: Adrenergic Mechanisms, *in* Applied Pharmacology, Chap. 16, Philadelphia, W. B. Saunders, 1976.

Patil, P. N., Miller, D. D., and Trendelenburg, U.: Molecular Geometry and Adrenergic Drug Activity, Pharmacol. Rev. 26:323, 1975.

Triggle, D. J.: Adrenergics: Catecholamines and Related Agents, *in* Wolff, M. E. (ed.): Burger's Medicinal Chemistry, 4th ed., Part III, p. 225, New York, John Wiley & Sons, 1981.

12

cholinergic drugs and related agents

George H. Cocolas

Few systems, if any, have been studied as extensively as those innervated by neurons that release acetylcholine at their endings. Since the classical studies of Dale,[1] who described the actions of esters and ethers of choline on isolated organs and their relationship to muscarine, pharmacologists, physiologists, chemists, and biochemists have applied their knowledge to understand the actions of the cholinergic nerve and its neurotransmitter. This chapter includes the drugs and chemicals that act on cholinergic nerves or the tissues that they innervate to either mimic or block the ac-

$$(CH_3)_3 \overset{+}{-} N - \overset{\alpha}{C}H_2 - \overset{\beta}{C}H_2 - O - \overset{\overset{O}{\parallel}}{C} - CH_3$$

Acetylcholine

tion of acetylcholine. Drugs that mimic the action of acetylcholine do so either by acting directly on the cholinergic receptors in the tissue or by inhibiting acetylcholinesterase, which prolongs the action of the neurotransmitter, acetylcholine. An understanding of acetylcholine-receptor function has been greatly augmented by studies on the electric organs from the marine ray Torpedo californica and the electric eel Electrophorus electricus.[2,3] Acetylcholine receptor has been isolated as a membrane bound protein from *Torpedo* and *Electrophorus* electroplax tissues; it displays all the properties of a nicotinic cholinergic receptor. The isolated protein complex in its native lipid environment retains many of the properties characteristic of the intact electroplax cells.[4] These include binding of cholinergic agonists (acetylcholine) and antagonists

(d-tubocurarine and α-neurotoxins), ion-permeability changes to physiologically significant cations (K^+, Na^+, Ca^{+2}), and pharmacological desensitization in response to prolonged exposure to agonists.[5]

Cholinergic nerves are found in the peripheral and the central nervous system of humans. Synaptic terminals in the cerebral cortex, corpus striatum, hippocampus, and several other regions in the central nervous system are rich in acetylcholine and in the enzymes that synthesize and hydrolyze this neurotransmitter. There is strong evidence that acetylcholine serves as a neurotransmitter in the central nervous system and although its function in the brain and brain stem is not clear, it has been implicated in memory and behavioral activity in humans.[6] The peripheral nervous system consists of those nerves outside the cerebrospinal axis and includes the somatic nerves and the autonomic nervous system. The somatic nerves are made up of a sensory (afferent) nerve and a motor (efferent) nerve. The motor nerves arise from the spinal cord and project uninterrupted throughout the body to all skeletal muscle. Acetylcholine mediates transmission of impulses from the motor nerve to skeletal muscle (i.e., neuromuscular junction). The neurotransmitter in sensory neurons is unknown.

The autonomic nervous system is composed of two divisions: the sympathetic and parasympathetic. Acetylcholine serves as a neurotransmitter at both sympathetic and parasympathetic preganglionic nerve endings, postganglionic nerve fibers in the parasympathetic, and some postganglionic fibers (e.g., salivary and sweat glands) in the sympathetic division of the autonomic nervous system. The autonomic system regulates the activities of smooth muscle and glandular secretions which function, as a rule, below the level of consciousness, e.g., respiration, circulation, digestion, body temperature, metabolism, etc. The two divisions have contrasting effects on the internal environment of the body. The sympathetic division frequently discharges as a unit, especially during conditions of rage or fright, and expends energy. The parasympathetic is organized for discrete and localized discharge and acts to store and conserve energy.

Drugs and chemicals that cause the parasympathetic system to react are termed *parasympathomimetics*, while those blocking the actions are called *parasympatholytics*. Agents that mimic the sympathetic system are *sypathomimetics* and those that block the actions are *sympatholytics*. Another classification used to describe drugs and chemicals acting on

the nervous system or the structures that the fibers innervate is based on the neurotransmitter released at the nerve ending. Drugs acting on the autonomic nervous system are divided into *adrenergic* for those postganglionic sympathetic fibers that release norepinephrine and epinephrine, and *cholinergic* for the remaining fibers in the autonomic nervous system and the motor fibers of the somatic nerves that release acetylcholine. Cholinergic drugs are also categorized according to the types of structures they affect. Thus, agents that act primarily on smooth muscle and stimulate secretory glands mimic the action of the alkaloid muscarine and are called *muscarinic*. Chemicals that stimulate skeletal muscle and autonomic ganglia are defined as *nicotinic* because their action resembles the action of nicotine.

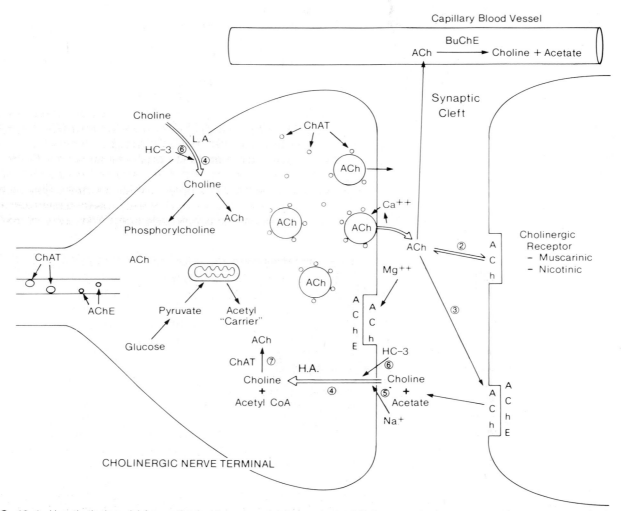

FIG. 12–1. *Hypothetical model for synthesis, storage, and release of acetylcholine. 1. Newly formed ACh is preferentially released under influence of Ca^{++} but inhibited by Mg^{++}. 2. ACh crosses synaptic cleft and reacts reversibly with cholinergic receptors or is (3) hydrolyzed by AChE to choline and acetate. 4. Choline is taken up by the neuron by a high-affinity (H.A.) or low-affinity (L.A.) uptake system. 5. Na^{+}-dependent, high-affinity uptake system of choline stimulates ACh synthesis. 6. Hemicholinium (HC3) specifically blocks both high and low uptake systems. 7. ACh is synthesized by ChAT and stored in synaptic vesicles. 8. Acetyl CoA is released from the mitochondria using acetate formed from glucose in the cytosol. 9. Choline from the low-affinity uptake system forms ACh (not released on nerve stimulation) and phosphorylcholine.*

CHOLINERGIC NEUROCHEMISTRY

Cholinergic neurons synthesize, store, and release acetylcholine (Fig. 12-1). The neurons also form choline acetyltransferase (ChAT) and acetylcholinesterase (AChE). These enzymes are synthesized in the soma of the neuron and transported down the axon through microtubules to the nerve terminal. Acetylcholine is prepared at the nerve ending by a transfer of an acetyl group from acetyl CoA to choline. The reaction is catalyzed by choline acetyltransferase. Cell-fractionation studies show that much of the acetylcholine is contained in synaptic vesicles at the nerve ending but that some is also free in the cytosol. The major source of choline for acetylcholine synthesis in vivo is hydrolysis of acetylcholine in the synapse. The choline formed is recaptured by the presynaptic terminal by a high-affinity uptake system under the influence of sodium ions[7,8] and is used to synthesize the acetylcholine released from the synaptic vesicles by the nerve-action potential. Free acetylcholine and phosphorylcholine found in the cytosol are prepared from choline brought into the neuron by the low-affinity uptake system but are not active in the nerve transmission process.

A number of quaternary ammonium bases act as competitive inhibitors of choline uptake. Hemicholinium (HC-3), a bis quaternary cyclic hemiacetal, and the triethyl analog of choline, 2-hydroxyethyltriethylammonium, act at the presynaptic membrane to inhibit the high-affinity uptake system of choline into the neuron. These compounds cause a delayed paralysis at repetitively activated cholinergic synapses and produce respiratory paralysis in test animals. The delayed block is due to the depletion of stored acetylcholine and may be reversed by choline. The acetyl group used for the synthesis of acetylcholine is obtained by conversion of glucose to pyruvate in the cytosol of the neuron and eventual formation of acetyl CoA. Owing to the impermeability of the mitochondrial membrane to acetyl CoA, this substrate is brought into the cytosol by the aid of an acetyl "carrier."

Hemicholinium (HC-3)

$$HO-CH_2CH_2-\overset{+}{N}(C_2H_5)_3$$

2-Hydroxyethyltriethylammonium

The synthesis of acetylcholine from choline and acetyl CoA is catalyzed by choline acetyltransferase.

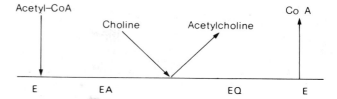

FIG. 12-2. *Ordered synthesis of acetylcholine by choline acetyltransferase.*

Transfer of the acetyl group from acetyl CoA to choline is an ordered reaction of the Theorell-Chance type, in which acetyl CoA first binds to the enzyme, forming a complex (EA) that then binds choline. The acetyl group is transferred and the acetylcholine that is formed dissociates from the enzyme active site. CoA is then released from the enzyme complex, EQ, to regenerate the free enzyme. The scheme is diagrammed in Figure 12-2. Choline acetyltransferase is inhibited in vitro by *trans*-N-methyl-4-(1-naphthylvinyl)pyridinium iodide;[9] however, its inhibitory activity in whole animals is unreliable.[10]

trans–N–Methyl–4(1–naphthylvinyl)pyridinium iodide

Newly formed acetylcholine is preferentially released from the presynaptic membrane when a nerve-action potential invades a presynaptic nerve terminal.[11,12] The release of acetylcholine results from a depolarization of the nerve terminal by the action potential, which alters membrane permeability to calcium ions. Calcium enters the nerve terminal and causes the release of the contents of several synaptic vesicles containing acetylcholine by exocytosis into the synaptic cleft. This burst or quantal release of acetylcholine causes depolarization of the postsynaptic membrane. The number of quanta of acetylcholine released may be as high as several hundred at a neuromuscular junction, with each quantum containing between 12,000 to 60,000 molecules. Acetylcholine is also released spontaneously in small amounts from presynaptic membranes. This small amount of neurotransmitter serves to maintain muscle tone by acting on the cholinergic receptors on the postsynaptic membrane.

The released acetylcholine either may diffuse out of the cleft and be hydrolyzed by acetylcholinesterase into choline and acetate or may bind to receptors in the postsynaptic membrane. Some of these receptors that bind acetylcholine may undergo a conformational change that opens ionic channels, causing an increase in conduction of the postsynaptic membrane. The open channels allow Na^+ to enter the cell and K^+ to flow out of the cell as these ions seek to decrease their electrochemical gradients. More Na^+ enters the cell than K^+ leaves it, resulting in a net influx of positive charges and depolarization of the membrane. When depolarized to a threshold potential (e.g., -50 mV) the membrane generates an action potential that propagates the stimulus.

The activation of cholinergic receptors by acetylcholine produces varied effects in different tissues. Skeletal and smooth muscle is stimulated by acetylcholine to cause contraction. The acinar cells of the exocrine pancreas are stimulated by the application of acetylcholine. Each of these target tissues is stimulated by the increased influx of Na^+ over K^+ to cause tissue depolarization. The action of acetylcholine on heart muscle through the vagus causes an increase in the K^+ permeability of atrial muscle membranes resulting in hyperpolarization and a slowing down of the cell-to-cell conduction of the action potential.

After acetylcholine has been released into the synaptic cleft, its concentration decreases rapidly. It is generally accepted that there is enough acetylcholinesterase at nerve endings to hydrolyze into choline and acetate any acetylcholine that has been liberated. The acetylcholinesterase activity from rat intercostal muscle is able to hydrolyze about 2.7×10^8 acetylcholine molecules in 1 msec; this far exceeds the 3×10^6 molecules released there by one nerve impulse.[13] In skeletal muscle the greater part of acetylcholinesterase found at the end-plate region is located postsynaptically. Acetylcholinesterase is present predominantly on presynaptic membranes in sympathetic ganglia.

The major sites of activity of acetylcholine are as follows: (1) the postganglionic parasympathetic (muscarinic) receptor, (2) the autonomic ganglia, and (3) the skeletal neuromuscular junction. The first is blocked by atropine, the second by hexamethonium,

and the third by decamethonium or curare. The above data might imply that the receptor sites are different, but the agonist acetylcholine may adopt a specific conformation to interact effectively with the appropriate receptor site. On the other hand, it is possible that all the receptor sites are very similar but that the immediate adjacent areas are different.

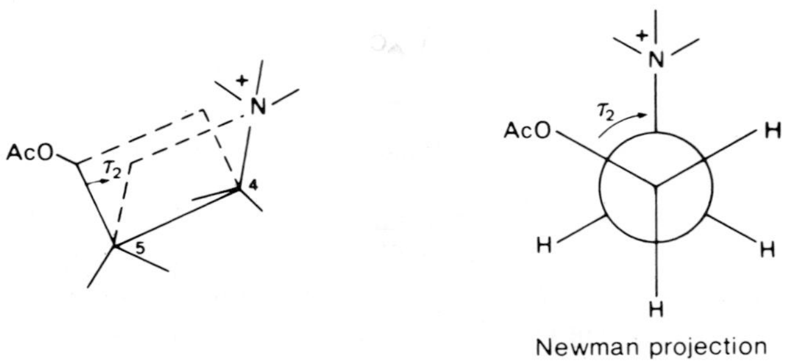

CHOLINERGIC AGONISTS

CHOLINERGIC STEREOCHEMISTRY

Three techniques have been used to study the conformational properties of acetylcholine and other cholinergic chemicals: roentgenographic (x-ray) crystallography, nuclear magnetic resonance, and molecular orbital calculations. Each of these methods may report the spatial distribution of atoms in a molecule in terms of torsion angles. However, they provide only circumstantial evidence about the conformation of the chemical as it acts on the biological receptor.

A torsion angle is defined as the angle formed between two planes as, for example, by the O-C5-C4-N atoms in acetylcholine. The angle between the oxygen and nitrogen atoms is best depicted by means of Newman projections (Fig. 12-3).

A torsion angle has a positive sign when the bond to the front atom is rotated to the right to eclipse the bond of the rear atom. The spatial orientation of acetylcholine is described by four torsion angles (Fig. 12-4).

The conformation of the choline moiety of acetylcholine has drawn the most attention in attempting to relate structure and pharmacological activity. The torsion angle ($\tau 2$) determines the spatial orientation of the cationic head of acetylcholine to the ester group. Roentgenographic diffraction studies have shown that the torsion angle ($\tau 2$) on acetylcholine has a value of $+77°$. Many compounds containing a choline component (i.e., $O\text{-}C\text{-}C\text{-}N^+(CH_3)_3$) have a preferred synclinal (gauche) conformation with $\tau 2$ values ranging from 68–89° (Table 12-1). Intermolecular packing forces in the crystal, as well as electrostatic interactions between charged nitrogen group and the ether oxygen of

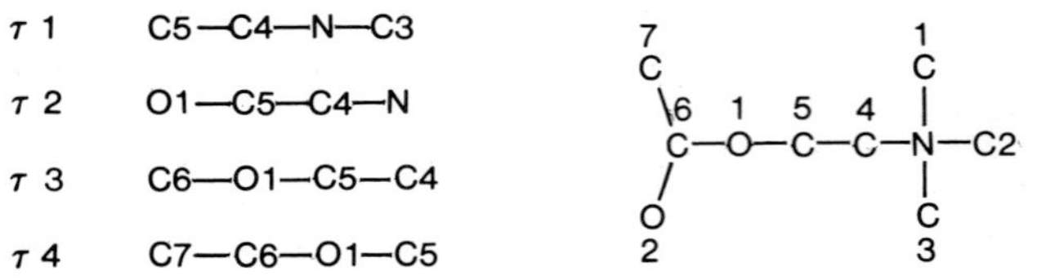

FIG. 12-3. *Gauche conformation of acetylcholine.*

Newman projection

$\tau 1$	C5—C4—N—C3
$\tau 2$	O1—C5—C4—N
$\tau 3$	C6—O1—C5—C4
$\tau 4$	C7—C6—O1—C5

FIG. 12-4. *Torsion angles of acetylcholine.*

the ester group, are probably the two dominant factors that lead to a preference for the synclinal conformation in the crystal state. Some choline esters display an antiperiplanar *(trans)* conformation between the onium and ester groups. For example, carbamoylcholine chloride ($\tau2$ +178°) is stabilized in this conformation by several hydrogen bonds,[14] acetylthiocholine iodide ($\tau2$ +171°) is in this conformation because of the presence of the more bulky and less electronegative sulfur atom, and (+) *trans*-1S,2S-acetoxycyclopropyl trimethylammonium iodide ($\tau2$ +137°) is fixed in this conformation by the rigidity of the cyclopropyl ring.

NMR spectroscopy of cholinergic molecules in solution is more limited than crystallography in delineating the conformation of compounds and is restricted to determining the torsion angle O1-C5-C4-N. Most of the NMR data are in agreement with the results of roentgenographic diffraction studies. NMR studies indicate that acetylcholine and methacholine are apparently not in their most stable *trans* conformation but exist in one of two gauche conformers[15] (Fig. 12-5). This agreement with the roentgenographic data may be a result of strong intramolecular interactions that stabilize these molecules and are not disturbed by solvents.[16]

Molecular orbital calculations based on the principles of quantum mechanics may be used to determine energy minima of rotating bonds and, therefore, to establish preferred conformations for the molecule. The measurements consider the molecules only in an isolated state and therefore do not take into account solvent interactions. Calculations[17] using the Hückel mo-

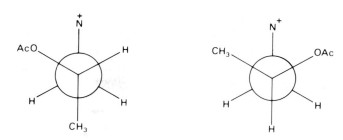

FIG. 12–5. *Gauche conformers of acetyl β-methacholine.*

lecular orbital method have found that acetylcholine has an energy minimum for the $\tau2$ torsion angle at about 80° and that the preferred conformation for the entire molecule of acetylcholine corresponds closely to that found in the crystal state.

The exact conformation adopted by acetylcholine to exert its cholinergic activity in vivo is not known. Studies of compounds that have significant muscarinic activity and contain semirigid or rigid structures have been used to suggest the active conformation of acetylcholine. The muscarinic receptor is stereoselective in its interaction with cholinergic agents. From this information a hypothesis about the conformational properties of acetylcholine at the muscarinic receptor has been developed.

The parasympathomimetic effects of muscarine were first reported in 1869,[18] but its structure was not elucidated until 1957.[19] Muscarine has four geometric isomers: muscarine, epimuscarine, allomuscarine, and epiallomuscarine (Fig. 12-6), none of which has a center or plane of symmetry. Each geometric isomer can exist as an enantiomeric pair.

The activity of muscarine is highly specific and resides primarily in the naturally occurring (+) muscarine. It is essentially free of nicotinic activity and apparently has the optimal stereochemistry to act on the muscarinic receptor. Synthetic molecules having a substituent on the carbon atom that corresponds to the β-carbon atom of acetylcholine show great differences in muscarinic activity between their isomers.

cis 2-Methyl-4-trimethylammonium-1,3-dioxolane

Acetyl (+) S β-methylcholine, (+) *cis*-2S-methyl-4R-trimethyl ammonium-1,3-dioxolane, (+) *trans*-1S,2S-acetoxycyclopropyl trimethylammonium, and naturally occurring (+) 2S,3R,5S-muscarine are more potent than their enantiomers and have very high ratios of activity between the S and R isomers (Table 12-2).

TABLE 12–1

CONFORMATIONAL PROPERTIES OF SOME CHOLINERGIC AGENTS*

Compound	O1-C5-C4-N Torsion Angle
Acetylcholine bromide	+ 77
Acetylcholine chloride	+ 85
(+) -2S,3R,5S-muscarine iodide	+ 73
Methylfurmethide iodide	+ 83
(+) Acetyl Sβ-methylcholine iodide	+ 85
(−) Acetyl Rα-methylcholine iodide	
Crystal form A	+ 89
Crystal form B	−150
(+) *cis* 2S-methyl-4R-trimethylammonium-1,3-dioxolane iodide	+ 68
(+) *trans* 1S,2S-acetoxycyclopropyltrimethylammonium iodide	+137
Carbamoylcholine bromide	+178
Acetylthiocholine bromide	+171
Acetyl Rα Sβ-dimethylcholine iodide (Erythro)	+ 76

* From Shefter, E., *in* Triggle, D. J., Moran, J. F., and Barnard, E. A., (eds.): Cholinergic Ligand Interactions, New York, Academic Press, 1971.

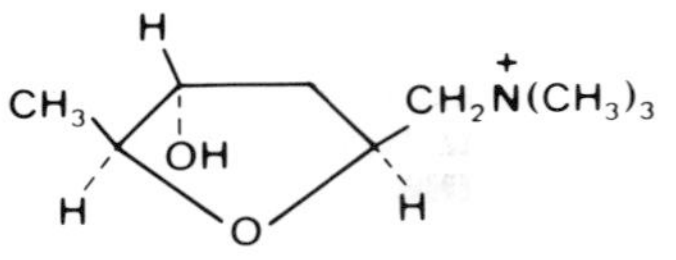

Muscarine

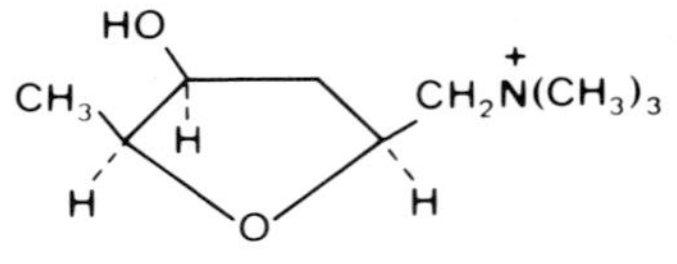

Epimuscarine

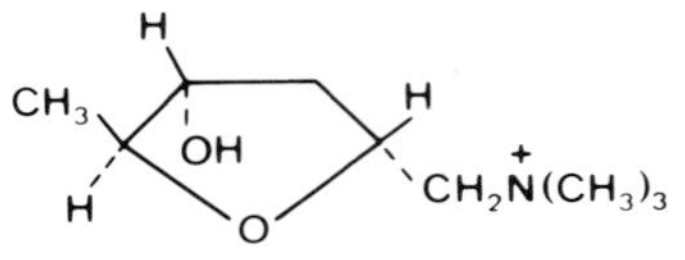

Allomuscarine

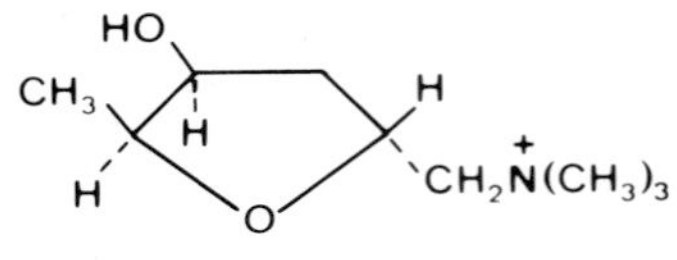

Epiallomuscarine

FIG. 12–6. *Geometric isomers of muscarine.*

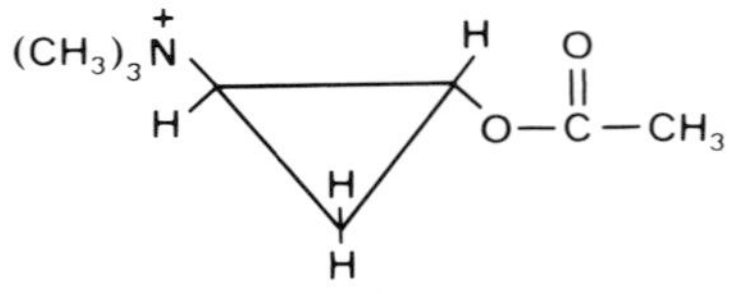

trans 2–Acetoxycyclopropyl Trimethylammonium

Each of these more active isomers has the same configuration at the carbon atom adjacent to the ester group. Although acetylcholine does not have an asymmetric center, its preferred conformation gives it the characteristics of an asymmetric molecule and perhaps acts in this orientation during its neurotransmit-

TABLE 12–2

EQUIPOTENT MOLAR RATIOS OF ISOMERS ON GUINEA-PIG ILEUM; RATIOS RELATIVE TO ACETYLCHOLINE

Compound	Guinea Pig Ileum	S to R Ratio
(+) Acetyl Sβ-methylcholine Chloride	1.0*	240
(−) Acetyl Rβ-methylcholine Chloride	24.0*	
(+) 2S, 3R, 5S-Muscarine Iodide	0.33†	394
(−) 2R, 3S, 5R-Muscarine Iodide	130†	
(+) cis 2S-Methyl-4R-trimethylammonium-1,3-dioxolane Iodide	6.00‡	100
(−) cis 2R-Methyl-4S-trimethylammonium-1,3-dioxolane Iodide	0.06‡	
(+) trans 1S,2S-Acetoxycyclopropyltrimethylammonium Iodide	0.88§	517
(−) trans 1R,2R-Acetoxycyclopropyltrimethylammonium Iodide	455§	

   * Beckett, A. H., Harper, N. J., Clitherow, J. W., and Lesser, E., Nature 189:671, 1961.
   † Waser, Pl: Pharmacol. Rev. 13:465, 1961.
   ‡ Bellean, B., and Puranen, J.: J. Med. Chem. 6:235–328, 1963.
   § Armstrong, P. D., Cannon, J. G., and Long, J. P.: Nature 220:65–66, 1968.

ter role. A similar observation may be made of (+) acetyl S β-methylcholine, (+) *cis*-2S-methyl-4R-trimethylammonium-1,3-dioxolane and (+) *trans*-1S,2S-acetoxycyclopropyl trimethylammonium, all of which have an S configuration at the carbon atom adjacent to the ester group. Each of these active muscarinic molecules may be deployed essentially free of steric interference on the receptor, in the same manner as acetylcholine and (+) muscarine. Their S to R ratios (Table 12-2) show the great degree of stereoselectivity of the muscarinic receptor in guinea-pig ileum for the configuration at the carbon adjacent to the ester group.

The nicotinic receptor is not considered as highly stereoselective as its muscarinic counterpart.

STRUCTURE–ACTIVITY RELATIONSHIPS OF CHOLINERGIC AGONISTS

Acetylcholine is a relatively simple molecule. Both its chemistry and its ease of testing for activity have allowed numerous chemical derivatives to be made and studied. Alterations on the molecule may be divided into three categories: the onium group, the ester function, and the choline moiety.

The onium group is essential for the intrinsic activity and contributes to the affinity of the drug to the receptors, partially through the binding energy and partially because of its action as a detecting and directing group. The trimethylammonium group is the optimal functional group for activity, although some significant exceptions are known, e.g., pilocarpine, arecoline, nicotine and oxytremorine. Phosphonium, sulfonium, arsenonium isosteres, or substituents larger than methyl on the nitrogen increase the size of the onium moiety, resulting not only in a diffusion of the charge and steric interference to proper drug-receptor interaction but also in a decrease in affinity (Table 12-3).

The ester group in acetylcholine contributes to the binding of the compound to the parasympathetic terminal synapse, probably because of its hydrogen bond-forming capacity. A comparison of the cholinergic activity of a series of alkyl trimethylammonium compounds (R-$\overset{+}{N}$(CH$_3$)$_3$; R = C$_1$-C$_9$) shows that n-amyl-trimethylammonium[20] which may be considered to have a similar size and mass as acetylcholine, is about one magnitude weaker as a muscarinic agonist. The presence of the acetyl group in acetylcholine is not as critical as is the size of the molecule. In studying a series of n-alkyl trimethylammonium salts it was noted[21] that for maximal muscarinic activity the quaternary ammonium group should be followed by a chain of 5 atoms; this was referred to as the *five-atom rule.* For the receptors for acetylcholine at the ganglionic synapse, compounds with a chain length of four atoms are more active than similar compounds with a chain length of five atoms.

Beckett[22] observed that a trimethylammonium group and ether oxygen were present in most active muscarinic agonists. It was proposed that the muscarinic receptor contained an anionic binding site (*Site 1,* Fig. 12-7) that accommodates the quaternary group of muscarine approximately 3.0 Å from a hydrogen bonding site (*Site 2,* Fig. 12-7) for the ring oxygen of muscarine or the other oxygen of the choline moiety. A distance of 5.7 Å separates the quaternary nitrogen from the alcohol group of muscarine, which is the location

of a subsidiary site (*Site 3,* Fig. 12-7). Site 3 can interact with the carbonyl group of acetylcholine, the ether oxygen of the dioxolane, and the double bond of furan.

TABLE 12-3

ACTIVITY OF ACETOXYETHYL ONIUM SALTS: EQUIPOTENT MOLAR RATIOS RELATIVE TO ACETYLCHOLINE*

CH$_3$COOCH$_2$CH$_2^-$	Cat blood pressure	Intestine	Frog heart
$\overset{+}{N}$Me$_3$	1	1 (Rabbit)	1
$\overset{+}{N}$Me$_2$H	50	40	50
$\overset{+}{N}$MeH$_2$	500	1,000	500
$\overset{+}{N}$H$_3$	2,000	20,000	40,000
$\overset{+}{N}$Me$_2$Et	3	2.5 (Guinea pig)	2
$\overset{+}{N}$MeEt$_2$	400	700	1,500
$\overset{+}{N}$Et$_3$	2,000	1,700	10,000†
$\overset{+}{P}$Me$_3$	13	12 (Rabbit)	12
$\overset{+}{A}$sMe$_3$	66	90	83
$\overset{+}{S}$Me$_2$	50	30 (Guinea pig)	96

Size of quaternary atom:‡

$$\underset{C \xleftarrow{\ d'\ } C}{\overset{\overset{+}{X}}{\nearrow\ \ \nwarrow}}$$

N	d = 1.47 Å	d′ = 2.4 Å
P	1.87	3.05
S	1.82	—
As	1.98	3.23

* Welsh, A. D., and Roepke, M. H.: J. Pharmacol. Exp. Ther. 55:118, 1935; Stehle, K. L., Melville, K. J., and Oldham, F. K.: J. Pharmacol. Exp. Ther. 56:473, 1936; Holton, P., and Ing, H. R.: Brit. J. Pharmacol. 4:190, 1949; Ing, H. R., Kordik, P., and Tudor Williams, D. P. H.: Brit. J. Pharmacol. 7:103, 1952.

† Reduces effect of acetylcholine

‡ From Barlow, R. B.: Introduction to Chemical Pharmacology, London, Methuen and Co. Ltd., 1964.

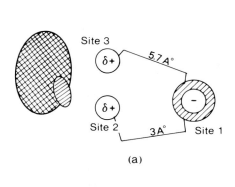

FIG. 12-7. Hypothetical structure of the muscarinic receptor (adapted from Michelson, M.J., and Zeimal, E.V. (eds.), ''Acetycholine,'' Oxford Pergamon Press, 1973.) A. Diagram of distribution of charges and hydrophobic portions (shaded) and large flat area for van der Waal's interaction (cross-hatched). B. Diagram of interaction of muscarine with muscarinic receptor. C. Diagram of interaction of acetyl β-methylcholine with muscarinic receptor.

The ether oxygen appears to be of primary importance for high muscarinic activity, because both choline ethyl ether and β-methylcholine ethyl ether have high muscarinic activity.[23,24] The attachment of the choline part of acetylcholine to the receptor seems to be related to that of (+) muscarine, or vice versa.

CH$_3$CH$_2$—O—CH$_2$CH$_2$$\overset{+}{N}$(CH$_3$)$_3$
Choline Ethyl Ether

CH$_3$CH$_2$—O—$\overset{\overset{\displaystyle CH_3}{|}}{C}$H—CH$_2$$\overset{+}{N}$(CH$_3$)$_3$
β–Methylcholine Ethyl Ether

Either shortening or lengthening the chain of atoms that separates the ester group from the onium moiety reduces muscarinic activity. Alpha-substitution on the choline moiety decreases both nicotinic and muscarinic activity, but the muscarinic activity is decreased to a greater extent. Nicotinic activity is decreased to a greater degree by substitution on the β-carbon. Therefore, acetyl α-methylcholine has nicotiniclike activity but little muscariniclike activity, and acetyl β-methylcholine (methacholine) has muscariniclike properties but little nicotinic activity. Hydrolysis by acetylcholinesterase is affected more by substitutions on the β-carbon than on the alpha. The hydrolysis rate of racemic acetyl β-methylcholine is about 50 percent of that of acetylcholine; racemic acetyl α-methylcholine is hydrolyzed about 90 per cent as fast.

PRODUCTS

Acetylcholine Chloride. Acetylcholine exerts a powerful stimulation of the parasympathetic nervous system. Attempts have been made to use it as a cholinergic agent,[25,26] but its duration of action is too short for sustained effects, owing to rapid hydrolysis by esterases and lack of specificity when administered for systemic effects. Acetylcholine is a cardiac depressant and an effective vasodilator. The stimulation of the vagus and the parasympathetic nervous system produces a tonic action on smooth muscle and induces a flow from the salivary and lacrimal glands. Its cardiac-depressant effect results from a negative chronotropic effect that causes a decrease in heart rate and from a negative inotropic action on heart muscle that produces a decrease in the force of myocardial contractions. The vasodilator action of acetylcholine is observed to be primarily on the arteries and the arterioles, with distinct effect on the peripheral vascular system. Bronchial constriction is a characteristic side-effect when the drug is given systemically.

CH$_3$—$\overset{\overset{\displaystyle CH_3}{\overset{+}{|}}}{\underset{\underset{\displaystyle CH_3}{|}}{N}}$—CH$_2CH_2$—O—$\overset{\overset{\displaystyle O}{||}}{C}$—CH$_3$ Cl$^-$
Acetylcholine Chloride

One of the most effective antagonists to the action of acetylcholine is atropine. Atropine blocks out the effect of acetylcholine on cardiac muscle and its production of peripheral vasodilatation (i.e., muscarinic effects) but does not affect the skeletal muscle contraction (i.e., nicotinic effect) that is produced. Acetylcholine chloride is a hygroscopic powder that is available in an admixture with mannitol to be dissolved in sterile water for injection shortly before use. It works as a short-acting miotic when introduced into the anterior chamber of the eye and is especially useful after cataract surgery during the placement of sutures. When applied topically to the eye it has little therapeutic value because of poor corneal penetration and rapid hydrolysis by acetylcholinesterase.

Methacholine Chloride U.S.P., acetyl β-methylcholine chloride, (2-hydroxypropyl)trimethylammonium chloride acetate.

CH$_3$—$\overset{\overset{\displaystyle CH_3}{\overset{+}{|}}}{\underset{\underset{\displaystyle CH_3}{|}}{N}}$—CH$_2$$\overset{\overset{\displaystyle CH_3}{|}}{C}$H—O—$\overset{\overset{\displaystyle O}{||}}{C}$—CH$_3$ Cl$^-$
Methacholine Chloride

Methacholine is the acetyl ester of β-methylcholine. Unlike acetylcholine, methacholine has sufficient stability in the body to give sustained parasympathetic stimulation, and this action is accompanied by little (1/1000 that of acetylcholine) or no "nicotinic effect." It exerts a depressant action on the cardiac auricular mechanism (which is blocked by quinidine), a stimulation of gastrointestinal peristalsis, and a general vasodilation followed by a fall in blood pressure. All of these effects are rapidly and completely blocked by atropine but are intensified and prolonged by physostigmine and neostigmine, which inhibit acetylcholinesterase.

Methacholine can exist as S and R enantiomers. Although the chemical is used as the racemic mixture, its muscarinic activity resides principally in the S isomer. The S to R ratio of muscarinic potency for these enantiomers is 240 to 1.

(+) Acetyl S β-methylcholine is hydrolyzed by acetylcholinesterase, whereas the R(−) isomer is not. (−) Acetyl R β-methylcholine is a weak competitive inhibitor (K$_i$ 4 × 10^{-4} M.) of acetylcholinesterase obtained from the electric organ of the eel (Electrophorus electricus). The hydrolysis rate of the S($+$) isomer is about 54 percent that of acetylcholine. This rate probably compensates for any decreased association (affinity) due to the β-methyl group with the muscarinic receptor site and may account for the fact that acetylcholine and ($+$) acetyl β-methylcholine have equimolar muscarinic potencies in vivo. ($-$) Acetyl R β-methylcholine weakly inhibits acetylcholinesterase and slightly reinforces the muscarinic activity of the S($+$) isomer in the racemic mixture of acetyl β-methylcholine.

In the hydrolysis of the acetyl α- and β-methylcholines, the greatest stereochemical inhibitory effects occur when the choline is substituted in the β-position. This also appears to be true of organophosphorus inhibitors (see below). The anionic site contributes to only a small degree to the stereospecificity of acetylcholinesterase. The R($-$) and S($+$) isomers of acetyl α-methylcholine are hydrolyzed at 78 percent and 97 percent of the rate of acetylcholine.

Methacholine chloride occurs as colorless or white crystals or as a white crystalline powder. It is odorless or has a slight odor and is very deliquescent. It is freely soluble in water, alcohol, or chloroform, and its aqueous solution is neutral to litmus and has a bitter taste. It is rapidly hydrolyzed in basic solution. Solutions are relatively stable to heat and will keep for at least 2 or 3 weeks when refrigerated to delay growth of molds.

Methacholine chloride may be administered subcutaneously for paroxysmal atrial tachycardia. The salutory effects may result from a transient stimulation of the vagal components, which results in the lowering of the heartbeat and the reestablishment of normal cardiac rhythmn.

Carbachol U.S.P., choline chloride carbamate. The pharmacological activity of carbachol is similar to that of acetylcholine. It is an ester of choline and thus possesses both muscarinic and nicotinic properties. Carbachol stimulates the autonomic ganglia and causes contraction of skeletal muscle, but it differs from a true muscarinic agent in that it does not have cardiovascular activity. Carbachol is used to reduce the intraocular tension of glaucoma when response cannot be obtained with pilocarpine, neostigmine, or methacholine. Penetration of the cornea is enhanced by the use of a wetting agent in the ophthalmic solution. In addition to its topical use for glaucoma, it is

Carbachol Chloride

used during ocular surgery where a more prolonged miosis is required than that which may be obtained with acetylcholine chloride.

Carbachol differs chemically from acetylcholine in its stability to hydrolysis. The carbamyl group of carbachol decreases the electrophilicity of the carbonyl and thus can form resonance structures more easily than acetylcholine. The result is that carbachol is less susceptible to hydrolysis and therefore more stable than acetylcholine in aqueous solutions.

The pharmacologic properties of carbachol may be the result of its ability to cause release of acetylcholine from cholinergic nerve endings. Carbachol also produces a cholinergic response by acting as an inhibitor of acetylcholinesterase. Carbachol acts as a semireversible inhibitor of acetylcholinesterase and serves to prolong the duration of acetylcholine at the neuromuscular junction.

Bethanechol Chloride U.S.P., Urecholine® Chloride, β-methylcholine carbamate chloride, (2-hydroxypropyl)trimethylammonium chloride carbamate, carbamylmethylcholine chloride. Bethanechol has similar pharmacological properties as methacholine. Both are esters of β-methylcholine and have feeble nicotinic activity. Bethanechol is more slowly inactivated than methacholine in vivo by acetylcholinesterase.

Bethanechol Chloride

Bethanechol is a carbamyl ester and would be expected to have the same stability in aqueous solutions as carbachol.

The main use of bethanechol chloride is in the relief of urinary retention and abdominal distention following surgery. The drug is used orally and by subcutaneous injection. It must never be administered by intramuscular or intravenous injection. Administration of the drug is associated with low toxicity and no serious side-effects, although it should be used with caution in asthmatic patients and, when used for glaucoma, does produce frontal headaches from the constriction of the sphincter muscle in the eye and from ciliary muscle spasms. Its duration of action is about one hour.

Bethanechol chloride occurs as a white crystalline solid with an aminelike odor. It is soluble in water 1:1 and 1:10 in alcohol but is nearly insoluble in most organic solvents. An aqueous solution has a pH of 5.5 to 6.3.

Aceclidine Hydrochloride, 3-quinuclidinol acetate hydrochloride, is a parasympathomimetic drug under clinical investigation. It is being used in the treatment of glaucoma.

Aceclidine Hydrochloride

Pilocarpine Hydrochloride U.S.P., pilocarpine monohydrochloride, is the hydrochloride of an alkaloid obtained from the dried leaflets of *Pilocarpus jaborandi* or *P. microphyllus* where it occurs to the extent of about 0.5 percent together with other alkaloids with a total of 1 percent.

Pilocarpine Hydrochloride

It occurs as colorless, translucent, odorless, faintly bitter crystals that are soluble in water (1:0.3), alcohol (1:3) and chloroform (1:360). It is hygroscopic and affected by light; its solutions are acid to litmus and may be sterilized by autoclaving. Alkalies saponify its ester group to give the corresponding inactive hydroxy acid (pilocarpic acid). Base-catalyzed epimerization at the ethyl group position occurs to an appreciable extent and is another major pathway of degradation.[27] Both routes result in loss of pharmacologic activity.

Pilocarpine has a physostigmine-like action but appears to act by direct cell stimulation rather than by disturbance of the cholinesterase-acetylcholine relationship as is the case with physostigmine.

Recent evidence supports the view that pilocarpine mimics the action of muscarine to stimulate ganglia through receptor site occupation similar to that of acetylcholine.[28] Its over-all molecular architecture and the interatomic distances of its functional groups in certain conformations are similar to those of muscarine, i.e., about 4 Å from the tertiary N—CH$_3$ nitrogen to the ether oxygen or carbonyl oxygen, and are compatible with this concept.

Outstanding pharmacologic effects are the production of copious sweating, salivation and gastric secretion. Pilocarpine causes pupillary constriction (miosis) and a spasm of accommodation. These effects are valuable in the treatment of glaucoma. The resulting pupil constriction and spasm of the ciliary muscle serve to reduce intraocular tension by establishing a better drainage of ocular fluid through the canal of Schlemm located near the corner of the iris and cornea. Pilocarpine is used as a 0.5 to 6 percent solution (i.e., of the salts) in treating glaucoma. Secretion in the respiratory tract is noted following therapeutic doses, and therefore the drug sometimes is used as an expectorant.

Pilocarpine Nitrate U.S.P., pilocarpine mononitrate. This salt occurs as shining, white crystals which are not hygroscopic but are light-sensitive. It is soluble in water (1:4), soluble in alcohol (1:75) and insoluble in chloroform and in ether. Aqueous solutions are slightly acid to litmus and may be sterilized in the autoclave. The alkaloid is incompatible with alkalies, iodides, silver nitrate and the usual alkaloidal precipitants.

INDIRECT-ACTING CHOLINERGIC AGONISTS—CHOLINESTERASE INHIBITORS

Termination of the action of acetylcholine at the synaptic junction is caused primarily by destruction of the neurotransmitter by acetylcholinesterase. Acetylcholinesterase (AChE) catalyzes the hydrolysis of acetylcholine to choline and acetic acid. Inhibition of acetylcholinesterases prolongs the life of the neurotransmitter in the junction and produces pharmacologic effects similar to those observed when acetylcholine is administered. Anticholinesterases have been used in the treatment of myasthenia gravis, atony in the gastrointestinal tract, and glaucoma. They have also been used as nerve gases and insecticides.

Butyrylcholinesterase (BuChE, psuedocholinesterase) is located in human plasma. Although its biological function and purpose in humans are not clear, BuChE has catalytic properties similar to those of acetylcholinesterase. Acetylcholinesterase isolated from red blood cells has the same substrate properties as the junctional esterase and hydrolyzes only acetyl and propionyl esters. The substrate specificity for butyrylcholinesterase is much more liberal (Table 12-4.)

Three different chemical groupings, acetyl, carbamyl, and phosphoryl, may react with the esteratic site of AChE. Although the chemical reactions are similar, the kinetic parameters for each of these types of substrates are different and result in the differences

TABLE 12–4

HYDROLYSIS OF VARIOUS SUBSTRATES BY AChE AND BuChE*

| Enzyme Substrate | AChE | | BuChE | |
	Source	Relative Rate†	Source	Relative Rate†
Acetylcholine	human or bovine RBC	100	human or horse plasma	100
Acetylthiocholine	bovine RBC	149	horse plasma	407
Acetyl β-methylcholine	bovine RBC	18	horse plasma	0
Propionylcholine	human RBC	80	horse plasma	170
Butyrylcholine	human RBC	2.5	horse plasma	250
Butyrylthiocholine	bovine RBC	0	horse plasma	590
Benzoylcholine	bovine RBC	0	horse plasma	67
Ethyl acetate	human RBC	2	human plasma	1
3,3-Dimethylbutyl acetate	human RBC	60	human plasma	35
2-Chloroethyl acetate	human RBC	37	human plasma	10
Iso-amyl acetate	human RBC	24	horse plasma	7
Iso-amyl propionate	human RBC	10	horse plasma	13
Iso-amyl butyrate	human RBC	1	horse plasma	14

* Adapted from Heath, D. F.: Organophosphorus Poisons—Anticholinesterases and Related Compounds, 1st ed, Pergamon Press, New York, 1961.
† Relative rates at approximately optimal substrate concentration; rate with acetylcholine = 100.

between the toxicity and utility of these materials as drugs or chemicals.

The initial step in the hydrolysis of ACh by AChE is a reversible enzyme-substrate complex formation. The association rate (k_{+1}) and dissociation rate (k_{-1}) are relatively large. The enzyme-substrate complex, E·ACh, may also form an acetyl-enzyme intermediate at a rate (k_2) that is slower than either the association or dissociation rates. Choline is released from this complex with the formation of the acetyl enzyme intermediate, E-A. This intermediate is then hydrolyzed to regenerate the free enzyme and acetic acid. The acetylation rate, k_2, is the slowest step in this sequence and is rate limiting (see below).

The active site of AChE consists of an anionic and an esteratic site. The anionic site is formed by the gamma carboxylate group (COO−) of a glutamic acid residue.[29] Also present in the active site of AChE are an acidic group, HA, pKa 9.2, believed to be a tyrosine residue,[30] two imidazole groups from histidine residues Im_1 and Im_2 with pKa values of 6.3 and 5.5, respectively, and a serine residue. Under physiological conditions free serine residues do not catalyze hydrolysis of esters. However, in conjunction with an imidazole

group, Im_2, the serine hydroxyl in the esteratic site of AChE is activated to become a strong nucleophile.

The acetyl ester is hydrolyzed by a mechanism that involves an imidazole group Im_2 of histidine acting as a general base catalyst. Im_2 accepts a proton from a serine hydroxyl group located at the esteratic site, creating a strong nucleophile. The activated serine residue then initiates the hydrolysis mechanism by attack on the carbonyl group of ACh. Choline is released, leaving the acetylated serine residue on the enzyme, E-A. The enzyme undergoes a conformational change, which brings the acetylated serine in close proximity to a second imidazole residue that catalyzes the hydrolysis of the acetylated enzyme (Fig. 12-8). The rate of this reaction is indicated by the k_3 rate constant (i.e., deacylation rate).

Carbamates such as in carbachol are also able to serve as substrates for acetylcholinesterase, which they carbamylate. The rate of carbamylation (k_2) is slower than the rate of acetylation. The hydrolysis (k_3, decarbamylation) of the carbamyl-enzyme intermediate is 10^7 times slower than its acetyl counterpart. Therefore, carbamyl esters acting as substrates for AChE form enzyme-carbamyl (E-C) intermediates that hydrolyze slowly and serve to limit the optimal functional capacity of AChE and may be considered to be semireversible inhibitors of AChE.

$$E + CX \xrightleftharpoons[k_{-1}]{k_{+1}} E \cdot CX \xrightarrow{k_2} E - C \xrightarrow{k_3} E + C$$

where CX = carbamylating substrate

In the mechanism above, k_3 is the rate-limiting step. The rate of k_2 depends not only on the nature of the alcohol moiety of the ester but also on the type of carbamyl ester. Esters of carbamic acid, i.e.,

$$R\text{-}O\text{-}\overset{\overset{\displaystyle O}{\|}}{C}\text{-}NH_2,$$

are better carbamylating agents of AChE than the methylcarbamyl, i.e.,

$$R\text{-}O\text{-}\overset{\overset{\displaystyle O}{\|}}{C}\text{-}NHCH_3,$$

and dimethylcarbamyl, i.e.,

$$R\text{-}O\text{-}\overset{\overset{\displaystyle O}{\|}}{C}\text{-}N(CH_3)_3,$$

$$E + ACh \xrightleftharpoons[k_{-1}]{k_{+1}} E \cdot ACh \xrightarrow[\text{choline}]{k_2} E - A \xrightarrow[H_2O]{k_3} E + CH_3COOH$$

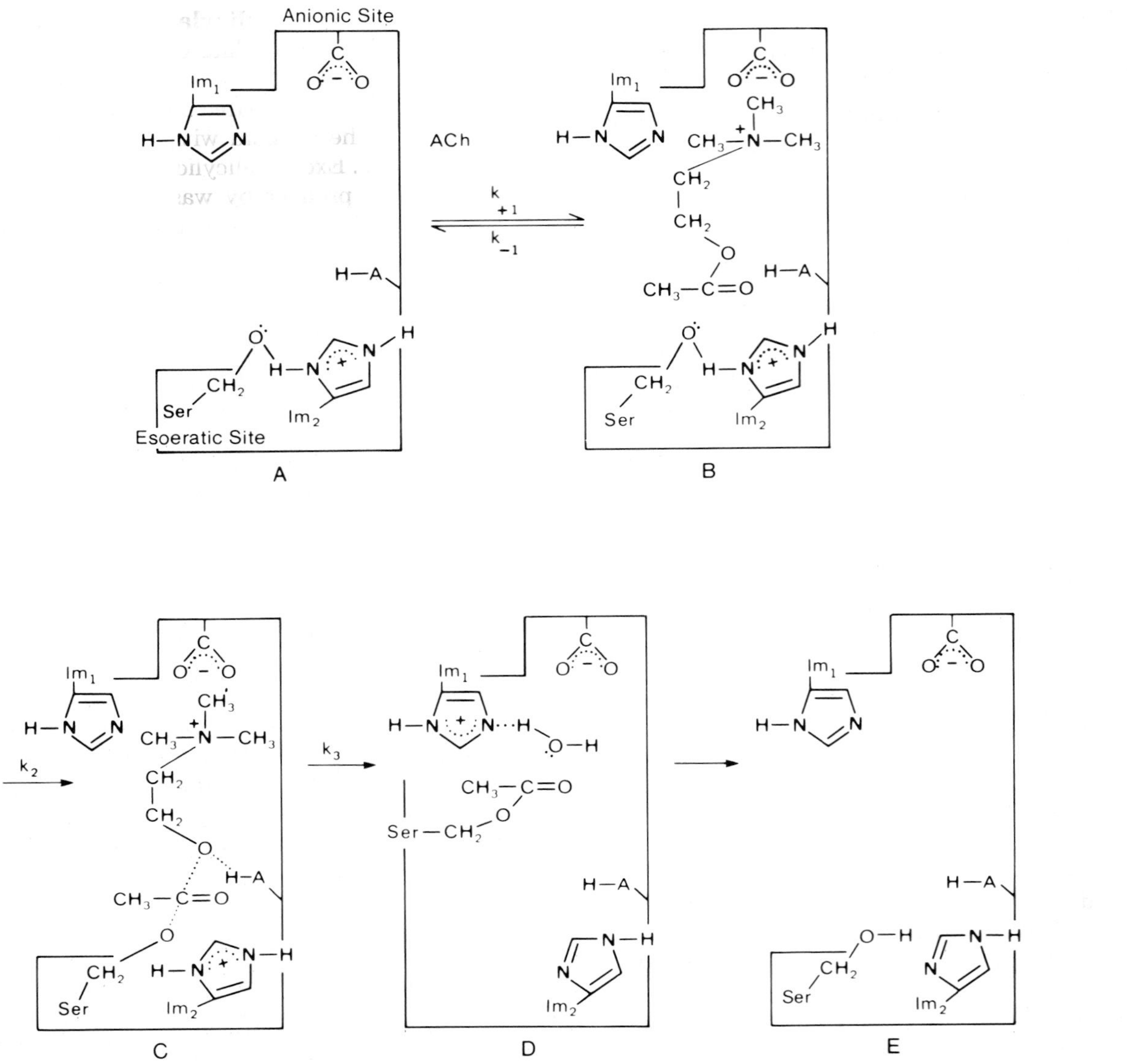

FIG. 12–8. *Mechanism of hydrolysis of ACh at active site of AChE, A. Activated esteratic site; B. ACh-AChE reversible complex; C. Acetylation; D. Deacetylation; E. Free enzyme. (Adapted from Krupka, R.M., Biochemistry 5, 1988 (1966).)*

analogs. They also decarbamylate from the enzyme faster.[31]

Organophosphate esters of selected compounds are also able to esterify the serine residue in the active site of AChE (Fig. 12-8). The hydrolysis rate (k_3) of the phosphorylated serine is extremely slow, and hydrolysis to the free enzyme and phosphoric-acid derivative is so limited that the inhibition is considered irreversible. These organophosphorus compounds are used as nerve gases in warfare, as agricultural insecticides, and for treatment of glaucoma.

It is possible for chemicals to inhibit acetylcholinesterase without interacting specifically with the ester-

where PX = phosphorylating substrate

atic site of the enzyme. Acetylcholinesterase as isolated from the electric eel is a macroanion. It possesses three distinct anionic sites[32] and perhaps other nonspecific anionic loci. Compounds containing an onium group within a large, flat, aromatic ring structure appear to be potentially good reversible inhibitors of the enzyme. The dissociation constant ($K_s = k_{-1}/k_{+1}$) is

TABLE 12–5

INHIBITION CONSTANTS FOR ANTICHOLINESTERASE POTENCY OF ACETYCHOLINESTERASE INHIBITORS

Reversible and semireversible inhibitors	$K_i(M.)$
Ambenonium	4.0×10^{-8}
Carbachol	1.0×10^{-4}
Demecarium	1.0×10^{-10}
Edrophonium	3.0×10^{-7}
Neostigmine	1.0×10^{-7}
Physostigmine	1.0×10^{-8}
Pyridostigmine	4.0×10^{-7}

Irreversible Inhibitors	$k_2(mole/min.)$
Isoflurophate	1.9×10^4
Echothiophate	1.2×10^5
Paraoxon	1.1×10^6
Sarin	6.3×10^7
Tetraethylpyrophosphate	2.1×10^8

of the magnitude of 10^{-4}M. Reversible inhibitors with dissociation constant values (K_i) of 10^{-7}M. have a 1000-fold greater affinity for the enzyme and, therefore, prevent acetylcholine from forming the enzyme-substrate complex that precedes the hydrolysis of the compound. Table 12-5 shows the relative potencies of several types of acetylcholinesterase inhibitors.

PRODUCTS

Physostigmine U.S.P. is an alkaloid usually obtained from the dried ripe seed of *Physostigma venenosum*. It occurs as a white, odorless, microcrystalline powder that is slightly soluble in water and freely soluble in alcohol, chloroform and the fixed oils. This alkaloid as the free base is quite sensitive to heat, light, moisture and bases and readily undergoes decomposition. When used topically to the conjunctiva it is better tolerated than its salts. Its lipid solubility properties permit adequate absorption from the appropriate ointment bases.

Physostigmine is a competitive inhibitor of acetylcholinesterase when acetylcholine is simultaneously present. The mechanism proposed is one of a reversible competition for the active site on the enzyme. A noncompetitive inhibition is observed when the enzyme is preincubated with physostigmine.

Physostigmine Salicylate U.S.P., eserine salicylate. This is the salicylate of an alkaloid usually obtained from the dried ripe seed of *Physostigma venenosum*. It may be prepared by neutralizing an ethereal solution of the alkaloid with an ethereal solution of salicylic acid. Excess salicylic acid is removed from the precipitated product by washing it with ether. The salicylate is less deliquescent than the sulfate.

Physostigmine Salicylate

It occurs as a white, shining, odorless crystal, or white powder that is soluble in water (1:75), alcohol (1:16) or chloroform (1:6) but is much less soluble in ether (1:250). Upon prolonged exposure to air and light, the crystals turn red in color. The red color may be removed by washing the crystals with alcohol, although this causes loss of the compound as well. Aqueous solutions are neutral or slightly acidic in reaction and take on a red coloration after a period of time. The coloration may be taken as an index of the loss of activity of physostigmine solutions. To guard against decomposed solutions, only enough should be made at one time for about a week's use. If it is necessary to sterilize the solution, it can be done by bacteriologic filtration. Physostigmine salicylate solutions are incompatible with the usual alkaloidal reagents, with alkalies and with iron salts. It is also incompatible because of the salicylate ion with benzalkonium chloride and related wetting agents. Physostigmine in solution is hydrolyzed to methylcarbamic acid and eseroline, which is not an inhibitor of acetylcholinesterase. Eseroline is oxidized readily to a red compound rubreserine[33] and is then converted to eserine blue and eserine brown. The addition of sulfite or ascorbic acid prevents the oxidation of the phenol to rubreserine. However, hydrolysis does take place and the physostigmine is inactivated. Solutions are most stable at pH 6 and never should be sterilized by heat.

Physostigmine   Eseroline   Rubreserine

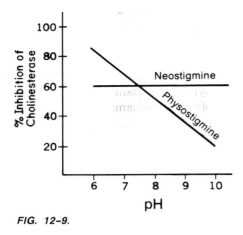

FIG. 12–9.

Physostigmine is a relatively poor carbamylating agent of acetylcholinesterase and is often considered a reversible inhibitor of the enzyme. It has a K_i value (i.e., k_{-1}/k_{+1}) on the order of 10^{-8}M. and is considered as a potent inhibitor of the enzyme. Its cholinesterase-inhibiting properties vary with pH (Fig. 12-9). Protonated physostigmine has a pKa of about 8 and as the pH is lowered more is in the protonated form. The inhibitory action is enhanced at lower pH's as shown in Figure 12-9; thus it is obvious that the protonated or salt form makes a marked contribution to its activity.

The ophthalmic effect (miotic) of physostigmine and related compounds is due to contraction of the ciliary body. This promotes drainage through the canal of Schlemm and thereby decreases intraocular pressure. For this reason, physostigmine is used in the treatment of glaucoma by direct instillation of a 0.1 to 1 percent solution in the eye. It is directly antagonistic to atropine in the eye and is sometimes used to help restore the pupil to normal size following atropine dilatation. Physostigmine also causes stimulation of the intestinal musculature and because of this is used in conditions of depressed intestinal motility. In gaseous distention of the bowel due to a number of causes, physostigmine often aids in the evacuation of gas as well as restoring normal bowel movement. It is administered by injection for this purpose. Much research has been done to find synthetic drugs with a physostigmine-like action. This has resulted in compounds of the neostigmine type which, at least for intestinal stimulation, are superior to physostigmine.

Physostigmine Sulfate U.S.P. occurs as a white, odorless, microcrystalline powder that is deliquescent in moist air. It is soluble in water 1:4, 1:0.4 in alcohol and 1:1,200 in ether. It has the advantage over the salicylate salt in that it is compatible in solution with benzalkonium chloride and related compounds.

Neostigmine Bromide U.S.P., Prostigmin® Bromide, (*m*-hydroxyphenyl)trimethylammonium bromide dimethylcarbamate, dimethylcarbamic ester of 3-hydroxyphenyltrimethylammonium bromide. A method of preparation is from dimethylcarbamyl chloride and the potassium salt of 3-hydroxyphenyldimethylamine. Methyl bromide readily adds to the tertiary amine, forming the stable quaternary ammonium salt (see formula for neostigmine bromide). It occurs as an odorless, white, crystalline powder having a bitter taste. It is soluble in water (1:0.5) and is soluble in alcohol. The crystals are much less hygroscopic than are those of neostigmine methylsulfate and thus may be used in tablets. Solutions are stable and may be sterilized by boiling. Aqueous solutions are neutral to litmus.

Neostigmine Bromide

Use of physostigmine as prototype of an indirect parasympathomimetic drug led to the development of stigmine in which a trimethylamine group was placed *para* to a dimethyl carbamate group in benzene. However, activity was obtained when these groups were placed *meta* to each other; neostigmine, a more active and useful drug, was obtained. Although physostigmine contains a methyl carbamate grouping, greater stability toward hydrolysis was obtained with a dimethyl carbamate group as in neostigmine.[34,35] The *meta* substituent makes the compound a better inhibitor of acetylcholinesterase than the *para*.

Of neostigmine that reaches the liver, 98 percent is metabolized in 10 minutes. Its transfer from plasma to liver cells and then to bile is probably passive in character. Since cellular membranes permit the passage of plasma proteins synthesized in liver into the bloodstream through capillary walls or lymphatic vessels, they may not present a barrier to the diffusion of quaternary amines such as neostigmine. Possibly the rapid hepatic metabolism of neostigmine provides a downhill gradient for the continual diffusion of this compound.[36] A certain amount may be hydrolyzed slowly by plasma cholinesterase.

Neostigmine has a mechanism of action quite similar to that of physostigmine. Neostigmine effectively inhibits cholinesterase at about 10^{-6} M concentration. Its activity does not vary with pH and at all ranges exhibits similar cationic properties (see Fig. 12-9). There may be a direct action of the drug on tissues innervated by cholinergic nerves, but this has not yet been confirmed.

The uses of neostigmine are similar to those of physostigmine, but they differ in that there are greater miotic activity, fewer and less unpleasant local and systemic manifestations and greater stability. Most frequent application is to prevent atony of the intestinal, skeletal and bladder musculature. An important use is in the treatment of myasthenia gravis. The bromide is used orally.

Neostigmine Methylsulfate U.S.P., Prostigmin® Methylsulfate, (*m*-hydroxyphenyl)trimethylammonium methylsulfate dimethylcarbamate, dimethylcarbamic ester of 3-hydroxyphenyltrimethylammonium methylsulfate. Neostigmine is prepared as in the method previously described, and the quaternary ammonium salt is made with methylsulfate. This compound is an odorless, white crystalline powder with a bitter taste. It is very soluble in water and is soluble in alcohol. Solutions are stable and can be sterilized by boiling. The compound is too hygroscopic for use in a solid form and thus always is used in injection. Aqueous solutions are neutral to litmus.

Neostigmine Methylsulfate

The methylsulfate is used postoperatively as a urinary stimulant, in the diagnosis and treatment of myasthenia gravis, and as an antiarrhythmic agent to terminate supraventricular tachycardia in patients who fail to respond to vagal stimulation.

Pyridostigmine Bromide U.S.P., Mestinon® Bromide, 3-hydroxyl-1-methylpyridinium bromide dimethylcarbamate, pyridostigminium bromide. This occurs as a white, hygroscopic crystalline powder having an agreeable characteristic odor. It is freely soluble in water, alcohol and in chloroform.

Pyridostigmine Bromide

Pyridostigmine bromide is about one-fifth as toxic as neostigmine. It appears to function in a manner like that of neostigmine but is said really to inactivate

pseudocholinesterase rather than cholinesterase. This agent is used primarily to treat myasthenia gravis. It has a longer period of duration and less muscarinic effect on the gastrointestinal tract.

Demecarium Bromide U.S.P., Humorsol®, (*m*-hydroxyphenyl)trimethylammonium bromide deca-methylene*bis*[methylcarbamate], is the diester of (*m*-hydroxyphenyl)trimethylammonium bromide with decamethylene *bis*(methylcarbamic acid) and thus is comparable to a *bis*-prostigmine molecule.

Demecarium Bromide

It occurs as a slightly hygroscopic powder that is freely soluble in water or alcohol. Aqueous solutions are neutral, stable and may be sterilized by heat. Its efficacy and toxicity are comparable to those of other potent anticholinesterase inhibitor drugs. It is a long-acting miotic used to treat wide-angle glaucoma and accommodative esotropia. Maximal effect occurs hours after administration and the effect may persist for days.

Ambenonium Chloride U.S.P., Mytelase® Chloride, [oxalyl*bis*(iminoethylene)]*bis*[(*o*-chlorobenzyl)diethylammonium dichloride]. This compound is a white, odorless powder, soluble in water and in alcohol, slightly soluble in chloroform, and practically insoluble in ether and in acetone. Ambenonium chloride is a cholinergic drug used in the treatment of myasthenia gravis. This condition is characterized by a pathologic exhaustion of the voluntary muscles caused by an impairment of neuromuscular transmission. There is strong suspicion that myasthenia gravis may be caused by an autoimmune mechanism that requires an increase in acetylcholine in the neuromuscular junction to sustain normal muscle activity.

Ambenonium Chloride

Ambenonium chloride acts by suppressing the activity of acetylcholinesterase. It possesses a relatively prolonged duration of action and causes fewer side-effects in the gastrointestinal tract than the other anticholinesterase drugs. The dosage requirements vary considerably, and the dosage must be individualized according to the response and tolerance of the patient.

Edrophonium Chloride U.S.P., Tensilon®, ethyl(*m*-hydroxphenyl)dimethylammonium chloride. Edrophonium chloride is a reversible anticholinesterase agent prepared by treating 3-dimethylaminophenol with ethyl iodide and converting the quaternary ammonium salt to the chloride with silver oxide and hydrochloric acid. It is a specific anticurare agent and acts within one minute to alleviate overdose of *d*-tubocurarine, dimethyl *d*-tubocurarine, or gallamine triethiodide. The drug also is used to terminate the action of any one of the above drugs when the physician so desires. However, it is of no value in terminating the action of the depolarizing blocking agents such as decamethonium, succinylcholine, etc., because it acts in a competitive manner.

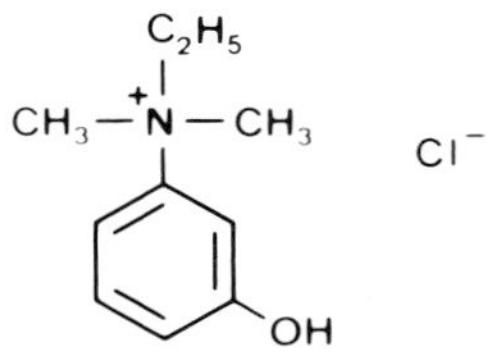

Edrophonium Chloride

Edrophonium chloride is related structurally to neostigmine methylsulfate and because of this has been tested as a potential diagnostic agent for myasthenia gravis. It has been found to bring about a rapid increase in muscle strength without significant side-effects. It is also used as an antiarrhythmic agent to terminate supraventricular arrhythmias that cannot be controlled by vagal stimulation alone.

IRREVERSIBLE INHIBITORS

Both AChE and BuChE can be inhibited irreversibly by a group of phosphate esters that are highly toxic (MLD for humans is 0.1 to 0.001 mg./kg.) These chemicals are nerve poisons and have been used in warfare and as agricultural insecticides. They permit ACh to accumulate at nerve endings and produce an exacerbation of ACh-like actions. The compounds belong to a class of organophosphorus esters. A general type formula of such compounds is as follows:

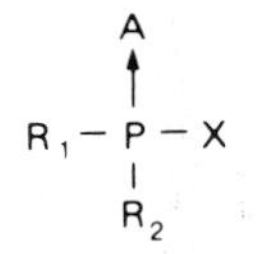

where R_1 = alkoxyl
R_2 = alkoxyl, alkyl, or tertiary amine
X = a good leaving group (e.g. F, CN, thiomalate *p*-nitrophenoxy)

A is usually oxygen or sulfur but may also be selenium. When A is other than oxygen, biological activation is required before the compound becomes effective as an inhibitor of cholinesterases. Phosphorothionates ($R_1R_2P(S)X$) have much poorer electrophilic character than their oxygen analogs and are much weaker hydrogen-bond forming molecules because of the sulfur atom.[37] Their anticholinesterase activity is 10^5-fold weaker than their oxygen analogs. X is the "leaving group" when the molecule reacts with the enzyme. Typical leaving groups include fluoride, nitrile, and *p*-nitrophenoxy. The R groups may be alkyl, alkoxy, aryl, aryloxy, or amino. The R moiety imparts lipophilicity to the molecule and contributes to its absorption through the skin. Inhibition of AChE by organophosphorus compounds takes place in two steps, association of enzyme and inhibitor and the phosphorylation step, completely analogous to acylation by the substrate (Fig. 12-10). Stereospecificity is mainly due to interactions of enzyme and inhibitor at the esteratic site.

Parathion $(EtO)_2P(S)O(p\text{-}NO_2C_6H_4)$ is inactive for cholinesterase in vitro and in vivo; its metabolite $(EtO)_2P(O)O(p\text{-}NO_2C_6H_4)$ is very active and is also inactivated by the liver. As is the case with some other biologically active substances, the route of administration may influence the quantitative effects. In the case of some cholinesterase inhibitors, the availability of the drug for metabolism by the liver is a major factor

TABLE 12–6

INFLUENCE OF THE ROUTE OF ADMINISTRATION ON TOXICITY*†

		L.D.$_{50}$, mg./kg. Body Weight (95% Fiducial Limits)			
		Hepatic Routes		Peripheral Routes	
Compound	***Molecular Weight***	Intra-peritoneal	Oral	Sub-cutaneous	Intra-venous
Physostigmine salicylate	413.45	ca. 1.0	5.50	1.12	0.46
Neostigmine methylsulfate	334.39	0.62	>5.0	0.66	0.47
Paraoxon	275.21	2.29	12.80	ca. 0.6	0.59
Parathion	291.27	15.1	25.7	21.4	17.4

*24-Hour median lethal doses of cholinesterase inhibitors following administration by different routes in female mice.
†Natoff, I. L.: J. Pharm. Pharmacol. 19:612, 1967.

in its toxicity. The data given in Table 12-6 tend to support these conclusions.[38]

The serine residue in the esteratic site forms a stable phosphoryl ester with the organophosphorus inhibitors. This stability permits labelling studies[39] to be carried out on this and other enzymes (e.g., trypsin, chymotrypsin) that have the serine hydroxyl as part of their active site.

Although insecticides and nerve gases are irreversible inhibitors of cholinesterases by forming a phosphorylated serine in the esteratic site of the enzyme, it is possible to reactivate the enzyme if action is taken soon after the phosphorylation has occurred. There are a number of compounds that can provide a nucleophilic attack on the phosphorylated enzyme and cause regeneration of the free enzyme. Substances such as choline, hydroxylamine, and hydroxamic acid have led to the development of more effective cholinesterase reactivators such as nicotinic hydroxamic acid and pyridine-2-aldoxime methiodide (2-PAM). A proposed mode of action for the reactivation of cholinesterase (inactivated by isoflurophate) by 2-PAM is shown in Figure 12-10.

Cholinesterases that have been exposed to phosphorylating agents (e.g., Sarin) become refractory to

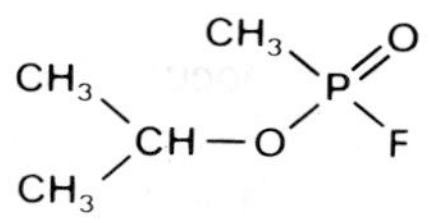

Sarin

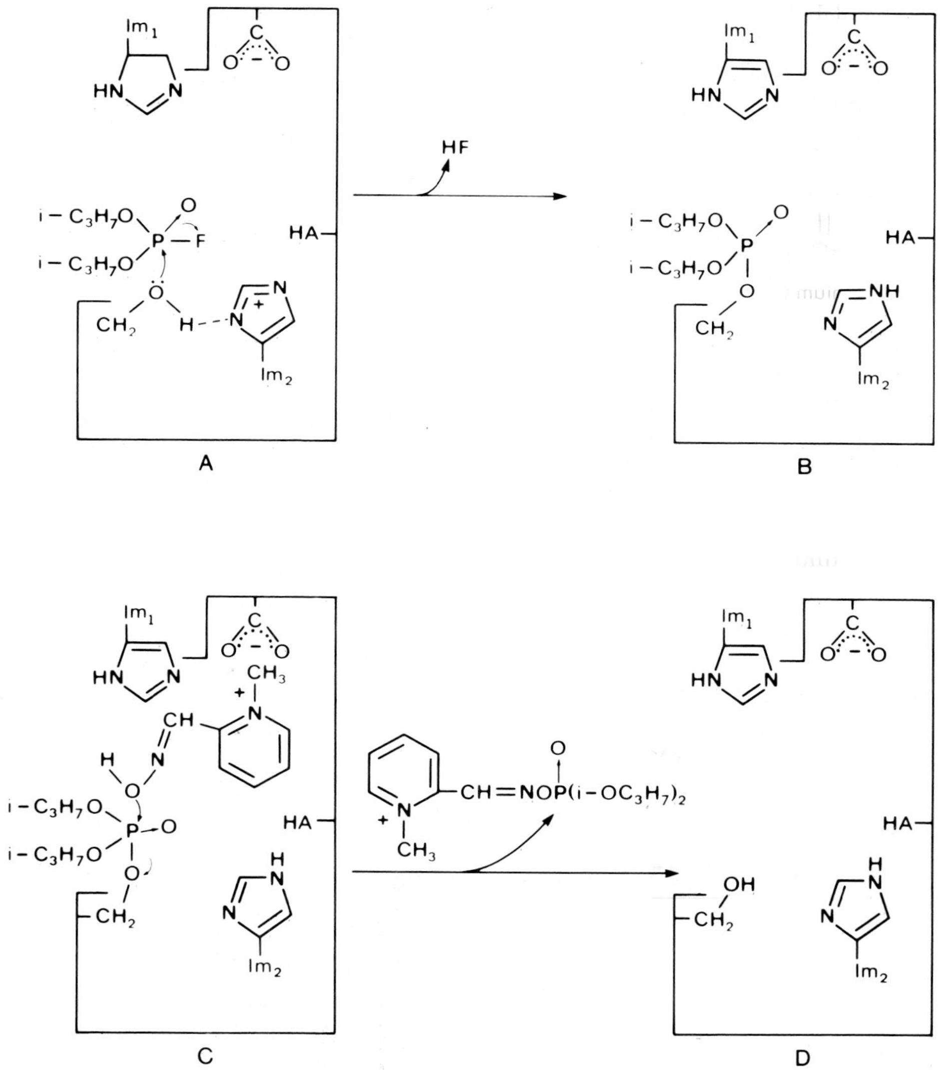

FIG. 12–10. *Phosphorylation and reactivation of cholinesterase. A. Phosphorylation of serine residue by isoflurophate. B. Phosphorylated serine residue in esteratic site. C. Nucleophilic attack on phosphorylated serine residue by 2-PAM. D. Removal of the phosphorylated 2-PAM to generate free enzyme.*

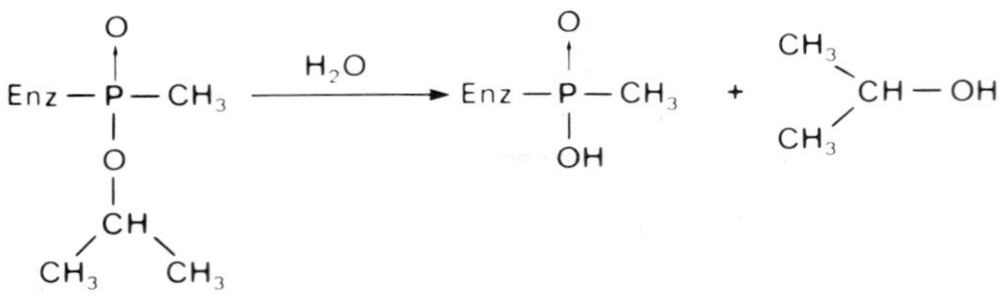

FIG. 12-11. *Aging of cholinesterase phosphorylated with sarin.*

reactivation by cholinesterase reactivators. The process is called aging and occurs both in vivo and in vitro to AChE and BuChE. Aging occurs by dealkylation of the phosphorylated moiety that is attached to the serine residue in the esteratic site of the enzyme (Fig. 12-11).

Some of the phosphate esters are used as insecticidal agents and must be handled with extreme caution because they also are very toxic to humans. Toxic symptoms are nausea, vomiting, excessive sweating, salivation, miosis, bradycardia, low blood pressure and respiratory difficulty that is the usual cause of death.

The organophosphate insecticides of low toxicity such as malathion generally cause poisoning only by ingestion of relatively large doses. On the other hand, parathion or methylparathion cause poisoning by inhalation or dermal absorption. Because these compounds are so long-acting, they are cumulative and serious toxic manifestations may result following a number of small exposures to them.

DFP and possibly related compounds also can phosphorylate the OH of the serine residue found as a functional group at the active site in other enzymes such as trypsin, alphachymotrypsin, etc. The derivative from DFP is stable to proteolytic enzymes and thus is found on the smaller polypeptides obtained by the degradation of the parent protein.

PRODUCTS

Isoflurophate U.S.P., Floropryl®, diisopropyl phosphofluridate, DFP, is a colorless liquid, soluble in water to the extent of 1.54 percent at 25° C to give a pH of 2.5. It is soluble in alcohol and to some extent in peanut oil. It is stable in the latter for a period of 1 year but decomposes in water in a few days. Solutions in peanut oil can be sterilized by autoclaving. The compound should be stored in hard glass, since continued contact with soft glass is said to hasten decomposition, as evidenced by a discoloration.

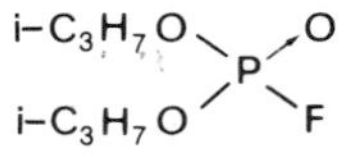

Diisopropyl Fluorophosphate

It must be handled with extreme caution. Avoid contact with eyes, nose, mouth and even the skin, because it can be absorbed readily through intact epidermis and more so through mucous tissues, etc.

Because DFP irreversibly[40] inhibits cholinesterase, its activity lasts for days or even weeks. During this period new cholinesterase may be synthesized in plasma, the erythrocytes and other cells.

A combination of atropine sulfate and magnesium sulfate has been found to give protection in rabbits against the toxic effects of DFP. One counteracts the muscarine, and the other the nicotine effect of the drug.[41] DFP has been used clinically in the treatment of glucoma.

Echothiophate Iodide U.S.P., Phospholine® Iodide, S-ester of (2-mercaptoethyl)trimethylammonium iodide with O,O-diethyl phosphorothioate. This occurs as a white, crystalline, hygroscopic solid that has a slight mercaptanlike odor. It is soluble in water 1:1, and 1:25 in dehydrate alcohol; aqueous solutions have a pH of about 4 and are stable at room temperature for about one month.

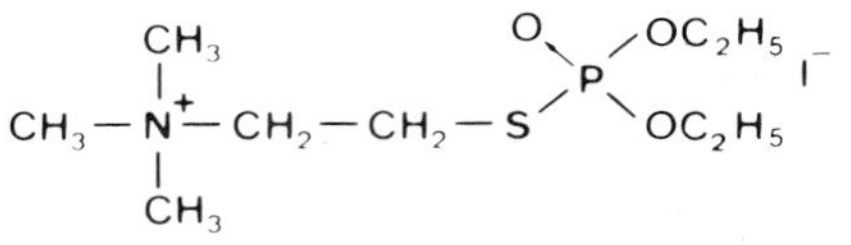

Echothiophate Iodide

Echothiophate iodide is a long-lasting cholinesterase inhibitor of the irreversible type such as isofluriphate. However, unlike the latter, it is a quaternary salt, and thus, when applied locally, its distribution in tissues is limited, which can be very desirable. It is used as a long-acting anticholinesterase agent in the treatment of glaucoma.

Hexaethyltetraphosphate (HETP) and **Tetraethylpyrophosphate (TEPP).** These two substances are compounds that also show anticholinesterase activity. HETP was developed by the Germans during World War II and is used as an insecticide against aphids. It has been reported that some HETP being sold in the United States does not have this structure but is a mixture, in which the active constituent is tetraethylpyrophosphate. When used as insecticides, these compounds have the advantage of being hydrolyzed rapidly to the relatively nontoxic water-soluble compounds phosphoric acid and ethyl alcohol. Fruit trees or vegetables sprayed with this type of compound retain no harmful residue after a period of a few days or weeks, depending on the weather conditions. The disadvantage of their use comes from their very high toxicity, which results mainly from their anticholinesterase activity. Workers spraying with these agents should use extreme caution that none of the vapors are breathed and that none of the vapor or liquid comes in contact with the eyes or skin.

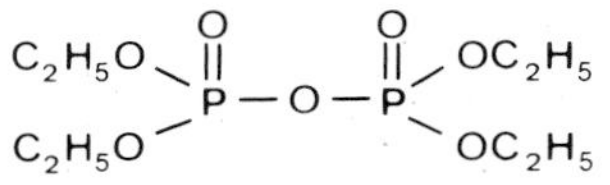

Tetraethylpyrophosphate

Malathion, diethyl 2(dimethoxyphosphinothioyl-thio)succinate is a water-insoluble phosphodithiolate ester that has been used as an agricultural insecticide. Malathion is a poor inhibitor of cholinesterases. Its effectiveness as a safe insecticide is due to the different rates at which humans and insects metabolize the chemical. Microsomal oxidation, which causes desulfuration, is slow in forming the phosphothiolate, which is 10,000 times more active than the phosphodithiolate (malathion) as a cholinesterase inhibitor. Insects detoxify the phosphothiolate by a phosphatase-forming dimethylphosphorothiolate that is inactive as

an inhibitor. Humans, however, rapidly hydrolyze malathion through a carboxyesterase enzyme yielding malathion acid, a yet poorer inhibitor of acetylcholinesterase. Phosphatases and carboxyesterases further metabolize malathion acid to dimethylphosphothiolate. The metabolic reactions are shown in Figure 12-12.

Parathion, Thiophos, O,O-diethyl-O-*p*-nitrophenyl phosphorothioate. This compound is a yellow liquid that is freely soluble in aromatic hydrocarbons, ethers, ketones, esters and alcohols but practically insoluble in water, petroleum ether, kerosene and the usual spray oils. It is decomposed at pH's higher than 7.5. Parathion is used as an agricultural insecticide. It is highly toxic, the effects being cumulative. Special precautions are necessary to prevent skin contamination or inhalation. Parathion is a relatively weak inhibitor of cholinesterase. There are enzymes present in liver microsomes and insect tissues that convert parathion ($pI_{50} < 4$) to paraoxon (a more potent inhibitor of cholinesterase, $pI_{50} > 8$).[42] Parathion is also metabolized by liver microsomes to yield *p*-nitrophenol and diethylphosphate, which is inactive as a cholinesterase inhibitor.[43]

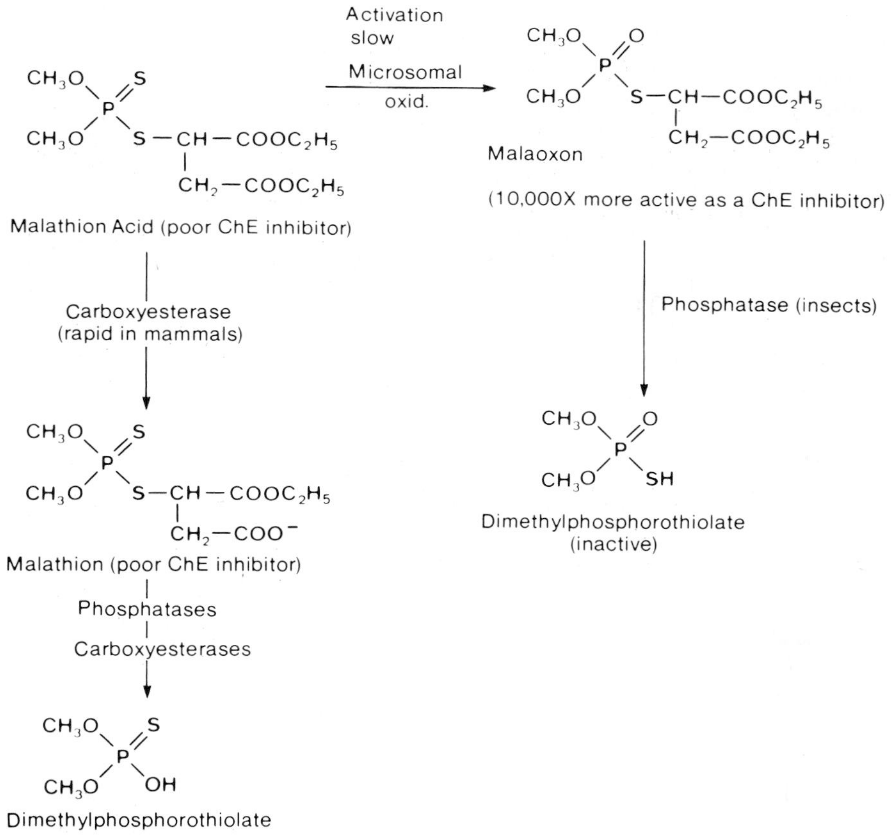

FIG. 12–12. *Metabolism of malathion in mammals and insects.*

Parathion
(low anti-AChE activity)

Paraoxon
(high anti-AChE activity)

Octamethylpyrophosphoramide, OMPA, Pestox III, Schradan, *bis*(bisdimethylaminophosphorous) anhydride. This compound is a viscous liquid that is miscible with water and soluble in most organic solvents. It is not hydrolyzed by alkalies or water but is hydrolyzed by acids. It is used as a systemic insecticide for plants, being absorbed by the plants without appreciable injury, but insects feeding on the plant are incapacitated.

OMPA is a weak inhibitor of cholinesterases in vitro. In vivo it is metabolized to the very strong inhibitor hydroxymethyl OMPA.[44] Hydroxymethyl OMPA is not stable and is further metabolized to the N-methoxide, which is a weak inhibitor of cholinesterase.[45]

Pralidoxime Chloride U.S.P., Protopam® Chloride, 2-formyl-1-methylpyridinium chlorideoxime, 2-PAM chloride, 2-pyridine aldoxime methyl chloride. It occurs as a white, nonhygroscopic crystalline powder that is soluble in water, 1 g. in less than 1 ml.

Pralidoxime chloride is used as an antidote for poisoning by parathion and related pesticides. It may be effective against some phosphates which have a quaternary nitrogen. It also is an effective antagonist for some carbamates such as neostigmine methylsulfate and pyridostigmine bromide.

In addition, pralidoxime effects depolarization at the neuromuscular junction, it is anticholinergic, cholinomimetic, it inhibits cholinesterase, potentiates the depressor action of acetylcholine in nonatropinized animals and potentiates the pressor action of acetylcholine in atropinized animals.

Pralidoxime Chloride

The mode of action of pralidoxime is described in Fig. 12-10.

The biological half-life of 2-PAM chloride in man is about 2 hours and its effectiveness is a function of its concentration in plasma that reaches a maximum in 2 to 3 hours after oral administration. Concentrations of 4 and 8 μg. per ml. of 2-PAM chloride in the blood plasma of rats significantly decrease the toxicity of sarin by factors of 2 and 2.5 respectively.[46]

Pralidoxime chloride, a quaternary ammonium compound, is most effective by intramuscular, subcutaneous or intravenous administration. Treatment of poisoning by an anticholinesterase will be most effective if given within a few hours. Little will be accomplished if the drug is used more than 36 hours after parathion poisoning has occurred.

OMPA
(weak cholinesterase inhibitor)

hydroxymethyl OMPA
(strong cholinesterase inhibitor)

OMPA–N–methoxide
(weak cholinesterase inhibitor)

TABLE 12–7

CHOLINERGIC AGENTS

Name Proprietary Name	Preparations	Application	Usual Adult Dose*	Usual Dose Range*	Usual Pediatric Dose*
Acetylcholine Chloride	Sterile powder		0.5 to 2 ml. of a freshly prepared 1 percent solution instilled in the anterior chamber of the eye		
Methacholine Chloride U.S.P. *Mecholyl Chloride*			S.C., initial, 10 mg.; then 25 mg. may be given 10 to 30 minutes later	10 to 40 mg.	
Carbachol U.S.P. *Isopto Carbachol, Carcholine*	Carbachol Ophthalmic Solution U.S.P.	Topically to the conjunctiva, 1 drop of a 1.5 percent solution 2 to 3 times daily			
	Carbachol Intraocular Solution U.S.P.		Intraocular irrigation, 0.5 ml. of a 0.01 percent solution instilled into the anterior chamber		
Bethanechol Chloride U.S.P. *Urecholine Chloride*	Bethanechol Chloride Injection U.S.P. Bethanechol Chloride Tablets U.S.P.		Oral, 10 to 30 mg. 3 times daily; S.C., 2.5 mg. 3 times daily	Oral, 30 to 120 mg. daily; S.C., 2.5 to 30 mg. daily	
Physostigmine U.S.P.	Physostigmine Salicylate Ophthalmic Ointment U.S.P.		Topical, to the conjunctiva, 1 cm. of a 0.25 percent ointment 1 to 3 times daily		
Physostigmine Salicylate U.S.P.	Physostigmine Salicylate Ophthalmic Solution U.S.P.		Topical, to the conjunctiva, 1 drop of a 0.25 or 0.5 percent solution 2 or 3 times daily		
Physostigmine Sulfate U.S.P. *Isopto Eserine*	Physostigmine Salicylate Injection U.S.P. Physostigmine Sulfate Ophthalmic Ointment U.S.P.		Topical, to the conjunctiva, 1 cm. of a 0.25 percent ointment 1 to 3 times daily		
Neostigmine Bromide U.S.P. *Prostigmin Bromide*	Neostigmine Bromide Tablets U.S.P.		15 to 45 mg. 3 to 6 times daily	15 to 375 mg. daily	330 μg. per kg. of body weight or 10 mg. per square meter of body surface, 6 times daily
Neostigmine Methylsulfate U.S.P. *Prostigmin Methylsulfate*	Neostigmine Methylsulfate Injection U.S.P.		Antidote to curare principles—I.V., 500 μg. to 2 mg. repeated as necessary (may be administered in combination with 600 μg. to 1.2 mg. of Atropine Sulfate Injection); cholinergic—I.M. or S.C., 250 to 500 μg. 4 to 6 times daily as necessary	250 μg. to 5 mg. daily	
Pyridostigmine Bromide U.S.P. *Mestinon Bromide*	Pyridostigmine Bromide Syrup U.S.P. Pyridostigmine Bromide Tablets U.S.P.		60 to 180 mg. 3 to 6 times daily	60 mg. to 1.5 g. daily	1.2 mg. per kg. of body weight or 33 mg. per square meter of body surface, 6 times daily

*See *U.S.P. D.I.* for complete dosage information.

(Continued)

TABLE 12-7

CHOLINERGIC AGENTS

Name Proprietary Name	Preparations	Application	Usual Adult Dose*	Usual Dose Range*	Usual Pediatric Dose*
Demecarium Bromide U.S.P. *Humorsol*	Demecarium Bromide Ophthalmic Solution U.S.P.	Topical, to the conjunctiva, of a 0.125 to 0.25 percent solution once or twice daily			
Ambenonium Chloride U.S.P. *Mytelase*	Ambenonium Chloride Tablets U.S.P.		Initial, 5 mg., gradually increasing as required up to 5 to 25 mg. 3 or 4 times daily	5 to 50 mg.	
Isoflurophate U.S.P. *Floropryl*	Isoflurophate Ophthalmic Ointment U.S.P.	Topically to the conjunctiva, as a 0.025 percent ointment once every 3 days to 3 times daily			
Echothiophate Iodide U.S.P. *Phospholine Iodide*	Echothiophate Iodide for Ophthalmic Solution U.S.P.	Topically to the conjunctiva, 1 drop of a 0.03 to 0.25 percent solution 1 or 2 times daily			
Edrophonium Chloride U.S.P. *Tensilon Chloride*	Edrophonium Chloride Injection U.S.P.	Antidote to curare principles; diagnostic aid (myasthenia gravis)	Antidote—I.V., 10 mg., repeated if necessary; diagnostic—I.V., 2 mg. followed by 8 mg. if no response in 45 seconds; I.M., 10 mg.	Antidote—1 to 40 mg. in one episode; diagnostic—I.V., 2 to 10 mg. per test; I.M., 10 to 12 mg. in one episode	Diagnostic, I.M.— children 34 kg. of body weight and under, 2 mg.; children over 34 kg. of body weight, 5 mg.; I.V.— infants, 500 µg.; children 34 kg. of body weight and under, 1 mg.; if no response after 45 seconds, 1 mg. every 30 to 45 seconds, up to 5 mg.; children over 34 kg. of body weight, 2 mg.; if no response after 45 seconds, 1 mg every 30 to 45 seconds, up to 10 mg.
Pilocarpine Hydrochloride U.S.P. *Almocarpine, Pilocel, Isopto Carpine, Pilocar, Pilomiotin*	Pilocarpine Hydrochloride Ophthalmic Solution U.S.P.	Topically to the conjunctiva, 1 drop to 0.25 of a 10 percent solution 1 to 6 times daily			
Pilocarpine U.S.P.	Pilocarpine Ocular System U.S.P.		Topical, to the conjunctiva, 1 ocular system delivering 20 or 40 µg. per hour, once every 7 days		
	Pilocarpine Nitrate Ophthalmic Solution U.S.P.				
Pilocarpine Nitrate U.S.P. *P.V. Carpine Liquifilm*		Topically to the conjunctiva, 1 drop of a 0.5 to 6 percent solution 1 to 4 times daily			

* See *U.S.P. D.I.* for complete dosage information.

CHOLINERGIC BLOCKING AGENTS

There are three types of peripheral cholinergic receptors that respond to acetylcholine. These are located at parasympathetic postganglionic nerve endings in smooth muscle, at the sympathetic and parasympathetic ganglia, and at the neuromuscular junctions in skeletal muscle. These receptors are all stimulated exclusively by acetylcholine, yet there are different drugs used as antagonists for each of these receptors. Thus, atropine is an effective blocking agent at parasympathetic postganglionic terminals, hexamethonium blocks transmission at autonomic ganglia, and *d*-tubocurarine blocks the effect of acetylcholine on skeletal muscle. The specificity of action of these blocking agents may be attributed to a number of factors which play a role in the accessibility of the blocking agent to the site. They include the anatomical characteristics of the receptor site and the affinity of the blocking agent for the receptor. Cholinergic synapses in the central nervous system are also subject to the action of anticholinergic drugs.

Anticholinergic action by drugs and chemicals is apparently dependent upon their ability to reduce the number of free receptors that can interact with acetylcholine. The theories of Clark and Stephenson[47] and Ariens[48] have explained the relationship between drug-receptor interactions and the observed biological response (see chap. 2). These theories indicate that the quantity of a drug-receptor complex formed at a given time depends upon the affinity of the drug for the receptor and that a drug that acts as an agonist must also possess another property called *efficacy* or *intrinsic activity*. Another explanation of drug-receptor interactions, the Paton rate theory,[49] defines a biological stimulus as being proportional to the rate of drug-receptor interactions (see chap. 2).

Both of these theories are compatible with the concept that a blocking agent that has a high affinity for the receptor may serve to decrease the number of available free receptors and decrease the efficiency of the endogenous neurotransmitter.

PARASYMPATHETIC POSTGANGLIONIC BLOCKING AGENTS

These blocking agents are also known as *anticholinergics, parasympatholytics,* or *cholinolytics.* One might be more specific in stating that members typical of this group are antimuscarinic because they block the action of muscarine.

Endogenous neurotransmitters, including acetylcholine, are relatively small molecules. It was noted by Ariens[48] that competitive reversible antagonists generally are larger molecules capable of additional binding to the receptor surface. The most potent anticholinergic drugs are derived from muscarinic agonists that contain one and sometimes two large or bulky groups. Ariens suggested that molecules that act as competitive reversible antagonists were capable of binding to the active site of the receptor but had an additional binding site that increased receptor affinity but did not contribute to the intrinsic activity (efficacy) of the drug. Consistent with this hypothesis, Bebbington and Brimblecombe[50] proposed that there is a relatively large area lying outside the agonist receptor-binding site where van der Waals interactions can take place (see Fig. 12-7) between the antagonist and the receptor area.

STRUCTURE–ACTIVITY RELATIONSHIPS

A wide variety of compounds possess anticholinergic activity. The development of such compounds has been largely empiric and based principally on atropine as the prototype. Nevertheless, structural permutations have resulted in compounds that do not have obvious relationships to the parent molecule. The following classification will serve to delineate the major chemical types that are encountered:

1. Solanaceous alkaloids and synthetic analogs
2. Synthetic aminoalcohol esters
3. Aminoalcohol ethers
4. Aminoalcohols
5. Aminoamides
6. Miscellaneous
7. Papaveraceous alkaloids and their synthetic analogs*

The chemical classification of anticholinergics acting on parasympathetic postganglionic nerve endings is complicated somewhat by the fact that some agents, especially the quaternary ammonium derivatives, act on the ganglia that have a musarinic component to their stimulation pattern and, at high doses, act at the neuromuscular junction in skeletal muscle.

There are several ways in which the structure–activity relationship could be considered, but in this discussion we shall follow, in general, the considerations of Long *et al.,*[51] who based their postulations on the

* Although these are not anticholinergics acting at the postganglionic nerve endings of the parasympathetic system, they are included here as a matter of convenience because of their deployment as antispasmodics.

1-hyoscyamine molecule as being one of the most active anticholinergics and, therefore, having an optimal arrangement of groups.

Anticholinergic compounds may be considered as chemicals that have some similarity to acetylcholine but contain additional substituents which enhance their binding to the cholinergic receptor.

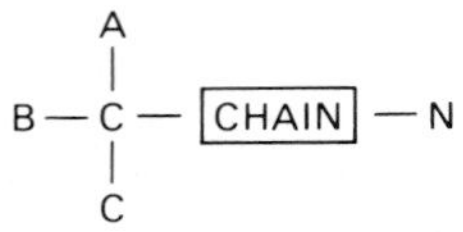

As depicted above, an anticholinergic may contain a quaternary ammonium function or possess a tertiary amine that is protonated in the biophase to form a cationic species. The nitrogen is separated from a pivotal carbon atom by a chain that may include an ester, ether, or hydrocarbon moiety. The substituent groups A and B contain at least one aromatic moiety capable of van der Waals interactions to the receptor surface and one cycloaliphatic or other hydrocarbon moiety for hydrophobic bonding interactions. C may be hydroxyl or carboxamide to undergo hydrogen bonding with the receptor.

The cationic head

Most authors consider that the anticholinergic molecules have a primary point of attachment to cholinergic sites through the *cationic head*, i.e., the positively charged nitrogen. In the case of quaternary ammonium compounds there is no question of what is implied, but in the case of tertiary amines one assumes, with good reason, that the cationic head is achieved by protonation of the amine at physiologic pH. The nature of the substituents on this cationic head is critical insofar as a mimetic response is concerned. Steric factors that cause a diffusion of the onium charge or produce less than an optimal drug-receptor interaction result in a decrease of mimetic properties and allow the drug to act as an antagonist because of other bonding interactions. It is undoubtedly true that a cationic head is better than none at all; it is possible to obtain a typical competitive block *without*

Benzilylcarbocholine

a cationic head. Ariens and his coworkers[48] have shown that *carbocholines*, e.g., benzilylcarbocholine, show a typical competitive action with acetylcholine although they are less effective than the corresponding compounds possessing a cationic head.

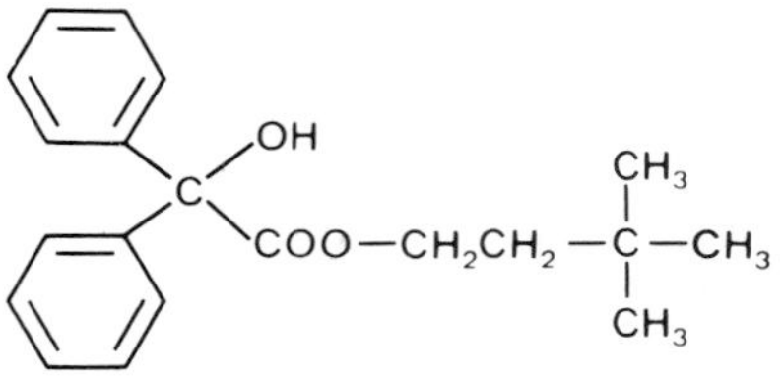

The hydroxyl group

Although not a requisite for activity, a suitably placed alcoholic hydroxyl group in an anticholinergic usually enhances the activity over a similar compound without the hydroxyl group. The position of the hydroxyl group with respect to the nitrogen appears to be fairly critical with the diameter of the receptive area being estimated at about 2 to 3 Å. It is assumed that the hydroxyl group contributes to the strength of binding, probably by hydrogen bonding to an electron-rich portion of the receptor surface.

The esteratic group

Many of the highly potent compounds possess an ester grouping, and it may be a necessary feature for the most effective binding. This is reasonable in view of the fact that the agonist (i.e., acetylcholine) possesses a similar function for binding to the same site. That an esteratic function is not necessary for activity is amply illustrated by the several types of compounds not possessing such a group (e.g., ethers, aminoalcohols, etc.). However, by far the greater number of active compounds possess this grouping. It is possible that it attaches to the receptor area at a positive site, similarly to acetylcholine, and may be necessary for maximal blocking activity.

Cyclic substitution

It will be apparent from an examination of the active compounds discussed in the following sections that at least one cyclic substituent (phenyl, thienyl, etc.) is a feature of the molecule. Aromatic substitution seems to be the most used in connection with the acidic moiety in esters. However, it will be noted that virtually all of the acids employed are of the aryl-substituted acetic acid variety. Use of aromatic acids per se leads to low activity as anticholinergics but with potential activity as local anesthetics. The question of the superiority of the cyclic species used (i.e., phenyl, thienyl, cyclohexyl, etc.) appears not to have been explored in depth, although phenyl rings seem to predominate. Substituents on the aromatic rings seem to contribute little to activity.

In connection with the apparent need for a cyclic group it is instructive to consider the postulations of Ariëns[48] in this respect. He points out that the "mimetic" molecules, richly endowed with polar groups, undoubtedly require a complementary polar receptor area for effective binding. As a consequence, it is implied that a relatively nonpolar area surrounds such sites. Thus, by increasing the binding of the molecule in this peripheral area by means of introducing flat nonpolar groups (e.g., aromatic rings) it should be possible to achieve compounds with excellent affinity but not possessing intrinsic activity. That this postulate is consistent with most anticholinergics, whether they possess an ester group or not, is quite obvious.

Stereochemical requirements

The stereochemistry of acetylcholine and other cholinergic agonists has been discussed earlier in this chapter. The results are summarized in Table 12-2 and show quite conclusively the large degree of stereoselectivity of the cholinergic receptor for the S over the R configuration in the agonist molecule. In contrast, the benzilate esters of the isomeric β-methylcholines show only a small difference in competitive antagonistic activity (R to S ratio = 1.2), indicating that the stereochemical requirement for antagonism is small in the alcohol portion of the molecule. If the stereochemical difference is removed from the choline moiety and introduced into the acidic portion once again a significant difference (R to S ratio = 100:1) in activity is noted in the R and S forms of cyclohexylphenylglycolate esters of choline. These observations are further reinforced by Ellenbroek's findings[52] on the comparative blocking activities of the four possible stereoisomers of the cyclohexylphenylglycolate esters of β-methylcholine as summarized in Table 12-8.

Similar relationships have been noted for tropic-acid esters. Because of the Cahn rules for describing

TABLE 12-8

EFFECT OF STEREOISOMERS ON ANTICHOLINERGIC ACTIVITY

Compound	pA₂*	R to S ratio
Benzilylcholine	8.6	
R-Hyoscyamine (R-Tropyltropeine)	6.9	
RS-Hyoscyamine (Atropine)	8.7	115 (S to R)
RS-Hyoscyamine (S-Tropyltropeine)	9.0	
Benzilyl Rβ-methylcholine	8.1	
Benzilyl Sβ-methylcholine	8.0	
R-Cyclohexylphenylglycolylcholine	10.4	100
S-Cyclohexylphenylglycolylcholine	8.4	
R-Cyclohexylphenylglycolyl Rβ-methylcholine	8.9	100
S-Cyclohexylphenylglycolyl Rβ-methylcholine	6.9	
R-Cyclohexylphenylglycolyl Sβ-methylcholine	8.3	40
S-Cyclohexylphenylglycolyl Sβ-methylcholine	6.6	
R-Tropylcholine	6.5 (S to R)	15.2 (S to R)
S-Tropylcholine	7.7	

*pA₂ = the logarithm of the reciprocal molar concentration of antagonist that requires a doubling of the concentration of agonist to compensate for the action of the agonist.
(Data from Ellenbroek, B. W. J., Nivard, F. J. R., van Rossum, J. M., and Ariens, E. J.: J. Pharm. Pharmacol. 17:393, 1967; Rama Sastry, B. V., and Cheng, H. C.: J. Pharmacol Exp. Ther., 202:105, 1977.)

absolute configuration S, tropic acid has the same spatial deployment of its aromatic, carboxyl, and hydroxy moieties as R-cyclohexylphenylglycolic acid. Hence hyoscyamine has an S to R ratio of 115 and tropylcholine an S to R ratio of 15.2. One is thus drawn to the conclusion that, for blocking activity, the structural requirements are low for the aminoalcohol portion and high for the acidic portion of cholinergic parasympathetic postganglionic blocking agents.

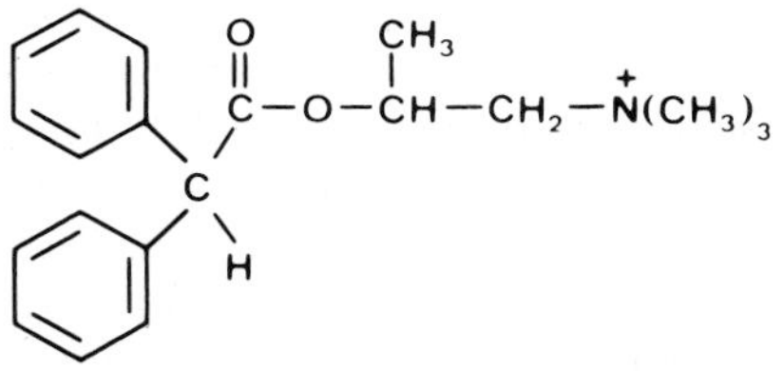

Benzilyl beta–methylcholine

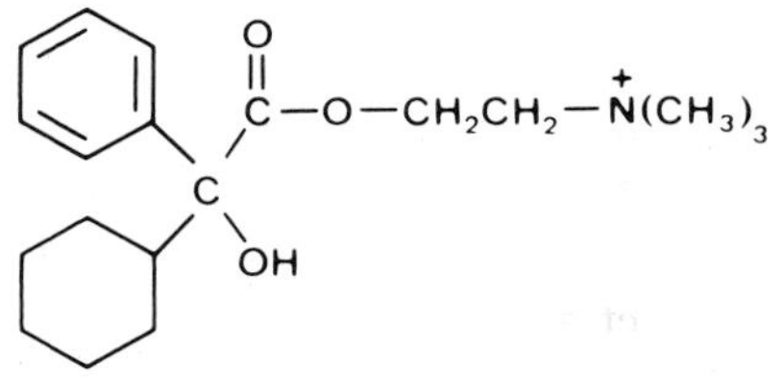

Cyclohexylphenylglycolyl choline

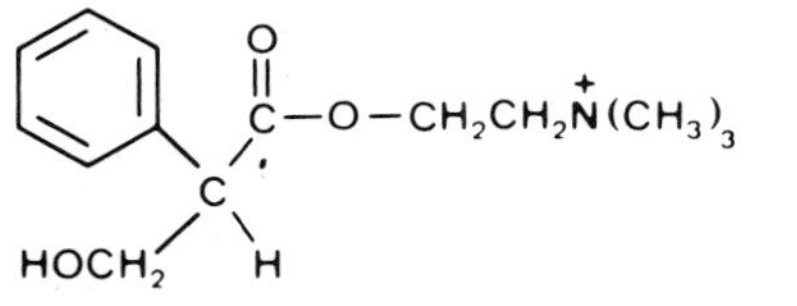

Tropylcholine

From another viewpoint there is a body of evidence[53] that suggests that muscarinic agonist and antagonist drugs do not interact with the same receptor. Competitive antagonism could result if the receptors were quite distinct, but the presence of an antagonist near the acetylcholine receptor could modify the latter in such a way that the affinity of an agonist would be reduced.[54]

THERAPEUTIC ACTIONS

Because organs controlled by the autonomic nervous system are doubly innervated by both the sympathetic and the parasympathetic systems, it is believed that there is a continual state of dynamic balance between the two systems. Theoretically, one should achieve the same end-result by stimulation of one of the systems or by blockade of the other and, indeed, in some cases this is true. Unfortunately, in most cases there is a limitation on this type of generalization, and the results of antimuscarinic blocking of the parasympathetic system are no exception. However, there are three predictable and clinically useful results from blocking the muscarinic effects of acetylcholine. These are:

1. *Mydriatic effect* (dilation of pupil of the eye) and *cycloplegia* (a paralysis of the ciliary structure of the eye, resulting in a paralysis of accommodation for near vision).
2. *Antispasmodic effect* (lowered tone and motility of the gastrointestinal tract and the genitourinary tract).
3. *Antisecretory effect* [reduced salivation *(antisialagogue)*, reduced perspiration *(anhidrotic)* and reduced acid and gastric secretion].

These three general effects of parasympatholytics can be expected in some degree from any of the known drugs, although in some cases it is necessary to administer rather heroic doses to demonstrate the effect. The mydriatic and cycloplegic effects, when produced by topical application, are not subject to any great undesirable side-effects due to the other two effects, because of limited systemic absorption. This is not the case with the systemic antispasmodic effects obtained by oral or parenteral administration. It is generally understood that drugs having effective blocking action on the gastrointestinal tract are seldom free of undesirable side-effects on the other organs. The same is probably true of the antisecretory effects. Perhaps the most commonly experienced obnoxious effects from the oral use of these drugs under ordinary conditions is dryness of the mouth, mydriasis and urinary retention.

Mydriatic and cycloplegic drugs are generally prescribed or used in the office by ophthalmologists. The principal purpose is for refraction studies in the process of fitting glasses. This permits the physician to examine the eye retina for possible discovery of abnormalities and diseases as well as to provide controlled conditions for the proper fitting of glasses. Because of the inability of the iris to contract under the influence of these drugs, there is a definite danger to the patient's eyes during the period of drug activity unless they are protected from strong light by the use of dark glasses. These drugs also are used to treat inflammation of the cornea (keratitis), inflammation of the iris and the ciliary organs (iritis and iridocyclitis), and inflammation of the choroid (choroiditis). Interestingly, a dark-colored iris appears to be more difficult to dilate than a light-colored one and may require more concentrated solutions. A caution in the use of mydriatics is advisable because of their demonstrated effect in raising the intraocular pressure. The pressure rises because pupil dilation tends to cause the iris to restrict drainage of fluid through the canal of Schlemm by crowding the angular space, thus leading to increased intraocular pressure. This is particularly the case with glaucomatous conditions which should be under the care of a physician.

It is well to note at this juncture that atropine is used widely as an antispasmodic because of its marked depressant effect on parasympathetically innervated smooth muscle. Indeed, atropine is the standard by which other similar drugs are measured. It is to be noted also that the action of atropine is a blocking action on the transmission of the nerve impulse, rather than a depressant effect directly on the musculature. Therefore, its action is termed *neurotropic* in contrast with the action of an antispasmodic such as papaverine, which appears to act by depression of the muscle cells and is termed *musculotropic*. Papaverine is the standard for comparison of musculotropic antispasmodics and, while not strictly a parasympatholytic, will be treated together with its synthetic analogs later in this chapter. The synthetic antispasmodics appear to combine neurotropic and musculotropic effects in greater or lesser measure, together with a certain amount of ganglion-blocking activity in the case of the quaternary derivatives.

Because of the widespread use of anticholinergics in the treatment of various gastrointestinal complaints, it is desirable to examine the pharmacologic basis on which this therapy rests. Smooth-muscle spasm, hypermotility and hypersecretion, individually or in combination, are associated with many painful ailments of the gastrointestinal tract. Among these are peptic ulcer, pylorospasm, cardiospasm and functional diarrhea. Although the causes have not been clearly defined, there are many who feel that emotional stress is the underlying common denominator to all of these conditions rather than a simple malfunction of the cholinergic apparatus. On the basis of Selye's original work on stress and Cannon's classic demonstration of the disruptive effects on normal digestive processes of anger, fear and excitement, stress is considered as being causative. The excitatory (parasympathetic) nerve of the stomach and the gut is intimately associated with the hypothalamus (the so-called "seat of feelings") as well as with the medullary and the sacral portions of the spinal cord. It is believed that emotions arising or passing through the hypothalamic area can transmit definite effects to the peripheral neural pathways such as the vagus and other parasympathetic and sympathetic routes. This is commonly known as a *psychosomatic reaction*. The stomach appears to be influenced by emotions more readily and more extensively than any other organ, and it does not strain the imagination to establish a connection between emotional effects and malfunction of the gastrointestinal tract. Individuals under constant stress are thought to develop a condition of "autonomic imbalance" due to repeated overstimulation of the parasympathetic pathways. The result is little rest and gross overwork on the part of the muscular and the secretory cells of the stomach and other viscera.

One of the earlier hypotheses advanced for the formation of ulcers proposed that strong emotional stimuli could lead to a spastic condition of the gut with accompanying anoxia of the mucosa due to prolonged vasoconstriction. The localized ischemic areas, combined with simultaneous hypersecretion of hydrochloric acid and pepsin, could then provide the groundwork for peptic ulcer formation by repeated irritation of the involved mucosal areas. Lesions in the protective mucosal lining would, of course, then permit the normal digestive processes to attack the tissue of the organ. Hydrochloric acid is considered as the causative agent because it is known that ulcer patients secrete substantially higher quantities of the acid than do normal people and also that ulcers can be induced in dogs with normal stomachs if the gastric acidity level is raised to the level of that found in ulcer patients. Nervous influence is thought to be basic to the hypersecretion of acid resulting in duodenal ulcers, whereas humoral or hormonal influences are believed to be responsible for excessive secretion in the case of gastric ulcers.[55]

The condition of overstimulation of the parasympathetic nervous supply (vagus) to the stomach is sometimes termed *parasympathotonia*. Reduction of this overstimulated condition can be achieved by surgery (surgical vagotomy) or by the use of anticholinergic drugs (chemical vagotomy), resulting in inhibition of both secretory and motor activity of the stomach. Although anticholinergic drugs can exert an antimotil-

ity effect, there is some question as to whether they can correct disordered motility or counteract spontaneous "spasms" of the intestine. In addition, although these drugs can (in adequate dosage) diminish the basal secretion of acid, there is said to be little effect on the acid secreted in response to food or to insulin hypoglycemia.

After the initial surge of chemical modification of the belladonna alkaloids, which began in the 1920's and ended in the early 1950's, few antispasmodics have been presented to the clinician to aid in the therapy of ulcers. It is suggested that neither the belladonna alkaloids nor the synthetic spasmolytics have achieved the degree of selectivity of cholinolytic action to have either class as the preferred one for the management of gastrointestinal ailments.[56] For the present, the most rational therapy seems to be a combination of a nonirritating diet to reduce acid secretion, antacid therapy, reduction of emotional stress, and administration of anticholinergic drugs. Most of the anticholinergic drugs on the market are offered either as the chemical alone or in combination with a central nervous system depressant such as phenobarbital or with one of the tranquilizers in order to reduce the central nervous system contribution to parasympathetic hyperactivity.

There does not seem to be an advantage to the use of combination products of antispasmodics with phenobarbital or phenothiazine tranquilizers because of the difficulty in balancing the effects of the two different CNS drugs with the anticholinergic agent. If the two drugs must be given together, it is suggested that they be given in separated preparations to permit better control of CNS effects.[57] Some clinical findings tend to show that phenobarbital is preferable to the tranquilizers. Whereas combinations of anticholinergics with sedatives are considered rational, there is not complete agreement on combinations with antacids. This is based on the fact that anticholinergic drugs affect primarily the fasting phases of gastrointestinal secretion and motility and are most efficient if administered at bedtime and well before mealtimes. Antacids neutralize acid largely present in the between-meal, digestive phases of gastrointestinal activity and are of more value if given after meals. H_2-receptor antagonists of histamine decrease hydrochloric acid secretion and are especially useful for those patients who suffer from the nocturnal secretory activity of the stomach (see Chap. 16).

In addition to the antisecretory effects of anticholinergics on hydrochloric acid and gastric secretion described above there have been some efforts to employ them as *antisialagogues* (to suppress salivation) and *anhidrotics* (to suppress perspiration).

Paralysis agitans or parkinsonism, first described by the English physician James Parkinson in 1817, is another condition that is often treated with the anticholinergic drugs. It is characterized by tremor, "pill rolling," cog-wheel rigidity, festinating gait, sialorrhea and masklike facies. Fundamentally, it represents a malfunction of the extrapyramidal system.[58] Skeletal muscle movement is controlled to a great degree by patterns of excitation and inhibition resulting from the feedback of information to the cortex and is mediated through the pyramidal and extrapyramidal pathways. The basal ganglia structures, such as the pallidum, corpus striatum, and substantia nigra serve as data processors for the pyramidal pathways and also the structures through which the extrapyramidal pathways pass on their way from the spinal cord to the cortex. Lesions of the pyramidal pathways cause spasticity, weakness, and exaggerated tendon reflexes. Interruption of the extrapyramidal pathways leads to a persistent increase in muscle tone, resulting in an excess of spontaneous involuntary movements along with changes in the reflexes. It is apparent, therefore, that the basal ganglia are functional in maintaining normal motor control. In parkinsonism there is a degeneration of the substantia nigra and corpus striatum, which involve controlled integration of muscle movement.

In the last decade or so the biochemistry of the pathologic state of parkinsonism has been characterized as (1) having decreased levels of dopamine and its chief metabolite, homovanillic acid, in the basal ganglia; (2) a reduced activity of aromatic L-amino acid decarboxylase, the enzyme which coverts L-dopa to dopamine; and (3) a decrease in melanin pigmentation in the substantive region. Despite this information the clinical picture remains as it has for many years. The disease state is apparently not reversible and chemotherapy is, of necessity, palliative.

The usefulness of the belladonna group of alkaloids was an empiric discovery by Charcot. The several synthetic preparations were developed in an effort to retain the useful antitremor and antirigidity effects of the belladonna alkaloids while at the same time reducing the undesirable side-effects. Incidentally, it was also discovered that antihistamine drugs (e.g., diphenhydramine) sometimes reduced tremor and rigidity. The antiparkinsonlike activity of the antihistamines has been attributed to their anticholinergic effects. The activity is confined to those drugs that can pass the blood–brain barrier, i.e., tertiary amines, not quaternary ammonium compounds.

Acetylcholine is widely found in the brain. Its concentration in the areas that control movement and behavior is higher than in the cortex. Large amounts of choline acetyltransferase and acetylcholinesterase are also found in the caudate nucleus. Therefore, all the components required for the synthesis and distribu-

tion of acetylcholine acting as a neurotransmitter are present here, although there remains no unequivocal proof that it functions as such in the brain. Nevertheless, the present assumptions are that ACh acts as a neurotransmitter in the central nervous system and that anticholinergics can block its action as they do in the peripheral nervous system. One reason for this statement is data from the investigations on oxytremorine.[59]

Injections of tremorine (1,4-dipyrrolidino-2-butyne) or its active metabolite oxotremorine [1-(2-pyrrolidono)-4-pyrrolidino-2-butyne] have been shown to increase the brain acetylcholine level in rats up to 40 percent.[60] This increase coincides roughly with the onset of tremors similar to those observed in parkinsonism. The mechanism of acetylcholine increase in rats is uncertain but has been shown not to be due to acetylcholinesterase inhibition or to activation of choline acetylase. The tremors are stopped effectively by administration of the tertiary amine type anticholinergic but not by the quaternaries.

Tremorine: R = H$_2$
Oxotremorine: R = O

Although many compounds have been introduced for treatment of parkinsonism, there is apparently a real need for compounds that will provide more potent action with fewer side-effects and, also, will provide a wide assortment of replacements for those drugs that seem to lose their efficacy with the passage of time.

The most significant advance in the treatment of parkinsonism stems from the discovery of the utility of L-dopa in managing the disease. This amino acid which is believed to act as a source of dopamine, known to be deficient in the patient afflicted with parkinsonism, initially was given in rather large doses with a concomitant increase in side-effects. More recent studies with combinations of L-dopa with a decarboxylase inhibitor (e.g., carbidopa) have in many cases allowed reduction of the dose to about one-fourth that required in the absence of the inhibitor. In particular, nausea and vomiting (caused by dopamine stimulation of the medullary vomiting center) have been sharply reduced although the mechanism by which beneficial activity is produced is not clear.

Apomorphine is also being intensively examined as a possible parkinsonlytic mainly because it may be considered to be a dopamine congener with the dopamine structure locked in a rigid conformation.

Dopamine

Apomorphine

A very interesting observation seems to indicate a link between the anticholinergic drugs and dopamine in the therapy of parkinsonism. It has been shown[61] that several antiparkinsonism drugs (e.g., benztropine, trihexyphenidyl, orphenadrine, diphenhydramine) were able to inhibit the uptake of catecholamines into synaptosomes from the corpus striatum. Subsequently atropine, benztropine and other antiparkinsonism drugs were shown to reduce the uptake of dopamine in rat striata.[62] The reduced uptake of dopamine presumably results in a potentiation of its effect and serves to ameliorate the symptoms of parkinsonism.

SOLANACEOUS ALKALOIDS AND SYNTHETIC ANALOGS

The solanaceous alkaloids represented by (−)−hyoscyamine, atropine [(±)−hyoscyamine] and scopolamine (hyoscine) are the forerunners of the class of parasympatholytic drugs. These alkaloids are found principally in henbane *(Hyoscyamus niger)*, deadly nightshade *(Atropa belladonna)* and jimson weed *(Datura stramonium)*. There are certain other alkaloids that are members of the solanaceous group (e.g., apoatropine, noratropine, belladonnine, tigloidine, meteloidine) but are not of sufficient therapeutic value to be considered in this text.

The crude drugs containing these alkaloids have been used since early times for their marked medicinal properties, which depend largely on inhibition of the parasympathetic nervous system and stimulation of the higher nervous centers. Belladonna, probably as a consequence of the weak local anesthetic activity of atropine, has been used topically for its analgesic effect on hermorrhoids, certain skin infections and various itching dermatoses. The application of sufficient amounts of belladonna or of its alkaloids results in mydriasis. Internally, the drug causes diminution of secretions, increases the heart rate (by depression of the vagus nerve), depresses the motility of the gastrointestinal tract and acts as an antispasmodic on various smooth muscles (ureter, bladder and biliary tract). In addition, it stimulates the respiratory center

directly. The very multiplicity of actions exerted by the drug causes it to be looked upon with some disfavor, because the physician seeking one type of response unavoidably obtains the others. The action of scopolamine-containing drugs differs from those containing hyoscyamine and atropine in that there is no central nervous system stimulation, and a narcotic or sedative effect predominates. The use of this group of drugs is accompanied by a fairly high incidence of reactions due to individual idiosyncrasies, death from overdosage usually resulting from respiratory failure. A complete treatment of the pharmacology and the uses of these drugs is not within the scope of this text, and the reader is referred to the several excellent pharmacology texts which are available. However, the introductory pages of this chapter have reviewed briefly some of the more pertinent points in connection with the major activities of these drug types.

STRUCTURAL CONSIDERATIONS

All of the solanaceous alkaloids are esters of the bicyclic aminoalcohol, 3-hydroxytropane, or of related aminoalcohols.

The structural formulas below show the piperidine ring system in the commonly accepted chair conformation because this form has the lowest energy requirement. However, the alternate boat form can exist under certain conditions, because the energy barrier is not great. Inspection of the 3-hydroxytropane formula also indicates that, even though there is no optical activity because of the plane of symmetry, two stereoisomeric forms (tropine and pseudotropine) can exist because of the rigidity imparted to the molecule through the ethane chain across the 1,5-positions. In tropine the axially oriented hydroxyl group, *trans* to the N-bridge, is designated as *alpha,* and the alternate *cis* equatorially oriented hydroxyl group is *beta.* The aminoalcohol derived from scopolamine, namely *scopine,* has the axial orientation of the 3-hydroxyl group but, in addition, has a *beta*-oriented epoxy group bridged across the 6,7-positions as shown. Of the several different solanaceous alkaloids known, it has already been indicated that (−)-hyoscyamine, atropine and scopolamine are the most important. Their structures are indicated, but it can be pointed out that antimuscarinic activity is associated with all of the solanaceous alkaloids that possess the tropinelike axial orientation of the esterfied hydroxyl group. It will be noted in studying the formulas below that tropic acid is, in each case, the esterifying acid. Tropic acid contains an easily racemized asymmetric carbon atom, the moiety accounting for optical activity in these compounds in the absence of racemization. The proper enantiomorph is necessary for high antimuscarinic activity, as illustrated by the potent (−)-hyoscyamine in comparison with the weakly active (+)-hyoscyamine. The racemate, atropine, has an intermediate activity. The marked difference in antimuscarinic potency of the optical enantiomorphs apparently does not extend to the action on the central nervous system, inasmuch as both seem to have the same degree of activity.[63]

The solanaceous alkaloids have been modified by preparing other esters of 3α-tropanol or quaternizing the nitrogen in tropanol or scopine with a methyl halide. These compounds were some of the initial attempts to separate the varied actions of atropine and scopolamine. It should be pointed out that few aminoalcohols have been found that impart the same degree of neurotropic activity as that exhibited by the ester formed by combination of tropine with tropic acid. Similarly, the tropic-acid portion is highly specific for the anticholinergic action, and substitution by other acids results in decreased neurotropic potency, although the musculotropic action may increase. The earliest attempts to modify the atropine molecule retained the tropine portion and substituted various acids for tropic acid.

Besides changing the acid residue, other changes have been directed toward the quaternization of the nitrogen. Examples of this type of compound are methscopolamine bromide, homatropine methylbromide, and anisotropine methylbromide. Quaternization of the tertiary amine produces variable effects in terms of increasing potency. Decreases in activity are apparent in comparing atropine with methylatropine (no longer used) and scopolamine with methscopolamine. Ariens *et al.*[64] ascribe decreased activity, especially where the groups attached to nitrogen are larger than methyl, to a possible decrease in affinity for the anionic site on the cholinergic receptor. This de-

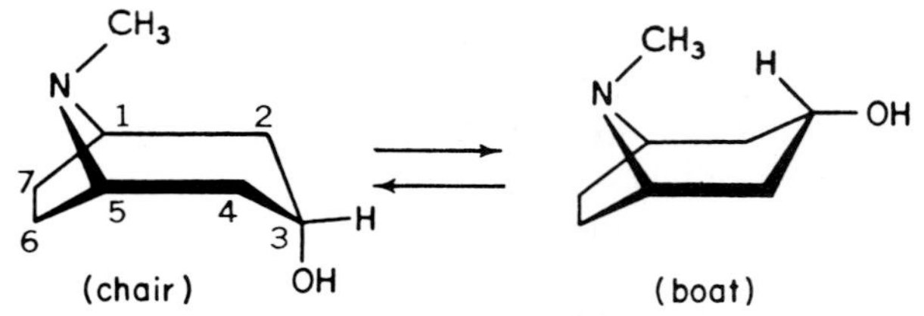

TROPINE
(3α-Hydroxytropane or
3α-tropanol)

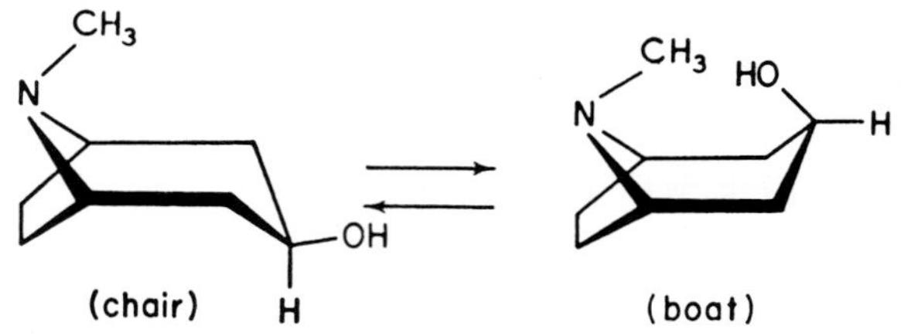

PSEUDOTROPINE
(3β-Hydroxytropane or
3β-tropanol)

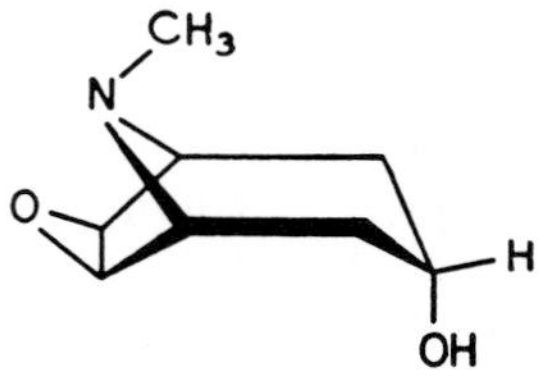

SCOPINE
(6:7β-Epoxy-3α-hydroxytropane
or 6:7β-Epoxy-3α-tropanol)

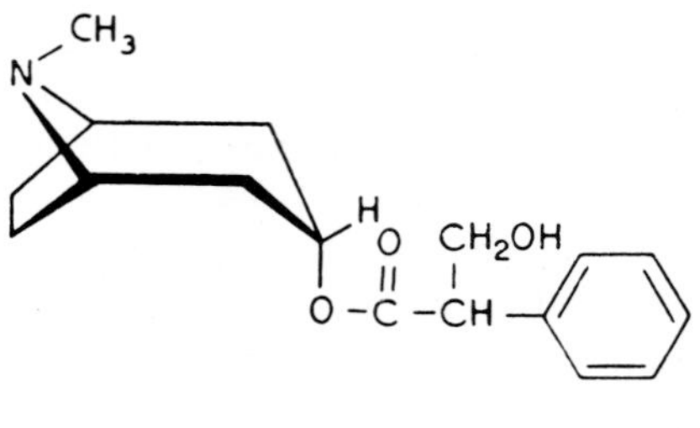

ATROPINE
(or Hyoscyamine)

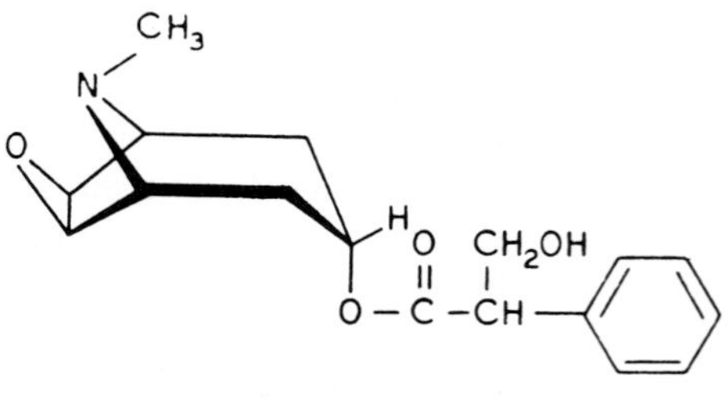

SCOPOLAMINE
(or Hyoscine)

creased affinity they attribute to a combination of greater electron repulsion by such groups and greater steric interference to the approach of the cationic head to the anionic site. In general, however, the effect of quaternization is much greater in reduction of parasympathomimetic action than of parasympatholytic action. This may be due partially to the additional blocking at the parasympathetic ganglion induced by quaternization, which could serve to offset the decreased affinity at the postgaglionic site. However, it also is to be noted that quaternization increases the curariform activity of these alkaloids and aminoesters, a usual consequence of quaternizing alkaloids. Another disadvantage in converting an alkaloidal base to the quaternary form is that the quaternized base is more poorly absorbed through the intestinal wall, with the consequence that the activity becomes erratic and, in a sense, unpredictable. The reader will find Brodie and Hogben's[65] comments on the absorption of drugs in the dissociated and the undissociated states of considerable interest, although space limitations do not permit expansion on the topic in this text. Briefly, however, they point out that bases (such as alkaloids) are absorbed through the lipoidal gut wall only in the undissociated form, which can be expected to exist in the case of a tertiary base, in the small intestine. On the other hand, quaternary nitrogen bases cannot revert to an undissociated form even in basic media and, presumably, would have difficulty passing through the gut wall. That quaternary compounds can be absorbed indicates that other less efficient mechanisms for absorption probably prevail. The comments of Cavallito[20] are interesting in this respect. Asher,[3] in connection with a long-term clinical study on anticholinergic compounds, states that "Observations concerning the synthetic tertiary amine derivatives were deleted since it was found that, in general, these drugs were quite weak when compared clinically with the drugs of the quaternary ammonium series."

PRODUCTS

Atropine U.S.P. Atropine is the tropine ester of racemic tropic acid (see above) and is optically inactive. It possibly occurs naturally in various *Solanaceae,* although some claim with justification that whatever atropine is isolated from natural sources results from racemization of (−)-hyoscyamine during the isolation process. Conventional methods of alkaloid isolation are used to obtain a crude mixture of atropine and hyoscyamine from the plant material.[66] This crude mixture is racemized to atropine by refluxing in chloroform or by treatment with cold dilute alkali. Be-

cause atropine is made by the racemization process, an official limit is set on the hyoscyamine content by restricting atropine to a maximum levorotation under specified conditions.

Synthetic methods for preparing atropine take advantage of Robinson's synthesis, employing modifications to improve the yield of tropinone. Tropinone may be reduced under proper conditions to tropine, which is then used to esterify tropic acid. Other acids may be used in place of tropic acid to form analogs, and numerous compounds of this type have been prepared which are known collectively as *tropëines.* The most important one, homatropine is considered later in this section.

Atropine occurs in the form of optically inactive, white, odorless crystals possessing a bitter taste. It is not very soluble in water (1:460; 1:90 at 80°) but is more soluble in alcohol (1:2; 1:1.2 at 60°). It is soluble in glycerin (1:27), in chloroform (1:1) and in ether (1:25).* Saturated aqueous solutions are alkaline in reaction (approximate pH = 9.5). The free base is useful when nonaqueous solutions are to be made, such as in oily vehicles and ointment bases.

Atropine Sulfate U.S.P., Atropisol®, is prepared by neutralizing atropine in acetone or ether solution with an alcoholic solution of sulfuric acid, care being exercised to prevent hydrolysis.

The salt occurs as colorless crystals or as a white crystalline powder. It is efflorescent in dry air and should be protected from light to prevent decomposition.

Atropine sulfate is freely soluble in water (1:0.5), in alcohol (1:5; 1:2.5 at boiling point) and in glycerin (1:2.5). Aqueous solutions of atropine are not very stable, although it has been stated[67] that solutions may be sterilized at 120° (15 lb. pressure) in an autoclave if the pH is kept below 6. Sterilization probably is best effected by the use of aseptic technique and a bacteriologic filter. The above reference suggests that no more than a 30-day supply of an aqueous solution should be made, and, for small quantities, the best procedure is to use hypodermic tablets and sterile distilled water. Kondritzer and his co-workers[68,69] have studied the kinetics of alkaline and proton-catalyzed hydrolyses of atropine in aqueous solution. The region of maximum stability lies between pH 3 and approximately 5. They also have proposed an equation to predict the half-life of atropine undergoing hydrolysis at constant pH and temperature.

The action of atropine or its salts is the same. It produces a mydriatic effect by paralyzing the iris and the ciliary muscles and for this reason is used by the

---

* In this chapter a solubility expressed as 1:460 indicates that 1 g. is soluble in 460 ml. of the solvent at 25°. Solubilities at other temperatures will be so indicated.

oculist in iritis and corneal inflammations and lesions. Its use is rational in these conditions because one of the first rules in the treatment of inflammation is rest, which, of course, is accomplished by the paralysis of muscular motion. Its use in the eye (0.5% to 1% solutions or gelatin disks) for fitting glasses is widespread. Atropine is administered in small doses before general anesthesia to lessen oral and air-passage secretions and, where morphine is administered with it, it serves to lessen the respiratory depression induced by morphine.

Atropine causes restlessness, prolonged pupillary dilation and loss of visual accommodation and furthermore gives rise to arrhythmias such as atrioventricular dissociation, ventricular extrasystoles and even ventricular fibrillation. Even though there has been a gradual replacement of ether as a general anesthetic with halothane, thereby eliminating problems with respiratory secretions caused by ether and thus requiring atropine, surgeons and anesthesiologists today continue to use it as an anesthetic premedicant to "dry up secretions" and to prevent vagal reflexes.[70]

Its ability to dry secretions also has been utilized in the so-called "rhinitis tablets" for symptomatic relief in colds. In cathartic preparations, atropine or belladonna has been used as an antispasmodic to lessen the smooth-muscle spasm (griping) often associated with catharsis.

In a recent application of atropine it has been found that this drug may be used in treatment of some types of arrhythmias. Atropine increases the heart rate by blocking the effects of acetylcholine on the vagus. In this context it is used to treat certain reversible bradyarrhythmias that may accompany acute myocardial infarction. It is also used as an adjunct to anesthesia to protect against bradycardia, hypotension and even cardiac arrest induced by the skeletal muscle relaxant succinylcholine chloride.

Another use for atropine sulfate has emerged following the development of the organic phosphates which are potent inhibitors of acetylcholinesterase. Atropine is a specific antidote to prevent the "muscarinic" effects of acetylcholine accumulation such as vomiting, abdominal cramps, diarrhea, salivation, sweating, bronchoconstriction and excessive bronchial secretions.[77] It is used intravenously but does not protect against respiratory failure due to depression of the respiratory center and the muscles of respiration.

Hyoscyamine U.S.P. is a levorotatory alkaloid obtained from various solanaceous species. One of the commercial sources is Egyptian henbane *(Hyoscyamus muticus)*, in which it occurs to the extent of about 0.5 percent. One method for extraction of the alkaloid utilizes *Duboisia* species.[72] Usually, it is prepared from the crude drug in a manner similar to that used for

atropine and is purified as the oxalate. The free base is obtained easily from this salt.

It occurs as white needles which are sparingly soluble in water (1:281), more soluble in ether (1:69) or benzene (1:150) and very soluble in chloroform (1:1) or alcohol. It is official as the sulfate and hydrobromide. The principal reason for the popularity of the hydrobromide has been its nondeliquescent nature. The salts have the advantage over the free base in being quite water-soluble.

As mentioned previously, hyoscyamine is the levoform of the racemic mixture which is known as atropine and, therefore, has the same structure. The dextro-form does not exist naturally but has been synthesized. Comparison of the activities of (−)-hyoscyamine, (+)-hyoscyamine and the racemate (atropine) was carried out by Cushny in 1903, wherein he found a greater peripheral potency for the (−)-isomer and twice the potency of the racemate. All later studies have essentially borne out these observations, namely, that the (+)-isomer is only weakly active and that the (−)-isomer is, in effect, the active portion of atropine. Inspection of the relative doses of Atropine Sulfate U.S.P. and Hyoscyamine Sulfate U.S.P. illustrates the difference very nicely. The principal criticism offered against the use of hyoscyamine sulfate exclusively is that it tends to racemize to atropine sulfate rather easily in solution so that atropine sulfate, then, becomes the more stable of the two. All of the isomers behave very much the same with respect to the central nervous system. A preparation containing the levorotatory alkaloids of belladonna but consisting principally of (−)-hyoscyamine malate is on the market under the trade name of Bellafoline. It has been promoted extensively on the basis of less central activity and greater peripheral activity than atropine possesses.

Hyoscyamine is used to treat disorders of the urinary tract more so than any other antispasmodic, although there is no evidence that it has any advantages over the other belladonna preparations and the synthetic anticholinergics. It is used to treat spasms of the bladder and in this manner serves as a urinary stimulant. It is used together with a narcotic to counteract the spasm produced by the narcotic when the latter is used to relieve the pain of urethral colic. Hyoscyamine preparations are also used in therapy of peptic ulcers as antispasmodics.

Hyoscyamine Hydrobromide U.S.P. This levorotatory salt occurs as white, odorless crystals or as a crystalline powder which is affected by light. It is not deliquescent. The salt is freely soluble in water, alcohol and chloroform but only slightly soluble in ether. The solutions, when freshly prepared, are neutral to litmus.

The uses are virtually the same as those cited for atropine and hyoscyamine, although it is believed that there is less central effect than with atropine.

Hyoscyamine Sulfate U.S.P., Levsin® Sulfate. This salt is a white, odorless, crystalline compound of a deliquescent nature. It is affected by light. It is soluble in water (1:0.5) and alcohol (1:5) but almost insoluble in ether. Solutions of hyoscyamine sulfate are acidic to litmus.

This drug is used as an anticholinergic in the same manner and for the same uses as atropine and hyoscyamine (q.v.), but possesses the disadvantage of being deliquescent.

Scopolamine, hyoscine. This alkaloid is found in various members of the *Solanaceae* (e.g., *Hyoscyamus niger, Duboisia myoporoides, Scopolia* sp. and *Datura metel*). It usually is isolated from the mother liquor remaining from the isolation of hyoscyamine.

The name *hyoscine* is the older name for this alkaloid, although *scopolamine* is more popular in this country. Scopolamine is the levo-component of the racemic mixture which is known as *atroscine*. Scopolamine is racemized readily when subjected to treatment with dilute alkali in the same way as is (−)-hyoscyamine (q.v.)

The alkaloid occurs in the form of a levo-rotatory, viscous liquid which is only slightly soluble in water but very soluble in alcohol, chloroform or ether. It forms crystalline salts with most acids, the hydrobromide being the most stable and the most popularly accepted. An aqueous solution of the hydrobromide, containing 10 percent of mannitol (Scopolamine Stable), is said to be less prone to decomposition than unprotected solutions.

Scopolamine Hydrobromide U.S.P., hyoscine hydrobromide. This salt occurs as white or colorless crystals or as a white granular powder. It is odorless and tends to effloresce in dry air. It is freely soluble in water (1:1.5), soluble in alcohol (1:20), only slightly soluble in chloroform and insoluble in ether.

Scopolamine gives the same type of depression of the parasympathetic nervous system as does atropine but it differs markedly from atropine in its action on the higher nerve centers. Both drugs readily cross the blood–brain barrier and even in therapeutic doses cause confusion, particularly in the elderly.

Whereas atropine stimulates the central nervous system, causing restlessness and talkativeness, scopolamine acts as a narcotic or sedative. In this capacity, it has found a use in the treatment of parkinsonism, although its value is depreciated by the fact that the effective dose is very close to the toxic dose. A sufficiently large dose of scopolamine will cause an individual to sink into a restful, dreamless sleep for a period of some 8 hours, followed by a period of approximately

the same length in which the patient is in a semiconscious state. During this time, the patient does not remember events that take place. When scopolamine is administered with morphine, this temporary amnesia is termed "twilight sleep."

Homatropine Hydrobromide U.S.P., Homatrocel®, 1αH,3αH-tropan-3α-ol mandelate (ester) hydrobromide. It may be prepared by evaporating tropine (obtained from tropinone) with mandelic and hydrochloric acids. The hydrobromide is obtained readily from the free base by neutralizing with hydrobromic acid. The hydrochloride may be obtained in a similar manner.

The hydrobromide occurs as white crystals, or as a white, crystalline powder which is affected by light. It is soluble in water (1:6) and in alcohol (1:40), less soluble in chloroform (1:420) and insoluble in ether.

Solutions are incompatible with alkaline substances, which precipitate the free base, and also with the common alkaloidal reagents. As in the case of atropine, solutions are sterilized best by filtration through a bacteriologic filter, although it is claimed that autoclaving has no deleterious effect.[73]

It is used topically in therapy to paralyze the ciliary structure of the eye (cycloplegia) and to effect mydriasis. It behaves very much like atropine but is weaker and less toxic. In the eye, it acts more rapidly but less persistently than atropine. The dilatation of the pupil takes place in about 15 to 20 minutes, and the action subsides in about 24 hours. By utilizing a miotic, such as physostigmine (q.v.), it is possible to restore the pupil to normality in a few hours. The drug is used in concentrations of 1 to 2 percent in aqueous solution or in the form of gelatin disks (lamellae).

Homatropine Hydrobromide

Homatropine Methylbromide U.S.P., Novatropine®, Mesopin®, 3α-hydroxy-8-methyl-1αH,5αH-tropanium bromide mandelate. This compound is the tropine methylbromide ester of mandelic acid. It may be prepared from homatropine by treating it with methyl bromide, thus forming the quaternary compound.

It occurs as a white, odorless powder having a bitter taste. It is affected by light. The compound is readily soluble in water and in alcohol but is insoluble in ether. The pH of a 1 percent solution is 5.9 and of a 10

Homatropine Methylbromide

percent solution is 4.5. Although a solution of the compound yields a precipitate with alkaloidal reagents, such as mercuric-potassium-iodide test solution, the addition of alkali hydroxides or carbonates does not cause a precipitate as is the case with non-quaternary nitrogen salts (e.g., atropine, homatropine).

Homatropine methylbromide is said to be less stimulating to the central nervous system than atropine, while retaining virtually all of its parasympathetic depressant action. It is used orally, in a manner similar to atropine, to reduce oversecretion and relieve gastrointestinal spasms.

Methscopolamine Bromide U.S.P., Pamine® Bromide, scopolamine methylbromide, $6\beta,7\beta$-epoxy-3α-hydroxy-8-methyl-1αH,5αH-tropanium bromide

Methscopolamine Bromide

($-$)-tropate. This compound may be made by treating either scopolamine or norscopolamine with methyl bromide.[74]

It is a crystalline, colorless compound, freely soluble in water, slightly soluble in alcohol and insoluble in acetone and chloroform. The drug is a potent parasympatholytic and is distinguished especially by its ability to inhibit the secretion of acid gastric juice through a depression of the vagus innervation of the stomach. This is in some contrast to methantheline bromide wherein the principal activity seems to be toward inhibition of the motility of the gastrointestinal tract. The effect of methscopolamine bromide is claimed to be specifically on the parasympathetic nervous system, although it is to be noted that blocking of the sympathetic system will occur with very large doses. According to Kirsner and Palmer,[75] methscopolamine bromide is one of the more effective antisecretory drugs, although they point out that it, too, has atropinelike side-effects when administered in large doses. The drug is also promoted for use as an antisialagogue and anhidrotic.

Perhaps the principal use of the drug is in the medical management of peptic ulcer, gastric hyperacidity and gastric hypermotility. Because of its atropinelike effect on secretions, it is of use in excessive salivation and sweating. Dryness of the mouth and blurred vision are the most common side-effects encountered. It is supplied in the form of 2.5-mg. tablets, with or without 15 mg. of phenobarbital, or in a protracted action form with 7.5 mg. per capsule. The usual form for injection is a solution containing 1 mg. per ml.

TABLE 12-9

ATROPINE AND RELATED COMPOUNDS

Name Proprietary Name	Preparations	Category	Application	Usual Adult Dose*	Usual Dose Range*	Usual Pediatric Dose*
Atropine U.S.P.		Anticholinergic				
Atropine Sulfate U.S.P. *Atropisol, Isopto Atropine*	Atropine Sulfate Injection U.S.P.	Anticholinergic; antidote to cholinesterase inhibitors		Anticholinergic—parenteral, 400 to 600 μg. 4 to 6 times daily; antidote to cholinesterase inhibitors—I.V., 2 to 4 mg. initially, followed by I.M., 2 mg. repeated every 5 to 10 minutes until muscarinic symptoms disappear or signs of atropine toxicity appear	300 μg. to 50 mg. daily	Anticholinergic—S.C., 10 μg. per kg. of body weight or 300 μg. per square meter of body surface, up to 44 μg. per dose, 4 to 6 times daily; antidote to cholinesterase inhibitors—I.V. or I.M., 1 mg. initially, followed by 500 μg. to 1 mg. every 10 to 15 minutes until signs of atropine toxicity appear

* See *U.S.P. D.I.* for complete dosage information.

(Continued)

TABLE 12-9

ATROPINE AND RELATED COMPOUNDS (Continued)

Name Proprietary Name	Preparations	Category	Application	Usual Adult Dose*	Usual Dose Range*	Usual Pediatric Dose*
	Atropine Sulfate Ophthalmic Ointment U.S.P.	Anticholinergic (ophthalmic)	Topical, to the conjunctiva, 0.3 to 0.5 cm. of a 1 percent ointment 1 to 3 times daily			
	Atropine Sul- Ophthalmic Solution U.S.P.	Anticholinergic (ophthalmic)	Topically to the conjunctiva, 1 drop of a 1 percent solution 1 to 3 times daily			
	Atropine Sulfate Tables U.S.P.	Anticholinergic		300 to 600 µg. 3 or 6 times daily	300 µg. to 8 mg. daily	
Hyoscyamine U.S.P. Cystospaz	Hyoscyamine Tablets U.S.P.	Anticholinergic		250 µg. 4 times daily		
Hyoscyamine Hydrobromide U.S.P.		Anticholinergic		250 to 500 µg.		
Hyoscyamine Sulfate U.S.P. Levsin	Hyoscyamine Sulfate Tablets U.S.P.	Anticholinergic			125 to 250 µg. 3 or 4 times daily	
Scopolamine Hydrobromide U.S.P. Isopto Hyoscine	Scopolamine Hydrobromide Injection U.S.P.	Anticholinergic		Parenteral, 320 µg. to 650 µg. as a single dose		S.C. 6 µg. per kg. of body weight or 200 µg. per square meter of body surface, as a single dose
	Scopolamine Hydrobromide Ophthalmic Solution U.S.P.	Anticholinergic (ophthalmic)	Topically to the conjunctiva, 1 drop of a 0.25 percent solution 1 to 3 times daily			
	Scopolamine Hydrobromide Tablets U.S.P.	Anticholinergic		400 to 800 µg.		6 µg. per kg. of body weight or 200 µg. per square meter of body surface, as a single dose
	Scopolamine Hydrobromide Ophthalmic Ointment U.S.P. Scopolamine Hydrobromide Ophthalmic Solution U.S.P.	Anticholinergic (ophthalmic)	Topical, to the conjunctiva, 0.3 to 0.5 cm. of a 0.2 percent ointment 1 to 3 times daily			
Homatropine Hydrobromide U.S.P. Homatrocel, Isopto Homatropine	Homatropine Hydrobromide Ophthalmic Solution U.S.P.	Anticholinergic (ophthalmic)	Topically to the conjunctiva, 1 drop of a 2 or 5 percent solution, 2 or 3 times daily			
Homatropine Methylbromide U.S.P.		Anticholinergic				
	Homatropine Methylbromide Tablets U.S.P.			2.5 to 5 mg. 4 times daily		
Methscopolamine Bromide U.S.P. Pamine Bromide	Methscopolamine Bromide Injection U.S.P. Methscopolamine Bromide Tablets U.S.P.	Anticholinergic		Oral, 2.5 mg. 1/2 hour before meals and 2.5 to 5 mg. at bedtime		

* See U.S.P. D.I. for complete dosage information.

SYNTHETIC AMINOALCOHOL ESTERS

It is generally agreed that the solanaceous alkaloids are potent parasympatholytics but that they have the undesirable property of producing a wide range of effects through their nonspecific blockade of autonomic functions. Thus, efforts to use the antispasmodic effect of the alkaloids most often results in a side-effect of dryness of the mouth. For this reason synthesis of compounds possessing specific cholinolytic actions was a very desirable field of study. Perhaps few prototype drugs were as avidly dissected in the minds of researchers as was atropine in attempts to modify its structure to separate the numerous useful activities of the prototype, i.e., antispasmodic, antisecretory, mydriatic, and cycloplegic. The majority of research was carried out in the pre- and post-World War II era. The ideal specificity of action may be an unattainable goal in view of the mode of action (competitive blockade) and present lack of information about the differences (if any) among the parasympathetic postganglionic cholinergic receptors.

Efforts at synthesis started with rather minor deviations from the atropine molecule, but a review of the commonly used drugs today indicates a marked departure from the rigid tropane aminoalcohol and tropic-acid residue. An examination of the structures of the present antispasmodics shows that the acid portion has been designed to provide a large hydrophobic moiety rather than the stereospecific requirement of S-tropic acid in (−) hyoscyamine that was once considered important. One of the major developments in the field of aminoalcohol esters was the successful introduction of the quaternary ammonium derivatives as contrasted with the tertiary amine-type esters synthesized originally. Although there are some effective tertiary amine esters in use today the quaternaries, as a group, represent the most popular type and appear to be slightly more potent than their tertiary counterparts despite some drawbacks.

It has already been pointed out that the stereochemical arrangement in the rigid atropine molecule lends itself to high activity, presumably because of a good fit of its prosthetic groups with the receptor site. Therefore, one might come to the conclusion that any deviation from this arrangement might reduce the activity substantially, if not remove it completely. However, early studies employing the empiric idea of structural dissection (so successful with local anesthetics) led to the conclusion that, even though atropine did seem to have a highly specific action, the tropine portion was nothing more than a highly complex aminoalcohol and was susceptible to simplification. The accompanying formula shows the portion of the atropine molecule (enclosed in the curved dotted line) believed to be responsible for its major activity. This group is sometimes called the "spasmophoric" group and compares with the "anesthesiophoric" group obtained by similar dissection of the cocaine molecule (q.v.). The validity of this conclusion has been amply borne out by the many active compounds having only a simple diethylaminoethyl residue replacing the tropine portion.

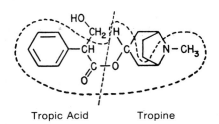

Tropic Acid   Tropine

Eucatropine may be considered as a conservative approach to the simplification of the aminoalcohol portion, in that the bicyclic tropine has been replaced by a monocyclic aminoalcohol and, in addition, mandelic acid replaces tropic acid.

One of the earliest compounds to be prepared utilizing a simplified noncyclic aminoalcohol was amprotropine which was prepared by Fromherz[76] in 1933 and for many years was widely used as a gastrointestinal antispasmodic but has been displaced by much more active compounds. In this particular case, the tropic acid residue was retained, but the bulk of

Amprotropine

research on antispasmodics of this nature has been directed toward compounds in which both the acid and the aminoalcohol portions have been modified. Acids formally related to phenylacetic acid, particularly with a hydroxy function on the carbon adjacent to the carbonyl (e.g., mandelic and benzilic), were shown early to be among the most highly active acids to be employed. Table 12-10 depicts the general situation with respect to substitution and spasmolytic potency of several different compounds. It will be noted that, starting with a simple acetyl ester (which is spasmogenic) the activity increases with increasing aromatic substitution. An enhancing effect is apparent when hydroxylation of the acetyl carbon is employed, although the dangers of broad generalizations are noted in the decreased activity of I vs. H. Likewise, two phenyl groups appear to be maximal, inasmuch as a sharp drop in activity is noted when three phenyls are employed. This is caused possibly by steric hindrance in the triphenylacetyl moiety. Comparison of compounds E and H would indicate also that enhancement of action results from bonding the phenyl groups together into the fluorene moiety, a compound which

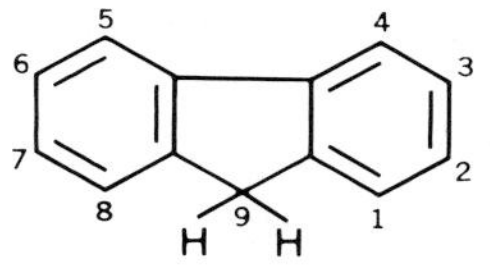

Fluorene

enjoyed some commercial success under the trade name of Pavatrine. However, it was withdrawn from the market some years ago because it has been far surpassed by other agents marketed by the same company.

It is evident that the acid portion (corresponding to tropic acid) should be somewhat bulky in nature, especially when the aminoalcohol portion is a simple one. This is an indication for the need of at least one portion of the molecule to have the space-occupying, umbrellalike shape which leads to firm binding at the receptor site area.

Although simplification of the aminoalcohol portion of the atropine prototype has been a guiding principle in most research, it is worth noting that many of the presently used anticholinergics still include a cyclic aminoalcohol moiety. It is interesting to note that the aminoalcohol-ester anticholinergics

are used primarily as antispasmodics or mydriatics and that cholinolytic compounds that are classed as aminoalcohol or aminoalcohol ether analogs of atropine are, with few exceptions, employed as antiparkinsonism drugs.

Another important feature to be found in many of the synthetic anticholinergics used as antispasmodics is that they contain a quaternary nitrogen, presumably to enhance activity. The initial synthetic quaternary compound methantheline bromide has served as a forerunner for many others. These compounds combine anticholinergic activity of the antimuscarinic type with some ganglionic blockade to reinforce the parasympathetic blockade. However, it must be noted that quaternization also introduces the possibility of blockade of voluntary synapses (curariform activity) as well; this can become evident with sufficiently high doses.

Products

Many active compounds have been discussed above, most of which have been on the market. Those currently in use are described in the following monographs.

Clidinium Bromide U.S.P., Quarzan® Bromide, 3-hydroxy-1-methylquinuclidinium bromide benzilate. The preparation of this compound is described by Sternbach and Kaiser.[77,78] It occurs as a white or nearly white, almost odorless, crystalline powder which is optically inactive. It is soluble in water and in

TABLE 12–10*

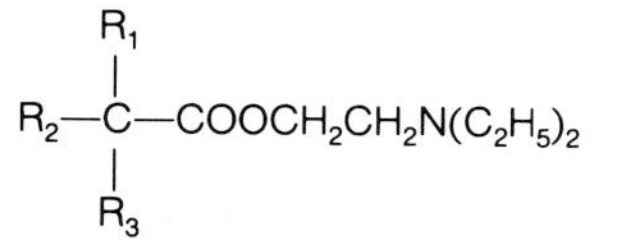

$$R_2 \underset{\overset{|}{R_3}}{\overset{\overset{\displaystyle R_1}{|}}{-}} C - COOCH_2CH_2N(C_2H_5)_2$$

	Structure			Spasmolytic Potency†	
Compound	R_1	R_2	R_3	Acetylcholine pA₂‡	Relative Potency %
A	H	H	H	Stimulates	
B	H	H	OH	4.0–4.3	1–2
C	Phenyl	H	H	5.0–5.3	10–20
D	Phenyl	H	OH	5.3–5.7	20–50
E§	Phenyl	Phenyl	H	6.0	100
F‖	Phenyl	Phenyl	OH	7.6	4,000
G	Phenyl	Phenyl	Phenyl	5.0	10
H¶	Fluorene-9-carboxylic			6.8	600
I	Fluorene-9-hydroxy-9-carboxylic			6.7	500

* Adapted from a table by Lands, A. M., *et al.*: J. Pharmacol. Exp. Ther. 100:19, 1950.
† All esters were tested as the hydrochlorides on rabbit small intestine (isolated segments).
‡ Logarithm of the reciprocal of the E.D.₅₀.
§ Trasentine.   ‖ WIN 5606.   ¶ Pavatrine.

alcohol but only very slightly soluble in ether and in benzene.

Clidinium Bromide

This anticholinergic agent is marketed alone and in combination with the minor tranquilizer chlordiazepoxide (Librium®), the resultant product being known as Librax®. The rationale of the combination for the treatment of gastrointestinal complaints is the use of an anxiety-reducing agent together with an anticholinergic based on the recognized contribution of anxiety to the development of the diseased condition. It is suggested for peptic ulcer, hyperchlorhydria, ulcerative or spastic colon, anxiety states with gastrointestinal manifestations, nervous stomach, irritable or spastic colon, etc. The combination capsule contains 5 mg. of chlordiazepoxide hydrochloride and 2.5 mg. of clidinium bromide. It is, of course, contraindicated in glaucoma and other conditions that may be aggravated by the parasympatholytic action, such as prostatic hypertrophy in the elderly male which could lead to urinary retention. The usual recommended dose for adults is 1 or 2 capsules 4 times a day before meals and at bedtime.

Cyclopentolate Hydrochloride U.S.P., Cyclogyl® Hydrochloride, 2-(dimethylamino)ethyl 1-hydroxy-α-phenylcyclopentaneacetate hydrochloride. This compound, together with a series of closely related compounds, was synthesized by Treves and Testa.[79] It is a crystalline, white, odorless solid which is very soluble in water, easily soluble in alcohol and only slightly soluble in ether. A 1 percent solution has a pH of 5.0 to 5.4.

Cyclopentolate Hydrochloride

It is used only for its effects on the eye, where it acts as an ophthalmic parasympatholytic. It produces cycloplegia and mydriasis quickly when placed in the eye. Its primary field of usefulness is in refraction studies. However, it can be used as a mydriatic in the management of iritis, iridocyclitis, keratitis and choroiditis. Although it does not seem to affect intraocular tension significantly it is desirable to be very cautious with patients with high intraocular pressure and also with elderly patients with possible unrecognized glaucomatous changes.

The drug has one-half the antispasmodic activity of atropine and has been shown to be nonirritating when instilled repeatedly into the eye. If not neutralized after the refraction studies, the effect is usually gone in 24 hours. Neutralization with a few drops of pilocarpine nitrate solution, 1 to 2 percent, often results in complete recovery in 6 hours.

It is supplied as a ready-made ophthalmic solution in concentrations of either 0.5 or 1 percent, and also in the form of a gel for better application to the eye.

Dicyclomine Hydrochloride U.S.P., Bentyl® Hydrochloride, 2-(diethylamino)ethyl [bicyclohexyl]-1-carboxylate hydrochloride. The synthesis of this drug is described by Tilford and his co-workers.[80] In common with similar salts, this drug is a white crystalline compound that is soluble in water.

Dicyclomine Hydrochloride

It is reported to have one-eighth of the neurotropic activity of atropine and approximately twice the musculotropic activity of papaverine. Again, this preparation has minimized the undesirable side-effects associated with the atropine-type compounds. It is used for its spasmolytic effect on various smooth-muscle spasms, particularly those associated with the gastrointestinal tract. It is also useful in dysmenorrhea, pylorospasm and biliary dysfunction.

The drug, introduced in 1950, is marketed in the form of capsules, with or without 15 mg. of phenobarbital, and also in the form of a syrup, with or without phenobarbital. For parenteral use (intramuscularly) it is supplied as a solution containing 20 mg. in 2 ml.

Eucatropine Hydrochloride U.S.P., euphthalmine hydrochloride, 1,2,2,6-tetramethyl-4-piperidyl mandelate hydrochloride. This compound possesses the aminoalcohol moiety characteristic of one of the early local anesthetics, i.e., *beta*-eucaine, but differs in the acidic portion of the ester by having a mandelate instead of a benzoate. The salt is an odorless, white, granular powder, providing solutions that are neutral to litmus. It is very soluble in water, freely soluble in alcohol and chloroform but almost insoluble in ether.

Eucatropine Hydrochloride

The action of eucatropine closely parallels that of atropine although it is much less potent than the latter. It is used topically in a 0.1-ml. dose as a mydriatic in 2 percent solution or in the form of small tablets. However, the use of concentrations of from 5 to 10 percent is not uncommon. Dilation, with little impairment of accommodation, takes place in about 30 minutes, and the eye returns to normal in from 2 to 3 hours.

Glycopyrrolate U.S.P., Robinul®, 3-hydroxy-1,1-dimethylpyrrolidinium bromide α-cyclopentylmandelate. The drug occurs as a white crystalline powder that is soluble in water or alcohol but is practically insoluble in chloroform or ether.

Glycopyrrolate

Glycopyrrolate is a typical anticholinergic and possesses, at adequate dosage levels, the atropinelike effects characteristic of this group. It has a spasmolytic effect on the musculature of the gastrointestinal tract as well as the genitourinary tract. It diminishes gastric and pancreatic secretions and diminishes the quantity of perspiration and saliva. Its side-effects are typically atropinelike also, i.e., dryness of the mouth, urinary retention, blurred vision, constipation, etc.[81] Because of its quaternary ammonium character it rarely causes central nervous system disturbances, although, in sufficiently high dosage, it can bring about ganglionic and myoneural junction block.

The drug is used as an adjunct in the management of peptic ulcer and other gastrointestinal ailments associated with hyperacidity, hypermotility and spasm. In common with other anticholinergics its use does not preclude dietary restrictions or use of antacids and sedatives if these are indicated.

Mepenzolate Bromide U.S.P., Cantil®, 3-hydroxy-1,1-dimethylpiperidinium bromide benzilate. This compound may be prepared by the method of Biel *et al.*[82] utilizing the transesterification reaction with 1-methyl-3-hydroxypiperidine and methyl benzilate. The resulting base is quaternized with methyl bromide to give a white, crystalline product which is water-soluble.

Mepenzolate Bromide

It has an activity of about one-half that of atropine in reducing acetylcholine-induced spasms of the guinea pig ileum, although some reports rate it as equal to atropine in effectiveness and duration of action. It is specifically promoted for a claimed "markedly selective action on the colon." The selective action on colonic hypermotility is said to relieve pain cramps and bloating and to help curb diarrhea. The evidence for this specific action is conflicting. Bachrach[2] for example, in a survey of the anticholinergic literature questions the specificity of these drugs "for any particular gastrointestinal organ, function, or segment of the gastrointestinal tract."

Methantheline Bromide U.S.P., Banthine® Bromide, diethyl(2-hydroxyethyl)methylammonium bromide xanthene-9-carboxylate. Methantheline may be prepared according to the method outlined by Burtner and Cusic[83] although this reference does not show the final formation of the quaternary salt. The compound from which the quaternary salt is prepared was in the series of esters from which aminocarbofluorene was selected as the best spasmolytic agent.

Methantheline Bromide

It is a white, slightly hygroscopic, crystalline salt which is soluble in water to produce solutions with a pH of about 5. Aqueous solutions are not stable and hydrolyze in a few days. The bromide form is preferable to the very hygroscopic chloride.

This drug, introduced in 1950, is a potent anticholinergic agent and acts at the nicotinic sites on the *ganglia* of the sympathetic and the parasympathetic systems, as well as at the myoneural junction of the postganglionic cholinergic fiber. Methantheline has no action at the effector site of the sympathetic system.

Among the conditions for which methantheline is indicated are gastritis, intestinal hypermotility, bladder irritability, cholinergic spasm, pancreatitis, hyperhidrosis and peptic ulcer, all of which are manifestations of parasympathotonia. The last indication (peptic ulcer) has been responsible for much of the publicity accorded the drug. The parasympathetic system is represented in its gastric innervation by the vagus nerve, and, prior to the introduction of such a drug as methantheline, the surgical procedure of vagotomy had been shown to give relief to peptic ulcer patients. The drug is, in effect a nonsurgical vagotomy that can be withdrawn whenever desired, in common with other quaternary anticholinergic agents.

Side-reactions are atropine-like (mydriasis, cycloplegia, dryness of mouth), and the drug is contraindicated in glaucoma. High overdosage may bring about a curare-like action, a not too surprising fact when it is considered that acetylcholine is the mediating factor for neural transmission at the somatic myoneural junction. This side-effect can be counteracted with neostigmine methylsulfate.

The drug is marketed as 50-mg. tablets for oral use with or without phenobarbital (15 mg.), and for parenteral use, 50-mg. ampules are supplied.

Oxyphencyclimine Hydrochloride U.S.P., Daricon®, Vistrax®, (1,4,5,6-tetrahydro-1-methyl-2-pyrimidinyl)methyl α-phenylcyclohexaneglycolate monohydrochloride. The synthesis of this compound is described by Faust *et al.*[84] The product is a white, crystalline compound which is sparingly soluble in water (1.2 g. per 100 ml. at 25° C.). It has a bitter taste.

Oxyphencyclimine Hydrochloride

This compound, introduced in 1958, is promoted as a peripheral anticholinergic-antisecretory agent with little or no curarelike activity and with little or no ganglionic blocking activity. That these activities are absent is probably due to the tertiary character of the compound, which is in somewhat marked contrast with the quaternaries that have dominated the anticholinergic scene for the most part and potentiate anticholinergic activity due to a coupling of antimuscarinic action with ganglion-blocking action. Also, the tertiary character of the nitrogen should promote its intestinal absorption, as previously outlined. Another feature of the compound is its relatively long duration of action (12 hours) which is suggested as being in some way related to the amidine-type structure to be found in the aminoalcohol portion of the molecule.[85] Perhaps that most significant activity of this compound is its marked ability to reduce both the volume and the acid content of the gastric juices[86], a desirable action in view of the more recent hypotheses pertaining to peptic ulcer therapy. Another important feature of this compound is its low toxicity in comparison with many of the other available anticholinergics.

Oxyphencyclimine is suggested for use in peptic ulcer, pylorospasm and functional bowel syndrome. It is contraindicated, as are other anticholinergics, in patients with prostatic hypertrophy and glaucoma.

Piperidolate Hydrochloride U.S.P., Dactil®, 1-ethyl-3-piperidyl diphenylacetate hydrochloride. This compound is synthesized according to the method of Biel and co-workers[82] utilizing the conventional acylation of the appropriate aminoalcohol with diphenylacetyl chloride in the presence of triethylamine. The hydrochloride exists as white crystals which are water-soluble.

Piperidolate Hydrochloride

The principal activity of this compound, introduced in 1954, seems to be antimuscarinic in nature with little or no action on ganglia or voluntary-muscle innervations. Its central action is negligible. Its antimuscarinic activity is about one one-hundredth that of atropine. The specificity as a spasmolytic for the smooth musculature of the gastrointestinal tract is said to be its main action, and it is termed a "visceral eutonic" (an agent producing normal tone of a viscus) in the promotional literature. There is claimed to be little or no effect on gastric secretion and, in therapeutic doses, it seems to have little action on the biliary tract musculature. The rapid effect on gastrointestinal motility (within 10 to 20 minutes) is attributed to a local anesthetic effect.

Its clinical usefulness has been as an adjunctive for management of functional gastrointestinal disorders characterized by spasm and hypermotility associated with pain. The upper gastrointestinal tract seems to be affected more by the drug than the lower tract and, whereas it is useful for gastroduodenal spasm, pylorospasm and cardiospasm, it is of little value for colonic spasm. It is *not* intended for use in peptic ulcer. The drug is also promoted for relief of spasm of biliary sphincter and biliary dyskinesia.

Poldine Methylsulfate U.S.P., 2-(hydroxymethyl)-1,1-dimethylpyrrolidinium methylsulfate benzilate. This compound occurs as a water-soluble, creamy-white crystaline powder.

Poldine Methylsulfate

It has, qualitatively, the same atropinelike actions as other anticholinergics both as to the therapeutically desirable effects and the undesirable side-effects.[81] It is promoted for the same purposes as, for example, glycopyrrolate and has the same side-effects and precautions concerning its use. Its principal use is as an adjunct in the management of peptic ulcer and related conditions.

It is marketed as 4-mg. tablets with or without 15 mg. of butabarbital sodium.

Propantheline Bromide U.S.P., Pro-Banthine® Bromide, (2-hydroxyethyethyl)diisopropylmethylammonium bromide xanthene-9-carboxylate.

Propantheline Bromide

The method of preparation of this compound is exactly analogous to that used for methantheline bromide (q.v.).

It is a white, water-soluble, crystalline substance with properties quite similar to those of methantheline.

Its chief difference from methantheline is in its potency, which has been estimated variously as being from 2 to 5 times as great. This greater potency is reflected in its smaller dose. For example, instead of a 50-mg. initial dose, a 15-mg. initial dose is suggested

for propantheline bromide. It is available in 15-mg. sugar-coated tablets and in the form of a powder (30 mg.) for preparing parenteral solutions.

AMINOALCOHOL ETHERS

The aminoalcohol ethers thus far introduced have been used as antiparkinsonism drugs rather than as conventional anticholinergics (i.e., as spasmolytics, mydriatics, etc.). In general, they may be considered as closely related to the antihistaminics and, indeed, do possess antihistaminic properties of a substantial order. In turn, the antihistamines possess anticholinergic activity and have been used as antiparkinsonism agents. Comparison of chlorphenoxamine and orphenadrine with the antihistaminic diphenhydramine illustrates the close similarity of structure. The use of diphenhydramine in parkinsonism has been cited earlier. Benztropine may also be considered as a structural relative of diphenhydramine, although the

Diphenhydramine

aminoalcohol portion is tropine and, therefore, more distantly related than chlorphenoxamine and orphenadrine. In the structure of benztropine a 3-carbon chain intervenes between the nitrogen and oxygen functions, whereas in the others a 2-carbon chain is evident. However, the rigid ring structure possibly orients the nitrogen and oxygen functions into more nearly the 2-carbon chain interprosthetic distance than is apparent at first glance. This, combined with the flexibility of the alicyclic chain, would help to minimize the distance discrepancy.

PRODUCTS

Benztropine Mesylate U.S.P., Cogentin® Methanesulfonate, 3α-(diphenylmethoxy)-1αH,5αH-tropane methanesulfonate. The compound occurs as a white, colorless, slightly hygroscopic, crystalline powder. It is very soluble in water, freely soluble in alcohol and very slightly soluble in ether. The pH of aqueous solutions is about 6. It is prepared by the method of Phillips[87] by interaction of diphenyldiazomethane and tropine.

Benztropine Mesylate

Benztropine mesylate combines anticholinergic, antihistaminic and local anesthetic properties of which the first is the applicable one in its use as an antiparkinsonism agent. It is about as potent as atropine as an anticholinergic and shares some of the side-effects of this drug, such as mydriasis, dryness of mouth, etc. Of importance, however, is the fact that it does not produce central stimulation but, on the contrary, exerts the characteristic sedative effect of the antihistamines and, for this reason, patients using the drug should not engage in jobs that require close and careful attention.

Tremor and rigidity are relieved by benztropine mesylate, and it is of particular value for those patients who cannot tolerate central excitation (e.g., aged patients). It also may have a useful effect in minimizing drooling, sialorrhea, masklike facies, oculogyric crises and muscular cramps.

The usual caution that is exercised with any anticholinergic in glaucoma, prostatic hypertrophy, etc., is observed with this drug.

Chlorphenoxamine Hydrochloride U.S.P., Phenoxene®, 2-[(p-chloro-α-methyl-α-phenylbenzyl)oxy]-N,N-dimethylethylamine hydrochloride. It occurs in the form of colorless needles which are soluble in water. Aqueous solutions are stable.

Chlorphenoxamine Hydrochloride

This drug was originally introduced in Germany as an antihistaminic. However, it is stated that this close relative of diphenhydramine (Benadryl®) has its antihistaminic potency lowered by the *para*-Cl and the α-methyl group present in the molecule. At the same time, the anticholinergic action is increased. The drug has an oral LD_{50} of 410 mg. in mice, indicating a substantial margin of safety. It was introduced to U.S. medicine in 1959.

It is indicated for the symptomatic treatment of all types of Parkinson's disease and is said to be especially useful when rigidity and impairment of muscle contraction are evident. It is not as useful against tremor, and combined therapy with other agents may be necessary. The drug has proved to be very useful either on its own or as a replacement for orphenadrine (effect tends to wear off) in counteracting akinesia, adynamia, mental sluggishness and lack of mobility in patients with paralysis agitans.

Orphenadrine Citrate U.S.P., Norflex®, N,N-dimethyl-2-[(o-methyl-α-phenylbenzyl)oxyl]ethylamine citrate (1:1). This compound is synthesized according to the method in the patent literature.[88] It occurs as a white, bitter-tasting crystalline powder. It is sparingly soluble in water, slightly soluble in alcohol, and insoluble in chloroform, in benzene and in ether. The hydrochloride salt is marketed as Disipal®.

Although this compound, introduced in 1957, is closely related to diphenhydramine structurally, it has a much lower antihistaminic activity and a much higher anticholinergic action. Likewise, it lacks the sedative effects characteristic of diphenhydramine. Pharmacologic testing indicates that it is not primarily a peripherally acting anticholinergic because it has only weak effects on smooth muscle, the eye and on secretory glands. However, it does reduce voluntary muscle spasm by a central inhibitory action on cerebral motor areas, a central effect similar to that of atropine.

Orphenadrine Citrate

The drug is used for the symptomatic treatment of Parkinson's disease. Although it is not effective in all patients, it appears from the literature that about one-half of the patients are benefited. It relieves rigidity better than it does tremor, and in certain cases it may accentuate the latter. The drug combats mental sluggishness, akinesia, adynamia and lack of mobility, but this effect seems to be diminished rather rapidly on prolonged use. It is best used as an adjunct to the other agents such as benztropine, procyclidine, cycrimine and trihexyphenidyl in the treatment of paralysis agitans.

The drug has a low incidence of side-effects, which are the usual ones for this group, namely, dryness of mouth, nausea, mild excitation, etc.

The development of amino alcohols as parasympatholytics took place in the 1940's. It was soon established, however, that these antispasmodics were equally efficacious in parkinsonism.

AMINOALCOHOLS

All the useful compounds have had the general characteristic of possessing rather bulky groups around the hydroxyl function, together with a cyclic amino function. This is reminiscent of the bulky groups in the acids that were found to be desirable in the aminoester type of anticholinergic (q.v.). It serves to emphasize the fact that the group, per se, is not a necessary adjunct to cholinolytic activity, provided that other polar groupings such as the hydroxyl can substitute as a prosthetic group for the carboxyl function. Another structural feature common to all aminoalcohol anticholinergics, with the notable exception of hexocyclium, is the γ-aminopropanol arrangement with 3 carbons intervening between the hydroxyl and amino functions. All of the aminoalcohols used for paralysis agitans are tertiary amines and, because the desired locus of action is central, quaternization of the nitrogen destroys the antiparkinsonism properties. However, quaternization of these aminoalcohols has been utilized to enhance the anticholinergic activity to produce antispasmodic and antisecretory compounds such as hexocyclium, mepiperphenidol, tricyclamol chloride, and tridihexethyl chloride. The marked difference in activity by simple quaternization is shown vividly by comparison of procyclidine with its methochloride, tricyclamol chloride. The former is a useful drug in parkinsonism, but the latter has very little value. However, there is not such a great disparity in their action as spasmolytics on smooth muscle, both being active, but with the greater activity being found in the quaternary ammonium form.

PRODUCTS

Biperiden U.S.P., Akineton®, α-5-norbornen-2-yl-α-phenyl-1-piperidinepropanol. The drug consists of a white, practically odorless, crystalline powder. It is practically insoluble in water and only sparingly soluble in alcohol although it is freely soluble in chloroform. Its preparation is described by Haas and Klavehn.[89]

Biperiden, introduced in 1959, has a relatively weak visceral anticholinergic but a strong nicotinolytic action in terms of its ability to block nicotine-induced convulsions. Therefore, its neurotropic action is rather low on intestinal musculature and blood vessels, but it

Biperiden

has a relatively strong musculotropic action, about equal to papaverine, in comparison with most synthetics. Its action on the eye, although mydriatic, is much less than that of atropine. These weak anticholinergic effects serve to add to its usefulness in Parkinson's syndrome by minimizing side-effects.

The drug is used in all types of Parkinson's disease (postencephalitic, idiopathic, arteriosclerotic) and helps to eliminate akinesia, rigidity and tremor. It is also used in drug induced extrapyramidal disorders by eliminating symptoms and permitting continued use of tranquilizers. Biperiden is also of value in spastic disorders not related to parkinsonism, such as multiple sclerosis, spinal cord injury and cerebral palsy. It is contraindicated in all forms of epilepsy.

It is usually taken orally in tablet form but the free base form is official to serve as a source for the preparation of Biperiden Lactate Injection U.S.P., which is a sterile solution of biperiden lactate in water for injection prepared from biperiden base with the aid of lactic acid. It usually contains 5 mg. per ml.

Biperiden Hydrochloride U.S.P., Akineton® Hydrochloride, α-5-norbornen-2-yl-α-phenyl-1-piperidinepropanol hydrochloride. It is a white optically inactive, crystalline, odorless powder which is slightly soluble in water, ether, alcohol and chloroform and sparingly soluble in methanol.

Biperiden hydrochloride has all of the actions described for biperiden above. The hydrochloride is used for tablets, because it is better suited to this dosage form than is the lactate salt. As with the free base and the lactate salt, xerostomia (dryness of the mouth) and blurred vision may occur.

Cycrimine Hydrochloride U.S.P., Pagitane® Hydrochloride, α-cyclopentyl-α-phenyl-1-piperidinepropanol hydrochloride. This drug is made by the procedure of Denton et al.[90] It occurs as a white, odorless, bitter solid which is sparingly soluble in alcohol (2:100), and only slightly soluble in water (0.6:100). A 0.5 percent solution in water is slightly acidic (pH 4.9-5.4).

The drug has been introduced as an aid in the treatment of paralysis agitans (Parkinson's disease). Cycrimine is a potent antispasmodic of the neurotropic type with an activity of about one-fourth to one-half that of atropine sulfate. It has little or no effect

Cycrimine Hydrochloride

against spasms induced by histamine. It is slightly more toxic than atropine sulfate.

The drug is supplied as 1.25-mg. and 2.5-mg. tablets. Dosage is quite variable and is adjusted to the individual. The postencephalitic group of patients was reported to be able to tolerate the largest doses (up to 30 and 50 mg. per day), whereas the arteriosclerotic and idiopathic types exhibited adverse effects with the larger doses. According to Magee and DeJong[91] the best procedure is to start with a small initial dose and increase the dose by 2.5-mg. (or smaller) increments to the point of tolerance. The side-effects are the usual atropine-like ones encountered with this group of compounds, namely, drying of the mouth, blurring of vision and epigastric distress.

Procyclidine Hydrochloride U.S.P., Kemadrin®, α-cyclohexyl-α-phenylpyrrolidinepropanol hydrochloride. This compound is prepared by the method of Adamson[92] or Bottorff[93] as described in the patent literature. It occurs as white crystals which are moderately soluble in water (3:100). It is more soluble in alcohol or chloroform and is almost insoluble in ether.

Procyclidine Hydrochloride

Although procyclidine, introduced in 1956, is an effective peripheral anticholinergic and indeed, has been used for peripheral effects similarly to its methochloride (i.e., tricyclamol chloride), its clinical usefulness lies in its ability to relieve spasticity of voluntary muscle by its central action. Therefore, it has been employed with success in the treatment of Parkinson's syndrome.[94] It is said to be as effective as cycrimine and trihexyphenidyl and is used for reduction of muscle rigidity in the postencephalitic, the arteriosclerotic and the idiopathic types of the disease. Its effect on tremor is not predictable and probably should be supplemented by combination with other similar drugs.

The toxicity of the drug is low, but side-effects are noticeable when the dosage is high. At therapeutic dosage levels dry mouth is the most common side-effect. The same care should be exercised with this drug as with all other anticholinergics when administered to patients with glaucoma, tachycardia or prostatic hypertrophy.

Tridihexethyl Chloride U.S.P., Pathilon® Chloride, (3-cyclohexyl-3-hydroxy-3-phenylpropyl)triethylammonium chloride. The preparation of this compound as the corresponding bromide is described by Denton and Lawson.[95] It occurs in the form of a white, bitter, crystalline powder possessing a characteristic odor. The compound is freely soluble in water and alcohol, the aqueous solutions being nearly neutral in reaction.

Although this drug, introduced in 1958, has ganglion-blocking activity, it is said that its peripheral atropinelike activity predominates; therefore, its therapeutic application has been based on the latter activity. It possesses the antispasmodic and the antisecretory activities characteristic of this group but, because of its quaternary character, is valueless in the Parkinson syndrome.

Tridihexethyl Chloride

The drug is useful for adjunctive therapy in a wide variety of gastrointestinal diseases such as peptic ulcer, gastric hyperacidity and hypermotility, spastic conditions such as spastic colon, functional diarrhea, pylorospasm and other related conditions. Because its action is predominately antisecretory it is most effective in gastric hypersecretion rather than in hypermotility and spasm. It is best administered intravenously for the later conditions.

The side-effects usually found with effective anticholinergic therapy occur with the use of this drug. These are dryness of mouth, mydriasis, etc. As with other anticholinergics, care should be exercised when administering the drug in glaucomatous conditions, cardiac decompension and coronary insufficiency. It is contraindicated in patients with obstruction at the bladder neck, prostatic hypertrophy, stenosing gastric and duodenal ulcers or pyloric or duodenal obstruction.

The drug may be administered orally or parenterally. Oral therapy is preferable. The drug is supplied in 25-mg. tablets and as powder for injection (10 mg. in 1 ml.).

Trihexyphenidyl Hydrochloride U.S.P., Artane® Hydrochloride, Tremin® Hydrochloride, Pipanol®, α-cyclohexyl-α-phenyl-1-piperidinepropanol hydrochloride. This compound was synthesized by Denton and his co-workers.[96] It occurs as a white, odorless crystalline compound that is not very soluble in water (1:100). It is more soluble in alcohol (6:100) and chloroform (5:100) but only slightly soluble in ether and benzene. The pH of a 1 percent aqueous solution is about 5.5 to 6.0.

Trihexyphenidyl Hydrochloride

Introduced in 1949, it is approximately one-half as active as atropine as an antispasmodic, but is claimed to have milder side-effects, such as mydriasis, drying of secretions and cardioacceleration. It has a good margin of safety, although it is about as toxic as atropine. It has found a place in the treatment of parkinsonism and is claimed also to provide some measure of relief from the mental depression often associated with this condition. However, it does exhibit some of the side-effects typical of the parasympatholytic-type preparation, although it is said that these often may be eliminated by adjusting the dose carefully. According to Pinder[58] it is the drug of choice for anticholinergic therapy of parkinsonism.

AMINOAMIDES

The aminoamide type of anticholinergic, from a structural standpoint, represents the same type of molecule as the aminoalcohol group with the important exception that the polar amide group replaces the corresponding polar hydroxyl group. Aminoamides retain the same bulky structural features as are found at one end of the molecule or the other in all of the active anticholinergics. Isopropamide is the only drug of this class currently in use.

Another amide-type structure is that of tropicamide, formerly known as bistropamide, a compound having some of the atropine features.

PRODUCTS

Isopropamide Iodide U.S.P., Darbid®, (3-carbamoyl-3,3-diphenylpropyl)diisopropylmethylammonium iodide. This compound may be made according to the method of Janssen and his co-workers.[97] It occurs as a

white to pale yellow crystalline powder with a bitter taste and is only sparingly soluble in water but is freely soluble in chloroform and alcohol.

Isopropamide Iodide

This drug, introduced in 1957, is a potent anticholinergic producing atropine-like effects peripherally. Even with its quaternary nature it does not cause sympathetic blockade at the ganglionic level except in high-level dosage. Its principal distinguishing feature is its long duration of action. It is said that a single dose can provide antispasmodic and antisecretory effects for as long as 12 hours.

It is used as adjunctive therapy in the treatment of peptic ulcer and other conditions of the gastrointestinal tract associated with hypermotility and hyperacidity. It has the usual side-effects of anticholinergics (dryness of mouth, mydriasis, difficult urination) and is contraindicated in glaucoma, prostatic hypertrophy, etc.

Tropicamide U.S.P., Mydriacyl®, N-ethyl-2-phenyl-N-(4-pyridylmethyl)hydracrylamide. The preparation of this compound is described in the patent literature.[98] It occurs as a white or practically white, crystalline powder which is practically odorless. It is only slightly soluble in water but is freely soluble in chloroform and in solutions of strong acids. The pH of ophthalmic solutions ranges between 4.0 and 5.0, the acidity being achieved with nitric acid.

Tropicamide

This drug is an effective anticholinergic for ophthalmic use where mydriasis is produced by relaxation of the sphincter muscle of the iris, allowing the adrenergic innervation of the radial muscle to dilate the pupil. Its maximum effect is achieved in about 20 to 25 minutes and lasts for about 20 minutes, with complete recovery being noted in about 6 hours. Its action is more rapid in onset and wears off more rapidly than that of most other mydriatics. To achieve mydriasis either the 0.5 or 1.0 percent concentration may be used, although cycloplegia is achieved only with the stronger solution. Its uses are much the same as those described in general for mydriatics earlier, but opinions differ as to whether the drug is as effective as homatropine, for example, in achieving cycloplegia. For mydriatic use, however, in examination of the fundus and treatment of acute iritis, iridocyclitis and keratitis it is quite adequate, and, because of its shorter duration of action, it is less prone to initiate a rise in intraocular pressure than are the more potent longer-lasting drugs. However, as with other mydriatics, pupil dilation can lead to increased intraocular pressure. In common with other mydriatics it is contraindicated in cases of glaucoma, either known or suspected, and should not be used in the presence of a shallow anterior chamber. Thus far, allergic reactions and/or ocular damage have not been observed with this drug.

MISCELLANEOUS

The miscellaneous group contains three useful compounds. Each of them has the typical bulky group that is characteristic of the usual anticholinergic molecule. In the one case it is represented by the diphenylmethylene moiety (e.g., diphemanil), in the second by a phenothiazine (e.g., ethopropazine) and in the third by a thioxanthene structure (e.g., methixene).

Diphemanil Methylsulfate U.S.P., Prantal® Methylsulfate, 4-(diphenylmethylene)-1,1-dimethylpiperidinium methylsulfate. This compound may be prepared by two alternative syntheses as outlined by Sperber and co-workers.[99] It was introduced in 1951.

The drug is a white, crystalline, odorless compound that is sparingly soluble in water (50 mg. per ml.), alcohol and chloroform. The pH of a 1 percent aqueous solution is between 4.0 and 6.0.

The methylsulfate salt was chosen as the best because the chloride is hygroscopic and because the bromide and iodide ions have exhibited toxic manifestations under clinical usage.

Diphemanil Methylsulfate

As mentioned previously, it is a potent cholinergic blocking agent. In the usual dosage range it acts as an effective parasympatholytic by blocking nerve impulses at the parasympathetic ganglia but does not invoke a sympathetic ganglionic blockade. It is claimed to be highly specific in its action upon those innervations that have to do with gastric secretion and motility. Although this drug is capable of producing atropine-like side-effects, these are not a problem because in the doses used they are reported to occur very rarely. The highly specific nature of its action on the gastric functions makes it useful in the treatment of peptic ulcer, and its lack of atropine-like effects makes this use much less distressing than is the case with some of the other similarly used drugs. In addition to its action in gastric hypermotility, it is valuable in hyperhidrosis in low doses (50 mg. twice daily) or topically.

The drug is not well absorbed from the gastrointestinal tract, particularly in the presence of food, so it is desirable to administer the oral doses between meals. In addition to the regular tablet form the drug also is supplied in a so-called "repeat action" tablet that has an enteric-coated tablet embedded in an ordinary tablet and gives about 8 hours of activity.

Ethopropazine Hydrochloride U.S.P., Parsidol®, 10-[2-(diethylamino)propyl]phenothiazine monohydrochloride. The compound is prepared in a number of ways, among which is the patented method of Berg and Ashley.[65] It occurs as a white crystalline compound with a poor solubility in water at 20° C. (1:400) but greatly increased solubility at 40° C. (1:20). It is soluble in ethanol and chloroform but almost insoluble in ether, benzene and acetone. The pH of an aqueous solution is about 5.8.

Ethopropazine Hydrochloride

This phenothiazine was introduced to therapy in 1954. It has similar pharmacologic activities and has been found to be especially useful in the symptomatic treatment of parkinsonism. In this capacity it has value in controlling rigidity, and it also has a favorable effect on tremor, sialorrhea and oculogyric crises. It is often used in conjunction with other parkinsonolytics for complementary activity.

Side-effects are common with this drug but not usually severe. Drowsiness and dizziness are the most common side-effects at ordinary dosage levels, and as the dose increases xerostomia, mydriasis, etc., become evident. It is contraindicated in conditions such as glaucoma because of its mydriatic effect.

TABLE 12-11

SYNTHETIC CHOLINERGIC BLOCKING AGENTS

Name Proprietary Name	Preparations	Category	Application	Usual Adult Dose*	Usual Dose Range*	Usual Pediatric Dose*
Clidinium Bromide U.S.P.	Clidinium Bromide Capsules U.S.P.	Anticholinergic		2.5 to 5.0 mg. 3 or 4 times daily	10 to 20 mg. daily	
Cyclopentolate Hydrochloride U.S.P. *Cyclogyl*	Cyclopentolate Hydrochloride Ophthalmic Solution U.S.P.	Anticholinergic (ophthalmic)	Topically to the conjunctiva, 1 drop of a 1 or 2 percent solution repeated once in 5 minutes			
Dicyclomine Hydrochloride U.S.P. *Bentyl*	Dicyclomine Hydrochloride Capsules U.S.P. Dicyclomine Hydrochloride Injection U.S.P.	Anticholinergic		10 to 20 mg. 3 or 4 times daily I.M., 20 mg. every 4 to 6 hours	Up to 120 mg. daily	
	Dicyclomine Hydrochloride Syrup, U.S.P.			10 to 20 mg. 3 or 4 times daily	Up to 120 mg. daily	Infants—5 mg. 3 or 4 times daily; children—10 mg. 3 or 4 times daily
	Dicyclomine Hydrochloride Tablets U.S.P.			10 to 20 mg. 3 or 4 times daily	Up to 120 mg. daily	
Eucatropine Hydrochloride U.S.P.	Eucatropine Hydrochloride Ophthalmic Solution U.S.P.	Anticholinergic (ophthalmic)				
Glycopyrrolate U.S.P. *Robinul*	Glycopyrrolate Injection U.S.P. Glycopyrrolate Tablets U.S.P.	Anticholinergic		Oral, 1 mg. 3 times daily; I.M., I.V., or S.C., 100 to 200 µg. at 4-hour intervals 3 or 4 times daily	1 to 2 mg.	
Mepenzolate Bromide U.S.P. *Cantil*	Mepenzolate Bromide Solution U.S.P. Mepenzolate Bromide Tablets U.S.P.	Anticholinergic		25 mg. 4 times daily	25 to 50 mg.	
Methantheline Bromide U.S.P. *Banthine*	Sterile Methantheline Bromide U.S.P. Methantheline Bromide Tablets U.S.P.	Anticholinergic		Oral, 50 mg. 4 times daily; I.M. or I.V., 50 mg. 4 times daily	50 to 100 mg.	
Oxyphencyclimine Hydrochloride U.S.P. *Daricon, Enarax, Vistrax*	Oxyphencyclimine Hydrochloride Tablets U.S.P.	Anticholinergic		10 mg. 2 times daily	10 to 50 mg. daily	
Piperidolate Hydrochloride U.S.P. *Dactil*	Piperidolate Hydrochloride Tablets U.S.P.	Anticholinergic		50 mg.		
Poldine Methylsulfate U.S.P. *Nacton*	Poldine Methylsulfate Tablets U.S.P.	Anticholinergic		4 mg. 3 or 4 times daily	2 to 4 mg.	
Propantheline Bromide U.S.P. *Pro-Banthine Bromide*	Sterile Propentheline Bromide U.S.P. Propantheline Bromide Tablets U.S.P.	Anticholinergic		I.M., of I.V., 15 to 30 mg. every 6 hours 15 mg. 3 times daily and 30 mg. at bedtime	Up to 240 mg. daily Up to 120 mg. daily	375 µg. per kg. of body weight or 10 mg. per square meter of body surface, 4 times daily

* See *U.S.P. D.I.* for complete dosage information.

(Continued)

TABLE 12-11

SYNTHETIC CHOLINERGIC BLOCKING AGENTS (CONTINUED)

Name Proprietary Name	Preparations	Category	Application	Usual Adult Dose*	Usual Dose Range*	Usual Pediatric Dose*
	Propantheline Bromide Extended-Release Tablets U.S.P.			30 mg. every 12 hours	Up to 120 mg. daily	
Benztropine Mesylate U.S.P. *Congentin Methanesulfate*	Benztropine Mesylate Injection U.S.P. Benztropine Mesylate Tablets U.S.P.	Antiparkinsonian		I.M., or I.V., 1 or 2 mg. 1 or 2 times daily 1 or 2 mg. 1 or 2 times daily	Up to 6 mg. daily Up to 6 mg. daily	
Chlorphenoxamine Hydrochloride U.S.P. *Phenoxene*	Chlorphenoxamine Hydrochloride Tablets U.S.P.	Skeletal-muscle relaxant		50 to 100 mg. 3 to 4 times a day	150 to 400 mg. daily	
Orphenadrine Citrate U.S.P. *Norflex*	Orphenadrine Citrate Injection U.S.P. Orphenadrine Citrate Tablets U.S.P.	Skeletal-muscle relaxant (antihistaminic)		I.M., or I.V., 60 mg. every 12 hours as needed Oral, 100 mg. twice daily		
Biperiden U.S.P. *Akineton*	Biperiden Lactate Injection U.S.P.	Anticholinergic		I.M. or I.V., 2 mg. of biperiden as the lactate which may be repeated every $\frac{1}{2}$ hour until relief is obtained, but no more than 4 consecutive doses should be given in a 24-hour period		
Biperiden Hydrochloride U.S.P. *Akineton Hydrochloride*	Biperiden Hydrochloride Tablets U.S.P.	Anticholinergic		Oral, 2 mg. 3 or 4 times daily		
Cycrimine Hydrochloride U.S.P. *Pagitane Hydrochloride*	Cycrimine Hydrochloride Tablets U.S.P.	Anticholinergic		Oral, 1.25 mg. 3 times daily	Up to 5 mg.	
Procyclidine Hydrochloride U.S.P. *Kemadrin*	Procyclidine Hydrochloride Tablets U.S.P.	Skeletal-muscle relaxant		Oral, 2 or 2.5 mg. 3 times daily, the dosage being adjusted as needed and tolerated or until the total dose reaches 20 to 30 mg. divided into 3 or 4 doses		
Tridihexethyl Chloride U.S.P. *Pathilon*	Tridihexethyl Chloride Injection U.S.P. Tridihexethyl Chloride Tablets U.S.P.	Anticholinergic		Oral, 25 mg. 3 times daily and 50 mg. at bedtime; parenteral, 10 to 20 mg. every 6 hours	25 to 75 mg. 1 to 4 times daily	
Trihexyphenidyl Hydrochloride U.S.P. *Artane, Pipanol, Tremin*	Trihexyphenidyl Hydrochloride, Elixir U.S.P. Trihexyphenidyl Hydrochloride Tablets U.S.P.	Antiparkinsonian		Initial, 1 to 2 mg. the first day, with increases of 2 mg. a day every 3 to 5 days until optimal effects are obtained; or 10 to 15 mg. usually divided into 3 or 4 doses		
Isopropamide Iodide U.S.P. *Darbid*	Isopropamide Iodide Tablets U.S.P.	Anticholinergic		5 mg. twice daily	10 to 20 mg. daily	

*See *U.S.P. D.I.* for complete dosage information.

(Continued)

TABLE 12-11

SYNTHETIC CHOLINERGIC BLOCKING AGENTS (CONTINUED)

Name Proprietary Name	Preparations	Category	Application	Usual Adult Dose*	Usual Dose Range*	Usual Pediatric Dose*
Tropicamide U.S.P. *Mydriacyl*	Tropicamide Ophthalmic Solution U.S.P.	Anticholinergic (ophthalmic)	Topically to the conjunctiva, 1 drop of a 1 percent solution, repeated in minutes			Topical, to the conjunctiva, 1 drop of a 0.5 or 1 percent solution repeated once in 5 minutes
Ethopropazine Hydrochloride U.S.P. *Parsidol*	Ethopropazine Hydrochloride Tablets U.S.P.	Antiparkinsonian		Initial, 50 mg. 1 or 2 times daily, the dose being gradually increased as necessary; maintenance, 100 to 150 mg. 1 to 4 times a day	50 to 600 mg. daily	
Diphemanil Methylsulfate U.S.P. *Prantal Methylsulfate*	Diphemanil Methylsulfate Tablets U.S.P.	Anticholinergic		100 mg. every 4 to 6 hours	50 to 200 mg.	

* See *U.S.P. D.I.* for complete dosage information.

PAPAVERINE AND RELATED COMPOUNDS

Previously, it was pointed out that papaverine is in actuality not a parasympatholytic. However, it exerts an antispasmodic effect and for that reason is customarily considered together with the solanaceous alkaloids. Papaverine does not interfere with the induction of the stimulus but rather with the response in the effector system. Because of its nonspecific action (i.e., with respect to the acetylcholine receptor) it is often called a nonspecific antagonist. This is sometimes referred to as a musculotropic type of spasmolysis, in contrast with the so-called neurotropic action of atropine and its congeners.

Papaverine interferes with the mechanism of muscle contraction by inhibiting the enzmye phosphodiesterase in smooth-muscle cells (Fig. 12-13). Cyclic AMP is formed by the action of the enzmye adenylate cyclase on the cellular nucleotide adenosine triphosphate (ATP). In turn the cyclic AMP formed by this action is broken down by the cellular enzyme cyclic nucleotide phosphodiesterase (PDE).[101] Papaverine is a potent inhibitor of vascular smooth muscle PDE[102] and there is a significant elevation of cyclic AMP following the administration of papaverine. The inhibition of PDE and elevation of cyclic AMP were shown to precede and be associated with smooth-muscle relaxation. The inhibition of PDE and the subsequent increase of cyclic AMP do not inactivate the contractile elements of the muscle because it is still possible to obtain a response after papaverine under certain conditions.

Regardless of the type of smooth muscle, papaverine acts as a spasmolytic, although its effectiveness is greater in some muscles than in others. It relaxes the smooth musculature of the larger blood vessels, especially coronary, systemic peripheral, and pulmonary arteries. Perhaps also, by its vasodilating action on cerebral blood vessels, papaverine increases cerebral blood flow and decreases cerebral vascular resistance. At the same time oxygen consumption is unaltered. These effects perhaps explain the benefits reported from the drug in cerebral vascular encephalopathy. Papaverine is devoid of the atropinelike effects on the central nervous system. The absence of such effects is a desirable characteristic of papaverine-type compounds, but unfortunately these compounds do not compare in potency to the atropine congeners.

Papaverine (see formula below) is the principal naturally occurring member of this group that is of any therapeutic consequence as an antispasmodic.

Papaverine Hydrochloride U.S.P., 6,7-dimethoxy-1-veratrylisoquinoline hydrochloride. This alka-

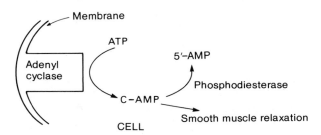

FIG. 12-13. Mechanism of antispasmodic activity of papaverine.

loid was isolated first by Merck (1848) from opium, in which it occurs to the extent of about 1 percent. Its structure was elucidated by the classic researches of Goldschmiedt, and its synthesis was effected first by Pictet and Gam in 1909.

Papaverine
(R = OCH₃)

Tetrahydropapaverine
(R = OCH₃)

Bis-β-phenyl-
ethylamine
(R = H)

Previous to World War II, papaverine had been obtained in sufficient quantities from natural sources. However, as a result of the war, the United States found itself early in 1942 without a source of opium and, therefore, of papaverine. Consequently, the commercial synthesis of papaverine took on a new significance, and methods soon were developed to synthesize the alkaloid on a large scale.[100]

Papaverine itself occurs as an optically inactive white, crystalline powder. It possesses one basic nitrogen and forms salts quite readily. The most important salt is the hydrochloride, which is official. The hydrochloride occurs as white crystals or as a crystalline, white powder. It is odorless and has a slightly bitter taste. The compound is soluble in water (1:30), alcohol (1:120) or chloroform. It not soluble in ether. Aqueous solutions are acid to litmus and may be sterilized by autoclaving. Unless properly handled and stored, extemporanous solutions of papaverine salts deteriorate rapidly.

Because of the antispasmodic action of papaverine on blood vessels, it has become extremely valuable for relieving the arterial spasm associated with acute vascular occlusion. It is useful in the treatment of peripheral, coronary and pulmonary arterial occlu-

sions. Administration of an antispasmodic is predicated on the concept that the lodgement of an embolus causes an intense reflex vasospasm. This vasospasm affects not only the artery involved but also the surrounding blood vessels. Relief of this neighboring vasospasm is imperative in order to prevent damage to these vessels and to limit the area of ischemia. Thus, it appears to increase collateral circulation in the affected area rather than to act on the occluded vessel.

Other than its antispasmodic action on the vascular system, it is used for bronchial spasm and visceral spasm. In the latter type of spasm, it is not advisable to administer morphine simultaneously because it opposes the relaxing action of papaverine.

Because papaverine is a musculotropic drug, it has provided the starting point for synthetic analogs in which it has been hoped that a neurotropic activity could be combined with its musculotropic action. This combination of activities would be desirable, if possible, without the introduction of any atropine-like side-effects. Comparing the results of this research with those of atropine analogs, it appears that the use of the latter has proved to be more successful.

One of the especially easily altered peripheral groups of papaverine is the methoxyl group, of which there are four. These have been changed to various alkoxyl groups, among which are the ethoxyl and the methylenedioxy groups. This research has shown that, as far as papaverine itself is concerned, the most active compound is one in which there are 4 ethoxyls replacing the 4 methoxyls. This is the commercially available compound, ethaverine, which has 3 times the activity of papaverine but causes serious liver toxicity. Dioxyline has a combination of 1 ethoxyl group with 3 methoxyl groups. It is not entirely certain that the alkoxyl groups are necessary for activity, although they seem to be present in most of the accepted compounds. Activity is known to reside in both 1-phenyl-3-methyl-isoquinoline (III) and 1-benzyl-3-methyl-isoquinoline (IV) but, on the other hand, spaso*genic* properties are found in 1-phenyl-3-methyl-6,7-methylenedioxy-3,4-dihydroisoquinoline (V).

Blicke[103] has pointed out that compounds based on a close similarity to papaverine very likely would have some of the defects peculiar to papaverine. Among these defects are the low water-solubility of the salts and the tendency of the salts to produce acidic solutions by hydrolysis because of a feebly basic nitrogen.

Recognizing these limitations and having observed that tetrahydropapaverine showed qualitatively the same type of action as papaverine, workers began to investigate the open chain models of tetrahydropa-

TABLE 12–12

Compound	Structure			Name
	R_1	R_2	R_3	
I	$-OC_2H_5$	$-H$	$-CH_2$⟨ring⟩$-OC_2H_5$, OC_2H_5	Ethaverine
II	$-OCH_3$	$-CH_3$	$-CH_2$⟨ring⟩$-OC_2H_5$, OCH_3	Dioxyline
III	$-H$	$-CH_3$	⟨phenyl⟩	1-Phenyl-3-methylisoquinoline
IV	$-H$	$-CH_3$	$-CH_2$⟨phenyl⟩	1-Benzyl-3-methylisoquinoline
V	O–CH₂–O (methylenedioxy)	$-CH_3$	⟨phenyl⟩	1-Phenyl-3-methyl-6,7-methylene-dioxy-3,4-dihydroisoquinoline

paverine. Inspection of the following formulas shows the logical progression from papaverine to the *bis-β-phenylethylamine* type of compound.

Rosenmund and his co-workers[104,105,106] have been among the most active in carrying out this type of permutation and early demonstrated that *bis-β-phenylethylamine* itself had a slight but unmistakable activity. From a study of a great number of these compounds, *bis*(γ-phenylpropyl)ethylamine was selected as the best all around compound. It has been recognized as alverine and it is said to be 2.3 times as active as papaverine.

As early as 1933, it was known that both saturated and unsaturated acyclic amines had spasmolytic properties. In addition, they had sympathomimetic properties. The best compound in this group was 2-methyl-amino-6-methyl-5-heptene which is commercially obtainable under the generic name of isometheptene (N). According to Issekutz,[107] this compound has a direct paralyzing effect on smooth muscle of the intestine in a manner similar to papaverine and also stimulates sympathetic nerve endings to thus inhibit intestinal functions.

In conclusion, it is well to point out that some of the sympathomimetic amines possess specialized antispasmodic properties toward the bronchi and are used as bronchodilators. Among this group, we find ephedrine, isoproterenol and epinephrine. However, the mechanism of action here is not a parasympatholytic or muscle-depressant action but may be characterized as an overstimulation of the sympathetic system which simulates in many ways the paralysis of the parasympathetic system.

Ethaverine Hydrochloride, Isovex®, Neopavrin®, 6,7-diethoxyl-1-(3,4-diethoxybenzyl)isoquinoline hydrochloride. This well-known derivative of papaverine is synthesized in exactly the same way as papaverine but intermediates that bear ethoxyl groups instead of methoxyl groups are utilized.[108]

The hydrochloride is soluble to the extent of 1 g. in 40 ml. of water at room temperature. The aqueous solutions are acidic in reaction, with a 1 percent solution having a pH of 3.6 and 0.1 percent solution having a pH of 4.6.

The pharmacologic action of ethaverine is quite similar to that of papaverine, although its effect is

TABLE 12–13

PAPAVERINE AND RELATED COMPOUNDS

Name Proprietary Name	Preparations	Category	Usual Adult Dose*	Usual Dose Range*
Papaverine Hydrochloride U.S.P. *Cerespan, Pavabid, Vasospan*	Papaverine Hydrochloride Injection U.S.P. Papaverine Hydrochloride Tablets U.S.P.	Smooth-muscle relaxant	Oral, 150 mg.; I.M., 30 mg.; I.V., 120 mg.	Oral, 100 to 300 mg.; I.M., 30 to 60 mg.
Ethaverine Hydrochloride *Ethaquin, Laverin, Isovex*	Ethaverine Hydrochloride Tablets Ethaverine Hydrochloride Injection Ethaverine Hydrochloride Elixir	Smooth-muscle relaxant	100 mg.	100 to 200 mg. three times a day
Dioxyline Phosphate *Paveril Phosphate*	Dioxyline Phosphate Tablets	Smooth-muscle relaxant		100 to 400 mg. 3 or 4 times daily

* See *U.S.P. D.I.* for complete dosage information.

said to be longer in duration. It is used in peripheral and cerebral vascular insufficiency associated with arterial spasm in doses of 100 mg.

Dioxyline Phosphate, Paveril® Phosphate, 1-(4-ethoxy-3-methoxybenzyl)-6,7-dimethoxy-3-methylisoquinoline phosphate. This may be prepared according to the usual Bischler-Napieralski isoquinoline synthesis followed by dehydrogenation.[109]

This compound is related quite closely to papaverine and gives the same type of antispasmodic action as papaverine, with less toxicity. By virtue of the lesser toxicity, it can be given in larger doses than papaverine if desired, although usually the same dosage regimen can be followed as with the natural alkaloid.

The drug is useful for mitigating the reflex vasospasm that already has been described for papaverine during peripheral, pulmonary or coronary occlusion. The indications are the same as for papaverine.

GANGLIONIC BLOCKING AGENTS

Autonomic ganglia have been the object of interest for many years for the study of the interactions occurring between drugs and nervous tissues. The first important account[110] was given by Langley and described the stimulating and blocking actions of nicotine on sympathetic ganglia. It was found that small amounts of nicotine stimulated ganglia and then produced a blockade of ganglionic transmission, thus causing a suppression of the various end-organs to electrical stimulation of preganglionic nerves. From these experiments Langley was able to outline the general pattern of innervation of organs by the autonomic nervous system. *Parasympathetic* ganglia are usually located near the organ they innervate and have preganglionic fibers that stem from the cervical and thoracic regions of the spinal cord. *Sympathetic* ganglia consist of 22 pairs that lie on either side of the

vertebral column to form lateral chains. These ganglia are connected both to each other by nerve trunks and also to the lumbar or sacral regions of the spinal cord. The transmission of impulses in autonomic ganglia may be described as similar to the neurohumoral processes that occur at almost all nerve endings. The released acetylcholine is normally not taken up by the nerve endings but hydrolyzed by acetylcholinesterase, and about half the choline formed is absorbed immediately into the nerve by an active process that is blocked by hemicholinium-3.

Nicotine

Using the sympathetic cervical ganglion as a model, it has been found that transmission in the autonomic ganglion is more complex than formerly believed. Traditionally, stimulation of autonomic ganglia by acetylcholine has been considered as the nicotinic action of the neurotransmitter. It is now understood that stimulation by acetylcholine produces a triphasic response in sympathetic ganglia. There is an initial excitatory postsynaptic potential (EPSP) with a latency of 1 millisecond, followed by an inhibitory postsynaptic potential (IPSP) with a latency of 35 milliseconds and, finally, a slowly generating EPSP with a latency of several hundred milliseconds. The initial EPSP is blocked by conventional competitive nondepolarizing ganglionic blocking agents such as hexamethonium and is considered the primary pathway for ganglionic transmission.[111] The slowly generating or late EPSP is blocked by atropine but not by the traditional ganglionic blocking agents. This receptor has muscarinic properties, because methacholine causes generation of the late EPSP without causing the initial spike characteristic of acetylcholine. Atropine also blocks the late EPSP produced by methacholine. In addition to these two cholinergic pathways, the cervical sympathetic ganglion was found to have a neuron that contains a catecholamine.[112] These neuronal cells identified initially by fluorescence-histochemistry studies and shown to be smaller than the postganglionic neurons are now referred to as small-intensity fluorescent cells or SIF cells. Dopamine has been identified as the fluorescent catecholamine in the SIF cells that are common to many other sympathetic ganglia. Dopamine apparently mediates an increase in cyclic AMP, which causes hyperpolarization of postganglionic neurons (Fig. 12-14). The IPSP phase of the transmission of sympathetic ganglia following acetylcholine administration can be blocked by both atropine and α-adrenergic blocking agents.[113]

If a similar nontraditional type of ganglionic transmission occurs in the parasympathetic ganglia, it has not yet been made evident.

With the anatomical and physiological differences between sympathetic and parasympathetic ganglia, it should be no surprise that ganglionic agents may show some selectivity between the two types of ganglia. Although we do not have drug classifications such as "parasympathetic ganglionic blockers" and "sympathetic ganglionic blockers," we do find that certain ganglia have a predominant effect over certain organs and tissues and that a nondiscriminant blockade of autonomic ganglia results in a change in the effect of the autonomic nervous system on that organ (Table 12-14). Nevertheless there are drugs that have some selective action. Garrett[114] has shown that tetraethyl-ammonium salts are nondiscriminating, whereas hexamethonium shows some selective blocking action. To date none of the commonly known ganglionic blockers

TABLE 12-14

RESULTS OF GANGLIONIC BLOCKERS ON ORGANS*

Organ	Predominant System	Results of Ganglionic Blockade
Cardiovascular system		
Heart	Parasympathetic	Tachycardia
Arterioles	Sympathetic	Vasodilation
Veins	Sympathetic	Dilatation
Eye:		
Iris	Parasympathetic	Mydriasis
Ciliary muscle	Parasympathetic	Cycloplegia
G.I. tract	Parasympathetic	Relaxation
Urinary bladder	Parasympathetic	Urinary retention
Salivary glands	Parasympathetic	Dry mouth
Sweat glands	Sympathetic†	Anhidrosis

*Adapted from Goth, A.: Medical Pharmacology, 9th ed., St. Louis, C. V. Mosby Co., 1978
†Neurotransmitter is acetylcholine.

has been identified as having a selective blockade of parasympathetic ganglia.

Van Rossum[115,116] has reviewed the mechanisms of ganglionic synaptic transmission, the mode of action of ganglionic stimulants, and the mode of action of ganglionic blocking agents. He has conveniently classified the blocking agents in the following manner:

Depolarizing Ganglionic Blocking Agents. These blocking agents are actually ganglionic stimulants. Thus, in the case of nicotine, it is well known that small doses give an action similar to that of the natural neuroeffector, acetylcholine, an action known as the "nicotinic effect of acetylcholine." However, larger amounts of nicotine bring about a ganglionic block, characterized initially by depolarization followed by a typical competitive antagonism. In order

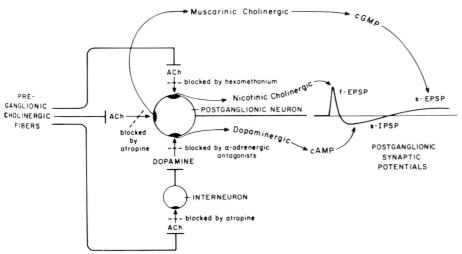

FIG. 12-14. Synaptic transmission in the mammalian superior cervical ganglion and the proposed role of the different receptors to the triphasic pattern of stimulation and selective blockade. (From Greengard, P., and Kebabian, J.W., Fed. Proc. 33:1059, 1974.)

to conduct nervous impulses the cell must be able to carry out a polarization and depolarization process, and if the depolarized condition is maintained without repolarization, it is obvious that no nerve conduction occurs. Acetylcholine itself, in high concentrations, will bring about an autoinhibition. There are a number of compounds which cause this type of ganglionic block but they are not of therapeutic significance. However, the remaining classes of ganglionic blocking agents have therapeutic utility.

Nondepolarizing Competitive Ganglionic Blocking Agents. Compounds in this class possess the necessary affinity to attach to the nicotinic receptor sites that are specific for acetylcholine but lack the intrinsic activity necessary for impulse transmission, i.e., they are unable to effect depolarization of the cell. Under experimental conditions, in the presence of a fixed concentration of blocking agent of this type, a large enough concentration of acetylcholine can offset the blocking action by competing successfully for the specific receptors. When such a concentration of acetylcholine is administered to a ganglion preparation, it appears that the intrinsic activity of the acetylcholine is as great as it was when no antagonist was present, the only difference being in the larger concentration of acetylcholine required. It is evident, then, that such blocking agents are "competitive" with acetylcholine for the specific receptors involved and either the agonist or the antagonist can displace the other if present in sufficient concentration. Drugs falling into this class are tetraethylammonium salts, hexamethonium, and trimethaphan. Mecamylamine possesses a competitive component in its action but is also noncompetitive—a so-called "dual antagonist."

Nondepolarizing Noncompetitive Ganglionic Blocking Agents. These blocking agents produce their effect, not at the specific acetylcholine receptor site, but at some point further along the chain of events that is necessary for transmission of the nervous impulse. When the block has been imposed, increase of the concentration of acetylcholine has no effect, and, thus, apparently acetylcholine is not acting competitively with the blocking agent at the same receptors. Theoretically, a pure noncompetitive blocker should have a high specific affinity to the noncompetitive receptors in the ganglia, and it should have very low affinity for other cholinergic synapses, together with no intrinsic activity. Mecamylamine, as mentioned before, has a noncompetitive component but is also a competitive blocking agent.

The first ganglionic blocking agents employed in therapy were tetraethylammonium chloride and bromide (I).* Although one might assume that curariform

*Compounds referred to by Roman numerals in this section are in Table 12-14.

activity would be a deterrent to their use, it has been shown that the curariform activity of the tetraethyl compound is less than 1 percent of that of the corresponding tetramethylammonium compound. A few years after the introduction of the tetraethylammonium compounds Paton and Zaimis[117] investigated the usefulness of the *bis*-trimethylammonium polymethylene salts:

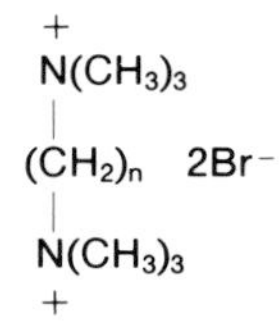

$$\begin{array}{l} \overset{+}{N}(CH_3)_3 \\ | \\ (CH_2)_n \quad 2Br^- \\ | \\ N(CH_3)_3 \\ + \end{array} \quad \begin{array}{l} n = 5\ or\ 6,\ active\ as\ ganglionic \\ blockers\ (feeble\ curariform \\ activity) \\ \\ n = 9\ to\ 12,\ weak\ ganglionic \\ blockers\ (strong\ curariform \\ activity) \end{array}$$

As shown above, their findings indicate that there is a critical distance of about 5 to 6 carbon atoms between the onium centers for good ganglion blocking action. Interestingly enough, the pentamethylene and the hexamethylene compounds are effective antidotes for counteracting the curare effect of the decamethylene compound. Hexamethonium (II), as the bromide and the chloride, emerged from this research as a clinically useful product.

Trimethaphan camphorsulfonate (III), a monosulfonium compound, bears some degree of similarity to the quaternary ammonium types because it, too, is a completely ionic compound. Although it produces a prompt ganglion blocking action on parenteral injection, its action is short, and it is used only for controlled hypotension during surgery. Almost simultaneously with the introduction of chlorisondamine (now long removed from the market), announcement was made of the powerful ganglionic blocking action of mecamylamine (IV), a secondary amine *without* quaternary ammonium character. As expected, the latter compound showed uniform and predictable absorption from the gastrointestinal tract as well as a longer duration of action. The action was similar to that of hexamethonium.

Drugs of this class have a limited usefulness as diagnostic and therapeutic agents in the management of peripheral vascular diseases (e.g., thromboangiitis obliterans, Raynaud's disease, diabetic gangrene, etc.). However, the principal therapeutic application has been in the treatment of hypertension through blockade of the sympathetic pathways. Unfortunately, the action is not specific, and the parasympathetic ganglia, unavoidably, are blocked simultaneously to a greater or lesser extent, causing visual disturbances, dryness of the mouth, impotence, urinary retention, constipation and the like. Constipation, in particular, probably due to unabsorbed drug in the intestine

TABLE 12–15

STRUCTURES OF GANGLIONIC BLOCKING AGENTS

Compound	Structure	Name
I	$(C_2H_5)_4\overset{+}{N}$ X^-	Tetraethylammonium Chloride (X = Cl) Tetraethylammonium Bromide (X = Br)
II	$(CH_3)_3\overset{+}{N}-(CH_2)_6-\overset{+}{N}(CH_3)_3$ $2X^-$	Hexamethonium Chloride (X = Cl) Hexamethonium Bromide (X = Br)
III		Trimethaphan Camphorsulfonate
IV		Mecamylamine Hydrochloride

(poor absorption), has been a drawback because the condition can proceed to a paralytic ileus if extreme care is not exercised. For this reason, cathartics or a parasympathomimetic (e.g., pilocarpine nitrate) are frequently administered simultaneously. Another serious side-effect is the production of orthostatic (postural) hypotension, i.e., dizziness when the patient stands up in an erect position. Prolonged administration of the ganglionic blocking agents results in their diminished effectiveness due to a build-up of tolerance, although some are more prone to this than others. Because of the many serious side-effects, this group of drugs has been largely abandoned by researchers seeking effective hypotensive agents.

In addition to the side-effects mentioned above, there are a number of contraindications to the use of these drugs. For instance, they are all contraindicated in disorders characterized by severe reduction of blood flow to a vital organ (e.g., severe coronary insufficiency, recent myocardial infarction, retinal and cerebral thrombosis, etc.) as well as situations where there have been large reductions in blood volume. In the latter case, the contraindication is based on the fact that the drugs block the normal vasoconstrictor compensatory mechanisms necessary for homeostasis. A potentially serious complication, especially in older male patients with prostatic hypertrophy, is urinary retention. These drugs should be used with care or not

at all in the presence of renal insufficiency, glaucoma, uremia and organic pyloric stenosis.

PRODUCTS

Trimethaphan Camsylate U.S.P., Arfonad®, (+)-1,3-dibenzyldecahydro-2-oxoimidazo[4,5-c]thieno[1,2-α]thiolium 2-oxo-10-bornanesulfonate (1:1). The drug consists of white crystals or is a crystalline powder with a bitter taste and a slight odor. It is soluble in water and alcohol but only slightly soluble in acetone and ether. The pH of a 1 percent aqueous solution is 5.0 to 6.0.

This ganglionic blocking agent is used only for certain neurosurgical procedures where excessive bleeding obscures the operative field. Certain craniotomies are included among these operations. The action of the drug is a direct vasodilation, and because of its transient action, it is subject to minute-by-minute control. On the other hand, this type of fleeting action makes it useless for hypertensive control. In addition, it is ineffective when given orally, and the usual route of administration is intravenous.

Mecamylamine Hydrochloride U.S.P., Inversine®, N,2,3,3-tetramethyl-2-norbornanamine hydrochloride. The drug occurs as a white, odorless, crystalline powder. It has a bittersweet taste. It is freely

TABLE 12-16

GANGLIONIC BLOCKING AGENTS

Name Proprietary Name	Preparations	Category	Usual Adult Dose*	Usual Dose Range*
Trimethaphan Camsylate U.S.P. *Arfonad*	Trimethaphan Camsylate Injection U.S.P.	Antihypertensive	I.V. infusion, 500 mg. in 500 ml. of 5 percent Dextrose Injection at a rate adjusted to maintain blood pressure at the desired level	200 µg. to 5 mg. per minute
Mecamylamine Hydrochloride U.S.P. *Inversine*	Mecamylamine Hydrochloride Tablets U.S.P.	Antihypertensive	Initial, 2.5 mg. twice daily, increased by 2.5 mg. increments at intervals of not less than 2 days as required; maintenance, 7.5 mg. 3 times daily	2.5 to 60 mg. daily

*See *U.S.P. D.I.* for complete dosage information.

soluble in water and chloroform, soluble in isopropyl alcohol, slightly soluble in benzene and practically insoluble in ether. The pH of a 1 percent aqueous solution ranges from 6.0 to 7.5, and the solutions are stable to autoclaving.

This secondary amine has a powerful ganglionic blocking effect which is almost identical with that of hexamethonium. It has an advantage over most of the ganglionic blocking agents in that it is readily and smoothly absorbed from the gastrointestinal tract. This makes it quite suitable for oral administration. It has a longer duration of action than hexamethonium, and the same effect can be obtained with lower doses. Although tolerance is built up to the drug on prolonged administration, this effect is less pronounced than that with hexamethonium and pentolinium. As with other ganglionic blocking agents, this drug is capable of producing the undesirable side-effects associated with parasympathetic blockade, although they are of less intensity than with most of the others. It is probably the drug of choice among the ganglion blockers.

It is used for the treatment of moderate to severe hypertension and is occasionally effective in malignant hypertension. The dosage is highly individualized and depends on the severity of the condition and the patient response.

NEUROMUSCULAR BLOCKING AGENTS

The possible existence of a junction between muscle and nerve was raised as early as 1856 when Claude Bernard observed that the site of action of curare was neither the nerve nor the muscle. Since that time it has been agreed that acetylcholine mediates transmission at the neuromuscular junction by a sequence of events that has been described earlier in this chapter. The neuromuscular junction consists of the axon impinging onto a specialized area of the muscle known as the muscle end-plate (Fig. 12-15). The axon is covered with a myelin sheath containing nodes of Ranvier but is bare at the ending. The nerve terminal is separated from the end-plate by a gap of 200 Å. The subsynaptic membrane of the end-plate contains the cholinergic receptor, the ion conducting channels (which are opened under the influence of acetylcholine), and the acetylcholinesterase.

One of the anatomical differences between the neuromuscular junction and other acetylcholine-responsive sites is the absence in the former of a membrane barrier or sheath that envelopes the ganglia or constitutes the blood–brain barrier. This is of significance in the accessibility of the site of action to drugs and is of particular importance with quaternary ammonium compounds, because they pass through living membranes with considerably greater difficulty and selectivity than do compounds that can exist in a nonionized species. The essentially bare nature of the myoneural junction permits ready access by quaternary ammonium compounds. In addition, compounds with considerable molecular dimensions are also accessible to the receptors in the myoneural junction. As a result of these properties, variations in the chemical structure of quaternaries have little influence on the

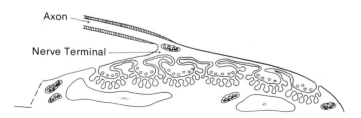

FIG. 12-15. *Semidiagrammatic representation of the neuromuscular junction at a fast twitch fiber (extensor digitorum longus) of the rat. The axon is covered with a myelin sheath but is bare at the actual end-plate region. The bulblike nerve terminal expansions are apposed by regular and deep junctional folds. (Reproduced with permission from Ellisman et al.: J. Cell Biol. 68:752–754, 1976.)*

potential ability of the molecule to reach the cholinergic receptor in the neuromuscular junction.

Agents that block the transmission of acetylcholine at the motor end-plate are called *neuromuscular blocking agents.* The therapeutic use of these compounds is primarily as adjuvants in surgical anesthesia to obtain relaxation of skeletal muscle. These drugs are also used in various orthopedic procedures such as alignment of fractures and correction of dislocations.

The therapeutically useful compounds in this group are sometimes referred to as possessing "curariform" or "curarimimetic" activity in reference to the original representatives of the class which were obtained from curare. Since then, synthetic compounds have been prepared with a similar activity. Although all of the compounds falling into this category, natural and synthetic alike, bring about substantially the same end-result, i.e., voluntary-muscle relaxation, there are some significant differences in the mechanisms whereby this is brought about. Thus, the following types of neuromuscular junction blockers have been noted.

Depolarizing Blocking Agents. Drugs in this category are known to bring about a depolarization of the membrane of the muscle end-plate. This depolarizing is quite similar to that produced by acetylcholine itself at ganglia and neuromuscular junctions (i.e., its so-called "nicotinic" effect), with the result that the drug, if it is in sufficient concentration, eventually will produce a block. It has been known for years that either smooth or voluntary muscle, when challenged repeatedly with a depolarizing agent, will eventually become insensitive. This phenomenon is known as *tachyphylaxis* or *desensitization* and is convincingly demonstrated under suitable experimental conditions with repeated applications of acetylcholine itself, the results indicating that within a few minutes the end-plate becomes insensitive to acetylcholine. The previous statements may imply that a blocking action of this type is quite clear-cut, but under experimental conditions it is not quite so clear and unambiguous because a block that initially begins with depolarization may regain the polarized state even before the block. Furthermore, a depolarization induced by increasing the potassium ion concentration does not prevent impulse transmission. For these and other reasons it is probably best to consider the blocking action as a desensitization until a clearer picture emerges. The drugs falling into this classification are decamethonium and succinylcholine.

Competitive Blocking Agents. There is no depolarization accompanying the block by these agents. The term *antidepolarizing* has been also suggested[118] on the grounds that these compounds all have the common property of actively preventing depolarization of the postjunctional membrane, not only by acetylcholine but also by the depolarizing neuromuscular blocking agents. It is thought that these agents successfully compete with acetylcholine for the receptor sites. Thus, by decreasing the effective acetylcholine-receptor combinations the end-plate potential becomes too small to initiate the propagated action potential. The action of these drugs is quite analogous to that of atropine at the muscarinic receptor sites of acetylcholine. Many experiments suggest that the agonist (acetylcholine) and the antagonist compete on a one-to-one basis for the end-plate receptors. Drugs falling into this classification are tubocurarine, dimethyltubocurarine and gallamine.

Mixed Blocking Agents. It has already been intimated that pure classifications of the blocking agents may be difficult. Because of this, some authorities believe that there are mixed types of blockers which possess both depolarizing and competitive components in the blocking action. Indeed, decamethonium and succinylcholine, while commonly classed as producing a depolarizing block, show evidence of some typical competitive action as well. This is particularly so in a variety of experiments on mammals including monkey, dog, rabbit, and guinea pig. It was found that decamethonium and succinylcholine initially cause a depolarization action on the muscle but during the blocking process the mechanism changes into that of a competitive type of blockade. This has been termed a *dual block* and is believed to be caused by an intracellular accumulation of the blocking agents.

Bovet[119] has classed antagonists at the neuromuscular junction according to their apparent morphological characteristics. He identified a class of long thin molecules as *leptocurares* and a group of more bulky molecules as *pachycurares.* The classification parallels considerably the mechanistic classification of these agents. Leptocurares are mechanistically depolarizing neuromuscular blocking agents and are represented by molecules such as decamethonium and succinylcholine. Pachycurares are nondepolarizing (competitive) blocking agents and are represented by d-tubocurarine. The morphological differences between leptocurares and pachycurares allow for a tentative explanation of mechanistic differences causing neuromuscular blockade in that the former type of agent possesses a thin depolarizing group that is able to penetrate the muscle end-plate. For maximal depolarizing activity the onium head should bear substituents of minimal steric bulk. It is established that a change from a depolarizing to a competitive action of neuromuscular blockade is accompanied by a progressive increase in size of the substituents on the onium centers although potency decreases at the same time.

CURARE AND CURARE ALKALOIDS

Originally *curare* was a term used to describe collectively the very potent arrow poisons used since early times by the South American Indians. The arrow poisons were prepared from numerous botanic sources and often were mixtures of several different plant extracts. Some were poisonous by virtue of a convulsant action and others by a paralyzant action. It is only the latter type that is of value in therapeutics and is ordinarily spoken of as "curare."

Chemical investigations of the curares were not especially successful because of the difficulties attendant on the obtaining of authentic samples of curare with definite botanic origin. It was only in 1935 that King was able to isolate a pure crystalline alkaloid, which he named d-tubocurarine chloride, from a curare of doubtful botanic origin[120] was shown to possess, in great measure, the paralyzing action of the original curare. Wintersteiner and Dutcher,[121] in 1943, also isolated the same alkaloid. However, they showed that the botanic source was *Chondodendron tomentosum* (Fam. *Menispermaceae*) and thus provided a known source of the drug.

Following the development of quantitative bioassay methods for determining the potency of curare extracts, a purified and standardized curare was developed and marketed under the trade name of Intocostrin® (Purified Chondodendron Tomentosum Extract), the solid content of which consisted of almost one-half (+)-tubocurarine solids. Following these essentially pioneering developments, (+)-tubocurarine chloride and dimethyltubocurarine iodide have appeared on the market as pure entities.

PRODUCTS

Tubocurarine Chloride U.S.P., (+)-tubocurarine chloride hydrochloride pentahydrate. This alkaloid is prepared from crude curare by a process of purification and crystallization.

Tubocurarine chloride occurs as a white or yellowish-white to grayish-white, odorless crystalline powder, which is soluble in water. Aqueous solutions of it are stable to sterilization by heat.

The structural formula for (+)-tubocurarine (see structure below) was long thought to be represented as in Ia. Through the work of Everette *et al.*[122] the structure is now known to be that of Ib. The monoquaternary nature of Ib thus revealed has caused some reassessment of thinking concerning the theoretical basis for the blocking action since all previous assumptions had assumed a diquaternary structure (i.e., Ia). Nevertheless, this does not negate the earlier conclusions that a diquaternary nature of the molecule provides better blocking action than does a monoquaternary (e.g., compare the potency of Ib with dimethyl tubocurarine iodide and note the approximately fourfold difference). It may also be of interest that (+)-isotubocurarine chloride (Ic)[124] prepared by monomethylation of (+)-tubocurarine, provides a compound with twice the activity of Ib in the particular test employed. Another point of interest may be that (−)-tubocurarine, enantiomer to Ib, has an activity estimated at one-twentieth to one-sixtieth that of Ib. The (−)-enantiomer has been isolated and tested for muscle-relaxant activity only once[125] and the results probably need to be validated.

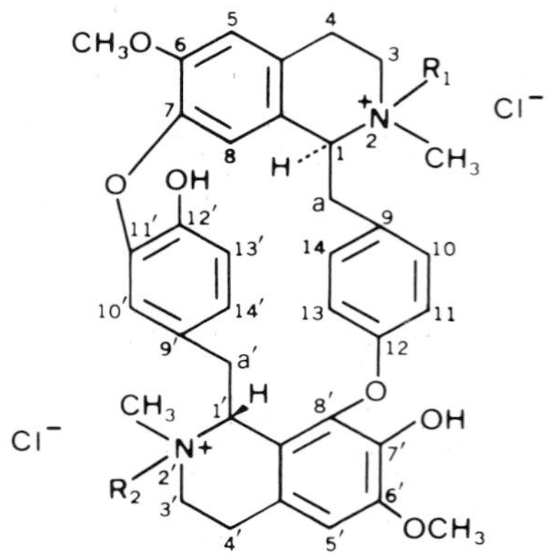

I a) $R_1 = R_2 = CH_3$

I b) $R_1 = H; R_2 = CH_3$

I c) $R_1 = CH_3; R_2 = H$

Tubocurarine is of value for its paralyzing action on voluntary muscles, the site of action being the neuromuscular junction. Its action is inhibited or reversed by the administration of acetylcholinesterase inhibitors such as neostigmine or by edrophonium chloride (Tensilon® Chloride). Such inhibition of its action is necessitated in respiratory embarrassment due to overdosage. It is often necessary to use artificial respiration as an adjunct until the maximum curare action has passed. The drug is inactive orally because of inadequate absorption through lipoidal membranes in the gastrointestinal tract and, when used therapeutically, is usually injected intravenously.

Tubocurarine, in the form of a purified extract, was first used in 1943 as a muscle relaxant in shock therapy of mental disorders. By its use the incidence of bone and spine fractures and dislocations resulting from convulsions due to shock was reduced markedly. Following this, it was employed as an adjunct in general anesthesia to obtain complete muscle relaxation, a usage that persists to this day. Prior to its use, satisfactory muscle relaxation in various surgical procedures (e.g., abdominal operations) was obtainable only with "deep" anesthesia using the ordinary general anesthetics. Tubocurarine permits a lighter plane of anesthesia with no sacrifice in the muscle relaxation so important to the surgeon. A reduced dose of tubocurarine is administered with ether, because ether itself has a curarelike action.

Another recognized use of tubocurarine is in the diagnosis of myasthenia gravis, because, in minute doses, it causes an exaggeration of symptoms by accentuating the already deficient acetylcholine supply. It has been experimented with to a limited extent in the treatment of spastic, hypertonic and athetoid conditions, but one of its principal drawbacks has been its relatively short duration of activity. When used intramuscularly its action lasts longer than when given by the intravenous route although this characteristic has not made it useful in the above conditions. Tubocurarine is frequently used with intravenous thiopental sodium anesthesia. Care should be used in the selection of the appropriate solution for this purpose since there are two available concentrations (i.e., 3 mg. per ml. and 15 mg. per ml.) of tubocurarine chloride injection. Most anesthesiologists employ the 15 mg. per ml. concentration injection and, although they note a transient cloudiness due to precipitation of the free barbiturate the condition clears up within minutes and, apparently, causes no problems.

Metocurine Iodide U.S.P., Metubine® Iodide, (+)-O,O′-dimethylchondrocurarine diiodide. This drug is prepared from natural crude curare by extracting the curare with methanolic potassium hydroxide. When the extract is treated with an excess of methyl iodide the (+)-tubocurarine is converted to the diquaternary dimethyl ether and crystallizes out as the iodide (see tubocurarine chloride). Other ethers besides the dimethyl ether also have been made and tested. For example, the dibenzyl ether was one-third as active as tubocurarine chloride and the diisopropyl compound had only one-half the activity. This is compared with the dimethyl ether which has approximately 4 times the activity of tubocurarine chloride. It is only moderately soluble in cold water but more so in hot water. It is easily soluble in methanol but insoluble in the water-immiscible solvents. Aqueous solutions have a pH of from 4 to 5, and the solutions are stable unless exposed to heat or sunlight for long periods of time.

The pharmacologic action of this compound is the same as that of tubocurarine chloride, namely, a competitive blocking effect on the motor end-plate of skeletal muscles. However, it is considerably more potent than the latter and has the added advantage of exerting much less effect on the respiration. The effect on respiration is not a significant factor in therapeutic doses. Accidental overdosage is counteracted best by forced respiration.

The drug is used for much the same purposes as tubocurarine chloride but in a smaller dose. The dose ranges from 2 to 8 mg. The exact dosage is governed by the physician and depends largely on the depth of surgical relaxation.

It is marketed in the form of a parenteral solution in ampules.

SYNTHETIC COMPOUNDS WITH CURARIFORM ACTIVITY

Over 100 years have passed since Crum-Brown and Fraser[125] described the curarimimetic properties induced in several tertiary alkaloids by quaternization with methyl sulfate. Their conclusion was that the quaternized forms of the tertiary alkaloids all had a more uniform pharmacologic activity (i.e., curarimimetic) than did the original tertiary forms which in many cases (e.g., atropine, strychnine, morphine) had widely different and characteristic activities. Their findings are sometimes known as the *rule of Crum-Brown and Fraser.* Since that time innumerable quaternary salts have been investigated in an effort to find potent, easily synthesized curarimimetics. It has been found that the curarelike effect is a common property of all "onium" compounds. In the order of decreasing activity they are:

$$(CH_3)_4N^+ > (CH_3)_3S^+$$
$$(CH_3)_4P^+ > (CH_3)_4As^+ > (CH_3)_4Sb^+$$

Even ammonium, potassium and sodium ions and other ions of alkali metals have been shown to exhibit a certain amount of curare action. Thus far, however, it has been impossible to establish any quantitative relationships between the magnitude or the mobility of the cation and the intensity of action. One of the exceptions to the rule that "onium" compounds are necessary for curarelike activity has been the demonstrated activity of the *Erythrina* alkaloids which are known to contain a tertiary nitrogen. Indeed, they seem to lose their potency when the nitrogen is quaternized. Whether or not two quaternary groups are necessary for maximum activity has led, through numerous studies, to the conclusion that the presence of two or more such quaternary groups permits higher activity by virtue of a more firm attachment at the site of action.[126,127]

Curare, until relatively recent times, remained the only useful curarizing agent, and it, too, suffered from a lack of standardization. The original pronouncement in 1935 of the structure of (+)-tubocurarine chloride, unchallenged for 35 years, led other workers to hope for activity in synthetic substances of less complexity. The quaternary ammonium character of the curare alkaloids, coupled with the known activity of the various simple "onium" compounds, hardly seemed to be coincidental, and it was natural for research to follow along these lines.

One of the first approaches to the synthesis of this type of compound was based on the assumption that the highly potent effect of tubocurarine chloride was a function of some optimum spacing of the two quaternary nitrogens. Indeed, the bulk of the experimental work tended to suggest an optimum distance of 12 to 15Å between quaternary nitrogen atoms in most of the bis-quaternaries for maximum curariform activity. However, other factors could modify this situation.[128,129] Bovet and his co-workers[130,131,132] were the first to develop synthetic compounds of significant potency through a systematic structure activity study based on (+)-tubocurarine as a model. One of their compounds, after consideration of potency-side-effect ratios was marketed in 1951 as Flaxedil® (gallamine triethiodide).

In 1948, another series of even simpler compounds was described independently by Barlow and Ing[133] and by Paton and Zaimis.[134] These were the *bis*-trimethyl-ammonium polymethylene salts (formulas above), and certain of them were found to possess a potency greater than that of (+)-tubocurarine chloride itself. Both groups concluded that the decamethylene compound was the best in the series and that the shorter chain lengths exhibited only feeble activity. Further investigations by Barlow and Zoller[135] revealed a second maximum of potency of 14 to 18 methylene groups in the *bis*-trimethylammonium polymethylene salt series. In particular the compound with 16 methylene groups proved to be 10 times more potent than decamethonium. Figure 12–16 shows the conclusions of Barlow and Ing and Barlow and Zoller with respect to

these compounds. The commercially obtainable preparation known as decamethonium (represented by the formula above where n = 10, salt may be I or Br) represents the decamethylene compound. An interesting finding is that the shorter-chain compounds, such as the pentamethylene and the hexamethylene compounds, are effective antidotes for counteracting the blocking effect of the decamethylene compound.

Cavallito *et al*[127] introduced another type of quaternary ammonium compound with high curarelike activity. This type is represented by the formula below and may be designated as ammoniumalkylaminobenzoquinones. The distance between the "onium" centers is the same as in the other less active "onium" compounds without the quinone structure, and for this reason, they speculate that the quinone itself may be involved in the activity. This appears to be reasonable in view of the fact that even the monoquaternary compounds and the corresponding nonquaternized amines show a significant curarelike activity. One compound (benzoquinonium chloride) was selected from this study and was marketed for several years as Mytolon® Chloride (n = 3, R_1 = C_2H_5, R_2 = $CH_2C_6H_5$) but since has been withdrawn because of its anticholinesterase side-effects.

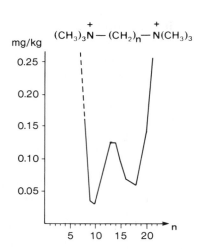

R_1 = methyl, ethyl, etc.

R_2 = benzyl, methyl

n = 2, 3, 4, 5

Ammonium alkylaminobenzoquinones

One of the most interesting pathways that research on neuromuscular junction blocking agents took was that which culminated in the widely used agent, succinylcholine chloride, a dicholine ester of succinic acid. It is rather surprising to find that this compound had been examined pharmacologically as early as 1906 and that the muscle-relaxant properties were not noticed until Bovet's pioneering study using (+)-tubocurarine as a model for inter-onium distances.[136] Others soon confirmed Bovet's observations and, in a commentary on the frequent outcomes of structure-activity studies, succinylcholine dichloride has withstood the test of time and is still the drug of choice as a depolarizing blocking agent. In retrospect, it may be looked upon as a "destabilized" decamethonium as pointed out by Ariëns *et al*[137] since it is metabolically disposed of by the action of cholinesterases whereas decamethonium persists since it cannot be similarly metabolized.

$$\overset{+}{(CH_3)_3N} - (CH_2)_n - \overset{+}{N(CH_3)_3}$$

FIG. 12–16. *Curarizing dose in the series of polymethylene-bis-trimethylammonium compounds. Cat, m. tibialis.*[133–135]

$$CH_2-COOCH_2CH_2\overset{+}{N}(CH_3)_3$$
$$CH_2-COOCH_2CH_2\underset{+}{N}(CH_3)_3 \quad 2Cl^-$$

Succinylcholine Dichloride

$$NH-COOCH_2CH_2\overset{+}{N}(CH_3)_3$$
$$(CH_2)_6 \qquad\qquad 2Cl^-$$
$$NH-COOCH_2CH_2\overset{+}{N}(CH_3)_3$$

Imbretil

Although new structural entities to replace succinylcholine are not envisioned, it is, nevertheless, interesting to consider the relative blocking activities of the dicholine esters of maleic and fumaric acid (cis- and trans-isomers, respectively). These were prepared by McCarthy et al[138] with the objective of determining whether succinylcholine acts at the receptor in the

$$CH_2-COOCH_2CH_2\overset{+}{N}(CH_3)_3$$
$$CH_2-COOCH_2CH_2\underset{+}{N}(CH_3)_3 \quad 2Cl^-$$

"Eclipsed"

$$(CH_3)_3\overset{+}{N}CH_2CH_2OOC-CH_2$$
$$CH_2-COOCH_2CH_2\underset{+}{N}(CH_3)_3 \quad 2Cl^-$$

"Staggered"

"eclipsed" or "staggered" conformation. With the same objective in mind, Burger and Bedford[139] and McCarthy et al.[140] have prepared dicholine esters of the cis- and trans-cyclopropane dicarboxylic acids. In all cases examined it is apparent that the "staggered" conformation is the most effective, which tends to reinforce the concept that the binding points for blocking agents are spaced approximately 12 to 14 Å apart as was suspected by the earliest workers.

Other structural studies that might have been anticipated as a consequence of the knowledge gained from the successes with succinylcholine would be the extension of the α,ω-dicholine ester concept to other diesters related to carbachol (i.e., choline carbamate). Although the direct analog of succinylcholine, i.e. the dicarbaminoyl choline ester, did have blocking activity, it was best in the ester where 6 methylene groups had to be interspersed between the nitrogen atoms of 2 carbachol moieties. This led to a clinically useful drug known as imbretil which seems to have an initial depolarizing action followed by a block which was typical of curarimimetics and was reversible with neostigmine. Although this blocking agent has been used abroad it has not been marketed in the U.S.

Another attempt to obtain a longer-lasting agent than succinylcholine was that of Phillips[141] in preparing the amide analogs to it together with a number of closely related compounds. The blocking activity engendered was a disappointment although it was noted that these compounds exerted a powerful action in prolonging the succinylcholine block when both were administered together.

Finally, it should be mentioned that another neuromuscular blocking agent to appear on the American scene has been pancuronium bromide, a bisquaternary derived from a steroidal framework. It is claimed to have about 5 times the potency of (+)-tubocurarine in man as a nondepolarizing agent. Although it has only 8 to 9 atoms interspersed between the two onium heads, it appears to have unusually high potency even though it does not conform to the usually expected inter-onium distances. However, it is probably safe to say that the steroidal skeleton makes little contribution to the activity as contrasted to its role in estrogenic and androgenic activities.

STERIC FACTORS

It is rather surprising that steric factors connected with neuromuscular junction (NMJ) blockade have not been examined more, in the face of the 20 to 60 times greater activity of the (+)-tubocurarine isomer over the (−)-enantiomer.[124] Studies that bear on this problem have been forthcoming from Soine et al.[142,143,144] as well as from the Stenlake group.[145,146,147] These findings indicate that, when a monoquaternary species is under consideration it appears almost inevitable that the (S)-configurational species is the most potent. On the other hand, it seems equally correct that bisquaternary species, even compounds derived from previously tested monoquaternary species, now show a decided R-configurational preference for blocking activity—directly opposite to that of the monoquaternaries. In view of the evidence accumulated to date it appears that there is a definite (ca. 2:1) superiority of the S-configuration over the R-configuration in the monoquaternary forms, even extending to (+)-tubocurarine itself.[132] On the other hand, it seems equally evident that, in bisquaternary forms, those with an R-configuration for the carbon adjacent to the quaternary moiety are destined to be more active. The reasons for these differences are not immediately apparent.

PRODUCTS

Decamethonium Bromide U.S.P., Syncurine®, decamethylene-*bis*-(trimethylammonium dibromide). This compound is prepared according to the method of Barlow and Ing.[133]

It is a colorless, odorless crystalline powder. It is soluble in water and alcohol, the solubility increasing with the temperature of the solvent. The compound is insoluble in chloroform and ether. It appears to be stable to boiling for at least 30 minutes in physiologic saline, either in the dark or in the sunlight. Twenty percent sodium hydroxide causes a white precipitate which is soluble on heating but reappears on cooling. Solutions are compatible with procaine hydrochloride and sodium thiopental.

The drug is used as a skeletal-muscle relaxant, especially in combination with the anesthetic barbiturates. Decamethonium is about 5 times as potent as (+)-tubocurarine and is used in a dose of 500 μg. to 3 mg. The antidote to overdosage is hexamethonium bromide or pentamethonium iodide.

Gallamine Triethiodide U.S.P., Flaxedil® Triethiodide, [*v*-phenenyl*tris*(oxyethylene)]*tris*[triethylammonium] triiodide. This compound is prepared by the method of Bovet *et al.*[132] and was introduced in 1951. It is a slightly bitter, amorphous powder. It is very soluble in water but is only sparingly soluble in alcohol. A 2 percent aqueous solution has a pH between 5.3 and 7.0.

Gallamine Triethiodide

Pharmacologically, it is a relaxant of skeletal muscle by blocking neuromuscular transmission. For this reason it is used as a muscular relaxant for both surgical and nonsurgical procedures. These have been mentioned in the general discussion of curare. It has an advantage over (+)-tubocurarine in that it exerts little or no effect on the autonomic ganglia, and it is readily miscible with the thiobarbiturate solutions used in anesthesia.

The drug is contraindicated in patients with myasthenia gravis, and it should also be borne in mind that the drug action is cumulative, as with curare. The antidote for gallamine triethiodide is neostigmine.

Succinylcholine Chloride U.S.P., Anectine®, Sucostrin®, choline chloride succinate (2:1). This compound may be prepared by the method of Phillips.[148]

It is a white, odorless, crystalline substance which is freely soluble in water to give solutions with a pH of about 4. It is stable in acidic solutions but unstable in alkali. The aqueous solutions should be refrigerated to ensure stability.

Succinylcholine is characterized by a very short duration of action and a quick recovery because of its rapid hydrolysis following injection. It brings about the typical muscular paralysis caused by a blocking of nervous transmission at the myoneural junction. Large doses may cause a temporary respiratory depression in common with other similar agents. Its action, in contrast with that of (+)-tubocurarine, is not antagonized by neostigmine, physostigmine or edrophonium chloride. These anticholinesterase drugs actually prolong the action of succinylcholine, and on this basis it is believed that the drug probably is hydrolyzed by cholinesterases. The brief duration of action of this curarelike agent is said to render an antidote unnecessary if the other proper supportive measures are available. However, succinylcholine has a disadvantage in that its action cannot be terminated promptly by the usual antidotes. This difficulty has led to further research, in an effort to overcome it.

It is used as a muscle relaxant for the same indications as other curare agents. It may be used for either short or long periods of relaxation, depending on whether one or several injections are given. In addition, it is suitable for the continuous intravenous drip method.

Succinylcholine chloride should not be used with thiopental sodium because of the high alkalinity of the latter or, if used together, should be administered immediately following mixing. However, separate injection is preferable.

Hexafluorenium Bromide U.S.P., Mylaxen®, hexamethylene-1,6-*bis*(9-fluorenyldimethylammonium) dibromide. This compound occurs as a white, water-soluble crystalline material. Aqueous solutions are stable at room temperature but are incompatible with alkaline solutions. Hexafluorenium bromide is used clinically to modify the dose and extend the duration of action of succinylcholine chloride.

Hexafluorenium Bromide

In most cases the dose of succinylcholine chloride can be lowered to about one-fifth the usual total dose.

The combination produces profound relaxation and facilitates difficult surgical procedures. Its principal mode of action appears to be suppression of the enzymatic hydrolysis of succinylcholine chloride, thereby prolonging its duration of action.

Pancuronium Bromide, Pavulon®, 2β,16β-dipiperidino-5α-androstane-3α,17β-dioldiacetatedimethobromide.

This new blocking agent is soluble in water and is marketed in concentrations of 1 mg. per ml. or 2 mg. per ml. for intravenous administration. It has been shown to be a typical nondepolarizing blocker with a potency approximately 5 times that of (+)-tubocurarine chloride and a duration of action approximately equal to the latter. Studies indicate that it has little or no histamine-releasing potential or ganglion-blocking activity and that it has little effect on the circulatory

Pancuronium Bromide

system except for causing a slight rise in the pulse rate. As one might expect, it is competitively antagonized by acetylcholine, anticholinesterases and potassium ion whereas its action is increased by inhalation anesthetics such as ether, halothane, enflurane and methoxyflurane. The latter enhancement in activity is

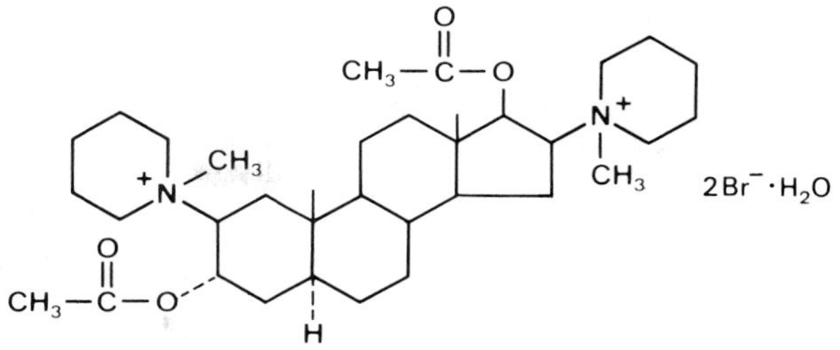

TABLE 12–17

NEUROMUSCULAR JUNCTION BLOCKING AGENTS

Name / Proprietary Name	Preparations	Category	Usual Adult Dose*	Usual Dose Range*	Usual Pediatric Dose*
Tubocurarine Chloride U.S.P.	Tubocurarine Chloride Injection U.S.P.	Skeletal-muscle relaxant	Initial, I.M. or I.V., 100 to 300 μg. per kg. of body weight, not exceeding 27 mg., then 25 to 100 μg. per kg. repeated as necessary	1 to 300 μg. per kg.	
Metocurine Iodide U.S.P.	Metocurine Iodide Injection, U.S.P.	Skeletal-muscle relaxant		I.V., initial, 1.5 to 8 mg. given over a 60-second period; maintenance, 500 μg. to 1 mg. every 25 to 90 minutes	
Decamethonium Bromide U.S.P. *Syncurine*	Decamethonium Bromide Injection U.S.P.	Skeletal-muscle relaxant		I.V., 40 to 60 μg. per kg. of body weight	Children—I.V. or I.M., 50 to 80 μg. per kg. of body weight
Gallamine Triethiodide U.S.P. *Flaxedil Triethiodide*	Gallamine Triethiodide Injection U.S.P.	Skeletal-muscle relaxant	I.V., 1 mg. per kg. of body weight, not exceeding 100 mg. per dose, repeated at 30- to 40-minute intervals if necessary	500 μg. to 1 mg. per kg.	Children less than 5 kg.—use is not recommended; over 5 kg.—see Usual Dose
Succinylcholine Chloride U.S.P. *Anectine, Sucostrin*	Succinylcholine Chloride Injection U.S.P.	Skeletal-muscle relaxant	I.V., 20 to 80 mg. I.V. infusion, 1 g. in 500 to 1000 ml. of 5 percent Dextrose Injection, Sodium Chloride Injection, or Sodium Lactate Injection at a rate of 500 μg. to 10 mg. per minute. I.M., up to 2.5 mg. per kg. of body weight, not exceeding a total dose of 150 mg.	I.V., 10 to 80 mg.	I.V., 1 to 2 mg. per kg. of body weight; I.M., see Usual Dose
	Sterile Succinylcholine Chloride U.S.P.		I.V. infusion, 1 g. in 500 to 1000 ml. of 5 percent Dextrose Injection, Sodium Chloride Injection, or Sodium Lactate Injection at a rate of 500 μg. to 10 mg. per minute		
Hexafluorenium Bromide U.S.P. *Mylaxen*	Hexafluorenium Bromide Injection U.S.P.	Potentiator (succinylcholine chloride)	I.V., initial, 400 μg. per kg. of body weight; maintenance, 100 to 200 μg. per kg. of body weight.		

* See *U.S.P. D.I.* for complete dosage information.

especially important to the anesthetist since the drug is frequently administered as an adjunct to the anesthetic procedure in order to relax the skeletal muscle. Perhaps the most frequent adverse reaction to this agent is the occasional prolongation of the neuromuscular block beyond the usual time course, a situation which can usually be controlled with neostigmine or by manual or mechanical ventilation since respiratory difficulty is a prominent manifestation of the prolonged blocking action.

As indicated, the principal use of pancuronium bromide is an adjunct to anesthesia to induce relaxation of skeletal muscle but it is employed to facilitate the management of patients undergoing mechanical ventilation. It should be administered only by experienced clinicians equipped with facilities for applying artificial respiration, and the dosage should be carefully adjusted and controlled.

REFERENCES

1. Dale, H. H.: J. Pharmacol, and Exp. Ther. 6:147, 1914.
2. Kasai, M., and Changeaux, J.-P.,: Biol. 6:1, 1971.
3. Popot, J. L., Sugiyama, H., and Changeaux, J.-P.: J. Mol. Biol. 106:469, 1976.
4. Moreau, M., and Changeaux, J.-P.: J. Mol. Biol. 106:457, 1976.
5. Sugiyama, H., Popot, J. L., and Changeaux, J.-P.: J. Mol. Biol. 16:485, 1976.
6. Karczmar, A. G., Jenden, D. J. (ed.): in Cholinergic Mechanisms and Psychopharmacology, New York, Plenum Press, 1977.
7. Haga, T., and Nada, H.: Biochem. Biophys, Acta 291:564, 1973.
8. Yamamura, H., and Snyder, S. H.: Neurochem. 21:1355, 1973.
9. Cavallito, C. J., Yun, H. S., Crispin-Smith, T., and Foldes, F. F.: J. Med. Chem. 12:134, 1969.
10. Aquilonius, S. M., Frankenberg, L., Stensio, K. E., and Winbladh, B.: Acta Pharmacol. Toxicol. 30:129, 1971.
11. Guyenet, Pl., Lefresne, P., Rossier, J., Beaujouan, J. C., and Glowinski, J.: Mol. Pharmacol. 9:630, 1973.
12. Guyenet, P., Lefresne, P., Rossier, J., Beaujouan, J. C., and Glowinski, J.: Brain Res. 62:523, 1973.
13. Namba, T., and Grob, D.: J. Neurochem. 15:1445, 1968.
14. Barrans, Y., and Clastre, J.: C. R. Acad. Sci.: (C) 270:306, 1970.
15. Partington, P., Feeney, J., and Burgen, A. S. V.: Mol. Pharmacol. 8:269, 1972.
16. Casey, A. F.: Prog. in Med. Chem. 11:1, 1975.
17. Kier, L. B.: Mol. Pharmacol. 3:487, 1967.
18. Schmeideberg, O., and Koppe, R.: Das Muscarine, das Giftige Alka id des Fielgenpiltzes, 1869, Vogel, Leipzig.
19. Hardegger, E., and Lohse, F.: Helv. Chim. Acta 40:2383, 1957.
20. Ariens, E. J., and Simonis, A. M., in deJong. H. (ed.): Quantitative Methods in Pharmacology, pp. 286–311, North Holland, Amsterdam, 1961.
21. Ing, H. R.: Science 109:264, 1949.
22. Beckett, A. H., Harper, N. J., Clitherow, J. W., and Lesser, E.: Nature 189:671, 1961.
23. Beckett, A. H., et al.: J. Pharm. Pharmacol. 15:362, 1963.
24. Wilson, I. B.: Ann. N. Y. Acad. Sci. 135:177, 1968.
25. Welsh, H. H., and Taub, R.: Science, 112:47, 1950.
26. Schueler, F. W., and Keasling, H. H.: Am. Sci. 133:512, 1951.
27. Nunes, M. A., and Brochmann-Hanssen, E. J.: J. Pharm. Sci. 63:716, 1974.
28. Jones, A.: J. Pharmacol. Exp. Ther. 141:195, 1963.
29. Englehard, N., Prchal, K., and Nenner, M.: Angew. Chem., Int. Ed. 6:615, 1967.
30. Bergmann, F.: Adv. Catalysis 10:131, 1958.
31. Wilson, I. B., Harrison, M. A., and Ginsberg, S.: J. Biol. Chem. 236:1498, 1961.
32. Belleau, B. DiTullio, V., and Tsai, Y. H.: Mol. Pharmacol. 6:41, 1970.
33. Ellis, S.: J. Pharm. Expt'l Ther. 79:309, 1943.
34. O'Brien, R. D., et al.: Mol. Pharmacol. 2:593, 1966; O'Brien, R. D., et al.: ibid. 4:121, 1968.
35. Aeschlimann, J. A., and Reinert, M.: J. Pharmacol. Exp. Ther. 43:413, 1931.
36. Calvey, T. H.: Biochem. Pharmacol. 16:1989, 1967.
37. Heath, D. F.: Organophosphorus Poisons, Pergamon Press, Oxford, 1961.
38. Natoff, I. L.: J. Pharm. Pharmacol. 19:612, 1967.
39. Oosterban, R. A., and Cohen, J. A.: in Googwin, T. W., Harris, I. J., and Hartley, B. S. (eds.): Structure and Activity of Enzymes, 87, New York Academic Press, 1964.
40. Tenn, J. G., and Toumarelli, R. C.: Am. J. Opthal. 35:46, 1952.
41. McNammara, P., et al.: J. Pharmacol, Exp. Ther. 87:281, 1946.
42. Diggle, W. M., and Gage, J. C.: Biochem. J. 49:491, 1951; Metcalf, R. L., and March, R. B.: Ann. Entomol. Soc. Amer. 46:63, 1953; Gage, J. C.: Biochem. J. 426, 1953.
43. Nakatsugawa, T., Tolman, N. M., and Dahm, P. A.: Biochem. Pharmacol. 17:1517, 1968.
44. O'Brien, R. D.: J. Agr. Food Chem. 11:163, 1963.
45. Mounter, L. A., and Cheatham, R. M.: Enzymol. 25:215, 1963.
46. Zvirblis, P., and Kondritzer, A.: J. Pharmacol. Exp. Ther. 157:432, 1967.
47. Stephenson, R. P.: Brit. J. Pharmacol. 11:378, 1956.
48. Ariens, E. J.: Adv. in Drug Res. 3:235, 1966.
49. Paton, W. D. M.: Proc. Roy. Soc. B154:21, 1961.
50. Bebbington, A., and Brimblecombe, R. W.: Adv. Drug. Res. 2:143, 1965.
51. Long, J. P., et al.: J. Pharmacol. Exp. Ther. 117:29, 1956.
52. Ellenbroek, B. W.: J. Pharm. Pharmacol. 17:393, 1965.
53. Abramson, F. B., Barlow, R. B., Mustafa, M. G., and Stephenson, R. P.: Brit. J. Pharmacol. 37:207, 1969; Brimblecombe, R. W., and Inch, T. D.: J. Pharm. Pharmacol. 22:881, 1970; Brimblecombe, R. W., Green, D. M., and Inch, T. D., J. Pharm. Pharmacol. 22:951, 1970.
54. Goldstein, A., Aronow, L., and Kalman, S., Principals of Drug Action, Harper & Row, 1968.
55. Fordtran, J. S., in Gastrointestinal Disease, p. 163, W. B. Saunders, Philadelphia, 1973.
56. Daniel, E. E., in Gastroenterology, p. 101, A. Bogoch (ed.); New York, McGraw-Hill, 1973.
57. AMA Drug Evaluations, 3rd ed., p. 1031, Littleton, Mass., Publishing Sciences Group, Inc., 1977.
58. Pinder, R. M.: Progress in Med. Chem. 9:191, 1973.
59. Brimblecome, R. W.: Drug Actions on Cholinergic Systems, p. 133, Baltimore, University Park Press, 1974.
60. Holmstedt, B., et al.: Biochem. Pharmacol. 14:189, 1965.
61. Coyle, J. T., and Snyder, S. J: Science 166:899, 1969.
62. Farnebo, L. O., et al.: J. Pharm. Pharmac. 22:733, 1970.
63. Gyermek, L., and Nador, K.: J. Pharm. Pharmacol 9:209, 1957.
64. Ariens, E. J., Simonis, A. M., and Van Rossum, J. M., in Ariens, E. J. (ed.): Molecular Pharmacology, p. 205, New York, Academic Press, 1964.
65. Brodie, B., and Hogben, C. A. M.: J. Pharm Pharmacol. 9:345, 1957.
66. Chemnitius, F: J. prakt. Chem. 116:276, 1927; see also Hamerslag, F.: The Chemistry and Technology of Alkaloids, p. 264, New York, Van Nostrand, 1950.
67. J. Am. Pharm. A. (Pract. Ed.) 8:377, 1947.
68. Zvirblis, P., et al.: J. Am. Pharm. A. (Sci. Ed.) 45:450, 1956.
69. Kondritzer, A. A., and Zvirblis, P.: J. Am. Pharm. A. (Sci. Ed.) 46:531, 1957.
70. Anesthesia 33:133, 1978.
71. Rodman, M. J.: Am. Prof. Pharm. 21:1049, 1955.
72. Ralph, C. S., and Willis, J. L.: Proc Roy. Soc. (N. S. Wales) 77:99, 1944.
73. Pittenger, P. S., and Krantz, J. C.: J. Am. Pharm. A. 17:1081, 1928.
74. U. S. Patent 2,753,288 (1956).
75. Kirsner, J., and Palmer, W.: J.A.M.A. 151:798, 1953.

76. Fromherz, K.: Arch. exp. Path. Pharmakol. 173:86, 1933.
77. Sternbach, L. H., and Kaiser, S.: J. Am. Chem. Soc. 74:2219, 1952.
78. U.S. Patent 2,648,667 (1953).
79. Treves, G.R., and Testa, F.C.: J. Am. Chem. Soc. 74:46, 1952.
80. Tilford, C. H., et al.: J. Am. Chem. Soc. 69:2902, 1947.
81. The Medical Letter 4:30, 1962.
82. Biel, J. H., et al.: J. Am. Chem. Soc. 77:2250, 1955; see also Long, J. P., and Keasling, H. K.: J. Am. Pharm. A. (Sci. Ed.) 43:616, 1954.
83. Burtner, R. R., and Cusic, J. W.: J. Am. Chem. Soc. 65:1582, 1943.
84. Faust, J. A., et al.: J. Am. Chem. Soc. 81:2214, 1959.
85. Nash, J. B., et al.: J. Pharmacol. Exp. Ther. 122:56A, 1958.
86. Steigmann, F., et al.: Am. J. Gastroent. 33:109, 1960.
87. U. S. Patent 2,595,405 (1952).
88. U. S. Patent 2,567,351 (1951).
89. Haas, H., and Klavehn, W.: Arch. exp. Path. Pharmakol, 226:18, 1955.
90. Denton, J. J., et al.: J. Am. Chem. Soc. 72:3795, 1950.
91. Magee, K., and DeJong, R.: J. Am. Med. Assoc. 153:715, 1953.
92. U. S. Patent 2,891,890 (1959); see also Adamson et al.: J. Chem. Soc. 52, 1951.
93. U. S. Patent 2,826,590 (1958).
94. Schwab, R. S., and Chafetz, M. E.: Neurology 5:273, 1955.
95. Denton, J. J., and Lawson, V. A.: J. Am. Chem. Soc. 72:3279, 1950, see also U. S. Patent 2,698, 325, 1954.
96. Denton, J. J., et al.: J. Am.: Chem. Soc. 71:2053, 1949.
97. Janssen, P., et al.: Arch intern. pharmacodyn. 103:82, 1955.
98. U. S. Patent 2,726,245 (1955).
99. Sperber, N., et al.: J. Am. Chem. Soc. 73:5101, 1951.
100. Caviezel, R., Eichenberger, E., Kunzle, F., and Schmutz, J.: Pharm. Acta Helv. 33:459, 1958.
101. Sutherland, E. W.: Science 177:401, 1972.
102. Kukowetz, W. R., and Poch, G.: Nauyn Schmeideberg's Arch. Pharmak. 267:189, 1970.
103. Blicke, F. F.: Ann. Rev. Biochem. 13:549, 1944.
104. Buth, W., et al.: Ber. deutsch. chem. Ges. 72:19, 1939.
105. Kulz, F., et al.: Ber. deutsch. chem. Ges. 72:2161, 1939.
106. Kulz, F., and Rosenmund, K. W.: Klin. Wschr. 17:345, 1938.
107. Issekutz, B. V., Jr.: Arch exp. Path. Pharmakol. 177:388, 1935.
108. Weijlard J., et al.: J. Am. Chem. Soc. 71:1889, 1949.
109. U. S. Patent 2,728,769, 1955.
110. International Encyclopedia of Pharmacology and Therapeutics. Section 12: Ganglionic Blocking and Stimulating Agents Ed. Karczmar, A. G., Vol. 1, Pergamon Press 1966, Oxford.
111. Greenspan, P. and Kebabian, J. W., Fed. Proc., 33:1059, 1974.
112. Volle, R. L., and Hancock, J. C.: Fed. Proceedings 29:1913, 1970.
113. Greegard, P., and Kebabina, J. W.: Fed. Proc 33:1059, 1974.
114. Garrett, J.: Arch. Intern. Pharmacodyn. 144:381, 1963.
115. Van Rossum, J. M.: Int. J. Neuropharmacol. 1:97, 1962.
116. ———: Int. J. Neuropharmacol. 1:403, 1962.
117. Paton, W. D. M., and Zaimis, E. J.: Brit. J. Pharmacol. 4:381, 1949.
118. Taylor, D. B., and Nedergaard, O. A.: Physiol. Rev. 45:523, 1964.
119. Bovet, D.: Ann. N. Y. Acad. Sci. 54:407, 1951.
120. King, H.: J. Chem. Soc. 1381, 1935; see also 265, 1948.
121. Wintersteiner, O., and Dutcher, J. D.: Science 97:467, 1943.
122. Everett, A. J., et al.: Chem. Comm. p. 1020, 1970.
123. Soine, T. O., and Naghaway, J.: J. Pharm. Sci. 63:1643, 1974.
124. King, H.: J. Chem. Soc., p. 936, 1947.
125. Brown, A. C., and Fraser, T.: Tr. Roy. Soc. Edinburgh 25:151, 693, 1868-1869.
126. Phillips, A. P., and Castillo, J. C.: J. Am. Chem. Soc. 73:3949, 1951.
127. Cavallito, C. J., et al.: J. Am. Chem. Soc. 72:2661, 1950.
128. ———: J. Am. Chem. Soc. 76:1862, 1954.
129. Macri, F. J.: Proc. Soc. Exp. Biol. Med 85:603, 1954.
130. Bovet, D., Courvoisier, S., Ducrot, R., and Horclois, R.: Compt. rend. Acad. sci. 223:597, 1946.
131. Bovet, D., Courvoisier, S., and Ducrot, R.: Compt. rend. Acad. sci. 224:1733, 1947.
132. Bovet, D., Depierre, F., and de Lestrange, Y.: Compt. rend. Acad. sci. 225:74, 1947.
133. Barlow, R. B., and Ing, H. R.: Nature (London) 161:718, 1948.
134. Paton, W. D. M., and Zaimis, E. J.: Nature (London) 161:718, 1948.
135. Barlow, R. B., and Zoller, A.: Brit. J. Pharmacol. 23:131, 1964.
136. Bovet, D.: Ann, N. Y. Acad Sci. 54:407, 1951.
137. Ariens, E. J.: Molecular Pharmacology, a Basis for Drug Design, in Jucker, E. (ed.): Progress in Drug Research, vol. 10, p. 514, Basel, Birkhauser, 1966.
138. McCarthy, J. F., et al.: J. Pharm. Sci. 52:1168, 1963.
139. Burger, A., and Bedford, G. R.: J. Med. Chem. 6:402, 1963.
140. McCarthy, J. F., et al.: J. Med. Chem. 7:72, 1964.
141. Phillips, A. P.: J. Am. Chem. Soc. 74:4320, 1952.
142. Erhardt, P. W., and Soine, T. O.: J. Pharm. Sci. 64:53, 1975.
143. Genenah, A. A., Soine, T. O., and Shaath, N. A.: J. Pharm. Sci. 64:62, 1975.
144. Soine, T. O., Hanley, W. S., Shaath, N. A., and Genenah, A. A.: J. Pharm. Sci. 64:67, 1975.
145. Stenlake, J. B., Williams, W. D., Dhar, N. C., and Marshall, I. G.: Europ. J. Med. Chem.–Chimica Therapeutica 9:233, 1974.
146. ——— ibid., 239.
147. ——— ibid., 243.
148. Phillips, A. P.: J. Am. Chem. Soc. 71:3264, 1949.

SELECTED READINGS

Ariens, E. J.: Receptor Theory and Structure Action Relationships, Adv. Drug Res. 3:235, 1966.
Barlow, R. B.: Introduction to Chemical Pharmacology, ed. 2, London, Methuen & Co. Ltd., 1964.
Barrett, E. F., and Magleby, K. L.: Physiology of Cholinergic Transmission, in Goldberg, A. M., and Hanin, I. (eds.): Biology of Cholinergic Function, p. 29, New York, Raven Press, 1976.
Bebbington, A., and Brimblecombe, R. W.: Muscarinic Receptors in the Peripheral and Central Nervous Systems, Adv. Drug. Res. 2:143, 1965.
Brimblecombe, R. W.: Review of Tremorganic Agents, in Waser, P. G. (ed.): Cholinergic Mechanisms, p. 405, New York, Raven Press, 1975.
Brimblecombe, R. W.: Drug Actions on Cholinergic Systems, London, University Park Press, 1974.
Casy, A. F.: Stereochemical Aspects of Parasympathomimetics an Their Antagonists: Recent Developments, Prog. Med. Chem 11:1, 1975.
Enranko, O.: Small Intensity Fluorescent (SIF) Cells and Nerv Transmission in Sympathetic Ganglia, Ann. Rev. Pharma Toxicol. 18:417, 1978.
Goldberg, A. M., and Hanin, I. (eds.): Biology of Cholin Function, New York, Raven Press, 1976.
Miyamoto, M. D.: The Actions of Cholinergic Drugs on Nerve Terminals, Pharmacol. Rev. 29:221, 1978.
Natoff, I. L.: Organophosphorus Pesticides: Pharmacology Med. Chem. 8:1, 1971.
O'Brien, R. D.: Design of Organophosphate and Car Inhibitors of Cholinesterases, in Ariens, E. J. (ed.): Dru 2:162, 1971.
Rama Sastry, B. V.: Stereoisomerism and Drug Acti Nervous System, Ann. Rev. Pharmacol. Toxicol. 13
Siegel, G. J., et al.: Basic Neurochemistry, ed. 2, Bo Brown & Co., 1976.
Taylor, P.: Neuromuscular Blocking Agents, in Goodm Gilman, A. (eds.): The Pharmacological Basis of ed. 6, p. 220. New York, Macmillan, 1980.
Waser, P. (ed.): Cholinergic Mechanisms, New York 1975.
Wilson, I. B., and Froede, H. C.: The Design of Irreversibly Blocked AChE, in Ariens, E. J. (ed 2:213, 1971.
Zaimis, E. (ed.): Neuromuscular Junction. Berlin, 1976.

l.

us
ol.

rgic

otor

Prog.

bamate
Design

n in the
53, 1973.
on, Little

, L. S. and
herapeutics,

Raven Press,

activators for
: Drug Design

pringer-Verlag,

13
diuretics

T. C. Daniels

E. C. Jorgensen

Prior to a study of the diuretic compounds it is desirable to review pertinent aspects of renal physiology and biochemistry. The kidney is the organ mainly responsible for maintaining an internal environment compatible with life processes. Its primary function is the regulation of the volume and composition of the body fluids, which it accomplishes by the elimination of variable amounts of water and selective ions such as Na^+, K^+, H^+, Cl^-, HPO_4^{--} and SO_4^{--}. The extracellular fluids, which comprise about 15 percent of normal body weight, are influenced directly by changes in kidney function. The fluid within the cell (intracellular) is under osmotic equilibrium with extracellular fluid, and changes in extracellular fluid composition lead to changes in internal cellular fluid composition and function. A diuretic substance increases the excretion of urine by the kidney, thereby decreasing body fluids, especially the extracellular fluids.[1]

The pH of the body fluids is maintained by the excretion of anions such as HPO_4^{--} and, through the mediation of carbonic anhydrase, the synthesis from carbon dioxide and water of carbonic acid, which dissociates to H^+ and HCO_3^- ions. The renal tubular cells are able to synthesize ammonia by the deamination of amino acids, and in this way also the acid-base balance is maintained.

The kidney also serves as an excretory organ for the elimination of water-soluble substances present in excess of body needs. Thus, urinary excretion controls plasma concentrations of many nonelectrolytes that are end-products of normal body metabolism, such as urea and uric acid, as well as metabolically solubilized derivatives of foreign molecules, such as glucuronides and sulfate esters of phenols.

The main functional unit is the nephron, and there are approximately one million nephrons in each kidney. It will be noted (Fig. 13-1) that each nephron has three functional parts:

1. The renal corpuscle, consisting of a cluster or tuft of capillaries, known as the glomerulus, which is enclosed in *Bowman's capsule*. The blood enters the nephron through the afferent arteriole under high capillary pressure, and portions of the dissolved substances are filtered through the walls of the capillaries and the epithelium of Bowman's capsule into the lumen of the capsule.

2. The renal tubule, which in turn may be divided into three segments: (a) the *proximal convoluted tubule,* (b) the *loop of Henle* and (c) the *distal convoluted tubule.*

3. The *collecting tubule,* which leads to the renal pelvis, the ureter and, finally, to a larger collecting duct emptying into the bladder, where the urine is stored.

The glomerular filtrate which enters Bowman's capsule has the same general composition as the blood plasma, except that substances with a molecular weight of 67,000 or more do not pass through the filtering membrane. In this way, serum albumin and globulin are retained and are not present in the ultrafiltrate of the normal kidney.

The glomerular filtrate in the normal adult is formed at the rate of approximately 120 ml. per minute, or more than 7 liters per hour. Since the total extracellular fluid in the average adult amounts to approximately 12.5 liters, it will undergo complete filtration in less than 2 hours. As the filtrate passes down the renal tubule (nephron), the epithelial cells of the tubule reabsorb most of the water and solutes (over 99%), returning them to the bloodstream. Thus, it requires more than 100 ml. of glomerular filtrate to produce 1 ml. of urine. Approximately 70 per cent of the

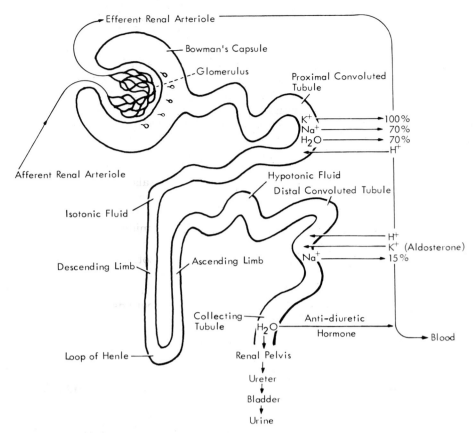

FIG. 13-1. *The nephron—the functional unit of the kidney.*

water and sodium ions and essentially all of the potassium ions are reabsorbed in the proximal tubules (Fig. 13-1). The fluid entering the loop of Henle is isotonic.[2] It becomes increasingly more concentrated as it passes through the descending limb, either by the loss of water,[3] or by the addition of solutes from the capillaries. The fluid passing through the ascending limb becomes increasingly more dilute, as about half of the remaining sodium and chloride ions are reabsorbed. Most of the remaining sodium ions may be reabsorbed in the distal tubule by a cation exchange mechanism under the control of the adrenal cortex hormone, aldosterone. Sodium from the renal tubular fluid is exchanged for potassium ions and, to a lesser extent, hydrogen ions from the blood. Reabsorption of water from the hypotonic fluid of the distal convoluted tubule takes place primarily under the influence of the antidiuretic hormone, vasopressin (ADH), of the posterior pituitary.

Most of the clinically useful diuretics increase the excretion of sodium ion (as chloride or bicarbonate) by decreasing reabsorption of the ion in the renal tubules. Any compound capable of interfering with the reabsorption of water and solutes from the glomerular filtrate will give a diuretic response.

There are four recognized sites of diuretic action in the renal tubule: a proximal site, two sites (medullary and cortical) in the thick ascending limb of the loop of Henle, and the far distal convoluted tubule sodium—potassium exchange site (see Fig. 13-1). The electrolyte excretion pattern given by a diuretic is determined by its sites of action in the renal tubule. Thus, diuretic compounds of quite diverse chemical structures such as mersalyl, ethacrynic acid, and furosemide act at the same sites in the renal tubule and produce a similar pattern of electrolyte excretion. On the other hand, two dichlorophenoxy acetic acid diuretics, ethacrynic acid and ticrynafen, although closely related structurally, act at different sites in the tubule and produce different diuretic profiles.

About half of the sodium ion remaining after passage through the proximal tubule is reabsorbed during passage through the loop of Henle. Some extremely powerful diuretics, such as furosemide and ethacrynic acid, block sodium transport in the ascending limb of the loop of Henle as well as at more proximal sites, and are sometimes referred to as "loop diuretics."

As a result of the inhibition of sodium reabsorption in the proximal tubule and loop of Henle, more sodium ion is delivered to the aldosterone-governed sodium-potassium exchange site in the distal tubule, leading to increased potassium secretion. The potassium ion loss may be appreciable, and hypokalemia is a major potential complication in the use of thiazide and loop diuretics. For this reason a potassium-sparing diuretic, such as amiloride, spironolactone or triamterene, may be used in combination with the more potent, but potassium-depleting, diuretics.[4] Since the mercurials inhibit both sodium reabsorption and potassium secretion, potassium depletion is not a problem with mercurial diuretic therapy.

The exact mechanism for the reabsorption of the normal electrolytes in the renal tubules remains to be determined, but it is generally believed that the electrolyte transport process may involve a "carrier molecule" to which the electrolyte has been attached through ion exchange. A number of specific transport mechanisms may be involved. It is also evident that an increase in the rate of glomerular filtration due to increased blood pressure or blood flow will increase urine formation. Likewise, inhibition of the antidiuretic hormone (vasopressin, ADH) of the posterior pituitary will produce diuresis. There are no diuretics currently in use which act primarily by these mechanisms.

The diuretics are generally employed for the treatment of all types of edema and, therefore, may be properly regarded as life-saving drugs. The main diseases associated with edema are congestive heart failure, premenstrual tension, edema of pregnancy, renal edema and cirrhosis with ascites. Diuretics are also used for the treatment of edema induced by the administration of ACTH and the other corticosteroids. Mild hypertension is normally controlled by use of a diuretic alone, while moderate to severe cases of hypertension may require a diuretic in combination with other agents (see Chap. 14).

Although there are a relatively large number of safe and efficient diuretics, there are none without inherent problems. The major deficiencies and side-effects are the following: (1) hypokalemia, (2) hyperglycemia with possible onset of diabetes, and (3) hyperuricemia with possible onset of gout.

With the recent introduction of highly active compounds, the diuretics may be divided into compounds with either "low-ceiling" or "high-ceiling" activity. The two classes differ by their dose-response curves. "Low-ceiling" compounds (e.g. thiazides, chlorthalidone and quinethazone) give dose-response curves that plateau at a certain point beyond which an increase in dose fails to increase the diuretic effect. In contrast, the "high-ceiling" diuretics, such as ethacry-

nic acid, furosemide and bumetanide give a dose-response curve that is linear over a wide range. The "low-ceiling" diuretics have a longer duration of action than the presently available "high-ceiling" compounds; for this reason, they are more desirable for use in long-term therapy.

The presently used diuretics may be conveniently divided into the following classes:

1. Water and osmotic agents
2. Acidifying salts
3. Mercurials
4. Phenoxyacetic acids
5. Purines and related heterocyclic compounds
6. Sulfonamides
   A. Inhibitors of carbonic anhydrase
   B. Benzothiadiazines ("thiazides") and related heterocyclic compounds
   C. Sulfamyl benzoic acid derivatives and related compounds
7. Endocrine antagonists

WATER AND OSMOTIC AGENTS

Any compound that is poorly reabsorbed by the renal tubules and is present in a concentration in excess of the concentration of electrolytes and dissolved substances in the body fluids will cause water and electrolytes to pass into the more concentrated solution and be excreted. Thus, diuresis and a mobilization of edema fluid take place.

The ingestion of large amounts of water and the resultant rapid dilution of blood increases urinary excretion by inhibiting the antidiuretic hormone (ADH), but the electrolyte concentration is not affected. There is no net loss of water from the tissues, since it is retained in proportion to the concentration of the electrolytes present.

In edematous states sodium salts are retained, and this produces an increase in the body fluids; therefore, low-salt or salt-free diets are used to restrict sodium intake. The substitution of potassium salts for sodium has little value and may be harmful by developing hyperkalemia.

A number of nonelectrolytes, such as urea and certain sugars, may be used as osmotic diuretics.

Urea U.S.P., Ureaphil®, carbamide. Urea is poorly reabsorbed by the renal tubules and therefore serves as an osmotic diuretic. When administered in large amounts, electrolytes are excreted, and a mobilization of edema fluid takes place. Oral administration is rarely used in the treatment of cardiac edema and ne-

$$H_2N-\overset{\overset{\displaystyle O}{\|}}{C}-NH_2$$

Urea

phrosis, since this requires doses of up to 20 g. 2 to 5 times daily.

Sterile lyophilized urea is available for the preparation of solutions containing 4 or 30 percent of urea in 10 percent invert sugar solutions. The 30 percent solution is used to control cerebral edema, or for the symptomatic relief of headache and vomiting due to increased intracranial pressure. It is also used in conjunction with surgical treatment of narrow angle closure glaucoma. The 4 percent urea and 10 percent invert sugar solution may be employed as a diuretic. A dose of 1 g. of urea per kg. of body weight reduces intracranial and intraocular pressure and produces diuresis. The lyophilized urea and invert sugar solutions should be freshly prepared for intravenous use.

Usual dose—intravenous infusion, 1 g. to 1.5 g. per kg. of body weight daily, as a 30 percent solution in Dextrose Injection at a rate not exceeding 4 ml. per minute.

Usual dose range—up to a maximum of 120 g. daily.

Usual pediatric dose—under 2 years of age: 100 mg. per kg. of body weight; over 2 years of age: 500 mg. to 1.5 g. per kg. or 35 g. per square meter of body surface.

Occurrence
Sterile Urea U.S.P.

Glucose, sucrose and mannitol have been used as osmotic diuretics. For this purpose they are administered intravenously and must be used in large doses in the order of 50 ml. of a 50 percent solution.

Mannitol U.S.P., D-Mannitol, is a hexahydroxy alcohol which is essentially not metabolized. It is filtered by the glomerulus, but only negligible amounts are reabsorbed by the tubules. Therefore, it is used as a diagnostic agent to measure glomerular filtration

$$HOCH_2CH-\overset{\overset{\displaystyle OH}{|}}{CH}-\overset{\overset{\displaystyle OH}{|}}{CH}-CHCH_2OH$$
$$\underset{\underset{\displaystyle OH}{|}}{} \quad \underset{\underset{\displaystyle OH}{|}}{}$$

D-Mannitol

rates (see Chap. 22) and to produce osmotic diuresis in various edematous states, including those which fail to respond to thiazide preparations. It is administered intravenously, the normal adult dose being 50 to 100 g. within a 24-hour period; the maximum recommended dose is 200 g. Approximately 80 percent is excreted in 12 hours.

The usual solution for injection contains 12.5 g. in 50 ml.

Usual dose—I.V., 200 mg. per kg. of body weight in a 15 to 25 percent solution, administered in 3 to 5 minutes.

Usual dose range—12.5 to 200 g. daily.

Usual pediatric dose—dosage is not established for children under 12 years of age.

Occurrence
Mannitol Injection U.S.P.
Mannitol and Sodium Chloride Injection U.S.P.

The osmotic diuretics have obvious disadvantages and have been largely replaced by more effective agents, except for special indications, such as the early treatment of acute reduction in renal blood flow, the reduction of intraocular pressure prior to eye surgery for glaucoma, or the enhancement of urinary excretion of intoxicants such as barbiturates.

ACIDIFYING SALTS

Acidifying salts, such as ammonium chloride and ammonium nitrate, elicit a weak diuretic response by producing an excess of the anion (Cl^-, NO_3^-) in the glomerular filtrate. This is made possible by the ammonium cation conversion to urea, which is a neutral compound. Accompanying the excess anion, there is an increase in Na^+ output, but a greater increase in H^+ concentration leading to an acidification of the urine. There remains an excess of hydrogen ions, leading to systemic acidosis. The same general mechanism is involved when calcium chloride or calcium nitrate is administered, except that the calcium cation (Ca^{++}) is depleted by deposition in the bone or is excreted as the phosphate.

The use of acidifying salts alone for diuresis is quite unsatisfactory, for the kidney is able to develop rapidly a compensating mechanism to neutralize the acids formed in the glomerular filtrate by increasing the formation of ammonia. Thus, in a comparatively short time (1 to 2 days) the acidifying salts lose their ability to produce acidosis and no longer serve effectively as diuretics. Other disadvantages of the acidifying salts are: gastric irritation, which may lead to anorexia, nausea and vomiting; the ammonium salts may produce hyperammonemia; and acidosis may lead to renal insufficiency.

At present, the acidifying agents (principally ammonium chloride) are used primarily in conjunction with the mercurial diuretics which they potentiate. In patients who cannot tolerate ammonium chloride, L-lysine monohydrochloride may be used as an acidifying agent.

MERCURIALS

Prior to 1950, the mercurial compounds were the only effective diuretics available for use. Although they have since been largely replaced by newer orally effective agents, they are still regarded as useful for the treatment of severe edematous states. Under optimal conditions, they are approximately four times as effective as the thiazides in increasing the excretion of sodium (as chloride).

The mercurial diuretics have a number of undesirable properties. They are not well absorbed from the gastrointestinal tract, but when administered parenterally usually give a rapid and reliable onset of diuresis. Toxic reactions are comparatively uncommon, but local irritation, necrosis, hypersensitivity reactions and electrolyte disturbances may occur. Mercurialism also may develop following prolonged use, especially with the mercurials used orally.

The continued use of the mercurial diuretics may be attributed to the following desirable properties: with the exception of the new potassium-sparing diuretics, including the aldosterone antagonists, the organic mercurials produce less potassium loss than most other classes of diuretics and supplementary potassium administration is normally unnecessary; they do not significantly alter the excretion of potassium, ammonium, bicarbonate or phosphate ions and therefore may be used without producing a marked disturbance of the electrolyte balance of the body fluids; carbohydrate metabolism is unaltered and thus they are free from the danger of producing hyperglycemia and the onset of diabetes; finally, uric acid elimination is unchanged, which avoids the possibility of hyperuricemia.

Mercurous chloride was first used as a diuretic by Paracelsus in the 16th century, and its use alone or in combination with other agents continued until comparatively recent times. The organic mercurial diuretics originated in 1919 following the observation that a new compound (merbaphen, novasural) introduced for

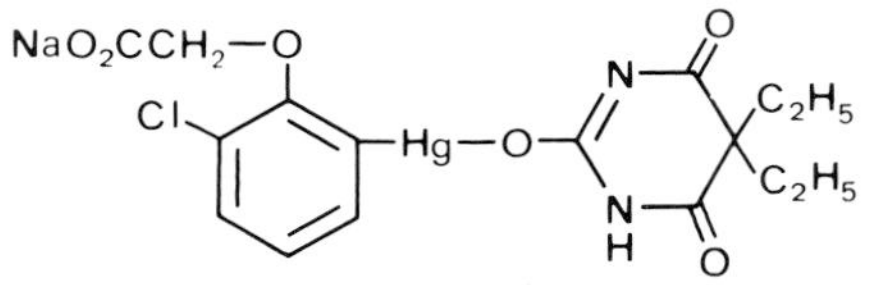

the treatment of syphilis elicited a pronounced diuretic response. However, merbaphen, with the mercury attached directly to the aromatic ring, was too toxic for general use as a diuretic, and it was later found that related compounds with the mercury attached to an aryl group were likewise too toxic.[1]

Structural variations led in 1924 to the less toxic mersalyl.

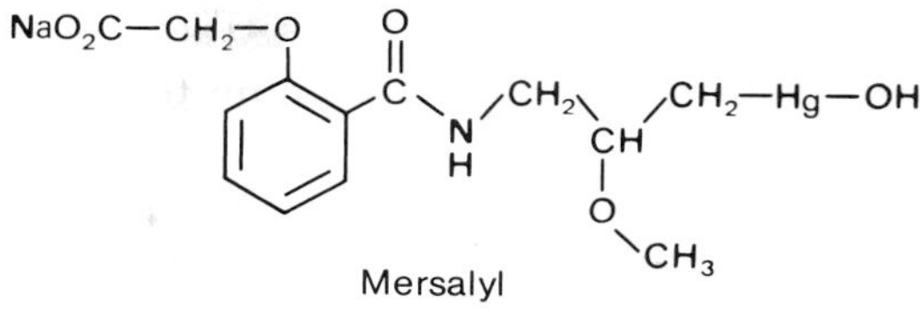

All of the mercurials in clinical use at present are close structural analogs, in which an alkoxymercuripropyl group is attached to a mono- or dicarboxylic acid or the amide derivative of an acid.[5] The following compound illustrates the general structural features of the mercurial diuretics:

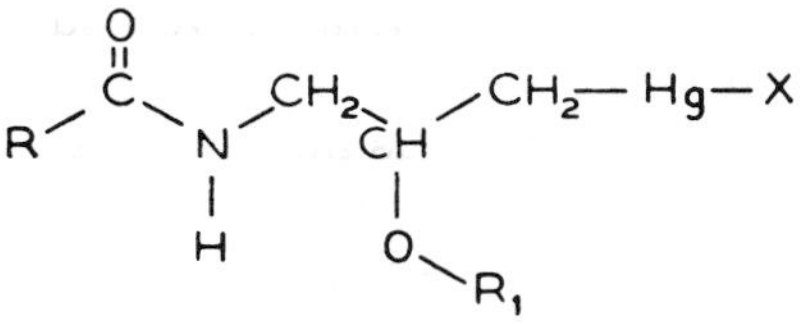

R = aliphatic, alicyclic, aromatic, heterocyclic groups, usually carrying a carboxyl function, and attached to the 3 carbon chains through amide, urea, ether or carbon-carbon linkages.
R_1 = —H, —CH$_3$, —C$_2$H$_5$ or —CH$_2$CH$_2$—OCH$_3$
X = —OH, —Cl, —Br, —O$_2$CCH$_3$, —SCH$_2$—CO$_2$H, —SCH$_2$(CHOH)$_4$CH$_2$OH

The general method for the synthesis of the mercurial diuretics involves the mercuration of an alkene:

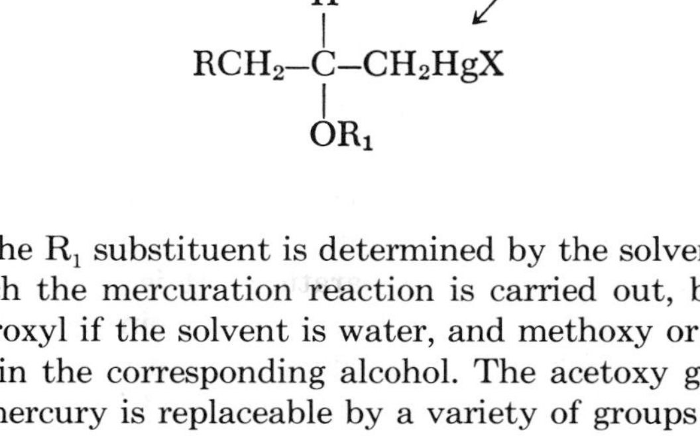

The R_1 substituent is determined by the solvent in which the mercuration reaction is carried out, being hydroxyl if the solvent is water, and methoxy or ethoxy in the corresponding alcohol. The acetoxy group on mercury is replaceable by a variety of groups (X), theophylline being most often used.

STRUCTURE-ACTIVITY RELATIONSHIPS

Diuretic activity for the organic mercurials requires a hydrophilic group (e.g., RCONH—) attached not less than three carbon atoms distant from the mercury.[6] Compounds with a shorter chain show little or no activity.

Each of the 3 groups (R, R_1 and X) influences the diuretic activity and the toxicity of the molecule. Of the 3 variants (R) has the greatest and (R_1) the least influence.[5] The (X) group does not increase the activity of the molecule per se, but when theophylline represents (X) there is improved absorption from the site of injection and an enhanced diuretic response due to a potentiating effect. Theophylline also lowers the tissue irritation and therefore is commonly used in combination with the mercurial diuretics.[6,7]

MODE OF ACTION

The mechanism of action of the mercurial diuretics remains to be established, but the primary site of action is in the proximal renal tubules.[8] Their action is believed to be due primarily to a combination of mercury ions with sulfhydryl groups attached to the renal enzymes responsible for the production of energy necessary for tubular reabsorption. When these specific enzymes are blocked, there is a marked increase in the amount of sodium chloride and water excreted, due to interference with the reabsorption process. Administration of dimercaprol (BAL) or other related *vicinal* dithiols with mercurial diuretics prevents blocking of the enzymes, and diuresis does not occur, because the dithiols have a greater affinity for ionic mercury than does the enzyme.

There is uncertainty as to whether the organic mercurials act in the mono- or the divalent form, that is, as intact molecules (R—Hg+) or, by a splitting of the molecule to give mercuric ions (Hg^{++}), at the site of action.[9] In support of the mercuric ion postulate, it has been observed that all active mercurial diuretics are acid-labile, yielding mercuric ion. Acid-stable compounds were inactive as diuretics.[1] If mercuric ions are responsible for the diuretic response, it must be assumed that the undissociated molecule contributes to the lower toxicity and serves to carry the mercury to the site of action where the splitting occurs. If the intact molecule is the active species, the hydrophilic group, separated by three carbon atoms from the mercury, may serve to reinforce binding to the enzyme receptor. The two proposed mechanisms of reaction, together with the blocking effect of BAL, are shown below.

A. Action due to intact molecule

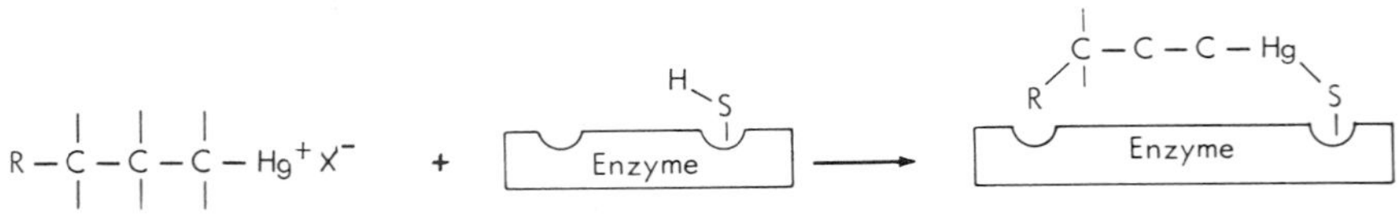

B. Action due to mercuric ion

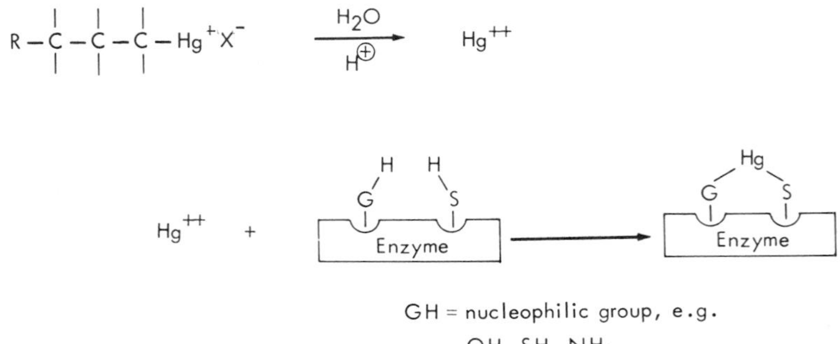

GH = nucleophilic group, e.g.
OH, SH, NH_2

C. Combination of BAL with mercuric ion

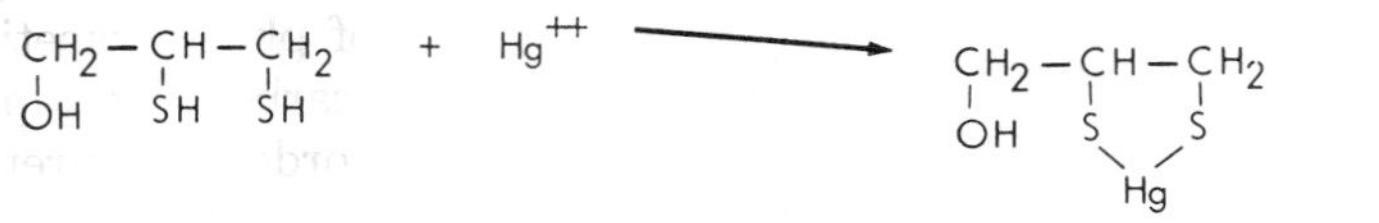

Stable Cyclic Mercurial

TABLE 13-1

MERCURIAL DIURETICS

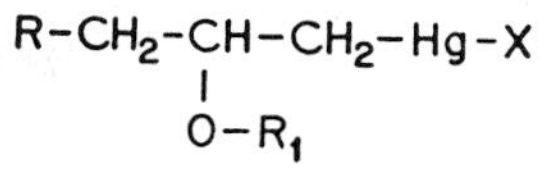

$$R-CH_2-CH-CH_2-Hg-X$$
$$\underset{O-R_1}{|}$$

Generic Name Proprietary Name	R	R₁	X	Administered
Sodium Mercaptomerin U.S.P. *Thiomerin*	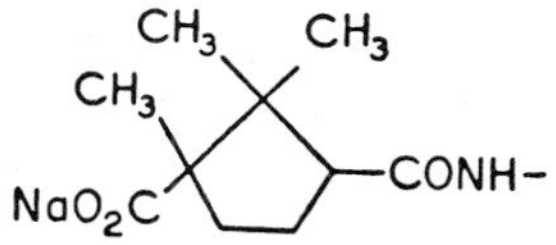 (structure: 1,2,2-trimethylcyclopentane with NaO₂C and CONH—)	—CH₃	—S—CH₂—CO₂Na	S.C.
Chlormerodrin *Neohydrin*	H₂N—CO—NH—	—CH₃	—Cl	Orally
Mercurophylline *Mercupurin*	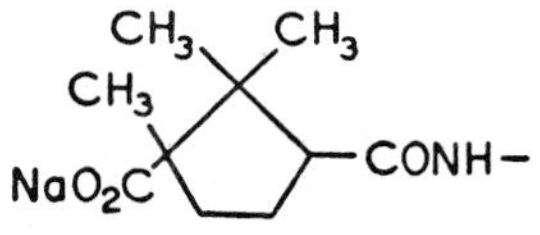 (structure: 1,2,2-trimethylcyclopentane with NaO₂C and CONH—)	—CH₃	Theophylline	I.M. and Orally
Mersalyl *Salyrgan*	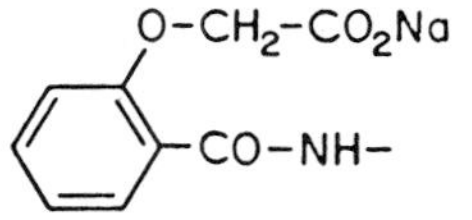 (structure: benzene ring with O—CH₂—CO₂Na and CO—NH—)	—CH₃	Theophylline	I.M. and I.V.

Table 13-1 gives the structural features and the modes of administration of the mercurial diuretics.

PRODUCTS

Mercaptomerin Sodium U.S.P., Thiomerin® Sodium, [[[3-(1-carboxy-1,2,2-trimethylcyclopentane-3-carboxamido)-2-methoxypropyl]thio]mercuri]acetate disodium salt.

Mercaptomerin was first introduced in 1946 and is the most widely used mercurial diuretic. Because of its low local irritation effect, it may be administered subcutaneously. Mercaptomerin is heat- and light-sensitive; therefore, it should be stored in lightproof containers under refrigeration.

This compound occurs as a white hygroscopic powder or amorphous solid that is freely soluble in water and soluble in alcohol. A 2 percent solution is neutral to litmus.

Usual dose—I.M. or S.C., 25 to 250 mg. daily.

Usual pediatric dose—I.M., the following amounts, or 125 mg. per square meter of body surface, are given from 1 to 2 times a week up to once daily: under 3 kg. of body weight: 16 mg.; 3 to 7 kg.: 31 mg.; 8 to 15 kg.: 63 mg.; 16 to 25 kg.: 94 mg.; and over 25 kg.: 125 mg.

Occurrence

Mercaptomerin Sodium Injection U.S.P.

Chlormerodrin, 3-chloro-(2-methoxy-3-ureido-propyl)mercury.

Chlormerodrin was first introduced in 1952 as an orally effective mercurial diuretic. It has since been largely replaced by more effective nonmercurial diuretics, but is still available for oral use in 18.3-mg. tablets. Chlormerodrin containing radioactive mercury, either ¹⁹⁷Hg or ²⁰³Hg, is used as a diagnostic aid in tumor localization (see Chap. 22).

PHENOXYACETIC ACIDS

The diuretic properties of the mercurials are postulated to be dependent on the blocking of an essential sulfhydryl group associated with the transport mechanism responsible for the reabsorption of electrolytes in the kidney tubules. Since the α,β-unsaturated ketones react readily and reversibly with sulfhydryl groups, derivatives of phenoxyacetic acid containing the α,β-unsaturated carbonyl group were studied and found to show a high order of diuretic activity.[10] Subsequently it was found that chloro-substituted phenoxyacetic acids in which the double bond of the α,β-unsaturated

$$RSH + R' - CH = CH - \overset{\overset{\displaystyle O}{\displaystyle \|}}{C} - R'' \rightleftharpoons R' - \underset{\underset{\displaystyle R}{\underset{\displaystyle |}{S}}}{\overset{}{CH}} - CH_2 - \overset{\overset{\displaystyle O}{\displaystyle \|}}{C} - R''$$

Sulfhydryl α,β-Unsaturated   Addition Product
Compound   Ketone

carbonyl group had been reduced retained diuretic activity, thus bringing into question the significance of sulfhydryl binding as a primary mode of action.[11]

STRUCTURE-ACTIVITY RELATIONSHIPS[1,10]

For maximum activity one position in the aromatic ring *ortho* to the unsaturated ketone must be substituted with a halogen or methyl group. Disubstitution in the 2,3 positions increases the activity; additional substitution in the ring may lower the activity. It is of interest that the first organic mercurial diuretic (merbaphen) was a derivative of a chlorophenoxyacetic acid, which suggests that such acids may serve to increase the activity of the functional groups of these two classes of diuretics (i.e., labile mercury bonding, reactive α,β-unsaturated carbonyl groups) and contribute physical properties (hydrophilic-lipophilic balance) to the intact molecules favoring their transport to receptor sites for interaction and blocking of the essential sulfhydryl transport mechanism.

The structure of the α,β-unsaturated carbonyl group influences the activity. Maximum activity is shown if the β-position of the unsaturated ketone is unsubstituted. Higher alkyl groups substituted for ethyl in the α-position lower the activity. Reduction of the double bond diminishes, but does not abolish, activity.[11]

The unsaturated ketone group must be *para* to the oxyacetic acid group for maximum activity. The *ortho* and *meta* isomers are much less active.

All the compounds in the series that gave a significant diuretic response were also highly reactive in vitro with sulfhydryl-containing compounds. Ethacrynic acid was the most active member of the series.

Ethacrynic Acid U.S.P., Edecrin®, [2,3-dichloro-4-(2-methylenebutyryl)phenoxy]acetic acid.

Ethacrynic Acid

Ethacrynic acid is a potent saluretic agent which may produce a marked diuretic response in cases of refractory edema. Increased diuresis begins within 30 minutes of an oral dose of ethacrynic acid, or within 5 minutes after an intravenous injection of ethacrynate sodium. Duration of action following oral administration is 6 to 8 hours, with peak diuresis at 2 hours. Under optimal conditions it has a natriuretic effect at least five times greater than the thiazides[12] and is equivalent in activity to the mercurials.[13] Unlike the mercurials, the diuretic activity is not influenced by metabolic alkalosis or acidosis. Like the thiazides, it may lower uric acid excretion and cause hyperuricemia but (unlike the thiazides) it has little influence on carbohydrate metabolism and blood glucose levels.[1] Ethacrynic acid inhibits sodium reabsorption primarily at two sites in the nephron:[13] (1) in the ascending limb of the loop of Henle, and (2) in the distal tubule where urinary dilution occurs. It is commonly referred to as a "loop diuretic." Chloride excretion is increased to a greater extent than sodium excretion, which may lead to systemic alkalosis.

Ethacrynic acid may produce hypoalbuminemia, hypochloremia, hyponatremia, hypokalemia, and metabolic alkalosis.[14] Transient and permanent deafness has been reported following treatment with ethacrynic acid in patients with uremia. The cause of deafness is unknown, but the drug should be used with caution in uremic patients.[15]

Ethacrynic acid is especially useful in the treatment of refractory edema and may be used in combination with a potassium-sparing diuretic.

Ethacrynic acid is a white crystalline powder, slightly soluble in water, but soluble in alcohol, chloroform or benzene. Ethacrynate sodium is soluble in water at 25° to the extent of about 7 percent. The solution at pH 7 is stable at room temperature for short periods, but stability decreases with increased temperature and pH. Neutral solutions of 5 percent Dextrose Injection or of Sodium Chloride Injection are used to prepare solutions for intravenous administration. A precipitate will form if the pH of the diluent is below 5, and the resulting hazy or opalescent preparation should not be used.

Usual dose—oral, 50 to 100 mg. once or twice daily; intravenous, ethacrynate sodium equivalent to 500 μg. to 1 mg. of ethacrynic acid per kg. of body weight.

Usual dose range—50 to 400 mg. daily.

Usual pediatric dose—infants: dosage is not established; children: 25 mg. initially with stepwise increments of 25 mg. until effect is achieved; maintain on alternate-day therapy with rest periods.

Occurrence
Ethacrynic Acid Tablets U.S.P.
Ethacrynate Sodium for Injection U.S.P.

Ticrynafen, Selacryn®, [2,3-dichloro-4-(2-thienyl-carbonyl)phenoxy]acetic acid.

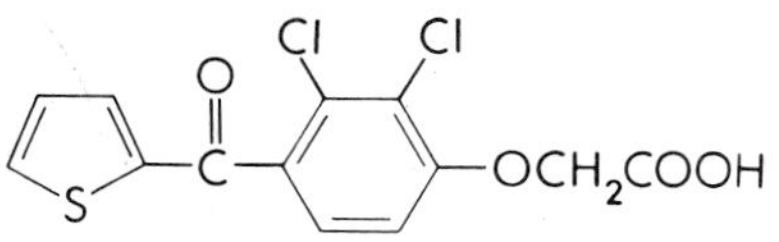

Ticrynafen

Ticrynafen was prepared as a heterocyclic derivative of the phenoxyacetic acids.[16] Despite its structural similarity to ethacrynic acid its diuretic action more closely resembles that of the thiazides. The unique additional property of ticrynafen is its uricosuric activity due to inhibition of urate reabsorption in the proximal tubules. The agent was used for several years in France under the generic name of tienilic acid and was introduced in the United States in 1979 as ticrynafen. Owing to reports of liver toxicity, the drug was withdrawn in 1980.

PURINES AND RELATED HETEROCYCLIC COMPOUNDS

Purine and pyrimidine bases are constituents of the nucleic acids. The purine bases, which consist of fused pyrimidine and imidazole rings, occur in nature primarily as oxidized derivatives. The 2,6-dihydroxypurine is xanthine, and the N-methylated xanthines, by virtue of their widespread occurrence in plant materials used by man, especially the traditional beverages (tea, coffee, cocoa), have long been recognized for their diuretic properties. The 3 most common naturally occurring xanthines are caffeine, theobromine and theophylline, and the diuretic potency increases in the order named.

These compounds are neither potent nor reliable diuretics, and their principal use has been in conjunction with the mercurial diuretics. The mode of action of the purine (xanthine) diuretics is not known, but they appear to give much the same type of action as the mercurials. When they are given concurrently with water diuresis, the urinary concentration of both sodium and chloride is increased, probably by decreased tubular reabsorption.[9]

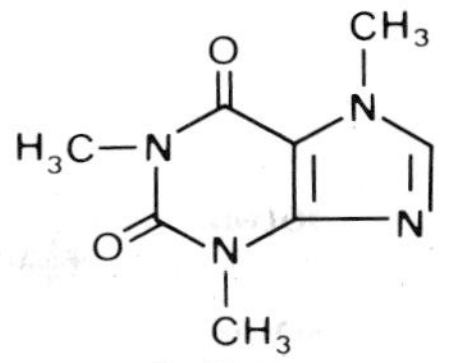

Caffeine

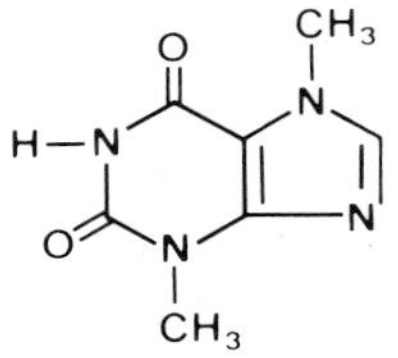

Theobromine

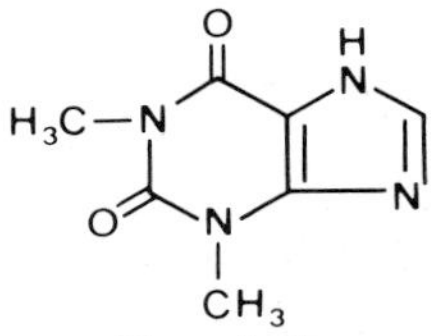

Theophylline

Theophylline U.S.P., Elixophyllin®, 1,3-dimethyl-xanthine. Theophylline is the most active diuretic of the xanthine alkaloids, but because of its relatively low activity and other pharmacologic effects (e.g., cardiovascular effects, bronchial muscle relaxation), it is seldom used alone. It is an important constituent of several mercurial diuretics in which it is used to reduce tissue-irritating effects, improve absorption and enhance the diuretic response.

Occurrence
Theophylline Tablets U.S.P.

Aminophylline U.S.P., theophylline ethylenediamine. Aminophylline may be used as a diuretic but it has largely been replaced for this purpose by orally more effective drugs. Aminophylline is effective as a diuretic only when given by the intravenous route. In addition to its diuretic effect, aminophylline is useful

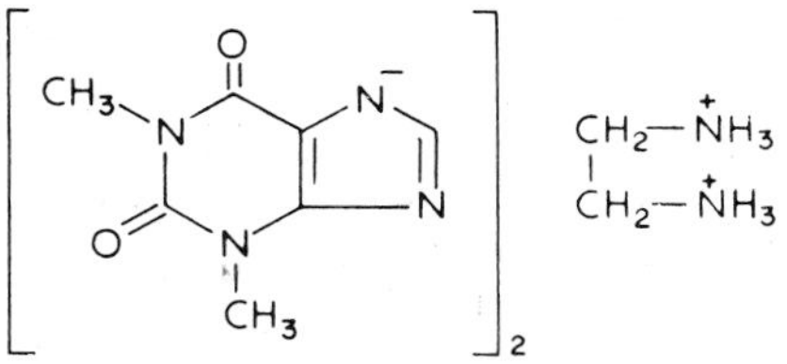

Aminophylline

as a peripheral vasodilator, a myocardial stimulant for the relief of pulmonary edema and as an antiasthmatic agent. Table 13-2 lists some xanthine derivatives, most of these consisting of salts possessing a higher solubility than the parent xanthine.

PTERIDINES

The pteridine ring system consists of fused pyrimidine and pyrazine rings. Derivatives of this heterocyclic system show unique diuretic properties.

TABLE 13–2

PARTIAL LIST OF PURINE (XANTHINE) DERIVATIVES AND COMBINATIONS

| Generic Name | Proprietary Name | Dosage Forms | | | Average Dose Range* mg. |
		Injection	Suppositories	Tablets	
Theophylline U.S.P.	Elixophyllin			X	100–200
Aminophylline U.S.P.	Lixaminol	X	X	X	200–500
Theophylline Sodium Acetate				X	200–300
Theophylline Sodium Glycinate U.S.P.	Glynazan	Aerosol Elixir	X	X	300–1,000
Theophylline Calcium Salicylate				X	250–300
Oxtriphylline U.S.P.	Choledyl			X	200–400
Dyphylline	Neothylline			X	100–200
Theobromine	—			X	300–500
Theobromine Calcium Gluconate				X	500–1,500

* See *U.S.P. D.I.* for complete dosage information.

Triamterene U.S.P., Dyrenium®, 2,4,7-triamino-6-phenylpteridine. Triamterene is a synthetic pteridine derivative which is orally effective in increasing urinary excretion of sodium and chloride ions, without increasing the excretion of potassium ions. Thus, it is one of the few potassium-sparing diuretics. Triamterene acts by interfering with the processes of cation exchange in the distal renal tubule by a mechanism other than antagonism of aldosterone.

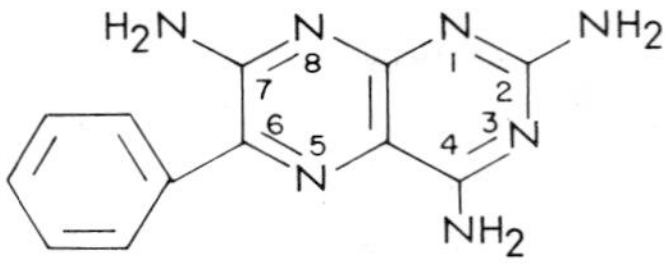

Triamterene

An increase in blood urea nitrogen levels has been observed with triamterene. This is believed to be due to a reduced glomerular filtration rate, and, therefore, use of the drug is contraindicated in the presence of renal disease and hepatitis.

Triamterene may be used alone or as an adjunct to long-term thiazide therapy to improve diuresis and prevent excessive potassium loss. Use of potassium chloride is not necessary with the appropriate combination of the two drugs.

The usual adult starting dose of triamterene is 100 mg. twice daily after meals. Maintenance therapy is usually 100 mg. daily or every other day, with the peak diuretic effect occurring 2 to 8 hours after administration.

Usual dose—100 mg. twice daily.

Usual dose range—100 mg. every other day to a maximum of 300 mg. daily.

Usual pediatric dose—1 to 2 mg. per kg. of body weight or 30 to 60 mg. per square meter of body surface twice daily, up to a maximum of 300 mg. daily.

Occurrence
Triamterene Capsules U.S.P.

AMINOPYRAZINES

Like the structurally related triamterene, an aminopyrazine derivative, amiloride, is a potent saluretic that prevents potassium depletion by direct blockade of tubular sodium–potassium exchange.

Amiloride, N-amidino-3,5-diamino-6-chloropyrazine carboxamide hydrochloride.

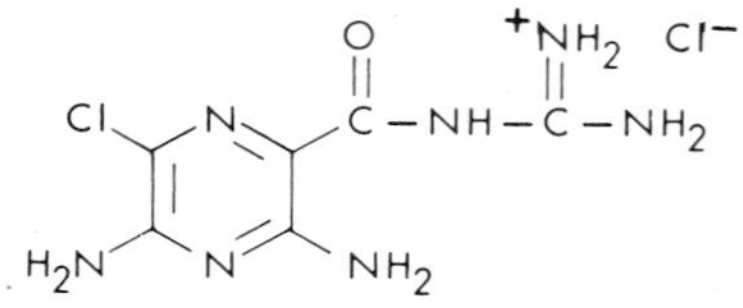

Amiloride

Amiloride is a long-acting, potassium-sparing saluretic that potentiates the diuretic effects of the thiazides. In clinical trials the maximal diuretic response was produced in 4 to 6 hours, with a duration of action from 10 to 12 hours. The usual oral dose ranges from 5 to 40 mg. daily for this experimental drug.

3-AMINOPYRAZOLIN-5-ONES

The structural element $-N-C\equiv-N-$, which is present in the diuretic purines, pteridines, and pyrazines, served as a guide in the discovery of a new class of potent high-ceiling diuretics, the 3-aminopyrazolin-5-ones.[17] Muzolimine has been selected for clinical evaluation.

Muzolimine, 3-amino-1-(3,4-dichloro-α-methylbenzyl)pyrazolin-5-one.

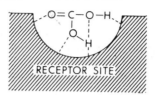

Muzolimine

Muzolimine is a high-ceiling saluretic agent with an electrolyte-excretion pattern and site of action like that of furosemide. Like furosemide it is amphoteric, forming soluble salts with acids or, in the enol form, with bases. The lipophilic nature of muzolimine is believed to account for its long duration of action.

SULFONAMIDES

INHIBITORS OF CARBONIC ANHYDRASE

Sulfanilamide was first introduced as an antibacterial agent in 1936–37. Soon thereafter it was noted that the drug altered the electrolyte balance, causing systemic acidosis due to increased excretion of bicarbonate.[18] In 1940 it was established that the electrolyte disturbance was due to the inhibitory effect of sulfanilamide on the enzyme carbonic anhydrase.[19] This observation, together with the fact that sulfanilamide was a comparatively weak carbonic anhydrase inhibitor, stimulated a search for more active compounds. Early studies revealed that carbonic anhydrase inhibition was limited to those compounds in which the amide nitrogen is free. The mono- and disubstituted derivatives of the sulfamyl ($-SO_2NH_2$) group are inactive. It has been proposed that the structural similarity of the unsubstituted sulfamyl group and carbonic acid permits competitive binding by the sulfonamide at the active site of the carbonic anhydrase enzyme.

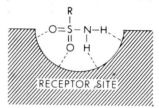

In 1950 Roblin and Clapp[20] synthesized a series of heterocyclic sulfonamides to test as carbonic anhydrase inhibitors. These workers set out to prepare more acidic sulfonamides in the hope that the more highly ionized compounds would bind more strongly to the carbonic anhydrase enzyme and thus be more active inhibitors. It was found that the degree of dissociation of the compounds roughly paralleled the carbonic anhydrase inhibition, with activity ranging as high as 2,500 times greater than sulfanilamide. From the new series of compounds, acetazolamide was selected for clinical trial and later was shown to be an effective diuretic agent.

The mode of action of the carbonic anhydrase inhibitors appears to be reasonably well established. Carbonic anhydrase is known to catalyze the hydration of carbon dioxide (produced metabolically in the renal tubules) to carbonic acid and likewise its reverse dissociation to carbon dioxide and water. The carbonic acid formed ionizes to give bicarbonate and hydrogen ions, as illustrated in the following equations:

$$HOH + CO_2 \underset{\text{Inhibition blocks this step}}{\overset{\text{Carbonic anhydrase}}{\rightleftharpoons}} H_2CO_3 \rightleftharpoons H^+ + HCO_3^-$$

The hydrogen ions formed exchange for sodium ions and, to a lesser extent, for potassium ions in the renal tubules, or they may combine with bicarbonate ions to produce carbonic acid and carbon dioxide, thereby propagating the cycle. Inhibition of carbonic anhydrase reduces the concentration of hydrogen ions in the renal tubules and leads to increased excretion of sodium and bicarbonate ions (decreased reabsorption), thereby producing diuresis. There may also be a significant loss of potassium, leading to hypokalemia; chloride excretion is not greatly altered. The normally acidic urine becomes alkaline, and hydrogen ion is retained, which may lead to systemic acidosis. When acidosis occurs, the carbonic anhydrase inhibitors are no longer effective as diuretics, and administration of the drug must be interrupted until the acid-base balance has returned to normal.

The ability of the carbonic anhydrase inhibitors to develop systemic acidosis makes this class of drugs useful as adjuncts to anticonvulsant therapy in epilepsy, since acidosis reduces, and may prevent, epileptic seizures.

The most important clinical use for the carbonic anhydrase inhibitors is in the treatment of glaucoma. Acetazolamide and other carbonic anhydrase inhibitors produce a partial depression of aqueous humor formation, thus reducing the high intraocular pressure associated with this disease.

Acetazolamide U.S.P., Diamox®, 5-acetamido-1,3,4-thiadiazole-2-sulfonamide.

Acetazolamide was introduced in 1953 as the first member of the series of carbonic anhydrase inhibitors. It is absorbed following oral administration to give peak levels in the blood plasma in about 2 hours, and the duration of its diuretic action is from 8 to 12 hours. The compound is well tolerated and may be used alone in mild or moderate cases of edema or in conjunction with a mercurial diuretic. When used alone, it may rapidly lose its effectiveness due to systemic acidosis, in which case interruption of therapy is necessary until the acid-base balance is restored. In addition to its diuretic effect, acetazolamide is a useful agent for the treatment of glaucoma and epilepsy. It occurs as a white to faintly yellowish-white, crystalline powder that is slightly soluble in water or alcohol. The usual range in the treatment of epilepsy is 375 mg. to 1 g. once a day. When used together with other anticonvulsants, the initial dose should not exceed 250 mg. The drug is available as 250-mg. tablets and as a

syrup containing 50 mg. per ml. Sterile Acetazolamide Sodium U.S.P. is available for intravenous administration when the oral route is impractical.

Methazolamide

Methazolamide U.S.P., Neptazane®, N-(4-methyl-2-sulfamoyl-Δ^2-1,3,4-thiadiazolin-5-ylidene)-acetamide, 5-acetylimino-4-methyl-Δ^2-1,3,4-thiadiazoline-2-sulfonamide. Methazolamide is a more active carbonic anhydrase inhibitor than the parent compound, acetazolamide. It is recommended for adjunctive treatment of chronic simple (open angle) glaucoma. A fall in intraocular pressure occurs within 2 to 4 hours of oral administration with a peak effect in 6 to 8 hours, and duration of effect of 10 to 18 hours. The drug is available as 50-mg. tablets for oral administration. Like acetazolamide, it can produce an electrolyte imbalance leading to acidosis.

Ethoxzolamide

Ethoxzolamide U.S.P., Cardrase®, 6-ethoxybenzothiazole-2-sulfonamide. Ethoxzolamide is approximately twice as active as a carbonic anhydrase inhibitor as acetazolamide. Following oral administration, maximum plasma levels are attained in approximately 2 hours, and the duration of action is from 8 to 12 hours. Like acetazolamide, the drug is used as an adjunct in the treatment of glaucoma and epilepsy. The properties of ethoxzolamide closely resemble those of acetazolamide.

Dichlorphenamide U.S.P., Daranide®, 1,2-dichloro-3,5-disulfamylbenzene. Dichlorphenamide is a carbonic anhydrase inhibitor recommended primarily as an adjunct for the treatment of glaucoma. Like other drugs of this class, it reduces intraocular pressure by inhibiting aqueous humor formation. Normally, it is used in conjunction with miotic agents such as pilocarpine, physostigmine, etc., and is claimed to be effective when other therapy, including

Acetazolamide

Dichlorphenamide

TABLE 13–3

CARBONIC ANHYDRASE INHIBITORS

Name Proprietary Name	Preparations	Usual Adult Dose*	Usual Dose Range*	Usual Pediatric Dose*
Acetazolamide U.S.P. Diamox	Acetazolamide Tablets U.S.P.	250 mg. 2 to 4 times daily	250 mg. to 1 g. daily	3 to 10 mg. per kg. of body weight or 100 to 300 mg. per square meter of body surface, 3 times daily
	Sterile Acetazolamide Sodium U.S.P.	I.M. or I.V., the equivalent of 500 mg. of acetazolamide, repeated in 2 to 4 hours	500 mg. to 1 g. daily	I.M., 5 mg. per kg. of body weight or 150 mg. per square meter of body surface, once daily
Methazolamide U.S.P. Neptazane	Methazolamide Tablets U.S.P.	50 to 100 mg. 2 or 3 times daily	100 to 300 mg. daily	
Ethoxzolamide U.S.P. Cardrase, Ethamide	Ethoxzolamide Tablets U.S.P.	62.5 to 250 mg. 2 to 4 times daily	62.5 mg. to 1 g. daily	
Dichlorphenamide U.S.P. Daranide, Oratrol	Dichlorphenamide Tablets U.S.P.	Initial, 100 to 200 mg., then 100 mg. twice daily until desired response has been obtained; maintenance, 25 to 50 mg. 1 to 3 times daily		

* See U.S.P. D.I. for complete dosage information.

miotics, has failed or is poorly tolerated. Since it is a carbonic anhydrase inhibitor, it is able to produce a disturbance of the acid-base balance, leading to systemic acidosis, but this is not usually experienced in the dosage recommended.

BENZOTHIADIAZINE ("THIAZIDES") AND RELATED HETEROCYCLIC COMPOUNDS

In a study of aromatic sulfonamides as diuretics, Novello and Sprague[21] observed an unexpected high order of activity in the benzene-1,3-disulfonamides, and that activity was enhanced by certain substituents on the ring; chloro, amino and acylamino groups gave a marked increase in the activity, as did the methyl substituent. Higher alkyl groups decreased activity. 1,4-Disulfonamides were less active. In a further study of the chemistry of the benzenedisulfonamides it was observed that when the acylamino group occupied a position *ortho* to an unsubstituted sulfamyl (—SO_2NH_2) group, the compound could be cyclized to give a new type of diuretic of still greater interest. Chlorothiazide is formed by ring closure (elimination of water) of 3-chloro-4,6-disulfamylformanilide.

By employing other acylated amines, analogs substituted in the 3-position are obtained.

Chlorothiazide was first introduced in 1958, and since that time several closely related analogs have been released for use. The heterocyclic benzothiadiazines are potent orally effective diuretics. Their potency approaches that of a parenteral organomercurial diuretic. Unlike most diuretic drugs, tolerance is not a problem. The benzothiadiazines are closely related chemically, but it has been observed that minor changes in structure may have a marked influence on activity. Thus, saturation of the thiadiazine ring of chlorothiazide gives dihydrochlorothiazide, which is approximately 10 times more active and less toxic than the parent compound.[22]

Mode of action

The diuretic effect of the benzothiadiazines is due largely to their ability to inhibit the renal tubular reabsorption of sodium and chloride ions and, to a lesser extent, potassium and bicarbonate ions.[23] This action occurs for the most part in the ascending limb of the

3-Chloro-4,6-disulfamylformanilide   Chlorothiazide

loop of Henle and in the upper segment of the distal convoluted tubule (see Fig. 13-1). The thiazides also decrease carbonic anhydrase activity in the distal tubule. Potassium loss produced by the thiazides is greater than that produced by the carbonic anhydrase inhibitors. The inhibition of sodium reabsorption by the thiazides at more proximal sites results in more sodium reaching the aldosterone-mediated sodium-potassium ion exchange site in the distal tubule. The larger amount of sodium reabsorbed at this site increases the ion-exchange secretion and loss of potassium ion.

Evaluation of the benzothiadiazines

The several benzothiadiazines in clinical use differ from the parent compound, chlorothiazide, primarily in their activity, toxicity and duration of action. The following observations, based on accumulated clinical experience have been made[24]: (1) Patients resistant to one benzothiadiazine derivative are likely to be resistant to others. (2) Each drug has a "ceiling dose" above which sodium loss does not increase. However, doses much lower than the "ceiling dose" are usually adequate. (3) The diuretic potency varies markedly, as reflected in the average dose for each of the compounds, which varies from 2 mg. to 500 mg. daily. (4) The doses necessary to produce equivalent loss of sodium and water may also produce comparable loss of potassium and bicarbonate. (5) To avoid hypokalemia the supplemental administration of potassium or concurrent administration of a potassium-sparing diuretic is recommended for all members of the series. Due to the observed increase in the incidence of small-bowel ulceration and stenosis following thiazide and potassium therapy,[25,26] the Food and Drug Administration has ordered certain preparations that contain potassium salts to go on prescription-order status. The regulation applies to any capsule and coated or uncoated tablet that supplies 100 mg. or more of potassium per unit dose, and to liquid preparations that contain potassium salts and supply 20 mg. or more potassium per ml. (6) The side-effects, which may be potentially serious, are common to all of the benzothiadiazines.

The side-effects of the benzothiadiazines may include nausea, anorexia and headaches, hyperuricemia and, in rare instances, leukopenia and rash. The hyperuricemia, which may precipitate attacks of gouty arthritis, is due to the fact that uric acid and the weakly acidic thiazides compete for the same limited-capacity secretion system. The excretion of uric acid may therefore be reduced following thiazide administration. Gastrointestinal distress, jaundice, photosensitization, acute glomerulonephritis and pancreatitis have been observed. They decrease the responsiveness to the catechol amines and increase the skeletal muscle paralysis of (+)-tubocurarine. The thiazides alter carbohydrate metabolism. Hyperglycemia and glycosuria may occur and precipitate onset of diabetes mellitus.

The thiazides, like other diuretics, are employed primarily for the treatment of edemas of both pathologic or drug-induced origin. They have the important advantage of potentiating the effect of antihypertensive agents such as reserpine, hydralazine and the ganglionic blocking agents, thus serving as useful adjuncts for the treatment of hypertension. Employed alone, the benzothiadiazines are useful antihypertensive agents, but they are used more commonly in combination with other antihypertensives.

The mechanism of the antihypertensive effect is not known but is generally assumed to be related to the diuretic and natriuretic properties. Interestingly, removal of the sulfamyl group gives compounds de-

TABLE 13-4

CHLOROTHIAZIDE AND ANALOGS

Generic Name	Proprietary Name	R	R_1
Chlorothiazide U.S.P.	Diuril	—Cl	H
Benthiazide U.S.P.	Exna, Aquatag	—Cl	$-CH_2-S-CH_2-$ (phenyl)

TABLE 13–5

HYDROCHLOROTHIAZIDE AND ANALOGS

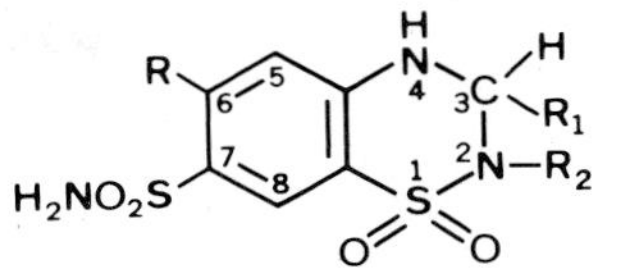

Generic Name	Proprietary Name	R	R$_1$	R$_2$
Hydrochlorothiazide U.S.P.	Hydrodiuril, Esidrix, Oretic	—Cl	—H	—H
Hydroflumethiazide U.S.P.	Saluron, Diucardin	—CF$_3$	—H	—H
Bendroflumethiazide U.S.P.	Naturetin	—CF$_3$	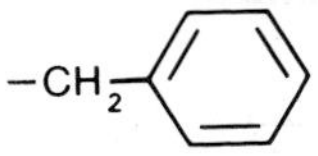 —CH$_2$-phenyl	—H
Trichlormethiazide U.S.P.	Naqua, Metahydrin	—Cl	—CHCl$_2$	—H
Methyclothiazide U.S.P.	Enduron, Aquatensen	—Cl	—CH$_2$Cl	—CH$_3$
Polythiazide U.S.P.	Renese	—Cl	—CH$_2$—S—CH$_2$—CF$_3$	—CH$_3$
Cyclothiazide U.S.P.	Anhydron	—Cl	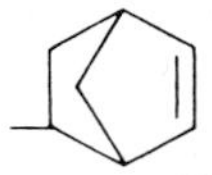 norbornenyl	—H

void of diuretic properties, yet retaining the antihypertensive activity. Thus, 7-chloro-3-methyl-2H-1,2,4,-benzothiadiazine 1,1-dioxide (diazoxide, Hyperstat®, see Chap. 14) has been reported[27] to be an effective antihypertensive agent without diuretic properties; other useful drugs may result from this observation.

The benzothiadiazine diuretics may be divided conveniently into two general types: (1) the analogs of chlorothiazide (Table 13-4) and (2) the analogs of hydrochlorothiazide (Table 13-5). They differ only with respect to the 3,4-positions of the thiadiazine ring, which is unsaturated in the chlorothiazide group.

Structure-activity Relationships of Benzothiadiazines

Optimum diuretic activity has thus far been associated with the following structural features: (1) The benzene ring must have a sulfamyl group, preferably unsubstituted, at position-7 and a halogen or halogenlike group (e.g., CF$_3$) at position-6. (2) Saturation of the 3,4-double bond generally produces increased activity. (3) Lipophilic substituents at position-3 enhance activity, as do lower alkyl groups, such as methyl at position-2. (4) Position-1 of the heterocyclic ring may be

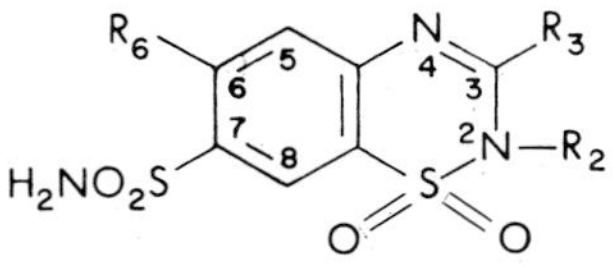

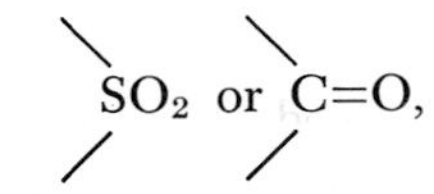

but higher activity is associated with the sulfur heterocycle.

Products

Chlorothiazide U.S.P., Diuril®, 6-chloro-2H-1,2,4-benzothiadiazine-7-sulfonamide 1,1-dioxide.

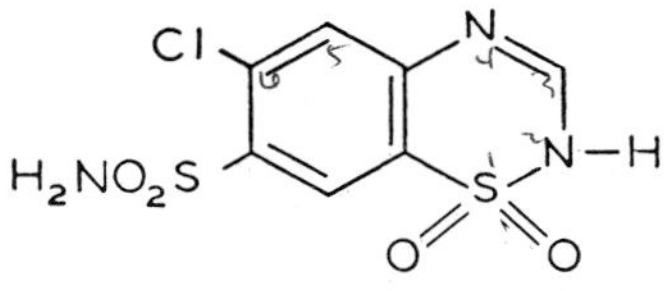

Chlorothiazide

Chlorothiazide was first introduced in 1958 and is the original benzothiadiazine diuretic. It gives primarily a saluretic effect but may also produce significant loss of potassium and bicarbonate. To prevent the development of hypokalemia, adequate amounts of potassium chloride or a potassium-sparing diuretic normally are given with chlorothiazide or with other thiazide diuretics. Hypochloremic alkalosis also may develop, and this may be treated by a temporary discontinuance of the drug or by giving appropriate amounts of ammonium chloride. Chlorothiazide is a potent diuretic and is used for the treatment of all types of edemas. It is also used alone and as an adjunct in the management of hypertension. Unlike other antihypertensive agents, chlorothiazide and other benzothiadiazine analogs lower blood pressure only in hypertensive and not in normotensive individuals. The onset of action is comparatively rapid (approximately 2 hours) and lasts from 6 to 12 hours. The drug normally maintains its effectiveness with prolonged administration.

Chlorothiazide shows no effect on intraocular pressure or on the rate of aqueous humor formation. Unlike the related carbonic anhydrase inhibitors, it is not used in the treatment of glaucoma.

Chlorothiazide Sodium for Injection U.S.P., Diuril® Sodium, is available for parenteral administration.

Benzthiazide U.S.P., Exna®, Aquatag®, 3-[(benzylthio)methyl]-6-chloro-2H-1,2,4-benzothiadiazine-7-sulfonamide 1,1-dioxide. Benzthiazide is a 3-substituted chlorothiazide and is significantly more active than the parent compound. It is reported to be approximately 85 percent as active as hydrochlorothiazide and to give a similar electrolyte excretion pattern. Hypokalemia and other electrolyte imbalances may occur on prolonged use.

Benzthiazide

Hydrochlorothiazide U.S.P., HydroDiuril®, Esidrix®, Oretic®, Thiuretic®, 6-chloro-3,4-dihydro-2H-1,2,4-benzothiadiazine-7-sulfonamide 1,1-dioxide.

Reduction of the 3,4-position of the thiadiazine ring increases activity of the benzothiadiazines by approximately 10 times. Hydrochlorothiazide is the parent compound belonging to the class of reduced benzothiadiazines. Qualitatively, the diuretic and metabolic properties of hydrochlorothiazide are similar to those of chlorothiazide. Although a lower dose is used, at the maximal therapeutic dosage all thiazides are approximately equal in their diuretic potency. Onset of diuresis occurs in 2 hours, the effect peaks at 4 hours, and action persists for 6 to 12 hours. The drug is effec-

Hydrochlorothiazide

tive in edemas associated with congestive heart failure, hepatic cirrhosis, steroid therapy, and various forms of renal dysfunction. It is useful alone in management of mild hypertension, or in combination with other classes of antihypertensive agents in cases of more severe hypertension.

Hydroflumethiazide U.S.P., Saluron®, Diucardin®, 3,4-dihydro-6-(trifluromethyl)-2H-1,2,4-benzothiadiazine-7-sulfonamide 1,1-dioxide. Hydroflumethiazide differs in structure from hydrochlorothiazide by having a trifluoromethyl group substituted for chlorine in the 6-position. The activity and the electrolyte excretion pattern are similar to those of hydrochlorothiazide, and the two drugs are roughly equivalent.

Hydroflumethiazide

Bendroflumethiazide U.S.P., Naturetin®, 3-benzyl-3,4-dihydro-6-(trifluoromethyl)-2H-1,2,4-benzothiadiazine-7-sulfonamide 1,1-dioxide. Bendroflume-

Bendroflumethiazide

thiazide is one of the more potent diuretic and antihypertensive agents available for use in terms of the dose required to produce the ceiling diuretic response characteristic for all thiazides. It incorporates in its structure a reduced thiadiazine ring and a benzyl substitution on the 3-position, both of which enhance the activity. The trifluoromethyl group is substituted for chlorine in the 6-position. The benzyl substitution on the 3-position enhances activity and gives a longer duration of action (approximately 18 hours), but this may not be important clinically.

Qualitatively, bendroflumethiazide is similar to hydrochlorothiazide. For long-term therapy, potassium chloride or a potassium-sparing diuretic is recommended as a supplement to avoid hypokalemia.

Trichlormethiazide U.S.P., Naqua®, Metahydrin®, 6-chloro-3-(dichloromethyl)-3,4-dihydro-2H-1,2,4-benzothiadiazine-7-sulfonamide 1,1-dioxide. Trichlormethiazide differs in structure from hydrochlorothiazide by the substitution of a dichloromethyl group for hydrogen in the 3-position, which increases the diuretic potency by approximately 10 times. The compound is excreted more slowly than hydrochlorothiazide, but the difference in duration of action has not been shown to be clinically important. Similar diuretic responses are reported following the administration of 8 mg. of trichlormethiazide and 75 mg. of hydrochlorothiazide.

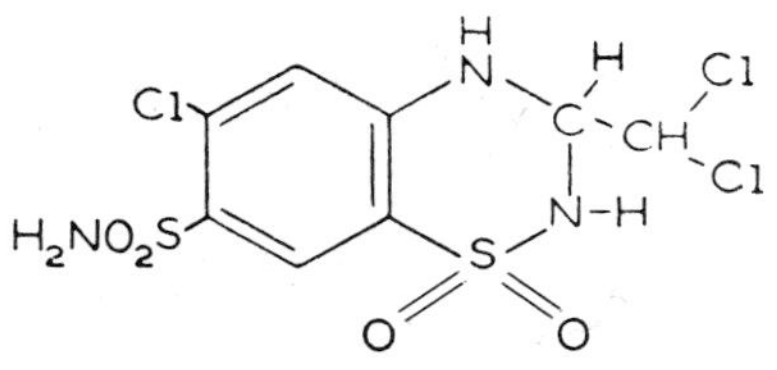

Trichlormethiazide

Methyclothiazide U.S.P., Enduron®, Aquatensen®, 6-chloro-3-(chloromethyl)-3,4-dihydro-2-methyl-2H-1,2,4-benzothiadiazine-7-sulfonamide 1,1-diox-

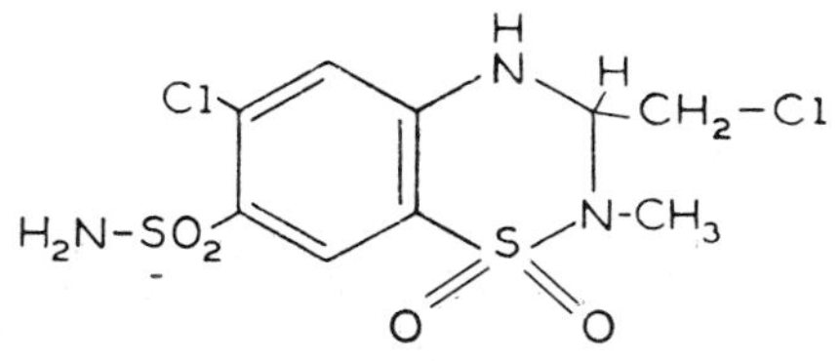

Methyclothiazide

ide. Methyclothiazide is the first of the benzothiadiazines to be substituted in the 2-position. It is well

absorbed and develops a diuretic response within 2 hours following administration and a maximum response at about 6 hours. The diuretic response continues for 24 hours or more; therefore, a continuous therapeutic effect may be obtained from a single daily dose. The predominant effects are diuresis, chloruresis and natriuresis. Urinary pH is not significantly altered.

Methyclothiazide is a potent oral diuretic, and, like other benzothiadiazines, it potentiates the effects of ganglionic blocking and other antihypertensive agents.

Polythiazide U.S.P., Renese®, 6-chloro-3,4-dihydro-2-methyl-3-[[(2,2,2-trifluoroethyl)thio]methyl]-2H-1,2,4-benzothiadiazine-7-sulfonamide 1,1-dioxide.

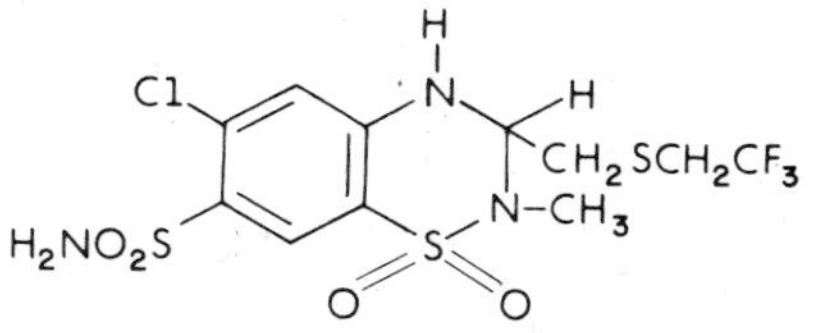

Polythiazide

The principal effect of polythiazide is on the renal excretion of sodium and chloride with a lesser effect on the excretion of potassium and bicarbonate. Substitution of both a methyl group in the 2-position and a trifluoroethylthiomethyl group in the 3-position results in an equivalent increase in urinary sodium excretion at about one-tenth the dose of hydrochlorothiazide. The drug is excreted slowly, due to binding by plasma proteins and reabsorption by the distal tubules. This may be responsible for its long duration of action. Sodium excretion levels have been reported to be elevated 72 hours after administration of 4 mg. of polythiazide.

Cyclothiazide U.S.P., Anhydron®, 6-chloro-3,4-dihydro-3-(5-norbornen-2-yl)-2H-1,2,4-benzothiadiazine-

Cyclothiazide

7-sulfonamide 1,1-dioxide. Cyclothiazide is a potent orally effective diuretic with a lipophilic terpene substituent in the 3-position of the benzothiadiazine ring.

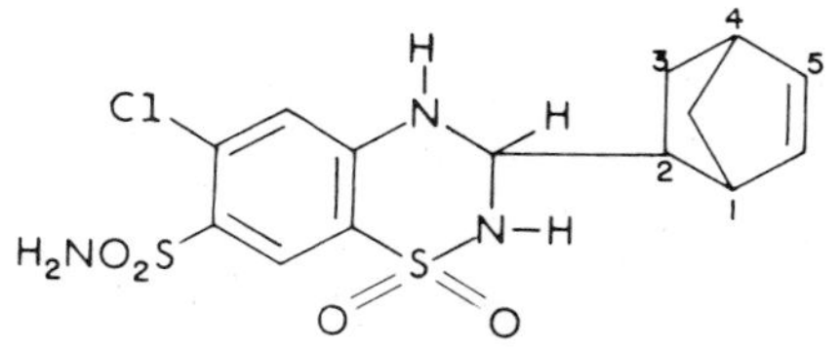

TABLE 13–6

THIAZIDE DIURETICS

Name Proprietary Name	Preparations	Category	Usual Adult Dose*	Usual Dose Range*	Usual Pediatric Dose*
Chlorothiazide U.S.P. Diuril	Chlorothiazide Oral Suspension U.S.P. Chlorothiazide Tablets U.S.P. Chlorothiazide Sodium for Injection U.S.P.	Diuretic	500 mg. to 1 g. once or twice daily I.V., 500 mg. to 1 g. once or twice daily	500 mg. to 2 g. daily	10 mg. per kg. of body weight or 300 mg. per square meter of body surface, twice daily
Benzthiazide U.S.P. Exna, Aquatag	Benzthiazide Tablets U.S.P.	Diuretic; antihypertensive		Diuretic—initial, 50 to 200 mg. daily; maintenance, 50 to 150 mg. daily; antihypertensive—initial, 25 to 100 mg. twice daily; maintenance, adjust to the response of the patient with a maximal dose of 50 mg. 3 times daily	
Hydrochlorothiazide U.S.P. HydroDiuril, Esidrix, Oretic, Thiuretic	Hydrochlorothiazide Tablets U.S.P.	Diuretic	25 to 100 mg. once or twice daily		1 mg. per kg. of body weight or 30 mg. per square meter of body surface twice daily
Hydroflumethiazide U.S.P.	Hydroflumethiazide Tablets U.S.P.	Antihypertensive; diuretic	25 to 100 mg. once or twice daily	1 mg. per kg. of body weight or 30 mg. per square meter of body surface once daily	
Bendroflumethiazide U.S.P. Naturetin	Bendroflumethiazide Tablets U.S.P.	Diuretic; antihypertensive		Diuretic—initial, 5 to 20 mg. daily; maintenance, 2.5 to 5 mg. daily; antihypertensive—initial, 5 to 20 mg. daily; maintenance, 2.5 to 15 mg. daily	
Trichlormethiazide U.S.P. Naqua, Metahydrin	Trichlormethiazide Tablets U.S.P.	Diuretic; antihypertensive		Initial, 1 to 4 mg. twice daily, then once every other day; or once daily for 3 to 5 days each week	
Methyclothiazide U.S.P. Enduron	Methyclothiazide Tablets U.S.P.	Diuretic; antihypertensive	2.5 to 10 mg. once daily, once every other day, or once daily for 3 to 5 days each week		50 to 200 μg. per kg. of body weight or 1.5 to 6 mg. per square meter of body surface once daily
Polythiazide U.S.P. Renese	Polythiazide Tablets U.S.P.	Diuretic; antihypertensive	1 to 4 mg. once daily, once every other day, or once daily for 3 to 5 days each week		20 to 8 μg. per kg. of body weight or 500 μg. per square meter of body surface once daily
Cyclothiazide U.S.P. Anhydron	Cyclothiazide Tablets U.S.P.	Diuretic; antihypertensive		Diuretic—initial, 1 to 2 mg. daily; maintenance, 1 to 2 mg. every other day or 2 or 3 times weekly; antihypertensive—2 mg. 1 to 3 times daily	

* See *U.S.P. D.I.* for complete dosage information.

QUINAZOLINONE DERIVATIVES

Quinethazone U.S.P., Hydromox®, 7-chloro-2-ethyl-1,2,3,4 - tetrahydro - 4 - oxo - 6 - quinazolinesulfonamide. Quinethazone differs structurally from the benzothiadiazines in having the ring sulfone (S-dioxide) replaced with the carbonyl group. The drug is a potent, long-acting, orally effective diuretic. The compound has the same order of potency as hydrochlorothiazide, with a duration of action between 18 and 24 hours.

Quinethazone acts like the structurally related thiazides by inhibition of the mechanisms for sodium and chloride reabsorption in the proximal tubule, and to a lesser extent in subsequent renal tubular regions. Like the thiazides, excess potassium may be lost due to increased exchange of potassium and sodium in the distal tubule. As with the thiazides, increases of serum uric acid and precipitation of gout may occur, as well as decreased glucose tolerance, hyperglycemia, and the aggravation or initiation of diabetes mellitus.

Quinethazone

Usual dose range—50 to 200 mg. daily.

Occurrence
Quinethazone Tablets U.S.P.

Metolazone, Zaroxolyn®, Diulo®, 7-chloro-2-methyl-3-*o*-tolyl-1,2,3,4-tetrahydro-4-oxo-6-quinazolinesulfonamide.

Metolazone

Metolazone, like quinethazone, is a quinazolinone derivative which replaces the sulfonyl, —SO$_2$—, group of the heterocyclic ring of the benzothiadiazines with the —CO— group. The presence of the highly lipophilic methyl and *o*-tolyl substituents greatly enhances diuretic potency. The drug was introduced in 1974 as an orally effective diuretic, which closely resembles the benzothiadiazines in its mode of action, therapeutic applications and side-effects. Diuresis begins within one hour of oral administration, peaks at two hours, and may persist for 12 to 24 hours. The prolonged action is due to protein binding and enterohepatic recycling. The oral dose, given once daily, ranges from 2.5 to 20 mg.

SULFAMYL BENZOIC ACID DERIVATIVES AND RELATED COMPOUNDS

All of the useful benzenesulfonamide diuretics, including the thiazides and related bicyclic compounds such as quinethazone, have a chlorine atom or trifluoro-

methyl group (a pseudohalogen) in the position *ortho* to the sulfamyl (—SO$_2$NH$_2$) group. In addition, they possess an electronegative group, such as —CO— or —SO$_2$—, *meta* to the sulfamyl group or in this position as part of a condensed ring. Furosemide, a 4-chloro-3-sulfamylbenzoic acid derivative, fits this general description, as do the structurally related 4-chloro-3-sulfamylbenzenecarboxamides, xipamide and clorexolone, and likewise the 4-chloro-3-sulfamyldiphenylketone, chlorthalidone. However, it has been shown[28] that the *ortho* halogen may be replaced by a wide variety of lipophilic substituents, with retention or enhancement of diuretic activity. Bumetanide, said to be many times more active than furosemide, is such a 4-phenoxy-3-sulfamylbenzoic acid.

Furosemide U.S.P., Lasix®, 4-chloro-N-furfuryl-5-sulfamoylanthranilic acid.

Furosemide

Furosemide is a potent high-ceiling saluretic agent that produces a rapid diuretic response of comparatively short duration (6 to 8 hours). It inhibits the reabsorption of sodium throughout the renal tubules, including the loop of Henle (a "loop diuretic"), which may account for its high potency and effectiveness in cases of reduced glomerular filtration in which the thiazides and other diuretics fail. In ceiling dosages furosemide shows 8 to 10 times the saluretic effect of the thiazides. Like the thiazides, furosemide promotes potassium excretion and is commonly used with potassium supplementation or a potassium-sparing diuretic. Other side-effects may include hypochloremic alkalosis, hyperuricemia, and hyperglycemia. Furosemide has a blood-pressure-lowering effect similar to that of the thiazides. In addition to the 20- and 40-mg. tablets for oral use, the drug is available as a sterile solution in 2-ml. ampules, each containing 20 mg.

Usual dose—20 to 80 mg. once daily initially.
Usual dose range—Up to 600 mg. daily.

Occurrence
Furosemide Injection U.S.P.
Furosemide Tablets U.S.P.

Xipamide is a new antihypertensive, natriuretic, kaliuretic, nonloop diuretic currently under clinical investigation. It is reported to have no effect on the proximal tubules of the kidney[29] and yet has a diuretic potency equal to that of furosemide. The time courses of the diuretic, natriuretic and kaliuretic effects are

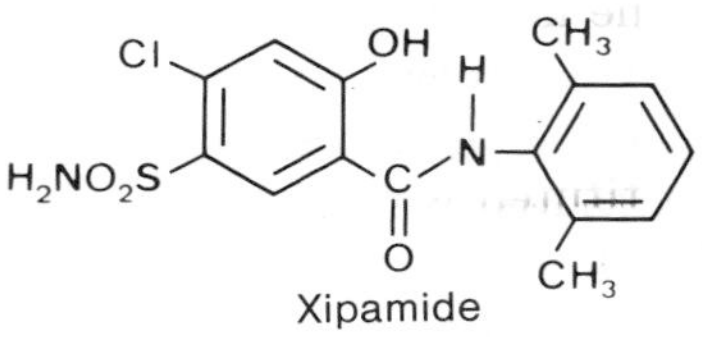

Xipamide

quite similar to hydrochlorthiazide but differ from those of furosemide. Xipamide has a slower onset of action and a longer duration. The peak diuretic effect is given in 4 to 6 hours and the duration of action is 12 hours or more.

Clorexolone, Nefrolan®, 5-chloro-2-cyclohexyl-1-oxo-6-sulfamylisoindoline. Clorexolone is a potassium-sparing long-acting diuretic, chemically related to chlorthalidone. Its electrolyte excretion pattern is

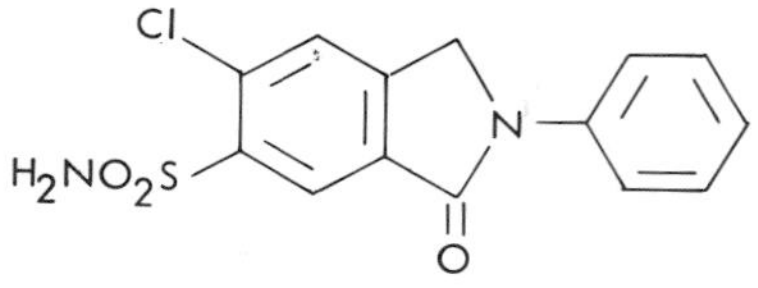

Clorexolone

similar to that of the thiazides, except that at therapeutic dosage levels of 25 to 100 mg. per day it produces a pronounced diuresis without causing hypokalemia. The natriuretic effect persists for more than 48 hours. Clorexolone increases the blood serum uric acid levels but appears to have no significant effect on carbohydrate metabolism and is, therefore, less likely to cause hyperglycemia.[13] The compound, which is currently under clinical study, is approximately two and one-half times more active than hydrochlorothiazide.[30]

Chlorthalidone U.S.P., Hygroton®, 2-chloro-5-(1-hydroxy-3-oxo-1-isoindolinyl)benzenesulfonamide. Chlorthalidone contains a sulfamyl and carboxamide group in each ring of the benzophenone system (II), but exists primarily in the tautomeric lactam form (I), in which the heterocyclic nucleus may be named as an isoindoline or as a phthalimidine. Chlorthalidone is a potent, long-acting, orally effective diuretic and antihypertensive agent. The compound is unique in that it is the only diuretic making use of the phthalimidine ring system but is clearly related structurally to other diuretic sulfonamides. Average therapeutic doses are reported to give primarily a saluretic effect with minimal loss of potassium and bicarbonate. The mode of action of chlorthalidone is not established, but the electrolyte excretion pattern is similar to that given by the benzothiadiazines. Chlorthalidone has no effect on either renal circulation or glomerular filtration, and the diuretic response is believed to be due to interfer-ence with the renal tubular reabsorption of sodium and chloride, thereby promoting loss of salt and water. Compared on a weight basis (orally), it is 1.8 times as potent as meralluride intramuscularly and gives a duration of action up to 60 hours. The compound is concentrated in the kidney, and a large portion of the drug is eliminated unchanged.

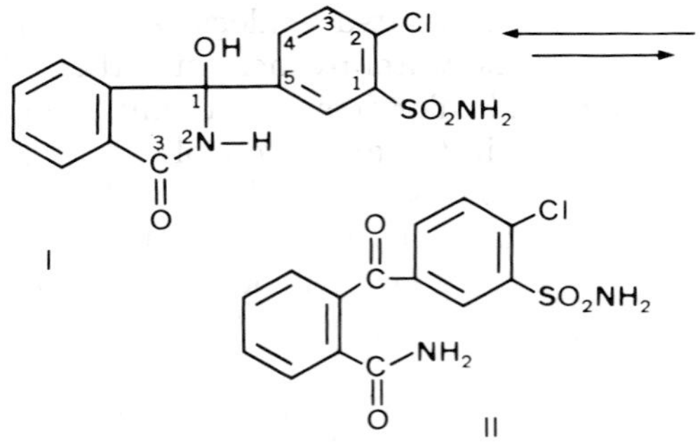

Chlorthalidone

Usual dose—25 to 100 mg. once daily, once every other day, or once daily for 3 days out of a week.

Usual pediatric dose—2 mg. per kg. of body weight or 60 mg. per square meter of body surface, 3 times a week.

Occurrence
Chlorthalidone Tablets U.S.P.

Bumetanide, Burinex®, 3-sulfamyl-4-phenoxy-5-*n*-butylaminobenzoic acid.

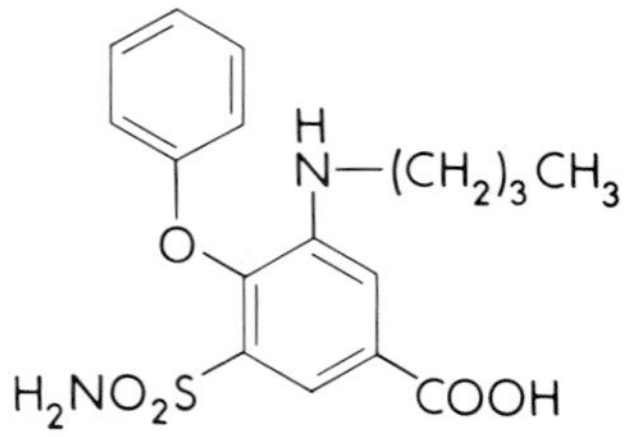

Bumetanide

Bumetanide is a highly potent short-acting diuretic which resembles the "loop diuretics" furosemide and ethacrynic acid in its inhibition of sodium and chloride reabsorption in the ascending limb of the loop of Henle, in addition to sites in the proximal tubule. The diuretic effect is maximal at 2 hours after oral administration and complete by 4 hours. Intravenous administration produces a maximal diuresis within 30 minutes. The usual oral dose of bumetanide, which is currently under clinical study, is 1 mg. once or twice daily; the usual intravenous dose is 0.5 mg.

ENDOCRINE ANTAGONISTS

Aldosterone is a potent antidiuretic hormone secreted by the adrenal cortex. In congestive heart failure, nephrosis and other pathologies associated with edema, there is increased secretion of aldosterone which is believed to be responsible for the retention of salt and water. The biologically active corticosteroids, such as cortisone, hydrocortisone, deoxycorticosterone, aldosterone, and closely related analogs tend to increase retention of salt and water by increasing the reabsorption of sodium and chloride and to promote the excretion of potassium in the distal convoluted tubule (see Fig. 13-1). The corticosteroids differ widely in their activity to produce retention of salt and water, but the most potent of the compounds is aldosterone, which is at least 1,000 times more active than hydrocortisone and is believed to play an important role in maintaining the normal electrolyte balance.

Aldosterone occurs as a hemiacetal in equilibrium with a hydroxy aldehyde form, as shown in the following structures:

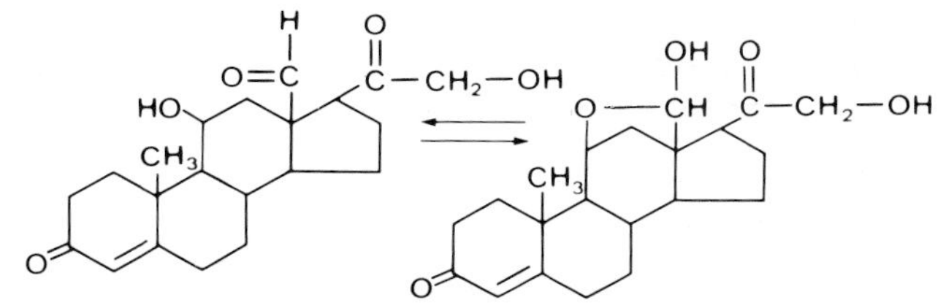

Hydroxy Aldehyde Form   Hemiacetal Form

Aldosterone

Antagonists of aldosterone decrease the amount of sodium and chloride reabsorbed by the renal tubules, thereby promoting diuresis—a process of competitive inhibition. Aldosterone antagonists in combination with other diuretics are able to restore the electrolyte balance and are potentially most useful agents. Spironolactone is the only aldosterone inhibitor available for use at the present time.

Spironolactone U.S.P., Aldactazide®, 17-hydroxy-7α-mercapto-3-oxo-17α-pregn-4-ene-21-carboxylic acid γ-lactone 7-acetate.

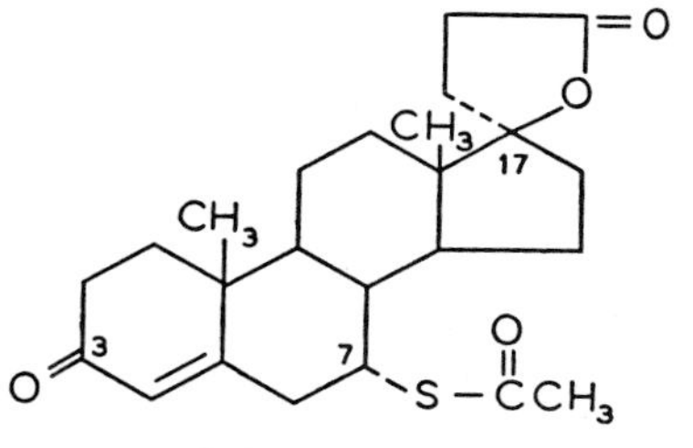

Spironolactone

Spironolactone is a synthetic steroid in which the side chain on the C-17 carbon of 4-androsten-3-one is replaced by a 5-membered lactone ring, and the 7α-position is substituted with an acetylthio group. Spironolactone has some structural similarity to aldosterone and blocks the latter's effect of promoting reabsorption of sodium and loss of potassium in the distal renal tubules. Unlike many other diuretics, spironolactone does not produce loss of potassium. The compound is sometimes effective when used alone, but a slow onset of action (3 to 7 days) and variable diuresis cause it to be used normally in combination with other diuretics, such as the mercurials and the benzothiadiazines. Spironolactone has been shown to be tumor-producing in chronic toxicity studies in rats.

Usual dose—25 mg. 2 to 4 times daily.

Usual dose range—50 to 400 mg. daily.

Usual pediatric dose—20 to 60 mg. per square meter of body surface, 3 times daily.

Occurrence
Spironolactone Tablets U.S.P.

MISCELLANEOUS COMPOUNDS UNDER INVESTIGATION

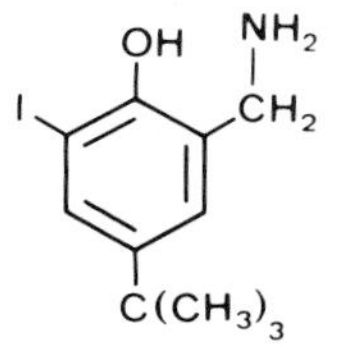

This unique class of compounds has been extensively studied for saluretic activity. From many structurally related compounds the substituted 2-aminomethylphenol MK-447 was selected as the most promising for pharmacological and clinical evaluation. It is a potent high-ceiling diuretic agent with significant antihypertensive and anti-inflammatory activity.[31] MK-447 is reported to be more effective than furosemide as a diuretic agent. Its antihypertensive activity is reduced by indomethacin which suggests the hypotensive effect may be prostaglandin-mediated. Depression of PGG_2 levels has been suggested as a possible explanation for its anti-inflammatory effect. MK-447 has low toxicity and a moderate duration of action. Since it has structural similarity to that of thyroxine, it was examined for stability and found not to undergo deiodination either in vitro or in vivo. MK-447 is rapidly absorbed and excreted. In humans the major metabolite is the N-glucuronide.

During recent years a large number of new compounds have been evaluated for their diuretic action, and a small percentage of these have been introduced as therapeutic agents. The ideal diuretic agent, one that effectively promotes sodium and water excretion without producing an electrolyte imbalance or disturbing carbohydrate metabolism and uric acid elimination, remains to be discovered. However, advances in renal physiology and biochemistry have led to a better understanding of the basic problem, and advances in pharmacology and medicinal chemistry have increased understanding of the modes of diuretic action and the rational design of more selective agents. Although many highly effective diuretics are now available, new agents which more nearly approach the ideal may be anticipated.

REFERENCES

1. Sprague, J. M.: Diuretics *in* Rabinowitz, J. L., and Myerson, R. M. (eds.): Topics in Medicinal Chemistry, vol. 2, New York, Interscience, John Wiley & Sons, 1968.
2. Early, L. E.: New Eng. J. Med. 276:966, 1967.
3. Kleeman, C. R., and Fichman, M. P.: New Eng. J. Med. 277:1300, 1967.
4. Early, L. E., and Orloff, J.: Ann. Rev. Med. 15:149, 1964.
5. Sprague, J. M.: Ann. N.Y. Acad. Sci. 71:328, 1958.
6. Kessler, R. H., Lozano, R., and Pitts, R. F.: J. Clin. Invest. 36:656, 1957.
7. Friedman, H. L.: Ann. N. Y. Acad. Sci. 65:461, 1957.
8. Vander, A. J., Malvin, R. L., Wilde, W. S., and Sullivan, L. P.: Am. J. Physiol. 195:558, 1958.
9. Mudge, G. H., and Weiner, I. N.: Ann. N.Y. Acad. Sci. 71:344, 1958.
10. Schultz, E. M., *et al.*: J. Med. Pharm. Chem. 5:660, 1962.
11. Cragoe, E. J., Jr., *et al.*: J. Med. Chem. 18:225, 1975.
12. Early, L. E., *et al.*: J. Clin. Invest. 43:1160, 1964.
13. Hutcheon, D. E.: Am. J. M. Sci. 253:620, 1967.
14. Kirkendall, W. M., and Stern, J. H.: Am. J. Cardiol. 22:162, 1968.
15. Schwartz, F. D.: Lancet 1:77, 1969.
16. Thullier, G., *et al.*: Eur. J. Med. Chem. 9:625, 1975.
17. Horstman, K., *et al. in* E.J. Cragoe, Jr. (ed.): Diuretic Agents, p. 125, Amer. Chem. Soc. Symposium Series No. 83, Washington, D.C., 1978.
18. Strauss, M. B., and Southworth, H.: Bull. Johns Hopkins Hosp. 63:41, 1938.
19. Mann, T., and Keilin, D.: Nature 146:164, 1940.
20. Roblin, R. O., Jr., and Clapp, J. W.: J. Am. Chem. Soc. 72:4289, 1950.
21. Novello, F. C., and Sprague, J. M.: J. Am. Chem. Soc. 79:2028, 1957.
22. Friend, D. G.: Clin. Pharmacol. Therap. 1:5, 1960.
23. Beyer, K. H.: Ann. N.Y. Acad. Sci. 71:363, 1958.
24. The Medical Letter 2 (No. 15):57, 1960; 3 (No. 9):36, 1961.
25. Boley, S. J., *et al.*: J.A.M.A. 192:93–98, 1965.
26. Abbruzzese, A. A., and Gooding, C. A.: J.A.M.A. 192:111–112, 1965.
27. Rubin, A. A., *et al.*: Science 133:2067, 1961.
28. Feit, P.W.: J. Med. Chem. 14:432, 1971.
29. Leary, W.P. and Asmal, A.C.: Curr. Therap. Res. 27:16, 1980.
30. Russell, R. R., *et al.*: Clin. Pharmacol. Therap 10:265, 1969.
31. Smith, R.L.: Ann. Reports Med. Chem. 13:64, 1978.

SELECTED READINGS

Cragoe, E.J., Jr. (ed): Diuretic Agents, Amer. Chem. Soc. Symposium Series 83, Washington D.C., 1978.

de Stevens, George: Diuretics: Chemistry and Pharmacology, Medicinal Chemistry, vol. 1, New York, Academic Press, 1963.

Early, L. E.: Current views on the concepts of diuretic therapy, New Eng. J. Med. 276:966, 1967.

Grollman, A. (ed.): New diuretics and antihypertensive agents, Ann. N.Y. Acad. Sci. 88:771–1020, 1960.

Hess, H.-J.: Diuretic Agents, *in* Cain, C. K. (ed.): Ann. Rep. Med. Chem., 1967, New York, Academic Press, 1968.

Hutcheon, D. E.: The pharmacology of the established diuretic drugs, Am. J. M. Sci. 253:620, 1967.

Kleeman, C. R., and Fickman, M. P.: The regulation of renal water metabolism, New Eng. J. Med. 277:1330, 1967.

Pitts, R. F., *et al.*: Chlorothiazide and other diuretics, Ann. N.Y. Acad. Sci. 71:371–478, 1958.

Sprague, J. M.: Diuretics, *in* Rabinowitz, J. L., and Myerson, R. M. (eds.): Topics in Med. Chem., vol. 2, New York, Wiley-Interscience, 1968.

Topliss, J. G.: Diuretics, *in* Burger, A. (ed.): Medicinal Chemistry, ed. 3, p. 976, New York, Wiley-Interscience, 1970.

14

cardiovascular agents

George H. Cocolas

Cardiovascular agents are used for their action on the heart or other parts of the vascular system to modify the total output of the heart or the distribution of blood to certain parts of the circulatory system. These include cardiotonic drugs (see Chap. 18); antihypertensive; antianginal agents and vasodilators; drugs that modify cardiac rhythm; antihyperlipidermic agents; and sclerosing agents. Other drugs that do not act directly on the cardiovascular system but are of considerable value in the treatment of cardiac disease are the diuretics (see Chap. 13) and anticoagulants.

This chapter is concerned with drugs that have a direct action on the cardiovascular system and drugs that affect certain constituents of blood. The latter include anticoagulants hypoglycemic agents, thyroid hormones, and the antithyroid drugs.

ANTIANGINAL AGENTS AND VASODILATORS

Angina pectoris (angina) may be treated with organic nitrates and β-adrenergic blocking agents (see Chap. 11) and is characterized by a severe constricting pain in the chest, often radiating from the precordium to the left shoulder and down the arm. The cause of angina is generally attributed to ischemia of heart muscle as a result of coronary artery disease. The syndrome has been described since 1772, but it was not until 1867 that amyl nitrite was introduced for the symptomatic relief of angina pectoris.[1] It was believed at that time that anginal pain was precipitated by an increase in blood pressure and that the use of amyl nitrite reduced both blood pressure and, concomitantly, the work that was needed to be done by the heart. Later it generally became accepted that nitrites relieved angina pectoris by dilating the coronary arteries and that changes in the work of the heart were of only secondary importance. However, it is now understood that the coronary blood vessels in the atherosclerotic heart are already dilated and that use of ordinary doses of dilator drugs does not significantly increase blood supply to the heart; instead, relief from anginal pain is caused by a reduction of cardiac consumption of oxygen.[2]

Although vasodilators are used primarily in the treatment of angina, a more sophisticated understanding of the hemodynamic response to these agents has broadened their clinical usefulness to other cardiovascular conditions. Because of their ability to reduce peripheral vascular resistance, there is a rapidly growing interest in using organic nitrates to improve cardiac output in some patients with congestive heart failure.

The coronary circulation supplies blood to the myocardial tissues to maintain cardiac function. It is capable of reacting to the changing demands of the heart by dilation of its blood vessels to provide sufficient oxygen and other nutrients and to remove metabolites. Myocardial metabolism is almost exclusively aerobic, which makes blood flow critical to the support of metabolic processes of the heart. This demand is met effectively by the normal heart, because it extracts a relatively large proportion of the oxygen delivered to it by the coronary circulation. The coronary blood flow is strongly dependent upon myocardial metabolism, which in turn is affected by work done by the heart and the efficiency of the heart. The coronary system normally has a reserve capacity that allows it to respond by vasodilation to satisfy the needs of the heart during strenuous activity by the body.

Coronary atherosclerosis, one of the more prevalent cardiovascular diseases, develops with increasing age and may lead to a reduction of the reserve capacity of the coronary system. It most often results in multiple stenoses and makes it difficult for the coronary system to meet adequately the oxygen needs of the heart that occur during physical exercise or emotional duress. The insufficiency of the coronary blood flow (myocardial ischemia) in the face of increased oxygen demand produces angina pectoris.

The principal goal in the prevention and relief of angina is to limit the oxygen requirement of the heart so that the amount of blood supplied by the stenosed arteries is adequate. Nitrate esters such as nitroglycerin lower arterial blood pressure and, in turn, reduce the work of the left ventricle. This action is produced by the powerful vasodilating effect of the nitrates acting directly on the arterial system and, to an even greater extent, on the venous system. The result is a reduction of cardiac filling pressure and ventricular size. This reduces the work required of the ventricle and decreases the oxygen requirements, allowing the coronary system to satisfy the oxygen demands of myocardial tissue and relieve anginal pain.

INTERMEDIARY MYOCARDIAL METABOLISM

Normal myocardial metabolism is aerobic, and the rate of oxygen utilization parallels the amount of adenosine triphosphate (ATP) synthesized by the cells.[3] Free fatty acids are the principal fuel for myocardial tissue, but lactate, acetate, acetoacetate, and glucose are also oxidized to CO_2 and water. A large volume of the myocardial cell consists of mitochondria in which two carbon fragments from free fatty-acid breakdown are metabolized through the Krebs cycle. The reduced flavin and nicotinamide dinucleotides formed by this metabolism are reoxidized by the electron transport chain because of the presence of oxygen (Fig. 14–1). In the hypoxic or ischemic heart, the lack of oxygen inhibits the electron transport-chain function and causes an accumulation of reduced flavin and nicotinamide coenzymes. As a result, fatty acids are converted to lipids rather than being oxidized. To compensate for this, glucose utilization and glycogenolysis increase, but the resulting pyruvate cannot be oxidized. A great loss of efficiency occurs as a result of

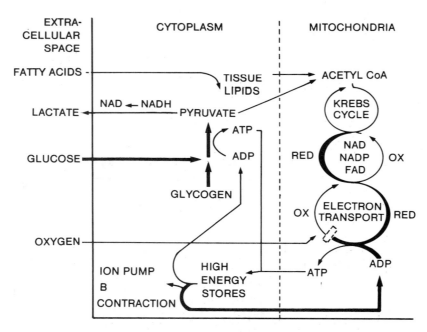

FIG. 14-1. *Energy pathways in the hypoxic myocardium. With oxygen deprivation there is relative blockade of the electron transport system, so that oxidative phosphorylation is inhibited, and high-energy phosphate stores may decline. Reduced flavin and adenine coenzymes accumulate so that citric acid cycle degradation of acetyl-CoA cannot proceed at a normal rate. Under these conditions, the utilization of glucose and glycogen increases, and pyruvate accumulates. The mitochondrial and cytoplasmic NAD to NADH ratios decline, and the lactate–pyruvate reaction becomes reversed. Fatty acids are deposited in tissue lipids. Increased glycolysis provides an inadequate compensatory mechanism for the formation of ATP. The size of labels and prominence of arrows in this figure are meant only to signify relative changes and do not have precise quantitative significance. By permission from Scheuer, J., Am. J. Cardiol. 19:385, 1967.*

the change of myocardial metabolism from aerobic to anaerobic pathways. Normally 36 moles of ATP are formed from the oxidation of 1 mole of glucose, but only 2 moles are formed from its glycolysis. This great loss of high-energy stores during hypoxia thus limits the functional capacity of the heart during stressful conditions and is reflected by the production of anginal pain.

MECHANISM OF SMOOTH MUSCLE VASODILATION

Relaxation of smooth muscle may be the result of one or more biological mechanisms. There appear to be receptors on smooth-muscle membranes which are activated by β-adrenergic agonists, e.g., isoproterenol; these, in turn, activate adenyl cyclase to increase the level of cyclic adenoside monophosphate (c-AMP) in the cell. The increased level of c-AMP is associated with smooth-muscle relaxation. Drugs such as papaverine (see Chap. 12) and theophylline (see Chap. 13) also function to relax smooth muscle through a c-AMP-mediated mechanism. These drugs inhibit phosphodiesterase and reduce the rate of conversion of c-AMP to 5'-AMP in the cell. The mechanism for vasodilation by organic nitrates proceeds according to the scheme in Figure 14–2. Organic nitrates react with reduced thiol groups in the vascular smooth-muscle

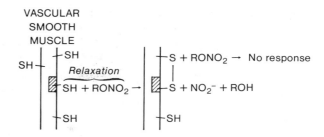

FIG. 14-2. *Schematic diagram of the reaction of organic nitrates with thiol groups in the vascular receptor. Adapted from Needleman, P. and Johnson, E.M. Jr.: J. Pharmacol. Exp. Ther. 184:709, 1973.*

receptor, leading to the formation of a disulfide link and the release of inorganic nitrite. This reaction converts the receptor to the disulfide form, which has a lower affinity for organic nitrates.[4] The organic nitrates have specific receptors on the membrane because these receptors, on continual exposure to nitrates, become tolerant to further action of organic nitrates but not to other vasodilators such as isoproterenol, papaverine, etc. Reversal of tolerance may be achieved by dithiothreitol, apparently by reduction of thiol groups on the membrane. Each of these vasodilatory mechanisms can be blocked by ethacrynic acid. Ethacrynic acid also can block the vasodilating effects of sodium nitroprusside. The ability of ethacrynic acid to block various types of structurally different

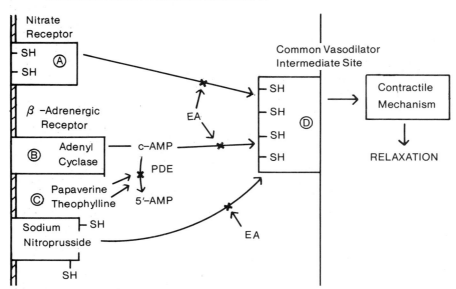

FIG. 14-3. *Interrelationship between vasodilators and tissue components. Specific receptor sites for nitrates (A), beta-adrenergic agonists (B), and sodium nitroprusside (C) mediate vasodilation through a common vasodilator site (D). Papaverine and theophylline inhibit phosphodiesterase conversion of C-AMP to 5'-AMP. Ethacrynic acid (EA) blocks vasodilation by nitrates, papaverine, theophylline and sodium nitroprusside possibly at the common vasodilator site. Adapted from Needleman, P., et al.: J. Pharmacol. Exp. Ther. 187:329, 1973, © 1973, American Society for Pharmacology and Experimental Therapeutics.*

vasodilatory agents suggests that this compound may be acting at a common site that affects muscle relaxation[5] (Fig. 14-3).

ESTERS OF NITROUS AND NITRIC ACIDS

Inorganic acids, like organic acids, will form esters with an alcohol. Pharmaceutically, the important ones are the bromide, chloride, nitrite and nitrate. The chlorides and the bromides are thought of conventionally as chloro- or bromo- compounds. Hydrogen cyanide forms the organic cyanides or nitriles (R—CN). Sulfuric acid forms organic sulfates, of which methyl sulfate and ethyl sulfate are the most common.

Nitrous acid, HNO_2, esters may be formed readily from an alcohol and nitrous acid. The usual procedure is to mix sodium nitrite, sulfuric acid and the alcohol. Organic nitrites are generally very volatile liquids that are only slightly soluble in water but soluble in alcohol. Preparations containing water are very unstable, due to hydrolysis.

The organic nitrates and nitrites and the inorganic nitrites have their primary utility in the prophylaxis and treatment of angina pectoris. They have a more limited application in treating asthma, gastrointestinal spasm and certain cases of migraine headache. Nitroglycerin (glyceryl trinitrate) was one of the first members of this group to be introduced into medicine and still remains an important member of the group. By varying the chemical structure of the organic nitrates, differences in speed of onset, duration of action and potency can be obtained. It is interesting to note that although the number of nitrate ester groups may vary from 2 to 6 or more, depending on the compound, there is no direct relationship between the number of nitrate groups and the level of activity. It appears that the higher the oil:water partition coefficient, the greater the potency. The orientation of the groups within the molecule may also affect potency.

TABLE 14-1

RELATION BETWEEN SPEED AND DURATION OF ACTION OF SODIUM NITRITE AND CERTAIN INORGANIC ESTERS

Compound	Action Begins (Minutes)	Maximum Effect (Minutes)	Duration of Action (Minutes)
Amyl Nitrite	$1/4$	$1/2$	1
Nitroglycerin	2	8	30
Isosobide Dinitrate	3	15	60
Sodium Nitrite	10	25	60
Erythrityl Tetranitrate	15	32	180
Pentaerythritol Tetranitrate	20	70	330

MECHANISM OF ANTIANGINAL ACTION OF NITRATES

The mechanism of action of short-acting sublingual nitrates in the relief of angina pectoris is complex. Although the short-acting sublingual nitrates relax vascular smooth muscle and dilate the coronary arteries of normal humans, there is little improvement of coronary blood flow when these chemicals are administered to individuals with coronary-artery disease. Nitroglycerin is an effective antianginal agent because it causes redistribution of coronary blood flow to the ischemic regions of the heart and also reduces myocardial oxygen demand. This latter effect is produced by a reduction of venous tone due to the nitrate vasodilating effect and a pooling of blood in the peripheral veins that results in a reduction in ventricular volume, stroke volume, and cardiac output. It also causes reduction of peripheral resistance during myocardial contractions. The combined vasodilatory effects cause a decrease in cardiac work and reduce oxygen demand.

METABOLISM OF ORGANIC NITRATES

Organic nitrates are metabolized rapidly on oral administration. The metabolism of organic nitrate occurs in the presence of reduced glutathione (GSH) and is catalyzed by hepatic glutathione–organic nitrate reductase. One molecule of nitroglycerin reacts with two GSH to release one inorganic nitrate ion to form either 1,2- or 1,3-glyceryl dinitrate. This is further metabolized in the liver by the same enzyme but at a slower rate to form a glyceryl mononitrate, which is the major urinary metabolite. The outcome of this metabolism is the conversion of a potent lipid-soluble vasodilator substance to a water-soluble metabolite that has lower biological potency and is readily excreted in the urine.

Buccal absorption reduces the immediate hepatic destruction of the organic nitrates because only 15 percent of the cardiac output is delivered to the liver; this allows a transient but effective circulating level of the intact organic nitrate before it is inactivated.[6]

PRODUCTS

Amyl Nitrite U.S.P., isopentyl nitrite. Amyl nitrite [$(CH_3)_2CHCH_2CH_2ONO$], is a mixture of isomeric amyl nitrites but is principally isoamyl nitrite. It may be prepared from amyl alcohol and nitrous acid by several procedures. It usually is dispensed in ampul form and used by inhalation, or orally in alcohol solution. Currently, it is recommended in treating cyanide poisoning; although not the best, it does not require intravenous injections.

Amyl nitrite is a yellowish liquid having an ethereal odor and a pungent taste. At room temperature it is volatile and inflammable. Amyl nitrite vapor forms an explosive mixture in air or oxygen. Inhalation of the vapor may involve definite explosion hazards if a source of ignition is present, as both room and body temperatures are within the flammability range of amyl nitrite mixtures with either air or oxygen. It is nearly insoluble in water but is miscible with organic solvents. The nitrite also will decompose into valeric acid and nitric acid.

Nitroglycerin, glyceryl trinitrate, is the trinitrate ester of glycerol and is official in tablet form in the *U.S.P.* It is prepared by carefully adding glycerin to a mixture of nitric and fuming sulfuric acids. This reaction is exothermic and the reaction mixture must be cooled to 10° to 20°.

The ester is a colorless oil with a sweet, burning taste. It is only slightly soluble in water, but it is soluble in organic solvents.

$$\begin{array}{ccc}
CH_2OH & & CH_2ONO_2 \\
| & \xrightarrow{3HNO_3} & | \\
CHOH & \xrightarrow{H_2SO_4} & CHONO_2 \quad + \quad 3H_2O \\
| & & | \\
CH_2OH & & CH_2ONO_2
\end{array}$$

Nitroglycerin

Nitroglycerin is used extensively as an explosive in dynamite. A solution of the ester, if spilled or allowed to evaporate, will leave a residue of nitroglycerin. To prevent an explosion, the ester must be decomposed by the addition of alkali. It has a strong vasodilating action and, since it is absorbed through the skin, is prone to cause headaches among workers associated with its manufacture. In medicine, it has the action typical of nitrites but its action is developed more slowly and is of longer duration. Of all the known coronary vasodilator drugs, nitroglycerin is the only one capable of stimulating the production of coronary collateral circulation and the only one able to prevent experimental myocardial infarction by coronary occlusion.

Previously, the nitrates were thought to be hydrolyzed and reduced in the body to nitrites, which then lowered the blood pressure. However, this is not the case.[2] The action depends on the intact molecule.

Nitroglycerin-tablet instability was reported in molded sublingual tablets.[8] The tablets, although uniform when manufactured, lost potency both because of volatilization of nitroglycerin into the surrounding materials in the container and intertablet migration of the active ingredient. Nitroglycerin may be stabilized in molded tablets by incorporating a "fixing" agent such as polyethylene glycol 400 or polyethylene glycol 4000.[9]

Diluted Erythrityl Tetranitrate U.S.P., Cardilate®, erythrol tetranitrate, tetranitrol, is the tetranitrate ester of erythritol and nitric acid, and it is prepared in a manner analogous to that used for nitroglycerin. The result is a solid, crystalline material. This ester also is very explosive and is diluted with lactose or other suitable inert diluents to permit safe handling; it is slightly soluble in water and is soluble in organic solvents.

$$\begin{array}{c}
H \\
| \\
HCONO_2 \\
| \\
HCONO_2 \\
| \\
HCONO_2 \\
| \\
HCONO_2 \\
| \\
H
\end{array}$$

Erythrityl Tetranitrate

Erythrityl tetranitrate requires slightly more time than nitroglycerin to develop its action and this is of longer duration. It is useful where a mild, gradual and prolonged vascular dilation is wanted and is used in the treatment of, and as a prophylaxis against, attacks of angina pectoris and to reduce blood pressure in arterial hypertonia.

Erythrityl tetranitrate produces a reduction of cardiac preload as a result of pooling of blood on the venous side of the circulatory system by its vasodilating action. This action results in a reduction of blood pressure on the arterial side during stressful situations and is an important factor in preventing the precipitation of anginal attacks.

Diluted Pentaerythritol Tetranitrate U.S.P., Peritrate®, Pentritol®, 2,2-*bis*(hydroxymethyl)-1,3-propanediol tetranitrate. This compound is a white, crystalline material with a melting point of 140°. It is insoluble in water, slightly soluble in alcohol and readily soluble in acetone. The drug is a nitric acid ester of the tetrahydric alcohol, pentaerythritol, and is a powerful explosive. For this reason it is diluted with lactose or mannitol or other suitable inert diluents to permit safe handling.

$$\begin{array}{ccc}
O_2NOCH_2 & & CH_2ONO_2 \\
& \diagdown C \diagup & \\
O_2NOCH_2 & & CH_2ONO_2
\end{array}$$

Pentaerythritol Tetranitrate

It relaxes smooth muscle of smaller vessels in the coronary vascular tree. It is used prophylactically to reduce the severity and frequency of anginal attacks. It is usually administered in sustained-release preparations to increase its duration of action. The mechanism of action is similar to that of erythrityl tetranitrate.

Diluted Isosorbide Dinitrate U.S.P., Isordil®, Sorbitrate®, 1,4:3,6-dianhydro-D-glucitol dinitrate, occurs as a white, crystalline powder. Its water-solubility is about 1 mg. per ml.

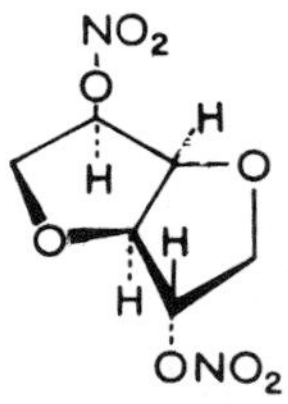

Isosorbide Dinitrate

Isosorbide dinitrate is probably effective in treatment of acute anginal attacks and for prophylaxis when given sublingually. The drug is less likely to be effective when given orally. When given sublingually, the effect begins in about 2 minutes, with a shorter duration of action than when given orally.

MISCELLANEOUS VASODILATORS

Nicotinyl Alcohol Tartrate, Roniacol®, β-pyridyl-carbinol bitartrate or 3-pyridinemethanol tartrate (the alcohol corresponding to nicotinic acid).

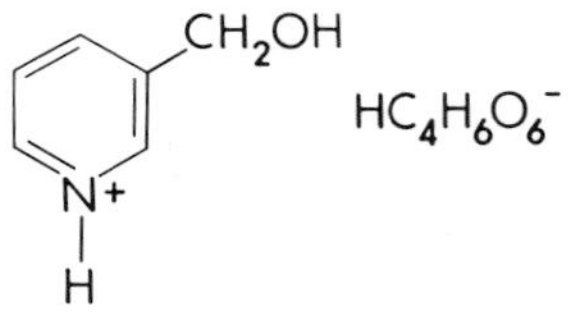

Nicotinyl Alcohol Tartrate

The free amine-alcohol is a liquid having a boiling point of 145°. It forms salts with acids. The bitartrate is crystalline and is soluble in water, alcohol and ether. An aqueous solution has a sour taste, partly due to the bitartrate form of the salt.

In 1950, it was introduced as a vasodilator, following the lead that nicotinic acid is a vasodilator. The action of the drug is peripheral vasodilation similar to that of nicotinic acid. There is a direct relaxing effect on peripheral blood vessels, producing a longer action

with less flushing than does nicotinic acid. It is given orally in tablets or as an elixir. Medicinal use includes the treatment of vascular spasm, Raynaud's disease, Buerger's disease, ulcerated varicose veins, chilblains, migraine, Ménière's syndrome and most conditions requiring a vasodilator.

Another use is in the treatment of dermatitis herpetiformis. This came about because both sulfapyridine and niacin were found to be effective.

The usual dose is 50 to 300 mg.

Dipyridamole, Persantine®, 2,2′,2″,2‴-{[4,8-dipiperidino(5,4-d)pyrimidino-2,6-diyl]dinitrilo}tetraethanol, is used for coronary and myocardial insufficiency. It is a yellow, crystalline powder, with a bitter taste. It is soluble in dilute acids, methanol or chloroform.

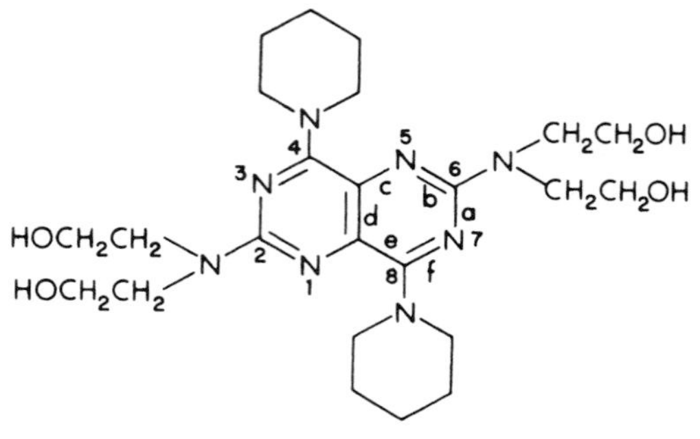

Dipyridamole

Dipyridamole is a long acting coronary vasodilator. Although its vasodilating action is selective on the coronary system its mechanism for causing vasodilitation of the coronary system is not clear. It is indicated for long term therapy of chronic angina pectoris.

The recommended oral dose is 50 mg. 2 or 3 times daily before meals. Optimum response may not be apparent until the third or fourth week of therapy. Dipyridamole is available in 25, 50, and 75 mg. sugar-coated tablets.

Cyclandelate, Cyclospasmol®, 3,5,5-trimethylcyclohexyl mandelate. This compound was introduced in 1956 for use especially in peripheral vascular disease in which there is vasospasm. It is a white to off-white crystalline powder, practically insoluble in water and readily soluble in alcohol and in other organic solvents. Its actions are similar to those of papaverine.

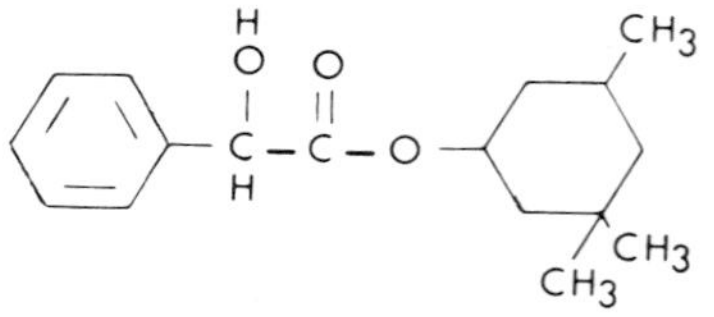

Cyclandelate

TABLE 14-2

ANTIANGINAL AGENTS AND VASODILATORS

Name Proprietary Name	Preparations	Category	Usual Adult Dose*	Usual Dose Range*
Amyl Nitrite U.S.P.	Amyl Nitrite Inhalant U.S.P.	Vasodilator	Inhalation, 300 µl., as required	Up to 10 mg. daily
	Nitroglycerin Sublingual Tablets U.S.P.	Anti-anginal	Sublingual, 150 to 600 µg. as a sublingual tablet repeated at 5-minute intervals as needed for relief of anginal attack	
	Nitroglycerin Extended-Release Capsules U.S.P.		Oral, 2.5, 6.5, or 9.0 mg. as an extended-release capsule every 12 hours, the dosage being increased every 8 hours if needed and tolerated	
	Nitroglycerin Extended-Release Tablets U.S.P.		Oral, 1.3, 2.6, or 6.5 mg. as an extended-release tablet every 12 hours, the dosage being increased to every 8 hours as needed and tolerated	
	Nitroglycerin Ointment U.S.P.		Topical, to the skin, 2.5 to 5 cm. (1 to 2 inches) of ointment as squeezed from the tube, every 3 to 4 hours as needed during the day and at bedtime	Up to 12.5 cm. (5 inches) of ointment as squeezed from the tube, per application
Diluted Erythrityl Tetranitrate U.S.P. *Cardilate*	Erythrityl Tetranitrate Tablets U.S.P. Erythrityl Tetranitrate Chewable Tablets U.S.P.	Vasodilator	Oral, 10 mg. as an oral (or chewable) tablet 4 times daily, the dosage being adjusted as needed or tolerated	Up to 100 mg. daily
Diluted Pentaerythritol Tetranitrate U.S.P. *EI-PETN, Pentritol*	Pentaerythritol Tetranitrate Tablets U.S.P.	Vasodilator	Oral, 10 to 20 mg. 4 times daily, the dosage being adjusted as needed and tolerated	Up to 160 mg. daily
	Pentaerythritol Tetranitrate Extended-Release Capsules U.S.P. Pentaerythritol Tetranitrate Extended-Release Tablets U.S.P.		Oral, 30 to 80 mg. as an extended-release capsule or tablet twice daily	Up to 160 mg. daily
Diluted Isosorbide Dinitrate U.S.P. *Isordil, Sorbitrate*	Isosorbide Dinitrate Tablets U.S.P.	Anti-anginal	Oral, 10 mg. 4 times daily, adjusting the dosage as needed and tolerated	
	Isosorbide Dinitrate Extended-Release Capsules U.S.P.		40 mg. as an extended-release capsule every 12 hours, the dosage being increased up to 40 mg. every 6 hours as needed and tolerated	
	Isosorbide Dinitrate Chewable Tablets U.S.P.		Oral 5 to 10 mg. chewed well, every 2 to 3 hours, the dosage being adjusted as needed and tolerated	
	Isosorbide Dinitrate Extended-Release Tablets U.S.P.		Oral, 40 mg. as an extended-release tablet every 12 hours, the dosage being increased up to 40 mg. every 6 hours as needed and tolerated	
	Isosorbide Dinitrate Sublingual Tablets U.S.P.		Sublingual, 5 to 10 mg. as a sublingual tablet every 2 to 3 hours as needed	
Cyclandelate *Cyclospasmol*		Vasodilator	100 to 400 mg. before meals and at bedtime	1.2 to 1.6 g. daily
Dipyridamole *Persantin*		Vasodilator	50 mg. 3 times daily	
Nicotinyl Alcohol Tartrate *Roniacol*		Vasodilator	50 to 100 mg. 3 times daily	

* See *U.S.P. D.I.* for complete dosage information.

When cyclandelate is effective, the improvement in peripheral circulation usually occurs gradually and treatment often must be continued over long periods. At the maintenance dose of 100 mg. four times daily, there is little incidence of serious toxicity. At higher doses, as high as 400 mg. four times daily, which may be needed initially, there is a greater incidence of unpleasant side-effects such as headache, dizziness and flushing. It must be used with caution in patients with glaucoma. The oral dosage forms are 200-mg. capsules and 100-mg. tablets.

ANTIARRHYTHMIC DRUGS

Cardiac arrhythmias are caused by a disturbance in the conduction of the impulse through the myocardial tissue, by disorders of impulse formation, or by a combination of these factors. The antiarrhythmic agents used most commonly affect impulse conduction by altering conduction velocity and the duration of the refractory period and by depressing spontaneous diastolic depolarization, causing a reduction of automaticity by ectopic foci.

There are many pharmacological agents currently available for the treatment of cardiac arrhythmias. Agents such as oxygen, potassium, and sodium bicarbonate relieve the underlying cause of some arrhythmias. Other agents, such as digitalis, propranolol, phenylephrine, edrophonium, and neostigmine, act on heart muscle or on the autonomic nerves to the heart and alter their influence on the cardiovascular system. Finally, there are drugs that alter the electrophysiologic mechanisms causing arrhythmias. This last group is discussed in this chapter.

Within the last two decades research on normal cardiac tissues and, in the clinical setting, on patients with disturbances of rhythm and conduction has brought to light information regarding the genesis of cardiac arrhythmias and the mode of action of antiarrhythmic agents. In addition to this, laboratory tests have been developed to measure blood levels of antiarrhythmic drugs such as phenytoin, disopyramide lidocaine, procainamide, and quinidine to help evaluate the pharmacokinetics of these agents. As a result it is possible to maintain steady-state plasma levels of these drugs that allow the clinician to use these agents more effectively and with greater safety. No other clinical intervention has been more effective in reducing mortality and morbidity in coronary-care units.[10]

CARDIAC ELECTROPHYSIOLOGY

The heart depends on the synchronous integration of electrical-impulse transmission and myocardial-tissue response to carry out its function as a pump. When the impulse is released from the sinoatrial node, excitation of the heart tissue takes place in an orderly manner by a spread of the impulse throughout the specialized automatic fibers in the atria, the A-V node, and, finally, the Purkinje fiber network in the ventricles. This spreading of impulses produces a characteristic electrocardiographic pattern that can be equated to predictable myocardial cell-membrane potentials and Na^+ and K^+ fluxes in and out of the cell.

A single fiber in the ventricle of an intact heart during the diastolic phase (phase 4, Fig. 14–4) has a membrane potential (resting potential) of −90 millivolts (mv). This potential is created by differential concentrations of K^+ and Na^+ in the intracellular and extracellular fluid. An active-transport system (pump) on the membrane is responsible for concentrating the K^+ inside the cell and maintaining higher concentrations of Na^+ in the extracellular fluid. Diastolic depolarization is caused by a decreased potassium ionic current into the extracellular tissue and a slow inward leakage of Na^+ until the threshold potential (−60 to −55 mv) is reached. At this time there is a sudden increase in the inward sodium current, and a self-propagated wave occurs to complete the membrane-depolarization process. Pacemaker cells possess this property, which is termed *automaticity*. This maximal

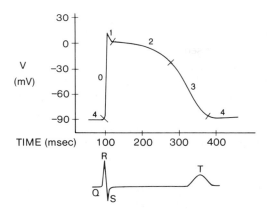

FIG. 14-4. *Diagrammatic representation of the membrane action potential as recorded from a Purkinje fiber and an electrogram recorded from an isolated ventricular fiber. The membrane resting potential is 90 mV with respect to the exterior of the fiber. At the point of depolarization there is a rapid change (phase 0) to a more positive value. The phases of depolarization and repolarization are indicated by the numbers 0, 1, 2, 3, 4. It should be noted that phases 0 and 3 of the membrane action potential correspond in time to the inscription of the QRS and T waves, respectively, of the local electrogram. From Lucchesi, B.R. in Antonuccio, M. (ed.): Cardiovascular Pharmacology, p. 270, New York, Raven Press, 1977.*

rate of depolarization (MRD) is represented by phase 0 or the spike-action potential (Fig. 14-4).

The form, duration, resting-potential level, and amplitude of the action potential are characteristic for different types of myocardial cells. The rate of rise of the response (phase 0) is related to the level of the membrane potential at the time of stimulation and has been termed *membrane responsiveness*. Less negative potentials produce smaller slopes of phase 0 and are characterized by slower conduction times. The phase-0 spike of the SA node corresponds to the inscription of the P wave on the electrocardiogram (Fig. 14-5). Repolarization is divided into three phases. The greatest amount of repolarization is represented by phase 3, in which there is a passive flux of K^+ ions out of the cell. Phase-1 repolarization is caused by influx of chloride ions. During phase 2, a small inward movement of calcium ions occurs through a slow-channel mechanism that is believed to be important in the process of coupling excitation with contraction.[11] The process of repolarization determines the duration of the action potential and is represented by the QT interval. The action-potential duration is directly related to the refractory period of cardiac muscle.

MECHANISMS OF ARRHYTHMIAS

The current understanding of the electrophysiological mechanisms responsible for the origin and perpetuation of cardiac arrhythmias is that they are due to either (1) altered impulse formation, i.e. change in automaticity, (2) altered conduction, or (3) both, acting simultaneously from different locations of the heart. As described above, the generation of cardiac impulses in the normal heart is usually confined to specialized tissues that spontaneously depolarize and initiate the action potential. These cells are located in the right atrium and are referred to as the *sinoatrial node* or the *pacemaker cells*. Although the spontaneous electrical depolarization of the sinoatrial pacemaker cells is independent of the nervous system, these cells are innervated by both sympathetic and parasympathetic fibers, which may cause an increase or decrease of the heart rate, respectively. Other special cells in the normal heart, which possess the property of automaticity, may influence cardiac rhythm when the normal pacemaker is suppressed or when pathologic changes occur in the mycocardium to make these cells the dominant source of cardiac rhythm, i.e., ectopic pacemakers. Automaticity of subsidiary pacemakers may develop when myocardial-cell damage occurs because of infarction or from digitalis toxicity, excessive vagal tone, excessive catecholamine release from sympathomimetic nerve fibers to the heart, or even high catecholamine plasma levels. The development of automaticity in specialized cells, such as that found in special atrial cells, certain atrioventricular nodal cells (A-V node), bundle of His, and Purkinje fibers, may lead to cardiac arrhythmias. Because production of ectopic impulses often is due to a defect in the spontaneous phase-4 diastolic depolarization (i.e., "T-wave"), drugs that are able to suppress this portion of the cardiac stimulation cycle are effective agents for these types of arrhythmias.

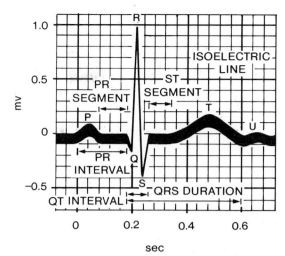

FIG. 14-5. *Normal electrocardiogram. From Ganong, W.F. "Review of Medical Physiology," 9th Ed. Lange Medical Publications 1979. San Francisco CA.*

Arrhythmias are also caused by disorders in the conduction of impulses and changes in the refractory period of the myocardial tissue. Pharmacological intervention is based on these two properties. The Purkinje fibers branch into a network of interlacing fibers, particularly at their most distant positions. This creates a number of pathways in which a unidirectional block in a localized area may establish circular (circus) micro- or macrocellular impulse movements that reenter the myocardial fibers and create an arrhythmia (Fig. 14-6). Unidirectional block results from localized myocardial disease (infarcts) or from a change in dependence of the tissue to sodium-ion fluxes that causes a longer conduction time and allows the tissue to repolarize to propagate the retrograde impulse.

CLASSES OF ANTIARRHYTHMIC DRUGS

Classification of antiarrhythmic agents into distinct categories is difficult because many drugs have more than one possible mode of action. Earlier reviews[12] have attempted to group the drugs into pharmacological categories based on their electrophysiologic properties on cardiac muscle. Below is a modified grouping of antiarrhythmic agents.

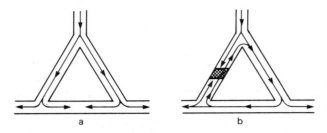

FIG. 14-6. *Reentry mechanism of Purkinje fibers. A. Normal conduction of impulses through triangular arrangement of cardiac fibers. B. Unidirectional block on left arm of triangular section allows impulse to reenter the regional conducting system and recycle.*

Classes of antiarrhythmic drugs

Group I. Depresses electrophysiological properties of myocardial cells
    Quinidine
    Procainamide
    Disopyramide
Group II. Facilitates conduction and shortens refractory period while depressing automaticity
    Lidocaine
    Phenytoin
Group III. β-adrenergic blocking agents
    Alprenolol   Practolol
    Metoprolol   Propranolol
    Pindolol   Sotalol
Group IV. Selective calcium antagonists
    Nifedipine
    Verapamil
Group V. Miscellaneous agents
    Atropine
    Edrophonium
    Neostigmine
    Bretylium
    Aprindine

Group I. Membrane-depressant drugs. Antiarrhythmic drugs that block the fast inward sodium current during the depolarization of the cardiac cell membrane are in this category. These drugs cause a decrease in the maximal rate of rise of the action potential (MRD) or a change in the membrane responsiveness. The end results of either of these two electrophysiological changes on the heart are a decrease in the conduction velocity and an increase in the threshold of excitability and effective refractory period of myocardial tissue. These actions are effective in aborting re-entrant arrhythmias in heart muscle. Drugs that fall into this category are quinidine, procainamide, and disopyramide. These drugs also decrease the slope of phase-4 depolarization and serve to control arrhythmias due to enhanced automaticity.

Group II. Agents that increase conduction and decrease refractoriness. The actions of lidocaine and phenytoin represent this group but also have membrane-depressant properties similar to Group-I antiarrhythmic agents. These agents depress phase-4 slopes of spontaneously depolarizing fibers in very low concentrations and thus are effective in stopping arrhythmias due to enhanced automaticity of ectopic pacemakers. However, some of the other electrophysiological effects observed with quinidine, such as decreased membrane responsiveness and conduction velocity or elevation of the electrical threshold to excitability, were not found to be caused by lidocaine or phenytoin. Both these drugs also have been reported often to enhance membrane responsiveness and conduction velocity. Their antiarrhythmic mechanism is based on their property to increase the relative duration of the functional refractory period by shortening the action-potential duration.

Group III. Beta-adrenergic blocking agents. Beta-adrenergic blocking drugs have the property of causing membrane-stabilizing or depressant effects on myocardial tissue. However, their antiarrhythmic properties are considered to be principally due to inhibition of adrenergic stimulation to the heart. The principle electrophysiologic property of these β-blocking agents is reduction of the phase-4 slope of potential sinus or ectopic pacemaker cells so that the heart rate decreases and ectopic tachycardias are either slowed down or converted to sinus rhythm.

Group IV. Selective calcium antagonists. The prototype drug in this group category is verapamil, an agent that has been used outside the United States since 1966 and has been in clinical trials for several years in the United States for the abolition of out-of-phase extra systoles. This drug selectively blocks entry of calcium into the myocardial cell and serves to depress AV conduction as well as block excitation–contraction coupling.

Group V. Miscellaneous antiarrhythmic agents. There are several agents, traditionally considered under cholinergic drug categories, that have found their way into the armament of cardiologists as antiarrhythmic agents. Atropine sulfate is used to treat certain reversible bradycardias accompanying myocardial infarctions. Atropine increases the heart rate by blocking the effects of acetylcholine on the heart. It is also used preoperatively to prevent sinus bradycardias that may be produced by certain drugs or surgical procedures. Edrophonium is an effective cholinesterase inhibitor used to terminate supraventricular tachycardias. Because of its reversible action on the esterase enzyme, the action, which is rapid in onset, is only of brief duration. Neostigmine is another cholinesterase inhibitor that has been used in the same manner as edrophonium. Bretylium tosylate is an adrenergic neuron-blocking agent that has been used as an antiarrhythmic agent. This drug has membrane-stabilizing properties, although some investigators believe its action is due to its antiadrenergic action. Bretylium is useful in terminating ventricular tachyarrhythmias of the re-entrant type, because of its ability to increase the duration of the action potential and functional refractory period. Aprindine is a long-acting antiarrhythmic agent that is effective orally and I.V. It is now in clinical trials and used to treat ventricular tachycardias.

PRODUCTS

Quinidine Sulfate U.S.P., quinidinium sulfate, is the sulfate of an alkaloid obtained from various species of Cinchona and their hybrids. It is a dextrorotary diastereoisomer of quinine. The salt crystallizes from water as the dihydrate in the form of fine, needlelike white crystals. Quinidine contains a hydroxymethyl group that serves as a link between a quinoline ring and a quinuclidine moiety. The structure contains two basic nitrogens, of which the quinuclidine nitrogen is the stronger base (pKa 10). Quinidine sulfate has a bitter taste and is light sensitive. Aqueous solutions are nearly neutral or slightly alkaline. It is soluble to the extent of one percent in water and is more highly soluble in alcohol or chloroform.

Quinidine

Quinidine is considered to be the prototype of antiarrhythmic drugs. Despite its long-term use, few structure–activity studies have been made to relate effectively the antiarrhythmic activity of quinidine to its chemical structure. Radioautographic studies show that quinidine binds to the lipoprotein of the cell membrane, which apparently hinders membrane cation transport.[13] In the resting phase quinidine diminishes Na^+ entry into the cell, causes depression of phase-4 diastolic depolarization, and shifts the intracellular threshold potential towards 0. These combined actions diminish spontaneous frequency of pacemaker tissues, depress the automaticity of ectopic foci, and, to a lesser extent, reduce impulse formation in the sinoatrial node. This last action results in bradycardia. During the spike-action potential quini-

dine decreases transmembrane permeability to passive influx of Na^+, causing a slowing down of the process of phase-0 depolarization which decreases conduction velocity. This is shown as a prolongation of the QRS complex of electrocardiograms. Quinidine also causes a prolongation of action potential duration which results in a proportionate increase in the QT interval. Quinidine is used to treat supraventricular and ventricular ectopic arrhythmias, such as atrial and ventricular premature beats, atrial and ventricular tachycardia, atrial flutter, and atrial fibrillation.

When quinidine is administered intramuscularly the peak effect, as measured by the prolongation of the QT interval on the electrocardiogram, is 1 to 1½ hours; given orally in the same dose, the peak effect occurs in 1 to 2 hours.[14] The duration of effect is about the same in both cases, but the intramuscular injection gives a greater peak effect. From 2 hours after administration the activity curves fall off at almost the same rate. When compared with the same oral dose of quinine, the cardiac response is qualitatively the same but of lesser magnitude and shorter duration. A study of the effect of gastric acidity on the absorption of quinidine sulfate from the gastrointestinal tract showed a consistently higher plasma level in the group with achlorhydria than in the normal subjects, but the difference between the mean values for each group was not statistically significant.[53]

Quinidine that has been absorbed from the gastrointestinal tract or from the site of an intramuscular injection is bound 80 percent to serum albumin.[10] The drug is taken up quickly from the blood stream by body tissues, so a significant concentration gradient is established within a few minutes. It is metabolized by the liver and has a half-life of 2 to 3 hours. Two metabolic products have been identified: 2-hydroxyquinidine, which had 30 to 40 percent of the activity of quinidine, and an inactive metabolite, which had a hydroxyl group on the α-carbon of the quinuclidine ring.

Quinidine Gluconate U.S.P., Quinaglute®, quinidinium gluconate. This occurs as an odorless, white powder with a very bitter taste. In contrast with the sulfate salt, it is freely soluble in water. This is important, because there are emergencies when the condition of the patient and the need for a rapid response may make the oral route of administration inappropriate. The high water-solubility of the gluconate salt along with a low irritant potential makes it of value when an injectable form is needed in these emergencies. Quinidine salts have been given intravenously for a prompt response, but this route is rather risky, so that the intramuscular route is usually used when the oral route is inadvisable. Quinidine dihydrochloride has been used parenterally, but this causes painful inflammatory induration at the injection site.[55] Quinidine hydrochloride dissolved in water with urea and antipyrine as solubilizers has been used successfully, but within a few months the solution turns brown, and crystallization occurs.[52] Quinidine sulfate in propylene glycol has been used satisfactorily.[17]

Quinidine gluconate forms a stable aqueous solution. When used for injection, it usually contains 80 mg. per ml., equivalent to 50 mg. of quinidine or 60 mg. of quinidine sulfate.

Quinidine Polygalacturonate, Cardioquin®. This is formed by reacting quinidine and polygalacturonic acid in a hydroalcoholic medium. It contains the equivalent of approximately 60 percent quinidine. This salt is only slightly ionized and slightly soluble in water, but studies have shown that although equivalent doses of quinidine sulfate give higher peak blood levels earlier, a more uniform and sustained blood level is achieved with the polygalacturonate salt.[56]

In many patients, the local irritant action of quinidine sulfate in the gastrointestinal tract causes pain, nausea, vomiting and especially diarrhea and often precludes oral use in adequate doses. It has been reported that in studies with the polygalacturonate salt no evidence of gastrointestinal distress was encountered. It is available as 275-mg. tablets. Each tablet is the equivalent of 200 mg. of quinidine sulfate or 166 mg. of free alkaloid.

Procainamide Hydrochloride U.S.P., Pronestyl® Hydrochloride, procainamidium chloride, *p*-amino-N-[2-(diethylamino)ethyl]benzamide monohydrochloride. It is the amide form of procaine hydrochloride in that the amide group ($\cdot CO \cdot NH$) replaces the ester group ($CO \cdot O$). The amide occurs as a white to tan crystalline powder, soluble in water but insoluble in alkaline solutions. Its aqueous solutions have a pH of about 5.5. Hydrolysis in water to the corresponding acid and amine is less likely than with procaine hydrochloride. This stability permits its use orally.

Procainamide Hydrochloride

A kinetic study of the acid-catalyzed hydrolysis of procainamide has shown it to be unusually stable to hydrolysis in the pH range 2 to 7, even at elevated temperatures.[57]

Procainamide appears to have all of the electrophysiological effects of quinidine. It diminishes automaticity, decreases conduction velocity, increases action-potential duration and thereby the refractory period of myocardial tissue. Clinicians have favored the use of procainamide for ventricular tachycardias and quinidine for atrial arrhythmias, even though the two drugs are effective in either type of disorder.

Procaine hydrochloride was at one time used for treating arrhythmias. However, it was not active orally and had to be given intravenously. Procaine is hydrolyzed by plasma esterases to the acid and diethylaminoethanol. The aminoalcohol is still effective, but it has hypotensive effects and is quickly removed from the bloodstream. A study of the disposition of procainamide showed 50 to 60 percent of the drug to be excreted unchanged in the urine following an I.M. dose. Procainamide is only slowly hydrolyzed by plasma enzymes. It is acetylated at the N-4 nitrogen, forming N-acetylprocainamide. The metabolite has been reported[20, 21] to be equipotent to procainamide as an antiarrhythmic agent in humans. Unlike quinidine, procainamide is only minimally bound to plasma proteins. Between 75 and 95 percent of the drug is absorbed from the gastrointestinal tract. Plasma levels appear 20 to 30 minutes after administration and peak in about one hour.[22]

After the steady-state is reached the plasma levels in man decrease at 10 to 20 percent per hour.[61] A study of the intramuscular use of the drug showed that the efficacy by this route was similar to that for oral or intravenous doses.[62] Appreciable serum levels were achieved in 5 minutes with the peak at 15 to 60 minutes; significant amounts were present after 6 hours. Higher serum levels and slower rate of decline were noted in patients with renal insufficiency.

Lidocaine Hydrochloride U.S.P., Xylocaine® Hydrochloride, 2-(diethylamino)-2',6'-acetoxylidide monohydrochloride. This drug, which was initially conceived as a derivative of gramine and introduced as a local anesthetic, is now being used intravenously as a rapid-acting agent to suppress ventricular arrhythmias.

Lidocaine Hydrochloride

Lidocaine has a different effect on the electrophysiological properties of myocardial cells than do procainamide and quinidine. It depresses diastolic depolarization and automaticity in the Purkinje-fiber network and increases the functional refractory period relative to action-potential duration as do procainamide and quinidine. However, it differs from the latter two drugs in that it does not decrease but even may enhance conduction velocity and it increases membrane responsiveness to stimulation. There is less data available on the subcellular mechanisms responsible for the antiarrhythmic actions of lidocaine than on the more established drug quinidine. It has been proposed that lidocaine has little effect on membrane cation exchange of the atria. Sodium-ion entrance into ventricular cells during excitation is not influenced by lidocaine, because it does not alter conduction velocity in this area. Lidocaine does depress Na^+ influx during diastole, as do all other antiarrhythmic drugs, to diminish automaticity in myocardial tissue. It also alters membrane responsiveness in Purkinje fibers, allowing an increase of conduction velocity and ample membrane potential at the time of excitation.[25]

Lidocaine administration is limited to the parenteral route and usually is given intravenously, although adequate plasma levels are achieved after intramuscular injections. Lidocaine is not bound to any extent to plasma proteins and is concentrated in the tissues. It is rapidly metabolized by the liver (Fig. 14–7). The first step is deethylation[26] with the formation of monoethylglycinexylide, followed by hydrolysis of the amide.[27] Metabolism is rapid, so the half-life of a single injection ranges from 15 to 30 minutes. Lidocaine is a popular drug because of its rapid action and its relative freedom from toxic effects on the heart, especially in the absence of hepatic disease. Monoethylglycinexylide, the initial metabolite of lidocaine, is an effective antiarrhythmic agent; however, its rapid hydrolysis by microsomal amidases prevents its use in humans. An analog, tocainide, which contains a methyl group *alpha* to the amide function, has a much slower rate of metabolism and is presently being investigated as an oral antiarrhythmic agent.

Precautions must be taken that lidocaine-hydrochloride solutions containing epinephrine salts are not used as cardiac depressants. Such solutions are intended only for local anesthesia and are not used intravenously. The aqueous solutions without epinephrine may be autoclaved several times, if necessary.

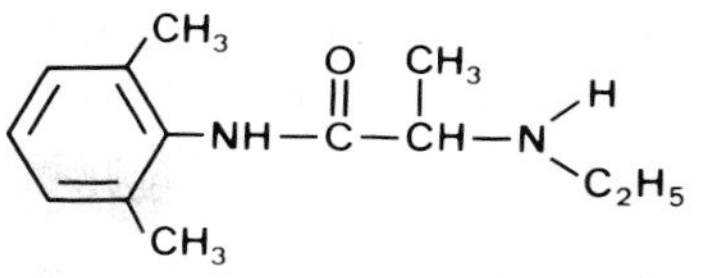

Tocainide

Phenytoin Sodium U.S.P., diphenylhydantoin sodium, Dilantin®, 5,5-diphenylhydantoin, 5,5-diphenyl-2,4-imidazolidinedione. This drug has been used for decades in the control of grand-mal types of epileptic seizures. It is structurally analogous to the barbiturates but does not possess the extensive sedative properties characteristic of the barbiturates. The compound is available as the sodium salt. Solutions for parenteral administration contain 40 percent propylene glycol and 10 percent alcohol to dissolve the sodium salt.

Phenytoin's cardiovascular effects were uncovered during observation of toxic manifestations of the drugs in patients being treated for seizure disorders. Phenytoin was found to cause bradycardia, prolong the PR interval, and produce T-wave abnormalities on electrocardiograms. Today phenytoin's greatest clinical use is in the treatment of digitalis-induced arrhythmias.[28] Its action is similar to that of lidocaine. It causes a depression of ventricular automaticity produced by digitalis, without adverse intraventricular conduction. Because it also reverses the prolongation of AV conduction by digitalis, phenytoin is useful in supraventricular tachycardias caused by digitalis intoxication.

Phenytoin is located in high amounts in the body tissues, especially fat and liver, leading to large gradients between the drug in tissues and the plasma concentrations. It is metabolized in the liver.

Bretylium Tosylate, Bretylol®, (*o*-bromobenzyl ethyl)dimethylammonium *p*-toluene sulfonate, is a white crystalline powder with an extremely bitter taste. The chemical is freely soluble in water and alcohol. Bretylium tosylate is an adrenergic neuronal

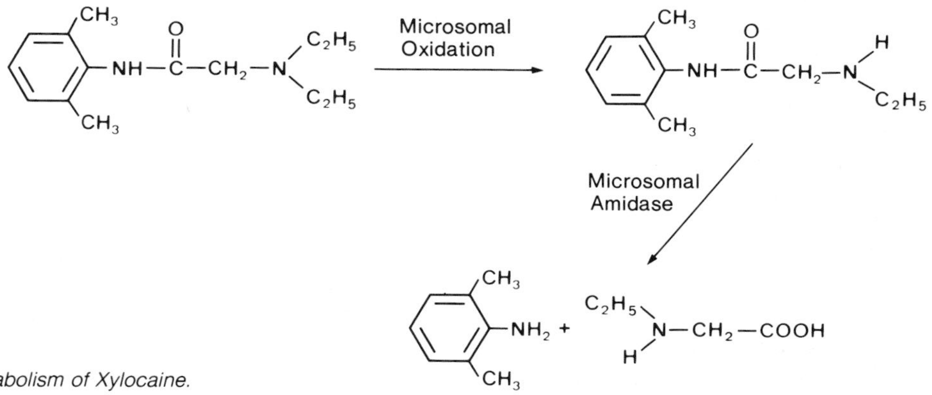

FIG. 14-7. Metabolism of Xylocaine.

blocking agent that accumulates selectively into the neurons and displaces norepinephrine. Because of this property, bretylium was used initially under the trade name of Darenthin® as an antihypertensive agent. It caused postural decrease in arterial pressure.[29] This use was discontinued because of the rapid development of tolerance, erratic oral absorption of the quaternary ammonium compound, and persistent pain in the parotid gland on prolonged therapy. At the present time bretylium is reserved for use in ventricular arrhythmias that are resistant to other therapy. Bretylium does not suppress phase-4 depolarization, a common action of other antiarrhythmic agents. It prolongs the effective refractory period relative to the action-potential duration but does not affect conduction time. Since bretylium does not have properties similar to those of the other antiarrhythmic agents, it has been suggested that its action is due to its adrenergic neuronal blocking properties; however, the antiarrhythmic properties of the drug are not affected by administration of reserpine. Bretylium is also a local anesthetic, but it has not been possible to demonstrate such an effect on atria of experimental animals except at very high concentrations.[30] Therefore, the precise mechanism of the antiarrhythmic action of bretylium remains to be resolved.

Bretylium Tosylate

Disopyramide Phosphate U.S.P., Norpace®, α-[2(diisopropylamino)ethyl]-α-phenyl-2-pyridine-acetamide phosphate, is a new oral and intravenous antiarrhythmic agent.[31] It is quite similar to quinidine and procainamide in its electrophysiological properties, in that it decreases phase-4 diastolic depolarization, decreases conduction velocity, and also has vagolytic properties.[32] It is used clinically in the treatment

Disopyramide Phosphate

of refractory, life-threatening ventricular tachyarrhythmias. Oral administration of the drug produces

peak plasma levels within 2 hours. The drug is bound approximately 50 percent to plasma protein and has a half-life of 6.7 hours in humans. More than 50 percent of the drug is excreted unchanged in the urine. Therefore, patients with renal insufficiency should be carefully monitored for evidence of overdose. Disopyramide commonly exhibits side-effects of dry mouth, constipation, urinary retention, and other cholinergic-blocking actions, because of its structural similarity to anticholinergic drugs.

Verapamil, 5-[(3,4-dimethoxyphenethyl)methyl-amino]-2-(3,4-dimethoxyphenyl)-2-isopropylvaleronitrile. Verapamil,[33] a papaverine derivative, was first

Verapamil

introduced as a coronary vasodilator in Germany in 1962. Subsequently it was found to have antiarrhythmic properties and has been used outside of the United States as an agent in the treatment of re-entrant AV junctional tachyarrhythmias. Verapamil is absorbed on oral administration and is bound significantly to plasma proteins. It has a half-life of 70 minutes. The drug is extensively metabolized in the body forming N- or O-demethyl metabolites, which have little biological activity. The metabolites are excreted in the urine and bile. Verapamil has the property of blocking the slow inward calcium current in cardiac fibers, which causes a slowing down of AV conduction and the sinus rate. The effect on calcium by verapamil also results in a block of excitation–contraction coupling and induces a negative inotropic effect on myocardial tissue.

Verapamil is contraindicated in the treatment of patients who are using β-adrenergic blocking agents. The combination of the two drugs causes a profound depression of ventricular function.

ANTIHYPERTENSIVE AGENTS

Hypertension has been defined as a measurement of the systolic or diastolic blood pressure above 160/95 mm Hg. in an adult under age 60. Despite this common signal, hypertension is not a single disease. Different types of hypertension vary in their hemodynamic and biochemical characteristics.[34] Essential hypertension is the most common form. Although advances have been made on the identification and con-

TABLE 14–3

ANTIARRHYTHMIC AGENTS

Name Proprietary Name	Preparations	Usual Adult Dose*	Usual Dose Range*	Usual Pediatric Dose*
Quinidine Sulfate U.S.P. *Quinora, Quinidex*	Quinidine Sulfate Capsules U.S.P. Quinidine Sulfate Tablets U.S.P.	Initial—oral, 200 to 800 mg. every 2 to 3 hours up to 5 times daily, as needed and tolerated	Up to 4 g. daily	6 mg. per kg. of body weight or 180 mg. per square meter of body surface, 5 times daily
Quinidine Gluconate U.S.P.	Quinidine Gluconate Injection U.S.P.	I.M., 600 mg., then 400 mg. repeated up to 12 times daily as necessary; I.V. infusion, 800 mg. in 40 ml. of 5 percent Dextrose Injection at the rate of 1 ml. per minute	Up to 5 g. daily	
Quinidine Polygalacturonate *Cardioquin*	Quinidine Polygalacturonate Tablets U.S.P.	Conversion of atrial and ventricular arrhythmias—oral, 275 to 825 mg. initially; then 275 to 825 mg. every 3 to 4 hours for 3 or 4 doses, with subsequent doses being increased by 137.5 to 275 mg. every third or fourth dose until rhythm is restored or toxic effects occur; maintenance—oral, 275 mg. 2 or 3 times daily as needed and tolerated		
Procainamide Hydrochloride U.S.P. *Pronestyl*	Procainamide Hydrochloride Capsules U.S.P. Procainamide Hydrochloride Tablets U.S.P.	Atrial arrhythmias—initial, 1.25 g. followed in 1 hour by 750 mg. if necessary, then 500 mg. to 1 g. every 2 to 3 hours as necessary or as tolerated; maintenance, 500 mg. to 1 g. 4 to 6 times daily Ventricular arrhythmias—1 g. initially, then 250 to 500 mg. every 3 hours as needed and tolerated	Up to 6 g. daily	12.5 mg. per kg. of body weight or 375 mg. per square meter of body surface, 4 times daily
	Procainamide Hydrochloride Injection U.S.P.	I.M., 500 mg. to 1 g. 4 times daily; I.V. infusion, 500 mg. to 1 g. at a rate of 25 to 50 mg. per minute followed by 2 to 6 mg. per minute to maintain therapeutic levels		
Disopyramide Phosphate U.S.P. *Norpace*	Disopyramide Phosphate Capsules U.S.P.	100 to 200 mg. four times daily	400 to 800 mg. daily	
Phenytoin Sodium U.S.P. *Dilantin*	Phenytoin Sodium Injection U.S.P.	Antiarrhythmic—I.V., 50 to 100 mg. every 10 to 15 minutes as necessary, but not to exceed a total dose of 15 mg. per kg. of body weight		
Bretylium Tosylate *Bretylol*	Bretylium Tosylate Injection	5 mg. per kg. repeated in 1 to 2 hours	20 to 40 mg. per kg. daily	
Lidocaine Hydrochloride U.S.P. *Xylocaine*	Lidocaine Hydrochloride Injection U.S.P.	Cardiac depressant (without epinephrine)—I.V., 50 to 100 mg.; may be repeated in 5 minutes (up to 300 mg. during a 1-hour period); I.V. infusion, 1 to 4 mg. per minute	Cardiac depressant (without epinephrine)—50 to 300 mg. during a 1-hour period	

*See *U.S.P. D.I.* for complete dosage information.

trol of essential hypertension, the etiology of this form of hypertension has not yet been resolved. The cause of renal hypertension is known. Renal hypertension can be created by experimentally causing renal artery stenosis in animals. Renal artery stenosis may also occur in pathologic conditions of the kidney such as nephritis, renal artery thrombosis, renal artery infarctions, or other conditions that have restricted blood flow through the renal artery. Hypertension may also originate from pathologic states in the central nervous system such as malignancies. Tumors in the adrenal medulla that cause release of large amounts of cate-

cholamines create a hypertensive condition known as *pheochromocytoma*. Excessive secretion of aldosterone by the adrenal cortex, often because of adenomas, also produces hypertensive disorders.

Arterial blood pressure is regulated by a number of physiological factors such as heart rate, stroke volume, peripheral vascular network resistance, blood vessel elasticity, blood volume, and viscosity of blood. Endogenous chemicals also play an important part in the regulation of arterial blood pressure. The peripheral vascular system is greatly influenced by the sympathetic–parasympathetic balance of the autonomic nervous system whose control originates in the central nervous system. Enhanced adrenergic activity is recognized as a principal contributor to essential hypertension.

Renal hypertension is controlled by the renin–angiotensin system. The renin–angiotensin system consists of peptidases and their substrates (found in blood and bound to tissue) that take part in the maintenance of blood pressure in low volume or low sodium states. Renal hypertension is created by the release of renin, an enzyme found primarily in the kidney, which forms a decapeptide, angiotensin I, from angiotensinogen, an α-2-globulin circulating in the blood. A converting enzyme splits two amino-acid residues from angiotensin I and forms the octapeptide, angiotensin II. Angiotensin II causes vasoconstriction and stimulates the synthesis of aldosterone in the adrenal cortex which, when released, enhances a retention of sodium ions that results in a rise in blood volume and an increase in blood pressure. The increase in blood pressure causes inhibition of renin release and limits the formation of angiotensin II. Angiotensin II is inactivated by peptidases in plasma and tissues. (Fig. 14–8)

Chemotherapy with antihypertensive agents evolved rapidly between 1950 and 1960. During that time a number of empiric discoveries were made that resulted in the marketing of drugs for the treatment and control of hypertensive disease. The first drugs of value were α-adrenergic blocking agents. These were intended to block the action of catecholamines with the expectation that contraction of the smooth muscle of the vascular walls would be blocked. These drugs had their limitations, because the duration of action was far too short and side-effects precluded long-term therapy. The clinical importance of these drugs lies in the value of the treatment of peripheral vascular disease and for diagnosis and short-term treatment of pheochromocytoma (see Chap. 11). Another type of chemical sympathectomy has been applied by the use of ganglionic blockers. Although ganglionic blockade is not selective, sympathetic control of the vascular smooth-muscle tone results in a clinically useful lowering of blood pressure from the use of these blockers (see Chap. 12).

The cholinergic agents, which act as antagonists to the adrenergic agents, cause peripheral vasodilatation and could possibly serve as hypotensive drugs. However, they produce a sharp fall in blood pressure that is of short duration, which limits their clinical use for prolonged antihypertensive therapy. Several other classes of drugs have been developed as antihypertensives and are discussed in other chapters. Diuretic agents such as the thiazides either alone or in combination with other drugs have proved to be very useful in reducing blood volume in the control of hypertension (see Chap. 13). Agents such as propranolol and metoprolol which block β-adrenergic receptors are some of the more recent additions to the chemotherapeutic tools that are used to treat patients with hypertension (see Chap. 11).

Chemicals that act to inhibit angiotensin-converting enzyme serve as effective angiotensin antagonists and may be useful potentially as a new class of antihypertensive agents. A promising investigational drug, captopril, (D-3-mercapto-2-methylpropanoyl)-L-proline, has been studied as a prototype inhibitor of this enzyme.

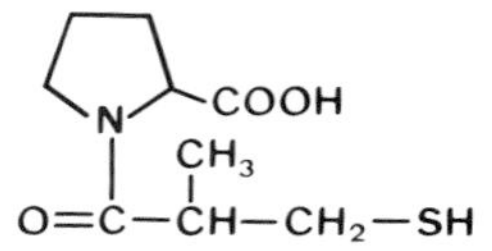

Captopril

AGENTS AFFECTING CONCENTRATION OF NEUROTRANSMITTERS AT NERVE ENDING

Rauwolfia Alkaloids

Folk remedies prepared from the species of Rauwolfia, a plant genus belonging to the Apocynaceae family, have been reported as early as 1563.[11] The root of the species *Rauwolfia serpentina* has been used for centuries as an antidote to stings and bites of insects, to reduce fever, as a stimulant to uterine contractions, for insomnia, and particularly for the treatment of insanity. Its use in hypertension was recorded in the

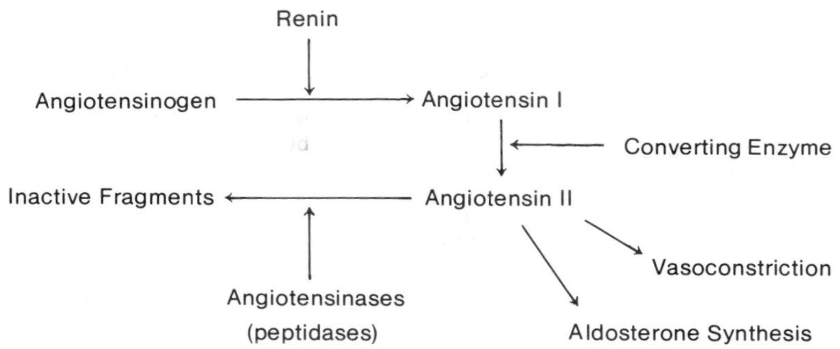

FIG. 14-8. Renin–angiotensin system.

Indian literature in 1918, but it was not until 1949 that hypotensive properties of *Rauwolfia* appeared in the Western literature.[36] *Rauwolfia* preparations were introduced in psychiatry in the treatment of schizophrenia in the early 1950s following confirmation of the folk-remedy reports on their use in mentally deranged patients. By the end of the 1960s, however, the drug had been replaced by more efficacious neurotropic agents. Reserpine and its preparations still remain useful in control of mild essential hypertension.

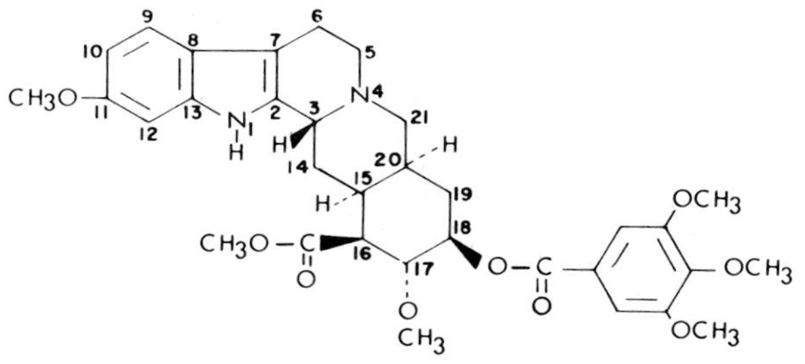

Reserpine

Chemical investigations of the active components of *R. serpentina* roots have yielded a number of alkaloids (e.g., ajmaline, ajmalicine, ajmalinine, serpentine, serpentinine, and others). Reserpine, which is the major active constituent of *Rauwolfia,* was isolated in

1952 by Muller *et al.*[37] and was found to be a much weaker base than the alkaloids mentioned above. Reserpinoid alkaloids are yohimbinelike bases that have an additional functional group on carbon-18. Only three naturally occurring alkaloids possess reserpine-like activity strong enough for use in treating hypertension: reserpine, deserpidine, and rescinnamine. Of these, only reserpine is official in the *U.S.P.*

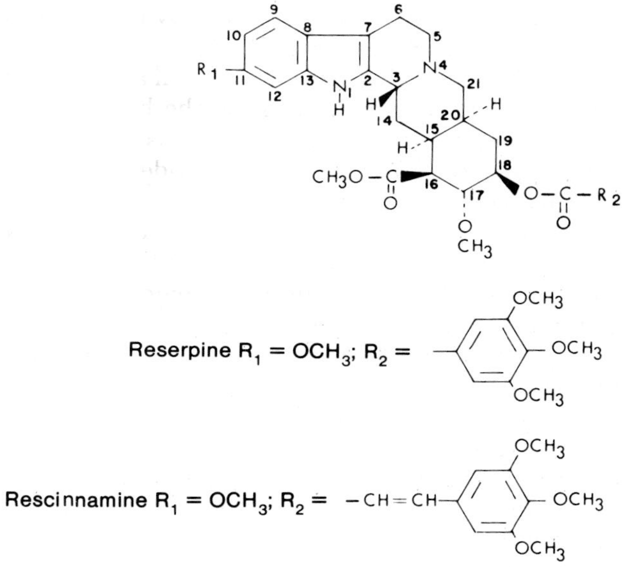

Reserpine $R_1 = OCH_3$; $R_2 =$

Rescinnamine $R_1 = OCH_3$; $R_2 = -CH=CH-$

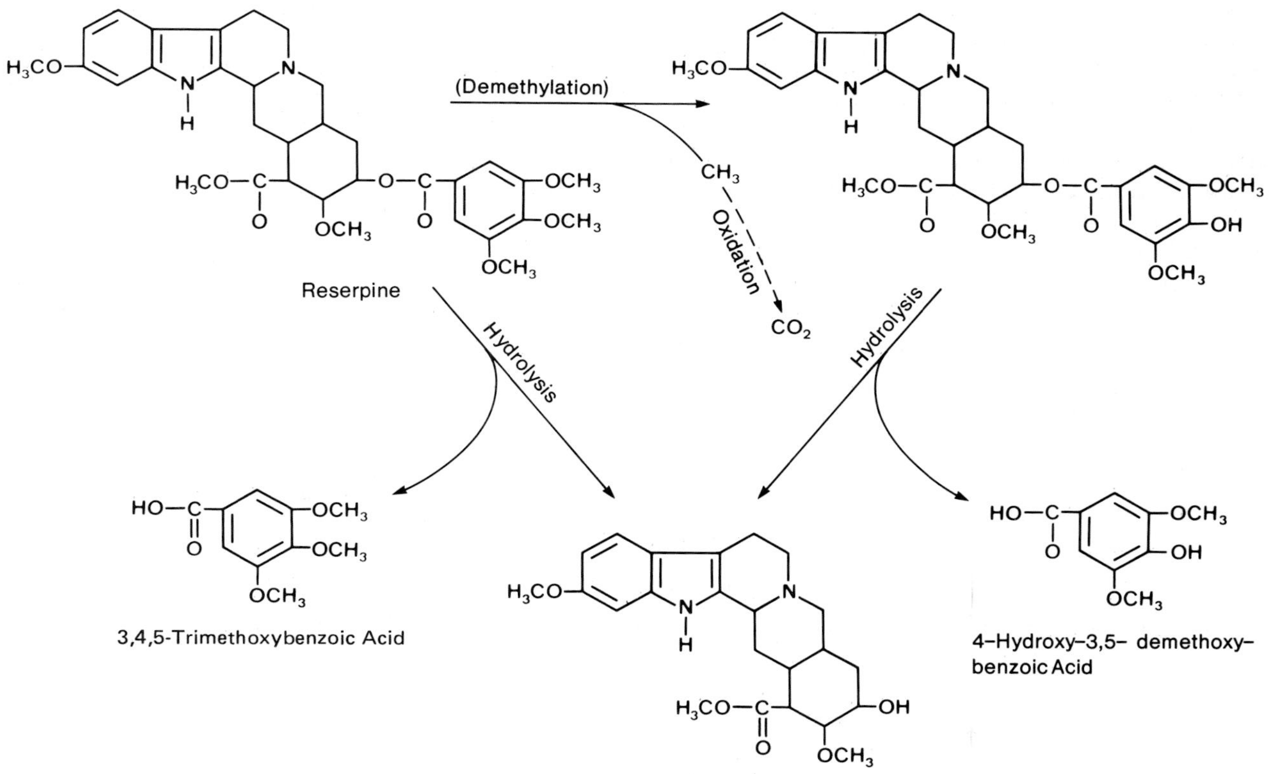

FIG. 14-9. *Metabolism of reserpine. (From J.M. Rand and H. Jurevics in "Antihypertensive Agents," F. Gross (ed). Springer-Verlag, 1977.)*

Reserpine is absorbed rapidly following oral administration. Fat tissue accumulates reserpine slowly, with a maximal level being reached between 4 and 6 hours. After 24 hours there are small amounts of reserpine in the liver and fat but none in the brain or other tissues. Reserpine is metabolized by the liver and intestine to methyl reserpate and 3,4,5-trimethoxybenzoic acid (Fig. 14–9).

The effects of reserpine do not correlate well with the tissue levels of the drug. It was observed that the pharmacological effects of reserpine were still manifest in animals at a time when reserpine could no longer be detected in the brain.[38] Subsequent to this observation it was found that reserpine causes depletion of catecholamines from postganglionic sympathetic nerves and the adrenal medulla. Both catecholamines and serotonin are depleted from the brain.[39, 40] Even though reserpine has central-nervous-system activity, its hypertensive action is primarily due to depletion of catecholamines from the peripheral sympathetic nerves.

Reserpine causes depletion of the catecholamines at the nerve ending by inhibiting the ATP–Mg^{++}-dependent uptake mechanism into the neuronal granules (uptake 2). The catecholamines that are not taken up into the granule are metabolized by mitochondrial monamine oxidase, causing the reduction of amine content. (Fig. 14–10).

Powdered Rauwolfia Serpentina U.S.P., Raudixin®, Rauserpa®, Rauval®, is the powdered whole root of *Rauwolfia serpentina* (Benth). It is a light-tan to light-brown powder, sparingly soluble in alcohol and only slightly soluble in water. It contains the total alkaloids, of which reserpine accounts for about 50 percent of the total activity. Orally, 200 to 300 mg. is roughly equivalent to 500 μg. of reserpine. It is used in the treatment of mild, labile hypertension or in combination with other hypotensive agents in severe hypertension.

Reserpine U.S.P., Serpasil®, Reserpoid®, RauSed®, Sandril®. This is a white to light-yellow crystalline alkaloid practically insoluble in water, obtained from various species of *Rauwolfia*. In common with other compounds with an indole nucleus, it is susceptible to decomposition by light and oxidation, especially when in solution. In the dry state discoloration occurs rapidly when exposed to light, but the loss in potency is usually small.[17] In solution there may be breakdown when exposed to light, especially in clear glass containers, with no appreciable color change; thus, color change cannot be used as an index of the amount of decomposition.

There are several possible points of breakdown in the reserpine molecule. Hydrolysis may occur at C-16 and C-18.[18] Reserpine is stable to hydrolysis in acid media, but in alkaline media the ester group at C-18 may be hydrolyzed to give methyl reserpate and trimethoxybenzoic acid (after acidification). If, in addition, the ester group at C-16 is hydrolyzed, reserpic acid (after acidification) and methyl alchol are formed. Citric acid helps to maintain reserpine in solution and in addition stabilizes the alkaloid against hydrolysis.

Storage of solutions in daylight causes epimerization at C-3 to form 3-isoreserpine. In daylight, oxidation (dehydrogenation) also takes place, 3-dehydroreserpine being formed. It is green in solution, but, as the oxidative process progresses, the color disappears and, finally, a strongly orange color appears. Oxidation of solutions takes place in the dark at an increasing rate with increased amounts of oxygen and at an even faster rate when exposed to light. Sodium meta-

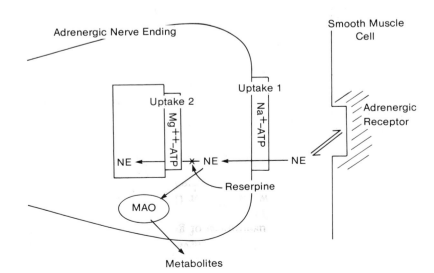

FIG. 14-10. *Action of reserpine at adrenergic nerve ending.*

bisulfite will stabilize the solutions if kept protected from light, but when exposed to light it actually oxidizes the reserpine so that the solutions are less stable than if the metabisulfite were absent. Nordihydroguaiaretic acid (NDGA) aids in stabilizing solutions when protected from light, but in daylight the degradation is retarded only slightly. Urethan in the solution stabilizes it in normally filled ampules but affords no protection in daylight.

Reserpine is effective orally and parenterally in the treatment of hypertension. After a single intervenous dose the onset of antihypertensive action usually begins in about 1 hour. After intramuscular injection the maximum effect occurs within approximately 4 hours and lasts about 10 hours. When given orally, the maximum effect occurs within about 2 weeks and may persist up to 4 weeks after the final dose. When used in conjunction with other hypotensive drugs in the treatment of severe hypertension the daily dose varies from 100 to 250 μg.

ADRENERGIC NEURON BLOCKING AGENTS

These agents inhibit the transmission of neuronal activity across the junction between sympathetic nerve terminals and the adrenergic receptor on the postjunctional membrane. They act presynaptically to prevent the release of norepinephrine that occurs as a result of nerve stimulation. The mechanism by which neuronal blocking agents inhibit the release of norepinephrine has not been resolved.

Blockade of the sympathetic system leads to hypotension, especially in hypertensive people. The hypotensive effect is more pronounced when patients are in an erect position than when they are supine. The variation in hypotensive effect because of body position is due to a reduction in the capacity of vasoconstrictor nerves to mediate the usual reflex compensations when erect posture is assumed. Neuronal blocking agents may enhance the response of the effector organs to injected norepinephrine. This effect is due to the property of the neuronal blocking agents inhibiting uptake 1 at the adrenergic nerve terminal.[43] Presumably this inhibition leads to an accumulation of norepinephrine in the area of the adrenergic receptor on the postjunctional membrane causing the enhanced response to the injected norepinephrine.[44]

Although several other neuronal blocking agents are known (e.g., bretylium, xylocholine) only trimethophan and guanethidine are marketed today for their antihypertensive properties. Guanethidine is taken up by the amine pump located on the neuronal membrane and retained in the nerve, displacing norepinephrine from its storage sites in the neuronal granules. The displaced norepinephrine is metabolized to homovanillic acid by mitochondrial monoamine oxidase, depleting the nerve ending of the neurotransmitter. The usefulness of guanethidine resides in the fact that although it is taken up by the nerve, it has a relatively low affinity for the amine pump and gives concentrations in the nerve endings that are too low to cause general depression of the neuronal membrane to stimulation[21] by the nerve-action potential. Guanethidine stored in the granules is released by the nerve-action potential but has a very low intrinsic activity for the adrenergic receptors on the postjunctional membrane. Moderate doses for a prolonged period of time or large doses of guanethidine may produce undesirable side-effects by causing neuromuscular blockade and adreneric nerve-conduction blockade.

Bretylium tosylate, now used as an antiarrythmic agent, was initially used as an antihypertensive agent because of its neuronal blocking properties. Its use has been discontinued because of orthostatic hypotension, the development of rapid tolerance to the drug, pain in the parotid glands, and a history of erratic oral absorption.

Guanethidine Sulfate U.S.P., Ismelin® Sulfate, [2-(hexahydro-1(2H)azocinyl)ethyl]guanidine sulfate, is a white crystalline material that is very soluble in water. It was one of a series of guanidine compounds prepared in the quest for potent antitrypanosomal agents. There is an absence of central nervous system effects, such as depression, from guanethidine, because the drug is highly polar and does not pass the blood–brain barrier easily. Guanethidine is chemically unrelated to antihypertensive agents previously introduced and produces a gradual, prolonged fall in blood pressure. Usually 2 to 7 days of therapy are required before the peak effect is reached, and usually this peak effect is maintained for 3 or 4 days; then, if the drug is discontinued, the blood pressure returns to pretreatment levels over a period of 1 to 3 weeks. Because of this slow onset and prolonged duration of action, only a single daily dose is needed.

Guanethidine Sulfate

Guanethidine is metabolized by microsomal enzymes to 2-(6-carboxyhexylamino)ethylguanidine and guanidine N-oxide (Fig. 14–11). Both metabolites have very weak antihypertensive properties.

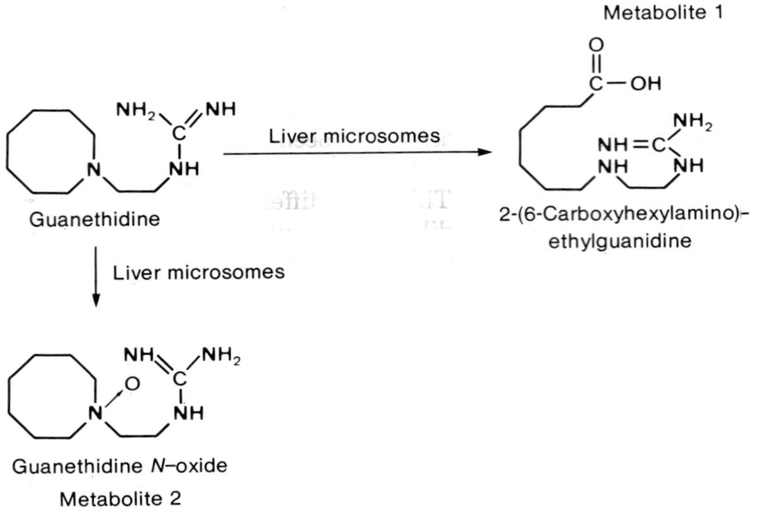

FIG. 14-11. Metabolism of guanethidine.

DRUGS ACTING DIRECTLY ON SMOOTH MUSCLE

Reduction of arterial smooth-muscle tone may occur by many mechanisms, such as reduction in sympathetic tone, stimulation of β-adrenergic receptors, or even direct action on the vasculature without interference with the autonomic innervation. Drugs acting on the arteriolar smooth muscle also have the property of increasing sympathetic reflex activity, causing an increase in heart rate and cardiac output as well as stimulating renin release from the kidney, which increases sodium retention and expansion of plasma volume. As a result, it is common to coadminister saluretics and β-adrenergic blocking agents with these agents.

PRODUCTS

Diazoxide U.S.P., Hyperstat I.V.® This drug is used as the sodium salt of 7-chloro-3-methyl-2H-1,2,4-benzothiadiazine 1,1-dioxide.

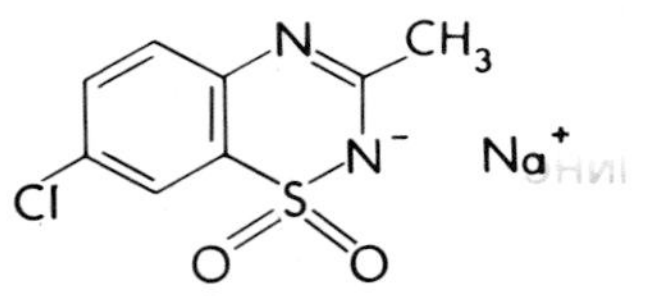

Sodium Diazoxide

Diazoxide is a direct vasodilating drug that lowers peripheral vascular resistance, increases cardiac output, and does not compromise renal blood flow.

This is a des-sulfamoyl analog of the benzothiazine diuretics and has a close structural similarity to chlorothiazide. It was intentionally developed in order to increase the antihypertensive action of the thiazides and minimize the diuretic effect.

It is used by intravenous injection as a rapidly acting antihypertensive agent for the emergency reduction of blood pressure in hospitalized patients with accelerated or malignant hypertension. Over 90 percent is bound to serum protein, and caution should be exercised when it is used in conjunction with other protein-bound drugs that may be displaced by diazoxide. The injection is given rapidly by the intravenous route to ensure maximal effect. The initial dose is usually 300 mg. (5 mg./kg.), with a second dose given if the first injection does not elicit a satisfactory lowering of blood pressure within 30 minutes. Further doses may be given at 4- to 24-hour intervals if needed. Oral antihypertensive therapy is begun as soon as possible.

The injection has a pH of about 11.5, which is necessary to convert the drug to its soluble sodium salt. There is no significant chemical decomposition after storage at room temperature for two years. When the solution is exposed to light, it darkens.

Diazoxide acts on arteriolar smooth muscle to cause relaxation. The vasodilatation is independent of a direct stimulation of α-adrenergic receptors. The mechanism causing relaxation of the smooth muscles is still under question; however, diazoxide probably acts by depleting an intracellular calcium pool or by inhibiting the release of calcium and thereby inhibiting the contractile process.[46]

Hydralazine Hydrochloride U.S.P., Apresoline® Hydrochloride, 1-hydrazinophthalazine monohydrochloride, originated from the work of a chemist[47] attempting to produce some unusual chemical compounds and from the observation[48] that this compound had antihypertensive properties. It occurs as yellow crystals and is soluble in water to the extent of about 3 percent. A 2 percent aqueous solution has a pH of 3.5 to 4.5.

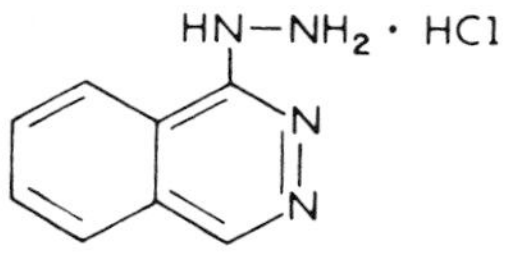

Hydralazine Hydrochloride

Hydralazine is useful in the treatment of moderate to severe hypertension. It is often used in conjunction with less potent antihypertensive agents, because when used alone in adequate doses there is a frequent occurrence of side-effects. In combinations it can be used in lower and safer doses. Its action appears to be centered on the smooth muscle of the vascular walls, with a decrease in peripheral resistance to blood flow. This results in an increased blood flow through the peripheral blood vessels. Also of importance is its unique property of increasing renal blood flow, an important consideration in patients with renal insufficiency.

Absorption of hydralizine hydrochloride taken orally is rapid and nearly complete. The maximal hypotensive effect is demonstrable within one hour. The drug is excreted rapidly by the kidney and within 24 hours 75 percent of the total amount administered appears in the urine as metabolites or unchanged drug. Hydralazine undergoes benzylic oxidation, glucuronide formation, and N-acetylation by the microsomal enzymes in the tissues (see Fig. 14–12). Acetylation appears to be a major determinant of the rate of hepatic removal of the drug from the blood and therefore of systemic availability.[49] Rapid acetylation results in a highly hepatic extraction ratio from blood and a greater first-pass elimination.[50]

Sodium Nitroprusside U.S.P., Nipride®, sodium nitroferricyanide, pentacyanonitrosylferrate (2-) diso- dium, $Na_2[Fe(CN)_5NO]$, is one of the most potent blood-pressure-lowering drugs. Its use is limited to hypertensive emergencies because of its short duration of action. The effectiveness of sodium nitroprusside as an antihypertensive has been known since 1928[51] but it was not until 1955 that its efficacy as a drug was established.[52] This drug differs from other vasodilators in that vasodilatation occurs in both venous and arterial vascular beds. Sodium nitroprusside is a reddish brown water-soluble powder. The rapid hypotensive effect of the chemical is due to the intact molecule when administered parenterally; however, it decomposes rapidly on oral administration to products that are claimed to be responsible for the small amount of oral activity observed.[53] Hydrocyanic acid (HCN), cyanogen [$(CN)_2$], and thiocyanogen [$(SCN)_2$] are decomposition products that have been suspected to be contributing to the oral hypotensive effects of sodium nitroprusside.

Minoxidil, Loniten®, 2,4-diamino-6-piperidinopyrimidine-3-oxide, was developed as a result of isosteric replacement by triaminopyrimidine of a triaminotriazine moiety. The triaminotriazines were initially observed to be potent vasodilators in cats and dogs following their formation of N-oxides in these animals. The triazines were inactive in humans because of people's inability to form N-oxide metabolites of the triazines; this led to the discovery of minoxidil.

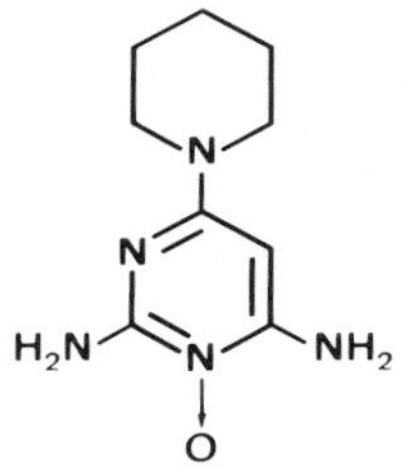

Minoxidil

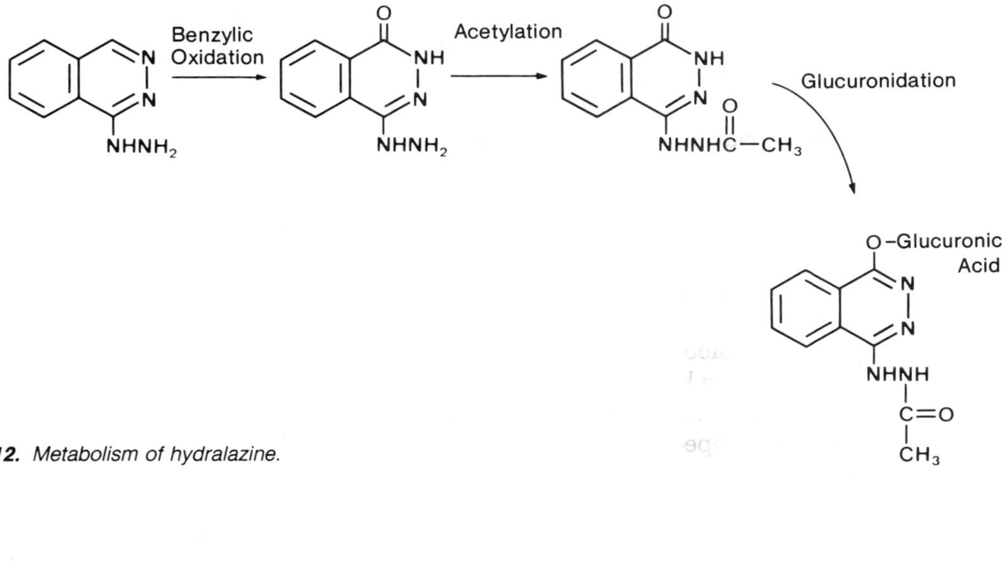

FIG. 14-12. *Metabolism of hydralazine.*

The antihypertensive properties of minoxidil are similar to those of hydralazine in that minoxidil is able to decrease arteriolar vascular resistance. Minoxidil exerts its vasodilator action by a direct effect on arteriolar smooth muscle and appears to have no effect on the central nervous system or on the adrenergic nervous system in animals.[54] The serum half-life of the drug is 4.5 hours, and the antihypertensive effect may last up to 24 hours.

Minoxidil is used for severe hypertension that is difficult to control by other antihypertensive agents. The drug has some characteristic side-effects of direct vasodilator drugs. It causes sodium and water retention and may require coadministration of a diuretic. Minoxidil also causes reflex tachycardia, which can be controlled by use of a β-adrenergic blocking agent.

Prazosin Hydrochloride, Minipress®, 1-(4-amino-6,7-dimethoxy-2-quinazolinyl)-4-(2-furoyl)piperazine monohydrochloride. The antihypertensive effects of this drug are due to its peripheral vasodilation either by direct action on the smooth muscle to cause their relaxation and interference with peripheral sympathetic function, or by its α-blocking action of adrenergic receptors. It has been suggested that the direct action of prazosin may be due to its ability to inhibit phosphodiesterase and a concurrent increase in c-AMP levels in muscle; however, it is not clear how it acts on the α-adrenergic receptors to decrease sympathetic function.[45]

Prazosin Hydrochloride

Prazosin is readily absorbed, and plasma concentrations reach a peak at about 3 hours after administration. Plasma half-life is between 2 and 3 hours. Prazosin is highly bound to plasma protein; however, it does not cause adverse drug reactions with drugs that might be displaced from their protein-binding sites,

e.g., cardiac glycosides. Prazosin may cause severe orthostatic hypotension because of its α-adrenergic blocking action, which prevents the reflex venous constriction that is activated when an individual changes position. At least 10 percent of the patients reported dizziness as a side-effect.

MISCELLANEOUS DRUGS

Methyldopa U.S.P., Aldomet®, α-methyldihydroxyphenylalanine, ($-$)-3-(3,4-dihydroxyphenyl)-2-methylalanine, is one of the more widely used antihypertensive agents. Its hypotensive properties were discovered during a systematic study of aromatic L-amino acid decarboxylase inhibitors. Methyldopa is not a potent inhibitor of the decarboxylase enzyme but in fact a poor substrate. Its hypotensive action is attributed to its conversion to α-methylnorepinephrine in adrenergic neurons in the central nervous system, where it accumulates in the storage granules and acts as a false transmitter upon receptor neurons. When α-methylnorepinephrine is released, it acts on adrenergic receptors of the central blood pressure-regulating mechanism to exert a peripheral depressant effect.[55] α-Methyldopamine, a metabolic intermediate of α-methyldopa, is apparently not an active hypotensive agent since blockade of dopamine β-hydroxylase in brains of experimental animals prevents the hypotensive effect of methyldopa. Earlier studies had suggested that the hypotensive action of α-methyldopa was due to the peripheral properties of the drug as a decarboxylase inhibitor or as a false transmitter. Although the drug is metabolized by peripheral neurons, this latter hypothesis does not seem valid because concomitant administration of peripherally active inhibitors of L-amino acid decarboxylase does not influence the antihypertensive effects of α-methyldopa.[56] Furthermore, there is little difference between norepinephrine and α-methylnorepinephrine in their potency on adrenergic receptors.[57]

In addition to the depression of vasomotor centers in the central nervous system, there is evidence that methyldopa suppresses release of renin by the kidney.[58]

Methyldopa α-Methyldopamine α-Methylnorepinephrine

Methyldopa is recommended for patients with high blood pressure that is not responsive to diuretic therapy alone.

Methyldopate Hydrochloride U.S.P., Aldomet® Ester Hydrochloride, (−)-3-(3,4-dihydroxyphenyl)-2-methylalanine ethyl ester hydrochloride. Methyldopa, suitable for oral use, is a zwitterion and is not soluble enough for parenteral use. This problem was solved by making the ester, leaving the amine free to form the water-soluble hydrochloride salt. It is supplied as a stable, buffered solution, protected with antioxidants and chelating agents.

Methyldopate Hydrochloride

Clonidine Hydrochloride, Catapres®, 2-[(2,6-dichlorophenyl)imino]-imidazoline monohydrochloride, was the first antihypertensive known to act on the central nervous system. It was synthesized in 1962 as a derivative of the known α-sympathomimetic drugs, naphazoline and tolazoline, as a potential nasal vasoconstrictor, but instead it has proven to be an effective drug in the treatment of mild to severe hypertension.

Clonidine Hydrochloride

Clonidine acts by both peripheral and central mechanisms in the body to affect blood pressure. It stimulates the peripheral α-adrenergic receptors to produce vasoconstriction, producing a brief period of hypertension. Clonidine acts centrally to inhibit the sympathetic tone and cause hypotension that is of much longer duration than the initial hypertensive effect. Administration of clonidine thus produces a biphasic change in blood pressure, beginning with a brief

hypertensive effect and followed by a hypotensive effect that persists for about 4 hours. This biphasic response is altered by dose only, in that larger doses produce a greater hypertensive effect and delay the onset of the hypotensive properties of the drug. Clonidine acts on *alpha* adrenoreceptors located in the hindbrain to produce its hypotensive action.[59] Clonidine also acts centrally to cause bradycardia and to reduce plasma levels of renin. Sensitization of baroreceptor pathways in the central nervous system appears to be responsible for the bradycardia transmitted by way of the vagus nerve. However, the central mechanism that results in plasma renin decrease is not known. The hypotensive properties of clonidine in animals can be blocked by applying α-adrenergic blocking agents directly to the brain.[60]

Clonidine has advantages over antihypertensive drugs such as guanethidine and prazosin, in that it seldom produces orthostatic hypotension side-effects. However, clonidine does have some sedative properties that are undesirable; it causes constipation and dryness of the mouth.

Clonidine is distributed throughout the body, with the highest concentrations found in the organs of elimination; kidney, gut, and liver. Brain concentrations are low but higher than plasma concentrations. The high concentration in the gut is due to an enterohepatic cycle wherein clonidine is secreted into the bile in rather high concentrations. The half-life in humans is about 20 hours. Clonidine is metabolized by the body to form two major metabolites, *p*-hydroxyclonidine and its glucuronide. *p*-Hydroxyclonidine does not cross the blood-brain barrier and has no hypotensive effect in people.

Pargyline Hydrochloride U.S.P., Eutonyl®, N-methyl-N-2-propynylbenzylamine hydrochloride. This drug, introduced in 1963, is a nonhydrazine monoamine oxidase (MAO) inhibitor that is effective in lowering systolic and diastolic blood pressure without depressing the patient.

Pargyline was discovered by Swett[61] as an MAO inhibitor and antihypertensive following observations that tuberculostatic hydrazides produced hypotension in patients and that these drugs were inhibitors of MAO. Pargyline is a therapeutic anomaly, because in-

Metabolism of Clonidine

TABLE 14–4

ANTIHYPERTENSIVE AGENTS

Name Proprietary Name	Preparations	Usual Adult Dose*	Usual Dose Range*	Usual Pediatric Dose*
Prazosin Hydrochloride *Minipress*	Prazosin Capsules	1 mg.	1 to 3 mg. daily	
Clonidine Hydrochloride *Catapress*	Clonidine Hydrochloride Tablets	0.1 mg.	0.1 to 0.3 mg. daily	
Diazoxide U.S.P. *Hyperstat*	Diazoxide Injection U.S.P.	300 mg.	0.3 to 1.2 g. daily	
Minoxidil *Loniten*	Minoxidil Tablets	0.2 mg. per kg.	10 to 40 mg. daily	
Powdered Rauwolfia Serpentina U.S.P. *Raudixin, Rauserpa, Rauval, Hyperloid, Rauja, Raulin, Veniber, Wolfina*	Rauwolfia Serpentina Tablets U.S.P.	Initial, 200 mg. daily for 1 to 3 weeks; maintenance 50 to 300 mg. daily		
Reserpine U.S.P. *Serpasil, Reserpoid, Rau- Sed, Sandril, Lemiserp, Resercen, Rolserp, Sertina, Vio-Serpine*	Reserpine Injection U.S.P. Reserpine Tablets U.S.P. Reserpine Elixir U.S.P.	I.M., 500 μg. to 1 mg., followed by 2 to 4 mg. 8 times daily as necessary Initial, 500 μg. once daily; maintenance, 100 to 250 μg. once daily	500 μg. to 32 mg. daily	70 μg. per kg. of body weight or 2 mg. per square meter of body surface, once or twice daily
Guanethidine Sulfate U.S.P. *Ismelin*	Guanethidine Sulfate Tablets U.S.P.	Ambulatory patients—initial, oral, 10 or 12.5 mg. once daily, the daily dosage being increased by 10 to 12.5 mg. at 5- to 7- day intervals if necessary for control of blood pressure		Oral, 200 μg. per kg. of body weight or 6 mg. per square meter of body surface once daily, the daily dosage being increased by 200 μg. per kg. of body weight or 6 mg. per square meter of body surface at 7- to 10-day intervals if necessary for control of blood pressure
Hydralazine Hydrochloride U.S.P. *Apresoline*	Hydralazine Hydrochloride Injection U.S.P. Hydralazine Hydrochloride Tablets U.S.P.	I.M., or I.V., 20 to 40 mg. repeated as necessary Oral, 10 mg. 4 times daily for the first 2 to 4 days, 25 mg. 4 times daily for the balance of the first week, and 50 mg. 4 times daily for the second and subsequent weeks, the dosage being adjusted to the lowest effective level	Up to 400 mg. daily	I.M. or I.V., 1.7 to 3.5 mg. per kg. of body weight or 50 to 100 mg. per square meter of body surface daily, divided into 4 to 6 doses Oral, 750 mg. per kg. of body weight or 25 mg. per square meter of body surface daily, divided into 4 doses, the dosage being increased gradually over 3 to 4 weeks as needed
Methyldopa U.S.P. *Aldomet*	Methyldopa Tablets U.S.P.	Initial—oral, 250 mg. 2 to 3 times daily for 2 days, the dosage then being adjusted, preferably at intervals of not less than 2 days until the desired response is obtained; maintenance— oral, 500 mg. to 2 g. daily, divided into 2 to 4 doses		Oral, initially, 10 mg. per kg. of body weight or 300 mg. per square meter of body surface, divided into 2 to 4 doses, the dosage then being adjusted, preferably at intervals of not less than 2 days, until the desired response is obtained, but not exceeding 65 mg. per kg. of body weight or 3 g. daily, whichever is less
Methyldopate Hydrochloride U.S.P. *Aldomet Ester*	Methyldopate Hydrochloride Injection U.S.P.	I.V. infusion 250 to 500 mg. in 100 ml. of 5 percent Dextrose Injection over a period of 30 to 60 minutes every 6 hours as necessary	Up to 1 g. daily every 6 hours	I.V. infusion, 5 to 10 mg. per kg. of body weight in 5 percent Dextrose Injection, administered slowly over a 30 to 60 minute period, every 6 hours if necessary, but not exceeding 65 mg. per kg. of body weight or 3 g. daily, whichever is less

*See *U.S.P. D.I.* for complete dosage information.

$$CH_2-N-CH_2C\equiv CH$$

with CH_3 on nitrogen

· HCl

Pargyline Hydrochloride

hibition of MAO would decrease catecholamine metabolism in the neuron and liver. One hypothesis is that blood pressure is lowered by the accumulation of a false transmitter substance, octopamine, which is a less effective pressor substance than norepinephrine.[62]

OH

$$CH-CH_2-NH_2$$

OH

Octopamine

ANTIHYPERLIPIDEMIC AGENTS

The major cause of death in the Western world today is attributed to vascular disease,[63] of which the most prevalent form is atherosclerotic heart disease. Although many causative factors of this disease are recognized, i.e., smoking, stress, diet, etc., atherosclerotic

disease can be treated through medication or surgery. There is, however, no specific means of prognosis of an impending coronary episode.

Hyperlipidemia is the most prevalent indicator to susceptibility to atherosclerotic heart disease; it is a term used to describe elevated plasma levels of lipids that are usually in the form of lipoproteins. Hyperlipidemia may be caused by an underlying disease involving the liver, kidney, pancreas, or thyroid; or it may not be attributable to any recognizable disease. Within recent years lipids have been indicated in the development of atherosclerosis in humans. Atherosclerosis may be defined as degenerative changes in the intima of medium and large arteries. The degeneration includes the accumulation of lipids, complex carbohydrates, blood, and blood products and is accompanied by the formation of fibrous tissue and calcium deposition on the intima of the blood vessels. These deposits or "placques" decrease the lumen of the artery, reduce its elasticity, and may create foci for thrombi and subsequent occlusion of the blood vessel.

LIPOPROTEIN METABOLISM

Lipoproteins are macromolecules consisting of lipid, i.e., cholesterol, triglycerides, etc., noncovalently bound with protein and carbohydrate. These combinations serve to solubilize the lipids and prevent them from forming insoluble aggregates in the plasma. The various lipoproteins found in plasma can be separated

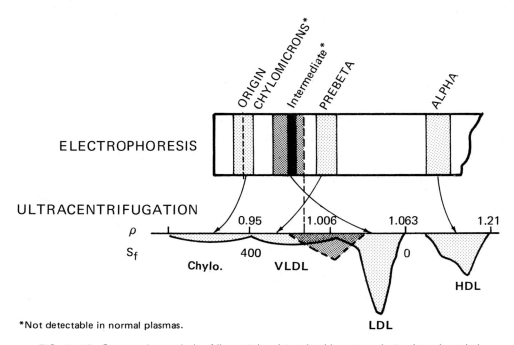

*Not detectable in normal plasmas.

FIG. 14-13. Comparative analysis of lipoproteins determined by paper electrophoresis and ultracentrifugation. From Morganroth, J. and Levy, R.I. in "Drugs in Cardiology, Vol. 1, Part 1" Donoso, E., Ed., Stratton Medical Book Corp., New York, N.Y., 1975.

by ultracentrifugal techniques into chylomicrons, very low density lipoproteins (VLDL), intermediate density lipoproteins (IDL), low density lipoproteins (LDL), and high density lipoproteins (HDL). These correlate with the electrophoretic separations of the lipoproteins as follows: chylomicrons, pre*beta* lipoprotein (VLDL); broad *beta* lipoprotein (IDL), *beta* lipoprotein (LDL), and *alpha* lipoprotein (HDL) (see Fig. 14–13).

Chylomicrons contain 90 percent triglycerides by weight and originate from exogenous fat from the diet. They are the least dense of the lipoproteins and migrate the least under the influence of an electric current. Chylomicrons are normally absent in plasma after 12 to 24 hours of fasting. VLDL are composed of about 60 percent triglycerides, 12 percent cholesterol, and 18 percent phospholipids. VLDL originates in the liver from free fatty acids. Although VLDL can be isolated from plasma, it is catabolized rapidly into IDL which is further degraded into LDL. Normally, IDL is also rapidly catabolized to LDL, but is not usually isolated from plasma. LDL consists of 50 percent cholesterol and 10 percent triglycerides. In normal subjects this lipoprotein accounts for about 65 percent of the plasma cholesterol and is of major concern in hyperlipidemic-related disease states. LDL is formed from the intravascular catabolism of VLDL. HDL is composed of 25 percent cholesterol and 50 percent protein and accounts for about 17 percent of the total plasma cholesterol in plasma.

It is now recognized that lipid disorders are related to problems in lipoprotein metabolism[64] that create conditions of hyperlipoproteinemia. The hyperlipoproteinemias have been classified into five types, each of which is treated differently (Table 14-5).

The abnormal lipoprotein pattern characteristic of type I is caused by a decrease in the activity of lipoprotein lipase, an enzyme that normally hydrolyzes the triglycerides present in chylomicrons and clears the plasma of this lipoprotein fraction. Because triglycerides that are found in chylomicrons come primarily from exogenous sources, this type of hyperlipoproteinemia may be treated by decreasing the intake of dietary fat. There are no drugs presently available that can be used to counteract type-I hyperlipidemia effectively.

Type II hyperlipoproteinemia has been divided into type IIa and IIb. Type IIa is characterized by elevated levels of LDL (β-lipoproteins) and normal levels of triglycerides. This subtype disorder is very common and may be caused by disturbed catabolism of LDL. Type IIb differs from type IIa in that this hyperlipidemia has elevated VLDL levels in addition to LDL. Type II hyperlipoproteinemia is often clearly familial and frequently inherited as an autosomal dominant abnormality with complete penetrance and expression in infancy.[65] Patients have been treated by use of diet restrictions of cholesterol and also by a low intake of saturated fats. This type of hyperlipoproteinemia responds to some form of chemotherapy. The combined therapy may bring LDL levels back to normal.

Type III is a rare disorder that is characterized by a broad band of β-lipoprotein. Like the type II, it is also familial. Patients respond favorably to diet and drug therapy. In type IV, hyperlipoproteinemia levels of VLDL are elevated. Because this type of lipoprotein is rich in triglycerides, plasma triglycerides are elevated. The metabolic defect that causes type IV is still unknown; however, this form of hyperlipidemia

TABLE 14–5

CHARACTERIZATION OF HYPERLIPOPROTEINEMIA TYPES

| Hyperlipo-proteinemia | Abnormality | | Appearance of Plasma* | Triglycerides | Cholesterol (total) |
	Electrophoresis	Ultracentrifuge			
I	Massive chylomicronemia		Clear. On top creamy layer of chylomicronemia	Massively elevated	Slightly to moderately elevated
IIa	β-lipoproteins elevated	LDL increased	Clear	Normal	Heavily elevated
IIb	Pre-β-lipoproteins elevated	LDL + VLDL increased	Slightly turbid	Slightly elevated	Heavily elevated
III	Broad β-lipoprotein band	VLDL/LDL of abnormal composition	Slightly turbid–turbid	Elevated	Elevated
IV	Pre-β-lipoproteins elevated	VLDL increased	Turbid	Moderately heavily elevated	Normal to elevated
V	Pre-β-lipoproteins elevated, chylomicronemia present	VLDL increased, chylomicronemia present	Turbid, on top chylomicronemia	Massively elevated	Slightly elevated

*After having been kept standing at 4°C for 25 hr.
Adapted from Witte, E. C.: Prog. Med. Chem. 11:199, 1975.

responds to diet and drug therapy. Type V hyperlipoproteinemia has high levels of chylomicrons and VLDL, resulting in high levels of plasma triglycerides. The biochemical defect of type V hyperlipoproteinemia is not understood. Clearance of dietary fat is impaired, and a reduction of dietary fat is indicated along with drug therapy.

The coronary-drug project

In 1966 a study was initiated in the United States to ascertain the efficacy of four hypolipidemic agents in the long-term treatment of men who had a history of several myocardial infarctions. The intent was to determine if long-term therapy would increase the survival of these individuals. The drugs chosen were nicotinic acid, conjugated estrogens, dextrothyroxine, and clofibrate. The results were such that there was no evidence to indicate that any of these drugs could be recommended for long-term therapy. The study on estrogens was discontinued one year into the study when it was shown that those individuals receiving conjugated estrogens (5 mg. per day) had an excessive number of nonfatal myocardial infarctions, pulmonary embolisms, and instances of thrombophlebitis compared with that recorded in the placebo group. About three years into the study, D-thyroxine was discontinued when data showed that 14.8 percent of the patients in this group died compared to 12.5 percent in the placebo group. Nicotinic acid and clofibrate were administered until the project's end in 1975. The data from both of these drugs did not provide evidence to allow clinicians to recommend long-term use of clofibrate or nicotinic acid to survivors of myocardial infarctions or coronary heart disease.[66]

Within recent years cholesterol has been believed to play an important role in the development of atherosclerosis in man. In patients with atherosclerosis, the fatty deposits in the blood vessels are high in cholesterol, either free or as esters. Further impetus has been given to the indictment of cholesterol by the fact that the incidence of atherosclerosis in Americans is significantly higher than in persons in other countries where the national diet is lower in cholesterol and saturated fats.[67]

The range of serum cholesterol in normal individuals is 190 to 250 mg. percent, with approximately 30 percent in the free state and 70 percent as cholesterol esters.[43] Three basic methods have been used in attempts to bring serum cholesterol levels within this range. These are: (1) diminish cholesterol absorption from the gastrointestinal tract; (2) increase the metabolism and the biliary excretion of cholesterol; and (3) inhibit endogenous liver synthesis of cholesterol.

Dihydrocholesterol, sitosterol and stigmasterol have been used, in the expectation that, because these sterols were not absorbed, the enzymatic reabsorption of cholesterol would be blocked.[68] It was found that dihydrocholesterol was itself absorbed. In tests using sitosterol and stigmasterol in the diet of experimental animals, there was found a reduction in serum cholesterol for several weeks, but then the levels returned to the original values. Thus it appears that there was an increased endogenous production of cholesterol to counterbalance the early inhibition (see Chap. 18).

Since fat is the major source of the precursors of cholesterol formed in the body, it seemed probable that modification of the fat content of the diet may modify cholesterol synthesis. This is indeed the case. The use of unsaturated fats from vegetable sources appears to lead to a decreased serum cholesterol, and evidence is accumulating to indicate there may also be an effect on the arterial concentration of cholesterol.

PRODUCTS

Clofibrate U.S.P., Atromid-S®, ethyl 2-(p-chlorophenoxy)-2-methylpropionate. Clofibrate is a stable, colorless to pale-yellow liquid with a faint odor and a characteristic taste. It is soluble in organic solvents, but insoluble in water.

Clofibrate

Clofibrate is prepared by a Williamson synthesis condensing p-chlorophenol with ethyl α-bromoisobutyrate[69] or by the interaction of a mixture of p-chlorophenol and chloroform in the presence of excess potassium hydroxide. The acid obtained by either of these methods is esterified to give clofibrate. Both acid and ester are active; however, the latter is preferred for medicinal use. Clofibrate is hydrolyzed rapidly to 2-p-chlorophenoxy-2-methylpropionic acid by esterases in vivo and circulates in blood bound to serum albumin. The acid has been investigated as a hypolipidemic agent. It is absorbed more slowly and to a lesser extent than the ester. The aluminum salt of the acid gave even lower blood levels than p-chlorophenoxy-2-methylpropionic acid.[70]

Clofibrate is the drug of choice in the treatment of type III hyperlipoproteinemias and may be useful also, to a lesser extent, in type IIa and IV hyperlipoproteinemias. The drug is not effective in types I and IIa. Clofibrate affects lipoprotein production. It causes reduction of VLDL by inhibiting its synthesis and increasing the clearance of this lipoprotein. It lowers serum triglyceride levels in the serum much more so than those of cholesterol and also decreases free fatty acids and phospholipids. The lowering of cholesterol may be the result of two mechanisms. Clofibrate inhibits the incorporation of acetate into the synthesis of cholesterol between the acetate and mevalonate step. Clofibrate also regulates cholesterol synthesis in the liver by inhibiting the microsomal reduction of 3-hydroxy-3-methylglutaroyl-CoA (HMG-CoA) catalyzed by HMG-CoA reductase. Clofibrate may lower plasma lipids by other means than impairment of cholesterol biosynthesis.

Clofibrate is well tolerated by most patients, the most common side-effects being nausea and, to a lesser extent, other gastrointestinal distress. The dosage of anticoagulants, if used in conjunction with this drug, should be reduced by one-third to one-half, depending on the individual response, so that the prothrombin time may be kept within the desired limits.

Usual adult dose—1.5 to 2 g. daily in 2 to 4 divided doses.

Usual pediatric dose—use in infants and children has not been established.

Occurrence
Clofibrate Capsules U.S.P.

Dextrothyroxine Sodium U.S.P., Choloxin®, sodium O-(4-hydroxy-3,5-diiodophenyl)-3,5-diiodo-D-tyrosine monosodium salt hydrate, sodium D-3,3′,5,5′-tetraiodothyronine. This compound occurs as light yellow to buff-colored powder. It is stable in dry air, but discolors on exposure to light; for this reason it should be stored in light-resistant containers. It is very slightly soluble in water, slightly soluble in alcohol and insoluble in acetone, in chloroform and in ether.

The hormones secreted by the thyroid gland have marked hypocholesterolemic activity along with their other well-known actions. With the finding that not all active thyroid principles possessed the same degree of physiologic actions, a search was made for congeners that would cause a decrease in serum cholesterol without other effects such as angina pectoris, palpitation and congestive failure. D-Thyroxine has resulted from this search. However, at the dosage required, the L-thyroxine contamination must be minimal, otherwise it will exert its characteristic actions. One route to optically pure (at least 99% pure) D-thyroxine is the use of an L-aminoacid oxidase from snake venom which acts only on the L-isomer and makes separation possible.

The mechanism of action of D-thyroxine appears to be stimulation of oxidative catabolism of cholesterol in the liver. The catabolic products are bile acids which are conjugated with glycine or taurine and ex-

Dextrothyroxine Sodium

creted via the biliary route into the feces. Cholesterol biosynthesis is not inhibited by the drug and abnormal metabolites of cholesterol do not accumulate in the blood. There is also a decrease in serum levels of triglycerides, but this is less consistent than the decrease of cholesterol. This is probably due to its preferential effect of increasing LDL catabolism. Because of this feature D-thyroxine is preferred in the treatment of type II hyperlipoproteinemias.

D-Thyroxine potentiates the action of anticoagulants such as warfarin or dicumarol; thus, dosage of the anticoagulants should be reduced by one-third if used concurrently and then further modified, if necessary, to maintain the prothrombin time within the desired limits. Also, it may increase the dosage requirements of insulin or of oral hypoglycemic agents if used concurrently with them.

Usual dose—initial, 1 to 2 mg. daily; maintenance, 4 to 8 mg. daily.

Usual dose range—1 to 8 mg. daily.

Occurrence
Dextrothyroxine Sodium Tablets U.S.P.

Cholestyramine Resin U.S.P., Cuemid®, Questran®, is the chloride form of a strongly basic anion-exchange resin. It is a styrene copolymer with divinylbenzene with quaternary ammonium functional groups. It has an affinity for bile salts so that the ingested resin combines with bile salts in the intestinal tract, leading to their increased fecal excretion.[71] In the process the chloride ion is exchanged for the bile salt anion. This makes the resin useful in pruritis resulting from partial biliary obstruction, a condition that leads to increased serum bile salt levels. Cholestyramine resin is also useful in lowering plasma lipids. By reducing the amounts of bile acids that are reabsorbed there results an increased catabolism of cholesterol to bile acids in the liver. Although the biosynthesis of cholesterol is increased, it appears that the rate of catabolism is greater, resulting in a net decrease in plasma cholesterol levels by affecting LDL clearance.

Cholestyramine

Cholestyramine resin does not bind with drugs that are neutral or with those that are amine salts; however, it is possible that acidic drugs (in the anion form) could be bound. For example, in animal tests it was found the absorption of aspirin given concurrently with the resin was only moderately depressed during the first 30 minutes.

Cholestyramine is the drug of choice for type IIa hyperlipoproteinemia. When used in conjunction with a controlled diet, it reduces β-lipoproteins. The drug is an insoluble polymer and thus probably one of the safest because it is not absorbed from the gastrointestinal tract to cause systemic toxic effects.

Category—ion-exchange resin (bile salts).

Usual dose—4 g. three times daily.

Usual dose range—10 to 16 g. daily.

Usual pediatric dose—children under 6 years of age, dosage is not established; over 6 years of age, 80 mg. per kg. of body weight or 2.35 g. per square meter of body surface, 3 times daily, or see Usual dose.

Colestipol Hydrochloride, Colestid®, is a high molecular-weight, insoluble, granular copolymer of tetraethylenepentamine and epichlorohydrin. It functions as anion-exchange, resin-sequestering agent in a manner similar to that of cholestyramine. Colestipol reduces cholesterol levels without affecting triglycerides and seems to be especially effective in treatment of type II hyperlipoproteinemias.

Colestipol

Usual dose—7.5 g. 2 to 4 times daily.

Usual dosage range—15 to 30 g. daily.

Niacin U.S.P., nicotinic acid, 3-pyridinecarboxylic acid, is effective in the treatment of all types of hyperlipoproteinemias with the exception of type I at doses above those given as a vitamin supplement. The drug reduces VLDL synthesis and subsequently its plasma products, IDL and LDL. Plasma triglyceride levels are reduced because of the decreased VLDL production. Cholesterol levels are lowered, in turn, because of the decreased rate of LDL formation from VLDL. Although niacin is the drug of choice for type II hyperlipoproteinemias, its use is limited because of the vasodilating side-effects. Flushing occurs in practically all patients but generally subsides when the drug is discontinued.

The basis of the hypolipidemic effects of niacin may be due to its ability to inhibit lipolysis, i.e., prevent the release of free fatty acids (FFA) and glycerol from fatty tissues. As a consequence there is a reduced reserve of FFA in the liver, and diminution of lipoprotein biosynthesis occurs, which results in a reduction of the production of VLDL. The decreased formation of lipoproteins leads to a pool of unused cholesterol normally incorporated in VLDL. This excess cholesterol is then excreted through the biliary tract.

Niacin (nicotinic acid) may be administered as aluminum nicotinate, Nicalex®. This is a complex of aluminum hydroxy nicotinate and niacin. The aluminum salt is hydrolyzed to aluminum hydroxide and niacin in the stomach. There seems to be no advantage of the aluminum salt over the free acid. The incidence of hepatic reaction appears more prevalent than with niacin.

Nicotinic acid has been esterified with the purpose of prolonging its hypolipidemic effect. Pentaerythritol tetranicotinate has been more effective experimentally than niacin in reducing cholesterol in rabbits. Sorbitol and myoinositol hexanicotinate polyesters have been employed in the treatment of patients with atherosclerosis obliterans.

The Coronary Drug Project investigation of niacin for long-term management against recurrent myocardial infarctions showed that the drug did not reduce total mortality of the patients. Niacin did appear to have an effect against recurrent nonfatal myocardial infarction, but the drug increased the incidence of cardiac arrhythmias and was concluded to be a poor risk as a long-term drug for such patients.

The usual maintenance dose of niacin is 3 to 6 grams per day given in three divided doses. The drug is usually given at mealtimes to reduce gastric irritation that often accompanies the large doses.

β-Sitosterol is a plant sterol whose structure is identical to that of cholesterol except for the substituted ethyl group on C-24 of its side chain. Although the mechanism of its hypolipidemic effect is not clearly understood, it is suspected that the drug inhibits the absorption of dietary cholesterol from the gastrointestinal tract. Sitosterols are poorly absorbed from the mucosal lining and appear to compete with cholesterol for absorption sites in the intestine.

β-Sitosterol

Usual dose—3 g.

Usual dosage range—3 to 18 g. daily.

Probucol, Lorelco®, DH-581, 4,4'-[(1-methylethyl-idene)*bis*(thio)]*bis*[2,6-*bis*(1,1-dimethylethyl)]phe-nol, is a novel hypolipidemic agent. It causes reduction of both liver- and serum-cholesterol levels by an as yet undisclosed mechanism. However, it does not inhibit the latter stages of cholesterol synthesis as did triparanol (Mer-29). It is effective in reducing levels of LDL and is used in those hyperlipoproteinemias that are characterized by elevated LDL levels.

Usual dose—500 mg. twice daily.

Usual pediatric dose—safety and effectiveness in children not established.

Probucol

ANTICOAGULANTS

In 1905 Morawitz introduced a theory of blood clotting based on the existence of four factors: thromboplastin (thrombokinase), prothrombin, fibrinogen, and ionized calcium. The clotting sequence proposed was that when tissue damage occurred thromboplastin entered the blood from the platelets and reacted with prothrombin in the presence of calcium to form thrombin. Thrombin then reacted with fibrinogen to form insoluble fibrin that enmeshed red blood cells to create a clot. The concept remained unchallenged for almost 50 years but has been modified now to accommodate the discovery of numerous additional factors that enter into the clotting mechanism (Table 14-6).

MECHANISM OF BLOOD COAGULATION

The process of blood coagulation as it is understood today (Fig. 14-14) involves a series of steps that occurs in a cascade fashion and terminates in the formation of a fibrin clot. Blood coagulation occurs by activation of either an *intrinsic* pathway (a relatively slow process of clot formation or the *extrinsic* pathway, which has a much faster rate of fibrin formation. Both pathways merge into a common pathway for the conversion of prothrombin to thrombin and subsequent transformation of fibrinogen to the insoluble strands of fibrin. Lysis of intravascular clots occurs through a plasminogen–plasmin system that consists of plasminogen, plasmin, urokinase, tissue lyokinase, natural tissue activators, and also some undefined inhibitors of the plasminogen–plasmin system.

The intrinsic pathway refers to the system for coagulation that occurs from the interaction of factors circulating in the blood. It is activated when blood comes into contact with a damaged vessel wall or a foreign substance. Each of the plasma coagulation factors (Table 14-6), with the exception of Factor III (tissue thromboplastin), circulates as an inactive proenzyme. Except for fibrinogen, which precipitates as fibrin, these factors are activated usually by enzymatic removal of a small peptide in the cascade of reactions that make up the clotting sequence (Fig. 14-14). The extrinsic clotting system refers to the mechanism by which thrombin is generated in plasma after the addition of tissue extracts. When various tissues such as brain or lung (containing thromboplastin) are added to blood, a complex between thromboplastin and factor VII in the presence of calcium ions activates factor X, thereby bypassing the time-consuming steps of the intrinsic pathway which form factor X. The intrinsic and extrinsic pathways interact in vivo. Small amounts of thrombin formed early following stimulation of the extrinsic pathway accelerate clotting by the intrinsic pathway by activating factor VIII. Thrombin also speeds up the clotting rate by activation of factor V located in the common pathway. Thrombin then converts the soluble protein, fibrinogen, into an insoluble fibrin gel and also activates factor XIII, which stabilizes the fibrin gel in the presence of calcium by inducing the formation of cross-linking between the chains of the fibrin monomer.

TABLE 14-6

THE ROMAN NUMERICAL NOMENCLATURE OF BLOOD-CLOTTING FACTORS AND SOME COMMON SYNONYMS

Factor	Synonyms
I	Fibrinogen
II	Prothombin
III	Thromboplastin; tissue factor
IV	Calcium
V	Proaccelerin; accelerator globulin; labile factor
VI	(This number is not now used)
VII	Proconvertin; stable factor; autoprothombin I, SPCA
VIII	Antihaemophilic factor; antihaemophilic globulin; platelet cofactor I; antihaemophilic factor A
IX	Plasma thromboplastin component (PTC); Christmas factor; platelet cofactor II; autoprothrombin II; antihaemophilic factor B
X	Stuart-Power factor, Stuart Factor, autoprothombin III
XI	Plasma thromboplastin antecedent (PTA); Antihaemophilic Factor C
XII	Hageman factor
XIII	Fibrin stabilizing factor; Fibrinase Laki-Lorand factor

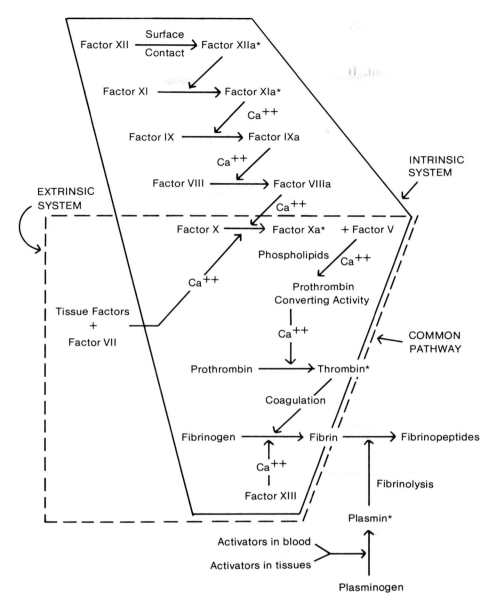

FIG. 14-14. *Scheme of blood coagulation and fibrinolysis. Reactions enclosed by the solid line are of the "intrinsic system," while those enclosed by the dotted line are of the "extrinsic system." The asterisk denotes a serine protease.*

ANTICOAGULANT MECHANISMS

The biosynthesis of prothrombin is dependent upon an adequate supply of vitamin K. A deficiency of vitamin K results in the formation of a defective prothrombin molecule. The defective prothrombin is antigenically similar to normal prothrombin but has reduced calcium-binding ability and no biological activity. In the presence of calcium ions, normal prothrombin adheres to the surface of phospholipid vesicles and greatly increases its activity in the clotting mechanism. The defect in the abnormal prothrombin

has been located in the N-terminal portion, in which the second carboxyl residue has not been added to the γ-carbon atom of some glutamic-acid residues on the prothrombin molecule to form γ-carboxylglutamic acid.[73] The administration of vitamin K-antagonists results in the decreased synthesis of a biologically active prothrombin molecule and results in an increase in the clotting time of blood in humans. The reaction involved in forming an active prothrombin is a vitamin K-dependent carboxylase located in the microsomal fraction of liver cells. The enzyme uses bicarbonate as the source of the carboxyl group (Fig. 14-

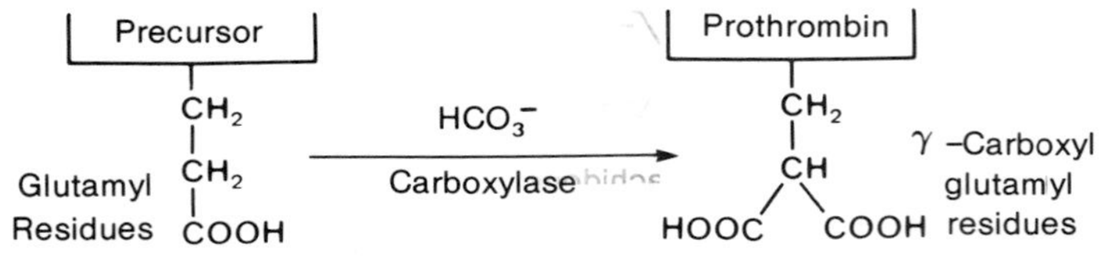

FIG. 14-15. Vitamin-K dependent carboxylase system.[74]

15). The role of vitamin K in the carboxylase reaction has not been resolved. It has been suggested that vitamin K acts to activate or carry bicarbonate or carbon dioxide in the carboxylation reaction. The increased clotting time is a consequence of the lack of prothrombin with γ-glutamyl residues to bind to phospholipid bilayer surfaces in the presence of Ca^{++}.

PLATELET AGGREGATION AND INHIBITORS

Blood platelets play a pivotal role in hemostasis and thrombus formation. Actually they have two roles in the cessation of bleeding, a hemostatic function in which platelets, through their mass, cause a physical occlusion of openings in blood vessels and a thromboplastic function in which the chemical constituents of the platelets take part in the blood-coagulation mechanism. The circulatory system is self-sealing because of the clotting properties of blood. On the other hand, the pathologic formation of clots within the circulatory system creates a potentially serious clinical situation that must be dealt with through the use of anticoagulants.

Platelets do not adhere to intact endothelial cells but do to subendothelial tissues exposed by injury, to cause hemostasis. Platelets bind to collagen in the vessel wall and trigger other platelets to adhere to them. This adhesiveness is accompanied by a change in shape of the platelets and may be caused by mobilization of calcium bound to the platelet membrane. The growth of the platelet mass is dependent on ADP released by the first few adhering cells and enhances the aggregation process. The shape change and aggregation of platelets is termed *phase I aggregation* and is reversible. A secondary phase (phase II) immediately follows with additional platelet aggregation. In phase II the platelets will undergo a secretory process during which enzymes, such as cathepsin and acid hydrolases along with fibrinogen, are released from α-granules in the platelets, and ADP, ATP, serotonin, and calcium are released from dense bodies in the platelets. The dense bodies are likened to the storage granules associated with adrenergic neurons. The selective process that releases the contents of dense bodies is called *release I* and the process that releases the contents of α-granules is referred to as *release II*. Increased levels of c-AMP inhibit platelet aggregation. c-AMP activates

specific dependent kinases which form protein-phosphate complexes that chelate calcium ions. The reduced levels of calcium inhibit aggregation (Fig. 14-16). Inhibitors of platelet aggregation can increase c-AMP levels by either stimulating adenyl cyclase or inhibiting phosphodiesterase.[75] Substances such as glucagon, adenosine, isoprenaline, and PGE_1 increase c-AMP levels and inhibit platelet aggregation. Drugs such as theophylline, aminophylline, dipyramidole, papaverine, and adenosine inhibit phosphodiesterase and also aggregation of platelets. Epinephrine, collagen, serotonin and PGE_2 inhibit adenyl cyclase and stimulate platelet aggregation.[76] The role of platelets in arterial thrombosis is similar to their role in hemostasis. The factors contributing to venous thrombosis are circulatory stasis, excessive generation of thrombin formation of fibrin, and, to a lesser extent than in the artery, platelet aggregation.

A number of different drugs and biochemicals have antithrombotic properties because of their ability to inhibit platelet aggregation. PGE_1 is one of the most potent inhibitors of platelet aggregation known. Its effect is attributed to stimulation of adenyl cyclase and subsequent accumulation of c-AMP.[77] In addition, studies with prostaglandin precursors and prostaglandin synthetase inhibitors suggest that prostaglandin formation is involved in the platelet-release reaction and subsequent aggregation. When platelets are stimulated with a variety of release inducers (such as ADP, collagen, and thrombin), arachidonic acid made

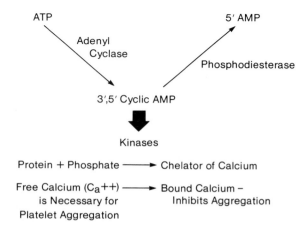

FIG. 14-16. Role of adenosine 3', 5'-cyclic monophosphate (cyclic AMP) in inhibition of platelet aggregation.

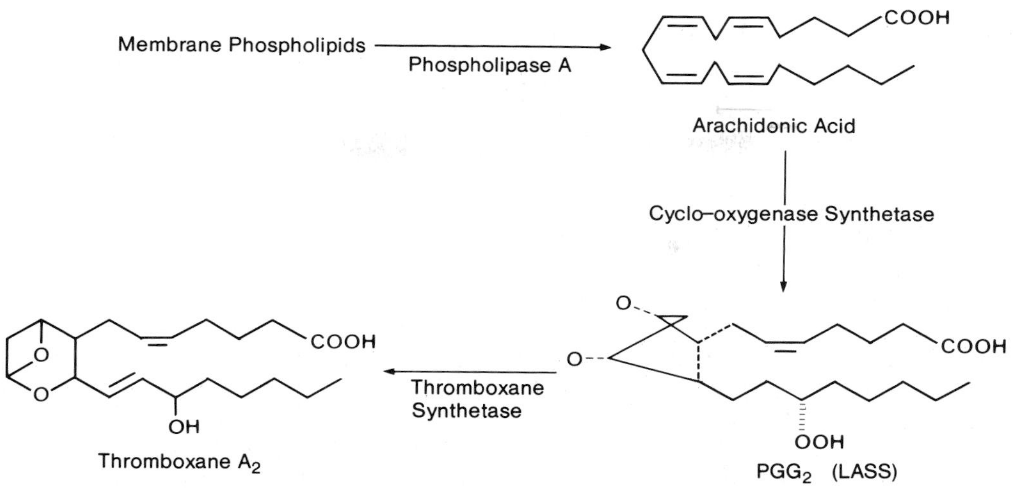

Membrane Phospholipids ⟶ Phospholipase A ⟶ Arachidonic Acid

Cyclo–oxygenase Synthetase

Thromboxane Synthetase

Thromboxane A$_2$

PGG$_2$ (LASS)

available by hydrolysis of membrane phospholipid is converted by cyclo-oxygenase to a labile cyclic endoperoxide, PGH$_2$ and PGG$_2$, the latter of which is a precursor of thromboxane A$_2$. Thromboxane A$_2$ can directly induce the release-I reaction. The endoperoxides, PGH$_2$ and PGG$_2$ have been referred to as *labile aggregation-stimulating substance* (LASS).[78] Aspirin, sulfinpyrazone, and indomethacin have an inhibitory effect on platelet aggregation. They inhibit cyclo-oxygenase, the enzyme that controls the formation of PG endoperoxides and increases the tendency for platelets to aggregate. Aspirin also inhibits the platelet-release reaction; it antagonizes phase II of ADP or epinephrine-induced aggregation as well as collagen- and thrombin-induced aggregation. Dipyridamole inhibits adenosine deaminase and adenosine uptake by platelets. As a result, the increased plasma concentrations of adenosine inhibit ADP-induced aggregation of platelets (Table 14-7).

A more recent development in prostaglandin effects on clotting has shown prostacyclin to be a very potent inhibitor of platelet aggregation. It causes an increase in c-AMP levels in blood platelets. A hypothesis by Vane[79] proposes that prostacyclin is formed from endoperoxides, PGG$_2$ or PGH$_2$, which are released from platelets when they collide with normal endothelial cells in blood vessels. The endothelial cells then release an enzyme that converts the endoperoxides into prostacyclin, which acts to prevent platelet aggregation. Damaged endothelial cells do not have the enzyme and so thromboxane A$_2$ is formed, enhancing clot formation.

TABLE 14–7

INDIVIDUAL CHARACTERISTICS OF COMMON ORAL ANTICOAGULANTS USED IN CLINICAL PRACTICE

Drug	Trade Name	Initial Dose (mg.) *	Maintenance Dose (mg.)	Time to Reach Therapeutic Level (Hours)	Time to Return to Near-Normal "Pro Time" (Days)
Coumarin Compounds					
Bishydroxycoumarin	Discoumarol	200–300	25–200	36–72	5–6
Warfarin	Coumadin, Panwarfarin, Prothromadin	40–60	2–10	36–48	3$\frac{1}{2}$–4$\frac{1}{2}$
Ethyl biscoumacetate	Tromexan	900–1200	150–900	18–36	1$\frac{1}{2}$–2$\frac{1}{2}$
Phenprocoumen	Liquamar	18–24	1–6	30–48	7–10
Indandione Compounds					
Phenindione	Danilone, Dindevan, Hedulin	200–300	25–100	36–48	3–4
Anisindione	Miraden	200–300	25–250	36-60	3–4

*When the traditional "loading dose" technique is used.
(Deykin, D.: N. Engl. J. Med. 283:691, 1970.)

Prostacyclin

Retardation of clotting is important in blood transfusions, to avoid thrombosis after operation or from other causes, to prevent recurrent thrombosis in phlebitis and pulmonary embolism and to lessen the propagation of clots in the coronary arteries. This retardation may be accomplished by agents that inactivate thrombin (heparin) or those substances that prevent the formation of prothrombin in the liver—the coumarin derivatives and the phenylindanedione derivatives.

Although heparin is a useful anticoagulant, it has limited applications. Many of the anticoagulants in use today were developed following the discovery of dicumarol, an anticoagulant that is present in spoiled sweet clover. These compounds are orally effective, but there is a lag period of 18 to 36 hours before they significantly increase the clotting time. Heparin, in contrast, produces an immediate anticoagulant effect following intravenous injection. A major disadvantage of heparin is that the only effective therapeutic route is parenteral.

Dicumarol and related compounds are not vitamin K antagonists in the classic sense. They appear to act by interfering with the function of vitamin K in the liver cells which are the sites of synthesis of the clotting factors, including prothrombin. This lengthens the clotting time by decreasing the amount of biologically active prothrombin in the blood.

The discovery of dicumarol and related compounds as potent reversible* competitors of vitamin K-coagulant-promoting properties led to the development of antivitamin K-compounds such as phenindione, which was designed in part according to metabolite-antimetabolite concepts. The active compounds of the phenylindanedione series are characterized by a phenyl, a substituted phenyl or a diphenylacetyl group in the 2-position. Another requirement for activity is a keto group in the 1- and 3-positions, one of which may form the enol tautomer. A second substituent, other than hydrogen, at the 2-position prevents this keto-enol tautomerism and the resulting compounds are ineffective as anticoagulants.

The activity of dicumarol and related compounds and the phenindione types can be reversed by the proper amounts of vitamin K_1†, menadione, etc.

Out of hundreds of active anticoagulants the following are accepted for clinical use.

PRODUCTS

Protamine Sulfate U.S.P. has an anticoagulant effect, but it counteracts the action of heparin if used in the proper amount and is used as an antidote for the latter in cases of overdosage. It is administered intravenously in a dose depending on the circumstances.

Usual adult dose—I.V., 1 mg. of protamine sulfate for every 90 U.S.P. Units of beef lung heparin sodium or for every 115 U.S.P. Units of porcine intestinal mucosa heparin sodium to be neutralized, administered slowly in 1 to 3 minutes, up to a maximum of 50 mg. in any 10-minute period.

Usual pediatric dose—see Usual Adult Dose.

Occurrence
Protamine Sulfate Injection U.S.P.
Protamine Sulfate for Injection U.S.P.

Dicumarol U.S.P., 3,3′-methylene*bis*(4-hydroxycoumarin), is a white or creamy-white, crystalline powder with a faint, pleasant odor and a slightly bitter taste. It is practically insoluble in water or alcohol, slightly soluble in chloroform and is dissolved readily by solutions of fixed alkalies. The effects after administration require 12 to 72 hours to develop and persist for 24 to 96 hours after discontinuance.

Dicumarol

Dicumarol is used alone or as an adjunct to heparin in the prophylaxis and treatment of intravascular clotting. It is employed in postoperative thrombophlebitis, pulmonary embolus, acute embolic and thrombotic occlusion of peripheral arteries and recurrent idiopathic thrombophlebitis. It has no effect on an already formed embolus but may prevent further intravascular clotting. Since the outcome of acute coronary thrombosis is largely dependent on extension of the clot and formation of mural thrombi in the heart chambers with subsequent embolization, dicumarol has been used in this condition. It also has been ad-

---

* At high levels dicumarol is not reversed by vitamin K.

† Vitamin K_1 is considerably more effective than menadione.

ministered to arrest impending gangrene after frostbite. The dose, after determination of the prothrombin clotting time, is 200 to 300 mg., depending on the size and the condition of the patient, the drug being given orally in the form of capsules or tablets. On the second day and thereafter, it may be given in amounts sufficient to maintain the prothrombin clotting time at about 30 seconds. If hemorrhages should occur, 50 to 100 mg. of menadione sodium bisulfite is injected, supplemented by a blood transfusion.

Warfarin Sodium U.S.P., Coumadin® Sodium, Panwarfin®, 3-(α-acetonylbenzyl)-4-hydroxycoumarin sodium salt, is a white, odorless, crystalline powder, having a slightly bitter taste; it is slightly soluble in chloroform, soluble in alcohol or water. A 1 percent solution has a pH of 7.2 to 8.5.

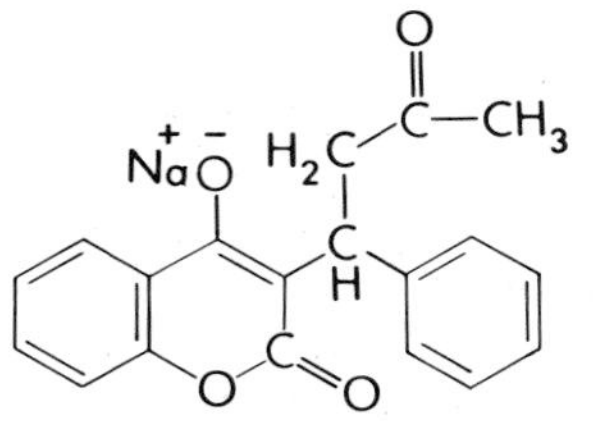

Warfarin Sodium

By virtue of its great potency warfarin at first was considered unsafe for use in humans and was utilized very effectively as a rodenticide, especially against rats. However, when used in the proper dosage level, it can be used in humans, especially by the intravenous route.

Warfarin Potassium U.S.P., Athrombin-K®, 3-(α-acetonylbenzyl)-4-hydroxycoumarin potassium salt. Warfarin potassium is readily absorbed after oral administration, with a therapeutic hypoprothrombinemia being produced in 12 to 24 hours after administration of 40 to 60 mg. This salt is therapeutically interchangeable with warfarin sodium.

Phenprocoumon U.S.P., Liquamar®, 3-(α-ethylbenzyl)-4-hydroxycoumarin. This drug has been shown to possess marked and prolonged anticoagulant activity.

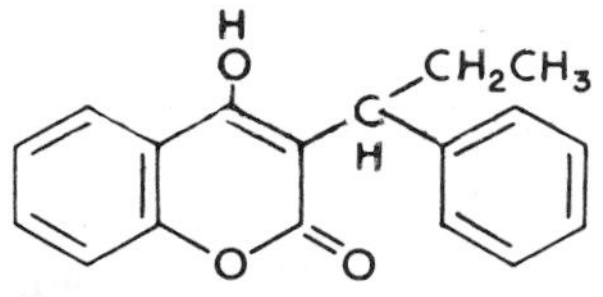

Phenprocoumon

Phenindione U.S.P., Hedulin®, Danilone®, 2-phenyl-1,3-indandione, is an oral anticoagulant specifically designed to function as an antimetabolite for vitamin K. It is a pale-yellow crystalline material that is slightly soluble in water but very soluble in alcohol. It is more prompt-acting than dicumarol. The rapid elimination is presumed to make the drug safer than others of the class.

TABLE 14–8

ANTICOAGULANTS

Name Proprietary Name	Preparations	Usual Adult Dose*	Usual Dose Range*
Dicumarol U.S.P.	Dicumarol Capsules U.S.P. Dicumarol Tablets U.S.P.	25 to 200 mg. once daily, as indicated by prothrombin-time determinations	
Warfarin Sodium U.S.P. *Coumadin, Panwarfin*	Warfarin Sodium for injection U.S.P. Warfarin Sodium Tablets U.S.P.	Oral, I.M. or I.V., 10 to 15 mg. daily for 2 to 3 days, then 2 to 10 mg. daily, as indicated by prothrombin-time determination	
Warfarin Potassium U.S.P. *Athrombin-K*	Warfarin Potassium Tablets U.S.P.	Oral, 10 to 15 mg. daily for 2 or 3 days, then 2 to 10 mg. daily, as indicated by prothrombin-time determinations	
Phenprocoumon U.S.P. *Liquamar*	Phenprocoumon Tablets U.S.P.		Initial, 21 mg. the first day, 9 mg. the second day, and 3 mg. the third day; maintenance, 1 to 4 mg. daily, according to prothrombin level
Phenindione U.S.P. *Hedulin*	Phenindione Tablets U.S.P.	Oral, 750 μg. to 6 mg. daily, as indicated by prothrombin-time determinations	
Anisindione U.S.P. *Miradon*	Anisindione Tablets	Initial, 300 mg. the first day, 200 mg. the second day, 100 mg. the third day; maintenance, 25 to 250 mg. daily	

*See *U.S.P. D.I.* for complete dosage information.

Phenindione

Anisindione U.S.P., Miradon®, 2-(*p*-methoxy-phenyl)-1,3-indandione, 2-(*p*-anisyl)-1,3-indandione, is a *p*-methoxy congener of phenindione. It is a white crystalline powder, slightly soluble in water, tasteless, and well absorbed after oral administration.

Anisindione

In instances where the urine may be alkaline, an orange color may be detected. This is due to metabolic products of anisindione and is not hematuria.

SCLEROSING AGENTS

Several different kinds of irritating agents have been used for the obliteration of varicose veins. These are generally called sclerosing agents and include invert sugar solutions, dextrose, ethyl alcohol, iron salts, quinine and urea hydrochloride, fatty acid salts (soaps) and certain sulfate esters. Many of these preparations contain benzyl alcohol which acts as a bacteriostatic agent and relieves pain after injection.

PRODUCTS

Morrhuate Sodium Injection U.S.P., is a sterile solution of the sodium salts of the fatty acids of cod-liver oil. The salt (a soap) was introduced first in 1918 as a treatment for tuberculosis and, in 1930, it was reported to be useful as a sclerosing agent. Morrhuate sodium is not a single entity, although morrhuic acid has been known for years. Morrhuate sodium is a mixture of the sodium salts of the saturated and unsaturated fatty acids from cod-liver oil.

The preparation of the free fatty acids of cod-liver oil is carried out by saponification with alkali and then acidulation of the resulting soap. The free acids are dried over anhydrous sodium sulfate before being

dissolved in an equivalent amount of sodium hydroxide solution. Morrhuate sodium is obtained by careful evaporation of this solution. The result is a pale-yellowish, granular powder having a slight fishy odor.

Commercial preparations are usually 5 percent solutions, which vary in properties and in color from light yellow to medium yellow to light brown. They are all liquids at room temperature and have congealing points that range from $-11°$ to $7°$. A bacteriostatic agent, not to exceed 0.5 percent, and ethyl or benzyl alcohol to the extent of 3 percent, may be added.

Usual dose—intravenous, by special injection, 1 ml. to a localized area.

Usual dose range—500 μl. to 5 ml.

SYNTHETIC HYPOGLYCEMIC AGENTS

The discovery that certain organic compounds will lower the blood sugar level is not a recent one. In 1918 guanidine was shown to lower the blood sugar level. The discovery that certain trypanosomes need much glucose and will die in its absence was followed by the discovery that galegine lowered the blood sugar level and was weakly trypanocidal. This led to the development of a number of very active trypanocidal agents such as the bisamidines, diisothioureas, bisguanidines, etc. Synthalin (trypanocidal at 1:250,000,000) and pentamidine are outstanding examples of very active trypanocidal agents. Synthalin lowers the blood sugar level in normal, depancreatized and completely alloxanized animals. This may be due to a reduction in the oxidative activity of mitochondria resulting from inhibition of the mechanisms which simultaneously promote phosphorylation of adenosine diphosphate and stimulate oxidation by nicotinamide adenine dinucleotide (NAD) in the citric acid cycle. Hydroxystilbamidine Isethionate U.S.P. is used as an antiprotozoal agent.

Galegine

In 1942, *p*-aminobenzenesulfonamidoisopropylthiadiazole (an antibacterial sulfonamide) was found to produce hypoglycemia. These results stimulated the research for the development of synthetic hypoglycemic agents, several of which are in use today.

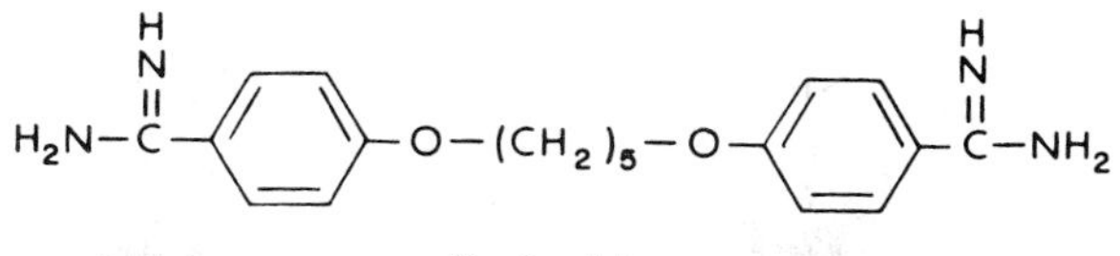

Pentamidine

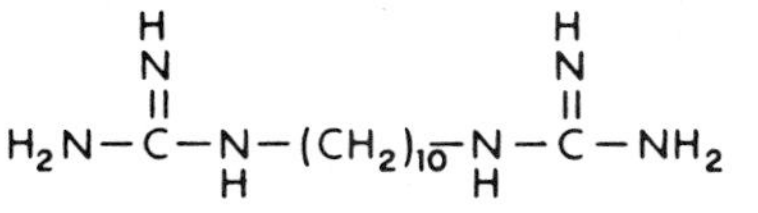

Synthalin

Sulfonylureas became widely available in 1955 for treatment of nonketosis-prone mild diabetes and are still the drugs of choice. A second class of compounds, the biguanides, in the form of a single drug, phenformin, was used since 1957. However, it has been withdrawn from the U.S. market recently because of its toxic effect. Phenformin causes lactic acidosis in which fatalities have been reported.

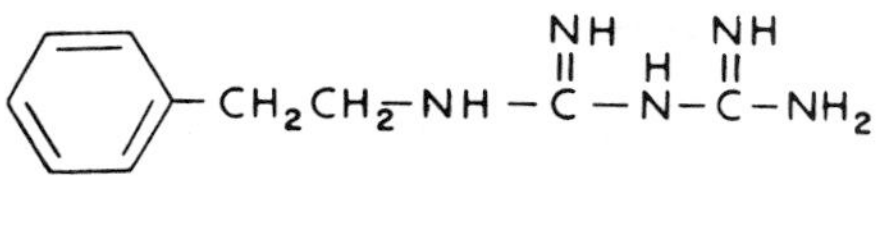

Phenformin

SULFONYLUREAS

The sulfonylureas may be represented by the following general structure:

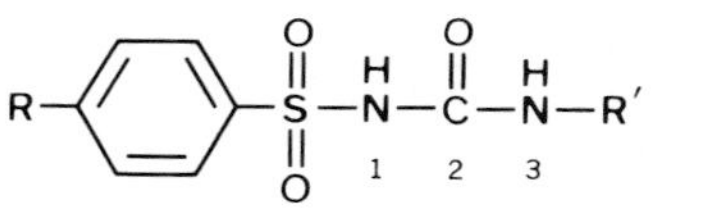

These are urea derivatives with an arylsulfonyl group in the 1-position and an aliphatic group at the 3-position. The aliphatic group, R', confers lipophilic properties to the molecule. Maximal activity results when R' consists of 3 to 6 carbon atoms as in chlorpropamide, tolbutamide and acetohexamide. Aryl groups at R' generally give toxic compounds. The R group on the aromatic ring primarily influences the duration of action of the compound. Tolbutamide disappears quite rapidly from the bloodstream through being metabolized to the inactive carboxy compound which is rapidly excreted. On the other hand, chlorpropamide is metabolized more slowly and persists in the blood for a much longer time.

The mechanism of action of the sulfonylureas is to increase the release of insulin from the functioning *beta* cells of the intact pancreas. In the absence of the pancreas, they have no significant effect on blood glu-

cose. They may have other actions, such as inhibition of glycogenolysis in the liver, but these are still uncertain. This group of drugs is of most value in the diabetic patient whose disease had its onset in adulthood. Accordingly, the group of sulfonylureas is not indicated in the juvenile-onset diabetic.

PRODUCTS

Tolbutamide U.S.P., Orinase®, 1-butyl-3-(*p*-tolylsulfonyl)urea, occurs as a white crystalline powder that is insoluble in water and soluble in alcohol or aqueous alkali. It is stable in air.

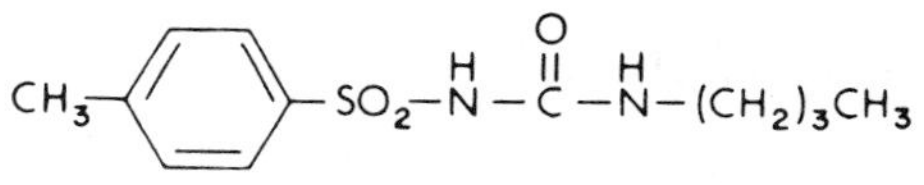

Tolbutamide

Tolbutamide is absorbed rapidly in responsive diabetic patients. The blood sugar level reaches a minimum after 5 to 8 hours. It is oxidized rapidly in vivo to 1-butyl-3-(*p*-carboxyphenyl)sulfonylurea, which is inactive. The metabolite is freely soluble at urinary pH; however, if the urine is strongly acidified, as in the use of sulfosalicylic acid as a protein precipitant, a white precipitate of the free acid may be formed.

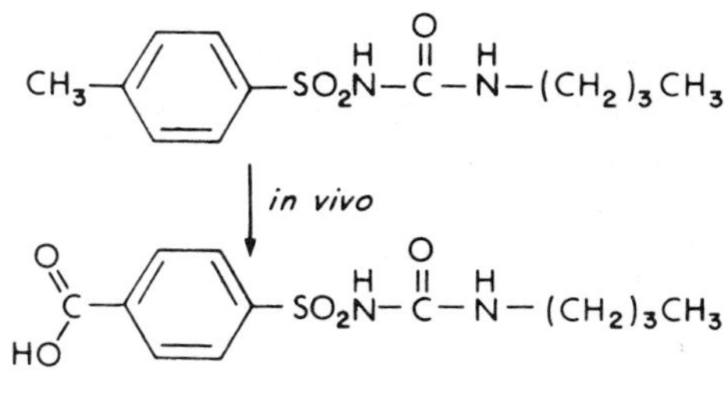

Tolbutamide should be used only where the diabetes patient is an adult or shows maturity onset in character, and the patient should adhere to dietary restrictions.

Tolbutamide Sodium U.S.P., Orinase® Diagnostic, 1-butyl-3-(*p*-tolylsulfonyl)urea monosodium salt. Tolbutamide sodium is a white crystalline powder, freely soluble in water, soluble in alcohol and in chloroform and very slightly soluble in ether.

This water-soluble salt of tolbutamide is used intravenously for the diagnosis of mild diabetes mellitus and of functioning pancreatic islet cell adenomas. The sterile dry powder is dissolved in sterile water for injection to make a clear solution which then should be administered within one hour. The main route of breakdown is to butylamine and sodium *p*-toluenesulfonamide.

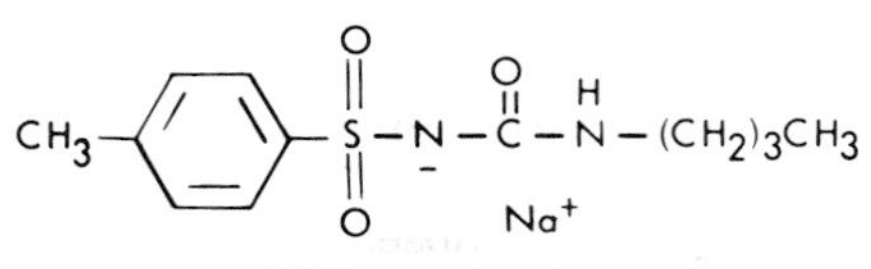

Tolbutamide Sodium

Chlorpropamide U.S.P., Diabinese®, 1-[(*p*-chlorophenyl)sulfonyl]-3-propylurea. Chlorpropamide is a white crystalline powder, practically insoluble in water, soluble in alcohol and sparingly soluble in chloroform. It will form water-soluble salts in basic solutions. This drug is more resistant to conversion to inactive metabolites than is tolbutamide and, as a result, has a much longer duration of action. One study showed that about half of the drug is excreted as metabolites, the principal one being hydroxylated in the 2-position of the propyl side chain.[23] After control of the blood sugar levels the maintenance dose is usually on a once-a-day schedule.

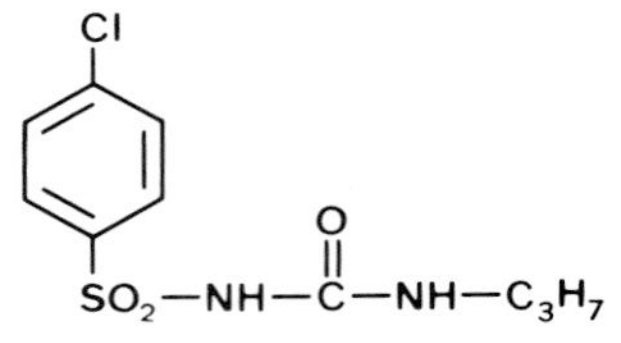

Chlorpropamide

Tolazamide U.S.P., Tolinase®, 1-(hexahydro-1H-azepin-1-yl)-3-(*p*-tolylsulfonyl)urea. This agent is an analog of tolbutamide and is reported to be effective, in general, under the same circumstances where tolbutamide is useful. However, tolazamide appears to be more potent than tolbutamide, and is nearly equal in potency to chlorpropamide. In studies with radioactive tolazamide, investigators found that 85 percent of an oral dose appears in the urine as metabolites which are more soluble than tolazamide itself.

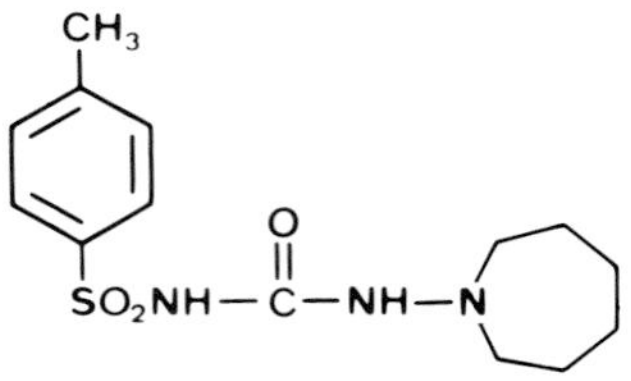

Tolazamide

Acetohexamide U.S.P., Dymelor®, 1-[(*p*-acetylphenyl)sulfonyl]-3-cyclohexylurea. Acetohexamide is chemically and pharmacologically related to tolbutamide and chlorpropamide. Like the other sulfonylureas, acetohexamide lowers the blood sugar, primarily by stimulating the release of endogenous insulin.[80]

TABLE 14-9

SYNTHETIC HYPOGLYCEMIC AGENTS

Name Proprietary Name	Preparations	Usual Adult Dose*
Tolbutamide U.S.P. *Orinase*	Tolbutamide Tablets U.S.P.	Initial, 500 mg. once or twice daily, adjusted gradually according to patient response or until the total daily dose reaches 3 g.
Tolbutamide Sodium U.S.P. *Orinase Diagnostic*	Sterile Tolbutamide Sodium U.S.P.	I.V., the equivalent of 1 g. of tolbutamide over a 2- to 3-minute period
Chlorpropamide U.S.P. *Diabinese*	Chlorpropamide Tablets U.S.P.	Oral, 100 to 250 mg. once daily initially, the dosage being increased by 50 to 125 mg. at one-week intervals until diabetic control is obtained or until the total daily dose reaches 750 mg.
Tolazamide U.S.P. *Tolinase*	Tolazamide Tablets U.S.P.	Oral, 100 to 250 mg. once daily initially, the dosage being adjusted gradually until diabetic control is obtained or until the total daily dose reaches 1 g.
Acetohexamide U.S.P. *Dymelor*	Acetohexamide Tablets U.S.P.	Oral, 250 mg. once daily initially, the dosage being adjusted gradually until diabetic control is obtained or until the total daily dose reaches 1.5 g.

*See *U.S.P. D.I.* for complete dosage information.

Acetohexamide

Acetohexamide is metabolized in the liver to a reduced form—the α-hydroxyethyl derivative. This metabolite, the main one in humans, possesses hypoglycemic activity. Acetohexamide is intermediate between tolbutamide and chlorpropamide in potency and duration of effect on blood sugar levels.

THYROID HORMONES

Desiccated, defatted thyroid substance has been used for many years as replacement therapy in thyroid gland deficiencies. The efficacy of the whole gland is now known to depend on its thyroglobulin content. This is an iodine-containing globulin. Thyroxine was obtained as a crystalline derivative by Kendall of the Mayo Clinic in 1916. It showed much the same action as the whole thyroid substance. Later thyroxine was synthesized by Harington and Barger in England. Later studies showed that an even more potent iodine-containing hormone existed, which is now known as triiodothyronine. There is now evidence that thyroxine may be the storage form of the hormone, while triiodothyronine is the circulating form. Another point of view is that, in the blood, thyroxine is more firmly bound to the globulin fraction than is triiodothyronine, which can then enter the tissue cells.

PRODUCTS

Levothyroxine Sodium U.S.P., Synthroid® Sodium, Letter®, Levoroxine®, Levoid®, O-[4-hydroxy-3,5-diiodophenyl]-3,5-diiodo-2-tyrosine monosodium salt, hydrate. This compound is the sodium salt of the *levo* isomer of thyroxine, which is an active physiologic principle obtained from the thyroid gland of domesticated animals used for food by man. It is also prepared synthetically. The salt is a light yellow, tasteless, odorless powder. It is hygroscopic but stable in dry air at room temperature. It is soluble in alkali hydroxides, 1:275 in alcohol and 1:500 in water to give a pH of about 8.9.

Levothyroxine sodium is used in replacement therapy of decreased thyroid function (hypothyroidism). In general, 100 μg. of levothyroxine sodium is clinically equivalent to 30 to 60 mg. of Thyroid U.S.P.

Levothyroxine Sodium

Usual adult dose—initial, mild hypothyroidism, oral 50 to 100 μg. as a single daily dose, with increments of 50 to 100 μg. at 2-week intervals until the desired result is obtained; I.V. or I.M., 100 to 300 μg. as a single daily dose.

Usual pediatric dose—children less than 1 year of age—oral, 25 to 50 μg. daily or a single daily dose for the first year of life; children over 1 year of age—3 to 5 μg. per kg. of body weight until the adult dose is reached; I.V. or I.M., daily dose equal to 75 percent of the usual oral pediatric dose.

Occurrence
Levothyroxine Sodium Tablets U.S.P.

Liothyronine Sodium U.S.P., Cytomel®, O-(4-hydroxy-3-iodophenyl)-3,5-diiodo-L-thyroxine monosodium salt, is the sodium salt of L-3,3′,5-triiodothyronine. It occurs as a light tan, odorless, crystalline powder slightly soluble in water or alcohol and has a specific rotation of +18° to +22° in acid (HCl) alcohol.

Liothyronine Sodium

Liothyronine occurs in vivo together with levothyroxine; it has the same qualitative activities as thyroxine but is more active. It is absorbed readily from the gastrointestinal tract, is cleared rapidly from the bloodstream and is bound more loosely to plasma proteins than is thyroxine, probably due to the less acidic phenolic hydroxyl group.

Its uses are the same as those of levothyroxine, including treatment of metabolic insufficiency, male infertility and certain gynecologic disorders.

Usual adult dose—mild hypothyroidism, initial: oral, 25 μg. daily, with increments of 12.5 or 25 μg. every 1 or 2 weeks until the desired result is obtained; maintenance: oral, 25 to 100 μg. daily.

Occurrence
Liothyronine Sodium Tablets U.S.P.

ANTITHYROID DRUGS

When hyperthyroidism exists (excessive production of thyroid hormones), the condition usually requires surgery, but prior to surgery the patient must be prepared by preliminary abolition of the hyperthyroidism through the use of antithyroid drugs. Thiourea and related compounds show an antithyroid activity, but they are too toxic for clinical use. The more useful drugs are 2-thiouracil derivatives and a closely related 2-thioimidazole derivative. All of these appear to have a similar mechanism of action, i.e., prevention of the iodination of the precursors of thyroxine and triiodothyronine. The main difference in the compounds lies in their relative toxicities.[81]

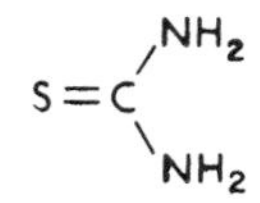

Thiourea 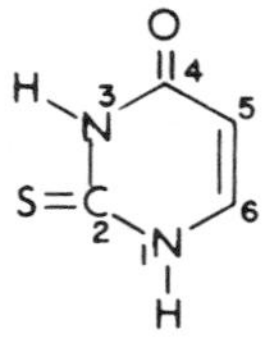   2-Thiouracil

These compounds are well absorbed after oral administration and are excreted in the urine.

The 2-thiouracils, 4-keto-2-thiopyrimidines, are undoubtedly tautomeric compounds and can be represented as follows:

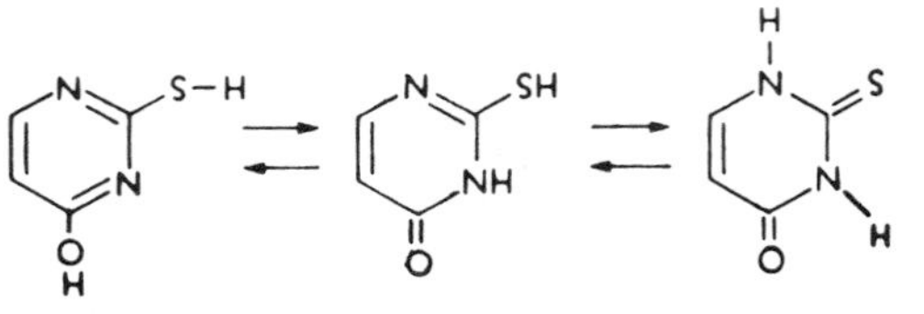

Some 300 related structures have been evaluated for antithyroid activity, but, of these, only the 6-alkyl-2-thiouracils and closely related structures possess useful clinical activity. The most serious side-effect of thiouracil therapy is agranulocytosis.

PRODUCTS

Propylthiouracil U.S.P., Propacil®, 6-propyl-2-thiouracil. Propylthiouracil is a stable, white crystalline powder with a bitter taste. It is slightly soluble in water but is readily soluble in alkaline solutions (salt formation).

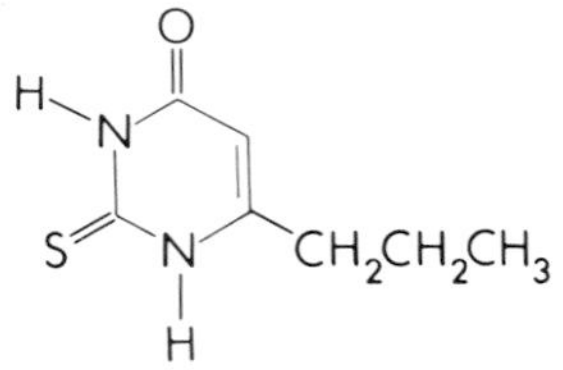

Propylthiouracil

This drug is useful in the treatment of hyperthyroidism. There is a delay in appearance of its effects, because propylthiouracil does not interfere with the activity of thyroid hormones already formed and stored in the thyroid gland. This lag period may vary from several days to weeks, depending on the condition of the patient. The need for three equally spaced doses during a 24-hour period is often stressed, but there is now evidence that a single daily dose is as effective as multiple daily doses in the treatment of most hyperthyroid patients.[82]

Methimazole, U.S.P., Tapazole®, 1-methylimidazole-2-thiol, occurs as a white to off-white crystalline

TABLE 14–10

ANTITHYROID DRUGS

Name Proprietary Name	Preparations	Usual Adult Dose*	Usual Pediatric Dose*
Propylthiouracil U.S.P. *Propacil*	Propylthiouracil Tablets U.S.P.	Initial—oral, 300 to 1,200 mg. daily, divided into 3 doses at eight-hour intervals or 4 doses at six-hour intervals, until patient becomes euthyroid; maintenance—oral, 50 to 800 mg. daily in 2 to 4 divided doses	Initial—children 6 to 10 years of age—oral, 50 to 200 mg. daily in 2 or 3 divided doses; children 10 years of age and over—150 to 600 mg. daily divided into 3 doses at eight-hour intervals; maintenance—determined by response
Methimazole U.S.P. *Tapazole*	Methimazole Tablets U.S.P.	Initial—mild hyperthyroidism: oral, 15 mg. daily, divided into 3 doses at eight-hour intervals; maintenance, 5 to 30 mg. daily, in 2 or 3 divided doses	Initial—oral, 400 μg. per kg. of body weight daily, divided into 3 doses at eight-hour intervals

*See *U.S.P. D.I.* for complete dosage information.

powder with a characteristic odor and is freely soluble in water. A 2 percent aqueous solution has a pH of 6.7 to 6.9. It should be packaged in well-closed, light-resistant containers.

Methimazole

Methimazole is indicated in the treatment of hyperthyroidism. It is more potent than propylthiouracil. The side-effects are similar to those of propylthiouracil. As with other antithyroid drugs, patients using this drug should be under medical supervision. Also, similar to the other antithyroid drugs, methimazole is most effective if the total daily dose is subdivided and given at 8-hour intervals.

REFERENCES

1. Robinson, B. F.: Adv. in Drug Res. 10:93, 1975
2. Aronow, W. S.: Am. Heart J. 84:273, 1972.
3. Sonnenblick, E. et al.: Am. J. Cardiol. 22:328, 1968.
4. Needleman, P., and Johnson E. M., Jr.: J. Pharmacol. Exp. Ther. 184:709, 1973.
5. Needleman, P., and Johnson, E. M., Jr., in Needleham, P. (ed.): Organic Nitrates, p. 97, Berlin, Springer-Verlag, 1975.
6. Needleman, P., Ann. Rev. Pharmacol. Toxicol., 16:81, 1976.
7. Krantz, J. C., et al.: J. Pharmacol. Exp. Ther. 70:323, 1940.
8. Fusari, S. A.: J. Pharm. Sci. 62:123, 1973.
9. Fusari, S. A.: J. Pharm. Sci. 62:2012, 1973.
10. Mason, D. T., et al.: Cardiovascular Drugs 1:75, 1977.
11. Bealer, G. W., and Reuter, H. Physiology (Lond.) 207:191, 1970.
12. Gettes, L. S., et al.: in Eliot, R. S., (ed.): Cardiac Emergencies, p. 245, Futura, Mount Kisco, N.Y., 1977
13. Hoffman, B. F., and Bigger, J. T.: Antiarrhythmic Drugs, in DiPalma, (ed.): Drill's Pharmacology in Medicine, 4th ed., New York, McGraw-Hill, 1971.
14. Riseman, J. E. F., et al.: Arch. Intern. Med. 71:460, 1943.
15. Mankin, J. W.: J. Lab. Clin. Med. 41:929, 1953.
16. Riseman, J. E. G., et al.: Am. Heart J. 22:219, 1941.
17. Gold H., et al.: J.A.M.A., 145:637, 1951; Brass, H.: J. Am. Pharm. A. (Pract. Ed.) 4:310, 1943.
18. Halpern, A., et al.: Antibiotics & Chemother. 9:97, 1959.
19. Marcus, A. D., and Taraszka, A. J.: J. Am. Pharm. A. (Sci. Ed.) 46:28, 1957.
20. Atkinson, A. J., et al.: Clin. Pharmacol. Ther. 21:575, 1977.
21. Elson, J., et al.: Clin. Pharmacol. Ther. 17:134, 1975.
22. Koch-Weser, J., Klein S.: J.A.M.A. 215:1454, 1971.
23. Brodie, B. B., et al.: J. Pharmacol. Exp. Ther. 102:5, 1951.
24. Bellet, S., et al.: Am. J. Med. 13:145, 1952.
25. Bigger, J. T., and Jaffe, C. C.: Am. J. Cardiol. 27:82, 1971.
26. Hollunger, G.: Acta Pharmacol. Toxicol. 17:356, 1960.
27. Hollunger, G.: Acta Pharmacol, Toxicol. 17:374, 1960.
28. Helfant, R. H., et al.: Am. Heart J. 77:315, 1969.
29. Boura, A. L. A., et al.: Lancet 2:17, 1959.
30. Papp, J. G., and Vaughan-Williams, E. M.: Brit. J. Pharmacol. 37:380, 1969.
31. Vismara, L. A., and Mason, D. T.: Clin. Pharm. Ther. 16:330, 1974.
32. Befeler, B. et al.: Amer. J. Cardiol. 35:282, 1975.
33. Rosen, M. R., Wit, A. L. and Hoffman, B. F.: Am. Heart J. 89:665, 1975.
34. Dustan, H. P., Tarzai, R. C. and Bravo, E. L.: Am. J. Med. 56:610, 1972.
35. Bein, H. J.: Experientia 9:107, 1973.
36. Vakil, R. J.: Brit. Heart J. 11:350, 1949.
37. Muller, J. M., Schlittler, E., and Brin, H. J.: Experientia 8:338, 1952.
38. Hess, S. M., Shore, P. A., and Brodie, B. B., J. Pharmacol. Exp. Ther., 118:54, 1956.
39. Kirshner, N.: J. Biol. Chem. 237:2311, 1962.
40. Euler, U. S. von, Lishajko, F.: Int. J. Neuropharmacol. 2:127, 1963.
41. Leyden, A. F., et al.: J. Am. Pharm. A. (Sci. Ed.) 45:771, 1956.
42. Weis-Fogh. O.: Pharm. acta helv. 35:442, 1960.
43. Iversen, L. L.: The Uptake and Storage of Noradrenaline in Sympathetic Nerves, Cambridge, Cambridge University Press, 1967.
44. Trendelenberg, U.: Pharm. Rev. 15:225, 1966.
45. Change, C. C., Chang, J. C., and Su, C. Y.: Brit. J. Pharmacol. 30:213, 1967.
46. Handbook of Exp. Pharmacologie, Gross, F., (ed.): Antihypertensives, p. 397, vol. 39, Heidelberg, Springer-Verlag, 1977.
47. Druey, J., and Ringier, B. H.: Helv. Chim. Acta 34:195, 1951.
48. Gross, F., Druey J., and Meier, R.: Experientia 6:19, 1950.
49. Zacest, R., and Koch-Weser, J.: Clin. Pharmacol. Ther. 13:420, 1972.
50. Gibaldi, M., Boyer, R. N., and Feldman, S. J.: Pharm. Sci. 60:1338, 1971.
51. Johnson, C. C.: Proc. Soc. Exp. Biol. Med. 26:102, 1928.
52. Corcoran, A. C., et al.: Circulation 2:188, 1955.
53. Schlitter, E.: Chemistry of Antihypertensive Agents, in Gross, F. (ed.): Handbuch Exp. Pharmacologic, vol. 39, p. 1, Heidelberg, Springer-Verlag, 1977.
54. DuCharme, D. W., et al.: Pharmacol. Exp. Ther. 184:662, 1973.
55. Heise, A., and Kroneberg, G.: Nauyn-Schmeideberg's Arch. Pharmacol. 279:285, 1973.
56. Sjoerdsma, A., et al.: Circulation 28:492, 1963.
57. Nies, A. S., and Shang, D. G.: Clin. Pharmacol. Exp. Ther. 14:823, 1973.
58. Halushka, P. V., and Keiser, H. R.: Cir. Res. 35:458, 1974.
59. Koblinger, W., and Klupp, H.: Recent Adv. in Hypertension 1:53, 1975.
60. Starke, K., and Montel, H.: Neuropharmacol. 12:1073, 1973.
61. Swett, L. R., et al.: N.Y. Acad. Sci. 107:891, 1963.
62. Cohen, R. A., et al.: Ann. Int. Med. 65:347, 1966.
63. Kritchevsky, D. (ed.): Hypolipidemic Agents, Berlin, Springer-Verlag, 1975.
64. Levy, R. L., and Rifkind, B. M.: Cardiovascular Drugs, vol. 1, p. 1, 1977.
65. Levy, R. L.: Ann. Rev. Pharmacol. Toxicol. 47:499, 1977.
66. Stamler, J.: Adv. Exp. Med. Biol. 82:52, 1977.
67. Oaks, W., et al.: Arch. Intern. Med. 104:527, 1959.
68. Curran, G. L.: Am. Pract. 7:1412, 1956.
69. Gilman, H., and Wilder, G. R.: J. Am. Chem. Soc. 82:1166, 1950.
70. Mannisto, P. T., et al.: Acta Pharmacol. Toxicol. 36:353, 1975.
71. Lindenbaum, S., and Higuchi, T: J. Pharm. Sci. 64:1887, 1975.
72. Morawitz, P.: Ergebnisse der Physiol. biologishen Chemie und Expt'l. Pharmacol. 4:307, 1905.
73. Stenflo, J., et al.: Proc. Natl. Acad. Sci., U.S.A., 71:2730, 1974.
74. Jackson, C. M., and Suttie, J. W.: Prog. Hematol. 10:333, 1978.
75. Triplett, D. A. (ed.): Platelet Function, Chicago, Am. Soc. Clin. Pathol., 1978.
76. Hamberg, M., et al.: Proc. Natl. Acad. Sci, U.S.A. 11:345, 1974.
77. Murer, E. D., and Day, H. J.: Platelets and Thromboses, in Sherry S., and Scriubine, A. (eds.): P, 1, Baltimore, University Park Press, 1974.
78. Willis, A. L.: Prostaglandins 8:453, 1974.
79. Tateson, J. B., Moncada, S., and Vane, J. R.: Prostaglandins 13:389, 1977.
80. Council on Drugs: J.A.M.A. 191:127, 1965.

81. McClintock, J. C., *et al.:* Surg. Cynec. Obstet. 112:653, 1961.
82. Greer, M. A., *et al.* New Eng. J. Med. 272:888, 1965.

SELECTED READINGS

Antianginal Agents and Vasodilators

Charlier, R. (ed.): Antianginal Drugs, Berlin, Springer-Verlag, 1971.

Dobbs, W., and Povalski, H. J.: Coronary Circulation, Angina Pectoris, and Antianginal Agents, *in* Antonaccio, M. J. (ed.): Cardiovascular Pharmacology, p. 461, New York, Raven Press, 1977.

Needham, P. (ed.): Organic Nitrates, Berlin, Springer-Verlag, 1975.

Parratt, J. R.: Pharmacological Approaches to the Therapy of Angina, Adv. in Drug Res. 9:103, 1974.

Robinson, B. F.: Mechanisms in Angina Pectoris in Relation to Drug Therapy, Adv. in Drug Res. 10:93, 1975.

Willis, W. H., *et al.:* Nitrates and Nitrites in the Treatment of Coronary Artery Disease, *in* Donoso, E. (ed.): Drugs in Cardiology, vol. I, Part I, p. 157, New York, Stratton Intercontinental Medical Book Corporation, 1975.

Antiarrhythmic Agents

Bassett, A. L., and Wit, A. L.: Electrophysiology of Antiarrhythmic Drugs Prog. in Drug Res. 17:34 (1973).

Caracta, A. R., and Damato, A. N.: Procainamide, *in* Donoso, E. (ed.): Drugs in Cardiology, vol. I, Part I, p. 1, New York, Stratton Intercontinental Medical Book Corporation, 1975.

Cohen, S. I.: Current Concepts in Quinidine Therapy, *In* Donoso, E. (ed.): Drugs in Cardiology, vol. I, Part I, p. 17, New York, Stratton Intercontinental Medical Book Corporation, 1975.

Cohen, L. S.: Diphenylhydantoin Sodium (Dilantin), *in* Donoso, E. (ed.): Drugs in Cardiology, vol. I, Part 1, p. 17, New York, Stratton Intercontinental Medical Book Corporation, 1975.

Dreifus, L. W., *et al.:* New Antiarrhythmic Drugs, *in* Donoso, E. (ed.): Drugs in Cardiology, vol. I, Part 1, p. 106, New York, Stratton Intercontinental Medical Book Corporation, 1975.

Romhilt, D. W., and Fowler, N. O.: Bretylium Tosylate, *in* Donoso, E. (ed.): Drugs in Cardiology, vol. I, Part 1, p. 80, New York, Stratton Intercontinental Medical Book Corporation, 1975.

Sasyniuk, B. I., and Ogilvie, R. I.: Antiarrhythmic Drugs: Electrophysiological and Pharmacokinetic Considerations, Ann. Rev. Pharmacol. 15:131, 1977.

Vaughan Williams, E. M.: Electrophysiological Basis for Rational Approach to Antidysrhythmic Drug Therapy, Adv. in Drug. Res. 9:69, 1974.

Antihypertensive Agents

Furet, C. I.: The Biochemistry of Guanethidine, Adv. in Drug. Res., 4:133, 1967.

Gross, F.: Antihypertensive Agents, *in* Handbook of Experimental Pharmacol., Berlin, Springer-Verlag, 1977.

Stitzel, R. E.: The Biological Fate of Reserpine, Pharmacol. Rev. 28:179, 1976.

Tarazin, R. C., and Gifford, R. W.: Drug Treatment of Hypertension, *in* Donoso, E. (ed.): Drugs in Cardiology, vol. I, Part 2, p. 1, New York, Stratton Intercontinental Medical Book Corporation, 1975.

Wilhelm, M., and Stevens, G. D.: Antihypertensive Agents, Prog. in Drug Res., 20:197, 1976.

Antihyperlipidemic Agents

Beneze, W. L.: Hypolipidemic Agents, *in* Kritchevsky, D. (ed.): Hypolipidemic Agents, p. 349, Berlin, Springer-Verlag, 1975.

Eisenberg, S., and Levy, R. I.: Lypoproteins and Lipoprotein Metabolism, *in* Kritchevsky, D. (ed.): Hypolipidemic Agents, p. 191, Berlin, Springer-Verlag, 1975.

Felts, J. M., and Rudel, L. L.: Mechanisms in Hyperlipidemic, *in* Kritchevsky, D. (ed.): Hypolipidemic Agents, p. 151, Berlin, Springer-Verlag, 1975.

Howe, R.: Hypolipidemic Agents, Adv. in Drug Res. 9:7, 1974.

Levy, R. I.: The Effect of Hypolipidemic Drugs on Plasma Lipoproteins, Ann, Rev. Pharmacol., 17:499, 1977.

Witte, E-C.: Antihyperlipidaemic Agents, Prog. in Med. Chem., 11:199, 1975.

Goldstein, J. L., and Brown, M. S.: The Low-Density Lipoprotein Pathway and Its Relation to Atherosclerosis, Ann. Rev. Biochem., 46:897, 1977.

Kritchevsky, D., *et al.:* Drugs, Lipid Metabolism and Athersclerosis, Adv. Exp. Med. Biol., 109: 1978.

Anticoagulants

Biggs, R. (ed.): Human Blood Coagulation, Haemostasis and Thrombosis, 2nd. ed., London, Blackwell Scientific Publications, 1976.

Jackson, G. M., and Suttie, J. W.: Recent Developments in Understanding the Mechanism of Vitamin K and Vitamin K-Antagonists Drug Action and the Consequences of Vitamin K Action in Blood Coagulation, Prog. in Hematol., 10:333, 1978.

Scriabine, A.: Platelets and Platelet Aggregation Inhibitors, *in* Antonaccio, M. J. (ed.): Cardiovascular Pharmacology, p. 429, New York, Raven Press, 1977.

Triplett, D. A.: Platelet Function, Chicago, American Society of Clinical Pathologists, 1978.

15

local anesthetic agents

Charles M. Darling

Local anesthesia may be defined as the loss of sensation or the loss of motor function in a circumscribed area of the body. Local anesthesics are drugs that produce this condition by causing a block of nerve conduction. They must be applied locally to the nerve tissue in appropriate concentrations. To be useful clinically, the action should always be reversible.

The local anesthetic agents are useful chemical tools for the temporary relief of localized pain in dentistry and minor surgical procedures, as well as for producing a state of nonresistance (e.g., spinal anesthesia) without general anesthesia. Many over-the-counter agents are used topically for temporary relief of pain and itching due to minor burns, insect bites, allergic responses, hemorrhoids, and other minor conditions.

An ideal local anesthetic has not been discovered; but several desirable properties may be stated[1]:

1. Nonirritating to tissue and not causing permanent damage. Most clinically available agents fulfill this requirement.
2. Low systemic toxicity because it is eventually absorbed from its site of application. Most local anesthetics are rapidly metabolized after absorption.
3. Effective whether injected into the tissue or applied locally to skin or mucous membranes. Intact skin is resistant generally to the action of most local anesthetics and requires high concentrations for prolonged periods of time. This is probably due to a slow rate of penetration and a rapid vascular diffusion following penetration. Thus, insufficient drug accumulates at the nerve endings in the dermis.
4. Rapid onset of anesthesia and a short duration of action. The action must last long enough to allow time for the contemplated surgery, yet not so long as to entail an extended period of recovery.

Historical development

Throughout history people have sought means to escape pain. History records the more popular methods[2] to be psychical influences, acupuncture, nerve compression, freezing, and drugs. Though the popularity of each method has been somewhat cyclical through the centuries, each persists today in one form of practice or another.

The modern-day local anesthetic drugs owe their origin to the isolation of cocaine in 1860. By nature humans are experimenters in the environment, always seeking to identify natural phenomena observed through the senses. In earlier times people sought to explore the smell and taste of all plants endemic to their region. Sometimes the results were catastrophic (Jamestown Weed[3]), sometimes valuable. As they explored other regions, they learned from the experiences of the natives of distant regions.

The natives of Peru learned to chew the leaves of the coca bush to stimulate the general feeling of well-being and to prevent hunger. Although no early literature citation is available to substantiate it, the suggestion is often made[2] that the Incas produced local anesthesia for surgical procedures by chewing a cud of the leaves and allowing the saliva to drop upon the site of incision.

The question of priority of discovery[2] is not pertinent to the present discussion; however, the use of a specific chemical substance to produce local anesthesia evolved slowly from the isolation of the crystalline alkaloid, cocaine, from coca leaves in 1860. A seemingly necessary historical component was the invention of the hypodermic syringe, which was introduced to the United States in 1856 and was used on a large scale for the first time during the Civil War.[4] Neither revolutionary concept can be assigned a historical precedence, because the fundamental observations were recorded over a span of centuries. Therefore, the modern use of local anesthetics evolved from the mid-19th century isolation of cocaine and the concomitant development of the hypodermic syringe.

In the late 19th century, cocaine was used as a topical anesthetic agent in ophthalmology, for the production of peripheral-nerve blockade, and in spinal anesthesia.[5] From these early experiments, a number of adverse effects were observed. Cocaine has a high acute toxicity and addicting properties that served to stimulate efforts to find effective and less toxic local anesthetic agents.

Although the structure of cocaine was not identified until 1924, earlier work had shown that it contained a benzoic acid ester moiety. This led to the preparation of the ester, ethyl *p*-aminobenzoate (benzocaine), in 1890. The low aqueous solubility of benzocaine limited its usefulness as an injectable agent and it was neglected for many years. It is recognized today, however, as an effective topical anesthetic agent for the production of surface anesthesia of mucous membranes.

During the next half-century, most compounds synthesized as local anesthetics were derivatives of benzoic acid. But unlike benzocaine, the molecules contained a basic amino group from which water-soluble salts could be prepared. A few of these agents are still in wide clinical use as local anesthetic agents: procaine, 1905; dibucaine, 1929; and tetracaine, 1930. Also, these synthetic derivatives, in contrast to cocaine, caused no contraction of the blood vessels and were rapidly absorbed from the site of injection. However, in 1903, the practice was begun of adding a potent vasoconstrictor, epinephrine, to the solutions to prolong the duration of local anesthesia.

In the mid-1930's, the serendipitous discovery of the local anesthetic effect of an isomer of the alkaloid, gramine, gave new direction to the search for new agents. Isogramine, 2-(dimethylaminomethyl)indole, was observed to produce local anesthesia to the tongue when tasted. The discovery of activity in an indole-containing structure represented a significant departure from the benzoic acid series.[2]

During the following decade, numerous compounds were synthesized and investigated for useful activity. This work culminated in the development of lidocaine, which has structural characteristics of the indole, isogramine. For example, each contains a nitrogen atom

Isogramine   Lidocaine

attached to an sp^2 carbon atom. An aromatic group is attached to one side of this moiety and a tertiary aminoalkyl group is attached to the other side.

Essentially all of the clinically useful local anesthetic agents are of the cocaine (benzoic acid derivative) or isogramine (anilide or so-called reversed amide) lineage. The more recently introduced agents are of the latter type.

NERVOUS TISSUE

The function of the nervous tissue is based on two fundamental properties. The first property is *irritability,* which refers to the ability to react to various stimulating agents. The second is *conductivity,* which is the ability to transmit the excitations. Both properties may be blocked by local anesthetic drugs.

The peripheral nerves are composed of fascicles (bundles) of nerve fibers of varying thickness, held together by connective tissue (Fig. 15-1). It is customary to classify nerve fibers according to their diameter. The speed of impulse transmission and the magnitude of the action potential are proportional to the size of the diameter. Fiber diameters cover a wide and continuous range from large myelinated (A fibers) through medium (B fibers) to small unmyelinated (C fibers).

The outer layer of connective tissue, *epineurium,* is made up of longitudinally arranged connective tissue cells and collagenous fibers interspersed with fat cells. Each of the smaller fascicles of a nerve is itself enclosed in a membrane of dense, concentric layers of connective tissue, called *perineurium.* The individual nerve fibers *(axons)* are interspersed within longitudinally arranged strands of collagenous fibers called *endoneurium.* Blood vessels run throughout the various sheaths, which are essentially lipid in nature. The molecular structure of these sheaths is unknown.[5]

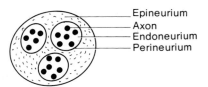

FIG. 15–1. *Diagram showing the cross-sectional view of the parts of a peripheral nerve.*

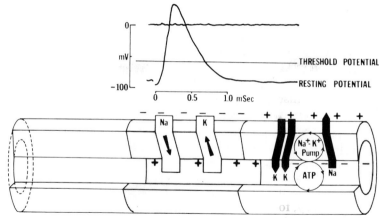

FIG. 15–2. *Relationship between membrane-action potential and ionic flux across the nerve membrane. From Covino, B.G., and Vassallo, H.G.: Local Anesthetics, Mechanism of Action and Clinical Use, p. 1. New York, Grune & Stratton, 1976. By permission.*

Each nerve fiber (axon) has its own cell membrane, the *axolemma,* which surrounds the cytoplasm *(axoplasm)* and serves as a barrier between the axoplasm and the endoneurium. The composition of the membrane is highly lipid with some protein.

During the period of nerve inactivity, the potential difference (resting potential) across the cell membrane is of the order of 50 to 100 mV inside negative (Fig. 15-2). Although this potential difference may seem rather small in absolute magnitude, it occurs across an interface of such extreme thinness that the resultant electric field approaches 100,000 V/cm.[6] It seems reasonable to suppose that such a high field strength is important to the orientation of molecules in the membrane structure and to the transport of ions across the lipid membrane.

Figure 15-2 illustrates the relationship between changes in membrane action potential and ionic flux across the nerve membrane when excitation occurs. An initial phase of depolarization is observed during which the electrical potential within the nerve cell becomes progressively less negative owing to the inflow of potassium ions. While the membrane permits potassium ions to flow back and forth with relative freedom, it resists movement of the sodium ions from the extracellular fluid to the inside of the cell until depolarization reaches a critical level.

When the action potential reaches a critical level (the threshold potential or firing level), the resistance to sodium ion (resistance is the inverse expression of conductance) decreases, allowing sodium ions to flow into the cell along its concentration gradient. As sodium ions flow in, the membrane becomes more and more depolarized; this further decreases the resistance to movement of sodium ions. A result of this self-perpetuating process is an extremely rapid phase of depolarization. The process is self-limited and should be proportional to the original extracellular and intracellular concentrations of sodium ion. However, the explosive response to depolarization reduces the membrane potential not merely to zero but over-shoots zero such that the inside of the membrane becomes positively charged with respect to the outside. At the peak of the action potential, the interior has a positive electrical potential of approximately +40 mV relative to the exterior of the cell. The peak of the action potential would be even greater were it not for the increased flow of potassium ions along its concentration gradient from inside to outside the cell.

At the conclusion of the depolarization phase, repolarization begins (refractory phase) utilizing the so-called *sodium–potassium pump,* which essentially "pumps" the sodium ions from inside to outside the cell and the potassium ions from outside to inside. This metabolic pump requires energy and is thought to be catalyzed by a specific adenosinetriphosphatase.[7] The electrical potential within the cell again becomes progressively more negative with respect to the outside until the resting potential of -60 to -90 mV is reestablished. Normally the entire process of depolarization and repolarization occurs within 1 millisecond.

MECHANISM OF ACTION

Several hypotheses on the mechanism of action of local anesthetic drugs have been examined.[1,8] It is generally accepted that bound calcium ions play a critical role in the generation of the nerve impulse. Possibly, the initial phase of depolarization (caused by a nerve impulse) results in the removal of calcium ions from sites in the nerve membrane leading to increased permeability to sodium ions, analogous to the opening of a pore through which sodium ions flow. Local anesthetics may inhibit the release of bound calcium ions or replace calcium ions at the site and thus stabilize the membrane to depolarization. The nerve impulse is not transmitted.

The association of local anesthetic molecules within the membrane may affect permeability by increasing the degree of disorder of the lipids that constitute the nerve membrane. Alternatively, the local anesthetic molecules may increase the surface pressure of the lipid layer and essentially close the pores through which ions move. The result of either alternative would restrict the openings to the sodium-ion channels and cause a general decrease in permeability, the fundamental change necessary for the generation of the action potential and the propagation of an impulse.

Takman has proposed a classification of local anesthetic agents based on site of action.[9] In this classification, the four classes are agents acting (1) at receptor sites on the external surface of the nerve membrane, (2) at receptor sites on the internal surface of the nerve membrane, (3) by a receptor-independent physicochemical mechanism, and (4) by a combination of a receptor and receptor-independent mechanism. The first class (1) of agents includes two guanidine-type molecules possessing local anesthetic potency greater than that of other known compounds. They are thought to produce local anesthesia by blocking the opening of the sodium-ion channel on the external surface of the nerve membrane.

Tetrodotoxin, obtained from the ovaries and other organs of the puffer fish *(Tetraodontidae)*, and saxitoxin, produced by certain marine dinoflagellates

Tetrodotoxin

Saxitoxin

(whose rapid growth causes discoloration of the ocean, hence the term *red tide*) that contaminate shellfish and cause paralytic shellfish poisoning in humans, are representatives of this group. Both highly hydrophilic substances are extremely toxic to humans and are not available for clinical use. However, tetrodotoxin is of experimental interest because it selectively blocks the increase in sodium conductance without affecting steady state conductances.

The second class (2) includes those agents acting at receptor sites near openings of the sodium-ion channels on the internal axoplasmic surface of the membrane. Studies to determine the site of action and whether the cationic or neutral form is the active species have been reviewed.[5,8] Quaternary ammonium salts of some local anesthetics are representative of this class and were shown to block conduction when applied internally by perfusion. The same drugs applied externally did not gain access to the inside of the cell and were ineffective in blocking conduction. It is interesting to note that high concentrations of tetrodotoxin applied internally for long periods had no effect on sodium-ion conductance. Because their probability of diffusion across lipid membranes is quite low, quaternary ammonium salts of local anesthetic drugs are not likely to have clinical value.

The compounds in the third class (3) are noncationic at physiological pH; these include benzocaine, alcohols, and barbiturates. Members of the latter two groups are not useful clinically as local anesthetics, but they are of value as experimental tools in the study of the mechanisms of action. Like other local anesthetic drugs, alcohols (e.g., *n*-butyl alcohol and benzyl alcohol) in low concentrations (higher concentrations result in irreversible nerve damage), and barbiturates block conduction reversibly by decreasing sodium-ion conductance.

Benzocaine, the only clinically useful member of this class, is known to exist in the uncharged form at physiological pH, and its activity is not affected by changes in pH. Thus, its local anesthetic action is due to physicochemical properties other than ionic bonding. And Takman suggests that compounds of this type exert their action by a receptor-independent physicochemical mechanism. They may restrict sodium-ion conductance either by increasing the degree of disorder of the lipids of the nerve membranes or by increasing the surface pressure of the lipid layer.

The fourth class (4) of compounds includes by far the most clinically useful local anesthetic agents. These agents are believed to act by a combination of receptor and receptor-independent mechanisms. A mechanistic interpretation by Ritchie[8] of the results reported in the literature is not in disagreement with this classification.

Agents of this class are tertiary amines in which the neutral form exists in equilibrium with the cationic conjugate-acid form at physiological pH. The suggestion has been made that these agents act at the inside surface of the nerve membrane in the cationic form. The neutral form then more readily passes through the lipid membrane to gain access to the inside surface. If this is the case, the pH of the axoplasm determines the active form of the drug in the region of the receptor. Because some reported results with mammalian C fibers do not support this view, Ritchie[8] suggested that the receptors with which local anesthetics interact may be within the membrane. The receptors may even be near the outer surface of the membrane but in such a position that externally applied drugs cannot reach them.

In summary, drugs produce local anesthesia by inhibiting membrane conductance of sodium ions. The evidence suggests that the inhibition may be produced at more than one site. Tetrodotoxin and saxitoxin interact with receptors on the outer surface, while quaternary ammonium salts interact with receptors on the inner surface of the nerve membrane. The clinically useful local anesthetic agents (members of classes 3 and 4, above) interact with receptor site(s) on the inside surface of the membrane or those within the membrane itself. The precise mechanism of action at the molecular level is not known.

STRUCTURE–ACTIVITY RELATIONSHIPS

As indicated in the above classification, a great variety of chemical structures may produce conduction blockade, and there appears to be no marked structural specificity. It would be difficult to define precisely the relationships between chemical structure and local anesthetic activity if all compounds were included without regard to site of action. Therefore, the following discussion of structure–activity relationships (SAR) is restricted to chemical classes of clinically useful local anesthetic agents (3 and 4 above), in which the salient features are quite similar.

The origin of the modern local anesthetic agents can be traced to the independent discoveries of two distinctly different alkaloids, cocaine and isogramine.

Cocaine   Isogramine

Cocaine is an aminoalkyl ester of benzoic acid; isogramine is a 2-(aminoalkyl)indole. Insufficient data is available to make definitive statements about the relationship of the common structural features of the two molecules to local anesthetic activity.

The structures of clinically useful agents contain a centrally located sp^2 hybridized carbon atom to which alkyl and aryl groups are attached either directly or through a heteroatom. When the aryl group is attached directly to the sp^2 carbon atom (cocaine type), activity appears to be relatively insensitive to the mode of attachment of the alkyl group. But when the alkyl group is attached directly to the sp^2 carbon atom (isogramine type), activity appears to require the attachment of an aryl group through a nitrogen bridge. The implications of these structural differences await further study. But apparently a positive mesomeric effect on the sp^2 carbon atom has a favorable effect on activity and was suggested[10] as leading to a greater affinity for the binding site at the receptor.

BENZOIC ACID DERIVATIVES

The benzoic-acid derivatives are synthetic compounds derived from the structure of cocaine and may be represented as follows.

$$Aryl - \overset{\overset{O}{\|}}{C} - X - Aminoalkyl$$

Aryl group. The clinically useful members of this series possess an aryl radical attached directly to the carbonyl group or attached through a vinyl group. Although alicyclic and aryl aliphatic carboxylic-acid esters are active, conjugation of the aromatic group with the carbonyl enhances local anesthetic activity. Substitution of the aryl group with substituents that increase the electron density of the carbonyl oxygen enhance activity. Favorable substituents include alkoxy, amino, and alkylamino groups in the *para* or *ortho* positions. However, these substituents also alter the liposolubility of the molecule, the effect of which will be treated separately (Coefficient of Distribution).

Aryl aliphatic radicals, which contain an alkylene group between the aryl radical and carbonyl group, result in compounds that have not found clinical use. In these compounds the mesomeric effect of the aryl radical does not extend to the carbonyl group.

Bridge, X. The bridge, X, may be carbon, oxygen, nitrogen, or sulfur. In an isosteric-procaine series, conduction–anesthetic potency[10] decreased in the following order: sulfur, oxygen, carbon, nitrogen. These modifications largely determine the chemical class to which each derivative belongs; they also affect duration of action and relative toxicity. In general, amides (X = N) are more resistant to metabolic hydrolysis than esters (X = O). Thioesters (X = S) may cause dermatitis.

Aminoalkyl group. The aminoalkyl group is not necessary for local anesthetic activity, but it is used to form water-soluble salts. For example, benzocaine (X = O; aminoalkyl = C_2H_5) is a local anesthetic, but the aqueous solutions of it must be highly acidic.

Because of its capacity for salt formation, the amino function is considered the hydrophilic part of the local-anesthetic molecule. Tertiary amines result in more useful agents. The secondary amines appear to be longer acting but they are more irritating; primary amines are not very active and cause irritation.

The alkyl groups (including the intermediate chain linked to X) primarily influence relative lipid solubility (distribution coefficient) as discussed below.

LIDOCAINE DERIVATIVES (ANILIDES)

Lidocaine derivatives are essentially anilide progenies of isogramine with the following general structural characteristics.

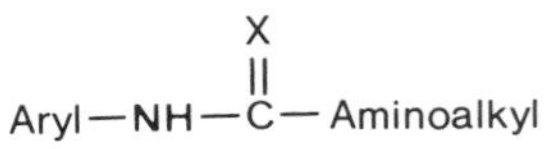

$$\text{Aryl} - \text{NH} - \overset{\overset{\displaystyle X}{\|}}{C} - \text{Aminoalkyl}$$

Aryl group. The clinically useful local-anesthetic agents of this type possess a phenyl group attached to the sp^2 carbon atom through a nitrogen bridge. Substitution of the phenyl with a methyl group in the 2- or 2- and 6-position enhances activity. The amide bond is more stable to hydrolysis than the ester bond. In addition, the methyl substituent(s) provide(s) steric hindrance to hydrolysis of the amide bond and increase(s) the coefficient of distribution.

Substituent X. In the general structure above, X may be carbon (isogramine), oxygen (lidocaine), or nitrogen (phenacaine). While the lidocaine series (X = O) has provided more useful products, insufficient data is available to state the relative contribution of X to activity.

Aminoalkyl group. As found for the benzoic-acid derivatives, the amino function has the capacity for salt formation and is considered the hydrophilic portion of the molecule. Tertiary amines are more useful clinically because the primary and secondary amines are more irritating to tissue.

Coefficient of Distribution. In general, nerve membranes consist primarily of lipids. Local anesthetic agents are injected near the site of action and must be capable of penetrating the nerve membrane. Increasing the lipid solubility of a series of compounds should result in facilitated penetration of nerve membranes. In in vitro experiments involving very simple systems such as isolated nerves, the potency of local anesthetic compounds is directly proportional to distribution coefficients.

The in vivo system is much more complicated, and often, within a congeneric series, increases in partition coefficients result in increasing potency to a maximum, after which the activity declines. Unfortunately, toxicity also increases concurrently.

Each structural component (aryl and aminoalkyl) contributes to the lipid solubility of the molecule and may be varied to produce derivatives of increasing distribution coefficient. Substitution of the aryl radical by alkyl, alkoxy, and alkylamino groups leads to homologous series in which partition coefficients increase with increasing number of methylene groups ($-CH_2-$). A review[10] of these series revealed that local anesthetic activity peaked with the C_4-, C_5-, or C_6- homologs, depending on the particular nature of the series being considered.

Similarly, variations in the aminoalkyl portion of the molecule lead to increases in activity and toxicity with increasing carbon number. The branching of N-alkyl groups is often accompanied by an intensification of activity. The aminoalkyl group may be part of an aliphatic heterocyclic ring.

The tertiary amino group may be diethylamino, piperidino, or pyrrolidino, leading to the products that exhibit essentially the same degree of activity. The more hydrophilic morpholino group usually leads to diminished potency. However, this weakening of the effect can be compensated fully by the introduction of additional lipophilic groups in other parts of the molecule.[11]

A given degree of lipid solubility does not assure local anesthetic activity, nor does a high degree of potency assure clinical utility. In the selection of a particular member of a series for clinical use, the relative potency must be weighed against the degree of toxicity.

Most of the clinically useful local anesthetic agents exhibit pKa values in the range between 8.0 and 9.5. The implication of this is that compounds with higher pKa values are ionized 100 percent at physiological pH and so have difficulty reaching the biophase. Substances having a lower pKa value are not sufficiently ionized and are less effective in practice, even though they reach the biophase.[10]

PRODUCTS

The numerous local anesthetic products for which official monographs have been prepared are discussed under two major headings, Benzoic-Acid Derivatives and

Lidocaine Derivatives (Anilides). If appropriate, the products are grouped by specific chemical class within each heading.

Except for cocaine, the clinically useful local anesthetics do not cause vasoconstriction. In clinical practice, therefore, a solution of a local anesthetic often contains epinephrine, norepinephrine, or a suitable synthetic congener such as phenylephrine. The vasoconstrictor serves a dual purpose. By decreasing the rate of absorption, it not only localizes the anesthetic at the desired site but also limits the rate at which the anesthetic is absorbed into the circulation. Thus, metabolism can reduce the systemic toxicity of the local anesthetic. In essence, then, the vasoconstrictor prolongs the action and lowers the systemic toxicity of local anesthetic agents.

BENZOIC-ACID DERIVATIVES

Cocaine U.S.P., (−)3-(benzoyloxy)-8-methyl-8-azabicyclo[3·2·1]octane-2-carboxylic acid methyl ester, [1(R), 2(R), 3(S)], methylbenzoylecgonine.

Cocaine is an alkaloid obtained from the leaves of *Erythroxylon coca* Lamarck and other species of *Erythroxylon* (Fam. Erythroxylaceae). It occurs to the extent of approximately 1 percent in South American leaves and to the equivalent (as other derivatives) of about 2 percent in Java leaves. It was first isolated as an amorphous mixture by Gaedcke (1855) who thought it was an alkaloid related to caffeine which he called erythroxyline. Later, Nieman (1860) obtained from coca leaves a crystalline alkaloid to which he gave the name cocaine.

Commercial production involves total extraction of bases followed by acid hydrolysis of the ester alkaloids to obtain the total content of (−)-ecgonine. After purification of the ecgonine, cocaine is synthesized by esterification with methanol and benzoic acid (Fig. 15-3).

The ecgonine portion of the cocaine molecule contains four asymmetric carbon atoms. Two of these (C_1 and C_5) are intramolecularly compensated; therefore, only eight optically active isomers (four racemates) exist. In cocaine, the benzoyloxy (C_3) and methoxycarbonyl (C_2) groups are *cis* to the nitrogen bridge. In

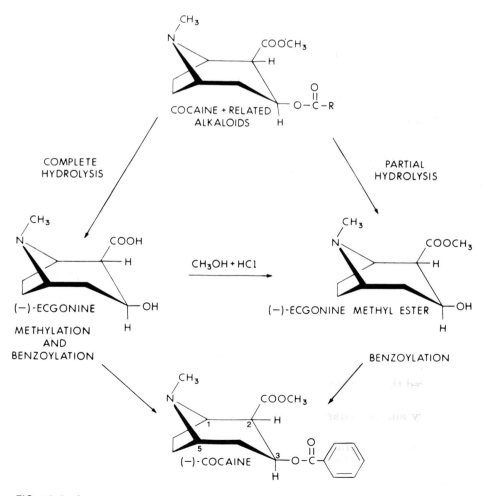

FIG. 15-3. *Commercial preparation of cocaine.*

(+)-pseudococaine, which is also active, the methoxycarbonyl is *trans*.[11]

Cocaine occurs as levorotatory, colorless crystals or as a white, crystalline powder that numbs the lips and the tongue when applied topically. It is slightly soluble in water (1:600), more soluble in alcohol (1:7), and quite soluble in chloroform (1:1) and in ether (1:3.5). The crystals are fairly soluble in olive oil (1:12) but less soluble in mineral oil (1:80 to 1:100). Because of its solubility characteristics, it is used principally where oily solutions or ointments are indicated. Cocaine is basic and readily forms crystalline, water-soluble salts with organic and inorganic acids.

Cocaine Hydrochloride U.S.P. The salt occurs as colorless crystals or as a white crystalline powder that is soluble in water (1:0.5), alcohol (1:3.5), chloroform (1:15), or glycerin. It has a pKa of 8.4.

Aqueous solutions of cocaine hydrochloride are stable if not subjected to elevated temperatures or stored for prolonged periods. Bacteriologic filtration is a better method of sterilization than autoclaving. At pH values near 7, solutions of the salt are very unstable and should be prepared freshly and not be autoclaved. Also, the solutions are incompatible with alkalis, the usual alkaloidal precipitants, silver nitrate, sodium borate, calomel, and mercuric oxide.

Koller, in 1884, reported[2] the first practical application of cocaine as a topical anesthetic in the eye. Its toxicity has prevented cocaine from being used for anything other than topical anesthesia and, even in this capacity, its use is limited for fear of causing systemic reactions and addiction.

Cocaine does not penetrate the intact skin but is readily absorbed from mucous membranes, as is evidenced by the quick response obtained by addicts when cocaine ("snow") is snuffed into the nostrils. Cocaine blocks the uptake of catecholamines at adrenergic nerve endings and is thus a potent vasoconstrictor. This accounts for the ulceration of the nasal septum after cocaine has been "snorted" for long periods in high doses.[12]

As a local anesthetic, cocaine is applied topically to mucous membranes such as those of the eye, the nose, and the throat in solutions of 2 to 5 percent. The anesthesia produced by such concentrations lasts approximately a half hour. Its use in ophthalmology is marred by the fact that, not infrequently, cocaine causes corneal damage with resultant opacity. For nose and throat work, concentrations higher than 10 percent are rarely used, 1 to 4 percent being a common concentration range. For external use, apply topically to mucous membranes, as a 2 to 20 percent solution.

Occurrence
Cocaine Hydrochloride Tablets for Topical Solution
U.S.P.

Hexylcaine Hydrochloride U.S.P., Cyclaine® Hydrochloride, 1-(cyclohexylamino)-2-propanol benzoate (ester) hydrochloride. The hydrochloride salt is soluble in water (1:8) and is freely soluble in alcohol and in chloroform. Solutions are stable to boiling and autoclaving. A 1 percent solution has a pH of 4.4.

Hexylcaine Hydrochloride

Hexylcaine has about the same toxicity as procaine; topical anesthesia (1 to 5%) is similar to cocaine and butacaine; for nerve-block anesthesia (1 to 2%), its toxicity is between that of butacaine and tetracaine. It is employed for spinal anesthesia in a 1 to 2.5 percent solution containing 10 percent glucose.

Meprylcaine Hydrochloride U.S.P., Oracaine® Hydrochloride, 2-methyl-2-(propylamino)-1-propanol benzoate (ester) hydrochloride. The hydrochloride occurs as a white, crystalline solid that is freely soluble in water, alcohol, or chloroform. A 2 percent solution has a pH of 5.7, and it has been reported that such a solution can be sterilized by autoclaving without decomposition.

Meprylcaine Hydrochloride

Meprylcaine is used primarily in dentistry in a 2 percent solution containing epinephrine (1:50,000) as an infiltration and nerve-block anesthetic. It is substantially more potent and more rapidly hydrolyzed in human serum than procaine.

Isobucaine Hydrochloride U.S.P., 2-(isobutylamino)-2-methyl-1-propanol benzoate (ester) hydrochloride. The salt is a white, crystalline solid that is freely soluble in water, alcohol, or chloroform. The pH of a 2 percent solution is about 6.

Isobucaine Hydrochloride

Structurally, isobucaine differs from meprylcaine in having an N-isobutyl group instead of an N-propyl group. Like meprylcaine, it is more toxic and exhibits a shorter duration of action than does procaine. It is used primarily in dentistry in a 2 percent solution containing epinephrine (1:65,000) for infiltration and nerve block.

Cyclomethycaine Sulfate U.S.P., Surfacaine® Sulfate, 3-(2-methylpiperidino)propyl 4-cyclohexyloxybenzoate bisulfate (1:1). The bisulfate occurs as a white, crystalline powder and is sparingly soluble in water or alcohol and slightly soluble in chloroform.

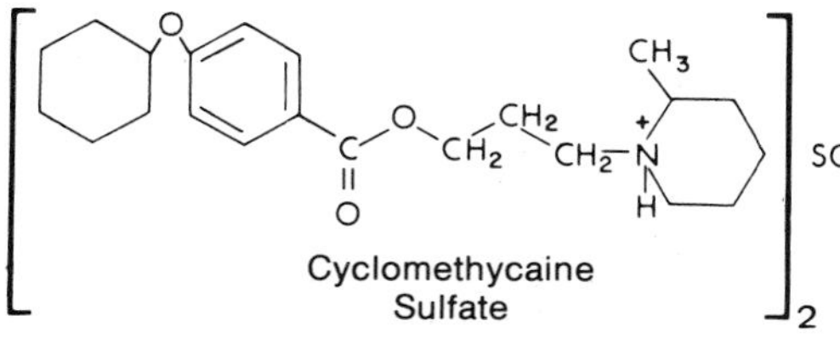

Cyclomethycaine
Sulfate

Cyclomethycaine is an effective local anesthetic on damaged or diseased skin and on rectal mucous membrane. It is useful topically on burns (sunburn), abrasions, and mucosa of rectum and genitourinary tract, and is applied most commonly in 0.25 to 1 percent ointments, jellies, creams, or suppositories. It is not recommended for use on mucous membranes of the upper-respiratory system or eye. As with all topical anesthetic agents, it carries a slight sensitizing potential.

Piperocaine Hydrochloride, Metycaine® Hydrochloride. Piperocaine differs structurally from cyclomethycaine in that the 4-cyclohexyloxy group of cyclomethycaine is absent. It is soluble in water (1:1.5) and in alcohol (1:4.5). Aqueous solutions are faintly acid and are stable to sterilization by autoclaving. Piperocaine is recommended for application to the eye in 2 to 4 percent solutions; to the nose and throat in 2 to 10 percent solutions; for infiltration in 0.5 to 1.0 percent solutions; for nerve block in 0.5 to 2 percent solutions; and for spinal anesthesia in 1.5 percent solutions, with the maximal quantity of the drug limited to 1.65 mg. per kg. of body weight.

Structural relationships of the benzoic-acid ester local anesthetics are shown in Table 15-1.

AMINOBENZOIC-ACID DERIVATIVES

Mixtures of sulfonamides and agents of this chemical class of local anesthetics should be avoided because of a potential pharmacological incompatibility. A hydrolysis product of these agents is *para*-aminobenzoic acid (PABA). Sulfonamides are considered competitive inhibitors of the incorporation of PABA (Chap. 5) in the biosynthesis of dihydrofolate.

TABLE 15-1

BENZOIC ACID DERIVATIVES

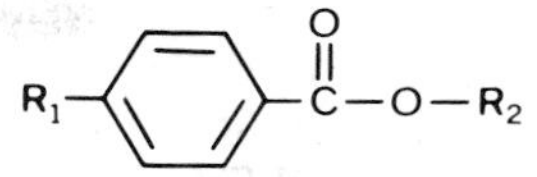

Proprietary Name	Generic Name	R₁	R₂
Cocaine		H	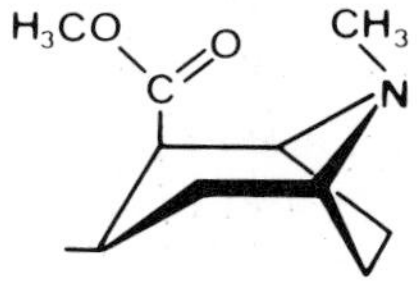
Hexylcaine	Cyclaine	H	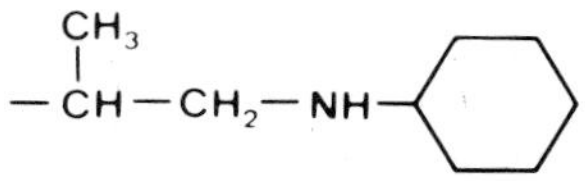
Meprylcaine	Oracaine	H	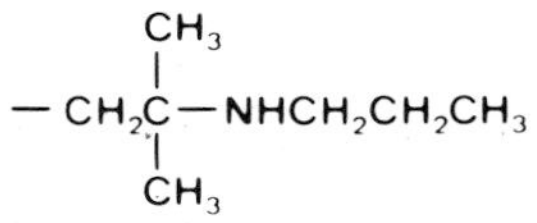
Isobucaine		H	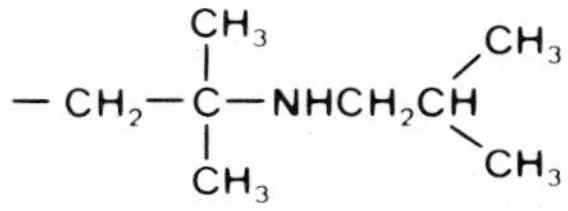
Cyclomethycaine	Surfacaine	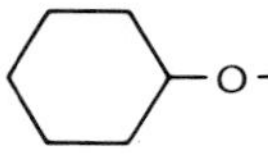	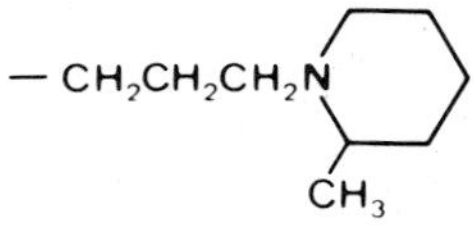
Piperocaine	Metycaine	H	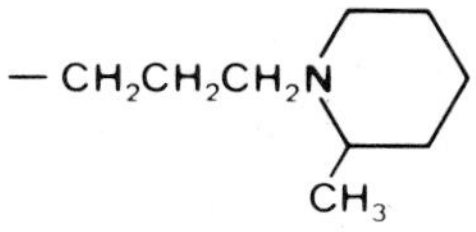

Benzocaine U.S.P., Americaine®, Anesthesin®, ethyl *p*-aminobenzoate. Benzocaine occurs as a white, odorless, crystalline powder, stable in air. It is soluble in alcohol (1:5), ether (1:4), chloroform (1:2), glycerin, or propylene glycol, and is insoluble in water (1:2500).

Unlike the other clinically employed local anesthetics, benzocaine and its homologs do not possess an aliphatic amino group that can be used for salt formation. The free electrons on the aromatic nitrogen are delocalized by the ring, and protonation at this position takes place less readily. Thus, the formation of water-soluble salts of benzocaine and its homologs is not possible and it is unsuited for injection.

Benzocaine

Benzocaine is sufficiently absorbed through abraded surfaces and mucous membranes to relieve the pain associated with ulcers, wounds, and inflamed mucous surfaces. It acts only as long as it is in contact with the skin or mucosal surface. It is used in ointment and cream preparations in concentrations up to 20 percent and is nonirritant and nontoxic.

Butamben U.S.P., Butesin®, butyl *p*-aminobenzoate. Butamben is the butyl ester analog of benzocaine, but it is claimed to be more effective and more toxic. Butamben picrate, a complex thought to contain two moles of butamben per mole of picric acid, is a yellow, amorphous powder. It is used in the eye as a saturated aqueous solution and for burns and denuded areas of the skin as a 1 percent ointment. A disadvantage is that the yellow color stains the skin and clothing. The amyl ester is called Ultracain®.

Procaine Hydrochloride U.S.P., Novocain® Hydrochloride, 2-diethylaminoethyl *p*-aminobenzoate monohydrochloride. The odorless, white, crystalline procaine hydrochloride is stable in air and soluble in water (1:1) and in alcohol (1:30). It was developed by Einhorn in 1906 and is one of the oldest and most used of the synthetic local anesthetics.

Procaine Hydrochloride

Procaine is most stable at pH 3.6 and becomes less stable as the pH is increased or decreased from this value.[14] Storage of buffered solutions at room temperature resulted in the following amounts of hydrolysis:

pH	Amount of Hydrolysis
3.7–3.8	0.5 to 1% in 1 year
4.5–5.5	1.0 to 1.5% in 1 year
7.5	1% in 1 day

Dosage forms are generally regarded as being satisfactory for use, as long as not more than 10 percent of the active ingredient has been lost and there is no increase in toxicity. The following times for a 10 percent loss in potency at 20°C are based on kinetic studies of procaine solutions.

pH	Time in days
3.6	2,300
5.0	1,200
7.0	7

It has also been calculated that the hydrolysis rate increases 3.1 times for each rise of 10° in the range of 20° to 70°.

The following data show the effect on buffered 2 percent procaine hydrochloride solutions of autoclaving at 15 pounds pressure for 2 hours.[15]

pH Before and After Autoclaving	Percent of Original Assay
2.4	97.5
2.6	97.9
2.8	98.1
3.0	98.4
3.2	98.5
3.4	98.5
3.6	98.3
3.8	98.2
4.0	97.8

Early studies suggested that neutral or slightly alkaline solutions of procaine had certain physiologic advantages over acid solutions in that there is less pain on injection, there is less tissue damage, and, most important, the rate of onset of anesthesia is more rapid and a smaller quantity of procaine is required.[16] However, neutral and alkaline solutions are very unstable and cannot be sterilized by autoclaving. The problem is even greater if epinephrine is to be used, because it is less stable than procaine in alkaline solutions.

The procaine molecule is a prototype of primary aromatic amines that are subject to oxidative decomposition. This decomposition can be controlled by nitrogen flushing of the solutions and by the addition of an antioxidant.

Another characteristic of the local anesthetics having a primary aromatic group is the reaction with glucose for the formation of procaine N-glucoside. There is no significant change in the clinical results, but there is the possibility of interference with the assay. Procaine also forms a soluble complex with sodium carboxymethyl cellulose.[17] However, sodium chloride displaces it. Again, there probably is no change in pharmacological action, but there may be an analytical problem.

Procaine hydrochloride is not effective on intact skin or mucous membranes but acts promptly when used by infiltration. The action may be prolonged by the concurrent administration of epinephrine or other vasoconstrictors that slow its release into the bloodstream, where procaine is rapidly inactivated by hydrolysis. Pseudocholinesterases are a group of plasma and tissue enzymes that catalyze the hydrolysis of esters, including procaine.

The toxicity of procaine is greatly decreased if absorption is restricted by vasoconstrictors, allowing the serum hydrolysis rate to exceed the rate of release from the tissues. Procaine induces a relatively high incidence of allergic reactions, and sensitive individuals usually also react to other aminobenzoates. For these patients, it is advisable to use an agent from a different chemical class. Unfortunately, tests for sensitivity are not reliable for predicting allergic responses.

Procaine is used to form an insoluble salt with penicillin G. The low solubility (7:1000) accounts for the prolonged action of this penicillin salt. Penicillin G is slowly released from the intramuscular depot.

Procainamide Hydrochloride U.S.P., the amide derivative of procaine, is a cardiac depressant (Chap. 14).

Chloroprocaine Hydrochloride U.S.P., Nesacaine® Hydrochloride, 2-diethylaminoethyl 4-amino-2-chlorobenzoate monohydrochloride. The salt, a white, crystalline powder, is stable in air, and its solutions are acid to litmus. It is soluble in water (1:22) and slightly soluble in 95 percent alcohol (1:100).

Chloroprocaine Hydrochloride

Chloroprocaine differs structurally from procaine in having a chlorine substituent in the 2-position of the aromatic ring. The electron-withdrawing chlorine atom destabilizes the ester group to hydrolysis. Chloroprocaine is hydrolyzed by plasma more than four times faster than procaine. It is more rapid in onset of action and more potent than procaine.

Tetracaine U.S.P., Pontocaine®, 2-dimethyl-aminoethyl 4-(butylamino)benzoate. Tetracaine, as the free base, occurs as a white to light-yellow waxy solid that must be protected from light. It is very slightly soluble in water and soluble in alcohol and in lipid substances.

Tetracaine Hydrochloride

Of the ester derivatives of procaine, tetracaine is one of the most easily absorbed drugs. Absorption of tetracaine from mucous membranes is rapid and simu-

lates intravenous injection closely,[13] whereas procaine is more slowly absorbed. The variation in absorption may be attributed to the difference in lipid solubility. The presence of a nonpolar n-butyl group on the aromatic nitrogen atom probably accounts for the greater lipid solubility of tetracaine. It is more slowly hydrolyzed in plasma than procaine. The free base is used to prepare ointments (0.5%) for topical and ophthalmic application.

Tetracaine Hydrochloride U.S.P. is a water-soluble salt (1:7) that occurs as a fine, white, crystalline powder and has a slightly bitter taste (pKa 8.5). Its solutions are more stable to hydrolysis than are procaine hydrochloride solutions and they may be sterilized by boiling.

Butacaine Sulfate U.S.P., Butyn® Sulfate, 3-(di-n-butylamino)-1-propanol p-aminobenzoate sulfate. The salt occurs as a white, crystalline powder that should be protected from light. It is freely soluble in water (ca. 1:1); the hydrochloride salt is less soluble. Structurally, butacaine differs from procaine in that the intermediate chain is increased to three carbons and the tertiary amino group has two n-butyl instead of two ethyl groups. Solutions may be sterilized by boiling and are effective on mucous membranes and in the eye. The rate of hydrolysis in plasma is comparable to that of tetracaine.

Butacaine Sulfate

Benoxinate Hydrochloride U.S.P., Dorsacaine® Hydrochloride, 2-diethylaminoethyl 4-amino-3-n-butoxybenzoate hydrochloride. The salt occurs as a white, crystalline powder and has a salty taste. It is very soluble in water and in chloroform and it is soluble in alcohol. The powder is stable in air and not affected by light or heat. Aqueous solutions have a pH of 4.5 to 5.2.

Benoxinate Hydrochloride

The chemical properties are similar to those of procaine except that the 3-butoxy group appears to stabilize the molecule to hydrolysis. This is in marked contrast to the 2-chloroprocaine, which is less stable to hydrolysis than procaine. The rate of hydrolysis is pH dependent, with the slowest rate at a pH of about 4.[18]

Benoxinate is employed primarily in ophthalmology as a 0.4 percent solution. Applications do not cause any significant irritation, constriction, or dilation of the pupil, any noticeable light sensitivity, or symptoms indicating that absorption into the system has taken place.

Propoxycaine Hydrochloride U.S.P., Blockain® Hydrochloride, 2-diethylaminoethyl 4-amino-2-propoxybenzoate monohydrochloride. This salt is a white, crystalline solid that is freely soluble in water and soluble in alcohol. The pH of a 2 percent solution is 5.4, which may be diluted and buffered at neutral pH without precipitation.

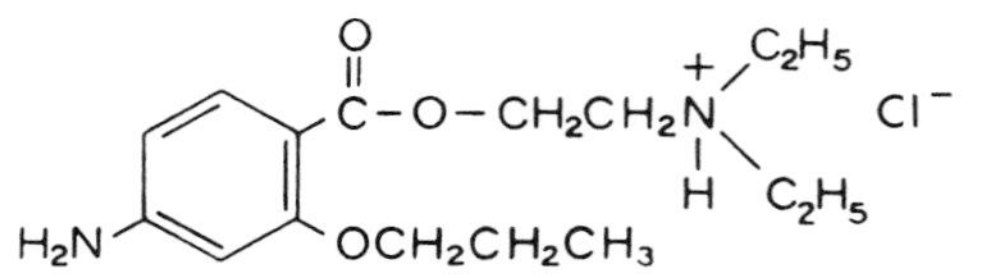

are minimal when the alkoxy substituent is in the 3-position.

Propoxycaine has a quicker onset and longer duration of action and is considerably more potent than procaine. Its greater lipophilic character probably accounts for these advantages. It is administered by injection for nerve block and infiltration anesthesia as a 0.5 percent solution without vasoconstrictors.

Proparacaine Hydrochloride U.S.P., Alcaine® Hydrochloride, Ophthaine® Hydrochloride, Ophthetic® Hydrochloride, 2-diethyaminoethyl 3-amino-4-propoxybenzoate monohydrochloride. The salt is a white to off-white, crystalline powder that is soluble in water (1:30) and in warm alcohol (1:30). Solutions are neutral to litmus and discolor in the presence of air. Discolored solutions should not be used.

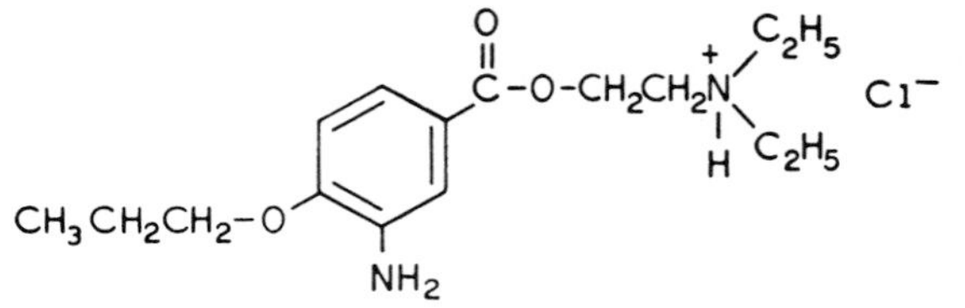

Propoxycaine Hydrochloride

Proparacaine Hydrochloride

Propoxycaine is not stable to autoclaving. The 2-propoxy group appears to labilize the ester group in much the same way as the 2-chloro group does in chloroprocaine. This is in contrast to the apparent stabilizing effect of a butoxy group placed in the 3-position (benoxinate). The inductive and steric effects of a 2-alkoxy substituent, which favor hydrolysis, apparently have a greater influence than the stabilizing positive-mesomeric (resonance) effect. These effects

Proparacaine is a positional isomer of propoxycaine. Proparacaine is less soluble, slightly more potent, and more toxic than propoxycaine. Although it is too toxic for use as an injection anesthetic, proparacaine is suitable for use topically in ophthalmology.

Table 15-2 shows the structural relationships of the local anesthetics that are esters of *p*-aminobenzoic acid.

TABLE 15-2

p-AMINOBENZOIC ACID DERIVATIVES

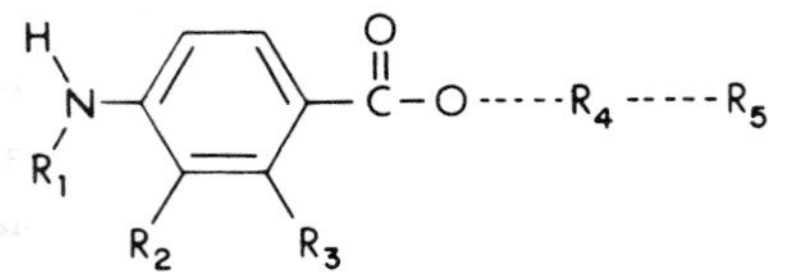

Generic Name	Proprietary Name	R_1	R_2	R_3	R_4	R_5
Benzocaine	Americaine	H	H	H	$-C_2H_5$	———
Butamben	Butesin	H	H	H	$-CH_2CH_2CH_2CH_3$	———
Procaine	Novocain	H	H	H	$-CH_2CH_2-$	$-N(C_2H_5)_2$
Chloroprocaine	Nesacaine	H	H	Cl	$-CH_2CH_2-$	$-N(C_2H_5)_2$
Tetracaine	Pontocaine	Bu	H	H	$-CH_2CH_2-$	$-N(CH_3)_2$
Butacaine	Butyn	H	H	H	$-CH_2 2CH_2CH_2-$	$-N(n-C_4H_9)_2$
Benoxinate	Dorsacaine	H	BuO—	H	$-CH_2CH_2-$	$-N(C_2H_5)_2$
Propoxycaine	Blockain	H	H	PrO—	$-CH_2CH_2-$	$-N(C_2H_5)_2$

ANILIDES

As a chemical class of local anesthetics, the currently used anilides were derived from lidocaine. Agents of this class are more stable to hydrolysis even under the conditions of prolonged autoclaving. They are more potent, have a lower incidence of side-effects, and induce less local irritation than the procaine-type local anesthetic agents. Also, there seems to be no cross-sensitization between the anilides and the benzoic-acid derivatives. Thus, the anilides are valuable alternatives for those individuals sensitive to the procaine-type local anesthetics.

Additionally, the anilides are effective with or without vasoconstrictors and are agents of choice in those instances where the dentist or patient may be sensitive to epinephrine and its congeners.

In contrast to the esters of benzoic-acid derivatives, the anilides undergo enzymatic degradation not in plasma but primarily in the liver. The major route of metabolism of the anilides is enzymatic hydrolysis of the amide bond.

The difference in site of metabolism is of clinical relevance. Patients possessing a genetic deficiency in the enzyme, pseudocholinesterase, are unable to hydrolyze ester-type local anesthetic agents at a normal rate and show reduced tolerance to these drugs. For these patients, an anilide agent may be preferable.

On the other hand, the anilides are metabolized mainly by hepatic microsomal enzymes, and patients with liver disease may show reduced tolerance to this type of agent. In such cases, the benzoic-acid derivatives provide suitable alternative agents.[5]

Structural relationships of the anilide-type local anesthetics are shown in Table 15-3.

Lidocaine U.S.P., Xylocaine®, 2-(diethylamino)-N-(2,6-dimethylphenyl)acetamide, 2-(diethylamino)-2′,6′-acetoxylidide. This compound may be considered the anilide of 2,6-dimethylaniline and N-diethylglycine. It occurs as a white to slightly yellow crystalline powder that has a characteristic odor and is stable in air. It is practically insoluble in water and very soluble in alcohol and lipid materials.

The hydrochloride salt (Lidocaine Hydrochloride U.S.P.) occurs as a white, odorless, crystalline powder that is very soluble in water and in alcohol. The salt has a pKa of 8.0.

Lidocaine Hydrochloride

TABLE 15-3

ANILIDES

Generic Name	Proprietary Name	R_1	R_2
Lidocaine	Xylocaine	CH_3	$-CH_2-N(C_2H_5)_2$
Mepivacaine	Carbocaine	CH_3	(2-methylpiperidin-1-yl, N-CH_3)
Prilocaine	Citanest	H	$-CH(CH_3)-NH-CH_2CH_2CH_3$
Bupivacaine	Marcaine	CH_3	(2-methylpiperidin-1-yl, N-CH_2CH_2CH_2CH_3)
Etidocaine	Duranest	CH_3	$-CH(C_2H_5)-N(C_2H_5)(CH_2CH_2CH_3)$

As opposed to ester-type local anesthetics, lidocaine is extremely resistant to hydrolysis. This is not unexpected because, in addition to the relative stability of the amide bond, the 2,6-dimethyl substituents provide steric hindrance to attack of the carbonyl. Solutions of the drug may be autoclaved and may be stored for months at room temperature without significant decomposition.

Lidocaine is about twice as potent as procaine and approximately 1.5 times as toxic. Systemic side-reactions and local irritant effects are few. Lidocaine appears to be relatively free of sensitizing reactions characteristic of the p-aminobenzoates. Because there seems to be no cross-sensitization between the anilide and the benzoic-acid types of local anesthetics, lidocaine is an agent of choice in individuals sensitive to procaine.

Also, lidocaine is effective with or without a vasoconstrictor. Hence, it is a valuable alternative for use in those individuals who are allergic to epinephrine and its congeners. The free base and its salts are effective topical local anesthetics. The hydrochloride is used for infiltration, peripheral-nerve block and epidural anesthesia.

Lidocaine is also an effective cardiac depressant and may be administered intravenously in cardiac surgery and life-threatening arrhythmias (see Chap. 14).

Mepivacaine Hydrochloride U.S.P., Carbocaine® Hydrochloride, ($\pm$)-N-(2,6-dimethylphenyl)-1-methyl-2-piperidinecarboxamide hydrochloride. The salt occurs as a white, crystalline solid that is freely soluble in water. The pH of a 2 percent solution is about 4.5.

Like lidocaine and the other anilides, mepivacaine is highly resistant to hydrolysis, and its solutions can be autoclaved without appreciable decomposition.

Mepivacaine Hydrochloride

Mepivacaine is used as the racemic mixture, because the two optical isomers were found[19] to have essentially the same toxicity and potency; the toxicity and potency of either are comparable to that of lidocaine. Its duration of action is considerably longer than that of lidocaine, even without a vasoconstrictor. Thus, mepivacaine hydrochloride is particularly useful when epinephrine and its congeners are contraindicated.

The effectiveness of mepivacaine as a topical local anesthetic has not been established.

Prilocaine Hydrochloride U.S.P., Citanest® Hydrochloride, N-(2-methylphenyl)-2-(propylamino)propanamide hydrochloride. The salt occurs as a white, crystalline powder, having a bitter taste. It is freely soluble in water and in alcohol.

Prilocaine (originally named propitocaine) is similar in stability, potency, toxicity, and duration of action to the other anilides. Its duration of action is intermediate between that of lidocaine and mepivacaine. Except for methemoglobinemia, its side-effects are similar to those observed with other anilides.

A primary metabolite of prilocaine is *o*-toluidine (2-methylaniline), which apparently is metabolized further to products that cause methemoglobinemia. Metabolic products of aniline (i.e., phenylhydroxylamine and nitrosobenzene) are known to produce methemoglobinemia.[20] Apparently 2,6-dimethylaniline (2,6-xylidine), the hydrolysis product of lidocaine and certain other anilides, is resistant to N-oxidation, and use of these anilides does not result in methemoglobinemia. This side-effect is not observed with other local anesthetic agents when used in normal dose ranges. However, when procaine was given intravenously in large doses to produce general anesthesia, methemoglobinemia was observed. It was suggested that aniline was formed by decarboxylation of PABA, a hydrolysis product of procaine.[21]

Prilocaine Hydrochloride

Prilocaine hydrochloride solutions are used without the addition of vasoconstrictors; therefore they are useful when such agents are contraindicated.

Bupivacaine Hydrochloride U.S.P., Marcaine® Hydrochloride. This local anesthetic differs from mepivacaine only in that the basic nitrogen contains a butyl instead of a methyl group. Its properties are very similar to those of mepivacaine.

Bupivacaine Hydrochloride

The duration of action of bupivacaine is 2 to 3 times longer than that of lidocaine or mepivacaine and 20 to 30 percent longer than that of tetracaine. Its potency is comparable to that of tetracaine but is about 4 times that of mepivacaine and lidocaine.

Etidocaine Hydrochloride, Duranest® Hydrochloride, was clinically introduced in 1972. Structurally, it is closely related to lidocaine. Its physicochemical properties and pharmacological actions are similar to those of lidocaine, but it possesses greater anesthetic potency and a longer duration of action.

Etidocaine Hydrochloride

MISCELLANEOUS

Phenacaine Hydrochloride U.S.P., Holocaine® Hydrochloride, N,N'-*bis*(4-ethoxyphenyl)acetamidine monohydrochloride monohydrate. The salt occurs as white crystals having a slightly bitter taste and producing a numbness of the tongue. It is stable in air and sparingly soluble in water (1:50), but freely soluble in alcohol. Aqueous solutions may be sterilized by boiling but they are very sensitive to alkalis. The alkalinity of a soft glass container may be sufficient to cause precipitation of the free base from solution.

Phenacaine Hydrochloride

Phenacaine is structurally related to the anilides in that an aromatic ring is attached to an sp^2 carbon through a nitrogen bridge. Its local anesthetic properties were reported in 1897; this preceded the discovery of the action of isogramine. However, the appearance of this unique structural feature appears to be an isolated incident and did not exert a significant influence on the subsequent evolution of local anesthetic development until the discovery of isogramine.

Another isolated historical incident apparently had no effect on the evolution of the anilide-type local anesthetic agents. The anilide, 5-(diethylaminoacetyl-amino)-2-hydroxybenzoic acid methyl ester (Nirvanine), was prepared as a water-soluble (salt) derivative of orthoform. It failed to stir the interest of clinicians because it was less active and more irritating than cocaine.

Phenacaine hydrochloride is slightly irritating, causing some discomfort to precede anesthesia. It is more toxic than cocaine and cannot be used for injection. However, it is fast acting and very effective on mucous membranes. Thus, because of toxicity, it is used primarily in ophthalmology in 1 percent solutions or 1 to 2 percent ointments.

Diperodon U.S.P., Diothane®, 3-piperidino-1,2-propanediol dicarbanilate monohydrate. Diperodon occurs as a white to cream-colored powder having a characteristic odor. The free base is insoluble in water and relatively unstable to heat, but aqueous solutions of the hydrochloride salt at pH 4.5 to 4.7 are stable and may be autoclaved. Structurally, it is related to the anilides in that an aromatic ring is attached to an sp^2 carbon via a nitrogen bridge. Thus a hydrolysis product is aniline and methemoglobinemia is a potential toxic side-effect.

After intravenous injection (as the hydrochloride salt), diperodon possesses comparable toxicity to cocaine. It is recommended for the relief of pain and

Diperodon

irritation in abrasions of the skin and mucous membranes, especially hemorrhoids.

Dimethisoquin Hydrochloride U.S.P., Quotane® Hydrochloride, 3-butyl-1-[2-(dimethylamino)ethoxy]-isoquinoline monohydrochloride. The salt occurs as a white to off-white, crystalline powder that is soluble in water (1:8), in alcohol (1:3), and in chloroform (1:2).

Dimethisoquin Hydrochloride

Structurally, dimethisoquin may be considered to be related to the benzoic-acid derivatives in that the phenyl ring is attached directly to an sp^2 carbon atom and the basic side chain is attached via an oxygen bridge.

Dimethisoquin is a safe, effective compound for general application as a topical anesthetic. It is available as a water-soluble ointment (0.5%) for dry dermatologic conditions and in lotion form for moist skin surfaces.

Pramoxine Hydrochloride U.S.P., Tronothane® Hydrochloride, 4-[3-(4-butoxyphenoxy)propyl]morpholine hydrochloride. The salt occurs as a white to off-white, crystalline solid that may have a slight aromatic odor. It is freely soluble in water or alcohol. The pH of a 1 percent solution is about 4.5.

Pramoxine Hydrochloride

Pramoxine is too irritating for ophthalmic use but is an effective topical local anesthetic of a low-sensitizing index with few toxic reactions. It is used for relief of pain and itching due to insect bites, minor wounds and lesions, and hemorrhoids.

Dyclonine Hydrochloride U.S.P., Dyclone® Hydrochloride, 4'-butoxy-3-piperidinopropiophenone hydrochloride. The salt occurs as a white, crystalline powder that may have a slight odor. It is soluble in water and in alcohol and stable in acidic solutions, if not autoclaved.

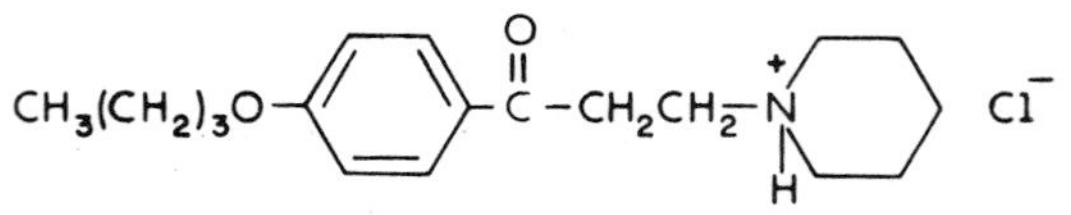

Dyclonine Hydrochloride

Dyclonine is a tissue irritant if injected but is an effective topical anesthetic for use on skin and mucous membranes. It is effective for anesthetizing mucous membranes prior to endoscopic and proctoscopic procedures. Anesthesia usually occurs in 5 to 10 minutes after application and persists for 20 minutes to 1 hour.

Dibucaine U.S.P., Nupercaine®, 2-butoxy-N-(2-diethylaminoethyl)cinchoninamide. The free base occurs as a white to off-white powder having a slight, characteristic odor. Chemically, the drug is an amide that darkens on exposure to light; it is slightly soluble in water and soluble in chloroform and in ether.

The lesser soluble free base is used to prepare cream, ointment, and rectal suppository dose forms for topical application. The more soluble hydrochloride salt (1:2 in water) is suitable for the preparation of the injectable dose form (Dibucaine Hydrochloride U.S.P., Nupercaine® Hydrochloride). Aqueous solutions of the salt may be sterilized by autoclaving, but the pH should not be above 6.2.

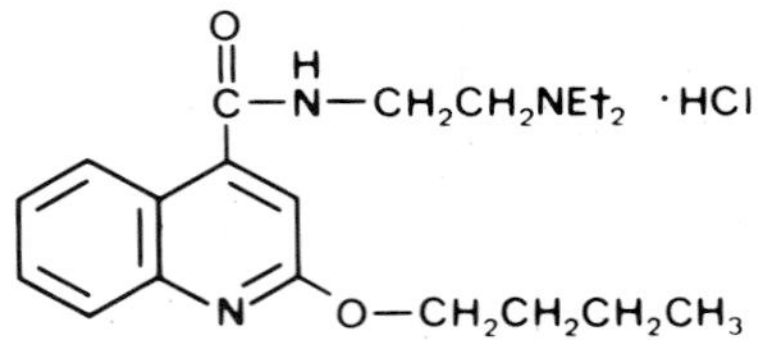

Dibucaine Hydrochloride

Because dibucaine is an amide, it is not hydrolyzed to any measurable extent in serum. It has been shown[13] to be metabolized rather slowly, which explains its high toxicity. It is the most potent, most toxic, and longest acting of the commonly employed local anesthetics.

The official titles of the local anesthetics listed in the *United States Pharmacopeia XX–National Formulary XV* are designated simply as U.S.P. in Table 15-4; the proprietary name(s), official preparations, application, usual dose, and usual dose range are also given.

TABLE 15–4

LOCAL ANESTHETICS

Name Proprietary Name	Preparations	Application	Usual Adult Dose*	Usual Dose Range*
Cocaine Hydrochloride U.S.P.	Cocaine Hydrochloride Tablets for Topical Solution U.S.P.	Topical, 2 to 20 percent solution to mucous membranes		
Hexylcaine Hydrochloride U.S.P. *Cyclaine*	Hexylcaine Hydrochloride Injection U.S.P. Hexylcaine Hydrochloride Topical Solution U.S.P.		Topical or by injection according to site and condition	
Meprylcaine Hydrochloride U.S.P. *Oracaine*	Meprylcaine Hydrochloride and Epinephrine Injection U.S.P.	Dental		
Isobucaine Hydrochloride U.S.P.	Isobucaine Hydrochloride and Epinephrine Injection U.S.P.	Dental		
Cyclomethycaine Sulfate U.S.P. *Surfacaine*	Cyclomethycaine Sulfate Cream U.S.P. Cyclomethycaine Sulfate Jelly U.S.P. Cyclomethycaine Sulfate Ointment U.S.P. Cyclomethycaine Sulfate Suppositories U.S.P.	Topical, 0.5 percent cream, 0.75 percent jelly, or 1 percent ointment to the skin	Rectal, 10 mg.	
Dibucaine U.S.P. *Nupercaine*	Dibucaine Cream U.S.P. Dibucaine Ointment U.S.P. Dibucaine Suppositories U.S.P.	Topical, 0.5 percent cream or 1 percent ointment several times daily	Rectal 2.5 mg. suppository several times daily	

*See *U.S.P. D.I.* for complete dosage information.

(Continued)

TABLE 15-4

LOCAL ANESTHETICS

Name Proprietary Name	Preparations	Application	Usual Adult Dose*	Usual Dose Range*
Dibucaine Hydrochloride U.S.P. *Nupercaine Hydrochloride*	Dubucaine Hydrochloride Injection U.S.P.			
Benzocaine U.S.P.	Benzocaine Cream U.S.P. Benzocaine Ointment U.S.P.	Topical, 1 to 20 percent aerosol, cream, or ointment to the skin		
Butamben U.S.P. *Butesin*		Topical		
Procaine Hydrochloride U.S.P. *Novocain*	Procaine Hydrochloride Injection U.S.P. Sterile Procaine Hydrochloride U.S.P. Procaine Hydrochloride and Epinephrine Injection U.S.P. Procaine and Phenylephrine Hydrochlorides Injection U.S.P. Procaine and Tetracaine Hydrochlorides and Levonordefrin Injection U.S.P.	Dental Dental Dental		
Chloroprocaine Hydrochloride U.S.P. *Nesacaine*	Chloroprocaine Hydrochloride Injection U.S.P.		Infiltration, 100 ml. of 0.5 percent solution; peripheral nerve block, 50 ml. of a 1 percent solution	
Tetracaine U.S.P. *Pontocaine*	Tetracaine Ointment U.S.P. Tetracaine Ophthalmic Ointment U.S.P.	Topical, 0.5. percent ointment to the conjunctiva to induce transient loss of corneal sensitivity		
Tetracaine Hydrochloride U.S.P. *Pontocaine Hydrochloride*	Tetracaine Hydrochloride Injection U.S.P. Sterile Tetracaine Hydrochloride U.S.P. Tetracaine Hydrochloride Ophthalmic Solution U.S.P. Tetracaine Hydrochloride Topical Solution U.S.P. Tetracaine Hydrochloride Cream U.S.P.	Topically to the conjunctiva, 0.05 to 0.1 ml. of a 0.5 or 1 percent solution Topically to the nose and throat 1 to 2 ml. of a 0.25 to 2 percent solution Topical, cream containing the equivalent of 1 percent of tetracaine	Subarachnoid, 0.5 to 2 ml. of a 0.5 percent solution in spinal fluid	
Butacaine Sulfate U.S.P. *Butyn*	Butacaine Sulfate Solution U.S.P.	Topical, 2 percent solution		
Benoxinate Hydrochloride U.S.P. *Dorsacaine*	Benoxinate Hydrochloride Ophthalmic Solution U.S.P.	Topical, to the conjunctiva, 50 to 200 μl. of a 0.4 percent solution		
Propoxycaine Hydrochloride U.S.P. *Blockain*	Propoxycaine and Procaine Hydrochlorides and Levonordefrin Injection U.S.P. Propoxycaine and Procaine Hydrochlorides and Norepinephrine Bitartrate Injection U.S.P.			

* See *U.S.P. D.I.* for complete dosage information.

(Continued)

TABLE 15–4

LOCAL ANESTHETICS

Name Proprietary Name	Preparations	Application	Usual Adult Dose*	Usual Dose Range*
Proparacaine Hydrochloride U.S.P. *Ophthaine*	Proparacaine Hydrochloride Ophthalmic Solution U.S.P.	Topically to the conjunctiva, 0.05 ml. of a 0.5 percent solution, repeated at 5- to 10-minute intervals if necessary		
Lidocaine U.S.P. *Xylocaine*	Lidocaine Topical Aerosol U.S.P. Lidocaine Ointment U.S.P.	Topically to mucous membranes, as a 2.5 to 5 percent ointment		
Lidocaine Hydrochloride U.S.P. *Xylocaine Hydrochloride*	Lidocaine Hydrochloride Injection U.S.P. Lidocaine Hydrochloride and Epinephrine Injection U.S.P. Lidocaine Hydrochloride Jelly U.S.P.	Topically to mucous membranes as a 2 percent jelly		
Mepivacaine Hydrochloride U.S.P. *Carbocaine*	Mepivacaine Hydrochloride Injection U.S.P. Mepivacaine Hydrochloride and Levonordefrin Injection U.S.P.			
Prilocaine Hydrochloride U.S.P. *Citanest*	Prilocaine Hydrochloride Injection U.S.P.			Therapeutic nerve block, 3 to 5 ml. of a 1 to 2 percent solution; infiltration, 20 to 30 ml. of a 1 or 2 percent solution; regional anesthesia, peridural and caudal, 15 to 20 ml. of a 3 percent solution or 20 to 30 ml. of a 1 or 2 percent solution; for infiltration and nerve block in dentistry, 0.5 to 5.0 ml. of a 4 percent solution
Diperodon U.S.P. *Diothane*	Diperodon Ointment U.S.P.	Topical or intrarectal, a 1 percent ointment 3 or 4 times daily		
Phenacaine Hydrochloride U.S.P.		To the conjunctiva, 1 or 2 percent ointment or a 1 percent solution		
Dimethisoquin Hydrochloride U.S.P. *Quotane*	Dimethisoquin Hydrochloride Lotion U.S.P. Dimethisoquin Hydrochloride Ointment U.S.P.	Topical, to the skin, 0.5 percent lotion or ointment 2 to 4 times daily		
Pramoxine Hydrochloride U.S.P. *Tronothane*	Pramoxine Hydrochloride Cream U.S.P. Pramoxine Hydrochloride Jelly U.S.P.	Topical, 1 percent cream or jelly every 3 to 4 hours		
Dyclonine Hydrochloride U.S.P. *Dyclone*	Dyclonine Hydrochloride Topical Solution U.S.P.	Topical, to mucous membranes, 0.5 to 1 percent solution		

* See *U.S.P. D.I.* for complete dosage information.

REFERENCES

1. Ritchie, J. M., and Cohen, P. J.: Cocaine, Procaine and Other Synthetic Local Anesthetics, *in* Goodman, L. S., and Gilman, A. (eds.): The Pharmacological Basis of Therapeutics, 5th ed., p. 379, New York, Macmillan, 1975.
2. Liljestrand, G.: The Historical Development of Local Anesthetics, *in* Lechat, P. (ed.): Local Anesthetics; International Encyclopedia of Pharmacology and Therapeutics, Section 8, vol. 1, p. 1, New York, Pergamon Press, 1971.
3. Claus, E. P., Tyler, V. E., and Brady, L. R.: Pharmacognosy, 6th ed., p. 238, Philadelphia, Lea & Febiger, 1970.
4. Ray, O. S.: Drugs, Society, and Human Behavior, p. 18, Saint Louis, C. V. Mosby Company, 1972.
5. Covino, B. G., and Vassallo, H. G.: Local Anesthetics, Mechanisms of Action and Clinical Use, p. 1, New York, Grune & Stratton, 1976.
6. Brinley, F. J., and Mullins, L. J.: Ann. N.Y. Acad. Sci. 242:406, 1974.
7. Askari, A., Ed.: Properties and Functions of $(Na^+ + K^+)$ — Activated Adenosinetriphosphatase, Ann. N.Y. Acad. Sci. 242:1–741, 1974.
8. Ritchie, J. M.: The Mechanism of Action of Local Anesthetic Agents, *in* Lechat, P. (ed.); Local Anesthetics, *International Encyclopedia* of *Pharmacology* and *Therapeutics,* Section 8, vol. 1, p. 131, New York, Pergamon, 1971.
9. Takman, B.: Br. J. Anaesth. (Suppl.) 47:183, 1975.
10. Büchi, J., and Perlia, X.: Structure-Activity Relations and Physicochemical Properties of Local Anesthetics, *in* Lechat, P. (ed.): Local Anesthetics, International Encyclopedia of Pharmacology and Therapeutics, Section 8, vol. 1, p. 39, New York, Pergamon, 1971.
11. Büchi, J., and Perlia, X.: Design of Local Anesthetics, *in* Ariens, E. J. (ed.): Drug Design, vol. 11-III, p. 243, New York, Academic Press, 1972.
12. Cohen S.: (J.A.M.A.) 231:74, 1975.
13. Hansson, E.: Absorption, Distribution, Metabolism and Excretion of Local Anesthetics, *in* Lechat, P. (ed.): Reference 2, p. 239.
14. Terp, P.: Acta Pharmacol. Toxicol. 5:353 1949, through Chem. Abstr. 44:6576c, 1950.
15. Bullock, K., and Cannell, J. S.: Quart J. Pharm. Pharmacol. 14:241, 1941.
16. Bullock, K.: Quart. J. Pharm. Pharmacol. 11:407, 1938.
17. Kennon, L., and Higuchi T.: J. Am. Pharm. A (Sci. Ed.) 45:157, 1956.
18. Willi, A. V.: Pharm. acta Helvet. 33:635, 1958.
19. Sadove, M., and Wessinger, G.D.: J. Int. Coll. Surg. 34:573, 1960.
20. Goldstein, A., Aronow, L., and Kalman, S. M.: Principles of Drug Action, p. 208, New York, Harper & Row, 1968.
21. Luduena, F. P.: Toxicity and Irritancy of Local Anesthetics, *in* Reference 2, p. 319.
22. Gutmann, G.: Deut. Med. Wochschr. 23:165, 1897.

SELECTED READINGS

Adriani, J., and Maraghi, M.: The Pharmacologic Principles of Regional Pain Relief, Ann. Rev. Pharmacol. Toxicol. 17:223, 1977.
Büchi, J., and Perlia, X.: The Design of Local Anesthetics, *in* Ariens, E. J. (ed.): Drug Design, vol. 11-III, p. 243, New York, Academic Press, 1972.
Covino, B. G., and Vassallo, H. G.: Local Anesthetics, Mechanisms of Action and Clinical Use, New York, Grune & Stratton, 1976.
Lechat, P. (ed.): Local Anesthetics, International Encyclopedia of Pharmacology and Therapeutics, New York, Pergamon, 1971.
Lofgren, N.: Studies on Local Anesthetics, Stockholm, University of Stockholm, 1948.
Ritchie, J. M., and Cohen, P. J.: Local Anesthetics, *in* Goodman, L. S., and Gilman, A. (eds.): The Pharmacological Basis of Therapeutics, 5th ed., p. 379, New York, Macmillan, 1975.
Takman, B. H., *et al.*: Local Anesthetics, *in* Foye, W. O. (ed.): Principles of Medicinal Chemistry 2nd ed., p. 339, Philadelphia, Lea & Febiger, 1981.
Takman, B. H., and Camougis, G.: Local Anesthetics, *in* Burger, A. (ed.): Medicinal Chemistry, ed. 3, p. 1607, New York, Wiley-Interscience, 1970.

16

histamine and antihistaminic agents

Charles M. Darling

HISTAMINE

Histamine (β-imidazolylethylamine or 1H-imidazole-4-ethanamine) was synthesized[1] in 1907 before its presence in tissues was recognized. It is widespread in nature, being found in ergot and other plants and in all organs and tissues of the human body. Its physiological importance is underscored by its classification as an autacoid, a word derived from the Greek *autos* (self) and *akos* (medicinal agent or remedy).

Noting species differences, Dale and Laidlaw[2,3,4] outlined the principal pharmacological activities of histamine from 1910 to 1919. After more than half a century of painstaking study, several physiological roles have been proposed for histamine.[5]

Histamine probably plays a basic role in the beginning of the inflammatory response of tissue to injury by dilating the capillaries and increasing their permeability. The usefulness of this effect, which usually leads to a "walling off" of the area of injury, may be counteracted by the effect of excess histamine released during the process. A conclusion was stated[6] that the inflammatory process would possibly develop faster and more favorably to the subject if an excess of histamine were not released.

The role of histamine in immediate hypersensitivity (antigen–antibody) reactions is complicated by release of other autacoids during such reactions. However, sufficient data[5] has been collected to suggest that histamine inhibits its own release and augments eosinophil chemotaxis.

Histamine exerts a variety of actions on the cardiovascular system.[7] Although the evidence is indirect, histamine continues to be implicated in the regulation of the microcirculation. The available quantitative evidence seems to indicate the presence of at least two distinct types of histamine receptors.

The increased synthesis of histamine in many tissues undergoing rapid growth or repair has been construed as evidence that "nascent" histamine may play a role in anabolic processes.[8] Histamine is distributed unevenly in the brain, as are other biogenic amines. Evidence on histamine as a central neurotransmitter has been reviewed.[5,9] The role of histamine in gastric secretion has been reviewed[10] and will be discussed here in a later section. Other exocrine glands remain to be investigated for their quantitative responses to histamine.

In general, histamine appears to be necessary for many physiological processes. Possibly, histamine has a homeostatic role and the histamine-forming capacity of various tissues is responsive to alterations in the concentration of the amine. However, certain conditions may lead to excessive production or release of the autacoid resulting in undesirable effects, e.g., shock. Less drastic actions of histamine are implicated in other diseases, such as peptic ulceration and asthmatic conditions. Therefore these and other actions of histamine have been studied for the purpose of developing drugs that will block selectively certain actions of histamine.

A review[11] of the reported evidence clearly indicates that there exists in the body and fluids of mammals one or more substances with antihistaminic activity. Natural antihistamine substance (NAS) appears to be

583

highly potent, but neither the mechanism of action nor the chemical structure is known.

HISTAMINE RECEPTORS

The recent discovery of H_2-antagonists has confirmed that more than one type of histamine receptor exists in biological systems. These antagonists have been shown to block gastric-acid secretion effectively. The stage is set not only to revolutionize the treatment of gastric ulcers but also for the birth of a new era of histamine antagonists that may elucidate the role of histamine in neurotransmission, microcirculatory control, inflammation, allergic disorders, and other phenomena. The setting of the stage began in the early years of this century.

Before the identification of histamine in the body, the testing of simple extracts from most organs and tissues of the body led to the use of such unspecific terms as *depressor substance, vasodilation, anaphylatoxin, H-substance,* and others. Although histamine had been identified chemically in tissue extracts, many were skeptical that histamine was a natural constituent of the body until Best and co-workers established its presence beyond doubt in 1927.[12] Unlike anticholinergic or antiadrenergic agents, some of which are found in nature, there were no potent antihistaminic drugs to be found in nature to serve as a model for developing synthetic antihistaminic drugs. This explains why the study of antihistaminic activity was delayed until the serendipitous discovery of synthetic antihistaminics. Research on antihistaminic drugs was initiated in France in 1933. While screening chemicals for other activities, Fourneau and Bovet[13] observed that certain aryl ethers containing a basic side chain protected guinea pigs against lethal doses of histamine. This initiated worldwide interest in the synthesis and study of related antihistaminic compounds during the next four decades. A review[14] of the chemistry of the more active antagonists reveals the surprising fact that these agents are structurally unrelated to histamine. Many of their physical and chemical properties differ greatly from those of histamine. With various degrees of potency, these classical antihistaminics inhibit the blood-pressure-lowering, gut-stimulation, and undesirable side-effects of histamine and its analogs. However, the gastric-stimulating effects of histamine are either unaffected or augmented by them.

Early on it was recognized that not only did the antihistaminics fail to antagonize all of the pharmacological actions of histamine but they also exhibited other pharmacological actions, e.g., anticholinergic and local anesthetic. This presented a challenge to establish the existence of receptors specific for histamine. A review[16] of the evidence for specific histamine receptors indicates that the postulated competitive antagonism of certain actions of histamine by the antihistaminics was supported by data collected in the late 1940's. The reader interested in the question of how to define an antihistaminic and its degree of activity is referred to a recent review.[21]

Acceptance of the proposal in 1948[17] that there exist at least two adrenergic receptors gave impetus to the search for more than one type of histamine receptor. In 1962, Lin and co-workers[18] suggested that, because the antihistaminics inhibited some but not all of the actions of histamine, there must be more than one type of histaminergic receptor. The concept was formalized in 1966[19] by defining H_1-receptors as those that are specifically antagonized by low concentrations of antihistaminic drugs. The other actions of histamine then are unlikely to be mediated by H_1-receptors. The wide variation in relative agonist activities of several histamine analogs further supported the postulate that histamine receptors could be differentiated into at least two classes.

Of necessity, the definition of H_2-receptors awaited the discovery of a specific antagonist. In 1972, Black and associates[20] provided additional evidence for the postulate of more than one type of histaminergic receptor. They observed that 2-methylhistamine (Fig. 16-1) had significantly more activity on the tissues containing H_1-receptors (e.g., guinea-pig ileum), whereas 4-methylhistamine had considerably more activity on guinea-pig atrium (no H_1-receptors). They also reported that a new compound, burimamide, acted as a specific competitive antagonist of histamine on those tissues refractory to block by the antihistaminics. They therefore proposed that the receptors mediating these responses to histamine be termed H_2-receptors.

Briefly, the pharmacological distinction of histamine receptors rests upon the actions of histamine and its antagonists. Histamine stimulates the contraction of smooth muscle from various organs such as the gut and bronchi. Because low concentrations of antihistaminic drugs suppress this effect, the pharmacological receptors that mediate this response are referred to as H_1-receptors and the drugs are said to be H_1-antagonists. Histamine also stimulates the secretion of acid by the stomach, increases the heart rate, and inhibits contractions in the rat uterus. Because the classical antihistaminics do not antagonize these effects, the receptors that mediate these actions are said to be H_2-receptors. Likewise, drugs that inhibit these responses to histamine are classified as H_2-antagonists.

Therapeutic areas that have evolved from research on antihistaminic drugs include tranquilizers, antipsy-

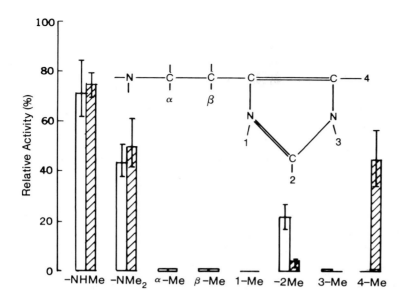

FIG. 16–1. *Relative activity of monomethylhistamines. Bar graph showing activity of monomethyl derivatives of histamine relative to histamine (equal to 100), estimated on both ileum (white columns) and atrium (hatched columns). Each bar indicates mean relative activity (with 95% confidence limits). The action on ileum is considered to involve H_1 receptors, whereas that on atrium involves activation of H_2 receptors. Reprinted by permission from Nature 236: 385. Copyright © 1972, Macmillan Journals Limited.*

chotics, antidepressants, antiparkinsonism agents, anti-emetics and others.[22] And now the events of the seventh decade of this century have set the stage for new discoveries to evolve from studies of the natural role of histamine in biological systems.

BIOSYNTHESIS AND METABOLISM

Even though injected histamine can be taken up by various tissues, ingested histamine, or that formed by bacteria in the gastrointestinal tract, does not appear to contribute significantly to the endogenous pool. The major source of histamine in the body appears to be the decarboxylation of the naturally occurring amino acid, histidine. The reaction is catalyzed by histidine decarboxylase, an enzyme that requires the coenzyme, pyridoxal phosphate (pyridoxine). A general mechanism for amino-acid decarboxylation catalyzed by pyridoxal phospate has been proposed.[23] Part of the driving force for the decarboxylation of the chelated Schiff base may be the resonance stabilization of the imine chelate that is formed. The resulting structure is hydrolyzed, yielding the free amine and the coenzyme in its original form. The model illustrates

the decarboxylation of histidine to form histamine. The histamine-forming capacity of some tissues is remarkably high.

The chief sites of histamine storage are the mast* cells and the basophils, which are the blood-circulating counterpart of the fixed-tissue mast cells. In non-mast-cell sites, histamine usually is undergoing a rapid turnover and is released rather than stored. Very little of the released histamine is excreted unchanged. In humans, most of it is excreted as the polar metabolites shown in Figure 16-2. The numerical values represent the percent recovery of histamine and metabolites in the urine in 12 hours following intradermal ^{14}C-histamine in human males.[24]

Recent reviews of the biogenesis[25] and metabolism and excretion[26] of histamine underscore the fact that nature has provided, on the one hand, some tissues with a high histamine-forming capacity and other tissues (mast cells and basophils) a great capacity to store histamine. On the other hand, the body has a strong defense against the unwanted effects of excessive histamine. The implication of this is that hista-

---

Mast is the German word for *fattening* or *forced-feeding*. The cell was recognized as a well-fed connective tissue cell.

Histidine Decarboxylation

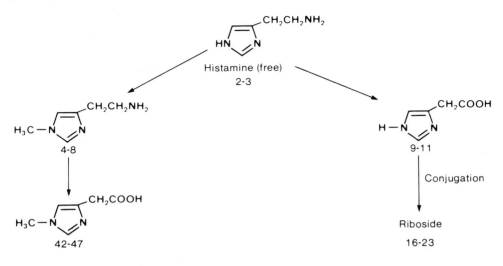

FIG. 16-2. *Metabolism of histamine.*

mine is very important to many physiologic and pathologic conditions. But our present knowledge still lacks experimental evidence of the role of histamine in these processes.

Histamine Phosphate U.S.P., 1H-imidazole-4-ethanamine phosphate (1:2), 4-(2-aminoethyl)imidazole *bis*(dihydrogen phosphate). The salt is stable in air but affected by light, and it is freely soluble in water (1:4). The chief use of histamine clinically is to diagnose impairment of the acid-producing cells (parietal cells) of the stomach. It is the most powerful gastric-secretory-stimulant that is available. The absence

Histamine Phosphate

of acid secretion after injection is considered proof that the gastric acid-secreting glands are nonfunctional, a condition (called *achlorhydria*) that is particularly symptomatic of pernicious anemia.

The wheal caused by intradermal injection of a 1:1000 solution has been suggested as a diagnostic test of local circulation. In normal individuals, the wheal appears in about 2.5 minutes, and any delay is considered a sign of vascular disease.

Usual dose (U.S.P. XIX)—subcutaneous, 27.5 μg. (the equivalent of 10 μg. of histamine base) per kg. of body weight.

Usual dose range (U.S.P. XIX)—10 to 40 μg. per kg.

Occurrence
Histamine Phosphate Injection U.S.P.

Betazole Hydrochloride U.S.P., Histalog® Hydrochloride, 1H-pyrazole-3-ethanamine dihydrochlo-

ride, 3-(2-aminoethyl)pyrazole dihydrochloride. The salt of this analog of histamine is a water-soluble, white, crystalline, nearly odorless powder. The pH of a 5 percent solution is about 1.5.

Betazole Hydrochloride

Betazole is less potent than histamine, but retains the ability to stimulate gastric secretion with much less tendency to produce the other effects that are usually observed after the use of histamine. The gastric secretory response to a dose of 5 mg. of betazole is comparable to the response to 10 μg. of histamine base.

Usual dose (U.S.P. XIX)—subcutaneous or intramuscular, 50 mg.

Usual dose range (U.S.P. XIX)—40 to 60 mg.

Occurrence
Betazole Hydrochloride Injection U.S.P.

ANTIHISTAMINIC AGENTS

The discovery of the H_2-antagonist burimamide in the early 1970's opened a new era in the history of the attempt to explain histamine-related physiological processes. Only history can record the importance of this discovery toward delineation of the role of histamine, as well as that of other autacoids, in homeostasis and disease. For a historical record of the progress thus far, we draw the reader's attention, in particular, to the literature surveys edited by Rocha e Silva[28] and Fordtran and Grossman.[29]

Even though the general term *antihistaminic* implies the inhibition of the action of histamine, the physicochemical properties, pharmacological activities, and clinical uses of the H_1-antagonists (classical antihistamines) differ greatly from those of the H_2-antagonists. Therefore, the H_1-antagonists and H_2-antagonists are discussed separately.

H_1-ANTAGONISTS
Mode of action

H_1-antagonists may be defined as those drugs which, in low concentrations, competitively inhibit the action of histamine on tissues containing H_1-receptors. H_1-antagonists are evaluated in terms of their ability to inhibit histamine-induced spasms in an isolated strip of guinea-pig ileum. Also, antihistamines may be evaluated in vivo in terms of their ability to protect animals against the lethal effects of histamine aerosols.

To distinguish competitive antagonism of histamine from other modes of action, the index pA is applied to in vitro assays. The index pA_2 for example, is defined as the inverse of the logarithm of the molar concentration of the antagonist, which reduces the response of a double dose of the agonist to that of a single one. The more potent H_1-antagonists exhibit a pA_2 value significantly greater than 6. While there are many pitfalls[30] to be avoided in the interpretation of structure–activity relationship (SAR) studies, the following example is illustrative in distinguishing competitive antagonism. It has been shown that pA_2 values for pyrilamine (mepyramine) antagonism ranged from 9.1 to 9.4 using human bronchii and guinea-pig ileum.[31] By contrast, the pA_2 value in guinea-pig atria (H_2-receptor) was 5.3[32] Thus, it may be concluded that pyrilamine is a weak, noncompetitive inhibitor of histamine at the atrial receptors and a competitive inhibitor at H_1-receptors.

For the most part, the therapeutic usefulness of the H_1-antagonists is in the symptomatic treatment of allergic disease. This action is clearly attributable to their antagonism of the action of histamine. In addition, the central properties of some of the agents are of clinical value, particularly in suppressing motion sickness.

H_1-antagonists are most useful in seasonal rhinitis (hay fever, pollinosis). The drugs best relieve the symptoms of these allergic diseases (sneezing, rhinorrhea, and itching of eyes, nose, and throat) at the beginning of the season when pollen counts are low.

Despite popular belief, antihistamines are of little value in treating the common cold.[32] Their weak anti-cholinergic effect may lessen rhinorrhea. Certain of the allergic dermatoses and the urticarial lesions of systemic allergies respond favorably to H_1-antagonists. However, the drugs are of little or no value in diseases such as systemic anaphylaxis and bronchial asthma in which autacoids other than histamine are important.

Structure–activity relationships

From a study of the activity of the thousands of compounds synthesized and screened for antihistamine activity, the structural requirements for optimal activity may be suggested. A more detailed discussion is available.[14,31]

Most compounds that antagonize the action of histamine at H_1-receptor sites may be described by the

2– Aminoethyl side chain

general structure where Ar is aryl (including phenyl, substituted phenyl, and heteroaryl groups such as 2-pyridyl); and Ar′ is a second aryl or arylmethyl group. Chemical classification of these agents is usually based on the unit X, which may be saturated carbon–oxygen (aminoalkyl ethers), nitrogen (ethylenediamines), or carbon (propylamines). Tricyclic derivatives, in which the two aromatic rings are bridged (e.g., phenothiazines), comprise additional chemical classes.

In general, the terminal N atom should be a tertiary amine for maximal activity. In contrast with many anticholinergics and local anesthetics, the dimethylamine derivatives appear to have a better therapeutic index. However, the terminal N atom may be part of a heterocyclic structure, as in antazoline and in chlorcyclizine, and may still result in a compound of high antihistaminic potency.

Extension or branching of the 2-aminoethyl side chain results in a less active compound. However, promethazine has a greater therapeutic index than its nonbranched counterpart.

Several properties of H_1-antagonists have been studied in an effort to relate them to activity.[14,31,33] While physicochemical and steric parameters appear to be important to H_1-antagonist action (especially lipophilic characteristics), no direct correlation has been established between a property and the antihistaminic effect.

The relationships of structure to overlapping actions (H_1-antagonist, anticholinergic, and local anesthetic) have been analyzed.[21]

Aminoalkyl ethers

Extensive studies of the qualitative and quantitative structure–activity relationships indicate that the critical length of the 2-aminoalkyl side chain shows more flexibility in the aminoalkylethers than in the ethylenediamines. However, the most active compounds have a chain length of two carbon atoms. Quaternization of the side chain nitrogen does not always result in less active compounds.

The drugs in this group possess significant anticholinergic activity, which enhances the H_1-blocking action on exocrine secretions. This action is not surprising in view of the structural similarity to the aminoalcohol ethers that exhibit cholinergic-blocking activity. Drowsiness is a side-effect common to the tertiary-aminoalkyl ethers. While this side-effect is exploited in over-the-counter sleeping aids, it may interfere with the patient's performance of tasks requiring mental alertness. The incidence of gastrointestinal side-effects is low. Structural relationships of the aminoalkyl ether derivatives are shown in Table 16-1, while the official preparations and dosages are given later (see Table 16-6).

Products

Diphenhydramine Hydrochloride U.S.P., Benadryl®, 2-(diphenylmethoxy)-N,N-dimethylethanamine hydrochloride. The oily, lipid-soluble free base is available as the bitter-tasting hydrochloride salt, which is a stable, white, crystalline powder, soluble in water (1:1), alcohol (1:2), and chloroform (1:2). The salt has a pKa value of 9, and a 1 percent aqueous solution has a pH of about 5.

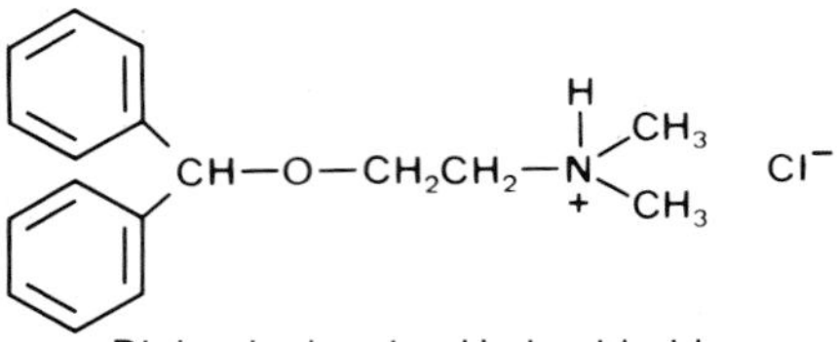

Diphenhydramine Hydrochloride

In the usual dose range of 25 to 400 mg., diphenhydramine is not a highly active H_1-antagonist; it has anticholinergic and sedative properties. Conversion to a quaternary ammonium salt does not alter the antihistaminic action greatly but does increase the anticholinergic action.

As an antihistaminic agent, diphenhydramine is recommended in various allergic conditions and, to a lesser extent, as an antispasmodic. It is administered either orally or parenterally in the treatment of urticaria, seasonal rhinitis (hay fever), and some dermatoses. The most common side-effect is drowsiness, and the concurrent use of alcoholic beverages and other CNS depressants should be avoided.

Dimenhydrinate U.S.P., Dramamine®, 8-chlorotheophylline 2-(diphenylmethoxy)-N,N-dimethylethylamine compound. The 8-chlorotheophyllinate (theoclate) salt of diphenhydramine is a white, crystalline, odorless powder that is slightly soluble in water and freely soluble in alcohol and chloroform.

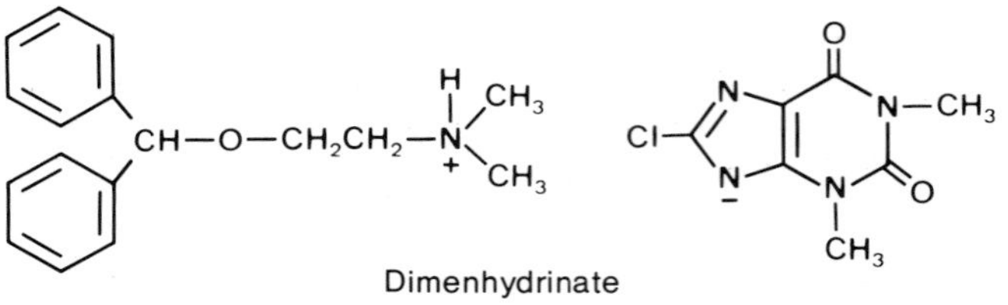

TABLE 16-1

AMINOALKYL ETHERS

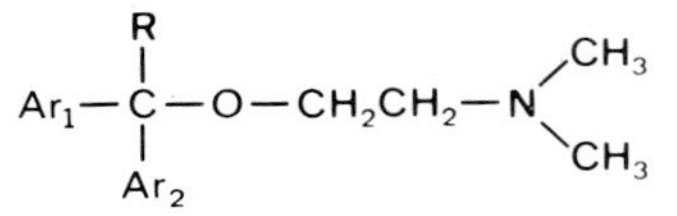

Generic Name		Ar₁	Ar₂	R
Diphenhydramine	Benadryl	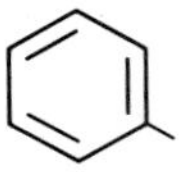	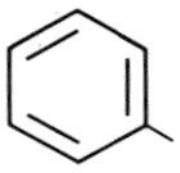	H
Bromodiphenhydramine	Ambodryl	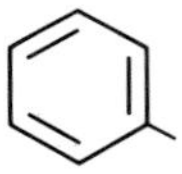	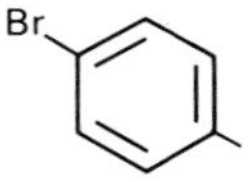	H
Doxylamine	Decapryn			CH₃
Carbinoxamine	Clistin	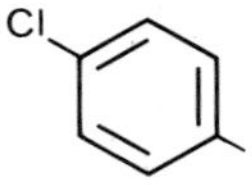	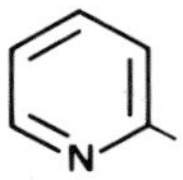	H
Clemastine	Tavist	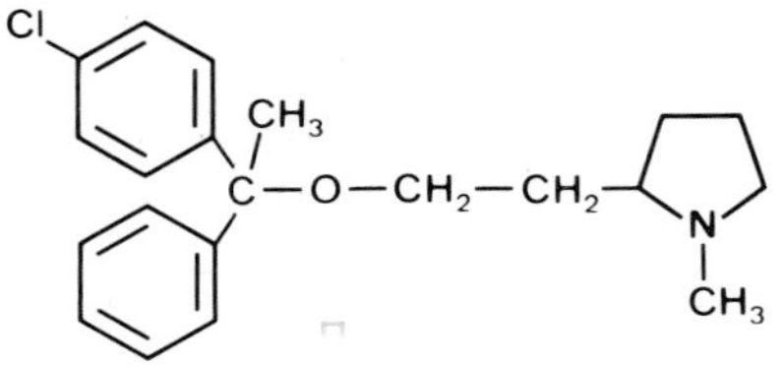		

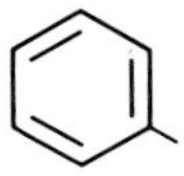

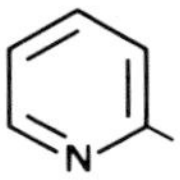

Dimenhydrinate

Dimenhydrinate is recommended for the nausea of motion sickness and for hyperemesis gravidarum (nausea of pregnancy). For the prevention of motion sickness, the dose should be taken at least one-half hour before beginning the trip. The cautions listed for diphenhydramine should be observed.

Bromodiphenhydramine Hydrochloride U.S.P., Ambodryl® Hydrochloride, 2-[(4-bromophenyl)phenylmethoxy]-N,N-dimethylethanamine hydrochloride. The hydrochloride salt is a white to pale-buff, crystalline powder that is freely soluble in water and in alcohol.

Relative to diphenhydramine, bromodiphenhydramine is more lipid soluble and was found to be twice as effective in protecting guinea pigs against the lethal effects of histamine aerosols.

Doxylamine Succinate U.S.P., Decapryn® Succinate, 2-[α-(2-dimethylaminoethoxy)-α-methylbenzyl]-pyridine bisuccinate. The acid succinate salt (bisuccinate) is a white to creamy-white powder that has a characteristic odor and is soluble in water (1:1), alcohol (1:2), and chloroform (1:2). A one percent solution has a pH of about 5.

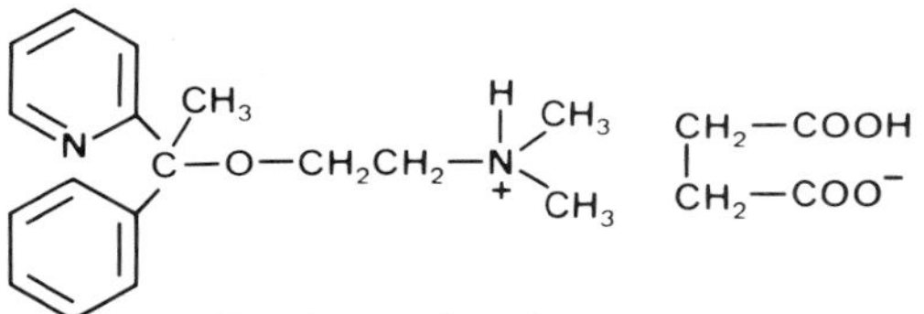

Doxylamine Succinate

Doxylamine succinate is comparable in potency to diphenhydramine. It is a good nighttime hypnotic when compared to secobarbital.[34] Concurrent use of alcohol and other CNS depressants should be avoided.

Carbinoxamine Maleate U.S.P., Clistin® Maleate, (d,l)-2-{p-chloro-α-[2-(dimethylamino)ethoxy]-benzyl}pyridine bimaleate. The oily, lipid-soluble free

Carbinoxamine Maleate

base is available as the bitter-tasting bimaleate salt, a white crystalline powder that is very soluble in water and freely soluble in alcohol and in chloroform. The pH of a 1 percent solution is between 4.6 and 5.1.

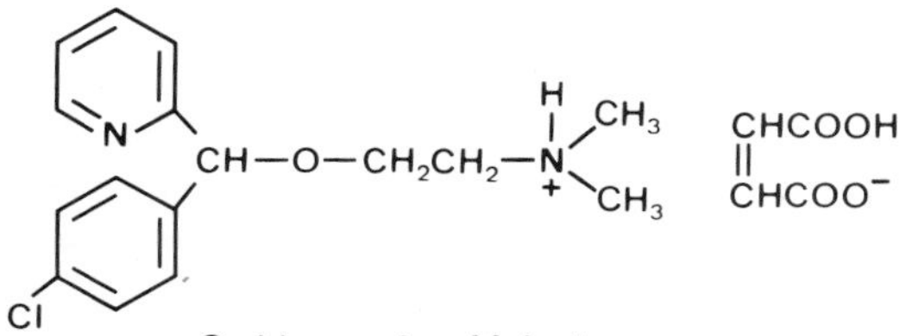

Carbinoxamine differs structurally from chlorpheniramine only in that an oxygen atom separates the asymmetric carbon atom from the aminoethyl side chain. The more active *levo* isomer of carbinoxamine has been shown to have the S absolute configuration[35] and to be superimposable upon the more active *dextro* isomer (S configuration[36]) of chlorpheniramine.

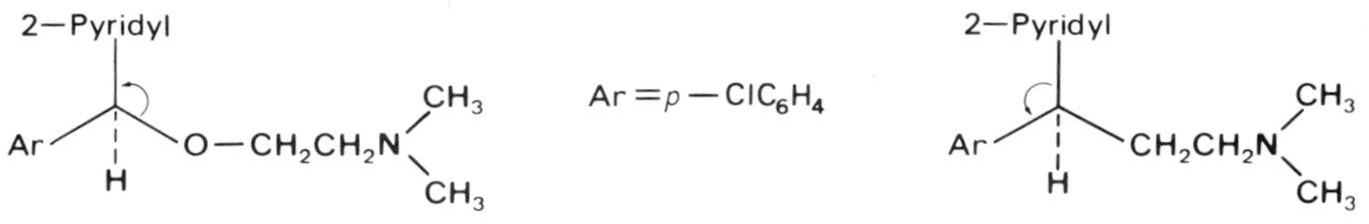

Carbinoxamine is a potent antihistaminic that lacks pronounced sedative effects and is available as the racemic mixture.

Clemastine Fumarate, Tavist®, 2-{2-[1-(4-chlorophenyl)-1-phenylethoxy]ethyl}-1-methylpyrrolidine hydrogen fumarate (1:1). Dextrorotatory clemastine has two chiral centers, each of which is of the R absolute configuration. A comparison of the activities of the antipodes indicates that the asymmetric center close to the side chain nitrogen is of lesser importance to antihistaminic activity.[31]

This member of the ethanolamine series is characterized by a long duration of action, with an activity that reaches a maximum in 5 to 7 hours and persists for 10 to 12 hours.

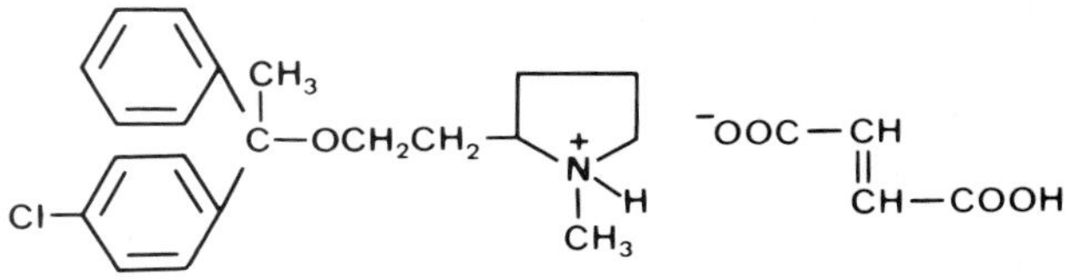

Clemastine Fumarate

It is well absorbed when administered orally and is excreted primarily in the urine. The side-effects are those usually encountered with this series of antihistamines. Clemastine is closely related to chlorphenoxamine, which is used for its central cholinergic-blocking activity.

Ethylenediamines

As a chemical class, the ethylenediamines contain the oldest useful antihistamines. They are highly effective H_1-antagonists with a relatively high incidence of CNS depressant and gastrointestinal side-effects.[32]

(S) Carbinoxamine (levo)

$Ar = p - ClC_6H_4$

(S) Chlorpheniramine (dextro)

The piperazine-, imidazoline-, and phenothiazine-type antihistamines contain the ethylenediamine moiety. These agents are discussed separately because they exhibit significantly different pharmacological properties.

In most drug molecules, the presence of a nitrogen atom appears to be a necessary condition for the formation of a stable salt with mineral acids. The aliphatic amino group in the ethylenediamines is sufficiently basic for salt formation, but the nitrogen atom to which an aromatic ring is bonded is considerably less basic. The free electrons on the aryl nitrogen are delocalized by the aromatic ring. A general resonance structure depicting this electron delocalization is as follows:

Because there is decreased electron density on nitrogen, it is less basic and protonation at this position takes place less readily.

TABLE 16–2

ETHYLENEDIAMINE DERIVATIVES

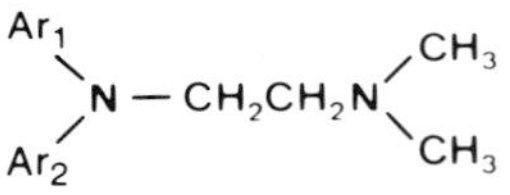

Generic Name	Proprietary Name	Ar₁	Ar₂
Pyrilamine		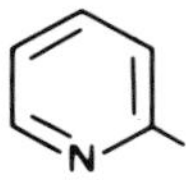	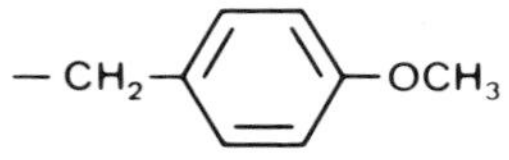 $-CH_2-$ ⬡ $-OCH_3$
Tripelennamine	Pyribenz-amine		$-CH_2-$ ⬡
Methapyrilene	Histadyl	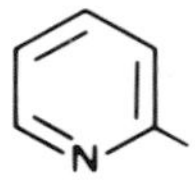	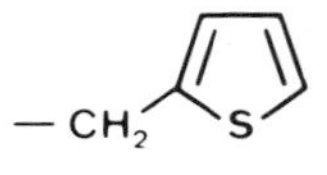 $-CH_2-$ ⬠S
Thonzylamine		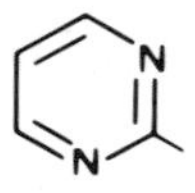	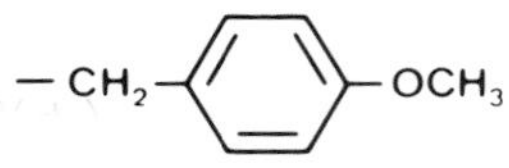 $-CH_2-$ ⬡ $-OCH_3$

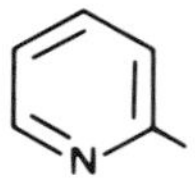

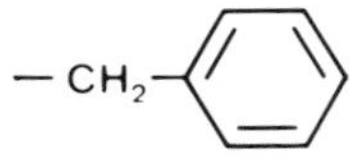

Table 16-2 shows structural relationships of the ethylenediamines (see Table 16-6 for the official preparations and dosages).

Tripelennamine Citrate U.S.P., Pyribenzamine® Citrate, PBZ, 2-{benzyl[2-(dimethylamino)ethyl]amino}pyridine dihydrogen citrate (1:1). The oily free base is available as the less bitter monocitrate salt, which is a white crystalline powder freely soluble in water and in alcohol. A 1 percent solution has a pH of 4.25. For oral administration in liquid dose forms, the citrate salt is less bitter and thus more palatable than the hydrochloride. Because of the difference in molecular weights, the doses of the two salts must be equated: 30 mg. of the citrate salt are equivalent to 20 mg. of the hydrochloride salt.

Tripelennamine Hydrochloride U.S.P., is a white crystalline powder that darkens slowly on exposure to light. The salt is soluble in water (1:0.77) and in alcohol (1:6). It has a pKa of about 9, and a 0.1 percent solution has a pH of about 5.5.

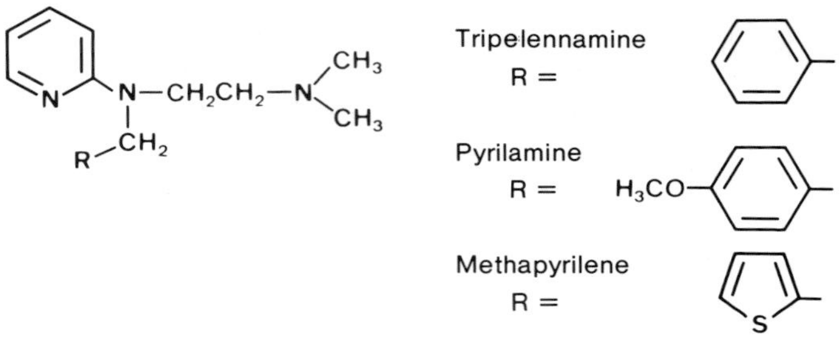

Tripelennamine, the first ethylenediamine developed in American laboratories, is well absorbed when given orally. On the basis of clinical experience, it appears to be as effective as diphenhydramine and may have the advantage of fewer and less severe side-reactions. However, drowsiness may occur and may impair ability to perform tasks requiring alertness. The concurrent use of alcoholic beverages should be avoided.

Pyrilamine Maleate U.S.P., 2-[(2-dimethylaminoethyl)(*p*-methoxybenzyl)amino]pyridine bimaleate (1:1), mepyramine. The oily free base is available as the acid maleate salt, which is a white crystalline powder having a faint odor and a bitter, saline taste. The salt is soluble in water (1:0.4) and freely soluble in alcohol. A ten percent solution has a pH of approximately 5. At a pH of 7.5 or above, the oily free base begins to precipitate.

Pyrilamine differs structurally from tripelennamine by having a methoxy group in the *para* position of the benzyl radical. It differs from its more toxic and less potent precursor, phenbenzamine (Antergan®) by having a 2-pyridyl group on the nitrogen atom in place of a phenyl group.

Clinically, pyrilamine and tripelennamine are considered to be among the less potent antihistaminics. They are highly potent, however, in antagonizing histamine-induced contractions of guinea-pig ileum.[14] Because of the pronounced local anesthetic action, the drug should not be chewed, but taken with food.

Methapyrilene Hydrochloride U.S.P., Histadyl® Hydrochloride, 2-[(2-dimethylaminoethyl)(2-thenyl)-amino]pyridine monohydrochloride. The oily free base is available as the bitter-tasting monohydrochloride salt, which is a white crystalline powder that is soluble in water (1:0.5), in alcohol (1:5), and in chloroform (1:3). Its solutions have a pH of about 5.5. It differs structurally from tripelennamine in having a 2-thenyl (thiophene-2-methylene) group in place of the benzyl group. The thiophene ring is considered isosteric with the benzene ring, and the isosteres exhibit similar activity.[37]

A study of the solid-state conformation of methapyrilene hydrochloride showed that the *trans* conformation is preferred for the two ethylenediamine nitrogen atoms. The Food and Drug Administration has declared methapyrilene a potential carcinogen in 1979, and all products containing it have been recalled.

Methapyrilene Fumarate U.S.P., is also available.

Thonzylamine Hydrochloride, 2-[(2-dimethyl-aminoethyl)(p-methoxybenzyl)amino]pyrimidine hydrochloride. The hydrochloride is a white, crystalline powder, soluble in water (1:1), alcohol (1:6), and chloroform (1:4). A 2 percent aqueous solution has a pH of 5.5. It is similar in activity to tripelennamine but is claimed to be less toxic.

The usual dose is 50 mg. up to 4 times daily.

Propylamine derivatives

The saturated members of this group are sometimes referred to as the *pheniramines*. The pheniramines are chiral molecules and the halogen-substituted derivatives have been resolved by crystallization of salts formed with *d*-tartaric acid.[14] The antihistamines in this group are among the most active H₁-antagonists. They are not so prone to produce drowsiness, but a significant proportion of patients do experience this effect.

In the unsaturated members, it has been suggested that a coplanar aromatic double-bond system (ArC=CH—CH₂N) is an important factor for antihistaminic activity. The pyrrolidino group is the side chain tertiary amine in the more active compounds.

The structural relationships of the propylamine derivatives are shown in Table 16-3. (For the official preparations and dosages, see Table 16-6.)

TABLE 16-3

PROPYLAMINE DERIVATIVES

Generic Name	Proprietary Name	Ar₁	Ar₂
SATURATED			
Pheniramine	Trimeton		
Chlorpheniramine Dexchlorpheniramine	Chlortrimeton Polaramine		
Brompheniramine Dexbrompheniramine	Dimetane Disomer		
UNSATURATED			
Pyrrobutamine	Pyronil		
Triprolidine	Actidil		

Pheniramine Maleate, Trimeton®, Inhiston® Maleate, 2-[α-(2-dimethylaminoethyl)benzyl]pyridine bimaleate. This salt is a white-crystalline powder, having a faint aminelike odor, that is soluble in water (1:5) and is very soluble in alcohol.

Pheniramine R = H
Chlorpheniramine R = Cl
Brompheniramine R = Br

This drug is the least potent member of the series and is marketed as the racemate. The usual adult dose is 20 to 40 mg. 3 times daily.

Chlorpheniramine Maleate U.S.P., Chlor-Trimeton® Maleate, (±)2-[p-chloro-α-(2-dimethylaminoethyl)benzyl]pyridine bimaleate. The bimaleate salt is a white crystalline powder that is soluble in water (1:3.4), in alcohol (1:10), and in chloroform (1:10). It has a pKa of 9.2, and an aqueous solution has a pH between 4 and 5.

Chlorination of pheniramine in the *para* position of the phenyl ring gave a 10-fold increase in potency with no appreciable change in toxicity. Most of the antihistaminic activity resides with the *dextro* enantiomorph (see dexchlorpheniramine below). The usual dose is 2 to 4 mg. three or four times a day.

Dexchlorpheniramine Maleate U.S.P., Polaramine® Maleate. Dexchlorpheniramine is the dextrorotatory enantiomer of chlorpheniramine. In vitro and in vivo studies of the enantiomorphs of chlorpheniramine showed that the antihistaminic activity exists predominantly in the *dextro* isomer.[14] As mentioned previously, the *dextro* isomer has been shown[36] to have the S configuration, which is superimposable upon the S configuration of the more active levorotatory enantiomorph of carbinoxamine.

Brompheniramine Maleate U.S.P., Dimetane® Maleate, (±)2-[p-bromo-α-(2-dimethylaminoethyl)-benzyl]pyridine bimaleate. This drug differs from chlorpheniramine by the substitution of a bromine atom for the chlorine atom. Its actions and uses are similar to those of chlorpheniramine.

Dexbrompheniramine Maleate U.S.P., Disomer®. Like the chlorine congener, the antihistaminic activity exists predominantly in the *dextro* isomer and is of comparable potency.

Pyrrobutamine Phosphate U.S.P., Pyronil® Phosphate, (E)-1-[4-(4-chlorophenyl)-3-phenyl-2-butenyl]pyrrolidine diphosphate.*

Pyrrobutamine Phosphate

E* (trans) isomer

The diphosphate occurs as a white crystalline powder that is soluble to the extent of 10 percent in warm water. Pyrrobutamine was investigated originally as the hydrochloride salt, but the diphosphate was found to be absorbed more readily and completely. Clinical studies indicate that it is long acting with a comparatively slow onset of action.

The feeble antihistaminic properties of several analogs point to the importance of having a planar ArC=CH—CH₂N unit and a pyrrolidino group as the side chain tertiary amine.[14]

Triprolidine Hydrochloride U.S.P., Actidil® Hydrochloride, (E)-2-[3-(1-pyrrolidinyl)-1-*p*-tolylpropenyl]pyridine monohydrochloride. Triprolidine hydrochloride occurs as a white crystalline powder having no more than a slight, but unpleasant, odor. It is soluble in water and in alcohol and its solutions are alkaline to litmus.

Triprolidine Hydrochloride

The activity is confined mainly to the geometric isomer in which the pyrrolidinomethyl group is *trans* to the 2-pyridyl group. Recent pharmacological studies[39] confirm the high activity of triprolidine, and the superiority of E- over corresponding Z-isomers as H₁-antagonists. At guinea-pig ileum sites, the affinity of triprolidine for H₁-receptors was more than 1000 times the affinity of its Z-partner.

The relative potency of triprolidine is of the same order as that of dexchlorpheniramine. The peak effect occurs in about 3 1/2 hours after oral administration, and the duration of effect is about 12 hours.

Phenothiazine derivatives

Beginning in the mid-1940's, a number of antihistaminic drugs have been discovered as a result of bridging the aryl units of agents related to the ethylenediamines. The search for effective antimalarials led to the investigation of phenothiazine derivatives in which the bridging entity is sulphur. In subsequent testing, the phenothiazine class of drugs was discovered to have not only antihistaminic activity but also a pharmacological profile of its own considerably different from that of the ethylenediamines. Thus began the era of the useful psychotherapeutic agent.[40]

The structural relationships of the phenothiazines exhibiting antihistaminic action are shown in Table 16-4. (For the official preparations and dosages, see Table 16-6.)

* Following IUPAC[38] stereochemical nomenclature, the choice of substituents used to denote configuration about an alkenic double bond is governed by the sequence rule. When the groups are *trans* use the prefix E, from the German word *entgegen* meaning "opposite"; when *cis* use Z from the German word *zusammen* meaning "together."

TABLE 16-4

PHENOTHIAZINE DERIVATIVES

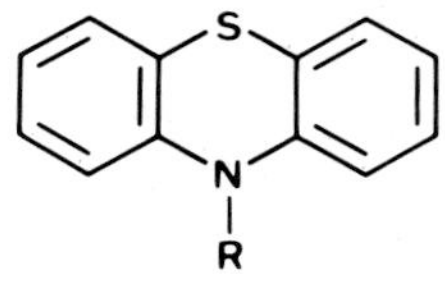

Generic Name	Proprietary Name	R
Promethazine	Phenergan	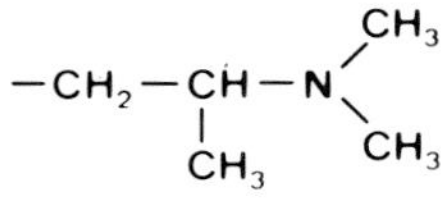 $-CH_2-CH-N(CH_3)_2$ with CH₃
Trimeprazine	Temaril	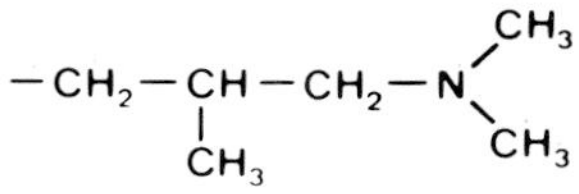 $-CH_2-CH-CH_2-N(CH_3)_2$ with CH₃
Methdilazine	Tacaryl	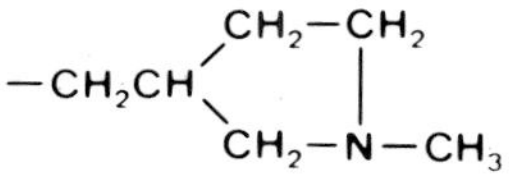 pyrrolidinyl structure

Promethazine Hydrochloride U.S.P., Phenergan® Hydrochloride, (±)10-(2-dimethylaminopropyl)phenothiazine monohydrochloride. The salt occurs as a white to faint yellow crystalline powder that is very soluble in water, in hot absolute alcohol, and in chloroform. Its aqueous solutions are slightly acid to litmus.

Promethazine is moderately potent by present-day standards with prolonged action and pronounced sedative side-effects. In addition to its antihistaminic

Promethazine Hydrochloride

action, it possesses an antiemetic effect, a tranquilizing action, and a potentiating action on analgesic and sedative drugs. In general, lengthening of the side chain and substitution of lipophilic groups in the 2-position of the aromatic ring results in compounds with decreased antihistaminic activity and increased psychotherapeutic properties.

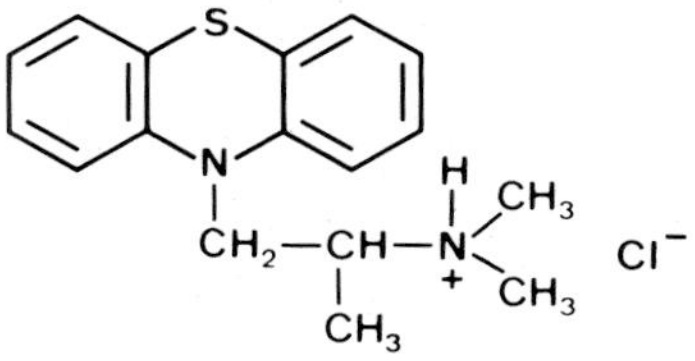

Enantiomers of promethazine have been resolved and shown to have similar antihistaminic and other pharmacologic properties.[41] This is in contrast to studies of the pheniramines and carbinoxamine compounds in which the chiral center is closer to the aromatic feature of the molecule. Asymmetry appears to be of lesser influence on antihistaminic activity when the chiral center lies near the positively charged side chain nitrogen.

The antihistaminic phenothiazines may cause drowsiness and so may impair the ability to perform tasks requiring alertness. Concurrent administration of alcoholic beverages and other CNS depressants should be avoided.

Trimeprazine Tartrate U.S.P., Temaril® Tartrate, (±)-10-[3-dimethylamino-2-methylpropyl]phenothiazine tartrate. The salt occurs as a white to off-white crystalline powder that is freely soluble in water and soluble in alcohol. Its antihistaminic action is reported to be from 1 1/2 to 5 times that of promethazine. Clinical studies have shown it to have a pronounced antipruritic action. This action may be unrelated to its histamine-antagonizing properties.

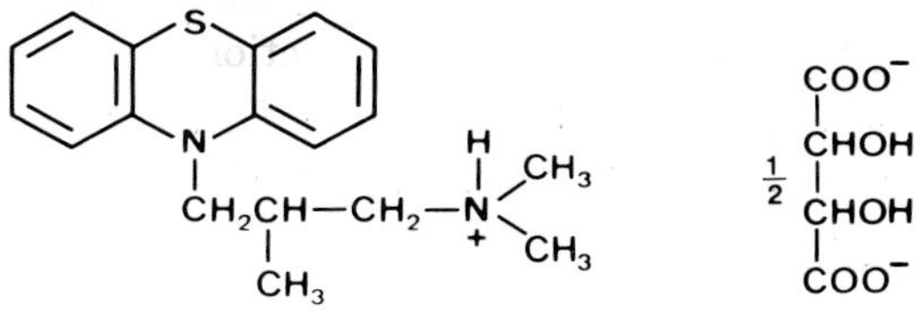

Trimeprazine Tartrate

Methdilazine U.S.P., Tacaryl®, 10-[(1-methyl-3-pyrrolidinyl)methyl]phenothiazine. This compound occurs as a light tan crystalline powder that has a characteristic odor, and is practically insoluble in water. Methdilazine, as the free base, is used in chewable tablets, because its low solubility in water contributes to its tastelessness. Some local anesthesia of the buccal mucosa may be experienced if the tablet is chewed and not swallowed promptly.

Methdilazine Hydrochloride U.S.P., Tacaryl® Hydrochloride, 10-[(1-methyl-3-pyrrolidinyl)methyl]phenothiazine monohydrochloride. The hydrochloride

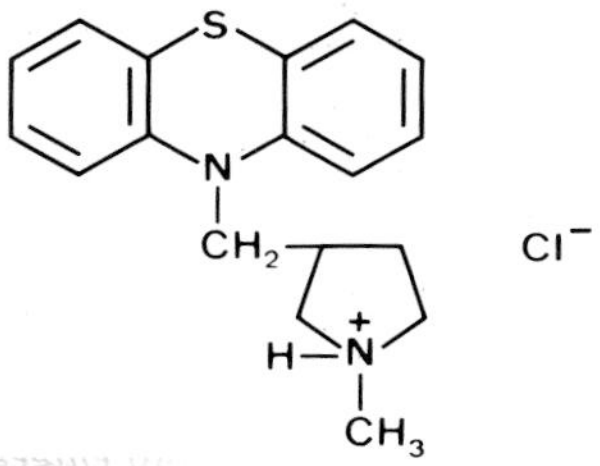

Methdilazine Hydrochloride

salt also occurs as a light tan crystalline powder having a slight characteristic odor. However, the salt is freely soluble in water and in alcohol.

The activity is similar to that of methdilazine and is administered orally for its antipruritic effect.

Piperazine derivatives

The activity of the piperazine-type antihistaminics (cyclic ethylenediamines) is characterized by a slow onset and long duration of action. These are moderately potent antihistaminics with a lower incidence of drowsiness. However, warning of the possibility of some dulling of mental alertness is advised.

As a group, these agents are useful as antiemetics as well as antihistamines. They have exhibited a strong teratogenic potential, inducing a number of malformations in rats. Norchlorcyclizine, a metabolite of these piperazines, was proposed to be responsible for the teratogenic effects of the parent drugs.[42]

Table 16-5 provides a structural comparison of the

TABLE 16–5

PIPERAZINE DERIVATIVES AND MISCELLANEOUS COMPOUNDS

Generic Name	Proprietary Name	Structure	Generic Name	Proprietary Name	Structure
PIPERAZINE DERIVATIVES					
Cyclizine	Marezine	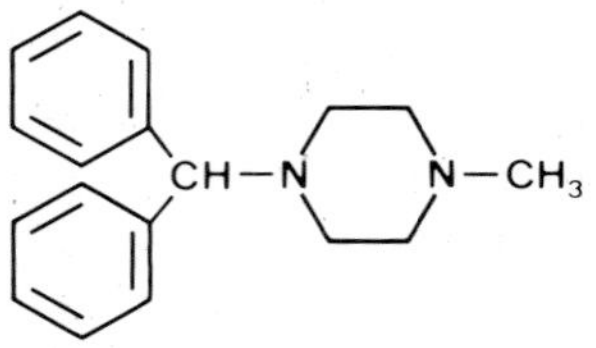	Meclizine	Bonine	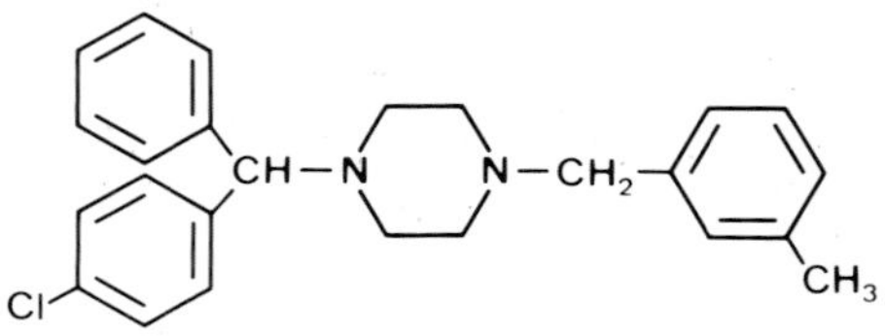
Chlorcyclizine		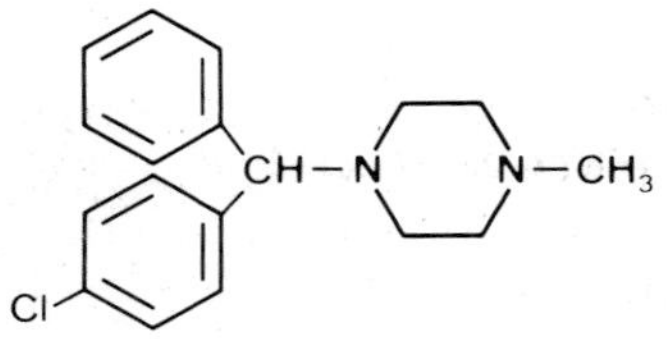	Buclizine	Softran	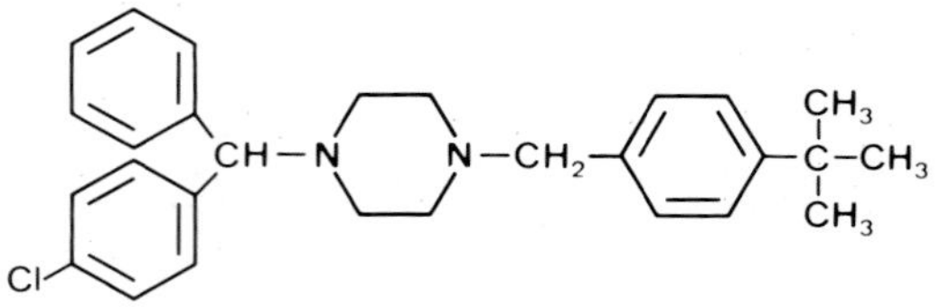
MISCELLANEOUS COMPOUNDS					
Diphenylpyraline	Hispril, Diafen	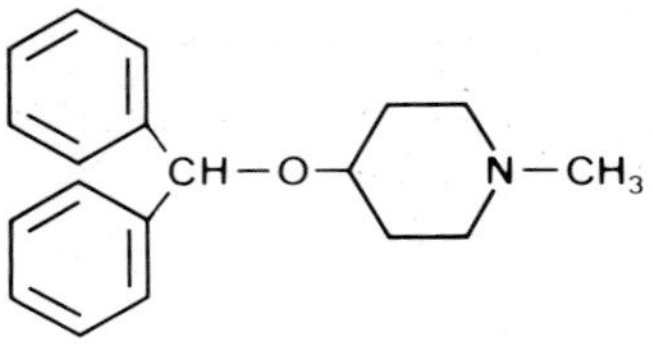	Antazoline		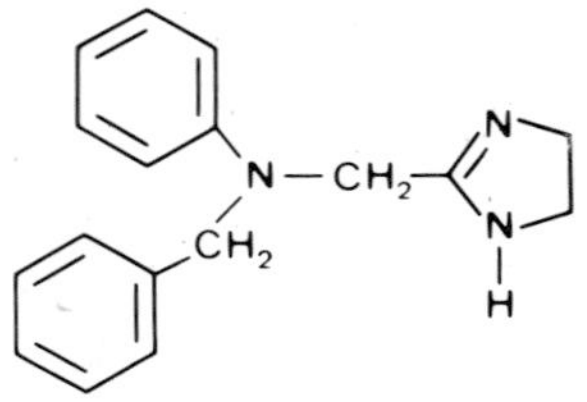
Phenindamine		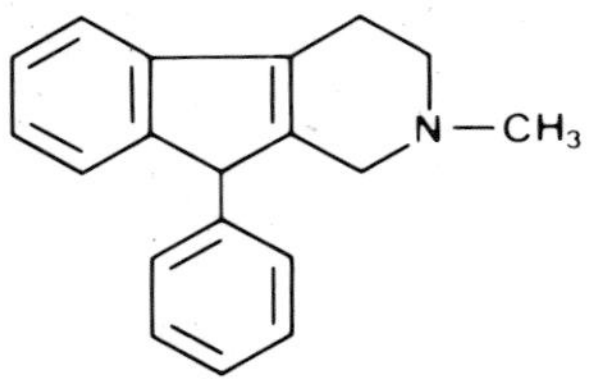			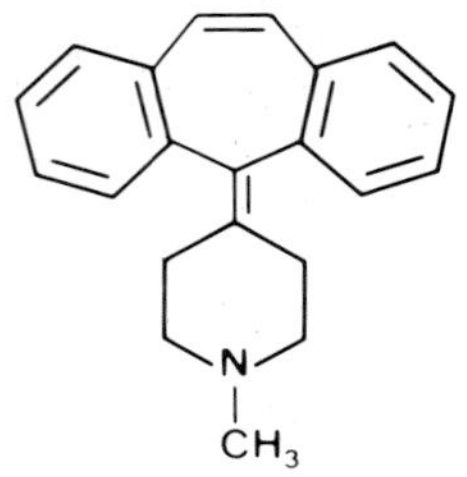
Dimethindene	Forhistal		Cyproheptadine	Periactin	
		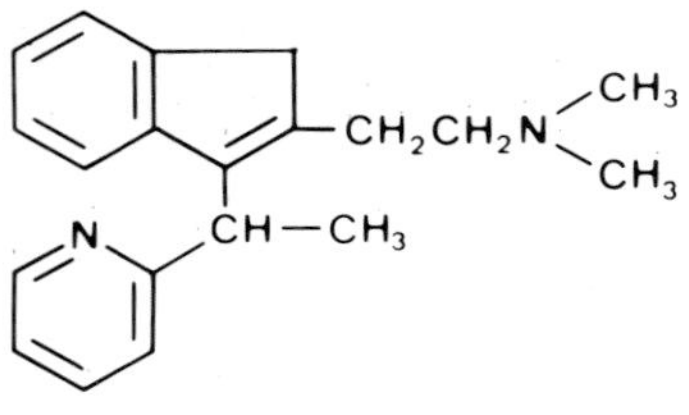	Azatadine	Optimine	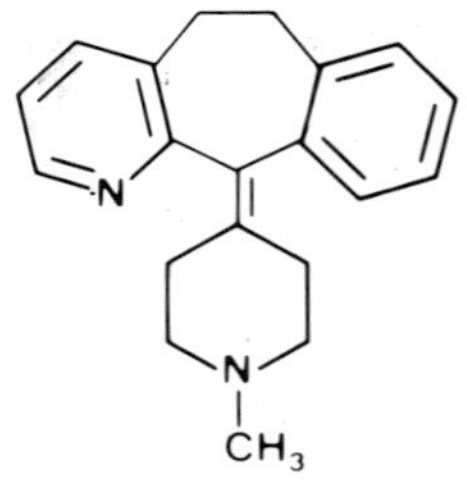

(Continued)

piperazine derivatives. (For official preparations and dosages see Table 16-6).

Cyclizine Hydrochloride U.S.P., Marezine® Hydrochloride, 1-(diphenylmethyl)-4-methylpiperazine monohydrochloride. This drug occurs as a light-sensitive, white crystalline powder having a bitter taste. It is slightly soluble in water (1:115), alcohol (1:115), and chloroform (1:75). It is used primarily in the prophylaxis and treatment of motion sickness.

The lactate salt (Cyclizine Lactate Injection U.S.P.) is used for intramuscular injection because of the limited water-solubility of the hydrochloride. The injection should be stored in a cold place, because a slight yellow tint may develop if stored at room temperature for several months. This does not indicate a loss in biological potency.

Chlorcyclizine Hydrochloride U.S.P., 1-(*p*-chloro-α-phenylbenzyl)-4-methylpiperazine monohydrochloride. This salt, a light-sensitive, white crystalline powder, is soluble in water (1:2), in alcohol (1:11), and in chloroform (1:4). A 1 percent solution has a pH between 4.8 and 5.5.

Disubstitution or substitution of halogen in the 2- or 3-position of either of the benzhydryl rings results in a much less potent compound.

Chlorcyclizine is indicated in the symptomatic relief of urticaria, hay fever, and certain other allergic conditions.

Meclizine Hydrochloride U.S.P., Bonine®, Antivert® Hydrochloride, 1-(*p*-chloro-α-phenylbenzyl)-4-(*m*-methylbenzyl)piperazine dihydrochloride. Meclizine hydrochloride is a tasteless, white or slightly yellowish crystalline powder that is practically insoluble in water (1:1000). It differs from chlorcyclizine by having an N-*m*-methylbenzyl group in place of the N-methyl group.

Although it is a moderately potent antihistaminic, meclizine is used primarily as an antinauseant in the prevention and treatment of motion sickness and in the treatment of nausea and vomiting associated with vertigo and radiation sickness.

Buclizine Hydrochloride, Softran®, 1-(*p*-chloro-α-phenylbenzyl)-4-(*p*-*tert*-butylbenzyl)piperazine dihydrochloride. The salt occurs as a white to slightly yellow crystalline powder that is insoluble in water.

The highly lipid-soluble buclizine has central-nervous-system depressant, antiemetic, and antihistaminic properties. The salt is available in 25 mg. and 50 mg. tablets for oral administration. The usual dose is 50 mg. 1 to 3 times daily.

MISCELLANEOUS COMPOUNDS

The miscellaneous group of compounds includes those agents that exhibit useful antihistaminic activity but do not fit conveniently into a chemical class. See Table 16-5 for structural comparison of these agents and Table 16-6 for the official preparations and dosages.

TABLE 16-6

ANTIHISTAMINIC AGENTS

Name Proprietary Name	Preparations	Usual Adult Dose*	Usual Dose Range*	Usual Pediatric Dose*
Diphenhydramine Hydrochloride U.S.P. *Benadryl*	Diphenhydramine Hydrochloride Capsules U.S.P. Diphenhydramine Hydrochloride Elixir U.S.P. Diphenhydramine Hydrochloride Tablets U.S.P.	Oral, 25 to 50 mg. 3 or 4 times daily	Up to 400 mg. daily	Use in premature and newborn infants is not recommended. 1.25 mg. per kg. of body weight or 37.5 mg. per square meter of body surface, 4 times daily, not to exceed 300 mg. daily
(Continued)				

TABLE 16–6

ANTIHISTAMINIC AGENTS

Name Proprietary Name	Preparations	Usual Adult Dose*	Usual Dose Range*	Usual Pediatric Dose*
	Diphenhydramine Hydrochloride Injection U.S.P.	I.M. or I.V., 10 to 50 mg.	Up to a maximum of 400 mg. daily	Use in premature and newborn infants is not recommended. I.M. or I.V., 1.25 mg. per kg. of body weight or 37.5 mg. per square meter of body surface, 4 times daily, not to exceed 300 mg. daily
Dimenhydrinate U.S.P. *Dramamine*	Dimenhydrinate Syrup U.S.P. Dimenhydrinate Tablets U.S.P.	Oral, 50 mg. every 4 hours as needed		Use in premature and newborn infants is not recommended. 1.25 mg. per kg. of body weight or 37.5 mg. per square meter of body surface, every 6 hours as needed, not to exceed 150 mg. daily
	Dimenhydrinate Injection U.S.P.	I.M. 50 mg. up to every 4 hours as needed; I.V. 50 mg. in 10 ml. of sodium chloride injection administered slowly over a period of at least 2 minutes, up to every 4 hours as needed		I.M., 1.25 mg. per kg. of body weight or 37.5 mg. per square meter of body surface, every 6 hours as needed not to exceed 300 mg. daily I.V., 1.25 mg. per kg. of body weight or 37.5 mg. per square meter of body surface, in 10 ml. of sodium chloride injection, administered slowly over a period of at least 2 minutes, every 6 hours as needed, not to exceed 300 mg. daily
	Dimenhydrinate Suppositories U.S.P.	Rectal, 100 mg. 1 or 2 times a day as needed		Pediatric dose has not been established
Bromodiphenhydramine Hydrochloride U.S.P.	Bromodiphenhydramine Hydrochloride Capsules U.S.P. Bromodiphenhydramine Hydrochloride Elixir U.S.P.	Oral, 25 mg. 3 times daily	Up to 150 mg. daily	Use in premature and full-term neonates is not recommended. Oral, 6.25 to 25 mg. 3 or 4 times a day as needed
Doxylamine Succinate U.S.P. *Decapryn*	Doxylamine Succinate Syrup U.S.P. Doxylamine Succinate Tablets U.S.P.	Oral, 12.5 to 25 mg. every 4 to 6 hours as needed		Use in premature and full-term neonates is not recommended. Oral, 500 μg. per kg. of body weight or 15 mg. per square meter of body surface every 4 to 6 hours as needed, not to exceed 75 mg. daily
Carbinoxamine Maleate U.S.P. *Clistin*	Carbinoxamine Maleate Elixir U.S.P. Carbinoxamine Maleate Tablets U.S.P.	Oral, 4 to 8 mg. 3 or 4 times daily		Use in premature or full-term neonates is not recommended. Children 1 to 3 years—oral, 2 mg. 3 or 4 times daily, age 4 to 6 years—oral 2 to 4 mg. 3 or 4 times daily; age 7 and over—oral, 4 mg. 3 or 4 times daily
	Carbinoxamine Maleate, Extended-release Tablets U.S.P.	Oral, 8 or 12 mg. every 8 to 12 hours		
Pyrilamine Maleate U.S.P.	Pyrilamine Maleate Tablets U.S.P.	Oral, 25 to 50 mg. every 6 to 8 hours		Children 6 years of age and over—oral, 12.5 to 25 mg. every 6 to 8 hours as needed
Tripelennamine Citrate U.S.P. *Pyribenzamine*	Tripelennamine Citrate Elixir U.S.P.	Oral, the equivalent of 25 to 50 mg. of tripelennamine hydrochloride every 4 to 6 hours as needed	Up to the equivalent of 600 mg. of tripelennamine hydrochloride daily	The equivalent of 1.25 mg. per kg. of body weight or 37.5 mg. of tripelennamine hydrochloride per square meter of body surface every 6 hours not to exceed 300 mg. daily; use in premature or full term neonates not recommended
Tripelennamine Hydrochloride U.S.P. *Pyribenzamine*	Tripelennamine Hydrochloride Tablets U.S.P.	Oral, 25 to 50 mg. every 4 to 6 hours as needed	Up to 600 mg. daily	1.25 mg. per kg. of body weight or 37.5 mg. per square meter of body surface every 6 hours as needed; not to exceed 300 mg. daily; use in premature or full-term neonates is not recommended
	Tripelennamine Hydrochloride Extended-release Tablets U.S.P.	Oral, 100 mg. every 8 to 12 hours as needed	Up to 600 mg. daily	Not recommended for children under 6 years of age. Oral, 50 mg. every 8 to 12 hours as needed
Methapyrilene Fumarate U.S.P.†	Methapyrilene Fumarate Syrup U.S.P.			

*See *U.S.P. D.I.* for complete dosage information.

†The official title and dosage forms of this compound were obtained from the *U.S.P. XX/N.F.XV,* but at the time of this printing, the U.S.P. Dispensing Information did not include this compound. Therefore, the usual dose, dose range, and pediatric dose were obtained from *U.S.P.XIX* or *N.F.XIV* if the dosage form was listed.

TABLE 16-6

ANTIHISTAMINIC AGENTS

Name Proprietary Name	Preparations	Usual Adult Dose*	Usual Dose Range*	Usual Pediatric Dose*
Methapyrilene Hydrochloride U.S.P.† *Histadyl*	Methapyrilene Hydrochloride Capsules U.S.P. Methapyrilene Hydrochloride Injection U.S.P.			
Chlorpheniramine Maleate U.S.P. *Chlor-Trimeton*	Chlorpheniramine Maleate Extended-Release Capsules U.S.P. Chlorpheniramine Maleate Syrup U.S.P. Chlorpheniramine Maleate Tablets U.S.P. Chlorpheniramine Maleate Extended-Release Tablets U.S.P. Chlorpheniramine Maleate Injection U.S.P.	Oral, 8 or 12 mg. every 8 to 12 hours as needed Oral, 4 mg. every 4 to 6 hours I.M., I.V., or S.C., 5 to 20 mg. as a single dose	Up to 40 mg. daily	Not recommended for use in children under 7 years of age. Age 7 years or over—oral, 8 mg. every 12 hours as needed Use in premature and full-term neonates is not recommended. Oral, 87.5 μg. per kg. of body weight or 2.5 mg. per square meter of body surface every 6 hours; children up to 12 years of age—oral, 1 to 2 mg. 3 or 4 times daily Not recommended for use in premature or full-term neonates. I.M., I.V., or S.C., 87.5 μg. per kg. of body weight or 2.5 mg. per square meter of body surface every 6 hours
Dexchlorpheniramine Maleate U.S.P. *Polaramine*	Dexchlorpheniramine Maleate Syrup U.S.P. Dexchlorpheniramine Maleate Tablets U.S.P. Dexchlorpheniramine Maleate Extended-Release Tablets U.S.P.	Oral, 2 mg. 3 or 4 times daily Oral, 4 or 6 mg. every 8 to 12 hours as needed		Children up to 12 years of age—oral, 500 μg. to 1 mg. 3 or 4 times daily as needed. Not recommended for use in premature or full-term neonates
Brompheniramine Maleate U.S.P. *Dimetane*	Brompheniramine Maleate Elixir U.S.P. Brompheniramine Maleate Injection U.S.P. Brompheniramine Maleate Tablets U.S.P. Brompheniramine Maleate Extended-Release Tablets U.S.P.	Oral, 4 mg. 3 or 4 times daily; I.M., I.V., or S.C., 10 mg. twice daily, up to 40 mg. daily Oral 8 or 12 mg. every 8 to 12 hours		Not recommended for use in premature or full-term neonates. Children 2 to 6 years—oral, 1 mg. 3 or 4 times daily as needed; children 6 to 12 years—oral, 2 mg. 3 or 4 times daily as needed; children up to 12 years—I.M., I.V., or S.C., 125 μg. per kg. of body weight or 3.75 mg. per square meter of body surface 3 or 4 times daily Not recommended for children under 6 years of age. Six years and over—oral, 8 or 12 mg. every 12 hours
Dexbrompheniramine Maleate U.S.P.† *Disomer*	Dexbrompheniramine Maleate Tablets			
Pyrrobutamine Phosphate U.S.P.† *Pyronil*				
Triprolidine Hydrochloride U.S.P. *Actidil*	Triprolidine Hydrochloride Syrup U.S.P. Triprolidine Hydrochloride Tablets U.S.P.	Oral, 2.5 mg. 2 or 3 times daily	Up to 10 mg. daily	Not recommended for use in premature or full-term neonates. Children up to 2 years of age—oral, 625 μg. 2 or 3 times daily; two years and over—oral, 1.25 mg. 2 or 3 times daily
Promethazine Hydrochloride U.S.P.† *Phenergan*	Promethazine Hydrochloride Syrup U.S.P. Promethazine Hydrochloride Tablets U.S.P. Promethazine Hydrochloride Injection U.S.P.	12.5 mg. every 4 to 6 hours as necessary, or 25 mg. once daily at bedtime I.M. or I.V., 12.5 to 25 mg. every 4 to 6 hours as needed	Up to 150 mg. daily Up to 150 mg. daily	Oral, 125 μg. per kg. of body weight every 4 to 6 hours, or 500 μg. per kg. at bedtime as needed
Trimeprazine Tartrate U.S.P.† *Temaril*	Trimeprazine Tartrate Syrup U.S.P. Trimeprazine Tartrate Tablets U.S.P.	The equivalent of 2.5 mg. of trimeprazine 4 times daily	10 to 80 mg. daily	6 months to 2 years of age—the equivalent of 1.25 mg. of trimeprazine 1 to 4 times daily as necessary, not exceeding 5 mg. daily; 3 to 6 years of age—2.5 mg. 1 to 4 times daily as necessary, not exceeding 10 mg. daily; 7 to 12 years of age—2.5 to 5 mg. 1 to 3 times daily as necessary, not exceeding 15 mg. daily

*See *U.S.P. D.I.* for complete dosage information.

†The official title and dosage forms of this compound were obtained from the *U.S.P. XX/N.F.XV,* but at the time of this printing, the U.S.P. Dispensing Information did not include this compound. Therefore, the usual dose, dose range, and pediatric dose were obtained from *U.S.P.XIX* or *N.F.XIV* if the dosage form was listed.

TABLE 16-6

ANTIHISTAMINIC AGENTS

Name Proprietary Name	Preparations	Usual Adult Dose*	Usual Dose Range*	Usual Pediatric Dose*
Methdilazine U.S.P.† Tacaryl	Methdilazine Tablets U.S.P.	7.2 mg. (equivalent of 8 mg. of methdilazine hydrochloride) 2 to 4 times daily		
Methdilazine Hydrochloride U.S.P. Tacaryl	Methdilazine Hydrochloride Syrup U.S.P. Methdilazine Hydrochloride Tablets U.S.P.		8 mg. 2 to 4 times daily	
Cyclizine U.S.P. Marezine	Cyclizine Lactate Injection U.S.P.	I.M., 50 mg., every 4 to 6 hours as needed		I.M., 1 mg. per kg. of body weight or 33 mg. per square meter of body surface 3 times daily as needed
Cyclizine Hydrochloride U.S.P. Marezine	Cyclizine Hydrochloride Tablets U.S.P.	50 mg. every 4 to 6 hours as needed	Up to 200 mg. daily	Oral, 1 mg. per kg. of body weight or 33 mg. per square meter of body surface, 3 times daily; children 6 to 12 years of age—oral, 25 mg. every 4 to 6 hours as needed
Chlorcyclizine Hydrochloride U.S.P.†		50 mg. 1 to 4 times daily	25 to 100 mg.	
Meclizine Hydrochloride U.S.P. Bonine, Antivert	Meclizine Hydrochloride Tablets U.S.P. Meclizine Hydrochloride Chewable Tablets U.S.P.	Oral, 25 to 50 mg. one hour before travel, repeated every 24 hours as needed		Pediatric dosage has not been established
Buclizine Hydrochloride U.S.P. Softran	Buclizine Hydrochloride Tablets U.S.P.	Oral, 50 mg. every 4 to 6 hours as needed	Up to 150 mg. daily	Pediatric dosage has not been established
Diphenylpyraline Hydrochloride U.S.P. Hispril, Diafen	Diphenylpyraline Hydrochloride Tablets U.S.P.	Oral, 2 mg. every 4 to 6 hours as needed		Use is not recommended in premature and full-term neonates. Children 2 to 6 years of age—oral, 1 to 2 mg. 2 times daily; children 6 years of age and over—oral, 2 mg. 3 times daily as needed
	Diphenylpyraline Hydrochloride Extended-Release Capsules U.S.P.	Oral, 5 mg. every 12 hours		Not recommended for use in children under 6 years of age. Children 6 years of age and over—oral, 5 mg. once a day as needed
Dimethindene Maleate U.S.P. Forhistal	Dimethindene Maleate Tablets U.S.P.	Oral, 1 to 2 mg. 1 to 3 times daily		Children 7 years of age and over—oral, 1 to 2 mg. 1 to 3 times daily as needed
	Dimethindene Maleate Extended-Release Tablets U.S.P.	Oral, 2.5 mg. 1 or 2 times daily as needed		Children 7 years of age and over—oral, 2.5 mg. 1 or 2 times daily as needed
Antazoline Phosphate U.S.P.†		Application, 1 or 2 drops of a 0.5 percent solution in each eye every 3 or 4 hours		
Cyproheptadine Hydrochloride U.S.P. Periactin	Cyproheptadine Hydrochloride Syrup U.S.P. Cyproheptadine Hydrochloride Tablets U.S.P.	Oral, 4 mg. 3 or 4 times daily	Up to 500 μg. per kg. of body weight daily	Use is not recommended in premature or full-term neonates. Oral, 125 μg. per kg. of body weight or 4 mg. per square meter of body surface, 2 times daily as needed. Children 2 to 6 years of age—oral, 2 mg. 2 or 3 times daily, not to exceed 12 mg. daily; children 7 to 14 years of age—oral, 4 mg. 2 or 3 times daily, not to exceed 16 mg. daily
Azatadine Maleate U.S.P. Optimine	Azatadine Maleate Tablets U.S.P.	Oral, 1 to 2 mg. twice daily as needed		Use is not recommended in premature or full-term neonates. Children 6 to 12 years of age—oral, 500 μg. to 1 mg. twice a day as needed

*See *U.S.P. D.I.* for complete dosage information.
†The official title and dosage forms of this compound were obtained from the *U.S.P. XX/N.F.XV*, but at the time of this printing, the U.S.P. Dispensing Information did not include this compound. Therefore, the usual dose, dose range, and pediatric dose were obtained from *U.S.P.XIX* or *N.F.XIV* if the dosage form was listed.

Diphenylpyraline Hydrochloride U.S.P., Hispril®, Diafen®, 4-diphenylmethoxy-1-methylpiperidine hydrochloride. The salt occurs as a white or slightly off-white crystalline powder that is soluble in water or alcohol. Diphenylpyraline is structurally related to diphenhydramine with the aminoalkyl side chain incorporated in a piperidine ring.

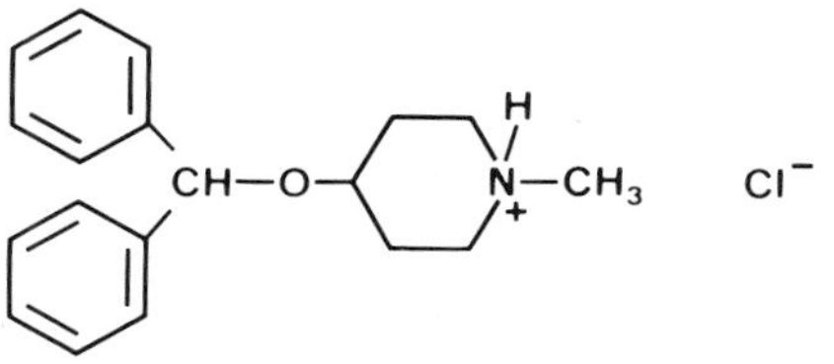

Diphenylpyraline Hydrochloride

Diphenylpyraline is a potent antihistaminic and the usual dose is 2 mg. 3 or 4 times daily. The hydrochloride is available as 2 mg. tablets and 5 mg. sustained-release capsules.

Phenindamine Tartrate, 2,3,4,9-tetrahydro-2-methyl-9-phenyl-1H-indeno[2,1-c]pyridine bitartrate. The hydrogen tartrate occurs as a creamy white powder, usually having a faint odor, and sparingly soluble in water (1:40). A 2 percent aqueous solution has a pH of about 3.5. It is most stable in the pH range of 3.5 to 5.0 and is unstable in solutions of pH 7 or higher. Oxidizing substances or heat may cause isomerization to an inactive form.

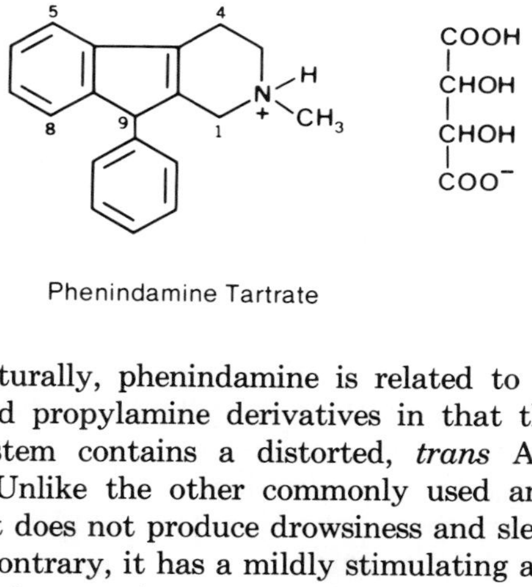

Phenindamine Tartrate

Structurally, phenindamine is related to the unsaturated propylamine derivatives in that the rigid ring system contains a distorted, *trans* Ar-C=C-CH$_2$N. Unlike the other commonly used antihistamines, it does not produce drowsiness and sleepiness; on the contrary, it has a mildly stimulating action in some patients and may cause insomnia when taken just before bedtime.[43]

Dimethindene Maleate U.S.P., Forhistal® Maleate, (±)2-{1-[2-(2-dimethylaminoethyl)inden-3-yl]ethyl}pyridine bimaleate (1:1). The salt occurs as a white to off-white crystalline powder that has a characteristic odor and is sparingly soluble in water. This potent antihistaminic agent may be considered as a

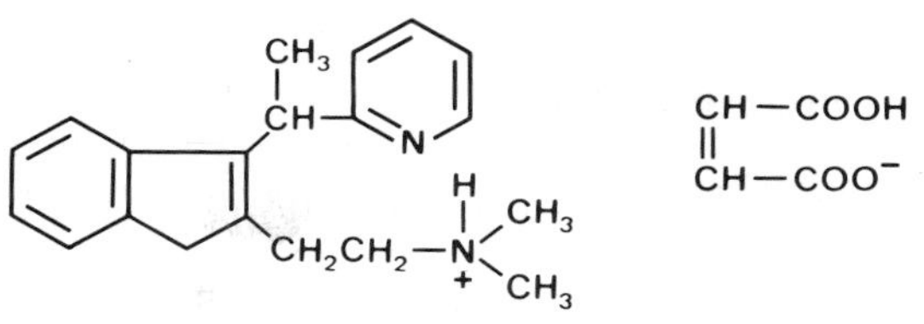

Dimethindene Maleate

derivative of the unsaturated propylamines. The principal side-effect is some degree of sedation or drowsiness. The antihistaminic activity resides mainly in the levorotatory isomer.[14]

Antazoline Phosphate U.S.P., 2-[(N-benzylanilino)methyl]-2-imidazoline dihydrogen phosphate. The salt occurs as a bitter-tasting, white to off-white crystalline powder that is soluble in water. It has a pKa of 10.0 and a 2 percent solution has a pH of about 4.5. Antazoline, similarly to the ethylenediamines, contains an N-benzylanilino group linked to a basic nitrogen through a 2-carbon chain.

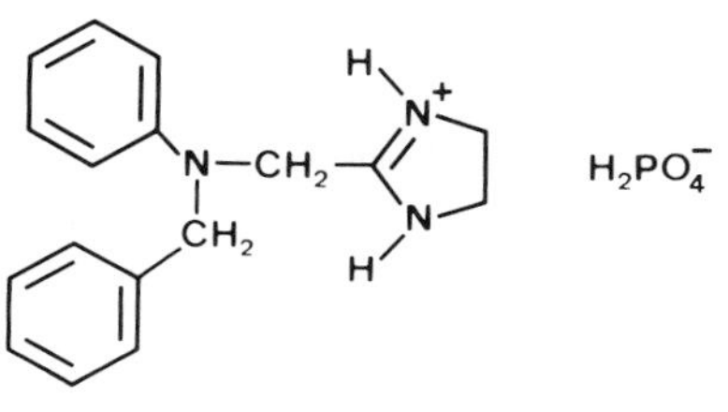

Antazoline Phosphate

Antazoline is less active than most of the other antihistaminic drugs, but it is characterized by the lack of local irritation. The more soluble phosphate salt is applied topically to the eye in a 0.5 percent solution. The less soluble hydrochloride is given orally. In addition to its use as an antihistamine, antazoline has over twice[44] the local anesthetic potency of procaine and also exhibits anticholinergic actions.

Cyproheptadine Hydrochloride U.S.P., Periactin® Hydrochloride, 4-(5H-dibenzo-[a,d]cyclohepten-5-ylidene)-1-methylpiperidine hydrochloride sesquihydrate. The salt is slightly soluble in water and sparingly soluble in alcohol.

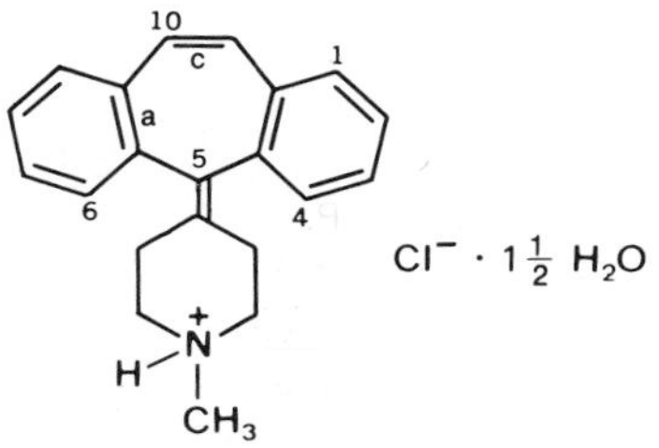

Cyproheptadine Hydrochloride

Cyproheptadine possesses both an anthistamine and an antiserotonin activity and is used as an antipruritic agent. Sedation is the most prominent side-effect, and this is usually brief, disappearing after 3 or 4 days of treatment.

This dibenzocycloheptene may be regarded as a phenothiazine analog in which the sulfur atom has been replaced by an isosteric vinyl group and the ring nitrogen replaced by an sp² carbon atom.

Azatadine Maleate U.S.P., Optimine® Maleate, 6,11-dihydro-11-(1-methylpiperid-4-ylidene)-5H-benzo[5,6]cyclohepta[1,2-b]pyridine dimaleate. Azatadine is an aza isostere of cyproheptadine in which the 10,11-double bond is reduced.

Azatadine Maleate

In early testing azatadine exhibited more than three times the potency of chlorpheniramine in the isolated guinea-pig ileum screen and more than seven times the oral potency of chlorpheniramine in protection of guinea pigs against a double lethal dose of intravenously administered histamine.[45]

It is a potent, long-acting antihistaminic with low sedation liability. The usual dose is 1 to 2 mg. twice daily. Azatadine is available in 1 mg. tablets.

INHIBITION OF HISTAMINE RELEASE

Interest has been generated in the suppression of release of autacoids as a therapeutic approach to the treatment of hypersensitivity.[32] The drug that has focused attention on this possibility is cromolyn.[46]

Cromolyn Sodium U.S.P., Intal®, disodium 1,3-*bis*(2-carboxychromon-5-yloxy)-2-hydroxypropane. The salt is a hygroscopic, white, hydrated crystalline powder that is soluble in water (1:10). It is tasteless at first, but leaves a very slightly bitter after-taste. The pKa of cromolyn is 2.0. Cromolyn belongs to a completely novel class of compounds and bears no structural relationship to other commonly used antiasthmatic compounds. Unlike its naturally occurring predecessor (khellin), cromolyn is not a smooth-muscle relaxant or a bronchodilator.

Cromolyn Sodium

Cromolyn inhibits release of the potent bronchial spasmogens, histamine and slow-reacting substance in anaphylaxis (SRS-A, leukotriene C), from human lung during allergic responses. Apparently, its action is on the pulmonary mast cells after the sensitization stage but before the antigen challenge. It does not seem to interfere with the antigen–antibody reaction, but it seems to suppress the responses to this reaction. The benefits of the drug are exclusively prophylactic. It is of no value after an asthmatic attack has begun.

The product is available as micronized particles mixed with an equal weight of lactose as an inert excipient. Emphasize to the patient that the drug is not absorbed when swallowed and is not effective by this route of administration. The powder is contained in a single-dose hard gelatin capsule designed for use in a turbo-inhaler.

Usual adult dose—Oral inhalation, 20 mg. (1 capsule) four times a day, up to 160 mg. (8 capsules) daily. Use in children up to 5 years of age is not recommended. Children 5 years of age and over, same as usual adult dose.

Dosage form—Cromolyn Sodium for Inhalation U.S.P.

H₂-ANTAGONISTS

While centrally acting agents, such as the sedatives and antianxiety agents, remain important adjuncts in the therapy of peptic-ulcer disease, they are not included in this section because they do not affect gastric secretions directly. Carbenoxolone sodium has been shown[47] to increase the rate of healing of ulcers by increasing the secretion of gastric mucoproteins. However, carbenoxolone is omitted here because its mechanism of action is different from that of agents affecting gastric-acid secretion.

In a very short time, H₂-antagonists have become an important alternative in the therapy of peptic ulcers. To place this category of drugs in proper perspective, the discussion follows the general outline below.

Mechanisms of inhibition of pepsin activity

1. Chemical complexation
2. pH control

a. Antacid

b. Antisecretory

The discussion emphasizes the advantages and disadvantages of the *mechanisms* involved and not the pros and cons of specific agents. When a side-effect is recognized as being a consequence of the desired mechanism of action, the search for a more specifically-acting agent among structural congeners is usually fruitless. Structural manipulation, in this case, can eliminate the side-effect only if it results in an agent that acts by a different mechanism.

Aside from the degree of protection *versus* the degree of insult, there seems to be general agreement that the common denominator in the etiology of peptic ulceration is the presence of the active proteolytic enzyme, pepsin. Therefore the mechanisms used to treat and prevent peptic-ulcer disease are mechanisms of pepsin inhibition.

Chemical complexation

The sulfate esters and sulfonate derivatives of polysaccharides and lignin form chemical complexes with the enzyme, pepsin. These complexes have no proteolytic activity. Because polysulfates and polysulfonates are poorly absorbed from the gastrointestinal tract, specific chemical complexation appears to be a desirable mechanism of pepsin inhibition. Unfortunately, these polymers are also potent anticoagulants.

pH Control

Pepsin activity is pH dependent. While some members of the prostaglandin E series have exhibited some inhibitory action on gastric secretion,[48] insufficient information is presently available to predict the usefulness of prostaglandins in the therapy of peptic-ulcer disease. Currently available drugs that inhibit pepsin activity affect the pH of gastric contents by neutralization (antacids) or inhibition of gastric-acid secretion (antimuscarinic agents and H_2-antagonists.

In Figure 16-3, the solid line illustrates the pH dependency[49] of human pepsin activity incubated at 37°C. Maximal peptic activity is obtained at a pH of 1.5 to 2.5 (the actual optimal pH varies slightly with the method used). Seventy percent of maximal activity occurs in the pH range of 2.5 to 5, with almost no activity above pH 5.

The dotted line in Figure 16-3 illustrates the stability of the enzyme at various pH values. At pH levels up to 7.5, pepsin is stable and lowering of pH restores maximal pepsin activity. However, at pH levels above 7.5, the enzyme is irreversibly inactivated (alkaline denaturation).

The best value at which the pH of the gastric contents should be controlled in the treatment of peptic-ulcer disease is not established. There is general agreement that antacids relieve pain of ulcers, but agreement falters over the question whether antacids are beneficial in the healing of the lesion.[50] Recently, two

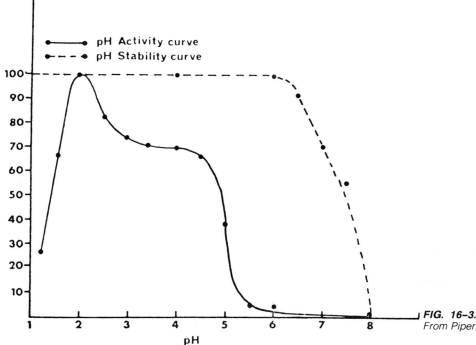

FIG. 16-3. The pH stability and pH activity of human pepsin. From Piper, D.W., and Fenton, B.H.: Gut 6: 506, 1965.

randomized, double-blind, multicenter trials[51,52] showed that intensive antacid therapy is superior to placebo in healing duodenal ulcer. In each trial, 210 ml. of an Al-Mg antacid was administered daily in seven divided doses. The in vitro buffering capacity of each 30 ml. dose was 123 mEq. of hydrochloric acid. Possibly, peak pH values were sufficiently high to inactivate pepsin irreversibly.

Antacid mechanism

Although the question of the best pH level to which the gastric contents should be buffered remains unanswered, the mechanism of antacid action is generally agreed to be neutralization of acid rather than chemical complexation with pepsin. Also, the antacid mechanism is desirable whether the attack factor in peptic ulceration is acid or acid-pepsin. However, larger doses of antacids would be required if the attack factor is acid-pepsin.[50]

Without regard to the side-effects of specific agents, several disadvantages (theoretically at least) are associated with the antacid mechanism.

One complicating factor is the uncertainty of the dose interval. Except with intragastric drip, continuous buffering is difficult to achieve, and practical considerations force a compromise in which buffering is achieved discontinuously. On the one hand, the rate and quantity of acid secretion varies with the individual's thoughts of food (the cephalic phase, in which the secretions are richer in pepsin), eating habits, and rate of gastric emptying. On the other hand, antacids promote gastric emptying, which limits the duration of antacid action.

A second problem with the titration of gastric contents involves a reflex mechanism. The concept of acid rebound is of historical interest, but as a clinical problem, it is not universally accepted and apparently is difficult to document.[53] Theoretically, acid rebound is a potential problem because pH of the gastric contents influences the release of gastrin (a potent gastric-stimulatory hormone). When the acidity of the gastric contents increases to a pH of 2.0, the gastrin mechanism for stimulating gastric secretion becomes totally blocked.[54] However, a rise in pH above 3 causes a release of gastrin. Therefore, the antacid mechanism indirectly stimulates acid secretion.

A third built-in difficulty associated with the antacid mechanism is related to the inhibition of secretion of intestinal factors. Excess acid in the duodenum stimulates the release of the hormone secretin, which stimulates the pancreas to release alkaline juices rich in digestive enzymes. The effect of antacid therapy on this complicated reflex mechanism has not been assessed. However, the potential does exist that antacids may interfere with release of intestinal hormones. Thus, they may indirectly prevent a slowing of gastric secretions, decrease the protection of the alkaline pancreatic juices to the duodenum, and adversely affect digestion of food products.

In addition, a number of problems, including systemic alkalosis, diarrhea, constipation, and a high sodium content, may be associated with the use of specific antacid products.

Antisecretory mechanisms–antimuscarinic agents

The older, more familiar method of controlling gastric-acid secretion involves blocking the vagus nerve. The neurogenic mechanism regulating gastric secretions is mediated through the parasympathetic fibers of the vagus nerves. Therefore, antimuscarinic agents have been used extensively as adjuncts to the treatment of peptic ulcer.

Except for CNS effects, atropine is relatively specific for muscarinic sites. The synthetic quaternary ammonium drugs generally exhibit varying degrees of nicotinic-blocking activity. However, both types of agents are highly potent vagal blockers, the action of which is sometimes referred to as *chemical vagotomy*.

The disadvantages of the antimuscarinic mechanism of controlling gastric-acid secretion are related to two important factors. For one, the antimuscarinic agents are not specific for vagal muscarinic receptors. These drugs block other muscarinic sites as well, and the resulting actions are considered to be side-effects inherent in the mechanism of action. Additionally, other side-effects may occur that are specifically related to the drug, e.g., CNS-stimulating actions of atropine.

Secondly, the antimuscarinic agents only partially inhibit gastric secretions. The greater volume of gastric secretion occurs during the gastric and intestinal phases, which are primarily under hormonal (gastrin) control. Further, the drugs reduce gastric secretion, provided the dose is increased to the limits of tolerance of side-effects. At doses sufficient to block vagally stimulated gastric secretions totally, intolerable side-effects occur.

An additional theoretical factor is delayed gastric emptying caused by antimuscarinic action. The desirability of this action, which would prolong the neutralizing effect of antacids, might be counteracted by equally prolonged production of acid because of the continued presence of food in the stomach. (The antimuscarinic agents and their actions are discussed in Chap. 12.)

In spite of the fact that antacids provide only one of several therapeutic measures in the treatment of peptic ulcer, they are among the first agents employed. Antimuscarinic agents have been used extensively as adjuncts to dietary and antacid treatment. The value of these two mechanisms in the treatment and prevention of peptic-ulcer disease is difficult to assess. However, the disadvantages of these mechanisms have demonstrated the need for the development of an alternative mechanism. The H_2-antagonists provide that alternative. It is too early to assess the value and impact of the H_2-antagonists on the therapy of this disease. The future will disclose the disadvantages, if any, of this alternative mechanism.

Antisecretory mechanisms—H₂-antagonists

Beginning with the discovery of the pharmacologic actions of histamine, many hypotheses have been proffered concerning the role of histamine in gastric secretion.[55] The most enduring hypothesis is the concept that histamine is the final common, local mediator of the parietal cells, by which it is meant that all other stimulants of gastric secretion act through histamine. While it has served very valuably to stimulate much highly productive research, the final-common-mediator hypothesis is no longer tenable. This certainly does not mean that histamine has no role in the normal physiology of gastric secretion.

The final-common-mediator hypothesis is not consistent with the patterns of action of H_2-antagonists and antimuscarinic agents. The hypothesis that histamine is the final common mediator cannot account for the observation that antimuscarinic agents inhibit the action of gastrin and histamine on gastric secretion. Likewise, the hypothesis that acetylcholine is the common mediator cannot account for the observation that H_2-antagonists inhibit the action of gastrin and cholinergic agents on gastric-acid secretion.[56]

The recent discovery of H_2-antagonists opened another door to the secrets of gastric secretory mechanisms. Of the three most potent known gastric secretagogues (gastrin, histamine, acetylcholine), two have yielded to useful competitive antagonists. The tools are now available to study their normal physiologic roles and their potentiating interactions.

A *final-common-potentiator hypothesis* has been proposed[56] that is a modified version of the previous final-common-mediator hypothesis. The hypothesis states simply that H_2-antagonists inhibit the direct action of histamine on acid secretion and inhibit the potentiating action of histamine on acid secretion stimulated by gastrin or acetylcholine. According to the modified hypothesis, secretagogues have two efficacies: (1) an intrinsic efficacy, which refers to the maximal response it can produce in the absence of other agents, and (2) a potentiating efficacy, which refers to the magnitude of its response in the presence of a second agent that potentiates its action.

Histamine is assumed to have both intrinsic efficacy and potentiating efficacy, whereas gastrin and acetylcholine have only potentiating efficacy. This means that in the absence of histamine, neither gastrin nor cholinergic agents would increase acid secretion. Only histamine can increase acid secretion. Gastrin or acetylcholine increases acid secretion by potentiating the effect of histamine.

Thus, an H_2-antagonist inhibits local secretion and gastric secretion stimulated by gastrin and cholinergic agents, by blocking the intrinsic and potentiating actions of histamine. An antimuscarinic agent (e.g., atropine) suppresses histamine-stimulated gastric secretion by blocking the potentiating action of acetylcholine.

Some in vitro studies[57] using parietal cells isolated from canine fundic mucosa, partially support the hypothesis. Using indices of response to stimulation other than secretion, the data suggest that the parietal cell contains receptors for each of the three secretagogues. Each secretagogue exhibited an intrinsic efficacy that could be blocked only by its specific antagonist. In other words, the intrinsic efficacy of histamine was blocked only by an H_2-antagonist, that of carbachol (cholinergic agent) was blocked only by atropine, and that of gastrin was not blocked by either an H_2-antagonist or atropine.

Also, each secretagogue markedly potentiated the action of each of the other two. Again, H_2-blockers inhibited the potentiating action only of histamine, and atropine inhibited only the potentiating effect of carbachol.

Structure–activity relationships

A review[58] of the characterization and development of cimetidine as a histamine H_2-receptor antagonist reveals a classic medicinal-chemistry approach to problem-solving. Beginning with the study of the relative potencies of the methylhistamines (see Fig. 16-1), hundreds of compounds have been synthesized and their actions studied leading to the development of cimetidine. From these studies several structure-activity relationships may be stated, but they must be considered tentative pending further study.

H_2–Antagonist*

* Numbering of the ring in the general structure of an H_2-Antagonist is according to the system followed in the literature.

The relationships may be separated into three parts: ring and substituents, side chain, and the terminal N-group.

Ring

The N^τ-H tautomer, as the predominant species, seems to be necessary for maximal H$_2$-antagonist* activity. The predominating species of the imidazole ring is not protonated at physiological pH (7.4). Therefore, histamine and the H$_2$-antagonists are capable of undergoing 1,3-prototropic tautomerism, and the N^τ- H tautomer is the more prevalent in both.

Electron-donating groups (e.g., methyl) favor the nearer N-H tautomer whereas electron-withdrawing groups do not. Hence the N^τ-H tautomer, drawn above, is the prevalent tautomer when the chain contains an electron-withdrawing group. Also, when the above R group is methyl (electron-donating), the N^τ-H tautomer is favored; thus, the tautomeric effect of an electron-withdrawing chain is reinforced by a 4-methyl substituent.

The structure of burimamide favors the N^π-H tautomer, whereas that of metiamide favors the N^τ-H tautomer. Metiamide is about 5 times more potent than burimamide.

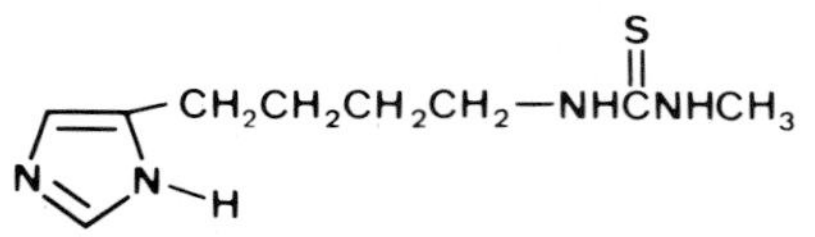

$$\text{CH}_2\text{CH}_2\text{CH}_2\text{CH}_2-\text{NHCNHCH}_3 \quad (\overset{S}{\overset{\|}{})}$$

Burimamide

Chain

For optimal activity the ring should be separated from the N-group by the equivalent of a four-carbon chain. A shorter chain drastically lowers antagonist activity. As discussed above, the chain should contain an electron-withdrawing substituent. The more active compounds contain an isosteric thioether ($-\text{S}-$) link in place of a methylene group ($-\text{CH}_2-$).

N-group

The terminal N-group should be a polar, nonbasic substituent for maximal antagonist activity. A positively charged group should bind more tightly to the receptor but it seems to permit the molecule to mimic histamine and act as an agonist. For example, a guanidine group, which is protonated (positively charged) at

---

* Numbering of the ring in the general structure of an H$_2$-antagonist is according to the system followed in the literature.

---

physiological pH, results in a compound that is a weak antagonist and a partial agonist.

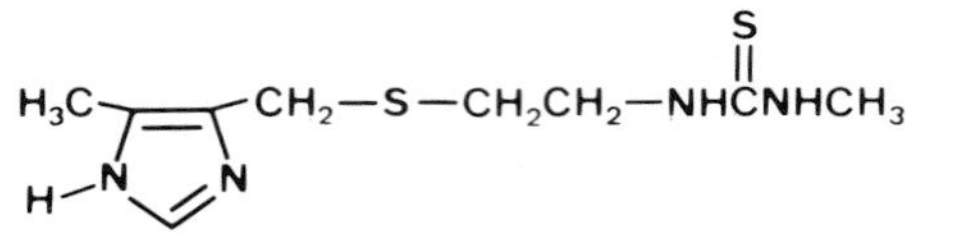

$$\text{H}_3\text{C}-\cdots-\text{CH}_2-\text{S}-\text{CH}_2\text{CH}_2-\text{NHCNHCH}_3 \quad (\overset{S}{\overset{\|}{})}$$

Metiamide

Metiamide, containing a polar, nonbasic thiourea group, was shown to be highly effective clinically in reducing gastric-acid secretion. However, granulocytopenia was observed in some of the patients. Because the thiourea group was possibly related to this effect, other polar but nonthiourea groups were investigated.

The corresponding cyanoguanidine substituent is a very polar group but predominantly nonionized at physiological pH. The resulting compound, cimetidine, is at least as active as metiamide and apparently lacks the side-effect (granulocytopenia).

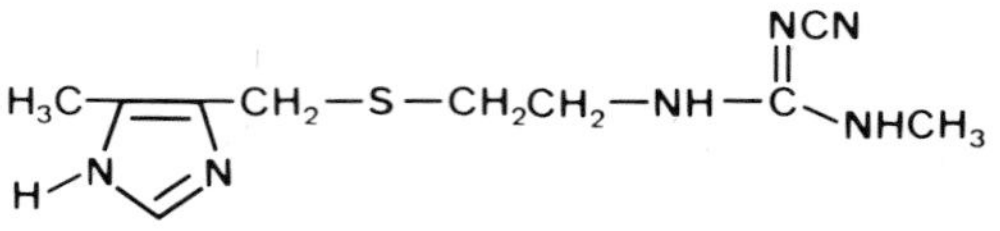

$$\text{H}_3\text{C}-\cdots-\text{CH}_2-\text{S}-\text{CH}_2\text{CH}_2-\text{NH}-\overset{\text{NCN}}{\overset{\|}{\text{C}}}-\text{NHCH}_3$$

Cimetidine

In a structural comparison of the H$_1$-antagonists and H$_2$-antagonists, there is a marked chemical distinction between the requirements for antagonist activity.[59] The H$_1$-antagonists possess aryl groups that need not have a structural relationship to the imidazole ring of histamine, but which do confer considerable lipophilicity to the molecule. They resemble histamine in possessing a side chain group (usually ammonium), which is positively charged at physiological pH.

In marked contrast, the H$_2$-antagonists are hydrophilic molecules. They bear a structural resemblance to histamine in having an imidazole ring capable of undergoing 1,3-prototropic tautomerism. They differ from histamine in the side chain group which, though polar, is uncharged. Thus they do not mimic the stimulant actions of histamine. Their low lipophilicity probably limits access to the central nervous system and avoids some of the side-effects associated with the use of the antihistaminic drugs.

These substantial chemical differences probably account for the considerable degree of selectivity exhibited by the respective antagonists in distinguishing the two types of receptors. Thus, H$_1$-receptor recognition seems to be determined by the lipophilic character (hydrophobic bonding) and the side chain ammonium group (ionic bonding). At H_2-receptors, recognition seems to be determined by the imidazole ring.

While the current interest clinically is in the therapy of peptic-ulcer disease, H_2-antagonists of the future may well be studied as therapeutic entities for other diseases associated with H_2-agonist activity. The focus of attention here is on the products used in the treatment and prevention of peptic-ulcer disease by a mechanism associated with gastric-acid secretion.

Cimetidine U.S.P., Tagamet®, N''-cyano-N-methyl-N'-[2-(5-methylimidazol -4- yl)methylthioethyl]guanidine. The drug is a colorless crystalline solid that is slightly soluble in water (1.14% at 37°C). The solubility is greatly increased with the addition of dilute acid to protonate the imidazole ring (apparent pKa of 7.09).[58] At pH 7, aqueous solutions are stable for at least 7 days. Cimetidine is a relatively hydrophilic molecule having an octanol–water partition coefficient of 2.5.

Usual adult dose: Duodenal ulcer—oral, 300 mg. four times a day, with meals and at bedtime. Parenteral, the equivalent of 300 mg. of cimetidine (as the hydrochloride salt) every 6 hours. The prescribing limit is 2400 mg. daily by either route of administration. For prophylaxis of recurrent ulcer, the recommended oral dose in 400 mg. at bedtime.

Pathological hypersecretory conditions—oral, 300 mg. four times a day, with meals and at bedtime, as long as clinically indicated.

Usual pediatric dose—oral, 5 to 10 mg. per kg. of body weight four times a day, with meals and at bedtime.

Dosage forms—Cimetidine Tablets U.S.P., Cimetidine Hydrochloride Injection U.S.P., and Cimetidine Hydrochloride Oral Solution U.S.P.

REFERENCES

1. Windaus, A., and Vogt, W.: Ber. deutsch. chem. Ges. 40:3691, 1907.
2. Dale, H. H., and Laidlaw, P. P.: J. Physiol. (London) 41:318, 1910.
3. *Ibid.* 43:182, 1911.
4. *Ibid.* 52:355, 1919.
5. Beaven, M. A.: Monographs in Allergy. Histamine: Its Role in Physiological and Pathological Processes, vol. 13, p. 22, New York, S. Karger, 1978.
6. Stern, P.: The Relation of Histamine to Inflammation, *in* Rocha e Silva, M. (ed.): Handbook of Experimental Pharmacology, vol. XVIII/1, p. 892, New York, Springer-Verlag, 1966.
7. Altura, B. M., and Halevy, S.: Cardiovascular Actions of Histamine, *in* Rocha e Silva, M. (ed.): Handbook of Experimental Pharmacology, vol. XVIII/2, p. 1, New York, Springer-Verlag, 1978.
8. Kahlson, G., and Rosengren, E.: *Physiol. Rev.* 48:155, 1968.
9. Schwartz, J. C.: Life Sci. 17:503, 1975.
10. Johnson, L. R.: Histamine and Gastric Secretion, *in* Rocha e Silva, M. (ed.): Handbook of Experimental Pharmacology, vol. XVIII/2, p. 41, New York, Springer-Verlag, 1978.
11. Pelletier, G.: Naturally Occurring Antihistaminics in Body Tissues, *in* Rocha e Silva, M. (ed.): Handbook of Experimental Pharmacology, vol. XVIII/2, p. 369, New York, Springer-Verlag, 1978.
12. Best, C. H., et al.: J. Physiol. (London), 62:397, 1927.
13. Fourneau, E., and Bovet, D.: Arch. Int. Pharmacodyn. 46:178, 1933.
14. Casy, A. F.: Chemistry of Anti-H_1 Histamine Antagonists, *in* Rocha e Silva, M. (ed.): Handbook of Experimental Pharmacology, vol. XVIII/2, p. 175, New York, Springer-Verlag, 1978.
15. Witiak, D. T.: Antiallergenic Agents, *in* Burger, A. (ed.): Medicinal Chemistry, 3rd ed., p. 1643, New York, Wiley-Interscience, 1970.
16. Paton, D. M.: Receptors for Histamine, *in* Schachter, M.: Histamine and Antihistamines, p. 3, New York, Pergamon Press, 1973.
17. Ahlquist, R. P.: Am. J. Physiol. 153:586, 1948.
18. Lin. T. M., et al.: Ann. N.Y. Acad. Sci. 99:30, 1962.
19. Ash, A. S. F., and Schild, H. O.: Brit. J. Pharmacol. Chemother. 27:427, 1966.
20. Black, J. W., et al.: Nature 236:385, 1972.
21. Rocha e Silva, M., and Antonio, A.: Bioassay of Antihistaminic Action, *in* Rocha e Silva, M. (ed.): Handbook of Experimental Pharmacology, vol. XVIII/2, p. 381, New York, Springer-Verlag, 1978.
22. Biel, J. H., and Martin, Y.C.: Organic Synthesis as a Source of New Drugs, *in* Gould, R.F.: Drug Discovery, Advances in Chemistry Series No. 108, p. 81, Washington, D.C., American Chemical Society, 1971.
23. Metzler, D. E., Ikawa, M., and Snell, E.E.: J. Am. Chem. Soc. 76:648, 1954.
24. Schayer, R. W., and Cooper, J. A. D.: J. Appl. Physiol 9:481, 1956.
25. Schayer, R. W.: Biogenesis of Histamine, *in* Rocha e Silva, M. (ed.): Handbook of Experimental Pharmacology, vol. XVIII/2, p. 109, New York, Springer-Verlag, 1978.
26. Wetterqvist, H.: Histamine Metabolism and Excretion, in Rocha e Silva, M. (ed.): Handbook of Experimental Pharmacology, vol. XVIII/2, p. 131, New York, Springer-Verlag, 1978.
27. Herbert, V.: Drugs Effective in Megaloblastic Anemias, *in* Goodman, L. S., and Gilman, A. (ed.): The Pharmacological Basis of Therapeutics, 5th ed., p. 1324, New York, Macmillan, 1975.
28. Rocha e Silva, M. (ed.): Handbook of Experimental Pharmacology, vol. XVIII/2, New York, Springer-Verlag, 1978.
29. Fordtran, J. S., and Grossman, M. I. (eds.): Third Symposium on Histamine H_2-Receptor Antagonists: Clinical Results with Cimetidine. Gastroenterology 74(2):339, 1978.
30. van den Brink, F. G., and Lien, E. J.: Competitive and Noncompetitive Antagonism, *in* Reference, 28, p. 333.
31. Nauta, W. Th., and Rekker, R. F.: Structure-Activity Relationships of H_1-Receptor Antagonists, *in* reference 28, p. 215.
32. Douglas, W. W.: Histamine and Antihistamines, *in* Goodman, L.S., and Gilman, A. (eds.): The Pharmacological Basis of Therapeutics, 5th ed., p. 590, New York, Macmillan, 1975.
33. Ganellin, C. R.: Histamine Receptors, *in* Hess, H.J.: Ann. Repts. Med. Chem. 14:91, 1979.
34. Sjoquist, F., and Lasagna, L.: Clin. Pharmacol. Ther. 8:48, 1967.
35. Barouh, V., et al.: J. Med. Chem. 14:834, 1971.
36. Shafi'ee, A, and Hite, G.: J. Med. Chem. 12:266, 1969
37. Nobles, W. L., and Blanton, C. D.: J. Pharm. Sci. 53:115, 1964.
38. IUPAC. Tentative rules for the nomenclature of organic chemistry. Section E. Fundamental Stereochemistry. J. Org. Chem. 35:2849, 1970.
39. Ison, R. R., Franks, F. M., and Soh, K. S.: J. Pharm. Pharmacol. 25:887, 1973.
40. Zirkle, Charles L.: To Tranquilizers and Antidepressants from Antimalarials and Antihistamines, *in* Clarke, F.H. (ed.): How Modern Medicines are Discovered, p. 55, New York, Futura, 1973.
41. Toldy, L., et al.: Acta Chim. Acad. Sci. Hungary 19:273, 1959.
42. King, C. T. G., Weaver, S. R., and Narrod, S. A.: J. Pharmacol. Exp. Ther. 147:391, 1965.

43. Criep, L. H.: Lancet 68:55, 1948.
44. Landau, S. W., Nelson, W. A., and Gay, L. N.: J. Allergy 22:19, 1951.
45. Villani, *et al.*: J. Med. Chem. 15:750, 1972.
46. Intal® Cromolyn Sodium, A Monograph: Fisons Corporation, Bedford, Massachusetts, 1973.
47. Baron, J. H., and Sullivan, F.M. (eds.): Symposium, Carbenoxolone Sodium, New York, Appleton-Century-Crofts, 1972.
48. Bass, P.: Gastric Antisecretory and Antiulcer Agents, *in* Harper, N. J., and Simmonds, A. B. (eds.): Advances in Drug Research, vol. 8, p. 205, New York, Academic Press, 1974.
49. Piper, D. W., and Fenton, B. H.: Gut 6:506, 1965.
50. Harvey, S. C.: Gastric Antacids and Digestants, *in* Goodman, L. S., and Gilman, A. (eds.): The Pharmacological Basis of Therapeutics, 5th ed., p. 960, New York, Macmillan, 1975.
51. Peterson, W. L., *et al.*: New Eng. J. Med. 297:341, 1977.
52. Ippoliti, A. F., *et al.*: Gastroenterology 74:393, 1978.
53. Perevia-Lima, J., and Hollander, F.: Gastroenterology 37:145, 1959.
54. Guyton, A. C.: *Textbook of Medical Physiology,* 5th ed., p. 874, Philadelphia, W. B. Saunders, 1976.
55. Johnson, L. R.: Histamine and Gastric Secretion, *in* Rocha e Silva, M. (ed.): Handbook of Experimental Pharmacology, vol. XVIII/2, p. 41, New York, Springer-Verlag, 1978.
56. Gardner, J. D., *et al.*: Gastroenterology 74:348, 1978.
57. Soll, A. H.: Gastroenterology 74:355, 1978.
58. Brimblecombe, R. W., *et al.*: Characterization and Development of Cimetidine as a Histamine H$_2$ Receptor Antagonist, *in* Fordtran, J. S., and Grossman, M. I. (eds.): Third Symposium on Histamine H$_2$ Receptor Antagonists: Clinical Results with Cimetidine, Gastroenterology, 74:339, 1978.
59. Durant, G. J., Emmett, J. C., and Ganellin, C. R.: The Chemical Origin and Properties of Histamine H$_2$ Receptor Antagonists, *in* Burland, W. L., and Simkins, M. A. (eds.): Cimetidine: Proceedings of the Second International Symposium on Histamine H$_2$ Receptor Antagonists, p. 1, New York, Elsevier North-Holland, 1977.

SELECTED READINGS

Bass, P.: Gastric Antisecretory and Antiulcer Agents, *in* Harper, N. J., and Simmonds, A. B. (eds.): Advances in Drug Research, vol. 8, p. 205, New York, Academic Press, 1974.

Beaven, M. A.: Histamine, Its Role in Physiological and Pathological Processes, Monographs in Allergy, vol. 13, New York, S. Karger, 1978.

Burland, W. L., and Simkins, M. A. (eds.): Cimetidine, Proceedings of the Second Int. Symposium on Histamine H$_2$—Receptor Antagonists, New York, Excerpta Medica/Elsevier, 1977.

Fordtran, J. S., and Grossman, M. I. (eds.): Third Symposium on Histamine H$_2$-Receptor Antagonists, Clinical Results with Cimetidine, Gastroenterology, 74, No. 2, Part 2:339, 1978.

Rocha e Silva, M. (ed.): Histamine and Antihistaminics, Handbook of Experimental Pharmacology, vol. XVIII/1, New York, Springer-Verlag, 1966.

Rocha e Silva, M. (ed.): Histamine II and Antihistaminics, Handbook of Experimental Pharmacology, vol. XVIII/2, New York, Springer-Verlag, 1978.

Schachter, M. (ed.): Histamine and Antihistamines, International Encyclopedia of Pharmacology and Therapeutics, Section 74, vol. 1, New York, Pergamon, 1973.

Thompson, J.H.: Gastrointestinal Disorders—Peptic Ulcer Disease, *in* Rubin, A.A. (ed.): Search for New Drugs, Medicinal Res. Series, vol. 6, p. 115, New York, Dekker, 1972.

Witiak, D.T.: Antiallergenic Agents, *in* Burger, A. (ed.): Medicinal Chemistry, ed. 3, p. 1643, New York, Wiley-Interscience, 1970.

Witiak, D.T. and Cavestri, R.C.: Antiallergenic Agents, *in* Foye, W.O. (ed.): Principles of Medicinal Chemistry, ed. 2, p. 473, Philadelphia, Lea & Febiger, 1981.

17

Robert E. Willette | analgesic agents

The struggle to relieve pain began with the origin of humanity. Ancient writings, both serious and fanciful, dealt with secret remedies, religious rituals, and other methods of pain relief. Slowly, there evolved the present, modern era of synthetic analgesics.*

Tainter[1] has divided the history of analgesic drugs into 4 major eras, namely:

1. The period of discovery and use of naturally occurring plant drugs.
2. The isolation of pure plant principles, e.g., alkaloids, from the natural sources and their identification with analgesic action.
3. The development of organic chemistry and the first synthetic analgesics.
4. The development of modern pharmacologic techniques, making it possible to undertake a systematic testing of new analgesics.

The discovery of morphine's analgesic activity by Sertürner, in 1806, ushered in the second era. It continues today only on a small scale. Wöhler introduced the third era indirectly with his synthesis of urea in 1828. He showed that chemical synthesis could be used to make and produce drugs. In the third era, the first synthetic analgesics used in medicine were the salicylates. These originally were found in nature (methyl salicylate, salicin) and then were synthesized by chemists. Other early, man-made drugs were acetanilid (1886), phenacetin (1887) and aspirin (1899).

These early discoveries were the principal contributions in this field until the modern methods of pharmacologic testing initiated the fourth era. The effects

of small structural modifications on synthetic molecules now could be assessed accurately by pharmacologic means. This has permitted systematic study of the relationship of structure to activity during this era. The development of these pharmacologic testing procedures, coupled with the fortuitous discovery of meperidine by Eisleb and Schaumann,[2] has made possible the rapid strides in this field today.

The consideration of synthetic analgesics, as well as the naturally occurring ones, will be facilitated considerably by dividing them into 2 groups: morphine and related compounds and the antipyretic analgesics.

It should be called to the reader's attention that there are numerous drugs which, in addition to possessing distinctive pharmacologic activities in other areas, may also possess analgesic properties. The analgesic property exerted may be a direct effect or may be indirect but is subsidiary to some other more pronounced effect. Some examples of these, which are discussed elsewhere in this text, are: sedatives (e.g., barbiturates); muscle relaxants (e.g., mephenesin, methocarbamol); tranquilizers (e.g., meprobamate), etc. These types will not be considered in this chapter.

MORPHINE AND RELATED COMPOUNDS

The discovery of morphine early in the 19th century and the demonstration of its potent analgesic properties led directly to the search for more of these potent principles from plant sources. In tribute to the remarkable potency and action of morphine, it has remained alone as an outstanding and indispensable analgesic from a plant source.

---

* An analgesic may be defined as a drug bringing about insensibility to pain without loss of consciousness. The etymologically correct term "analgetic" may be used in place of the incorrect but popular "analgesic."

607

It is only since 1938 that synthetic compounds rivaling it in action have been found, although many earlier changes made on morphine itself gave more effective agents.

Modifications of the morphine molecule will be considered under the following headings:

1. Early changes on morphine prior to the work of Small, Eddy and their co-workers.
2. Changes on morphine initiated in 1929 by Small, Eddy and co-workers[3] under the auspices of the Committee on Drug Addiction of the National Research Council and extending to the present time.
3. The researches initiated by Eisleb and Schaumann[2] in 1938, with their discovery of the potent analgesic action of meperidine, a compound departing radically from the typical morphine molecule.
4. The researches initiated by Grewe, in 1946, leading to the successful synthesis of the morphinan group of analgesics.

Early morphine modifications

Morphine is obtained from **opium,** which is the partly dried latex from incised unripe capsules of *Papaver somniferum.* Opium contains numerous alkaloids (as meconates and sulfates), of which morphine, codeine, noscapine (narcotine) and papaverine are therapeutically the most important, and thebaine, which has convulsant properties but is an important starting material for many other drugs. Other alkaloids, such as narceine, also have been tested medicinally but are not of great importance. The action of opium is due principally to its morphine content. As an analgesic, opium is not as effective as morphine because of its slower absorption, but it has a greater constipating action and is thus better suited for antidiarrheal preparations (e.g., paregoric). Opium, as a constituent of Dover's powders and Brown Mixture, also exerts a valuable expectorant action that is superior to that of morphine.

Two types of basic structures usually are recognized among the opium alkaloids, i.e., the *phenanthrene* (morphine) type and the *benzylisoquinoline* (papaverine) type (see below).

The pharmacologic actions of the two types of alkaloids are dissimilar. The morphine group acts principally on the central nervous system as a depressant and stimulant, whereas the papaverine group has little effect on the nervous system but has a marked antispasmodic action on smooth muscle.

Clinically, the depressant action of the morphine group is the most useful property, resulting in an increased tolerance to pain, a sleepy feeling, a lessened perception to external stimuli, and a feeling of well-

being (euphoria). Respiratory depression, central in origin, is perhaps the most serious objection to this type of alkaloid, aside from its tendency to cause addiction. The stimulant action is well illustrated by the convulsions produced by certain members of this group (e.g., thebaine).

Phenanthrene Type
(Morphine, R & R′ = H)

Benzyl-Isoquinoline Type
(Papaverine)

Prior to 1929, the derivatives of morphine that had been made were primarily the result of simple changes on the molecule, such as esterification of the phenolic or alcoholic hydroxyl group, etherification of the phenolic hydroxyl group and similar minor changes. The net result had been the discovery of some compounds with greater activity than morphine but also greater toxicities and addiction tendencies. No compounds had been found that did not possess in some measure the addiction liabilities of morphine.*

Some of the compounds that were in common usage prior to 1929 are listed in Table 17-1, together with some other more recently introduced ones. All have the morphine skeleton in common.

Among the earlier compounds is codeine, the phenolic methyl ether of morphine which also had been obtained from natural sources. It has survived as a good analgesic and cough depressant, together with the corresponding ethyl ether which has found its principal application in ophthalmology. The diacetyl

---

* The term "addiction liability," or the preferred term "dependence liability," as used in this text, indicates the ability of a substance to develop true addictive tolerance and physical dependence and/or to suppress the morphine abstinence syndrome following withdrawal of morphine from addicts.

TABLE 17–1

SYNTHETIC DERIVATIVES OF MORPHINE

Compound Proprietary Name	R	R'	R''	Principal Use
Morphine	H	H	(CH=CH₂ with H, OH)	Analgesic
Codeine	CH_3	H	Same as above	Analgesic and to depress cough reflex
Ethylmorphine *Dionin*	C_2H_5	H	Same as above	Ophthalmology
Diacetylmorphine (Heroin)	CH_3CO	H	$-O-C(=O)-CH_3$ (with H)	Analgesic (prohibited in U.S.)
Hydromorphone (Dihydromorphinone) *Dilaudid*	H	H	($-CH_2-CH_2-C=O$)	Analgesic
Hydrocodone (Dihydrocodeinone) *Dicodid*	CH_3	H	Same as above	Analgesic and to depress cough reflex
Oxymorphone (Dihydrohydroxy-morphinone)	H	OH	Same as above	Analgesic
Oxycodone (Dihydrohydroxy-codeinone)	CH_3	OH	Same as above	Analgesic and to depress cough reflex
Dihydrocodeine *Paracodin*	CH_3	H	($-CH_2-CH_2-CH(H)-OH$)	Depress cough reflex
Dihydromorphine	H	H	Same as above	Analgesic
Methyldihydro-morphinone *Metopon*	H	H	($-CH_2-CH_2-$ ... O CH_3 O)	Analgesic

derivative of morphine, heroin, has been known for a long time; it has been banished for years from the United States and is being used in decreasing amounts in other countries. It is the most widely used illicit drug used by narcotic addicts. Among the reduced compounds were dihydromorphine and dihydrocodeine and their oxidized congeners, dihydromorphinone (hydromorphone) and dihydrocodeinone (hydrocodone). Derivatives of the last two compounds possessing a hydroxyl group in position 14 are dihydrohydroxymorphinone, or oxymorphone, and dihydrohydroxycodeinone, or oxycodone. These represent the principal compounds that either had been on the market or had been prepared prior to the studies of Small, Eddy and co-workers.† It is well to note that no really systematic effort had been made to investigate the structure-activity relationships in the molecule, and only the easily changed peripheral groups had been modified.

Morphine modifications initiated by the researches of Small and Eddy

The avowed purpose of Small, Eddy and co-workers,[3] in 1929, was to approach the morphine problem from the standpoint that:

1. It might be possible to separate chemically the addiction property of morphine from its other more salutary attributes. That this could be done with some addiction-producing compounds was shown by the development of the nonaddictive procaine from the addictive cocaine.
2. If it were not possible to separate the addictive tendencies from the morphine molecule, it might be possible to find other synthetic molecules without this undesirable property.

Proceeding on these assumptions, they first examined the morphine molecule in an exhaustive manner. As a starting point, it offered the advantages of ready availability, proven potency, and ease of alteration. In addition to its addictive tendency, it was hoped that other liabilities, such as respiratory depression, emetic properties, and gastrointestinal tract and circulatory disturbances, could be minimized or abolished as well. Since early modifications of morphine (e.g., acetylation or alkylation of hydroxyls, quaternization of the nitrogen, and so on) caused variations in the addictive potency, it was felt that the physiologic effects of morphine could be related, at least in part, to the peripheral groups.

It was not known if the actions of morphine were primarily a function of the peripheral groups or of the structural skeleton. This did not matter, however, because modification of the groups would alter activity

---

† The only exception is oxymorphone; this was introduced in the U. S. in 1959 but is mentioned here because it obviously is closely related to oxycodone.

in either case. These groups and the effects on activity by modifying them are listed in Table 17-2. The results of these and earlier studies[4] have not, in all cases, shown quantitatively the effects of simple modifications on the analgesic action of morphine. However, they do indicate in which direction the activity is apt to go. The studies are far more comprehensive than Table 17-2 indicates, and the conclusions depend on more than one pair of compounds in most cases.

Unfortunately, these studies on morphine did not provide the answer to the elimination of addiction potentialities from these compounds. In fact, the studies

suggested that any modification bringing about an increase in the analgesic activity caused a concomitant increase in addiction liability.

The second phase of the studies, engaged in principally by Mosettig and Eddy,[3] had to do with the attempted synthesis of substances with central narcotic and, especially, analgesic action. It is obvious that the morphine molecule contains in its makeup certain well-defined types of chemical structures. Among these are the phenanthrene nucleus, the dibenzofuran nucleus and, as a variant of the latter, carbazole. These synthetic studies, although extensive and inter-

TABLE 17-2

SOME STRUCTURAL RELATIONSHIPS IN THE MORPHINE MOLECULE

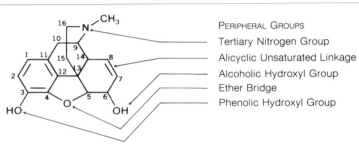

PERIPHERAL GROUPS
— Tertiary Nitrogen Group
— Alicyclic Unsaturated Linkage
— Alcoholic Hydroxyl Group
— Ether Bridge
— Phenolic Hydroxyl Group

Peripheral Groups of Morphine	Modification (On Morphine Unless Otherwise Indicated)	Effects on Analgesic Activity* (Morphine or Another Compound as Indicated = 100)
Phenolic Hydroxyl	—OH→—OCH₃ (codeine)	15
	—OH→—OC₂H₅ (ethylmorphine)	10
	—OH→—OCH₂CH₂—N◯O (pholcodine)	1
Alcoholic Hydroxyl	—OH→—OCH₃ (heterocodeine)	500
	—OH→—OC₂H₅	240
	—OH→—OCOCH₃	420
	—OH→=O (morphinone)	37
	†—OH→=O (dihydromorphine to dihydromorphinone)	600 (dihydromorphine vs. dihydromorphinone)
	†—OH→=O (dihydrocodeine to dihydrocodeinone)	390 (dihydrocodeine vs. dihydrocodeinone)
	†—OH→—H (dihydromorphine to dihydrodesoxymorphine-D)	1000 (dihydromorphine vs. dihydrodesoxymorphine-D)
Ether Bridge	‡=C—O—CH—→=C—OH HCH— (dihydrodesoxymorphine-D to tetrahydrodesoxymorphine)	13 (dihydrodesoxymorphine-D vs. tetrahydrodesoxymorphine)
Alicyclic Unsaturated Linkage	—CH=CH—→—CH₂CH₂— (dihydromorphine)	120
	†—CH=CH—→—CH₂CH₂—(codeine to dihydrocodeine)	115 (codeine vs. dihydrocodeine)
Tertiary Nitrogen	N—CH₃ → N—H (normorphine)	5

(Continued)

* Percent ratio of the E.D.$_{50}$ of morphine (or other compound indicated) to the E.D.$_{50}$ of the compound as determined in mice. These conclusions have been adapted from data in references 3 and 4. For a wealth of additional tabular material the reader is urged to consult the original references.

† These represent cases in which, for various reasons, a direct comparison with morphine itself cannot be made. The alternative has been to compare the effect of modifying the group in a pair of compounds where the changes can be made. It is felt that the direction of change in analgesic activity at least can be determined in this way.

‡ See, however, discussion of N-methylmorphinan later in this chapter.

§ Although many of these derivatives possess morphine antagonism, it has been shown that many of them also possess analgesic activity in their own right. Indeed, the ability to antagonize morphine in the rat is used as a screening method to assure low addiction potential in man.[44]

Not included in the studies of Small et al. See reference 8.

TABLE 17-2

SOME STRUCTURAL RELATIONSHIPS IN THE MORPHINE MOLECULE (Continued)

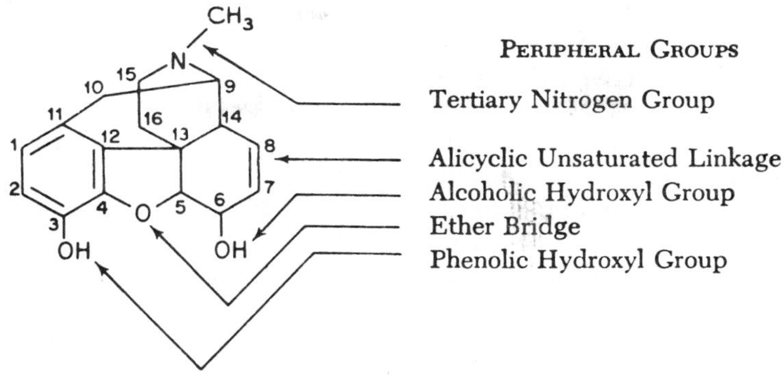

PERIPHERAL GROUPS

Tertiary Nitrogen Group

Alicyclic Unsaturated Linkage
Alcoholic Hydroxyl Group
Ether Bridge
Phenolic Hydroxyl Group

Peripheral Groups of Morphine	Modification (On Morphine Unless Otherwise Indicated)	Effects on Analgesic Activity* (Morphine or Another Compound as Indicated = 100)
	$N-CH_3 \rightarrow N-CH_2CH_2-C_6H_5$	1400
	§ $N-CH_3 \rightarrow N-R$	Reversal of activity (morphine antagonism); R = propyl, isobutyl, allyl, methallyl
	$N-CH_3 \rightarrow N^+(CH_3)_2\ Cl^-$	1 (strong curare action)
	Opening of nitrogen ring (morphimethine)	Marked decrease in action
Nuclear Substitution	Substitution of:	
	—NH$_2$ (most likely at position 2)	Marked decrease in action
	—Cl or —Br (at position 1)	50
	—OH (at position 14 in dihydromorphinone)	250 (dihydromorphinone vs. oxymorphone)
	—OH (at position 14 in dihydrocodeinone)	530 (dihydrocodeinone vs. oxycodone)
	#—CH$_3$ (at position 6)	280
	#—CH$_3$ (at position 6 in dihydromorphine)	33 (dihydromorphine vs. 6-methyldihydromorphine)
	#—CH$_3$ (at position 6 in dihydrodesoxymorphine-D)	490 (dihydrodesoxymorphine-D vs. 6-methyldihydrodesoxymorphine)
	#=CH$_2$ (at position 6 in dihydrodesoxymorphine-D)	600 (dihydrodesoxymorphine-D vs. 6-methylenedihydrodesoxymorphine)

(Continued)

* Percent ratio of the E.D.$_{50}$ of morphine (or other compound indicated) to the E.D.$_{50}$ of the compound as determined in mice. These conclusions have been adapted from data in references 3 and 4. For a wealth of additional tabular material the reader is urged to consult the original references.

† These represent cases in which, for various reasons, a direct comparison with morphine itself cannot be made. The alternative has been to compare the effect of modifying the group in a pair of compounds where the changes can be made. It is felt that the direction of change in analgesic activity at least can be determined in this way.

‡ See, however, discussion of N-methylmorphinan later in this chapter.

§ Although many of these derivatives possess morphine antagonism, it has been shown that many of them also possess analgesic activity in their own right. Indeed, the ability to antagonize morphine in the rat is used as a screening method to assure low addiction potential in man.[44]

Not included in the studies of Small et al. See reference 8.

esting, failed to provide significant findings and will not be discussed further in this text.

One of the more useful results of the investigations was the synthesis of 5-methyldihydromorphinone* (see Table 17-1). Although it possessed addiction liabilities, it was found to be a very potent analgesic with a minimum of the undesirable side-effects of morphine, such as emetic action and mental dullness.

Later, the high degree of analgesic activity demonstrated by morphine congeners in which the alicyclic ring is either reduced or methylated (or both) and the alcoholic hydroxyl at position 6 is absent has prompted the synthesis of related compounds possessing these features. These include 6-methyldihydromorphine and its dehydrated analog 6-methyl-Δ^6-desoxymorphine or methyldesorphine,[6] both of which

_____

* The location of the methyl substituent was originally assigned to position 7.[5]

have shown high potency. Also of interest were compounds reported by Rapoport and his co-workers[7]: morphinone; 6-methylmorphine; 6-methyl-7-hydroxy-, 6-methyl- and 6-methylenedihydrodesoxymorphine. In analgesic activity in mice, the last-named compound proved to be 82 times more potent, milligram for milligram, than morphine. Its therapeutic index ($T.I._{50}$) was 22 times as great as that of morphine.[8]

The structure-activity relationships of 14-hydroxymorphine derivatives have been reviewed recently, and several new compounds have been synthesized.[9] Of these, the dihydrodesoxy compounds possessed the highest degree of analgesic activity. Also, esters of 14-hydroxycodeine derivatives have shown very high activity.[10] For example, in rats, 14-cinnamyloxycodeinone was 177 times more active than morphine.

In 1963, Bentley and Hardy[11] reported the synthesis of a novel series of potent analgesics derived from the opium alkaloid thebaine. In rats the most active members of the series (I, $R_1 = H$, $R_2 = CH_3$, $R_3 = $ isoamyl; and I, $R_1 = COCH_3$, $R_2 = CH_3$, $R_3 = n\text{-}C_3H_7$) were found to be several thousand times stronger than morphine.[12] These compounds exhibited marked differences in activity of optical isomers, as well as other interesting structural effects. It was postulated that the more rigid molecular structure might allow them to fit the receptor surface better. Extensive structural and pharmacologic studies have been reported.[13] Some of the N-cyclopropylmethyl compounds are the most potent antagonists yet discovered and are currently being studied very intensively.

As indicated in Table 17-2, replacement of the N-methyl group in morphine by larger alkyl groups not only lowers analgesic activity but confers morphine antagonistic properties on the molecule (see below). In direct contrast to this effect, the N-phenethyl derivative has 14 times the analgesic activity of morphine. This enhancement of activity by N-aralkyl groups has wide application, as will be shown later.

It has been observed that the morphine antagonists, such as nalorphine, are also strong analgesics.[14] The similarity of the ethylenic double bond and the cyclopropyl group has prompted the synthesis of N-cyclopropylmethyl derivatives of morphine and its derivatives.[15] This substituent confers strong narcotic antagonistic activity in most cases, with variable effects on analgesic potency. The dihydronormorphinone derivative had only moderate analgesic activity.

Morphine modifications initiated by the Eisleb and Schaumann research

In 1938 Eisleb and Schaumann[2] reported the fortuitous discovery that a simple piperidine derivative, now known as meperidine, possessed analgesic activity. It was prepared as an antispasmodic, a property it shows as well. As the story is told, during the pharmacologic testing of meperidine in mice, it was observed to cause the peculiar erection of the tail known as the Straub reaction. Because the reaction is characteristic of morphine and its derivatives, the compound then was tested for analgesic properties and was found to be about one-fifth as active as morphine. This discovery led not only to the finding of an active analgesic but, far more important, it served as a stimulus to research workers. The status of research in analgesic compounds with an activity comparable with that of morphine was at a low ebb in 1938. Many felt that potent compounds could not be prepared, unless they were very closely related structurally to morphine. However, the demonstration of high potency in a synthetic compound that was related only distantly to morphine spurred the efforts of various research groups.[16,17]

The first efforts, naturally, were made upon the meperidine type of molecule in an attempt to enhance its activity further. It was found that replacement of the 4-phenyl group by hydrogen, alkyl, other aryl, aralkyl and heterocyclic groups reduced analgesic activity. Placement of the phenyl and ester groups at the 4-position of 1-methylpiperidine also gave optimum activity. Several modifications of this basic structure are listed in Table 17-3.

Among the simplest changes to increase activity is the insertion of a m-hydroxyl group on the phenyl ring. It is in the same relative position as in morphine. The effect is more pronounced on the keto compound (A-4) than on meperidine (A-1). Ketobemidone is equivalent to morphine in activity and was widely used.

More significantly, Jensen and co-workers[18] discovered that replacement of the carbethoxyl group in meperidine by acyloxyl groups gave better analgesic, as well as spasmolytic, activity. The "reversed" ester of meperidine, the propionoxy compound (A-6), was the most active, being 5 times as active as meperidine. These findings were validated and expanded upon by Lee et al.[19] In an extensive study of structural modifications of meperidine, Janssen and Eddy[20] concluded that the propionoxy compounds were always more ac-

TABLE 17–3

COMPOUNDS RELATED TO MEPERIDINE

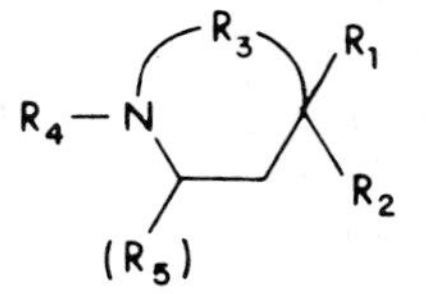

(R_5 = H except in trimeperidine, where it is CH_3)

Com-pound	Structure				Name (If Any)	Analgesic Activity* (Meperidine = 1)
	R_1	R_2	R_3	R_4		
A-1	$-C_6H_5$	$-COOC_2H_5$	$-CH_2CH_2-$	$-CH_3$	Meperidine	1.0
A-2	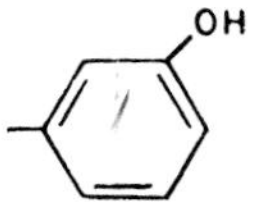 (phenol, OH)	$-COOC_2H_5$	$-CH_2CH_2$	$-CH_3$	Bemidone	1.5
A-3	$-C_6H_5$	$-COOCH(CH_3)_2$	$-CH_2CH_2-$	$-CH_3$	Properidine	15
A-4	$-C_6H_5$	$-\overset{\|}{\underset{O}{C}}-C_2H_5$	$-CH_2CH_2-$	$-CH_3$		0.5
A-5	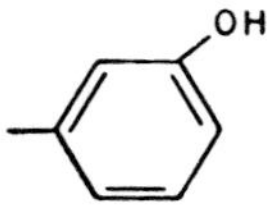 (phenol, OH)	$-\overset{\|}{\underset{O}{C}}-C_2H_5$	$-CH_2CH_2-$	$-CH_3$	Ketobemidone	6.2
A-6	$-C_6H_5$	$-O-\overset{\|}{\underset{O}{C}}-C_2H_5$	$-CH_2CH_2-$	$-CH_3$		5
A-7	$-C_6H_5$	$-O-\overset{\|}{\underset{O}{C}}-C_2H_5$	$-CH_2\overset{CH_3}{\overset{\|}{CH}}-$	$-CH_3$	Alphaprodine / Betaprodine	5 / 14
A-8	$-C_6H_5$	$-O-\overset{\|}{\underset{O}{C}}-C_2H_5$	$-CH_2\overset{CH_3}{\overset{\|}{CH}}-$	$-CH_3 (R_5 = CH_3)$	Trimeperidine	7.5
A-9	$-C_6H_5$	$-COOC_2H_5$	$-CH_2CH_2-$	$-CH_2CH_2C_6H_5$	Pheneridine	2.6
A-10	$-C_6H_5$	$-COOC_2H_5$	$-CH_2CH_2-$	$-CH_2CH_2-C_6H_4-NH_2$ 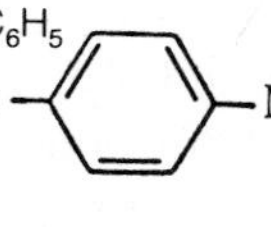	Anileridine	3.5
A-11	$-C_6H_5$	$-COOC_2H_5$	$-CH_2CH_2-$	$-(CH_2)_3-NH-C_6H_5$	Piminodine	55†
A-12	$-C_6H_5$	$-O-\overset{\|}{\underset{O}{C}}-C_2H_5$	$-CH_2CH_2-$	$-CH_2CH_2\overset{\|}{\underset{O-\overset{\|}{\underset{O}{C}}-C_2H_5}{CH}}C_6H_5$		1880†
A-13	$-C_6H_5$	$-COOC_2H_5$	$-CH_2CH_2-$	$-CH_2CH_2\overset{\|}{\underset{CN}{C}}(C_6H_5)_2$	Diphenoxylate	None

*Ratio of the E.D.$_{50}$ of meperidine to the E.D.$_{50}$ of the compound in mg./kg. administered subcutaneously in mice, based on data in references 4, 27, 33, 34.
† In rats. See reference 21.
‡ In rats. See reference 28.

(Continued)

TABLE 17–3

COMPOUNDS RELATED TO MEPERIDINE

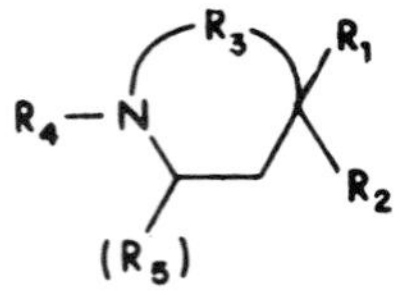

(R$_5$ = H except in trimeperidine,
where it is CH$_3$)

Com-pound	Structure				Name (If Any)	Analgesic Activity* (Meperidine = 1)
	R$_1$	R$_2$	R$_3$	R$_4$		
A-14	—C$_6$H$_4$-p-Cl	—OH	—CH$_2$CH$_2$—	—CH$_2$CH$_2$C(C$_6$H$_5$)$_2$ $\mid$ CN(CH$_3$)$_2$ $\parallel$ O	Loperamide	None
A-15	—C$_6$H$_5$	—COOC$_2$H$_5$	—CH$_2$CH$_2$CH$_2$—	—CH$_3$	Ethoheptazine	1
A-16	—C$_6$H$_5$	—O—CC$_2$H$_5$ ($\parallel$ O)	—CH— (CH$_3$)	—CH$_3$	Prodilidine	0.3
A-17	—H	—N—CC$_2$H$_5$ ($\parallel$ O) (C$_6$H$_5$)	—CH$_2$CH$_2$—	—CH$_2$CH$_2$C$_6$H$_5$	Fentanyl	940
A-18	—COOCH$_3$	—N—CC$_2$H$_5$ ($\parallel$ O) (C$_6$H$_5$)	—CH$_2$CH— (CH$_3$)	—CH$_2$CH$_2$C$_6$H$_5$	R 34,995	8400‡

(Continued)

*Ratio of the E.D.$_{50}$ of meperidine to the E.D.$_{50}$ of the compound in mg./kg. administered subcutaneously in mice, based on data in references 4, 27, 33, 34.
† In rats. See reference 21.
‡ In rats. See reference 28.

tive, usually about 2-fold, regardless of what group was attached to the nitrogen.

Lee[21] had postulated that the configuration of the propionoxy derivative (A-6) more closely resembled that of morphine, with the ester chain taking a position similar to that occupied by carbons 6 and 7 in morphine. His speculations were based on space models and certainly did not reflect the actual conformation of the nonrigid meperidine. However, he did arrive at the correct assumption that introduction of a methyl group into position 3 of the piperidine ring in the propionoxy compound would yield two isomers, one with activity approximating that of desomorphine and the other with lesser activity. One of the two diastereoisomers (A-7), betaprodine, has an activity in mice of about 9 times that of morphine and 3 times that of A-6. Beckett *et al.*[22] have established it to be the *cis* (methyl/phenyl) form. The *trans* form, alphaprodine, is twice as active as morphine. Resolution of the racemates shows one enantiomer to have the predominant activity. In man, however, the sharp differ-

ences in analgesic potency are not so marked. The *trans* form is marketed as the racemate. The significance of the 3-methyl has been attributed to discrimination of the enantiotropic edges of these molecules by the receptor. This is even more dramatic in the 3-allyl and 3-propyl isomers, where the α-*trans* forms are considerably more potent than the β-isomers, indicating 3-carbon substituents are not tolerated in the axial orientation. The 3-ethyl isomers are nearly equal in activity, further indicating that two or less carbons are more acceptable in the drug-receptor interaction.[23]

Until only the last few years it appeared that a small substituent, such as methyl, attached to the nitrogen was optimal for analgesic activity. This was believed to be true not only for the meperidine series of compounds but for all the other types as well. It is now well established that replacement of the methyl group by various aralkyl groups can increase activity markedly.[20] A few examples of this type of compound in the meperidine series are shown in Table 17-3. The phenethyl derivative (A-9) is seen to be about 3 times as active as meperidine (A-1). The *p*-amino congener, anileridine (A-10) is about 4 times more active. Piminodine, the phenylaminopropyl derivative (A-11), has 55 times the activity of meperidine in rats and in clinical trials is about 5 times as effective in man as an analgesic.[24] The most active meperidine-type compounds to date are the propionoxy derivative (A-12), which is nearly 2,000 times as active as meperidine, and the N-phenethyl analog of betaprodine, which is over 2,000 times as active as morphine.[22] Diphenoxylate (A-13), a structural hybrid of meperidine and methadone types, lacks analgesic activity although it reportedly suppresses the morphine abstinence syndrome in morphine addicts.[25,26] It is quite effective as an intestinal spasmolytic and is used for the treatment of diarrhea. Several other derivatives of it have been studied.[27] The pyrrolidine oxyamide derivative, diphenoximide, is currently being investigated as a heroin detoxification agent. The related *p*-chloro analog loperamide (A-14) has been shown to bind to opiate receptors in the brain but not to penetrate the blood–brain barrier sufficiently to produce analgesia.[28]

Another manner of modifying the structure of meperidine with favorable results has been the enlargement of the piperidine ring to the 7-membered hexahydroazepine (or hexamethylenimine) ring. As was the case in the piperidine series, the most active compound was the one containing a methyl group on position 3 of the ring adjacent to the quaternary carbon atom in the propionoxy derivative, that is, 1,3-dimethyl-4-phenyl-4-propionoxyhexahydroazepine, to which the name proheptazine has been given. In the study by Eddy and co-workers, previously cited, proheptazine was one of the more active analgesics included and had one of the highest addiction liabilities. The higher ring homolog of meperidine, ethoheptazine, has been marketed. Though originally thought to be inactive,[29] it is less active than codeine as an analgesic in man and has the advantages of being free of addiction liability and having a low incidence of side-effects.[30] Because of its low potency it is not very widely used.

Contraction of the piperidine ring to the 5-membered pyrrolidine ring has also been successful. The lower ring homolog of alphaprodine, prodilidene (A-16), is an effective analgesic, 100 mg. being equivalent to 30 mg. of codeine, but because of its potential abuse liability has not been marketed.[31]

A more unusual modification of the meperidine structure may be found in fentanyl (A-17), in which the phenyl and the acyl groups are separated from the ring by a nitrogen. It is a powerful analgesic, 50 times stronger than morphine in man, with minimal side-effects.[32] Its short duration of action makes it well suited for use in anesthesia.[33] It is marketed for this purpose in combination with a neuroleptic, droperidol. The *cis*-(−)-3-methyl analog with an ester group at the 4-position like meperidine (A-18) was found to be 8,400 times more potent than morphine as an analgesic. In addition, it has shown the highest binding affinity to isolated opiate receptors than other compounds tested.[28]

It should be recalled by the reader that when the nitrogen ring of morphine is opened, as in the formation of morphimethines, the analgesic activity virtually is abolished. On this basis, the prediction of whether a compound would or would not have activity without the nitrogen in a ring would be in favor of lack of activity or, at best, a low activity. The first report indicating that this might be a false assumption was based on the initial work of Bockmuehl and Ehrhart[34] wherein they claimed that the type of compound represented by B-1 in Table 17-4 possessed analgesic as well as spasmolytic properties. The Hoechst laboratories in Germany followed up this lead during World War II by preparing the ketones corresponding to these esters. Some of the compounds they prepared with high activity are represented by formulas B-2 through B-7. Compound B-2 is the well-known methadone. In the meperidine and bemidone types, the introduction of a *m*-hydroxyl group in the phenyl ring brought about slight to marked increase in activity, whereas the same operation with the methadone-type compound brought about a marked decrease in action. Phenadoxone (B-8), the morpholine analog of methadone, has been marketed in England. The piperidine analog, dipanone (q.v.), was under study in this country after successful results in England.

Methadone was first brought to the attention of American pharmacists, chemists and allied workers by the Kleiderer report[35] and by the early reports of Scott and Chen.[36] Since then, much work has been done on this compound, its isomer known as isomethadone, and allied compounds. The report by Eddy, Touchberry and Lieberman[37] covers most of the points concerning the structure-activity relationships of methadone. It was demonstrated that the *levo* isomer (B-3) of methadone (B-2) and the *levo* isomer of isomethadone (B-4) were twice as effective as their racemic mixtures. It is also of interest that all structural derivatives of methadone demonstrated a greater activity than the corresponding structural derivatives of isomethadone. In other words, the superiority of methadone over isomethadone seems to hold even through the derivatives. Conversely, the methadone series of compounds was always more toxic than the isomethadone group.

More extensive permutations, such as replacement of the propionyl group (R₃ in B-2) by hydrogen, hydroxyl or acetoxyl, led to decreased activity. In a series of amide analogs of methadone, Janssen and Jageneau[38] synthesized racemoramide (B-12), which is more active than methadone. The (+)-isomer, dextromoramide (B-13), is the active isomer and has been marketed. A few of the other modifications that have been carried out, together with the effect on analgesic activity relative to methadone, are described in Table 17-4, which comprises most of the methadone congeners that are or were on the market. It can be assumed that much deviation in structure from these examples will result in varying degrees of activity loss.

Particular attention should be called to the two phenyl groups in methadone and the sharply decreased action resulting by removal of one of them. It is believed that the second phenyl residue helps to lock the $-COC_2H_5$ group of methadone in a position to simulate again the alicyclic ring of morphine, even though the propionyl group is not a particularly rigid group. However, in this connection it is interesting to note that the compound with a propionoxy group in place of the propionyl group (R_3 in B-2) is without significant analgesic action.[17] In direct contrast with this is (+)-propoxyphene (B-14), which is a propionoxy derivative with one of the phenyl groups replaced by a benzyl group. In addition, it is an analog of isomethadone (B-4), making it an exception to the rule. This compound is lower than codeine in analgesic activity, possesses few side-effects and has a limited addiction liability.[39] Replacement of the dimethylamino group in (+)-propoxyphene with a pyrrolidyl group gives a compound that is nearly three-fourths as active as methadone and possesses morphinelike

TABLE 17-4

COMPOUNDS RELATED TO METHADONE*

Compound	R₁	R₂	R₃	R₄	Name	Isomer, Salt	Analgesic Activity† (Methadone = 1)
	R_1	R_2	R_3	R_4			
B-1	$-C_6H_5$	$-C_6H_5$	$-COO-Alkyl$	$-CH_2CH_2N(CH_5)_2$		——	0.17
B-2	$-C_6H_5$	$-C_6H_5$	$-C(=O)-C_2H_5$	$-CH_2CH(CH_3)N(CH_3)_2$	Methadone	(±)-HCl	1.0
B-3		Same as in B-2			Levanone	(−)-bitartr.	1.9
B-4	$-C_6H_5$	$-C_6H_5$	$-C(=O)-C_2H_5$	$-CH(CH_3)CH_2N(CH_3)_2$	Isomethadone	(±)-HCl	0.65
B-5	$-C_6H_5$	$-C_6H_5$	$-C(=O)-C_2H_5$	$-CH_2CH_2N(CH_3)_2$	Normethadone	HCl	0.44
B-6	$-C_6H_5$	$-C_6H_5$	$-C(=O)-C_2H_5$	$-CH_2CH(CH_3)N\langle\text{piperidyl}\rangle$	Dipanone	(±)-HCl	0.80
B-7	$-C_6H_5$	$-C_6H_5$	$-C(=O)-C_2H_5$	$-CH_2CH_2N\langle\text{piperidyl}\rangle$	Hexalgon	HBr	0.50
B-8	$-C_6H_5$	$-C_6H_5$	$-C(=O)-C_2H_5$	$-CH_2CH(CH_3)N\langle\text{morpholinyl}\rangle$	Phenadoxone	(±)-HCl	1.4

* Table adapted from Janssen, P. A. J.: Synthetic Analgesics, Part i, New York, Pergamon Press, 1960.
† Ratio of the E.D.$_{50}$ of methadone to the E.D.$_{50}$ of the compound in mg./kg. administered subcutaneously to mice as determined by the hot-plate method.

TABLE 17–4

COMPOUNDS RELATED TO METHADONE* *(Continued)*

Com-pound	Structure				Name	Isomer, Salt	Analgesic Activity† (Methadone = 1)
	R_1	R_2	R_3	R_4			
B-9	—C_6H_5	—C_6H_5	—CHC$_2$H$_5$, O—CCH$_3$ ‖ O	—CH$_2$CHN(CH$_3$)$_2$, CH$_3$	Alphacetylmethadol	α, ($\pm$)-HCl	1.3
B-10	Same as in B-9				Betacetylmethadol	β, ($\pm$)-HCl	2.3
B-11	—C_6H_5	—C_6H_5	—COOC$_2$H$_5$	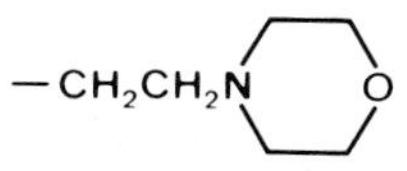 —CH$_2$CH$_2$N⟨O⟩	Dioxaphetyl butyrate	HCl	0.25
B-12	—C_6H_5	—C_6H_5	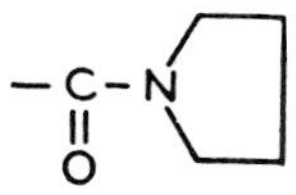 —C‖O—N⟨⟩	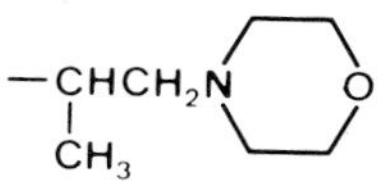 —CHCH$_2$N⟨O⟩, CH$_3$	Racemoramide	($+$)-base	3.6
B-13	Same as in B-12				Dextromoramide	($+$)-base	13
B-14	—C_6H_5	—CH$_2$C$_6$H$_5$	O—C—C$_2$H$_5$ ‖ O	—CHCH$_2$N(CH$_3$)$_2$, CH$_3$	Propoxyphene	($+$)-HCl	0.21

(Continued)

* Table adapted from Janssen, P. A. J.: Synthetic Analgesics, Part i, New York, Pergamon Press, 1960.

† Ratio of the E.D.$_{50}$ of methadone to the E.D.$_{50}$ of the compound in mg./kg. administered subcutaneously to mice as determined by the hot-plate method.

properties. The ($-$)-isomer of alphacetylmethadol (B-9), known as LAAM, is being intensely investigated as a long-acting substitute for methadone in the treatment of addicts.[40]

Morphine modifications initiated by Grewe

Grewe, in 1946, approached the problem of synthetic analgesics from another direction when he synthesized the tetracyclic compound which he first named morphan and then revised to N-methylmorphinan. The relationship of this compound to morphine is obvious.

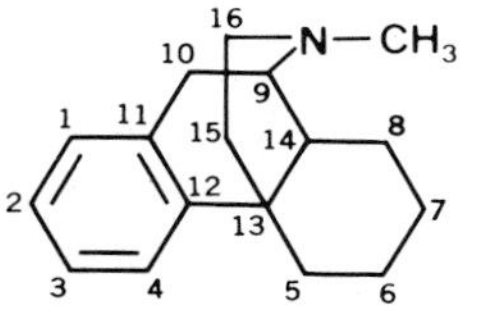

N-Methylmorphinan

N-Methylmorphinan differs from the morphine nucleus in the lack of the ether bridge between carbon atoms 4 and 5. Because this compound has been found to possess a high degree of analgesic activity, it suggests the nonessential nature of the ether bridge. The 3-hydroxyl derivative of N-methylmorphinan (racemorphan) was on the market and had an intensity and duration of action that exceeded that of morphine. The original racemorphan was introduced as the hydrobromide and was the ($\pm$)- or racemic form as obtained by synthesis. Since then, realizing that the levorotatory form of racemorphan was the analgesically active portion of the racemate, the manufacturers have successfully resolved the ($\pm$)-form and have marketed the *levo*-form as the tartrate salt (levorphanol). The *dextro*-form has also found use as a cough depressant (see dextromethorphan). The ethers and acylated derivatives of the 3-hydroxyl form also exhibit considerable activity. The 2- and 4-hydroxyl isomers are, not unexpectedly, without value as analgesics. Likewise, the N-ethyl derivative is lacking in activity and the N-allyl compound, levallorphan, is a potent morphine antagonist.

Eddy and co-workers[41] have reported on an extensive series of N-aralkylmorphinan derivatives. The effect of the N-aralkyl substitution was more dramatic in this series than it was in the case of morphine or meperidine. The N-phenethyl and N-*p*-aminophenethyl analogs of levorphanol were about 3 and 18 times, respectively, more active than the parent compound in analgesic activity in mice. The most potent member of the series was its N-β-furylethyl analog, which was nearly 30 times as active as levorphanol or 160 times as active as morphine. The N-acetophenone analog, levophenacylmorphan, was once under clinical investigation. In mice, it is about 30 times more active than morphine, and in man a 2-mg. dose is equivalent to 10 mg. of morphine in its analgesic response.[42] It has a much lower physical dependence liability than morphine.

The N-cyclopropylmethyl derivative of 3-hydroxymorphinan (cyclorphan), was reported to be a potent morphine antagonist capable of precipitating morphine withdrawal symptoms in addicted monkeys, indicating that it is nonaddicting.[15] Clinical studies have indicated that it is about 20 times stronger than morphine as an analgesic but has some undesirable side-effects, primarily hallucinatory in nature. However, the N-cyclobutyl derivative, butorphanol, possesses mixed agonist–antagonist properties and has been marketed as a potent analgesic.

Inasmuch as removal of the ether bridge and all the peripheral groups in the alicyclic ring in morphine did not destroy its analgesic action, May and co-workers[43] synthesized a series of compounds in which the alicyclic ring was replaced by one or two methyl groups. These are known as benzomorphan derivatives, or, more correctly, as benzazocines. They may be represented by the following formulae:

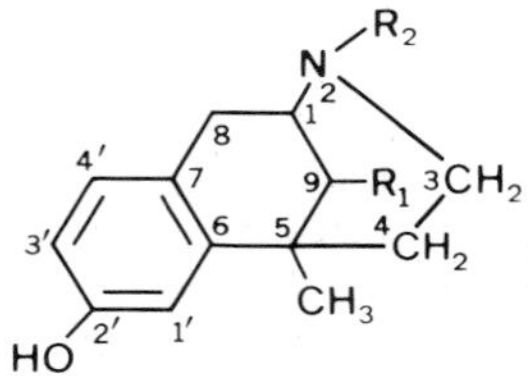

The trimethyl compound (II, $R_1 = R_2 = CH_3$) is about 3 times more potent than the dimethyl (II, $R_1 = H$, $R_2 = CH_3$). The N-phenethyl derivatives have almost 20 times the analgesic activity of the corresponding N-methyl compounds. Again, the more potent was the one containing the two ring methyls

(II, $R_1 = CH_3$, $R_2 = CH_2CH_2C_6H_5$). Deracemization proved the *levo*-isomer of this compound to be more active, being about 20 times as potent as morphine in mice. The (±)-form, phenazocine, was on the market but has been removed.

May and his co-workers[44] have demonstrated an extremely significant difference between the two isomeric N-methyl benzomorphans in which the alkyl in the 5-position is *n*-propyl (R_1) and the alkyl in the 9-position is methyl (R_2). These have been termed the α-isomer and the β-isomer and have the groups oriented as indicated. The isomer with the alkyl *cis* to the phenyl has been shown to possess analgesic activity (in mice) equal to that of morphine but has little or no capacity to suppress withdrawal symptoms in addicted monkeys. On the other hand, the *trans*-isomer has one of the highest analgesic potencies among the benzomorphans but is quite able to suppress morphine withdrawal symptoms. Further separation of properties is found between the enantiomers of the *cis*-isomer. The (+)-isomer has weak analgesic activity but a high physical dependence capacity. The (−)-isomer is a stronger analgesic without the dependence capacity, and possesses antagonistic activity.[45] The same was found true with the 5,9-diethyl and 9-ethyl-5-phenyl derivatives. The (−)-*trans*-5,9-diethyl isomer was similar except it had no antagonistic properties. This demonstrates that it is possible to divorce analgesic activity comparable to morphine from addiction potential. The fact that N-methyl compounds have shown antagonistic properties is of great interest as well. The most potent of these is the benzomorphan with an α-methyl and β-3-oxoheptyl group at position 9. The (−)-isomer shows greater antagonistic activity than naloxone and is still three times more potent than morphine as an analgesic.[46]

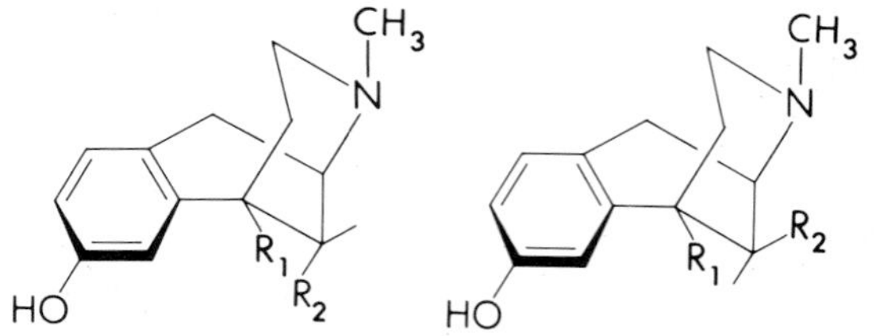

α-isomer (cis)   β-isomer (trans)

An extensive series of the antagonist-type analgesics in the benzomorphans has been reported.[47] Of these, pentazocine (II, $R_1 = CH_3$, $R_2 = CH_2CH = C(CH_3)_2$) and cyclazocine (II, $R_1 = CH_3$, $R_2 = CH_2$—cyclopropyl) have proved to be the most interesting. Pentazocine has about half the analgesic activity of morphine, with a lower incidence of side-effects.[48] Its addiction liability is much lower, approximating that of propoxyphene.[49] It is currently available in parenteral and tablet form. Cyclazocine is a strong morphine antagonist, showing about 10 times the analgesic activity of morphine.[50] It is currently being investigated as an analgesic and for the treatment of heroin addiction.

It was mentioned previously that replacement of the N-methyl group in morphine by larger alkyl groups lowered analgesic activity. In addition, these compounds were found to counteract the effect of morphine and other morphinelike analgesics and are thus known as *narcotic antagonists*. The reversal of activity increases from ethyl to propyl to allyl, with the cyclopropylmethyl usually being maximal. This property was found to be true not only in the case of morphine but with other analgesics as well. N-Allyl-normorphine (nalorphine) was the first of these but has been taken off the market due to side-effects. Levallorphan, the corresponding allyl analog of levorphanol, and naloxone, N-allylnoroxymorphone, are the two narcotic antagonists presently on the market. Naloxone appears to be a pure antagonist with no morphine- or nalorphine-like effects. It also blocks the effects of other antagonists. These drugs are used to prevent, diminish or abolish many of the actions or the side-effects encountered with the narcotic analgesics. Some of these are respiratory and circulatory depression, euphoria, nausea, drowsiness, analgesia and hyperglycemia. They are thought to act by competing with the analgesic molecule for attachment at its, or a closely related, receptor site. As indicated previously, the observation that some narcotic antagonists, which are devoid of addiction liability, are also strong analgesics has spurred considerable interest in them.[14] The N-cyclopropylmethyl compounds mentioned are the most potent antagonists, but appear to produce psychotomimetic effects and may not be useful as analgesics. However, one of these, buprenorphine, has shown an interesting profile and has been introduced in Europe and is currently under study in the United States as a potent analgesic.[51] It has also been studied as a possible treatment for narcotic addicts.[52]

Very intensive efforts are under way to develop narcotic antagonists that can be used to treat narcotic addiction.[53] The continuous administration of an antagonist will block the euphoric effects of heroin, thus aiding rehabilitation of an addict. The cyclopropylmethyl derivative of naloxone, naltrexone, is the antagonist that is most widely being studied. The oral dose of 100 to 150 mg. three times a week is sufficient to block several usual doses of heroin.[54] Long-acting preparations are also under study.[55]

Much research, other than that described in the foregoing discussion, has been carried out by the systematic dissection of morphine to give a number of interesting fragments. These approaches have not pro-

duced important analgesics yet; therefore, they are not discussed in this chapter. However, the interested reader may find a key to this literature from the excellent reviews of Eddy,[4] Bergel and Morrison,[17] and Lee.[21]

STRUCTURE-ACTIVITY RELATIONSHIPS

Several reviews on the relationship between chemical structure and analgesic action have been published.[4,25,56-64] Only the major conclusions will be considered here, and the reader is urged to consult these reviews for a more complete discussion of the subject.

From the time Small and co-workers started their studies on the morphine nucleus to the present, there has been much light shed on the structural features connected with morphinelike analgesic action. In a very thorough study made for the United Nations Commission on Narcotics in 1955, Braenden and co-workers[57] found that the features possessed by all known morphine-like analgesics were:

1. A tertiary nitrogen, the group on the nitrogen being relatively small.
2. A central carbon atom, of which none of the valences is connected with hydrogen.
3. A phenyl group or a group isosteric with phenyl, which is connected to the central carbon atom.
4. A 2-carbon chain separating the central carbon atom from the nitrogen for maximal activity.

From the foregoing discussion it is evident that a number of exceptions to these generalizations may be found in the structures of compounds that have been synthesized in the last several years. Eddy[25] has discussed the more significant exceptions.

In regard to the first feature mentioned above, extensive studies of the action of normorphine have shown it to possess analgesic activity in the order of morphine. In man, it is about one-fourth as active as morphine when administered intramuscularly but was slightly superior to morphine when administered intracisternally. On the basis of the last-mentioned effect, Beckett and his co-workers[65] postulated that N-dealkylation was a step in the mechanism of analgesic action. This has been questioned.[66] Additional studies indicate that dealkylation does occur in the brain, although its exact role is not clear.[67] It is clear, from the previously discussed N-aralkyl derivatives, that a small group is not necessary.

Several exceptions to the second feature have been synthesized. In these series, the central carbon atom has been replaced by a tertiary nitrogen. They are related to methadone and have the following structures:

III

IV

Diampromide (III) and its related anilides have potencies that are comparable to those of morphine;[68] however, they have shown addiction liability and have not appeared on the market. The closely related cyclic derivative fentanyl (A-17, Table 17-3), is used in surgery. The benzimidazoles, such as etonitazene (IV), are very potent analgesics, but show the highest addiction liabilities yet encountered.[69]

Possibly an exception to feature 3, and the only one that has been encountered, may be the cyclohexyl analog of A-6 (Table 17-3), which has significant activity.

Eddy[25] mentions two possible exceptions to feature 4 in addition to fentanyl.

As a consequence of the many studies on molecules of varying types that possess analgesic activity, it became increasingly apparent that activity was associated not only with certain structural features but also with the size and the shape of the molecule. The hypothesis of Beckett and Casy[70] has dominated thinking for a number of years in the area of stereochemical specificity of these molecules. They noted initially that the more active enantiomers of the methadone and thiambutene type analgesics were related configurationally to R-alanine. This suggested to them that a stereoselective fit at a receptor could be involved in analgesic activity. In order to depict the dimensions of an analgesic receptor, they selected morphine (because of its semirigidity and high activity) to provide them with information on a complementary receptor. The features that were thought to be essential for proper receptor fit were:

1. A basic center able to associate with an anionic site on the receptor surface;
2. A flat aromatic structure, coplanar with the basic center, allowing for van der Waal's bonding to a flat surface on the receptor site to reinforce the ionic bond; and
3. A suitably positioned projecting hydrocarbon moiety forming a 3-dimensional geometric pattern with the basic center and the flat aromatic structure.

These features were selected, among other reasons, because they are present in N-methylmorphinan which may be looked upon as a "stripped down" morphine, i.e., morphine without the characteristic peripheral groups (except for the basic center). Inasmuch as N-methylmorphinan possessed significant activity of the morphine type, it was felt that these three features were the fundamental ones determining activity and that the peripheral groups of morphine acted essentially to modulate the activity.

In accord with the above postulations, Beckett and Casy,[70] proposed a complementary receptor site (Fig. 17-1) and suggested ways[71,72] in which the known active molecules could be adapted to it. Subsequent to their initial postulation it was demonstrated that natural (−)-morphine was related configurationally to methadone and thiambutene, a finding that lent weight to the hypothesis. Fundamental to their pro-

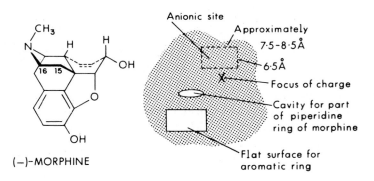

FIG. 17-1. *Diagram of the surface of the analgesic receptor site with the corresponding lower surface of the drug molecule. The 3-dimensional features of the molecule are shown by the bonds: —, - - -, and — — which represent in front of, behind, and in the plane of the paper, respectively. (Gourley, D. R. H., in Jucker, E. (ed.): Progress in Drug Research, vol. 7, p. 36, Basel, Birkhauser, 1964)*

posal, of course, was that such a receptor was essentially inflexible and that a lock-and-key type situation existed. Recently, the unnatural (+)-morphine was synthesized and shown to be inactive.[73]

Although the above hypothesis appeared to fit the facts quite well and was a useful hypothesis for a number of years, it now appears that certain anomalies exist which cannot be accommodated by it. For example, the more active enantiomer of α-methadol is not related configurationally to R-alanine, in contrast to the methadone and thiambutene series. This is also true for the carbethoxy analog of methadone (V) and for diampromide (III) and its analogs. Another factor that was implicit in considering a proper receptor fit for the morphine molecule and its congeners was that the phenyl ring at the 4-position of the piperidine moiety should be in the axial orientation for maximum activity. The fact that structure VI has only an equatorial phenyl group, yet possesses activity equal to that of morphine would seem to cast doubt on the necessity for axial orientation as a receptor-fit requirement.

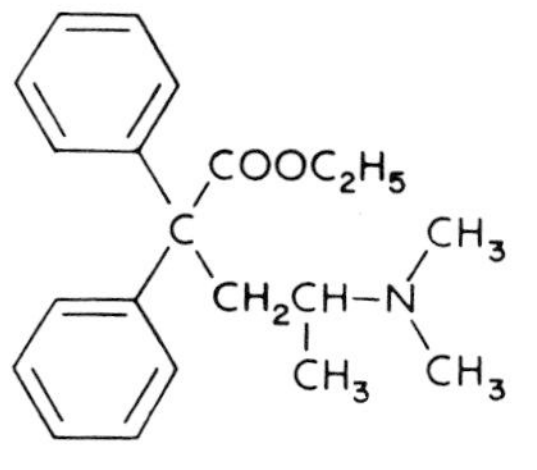

V

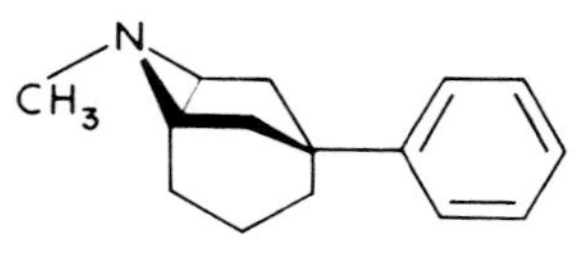

VI

In view of the difficulty of accepting Beckett and Casy's hypothesis as a complete picture of analgesic-receptor interaction, Portoghese[74,76] has offered an alternative hypothesis. This hypothesis is based in part on the established ability of enzymes and other types of macromolecules to undergo conformational changes[77,78] on interaction with small molecules (substrates or drugs). The fact that configurationally unrelated analgesics can bind and exert activity is interpreted as meaning that more than one mode of binding may be possible at the same receptor. Such different modes of binding may be due to differences in positional or conformational interactions with the receptor. The manner in which the hypothesis can be adapted to the methadol anomaly is illustrated in

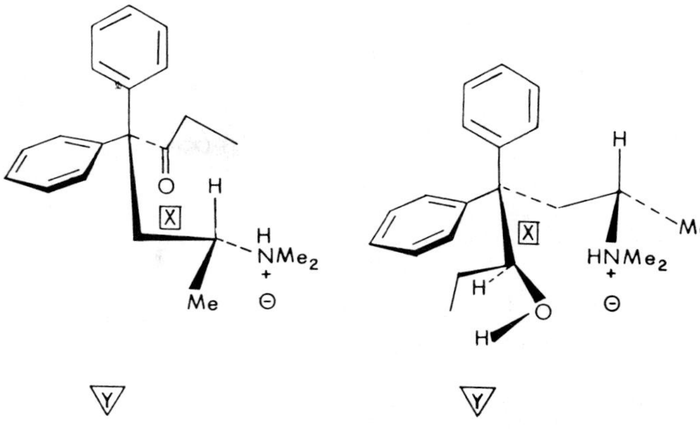

FIG. 17-2. *An illustration of how different polar groups in analgesic molecules may cause inversion in the configurational selectivity of an analgesic receptor. A hydrogen bonding moiety is denoted by x. Y represents a site which is capable of being hydrogen bonded.*

Figure 17-2. Portoghese, after considering activity changes in various structural types (i.e., methadones, meperidines, prodines, etc.) as related to the identity of the N-substituent, noted that in certain series there was a parallelism in the direction of activity when identical changes in N-substituents were made. In others there appeared to be a nonparallelism. He has interpreted parallelism and nonparallelism, respectively, as being due to similar and to dissimilar modes of binding. As viewed by this hypothesis, while it is still a requirement that analgesic molecules be bound in a fairly precise manner, it nevertheless liberalizes the concept of binding in that a response may be obtained by two different molecules binding stereoselectively in two different precise modes at the same receptor. A schematic representation of such different possible binding modes is shown in Figure 17-3. This

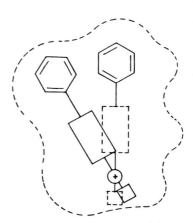

FIG. 17-3. *A schematic illustration of two different molecular modes of binding to a receptor. The protonated nitrogen is represented by ⊕. The square denotes an N-substituent. The anionic sites lies directly beneath ⊕.*

representation will aid in visualizing the meaning of *similar* and *dissimilar* binding modes. If two different analgesiophores* bearing identical N-substituents are positioned on the receptor surface so that the N-substituent occupies essentially the same position, a similar pharmacologic response may be anticipated. Thus, as one proceeds from one N-substituent to another the response should likewise change, resulting in a parallelism of effect. On the other hand, if two different analgesiophores are bound to the receptor so that the N-substituents are not arranged identically, one may anticipate nonidentical responses on changing the N-substituent, i.e., a nonparallel response. From the preceding statements, as well as the diagram, it is not to be inferred that the analgesiophore necessarily will be bound in the identical position within a series. They do, however, suggest that, in series with parallel activities, the pairs being compared will be bound identically to produce the parallel effect. Interestingly, when binding modes are similar he has been able to demonstrate the existence of a linear free energy relationship. There also is the possibility that more than one receptor is involved.

Considerable evidence is now available to demonstrate that multiple receptors exist. Martin has characterized and named these by responses from probe molecules, i.e., μ (mu) receptors for morphine specific effects, σ (sigma) for cyclazocine, and κ (kappa) for ketocyclazocine.[79] Various combinations of these in different tissues could be responsible for the varying effects observed.[80,81]

Although this hypothesis is new, it appears to embrace virtually all types of analgesic molecules presently known,† and it will be interesting to see whether it is of further general applicability as other molecules with activity are devised.

Another of the highly important developments in structure-activity correlations has been the development of highly active analgesics from the N-allyl type derivatives that once were thought to be only morphine antagonists and devoid of analgesic properties. Serendipity played a major role in this discovery: Lasagna and Beecher,[82] in attempting to find some "ideal" ratio of antagonist (N-allylnormorphine, nalorphine) to analgesic (morphine) so as to maintain the desirable effects of morphine while minimizing the undesirable ones, discovered that nalorphine was, milligram for milligram, as potent an analgesic as morphine. Unfortunately, nalorphine has depersonalizing

and psychotomimetic properties which preclude its use clinically as a pain reliever. However, the discovery led to the development of related derivatives such as pentazocine and cyclazocine. Pentazocine has achieved some success in providing an analgesic with low addiction potential, although it is not totally free of some of the other side-effects of morphine. The pattern of activity in these and other N-allyl and N-cyclopropylmethyl derivatives indicates that most potent antagonists possess psychotomimetic activity, whereas the weak antagonists do not. It is from this latter group that useful analgesics, such as pentazocine, butorphanol, and nalbuphine, have been found. The latter two possess N-cyclobutylmethyl groups.

What structural features are associated with antagonistlike activity has become uncertain. The N-allyl and dimethylallyl substituent does not always confer antagonist properties. This is true in the meperidine and thevinol series. Demonstration of antagonistlike properties by specific isomers of N-methyl benzomorphans has raised still further speculation. The exact mechanisms by which morphine and the narcotic antagonists act are not clearly defined, and a great amount of research is presently being carried on. Recent reviews and symposia may be consulted for further discussions of these topics.[53,83-85]

A further problem also is demonstrated in the testing for analgesic activity. As noted above, the analgesic activity of the antagonists was not apparent from animal testing but was observed only in man. Screening in animals can be used to assess the antagonistic action, which indirectly indicates possible analgesic properties in humans.[86]

It has been customary in the area of analgesic agents to attribute differences in their activities to structurally related differences in their receptor interactions. This rather universal practice continues in spite of early warnings and recent findings. It now appears clear that much of the differences in relative analgesic potencies can be accounted for on the basis of pharmacokinetic or distribution properties.[84] For example, a definite correlation was found between the partition coefficients and the intravenous analgesic data for 17 agents of widely varying structures.[87] Usual test methods do not help define which structural features are related to receptor and which to distribution phenomena. Studies directed toward making this distinction are using the measurement of actual brain and plasma levels[88,89] or direct injection into the ventricular area,[87] the measurement of ionization potentials and partition coefficients,[90] and the application of molecular orbital theories and quantum mechanics.[85,91-93] These are providing valuable insight in regard to the designing of new and more successful agents. All of the work described above had strongly

---

* The analgesic molecule less the N-substituent, i.e., the portion of the molecule giving the characteristic analgesic response.

† Two possible exceptions are 4-propionoxy-4-cyclohexyl-1-methylpiperidine and 1-tosyl-4-phenyl-4-ethylsulfone piperidine. (Helv. chim. acta 36:819, 1953)

suggested for years the existence of specific binding sites or *receptors* in brain and other tissue. The demonstration of the high degree of steric and structural specificity in the action of the opiates and their antagonists led many investigators to search for such receptors.[94,95] Thus in 1971, Goldstein and co-workers demonstrated stereospecific binding in brain homogenates.[96] This was quickly followed by refinements and further discoveries by Simon, Terenius, and Snyder.[97-99] These receptor binding studies have now become a routine assay for examining structure–activity relationships.

In addition to the binding studies, considerable attention continued on the use of in vitro models, in particular the isolated guinea-pig ileum, rat jejunum, and mouse vas deferens.[100] While working with these preparations, Hughes was the first to discover the existence of an endogenous factor from pig brains that possessed opiatelike properties.[101] This factor, given the name *enkephalin,* was found to consist of two pentapeptides, called *methonine-* or *met-enkephalin* and *leucine-* or *leu-enkephalin.* These two enkephalins have subsequently been shown to exist in all animals including humans and to possess all morphinelike properties. It was also observed that they exist as segments of a pituitary hormone, the 91 amino acid β-lipotropin, which is cleaved selectively to release specific segments that have now been found to have functions within the body. Thus, segment 61 to 65 is met-enkephalin, 61 to 76 is α-endorphin, 61 to 77 is γ-endorphin, and probably the most important, 61 to 91 is β-endorphin.

H-Glu-Leu-Thr-Gly-Gln-Arg-Leu-Arg-Gln-Gly-Asp-Pro-Asn-Ala
Leu-Ser-His-Glu-Leu-Ala-Asn-Pro-Gly-Glu-Gly-Asp-Asn-Ala-Gly
Leu-Ala-Asp-Leu-Val-Ala-Ala-Glu-Lys-Lys-Asp-Glu-Gly-Pro-Tyr
Arg-Lys-Asp-Lys-Pro-Pro-Ser-Gly-Trp-Asg-Phe-His-Glu-Met-Arg

Methionine Enkephalin
Tyr 61
Gly-Gly
Met-Phe 65
Thr-Ser-Glu-Lys-Ser 70
Val-Leu-Pro-Thr-Gln
Thr-Leu-Phe-Lys-Asn 76 80
Asn-Lys-Ile-Ile-Ala 85
Ala-Tyr-Lys-Lys-Gly-Glu-OH 91

β-Endorphin

β-Lipotropin, β-Endorphin, Methionine Enkephalin Relationships

The last endorphin (short for *end*ogenous *morphine*) has 20 times the analgesic potency of morphine when injected into rat brain. It has also been shown that these substances can produce tolerance and dependence. Although there have been numerous suggestions made concerning possible physiological function, e.g., neuromodulators or transmitters for natural pain relief, schizophrenia, etc., there is still no proof for

such roles. It appears that this proof may just be a question of time.[102,103]

It is obvious that all of these techniques will lead to new concepts and understanding of the processes of analgesia, tolerance and dependence. It is hoped that learning how these mechanisms operate will aid in the design and development of better analgesics.

PRODUCTS*

Morphine. This alkaloid was isolated first in 1803 by Derosne, but the credit for isolation generally goes to Serturner (1803) who first called attention to the basic properties of the substance. Morphine, incidentally, was the first plant base isolated and recognized as such. Although intensive research was carried out with respect to the structure of morphine, it was only in 1925 that Gulland and Robinson[105] postulated the currently accepted formula. The total synthesis of morphine finally was effected by Gates and Tschudi[106] in 1952, thus confirming the Gulland and Robinson formula.

Morphine is obtained only from the opium poppy, *Papaver somniferum,* either from opium, the resin obtained by lancing the unripe pod, or from poppy straw. The latter process is being favored as it helps to eliminate illicit opium from which heroin is readily produced. It occurs in opium in amounts varying from 5 to 20 percent (*U.S.P.* requires not less than 9.5%). It is isolated by various methods, of which the final step is usually the precipitation of morphine from an acid solution by using excess ammonia. The precipitated morphine then is recrystallized from boiling alcohol.

The free alkaloid occurs as levorotatory, odorless, white, needlelike crystals possessing a bitter taste. It is almost insoluble in water (1:5,000,† 1:1,100 at boiling point), ether (1:6,250) or chloroform (1:1,220). It is somewhat more soluble in ethyl alcohol (1:210, 1:98 at boiling point). Because of the phenolic hydroxyl group, it is readily soluble in solutions of alkali or alkaline earth metal hydroxides.

Morphine is a mono-acidic base and readily forms water-soluble salts with most acids. Thus, because morphine itself is so poorly soluble in water, the salts are the preferred form for most uses. Numerous salts have been marketed, but the ones in use are principally the sulfate and, to a lesser extent, the hydrochlo-

---

* In General Circular No. 253, March 10, 1960, the Treasury Department, Bureau of Narcotics, Washington D.C. 20525 has published an extensive listing of narcotics of current interest in the drug trade. This listing will be much more extensive than the following monographic coverage of compounds primarily of interest to American pharmacists.

† In this chapter a solubility expressed as (1:5,000) indicates that 1 g. is soluble in 5,000 ml. of the solvent at 25°. Solubilities at other temperatures will be so indicated.

ride. Morphine acetate, which is freely soluble in water (1:2.5) but is relatively unstable, has been used to a limited extent in liquid antitussive combinations.

Many writers have pointed out the "indispensable" nature of morphine, based on its potent analgesic properties toward all types of pain. It is properly termed a narcotic analgesic. However, because it causes addiction so readily, it should be used only in those cases where other pain-relieving drugs prove to be inadequate. It controls pain caused by serious injury, neoplasms, migraine, pleurisy, biliary and renal colic and numerous other causes. It often is administered as a preoperative sedative, together with atropine to control secretions. With scopolamine, it is given to obtain the so-called "twilight sleep." This effect is used in obstetrics, but care is exercised to prevent respiratory depression in the fetus. It is worthy of note that the toxic properties of morphine are much more evident in young and old people.

Morphine Hydrochloride. This salt may be prepared by neutralizing a hot aqueous suspension of morphine with diluted hydrochloric acid and then concentrating the resultant solution to crystallization. It is no longer commercially available.

It occurs as silky, white, glistening needles or cubical masses or as a crystalline, white powder. The hydrochloride is soluble in water (1:17.5, 1:0.5 at boiling point), alcohol (1:52, 1:46 at 60°) or glycerin, but it is practically insoluble in ether or chloroform. Solutions have a pH of approximately 4.7 and may be sterilized by boiling.

Its uses are the same as those of morphine.

The usual oral and subcutaneous dose is 15 mg. every 4 hours as needed, with a suggested range of 8 to 20 mg.

Morphine Sulfate U.S.P. This morphine salt is prepared in the same manner as the hydrochloride, i.e., by neutralizing morphine with diluted sulfuric acid.

It occurs as feathery, silky, white crystals, as cubical masses of crystals or as a crystalline, white powder. Although it is a fairly stable salt, it loses water of hydration and darkens on exposure to air and light. It is soluble in water (1:16, 1:1 at 80°), poorly soluble in alcohol (1:570, 1:240 at 60°) and insoluble in chloroform or ether. Aqueous solutions have a pH of approximately 4.8 and may be sterilized by heating in an autoclave.

This common morphine salt is used widely in England by oral administration for the management of pain in cancer patients. It has largely replaced Brompton's mixture, a combination of heroin and cocaine in chloroform water. In the United States this preparation has become mistakenly popular, substituting morphine sulfate for the heroin. Moreover, Twycross

has advised that the stimulant cocaine is contraindicated because it interferes with sleep.[107]

Codeine U.S.P. Codeine is an alkaloid which occurs naturally in opium, but the amount present is usually too small to be of commercial importance. Consequently, most commercial codeine is prepared from morphine by methylating the phenolic hydroxyl group. The methylation methods usually are patented procedures and make use of reagents such as diazomethane, dimethyl sulfate and methyl iodide. Newer methods are based on its synthesis from thebaine, which makes it possible to use *P. bracteatum* as a natural source (see beginning of chapter).

It occurs as levorotatory, colorless, efflorescent crystals or as a white, crystalline powder. It is light-sensitive. Codeine is slightly soluble in water (1:120) and sparingly soluble in ether (1:50). It is freely soluble in alcohol (1:2) and very soluble in chloroform (1:0.5).

Codeine is a mono-acidic base and readily forms salts with acids, the most important salts being the sulfate and the phosphate. The acetate and the methylbromide derivatives have been used to a limited extent in cough preparations. The free base is used little as compared with the salts, its greatest use being in Terpin Hydrate and Codeine Elixir U.S.P.

The general pharmacologic action of codeine is similar to that of morphine but, as previously indicated, it does not possess the same degree of analgesic potency. Lasagna[108] comments on the status of the drug as follows:

> "Despite codeine's long use as an analgesic drug, it is amazing how little reliable information there is about its efficacy, particularly by the parenteral route."

There are studies that indicate that 30 to 120 mg. of codeine are considerably less efficient parenterally than 10 mg. of morphine and the usual side-effects of morphine—respiratory depression, constipation, nausea, etc.—are apparent. Codeine is less effective orally than parenterally, and it has been stated by Houde and Wallenstein[109] that 32 mg. of codeine is about as effective as 650 mg. of aspirin in relieving terminal cancer pain. However, it also has been recognized that combinations of aspirin and codeine act additively as analgesics, thus giving some support to the common practice of combining the two drugs.

Codeine has a reputation as an antitussive, depressing the cough reflex, and is used in many cough preparations. It is one of the most widely used morphine-like analgesics. It is considerably less addicting than morphine and in the usual doses respiratory depression is negligible, although an oral dose of 60 mg. will cause such depression in a normal person. It is probably true that much of codeine's reputation as an antitussive rests on subjective impressions rather than on objective studies. The average 5-ml. dose of Terpin Hydrate and Codeine Elixir contains 10 mg. of codeine. This preparation and many like it have been sold over the counter as exempt narcotic preparations. However, abuse or misuse of these preparations has led to their being placed on a prescription-only status in many states.

A combination of codeine and papaverine (Copavin®) was advocated by Diehl[110] for the prophylaxis and treatment of common colds. When administered at the first signs of a cold, it was claimed to have aborted the cold in a significant percentage of the cases.

Codeine Phosphate U.S.P. This salt may be prepared by neutralizing codeine with phosphoric acid and precipitating the salt from aqueous solution with alcohol.

Codeine phosphate occurs as fine, needle-shaped, white crystals or as a white, crystalline powder. It is efflorescent and is sensitive to light. It is freely soluble in water (1:2.5, 1:0.5 at 80°) but less soluble in alcohol (1:325, 1:125 at boiling point). Solutions may be sterilized by boiling.

Because of its high solubility in water as compared with the sulfate, this salt is used widely. It is often the only salt of codeine stocked by pharmacies and is dispensed, rightly or wrongly, on all prescriptions calling for either the sulfate or the phosphate.

Codeine Sulfate U.S.P. Codeine sulfate is prepared by neutralizing an aqueous suspension of codeine with diluted sulfuric acid and then effecting crystallization.

It occurs as white crystals, usually needlelike, or as a white, crystalline powder. The salt is efflorescent and light-sensitive. It is soluble in water (1:30, 1:6.5 at 80°), much less soluble in alcohol (1:1,280) and insoluble in ether or chloroform.

This salt of codeine is prescribed frequently but is not as suitable as the phosphate for liquid preparations. Solutions of the sulfate and the phosphate are incompatible with alkaloidal reagents and alkaline substances.

Ethylmorphine Hydrochloride, dionin. This synthetic compound is analogous to codeine, but instead of being the methyl ether it is the ethyl ether. Ethylmorphine may be prepared by treating an alkaline alcoholic solution of morphine with diethyl sulfate. The hydrochloride is obtained from the free base by neutralizing it with diluted hydrochloric acid.

The salt occurs as a microcrystalline, white or faintly yellow, odorless powder. It has a slightly bitter taste. It is soluble in water (1:10) and in alcohol (1:25) but only slightly soluble in ether and in chloroform.

The systemic action of this morphine derivative is

intermediate between those of codeine and morphine. It has analgesic qualities and sometimes is used for the relief of pain. As a depressant of the cough reflex, it is as effective as codeine and, for this reason, is found in some commercial cough syrups. However, the chief use of this compound is in ophthalmology. By an irritant dilating action on vessels, it stimulates the vascular and lymphatic circulation of the eye. This action is of value in chemosis (excessive edema of the ocular conjunctiva), and the drug is termed a *chemotic.*

Diacetylmorphine Hydrochloride, heroin hydrochloride, diamorphine hydrochloride. Although heroin is 2 to 3 times more potent than morphine as an analgesic, its sale and use are prohibited in the United States because of its intense addiction liability. It is available in some European countries where it has a limited use as an antitussive and as an analgesic in terminal cancer patients. Because of its superior solubility over morphine sulfate, arguments have been raised for its availability. However, the other more potent analgesics described here have significant advantages in being more stable and longer acting. It remains as one of the most widely used narcotics for illicit purposes and still places major economic burdens on our society.

Hydromorphone, dihydromorphinone. This synthetic derivative of morphine is prepared by the catalytic hydrogenation and dehydrogenation of morphine under acidic conditions, using a large excess of platinum or palladium.

The free base is similar in properties to those of morphine, being slightly soluble in water, freely soluble in alcohol and very soluble in chloroform.

This compound is of German origin, and was introduced in 1926. It is a substitute for morphine (5 times as potent) but has approximately equal addicting properties and a shorter duration of action. It possesses the advantage over morphine of giving less daytime sedation or drowsiness. It is a potent antitussive and is often used for coughs that are difficult to control.

Hydromorphone Hydrochloride U.S.P., Dilaudid®, dihydromorphinone hydrochloride. Hydromorphone hydrochloride occurs as a light-sensitive, white, crystalline powder which is freely soluble in water (1:3), sparingly soluble in alcohol and practically insoluble in ether. It is used in about one-fifth the dose of morphine for any of the indications of morphine.

Hydrocodone Bitartrate U.S.P., Dicodid®, Codone®, dihydrocodeinone bitartrate. This drug is prepared by the catalytic rearrangement of codeine or by hydrolyzing dihydrothebaine. It occurs as fine, white crystals or as a white, crystalline powder. It is soluble in water (1:16), slightly soluble in alcohol and insoluble in ether. It forms acidic solutions and is affected by light. The hydrochloride is also available.

Hydrocodone has a pharmacologic action midway between those of codeine and morphine, with 15 mg. being equivalent to 10 mg. of morphine in analgesic power. Although it has been shown to possess more addiction liability than codeine, it has been said to give no evidence of dependence or addiction when used for a long time. Its principal advantage is in the lower incidence of side-effects encountered with its use. It is more effective than codeine as an antitussive and is used primarily for this purpose. It is on the market in many cough preparations as well as in tablet and parenteral forms. It has also been marketed in an ion-exchange resin complex form under the trade name of Tussionex®. The complex has been shown to release the drug at a sustained rate and is said to produce effective cough suppression over a 10- to 12-hour period.

Although this drug found extensive use in antitussive formulations for many years, recently it has been placed under more stringent narcotic regulations, and it is being replaced gradually by codeine or dextromethorphan in most over-the-counter cough preparations.

Oxymorphone Hydrochloride U.S.P., Numorphan®, (−)-14-hydroxydihydromorphinone hydrochloride. Oxymorphone, introduced in 1959, is prepared by cleavage of the corresponding codeine derivative. It is used as the hydrochloride salt, which occurs as a white, crystalline powder freely soluble in water and sparingly soluble in alcohol. In man, oxymorphone is as effective as morphine in one-eighth to one-tenth the dosage, with good duration and a slightly lower incidence of side-effects.[111] It has high addiction liability. It is used for the same purposes as morphine, such as control of postoperative pain, pain of advanced neoplastic diseases as well as other types of pain that respond to morphine. Because of the risk of addiction it should not be employed for relief of minor pains that can be controlled with codeine. It is also well to note that it has poor antitussive activity and is not used as a cough suppressant.

It may be administered orally, parenterally (intravenously, intramuscularly or subcutaneously) or rectally and for these purposes is supplied as a solution for injection (1.0 and 1.5 mg. per ml.), suppositories (2 and 5 mg.) and in tablets (10 mg.).

Nalbuphine Hydrochloride, Nubain®, N-cyclobutylmethyl-14-hydroxy-N-nordihydromorphinone hydrochloride, N-cyclobutylmethylnoroxymorphone hydrochloride, was introduced in 1979 as a potent analgesic of the agonist-antagonist type with no to low abuse liability. It is somewhat less potent as an analgesic than its parent oxymorphone but shares some of the antagonist properties of the closely related but pure antagonists naloxone and naltrexone. Nalbuphine hydrochloride occurs as a white to off-white crystalline powder that is soluble in water and sparingly soluble in alcohol. It is prepared from cyclobutylmethyl bromide and noroxycodone followed by cleavage of the O-methyl group.

This new analgesic shows a very rapid onset with a duration of action of up to 6 hours. It has relatively low abuse liability, being judged to be less than that of codeine and propoxyphene. The injection was therefore introduced without narcotic controls, although caution is urged for long-term administration or use in emotionally disturbed patients. Abrupt discontinuation after prolonged use has given rise to withdrawal signs. Usual doses cause respiratory depression comparable to that of morphine, but no further decrease is seen with higher doses. It has fewer cardiac effects than pentazocine and butorphanol. The most frequent adverse effect is sedation and, as with most other CNS depressants and analgesics, caution should be urged when administered to ambulatory patients who may need to drive a car or operate machinery.

Nalbuphine was marketed initially as an injectable (10 mg. per ml.), but a tablet form is expected to be available soon. The usual dose is 10 mg. administered subcutaneously, intramuscularly, or intravenously at 3- to 6-hour intervals, with a maximal daily dose of 160 mg.

Oxycodone Hydrochloride, dihydrohydroxycodeinone hydrochloride. This compound is prepared by the catalytic reduction of hydroxycodeinone, the latter compound being prepared by hydrogen peroxide (in acetic acid) oxidation of thebaine. This derivative of morphine occurs as a white, crystalline powder which is soluble in water (1:10) or alcohol. Aqueous solutions may be sterilized by boiling. Although this drug is almost as likely to cause addiction as morphine, it is sold in the United States in Percodan® and Tylox® as a mixture of its hydrochloride and terephthalate salts in combination with aspirin, phenacetin and caffeine.

It is used as a sedative, an analgesic and a narcotic. Because it is believed to exert a physostigmine-like action, it is used externally in the eye in the treatment of glaucoma and related ocular conditions. To depress the cough reflex, it is used in 3- to 5-mg. doses and as an analgesic in 5- to 10-mg. doses. For severe pain, a dose of 20 mg. is given subcutaneously.

Dihydrocodeine Bitartrate. Dihydrocodeine is obtained by the reduction of codeine. The bitartrate salt occurs as white crystals which are soluble in water (1:4.5) and only slightly soluble in alcohol. Subcutaneously, 30 mg. of this drug is almost equivalent to 10 mg. of morphine as an analgesic, giving more prompt onset and negligible side-effects. It has addiction liability. It is available only as a cough preparation (Cophene-S®). As an antitussive, the usual dose is 10 to 30 mg.

Normormorphine. This drug may be prepared by N-demethylation of morphine.[112] It is still undergoing investigation and evaluation of its pharmacologic properties. In man, by normal routes of administration, it is about one-fourth as active as morphine in producing analgesia but has a much lower physical dependence capacity. Its analgesic effects are nearly equal by the intraventricular route. It does not show the sedative effects of morphine in single doses but does so cumulatively. Normorphine suppresses the morphine abstinence syndrome in addicts, but after its withdrawal it gives a slow onset and a mild form of the abstinence syndrome.[113] It has been considered for possible use in the treatment of narcotic addiction.

Concentrated Opium Alkaloids, Pantopon®, consists of a mixture of the total alkaloids of opium. It is free of nonalkaloidal material, and the alkaloids are said to be present in the same proportions as they occur naturally. The alkaloids are in the form of the hydrochlorides, and morphine constitutes 50 percent of the weight of the material.

This preparation is promoted as a substitute for morphine, the claim being that it is superior to the latter, due to the synergistic action of the opium alkaloids. This synergism is said to result in less respiratory depression, less nausea and vomiting and an antispasmodic action on smooth muscle. According to several authorities, however, the superiority to morphine is overrated, and the effects produced are comparable with the use of an equivalent amount of morphine. The commercial literature suggests a dose of 20 mg. of Pantopon to obtain the same effect as is given by 15 mg. of morphine.

Solutions prepared for parenteral use may be slightly colored, a situation which does not necessarily indicate decomposition.

Apomorphine Hydrochloride U.S.P. When morphine or morphine hydrochloride is heated at 140° under pressure with strong (35%) hydrochloric acid, it loses a molecule of water and yields a compound known as apomorphine.

The hydrochloride is odorless and occurs as minute, glistening, white or grayish-white crystals or as a white powder. It is light-sensitive and turns green on exposure to air and light. It is sparingly soluble in water (1:50, 1:20 at 80°) and in alcohol (1:50) and is very slightly soluble in ether or chloroform. Solutions are neutral to litmus.

The change in structure from morphine to apomorphine causes a profound change in its physiologic action. The central depressant effects of morphine are

Apomorphine

much less pronounced, and the stimulant effects are enhanced greatly, thereby producing emesis by a purely central mechanism. It is administered subcutaneously to obtain emesis. It is ineffective orally. Apomorphine is one of the most effective, prompt (10 to 15 minutes) and safe emetics in use today. However, care should be exercised in its use because it may be depressant in already depressed patients.

Meperidine Hydrochloride U.S.P., Demerol® Hydrochloride, ethyl 1-methyl-4-phenylisonipecotate hydrochloride, ethyl 1-methyl-4-phenyl-4-piperidinecarboxylate hydrochloride. This is a fine, white, odorless, crystalline powder that is very soluble in water, soluble in alcohol and sparingly soluble in ether. It is stable in the air at ordinary temperature, and its aqueous solution is not decomposed by a short period of boiling. The free base may be made by heating benzyl cyanide with bis(β-chloroethyl)methylamine, hydrolyzing to the corresponding acid and esterifying the latter with ethyl alcohol.[2]

Meperidine first was synthesized in order to study its spasmolytic character, but it was found to have analgesic properties in far greater degree. The spasmolysis is due primarily to a direct papaverine-like depression of smooth muscle and, also, to some action on parasympathetic nerve endings. In therapeutic doses, it exerts an analgesic effect which lies between those of morphine and codeine, but it shows little tendency toward hypnosis. It is indicated for the relief of pain in the majority of cases for which morphine and other alkaloids of opium generally are employed, but it is especially of value where the pain is due to spastic conditions of intestine, uterus, bladder, bronchi, and so on. Its most important use seems to be in lessening the severity of labor pains in obstetrics and, with barbiturates or tranquilizers, to produce amnesia in labor. In labor, 100 mg. is injected intramuscularly as soon as contractions occur regularly, and a second dose may be given after 30 minutes if labor is rapid or if the cervix is thin and dilated (2 to 3 cm. or more). A third dose may be necessary an hour or two later, and at this stage a barbiturate may be administered in a small dose to ensure adequate amnesia for several hours. Meperidine possesses addiction liability. There

TABLE 17-5

MORPHINE AND RELATED COMPOUNDS

Name Proprietary Name	Preparations	Category	Usual Adult Dose*	Usual Dose Range*	Usual Pediatric Dose*
Morphine Sulfate U.S.P.	Morphine Sulfate Injection U.S.P.	Narcotic analgesic	Oral, I.M. or S.C., 5 to 20 mg. every 4 hours as needed	12 to 120 mg. daily	S.C., 100 to 200 μg. per kg. of body weight, every 4 hours as needed, up to a maximum of 15 mg. per dose
Codeine U.S.P.	Terpin Hydrate and Codeine Elixir U.S.P.	Analgesic (narcotic); antitussive	Antitussive, 5 to 10 mg. every 4 hours		
Codeine Phosphate U.S.P.	Codeine Phosphate Injection U.S.P. Codeine Phosphate Tablets U.S.P.	Narcotic analgesic; antitussive	Analgesic—I.M., I.V., S.C. or oral, 15 to 60 mg. every 4 hours as needed		Analgesic—I.M., I.V., S.C. or oral, 500 μg. per kg. of body weight or 16.7 mg. per square meter of body surface every 4 hours as necessary
Codeine Sulfate U.S.P.	Codeine Sulfate Tablets U.S.P.	Analgesic (narcotic); antitussive	See Codeine Phosphate Tablets U.S.P.		See Codeine Phosphate Tablets U.S.P.
Ethylmorphine Hydrochloride *Dionin*		Antitussive	5 to 15 mg. 3 or 4 times daily		Children, 1 mg. per kg. of body weight in 4 to 6 divided doses
Hydromorphone Hydrochloride U.S.P. *Dilaudid, Hymorphan*	Hydromorphone Hydrochloride Injection U.S.P. Hydromorphone Hydrochloride Tablets U.S.P.	Analgesic (narcotic)	Oral and S.C., 2 mg. every 4 hours as necessary	1 to 4 mg.	
Hydrocodone Bitartrate U.S.P. *Dicodid, Codone*	Hydrocodone Bitartrate Tablets U.S.P.	Antitussive	5 to 10 mg. 3 or 4 times daily as necessary	5 to 50 mg. daily	
Oxymorphone Hydrochloride U.S.P. *Numorphan*	Oxymorphone Hydrochloride Injection U.S.P. Oxymorphone Hydrochloride Suppositories U.S.P.	Analgesic (narcotic)	S.C. and I.M., 1.0 to 1.5 mg. every 4 to 6 hours as needed; I.V., 500 μg. initially, repeated in 4 to 6 hours, if necessary; rectal, 2 or 5 mg. every 4 to 6 hours		
Apomorphine Hydrochloride U.S.P.	Apomorphine Hydrochloride Tablets U.S.P.	Emetic	S.C., 5 mg.		

*See *U.S.P. D.I.* for complete dosage information.

is a development of psychic dependence in those individuals who experience a euphoria lasting for an hour or more. The development of tolerance has been observed, and it is significant that meperidine can be successfully substituted for morphine in addicts who are being treated by gradual withdrawal. Furthermore, mild withdrawal symptoms have been noted in certain persons who have become purposely addicted to meperidine. The possibility of dependence is great enough to put it under the federal narcotic laws. Nevertheless, it remains as one of the more widely used analgesics.

Alphaprodine Hydrochloride U.S.P., (±)-1,3-dimethyl-4-phenyl-4-piperidinol propionate hydrochloride. This compound is prepared according to the method of Ziering and Lee.[114]

It occurs as a white, crystalline powder, which is freely soluble in water, alcohol and chloroform but insoluble in ether.

The compound is an effective analgesic, similar to meperidine and has been found to be of special value in obstetric analgesia. It appears to be quite safe for use in this capacity, causing little or no depression of respiration in either mother or fetus.

Anileridine U.S.P., Leritine®, ethyl 1-(*p*-aminophenethyl)-4-phenylisonipecotate. It is prepared by the method of Weijlard *et al.*[115] It occurs as a white to yellowish-white, crystalline powder that is freely soluble in alcohol but only very slightly soluble in water. It is oxidized on exposure to air and light. The injection is prepared by dissolving the free base in phosphoric acid solution.

Anileridine is more active than meperidine and has the same usefulness and limitations. Its dependence capacity is less and it is considered a suitable substitute for meperidine.

Anileridine Hydrochloride U.S.P., Leritine® Hydrochloride, ethyl 1-(*p*-aminophenethyl)-4-phenylisonipecotate dihydrochloride. It is prepared as cited above for anileridine except that it is converted to the dihydrochloride by conventional procedures. It occurs as a white or nearly white, crystalline, odorless powder which is stable in air. It is freely soluble in water, sparingly soluble in alcohol and practically insoluble in ether and chloroform.

This salt has the same activity as that cited for anileridine (see above).

Diphenoxylate Hydrochloride U.S.P., Colonil®, Diphenatol®, Enoxa®, Lofene®, Loflo®, Lomotil®, Lonox®, Lo-Trol®, Nor-nil®, ethyl 1-(3-cyano-3,3-diphenylpropyl)-4-phenylisonipecotate monohydrochloride. It occurs as a white, odorless, slightly water-soluble powder with no distinguishing taste.

Although this drug has a strong structural relationship to the meperidine-type analgesics it has very little, if any, such activity itself. Its most pronounced activity is its ability to inhibit excessive gastrointestinal motility, an activity reminiscent of the constipating side-effect of morphine itself. Investigators have demonstrated the possibility of addiction,[25,26] particularly with large doses, but virtually all studies using ordinary dosage levels show nonaddiction. Its safety is reflected in its classification as an exempt narcotic, with, however, the warning that it may be habit forming. To discourage possible abuse of the drug, the commercial product (Lomotil®) once contained a subtherapeutic dose (25 μg.) of atropine sulfate in each 2.5-mg. tablet and in each 5 ml. of the liquid which contains a like amount of the drug. Atropine has now been removed because of unwarranted side-effects.

It is indicated in the oral treatment of diarrheas resulting from a variety of causes. The usual initial adult dose is 5 mg. 3 or 4 times a day, with the maintenance dose usually being substantially lower and being individually determined. Appropriate dosage schedules for children are available in the manufacturer's literature.

The incidence of side-effects is low, but the drug should be used with caution, if at all, in patients with impaired hepatic function. Similarly, patients taking barbiturates concurrently with the drug should be observed carefully, in view of reports of barbiturate toxicity under these circumstances.

Loperamide Hydrochloride U.S.P., Imodium®, 4-(4-*p*-chlorophenyl-4-hydroxypiperidino)-N,N-dimethyl-2,2-diphenylbutyramide hydrochloride. This hybrid of a methadonelike and meperidine molecule is closely related to diphenoxylate, being more specific, potent, and longer acting. It acts as an antidiarrheal by a direct effect on the circular and longitudinal intestinal muscles. Following oral administration it reaches peak blood levels within 4 hours and has a very long plasma half-life (40 hours). Tolerance to its effects have not been observed.[116] Although it has shown minimal CNS effects, it has been controlled under schedule V.

Loperamide is available as 2 mg. capsules (Loperamide Hydrochloride Capsules U.S.P.) for treatment of acute and chronic diarrhea. Dosage recommended is 4 mg. initially, with 2 mg. following each loose stool for a maximum of 16 mg. per day.

Ethoheptazine Citrate, Zactane Citrate®, ethyl hexahydro-1-methyl-4-phenyl-1H-azepine-4-carboxylate dihydrogen citrate, 1-methyl-4-carbethoxy-4-phenylhexamethylenimine citrate. It is effective orally against moderate pain in doses of 50 to 100 mg., with minimal side-effects. Parenteral administration is limited, due to central stimulating effects. It appears to have no addiction liability, but toxic reactions have occurred with large doses. A double blind study in man rated 100 mg. of the hydrochloride salt equivalent to 30 mg. of codeine, and found that the addition of 600 mg. of aspirin increased analgesic effectiveness.[29] In another study, 150 mg. was found to be equal to 65 mg. of propoxyphene, both being better than placebo.[117] It is available as a 75-mg. tablet and in combination with 600 mg. of aspirin (Zactirin®).

Fentanyl Citrate U.S.P., Sublimaze®, N-(1-phenethyl-4-piperidyl)propionanilide citrate. This compound occurs as a crystalline powder, soluble in water (1:40) and methanol, and sparingly soluble in chloroform.

This novel anilide derivative has demonstrated analgesic activity 50 times that of morphine in man.[32] It has a very rapid onset (4 minutes) and short duration of action. Side-effects similar to those of other potent analgesics are common—in particular, respiratory depression and bradycardia. It is used primarily as an adjunct to anesthesia. For use as a neuroleptanalgesic in surgery, it is available in combination with the neuroleptic droperidol as Innovar®. It has dependence liability.

Methadone Hydrochloride U.S.P., Dolophine®, Westadone®, 6-(dimethylamino)-4,4-diphenyl-3-heptanone hydrochloride. It occurs as a white, crystalline powder with a bitter taste. It is soluble in water, freely soluble in alcohol and chloroform, and insoluble in ether.

Methadone is synthesized in several ways. The method of Easton and co-workers[118] is noteworthy in that it avoids the formation of the troublesome isomeric intermediate aminonitriles. The analgesic effect and other morphine-like properties are exhibited chiefly by the (−)-form. Aqueous solutions are stable and may be sterilized by heat for intramuscular and intravenous use. Like all amine salts, it is incompatible with alkali and salts of heavy metals. It is somewhat irritating when injected subcutaneously.

The toxicity of methadone is 3 to 10 times greater than that of morphine, but its analgesic effect is twice that of morphine and 10 times that of meperidine. It has been placed under federal narcotic control because of its high addiction liability.

Methadone is a most effective analgesic, used to alleviate many types of pain. It can replace morphine for the relief of withdrawal symptoms. It produces less sedation and narcosis than does morphine and appears to have fewer side-reactions in bed-ridden patients. In spasm of the urinary bladder and in the suppression of the cough reflex, methadone is especially valuable.

The *levo*-isomer, levanone, is said not to produce euphoria or other morphinelike sensations and has been advocated for the treatment of addicts.[119] Methadone itself is being used quite extensively in addict treatment, although not without some controversy.[120] It will suppress withdrawal effects and is widely used to maintain former heroin addicts during this rehabilitation. Large doses are often used to "block" the effects of heroin during treatment.

The use of methadone in treating addicts is subject to F.D.A. regulations that require special registration of physicians and dispensers. Methadone is available, however, for use as an analgesic under the usual narcotic requirements.

Levo-alpha-acetylmethadol, (−)-α-6-(dimethylamino)-4,4-diphenyl-3-heptyl acetate hydrochloride, methadyl acetate, LAAM. It occurs as a white, crystalline powder that is soluble in water, but dissolves with some difficulty. It is prepared by hydride reduction of (+)-methadone followed by acetylation.

Of the four possible methadol isomers, the 3S,6S-isomer LAAM has the unique characteristic of producing long-lasting narcotic effects. Extensive metabolism studies have shown that this is due to its N-demethylation to give (−)-α-acetylnormethadol, which is more potent than its parent LAAM and possesses a long half-life.[121] This is further accentuated by its demethylation to the dinor metabolite, which has similar properties.[121,122]

Because of the need to administer methadone daily, which leads to inconvenience to the maintenance patient and illicit diversion, the long-acting LAAM is being actively investigated as an addict maintenance drug to replace methadone. Generally, a 70-mg. dose three times a week is sufficient for routine maintenance.[40,123] The drug is undergoing extensive clinical trials.

It is of interest to note that the racemate of the nor metabolite, noracylmethadol, was once studied in the clinic as a potential analgesic.[124]

Propoxyphene Hydrochloride U.S.P., Darvon®, Dolene®, Doloxene®, D-orafen®, Progesic Compound-65®, Proxagesic®, Proxene®, SK 65®, (+)-α-4-dimethylamino-1,2-diphenyl-3-methyl-2-butanol propionate hydrochloride. This drug was introduced into therapy in 1957. It may be prepared by the method of Pohland and Sullivan.[125] It occurs as a bitter, white, crystalline powder which is freely soluble in water, soluble in alcohol, chloroform and acetone but practically insoluble in benzene and ether. It is the α-(+)-isomer, the α-(−)-isomer and β-diastereoisomers being far less potent in analgesic activity. The α-(−)-isomer, *levo*-propoxyphene, is an effective antitussive (see below).

In analgesic potency, propoxyphene is approximately equal to codeine phosphate and has a lower incidence of side-effects. It has no antidiarrheal, antitussive or antipyretic effect, thus differing from most analgesic agents. It is able to suppress the morphine abstinence syndrome in addicts but has shown a low level of abuse because of its toxicity. It is not very effective in deep pain and appears to be no more effective in minor pain than aspirin. Its widespread use in dental pain seems justified, since aspirin is reported to be relatively ineffective. It was recently classified as a narcotic and controlled under federal law. It does give some euphoria in high doses and has been abused. It has been responsible for numerous overdosage deaths. Refilling of the drug should be avoided if misuse is suspected.

It is available in several combination products with aspirin (e.g., Darvon w/A.S.A.®) or acetaminophen (e.g., Dolene A.P.®, Wygesic®).

Propoxyphene Napsylate U.S.P., Darvon-N®, (+)-α-4-dimethylamino-1,2-diphenyl-3-methyl-2-butanol propionoate (ester) 2-naphthylenesulfonate (salt). It is very slightly soluble in water, but soluble in alcohol, chloroform and acetone.

The napsylate salt of propoxyphene was introduced shortly before the patent on Darvon® expired. As an insoluble salt form it is claimed to be less prone to abuse because it can not be readily dissolved for injection, and upon oral administration gives a slower, less pronounced peak blood level.

Because of its mild narcoticlike properties it was being intensely investigated as an addict maintenance drug to be used in place of methadone. It was hoped that it would offer the advantage of providing an easier withdrawal and serve as an addict detoxification drug. Unfortunately, toxicity at higher doses has limited this application.

It is available in combination with aspirin and acetaminophen, Darvocet–N®.

Levorphanol Tartrate U.S.P., Levo-Dromoran®

Tartrate, (−)-3-hydroxy-N-methylmorphinan bitartrate. The basic studies in the synthesis of this type of compound were made by Grewe, as already pointed out above. Schnider and Grüssner synthesized the hydroxymorphinans, including the 3-hydroxyl derivative, by similar methods. The racemic 3-hydroxy-N-methylmorphinan hydrobromide (racemorphan, (±)-Dromoran®) was the original form in which this potent analgesic was introduced. This drug is prepared by resolution of racemorphan. It should be noted that the *levo* compound is available in Europe under the original name Dromoran®. As the tartrate, it occurs in the form of colorless crystals. The salt is sparingly soluble in water (1:60) and is insoluble in ether.

The drug is used for the relief of severe pain and is in many respects similar in its actions to morphine except that it is from 6 to 8 times as potent. The addiction liability of levorphanol is as great as that of morphine, and, for that reason, caution should be observed in its use. It is claimed that the gastrointestinal effects of this compound are significantly less than those experienced with morphine. Naloxone (q.v.) is an effective antidote for overdosage. Levorphanol is useful for relieving severe pain originating from a multiplicity of causes, e.g., inoperable tumors, severe trauma, renal colic, biliary colic. In other words, it has the same range of usefulness as morphine and is considered an excellent substitute. It is supplied in ampules, in multiple-dose vials and in the form of oral tablets. The drug requires a narcotic form.

Butorphanol Tartrate U.S.P., Stadol®, (−)-N-cyclobutylmethyl-3,14-dihydroxymorphinan bitartrate. This new, potent analgesic occurs as a white, crystalline powder soluble in water and sparingly soluble in alcohol. It is prepared from the dihydroxy-N-normorphinan obtained by a modification of the Grewe synthesis. It is the cyclobutyl analog of levorphanol and levallorphan, being equally potent as an analgesic as the former and somewhat less active as an antagonist than the latter.

Butorphanol

The onset and duration of action of the drug is comparable to that of morphine, but it has the advantages of showing a maximal ceiling effect on respiratory depression and a greatly reduced abuse liability. The injectable form was marketed without narcotic controls; however, this product and the tablets are being placed in schedule IV because of suspected misuse and lack of recognition of its potential abuse liability. The drug has also been used illegally in the doping of race horses.

Butorphanol shares the adverse hemodynamic effects of pentazocine, causing increased pressure in specific arteries and on the heart work-load. It should, therefore, be used with caution and only with patients hypersensitive to morphine in the treatment of patients with myocardial infarction or other cardiac problems. Other adverse effects include a high incidence of sedation and, less frequently, nausea, headache, vertigo, and dizziness.[126]

It is available as a parenteral for intramuscular and intravenous administration in a dose of 1 or 2 mg. every 3 to 4 hours, with a maximal single dose of 4 mg. A tablet form has been studied in clinical trials and is under review for marketing.

Pentazocine U.S.P., Talwin®, 1,2,3,4,5,6-hexahydro-*cis*-6,11-dimethyl-3-(3-methyl-2-butenyl)-2,6-methano-3-benzazocin-8-ol, *cis*-2-dimethylallyl-5,9-dimethyl-2'-hydroxy-6,7-benzomorphan. It occurs as a white, crystalline powder which is insoluble in water and sparingly soluble in alcohol. It forms a poorly soluble hydrochloride salt but is readily soluble as the lactate.

Pentazocine in a parenteral dose of 30 mg. or an oral dose of 50 mg. is about as effective as 10 mg. of morphine in most patients. There is now some evidence that the analgesic action resides principally in the (−)-isomer, with 25 mg. being approximately equivalent to 10 mg. of morphine sulfate.[127] Occasionally, doses of 40 to 60 mg. may be required. Pentazocine's plasma half-life is about 3½ hours.[128] At the lower dosage levels, it appears to be well tolerated, although some degree of sedation occurs in about one-third of those persons receiving it. The incidence of other morphinelike side-effects is as high as with morphine and other narcotic analgesics. In patients who have been receiving other narcotic analgesics, large doses of pentazocine may precipitate withdrawal symptoms. It shows an equivalent or greater respiratory depressant activity. Pentazocine has given rise to a few cases of possible dependence liability. It was recently placed under control and its abuse potential should be recognized and close supervision of its use maintained. Levallorphan cannot reverse its effects, although naloxone can, and methylphenidate is recommended as an antidote for overdosage or excessive respiratory depression.

Pentazocine as the lactate is available in vials containing the equivalent of 30 mg. of base per ml., buffered to pH 4 to 5. It should not be mixed with barbiturates. Tablets of 50 mg. (as the hydrochloride) are also available for oral administration.

Methotrimeprazine U.S.P., Levoprome®, (−)-10-[3-(dimethylamino)-2-methylpropyl]-2-methoxy-

TABLE 17-6

SYNTHETIC ANALGESICS

Name Proprietary Name	Preparations	Category	Usual Adult Dose*	Usual Dose Range*	Usual Pediatric Dose*
Meperidine Hydrochloride U.S.P. *Demerol*	Meperidine Hydrochloride Injection U.S.P. Meperidine Hydrochloride Tablets U.S.P.	Narcotic analgesic	I.M., S.C. or oral, 50 to 150 mg. every 3 or 4 hours		I.M., S.C. or oral, 1.1 to 1.76 mg. per kg. of body weight, not to exceed the adult dose, every 3 or 4 hours as necessary
	Meperidine Hydrochloride Syrup U.S.P.	Analgesic (narcotic)			
Alphaprodine Hydrochloride U.S.P.	Alphaprodine Hydrochloride Injection U.S.P.	Analgesic (narcotic)	S.C., 0.4 to 1.2 mg./kg., I.V., 0.4 to 0.6 mg./kg.	S.C., 20 to 60 mg.; I.V., 20 to 30 mg.	
Anileridine U.S.P. *Leritine*	Anileridine Injection U.S.P.	Analgesic (narcotic)	S.C. or I.M., 25 to 50 mg. of anileridine, as the phosphate, repeated every 6 hours, if necessary	S.C. or I.M., 25 to 75 mg.	
Anileridine Hydrochloride U.S.P. *Leritine*	Anileridine Hydrochloride Tablets U.S.P.	Analgesic (narcotic)	25 mg. of anileridine, as the dihydrochloride, repeated every 6 hours, if necessary	25 to 50 mg.	
Fentanyl Citrate U.S.P. *Sublimaze*	Fentanyl Citrate Injection U.S.P.	Narcotic analgesic	Induction—I.V., the equivalent of 50 to 100 µg. of fentanyl; may be repeated every 2 to 3 minutes until the desired effect is achieved; maintenance—I.V., the equivalent of 25 to 50 µg. of fentanyl as necessary; postoperative analgesia—I.M., the equivalent of 50–100 µg. of fentanyl every 1 to 2 hours as necessary.	25 to 100 µg.	Dosage is not established in children under 2 years of age
Methadone Hydrochloride U.S.P. *Dolophine, Westadone*	Methadone Hydrochloride Injection U.S.P. Methadone Hydrochloride Tablets U.S.P. Methadone Hydrochloride Tablets for Oral Solution U.S.P.	Narcotic abstinence syndrome suppressant; narcotic analgesic	Analgesic— I.M. or S.C., 2.5 to 10 mg. every 3 or 4 hours as necessary Narcotic abstinence syndrome suppressant—detoxification—I.M. or S.C., 15 to 40 mg. once daily, the dose gradually being reduced according to patient response	Analgesic—15 to 80 mg. daily Narcotic abstinence syndrome suppressant—up to 100 mg. daily	
	Methadone Hydrochloride Oral Solution U.S.P.		Analgesic, oral, 2.5 to 10 mg. every 3 or 4 hours as needed; suppresant (narcotic abstinence syndrome)—oral, 15 to 40 mg. once daily, the dosage being decreased according to patient response Narcotic abstinence syndrome suppressant—detoxification—40 mg. once daily, the dose gradually being reduced after 2 to 3 days of stabilization; maintenance, 40 to 120 mg. once daily	Up to 100 mg. daily Narcotic abstinence syndrome suppressant—15 to 120 mg. daily	Dosage has not been established
Propoxyphene Hydrochloride U.S.P.	Propoxyphene Hydrochloride Capsules U.S.P.	Analgesic (C-IV)	Oral, 65 mg. every 4 hours as necessary		Dosage has not been established

*See *U.S.P. D.I.* for complete dosage information.
(Continued)

TABLE 17-6

SYNTHETIC ANALGESICS

Name Proprietary Name	Preparations	Category	Usual Adult Dose*	Usual Dose Range*	Usual Pediatric Dose*
Darvon, Dolene, Boraphen, Maragesic, Proxene Propoxphene Napsylate U.S.P. Darvon-N	Propoxyphene Napsylate Oral Suspension U.S.P. Propoxyphene Napsylate Tablets U.S.P.	Analgesic (C-IV)	Oral 100 mg. every 4 hours, as needed		Dosage has not been established
Levorphanol Tartrate U.S.P. Levo-Dromoran	Levorphanol Tartrate injection U.S.P. Levorphanol Tartrate Tablets U.S.P.	Analgesic (narcotic) (C-II)	Oral and S.C., 2 mg.	1 to 3 mg.	
Pentazocine U.S.P. Talwin	Pentazocine Lactate Injection U.S.P. (with the aid of lactic acid)	Analgesic (C-IV)	I.M., I.V., or S.C., the equivalent of 30 mg. of pentazocine every 3 or 4 hours as needed	Up to the equivalent of 360 mg. of pentazocine daily	Dosage has not been established
Pentazocine Hydrochloride U.S.P. Talwin	Pentazocine Hydrochloride Tablets U.S.P.	Analgesic (C-IV)	Oral, the equivalent of 50 mg. of pentazocine every 3 or 4 hours, as needed. Dosage may be increased to 100 mg. if necessary, but the total daily dose should not exceed 600 mg.	Up to the equivalent of 600 mg. of pentazocine daily	Dosage has not been established
Methotrimeprazine U.S.P. Levoprome	Methotrimeprazine Injection U.S.P.	Analgesic	I.M., 10 to 30 mg. every 4 to 6 hours	5 to 40 mg.	

(Continued)

* See *U.S.P. D.I.* for complete dosage information.

phenothiazine. This phenothiazine derivative, closely related to chlorpromazine, possesses strong analgesic activity. An intramuscular dose of 15 to 20 mg. is equal to 10 mg. of morphine in man. It has not shown any dependence liability and appears not to produce respiratory depression. The most frequent side-effects are similar to those of phenothiazine tranquilizers, namely, sedation and orthostatic hypotension. These often result in dizziness and fainting, limiting the use of methotrimeprazine to nonambulatory patients. It is to be used with caution along with antihypertensives, atropine, and other sedatives. It shows some advantage in cases in which addiction and respiratory depression are problems.[129]

Nefopam, Acupan®, 5-methyl-1-phenyl-3,4,5,6-tetrahydro-[1H]-2,5-benzoxazocine. This rather novel an-

Nefopam

algesic represents a departure from traditional structure-activity relationships, but shows activity comparable to that of codeine. It gives very rapid onset due to rapid absorption with 60 mg. giving pain relief comparable to 600 mg. of aspirin. Side-effects were minimal.[130]

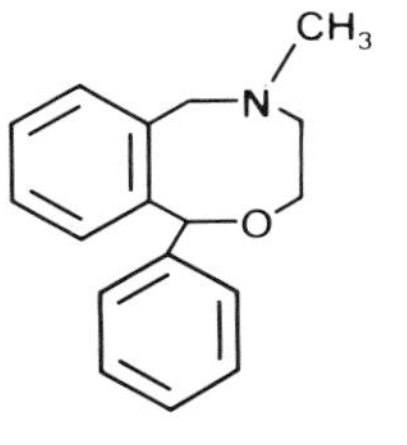

NARCOTIC ANTAGONISTS

Nalorphine Hydrochloride U.S.P., N-allylnormorphine hydrochloride. This morphine derivative may be prepared according to the method of Weijlard and Erickson.[112] It occurs in the form of white or practically white crystals that slowly darken on exposure to air and light. It is freely soluble in water (1:8) but is sparingly soluble in alcohol (1:35) and is almost insoluble in chloroform and ether. The phenolic hydroxyl group confers water-solubility in the presence of fixed alkali. Aqueous solutions of the salt are acid, having a pH of about 5.

Nalorphine has a direct antagonistic effect against morphine, meperidine, methadone and levorphanol. However, it has little antagonistic effect toward barbiturate or general anesthetic depression.

Perhaps one of the most striking effects is on the respiratory depression accompanying morphine overdosage. The respiratory minute volume is quickly returned to normal by intravenous administration of the drug. However, it does have respiratory depressant activity itself, which may potentiate the existing depression. It affects circulatory disturbances in a similar way, reversing the effects of morphine. Other effects of morphine are affected similarly. It is interesting to note that morphine addicts, when treated with the drug, exhibit certain of the withdrawal symptoms associated with abstinence from morphine. Thus, it was used as a diagnostic test agent to determine narcotic addiction. Chronic administration of nalorphine along with morphine prevents or minimizes the development of dependence on morphine. As pointed out earlier, it has been found to have strong analgesic properties but is not acceptable for such use, due to the high incidence of undesirable psychotic effects. Because of these properties and the availability of alternate antagonists, it was withdrawn from the market.

Levallorphan Tartrate U.S.P., Lorfan®, (−)-N-allyl-3-hydroxymorphinan bitartrate. This compound occurs as a white or practically white, odorless, crystalline powder. It is soluble in water (1:20), sparingly soluble in alcohol (1:60) and practically insoluble in chloroform and ether. Levallorphan resembles nalorphine in its pharmacologic action, being about 5 times more effective as a narcotic antagonist. It has been found also to be useful in combination with analgesics such as meperidine, alphaprodine and levorphanol to prevent the respiratory depression usually associated with these drugs.

Naloxone Hydrochloride U.S.P., Narcan®, N-allyl-14-hydroxynordihydromorphinone hydrochloride, N-allylnoroxymorphone hydrochloride is presently on the market as the agent of choice for treating narcotic overdosage. It lacks not only the analgesic activity shown by other antagonists but all of the other agonist effects. It is almost 7 times more active than nalorphine in antagonizing the effects of morphine. It shows no withdrawal effects after chronic administra-

tion. The duration of action is about 4 hours. It was briefly investigated for the treatment of heroin addiction. With adequate doses of naloxone, the addict does not receive any effect from heroin. It is given to an addict only after a detoxification period. Its long-term usefulness is currently limited, because of its short duration of action, thereby requiring large oral doses. Long-acting forms or alternate antagonists are being investigated.

Cyclazocine, cis-2-cyclopropylmethyl-5,9-dimethyl-2′-hydroxy-6,7-benzomorphan, is a potent narcotic antagonist that has shown analgesic activity in man in 1-mg. doses. It is presently being reinvestigated as a clinical analgesic. It does possess hallucinogenic side-effects at higher doses which may limit its usefulness as an analgesic. It was studied like naloxone, in the treatment of narcotic addiction. By voluntary treatment with cyclazocine, addicts are deprived of the euphorogenic effects of heroin. Its dependence liability is lower, and the effects of withdrawal develop more slowly and are milder. Tolerance develops to the side-effects of cyclazocine but not to its antagonist effects.[131] A usual maintenance dose of 4 mg. is obtained by gradually increasing doses. The effects are long-lasting and are not reversed by other antagonists such as nalorphine.

Naltrexone, N-cyclopropylmethyl-14-hydroxynordihydromorphinone, N-cyclopropylmethylnoroxymorphone, EN-1639. This naloxone analog is being actively investigated as the preferred agent for treating former opiate addicts. Oral doses of 50 mg. daily or 100 mg. three times weekly are sufficient to "block" or protect a patient from the effects of heroin. Its metabolism,[132] pharmacokinetics[133] and pharmacology[134] are being intensely studied due to tremendous governmental interest in developing new agents for the treatment of addiction.[54]

Several sustained-release or depot dosage forms of naltrexone are also being developed in order to avoid the recurrent decision on the part of the former addict as to whether a protecting dose of antagonist is needed.[55,135]

TABLE 17–7

NARCOTIC ANTAGONISTS

Name Proprietary Name	Preparations	Usual Adult Dose*	Usual Dose Range*	Usual Pediatric Dose*
Levallorphan Tartrate N.F. *Lorfan*	Levallorphan Tartrate Injection N.F.	I.V., 1 mg., repeated twice at 10- to 15-minute intervals, if necessary	500 μg. to 2 mg., repeated, if necessary	0.05 to 0.1 mg. in neonates to decrease respiratory depression
Naloxone Hydrochloride U.S.P. *Narcan*	Naloxone Hydrochloride Injection U.S.P.	Parenteral, 400 βg., repeated at 2- to 3-minute intervals as necessary		0.01 mg. as above

*See *U.S.P. D.I.* for complete dosage information.

There are several other narcotic antagonists that are also being investigated, e.g., diprenorphine[136] and oxilorphan.[137]

ANTITUSSIVE AGENTS

Cough is a protective, physiologic reflex that occurs in health as well as in disease. It is very widespread and commonly ignored as a mild symptom. However, in many conditions it is desirable to take measures to reduce excessive coughing. It should be stressed that many etiologic factors cause this reflex; and in a case where a cough has been present for an extended period of time, or accompanies any unusual symptoms, the person should be referred to a physician. Cough preparations are widely advertised and often sold indiscriminately; so it is the obligation of the pharmacist to warn the public of the inherent dangers.

Among the agents used in the symptomatic control of cough are those which act by depressing the cough center located in the medulla. These have been termed anodynes, cough suppressants and centrally acting antitussives. Until recently, the only effective drugs in this area were members of the narcotic analgesic agents. The more important and widely used ones are morphine, hydromorphone, codeine, hydrocodone, methadone and levorphanol, which were discussed in the foregoing section.

In recent years, several compounds have been synthesized that possess antitussive activity without the addiction liabilities of the narcotic agents. Some of these act in a similar manner through a central effect. In a hypothesis for the initiation of the cough reflex, Salem and Aviado[138] proposed that bronchodilation is an important mechanism for the relief of cough. Their hypothesis suggests that irritation of the mucosa initially causes bronchoconstriction, and this in turn excites the cough receptors. Many of these compounds are summarized in Table 17-8, together with the mechanism of action(s) attributed to them.

Chappel and von Seemann[139] have pointed out that most antitussives of this type fall into two structural groups. The larger group represented in Table 17-8 has those that bear a structural resemblance to methadone. The other group has large, bulky substituents on the acid portion of an ester, usually connected by means of a long, ether-containing chain to a *tertiary* amino group. The notable exceptions shown in Table 17-8 are benzonatate and sodium dibunate. Noscapine could be considered as belonging to the first group.

It should be pointed out that many of the cough preparations sold contain various other ingredients in addition to the primary antitussive agent. The more

TABLE 17-8

NON-NARCOTIC ANTITUSSIVE AGENTS

important ones include: antihistamines, useful when the cause of the cough is allergic in nature, although some antihistaminic drugs, e.g., diphenhydramine, have a central antitussive action as well; sympathomimetics, which are quite effective due to their bronchodilatory activity, the most useful being ephedrine, methamphetamine, phenylpropanolamine, homarylamine, isoproterenol and isoöctylamine; parasympatholytics, which help to dry secretions in the upper respiratory passages; and expectorants. It is not known if these drugs potentiate the antitussive action, but they usually are considered as adjuvant therapy.

The more important drugs in this class will be discussed in the following section. For a more exhaustive coverage of the field the reader is urged to consult the excellent review of Chappel and von Seemann.[139]

PRODUCTS

Some of the narcotic antitussive products have been discussed previously with the narcotic analgesics (q.v.).

Noscapine U.S.P., Tusscapine®, (−)-narcotine. This opium alkaloid was isolated, in 1817, by Robiquet. It is isolated rather easily from the drug by ether extraction. It makes up 0.75 to 9 percent of opium. Present knowledge of its structure is due largely to the researches of Roser.

Noscapine occurs as a fine, white or practically white, crystalline powder which is odorless and stable in the presence of light and air. It is practically insoluble in water, freely soluble in chloroform, and soluble in acetone and benzene. It is only slightly soluble in alcohol and ether.

With the discovery of its unique antitussive properties, the name of this alkaloid was changed from narcotine to noscapine. It was realized that it would not meet with widespread acceptance as long as its name was associated with the narcotic opium alkaloids. The selection of the name "noscapine" was probably due to the fact that a precedent existed in the name of (±)-narcotine, namely "gnoscopine."

Although noscapine had been used therapeutically as an antispasmodic (similar to papaverine), antineuralgic and antiperiodic, it had fallen into disuse. It had also been used in malaria, migraine and other conditions in the past in doses of 100 to 600 mg. Newer methods of testing for antitussive compounds were responsible for revealing the effectiveness of noscapine in this respect. In addition to its central action, it has been shown to exert bronchodilation effects.

Noscapine is an orally effective antitussive, approximately equal to codeine in effectiveness. It is free of the side-effects usually encountered with the narcotic antitussives and, because of its relatively low toxicity, may be given in larger doses in order to obtain a greater antitussive effect. Although it is an opium alkaloid, it is devoid of analgesic action and addiction liability. It is available in various cough preparations (e.g., Conar®).

Dextromethorphan Hydrobromide U.S.P., Romilar® Hydrobromide, (+)-3-methoxy-17-methyl-9α,13α,14α-morphinan hydrobromide. This drug is the O-methylated (+)-form of racemorphan left after the

resolution necessary in the preparation of levorphanol. It occurs as practically white crystals, or as a crystalline powder, possessing a faint odor. It is sparingly soluble in water (1:65), freely soluble in alcohol and chloroform and insoluble in ether.

It possesses the antitussive properties of codeine without the analgesic, addictive, central depressant and constipating features. Ten milligrams is suggested as being equivalent to a 15-mg. dose of codeine in antitussive effect.

It affords an opportunity to note the specificity exhibited by very closely related molecules. In this case, the (+)- and (−)-forms both must attach to receptors responsible for the suppression of cough reflex, but the (+)-form is apparently in a steric relationship such that it is incapable of attaching to the receptors involved in analgesic, constipative, addictive and other actions exhibited by the (−)-form. It has largely replaced many older antitussives, including codeine, in prescription and non-prescription cough preparations.

Levopropoxyphene Napsylate U.S.P., Novrad®, (−)-α-4-(dimethylamino)-3-methyl-1,2-diphenyl-2-butanol propionate (ester) 2-naphthalenesulfonate (salt). This compound, the *levo*-isomer of propoxyphene, does not possess the analgesic properties of the (+)-form but is equally effective as an antitussive, 50 mg. being equivalent to 15 mg. of codeine.[140] Side-effects are infrequent. Levopropoxyphene napsylate is also available in suspension form and has the advantage of being virtually tasteless.

Benzonatate U.S.P., Tessalon®, 2,5,8,11,14,17,20,-23,26-nona-oxaoctacosan-28-yl *p*-(butylamino)benzoate. This compound was introduced in 1956. It is a pale yellow, viscous liquid insoluble in water and soluble in most organic solvents. It is chemically related to *p*-aminobenzoate local anesthetics except that the aminoalcohol group has been replaced by a methylated polyethylene glycol group (see Table 17-8).

Benzonatate is said to possess both peripheral and central activity in producing its antitussive effect. It somehow blocks the stretch receptors thought to be responsible for cough. Clinically, it is not as effective as codeine but produces far fewer side-effects and has a very low toxicity. It is available in 50- and 100-mg. capsules ("perles") and ampules (5 mg./ml.).

Chlophedianol, ULO®, 1-*o*-chlorophenyl-1-phenyl-3-dimethylaminopropan-1-ol. This compound, which first was described as an antispasmodic, was found to be an effective antitussive agent.[141] It is useful in doses of 20 to 30 mg. given 3 to 5 times daily, with a duration of effect for a single dose lasting up to 5 hours. It has a low incidence of side-effects. It is available in several combinations (15 mg./ml.) (Acutuss®, Ulogesic®).

Caramiphen Edisylate, 2-diethylaminoethyl 1-

TABLE 17-9

ANTITUSSIVE AGENTS

Name Proprietary Name	Preparations	Usual Adult Dose*	Usual Dose Range*
Dextromethorphan Hydrobromide U.S.P. *Romilar*	Dextromethorphan Hydrobromide Syrup U.S.P.	15 to 30 mg. 1 to 4 times daily	
Levopropoxyphene Napsylate U.S.P. *Novrad*	Levopropoxyphene Napsylate Capsules U.S.P. Levopropoxyphene Napsylate Oral Suspension U.S.P.	50 to 100 mg. of levopropoxyphene, as the napsylate, every 4 hours	
Benzonatate U.S.P. *Tessalon*	Benzonatate Capsules U.S.P.	100 mg. 3 times daily	100 to 200 mg.

* See *U.S.P. D.I.* for complete dosage information.

phenylcyclopentane-1-carboxylate ethanedisulfonate. Caramiphen occurs in the form of water- and alcohol-soluble crystals. The antitussive activity of this compound is less than that of codeine. It has been shown to have both central and bronchodilator activity. The incidence of side-effects is lower than with the narcotic antitussives. It is currently marketed as a combination under the tradenames of Tuss-Ornade®, both in a liquid form and in a sustained-release form, and as De-Tuss®, Tuss-Ade®, Tuss-Liquid®, and Tusscaps®.

Carbetapentane Citrate, 2-[2-(diethylamino)-ethoxy]ethyl 1-phenylcyclopentanecarboxylate citrate. This salt is a white, odorless, crystalline powder, which is freely soluble in water (1:1), slightly soluble in alcohol and insoluble in ether. It is similar to caramiphen chemically and is said to be equivalent to codeine as an antitussive. Introduced in 1956, it is well tolerated and has a low incidence of side-effects. It is available as a syrup (7.5 mg./5 ml.) in combination with codeine (Tussar-2®) or alone (20 mg./5 ml. Cophene-X and -XP).

The tannate is also available (Rynatuss®) and is said to give a more sustained action.

THE ANTI-INFLAMMATORY ANALGESICS

The growth of this group of analgesics was related closely to the early belief that the lowering or "curing" of fever was an end in itself. Drugs bringing about a drop in temperature in feverish conditions were considered to be quite valuable and were sought after eagerly. The decline of interest in these drugs coincided more or less with the realization that fever was merely an outward symptom of some other, more fundamental, ailment. However, during the use of the several antipyretics, it was noted that some were excellent analgesics for the relief of minor aches and pains. These drugs have survived to the present time on the basis of the analgesic rather than the antipyretic effect. Although these drugs are still widely utilized for the alleviation of minor aches and pains, they are also employed extensively in the symptomatic treatment of rheumatic fever, rheumatoid arthritis and osteoarthritis. The dramatic effect of salicylates in reducing the inflammatory effects of rheumatic fever is time-honored, and, even with the development of the corticosteroids, these drugs are still of great value in this respect. It has been reported that the steroids are no more effective than the salicylates in preventing the cardiac complications of rheumatic fever.[142]

The analgesic drugs that fall into this category have been disclaimed by some as not deserving the term "analgesic" because of the low order of activity in comparison with the morphine-type compounds. Indeed, Fourneau has suggested the name "antalgics" to designate this general category and, in this way, to make more emphatic the distinction from the narcotic or so-called "true" analgesics. Two of the principal features distinguishing these minor analgesics from the narcotic analgesics are: the low activity for a given dose and the fact that higher dosage does not give any significant increase in effect.

Considerable research has continued in an effort to find new nonsteroidal anti-inflammatory agents. Long-term therapy with the corticosteroids is often accompanied by various side-effects. Efforts to discover new agents have been limited for the most part to structural analogs of active compounds due to a lack of knowledge about the causes and mechanisms of inflammatory diseases.[143] Although several new agents have been introduced for use in rheumatoid arthritis, aspirin appears to remain the agent of choice.

Of considerable interest is the observation that prostaglandins appear to play a major role in the inflammatory processes.[144] Of particular significance are reports that drugs such as aspirin and indomethacin inhibit prostaglandin synthesis in several tissues.[145] Furthermore, almost all classes of nonsteroidal anti-inflammatory agents strongly inhibit the conversion of arachidonic acid into prostaglandin E_2.[146,147] This has been shown to occur at the stage of conversion of arachidonic acid, released by the action of phospholipase A on damaged tissues, to the cyclic endoperoxides, PGG_2 and PGH_2, by prostaglandin synthetase. These are known to cause vasoconstriction and pain. They in turn are converted in part to PGE_2 and $PGF_{2\alpha}$, which can cause pain and vasodilation. This effect of the nonsteroidol anti-inflammatory agents parallels their relative potency in various tests and is stereospecific.[146] The search for specific inhibitors of prostaglandin synthesis has opened a new area of research in this field.

Discussion of these drugs will be facilitated by considering them in their various chemical categories.

SALICYLIC ACID DERIVATIVES

Historically, the salicylates were among the first of this group to achieve recognition as analgesics. Leroux, in 1827, isolated salicin, and Piria, in 1838, prepared salicylic acid. Following these discoveries, Cahours (1844) obtained salicylic acid from oil of wintergreen (methyl salicylate); and Kolbe and Lautermann (1860) prepared it synthetically from phenol. Sodium salicylate was introduced in 1875 by Buss, followed by the introduction of phenyl salicylate by Nencki, in 1886. Aspirin, or acetylsalicylic acid, was first prepared in 1853 by Gerhardt but remained obscure until Felix Hoffmann discovered its pharmacologic activities in 1899. It was tested and introduced into medicine by Dreser, who named it *aspirin* by taking the "a" from acetyl and adding it to "spirin," an old name for salicylic or spiric acid, derived from its natural source of spirea plants.

The pharmacology of the salicylates and related compounds has been reviewed extensively by Smith.[148,149] Salicylates, in general, exert their antipyretic action in febrile patients by increasing heat elimination of the body through the mobilization of water and consequent dilution of the blood. This brings about perspiration, causing cutaneous dilation. This does not occur with normal temperatures. The antipyretic and analgesic actions are believed to occur in the hypothalamic area of the brain. It is also thought by some that the salicylates exert their analgesia by their effect on water balance, reducing edema usually associated with arthralgias. Aspirin has been shown to be particularly effective in this respect.

For an interesting account of the history of aspirin and a discussion of its mechanisms of action, the reader should consult an article on the subject by Collier,[150] as well as the reviews by Smith,[148,149] and Nickander *et al.*[147]

The possibility of hypoprothrombinemia and concomitant capillary bleeding in conjunction with salicylate administration accounts for the inclusion of menadione in some salicylate formulations. However, there is some doubt as to the necessity for this measure. A more serious aspect of salicylate medication has been the possibility of inducing hemorrhage due to direct irritative contact with the mucosa. Alvarez and Summerskill have pointed out a definite relationship between salicylate consumption and massive gastrointestinal hemorrhage from peptic ulcer.[151] Barager and Duthie,[152] on the other hand, in an extensive study find no danger of increase in anemia or in development of peptic ulcer. Levy[153] has demonstrated with the use of labeled iron that bleeding does occur following administration of aspirin. The effects varied with the formulation. It is suggested by Davenport[154] that back-diffusion of acid from the stomach is responsible for capillary damage.

Because of these characteristics of aspirin, it has been extensively studied as an antithrombotic agent in the treatment and prevention of clinical thrombosis.[155] It is thought to act by its selective action on the synthesis of the prostaglandin-related thromboxane and prostacyclin, which are the counterbalancing factors involved in platelet aggregation and are released when tissue is injured. Although aspirin has now been approved for the prevention of transient ischemic attacks, indicators of an impending stroke, it is not recommended for patients who have suffered heart attacks.[156]

The salicylates are readily absorbed from the stomach and the small intestine, being quite dependent on the pH of the media. Absorption is considerably slower as the pH rises (more alkaline), due to the acidic nature of these compounds and the necessity for the presence of undissociated molecules for absorption through the lipoidal membrane of the stomach and the intestines. Therefore, buffering agents administered at the same time in *excessive* amounts will decrease the rate of absorption. In small quantities, their principal effect may be to aid in the dispersion of the salicylate into fine particles. This would help to increase absorption and decrease the possibility of gastric irritation due to the accumulation of large particles of the undissolved acid and their adhesion to the gastric mucosa. Levy and Haves[157] have shown that the absorption rate of aspirin and the incidence of gastric distress were a function of the dissolution rate of its particular dosage form. A more rapid dissolution rate of calcium and buffered aspirin was believed to account for faster absorption. They also established that significant variations exist in dissolution rates of different nationally distributed brands of plain aspirin tablets. This may account for some of the conflicting reports and opinions concerning the relative advantages of plain and buffered aspirin tablets. Lieberman and co-workers[158] have also shown that buffering is effective in raising the blood levels of aspirin. In a measure of the antianxiety effect of aspirin by means of electroencephalograms (EEG), differences between buffered, brand name, and generic aspirin preparations were found.[159]

Potentiation of salicylate activity by virtue of simultaneous administration of *p*-aminobenzoic acid or its salts has been the basis for the introduction of numerous products of this kind. Salassa and his co-workers have shown this effect to be due to the inhibition both of salicylate metabolism and of excretion in the urine.[160] This effect has been proved amply, provided that the ratio of 24 g. of *p*-aminobenzoic acid to 3 g. of salicylate per day is observed. However, there is no strong evidence to substantiate any significant elevation of plasma salicylate levels when a lesser quantity of *p*-aminobenzoic acid is employed.

The derivatives of salicylic acid are of two types (I and II (a, b)):

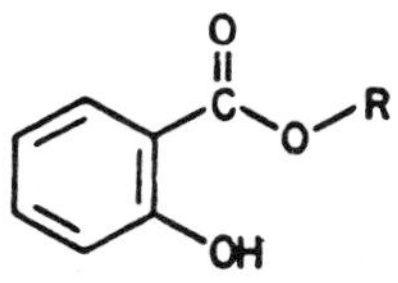

I

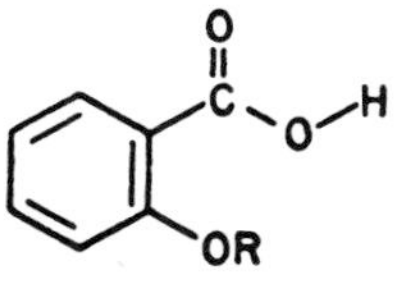

IIa 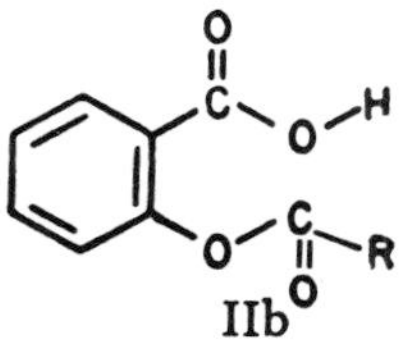 **IIb**

Type I represents those which are formed by modifying the carboxyl group (e.g., salts, esters or amides). Type II (a and b) represents those which are derived by substitution on the hydroxyl group of salicylic acid. The derivatives of salicylic acid were introduced in an attempt to prevent the gastric symptoms and the undesirable taste inherent in the common salts of salicylic acid. Hydrolysis of type I takes place to a greater extent in the intestine, and most of the type II compounds are absorbed into the bloodstream (see aspirin).

Compounds of type I

The alkyl and aryl esters of salicylic acid (type I) are used externally, primarily as counterirritants, where most of them are well absorbed through the skin. This type of compound is of little value as an analgesic.

A few inorganic salicylates are used internally when the effect of the salicylate ion is intended. These compounds vary in their irritation of the stomach. To prevent the development of pink or red coloration in the product, contact with iron should be avoided in the manufacture.

Sodium Salicylate U.S.P. may be prepared by the reaction, in aqueous solution, between 1 mole each of salicylic acid and sodium bicarbonate; upon evaporating to dryness, the white salt is obtained.

Generally, the salt has a pinkish tinge or is a white, microcrystalline powder. It is odorless or has a faint, characteristic odor, and it has a sweet, saline taste. It is affected by light. The compound is soluble in water (1:1), alcohol (1:10) and glycerin (1:4).

In solution, particularly in the presence of sodium bicarbonate, the salt will darken on standing (see salicylic acid). This darkening may be lessened by the addition of sodium sulfite or sodium bisulfite. Also, a color change is lessened by using recently boiled distilled water and dispensing in amber-colored bottles. Sodium salicylate forms a eutectic mixture with antipyrine and produces a violet coloration with iron or its salts. Solutions of the compound must be neutral or slightly basic to prevent precipitation of free salicylic acid. However, the U.S.P. salt forms neutral or acid solutions.

This salt is the one of choice for salicylate medication and usually is administered with sodium bicarbonate to lessen gastric distress, or it is administered in enteric-coated tablets. The use of sodium bicarbonate[161] is ill-advised, since it has been shown to decrease the plasma levels of salicylate and to increase the excretion of free salicylate in the urine.

Sodium Thiosalicylate, Arthrolate®, Nalate®, Thiodyne®, Thiolate®, Thiosul®, TH Sal®, is the sulfur or thio analog of sodium salicylate. It is more soluble and better absorbed, thereby requiring lower dosages. It is recommended for gout, rheumatic fever, and muscular pains in doses of 100 to 150 mg. every 3 to 6 hours for 2 days, and then 100 mg. once or twice daily.

Magnesium Salicylate U.S.P., Analate®, Causalin®, Lorisal®, Mobidin®, MSG-600®, Triact®, Magan®, is a sodium-free salicylate preparation that may be used in conditions in which sodium intake is restricted. It is claimed to produce less gastrointestinal upset. The dosage and indications are the same as for sodium salicylate.

Choline Salicylate, Arthropan®. This salt of salicylic acid is extremely soluble in water. It is claimed to be absorbed more rapidly than aspirin, giving faster peak blood levels. It is used in conditions where salicylates are indicated in a recommended dose of 870 mg. to 1.74 g. 4 times daily.

Other salts of salicylic acid that have found use are

those of ammonium, lithium and strontium. They offer no distinct advantage over sodium salicylate.

Carbethyl Salicylate, Sal-Ethyl Carbonate®, ethyl salicylate carbonate, is an ester of ethyl salicylate and carbonic acid and thus is a combination of a type I and type II compound.

Carbethyl Salicylate

It occurs as white crystals, insoluble in water and in diluted hydrochloric acid, slightly soluble in alcohol or ether and readily soluble in chloroform or acetone. The insolubility tends to prevent gastric irritation and makes it tasteless.

In action and uses it resembles aspirin and gives the antipyretic and analgesic effects of the salicylates. The pharmaceutic forms are powder, tablet and a tablet containing aminopyrine.

The usual dose is 1.0 g.

The salol principle

Nencki introduced salol in 1886 and by so doing presented to the science of therapy the "Salol Principle." In the case of salol, two toxic substances (phenol and salicylic acid) were combined into an ester which, when taken internally, will slowly hydrolyze in the intestine to give the antiseptic action of its components (q.v.). This type of ester is referred to as a "Full Salol" or "True Salol" when both components of the ester are active compounds. Examples are guaiacol benzoate, β-naphthol benzoate, and salol.

This "Salol Principle" can be applied to esters of which only the alcohol or the acid is the toxic, active or corrosive portion, and this type is called a "Partial Salol."

Examples of a "Partial Salol" containing an active acid are ethyl salicylate, and methyl salicylate. Examples of a "Partial Salol" containing an active phenol are creosote carbonate, thymol carbonate and guaiacol carbonate.

Although a host of the "salol" type of compounds have been prepared and used to some extent, none is presently very valuable in therapeutics, and all are surpassed by other agents.

Phenyl Salicylate, Salol. Phenyl salicylate occurs as fine, white crystals or as a white, crystalline powder with a characteristic taste and a faint, aromatic odor. It is insoluble in water (1:6700), slightly soluble in glycerin, soluble in alcohol (1:6), ether, chloroform, acetone or fixed and volatile oils.

Damp or eutectic mixtures form readily with many organic materials, such as thymol, menthol, camphor, chloral hydrate and phenol.

Salol is insoluble in the gastric juice but is slowly hydrolyzed in the intestine into phenol and salicylic acid. Because of this fact, coupled with its low melting point (41 to 43°), it has been used in the past as an enteric coating for tablets and capsules. However, it is not efficient as an enteric coating material, and its use has been superseded by more effective materials.

It also is used externally as a sun filter (10 % ointment) for sunburn prevention.

Salicylamide, *o*-hydroxybenzamide. This is a derivative of salicylic acid that has been known for almost a century and has found renewed interest. It is readily prepared from salicyl chloride and ammonia. The compound occurs as a nearly odorless, white, crystalline powder. It is fairly stable to heat, light and moisture. It is slightly soluble in water (1:500), soluble in hot water, alcohol (1:15) and propylene glycol, and sparingly soluble in chloroform and ether. It is freely soluble in solutions of alkalies. In alkaline solution with sodium carbonate or triethanolamine, decomposition takes place, resulting in a precipitate and yellow to red color.

Salicylamide

Salicylamide is said to exert a moderately quicker and deeper analgesic effect than does aspirin. Long-term studies on rats revealed no untoward symptomatic or physiologic reactions. Its metabolism is different from that of other salicylic compounds, and it is not hydrolyzed to salicylic acid.[148] Its analgesic and antipyretic activity is probably no greater than that of aspirin and possibly less. However, it can be used in place of salicylates and is particularly useful for those cases where there is a demonstrated sensitivity to salicylates. It is excreted much more rapidly than other salicylates, which probably accounts for its lower toxicity, and thus does not permit high blood levels.

The dose for simple analgesic effect may vary from 300 mg. to 1 g. administered 3 times daily; but for rheumatic conditions the dose may be increased to 2 to 4 g. 3 times a day. However, gastric intolerance may limit the dosage. The usual period of this higher dosage should not extend beyond 3 to 6 days. It is available in several combination products.

Aspirin U.S.P., Aspro®, Empirin®, acetylsalicylic acid. Aspirin was introduced into medicine by Dreser in 1899. It is prepared by treating salicylic acid, which was first prepared by Kolbe in 1874, with acetic anhydride.

The hydrogen atom of the hydroxyl group in salicylic acid has been replaced by the acetyl group; this also may be accomplished by using acetyl chloride with salicylic acid or ketene with salicylic acid.

Aspirin

Aspirin occurs as white crystals or as a white, crystalline powder. It is slightly soluble in water (1:300) and soluble in alcohol (1:5), chloroform (1:17) and ether (1:15). Also, it dissolves easily in glycerin. Aqueous solubility may be increased by using acetates or citrates of alkali metals, although these are said to decompose it slowly.

It is stable in dry air, but in the presence of moisture, it slowly hydrolyzes into acetic and salicylic acids. Salicylic acid will crystallize out when an aqueous solution of aspirin and sodium hydroxide is boiled and then acidified.

Aspirin itself is sufficiently acid to produce effervescence with carbonates and, in the presence of iodides, to cause the slow liberation of iodine. In the presence of alkaline hydroxides and carbonates, it decomposes, although it does form salts with alkaline metals and alkaline earth metals. The presence of salicylic acid, formed upon hydrolysis, may be confirmed by the formation of a violet color upon the addition of ferric chloride solution.

Aspirin is not hydrolyzed appreciably on contact with weakly acid digestive fluids of the stomach, but on passage into the intestine, is subjected to some hydrolysis. However, most of it is absorbed unchanged. The gastric mucosal irritation of aspirin has been ascribed by Garrett[162] to salicylic acid formation, the natural acidity of aspirin, or the adhesion of undissolved aspirin to the mucosa. He has also proposed the nonacidic anhydride of aspirin as a superior form for oral administration. Davenport[154] concludes that aspirin causes an alteration in mucosal cell permeability, allowing back-diffusion of stomach acid which damages the capillaries. A number of proprietaries (e.g.,

Bufferin®) employ compounds, such as sodium bicarbonate, aluminum glycinate, sodium citrate, aluminum hydroxide or magnesium trisilicate, to counteract this acid property. One of the better antacids is Dihydroxyaluminum Aminoacetate U.S.P. Aspirin has been shown to be unusually effective when prescribed with calcium glutamate. The more stable, nonirritant calcium acetylsalicylate is formed, and the glutamate portion (glutamic acid) maintains a pH of 3.5 to 5.

Preferably, dry dosage forms (i.e., tablets, capsules or powders) should be used, since aspirin is somewhat unstable in aqueous media. In tablet preparations, the use of acid-washed talc has been shown to improve the stability of aspirin.[163] Also, it has been found to break down in the presence of phenylephrine hydrochloride.[164] Aspirin in aqueous media will hydrolyze almost completely in less than 1 week. However, solutions made with alcohol or glycerin do not decompose as quickly. Citrates retard hydrolysis only slightly. Some studies have indicated that sucrose tends to inhibit hydrolysis. A study of aqueous aspirin suspensions has indicated sorbitol to exert a pronounced stabilizing effect.[165] Stable liquid preparations are available that use triacetin, propylene glycol or a polyethylene glycol. Aspirin lends itself readily to combination with many other substances but tends to soften and become damp with methenamine, aminopyrine, salol, antipyrine, phenol or acetanilid.

Aspirin is one of the most widely used compounds in therapy and, until recently, was not associated with untoward effects. Allergic reactions to aspirin now are observed commonly. Asthma and urticaria are the most common manifestations and, when they occur, are extremely acute in nature and difficult to relieve. Like sodium salicylate, it has been shown to cause congenital malformations when administered to mice.[166] Pretreatment with sodium pentobarbital or chlorpromazine resulted in a significant lowering of these effects.[167] Similar effects have been attributed to the consumption of aspirin in women and its use during pregnancy should be avoided. However, other studies indicate that no untoward effects are seen. The reader is urged to consult the excellent review by Smith for an account of the pharmacologic aspects of aspirin.[148,149]

Practically all salts of aspirin, except those of aluminum and calcium, are unstable for pharmaceutic use. These salts appear to have fewer undesirable side-effects and to induce analgesia faster than aspirin.

A timed-release preparation (Measurin®) of aspirin is available. It does not appear to offer any advantages over aspirin except for bedtime dosage.

Aspirin is used as an antipyretic, analgesic and antirheumatic, usually in powder, capsule, suppository or tablet form. Its use in rheumatism has been reviewed and it is said to be the drug of choice over all other salicylate derivatives.[168,169] There is some anesthetic action when applied locally, especially in powder form in tonsilitis or pharyngitis, and in ointment form for skin itching and certain skin diseases. In the usual dose, 52 to 75 percent is excreted in the urine, in various forms, in a period of 15 to 30 hours. It is believed that analgesia is due to the unhydrolyzed acetylsalicylic acid molecule.[148-150] A widely used combination is aspirin, phenacetin, or, as of recently, acetominophen and caffeine, known as APC.

Aluminum Aspirin, hydroxy*bis*(salicylato)aluminum diacetate. This salt of aspirin may be prepared by thoroughly mixing aluminum hydroxide gel, water and acetylsalicylic acid, maintaining the temperature below 65°. Aluminum aspirin occurs as a white to off-white powder or granules and is odorless or has only a slight odor. It is insoluble in water and organic solvents, is decomposed in aqueous solutions of alkali hydroxides and carbonates and is not stable above 65°. It offers the advantages of being free of odor and taste and possesses added shelf-like stability. It is available in a flavored form for children (Dulcet®).

Aluminum Aspirin

Calcium Acetylsalicylate, soluble aspirin, calcium aspirin. This compound is prepared by treating acetylsalicylic acid with calcium ethoxide or methoxide in alcohol or acetone solution. It is readily soluble in water (1:6) but only sparingly soluble in alcohol (1:80). It is more stable in solution than aspirin and is used for the same conditions.

Calcium Acetylsalicylate

Calcium aspirin is marketed also as a complex salt with urea, calcium carbaspirin (Calurin®), which is claimed to give more rapid salicylate blood levels and to be less irritating than aspirin, although no clear advantage has been shown.

The usual dose is 500 mg. to 1.0 g.

Salsalate, Arcylate®, Disalcid®, salicylsalicylic acid, is the ester formed between two salicylic-acid molecules to which it is hydrolyzed following absorption. It is said to cause less gastric upset than aspirin because it is relatively insoluble in the stomach and is not absorbed until it reaches the small intestine. Limited clinical trials[170-172] suggest that it is as effective as aspirin and that it may have fewer side-effects.[173] The recommended dose is 325 to 1000 mg. 2 or 3 times a day.

Flufenisal, acetyl-5-(4-fluorophenyl)salicylic acid, 5'-fluoro-4-hydroxy-3-biphenylcarboxylic acid acetate. Over the years several hundred analogs of aspirin have been made and tested in order to produce a compound that was more potent, longer acting and with less gastric irritation. By the introduction of a hydrophobic group in the 5-position, flufenisal appears to meet these criteria. In animal tests it is at least 4 times more potent. In humans, it appears to be about twice as effective with twice the duration.[174] Like other aryl acids it is highly bound to protein plasma as its deacylated metabolite. Further clinical trials are in progress.

THE N-ARYLANTHRANILIC ACIDS

One of the early advances in the search for non-narcotic analgesics was centered in the N-arylanthranilic acids. Their outstanding characteristic is that they are primarily nonsteroidal anti-inflammatory agents and, secondarily, that some possess analgesic properties.

Mefenamic Acid, Ponstel®, N-(2,3-xylyl)anthranilic acid, (a), occurs as an off-white, crystalline powder that is insoluble in water and slightly soluble in alcohol. It appears to be the first genuine antiphlogistic analgesic discovered since aminopyrine. Because it is believed that aspirin and aminopyrine owe their general-purpose analgesic efficacy to a combination of peripheral and central effects,[175] a wide variety of arylanthranilic acids were screened for antinociceptive (analgesic) activity if they showed significant anti-inflammatory action. It has become evident that the combination of both effects is a rarity among these compounds. The mechanism of analgesic action is believed to be related to its ability to block prostaglandin synthetase. No relationship to lipid-plasma distribution, partition coefficient, or pKa has been noted. The interested reader, however, will find additional information with respect to antibradykinin and anti-UV-erythema activities of these compounds together with speculations on a receptor site in the literature.[176]

It has been shown[177] that mefenamic acid in a dose of 250 mg. is superior to 600 mg. of aspirin as an anal-

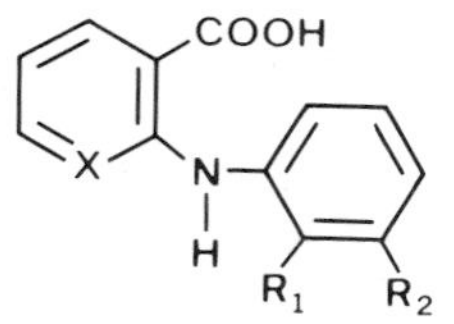

(a) $R_1 = CH_3$, $X = CH$, $R_2 = H$

(b) $R_1 = H$

$R_2 = CF_3$, $X = CH$

gesic and that doubling the dose gives a sharp increase in efficacy. A study[178] examining this drug with respect to gastrointestinal bleeding indicated that it has a lower incidence of this side-effect than has aspirin. Diarrhea, drowsiness and headache have accompanied its use. The possibility of blood disorders has prompted limitation of its administration to 7 days. It is not recommended for children or during pregnancy. It has been approved for use in the management of primary dysmenorrhea, which is thought to be caused by excessive concentrations of prostaglandins and endoperoxides.

Meclofenamate Sodium, Meclomen®, is sodium N-(2,6-dichloro-*m*-tolyl)anthranilate.

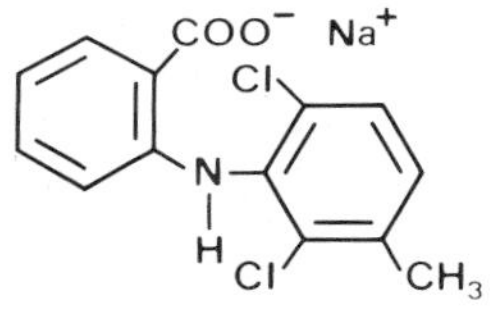

Meclofenamate Sodium

This drug is available in 50 and 100 mg. capsules for use in the treatment of acute and chronic rheumatoid arthritis. The most significant side-effects are gastrointestinal in nature, including diarrhea.

ARYLACETIC ACID DERIVATIVES

This group of anti-inflammatory agents is receiving the most intensive attention for new clinical candidates. As a group they have the characteristic of showing high analgesic potency in addition to their anti-inflammatory activity.

Indomethacin U.S.P., Indocin®, 1-(*p*-chlorobenzoyl)-5-methoxy-2-methylindole-3-acetic acid, occurs as a pale yellow to yellow-tan crystalline powder which is soluble in ethanol and acetone and practically insoluble in water. It is unstable in alkaline solution and sunlight. It shows polymorphism, one form melting at about 155° and the other at about 162°. It may

occur as a mixture of both forms with a melting range between the above melting points.

Since its introduction in 1965, it has been widely used as an anti-inflammatory analgesic in rheumatoid arthritis, spondylitis, and osteoarthritis, and to a lesser extent in gout. Although both its analgesic and anti-inflammatory activities have been well established, it appears to be no more effective than aspirin.[179]

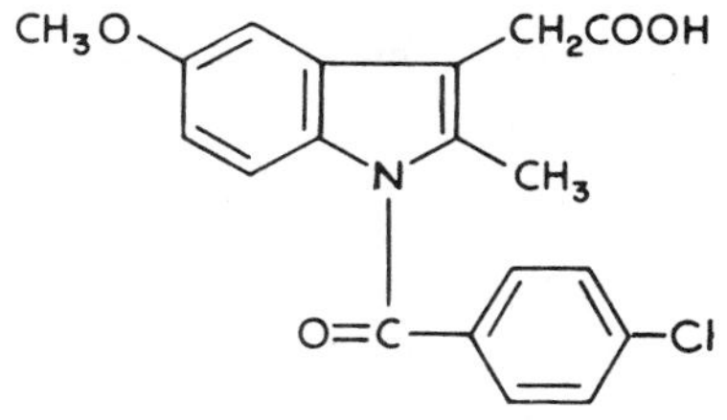

Indomethacin

The most frequent side-effects are gastric distress and headache. It has also been associated with peptic ulceration, blood disorders, and possible deaths. The side-effects appear to be dose-related and sometimes can be minimized by reducing the dose. It is not recommended for use in children because of possible interference with resistance to infection. Like many other acidic compounds, it circulates bound to blood protein, requiring caution in the concurrent use of other protein-binding drugs.

Indomethacin is recommended only for those patients by whom aspirin cannot be tolerated, and in place of phenylbutazone in long-term therapy, for which it appears to be less hazardous than corticosteroids or phenylbutazone.

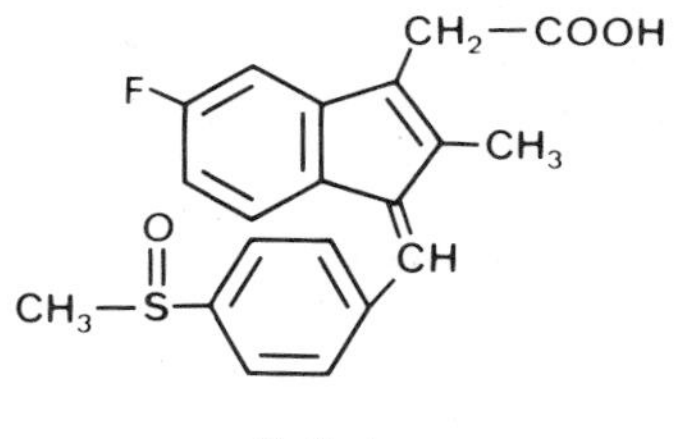

Sulindac

Sulindac U.S.P., Clinoril®, (Z)-5-fluoro-2-methyl-1-{[*p*-(methylsulfinyl)phenyl]methylene}-1H-indene-3-acetic acid, occurs as yellow crystals soluble in alkaline but insoluble in acidic solutions. The drug reaches peak blood levels within 2 to 4 hours and undergoes a complicated, reversible metabolism shown as follows:

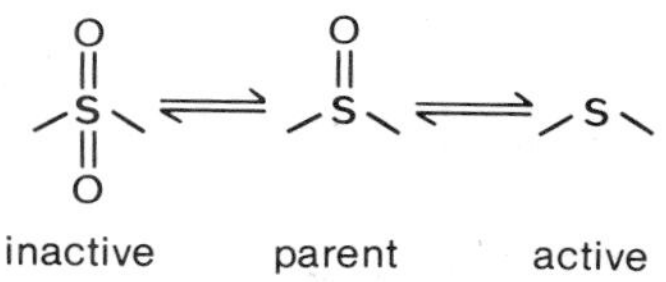

inactive   parent   active

The parent sulfinyl has a plasma half-life of 8 hours, with that of the active sulfide metabolite being 16.4 hours. The more polar and inactive sulfoxide is virtually the sole form excreted. The long half-life is due to extensive enterohepatic recirculation.[180] In in vitro studies, only the sulfide species inhibits prostaglandin synthetase. Although these forms are highly protein bound, the drug does not appear to affect binding of anticoagulants or hypoglycemics. Coadministration of aspirin is contraindicated because it reduces significantly the sulfide blood levels.

Careful monitoring of patients with a history of ulcers is recommended. Gastric bleeding, nausea, diarrhea, dizziness, and other adverse effects have been noted but with a lower incidence than with aspirin. Sulindac is recommended for rheumatoid arthritis, osteoarthritis, and ankylosing spondolitis in 150 to 200 mg. dose, twice daily.[181] It is available as tablets (150 and 200 mg.).

Tolmetin Sodium U.S.P., Tolectin®, McN-2559, 1-methyl-5-(p-toluoyl)pyrrole-2-acetate dihydrate sodium, is a newly introduced arylacetic acid derivative with a pyrrole as the aryl group. This drug is rapidly absorbed with a relatively short plasma half-life (1 hour). It is recommended for use in the management of acute and long-term rheumatoid arthritis. It shares similar but less frequent adverse effects as with aspirin. It does not potentiate coumarinlike drugs nor alter the blood levels of sulfonylureas or insulin. As with other drugs in this class, it is known to inhibit prostaglandin synthetase and lower prostaglandin-E blood levels.

Available as tablets (200 mg.), a dose of 400 mg. three times daily with a maximum of 2000 mg. is recommended. Clinical trials indicate a usual daily dose of 1200 mg. is comparable in relief to 3.9 grams of aspirin and 150 mg. of indomethacin per day.[182]

Tolmetin: $R_1 = CH_3$, $R_2 = H$
Zomepirac: $R_1 = Cl$, $R_2 = CH_3$

Zomepirac Sodium, Zomax®, McN-2783, 5-(4-chlorobenzoyl)-1,4-dimethyl-1*H*-pyrrole-2-acetate dihydrate sodium, is the chloro analog of tolmetin. It shows significantly longer plasma levels (7 hours),[183] thereby requiring less frequent dosing. In pain relief, 25 to 50 mg. is reported to give relief equivalent to 650 mg. of aspirin. In a study on cancer patients, oral doses of 100 to 200 mg. were as effective as moderate parenteral doses of morphine.[184] Zomepirac is marketed in 100 mg. tablets.

Ibuprofen U.S.P., Motrin®, 2-(4-*iso*butylphenyl)-propionic acid. This arylacetic acid derivative was introduced into clinical practice following extensive clinical trials. It appears comparable to aspirin in the treatment of rheumatoid arthritis, with a lower incidence of side-effects.[185] It has also been approved for use in primary dysmenorrhea.

Ibufenac R = H
Ibuprofen R = CH₃

Of interest in this series of compounds is that it was noted that potency was enhanced by introduction of the α-methyl group on the acetic acid moiety. The precursor ibufenac (R = H), which was abandoned due to hepatotoxicity, was less potent. Moreover, it was found that the activity resides in the S-(+)-isomer, not only in ibuprofen, but throughout the arylacetic acid series. Furthermore, it is these isomers that are the more potent inhibitors of prostaglandin synthetase.[146]

Namoxyrate, Namol®, 2-(4-biphenyl)butyric acid dimethylaminoethanol salt, is another phenylacetic acid derivative under investigation. Namoxyrate shows high analgesic activity, being about 7 times that of aspirin and nearly as effective as codeine. It has high antipyretic activity but appears to be devoid of anti-inflammatory activity. These effects are peripheral. The dimethylaminoethanol increases its activity by increasing intestinal absorption. The ester of these two components is less active.[186]

Namoxyrate

Naproxen U.S.P., Anaprox®, Naprosyn®, (+)-6-methoxy-α-methyl-2-naphthaleneacetic acid, occurs as white to off-white crystals that are sparingly soluble in acidic solutions, freely soluble in alkaline solutions, and highly soluble in organic or lipidlike solutions. Following oral administration, it is well absorbed, giving peak blood levels in 2 to 4 hours and a half-life of 13 hours. A steady-state blood level is usually achieved following 4 to 5 doses. This drug is very highly protein bound and displaces most protein-bound drugs. Dosages of these must be adjusted accordingly.

Naproxen

Naproxen is recommended for use in rheumatoid and gouty arthritis. It shows good analgesic activity, 400 mg. being comparable to 75 to 150 mg. of oral meperidine and superior to 65 mg. propoxyphene and 325 mg. of aspirin plus 30 mg. of codeine. A 220- to 330-mg. dose was comparable to 600 mg. of aspirin alone. It has been reported to produce dizziness, drowsiness, and nausea, with infrequent mentions of gastrointestinal-tract irritation. Like aspirin it inhibits prostaglandin synthetase and prolongs blood-clotting time. It is not recommended for pregnant or lactating women or in children under 16.[189]

Fenoprofen Calcium U.S.P., Nalfon®, α-methyl-3-phenoxybenzeneacetic acid dihydrate calcium, occurs as a white, crystalline powder that is slightly soluble in water, soluble in alcohol, and insoluble in benzene. It is rapidly absorbed orally, giving peak blood levels within 2 hours and has a short plasma half-life (3 hrs.). It is highly protein bound like the other acylacetic acids, and caution must be exercised when used concurrently with hydantoins, sulfonamides, and sulfonylureas. It shares many of the adverse effects common to this group of drugs, with gastrointestinal bleeding, ulcers, dyspepsia, nausea, sleepiness, and dizziness reported at a lower incidence than with aspirin. It inhibits prostaglandin synthetase.[187]

Fenoprofen

Available as capsules (200 and 300 mg.) and tablets (600 mg.) it is recommended for rheumatoid arthritis and osteoarthritis in divided doses four times a day for a maximum of 3200 mg. per day. It should be taken at least 30 minutes before or 2 hours after meals. It is not yet recommended for the management of acute flare-ups. Doses of 2.4 g. per day have been shown to be comparable to 3.9 g. per day of aspirin in arthritis. For pain relief, 400 mg. gave similar results to 650 mg. of aspirin.[188]

Piroxicam, Feldene®, CP 16171, 4-hydroxy-2-methyl-N-2-pyridyl-H-1,2-benzothiazine-3-carboximide 1,1-dioxide, represents a new class of acidic in-

hibitors of prostaglandin synthetase, although it does not antagonize PGE$_2$ directly.[190] This drug, recently introduced in Europe, is very long acting with a plasma half-life of 38 hours, thereby requiring a dose of only 20 to 30 mg. once daily. It is reported to give similar results to 25 mg. of indomethacin or 400 mg. of ibuprofen 3 times a day.[191,192]

Piroxicam

Several other arylacetic acid derivatives are under current clinical evaluation. These include ketoprofen, alclofenac, fenclofenac, pirprofen, and prodolic and bucloxic acids. Although only early reports are available, many of these appear to show superiority over indomethacin and aspirin. The reader may consult the reviews of Evens and Scherrer and Whitehouse for further details.

ANILINE AND p-AMINOPHENOL DERIVATIVES

The introduction of aniline derivatives as analgesics is based on the discovery by Cahn and Hepp, in 1886, that aniline (C-1)* and acetanilid (C-2) both have powerful antipyretic properties. The origin of this group from aniline has led to their being called "coal tar analgesics." Acetanilid was introduced by these workers because of the known toxicity of aniline itself. Aniline brings about the formation of methemoglobin, a form of hemoglobin that is incapable of functioning as an oxygen carrier. The acyl derivatives of aniline were thought to exert their analgesic and antipyretic effects by first being hydrolyzed to aniline and the corresponding acid, following which the aniline was oxidized to p-aminophenol (C-3). This is then excreted in combination with glucuronic or sulfuric acid.

The aniline derivatives do not appear to act upon the brain cortex; the pain impulse appears to be intercepted at the hypothalamus, wherein also lies the thermoregulatory center of the body. It is not clear if this is the site of their activity, because most evidence suggests that they act at peripheral thermoceptors. They are effective in the return to normal temperature of feverish individuals. Normal body temperatures are not affected by the administration of these drugs.

* See Table 17-10.

It is significant to note that, of the antipyretic-analgesic group, the aniline derivatives show little if any anti-inflammatory activity.

Table 17-10 shows some of the types of aniline derivatives that have been made and tested in the past. In general, any type of substitution on the amino group that reduces its basicity results also in a lowering of its physiologic activity. Acylation is one type of substitution that accomplishes this effect. Acetanilid (C-2) itself, although the best of the acylated derivatives, is toxic in large doses but when administered in analgesic doses is probably without significant harm. Formanilid (C-4) is readily hydrolyzed and too irritant. The higher homologs of acetanilid are less soluble and, therefore, less active and less toxic. Those derived from aromatic acids (e.g., C-5) are virtually without analgesic and antipyretic effects. One of these, salicylanilide (C-6), is used as a fungicide and antimildew agent. Exalgin (C-7) is too toxic.

The hydroxylated anilines *(o, m, p)*, better known as the aminophenols, are quite interesting from the standpoint of being considerably less toxic than aniline. The *para* compound (C-3) is of particular interest from two standpoints, namely, it is the metabolic product of aniline, and it is the least toxic of the three possible aminophenols. It also possesses a strong antipyretic and analgesic action. However, it is too toxic to serve as a drug and, for this reason, there were numerous modifications attempted. One of the first was the acetylation of the amine group to provide N-acetyl-*p*-aminophenol (acetaminophen) (C-8), a product which retained a good measure of the desired activities. Another approach to the detoxification of *p*-aminophenol was the etherification of the phenolic group. The best known of these are anisidine (C-9) and phenetidine (C-10), which are the methyl and ethyl ethers, respectively. However, it became apparent that a free amino group in these compounds, while promoting a strong antipyretic action, was also conducive to methemoglobin formation. The only exception to the preceding was in compounds where a carboxyl group or sulfonic acid group had been substituted on the benzene nucleus. In these compounds, however, the antipyretic effect also had disappeared. The above considerations led to the preparation of the alkyl ethers of N-acetyl-*p*-aminophenol of which the ethyl ether was the best and is known as phenacetin (C-11). The methyl and propyl homologs were undesirable from the standpoint of causing emesis, salivation, diuresis and other reactions. Alkylation of the nitrogen with a methyl group has a potentiating effect on the analgesic action but, unfortunately, has a highly irritant action on mucous membranes.

The phenacetin molecule has been modified by changing the acyl group on the nitrogen with some-

TABLE 17–10

SOME ANALGESICS RELATED TO ANILINE

Com-pound	R₁	R₂	R₃	Name
C-1	—H	—H	—H	Aniline
C-2	—H	—H	—C(=O)—CH₃	Acetanilid
C-3	—OH	—H	—H	p-Aminophenol
C-4	—H	—H	—C(=O)—H	Formanilid
C-5	—H	—H	—C(=O)—C₆H₅	Benzanilid
C-6	—H	—H	—C(=O)— (o-hydroxyphenyl)	Salicylanilide (not an analgesic, but is an antifungal agent)
C-7	—H	—CH₃	—C(=O)—CH₃	Exalgin
C-8	—OH	—H	—C(=O)—CH₃	Acetaminophen
C-9	—OCH₃	—H	—H	Anisidine
C-10	—OC₂H₅	—H	—H	Phenetidine
C-11	—OC₂H₅	—H	—C(=O)—CH₃	Phenacetin
C-12	—OC₂H₅	—H	—C(=O)—CHCH₃ (with OH)	Lactylphenetidin
C-13	—OC₂H₅	—H	—C(=O)—CH₂NH₂	Phenocoll
C-14	—OC₂H₅	—H	—C(=O)—CH₂OCH₃	Kryofine
C-15	—OC(=O)—CH₃	—H	—C(=O)—CH₃	p-Acetoxyacetanilid
C-16	—OC(=O)— (o-hydroxyphenyl)	—H	—C(=O)—CH₃	Phenetsal
C-17	—OCH₂CH₂OH	—H	—C(=O)—CH₃	Pertonal

times beneficial results. Among these are lactylphenetidin (C-12), phenocoll (C-13) and kryofine (C-14). None of these, however, is in current use.

Changing the ether group of phenacetin to an acyl type of derivative has not always been successful. *p*-

Acetoxyacetanilid (C-15) has about the same activity and disadvantages as the free phenol. However, the salicyl ester (C-16) exhibits a diminished toxicity and an increased antipyretic activity. Pertonal (C-17) is a somewhat different type in which glycol has been used to etherify the phenolic hydroxyl group. It is very similar to phenacetin. None of these is presently on the market.

With respect to the fate in man of the types of compounds discussed above, Brodie and Axelrod[193] point out that acetanilid and phenacetin are metabolized by two different routes. Acetanilid is metabolized primarily to N-acetyl-p-aminophenol, acetaminophen, and only a small amount to aniline, which they showed to be the precursor of phenylhydroxylamine, the compound responsible for methemoglobin formation. Phenacetin is mostly de-ethylated to acetaminophen, whereas a small amount is converted by deacetylation to p-phenetidine, also responsible for methemoglobin formation. With both acetanilid and phenacetin, the metabolite acetaminophen formed is believed to be responsible for the analgesic activity of the compounds.

Acetanilid, antifebrin, phenylacetamid, is the monoacetyl derivative of aniline, prepared by heating aniline and acetic acid for several hours.

It can be recrystallized from hot water and occurs as a stable, white, crystalline compound. It is slightly soluble in water (1:190) and easily soluble in hot water, acetone, chloroform, glycerin (1:5), alcohol (1:4) or ether (1:17).

Acetanilid is a neutral compound and will not dissolve in either acids or alkalies.

It is prone to form eutectic mixtures with aspirin, antipyrine, chloral hydrate, menthol, phenol, pyrocatechin, resorcinol, salol, thymol or urethan.

It is definitely toxic in that it causes formation of methemoglobin, affects the heart and may cause skin reactions and a jaundiced condition. Nevertheless, in the doses used for analgesia, it is a relatively safe drug. However, it is recommended that it be administered in intermittent periods, no period exceeding a few days.[194]

The analgesic effect is selective for most simple headaches and for the pain associated with many muscles and joints.

The usual dose is 200 mg.

A number of compounds related to acetanilid have been synthesized in attempts to find a better analgesic, as previously indicated. They have not become very important in the practice of medicine, for they have little to offer over acetanilid. The physical and chemical properties are also much the same. Eutectic mixtures are formed with many of the same compounds.

Phenacetin U.S.P., acetophenetidin, p-acetophenetidide, may be synthesized in several steps from p-nitrophenol.

It occurs as stable, white, glistening crystals, usually in scales, or a fine, white, crystalline powder. It is odorless and has a slightly bitter taste. It is very slightly soluble in water (1:1,300), soluble in alcohol (1:15) and chloroform (1:15), but only slightly soluble in ether (1:130). It is sparingly soluble in boiling water (1:85).

In general properties and incompatibilities, such as decomposition by acids and alkalies, it is similar to acetanilid. Phenacetin forms eutectic mixtures with chloral hydrate, phenol, aminopyrine, pyrocatechin or pyrogallol.

It is used widely as an analgesic and antipyretic, having essentially the same actions as acetanilid. It should be used with the same cautions because the toxic effects are the same as those of acetaminophen, the active form to which it is converted in the body. Some feel there is little justification for its continued use,[193] and it is presently restricted to prescription use only. In particular, a suspected nephrotoxic action[195] has been the basis for the present warning label requirements by the Food and Drug Administration, i.e., "This medication may damage the kidneys when used in large amounts or for a long period of time. Do not take more than the recommended dosage, nor take regularly for longer than 10 days without consulting your physician." Some recent evidence suggests that phenacetin may not cause nephritis to any greater degree than aspirin, with which it has been most often combined.[196] However, it has been strongly indicated as being carcinogenic in rats and associated with tumors in abusers of phenocetin.[197,198] It has been removed from many combination products and replaced either with additional aspirin, e.g., Anacin®, or with acetaminophen.

Acetaminophen U.S.P., Datril®, Tempra®, Tylenol®, N-acetyl-p-aminophenol, 4'-hydroxyacetanilide. This may be prepared by reduction of p-nitrophenol in glacial acetic acid, acetylation of p-aminophenol with acetic anhydride or ketene, or from p-hydroxyacetophenone hydrazone. It occurs as a white, odorless, slightly bitter crystalline powder. It is slightly soluble in water and ether, soluble in boiling water (1:20), alcohol (1:10), and sodium hydroxide T.S.

Acetaminophen has analgesic and antipyretic activities comparable to those of acetanilid and is used in the same conditions. Although it possesses the same toxic effects as acetanilid, they occur less frequently and with less severity; therefore, it is considered somewhat safer to use. However, the same cautions should be applied. The required Food and Drug Administration warning label reads: "Warning: Do not give to children under three years of age or use for more than 10 days unless directed by a physician."[195]

It is available in several nonprescription forms and, also, is marketed in combination with aspirin and caffeine (Trigesic®).

THE PYRAZOLONE AND PYRAZOLIDINEDIONE DERIVATIVES

The simple doubly unsaturated compound containing 2 nitrogen and 3 carbon atoms in the ring, and with the nitrogen atoms neighboring, is known as pyrazole. The reduction products, named as are other rings of 5 atoms, are pyrazoline and pyrazolidine. Several pyrazoline substitution products are used in medicine. Many of these are derivatives of 5-pyrazolone. Some can be related to 3,5-pyrazolidinedione.

Pyrazole   Pyrazoline   Pyrazolidine

5-Pyrazolone   3,5-Pyrazolidinedione

Ludwig Knorr, a pupil of Emil Fischer, while searching for antipyretics of the quinoline type, in 1884, discovered the 5-pyrazolone now known as antipyrine. This discovery initiated the beginnings of the great German drug industry that dominated the field for approximately 40 years. Knorr, although at first mistakenly believing that he had a quinoline-type compound, soon recognized his error, and the compound was interpreted correctly as being a pyrazolone. Within 2 years, the analgesic properties of this compound became apparent when favorable reports began to appear in the literature, particularly with reference to its use in headaches and neuralgias. Since then, it has retained some of its popularity as an analgesic, although its use as an antipyretic has declined steadily. Since its introduction into medicine, there have been over 1,000 compounds made in an effort to find others with a more potent analgesic action combined with a lesser toxicity. That antipyrine remains as one of the useful analgesics today is a tribute to its value. Many modifications of the basic compound have been made. The few derivatives and modifications on the

TABLE 17-11
DERIVATIVES OF 5-PYRAZOLONE

Compound Proprietary Name	R₁	R₂	R₃	R₄
Antipyrine (Phenazone)	$-C_6H_5$	$-CH_3$	$-CH_3$	$-H$
Aminopyrine (Amidopyrine)	$-C_6H_5$	$-CH_3$	$-CH_3$	$-N(CH_3)_2$
Dipyrone (Methampyrone)	$-C_6H_5$	$-CH_3$	$-CH_3$	$-NCH_2SO_3Na$ $\vert$ CH_3

market are listed in Tables 17-11 and 17-12. Phenylbutazone, although analgesic itself, was originally developed as a solubilizer for the insoluble aminopyrine. It is being used at present for the relief of many forms of arthritis, in which capacity it has more than an analgesic action in that it also reduces swelling and spasm by an anti-inflammatory action.

PRODUCTS

Antipyrine U.S.P., Felsol®, phenazone, 2,3-dimethyl-1-phenyl-3-pyrazolin-5-one. This was one of the first important drugs to be made (1887) synthetically.

Antipyrine and many related compounds are prepared by the condensation of hydrazine derivatives

TABLE 17-12
DERIVATIVES OF 3,5-PYRAZOLIDINEDIONE

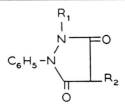

Compound Proprietary Name	R₁	R₂
Phenylbutazone Azolid, Butazolidin	$-C_6H_5$	$-C_4H_9$ (n)
Oxyphenbutazone Oxalid, Tandearil	$-C_6H_4(OH)$ (p)	$-C_4H_9$ (n)

with various esters. Antipyrine itself is prepared by the action of ethyl acetoacetate on phenylhydrazine and subsequent methylation.

It consists of colorless, odorless crystals or a white powder with a slightly bitter taste. It is very soluble in water, alcohol or chloroform, less so in ether, and its aqueous solution is neutral to litmus paper. However, it is basic in nature, which is due primarily to the nitrogen at position 2.

Locally, antipyrine exerts a paralytic action on the sensory and the motor nerves, resulting in some anesthesia and vasoconstriction, and it also exerts a feeble antiseptic effect. Systemically, it causes results that are very similar to those of acetanilid, although they are usually more rapid. It is readily absorbed after oral administration, circulates freely and is excreted chiefly by the kidneys without having been changed chemically. Any abnormal temperature is reduced rapidly by an unknown mechanism, usually attributed to an effect on the serotonin-mediated thermal regulatory center of the nervous system. It has a higher degree of anti-inflammatory activity than aspirin, phenylbutazone, and indomethacin. It also lessens perception to pain of certain types, without any alteration in central or motor functions, which differs from the effects of morphine. Very often it produces unpleasant and possibly alarming symptoms, even in small or moderate doses. These are giddiness, drowsiness, cyanosis, great reduction in temperature, coldness in the extremities, tremor, sweating and morbilliform or erythematous eruptions; with very large doses there are asphyxia, epileptic convulsions, and collapse. Treatment for such untoward reactions must be symptomatic. It is probably less likely to produce collapse than acetanilid and is not known to cause the granulocytopenia that sometimes follows aminopyrine.

Antipyrine has been employed in medicine less often in recent years than formerly. It is administered orally to reduce pain and fever in neuralgia, the myalgias, migraine, other headaches, chronic rheumatism and neuritis but is less effective than salicylates and more toxic. When used orally it is given as a 300-mg. dose. It sometimes is employed in motor disturbances, such as the spasms of whooping cough or epilepsy. Occasionally, it is applied locally in 5 to 15 percent solution for its vasoconstrictive and anesthetic effects in rhinitis and laryngitis and sometimes as a styptic in nosebleed.

The great success of antipyrine in its early years led to the introduction of a great many derivatives, especially salts with a variety of acids, but none of these has any advantage over the parent compound. Currently in use is the compound with chloral hydrate (Hypnal).

Aminopyrine, amidopyrine, aminophenazone, 2,3-dimethyl-4-dimethylamino-1-phenyl-3-pyrazolin-5-one. It is prepared from nitrosoantipyrine by reduction to the 4-amino compound followed by methylation.

It consists of colorless, odorless crystals that dissolve in water and the usual organic solvents. It has about the same incompatibilities as antipyrine.

It has been employed as an antipyretic and analgesic, as is antipyrine, but is somewhat slower in action. However, it seems to be much more powerful, and its effects last longer. The usual dose is 300 mg. for headaches, dysmenorrhea, neuralgia, migraine and other like disorders, and it may be given several times daily in rheumatism and other conditions that involve continuous pain.

One of the chief disadvantages of therapy with aminopyrine is the possibility of producing agranulocytosis (granulocytopenia). It has been shown that this is caused by drug therapy with a variety of substances, including mainly aromatic compounds, but particularly with aminopyrine; indeed, a number of fatal cases have been traced definitely to this drug. The symptoms are a marked fall in leukocytes, absence of granulocytes in the blood, fever, sore throat, ulcerations on mucous surfaces, and prostration, with death in the majority of cases from secondary complications. The treatment is merely symptomatic with penicillin to prevent any possible superimposed infection. The condition seems to be more or less an allergic reaction, because only a certain small percentage of those who use the drug are affected, but great caution must be observed to avoid susceptibility. Many countries have forbidden or greatly restricted its administration, and it has fallen more or less into disfavor.

Dipyrone, methampyrone, occurs as a white, odorless, crystalline powder possessing a slightly bitter taste. It is freely soluble in water (1:1.5) and sparingly soluble in alcohol.

It is used as an analgesic, an antipyretic and an antirheumatic. The recommended dose is 300 mg. to 1 g. orally and 500 mg. to 1 g. intramuscularly or subcutaneously.

Phenylbutazone U.S.P., Azolid®, Butazolidin®, Phenylzone-A®, 4-butyl-1,2-diphenyl-3,5-pyrazolidinedione. This drug is a white to off-white, odorless, slightly bitter-tasting powder. It has a slightly aromatic odor and is freely soluble in ether, acetone and ethyl acetate, very slightly soluble in water, and is soluble in alcohol (1:20).

According to the patents describing the synthesis of this type of compound, it can be prepared by condensing n-butyl malonic acid or its derivatives with hydrazobenzene to get 1,2-diphenyl-4-n-butyl-3,5-pyrazolidinedione. Alternatively, it can be prepared by

TABLE 17–13

ANTIPYRETIC ANALGESICS

Name Proprietary Name	Preparations	Category	Usual Adult Dose*	Usual Dose Range*	Usual Pediatric Dose*
Sodium Salicylate U.S.P.	Sodium Salicylate Tablets U.S.P.	Analgesic	Oral, 325 to 650 mg. every 4 hours, as needed; for arthritis—oral, 3.6 to 5.4 g. daily in divided doses		Oral, 1.5 g. per square meter of body surface daily in divided doses or for children up to 2 years of age—dosage to be individualized by physician; 2 to 4 years of age—162.5 mg. every 4 hours as needed; 4 to 6 years of age—243.8 mg. every 4 hours as needed; 6 to 9 years of age—325 mg. every 3 hours as needed; 9 to 11 years of age—406.3 mg. every 4 hours as needed; 11 to 12 years of age—487.5 mg. every 4 hours as needed
Aspirin U.S.P.	Aspirin Tablets U.S.P. Aspirin Capsules U.S.P.	Analgesic; antipyretic; antirheumatic	Oral, 325 to 650 mg. every 4 hours, as needed; for arthritis—oral, 3.6 to 5.4 g. daily in divided doses		
Indomethacin U.S.P. *Indocin*	Indomethacin Capsules U.S.P.	Antiinflammatory (nonsteroid)	Antirheumatic—oral, initially 25 or 50 mg. 2 to 4 times a day; if well tolerated, the dosage per day may be increased by 25 or 50 mg. at weekly intervals until a satisfactory response is obtained or up to a maximal dose of 200 mg. per day		Children up to 15 years of age—dosage has not been established
Acetaminophen U.S.P. *Tempra, Tylenol, Valadol, Datril*	Acetaminophen Elixir U.S.P. Acetaminophen Tablets U.S.P. Acetaminophen Capsules U.S.P. Acetaminophen Oral Suspension U.S.P. Acetaminophen Syrup U.S.P. Acetaminophen Chewable Tablets U.S.P.	Analgesic; antipyretic	Oral, 325 mg. to 650 mg. every 4 hours as necessary	325 mg. to 3.9 g. daily	Oral, 1.5 g. per square meter of body surface a day in divided doses; or for children up to 2 years of age—dosage to be individualized by physician; children 2 to 4 years of age—oral, 160 mg. every 4 hours as needed; children 4 to 6 years of age—oral, 240 mg. every 4 hours as needed; children 6 to 9 years of age—oral, 320 mg. every 4 hours as needed; children 9 to 11 years of age—oral, 400 mg. every 4 hours as needed; children 11 to 12 years of age—oral, 480 mg. every 4 hours as needed
	Acetaminophen Suppositories U.S.P.		Rectal, 325 to 650 mg. every 4 hours as needed		1.5 g. per square meter of body surface daily in divided doses; or for children up to 2 years of age—dosage to be individualized by physician; children 2 to 4 years of age—rectal, 160 mg. every 4 hours as needed; children 4 to 6 years of age—rectal, 240 mg. every 4 hours as needed; children 6 to 9 years of age—rectal, 320 mg. every 4 hours as needed; children 9 to 11 years of age—rectal, 400 mg. every 4 hours as needed; children 11 to 12 years of age—rectal, 480 mg. every 4 hours as needed

* See *U.S.P. D.I.* for complete dosage information.
(Continued)

TABLE 17-13

ANTIPYRETIC ANALGESICS

Name Proprietary Name	Preparations	Category	Usual Adult Dose*	Usual Dose Range*	Usual Pediatric Dose*
Phenylbutazone U.S.P. *Butazolidin, Azolid*	Phenylbutazone Tablets U.S.P.	Antirheumatic	Rheumatoid arthritis, rheumatoid spondylitis, osteoarthritis, psoriatic arthritis, and painful shoulder: initial—oral, 100 or 200 mg. 3 times daily; maintenance—oral, 100 mg. 1 to 4 times daily		Children up to 15 years of age—use is not recommended
Oxyphenbutazone U.S.P. *Tandearil, Oxalid*	Oxyphenbutazone Tablets U.S.P.	Antiarthritic; antiinflammatory (nonsteroid)	Acute local inflammation: oral, 100 mg. 4 times daily to 200 mg. 3 times daily for the first 2 or 3 days; then 100 mg. 3 times daily, with the average duration of therapy usually not exceeding 2 to 7 days		Children up to 15 years of age—use is not recommended
Sulindac U.S.P.	Sulindac Tablets U.S.P.		Oral, 150 to 200 mg. twice daily		Dosage has not been established
Tolmetin Sodium U.S.P. *Tolectin*	Tolmetin Sodium Tablets U.S.P.		Initial—oral, the equivalent of tolmetin 3 times daily, preferably including a dose in the morning and a dose at bedtime; maintenance—oral, the equivalent of 2 g. of tolmetin a day in divided doses	Up to the equivalent of 2 g. of tometin daily	Children up to 2 years of age—dosage has not been established; children 2 years of age and over—initial, oral, the equivalent of 20 mg. of tolmetin per kg. of body weight a day in divided doses; maintenance, oral, the equivalent of 15 to 30 mg. of tolmetin per kg. of body weight daily in divided doses
Naproxen U.S.P. *Naprosyn*	Naproxen Tablets U.S.P.		Oral, 250 mg. twice daily, morning and evening	Up to 750 mg. daily	Dosage has not been established
Fenoprofen Calcium U.S.P. *Nalfon*	Fenoprofen Calcium Capsules U.S.P. Fenoprofen Tablets U.S.P.		Orally, the equivalent of 300 to 600 mg. of fenoprofen 4 times daily	Up to the equivalent of 3.2 g. of fenoprofen daily	Dosage has not been established

(Continued)

*See *U.S.P. D.I.* for complete dosage information.

treating 1,2-diphenyl-3,5-pyrazolidinedione, obtained by a procedure analogous to the above condensation, with butyl bromide in 2*N* sodium hydroxide at 70° or with *n*-butyraldehyde followed by reduction utilizing Raney nickel catalyst.

The principal usefulness of phenylbutazone lies in the treatment of the painful symptoms associated with gout, rheumatoid arthritis, psoriatic arthritis, rheumatoid spondylitis and painful shoulder (peritendinitis, capsulitis, bursitis and acute arthritis of the joint). Because of its many unwelcome side-effects, this drug is not generally considered to be the drug of choice but should be reserved for trial in those cases that do not respond to less toxic drugs. It should be emphasized that, although the drug is an analgesic, it is not to be considered as one of the simple analgesics and is not to be used casually. The initial daily dosage in adults ranges from 300 to 600 mg., divided into 3 or 4 doses. The manufacturer suggests that an average initial daily dosage of 600 mg. per day administered for 1 week should determine whether the drug will give a favorable response. If no results are forthcoming in this time, it is recommended that the drug be discontinued to avoid side-effects. In the event of favorable response, the dosage is reduced to a minimal effective daily dose, which usually ranges from 100 to 400 mg.

The drug is contraindicated in the presence of edema, cardiac decompensation, a history of peptic ulcer or drug allergy, blood dyscrasias, hypertension and whenever renal, cardiac or hepatic damage is present. All patients, regardless of the history given, should be careful to note the occurrence of black or tarry stools which might be indicative of reactivation of latent peptic ulcer and is a signal for discontinuance of the drug. The physician is well advised to read the manufacturer's literature and warnings thoroughly before attempting to administer the drug. Among the precautions the physician should take with regard to the patient are to examine the patient periodically for toxic reactions, to check for increase in weight (due to water retention) and to make periodic blood counts to guard against agranulocytosis.

Oxyphenbutazone U.S.P., Oxalid®, Tandearil®, 4-butyl-1-(*p*-hydroxyphenyl)-2-phenyl-3,5-pyrazolidinedione. This drug is a metabolite of phenylbutazone and has the same effectiveness, indications, side-effects and contraindications. Its only apparent advantage is that it causes acute gastric irritation less frequently.

The pharmacology of these and other analogs has been reviewed extensively.[199]

REFERENCES

1. Tainter, M. L.: Ann. N. Y. Acad. Sci. 51:3, 1948.
2. Eisleb, O., and Schaumann, O.: Deutsche med. Wschr. 65:967, 1938.
3. Small, L. F., Eddy, N. B., Mosettig, E., and Himmelsbach, C. K.: Studies on Drug Addiction, Supplement No. 138 to the Public Health Reports, Washington, D.C., Supt. Doc., 1938.
4. Eddy, N. B., Halbach, H., and Braenden, O. J.: Bull. W.H.O. 14:353–402, 1956.
5. Stork, G., and Bauer, L.: J. Am. Chem. Soc. 75:4373, 1953.
6. U. S. Patent 2,831,531; through Chem. Abstr. 52:13808, 1958.
7. Rapoport, H., Baker, D. R., and Reist, H. N.: J. Org. Chem. 22:1489, 1957; Chadha, M. S., and Rapoport, H.: J. Am. Chem. Soc. 79:5730, 1957.
8. Okun, R., and Elliott, H. W.: J. Pharmacol. Exp. Ther. 124:255, 1958.
9. Seki, I., Takagi, H., and Kobayashi, S.: J. Pharm. Soc. Jap. 84:280, 1964.
10. Buckett, W. R., Farquharson, M. E., and Haining, C. G.: J. Pharm. Pharmacol. 16:174, 68T, 1964.
11. Bentley, K. W., and Hardy, D. G.: Proc. Chem. Soc. 220, 1963.
12. Lister, R. E.: J. Pharm. Pharmacol. 16:364, 1964.
13. Bentley, K. W., and Hardy, D. G.: J. Am. Chem. Soc. 89:3267, 1967.
14. Telford, J., Papadopoulos, C. N., and Keats, A. S.: J. Pharmacol. Exp. Ther. 133:106, 1961.
15. Gates, M., and Montzka, T. A.: J. Med. Chem. 7:127, 1964.
16. Schaumann, O.: Arch. exp. Path. Pharmakol. 196:109, 1940.
17. Bergel, F., and Morrison, A. L.: Quart. Revs. (London) 2:349, 1948.
18. Jensen, K. A., Lindquist, F., Rekling, E., and Wolffbrandt, C. G.: Dansk. tids. farm. 17:173, 1943; through Chem. Abstr. 39:2506, 1945.
19. Lee, J., Ziering, A., Berger, L., and Heineman, S. D.: Jubilee Volume—Emil Barell, p. 267, Basel, Reinhardt, 1946; J. Org. Chem. 12:885, 894, 1947; Berger, L., Ziering, A., and Lee, J.: J. Org. Chem. 12:904, 1947; Ziering, A., and Lee, J.: J. Org. Chem. 12:911, 1947.
20. Janssen, P. A. J., and Eddy, N. B.: J. Med. Pharm. Chem. 2:31, 1960.
21. Lee, J.: Analgesics: B. Partial structures related to morphine, *in* American Chemical Society: Medicinal Chemistry, vol. 1, pp. 438–466, New York, Wiley, 1951.
22. Beckett, A. H., Casy, A. F., and Kirk, G.: J. Med. Pharm. Chem. 1:37, 1959.
23. Bell, K. H., and Portoghese, P. S.: J. Med. Chem. 16:203, 589, 1973; *ibid.* 17:129, 1974.
24. Groeber, W. R., *et al.*: Obstet. Gynec. 14:743, 1959.
25. Eddy, N. B.: Chem. & Ind. (London), p. 1462, Nov. 21, 1959.
26. Fraser, H. F., and Isbell, H.: Bull. Narcotics 13:29, 1961.
27. Janssen, P. A. J., *et al.*: J. Med. Pharm. Chem. 2:271, 1960.
28. Stahl, K. D., *et al.*: Eur. J. Pharmacol. 46:199, 1977.
29. Blicke, F. F., and Tsao, E.: J. Am. Chem. Soc. 75:3999, 1953.
30. Cass, L. J., *et al.*: J.A.M.A. 166:1829, 1958.
31. Batterham, R. C., Mouratoff, G. J., and Kaufman, J. E.: Am. J. M. Sci. 247:62, 1964.
32. Finch, J. S., and DeKornfeld, T. J.: J. Clin. Pharmacol. 7:46, 1967.
33. Yelnosky, J., and Gardocki, J. F.: Tox. Appl. Pharmacol. 6:593, 1964.
34. Bockmuehl, M., and Ehrhart, G.: German Patent 711,069.
35. Kleiderer, E. C., Rice, J. B., and Conquest, V.: Pharmaceutical Activities at the I. G. Farbenindustrie Plant, Höchst-am-Main, Germany. Report 981, Office of the Publication Board, Dept. of Commerce, Washington, D.C., 1945.
36. Scott, C. C., and Chen, K. K.: Fed. Proc. 5:201, 1946; J. Pharmacol. Exp. Ther. 87:63, 1946.
37. Eddy, N. B., Touchberry, C., and Lieberman, J.: J. Pharmacol. Exp. Ther. 98:121, 1950.
38. Janssen, P. A. J., and Jageneau, A. H.: J. Pharm. Pharmacol. 9:381, 1957; 10:14, 1958. See also Janssen, P. A. J.: J. Am. Chem. Soc. 78:3862, 1956.
39. Cass, L. J., and Frederik, W. S.: Antibiot. Med. 6:362, 1959, and references cited therein.
40. Blaine, J., and Renault, P. (eds.): Rx 3 times/wk LAAM—Methadone Alternative, NIDA Research Monograph 8, DHEW, 1976.
41. Eddy, N. B., Besendorf, H., and Pellmont, B.: Bull. Narcotics, U.N. Dept. Social Affairs 10:23, 1958.
42. DeKornfeld, T. J.: Curr. Res. Anesth. 39:430, 1960.
43. Murphy, J. G., Ager, J. H., and May, E. L.: J. Org. Chem. 25:1386, 1960, and references cited therein.
44. Chignell, C. F., Ager, J. H., and May, E. L.: J. Med. Chem. 8:235, 1965.
45. May, E. L., and Eddy, N. B.: J. Med. Chem. 9:851, 1966.
46. Michne, W. F., *et al.*: J. Med. Chem. 22:1158, 1979.
47. Archer, S., *et al.*: J. Med. Chem. 7:123, 1964.
48. Cass, L. J., Frederik, W. S., and Teodoro, J. V.: J.A.M.A. 188:112, 1964.
49. Fraser, H. F., and Rosenberg, D. E.: J. Pharmacol. Exp. Ther. 143:149, 1964.
50. Lasagna, L., DeKornfeld, T. J., and Pearson, J. W.: J. Pharmacol. Exp. Ther. 144:12, 1964.
51. Houde, R. W.: Brit. J. Clin. Pharmacol. 7:297, 1979.
52. Mello, N. K., and Mendelson, J. H.: Science 207:657, 1980.
53. Martin, W. R.: Pharmacol. Rev. 19:463, 1967.
54. Julius, D., and Renault, P. (eds.): Narcotic Antagonists: Naltrexone, Progress Report, NIDA Research Monograph 9, DHEW, 1976.
55. Willette, R. E. (ed.): Narcotic Antagonists: The Search for Long-Acting Preparations, NIDA Research Monograph 4, DHEW, 1975.
56. deStevens, G. (ed.): Analgetics, New York, Academic Press, 1965.
57. Braenden, O. J., Eddy, N. B., and Halbach, H.: Bull. W.H.O. 13:937, 1955.
58. Leutner, V.: Arzneimittelforschung 10:505, 1960.
59. Janssen, P. A. J.: Brit. J. Anaesth. 34:260, 1962.
60. Beckett, A. H., and Casy, A. F., *in* Ellis, G. P., and West, G. B. (eds.): Progress in Medicinal Chemistry, vol. 2, pp. 43–87, London, Butterworth, 1962.
61. Mellet, L. B., and Woods, L. A., *in* Progress in Drug Research, vol. 5, pp. 156–267, Basel, Birkhäuser, 1963.
62. Casy, A. F., *in* Ellis, G. P., and West, G. B. (eds.): Progress in Medicinal Chemistry, vol. 7, pp. 229–284, London, Butterworth, 1970.
63. Lewis, J., Bently, K. W., and Cowan, A.: Ann. Rev. Pharmacol. 11:241, 1970.
64. Eddy, N. B., and May, E. L.: Science 181–407, 1973.

65. Beckett, A. H., Casy, A. F., and Harper, N. J.: Pharm. Pharmacol. 8:874, 1956.
66. Lasagna, L., and DeKornfeld, T. J.: J. Pharmacol. Exp. Ther. 124:260, 1958.
67. Fishman, J., Hahn, E. F., and Norton, B. I.: Nature 261:64, 1976.
68. Wright, W. B., Jr., Brabander, H. J., and Hardy, R. A., Jr.: J. Am. Chem. Soc. 81:1518, 1959.
69. Gross, F., and Turrian, H.: Experientia 13:401, 1957; Fed. Proc. 19:22, 1960.
70. Beckett, A. H., and Casy, A. F.: J. Pharm. Pharmacol. 6:986, 1954.
71. Beckett, A. H.: J. Pharm. Pharmacol. 8:848, 860, 1958.
72. ———: Pharm. J., p. 256, Oct. 24, 1959.
73. Jacquet, Y. F., et al.: Science 198:842, 1977.
74. Portoghese, P. S.: J. Med. Chem. 8:609, 1965.
75. ———: J. Pharm. Sci. 55:865, 1966.
76. ———: Acc. Chem. Res. 11:21, 1978.
77. Koshland, D. E., Jr.: Proc. First Intern. Pharmacol. Meeting 7:161, 1963, and references cited therein.
78. Belleau, B.: J. Med. Chem. 7:776, 1964.
79. Martin, W. R., et al: J. Pharmacol. Exp. Ther. 197:517, 1976.
80. Gilbert, P. E., and Martin, W. R.: J. Pharmacol. Exp. Ther. 198:66, 1976.
81. Beaumont, A., and Hughes, J.: Ann. Rev. Pharmacol. Toxicol. 19:245, 1979.
82. Lasagna, L., and Beecher, H. K.: J. Pharmacol. Exp. Ther. 112:356, 1954.
83. Soulairac, A., Cahn, J., and Charpentier, J. (eds.): Pain, New York, Academic Press, 1968.
84. Willette, R. E.: Am. J. Pharm. Educ. 34:662, 1970.
85. Barnett, G., Trsic, M., and Willette, R. E. (eds.): Quantitative Structure–Activity Relationships of Analgesics, Narcotic Antagonists, and Hallucinogens, NIDA Research Monograph 22, DHEW, 1978.
86. Archer, S., and Harris, L. S., in Jucker, E. (ed.): Progress in Drug Research, vol. 8, p. 262, Basel, Birkhäuser, 1965.
87. Kutter, E., et al.: J. Med. Chem. 13:801, 1970.
88. Portoghese, P. S., et al.: J. Med. Chem. 14:144, 1971.
89. ———: J. Med. Chem. 11:219, 1968.
90. Kaufman, J. J., Semo, N. M., and Koski, W. S.: J. Med. Chem., 18:647, 1975.
91. Kaufman, J. J., Kerman, E., and Koski, W. S.: Internat. J. Quantum Chem. 289, 1974.
92. Loew, G. H., and Berkowitz, D. S.: J. Med. Chem. 21:101, 1978.
93. ———: J. Med. Chem. 22:603, 1979.
94. Simon, E. J., and Hiller, J. B.: Ann. Rev. Pharmacol. Toxicol. 18:371, 1978.
95. Goldstein, A.: Life Sci. 14:615, 1974.
96. Goldstein, A., Lowney, L. I., and nad Pal, B. K.: Proc. Natl. Acad. Sci. U.S.A. 68:1742, 1971.
97. Simon, E. J., Hiller, J. M., and Edelman, I.: Proc. Natl. Acad. Sci. U.S.A. 70:1947, 1973.
98. Terenius, L.: Acta Pharmacol. Toxicol. 32:317, 1973.
99. Pert, C. B., and Snyder, S. H.: Science 179:1011, 1973.
100. Kosterlitz, H. W., and Watt, A. J.: Brit. J. Pharmacol. Chemother. 33:266, 1968.
101. Hughes, J.: Brain Res. 88:295, 1975; ibid.: Neurosci. Res. Program. Bull. 13:55, 1975.
102. Terenius, L.: Ann. Rev. Pharmacol. Toxicol. 18:189, 1978.
103. Goldstein, A.: Science 193:1081, 1976.
104. Kolanta, G. B.: Science 205:774, 1979.
105. Proc. Manchester Lit. Phil. Soc. 69–79, 1925.
106. Gates, M., and Tschudi, G.: J. Am. Chem. Soc. 74:1109, 1952; 78:1380, 1956.
107. Twycross, R. G.: Int. J. Clin. Pharmacol. 9:184, 1974.
108. Lasagna, L.: Pharmacol. Rev. 16:47, 1964.
109. Houde, R. W., and Wallenstein, S. L.: Minutes of the 11th Meeting, Committee on Drug Addiction and Narcotics, National Research Council, 1953, p. 417.
110. Diehl, H. S.: J.A.M.A. 101:2042, 1933.
111. Eddy, N. B., and Lee, L. E.: J. Pharmacol. Exp. Ther. 125:116, 1959.
112. Weijlard, J., and Erickson, A. E.: J. Am. Chem. Soc. 64:869, 1942.
113. Fraser, H. F., et al.: J. Pharmacol. Exp. Ther. 122:359, 1958; Cochin, J., and Axelrod, J.: J. Pharmacol. Exp. Ther. 125:105, 1959.
114. Ziering, A., and Lee, J.: J. Org. Chem. 12:911, 1947.
115. Weijlard, J., et al.: J. Am. Chem. Soc. 78:2342, 1956.
116. The Medical Letter 19:73, 1977.
117. Wang, R. I. H.: Eur. J. Clin. Pharmacol. 7:183, 1974.
118. Easton, N. R., Gardner, J. H., and Stevens, J. R.: J. Am. Chem. Soc. 69:2941, 1947. See also reference 34.
119. Freedman, A. M.: J.A.M.A. 197:878, 1966.
120. The Medical Letter 11:97, 1969.
121. Smits, S. E.: Res. Commun. Chem. Path. Pharmacol. 8:575, 1974.
122. Billings, R. E., Booher, R., Smits, S. E., Pohland, A., and McMahon, R. E.: J. Med. Chem. 16:305, 1973.
123. Jaffe, J. H., Senay, E. C., Schuster, C. R., Renault, P. F., Smith, B., and DiMenza, S.: J. A. M. A. 222:437, 1972.
124. Gruber, C. M., and Babtisti, A.: Clin. Pharmacol. Therap. 4:172, 1962.
125. Pohland, A., and Sullivan, H. R.: J. Am. Chem. Soc. 75:4458, 1953.
126. The Medical Letter 20:111, 1978.
127. Forrest, W. H., et al.: Clin. Pharmacol. Therap. 10(4):468, 1969.
128. Ehrnebo, M., Boreus, L. O., and Lönroth, V.: Clin. Pharmacol. Ther. 22:888, 1977.
129. The Medical Letter 9:49, 1967.
130. Klatz, A. L.: Curr. Ther. Res. 16:602, 1974; Workman, F. C., and Winter, L.: Curr. Ther. Res. 16:609, 1974.
131. Jasinski, D. R., Martin, W. R., and Sapira, J. D.: Clin. Pharmacol. Therap. 9:215, 1968.
132. Cone, E. J.: Tetrahedron Letters 28:2607, 1973; Chatterjie, N., et al.: Drug Metab. Disp. 2:401, 1974.
133. Batra, V. K., Sams, R. A., Reuning, R. H., and Malspeis, L.: Acad. Pharm. Sci. 4:122, 1974.
134. Blumberg, H., and Dayton, H. B., in Kosterlitz, H., and Villarreal, J. E. (eds.): Agonist and Antagonist Actions of Narcotic Analgesic Drugs, pp. 110–119, London, Macmillan, 1972.
135. Woodland, J. H. R., et al.: J. Med. Chem. 16:897, 1973.
136. Takemori, A. E. A., Hayashi, G., and Smits, S. E.: Eur. J. Pharmacol. 20:85, 1972.
137. Nutt, J. G., and Jasinsky, D. R.: Pharmacologist 15:240, 1973.
138. Salem, H., and Aviado, D. M.: Am. J. Med. Sci. 247:585, 1964.
139. Chappel, C. I., and von Seemann, C., in Ellis, G. P., and West, G. B. (eds.): Progress in Medicinal Chemistry, vol. 3, pp. 133–136, London, Butterworth, 1963.
140. Chernish, S. M.: Annals Allergy 21:677, 1963.
141. Chen, J. Y. P., Biller, H. F., and Montgomery, E. G.: J. Pharmacol. Exp. Ther. 128:384, 1960.
142. Five Year Report, Brit. Med. J. 2:1033, 1960.
143. Wong, S., in Heinzelman, R. V. (ed.): Annual Reports in Medicinal Chemistry, vol. 10, pp. 172–181, New York, Academic Press, 1975.
144. Collier, H. O. J.: Nature 232:17, 1971.
145. Vane, J. R.: Nature 231:232, 1971.
146. Shen, T. Y.: Angew, Chem. (Internat. Ed.) 11:460, 1972.
147. Nickander, R., McMahon, F. G., and Ridolfo, A. S.: Ann. Rev. Pharmacol. Toxicol. 19:469, 1979.
148. Smith, P. K.: Ann. N.Y. Acad. Sci. 86:38, 1960.
149. Smith, M. J. H., and Smith, P. K. (eds.): The Salicylates. A Critical Bibliographic Review, New York, Wiley, 1966.
150. Collier, H. O. J.: Sci. Am. 209:97, 1963.
151. Alvarez, A. S., and Summerskill, W. H. J.: Lancet 2:920, 1958.
152. Barager, F. D., and Duthie, J. J. R.: Brit. Med. J. 1:1106, 1960.
153. Leonards, J. R., and Levy, G.: Abstracts of the 116th Meeting of the American Pharmaceutical Association, p. 67, Montreal, May 17–22, 1969.
154. Davenport, H. W.: New Eng. J. Med. 276:1307, 1967.
155. Weiss, H. J.: Schweiz. med. Wschr. 104:114, 1974; Elwood, P. C., et al.: Brit. Med. J. 1:436, 1974.
156. Aspirin Myocardial Infraction Study Research Group: J.A.M.A. 243:661, 1980.

157. Levy, G., and Hayes, B. A.: New Eng. J. Med. 262:1053, 1960.
158. Lieberman, S. V., et al.: J. Pharm. Sci. 53:1486, 1492, 1964.
159. Pfeiffer, C. C.: Arch. Biol. Med. Exp. 4:10, 1967.
160. Salassa, R. M., Bollman, J. M., and Dry, T. J.: J. Lab. Clin. Med. 33:1393, 1948.
161. Smith, P. K., et al.: J. Pharmacol. Exp. Ther. 87:237, 1946.
162. Garrett, E. R.: J. Am. Pharm. A. (Sci. Ed.) 48:676, 1959.
163. Gold, G., and Campbell, J. A.: J. Pharm. Sci. 53:52, 1964.
164. Troup, A. E., and Mitchner, H.: J. Pharm. Sci. 53:375, 1964.
165. Blaug, S. M., and Wesolowski, J. W.: J. Am. Pharm. A. (Sci. Ed.) 48:691, 1959.
166. Obbink, H. J. K.: Lancet 1:565, 1964.
167. Goldman, A. S., and Yakovac, W. C.: Proc. Soc. Exp. Biol. Med. 115:693, 1964.
168. Anon.: Brit. Med. J. 2:131T, 1963.
169. The Medical Letter 8:7, 1966.
170. Liyanage, S. P., and Tambar, P. K.: Curr. Med. Res. Opin. 5:450, 1978.
171. Deodhar, S. D., et al.: Curr. Med. Res. Opin. 5:185, 1978.
172. Regaldo, R. G.: Curr. Med. Res. Opin. 5:454, 1978.
173. Leonards, J. R.: J. Lab. Clin. Med. 74:911, 1969.
174. Bloomfield, S. S., Barden, T. P., and Hille, R.: Clin. Pharmacol. Therap. 11:747, 1970.
175. Winder, C. V.: Nature 184:494, 1959.
176. Scherrer, R. A., in Scherrer, R. A., and Whitehouse, M. W. (eds.): Antiinflammatory Agents, p. 132, New York, Academic Press, 1974.
177. Cass, L. J., and Frederik, W. S.: J. Pharmacol. Exp. Ther. 139:172, 1963.
178. Lane, A. Z., Holmes, E. L., and Moyer, C. E.: J. New Drugs 4:333, 1964.
179. The Medical Letter 10:37, 1968.
180. Walker, R. W., et al.: Anal. Biochem. 95:579, 1979.
181. Brogden, R. N., et al.: Drugs 16:97, 1978, see also The Medical Letter 21:i, i979.
182. ———: Drugs 15:429, 1978.
183. O'Neill, P. J., et al.: J. Pharmacol. Exp. Ther. 209:366, 1979.
184. Wallenstein, S. L.: Unpublished report.
185. Dornan, J., and Reynolds, W.: Can. Med. Assoc. J. 110:1370, 1974.
186. Emele, J. F., and Shanaman, J. E.: Arch. Int. Pharmacodyn. Ther. 170:99, 1967.
187. Brogden, R. N., et al.: Drugs 13:241, 1977.
188. Chernish, S. M., et al.: Arth. Rheum. 22:376, 1979.
189. Brogden, R. N., et al.: Drugs 18, 241, 1979.
190. Wiseman, E. H.: R. Soc. Med. Int. Congr. Ser. 1:11, 1978.
191. Weintraub, M., et al.: J. Rheumatol. 4:393, 1979.
192. Balogh, Z., et al.: Curr. Med. Res. Opin. 6:148, 1979.
193. Brodie, B. B., and Axelrod, J.: J. Pharmacol. Exp. Ther. 94:29, 1948; 97:58, 1949. See also Axelrod, J.: Postgrad. Med. 34:328, 1963.
194. Bonica, J. J., and Allen, G. D., in Modell, W. (ed.): Drugs of Choice 1970-1971, p. 210, St. Louis, C. V. Mosby, 1970.
195. The Medical Letter 6:78, 1964.
196. Brown, D. M., and Hardy, T. L.: Brit. J. Pharmacol. Chemotherap. 32:17, 1968.
197. Tomatis, L., et al.: Cancer Res. 38:877, 1978.
198. Bengsston, V., Johansson, S., and Angervall, L.: Kidney Int. 13:107, 1978, see also Science 240:129, 1979.
199. Burns, J. J., et al.: Ann. N.Y. Acad. Sci. 86:253, 1960; Domenjoz, R.: Ann. N.Y. Acad. Sci. 86:263, 1960.

SELECTED READINGS

American Chemical Society, First National Medicinal Chemistry Symposium, pp. 15-49, 1948.
Anon.: Codeine and Certain Other Analgesic and Antitussive Agents: A Review, Rahway, Merck & Co., 1970.
Archer, S., and Harris, L. S.: Narcotic Antagonists, in Jucker, E. (ed.): Progress in Drug Research, vol. 8, pp. 262-320, Basel, Birkhäuser, 1965.
Arrigoni-Martelli, E.: Inflammation and Antiinflammatories, New York, Spectrum, 1977.
Barlow, R. B.: Morphine-like Analgesics, in Introduction to Chemical Pharmacology, pp. 39-56, New York, Wiley, 1955.
Beckett, A. H., and Casy, A. F.: The Testing and Development of Analgesic Drugs, in Ellis, G. P., and West, G. B. (eds.): Progress in Medicinal Chemistry, vol. 2, pp. 43-87, London, Butterworth, 1963.
Bergel, F., and Morrison, A. L.: Synthetic analgesics, Quart. Rev. (London) 2:349, 1948.
Berger, F. M., et al.: Non-narcotic drugs for the relief of pain and their mechanism of action, Ann. N.Y. Acad. Sci. 86:310, 1960.
Braenden, O. J., Eddy, N. B., and Hallbach, H.: Relationship between chemical structure and analgesic action, Bull. W.H.O. 13:937, 1955.
Braude, M. C., et al. (eds.): Narcotic Antagonists, New York, Raven Press, 1973.
Brümmer, T.: Die historische Entwicklung des Antipyrin und seiner Derivative, Fortschr. Therap. 12:24, 1936.
Casy, A. F.: Analgesics and Their Antagonists: Recent Developments, in Ellis, G. P., and West, G. B. (eds.): Progress in Medicinal Chemistry, vol. 7, pp. 229-284, London, Butterworth, 1970.
Chappel, C. I., and von Seemann, C.: Antitussive Drugs, in Ellis, G. P., and West, G. B. (eds.): Progress in Medicinal Chemistry, vol. 3, pp. 89-145, London, Butterworth, 1963.
Chen, K. K.: Physiological and pharmacological background, including methods of evaluation of analgesic agents, J. Am. Pharm. A. (Sci. Ed.) 38:51, 1949.
Clouet, D. H.: Narcotic Drugs: Biochemical Pharmacology, New York, Plenum Press, 1971.
Collins, P. W.: Antitussives, in Burger, A. (ed.): Medicinal Chemistry, ed. 3, pp. 1351-1364, New York, Wiley-Interscience, 1970.
Coyne, W. E.: Nonsteroidal Antiinflammatory Agents and Antipyretics, in Burger, A. (ed.): Medicinal Chemistry, ed. 3, pp. 953-975, New York, Wiley-Interscience, 1970.
deStevens, G. (ed.): Analgetics, New York, Academic Press, 1965.
Eddy, N. B.: Chemical structure and action of morphine-like analgesics and related substances, Chem. & Ind. (London), p. 1462, Nov. 21, 1959.
Eddy, N. B., Halbach, H., and Braenden, O. J.: Bull. W.H.O. 14:353-402, 1956; 17:569-863, 1957.
Evens, R. P.: Drug Therapy Reviews: Antirheumatic Agents, Am. J. Hosp. Pharm. 36:622, 1979.
Fellows, E. J., and Ullyot, G. E.: Analgesics: A. Aralkylamines, in American Chemical Society, Medicinal Chemistry, vol. 1, pp. 390-437, New York, Wiley, 1951.
Gold, H., and Cattell, M.: Control of Pain, Am. J. Med. Sci. 246(5):590, 1963.
Greenberg, L.: Antipyrine: A Critical Bibliographic Review, New Haven, Hillhouse, 1950.
Gross, M.: Acetanilid: A Critical Bibliographic Review, New Haven, Hillhouse, 1946.
Hellerbach, J., Schnider, O., Besendorf, H., Dellmont, B., Eddy, N. B., and May, E. L.: Synthetic Analgesics: Part II. Morphinans and 6,7-Benzomorphans, New York, Pergamon Press, 1966.
Jacobson, A. E., May, E. L., and Sargent, L. J.: Analgetics, in Burger, A. (ed.): Medicinal Chemistry, ed, 3, pp. 1327-1350, New York, Wiley-Interscience, 1970.
Janssen, P. A. J.: Synthetic Analgesics: Part I. Diphenylpropylamines, New York, Pergamon Press, 1960.
Janssen, P. A. J., and van der Eycken, C. A. M.: in Burger, A. (ed.): Drugs Affecting the Central Nervous System, pp. 25-85, New York, Dekker, 1968.
Lasagna, L.: The clinical evaluation of morphine and its substitutes as analgesics, Pharmacol. Rev. 16:47-83, 1964.
Lee, J.: Analgesics: B. Partial Structures Related to Morphine, in American Chemical Society, Medicinal Chemistry, vol. 1, pp. 438-466, New York, Wiley, 1951.
Martin, W. R.: Opioid antagonists, Pharmacol. Rev. 19:463-521, 1967.
Mellet, L. B., and Woods, L. A.: Analgesia and Addiction, in Progress in Drug Research, vol. 5, pp. 156-267, Basel, Birkhäuser, 1963.

Portoghese, P. S.: Stereochemical factors and receptor interactions associated with narcotic analgesics, J. Pharm. Sci. 55:865, 1966.

Reynolds, A. K., and Randall, L. O.: Morphine and Allied Drugs, Toronto, Univ. Toronto Press, 1957.

Salem, H., and Aviado, D. M.: Antitussive Agents, vols. 1–3 (Section 27 of International Encyclopedia of Pharmacology and Therapeutics), Oxford, Pergamon Press, 1970.

Scherrer, R. A., and Whitehouse, M. W.: Antiinflammatory Agents, New York, Academic Press, 1974.

Shen, T. Y.: Perspectives in nonsteroidal anti-inflammatory agents, Angew. Chem. (Internat. Ed.) 11:460, 1972.

Snyder, S. H.: Opiate Receptors and Internal Opiates, Sci. Am. 237:236–44, 1977.

Winder, C. A.: Nonsteroid Anti-inflammatory Agents, *in* Jucker, E. (ed.): Progress in Drug Research, vol. 10, pp. 139–203, Basel, Birkhäuser, 1966.

18

Dwight S. Fullerton

steroids and therapeutically related compounds

Steroids are widely distributed throughout the plant and animal kingdoms, and are formed by identical or nearly identical biosynthetic pathways in both plants and animals. Furthermore, because of their relatively rigid chemical structures, the steroids usually have easily predictable physical and chemical properties.

However, the similarity among the steroids ends with their fundamental chemical properties. The steroids have little in common therapeutically, except that as a group they are the most extensively used drugs in modern medicine. The major therapeutic classes of steroids are illustrated in Figure 18-1. The fact that minor changes in steroid structure can cause extensive changes in biological activity has been a continual source of fascination for medicinal chemists and pharmacologists for some three decades.

In this chapter, we will consider the steroids used in modern medicine. Some nonsteroidal compounds which have similar therapeutic uses will also be discussed, e.g., the diethylstilbestrol estrogens and the nonsteroidal chemical contraceptive agents.

Many general reviews on steroid chemistry, synthesis and analysis,[1-20] biochemistry and receptors,[21-59] pharmacology and therapy[60-68] and metabolism[69-76] have been published.* Additional reviews on particular classes of steroids will be cited in subsequent sections.

STEROID RECEPTORS AND X-RAY STUDIES

The greatest progress in steroid research in recent years has been in the area of steroid receptors. Several

---

* Witzmann, Rupert F.: Steroids. Keys To Life, New York, Reinhold, 1981, is especially recommended for a fascinating overview.

excellent books summarizing research data in these areas have been published.[27-32,34-37].

Considering the many diverse actions of even a single class of steroid hormones, e.g., the estrogens, it is not surprising that several receptors (or specific binding proteins) have been isolated and partially purified for each class. In view of the complex interrelationship of the many receptors involved, simple structure-activity relationships now need to be interpreted with much greater caution.

In spite of a great amount of research, we have only begun to understand the mechanism of many steroid-receptor interactions. The study of steroid receptors holds promise of many exciting discoveries for years to come. As King and Mainwaring[22] have so succinctly stated in their thorough review of steroid receptors, "Many scientific discoveries appear delightfully simple at first but, as further experiments are performed, the simplicity disappears and a phase of maximum confusion occurs . . . this is (hopefully) followed by an answer. . . ."

As noted by Baxter and Funder,[24] because of the presence of a variety of steroid-binding proteins in the plasma and cytoplasm that do not cause a physiological response (i.e., they are not true receptors), steroid-receptor studies are often very difficult. Previously identified "receptors" have been found not to be the true receptors at all. A general scheme for steroid hormone-receptor interactions is shown in Figure 18-2.

Not all steroids cause a response in the same way as steroid hormones, however. For example, see the discussion on the cardiac steroid receptor, Na^+, K^+-ATPase, later in this chapter. This receptor lies in the center of the cell membrane, having adenosine triphosphate (ATP) receptor sites on the inner surface of the cell membrane and steroid receptor sites on the outer surface. The cardiac steroids inhibit Na^+, K^+-

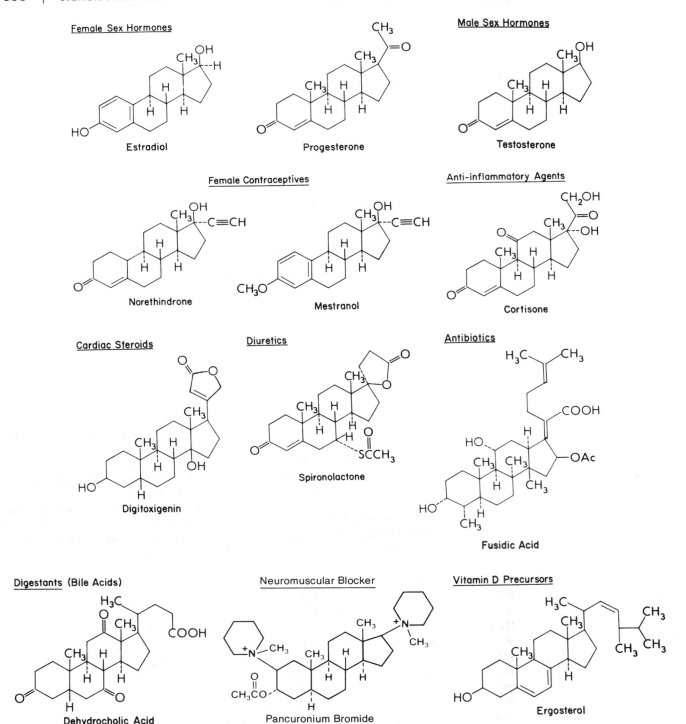

FIG. 18-1. *Representative examples of primary therapeutic classes of steroids.*

ATPase from its important actions of pumping Na^+ and K^+ during heart contractions and of nerve transmission.

The superbly illustrated Scientific American review on steroid hormone receptors by O'Malley and Schrader[23] is recommended. The excellent 1979 brief review of recent hormone-receptor research, mechanisms of hormone action, Kd, affinity, and receptor-related diseases by Baxter and Funder[24] is also highly recommended. More extensive reviews by Baxter, O'Malley, Jensen and others have also recently been published.[25–32] Roles of cyclic AMP in steroid hormone-receptor response have also been investigated.[28]

The use of site-specific (affinity and photo-affinity)

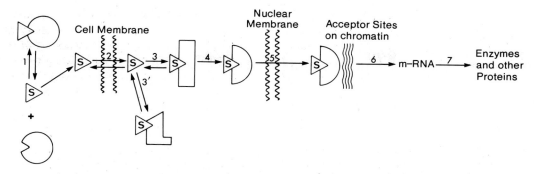

FIG. 18-2. *Steroid hormone action. (1) Steroid can be free or in equilibrium with cytoplasmic binding proteins which can be quite specific, e.g., uteroglobin. (2) Steroid (or active metabolite) enters the cell by an active or passive transport process. (3) Steroid binds to its receptor protein to form its steroid-receptor complex. (3′) Steroid can also bind to other cytoplasmic proteins, some of which can be quite specific. However these steroid-protein complexes are unable to cause a response at the chromatin acceptor sites. (4) The receptor undergoes a conformational change and is thereby activated. (5) The steroid-receptor complex goes into the nucleus and binds to acceptor sites on chromatin (DNA and its associated enzyme systems). Bruchovsky and Craven[40] have reported that the steroid-receptor complex of dihydrotestosterone may give up its dihydrotestosterone, which then acts upon the acceptor sites by itself. However, this mechanism is not likely for most steroid hormones. (6) The production of certain RNA's and m-RNA's is either increased or decreased, along with (7) the corresponding enzymes and other proteins which cause the steroid hormone's response (e.g., increased glucose production by glucocorticoids, development of secondary sex characteristics by estrogens or androgens). The role of the steroid in regulating protein biosynthesis has been hypothesized as being of the following types (elegantly discussed by King and Mainwaring[22]):*
A. The steroid itself is the regulating agent, but cannot reach its nuclear site of action without the aid of a transporter protein receptor. The transporting protein receptor may or may not be necessary to cause a response at the nuclear site of action.
B. The cytoplasmic binding proteins are the regulating agents, but need the steroid hormones to effect their transport into the nucleus. The steroid may or may not be necessary to cause a response at the nuclear site of action.
C. The steroid activates cytoplasmic or nuclear enzyme systems directly. In fact, the "nuclear site of action" in most cases may not be the chromosomes themselves, but instead regulatory enzymes. This is also true for the two possible roles discussed above.

labeling and chromatography in steroid-receptor isolation and study has been extensive.[23,33,35] Special (noncovalent binding) steroid hormone analog "tags" have also been very useful.[36,37]

X-ray crystal and computer graphic studies have begun to give exciting new insights into steroid structure–receptor relationships. Duax, Rohrer, and Weeks of the Medical Foundation of Buffalo have recently published *The Atlas of Steroid Structure* and an extensive review of their own and others' steroid roentgenographic (x-ray) studies.[38–42]

Medicinal chemists traditionally have assumed that there could be no relationship between the conformations of rigid molecules in crystals and their preferred conformations in solution with receptors. However, it is now clear from roentgenographic studies of steroids, prostaglandins, thyroid compounds, and many other drug classes that roentgenography can be a powerful tool in understanding drug action and in designing new drugs. The ways in which x-ray crystallography and computer graphics revealed how cardiac steroids interact with their receptors are discussed later in this chapter. The state of the art is the new computer graphics system of Professor Robert Lan-

gridge at the University of California at San Francisco. Drugs and their receptors can be drawn, fitted, turned, and twisted three dimensionally in full color with this remarkable system.[42a]

In 1980, Mornon and coworkers reported the first x-ray crystallographic, three-dimensional structure determination of a specific plasma progesterone-binding protein, uteroglobin.[77] Although uteroglobin is not a true receptor (in that it does not carry progesterone to a nuclear acceptor), its three-dimensional structure begins to give a graphic illustration of how steroid hormones bind to their receptors. No true steroid receptors have yet been isolated from cytoplasm in sufficient quantity and purity to permit their three-dimensional structures to be determined.

There is far more than just a single receptor per cell in tissues sensitive to a particular steroid. This is probably not surprising to those who have used Avogadro's number to determine how many molecules are taken in a single dose of any drug. Wolff and co-workers, for example, have found that there are about 65,000 glucocorticoid receptor sites per rat hepatoma cell, and about 110,000 affinity sites on the chromatin.[78] Baxter notes that, typically, there are 3,000 to

100,000 glucocorticoid receptors per cell.[24] Wallick and Schwartz have found that there are as many as 5,200,-000 digitalis receptor sites in a cell of a very digitalis steroid-sensitive tissue such as cat ventricle, but there are many fewer in a less sensitive tissue (e.g., 100,000 per guinea-pig atrium cell).[79]

Changes in steroid receptors and the number of receptors during development and puberty[80] and aging[81] have recently been reported, giving new insight into the roles of steroid hormones during life.

STEROID NOMENCLATURE, STEREOCHEMISTRY AND NUMBERING

As shown in Figure 18-3, nearly all steroids are named as derivatives of cholestane, androstane, pregnane or estrane. The standard system of numbering is illustrated with 5α-cholestane.

The absolute stereochemistry of the molecule and any substituents is shown with solid (β) and dashed (α) bonds. Most carbons have one β-bond and one α-bond, with the β-bond lying closer to the "top" or C-18 and C-19 methyl side of the molecule. Both α- and β-substituents may be axial or equatorial. This system of designating stereochemistry can best be illustrated using 5α-androstane.

The stereochemistry of the H at C-5 is always indicated in the name. Stereochemistry of other H atoms

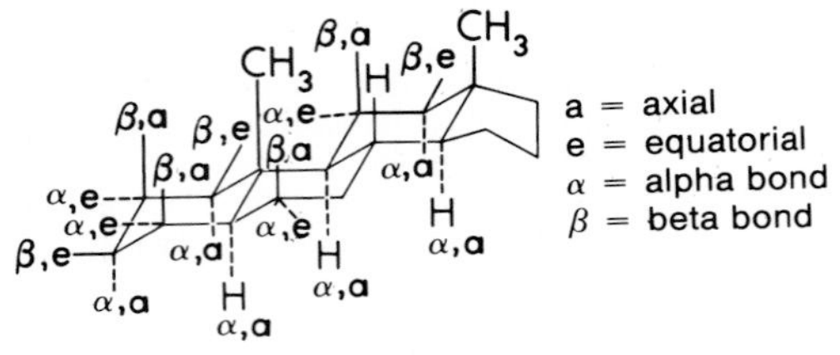

a = axial
e = equatorial
α = alpha bond
β = beta bond

5α-Androstane

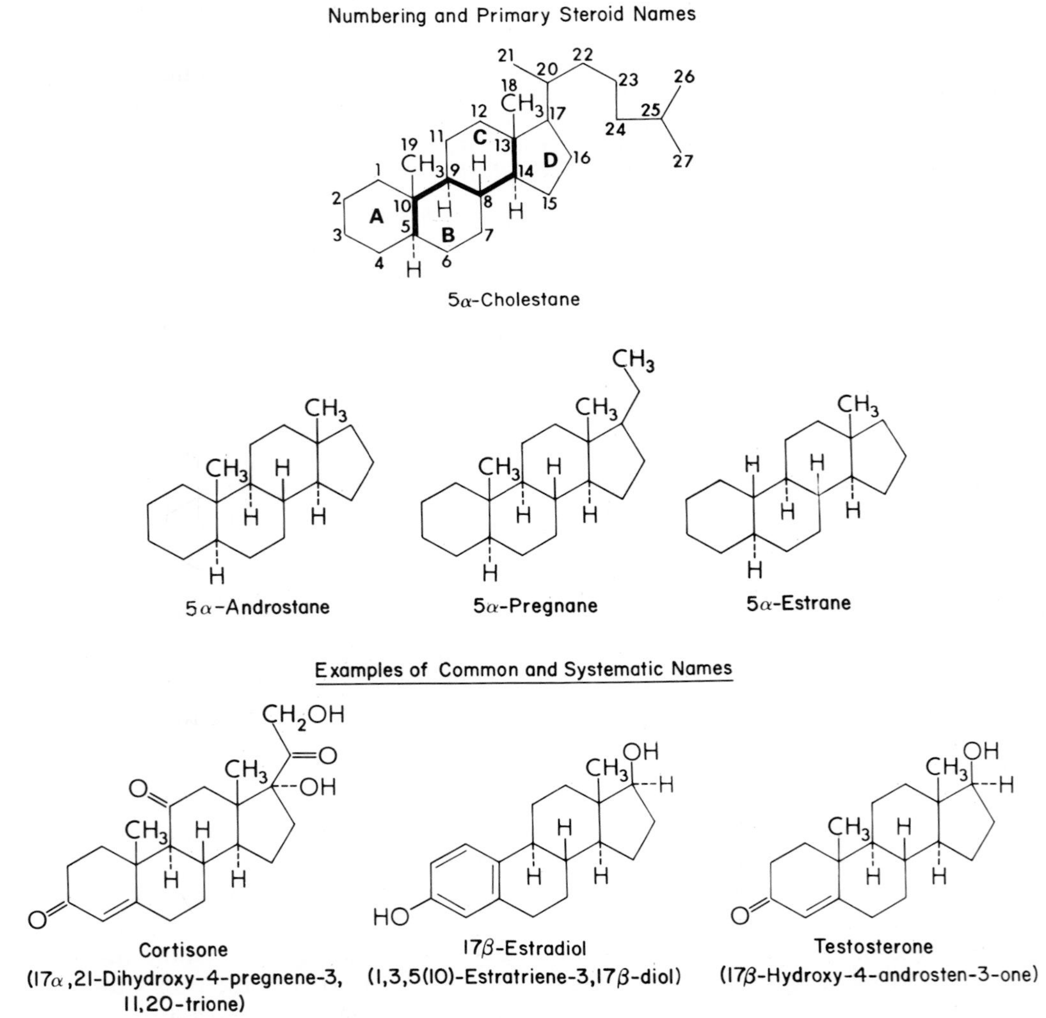

Numbering and Primary Steroid Names

5α-Cholestane

5α-Androstane   5α-Pregnane   5α-Estrane

Examples of Common and Systematic Names

Cortisone
(17α,21-Dihydroxy-4-pregnene-3, 11,20-trione)

17β-Estradiol
(1,3,5(10)-Estratriene-3,17β-diol)

Testosterone
(17β-Hydroxy-4-androsten-3-one)

FIG. 18-3.

is not indicated unless different from 5α-cholestane. Changing the stereochemistry of any of the ring-juncture or backbone carbons (shown in Fig. 18-3 with a heavy line on 5α-cholestane) greatly changes the shape of the steroid.

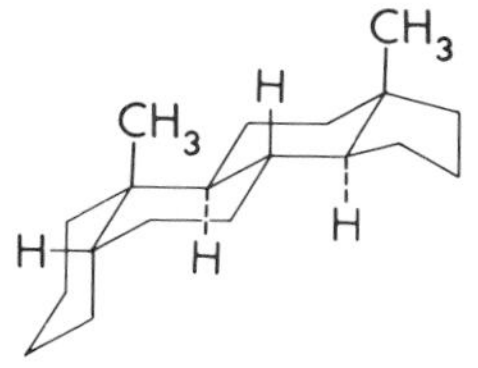

5β-Androstane 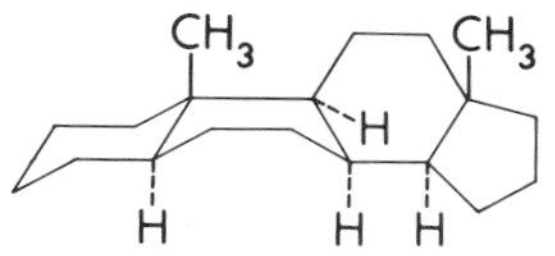   5α,8α-Androstane

Because of the immense effect that "backbone" stereochemistry has upon the shape of the molecule, IUPAC rules[72] require the stereochemistry at all backbone carbons to be clearly shown. That is, all *hydrogens* along the backbone must be drawn. When the stereochemistry is not known, a wavy line is used in the drawing, and the Greek letter xi (ξ) instead of α or β is used in the name. Methyls are always drawn as CH_3. Some authors also draw hydrogens at C-17.

The position of double bonds can be designated in any of the various ways shown below. Double bonds from carbon 8 may go toward C-9 or C-14; and those from C-20 may go toward C-21 or C-22. In such cases, both carbons are indicated in the name if the double bond is not between sequentially numbered carbons.

These principles of modern steroid nomenclature are applied to naming several common steroid drugs shown in Figure 18-3.

Such common names as "testosterone" and "cortisone" are obviously much easier to use than the long systematic names. However, substituents must always have their position and stereochemistry clearly indicated when common names are used; e.g., 17α-methyltestosterone, 9α-fluorocortisone.

The terms *cis* and *trans* are occasionally used in steroid nomenclature to indicate the backbone stereochemistry *between* rings. For example, 5α-steroids are A/B *trans;* and 5β-steroids are A/B *cis.* The terms *syn* and *anti* are used analogously to *trans* and *cis* for indicating stereochemistry in bonds *connecting* rings, e.g.,

the C-9:C-10 bond which connects rings A and C. The use of these terms is indicated below.

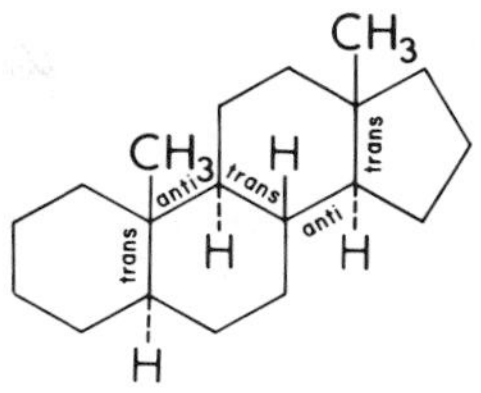

Other methods of indicating steroid stereochemistry and nomenclature occur in the early medical literature, but these methods are seldom used now.

5β-Steroids were sometimes called "normal," and 5α-steroids "allo"—a historical result of many 5β-steroids such as the bile acids being characterized before 5α-steroids. 5β-Cholestanol was also known as coprostanol, a name which was the result of many 5β-steroids being found in feces (Greek "kopros," or dung).

Steroid drawings sometimes appear with lines drawn instead of methyls (CH_3) (even though incorrect by IUPAC rules), and backbone stereochemistry is not indicated unless different from 5α-androstane, as follows.

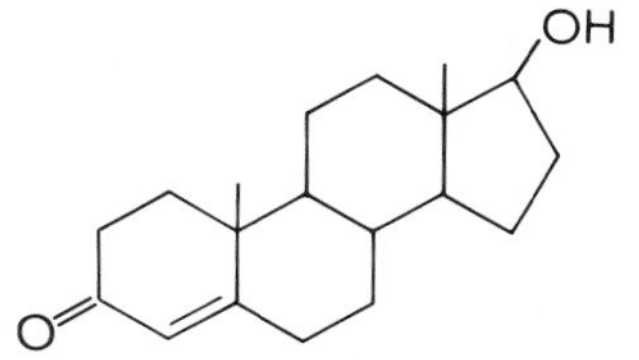

Testosterone 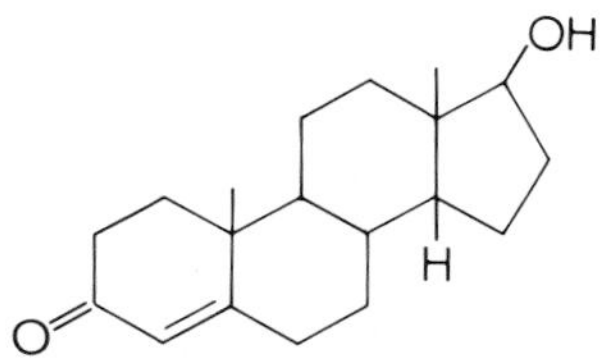   14β-Testosterone

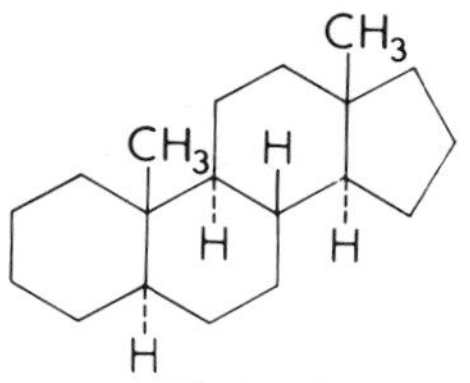

5α-Androstane

Finally, circles were sometimes used to indicate α-hydrogens, and dark dots to indicate β-hydrogens.

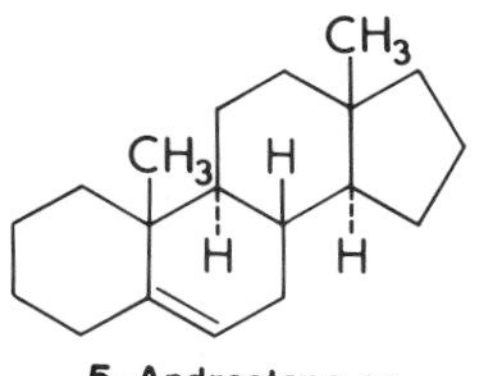

5-Androstene or
Δ⁵-Androstene or
Androst-5-ene

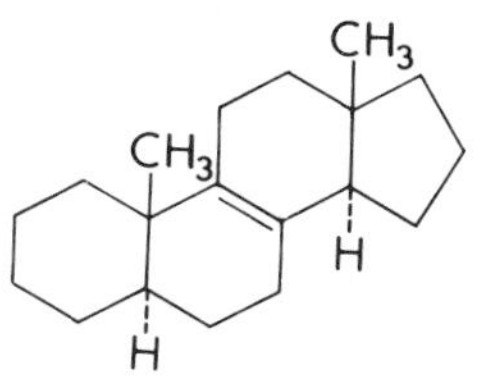

5α-Androst-8-ene or
5α-Δ⁸-Androstene

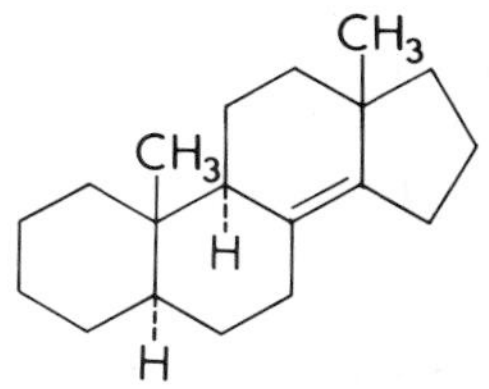

5α-Androst-8(14)-ene or
5α-Δ⁸⁽¹⁴⁾-Androstene

Testosterone   14 β-Testosterone   5 α-Androstane

STEROID BIOSYNTHESIS

Steroid hormones in mammals are biosynthesized from cholesterol, which in turn is made in vivo from acetyl coenzyme A. About one gram of cholesterol is biosynthesized per day in man, and an additional 300 mg. is provided in the diet. (The possible roles of cholesterol and diet in atherosclerosis will be discussed later.) A schematic outline of these biosynthetic pathways is shown in Figure 18-4, and the interested reader is referred to recent reviews[83-87] for additional details.

CHEMICAL AND PHYSICAL PROPERTIES OF STEROIDS

With few exceptions, the steroids are white crystalline solids. They may be in the form of needles, leaflets, platelets or amorphous particles depending upon the particular compound, solvent used in crystallization, and skill and luck of the chemist. Since the steroids have 17 or more carbon atoms, it is not surprising that they tend to be water-insoluble. Addition of hydroxyl or other polar groups (or decreasing carbons) increases water-solubility slightly as expected. Salts of course are the most water-soluble. Examples are shown in Table 18-1.

CHANGES TO MODIFY PHARMACOKINETIC PROPERTIES OF STEROIDS

As with many other compounds described in previous chapters, the steroids can be made more lipid-soluble or more water-soluble simply by making suitable ester derivatives of hydroxyl groups. Derivatives with increased lipid-solubility are often made to decrease the rate of release of the drug from intramuscular injection sites, i.e., in depot preparations. More lipid-soluble derivatives also have improved skin absorption properties, and so are preferred for dermatological preparations. Derivatives with increased water-solubility are needed for intravenous preparations. Since hydrolyzing enzymes are found throughout mamma-

lian cells, especially in the liver, converting hydroxyl groups to esters does not significantly modify the activity of most compounds. These principles of modifying pharmacokinetic properties have been discussed in detail by Ariens.[88]

Some steroids are particularly susceptible to rapid metabolism after absorption or rapid inactivation in the gastrointestinal tract before absorption. Often a simple chemical modification can be made to decrease these processes, and thereby increase the drug's half-life—or make it possible to be taken orally.

Examples of common chemical modifications are illustrated in Figure 18-5. Drugs such as testosterone cyclopentylproprionate and methylprednisolone sodium succinate (which are converted in the body to more active drugs) are called *prodrugs*.

R. E. Counsell and co-workers have given particular attention to the tissue distribution of steroids and its implication in drug design. For example, it has long been known that cholesterol is found in the highest concentration in the adrenal gland, and so 19-iodocholesterol ^{131}I is now used therapeutically for the diagnosis of various adrenal cortical diseases.[89-91] Radioactive

TABLE 18-1

	Solubility (g. / 100 ml.)		
	CHCl$_3$	EtOH	H$_2$O
Cholesterol	22	1.2	Insoluble
Testosterone	50	15	Insoluble
Testosterone Propionate	45	25	Insoluble
Dehydrocholic Acid	90	0.33	0.02
Estradiol	1.0	10	Insoluble
Estradiol Benzoate	0.8	8	Insoluble
Betamethasone	0.1	2	Insoluble
Betamethasone Acetate	10	3	Insoluble
Betamethasone NaPO$_4$ Salt	Insoluble	15	50
Hydrocortisone	1.0	2.5	0.01
Hydrocortisone Acetate	0.5	0.4	Insoluble
Hydrocortisone NaPO$_4$ Salt	Insoluble	1.0	75
Prednisolone	0.4	3	0.01
Prednisolone Acetate	1.0	0.7	Insoluble
Prednisolone NaPO$_4$ Salt	0.8	13	25

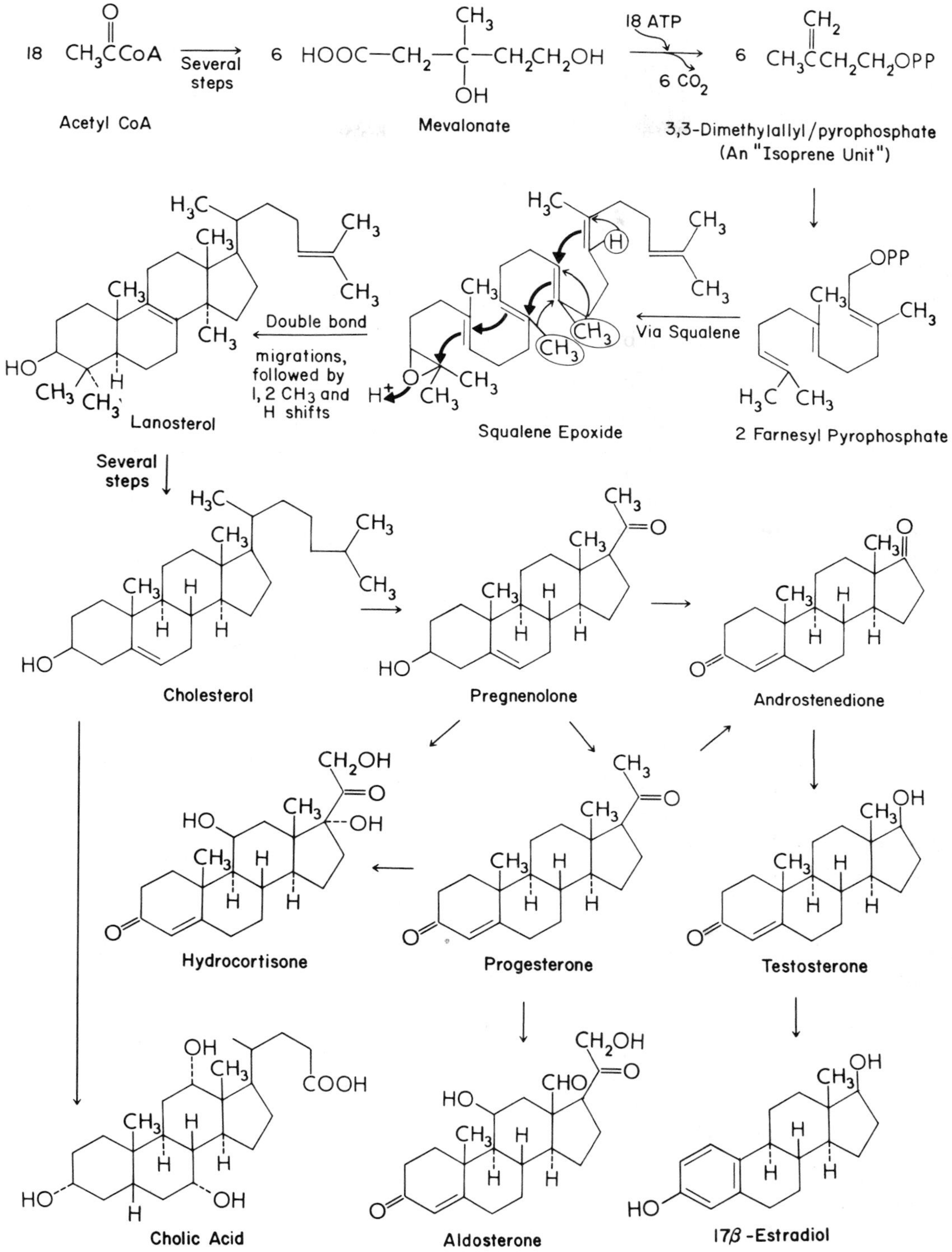

FIG. 18-4. A schematic outline of the biosynthesis of steroids.

steroids have been recognized for many years to bind most selectively to tissues which respond to them, and so labeled steroids have been used for many receptor and tissue studies.

Drugs with high affinity for the adrenals or other hormone-synthesizing tissues also have been studied as potential blockers of biosynthetic pathways, e.g., to block the biosynthesis of cholesterol in hyperlipidemia and heart disease, or the biosynthesis of excessive hormones from cholesterol in diseases of the adrenals.[91]

1. Increase Lipid-Solubility (Slower rate of release for depot preparation; increase skin absorption)

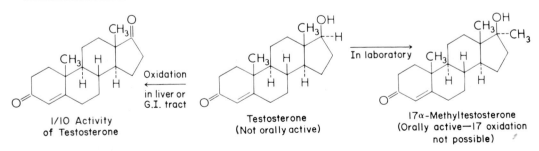

(I.M. dose: 10-25 mg.
2-3 times/week)

In Vivo
In laboratory

Testosterone Cyclopentylpropionate
(I.M. dose: 200-400 mg. every 4 weeks)

Triamcinolone

In laboratory

Triamcinolone Acetonide
(Active)

2. Increase Water-Solubility (Suitable for I.V. use)

Methylprednisolone
(Not water-soluble)

In vivo
In laboratory

Methylprednisolone Sodium Succinate
(Sufficiently water-soluble for I.V.)

3. Decrease Inactivation

1/10 Activity
of Testosterone

Oxidation
in liver or
G.I. tract

Testosterone
(Not orally active)

In laboratory

17α-Methyltestosterone
(Orally active—17 oxidation
not possible)

FIG. 18-5. *Common steroid modifications to alter therapeutic utility.*

SEX HORMONES

Although the estrogens and progesterone are usually called female sex hormones, and testosterone is called a male sex hormone, it should be noted that all these steroids are biosynthesized in *both* males and females. For example, an examination of the biosynthetic pathway in Figure 18-4 will reveal that progesterone serves as a biosynthetic precursor to cortisone and aldosterone and, to a lesser extent, to testosterone and the estrogens. Testosterone is one of the precursors of the estrogens. However, the estrogens and progesterone are produced in much larger amounts in females, as is testosterone in males. These hormones play profound roles in reproduction, in the menstrual cycle and in giving women and men their characteristic physical differences.

Furthermore, ovulation and the secretion of estrogens and progesterone in women, and spermatogenesis and the secretion of testosterone in men, are partially controlled by the same hormones. Of greatest importance are follicle-stimulating hormone (FSH) and luteinizing hormone (LH), both of which are released by the anterior lobe of the pituitary. Neither FSH nor LH is a steroid.

A larger number of synthetic or semisynthetic steroids having biological activities similar to those of progesterone have been made, and these are commonly called progestins. Several nonsteroidal compounds have also been found to have estrogenic activity. Although the estrogens and progestins have had their most extensive use in chemical contraceptive agents for women, their wide spectrum of activity has given them many therapeutic uses in both women and men.

Testosterone has been found to have two primary kinds of activities—androgenic (or male-physical-characteristic-promoting) and anabolic (or muscle-building). Many synthetic and semisynthetic androgenic and anabolic steroids have been prepared. A great deal of interest has focused on the preparation of anabolic agents, e.g., for use in aiding recovery from debilitating illness or surgery. However, the androgenic agents do have some therapeutic usefulness in women, e.g., in the palliation of certain sex-organ cancers.

In summary, it can be said that while many sex-hormone products have their greatest therapeutic uses in either women or in men, nearly all have some uses in both sexes. Nevertheless, the higher concentrations of estrogens and progesterone in women, and of testosterone in men, cause the development of the complementary reproductive systems and characteristic physical differences of women and men.

ESTROGENS AND PROGESTINS

The estrogens and progestins commonly used in medicine today are shown in Figures 18-6 and 18-7. Although most widely used as chemical contraceptive agents for women, these compounds are also indicated in a wide variety of physiologic and disease conditions.

Therapeutic uses

The estrogens have been used primarily for treatment of postmenopausal symptoms and as ovulation inhibitors for contraception (to be discussed later). In early 1976, there were extensive warnings about the dangers of excess estrogen use. These problems will be discussed in detail later in this chapter along with the safety of oral contraceptives. Estrogens have also been used for other conditions of estrogen insufficiency be-

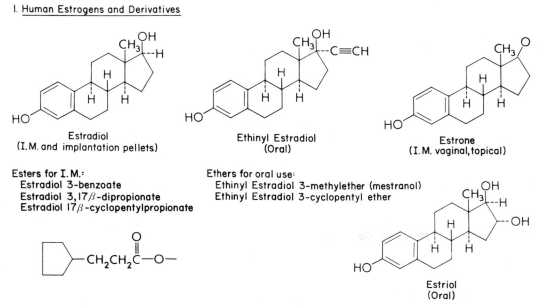

FIG. 18-6. *Natural and synthetic estrogens. (Continued on overleaf.)*

2.Equine Estrogens   (1.) Conjugated Estrogens: 50-65% Sodium Estrone Sulfate
(Oral, I.M.,topical,   20-35% Sodium Equilin Sulfate
vaginal)   plus nonestrogenic compounds

                    (2.) Esterified Estrogens: 70-85% Sodium Estrone Sulfate
                                           6.5-15% Sodium Equilin Sulfate
                                           plus nonestrogenic compounds

Estrone Sodium Sulfate   Equilin Sodium Sulfate   Equilenin

Other salt available:
Piperazine Estrone Sulfate

3. Synthetic Estrogens (Oral, I.M., topical, vaginal)

Diethylstilbestrol   Dienestrol

Chlorotrianisene   Benzestrol

FIG. 18-6. *Continued.*

sides menopause, including hypogenitalism, amenorrhea, and ovarian failure; however, the apparent risks of estrogen use are expected to greatly limit the doses of estrogens prescribed in the future.

The progestins are indicated in conditions characterized by progesterone insufficiency, such as amenorrhea or functional uterine bleeding. Progestins were previously used to prevent habitual abortions and as a

pregnancy test. However, the F.D.A. has strongly warned against these uses because the use of sex hormones during early pregnancy can seriously damage the fetus.[92] These side-effects were reviewed in 1979.[93]

Side-effects associated with low doses of progesterone are usually minimal and include nausea and "spotting". However, in higher doses, progesterone and especially the progestins which are 19-nortestos-

I. Progesterones and Derivatives

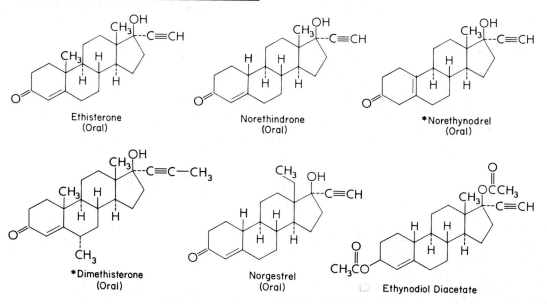

FIG. 18-7. *Natural and synthetic progestins.*

(*Available only in Contraceptive Products)

terone derivatives have many side-effects—often directly attributable to a minor androgenic action. These side-effects may include weight gain, congestion of the breasts, masculinization of the female fetus, increased fluid retention, etc. As with the estrogens, the possibility of hormone-dependent cancer or pregnancy must be excluded before estrogen or progestin therapy is begun. Regular physical examinations are essential during therapy.

Combinations of estrogens and progestins are also used to control excessive uterine bleeding, to stimulate redevelopment of the endometrium following curettage, and to treat amenorrhea.

Biosynthetic sources

The estrogens are normally produced in relatively large quantities in the ovaries and placenta, in lower amounts in the adrenals, and in trace quantities in the testes. About 50 to 350 μg. per day of estradiol are produced by the ovaries (especially the corpus luteum) during the menstrual cycle.[65] During the first months of pregnancy, the corpus luteum produces larger amounts of estradiol and other estrogens, whereas the placenta produces most of the circulating hormone in late pregnancy. During pregnancy the estrogen blood levels are up to 1000 times greater than during the menstrual cycle.[60]

Progesterone is produced in the ovaries, testes and adrenals. Much of the progesterone which is synthesized is immediately converted to other hormonal intermediates and is not secreted. (Refer to the biosynthetic pathway, Fig. 18-4.) The corpus luteum secretes the most progesterone, 20 to 30 mg. per day during the last or "luteal" stage of the menstrual cycle.[65] Normal men secrete about 1 to 5 mg. of progesterone daily.

The biosynthesis, mechanism of action and other effects of progesterone have recently been reviewed.[94]

Roles in the menstrual cycle

As shown in Figures 18-8 and 18-9, plasma concentrations of follicle-stimulating hormone (FSH), luteinizing hormone (LH), progesterone and estradiol vary throughout the menstrual cycle. The varying concentrations of these hormones and the events of the menstrual cycle are closely related, although not all the relationships are understood.

At the start of the cycle (with day one being the first day of menstruation), plasma concentrations of estradiol and other estrogens (see Fig. 18-8) and progesterone are low. FSH and LH stimulate several ovarian follicles to enlarge and begin developing more

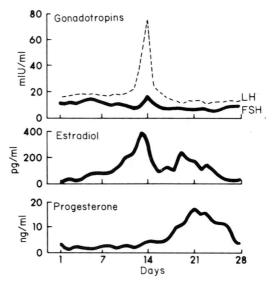

FIG. 18-8. *Hormone changes in the normal menstrual cycle. (Meyers, F. H., Jawetz, E., and Goldfien, A.: Review of Medical Pharmacology, ed. 4, New York, Lange, 1974. Used with permission.)*

rapidly than the others. After a few days only one follicle continues the development process to the final release of a mature ovum. The granulosa cells of the maturing follicles begin secreting estrogens, which then cause the uterine endometrium to thicken. Vaginal and cervical secretions increase. Estrogen, LH and FSH reach their maximum plasma concentrations at about the fourteenth day of the cycle. The sudden increase in LH causes the follicle to break open, releasing a mature ovum. Under the stimulation of LH, the follicle changes into the corpus luteum, which begins secreting progesterone as well as estrogen. The in-

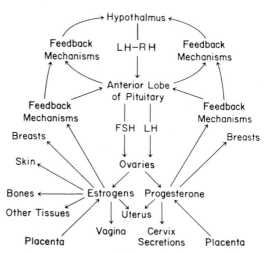

FIG. 18-9. *Tissue interrelationships of estrogens and progesterone. Abbreviations: FSH, follicle-stimulating hormone; LH, luteinizing hormone; LH–RH, luteinizing hormone–releasing hormone.*

creased concentrations of estrogens and progesterone inhibit the hypothalamus and the anterior pituitary by a feedback inhibition process. The estrogens and progesterone also stimulate the continued development of the uterine endometrium. If fertilization does not occur by about day 25, the corpus luteum begins to degenerate, slowing down its production of hormones. The concentrations of estrogens and progesterone become too low to maintain the vascularization of the endometrium, and menstruation results.

Yet it is easily recognized that this description of the menstrual process is at best incomplete. Although the interested reader is referred to recent reviews,[60-68] some very interesting questions (with particular relevance to chemical contraception) remain to be answered. For example, what causes the plasma concentrations of LH and FSH to peak so suddenly?

LH–RH—Roles in males and females

As shown in Figure 18-9, the hypothalamus releases a hormone called *luteinizing hormone–releasing hormone* (LH–RH). This peptide controls and regulates both male and female reproduction. Its isolation, purification, and structure determination were achieved in the early 1970's by two groups headed by R. Guillemin and A. V. Schalley, an accomplishment that earned them a Nobel Prize in 1977. LH–RH is a decapeptide (10 amino acids), and because of its simple structure and key roles in reproduction, over a thousand LH–RH analogs had been synthesized by early 1981.[95-102]

LH–RH stimulates the anterior pituitary gland to secrete LH and FSH (both complicated proteins) in males and females. LH causes testosterone production by the testes in males, which stimulates spermatogenesis (sperm production). The role of FSH in males is not as well known. LH and FSH in females regulate ovulation as described in the section above.

As a result of LH–RH's important roles in reproduction, LH–RH analogs have attracted wide interest and research effort as contraceptive agents.[51a-f] These approaches to fertility control will be discussed in the section on chemical contraception.

Other biological activities of estrogens and progesterone

In addition to having important roles in the menstrual cycle, the estrogens and, to a lesser extent, progesterone are largely responsible for the development of secondary sex characteristics in women at puberty.

The estrogens cause a proliferation of the breast ductile system, and progesterone stimulates development of the alveolar system. The estrogens also stimulate the development of lipid and other tissues which contribute to breast shape and function. Pituitary hormones and other hormones are also involved. Fluid retention in the breasts during the later stages of the menstrual cycle is a common effect of the estrogens. Interestingly, the breast engorgement which occurs after childbirth (stimulated by prolactin, oxytocin and other hormones) can be suppressed by administration of estrogen—probably due to feedback inhibition of the secretion of pituitary hormones.

Murad and Haynes nicely summarized the important role of the estrogens in puberty in young women: "The estrogens . . . go a long way toward accounting for that intangible attribute called femininity."[61] The estrogens directly stimulate the growth and development of the vagina, uterus and fallopian tubes, and in combination with other hormones play a primary role in sexual arousal and in producing the body contours of the mature woman. Pigmentation of the nipples and genital tissues, and stimulation of the growth of pubic and underarm hair (possibly with the help of small amounts of testosterone) are other results of estrogen action.

The physiologic changes at menopause emphasize the important roles of estrogens in the young woman. Breast and reproductive tissues atrophy, the skin loses some of its suppleness, coronary atherosclerosis and gout become potential health problems for the first time, and the bones begin to lose density due to decreased mineral content.

A very important role of progesterone during pregnancy is to depress the contractility of the uterus. In the third trimester, progesterone production decreases, estrogen production increases and the uterus becomes increasingly excitable in preparation for childbirth.

Metabolism of estrogens and progesterone

The metabolism of natural synthetic (e.g., mestranol) estrogens was reviewed in detail by Bolt in 1979.[76]

The three primary estrogens in women are 17β-estradiol, estrone and estriol. While 17β-estradiol is produced in greatest amounts, it is quickly oxidized (Fig. 18-10) to estrone,

Sodium Glucuronide
of Estriol

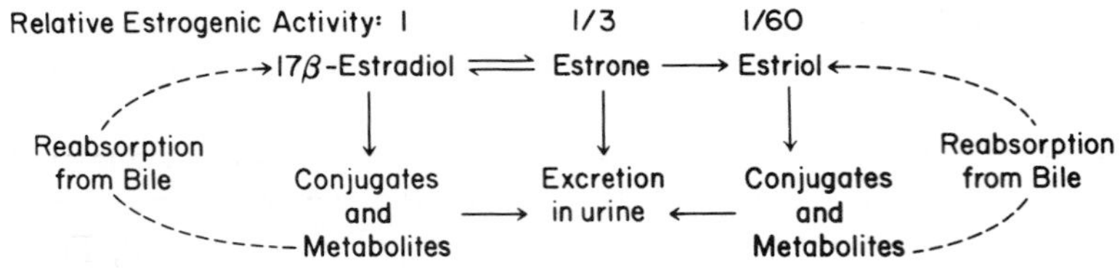

Relative Estrogenic Activity: 1   1/3   1/60

FIG. 18-10. *Interconversion and metabolism of natural estrogens.*

the estrogen found in highest concentration in the plasma. Estrone in turn is oxidized to estriol, the major estrogen found in human urine. During pregnancy, the placenta produces large amounts of estrone. However, in both pregnant and nonpregnant women the three primary estrogens are also metabolized to small amounts of other derivatives, e.g., 2-methoxyestrone and 16β-hydroxy-17β-estradiol. Only about 50 percent of therapeutically administered estrogens (and their various metabolites) are excreted in the urine during the first 24 hours. The remainder is excreted into the bile and reabsorbed so that several days are required for complete excretion of a given dose.

Conjugation appears to be very important in estrogen transport and metabolism. Although the estrogens are unconjugated in the ovaries, in the plasma and other tissues significant amounts of conjugated estrogens may predominate. Most of the conjugation takes place in the liver.

The primary estrogen conjugates found in plasma and urine are the combination of estrogen with glucuronic acid and, to a lesser extent, with sulfate. The conjugates are called glucuronides and sulfates, respectively. As the sodium salts, they are, of course, quite water-soluble. The sodium glucuronide of estriol and the sodium sulfate ester of estrone are shown below.

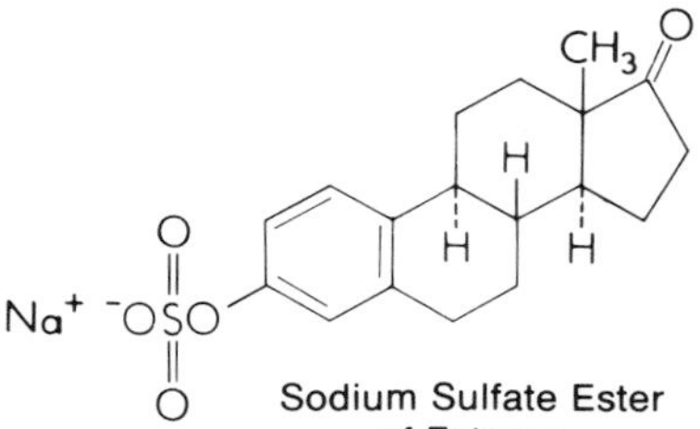

Sodium Sulfate Ester
of Estrone

As shown in Figure 18-4, progesterone can be biotransformed to many other steroid hormones and in that sense it has a great number of metabolic products. However, the principle excretory product of progesterone metabolism is 5β-pregnane-3α,20-diol and its conjugates.

The metabolism of progesterone is extremely rapid and, therefore, it is not effective orally. That fact has been a major stimulus in the development of the 19-nortestosterones with progesteronelike activity.

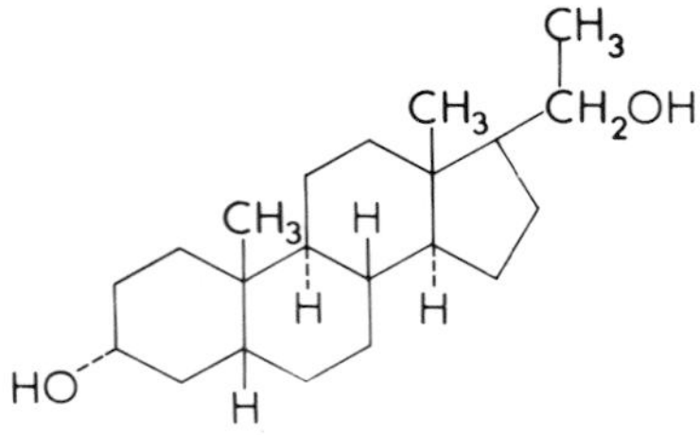

Structure-Activity relationships of the estrogens

The activity of the steroid and nonsteroid estrogens was explained by Schueler in 1946[103] as being due to a similarity in distance between hydrogen bonding groups, specifically the 3-OH and 17-OH of estradiol and the phenolic hydroxyls in DES. However the "critical distance" later cited by Schueler[104] as approximately 14.5 Å is incorrect. Unfortunately the error has persisted in the recent medicinal chemical literature.[105] The crystallographically observed distances between the terminal oxygens in DES and estradiol are actually 12.1 Å and 10.9 Å respectively.[42] In the crystal and in plasma, steroids are usually hydrated. Estradiol is no exception. As shown in the crystallographic diagrams below, two molecules of water have been found hydrogen-bonded to the 17-OH of estra-

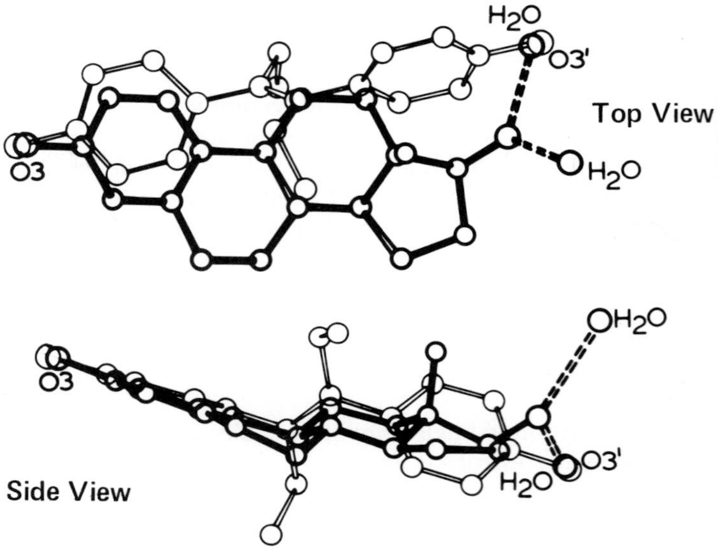

Estradiol (H$_2$O)$_2$ = dark lines
DES = light lines

diol. The distance from one water to the 3-OH is exactly 12.1 Å.[42] If a 12.1 Å distance is essential for receptor binding, this strongly suggests that water may play a significant role. DES and hydrated estradiol are superimposed in the "top and side view" crystallographic drawings below (courtesy of the Medical Foundation of Buffalo).[42]

As long as the Schueler relationship between hydrogen bonding groups is maintained, significant estrogenic activity remains in most cases. For example, the *cis*-isomer of DES has only one-tenth of the estrogenic activity of the *trans*-isomer. The *meso*-isomer of dihydro-DES (hexestrol, Fig. 18-11) is active. It can keep the two phenolic groups appropriately separated,

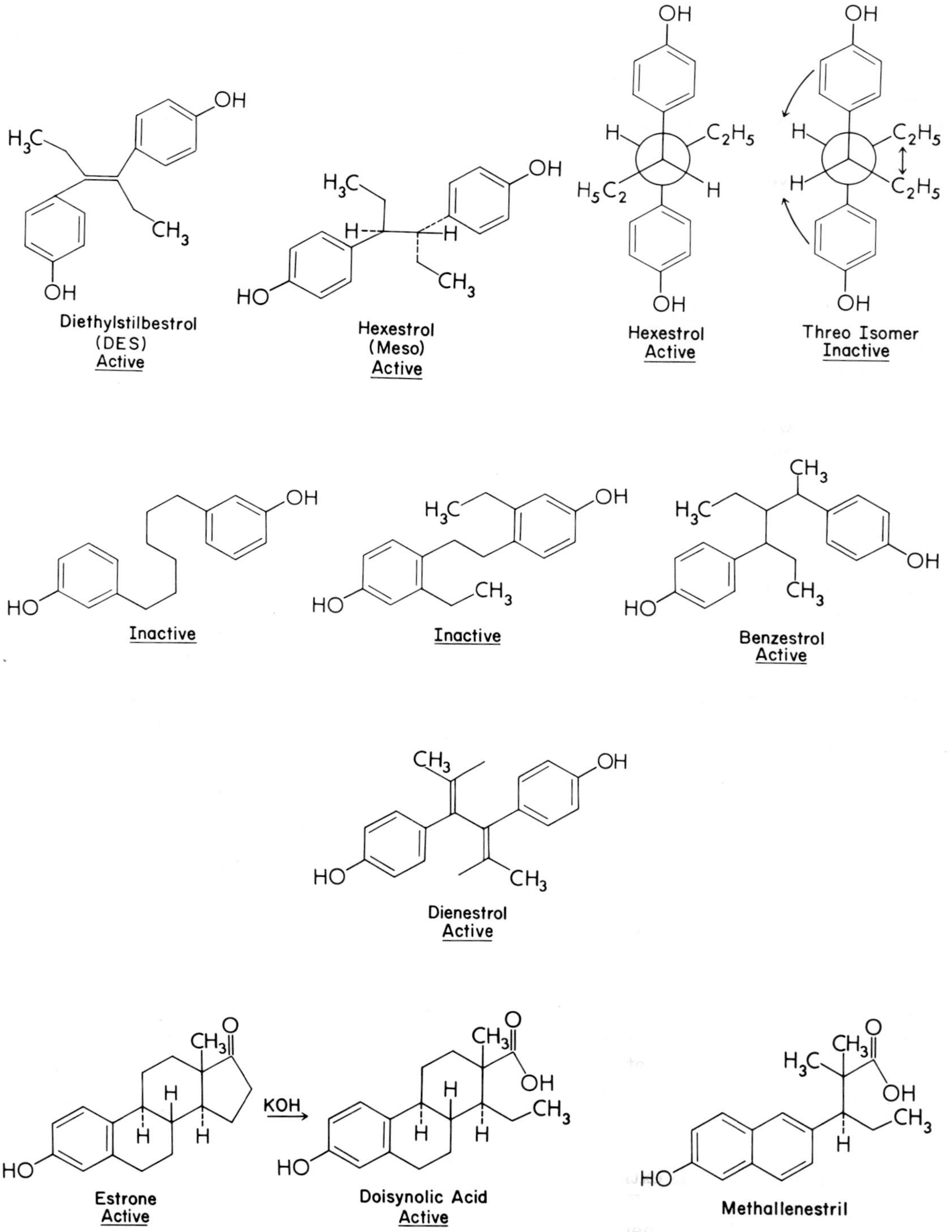

FIG. 18-11. *Synthetic estrogens: similarity to DES and estrone. Data from references 107, 108.*

but the *threo*-isomer cannot, due to steric repulsion of the ethyl groups.

Similarly, the two central double bonds in dienestrol (Fig. 18-11) keep the molecules fairly rigid and the phenolic hydroxyls separated. Benzestrol has three asymmetric carbons and can exist in eight possible diastereomeric forms, one of which is much more active than the others. Many other derivatives and modifications of DES have been made and studied,[106-110] some of which are shown in Figures 18-11 and 18-12.

Estradiol has three times and sixty times the estrogenic activity of estrone and estriol, respectively. However, estradiol is rapidly metabolized in the liver and by bacteria in the gastrointestinal tract (see Fig. 18-10).

Adding a 17α-alkyl group to estradiol blocks oxidation to estrone and in general greatly slows metabolic inactivation. In particular, 17α-ethinyl derivatives have been found to be highly estrogenic and orally active.

The nonsteroid estrogens also have good activity when taken orally. This property made them particularly attractive before steroidal estrogens were available at low prices. Today, however, steroid estrogens are easily made from diosgenin and can be obtained from the urine of horses.

Estrogen receptors

Estrogen receptors have been of considerable interest during the last few years. Many studies have shown that a number of proteins have selective and high affinity for estrogens, and progress in estrogen receptor research has been extensively reviewed.[21-32] The study of estrogen receptors has been stimulated by their apparent role in the carcinogenesis of several cancers.[111,50-52,54a]

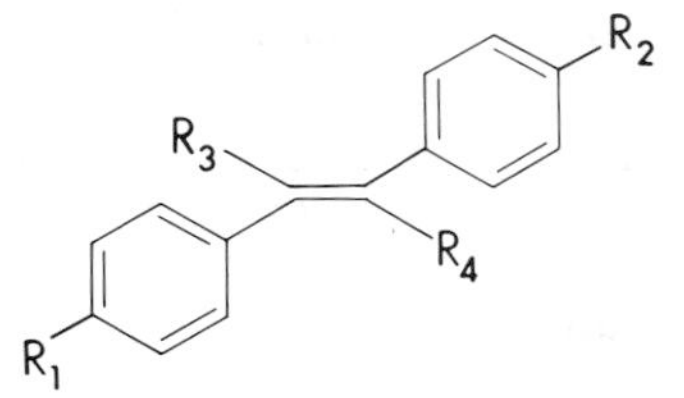

Several studies have found a general similarity of estrogen receptor protein, irrespective of the tissue or species studied. The general mechanism for estrogen action appears to follow Process 4 of Figure 18-2. The first step is the formation of a 4S hormone-protein complex in the cytoplasm. This complex then changes into a 5S complex, and it in turn is translocated into the nucleus. Cyclic AMP appears to play a major role.[28,55,56,57] O'Malley and co-workers have shown that once the receptor-estrogen complex is translocated into the nucleus, the complex acts upon the chromatin to increase template activity by increasing the number of initiation sites.[26,35,58,59]

Products

The estrogens are available in three groups of products: human estrogens and derivatives (obtained from degradation of sapogenins or cholesterol), equine estrogens (obtained from the urine of horses), and nonsteroidal estrogens.

The estrogens have been related with a number of serious side effects. These will be discussed in the section on oral contraceptives.

The human estrogens are available as a variety of C-3 and C-17 esters and ethers to increase their duration of action. 17α-Ethinyl estradiol is the most active orally and is widely used along with its 3β-methyl ether (mestranol) in oral contraceptives.

Equine estrogens contain equilenin and the sulfate ester salts of estrone and equilin. Conjugated Estrogens U.S.P. contains a larger amount of sodium estrone sulfate than Esterified Estrogens U.S.P., but the doses for both products are usually similar.

Both the steroidal and nonsteroidal estrogens are used for a number of medical indications, each with a specific dose range and schedule. For that reason, manufacturers' literature or reference texts with extensive dose and indication tables should be consulted when advising physicians.

Other indications (with specific doses) for most steroidal and nonsteroidal estrogens include: retarding progression of osteoporosis; senile vaginitis; abnormal uterine bleeding; hypogenitalism, amenorrhea; postmenopausal mammary carcinoma; and inoperable prostatic carcinoma.

General U.S.P. D.I. doses appear in the accompanying tables.

Major contraindications are thromboembolic disease (see discussion on oral contraceptives); primary carcinoma of the cervix, uterus, vagina, or breasts (ex-

R¹	R²	R³	R⁴	Equivalent Dose in μg.
OH	OH	Et	Et	0.3
OH	OCH₃	Et	Et	2.5
NH₂	OCH₃	Et	Et	1000
H	H	Et	Et	Inactive at 1000
OH	Br	Et	Et	100
OH	OH	CH₃	CH₃	20
OH	OH	CH₃	Et	0.5
OH	OH	CH₃CH₂CH₂—	Et	1-10
OH	OH	CH₃CH₂CH₂—	CH₃CH₂CH₂—	10-100

FIG. 18-12. *Stilbene derivatives: equivalent estrogen doses in rats. Data from references 109, 110.*

cept in postmenopausal women); and pregnancy (see discussion on oral contraceptives).

Patient information

All patients in the U.S. receiving a prescription for an estrogen product are now required to receive a patient package insert (PPI). Long, difficult to read PPI's have been the subject of intense discussion, as reviewed in 1980.[113] The American Council on Science and Health also publishes a very readable and accurate analysis of the benefits and risks of postmenopausal estrogen therapy.[114]

Estrone U.S.P, 3-hydroxyestra-1,3,5(10)-trien-17-one, is only one third as active as its natural precursor, estradiol (Fig. 18-10). As the salt of its 3-sulfate ester, estrone is the primary ingredient in Conjugated Estrogens U.S.P. and Esterified Estrogens U.S.P. Although originally obtained from the urine of pregnant mares (about 10 mg. per liter), estrone is now prepared from the Mexican yam, discussed later in this chapter. Assay is usually by ultra-violet light, using the maximum absorption 280 nm (EtOH). Radioimmunoassay procedures are also being developed for assay of estrone in plasma.

Piperazine Estrone Sulfate U.S.P., 3-sulfoxyestra-1,3,5(10)-trien-17-one piperazine salt. All the estrone 3-sulfate salts have the obvious pharmaceutical advantage of increased water-solubility (as one would predict from Table 18-1) and better oral absorption. Acids will not only convert the salts to the free 3-sulfate esters but will also cause some hydrolysis of the ester. This does not seem to adversely affect absorption, but precipitation of the free sulfate esters in acidic pharmaceutical preparations should be avoided. The dibasic piperazine molecule acts as a buffer, giving it somewhat greater stability.

Conjugated Estrogens U.S.P. and **Esterified Estrogens U.S.P.** These products are mixtures of steroidal estrogens and nonestrogenic materials extracted from the urine of horses, especially pregnant mares. **Conjugated Estrogens** contain 50 to 65 percent of sodium estrone sulfate and 20 to 35 percent of sodium equilin sulfate (based on the total estrogen content of the product). **Esterified Estrogens** have an increased amount of sodium estrone sulfate, 70 to 85 percent, often synthetically prepared from diosgenin and added to the urine extract. Although most commonly used to treat postmenopausal symptoms, the **Conjugated Estrogens** and **Esterified Estrogens** are used for the entire range of indications described previously.

Estradiol U.S.P., estra-1,3,5(10)-triene-3,17β-diol, is the most active of the natural steroid estrogens (Fig. 18-10). Although its 17β-OH group is vulnerable to bacterial and enzymatic oxidation to estrone (Fig. 18-10), it can be temporarily protected as an ester or permanently protected by adding a 17α-alkyl group (giving 17α-ethinyl estradiol and the 3-methyl ether, mestranol, the most commonly used estrogen in oral contraceptives). 3-Esters increase the duration of activity. These derivatives illustrate the principles of steroid modification shown in Figure 18-5. The increased oil-solubility of the 3- and 17β-esters (relative to estradiol) permits the esters to remain in oil at the injection site for extended periods of time. The commercially available estradiol esters are listed below and illustrated in Figure 18-6.

Estradiol Benzoate U.S.P.
Estradiol Valerate U.S.P.
Estradiol Cypionate U.S.P.

Ethinyl Estradiol U.S.P., 17α-ethinyl estradiol, has the great advantage over other estradiol products in that it is orally active. It is equal to estradiol in potency by injection, but 15 to 20 times more active orally. The 3-methyl ether of ethinyl estradiol is **Mestranol U.S.P.,** widely used in oral contraceptives.

Estriol U.S.P., possesses estrogenic activity and is reported to be orally active.

Diethylstilbestrol U.S.P., trans-α,α'-diethyl-4,4'-stilbenediol, DES, is the most active of the nonsteroidal estrogens (see Structure-Activity Relationships of the Estrogens), having about the same activity as estrone when given intramuscularly. The cis-isomer has only one-tenth the activity of the trans. The trans-isomer is also well absorbed orally and slowly metabolized, so it has been a popular estrogen for many medical purposes (see Therapeutic Uses). However, diethylstilbestrol must never be taken during pregnancy, except as an abortifacient (discussed later.) If taken during pregnancy, the drug increases the risk of cervical cancer in female offspring when they reach adulthood. The diphosphate ester, Diethylstilbestrol Diphosphate U.S.P., is used only for cancer of the prostate and is available for intravenous use. However, it has been reported that there may be an increased incidence of deaths from cardiovascular causes in men who received 5 mg. of DES daily for prolonged periods. The diphosphate salt has great water-solubility, as one would predict from Table 18-1. Diethylstilbestrol is extensively used in low doses as an aid to fatten cattle. Because DES has been implicated in cancer (albeit in higher doses), the United States Congress and F.D.A. began action in September, 1975, to ban DES in animal feed until further scientific studies are completed.

Note: all stilbene derivatives such as DES and dienestrol are light-sensitive and must be kept in light-resistant containers.

Dienestrol U.S.P., 4,4'-(1,2-diethylidene-1,2-eth-

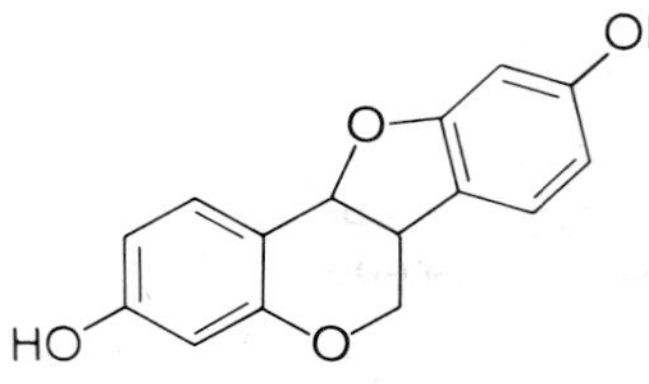

Coumestrol 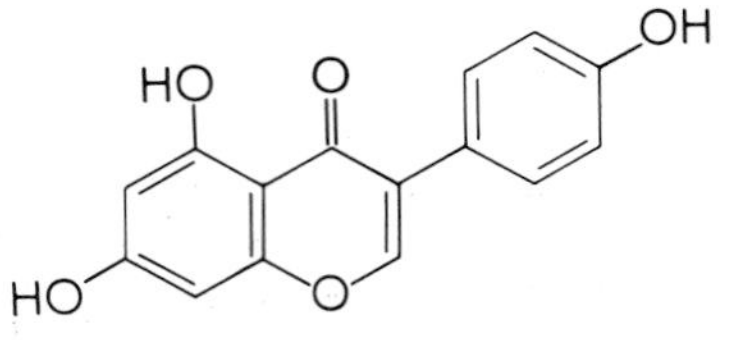   Genistein 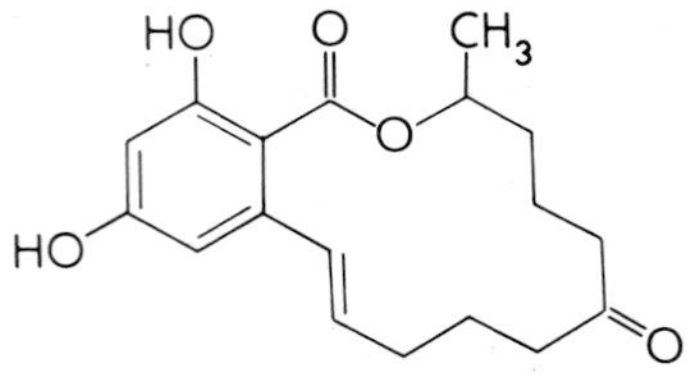   Zearalenone

anediyl)*bis*phenol, has about the same activity as DES when taken orally. The cream is used to treat atrophic vaginitis.

Benzestrol U.S.P., 4,4'-(1,2-diethyl-3-methyl-1,3-propanediyl)*bis*phenol, when drawn like DES in Figure 18-6, obviously resembles DES. Yet it has no double bonds such as have DES or dienestrol to keep the phenolic groups in a *trans* spatial arrangement. However, the adjacent ethyl groups do not prefer eclipsed conformation (much higher in energy than *trans*), thereby helping keep the phenolic groups *trans*. Benzestrol is used for all the usual indications for estrogens (see Therapeutic Uses).

Chlorotrianisene U.S.P., chloro*tris*-(*p*-methoxyphenyl)ethylene, is more active orally than by injection, and is thought to be converted to a more active form hepatically. When given by injection, it is quite a weak estrogen. It has good lipid-solubility and is slowly released from lipid tissues, thus giving it a relatively longer duration of action. The fat storage can also delay its onset of action.

Estrogens from plants. Several natural plant substances which differ from DES in structure are also potent estrogens. These include genistein, from a species of clover[115]; coumestrol, found in certain legumes[116]; and zearalenone, from a *Fusarium* fungus.[117] These and others have antifertility activity, as reviewed by Briggs.[119]

ANTIESTROGENS (OVULATION STIMULANTS)

Whereas estrogens have been very important in chemical contraception, estrogen antagonists (antiestrogens) have been of great interest as ovulation stimulants. While the term "antiestrogen" has been rather loosely applied to progestins and androgens, a few compounds have been found to have a direct effect in increasing FSH production by the hypothalamus. The mechanism is presumably a blocking of feedback inhibition of ovary-produced estrogens. The result is a greatly increased level of FSH and possibly LH, and, therefore, a stimulation of ovulation.

Clomiphene citrate is known to induce ovulation, and is believed to act on the hypothalamus. In tests with experimental animals, it has no effect in the absence of a functioning pituitary gland. Its great structural similarity to chlortrianisene can be seen below. A related compound, ethamoxytriphetol, is also strongly antiestrogenic, but not all F.D.A.-required studies have been completed.

Clomiphene causes a number of side-effects, especially enlargement of the ovaries. Abdominal discomfort should immediately be discussed with the physician. Other side-effects include nausea, visual disturbance, depression, breast soreness and increased nervous tension. Multiple births occur in about 10 percent of patients.

Alternatively, it would seem logical that ovulation could be stimulated by administering LH and FSH. However, animal preparations of LH and FSH either have not been effective (due to species differences) or have caused antigen-antibody reactions. A limited amount of human LH and FSH extracts has been obtained from human pituitary glands or from the urine of postmenopausal women. The extract is called human menopausal gonadotropin (HMG). As with clomiphene, the patient must have partially functioning

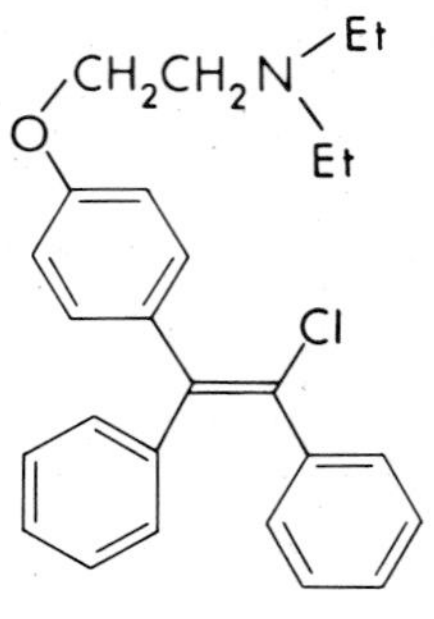

Clomiphene 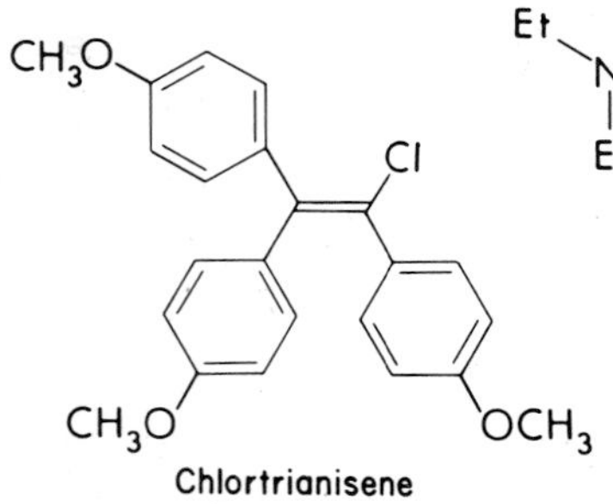   Chlortrianisene 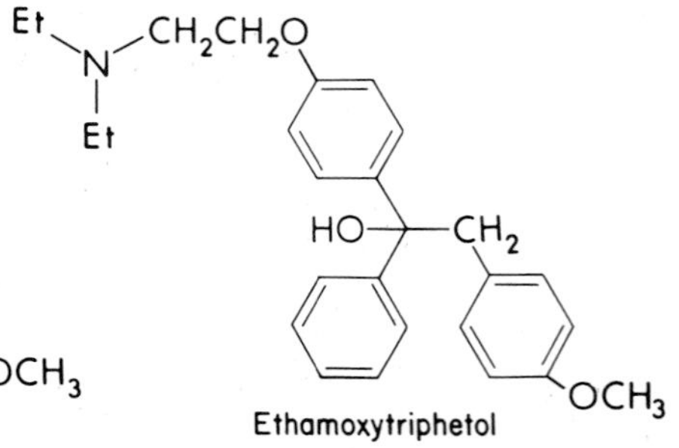   Ethamoxytriphetol

TABLE 18-2

ESTROGEN PRODUCTS (STEROIDAL ESTROGENS)

Name Proprietary Name	Preparations	Application	Usual Adult Dose*	Usual Dose Range*
Estrone U.S.P. *Theelin,* *Menformon,* *Urestrin*	Sterile Estrone Suspension U.S.P. Estrone Injection U.S.P.	Available also in vaginal suppositories	I.M., 1 mg. one or more times weekly as required; reduce to maintenance dose as soon as response is obtained.	200 μg. to 5 mg. weekly
Piperazine Estrone Sulfate U.S.P. *Ogen*	Piperazine Estrone Sulfate Tablets U.S.P. Piperazine Estrone Sulfate Vaginal Cream U.S.P.		1.5 mg. daily	750 μg. to 10 mg. daily
Conjugated Estrogens U.S.P. *Premarin,* *Menotabs*	Conjugated Estrogens Tablets U.S.P.	Available also as vaginal cream, and I.V. and I.M. dosage forms	1.25 to 2.5 mg. 1 to 3 times a day for 3 weeks of every month	300 μg. to 30 mg. daily
Esterified Estrogens U.S.P. *Amnestrogen,* *Menest,* *SK-Estrogens,* *Evex, Glyestrin*	Esterified Estrogens Tablets U.S.P.		1.25 to 2.5 mg. 1 to 3 times a day for 3 weeks of every month	300 μg. to 30 mg. daily
Estradiol U.S.P.	Sterile Estradiol Suspension U.S.P. Estradiol Pellets U.S.P. Estradiol Tablets U.S.P.			Implantation, 25 mg. repeated when necessary; I.M., 220 μg. to 1.5 mg. 2 or 3 times weekly. Oral, 200 to 500 μg. 1 to 3 times daily
Estradiol Benzoate U.S.P. *Progynon* *Benzoate*	Estradiol Benzoate Injection U.S.P.			I.M., initial, 1.0 to 1.66 mg. 2 or 3 times weekly for 2 or 3 weeks; maintenance, 330 μg. to 1.0 mg. twice weekly
Estradiol Valerate U.S.P. *DelEstrogen*	Estradiol Valerate Injection U.S.P.		I.M., 5 to 30 mg. every 2 weeks	5 to 40 mg. every 1 to 3 weeks
Estradiol Cypionate U.S.P. *Dep-Estradiol*	Estradiol Cypionate Injection U.S.P.		Initial, I.M., 1 to 5 mg. weekly for 2 to 3 weeks; maintenance, 2 to 5 mg. every 3 to 4 weeks	
Ethinyl Estradiol U.S.P. *Lynoral, Estinyl,* *Feminone*	Ethinyl Estradiol Tablets U.S.P.		50 μg. 1 to 3 times a day	20 μg. to 3 mg. daily
Mestranol U.S.P.				See Table 18-8 for doses in oral contraceptives

* See *U.S.P. D.I.* for complete dosage information.

TABLE 18-3

ESTROGEN PRODUCTS (NONSTEROIDAL ESTROGENS)

Name Proprietary Name	Preparations	Application	Adult Usual Dose*	Usual Dose Range*
Diethylstilbestrol U.S.P. *Stilbetin*	Diethylstilbestrol Tablets U.S.P.		Mammary carcinoma, 15 mg. once daily; carcinoma of prostate, 1 to 3 mg. once daily; estrogen, 200 μg. to 2 mg. once daily	Mammary carcinoma, 1 to 15 mg. or more daily; carcinoma of prostate, 1 to 5 mg. daily; estrogen, 100 μg. to 25 mg. daily

* See *U.S.P. D.I.* for complete dosage information.

(Continued)

TABLE 18-3

ESTROGEN PRODUCTS (NONSTEROIDAL ESTROGENS)

Name Proprietary Name	Preparations	Application	Adult Usual Dose*	Usual Dose Range*
	Diethylstilbestrol Injection U.S.P.		Carcinoma of prostate, I.M., 2 to 5 mg. twice a week; estrogen, I.M., 250 µg. to 1 mg. 2 or 3 times a week	Estrogen, 100 µg. twice a week to 10 mg. daily
	Diethylstilbestrol Suppositories U.S.P.		Vaginal, 100 µg. to 1 mg. once daily	
Dienestrol U.S.P. *Synestrol*	Dienestrol Cream U.S.P.	Vaginal, 5 g. of a 0.01 percent cream once or twice daily for 7 to 14 days, then once every 48 hours for 7 to 14 days	500 µg. daily	100 µg. to 1.5 mg.
Benzestrol U.S.P. *Chemestrogen*	Benzestrol Tablets U.S.P.		1 to 2 mg. daily	500 µg. to 5 mg.
Chlorotrianisene U.S.P. *Tace*	Chlorotrianisene Capsules U.S.P.		24 mg. daily	12 to 144 mg. daily, as determined by the practitioner for the condition being treated

* See *U.S.P. D.I.* for complete dosage information.

ovaries for HMG to be effective, and other causes of infertility should be excluded before HMG treatment. HMG should be used with caution because ovarian enlargement is quite common. Multiple births occur in up to 20 percent of the cases, and pregnancies followed by spontaneous abortions occur in 20 to 30 percent of the cases. HMG has also been used to treat obesity, but the F.D.A. has strongly warned against its use for that purpose.[118]

In general, it is strongly recommended that product literature or detailed general references such as *Facts and Comparisons* or the *Hospital Formulary* be consulted before dispensing either clomiphene citrate or HMG.

Clomiphene Citrate U.S.P., Clomid®, N,N-diethyl-2-[4-(2-chloro-1,2-diphenylethenyl)phenoxy]ethanamine, is given to stimulate ovulation in the usual dose of 50 mg. daily for five days starting on the fifth day of the menstrual cycle. If ovulation does not occur, the dose is increased to 100 mg. daily for five days in the next cycle. The patient should be warned to report any visual disturbances or abdominal pain to the physician. If menstruation does not occur at the end of the first full cycle following treatment, pregnancy tests should be conducted before additional clomiphene is taken. A careful physical examination prior to treatment is recommended, especially to determine the possible presence of ovarian cysts, since ovarian enlargement sometimes occurs.

Structure-Activity Relationships of the Progestins

The progestins are compounds with progestational activity, primarily including progesterone and 19-nortestosterones. Although the 19-nortestosterones do have androgenic side-effects, their primary activity is nevertheless progestational. Another reason for great interest in these progestins is that progesterone is not orally effective. Its plasma half-life is only about five minutes, and it is almost completely metabolized in one passage through the liver.[69-71,94]

It is known that addition of 17α-alkyl groups to testosterone blocks oxidation at C-17. However, 17α-methyltestosterone has only half the androgenic activity of parenterally administered testosterone, and the 17α-ethyl analog is nearly inactive. Adding the electron density of a triple bond as in 17α-ethinyl causes a marked increase in progestational activity, and simultaneously blocks metabolic or bacterial oxidation to the corresponding 17-ones. Thus, by adding a 17α-ethinyl or propinyl group to testosterone, one can simultaneously decrease anabolic activity and promote good progestational activity, and have an orally active compound as well. Table 18-4 illustrates the relative progestational activity of a number of progestins. As shown in Figure 18-7, a 17α-hydroxy or a 6α-methyl group significantly increases progestational activity. Metabolic inactivation is presumably reduced.

The 19-nor derivatives have also been found to have marked ovulation-inhibiting activity, which does not necessarily parallel progestational activity. The endometrial proliferation (Clauberg-McPhail) test is most often used to evaluate progestational activity, while antiovulation activity is determined by examining treated female rabbits for ovulation-rupture points in their ovaries. Other methods are discussed by Deghenghi and Manson.[3]

TABLE 18-4

COMPARATIVE PROGESTATIONAL ACTIVITY OF SELECTED PROGESTINS[118A]

	Relative Oral Activity	Activity SC
Progesterone	(nil)	1
17α-Ethinyltestosterone (Ethisterone)	1	0.1
17α-Ethinyl-19-nortestosterone (Norethindrone)	5–10	0.5–1
Norethynodrel	0.5–1	0.05–1
17α-Hydroxyprogesterone Caproate	2–10	4–10
Medroxyprogesterone Acetate	12–25	50
19-Norprogesterone	—	5–10
Norgestrel	—	3
Dimethisterone	12	—

The development of 19-norsteroids as contraceptive agents and the historic work of Pincus, Rock and Garcia will be discussed in the section on chemical contraception.

Progesterone receptors

Since progesterone is a biosynthetic precursor to other steroids and, in addition, is rapidly metabolized, the study of progesterone has been more difficult than the study of the estrogens. It is clear that relatively large amounts of progesterone are required to cause a biological response, and in most cases there is a synergistic effect with an estrogen.

It has been suggested that progesterone may inhibit enzymes involved in the maintenance of the uterine wall membrane potential, thus causing its known depressant effect on uterine contractility.

A number of progesterone receptors have been found in the uterine cells of many animal species. However, in nearly all cases the receptors have been found to be unresponsive to progesterone unless pretreated with estrogens. It appears that there are at least two cytoplasmic receptors for progesterone, and one or more nuclear receptors. There is evidence supporting the possibility that either the nuclear receptor or a nuclear receptor-DNA complex causes the observed physiologic responses of uterine-bound progesterone.

Preliminary binding studies have also been completed with progesterone and other tissues. For example, it has been found that ³H-progesterone accumulates in mammary gland tissue in three times greater concentrations than in plasma. Since there has been shown to be a very high correlation between protein binding and tissue response in the case of the steroids, these preliminary data suggest the existence of receptors within the mammary gland.

Products

The progestins are primarily used in oral contraceptive products for women and they are also used to treat a number of gynecological disorders: dysmenorrhea, endometriosis, amenorrhea and dysfunctional uterine bleeding. Estrogens are given simultaneously in most of these situations. Progestins have been used to prevent habitual abortions, but the F.D.A. has strongly warned against the use of steroids during pregnancy.[92] Large doses of progestins have also been given as a test for pregnancy, but the F.D.A. warning would seem to discourage this practice as well. Adverse effects of sex hormones given during pregnancy were reviewed in 1979.[93]

The doses appropriate for the various indications described above can vary significantly, and detailed manufacturers' literature or general references should be consulted prior to advising physicians.

General U.S.P. D.I. doses are listed in Table 18-5.

Progesterone U.S.P., pregn-4-en-3,20-dione, is so rapidly metabolized that it is not very effective orally, being only one-twelfth as active as intramuscularly. It can also be very irritating when given intramuscularly. Buccally it is only slightly more active than orally. Originally obtained from animal ovaries, it was prepared in ton quantities from diosgenin in the 1940's. This marked the start of the modern steroid industry, a fascinating history discussed later in this chapter. The discovery of 19-nortestosterones with progesterone activity made synthetic modified progestins of tremendous therapeutic importance.

Progesterone (and all other steroid 4-ene-3-ones) is light-sensitive and should be protected from light.

Hydroxyprogesterone Caproate U.S.P., 17α-hydroxypregn-4-en-3,20-dione hexanoate, is much more active and longer acting than progesterone (see Table 18-4), probably because the 17α-ester function hinders reduction to the 20-ol. It is given only intramuscularly. The hexanoate ester greatly increases oil-solubility, allowing it to be slowly released from depot preparations, as one would predict from Figure 18-5.

Medroxyprogesterone Acetate U.S.P., 17α-hydroxy-6α-methylpregn-4-en-3,20-dione 17α-acetate, adds a 6α-methyl group to the 17α-hydroxyprogesterone structure to greatly decrease the rate of reduction of the 4-ene-3-one system. The 17α-acetate group also decreases reduction of the 20-one, just as with the 17α-caproate. Medroxyprogesterone acetate is very active orally (Table 18-4), and has such a long duration of action intramuscularly that it cannot be routinely used intramuscularly for treating many menstrual disorders.

Norethindrone U.S.P. and **Norethynodrel U.S.P.,** 17α-ethinyl-19-nortestosterone, and its Δ⁵⁽¹⁰⁾ isomer, respectively, might appear at first glance to be

TABLE 18-5

PROGESTIN PRODUCTS

Name Proprietary Name	Preparations	Usual Adult Dose*	Usual Dose Range*
Progesterone U.S.P. *Lipo-Lutin*	Progesterone Injection U.S.P. Sterile Progesterone Suspension U.S.P.	I.M., 5 to 25 mg. once a day beginning 8 to 10 days before menstruation	5 to 50 mg. daily
Hydroxyprogesterone Caproate U.S.P. *Delalutin, Corlutin, Hylutin*	Hydroxyprogesterone Caproate Injection U.S.P.	Menstrual disorders, I.M., 375 mg. once a month; uterine cancer, I.M., 1 g. or more, repeated 1 or more times per week	375 mg. monthly to 7 g. weekly
Medroxyprogesterone Acetate U.S.P. *Provera*	Sterile Medroxyprogesterone Acetate Suspension U.S.P.	Endometriosis, I.M., 50 mg. once a week; uterine cancer, I.M., 400 mg. to 1 g. once a week	50 mg. to 1 g. weekly
	Medroxyprogesterone Acetate Tablets U.S.P.	Habitual and threatened abortion, 10 to 40 mg. daily; menstrual disorders, 2.5 to 20 mg. daily for 5 to 10 days, during the second half of the menstrual cycle	
Norethindrone U.S.P. *Norlutin*	Norethindrone Tablets U.S.P. Norethindrone and Ethanyl Estradiol Tablets U.S.P. Norethindrone and Mestranol Tablets U.S.P. Norethindrone Acetate Tablets U.S.P.	5 to 20 mg. once a day	5 to 40 mg. daily
Norethindrone Acetate U.S.P. *Norlutate*	Norethindrone Acetate and Ethinyl Estradiol Tablets U.S.P.†	1 to 2.5 mg. of norethindrone acetate and 20 to 50 µg. of ethinyl estradiol once daily for 20 or 21 days, beginning on the 5th day of the menstrual cycle	
Norethynodrel U.S.P.	Norgestrel Tablets U.S.P.	2.5 to 10 mg. once daily	2.5 to 30 mg. daily
Norgestrel U.S.P.	Norgestrel and Ethinyl Estradiol Tablets U.S.P.†	500 µg. of norgestrel and 50 µg. of ethinyl estradiol for 21 days, beginning on the 5th day of the menstrual cycle.	
Ethynodiol Diacetate U.S.P.	Ethynodiol Diacetate and Ethinyl Estradiol Tablets U.S.P.†	1 mg. of ethynodiol diacetate and 50 µg. of ethinyl estradiol once a day for 21 days, beginning on the 5th day of the menstrual cycle	
Dydrogesterone U.S.P. *Duphaston*	Dydrogesterone Tablets U.S.P.	10 to 20 mg. daily in divided doses	10 to 30 mg.

* See *U.S.P. D.I.* for complete dosage information.
† Oral contraceptive.

subtle copies of each other. One would predict that the $\Delta^{5(10)}$ double bond would isomerize in the stomach's acid to the Δ^3 position. In fact, however, the two drugs were simultaneously and independently developed so neither can be considered a copy of the other. Furthermore, norethindrone is about ten times more active than norethynodrel (Table 18-4), indicating that isomerization is not as facile in vivo as one might predict. Although they are less active than progesterone when given subcutaneously they have the important advantage of being orally active. The discovery of the potent progestin activity of 17α-ethinyltestosterone (ethisterone) and 19-norprogesterone preceded the development of these potent progestins. All are orally active, with the 17α-ethinyl group blocking oxidation to the less active 17-one. The rich electron density of the ethinyl group and the absence of the 19-methyl greatly enhance progestin activity. Both compounds

have become of great importance as progestin ingredients of oral contraceptives, although Norethindrone U.S.P. and Norethindrone Acetate U.S.P. are widely employed for all the usual indications of the progestins. Since these compounds retain the key feature of the testosterone structure—the 17β-OH—it is not surprising that they possess some androgenic side-effects. The related compound, **Norgestrel U.S.P.,** has an ethyl group instead of the C-13-methyl, but has similar biological properties. Norgestrel is used only in oral contraceptives. All these 19-nortestosterone derivatives will be discussed in the later section on chemical contraceptives.

Dydrogesterone U.S.P., 9β,10α-pregna-4,6-dien-3,20-dione, is a "retro" or C_{19}-iso steroid. It has good progestin activity, but no ovulation-inhibition (contraceptive) activity, and it is not as effective in treating some menstrual disorders as other progestins.

ANDROGENS AND ANABOLIC AGENTS

Although produced in small concentrations in females, testosterone and dihydrotestosterone are produced in much greater amounts in males. Testosterone has two important activities: *androgenic activity* (or male-sex-characteristic-promoting) and *anabolic activity* (or muscle-building). Compounds which have these two activities are generally called androgens and anabolic agents. Since it would be very useful to have drugs which were anabolic but not androgenic (e.g., to aid the recovery of severely debilitated patients), many compounds with increased anabolic activity have been synthesized. However, significant levels of androgenic activity have limited the therapeutic uses of all these compounds.

The commonly used androgenic and anabolic agents are shown in Figure 18-13. Several excellent reviews on androgens and anabolic agents have been published.[2,120-122]

Therapeutic uses

The primary use of androgens and anabolic agents is as androgen replacement therapy in men, either at maturity or in adolescence. The cause of testosterone deficiency may either be hypogonadism or hypopituitarism.

The use of the androgens and anabolic agents for their anabolic activity, or for uses other than androgen replacement, has been very limited due to their masculinizing actions. This has greatly limited their use in women and children. Although anabolic activity is often needed clinically, none of the products presently available has been found to be free of significant androgenic side-effects.

The masculinizing (androgenic) side-effects in females include hirsutism, acne, deepening of the voice, clitoral enlargement and depression of the menstrual cycle. Furthermore, the androgens and anabolic agents generally alter serum lipid levels and increase the probability of atherosclerosis, characteristically a disease of males and postmenopausal females.

Androgens in low doses are sometimes used in the treatment of dysmenorrhea and postpartum breast enlargement. However, the masculinizing effects of the androgens and anabolic agents, even in small doses, preclude their use in most circumstances. Secondary treatment of advanced or metastatic breast carcinoma in selected cases is generally considered to be the *only* indication for large-dose, long-term androgen therapy in women.

Androgens and anabolic agents are also used to treat certain anemias, osteoporosis, and to stimulate growth in postpuberal boys. In all cases, use of these agents requires caution.

Androgens and sports

The use of anabolic steroids by athletes began in the late 1940's[123] and is now widespread. Up to 80 percent of competitive weightlifters and about 75 percent of professional football players use these drugs,[124] along with a variety of other athletes. During the 1972 Olympic games in Munich, 68 percent of interviewed track and field athletes (not including long distance runners) had used anabolic steroids in preparation for the Olympics.[125]

Yet the understanding by athletes of the rational selection, risks, and risk/benefit ratios of the drugs they take appears to be quite limited. A local weightlifter, for example, noted to the author's Research Associate that he takes "1 ml. of testosterone injected each week, and 10 to 20 mg. of Dianabol® per day. I switch the Dianabol® to Anavar® two weeks before competition." (*Note:* this is a potentially dangerous drug regimen!) He did not know of any side-effects of the drugs and had never read a research report proving the drugs did any good.

Many studies have attempted to determine if taking anabolic steroids improves athletic performance.[126-138] However, some[132,135,136] failed to use controls (athletes who trained in an identical manner but who did not take anabolic steroids). Others failed to use placebos in at least a single blind research design (neither the treated nor control groups knowing which they were taking). In short, only ten studies[125-127,129-132,134,137-138] published by early 1980 are useful in delineating this interesting question.

Of these ten studies, four reported that anabolic steroids did increase athletic performance,[129,131,134,138] while six others found they did not.[125-127,130,133,137] It would be fair to say, therefore, that the benefit of anabolic steroids to athletic performance is uncertain.

The risks of using these drugs appear to outweigh their uncertain benefits. These risks include significantly depressed testosterone production that may not be reversible,[127,130-131,139-141] as well as jaundice, gynecomastia, testicular shrinkage, swelling of the prostate, infertility, oligospermia, and decreased libido.[123-124,139]

(Novich humorously suggests that the reduction in sexual activity leads to increased eating, which is the cause of the weight gain.[139]) Liver and bone damage are also possible.[124] The use of anabolic steroids by adolescent or prepubertal males and females is particularly dangerous.

Because of these risks, the International Olympic

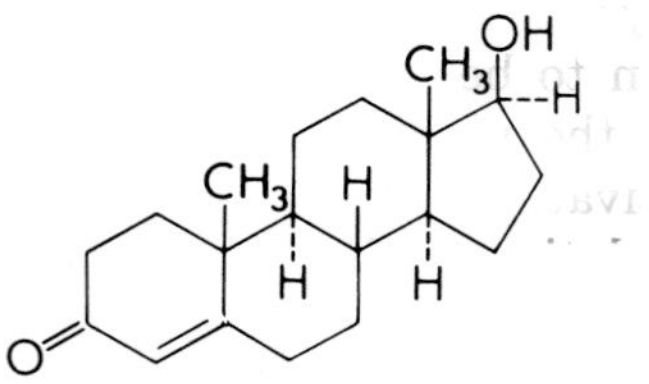

Testosterone
(1:1)

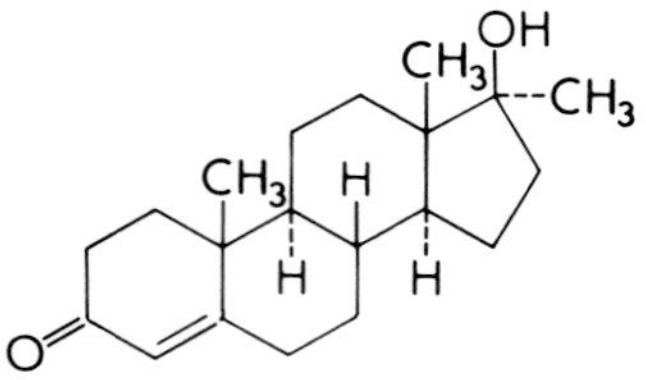

17α-Methyltestosterone
(1:1 but 1/2 as potent
as testosterone)

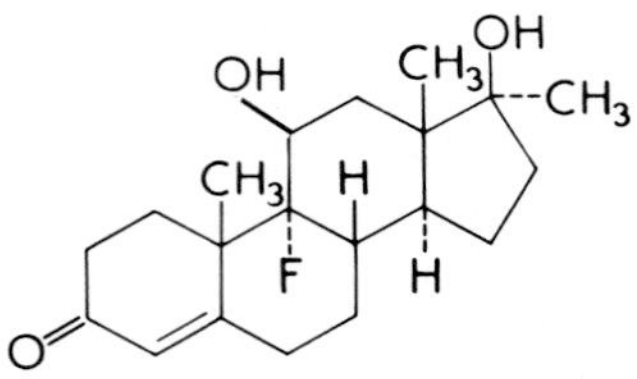

Fluoxymesterone
(1:1 to 2:1 and 5 to 10 times
more potent than testosterone)

17β -Esters Commercially Available:

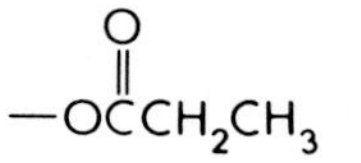

$-OCCH_2CH_3$ Testosterone Propionate

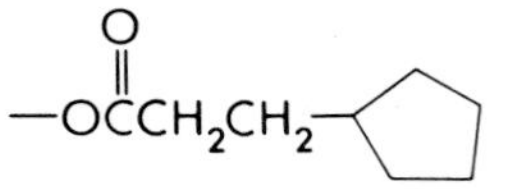

$-OCCH_2CH_2-$⬠   Testosterone Cyclopentylpropionate
(Cypionate)

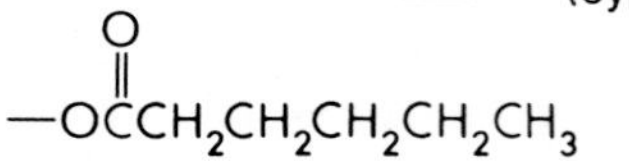

$-OCCH_2CH_2CH_2CH_2CH_3$ Testosterone Enanthate

All are I.M.— some available as implantation pellets

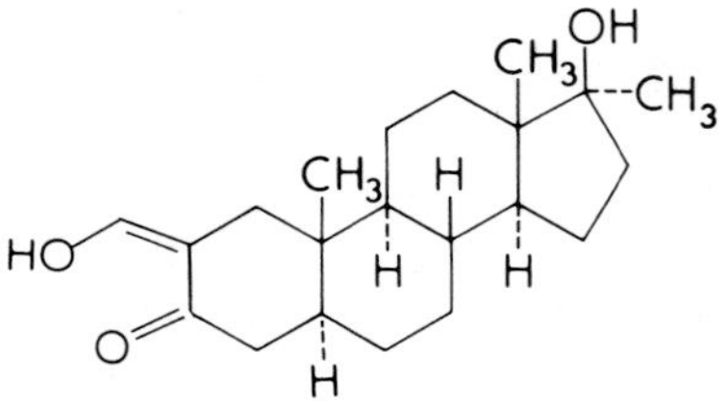

Oxymetholone
(2.5:1; 6:1 S.C.)

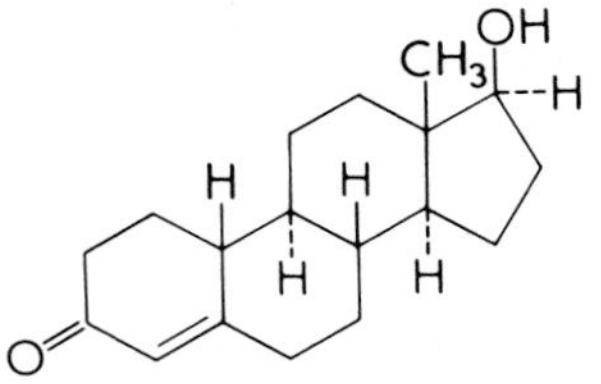

Nandrolone
(2.5:1 to 4:1)

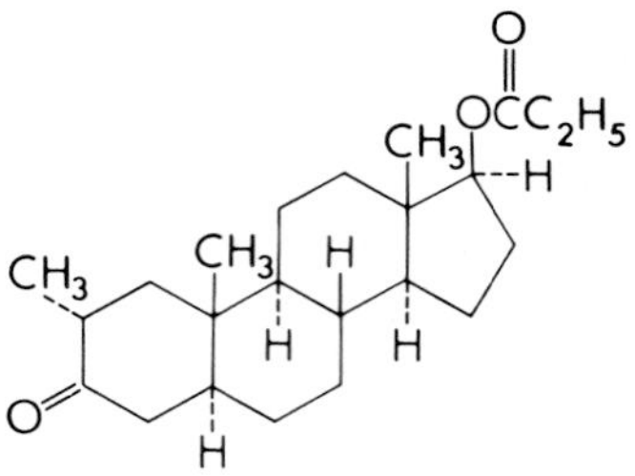

Dromostalone
(Propionate, 3:1 to 4:1)

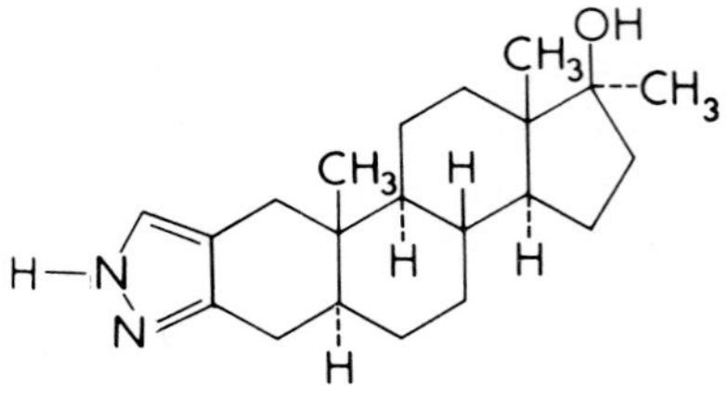

Stanozolol
(3:1 to 6:1)

17β -Esters Commercially Available:

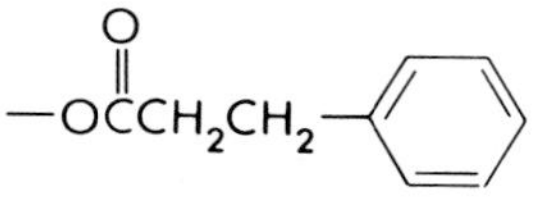

$-OCCH_2CH_2-$⬡   Nandrolone Phenpropionate

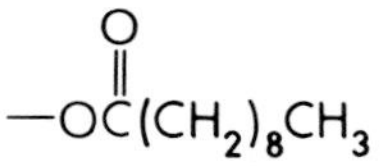

$-OC(CH_2)_8CH_3$ Nandrolone Decanoate

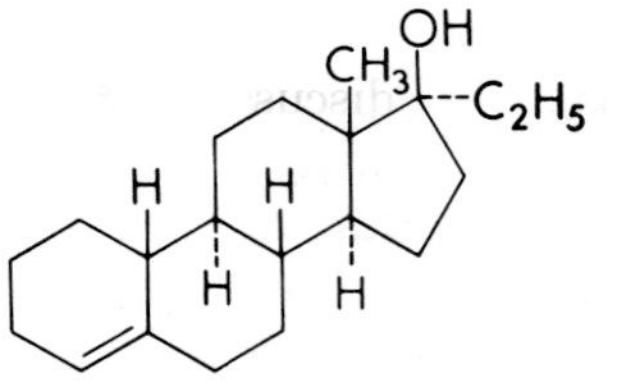

Ethylestrenol
(3:1)

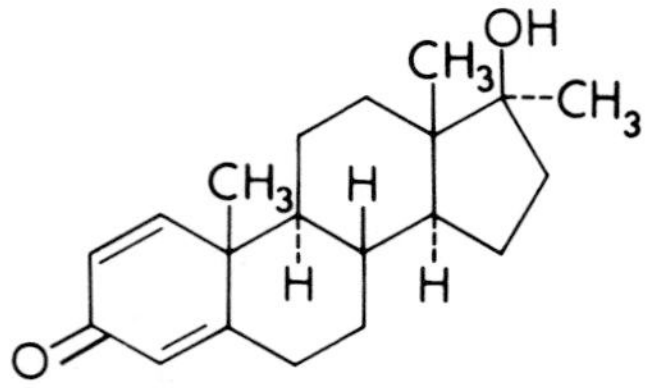

Methandrostenolone
(1:1)

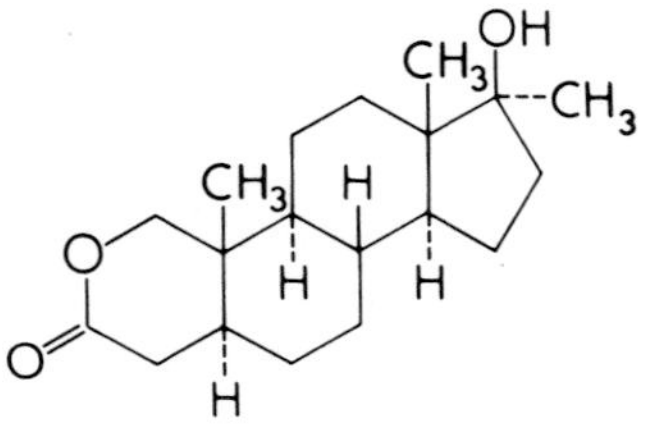

Oxandrolone

FIG. 18-13. *Androgens and anabolic agents (anabolic:androgenic ratio).*

Committee has banned all anabolic drugs. The top six winners in each Olympic event are to be tested for nontherapeutic drugs of all types.[142] The use of these drugs by athletes has also been criticized and discouraged by coaches,[143] physicians,[139] and other athletic governing bodies.[144]

Biosynthetic sources

As shown in Figure 18-4, testosterone can be synthesized via progesterone and androstenedione. Labeling experiments have also shown that it can be biosynthesized from androst-5-ene-3β,17β-diol,[70] not shown in Figure 18-4.

Testosterone is primarily produced by the Leydig cells of the testes. The ovaries and adrenal cortex also synthesize androstenedione and 5-androsten-17-one-3β-ol (dehydroepiandrosterone), which can be rapidly converted to testosterone in many tissues.[70]

Testosterone levels in the plasma of men are 5 to 100 times greater than the levels in the plasma of women, with about 4 to 12 mg. per day being produced in young men and 0.5 to 2.9 mg. per day in young women.[145,146]

Testosterone is produced in the testes in response to FSH and LH (interstitial cell-stimulating hormone, or ICSH) release by the anterior pituitary. Testosterone and dihydrotestosterone inhibit the production of LH and FSH by a feedback inhibition process. This is quite similar to the feedback inhibition by estrogens and progestins in FSH and LH production.

Biological activities

Testosterone and dihydrotestosterone cause pronounced masculinizing effects even in the male fetus. They induce the development of the prostate, penis and related sexual tissues.[120–122]

At puberty, the secretion of testosterone by the testes increases greatly, leading to an increase in facial and body hair, a deepening of the voice, an increase in protein anabolic activity and muscle mass, a rapid growth of long bones, and a loss of some subcutaneous fat. Spermatogenesis begins, and the prostate and seminal vesicles increase in activity. Sexual organs increase in size. The skin becomes thicker and sebaceous glands increase in number, leading to acne in many young people. The androgens also play important roles in male psychology and behavior.

Metabolism

Testosterone is rapidly converted to 5α-dihydrotestosterone in many tissues, and 5α-dihydrotestosterone is also secreted by the testes. In fact, 5α-dihydrotestosterone is known to be the active androgen in many tissues, e.g., in the prostate. The primary route for metabolic inactivation of testosterone and dihydrotestosterone is oxidation to the 17-one. The 3-one group is also reduced to the 3α- and 3β-ols. The products are shown in Figure 18-14. A few other metabolites have also been detected.[69,145]

Assay procedures

The most commonly utilized test to determine the androgenic and anabolic activity of various compounds is with castrated rats. After a period of treatment with the drug, the rats are sacrificed. The increased weight of the levator ani muscle relative to non-drug-treated control animals is used as a measure of anabolic activity.[69,147] The increased weight of the ventral prostate and seminal vesicles relative to controls is used as a measure of androgenic activity. The tests are inexpensive and easy to perform. Activities are always evaluated against testosterone- or methyltestosterone-treated animals as well as the controls.

However, the tests do have their limitations.[148–150] In particular, it has been noted that the levator ani muscle is sometimes more sensitive to androgens than is skeletal muscle. As a result, dogs and ovariectomized monkeys are studied in nitrogen balance experiments (with an increase in retained nitrogen being a measure of protein synthesis in the body).[147] Unfortunately, nitrogen balance assays also have limitations.[151]

Other assay procedures are sometimes employed, including measurement of the absorption of labeled α-aminoisobutyric acid in the levator ani, as well as determining the effectiveness of counteracting the catabolic effects of various drugs such as cortisone.

Structure-Activity studies

In his book *Androgens and Anabolic Agents*, Julius A. Vida[152] has summarized the structures and biological activity of over 500 different androgens and anabolic agents. The excellent discussion by Counsell and Brueggemier[2] also cites many compounds. One might suppose that the structure-activity relationships of these drugs have been well delineated. However, the structural requirements for selective anabolic activity are still unclear, and there is even uncertainty about the relationship of structure to androgenic activity.

Since bacterial and hepatic oxidation of the 17β-hydroxyl to the 17-one is the primary route of metabolic inactivation,[2,69] 17α-alkyl groups have been

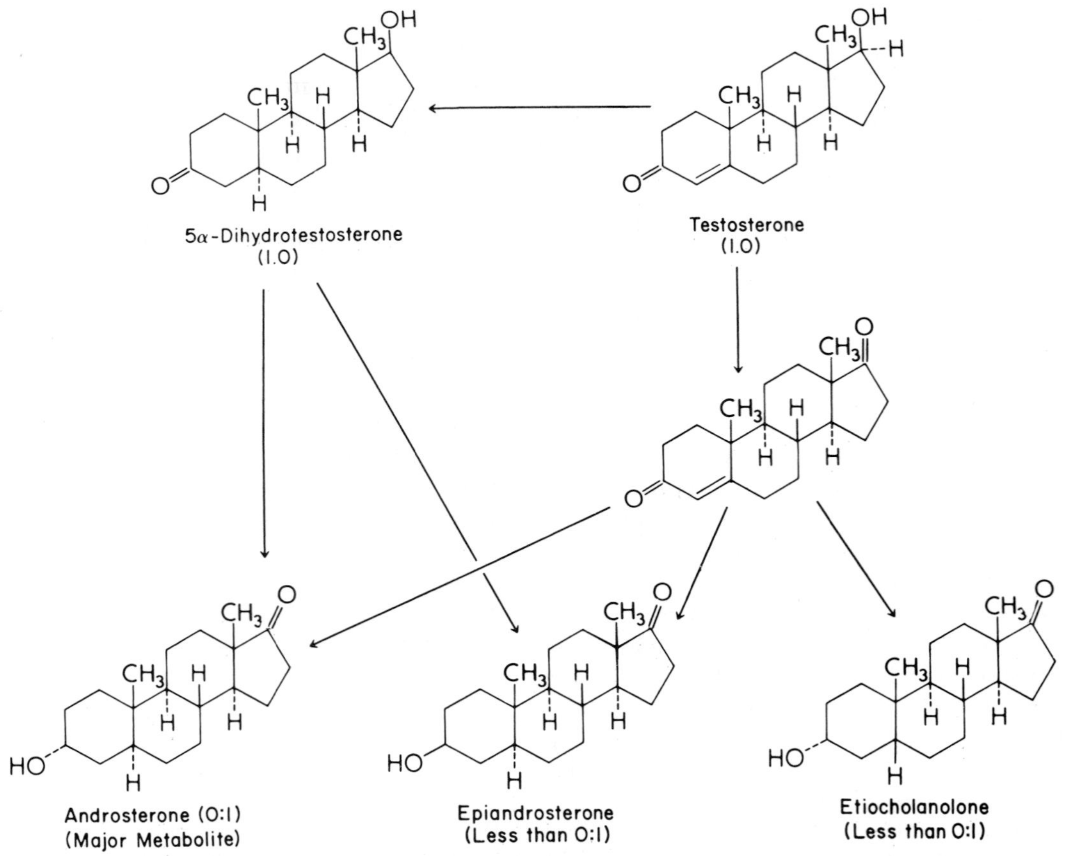

FIG. 18-14. *Metabolism of testosterone and 5α-dihydrotestosterone (relative androgenic activity).*

added. Even though 17α-methyltestosterone is only about half as active as testosterone, it can be taken orally. 17α-Ethyltestosterone has greatly reduced activity, as shown in Table 18-6. A disadvantage of the 17α-alkyl testosterones is that hepatic disturbances (and occasionally jaundice) may occur.

Table 18-6 illustrates other structure-activity effects of the androgen; for example, the greatly decreased activity of the 17β-ol isomer of testosterone.

Many hypotheses have been made to attempt to summarize the structure-activity relationships of all the known androgens, including proposals of Vida,[152] Wolff[154] and others.[2,152,155] Vida,[2,156] Counsell[152] and Klimstra[154] have published detailed discussions of the various hypotheses. They have also discussed the activity of the many substituted androstanes and testosterones studied to evaluate steric and electronic effects upon androgenic and anabolic activity.

TABLE 18-6

ANDROGENIC ACTIVITIES OF SOME ANDROGENS[153]

Compound	Micrograms Equivalent to an International Unit
Testosterone (17β-ol)	15
Epitestosterone (17α-ol)	400
17α-Methyltestosterone	25–30
17α-Ethyltestosterone	70–100
17α-Methylandrostane-3α,17β-diol	35
17α-Methylandrostane-3-one-17β-ol	15
Androsterone	100
Epiandrosterone	700
Androstane-3α,17β-diol	20–25
Androstane-3α,17α-diol	350
Androstane-3β,17β-diol	500
Androstane-17β-ol-3-one	20
Androstane-17α-ol-3-one	300
Δ5-Androstene-3α,17β-diol	35
Δ5-Androstene-3β,17β-diol	500
Androstanedione-3,17	120–130
Δ4-Androstenedione	120

Drugs with improved anabolic activity

Many drugs are available which have improved anabolic:androgenic activity ratios, but none is free of androgenic activity. This has greatly limited their therapeutic utility. Examples of drugs which have been found to have marked improvements in anabolic activity are illustrated in Figure 18-15, but these have

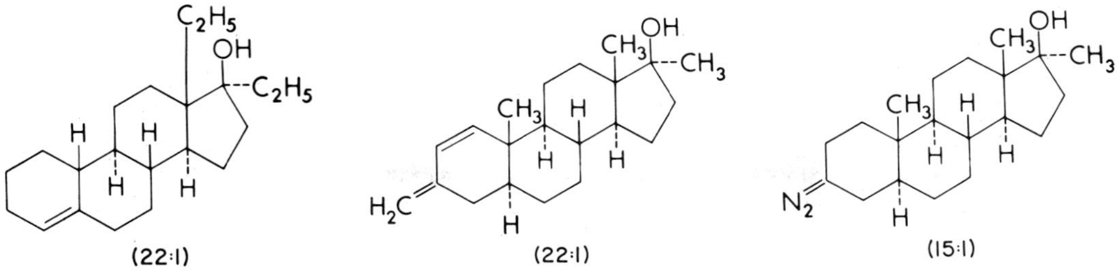

FIG. 18-15. *Experimental compounds with improved anabolic activity (anabolic activity:androgenic activity ratio).*

not been used clinically due to hepatic toxicity or other side-effects. For example, 19-nor steroids have been found to be quite anabolic, but their significant progestational activity has generally precluded their use.

Generalizations about the structural changes which enhance anabolic activity are difficult to make, but Vida,[152] Counsell and Klimstra[2,156] have presented detailed analyses. Albanese[115] has also made comparative studies of various anabolic agents in men and women. An examination of the compounds in Figure 18-15 shows that greater planarity and electron density in ring A seems to favor anabolic activity.

As with other compounds we have discussed, hydroxyl groups in the testosterones are often converted to the corresponding esters to prolong activity, or to provide some protection from oxidation.

Testosterone and dihydrotestosterone receptors

Human prostrate glands have a high affinity for binding of labeled testosterone and 5α-dihydrotestosterone. Two receptors, or selective binding proteins, have been isolated.[121,122,157,158]

Receptors from rat prostate have been much more extensively studied.[22] Testosterone enters the rat prostate cells and is metabolized almost immediately to 5α-dihydrotestosterone. The dihydrotestosterone is bound to a cytoplasmic receptor (Complex I) which is modified by a temperature-dependent process to a steroid-receptor complex called Complex II. Nuclear binding of Complex II and other steroid-protein complexes has been intensively studied by Fang and Liao and others.[22,121,122] It has been proposed by Ahmed and Wilson[159] that the complex binds to chromosomal acceptor sites (probably nonhistone acidic proteins). Their data indicates that this process activates protein phosphokinases, which in turn activate enzymes (such as RNA-polymerase) which facilitate transcription.

There are a number of significant puzzles concerning testosterone action.[121,122] Some species, such as the rat, have no testosterone receptors (or specific binding proteins) in the levator ani[160] or other androgen-sensitive muscle tissues.[161] Formation of 5α-dihydrotestosterone from testosterone in muscle tissue is negligible. It is known that the stimulation of protein synthesis in muscle tissues by androgens is slow,[162] so the process is clearly not a direct one. Growth hormone may also play a role in androgenic action.[163] In short, there is a need for many more additional studies—especially in vivo.

Recent research on androgen action and receptors has been reviewed in detail by Bardin and co-workers,[121] Mainwaring,[122] and Counsell and Brueggemier.[2]

Products

Therapeutic uses of the androgens and anabolic agents have been previously discussed. 17β-Esters and 17α-alkyl products are available for a complete range of therapeutic uses (see Fig. 18-5). These drugs are contraindicated in men with prostatic cancer; in men or women with heart disease, kidney disease or liver disease; and in pregnancy. Diabetics using the androgens and anabolic agents should be carefully monitored. A possible interaction of these drugs is with anticoagulants.[164] Female patients may develop virilization side-effects, and doctors should be warned that some of these effects may be irreversible, e.g., voice changes. Virtually all the anabolic agents presently commercially available have significant androgenic activity, so virilization is a potential problem with all women patients. The 17α-alkyl products may cause cholestatic hepatitis in some patients.

Doses and dosage schedules for specific indications can vary markedly (see Therapeutic Uses for indications), so specialized dose-indication references such as the *Hospital Formulary* or *Facts and Comparisons* should be consulted when advising physicians on doses. General U.S.P. D.I. doses are listed in Table 18-7.

All steroid 4-en-3-ones are light-sensitive and should be kept in light-resistant containers.

TABLE 18-7

ANDROGENS AND ANABOLIC AGENTS

Name Proprietary Name	Preparations	Usual Adult Dose*	Usual Dose Range*
Testosterone U.S.P. *Oreton, Neo-Hombreol (F)*	Testosterone Pellets U.S.P. Sterile Testosterone Suspension U.S.P.	Implantation, 300 mg.; I.M., 25 mg. twice weekly to once daily, depending on condition being treated	

Testosterone Cypionate U.S.P.
 	
   *Depo-Testosterone, Malogen
     CYP, Durandro,
     T-Ionate-P.A.*
	Testosterone Cypionate
	Injection U.S.P.
		I.M., 200 to 400 mg. once every 3 to 4 weeks	100 to 400 mg.

Testosterone Enanthate U.S.P.
 	
   *Delatestryl, Malogen L.A.,
     Testate, Testostroval-P.A.*
	Testosterone Enanthate
	Injection U.S.P.
		I.M., 200 to 400 mg. once a month	100 to 400 mg.

Testosterone Propionate U.S.P.
 	
   *Neo-Hombreol, Oreton
     Propionate*
	Testosterone Propionate
	Injection U.S.P.
	Testosterone Propionate Tablets
	U.S.P.
		Replacement therapy, I.M., 10 to 25 mg. 2 or 3 times a week; inoperable mammary cancer, I.M., 100 mg. 3 times a week	20 to 300 mg.
			weekly

Methyltestosterone U.S.P.
 	
   *Android, Metandren,
     Oreton Methyl, Testred,
     Neo-Hombreol (M)*
	Methyltestosterone Tablets
	U.S.P.
	Methyltestosterone Capsules
	U.S.P.
		Inoperable breast cancer, 100 mg. 2 times a day; replacement therapy, 5 to 20 mg. 2 times a day	5 to 200 mg. daily

Fluoxymesterone U.S.P.
 	
   *Halotestin, Ora-Testryl,
     Ultandren*
	Fluoxymesterone Tablets U.S.P.	Replacement therapy, 2 to 2.5 mg. 1 to 4 times a day; inoperable mammary cancer, 5 to 10 mg. 3 times a day	2 to 30 mg. daily

Methandrostenolone U.S.P. *Dianabol*	Methandrostenolone Tablets U.S.P.	2.5 to 5 mg. daily	
Oxymetholone U.S.P. *Adroyd, Anadrol, Anadrol-50*	Oxymetholone Tablets U.S.P.	5 to 10 mg. daily	5 to 50 mg.
Oxandrolone U.S.P. *Anavar*	Oxandrolone Tablets U.S.P.		Initial, 5 to 10 mg. daily; maintenance, 2.5 to 5 mg. daily
Nandrolone Decanoate U.S.P. *Deca-Durabolin*	Nandrolone Decanoate Injection U.S.P.	I.M., 50 to 100 mg. every 3 to 4 weeks	
Nandrolone Phenpropionate U.S.P. *Durabolin, Durabolin-50*	Nandrolone Phenpropionate Injection U.S.P.	I.M., 25 to 50 mg. weekly	
Stanozolol U.S.P. *Winstrol*	Stanozolol Tablets U.S.P.	2 mg. 3 times daily	

* See *U.S.P. D.I.* for complete dosage information.

Testosterone U.S.P., 17β-hydroxy-4-androsten-3-one, is a naturally occurring androgen in men, and in women where it serves as a biosynthetic precursor to estradiol. However, it is rapidly metabolized to relatively inactive 17-ones (Fig. 18-14), so it is not orally active. Testosterone 17β-esters are available in long-acting intramuscular depot preparations, illustrated in Figures 18-5 and 18-13, including the following.

Testosterone Cypionate U.S.P., testosterone 17β-cyclopentylpropionate.

Testosterone Enanthate U.S.P., testosterone 17β-heptanoate.

Testosterone Propionate U.S.P., testosterone 17β-propionate.

Methyltestosterone U.S.P., 17β-hydroxy-17α-methylandrost-4-en-3-one, is only about half as active as testosterone (when compared intramuscularly), but

it has the great advantage of being orally active (see Fig. 18-5). (Methyltestosterone given by the buccal route is about twice as active as oral.) Both testosterone and methyltestosterone have high androgenic activity, limiting their usefulness where good anabolic activity/low androgenic activity is desired.

Fluoxymesterone U.S.P., 9α-fluoro-11β,17β-dihydroxy-17α-methyltestosterone, is a highly potent, orally active androgen, about five to ten times more potent than testosterone. It can be used for all the indications discussed previously, but its great androgenic activity has made it useful primarily for treatment of the androgen-deficient male.

Methandrostenolone U.S.P., 17β-hydroxy-17α-methylandrosta-1,4-dien-3-one, is orally active and about equal in potency to testosterone.

Anabolic Agents include the commercially available androgens with improved anabolic activity (Fig. 18-13) and those which are still experimental (examples in Fig. 18-15). It should be emphasized that virtually all the commercial products have significant androgenic properties (ratios given in Fig. 18-13), so virilization in women and children can be expected. Many of the anabolic agents are orally active, as one would predict by noting a 17α-alkyl group in many of them (Fig. 18-13). Those without the 17α-alkyl (nandrolone and dromostalone) are only active intramuscularly. The commercially available anabolic agents include:

Oxymetholone U.S.P., 17α-methyl-17β-hydroxy-2-(hydroxymethylene)-5α-androstan-3-one.

Oxandrolone U.S.P., 17α-methyl-17β-hydroxy-2-oxa-5α-androstan-3-one.

Stanozolol U.S.P., 17α-methyl-17β-hydroxy-5α-androstan-3-one.

Nandrolone Decanoate U.S.P. and **Nandrolone Phenpropionate U.S.P.,** 17β-hydroxy-estr-4-en-3-one 17β-decanoate and 17β-(3'-phenyl)propionate.

ANTIANDROGENS

Five experimental compounds[165-171] (Fig. 18-16) have been intensively studied as androgen antagonists, or antiandrogens. Yoshioka and Goto's TSAA-291 is the newest of the group and may soon be commercially available. (The other four will probably not.) Estrogens have been used as antiandrogens, but their feminizing side-effects (e.g., loss of libido) have precluded their extensive use in men. Antiandrogens would be of therapeutic use in treating conditions of hyperandrogenism (e.g., hirsutism, acute acne, and premature baldness) or androgen-stimulated cancers, e.g., prostatic carcinoma. The ideal antiandrogen would be nontoxic, highly active, devoid of any hormonal activity, and would not decrease libido. Unfortunately the four compounds in Figure 18-16 have not met all these criteria completely,[85] although SCH 13521 and TSAA-291 have had some partial successes in clinical trials.[166,172] Mainwaring and co-workers have recently summarized progress in the field, including their own work on SCH 13521.[165] Spironolactone also has been found to have some antiandrogenic actions.[173] which is evidenced by its gynecomastia side-effects.

CHEMICAL CONTRACEPTIVE AGENTS

In a superb review of chemical contraception, John P. Bennett notes that it has taken about two million years for the world's population to reach 3 billion.[174] Only 40 more years will be needed to increase the population to 6 billion at current population growth rates. While some may disagree with such "doomsday" predictions, there is no doubt that the world's increasing population is a major concern.

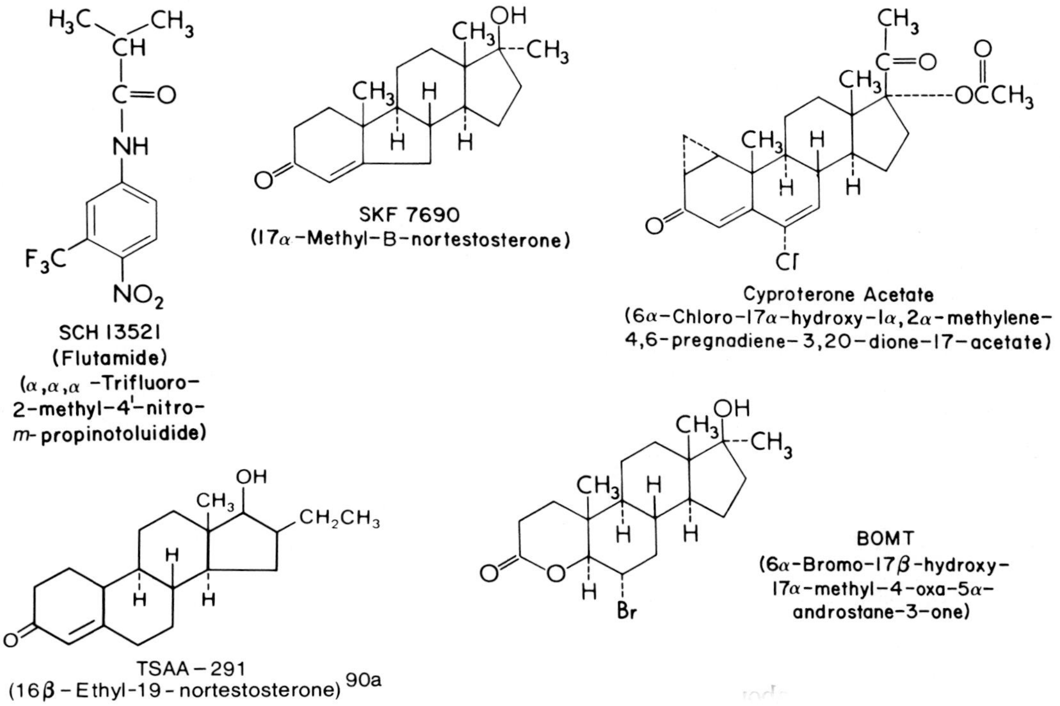

FIG. 18-16. *Antiandrogens.*

Political, cultural, and research cost barriers have enormously complicated the development of contraceptive agents in modern times. The reviews by Carl Djerassi,[175] former Research Director of Syntex and inventor of norethindrone, and by Lednicker[175a] are important reading. (Their "insiders' viewpoint" of the research competition during the 1950's and 1960's to develop steroid products is especially interesting.)

The need for birth control was most dramatically and effectively expressed from 1910 to 1950 by Margaret Sanger. This remarkable American woman made birth control information generally available in the United States, made the medical profession better aware of the needs of women, and also raised the funds necessary for the early research on oral contraceptives. Margaret Sanger is generally recognized as the "mother of birth control."

During the 1940's and 1950's, great progress was made in the development of intravaginal spermicidal agents. However, the most notable achievement in chemical contraception came in the early 1960's with the development of oral contraceptive agents—"the pill." Since that time, a number of postcoital contraceptives and abortifacients have been developed. Hormone-releasing intrauterine devices are also being tested. However, progress has been much slower in the development of male contraceptive agents.

In the following pages, each of these approaches to chemical contraception will be discussed. For additional information the interested reader is referred to several excellent clinical[176-179] and patient-oriented[180] guides on all methods of birth control, and to specific clinical reviews on the oral contraceptives.[181-189] Individual compounds have already been discussed with the estrogens and progestins.

OVULATION INHIBITORS AND RELATED HORMONAL CONTRACEPTIVES

History [174,175,190]

In the 1930's, several research groups found that injections of progesterone inhibited ovulation in rats, rabbits and guinea pigs.[190-193] Sturgis, Albright and Kurzrok, in the early 1940's, are generally credited with the concept that estrogens and/or progesterone could be used to prevent ovulation in women.[194-196] In 1955, Pincus[197] reported that progesterone given from day 5 to day 25 of the menstrual cycle would inhibit ovulation in women. During this time, Djerassi and Rosenkranz[198] of Syntex, and Colton[199] of G. D. Searle and Co. reported the synthesis of norethindrone and norethynodrel. These progestins possessed very high progestational and ovulation-inhibiting activity. Most of

the synthetic work was made possible by the development of the Birch reduction by Arthur J. Birch in 1950, and used by Birch to synthesize 19-nortestosterone itself.[200]

Extensive animal and clinical trials conducted by Pincus, Rock and Garcia confirmed in 1956 that Searle's norethynodrel and Syntex's norethindrone were effective ovulation inhibitors in women. In 1960 Searle marketed Enovid® (a mixture of norethynodrel and mestranol), and in 1962 Ortho marketed Ortho Novum® (a mixture of norethindrone and mestranol) under contract with Syntex. Norethynodrel and norethindrone have remained the most extensively used progestins in oral contraceptives, but several other useful agents have been developed. These will be discussed in the sections which follow.

Therapeutic classes and mechanism of action

The ovulation inhibitors and modern hormonal contraceptives fall into several major categories (Table 18-8), each with its own mechanism of contraceptive action.[173,175,179,201] Individual compounds have been discussed with the estrogens and progestins in the previous section.

1. Combination tablets

Although as noted earlier, Sturgis and Albright recognized in the early 1940's that either estrogens or progestins could inhibit ovulation, it was subsequently found that combinations were highly effective. Some problems such as breakthrough (midcycle) bleeding were also found to be reduced by the use of a combination of progestin and estrogen.

Although all the details of the process are still not completely understood, it is now believed that the combination tablets suppress the production of LH and/or FSH by a feedback inhibition process (see Fig. 18-9). Without FSH or LH, ovulation is prevented. The process is similar to the natural inhibition of ovulation during pregnancy due to the release of estrogens and progesterone from the placenta and ovaries. An additional effect comes from the progestin in causing the cervical mucus to become very thick, providing a barrier for the passage of sperm through the cervix. However, since pregnancy is impossible without ovulation, the contraceptive effects of thick cervical mucus or alterations in the lining of the uterus[201] (to decrease the probability of implantation of a fertilized ovum) would appear to be quite secondary. However, some authors have reported that occasionally ovulation may occur,[202,203] and thus the alterations of the cervical mucus and the endometrium may actually serve an

TABLE 18–8

COMPARISON OF ORAL CONTRACEPTIVE REGIMENS

(Day 1 = First day of menstruation)

(If menstruation does not occur, day 5 = 7th day since last active tablet was taken.)

1 2 3 4 5 6 7 8 9 10 11 12 13 14 15 16 17 18 19 20 21 22 23 24 25 26 27 28

No Tablets or Inactive Tablets	**Combination**	No Tablets or Inactive Tablets
	←————————————————————————————→	
	20 or 21 Tablets Each Containing Estrogen & Progestin	

Progestin Only

1 Tablet Every Day of the Year

Brand	Progestin		Estrogen		Dosage Cycle
Combinations					
Loestrin 1/20	Norethindrone Acetate	1 mg.	Ethinyl Estradiol	20 μg.	28-day*
Loestrin 1.5/30	Norethindrone Acetate	1.5 mg.	Ethinyl Estradiol	30 μg.	28-day*
Lo/Ovral	Norgestrel	0.3 mg.	Ethinyl Estrodiol	30 μg.	
Ovcon-35	Norethindrone	0.4 mg.	Ethinyl Estradiol	35 μg.	21, 28 day
Brevicon	Norethindrone	0.5 mg.	Ethinyl Estradiol	35 μg.	21, 28 day
Modicon	Norethindrone	0.5 mg.	Ethinyl Estradiol	35 μg.	21, 28 day
Norlestrin 1/50	Norethindrone Acetate	1 mg.	Ethinyl Estradiol	50 μg.	21, 28-day*†
Norlestrin 2.5/50	Norethindrone Acetate	2.5 mg.	Ethinyl Estradiol	50 μg.	21, 28-day*
Demulen	Ethynodiol Diacetate	1 mg.	Ethinyl Estradiol	50 μg.	21, 28-day†
Ovral	Norgestrel	0.5 mg.	Ethinyl Estradiol	50 μg.	21, 28-day†
Norinyl 1 + 50	Norethindrone	1 mg.	Mestranol	50 μg.	21, 28-day†
Ortho-Novum 1/50	Norethindrone	1 mg.	Mestranol	50 μg.	20, 21-day
Ovcon-50	Norethindrone	1 mg.	Ethinyl Estradiol	50 μg.	
Ortho-Novum 10 mg.	Norethindrone	10 mg.	Mestranol	60 μg.	20-day
Enovid 5 mg.	Norethynodrel	5 mg.	Mestranol	75 μg.	20-day
Ortho-Novum 1/80	Norethindrone	1 mg.	Mestranol	80 μg.	21-day
Norinyl 1 + 80	Norethindrone	1 mg.	Mestranol	80 μg.	21, 28-day†
Ortho-Novum 2 mg.	Norethindrone	2 mg.	Mestranol	100 μg.	20-day
Norinyl 2 mg.	Norethindrone	2 mg.	Mestranol	100 μg.	20-day
Enovid-E	Norethynodrel	2.5 mg.	Mestranol	100 μg.	20, 21-day
Ovulen	Ethynodiol Diacetate	1 mg.	Mestranol	100 μg.	20, 21, 28-day†
Progestin Only					
Micronor	Norethindrone	0.35 mg.			Continuous daily
Nor-Q.D.	Norethindrone	0.35 mg.			Continuous daily
Ovrette	Norgestrel	0.075 mg.			Continuous daily
Injectable Depot Hormonal Contraceptives					
Depo-Provera	Medroxyprogesterone Acetate				400 to 1,000 mg. initially, with maintenance dose as low as 400 mg. per month
Once-A-Month Oral Contraceptive					
—	Norethindrone Acetate 3-Cyclopentyl Enol Ether (Quinestrol)		Ethinyl Estradiol 3-Cyclopentyl Ether (Quingestanol)		
Hormone-Releasing Implants and IUD's					
—	Subcutaneous Silastic implants containing estrogens and/or progestins implanted in forearms				Reported effective up to one year per implant
—	Intravaginal Silastic rings containing progestins				Under study by Upjohn
Progestasert	Progesterone-releasing IUD				38 mg. dose in IUD lasts 1 year

* The 28-day regimen includes 7 tablets of 75 mg. ferrous fumarate.

† The 28-day regimen includes 6 to 7 inert tablets.

Table taken in part from Medical Letter, April 26, 1974, and used with permission.

important contraceptive function (especially, perhaps, when the patient forgets to take one of the tablets). During combination drug treatment, the endometrial lining develops sufficiently for withdrawal bleeding to occur about four or five days after taking the last active tablet of the series (Table 18-8).

The combination tablets are usually taken from the fifth to the twenty-fifth day of the menstrual cycle, sometimes preceded or followed by inert tablets (Table 18-8). With a few women, there may occasionally be little or no menstrual flow, so in any case, the active tablet cycle must begin seven or eight days following completion of the previous active tablet cycle.

It used to be a common practice to prescribe progestins to prevent spontaneous abortions in some women; to administer high doses of progestins as a test for pregnancy; and to "just continue taking" oral contraceptives for an additional month if the patient misses a menstrual period. However, the F.D.A. and independent investigators have recently strongly warned *against* the use of any steroidal hormones for any purpose *during* early pregnancy[92] because of possible damage to the fetus.[93,204-206] On the other hand, several studies have shown that there is no significant effect on progeny when women have taken "the pill" *before* becoming pregnant.[92,93,207-209]

2. Progestin only (mini pill)

The estrogen component of sequential and combination oral contraceptive agents has been related to some side-effects, with thromboembolism being a particular concern. One solution to this problem has been to develop new products with decreased estrogen content. In the case of the "mini pill," there is no estrogen at all.

Although higher doses of progestin are known to suppress ovulation, mini-pill doses of progestin are not sufficient to suppress ovulation in all women. Some studies have indicated that an increase in the viscosity of the cervical mucus (or sperm barrier) could account for much of the contraceptive effect, while other studies disagree.[174,181,184,201] Low doses of progestin have also been found to increase the rate of ovum transport and to disrupt implantation.[193] There is a good probability that most or all of these factors contribute to the overall contraceptive effect of the mini pill.

The progestin-only tablets are taken every day of the year. As with the other oral contraceptive products, if menstruation does not occur, the patient should contact her doctor immediately.

3. Injectable depot hormonal contraceptives

In principle, there is no reason why a long-acting depot preparation of a progestin or an estrogen-progestin combination could not be developed. Progress in

this approach was reviewed in 1976.[210] While there have been tests on a number of animals with various hormone preparations, only one drug (Depo-Provera®) has been sufficiently tested to merit temporary F.D.A. approval (later withdrawn). Other drugs still undergoing clinical or preclinical trials have been found to be effective contraceptive agents, but irregular menstrual cycles and menstrual "spotting" remain major problems.

Injectable medroxyprogesterone acetate was briefly approved by the F.D.A. as a depot hormonal contraceptive for women who cannot use other methods of contraception, e.g., in mental institutions and low socio-economic areas where patients probably would not follow the important dosage schedule of the oral contraceptive hormones.[176] However, the approval was stayed until its safety is more carefully evaluated.[94,211] Doses of 150 mg. are injected once every three months, with other methods of contraception recommended for the first month following the initial injection. Prolonged infertility after stopping use of the drug is common, and a package insert for prospective patients warns that permanent infertility is a possibility.[94,211]

4. Once-a-month and once-a-week oral contraceptives

The advantages of a once-a-month oral contraceptive are obvious, and good progress has been made in the development of such drugs. However, as reviewed by Diczfalusy, development of a long-acting, fertility-regulating agent is a difficult, costly, and time-consuming process.[211a] Berman and co-workers have reported that a small oral dose of ethinyl estradiol 3-cyclopentyl ether (quinestrol) and norethindrone acetate 3-cyclopentyl enol ether (quingestanol) is effective in humans when given once a month.[212] However, it has been found that full contraceptive protection is not achieved until the second month's dose has been taken. After that time, contraceptive efficiency is reported to be excellent.

R-2323 [the 13β-ethyl,9(10),11(12)-triene derivative of norethindrone] is now undergoing clinical trials in the United States as a once a week oral contraceptive. Initial clinical studies in Haiti and Chile showed a high rate of pregnancy (6–7 per 100 woman-years), but it is believed that poor patient compliance with directions was largely responsible.[213]

5. Hormone-releasing implants and intrauterine devices[214,215]

As mentioned previously, the low progestin doses of the "mini pill" seem to have a direct effect on the uterus and associated reproductive tract. It would, therefore, seem possible to lower the progestin dose even more if the drug was released in the reproductive tract itself. Several devices employing these concepts are now being studied in clinical trials and are expected to be on the market in the near future.[174,97] These intravaginal and intracervical devices have been reviewed by The World Health Organization.[224a]

In 1964, Folkman and Long[216] showed that chemicals can be released via diffusion through the walls of a silicone rubber capsule at a constant rate. A particularly attractive silicone rubber was found to be Silastic (Dow) which was nontoxic and apparently nonallergenic. During initial studies, capsules made of Silastic and containing estrogens and/or progestins have been implanted subcutaneously in the forearms of women patients.[217-219] These studies are still in progress, but it has been possible to obtain efficient contraception for one year with forearm transplants.

Similar studies have been begun with uterine-implanted Silastic capsules containing low doses of progestins.[220-222] It was envisioned that progestin-containing intrauterine devices (IUD's) would have some particular advantages over other IUD's. First, the progestin should decrease uterine contractility (thus decreasing the number of IUD's ejected). Second, it should decrease the vaginal bleeding sometimes associated with IUD's. Additional studies are in progress to evaluate these predictions.

Intravaginal Silastic rings containing low doses of progestins are also under study. Initial clinical studies have been promising.[223,224a]

The Progestasert® IUD (Progesterone Intrauterine Contraceptive System U.S.P.) has 38 mg. of microcrystalline progesterone dispersed in silicone oil. The dispersion is contained in a flexible polymer in the approximate shape of a 'T'.[225,226] The polymer acts as a membrane to permit 65 μg. of progesterone to be slowly released into the uterus each day for 1 year. Contrary to prediction, the progesterone-containing IUD has had some of the therapeutic problems of other IUD's, including a relatively low patient continuation rate, some septic abortions, and some perforations of uterus and cervix. Clinical studies[227] have produced the following data on Progestasert®, expressed as events per 100 women through 12 months of use.

	Parous	*Nulliparous*
Pregnancy	1.9	2.5
Expulsion	3.1	7.5
Medical removals	12.3	16.4
Continuation rate	79.1	70.9

6. Biodegradable sustained-release systems

A very interesting and exciting approach to fertility control uses biodegradable polymers and microparticles to release the estrogen or progestin. As reviewed by Benagiano and Gabelnick[227a] of the World Health Organization, the microparticles can be injected through regular needles. Release of the active drug occurs by erosion, diffusion, and cleavage of covalent bonds between the drug and polymer. Polymer matrices that have been investigated for these purposes include caprolactone, glutamic acid, lactic acid, and glycolic acid polymers.

Safety of postmenopausal estrogens

Estrogens have been used for over three decades for treatment of postmenopausal symptoms and for many other therapeutic uses, some appropriate and some inappropriate. These estrogens have included equine "conjugated" steroid estrogens, synthetic estrogens such as DES, and synthetic human steroid estrogens. In early 1976, the FDA issued strong warnings about the misuse and overuse of estrogens with postmenopausal women.[228]

In 1979, as reviewed in an FDA Bulletin, "FDA has reviewed three recent studies on estrogen use, and the agency affirms that menopausal and postmenopausal women who take estrogens have an increased risk of endometrial cancer."[229] The risk relative to postmenopausal women not taking estrogens varied from 4.5 to over 13. The first FDA Bulletin warning about the risk of endometrial cancer was issued in February, 1976,[228] focusing on the conjugated estrogens in particular. In May, 1976, the *Medical Letter* warned that all estrogens should be considered potentially carcinogenic.[130] The reports also examined in detail the widespread overuse and misuse of these products. The FDA has recommended cyclic administration of the lowest effective dose for the shortest possible time with appropriate monitoring for endometrial cancer.[228] The risks and benefits for postmenopausal estrogens have also been summarized by the American Council on Science and Health.[114]

It must be noted, however, that current evidence indicates that estrogens might help prevent or retard osteoporosis, an often debilitating condition with some postmenopausal women.[228a] As summarized in late 1980 by Gastel, Cornoni-Huntley, and Brody,[228b] a panel of experts was convened by the National Institute on Aging. They concluded that the use of estrogens does alleviate vasomotor symptoms and atrophy of the vaginal epithelium and might aid in preventing osteoporosis. However, they reaffirmed that these estrogens increase the incidence of endometrial cancer. The experts concluded that any candidate for postmenopausal estrogens should be given as much information as possible about both benefits and risks, and then with her physician reach an individual decision on whether to take these drugs or not.

Safety of oral contraceptives

Estrogens have also been used widely in oral contraceptive products since 1964. However, only recently has there been widespread agreement that the use of *any* estrogen product can involve significant risks to the patient.[230-232]

When the dose of estrogen is high, the risk can be significant, irrespective of the type of estrogen or product. Preliminary studies suggest that the risk can be minimized by lowering the estrogen dose. The results of these findings have been that: (1) the sequential contraceptive products with their high doses of estrogen have been removed from American markets; (2) many combination contraceptives containing less than 50 μg. estrogen per dose have recently been marketed (Table 18-8); (3) progestin only or mini-pill products have appeared (Table 18-8); (4) a few groups of women have been identified who should definitely not take oral contraceptives, (e.g., women over 40 and women who are moderate to heavy smokers); and (5) the use and misuse of estrogens in postmenopausal women is expected to be greatly reduced.

The safety of "the pill" has been one of the most intensively discussed subjects in the press.[215] Several excellent reviews, now partially out of date, have been published in the medical literature.[174,176,177,178] Current information can be obtained from reviews published in early 1976[228,230,232] and 1979.[175,185-186,189]

However, great caution must be used in interpreting these reviews. Low-estrogen-dose (i.e., less than 50 μg.) combination products and progestin-only products (Table 18-8) have been marketed only recently. Nearly all long-term studies on oral contraceptives have been on high-estrogen-dose products.[234-239] As a result, the usual practice of lumping all the contraceptive products together when describing serious side effects is inaccurate and extremely misleading.

Also it must be emphasized that the increased risks associated with taking oral contraceptives can be very deceptive unless compared with the actual incidence of the health problems themselves. In turn, one might also consider these risks relative to other common causes of death, (e.g., traffic accidents). Data for the United States and Minnesota are provided in Table 18-9. As can be seen from these data, the risk of death from myocardial infarction is much greater for men (who obviously do not take contraceptives) than for women.

Smoking greatly increases the risks. A 1979 study by Shapiro, Rosenberg, Slone, and co-workers[189] showed that the risk of myocardial infarction with oral contraceptive users (median age 43) was four times that of nonusers. However the risk for heavy smokers (more than 25 cigarettes per day) was 20 to 40 times greater.

The primary risk in using oral contraceptives is cardiovascular disease, especially myocardial infarction and thromboembolic disease. (Other side-effects have recently been reviewed.[185,186])

Jain has reported that the risk of death from myocardial infarction for "pill users" aged 30 to 39 is about 1.8 per 100,000 for nonsmokers; and 13.0 per 100,000 for heavy smokers (15 or more cigarettes per day). The risk for nonsmokers who did not use the "pill" was 1.2 per 100,000.[187] The risks increase significantly for women 40 and older. In short, the actual incidence of "pill-induced" cardiovascular death for nonsmoking young women appears to be quite small. Further, an analysis by Tietze has shown that in the United States, since adoption of the "pill" in the mid-1960's, deaths of women of reproductive age from cardiovascular disease have declined much more steeply than have death rates for men of the same age group.[188]

Most reports of side-effects have focused on women taking oral contraceptive products containing over 50 μg. estrogen per dose. Further studies are needed to delineate the frequency of these side-effects with low-estrogen-dose and progestin-only products. However, initial studies suggest the frequency of these side-effects is significantly decreased with low-estrogen-dose and progestin-only products.[220-222,238,239]

Unfortunately, the problem of minor (but annoying) midcycle bleeding (spotting) and the need for a precise dosing schedule increase as the dose of the oral contraceptive decreases. The spotting can often be

TABLE 18-9

INCIDENCE OF DEATH FROM VARIOUS CAUSES, BY SEX AND AGE, 1974

Cause of Death	Sex	Age Group (yrs.)	Death per 100,000* Minnesota	Death per 100,000* United States
Myocardial Infarction	F	20–34	0.22	3.3
	M	20–34	1.71	10.9
	F	35–44	4.10	20.2
	M	35–44	38.9	101.5
Cancer of Uterus and Cervix	F	20–34	0.45	7.8
	F	35–44	4.61	37.1
Breast Cancer	F	20–34	1.79	3.7
	F	35–44	28.7	12.0
Car Accident	F	20–34	13.6	40.5
	M	20–34	48.9	144.1
	F	35–44	6.66	17.6
	M	35–44	21.7	57.9

* Data Supplied by the National Center for Health Statistics, Rockville Maryland; the Bureau of Health Statistics, St. Paul, Minnesota; and the Traffic Safety Bureau, St. Paul, Minnesota. Data are for all people in the sex and age group, so does not reflect effects of other health factors, such as diet, exercise, smoking, medicines.

controlled by increasing slightly the dose of the oral contraceptive. However, the health risks of this choice, and the question of being able to remember to take the pill on time, are matters which must be left to the woman herself to decide. The selection of contraceptive method will be discussed in detail at the end of this section.

In addition, as has been noted previously in this chapter, the FDA has strongly warned against the use of any steroid hormones during pregnancy[92] because of possible damage to the fetus.[93,204-206] If the woman taking any oral contraceptive misses a period, she should immediately inform her physician. The effects of sex-hormone exposure during pregnancy have recently been reviewed.[93]

Finally, as will be discussed in the next section, the estrogen DES when used as a postcoital contraceptive has a number of additional serious side effects.

The question of which pill, and which contraceptive method, will be discussed later in this section.

OTHER METHODS OF CHEMICAL CONTRACEPTION

Postcoital contraceptives

Progress in the development of postcoital, or "morning-after" chemical contraceptive agents has been slow in spite of a relatively large amount of research. Although many studies have been conducted in animals, the potential dangers of experimental postcoital contraceptives (including ectopic pregnancy) have understandably limited the number of women volunteers for clinical studies.

Nevertheless, many compounds have been found to be effective postcoital contraceptives in animals. Many, if not most, are estrogenic, and possibly act by an alteration of the mechanisms of fertilized-egg transport into the uterus. Simultaneously, the timing of the development of the uterine endometrium is believed to be sufficiently altered to prevent implantation.

Compounds which have been found to be effective postcoital contraceptives in animal studies include steroid estrogens, stilbestrols and a variety of synthetic compounds which bear little similarity to other known hormonal agents. The interested reader is referred to the review by Bennett[174] for a complete discussion of these compounds.

The only postcoital contraceptive presently approved by the FDA, diethylstilbestrol (DES), illustrates the use of and problems of drugs with this potential use. Relatively high (25 mg. twice daily) doses must be given for five continuous days, starting not

later than 72 hours after coitus. The need for drug treatment soon after coitus is obvious in view of the proposed mechanism above, and several-day therapy is needed because fertilization can occur several days after coitus. The high doses that are required cause nausea and vomiting in many women. The "Copper T" has also been studied as a postcoital contraceptive and initial results have been promising.[239a]

DES babies

High doses of diethylstilbestrol can be teratogenic, and daughters of women who have had diethylstilbestrol during pregnancy were believed to have a high risk of vaginal or cervical cancer.[241-243] However, studies reported in 1979[240,241] of over 2,000 women exposed to DES in utero revealed very few cases of cancer, although a high percentage of the women had vaginal epithelial changes. Forsberg has proposed a mechanism for the carcinogenic action of DES.[242]

Abortifacients

History records many different compounds which have been tried as abortifacients—everything from plant extracts to rusty nail water. Many chemicals have been found to be very effective with animals, including metabolites, cytotoxic agents, 5-hydroxytryptamine, monoamine oxidase inhibitors, androgens, and others. Usually these same compounds also have been found to be toxic or mutagenic, or cause severe hemorrhaging along with the abortion.

However, one compound has recently been approved by the FDA to induce second trimester abortions—prostaglandin $PGF_{2\alpha}$.[244-246] $PGF_{2\alpha}$ and PGE_2 concentrations significantly increase in amniotic fluid prior to normal labor and childbirth.

$PGF_{2\alpha}$

Good surgical support is essential with $PGF_{2\alpha}$, since some clinicians report a high incidence of incomplete abortions that require dilatation and curettage. Furthermore, in those cases where the placenta is retained, severe hemorrhage requiring transfusion may result. The drug is approved only for intra-amniotic injection in the second trimester. Suction is a common (and probably safer) method of clinical abortion during the first trimester; and saline-induced abortions are sometimes used during the second trimester. However, the saline method has been associated with disseminated intravascular coagulation in the patient, a problem not reported with $PGF_{2\alpha}$.

Spermicides

"As early as the 19th Century B.C., the Egyptians were mixing honey, natron (sodium carbonate), and crocodile dung to form a vaginal contraceptive paste. . . . During the middle ages, rock salt and alum were frequently used as vaginal contraceptives."[247] The history of spermicidal agents is indeed a long one. Modern spermicidal agents, or "vaginal contraceptives," fall into three categories: surface-active, or sulfhydryl-binding agents; bactericides; and acids. These agents have recently been reviewed in the medical literature.[174,179,180,248] The new foaming tablet Encare Oval® has also been studied.[249]

In addition to the inherent spermicidal properties of the active agent, the efficiency of spermicidal products depends upon many more factors. They must be inserted high into the vagina (usually with an applicator). They must (perhaps inconveniently) be used just before intercourse and reused if intercourse is to be repeated.

Further, the formulation of these contraceptive products becomes almost as important as the active spermicidal agent itself. The product's formulation must permit diffusion into the cervix, since some spermatozoa may be released directly into it. The product must also have a reasonable stability in the vagina so that enough active spermicide remains after intercourse. Finally, the ideal vaginal contraceptive must be nontoxic and nonirritating to both partners.

The primary action of the surface-active agents is to reduce the surface tension at the sperm cell surface and cause a lethal osmotic imbalance. They may also inhibit oxygen uptake. The bactericides also may alter the surface properties of the sperm cells and after penetrating the cell membrane can disrupt metabolic processes. The acidic agents cause direct damage to the surface of the sperm cell membranes by denaturation of cell protein material. Examples of common spermicidal agents are shown in Table 18-10.

There are four primary types of vaginal contraceptive products: (1) creams, jellies and pastes which are squeezed from a tube or applicator; (2) suppositories; (3) foams (from aerosol pressurized containers or tablets); and (4) soluble films. Often the vaginal contraceptives are used in combination with another contraceptive method, e.g., diaphragm or "rhythm" method.

TABLE 18–10. EXAMPLES OF COMMONLY USED SPERMICIDES[247, 248]
(In jellies, creams, suppositories, foaming tablets, aerosol foams and soluble films—not all available in the United States)

1. *Surface-Active Agents* (also somewhat bactericidal)

   a. $CH_3\ (CH_2)_6CH_2$—⬡—$(OCH_2CH_2)_9OH$

   Nonoxynol-9, U.S.P.
   (Nonylphenoxypolyoxyethyleneethanol)
   (Delfen, Immolin, Emko, Because, Encare Oval)
   b. Others: p-Di-isobutylphenoxypolyethoxyethanol; Polyoxyethylenenonylphenol; Otoxynol, U.S.P. (Koromex); Dodecoethylene glycol monolaurate (Ramses 10 hour®)

2. *Bactericides*

   a. $C_6H_5HgOCCH_3$ (with O double-bonded above the C)
   Phenylmercuric Acetate
   (Lorophyn®)
   b. Others: Benzethonium Chloride, Methylbenzethonium Chloride, Phenylmercuric Borate

3. *Acids*
   a. Boric Acid
   b. Others: Tartaric Acid, Phenols, etc.

Both the American Medical Association and the *Birth Control Handbook* rank the foam products higher in contraceptive effectiveness than creams or jellies, although the creams or jellies may have some advantages when used with diaphragms. (The excellent Emory University pamphlet contains complete instructions to be used with these and other contraceptive products.) The soluble films, primarily used in Europe, are transparent, water-soluble films which are impregnated with a spermicidal agent and then inserted into the vagina prior to intercourse.

Recent research has also suggested that some vaginal spermicides may also provide significant protection against venereal disease transmission.[247]

Chemical contraceptives for men

Much less research has been done on the development of chemical contraceptives for men than for women. Current progress has been reviewed in 1978 by Barwin[250] and in 1980 by Jackson and Morris.[251] As stated by one of the research leaders of the pharmaceutical industry, "The chief reason . . . is the presumption that men would not use a chemical contraceptive through fears of psychological or clinical effects upon libido and masculinity. . . . I have never agreed with this forecast."[174] Further, the toxicologic and clinical studies required for the FDA approval of an entirely new kind of drug are rather immense, so a chemical contraceptive product for men is not expected in the near future.

Ideally, one would like to have a drug which would inhibit spermatogenesis (without being mutagenic), would not decrease libido, and would not have any other effect on testicular function, e.g., hormone function. Alternatively, drugs which would only affect the spermatozoa *after* formation would be of great interest, e.g., drugs which could block the fertilizing ability of sperm stored in the epididymis.

Examples of drugs which have been found to have these properties in animals (and for a few in man) are shown in Figure 18-17.

LH-RH AGONISTS AND ANTAGONISTS

As described earlier in this chapter, luteinizing hormone-releasing hormone (LH-RH, also called *gonadotropin-releasing hormone,* GnRH) controls and regulates both male and female reproduction. Once its structure was determined in the early 1970's to be a simple decapeptide, an intense effort was made to make analogs as fertility and antifertility agents. Already over 1000 have been made. Some have turned out to be agonists and others, antagonists of LH-RH.[97-100d] Extremely potent agonists and antagonists have been synthesized by varying the amino acids at positions 6 (Gly), 7 (Leu), and 9 (ProNEt).

To date, more agonists of LH-RH have been synthesized and studied than antagonists. Although research on LH-RH antagonists is about two years behind agonist studies, both groups have great potential as future antifertility agents.

LH-RH agonists cause their antifertility effects in females by causing overproduction of LH, a general disruption in FSH secretion, implantational failure, and an actual decrease in LH-receptor populations in the ovaries. Although most actions of LH-RH agonists appear to be upon the pituitary, there is evidence that they may able to cause some direct effects on ovarian tissues. These actions together account for LH-RH's potential as both precoital and postcoital contraceptives. In particular, it appears that, as LH-receptor populations decrease, progesterone levels drop and the pregnancy is terminated.

The mechanism for the antifertility effect of LH-RH agonists in males is not as well understood. It appears that LH-RH agonists may also decrease testosterone production, causing concern that these ana-

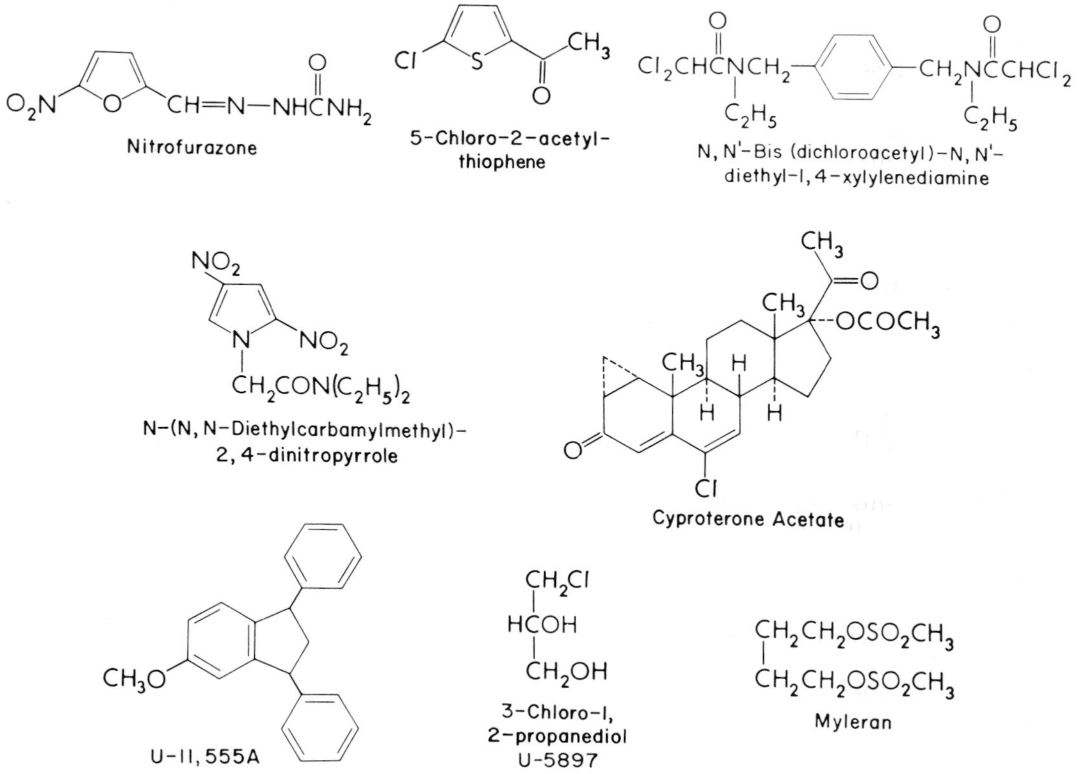

FIG. 18-17. Examples of chemical contraceptives for males. (Most have been tested only in animals. Some are quite toxic.)

logs may not hold as much promise for men as for women.

LH agonists have also been used to induce ovulation in women, and to stimulate spermatogenesis in men. However, in those cases where ovulation was induced, actual pregnancy was infrequent. Successes with men have also been mixed. Much more research is yet to be done.

Are LH-RH analogs the future replacements of steroid contraceptives, as many have speculated? Only time will tell, but their future looks bright indeed. The excellent reviews of their status as fertility and antifertility agents by Corbin, Sharpe, Vale, and Schally are recommended for further information.[97-100d]

RELATIVE CONTRACEPTIVE EFFECTIVENESS OF VARIOUS METHODS

Some caution is required in interpreting data on the effectiveness of contraceptive methods. Even the "best" method can lead to pregnancy if not used consistently and correctly. Even the generally least effective method is better than no contraceptive at all. Table 18-11 presents some data on numbers of pregnancies per method. The excellent article of Huff and Hernandez[248] on over-the-counter contraceptives is suggested for a comparative study.

SELECTION OF A CONTRACEPTIVE METHOD

It is obvious after reading the preceding pages that selection of a contraceptive method involves weighing the risk of pregnancy against the health risks of the particular contraceptive method. Along with the health risks and effectiveness of each method, the health professional and patient should also discuss convenience, method of use, and minor but possible bothersome side effects.

While there is an inherent risk of death due to complications of pregnancy, the usually quoted morbidity figures are quite out of date. For example, there were 25 pregnancy-related deaths per 100,000 pregnancies in Britain in 1964,[180] but less than 5 per 100,000 in Minnesota in 1973,[252] and less than four in Oregon in the mid-1970's.

The health professional should also clearly explain the decreased effectiveness of particular contraceptive methods if instructions for proper use are not carefully followed. Forgetting to take a low-dose "pill" at the same time every day, or not inserting a spermicidal foam within an hour of intercourse can greatly decrease contraceptive effectiveness.

There are also some groups of women for whom particular contraceptive methods involve too great a health risk to even be considered. Women with particular physiological problems may not be able to use IUD's. (An October 3, 1980 *Medical Letter* also noted the high risks of all IUD's in causing pelvic inflammatory disease and sometimes resulting sterility.) Women over 40, women with histories of thromboembolic disease, women with breast tumors, women who are DES babies, and women who might be pregnant should not use an oral contraceptive.

The advantages, risks, effectiveness, and methods of use should be described by the health professional. However, the final choice of a particular method should be left completely to the woman herself to decide.

TABLE 18–11

FAILURE RATE OF CONTRACEPTIVE METHODS*

| *(Data Based on Actual Users When Available)* | |
Method	Pregnancies/100 Woman-Years
Abortion	0
Tubal ligation	0.04
Vasectomy	0.15
Combination Oral Contraceptive	
High Estrogen dose (750 μg.)	0.05–0.1
Mid-Estrogen dose (50 μg.)	0.05–0.1
Low Estrogen dose (30–35 μg.)	0.03–0.7
Progestin Only, Oral (mini pill)	.63–1.58
Progestasert IUD	1.9–2.5
R-2323 (once-a-week oral)†‡	6–7
IUD	1.6–5
Forearm Implant, Progestin Releasing	Data not available
Intravaginal Ring, Progestin Releasing	Data not available
Condom and Spermicide	1–5
Spermicidal Foam	3–29
Diaphragm	9–33
Condom	3–28
Coitus Interruptus	18–22
Douche	33–60
Rhythm	1–47
No Contraceptive Method	60–80

* Table does not reflect relative safety, ease of use, etc.
† Clinical trials
‡ Number of pregnancies believed to be high because of poor patient compliance during early trials.

ADRENAL CORTEX HORMONES

The adrenal glands (which lie just above the kidneys) secrete over fifty different steroids, including precursors for other steroid hormones. However, the most important hormonal steroids produced by the adrenal cortex are aldosterone and hydrocortisone. Aldosterone is the primary *mineralocorticoid* in man, i.e., it causes significant salt retention. Hydrocortisone is the primary *glucocorticoid* in man, i.e., it has its primary effects on intermediary metabolism. The glucocorticoids have become very important in modern medicine, especially for their anti-inflammatory effects.

Medically important adrenal cortex hormones and synthetic mineralocorticoids and glucocorticoids are shown in Figure 18-18. Since salt-retention activity is usually undesirable, the drugs are classified by their salt-retention activities.

THERAPEUTIC USES

The adrenocortical steroids are used primarily for their glucocorticoid effects, including immunosuppression, anti-inflammatory activity and antiallergic activity.[4,60,64] The mineralocorticoids are used only for treatment of Addison's disease. Addison's disease is caused by chronic adrenocortical insufficiency and may be due either to adrenal or anterior pituitary failure. The anterior pituitary secretes ACTH, adrenocorticotropic hormone, a polypeptide which stimulates the adrenal cortex to synthesize steroids.

The symptoms of Addison's disease illustrate the great importance of the adrenocortical steroids in the body and, especially, the importance of aldosterone. These symptoms include increased loss of body sodium, decreased loss of potassium, hypoglycemia, weight loss, hypotension, weakness, increased sensitivity to insulin and decreased lipolysis. The roles of aldosterone in clinical physiology[5,253,254] and the uses of spironolactone (an important aldosterone antagonist) will be discussed in subsequent sections.

Hydrocortisone is also used during postoperative recovery following surgery for Cushing's syndrome—excessive adrenal secretion of glucocorticoids. Cushing's syndrome can be caused by bilateral adrenal hyperplasia or adrenal tumors and is treated by surgical removal of the tumors or resection of hyperplastic adrenals.

The use of glucocorticoids during recovery from surgery for Cushing's syndrome illustrates a very important principle of glucocorticoid therapy: *abrupt withdrawal of glucocorticoids may result in adrenal insufficiency*—showing clinical symptoms similar to Addison's disease. For that reason, patients who have been on long-term glucocorticoid therapy must have the dose *gradually* reduced. Furthermore, prolonged treatment with glucocorticoids can cause adrenal suppression, especially during times of stress. The symptoms are similar to those of Cushing's syndrome, for example, rounding of the face, hypertension, edema, hypokalemia, thinning of the skin, osteoporosis, diabetes, and even subcapsular cataracts. In doses of 45 mg.

per square meter of body surface area or more daily, growth retardation occurs in children.

The glucocorticoids are used in the treatment of collagen vascular diseases, including rheumatoid arthritis, disseminated lupus erythematosus, and dermatomyositis.

Although there is usually prompt remission of redness, swelling and tenderness by the glucocorticoids in rheumatoid arthritis, continued long-term use may lead to serious systemic forms of collagen disease. As a result, the glucocorticoids should be used infrequently in rheumatoid arthritis.

The glucocorticoids are used extensively topically, orally and parenterally to treat inflammatory conditions. They also usually produce relief from the discomforting symptoms of many allergic conditions—intractable hay fever, exfoliative dermatitis, generalized eczema, etc.

The glucocorticoids' lymphocytopenic actions make them very useful for treatment of chronic lymphocytic leukemia in combination with other antineoplastic drugs.

The glucocorticoids are also used in the treatment of congenital adrenal hyperplasias. These disorders are caused by an inability of the adrenals to carry out 11β-, 17α-, or 21-hydroxylations (see Fig. 18-19). The most common is a lack of 21-hydroxylase activity, which will result in decreased production of hydrocortisone and a compensatory increase in ACTH production. Furthermore, the resultant build-up of 17α-hydroxyprogesterone will lead to an increase of testosterone. When 11β-hydroxylase activity is low, large amounts of 11-deoxycorticosterone will be produced. Since 11-deoxycorticosterone is a potent mineralocorticoid, there will be symptoms of mineralocorticoid excess, including hypertension, etc. When 17α-hydroxylase activity is low, there will be decreased production of testosterone and estrogens as well as hydrocortisone.

The adrenocortical steroids are contraindicated or should be used with great caution in patients having: (1) peptic ulcer (in which the steroids may cause hemorrhage); (2) heart disease; (3) infections (the glucocorticoids suppress the body's normal infection-fighting processes); (4) psychoses (since behavorial disturbances may occur during steroid therapy); (5) diabetes (the glucocorticoids increase glucose production, so more insulin may be needed); (6) glaucoma; (7) osteoporosis; and (8) herpes simplex involving the cornea.

BIOSYNTHESIS

As shown in a simplified scheme in Figure 18-19, aldosterone and hydrocortisone are biosynthesized from

(Text continues on p. 700)

I. Mineralocorticoids (High Salt Retention)

Aldosterone
(Not Commercially Available)

Desoxycorticosterone (R=H)

Esters Available:
Desoxycorticosterone Acetate: R=COCH₃
Desoxycorticosterone Pivalate: R=COC(CH₃)₃

Fludrocortisone Acetate

2. Glucocorticoids with Moderate to Low Salt Retention

Cortisone (R=H)

Ester Available:
Cortisone Acetate: R=COCH₃

Hydrocortisone (R=H)
(or Cortisol)

Esters Available:
Hydrocortisone Acetate: R=COCH₃
Hydrocortisone Cypionate: R= COCH₂CH₂—⬠

Salts Available:
Hydrocortisone Sodium Phosphate: R=PO₃⁻(Na⁺)₂
Hydrocortisone Sodium Succinate:
R=COCH₂CH₂COO⁻ Na⁺

Prednisolone (R=H)

Salts Available:
Prednisolone Sodium Phosphate: R=PO₃⁻ (Na⁺)₂
Prednisolone Succinate: R=COCH₂CH₂COO⁻ Na⁺

Esters Available:
Prednisolone Acetate: R=Ac
Prednisolone Succinate: R=COCH₂CH₂COOH
Prednisolone Tebutate: R=COCH₂C(CH₃)₃

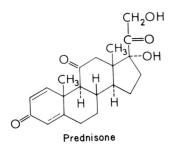

Prednisone

FIG. 18-18. *Natural and semisynthetic adrenal cortex hormones.*

3. <u>Glucocorticoids with Low Salt Retention</u>

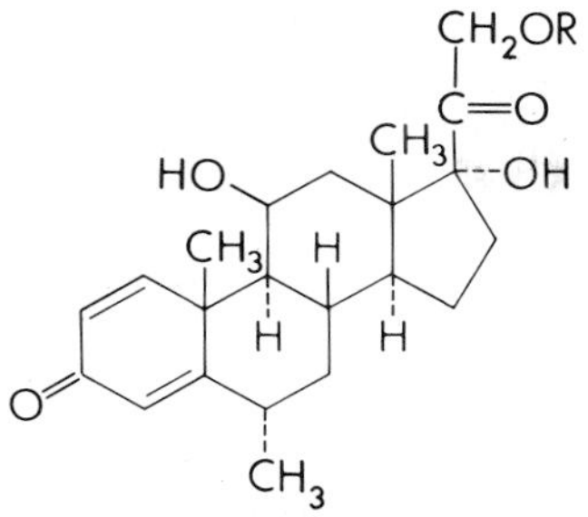

Methylprednisolone (R=H)

Ester Available:

Methylprednisolone Acetate: R=COCH₃

Salt Available:

Methylprednisolone Sodium Succinate:
R=COCH₂CH₂COO⁻ Na⁺

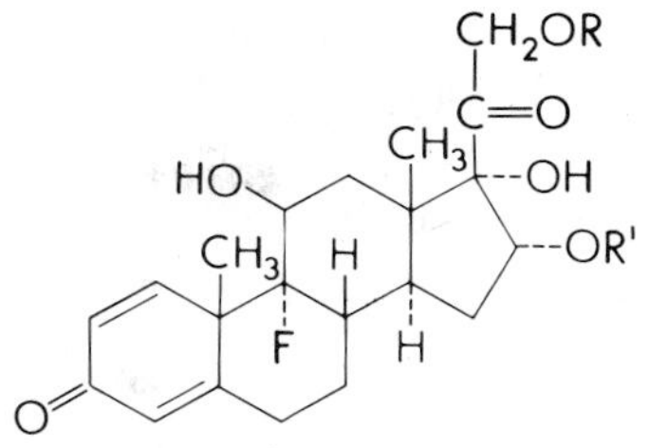

Triamcinolone

Ester Available:
Triamcinolone Diacetate: R,R'=COCH₃

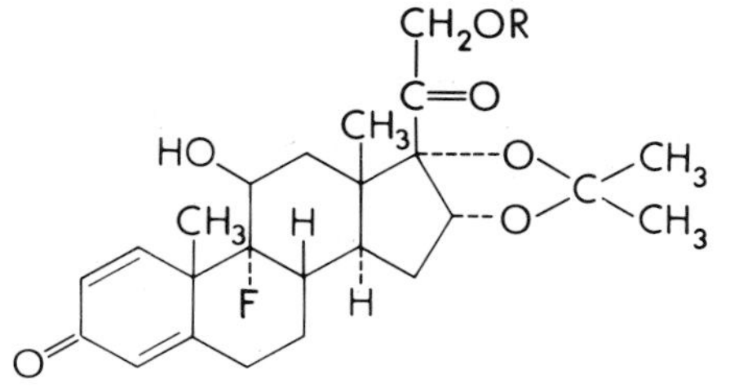

Triamcinolone Acetonide
Triamcinolone Hexacetonide R=COCH₂C(CH₃)₃

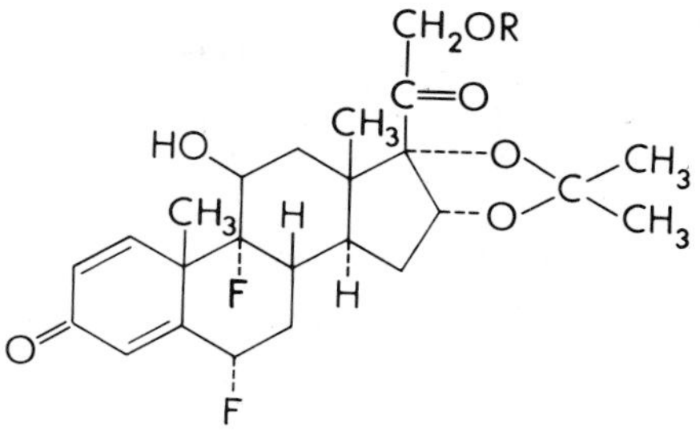

Fluocinolone Acetonide
(R=H)

Fluocinonide
(R=COCH₃)

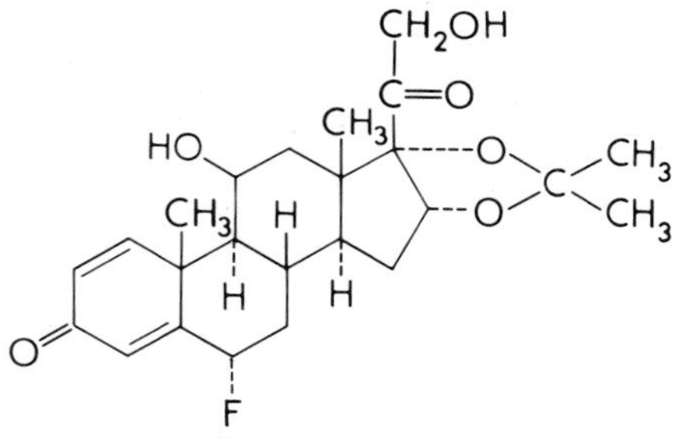

Flurandrenolone

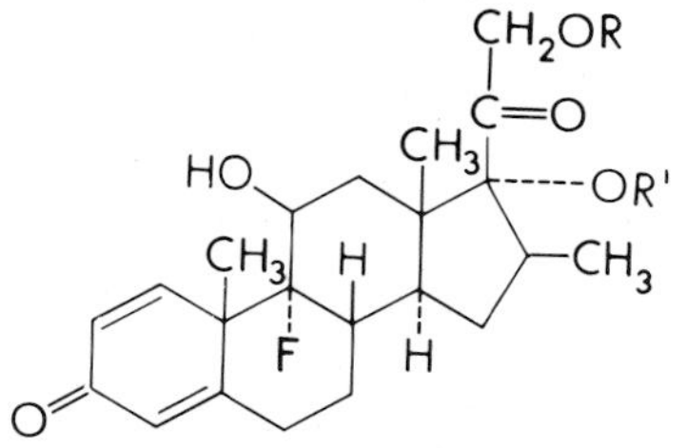

Betamethasone (R'=H)

Esters Available:
Betamethasone Acetate: R=COCH₃, R'=H
Betamethasone Valerate: R=H, R'=CO(CH₂)₃CH₃
Salt Available:
Betamethasone Sodium Phosphate: R=PO₄⁻ (Na⁺)₂,R'=H

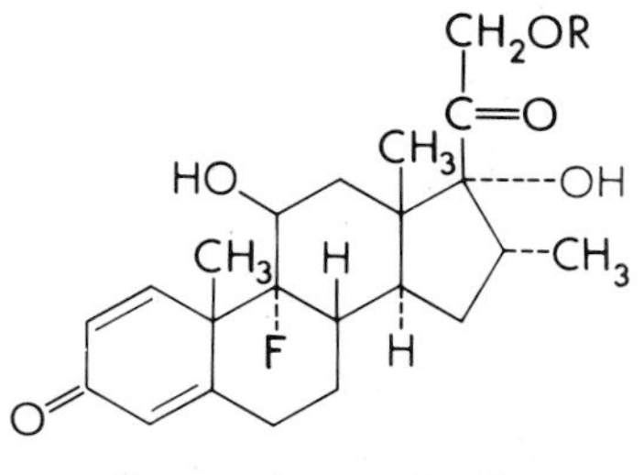

Dexamethasone (R=H)

Salt Available:

Dexamethasone Sodium Phosphate:
R=PO₄⁻ (Na⁺)₂

FIG. 18-18. (Cont.)

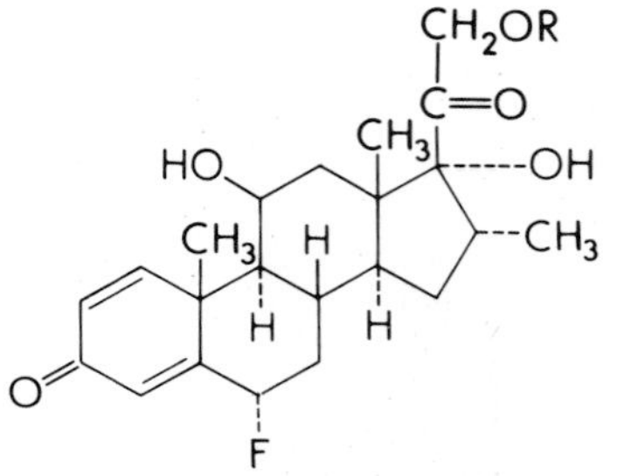

Paramethasone (R=H)

Ester Available:
Paramethasone Acetate: R=Ac

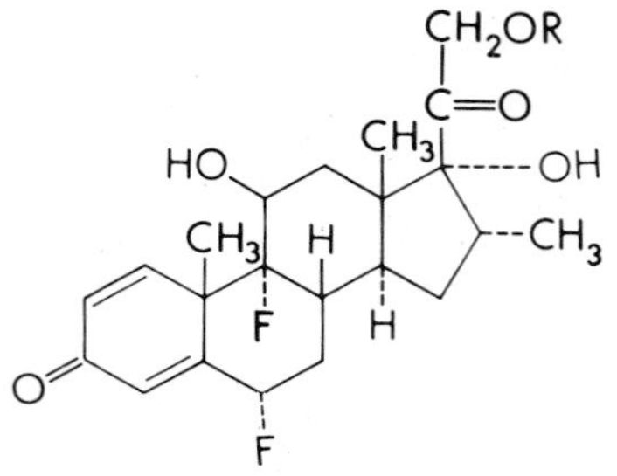

Flumethasone R=H

Ester Available:
Flumethasone Pivalate: R=COC(CH$_3$)$_3$

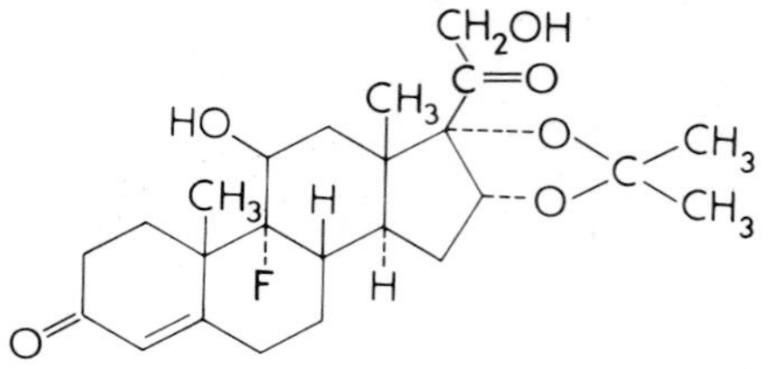

Halcinonide

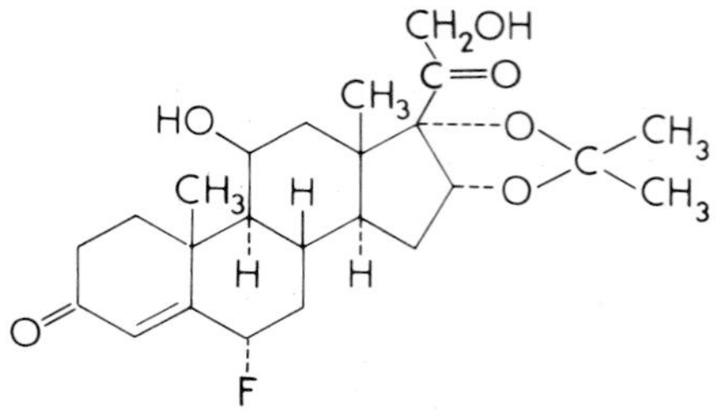

Flurandrenolide

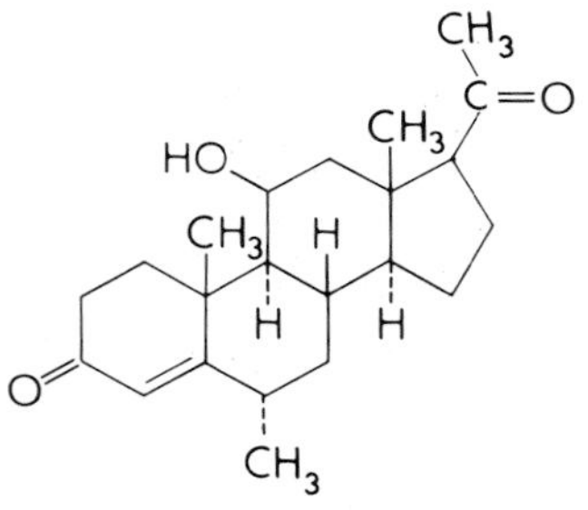

Medrysone

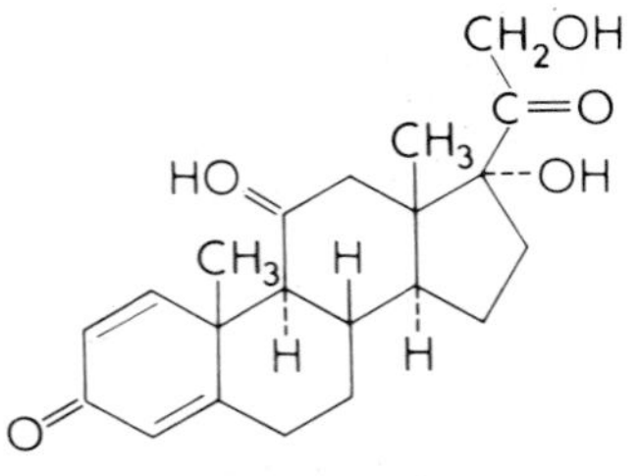

Meprednisone

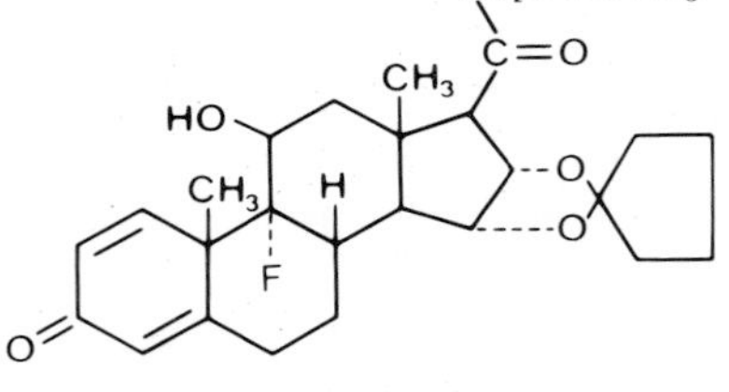

Amcinonide

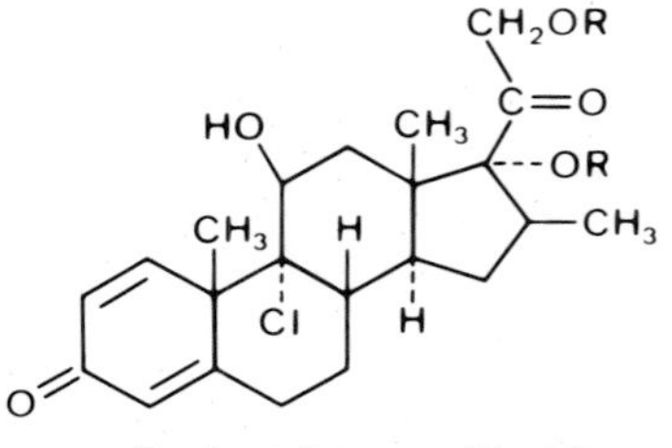

Beclomethasone (R=H)

Ester Available
Dipropionate: R=CCH$_2$CH$_3$

FIG. 18-18. (Cont.)

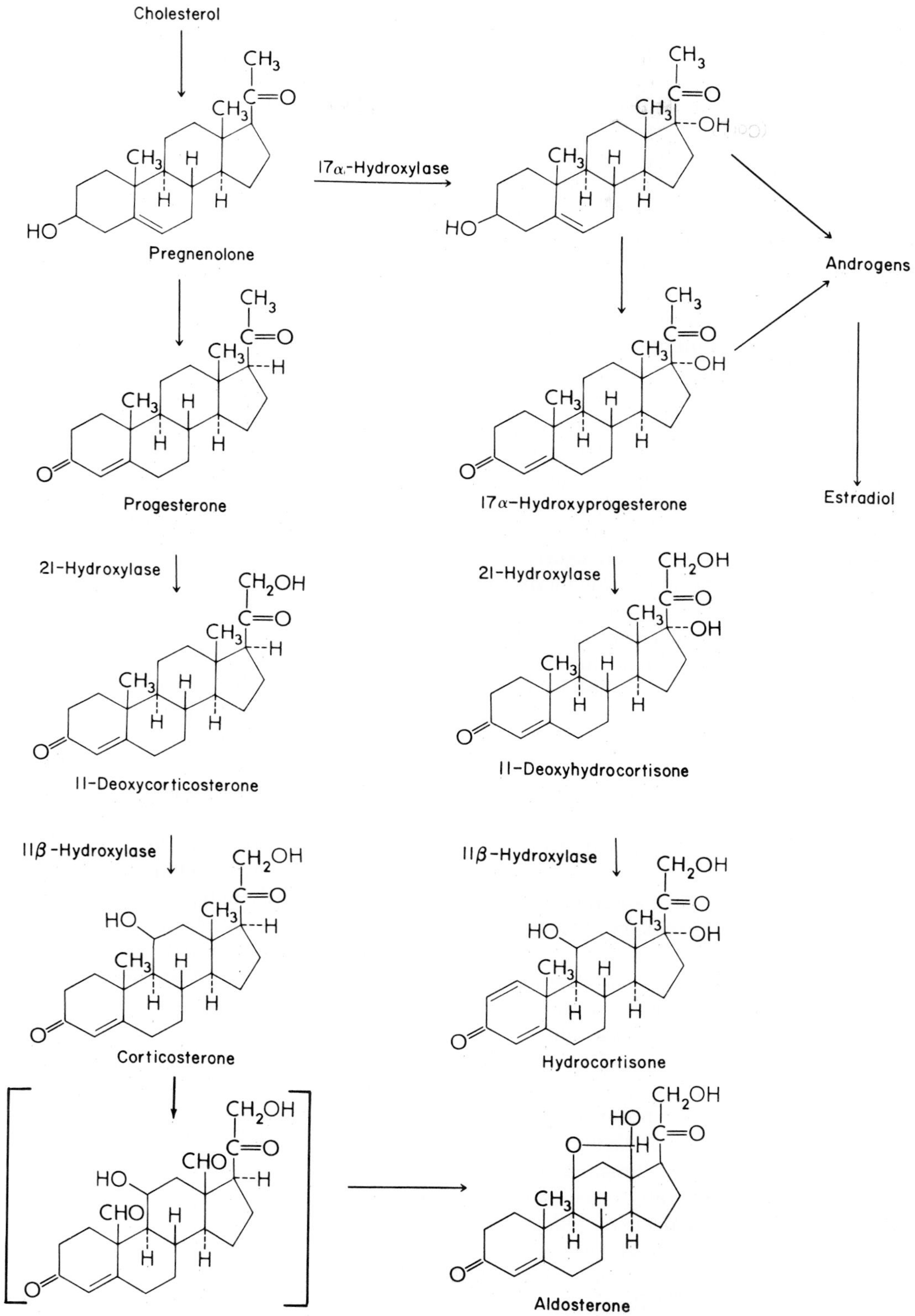

FIG. 18-19. *A simplified scheme of the biosynthesis of hydrocortisone and aldosterone. The biosynthetic pathways are more complex than presented here. There are several excellent reviews in the literature which present the pathways in their entirety.*

pregnenolone via a series of steps involving hydroxylations at C-17, C-11 and C-21. Deficiencies in any of the three hydroxylase enzymes are the cause of congenital adrenal hyperplasias, discussed in the previous section on therapeutic uses.

Although the details are not completely known, the polypeptide ACTH produced by the anterior pituitary enhances or is necessary for the conversion of cholesterol to pregnenolone. ACTH also stimulates the synthesis of hydrocortisone. (ACTH is discussed in Chap. 20.) Hydrocortisone then acts by feedback inhibition to suppress the formation of additional ACTH.

The release of the primary mineralocorticoid, aldosterone, is only slightly dependent upon ACTH. Aldosterone is an active part of the angiotensin-renin–blood-pressure cycle which controls blood volume. A decrease in blood volume stimulates the juxtaglomerular cells of the kidneys to secrete the enzyme renin. Renin, in turn, converts angiotensinogen to angiotensin, then angiotensin stimulates the adrenal cortex to release aldosterone. Aldosterone then causes the kidneys to retain sodium, and blood volume increases. When the blood volume has increased sufficiently, there is a decreased production of renin, until blood volume drops again. These physiologic mechanisms are clearly illustrated in booklets published by G. D. Searle and Co.[253,254]

BIOCHEMICAL ACTIVITIES

The adrenocortical steroids permit the body to adjust to environmental changes, to stress and to changes in the diet. As Sayers and Travis have succinctly stated in their fine review of the pharmacology of these steroids, "The adrenal cortex is the organ, *par excellence*, of homeostasis."[255] Aldosterone and, to a lesser extent, other mineralocorticoids maintain a constant electrolyte balance and blood volume, and the glucocorticoids have key roles in controlling carbohydrate, protein and lipid metabolism.

Aldosterone increases sodium reabsorption in the kidneys. Increased plasma sodium concentration will, in turn, lead to increased blood volume, since blood volume and urinary excretion of water are directly related to plasma sodium. Simultaneously, aldosterone increases potassium ion excretion. 11-Deoxycorticosterone is also quite active as a mineralocorticoid. Similar actions are exhibited with hydrocortisone and corticosterone, but to a much smaller degree.[255]

Aldosterone controls the movement of sodium ions in most epithelial structures involved in active sodium transport. Although aldosterone acts primarily on the distal convoluted tubules of the kidneys, it also acts on the proximal convoluted tubules and collecting ducts. Aldosterone also controls the transport of sodium in sweat glands, small intestine, salivary glands and the colon. In all these tissues, aldosterone enhances the inward flow of sodium ions and promotes the outward flow of potassium ions.

However, aldosterone (and other steroids with mineral activity) does not cause *immediate* changes in sodium excretion. There is a latent period after administration of any of the mineralocorticoids. This supports the view that aldosterone acts via stimulation of the synthesis of enzymes which, in turn, are actually responsible for active ion transport. This will be discussed in greater detail with the adrenocortical steroid receptors.

The glucocorticoids have many physiologic and pharmacologic actions. They control or influence carbohydrate, protein, lipid and purine metabolism. They also affect the cardiovascular and nervous systems and skeletal muscle.

Glucocorticoids stimulate glucose and glycogen synthesis by inducing the syntheses of required enzymes. They have a catabolic effect on muscle tissue, stimulating the formation and transamination of amino acids into glucose precursors in the liver. The catabolic actions in Cushing's syndrome are demonstrated by a wasting of the tissues, osteoporosis and reduced muscle mass. Lipid metabolism and synthesis are significantly increased in the presence of glucocorticoids, but the actions seem to be dependent on the presence of other hormones or cofactors in most cases.

Patients with Addison's disease exhibit muscle weakness and are easily fatigued. This may be due primarily to inadequate blood volume and aldosterone insufficiency, although changes in glucose availability may also be involved. A lack of adrenal cortex steroids also causes depression, irritability and even psychoses, reflecting significant effects on the nervous system.

Glucocorticoids decrease lymphocyte production and are generally immunosuppressive. Of great importance therapeutically is the fact that the glucocorticoids and ACTH decrease inflammation and the physiologic changes which occur to cause inflammation—edema, capillary dilatation, migration of phagocytes, capillary proliferation, deposition of collagen, etc.

METABOLISM

Cortisone and hydrocortisone are enzymatically interconvertible and so one finds metabolites from both. Most of the metabolic processes occur in the liver, with the metabolites excreted primarily in the urine. Although many metabolites have been isolated, the primary routes of catabolism are: (1) reduction of the

C-4 double bond to yield 5β-pregnanes; (2) reduction of the 3-one to give 3α-ols; (3) reduction of the 20-one to the corresponding 20α-ol. The two primary metabolites are tetrahydrocortisol and tetrahydrocortisone (shown below) and their conjugates.

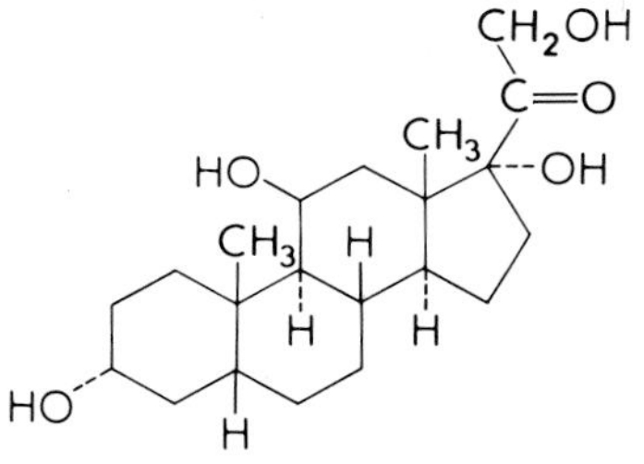

Tetrahydrocortisol

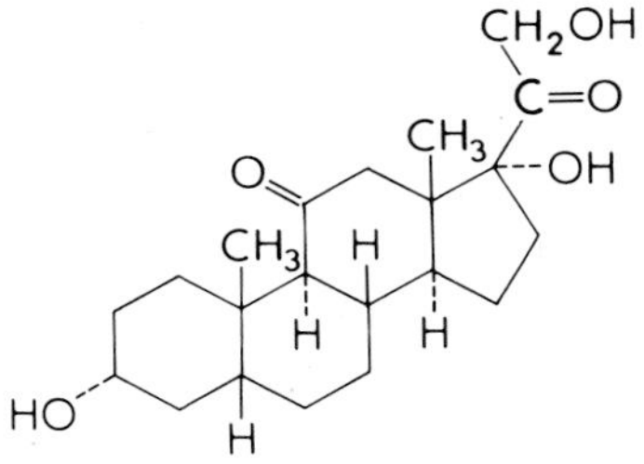

Tetrahydrocortisone

However, other metabolites include 20-ols and derivatives of side-chain oxidation and cleavage, as shown below.

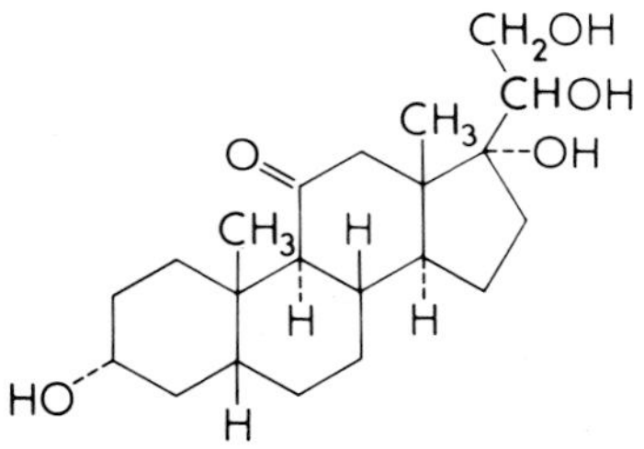

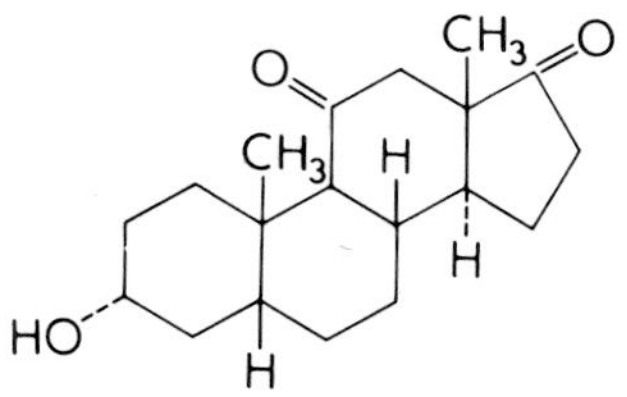

The C_{19} metabolites of the latter type are often androgenic.

GLUCOCORTICOID RECEPTORS

The glucocorticoids have been found to bind with great specificity to all tissues which elicit a "glucocorticoid response." The glucocorticoids are readily metabolized, which has made definitive studies particularly difficult. Further, plasma protein binding is quite high, leaving relatively little "plasma glucocorticoid" actually free to cause a physiologic response.

Tomkins, Baxter and others[24-27,256] have studied glucocorticoid receptors intensively and shown that the glucocorticoids react with them essentially as shown in Figure 18-2.

The glucocorticoids are known to stimulate the production or inhibition of a variety of enzymes, some of which are involved in the anabolic processes and others in catabolic processes of glucocorticoid activity. Tyrosine aminotransferase and other enzymes involved with transamination (necessary for amino acids to be converted to glucose and glycogen precursors) are induced by glucocorticoids.

The glucocorticoids appear to inhibit phosphofructokinase (which converts fructose 6-phosphate to fructose 1,6-diphosphate in glycolysis), thus decreasing glucose metabolism. They also may inhibit the conversion of pyruvate to acetyl CoA, thus "forcing" pyruvate to be used in glyconeogenesis. Both inhibition processes are likely, due to gene repression, i.e., repression of the formation of RNA templates needed to synthesize phosphofructokinase and other enzymes in glucose metabolism.

Glucocorticoid stimulation of lipid metabolism may be due to an increase in c-AMP formation (by inducing adenyl cyclase or inhibiting phosphodiesterase).

The roles of the glucocorticoids in acting as anti-inflammatory agents are still not well defined. Many different kinds of substances have been implicated in causing inflammation. These include histamine, serotonin, kinins, hyaluronic acid depolymerizers, acetylcholine, epinephrine, prostaglandins, antigen/antibody complexes, etc. It is known that the glucocorticoids' eosinopenic (reducing eosinophils) and hyperglycemic activities parallel their anti-inflammatory effectiveness. (Anti-inflammatory activity can be measured directly by using chemical irritants with experimental animals.) It is known that cortisone inhibits the release of inflammation-producing lysosomal enzymes; inhibits the ingestion of antigen/antibody complexes by white blood cells (which releases lysosomal enzymes); and reduces capillary permeability. How these effects are caused by glucocorticoids is still not known.

MINERALOCORTICOID RECEPTORS

Aldosterone is secreted in very small amounts, much smaller than other steroid hormones. Human plasma concentrations of aldosterone are only about 8 ng. per ml., with glucocorticoids up to 10^3 higher.

It now appears that aldosterone binds to cytoplasmic receptors of 8.5 S or 4.5 S size. This cytoplasmic receptor then undergoes a change into 3 S, which can be found in the nucleus. Finally, the 3 S receptor/steroid complex appears to act directly upon chromatin, specifically DNA. This over-all process has been illustrated in Figure 18-2. The DNA would then be stimulated to produce RNA templates for the synthesis of enzymes necessary for active Na$^+$ transport. The latent period of aldosterone action is then easy to rationalize.

STRUCTURE-ACTIVITY RELATIONSHIPS

Aldosterone cannot be produced in sufficient quantities and at a sufficiently low cost to make it a practical drug product. Cortisone and hydrocortisone have too much salt-retaining activity in the doses needed for some therapeutic purposes. For these two reasons, especially the latter, a great amount of effort has been made to design semisynthetic glucocorticoids and mineralocorticoids.

The structure-activity relationships expressed in Tables 18-12 and 18-13, along with other studies, can be summarized as shown below:

1. Substituents which significantly increase anti-inflammatory and glucocorticoid activity:
   1-dehydro (Δ^1)
   6α-fluoro
2. Substituents which significantly decrease mineralocorticoid activity:
   16α-hydroxy   16α- and 16β-methyl
   16α,17α-ketals
3. Substituents which significantly increase both glucocorticoid and mineralocorticoid activities:
   9α-fluoro   21-hydroxy   2α-methyl
   9α-chloro

The 11β-hydroxyl of hydrocortisone is believed to be of major importance in binding to the receptors; cortisone may be reduced in vivo to yield hydrocortisone as the active agent.[179] The increased activity of the 9α-fluoro derivative may be due to its electron-withdrawing inductive effect on the 11β-hydroxyl, making it more acidic and, therefore, binding better with the receptors. The 9α-fluoro also reduces oxidation of the 11β-OH to the less active 11-one.

The effect of introducing the Δ^1 double bond to significantly increase glucocorticoid activity and potency may be due to the resultant change in the shape of ring A. X-ray crystal studies by Duax, Weeks, Rohrer and Griffin have confirmed this presumption.[42]

TABLE 18–12

APPROXIMATE RELATIVE ACTIVITIES OF CORTICOSTEROIDS*

	Biological Half-Life (minutes)	Anti-inflammatory Activity	Topical Activity	Salt-Retaining Activity	Equivalent Dose (mg.)
Mineralocorticoids					
Aldosterone	—	0.2	0.2	800	—
11-Deoxycorticosterone	—	0	0	40	—
9α-Fluorohydrocortisone	—	10	5–40	800	2
Glucocorticoids					
Hydrocortisone	102	1	1	1	20
Cortisone	—	0.8	0	0.8	25
Prednisolone	200	4	4	0.6	5
Prednisone	—	3.5	0	0.6	5
6α-Methylprednisolone (Methyprednisolone)	—	5	5	0	4
16β-Methylprednisone (Meprednisone)	Used in eyes only; comparative data not available				
6α-Fluoroprednisolone (Fluprednisolone)	—	15	7	0	1.5
Triamcinolone Acetonide	300	5	5–100	0	4
Triamcinolone	100–200	—	1–5	—	—
6α-Fluorotriamcinolone Acetonide (Fluocinolone Acetonide)	—	—	over 40	—	—
Flurandrenolone Acetonide (Flurandrenolide)	—	—	over 20	—	—
Fluocinolone	—	—	over 40	—	—
Fluocinolone 21-Acetate (Fluocinonide)	—	—	over 40–100	—	—
Betamethasone	—	35	5–100	0	0.6
Dexamethasone (16 α-Isomer of Betamethasone)	200	30	10–35	0	0.75

* The data in this table are only approximate. Blanks indicate that comparative data are not available to the author or that the product is used only for one use, e.g., topically. Data were taken from several sources, and there is an inherent risk in comparing such data. However, the table should serve as a guide of relative activities. Readers who have access to data which may be useful for revising this table in future editions are encouraged to write to the author or editor. A valuable study that might also be consulted is Ringler, I.: Methods in Hormone Research, 3 (R.I. Dorfman, ed.), Part A, Academic Press, New York, 1964.

Data in Table 18–12 are abstracted from the following references, among others: Sciuchetti, L.A.: Pharm. Index 5:7, 1963. Fischer, D.A., and Panos, T.C.: Postgrad. Med. 39:650, 1966; references 31, 186, 187; The Hospital Formulary, American Society of Hospital Pharmacists, 1975, Medical Letter, *17*, 97 (1975).

TABLE 18–13

EFFECTS OF SUBSTITUENTS ON GLUCOCORTICOID ACTIVITY*

Clinical Antirheumatic Enhancement Factors

Functional Group	Factor	Functional Group	Factor
1-Dehydro	2.8	16α-Methyl	1.6
6-Dehydro	0.9†	6β-Methyl	1.3†
6α-Methyl	0.9†	16α, 17α-Isopropyl-idenedioxy	0.6†
6α-Fluoro	1.9	17α-Acetoxy	0.3†
9α-Fluoro	4.9	21-Deoxy	0.2†
16α-Hydroxy	0.3	21-Methyl	0.3†

Enhancement Factors for Various Functional Groups of Corticosteroids

Functional Group	Glycogen Deposition	Anti-inflammatory Activity	Effects on Urinary Sodium‡
9α-Fluoro	10	7–10	+ + +
9α-Chloro	3–5	**3	+ +
9α-Bromo	0.4**		+
12α-Fluoro	6–8§		+ +
12α-Chloro	4§		
1-Dehydro	3–4	3–4	—
6-Dehydro	0.5–0.7		+
2α-Methyl	3–6	1–4	+ +
6α-Methyl	2–3	1–2	— — —
16α-Hydroxy	0.4–0.5	0.1–0.2	— — —
17α-Hydroxy	1–2	4	—
21-Hydroxy	4–7	25	+ +
21-Fluoro	2	2	— —

* From Rodig, O.R., *in* Burger, A. (ed.): Medicinal Chemistry, Part II, ed. 3, New York, Wiley–Interscience, 1970. Used with permission.
† Two observations or less.
‡ + = retention; − = excretion.
** In 1-dehydrosteroids this value is 4.
§ In the presence of a 17α-hydroxyl group this value is <0.01.

TOPICAL POTENCY

The uses, problems, and effectiveness of topical corticosteroids were reviewed in 1979 and 1980.[264]

Although, as shown in Table 18-12, cortisone and prednisone are not active topically, most other glucocorticoids are active. Some compounds, such as triamcinolone and its acetonides, have striking activity topically. Skin absorption is favored by increased lipid-solubility of the drug (see Fig. 18-5). Absorption can also be greatly affected by extent of skin damage, concentration of the glucocorticoid, cream or ointment base used, and similar factors. One must not, therefore, assume from study of Table 18-12 that, for example, a 0.25 percent cream of prednisolone is necessarily exactly equivalent in anti-inflammatory potency to 1 percent hydrocortisone. Nevertheless, the table can serve as a preliminary guide. Furthermore, particular patients may seem to respond better to one topical anti-inflammatory glucocorticoid than another, irrespective of the relative potencies shown in Table 18-12.

Except for fludrocortisone, the topical corticosteroids do not cause effects of absorption when used on small areas of intact skin. However, when these compounds are used on large areas of the body, systemic absorption may occur—especially if the skin is damaged or if occlusive dressings are used. Fludrocortisone is more readily absorbed than other topical corticosteroids, so systemic problems can be expected more frequently with it. (Up to 20 to 40% of hydrocortisone given rectally may also be absorbed.)

When topically administered the topical glucocorticoids present relatively infrequent therapeutic problems, but it should be remembered that their anti-inflammatory action can mask symptoms of infection. Many physicians prefer not giving a topical anti-inflammatory steroid until after an infection is controlled with topical antibiotics. The immunosuppressive activity of the topical glucocorticoids can also prevent natural processes from curing the infection. Topical steroids may also actually cause any of several dermatoses in some patients.

Finally, as discussed before with the oral contraceptives, steroid hormones should not be used during pregnancy. If absolutely necessary to use the glucocorticoids topically during pregnancy, they should be limited to small areas of intact skin and used for a limited time.

PRODUCTS

The adrenal corticosteroid products are shown in Figure 18-18. The structures illustrate the usual changes (Fig. 18-5) made to modify solubility of the products—and, therefore, their therapeutic uses. In particular, the 21-hydroxyl can be converted to an ester to make it less water-soluble to modify absorption; or to a phosphate ester salt or hemisuccinate ester salt to make it more water-soluble and appropriate for intravenous use. The products also reflect the previously discussed structure-activity relationships change to increase anti-inflammatory activity or potency, or to decrease salt retention.

It must be emphasized again that patients who have been on long-term glucocorticoid therapy must have the dose *gradually* reduced. This "critical rule" and indications have been previously discussed under Therapeutic Uses. Dosage schedules and gradual reduction of dose schedules can be quite complex and specific for each indication. For that reason, specialized indication and dose references such as *Facts and Comparisons* and *The Hospital Formulary* should be consulted before advising physicians on dosages.

General U.S.P. D.I. doses are shown in Table 18-14.

Many of the glucocorticoids are available in topical dosage forms, including creams, ointments, aerosols, lotions and solutions. They are usually applied three to four times a day to well-cleaned areas of affected skin. (The patient should be instructed to apply them with well-washed hands as well). Ointments are usually prescribed for dry, scaly dermatoses. Lotions are well suited for weeping dermatoses. Creams are of general use for many other dermatoses. When applied to very large areas of skin or to damaged areas of skin, significant systemic absorption can occur. The use of an occlusive dressing can also greatly increase systemic absorption.

Desoxycorticosterone Acetate U.S.P., pregn-4-en-3,20-dione-21-ol 21-acetate, is a potent mineralocorticoid used only for the treatment of Addison's disease. It has essentially no anti-inflammatory (glucocorticoid) activity but has 100 times the salt-retention (mineralocorticoid) activity of hydrocortisone. (It has only one-thirtieth the activity of aldosterone.) Hydrocortisone or other glucocorticoids should be given simultaneously for patients with acute adrenal insufficiency. Its great salt-retaining activity can be expressed as edema and pulmonary congestion as toxic doses are reached. It is insoluble in water (as one would predict from Table 18-1). Since Addison's disease is essentially incurable, treatment continues for life. With a serum half-life of only 70 minutes, the drug is sometimes given in the form of subcutaneous pellets administered every 8 to 12 months. The duration of action of **Desoxycorticosterone Pivalate U.S.P.** when given intramuscularly in depot preparations is longer than the acetate, often being administered only once every four weeks.

Fludrocortisone Acetate U.S.P., 9α-fluoro-11β,17α,21-trihydroxypregn-4-en-3,20-dione 21-acetate, 9α-fluorohydrocortisone, is used only for the treatment of Addison's disease and for inhibition of endogenous adrenocortical secretions. As shown in Table 18-12, it has up to about 800 times the mineralocorticoid activity of hydrocortisone and about 11 times the glucocorticoid activity.[259-262] This compound was made originally as an intermediate in the total synthesis of Fried and Sabo,[263] but its potent activity stimulated the synthesis and study of the many fluorinated steroids shown in Figure 18-18. Although its great salt-retaining activity limits its use to Addison's disease, it has sufficient glucocorticoid activity so that in many cases of the disease additional glucocorticoids need not be prescribed.

Cortisone Acetate U.S.P., 21,17β-dihydroxypregn-4-en-3,11,20-dione 21-acetate, is a natural cortical steroid with good anti-inflammatory activity and low to moderate salt-retention activity. It is used for the entire spectrum of uses discussed previously under Therapeutic Uses—collagen diseases, especially rheumatoid arthritis; Addison's disease; severe shock; allergic conditions; chronic lymphatic leukemia; and many other indications. Cortisone acetate is relatively ineffective topically, in part because it must be reduced in vivo to hydrocortisone which is more active. Its plasma half-life is only about 30 minutes, compared to one and a half to three hours for hydrocortisone.

Hydrocortisone U.S.P., 11β,17α,21-trihydroxypregn-4-en-3,20-dione, is the primary natural glucocorticoid in man. Synthesis of 9α-fluorohydrocortisone during the synthesis of hydrocortisone has led to the array of semisynthetic glucocorticoids shown in Figure

TABLE 18-14

NATURAL AND SEMISYNTHETIC MINERALOCORTICOID HORMONES

Name Proprietary Name	Preparations	Usual Adult Dose*	Usual Dose Range*	Usual Pediatric Dose*
Desoxycorticosterone Acetate U.S.P. *Doca,* *Percorten Acetate*	Desoxycorticosterone Acetate Injection U.S.P.	I.M., 1 to 6 mg. once daily	1 to 10 mg. daily	I.M., 1 to 5 mg. once a day or 1.5 to 2 mg. per square meter of body surface, once a day
	Desoxycorticosterone Acetate Pellets U.S.P.	Subcutaneous implantation—125 mg. (1 pellet) per 500 μg. of daily I.M. dose of desoxycorticosterone acetate		
Desoxycorticosterone Pivalate U.S.P. *Percorten Pivalate*	Sterile Desoxycorticosterone Pivalate Suspension N.F.			I.M., 25 mg. for each 1 mg. of daily maintenance dose of desoxycorticosterone acetate injection, at 4-week intervals
Fludrocortisone Acetate U.S.P. *Florinef Acetate*	Fludrocortisone Acetate Tablets U.S.P.	100 μg. once daily	100 to 200 μg. (0.1 to 0.2 mg) a day	

*See *U.S.P. D.I.* for complete dosage information.

18-18, many of which have greatly improved anti-in-flammatory activity. Nevertheless, hydrocortisone, its esters and salts remain a mainstay of modern adreno-cortical steroid therapy—and the standard for comparison of all other glucocorticoids and mineralocorticoids (Table 18-12). It is used for all the indications previously mentioned. Its esters and salts illustrate the principles of chemical modification to modify pharmacokinetic utility shown in Figure 18-5. The

commercially available salts and esters (Fig. 18-18) include:

Hydrocortisone Acetate U.S.P.
Hydrocortisone Sodium Succinate U.S.P.
Hydrocortisone Cypionate U.S.P.
Hydrocortisone Sodium Phosphate U.S.P.
Hydrocortisone Valerate U.S.P.
Prednisolone U.S.P., Δ^1-hydrocortisone, $11\beta,17\alpha,$-21-trihydroxypregna-1,4-dien-3-20-dione, has less salt-

TABLE 18–15

ANTI-INFLAMMATORY GLUCOCORTICOIDS WITH MODERATE TO LOW SALT RETENTION

Name / Proprietary Name	Preparations	Application	Usual Adult Dose*	Usual Dose Range*	Usual Pediatric Dose*
Cortisone Acetate U.S.P. / *Cortone*	Cortisone Acetate Tablets U.S.P.		25 to 300 mg. a day as a single dose or in divided doses	10 to 400 mg. daily	175 g. to 2.5 mg. per kg. of body weight or 5 to 75 mg. per square meter of body surface, 4 times daily
	Sterile Cortisone Acetate Suspension U.S.P.		20 to 30 mg. a day	10 to 400 mg. daily	200 µg. to 1.25 mg. per kg. of body weight or 7 to 37.5 mg. per square meter of body surface once or twice daily
Hydrocortisone U.S.P. / *Cortef, Hydrocortone*	Hydrocortisone Enema U.S.P.		Rectal, 100 mg. each night for 21 days, or until clinical and proctological remission is obtained		
	Hydrocortisone Ointment U.S.P. / Hydrocortisone Lotion U.S.P. / Hydrocortisone Cream U.S.P.	Topically to the skin as a 0.25 percent ointment, 0.125 to 1 percent lotion, or 0.125 to 2.5 percent cream 2 to 4 times daily			
	Hydrocortisone Tablets U.S.P. / Sterile Hydrocortisone Suspension U.S.P.		5 to 60 mg. 4 times daily	20 to 240 mg. daily	140 µg. to 2 mg. per kg. of body weight or 4 to 60 mg. per square meter of body surface, 4 times daily
Hydrocortisone Acetate U.S.P. / *Cortril Suspension, Hydrocortone Acetate Suspension*	Hydrocortisone Acetate Ointment U.S.P. / Hydrocortisone Acetate Cream U.S.P. / Hydrocortisone Acetate Lotion U.S.P.	Topically to the skin, as a 0.5 to 2.5 percent ointment 1 to 4 times a day			
	Hydrocortisone Acetate Ophthalmic Suspension U.S.P.	Topically to the conjunctiva, 0.05 to 0.1 ml. of a 0.5 to 2.5 percent suspension 3 to 20 times daily			
	Hydrocortisone Acetate Ophthalmic Ointment U.S.P.	Topically to the conjunctiva, as a 0.5 to 1.5 percent ointment 1 to 4 times daily			
	Sterile Hydrocortisone Acetate Suspension U.S.P.		Intra-articular, intralesional, or soft-tissue injection, 5 to 75 mg., repeated at 2- to 3-week intervals		

*See *U.S.P. D.I.* for complete dosage information.
(Continued)

TABLE 18–15

ANTI-INFLAMMATORY GLUCOCORTICOIDS WITH MODERATE TO LOW SALT RETENTION (Continued)

Name Proprietary Name	Preparations	Application	Usual Adult Dose*	Usual Dose Range*	Usual Pediatric Dose*
Hydrocortisone Sodium Succinate U.S.P. *Solu-Cortef, A-hydrocort*	Hydrocortisone Sodium Succinate for Injection U.S.P.		I.M. or I.V., the equivalent of 100 to 500 mg. hydrocortisone at 1- to 10-hour intervals	100 mg. to 8 g. daily	I.M., 160 μg. to 1 mg. per kg. of body weight or 6 to 30 mg. per square meter of body surface, 1 or 2 times a day
Hydrocortisone Cypionate U.S.P. *Cortef Fluid*	Hydrocortisone Cypionate Oral Suspension U.S.P.		The equivalent of 15 to 30 mg. of hydrocortisone 3 or 4 times daily	The equivalent of 25 to 320 mg. of hydrocortisone daily	
Hydrocortisone Sodium Phosphate U.S.P. *Hydrocortone Phosphate*	Hydrocortisone Sodium Phosphate Injection U.S.P.		I.M., I.V., or S.C., the equivalent of cortisol, 15 to 240 mg. a day	25 mg. to 1 g. daily	I.M., 160 μg. to 1 mg. per kg. of body weight or 6 to 30 mg. per square meter of body surface, 1 or 2 times a day
Prednisolone U.S.P. *Delta-Cortef, Prednis, Sterane*	Prednisolone Tablets U.S.P. Prednisdone Cream U.S.P.		5 to 15 mg. 1 to 4 times daily	5 to 250 mg. daily	35 to 500 μg. per kg. of body weight or 1 to 15 mg. per square meter of body surface, 4 times daily
Prednisolone Acetate U.S.P. *Meticortelone Acetate, Nisolone, Savacort*	Sterile Prednisolone Acetate Suspension U.S.P.		Intra-articular, 12.5 to 25 mg. at each site every 1 to 3 weeks; I.M., 2 to 30 mg. twice daily		I.M., 40 to 250 μg. per kg. of body weight or 1.5 to 7.5 mg. per square meter of body surface, once or twice daily
Prednisolone Succinate U.S.P. *Meticortelone Soluble*	Available only as sodium salt (below) Prednisolone Sodium Succinate for Injection U.S.P.		I.M. or I.V., the equivalent of 2 to 30 mg. of prednisolone twice daily		I.M. or I.V., the equivalent of 40 to 250 μg. of prednisolone per kg. of body weight or 1.5 to 7.5 mg. per square meter of body surface, once or twice daily
Prednisolone Tebutate U.S.P. *Hydeltra-T.B.A.*	Sterile Prednisolone Tebutate Suspension U.S.P.				
Prednisolone Sodium Phosphate U.S.P. *Sodasone*	Prednisolone Sodium Phosphate Injection U.S.P.		I.M. or I.V., the equivalent of 10 to 50 mg. of prednisolone phosphate twice daily	10 to 400 mg. daily	I.M. or I.V., the equivalent of 40 to 250 μg. of prednisolone phosphate per kg. of body weight or 1.5 to 7.5 mg. per square meter of body surface once or twice daily
Hydeltrasol	Prednisolone Sodium Phosphate Ophthalmic Solution U.S.P.	Topical to the conjunctiva, the equivalent of 0.05 to 0.1 ml. of a 0.113 to 0.9 percent solution of prednisolone phosphate 2 to 20 times a day			
Prednisone U.S.P. *Delta-Dome, Orasone, Deltasone, Servisone, Meticorten, Paracort*	Prednisone Tablets U.S.P.		5 to 60 mg. a day as a single dose or in divided doses	5 to 250 mg. daily	35 to 500 μg. per kg. of body weight or 1 to 15 mg. per square meter of body surface 4 times daily

* See *U.S.P. D.I.* for complete dosage information.

retention activity than hydrocortisone (see Table 18-12), but some patients have more frequently experienced complications such as gastric irritation and peptic ulcers. Because of low mineralocorticoid activity, it cannot be used alone for adrenal insufficiency. Prednisolone is available in a variety of salts and esters to maximize its therapeutic utility (see Fig. 18-5):

Prednisolone Acetate U.S.P.

Prednisolone Succinate U.S.P.

Prednisolone Sodium Succinate for Injection U.S.P.

Prednisolone Sodium Phosphate U.S.P.

Prednisolone Tebutate U.S.P.

Prednisone U.S.P., Δ^1-cortisone, $17\alpha,21$-dihydroxypregna-1,4-dien-3,11,20-trione, has activity very similar to that of prednisolone, and because of its lower salt-retention activity is often preferred over cortisone or hydrocortisone.

GLUCOCORTICOIDS WITH LOW SALT RETENTION

Most of the key differences between the many glucocorticoids with low salt retention (Fig. 18-18) have been summarized in Tables 18-12 and 18-13. The tremendous therapeutic and, therefore, commercial importance of these drugs has been a stimulus to the proliferation of new compounds and their products. It is difficult to point out additional major unique features (not shown on Tables 18-12 and 18-13) of many of these compounds. Within a given anti-inflammatory or topical anti-inflammatory potency range (Table 18-12), compounds have similar side-effects and actions. Some individual patients may respond better to one compound (within a given potency range) than another, but in most cases generalizations cannot be made. Many compounds also are available as salts or esters to give the complete range of therapeutic flexibility illustrated in Figure 18-5. When additional pertinent information (other than that shown in Tables 18-12 and 18-14) is available, it will be given below.

Methylprednisolone U.S.P., $11\beta,17\alpha,21$-trihydroxy-6α-methylpregna-1,4-dien-3,20-dione.

Methylprednisolone Acetate U.S.P.

Methylprednisolone Hemisuccinate U.S.P.

Methylprednisolone Sodium Succinate U.S.P.

Triamcinolone U.S.P., 9α-fluoro-$11\beta,16\alpha,17\alpha,21$-tetrahydroxypregna-1,4-dien-3,20-dione.

Triamcinolone Acetonide U.S.P., triamcinolone-$16\alpha,17\alpha$-acetone ketal, $16\alpha,17\alpha$-[(1-methylethylidene)*bis*(oxy)]triamcinolone.

Triamcinolone Hexacetonide U.S.P., triamcinolone acetonide 21-[3-(3,3-dimethyl)butyrate].

Triamcinolone Diacetate U.S.P. The hexaceto-

nide is slowly converted to the acetonide in vivo and is given only by intra-articular injection. Only triamcinolone and the diacetate are given orally. When triamcinolone products are given intramuscularly, they are often given deeply into the gluteal region, since local atrophy may occur with shallow injections. The acetonide and diacetate may be given by intra-articular or intrasynovial injection, and the acetonide additionally may be given by intrabursal or sometimes by intramuscular or subcutaneous injection. A single intramuscular dose of the diacetate or acetonide may last up to three or four weeks. Plasma levels with intramuscular doses of the acetonide are significantly higher with the acetonide than triamcinolone itself.

Topically applied triamcinolone acetonide is a potent anti-inflammatory agent (see Table 18-12), about ten times more potent than triamcinolone.

Fluocinolone Acetonide U.S.P., 6α-fluorotriamcinolone acetonide, $6\alpha,9\alpha$-difluoro-$11\beta,16\alpha,17\alpha,21$-tetrahydroxypregna-1,4-dien-3,20-dione,$16\alpha,17\alpha$-acetone ketal, the 21-acetate of fluocinonide. Fluocinonide is about five times more potent than the acetonide in the vasoconstrictor assay.

Flurandrenolone, $16\alpha,17\alpha$-dihydroxy-6α-fluoropregn-4-en-3,20-dione. The $16\alpha,17\alpha$-acetone ketal is **Flurandrenolide U.S.P.** and has replaced flurandrenolone in clinical practice. Although a flurandrenolide tape product is available, it can stick to and remove damaged skin, so it should be avoided with vesicular or weeping dermatoses.

Betamethasone U.S.P., 9α-fluoro-$11\beta,17\alpha,21$-trihydroxy-16β-methylpregna-1,4-dien-3,20-dione.

Betamethasone Valerate U.S.P.

Betamethasone Acetate U.S.P.

Betamethasone Sodium Phosphate U.S.P.

Betamethasone Benzoate U.S.P.

Betamethasone Dipropionate U.S.P.

Beclomethasone Dipropionate U.S.P., 9 chloro-$11\beta,17,21$-trihydroxy-16β-methylpregna-1,4-diene-3,20-dione, 17,21-diproprionate, is available as a nasal inhaler and is used for seasonal rhinitis poorly responsive to conventional treatment.

Dexamethasone U.S.P., 9α-fluoro-$11\beta,17\alpha,21$-trihydroxy-16α-methylpregna-1,4-dien-3,20-dione, is essentially the 16α-isomer of betamethasone.

Dexamethasone Acetate U.S.P.

Dexamethasone Sodium Phosphate U.S.P.

Desoximetasone U.S.P.

Paramethasone Acetate U.S.P., 6α-fluoro-$11\beta,17\alpha,21$-trihydroxy-16α-methylpregna-1,4-dien-3-20-dione 21-acetate.

Flumethasone Pivalate U.S.P., $6\alpha,9\alpha$-difluoro-$11\beta,17\alpha,21$-trihydroxy-16α-methylpregna-1,4-dien-3,20-dione 21-pivalate.

Halcinonide, 21-chloro-9α-fluoro-$11\beta,16\alpha,17$-tri-

(Text continues on p. 711)

TABLE 18–16

ANTI-INFLAMMATORY GLUCOCORTICOIDS WITH LOW SALT RETENTION

Name Proprietary Name	Preparations	Application	Usual Adult Dose*	Usual Dose Range*	Usual Pediatric Dose*
Methylprednisolone U.S.P. *Medrol*	Methylprednisolone Tablets U.S.P.		4 mg. 4 times daily	2 to 48 mg. daily	117 µg. (0.117 mg.) to 1.2 mg. per kg. of body weight or 3.3 to 35 mg. per square meter of body surface a day as a single dose or in divided doses
Methylprednisolone Acetate U.S.P. *Depo-Medrol*	Methylprednisolone Acetate Cream U.S.P.	Topical, 0.25 or 1 percent cream 1 to 3 times daily			
	Sterile Methylprednisolone Acetate Suspension U.S.P.		Intra-articular, intra-lesional, or soft-tissue injection, 4 to 80 mg. repeated at 1-day to 2-week intervals, if necessary Intramuscular, 4 to 120 mg, repeated at 1-day to 2-week intervals if necessary	4 to 120 mg. of methyl-prednisolone acetate	Intramuscular, 500 µg. (0.5 mg.) to 1.5 mg. per kg. of body weight or 15 to 45 mg. per square meter of body surface, repeated at 1-day to 2-week intervals, if necessary
	Methylprednisolone Acetate for Enema U.S.P.		Rectal, 40 mg., 3 to 7 times a week	120 to 280 mg. weekly	500 µg. (0.5 mg.) to 1 mg. per kg. of body weight or 15 to 30 mg. per square meter of body surface, every one or two days
Methylprednisolone Sodium Succinate U.S.P. *Solu-Medrol*	Methylprednisolone Sodium Succinate for Injection U.S.P.		I.M. or I.V., the equivalent of methylprednisolone, 10 to 40 mg. repeated as needed	10 mg. to 1.5 g. daily	I.M. or I.V. the equivalent of methylprednisolone 500 µg. (0.5 mg.) to 1 mg. per kg. of body weight or 15 to 30 mg. per square meter of body surface a day
Triamcinolone *Aristocort, Kenacort*	Triamcinolone Tablets U.S.P.		Replacement therapy—oral, the equivalent of 4 to 6 mg. of triamcinolone a day as a single dose or in divided doses, in conjuction with a mineralocorticoid Disease therapy—oral, the equivalent of 4 to 48 mg. of triamcinolone a day as a single dose or in divided doses		Replacement therapy—117 µg. (0.117 mg.) per kg. of body weight or 3.3 mg. per square meter of body surface a day as a single dose or in divided doses in conjunction with a mineralocorticoid Disease therapy—416 µg. (0.416 mg.) to 1.7 mg. per kg. of body weight or 12.5 to 50 mg. per square meter of body surface as a single dose or in divided doses
Triamcinolone Diacetate U.S.P. *Aristocort Diacetate*	Sterile Triamcinolone Diacetate Suspension U.S.P. Triamcinolone Diacetate Syrup U.S.P.		See Triamcinolone Tablets	Oral, 4 to 30 mg. daily	See Triamcinolone Tablets
Triamcinolone Acetonide U.S.P. *Kenalog, Aristocort Acetonide, Aristoderm*	Triamcinolone Acetonide Cream U.S.P. Triamcinolone Acetonide Ointment U.S.P. Triamcinolone Acetonide Topical Aerosol U.S.P.	Topically to the skin, as a 0.025 to 0.5 percent cream or ointment 2 to 4 times daily			
	Triamcinolone Acetonide Dental Paste U.S.P.	Topically to the oral mucous membranes, as a 0.1 percent paste 2 to 4 times daily			

(Continued)

TABLE 18-16

ANTI-INFLAMMATORY GLUCOCORTICOIDS WITH LOW SALT RETENTION *(Continued)*

Name Proprietary Name	*Preparations*	*Application*	*Usual Adult Dose**	*Usual Dose Range**	*Usual Pediatric Dose**
	Sterile Triamcinolone Acetonide Suspension U.S.P.		Intra-articular, intrabursal, or tendon-sheath injection, 2.5 to 15 mg. Intradermal or intralesional up to 1 mg., repeated at weekly or less frequent intervals, if necessary Intramuscular, 40 to 80 mg. repeated at 4 week intervals, if necessary	Intra-articular or intrabursal, 2.5 to 80 mg.; I.M., 20 to 100 mg.	Children up to 6 years of age—use is not recommended Children 6 to 12 years of age—intra-articular, intrabursal, or tendon-sheath injection, 2.5 to 15 mg., repeated as needed Intramuscular, 40 mg. repeated at 4-week intervals if necessary or 30 to 200 μg. (0.03 to 0.2 mg.) per kg. of body weight or 1 to 6.25 mg. per square meter of body surface, repeated at 1- to 7-day intervals
Triamcinolone Hexacetonide U.S.P. *Aristospan*	Triamcinolone Hexacetonide Suspension U.S.P.		Intra-articular, 2 to 20 mg. at each site, once every 3 to 4 weeks; intralesional or sublesional, up to 500 μg. per square inch of affected skin, repeated as needed		
Fluocinolone Acetonide U.S.P. *Fluonid, Synalar*	Fluocinolone Acetonide Cream U.S.P. Fluocinolone Acetonide Ointment U.S.P. Fluocinolone Acetonide Topical Solution U.S.P.	Topically to the skin, as a 0.01 to 0.2 percent cream, or as a 0.025 percent ointment, or as a 0.01 percent solution, 3 or 4 times daily, as necessary			
Flurandrenolide U.S.P. *Cordran*	Flurandrenolide Cream U.S.P. Flurandrenolide Ointment U.S.P. Flurandrenolide Lotion U.S.P. Flurandrenolide Tape U.S.P.	Topically to the skin, as a 0.025 to 0.05 percent cream or ointment 2 or 3 times daily			
Betamethasone U.S.P. *Celestone*	Betamethasone Cream U.S.P. Betamethasone Syrup U.S.P. Betamethasone Tablets U.S.P.	Topical, 0.2 percent cream applied to the skin 2 or 3 times daily	Oral, 600 μg. (0.6 mg.) to 7.2 mg. a day as a single dose or in divided doses	600 μg. to 7.2 mg. daily	Oral, 17.5 to 250 μg. (0.017 to 0.25 mg.) per kg. of body weight or 500 μg. (0.5 mg.) to 7.5 mg. per square meter of body surface a day as a single dose or in divided doses
Betamethasone Acetate U.S.P. *Celestone Soluspan (with the sodium phosphate)*	Sterile Betamethasone Sodium Phosphate and Betamethasone Acetate Suspension U.S.P.		Intra-articular, 3 to 12 mg., repeated as needed Intrabursal, 6 mg. repeated as needed Intradermal or intralesional, 1.2 mg. per square centimeter of affected skin up to a total amount of 6 mg. repeated at 1-week intervals, if necessary Intramuscular, 500 μg. (0.5 mg.) to 9 mg. a day		

(Continued)

TABLE 18-16

ANTI-INFLAMMATORY GLUCOCORTICOIDS WITH LOW SALT RETENTION *(Continued)*

Name Proprietary Name	*Preparations*	*Application*	*Usual Adult Dose**	*Usual Dose Range**	*Usual Pediatric Dose**
Betamethasone Valerate N.F. *Valisone*	Betamethasone Valerate Topical Aerosol U.S.P. Betamethasone Valerate Cream U.S.P. Betamethasone Valerate Lotion U.S.P. Betamethasone Valerate Ointment U.S.P.	Topical, aerosol containing the equivalent of 0.15 percent of betamethasone, cream containing the equivalent of 0.01 or 0.1 percent of betamethasone, or lotion or ointment containing the equivalent of 0.1 percent of betamethasone to the affected area 1 to 3 times daily			
Betamethasone Sodium Phosphate U.S.P. *Celestone Soluspan (with the acetate)*	Sterile Betamethasone Sodium Phosphate and Betamethasone Acetate Suspension U.S.P.		See Betamethasone Acetate U.S.P.		
Betamethasone Dipropionate U.S.P.	Betamethasone Dipropionate Topical Aerosol U.S.P. Betamethasone Dipropionate Cream U.S.P. Betamethasone Dipropionate Lotion U.S.P. Betamethasone Dipropionate Ointment U.S.P.				
Dexamethasone U.S.P. *Decadron, Deronil, Dexameth, Gammacorten*	Dexamethasone Elixir U.S.P. Dexamethasone Tablets U.S.P. Dexamethasone Topical Aerosol U.S.P. Dexamethasone Ophthalmic Suspension U.S.P. Dexamethasone Gel U.S.P.		Oral, 750 μg. (0.75 mg.) to 9 mg. a day as a single dose or in divided doses		Oral, 23 μg. (0.023 mg.) to 330 μg. (0.33 mg.) per kg. of body weight or 670 μg. (0.67 mg.) to 10 mg. per square meter of body surface a day as a single dose or in divided doses
Dexamethasone Sodium Phosphate U.S.P. *Decadron Phosphate, Hexadrol Phosphate*	Dexamethasone Sodium Phosphate Cream U.S.P.	Topically to the skin, as the equivalent of a 0.1 percent cream 3 or 4 times daily			
	Dexamethasone Sodium Phosphate Injection U.S.P.		Intra-articular or soft-tissue injection, the equivalent of 200 μg. to 6 mg. of dexamethasone phosphate repeated at 3-day to 3-week intervals, if necessary	Intra-articular or soft-tissue injection, 200 μg. to 6 mg. at each site once every 3 days to 3 weeks; I.M. or I.V., 500 μg. to 80 mg. daily	I.M. or I.V., 6 to 40 μg. per kg. of body weight or 235 μg. to 1.25 mg. per square meter of body surface 1 or 2 times daily

(Continued)

TABLE 18–16

ANTI-INFLAMMATORY GLUCOCORTICOIDS WITH LOW SALT RETENTION *(Continued)*

Name Proprietary Name	Preparations	Application	Usual Adult Dose*	Usual Dose Range*	Usual Pediatric Dose*
	Dexamethasone Sodium Phosphate Inhalation Aerosol U.S.P.				
	Dexamethasone Sodium Phosphate Ophthalmic Ointment U.S.P.	Topically to the conjunctiva, as the equivalent of a 0.05 percent ointment 1 to 4 times daily			
	Dexamethasone Sodium Phosphate Ophthalmic Solution U.S.P.	Topically to the conjunctiva, 0.05 to 0.1 ml. or the equivalent of a 0.1 percent solution 3 to 20 times daily			
	Dexamethasone Acetate U.S.P.	Sterile Dexamethasone Acetate Suspension U.S.P.			
Paramethasone Acetate U.S.P. *Haldrone, Stemex*	Paramethasone Acetate Tablets U.S.P.		Oral, 2 to 24 mg. a day as a single dose or in divided doses		Oral, 58 to 800 μg. (0.058 to 0.8 mg.) per kg. of body weight or 1.67 to 25 mg. per square meter of body surface a day as a single dose or in divided doses
Flumethasone Pivalate U.S.P. *Locorten*	Flumethasone Pivalate Cream U.S.P.	Topical, 0.03 percent cream			
Fluprednisolone *Alphadrol*	Fluprednisolone Tablets			Initial, 1.5 to 18 mg. daily; maintenance, 1.5 to 12 mg. daily	
Halcinonide *Halog*	Halcinonide Cream	Topically to the skin, as a 0.1 percent cream 2 or 3 times daily			
Fluocinonide U.S.P. *Lidex*	Fluocinonide Cream U.S.P. Fluocinonide Ointment U.S.P. Fluocinonide Gel U.S.P.	Topically to the skin, as a 0.05 percent cream or ointment 3 or 4 times daily			
Medrysone U.S.P. *Medrocort*	Medrysone Ophthalmic Suspension U.S.P.	Topically to the conjunctiva, 0.05 ml. of a 1 percent suspension up to 6 times daily			
Meprednisone N.F. *Betapar*	Meprednisone Tablets U.S.P.		Oral, 4 to 60 mg. a day as a single dose or in divided doses		
Amcinonide *Cyclocort*	Amcinonide Cream	Topically to affected area, as a 0.1 percent cream 2 or 3 times daily			

*See *U.S.P. D.I.* for complete dosage information.

hydroxypregn-4-en-3,20-dione 16α,17α-ketal, is the first *chloro*-glucocorticoid yet marketed. As with several other glucocorticoids (Table 18-16), it is used only topically. In one double-blind study with beta-methasone valerate cream, halcinonide was found superior in the treatment of psoriasis. However, it can

be used for the usual range of indications previously described.

Medrysone U.S.P., 11β-hydroxy-6α-methylpregn-4-en-3,20-dione, is unique among the other corticosteroids shown on Figure 18-18 in that it does not have the usual 17α,21-diol system of the others. At present it is used only for treatment of inflammation of the eyes.

Meprednisone U.S.P., 16β-methylprednisone.

Fluorometholone U.S.P., 9α-fluoro-11β,17-dihydroxy-6α-methylpregna-1,4-diene-3,20-dione.

CARDIAC STEROIDS

Poison and heart tonic—these are the "split personalities" of the cardiac steroids which have troubled and fascinated physicians and chemists for several centuries. Plants containing the cardiac steroids have been used as poisons and heart drugs at least since 1500 B.C., with squill appearing in the Ebers Papyrus of ancient Egypt. Throughout history these plants or their extracts have variously been used as arrow poisons, emetics, diuretics and heart tonics. Toad skins containing cardiac steroids have been used as arrow poisons and even as aids for treating toothaches and as diuretics.

The poison–heart tonic dichotomy continues even today. Cardiac steroids are absolutely indispensable in the modern treatment of congestive heart failure. Nevertheless, their toxicity remains a serious problem. In their review on "Clinical Correlates of the Electrophysiologic Action of Digitalis," Dreifus and Watanabe[265] noted in 1972 that digitalis steroid toxicity may account for half of drug-induced in-hospital deaths, and that when the desired therapeutic response is attained, about 60 percent of the toxic dose has been administered. Advances in biopharmaceutics have fortunately reduced the incidence of toxicity with these drugs today.

The clinical importance and toxicity of the digitalis cardiac steroids have stimulated intensive chemical and clinical studies for several decades. More recently, the cardiac steroids have been found to be potent inhibitors of sodium- and potassium-dependent ATPase (Na^+,K^+-ATPase), an enzyme which has been used as a probe of the chemistry of cell membranes by many biochemists. Na^+,K^+-ATPase inhibition is also directly or indirectly involved in the therapeutic activity of the cardiac steroids.

Many major reviews of research on the cardiac steroids' chemistry,[265-269] glycoside structures,[270-271] clinical uses and problems,[272-276] structure-activity relationships,[277-279] and pharmacology[280-284] have been published. Na^+,K^+-ATPase has also been reviewed.[280-281,285-287] These reviews are suggested for additional study.

In spite of a great amount of research, the mechanisms of the cardiac steroids' therapeutic and toxic actions have still not been completely delineated.

The cardiac steroids actually include two groups of compounds—the cardenolides and the bufadienolides. The cardenolides, illustrated by digitoxigenin in Figure 18-20, have an unsaturated butyrolactone ring at C-17, while the bufadienolides have an α-pyrone ring. Both have very similar activities and are found in a variety of plant species. However, the bufadienolides are commonly called "toad poisons" because several are found in the skin secretions of various toad species.

The $5\beta,14\beta$-stereochemistry of the cardenolides and bufadienolides gives the molecules an interesting shape, caused by the resulting A/B *cis* and C/D *cis* ring junctures. This stereochemistry appears to be an important prerequisite for some but possibly not all cardiac steroid activity. This will be discussed later in this section.

By far the most historically and commercially important sources of cardiac steroids have been two species of *Digitalis—D. purpurea* and *D. lanata.* Whole-leaf digitalis preparations appeared in the London Pharmacopeia in the 1500's but inconsistent results and common fatalities caused their removal. In 1785, William Withering published his classic, *An Account of the Foxglove and Its Medical Uses,* noting that digitalis could be used to treat cardiac insufficiency with its associated dropsy (edema).[288] Nevertheless, it was not until the early 1900's that digitalis and the purified glycosides were commonly used for the treatment of heart disease.

The cardenolides and bufadienolides are also moderately cytotoxic to some cancer cell lines.[289]

CARDIAC STEROID GLYCOSIDES

The cardiac steroids are usually found in nature as the corresponding 3β-glycosides. One to four sugars are added to the steroid 3β-hydroxyl to form the glycoside structure. Although hundreds of cardiac steroid glycosides have been found in nature,[270,271] relatively few different cardenolide or bufadienolide aglycones have been found. For example, as shown in Table 18-17, only three make up the digitalis glycosides. (The substance which forms a glycoside with a sugar is called the aglycone. Thus, the cardenolides and bufadienolides are the *aglycones* or "genins" of the *cardiac glycosides.*) The structure of a representative cardenolide glycoside, lanatoside C, is shown in Figure 18-21.

The sugars found as part of the cardiac glycosides appear in Figure 18-22. Several have been found only in cardiac glycosides. The sugar groups may be selectively removed by enzymatic, acid or base-catalyzed hydrolysis.

The Cardenolides and Bufadienolides

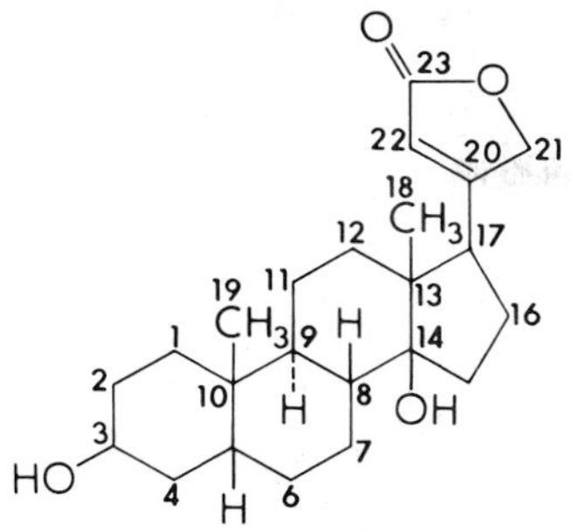

Digitoxigenin
(Cardenolide Prototype)
(Also in commercially
available products)

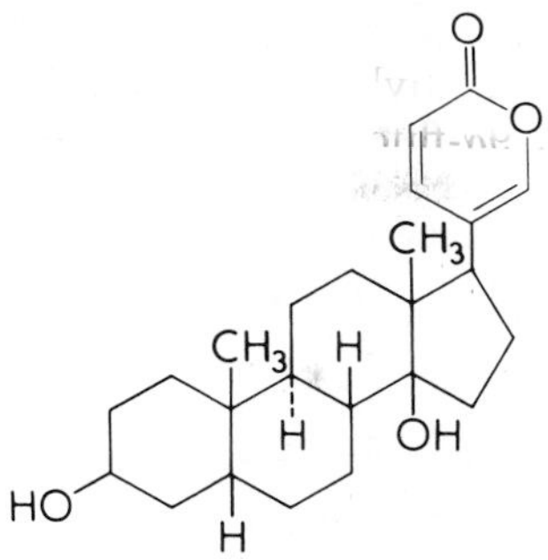

Bufalin
(Bufadienolide Prototype)

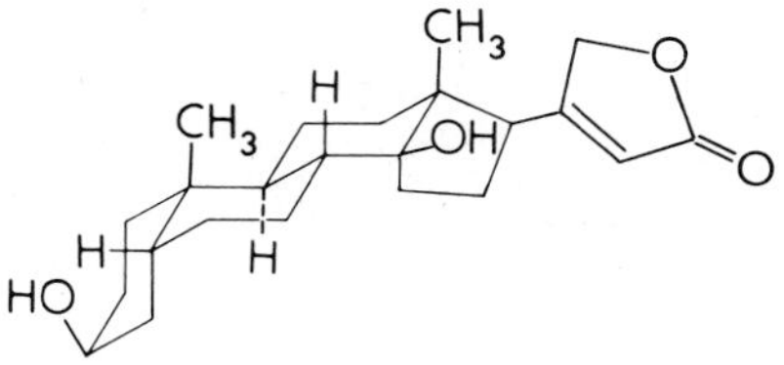

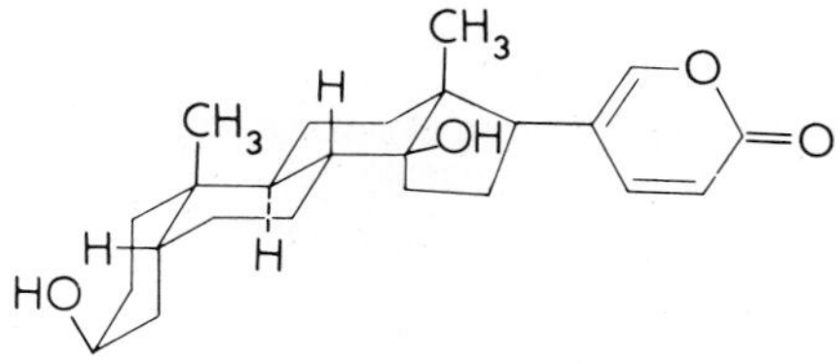

Cardenolide Aglycones
in Commercially Available Products

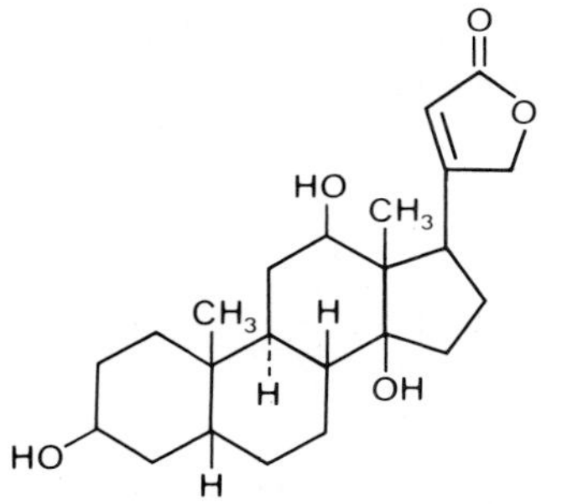

Digoxigenin

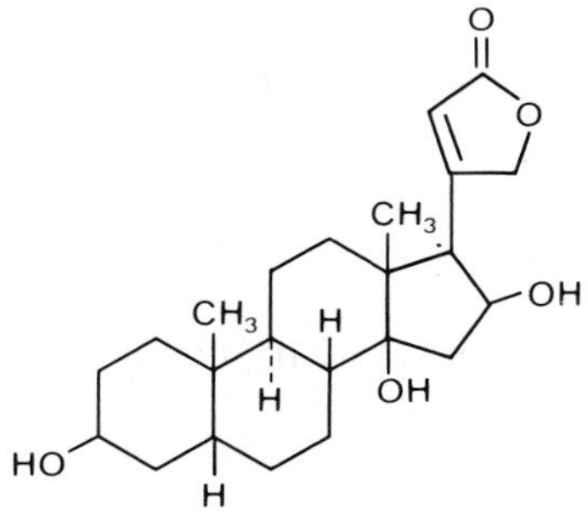

Gitoxigenin

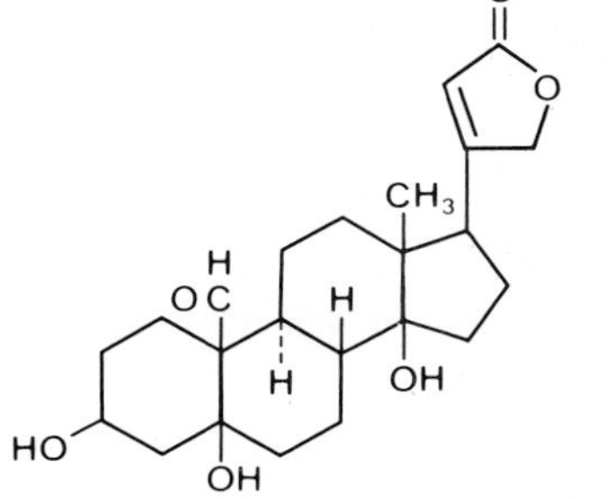

Strophanthidin

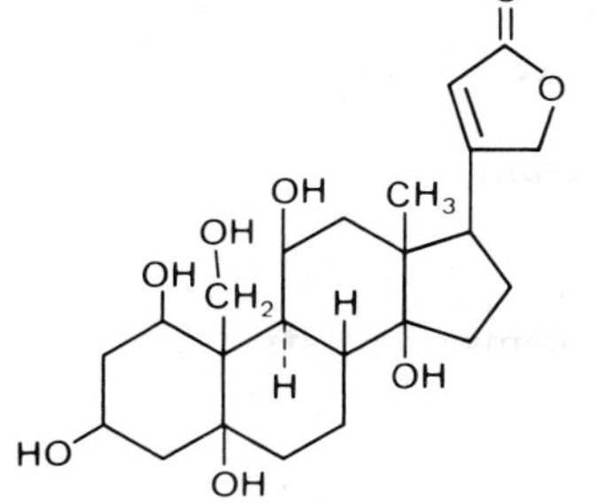

Ouabagenin

FIG. 18-20.

TABLE 18–17

CARDIAC GLYCOSIDES AND HYDROLYSIS PRODUCTS FROM COMMON SOURCES

Structure	Name
1. From *Digitalis purpurea* leaf	
Glucose-digitoxose$_3$-digitoxigenin	*Purpurea glycoside A*
Digitoxose$_3$-digitoxigenin	Digitoxin
Glucose-digitoxose$_3$-ditoxigenin	*Purpurea glycoside B*
Digitoxose$_3$-gitoxigenin	Gitoxin
2. From *Digitalis lanata* leaf	
Glucose-3-acetyldigitoxose-digitoxose$_2$-digitoxigenin	*Lanatoside A*
Glucose-digitoxose$_3$-digitoxigenin	Desacetyl lanatoside A
	(same as Purpurea glycoside A)
3-Acetyldigitoxose-digitoxose$_2$-digitoxigenin	Acetyldigitoxin
Digitoxose$_3$-digitoxigenin	Digitoxin
Glucose-3-acetyldigitoxose-digitoxose$_2$-gitoxigenin	*Lanatoside B*
Glucose-digitoxose$_3$-gitoxigenin	Desacetyl lanatoside B
	(same as Purpurea glycoside B)
3-Acetyldigitoxose-digitoxose$_2$-gitoxigenin	Acetylgitoxin
Digitoxose$_3$-gitoxigenin	Gitoxin
Glucose-3-acetyldigitoxose-digitoxose$_2$-digoxigenin	*Lanatoside C*
Glucose-digitoxose$_3$-digoxigenin	Desacetyl lanatoside C
3-Acetyldigitoxose-digitoxose$_2$-digoxigenin	Acetyldigoxin
Digitoxose$_3$-digoxigenin	Digoxin
3. From *Strophanthus gratus* seed	
Rhamnose Ouabagenin	Ouabain

CARDIAC EXCITATION AND CARDIAC INNERVATION

If one is to understand the actions of drugs upon the heart, an understanding of cardiac physiology and cardiac electrical impulses is essential. Only a very brief discussion will be presented here. The concise, well-illustrated discussion of William F. Ganong in *Review of Medical Physiology* (Lange Medical Publications, 1975) is recommended for further study.

Adrenergic sympathetic nerves which stimulate the heart can increase heart rate (a chronotropic effect) and increase the force of contraction (an inotropic

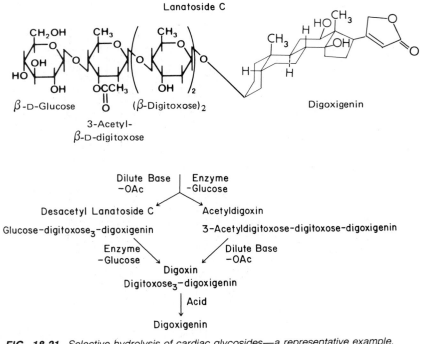

FIG. 18-21. *Selective hydrolysis of cardiac glycosides—a representative example.*

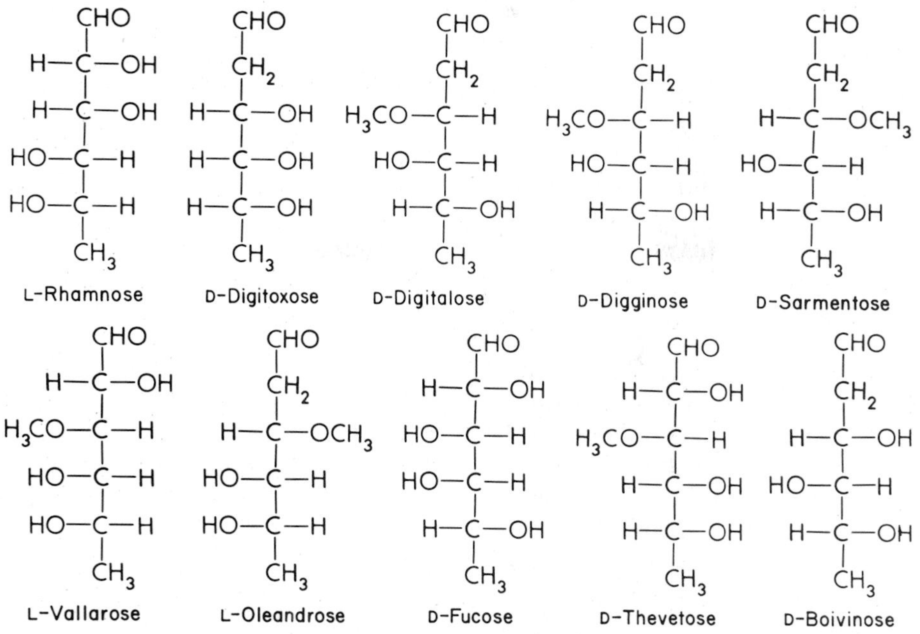
L-Rhamnose
D-Digitoxose
D-Digitalose
D-Digginose
D-Sarmentose
L-Vallarose
L-Oleandrose
D-Fucose
D-Thevetose
D-Boivinose

FIG. 18-22. Cardiac glycoside sugars.

effect). Since the digitalis steroids also have an inotropic effect, it is important to note that their mode of action is quite unlike that of the catecholamines, for example. Further, the heart is also innervated by cholinergic vagal fibers which decrease heart rate (also a chronotropic effect). When the vagal fibers are cut or chemically blocked with parasympatholytic drugs, the heart rate increases markedly.

Atrial contraction (atrial systole) precedes ventricular contraction (ventricular systole). Figure 18-23 shows the major parts of the cardiac electrical conduction system. The sino-atrial (SA) node is a specialized muscle which is self-excitatory and sends out depolarization impulses rhythmically. Depolarization initiated at the SA node spreads through the atria, reaching the atrioventricular (AV) node. The AV node is also a pacemaker tissue, and after a short delay, sends an excitation wave through the bundle of His, through the Purkinje system, to the ventricular muscle. Since the rate of conduction in all these tissues is different, the *direction* of stimulation in the heart can be controlled as well as the critical sequence of ventricles contracting after their respective atria. The direction and sequence of these waves of depolarization (i.e., impulses) can be recorded on electrocardiograms.

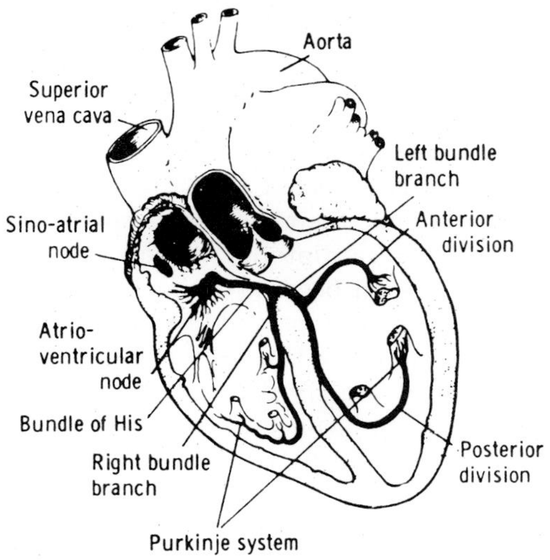
Aorta
Superior vena cava
Left bundle branch
Anterior division
Sino-atrial node
Atrio-ventricular node
Bundle of His
Right bundle branch
Posterior division
Purkinje system

FIG. 18-23. Major parts of the cardiac electrical conduction system of the heart. (From Goldman, M. L.: Principles of Clinical Electrocardiography, Los Altos, Cal., Lange, 1970)

CLINICAL AND PHYSIOLOGICAL ACTIONS

Although there is great uncertainty about the cardiac steroids' mechanism of inotropic action, their clinical and physiologic actions are well delineated. The cardiac steroids have a great number of effects upon the heart, but by far their most important clinical effect is to increase the force of myocardial contraction. This is commonly called the inotropic action of the cardiac steroids, and the resultant effects are increased cardiac output, diuresis and reduction of edema, and decreased heart size.

The importance of the inotropic action on the failing heart cannot be overemphasized. A failing heart does not pump blood efficiently enough to meet the body's needs for oxygen and nutrients. There is an increase in blood volume and subsequent edema caused by sodium and water retention by the kidneys. Sympathetic nervous system stimulation of the heart increases, leading to tachycardia, vasoconstriction and sweating. These are largely compensatory mechanisms of the body to counterbalance the effects of the inefficient and poorly pumping heart. Eventually, pulmonary edema results and, if left untreated, it results in death.

The cardiac steroids cause the heart to contract more strongly *and efficiently,* i.e., with a resulting reduction in heart size and, secondarily, a reduction in myocardial oxygen consumption. In addition, the cardiac steroids have another important effect on the failing heart—they slow the heart. This is the so-called chronotropic effect. Since (as mentioned above) ventricular tachycardia is common during heart failure, this dual action of the cardiac steroids is of immense help to the patient.

The chronotropic effect is the result of a number of actions of the cardiac steroids. First, the inotropic action causes cardiac output to increase, thus decreasing the compensatory sympathetic nervous stimulation indirectly. Second, digitalis increases the sensitivity of the heart to vagal stimulation. Third, atrioventricular (AV) node conductivity is decreased. During atrial tachycardia or fibrillation, this decrease will mean that fewer atrial impulses will activate the AV node. Fourth, the sino-atrial (SA) node is depressed by the cardiac steroids.

Secondary effects of the cardiac steroids include diuresis due to more normal cardiac output, not to any direct effect on the renal tubules, and a decrease in venous blood pressure as extracellular fluid decreases.

Additional details about these effects of the cardiac steroids can be found in the excellent discussion of Moe and Farah.[273]

Toxic symptoms of the cardiac steroids include gastrointestinal upset and nausea at lower toxic doses. Mental changes, visual disturbances, hypokalemia, abdominal pain and various arrhythmias result at higher toxic levels. In the extreme, the drugs cause ventricular fibrillation, systolic arrest and death.

CARDIOTONIC ACTIVITY AND TOXICITY

Before discussing the hypotheses concerning the mechanism of inotropic activity, it is important to review the history of cardiac steroid testing.

Until about 1970, it was generally believed that the toxic effects of the cardiac steroids were simply a physiologic extension of their therapeutic effects.

Stated simply, too much increase in cardiac muscle contractility was assumed to be ventricular arrest. Anesthetized cats were used to evaluate the therapeutic potential of natural and semisynthetic cardenolides and bufadienolides. Cardiac arrest (and death of the cat) caused by a slow infusion of the drug in precisely measured amounts was used as the determinant of cardiotonic activity.[270,271,290]

Two decades of structure-activity studies were based on toxicity data. At first glance it may seem absurd that efforts to design "better" cardiac drugs were based on measuring toxicity rather than efficacy. However, it must be noted that Brown, Stafford and Wright[291] found a striking parallel between toxicity in the cat, guinea pig and chick and the directly measured inotropic activity of some common cardenolides and cardenolide glycosides.

In the late 1960's, it was found that the cardiac steroids were potent inhibitors of Na^+,K^+-ATPase, and there was significant data to indicate that this inhibition may be the determinant of inotropic activity. Inhibition of Na^+,K^+-ATPase then became a much-used method of evaluating cardiotonic activity. However, in 1970, M. E. Wolff of the University of California reported finding a number of compounds which were highly Na^+,K^+-ATPase-inhibiting and toxic, but completely devoid of any inotropic activity.[292]

As shown in Figure 18-24, there are many compounds which have been found to have partially selective activities. Especially interesting are the structures of amrinone,[293] anthopleurin[294] (a 49 amino acid peptide obtained from a northwest sea anemone) and the progestins chlormadinone acetate and medroxyprogesterone acetate.[295,296,297] Amrinone has been of great interest, but its thrombocytopenia is viewed as a barrier to eventual marketing. Anthopleurin is a peptide and is expected to cause allergenic responses with some patients. The protestins' Na^+,K^+-ATPase inhibiting activity has just recently begun to be studied, but it is expected that their progestational activity would limit their use with heart patients.

A variety of "cardiotonic" agents have aroused considerable interest because of the claim (or hope) of decreased toxicity or improved pharmacokinetic properties that would make them safer to use. Several are shown in Fig. 18-25. Additional investigations are continuing, but no compound has yet been marketed in the U.S. Claims of decreased toxicity of actodigin have been questioned by Thomas and Schwartz,[300] and Cagin et al.[301] Fullerton, Rohrer, and co-workers found that actodigin genin behaved exactly like other genins in vitro.[302]

Inotropic activity is now measured directly.[275,303] For example, cat, rabbit or guinea pig hearts are

1. Inotropic and Na⁺,K⁺–ATPase-inhibiting
   Cardenolides
   Bufadienolides
   Cardenolide 3–bromoacetates

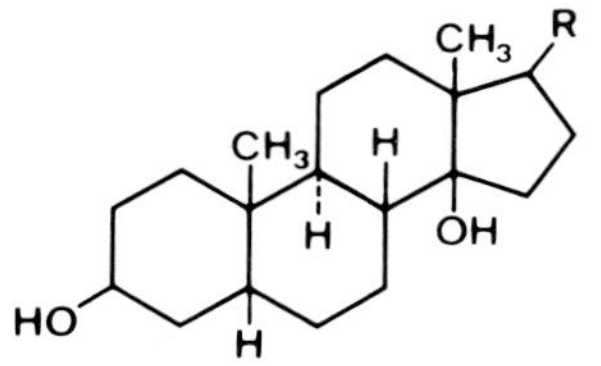

Active: R= COOCH₃

C≡N

CHO

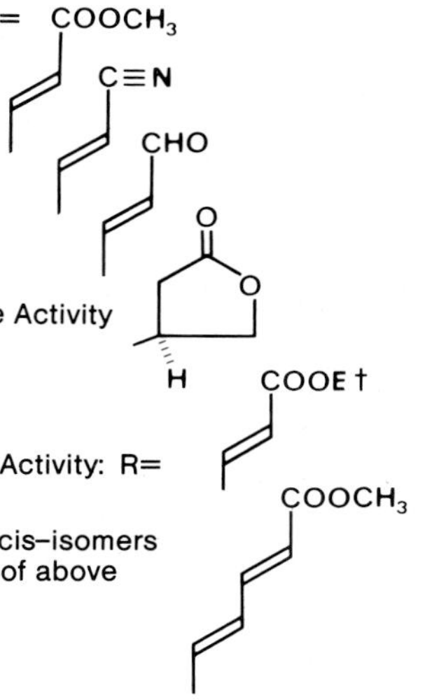

Moderate Activity

Marginal Activity: R=

COOE †

COOCH₃

Inactive: cis–isomers
of above

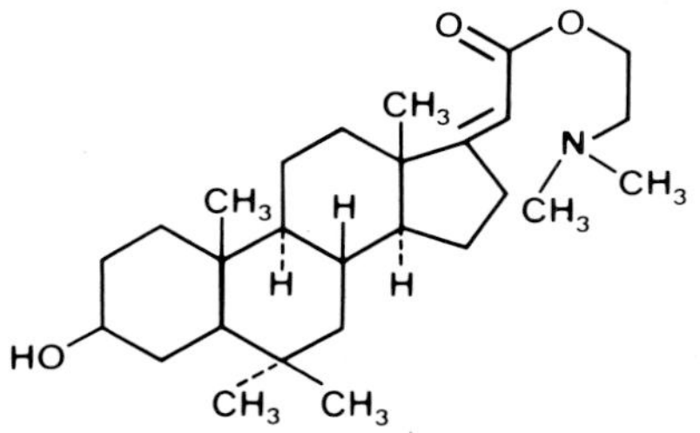

Cassaine

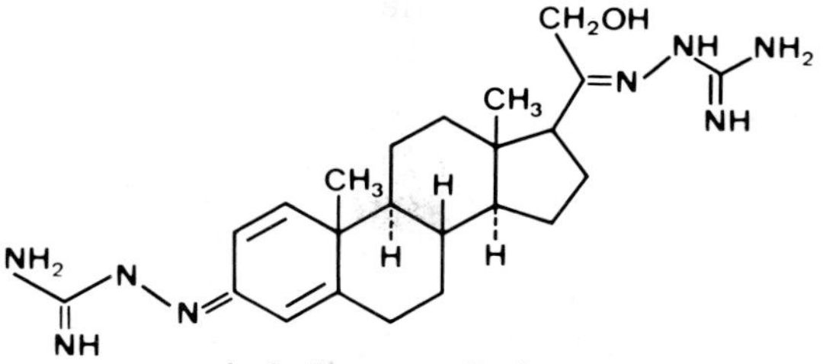

And other guanylhydrazones

* 2. Inotropic but not Na⁺,K⁺–ATPase-inhibiting
   Catecholamines (immediate but short-acting)
   Caffeine, veratrum alkaloids, etc. (undesirable
   side-effects or toxic at chronic dose levels)
   Anthopleurin

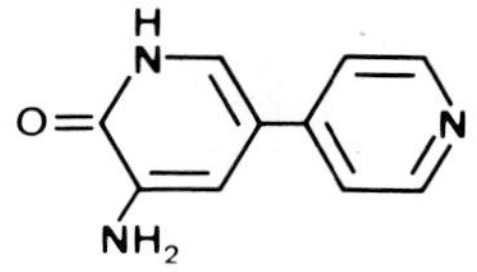

Amrinone

† 3. Na⁺,K⁺–ATPase-inhibiting but not inotropic
   Sodium azide
   –SH blocking reagents
   Mersalyl
   Fatty acids
   Disopropylfluorophosphate

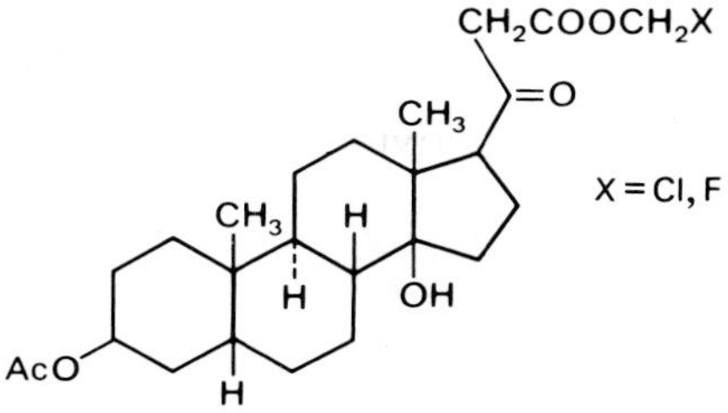

X = Cl, F

And other steroid alkylating agents[292]

4. Na⁺,K⁺–ATPase-inhibiting, inotropic
   Activity uncertain
   Chlormadinone acetate

*Other drugs, including chlorpromazine and quinidine, have been described as belonging to this group. However, T. M. Brodie and co-workers (Ann. N.Y. Acad. Sci. 242:527, 1974) have found chlorpromazine's inotropic effect is blocked by B-adrenergic blockers. Quinidine has also been considered to have a negative inotropic effect due to its general depressant effects on the myocardium.
†Some or all may be so enzyme- or membrane-destructive in vivo that any muscle contraction is not possible. For that reason, this category can not be used as solid evidence against the hypothesis that Na⁺,K⁺,–ATPase inhibition is directly related to inotropic action in vivo.

FIG. 18-24. *Examples of compounds with some digitalis activities.*

excised, and strips of the atria or ventricles are attached to recording devices in nutrient media. After the preparations stabilize (usually one hour or less), the cardiac steroids being tested are added to the infusion medium. Direct electronic stimulation of the heart tissue causes the tissues to contract, and the amount of contraction is directly measured on a recording device. Studies are also made utilizing electrocardiograms (ECG) and other recording devices with excised whole hearts.

Dog "heart-lung" preparations are also used to determine cardiac output as a measure of over-all cardiotonic effects.[219]

POSSIBLE MECHANISMS OF ACTION

The great toxicity of the cardiac steroids has been a major stimulus in studying their mechanisms of action. If inotropic activity and toxicity are caused by

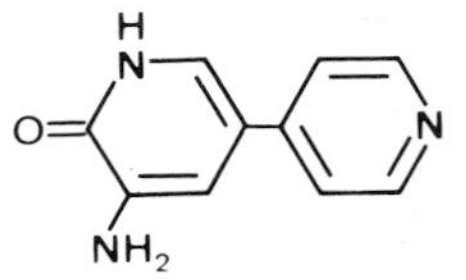

Amrinone[293]

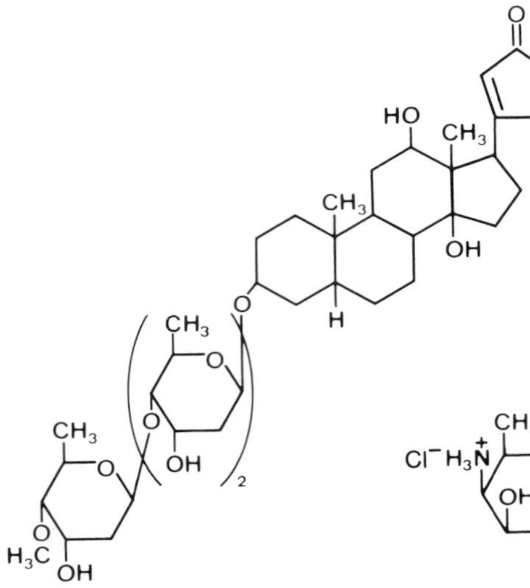

β–Methyldigoxin[298]

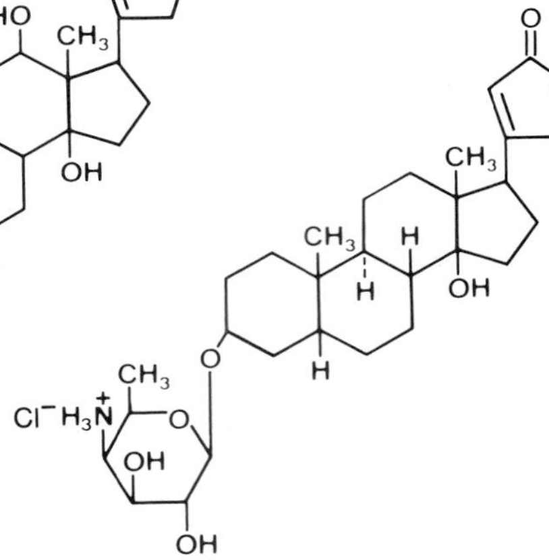

ASI–222[299]

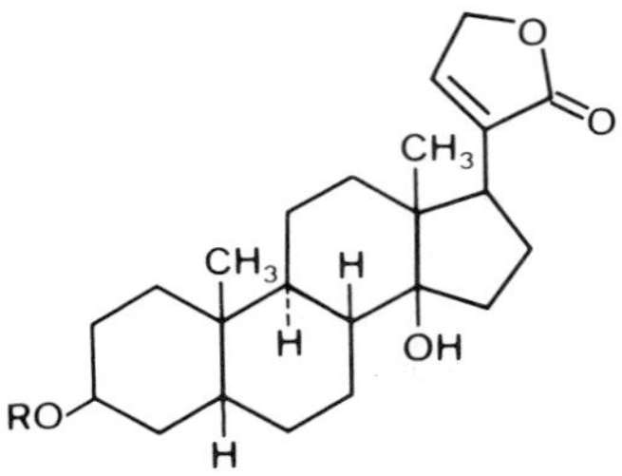

Actodigin (AY22,241)[300-302]
R=Glucose

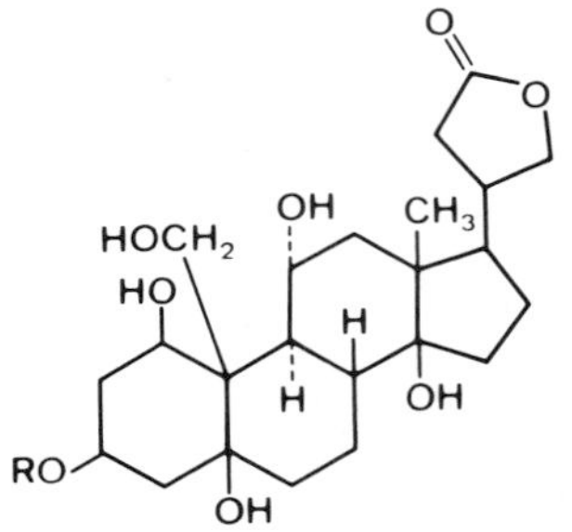

20,22–Dihydro–Ouabain[303]
R=Rhamnose

FIG. 18-25. *Compounds which might have decreased or more manageable toxicity.*

different mechanisms, then it should be possible to design safer inotropic steroids. Unfortunately, our understanding of the cardiac steroids' mechanisms of action has been impeded by the lack of direct inotropic testing data for many compounds. There are also many technical difficulties in measuring intracellular Ca^{++} movement, and in studying Na^+, K^+-ATPase.

It is known that cardiac steroids inhibit sodium- and potassium-dependent ATPase (transport ATPase, Na^+, K^+-ATPase).[285,286,287,305] This remarkable enzyme plays many key roles in the body (in heart contraction, in ion balance, and in nerve transmission). As elegantly reviewed by Schwartz,[304] Na^+, K^+-ATPase consists of at least two subunits (the α unit, molecular weight or mw about 100,000; and the β unit, mw about 45,000) and possibly a third and fourth (γ_1 and γ_2, each about mw 12,000). The α unit goes from the outside of the membrane all the way to the inside, a distance in some cells of 80 Å. The digitalis (ouabain) binding site is on the outer surface of the α unit, and the ATP binding site is on the inner surface.

Na^+, K^+-ATPase is responsible for maintaining the unequal distribution of Na^+ and K^+ ions across cell membranes. Na^+ is maintained in higher concentration in the extracellular fluid, while K^+ is in higher concentration inside the cell. When a wave of depolarization passes through the heart, there is a change in the permeability of the heart cell membranes. Na^+ quickly moves into the cell by passive diffusion and K^+ moves out. After the heart "beats," the process must be reversed, i.e., K^+ pumped against a concentration gradient into the cell, and Na^+ against a concentration gradient out of the cell. This process (shown in Fig. 18-26) is commonly called the "sodium

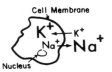

FIG. 18-26. *Movements of ions with the "Sodium pump." Sodium- and potassium-dependent ATPase catalyzes the pump. The hydrolysis of ATP to ADP provides the needed energy to move the ions against concentration gradients.*

pump," and is catalyzed by Na^+,K^+-ATPase. Since ions are being moved against concentration gradients, it is not surprising that energy is required. The hydrolysis of ATP provides the needed energy.

Na^+,K^+-ATPase operates in all cell membranes to maintain the unequal distribution of Na^+ and K^+ ions across the membrane. However, since there is rapid ion "pumping" between each beat of the heart, inhibition of Na^+,K^+-ATPase has the greatest effect on heart tissue.

There have been many discussions and even debates in the literature on the relationship between Na^+,K^+-ATPase inhibition, inotropic activity and toxicity.[200,203-205,210] The most fundamental question is: Is Na^+,K^+-ATPase inhibition the cause of inotropic activity? The most probable answer is "yes," but many questions remain to be answered. The possible relationships of changes in intracellular Na^+, K^+, Ca^{++}, and c-AMP with Na^+,K^+-ATPase inhibition have been reviewed in 1980 by Schwartz[304] and in 1978 by Akera and Brody.[305]

STRUCTURE-ACTIVITY RELATIONSHIPS

Until it was discovered that some toxic and Na^+,K^+-ATPase-inhibiting compounds were *not* inotropic, many relatively simple structure-activity relationships were "recognized" for the cardiac steroids. (Detailed reviews have been published by Thomas,[277,316] Gunter and Linde[279,] and by Flasch and Heinz.[280]) The lactone ring, the 20(22)-lactone double bond, the 14β-OH—all were considered "essential." As mentioned before, many of these concepts were based almost entirely on toxicity data.

THE GENIN PORTION

Many often-conflicting models have been proposed over the last two decades to describe structural and geometric features that govern the ability of a particular digitalis genin (cardenolide) or bufalin genin (bufadienolide) to inhibit Na^+,K^+-ATPase (or cause an inotropic response). Examples are shown in Fig. 18-27. Since hydrogenation of the side group carbon–carbon double bond was known to decrease activity significantly, an active role in Na^+,K^+-ATPase inhibition was generally envisioned for this double bond. (C-20 also changes from sp^2 to sp^3, greatly moving the side group carbonyl.)

Kupchan and co-workers[307,308] proposed a Michael reaction between a nucleophile on the enzyme (e.g., SH, NH_2) and ene-one part of the side group. Thomas and associates have suggested a similar mechanism

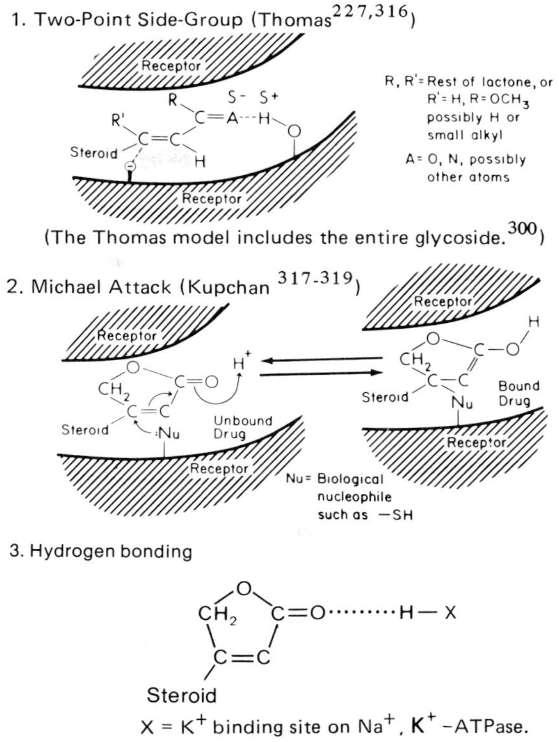

1. Two-Point Side-Group (Thomas[227,316])

R, R'= Rest of lactone, or R'= H, R= OCH_3 possibly H or small alkyl

A= O, N, possibly other atoms

(The Thomas model includes the entire glycoside.[300])

2. Michael Attack (Kupchan[317-319])

Nu= Biological nucleophile such as —SH

3. Hydrogen bonding

Steroid

X = K^+ binding site on Na^+, K^+–ATPase.

FIG. 18-27. *Proposed models of cardenolide-receptor binding.*

but did not include actual covalent binding.[277,300,306] Portius and Repke[310-312] thought that hydrogen bonding between the side group and the K^+ binding site on Na^+,K^+-ATPase determined the degree of Na^+,K^+-ATPase inhibition, and that the molecule's dipole was thus quite important. Other models have been proposed by Glynn[313] and by Hoffman.[314]

Fullerton, Rohrer, Ahmed, From and co-workers found that a number of genins they synthesized and tested did not fit these models.[315-317] This led them to a detailed study of cardenolide and bufadienolide genins with the use of the National Institutes of Health PROPHET computer system.[315,317]

Using PROPHET, the most stable conformations of each of a variety of genins were graphically superimposed with the cardenolide prototype, digitoxigenin. An example is shown in Figure 18-28. PROPHET was then used to measure distances between each superimposed analog and the corresponding atoms in digitoxigenin. They found[315] that the position of a particular genin's carbonyl oxygen (or nitrile nitrogen) relative to digitoxigenin's was a nearly perfect predictor of its activity (Fig. 18-29).

$$\log I_{50} = 0.44D - 6.36 \qquad r^2 = 0.98$$

where I_{50} = amount of genin required to inhibit 50 percent of a standard guinea-pig brain Na^+,K^+-ATPase preparation in vitro.

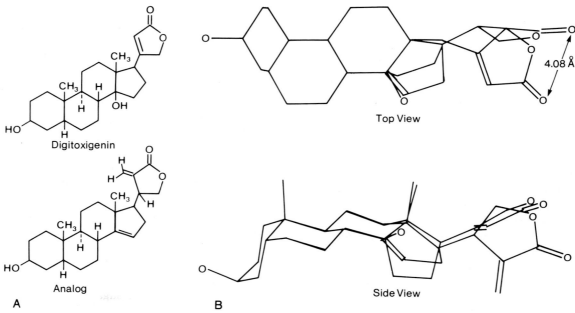

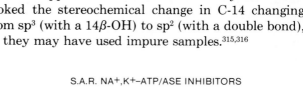

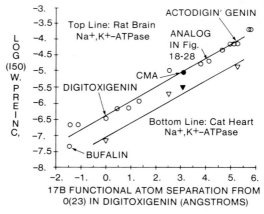

FIG. 18-28. *(A) Digitoxigenin and its 14-ene, 22-methylene analog. (B) PROPHET superposition of digitoxigenin and its analog, both in their most stable C17-side group conformations, showing distance between carbonyl oxygens; as shown in Fig. 18-29, as the distance between carbonyl oxygens increases, activity (NA⁺, K⁺-ATPase inhibition) decreases. With the superimposed pair above, it is 4.08 Å. (For additional drawings and discussion see Fullerton, Rohrer, et al.[315,317])*

"D" equals distance between carbonyl oxygens of the analog and digitoxigenin in angstroms. This relationship means that for every 2.2 Å the carbonyl oxygen (or nitrile N, for one compound) moves, activity changes by one order of magnitude (10 or 0.1).[315]

The "r^2" is the statistical correlation. (Perfect would be 1.00.) As shown in Figure 18-29, a corresponding relationship was found for genins tested in a cat heart Na⁺,K⁺-ATPase system.[318]

The genins graphed in Figure 18-29 included bufalin, cardenolide analogs with and without a 14β-OH, with acyclic side groups such as esters and a nitrile, and also included the progesterone derivative chlormadinone acetate (CMA) (see Fig. 18-7). From these studies the following conclusions may be made about digitalis genin-structure activity relationships.

A-B-C-D ring system. Both cardenolide/bufadienolide or pregnene ring systems appear to be able to fit the "digitalis" (ouabain) binding site on Na⁺,K⁺-ATPase. The activity-determining requirement appears to be primarily that the side group carbonyl oxygen (or nitrile N) be in the right spot relative to the A-B-C-D rings for maximal activity. It is not yet known what structural feature on Na⁺,K⁺-ATPase causes this requirement.

14β-OH. Not necessary for activity. Some compounds that did not have a 14β-OH have been found

to be more active than otherwise identical genins that did.[316] It appears that earlier studies may have overlooked the stereochemical change in C-14 changing from sp³ (with a 14β-OH) to sp² (with a double bond), or they may have used impure samples.[315,316]

S.A.R. NA+,K+-ATP/ASE INHIBITORS

FIG. 18-29. *Relationship of cardiac-steroid-genin-carbonyl-oxygen position to inhibition of Na+,K+-ATPase (Fullerton, Rohrer, Yoshioka, Kitatsuji, Deffo, From, and Ahmed, unpublished data) Circles: Genins studied with rat brain Na+,K+-ATPase*
Triangles: Genins studied with cat heart Na+,K+-ATPase. Triangles directly under the circles represent the same genin. CMA = Chlormadinone Acetate

Lactone ring. Not necessary, as first shown by Thomas and co-workers[306] (see Fig. 18-24).

Side group carbon–carbon double bond. It appears that this double bond may just be keeping the side group's carbonyl oxygen in the right place for most efficient Na^+,K^+-ATPase inhibition.

STRUCTURE AND PARTITION COEFFICIENT

As shown in Table 18-18, commercially available cardiac steroids differ markedly in their degree of absorption, half-life and time to maximal effect. In most cases this is due to polarity differences caused by the number of sugars at C-3 and the presence of additional hydroxyls on the cardenolide.

Nevertheless, it has always been difficult to visualize how apparently minor structural variations can cause major differences in partition coefficient and absorption. For example, lanatoside C and digoxin differ only by the presence of an additional sugar in lanatoside C (Table 18-17). One might expect that "one more sugar shouldn't make much difference." However, the $CHCl_3$/16 percent aqueous MeOH partition coefficients for the compounds are very different indeed: 16.2 for lanatoside C, 81.5 for digoxin, 96.5 for digitoxin and 10 for gitoxin.[322]

The compounds with increased lipid-solubility also are the slowest to be excreted. Additional hydroxyl groups in the more polar compounds provide additional sites for conjugate formation and other metabolic processes. In addition, it is commonly known that more lipid-soluble drugs tend to be excreted more slowly due to increased accumulation in lipid tissues.

THE SUGAR PORTION

Sugars at C-3 are not necessary for high inotropic and Na^+,K^+-ATPase inhibiting activities. Digitoxigenin, digoxigenin, and other genins are very active in vitro, but slightly less active than their β-D-digitoxose or α-L-rhamnose glycosides. In vivo, sugars attached to the C-3-OH protect the C-3-OH from being rapidly metabolized to the 3-one and then reduced to the less active 3α-OH isomer.

Most research on the roles of glycoside sugars has been done by Yoda and Yoda and co-workers.[319,320] They propose that the genin portion binds first to the receptor site on Na^+,K^+-ATPase. This is believed to cause a conformational change in the Na^+,K^+-ATPase, which activates the sugar-binding site. Finally, the sugar or sugars bind. All the naturally occurring glycosides bind reversibly but often dissociate so slowly that they appear to have bound irreversibly. The corresponding genins dissociate much faster.

Yoda and Yoda have also found that some sugars attached to the C-3-OH bind well (e.g., β-D-digitoxose and α-L-rhamnose) but others do not (e.g., 6-deoxyglucose). They believe that the C-3'-OH on the sugar directly attached must be axial to bind well, as with digitoxose. If it is equatorial, as with glucose, binding is poor. Further research will help delineate and explain fully these structural requirements for the sugar-binding site.

TABLE 18–18

CARDIAC GLYCOSIDE PREPARATIONS*

Agent	Gastrointestinal Absorption	Onset of Action † (min.)	Peak Effect (hr.)	Average Half-life ‡	Principal Metabolic Route (Excretory) Pathway	Average Digitalizing Dose		Usual Daily Oral Maintenance Dose §
						Oral**	I.V.***	
Ouabain	Unreliable	5–10	0.5–2	21 hr.	Renal; some G.I. excretion	—	0.3–0.5 mg.	—
Deslanoside	Unreliable	10–30	1–2	33 hr.	Renal	—	0.8 mg.	—
Digoxin	55–75%	15–30	1.5–5	36 hr.	Renal; some G.I. excretion	1.25–1.5 mg.	0.75–1.0 mg.	0.25–0.5 mg.
Digitoxin	90–100%	25–120	4–12	4–6 days	Hepatic; renal excretion of metabolites	0.7–1.2 mg.	1.0 mg.	0.1 mg.
Digitalis leaf	About 40%	—	—	4–6 days	Similar to digitoxin	0.8–1.2 g.	—	0.1 g.

*Table from Smith, T. W., and Haber, E.: New Eng. J. Med. 289:1063, 1973. Used by permission.
† For intravenous dose.
‡ For normal subjects (prolonged by renal impairment with digoxin, ouabain and deslanoside and probably by severe hepatic disease with digitoxin and digitalis leaf.
** Divided doses over 12 to 24 hours at intervals of 6 to 8 hours.
*** Given in increments for initial subcomplete digitalization; supplement with additional increments p.r.n.
§ Average for adult patients without renal or hepatic involvement; varies widely among patients and requires close medical supervision.

CLINICAL ASPECTS OF DIGITALIS THERAPY

In addition to knowing pharmacokinetic differences among the commonly available cardiac steroid products (Table 18-18), it is essential that the pharmacist and physician understand all aspects of digitalis therapy so that needless deaths due to digitalis overdoses may be avoided. Basic principles are presented here, but it is important to carefully study more complete discussions—e.g., the excellent article by Jelliffe[276] and the text by Niles, Melmon and Morelli.[274] The guide of Gerbino, "Digitalis Glycoside Intoxication—a Preventative Role for Pharmacists,"[321] is essential for all pharmacists to study.

The essential features of digitalis therapy are:

1. *A "Loading" or "Digitalizing" Dose.* The potent activity of the digitalis steroids combined with their potential for toxicity makes selection of individual doses more complicated than for almost any other drug. Most important, one must carefully consider the renal function of the patient, since much of the dose is excreted in the urine. In patients with normal renal function, the average half-life of digoxin is much shorter than digitoxin (Table 18-18). "Approximately 35 percent of digoxin in the body is excreted per day—but only 10 percent of the digitoxin. . . . However, no matter what loading dose is chosen, over a long period (exceeding five half-lives) the final concentration of drug in the body is determined by the daily maintenance dose. This has led some authors to advocate digitalization without a loading dose."[274]

2. *A Maintenance Dose.* As with the "loading dose" (if used at all), the maintenance dose must be carefully tailored to each individual patient. Average doses must not be used without accounting for individual patient variables—kidney function, age, potential drug interactions, presence of heart, thyroid or hepatic disease. (Gerbino lists several other factors which should be considered.[321]) Jelliffe has developed a computer program for selecting digitalis doses for individual patients—and in using the program, hospitals have reduced the incidence of toxic reactions by over 60 percent.[276]

   The pharmacist and the patient can perform key roles in detecting toxic symptoms in patients. Table 18-19 shows many common noncardiac symptoms of possible digitalis intoxication. Changes in heart rate or rhythm, often very obvious to the patient, are probably the most important signs of digitalis excess or deficiency.

3. *Avoiding or Controlling Drug Interactions.* Hypokalemia, such as that brought about by diuretics, or hyperkalemia can cause cardiac arrhythmias—

TABLE 18-19

NONCARDIAC SYMPTOMS OF DIGITALIS TOXICITY[321]

Symptoms	Frequency	Manifestations
Gastrointestinal	Most common	Anorexia, nausea, vomiting, diarrhea, abdominal pain, constipation
Neurological	Common	Headache, fatigue, insomnia, confusion, vertigo
	Uncommon	Neuralgias (especially trigeminal), convulsions, paresthesias, delirium, psychosis
Visual	Common	Color vision, usually green or yellow; colored halos around objects
	Uncommon	Blurring, shimmering vision
	Rare	Scotomata, micropsia, macropsia, amblyopias (temporary or permanent)
Miscellaneous	Rare	Allergic (urticaria, eosinophilia), idiosyncrasy, thrombocytopenia, gastrointestinal hemorrhage, and necrosis

From Gerbino.[321] Copyright © 1973, American Society of Hospital Pharmacists, Inc. All rights reserved. Used with permission.

arrhythmias which can also be caused by digitalis. Calcium and digitalis glycosides are synergistic in their actions on the heart.

Many drugs can affect absorption of digitalis, for example, cathartics and neomycin. Protein binding can be disturbed by coumarin anticoagulants, phenylbutazone and some sulfonamides. However, these drug interactions are much less common than those involving potassium or calcium.

4. *Prompt Treatment of Digitalis Toxicity,* if it occurs, by the physician. Most important, digitalis therapy must be stopped until symptoms are under control. Supplemental K^+ therapy is sometimes necessary as well.

BIOAVAILABILITY OF THE DIGITALIS GLYCOSIDES

The great potency of the digitalis glycosides and their high toxicity have made bioavailability information more important for these compounds than for almost any other drug in use today. Butler,[323] Smith[324] and Saki[325] have developed radioimmunoassay techniques to precisely measure extremely small digitalis serum concentrations. Radioimmunoassay kits also are available commercially[326,327] as well as nonradioactive "Emit" type systems. Many studies have shown a strong correlation between the one-hour dissolution rate of digoxin tablets and measured serum concentrations.[328-334] The F.D.A. has issued recommendations for in vivo bioavailability tests for manufacturers of digitalis glycoside tablets.[346] In addition, manufacturers are required to perform in vitro U.S.P. dissolution tests[336] on tablets from each batch. All marketed products must fall within the range of 55 to 95 percent for one-hour dissolution. Any product below 55 percent will be removed from the market and any above 95 percent will require an Investigational New Drug and New Drug Application to be filed before marketing. Jelliffe[276] and Smith and Haber[272] presented additional information on evaluating bioavailability of digitoxin and digoxin; this is important reading for anyone monitoring drug therapy of cardiac patients.

PRODUCTS

The most important information on the digitalis steroids appears in Table 18-18. Structural differences are shown in Figure 18-20 and Table 18-17. Any additional pertinent information will be given below. The previous sections on digitalis toxicity and therapy should be carefully studied as well.

Powdered Digitalis U.S.P. is the dried, powdered leaf of *Digitalis purpurea*. When digitalis is prescribed, powdered digitalis is to be dispensed. One hundred mg. is equivalent to one U.S.P. Digitalis Unit, used as a relative measure of activity in pigeon assays. Powdered digitalis contains digitoxin, gitoxin and gitalin, of which digitoxin is usually in highest concentration. Because of the significant presence of digitoxin, powdered digitalis has a slow onset of action and long half-life (see Table 18-18). The long half-life makes toxic symptoms more difficult to treat than with cardiotonic steroids with shorter half-lives.

Digoxin U.S.P., because of its moderately fast onset of action and relatively short half-life (Table 18-18), has become the most frequently prescribed digitalis steroid. It is a *Digitalis lanata* glycoside of *digoxigenin* (Fig. 18-20), 3β,12β,14β-trihydroxy-5β-card-20(22)-enolide. Digoxin was first isolated by Smith,[337] in 1930. It may be given orally, intravenously or intramuscularly (into deep muscle, followed by firm massage).

Digitoxin and digoxin are the most frequently prescribed digitalis steroids, and Jelliffe[276] has published a superb comparison of their properties including, in part, the following: digoxin is more rapidly excreted and therefore is also more rapidly cumulative in the presence of impaired renal function. Clinical changes due to changing maintenance dose are quickly observed. Digitoxin is excreted more slowly and cumulates more slowly. Since it is more slowly excreted, its kinetics are less affected by renal function. It has better absorption (bioavailability) than digoxin, and therefore probably is more reproducible. The rapid excretion of digoxin is certainly useful when

TABLE 18–20

DIGITALIS PRODUCTS*

Name Proprietary Name	Preparations
Digitalis U.S.P. (*Note:* When digitalis is prescribed, Powdered Digitalis U.S.P. is to be dispensed.)	The dried leaf of *Digitalis purpurea*—not used therapeutically until powdered
Powdered Digitalis U.S.P. *Digitora, Digifortis, Pil-Digis*	Digitalis Capsules U.S.P. Digitalis Tablets U.S.P. Digitalis Tincture
Digitalis Purpurea Glycosides *Digiglusin, Gitaligin*	Tablets Injection (Each tablet or 1 ml. of injection is equivalent to 1 U.S.P. Unit, i.e., 100 mg. of digitalis.)
Digoxin U.S.P. *Lanoxin*	Digoxin Elixir U.S.P. Digoxin Injection U.S.P. Digoxin Tablets U.S.P.
Digitoxin U.S.P. *Crystodigin, Purodigin*	Digitoxin Injection U.S.P. Digitoxin Tablets U.S.P.
Acetyldigitoxin U.S.P.	Acetyldigitoxin Tablets U.S.P.
Ouabain U.S.P.	Ouabain Injection U.S.P.
Lanatoside C *Cedilanid*	Tablets
Deslanoside U.S.P. *Cedilanid-D*	Deslanoside Injection U.S.P.

* For doses, see Table 18–18.

toxicity develops. However, if renal function should be impaired during long-term maintenance therapy, the risk of serious toxicity appears to be considerably greater for the patient receiving digoxin rather than digitoxin. Thus, Jelliffe suggests that digitoxin would be preferred with patients with potentially variable renal function.

Digitoxin U.S.P. is obtained from *Digitalis purpurea* and *Digitalis lanata,* as well as several other species of *Digitalis*. It was obtained in crystalline form in 1869 by Nativelle.[349] It is a glycoside of digitoxigenin (Fig. 18-20), 3β,14β-dihydroxy-5β-card-20(22)-enolide. The properties of digitoxin have been compared to digoxin above.

Acetyldigitoxin U.S.P. is obtained from the enzymatic hydrolysis of lanatoside A (Table 18-17).

Ouabain U.S.P., also called G-strophanthin, is a glycoside obtained from the seeds of *Strophanthus gratus* or the wood of *Acokanthera schimperi*. It is too poorly and unreliably absorbed to be used orally, but its extremely fast onset of action (Table 18-18) makes it useful for rapid digitalization in emergencies (e.g., nodal tachycardia, atrial flutter or acute congestive heart failure). Its synonym G-strophanthin makes it easily confused with strophanthin (or K-strophanthin), a glycoside obtained from *Strophanthus kombe.*

The aglycone of ouabain is ouabagenin, while the aglycone of strophanthin is strophanthidin.

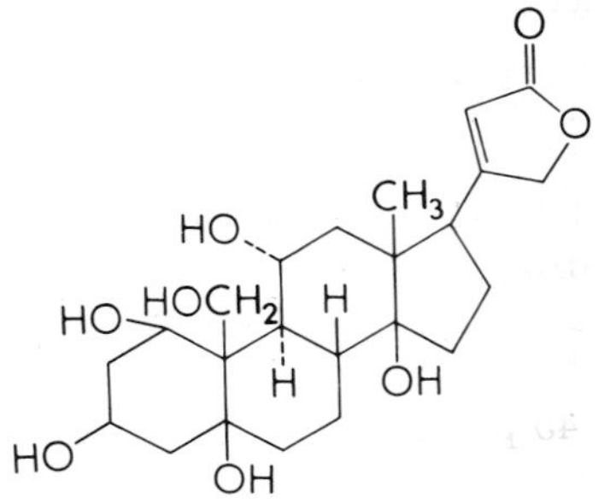

Ouabagenin

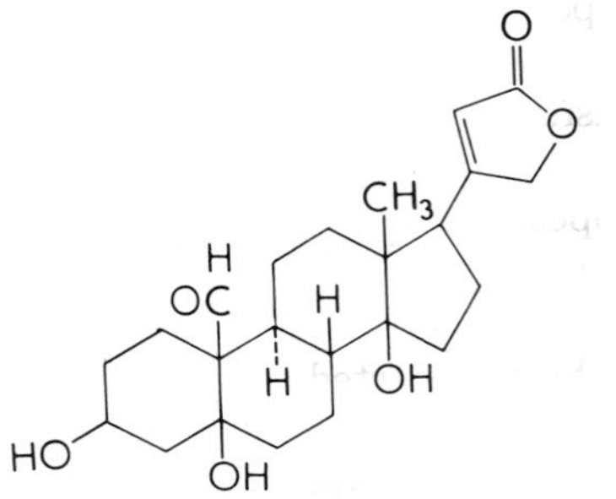

Strophanthidin

Lanatoside C is a digoxigenin glycoside obtained from the leaves of *Digitalis lanata*. It is poorly and irregularly absorbed from the gastrointestinal tract and has a variable metabolic half-life.

Deslanoside U.S.P. is a digoxigenin glycoside obtained from lanatoside C by alkaline deacetylation (Table 18-17). It is used only for rapid digitalization in emergency situations and may be given intravenously or intramuscularly.

STEROIDS WITH OTHER ACTIVITIES

As shown in Figure 18-1, there are a number of important steroids which don't fall into the previous classifications. (Vitamin D precursors are discussed in Chap. 21.) Since these compounds have diverse activities and uses, they will be presented individually in the monographs which follow. Products are listed in Table 18-21.

The reader is also reminded of 19-iodocholesterol[131]I compounds discussed in the section Changes to Modify Pharmacokinetic Properties.

PRODUCTS

Cholesterol U.S.P. is used as an emulsifying agent. Its biosynthesis and structure are shown in Figure 18-4, which also illustrates its essential role as a steroid hormone precursor. Cholesterol is the precursor of virtually all other steroid hormones.

It is important to note that significantly more cholesterol is biosynthesized in the body each day (about 1 to 2 grams) than is contained in the usual Western diet (about 300 mg.). Cholesterol has been implicated in coronary artery disease, but there is increasing evidence that high stress, low exercise, "junk" foods, smoking and genetics are possibly

TABLE 18-21

STEROIDS WITH OTHER ACTIVITIES

Name Proprietary Name	Preparations	Category	Usual Adult Dose*	Usual Dose Range*	Usual Pediatric Dose*
Cholesterol U.S.P.		Pharmaceutic aid (emulsifying agent)			
Spironolactone U.S.P. *Aldactone*	Spironolactone Tablets U.S.P.	Diuretic	25 mg. 2 to 4 times daily	50 to 400 mg. daily	20 to 60 mg. per square meter of body surface 3 times daily
Ox Bile Extract	Ox Bile Extract Tablets	Digestant or choleretic	300 mg. with water 3 times daily		
Dehydrocholic Acid U.S.P.	Dehydrocholic Acid Tablets U.S.P.	Choleretic	500 mg. 3 times daily	250 to 750 mg.	
Fusidic Acid *Fucidin*	Tablets, Solution for infusion	Antibiotic (gram-positive only)	500 mg. 3 times daily		20 to 40 mg. per kg. of body weight per day
Lanolin U.S.P. (Mixture of steroids, other fats and oils)		Water-in-oil emulsion ointment base			
Anhydrous Lanolin U.S.P. (Mixture of steroids, other fats and oils)		Absorbent ointment base			

*See *U.S.P. D.I.* for complete dosage information.

primary causes of heart disease. The famous heart surgeon, Dr. Michael DeBakey, has noted: "Much to the chagrin of many of my colleagues who believe in this polyunsaturated fat and cholesterol business, we have put our patients on no dietary programs. . . . About 80 percent of my sickest patients have cholesterol levels of normal people. . . ."[339] Friedman and Rosenman[340] have recently written a book on stress and exercise factors in heart disease which also reaches a similar conclusion. Passwater[339] has summarized this view as well. Nevertheless, there is a large volume of data which clearly shows that cholesterol does play a significant role in heart disease.

Cholesterol is found in most plants and animals. Brain and spinal cord tissues are rich in cholesterol. Gallstones are almost pure cholesterol. In fact, cholesterol was originally isolated from gallstones, by Paulleitier de Lasalle in about 1770. In 1815, Chevreul[341] showed that cholesterol was unsaponifiable and he called it cholesterin (*chole,* bile; *steros,* solid). In 1859, Berthelot[342] established its alcoholic nature, and since then it has been called cholesterol.

Cholesterol, lanosterol (structure shown in Fig. 18-4), fatty acids and their esters make up **Anhydrous Lanolin U.S.P.** and **Lanolin U.S.P.** Lanolin (or hydrous wool fat) is the purified, fat-like substance from the wool of sheep, *Ovis aries,* and contains 25 to 30 percent water. Anhydrous lanolin (or wool fat) contains not more than 0.25 percent water.

Spironolactone U.S.P. (Fig. 18-1), 17β-hydroxy-7α-acetylthio-17α-pregn-4-en-3-one-21-carboxylic acid γ-lactone, is an aldosterone antagonist of great medical importance because of its diuretic activity. Spironolactone is discussed in Chapter 13. The roles of aldosterone in ion and blood volume regulation have been discussed previously in this chapter, and well-illustrated booklets on aldosterone and spironolactone by G. D. Searle and Co.[253,254] are recommended for further information.

Dehydrocholic Acid U.S.P., 3,7,12-triketocholanic acid (Fig. 18-1), is a product obtained by oxidizing bile acids. The bile acids serve as fat emulsifiers during digestion. About 90 percent of the cholesterol not used for biosynthesis of steroid hormones is degraded to bile acids. All are 5β-steroid-3α-ols, giving rise to the "normal" designation discussed previously in Nomenclature, Stereochemistry and Numbering. As shown in Figure 18-30, cholesterol has part of its side chain oxidatively removed in the liver and two or more hydroxyls are added.[342] The resulting bile acids are then converted to their glycine or taurine conjugate salts which are secreted in the bile. After entering the large intestine, the conjugate salts are converted to cholic acid, desoxycholic acid and several other bile acids. Much of the bile acids are then reabsorbed, with

cholic acid having a biological half-life of about three days.[343]

The bile acids are anionic detergents which emulsify fats, fat-soluble vitamins and other lipids so that they may be absorbed. Dehydrocholic acid also stimulates the production of bile (choleretic effect). It is used following surgery on the gallbladder or bile duct to promote drainage, and for its lipid-solubilizing effects in certain manifestations of cirrhosis or steatorrhea. A related product, **Ox Bile Extract,** contains not less then 45 percent of cholic acid and is used for the same purposes as dehydrocholic acid.

Fusidic Acid (Fig. 18-1) and its sodium salt are used in Europe as antibiotics for gram-positive bacterial infections, particularly with patients who are penicillin-sensitive. It acts by inhibition of G-factor during protein biosynthesis.[344,345] It is also of interest because it appears that it is formed from an intermediate common to the biosynthesis of lanosterol during squalene epoxide cyclization.[346] Structure-activity studies have been reported by Godtfredsen, in 1966, and it was found that just about any minor structural modification of the molecule will result in significantly decreased activity.[347] **Cephalosporin P₁**[348] and **Helvolic Acid**[349] are steroids with structures very similar to fusidic acid, and both are antibiotics useful in some gram-positive bacterial infections. The clinical uses and properties of fusidic acid have been reviewed by Kucers.[350]

COMMERCIAL PRODUCTION OF STEROIDS

HISTORY

This chapter on steroids would not be complete without brief mention of the fascinating history of the steroid industry.[175,190,351] In the 1930's, steroid hormones had to be obtained by extraction of cow, pig and horse ovaries, adrenals and urine. The extraction process was not only inefficient, it was expensive. Progesterone was valued at over $80 per gram. However, by the late 1940's, progesterone was being sold for less than 50 cents a gram, and was available in ton quantities. The man who made steroid hormones cheaply and plentifully available is Russell E. Marker, the "founding father" of the modern steroid industry.

After leaving graduate school in 1925, Marker worked in a variety of areas in organic chemistry research. In 1935, he went to the Pennsylvania State University to begin studying steroids, turning his full attention to finding inexpensive starting materials for steroid hormone syntheses. In 1939, he correctly determined the structure of sarsasapogenin, a sapogenin (aglycone of a saponin, i.e., a glycoside which

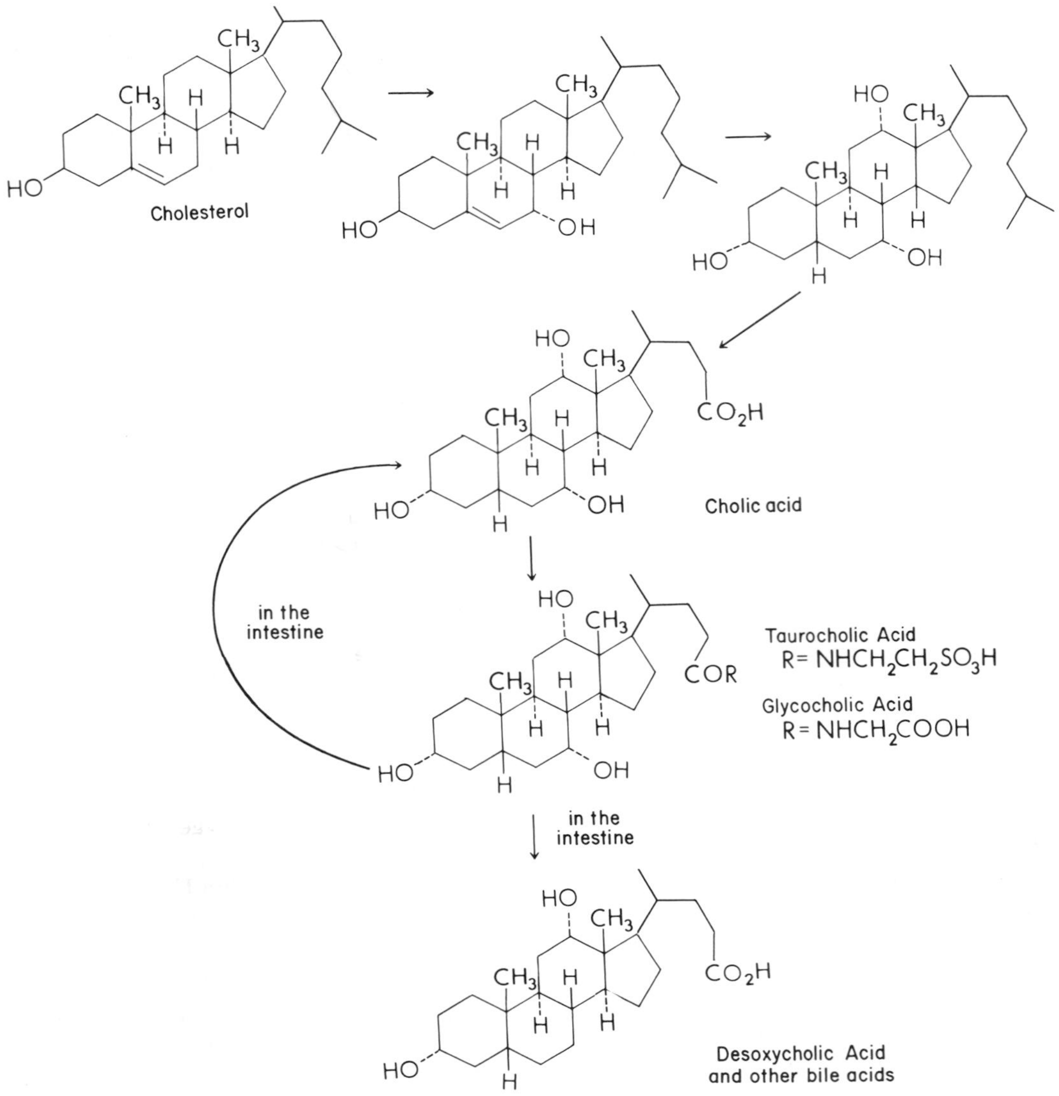

FIG. 18-30. *Metabolism of cholesterol to bile salts.*

foams in water) whose structure had been incorrectly published by many other chemists a few years earlier.[352]

Marker quickly developed a procedure (Fig. 18-31) to degrade the side chain of sarsasapogenin to yield a pregnane. Soon thereafter he degraded diosgenin (Δ^5-sarsasapogenin) to progesterone (Fig. 18-30) in excellent yield.

The commercial potential of the process was obvious to Marker. He immediately launched a series of plant-collecting expeditions from 1939 to 1942 to find a high-yield source of diosgenin, isolated previously from a *Dioscorea* species in Japan.[353] Over 400 species were collected (over 40,000 kg. of plant material) in Mexico and the American Southwest.

Two particularly high-yielding sources of diosgenin were found in Mexico—*Dioscorea composita* ("barbasco") and *D. macrostachya* ("cabeza de negro")—

commonly called "the Mexican yams." Although barbasco had five times the diosgenin content of cabeza, it was in generally inaccessible areas, and so Marker concentrated on cabeza. He knew he had a high-yield, low-cost source of progesterone but was unable to interest several American drug companies. In 1943, he returned to Mexico City and promptly made 3 kg. of progesterone (valued at $240,000) from cabeza. On January 21, 1944, Marker, Lehmann and Somlo incorporated Syntex Laboratories, and by 1951, Syntex was taking orders for 10-ton quantities of progesterone.[268]

However, in 1945, Marker, Somlo and Lehmann had a general "falling out," and Marker sold his 40 percent interest in Syntex to the other two partners. Syntex then brought in Rosenkranz, Djerassi and other chemists to continue the synthesis of hormones from diosgenin. In 1951 and 1953, Frank Colton of

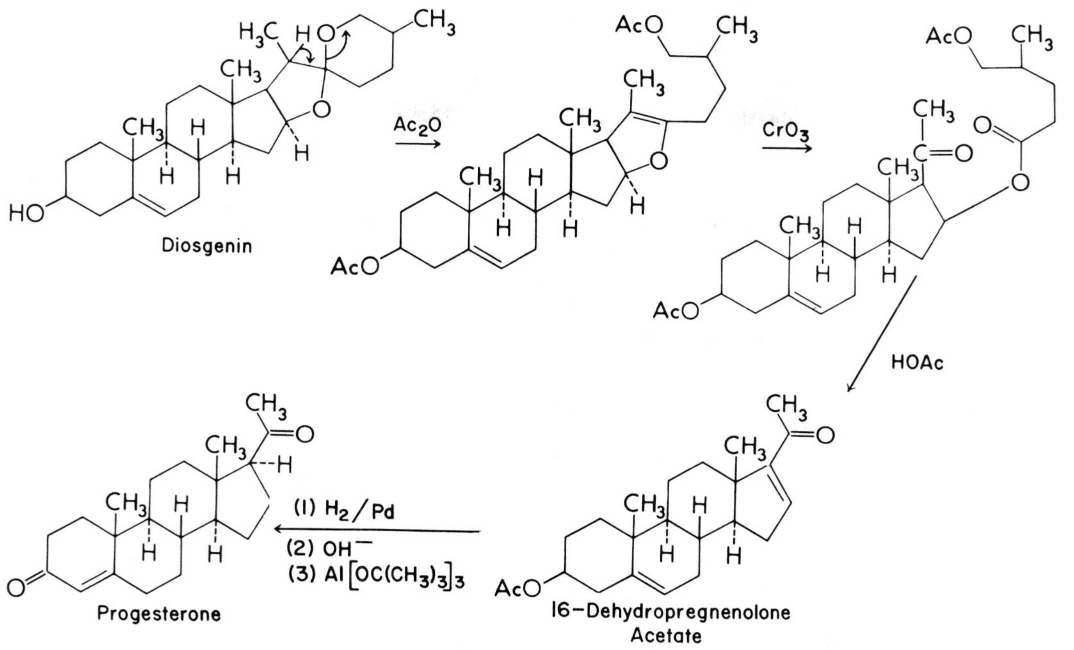

FIG. 18-31. *The Marker synthesis of progesterone from diosgenin.*

G. D. Searle and Co.[354] and Djerassi and Rosenkranz of Syntex Laboratories[355] synthesized norethynodrel and norethindrone, respectively, thus beginning the era of oral contraceptives which continues to this day.

During the 1950's virtually all the steroid hormones had been made from diosgenin by chemists in North America and Europe. The "Mexican yams" have been "nationalized" by Mexico, thus blocking export. Attempts to grow high-yield barbasco or cabeza in other countries have been generally unsuccessful. In 1951, Upjohn patented a process of converting progesterone to 11α-hydroxyprogesterone,[356] a useful intermediate in the synthesis of cortisone. Since that time, microorganisms have continued to play many key roles in the inexpensive commercial production of steroid drugs.

CURRENT METHODS

In 1976, the Mexican government raised the price of diosgenin by 250 percent. However, by this time a number of alternative synthetic routes had been developed to produce all the steroid hormones from other starting materials. After the diosgenin price rise, these procedures were actually cheaper, and use of diosgenin by most of the world's steroid companies stopped almost immediately.

Today nearly all steroid hormones are made from stigmasterol (an inexpensive component of soybean oil) or cholesterol (available in ton quantities from wool fat). Microbiological side-group cleaving processes of Sih[357] and Kraychy[358] are the basis for the routes shown in Figure 18-32. The 19-methyl group is usually removed by the method of Uberwasser[359] (oxidation to form an oxygen bridge at C-6); but Birch reduction with Li/NH_3, an industrially more difficult process, is also used. The Upjohn microbial process[367] for converting progesterone to 11α-hydroxyprogesterone is still used to make cortisone and hydrocortisone products. An overview of these processes has been published by Klimstra and Colton.[360] Some total synthetic routes are also used by some companies to make estrone.[361]

I would like to thank Paula Stewart for her first draft of the section on "Androgens and Sports." Vince Wilmarth provided invaluable assistance in updating tables to be consistent with the new U.S.P. The courtesy of Dr. Alan Corbin of Wyeth Laboratories and of Dr. John Baxter of the University of California in advising me on the LH-RH and hormone receptor sections is also deeply appreciated.

REFERENCES

1. Wolff, M. E. (ed.): Burger's Medicinal Chemistry, 4th ed, vol. 1, vol. 2, vol 3. New York, Wiley, 1979–1980.
2. Counsell, R. E., and Brueggemier, R.: The Male Sex Hormones, Chap. 28, vol. 2, p. 873, New York, Wiley, 1979.
3. Deghenghi, R., and Givner, M. L.: The Female Sex Hormones, Chap. 29, vol 2, p. 917, New York, Wiley, 1979.
4. Wolff, M. E.: The Anti-inflammatory Steroids, Chap. 63, vol. 3, p. 1273, New York, Wiley, 1980.
5. Stockigt, J. R.: Mineralocorticoid Hormones, Adv. Steroid Biochem, Pharmacol. in Briggs, M. H., and Christie, G. A., (eds.) 5:161, 1976.
6. Fieser, L. F., and Fieser, M.: Steroids, New York, Reinhold, 1967.
7. Johns, W. F. (ed.): Steroids, MIP International Review of Science, Baltimore, University Park Press, 1973.

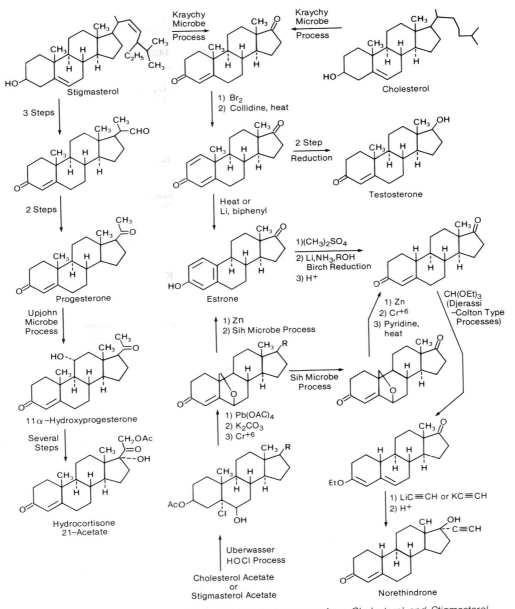

FIG. 18-32. *Commercial production of steroid hormones from Cholesterol and Stigmasterol.*

8. Clayton, R. B. (ed.): Steroids and Terpenoids, New York Academic Press, 1969.
9. Heftmann, E.: Steroid Biochemistry, New York, Academic Press, 1970.
10. Djerassi, C.: Steroid Reactions, San Francisco, Holden-Day, 1963.
11. Blickenstaff, R. T., *et al.:* Total Synthesis of Steroids, New York, Academic Press, 1964.
12. Fried, J., and Edwards, J. A.: Organic Reactions in Steroid Chemistry, vols. 1 and 2, New York, Van Nostrand-Reinhold, 1972.
13. Kirk, D. N.: Steroid Reaction Mechanism, New York, Hartshorn, 1969.
14. Akhrem, A. A., and Titov, Y. A.: Total Synthesis of Steroids, translated by J Schmorak, Israel Program for Scientific Translations, Plenum, New York, N.Y., 1970.
15. Charney, W., and Herzog, H. L.: Microbial Transformations of Steroids, New York, Academic Press, 1967.

16. Lizuka, H.: Microbial Transformation of Steroids and Alkaloids, Baltimore, University Park Press, 1967.
17. Grant, J. K., and Hall, P. E.: Ad. Steroid Biochem. Pharmacol. 1:419, 1970.
18. Heftmann, E.: Modern Methods of Steroid Analysis, New York, Academic Press, 1973.
19. Eik-Nes, K. B., and Horning, E. C.: Gas Phase Chromatography of Steroids, New York, Springer, 1968.
20. Kirkland, J. J., and Synder, L. R.: Introduction to Modern Liquid Chromatography, New York, Wiley-Interscience, 1974.
21. McKerns, K. W. (ed.): The Sex Steroids—Molecular Mechanisms, New York, Appleton, 1971.
22. King, R. J. G., and Mainwaring, W. I. P.: Steroid-Cell Interactions, Baltimore, University Park Press, 1974.
23. O'Malley, B. W., and Schrader, W. T.: Scientific American, 234:32, 1976.
24. Baxter, J. D., and Funder, J. W.: New Eng J. Med. 301:1149, 1979.

25. Baxter, J. D., and MacLeod, K.: *in* Bondy, P. K., and Rosenberg, L. E. (eds.): Duncan's Diseases of Metabolism, Philadelphia, W. B. Saunders, 1979.

26. O'Malley, B., and Birnbaumer, L. (eds.): Receptors and Hormone Action, vol. 2., New York, Academic Press, 1978.

27. Baxter, J. D., and Rousseau, G. G. (eds.): Glucocorticoid Hormone Action, New York, Springer-Verlag, 1979.

28. Cho-Chung, Y. S.: Life Sci., 24:1231, 1979.

29. Jensen, E. V.: Pharmacol. Rev., in press.

30. Segal, S. J., and Koide, S. S.: Pharmacol. Ther. 4:183, 1979.

31. McGuire, W. L. (ed.): Progesterone Receptors in Normal and Neoplastic Tissues, New York, Raven Press, 1977.

32. Gorski, J., and Gannon, F.: Ann. Rev. Physiology 38:425, 1976.

33. Lakoby, W. B., and Wilchek M. (eds.): Methods in Enzymology, vol. XLVI, Affinity Labeling, New York, Academic Press, 1977.

34. Pasqualini, J. R. (ed.): Receptors and Mechanism of Action of Steroid Hormones, Part I and Part II, New York, Dekker, 1978.

35. O'Malley, B.W., and Hardman, J. F.: Steroid Hormones, Methods in Enzymology, vol. XXXVI, New York, Academic Press, 1975.

36. Raynaud, J. P.: *in* McGuire, W. L. (ed.): Progesterone Receptors in Normal and Neoplastic Tissues, Raven Press, New York, 1977.

37. Ojasso, T., and Raynaud, J. P.: Cancer Res. 38:4186, 1978.

38. Duax, W. L., and Norton D. A. (eds.): Atlas of Steroid Structure, vol. I, New York, Plenum, 1975.

39. Duax, W. L. Weeks, C. M., and Rohrer, D. C.: Crystal structure of steroids: molecular conformation and biological function, Laurentian Hormone Conference Abstracts, 1975.

40. ———— Crystal structure of steroids: molecular conformation and biological function, *In* Recent Progress in Hormone Research 32:81, 1976.

41. ———— Crystal structure of steroids, *In* Eliel, E. L. and Allinger, N. (eds.), Topics in Stereochemistry (9:271, 1976.

42. Duax, W. L., Weeks, C. M., Rohrer, D. C., and Griffin, J. F.: Crystal and molecular structures of steroids: identification, analysis and drug design, Basle, Excerpta Medica 2:565, 1977.

42a. Langridge, R., Ferrin, T. E., Kuntz, I. D., and Connolly, M.: Science 211:661, 1981.

43. Thomas, J. A., and Singhal, R. L. (eds.): Molecular Mechanisms of Gonadal Hormone Action, Baltimore, University Park Press, 1974.

44. Krebs, H. A.: Metabolic Control Through Estrogen Action, *in* Advances in Enzyme Regulation, vol. 11, (George Weber, Editor), Pergamon Press, 1973 (proceedings of an October, 1972, conference).

45. Pasqualini, J. R.: Recent Advances in Steroid Biochemistry, New York, Pergamon Press, 1975.

46. Gorski, J. and Gannon, F., E. Knobil, R. R. Sonnenschein and E. S. Edelman, (Eds.): Ann. Rev. Physiology, 38:425, 1976.

47. Westphal, V.: Steroid-Protein Interactions, New York, Springer, 1971.

48. Jensen, E. V., and DeSombre, E. R.: Ann. Rev. Biochem. 41:789, 1972.

49. O'Malley, B. W. and Means, A. R.: Science 183:610, 1974.

50. Marx, J. L., *ibid*, 191:838, 1976.

51. Henderson, B. E., et al.: New Eng. J. Med 293:790, 1975.

52. Edmonson, H. A., Henderson, B. E., and Benton, B.: *ibid*, 294:470, 1976.

53. Li, J. L., Talley, D. L., Li, S. A., and Villee, C. A.: Cancer Research, 36:1127, 1976.

54. Harrison, R. W., and Toft, D. O.: Endrocrinology, 16:199, 1975.

54a. Lippman, M.: Life Sci. 18:143, 1976.

55. Natides, A. C., and Nielsen, S.: J. Biol Chem. 249:1866, 1974.

56. Kvinnsland, S.: Life Sci. 12:373, 1973.

57. Kvinnsland, S., et al.: J. Ster. Biochem. 6:1121, 1975.

58. O'Malley, B. W., et al.: J. Biol. Chem. 250:5175, 1975.

59. ———— J. Biol. Chem. 251:1960, 1976.

60. Thomas J. A., and Mawhinney, M. G.: Synopsis of Endocrine Pharmacology, Baltimore, University Park Press, 1973.

61. Gilman, A. G., Goodman, L. S., and Gilman, A. (eds.): The Pharmacological Basis of Therapeutics, 6th ed., New York, MacMillian, 1980.

62. Murad, F., and Haynes, R. C.: Estrogens and Progestins, Chap. 61, *ibid*, p. 1420.

63. Murad, F., and Haynes, R. C.: Androgens and Anabolic Steroids, Chap. 62,*ibid.*, p. 1448.

64. Haynes, R. C., and Murad F.: Adrenocortical Steroids, Chap. 63, *ibid.*, p. 1466.

65. Meyers, F. H., *et al.* (eds.): Review of Medical Pharmacology, ed. 4, Los Altos, Cal., Lange Medical Pub., 1974.

66. Azarnoff, D. L., (ed.): "Steroid Therapy," Philadelphia, W. B. Saunders, 1975.

67. Briggs, M. H.: Steroid Biochemistry and Pharmacology, New York, Academic Press, 1970.

68. Freedman, M. A., and Freedman, S. N.: Introduction to Steroid Biochemistry and Its Clinical Applications, New York, Harper, 1970.

69. Dorfman, R. I., and Unger, F.: Metabolism of Steroid Hormones, New York, Academic Press, 1965.

70. Salhanick, H. A., *et al.* Metabolic Effects of Gonadal Hormones and Contraceptive Steroids, New York, Plenum Press, 1969.

71. McKerns, K. W.: Steroid Hormones and Metabolism, New York, Appleton, 1969.

72. Fotherby, K. and James, F.: Metabolism of synthetic steroids, Ad. Steroid Biochem. Pharmacol., 3:67, 1972.

73. Hadd, H. E., and Blickenstaff, R. T.: Conjugates of Steroid Hormones, New York, Academic Press, 1969.

74. Bernstein, S., and Solomon, S.: Chemical and Biological Aspects of Steroid Conjugation, New York, Springer, 1970.

75. Bruchovsky, N., and Craven, S.: Biochem. Biophys. Res. Commun. 62:837, 1975.

76. Bolt, H. M.: Pharmacol. Ther. 4:155, 1979.

77. Mornon, J. P., *et al.*: J. Mol. Biol. 137, 1980.

78. Wolff, M. E., *et al.*: 1979 PROPHET Colloquoum, Airlie, Virginia, 1980.

79. Michael, L. H., *et al.*: Mol. Pharmacol. 16:135, 1979.

80. Greenstein, B. D.: J. Reprod. Fertil. 52:365, 1978.

81. Chang, W. C., and Roth, G. S.: J. Steroid, Biochem. 11:889, 1979.

82. IUPAC Commision: Nomenclature of Steroids, Steroids 13:278, 1969.

83. Dempsey, M. E.: Ann. Rev. Biochem. 43:967, 1974.

84. Clayton, R. B.: Quart. Rev. 19:201, 1965.

85. Appelgren, L. E.: Sites of Steroid Hormone Formation, Stockholm, Eigsp Press, 1967.

86. Cornforth, L. W.: Quart. Rev. 23:125, 1969.

87. Mulheirn, L. J., and Ramm, P. J.: Chem. Rev. 72:259, 1972.

88. Ariens, E. J. (ed.) Drug Design, vols. 1, 2, 3, and 4, New York, Academic Press, 1971–74.

89. Counsell, R. E., *et al.*: J. Med. Chem. 16:945, 1973.

90. Counsell, R. E., *et al.*: J. Nucl. Med. 14:777, 1973.

91. Lu, M. C., *et al.*: Med. Chem. 15:1284, 1972.

92. F.D.A. Drug Bulletin 5:1, 1975.

93. Briggs, M. H., and Briggs, M.: Sex Hormone Exposure During Pregnancy and Malformations, *in* Briggs, M. H., and Corbin, A.: (ed.): Adv. Steroid Biochem. Pharmacol. 7:51, 1979.

94. Aufrere, M. B., and Benson, H.: J. Pharm. Sci. 65:783, 1976.

95. Corbin, A., *et al.* Int. J. Gynec. Obstet. 16:359, 1979.

96. Mangan, C. E., *et al.*: Am. J. Obstet. Gynec. 134:860, 1979.

97. Corbin, A.: Contraceptive Technology *in* Briggs, M. H., and Corbin, A., (eds.): The Future, Adv. Steroid Biochem. Pharmacol. 7:1, 1979.

98. Corbin, A., and Bex, F. J.: a) Antifertility Effects of LHRH and Agonists, *in* McKerns, K. W. (ed.): Reproductive Processes and Contraception Plenum Press, 1980; b) LHRH and Analogues, Contraceptive and Contraceptive Potential, *in* Briggs, M, and Corbin, A. (ed.): Progress in Hormone Biochemistry and Pharmacology Montreal, Eden Press, 1980.

99. Corbin, A., *et al.*: Int. J. Fertil. 23:81, 1978.

100. Corbin, A., and Bex, F. J.: The Pharmacology of LH–RH Analogues, 179th ACS National Meeting Abstracts, Houston, Texas, March, 1980.

100a. Casper, R., *et al.*: The Effectiveness of a Superactive LRF Agonist As A Contraceptive Agent *in* Yen, S. *et al., ibid,* March, 1980.
100b. Schally, A. V., *et al.*: Present Status of LH-RH and Its Stimulatory and Inhibitory Analogs, *ibid,* March, 1980.
100c. Vale, W.: Regulation of Reproductive Functions by GnRH and Synthetic Agonists and Antagonists, *ibid,* March, 1980.
100d. Sharpe, R. M.: "Extra-pituitary Actions of LHRH and its Agonists, Nature 286:12, 1980.
101. Schally, A. V., *et al.*: Present Status of LH-RH and Its Stimulatory and Inhibitory Analogs, *ibid,* 1980.
102. Bell, M. R., *et al.*: Chemical Control of Fertility, Ann. Rep. Med. Chem., 14:168, 1979.
103. Schueler, F. W. Science 103:221, 1946.
104. Keasling, H. H., and Schueler, F. W.: J. Am. Pharm. Assn. 39:87, 1950.
105. Salerni, O. L.: Natural and Synthetic Organic Medicinal Compounds, p. 145, St. Louis, C. V. Mosby, 1976.
106. Solmessen, U. V.: Chem. Rev. 37:481, 1945.
107. Blanchard, E. W., and Stebbins, B. B.: Endocrinology 32:307, 1943.
108. Baker, B. R.: J. Am. Chem. Soc. 65:1572, 1943.
109. Rubin, M., and Wishinsky, H.: J. Am. Chem. Soc. 66:1948, 1944.
110. Dodds, E. C.: Nature 142:34, 1938.
111. McGuire, W. L. *et al.*: Current Topics in Exp. Endocrin., 3:93, 1978.
112. Facts and Comparison, St. Louis, Facts and Comparisons, Inc., 1975. (Suggested doses may change. Facts and Comparisons, Inc. is not responsible for any errors which appear in this text.)
113. Morris, L. A., *et al.*: American Pharmacy, NS20:318, 1980.
114. The American Council on Science and Health, Post-Menopausal Estrogen Therapy, 1979 (Available from The American Council on Science and Health, 199T Broadway, New York, N.Y., 10023).
115. Albanese, A. A.: N.Y.J. Med. 65:2116, 1965.
116. Heftmann, E.: Steroid Biochemistry, p. 141, New York, Academic Press, 1970.
117. Mirocha, C. J., *et al.*: Microbial Toxins 7:107, 1971.
118. F.D.A. Drug Bulletin 5:2, June, 1975.
118a. Kincl, F., and Dorfman, R.: Acta Endocrinol. 42 (Suppl 73): 3, 1963; and Steroids 2:521, 1963.
119. Briggs, M. H., and Christie, G. A.: (Estrogenic and Non-Estrogenic) Antifertility Substances in Plants, Adv. Steroid Biochem. Pharmacol. 6:xi, 1977.
120. Kocharian, C. D., *et al.*: Pharmacology and Therapeutics, Part B 1(2):149, 1975.
121. Bardin, C. W., *et al.*: The role of receptors in the anabolic action of androgens, *in* O'Malley, B. W., and Birnbaumer, L. (eds.): Receptors and Hormone Action, New York, Academic Press, 1978.
122. Mainwaring, W.: The Mechanism of Action of Androgens, Berlin, Springer-Verlag, 1977.
123. Hendershott, J.: Track and Field News, Part 5 22:3, 1969.
124. Wade, N.: Science 176:1399, 1972.
125. Loughton, S. J., and Ruhling, R. O.: J. Sports Med. and Phys. Fit. 17:285, 1977.
126. Casner, S., *et al.*: *ibid,* 11:98, 1971.
127. Fahey, T. D., and Brown, C. H.: Med. Sci. Sports 5:272, 1973.
128. Fowler, W. M., Jr., *et al.*: J. Appl. Physiol. 20:1038, 1965.
129. Freed, D. L., and Banks, A. J.: Brit. J. Sports Med. 9:78, 1975.
130. Hervey, G. R., *et al.*: Lancet 2:699, 1976.
131. Johnson, L. C., *et al.*: Med. Sci. Sports 4:43, 1972.
132. Johnson, L. C., and O'Shea, J. P.: Science 64:957, 1969.
133. Johnson, L. C., *et al.*: Med. Sci. Sports 7:287, 1975.
134. O'Shea, J. P.: Nutr. Report Internat. 4:363, 1971.
135. O'Shea, J. P., and Winkler, W.: *ibid,* 2:351, 1970.
136. Stamford, B. A., and Moffat, R.: J. Sports Med. and Phys. Fit. 14:191, 1974.
137. Stromme, S. B., *et al.*: Med. Sci. Sports 6:203, 1974.
138. Ward, P.: *ibid,* 5:227, 1973.
139. Novich, M. M.: N.Y. State J. Med. 73:2679, 1973.
140. Brooks, R. V.: *et al.,* J. Steroid Biochem. 11:913, 1979.
141. Harkness, R. A., *et al.*: Brit. J. Sports Med. 9:70, 1975.
142. Frischkorn, C. G., and Frischkorn, E. H.: J. Chromatogr. 151:33, 1978.
143. Wieder, B.: J. Sports Med. Phys. Fit. 13:131, 1973.
144. American College of Sports Medicine: Position statement on the use and abuse of anabolic-androgenic steroids in sports Med. Sci. Sports 9:xi-xii, Winter, 1977.
145. Prunty, F. T. G.: Brit. Med. J. 2:605, 1960.
146. Lipsett, M. B., and Korenman, S. G.: J.A.M.A. 190:757, 1964.
147. Kincl, F. A.: Methods Hormone Res. 4:21, 1965.
148. Nimni, M. E., and Geiger, E.: Proc. Soc. Exp. Bio. Med. 94:606, 1957.
149. Kochakian, C. D., and Tillotson, C.: Endocrinology 60:607, 1957.
150. Scow, R. O.: Endocrinology 51:42, 1952.
151. Potts, G. O., *et al.* Endocrinology 67:849, 1960.
152. Vida, J. A.: Androgens and Anabolic Agents, New York, Academic Press, 1969.
153. Compiled from data prepared by O. Gisvold, Chap. 25, Steroids, in the 6th edition of this text.
154. Wolff, M. E., Ho, W., and Kwok, R.: J. Med. Chem. 1:577, 1964.
155. Bowers, A., et al.: J. Med. Chem. 6:156, 1963.
156. Klimstra, P. D.: *in* Chemistry and Biochemistry of Steroids, Chap. 8, vol. 3, Los Altos, Cal., Geron-X, 1969.
157. Hansson, V., and Tueter, K. J.: Acta. Endo. (Kbh.) 155:148, 1971.
158. Mainwaring, W. I. P., and Milroy, E. G. P.: J. Endocrinology 57:371, 1973.
159. Ahmed, K., and Wilson, M.: J. Biol. Chem. 250:2370, 1975.
160. Mainwaring, W. I. P., and Mangan, F. R.: J. Endocrinology 59:121, 1973.
161. Eikness, K. B., *et al.*: J. Biol. Chem. 206:411, 1954.
162. Brinkman, A. O., *et al.* Ann. Endocrinology 31:789, 1970.
163. Westphal, V.: Steroid-Protein Interaction, Berlin, Springer-Verlay, 1971.
164. American Pharmaceutical Association: Evaluations of Drug Interactions, Washington, D.C., 1973-1975.
165. Mainwaring, W. I. P., *et al.*: Mol. Cell. Endocrin. 133, 1974.
166. Stoliar, B., and Albert, D. J.: J. Urology 111:803, 1974.
167. Saunders, H. C., Holden, K., and Kerwin, J. F.: Steroids 3:687, 1964.
168. Neumann, F., *et al.*: J. Endocrinology 35:363, 1966; Recent *et al.*: Neumann, F., Prog. Hormone Res. 26:337, 1970.
169. Boris, A., *et al.*: Endocrinology 88:1086, 1971.
170. Liao, S., *et al.*: Endocrinology 94:1205, 1974.
171. Goto, C., *et al.*: Chem. Pharm. Bull 25:1295, 1975.
172. Nakayama, R., *et al.*: Acta Endocrinol. Supp. 92: 229, 1979.
173. Bonne, C., and Raynaud, J. P.: Mol. Cell. Endocrin. 2:59, 1974.
174. Bennett, J. P.: Chemical Contraception, New York, Columbia University Press, 1974.
175. Djerassi, C.: The Politics of Contraception, New York, Norton, 1979.
175a. Lednicer, D. (ed.): Contraception, The Chemical Control of Fertility, New York, Marcel Deker, 1969.
176. The Medical Letter 16:37, 1974.
177. Emory University School of Medicine: Contraceptive Technology 1973-74, Atlanta, 1974.
178. Publications of the Population Information Program: Population Reports, Department of Medical and Public Affairs, The George Washington University Medical Center, 2001 S. Street, N.W., Washington, D.C., 20009.
179. Editors, American Pharmacy NS 19: July, 1979.
180. Cherniak, D., and Feingold, A.: Birth Control Handbook, 14th ed. P. O. Box 1000, Station G., Montreal, Quebec, Montreal Health Press, 1979.
181. Balin, H., *et al.*: Seminars in Drug Treatment 3:121, 1973.
182. Lehfeldt, H.: *in* Ralph M. Wynn, (ed.) Obstet. Gynec. Ann. 261-315, 1973.
183. Bingel, A. S., and Benoit, P. S.: J. Pharm. Sci. 62:179, 349, 1973.
184. Stevens, V. C., and Vorys, N.: Obstet. Gynec. Survery 22:781, 1967.
185. Durand, J. C., and Bressler, R.: Adv. Int. Med. 24:97, 1979.
186. Andrews, W. C.: Clin. Obstet. Gynec. 6:3, 1979.

187. Lain, A. K.: Studies in Family Planning, 8:50, 1977.

188. Tietze, C.: Family Planning Perspectives 11:80, 1979.

189. Shapiro, S., *et al.*: Lancet 1:743, 1979.

190. Syntex, A Corporation and A Molecule, Palo Alto, Cal., National Press, 1966.

191. Makepeace, A. W., *et al.*: Am. J. Physiol. 119–512, 1937.

192. Selye, H., *et al.*: Fertil. Steril. 22:735, 1971, and earlier references cited.

193. Dempsey, E. W.: Am. J. Physiol. 120:926, 1937.

194. Kurzrok, R.: J. Contracept. 2:27, 1937.

195. Albright, F.: Musser, J. H. (ed.): Internal Medicine Philadelphia, Lea & Febiger, 1945.

196. Sturgis, S. H., and Albright, F.: Endocrinology 26:68, 1940.

197. Pincus, G.: The Control of Fertility, New York, Academic Press, 1965, and cited references.

198. Djerassi, C., *et al.*: J. Chem. Soc. 76:4092, 1954.

199. Colton, F. B.: U.S. Patent 2,691,028, 1954 (applied May, 1953).

200. Birch, A. J.: Quart. Rev. (London) 12, 1958:4, 69, 1950.

201. Rudel, H. W., and Martinez-Manautou, J.: Oral Contraceptives, *in* Rabinowitz, J. L., and Myerson, R. M., (eds.): Topics in Medicinal Chemistry, p. 339, New York, Wiley-Interscience, 1967.

202. Behrman, S. J.: *in* Austin, C. R., and Perry, J. S., (eds.): a discussion of Mears, E.: Agents Affecting Fertility London, Churchill, 1965.

203. Goldzieher, J. W., *et al.*: J.A.M.A. 180:359, 1962.

204. Levy, E. P., *et al.*: Lancet 1:611, 1973.

205. Nora, A. H., and Nora, J. J.: Arch. Env. Health 30:17, 1975.

206. Gardner, L. I., *et al.*: Lancet 2:667, 1970.

207. Robinson, S. C.: Am. J. Obstet. Gynec. 109:354, 1971.

208. Banks, A. L.: Int. J. Fertil. 13:346, 1968.

209. Poland, B. J., and Ash, K. A.: Am. J. Obstet. Gynec. 116:1138, 1973.

210. Vecchio, T.H.: Long-Acting Injectable Contraceptives, *in* Briggs, M. H., and Christie, G. A. (eds.): Adv. Steroid Biochem. Pharmacol. 5:1, 1976.

211. F.D.C. Reports, Sept. 16, 1973.

211a. Diczfalusy, E.: J. Steroid Biochem. 11:443, 1979.

212. Berman, E.: J. Reprod. Med. 5:37, 1970.

213. Patient Information on R-2323 provided during 1976 Minnesota clinical trials.

214. Wheeler, R. G., *et al.*: Intrauterine Devices, New York, Academic Press, 1974.

215. Population Reports, Intrauterine Devices, Series B, January 1975, The George Washington University Medical Center, Washington, D.C.

216. Folkman, J., and Long, D. M.: J. Surg. Res. 4:139, 1964.

217. Croxatto, H., *et al.*: Am. J. Obstet. Gynec. 105:1135, 1969.

218. Coutinho, E. M.: J. Reprod. Fertil. 23:345, 1970.

219. Da Silva, A. R., and Coutinho, E. M.: Int. J. Fertil. 23:185, 1978.

220. Vickery, B. H., *et al.*: Fertil. Steril. 21:201, 1970.

221. Scommegna, A., *et al.*: Fertil. Steril. 21:201, 1970.

222. Doyle, L. L., and Clewe, T. H.: Am. J. Obstet. Gynec. 101:564, 1968.

223. Mishell, D. R., Jr.: Fertil. Steril. 21:99, 1970.

224. Mishell, D. R., Jr.: Obstet. Gynec. News 6:33, 1971.

224a. World Health Organization: J. Steroid Biochem. 11:461, 1979.

225. Pharriss, B. B., *et al.*: Fertil. Steril. 25:922, 1974, and cited references.

226. Pharriss, B. B.: Uterine Progesterone System, pp. 203–209, *in* Intrauterine Devices, reference 125 above.

227. Product insert for Progestasert, Alza, December, 1975.

227a. Benagiano, G., and Gabelnick, H. L.: J. Steroid Biochem. 11:449, 1979.

228. F.D.A. Drug Bulletin, February, 1976.

228a. Nordin, B. E.: Drugs 484, 1979.

228b. Gastel, B., Cornoni-Huntley, J. and Brody, J. A.: J. Fam. Pract. 11:851, 1980.

230. The Medical Letter, May 21, 1976.

231. Marx, J. L.: Science 191:838, 1976.

232. The Medical Letter, February 27, 1976.

233. Consumer Reports, May, 1970, and Vaugn, P.: The Pill on Trial, New York, Coward-McCann, 1970.

234. Inman, W. H. W., and Vessey, M. P.: Brit. Med. J. 2:193, 1968.

235. Vessey, M. P., and Doll, R.: Brit. Med. J. 2:651, 1969.

236. Mann, J. I., and Vessey, M. P.: Brit. Med. J. 2:241, 1975.

237. Mann, J. I., and Inman, W. H. W.: Brit. Med. J. 2:245, 1975.

238. F.D.A. Drug Bulletin, July-August, 1975.

239. Inman, W. H. W., *et al.*: Brit. Med. J. 2:203, 1970.

240. O'Brien, P. C., *et al.*: Obstet. Gynecol. 53:300, 1979.

241. Mangan, C. E., *et al.*: Am. J. Obstet. Gynecol. 134:860, 1979.

242. Forsberg, J. G.: Arch. Toxicol. (Suppl.) 2:263, 1979.

243. F.D.A. Drug Bulletin, May, 1973.

244. Gebhard, P., *et al.*: Pregnancy, Birth and Abortion, New York, Harper & Row, 1958.

245. The Medical Letter 16:89, 1974.

246. Population Reports, Prostaglandins, Series G, The George Washington University Medical Center, Washington, D.C., No. 1, April, 1973; No. 4, March, 1974; No. 5, July, 1974.

247. Population Reports, Barrier Methods, Series H, No. 3, January, 1975, Department of Medical and Public Affairs, The George Washington University, see reference 95, above.

248. Huff, J. E., and Hernandex, L.: Contraceptive Methods and Products in Handbook of Nonprescription Drugs, 6th ed., American Pharmaceutical Association, Washington, D.C., 1979.

249. Stone, S. C., and Cardinale, F.: Am. J. Obstet. Gynec. 133:635, 1979.

250. Barwin, B. N.: CMA Journal 119:757, 1978.

251. Jackson, H., and Morris, I. D.: Clin. Obstet. and Gynec. 6:129, 1979.

252. Data supplied by Bureau of Records, Minnesota Department of Health.

253. Aldosterone in Clinical Practice, G. D. Searle and Co., 1974, Available from G. D. Searle and Co.

254. Aldosterone in Clinical Medicine, MedCom Learning Systems, 1972. Available from G. D. Searle and Co.

255. Sayers, G., and Travis, R. H.: ACTH and Adrenocortical Steroids, *in* Goodman, L. S., and Gilman, A. (eds.): The Pharmacological Basis of Therapeutics, ed. 5, Chap. 72, New York, Macmillian, 1975.

256. Yamamoto, K. R., Stampfer, M. R., and Tomkins, G. M.: Proc. Nat. Acad. Sci. 71:3901, 1974, and cited references.

257. Tait, J. F., *et al.*: Clin. Invest. 40:72, 1961.

258. Gray, C. H., and Bacharach, A. L.: Hormones in Blood ed. 2, New York, Academic Press, 1970.

259. Robinson, H. M.: Bull. Sci. Med. Univ. Maryland 40:72, 1955.

260. Stuart, D.: Pharmindex 1:6, 1959.

261. Fried, J.: Ann. N.Y. Acad. Sci. 61:573, 1955.

262. Thorn, G. W., *et al.*: Ann. Int. Med. 43:979, 1955.

263. Fried, J., and Sabo, E. F.: J. Am. Chem. Soc. 75:2273, 1953.

264. Blank, H., *et al.*: Round Table Discussion, Cutis, 24: October, December, 1979; January, March, April, 1980.

265. Dreifus, L. S., and Watanabe, Y.: Seminars in Drug Treatment 2:147, 1972.

266. Sondheimer, F.: Chemistry in Britain, 454, October, 1965, and cited references.

267. May, P. J.: Terpenoids and steroids, Specialist Periodical Reports of the Chemical Society 1:404, 1971.

268. Fieser, L. F., and Fieser, M.: Steroids, ed. 2, New York, Reinhold, 1967.

269. Ode, R. H., Kamano, Y., Johns, W. F., and Pettit, G.: Cardenolides and Bufadienolides, *in* Steroids: MTP International Review, Organic Chemistry Series, vol. 8, Baltimore, University Park Press, 1973.

270. Chen, K. K.: J. Med. Chem. 13:1029, 1035, 1970, and cited references.

271. Reichstein, T., Naturwiss., 54:53, 1967.

272. Smith, T. W., and Haber, E.: New Eng. J. Med. 289:945, 1010, 1063, 1072, 1125, 1973.

273. Hoffman, B. F., and Bigger, J. T.: Digitalis and Allied Cardiac Glycosides, *in* Gilman, A. G., Goodman, L. S., and Gilman, A. (eds.): The Pharmacological Basis of Therapeutics, 6th ed., p. 729, New York, Macmillan, 1980.

274. Niles, A. S.: Cardiovascular Disorders, *in* Melmon, K. L., and Morrelli, H. F. (eds.): Clinical Pharmacology, New York, Macmillan, 1972.

275. Thorp, R. H., and Cobin, L. B.: Cardiac Stimulant Substances, New York, Academic Press, 1967.
276. Jelliffe, R. W.: Therapeutics 3:3, 1975.
277. Thomas, R.: Cardiac Drugs, Chap. 38, *in* Wolff, M. E. (ed.): Burger's Medicinal Chemistry New York, Wiley-Interscience, 1980.
278. Flasch, H. F., and Heinz, N.: Naunyn Schm. Arch. Pharmacol. 304:37, 1978.
279. Guntert, W., and Linde-, H. H.: Arch. Int. Pharm. Ther. 233:53, 1978.
280. Schwartz, A., and Adams, R. J.: Circ. Res. (Suppl. I) 46:I-154, 1980.
281. Akera, T., and Brody, T. M.: Pharmacol. Rev. 29:187, 1978.
282. Glynn, I. M., and Karlish, S. J. D.: Ann. Rev. Physiology 37:13, 1975.
283. Glynn, I. M.: *in* Fisch, C., and Surawicz, B. (eds.): Digitalis, New York, Grune & Stratton, 1969.
284. Chung, E. K.: Digitalis Intoxication, Baltimore, Williams & Wilkins, 1969.
285. Schwartz, A., et al.: Pharmacol. Rev. 27:3, 1975.
286. Askari, A.: Ann. N.Y. Acad. Sci. 242:5-740, 1974, and cited references.
287. Hokin, L. F., and Dahl, J. L.: Chap. 8 *in* Hokin, R. E., (ed.): Metabolic Pathways, vol. 6, Metabolic Transport, Academic Press, New York, 1972.
288. Withering, W.: *in* Shusten, L. (ed.): Readings in Pharmacology, Boston, Little, Brown & Co., 1962.
289. Kupchan, S. M., et al.: Bioorg. Chem. 1:13, 1971, and cited references.
290. Chen, K. K., and Henderson, F. G.: J. Med. Chem. 8:577, 1965, and cited references.
291. Brown, B. T., et al.: Brit. J. Pharmacol. 18:311, 1962.
292. Wolff, M. E., et al.: J. Med. Chem. 13:657, 1970, and cited references.
293. Hospital Formulary, December, 1979.
294. Norton, T. R.: J. Pharmacol. Exp. Ther., 205:683, 1978.
295. Schwartz, A.: *Nature,* in Press, 1980.
296. LaBella, F. S., et al.: in press, 1980.
297. Fullerton, D. S., Rohrer, D., Kitatsuji, E., Yoshioka, K., From, A., and Ahmed, K. Unpublished results.
298. Das, G., et al.: Clin. Pharmacol. Ther. 22:280, 1977.
299. Cook, L. S., et al.: J. Cardiovasc. Pharmacol. 1:551, 1979.
300. Thomas, R., et al.: Eur. J. Pharmacol. 53:227, 1979.
301. Cagin, N. A., et al.: Arch. Int. Pharmaco dyn. Ther. 226:263, 1977.
302. Fullerton, D. S., et al.: Mol. Pharmacol. 17:43, 1980.
303. Schwartz A.: Methods in Pharmacology 1:105, 1971.
304. Schwartz, A., and Adams, R. J.: Circ. Res. (Supp. I) 46:154, 1980.
305. Akera, T., and Brody, T. M.: Pharmacol. Rev. 29:187, 1978.
306. Thomas, R., et al.: J. Pharm. Sci. 63:1649, 1974.
307. Kupchan, S. M., et al.: Bioorg. Chem.: 1:24, 1971.
308. Kupchan, S. M., et al.: J. Org. Chem. 35:3539, 1970.
309. Jones, J. B., and Middleton, H. W.: Can. J. Chem. 48:3819, 1970.
310. Portius, H. L., and Repke, K.: Arzneimittel Forsch. 14:1073, 1964.
311. Repke, K.: *in* Wilbrandt, W., and Lindgren, P. (eds.): New Aspects of Cardiac Glycosides, Proc. 1st Int. Pharmacol. Meeting, 3:203, 1963.
312. Repke, K., et al.: Ann. N.Y. Acad. Sci. 242:737, 1974.
313. Glynn, I. M.: J. Physiol. 136:148, 1957.
314. Hoffman, L. F.: Am. J. Med. 41:666, 1966.
315. Fullerton, D. S., Yoshioka, K., Rohrer, D., From, A., and Ahmed, K.: Science 205:917, 1979.
316. Fullerton, D. S., Yoshioka, K., Rohrer, D., From, A., and Ahmed, K.: J. Med. Chem. 22:529, 1979.
317. Rohrer, D. C., et al.: Functional Receptor Mapping of Modified Cardenolides: Use of the PROPHET System, *in* Christofferson, R. E., and Olson, R. C. (eds.): Computer Assisted Drug Design, American Chemical Society, 1979.
318. Fullerton, D. S., Kitatsuji, E., Yoshioka, K., Rohrer, D., From, A., and Ahmed, K.: Unpublished results.
319. Yoda, A.: Ann. N.Y. Acad. Sci. 242:598, 1974.
320. Yoda, A.: Mol. Pharmacol. 12:399, 1976, and cited references.
321. Gerbino, P. P.: Am. J. Hosp. Pharm. 30:499, 1973.
322. White, W. F., and Gisvold, O.: J. Pharm. Sci. 41:42, 1952.
323. Butler, V. P., Jr.: Prog. Cardiovasc. Dis. 14 (6): May, 1972.
324. Smith, T. W.: Circulation 44:29, 1971; Smith, T. W.: J. Pharmacol. Exp. Ther. 175:352, 1970.
325. Saki, H. K., and Sakai, H.: Clin. Chem. 21:227, 1975.
326. Kubasik, N. P., et al.: Clin. Biochem. 7:206, 1974.
327. Lab. Management, May, 1973.
328. Lindenbaum, J., et al.: Lancet, Cresswell 1:1215, 1973.
329. Binnion, P. F., and Aristarco, M.: Clin. Pharmacol. Ther. 16(5):807, 1974.
330. Johnson, B. F., et al.: Lancet 1:1473, 1973.
331. Shaw, T. R. D., et al.: Brit. Med. J. 4:763, 1973.
332. Fraser, E. J., et al.: J. Pharm. Pharmacol. 25:968, 1973.
333. Fleckenstein, L., et al.: Clin. Pharmacol. Ther. 16(3):435, 1974.
334. Steiness, E., et al.: Clin. Pharmacol. Therp. 14(6):949, 1973.
335. Digoxin—The regulatory viewpoint, Circulation 44:395-398, 1974.
336. The United States Pharmacopeia, Rev. 20, Rockville, Md., United States Pharmacopeial Convention, 1979.
337. Smith, S.: J. Chem. Soc., 508, 1930; ibid, 23, 1931.
338. Nativelle, C. A.: J. Pharm. Chem. 9:225, 1869.
339. Passwater, R. A.: Dietary Cholesterol and Heart Disease, American Laboratory, Sept. 1972.
340. Friedman, M., and Rosenman, H. R.: Type A Behaviors and Your Heart, New York, Fawcett, 1975.
341. Chevreul, M.: Ann. Chim. Phys. (Ser. 1) 95:5, 1815.
342. Nair, P. P., and Kritchevsky, D. (eds): The Bile Acids, vols. 1 and 2, New York, Plenum Press, 1971–72, and cited references.
343. Lindstedt, S.: Acta Physiol. Scand. 40:1, 1957.
344. Lucas, L. J., and Lipman, F.: Ann. Rev. Biochem. 40:409, 1971.
345. Bodley, J. W., and Lin, L.: Biochem. 11:782, 1972, and cited references.
346. Mulheirn, L. J., and Caspi, E.: J. Biol. Chem. 246:2494, 1971, and cited references.
347. Godtfredsen, W. O., et al.: J. Med. Chem. 9:15, 1966.
348. Chou, T. S., et al.: Tetrahedron Letters 25:3341, 1969.
349. Iwasaki, S., et al.: Chem. Comm., 119, 1970.
350. Kucers, A.: The Use of Antibiotics, Philadelphia, J. B. Lippincott, 1972.
351. Lehmann, P. A., et al.: J. Chem. Ed. 50:195, 1973.
352. Fieser, L. F., and Fieser, M.: Steroids, pp. 816–825, New York, Reinhold, 1959.
353. Tsukamoto, T., et al.: J. Pharm. Soc. Jap. 56:931, 1936.
354. Colton, F. B.: U.S. Patents 2,655,518, 1952; 2,691,028, 1953; 2,725,378, 1953.
355. Djerassi, C., et al.: J. Am. Chem. Soc. 76:4092, 1954.
356. See Fieser, L. F., and Fieser, M.: Reference 6 above, pp. 672–678; and Peterson, D. H., et al.: J. Am. Chem. Soc. 74:5933, 1952; and *ibid,* 75:408, 1953.
357. Sih, C., et al.: J. Am. Chem. Soc. 87:2765, 1965; Patent described in Chem. Abst. 68:103880, 1968.
358. Kraychy, S., et al.: Applied Microbiology, 23:72, 1972; Patent described in Chem. Abst. 77:124777, 1972.
359. Ueberwasser, H., et al.: Helv. Chim. Acta. 46:344, 1963; *ibid.,* 46:361, 1963.
360. Klimstra, P. D., and Colton, F. B.: Chemistry of The Steroidal Contraceptives, Chap. 3, *in* Lednicer, D. (ed.): New York, Marcel Dekker, 1969.
361. Velluz, L., et al.: Compt. Rend. 257:3086; Ananchenko, S. M., et al.: Tetrahedron 18:1355, 1962.

19
carbohydrates

Jaime N. Delgado

Carbohydrates, usually called "sugars" (e.g., glucose, sucrose, starch and glycogen), were thought to be correctly represented by the generalized formula, $C_x(H_2O)_y$, and thus the term "carbohydrate" became extensively used. However, many compounds now classified as carbohydrates (2-deoxyribose, digitoxose, glucuronic and gluconic acids, the amino sugars) possess structures that cannot be represented by such a formula. On a functional group basis carbohydrates are characterized as polyhydroxy aldehydes or polyhydroxy ketones and their derivatives.

Carbohydrates are extensively distributed in both the plant and animal kingdoms. Chlorophyll-containing plant cells produce carbohydrates by photosynthesis which involves the fixation of CO_2 via reduction by H_2O and requires solar electromagnetic energy. Carbohydrates serve as a source of energy for plants and animals, and in the form of cellulose they function as the supporting structures of plants. In plants and microorganisms carbohydrates are metabolized through various pathways leading to amino acids, purines, pyrimidines, fatty acids, vitamins, etc. Together with other dietary components (such as proteins, lipids, minerals and vitamins) some carbohydrates are metabolically utilized by animals in many processes, degraded to acetyl CoA for the synthesis of lipids or oxidized to obtain ATP; and in plants they are used for the synthesis of other organic compounds. Most of the carbohydrate which is utilizable by the human consists of starch, glycogen, sucrose, maltose or lactose, whereas cellulose, xylans and pectins cannot be degraded by digestive processes because of the lack of the appropriate enzymes.

As the foregoing statements indicate, the biological importance of carbohydrates is readily obvious. Various textbooks of biochemistry provide complete discussions of the chemistry and metabolism of carbohy-drates.[1,2,3] Moreover, in medicinal chemistry it is recognized that many pharmaceutic products contain carbohydrates or modified carbohydrates as therapeutic agents or as pharmaceutic necessities. Certain antibiotics are carbohydrate derivatives. The streptomycins, neomycins, paromomycins, gentamicins and kanamycins are basic carbohydrates which have significant antimicrobial properties.[4] The cardioactive glycosides represent another class of medicinal agents possessing carbohydrate moieties which contribute to their therapeutic efficacy (see Chap. 18).[5]

Some knowledge of the interrelationships of carbohydrates with lipids and proteins in human metabolism is necessary for the study of the medicinal biochemistry of diabetes mellitus and the actions of antidiabetic agents. Accordingly, a brief discussion of these topics will be presented later in this chapter for purposes of emphasizing how some factors affecting carbohydrate metabolism also affect metabolic processes involving lipids and proteins.

CLASSIFICATION

A brief review of elementary characterizations of the more important carbohydrates is fundamental to the understanding of the structural and functional differences among the vast array of natural products which are classified as carbohydrates. The following summary is intended to delineate and exemplify the major classes and types of carbohydrates.

It is conventional to classify carbohydrates as *monosaccharides, oligosaccharides* and *polysaccharides,* depending on the number of sugar residues present per molecule. Furthermore, monosaccharides containing three carbon atoms are called *trioses,* those containing four carbon atoms are *tetroses,* whereas *pentoses, hexoses* and *heptoses* contain five, six and seven carbon atoms, respectively. On a functional group basis, monosaccharides having a potential aldehyde group in addition to hydroxyl functions are known as *aldoses* and those bearing a ketone function are *ketoses.* For example, glyceraldehyde is an aldotriose and dihydroxyacetone is a ketotriose, whereas glucose is an aldohexose and fructose is a ketohexose.

Diasaccharides, trisaccharides and tetrasaccharides are oligosaccharides. Sucrose, lactose, maltose, cellobiose, gentiobiose and melibiose are important disaccharides. Raffinose, melecotose and gentianose are trisaccharides; stachyose is a tetrasaccharide.

Monosaccharides existing in the form of heterocycles are classified with respect to the size of the ring system; i.e., the 6-membered ring structures considered to be related to pyran are called *pyranoses* and the 5-membered ring structures related to furan are called *furanoses.* This type of nomenclature can be applied to oligosaccharides and glycoside derivatives. Thus, maltose can be named 4-D-glucopyranosyl-α-D-glucopyranoside, lactose is 4-D-glucopyranosyl-β-D-galactopyranoside and sucrose is 1-α-D-glucopyranosyl-β-D-fructofuranoside. (Stereochemical classification of carbohydrates is considered briefly below as the basis for the aforementioned configurational designations.)

Most carbohydrate material in nature exists as high molecular weight polysaccharides which on hydrolysis yield monosaccharides or their derivatives. Glucose, mannose, galactose, arabinose, and glucuronic, galacturonic and mannuronic acids and some amino sugars occur as structural components of polysaccharides, glucose being the most common component.

Polysaccharides yielding only one variety of monosaccharide are called *homopolysaccharides,* and those yielding a mixture of different monosaccharides are known as *heteropolysaccharides.* Homopolysaccharides of importance include the starches, cellulose and glycogen which when hydrolyzed yield glucose. Heparin, hyaluronic acid and the immunochemically specific polysaccharide of type III pneumococcus are representative examples of heteropolysaccharides. Heparin's polymeric structure is composed of a repeating monomer of glucuronic acid 2-sulfate and glucosamine N-sulfate with an additional sulfate at C-6 (see the abbreviated structure for heparin, below). Hyal-

Abbreviated Structure for Heparin

uronic acid contains glucuronic acid and N-acetyl glucosamine units, and the type III pneumococcus polysaccharide on hydrolysis yields glucose and glucuronic acid. These heteropolysaccharides contain two different sugars in each component monomer. Much more complex polysaccharides contain more than two monosaccharides; e.g., gums and mucilages upon hydrolysis yield galactose, arabinose, xylose and glucuronic and galacturonic acids.

Research in the field of structure-activity relationships among polysaccharides continues to increase the understanding of the relationship between their conformations in solution and biological function. It has been noted that polysaccharides of the pyranose forms of glucose, galactose, mannose, xylose and arabinose have conformations that are restricted by steric factors. Such polysaccharides have been characterized and classified on the basis of conformation properties: type A, extended and ribbonlike; type B, helical and flexible; type C, rigid and crumpled; and type D, very flexible and extended. It is interesting to note that most support materials are categorized in type A and most matrix materials belong to type B. Cellulose and chitin form rigid structures, and these polysaccharides are the most important structural polysaccharides in nature. Matrix materials form gels, and this property is fundamental to their biological functions. It has been suggested that some matrix materials produce gels by forming double helices. Hyaluronic acid has been studied in this regard and its gelling properties appear to be dependent on double helix formation.[6] The foregoing summary and particularly the reference cited illustrate the significance of polysaccharides as the fibrous and matrix materials in support structures of plants and animal organisms.

Many different carbohydrates occur as components of glycoproteins. The term *glycoprotein* is used in a general sense and includes proteins that contain covalently bonded carbohydrate. Glycoproteins are widely distributed in animal tissues, and some have been found in plants and microorganisms. All plasma proteins except albumin, proteins of mucous secretions, some hormones (e.g., thyroglobulin, chorionic gonadotropin), certain enzymes (e.g., serum cholinesterase, deoxyribonuclease), components of cellular and extracellular membranes, and constituents of connective tissue are classified as glycoproteins. The bonding of the carbohydrate moiety to the peptide usually involves C-1 of the most internal sugar and a functional group of an amino acid within the peptide chain; e.g., the linkage of N-acetylglucosamine through a β-glycosidic bond to the amide group of asparagine. The metabolism of glycoproteins has been studied by numerous investigators, and the major studies have been recently reviewed.[7]

Glycolipids are carbohydrates containing lipids, and some are derivatives of sphingosine. The carbohydrates containing derivatives of ceramides are called *glycosphingolipids*. Under normal circumstances there is a steady state of balance between the synthesis and catabolism of glycosphingolipids in all cells. In the absence of any one of the hydrolases necessary for degradation, there is abnormal accumulation of intermediate metabolites, particularly in nervous tissue, which leads to various sphingolipodystrophies. There are three classes of glycosphingolipids: cerebrosides, gangliosides and ceramide oligosaccharides.

Lipopolysaccharides of gram-negative bacteria have been studied with emphasis on structural elucidation. The peripheral portions of the lipopolysaccharides, called O-antigens, are composed of various carbohydrates arranged as oligosaccharide-repeating units forming high molecular weight polysaccharides. Structural details differ with the serotype of the organism. Some somatic O-antigens are highly toxic to animals. The lipopolysaccharide of the *Enterobacteriaceae* is one of the most complex of all polysaccharides, if not the most complex carbohydrate known. This polysaccharide has a gross structure whose carbohydrate moiety, the outermost portion, consists of abequose, mannose, rhamnose, galactose and N-acetylglucosamine units; the lipid fraction includes glucosamine, phosphate, acetate and β-hydroxymyristic acid.

BIOSYNTHESIS

Photosynthesis proceeds in the chlorophyll-containing cells of plants. The photosynthetic process involves the absorption of radiant energy by chlorophyll, and the conversion of the absorbed light energy into chemical energy. This chemical energy is necessary for the reduction of CO_2 from the atmosphere to form glucose. The so-formed glucose may be metabolized by the plant cells, forming other carbohydrates, degraded to form precursors for the synthesis of other organic compounds, and oxidized as an energy source for the plant's physiology.

In higher plants sucrose is synthesized via the activated form of glucose, uridine diphosphoglucose (UDPG) and fructose.

Polysaccharide biosynthesis also requires UDPG. For illustration of polysaccharide formation, consider glycogenesis in hepatic tissue: a glycogen synthetase enzyme catalyzes the polymerization of glucose units from UDPG. The latter is obtained from the reaction between glucose-1-phosphate and uridine triphosphate (UTP). It is noteworthy that UDPG performs an important role in the formation of the glycosidic linkage fundamental to the structure of oligosaccharides and polysaccharides. Analogous UDP compounds involving other monosaccharides are utilized in the biosynthesis of polysaccharides containing these sugars. The biosynthesis of cellulose is supposed to occur through the guanine-containing analog of UDPG, guanosine diphosphoglucose (GDPG).

STEREOCHEMICAL CONSIDERATIONS

Basic organic chemistry textbooks cover the principles of stereoisomerism relevant to the study of carbohydrates.* The configurational and conformational aspects of carbohydrates have been recently reviewed by Bentley.[8] The stereochemistry of carbohydrates has presented many challenges to scientists, and there are several books[3,9] which treat this subject comprehensively; hence, here only a brief resumé is presented. Stoddard[9] reviewed stereochemical studies, including nomenclature, on the basis of conformational analysis. In addition to configurational designations (e.g., β-D-glucopyranose), italic letters are used to specify conformation: *C*, chair; *B*, boat; *S*, twist boat; *H*, half-chair; etc. As an illustration consider the structure of β-D-glucose: according to this system, conformation is defined by numerals indicating ring atoms lying above or below a defined reference plane; in structure 1 below for β-D-(+)-glucose the reference plane contains C-2, C-3, C-5 and O; C-4 is above the plane and C-1 is below; hence, this conformation is designated as $^4C_1$ (compare with α-D-glucose). These symbols were proposed by the British Carbohydrate Nomenclature Committee, and they seem to be receiving general usage.

Advances in studies of configurational and conformational features of carbohydrate structures have been facilitated by x-ray crystallography, nuclear magnetic resonance spectroscopy, and mass spectrometry combined with gas chromatography.

---

* Symbols *d* and *1* or (+) and (−) are used to designate sign of rotation of plane-polarized light, and the configuration is designated by the symbols in small capitals D and L; monosaccharides are designated as D or L on the basis of the configuration of the highest-numbered asymmetric carbon, the carbonyl being at the top: D if the —OH is on the right and the L if the —OH is on the left. (+)-Mannose, (−)-arabinose are assigned to the D-family because of their relation to D-(+)-glucose and D-(+)-glyceraldehyde. Thus, sugars configurationally related to D-glyceraldehyde are said to be members of the D-family, and those related to L-glyceraldehyde belong to the L-series.

β-D-(+)-Glucose

α-D-(+)-Glucose

INTERRELATIONSHIPS WITH LIPIDS AND PROTEINS

Various interrelationships of carbohydrates with proteins and lipids have been noted above as glycoproteins and glycolipids were characterized. Many other relationships exist among metabolic processes involving carbohydrates, lipids and proteins. Some of these relationships exemplify how regulation of metabolism is maintained via numerous mechanisms, e.g., feedback regulation and hormonal regulation.

The Krebs' citrate cycle requires acetyl CoA, and this requirement is satisfied by glycolysis and pyruvic acid decarboxylation, by the β-oxidation of fatty acids, by oxidation of glycerol via the glycolytic pathway, and by pyruvic acid from alanine transamination. Even more correlations are found within the steps of the Krebs' cycle. Oxaloacetate can be transformed into aspartate or into phosphoenolpyruvate and subsequently to carbohydrates. Such transformation of oxaloacetate into carbohydrates is the metabolic route of gluconeogenesis. Lactate from anaerobic glycolysis is the major starting material for gluconeogenesis; the lactate (via pyruvate and carboxylation of pyruvate) is transformed into the key intermediate phosphoenolpyruvate.

In anabolism, acetyl CoA which can originate from carbohydrates, proteins and lipids is utilized in the synthesis of important metabolites such as steroids and fatty acids.

Appreciation of the foregoing correlations between so many important metabolic and anabolic processes facilitates the understanding of how and why factors affecting certain processes directly also affect other processes indirectly.

Glucose must be activated via formation of UDP-glucose (UDPG) prior to utilization in glycogenesis. The enzyme glycogen-synthetase catalyzes the transformation of UDPG into glycogen. Glycogen catabolism to glucose proceeds through the action of phosphorylase-a which catalyzes phosphorolysis of glycogen, providing glucose-1-phosphate; the latter then enters glycolysis. At this point, the dynamism of hormonal regulation can be illustrated by referring to the following phenomena. If and when the blood glucose concentration decreases below the normal level, epinephrine (from the adrenal medulla) activates adenylcyclase which catalyzes formation of c-AMP (cyclic-3′,5′-adenosinemonophosphate). c-AMP is a general mediator of many hormone actions, and herein stimulates the activation of phosphorylase-b (inactive) providing phosphorylase-a (active). Thus, due to phosphorylase-a action the net effect promotes glycogen breakdown, leading to glucose-1-phosphate and an increase in blood glucose concentration. This accounts for epinephrine's hyperglycemic action and also explains how epinephrine agonists can affect carbohydrate metabolism, whereas opposite effects can be expected from epinephrine antagonism. Newton and Hornbrook[10] investigated the metabolic effects exerted by adrenergic agonists and antagonists and concluded that the order of potency was isoproterenol > norepinephrine > salbutamol, with respect to stimulation of rat liver adenylcyclase; similar order of potency is reported for increased rat liver phosphorylase activity, but epinephrine produced a greater maximal response than isoproterenol. These authors also report that the β-adrenergic antagonist propranolol blocked the effects of isoproterenol or epinephrine on adenylcyclase, whereas the α-adrenergic blockers ergotamine and phenoxybenzamine produced only partial inhibition. It is therefore noteworthy that the adrenergic metabolic receptor in the liver (rat) reacts to agonists and antagonists in parallelism with the responses of those receptors in other tissues which have been designated β-adrenergic receptors.

Abnormally low blood glucose levels also stimulate pancreatic α-cells to release the hormone glucagon, another hyperglycemic hormone. Glucagon, much like epinephrine, activates adenylcyclase, promoting c-AMP formation, etc., leading to enhancement of glycogen catabolism, but glucagon affects liver cells and epinephrine affects both muscle and liver cells.

Adrenocortical hormones (i.e., the glucocorticoids) affect carbohydrate metabolism by promoting gluconeogenesis and glycogen formation. Since gluconeogenesis from amino acids is enhanced, and since these hormones also inhibit protein synthesis in nonhepatic tissues, the precursor amino acids are made available for gluconeogenesis in the liver. Glucocorticoids stimulate synthesis of specific proteins in liver while inhibiting protein formation in muscle and other tissues. As protein catabolism continues in these tissues, the ultimate result is a net protein catabolic effect.

Sufficiently high blood glucose concentration stimulates pancreatic β-cells to secrete the hypoglycemic hormone insulin. Insulin exerts numerous biochemical actions,[11] affecting not only carbohydrate metabolism but also lipid and protein metabolism; glycogenesis, lipogenesis and protein synthesis are enhanced by insulin, whereas ketogenesis from fatty acids, glycogenolysis and lipolysis are processes which are suppressed by insulin. (Of course, insulin deficiency leads to the opposite effects on these processes.) It is clear that insulin modifies the reaction rates of many processes in its target cells, and highly specific insulin-receptor interactions have been implicated.[12] Insulin receptors on adipose and liver cells have been characterized and they appear to have uniform characteristics. Experimental evidence limits insulin action to the plasma membrane of target cells. Insulin-receptor interactions can lead to modulation of other hormone actions through mechanisms involving c-AMP-phosphodiesterase activation and inhibition of adenylcyclase activation. c-AMP-phosphodiesterase is responsible for catalyzing c-AMP hydrolysis which inactivates c-AMP; thus insulin activation of this phosphodiesterase results in reversal of the metabolic effects of hormones which act through c-AMP. On the other hand, it is of interest to note that many other compounds show capability of activating phosphodiesterase, e.g., c-GMP and nicotinic acid, and even more compounds demonstrate inhibitory activity, e.g., xanthine derivatives, papaverine and related isoquinoline compounds and some adrenergic amines. The foregoing processes have been recently reviewed.[13]

The specific biochemical effects exerted by insulin and glucagon are delineated in more detail in Chapter 20. Suffice it to say here that medicinal agents which promote insulin availability exert actions via insulin and also a variety of other effects. Consider the hypoglycemic sulfonylureas which stimulate insulin secretion also might act on phosphodiesterase inhibiting the inactivation of c-AMP; moreover, c-AMP has been implicated as a factor promoting insulin release; reportedly some sulfonylureas also reduce glucagon secretion.

The biguanide phenformin was removed from the pharmeceutical market in 1977 because of the serious side-effect called *lactic acidosis*.[14] The hypoglycemic action of phenformin involves various factors that promote glucose use. Although the exact molecular mechanism of action is unclear, it is known that phenformin promotes anaerobic glycolysis and exerts other effects. The action on anaerobic glycolysis is the effect that is responsible for excessive lactic-acid formation (from pyruvic-acid reduction) and the development of lactic acidosis.

Feedback regulation of enzyme-catalyzed reactions is another basic mechanism for the regulation of metabolism, i.e., allosteric inhibition of a key enzyme. Phosphofructokinase, the pacemaker enzyme of glycolysis, is inhibited allosterically by ATP, and through such modulation ATP suppresses carbohydrate catabolism. Of course there are other cases of feedback regulation. Atkinson's classic article[15] on phenomena associated with biological feedback control at the molecular level should be consulted in order to compare negative and positive feedback regulation. AMP, in contrast to the effect of ATP, can exert positive regulatory action on phosphofructokinase. The regulatory metabolite acting as modulator modifies the affinity of the enzyme for its substrates, and the terms positive and negative are used to indicate whether there is an increase or a decrease in affinity.

SUGAR ALCOHOLS

Sorbitol, glucitol, mannitol, galactitol and dulcitol are natural products which are so closely related to the carbohydrates that it is traditional to classify them as carbohydrate derivatives, i.e., sugar alcohols. These compounds are reduction products of the corresponding aldohexoses, glucose, mannose and galactose, respectively. Therefore, such sugar alcohols are characterized as hexahydroxy alcohols.

Sorbitol formation has recently been implicated as a factor contributing to complications of diabetes due to high glucose concentration in nerve and eye-lens cells. Cells that have excessive glucose tend to convert normally minor metabolic pathways to major processes; e.g., glucose reduction to form sorbitol. Sorbitol is not usually metabolized rapidly and cannot be effectively elminated by the cell; because it accumulates in high concentrations in lens cells of diabetic rodents, osmotic swelling of the lens cells occurs. This osmotic swelling of lens calls has been associated with the development of cataracts as a complication of diabetes. In diabetic rodents cataract formation can be prevented with agents that inhibit the aldose reductase enzyme that catalyzes glucose reduction to sorbitol.[16] Such aldose reductase inhibitors are therefore, under investigation as potential medicinal agents.[17] In this connection it should be noted that some hydantoins have demonstrated aldose reductase-inhibitory activity; for example, sorbinil is listed by the U.S.A.N. (United States Adopted Names) and U.S.P. Dictionary of Drug Names, U.S.P. Pharmacopeial Convention, Inc., 1980.

Sorbitol N.F is very water-soluble and produces sweet and viscous solutions. Hence it is used in the formulation of some food products, cosmetics and pharmaceuticals. Sorbitol Solution U.S.P. is a 70 percent w/w solution that contains at least 64 percent D-sorbitol, the balance being related sugar alcohols. Upon dehydration it forms tetrahydropyran and tetrahydrofuran derivatives, the fatty acid monoesters of which are the nonionic surface-active agents called Spans®. Alternatively, these dehydration products react with ethylene oxide to form the Tweens® which are also useful surfactants.

Mannitol U.S.P. is a useful medicinal. It acts as an osmotic diuretic and is administered intravenously. After intravenous infusion (in the form of a sterile 25 percent solution), it is filtered by glomeruli and passes unchanged through the kidneys into the urine; however, while in the proximal tubules, the loops of Henle, the distal tubules and the collecting ducts, mannitol increases the osmotic gradient against which these structures absorb water and solutes. Due to the foregoing osmotic effect, the urinary water, sodium and chloride ions are increased. Mannitol is also indicated as an irrigating solution in transurethral prostatic resection.

Mannitol is also widely used as an excipient in chew tablets. In contrast to sorbitol, it is nonhygroscopic. In addition, it has a sweet and cooling taste.

SUGARS

Dextrose U.S.P., D-(+)-glucopyranose, grape sugar, D-glucose, glucose. Dextrose is a sugar usually obtained by the hydrolysis of starch. It can be either α-D-glucopyranose or β-D-glucopyranose or a mixture of the two. A large amount of the dextrose of commerce, whether crystalline or syrupy, usually is obtained by the acid hydrolysis of corn starch, although other starches can be used.

Although some free glucose occurs in plants and animals, most of it occurs in starches, cellulose, glycogen and sucrose. It also is found in other polysaccharides, oligosaccharides and glycosides.

Dextrose occurs as colorless crystals or as a white, crystalline or granular powder. It is odorless and has a sweet taste. One g. of dextrose dissolves in about 1 ml.

of water and in about 100 ml. of alcohol. It is more soluble in boiling water and in boiling alcohol.

Aqueous solutions of glucose can be sterilized by autoclaving.

Glucose can be used as a ready source of energy in various forms of starvation. It is the sugar found in the blood of animals and in the reserve polysaccharide glycogen which is present in the liver and muscle. It can be used in solution intravenously to supply fluid and to sustain the blood volume temporarily. It has been used in the management of the shock which may follow the administration of insulin used in the treatment of schizophrenia. This, because a "hypoglycemia" results from the use of insulin in this type of therapy, and the "hypoglycemic" state can be reversed by the use of dextrose intravenously. When dextrose is used intravenously, its solutions (5 to 50%) usually are made with physiologic salt solution or Ringer's solution. The dextrose used for intravenous injection must conform to the *U.S.P.* requirements for dextrose.

Liquid Glucose N.F. is a product obtained by the incomplete hydrolysis of starch. It consists chiefly of dextrose (D-glucose, $C_6H_{12}O_6$), with dextrins, maltose and water. This glucose usually is prepared by the partial acid hydrolysis of cornstarch and, hence, the common name corn syrup and other trade names refer to a product similar to liquid glucose. The official product contains not more than 21 percent of water.

Liquid glucose is a colorless or yellowish, thick, syrupy liquid. It is odorless, or nearly so, and has a sweet taste. Liquid glucose is very soluble in water, but is sparingly soluble in alcohol.

Liquid glucose is used extensively as a food (sweetening agent) for both infants and adults. It is used in the massing of pills, in the preparation of pilular extracts and for other similar uses. It is not to be used intravenously.

Calcium Gluconate U.S.P. The gluconic acid used in the preparation of calcium gluconate can be prepared by the electrolytic oxidation of glucose as follows:

D-Glucose → (NaBr, CaCO₃, Carbon Electrodes) → Calcium Gluconate

Gluconic acid is produced on a commercial scale by the action of a number of fungi, bacteria and molds upon 25 to 40 percent solutions of glucose. The fermentation is best carried out in the presence of calcium carbonate and oxygen to give almost quantitative yields of gluconic acid. A number of organisms can be used, for example, *Acetobacter oxydans, A. aceti, A. rancens, B. gluconicum, A. xylinum, A. roseus* and *Penicillium chrysogenum*. The fermentation is complete in 8 to 18 days.

Calcium gluconate occurs as a white, crystalline or granular powder without odor or taste. It is stable in air. Its solutions are neutral to litmus paper. One g. of calcium gluconate dissolves slowly in about 30 ml. of water and in 5 ml of boiling water. Each ml. of a 10 percent solution represents 9.3 mg. of calcium. It is insoluble in alcohol and in many other organic solvents.

Calcium gluconate will be decomposed by the mineral acids and other acids that are stronger than gluconic acid. It is incompatible with soluble sulfates, carbonates, bicarbonates, citrates, tartrates, salicylates and benzoates.

Calcium gluconate fills the need for a soluble, nontoxic well-tolerated form of calcium that can be employed orally, intramuscularly or intravenously. Calcium therapy is indicated in conditions such as parathyroid deficiency (tetany), general calcium deficiency, and when calcium is the limiting factor in increased clotting time of the blood. It can be used both orally and intravenously.

Calcium Gluceptate, U.S.P., calcium glucoheptonate, is a sterile, aqueous, approximately neutral solution of the calcium salt of glucoheptonic acid, a homolog of gluconic acid. Each ml. of Calcium Gluceptate Injection U.S.P. represents 18 mg. of Ca. Its uses and actions are the same as those of calcium gluconate.

Ferrous Gluconate U.S.P., Fergon®, iron (2+) gluconate, occurs as a fine yellowish-gray or pale greenish-yellow powder with a slight odor like that of burnt sugar. One gram of this salt is soluble in 10 ml. of water; however, it is nearly insoluble in alcohol. A 5 percent aqueous solution is acid to litmus.

Ferrous gluconate can be administered orally or by injection for the utilization of its iron content.

Glucuronic Acid occurs naturally as a component of many gums, mucilages, hemicelluloses and in the mucopolysaccharide portion of a number of glycoproteins. It is used by animals and humans to detoxify such substances as camphor, menthol, phenol, salicylates and chloral hydrate. None of the above can be used to prepare glucuronic acid for commercial purposes. It is prepared by oxidizing the terminal primary alcohol group of glucose or a suitable derivative thereof, such as 1,2-isopropylidine-D-glucose. It is a

white, crystalline solid that is water-soluble and stable. It exhibits both aldehydic and acidic properties. It also may exist in a lactone form and as such is marketed under the name "Glucurone," an abbreviation of glucuronolactone.

D-Glucuronic Acid

An average of 60 percent effectiveness was obtained in the relief of certain arthritic conditions by the use of glucuronic acid. A possible rationale for the effectiveness of glucuronic acid in the treatment of arthritic conditions is based upon the fact that it is an important component of cartilage, nerve sheath, joint capsule tendon and joint fluid and intercellular cement substances. The dose is 500 mg. to 1.0 g. orally 4 times a day or 3 to 5 ml. of a 10 percent buffered solution given intramuscularly.

Fructose U.S.P., D-(−)-fructose, levulose, β-D-(−)-fructopyranose, is a sugar usually obtained by hydrolysis of aqueous solutions of sucrose and subsequent separation of fructose from glucose.* It occurs as colorless crystals or as a white or granular powder that is odorless and has a sweet taste. It is soluble 1:15 in alcohol and is freely soluble in water. Fructose is considerably more sensitive to heat and decomposition than is glucose and this is especially true in the presence of bases.

D-Fructose

Fructose (a 2-ketohexose) can be utilized to a greater extent than glucose by diabetics and by patients who must be fed by the intravenous route.

Lactose U.S.P., saccharum lactis, milk sugar, is a sugar obtained from milk. Lactose is a by-product of whey, which is the portion of milk that is left after the

---

* The crystalline form of fructose is the β-anomer having a 6-membered ring, but when dissolved in water it is converted not only to the α form but the α and β forms of fructofuranose are formed also. The fructofuranose forms were called "gamma" sugars.

TABLE 19-1

SUGAR PRODUCTS

Name	Preparations	Category	Application	Usual Adult Dose*	Usual Dose Range*	Usual Pediatric Dose*
Dextrose U.S.P.	Dextrose Injection U.S.P.	Fluid and nutrient replenisher		I.V. infusion, 1 liter		
	Dextrose and Sodium Chloride Injection U.S.P.	Fluid, nutrient, and electrolyte replenisher		I.V. infusion, 1 liter		
	Anticoagulant Citrate Dextrose Solution U.S.P.	Anticoagulant for storage of whole blood	For use in the proportion of 75 ml. of Solution A or 125 ml. of Solution B for each 500 ml. of whole blood			
	Anticoagulant Citrate Phosphate Dextrose Solution U.S.P.	Anticoagulant for storage of whole blood	For use in the proportion of 70 ml. of solution for each 500 ml. of whole blood			
Calcium Gluconate U.S.P.	Calcium Gluconate Injection U.S.P.	Calcium replenisher		I.V., 10 ml. of a 10 percent solution at a rate not exceeding 0.5 ml. per minute at intervals of 1 to 3 days	1 g., weekly to 15 g. daily	125 mg. per kg. of body weight or 3 g. per square meter of body surface, up to 4 times daily, diluted and given slowly
	Calcium Gluconate Tablets U.S.P.	Calcium replenisher		1 g. 3 or more times daily	1 to 15 g. daily	125 mg. per kg. of body weight or 3 g. per square meter of body surface, up to 4 times daily
Ferrous Gluconate U.S.P.	Ferrous Gluconate Capsules U.S.P. Ferrous Gluconate Tablets U.S.P.	Iron supplement		300 mg. 3 times daily	200 to 600 mg.	
Fructose U.S.P.	Fructose Injection U.S.P.	Fluid replenisher and nutrient		I.V. and S.C., as required		
	Fructose and Sodium Chloride Injection U.S.P.	Fluid replenisher, nutrient, and electrolyte replenisher		I.V. and S.C., as required		
Lactose U.S.P.		Pharmaceutic aid (tablet and capsule diluent)				
Sucrose N.F.	Compressible Sugar N.F. Confectioner's Sugar N.F.	Pharmaceutic aid (sweetening agent; tablet excipient)				

* See *U.S.P. D.I.* for complete dosage information.

fat and the casein have been removed for the production of butter and cheese. Cows' milk contains 2.5 to 3 percent of lactose, whereas that of other mammals contains 3 to 5 percent. Although common lactose is a mixture of the *alpha* and *beta* forms, the pure *beta* form is sweeter than the slightly sweet-tasting mixture.

Lactose occurs as white, hard, crystalline masses or as a white powder. It is odorless, and has a faintly sweet taste. It is stable in air, but readily absorbs odors. Its solutions are neutral to litmus paper. One g. of lactose dissolves in 5 ml. of water, and in 2.6 ml. of boiling water. Lactose is very slightly soluble in alcohol and is insoluble in chloroform and in ether.

Lactose is hydrolyzed readily in acid solutions to yield one molecule each of D-glucose and D-galactose. It reduces Fehling's solution.

Lactose is used as a diluent in tablets and powders and as a nutrient for infants.

β-Lactose when applied locally to the vagina brings

α-Lactose

Galactose   Glucose

β-Lactose

Galactose   Glucose

about a desirable lower pH. The lactose probably is fermented, with the production of lactic acid.

Maltose or malt sugar, 4-D-glucopyranosyl-α-D-glucopyranoside, is an end-product of the enzymatic hydrolysis of starch by the enzyme diastase. It is a reducing disaccharide that is fermentable and is hydrolyzed by acids or the enzyme maltose to yield 2 molecules of glucose.

Maltose is a constituent of malt extract and is used for its nutritional value for infants and adult invalids.

Malt Extract is a product obtained by extracting malt, the partially and artificially germinated grain of one or more varieties of *Hordeum vulgare* Linné (*Fam.* Gramineae). Malt extract contains maltose, dextrins, a small amount of glucose and amylolytic enzymes.

Malt extract is used in the brewing industry because of its enzyme content which converts starches to fermentable sugars. It also is used in infant feeding for its nutritive value and laxative effect.

The usual dose is 15 g.

Dextrins are obtained by the enzymatic (diastase) degradation of starch. These degradation products vary in molecular weight in the following decreasing order: amylodextrin, erythodextrin and achroodextrin. Lack of homogeneity precludes the assignment of definite molecular weights. With the decrease in molecular weight, the color produced with iodine changes from blue to red to colorless.

Dextrin occurs as a white, amorphous powder that is incompletely soluble in cold water but freely soluble in hot water.

Dextrins are used extensively as a source of readily digestible carbohydrate for infants and adult invalids.

They often are combined with maltose or other sugars.

Sucrose N.F., saccharum, sugar, cane sugar, beet sugar. Sucrose is a sugar obtained from *Saccharum officinarum* Linné (*Fam.* Gramineae), *Beta vulgaris* Linné (*Fam.* Chenopodiaceae), and other sources. Sugar cane (15 to 20% sucrose) is expressed, and the juice is treated with lime to neutralize the plant acids. The water-soluble proteins are coagulated by heat and are removed by skimming. The resultant liquid is decolorized by means of charcoal and concentrated. Upon cooling, the sucrose crystallizes out. The mother liquor, upon concentration, yields more sucrose and brown sugar and molasses.

Sucrose occurs as colorless or white crystalline masses or blocks, or as a white, crystalline powder. It is odorless, has a sweet taste, and is stable in air. Its solutions are neutral to litmus. One g. of sucrose dissolves in 0.5 ml. of water and in 170 ml. of alcohol.

Sucrose does not respond to the tests for reducing sugars, i.e., reduction of Fehling's solution and others. It is hydrolyzed readily, even in the cold, by acid solutions to give one molecule each of D-glucose and D-fructose. This hydrolysis also can be effected by the enzyme invertase. Sucrose caramelizes at about 210°.

Sucrose is used in the preparation of syrups and as a diluent and sweetening agent in a number of pharmaceutic products, e.g., troches, lozenges and powdered extracts. In a concentration of 800 mg. per ml., sucrose is used as a sclerosing agent.

Invert sugar, Travert®, is a hydrolyzed product of sucrose (invert sugar) prepared for intravenous use.

Xylose U.S.P. is used as a diagnostic aid in testing for intestinal absorptive capacity in the diagnosis of celiac disease.

α-D-Xylose

STARCH AND DERIVATIVES

Starch N.F., amylum, cornstarch, consists of the granules separated from the grain of *Zea mays* Linné (*Fam.* Gramineae). Corn, which contains about 75 percent dry weight of starch, is first steeped with sulfurous acid and then milled to remove the germ and the seed coats. It then is milled with cold water, and the starch is collected and washed by screens and flotation. Starch is a high molecular weight carbohydrate composed of 10 to 20 percent of a hot-water-soluble "amylose" and 80 to 90 percent of a hot-water-insoluble "amylopectin." Amylose is hydrolyzed completely to maltose by the enzyme β-amylase, whereas amylopectin is hydrolyzed only incompletely (60%) to maltose. The glucose residues are in the form of branched chains in the amylopectin molecule. The chief linkages of the glucose units in starch are α-1,4, since β-amylase hydrolyzes only *alpha* linkages and maltose is 4-D-glucopyranosyl-α-D-glucopyranoside.

Starch occurs as irregular, angular, white masses or as a fine powder, and consists chiefly of polygonal, rounded or spheroidal grains from 3 to 35 microns in diameter and usually with a circular or several-rayed central cleft. It is odorless and has a slight characteristic taste. Starch is insoluble in cold water and in alcohol.

Amylose gives a blue color on treatment with iodine, and amylopectin gives a violet to red-violet color.

Starch is used as an absorbent in starch pastes, as an emollient in the form of a glycerite and in tablets and powders.

Pregelatinized Starch N.F. This is starch which has been modified to make it suitable for use as a tablet excipient. It has been processed in the presence of water to rupture most of the starch granules and then dried.

CELLULOSE AND DERIVATIVES

Cellulose is the name generally given to a group of very closely allied substances rather than to a single entity. The celluloses are anhydrides of β-glucose, possibly existing as long chains that are not branched, consisting of 100 to 200 β-glucose residues. These chains may be cross-linked by residual valences (hydrogen bonds) to produce the supporting structures of the cell walls of plants. The cell walls found in cotton, pappi on certain fruits and other sources are the purest forms of cellulose; however, because they are cell walls, they enclose varying amounts of substances that are proteinaceous, waxy, or fatty. These, of course, must be removed by proper treatment in order to obtain pure cellulose. Cellulose from almost all other sources is combined by ester linkages, glycoside linkages and other combining forms with encrusting substances, such as lignin, hemicelluloses, pectins. These can be removed by steam under pressure, weak acid or alkali solutions, and sodium bisulfite and sulfurous acid. Plant celluloses, especially those found in wood, can be resolved into β-cellulose, which is soluble in 17.5 percent sodium hydroxide, and alkali-insoluble α-cellulose. The cellulose molecule can be depicted in part as shown below.

Purified Cotton U.S.P. is the hair of the seed of cultivated varieties of *Gossypium hirsutum* Linné, or of other species of *Gossypium* (*Fam.* Malvaceae), freed from adhering impurities, deprived of fatty matter, bleached and sterilized.

Microcrystalline Cellulose N.F. is purified, partially depolymerized cellulose prepared by treating alpha cellulose, obtained as a pulp from fibrous plant material, with mineral acids.

It occurs as a fine, white, odorless crystalline powder that is insoluble in water, in dilute alkalies and in most organic solvents.

Methylcellulose U.S.P., Syncelose®, Cellothyl®, Methocel®, is a methyl ether of cellulose whose methoxyl content varies between 26 and 33 percent. A 2 percent solution has a centipoise range of not less than 80 and not more than 120 percent of the labeled amount when such is 100 or less, and not less than 75 and not more than 140 percent of the labeled amount for viscosity types higher than 100 centipoises.

Methyl- and ethylcellulose ethers (Ethocel®) can be prepared by the action of methyl and ethyl chlorides or methyl and ethyl sulfates, respectively, on cellulose that has been previously treated with alkali. Purification is accomplished by washing the reaction product with hot water. The degree of methylation or ethylation can be controlled to yield products that vary in their viscosities when they are in solution. Seven viscosity types of methylcellulose are produced commercially and have the following centipoise values: 10, 15, 25, 100, 400, 1,500 and 4,000, respectively. Other intermediate viscosities can be obtained by the use of a blending chart. The ethylcelluloses have similar properties.

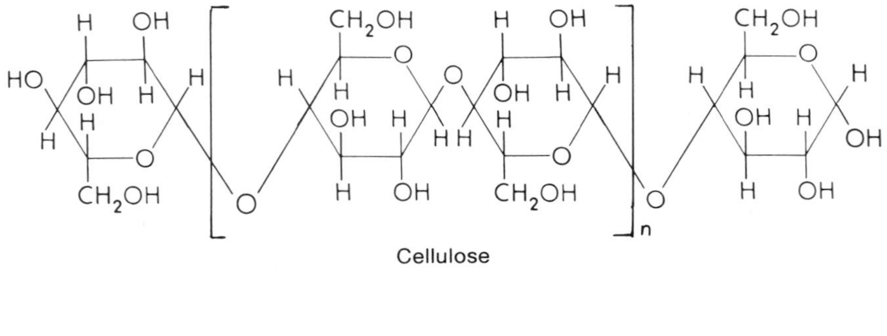

Cellulose

Methylated celluloses of a lower methoxy content are soluble in cold water, but, in contrast to the naturally occurring gums, they are insoluble in hot water and are precipitated out of solution at or near the boiling point. Solutions of powdered methylcellulose can be prepared most readily by first mixing the powder thoroughly with one-fifth to one-third of the required water as hot water (80° to 90°) and allowing it to macerate for 20 to 30 minutes. The remaining water then is added as cold water. With the increase in methoxy content, the solubility in water decreases until complete water-insolubility is reached.

Methylcellulose resembles cotton in appearance and is neutral, odorless, tasteless and inert. It swells in water and produces a clear to opalescent, viscous, colloidal solution. Methylcellulose is insoluble in most of the common organic solvents. On the other hand, aqueous solutions of methylcellulose can be diluted with ethanol.

Methylcellulose solutions are stable over a wide range of pH (2 to 12) with no apparent change in viscosity. The solutions do not ferment and will carry large quantities of univalent ions, such as iodides, bromides, chlorides and thiocyanates. However, smaller amounts of polyvalent ions, such as sulfates, phosphates, carbonates and tannic acid or sodium formaldehyde sulfoxylate, will cause precipitation or coagulation.

The methylcelluloses are used as substitutes for the natural gums and mucilages, such as gum tragacanth, gum karaya, chrondrus or quince seed mucilage. They can be used as bulk laxatives and in nose drops, ophthalmic preparations, burn preparations, ointments and like preparations. Although methylcellulose when used as a bulk laxative takes up water quite uniformly, tablets of methylcellulose have caused fecal impaction and intestinal obstruction. Commercial products include Hydrolose® Syrup, Anatex®, Cologel® Liquid, Premocel® Tablets and Valocall®. In general, methylcellulose of the 1,500 or 4,000 cps. viscosity type is the most useful as a thickening agent when used in 2 to 4 percent concentrations. For example, a 2.5 percent concentration of a 4,000 cps. type methylcellulose will produce a solution with a viscosity obtained by 1.25 to 1.75 percent of tragacanth.

Ethylcellulose N.F. is an ethyl ether of cellulose containing not less than 45 percent and not more than 50 percent of ethoxy groups and is prepared from ethyl chloride and cellulose. It occurs as a free-flowing, stable white powder that is insoluble in water, glycerin and propylene glycol but is freely soluble in alcohol, ethyl acetate or chloroform. Aqueous suspensions are neutral to litmus. Films prepared from organic solvents are stable, clear, continuous, flammable and tough.

Hydroxypropyl Methylcellulose 2208 U.S.P., propylene glycol ether of methyl cellulose, contains a degree of substitution of not less than 19 and not more than 24 percent as methoxyl groups (OCH_3), and not less than 4 and not more than 12 percent as hydroxypropyl groups (OC_3H_6OH). It occurs as a white, fibrous or granular powder that swells in water to produce a clear to opalescent, viscous, colloidal solution.

Oxidized Cellulose U.S.P., Oxycel®, when thoroughly dry contains not less than 16 nor more than 24 percent of carboxyl groups. Oxidized cellulose is cellulose in which a part of the terminal primary alcohol groups of the glucose residues have been converted to carboxyl groups. Therefore, the product is possibly a synthetic polyanhydrocellobiuronide. Although the *U.S.P.* accepts carboxyl contents as high as 24 percent, it is reported that products which contain 25 percent carboxyl groups are too brittle (friable) and too readily soluble to be of use. Those products which have lower carboxyl contents are the most desirable. Oxidized cellulose is slightly off-white in color, is acid to the taste and possesses a slight, charred odor. It is prepared by the action of nitrogen dioxide, or a mixture of nitrogen dioxide and nitrogen tetroxide, upon cellulose fabrics at ordinary temperatures. Because cellulose is a high molecular weight carbohydrate composed of glucose residues joined 1,4- to each other in their *beta* forms, the reaction must be as shown below on the cellulose molecule in part.

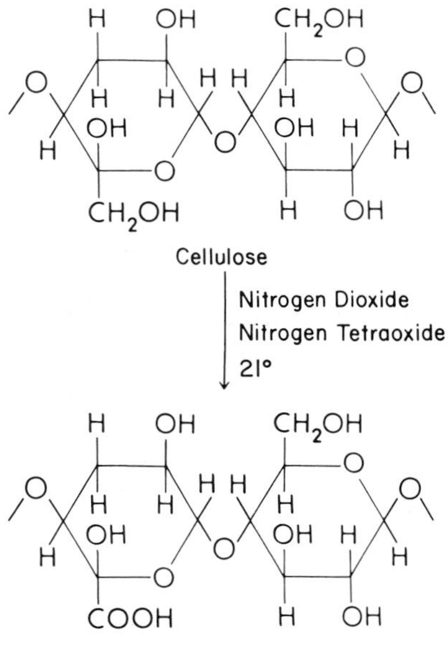

Cellulose

Nitrogen Dioxide
Nitrogen Tetraoxide
21°

Oxidized Cellulose

The oxidized cellulose fabric, such as gauze or cotton, resembles the parent substance. It is insoluble in water and in acids but is soluble in dilute alkalies. In weakly alkaline solutions, it swells and becomes translucent and gelatinous. When wet with blood, it becomes slightly sticky and swells, forming a dark brown, gelatinous mass. Oxidized cellulose cannot be sterilized by autoclaving. Special methods are needed to render it sterile.

Oxidized cellulose has noteworthy hemostatic properties. However, when it is used in conjunction with thrombin, it should be neutralized previously with a solution of sodium bicarbonate. It is used in various surgical procedures in much the same way as gauze or cotton, by direct application to the oozing surface. Except when used for hemostasis, it is not recommended as a surface dressing for open wounds. Oxidized cellulose implants in connective tissue, muscle, bone, serous and synovial cavities, brain, thyroid, liver, kidney and spleen were absorbed completely in varying lengths of time, depending on the amount of material introduced, the extent of operative trauma and the amount of blood present.

Carboxymethylcellulose Sodium U.S.P., CMC®, sodium cellulose glycolate, is the sodium salt of a polycarboxymethyl ether of cellulose, containing, when dried, 6.5 to 9.5 percent of sodium. It is prepared by treating alkali cellulose with sodium chloroacetate. This procedure permits a control of the number of $—OCH_2COO^-$ Na^+ groups that are to be introduced. The number of $—OCH_2COO^-$ Na^+ groups introduced is related to the viscosity of aqueous solutions of these products. CMC® is available in various viscosities, i.e., 5 to 2,000 centipoises in 1 percent solutions. Therefore, high molecular weight polysaccharides containing carboxyl groups have been prepared whose properties in part resemble those of the naturally occurring polysaccharides, whose carboxyl groups contribute to their pharmaceutic and medicinal usefulness.

Carboxymethylcellulose sodium occurs as a hygroscopic white powder or granules. Aqueous solutions

TABLE 19-2

PHARMACEUTICALLY IMPORTANT CELLULOSE PRODUCTS

Name Proprietary Name	Preparations	Category	Usual Adult Dose*	Usual Dose Range*
Purified Cotton U.S.P.		Surgical aid		
Purified Rayon U.S.P.		Surgical aid		
Microcrystalline Cellulose N.F.		Pharmaceutic aid (tablet diluent)		
Powdered Cellulose N.F.		Pharmaceutic aid (tablet diluent; adsorbant; suspending agent)		
Methylcellulose U.S.P.		Pharmaceutic aid (suspending agent; tablet excipient; viscosity-increasing agent)		
	Methylcellulose Ophthalmic Solution U.S.P.	Topical protectant (ophthalmic)		
	Methylcellulose Tablets U.S.P.	Cathartic	1 to 1.5 g. 2 to 4 times daily	1 to 6 g. daily
Ethylcellulose N.F.		Pharmaceutic aid (tablet binder)		
Hydroxypropyl Methylcellulose U.S.P.		Pharmaceutic aid (suspending agent; tablet excipient; viscosity-increasing agent)	Topically to the conjunctiva, 0.05 to 0.1 ml. of a 0.5 to 2.5 percent solution 3 or 4 times daily, or as needed, as artificial tears or contact lens solution	
	Hydroxypropyl Methylcellulose Ophthalmic Solution U.S.P.	Topical protectant (ophthalmic)		
Oxidized Cellulose U.S.P. *Oxycel*		Local hemostatic	Topically as necessary to control hemorrhage.	
Carboxymethylcellulose Sodium U.S.P.		Pharmaceutic aid (suspending agent; tablet excipient; viscosity-increasing agent)		
	Carboxymethylcellulose Sodium Tablets U.S.P.	Cathartic	1.5 g. 3 times daily	
Pyroxylin U.S.P.		Pharmaceutic necessity for Collodion U.S.P.		
Cellulose Acetate Phthalate N.F.		Pharmaceutic aid (tablet coating agent)		

* See *U.S.P. D.I.* for complete dosage information.

may have a pH between 6.5 and 8. It is easily dispersed in cold or hot water to form colloidal solutions that are stable to metal salts and pH conditions from 2 to 10. It is insoluble in alcohol and organic solvents.

It can be used as an antacid but is more adaptable for use as a nontoxic, nondigestible, unabsorbable, hydrophilic gel as an emollient-type bulk laxative. Its bulk-forming properties are not as great as those of methylcellulose; on the other hand, its lubricating properties are superior, with little tendency to produce intestinal blockage.

Pyroxylin U.S.P., soluble guncotton, is a product obtained by the action of nitric and sulfuric acids on cotton, and consists chiefly of cellulose tetranitrate $[C_{12}H_{16}O_6(NO_3)_4]$. The glucose residues in the cellulose molecule contain 3 free hydroxyl groups which can be esterified. Two of these 3 hydroxyl groups are esterified to give the official pyroxylin, and, therefore, it is really a dinitrocellulose or cellulose dinitrate which conforms to the official nitrate content.

Pyroxylin occurs as a light yellow, matted mass of filaments, resembling raw cotton in appearance, but harsh to the touch. It is exceedingly flammable and decomposes when exposed to light, with the evolution of nitrous vapors and a carbonaceous residue. Pyroxylin dissolves slowly but completely in 25 parts of a mixture of 3 volumes of ether and 1 volume of alcohol.

In the form of collodion and flexible collodion, it is used for coating purposes per se or in conjunction with certain medicinal agents.

Cellulose Acetate Phthalate N.F. is a partial acetate ester of cellulose which has been reacted with phthalic anhydride. One carboxyl of the phthalic acid is esterified with the cellulose acetate. The finished product contains about 20 percent acetyl groups and about 35 percent phthalyl groups. In the acid form it is soluble in organic solvents and insoluble in water. The salt form is readily soluble in water. This combination of properties makes it useful in enteric coating of tablets because it is resistant to the acid condition of the stomach but is soluble in the more alkaline environment of the intestinal tract.

HEPARIN

Heparin is a mucopolysaccharide composed of α-D-glucuronic acid and 2-amino-2-deoxy-α-D-glucose units; these monosaccharide units are partially sulfated and are linked in the polymeric form through $1 \rightarrow 4$ linkages, as indicated by the structure shown at the beginning of the chapter. Heparin is present in animal tissue of practically all types but mainly in lung and liver tissue.[18]

The chemistry and pharmacology of heparin have been reviewed by Ehrlich and Stivala.[19] This review comprehensively covers most topics pertinent to medicinal chemistry. Heparin is included among the A.M.A. *Drug Evaluations, 1980* anticoagulants.[20] Its greatest use has been in the prevention and arrest of thrombosis. (See Chap. 20, on the biochemical functions performed by thrombin, fibrinogen and fibrin in normal blood coagulation.)

The mechanism of anticoagulant action exerted by heparin has been investigated from various standpoints, and now it is recognized that the mechanism involves the plasma protein inhibitor of serine proteases, which is called *antithrombin III*. This naturally occurring inhibitor inactivates various critical clotting factors, which are enzymes designated as IXa, Xa, XIa, thrombin, and perhaps also XIIa, that have a serine residue within the reactive center. Antithrombin III interacts with and inhibits these factors irreversibly. Heparin interacts with antithrombin III and induces conformational changes that complement the interaction between antithrombin III and the aforementioned factors.[21]

Jaques[22] summarized recent studies that have shown that heparin is a biochemical representative of a class of compounds characterized as linear anionic polyelectrolytes. Such compounds demonstrate interesting specific reactions with biologically active proteins, forming stable complexes that change the bioactivity of these proteins. These complexations increase the negative charge of cell surfaces, including those of the blood-vessel walls. The increase in the negative charge of the vessel wall is considered to be a factor that contributes to the prevention of thrombosis by heparin and similar compounds.

Heparin also affects fibrinolysis. It seems to reduce the inhibition of antifibrinolysin and thus enhances fibrinolysis. Heparin effects on platelets have been studied, and heparin was shown to prevent conversion of degenerated platelets in solution from forming a gel; heparin inhibits platelet adhesion to intercellular cement; it also prevents platelet disintegration and release of phospholipids. Another major effect of heparin is on blood lipids. Heparin stimulates the release of lipoprotein lipase, an enzyme that catalyzes the hydrolysis of triglycerides associated with chylomicrons, and through this action promotes the clearing of lipemic plasma. Research on heparin has included the investigation of the possible effect of heparin on tumor growth and metastasis. Some studies show that heparin is a miotic inhibitor in Ehrlich's ascites tumor. Other investigations have produced negative data, and hence the question remains unanswered.[19]

Protamine has been characterized as a heparin antagonist, but it has the characteristic of prolonging clotting time on its own. Protamine (discussed also in Chap. 20) is basic enough to interact with heparin (which is acidic due to its O-sulfate and N-sulfate groups). When protamine and heparin interact, they neutralize the action of each other.

It appears that the reticuloendothelial system may be involved in the disposition of heparin; i.e., heparin may leave the plasma by uptake into the reticuloendothelial system. Recent data from kinetic studies of heparin removal from circulation of the minipig are consistent with this suggestion.[23]

Heparin is catabolized primarily in the liver by partial cleavage of the sulfate groups and is excreted by the kidneys primarily as a partially sulfated product. Up to 50 percent may be excreted unchanged when high doses are given. The partially desulfated product excreted in the urine has been shown to be one-half as active as heparin with respect to anticoagulant properties.

Heparitin sulfate is the polysaccharide found as a by-product in the preparation of heparin from lung and liver tissue. Heparitin sulfate has a lower sulfate content, and its glucosamine residues are partially acetylated and N-sulfated. Heparitin sulfate isolated from the aorta has negligible antithrombin activity.

Heparin Sodium U.S.P. Heparin may be prepared commercially from lung and liver, employing the procedure of Kuizenga and Spaulding[15] combined with suitable methods for purifying the isolated heparin. The sodium salt is a white, amorphous, hygroscopic powder that is soluble (1:20) in water, but poorly soluble in alcohol. A 1 percent aqueous solution has a pH of 5 to 7.5. It is relatively stable to heat and solutions may be sterilized by autoclaving.

Heparin is administered intravenously in two ways: (1) the intermittent dose method and (2) the continuous drip method. In the intermittent dose method, a dose of 50 mg. is repeated every 4 hours until a total of 250 mg. per day has been given. The continuous drip method is to be preferred; it consists of a slow infusion of a heparin-containing solution into the vein, adjusting the flow according to the observed clotting time. A solution containing 100 to 200 mg. of heparin in each 1,000 ml. of 5 percent dextrose or physiologic saline solution is used for the latter method.

The therapeutic use of subcutaneous heparin in low doses is under extensive investigation, and some of the recent reports have favorably evaluated this mode of administration.[24,25,26]

A common side-effect with heparin can be hemorrhage, but this can be minimized with the low dose regimen.[27]

Category—anticoagulant.

Usual dose—parenteral, the following amounts, as indicated by prothrombin-time determinations: I.V., 10,000 U.S.P. Heparin Units initially, then 5,000 to 10,000 Units every 4 to 6 hours; infusion, 20,000 to 40,000 Units per liter at a rate of 15 to 30 Units per minute; subcutaneous, 10,000 to 20,000 Units initially, then 8,000 to 10,000 Units every 8 hours.

Usual pediatric dose—I.V. infusion, 50 Units per kg. of body weight initially, followed by 100 Units per kg., added and absorbed every 4 hours.

Occurrence
Heparin Sodium Injection U.S.P.

GLYCOSIDES

Because a number of plant constituents yielded glucose and an organic hydroxide upon hydrolysis, the term "glucoside" was introduced as a generic term for these substances. The fact that a number of plant constituents yielded sugars other than glucose led to the suggestion of the less specific general term "glycoside." When the nature of the sugar residue is known, more specific terms can be used where desired, such as glucoside, fructoside, rhamnoside and others, respectively. The nonsugar portion of the glycoside generally is referred to as the aglycon or genin.

Two general types of glycosides are known, viz., the nitrogen glycosides and the conventional type glycoside. The conventional type glycoside has an acetal structure and can be illustrated by the simplest type in which methyl alcohol is the aglycon or organic hydroxide. Two forms of this as well as all other glycosides are possible, viz., *alpha* and *beta,* because of the asymmetry centering about carbon atom 1 of the sugar residue that contains the acetal structure. It is thought that all naturally occurring glycosides are of the *beta* variety because the enzyme emulsin, which cannot hydrolyze synthetic *alpha* glycosides, hydrolyzes naturally occurring glycosides. Some of the *beta* glycosides also are hydrolyzed by amygdalase, cellobiase, gentiobiase and the phenol-glycosidases. The *alpha* glycosides are hydrolyzed by maltase, mannosidase and trehalase.

Glycosides usually are hydrolyzed by acids and are relatively stable toward alkalies. Some glycosides are much more resistant to hydrolysis than others. For example, those glycosides that contain a 2-desoxy sugar (see cardiac glycosides) are easily cleaved by weak acids, even at room temperature. On the other hand, most of the glycosides containing the normal type sugars are quite resistant to hydrolysis, and of these, some may require rather drastic hydrolytic measures. The drastic treatment required for the hydrolysis of some glycosides causes chemical changes to take place in the aglycon portion of the molecule;

these changes present problems in the elucidation of their structures. On the other hand, those glycosides that are very easily hydrolyzed present problems in regard to isolation and storage. Examples of the latter are the cardiac glycosides.

Although most glycosides are stable to hydrolysis by bases, the structure of the aglycon may determine its base sensitivity; e.g., picrocrocin has a half-life of 3 hours in 0.007 N KOH at 30°.

The sugar component of glycosides may be a mono-, di-, tri- or tetrasaccharide. There is a wide variety of sugars found in the naturally occurring glycosides. Most of the unusual and rare sugars found in nature are components of glycosides.

The aglycons or nonsugar portions of glycosides are represented by a wide variety of organic compounds, as illustrated by the cardiac glycosides, the saponins, etc. (see Chap. 18).

Because of the complexity of the structures of the naturally occurring glycosides, no generalizations are possible with regard to their stabilities if the stabilities of the glycosidic linkages are excluded. It also follows that considerable deviations are met with in their solubility properties. Many glycosides are soluble in water or hydroalcoholic solutions because the solubility properties of the sugar residues exert a considerable effect. Some glycosides, such as the cardiac glycosides, are slightly soluble or insoluble in water. In these cases, the steroid aglycon is markedly insoluble in water and offsets the solubility properties of the sugar residues. Most glycosides are insoluble in ether. Some glycosides are soluble in ethyl acetate, chloroform or acetone.

Glycosides occur widely distributed in nature. They are found in varying amounts in seeds, fruits, roots, bark and leaves. In some cases, two or more glycosides are found in the same plant, e.g., cardiac glycosides and saponins. Glycosides often are accompanied by enzymes that are capable of synthesizing or hydrolyzing them. This phenomenon introduces problems in the isolation of glycosides because the disintegration of plant tissues, with no precautions to inhibit enzymatic activity, leads, in some cases, to partial or complete hydrolysis of the glycosides.

Most glycosides are bitter to the taste, although there are many that are not. Glycosides per se or their hydrolytic products furnish a number of drugs, some of which are very valuable. Some plants that contain the cyanogenetic type of glycoside present an agricultural problem. Cattle have been poisoned by eating plants which are rich in the cyanogenetic type of glycoside.

REFERENCES

1. Harper, H. A.: Review of Physiological Chemistry, 17th, ed., Los Altos, Calif., Lange Medical Pub., 1980.
2. Montgomery, R., et al.: Biochemistry, A Case-Oriented Approach, pp. 246–304, St. Louis, C. V. Mosby, 1974.
3. White, A., et al.: Principles of Biochemistry, 6th, ed. pp. 423–567, New York, McGraw-Hill, 1978.
4. Sensi, P., and Gialdroni-Grassi, G.: Antimycobacterial Agents, in Wolff, M. E., (ed.): Burger's Medicinal Chemistry, ed. 4th pp. 311–321, New York, John Wiley & Sons, 1979.
5. Thomas, R., et al.: J. Pharm. Sci. 63:1649, 1974.
6. Kirkwood, S.: Ann. Rev. Biochem. 43:401, 1974.
7. Spiro, R. G.: Ann. Rev. Biochem. 39:599, 1970.
8. Bentley, R.: Ann. Rev. Biochem. 41:953, 1972.
9. Stoddard, J. F.: Stereochemistry of Carbohydrates, New York, Wiley-Interscience, 1971.
10. Newton, N. E., and Hornbrook, K. R.: J. Pharmacol. Exp. Ther. 181:479, 1972.
11. Piles, S. J., and Parks, C. R.: Ann. Rev. Pharmacol. 14:365, 1974.
12. White, A., et al.: Principles of Biochemistry, ed. 6, pp. 1265–1279, New York, McGraw-Hill, 1978.
13. Amer, M. S., and Kreighbaum, W. E.: J. Pharm. Sci. 64:1, 1975.
14. Dept. of Health, Education and Welfare, Food and Drug Administration: FDA Drug Bulletin, vol. 7, number 3, August 1977; Science 203:1094, 1979.
15. Atkinson, D. E.: Science 150:1, 1965.
16. Kolata, G. B.: Science 203:1098, 1979.
17. Blank, B.: in Wolff, M. E., (ed.): Burger's Medicinal Chemistry, ed. 4, chap. 31, New York, John Wiley & Sons, 1979.
18. Kuizenga, M. H., and Spaulding, L. B.: J. Biol. Chem. 148:641, 1943.
19. Ehrlich, J., and Stivala, S. S.: J. Pharm. Sci. 52:517,1973.
20. A.M.A. Department of Drugs: Drug Evaluations, 4th ed., p. 1078, New York, John Wiley & Sons, 1980.
21. von Kaulla, K. N.: in Wolff, M. E. (ed.): Burger's Medicinal Chemistry, ed 4, Chap. 32, p. 1081, New York, John Wiley & Sons, Inc. 1979.
22. Jaques, L. B.: Science 206:528, 1979.
23. Harris, P. A., and Harris, K. L.: J. Pharm. Sci. 63:138, 1974.
24. Flemming, J. S., and MacNintch, J. E.: Ann. Rep. Med. Chem. 9:75, 1974.
25. Gallus, A. S., et al.: New Eng. J. Med. 288:545, 1973.
26. Skillman, J. J.: Surgery 75:114, 1974.
27. Herrman, R. G., and Lacefield, W. B.: Ann. Rep. Med. Chem. 8:73, 1973.

SELECTED READINGS

Harper, H. A.: Review of Physiological Chemistry, ed. 14, pp. 232–267, Los Altos, Cal., Lange Medical Pub., 1973.
Montgomery, R., et al.: Biochemistry, A Case-Oriented Approach, ed. 2, pp. 293–362, St. Louis, C.V. Mosby, 1977.
Morrison, R. T., and Boyd, R. N.: Organic Chemistry, ed. 3, pp. 1070–1132, Boston, Allyn & Bacon, 1973.
Rafelson, M.E., et al.: Basic Biochemistry, ed. 4, Chaps. 2, 4, New York, Macmillan, 1980.
Snell, E. E., et al. (eds.): Ann. Rev. Biochem. Vol. 44, Palo Alto, Calif., 1979. (Series containing many reviews relating to carbohydrates.)
Tipson, R. S., and Horton, D. (eds.): Advances in Carbohydrate Chemistry and Biochemistry, vol. 34, New York, Academic Press, 1979. (Series started in 1945, W. W. Pigman (ed.): Advances in Carbohydrate Chemistry.
White, A., et al.: Principles of Biochemistry, ed. 6, New York, McGraw-Hill, 1978.

20

amino acids, proteins, enzymes and peptide hormones

Jaime N. Delgado

It is well known that proteins are essential components of all living matter. As cellular components, proteins perform numerous functions. The chemical reactions fundamental to the life of the cell are catalyzed by proteins called enzymes. Other proteins are structural constituents of protoplasm and cell membranes. Some hormones are characterized as proteins or protein-like compounds because of their polypeptide structural features.

Protein chemistry is essential not only to the study of molecular biology in understanding how cellular components participate in the physiologic processes of organisms, but also to medicinal chemistry. An understanding of the nature of proteins is necessary for the study of those medicinal agents which are proteins or protein-like compounds and their physicochemical/biochemical properties relating to mechanisms of action. Also, in medicinal chemistry drug-receptor interactions are implicated in the rationalization of structure-activity relationships and in the science of rational drug design. Drug receptors are considered to be macromolecules, some of which seem to be proteins or protein-like.

This chapter reviews the medicinal chemistry of proteins, and also includes some discussion of those amino acids which are products of protein hydrolysis. Some amino acids (e.g., dopa) are useful therapeutic agents and their mode of action relates to amino acid metabolism. Some medicinals are amino acid antagonists and their biochemical effects relate to their therapeutic uses; hence, brief mention of some representative cases of amino acid antagonism will be made in appropriate context. Moreover, the hormones with protein-like structure are also discussed, with emphasis on their biochemical effects.

A study of medicinal chemistry cannot be made without including some enzymology, not only because many drugs affect enzyme systems and vice versa, but also because fundamental lessons of enzymology have been applied to the study of drug-receptor interactions. Accordingly, this chapter includes a section on enzymes.

AMINO ACIDS

Proteins are biosynthesized from α-amino acids, and when proteins are hydrolyzed, amino acids are obtained. Some very complex (conjugated) proteins yield other hydrolysis products in addition to amino acids. α-Amino acids are commonly characterized with the generalized structure*:

$$R-\overset{\overset{\displaystyle H}{|}}{\underset{\underset{\displaystyle NH_2}{|}}{C}}-COOH$$

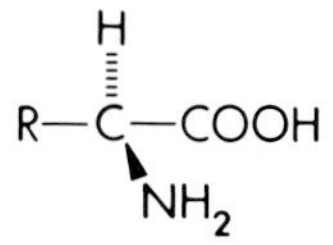

The most important amino acids are described in Table 20-1. Although the above structure for amino acids is widely used, physical, chemical and some biochemical properties of these compounds are more consistent with a dipolar ion structure.

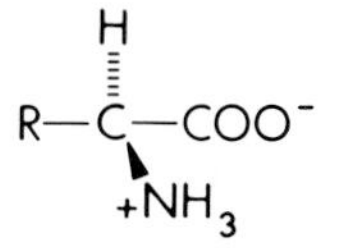

$$R-\overset{\overset{\displaystyle H}{|}}{\underset{\underset{\displaystyle ^+NH_3}{|}}{C}}-COO^-$$

---

* All α-amino acids, except glycine, are optically active since the R for the generalized structure represents some moiety other than H; the amino acids of proteins have the same absolute configuration as L-alanine, which is related to L-glyceraldehyde. (The D and L designations refer to configuration, rather than to optical rotation.)

TABLE 20–1

NATURALLY OCCURRING AMINO ACIDS

Name	Symbol	Formula
Glycine	Gly	H_2NCH_2COOH
Alanine	Ala	$CH_3CH(NH_2)COOH$
Valine	Val	$(CH_3)_2CHCH(NH_2)COOH$
Leucine	Leu	$(CH_3)_2CHCH_2CH(NH_2)COOH$
Isoleucine	Ileu	$CH_3CH_2CH(CH_3)CH(NH_2)COOH$
Serine	Ser	$HOCH_2CH(NH_2)COOH$
Threonine	Thr	$CH_3CH(OH)CH(NH_2)COOH$
Cysteine	CySH	$HSCH_2CH(NH_2)COOH$
Cystine	CySSCy	$(-SCH_2CH(NH_2)COOH)_2$
Methionine	Met	$CH_3SCH_2CH_2CH(NH_2)COOH$
Proline	Pro	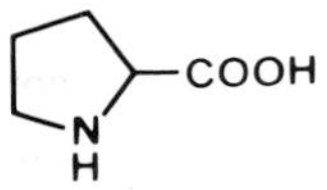
Hydroxyproline	Hypro	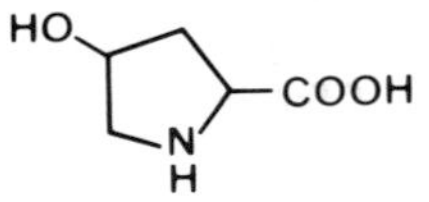
Phenylalanine	Phe	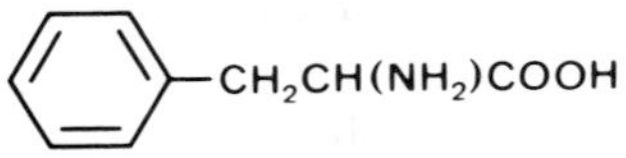
Tyrosine	Tyr	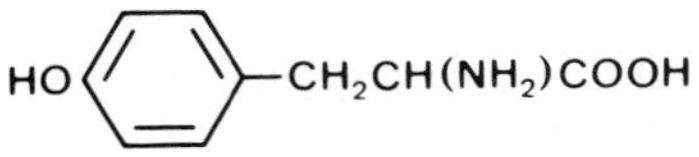
Tryptophan	Try	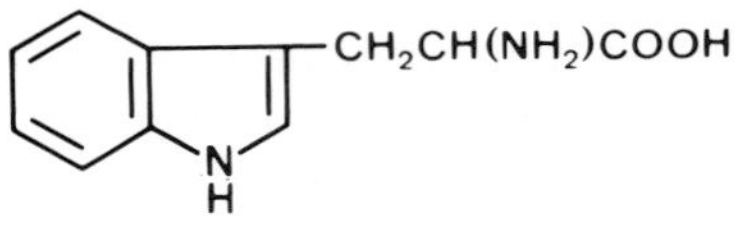
Aspartic acid	Asp	$HOOCCH_2CH(NH_2)COOH$
Glutamic acid	Glu	$HOOCCH_2CH_2CH(NH_2)COOH$
Lysine	Lys	$H_2NCH_2CH_2CH_2CH_2CH(NH_2)COOH$
Arginine	Arg	$H_2NC(=NH)NHCH_2CH_2CH_2$ $CH(NH_2)COOH$
Histidine	His	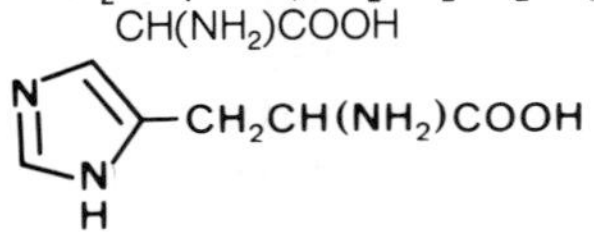

The relatively high melting point, solubility behavior and acid-base properties characteristic of amino acids can be accounted for on the basis of the dipolar ion structure (commonly called zwitterion). Amino acids in the dry solid state are dipolar ions (inner salts).

Amino acids when dissolved in water can exist as dipolar ions, and in this form would make no contribution to migration in an electric field. The concentration of the dipolar ion will vary depending on the pKa's of the amino acids and the hydronium ion con-

centration of the aqueous solution according to the following equilibrium:

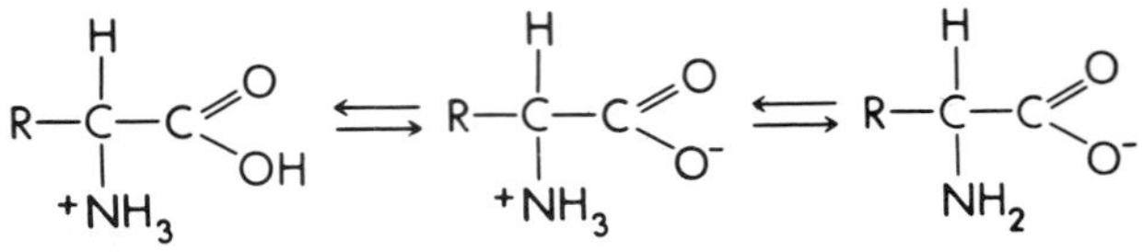

The hydronium ion concentration of the solution can be adjusted, and, if expressed in terms of pH, the pH at which the concentration of the dipolar form is maximal has been called the isoelectric point for the amino acid. (Since proteins are polymers of amino acids, they also have zwitterion character and isoelectric points.)

Glycine has $pKa_1 = 2.34$ for the carboxyl group and $pKa_2 = 9.6$ for the protonated amino group. The R groups of other amino acids change the pKa's slightly. The positive charge of I tends to repel a proton from the carboxyl group so that I is more strongly acidic than acetic acid (pKa = 4.76). The pKa_2 value for III is less than methylamine because of the electron-withdrawing effect of the carboxyl group (see structure below).

Table 20-1 demonstrates that most amino acids have complex side chains and that some amino acids have other functions (in addition to the α-carboxyl and α-amino groups) such as —OH, —NH₂, —CO₂H, —SH, phenolic —OH, guanidine, etc. These functions contribute to the physicochemical and biochemical properties of the respective amino acids or to their derivatives, including the proteins in which they are present. It has been customary to designate those amino acids that cannot be synthesized in the organism (animal) at a rate adequate to meet metabolic requisites as essential amino acids. According to White, et al.,[1] nutritionally essential amino acids (for man) are: arginine, histidine, isoleucine, leucine, lysine, methionine, phenylalanine, threonine, tryptophan and valine. At this point it is important to note that some of these essential amino acids participate in the biosynthesis of other important metabolites; e.g., histamine (from histidine); catecholamines and the thyroid hormones (from phenylalanine via tyrosine); serotonin from tryptophan; etc.

Amino acid antagonists have received the attention of many medicinal chemists. As antimetabolites these compounds interfere with certain metabolic processes and thus exert, in some cases, therapeutically useful pharmacologic actions, e.g., α-methyldopa as a dopa-decarboxylase inhibitor. Table 20-2 lists some other amino acid antagonists. The study of such antimetabolites as potential chemotherapeutic agents contin-

TABLE 20-2

SELECTED AMINO ACID ANTAGONISTS

Amino Acid Antagonist	Amino Acid Antagonized	Other Inhibitory Effects
D-Alanine	L-Alanine	Carboxypeptidase
D-Phenylalanine	L-Phenylalanine	D-Amino acid oxidase
α-Methyl-L-methionine	L-Methionine	D-Amino acid oxidase
α-Methyl-L-glutaric acid	L-Glutaric acid	Glutamic decarboxylase
Ethionine	Methionine	
α-Methyldopa	Dopa	Dopa decarboxylase
Allyl glycine	Methionine	Growth of E. coli
Propargylglycine	Methionine	Growth of E. coli
2-Amino-5-heptenoic acid	Methionine	Growth of E. coli
2-Thienylalanine	Phenylalanine	Growth of yeast
p-Fluorophenylalanine	Phenylalanine	Incorporation of phenylalanine into protein molecules
L-O-Methyl threonine	Isoleucine	Competitive incorporation of leucine into proteins
4-Oxalysine	Lysine	Growth of E. coli, L. casei, etc.
6-Methyltryptophan	Tryptophan	
5,5,5-Trifluoronorvaline	Leucine, Methionine	Growth of E. coli, etc.
3-Cyclohexene-I-glycine	Isoleucine	Inhibits E. coli
O-Carbamyl-L serine	L-Glutamine	Inhibits E. coli, S. lactis

ues. Research in cancer chemotherapy has involved experimentation with many antagonists of amino acids. Glutamine antagonists, azaserine and 6-diazo-5-oxonorleucine (DON), interfere with the metabolic processes which require glutamine and thus disrupt nucleic acid synthesis (glutamine is required for nucleic acid formation; glutamine is derived from glutamic acid). The phenomenon, lethal synthesis, involves the incorporation of the antimetabolite into protein structure or into the structure of some other macromolecule, and this unnatural macromolecule alters metabolic processes dependent on it. O-Methylthreonine competes with isoleucine for incorporation into protein molecules, whereas O-ethylthreonine is incorporated into t-RNA in *E. coli.*

Although all of the naturally occurring amino acids have been synthesized, and a number of them are available by the synthetic route, others are available more economically by isolation from hydrolyzed proteins. The latter are leucine, lysine, cystine, cysteine, glutamic acid, arginine, tyrosine, the prolines and tryptophan.

PRODUCTS

Some pharmaceutically important amino acids are listed in Table 20-3.

Aminoacetic Acid U.S.P., Glycocoll®, glycine, contains not less than 98.5 percent and not more than 101.5 percent of $C_2H_5NO_2$. It occurs as a white, odorless, crystalline powder, having a sweetish taste. It is insoluble in alcohol but soluble in water (1:4) to make a solution that is acid to litmus paper.

A 1.5 percent solution is preferred over the 2.1 percent isotonic solution for use as an irrigating solution during transurethral resection of the prostate gland. From 10 to 15 liters of the solution may be used during the surgical operation.

Methionine U.S.P., Amurex®, DL-2-amino-4-(methylthio)butyric acid, occurs as white, crystalline platelets or powder with a slight, characteristic odor; it is soluble in water (1:30), and a 1 percent solution has a pH of 5.6 to 6.1. It is insoluble in alcohol. In recent years, the racemic compound has been produced in ever-increasing quantities and at considerably reduced cost. The human body needs proteins that furnish methionine in order to prevent pathologic accumulation of fat in the liver, a condition which can be counteracted by administration of the acid or proteins that provide it. Methionine also has a function in the synthesis of choline, cystine, lecithin and, probably, creatine. Deficiency not only limits growth in rats but also inhibits progression of tumors.

In therapy, methionine has been employed in the treatment of liver injuries caused by poisons such as carbon tetrachloride, chloroform, arsenic and trinitrotoluene. While many physicians are enthusiastic about its value under such circumstances, this action has not been established satisfactorily.

Another use for methionine is as a urinary acidifier to help control the odor and dermatitis in incontinent patients caused by ammoniacal urine. It has been reported to be effective in both short- and long-term usage. Treatment must be continued for 3 or 4 days before the ammoniacal odor is eliminated.

Dihydroxyaluminum Aminoacetate U.S.P., Robalate®, basic aluminum glycinate, may be represented by the formula $H_2NCH_2COOAl(OH)_2$. It is a white, odorless, water-insoluble powder with a faintly sweet taste and is employed as a gastric antacid in the same way as aluminum hydroxide gel. Over the latter, it is claimed to have the advantages of more prompt, greater and more lasting buffering action. Also, it is said to have less astringent and constipative effects because of its smaller content of aluminum. However, all medical authorities are not yet satisfied that any of these claims are justified. The compound is furnished in powder, magma, or in tablets containing 500 mg.

Aminocaproic Acid U.S.P., Amicar®, 6-aminohexanoic acid, occurs as a fine, white, crystalline powder that is freely soluble in water, slightly soluble in alcohol and practically insoluble in chloroform.

TABLE 20-3

PHARMACEUTICALLY IMPORTANT AMINO ACIDS

Name Proprietary Name	Preparations	Category	Application	Usual Adult Dose*	Usual Dose Range*
Aminoacetic Acid U.S.P.	Aminoacetic Acid Irrigation U.S.P.	Irrigating Solution	Topically to the body cavities, as a 1.5 percent solution		
Methionine U.S.P. *Amurex, Odor- Scrip, Oradash, Uranap*	Methionine Capsules U.S.P. Methionine Tablets U.S.P.	Acidifier (urinary)		400 to 600 mg. daily	
Dihydroxyaluminum Aminoacetate U.S.P. *Robalate, Spenalate*	Dihydroxyaluminum Aminoacetate Magma U.S.P. Dihydroxyaluminum Aminoacetate Tablets U.S.P.	Antacid		500 mg. to 1 g. 4 times daily	500 mg. to 2 g.
Aminocaproic Acid U.S.P. *Amicar*	Aminocaproic Acid Injection U.S.P. Aminocaproic Acid Syrup U.S.P. Aminocaproic Acid Tablets U.S.P.	Hemostatic		Oral and I.V., initial, 5 g. followed by 1 to 1.25 g. every hour to maintain a plasma level of 13 mg. per 100 ml. No more than 30 g. per 24-hour period is recommended	
Acetylcysteine U.S.P. *Mucomyst*	Acetylcysteine Solution U.S.P.	Mucolytic agent			By inhalation of nebulized solution, 3 to 5 ml. of a 20 percent solution or 4 to 10 ml. of a 10 percent solution 3 or 4 times daily; by direct instillation, 1 to 2 ml. of a 10 or 20 percent solution every 1 to 4 hours
Levodopa U.S.P. *Larodopa, Levopa, Dopar, Bendopa*	Levodopa Capsules U.S.P. Levodopa Tablets U.S.P.	Antiparkin- sonian		Initial, 250 mg. 2 to 4 times a day, gradually increasing the total daily dose in increments of 100 to 750 mg. every 3 to 7 days as tolerated	500 mg. to 8 g. daily
Glutamic Acid Hydrochloride *Acidulin*	Glutamic Acid Hydrochloride Capsules	Acidifier (gastric)			Oral, 340 mg. to 1 g. 3 times daily before meals

* See *U.S.P. D.I.* for complete dosage information.

Aminocaproic acid is a competitive inhibitor of plasminogen activators such as streptokinase and urokinase. It is effective because it is an analog of lysine whose position in proteins is attacked by plasmin. To a lesser degree it also inhibits plasmin (fibrinolysin). Lowered plasmin levels lead to more favorable amounts of fibrinogen, fibrin and other important clotting components.

Aminocaproic acid has been used in the control of hemorrhage in certain surgical procedures. It is of no value in controlling hemorrhage due to thrombocytopenia or other coagulation defects or vascular disruption, e.g., bleeding ulcers, functional uterine bleeding, post-tonsillectomy bleeding, etc. Since it inhibits the dissolution of clots, it may interfere with normal mechanisms for maintaining the patency of blood vessels.

Aminocaproic acid is well absorbed orally. Plasma peaks occur in about two hours. It is excreted rapidly, largely unchanged.

Acetylcysteine U.S.P., Mucomyst®, is the N-acetyl derivative of L-cysteine. It is used primarily to reduce the viscosity of the abnormally viscid pulmonary secretions in patients with cystic fibrosis of the pancreas (mucoviscidosis) or various tracheobronchial and bronchopulmonary diseases.

Acetylcysteine is more active than cysteine and its mode of action in reducing the viscosity of mucoprotein solutions, including sputum, may be by opening the disulfide bonds in the native protein.

Acetylcysteine is most effective in 10 to 20 percent solutions with a pH of 7 to 9. It is used by direct instillation or by aerosol nebulization. It is available as a 20 percent solution of the sodium salt in 10- and 30-ml. containers. An opened vial of acetylcysteine must be covered, stored in a refrigerator and used within 48 hours.

Glutamic Acid Hydrochloride, Acidulin®, is essentially a pure compound that occurs as a white, crystalline powder soluble 1:3 in water and insoluble in alcohol. It has been used in place of glycine in the treatment of muscular dystrophies with rather unpromising results. It also is combined (8 to 20 g. daily) with anticonvulsants for the petit mal attacks of epilepsy, a use which appears to depend on change in pH of the urine.

The hydrochloride, which releases the acid readily, has been recommended under a variety of names for furnishing acid to the stomach in the achlorhydria of pernicious anemia and other conditions. The usual dosage range is 600 mg. to 1.8 g. taken during meals.

Levodopa U.S.P., Larodopa®, Dopar®, Levopa®, is (−)-3-(3,4-dihydroxyphenyl)-L-alanine. It occurs as a colorless, crystalline material. It is slightly soluble in water and insoluble in alcohol. Levodopa is a precursor of dopamine and has been found to be of value in the treatment of Parkinson's disease. Dopamine does not cross the blood-brain barrier and, therefore, is ineffective. Levodopa does cross the blood-brain barrier and presumably is metabolically converted to dopamine in the basal ganglia. The dose must be carefully determined for each patient.

Levodopa

Carbidopa U.S.P., Lalosyn®, Sinemet®, is a combination of carbidopa and levodopa. The former is the hydrazine analog of α-methyldopa, and it is an inhibitor of aromatic acid decarboxylation. Accordingly, when carbidopa and levodopa are administered in combination, carbidopa inhibits decarboxylation of peripheral levodopa but carbidopa does not cross the blood-brain barrier and hence does not affect the metabolism of levodopa in the central nervous system. Since carbidopa's decarboxylase-inhibiting activity is limited to extracerebral tissues, it makes more levodopa available for transport to the brain. Thus, carbidopa reduces the amount of levodopa required by approximately 75 percent.

Sinemet® is supplied as tablets in two strengths, Sinemet®-10/100, containing 10 mg. of carbidopa and

Carbidopa

100 mg. of levodopa, and Sinemet®-25/250, containing 25 mg. of carbidopa and 250 mg. of levodopa.

Management of acute overdosage with Sinemet® is fundamentally the same as management of acute overdosage with levodopa; however, pyridoxine is not effective in reversing the actions of Sinemet®.

PROTEIN HYDROLYSATES

In therapeutics, agents affecting volume and composition of body fluids include various classes of parenteral products. Idealistically, it would be desirable to have parenteral fluids available which would provide adequate calories, important proteins and lipids so as to mimic as closely as possible an appropriate diet. However, this is not the case. Usually, sufficient carbohydrate is administered intravenously in order to prevent ketosis, and in some cases it is necessary to give further sources of carbohydrate by vein so as to reduce the wasting of protein. Sources of protein are made available in the form of protein hydrolysates and these can be administered to favorably influence the balance.[2] The *United States Pharmacopeia XX* recognizes protein hydrolysates in the form of (intravenous) parenteral solutions. Protein hydrolysates are also used to supplement the diet in cases of protein deficiencies.

Complete hydrolysis of simple proteins yields mixtures of the component amino acids, whereas incomplete hydrolysis yields a mixture of the amino acids together with dipeptides and polypeptides. The hydrolysis can be conducted under alkaline, acidic or enzymic conditions. Under either alkaline or acidic conditions some of the amino acids undergo decomposition. When the hydrolysis is catalyzed by proteolytic enzymes (proteases), the process is slow and seldom complete. Raw materials for the process are purified proteins (e.g., casein and lactalbumin) or yeast, liver, beef and certain vegetables.

Protein deficiencies in human nutrition are sometimes treated with protein hydrolysates. The lack of adequate protein may result from a number of conditions, but the case is not always easy to diagnose. The deficiency may be due to insufficient dietary intake; temporarily increased demands as in pregnancy; impaired digestion or absorption; liver malfunction; increased catabolism or loss of proteins and amino acids as in fevers, leukemia, hemorrhage, after surgery, burns, fractures or shock.

Dietary protein deficiency, kwashiorkor, develops in children who are fed almost exclusively on gruel of a cereal such as plantain, taro or corn (most common). Corn contains only 2 grams of protein per 100 calories, whereas milk contrastingly contains 5.4 grams per 100 calories. Such protein deficiency is considered to be the world's major nutritional problem. The negative nitrogen balance of severe protein restriction drastically affects the liver which may lose up to 50 percent of its total nitrogen. Mitochondria, microsomes, cytoplasmic enzymes, and RNA but not DNA are all decreased. Muscle tissue also suffers nitrogen loss, but not to the same extent as liver tissue. The symptoms of kwashiorkor (growth retardation, anemia, hypoproteinemia, fatty liver) usually respond favorably to a high-protein diet containing meat and milk products.[3]

PRODUCTS

Protein Hydrolysate Injection U.S.P., Protein Hydrolysates (Intravenous), Aminogen®, Aminosal®, Travamin®. Protein hydrolysate injection is a sterile solution of amino acids and short-chain peptides which represent the approximate nutritive equivalent of the casein, lactalbumin, plasma, fibrin, or other suitable protein from which it is derived by acid, enzymatic, or other method of hydrolysis. It may be modified by partial removal and restoration or addition of one or more amino acids. It may contain dextrose or other carbohydrate suitable for intravenous infusion. Not less than 50 percent of the total nitrogen present is in the form of α-amino nitrogen. It is a yellowish to red-amber transparent liquid that has a pH of 4 to 7.

Parenteral preparations are employed for the maintenance of a positive nitrogen balance in cases where there is interference with ingestion, digestion or absorption of food. In such cases, the material to be injected must be nonantigenic and must not contain pyrogens or peptides of high molecular weight. Injection may result in untoward effects, such as nausea, vomiting, fever, vasodilatation, abdominal pain, twitching and convulsions, edema at the site of injection, phlebitis and thrombosis. Sometimes these reactions are due to inadequate care in cleanliness or too rapid administration.

Category—Fluid and nutrient replenisher.

Usual dose—I.V. infusion, 2 to 3 liters of a 5 percent solution once daily at a rate of 1.5 to 2 ml. per minute initially, then increased gradually as tolerated to 3 to 6 ml. per minute.

Usual dose range—2 to 8 liters daily.

Usual pediatric dose—infants; I.V. infusion, 2 to 3 g. of protein per kg. of body weight in a 4 to 7 percent solution once daily at a rate not exceeding 0.2 ml. per minute initially, then increased gradually as tolerated to 0.2 to 0.6 ml. per minute; children, I.V. infusion, 1 to 2 g. of protein per kg. in a 4 to 7 percent solution once daily at a rate not exceeding 0.2 ml. per minute initially, then increased gradually as tolerated to 1 to 3 ml. per minute.

PROTEINS AND PROTEIN-LIKE COMPOUNDS

The chemistry of proteins is very complex, and some of the most complex facets remain to be clearly understood. Protein structure is usually studied in basic organic chemistry and to a greater extent in biochemistry, but for the purposes of this chapter some of the more important topics will be summarized with emphasis on relationships to medicinal chemistry. Much progress has been made in the last 25 years in the understanding of the more sophisticated features of protein structure and its correlation with physicochemical and biological properties. With the total synthesis of ribonuclease in 1969, new approaches to the study of structure-activity relationships among proteins involve the synthesis of modified proteins.

Many types of compounds which are important in medicinal chemistry are structurally classified as proteins. Enzymes, antigens and antibodies are proteins.* Numerous hormones are low molecular weight proteins, hence relative to the foregoing they are called simple proteins. Fundamentally, all proteins are composed of one or more polypeptide chains; i.e., the primary organizational level of protein structure is the polypeptide (polyamide) chain composed of naturally occurring amino acids bonded to one another via amide linkages. An extended polypeptide chain can be visualized with the aid of Figure 20-1. The specific physicochemical and biological properties of proteins depend not only on the nature of the specific amino acids and on their sequence within the polypeptide chain, but also on conformational characteristics.

---

* The term *interferon* is generally applied to the antiviral proteins naturally produced by various cells. Since the characterization of interferon as an antiviral protein in 1957, much has been learned about the purification and characterization of human leukocyte interferon (e.g., purification by HPLC), antiviral properties of interferon from various species, cloning and expression of human interferons in bacteria, and monoclonal antibodies of human leukocyte interferon. The interested reader should refer to S. Pestka et al., Annual Reports in Medicinal Chemistry, Vol. 16, 229, 1981.

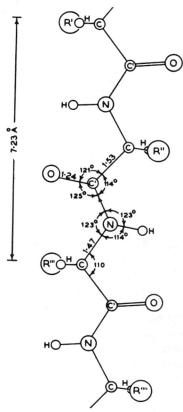

FIG. 20-1. *A diagrammatic representation of a fully extended polypeptide chain with the bond lengths and the bond angles derived from crystal structures and other experimental evidence.[4] (Corey and Pauling)*

CONFORMATIONAL FEATURES OF PROTEIN STRUCTURE

As indicated above, the polypeptide chain is considered to be the primary level of protein structure, and the folding of the polypeptide chains into a specific coiled structure is maintained through hydrogen-bonding interactions (intramolecular). (See Fig. 20-2.) The latter folding pattern is called the secondary level of protein structure. The intramolecular hydrogen bonds involve the partially negative oxygens of amide carbonyl groups and the partially positive hydrogens of the amide —NH (See below.) Additional factors contribute to the stabilization of such folded structures; e.g., ionic bonding between positively charged and negatively charged groups, and disulfide bonds.

The arrangement and interfolding of the coiled chains into layers determine the tertiary and higher levels of protein structure. Such final conformational character is determined by various types of interactions, primarily hydrophobic forces and to some extent hydrogen bonding and ion pairing. Hydrophobic forces are implicated in many biological phenomena associated with protein structure and interactions.[5] The side chains (R groups) of various amino acids have hydrocarbon moieties which are hydrophobic and they have minimal tendency to associate with water molecules, whereas water molecules are strongly associated through hydrogen bonding. Such hydrophobic R groups tend to get close to one another, with exclusion of water molecules, to form "bonds" between different segments of the chain or between different chains. These "bonds" are often termed hydrophobic bonds, hydrophobic forces, or hydrophobic interactions.

The study of protein structure has required several physicochemical methods of analysis. Ultraviolet spectrophotometry has been applied to the assessment of conformational changes that proteins undergo. Conformational changes can be investigated by the direct plotting of the difference in absorption between the protein under various sets of conditions. X-ray analysis has been very useful in the elucidation of the structures of several proteins, e.g., myoglobulin and lysozyme. Absolute determinations of conformation and helical content can be made by x-ray diffraction analysis. Optical rotation of proteins also has been studied fruitfully. It is interesting that the specific rotations of proteins are always negative, but changes in pH (when the protein is in solution) and conditions which promote denaturation (urea solutions, increased temperatures) tend to augment the negative optical rotation. For this reason it is rationalized that the changes in rotation are due to conformational changes (i.e., changes in protein structure at the secondary and higher levels of organization). Optical rotatory dispersion (ORD) also has been experimented with in the study of conformation alterations and conformational differences among globular proteins. Circular dichroism methodology also has been involved in structural studies. The shape and magnitude of rotatory dispersion curves and circular dichroism spectra are very sensitive to conformational alterations, thus the effects of enzyme inhibitors on conformation can be analyzed. Structural studies have included the investigation of the tertiary structures of proteins in high-frequency nuclear magnetic resonance (NMR).[6] NMR spectroscopy has been of some use in the study of interactions between drug molecules and proteins such as enzymes, proteolipids, etc. NMR has been applied to the study of binding of atropine analogs to acetylcholinesterase[7] and of interactions involving cholinergic ligands to housefly brain and torpedo electroplax.[8] Kato[9,10] has also investigated the binding of inhibitors (e.g., physostigmine) to acetylcholinesterase utilizing NMR spectroscopy.*

---

* C. M. Deber, *et al.,* have reviewed some modern approaches to the deduction of peptide conformation in solution: [13]C nuclear magnetic resonance, conformational energy calculations, and circular dichroism (C. M. Deber, *et al.,* Science, 9:106, 1976).

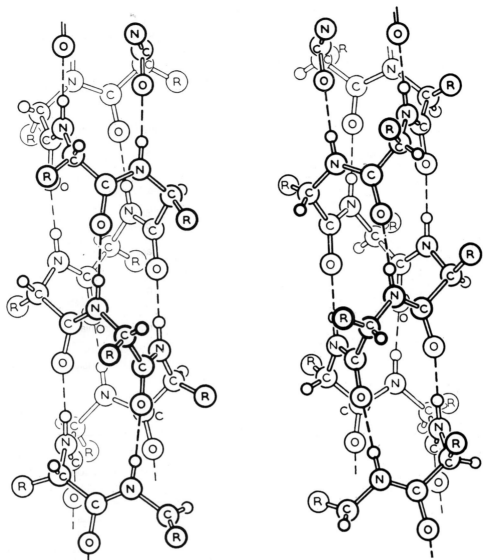

FIG. 20-2. *Left-handed and right-handed α-helices. The R and H groups on the α-carbon atom are in the correct position corresponding to the known configuration of the L-amino acids in proteins. (L. Pauling and R. B. Corey, unpublished drawings.)*

FACTORS AFFECTING PROTEIN STRUCTURE

Conditions which promote the hydrolysis of amide linkages affect protein structure, as noted above under Protein Hydrolysates.

The highly ordered conformation of a protein can be disorganized (without hydrolysis of the amide linkages), and in the process the protein's biological activity is obliterated. This process is customarily called denaturation, and it involves unfolding of the polypeptide chains, loss of the native conformation of the protein and disorganization of the uniquely ordered structure without the cleavage of covalent bonds. The rupture of native disulfide bonds is usually considered to be a more extensive and drastic change than denaturation. Criteria for the detection of denaturation involve: detection of previously masked —SH, imidazole, and —NH₂ groups; decreased solubility; increased susceptibility to the action of proteolytic enzymes; decreased diffusion; constant and increased viscosity of protein solution; loss of enzymatic activity if the protein is an enzyme; modification of antigenic properties.

For many years Eyring[11,12] has carried on studies of factors affecting protein structure and, therefore, biochemical processes. Eyring's studies,[13] involving interactions between general anesthetic molecules and proteins are fundamental to medicinal chemistry and pharmacology. The interested reader should consult the references cited; however, herein brief mention must be made in order to exemplify the importance of hydrophobic phenomena in mechanisms of drug action involving proteins or other macromolecules.

Eyring proposes that anesthetics affect the action of proteins necessary for central nervous system function. It is emphasized that there are certain proteins needed for the maintenance of consciousness. In order to function normally the protein must have a particular conformation. Anesthetic molecules are implicated as interacting with the hydrophobic regions of the protein, thus disrupting (unfolding) the conformation. These conformational changes in essential proteins affect their activities and function; hence, it is believed that these effects lead to blockade of synapses.[13] It is interesting to compare Eyring's proposals with those of Pauling (see Chap. 2) relating to mechanisms of general anesthetic action;* recently, the latter have been substantiated by Haberfield and Kivuls.[14]

PURIFICATION AND CLASSIFICATION

It may be said that it is old-fashioned to classify proteins according to the following system since so much progress has been made in the understanding of protein structure. Nevertheless, an outline of this system of classification is given because the terms used are still found in the pharmaceutical and medical literature. Table 20-4 includes the classification and characterization of simple proteins. Prior to the classification it must be insured that the protein material is purified to the extent practically possible, and this is a very challenging task. Several criteria are used to determine homomolecularity; e.g., crystallinity, constant solubility at a given temperature; osmotic pressure in different solvents; diffusion rate; electrophoretic mobility; dielectric constant; chemical assay; spectrophotometry; quantification of antigenicity. The methodology of purification is complex; procedures can involve various techniques of chromatography (column), electrophoresis, ultracentrifugation, etc. In some cases high-pressure liquid chromatography (hplc) has been applied to the separation of peptides; e.g., Folkers et al.[15] have reported the purification of some hypothalamic peptides using a combination of chromatographic methods including hplc.

Conjugated proteins contain a nonprotein structural component in addition to the protein moiety,

---

* Also compare: Di Paolo and Sandorfy: J. Med. Chem. 17:809, 1974.

TABLE 20-4

SIMPLE (TRUE) PROTEINS

Class	Characteristics	Occurrence
Albumins	Soluble in water, coagulable by heat and reagents	Egg albumin, lactalbumin, serum albumin, leucosin of wheat, legumelin of legumes
Globulins	Insoluble in water, soluble in dilute salt solution, coagulable	Edestin of plants, vitelline of egg, serum globulin, lactoglobulin, amandin of almonds, myosin of muscles
Prolamines	Insoluble in water or alcohol, soluble in 60–80% alcohol, not coagulable	Found only in plants, e.g., gliadin of wheat, hordein of barley, zein of corn and secalin of rye
Glutelins	Soluble only in dilute acids or bases, coagulable	Found only in plants, e.g., glutenin of wheat and oryzenin of rice
Protamines	Soluble in water or ammonia, strongly alkaline, not coagulable	Found only in the sperm of fish, e.g., salmine from salmon
Histones	Soluble in water, but not in ammonia, predominantly basic, not coagulable	Globin of hemoglobin, nucleohistone from nucleoprotein
Albuminoids	Insoluble in all solvents	In keratin of hair, nails and feathers; collagen of connective tissue; chondrin of cartilage; fibroin of silk and spongin of sponges

whereas simple proteins contain only the polypeptide chain of amino acid units. Nucleoproteins are conjugated proteins containing nucleic acids as structural components. Glycoproteins are carbohydrate-containing conjugated proteins (e.g., thyroglobulin). Phosphoproteins contain phosphate moieties (e.g., casein); lipoproteins are lipid-bearing; metalloproteins have some bound metal. Chromoproteins, such as hemoglobin or cytochrome, have some chromophoric moiety.

There are alternative classification systems. The system of choice varies with the purpose of the characterization study. For example, it is well known that hemoproteins include the hemoglobins, the cytochromes and the enzymes, catalase and peroxidase. The component protein of these is colorless, but the nonprotein moiety is colored and it is an iron-porphyrin. Hence, these conjugated proteins also fall into the chromoprotein classification.

PROPERTIES OF PROTEINS

The classification delineated in Table 20-4 is based on solubility properties. Fibrous proteins are water-insoluble and highly resistant to hydrolysis by proteolytic enzymes; the collagens, elastins and keratins are in this class. On the other hand, globular proteins (albumins, globulins, histones and protamines) are relatively water-soluble; they are also soluble in aqueous solutions containing salts, acids, bases or ethanol. Enzymes, oxygen-carrying proteins and protein hormones are globular proteins.

Another important characteristic of proteins is the amphoteric behavior. In solution proteins migrate in an electric field, and the direction and rate of migration is a function of the net electrical charge of the protein molecule which in turn depends on the pH of the solution. The isoelectric point is the pH value at which a given protein does not migrate in an electric field, and it is a constant for any given protein and can be used as an index of characterization. Proteins differ in rate of migration and also in their isoelectric points. Electrophoretic analysis is used to determine purity and for quantitative estimation since proteins differ in electrophoretic mobility at any given pH.

Being ionic in solution, proteins bind with cations and anions depending on the pH of the environment. In some cases complex salts are formed and precipitation takes place; e.g., trichloracetic acid is a precipitating agent for proteins and is used for deproteinizing solutions.

Proteins possess chemical properties characteristic of their component functional groups, but in the native state some of these groups are "buried" within the tertiary protein structure and may not readily react. Certain denaturation procedures can expose these functions and allow them to respond to the usual chemical reagents, e.g., an exposed —NH$_2$ group can be acetylated by ketene, —CO$_2$H can be esterified with diazomethane, etc.

COLOR TESTS, MISCELLANEOUS SEPARATION AND IDENTIFICATION METHODS

Proteins respond to the following color tests: (1) biuret, a pink to purple color with an excess of alkali and a small amount of copper sulfate; (2) ninhydrin, a blue color when boiled with ninhydrin (triketohydrindene hydrate) that is intensified by the presence of pyridine; (3) Millon's test for tyrosine, a brick-red color or precipitate when boiled with mercuric nitrate in an excess of nitric acid; (4) Hopkins-Cole test for tryptophan, a violet zone with a salt of glyoxylic acid and stratified over sulfuric acid; and (5) xanthoproteic test, a brilliant orange zone when a solution in concentrated nitric acid is stratified under ammonia.

Almost all so-called alkaloidal reagents will precipitate proteins in slightly acid solution.

The qualitative identification of the amino acids found in proteins, etc., has been simplified very greatly by the application of paper chromatographic techniques to the proper hydrolysate of proteins and related substances. End-member degradation techniques for the detection of the sequential arrangements of the amino acid residues in polypeptides (proteins, hormones, enzymes, etc.) have been developed to such a high degree with the aid of paper chromatography that very small samples of the polypeptides can be utilized. These techniques, together with statistical methods, have led to the elucidation of the sequential arrangements of the amino acid residues in oxytocin, vasopressin, insulin, hypertensin, glucagon, corticotropins, etc.

Ion-exchange chromatography has been applied to protein analysis and to the separation of amino acids. The principles of ion-exchange chromatography can be applied to the design of automatic amino acid analyzers with appropriate recording instrumentation. One- or two-dimensional thin-layer chromatography also has been used to accomplish separations not possible with paper chromatography. Another method for separating amino acids and proteins involves a two-dimensional analytical procedure, using electrophoresis in one dimension and partition chromatography in the other. The applicability of high-pressure liquid chromatography has been noted above.[15]

PRODUCTS

Gelatin U.S.P. is a protein obtained by the partial hydrolysis of collagen, an albuminoid found in bones, skins, tendons, cartilage, hoofs and other animal tissues. The products seem to be of great variety, and, from a technical standpoint, the raw material must be selected according to the purpose intended. The reason for this is that collagen usually is accompanied in nature by elastin and especially by mucoids, such as chondromucoid, which enter into the product in a small amount. The raw materials for official gelatin, and also that used generally for a food, are skins of calf or swine and bones. First, the bones are treated with hydrochloric acid to remove the calcium compounds and then are digested with lime for a prolonged period, which converts most other impurities to a soluble form. The fairly pure collagen is extracted with hot water at a pH of about 5.5, and the aqueous solution of gelatin is concentrated, filtered and cooled to a stiff gel. Calf skins are treated in about the same way, but those from hogs are not given any lime treatment. The product derived from an acid-treated precursor is known as Type A and exhibits an isoelectric point between pH 7 and 9, while that where alkali is used is known as Type B and exhibits an isoelectric point between pH 4.7 and 5. The minimum of gel strength officially is that a 1 percent solution, kept at 0° for 6 hours, must show no perceptible flow when the container is inverted.

Gelatin occurs in sheets, shreds, flakes or coarse powder. It is white or yellowish, has a slight but characteristic odor and taste and is stable in dry air but subject to microbial decomposition if moist or in solution. It is insoluble in cold water but swells and softens when immersed and gradually absorbs 5 to 10 times its own weight of water. It dissolves in hot water to form a colloidal solution; it also dissolves in acetic acid and in hot dilute glycerin. Gelatin commonly is bleached with sulfur dioxide, but that used medicinally must have not over 40 parts per million of sulfur dioxide. However, a proviso is made that for the manufacture of capsules or pills it may have certified colors added, may contain as much as 0.15 percent of sulfur dioxide and may have a lower gel strength.

Gelatin is used in the preparation of capsules and the coating of tablets and, with glycerin, as a vehicle for suppositories. It also has been employed as a vehicle for other drugs when slow absorption is required. When dissolved in water, the solution becomes somewhat viscous, and such solutions are used to replace the loss in blood volume in cases of shock. This is accomplished more efficiently now with blood plasma, which is safer to use. In hemorrhagic conditions, it sometimes is administered intravenously to increase the clotting of blood or is applied locally for the treatment of wounds.

The most important value in therapy is as an easily digested and adjuvant food. It fails to provide any tryptophan at all and is lacking notably in adequate amounts of other essential acids; approximately 60 percent of the total acids consist of glycine and the prolines. Nevertheless, when supplemented, it is very useful in various forms of malnutrition, gastric hyperacidity or ulcer, convalescence and general diets of the sick. It is specially recommended in the preparation of modified milk formulas for feeding infants.

Special Intravenous Gelatin Solution is a 6 percent sterile, pyrogen-free, nonantigenic solution in isotonic chloride, the gelatin being specially prepared from beef-bone collagen. It is odorless, clear, amber-colored, slightly viscous above 29°C but a gel at room temperature and has a pH of 6.9 to 7.4. It is employed as an infusion colloid to support blood volume in various types of shock and thus is a substitute for plasma and whole blood. It is contraindicated in kidney ailments and must be used with care in cardiac disease. Any typing of blood must be done before injection because gelatin interferes with proper grouping. The semisolid preparation is warmed to 50°C before use, and about 500 ml. is injected at the rate of not over 30

ml. per minute. It gives adequate protection for 24 to 48 hours.

Absorbable Gelatin Film U.S.P., Gelfilm®, is a sterile, nonantigenic, absorbable, water-insoluble gelatin film. The gelatin films are prepared from a solution of specially prepared gelatin-formaldehyde combination by spreading on plates and drying under controlled humidity and temperature. The film is available as light yellow, transparent, brittle sheets 0.076 to 0.228 mm. thick. Although insoluble in water, they become rubbery after being in water for a few minutes.

Absorbable gelatin film is used primarily in surgical closures and for repair of defects in such tissues as the dura mater and the pleura.

Absorbable Gelatin Sponge U.S.P., Gelfoam®, is a sterile, absorbable, water-insoluble, gelatin-base sponge that is a light, nearly white, nonelastic, tough, porous matrix. It is stable to dry heat at 150°C for 4 hours. It absorbs 50 times its own weight of water or 45 oxalated whole blood.

It is absorbed in 4 to 6 weeks when it is used as a surgical sponge. When applied topically to control capillary bleeding, it should be moistened with sterile isotonic sodium chloride solution or thrombin solution.

Nonspecific Proteins. The intravenous injection of foreign protein is followed by fever, muscle and joint pain, sweating and decrease and then increase in leukocytes; it even can result in serious collapse. The results have been used in the treatment of various infections, originally the chronic form. The method is presumed to be of value in acute and chronic arthritis, peptic ulcer, certain infections of the skin and eye, some vascular diseases, cerebrospinal syphilis, especially dementia paralytica, and other diseases. Since a fever is necessary in this system, the original program has developed into the use of natural fevers, such as malaria, of external heat and similar devices. However, the slightly purified proteins of milk still are rec-

TABLE 20–5

PHARMACEUTICALLY IMPORTANT PROTEIN PRODUCTS

Name Proprietary Name	Category
Gelatin U.S.P.	Pharmaceutic aid (encapsulating agent; suspending agent; tablet binder and coating agent)
Absorbable Gelatin Film U.S.P. *Gelfilm*	Local hemostatic
Absorbable Gelatin Sponge U.S.P. *Gelfoam*	Local hemostatic

ommended for some diseases; they are available commercially as Activin®, Caside®, Clarilac®, Bu-Ma-Lac®, Lactoprotein®, Mangalac®, Nat-i-lac®, Neolacmanese® and Proteolac®. Muscosol® is a purified beef peptone, and Omniadin® is a similar purified bacterial protein. Synodal® contains nonspecific protein with lipoids, animal fats and emetine hydrochloride and is designed for the treatment of peptic ulcer. One of the favorite agents of this class has been typhoid vaccine.

Venoms. Cobra (Naja) Venom Solution, from which the hemotoxic and proteolytic principles have been removed, has been credited with virtues due to toxins and has been injected intramuscularly as a non-narcotic analgesic in doses of 1 ml. daily. Snake Venom Solution of the water moccasin is employed subcutaneously in doses of 0.4 to 1.0 ml. as a hemostatic in recurrent epistaxis, thrombocytopenic purpura and as a prophylactic before tooth extraction and minor surgical procedures. Stypven® from the Russell viper is used topically as a hemostatic and as thromboplastic agent in Quick's modified clotting time test. Ven-Apis®, the purified and standardized venom from bees, is furnished in graduated strengths of 32, 50 and 100 bee-sting units. It is administered topically in acute and chronic arthritis, myositis and neuritis.

Nucleoproteins. The nucleoproteins previously mentioned are found in the nuclei of all cells and also in the cytoplasm. They can be deproteinized by a number of methods. Those compounds that occur in yeast usually are treated by grinding with a very dilute solution of potassium hydroxide, adding picric acid in excess and precipitating the nucleic acids with hydrochloric acid, leaving the protein in solution. The nucleic acids are purified by dissolving in dilute potassium hydroxide, filtering, acidifying with acetic acid and finally precipitating with a large excess of ethanol.

The nucleic acids prepared in some such manner differ in a few respects according to source, but they seem to be remarkably alike in chemical composition. They are slightly soluble in cold water, more readily soluble in hot water and easily soluble in dilute alkalies with the production of salts, from which they can be reprecipitated by acids. Neutral solutions of the sodium salts from thymonucleic acids set to a jelly on cooling, but those from yeast do not. All of them can be hydrolyzed to nucleotides, and these in turn to nucleosides and phosphoric acid. The nucleosides further hydrolyze to D-ribose or 2-deoxyribose and derivatives of pyrimidine: adenine and guanine, which are purines, and cytosine (6-amino-2-hydroxypyrimidine), uracil (2,6-dihydroxypyrimidine), thymine (5-methyluracil), and 5-methylcytosine. The nucleotides are known as adenylic acid, guanylic acid, cytidylic acid,

uridylic acid, thymylic acid, 5-methylcytidylic acid and their corresponding deoxy-congeners.

Adenylic acid (AMP) is found in muscle in the free state and in combination with additional phosphoric acid as adenyl diphosphate (ADP) and as adenyl triphosphate (ATP). During muscular exertion, the last compound is hydrolyzed enzymatically to adenylic acid or the diphosphate to furnish phosphoric acid and energy during metabolism, and regeneration of the triphosphate takes place in the muscle by further enzymatic action.

AMP

Both the nucleotides and the deoxynucleotides are the residues of the polynucleotide RNA, and DNA is a polydeoxynucleotide. DNA transmits genetic information in organisms that contain it. RNA carries instructions from the genes to the sites where it directs the assembly of proteins.

ATP

ENZYMES

Those proteins which have catalytic properties are called enzymes (i.e., enzymes are biological catalysts of protein nature).[*] Some enzymes have full catalytic reactivity per se; these are considered to be simple proteins because they do not have a nonprotein moiety. On the other hand, other enzymes are conjugated proteins, and the nonprotein structural components are necessary for reactivity. In some cases enzymes require metallic ions. Since enzymes are proteins or conjugated proteins, the general review of protein structural studies presented earlier in this chapter

---

[*] Important factors limiting rates of enzyme-catalyzed reactions have been critically evaluated by W. W. Cleland (Accounts of Chemical Research, 8:145, 1975).

(e.g., protein conformation and denaturation) is fundamental to the following topics. Conditions that effect denaturation of proteins usually have adverse effects on the activity of the enzyme.

General enzymology is discussed effectively in numerous standard treatises, and one of the most concise discussions appears in the classic work by Ferdinand.[16] Ferdinand includes reviews of enzyme structure and function, bioenergetics and kinetics, and appropriate illustrations with a total of 37 enzymes selected from the six major classes of enzymes. Accordingly, for additional basic studies of enzymology, the reader should refer to this classic monograph and to the literature cited below.

RELATION OF STRUCTURE AND FUNCTION

Koshland[17] has reviewed concepts concerning correlations of protein conformation and conformational flexibility of enzymes with enzyme catalysis. Enzymes do not exist initially in a conformation complementary to that of the substrate. The substrate induces the enzyme to assume a complementary conformation. This is the so-called "induced fit" theory. There is proof that proteins do possess conformational flexibility and undergo conformational changes under the influence of small molecules. It is emphasized that this does not mean that all proteins must be flexible; nor does it mean that conformationally flexible enzymes must undergo conformation changes when interacting with all compounds. Furthermore, a regulatory compound that is not directly involved in the reaction can exert control on the reactivity of the enzyme by inducing conformational changes, i.e., by inducing the enzyme to assume the specific conformation complementary to the substrate. (Conceivably, hormones as regulators function according to the foregoing mechanism of affecting protein structure.) So-called flexible enzymes can be distorted conformationally by molecules classically called inhibitors. Such inhibitors can induce the protein to undergo conformation changes disrupting the catalytic functions or the binding function of the enzyme. In this connection it is interesting to note how the work of Belleau and the molecular perturbation theory of drug action relate to Koshland's studies (see Chap. 2).

Evidence continues to support the explanation of enzyme catalysis on the basis of the "active site" (reactive center) of amino acid residues which is considered to be that relatively small region of the enzyme's macromolecular surface involved in catalysis. Within this site, the enzyme has strategically positioned functional groups (from the side chains of amino acid

units) which participate cooperatively in the catalytic action.[18]

Some enzymes have absolute specificity with respect to a single substrate, but in other cases enzymes catalyze a particular type of reaction that various compounds undergo. In the latter case the enzyme is said to have relative specificity. Nevertheless, when compared with other catalysts, enzymes are outstanding with respect to specificity for certain substrates.† Of course, the physical, chemical, conformational and configurational properties of the substrate determine its complementarity to the enzyme's reactive center. These factors therefore determine whether a given compound satisfies the specificity of a particular enzyme. Enzyme specificity must be a function of the nature, including conformational and chemical reactivity, of the reactive center, but when the enzyme is a conjugated protein with a coenzyme moiety, the nature of the coenzyme also contributes to specificity characteristics.

It seems that in some cases the active center of the enzyme is complementary to the substrate molecule in a strained configuration corresponding to the "activated" complex for the reaction catalyzed by the enzyme. The substrate molecule is attracted to the enzyme and is caused by the forces of attraction to assume the strained state, with conformational changes, which favors the chemical reaction; that is, the activation energy requirement of the reaction is decreased by the enzyme to such an extent as to cause the reaction to proceed at an appreciably greater rate than it would in the absence of the enzyme. If in all cases the enzyme were completely complementary in structure to the substrate, then no other molecule would be expected to compete successfully with the substrate in combination with the enzyme, which in this respect would be similar in behavior to antibodies. However, in some cases an enzyme complementary to a strained substrate molecule might attract more strongly to itself a molecule resembling the strained substrate molecule itself; e.g., the hydrolysis of benzoyl-L-tyrosylglycineamide was practically inhibited by an equal amount of benzoyl-D-tyrosylglycineamide. This illustration also might serve to illustrate a type of antimetabolite activity.

Several types of interactions contribute to the formation of enzyme-substrate complexes: attractions between charged (ionic) groups on the protein and the substrate; hydrogen bonding; hydrophobic forces (the

---

† A recent review of interpretations of enzyme reaction stereospecificity offers interesting reading pertaining to these considerations; see Hanson and Rose.[19] J. W. Cornforth's lecture, delivered when he received the 1975 Nobel Prize in Chemistry (a prize he shared with V. Prelog), on asymmetry and enzyme action is relevant to this discussion (Science, 193:121, 1976).

tendency of hydrocarbon moieties of side chains of amino acid residues to associate with the nonpolar groups of the substrate in a water environment); and London forces (induced-dipole interactions).

Many studies of enzyme specificity have been made on proteolytic enzymes (proteases). Configurational specificity can be exemplified with the case of aminopeptidase which cleaves L-leucylglycylglycine but does not affect D-leucylglycylglycine. D-Alanylglycylglycine is slowly cleaved by this enzyme. These phenomena illustrate the significance of steric factors; at the active center of aminopeptidase, a critical factor is a matter of closeness of approach that affects the kinetics of the reaction.

One can easily imagine how difficult it is to study the reactivity of enzymes on a functional group basis since the mechanism of enzyme action is so complex.[17] Nevertheless, it can be said that the —SH group probably is found in more enzymes as a functional group than are the other polar groups. It should be noted that in some cases, e.g., urease, the less readily available SH groups are necessary for biological activity and can not be detected by the nitroprusside test that is used to detect the freely reactive SH groups.

A free —OH group of the tyrosyl residue is necessary for the activity of pepsin. Both the —OH of serine and the imidazole portion of histidine appear to be necessary parts of the active center of certain hydrolytic enzymes such as trypsin and chymotrypsin and furnish the electrostatic forces involved in a proposed mechanism (shown below), in which E denotes enzyme, the other symbols being self-evident.*

These two groups, i.e., —OH and =NH, could be located on separate peptide chains in the enzyme so long as the specific 3-dimensional structure formed during activation of the zymogen brought them near enough to form a hydrogen bond. The polarization of the resulting structure would cause the serine oxygen to be the nucleophilic agent which attacks the carbonyl function of the substrate. The complex is stabilized by the simultaneous "exchange" of the hydrogen bond from the serine oxygen to the carbonyl oxygen of the substrate.

The intermediate acylated enzyme is written with the proton on the imidazole nitrogen. The deacylation reaction involves the loss of this positive charge simultaneously with the attack of the nucleophilic reagent (abbreviated Nu:H).

* Alternative mechanisms have been proposed[17]; esterification and hydrolysis have been extensively studied by M. L. Bender (J. Am. Chem. Soc. 79:1258, 1957; 80:5388, 1958; 82:1900, 1960; 86:3704, 5330, 1964). More recently, D. M. Blow has reviewed studies concerning the structure and mechanism of chymotrypsin (Accounts of Chemical Research, 9:145, 1976).

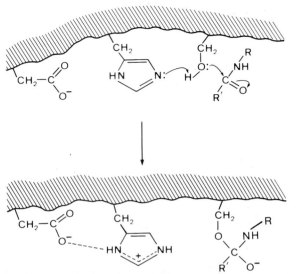

Generalized Mechanism of Protease Catalysis. Adapted from Chemical and Engineering News, Apr. 16, 1979, p. 23.

Roberts[20] effectively used nitrogen-15 (^{15}N) NMR to study the mechanism of protease catalysis. A schematic summary of the generalized mechanism is represented below. It is concluded that the tertiary N-1 nitrogen of the histidine unit within the reactive center of the enzyme deprotonates the hydroxyl of the neighboring serine unit and simultaneously the hydroxyl oxygen exerts nucleophilic attack on the carbonyl carbon of the amide substrate as depicted in the scheme. A tetrahedral intermediate is implicated, and the carboxylate group of the aspartate unit (the third functional group within the reactive center) stabilizes the developing imidazolium ion by a hydrogen bonding to the N-3 hydrogen. Finally, the decomposition of the anionic tetrahedral intermediate toward product formation (amine and acylated serine) is promoted by prior protonation of the amide nitrogen by the imidazolium group.

A possible alternative route to deacylation would involve the nucleophilic attack of the imidazole nitrogen on the newly formed ester linkage of the postulated acyl intermediate, leading to the formation of the acyl imidazole. The latter is unstable in water, hydrolyzing rapidly to give the product and regenerated active enzyme.

The reaction of an alkyl phosphate in such a scheme may be written in an entirely analogous fashion, except that the resulting phosphorylated enzyme would be less susceptible to deacylation through nucleophilic attack. The following diagrammatic scheme has been proposed to explain the function of the active thiol ester site of papain. This ester site is formed and maintained by the folding energy of the enzyme (protein) molecule.

Enzyme-catalyzed Hydrolysis of

$$
\begin{array}{c} O \\ \parallel \\ R\!-\!C\!-\!X \end{array}: \text{A Proposed}
$$

Generalized Mechanism

ZYMOGENS (PROENZYMES)

Zymogens, also called proenzymes, are enzyme precursors. These proenzymes are said to be activated when they are transformed to the enzyme. This activation usually involves catalytic action by some proteolytic enzyme. In some cases the activators merely effect a reorganization of the tertiary structure (conformation) of the protein so that the groups involved within the reactive center become functional, i.e., unmasked.

SYNTHESIS AND SECRETION OF ENZYMES

Exportable proteins (enzymes) such as amylase, ribonuclease, chymotrypsin(ogen), trypsin(ogen), insulin, etc. are synthesized on the ribosomes. They pass across the membrane of the endoplasmic reticulum into the cisternae and directly into a smooth vesicular structure which effects further transportation. They are finally stored in highly concentrated form within membrane-bound granules. These are called zymogen granules whose exportable protein content may reach a value of 40 percent of the total protein of the gland cell. In the above sequences the newly synthesized ex-

portable protein (enzymes) is not free in the cell sap. The stored exportable proteins are released into the extracellular milieu in the case of the digestive enzymes and into adjacent blood capillaries in the case of hormones. The release of these proteins is initiated (triggered) by specific inducers: for example, cholinergic agents (but not epinephrine) and Ca^{++} effect a discharge of amylase, lipase, etc. into the medium; increase in glucose levels stimulates the secretion of insulin, etc. This release of the reserve enzymes and hormones is completely independent of the synthetic process as long as the stores in the granules are not completely depleted. Energy-oxidative phosphorylation does not play an important role in these releases. Electron microscope studies indicate a fusion of the zymogen granule membrane with the cell membrane so that a direct opening of the granule into the extracellular lumen of the gland is formed.

CLASSIFICATION

There are various systems for the classification of enzymes, e.g., the International Union of Biochemistry (IUB) system. This system includes some of the termi-

The Action of Papain: A Proposed Scheme

nology which is used in the literature of medicinal chemistry, and in many cases the terms are self-explanatory: e.g., oxidoreductases; transferases (catalyze transfer of a group, such as methyltransferases); hydrolases (catalyze hydrolysis reactions, such as esterases and amidases); lyases (catalyze nonhydrolytic removal of groups leaving double bonds); isomerases; ligases. Other systems are sometimes used to classify and characterize enzymes, and the following terms are frequently encountered: lipases, peptidases, proteases, phosphatases, kinases, synthetases, dehydrogenases, oxidases, reductases, etc.

PRODUCTS

Pharmaceutically important enzyme products are listed in Table 20-6.

Pancreatin U.S.P., Panteric®, is a substance obtained from the fresh pancreas of the hog or of the ox and contains a mixture of enzymes, principally pancreatic amylase (amylopsin), protease and pancreatic lipase (steapsin). It converts not less than 25 times its weight of U.S.P. Potato Starch Reference Standard into soluble carbohydrates, and not less than 25 times its weight of casein into proteoses. Pancreatin of a higher digestive power may be brought to this standard by admixture with lactose, or with sucrose containing not more than 3.25 percent of starch, or with pancreatin of lower digestive power. Pancreatin is a cream-colored, amorphous powder having a faint, characteristic, but not offensive, odor. It is slowly but incompletely soluble in water and insoluble in alcohol. It acts best in neutral or faintly alkaline media, and excessive acid or alkali renders it inert. Pancreatin can be prepared by extracting the fresh gland with 25 percent alcohol or with water and subsequently precipitating with alcohol. Besides the enzymes mentioned, it contains some trypsinogen, which can be activated by enterokinase of the intestines, chymotrypsinogen, which is converted by trypsin to chymotrypsin, and carboxypeptidase.

Pancreatin is used largely for the predigestion of food and for the preparation of hydrolysates. The value of its enzymes orally must be very small because they are digested by pepsin and acid in the stomach, although some of them may escape into the intestines without change. Even if they are protected by enteric coatings, it is doubtful if they could be of great assistance in digestion.

Trypsin Crystallized U.S.P., is a proteolytic enzyme crystallized from an extract of the pancreas gland of the ox, *Bos taurus*. It occurs as a white to yellowish-white, odorless, crystalline or amorphous

TABLE 20-6

PHARMACEUTICALLY IMPORTANT ENZYME PRODUCTS

Name Proprietary Name	Preparations	Category	Application	Usual Adult Dose*	Usual Dose Range*
Pancreatin U.S.P. *Panteric,* *Viokase*	Pancreatin Capsules U.S.P. Pancreatin Tablets U.S.P.	Digestive aid		325 mg. to 1 g.	
Trypsin Crystallized U.S.P.	Trypsin Crystallized for Aerosol U.S.P.	Proteolytic enzyme		Aerosol, 125,000 U.S.P. Units in 3 ml. of saline daily	
Pancrelipase U.S.P. *Cotazym*	Pancrelipase Capsules U.S.P. Pancrelipase Tablets U.S.P.	Digestive aid			An amount of pancrelipase equivalent to 8,000 to 24,000 U.S.P. Units of lipolytic activity prior to each meal or snack, or to be determined by the practitioner according to the needs of the patient
Chymotrypsin U.S.P. *Chymar*	Chymotrypsin for Ophthalmic Solution U.S.P.	Proteolytic enzyme (for zonule lysis)	1 to 2 ml. by irrigation to the posterior chamber of the eye, under the iris, as a solution containing 75 to 150 Units per ml.		
Hyaluronidase for Injection U.S.P. *Alidase, Wydase*	Hyaluronidase Injection U.S.P.	Spreading agent		Hypodermoclysis, 150 U.S.P. Hyaluronidase Units	
Sutilains U.S.P. *Travase*	Sutilains Ointment U.S.P.	Proteolytic enzyme	Topical, ointment, 2 to 4 times daily		

*See *U.S.P. D.I.* for complete dosage information.

powder, and 500,000 U.S.P. Trypsin Units are soluble in 10 ml. of water of saline T.S.

Trypsin has been used for a number of conditions in which its proteolytic activities relieve certain inflammatory states, liquefy tenacious sputum, etc.; however, the many side-reactions encountered, particularly when it is used parenterally, militate against its use.

Pancrelipase U.S.P., Cotazym®. This preparation has a greater lipolytic action than do other pancreatic enzyme preparations. For this reason it is used to help control steatorrhea and in other conditions in which pancreatic insufficiency impairs the digestion of fats in the diet.

Chymotrypsin U.S.P., Chymar®. This enzyme is extracted from mammalian pancreas and is used in cataract surgery. A dilute solution is used to irrigate the posterior chamber of the eye in order to dissolve the fine filaments which hold the lens.

Hyaluronidase for Injection U.S.P., Alidase®, Wydase®, Premdase®, is a sterile, dry, soluble enzyme product prepared from mammalian testes and capable of hydrolyzing the mucopolysaccharide hyaluronic acid. It contains not more than 0.25 μg. of tyrosine for each U.S.P. Hyaluronidase Unit. Hyaluronidase in solution must be stored in a refrigerator. Hyaluronic acid, an essential component of tissues, limits the spread of fluids and other extracellular material, and, because the enzyme destroys this acid, injected fluids and other substances tend to spread farther and faster than normal when administered with this enzyme. Hyaluronidase may be used to increase the spread and consequent absorption of hypodermoclysis solutions, to diffuse local anesthetics, especially in nerve blocking, and to increase diffusion and absorption of other injected materials, such as penicillin. It also enhances local anesthesia in surgery of the eye and is useful in glaucoma because it causes a temporary drop in intraocular pressure.

Hyaluronidase is practically nontoxic, but caution must be exercised in the presence of infection, because the enzyme may cause a local infection to spread, through the same mechanism; it never should be injected in an infected area. Sensitivity to the drug is rare.

The activity of hyaluronidase is determined by measuring the reduction of turbidity that it produces on a substrate of native hyaluronidate and certain proteins, or by measuring the reduction in viscosity that it produces on a buffered solution of sodium or potassium hyaluronidate. Each manufacturer defines its product in turbidity or viscosity units, but they are not the same because they measure different properties of the enzyme.

Sutilains U.S.P., Travase®, is a proteolytic enzyme obtained from cultures of *B. subtilis* and is used to dissolve necrotic tissue occurring in second- and third-degree burns as well as in bed sores and ulcerating wounds.

Many substances are contraindicated during the topical use of sutilains. These include detergents and anti-infectives which have a denaturing action on the enzyme preparation. The antibiotics penicillin, streptomycin and neomycin do not inactivate sutilains. Mafenide acetate is also compatible with the enzyme.

Streptokinase-Streptodornase, Varidase®, is a mixture containing streptokinase and streptodornase. The former activates an enzyme in the blood that reacts on fibrin and brings about dissolution of blood clots and fibrinous exudates. The latter acts in a similar way to dissolve constituents of pus and has no effect on living cells. The mixture is used locally to remove dead tissue in surgery and before making skin grafts. It is recommended also in the treatment of hemothorax, hematoma, empyema, osteomyelitis, draining sinuses, tuberculous abscesses, infected wounds or ulcers, severe burns and other chronic suppurations. It is supplied in vials containing 100,000 units of streptokinase and 25,000 units of streptodornase.

Fibrinolysin and deoxyribonuclease are available in Elase®, which rapidly lyses fibrinous material in serum, clotted blood and purulent exudates but does not appreciably attack living tissue. It is used topically in surgical wounds, burns, chronic skin ulcerations, sinus tracts, abscesses, etc.

Papain U.S.P., Caroid®, Papase®, the dried and purified latex of the fruit of *Carica papaya* L. (*Fam.* Caricaceae), has the power of digesting protein in either acid or alkaline media; it is best at a pH of from 4 to 7, and at 65 to 90°C. It occurs as light brownish-gray to weakly reddish-brown granules or as a yellowish-gray to weakly yellow powder. It has a characteristic odor and taste and is incompletely soluble in water to form an opalescent solution. The commercial material is prepared by evaporating the juice, but the pure enzyme also has been prepared and crystallized. In medicine, it has been used locally in various conditions similar to those for which pepsin is employed. It has the advantage of activity over a wider range of conditions, but it is often much less reliable. Intraperitoneal instillation of a weak solution has been recommended to counteract a tendency to adhesions after abdominal operations, and several enthusiastic reports have been made about its value under these conditions. Papain has been reported to cause allergies in persons who handle it, especially those who are exposed to inhalation of the powder.

Bromelin is a somewhat similar proteolytic enzyme from the pineapple, *Ananas comosus* L. Merr., and can be prepared from the juice by precipitating with ammonium sulfate or by alcohol. Its activity is greatest at a pH of from 3 to 4. It has been suggested as an anthelmintic because it has the power of digesting living worms in a test tube.

Plant Protease Concentrate, Ananase®, is a mixture of proteolytic enzymes obtained from the pineapple plant. It is proposed for use in the treatment of soft tissue inflammation and edema associated with traumatic injury, localized inflammations and postoperative tissue reactions. The swelling that accompanies inflammation may possibly be caused by occlusion of the tissue spaces with fibrin. If this be true, sufficient amounts of Ananase® would have to be absorbed and reach the target area after oral administration to act selectively on the fibrin. This is yet to be firmly established and its efficacy as an anti-inflammatory agent is inconclusive. On the other hand, an apparent inhibition of inflammation has been demonstrated with irritants such as turpentine and croton oil (granuloma pouch technique).

Ananase® is available in 50,000-unit tablets for oral use.

Diastase, Taka®-Diastase, is derived from the action of a fungus, *Aspergillus oryzae* Cohn (*Eurotium O.* Ahlburg), on rice hulls or wheat bran. It is a yellow, hygroscopic, almost tasteless powder that is freely soluble in water and can solubilize 300 times its weight of starch in 10 minutes. It is employed in doses of 0.3 to 1.0 g. in the same conditions as malt diastase. Taka®-Diastase is combined with alkalies as an antacid in Takazyme®, with vitamins in Taka-Combex® and in other preparations.

HORMONES

Certain hormones are polypeptides and are classified as simple proteins when compared with the much more complex proteins, such as enzymes and conjugated proteins. These hormones include metabolites elaborated by the hypothalamus, as well as the pituitary hormones. Insulin and glucagon are polypeptides produced by the pancreas.

HORMONES FROM THE HYPOTHALAMUS

The physiologic and clinical aspects of hypothalamic-releasing hormones have been reviewed.[21-25.] Through these hormones, the central nervous system regulates other essential endocrine systems, including the pituitary, which in turn controls other systems, e.g., the thyroid.*

Thyroliberin (thyrotropin-releasing hormone,

TRH) is the hypothalamic hormone that is responsible for the release of the pituitary's thyrotropin (TSH). Thyrotropin stimulates the production of thyroxine and liothyronine by the thyroid. The latter thyroid hormones, by feedback regulation, inhibit the action of TRH on the pituitary. Thyroliberin is a relatively simple tripeptide that has been characterized as pyroglutamyl-histidyl-prolinamide. TRH possesses interesting biological properties. In addition to stimulating the release of thyrotropin, it promotes the release of prolactin. It also has some CNS effects that have been evaluated for antidepressant therapeutic potential, but so far the results of clinical studies are not considered to be conclusive.[26]

Gonadoliberin, as the name implies, is the gonadotropin-releasing hormone (GnRH). This hypothalamic hormone has been identified to be the decapeptide H-Glu-His-Trp-Ser-Tyr-Gly-Leu-Arg-Pro-Gly-NH$_2$. GnRH stimulates the release of lutropin (LH) and follitropin (FSH) by the pituitary. Gonadoliberin is considered to be of potential therapeutic importance in the treatment of hypogonadotropic infertility in both males and females.[22]

GRF, a hypothalamic growth-releasing factor also called somatoliberin, continues to be under intensive investigation. Its identification and biological characterization remain to be completed, but physiologic and clinical data support the existence of hypothalamic control of pituitary release of somatotropin.

Somatostatin is another very interesting hypothalamus hormone.[21] It is a tetradecapeptide possessing a disulfide linking two cysteine residues, 3–14, in the form of a 38-member ring. Somatostatin suppresses several endocrine systems. It inhibits the release of somatotropin and thyrotropin by the pituitary. It also inhibits the secretion of insulin and glucagon by the pancreas. Gastrin, pepsin, and secretin are intestinal hormones that are likewise affected by somatostatin. The therapeutic potential of somatostatin is discussed below in relation to the role of glucagon in the pathology of human diabetes.

Other hypothalamus hormones include the lutropin-release-inhibiting factor (LHIF), prolactin-releasing factor (PRF), prolactin-release-inhibiting factor (PIF), corticotropin-releasing factor (CRF), melanoliberin (MRF), and melanostatin (MIF).

As the foregoing discussion illustrates, the hypothalamus endocrine system performs many essential functions affecting other endocrine systems.[25] In turn, the thalamus and cortex exert control on the secretion of these (hypothalamic) factors. A complete review of

---

* Recent developments concerning the synthesis, structure-activity relationships, and potential clinical applications of the peptides of the hypothalamus have been reviewed. See A.F. Spatola, Annual Reports in Medicinal Chemistry, Vol. 16, 199, 1981.

this field is beyond the scope of this chapter, hence the interested reader should refer to the literature cited.[21-25]

PITUITARY HORMONES

As noted above, the anterior pituitary (adenohypophysis) is under control by hypothalamic regulatory hormones, and it secretes ACTH, GH, prolactin, etc.

Adrenocorticotropic hormone

The adrenocorticotropic hormone (adrenocorticotropin, ACTH, corticotropin) is a medicinal agent which has been the center of much research. In the late 1950's its structure was elucidated and the total synthesis was accomplished in the 1960's. Related peptides also have been synthesized and some of these possess similar physiologic action. Human ACTH has 39 amino acid units within the polypeptide chain. Full activity has been reported for synthetic peptides containing the first 20 amino acids. A peptide containing 24 amino acids has full steroidogenic activity without allergenic reactions. This is of practical importance since natural ACTH preparations sometimes produce clinically dangerous allergic reaction.

ACTH exerts its major action on the adrenal cortex promoting steroid synthesis by stimulating the formation of pregnenolone from cholesterol. An interaction between ACTH and specific receptors is implicated in the mechanism leading to stimulation of adenylcyclase and acceleration of steroid production. Thus c-AMP, the general mediator of many hormone actions, is involved, perhaps through enhancement of the synthesis of some protein associated with steroidogenesis. Other biochemical effects exerted by ACTH include stimulation of phosphorylase and hydroxylase activities. Glycolysis also is increased by this hormone. Enzyme systems which catalyze processes involving the production of NADPH are also stimulated. (It is noteworthy that NADPH is required by the steroid hydroxylations which take place in the over-all transformation of cholesterol to hydrocortisone, the major glucocorticoid hormone.)

c-AMP

Corticotropin Injection U.S.P., ACTH injection, adrenocorticotropin injection, is a sterile preparation of the principle or principles derived from the anterior lobe of the pituitary of mammals used for food by man. It occurs as a colorless or light straw-colored liquid, or soluble amorphous solid by drying such liquid from the frozen state. It exerts a tropic influence on the adrenal cortex. The solution has a pH range of 3.0 to 7.0 and is used for its adrenocorticotropic activity.

Repository Corticotropin Injection U.S.P., corticotropin gel, purified corticotropin, ACTH, purified, is corticotropin in a solution of partially hydrolyzed gelatin to be used intramuscularly for a more uniform and prolonged maintenance of activity.

Sterile Corticotropin Zinc Hydroxide Suspension U.S.P. is a sterile suspension of corticotropin, adsorbed on zinc hydroxide and contains not less than 45 and not more than 55 µg. of zinc for each 20 U.S.P. Corticotropin Units. Because of its prolonged activity due to slow release of corticotropin, an initial dose of 40 U.S.P. Units can be administered intramuscularly, followed by a maintenance dose of 20 Units, 2 or 3 times a week.

Cosyntropin, Cortrosyn®, is a synthetic peptide containing the first 24 amino acids of natural corticotropin. Cosyntropin is used as a diagnostic agent to test for adrenal cortical deficiency. Plasma hydrocortisone concentration is determined before and 30 minutes after the administration of 250 µg. of cosyntropin. Most normal responses result in an approximate doubling of the basal hydrocortisone concentration in 30 to 60 minutes. If the response is not normal, adrenal insufficiency is indicated. Such adrenal insufficiency could be due to either adrenal or pituitary malfunction, and further testing is required in order to distinguish between the two. Cosyntropin (250 µg.

Ser•Tyr•Ser•Met•Glu•His•Phe•Arg•Trp•Gly•Lys•Pro•Val•Gly•Lys
Lys
Glu•Ala•Gly•Asn•Pro•Tyr•Val•Lys•Val•Pro•Arg•Arg
Asp
Glu•Ser•Ala•Glu•Ala•Phe•Pro•Leu•Glu•Phe

Human ACTH

infused within 4 to 8 hours) or corticotropin (80 to 120 units daily for 3 to 4 days) is administered. Patients with functional adrenal tissue should respond to this dosage. Patients who respond accordingly are suspected of hypopituitarism and the diagnosis can be confirmed by other tests for pituitary function. On the other hand, little or no response is shown by patients who have Addison's disease.

Somatotropin

The growth hormone (GH) is another polypeptide elaborated by the anterior pituitary. The amino acid sequence of human gonadotropin (HGH) has been determined, and structural comparisons with bovine and ovine hormones have been made.[28] In addition to promoting body growth, this hormone exerts several other actions. Moreover, it should not be inferred that this polypeptide is the only factor which is known to promote growth. Other hormones also contribute to normal growth of the organism. Promotion of body growth is associated with skeletal development and protein anabolism, and in fact GH has an anabolic effect, promoting protein synthesis in liver and peripheral tissues. GH also causes acute hypoglycemia followed by elevated blood glucose concentration and perhaps glucosuria. GH stimulates glucagon secretion by the pancreas, increases muscle glycogen, augments release of fatty acids from adipose tissue and increases osteogenesis.

Prolactin

Three of the anterior pituitary hormones are polypeptides: ACTH, GH and prolactin. (The others are

TABLE 20-7

PHARMACEUTICALLY IMPORTANT ACTH PRODUCTS

Preparation Proprietary Name	Category	Usual Adult Dose*	Usual Dose Range*	Usual Pediatric Dose*
Corticotropin Injection U.S.P. Corticotropin for Injection U.S.P.	Adrenocorticotropic hormone; adrenocortical steroid (anti-inflammatory); diagnostic aid (adrenocortical insufficiency)	Adrenocorticotropic hormone—parenteral, 20 U.S.P. Units 4 times daily; adrenocortical steroid (anti-inflammatory)—parenteral, 20 U.S.P. Units 4 times daily; diagnostic aid (adrenocortical insufficiency)—rapid test—I.M. or I.V., 25 U.S.P. Units, with blood sampling in 1 hour; adrenocortical steroid output—I.V. infusion, 25 Units in 500 to 1,000 ml. of 5 percent Dextrose Injection over a period of 8 hours on each of 2 successive days, with 24-hour urine collection done on each day	Adrenocorticotropic hormone—40 to 80 Units daily; adrenocortical steroid (anti-inflammatory)—40 to 80 Units daily	Parenteral, 0.4 Unit per kg. of body weight or 12.5 Units per square meter of body surface, 4 times daily
Repository Corticotropin Injection U.S.P. *Acthar Gel, Cortrophin Gel*	Adrenocorticotropic hormone; adrenocortical steroid (anti-inflammatory); diagnostic aid (adrenocortical insufficiency)	Adrenocorticotropic hormone—I.M. or S.C., 40 to 80 U.S.P. Units every 24 to 72 hours; I.V. infusion, 40 to 80 U.S.P. Units in 500 ml. of 5 percent Dextrose Injection given over an 8-hour period, once daily; adrenocortical steroid (anti-inflammatory)—I.M. or S.C., 40 to 80 U.S.P. Units every 24 to 72 hours; I.V. infusion, 40 to 80 U.S.P. Units in 500 ml. of 5 percent Dextrose Injection given over an 8-hour period, once daily; diagnostic aid (adrenocortical insufficiency)—I.M., 40 U.S.P. Units twice daily on each of 2 successive days, with 24-hour urine collection done each day		Adrenocorticotropic hormone— parenteral, 0.8 Unit per kg. of body weight or 25 Units per square meter of body surface, per dose
Sterile Corticotropin Zinc Hydroxide Suspension U.S.P. *Cortrophin-Zinc*	Adrenocorticotropic hormone; adrenocortical steroid (anti-inflammatory); diagnostic aid (adrenocortical insufficiency)	Adrenocorticotropic hormone—I.M., initial, 40 to 60 U.S.P. Units daily, increasing interval to 48, then 72 hours; reduce dose per injection thereafter; maintenance, 20 Units daily to twice weekly; adrenocortical steroid (anti-inflammatory)—I.M., initial, 40 to 60 Units daily, increasing interval to 48, then 72 hours; reduce dose per injection thereafter; maintenance, 20 Units daily to twice weekly; diagnostic aid (adrenocortical insufficiency)—I.M., 40 U.S.P. Units on each of 2 successive 24-hour periods		
Cosyntropin *Cortrosyn*	Diagnostic aid (adrenocortical insufficiency)	I.M. or I.V., 250 μg.		Children 2 years of age or less, 0.125 mg.

*See *U.S.P. D.I.* for complete dosage information.

conjugated proteins, glycoproteins.) Prolactin (lacto-genic hormones, luteotropin, PRL) stimulates lactation of parturition. At one time it appeared that PRL might be identical with growth hormone but now it is established that human PRL and GH are discrete and separable.[17]

Follicle-stimulating hormone

Follicle-stimulating hormone (FSH) promotes the development of ovarian follicles to maturity. FSH also promotes spermatogenesis in testicular tissue. It is a glycoprotein whose carbohydrate component is considered to be associated with its activity.

Luteinizing hormone

Luteinizing hormone (LH) is another glycoprotein. It acts after the maturing action of FSH on ovarian follicles and stimulates production of estrogens and transforms the follicles into corpora lutea. LH also acts in the male of the species by stimulating the Leydig cells which produce testosterone.

Menotropins

Pituitary hormones prepared from the urine of post-menopausal women whose ovarian tissue does not respond to gonadotropin are available for medicinal use in the form of the product, menotropins (Pergonal®). The latter has FSH and LH gonadotropin activity in a 1:1 ratio. Menotropins is useful in the treatment of anovular women whose ovaries are responsive to pituitary gonadotropins but have a gonadotropin deficiency due to either pituitary or hypothalamus malfunction. Usually, menotropins is administered intramuscularly: initial dose of 75 I.U. of FSH and 75 I.U. of LH daily for 9 to 12 days, followed by 10,000 I.U. of chorionic gonadotropin one day after the last dose of menotropins.

Thyrotropin

The thyrotropic hormone, also called thyrotropin (TSH), is a glycoprotein consisting of two polypeptide chains. This hormone promotes production of thyroid hormones by affecting the kinetics of the mechanism whereby the thyroid concentrates iodide ions from the bloodstream, thereby promoting incorporation of the halogen into the thyroid hormones and release of hormones by the thyroid.

Neurohypophyseal hormones

The posterior pituitary (neurohypophysis) is the source of vasopressin, oxytocin, α- and β-melanocyte-stimulating hormones, and coherin. The synthesis, transport, and release of these hormones have recently been reviewed by Brownstein et al.[27] It is herein noted that vasopressin and oxytocin are synthesized and released by neurons of the hypothalamic–neurohypophyseal system. These peptide hormones and their respective neurophysin-carrier proteins are synthesized as structural components of separate precursor proteins, and these proteins appear to be partially degraded into smaller bioactive peptides in the course of transport along the axon.

Vasopressin and oxytocin are completely known structurally and have been synthesized. Actually, three closely related octapeptides have been isolated from mammalian posterior pituitary: oxytocin and arginine-vasopressin from most mammals, and lysine-vasopressin from pigs. The vasopressins differ from one another with respect to the nature of the eighth amino acid residues: arginine and lysine, respectively. Oxytocin has leucine at position 8 and its fourth amino acid is isoleucine instead of phenylalanine. (These hormones are actually secreted by the hypothalamus and are stored in the posterior pituitary.)

Vasopressin is also known as the pituitary antidiuretic hormone (ADH). This hormone can effect graded changes in the permeability of the distal portion of the mammalian nephron to water, resulting in either conservation or excretion of water; thus it modulates the renal tubular reabsorption of water. ADH has been shown to increase c-AMP production in several tissues. Theophylline, which promotes c-AMP by inhibiting the enzyme (phosphodiesterase) which catalyzes its hydrolysis, causes permeability changes similar to those due to ADH. c-AMP also effects similar permeability changes, hence, it is suggested that c-AMP is involved in the mechanism of action of ADH.

ADH is therapeutically useful in the treatment of diabetes insipidus of pituitary origin. It also has been used to relieve intestinal paresis and distention.

Oxytocin is appropriately named on the basis of its oxytocic action. Oxytocin exerts stimulant effects on the smooth muscle of the uterus and mammary gland. On the other hand, this hormone has a relaxing effect on vascular smooth muscle when administered in high doses. It is considered to be the drug of choice to induce labor and to stimulate labor in cases of intrapartum hypotonic inertia. Oxytocin also is used in inevitable or incomplete abortion after the twentieth week of gestation. It also may be used to prevent or control hemorrhage and to correct uterine hypotonic-ity. In some cases oxytocin is used to promote milk ejection; it acts by contracting the myoepithelium of the mammary glands. Oxytocin is usually administered parenterally via intravenous infusion, intravenous injection or intramuscular injection. Oxytocin citrate buccal tablets are also available, but the rate of absorption is unpredictable and buccal administration is less precise. Topical administration (nasal spray) two or three minutes before nursing to promote milk ejection is sometimes recommended.[29]

Oxytocin Injection U.S.P. is a sterile solution in water for injection of oxytocic principle prepared by synthesis or obtained from the posterior lobe of the pituitary of healthy, domestic animals used for food by man. The pH is 2.5 to 4.5; expiration date, 3 years.

Oxytocin preparations are widely used with or without amniotomy to induce and stimulate labor. Although injection is the usual route of administration, the sublingual route is extremely effective. Sublingual and intranasal spray (Oxytocin Nasal Solution U.S.P.) routes of administration also will stimulate milk let-down.

Vasopressin Injection U.S.P. is a sterile solution in water for injection of the water-soluble pressor principle of the posterior lobe of the pituitary of healthy domestic animals used for food by man, or prepared by synthesis. Each ml. possesses a pressor activity equal to 20 U.S.P. Posterior Pituitary Units, expiration date, 3 years.

Vasopressin Tannate, Pitressin® Tannate, is a water-insoluble tannate of vasopressin administered intramuscularly (1.5 to 5 pressor units daily) for its prolonged duration of action due to the slow release of vasopressin. It is particularly useful for patients who have diabetes insipidus, but it never should be used intravenously.

Felypressin, 2-phenylalanine-8-lysine vasopressin, has relatively small antidiuretic activity and little oxytocic activity. It has considerable pressor (i.e., vasoconstrictor) activity which, however, differs from that of epinephrine, i.e., following capillary constriction in the intestine it lowers the pressure in the vena portae, whereas epinephrine raises the portal pressure. Felypressin also causes an increased renal blood flow in the cat, whereas epinephrine brings about a fall in renal blood flow. Felypressin is 5 times more effective a vasopressor than lysine vasopressin and is recommended in surgery to minimize blood flow, especially in obstetrics and gynecology.

Lypressin is synthetic lysine-8-vasopressin, a polypeptide similar to the antidiuretic hormone. The lysine analog is considered to be more stable and it is rapidly absorbed from the nasal mucosa. Lypressin (Diapid®) is pharmaceutically available as a topical solution, spray, 50 pressor units (185 μg.) per ml. in 5-ml. containers. Usual dosage, topical (intranasal), one or more sprays applied to one or both nostrils one or more times daily.[29]

Melanocyte-stimulating hormone

The middle lobe of the pituitary secretes intermedin which increases the deposition of melanin by the melanocytes of the human skin; hence this principle is called melanocyte-stimulating hormone (MSH). It is important to note some endocrinologic correlations by referring to the effect of hydrocortisone inhibiting secretion of MSH; epinephrine and norepinephrine inhibit the action of MSH. Two peptides which have been isolated from the pituitary have been designated α-MSH and β-MSH. α-MSH contains the same amino acid sequence of the first 13 amino acids of ACTH. β-MSH has 18 amino acid units.

TABLE 20-8

NEUROHYPOPHYSEAL HORMONES: PHARMACEUTICAL PRODUCTS

Preparation Proprietary Name	*Category*	*Usual Adult Dose**	*Usual Dose Range**	*Usual Pediatric Dose**
Oxytocin Injection U.S.P. *Pitocin*	Oxytocic	I.M., 3 to 10 Units; I.V. infusion, 10 Units in 1 liter of 5 percent Dextrose Injection at a rate of 0.5 to 2 ml. per minute		
Vasopressin Injection U.S.P. *Pitressin*	Antidiuretic posterior pituitary hormone	I.M. or S.C., 5 to 10 Units 2 to 4 times daily as necessary	5 to 60 Units daily	I.M. or S.C., 2.5 to 10 Units 2 to 4 times daily as necessary
Vasopressin Tannate Injection *Pitressin Tannate Injection*	Antidiuretic posterior pituitary hormone		I.M., 2.5 to 5 units as required, usually every 1 to 3 days	Children, I.M., 1.25 to 2.5 units as required, usually every 1 to 3 days

* See *U.S.P. D.I.* for complete dosage information.

OTHER BRAIN PEPTIDES

Numerous peptides are elaborated in the brain, and some have been implicated as neurotransmitters. The endogenous opiates, the enkephalins and endorphins, are among the most interesting and are discussed in the chapter on analgesics (see Chap. 17). Miller and Cuatrecasas[30] reviewed comprehensively most of the literature concerning the biosynthesis and degradation, neuroendocrine effects, and structural pharmacology of these opioid peptides.

Substance-P has been appearing in the literature since the early 1930's. A review by Snyder and Innis[31a] includes discussions of Substance-P, neurotensin, vasoactive intestinal peptide (VIP), cholecystokinin, angiotensin-II, etc. Substance-P has been characterized as a peptide with the following sequence of amino acids: Arg-Pro-Lys-Pro-Gln-Gln-Phe-Phe-Gly-Leu-Met. It appears to function as a transmitter of signals carried by sensory nerves into the spinal cord and then relayed to the brain.[32]

Certain peptides usually considered to be of pituitary origin have been detected in the central nervous system; e.g., ACTH, melanocyte stimulating hormone (MSH), β-lipotropic hormone (β-LPH) and β-endorphin. Krieger and Liotta[32a] have summarized and reviewed critically the available data concerning synthesis, distribution, regulation, and function of these hormones, and they conclude that pituitary hormones originating in the brain may be involved in central coordination of responses independent of those affected by peripheral secretion of such pituitary hormones. On the other hand, those hormones that appear to enter the brain by possible retrograde portal blood flow may participate in short-loop feedback regulation of anterior pituitary function.

PLACENTAL HORMONES

Human chorionic gonadotropin

Human chorionic gonadotropin (HCG) is a glycoprotein synthesized by the placenta. Estrogens stimulate the anterior pituitary to produce placentotropin, which in turn stimulates HCG synthesis and secretion. HCG is produced primarily during the first trimester of pregnancy. It exerts effects which are similar to those of pituitary LH.

HCG is used therapeutically in the management of crytorchidism in prepubertal boys. Is is also used in women in conjunction with menotropins to induce ovulation when the endogenous availability of gonadotropin is not normal.

Human placental lactogen

Human placental lactogen (HPL) is also called human choriomammotropin and chorionic growth-hormone prolactin. This hormone exerts numerous actions. In addition to mammotropic and lactotropic effects, this hormone exerts somatotropic and luteotropic actions. It has been identified as a protein composed of 191 amino-acid units in single peptide chain with two disulfide bridges.[22] HPL resembles human somatotropin.

PANCREATIC HORMONES

Relationships between lipid and glucose levels in the blood and the general disorders of lipid metabolism found in diabetic subjects have received the attention of many chemists and clinicians. In order to understand diabetes mellitus, its complications and its treatment, one has to begin at the level of basic biochemistry regarding the pancreas and the ways carbohydrates are correlated with lipid and protein metabolism (see Chap. 19). The pancreas produces insulin as well as glucagon; β-cells secrete insulin and the α-cells secrete glucagon. Insulin will be considered first.

Insulin

Recent advances in the biochemistry of insulin have been reviewed with emphasis on proinsulin biosynthesis, conversion of proinsulin to insulin, secretion, insulin receptors, catabolism, effects by sulfonylureas, etc.[33] The existence of a precursor (proinsulin) in insulin formation has been demonstrated in isolated islets. Proinsulin is synthesized in islet endoplasmic reticulum and is stored in the secretory granules where the cleavage of proinsulin to insulin takes place. This cleavage, which is considered to be the rate-limiting step in insulin synthesis, is catalyzed by some protease. Proinsulin is a polypeptide containing 84 amino acids; upon activation as indicated above a segment of the chain containing 33 amino acids (chain C) is cleaved off, leaving chains A and B, having 21 and 30 amino acid residues, respectively. Chains A and B joined through two disulfide linkages constitute the insulin structure. (See Fig. 20-3.)

Modern x-ray crystallography has been successfully applied to the study of insulin. Now that the three-dimensional structure of insulin has been determined, it is possible both to understand that the high bioactivity of insulin depends on the integrity of the overall conformation (three-dimensional structure) and to identify a receptor-binding region consisting of A-1, 4, 5, 19, 21, and B-12, 16, 24 to 26. It should be noted that the crystal structure of insulin appears to be conserved in solution and during its receptor interaction.[22]

The insulins from various animal species differ to a minor degree with respect to certain amino acid residues.

Insulin comprises 1 percent of pancreatic tissue, and secretory protein granules contain about 10 percent insulin. These granules fuse with the cell membrane, with simultaneous liberation of insulin which enters the portal vein and passes through the liver. In the liver, large amounts are trapped and the remainder is delivered to the systemic circulation. The half-life of insulin in plasma is about 40 minutes.

In most cases exogenous insulin is weakly antigenic. No insulin antibodies were found in thousands of persons who had never received insulin. The release of insulin is probably triggered by certain levels of glucose in the blood or a metabolic product of glucose and the insulin levels in the blood. Secretin and ACTH can directly stimulate the secretion of insulin. Other factors such as glucagon cause an increase in plasma insulin probably via indirect mechanisms, i.e., release of glucose.

"Clinical" insulin that has been crystallized 5 times and then subjected to countercurrent distribution (2-butanol: 1% dichloroacetic acid in water) yields about 90 percent insulin-A, with varying amounts of insulin-B, together with other minor components. A and B differ by an amide group and have the same activity. End-member analysis, sedimentation and diffusion studies indicate a molecular weight of about 6,000. The value of 12,000 for the molecular weight of insulin containing trace amounts of zinc (obtained by physical methods) is probably a bimolecular association product through the aid of zinc.

The extensive studies of Sanger[34] and others have elucidated the amino acid sequence and structure of insulin. This breakthrough led researchers to pursue further chemical studies.

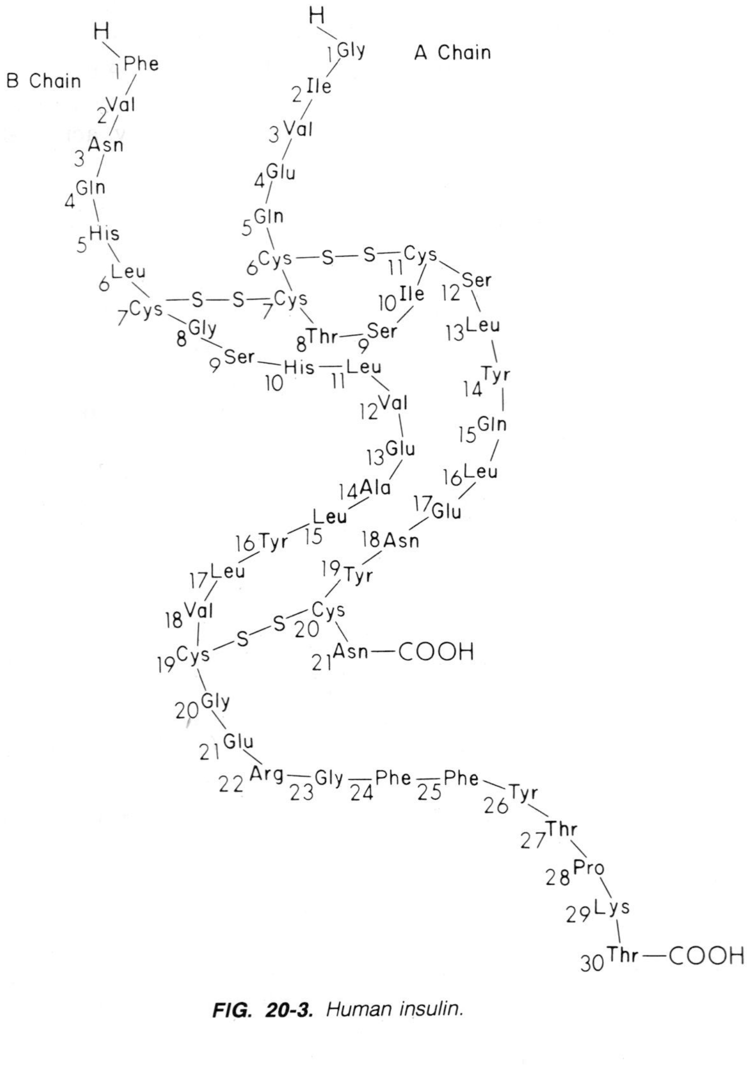

FIG. 20-3. Human insulin.

Recently the A and B chains of human, bovine and sheep insulin have been synthesized in a few weeks by the peptide synthesis on solid supports. The A and B chains have been combined to form insulin in 60 to 80 percent yields, with a specific activity comparable to that of the natural hormone.[35] This lends support to the suggestion that the A and B chains are synthesized in vivo separately and are subsequently combined to form insulin.

The total synthesis of human insulin has been reported by Rittel et al.[36,37] These workers were able to selectively synthesize the final molecule appropriately cross-linked by disulfide (—S—S—) groups in yields ranging between 40 to 50 percent, whereas earlier synthetic methods involved random combination of separately prepared A and B chains of the molecule.

Although insulin is readily available from natural sources (e.g., porcine and bovine pancreatic tissue), partial syntheses and molecular modifications have been developed as the basis for structure-activity relationship (SAR) studies. Such studies have revealed that amino-acid units cannot be removed from the insulin-peptide chain A without significant loss of hormonal activity. On the other hand, several amino acids of chain B are not considered to be essential for activity. Up to the first six and the last three amino-acid units can be removed without significant decrease in activity.[22]

Two insulin analogs, which differ from the parent hormone in that the NH_2 terminus of chain A (A^1) glycine has been replaced by L- and D-alanine respectively, have been synthesized by Cosmatos et al.[38] for SAR studies. The relative potencies of the L- and D-analogs reveal interesting SAR's. The L- and D-alanine analogs are 9.4 percent and 95 percent, respectively, as potent as insulin in glucose oxidation. The relative binding affinity to isolated fat cells is reported to be approximately 10 percent for the L- and 100 percent for the D-analog. Apparently, substitution on the α-carbon of A^1 glycine of insulin with a methyl in a particular configuration interferes with the binding; hence the resulting analog (the L-alanine analog) is much less active. Methyl substitution in the opposite configuration does not affect the binding nor the bioactivity.

It appears that molecular modifications of insulin on the amino groups lead to reduction of bioactivity, but modifications of the epsilon amino group of lysine number 29 on chain B (B-29) may yield active analogs. Accordingly, May et al.[39] synthesized N-epsilon-(+)-biotinyl insulin which was demonstrated to be equipotent with natural insulin. Complexes of this biotinyl-insulin derivative with avidin were also prepared and evaluated biologically; these complexes showed a potency decrease to 5 percent of that of insulin. Such complexes conjugated with ferritin are expected to be useful in the development of electron-microscope stains of insulin receptors. Insulin is inactivated in vivo by (1) an immunochemical system in the blood of insulin-treated patients, (2) reduction of the disulfide bonds (probably by glutathione) and (3) by insulinase (a proteolytic enzyme) that occurs in liver. Pepsin and chymotrypsin will hydrolyze some peptide bonds that lead to inactivation. It is inactivated by reducing agents such as sodium bisulfite, sulfurous acid and hydrogen.

Insulin has many effects on metabolic processes; these actions can be direct or indirect, and it is difficult if not impossible to establish which are primary actions and which are secondary effects. Insulin affects skeletal and heart muscle, adipose tissue, the liver, the lens of the eye, and perhaps leukocytes.

In muscle and adipose tissue insulin promotes transport of glucose and other monosaccharides across cell membranes; it also facilitates transport of amino acids, potassium ions, nucleosides and ionic phosphate. Insulin also activates certain enzymes, kinases and glycogen synthetase in muscle and adipose tissue. In adipose tissue insulin decreases the release of fatty acids induced by epinephrine or glucagon. c-AMP promotes fatty acid release from adipose tissue; therefore, it is possible that insulin decreases fatty acid release by reducing tissue levels of c-AMP. Insulin also facilitates the incorporation of intracellular amino acids into protein.

In the liver there is no barrier to the transport of glucose into cells, but, nevertheless, insulin influences liver metabolism, decreasing glucose output, decreasing urea production, lowering c-AMP and increasing potassium and phosphate uptake. It appears that insulin exerts induction of specific hepatic enzymes involved in glycolysis, while inhibiting gluconeogenic enzymes. Thus, insulin promotes glucose utilization via glycolysis by increasing the synthesis of glucokinase, phosphofructokinase and pyruvate kinase. Insulin decreases the availability of glucose from gluconeogenesis by suppressing pyruvate carboxylase, phosphoenolpyruvate carboxykinase, fructose-1,6-diphosphatase, and glucose-6-phosphatase.

Insulin effects on lipid metabolism also are important. In adipose tissue insulin has an antilipolytic action (i.e., an effect opposing the breakdown of fatty acid triglycerides). It also decreases the supply of glycerol to the liver. Thus, at these two sites, insulin decreases the availability of precursors for the formation of triglycerides. Insulin is necessary for the activation and synthesis of lipoprotein lipases, enzymes responsible for lowering very low-density lipoprotein (VLDL) and chylomicrons in peripheral tissue. Other effects due to insulin include stimulation of the synthesis of fatty acids (lipogenesis) in the liver.[40]

Classically, diabetes mellitus has been characterized as a deficiency of insulin. Various types of diabetes are recognized.* The juvenile diabetic has little detectable circulating insulin, and the pancreas does not respond to a glucose load. However, maturity-onset diabetes may show an abnormal response to glucose; because of the continued elevated glucose levels an individual may ultimately secrete more insulin than a normal subject. Vinik, Kalk and Jackson[41] noted that it is not known whether the initial lesion in diabetes is associated with excess insulin secretion or deficient insulin secretion. It seems as if both α- and β-cells of the pancreas are impaired in diabetes and that the glucoreceptor mechanism in both is damaged early, so that response of both insulin and glucagon to hyperglycemia is impaired. These authors continue to note that the early lesion may be an acquired or inherited selective insensitivity of both α- and β-cells to glucose.

Hyperlipidemia as a diabetic complication has been investigated from various viewpoints. Hyperlipidemia has been implicated in the development of atherosclerosis. Severe hyperlipidemia may lead to life-threatening attacks of acute pancreatitis. It also seems that severe hyperlipidemia causes xanthoma. Researchers also are attempting to elucidate the relationship between diabetes and endogenous hyperlipidemia

(hypertriglyceridemia).[40] Considering the effects of insulin on lipid metabolism as summarized earlier, one can rationalize that in adult-onset diabetes in which the patient may actually have an absolute excess of insulin in spite of the evidence of the glucose tolerance test, the effect of the excessive insulin on lipogenesis in the liver may be enough to increase the level of circulating triglycerides and of very low-density lipoproteins. In juvenile-onset diabetes with a deficiency of insulin, the circulating level of lipids may rise because too much precursor is available, with fatty acids and carbohydrate going to the liver.

There is current concern about the relation between the carbohydrate metabolic manifestations of diabetes and two types of vascular lesion: macroangiopathy, or athema; and microangiopathy which is more subtle and can be properly studied only with the electron microscope. Siperstein[40] has studied these lesions, and he states that these lesions (both types) are responsible for many of the complications of diabetes, including intercapillary glomerulosclerosis, premature atherosclerosis, retinopathy with its specific microaneurysms and retinitis proliferans, leg ulcers, and limb gangrene. Thus far it has been observed that the level of hyperglycemia is dissociated from the severity of microangiopathy; there is no relation between the severity of the carbohydrate metabolic abnormality and the basement membrane thickening of microangiopathy. On the other hand, Azarad[42] believes that the cause of microangiopathy is a disorder in carbohydrate metabolism, i.e., the lesions are a consequence of diabetes rather than a genetically linked association.

The several insulin preparations available as medicinal products are presented in Table 20-9 with pertinent characterizations, including onset and dura-

---

* Comprehensive reviews have recently covered the progress in our understanding of the clinical biochemistry of diabetes mellitus; e.g., diagnosis and classification, genetic and etiological factors, the biosynthesis of insulin and recombinant DNA techniques, the characterization of insulin receptors, inhibitors of fatty-acid oxidation, etc. See C.R. Rasmussen et al, Annual Reports in Medicinal Chemistry, Vol. 16, 173, 1981.

TABLE 20-9

INSULIN PREPARATIONS

Name	Particle Size (Microns)	Action	Composition	pH	Duration (Hours)
Insulin Injection* U.S.P.	. . .	Prompt	Insulin + $ZnCl_2$	2.5–3.5	5–7
Prompt Insulin Zinc Suspension* U.S.P.	2‡	Rapid	Insulin + $ZnCl_2$ + buffer	7.2–7.5	12
Insulin Zinc Suspension* U.S.P.	10–40 (70%) 2 (30%)‡	Intermediate	Insulin + $ZnCl_2$ + buffer	7.2–7.5	18–24
Extended Insulin Zinc Suspension* U.S.P.	10–40	Long-acting	Insulin + $ZnCl_2$ + buffer	7.2–7.5	24–36
Globin Zinc Insulin Injection* U.S.P.	. . .	Intermediate	§Globin + $ZnCl_2$ + insulin	3.4–3.8	12–18
Protamine Zinc Insulin Suspension† U.S.P.	. . .	Long-acting	‖Protamine + insulin + Zin	7.1–7.4	24–36
Isophane Insulin Suspension* U.S.P.	30	Intermediate	Protamine# $ZnCl_2$ insulin buffer	7.1–7.4	18–24

* Clear or almost clear.
† Turbid.
‡ Amorphous.
§ Globin (3.6 to 4.0 mg. per 100 U.S.P. Units of insulin) prepared from beef blood.
‖ Protamine (1.0 to 1.5 mg. per 100 U.S.P. Units of insulin) from the sperm or the mature testes of fish belonging to the Genus Oncorhynchus or Salmo.
Protamine (0.3 to 0.6 mg. per 100 U.S.P. Units of insulin) (q.v.).

tion of action. Amorphous insulin was the first form made available for clinical use. Further purification afforded crystalline insulin which is now commonly called "Regular Insulin." Insulin Injection U.S.P. is made from zinc insulin crystals. For some time, regular insulin solutions have been prepared at a pH of 2.8 to 3.5; if the pH were increased above the acidic range, particles would be formed. However, more highly purified insulin can be maintained in solution over a wider range of pH even when unbuffered. Neutral insulin solutions are found to have greater stability than acidic solutions; neutral insulin solutions maintain nearly full potency when stored up to 18 months at 5° and 25°. As noted in Table 20-9, the various preparations differ with respect to onset and duration of action. Many attempts have been made to prolong the duration of action of insulin; e.g., the development of insulin forms possessing less water-solubility than the highly soluble (in body fluids) regular insulin. Protamine insulin preparations proved to be less soluble and less readily absorbed from body tissue. Protamine zinc insulin suspensions proved to

be even more long-acting then protamine insulin; these are prepared by mixing insulin, protamine and zinc chloride with a buffered solution. Isophane insulin suspension incorporates some of the qualities of regular insulin injection and usually is sufficiently long-acting (although not as much as protamine zinc insulin) to protect the patient from one day to the next (the term "isophane" is derived from the Greek *iso* and *phane* meaning *equal* and *appearance,* respectively). Isophane insulin is prepared by the careful control of the ratio of protamine and insulin and the formation of a crystalline entity containing stoichiometric amounts of insulin and protamine. (Isophane insulin is also known as NPH; the code N indicates neutral pH, the P stands for protamine, and the H for Hegedorn, the developer of the product.) Long-acting insulin preparations (longer-acting than protamine zinc insulin) are pharmaceutically available. If the concentration of zinc chloride is increased to 10 times the amount needed for the formation of soluble zinc insulin, and if the buffer is changed from phosphate to acetate, the excess zinc ions complex with insulin to form a product which is much less soluble at pH 7.4. Two forms of the high-zinc insulin product can be prepared by adjusting the pH—one crystalline and one amorphous or microcrystalline; the crystalline form is much more insoluble and very long-acting. The amorphous form is more readily absorbed. The very long-acting crystalline form is available under the U.S.P. name Extended Insulin Zinc Suspension, whereas the shorter-acting amorphous form is available as Prompt Insulin Zinc Suspension. The slow acting variety is recognized by the *U.S.P.* as Insulin Zinc Suspension.[43]

The posology of the various insulins is summarized in Table 20-10.

Modern concepts regarding the therapeutics of diabetes mellitus have been reviewed by Maurer.[44] This review emphasizes that insulin therapy does not always prevent serious complications. Even diabetics who are considered to be well under insulin therapeutic control experience wide fluctuations in blood-glucose concentration, and it is hypothesized that these fluctuations eventually cause the serious complications of diabetes, e.g., kidney damage, retina degeneration, premature atherosclerosis, cataracts, neurological dysfunction, and a predisposition to gangrene. Hence, it is hoped that future diabetes therapeutics may include a so-called *artificial pancreas* (for implantation); such an apparatus could be a mechanical system capable of releasing insulin in appropriate amounts in response to blood-glucose requirements, or it could be a system of encapsulated cultured, living pancreatic islet cells. Another approach is the attempt to supplement the patient's defective pancreas by

TABLE 20-10

DOSAGE OF INSULIN PREPARATIONS

Preparation Proprietary Name	Usual Adult Dose*
Insulin Injection U.S.P. *Regular Iletin, Regular Insulin*	Diabetic hyperglycemia—S.C., as directed by the physician, 15 to 30 minutes before meals up to 3 or 4 times daily
Prompt Insulin Zinc Suspension U.S.P. *Semilente Iletin, Semilente Insulin*	S.C., as directed by the physician, once daily 30 to 60 minutes before breakfast. An additional dose may be necessary for some patients about 30 minutes before a meal or at bedtime
Globin Zinc Insulin Injection U.S.P.	S.C., as directed by the physician, usually once daily 30 to 60 minutes before breakfast; may also be used twice daily when needed
Insulin Zinc Suspension U.S.P. *Lente Iletin, Lente Insulin*	S.C., as directed by physician, once daily 30 to 60 minutes before breakfast. An additional dose may be necessary for some patients about 30 minutes before a meal or at bedtime
Isophane Insulin Suspension U.S.P. *NPH Iletin, NPH Insulin*	S.C., as directed by physician, once daily, 30 to 60 minutes before breakfast. An additional dose may be necessary for some patients about 30 minutes before a meal or at bedtime
Extended Insulin Zinc Suspension U.S.P. *Ultralente Iletin, Ultralente Insulin*	S.C., as directed by physician, once daily 30 to 60 minutes before breakfast
Protamine Zinc Insulin Suspension U.S.P. *Protamine, Zinc Iletin, Protamine Zinc Insulin*	S.C., as directed by physician, once daily 30 to 60 minutes before breakfast

* See *U.S.P. D.I.* for complete dosage information.

transplantation with a normally functioning pancreas from an appropriate donor.

Zinc and chromium have recently received attention because of their effects on maintaining normal carbohydrate metabolism.[44] Zinc is an integral component of the insulin structure, so a dietary deficiency of this trace element may reduce insulin availability. Chromium is believed to potentiate the actions of insulin at the level of its receptors. There is no definitive consensus on the therapeutic importance of these two trace elements, but many scientists believe that marginal chromium deficiency is an important cause of maturity-onset diabetes, and that foods rich in chromium are valuable in treating this malfunction. Chromium and zinc are usually available in generous amounts in unrefined foods such as whole-grain products and are almost absent in highly processed foods[44.]

Glucagon

Glucagon U.S.P. The hyperglycemic-glycogenolytic hormone elaborated by the α-cells of the pancreas is known as glucagon. It contains 29 amino acid residues in the sequence shown below. Glucagon has been isolated from the amorphous fraction of a commercial insulin sample (4% glucagon).

H·His·Ser·Gln*·Gly·Thr·Phe·Thr·Ser·Asp·Tyr·
Ser·Lys·Tyr·Leu·Asp·Ser·Arg·Arg·Ala·Gln.

Asp·Phe·Val·Gln·Tyr·Leu·Met·Asn†·Thr·OH

Recently there has been attention focused on glucagon as a factor in the pathology of human diabetes. According to Unger, Orci and Maugh,[45.] the following observations support this implication of glucagon: an elevation in glucagon blood levels (hyperglucagonemia) has been observed in association with every type of hyperglycemia; when secretion of both glucagon and insulin are suppressed, hyperglycemia is not observed unless the glucagon levels are restored to normal by the administration of glucagon; the somatostatin-induced suppression of glucagon release in diabetic animals and humans restores blood sugar levels to normal and alleviates certain other symptoms of diabetes.

A recent comprehensive review includes a summary of the biochemistry of somatostatin, insulin, and glucagon. Although somatostatin was first discovered in the hypothalamus, it is now understood that it is elaborated by the delta cells of the pancreas and elsewhere in the body, and that it regulates the output of both insulin and glucagon.‡

Somatostatin suppresses the release of both insulin and glucagon. (Somatostatin is the somatotropin-release-inhibiting factor from the hypothalamus, and it has been characterized as an oligopeptide containing 14 amino acid residues.) Although somatostatin can serve as a useful experimental tool in the study of glucagon and the etiology of diabetes, it is doubtful that it will prove to be a useful medicinal per se because it suppresses the release of other hormones in addition to glucagon and insulin, particularly the growth hormone.

Unger, Orci and Maugh propose that while the major role of insulin is regulation of the transfer of glucose from the blood to storage in insulin-responsive tissues, e.g., liver, fat and muscle, the role of glucagon is regulation of the liver-mediated mobilization of stored glucose. The principal consequence of high concentrations of glucagon is liver-mediated release into the blood of abnormally high concentrations of glucose, thus causing persistent hyperglycemia. It is therefore indicated that the presence of relative excess of glucagon is an essential factor in the development of diabetes.[45]

Glucagon's solubility is 50 μg. per ml. in most buffers between pH 3.5 and 8.5. It is soluble 1 to 10 mg. per ml. in the pH ranges 2.5 to 3.0 and 9.0 to 9.5. Solutions of 200 μg. per ml. at pH 2.5 to 3.0 are stable for at least several months at 4° if sterile. Loss of activity via fibril formation occurs readily at high concentrations of glucagon at room temperature or above at pH 2.5. The isoelectric point appears to be at pH 7.5 to 8.5. Because it has been isolated from commercial insulin its stability properties should be comparable with those of insulin.

As in the case of insulin and some of the other polypeptide hormones, glucagon-sensitive receptor sites in target cells bind glucagon. This hormone-receptor interaction leads to activation of membrane adenylcyclase which catalyzes c-AMP formation. Thus, intracellular c-AMP is elevated. The mode of action of glucagon in glycogenolysis is basically the same as the mechanism of epinephrine, i.e., via stimulation of adenylcyclase, etc. Subsequently, the increase in c-AMP results in activating the protein kinase which catalyzes phosphorylation of phosphorylase kinase → phospho-phosphorylase kinase. The latter is necessary for the activation of phosphorylase-b forming phosphorylase-a. Finally, phosphorylase-a catalyzes glycogenolysis, and this is the basis for the hyperglycemic action of glucagon. Although both glucagon and epinephrine exert hyperglycemic action via c-AMP, glucagon affects liver cells and epinephrine affects both muscle and liver cells.

Fain[46] recently reviewed the many phenomena associated with hormones, membranes, and cyclic

---

* Glutamine.
† Asparagine.
‡ See Sanders, H. J.: *Chem. and Engin. News*, p. 30, March 2, 1981.

nucleotides, including several factors activating glycogen phosphorylase in rat liver. These factors involve not only glucagon but also vasopressin and the catecholamines. Glucagon and β-catecholamines mediate their effects on glycogen phosphorylase via cyclic-AMP but may also involve other factors.

Glucagon exerts other biochemical effects. Gluconeogenesis in the liver is stimulated by glucagon, and this is accompanied by enhanced urea formation. Glucagon inhibits the incorporation of amino acids into liver proteins. Fatty acid synthesis is decreased by glucagon. Cholesterol formation also is reduced. On the other hand, glucagon activates liver lipases and stimulates ketogenesis. Ultimately, the availability of fatty acids from liver triglycerides is elevated, fatty acid oxidation increases acetyl CoA and other acyl CoA's, and ketogenesis is promoted. As glucagon effects elevation of c-AMP levels, release of glycerol and free fatty acids from adipose tissue is also increased.

Glucagon is therapeutically important. It is recommended for the treatment of severe hypoglycemic reactions caused by the administration of insulin to diabetic or psychiatric patients. Of course, this treatment is effective only when hepatic glycogen is available. Nausea and vomiting are the most frequently encountered reactions to glucagon.

Usual dose—parenteral, adults, 500 μg. to 1 mg. repeated in 20 minutes if necessary; pediatric, 25 μg. per kg. of body weight, repeated in 20 minutes if necessary.

PARATHYROID HORMONE

This hormone is a linear polypeptide containing 84 amino acid residues. It regulates the concentration of calcium ion in the plasma within the normal range in spite of variations in calcium intake, excretion and anabolism into bone. Also in the case of this hormone c-AMP is implicated as a secondary messenger. Parathyroid hormone activates adenylcyclase in renal and skeletal cells, and this effect promotes formation of c-AMP from ATP. The c-AMP increases the synthesis and release of the lysosomal enzymes necessary for the mobilization of calcium from bone.

Parathyroid Injection U.S.P. has been employed therapeutically as an antihypocalcemic agent for the temporary control of tetany in acute hypoparathyroidism. However, the *A.M.A. Drug Evaluations, 1977,* considers this preparation to be obsolete.

Usual dose—parenteral, 20 to 40 Units twice daily.
Usual dose range—40 to 300 Units daily.

HYPERTENSIN (ANGIOTENSIN)

Angiotensin I is a decapeptide which is activated by partial degradation to the octapeptide called angiotensin II.* The latter is a pressor hormone. Angiotensin I is released by the action of renin (a proteolytic enzyme from the kidneys) from angiotensinogen. Angiotensinogen is produced by the liver and contained in plasma. It has been said that angiotensin II is the most powerful pressor substance known. It is found in the blood of many humans with essential hypertension. Normal plasma is devoid of angiotensin II. All tissues have peptidase activity, particularly intestine and kidney tissues, which inactivate angiotensin II via hydrolysis. (Angiotensin also exerts a stimulating action on the adrenal cortex, thus promoting aldosterone release. Due to the latter, sodium ion retention results.)

Angiotensin amide, a synthetic polypeptide, has about twice the pressor activity of angiotensin II. It is pharmaceutically available as a lyophilized powder for injection (0.5 to 2.5 mg. diluted in 500 ml. of sodium chloride injection or 5% dextrose for injection) to be administered by continuous infusion. The pressor effect of angiotensin is due to an increase in peripheral resistance; it constricts resistance vessels but has little or no stimulating action on the heart and little effect on the capacitance vessels. Angiotensin has been utilized as an adjunct in various hypotensive states. It is mainly useful in controlling acute hypotension during administration of general anesthetics which sensitize the heart to the effects of catecholamines.

BRADYKININ AND KALLIDIN

These are potent vasodilators and hypotensive agents which have peptide structures. Bradykinin is a nonapeptide, whereas kallidin is a decapeptide. Bradykinin's amino acid sequence is: H·Arg·Pro·Pro·Gly·Phe·Ser·Pro·Phe·Arg·OH. Kallidin is lysylbradykinin; i.e., it has an additional lysine residue at the amino ("left") end of the chain. These two compounds are made available from kininogen, a blood globulin, upon hydrolysis. Trypsin, plasmin or the preoteases of certain snake venoms can catalyze the hydrolysis of kininogen.

---

* The enzyme system that catalyzes the transformation of angiotensin-I into angiotensin-II is called the *angiotensin-converting enzyme* (ACE), and certain compounds that inhibit the ACE system have recently proved to have antihypertensive properties; e.g., the novel antihypertensive agent, captopril, has recently been made pharmaceutically available. For a recent review of renin-angiotensin inhibition refer to Ondetti and Cushman, J. Med. Chem., 24: 355, 1981.

Bradykinin is one of the most powerful vasodilators known; 0.05 to 0.5 μg. per kg. intravenously can produce a decrease in blood pressure in all mammals so far investigated.

Although the kinins per se are not used as medicinals, kallikrein enzyme preparations which release bradykinin from the inactive precursor have been used in the treatment of Raynaud's disease, claudication, and circulatory diseases of the eyegrounds. (Kallikreins is the term used to designate the group of proteolytic enzymes which catalyze the hydrolysis of kininogen, forming bradykinin.)

THYROCALCITONIN

The thyroid produces a polypeptide containing 32 amino acids which inhibits calcium resorption from bone; changes in plasma phosphate usually parallel changes in plasma calcium. This hormone is known as thyrocalcitonin (TCT).

The clinical potential of TCT is in the treatment of osteoporosis and other bone disorders, in hypercalcemia of malignancy and in the treatment of infants with idiopathic hypercalcemia.

THYROTROPIN

Thyrotropin, Thytropar®, thyroid stimulating hormone, TSH, appears to be a glycoprotein (molecular weight 26,000 to 30,000) containing glucosamine, galactosamine, mannose and fucose, whose homogeneity is yet to be established. It is produced by the basophil cells of the anterior lobe of the pituitary gland. TSH enters the circulation from the pituitary, presumably traversing cell membranes in the process. After exogenous administration it is widely distributed and disappears very rapidly from circulation. Some evidence suggests that the thyroid may directly inactivate some of the TSH via an oxidation mechanism that may involve iodine. TSH thus inactivated can be reactivated by certain reducing agents. TSH regulates the production by the thyroid gland of thyroxine which stimulates the metabolic rate. Thyroxine feedback mechanisms regulate the production of TSH by the pituitary gland.

The decreased secretion of TSH from the pituitary is a part of a generalized hypopituitarism that leads to hypothyroidism. This type of hypothyroidism can be distinguished from primary hypothyroidism by the administration of TSH in doses sufficient to increase the uptake of radioiodine or to elevate the blood or plasma protein-bound iodine (PBI) as a consequence of enhanced secretion of hormonal iodine (thyroxine).

It is of interest that massive doses of vitamin A inhibit the secretion of TSH.

TSH is used as a diagnostic agent to differentiate between primary and secondary hypothyroidism. Its use in hypothyroidism due to pituitary deficiency has limited application; other forms of treatment are preferable.

Dose, intramuscular or subcutaneous, 10 International Units.

THYROGLOBULIN

Thyroglobulin, a glycoprotein, is composed of several peptide chains; it also contains 0.5 to 1 percent iodine and 8 to 10 percent carbohydrate in the form of two types of polysaccharides. The formation of thyroglobulin is regulated by thyrotropin (TSH). Thyroglobulin has no hormonal properties. It must be hydrolyzed to release the hormonal iodothyronines: thyroxine and liothyronine (see Chap. 14.)

PENTAGASTRIN

Pentagastrin, Peptavlon®, a physiologic gastric acid secretagogue, is the synthetic pentapeptide derivative N-*t*-butyloxycarbonyl-β-alanyl-L-tryptophyl-L-methionyl-L-aspartyl-L-phenylalanyl amide. It contains the C-terminal tetrapeptide amide (H·Try·Met·Asp·Phe·NH$_2$) which is considered to be the active center of the natural gastrins. Accordingly, pentagastrin appears to have the physiologic and pharmacologic properties of the gastrins, including: stimulation of gastric secretion, pepsin secretion, gastric motility, pancreatic secretion of water and bicarbonate, pancreatic enzyme secretion, biliary flow and bicarbonate output, intrinsic factor secretion, contraction of the gall bladder.

Pentagastrin is indicated as a diagnostic agent to evaluate gastric acid secretory function, and it is useful in testing for anacidity in patients with suspected pernicious anemia, atrophic gastritis or gastric carcinoma, hypersecretion in patients with suspected duodenal ulcer or postoperative stomal ulcers, and for the diagnosis of Zollinger-Ellison tumor.

Pentagastrin is usually administered subcutaneously; the optimal dose is 6 μg. per kg. Gastric acid secretion begins approximately 10 minutes after administration and peak responses usually occur within 20 to 30 minutes. The usual duration of action is from 60 to 80 minutes. Pentagastrin has a relatively short plasma half-life, perhaps under 10 minutes. The available data from metabolic studies indicate that pentagastrin is inactivated by the liver, kidney and tissues of the upper intestine.

Contraindications include hypersensitivity or idiosyncrasy to pentagastrin. It should be used with caution in patients with pancreatic, hepatic or biliary disease.

BLOOD PROTEINS

The blood is the transport system of the organism and thus performs important distribution functions. Considering the multitude of materials transported by the blood (e.g., nutriments, oxygen, carbon dioxide, waste products of metabolism, buffer systems, antibodies, enzymes and hormones), its chemistry is very complex. Grossly, approximately 45 percent consists of the formed elements that can be separated by centrifuging, and of these only 0.2 percent are other than erythrocytes. The 55 percent of removed plasma contains approximately 8 percent solids of which a small portion (less than 1%) can be removed by clotting to produce defibrinated plasma, which is called serum. Serum contains inorganic and organic compounds, but the total solids are chiefly protein, mostly albumin and the rest nearly all globulin. The plasma contains the protein fibrinogen which is converted by coagulation to insoluble fibrin. The separated serum has an excess of the clotting agent thrombin.

Serum globulins can be separated by electrophoresis into α-, β- and γ-globulins that contain most of the antibodies. The immunologic importance of globulins is well known. Many classes and groups of immunoglobulins are produced in response to antigens or even to a single antigen. The specificity of antibodies has been studied from various points of view, and recently Richards et al.[47] reported evidence which suggests that even though immune serums appear to be highly specific with respect to antigen binding, individual immunoglobulins may not only interact with a number of structurally diverse determinants, but may bind such diverse determinants to different sites within the combining region.

The importance of the blood coagulation process has been obvious for a long time. Coagulation mechanisms are well covered in several biochemistry texts,[48,49] hence herein a brief summary suffices. The required time for blood clotting is normally 5 minutes, and any prolongation beyond 10 minutes is considered abnormal. Thrombin, the enzyme responsible for the catalysis of fibrin formation, originates from the inactive zymogen, prothrombin; the prothrombin → thrombin transformation is dependent on calcium ions and thromboplastin. The fibrinogen → fibrin reaction catalyzed by thrombin involves: proteolytic cleavage (partial hydrolysis); polymerization of the fibrin monomers from the preceding step; actual clotting (hard clot formation). The final process forming the hard clot occurs in the presence of calcium ions and the enzyme fibrinase.[43]

Thrombin U.S.P. is a sterile protein substance prepared from prothrombin of bovine origin. It is used as a topical hemostatic due to its capability of clotting blood, plasma or a solution of fibrinogen without adding other substances. Thrombin also may initiate clotting when combined with gelatin sponge or fibrin foam.

For external use—topically to the wound, as a solution containing 100 to 2000 N.I.H. Units per ml. in Sodium Chloride Irrigation or Sterile Water for Injection or as a dry powder.

Hemoglobin

Erythrocytes contain 32 to 55 percent hemoglobin, about 60 percent water and the rest as stroma. The last can be obtained, after hemolysis of the corpuscles by dilution, through the process of centrifuging and is found to consist of lecithin, cholesterol, inorganic salts and a protein, stromatin. Hemolysis of the corpuscles, or laking as it sometimes is called, may be brought about by hypotonic solution, by fat solvents, by bile salts which dissolve the lecithin, by soaps or alkalies, by saponins, by immune hemolysins and by hemolytic serums, such as those from snake venom and numerous bacterial products.

Hemoglobin (Hb) is a conjugated protein, the prosthetic group being heme (hematin) and the protein (globin) which is composed of four polypeptide chains, usually in identical pairs. The total molecular weight is about 66,000 including four heme molecules. The molecule has an axis of symmetry and therefore is composed of identical halves with an over-all ellipsoid shape of the dimensions $55 \times 55 \times 70$Å.

Iron in the heme of hemoglobin (ferrohemoglobin) is in the ferrous state and can combine reversibly with oxygen to function as a transporter of oxygen.

$$\text{Hemoglobin} + \text{Oxygen (O}_2) \rightleftharpoons \text{Oxyhemoglobin}$$

In this process, the formation of a stable oxygen complex, the iron remains in the ferrous form because the heme moiety lies within a cover of hydrophobic groups of the globin. Both Hb and O_2 are magnetic, whereas HbO_2 is dimagnetic because the unpaired electrons in both molecules have become paired. When oxidized to the ferric state (methemoglobin or ferrihemoglobin) this function is lost. Carbon monoxide will combine with hemoglobin to form carboxyhemoglobin (carbonmonoxyhemoglobin) to inactivate it.

The stereochemistry of the oxygenation of hemoglobin is very complex and it has been investigated to some extent. Some evidence from x-ray crystallographic studies reveals that the conformations of the α and β chains are altered when their heme moieties complex with oxygen, thus promoting the complexation with oxygen. It is assumed that hemoglobin can exist in two forms, the relative position of the subunits in each form being different. In the deoxy form α and β subunits are bound to each other by ionic bonds in a compact structure that is less reactive toward oxygen than is the oxy form. Some ionic bonds are cleaved in the oxy form, relaxing the conformation. The latter conformation is more reactive to oxygen.[50]

REFERENCES

1. White, A., *et al.:* Principles of Biochemistry, 6th ed., p. 684, New York, McGraw-Hill, 1978.
2. Mudge, G. H., and Welt, L. H.: *in* Goodman, L. S., and Gilman, A. (eds.): The Pharmacological Basis of Therapeutics, 5th ed., p. 769, New York, Macmillan, 1975.
3. White, A., *et al.:* Principles of Biochemistry, 6th ed., pp. 1325–1326, New York, McGraw-Hill, 1978.
4. Corey, R. B., and Pauling, L.: Proc. Roy. Soc. London (ser. B) 141:10, 1953; see also Ad. Protein Chem., p. 147, 1957.
5. Tanford, C.: The Hydrophobic Effect: Formation of Micelles and Biological Membranes, 2nd ed., New York, John Wiley & Sons, 1979.
6. McDonald, C. C., and Phillips, W. D.: J. Am. Chem. Soc. 89:6332, 1967.
7. Kato, G., and Yung, J.: Mol. Pharmacol. 7:33, 1971.
8. Elefrawi, M. E., *et al.:* Mol. Pharmacol. 7:104, 1971.
9. Kato, G.: Mol. Pharmacol. 8:575, 1972.
10. ———: Mol. Pharmacol. 8:582, 1972.
11. Johnson, F. H., *et al.:* The Kinetic Basis of Molecular Biology, New York, John Wiley & Sons, 1954.
12. Eyring, H., and Eyring, E. M.: Modern Chemical Kinetics, New York, Rheinhold, 1963.
13. Eyring, H.: Am. Chem. Soc. National Meeting, Dallas, April, 1973; for abstract of paper see Chem & Engin. News, p. 17, April 30, 1973.
14. Haberfield, P., and Kivuls, J.: J. Med. Chem. 16:942, 1973.
15. Folkers, K., *et al.:* Biochem. Biophys. Res. Commun. 59:704, 1974.
16. Ferdinand, W.: The Enzyme Molecule, New York, John Wiley & Sons, 1976.
17. Koshland, D. E.: Sci. Am. 229:52, 1973; see also Ann. Rev. Biochem. 37:359, 1968.
18. Lowe, J. N., and Ingraham, L. L.: An Introduction to Biochemical Reaction Mechanisms, Englewood Cliffs, N.J., Prentice-Hall, 1974.
19. Hanson, K. R., and Rose, I. A.: Acc. Chem. Res. 8:1, 1975.
20. Chem. and Engin. News, p. 23, April 16, 1979.
21. White, A., *et al.:* Principles of Biochemistry, 6th ed., p. 1277, New York, McGraw-Hill, 1978.
22. Meinhoffer, J.: Peptide and Protein Hormones, *in* Wolff, M. E. (ed.): Burger's Medicinal Chemistry, 4th ed., Part II, p. 751, New York, John Wiley & Sons, 1979.
23. Turkin, D.: American Pharmacy NS 20:45, 1980.
24. Veber, D. F., and Saperstein, R.: Somatostatin, *in* Hess, H. J. (ed.): Annual Reports of Medicinal Chemistry, vol. 14, p. 209, New York, Academic Press, 1979.
25. White, W. F.: Ann. Rep. Med. Chem. 8:204, 1973.
26. Schally, A. V., Arimura, A., and Kastin, A. J.: Science 179:341–350, 1973.
27. Brownstein, M. J.: Science 207:373,1980.
28. Li, C. H., *et al.:* J. Protein Res. 4:151, 1972.
29. A.M.A. Department of Drugs: Drug Evaluations, 4th ed., pp. 791–792, New York, John Wiley & Sons, Inc., 1980.
30. Miller, R. J., and Cuatrecaseas, P.: Vitamins and Hormones 36:297, 1978.
31. Snyder, S. H., Innis, R. B.: Ann. Rev. Biochem. 48:755, 1979.
32. Marx, J. L.: Science 205:886, 1979.
32a. Krieger, D. T., and Liotta, A. S.: Science 205:366, 1979.
33. Chang, A. Y.: Ann. Rep. Med. Chem. 9:182, 1974.
34. For references to Sanger's studies, see Ann. Rev. Biochem. 27:58, 1958.
35. Katsoyannis, P. G.: Science 154:1509, 1966.
36. Rittel, W., *et al.:* Helv. chim. acta 67:2617, 1974.
37. Complex techniques lead to insulin synthesis, Chem. & Engin. News, April 28, 1975.
38. Cosmatos, A., *et al.:* J. Biol. Chem. 263:6586, 1978.
39. May, J. M., *et al.:* J. Biol. Chem. 253:686, 1978.
40. Report from the Geigy Symposium in Albuquerque, New Mexico: Diabetes Re-examined, Diabetology, Feb. 6, 1974.
41. Vinik, A. I., Kalk, W. J., and Janckson, W. P. U.: Lancet, pp. 485–486, March 23, 1974.
42. Azarad, E.: Nouvelle Presse Med. 2:3037, 1973.
43. Waife, S. O. (ed.): Diabetes Mellitus, 8th ed., pp. 37–44, Indianapolis, Eli Lilly & Co., 1980.
44. Maurer, A. C.: American Scientist 67:422, 1979.
45. Unger, R. J., Orci, L., and Maugh, T. H., II: Science 188:923, 1975.
46. Fain, J. N.: Receptors and Recognition, Series A, 6:3, 1978.
47. Richards, F. F., *et al.:* Science 187:130, 1975.
48. Harper, H. A.: Review of Physiological Chemistry, 17th ed., Los Altos, Cal., Lange Medical Pubs., 1980.
49. White, A., *et al.:* Principles of Biochemistry, 6th ed., pp. 916–928, New York, McGraw-Hill, 1978.
50. Montgomery, R., *et al.:* Biochemistry: A Case-oriented Approach, 2nd ed., pp. 78–80, St. Louis, C. V. Mosby, 1977.

SELECTED READINGS

Boyer, P. D. (ed.): The Enzymes, 3rd ed., New York, Academic Press, 1970.

Brockerhoff, H., and Jensen, R. G.: Lipolytic Enzymes, New York, Academic Press, 1974.

Ferdinand, W.: The Enzyme Molecule, New York, John Wiley & Sons, 1976.

Grollman, A. P.: Inhibition of Protein Biosynthesis, *in* Brockerhoff, H., and Jensen, R. G.: Lipolytic Enzymes, pp. 231–247, New York, Academic Press, 1974.

Haschemeyer, R. H., and de Harven, E.: Electron microscopy of enzymes, Ann. Rev. Biochem. 43:279, 1974.

Jenks, W. P.: Catalysis in Chemistry and Enzymology, New York, McGraw-Hill, 1969.

Lowe, J. N., and Ingraham, L. L.: An Introduction to Biochemical Reaction Mechanisms, Englewood Cliffs, N.J., Prentice-Hall, 1974. (This book includes elementary enzymology including mechanisms of coenzyme function.)

Meinhoffer, J.: Peptide and Protein Hormones, *in* Wolff, M.E. (ed.): Burger's Medicinal Chemistry, 4th ed., Part II, p. 751, New York, John Wiley & Sons, 1979.

Meienhofer, J.: Peptide Hormones of the Hypothalamus and Pituitary, *in* Heinzelman, R. V. (ed.): Annual Reports in Medicinal Chemistry, vol. 10, New York, American Chemical Society, 1975.

Mildvan, A. S.: Mechanism of Enzyme Action, Ann. Rev. Biochem. 43:357, 1974.

Pikes, S. J., and Parks, C. R.: The Mode of Action of Insulin, Ann. Rev. Pharmacol. 14:365, 1974.

Rafelson, M. E., *et. al.:* Basic Biochemistry, 4th ed., chaps. 3,4,8,11, New York, Macmillan, 1980.

Schaeffer, H. J.: Factors in the Design of Reversible and Irreversible Enzyme Inhibitors, *in* Ariens, E. J. (ed.): Drug Design, vol. 2, pp. 129–159, New York, Academic Press, 1971.

Tager, H. S., and Steiner, D. F., Peptide Hormones, Ann. Rev. Biochem. 43:509, 1974.

Waife, S. O. (ed.): Diabetes Mellitus, 8th ed., Indianapolis, Eli Lilly and Co., 1980.

Wilson, C. A.: Hypothalamic Amines and the Release of Gonadotrophins and Other Anterior Pituitary Hormones, *in* Simonds, A. B. (ed.): Adv. in Drug Res., New York, Academic Press, 1974.

21

vitamins and related compounds

Jaime N. Delgado

Vitamins have traditionally been considered to be "accessory food factors." Generally, vitamins are among those nutrients that the human organism cannot synthesize from other dietary components. Together with certain amino acids (i.e., essential amino acids), the vitamins constitute a total of 24 organic compounds that have been characterized as dietary essential.[1] Many vitamins function biochemically as precursors in the synthesis of coenzymes necessary in human metabolism; thus, vitamins perform essential functions. When they are not available in appropriate amounts, the consequences may lead to serious disease states.

Although there are relatively few therapeutic indications for vitamin pharmaceutical preparations, diseases caused by certain vitamin deficiencies do respond favorably to vitamin therapy. Additionally, there are products indicated for prophylactic use as dietary supplements. An optimal diet provides all of the necessary nutrients; however, in some cases of increased demands, vitamin and mineral supplementation is recommended.[2]

The medicinal chemistry of vitamins is fundamental not only to the therapeutics of nutritional problems but also to the understanding of the biochemical actions of other medicinal agents that directly or indirectly affect the metabolic functions of vitamins and coenzymes. Accordingly, this chapter includes a brief summary of basic biochemistry of vitamins, structure–activity relationships, physicochemical properties and some stability considerations, nutritional and therapeutic applications, and brief characterizations of representative pharmaceutical products.

In 1912 Funk described a substance that was present in rice polishings and in foods that cured polyneuritis in birds and beri-beri in humans. This substance was referred to as "vitamine" because it was charac-

terized as an amine and as a vital nutritional component. After other food factors were also noted to be vital nutritional components that were not amines and did not even contain nitrogen, Drummond suggested the modification that led to the term *vitamin*. In 1913 McCollum and Davis described a lipid-soluble essential food factor in butterfat and egg yolk, and two years later reference was made to a water-soluble factor in wheat germ. Thus, the terms *fat-soluble A* and *water-soluble B* were respectively applied to these food factors. Since then many other dietary components have been discovered to be essential nutritional components, i.e., vitamins. It is traditional to classify these compounds as either lipid-soluble or water-soluble vitamins. This classification is convenient, because members of each category possess important properties in common.

LIPID-SOLUBLE VITAMINS

The lipid-soluble vitamins include vitamins A, D, E, and K. These compounds possess other characteristics in common besides solubility. They are usually associated with the lipids of foods and are absorbed from the intestine with these dietary lipids. The lipid-soluble vitamins are stored in the liver and thus conserved by the organism, whereas storage of the water-soluble vitamins in most cases is not significant.

THE VITAMIN A'S

Vitamin A was first recognized as a vitamin by McCollum and Davies[3] in 1913 to 1915, but studies of the molecular mechanism of action of retinol in the visual process were not significantly productive until 1968 to 1972. The mechanism of action of vitamin A in physiological processes other than vision has been very difficult to study. It has been difficult to identify a single biochemical change that is directly due to vitamin-A deficiency. Nevertheless, there is convincing evidence that vitamin A performs an important function in the biosynthesis of glycoproteins. Vitamin A is involved in sugar-transfer reactions in mammalian membranes.[4] Moreover, vitamin A has been demonstrated to control and direct differentiation of epithelial tissues; this has led to the suggestion that vitamin A has hormone-like properties. Investigations on the mechanism of this action have been stymied by difficulties in elucidating definitely the nature of the biochemically active form of the vitamins whether it is all *trans*-retinol or retinoic acid. This question remains under intensive investigation and is comprehensively reviewed by Chytil and Ong.[5]

Vitamin A (all *trans*-retinol) is biosynthesized in animals from plant pigments called *carotenoids,* which are terpenes composed of isoprenoid units; e.g., β–carotene is the precursor of retinol (vitamin A).

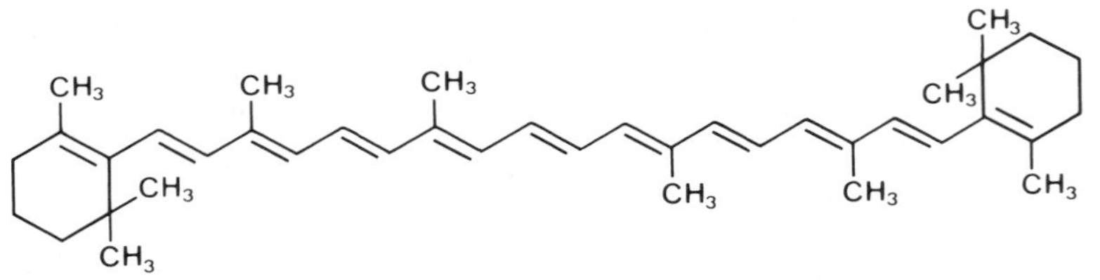

The stereochemistry of vitamin A and related compounds is very complex, and a complete stereochemical analysis is beyond the scope of this chapter. A brief summary of some stereochemical features is presented here as the basis for the characterization of the biochemical actions exerted by this vitamin. The study of the structural relationships among vitamin A and its stereoisomers has been complicated by the common use of two numbering systems, as exemplified by the vitamin A (all *trans*; retinol) and neovitamin A ($\Delta^4 cis$ or 11-mono-*cis* vitamin A).

For steric reasons the number of isomers of vitamin A most likely to occur would be limited. These are all-*trans*, 9-*cis* (Δ^3-*cis*), 13-*cis* (Δ^5-*cis*) and the 9,13-di-*cis*. A *cis* linkage at double bond 7 or 11 encounters steric hindrance. The 11-*cis* isomer is twisted as well as bent at this linkage; nevertheless, this is the only isomer that is active in vision.

Most liver oils contain vitamin A and neovitamin A in the ratio of 2 to 1.

The biological activity[7] of the isomers of vitamin A acetate, in terms of U.S.P Units per gram, are as follows: vitamin A, all *trans*, 2,907,000; neovitamin A, 2,190,111; Δ^3-*cis*, 634,000; $\Delta^{3,5}$-di-*cis*, 688,000 and $\Delta^{4,6}$-di-*cis*, 679,000. In the case of the isomers of vitamin-A aldehyde,[7] the following values have been reported: all *trans*, 3,050,000; neo (Δ^5-*cis*), 3,120,000; Δ^3-*cis*, 637,000; $\Delta^{3,5}$-di-*cis*, 581,000; and $\Delta^{4,6}$-di-*cis*, 1,610,000.

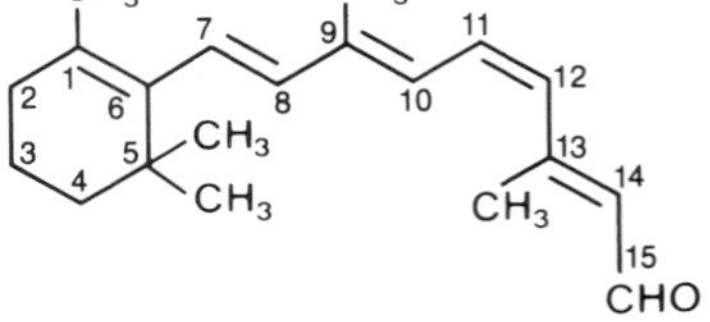

β – Carotene

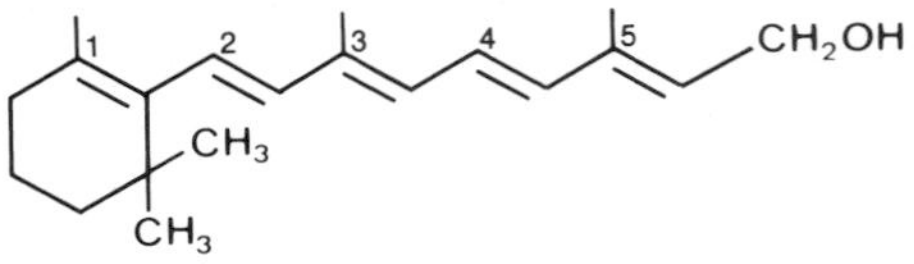

Vitamin A (Retinol)

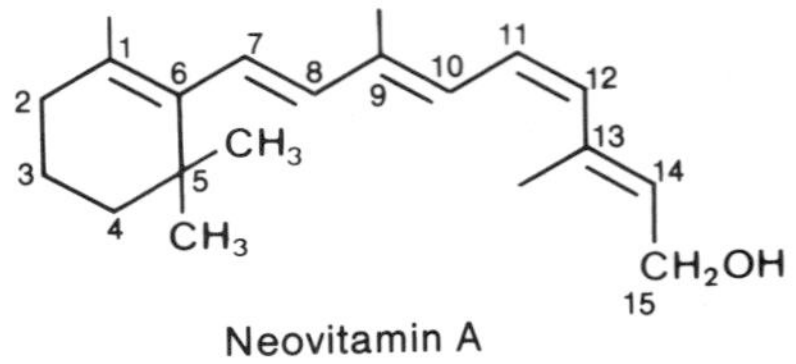

Neovitamin A

The intestinal mucosa is the main site of β-carotene transformation[1] to retinal but the enzyme that catalyzes the transformation also occurs in hepatic tissue. The enzyme is an iron-containing dioxygenase responsible for formation of two retinal molecules from one β-carotene molecule; subsequently, the retinal undergoes NADH or NADPH reduction to retinol. Retinoic acid, the corresponding carboxylic acid, promotes development of bone and soft tissues and sperm production, but it does not participate in the visual process. Retinoic acid is found in the bile in the glucuronide form.

Retinol ⇌ Retinal ⟶ Retinoic Acid

Neoretinene b (Retinal)

Disregarding stereochemical variations, a number of compounds with structures corresponding to vitamin A, its ethers and its esters have been prepared.[9-11] These compounds, as well as synthetic vitamin A acid, possess biological activity.

Although fish-liver oils were used for their vitamin A content, purified or concentrated forms of vitamin A are of great commercial significance. These are prepared in three ways: (1) saponification of the oil and concentration of the vitamin A in the nonsaponifiable matter by solvent extraction, the product is marketed as such; (2) molecular distillation of the nonsaponifiable matter, from which the sterols have previously been removed by freezing, giving a distillate of vitamin A containing 1,000,000 to 2,000,000 I.U. per gram; (3) subjecting the fish oil to direct molecular distillation to recover both the free vitamin A and vitamin A palmitate and myristate.

Pure crystalline vitamin A occurs as pale yellow plates or crystals. It melts at 63 to 64° and is insoluble in water but soluble in alcohol, the usual organic solvents and the fixed oils. It is unstable in the presence of light and oxygen and in oxidized or readily oxidized fats and oils. It can be protected by the exclusion of air and light and by the presence of antioxidants.

Like all substances that have a polyene structure, vitamin A gives color reactions with many reagents, most of which are either strong acids or chlorides of polyvalent metals. An intense blue color (Carr-Price) is obtained with vitamin A in dry chloroform solution upon the addition of a chloroform solution of antimony trichloride. This color reaction has been studied extensively and is the basis of a colorimetric assay for vitamin A.[12]

The chief source of natural vitamin A is fish-liver oils, which vary greatly in their content of this vitamin (Table 21–1). It occurs free and combined as the biologically active esters, chiefly of palmitic and some myristic and dodecanoic acids. It also is found in the livers of animals, especially those which are herbivorous. Milk and eggs are fair sources of this vitamin. The provitamins A, e.g., *beta, alpha,* and *gamma* carotenes and cryptoxan-thin, are found in green parts of plants, carrots, red palm oil, butter, apricots, peaches, yellow corn, egg yolks and other similar sources. The carotenoid pigments are utilized poorly by humans, whereas animals differ in their ability to utilize these compounds. These carotenoid pigments are provitamins A because they are converted to the active vitamin A. For example, β-carotene has been shown to be absorbed intact by the intestinal mucosa, then cleaved to retinal by β-carotene-15,15-dioxygenase which requires molecular oxygen.[13] β-Carotene can give rise to 2 molecules of retinal, whereas in the other 3 carotenoids only 1 molecule is possible by this transformation. These carotenoids have only 1 ring (see formula for β-carotene) at the end of the polyene chain that is identical with that found in β-carotene and is necessary and found in vitamin A.

Oils or lipids enhance the absorption of carotene, which is poorly absorbed (10%) when ingested in dry vegetables.

The conjugated double bond systems found in vitamin A and β-carotene are necessary for activity, for when these compounds are partially or completely reduced, activity is lost. The ester and methyl ethers of vitamin A have a biological activity on a molar basis equal to vitamin A. Vitamin A acid is biologically active but is not stored in the liver.

Vitamin A often is called the "growth vitamin" because a deficiency of it in the diet causes a cessation of growth in young rats. A deficiency of vitamin A is manifested chiefly by a degeneration of the mucous membranes throughout the body. This degeneration is evidenced to a greater extent in the eye than in any other part of the body and gives rise to a condition known as xerophthalmia. In the earlier stages of vitamin A deficiency, there may develop a night blindness (nyctalopia) which can be cured by vitamin A. Night blindness can be defined as the inability to see in dim light.

"Dark adaptation" or "visual threshold" is a more suitable description than "night blindness" when applied to many subclinical cases of vitamin A deficiency. The visual threshold at any moment is just that light intensity required to elicit a visual sensation. Dark adaption is the change which the visual threshold undergoes during a stay in the dark after an exposure to light. This change may be very great. After exposure of the eye to daylight, a stay of 30 minutes in the dark results in a decrease in the threshold by a factor of a million. This phenomenon is used as the basis to detect subclinical cases of vitamin A deficiencies. These tests vary in their technique, but, essentially, they measure visual dark adaptation after exposure to bright light and compare it with the normal.[14]

Advanced deficiency of vitamin A gives rise to a

TABLE 21–1

VITAMIN A CONTENT OF SOME FISH-LIVER OILS

Source of Oil	Animal	Potency (I.U./g.)
Halibut, liver	*Hippoglossus hippoglossus*	60,000
Percomorph, liver	Percomorph fishes (mixed oils)	60,000
Shark, liver	*Galeus zygopterus*	25,500
Shark, liver	*Hypoprion brevirostris* and other varieties	16,500
Burbot, liver	*Lota maculosa*	4,880
Cod, liver	*Gadus morrhua*	850

dryness and scaliness of the skin, accompanied by a tendency to infection. Characteristic lesions of the human skin due to vitamin A deficiency usually occur in sexually mature persons between the ages of 16 and 30 and not in infants. These lesions appear first on the anterolateral surface of the thigh and on the posterolateral portion of the upper forearms and later spread to adjacent areas of the skin. The lesions consist of pigmented papules, up to 5 mm. in diameter, at the site of the hair follicles.

Vitamin A regulates the activities of osteoblasts and osteoclasts, influencing the shape of the bones in the growing animal. The teeth also are affected. In vitamin A deficiency states, a long overgrowth occurs. Overdoses of vitamin A in infants for prolonged periods of time led to irreversible changes in the bones, including retardation of growth, premature closure of the epiphyses and differences in the lengths of the lower extremities. Thus, a close relationship exists between the functions of vitamins A and D with regard to cartilage, bones and teeth.[15]

The tocopherols exert a sparing and what appears to be a synergistic action[16] with vitamin A.

Blood levels of vitamin A decrease very slowly, and a decrease in dark adaptation was observed in only 2 of 27 volunteers (maintained on a vitamin A-free diet) after 14 months, at which time blood levels had decreased from 88 I.U. per 100 ml. of blood to 60 I.U.

Vitamin A performs numerous biochemical functions; it has been demonstrated that vitamin A promotes the production of mucus by the basal cells of the epithelium, whereas in its absence keratin can be formed. Vitamin A performs a function in the biosynthesis of glycogen and some steroids, and increased quantities of coenzyme Q are found in the livers of vitamin-deficient rats. Significantly, the most well-known action of vitamin A is its function in the chemistry of vision.

The molecular mechanism of action of vitamin A in the visual process has been under investigation for many years. Wald in 1968 and Morton in 1972 characterized this mechanism of action. The chemistry of vision was comprehensively reviewed in *Accounts* of *Chemical Research* (1975) by numerous investigators. These reviews include theoretical studies of the visual chromophore, characterization of rhodopsin in synthetic systems, dynamic processes in vertebrate rod visual pigments and their membranes, and the dynamics of the visual protein opsin.[17-21]

Vitamin A (all *trans* retinol) undergoes isomerization to the 11-*cis* form in the liver. This transformation is catalyzed by a retinol isomerase. Subsequently, 11-*cis* retinol interacts with a protein called retinol-binding protein (RBP) to form a complex that is transported to the retina photoreceptor cells, which contain specific receptors for the RBP–retinol complex.

The retina has been considered[20,22] to be a double sense organ in which the rods are concerned with colorless vision at low light intensities and the cones with color vision at high light intensities. A dark-adapted, excised retina is rose-red in color, when it is exposed to light, its color changes to chamois, to orange, to pale yellow; finally, upon prolonged irradiation, it becomes colorless. The rods contain photosensitive visual purple (rhodopsin) which, when acted upon by light of a definite wavelength, is converted to visual yellow and initiates a series of chemical steps necessary to vision. Visual purple is a conjugated, carotenoid protein having a molecular weight of about 40,000 and 1 prosthetic group per molecule. It has an absorption maximum of about 510 nm. The prosthetic group is retinene (neoretinene b or retinal) which is joined to the protein through a protonated Schiff's base linkage. The function of retinene in visual purple is to provide an increased absorption coefficient in visible light and thus sensitize the protein which is denatured. This process initiates a series of physical and chemical steps necessary to vision. The protein itself differs from other proteins by having a lower energy of activation, which permits it to be denatured by a quantum of visible light. Other proteins require a quantum of ultraviolet light to be denatured. The bond between the pigment and the protein is much weaker when the

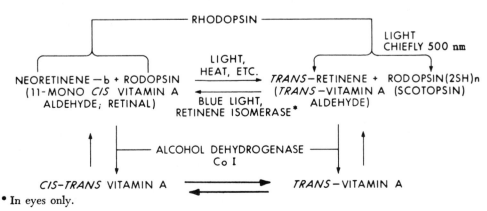

* In eyes only.

protein is denatured than when it is native. The denaturation process of the protein is reversible and takes place more readily in the dark to give rise, when combined with retinene, to visual purple. The effectiveness of the spectrum in bleaching visual purple runs fairly parallel with its absorption spectrum (510 nm.) and with the sensibility distribution of the eye in the spectrum at low illuminations. It has been calculated that for man to see a barely perceptible flash of light, in a dark-adapted eye there need be transformed photochemically only 1 molecule of visual purple in each 5 to 14 rod cells. In vivo, visual purple is constantly reformed as it is bleached by light, and under continuous illumination, an equilibrium between visual purple,* visual yellow† and visual white‡ is maintained. If an animal is placed in the dark, the regeneration of visual purple continues until a maximum concentration is obtained. Visual purple in the eyes of an intact animal may be bleached by light and regenerated in the dark an enormous number of times.

Visual purple occurs in all vertebrates. It is not distributed evenly over the retina. It is missing in the fovea, and in the regions outside of the fovea its concentration undoubtedly increases to a maximum in the region about 20° off center, corresponding to the high density of rods in this region. Therefore, to see an object best in the dark, one should not look directly at it.

The diagram shown above represents some of the changes that take place in the visual cycle involving the rhodopsin system in which the 11-mono-*cis* isomer of vitamin A is functional in the aldehyde form.[23]

Temperature controlled studies led to results that are depicted in the right-hand column.

Pure vitamin A has the activity of 3,500,000 I.U. per gram. Moderate to massive doses of vitamin A have been used in pregnancy, lactation, acne, abortion of colds, removal of persistent follicular hyperkeratosis of the arms, persistent and abnormal warts, corns and calluses and similar conditions. Phosphatides or the tocopherols enhance the absorption of vitamin A. Vitamin A applied topically appears to reverse the impairment of wound healing by corticoids.

Vitamin A U.S.P. is a product that contains retinol (vitamin A alcohol) or its esters from edible fatty acids chiefly acetic and palmitic acids, and whose activity is not less than 95 percent of the labeled amount; 0.3 μg. of vitamin A alcohol (retinol) equals 1 U.S.P. Unit.

Vitamin A occurs as a yellow to red, oily liquid; it is nearly odorless or has a fishy odor and is unstable to

---

*Protein combined with retinene.
†Protein denatured plus free retinene.
‡Protein plus vitamin A.

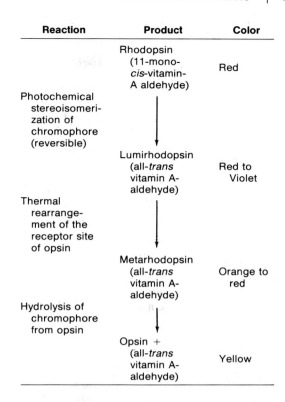

Reaction	Product	Color
Photochemical stereoisomerization of chromophore (reversible)	Rhodopsin (11-mono-*cis*-vitamin-A aldehyde)	Red
Thermal rearrangement of the receptor site of opsin	Lumirhodopsin (all-*trans* vitamin A-aldehyde)	Red to Violet
Hydrolysis of chromophore from opsin	Metarhodopsin (all-*trans* vitamin A-aldehyde)	Orange to red
	Opsin + (all-*trans* vitamin A-aldehyde)	Yellow

air and light. It is insoluble in water or glycerin and is soluble in absolute alcohol, vegetable oils, ether and chloroform.

Tretinoin U.S.P., Retin-A®, retinoic acid. This is the acid analog of all *trans* vitamin A. It is used, not for vitamin A activity, but as a keratolytic in the topical treatment of acne vulgaris. It is usually applied as a 0.05 percent polyethylene glycol ethyl alcohol solution once daily. Within 48 hours after the first application the skin may become red and begin to peel. It may be necessary to continue the daily application for 3 to 4 months.

All *trans* Retinoic Acid

***Beta*-carotene,** Solatene®, is a carotenoid pigment, occurring naturally in green and yellow vegetables, which is pharmaceutically available and indicated for the treatment of photosensitivity in patients with erythoopoietic protoporphyria.

VITAMIN A₂

Vitamin A_2 is found in vertebrates which live or at least begin their lives in fresh water. Vitamin A_2 exhibits chemical, physical and biological properties very similar to those of vitamin A. It has the structural formula shown below.

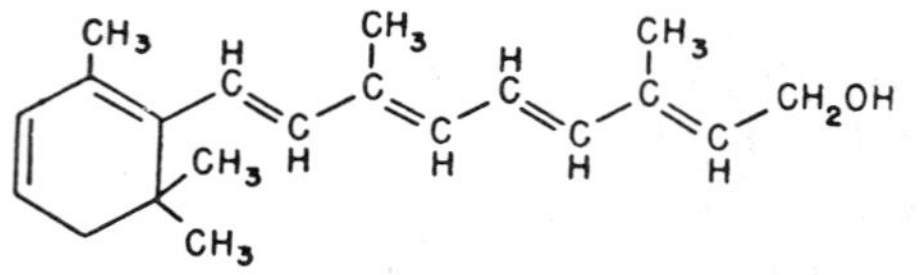

Vitamin A₂ (all *trans*)
3-Dehydroretinol or Dehydroretinol

Vitamin A_2 has a biological potency of 1,300,000 U.S.P. Units per gram which is approximately 40 percent of the activity of crystalline vitamin A acetate.

Three additional visual pigments are known: (1) iodospin, composed of cone opsin and retinene₁; (2) porphyropsin, composed of retinene₂ and rod opsin; (3) cyanopsin, composed of retinene₂ and cone opsin.

Only neoretinene b (retinal) (Δ^4- or 11-mono-*cis*) can combine with opsin (scotopsin) to form rhodopsin. The isomerization of *trans*-retinene may take place in the presence of blue light. However, vision continues very well in yellow, orange and red light in which no isomerization takes place. The neoretinene b (retinal) under these circumstances is replaced by an active form of vitamin A from the bloodstream which, in turn, obtains it from stores in the liver. The isomerization of *trans*-vitamin A in the body to *cis-trans*-vitamin A seems to keep pace with long-term processes such as growth, since vitamin A, neovitamin A and neoretinene b (retinal) are equally active in growth tests in rats.

The sulfhydryl groups (2 for each retinene molecule isomerized) exposed on the opsin the transmission of impulses in the phenomenon of vision.

The assumption has been made that the mechanism of action of vitamin A might be similar to that of the steroid hormones (see the sections on steroids and steroid-hormone receptors), and this assumption has led to the characterization of two intracellular-binding proteins, one specific for retinal and the other specific for retinoic acid.[5]

The available experimental data do not provide complete evidence that these two proteins are in fact receptors analogous to steroid-hormone receptors, but there is convincing evidence that these two proteins mediate important aspects of vitamin-A function. The existence of a protein that specifically binds retinoic acid substantiates the implication of retinoic acid as a physiological form of vitamin A.

For many years attempts have been made to demonstrate relationships between vitamin A and cancer. Recently, it has been demonstrated that dietary deficiency of vitamin A may lead to increased incidence of spontaneous and carcinogen- or virus-induced epithelial metaplasias and tumors in experimental animals and possibly in humans. It also has been observed that in several human tumors from lung, breast, skin, and stomach, cellular retinoic acid-binding protein is detectable only in the tumor and not in the normal tissue.[5]

THE VITAMIN D'S

The term *vitamin D* was originally applied to the material obtained by irradiation of yeast ergosterol. This material was later analyzed to be a mixture of calciferol isomers and some steroids. Upon purification and further characterization, calciferol proved to possess significant antirachitic properties and became known as vitamin D_2. Analogously, 7-dehydrocholesterol, which originates from cholesterol in human metabolism, normally undergoes transformation to cholecalciferol, vitamin D_3, upon exposure of the skin to sunlight (see the structures represented below).

Recent advances[24] in the understanding of the biochemistry of the vitamin D's have led to the classification of vitamin D's physiological function as an endocrine system; vitamin D_3 (cholecalciferol) actually appears to function as a hormone precursor. According to this classification, cholecalciferol is not a vitamin because it normally can be synthesized from cholesterol in human metabolism.

Cholecalciferol does not perform its functions directly. It must be metabolically transformed in the liver to 25-hydroxycholecalciferol and then in the kidney to 1,25-dihydroxycholecalciferol. The latter metabolite is considered to be the physiologically active form. Subsequently, 24-hydroxylation proceeds in the kidney, and this initiates inactivation. The production of the hormone 1,25-dihydroxycholecalciferol is regulated by the body's need for calcium and phosphate ions. 1,25-Dihydroxycholecalciferol can bring about the appearance of the 24-hydroxylase system that catalyzes its metabolic inactivation. The need for calcium stimulates parathyroid-hormone secretion. The parathyroid hormone, in turn, suppresses the 24-hydroxylase and stimulates the 1-hydroxylase system. When phosphate availability is below normal the 1-hydroxylase is stimulated and the 24-hydroxylase undergoes suppression.

1,25-Dihydroxycholecalciferol promotes Ca^{++} intestinal absorption and mobilization of Ca^{++} from bone. The mechanism of action promoting Ca^{++} transport in the intestine involves formation of a calcium-binding protein. 1,25-Dihydroxycholecalciferol promotes availability of this protein. A calcium-dependent ATPase, Na^+, and the calcium-binding protein are necessary for the intestinal Ca^{++}-transport process. 1,25-Dihydroxycholecalciferol also promotes intestinal phosphate absorption, mobilization of Ca^{++} and phosphate from bone, and renal reabsorption of Ca^{++} and phosphate.

The vitamin D's are important in the therapeutics of hypoparathyroidism and of vitamin-D deficiency.[2] Ergocalciferol, cholecalciferol, and dihydrotachysterol are recognized by the *U.S.P.* Although dihydrotachysterol has relatively weak antirachitic activity, it is effective and quicker acting in increasing serum-Ca^{++} concentration in parathyroid deficiency. Dihydrotachysterol has a shorter duration of action; hence, it has less potential of toxicity due to hypercalcemia.

Products

Ergocalciferol U.S.P., irradiated ergosta-5,7,22-trien-3β-ol, vitamin D$_2$, calciferol. The history and preparation of this vitamin have been described.

Vitamin D$_2$ is a white, odorless, crystalline compound that is soluble in fats and in the usual organic solvents, including alcohol. It is insoluble in water. Vitamin D$_2$ is oxidized slowly in oils by oxygen from the air, probably through the fat peroxides that are formed. Vitamin A is much less stable under the same conditions.

Pure vitamin D$_2$ protects rats from rickets in daily doses of 15 μg. However, it was soon shown that, rat unit for rat unit, vitamin D$_2$ or irradiated ergosterol was not as effective as cod-liver oil for the chick. Therefore, the vitamin D of cod-liver oil appears to differ from vitamin D$_2$.

One microgram equals 40 U.S.P. Units.

Cholecalciferol U.S.P., activated 5,7-cholesta-dien-3β-ol, vitamin D$_3$, activated 7-dehydrocholesterol, occurs as white, odorless crystals that are soluble in fatty oils, alcohol, and many organic solvents. It is insoluble in water.

Vitamin D$_3$ also occurs in tuna- and halibut-liver oils. Vitamin D$_3$ has the same activity as vitamin D$_2$ in rats. Vitamin D$_3$ is more effective for the chick; however, both vitamins have equal activity for humans.

Vitamin D$_3$ exhibits stability comparable to that of vitamin D$_2$.

Epimerization of the —OH at C-3 in vitamin D$_2$ or D$_3$ or conversion of the —OH at C-3 to a ketone group diminishes the activity greatly but does not completely destroy it. Ethers and esters that cannot be cleaved in the body have no vitamin-D activity. Inversion of the hydrogen at C-9 in ergosterol and other 7-dehydrosterols prevents the normal course of irradiation.

Dihydrotachysterol U.S.P., Hytakerol®, A. T. 10, dihydrotachysterol$_2$, 9,10-*seco*-5,7,22-ergastatrien-3-β-ol. Tachysterol$_2$, represented below, is a by-product from ergosterol irradiation. Reduction of tachysterol$_2$ led to dihydrotachysterol$_2$.

Dihydrotachysterol occurs as colorless or white crystals or a white, crystalline, odorless powder. It is soluble in alcohol, freely soluble in chloroform, sparingly soluble in vegetable oils, and practically insoluble in water.

Dihydrotachysterol has slight antirachitic activity. It causes an increase of the calcium concentration in the blood, an effect for which tachysterol is only one-tenth as active.

Ergosterol → Vitamin D$_2$ (Ergocalciferol)

7–Dehydrocholesterol → Vitamin D$_3$ (Cholecalciferol)

Tachysterol₂

Dihydrotachysterol₂

25-Hydroxydihydrotachysterol$_3$, prepared recently, has weak antirachitic activity, but it is a more important bone-mobilizing agent and is more effective than dihydrotachysterol$_3$. Also, it is more effective increasing intestinal calcium transport and bone mobilization in thyroparathyroidectomized rats. Its activity suggests that it may be the drug of choice in the treatment of hypoparathyroidism and similar bone diseases.[25]

VITAMIN E

Since the early 1920's it has been known that rats fed only cow's milk are not able to produce offspring. The principle from wheat germ that can rectify this deficiency in both male and female rats was named *vitamin E*. When the compound known as vitamin E was isolated in 1936, it was named tocopherol. Since then several other closely related compounds have been discovered from natural sources, and this family of natural products took the generic name *tocopherols*. The most well-known tocopherols include α-tocopherol (vitamin E), which has the greatest biological activity; β-tocopherol; γ-tocopherol; and Δ-tocopherol.

α−Tocopherol(5,7,8−trimethyltocol)

The structure represented above shows that the tocopherols are various methyl-substituted tocol derivatives. β-Tocopherol is 5,8-dimethyltocol; the γ-compound is 7,8-dimethyltocol; Δ-tocopherol is 8-methyltocol.

The tocopherols are diterpenoid natural products biosynthesized from a combination of four isoprenoid units; geranylgeranyl pyrophosphate is the key intermediate that leads to these compounds.[26]

Vitamin E U.S.P. may consist of *d*- or *dl*- α-tocopherol or their acetates or their succinates, 97.0 to 100 precent pure. It also may be mixed tocopherols concentrate containing not less than 33 percent of total tocopherols of which not less than 50 percent is *dl*- or *d*-α-tocopherol and is obtained from edible vegetable oils that may be used as diluents when needed. It also may be a 25 percent *dl*- or *d*-α-tocopheryl acetate in concentrate, the vehicle being an edible vegetable oil.

The tocopherols and their acetates are light yellow, viscous, odorless oils that have an insipid taste. They are insoluble in water and soluble in alcohol, organic solvents and fixed oils. They are stable in air for reasonable periods of time, but are oxidized slowly by air. They are oxidized readily by ferric salts, mild oxidizing agents and by air in the presence of alkali. They are inactivated rapidly by exposure to ultraviolet light; however, not all samples behave alike in this respect because traces of impurities apparently affect the rate of oxidation very much. The tocopherols have antioxidant properties for fixed oils in the following decreasing order of effectiveness: δ-, γ-, β- and α-.[27] In the process of acting as antioxidants, the tocopherols are destroyed by the accumulating fat peroxides that are decomposed by them. They are added to Light Mineral Oil N.F. and Mineral Oil U.S.P. because of their antioxidant property. The tocopherols can be converted to the acetates and benzoates, respectively, which are oils and are as active as the parent compounds and have the advantage of being more stable toward oxidation.

l-α-Tocopherol is absorbed from the gut more rapidly than the *d*-form; however, the absorption of the mixture of *d*- and *l*-α-tocopherol was considerably higher (about 55% av.) than was to be expected from the data obtained after administration of the single compounds. No marked differences were noted in the distribution in various tissues and the metabolic degradation of *d*- and *l*-α-tocopherols.[28] The liver is an important storage site.

The tocopherols are especially abundant in wheat germ, rice germ, corn germ, other seed germs, lettuce, soya and cottonseed oils. All green plants contain some tocopherols, and there is some evidence that some green leafy vegetables and rose hips contain more than wheat germ. It probably is synthesized by leaves and translocated to the seeds. All 4 tocopherols have been found in wheat-germ oil; α-, β-, and γ-tocopherols have been found in cottonseed oil. Corn oil contains predominantly γ-tocopherol and thus furnishes a convenient source for the isolation of this, a difficult member of the tocopherols to prepare. δ-Tocopherol is 30 percent of the mixed tocopherols of soya bean oil.

d-α-Tocopherol is about 1.36 times as effective as dl-α-tocopherol in rat antisterility bio-assays. β-Tocopherol is about one-half as active as α-tocopherol, and the γ- and δ-tocopherols are only one one-hundredths as active as α-tocopherol. The esters of the tocopherol, such as the acetate, propionate and butyrate, are more active than the parent compound.[29] This is also true of the phosphoric acid ester of (±)-δ-tocopherol when it is administered parenterally.[30] The ethers of the tocopherols are inactive. The oxidation of the tocopherols to their corresponding quinones also leads to inactive compounds. Replacement of the methyl groups by ethyl groups leads to decreased activity. The introduction of a double bond in the 3,4 position of α-tocopherol reduces its activity by about two-thirds. Reduction of the size of the long alkyl side chain or the introduction of double bonds in this side chain markedly reduces activity.

Most of the gastrointestinal absorption of vitamin E occurs through the mucosa and the lymphatic system. It has been demonstrated that bile performs an important function in promoting tocopherol absorption. The tocopherols in lymph are associated with chylomicrons and very low-density lipoproteins (VLDL). Circulating tocopherols are also associated mainly with the blood low-density lipoproteins. The tocopherols are readily and reversibly bound to most tissues including adipose tissue, and the vitamin is thus stored.

For decades there has been significant interest in investigating the biochemical functions of vitamin E, but it is still difficult to explain many of the biochemical derangements caused by vitamin-E deficiency in animals. There seems to be general agreement that one of the primary metabolic functions of the vitamin is that of an antioxidant of lipids, particularly unsaturated fatty acids. This function preventing lipid oxidation does not, however, explain all the biochemical abnormalities due to vitamin-E deficiency. Moreover, vitamin E is not the only in vivo antioxidant. Two enzyme systems, glutathione reductase and o-phenylenediamine peroxidase, also function in this capacity.[31]

It has been postulated that vitamin E has a role in the regulation of protein synthesis. Other actions of this vitamin have also been investigated, for example,

effects on muscle creatine kinase and liver xanthine oxidase. It has been noted that vitamin-E deficiency leads to an increase in the turnover of creatine kinase. There is also an increase in liver xanthine-oxidase activity in vitamin-E-deficient animals, and this increase is due to an increase in de-novo synthesis.[31]

Although it has been difficult to establish clinical correlates of vitamin-E deficiency in humans; Bieri and Farrell[31] have summarized some useful generalizations and conclusions. These workers have noted that the infant, especially the premature infant, is susceptible to tocopherol deficiency due to ineffective transfer of the vitamin from placenta to fetus, and that growth in infants requires greater availability of the vitamin. On the other hand, in adults the tocopherol storage depots provide adequate availability that is not readily depleted, but intestinal malabsorption syndromes, when persistent, can lead to depletion of the storage depots. Children with cystic fibrosis suffer from severe vitamin-E deficiency due to malabsorption. Tropical sprue, celiac disease, gastrointestinal resections, hepatic cirrhosis, biliary obstruction, and excessive ingestion of mineral oil also may cause long-term malabsorption.

Vitamin-E therapeutic indications include the clinical conditions characterized by low serum-tocopherol levels, increased fragility of red blood cells to hydrogen peroxide, or conditions that require additional amounts. The latter case can be exemplified with individuals who consume excessive amounts of polyunsaturated fatty acids (more than 20 g. per day over normal diet).[32]

It has been claimed that vitamin E could be of therapeutic benefit in ischemic heart disease, but evidence against this claim continues to accumulate. It also has been suggested that megadoses of tocopherol be used in the treatment of peripheral vascular disease with intermittent claudication. Although some studies support this proposal, experts in the field state that further clinical studies are necessary in order to make a definitive recommendation. Nevertheless, it continues to be popular and controversial to consider the beneficial effects of vitamin E and other vitamins in large (mega) dietary supplements, and investigations of megavitamin-E therapy for cardiovascular disease continue to appear in the literature.[31]

The eminent vitamin biochemist, R. J. Williams has emphasized that

"lipid peroxidation, the formation of harmful peroxides, from the interaction between oxygen and highly unsaturated fats (polyunsaturates) needs to be controlled in the body. Both oxygen and the polyunsaturated lipids are essential to our existence, but if the protection against peroxidation is inadequate, serious damage to various body proteins may result. Vitamin

E is thought to be the leading agent for the prevention of peroxidation and the free radical production that is associated both with it and with radiation."[33] Williams also notes that although exact mechanisms of action of these antioxidants are not yet known, "providing plenty of vitamin E and ascorbic acid—both harmless antioxidants—is indicated as a possible means of preventing premature aging, especially if one's diet is rich in polyunsaturated acids."

Regarding the above implication of unnecessary peroxidation of unsaturated lipids, it is interesting to note that very recently it has been postulated that atherosclerosis appears to be due to a deficiency of prostacyclin, and that this deficiency is caused by inhibition of prostacyclin synthetase by lipid peroxides or by free radicals that are likely to be generated during hyperlipidemia. Although there is no direct evidence that in experimental or human atherosclerosis lipid peroxidation is the earliest sign of the disease state, lipid peroxides have been found in arteries from atherosclerotic patients and in ceroid atheromatic plaques and, at the same time, prostacyclin is hardly generated in human atheromatic plaques.[34]

THE VITAMIN K'S

Vitamin K biological activity was first discovered in 1929, and vitamin K was finally identified structurally in 1931 as represented below. The term *vitamin K* was applied to the vitamin isolated from alfalfa, and a similar principle from fish meal was named vitamin K_2. Many other closely related compounds possess vitamin-K activity; e.g., menadione (2-methyl-1,4-naphthoquinone) is as active as vitamin K on a molar basis.

Vitamin K is a naphthoquinone derivative containing diterpenoid units biosynthesized by the intermediate geranylpyrophosphate.[35] Alfalfa, chestnut leaves, and spinach are excellent sources of vitamin K, which also occurs in hog liver fat, hempseed, tomatoes, kale, and soy bean oil.

Vitamin-K-deficiency disease, which is known to lead to fatal hemorrhage, has been recognized since 1929, but the molecular mechanism of action of this vitamin continues to attract the attention of many researchers. The isolation and identification of γ-carboxyglutamic acid in bovine prothrombin in 1974 required 45 years of active research on the biochemistry of vitamin K. This discovery is considered to be a milestone because it led to better understanding of vitamin K and blood coagulation. It is already understood that vitamin K is a component of a membrane-bound carboxylase enzyme system and functions in the post-translational carboxylation of some peptide-bound glutamate residues in a number of so-called vitamin-K-dependent proteins that are necessary for normal blood coagulation. The vitamin-K-dependent carboxylase system includes a specialized microsomal election transport system coupled to a carbon dioxide fixation reaction. Although the reaction does not require ATP, it uses the energy from the oxidation of reduced vitamin K to execute the carboxylation of glutamic acid.[36]

Although the molecular involvement of vitamin K in blood coagulation requires further clarification, some conclusions regarding its mechanism can be summarized. Vitamin K undergoes reduction of the quinone ring as a preliminary step. This reduction to vitamin-K hydroquinone is effected by NADH. Vitamin-K hydroquinone is necessary for the carboxylation of glutamic acid to form carboxylglutamic acid, which participates in the complexation (chelation) of the necessary Ca^{++}. It is, therefore, the gen-

Vitamin K_1
(2-Methyl-3-phytyl-1,4-naphthoquinone)

$n = 4 =$ Vitamin $K_2(30)$
$n = 5 =$ Vitamin $K_2(35)$

eral consensus that vitamin K, through its hydroquinone metabolite, participates in the formation of specific Ca^{++}-binding sites on prothrombin and that the strong binding sites for Ca^{++} involve adjacent γ-carboxylglutamic acid residues within the prothrombin structure. It is also significant that γ-carboxyglutamic acid has been found as a residue in all vitamin-K-dependent clotting factors. Moreover, it is interesting to note that the anticoagulant coumarin derivatives and related compounds (see Chap. 14) interfere with the carboxylase system and thus interfere with the γ-carboxylation of the glutamic-acid residues of the prothrombin structure.

The therapeutic use of vitamin K as a systemic hemostatic agent is based on the critical function that the vitamin performs in blood coagulation. Vitamin K_1 (phytonadione) is effective both in the treatment of hypoprothrombinemia due to dietary deficiency of the vitamin or malabsorption and in bleeding due to oral anticoagulants (e.g., coumadin derivatives). Phytonadione exerts prompt and prolonged action. It can be administered orally, subcutaneously, or intramuscularly; in emergencies it can be given by slow intravenous injection.

PRODUCTS

Phytonadione U.S.P., Mephyton®, Konakion®, 2-methyl-3-phytyl-1,4-naphthoquinone, Vitamin K_1, is described as a clear, yellow, very viscous, odorless or nearly odorless liquid.

Pure vitamin K_1 is a yellow, crystalline solid that melts at 69° C. It is insoluble in water, slightly soluble in alcohol, soluble in vegetable oils and in the usual fat solvents. It is unstable toward light, oxidation, strong acids and halogens. It easily can be reduced to the corresponding hydroquinone, which, in turn, can be esterfied. A large number of compounds have been tested for their antihemorrhagic activity.

Significant biological activity is manifested in compounds with the following structure when:

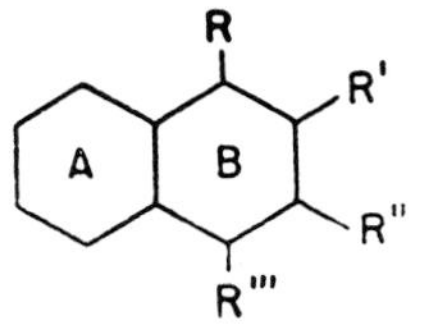

1. Ring A is aromatic or hydro-aromatic.
2. Ring A is not substituted.
3. Ring B is aromatic or hydro-aromatic.
4. R equals OH, CO, OR, OAc (the R in OR equals methyl or ethyl).
5. R' equals methyl.

6. R'' equals H, sulfonic acid, dimethylamino or an alkyl group containing 10 or more carbon atoms. A double bond in the β, γ position of this alkyl group enhances potency, whereas, if the double bond is further removed, it exerts no effect. Isoprenoid groups are more effective than straight chains. In the case of the vitamin $K_{2(30)}$ type compounds the 6',7'-mono-*cis* isomer is significantly less active than the all-*trans* or the 18',19'-mono-*cis* isomer. This also was true of the vitamin $K_{2(20)}$ isoprenolog. A vitamin $K_{2(25)}$ isoprenolog was 20 percent more active than vitamin $K_{1(37)}$.
7. R''' equals H, OH, NH_2, CO, OR, Ac (the R in OR equals methyl or ethyl).

Decreased antihemorrhagic activity is obtained when:

1. Ring A is substituted.
2. R' is an alkyl group larger than a methyl group.
3. R'' is a hydroxyl group.
4. R'' contains a hydroxyl group in a side chain.

It is interesting to note that, if ring A is benzenoid in character, the introduction of sulfur in place of a —CH=CH— in this ring in 2-methylnaphthoquinone permits the retention of some antihemorrhagic activity. This might indicate that, in the process of exerting vitamin K activity, the benzenoid end of the molecule must fit into a pocket carefully tailored to it. That the other end is not so closely surrounded is shown by the retention of activity on changing the alkyl group in the 2-position.

Although marked antihemorrhagic activity is found in a large number of compounds, the possibility exists that they may be converted in the body to a vitamin K_1 type compound. The esters of the hydroquinones may be hydrolyzed, and the resulting hydroquinone may be oxidized to the quinone. The methyl tetralones, which are very active, possibly could be dehydrogenated to the methylnaphthols, which are hydroxylated, and the latter product converted to the biologically equivalent quinone. Compounds with a dihydrobenzenoid ring (such as 5,8-dihydrovitamin K_1) appear to be moderately easily dehydrogenated, whereas the corresponding tetrahydrides are resistant to such a change.

The only known function of vitamin K in higher animals is to maintain adequate plasma levels of the protein prothrombin (factor II), and three other essential clotting factors: VII (proconvertin). IX (autoprothrombin II) and X (Stuart-Prower Factor). It follows that any condition which does not permit the full utilization of the antihemorrhagic agents or the production of prothrombin would lead to an increase in the amount of time in which the blood will clot or to hemorrhagic conditions. Some of these conditions are:

(1) faulty absorption caused by a number of conditions, e.g., obstructive jaundice, biliary fistulas, intestinal polyposis, chronic ulcerative colitis, intestinal fistula, intestinal obstruction and sprue; (2) damaged livers or primary hepatic diseases, such as atrophy, cirrhosis or chronic hepatitis; (3) insufficient amounts of bile or abnormal bile in the intestinal tract and (4) insufficient amounts of vitamin K.

Bile of a normal composition is necessary to facilitate the absorption of vitamin K from the intestinal tract. The bile component principally concerned in the absorption and transport of fat-soluble vitamin K from the digestive tract is thought to be deoxycholic acid. The molecular compound of vitamin K with deoxycholic acid was effective upon oral administration to rats with biliary fistula.

Vitamin K is administered in conjunction with bile salts or their derivatives in pre- and postoperative jaundiced patients to bring about and maintain a normal prothrombin level in the blood.

In the average infant, the birth values of prothrombin content are adequate, but during the first few days of life they appear to fall rapidly, even dangerously low, and then slowly recover spontaneously. This transition period was and is a critical one because of the numerous sites of hemorrhagic manifestations, traumatic or spontaneous, that may prove serious if not fatal. This condition now is recognized as a type of alimentary vitamin K deficiency. The spontaneous recovery is due perhaps to the establishment of an intestinal flora capable of synthesizing vitamin K after ingestion of food. However, administration of vitamin K orally effects a prompt recovery.

Vitamin K can be used to diagnose liver function accurately. The intramuscular injection of 2 mg. of 2-methyl-1,4-naphthoquinone has led to response in prothrombin index in patients with jaundice of extrahepatic origin but not in patients with jaundice of intrahepatic origin, e.g., cirrhosis.

Vitamin K_1 acts more rapidly (effect on prothrombin time) than menadione within 2 hours after intravenous administration. However, no difference could be detected after 2 hours.[37]

The menadiones are much less active than vitamin K_1 in normalizing the prolonged blood-clotting times caused by dicumarol and related drugs.[37]

Vitamin K_1 is the drug of choice for humans because of its low toxicity. Its duration of action is longer than that of menadione and its derivatives. Vitamin K should not be administered to patients receiving warfarin or coumarin anticoagulants.

Menadione U.S.P., 2-methyl-1,4-naphthoquinone, menaphthone, thyloquinone. Menadione can be prepared very readily by the oxidation of 2-methylnaphthalene with chromic acid. It is a bright yellow, crystalline powder and is nearly odorless. It is affected by sunlight. Menadione is practically insoluble in water; it is soluble in vegetable oils, and 1 g. of it is soluble in about 60 ml. of alcohol. The *N.F.* has a caution that menadione powder is irritating to the respiratory tract and to the skin, and an alcoholic solution has vesicant properties.

On a mole for mole basis, menadione is equal to vitamin K_1 in activity and can be used as a complete substitute for this vitamin. It is effective orally, intravenously and intramuscularly. If given orally to patients with biliary obstruction, bile salts or their equivalent should be administered simultaneously in order to facilitate absorption. It can be administered intramuscularly in oil when the patient cannot tolerate an oral product, has a biliary obstruction or where a prolonged effect is desired.

[14]C-labeled menadiol diacetate in small physiologic doses is converted in vivo to a vitamin $K_{2(20)}$, and the origin of the side chain probably is via mevalonic acid. This suggests that menadione may be an intermediate or a provitamin K.[37]

Menadione in oil is three times more effective than a menadione suspension in water. More of menadione than of vitamin K_1 is absorbed orally, but 38 percent of the former is excreted by the kidney in 24 hours whereas only very small amounts of the latter are excreted by this route in 24 hours. In rats menadione in part is reduced to the hydroquinone and excreted as the glucuronide 19 percent and the sulfate 9.3 percent.

Menadione Sodium Bisulfite U.S.P., Hykinone®, 2-methyl-1,4-napthoquinone sodium bisulfite, menadione bisulfite, is prepared by adding a solution of sodium bisulfite to menadione.

Menadione

Menadione Sodium Bisulfite

Menadione sodium bisulfite occurs as a white, crystalline, odorless powder. One gram of it dissolves in about 2 ml. of water, and it is slightly soluble in alcohol. It decomposes in the presence of alkali to liberate the free quinone.

Menadiol Sodium Diphosphate U.S.P., Synkayvite®, Kappadione®, tetrasodium 2-methyl-1,4-naphthalenediol *bis*(dihydrogen phosphate), tetrasodium 2-methylnaphthohydroquinone diphosphate, is a white hygroscopic powder, very soluble in water, giving solutions that have a pH of 7 to 9. It is available in ampules for use subcutaneously, intramuscularly or intravenously and in tablets for oral administration.

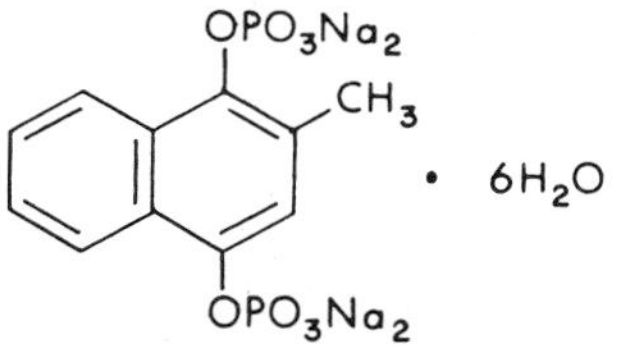

Menadiol Sodium Diphosphate

Menadione bisulfite and menadiol diphosphate have been shown to produce hemolytic symptoms (reticulocytosis, increase in Heinz bodies) in newborn premature infants when given in excessive doses (more than 5 to 10 mg. per kg.). In severe cases overt hemolytic anemia with hemoglobinuria may occur. The increased red cell breakdown may lead to hyperbilirubinemia and kernicterus.

These compounds may interfere with bile pigment secretion also. Newborns with a congenital defect of glucose-6-phosphate dehydrogenase can react with severe hemolysis even with small doses of menadione derivatives. However, small nonhemolyzing doses can be used in the newborn, and combination with vitamin E is not considered essential.[38]

WATER-SOLUBLE VITAMINS

Although these vitamins are structurally diverse, they are characterized as a general class on the basis of water solubility in order to distinguish them from the

TABLE 21-2

LIPID-SOLUBLE VITAMINS

Name Proprietary Name	Preparations	Category	Usual Adult Dose*	Usual Dose Range*	Usual Pediatric Dose*
Vitamin A U.S.P. *Acon, Aquasol A, Dispatabs, Homagenets-A oral, Testavol-S, Vi-Dom-A, Vio-A, Alphalin, Anatola, Super A Vitamin*	Vitamin A Capsules U.S.P.	Vitamin A (antixerophthalmic)	Prophylactic—1.5 mg. (5,000 U.S.P. Vitamin A Units) once daily; therapeutic—3 to 15 mg. (10,000 to 50,000 Units) once daily	Prophylactic—1.5 to 2.4 mg. (5,000 to 8,000 Units) once daily; therapeutic—3 to 150 mg. (10,000 to 500,000 Units) once daily	Prophylactic—the following amounts once daily; infants up to 1 year of age—450 μg. (1,500 Units); 1 to 3 years—600 μg. (2,000 Units); 3 to 6 years—750 μg. (2,500 Units); 6 to 10 years—1.05 mg. (3,500 Units); 10 to 12 years—1.35 mg. (4,500 Units); 12 years and older—see Usual Dose. Therapeutic—see Usual Dose
Ergocalciferol U.S.P. *Deltalin, Drisdol*	Ergocalciferol Capsules U.S.P. Ergocalciferol Solution U.S.P. Ergocalciferol Tablets U.S.P.	Vitamin D (antirachitic)	Rickets, Prophylactic—10 μg. (400 U.S.P. Vitamin D Units) once daily; therapeutic, deficiency rickets—300 μg. to 1.25 mg. (12,000 to 50,000 Units) once daily; refractory rickets—1.25 to 25 mg. (50,000 to 1,000,000 Units) once daily; hypocalcemic tetany—1.25 to 10 mg. (50,000 to 400,000 Units) once daily		
Cholecalciferol U.S.P.	Decavitamin Capsules U.S.P. Decavitamin Tablets U.S.P.	Vitamin D (antirachitic)	1 dosage unit daily		
Dihydrotachysterol U.S.P. *Hytakerol*	Dihydrotachysterol Tablets U.S.P.	Antihypocalcemic	Initial, 800 μg. to 2.4 mg. once daily; maintenance, 200 μg. weekly to 1 mg. daily		

* See *U.S.P. D.I.* for complete dosage information.
(Continued)

TABLE 21-2

LIPID-SOLUBLE VITAMINS *(Continued)*

Name Proprietary Name	Preparations	Category	Usual Adult Dose*	Usual Dose Range*	Usual Pediatric Dose*
Vitamin E U.S.P. *E-Ferol, E-Ferol Succinate, Eprolin, Epsilan-M, Ecofrol, Tokols*	Vitamin E Capsules U.S.P.	Vitamin E supplement		Prophylactic—from 5 to 30 International Units of Vitamin E; therapeutic—to be determined by the practitioner according to the needs of the patient	
Phytonadione U.S.P. *AquaMephyton, Konakion, Mephyton*	Phytonadione Injection U.S.P.	Vitamin K (prothrombogenic)	Parenteral, 2.5 to 25 mg., repeated in 6 to 8 hours, if necessary	2.5 to 50 mg. daily	Hemorrhagic disease of the newborn, prophylactic—I.M., 500 μg. to 1 mg.; thera-peutic—I.M. or S.C., 1 mg. Other prothrombin deficien-cies—infants, parenteral, 2 mg.; older infants and children, 5 to 10 mg.
	Phytonadione Tablets U.S.P.		2.5 to 25 mg., repeated in 12 to 48 hours, if necessary	1 to 50 mg. daily	Prothrombin deficiencies—infants, 2 mg.; older infants and children, 5 to 10 mg.
Menadione U.S.P.	Menadione Injection U.S.P. Menadione Tablets U.S.P.	Source of vitamin K	Oral and I.M., 2 mg. daily	2 to 10 mg.	
Menadione Sodium Bisulfite U.S.P. *Hykinone*	Menadione Sodium Bisulfite Injection U.S.P.	Source of vitamin K	I.V. and S.C., 2 mg. daily		
Menadiol Sodium Diphosphate U.S.P. *Kappadione, Synkayvite*	Menadiol Sodium Diphosphate Injection U.S.P. Menadiol Sodium Diphosphate Tablets U.S.P.	Source of vitamin K	Oral, I.M., I.V., or S.C., 3 to 6 mg. daily	5 to 75 mg. daily	

* See *U.S.P. D.I.* for complete dosage information.

lipid-soluble vitamins. This class includes the β-complex vitamins and ascorbic acid (vitamin C). The term *β-complex vitamins* usually refers to thiamine, riboflavin, pyridoxine, nicotinic acid, pantothenic acid, biotin, cyanocobalamine, and folic acid. Dietary deficiencies of any one of the B vitamins are commonly complicated by deficiencies of more than one member of the group, hence treatment with β-complex preparations is usually indicated.

THIAMINE (VITAMIN B₁)

Thiamine was the first water-soluble vitamin to be discovered in 1926, but the complete determination of its structure and synthesis were not accomplished until 1936.

Many natural foods provide adequate amounts of this vitamin. The germ of cereals, brans, egg yolks, yeast extracts, peas, beans, and nuts usually provide enough thiamine to satisfy adult requirements. It is not economically practical to isolate the crystalline vitamin from natural sources on a commercial scale, hence the commercially available thiamine is prepared synthetically.

Thiamine is biologically synthesized from the pyrimidine derivative 4-amino-5-hydroxymethyl-2-methyl pyrimidine methylpyrimidine and 5-(β-hydroxyethyl)-4-methylthiazole. These two precursors are converted to phosphate derivatives under kinase catalysis, which requires ATP. The respective phosphate derivatives then interact to form thiamine phosphate. The latter reaction is catalyzed by thiamine-phosphate pyrophosphorylase.

In higher mammalian organisms, thiamine is transformed to the coenzyme thiamine pyrophosphate by direct pyrophosphate transfer from ATP. This coenzyme performs important metabolic functions, e.g., as codecarboxylase in the decarboxylation of pyruvate to form acetyl-coenzyme A.

In the decarboxylation of pyruvate, the coenzyme interacts with pyruvic acid to form so-called *active aldehyde* as shown below.

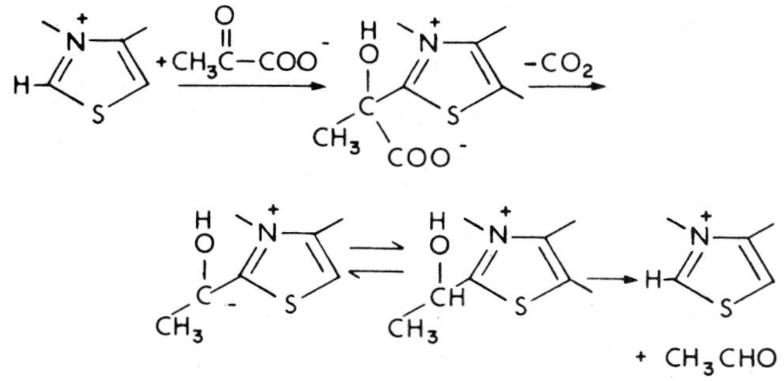

Active Aldehyde

The "active-aldehyde" intermediate then interacts with thioctic acid to form acetyl-thioctate, which is responsible for acetylating CoA-SH to form acetyl CoA.

Thiamine hydrochloride is unstable in aqueous solutions of pH greater than 5. Under these conditions it undergoes decomposition and inactivation. Thiamine is also susceptible to oxidation. It is readily oxidized by exposure to the atmosphere or by oxidizing reagents such as hydrogen peroxide, permanganate, or alkaline potassium ferricyanide. This oxidation forms thiochrome as represented below.

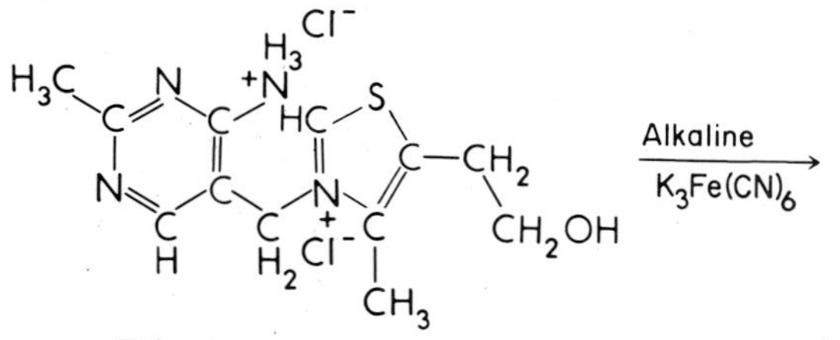

Thiamine Hydrochloride

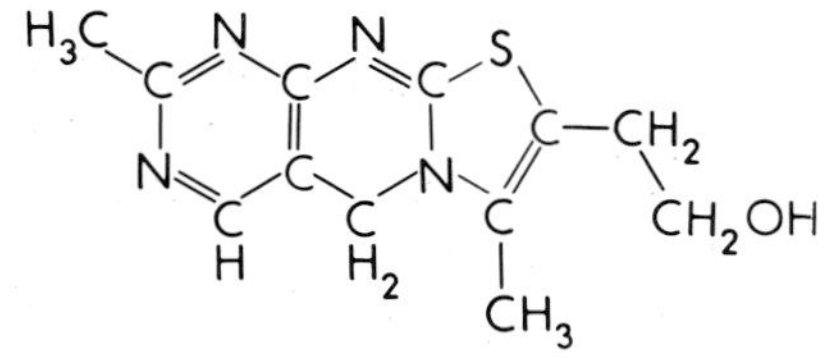

Thiochrome

Thiochrome exhibits a vivid blue fluorescence; hence, this reaction is the basis for the quantitative fluorometric assay of thiamine in the *U.S.P.*

Thiamine* Hydrochloride U.S.P., thiamine monohydrochloride, thiamine chloride, vitamin B₁ hy-

---

*The structure of thiamine hydrochloride has been revised; see Ganellin, G.R.: Chemistry and structure–activity analysis, *in* Roberts, G.C.K. (ed.): Drug Action at the Molecular Level, Baltimore, University Park Press, 1977.

drochloride, vitamin B₁, aneurine hydrochloride occurs as small, white crystals or as a crystalline powder; it has a slight, characteristic yeastlike odor. The anhydrous product, when exposed to air, will absorb rapidly about 4 percent of water. One gram is soluble in 1 ml. of water and in about 100 ml. of alcohol. It is soluble in glycerin. An aqueous solution, 1 in 20, has a pH of 3. Aqueous solutions 1:100 have a pH of 2.7 to 3.4.

Thiamine hydrochloride is sensitive toward alkali. The addition of 3 moles of sodium hydroxide per mole of thiamine hydrochloride reacts as shown below.

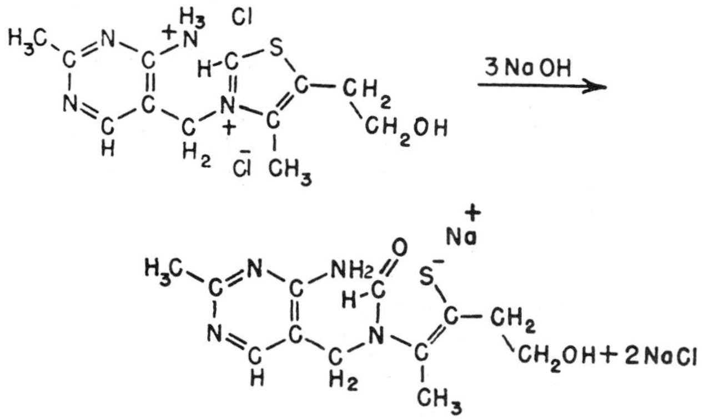

Thiamine Hydrochloride Degradation

Thiamine Mononitrate U.S.P., thiamine nitrate, vitamin B₁ mononitrate, is a colorless compound that is soluble in water 1:35 and slightly soluble in alcohol. Two percent aqueous solutions have a pH of 6.0 to 7.1. This salt is more stable than the chloride hydrochloride in the dry state, is less hygroscopic, and is recommended for multivitamin preparations and the enrichment of flour mixes.

PANTOTHENIC ACID

During the decade of the 1930's, R. J. Williams and his collaborators recognized, isolated, and synthesized pantothenic acid. Because its occurrence is so widespread, it was called *pantothenic acid* from the Greek, meaning "from everywhere."

This vitamin is synthesized by most green plants and microorganisms. The precursors are γ-ketoisovaleric acid and β-alanine.[1] The latter originates from the decarboxylation of aspartic acid. γ-Ketoisovaleric acid is converted to ketopantoic acid by N⁵,N¹⁰-methylenetetrahydrofolic acid, and then, upon reduction, pantoic acid is formed. Finally, pantoic acid and β-alanine react by amide formation to form pantothenic acid.

The metabolic functions of pantothenic acid in human biochemistry are mediated through the synthesis of coenzyme-A; i.e., this vitamin is a structural component of coenzyme-A, which is necessary for many important metabolic processes. CoA is involved in the activation of fatty acids prior to β-oxidation; this activation requires ATP to form the respective fatty acyl CoA derivatives. It also participates in fatty acid β-oxidation in the final step, forming acetyl-coenzyme-A. The latter also is formed from pyruvate decarboxylation. In pyruvate decarboxylation CoA participates in collaboration with thiamine-pyrophosphate and lipoic acid, two other important coenzymes. Thiamine-pyrophosphate is the actual decarboxylating coenzyme that functions with lipoic acid to form acetyldihydrolipoic acid from the decarboxylation of pyruvate. CoA then accepts the acetyl group from acetyldihydrolipoic acid to form acetyl CoA. Acetyl CoA participates as an acetyl donor in many processes and is also the precursor in important biosyntheses; i.e., fatty acids and steroids are synthesized from acetyl CoA.

ties are very similar to those of Calcium Pantothenate U.S.P.

Panthenol, the alcohol analog of pantothenic acid, exhibits both qualitatively and quantitatively the vitamin activity of pantothenic acid. It is considerably more stable than pantothenic acid in solutions with pH values of 3 to 5, but of about equal stability at pH 6 to 8. It appears to be more readily absorbed from the gut, particularly in the presence of food.

NICOTINIC ACID

Nicotinic acid (niacin) was first prepared by the oxidation of the alkaloid, nicotine, but not until 1913 was it isolated from yeast and recognized as an essential food factor (refer to the structures below). In 1934–1935 nicotinamide was obtained from the hydrolysis of a

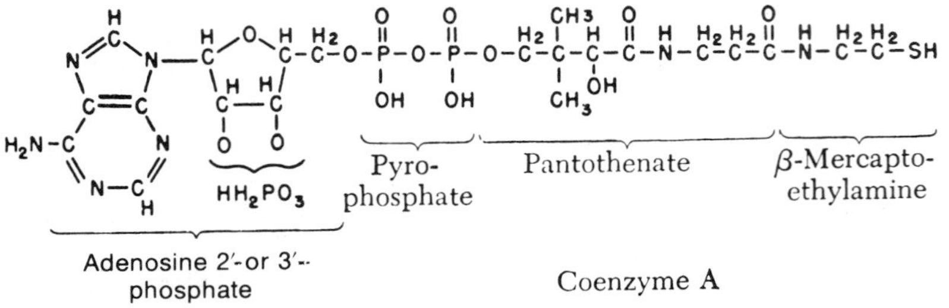

Coenzyme A

Clinical cases of pantothenic-acid deficiency do not commonly develop, unless they arise in combination with deficiencies of the other B vitamins.[2] Accordingly, pantothenic acid is usually included in multiple-vitamin preparations. The calcium salt is commonly used in pharmaceutical preparations. Panthenol, the alcohol derivative, is another form commonly used.

Products

Calcium Pantothenate U.S.P., calcium D-pantothenate, is a slightly hygroscopic, white, odorless, bitter-tasting powder that is stable in air. It is insoluble in alcohol, soluble 1:3 in water, and aqueous solutions have a pH of about 9 and $[\alpha]_D = +25°$ to $+27.5°C$. Autoclaving calcium pantothenate at 120°C for 20 minutes may cause a 10 to 30 percent decomposition. Some of the phosphates of pantothenic acid that occur naturally in coenzymes are quite stable to both acid and alkali, even upon heating.[41]

Racemic Calcium Pantothenate U.S.P. is recognized to provide a more economical source of this vitamin. Other than containing not less than 45 percent of the dextrorotatory biologically active form, its proper-

coenzyme isolated from horse red blood cells. This coenzyme was later named *coenzyme II* and is now more commonly called nicotinamide-adenine-dinucleotide-phosphate (NADP).

Generous sources of this vitamin include pork, lamb, and beef livers; hog kidneys; yeasts; pork; beef tongue; hearts; lean meats; wheat germ; peanut meal; and green peas.

Nicotinic acid can be synthesized by almost all plants and animals. Tryptophan can be metabolized to a nicotinic-acid nucleotide in animals, but the efficiency of this multi-step process varies from species to species. Plants and many microorganisms synthesize this vitamin through alternative routes using aspartic acid.

In the human, nicotinic acid reacts with 5-phosphoribosyl-1-pyrophosphate to form nicotinic-acid mononucleotide, which then reacts with ATP to produce desamido-NAD (the intermediate dinucleotide with the nicotinic-acid moiety). Finally, the latter intermediate is converted to NAD (nicotin-amide-adenine-dinucleotide, originally called *coenzyme I*) by transformation of the carboxyl of the nicotinic-acid moiety to the amide by glutamine. This final step is catalyzed by NAD synthetase; NADP is produced from NAD by ATP under kinase catalysis.[1]

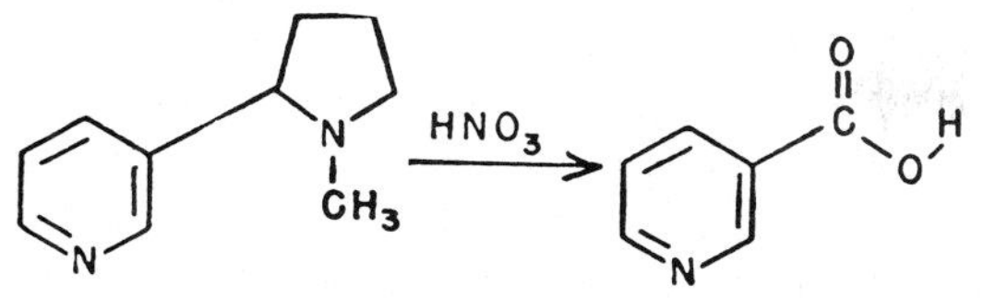

Nicotine   Nicotinic Acid

NAD and NADP participate as oxidizing coenzymes for many (more than 200) dehydrogenases. Some dehydrogenases require NAD and others require NADP; some dehydrogenases function with either. The following generalized representation illustrates the function of these coenzymes in metabolic oxidations and reductions. The abbreviation [NAD$^+$] emphasizes the electrophilicity of the pyridine C_4 moiety (which is the center of reactivity) and the substrate designated as

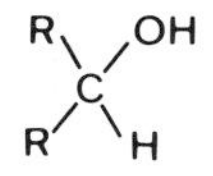

could be a primary or secondary alcohol. Arrow *a* symbolizes the function of NAD as oxidant in the hydride transfer from the substrate to the coenzyme forming NADH, reduced coenzyme. The hydroxyl of the substrate is visualized as undergoing deprotonation concertedly by either water or the pyridine nitrogen of NADH. Arrow *b* shows concerted formation of the carbonyl-π bond of the oxidation product. Arrow *c* symbolizes the reverse hydride transfer from reduced coenzyme, NADH, to the carbonyl carbon; and concertedly, as the carbonyl oxygen undergoes protonation, the reduction of the carbonyl group forms the corresponding alcohol. Thus, NAD and NADP function as hydride acceptors, while NADH and NADPH are hydride donors. Although the foregoing is a simplistic representation, it illustrates the dynamism of such oxidation–reduction reactions effected by these coenzymes under appropriate dehydrogenase catalysis.

As noted above, nicotinic acid (also known as niacin) can be made available in the nucleotide form from the amino acid tryptophan. Thus, the human and other mammals can synthesize the vitamin, provided there is appropriate dietary availability of trytophan. (Thus, it appears that nicotinic acid is not a true vitamin according to the classical definition of the term.)

Serious deficiency of niacin or tryptophan may lead to pellagra. Psychoses sometimes develop in severe cases of pellagra. Niacin or niacinamide (nicotinamide) are effective in the prevention or treatment of pellagra, but the acid may cause vasodilation when administered in high doses. Niacin is also effective therapeutically as a lipid-lowering agent in the treatment of certain cases of hyperlipidemia.[2]

Niacin U.S.P., nicotinic acid, 3-pyridinecarboxylic acid occurs as white crystals or as a crystalline powder. It is odorless or it may have a slight odor. One gram of nicotinic acid dissolves in 60 ml. of water. It is

Generalized Representation of The Hydride Transfer Reaction:

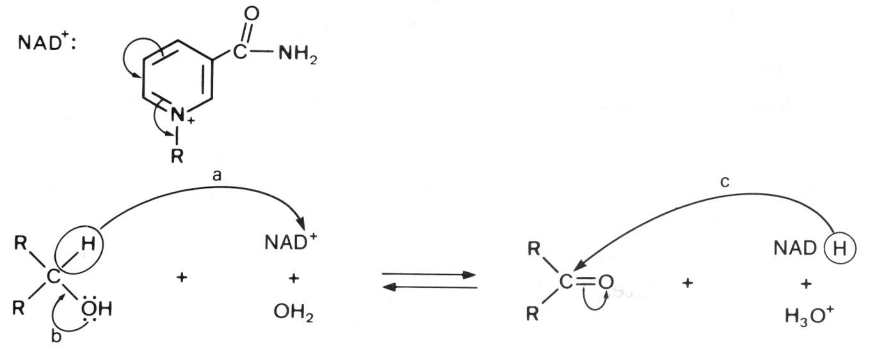

Ethanol Oxidation is Schematically Illustrated Below:

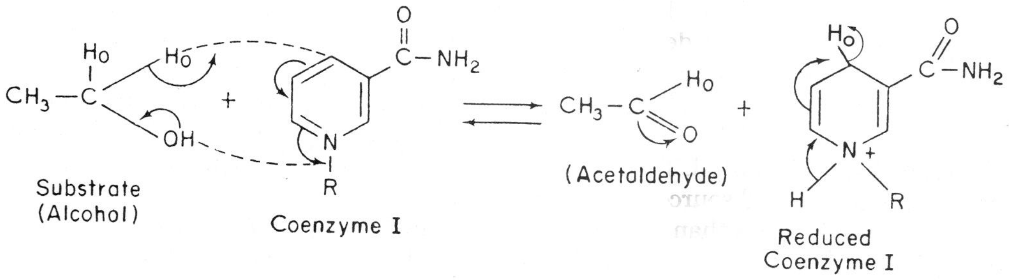

Substrate (Alcohol)   Coenzyme I   (Acetaldehyde)   Reduced Coenzyme I

freely soluble in boiling water, in boiling alcohol and in solutions of alkali hydroxides and carbonates but is almost insoluble in ether. A 1 percent aqueous solution has a pH of 6.

Nicotinic acid is stable under normal storage conditions. It sublimes without decomposition.

Niacinamide U.S.P., nicotinamide, nicotinic acid amide. Nicotinamide is prepared by the amidation of esters of nicotinic acid or by passing ammonia gas into nicotinic acid at 320°C.

Ethyl Nicotinate   Nicotinamide

Nicotinamide is a white, crystalline powder that is odorless or nearly so and has a bitter taste. One gram is soluble in about 1 ml. of water, 1.5 ml. of alcohol and in about 10 ml. of glycerin. Aqueous solutions are neutral to litmus. For occurrence, action and uses see nicotinic acid.

Niacinamide hydrochloride recently has been made available. It is more stable in solution and more compatible with thiamine chloride in solution.

RIBOFLAVIN (VITAMIN B₂)

Although the isolation of crystalline riboflavin was not accomplished until 1932, interest in this compound as a pigment dates back to 1881 in connection with the color in the whey of milk. In 1932 riboflavin was isolated as a coenzyme–enzyme complex from yeast by Warburg and Christian, and this complex was designated as *yellow oxidation ferment*.

Riboflavin is synthesized by all green plants and by most bacteria and fungi. It is known that the precursor is a guanosine-phosphate derivative, but the exact synthetic steps leading to the vitamin are not completely understood.

In higher mammals riboflavin is readily absorbed from the intestine and is distributed to all tissues. It is the precursor in the biosynthesis of the coenzymes FMN (flavin mononucleotide) and FAD (flavin adenine dinucleotide). The metabolic functions of this vitamin involve these two coenzymes, which participate in numerous vital oxidation–reduction processes. FMN, which is riboflavin-5′-phosphate, is produced from the vitamin and ATP under flavokinase catalytic action. FAD originates from an FMN and ATP reaction that involves reversible dinucleotide formation

catalyzed by flavin nucleotide pyrophosphorylase. These coenzymes function in combination with a number of enzymes as coenzyme–enzyme complexes often characterized as *flavoproteins*. The riboflavin moiety of the complex is considered to be a hydrogen-transporting agent (carrier) functioning with O_2, cytochromes, etc., as hydrogen acceptors; the hydrogen donors may be NADH, NADPH, or some suitable substrate.

Riboflavin
6,7-Dimethyl-9-(D-1′-ribityl)isoalloxazine

Riboflavin U.S.P. (riboflavine, lactoflavin, vitamin B₂, vitamin G) is a yellow to orange-yellow crystalline powder with a slight odor. It is soluble in water 1:3,000 to 1:20,000 ml., the variation in solubility being due to differences in internal crystalline structure, but it is more soluble in an isotonic solution of sodium chloride. A saturated aqueous solution has a pH of 6. It is less soluble in alcohol and insoluble in ether or chloroform. Benzyl alcohol (3%), gentistic acid (3%), urea in varying amounts, and niacinamide are used to solubilize riboflavin when relatively high concentrations of this factor are needed for parenteral solutions. Gentistic ethanol amide and sodium 3-hydroxy-2-naphthoate are also effective solubilizing agents for riboflavin.

When dry, riboflavin is not appreciably affected by diffused light; however, as previously mentioned, it deteriorates in solution in the presence of light, and this deterioration is very rapid in the presence of alkalies. This deterioration may be retarded by buffering on the acid side.

PYRIDOXINE

In 1935 the term *vitamin B₆* was applied to the principle that cured a dermatitis in rats fed a vitamin-free diet supplemented with thiamine and riboflavin. Three years later vitamin B₆ was isolated from rice paste and yeast. By 1939 the structure elucidation and synthesis of the vitamin were accomplished. This vita-

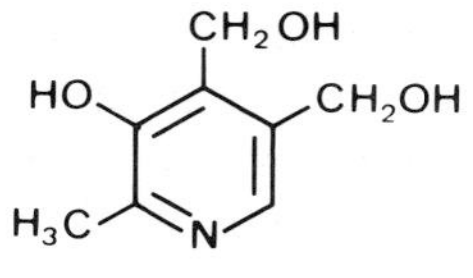

CH₂OH ... Pyridoxine

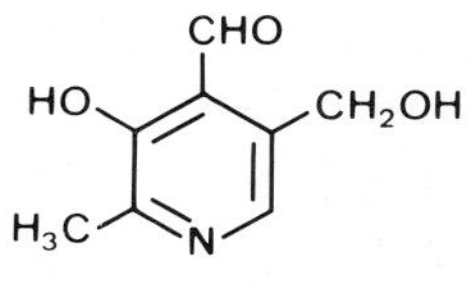

CHO ... Pyridoxal

CH₂NH₂ ... Pyridoxamine

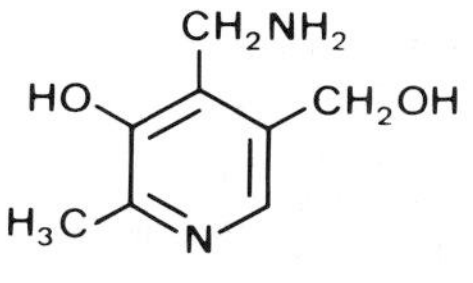

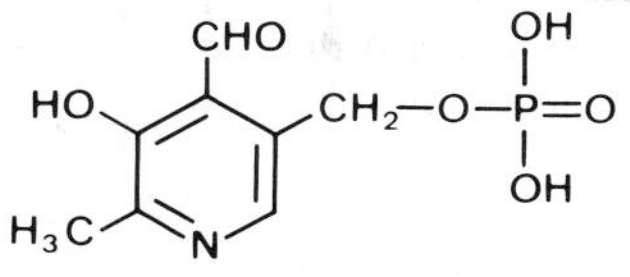

Pyridoxal phosphate

min is also known as *pyridoxine* or *pyridoxol.* Two additional chemical forms, pyridoxal and pyridoxamine, have also been isolated from natural sources. These three compounds are metabolically and functionally interrelated.

Pyridoxine is available from whole-grain cereals, peanuts, corn, meat, poultry, and fish. It is synthesized by plants and certain microorganisms from glycolysis intermediates, glyceraldehyde, dihydroxyacetone, and pyruvic acid.

In the human, pyridoxine first undergoes hepatic phosphorylation catalyzed by a specific kinase and then is oxidized to the aldehyde state by a flavoenzyme. Thus, the vitamin is the precursor necessary for the production of pyridoxal-5-phosphate, which is a coenzyme[42,43] that performs many vital functions in human metabolism. This coenzyme functions in transaminations and decarboxylations that amino acids generally undergo; e.g., it functions as a cotransaminase in the transamination of alanine to form pyruvic acid, as a codecarboxylase in the decarboxylation of dopa to form dopamine, etc. Other biological transformations[42,43] of amino acids in which pyridoxal can function are racemization, elimination of the α-hydrogen together with a β-substituent (i.e., OH or SH) or a γ-substituent, and probably the reversible cleavage of β-hydroxyamino acids to glycine and carbonyl compounds.

An electromeric displacement of electrons from bonds a, b or c would result in the release of a cation (H, R′, or COOH) and subsequently lead to the variety of reactions observed with pyridoxal. The extent to which one of these displacements predominates over others depends on the structure of the amino acid and the environment (pH, solvent, catalysts, enzymes, etc.). When the above mechanism applies in vivo, the pyridoxal component is linked to the enzyme through the phosphate of the hydroxymethyl group.

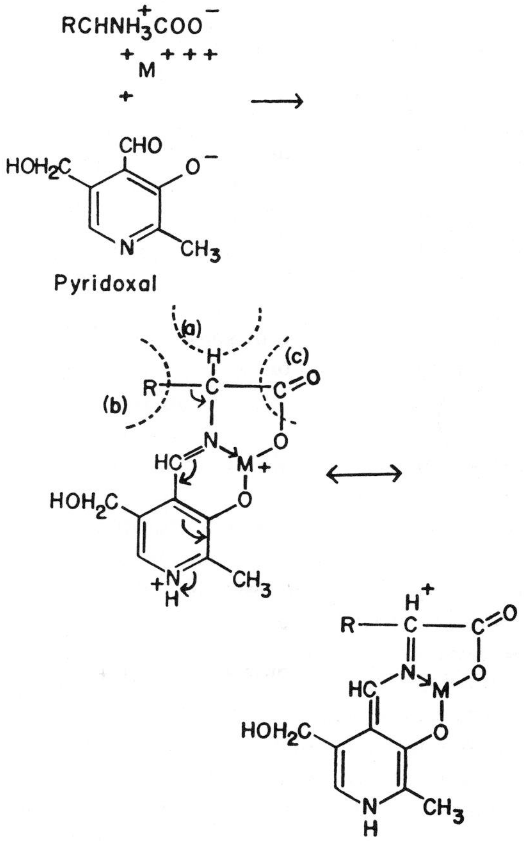

Metals such as iron and aluminum that markedly catalyze nonenzymatic transaminations in vitro probably do so by promoting the formation of the Schiff base and maintaining planarity of the conjugated system through chelate ring formation which requires the presence of the phenolic group. This chelated metal ion also provides an additional electron-attracting group that operates in the same direction as the heterocyclic nitrogen atom (or nitro group), thus increasing the electron displacements from the alpha carbon atom as shown here.

Pyridoxine metabolism in humans leads mainly to the corresponding carboxylic acid, 4-pyridoxic acid, which is the product from pyridoxal aldehyde oxidation.

4–Pyridoxic Acid

Because of inadequate diets some infants suffer from severe vitamin B_6 deficiencies that can lead to epilepticlike convulsive seizures, and the convulsions can be controlled by treatment with pyridoxine.[1] It is believed that the convulsions are due to a below-normal availability of the CNS neurohormone, γ-aminobutyric acid (GABA), from glutamic-acid decarboxylation, which is effected by the coenzyme pyridoxal-5-phosphate.[43a]

It also should be noted that certain hydrazine derivatives, when administered therapeutically (e.g., isoniazid), can induce a deficiency of the coenzyme (pyridoxal-5-phosphate) by inactivation through the mechanism of hydrazone formation with the aldehyde functional group.

Another hydrazine derivative, hydralazine, when administered in high doses to control hypertension, can cause similar B_6 deficiency, conceivably through a similar mechanism involving hydrazone formation.

Hypochromic anemias due to familial-type pyridoxine dependency respond to pyridoxine therapy. Similarly, this vitamin has been useful in the treatment of hypochromic or megaloblastic anemias that are not due to iron deficiency and do not respond to other hematopoietic agents.[2]

A review by Rose[44] summarizes studies on the effects of certain hormones on vitamin-B_6 nutrition in humans, on the biochemical interrelationship between steroid hormones and pyridoxal phosphate-dependent enzymes, and on the role of vitamin B_6 in regulating hypothalamus–pituitary functions. Some of these studies have important clinical implications that are noteworthy. The use of estrogen-containing oral contraceptives has been investigated as a factor leading to an abnormality of tryptophan metabolism. This abnormality resembles a dietary vitamin-B_6 deficiency and responds favorably to treatment with the vitamin. For some time there has been clinical interest in the relationship between certain hormones and vitamin-B_6 function, because abnormal urinary excretions of tryptophan metabolites were observed during pregnancy and in patients with hyperthyroidism.

Estrogens and tryptophan metabolism have been studied, because estrogen administration leads in some cases to the excretion of abnormally large amounts of xanthurenic acid, a metabolic product from tryptophan. This metabolic malfunction has been related to the inhibitory effect of estrogen sulfate conjugates on another pathway of tryptophan metabolism, the transamination of the kynurenine from tryptophan. Consequently, xanthurenic-acid formation appears to be abnormally increased owing to this estrogen effect or to B_6 deficiency.

In vitro studies have been conducted to determine the effect of estrogens on kynurenine aminotransferase, which catalyzes the B_6-dependent transamination of kynurenine to kynurenic acid. Some estrogen-ester conjugates, e.g., estradiol disulfate and diethylstilbestrol sulfate, interfere with this transamination, apparently by reversible inhibition of the aminotransferase apoenzyme. It appears that the estrogen sulfate competes with pyridoxal-5-phosphate for the interaction with the apoenzyme. In contrast, free estradiol and estrone do not possess this inhibitory property.

Some women suffer from mental depression when taking estrogen-containing oral contraceptives, and this depression could be due to another malfunction in tryptophan metabolism leading to 5-hydroxy tryptamine (serotonin). There is some evidence that the decarboxylation of 5-hydroxytryptophan is inhibited (in vitro) by estrogen conjugates competing with pyridoxal phosphate for the decarboxylase apoenzyme.

Other endocrine systems are interrelated. Both corticosteroids and thyroid hormones may increase the requirement for pyridoxine and affect pyridoxal-5-phosphate-dependent metabolic processes. Moreover, there appear to be associations between vitamin B_6 and anterior pituitary hormones. These associations seem to involve the hypothalamus, 5-hydroxytryptamine, and dopamine. The latter two neurotransmitters are synthesized by metabolic processes that require pyridoxal-5-phosphate.

Pyridoxine Hydrochloride U.S.P., 5-hydroxy-6-methyl-3,4-pyridinedimethanol hydrochloride (vitamin B_6 hydrochloride, rat antidermatitis factor). In 1935, P. Gyorgy showed that "rat pellagra" was not the same as human pellagra but that it resembled a particular disease of infancy known as "pink disease" or acrodynia. This "rat acrodynia" is characterized by a symmetric dermatosis affecting first the paws and the tips of the ears and the nose. These areas become swollen, red and edematous, with ulcers developing frequently around the snout and on the tongue. Thickening and scaling of the ears is noted, and there is a loss of weight, with fatalities occurring in from 1 to 3 weeks after the appearance of the symptoms. Gyorgy was able to cure the above conditions with a supplement obtained from yeast which he called "vitamin B_6." In 1938, this factor was isolated from rice paste and yeast in a crystalline form in a number of laboratories. A single dose of about 100 μg. produced healing in 14 days in a rat having severe vitamin-B_6-deficiency symptoms.

Chemical tests, electrometric titration determinations and absorption spectrum studies gave clues as to its composition. These were substantiated by the synthesis of vitamin B_6 (1938 and 1939).

Pyridoxine hydrochloride is a white, odorless, crystalline substance that is soluble 1:5 in water, and 1:100 in alcohol and insoluble in ether. It is relatively stable to light and air in the solid form and in acid solutions at a pH of not greater than 5, at which pH it can be autoclaved at 15 lbs. at 120°C for 20 to 30 minutes. Pyridoxine is unstable when irradiated in aqueous solutions at pH 6.8 or above. It is oxidized readily by hydrogen peroxide and other oxidizing agents. Pyridoxine is stable in mixed vitamin preparations to the same degree as riboflavin and nicotinic acid. A 1 percent aqueous solution has a pH of 3.

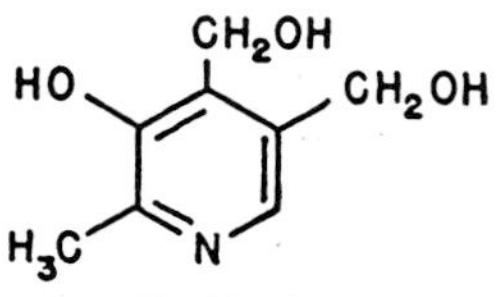

Pyridoxine

The pKa_1 values for pyridoxine, pyridoxal and pyridoxamine are 5.00, 4.22 and 3.40, respectively, and their pKa_2 values are 8.96, 8.68 and 8.05, respectively.

THE COBALAMINS

Vitamin B_{12} (cyanocobalamin) occurs in nature as a cofactor that originally was isolated as cyanocobala-min and vitamin B_{12b} (hydroxocobalamin, q.v.). In April 1948, Rickes et al.[45] isolated from clinically active liver fractions minute amounts of a red, crystalline compound that was also highly effective in promoting the growth of *L. lactis*. This compound was called vitamin B_{12} and in single doses, as small as 3 to 6 μg., it produced positive hematologic activity in patients having addisonian pernicious anemia. Evidence indicates that its activity is comparable with that of Castle's extrinsic factor and that it can be stored in liver.

Vitamin B_{12} is found in commercial fermentation processes of antibiotics, such as *Streptomyces griseus, S. olivaceus, S. aureofaciens,* sewage, milorganite, and others. Some of these fermentations furnish a commercial source of vitamin B_{12}.

In the biosynthesis of the coenzymes[46] derived from vitamin B_{12}, the cobalt is reduced from a trivalent to a monovalent state before the organic anionic ligands are attached to the structure. The two types of cobamides that participate as coenzymes in human metabolism are the adenosylcobamides and the methylcobamides. These coenzymes perform vital functions in methylmalonate-succinate isomerization and in methylation of homocysteine to methionine. Methylcobalamin is the major form of the coenzyme in the plasma, whereas 5-deoxyadenosylcobalamine is the major form in the liver and in other tissues. The enzyme system methylmalonyl-CoA mutase requires 5′-deoxyadenosylcobamide, and this enzyme system catalyzes the methylmalonyl-CoA transformation to succinyl CoA, which is the major pathway of propionyl-CoA metabolism. Propionyl CoA from lipid metabolism has to be processed through this pathway via succinyl CoA in order to enter the Krebs citric-acid cycle to be either converted to γ-oxaloacetate leading to gluconeogenesis or oxidized aerobically to CO_2 with production of ATP. The methylation of homocysteine to form methionine requires methylcobalamin, and it

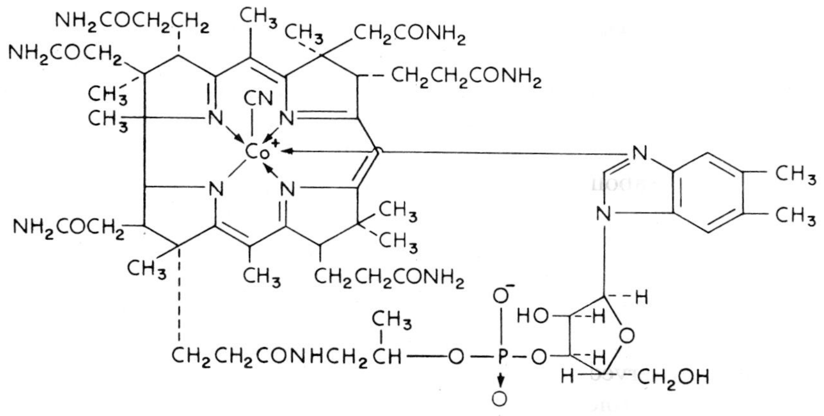

Cyanocobalamin

is catalyzed by a transmethylase that is also dependent on 5-methyltetrahydrofolic acid and reduced FAD.[46]

Herbert and Das[46] recently reviewed the biochemical roles of vitamin B_{12} and folic acid in hemato- and other cell poiesis; they emphasize that vitamin B_{12} and folic acid are essential for normal growth and proliferation of all human cells. It is further noted that vitamin B_{12} also has a function in the maintenance of myelin throughout the nervous system. It is known that deficiency of either vitamin leads to megaloblastic anemia involving below-normal DNA synthesis in all proliferating cells in the body. It has been postulated that vitamin-B_{12} deficiency is largely a conditioned folic-acid deficiency due to below-normal transformation of 5-methyl-THF to THF by the B_{12}-dependent homocysteine methyl-transferase reaction and defective cellular uptake of 5-methyl-THF in vitamin-B_{12} deficiency. This so-called *methyl-THF trap* hypothesis seems to rationalize a mechanism for the pathogenesis of megaloblastic anemia in vitamin-B_{12} deficiency, and recent investigations continue to provide evidence that supports this rationalization.

PRODUCTS

Cyanocobalamin U.S.P., vitamin B_{12}, is a cobalt-containing substance usually produced by the growth of suitable organisms or obtained from liver. It occurs as dark red crystals or as an amorphous or crystalline powder. The anhydrous form is very hygroscopic and may absorb about 12 percent of water. One gram is soluble in about 80 ml. of water. It is soluble in alcohol but insoluble in chloroform and in ether.

Vitamin B_{12} loses about 1.5 percent of its activity per day when stored at room temperature in the presence of ascorbic acid; whereas, vitamin B_{12b} is very unstable (completely inactivated in one day). This loss in activity is accompanied by a release of cobalt and a disappearance of color. The greater stability of vitamin B_{12} is attributed to the increased strength of the bond between cobalt and the benzimidazole nitrogens by cyanide. Unusual resonance energy is imputed to the cobalt-cyanide complex, giving a positive charge to the cobalt atom and thereby strengthening the Co-N bond. The protective action of certain liver extracts of vitamin B_{12b} toward ascorbic acid and its sodium salt is, no doubt, due to the presence of copper and iron. Iron salts will protect vitamin B_{12b} in 0.001 percent concentration. Catalysis of the oxidative destruction of ascorbate by iron is well known. On exposure to air, liver extracts containing B_{12} lose most of the B_{12} activity in 3 months. The most favorable pH for a mixture of cyanocobalamin and ascorbic acid appears to be 6 to 7. Niacinamide can stabilize aqueous parenteral solutions of cyanocobalamin and folic acid at a pH of 6 to 6.5. However, it is unstable in B complex solution. Cyanocobalamin is stable in solutions of sorbitol and glycerin but not in dextrose or sucrose.

Aqueous solutions of vitamin B_{12} are stable to autoclaving for 15 minutes at 121°C. It is almost completely inactivated in 95 hours by 0.015 N sodium hydroxide or 0.01N hydrochloric acid. The optimum pH for the stability of cyanocobalamin is 4.5 to 5.0. Cyanocobalamin is stable in a wide variety of solvents.

Hydroxocobalamin U.S.P., cobinamide, dihydroxide, dihydrogen phosphate (ester), mono(inner salt), 3'-ester with 5,6-dimethyl-1-α-D-ribofuranosylbenzimidazole, vitamin B_{12b}, is cyanocobalamin in which the CN group is replaced by an OH group. It occurs as dark red crystals or as a red crystalline powder that is sparingly soluble in water or alcohol and practically insoluble in the usual organic solvents.

Under the usual conditions, in the absence of cyanide ions only the hydroxo form of cobalamin is isolated from natural sources. It has good depot properties but is less stable than cyanocobalamin.

Cyanocobalamin Co 57 Capsules U.S.P. contain cyanocobalamin in which some of the molecules contain radioactive cobalt (Co 57). Each μg. of this cyanocobalamin preparation has a specific activity of not less than 0.5 microcurie.

The *U.S.P.* cautions that in making dosage calculations one should correct for radioactive decay. The radioactive half-life of Co 57 is 270 days.

Cyanocobalamin Co 57 Solution U.S.P. has the same potency, dosage and use as described under Cyanocobalamin Co 57 Capsules U.S.P. It is a clear, colorless to pink solution that has a pH range of 4.0 to 5.5.

Cyanocobalamin Co 60 Capsules U.S.P. is the counterpart of Cyanocobalamin Co 57 Capsules in potency, dosage and use. It differs only in its radioactive half-life, which is 5.27 years.

Cyanocobalamin Co 60 Solution U.S.P. has the same potency, dosage and use as Cyanocobalamin Co 60 Capsules. It is a clear, colorless to pink solution that has a pH range of 4.0 to 5.5.

The above four preparations must be labeled "Caution—Radioactive Material" and "Do not use after 6 months from date of standardization."

Cobalamin Concentrate U.S.P., derived from Streptomyces cultures or other cobalamin-producing microorganisms, contains 500 μg. of cobalamin per g. of concentrate.

A cyanocobalamin zinc tannate complex can be used as a repository form for the slow release of cyanocobalamin when it is administered by injection.

FOLIC ACID

In the early 1940's R. J. Williams *et al.*[47] reported the term *folic acid* in referring to a vitamin occurring in leaves and foliage of spinach. Since then, folic acid has been found in whey, mushrooms, liver, yeast, bone marrow, soybeans, and fish meal. The structure (see below) has been proved by synthesis in many laboratories; *e.g.*, see Waller *et al.*[48]

Folic Acid

Folic acid is a pteridine derivative (rings A and B constitute the pteridine heterocyclic system) synthesized by bacteria from guanosine triphosphate, *p*-aminobenzoic acid and glutamic acid. Accordingly, the structure of folic acid is composed of three moieties: the pteridine moiety derived from guanosine triphosphate, the *p*-aminobenzoic-acid and glutamic-acid moieties. (It is interesting to relate that antibacterial sulfonamides (see Chap. 5) compete with *p*-aminobenzoic acid and thus interfere with bacterial folic-acid synthesis.) Of course, the human is not able to synthesize folic acid.

In the human, dietary folic acid must be reduced metabolically to tetrahydrofolic acid (THF) in order to exert its vital biochemical actions. This reduction, which proceeds through the intermediate dihydrofolic acid, is catalyzed by a reductase. This reductase enzyme system has been implicated as the catalyst in both reaction steps, folic-acid reduction and dihydrofolic-acid reduction. The coenzyme THF is converted to other cofactors by formulation of the N-10 nitrogen. These coenzymes, derived from folic acid, participate in the biosynthesis of nucleic acids by performing essential functions as formyl-group carriers in several "one-carbon" transfers. The most critical "one-carbon" transfer that is involved in DNA synthesis requires N^5, N^{10}-methylene-THF as the methylating coenzyme responsible for converting uridylic acid to thimydilic acid. It is interesting to note that some folic-acid antagonists useful in cancer chemotherapy (e.g., methotrexate, see Chap. 8) interfere with DNA synthesis by inhibiting this methylation step.[49]

There is a fundamental relationship between folic-acid metabolism and vitamin B_{12}. The reduction of methylene-THF to 5-methyl-THF is essentially irreversible, hence there is only one pathway for the regeneration of THF from 5-methyl-THF. The THF is regenerated by the B_{12}-dependent methyl-group transfer from 5-methyl-THF to homocysteine. This biochemical interrelationship has been implicated in the etiology of megaloblastic anemia. (See corresponding discussion under vitamin B_{12}.[46])

Folic Acid U.S.P., Folacine®, Folvite®, N-{[(2-amino-4-hydroxy-6-pteridinyl)methyl]amino}benzoyl-glutamic acid, pteroylglutamic acid. Folic acid occurs as a yellow or yellowish-orange powder that is only slightly soluble in water (1 mg. per 100 ml.). It is insoluble in the common organic solvents. The sodium salt is soluble (1:66) in water.

Aqueous solutions of folic acid or its sodium salt are stable to oxygen of the air, even upon prolonged standing. These solutions can be sterilized by autoclaving at a pressure of 15 pounds per square inch in the usual manner. Folic acid in the dry state and in very dilute solutions is decomposed readily by sunlight or ultraviolet light. Although folic acid is unstable in acid solutions, particularly below a pH of 6, the presence of liver extracts has a stabilizing effect at lower pH levels than is otherwise possible. Iron salts do not materially affect the stability of folic-acid solutions. The water-soluble vitamins that have a deleterious effect on folic acid are listed in their descending order of effectiveness as follows: riboflavin, thiamine hydrochloride, ascorbic acid, niacinamide, pantothenic acid, and pyridoxine. This deleterious effect may be overcome, to a considerable degree, by the inclusion of approximately 70 percent of sugars in the mixture.

Folic acid in foods is destroyed more readily by cooking than are the other water-soluble vitamins. These losses range from 46 percent in halibut to 95 percent in pork chops and from 69 percent in cauliflower to 97 percent in carrots.

Leucovorin Calcium U.S.P., calcium folinate, calcium N-[*p*-{(-amino-5-formyl-5,6,7,8-tetrahydro-4-hydroxy-6-pteridinyl)methyl}-amino] benzoylglutamate, calcium 5-formyl-5,6,7,8-tetrahydrofolate, occurs as a yellowish-white or yellow, odorless, microcrystalline powder that is insoluble in alcohol and very soluble in water.

ASCORBIC ACID

The historical significance of vitamin C was so eloquently summarized by the eminent medicinal chemist and pharmacist, the late Professor Ole Gisvold, that the following direct quotation from the seventh edition of this textbook is an appropriate introduction to the significance of ascorbic acid in medicinal chemistry and basic biochemistry:

The disease scurvy, which now is known as a condition due to a deficiency of ascorbic acid in the diet, has considerable historical significance.[50] For example, in the war between Sweden and Russia (most likely the march of Charles XII into the Ukraine in the winter of 1708–1709) almost all of the soldiers of the Swedish army became incapacitated by scurvy. But further progress of the disease was stopped by a tea prepared from pine needles. The Iroquois Indians cured Jacques Cartier's men in the winter of 1535–1536 in Quebec by giving them a tea brewed from an evergreen tree. Many of Champlain's men died of scurvy when they wintered near the same place in 1608–1609. During the long siege of Leningrad, lack of vitamin C made itself particularly felt, and a decoction made from pine needles played an important role in the prevention of scurvy. It is somewhat common knowledge that sailors on long voyages at sea were subject to the ravages of scurvy. The British used supplies of limes to prevent this, and the sailors often were referred to as "limeys."

Holst and Frolich,[51] in 1907, first demonstrated that scurvy could be produced in guinea pigs. A comparable condition cannot be produced in rats.

Although Waugh and King[52] (1932) isolated crystalline vitamin C from lemon juice and showed it to be the antiscorbutic factor of lemon juice, Szent-Gyorgyi[53] had isolated the same substance from peppers in 1928, in connection with his biological oxidation–reduction studies. At the time, he failed to recognize its vitamin properties and reported it as a hexuronic acid because some of its properties resembled those of sugar acids. Hirst et al.,[54] suggested that the correct formula should be one of a series of possible tautomeric isomers and offered basic proof that the formula now generally accepted is correct. The first synthesis of L-ascorbic acid (vitamin C) was announced almost simultaneously by Haworth and Reichstein,[55] in 1933. Since that time, ascorbic acid has been synthesized in a number of different ways.

This vitamin is now better known as ascorbic acid because of its acidic character and its effectiveness in the treatment and prevention of scurvy. The acidic character is due to the two enolic hydroxyls; the C-3 hydroxyl has the pKa value of 4.1, and the C-2 hydroxyl has a pKa of 11.6. The monobasic sodium salt is the usual salt form, e.g., Sodium Ascorbate U.S.P.

Ascorbic Acid

It is well known that ascorbic acid can be synthesized by nearly all living organisms, plants, and animals; but primates, guinea pigs, bats, and some other species are not capable of producing this vitamin. The consensus is that organisms that cannot synthesize ascorbic acid lack the liver microsomal enzyme L-gulonolactone oxidase, which catalyzes the terminal step of the biosynthetic process. Sato and Udenfriend[56] summarized recent studies of the biosynthesis of ascorbic acid in mammals and the biochemical and genetic basis for the incapability of some species to synthesize the vitamin. Because people are one of the few animal species that cannot synthesize ascorbic acid, the vitamin has to be available as a dietary component.

Ascorbic acid performs important metabolic functions, as evidenced by the severe manifestations of its deficiency in humans. It has been demonstrated that this vitamin is involved in metabolic hydroxylations in numerous important metabolic processes. Ascorbic acid has also been implicated as an important factor in other critical oxidation–reduction processes in human metabolism.[1]

Although it is well known that ascorbic acid is an effective reducing agent and antioxidant, the biochemical functions of this vitamin are not well understood. It is controversial to consider ascorbic acid to be an antiviral agent, but some scientists still argue that ascorbic acid is an effective cure or preventative of "common colds."[57] A recent study provides some evidence that ascorbic acid appears to help the organism recover from viral infections through an indirect mechanism on the body's immune system.[57a] Ascorbic acid has also received attention as a possible anticancer agent. Recently, it has been demonstrated in cell-culture studies that ascorbic acid, both alone and in combination with copper ions, is selectively toxic to melanoma cancer cells.[57b]

Aqueous solutions of ascorbic acid are not very stable. The ascorbic acid in such preparations undergoes oxidation, particularly under aerobic conditions. Oxidation to dehydroascorbic acid is followed by hydrolytic cleavage of the lactone. The effect of pH on the aerobic degradation of ascorbic acid aqueous solutions has been studied by various investigators. Rogers and Yacomeni[57c] concluded that the degradation rate shows a maximum near pH 4 and a minimum near pH 5.6. It was also noted that if a preparation of ascorbic acid develops acidity on storage and if its initial pH is between 5 and 5.6, the rate of degradation will increase as the pH decreases; hence, an initial pH in the range of 5.6 to 6 is recommended.

Ascorbic Acid U.S.P., Cevitamic Acid®, Cebione®, vitamin C, L-ascorbic acid. Ascorbic acid occurs as white or slightly yellow crystals or powder. It is odorless, and on exposure to light it gradually darkens. One gram of ascorbic acid dissolves in about 3 ml. of water and in about 30 ml. of alcohol. A 1 percent aqueous solution has a pH of 2.7.

Sodium Ascorbate U.S.P. is a white, crystalline

powder that is soluble 1:1.3 in water and is insoluble in alcohol.

Ascorbic Acid Injection U.S.P. is a sterile solution of sodium ascorbate that has a pH of 5.5 to 7.0. It is prepared from ascorbic acid with the aid of sodium hydroxide, sodium carbonate, or sodium bicarbonate. It may be used for intravenous injection, whereas ascorbic acid is too acidic for this purpose.

Ascorbyl Palmitate N.F., ascorbic acid palmitate (ester), is the C-6 palmitic-acid ester of ascorbic acid. It occurs as a white to yellowish-white powder that is very slightly soluble in water and in vegetable oils. It is freely soluble in alcohol. Ascorbic acid has antioxidant properties and is a very effective synergist for the phenolic antioxidants such as propylgallate, hydroquinone, catechol, and nordihydroguaiaretic acid (NDGA) when they are used to inhibit oxidative rancidity in fats, oils, and other lipids. Long-chain, fatty-acid esters of ascorbic acid are more soluble and suitable for use with lipids than is ascorbic acid.

BIOTIN

Biotin was discovered, isolated, and identified structurally in the 1930's. Since then it has been noted that small amounts of biotin can be detected in almost all higher animals. The highest concentrations have been discovered in liver, kidney, eggs and yeast, as a water-insoluble complex. Considerable quantities are found both free and in the complex form in vegetables, grains, and nuts. Alfalfa, string beans, spinach, and grass are fair sources of this vitamin.

Microorganisms synthesize biotin from the fatty acid, oleic acid. The biosynthetic process involves numerous complex reactions that remain to be better understood. The final reaction step requires formation of the sulfur heterocycle, but the source of the sulfur is not yet known.

Although this vitamin is known to perform essential metabolic functions in the human, the minimal nutritional requirement has not been established because it has been difficult to quantify the amounts of the vitamin made available by intestinal microorganisms. Nevertheless, deficiency states may develop owing to prolonged feeding of large quantities of raw egg white. Raw egg white contains avidin, a protein that complexes biotin and minimizes its absorption from the gastrointestinal tract.

Biotin performs vital metabolic functions in important carboxylation processes in the form of carboxy-biotin, which is in combination with a carboxylase as represented below.

$$\text{Biotin-enzyme} + HCO_3^- \xrightleftharpoons{Mg^{++}} CO_2\text{-biotin enzyme} + ADP + Pi.$$

The oxygen for ATP cleavage is derived from bicarbonate and appears in the Pi.

CO$_2$-Biotin enzyme

Purified preparations of acetyl-CoA carboxylase contained biotin (1 mole of biotin per 350,000 g. of protein (enzyme)). It catalyzed the first step in palmitate synthesis as follows:

$$CH_3COSCoA + HCO_3^- + ATP \xrightleftharpoons{Mg^{++}} {}^-OOCCH_2COSCoA + ATP + Pi.$$

Other enzymes with which biotin appears to be intimately associated in carboxylation are *beta*-methylerotonyl-CoA carboxylase, propionyl-CoA carboxylase, pyruvate carboxylase and methylmalonyloxalacetic transcarboxylase.

Biotin also is joined in an amide linkage to the *epsilon* amino group of a lysine residue of carbamyl phosphate synthetase (CPS) to form biotin-CPS which participates with 2 ATP, HCO$_3^-$ and glutamine in the synthesis of carbamyl phosphate. This takes place stepwise as follows:

(1) Biotin CPS + ATP + HCO$_3^-$ ⇌ carbonic phosphoric anhydride biotin CPS (CPA biotin CPS) + ADP;

(2) CPA biotin CPS ⇌ −OOC biotin CPS + Pi;

(3) −OOC biotin CPS + glutamine ⇌ H$_2$NOC biotin CPS + ATP ⇌ biotin CPS + carbamyl-phosphate + ADP.

Carbamyl phosphate can participate in amino acid metabolism and some nucleic acid syntheses.

MISCELLANEOUS CONSIDERATIONS

Some dietary components are difficult to characterize as essential nutritional factors in human metabolism because the organism has the necessary chemistry to produce these compounds from other dietary components. (Consider vitamin D and nicotinic acid, which have been discussed above.) Vitamin D and nicotinic acid are, however, generally considered among the classical vitamins. Moreover, there is no clear consensus on the necessity for inositol, choline, and *p*-aminobenzoic acid. Nevertheless, it is well known that such dietary components do perform important metabolic functions, hence a brief characterization of these should be noted.

Inositol, 1,2,3,5/4,6-cyclohexanehexol, *i*-inositol, *meso*-inositol (myo-inositol) (mouse anti-alopecia factor), is prepared from natural sources, such as corn steep liquors, and is available in limited commercial quantities. It is a white, crystalline powder and is soluble in water 1:6 and in dilute alcohol. It is slightly soluble in alcohol, the usual organic solvents and in fixed oils. It is stable under normal storage conditions.

Inositol is one of nine different *cis-trans* isomers of hexahydroxycyclohexane and usually is assigned the following configuration:

Inositol

Inositol has been found in most plants and animal tissues. It has been isolated from cereal grains, other plant parts, eggs, blood, milk, liver, brain, kidney, heart muscle and other sources. The concentration of inositol in leaves reaches a maximum shortly before the time that the fruit ripens. Good sources of this factor are fruits, especially citrus fruits,[58] and cereal grains.

Inositol occurs free and combined in nature. In plants, it is present chiefly as the well-known phytic acid which is inositol hexaphosphate. It is also present in the phosphatide fraction of soybean as a glycoside. In animals, much of it occurs free.[59]

Inositol in the form of phosphoinositides is almost as widely distributed as inositol, and these forms are, in some cases, more active metabolically. Phosphatidylinositol (monophosphoinositide) is the most widely distributed of the inositides and the chief fatty acid residue is stearic acid.

Phosphatidylinositol

Inositol lipids occur in all mammalian tissues which have been investigated, and phosphatidylinositol is present in many tissues as about 2 to 8 percent of the total lipid phosphorus.

The phosphoinositides may be involved in the transport of certain cations and have as yet undetermined functions.

Eastcott,[60] in 1928, showed that bios I was the well-known, naturally occurring optically inactive inositol. In 1941, Woolley[61] showed that mice maintained on an inositol-deficient diet ceased growing, lost their hair and finally developed a severe dermatitis. In rats, a denudation about the eyes, called "spectacle eye," takes place in the absence of inositol in the diet. These symptoms also are accompanied by the development of a special type of fatty liver containing large amounts of cholesterol.

Inositol has been shown to be a growth factor for a wide variety of human cell lines in tissue culture. It is considered a characteristic component of seminal fluid, and the content is an index of the secretory activity of the seminal vesicles.

Evidence is accumulating to indicate that inositol will reduce elevated blood cholesterol levels. This, in turn, may prevent or mitigate cholesterol depositions in the intima of blood vessels in humans and animals and, therefore, be of value in atherosclerosis.

Inositol has also been considered as a lipotropic agent. Because the human can synthesize inositol, the need for inositol as a nutritional requirement has not been proved.[2]

Methionine U.S.P. An adequate diet should provide the methionine necessary for normal metabolism in the human. Methionine is considered to be an essential amino acid in people. It is the precursor in the biosynthesis of adenosyl methionine, which is an important methylating coenzyme involved in a variety of methylations, for example, in N-methylation of norepinephrine to form epinephrine and O-methylation of catecholamines catalyzed by catechol-O-methyltransferases. Adenosylmethionine also participates in the methylation of phosphatidylethanolamine to form phosphatidylcholine, but this pathway is not efficient enough to provide all the choline required by higher animals, hence adequate dietary availability of choline is necessary.[2]

Choline is a component of lipoproteins and a constituent of cell membranes. Choline is necessary for acetylcholine synthesis. The A.M.A. *Drug Evaluations*, 1977, notes that although choline has been studied as a lipotropic agent, it has not been definitively demonstrated to alleviate fatty infiltration of the liver, cirrhosis, in humans.

p-**Aminobenzoic Acid U.S.P. (PABA)** has been mentioned as a biosynthetic component of folic acid in bacteria, but it is well known that higher mammalian organisms cannot synthesize folic acid from its precursors. Nevertheless, it seems that PABA performs certain metabolic functions in some animals. In the early

1950's PABA was reported to be an essential factor in the normal growth and life of the chick.

Since these original developments in this field, various claims[62] have been made for the chromotrichial value of p-aminobenzoic acid in rats, mice, chicks, minks and humans. The problem of nutritional achromotrichia is a complex one that may involve several vitamin or vitamin-like factors and is complicated by the synthesis and absorption from the intestinal tract of a number of factors produced by bacteria.

PABA is a white, crystalline substance that occurs widely over the plant and animal kingdom. It occurs both free and combined[63] and has been isolated[64] from yeast, of which it is a natural constituent. It is soluble 1:170 in water, 1:8 in alcohol and freely soluble in alkali.

p-Aminobenzoic acid is thought to play a role in melanin formation and to influence or catalyze tyrosine activity.[65] It inhibits oxidative destruction of epinephrine and stilbestrol, counteracts the graying of fur attributable to hydroquinone in cats and mice, exhibits antisulfanilamide activity and counteracts the toxic effects of carbarsone and other pentavalent phenylarsonates.[66]

PABA when given either parenterally or in the diet to experimental animals, will protect them against otherwise fatal infections of epidemic or murine typhus, Rocky Mountain spotted fever and tsutsugamushi disease.[67] These diseases have been treated clinically with most encouraging results by maintaining blood levels of 10 to 20 mg. percent for Rocky Mountain spotted fever and tsutsugamushi diseases. The mode of action of p-aminobenzoic acid in the treatment of the above diseases appears to be rickettsiostatic rather than rickettsicidal, and the immunity mechanisms of the host finally overcome the infection.

PABA appears to function as a coenzyme in the conversion of certain precursors to purines.[68] p-Aminobenzoic acid has been suggested as an effective sun screen as a 5 percent solution in 55 to 75 percent ethyl alcohol on excessive sunlight-exposed areas of the skin.[69]

The historical significance of the effect of PABA on the antimicrobial action of sulfonamides and sulfones has been reviewed by Anand.[70]

REFERENCES

1. White, A., et al.: Principles of Biochemistry, 6th ed., p. 1320, 1333, 1362, New York, McGraw-Hill, 1978.
2. American Medical Association: Drug Evaluations, 3rd ed., p. 175, Littleton, Mass., 1977.
3. McCollum, E. V., and Davis, M.: J. Biol Chem. 19:245, 1914; 21:179, 1915.
4. De Luca, L. M.: Vitamins and Hormones 35:1, 1977.
5. Chytil, F., and Ong, D. E.: Vitamins and Hormones 36:1, 1978.
6. Robeson, C. D., et al.: J. Am. Chem. Soc. 77:4111, 1955.
7. Snell, E. E., et al.: J. Am. Chem. Soc. 77:4134, 4136, 1955.
8. Hanze, A. R., et al.: J. Am. Chem Soc. 68:1389, 1946. Milas, N. A.: U. S. Patents 2,369,156, 2,369,168, 2,382,085, 2,382,086; Isler, O., et al.: Experientia 2:31, 1946; Karrer, P., et al.: Helv. Chim. acta 29:704, 1946; Milas, N. A., and Harrington, T. M.: J. Am. Chem. Soc. 69:2248, 1947; Oroshnik, W.: J. Am. Chem. Soc. 67:1627, 1945; Isler, O., et al.: Helv. Chim. acta 30:1911, 1947.
9. Milas, N. A.: Science 103:581, 1946.
10. Arens, J. F., and van Dorp, D. A.: Nature (London) 157:190, 1946. ———: Rec. trav. chim. 65:338, 1946.
11. Isler, O., et al.: Hev. Chim acta 30:1911, 1947.
12. Carr, F. H., and Price, E. A.: Biochem. J. 20:497, 1926.
13. Goodman, P. S., et al.: J. Biol. Chem. 242:3543, 1967.
14. Pett, L. B.: J. Lab. Clin. Med. 25:149, 1939; Hecht, S., and Mandelbaum, J.: J.A.M.A. 112:1910, 1939.
15. McLean, F., and Budy, A.: Vitamins and Hormones 21:51, 1963.
16. Green, J.: Vitamins and Hormones 20:485, 1962.
17. Kliger, D. S., and Menger, E. L.: Acc. Chem. Res. 8:81, 1975.
18. Hubbell, W. L.: Acc. Chem. Res. 8:85, 1975.
19. Honig, B., et al.: Acc. Chem. Res. 8:92, 1975.
20. Abrahamson, E. W.: Acc. Chem. Res. 8:101, 1975.
21. Williams, T. P.: Acc. Chem. Res. 8:107, 1975.
22. Hecht, S.: Am. Sci. 32:159, 1944.
23. White, A., et al.: Principles of Biochemistry, 6th ed., pp. 1173–1178, New York, McGraw-Hill, 1978.
24. DeLuca, H. F.: J. Lab. Clin. Med. 87:7, 1976.
25. Suda, T., et al.: Biochem. 9:1651, 1970.
26. Geisman, T. A., and Crout, D.H.G.: Organic Chemistry of Secondary Plant Metabolism, p. 292, San Francisco, Freeman-Cooper Co., 1969.
27. Stern, M. A., et al.: J. Am. Chem. Soc. 69:869, 1947.
28. Weber, F., et al.: Biochem. Biophys. Res. Commun. 14:186, 1964.
29. Demole, V., et al.: Helv. chim. acta 22:65, 1939.
30. Karrer, P., and Bussmann, G.: Helv. chim. acta 23:1137, 1940.
31. Bieri, J. G., and Farrell, P. M.: Vitamins and Hormones 34:31, 1976.
32. American Medical Association: Drug Evaluations, 3rd ed., pp. 186–187, Littleton, Mass., 1977.
33. Williams, R. J.: Nutrition Against Disease, p. 148, New York, Pitman Publ. Corp., 1971; see also Williams, R. J., and Kalita, D. K.: A Physician's Handbook on Orthomolecular Medicine, p. 64, New Canaan, Conn., Keats Publ., Inc., 1977; and for a general review see Williams, R. J.: Physician's Handbook of Nutritional Science, Springfield, Ill., Charles C Thomas, 1978.
34. Gryglewski, R. J.: Trends in Pharmacol. Sciences 1:164, 1980.
35. Geisman, T. A., and Crout, D.H.G.: Organic Chemistry of Secondary Plant Metabolism, p. 291–311, San Francisco, Freeman-Cooper Co., 1969.
36. Olson, R. E., and Suttie, J. W.: Vitamins and Hormones 35:59, 1977.
37. See Chemistry and Biochemistry of the K Vitamins, in Vitamins and Hormones 17:531, 1959.
38. Gyorgy, P.: Vitamins and Hormones 20:600, 1962.
39. Fieser, L. F.: J. Biol. Chem. 133:391; 1940.
40. White, A., et al.: Principles of Biochemistry, 6th ed., p. 1334, New York, McGraw-Hill, 1978.
41. King, T. E., and Strong, F. M.: Science 112:562, 1950.
42. Braunstein, A. E.: Enzymes 2:115, 1960.
43. Ikawa, K., and Snell, E.: J. Am. Chem. Soc. 76:637, 1954.
43a. Isaacson, E. I., and Delgado, J. N.: Anticonvulsants, in Wolff, M. E. (ed.): Burger's Medicinal Chemistry, ed. 4, Part 3, New York, John Wiley & Sons, 1981.
44. Rose, D. P.: Vitamins and Hormones 36:53, 1978.
45. Rickes, E. L., et al.: Science 107:397, 1948; Smith: Nature (London) 162:144, 1948; Ellis, et al.: J. Pharm. Pharmacol. 1:60, 1949.
46. Herbert, V., and Das, K. C.: Vitamins and Hormones 34:1, 1976.
47. Williams, R. J., et al.: J. Am. Chem. Soc. 63:2284, 1941; 66:267, 1944.

48. Waller, C. W., *et al.*: J. Am. Chem. Soc. 70:19, 1948; see also Angier, R. G., *et al.*: Science 103, 667, 1946.
49. Montgomery, J. A.: Drugs for Neoplastic Diseases, *in* Wolff, M. E. (ed.): Burger's Medicinal Chemistry, ed. 4, pp. 602–604, New York, John Wiley & Sons, 1979.
50. Schick, B.: Science 98:325, 1943.
51. Holst, A., and Frolich, T.: J. Hyg. 7:634, 1907.
52. Waugh, W. A., and King, C. C.: Science 75:357, 630, 1932, and J. Biol. Chem. 97:325, 1932; Svirbely, J. L., and Szent-Gyorgyi, A.: Nature (London) 129:576, 609, 1932, and Biochem. J. 26:865, 1932, and 27:279, 1933; Tillmans, J. *et al.*: Biochem. Z. 250:312, 1932.
53. Szent-Gyorgyi, A.: Biochem. J. 22:1387, 1928.
54. Hirst, E. L., *et al.*: J. Soc. Chem. Ind. 2:221, 482, 1933; Cox, E. G., and Goodwin, T. H.: J. Chem. Soc., 769, 1936, and Nature 130:88, 1932.
55. Haworth, W. N., *et al.*: J. Chem. Soc., 1419, 1933; Reichstein, T.: Helv. Chim. Acta 16:1019, 1933.
56. Sato, P., and Udenfriend, S.: Vitamins and Hormones 36:33, 1978.
57. Williams, R. J.: Physician's Handbook of Nutritional Science, p. 94, Springfield, IL, Charles C Thomas, 1978; Avery G. S.: Drug Treatment: Principles and Practice of Clinical Pharmacology and Therapeutics, p. 865, Littleton, MA, Publ. Sciences Group, Inc., 1976.
57a. Bram, S., *et al.*: Nature 284:629, 1980.
57b. Manzella, J. P., and Roberts, N. J.: J. Immunology 123:1940, 1979.
57c. Rogers, A. R., and Yacomeni, J. A.: J. Pharm Pharmacol. 23:2185, 1971.
58. Nelson, E. K., and Keenan, G. L.: Science 77:561, 1933.
59. Anderson, R. J., *et al.*: J. Biol. Chem. 125:299, 1938.
60. Eastcott, E. V.: J. Physiol. Chem. 28:1180, 1928.
61. Woolley, D. W.: Science 92:384, 1940, and J. Biol. Chem. 136:113, 1940; Martin, G. J., and Ansbacher, S.: Proc. Soc. Exp. Biol. Med. 48:118, 1941.
62. Emerson, G. A.: Proc. Soc. Exp. Biol. Med. 47:448, 1941.
63. Diamond, N. S.: Science 94:420, 1941.
64. Rubbo, S. D., and Gillespie, J. M.: Nature (London) 146:838, 1940.
65. Wisansky, W. A., *et al.*: J. Am. Chem. Soc. 63:1771, 1941.
66. Sandground, J. H., and Hamilton, C. R.: J. Pharmacol. Exp. Ther. 78:109, 1943.
67. Am. Prof. Pharm. 13:451, 1947.
68. Shive, W. *et al.*: J. Am. Chem. Soc. 69:725, 1947.
69. Avery, G. S.: Drug treatment: Principles and Practice of Clinical Pharmacology and Therapeutics, p. 345, Littleton, Mass., Publ. Sciences, Group, Inc., 1976.
70. Anand, N.: Sulfonamides and Sulfones, *in* (ed.): Burger's Medicinal Chemistry, 4th ed., Part 2, Wolff, M. E. p. 1, New York, John Wiley & Sons, 1979.

SELECTED READINGS

American Medical Association (Department of Drugs): AMA Drug Evaluations, Chap. 13 and 14, Littleton, MA, Publishing Sciences Group, 1977.
Harper, H. A.: Review of Physiological Chemistry, ed. 17, Chap. 7, Los Altos, CA, Lange Medical Publications, 1980.
Napoli, J.L., and DeLuca, H. F.: Blood calcium regulators, *in* Wolff, M. D. (ed.): Burger's Medicinal Chemistry, ed. 4, Chap. 26, p. 705, New York, John Wiley & Sons, 1979.
Sanders, H. J.: Nutrition and health, Chem. and Engin. News, Mar. 26:27, 1979.
Suttie, J. W. (ed.): Vitamin K Metabolism and Vitamin K-dependent Proteins, Baltimore, University Park Press, 1980.
Vitamins and Hormones, a series of annual volumes containing up-to-date reviews and current bibliographies, New York, Academic Press.
White, A., *et al.*: Principles of Biochemistry, ed. 6, Chap. 49, 50, 51, New York, McGraw-Hill, 1978.

22

miscellaneous organic pharmaceuticals

Robert F. Doerge

DIAGNOSTIC AGENTS

Diagnostic agents are used to detect impaired function of the body organs or to recognize abnormalities in tissue structure. Usually, these agents find no other use in medicine; however, a few are also valuable therapeutic agents. Factors that often determine the usefulness of a diagnostic agent are its solubility, mode and rate of excretion, metabolism, chemical configuration (e.g., color) and chemical composition (e.g., iodine).

Compounds used in diagnosis generally are divided into two classes. First, there are the many clinical diagnostic chemicals used to determine normal and pathologic products in urine, blood, feces and other body fluids or excrement. Also in the first group are the serologic solutions and tissue-staining dyes necessary in microscopic examination. Second, there is the group that is being discussed here, which finds application directly to or in the body and is most often described by the use of the term "diagnostic agent." These agents are conveniently arranged into three groups: (1) radiopaque substances; (2) compounds for testing functional capacity; (3) compounds modifying a physiologic action.

RADIOPAQUE DIAGNOSTIC AGENTS

Radiopaque diagnostic agents include both inorganic and organic compounds. These compounds have the property of casting a shadow on x-ray film and are also useful in fluoroscopic examination. Barium Sulfate U.S.P. is the only inorganic compound used. Suspensions are generally used in roentgenographic (x-ray) examination of the gastrointestinal tract (given orally or by enema). Barium Sulfate for Suspension U.S.P. may contain dispersing or suspending agents; in addition, coloring agents, flavors, fluidizing agents, and preservatives may be added. Organic iodinated compounds are usually considered more useful; they are more opaque and are used most in x-ray studies.

Iodine was observed to contribute opacity to x-ray in 1924 and was studied more fully by Binz, in 1935.[1] Useful iodinated compounds contain iodine in a strong covalent linkage and do not release iodide ions readily. However, their use in conditions of thyroid disease or tuberculosis should be with caution. The iodinated compounds are used primarily by two techniques: systemic and retrograde.

In the systemic procedure, the agent is administered orally or intravenously and is used to examine the kidney (urography) or liver (cholecystography). The contrast medium is used in the roentgenographic visualization of accessible parts of the body, such as renal cavities, ureters, biliary tract, blood vessels, the heart and the large vessels. The patient is given a preliminary test to determine individual sensitivity by instillation of a small amount into the conjunctival sac, then a cathartic is given the night before the injection and food and liquid are withheld for at least 18 hours previous to prevent blurring of the pictures. The solution is warmed to 98°F. and injected slowly into the vein; the patient is kept under careful observation. When renal functioning is normal, good exposures are obtained in from 5 to 15 minutes. Some iodinated compounds will concentrate in the kidney or the bladder and others in the liver or the gallbladder.

The retrograde method is the introduction of the diagnostic agent by mechanical means. An iodinated compound may be introduced into the urethra, the bladder, the vagina, the lower bowel, the ulcer area or varicose veins, for example. For retrograde pyelography, the solution is diluted with normal saline to about 15 percent and allowed to flow by gravity through a catheter that has been inserted into the ureteral orifice by means of a cystoscope, about 20 ml. being required. For the visualization of the blood vessels or the heart, a solution of special concentration (usually about 70%) is used, and the technique is more complicated. In all of these methods, mildly toxic reactions are quite frequent, but serious ones are encountered rarely, provided that the patient has been tested for susceptibility, that the injections are not repeated too often and that contraindicating diseases are not present. The most serious reactions are cyanosis and a fall in blood pressure, which lasts for less than 1 hour and can be overcome by epinephrine.

Requirements of a satisfactory radiopaque are as follows:

1. Adequate radiopacity. This usually requires an iodine content of 50 percent or more.
2. The solution should be capable of selective concentration in certain structures, such as gallbladder or kidney.
3. The solution should be retained in the area or organ long enough for x-ray visualization; then, it should be excreted rapidly with no toxic effects.
4. High solubility is desirable, often in the range of 40 percent.
5. It should be stable under the conditions of use (resist change in vivo) as well as during storage prior to use.
6. The compound should have a low toxicity, with a minimum of pharmacodynamic activity.

Contrast media may be divided arbitrarily into those which are water-soluble and those which are not. This serves to divide them also according to their general use. The water-soluble group is used mainly for urography and also for angiography. Angiography is the term generally used to define visualization of the blood vessels using a contrast medium. It is also used to designate visualization of the heart, lymph and bile ducts. The water-insoluble group is used mainly for cholecystography, with some use in bronchography and myelography.

WATER-SOLUBLE CONTRAST MEDIA

Iodohippurate Sodium I 131 Injection U.S.P., Sodium o-iodohippurate. This is a sterile solution containing sodium o-iodohippurate in which a portion of the molecules contains radioactive iodine ^{131}I in the structure.

Sodium Iodohippurate

Category—Diagnostic aid (renal function determination)..

Usual dose—renogram, I.V., the equivalent of 1 to 3 μCi; scanning, I.V., the equivalent of 200 to 300 μCi.

Diatrizoic Acid U.S.P., 3,5-*bis*(acetylamino)-2,4,6-triiodobenzoic acid. This is the parent acid for the sodium and meglumine salts.

Diatrizoic Acid

Diatrizoate Sodium U.S.P., Hypaque®, Sodium 3,5-*bis*(acetylamino)-2,4,6-triiodobenzoate. The sodium salt of diatrizoic acid is used because of its high water-solubility. A 50 percent aqueous solution is essentially neutral in reaction. The solutions may be sterilized by autoclaving but, in common with most iodinated compounds, should be protected from light. Diatrizoate Sodium Solution U.S.P. is for oral administration and many contain an antimicrobial agent. Diatrizoate Sodium Injection U.S.P. may be buffered and may contain edetate disodium or edetate calcium disodium as a chelating agent. If it is intended for intravascular administration, it must not contain an antimicrobial agent.

The 50 percent solution which is commonly used in urographic studies may become cloudy or form a precipitate when stored at low temperatures. When warmed to 25°C., the solution should be free of haze or crystals.

This diagnostic agent is also available with a coloring agent and a surfactant for making solutions for oral administration or to be given as an enema. This powder is not intended for use in preparing solutions for parenteral use.

A mixture of sodium and methylglucamine diatrizoate salts in various ratios is used in angiography and urography. At body temperature the solutions should be clear, but at room temperature or below crystals may form. These solutions are for use in cases which present difficult diagnostic problems, and by persons specially trained in their use.

Diatrizoate Meglumine U.S.P., Cardiografin®, Gastrografin®, Renografin®, Hypaque® Meglumine,

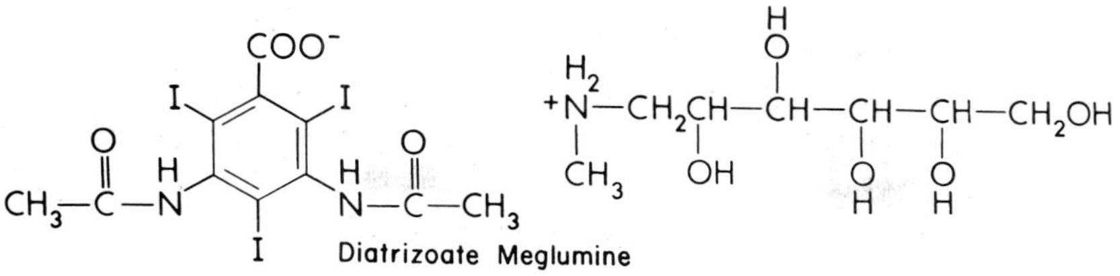

Diatrizoate Meglumine

is the N-methylglucamine salt of 3,5-diacetamido-2,4,-6-triiodobenzoic acid. The solutions that are available commercially usually contain a citrate buffer and a chelating agent. All solutions should be protected from light. At body temperature the solutions should be clear and free of any crystals.

The action and uses of this salt are similar to those of the sodium salt; however, the methylglucamine salt has the advantage of not introducing large amounts of the sodium ion into the bloodstream. It has been used most extensively for intravenous excretory urography, but it is also useful in visualization of the cardiovascular system. When given orally it is only slightly absorbed and may be used instead of suspensions of barium sulfate for visualization of the gastrointestinal tract.

Iothalamic Acid U.S.P., 3-(acetylamino)-2,4,6-triiodo-5-[(methylamino)carbonyl]benzoic acid. Iothalamic acid was synthesized as part of a research project directed toward the development of contrast agents with a higher water-solubility and lower incidence of toxic reactions than reported for known agents.[2] The N-methylcarbamoyl group replaces one of the acetamido groups of diatrizoic acid. Iothalamic acid is the parent acid for the preparation of the sodium and meglumine salts in the Iothalamate Sodium Injection U.S.P. and Iothalamate Meglumine Injection U.S.P., respectively.

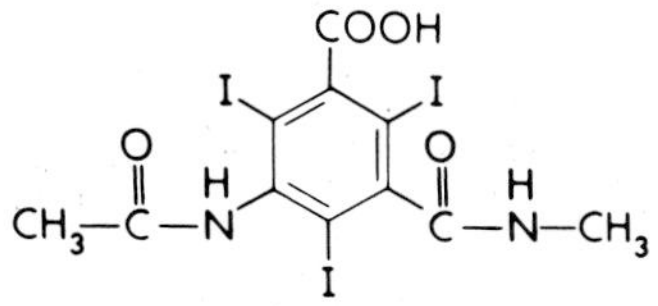

Iothalamic Acid

Iothalamate Sodium Injection U.S.P. is a 66.8 percent solution with a buffer and a chelating agent present. The injection is a clear, pale yellow, slightly viscous liquid with pH 6.8 to 7.5.

Iothalamate Meglumine Injection U.S.P. is the N-methylglucamine salt of iothalamic acid. Solutions of 30 and 60 percent concentrations are commercially available. They contain a phosphate buffer and a chelating agent. The solutions are sensitive to light and must be protected.

Iodipamide U.S.P., 3,3'-(adipoyldiimino)-*bis* [2,4,-6-triiodobenzoic acid.]

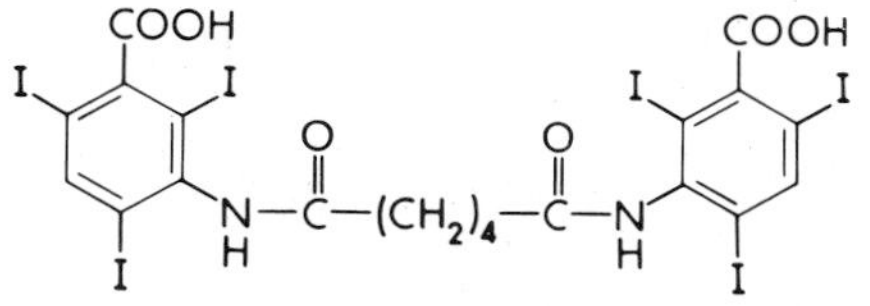

Iodipamide

Iodipamide is the parent acid for the preparation of the meglumine salt. The free acid has a pKa of 3.5. The meglumine salt is highly water-soluble with the usual concentration of the injection being 52 percent.

Iodipamide Meglumine Injection U.S.P., Cardiografin® Meglumine, with the increased molecular weight is excreted in the feces with only about 10 percent of a dose being excreted in the urine. This preparation is given intravenously to visualize the gallbladder and biliary ducts. It is used for patients who cannot tolerate oral products or intraductal injection. The injection may contain a chelating agent and a phosphate buffer.

Metrizamide, Amipaque®, 2-[3-acetamido-2,4,6-triiodo-5-(N-methylacetamido)-benzamido]-2-deoxy-D-glucopyranose is the amide of metrizoic acid and glucoseamine.

This highly water-soluble, nonionic radiopaque agent is an important advance in the development of contrast agents for use in myelography. It is supplied as a lyophilized powder that is reconstituted using a bicarbonate buffer solution just prior to use. A solution isotonic with cerebrospinal fluid contains 166 mg.

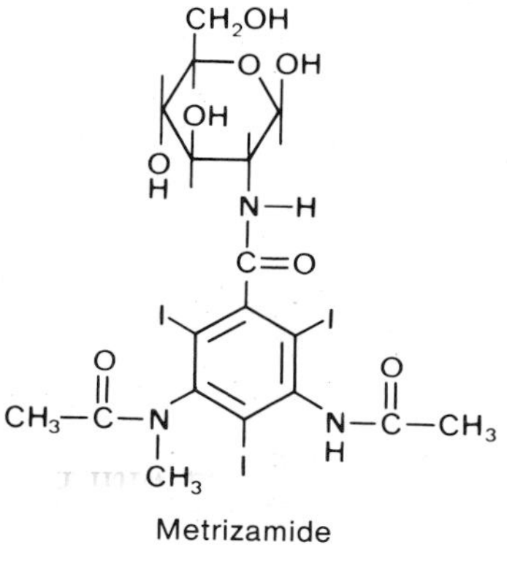

Metrizamide

of iodine per ml. The molecule contains 42 percent iodine. The solution is completely miscible with the cerebrospinal fluid and possesses radio-opacity adequate to demonstrate the spinal cord and nerve roots without being so radiopaque that it obscures details. The material does not have to be withdrawn from the subarachnoid space; 70 percent of the dose is absorbed into the bloodstream and excreted unchanged in the urine during the first 24 hours after lumbar injection.

Methiodal Sodium U.S.P., Skiodan® Sodium, is sodium monoiodomethanesulfonate, ICH_2SO_3Na, having an iodine content of about 52 percent. It is a white, crystalline, odorless powder which has a mild saline taste and a sweetish after-taste. It is soluble in water (7:10), forming a solution neutral to litmus (pH 6 to 8), and is only slightly soluble in alcohol. Solubility in organic solvents is negligible. The salt is prone to decompose in light, turning to a yellow color (iodine). Both the solid compound and its water solutions should be kept protected from light.

Methiodal sodium is useful both by intravenous injection and by retrography. After injection the urine concentration is 4 to 6 percent, and 75 percent is excreted in 3 hours.

Ipodate Sodium U.S.P., Oragrafin® Sodium, sodium 3-[[(dimethylamino)methylene]amino]-2,4,6-triiodohydrocinnamate, occurs as a water-soluble white to off-white odorless powder with a weakly bitter taste. This compound is stable in the dry form, and aqueous solutions are stable except at elevated temperatures. Both the dry material and aqueous solutions must be protected from light.

Ipodate Sodium

Sodium ipodate is given orally, as capsules, in cholecystography and in cholangiography. Maximal concentration in the hepatic and biliary ducts occurs in 1 to 3 hours in most patients and persists for about 45 minutes.

Ipodate Calcium U.S.P., Oragrafin® Calcium, calcium 3-[[(dimethylamino)methylene]amino]-2,4,6-triiodohydrocinnamate, differs from the sodium salt only in that it is almost insoluble in water and is supplied as granules with flavored sucrose. It may be administered by using an aqueous suspension.

WATER-INSOLUBLE CONTRAST MEDIA

Cholecystopexy is any gallbladder disease, and, to aid in diagnosing the disease, a compound is desirable that is opaque to x-rays and will be concentrated in vivo in the gallbladder and the bile duct. Usually, these agents are taken orally after a fat-free meal, and then some hours later (12) or the next day, with no other intake of food, the x-ray or fluoroscopic examination is made.

These diagnostic agents generally are only slightly soluble in water. They most often are used as the free organic acid. The formula below summarizes the structural modifications in this group.

PRODUCTS

Iopanoic Acid U.S.P., Telepaque®, is 3-amino-α-ethyl-2,4,6-triiodohydrocinnamic acid, a cream-colored solid which contains 66.68 percent iodine. It is insoluble in water but soluble in dilute alkali and 95 percent alcohol, as well as in other organic solvents.

Iopanoic Acid

In a study of derivatives, it was observed that the optimum visualization of the gallbladder was obtained when the number of carbon atoms in the alkanoic acid side chain approached five. Comparative studies with iodoalphionic acid showed iopanoic acid to be 1¼ times as effective. Also, it is about ¾ as toxic as iodoalphionic acid.

Iopanoic acid taken orally is well tolerated by the gastrointestinal tract and gives no impairment of hepatic or renal function. It is excreted in the feces and to a slight extent in the urine.

Propyliodone U.S.P., is propyl 3,5-diiodo-4-oxo-1(4H)pyridineacetate and is used as a sterile oil suspension for instillation into the trachea prior to bronchography. It occurs as a white, crystalline powder which is practically insoluble in water.

Propyliodone

Tyropanoate Sodium U.S.P., Bilopaque®, sodium 3-butyramido-α-ethyl-2,4,6-triiodohydrocinnamate.

Tyropanoate Sodium

This compound was introduced in the United States in 1972. It is an acylated iopanoic acid used as the sodium salt. It is used orally for cholecystography and cholangiography. Adverse reactions, nausea, vomiting and diarrhea, are infrequent. The usual oral dose is four 750-mg. capsules (Tyropanoate Sodium Capsules U.S.P.).

Iocetamic Acid U.S.P., Cholebrine®, N-acetyl-N-(3-aminotriiodophenyl)-β-aminoisobutyric acid.

Iocetamic Acid

Iocetamic acid is administered orally 10 to 15 hours before x-ray films are to be taken of the gallbladder. This compound, although it localizes in the biliary tract, is eliminated primarily via the renal route, with only a small proportion being eliminated in the feces. According to reports, the compound is well tolerated. The usual dose is four to six 750-mg. tablets.

Ethiodized Oil Injection U.S.P., Ethiodol®, is an iodine addition product of the ethyl esters of the fatty acids from poppyseed oil and contains about 37 percent of bound iodine. Since oleic acid is about 28 percent and linoleic acid is about 58 percent of the fatty acids derived from poppyseed oil, this would indicate that the main components of ethiodized oil are ethyl 9,10-diiodostearate and ethyl 9,10,12,13-tetraiodostearate. This product has about one-fifth the viscosity of the iodized oils (glyceryl esters) that were formerly used. This makes the injections easier to administer and makes the procedure more comfortable for the patient.

Iophendylate U.S.P., Pantopaque®, ethyl 10-(iodophenyl)undecanoate.

Iophendylate

It is a mixture of the isomers of ethyl iodophenylundecylate and consists chiefly of ethyl 10-(iodophenyl)undecanoate. It is a pale yellowish, odorless, viscous liquid. It is only slightly soluble in water but is fully soluble in most organic solvents.

AGENTS FOR KIDNEY FUNCTION TEST

Aminohippuric Acid U.S.P., ρ-aminohippuric acid. This is a white crystalline powder which discolors on exposure to light.

Aminohippuric Acid

It is soluble to the extent of 1 in 100 in water or alcohol, and is readily soluble in acids or bases with salt formation occurring.

Aminohippurate Sodium Injection U.S.P. is prepared by treating the free acid with an equivalent amount of sodium hydroxide and adjusting the pH to 7.0 to 7.2 with citric acid. This solution is used without isolating the sodium salt. The acid is prepared from p-nitrobenzoyl chloride and glycine. The p-nitro acid is isolated and reduced.

Solutions of sodium p-aminohippurate are sensitive to light.[3] The addition of 0.1 percent of sodium bisulfite markedly retards the darkening of solutions in ampules and prevents discoloration for at least 2 weeks in direct sunlight and 3 years in the dark or in diffused sunlight if the solution and ampules are nitrogen-purged before filling. Dextrose should not be included in the solutions.

The sodium salt is excreted by the tubular epithelium of the kidney and by the glomerulus, thus serving as a means for measuring the effective renal plasma flow and for determining the functional capacity of the tubular excretory mechanism.

Category—diagnostic aid (renal function determination).

Usual dose—I.V., 2 g.

Indigotindisulfonate Sodium U.S.P., sodium 5,5′-indigotindisulfonate, indigo carmine, occurs as a blue powder or crystal with a copper luster and is prepared from indigotin by sulfonation. This is an exam-

TABLE 22-1

RADIOPAQUE DIAGNOSTIC AGENTS

Name Proprietary Name	Preparations	Usual Adult Dose*	Usual Dose Range*	Usual Pediatric Dose*
Diatrizoate Sodium U.S.P. *Hypaque Sodium*	Diatrizoate Sodium Injection U.S.P.	Cholangiography—10 to 15 ml. of a 25 to 50 percent solution; excretory urography—I.V., 30 ml. of a 50 percent solution; retrograde pyelography—unilateral, 6 to 10 ml. of a 20 percent solution; hysterosalpingography—8 ml. of a 50 percent solution	Cholangiography—10 to 100 ml. of a 25 to 50 percent solution; excretory urography—20 to 60 ml. of a 50 percent solution; hysterosalpingography—6 to 10 ml. of a 50 percent solution; retrograde pyelography—unilateral, 6 to 15 ml. of a 20 percent solution	Excretory urography—Infants, 5 ml. of a 5 percent solution; children, 6 to 20 ml. of a 50 percent solution; retrograde urography—under 5 years of age; unilateral, 1.5 to 3.0 ml. of a 20 percent solution; over 5 years of age, unilateral, 4 to 5 ml. of a 20 percent solution
	Diatrizoate Sodium Solution U.S.P.	Gastrointestinal tract—90 to 180 ml. of a 25 to 41.7 percent solution		Gastrointestinal tract—30 to 75 ml. of a 20 to 41.7 percent solution
Diatrizoate Meglumine U.S.P. *Cardiografin, Gastrografin, Renografin, Hypaque Meglumine*	Diatrizoate Meglumine Injection U.S.P.	Angiocardiography—I.V. or intra-arterial, 25 to 50 ml. of a 76 to 85 percent solution; aortography—intra-arterial, 15 to 40 ml. of a 76 percent solution; cerebral angiography—intra-arterial, 10 ml. of a 60 percent solution; excretory urography—I.V., 20 to 60 ml. of a 60 to 76 percent solution; peripheral arteriography—intra-arterial, 10 to 40 ml. of a 60 to 76 percent solution; retrograde pyelography—unilateral, 15 ml. of a 30 percent solution; venography—I.V., 10 to 20 ml. of a 60 percent solution		Angiocardiography—under 5 years of age, 10 to 20 ml. of a 76 percent solution; 5 to 10 years of age, 20 to 30 ml. of a 76 percent solution; excretory urography—the following amounts of a 60 to 76 percent solution: under 6 months of age, 4 ml.; 6 to 12 months of age, 6 ml.; 1 to 2 years of age, 8 ml.; 2 to 5 years of age, 10 ml.; 5 to 7 years of age, 12 ml.; 8 to 10 years of age, 14 ml.; 11 to 15 years of age, 16 ml.
Iothalamic Acid U.S.P. *Angio-Conray, Conray-400*	Iothalamate Sodium Injection U.S.P.	Angiocardiography—intra-arterial or I.V., 40 to 50 ml. of a 66.8 percent solution; aortography—intra-arterial or I.V., the following amounts: I.V. aortography, 1 ml. per kg. of body weight, up to a maximum of 80 to 100 ml. of a 66.8 percent solution per injection; renal aortography, 10 to 25 ml. of a 66.8 percent solution; translumbar aortography, 20 ml. of a 66.8 percent solution; excretory urography—I.V., 25 ml. of a 66.8 percent solution	Angiocardiography—40 to 50 ml.; aortography—10 to 100 ml.; urography—25 to 60 ml.	Angiocardiography—intra-arterial or I.V., 0.5 to 1.0 ml. of a 66.8 percent solution per kg. of body weight; excretory urography—I.V., 0.5 ml. per kg.
Conray	Iothalamate Meglumine Injection U.S.P.	Cerebral angiography—intra-arterial, 6 to 10 ml. of a 60 percent solution; excretory urography—I.V., 30 ml. of a 60 percent solution; peripheral arteriography—intra-arterial, 20 to 40 ml. of a 60 percent solution; peripheral pyelography—I.V., 4.4 ml. of a 30 percent solution per kg. of body weight	Cerebral angiography—6 to 50 ml. of a 60 percent solution; excretory urography—25 to 60 ml. of a 60 percent solution; peripheral pyelography—up to 300 ml. of a 30 percent solution	0.5 ml. of a 60 percent solution per kg. of body weight
Iodipamide U.S.P. *Cholografin Meglumine*	Iodipamide Meglumine Injection U.S.P.	Cholangiography and cholecystography—I.V., 20 ml. over a period of 10 minutes. Do not repeat within 24 hours		0.3 to 0.6 ml. per kg. of body weight
Methiodal Sodium U.S.P. *Skiodan Sodium*	Methiodal Sodium Injection U.S.P.	I.V., 20 g. in 50 ml.	10 to 30 g.	
Ipodate Sodium U.S.P. *Oragrafin Sodium*	Ipodate Sodium Capsules U.S.P.	Cholecystography—3 g. 10 to 12 hours before examination	3 to 6 g.	

*See *U.S.P. D.I.* for complete dosage information.
(Continued)

TABLE 22–1

RADIOPAQUE DIAGNOSTIC AGENTS *(Continued)*

Name Proprietary Name	*Preparations*	*Usual Adult Dose**	*Usual Dose Range**	*Usual Pediatric Dose**
Ipodate Calcium U.S.P. *Oragrafin Calcium*	Ipodate Calcium for Oral Suspension U.S.P.	Cholecystography—3 g. 10 to 12 hours before examination	3 to 6 g.	
Iopanoic Acid U.S.P. *Telepaque*	Iopanoic Acid Tablets U.S.P.	Cholecystography—3 g.	3 to 6 g.	
Iocetamic Acid U.S.P. *Cholebrine*	Iocetamic Acid Tablets U.S.P.	Cholecystography—3 g.	3 to 6 g.	
Propyliodone U.S.P. *Dionosol Oil*	Sterile Propyliodone Oil Suspension U.S.P.	Bronchography—intratracheal 0.75 to 1 ml. of a 60 percent oil suspension for each year of age, up to a maximum of 12 to 18 ml.		
	Ethiodized Oil Injection U.S.P. *Ethiodol*	Hysterosalpingography—by special injection, initial, 5 ml. followed by increments of 2 ml. until tubal patency is established or patient's limit of tolerance is reached; lymphography—by special injection, lower extremity, 6 to 8 ml. per extremity, at a rate of 0.1 to 0.2 ml. per minute; upper extremity, 2 to 4 ml. per extremity, at a rate of 0.1 to 0.2 ml. per minute		Lymphography—1 ml. to a maximum of 6 ml.
Iophendylate U.S.P. *Pantopaque*	Iophendylate Injection U.S.P.	Myelography—intrathecal or by special injection, 3 to 12 ml.		

* See *U.S.P. D.I.* for complete dosage information.

ple of solubilizing a compound with sodium sulfonate groups. It is soluble in water (1:100), is slightly soluble in alcohol and almost insoluble in other organic solvents. It is affected by light, but its solutions may be sterilized by autoclaving.

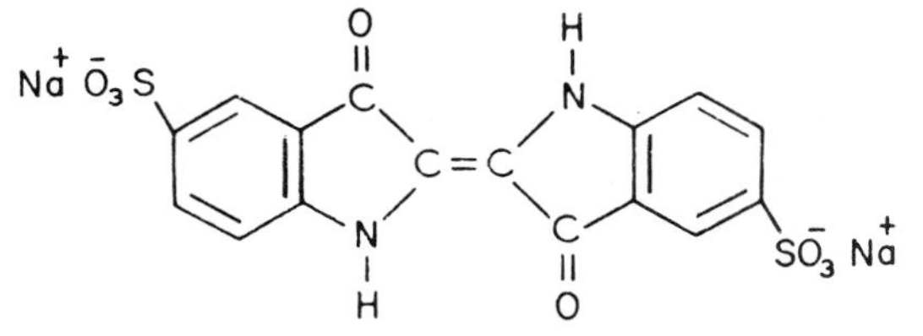

Indigotindisulfonate Sodium

The dye is used in the laboratory as a coloring agent, stain and reagent. It is used to determine renal function and to locate the ureteral orifices. Normally, it appears in the urine in 10 minutes, and about 10 percent of it is eliminated during the first hour.

Category—diagnostic acid (cystoscopy).

Usual dose—I.V., 40 mg.

Occurrence
Indigotindisulfonate Sodium Injection U.S.P.

Phenolsulfonphthalein U.S.P., α-hydroxy-α,α-*bis*(*p*-hydroxyphenyl)-*o*-toluenesulfonic acid γ-solutone, PSP, phenol red, is a red, crystalline powder that is stable in air. It is soluble in water (1:1,300), in alcohol (1:350) and almost insoluble in ether. It dissolves readily in bases. The compound may be considered as a derivative of phenolphthalein in which the CO group is replaced by an SO_2 group.

This compound, pKa 7.9, is used in the laboratory as an acid-base indicator using a 0.02 to 0.05 percent alcohol solution. At pH 6.8 it is yellow, and at pH 8.4 it is red. The dye is employed medicinally as a diagnostic agent for determining renal function. For this purpose, the monosodium salt is injected intrave-

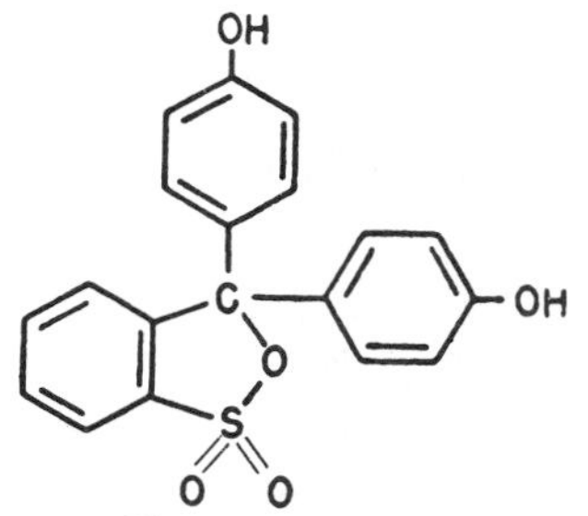

Phenolsulfonphthalein

nously or intramuscularly, and the amount of phenolsulfonphthalein excreted in the urine is measured quantitatively. When kidney function is normal, the dye is excreted in a shorter time interval than when kidney function is impaired.

Category—diagnostic aid (renal function).

Usual dose—I.M. or I.V., 6 mg.

Occurrence

Phenolsulfonphthalein Injection U.S.P.

AGENTS FOR LIVER FUNCTION TEST

Sulfobromophthalein Sodium U.S.P., Bromsulphalein® Sodium, disodium 4,5,6,7-tetrabromo-3′,3″-disulfophenolphthalein, disodium phenoltetrabromophthalein disulfonate, is a white, crystalline, hygroscopic powder that has a bitter taste and is odorless. It is soluble in water but is insoluble in alcohol and acetone.

The bromine atoms in the compound cause it to be removed from the blood almost entirely by way of the liver. The introduction of sulfonic acid groups into compounds of this type increases the toxicity and greatly increases the water-solubility. The compound is injected intravenously, as a 5 percent solution, and the amount remaining in the blood after a certain time interval is determined colorimetrically. The rate at which the dye is removed from the blood is a measure of the hepatic function. The concentration of the dye in the bloodstream is measured at the end of one hour and at regular time intervals thereafter in order to determine the rate of clearance.

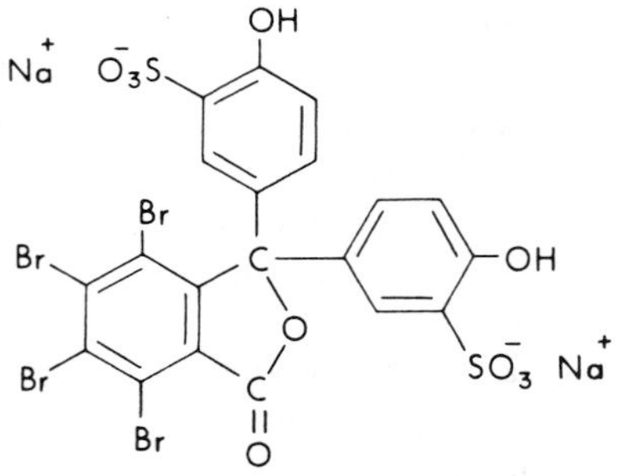

Sulfobromophthalein Sodium

Category (Injection)—diagnostic aid (hepato-biliary function determination).

Usual dose—I.V., 5 mg. per kg. of body weight not exceeding 500 mg.

Usual dose range—2 to 5 mg. per kg.

Occurrence

Sulfobromophthalein Sodium Injection U.S.P.

Rose Bengal, tetraiodotetrachlorofluorescein, is made by reacting tetrachlorophthalic anhydride with resorcinol and iodinating the resulting product. It is used as a test for liver function. The liver almost exclusively removes the dye from the bloodstream. From 100 to 150 mg. of the dye is injected intravenously in sterile saline. A normally functioning liver will remove 50 percent of the dye within 2 minutes. The dye is photosensitive, so the dye, its solutions and the patients receiving it should be protected from light.

This compound is also available as ^{131}I-labeled tetraiodotetrachlorofluorescein in sterile, neutral solution. A small amount of the radioactive dye is injected intravenously, then the rates of clearance from the blood by the liver and excretion into the small intestine are determined. The clearance and excretion rates are determined using standard radioisotope counting equipment. The usual intravenous dose is the equivalent of 5 to 25 microcuries. The usual I.V. dose is the equivalent of 1 to 4 microcuries.

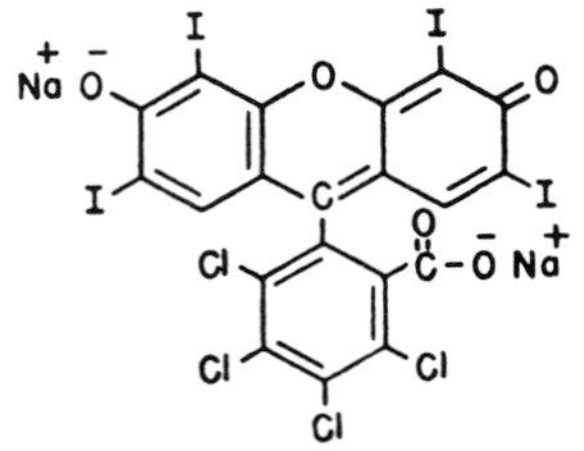

Rose Bengal

Occurrence

Rose Bengal Sodium I 131 Injection U.S.P.

MISCELLANEOUS DIAGNOSTIC AGENTS

Fluorescein Sodium U.S.P., resorcinolphthalein sodium, soluble fluorescein, is an orange, odorless, hygroscopic powder. It is soluble in water and sparingly soluble in alcohol.

The disodium salt forms highly fluorescent solutions when dissolved in water. The acidified solution has practically no fluorescence.

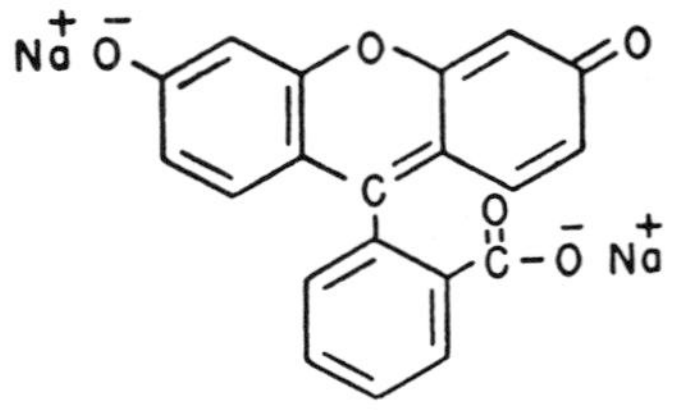

Fluorescein Sodium

Fluorescein sodium is used as an ophthalmologic diagnostic agent. For this purpose, an ophthalmic strip impregnated with the dye is used. Diseased or abraded areas of the cornea, such as corneal ulcers, are stained green by the solution. Foreign bodies appear with a green ring around them, while the normal cornea is not stained.

Fluorescein Sodium Injection U.S.P. is used to determine circulation time. The usual intravenous dose is 500 mg.; the usual dose range is 500 to 1250 mg.; and the usual intravenous pediatric dose is 15.4 mg. per kg. of body weight.

Occurrence
Fluorescein Sodium Ophthalmic Strips U.S.P.
Fluorescein Sodium Injection U.S.P.

Evans Blue U.S.P. is a complex azo dye known chemically as 4,4'-*bis*[7-(1-amino-8-hydroxy-2,4-disulfo)naphthylazo]-3,3'-bitolyl tetrasodium salt.

It exists as blue crystals having a bronze to green luster and is soluble in water, alcohol, acids and alkalies. The aqueous solutions are quite stable and may be autoclaved. Saline solutions are less stable and should not be autoclaved.

Evans Blue

Evans blue dye when injected into the bloodstream combines firmly with the plasma albumin. The color developed is directly proportional to its concentration. Spectral absorption is greatest at about 610 nm. where the photometric determination is made, and, by means of color intensity, the total blood volume may be found. This is used as a guide in replacement therapy, in shock and in hemorrhage.

Category—diagnostic aid (blood volume determination).

Usual dose—intravenous, the equivalent of 22.6 mg. of dried Evans Blue.

Occurrence	Percent Evans Blue
Evans Blue Injection U.S.P.	0.45

Indocyanine Green U.S.P., Cardio-Green®. This is a dark green to black powder which forms deep emerald-green solutions. The solutions are not stable over long periods, thus they must be made just prior to administration.

Sterile Indocyanine Green U.S.P. is indocyanine green suitable for parenteral use. The usual intravenous dose for cardiac output determination is 5 mg. in 1 ml., repeated as necessary, and for hepato-biliary function determination is 500 μg. per kg. of body weight. The usual dose range for cardiac output determination is 5 to 25 mg.; the total dose should be less than 2 mg. per kg. of body weight. The usual pediatric dose for cardiac output determination in infants is 1.25 mg. in 1 ml., and in children is 2.5 mg. in 1 ml., repeated as necessary.

Occurrence
Sterile Indocyanine Green U.S.P.

Chlormerodrin Hg 197 Injection U.S.P., Neohydrin-197®, and **Chlormerodrin Hg 203 Injection U.S.P.,** Neohydrin-203®, are available in sterile solution as radioactive diagnostic aids, employed for locating lesions of the brain and anatomic or functional defects of the kidneys. Mercury-197 has a shorter half-life (65 hours) than mercury-203 (46.6 days) and also a lower gamma radiation energy. Chlormerodrin Hg 203 delivers less than one-half the total body radiation of radioiodinated (^{131}I) serum albumin, and is said to be diagnostically superior in locating brain lesions. Chlormerodrin Hg 197 solution contains about 1,000 microcuries per ml.; chlormerodrin Hg 203 solution contains about 250 microcuries per ml.

Chlormerodrin Hg 197 Injection U.S.P. is used as a diagnostic aid in renal scanning with the usual intravenous dose being the equivalent of 100 to 150 microcuries. Chlormerodrin Hg 203 Injection U.S.P. is used as a diagnostic aid in tumor localization with the usual intravenous dose being 10 microcuries per kg. of body weight.

Azuresin, Diagnex® Blue, azure A carbacrylic resin, is a carbacrylic resin-dye combination that is used to diagnose achlorhydria. In the presence of acid in the gastric juice, the dye is released and absorbed from the upper intestine and then promptly excreted in the urine where it can be determined colorimetrically.

Each test unit contains two 250-mg. tablets of caffeine and sodium benzoate to be taken to stimulate gastric secretion. Histamine phosphate or betazole hydrochloride may be used in place of the caffeine and sodium benzoate.

Category—diagnostic aid (gastric secretion).

Usual dose—2 g. preceded by 500 mg. of caffeine and sodium benzoate.

Metyrapone U.S.P., Metopirone®, 2-methyl-1,2-di-3-pyridyl-1-propanone, occurs as a white to off-white crystalline powder. It has a characteristic odor. It should be protected from heat and light because of its low melting point and its light-sensitivity.

Metyrapone

Metyrapone possesses the property of selective inhibition in vivo of hydroxylation of the three principal adrenocorticoid hormones, hydrocortisone, corticosterone and aldosterone.[4] Thus, it finds use as a diagnostic tool to determine residual pituitary function in patients with hypopituitarism and, also, to evaluate a patient's ability to withstand surgery and other stresses.

Metyrapone is available as 250-mg. tablets of the base, and as ampules with each ml. containing 100 mg. of the bitartrate salt which is equivalent of 43.8 mg. of the base.

Category—diagnostic aid (hypothalamico-pituitary function determination).

Usual dose—750 mg. every 4 hours for six doses.

Usual pediatric dose—15 mg. per kg. of body weight every 4 hours for six doses.

Occurrence
Metyrapone Tablets U.S.P.
Metyrapone Tartate Injection U.S.P.

MISCELLANEOUS GASTROINTESTINAL AGENTS

This is a heterogeneous group of drugs, most of which are used as laxatives or cathartics. If properly used they serve a useful purpose in easing defecation in patients with hemorrhoids, hernias or hypertensive disorders. They are useful in emptying the lower intestinal tract before x-ray examination or surgery.

Mineral Oil U.S.P., liquid paraffin, white mineral oil, heavy liquid petrolatum. Mineral oil is a mixture of liquid hydrocarbons obtained from petroleum. The hydrocarbons usually present range in carbon content from C_{18} to C_{24}. Mineral oil has a specific gravity range of 0.845 to 0.905 and a kinematic viscosity at 40.0°C of not less than 34.5 centistokes. Heavy Russian-type mineral oils may have higher viscosities.

Although mineral oil is composed of hydrocarbons of marked stability, some oils, particularly those less highly refined, on exposure to light and air develop a kerosene odor and taste. This is believed to be due to peroxide formation. The *U.S.P.* allows the addition of an antioxidant to prevent peroxide formation. A concentration of 10 p.p.m. of *dl*-α-tocopherol may be used. There is no official test prescribed for measuring the stability of a mineral oil. A shelf-life test[5] has been developed based upon heating the oil for 2 to 15 minutes at 300° F. and testing for peroxide formation with an acetone solution of ferrous thiocyanate. Those oils that remain free of peroxide formation for 15 minutes have an estimated shelf-life of at least a year.

Mineral oil has been used widely as an intestinal lubricant and laxative and for softening the contents of the lower intestine in the treatment of hemorrhoids and other rectal disturbances. Oils of higher viscosity are desirable because they are less likely to leak out from the lower bowel. Petrolatum is sometimes added further to prevent such leakage. Mineral oil also has been used as a noncaloric oil in obesity diets. Mineral oil used near mealtime interferes with the absorption of vitamins A, D and K from the digestive tract[6] and, therefore, interferes with the utilization of calcium and phosphorus, leaving the user liable to deficiency diseases; when used during pregnancy it predisposes to hemorrhagic diseases of the newborn. Mineral oil should be prescribed for limited periods and be administered only at bedtime. A study has shown that mineral oil in doses up to 30 ml. taken at bedtime over long periods of time did not have any effect on the vitamin A concentration of the blood nor were any other deleterious effects noted. It should be given to infants only upon the advice of a physician. The usual dose is 15 to 60 ml. once daily preferably at bedtime.

Castor Oil U.S.P. is the fixed oil expressed from the seed of *Ricinus communis* Linné (Fam. *Euphorbiaceae*). Due to the presence of the glyceride of ricinoleic acid (80%), the oil is used as a laxative. It is the only fixed oil that is soluble in alcohol, so it is added to collodion to increase the flexibility. Solubility in alcohol is due to the presence of hydroxyl groups in the ricinolein.

Castor oil is quite different in solubility from other fatty oils.[7] It tends to dissolve in oxygenated solvents (alcohols) and be insoluble in hydrocarbon-type solvents (benzin), which is opposite to other vegetable oils. It is miscible with dehydrated alcohol, glacial acetic acid, chloroform or ether.

The usual dose range is 15 to 60 ml. The usual dose for infants is 1 to 5 ml., and for children is 5 to 15 ml. or 15 ml. per square meter of body surface.

Phenolphthalein U.S.P., 3,3-*bis*(p-hydroxyphenyl)phthalide, is a white or faintly yellowish-white, crystalline powder. It is soluble in alcohol (1:15), in ether (1:100) and in dilute bases but is almost insoluble in water.

Phenolphthalein

Phenolphthalein, in addition to being used as a laxative, is one of the most commonly used indicators for the titration of weak acids with alkali.

Phenolphthalein is used as a mild, tasteless laxative in the treatment of constipation. The colorless or almost colorless U.S.P. product has only about one-third the laxative action of yellow phenolphthalein, a more impure product. It was thought for some time that the greater laxative action of the yellow product was due to hydroxyanthraquinones which were presumed to have been formed during the synthesis. However, recent work has shown that hydroxyanthraquinones are not present in yellow phenolphthalein nor is the laxative action of Phenolphthalein U.S.P. increased by adding hydroxyanthraquinones to it.

It may be combined with other drugs, such as agar-agar or mineral oil. Phenolphthalein is not well absorbed from the intestinal tract and has a low toxicity. It is found in many of the commercial laxative preparations.

Category—cathartic.
Usual dose—60 mg.

Occurrence
Phenolphthalein Tablets U.S.P.

Bisacodyl U.S.P., Dulcolax®, is 4,4′-(2-pyridyl-methylene)diphenol diacetate (ester). It occurs as tasteless crystals which are practically insoluble in water and alkaline solutions. It is soluble in acids and organic solvents.

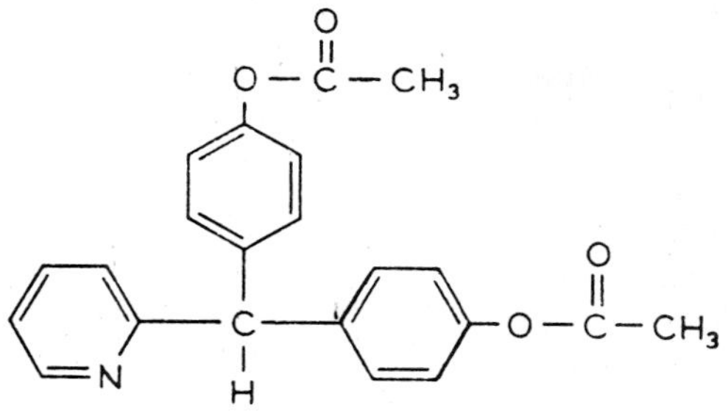

Bisacodyl

Bisacodyl appears to act directly on the colonic and rectal mucosa with little effect on the small intestine. It is recommended for use in constipation and in the preparation of patients for surgery or radiography. It is supplied as enteric-coated 5-mg. tablets and as 10-mg. suppositories which may be stored at normal room temperature.

The tablets must be swallowed whole, not chewed or crushed, and should not be taken within one hour of antacids. These precautions are necessary so that the enteric coating is not disturbed until after the drug leaves the stomach. If released in the stomach, the drug may cause vomiting.

Category—cathartic.
Usual dose—oral and rectal, 10 mg.
Usual dose range—oral, 10 to 15 mg. daily.

Usual pediatric dose—rectal, under 2 years of age, 5 mg., and over 2 years of age, 10 mg.; oral, 300 μg. per kg. of body weight or 8 mg. per square meter of body surface.

Occurrence
Bisacodyl Suppositories U.S.P.
Bisacodyl Tablets U.S.P.

Danthron U.S.P., Dorbane®, is 1,8-dihydroxyanthraquinone. It is structurally related to the anthraquinone derivatives found in Cascara sagrada and other vegetable cathartics.

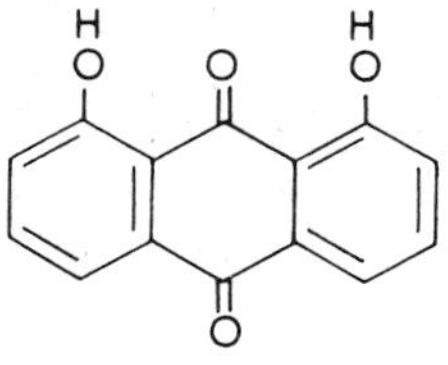

Danthron

Danthron is administered orally at bedtime. It is frequently used in combination with a fecal softening agent, dioctyl sodium sulfosuccinate.

Category—cathartic.
Usual dose—75 to 150 mg.

Occurrence
Danthron Tablets U.S.P.

Metoclopramide Hydrochloride, Reglan®, 4-amino-5-chloro-N-[(2-diethylamino)ethyl]-2-methoxy-benzamide hydrochloride. This agent is used by injection to increase the motility of the upper gastrointestinal tract. It is useful in facilitating intubation of the small bowel and in stimulating gastric emptying and intestinal transit of barium sulfate.

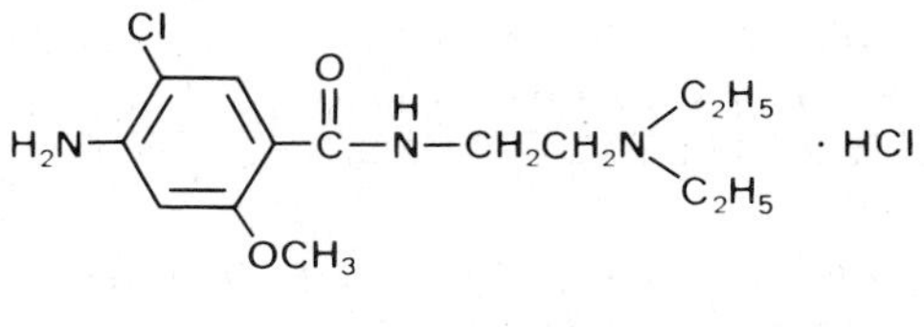

Metoclopramide Hydrochloride

Simethicone U.S.P., polydimethylsiloxane, is a mixture of linear siloxane polymers of the general formula below.

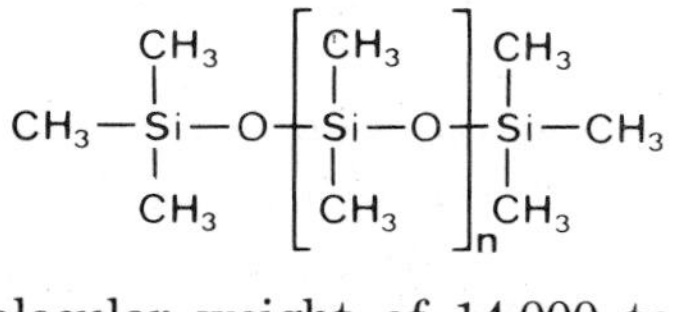

It has a molecular weight of 14,000 to 21,000. The clinical use is based on its antifoam properties. It forms a film of low surface tension and causes collapse of foam bubbles. It is physiologically inert, is not absorbed, does not interfere with gastric secretions or interfere with absorption of nutrients, and is excreted unchanged. It is used orally in the treatment of flatulence caused by entrapped gas in the stomach and gastrointestinal tract.

Usual adult dose—40 to 100 mg. four times daily.
Usual dose range—40 to 500 mg. daily.

Occurrence
Simethicone Oral Suspension U.S.P.
Simethicone Tablets U.S.P.

Docusate Sodium U.S.P., Colace®, Doxinate®, dioctyl sodium sulfosuccinate sodium 1,4-*bis*[2-ethylhexyl)sulfosuccinate, is a white, waxlike plastic solid with an odor suggestive of octyl alcohol. It is soluble in alcohol, glycerin, petroleum benzin, and other organic solvents. It dissolves slowly in water (1:70).

Docusate sodium is an excellent wetting agent, exceeding most other agents in this ability. An important therapeutic use is in the treatment of constipation and fecal impaction. It is relatively inert pharmacologically and is considered to act primarily through its surfactant action by permitting fluids to penetrate and soften the fecal mass.

Usual adult dose—50 to 100 mg. once or twice daily.
Usual dose range—50 to 500 mg. daily.
Usual pediatric dose—1.25 mg. per kg. of body weight or 37.5 mg. per square meter of body surface four times daily.

Occurrence
Docusate Sodium Capsules U.S.P.
Docusate Sodium Solution U.S.P.
Docusate Sodium Syrup U.S.P.
Docusate Sodium Tablets U.S.P.

Docusate Calcium U.S.P., Surfak®, calcium 1,4-*bis*(2-ethylhexyl) sulfosuccinate, is a white gelatinous solid, slightly soluble in water and freely soluble in alcohol and glycerin. It is similar in action to the sodium salt and is especially useful in conditions in which the sodium ion is contraindicated.

Usual dose—240 mg.

Occurrence
Docusate Calcium Capsules U.S.P.

ANTIRHEUMATIC GOLD COMPOUNDS

Gold and its compounds have been used since early times in the treatment of various diseases, including syphilis, tuberculosis and cancer. However, they have not proved to be effective therapeutic agents for these conditions. At present they are used for the treatment of lupus erythematosus and rheumatoid arthritis. Gold compounds are among the most toxic of all the metal compounds. Toxic manifestations involve skin, renal and hematologic reactions.

Gold Sodium Thiomalate U.S.P., Myochrysine®, (disodium mercaptosuccinato)-gold. This compound occurs as a white or yellowish-white powder that is almost insoluble in alcohol and ether but very soluble in water. A 5 percent aqueous solution has a pH of about 6. Solutions should be protected from light and not be used if they have darkened to more than a pale yellow. It is used for the treatment of rheumatoid arthritis.

Gold Sodium Thiomalate

Usual adult dose—I.M., initial, 10 mg. the first week, 25 mg. the second week, then 25 to 50 mg. once a week, up to a total dose of 1 g.; maintenance, 25 to 50 mg. every 2 weeks for 2 to 20 weeks, then 25 to 50 mg. every 3 to 4 weeks.
Usual pediatric dose—I.M., 1 mg. per kg. of body weight, not to exceed 50 mg. per dose.

Occurrence
Gold Sodium Thiomalate Injection U.S.P.

Aurothioglucose U.S.P., Solganal®, (1-thio-D-glucopyranosato)gold. Aurothioglucose is a water-soluble, oil-insoluble compound containing about 50 percent of gold.

It occurs as yellow crystals with a slight mercaptanlike odor. It decomposes in water solution, so is used as a suspension in an anhydrous vegetable oil.

Aurothioglucose

Usual adult dose—I.M., 10 mg. the first week, 25 mg. the second and third weeks, then 50 mg. once a week until a total dose of 800 mg. to 1 g. has been given; thereafter, 25 to 50 mg. every 3 to 4 weeks.
Usual adult limit—up to 50 mg. per week.
Usual pediatric dose—up to 6 years of age, dosage has not been established; 6 to 12 years, I.M. 2.5 mg. the first week, 6.25 mg. the second and third weeks, then 12.5 mg. every week until a total dose of 200 to

250 mg. has been given; thereafter, 6.25 to 12.5 mg. every 3 to 4 weeks.

Occurrence
Sterile Aurothioglucose Suspension U.S.P.

ALCOHOL DETERRENT AGENTS

Disulfiram U.S.P., Antabuse®, tetraethylthiuram disulfide, *bis*(diethylthiocarbamyl)disulfide, TTD. Since the discovery that this compound causes nausea, pallor, copious vomiting and other unpleasant symptoms when alcohol is ingested after its use, it has been proposed as a treatment for alcoholism. Up to 6 g. of the drug has been tolerated without symptoms if alcohol is not taken. However, if alcohol is taken in appreciable quantities after disulfiram, dizziness, palpitation, unconsciousness and even death may result.

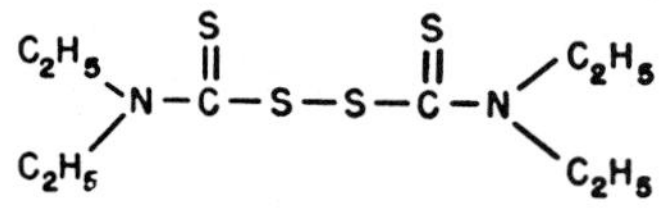

Tetraethylthiuram Disulfide

It has been observed that individuals ingesting alcohol after disulfiram have a blood acetaldehyde level 5 to 10 times greater than that obtained when the same amount of alcohol is ingested by untreated persons. The breath has a noticeable aldehyde odor. The intravenous infusion of acetaldehyde to give the same blood level produces similar symptoms of approximately the same intensity. The mode of action of disulfiram apparently involves inhibition of enzymes which oxidize acetaldehyde and thus allow high concentrations of acetaldehyde to be built up in the body. The compound is insoluble in water but freely soluble in alcohol, benzene and carbon disulfide.

The initial usual adult dose is up to 500 mg. once daily for one or two weeks; the maintenance dose is 250 mg. once daily.

PSORALENS

The psoralens are furocoumarins which are widely distributed in nature. Plants containing these psoralens have been used since antiquity to produce pigmentation. The probable mechanism of action is the concentration of the psoralen in the melanocytes, which when activated by ultraviolet irradiation initiates melanin production. After an oral dose the skin becomes photosensitive in about 1 hour, reaches a peak in sensitivity in 2 hours and the effect wears off in 8 hours.[9]

Methoxsalen U.S.P., Meloxine®, 8-methoxypsoralen, is obtained from the fruit of *Ammi majus*. Methoxsalen increases the normal response of the skin to ultraviolet radiation. Overdoses or excessive exposure early in the treatment may cause severe burning.

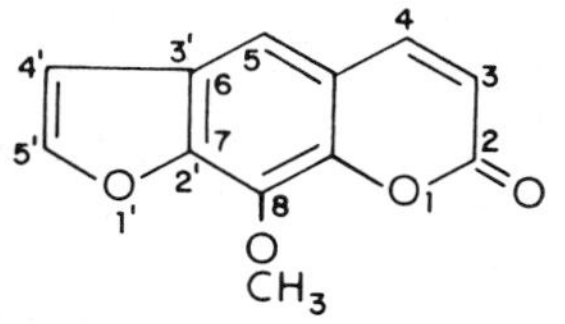

Methoxsalen

Methoxsalen is used topically as a pigmenting agent. It is applied to the lesion as a 1 percent solution prior to exposure to sunlight or a long-wave ultraviolet light source. The irradiation must be carefully controlled to avoid the development of severe erythema and blistering. Removal of the solution after controlled irradiation is advisable.

Trioxsalen U.S.P., Trisoralen®, 2,5,9-trimethyl-7H-furo[3,2-g][1]benzopyran-7-one, 4,5′,8-trimethylpsoralen. This is a synthetic psoralen with much the same uses as methoxsalen, but it is reported to be more potent and less toxic.

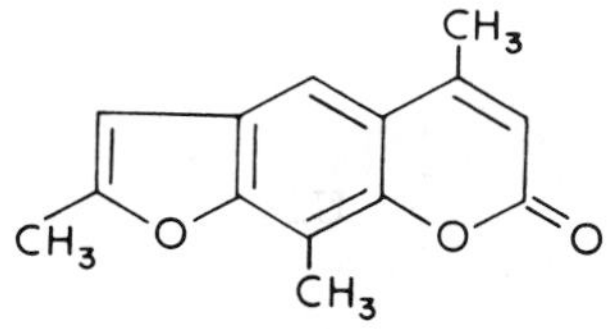

Trioxsalen

Category—oral pigmenting agent.

Usual dose—10 mg. once daily, 2 to 4 hours before exposure to sunlight or ultraviolet light.

Usual dose range—5 to 10 mg. daily, up to a maximum total dose of 140 mg.

Usual pediatric dose—use is not recommended in children 12 years of age and under; over 12 years of age, see Usual Dose.

Occurrence
Trioxsalen Tablets U.S.P.

SUNSCREEN AGENTS

Sunscreen agents are applied topically to the skin to prevent sunburn. These agents screen out the part of the ultraviolet spectrum that is responsible for sunburn. This is generally accepted as being 280 to 315 nm. By proper selection of agents the irradiation can be completely screened to prevent any skin exposure or the irradiation can be partially screened out so that the suntan can develop without burning.

Aminobenzoic Acid U.S.P., Pabanol®, Presun®, *p*-aminobenzoic acid. This is generally recognized as one of the most effective sunscreen agents. Five percent solutions in 55 to 70 percent alcohol are widely used. However, this material washes off easily and must be reapplied every 2 hours and after swimming.

Dioxybenzone U.S.P., 2,2′-dihydroxy-4-methoxybenzophenone. This sunscreen is used in 10 percent concentration, usually in a topical cream. It is effective when freshly applied but must be frequently reapplied, especially after swimming.

Oxybenzone U.S.P., 2-hydroxy-4-methoxybenzophenone. This compound is closely related to dioxybenzone and the two are frequently used in combination (Solbar®).

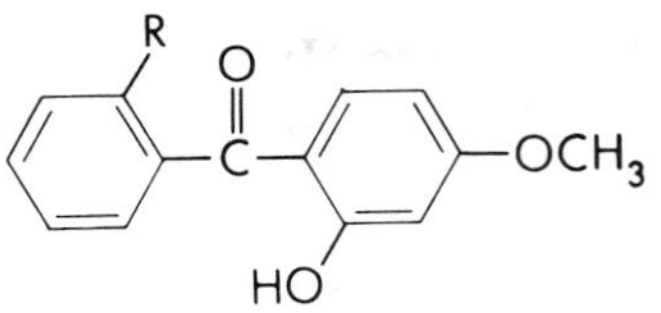

Dioxybenzone
R=OH

Oxybenzone
R=H

URICOSURIC AGENTS

Most purine derivatives in the diet are converted to uric acid and in humans are excreted as such. Gout is characterized by an error in the metabolism of uric acid; there is an elevation of serum urate and crystals form in the cartilages. Colchicine is used for acute attacks and uricosuric agents are used to aid in the excretion of the elevated levels of urates.

Colchicine U.S.P. occurs as a pale yellow powder, which is soluble in water, freely soluble in alcohol and in chloroform and is slightly soluble in ether. It darkens on exposure to light. This drug is used for the relief of acute attacks of gout. It is also useful for the

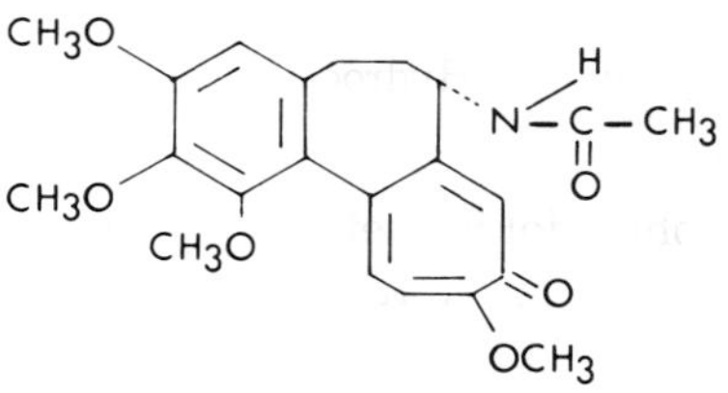

Colchicine

prevention of acute gout when there is frequent recurrence of the attacks. The exact mechanism is not yet established, but it is known not to have any effect in uric acid metabolism. It is considered here with the uricosuric agents as a matter of convenience.

Occurrence
Colchicine Tablets U.S.P.
Colchicine Injection U.S.P.

Probenecid U.S.P., Benemid®, is *p*-(dipropylsulfamoyl)benzoic acid. It is a white, nearly odorless crystalline powder. It is soluble in dilute alkali (salt formation) but is practically insoluble in water and dilute acids.

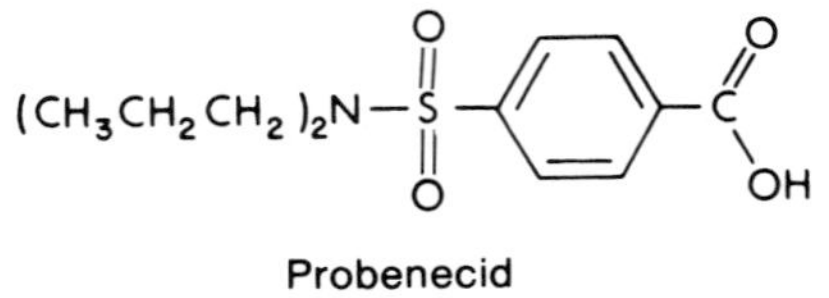

Probenecid

Uric acid is normally excreted through the glomeruli and reabsorbed by the tubules in the kidney. Probenecid acts by interfering with this tubular reabsorption. Probenecid also inhibits excretion of compounds such as aminosalicylic acid, penicillin and sulfobromophthalein. It is rapidly absorbed after oral administration, then is metabolized and excreted slowly, mainly as the glucuronate conjugate.

Category (tablets)—uricosuric.

Usual dose—250 mg. twice daily for 1 week, then 500 mg. twice daily.

Occurrence
Probenecid Tablets U.S.P.

Allopurinol U.S.P., Zyloprim®, 1H-pyrazolo[3,4-d]pyrimidin-4-ol, 4-hydroxypyrazolo[3,4-d]pyrimidine. Allopurinol, an isostere of 6-hydroxypurine or hypoxanthine, is an off-white powder which is insoluble in water but soluble in solutions of fixed alkali hydroxides. It has a pKa of about 9.4. It blocks the formation of uric acid by inhibiting xanthine oxidase, the enzyme responsible for the biotransformation of hypoxanthine to xanthine and of xanthine to uric acid. Thus, it is useful in the control of uric acid levels associated with gout and other conditions. Allopurinol is metabolized to the corresponding xanthine isostere, alloxanthine, which in turn contributes to the inhibition of xanthine oxidase.

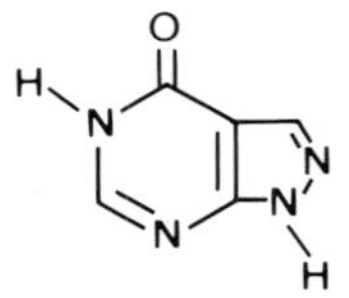

Allopurinol 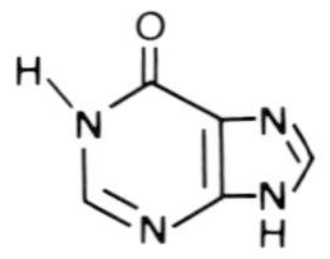   Hypoxanthine

Allopurinol also inhibits the enzymatic oxidation of mercaptopurine, which is used as an antineoplastic antimetabolite. When the two compounds are co-administered, there may be as much as a 75 percent reduction in the dose requirement of mercaptopurine. Salicylates do not interfere with the action of allopurinol, in contrast to their interference with the activity of other uricosuric agents.

The dosage of allopurinol required to lower serum uric acid to normal or near-normal levels varies with the severity of the disease. Divided daily doses are advisable because of the short biological half-life of the drug. While the drug is being administered, fluid intake should be adequate to produce a daily urinary output of at least 2 liters and it is desirable that the urine be maintained at a neutral or slightly alkaline pH value in order to increase the solubility of the drug and of hypoxanthine.

Hypoxanthine   Xanthine   Uric Acid

Category (tablets)—xanthine oxidase inhibitor.
Usual dose—100 to 200 mg. 2 or 3 times daily; or 300 mg. as a single daily dose.

Occurrence
Allopurinol Tablets U.S.P.

Sulfinpyrazone U.S.P., Anturane®, 1,2-diphenyl-4-[2-(phenylsulfinyl)ethyl]-3,5-pyrazolidinedione. Unlike the closely related phenylbutazone, which is used as an anti-inflammatory agent, this is a potent uricosuric agent and is used primarily for the prevention of attacks of acute gouty arthritis. It has only a weak analgesic and anti-inflammatory action, so pain relief must be obtained by administration of phenylbutazone or other analgesics. Salicylates, however, are contraindicated, as they antagonize its uricosuric action. Gastric distress is the most common side-effect and,

Sulfinpyrazone

like the other pyrazolones, it should be taken with milk or food.

Occurrence
Sulfinpyrazone Tablets U.S.P.
Sulfinpyrazone Capsules U.S.P.

ANTIEMETIC AGENTS

Only miscellaneous antiemetic agents will be considered here. Other classes of compounds include the sedatives and hypnotics, the antihistamines and the phenothiazine tranquilizers; these are considered elsewhere in the text. Nausea and vomiting often accompany many disease conditions. The proper approach is to determine the cause and, if possible, correct it. Certain types of therapy cause nausea and vomiting and here the use of antiemetics may be included as part of the treatment.

Trimethobenzamide Hydrochloride U.S.P., Tigan®, N-{4-[2-(dimethylamino)ethoxy]benzyl}-3,4,5-trimethoxybenzamide monohydrochloride. This drug is reported to block the emetic mechanism without undesirable side-effects. It is useful in nausea and vomiting associated with pregnancy, radiation therapy, drug administation and travel sickness. Effects appear within 20 to 40 minutes after administration and last for 3 to 4 hours.

Trimethobenzamide Hydrochloride

Usual adult dose—oral, 250 mg., I.M. or rectal, 200 mg. 3 or 4 times daily as needed.
Usual pediatric dose—oral or rectal, 15 mg. per kg. of body weight a day as needed, divided into 3 or 4 doses; children less than 15 kg.—rectal, 100 mg. 3 or 4 times daily as needed; children 15 to 45 kg.—oral or rectal, 100 to 200 mg. 3 or 4 times daily as needed.

Occurrence
Trimethobenzamide Hydrochloride Capsules U.S.P.
Trimethobenzamide Hydrochloride Injection U.S.P.

Diphenidol, Vontrol®, α,α-diphenyl-1-piperidinebutanol. Diphenidol is useful in the control of vertigo and of nausea and vomiting. The soluble hydrochloride salt is used in the injectable forms and the tablets; the pamoate is used in the suspension and the free base in the suppositories.

Diphenidol

The adult dosage is 25 to 50 mg. orally or rectally 4 times daily; for acute symptoms, 20 to 40 mg. may be given 4 times daily by deep intramuscular injection.

DEPIGMENTING AGENTS

Hydroquinone U.S.P., 1,4-benzenediol, occurs as fine white needles that darken on exposure to light and air. It is freely soluble in water, in alcohol, or in ether. Hydroquinone and its derivative, the monobenzyl ether, possess the property of causing depigmentation of the skin. Hydroquinone is generally used in a 2 percent concentration in a cream.

Hydroquinone
R=H

Monobenzone
R= —CH₂—

For external use—topically as a 2 to 4 percent cream to the affected area once or twice daily at 12-hour intervals. Do not use near eyes, or on open cuts, sunburned or irritated skin, or prickly heat. Notify physician if skin rash or irritation occurs.

Occurrence
Hydroquinone Cream U.S.P.

Monobenzone U.S.P., Benoquin®, *p*-(benzyloxy)-phenol. This occurs as a white, crystalline, odorless powder that is freely soluble in alcohol but insoluble in water. The activity of monobenzone was discovered when people using rubber gloves containing it as an antioxidant developed areas of hypopigmentation. With proper precautions it is used for the treatment of hyperpigmentation due to an increased amount of melanin in the skin.

For external use—topically as a 20 percent ointment once or twice daily.

Occurrence
Monobenzone Ointment U.S.P.

CHELATING AGENTS

Complexation is a general term encompassing many different types of interactions among substances. A specific type of complexation involving metal ions is known as *chelation*. Under appropriate conditions the metal ion may be released or exchanged with other ions.

In the chelated form the complexing agent is known as the ligand. This portion of the molecule contains two or more groups which bind to the metal. The terms polydentate, bidentate and tridentate are used to describe ligands containing several, two or three metal-binding sites, respectively.

Sequestration is another term relating to complexation. It involves the process in which a complexing agent and a metal ion form a water-soluble complex. The complexing agent may be either of a chelating or a nonchelating type. As an example, both sodium metaphosphate and disodium ethylenediaminetetraacetate complex (sequester) calcium ions in aqueous solution. However, only the latter agent forms the complex through chelation.

A metal ion when complexed assumes physical, chemical and biological properties which may be different from its ionic state. For example: the solubility in water and other solvents may change; the presence of a usual precipitating agent may not produce a precipitate; the reactivity with certain enzyme sites in the body may be greatly diminished. These various properties of the metal in the chelated form are dependent upon such factors as the particular chelating agent, pH, temperature and presence of other metal ions.

Chelation has several important pharmaceutical applications. These are: (1) chelating agents are used for the removal of unwanted or excess metal ions in the body; (2) chelates may provide a more efficacious form for the administration of certain metal ions; (3) chelating agents in formulations may prevent certain undesired reactions that would otherwise be caused by trace quantities of metal ions; and (4) chelating agents serve as reagents in titrations, extractions and other procedures used in the separation and analysis of various metal ions. Products having applications in the first three categories are described below.

PRODUCTS

Edetic Acid N.F., Versene®, Sequestrene® AA, ethylenediaminetetraacetic acid, EDTA. This compound and its alkali salts form chelates with alkaline earth ions and heavy metal ions. The reaction of EDTA with calcium is shown on the next page.

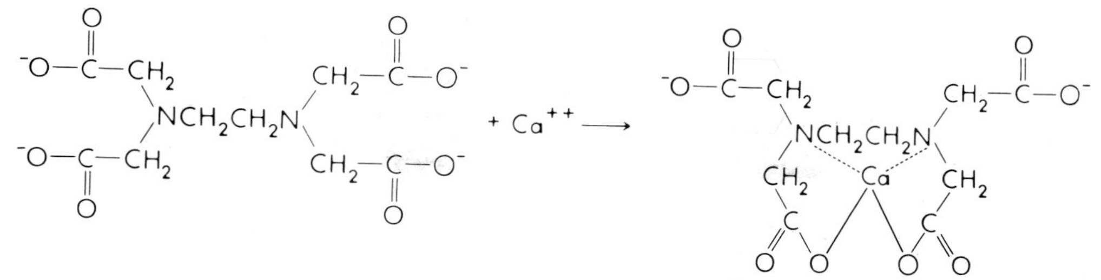

From the reaction it may be seen that the complex contains a 1:1 molar ratio of EDTA to calcium ion. Other metal ions also bond similarly at the same sites on EDTA to form 1:1 complexes.

An equilibrium reaction is associated with each of the metal ions which form a chelate with EDTA. The general equation for this equilibrium is:

$$EDTA^{-4} + M^{+n} \rightleftharpoons MEDTA^{n-4}$$

The equilibrium constant for the reaction may be expressed as

$$K = \frac{[MEDTA^{n-4}]}{[M^{+n}][EDTA^{-4}]}$$

This constant is known as the stability constant of the complex. For the given conditions of temperature and pH, each metal ion has a characteristic stability constant. As these values are extremely large, they are widely used in the logarithmic form and are shown as log K values. The stability constants for EDTA chelates of various metal ions are given in Table 22–1.

Stability constants provide a basis for explaining and predicting, both qualitatively and quantitatively, the effects produced by chelation. As an example, the metal complex having the greatest stability constant will form when EDTA is placed in a solution containing more than one type of metal ions. Also, in a related manner, the metal complex having the greatest stability constant will form when an EDTA-metal chelate is placed in a solution containing other metal ions. This exchange of a metal ion to give a more stable complex is illustrated in the following reaction:

$$CaEDTA^{-2} + Pb^{+2} \rightarrow PbEDTA^{-2} + Ca^{+2}$$

Relatively small quantities (0.1 to 1%) of EDTA and its salts are added as stabilizing agents to pharmaceutical, cosmetic and food preparations. Frequently these preparations contain substances which are catalytically oxidized by iron, copper, and manganese ions. This results in loss of potency and undesirable changes in color and flavor. Trace quantities of these metals may be introduced into the preparation through the water supply and through contact with metallic processing equipment. The chelates which are formed upon the addition of EDTA do not produce

degradation of the components. Examples of substances which may be stabilized in this manner with EDTA include penicillin, ascorbic acid, epinephrine, salicylic acid, and unsaturated fatty acids. EDTA also inactivates alkaline earth metal ions through the formation of stable, water-soluble complexes. These ions may otherwise interact with various substances (e.g., soaps, quaternary ammonium compounds and dyes) contained in liquid preparations to cause such changes as turbidity, precipitation, and loss of potency.

Edetate Disodium U.S.P., Endrate®, Disotate®, disodium ethylenediaminetetraacetate. This is a white, crystalline powder, soluble in water. It is used intravenously as an antidote for hypercalcemia and digitalis poisoning. In the latter treatment, EDTA acts by lowering the calcium ion level in the blood to provide protection against the occurrence of ventricular arrhythmias. Also, it is used topically in concentrations of 0.35 percent to 1.85 percent to remove corneal calcium deposits that impair vision or cause pain.

Edetate Calcium Disodium U.S.P., Calcium Disodium Versenate®, calcium disodium ethylenediaminetetraacetate. This is a white, crystalline powder, freely soluble in water. The primary use of this agent is in the treatment of lead poisoning. It does not appear to be of value for treating other metal poisonings. The complex formed with lead is stable, water-soluble and readily excreted by the kidneys. Hypocalcemia is not produced by the drug because of its calcium content. It is available in a 20 percent aqueous solution.

TABLE 22–2

STABILITY CONSTANTS OF METAL-EDTA COMPLEXES

M^{+n}	log K (at 20°C.)
Mg^{+2}	8.7
Ca^{+2}	11.0
Mn^{+2}	14.0
Fe^{+2}	14.3
Al^{+3}	16.1
Co^{+2}	16.3
Zn^{+2}	16.5
Pb^{+2}	18.0
Ni^{+2}	18.6
Cu^{+2}	18.8
Fe^{+3}	25.1

Dimercaprol U.S.P., British Anti-Lewisite, 2,3-di-mercapto-1-propanol, BAL.

$$CH_2-CH-CH_2-OH$$
$$| \quad \quad |$$
$$SH \quad \quad SH$$

Dimercaprol

Dimercaprol was introduced originally as an antidote for poisoning by Lewisite and other organic arsenical war gases. Further studies led to its present important use as an antidote given intramuscularly for arsenic, mercury and gold poisoning. In this treatment, competition for the toxic metal ions occurs between the SH groups of the drug and various SH groups in the tissues. The drug forms stable water chelates of these toxic metals which are excreted in the urine.

Penicillamine U.S.P., Cuprimine®, D-3-mercaptovaline.

$$\begin{array}{c} CH_3 \quad H \\ | \quad \quad | \\ -HS-C---C-COOH \\ | \quad \quad | \\ CH_3 \quad NH_2 \end{array}$$

Penicillamine

This is a white, crystalline powder, soluble in water, slightly soluble in alcohol, and insoluble in ether and chloroform. This agent is formed as a degradation product in the acid hydrolysis of penicillin and is devoid of antibacterial activity. Its therapeutic use is based on its ability to form water-soluble chelates with several metal ions when given orally. Its primary use is in the treatment of Wilson's disease for removing excess serum copper. Also, it is used as an oral agent for the long-term treatment of lead poisoning.

TABLE 22–3

CHELATING AGENTS

Name Proprietary Name	Preparations	Category	Usual Adult Dose*	Usual Dose Range*	Usual Pediatric Dose*
Edetic Acid N.F. *Versene, Sequestrene AA*		Pharmaceutic aid (metal complexing agent)			
Edetate Disodium U.S.P. *Endrate, Disotate*	Edetate Disodium Injection U.S.P.	Pharmaceutic aid (chelating agent) Metal complexing agent	I.V. infusion, 50 mg. per kg. of body weight in 500 ml. of 5 percent Dextrose Injection or Sodium Chloride Injection over a period of 3 to 4 hours, once a day	Up to a maximum of 3 g. daily	I.V. infusion, 40 mg. per kg. of body weight of a maximum of 3 percent edetate disodium in 5 percent Dextrose Injection or Sodium Chloride Injection over a period of 3 to 4 hours, once a day
Edetate Calcium Disodium U.S.P. *Calcium Disodium Versenate*	Edetate Calcium Disodium Injection U.S.P.	Metal complexing agent	I.M., 1 g. in 0.5 percent Procaine Hydrochloride Injection 2 times a day; I.V. infusion, 1 g. in 250 to 500 ml. of Sodium Chloride Injection or 5 percent Dextrose Injection over a period of 1 to 2 hours, 2 times a day	Not exceeding 750 mg. per kg. of body weight per course of treatment	I.M., up to 35 mg. per kg. of body weight or 850 mg. per square meter of body surface in 0.5 percent Procaine Hydrochloride Injection, 2 times a day; I.V. infusion, up to 35 mg. per kg. or 850 mg. per square meter as a 0.2 to 0.4 percent solution in Sodium Chloride Injection or 5 percent Dextrose Injection over a period of 1 to 2 hours, 2 times a day
Dimercaprol U.S.P.	Dimercaprol Injection U.S.P.	Antidote to arsenic, gold and mercury poisoning; metal complexing agent	I.M., 2.5 mg. per kg. of body weight 4 to 6 times a day on the first two days, then 2 times a day for the next 8 days, if necessary	2.5 to 5 mg. per kg.	I.M., 2.5 to 3 mg. per kg., 6 times a day on the first day, 4 times a day on the second day, 2 times a day on the third day, then once a day for the next 10 days, if necessary
Penicillamine U.S.P. *Cuprimine*	Penicillamine Capsules U.S.P. Penicillamine Tablets U.S.P.	Metal complexing agent	250 mg. 4 times a day	250 mg. to 2 g. daily	Infants over 6 months and young children—250 mg. as a single dose; older children—see Usual Adult Dose

* See *U.S.P. D.I.* for complete dosage information.

TABLE 22-4

IRON COMPLEXES

Preparations Proprietary Name	Category	Usual Adult Dose*	Usual Dose Range*	Usual Pediatric Dose*
Iron Dextran Injection U.S.P. *Imferon*	Hematinic	I.M. or I.V., the equivalent of 25 mg. of iron on the first day. If no adverse reactions are noted, administer as follows: I.M., the equivalent of 100 to 250 mg. of iron once a day; I.V., the equivalent of 25 to 100 mg. of iron once a day	25 to 250 mg. daily	I.M., infants under 4.5 kg. of body weight—up to 25 mg. once a day; children under 9 kg.—up to 50 mg. once a day; under 50 kg.—up to 100 mg. once a day
Iron Sorbitex Injection U.S.P. *Jectofer*	Iron supplement	I.M., the equivalent of 100 mg. of iron once daily	100 to 200 mg. daily	

*See *U.S.P. D.I.* for complete dosage information.

Iron Dextran Injection U.S.P., Imferon®. This is a sterile, colloidal solution of ferric hydroxide complexed with partially hydrolyzed dextran of low molecular weight in water for injection. The iron complex has a molecular weight of about 180,000 and contains about 1 percent total iron. It may contain up to 0.5 percent phenol as a preservative. It is a dark brown, slightly viscous liquid with a pH range of 5.2 to 6.5.

This agent is administered intramuscularly or intravenously for the treatment of confirmed iron-deficiency anemia. It is intended for use only in cases in which oral administration of iron is ineffective or impractical. Severe anaphylactic reactions, pain and temporary staining of the skin at the site of injection may occur. These are side-effects common to parenteral iron administration. Each ml. contains the equivalent of 50 mg. elemental iron.

Iron Sorbitex Injection U.S.P., Jectofer®. This is a sterile, aqueous solution of a complex of iron, sorbitol and citric acid, stabilized with dextrin and an excess of sorbitol. It is a dark brown, clear liquid, having a pH range of 7.2 to 7.9.

The drug is administered only intramuscularly for the same indications as given for iron dextran injection, above. Each ml. contains the equivalent of 50 mg. elemental iron.

Deferoxamine Mesylate U.S.P., Desferal®, DFOM, is obtained from *Streptomyces pilosus*.

Deferoxamine Mesylate

As the mesylate (methanesulfonate) salt the agent is soluble 25 percent in water. The hydrochloride salt is only soluble 5 percent in water.

The drug has a very high and specific binding capacity for iron. It readily chelates the ferric ion and has some affinity for the ferrous ion. Ferrioxamine, a 1:1 chelate, forms with the ferric ion. This chelate is red-colored, stable, water-soluble and readily excreted by the kidneys.

Ferrioxamine

When administered parenterally, the drug combines with ionic iron in the blood. Also, it can remove iron from proteins such as transferrin in blood serum and ferritin in the tissues. However, DFOM does not bind the iron of hemoglobin, nor does it produce anemia during therapy. The drug is given intramuscularly (route of choice) or intravenously for the treatment of acute iron poisoning. It is not a substitute, but rather an adjunct, to other therapeutic measures used in treating iron intoxication. Effectiveness is greatest when therapy is initiated within a short interval following iron ingestion. It is available in an ampul containing 500 mg. lyophilized powder.

When given intramuscularly the initial dose is 1 g. dissolved in 2 ml. of sterile water for injection. This is followed by 0.5 g. every 4 to 6 hours, depending upon the clinical response. For intravenous use 0.5 g. is dissolved in 2 ml. of sodium chloride injection or in dextrose injection, and administered at a rate not exceeding 15 mg. per kg. per hour. The reconstituted solution may be stored under sterile conditions at room temperature for not longer than two weeks.

REFERENCES

1. Binz, A.: Agnew, Chem. 48:425, 1935.
2. Hoey, G. B., *et al.*: J. Med. Chem. 6:24, 1963.
3. Whittet, T. D., and Robinson, A. E.: Pharm. J. 1964:39 (July 11).
4. Coppage, W. S., Jr.: J. Clin. Invest. 38:2101, 1959.
5. Golden, M. J.: J. Am. Pharm. A. (Sci. Ed.) 34:76, 1945.
6. Cataline, F. L., Jeffries, S. F., and Reinish, F.: J. Am. Pharm. A. (Sci. Ed.) 34:33, 1945.
7. Gilvert, E. E.: J. Chem. Ed. 18:338, 1941.
8. Hubacher, M. H., and Doernberg, S.: J. Am. Pharm. A. (Sci. Ed.) 37:261, 1948.
9. Becker, S. W.: J.A.M.A. 173:1483, 1960.

SELECTED READINGS

Almen, T: Contrast agent design, J. Theoret. Biol. 24:216, 1969.
Chaberek, S., and Martel, A.E.: Organic Sequestering Agents, New York, Wiley, 1959.
Chenoy, N.C.: Radiopaques—a review, Pharm. J. 194:663, 1965.
Dwyer, F.P., and Mellor, D.P.: Chelating Agents and Metal Chelates, New York, Academic Press, 1964.
Hermes, H., and Taenzer, V.: Design of X-Ray Contrast Media, *in* Ariens, E.J. (ed.): Drug Design, vol. 6, p. 261, New York, Academic Press, 1975.
Hoppe, J.O.: X-ray contrast media, Med. Chem. 6:290, 1963.
Shockman, A.T.: Radiologic diagnostic agents, Topics in Med. Chem. 1:381, 1967.
vanHam, G.W., and Herzog, W.P.: The Design of Sunscreen Preparations, *in* Ariens, E.J. (ed.): Drug Design, vol. 4, p. 193, New York, Academic Press, 1974.

pharmaceutic aids

SOLVENTS AND VEHICLES

Light Mineral Oil U.S.P., light liquid paraffin, light white mineral oil. Light mineral oil is a mixture of liquid hydrocarbons obtained from petroleum. Mineral oils for pharmaceuticals are purified and freed from sulfur compounds, unsaturated hydrocarbons and solid paraffins. The *U.S.P.* permits the addition of an antioxidant to prevent the development of oxidative rancidity. Vitamin E *(dl-α-tocopherol)* is often used in 10 p.p.m. Often, unsaturated aliphatic compounds are present in minute amounts and will be oxidized to low molecular weight aldehydes and acids which affect both taste and odor.

Petrolatum U.S.P., petrolatum jelly, yellow petrolatum. Petrolatum is a purified, semisolid mixture of hydrocarbons obtained from petroleum. Petrolatum is a colloidal dispersion of aliphatic liquid hydrocarbons (C_{18} to C_{24}) in solid hydrocarbons (C_{25} to C_{30}), the disperser being a plastic material known as protosubstance, consisting of noncrystalline, naturally occurring, branched chain paraffinic-type hydrocarbons (C_{25} to C_{30}). Without this, a mixture of mineral oil and paraffin would not be stable; the oil would leak out or "sweat." The fact that petrolatum does not produce an oily stain on paper indicates the presence of the oil in the inner phase of the dispersion.

Petrolatum is yellowish to light amber in color. It has not more than a slight fluorescence even after being melted, and it is transparent in thin layers. It is free or nearly free from odor and taste. It is soluble in benzene, chloroform, ether, petroleum, benzin, carbon disulfide, solvent hexane or in most fixed and volatile oils. It is partly soluble in acetone; the protosubstance, which is precipitated out, may be dissolved by the addition of amyl acetate. Petrolatum is insoluble in alcohol or water.

The melting point ranges from 38° to 60°C, and the specific gravity between 0.815 and 0.880 at 60°C. These should not be confused with consistency, measured by a penetrometer, which characterizes the firmness of texture of a petrolatum. Consistency is dependent on the microscopic fibers which make up a petrolatum. If these are tough and stiff, a product very firm in consistency results. Fibers also may vary in length. If they are too short, the product will be soupy at summer temperatures. If they are too long, the petrolatum is too tacky. Medium length is preferred for most pharmaceutical products. Petrolatums of soft consistency generally are used when ease of spreading is desired. Those of medium consistency are used most widely for ointments and cosmetic creams. Those of hard consistency are used in lipsticks and when considerable mineral oil is to be incorporated.

Petrolatum is miscible with a relatively small proportion of water. However, the addition of small amounts of such ingredients as cetyl alcohol, lanolin and cholesteryl esters may increase the water-absorption properties to nearly 10 times the weight of the petrolatum base. It is used as an ointment base.

White Petrolatum U.S.P., white petroleum jelly. White Petrolatum is a purified mixture of semisolid hydrocarbons obtained from petroleum and is wholly or nearly decolorized. This differs from Petrolatum U.S.P. only in respect to color. It is white or faintly yellowish and transparent in thin layers. This type is preferred as a household topical dressing.

White petrolatum is widely used as an oleaginous ointment base.

Occurrence
Hydrophilic Petrolatum U.S.P.
Hydrophilic Ointment U.S.P.
Petrolatum Gauze U.S.P.
White Ointment U.S.P.

Oleyl Alcohol N.F. This is a mixture of unsaturated and saturated high molecular weight fatty alcohols consisting chiefly of $CH_3(CH_2)_7CH=CH(CH_2)_7CH_2OH$. The alcohol exists as a pale yellow liquid having a faint characteristic odor and bland taste. It is soluble in alcohol, fixed oils and in mineral oil.

Oleyl alcohol increases the water-adsorption capacity of ointments. This value refers to the quantity of water which may be blended into an ointment without an appreciable loss of consistency. Also, oleyl alcohol gives the quality of "softness" to ointments.

Cetyl Alcohol N.F., 1-hexadecanol, palmityl alcohol. Cetyl alcohol is a mixture of solid alcohols consisting chiefly of cetyl alcohol, $CH_3(CH_2)_{14}CH_2OH$. It is a component of spermaceti and at one time was obtained from this source. The alcohol is prepared now by hydrogenating a mixture of fatty acids having a high percentage of palmitic acid. This saturated alcohol is marketed in various forms, namely, white flakes, granules, cubes or castings. It is an unctuous material having a slight odor and a bland, mild taste. It is soluble in alcohol, ether and mineral and vegetable oils and insoluble in water.

Similarly to oleyl alcohol, it is used in ointments for its water-absorbing property. Because of its existence as a solid at room temperature it increases the "firmness" of ointments.

Stearyl Alcohol N.F., stenol. This consists of at least 90 percent stearyl alcohol, $CH_3(CH_2)_{16}CH_2OH$. This alcohol is similar to cetyl alcohol in appearance and properties.

It is similar in use to cetyl and oleyl alcohol. Among these alcohols, stearyl alcohol gives the greatest increase in firmness.

Polyethylene Glycol N.F., Carbowax®, is a linear polymer of ethylene oxide and water, having the general formula $HOCH_2(CH_2OCH_2)_nCH_2OH$ in which n equals the average number of xyethylene groups.

This is a series of polyethylene glycols having average molecular weights ranging from 200 to 6000 (see Table A-1). Those in the series from 200 to 700 are liquids, while those above 1000 are waxlike, unctuous solids. All of the agents are soluble in water and form clear solutions. They are soluble in many organic solvents, such as aliphatic alcohols, ketones, esters and aromatic hydrocarbons, and are insoluble in aliphatic hydrocarbons. The liquid polyethylene glycols are less hygroscopic than glycerin and other simple glycols. A relationship exists between the molecular weight and many of the physical properties. The following properties are decreased upon an increase in the molecular weight: solubility in water, solubility in organic solvents, vapor pressure and hygroscopicity. Various combinations of these properties may be obtained by blending two or more of the agents. As additional properties, these polymers are nonirritating and have a low order of toxicity. Their pharmaceutical use is primarily as anhydrous bases for external preparations including ointments and suppositories.

Paraffin N.F., petrolatum wax. Paraffin is a purified mixture of solid hydrocarbons obtained from petroleum. It is a white, translucent solid of crystalline structure composed of C_{24} to C_{30} hydrocarbons. Its solubility is similar to that of petrolatum. Its congealing range is 47° to 65°C. It is used in pharmacy mainly to raise the melting point of ointment bases.

The paraffins of commerce are, mainly, interlaced plate-type crystals representing straight chain hydrocarbons. Within recent years, paraffinic fractions which consist chiefly of branched chain hydrocarbons have been separated. Physically, they are composed of minute interlacing needles; they are known as microcrystalline waxes. They are plastic; they have a higher melting point, and they are tougher and more flexible than regular paraffin. They are used in polishes, paper coatings and laminated boards.

Microcrystalline Wax N.F. is obtained from petroleum and has a much finer crystalline structure than paraffin. Its melting range is from 54° to 102°C. The main pharmaceutical uses are in cosmetics, ointment bases and as a component of some of the wax-fat coatings used in sustained-release products.

Yellow Wax N.F. Waxes are composed primarily of esters of high molecular-weight monohydric alcohols and high molecular-weight fatty acids. The alcohols range from C_{24} to C_{32}, and the acids are C_{16} and above. Yellow Wax N.F. or purified beeswax is mainly ceryl myristate. Beeswax may contain 10 to 14 percent hydrocarbons, with those of about C_{31} predominating. The main pharmaceutical use of Yellow Wax N.F. is

TABLE A–1

POLYETHYLENE GLYCOLS

$HOCH_2(CH_2OCH_2)_nCH_2OH$

Average Molecular Weight	n	Molecular Weight Range	Physical State	Water-Solubility at 20°C. (Percent by Weight)
300	5–5.75	285–315	Liquid	Complete
400	7.2–8.1	380–420	Liquid	Complete
600	11.5–12.9	570–630	Liquid	Complete
1500	28–35	1300–1600	Solid	—
1540	28–36	1296–1648	Solid	70
4000	67–83	3000–3700	Solid	62
6000	157–203	7000–9000	Solid	50

to stiffen ointments that contain a large amount of liquid.

White Wax N.F. is beeswax that has been decolorized by heating with activated charcoal and filtering while hot. It has the same general uses as Yellow Wax N.F.

Squalene N.F. is a hydrocarbon obtained by the complete reduction of squalene. This is usually obtained from shark-liver oil.

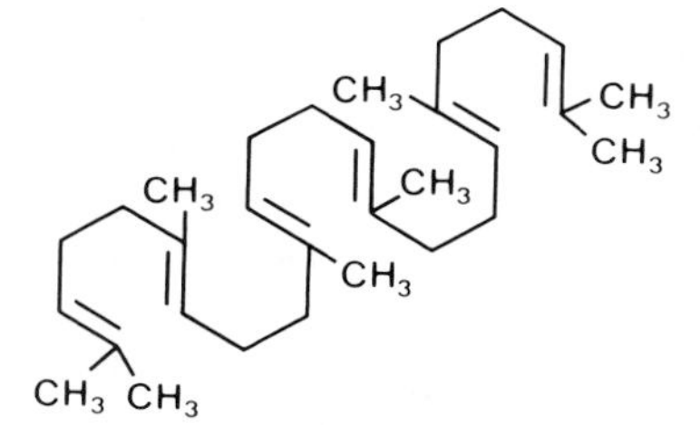

Squalane

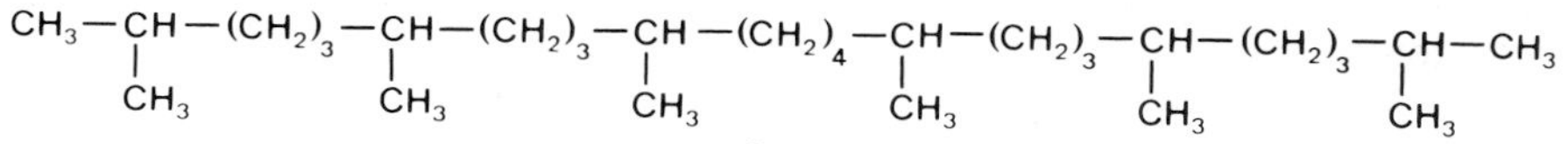

Squalane

It is a liquid at room temperature, is soluble in ether and other hydrocarbon solvents, but is only slightly soluble in alcohol or acetone. It is used in suppository bases and as a skin lubricant in cosmetics.

Oleic Acid N.F. is a nearly colorless liquid at room temperature. When exposed to air during storage it develops a deeper color and a rancid odor. The fatty acids derived from olive oil are about 85 percent oleic acid.

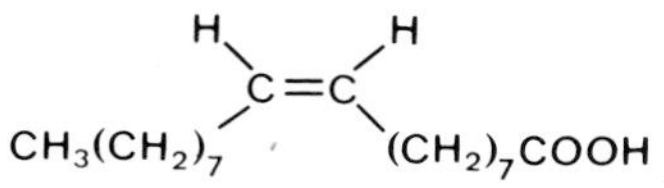

Oleic Acid

Oleic acid is used as a solvent in ointment bases. It may also be converted to a soap, such as triethanolamine oleate, and used in lotion vehicles.

Ethyl Oleate N.F. is the ester of ethyl alcohol and oleic acid. Other closely related fatty-acid ethyl esters may also be present. The product has a saponification value between 177 and 188. It is used as a solvent in pharmaceutical and cosmetic lotions.

Isopropyl Myristate N.F. is a mixture composed principally of the isopropyl ester of myristic acid with lesser amounts of the isopropyl esters of other saturated fatty acids. It is used in cosmetics and pharmaceuticals for its emollient and solvent properties.

Methyl Alcohol N.F., methanol, was originally made by the destructive distillation of wood, hence the name *wood alcohol*. It is a flammable, poisonous liquid. Its main pharmaceutical use is as an extracting agent; all traces of it must be removed from the final product.

Glycerin U.S.P., glycerol, propanetriol, trihydroxypropane. Glycerin, an important pharmaceutical for many years, was isolated in 1779; its structure was determined in 1835. For over 100 years, the principal source of glycerin was the hydrolysis of fats. In 1938, a method was devised to produce glycerin from propylene, a petroleum product. Glycerin is a clear, colorless, viscous liquid with a faint odor and a sweet taste. It is highly hygroscopic and is miscible with alcohol or water but insoluble in organic solvents.

The applications of glycerin in pharmacy are extremely varied, and its uses in industry alone would require a book for complete discussion. The solvent and the preservative properties of glycerin are used widely, while the consistency and sweet taste make it suitable for use in cough remedies. As an emollient and demulcent, it is an ingredient in lotions and hand creams. In prescription compounding, it is used frequently as a stabilizer, to retard precipitation by decreasing ionization in many cases and as a softening agent.

Occurrence
Glycerin Ophthalmic Solution U.S.P.
Glycerin Oral Solution U.S.P.
Glycerin Otic Solution U.S.P.
Glycerin Suppositories U.S.P.

Propylene Glycol U.S.P., 1,2-propanediol, is a hygroscopic, viscous liquid, with a slight, unpleasant taste. It is miscible with water in all proportions. The commercial product is made

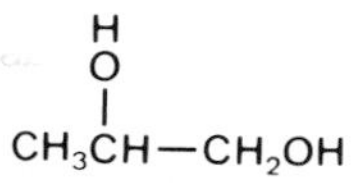

Propylene Glycol

synthetically and is a racemic mixture. It finds use as a substitute for glycerin in pharmaceutical products in which taste is not a consideration. It is a good solvent for many medicinal agents; it possesses emollient properties when used in cosmetic preparations.

Chloroform N.F., trichloromethane. Chloroform ($CHCl_3$) was first synthesized by Leibig in 1831; it was introduced as an obstetric anesthetic in 1847 by Simpson, an Edinburgh surgeon, within a year of the introduction of ether as an anesthetic in this country.

Chloroform is a colorless, mobile liquid of ethereal odor and sweet taste. It has a boiling point of 61°C and a specific gravity of 1.475. It is soluble in about 200 volumes of water and miscible with most organic solvents and oils. Its heated vapors burn with a green flame; the liquid itself is not inflammable. In the presence of air, sunlight or open flames, it is oxidized to phosgene,

$$Cl-C=O$$
$$|$$
$$Cl$$

a highly reactive gas which hydrolyzes in the lung tissues to hydrogen chloride, producing pulmonary edema. Other impurities that should be absent from chloroform are chlorine and chlorinated decomposition products, acids, aldehydes, ketones and readily carbonizable substances. Chloroform should be stored in airtight, light-resistant containers at a temperature not above 30°C. When corks are used, they should be covered with tin foil.

Chloroform serves as a solvent for fats, resins and some plastics and is an excellent extractant for alkaloids and other soluble medicinal agents in their manufacture and assay.

Carbon Tetrachloride N.F., tetrachloromethane, is a clear, colorless, noninflammable liquid. Its solvent properties are similar to those of chloroform. It is miscible with alcohol, ether, benzene, and other organic solvents, but it is practically insoluble in water. It should be stored in tight, light-resistant containers.

Acetone U.S.P., 2-propanone, dimethyl ketone, was observed first in the distillate from wood in 1661 by Robert Boyle; it was isolated from the heating of acetates by Macaire and Marcet in 1823. A small amount is present in normal blood and urine; this may be increased greatly in diabetes, apparently by decarboxylation of acetoacetic acid.

$$CH_3-C-CH_3$$
$$||$$
$$O$$

Acetone

Commercial acetone is usually of a high degree of purity, seldom less than 99 percent, the remainder being practically all water. It is a transparent, colorless, mobile liquid with a characteristic odor, boiling at about 56°C but volatile and inflammable at much lower temperatures. It has a specific gravity of about 0.79 and is miscible with water, alcohol or other solvents.

Although acetone is not used in medicine, it is of immense value in industry, chiefly as a solvent for fats, waxes, oils, varnishes, lacquers, rubber and like materials. In addition, it is employed in the manufacture of a great number of substances, including chloroform, explosives, varnish removers, plastics, rayon and medicinals.

Amylene Hydrate, N.F., 2-methyl-2-butanol, *t*-pentyl alcohol, is a colorless liquid with a characteristic odor and a burning taste.

$$CH_3-CH_2-\underset{\underset{CH_3}{|}}{\overset{\overset{CH_3}{|}}{C}}-OH$$

Amylene Hydrate

It has been used as a solvent for tribromoethanol, the solution being used as a basal anesthetic.

FLAVORS

Anethole N.F., *p*-propenylanisole, paramethoxypropenylbenzene. Anethole is parapropenylanisole. It is obtained from anise oil and other sources, or is prepared synthetically. This aromatic ether is isolated from several volatile oils by fractionating, chilling and crystallizing. It may also be prepared synthetically.

$$\underset{OCH_3}{\overset{\overset{\displaystyle H}{\underset{\displaystyle}{C}}=\overset{\displaystyle CH_3}{\underset{\displaystyle H}{C}}}{\bigcirc}}$$

Anethole

This ether is a colorless or faintly yellow liquid which congeals at about 20° to 23°C. It has the aromatic odor of anise oil, a sweet taste and the characteristic of being affected by light. It is soluble in organic solvents but is insoluble in water.

Benzaldehyde N.F., artificial oil of bitter almond, is found naturally combined in some glycosides, such as amygdalin of bitter almonds and other rosaceous kernels; commercial production is by synthesis.

Benzaldehyde

Benzaldehyde is a colorless, strongly refractive liquid, having an odor resembling that of bitter almond oil and a burning, aromatic taste. It dissolves in about 350 volumes of water and is miscible with alcohol, ether or fixed or volatile oils. It is heavier than water, having a specific gravity of about 1.045. If made from glycosides, it may contain some hydrocyanic acid; if made from benzal chloride, it may contain chlorinated products. It is assayed by reaction with hydroxylamine hydrochloride and titration of the released hydrogen chloride.

One remarkable property of benzaldehyde is its great tendency to auto-oxidize. While it is affected much less easily by oxidizing agents than aliphatic aldehydes are, oxygen of the air is absorbed rather rapidly to form a peroxide, which decomposes to give benzoic acid.

$$C_6H_5CHO + O_2 \rightarrow C_6H_5CO_3H$$
$$C_6H_5CHO + C_6H_5CO_3H \rightarrow 2C_6H_5COOH$$

Consequently, benzaldehyde must be stored in well-filled, tight, light-resistant containers. The presence of a small amount of an antioxidant, such as hydroquinone, will make the commercial article more stable.

It is used almost entirely as a flavor and in the preparation of synthetics, such as triphenylmethane dyes, cinnamic aldehyde and acid and unsaturated esters.

Vanillin N.F., 4-hydroxy-3-methoxy-benzaldehyde, occurs naturally in a large number of plants, including vanilla beans, but it is prepared more conveniently and economically from lignin waste in the manufacture of paper or from eugenol of clove oil. In the latter process, the eugenol is converted by alkalies to isoeugenol, acetylated, oxidized and finally hydrolyzed.

Vanillin occurs as fine, white to slightly yellow crystals, usually needlelike, having an odor and taste suggestive of vanilla. It is soluble in water (1:100), in glycerin (1:20) and in other organic solvents. Because of the phenolic group, it dissolves also in fixed alkali hydroxides and gives a blue color with ferric chloride. Solutions of vanillin in water are acid and give a white precipitate with lead subacetate. Vanillin is rather volatile, easily oxidized and affected by light, so that it must be stored in tight, light-resistant containers.

Ethyl Vanillin N.F., 3-ethoxy-4-hydroxy-benzaldehyde, is synthesized and occurs as fine white or slightly yellowish crystals. It has the same general physical and chemical characteristics as vanillin, but

Vanillin R = CH₃
Ethyl Vanillin R = C₂H₅

ethyl vanillin possesses a more delicate and a more intense vanilla odor and taste.

Ethyl Acetate N.F., acetic ether, vinegar naphtha, is obtained by the slow distillation of a mixture of ethyl alcohol, acetic acid and sulfuric acid. It is a transparent, colorless liquid, with a fragrant, refreshing, slightly acetous odor and a peculiar, acetous, burning taste. The ester is miscible with ether, alcohol and fixed and volatile oils.

At present, pharmaceutically it is used to impart a pleasant odor and flavor; it has wider application in industry as a solvent.

Methyl Salicylate N.F., wintergreen oil, sweet birch oil. Methyl salicylate is produced synthetically or is obtained by maceration and subsequent distillation with steam from the leaves of *Gaultheria procumbens* Linné (Fam. *Ericaceae*) or from the bark of *Betula lenta* Linné (Fam. *Betulaceae*).

It is prepared synthetically through the esterification of salicylic acid and methyl alcohol in the presence of sulfuric acid.

Methyl salicylate is a colorless, yellowish or reddish, fragrant, oily liquid, slightly soluble in water and soluble in alcohol. It is labeled to indicate whether it was prepared synthetically or distilled from natural sources.

Only the carboxyl group of salicylic acid has reacted with methyl alcohol. Therefore, the hydroxyl group reacts with ferric chloride T.S. to produce a violet color. It is readily saponified by alkalies and reacts like the other salicylates.

It is used most often as a flavoring agent and in external medication as a rubefacient in liniments. In water and hydroalcoholic solutions it is absorbed rapidly, thus penetrating deeply into the tissue and exerting also a systemic action. It is generally applied as a lotion in 10 to 25 percent concentration.

Internal use is limited to small quantities due to its toxic effects in large doses. The average lethal dose is 10 ml. for children and 30 ml. for adults. In veterinary practice, it finds some use as a carminative.

AEROSOL PROPELLANTS

Compounds in common use are the chlorofluoromethanes and the chlorofluoroethanes. Difluorodichloromethane (CF_2Cl_2) is an example. Because of their low boiling point ($-30°$), noninflammability and unexpected freedom from toxicity, Midgley and Henne introduced the Freon group as refrigerants in 1930. Tests by the U.S. Bureau of Mines have shown them to be nontoxic at concentrations of 20 percent in air for exposures as long as 8 hours.

The aerosols consist of a solution of active ingredients and of a propellant in a sealed container with a specially designed valve and a standpipe.

Therapeutic application of aerosols is a well-established medical practice. Current medical use is restricted to products that are designed for inhalation. Compounds recognized by the *N.F.* are:

Dichlorodifluoromethane N.F.
Dichlorotetrafluoroethane N.F.
Trichloromonofluoromethane N.F.

SYNTHETIC SWEETENING AGENTS

Saccharin N.F., 1,2-benzisothiazolin-3-one 1,1-dioxide. Saccharin occurs as white crystals, or a white crystalline powder. It is odorless or nearly so, but has a very pronounced sweet taste. In dilute solution, it is about 500 times as sweet as sucrose.

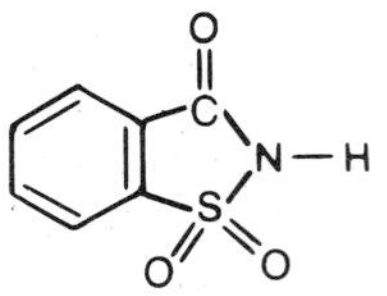

Saccharin

Saccharin is relatively stable in solution from pH 3.3 to 8.0. The following chart shows the percent of unchanged saccharin remaining after autoclaving 0.35 percent aqueous solutions for 1 hour. These data indicate that under the usual conditions there would be only a minor loss from hydrolysis.

Solvent	pH	100°C.	125°C.	150°C.
H$_2$O	2.0	97.1	91.5	81.4
Buffer	3.3	100	99	98.1
Buffer	7.0	99.7	99.7	98.4
Buffer	8.0	100	100	100

Since it does not enter into the body's metabolism, it is employed as a sweetening agent in the diets of diabetics and others who need to restrict their intake of calories.

Saccharin Sodium U.S.P., sodium 1,2-benzisothiazolin-3-one 1,1-dioxide. The compound occurs as a white, crystalline powder. It is soluble in water (1:1.5) and in alcohol (1:50). Like saccharin, the sodium salt is about 500 times as sweet as sugar; and, since the salt is much more soluble in water, this is the form in which it usually is employed as a sweetening agent.

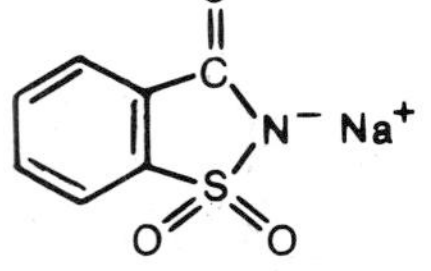

Sodium Saccharin

Saccharin Calcium U.S.P., calcium 1,2-benziso-thiazolin-3-one 1,1-dioxide. This compound occurs as white crystals or as a white crystalline powder. One gram is soluble in 1.5 ml. of water.

ACIDIFYING AGENTS

Acetic Acid N.F., ethanoic acid, contains 36 to 37 percent of CH_3COOH. Acetic acid and its salts and esters are distributed widely in nature. The acid has been known for over 100 years, first in the form of vinegar. Acetic acid has been obtained from many sources but is now produced from ethanol by oxidation and from acetylene by hydration to acetaldehyde, which then is oxidized.

Acetic acid is a corrosive, colorless, liquid with a pungent odor and a sharp, acid taste. It is nontoxic and of little therapeutic value. In most cases the acid is used in diluted form where a weak and innocuous acid is required.

Acetates usually employed in pharmacy are Sodium Acetate U.S.P., Potassium Acetate U.S.P., and solutions of aluminum acetate and subacetate.

Sodium acetate and potassium acetate are the common salts. In solution their alkalinity is due to hydrolysis where the acetate ion functions as a proton acceptor. The alkalinity is utilized with preparations of theobromine and theophylline.

Most acetates are very soluble in water. The only one not very soluble is silver acetate. Acetates are stable in solution. Ethyl Acetate N.F. is used as a flavor.

Occurrence	Percent Acetic Acid
Glacial Acetic Acid U.S.P.	99.5
Acetic Acid N.F.	36
Diluted Acetic Acid N.F.	6
Aluminum Subacetate Topical Solution U.S.P. .	16
Aluminum Acetate Topical Solution U.S.P.	1.5
Acetic Acid Irrigation U.S.P.	0.25

Lactic Acid U.S.P., 2-hydroxypropionic acid, α-hydroxypropionic acid, is a mixture of lactic acid,

CH₃CHOHCOOH, and lactic anhydride, CH₃-CHOHCO·OCH(CH₃)COOH. In 1780, Scheele discovered this product of bacterial fermentation; it is found in products such as sour milk, cheese, buttermilk, wine and sauerkraut.

Although lactic acid may be synthesized, it is produced commercially by the fermentation of molasses, whey or corn sugar, with either *Lactobacillus delbruckii* or *L. bulgaricus*. Because of an asymmetric carbon atom, lactic acid may exist in three forms; it usually is supplied in the DL form. The acid is a clear, slightly yellow, odorless, syrupy liquid, miscible with water, alcohol, or ether but immiscible with chloroform. A 0.1 N solution has a pH of 2.4.

The free acid seldom is used because it is caustic in concentrated form. It is added to infant formulas to aid digestion and to decrease the tendency of regurgitation, and it is employed as a spermicidal agent in contraceptives. It is used to prepare lactates of minerals which are intended for internal administration, e.g., Calcium Lactate U.S.P., Sodium Lactate Injection U.S.P. By the use of *Lactobacillus acidophilus,* an acidic intestinal flora which produces lactic acid by fermentation is developed.

Tartaric Acid N.F., dihydroxysuccinic acid, is a dihydroxy dicarboxylic acid that was first isolated by Scheele. It is obtained from tartar, a crystalline deposit of crude potassium bitartrate occurring in wine.

$$\begin{array}{c} H \\ | \\ HO-C-COOH \\ | \\ HO-C-COOH \\ | \\ H \end{array}$$

Tartaric Acid

The acid contains 2 asymmetric carbon atoms, and it exists in racemic, *levo, dextro* and *meso* forms. In the wine industry, a crude form of potassium bitartrate, called argol, is produced. This is treated with a calcium salt to form insoluble calcium tartrate, which is acidified with sulfuric acid to yield insoluble calcium sulfate and soluble tartaric acid.

The acid occurs as large, colorless crystals or as a fine, white, crystalline powder that is practically 100 percent pure. It is stable in air, has an acid taste and is odorless. Its pKa values of 3.0 and 4.3 make it useful as a buffer in the pH range of 2.5 to 5.0. It is soluble in water (1:0.75) or in alcohol (1:3), but it is insoluble in most organic solvents. The properties are typical of those of organic acids, and it forms a precipitate with salts of potassium, calcium, barium, strontium, lead,

silver or copper. Potassium bitartrate is one of the few relatively water-insoluble (1:165) potassium salts.

Tartrates in general are insoluble or very slightly soluble. Sodium tartrate is soluble, Potassium Sodium Tartrate U.S.P. is very soluble, potassium and ammonium dissolve with difficulty, and the others are insoluble. Tartrates and tartaric acid are sequestering agents similar to the citrates. Soluble complex ions are formed in the presence of excess tartrate. Examples of this property are the solubilizing of cupric ions by Rochelle salt in Fehling's solution, the soluble antimony salt, Antimony Potassium Tartrate U.S.P. and bismuth potassium tartrate.

The hydrogen of the alcoholic OH is more active than the OH of ethanol and is easily replaced in basic media by ions of copper, bismuth, antimony or iron. Thus, there is formed a complex that results in a soluble compound. This sequestering property of tartrates is very pronounced. With ferric ions a soluble ferritartaric acid is formed.

Tartaric acid is used in preparing many useful tartrates, and, as the free acid, it is employed in effervescent salts and refrigerant drinks. The acid is not metabolized as are most organic acids, but it is excreted in the urine and, therefore, increases the acidity of the system. However, most of it is retained within the intestine and functions by osmotic pressure as a saline laxative. Several salts are used as saline cathartics, i.e., Potassium Sodium Tartrate U.S.P., and potassium bitartrate.

Citric Acid U.S.P. is a tribasic acid that is distributed widely in nature. The official form is of not less than 99.5 percent purity calculated on the anhydrous basis.

$$\begin{array}{c} H \\ HC-COOH \\ | \\ HOC-COOH \\ | \\ HC-COOH \\ H \end{array}$$

Citric Acid

Commercially, the acid may be produced from limes or lemons, from the residue from pineapple canning and from the fermentation of beet molasses. Juice from the citrus fruits is treated with chalk, and the precipitated calcium citrate is acidified with sulfuric acid. Calcium sulfate is filtered off and the citric acid is recovered from the filtrate. The fermentation process accounts for the largest amount of citric acid and may be carried out with any one of over nineteen varieties of fungi *(Citromyces, Aspergillus, Penicillium)* to give liquor concentrations of from 10 to 15 percent citric acid.

The acid occurs as a white, crystalline powder or as large, colorless, translucent crystals. It is efflorescent in air, odorless, sour-tasting and is soluble in water (1:0.5), alcohol (1:2) or ether (1:30) and insoluble in other organic solvents. An aqueous solution of citric acid is unstable because it undergoes slow decomposition. Its chemical reactions are those characteristic of organic acids. Salts are formed readily with all hydroxides, and they produce alkaline, aqueous solutions (sodium, potassium, calcium, magnesium). Citric acid effervesces carbonates, and this property is employed widely in effervescent salts.

The citrates of the alkali metals are soluble in water, whereas most other citrate salts are insoluble in water. However, the insoluble citrates may be solubilized by an excess of citric acid or citrate ion (usually furnished by sodium citrate). The citrate ion often is used as a sequestering agent with ions of metals, such as magnesium, manganese, calcium, ferric, bismuth, strontium, barium, copper and silver. The metal ion is held in solution in a complex anion form that is soluble and yet prevents the metal ion from exhibiting its usual properties. This principle is utilized in iron preparations, Benedict's solution and Anticoagulant Sodium Citrate Solution U.S.P.

Fumaric Acid N.F., is made commercially by fermentation of glucose. Its pKa values

$$H-C-COOH$$
$$\|$$
$$HOOC-CH$$

Fumaric Acid

of 3.0 and 4.4 make it useful for many of the same applications as is tartaric acid.

Therapeutic application of free citric acid is rare, but for a variety of reasons it has extensive pharmaceutical use. Many salts of citric acid are available, and they are used for both the metal ion and the citrate portion. Examples are Sodium Citrate U.S.P. and Potassium Citrate N.F.

Trichloroacetic Acid U.S.P. is one of the three chlorinated acetic acids. It has been known since 1838 and is prepared by oxidizing chloral hydrate with either nitric acid or permanganate.

$$CCl_3CHO \xrightarrow{(O)} CCl_3COOH$$

The acid occurs as colorless, deliquescent crystals that have a characteristic odor. Usually, the introduction of a halogen atom into an acid increases the acidic properties, and this is true of acetic acid; the acid properties increase with increase in the number of chlorine atoms. Trichloroacetic acid is a stronger acid (as strong as hydrochloric acid) than acetic acid and is very corrosive to the skin. It is soluble in water (1:0.1) or most organic solvents.

It is astringent, antiseptic and caustic; the caustic properties are the most useful, as in treating forms of keratosis, such as moles and warts. Solutions of trichloroacetic acid are very efficient protein precipitants.

Category—caustic.

For external use—topically, to the skin.

ALCOHOL DENATURANTS

Denatonium Benzoate N.F., benzyldiethyl[(2,6-xylylcarbamoyl)methyl]ammonium benzoate. This compound occurs as a white, crystalline powder with an intensely bitter taste. It is soluble in water and freely soluble in alcohol and in chloroform. Because of its intensely bitter taste and good solubility in alcohol and in water it can replace brucine as a denaturant for ethyl alcohol.

Denatonium Benzoate

Sucrose Octaacetate N.F. is a white, practically odorless powder with an intensely bitter taste. It is soluble in alcohol and in ether but very slightly soluble in water (1 in 1100). It is made by complete acetylation of sucrose.

Sucrose Octaacetate

Methyl Isobutyl Ketone N.F., 4-methyl-2-pentanone, isopropylacetone. This is a colorless liquid with a faint ketonic and camphorlike odor. It is miscible with alcohol and other organic solvents.

$$(CH_3)_2CHCH_2-C-CH_3$$
$$\|$$
$$O$$

Methyl Isobutyl Ketone

These properties make it useful as a component of S.D.A. Formula 23-H (Internal Revenue Service, U.S. Treasury Department), which is used in Rubbing Alcohol U.S.P.

TABLET LUBRICANTS

Stearic Acid N.F. Stearic Acid is a mixture of solid acids obtained from fats, and consists chiefly of stearic acid [$CH_3(CH_2)_{16}COOH$] and palmitic acid [$CH_3(CH_2)_{14}COOH$]. The stearic acid is obtained by the saponification of fats (beef tallow).

Stearic Acid N.F. is a solid, white, waxlike, crystalline substance having practically no taste or odor; it is insoluble in water but soluble in organic solvents.

It is used as a tablet lubricant and to prepare stearates and in ointments, suppositories, creams and cosmetic products. The potassium and sodium salts are of special interest, since these are the common soaps. Esters, such as glycol stearate, are used in ointments, and butyl stearate is satisfactory as an enteric coating for tablets.

Purified Stearic Acid N.F. This is a more pure form of stearic acid. The stearic acid content is not less than 90 percent and most of the balance is palmitic acid. The uses are the same for both grades.

Hydrogenated Vegetable Oil N.F. This product is made by the hydrogenation of vegetable oils and consists primarily of the triglycerides of stearic and palmitic acid. It is a fine, white powder at room temperature. It is used as a tablet lubricant.

Sodium Stearate N.F. This is a mixture of varying proportions of sodium stearate and sodium palmitate. It is prepared by neutralizing Stearic Acid N.F. with sodium carbonate.

$$CH_3(CH_2)_nCOO^- \; Na^+$$

Sodium Palmitate n = 14
Sodium Stearate n = 16

Calcium Stearate N.F. is the calcium salt of stearic and palmitic acids in varying proportions. It is a fine, white to off-white, unctuous powder with a slight characteristic odor. It is insoluble in water, in alcohol, and in ether. Its primary use is in tablet formulation as a solid lubricant for the granules during the tableting process.

Zinc Stearate U.S.P. is a light, fluffy, fine, white powder which is insoluble in organic solvents and in water.

It is used in ointments and powders as a mild astringent and antiseptic. It is also used as a tablet lubricant in the tableting process. The salt is a water-repellent powder and was used widely to replace talcum powder until it was found that inhalation may cause pulmonary inflammation.

Magnesium Stearate N.F. is the magnesium salt of stearic and palmitic acids in varying proportions. It is a fine, white, unctuous powder with a faint characteristic odor. Its properties and uses are similar to those of Calcium Stearate N.F.

Aluminum Monostearate N.F. occurs as a white to off-white bulky powder. It is insoluble in water, alcohol and ether. It is prepared by mixing solutions of a soluble aluminum salt and sodium stearate. Aluminum monostearate is used in Sterile Penicillin G Procaine with Aluminum Stearate Suspension U.S.P. for its action as a dispersing agent and its thixotropic properties. It forms a gel which becomes a free-flowing solution when shaken gently before use.

ANTIOXIDANTS

Butylated Hydroxyanisole N.F., *tert*-butyl-4-methoxyphenol. Butylated hydroxyanisole occurs as a white or slightly yellow, waxy solid that is insoluble in water but freely soluble in alcohol, in propylene glycol, in ether and in many lipids.

Butylated hydroxyanisole, when combined with other antioxidants, offers a means of protection against oxidative rancidity in lipid materials.

Butylated Hydroxyanisole

Butylated Hydroxytoluene N.F., 2,6-di-*tert*-butyl-*p*-cresol, occurs as a white, crystalline solid that is insoluble in water and in propylene glycol but is freely soluble in alcohol and more soluble in lipids than is butylated hydroxyanisole. Butylated hydroxytoluene is used like butylated hydroxyanisole.

Butylated Hydroxytoluene

Monothioglycerol N.F., 3-mercapto-1,2-propanediol, is a light yellow, viscous liquid with a characteristic odor. It is slightly soluble in water.

$$CH_2-CH-CH_2-SH$$
$$||$$
$$OHOH$$

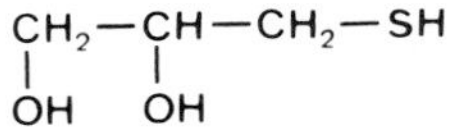

Monothioglycerol

Propyl Gallate N.F., propyl ester of 3,4,5-trihydroxybenzoic acid, is a crystalline material only slightly soluble in water but more soluble in fats and oils. It is generally used in combination with other antioxidants to protect lipid materials from oxidative rancidity.

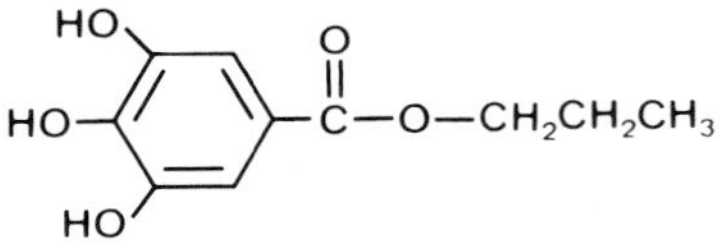

Propyl Gallate

Sodium Formaldehyde Sulfoxylate N.F., consists of white crystals or pieces that are soluble in water but only sparingly so in alcohol. It is almost odorless when freshly prepared but quickly develops a characteristic garliclike odor. It is decomposed rapidly by acids and reduces even very mild oxidizing agents. It has been used as an antioxidant in liquid dose forms of some of the phenothiazine derivatives.

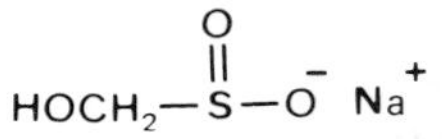

SYNTHETIC SUSPENDING AGENTS

Carbomer N.F., Carbopol®, is a synthetic, high molecular-weight, cross-linked polymer of acrylic acid containing between 56 and 68 percent carboxylic-acid groups.

Several types are available, ranging in average molecular weight from approximately 250,000 to 4,000,000. A pharmaceutical grade has an approximate molecular weight of 3,000,000. It is a fluffy, white, hygroscopic powder with a slight, characteristic odor. It reacts with alkali hydroxides, amines, or amino acids to form salts that possess excellent thickening, suspending, and emulsifying properties. The choice of neutralizing agent determines the hydrophilic and lipophilic properties of the salt that is formed. Maximal viscosity is reached at pH 5.6 and is maintained until about pH 10.

SURFACTANTS AND EMULSIFYING AIDS

POLYSORBATES

The polysorbates are a mixture of sorbitol and its anhydrides (sorbitans), copolymerized with ethylene oxide and monoesterified with a fatty acid. The following general formula shows one of the possible sorbitans reacted with ethylene oxide so that $w+x+y+z$ equals approximately 20; this, in turn, is esterified at a hydroxyl group with a fatty acid to give a polysorbate.

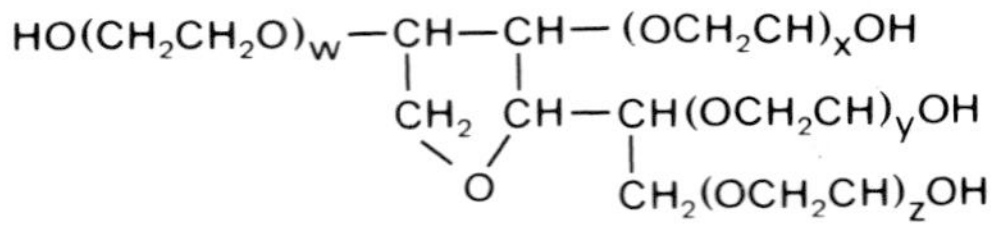

The presence of the polyoxyethylene groups in the molecules leads to a predominance of hydrophilic properties as compared to the sorbitan fatty-acid esters.

The polysorbates are used as solubilizing and emulsifying agents. A blend of more than one surfactant may be used to obtain an emulsion system having the desired hydrophile–lipophile balance. The polysorbates are listed in Table A–2 with some of their properties.

SORBITAN FATTY-ACID ESTERS AND RELATED COMPOUNDS

These are derivatives of sorbitol and its mixed anhydrides or sorbitans that have been esterified partially with an appropriate quantity and type of fatty acid.

TABLE A–2

POLYOXYETHYLENE (20) SORBITAN FATTY-ACID ESTERS

Title	Chemical Name	Product Name	HLB Value	Form
Polysorbate 20 N.F.	Polyoxyethylene 20 Sorbitan Monolaurate	Tween 20	16.7	Liquid
Polysorbate 40 N.F.	Polyoxyethylene 20 Sorbitan Monopalmitate	Tween 40	15.6	Liquid
Polysorbate 60 N.F.	Polyoxyethylene 20 Sorbitan Monostearate and Monolaurate (mixture)	Tween 60	14.9	Liquid
Polysorbate 80 N.F.	Polyoxyethylene 20 Sorbitan Monooleate	Tween 80	15	Liquid

TABLE A–3

SORBITAN FATTY-ACID ESTERS

Title	Product Name	HLB Value	Form
Sorbitan Monolaurate N.F.	Span 20	8.6	Liquid
Sorbitan Monopalmitate N.F.	Span 40	6.7	Solid
Sorbitan Monostearate N.F.	Span 60	4.7	Solid
Sorbitan Monooleate N.F.	Span 80	4.3	Liquid

They possess greater lipophilic properties than the polysorbates so that their hydrophile–lipophile balance is lower than that of the polysorbates. The names and some of their properties are listed in Table A–3. These compounds are used as emulsifying agents for preparing water/oil (w/o) emulsions. They may be blended with polysorbates so that, depending on the proportions used, either w/o or o/w emulsions may be prepared.

Polyoxyl 40 Stearate N.F., Myrj® 52, polyethylene glycol monostearate, polyoxyethylene 40 stearate, stearethate 40, PEG–40 stearate. This is a mixture of the monostearate and distearate esters of mixed polyoxyethylene diols and the corresponding free glycols, the average polymer length being equivalent to about 40 oxyethylene units. The following structure, where n has a value of approximately 40, is representative of this product.

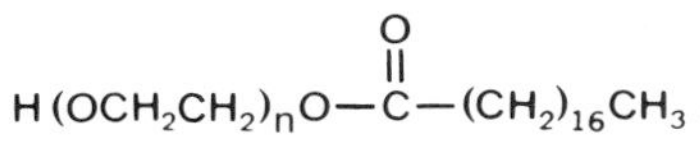

$$H(OCH_2CH_2)_nO-\overset{\displaystyle O}{\overset{\displaystyle \|}{C}}-(CH_2)_{16}CH_3$$

It is a waxy, nearly colorless solid having a congealing range between 38° and 46°C. It is soluble in water, alcohol, ether and acetone and is insoluble in mineral and vegetable oils. The HLB value is 16.9.

Like other agents of this type, it is useful as an emulsifying agent in the preparation of o/w emulsions.

Polyoxyethylene 50 Stearate N.F., Myrj® 53, PEG–50 stearate. This is a mixture of the monostearate and distearate esters of mixed polyoxyethylene diols and the corresponding free diols, with an average polymer length equivalent to about 50 oxyethylene units.

It occurs as a soft, cream-colored, waxy solid having a faint fatlike odor and melting at about 45°C. It is soluble in water and in isopropyl alcohol and insoluble in mineral oil. The HLB value is 17.9.

Polyoxyl 10 Oleyl Ether N.F. Polyoxyl 10 Oleyl Ether is a mixture of the monoleyl and dioleyl ethers of mixed polyoxyethylene diols, the average polymer length being equivalent to about 10 oxyethylene units.

Polyoxyl 20 Cetostearyl Ether N.F. is a mixture of the monocetostearyl and dicetostearyl (cetyl and stearyl mixture) ethers of mixed polyoxyethylene diols, the average polymer length being equivalent to about 20 oxyethylene units. It is also used as an emulsifier and surfactant.

Octoxynol 9 N.F., octylphenoxypolyethoxyethanol. This is an anhydrous mixture of related ethers of the following general structure in which n varies from 5 to 15.

It is a pale yellow liquid with a faint odor and bitter taste. It is miscible with water, alcohol or acetone.

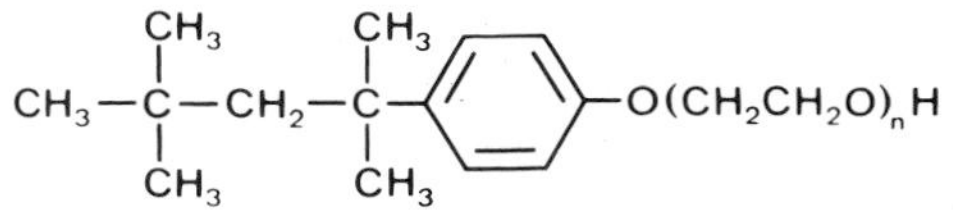

It is soluble in benzene or toluene but is insoluble in solvent hexane. It is used as a surfactant.

Nonoxynol 10 N.F., nonylphenoxypolyethoxyethanol, is a similar mixture of related ethers of the following general structure in which n varies from 6 to 16.

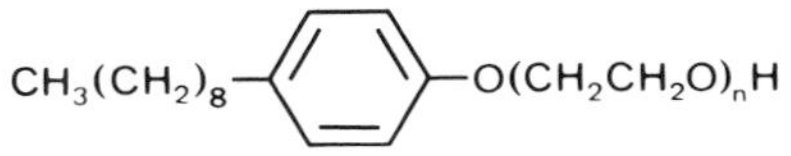

Its uses are similar to those of Octoxynol 9.

Tyloxapol U.S.P., *p*-isooctylpolyoxyethylene phenol formaldehyde polymer, has the following general structure in which m is 8 to 10 and n is not more than 5.

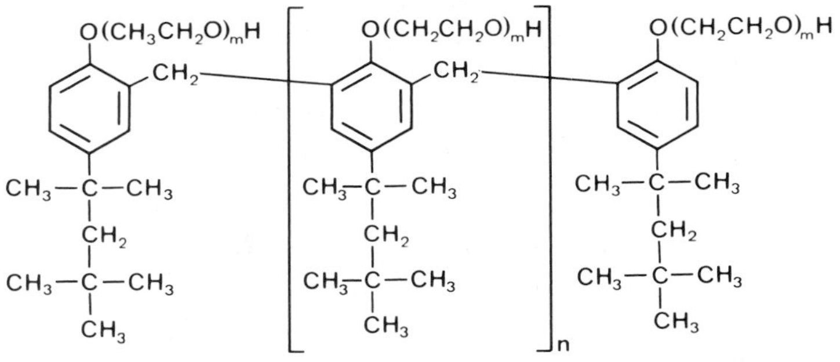

It is a liquid freely soluble in water and soluble in benzene or chlorinated hydrocarbon solvents. Precautions must be taken to avoid contact with metals. Tyloxapol is used as a detergent.

Tromethamine U.S.P., 2-amino-2-hydroxymethyl-1,3-propanediol, *tris*(hydroxymethyl)aminomethane, trisamine, tris buffer, is a monoacidic base (pKa 8.1 for the protonated amine). It is widely used as an emulsifying agent for cosmetic creams and lotions. It is also

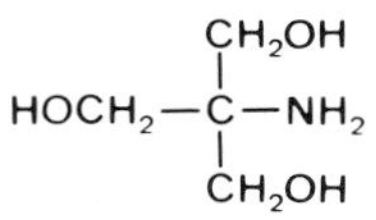

Tromethamine

used with its conjugate acid as a buffer with a useful range of pH 7 to 9. Tromethamine for Injection U.S.P. is used intravenously to correct systemic acidosis.

Meglumine U.S.P., N-methylglucamine. This is a

Meglumine

monoacidic base (pKa about 8 for the protonated amine). It forms salts with acids and chelates metal ions. Salts with alkyl aryl sulfonic acids are useful surface active agents.

A pharmaceutical use is the formation of highly water-soluble salts with triiodobenzoic acids used as radiopaque agents, e.g., diatrizoate meglumine.

Trolamine N.F., triethanolamine. This is a mixture consisting largely of triethanolamine, $N(CH_2CH_2OH)_3$, with smaller amounts of the primary and the secondary compounds, $NH_2CH_2CH_2OH$ and $NH(CH_2CH_2OH)_2$, respectively. The substance is a colorless to pale yellow, viscous, hygroscopic liquid having a slightly ammoniacal odor. It is miscible with water or alcohol and is soluble in chloroform. The pH of a 25 percent solution is 11.2.

Triethanolamine combines with fatty acids to form amine soaps with good emulsifying and detergent properties. These compounds are water-soluble and are less alkaline than the corresponding sodium, potassium or ammonium salts.

Monoethanolamine N.F., 2-aminoethanol, $NH_2CH_2CH_2OH$, is a viscous, colorless, very hygroscopic liquid with an ammoniacal odor. It has about the same base strength as ammonia. As compared to triethanolamine, it may be more advantageous for the preparation of soaps because it is more basic and has a lower combining weight. Also, it is useful as a general alkalizing agent. The use of this agent in thimerosal solutions is based on its alkalizing action as these solutions are most stable at an alkaline pH.

Glyceryl Monostearate N.F., monostearin.

Glyceryl Monostearate

It is obtained from mono- and diglycerides of saturated fatty acids. The main components are glyceryl monostearate and glyceryl monopalmitate.

This agent occurs as a white, waxlike solid or as white, waxlike beads or flakes. It has a slight, agreeable, fatty odor and taste. It is affected by light.

It melts above 55°C and is soluble in hot organic solvents, such as alcohol, mineral oil and acetone. It is insoluble in water, but readily dispersible in hot water with the aid of soap or other surface-active agents.

Glyceryl monostearate has a strong lipophilic property as indicated by its HLB value of 3.5. It serves as an emulsifying agent for w/o emulsions. When mixed with a suitable surfactant it forms o/w emulsions. The combination of glyceryl monostearate with a soap or other surfactant is available as a commercial product known as glyceryl monostearate, self-emulsifying. In addition to its emulsifying action, glyceryl monostearate increases the consistency of preparations.

Propylene Glycol Monostearate N.F., 1,2-propanediol monostearate.

Propylene Glycol Monostearate

This is a mixture of the propylene glycol mono- and diesters of stearic and palmitic acids, consisting chiefly of propylene glycol monostearate and propylene glycol monopalmitate.

It occurs as a white, waxlike solid or as beads or flakes. It is insoluble in water, but dispersible in hot water with the aid of soap or other surface-active agents.

Similar to glyceryl monostearate, this agent has a strong lipophilic property and is used as an emulsifying agent for w/o emulsions.

Sodium Lauryl Sulfate N.F., Duponol® C, is a mixture of sodium alkyl sulfates consisting chiefly of sodium lauryl sulfate $[CH_3(CH_2)_{10}CH_2OSO_3Na]$. This is prepared by sulfating long-chain alcohols and neutralizing to form the sodium salts. The alcohols are derived by reduction of coconut oil and other fatty glycerides by high-pressure hydrogenation with copper-chromium oxide catalyst.

It occurs as white or light-yellow crystals or flakes having a slight coconut fatty odor. It is soluble in water (1:10). The official form permits up to 8 percent of combined sodium sulfate and sodium chloride.

It is used for its foaming and cleansing activity in shampoos and dental preparations. It serves as an emulsifier in preparing water-miscible ointment bases for pharmaceutic and cosmetic preparations.

pKa's of drugs and reference compounds

Name	pKa*	Reference	Name	pKa*	Reference
Acenocoumarol	4.7	1	Aniline	4.6	6
Acetaminophen	9.9 (phenol)	2	Anisindione	4.1	27
Acetanilid	0.5	3	Antazoline	10.0	28
Acetarsone	3.7 (acid)	4	Antifebrin	1.4	29
	7.9 (phenol)		Antipyrine	2.2	29
	9.3 (acid)		Apomorphine	7.0	30
Acetazolamide	8.8 (acetamido)	5	Aprobarbital	7.8	10
Acetic Acid	4.8	6	Arecoline	7.6	31
α-Acetylmethodol	8.3	7	Arsthinol	9.5 (phenol)	4
Acetylpromazine	9.3	8	Ascorbic Acid	4.2	18
N⁴-Acetylsulfadiazine	6.1	9		11.6	
N⁴-Acetylsulfametiazole	5.2	9	Aspirin	3.5	7
N⁴-Acetylsulfamethoxypyridazine	6.9	9	Atropine	9.7	30
N⁴-Acetylsulfapyridine	8.2	9	Barbital	7.8	32
N⁴-Acetylsulfisoxazole	4.4	9	Barbituric Acid	4.0	32
Allobarbital	7.5	10	Bemegride	11.2	33
Allopurinol	9.4	11	Bendroflumethiazide	8.5	34
Allylamine	10.7	12	Benzilic Acid	3.0	12
Allylbarbituric Acid	7.6	13	Benzocaine	2.8	30
Alphaprodine	8.7	14	Benzoic Acid	4.2	7
Alprenolol	9.6	15	Benzphetamine	6.6	35
Amiloride	8.7	15	Benzquinamide	5.9	36
Amantadine	10.8	16	Benzylamine	9.3	12
p-Aminobenzoic Acid	2.4 (amine)	17	Biscoumacetic Acid	3.1	1
	4.9			7.8 (enol)	174
Aminocaproic Acid	4.4	18	Bromodiphenhydramine	8.6	37
	10.8		p-Bromophenol	9.2	38
6-Aminopenicillanic Acid	2.3 (carboxyl)	19	Bromothen	8.6	28
	4.9 (amine)		8-Bromotheophylline	5.5	13
Aminopterin	5.5 (heterocyclic ring)	20	Brucine	8.0	6
Aminopyrine	5.0	6	Bupivacaine	8.1	39
Aminosalicylic Acid	1.7 (amine)	17	Butabarbital	7.9	15
	3.9		Butethal	8.1	40
Amitriptyline	9.4	21	Butylparaben	8.4	41
Ammonia	9.3	12	Butyric Acid	4.8	12
Amobarbital	8.0	22	Caffeine	14.0	42
Amoxicillin	2.4 (carboxyl)	23		0.6 (amine)	
	7.4 (amine)		Camphoric Acid	4.7	12
	9.6 (phenol)		Carbinoxamine	8.1	43
Amphetamine	9.8	24	Carbonic Acid	6.4 (1st)	6
Amphotericin B	5.7 (carboxyl)	25		10.4 (2nd)	
	10.0 (amine)		Cefazolin	2.3	44
Ampicillin	2.7	26	Cephalexin	3.6	45
	7.3 (amine)		Cephaloglycin	2.5	45

* pKa given for protonated amine.
(Continued)

Name	pKa*	Reference	Name	pKa*	Reference
Cephaloridine	3.4	46	Dihydrocodeine	8.8	3
Cephalothin	2.4	45	3,5-Diiodo-L-tyrosine	2.5 (amine)	32
Cephradine	2.6 (carboxyl)	47		6.5	
	7.3 (amine)			7.5 (phenol)	
Chlorambucil	5.8	48	Dimethylamine	10.7	12
Chlorcyclizine	7.8	37	p-Dimethylaminobenzoic Acid	5.1	73
Chlordiazepoxide	4.8	49	p-Dimethylaminosalicylic Acid	3.8	73
Chlorindione	3.6	27	Dimethylbarbituric Acid	7.1	12
Chloroquine	8.1	50	Dimethylhydantoin	8.1	3
	9.9		2,4-Dinitrophenol	4.1	74
8-Chlorotheophylline	5.3	51	Diperodon	8.4	75
Chlorothiazide	6.7	52	Diphenhydramine	9.0	37
	9.5		Diphenoxylate	7.1	76
Chlorpheniramine	9.2	37	Doxorubicin	8.2	77
Chlorphentermine	9.6	35		10.2	
Chlorpromazine	9.2	53	Doxycycline	3.4	78
Chlorpromazine Sulfoxide	9.0	44		7.7	
Chlorpropamide	5.0	54		9.3	
Chlorprothixene	8.4	55	Doxylamine	9.2	37
Chlortetracycline	3.3	56	Droperidol	7.6	79
	7.4		Ephedrine	9.6	7
	9.3		Epinephrine	8.7 (phenol)	80
Cinchonidine	4.2 (1st)	3		9.9 (amine)	
	8.4 (2nd)		Equilenin	9.8	61
Cinchonine	4.0 (1st)	30	Ergotamine	6.3	81
	8.2 (2nd)		Erythromycin	8.8	82
Cinnamic Acid	4.5	12	Erythromycin Estolate	6.9	83
Citric Acid	3.1 (1st)	57	17α-Estradiol	10.7	61
	4.8 (2nd)		Estriol	10.4	61
	6.4 (3rd)		Ethacrynic Acid	3.5	18
Clindamycin	7.5	58	Ethambutol	6.6	84
Clofibrate	3.0 (acid)	15		9.5	
Clonazepam	10.5 (1-position)	59	Ethanolamine	9.5	72
	1.5 (4-position)		Ethopropazine	9.6	28
Clonidine	8.0	60	Ethosuximide	9.3	15
Cloxacillin	2.7	19	p-Ethoxybenzoic Acid	4.5	73
Cobefrin	8.5	24	p-Ethoxysalicylic Acid	3.2	73
Cocaine	8.4	3	Ethylamine	10.7	12
Codeine	7.9	42	Ethylbarbituric Acid	4.4	12
Colchicine	1.7	18	Ethyl Biscoumacetate	3.1	1
o-Cresol	10.3	17	Ethylenediamine	6.8 (1st)	72
m-Cresol	10.1	17		9.9 (2nd)	
p-Cresol	10.3	61	Ethylparaben	8.4	41
Cyanic Acid	3.8	12	Ethylphenylhydantoin	8.5	85
Cyanopromazine	9.3	62	Etidocaine	7.7	39
Cyclizine	8.2	63	β-Eucaine	9.4	3
Cyclopentamine	3.5	64	Fenfluramine	9.1	35
Cyclopentolate	7.9	65	Fenoprofen	4.5	86
Dantrolene	7.5	66	Flucytosine	10.7 (amide)	87
Debrisoquin	11.9	67		2.9 (amine)	
Dehydrocholic Acid	5.0	68	Flufenamic Acid	3.9	74
Demeclocycline	3.3	56	Flunitrazepam	1.8	59
	7.2		p-Fluorobenzoic Acid	4.2	38
	9.3		Fluorouracil	8.0	88
Desipramine	10.2	21		13.0	
Dextromethorphan	8.3	43	Fluphenazine	8.1 (1st)	53
Diatrizoic Acid	3.4	69		9.9 (2nd)	
Diazepam	3.3	49	Fluphenazine Enanthate	3.5	89
Dibucaine	8.5	70		8.2	
Dichloroacetic Acid	1.3	42	Flurazepam	8.2	90
Dicloxacillin	2.8	26		1.9	
Dicumarol	4.4 (1st)	71	Formic Acid	3.7	12
	8.0 (2nd)		Fumaric Acid	3.0 (1st)	42
Diethanolamine	8.9	72		4.4 (2nd)	
Diethylamine	11.0	12	Furaltadone	5.0	91
p-Diethylaminobenzoic Acid	6.2	73	Furosemide	4.7	92
p-Diethylaminosalicylic Acid	3.8	73	Gallic Acid	4.2	38

* pKa given for protonated amine.
(Continued)

Name	pKa*	Refer-ence	Name	pKa*	Refer-ence
Glibenclamide	6.5	54	Levodopa (cont.)	9.7 (1st phenol)	113
Gluconic Acid	3.6	18		13.4 (2nd phenol)	
Glucuronic Acid	3.2	93	Levomepromazine	9.2	64
Glutamic Acid	4.3	12	Levorphanol	8.9	14
Glutarimide	11.4	16	Levulinic Acid	4.6	12
Glutethimide	11.8	94	Lidocaine	7.9	114
Glycerophosphoric Acid	1.5 (1st)	42	Lincomycin	7.5	115
	6.2 (2nd)		Liothyronine	8.4 (phenol)	116
Glycine	2.4	42	Lorazepam	11.5	117
	9.8 (amine)			1.3	
Glycollic Acid	3.8	12	Malamic Acid	3.6	12
Guanethidine	11.9	67	Maleic Acid	1.9	12
Guanidine	13.6	3	Malic Acid	3.5 (1st)	57
Haloperidol	8.3	95		5.1 (2nd)	
Heroin	7.8	14	Malonic Acid	2.8	12
Hexachlorophene	5.7	74	Mandelic Acid	3.8	97
Hexetidine	8.3	96	Mecamylamine	11.2	7
Hexobarbital	8.3	85	Meclizine	3.1	118
Hexylcaine	9.1	70		6.2	
Hippuric Acid	3.6	97	Medazepam	6.2	119
Histamine	9.9 (side chain)	98	Mefenamic Acid	4.3	74
	6.0 (imidazole)		Mepazine	9.3	53
Homatropine	9.7	3	Meperidine	8.7	14
Hydantoin	9.1	17	Mephentermine	10.3	35
Hydralazine	0.5 (ring N)	99	Mephenytoin	8.1	120
	6.9 (hydrazine)		Mephobarbital	7.7	85
Hydrochlorothiazide	7.0	18	Mepivacaine	7.6	39
	9.2		Mercaptopurine	7.8	121
Hydrocortisone Hemisuccinate Acid	5.1	100	Metaproterenol	8.8	15
Hydroflumethiazide	8.9	101	Methacycline	3.5	15
	10.5			7.6	
Hydrogen Peroxide	11.3	12		9.2	
Hydromorphine	7.8	3	Methadone	8.3	14
Hydroxyamphetamine	9.6	24	Methamphetamine	9.5	43
p-Hydroxybenzoic Acid	4.1	38	Methapyrilene	3.7	3
o-Hydroxycinnamic Acid	4.7	38		8.9 (side chain)	
m-Hydroxycinnamic Acid	4.5	38	Methaqualone	2.5	122
p-Hydroxycinnamic Acid	4.4	38	Metharbital	8.2	85
Hydroxylamine	6.0	6	Methazolamide	7.3	18
p-Hydroxysalicylic Acid	3.2	73	Methenamine	4.9	29
Hydroxyzine	1.8	102	Methicillin	2.8	19
Ibuprofen	5.2	103	Methopromazine	9.4	62
Idoxuridine	8.3	104	Methotrexate	4.8	124
Imidazole	7.0	72		5.5	
Imipramine	9.5	21	Methohexital	8.3	123
Indomethacin	4.5	18	Methoxamine	9.2	64
Indoprofen	5.8	105	Methoxyacetic Acid	3.5	12
Iodipamide	3.5	106	o-Methoxybenzoic Acid	4.2	38
Iophenoxic Acid	7.5	1	m-Methoxybenzoic Acid	4.2	38
Isocarboxazid	10.4	107	o-Methoxycinnamic Acid	4.7	38
Isomethadone	8.1	14	m-Methoxycinnamic Acid	4.5	38
Isoniazid	10.8 (pyridine)	108	p-Methoxycinnamic Acid	4.9	38
	11.2 (hydrazide)		Methyclothiazide	9.4	125
Isophthalic Acid	3.6	12	Methylamine	10.6	12
Isoproterenol	8.7 (amine)	109	1-Methylbarbituric Acid	4.4	17
	9.9 (phenol)		Methyldopa	2.2	126
Kanamycin	7.2	15		10.6 (amine)	
Ketamine	7.5	110		9.2 (1st phenol)	
Lactic Acid	3.9	18		12.0 (2nd phenol)	
Leucovorin	3.1	18	N-Methylephedrine	9.3	127
	4.8		Methylergonovine	6.7	81
	10.4 (phenol)		N-Methylglucamine	9.2	126
Levallorphan Tartrate	6.9	111	Methylhexylamine	10.5	64
Levarterenol	8.7 (phenol)	112	Methylparaben	8.4	41
	9.7 (amine)		Methylphenidate	8.8	128
Levodopa	2.3 (carboxyl)	113	Methylprednisolone-21-phosphate	2.6	129
	8.7 (amine)			6.0	

* pKa given for protonated amine.
(Continued)

Name	pKa*	Reference	Name	pKa*	Reference
Methylpromazine	9.4	62	Perphenazine	7.8	21
Methyprylon	12.0	130	Phenacetin	2.2	29
Methysergide	6.6	81	Phenadoxane	6.9	14
Metopon	8.1	14	Phendimetrazine	7.6	35
Metoprolol	9.7	15	Phenethicillin	2.7	19
Metronidazole	2.6	131	Phenformin	11.8	140
Miconazole	6.9	132	Phenindamine	8.3	37
Minocycline	2.8	15	Phenindione	4.1	27
	5.0		Pheniramine	9.3	37
	7.8		Phenmetrazine	8.5	35
	9.5		Phenobarbital	7.5	13
Molindone	6.9	133	Phenol	9.9	7
Monochloroacetic Acid	2.9	42	Phenolsulfonphthalein	7.9	18
Morphine	8.0	32	Phenoxyacetic Acid	3.1	12
	9.6 (phenol)		Phentermine	10.1	35
Nafcillin	2.7	26	Phenylbutazone	4.4	141
Nalidixic Acid	6.0 (amine)	134	Phenylbutazone (isopropyl analog)	5.5	142
	1.0		Phenylephrine	9.8 (amine)	143
Nalorphine	7.8	14		8.8 (phenol)	
Naphazoline	3.9	64	Phenylethylamine	9.8	24
1-Naphthol	9.2	38	Phenylpropanolamine	9.4	109
2-Naphthol	9.4	38	Phenylpropylmethylamine	9.9	24
Naproxen	4.2	135	Phenyramidol	5.9	18
Narcotine	5.9	3	Phenytoin	8.3	144
Nicotine	3.1	136	o-Phthalamic Acid	3.8	12
	8.0		Phthalic Acid	2.9	6
Nicotine Methiodide	3.2	136	Phthalimide	7.4	12
Nicotinic Acid	4.8	18	Physostigmine	2.0	3
Nitrazepam	10.8	117		8.1	
	3.2		Picolinic Acid	5.3	38
o-Nitrobenzoic Acid	3.2	38	Picric Acid	0.4	42
m-Nitrobenzoic Acid	3.6	38	Pilocarpine	1.6	3
p-Nitrobenzoic Acid	3.7	38		7.1	
Nitrofurantoin	7.2	91	Piperazine	5.7	30
Nitrofurazone	10.0	137		10.0	
Nitromethane	11.0	12	Piperidine	11.2	6
o-Nitrophenol	7.2	38	Pirbuterol	3.0 (pyridine)	145
m-Nitrophenol	8.3	38		7.0 (pyridol)	
p-Nitrophenol	7.1	38		10.3 (amine)	
8-Nitrotheophylline	2.1	51	Plasmoquin	3.5	146
Norhexobarbital	7.9	123		10.1	
Norketamine	6.7	110	Polymyxin B	8.9	15
Norparamethadione	6.1	85	Prazepam	3.0	147
Nortrimethadione	6.2	85	Prilocaine	7.9	39
Noscapine	6.2	18	Probarbital	8.0	22
Novobiocin	4.3	18	Probenecid	3.4	1
	9.1		Procainamide	9.2	148
Ornidazole	2.6	131	Procaine	9.0	149
Orphenadrine	8.4	15	Procarbazine	6.8	150
Oxamic Acid	2.1	12	Prochlorperazine	3.6	64
Oxazepam	1.8	138		7.5	
	11.1		Promazine	9.4	53
Oxyphenbutazone	4.5	93	Promethazine	9.1	64
	10.0 (phenol)		Propicillin	2.7	19
Oxytetracycline	3.3	56	Propiomazine	6.6	151
	7.3		Propionic Acid	4.9	42
	9.1		Propranolol	9.5	15
Pamaquine	8.7	3	i-Propylamine	10.6	12
Papaverine	5.9	42	n-Propylamine	10.6	12
Penicillamine	1.8 (carboxyl)	18	Propylhexedrine	10.5	64
	7.9 (amino)		Propylparaben	8.4	41
	10.5 (thiol)		Propylthiouracil	7.8	152
Penicillin G	2.8	42	Pseudoephedrine	9.9	35
Penicillin V	2.7	19	Pyrathiazine	8.9	64
Penicilloic Acid	5.2	139	Pyrazinamide	0.5	18
Pentachlorophenol	4.8	74	Pyridine	5.2	6
Pentobarbital	8.0	32	Pyridoxine	2.7	153

* pKa given for protonated amine.
(Continued)

Name	pKa*	Reference	Name	pKa*	Reference
Pyridoxine (cont.)	5.0 (amine)	153	Talbutal	7.8	10
	9.0 (phenol)	154	Tartaric Acid	3.0 (1st)	57
Pyrilamine	4.0	3		4.3 (2nd)	
	8.9		Tetracaine	8.5	70
Pyrimethamine	7.2	1	Tetracycline	3.3	56
Pyrimethazine	9.4	53		7.7	
Pyrrobutamine	8.8	37		9.5	
Pyruvic Acid	2.5	18	Thenyldiamine	3.9	3
Quinacrine	8.0	3		8.9	
	10.2		Theobromine	8.8	32
Quinidine	4.2	3		0.7 (amine)	
	8.3		Theophylline	8.8	32
Quinine	4.2	3		0.7 (amine)	
	8.8		Thiamine	4.8	153
Reserpine	6.6	42		9.0	
Resorcinol	6.2	12	Thiamylal	7.3	123
Riboflavin	1.7	3	Thioacetic Acid	3.3	12
	10.2		Thioglycolic Acid	3.6	12
Rifampin	1.7 (C–8 phenol)	155	Thiopental	7.5	123
	7.9 (piperazine N)		Thiopropazate	3.2	64
Saccharic Acid	3.0	12		7.2	
Saccharin	1.6	42	Thioridazine	9.5	53
Salicylamide	8.1	32	Thiouracil	7.5	152
Salicylic Acid	3.0	6	Thonzylamine	8.8	37
	13.4 (phenol)		L-Thyronine	9.6 (phenol)	116
Scopolamine	7.6	3	L-Thyroxine	2.2 (carboxyl)	162
Secobarbital	8.0	156		6.7 (phenol)	
Serotonin	4.9	18		10.1 (amine)	
	9.8		Tolazamide	5.7	15
Sorbic Acid	4.8	12	Tolazoline	10.3	163
Sotalol	9.8 (amine)	157	Tolbutamide	5.3	32
	8.3 (sulfonamide)		p-Toluidine	5.3	7
Spectinomycin	7.0	18	Trichloroacetic Acid	0.9	42
	8.7		Triethanolamine	7.8	72
Strychnine	2.5	3	Triethylamine	10.7	29
	8.2		Trifluoperazine	4.1	53
Succinic Acid	4.2 (1st)	57		8.4	
	5.6 (2nd)		Triflupromazine	9.4	64
Succinimide	9.6	158	Trimethobenzamide	8.3	164
Succinuric Acid	4.5	12	Trimethoprim	7.2	165
Sulfacetamide	5.4	159	Trimethylamine	9.8	12
	1.8		Tripelennamine	9.0	37
Sulfadiazine	6.5	32	Triprolidine	6.5	166
Sulfadimethoxine	6.7	159	Troleandomycin	6.6	18
	2.0 (amine)		Tromethamine	8.1	72
Sulfadimethoxytriazine	5.0	9	Tropacocaine	9.7	30
Sulfaethidole	5.4	32	Tropic Acid	4.1	12
Sulfaguanidine	2.8	32	Tropicamide	5.2	167
Sulfamerazine	7.1	42	Tropine	10.4	12
Sulfameter	6.8	18	Tubocurarine Chloride	7.4	168
Sulfamethazine	7.4	32	Tyramine	9.5 (phenol)	109
Sulfamethizole	5.4	32		10.8 (amine)	
Sulfamethoxypyridazine	7.2	9	Urea	0.2	169
Sulfanilamide	10.4	160	Uric Acid	5.4	170
Sulfanilic Acid	3.2	6		10.3	
Sulfaphenazole	6.5	159	Valeric Acid	4.8	42
	1.9 (amine)		Vanillic Acid	4.5	12
Sulfapyridine	8.4	42	Vanillin	7.4	17
Sulfasalazine	0.6 (amine)	161	Vinbarbital	8.0	22
	2.4 (carboxyl)		Vinblastine	5.4	171
	9.7 (sulfonamide)			7.4	
	11.8 (phenol)		Viomycin	8.2	18
Sulfathiazole	7.1	42		10.3	
Sulfinpyrazone	2.8	93		12.0	
Sulfisomidine	7.5	159	Warfarin	5.1	172
	2.4 (amine)		Xipamide	4.8 (phenol)	173
Sulfisoxazole	5.0	32		10.0 (sulfonamide)	

*pKa given for protonated amine.
(Continued)

REFERENCES

1. Anton, A. H.: J. Pharmacol. Exp. Ther. 134:291, 1961.
2. Rhodes, H. J., *et al.*: J. Pharm. Sci. 64, 1387, 1975.
3. Perrin, D. D.: Dissociation Constants of Organic Bases, London, Butterworths, 1965.
4. Hiskey, C. F., and Cantwell, F. F.: J. Pharm. Sci. 57:2105, 1968.
5. Coleman, J. E.: Ann. Rev. Pharmacol. 15: 238, 1975.
6. Kolthoff, I. M., and Stenger, V. A.: Volumetric Analysis, vol. 1, ed. 2, New York, Interscience Publishers, Inc., 1942.
7. Schanker, L. S., *et al.*: J. Pharmacol. Exp. Ther. 120:528, 1957.
8. Liu, S., and Hurwitz, A.: J. Colloid Interface Sci. 60:410, 1977, through Chem. Abstr. 87: 44189u, 1977.
9. Scudi, J. W., and Plekss, O. J.: Pro. Soc. Exptl. Biol. Med. 97:639, 1958.
10. Carstensen, J. T., *et al.*: J. Pharm. Sci. 53:1547, 1964.
11. Gressel, P. D., and Gallelli, S. F.: J. Pharm. Sci. 57, 335, 1968.
12. Washburn, E. W. (ed. in chief), International Critical Tables, New York, McGraw-Hill, 6:261, 1929.
13. Maulding, H. V., and Zoglio, M. A.: J. Pharm. Sci. 60:311, 1971.
14. Beckett, A. H.: J. Pharm. Pharmacol. 8:851, 1956.
15. Heel, R. C., and Avery, G. S., *in* Avery, G.S. (ed.): Drug Treatment, ed. 2, pp. 1212–1222, Sidney, ADIS, 1980.
16. Albert, A.: Selective Toxicity, ed. 4, p. 281, London, Methuen, 1968.
17. Kortüm, G., *et al.*: Dissociation Constants of Organic Acids, London, Butterworths, 1961.
18. Windholz, M. (ed.): The Merck Index, 9th ed., Rahway, N.J., Merck & Co., 1976.
19. Rapson, H. D. C., and Bird, A. E.: J. Pharm. Pharmacol 15:226T, 1963.
20. Baker, R. B., and Jordan, J. H.: J. Pharm. Sci. 54:1741, 1965.
21. Green, A. L.: J. Pharm. Pharmacol. 19:10, 1967.
22. Krahl, M. E.: J. Phys. Chem. 44: 449, 1940.
23. Rolinson, G. N.: J. Infect. Dis. 129:S143, 1974.
24. Leffler, E. B., *et al.*: J. Am. Chem. Soc. 73:2611, 1951.
25. Asher, I. M., *et al.*: Analyt. Profiles 6:11, 1977.
26. Hou, J. P., and Poole, J. W.: J. Pharm. Sci. 58:1150, 1969.
27. Stella, V. J., and Gish, R.: J. Pharm. Sci. 68:1047, 1979.
28. Marshall, P. B.: Brit. J. Pharmacol. 10:270, 1955.
29. Evstratova, K. I., *et al.*: Farmatsiya 17:33, 1968; through Chem. Abstr, 69:9938a, 1968.
30. Kolthoff, J. M.: Biochem. Z. 162:289, 1925; through Trans. Faraday Soc. 39:338, 1945.
31. Burgen, A. S. V.: J. Pharm. Pharmacol. 16:638, 1964.
32. Ballard, B. B., and Nelson, E.: J. Pharmacol. Exp. Ther. 135:120, 1962.
33. Peinhardt, G.: Pharmazie 32:725, 1977; through Chem. Abstr. 88:141587a, 1978.
34. Agren, A., and Bäck, T.: Acta. Pharm. Suec. 10:225, 1973.
35. Vree, T. B., *et al.*: J. Pharm. Pharmacol. 21:774, 1969.
36. Wiseman, E. H., *et al.*: Biochem. Pharmacol. 13:1421, 1964.
37. Lordi, N. G., and Christian, J. E.: J. Am. Pharm. A. 45:300, 1956.
38. Conners, K. A., and Lipari, J. M.: J. Pharm. Sci. 65:380, 1976.
39. de Jong, R. H.: J. Am. Med. Assoc. 238:1284, 1977.
40. Fincher, J. H., *et al.*: J. Pharm. Sci. 55:24, 1966.
41. Tammilehto, S., and Büchi, J.: Pharm. Acta Helvet. 43:726, 1968.
42. Martin, A. N., *et al.*: Physical Pharmacy, ed. 2, p. 194, Philadelphia, Lea & Febiger, 1969.
43. Borodkin, S., and Yunker, M. H.: J. Pharm. Sci. 59:481, 1970.
44. Nightengale, C. H., *et al.*: J. Pharm. Sci. 64:1907, 1975.
45. Streng, W. H.: J. Pharm. Sci. 67:667, 1978.
46. Flynn, E. H. (ed.): Cephalosporins and Penicillins, New York, Academic Press, p. 316, 1972.
47. Florey, K.: Analyt. Profiles 5:36, 1976.
48. Linford, J. H.: Biochem. Pharmacol. 12:321, 1963.
49. Van der Kleijn, E.: Arch. Int. Pharmacodyn. 179:242, 1969.
50. Hong, P. D.: Analyt. Profiles 5:69, 1976.
51. Meyer, M. C., and Guttman, D. E.: J. Pharm. Sci. 57:245, 1968.
52. Resetarits, D. E., and Bates, T. R.: J. Pharm. Sci. 68:126, 1979.
53. Sorby, D. L., *et al.*: J. Pharm. Sci. 58:788, 1966.
54. Crooks, M. J., and Brown, K. F.: J. Pharm. Pharmacol 26:305, 1974.
55. Rudy, B. C., and Senkowski, B. Z.: Analyt. Profiles 2:75, 1973.
56. Benet, L. Z., and Goyan, J. E.: J. Pharm. Sci. 55:983, 1965.
57. Pitman, I. H., *et al.*: J. Pharm. Sci. 57:239, 1968.
58. Taraszka, M. J.: J. Pharm. Sci. 60:946, 1971.
59. Kaplan, S. A., *et al.*: J. Pharm. Sci. 63:527, 1974.
60. Timmermars, P. B. M. W. M., and van Zweiten, P. A.: Arzn. Forsch. 28:1676, 1978.
61. Hurwitz, A. R., and Liu, S. T.: J. Pharm. Sci. 66:626, 1977.
62. Hulshoff, A., and Perrin, J.: Pharm. Acta. Helv. 51:67, 1976.
63. Barlow, R. B.: Introduction to Chemical Pharmacology, ed. 2, p. 357. New York, Wiley, 1964.
64. Chatten, L. G., and Harris, L. E.: Anal. Chem. 34:1499, 1962.
65. Wang, E. S. N., and Hammerlund, E. R.: J. Pharm. Sci. 59:1561, 1970.
66. Vallner, J. J., *et al.*: J. Pharm. Sci. 65:873, 1976.
67. Hengstmann, J. H., *et al.*: Analyt. Chem. 46:35, 1974.
68. Jons, W. H., and Bates, T. R.: J. Pharm. Sci. 59:329, 1970.
69. Langecker, A. A., *et al.*: Arch. Exp. Pathol. Pharmacol. 222:584, 1954.
70. Truant, A. P., and Takman, B.: Anesth. Analg. 38:478, 1959.
71. Cho, M. J., *et al.*: J. Pharm. Sci. 60:197, 1971.
72. Bates, R. G.: Ann. N.Y. Acad. Sci. 92:341, 1961.
73. Pothisiri, P., and Carstensen, J. T.: J. Pharm. Sci. 64:1933, 1975.
74. Terrada, H., *et al.*: J. Med. Chem. 17:330, 1974.
75. Cohen, J. L.: Analyt. Profiles 6:108, 1978.
76. Peeters, J. J.: J. Pharm. Sci. 67:129, 1978.
77. Sturgeon, R. J., and Schulman, S. G.: J. Pharm. Sci. 66:959, 1977.
78. Jaffe, J. M., *et al.*: J. Pharmacokin. Biopharm. 1:281, 1973.
79. Janicki, C. A., and Gilpin, R. K.: Analyt. Profiles 7:181, 1978.
80. Marten, R. B.: J. Phys. Chem. 75:2659, 1971.
81. Maulding, H. V., and Zoglio, M. A.: J. Pharm. Sci. 59:700, 1970.
82. Garrett, E. R., *et al.*: J. Pharm. Sci. 59:1449, 1970.
83. Mann, J. M.: Analyt. Profiles 1:107, 1972.
84. Shepherd, R. G., *et al.*: Ann. N.Y. Acad. Sci. 135:698, 1966.
85. Butler, T. C.: J. Am. Pharm. A. 44:367, 1955.
86. Ward, C. K., and Schirmer, R. E.: Analyt. Profiles 6:165, 1977.
87. Waysek, E. H., and Johnson, J. H.: Analyt. Profiles 5:128, 1976.
88. Rudy, B. C., and Senkowski, B. Z.: Analyt. Profiles 2:234, 1973.
89. Florey, K.: Analyt. Profiles 2:254, 1973.
90. Rudy, B. C., and Senkowski, B. Z.: Analyt. Profiles 3:321, 1974.
91. Buzard, J. A., *et al.*: Am. J. Physiol, 201:492, 1961.
92. McCallister, J. B., *et al.*: J. Pharm. Sci. 59:1288, 1970.
93. Perel, J. M., *et al.*: Biochem. Pharmacol. 13:1305, 1964.
94. DeLuca, P. P., *et al.*: J. Pharm. Sci. 62:1321, 1973.
95. Janssen, P. A. J., *et al.*: J. Med. Pharm. Chem. 1:282, 1959.
96. Satzinger, G., *et al.*: Analyt. Profiles 7:284, 1978.
97. Parrott, E. L., and Saski, W.: Exper. Pharm. Technol., p. 255. Minneapolis, Burgess, 1965.
98. Paiva, T. B., *et al.*: J. Med. Chem. 13:690, 1970.
99. Naik, *et al.*: J. Pharm. Sci. 65:275, 1976.
100. Garrett, E. R.: J. Pharm. Sci. 51:445, 1962.
101. Smith, R. B., *et al.*: J. Pharm. Sci. 65:1209, 1976.
102. Florey, K.: Analyt. Profiles 7:325, 1978.
103. Upjohn product information 1975.
104. Prusoff, W. H.: Pharmacol. Rev. 19:223, 1967.
105. Fucella, L. M., *et al.*: Clin. Pharmacol. Therap. 17:277, 1975.
106. Neudent, W., and Röpke, H.: Chem. Ber. 87:666, 1954.
107. Rudy, B. C., and Senkowski, B. Z.: Analyt. Profiles 2:307, 1973.
108. Brewer, G. A.: Analyt. Profiles 6:198, 1978.
109. Lewis, G. G.: Brit. J. Pharmacol. 9:488, 1954.
110. Cohen, M. L., and Trevor, A. J.: J. Pharmacol. Exp. Ther. 189:354, 1974.
111. Rudy, B. C., and Senkowski, B. Z.: Analyt. Profiles 2:354, 1973.
112. Kappe, T., and Armstrong, M. D.: J. Med. Chem. 8:371, 1965.
113. Gorten, J. E., and Jameson, R. F.: J. Chem. Soc. (A), 2615, 1968.
114. Narahashi, T. I., *et al.*: J. Pharmacol. Exp. Ther. 171:32, 1970.
115. Hoerksema, H.: J. Am. Chem. Soc. 86:4223, 1964.
116. Smith, R. L.: Med. Chem. 2:477, 1964.
117. Barrett, J., *et al.*: J. Pharm. Pharmacol. 25:389, 1973.
118. Persson, B. A., and Schill, G.: Acta. Pharm. Suec. 3:291, 1966.
119. le Pettit, G. F.: J. Pharm. Sci. 65:1095, 1976.

120. Sandoz Pharmaceuticals brochure, 1962.
121. Fox, J. J., *et al.*: J. Amer. Chem. Soc. 80:1672, 1958.
122. Zalipsky, J. J., *et al.*: J. Pharm. Sci. 65:461, 1976.
123. Bush, M. T., *et al.*: Clin. Pharmacol. Therap. 7:375, 1966.
124. Liegler, D. G., *et al.*: Clin. Pharmacol. Therap. 10:849, 1969.
125. Raihle, J. A.: Analyt. Profiles 5:320, 1976.
126. Balasz, L., and Pungor, E.: Mikrochim. Acta 1962:309; through Chem. Abstr. 56:13524g, 1962.
127. Halmekoski, J., and Hannikainen, H.: Acta Pharma. Suec. 3:145, 1966.
128. Siegel, S., *et al.*: J. Am. Pharm. A. 48:431, 1959.
129. Flynn, G. L., and Lamb, D. J.: J. Pharm. Sci. 59:1436, 1970.
130. Rudy, B. C., and Senkowski, B. Z.: Analyt. Profiles, 2:376, 1973.
131. Schwartz, D. E., and Jeunet, F.: Chemother. 22:19, 1976.
132. Peetere, J.: J. Pharm. Sci. 67:129, 1978.
133. Dudzinski, J., *et al.* J. Pharm. Sci. 62:624, 1973.
134. Storoscik, R., *et al.*: Acta Pol. Pharm. 28:601, 1971; through Chem. Abstr. 76:158322k, 1972.
135. Chowhan, Z. T.: J. Pharm. Sci. 67:1258, 1978.
136. Barlow, R. B., and Hamilton, J. T.: Brit. J. Pharmacol. 18:543, 1962.
137. Sanders, H. J., *et al.*: Ind. & Eng. Chem. 47:358, 1955.
138. Shearer, C. M., and Pilla, C. R.: Analyt. Profiles 3:452, 1974.
139. Frazakenly, G. V., and Jackson, E. G.: J. Pharm. Sci. 57:335, 1968.
140. Elpern, B.: Ann. N.Y. Acad. Sci. 148:579, 1968.
141. Stella, V. J., and Pipkin, J. D.: J. Pharm. Sci. 65:1160, 1976.
142. Dayton, P. G., *et al.*: Fed. Proc. 18:382, 1959.
143. Riegelman, S., *et al.*: J. Pharm. Sci. 51:129, 1962.
144. Agarwal, S. P., and Blake, M. I.: J. Pharm. Sci. 57:1434, 1958.
145. Bansal, P. C., and Monkhouse, P. C.: J. Pharm. Sci. 66:820, 1977.
146. Christophers, S. R.: Ann. Trop. Med. 31:43, 1937.
147. Dox, T., *et al.*: Iyakirkin Kenkyn 9:205, 1978: through Chem. Abstr. 88:141601a, 1978.
148. Poet, R. B., and Kadin, H.: Analyt. Profiles 4:360, 1975.
149. Strobel, G. B., and Bianchi, C. P.: J. Pharmacol. Exp. Ther. 172:5, 1970.
150. Rucki, R. J.: Analyt. Profiles 5:415, 1976.
151. Cromble, K. B., and Cullen L. F.: Analyt. Profiles 2:452, 1973.
152. Garrett, E. R., and Weber, D. J.: J. Pharm. Sci. 59:1389, 1970.
153. Carlin, H. S., and Perkins, A. J.: Am. J. Hosp. Pharm. 25:271, 1968.
154. Snell, E. E.: Vitamins and Hormones 16:84, 1958.
155. Gallo, A. G., and Radaelli, P.: Analyt. Profiles 5:483, 1976.
156. Knochel, J. P., *et al.*: J. Lab. Clin. Med. 65:361, 1965.
157. Garrett, E. R., and Schnelle, K.: J. Pharm. Sci. 60:836, 1971.
158. Conners, K. A.: Textbook of Pharmaceutical Analysis, p. 475, New York, John Wiley & Sons, 1967.
159. Suzuki, A., *et al.*: J. Pharm. Sci. 59:651, 1970.
160. Brueckner, A. H.: Yale J. Biol. Med. 15:813, 1943.
161. Nygard, B., *et al.*: Acta Pharm. Suec. 3:313, 1966.
162. Post, A., and Warren, R. J.: Analyt. Profiles 5:241, 1976.
163. Shore, R. A., *et al.*: J. Pharmacol. Exp. Ther. 119:361, 1957.
164. Blessel, K. W., *et al.*: Analyt. Profiles 2:564, 1973.
165. Kaplan, S. A.: J. Pharm. Sci. 59:358, 1970.
166. De Angelis, R. L., *et al.*: J. Pharm. Sci. 66:842, 1977.
167. Blessel, K. W., *et al.*: Analyt. Profiles 3:577, 1974.
168. Papastephanou, C.: Analyt. Profiles 7:492, 1978.
169. McLean, W. M., *et al.*: J. Pharm. Sci. 56:1614, 1967.
170. White, A., *et al.*: Principles of Biochemistry, p. 184, New York, McGraw-Hill, 1968.
171. Neuss, N., *et al.*: J. Am. Chem. Soc. 81:4754, 1959.
172. Hiskey, C. F., *et al.*: J. Pharm. Sci. 51:43, 1962.
173. Hempelmann, F. W.: Arzn. Forsch. 27:2140, 1977.
174. Burns, J. J.: J. Am. Chem. Soc. 75:2345, 1953.

Index

index